CHILTON.

CHRYSLER

DIAGNOSTIC SERVICE
2005 Edition

THOMSON

DELMAR LEARNING

Australia • Canada • Mexico • Singapore • Spain • United Kingdon • United States

THOMSON

DELMAR LEARNING

Chilton®

Chrysler Diagnostic Service

2005 Edition

Vice President, Technology and Trades SBU:

Alar Elken

Executive Director, Professional Business Unit:

Gregory L. Clayton

Publisher, Professional Business Unit:

David Koontz

Marketing Director:

Beth A. Lutz

Production Director:
Mary Ellen Black

Marketing Specialist:
Brian McGrath

Marketing Coordinator:
Marissa Mariella

Production Editor:
Elizabeth Hough

Editorial Assistant:
Christine Wade

Editor:
Timothy A. Crain

Publishing Assistant:
Paula Baillie

Cover Design:
Melinda Possinger

ISBN: 1-4180-0550-9

NOTICE TO THE READER

TABLE OF CONTENTS

WHEN USING THIS MANUAL

Manufacturer and Model Coverage

This manual does not cover every Daimler Chrysler model that is currently available on the market. Rather, the Chilton editorial staff makes judicious decisions as to which models warrant coverage, based on which vehicles are serviced by most technicians.

Model Year Information

This manual is published toward the end of the year prior to the edition year. Every effort is made to gather current data from the Original Vehicle Manufacturers (OEMs) when they publish it. Different OEMs choose to release their new model information at different times of the year. Indeed, the same OEM can publish information early one season and late the next season. As a result, not all models are equally current when each edition of this manual is published.

Although information in this manual is based on industry sources and is as complete as possible at the time of publication, some vehicle manufacturers may make changes which cannot be included here. Information on late models may not be available in some circumstances. While striving for total accuracy, the publisher cannot assume responsibility for any errors, changes, or omissions that may occur in the compilation of this data.

Safety Notice

Proper service and repair procedures are vital to the safe, reliable operation of all motor vehicles, as well as the personal safety of those performing the repairs. This manual outlines procedures for diagnosing and serving vehicles using safe, effective methods. The procedures may contain many NOTES and CAUTIONS which should be followed along with standard safety procedures to reduce the possibility of personal injury or improper service which could damage the vehicle or compromise its safety.

Diagnostic procedures, tools, parts, and technician skill and experience vary widely. It is not possible to anticipate all conceivable ways or conditions under which vehicles may be serviced, or to provide cautions for all possible hazards that may result. Standard and accepted safety precautions and equipment should be used when handling toxic or flammable substances, and safety goggles or other protection should be used during any process that may cause sparking, material removal or projectiles.

Some procedures require the use of tools specially designed for a specific purpose. Before substituting another tool or procedure, you must be completely satisfied that neither your personal safety, nor the performance of the vehicle will be endangered.

Special Tools

Special tools are recommended by the vehicle manufacturer to perform specific jobs. When necessary, special tools may be referred to in the text by part number. These tools may be purchased, under the appropriate part number, from your local dealer or regional distributor, or an equivalent tool can be purchased locally from a tool supplier or parts outlet. Before substituting any tool for the one recommended, read the previous Safety Notice.

ACKNOWLEDGEMENT

The publisher would like to express appreciation to Daimler Chrysler for its assistance in producing this publication. No further reproduction or distribution of the material in this manual is allowed without the expressed written permission of DaimlerChrysler and the publisher.

Understanding On-Board Diagnostics

Introduction

OBD II OVERVIEW

The OBD II system was developed as a step toward compliance with California and Federal regulations that set standards for vehicle emission control monitoring for all automotive manufacturers. The primary goal of this system is to detect when the degradation or failure of a component or system will cause emissions to rise by 50%. Every manufacturer must meet OBD II standards by the 1996 model year. Some manufacturers began programs that were OBD II mandated as early as 1992, but most manufacturers began an OBD II phase-in period starting in 1994.

The changes to On-Board Diagnostics influenced by this new program include:

- Common Diagnostic Connector
- Expanded Malfunction Indicator Light Operation
- Common Trouble Code and Diagnostic Language
- Common Diagnostic Procedures
- New Emissions-Related Procedures, Logic and Sensors
- Expanded Emissions-Related Monitoring

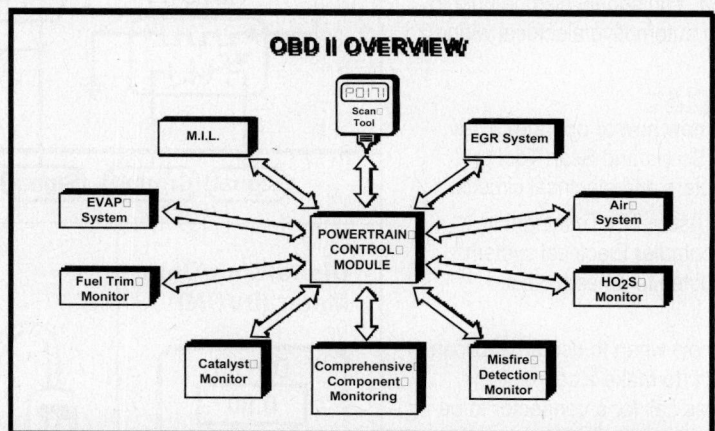

COMMON TERMINOLOGY

OBD II introduces common terms, connectors, diagnostic language and new emissions-related monitoring procedures. The most important benefit of OBD II is that all vehicles will have a common data output system with a common connector. This allows equipment Scan Tool manufacturers to read data from every vehicle and pull codes with common names and similar descriptions of fault conditions. In the future, emissions testing will require the use of an OBD II certifiable Scan Tool.

TECHNICIAN REQUIREMENTS

As an automotive repair technician, you should have a basic understanding of how to use the hand tools and meters necessary to effectively use the information in this OBD II manual.

■ **NOTE:** *Lack of basic knowledge of the Powertrain when performing test procedures could cause incorrect diagnosis or damage to Powertrain components. Do not attempt to diagnose a Powertrain problem without having this basic knowledge.*

ELECTRICITY AND ELECTRICAL CIRCUITS

You should understand basic electricity and know the meaning of voltage (volts), current (amps), and resistance (ohms). You should be able to identify a *Series* circuit as well as a *Parallel* circuit in an automotive wiring diagram. Refer to the examples in the Graphic to the right.

You should understand what happens in an electrical circuit with an open circuit or a shorted wire, and you should be able to identify an open or shorted circuit condition using a DVOM. You should also be able to read and understand an automotive electrical wiring diagram.

CIRCUIT TESTING TOOLS

You should have (and know how to operate) a 12v Test Light, DVOM, Lab Scope and Scan Tool to diagnose vehicle computers and electrical circuits.

You should know not to use a 12v Test Light to diagnose the Engine Controller Electrical system unless specifically instructed to do so by test procedures.

You should have and know when to use an applicable aftermarket connector kit (to make a connection) whenever test procedures call for a connector to be probed in order to make a measurement.

ELECTRICAL CIRCUITS

When you encounter a wiring problem during testing, and need to refer to electrical circuit information, you should be comfortable with this type of information:

- Wiring schematics (including circuit numbers and colors)
- Electrical component connector, splice and ground locations
- Wiring repair procedures and wiring repair parts information

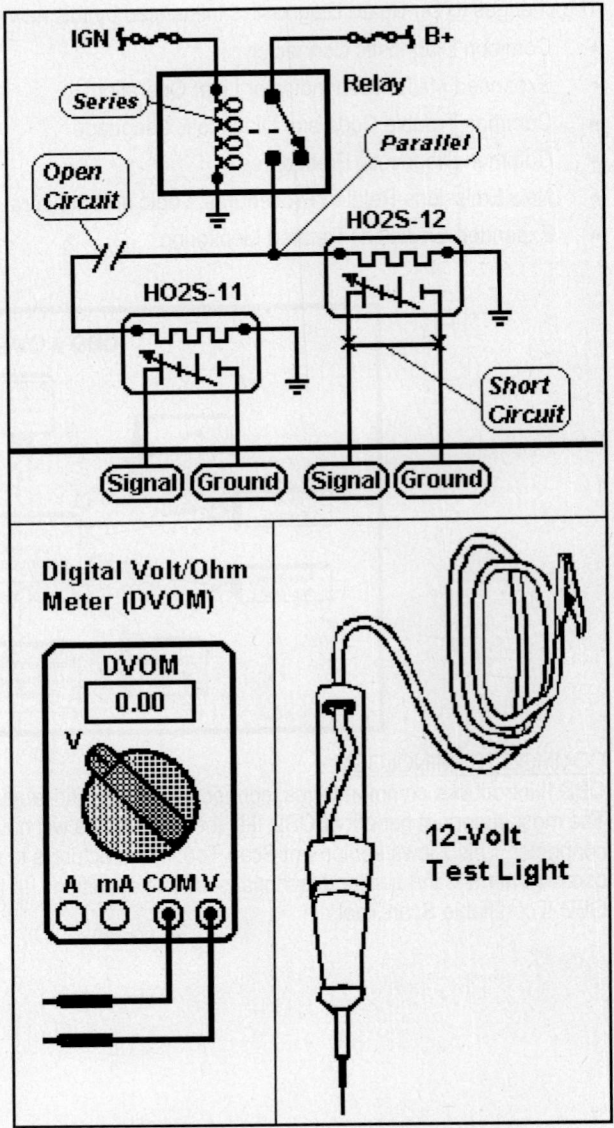

OBD II System

History of OBD Systems

INTRODUCTION

Starting in 1978, several vehicle manufacturers introduced a new type of control for several vehicle systems and computer control of engine management systems. These computer-controlled systems included programs to test for problems in the engine mechanical area, electrical fault identification and tests to help diagnose the computer control system. Early attempts at diagnosis involved expensive and specialized diagnostic testers that hooked up externally to the computer in series with the wiring connector and monitored the input/output operations of the computer.

By early 1980, vehicle manufacturers had designed systems in which the onboard computer incorporated programs to monitor selected components, and to store a trouble code in its memory that could be retrieved at a later time. These trouble codes identified failure conditions that could be used to refer a technician to diagnostic repair charts or test procedures to help pinpoint the problem area.

EVOLUTION OF DAIMLER CHRYSLER COMPUTERIZED ENGINE CONTROLS

The evolution of Computerized Engine Controls on Chrysler vehicles equipped with fuel injection is highlighted in the Graphic below.

Computerized Engine Controls Evolution Graphic - Chrysler & Jeep

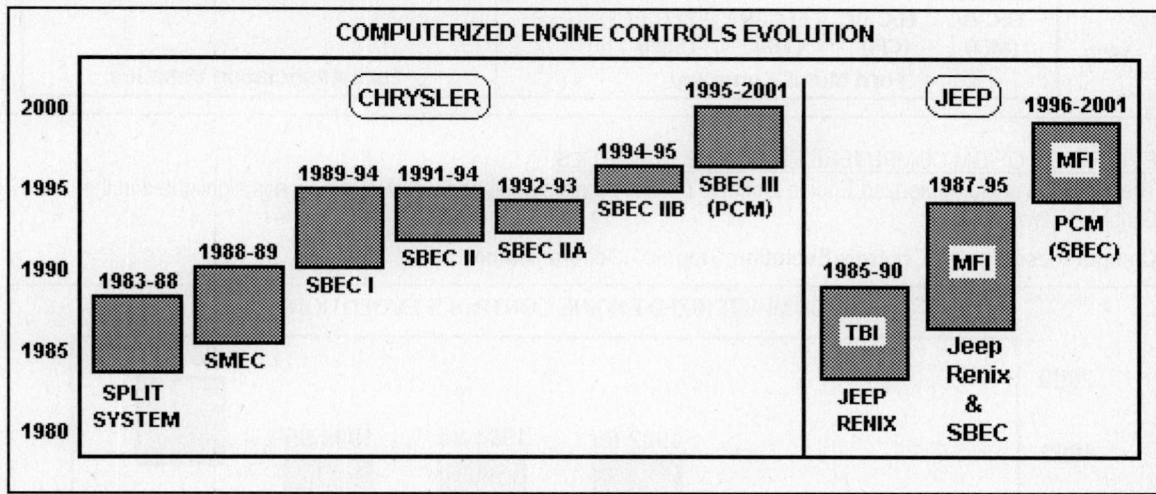

EVOLUTION OF FORD MOTOR COMPANY COMPUTERIZED ENGINE CONTROLS

The evolution of Computerized Engine Controls on Ford vehicles equipped with fuel injection is highlighted in the Graphic below.

Computerized Engine Controls Evolution Graphic - Ford Motor Company

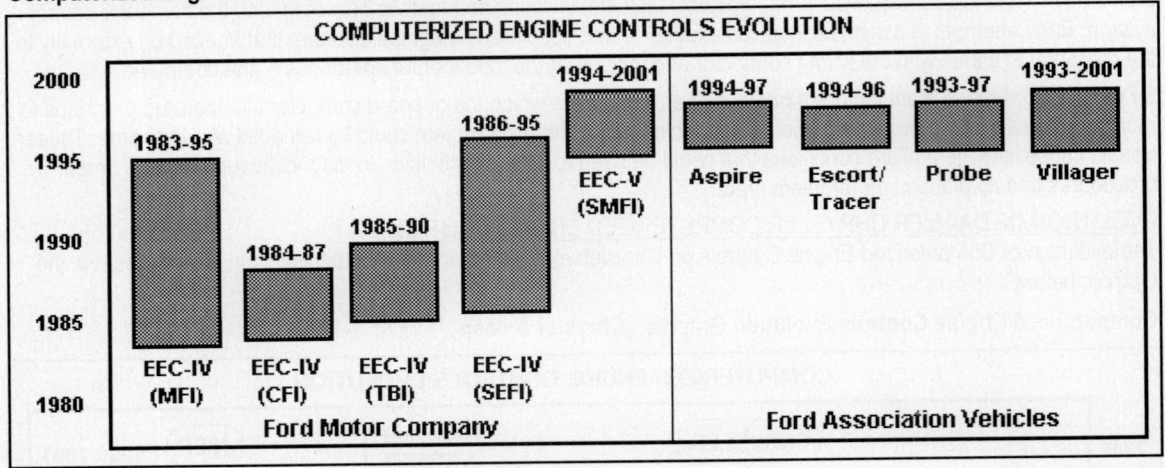

EVOLUTION OF GM COMPUTERIZED ENGINE CONTROLS

The evolution of Computerized Engine Controls on GM vehicles equipped with fuel injection is highlighted in the Graphic below.

Computerized Engine Controls Evolution Graphic - General Motors

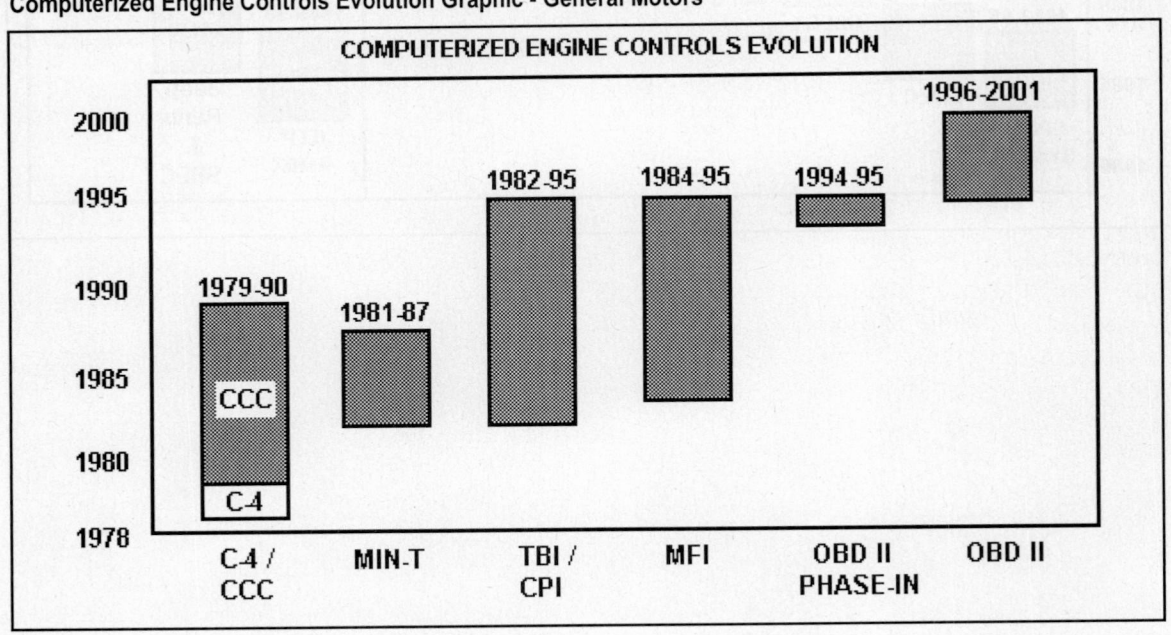

Computer Diagnostics

INTRODUCTION

General diagnostics of computers fall into these two categories:

- External Onboard Diagnostics
- Internal Onboard Diagnostics

The first level of diagnostics uses an external tool that taps into the computer and runs a series of diagnostic tests. This method of diagnostics was popular in the 1970's and was used in the 1980's on many European vehicles. The second level incorporates diagnostics into the circuit board of the computer and is known in the industry as "On-Board" diagnostics because the diagnostics are on the computer circuit board.

In 1980 General Motors incorporated an On-Board Computer Program where the "check engine" light came on to inform the vehicle owner that there was a fault in the computer system. The light was turned on when a diagnostic code was set to alert the driver that service was needed on the vehicle.

California formed a government agency, the California Air Resources Board (CARB) to monitor the air quality and establish regulations to reduce air pollution. The California Health and Safety Code authorized the Air Resources Board to adopt motor vehicle emissions standards and in-use performance standards that it finds necessary, cost effective and technologically feasible. In 1988, CARB required that all vehicles sold in California incorporate a system with an On-Board Diagnostic program where a "check engine" light would come on to notify the vehicle owner of a potential failure of computer sensors and/or their systems. This system is known as On-Board Diagnostics First Generation and is now referred to as OBD I.

PROBLEMS WITH OBD I SYSTEMS

One of the problems with OBD I was that the code retrieval methods varied from manufacturer to manufacturer and there was no consistency between systems. Most manufacturers looked at similar computer sensors and circuits, but codes were inconsistent and difficult to identify and define. Some manufacturers require special tools to retrieve trouble codes or required special test procedures for these tools which self-tested circuits and systems or energized the components for testing in the field.

SCAN TOOL INTRODUCTION

Domestic vehicle manufacturers (Chrysler, Ford and GM) designed their computers to have an accessible data line where a diagnostic tester could retrieve data on sensors and the status of operation for components.

These testers became known in the automotive repair industry as "Scan Tools" because they scanned the data on the computers and provided information for the technician.

Ford Motor Company developed a tester that would access codes, activate sensors and perform limited tests and adjustments, however they did not incorporate data stream features until 1988.

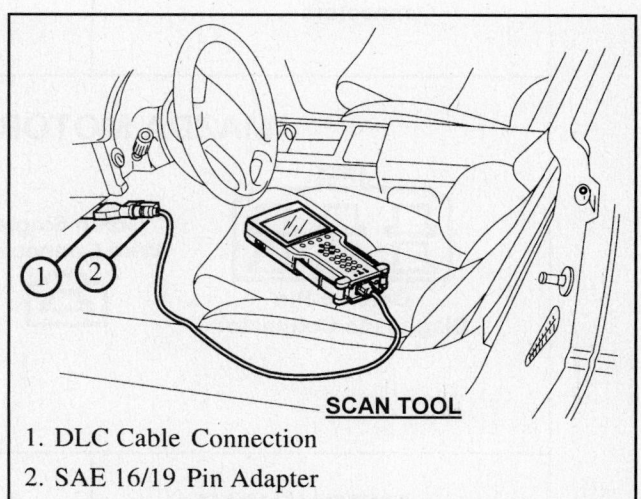

SCAN TOOL

1. DLC Cable Connection
2. SAE 16/19 Pin Adapter

OBD I SYSTEM CONNECTORS

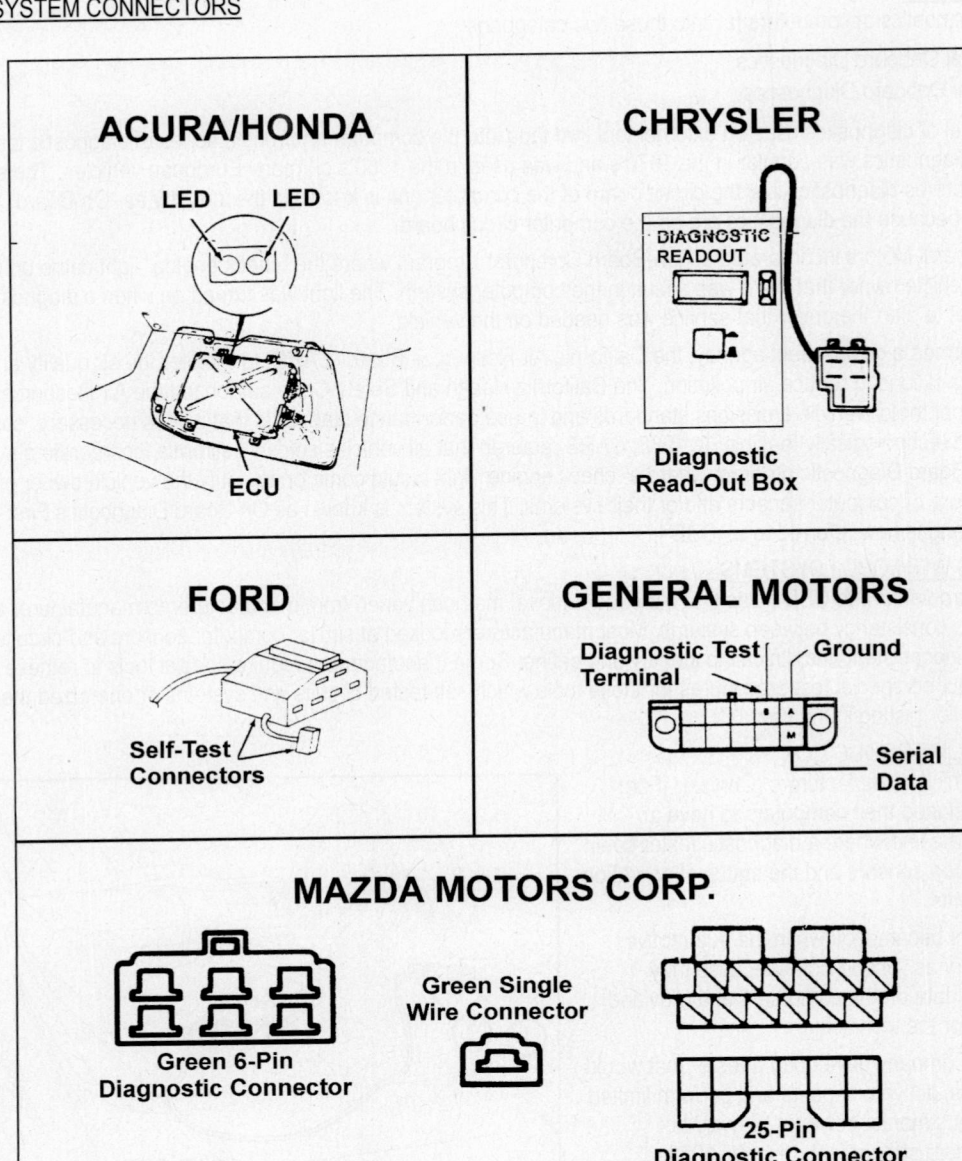

ACURA/HONDA

LED   LED

ECU

CHRYSLER

DIAGNOSTIC READOUT

Diagnostic Read-Out Box

FORD

Self-Test Connectors

GENERAL MOTORS

Diagnostic Test Terminal — Ground

Serial Data

MAZDA MOTORS CORP.

Green 6-Pin Diagnostic Connector

Green Single Wire Connector

25-Pin Diagnostic Connector

MITSUBISHI

Diagnostic Terminal

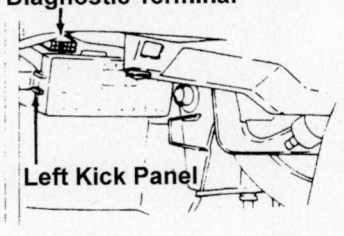

Left Kick Panel

NISSAN MOTOR CO.

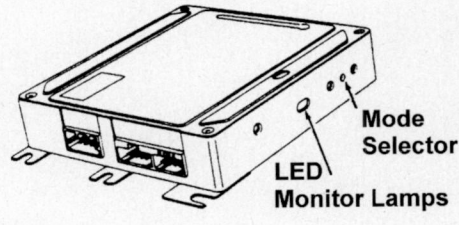

Mode Selector

LED Monitor Lamps

Government Regulations

INTRODUCTION

The California Air Resources Board (CARB) conducted research on OBD I vehicle emissions and the study resulted in the following conclusions:

- The research found a significant number of pre-1988 vehicles with degraded emissions components. These components were not failing outright, but deterioration increased emissions levels. This problem did not usually set codes alerting the vehicle owner or technician that there was a problem, therefore the condition was not perceived as a problem in the field. However, CARB viewed this as a problem due to the increased emission levels.

- Vehicle testing programs found failures in Canister Purge systems and Secondary Air Management systems. Many of these failures occurred under road load conditions and were not quickly or easily detectable in the service bay. These failures resulted in increased emissions.

- Catalytic Converters were failing and vehicles were being driven with deteriorated catalysts. A leading cause of this failure was engine misfire.

- The On-Board Monitoring Systems did not detect fuel system faults that were responsible for increasing emissions even though fuel systems were deteriorated enough to have excessive emissions.

- The monitoring systems did not detect oxygen sensors that were "lazy" or slow in response. This condition was found to result in an increase in emissions levels.

- EGR monitoring did not verify if the system was operating within a range that could result in an increase of emissions. There was a need to monitor the flow of EGR gases through the system in order to verify the EGR passages were not clogged.

- Codes were different for each manufacturer and this was confusing for a technician working on different vehicles.

DEVELOPMENT OF OBD I STANDARDS

CARB reviewed the system of monitoring Engine Control sensors and systems developed by Chrysler Motors and General Motors. They incorporated this concept into their regulations.

The result was that the California regulations required that all vehicle manufacturers develop a set of diagnostics that would incorporate a system where codes and data are made available through a Scan Tool accessible to every technician.

These California standards, originally published in October of 1988, generally apply to 1994 and later passenger cars, light duty trucks and medium duty vehicles. Similar diesel and alternative fuel vehicle regulations took affect in 1996. After 1988, California made the decision to accept the Federal (EPA) OBD II regulations.

FEDERAL TEST PROCEDURE

OBD II requires that the on-board computer monitors vehicle emissions and in some cases perform an "Active" diagnostic test of those systems. These tests were developed by the EPA and are a reflection of the Federal Test Procedure (FTP).

The FTP is a series of programmed tests where a vehicle is driven through specific drive cycles while emissions are being monitored. These tests are conducted at various mileage levels and test emissions under very specific conditions. The amount of fuel in the gas tank is monitored, the type of fuel and octane level are all controlled. These tests are conducted on a dynamometer and are performed under hot and cold vehicle conditions. They are conducted under EPA supervision and are required to certify a vehicle for sale in the USA.

The OBD II system was designed to monitor these same systems and a Malfunction Indicator Lamp (MIL) must illuminate if a system or component either fails or deteriorates to a point where the vehicle emissions could rise beyond 1.5 times the FTP standard.

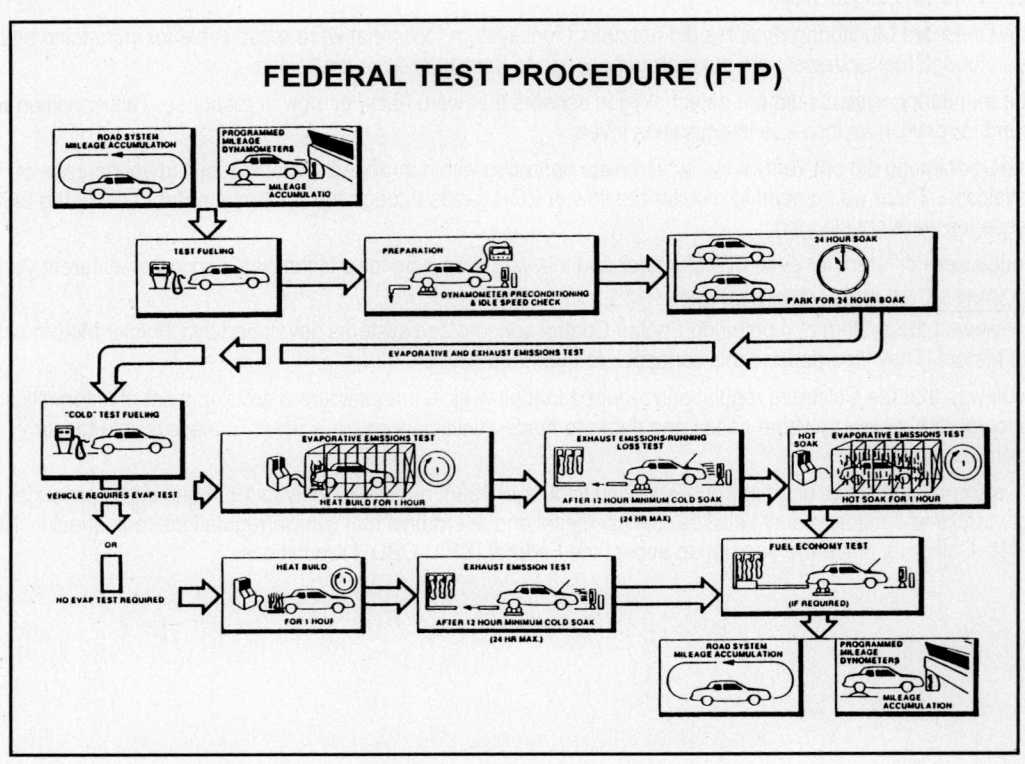

CARB Regulations

SUMMARY OF CARB REGULATIONS

- All vehicles are equipped with a Malfunction Indicator Light (MIL) that will remain "on" when certain faults are detected by the controller. If the MIL is "on", the program should be able to turn it "off" if the fault does not reappear on two consecutive trips. Codes can be erased from memory if the fault does not occur for 40 warmup periods.

- The catalytic converter closest to the engine (if more than one converter is on the vehicle) must be monitored to see if it has deteriorated.

- The oxygen sensor must be monitored for the output voltage and the response rate. The monitor should include the rich/lean transitions (cross counts or oxygen sensor switch rate) and should check the oxygen sensor transition from rich to lean and lean to rich. The oxygen sensor heater should be monitored for proper operation.

- The diagnostic system must monitor the engine for misfire and identify the specific cylinders misfiring. This system must include identification informing the technician that the catalyst is deteriorated, damaged, or has failed a Federal Test Procedure type drive cycle. If the engine is misfiring at a level that could cause damage to the converter catalyst, the MIL should flash when these conditions are present and the controller should switch to a backup program designed to reduce the level of misfire.

- The fuel delivery system should be monitored for its ability to control fuel. It should determine if the system has a condition that is over the range for optimal fuel control. This system is referred to as fuel trim monitor.

- The EGR system should be monitored for both low and high flow rates as well as verify that solenoids controlling the EGR are working properly.

- The EVAP System should be monitored for HC flow changes based on fuel tank fill level and the monitor should calculate for these differences.

- The Secondary Air Management system should be monitored for the control valve operation and for the airflow throughout the system.

- Any emissions-related component or system not otherwise described that provides input directly or indirectly to the processor should be monitored. This includes all input and output components.

EPA Regulations

The Federal Environmental Protection Act (EPA) also established diagnostic regulations, but in an effort to simplify the development process for vehicle manufacturers, the EPA decided to comply with the California OBD II regulations from 1994 to 1998.

SAE Forms Committees to Develop Standards

The Society of Automotive Engineers formed committees to help its member engineers coordinate efforts and develop a second generation of diagnostics. The new diagnostics, known as OBD II, would standardize the diagnostic connector, diagnostic data retrieval and code identification.

EPA Expands the SAE Standards

The EPA reviewed the SAE standards and added regulations requiring all manufacturers to meet Federal OBD II standards by 1996. To assist in development of expanded diagnostics, the EPA decided to use the California regulations as federal standards through 1998. Federal and California regulations established a two-year phase-in to take effect from 1994-1996 to allow for design and phase-in of expanded diagnostics.

Government (CARB & EPA) and SAE Regulations

OVERVIEW
The Society of Automobile Engineers (SAE), CARB and the EPA set the standards that relate to changes in the industry (terminology, common Scan Tool interface, etc.) while others set the diagnostic standard for how information is handled by vehicle controllers.

Industry Regulations

Government Mandated Regulations	SAE "Recommended" Compliance
J1930 - Industry Terminology Standardization	J2201 - Additional Guidelines for Generic Scan Tool Interface
J1978 - Standards for Generic OBD II Scan Tool Interface Protocol	J2190 - Enhanced Test Mode Standards
J2205 - Standards for Expanded Scan Tool Interface Protocol	J2008 - Guidelines for Repair Service Information (CARB Standards)
J2008 - Standards for Repair and Service Information (EPA Guidelines)	

On-Board Computer Regulations

Government Mandated Regulations	SAE "Recommended" Compliance
J2012 - Standards for a Diagnostic Trouble Code (DTC)	J2186 - CARB approved standards for anti-tamper procedures
J1962 - Standard 16-Pin diagnostic connector	J2178 - Scan Tool message strategy guidelines
J1979 - Standards for Diagnostic Test Modes	J2190 - Enhanced Test Mode Standards
J1850 - Scan Tool Communication guidelines for Class 'B' Data Interface	J1724 - Vehicle Electronic Identification Standards
J2186 - EPA mandated Anti-Tamper procedures	

CARB AND EPA REGULATIONS
The government agencies that set OBD regulations are CARB and the EPA. The tables below compare differences between the EPA and CARB requirements for OBD II.

Part One - Industry Regulations

CARB	EPA	Government Requirements
	X	1994 model year - all service information must conform to J1930.
X	X	Service manuals must publish a normal range for calculated load values and Mass Air Flow Rate at idle and at 2,500 RPM.
	X	The vehicle manufacturer is responsible for ensuring information is available even if the information is provided by an intermediary.
	X	The cost of repair information to the independent technician shall not exceed the lowest price that is available to a dealership.
	X	All other information available at a fair and reasonable price, otherwise it is considered not available (a fine of $25K per day could be applied).
	X	Electronic service information must be available by the 1998 model year.
X		Repair procedures must be available which allow effective diagnosis and repairs using a J1978 Generic Scan Tool and readily available repair tools.
X	X	J1978 Scan Tool compatibility - communication protocol. All serial data and enhanced tests must be available to a Scan Tool. Scan Tool must inform user which emissions systems are monitored. EPA added requirements that the VIN be accessible off the DLC.
	X	Requires Bi-directional diagnostic control of the computer be available on the Scan Tool meeting J2205 and J1979 standards.
	X	1996 model year - all service information must be in the J2008 format.
	X	Labeling requirements must meet J1877 and J1892 standards.

Part Two - On-Board Computer Regulations

CARB	EPA	Government Requirements
X	X	J2012 Diagnostic Trouble Codes - If the PCM detects a fault, it must set a code to identify a fault (uniform identification), include conditions that describe how the PCM reverts to default mode and erase the code after 40 warm up cycles with the MIL off.
X	X	J1962 Diagnostic Connector mounted on Instrument Panel driver's side of vehicle with standard pins for serial data, power and ground.
X	X	J1979 Diagnostic Test Mode Messages - defines standard messages for access to trouble codes, vehicle data stream and Freeze Frame data.
X	X	Scan Tool Interface for DLC - SAE J1850 serial data link required to access all emission-related data. Requires 3 byte headers (does not allow IBS or checksum).
X	X	Vehicle manufacturers must provide tampering deterrence for a PCM that is programmable (where the PROM is rewritten to change operating parameters). This must include write-protect standards for programmable memory and references J2186 - Data Link Security for write protect.
	X	Access to vehicle calibration data, odometer and keyless entry codes can be limited, but OEMs must provide "the best means available for providing non-dealer technicians with calibration data necessary to perform repairs".
X	X	Freeze Frame Data Stored with the first fault of any component or system and replace data if there is a subsequent fuel system or misfire fault. EPA added "airflow rate" to required Freeze Frame data.
X	X	Signal access to the required diagnostic data must be made available through the diagnostic connector using standard messages. Actual values should be identified separately from default or limp home values.
X	X	The vehicle must have only one Malfunction Indicator Light (MIL) for emission related problems. The MIL must remain "on" if a malfunction is detected and stays on until three trips indicate the fault is gone. The MIL must blink if a catalyst-threatening misfire condition is present. Note: A few manufacturers received exemptions from this standard for specific 1994-96 models.

Part Three - On-Board Computer Monitor Regulations

CARB	EPA	Government Requirements
X	X	Oxygen Sensor Monitor - It must check the output voltage and response rate for all oxygen sensors once per trip. The results of most recent oxygen sensor evaluation test must be available over the data link connector as serial data. The EPA added a requirement that the results of the most recent on-board monitoring data and test limits for all systems with specific Monitor evaluation tests must be made available.
X	X	Catalyst Monitor - it must verify the catalyst is functioning at steady state efficiency and that it does not deteriorate over 1.5X the standard. This test is done once per trip.
X	X	Misfire Monitor - It must run continuously under all conditions in order to identify a misfiring cylinder. On vehicles with SFI, it must cutoff fuel to a misfiring cylinder.
X		EGR System - It must monitor for both low and high EGR flow rate once per trip.
X		EVAP Purge Monitor - It must check the system function and for leaks once per trip.
X		Secondary AIR Management Monitor - It must perform a functional test of the AIR system and switching valves (a test for proper function and airflow once per trip).
X		Fuel System Monitor - It must check the ability of the controller to control fuel delivery.
X	X	Comprehensive Component Monitor - It must monitor all components or systems that send data to or receives data from the PCM. CARB requires a check for out-of-range signals and a functional response test of related outputs. CARB requires continuous monitoring while the EPA requires evaluation periodically once per "Drive Cycle".
	X	The system must monitor any deterioration or malfunction which occurs which can cause exhaust or evaporative emissions to increase 1.5X the Federal Test Standard.
X		Air Conditioning system must be monitored for loss of reactive refrigerant once per trip. Non-reactive refrigerant does not have to be monitored.

Explanation of SAE Standards

J1930 - Common Names for Components

J1930 established common nomenclature for emissions and computer-related components and systems. This includes common definitions, abbreviations and acronyms.

This standard is designed to provide the technician with a recognizable name for components that apply to all vehicles. This nomenclature has been determined to be beneficial for technicians who work on multiple lines of vehicles as well as vehicles from different manufacturers.

J2008 - Service Information Availability

J2008 requires that "all information" must be made available to "any person engaged" in the repair of the vehicle. The legislation is very specific and requires that "no such information on vehicle repair" may be withheld from any technician. J2008 is still being finalized and will continue to be interpreted by the vehicle manufacturer. It also sets standards for the organization of vehicle service information. This includes the data model, data type definition, graphics standards and electronic transmission of data.

EPA Guidelines on Repair Information

The EPA published guidelines that state that information availability requirements include 1994 model year vehicles, and that vehicle manufacturer must furnish to "any person" engaged in the repair or service of a motor vehicle with "all information" required to make emission related diagnosis and repairs. Includes, but is not limited to service manuals, technical service bulletins, vehicle recalls, engine control emissions system information, bi-directional control and training information.

None of this information may be withheld if provided directly or indirectly to the dealers. Information cost to independent technicians shall not exceed the lowest price of the same information to the dealerships. Other repair information must be made available at a "fair and reasonable" cost, otherwise it is considered unavailable.

J2205 - Expanded Diagnostic Protocol for Scan Tools

Some Scan Tools incorporate a protocol that allows the technician to access information not specifically required by OBD II standards. The information in the messages on this tool will be specified in factory service information and provided to the technician. Refer to the examples under Diagnostic Function in the Graphic to the right.
Source: OTC Scan Tool with a 1999 Pathfinder cartridge.

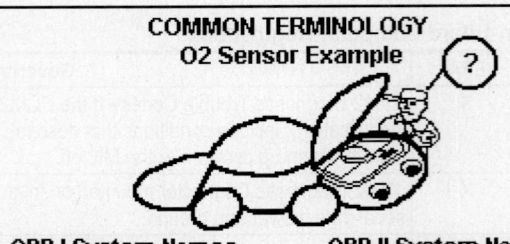

COMMON TERMINOLOGY
O2 Sensor Example

OBD I System Names	OBD II System Names
• O2 Sensor	• O2S-11 Bank 1 Sensor 1
• HEGO Sensor	• HO2S-11 Bank 1 Sensor 1
• EGO Sensor	• HO2S-12 Bank 1 Sensor 2
• LAMBDA Sensor	• HO2S-13 Bank 1 Sensor 3
• Feedback Sensor	• HO2S-21 Bank 2 Sensor 1
	• HO2S-22 Bank 2 Sensor 2

SCAN TOOL MENUS

Press:
> 1-OEM Tests
  2-OBD II

DIAGNOSTIC FUNCTION

Press:
> 1- Datastream
  2-Diagnostic Codes
  4-Record/Playback
  5-Special Test
  7-Monitor Setup

1-DATASTREAM

BARO	29.4" HG
BATT TEMP	49°F
ENGINE RPM	750
IAC DESIRED	37
IAC MOTOR	37

2-DIAGNOSTIC CODES

Press:
> 1-Read Codes
  2-Clear Codes
  3-Code History

EXPLANATION OF ONBOARD COMPUTER REGULATIONS

J1978 - Generic Scan Tool Usage

J1978 requires that all vehicle manufacturers make readily available to the automotive repair industry all data, codes and emissions-related information that can be accessed by a generic Scan Tool. The values for all trouble codes, sensors and components along with Freeze Frame data stored in the computer must be accessible for download to a Generic Scan Tool.

Once a Generic Scan Tool is connected to the 16-pin OBD II connector, it can retrieve certain data from the computer data stream, retrieve Freeze Frame data, read any 5-digit codes and clear these codes from memory.

The EPA expanded this regulation to include the ability to perform bi-directional diagnostic control. The EPA did not define "bi-directional", and the vehicle manufacturers requested that more specific standards be written and incorporated into J2205 after review by SAE committees.

J2178 sets the standards for how vehicle interface messages are displayed on the Generic Scan Tool.

J2201 - Generic Scan Tool Terminal Designation

J2201 sets additional guidelines for Generic Scan Tool interface and assigns the designation of terminals for voltage feed, ground and data transmission.

Generic Scan Tool Menu Example

The Scan Tool Parameter Identification (PID) Mode allows access to certain data values, analog and digital input and output signals, calculated values and system status information.

Generic Scan Tool Navigation

An example of how to navigate through the Scan Tool menus (Snap On example) to locate the Generic PID information is shown in the Graphic to the right.

There are 16 engine related parameters for this vehicle on a Generic OBD II Scan Tool. The parameters in the last frame of this example represent known good values.

Parameter ID (PID) Information

The proper sequence to follow to obtain a complete Generic PID list for this vehicle is shown in the Graphic.

1) Scroll through the main menu and line up the tilde (~) with the desired choice (in this case, GENERIC).
2) Scroll to CODES & DATA. Select it with the tilde (~).
3) Connect an OBD II K2 Adapter to the test connector to allow the tool to read OBD II Generic information.
4) Scroll through the menu and then line up the tilde (~) with the desired choice (DATA - NO CODES).

Source: Snap On Scan Tool with a 1999 cartridge.

SCAN TOOL MENUS

GM/SATURN (1980-1999)
CHRYSLER (1983-1999)
JEEP (1984-1999)
FORD (1981-1999)
~GENERIC OBD II

OBD II GENERIC SCREENS

MAIN MENU-EMISSIONS
[PRESS N FOR HELP]
~CODES & DATA MENU
CUSTOM SETUP

CONNECT OBD-II
K2 ADAPTER TO
16-PIN OBD II TEST
CONNECTOR.
NO REPAIR TIPS
AVAILABLE IN
GENERIC MODE.
PRESS Y TO CONTINUE

CODES & DATA MENU
CODES ONLY
~DATA (NO CODES)
O2 MONITORS
FREEZE FRAME
PENDING CODES

OBD II DATA
(CODES NOT AVAILABLE)

ENGINE RPM_____720
THROTTLE(%)_____16.4
FUEL SYS1_____OL
FUEL SYS2_____N/A
COOLANT(°F)_____117
MAP("Hg)_____29.00
IGN ADVANCE(°)___0.0
ST TRIM B1(%)_____0.00
LT TRIM B1(%)_____00.0
O2 B1-S1(V)_____0.470
TRIM B1-S1(%)____0.00
O2 B1-S2(V)_____0.510
TRIM B1-S2(%)____N/A
VEH SPEED(MPH)__0
ENG LOAD(%)_____17.0
MIL STATUS_____OFF

J1724 - Vehicle Electronic Identification

SAE has developed a recommended practice to provide electronic access to vehicle content information necessary to diagnose, service, test and repair passenger cars and light duty trucks. The SAE committee in charge of this area continues to look at a wide range of interpretations for this standard.

J1850 - Scan Tool Access to Emission Related Data

Access to emission related data must be made available on a standard Diagnostic Data Link (DDL) defined in the J1850 standard (Class 'A' data). There are also other systems on the vehicle that use PCM data. For example, the Climate Control Automatic Air Conditioning System uses signals from the ECT sensor to help determine when to operate the Air Conditioning and Electric Cooling Fan. SAE also developed standards for Vehicle Network and Multiplexing Data Communications (referred to as Class 'B' data). Class 'B' data communications use a system where data is transferred between one or more controllers (or Modules) to eliminate redundant sensors and other system duplication. The modules in this type of system form a multiplex of interactive systems.

Class 'B' Data Communication

Class 'B' data communications have to be able to perform all Class 'A' data functions. However, these two types of communication protocols differ from each other and usually do not communicate in the same format. Scan Tools that communicate with both formats will be available as this standard is defined further and may be made available from the vehicle manufacturer. This means that the Generic Scan Tool (GST) may not be able to access information from computers that control ABS, Air Conditioning, Steering and Suspension, Electronic Transmissions and other related systems.

J1962 - Common Diagnostic Connector

J1962 establishes a set of standards for the OBD II 16-Pin diagnostic connector. The 8 pins assigned by SAE include two pins for a Serial Data Link, two pins for an ISO 9141 Serial Data Link (European) and pins for battery power and ground.

J1979 - Diagnostic Test Mode Messages

Defines standard messages for access to trouble codes, vehicle data stream information and Freeze Frame data.

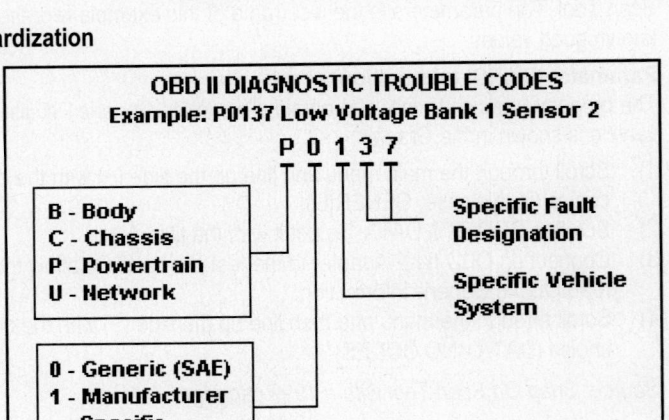

J2012 - Diagnostic Trouble Codes (DTC) Standardization

Diagnostic Trouble Code (DTC) is a term used to describe the method used when a vehicle computer detects a problem in a component or system that it is monitoring.

OBD I system trouble codes were one (1), two (2) or three (3) digit numbers. SAE J2012 set standards for trouble codes and definitions for emission-related systems.

OBD II codes use a five-digit code. OBD II codes begin with a letter followed by four numbers. Refer to the example in the Graphic to the right.

J2186 - Diagnostic Data Link Security Standards

Procedures used in tamper protection must discourage tampering yet allow for the service industry to reprogram or service the PCM as deemed necessary by a vehicle manufacturer. Legitimate service of the PROM will be referenced in later EPA regulations. This service can only be performed if the proper security codes are transmitted from the Scan Tool. However, normal Generic Scan Tool (GST) communications are not affected by this standard.

SAE J2186 defines EE Data Link Security standards for computers with electronically erasable and re-programmable PROMS. There are established standards intended to eliminate "hot rod" PROMS that could disable/defeat emissions-related control systems.

Engine operating parameters should not be changed without the use of manufacturer specialized tools, codes and procedures. CARB specified that any re-programmable computer coded system should include proven write-protection procedures and hardware. The Federal EPA requires that any re-programmable computer codes or operating parameters must be tamper resistant and must conform to SAE J2186 EE Data Link Security standards.

Tamper Protection Explanation

The possibility that On-Board programs may be tampered with from manufacturer specifications introduces the possibility of additional vehicle problems that would need to be diagnosed. Even before vehicles used computers, it was difficult to diagnose a driveability problem in a modified or performance vehicle. The potential of these changes could make it difficult for committees to balance between allowing individuals to install performance changes and protecting the repair industry from the changes.

There is a need for standards that would allow a PCM to identify a vehicle ID number and current calibration, whether the change is a factory update, or an aftermarket modification. This information could also be made accessible to a repair technician.

This would allow the technician to know at the start of a repair procedure that the diagnostic situation might not follow published diagnostic information. To avoid this situation, EPA proposed regulations that would force vehicle manufacturers to utilize complicated methods to deter unauthorized reprogramming. They would include executable routines that could have copyright protection with encrypted data and mandated electronic access by service facilities to an off-site computer maintained by the vehicle manufacturer. Access to an executable routine would be controlled by the vehicle manufacturer and made available only at one of their authorized dealers.

OBD II Warmup Cycle Definition

It is important to understand the meaning of the expression *warmup cycle*. Once the MIL is turned off and the fault that caused the code does not reappear, the OBD II code that was stored in PCM memory will be erased after 40 warmup cycles. The exceptions to this rule are codes related to a Fuel system or Misfire problem. These trouble codes require that 80 warmup cycles occur without the reoccurrence of the fault before the code will be erased from the PCM memory.

A warmup cycle is defined as engine operation (after a key off and engine cool-down period) in which engine temperature increases at least 40ºF and reaches at least 160ºF.

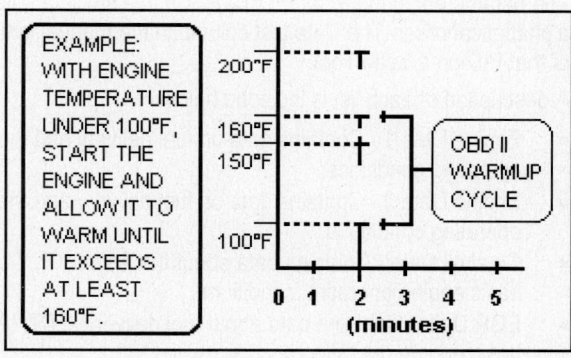

EXPLANATION OF SAE STANDARDS

J2190 - Enhanced Diagnostic Test Modes

This standard identifies test modes and diagnostics for issues not covered by the EPA and CARB regulations. They include Enhanced Test Modes (including an expanded diagnostic routine), and include the protocols required to establish the screens on the Scan Tool for these items:

- Request a diagnostic session
- Request trouble code related Freeze Frame data
- Request all diagnostic trouble codes
- Request status of Main Monitors, clear all test data
- Request diagnostic data, security access
- Disable or enable normal message transmission
- Request diagnostic data packets, test routine results
- Enter or exit diagnostic routines
- Substitute sensor values, substitute output controls
- Read or write to the PROM
- Messages from Enhanced Diagnostic Test Modes are available through an Enhanced Scan Tool, but may not be available through a Generic Scan Tool.

Scan Tool Enhanced Menu Example

An example of how to navigate through the Vetronix Scan Tool menus to locate the OEM PID information is shown in the Graphic. An example of first seven steps to follow is shown. Step (8) contains examples of PID data.

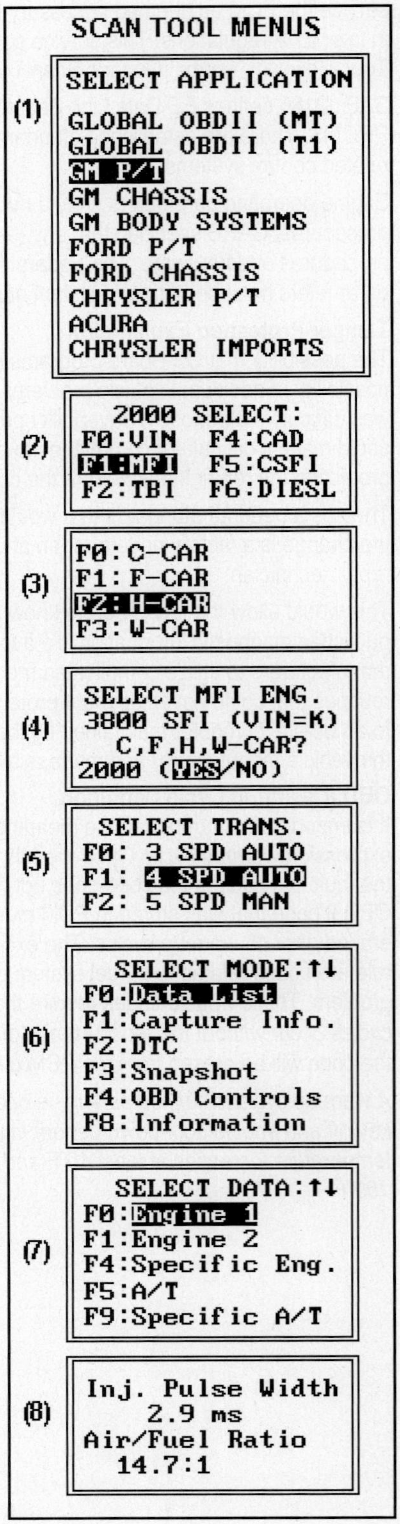

Parameter ID (PID) Information

The PID information for this vehicle is organized into various Data Lists (Engine Data 1, etc). Each PID is categorized into a particular list.

The parameters in the PCM PID Tables in this article are listed in alphabetical order. The Data List column in the manual indicates the location of that PID on a Scan Tool.

A description of each list is included below:

- Engine Data 1 - Contains data on fuel delivery and the basic engine operating conditions.
- Engine Data 2 - Contains data on fuel delivery and the basic engine operating conditions.
- Catalyst Data - Contains data about the A/C, CKP, CMP, KS, and the basic engine operating conditions.
- EGR Data - Contains data about fuel delivery, ECT, IAT, VTD, and the basic engine operating conditions.
- EVAP Data - Contains data that allows it to display parameters needed to verify EVAP system operation.
- HO2S Data - Contains data on the Oxygen sensor.
- Misfire Data - Contains data for Misfire diagnostics.
- Output Device Driver Data - Contains data specific to the ODD operation.

Source: Vetronix Scan Tool & 2000 Mass Storage Unit.

SCAN TOOL MENUS

(1)
```
SELECT APPLICATION
GLOBAL OBDII (MT)
GLOBAL OBDII (T1)
GM P/T
GM CHASSIS
GM BODY SYSTEMS
FORD P/T
FORD CHASSIS
CHRYSLER P/T
ACURA
CHRYSLER IMPORTS
```

(2)
```
2000 SELECT:
F0:VIN    F4:CAD
F1:MFI    F5:CSFI
F2:TBI    F6:DIESL
```

(3)
```
F0: C-CAR
F1: F-CAR
F2: H-CAR
F3: W-CAR
```

(4)
```
SELECT MFI ENG.
3800 SFI (VIN=K)
  C,F,H,W-CAR?
2000 (YES/NO)
```

(5)
```
SELECT TRANS.
F0: 3 SPD AUTO
F1: 4 SPD AUTO
F2: 5 SPD MAN
```

(6)
```
SELECT MODE:↑↓
F0:Data List
F1:Capture Info.
F2:DTC
F3:Snapshot
F4:OBD Controls
F8:Information
```

(7)
```
SELECT DATA:↑↓
F0:Engine 1
F1:Engine 2
F4:Specific Eng.
F5:A/T
F9:Specific A/T
```

(8)
```
Inj. Pulse Width
     2.9 ms
Air/Fuel Ratio
     14.7:1
```

Malfunction Indicator Lamp

INTRODUCTION

The CARB and Federal EPA regulations require that a Malfunction Indicator Lamp (MIL) be illuminated when an emissions related fault is detected and that a Diagnostic Trouble Code be stored in the vehicle controller (PCM) memory.

Most vehicle manufacturers provided the "Check Engine" light diagnostics required by the 1988 California regulations in time to meet this deadline.

OBD II regulations established changes in the "Check Engine" light operation. A new universal term identified this "light" as a Malfunction Indicator Light (MIL). However, the light on the dash may still be identified with the term "Check Engine" or "Service Engine Soon" for ease of customer understanding.

OBD II guidelines set tight conditions for activating and de-activating the MIL (lamp). This strict set of guidelines has resulted in multiple "levels" of diagnostics with different criteria and conditions for *when* an emissions-related fault will cause the MIL to activate and set a code. Also, there are other codes available that will not cause the PCM to activate the MIL. The guidelines established how quickly the onboard diagnostics must be able to identify a fault, set the trouble code in memory and activate the MIL (lamp).

REGULATIONS FOR CLEARING CODES AND CONTROLLING THE MIL

There are strict regulations for conditions to turn off the light and to clear trouble codes. In the past, some vehicle manufacturers had the technician remove battery voltage from the computer to clear the codes. These new regulations contain significant changes in how and when the controller turns off the MIL. The vehicle must be driven under specific conditions while the emission systems are monitored. Once a fault is detected, the system or component that failed must pass three consecutive tests (three trips) without failing before the MIL will be turned off. OBD II regulations include:

- A standard to regulate how quickly a computer must identify a fault, activate the MIL and set a trouble code.
- A standard to regulate criteria that can turn off the MIL when a fault is not present.
- A standard that establishes how long a trouble code remains in the computer memory once the problem has been repaired and the code is cleared.
- A standard to regulate what information must be available from the vehicle manufacturer that the repair technician can use to assist them in identifying the cause of a fault (i.e., Scan Tool and Freeze Frame data).

FAULTS NOT RELATED TO EMISSION CONTROL SYSTEMS

On an OBD II system, the MIL is not activated unless the computer determines that a failure in a component or system will affect the emissions levels of the vehicle. In effect, this means that **only emissions-related faults (codes)** will cause the PCM to activate the MIL. Be aware that some driveability-related problems not related to emission control components or systems can cause a code to set without the PCM activating the MIL. However, any trouble codes associated with the fault will still be set in memory for a technician to access with an OBD II certified Scan Tool.

KEY POINTS

Just like with OBD I systems, there can be trouble codes without activating the MIL, and there can be failure conditions on some systems not related to emission controls that do not set a trouble code. However, on OBD II systems, when diagnosing any driveability or emissions-related problems, all codes are considered "hard" codes. You should first read and record the codes and related data, then make the repairs. Once these steps are done, you can clear the trouble codes and related Freeze Frame data.

UNDERSTANDING MIL CONDITIONS

The three (3) possible MIL conditions are explained next.

Condition 1: MIL Off

This condition indicates that the PCM has not detected any faults in an emission-related component or system, or that the MIL power or control circuit is not working properly.

Condition 2: MIL On Steady

This condition indicates a fault in an emission-related component or system that could increase tailpipe emissions.

Condition 3: MIL Flashing

This condition indicates either a misfire or fuel system related fault that could cause damage to a catalytic converter.

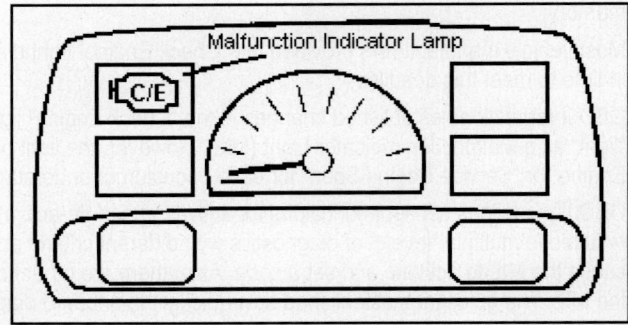

Note: *If a misfire condition exists with the MIL "on" steady, and the driver reaches a vehicle speed and load condition where the engine misfires at a level that could cause catalyst damage, the MIL will begin to flash. It will continue to flash until the engine speed and load conditions that caused that level of misfire subside. Then the MIL will return to the MIL "on" steady condition. This situation may result in a customer complaint as described next: "The MIL in my instrument cluster comes on and then flashes intermittently".*

ACTIONS OR CONDITIONS TO TURN OFF THE MIL

The PCM will turn off the MIL if any of the following actions or conditions occurs:

- The codes are cleared with a Generic or Proprietary Scan Tool
- Power to the PCM is removed (at the battery or with the PCM power fuse)
- A vehicle is driven on three consecutive trips **(including three warmup cycles)** and meets all of the particular code set conditions without the PCM detecting any faults

The PCM will set a code if a fault is detected that could cause tailpipe emissions to exceed 1.5 times the FTP Standard. However, the PCM will not de-activate the MIL until the vehicle has been driven on three consecutive trips with vehicle conditions similar to actual conditions present when the fault was detected. *This is not just three (3) vehicle startups and trips. It means three trips where certain engine operating conditions are met so that the OBD II Monitor that found the fault can "rerun" and pass that diagnostic test.*

Once the MIL is de-activated, the original code will remain in memory until forty warmup cycles are completed without the fault reappearing. A warmup cycle is defined as a trip where with an engine temperature change of at least 40°F, and where the engine temperature reaches at least 160°F.

SIMILAR CONDITIONS (FUEL TRIM AND MISFIRE CODES)

If a Fuel Control system (fuel trim) or misfire-related code is set, the vehicle must be driven under conditions similar to conditions present when the fault was detected before the PCM will de-activate the MIL (lamp). These "similar conditions: are described next:

- The vehicle must be driven with engine speed within 375 RPM of the engine speed stored in the Freeze Frame data when the code set
- The vehicle must be driven within engine load ± 10% of the engine load value stored in the Freeze Frame data when the code set
- The vehicle must be driven with engine temperature conditions similar to the temperature value stored in Freeze Frame data when the code set

Diagnostic Trouble Codes

INTRODUCTION

One of the key features in the OBD II system was an attempt to standardize the wording that describes a diagnostic trouble code or DTC (a term used to describe the method applied when the onboard controller recognizes and identifies a problem in one or more of the circuits or components that it monitors). As a point of review, keep in mind that the trouble codes used with OBD I systems consisted of codes identified with one (1), two (2), or three (3) digits. In effect, trouble codes were only identified with numbers.

DIAGNOSTIC TROUBLE CODES (5-DIGIT)

As previously discussed, SAE J2012 set standards for trouble codes and definitions for emission-related systems. OBD II trouble codes use a five-digit code, and these codes begin with a letter and are followed by four numbers. Since a letter is involved in the sequence, the correct way to read this type of code is with an OBD II certified Scan Tool.

The range of the code designations was designed to allow for future expansion and to allow for manufacturer specific usage on some systems. The illustration in the Graphic includes an explanation of OBD II Code Standardization for a DTC P0137.

UNIVERSAL CODE DESIGNATION EXPLANATION

The number in the thousandths position indicates that the trouble code is common to all manufacturers (a "P0" code).

Most vehicle manufacturers use this designation and then assign a common number and fault message to the problem. The code repair chart is not universal and service procedures will vary between the different vehicle manufacturers. However, the fault described in the code title is common to all systems on the vehicles (e.g., it was assigned a universal code designation). The first letter in the code identifies the system that controlled the device that failed (refer to the table below).

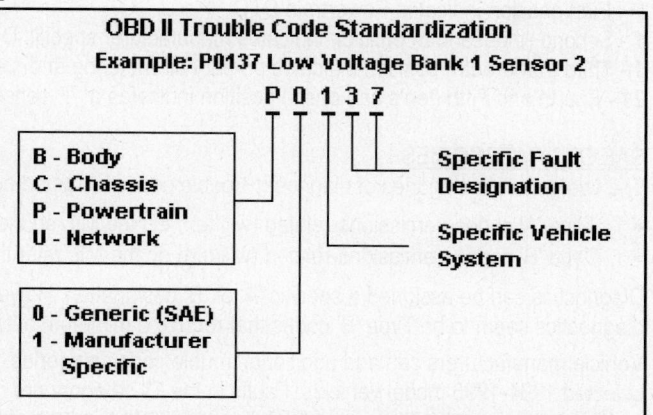

Code Description Table

System ID	System Description
B	Body Control System
C	Chassis Control System
P	Powertrain System
U	UART Data Link, Network Code

MANUFACTURER SPECIFIC DESIGNATION EXPLANATION

Vehicle manufacturers had some code conditions that are specific to the design of their individual system. Not all vehicle manufacturers have chosen to use P1xxx series codes due to differences in their basic systems, diagnostic strategy and their implementation.

These codes are designated as manufacturer specific codes (e.g., a "P1xxx" code), and each manufacturer can define the code and fault description for this designation. Although it was expected that each vehicle manufacturer would remain consistent across their product line, there has been considerable variation on P1xxx designations.

DTC NUMBERING EXPLANATION

The Number in the hundredth position indicates the specific vehicle system or subgroup that failed. This position should be consistent for P0xxx and P1xxx type trouble codes. An SAE committee established the numbers and systems listed below:

- **P0100** - Air Metering and Fuel System fault
- **P0200** - Fuel System (fuel injector only) fault
- **P0300** - Ignition System or Misfire fault
- **P0400** - Emissions Control System fault
- **P0500** - Idle Speed Control, Vehicle Speed Sensor fault
- **P0600** - Computer Output Circuit (relay, solenoid, etc.) fault
- **P0700** - Transaxle, Transmission faults

Note: The "ten's" and "one's" in the numbers indicate the part of the system at fault.

DTC NUMBERING EXAMPLE

DTC P1121 - GM Throttle Position Sensor Circuit Intermittent High Voltage:

P - First position indicates Powertrain DTC
1 - Second (thousandth) position indicates manufacturer specific DTC
1 - Third (hundredth) position indicates primary air metering and fuel system
21 - Fourth and Fifth (ten's and one's) position indicates a TP sensor fault

SAE DTC CATEGORIES

The two general categories of diagnostic trouble codes are listed below:

- Type 'A' codes - emissions-related (will turn On the MIL on the first failure)
- Type 'B' codes - emissions-related (will turn on the MIL after the second consecutive trip with a failure)

Diagnostics can be assigned a specific 'A' or 'B' designation. However, most vehicle manufacturer expanded diagnostics seem to be Type 'B' codes that require a minimum of two consecutive trips with a fault to activate the MIL.

Vehicle manufacturers can add additional trouble code categories. For example, GM has a 'D' category for a few selected 1994-1995 model vehicles. Faults in the 'D' category are non-emissions faults that will not cause tailpipe emissions to exceed 1.5 times the FTP standard. With this type of fault, the PCM does not activate any lamps or store any fault data in the Freeze Frame buffer (used for Type 'A' and Type 'B' faults).

COMMON CODE NAMES AND DESCRIPTIONS

OBD II guidelines set standards to universalize the Code Name and Description. These standards only apply to P0xxx codes. You need to be careful because there are several OBD II trouble codes where the same code number can have a different code title.

CODE VARIATION EXAMPLE

Note the use of the same code number for 2 different electrical faults in the table below.

DTC Number	Code Description & Conditions
P1641 (N/MIL) 1996-98 A, L, N & W Body: VIN M, X	**MIL Control Circuit Conditions** Key on, then the PCM received an improper voltage level on the MIL driver circuit (ODM 'A' output 1), condition met for 30 seconds. • Refer to the correct code repair chart.
P1641 (N/MIL) 1996 B Body: VIN P & W 1996 Y Body: VIN 5 & P	**Fan Control Relay 1 Control Circuit Conditions** No A/C or ECT codes set, engine speed over 600 rpm, then the PCM detected that the commanded state of the FC Relay 1 driver and Actual state did not match for 5 seconds. • Refer to the correct code repair chart.

DTC DESCRIPTOR DEFINITIONS

The SAE J2012 document further defines most circuit, component or system codes into the four basic categories explained next.

Circuit Malfunction - Indicates a fixed value or no response from the system. This descriptor can be used instead of a dual High/Low Voltage Code or used to indicate another failure mode.

Range/Performance - Indicates that the circuit is functional, but not operating normally. This descriptor may also indicate a stuck, erratic, intermittent or skewed value that could cause poor performance of an emission control circuit, component or system.

Low Input - Indicates that a signal circuit voltage, frequency or other measurement at a PCM input terminal is at or near zero. The test is made with the external circuit, component or system connected. The signal type is used in place of the word "input."

High Input - Indicates that a signal circuit voltage, frequency or other test measurement at a PCM input terminal is at or near full scale. This test is made with the external circuit, component, or system connected. Signal type is used in place of the word "input."

CONDITIONS TO CLEAR TROUBLE CODES

Diagnostic trouble codes are cleared from the PCM memory using several different methods (the actual method varies between vehicle manufacturers). An example of the Scan Tool navigation screens that appear on a 1999 Ford Windstar during a code clearing procedure is shown in the Graphic (Source: Snap On).

The list below contains a summary of a few of the methods that can be used to clear OBD II trouble codes. The actual conditions for each vehicle manufacturer must be determined and followed exactly.

- Regulations adopted with OBD II allow codes to be cleared by the PCM once 40 warmup cycles occur after the "last test failed" message clears and after 40 "last test passed" messages occur. Refer to Page 1-17 for the definition of a "warmup cycle".
- The Scan Tool can be used to clear any stored codes (and Freeze Frame data).
- On some vehicles, if battery voltage to the PCM is removed, the trouble codes, Freeze Frame data, "trip" or "drive cycle" status and I/M Readiness status will be lost. The battery voltage must be removed for 5 minutes or longer for this action to occur.

■ NOTE: *Do not clear the trouble codes unless the code repair chart diagnostic procedure instructs you to do so. Most manufacturers will clear Freeze Frame data (that can be used to diagnose the cause of the fault) at the same time a code is cleared. In effect, this step will result in the loss of the Freeze Frame data on most systems.*

Source: Snap On Scan Tool with a 1999 cartridge.

SCAN TOOL MENUS

SELECT 8th VIN CHAR.
VIN: --T--U--4----------
VEH: 1999 FORD VAN
ENG: 3.8L V6 EEC-V SEFI

SCROLL TO SELECT
THE SYSTEM:
~ENGINE & PCM
ABS
AIRBAG
GEM

SERVICE CODE MENU
KOEO SELF-TEST
~CLEAR CODES
MEMORY CODES

CLEAR CODES

THIS STEP WILL CLEAR
ALL TROUBLE CODES,
FREEZE FRAME DATA &
READINESS INFORMATION

ARE YOU SURE?
~ YES
NO

Diagnostic Routines

OBD I SYSTEM DIAGNOSTICS

One of the most important things to understand about the automotive repair industry is the fact that you have to continually learn new systems and new diagnostic routines (the test procedures designed to isolate a problem on a vehicle system). For OBD I and II systems, a diagnostic routine can be defined as a procedure (a series of steps) that you follow to find the cause of a problem, make a repair and then verify the problem is fixed.

CHANGES IN DIAGNOSTIC ROUTINES

In some cases, a new Engine Control system may be similar to an earlier system, but it can have more indepth control of vehicle emissions, input and output devices and it may include a diagnostic "monitor" embedded in the engine controller designed to run a thorough set of emission control system tests.

OBD I Diagnostic Flowchart

The OBD I Diagnostic Flowchart on this page can be used to find the cause of problems related to Engine Control system trouble codes or driveability symptoms detected on OBD I systems. It includes a step-by-step procedure to use to repair these systems. To compare this flowchart with the one used on OBD II systems, refer to the next page.

The steps in this flow chart should be followed as described below (from top to bottom).

- Do the Pre-Computer Checks.
- Check for any trouble codes stored in memory.
- Read the trouble codes - If trouble codes are set, record them and then clear the codes.
- Start the vehicle and see if the trouble code(s) reset. If they do, then use the correct trouble code repair chart to make the repair.
- If the codes do not reset, than the problem may be intermittent in nature. In this case, refer to the test steps used to find the cause of an intermittent fault (wiggle test).
- In no trouble codes are found at the initial check, then determine if a driveability symptom is present. If so, then refer to the approriate driveability symptom repair chart to make the repair. If the first symptom chart does not isolate the cause of the condition, then go on to another driveability symptom and follow that procedure to conclusion.
- If the problem is intermittent in nature, then refer to the special intermittent tests. Follow all available intermittent tests to determine the cause of this type of fault (usually an electrical connection problem).

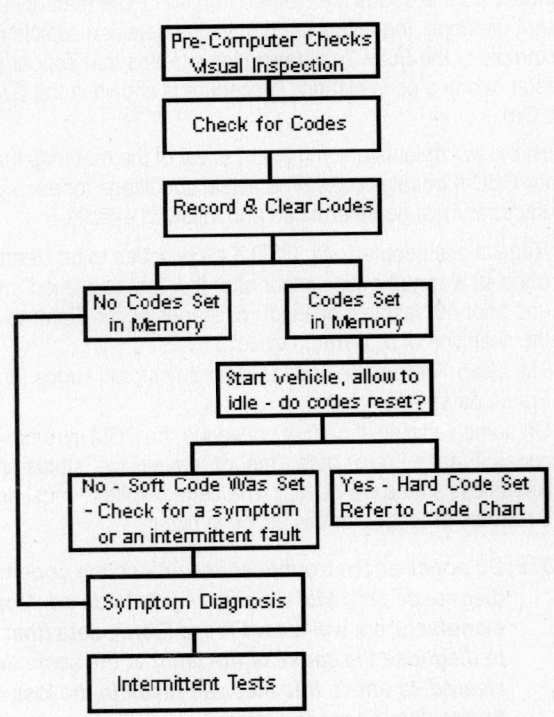

OBD I DIAGNOSTIC FLOWCHART

OBD II SYSTEM DIAGNOSTICS

The diagnostic approach used in OBD II systems is more complex than that of the one for OBD I systems. This complexity will effect how you approach diagnosing the vehicle. On an OBD II system, the onboard diagnostics will identify sensor faults (i.e., open, shorted or grounded circuits) as well as those that lose calibration. Another new test that arrived with OBD II is the rationality test (a test that checks whether the value for one input makes rational sense when compared against other sensor input values). The changes plus the use of OBD II Monitors have dramatically changed OBD II diagnostics.

The use of a repeatable test routine can help you quickly get to the root cause of a customer complaint, save diagnostic time and result in a higher percentage of properly repaired vehicles. You can use this Diagnostic Flow Chart to keep on track as you diagnose an Engine Control problem or a base engine fault on vehicles with OBD II.

OBD II Diagnostic Flowchart

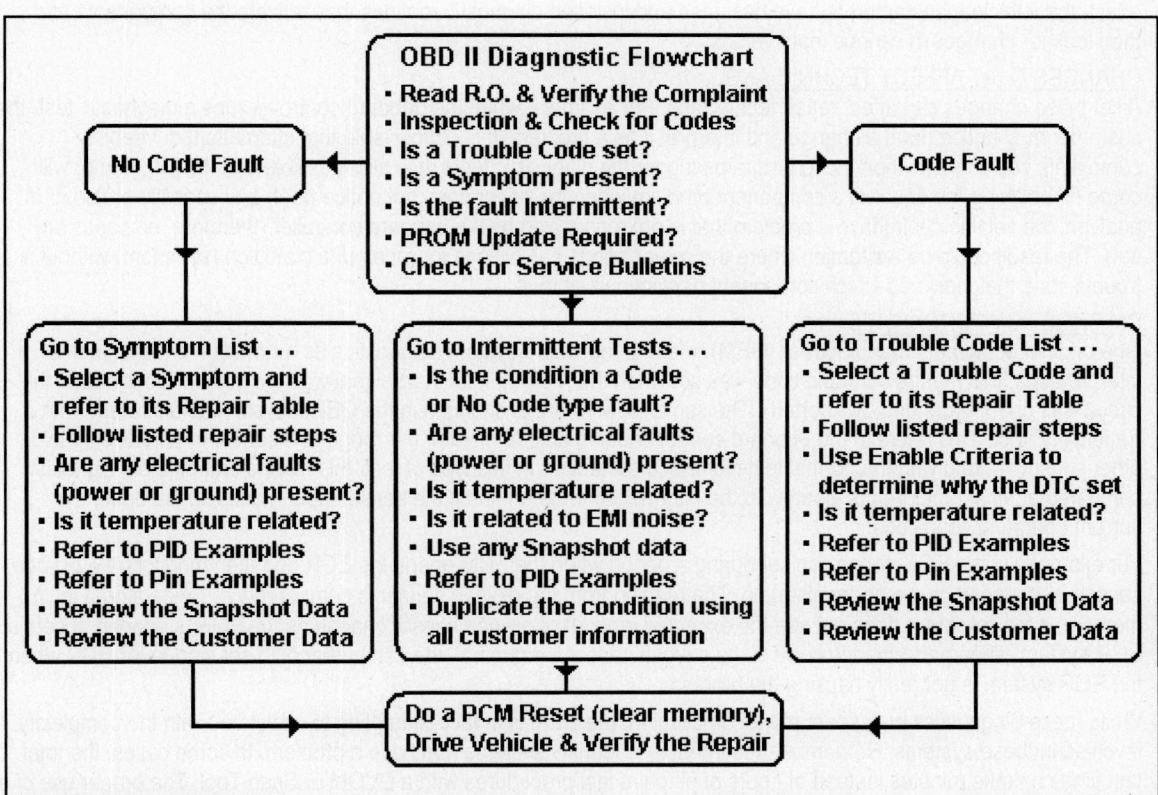

FLOW CHART STEPS

Here are some of the steps included in the Diagnostic Routine:

- Review the repair order and verify the customer complaint as described
- Perform a Visual Inspection of underhood or engine related items
- If the engine will not start, refer to No Start Tests
- If codes are set, refer to the trouble code list, select a code and use the repair chart
- If no codes are set, and a symptom is present, refer to the Symptom List
- Check for any related technical service bulletins (for both Code and No Code Faults)
- If the problem is intermittent in nature, refer to the special Intermittent Tests

Expanded Diagnostics

INTRODUCTION

The primary focus of OBD Expanded Diagnostics is to verify that all Emission Control systems continue to operate efficiently and that they do not deteriorate to a point where tailpipe emissions would increase to a point more than 1.5 times the FTP standard.

On a vehicle equipped with an OBD II system, all emissions faults must set diagnostic codes. However, instead of just identifying a circuit or component failure, this system will identify some problems where the component is deteriorating. These types of problems often occur before the vehicle driver notices a problem. If the fault is driveability related, the driver may not notice it while driving the vehicle. Some faults do not impact vehicle operation (and may not be noticed by the driver - they are not symptom-related faults).

All of these faults can be identified by an OBD II system when it monitors an emissions-related system and determines that it has deteriorated to a point where it would increase emissions beyond the FTP standard. Evaluation of these systems requires expanded programs that monitor all emissions-related components and systems for deterioration. In effect, the vehicle manufacturers have designed and installed diagnostic routines that activate the components and then look for changes in various input values.

CHANGES THAT AFFECT TECHNICIANS

All of these changes can affect repair technicians. For example, when the onboard controller runs a diagnostic test, the customer may notice (feel) a change and interpret it as a problem. It is conceivable that intermittent driveability complaints could result when these programs trigger the diagnostic tests. It is also possible that the MIL (lamp) will come on notifying the driver of a component problem when the driver does not notice (feel) a driveability problem. In addition, the vehicle could have a problem that is not recognized by the onboard controller (therefore, no codes are set). The result could be a situation where the owner brings a vehicle in for repair of a condition (symptom) without a trouble code that indicates which component or system is at fault.

DIAGNOSTIC APPROACHES

One original equipment manufacturer (OEM) refers to the OBD I system diagnostics as "normality" checks. In this interpretation, this means a trouble code was set when the "normal" electrical signal was too high or too low (i.e., the circuit was open, grounded, or shorted). The same OEM refers to changes in the OBD II system as a strategy with a "rationality" test. This refers to the onboard controller being able to monitor the range of a sensor in relationship to other sensors (along with the normal tests for electrical faults). In this type of test, this means that an input signal is compared against other inputs along with the information to determine if the sensor input makes sense under the current operating conditions.

For example, some EGR codes can set during a period when the PCM opens the EGR and then monitors the oxygen sensor response at cruise or deceleration. The reading from the oxygen sensor is compared to a range stored in memory. If the computer does not see the expected amount of oxygen sensor change as the EGR is opened, it sets an EGR system range/performance code. If the oxygen sensor is marginal, the computer could set a code for EGR when the EGR system is not really causing the problem.

While these diagnostics may seem more complicated, the OEM is in fact attempting to assist you with the complexity involved in these systems. Remember, the intent is to identify systems that have a problem. In some cases, the total test time can take minutes instead of hours of pinpoint test procedures with a DVOM or Scan Tool. The proper use of a DVOM, Scan Tool and Lab Scope must be understood to work on OBD II systems.

Expanded Diagnostics

SCAN TOOL INFORMATION

CARB regulations require that vehicle manufacturers make available to repair technicians procedures which allow effective emission related diagnostic and repair using a Generic Scan Tool. This regulation was developed into SAE J1978 that sets guidelines for a common Scan Tool to access On-Board information.

The actual information shown on a Scan Tool can vary between different vehicle systems. Each manufacturer emphasized certain programs and then displayed the information in their own format. SAE J2205 sets the standard for how the Scan Tool will interface with the computer and access the computer information. This standard was necessary because computers are interfacing in bi-directional formats on these vehicles.

There is a difference in the amount of information available on each brand of Scan Tool. Review how this information is accessed on your OBD II certified Scan Tool. The Scan Tool gets its power from the vehicle being tested and talks or "interfaces" with the vehicle diagnostic system or the diagnostic executive program.

TROUBLE CODE INFORMATION

The trouble code information on a Scan Tool includes:

- Current and History trouble codes
- MIL Requested Information ("MIL ON" data)
- Diagnostic test status (test run/test pass or fail)
- Last test pass or fail message
- Freeze Frame data for the 1st emission fault
- Some Scan Tools can display Failure Records

FREEZE FRAME INFORMATION

CARB and EPA regulations require that the controller store specific Freeze Frame (engine related) data when the first emission related fault is detected. The data stored in Freeze Frame can only be replaced by data from a trouble code with a higher priority (i.e., a trouble related to a Fuel system or Misfire Monitor fault).

The Freeze Frame has to contain data values that occurred at the time the code was set (these values are provided in standard units of measurements). As a result, OBD II systems record the data present at the time an emission related code is recorded and the MIL activated. This data can be accessed and displayed on a Scan Tool. Freeze Frame data is one frame or one instant in time. It records the data that set the code.

REQUIRED FREEZE FRAME DATA ITEMS

- Calculated load value, Engine Speed (rpm), Short and Long Term fuel trim values
- Fuel system pressure value (where applicable)
- Vehicle speed (MPH) & Closed / Open Loop status
- Engine coolant temperature and Intake manifold pressure
- Trouble Code that triggered the Freeze Frame
- If misfire code set - identify which cylinder is misfiring

```
( MAIN MENU (CHRYSLER) )

Press:
  OTHER SYSTEMS
  FUNCTIONAL TESTS
  CODES & DATA MENU
  CUSTOM SETUP
~ SYSTEM TESTS


( SYSTEM TESTS )

Press:
SYSTEM TESTS:
  EGR SYSTEMS TEST
  GENERATOR FIELD TEST
  INJ. KILL TEST
  MISFIRE COUNTERS
~ PURGE VAPORS TEST
  READ VIN


( PURGE VAPORS TEST )

Press:
* Y TO SWITCH BETWEEN
  NORM, FLOW & BLOCK
  PURGE STATUS___NORM
  ENGINE RPM_____736
  NO CODES IN THIS MODE
  UPSTRM O2S(V)____0.63
  DWNSTRM O2S(V)__0.14
  PURGE(mA)_____120
  ST ADAP(%)_____1.4
```

Expanded Diagnostics

TROUBLE CODE "TEST CONDITIONS"

Some vehicle emission control components and systems are "continuously" monitored by the Comprehensive Component, Fuel System and Misfire Monitors while some of the OBD II Main Monitors only run their diagnostic tests after certain test conditions or enable criteria have been met (e.g., the EGR system and EVAP system Monitors).

Key Point - Certain code "test conditions" must be met to "run" certain Monitors, and the conditions vary by vehicle and engine configuration. Also, the information related to each trouble code contains the actual conditions present when that particular code set.

INTRODUCTION CONTENTS

Chrysler Motors Systems

Chrysler Vehicle Performance

ABOUT THIS MANUAL

This manual was developed to provide you with information that explains the theory of operation, diagnosis and repair of various control systems on several late model Daimler/Chrysler Motors vehicles.

ASE Test Help
The information in this manual can be a valuable asset as you prepare to take the ASE A-8 or L-1 tests. It can also expand on your previous repair knowledge if you have already passed both of these certification tests.

Main Features
The main features of this product include an Introduction to Vehicles Performance as well as the individual sections that explain how to use Chrysler vehicle diagnostics along with a DVOM, Lab Scope and Scan Tool to test the Powertrain, Transmission, and Body Control Module input and output signals used on Daimler/Chrysler vehicle applications.

SECTION OVERVIEWS

Introduction Section - This section contains separate articles on these subjects: Engine Controls, Vehicle Identification, Problem Solving, Serial Data Communication, Flash Reprogramming, Powertrain Diagnostics, Chrysler DTC Test and OBD Systems.

Car Section - This section contains information specific to a 2002 Sebring including these subjects: Anti-Theft, Sentry Key Immobilizer, Body Controls, Electronic Ignition, EVAP system, Fuel Delivery system, Idle Air Control Motor, MAP, Oxygen Sensor, Transmission Controls, PID Data, Pin Voltage Tables and Wiring Diagram examples.

Ram Pickup Section - This section contains information specific to a 1998 Ram 1500 including these subjects: Electronic Ignition, EVAP system, Fuel Delivery system, Idle Air Control Motor, Speed Control, MAP, TPS, ECT, IAT, Oil Pressure Sensor, Oxygen Sensors, Transmission Controls, PID Data, Pin Voltage and Wiring Diagram examples.

Minivan Section - This section contains information specific to a 1999 Town & Country including these subjects: Charging System, Electronic Ignition, EGR & EVAP systems, Fuel Delivery system, Speed Control, Idle Air Control Motor, MAP, VSS, Oxygen Sensor, Body Control Module, PID Data, Pin Voltage Tables and Wiring Diagram examples.

SUV Section - This section contains information specific to a 1999 Jeep Cherokee and a 2002 Durango including these subjects: New Generation Controller (NGC), Electronic Ignition, EVAP systems, Fuel Delivery system, PCM controlled cooling fan and generator, Speed Control, Idle Air Control Motor, TPS, MAP, Oxygen Sensor, Transmission Controls, PID Data, Pin Voltage Tables and Wiring Diagram examples.

Diagnostic Help
Sections 3 through 6 contain articles that include "real world" test examples and results that can be used with a DVOM, Lab Scope or Scan Tool. This *vehicle specific information* can be very helpful during actual repair work and vehicle testing. The example repair tables developed for this manual combine theory with a practical hands-on testing approach.

Computerized Engine Controls

INTRODUCTION

The evolution of various Computerized Engine Controls (CEC) used on Daimler/Chrysler vehicles from 1983-2002 is shown in the Graphic below. To read about the fuel injection system and diagnostics for a particular vehicle, refer to one of the four vehicle sections.

CEC Evolution Graphic

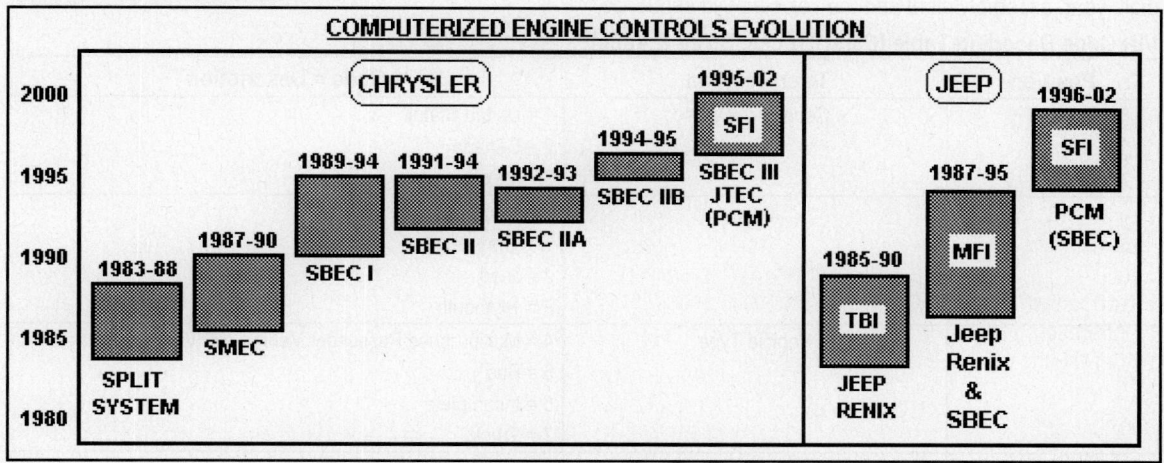

ELECTRONIC FUEL INJECTION

Daimler/Chrysler Vehicle Applications

SPLIT SYSTEM - The Split System (Logic Module and Power Module) was used on 1983-87 vehicles with a 2.2L I4 engine and high-pressure throttle body injection (TBI).

In 1984, the Split System was introduced on the 2.2L I4 turbocharged engine. This engine and fuel system was used through model year 1988.

SMEC - The Single Module Engine Controller (SMEC) system was used on 1988-89 models with 2.2L and 2.5L I4 TBI, and 1987½ -90 vehicles with a 3.0L V6 MFI engine.

SBEC I, IIA & IIB - The Single Board Engine Controller (SBEC) system was used on 1989-95 models equipped with both throttle body and multiport fuel injection. These systems were identified as IIA, IIB and IIC and were equipped with OBD I diagnostics.

SBEC III - The Single Board Engine Controller III (SBEC III) system was used on 1996-2002 vehicles with multiport fuel injection. These vehicles have OBD II systems.

Jeep Fuel Injection Systems

JEEP RENIX (TBI) - The first computerized engine control system was the Jeep Renix system used on 1985-90 models with 2.5L I4 engines (TBI).

JEEP RENIX (MFI) - The second computerized engine control system was the Jeep Renix system used on 1987-90 models with 4.0L I6 engines (MFI).

SBEC (MFI) - The third computerized engine control system was the Single Board Engine Controller (SBEC) system used on 1991-95 models with 2.5L, 4.0L and 5.2L engines (MFI). 1993-95 Grand Cherokee and Grand Wagoneer used the 5.2L engine.

SBEC III - The fourth computerized engine control system was the Single Board Engine Controller (SBEC) system used on 1996-2002 models with the 2.5L, 4.0L, 5.2L and 5.9L SFI engines. This system was designed as an OBD II system.

Vehicle Identification

VEHICLE IDENTIFICATION NUMBER

The vehicle identification number (VIN) is a seventeen digit legal identifier of the vehicle. It is located on a plate attached to the upper left corner of the instrument panel near the left windshield pillar. It can be seen through the windshield from outside the vehicle.

The VIN includes the country of origin, make, vehicle type, passenger safety equipment, car line, body, engine, check digit, year, assembly plant and the vehicle build sequence.

VIN Code Decoding Table (Cars, Trucks, SUVs & Vans)

Position	Interpretation	Code = Description
1	Country of Origin	1 = United States 2 = Canada 3 = Mexico
2	Make	B = Dodge C = Chrysler J = Jeep P = Plymouth
3	Vehicle Type	4 = Multipurpose Passenger Vehicle (MPV) 5 = Bus 6 = Incomplete 7 = Truck
4	Gross Vehicle Weight Rating	F = 4001-5000 H = 6001-7000 J = 7001-8000 K = 8001-9000 L = 9001-10,000 M = 10,001-14,000 W = Hydraulic Brakes
4	Passenger Safety (Neon)	E = Active Restraints, Driver & Passenger Airbags
5	Car or Vehicle Line	B = Ram Wagon/Ram Van C = Ram Cab Chassis/Ram Pickup 4x2 F = Ram Cab Chassis/Ram Pickup 4x4
6	Series	1 = 1500 2 = 2500 3 = 3500
7	Body Style	1 = Van 2 = Club Cab 3 = Quad Cab 5 = Wagon 6 = Conventional Cab/Cab Chassis 7 = 2DR Sport Utility 8 = 4DR Sport Utility
8	Engine	T = 5.2L V8 CNG W = 8.0L V10 MFI X = 3.9L V6 MFI Y = 5.2L V8 MFI Z = 5.9L V8 MFI - LDC N - 4.7L V8 MFI 2 = 5.2L V8 LPG 5 = 5.9L V8 MFI - HDC 6 = 5.9L V6 Diesel 7 = 5.9L V6 Diesel
9	Check Digit	-
10	Model Year	T ='96, V='97, W='98, X='99, Y='00, 1-'01, 2='02
11	Plant Location	D = Belvidere G = Saltillo J = St Louis North K = Pillette Rd Windsor L = Toledo #1 L = Lago Alberto S = Dodge City T = Toluca
12 through 17	Vehicle Build Sequence	- - - - - - -

VIN Code Graphic

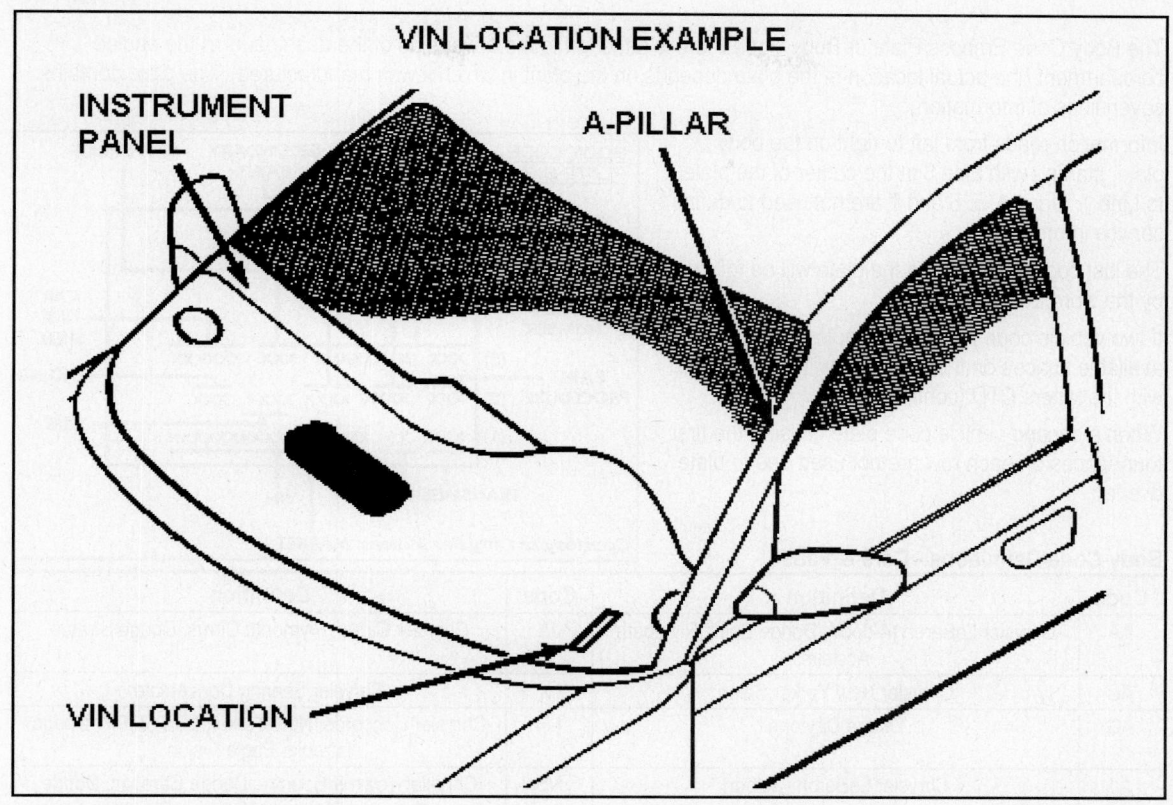

VEHICLE SAFETY CERTIFICATION LABEL

The vehicle safety certification label is attached to the driver side door on the side facing the B-pillar.

This label indicates the following:

- Date of Manufacturer (month and year)
- Gross vehicle weight rating
- Front and rear gross axle weight rating (GAWR)
- Vehicle identification number

This label also includes the month, day and hour of manufacturer.

Any communications or inquiries to Daimler/Chrysler regarding a particular vehicle should include the Month-Day-Hour and VIN information for this label.

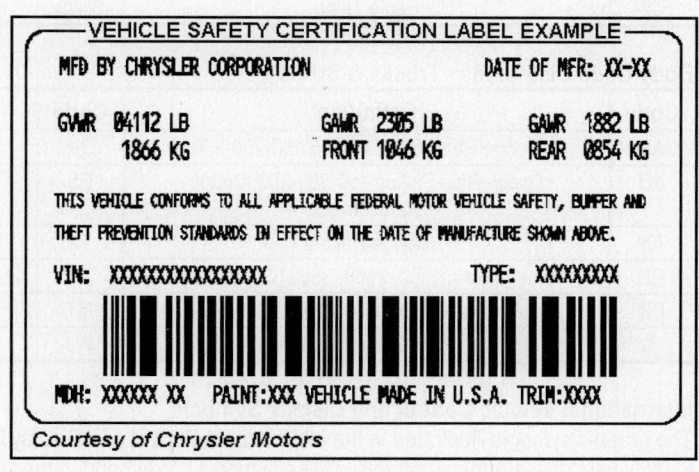

BODY CODE PLATE

The Body Code Emboss/Plate or Body Code Plate is attached to the driver side of the dash panel in the engine compartment (the actual location of the plate depends on the plant in which it was manufactured). The plate contains seven lines of information.

Information reads from left to right on the body plate, starting with Line 3 in the center of the plate to Line 1. Lines 4, 5, 6 and 7 are not used to define service information.

The last code imprinted on the plate will be followed by the word END.

If two vehicle code plates are required, the last available spaces on the first plate will be imprinted with the letters CTD (continued).

When a second vehicle code plate is used, the first four spaces on each row are not used due to plate overlap.

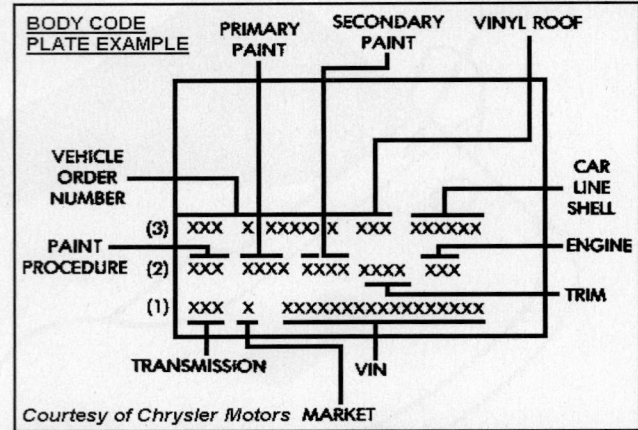

Courtesy of Chrysler Motors

Body Code Definitions - Cars & Vans

Code	Definition	Code	Definition
AA	Chrysler LeBaron (4-door), Dodge Spirit, Plymouth Acclaim	JA	Chrysler Cirrus, Plymouth Cirrus, Dodge Stratus
AC	Chrysler New Yorker Salon	JX	Chrysler Sebring Convertible
AG	Dodge Daytona	LH	Chrysler Concorde, New Yorker, LHS, 300M; Dodge Intrepid, Eagle Vision
AJ	Chrysler LeBaron (2-door)	NS	Chrysler Town & Country, Dodge Caravan, Dodge Grand Caravan, Plymouth Voyager, Plymouth Grand Voyager
AP	Dodge Shadow, Plymouth Sundance	PL	Dodge Neon, Plymouth Neon
AY	Chrysler Imperial & Fifth Avenue	PR	Prowler
FJ22	Chrysler Sebring, Dodge Avenger	SR27	Dodge Viper (Roadster)
FJ24	Eagle Talon	SR29	Dodge Viper (Coupe)

Body Code Definitions - Trucks & SUV's

Code	Definition	Code	Definition
AB	Dodge Ram Van & Wagon 150-250-350	DN	Dodge Durango
AD	Dodge Ram Pickup 150-250-350, Dodge Ramcharger	ES	Chrysler Grand Voyager & Grand Ram
AN	Dodge Dakota	TJ	Jeep Wrangler
BE	Dodge Ram 1500, 2500, 3500 (Quad Cab)	XJ	Jeep Cherokee
BR	Dodge Ram 1500, 2500, 3500	YJ	Jeep Wrangler
T	Dodge Ram Cab 1500, 2500, 3500	WJ	Jeep Grand Cherokee

International Vehicle Control and Display Symbols

The graphic symbols illustrated in the International Control and Display Symbols graphic on this page are used to identify various instrument controls. These symbols correspond to the controls and displays that are located on the instrument panel.

VECI LABEL

The underhood Vehicle Emissions Control Information (VECI) Label combines both emission control information and vacuum hose routing.

VECI labels are permanently attached and cannot be removed without defacing or destroying the information.

The VECI label is located in the engine compartment in various locations (e.g., it is in front of the radiator on 3.9L and 5.2L engines).

The label contains this information:

- Engine family and displacement
- Evaporative family
- Emission control schematic
- Certification application
- Engine timing specification
- Idle speed
- Spark plug type and gap

The 5.9L HDC-gas powered engine has two VECI labels. One label is located in front of the radiator and contains the vacuum hose routing. The other VECI label is attached to the air cleaner (driver's side) and contains the following information:

- Engine family and displacement
- Evaporative family
- Certification information
- Engine timing specification (if applicable)
- Idle speed (if applicable)
- Spark plug type and gap

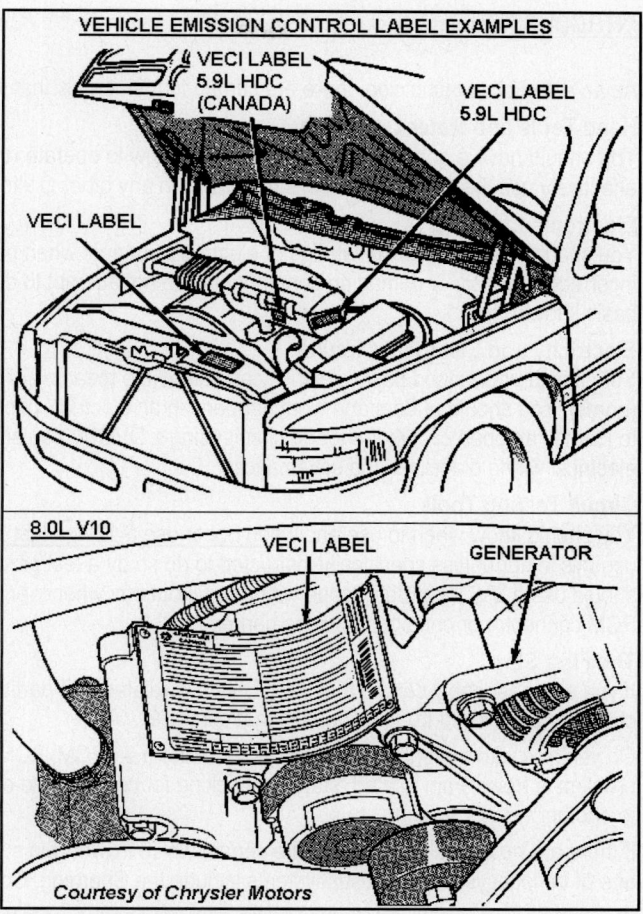

Courtesy of Chrysler Motors

The VECI label for the 8.0L V10 HDC-gas powered engine is also located in the engine compartment. It is attached to a riveted metal plate located to the right side of the generator.

Canadian Vehicles

There are unique labels for vehicles built for sale in the country of Canada. They include engines built for light duty cycle (LDC) and heavy duty cycle (HDC) use. Canadian labels are written in both the English and French languages. The VECI label is split into two different labels for all Canadian vehicles.

Technician Requirements

INTRODUCTION

As an automotive technician, there are certain fundamentals that you should understand:

Hand Tools and Meter Operation
You should have a good basic understanding of how to operate required hand tools and test meters in order to effectively use the information in this manual or in any other Chilton service repair manual or electronic repair media.

Electronic Controls
You should have a basic knowledge of electronic controls when performing test procedures to keep from making an incorrect diagnosis or damaging components. Do not attempt to diagnose an electronic control problem without this basic knowledge!

Electricity and Electrical Circuits
You should understand basic electricity and know the meaning of voltage (volts), current (amps), and resistance (ohms). You should understand what happens in an electrical circuit when it is open or shorted, and you should be able to identify an open circuit or shorted circuit using a DVOM. You should also be able to read and understand automotive electrical wiring diagrams and schematics.

Circuit Testing Tools
You should know when to use and <u>when not to use a 12-volt test light</u> during diagnosis of electronic controls (do not use this tester unless specifically instructed to do so by a test procedure). Instead of using a 12-volt test light, you should use a DVOM or Lab Scope with a breakout box whenever a diagnostic procedure calls for a measurement at a PCM connector or component wiring harness.

The First Step
If you are reasonably certain that the problem is related to a particular electronic control system, the first step is to check for any stored trouble codes in that controller.

On vehicles with more than one vehicle controller (i.e., PCM, BCM, MIC, TCM, etc.), if you are unsure whether the problem is Powertrain related, start by checking for codes in the other controllers to determine if the problem is related to another vehicle system.

If there are no codes set, and you are certain which Powertrain subsystem has a problem, you can start by checking one of the subsystems. The subsystems include the Charging, Cooling, Fuel, Ignition and Speed Control systems.

If a wiring problem is found during testing, you will need to refer to wiring diagrams in this manual, other manuals or electronic media. This may include:

- Wiring schematics (including circuit numbers and colors)
- Electrical component connector, splice and ground locations
- Wiring repair procedures and wiring repair parts information

The Last Step
Once you decide how to repair the vehicle, in addition to performing the repair, it is a good idea to clear any trouble codes that were set and to verify they do not reset.

An explanation of how to use the PCM Reset Step to clear codes is included in this section. To verify a repair, you should verify that the Check Engine Light is operational and goes out after the 4-second key-on bulb check. Then you need to duplicate the conditions present when the customer complaint occurred or when a trouble code set; these are the actual code conditions that caused a code to set. The individual code conditions are included later in this manual. You can use this information to find out how to drive a vehicle for repair verification.

Serial Data Communication

SCI RECEIVE & TRANSMIT CIRCUITS

The SCI Receive and Transmit circuits make up a communications network through which the PCM communicates with a Scan Tool. The SCI Receive (circuit) is the serial data communication "receive" circuit for OEM and Aftermarket Scan Tools. The PCM "receives" data from the Scan Tool via this circuit.

The SCI Transmit (circuit) is the serial data communication "transmit" circuit for OEM and Aftermarket Scan Tools. The PCM "transmits" data to the Scan Tool via this circuit.

Serial data refers to information transferred in a linear fashion over a single line, one bit at a time. During actual communication, the serial data captured from the Chrysler SCI transmit and receive circuits will appear similar to the examples in the Graphic below.

SCI Serial Data Examples (1999 Chrysler Town & Country)

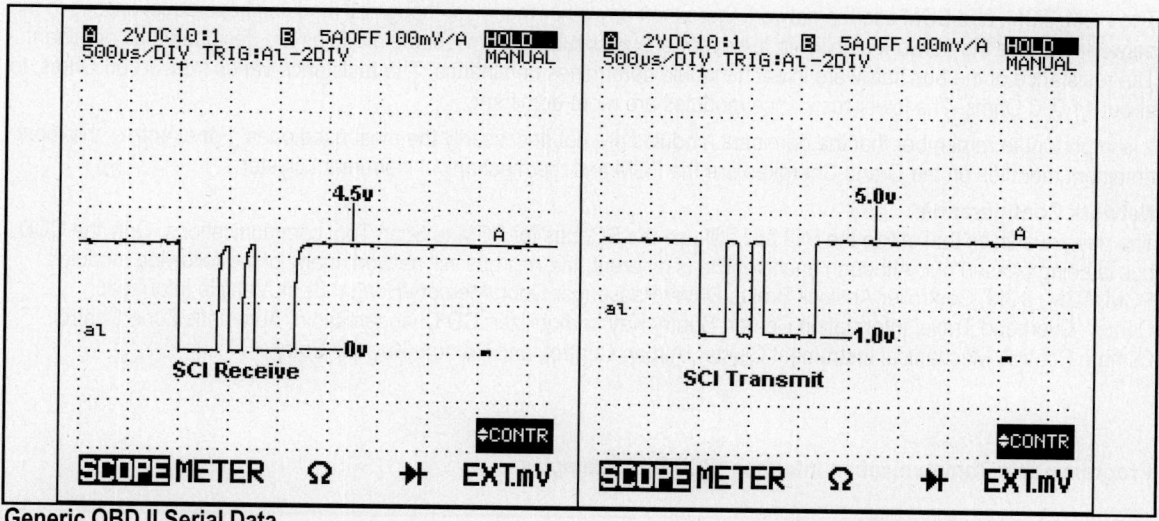

Generic OBD II Serial Data

The SCI data bus is also used by the PCM to transmit Generic OBD II data. When the PCM is transmitting information (data, codes, etc) to a Generic Scan Tool, it switches from the normal 0-5v square wave signal to a 0-12v square wave signal.

CCD BUS CIRCUITS

The Chrysler Collision Detection (CCD) bus is a network that allows communications between various vehicle modules and the DLC. The PCM exchanges MIL On/Off, engine speed and vehicle load information with various controllers and modules through this network. The CCD Bus utilizes a "Class 2" communications system where each bit of information can have one of two lengths: long or short. This allows multiple signals over a one wire. The network consists of a twisted pair of wires (CCD+ and CCD-) that connect to each module on the network.

Messages carried on the Class 2 data bus are also prioritized. If two messages attempt to establish communications on the data line at the same time, only the message with higher priority will continue. The device with the lower priority message must wait.

PCI BUS CIRCUITS

The Programmable Communication Interface (PCI) bus is a communications network that replaces the CCD bus network. The PCI bus was introduced on LH body vehicles, and has now expanded to most of the Chrysler line of vehicles.

Like other networks, the PCI bus uses variable pulsewidth signals to communicate digital messages between control modules. Unlike the CCD bus, the PCI network uses only 1 wire. When communicating, the signal toggles from near 0v to near 7.5v.

PCI Features
The PCI bus is a simpler network because it contains fewer wires, and does not rely on the twisting of the wires for shielding. Each module on the network supplies voltage and termination, which allows the rest of the network to continue functioning even if there is an open circuit leading to one module. Shorts to ground or voltage will cause the entire network to stop communicating. The Scan Tool communicates with the PCM on the SCI bus, so the PCI bus cannot cause Scan Tool communication faults with the PCM.

Dominant Modules
The PCM, TCM, and BCM use the network more than any other modules, though all modules have equal access to the network. To give the network stability in the event of bus circuit problems, some modules are designated as dominant. The resistance of the bus hardware in each module determines dominance. The resistance varies from 1,000 Ohms, to about 11,000 Ohms. The lower resistance modules are more dominant.

It is important to remember that the dominant modules are not necessarily the most used ones. For example, the most dominant modules on the Grand Cherokee are the PCM and the mechanical instrument cluster.

Network Configuration
The newer vehicles that utilize the PCI bus still use the SCI bus for PCM to Scan Tool communications. Only the CCD bus is being phased out. Although each vehicle is different, the PCI bus will network many of the following modules: PCM, TCM, BCM, Controller Antilock Brake, Driver/Passenger Door, Memory/Heated Seat, Vehicle Information Center, Overhead Travel Information Center, Sentry Key Immobilizer, CD Changer/Radio, Automatic Zone Control, Climate Control, Mechanical Instrument Cluster, Airbag Control, and Remote Keyless Entry.

Programmable Communication Interface (PCI) Bus Examples

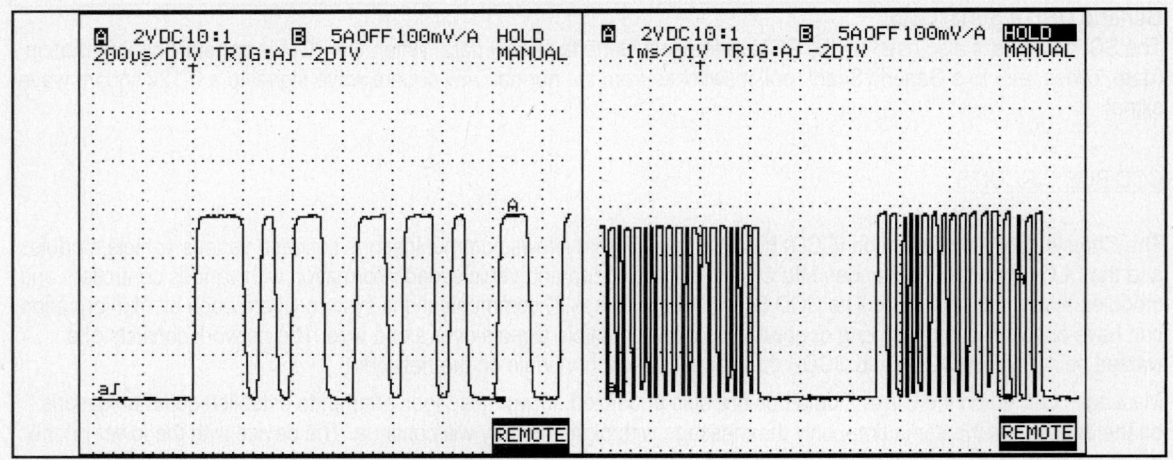

SCAN TOOL COMMUNICATION

For the Scan Tool to communicate with various vehicle stand-alone modules, it must be connected into the system. On OBD I systems, connect the Scan Tool to the underhood test connector. On OBD II systems, connect the Scan Tool to the DLC under the dash.

To access the PCM, connect the Scan Tool to the appropriate DLC. Turn the ignition on to allow the vehicle to identify itself to the Scan Tool. Once communication is established with the PCM, select a function (codes, data, actuator tests, system tests, etc) from the tool main menu. The Scan Tool performs these functions by communicating with the vehicle modules through the data bus circuits.

To read trouble codes without a Scan Tool, cycle the key from "off" to "on" three times within five (5) seconds. This test can establish whether the PCM is capable of responding to a request. It can also be used if the Scan Tool will not communicate.

The Scan Tool "talks" to the vehicle controllers on separate networks. The tool communicates with the PCM via the SCI Transmit and Receive circuits, and with other modules through the CCD or PCI data bus circuits.

OBD I SCI Schematic (1983-95 Vehicles)

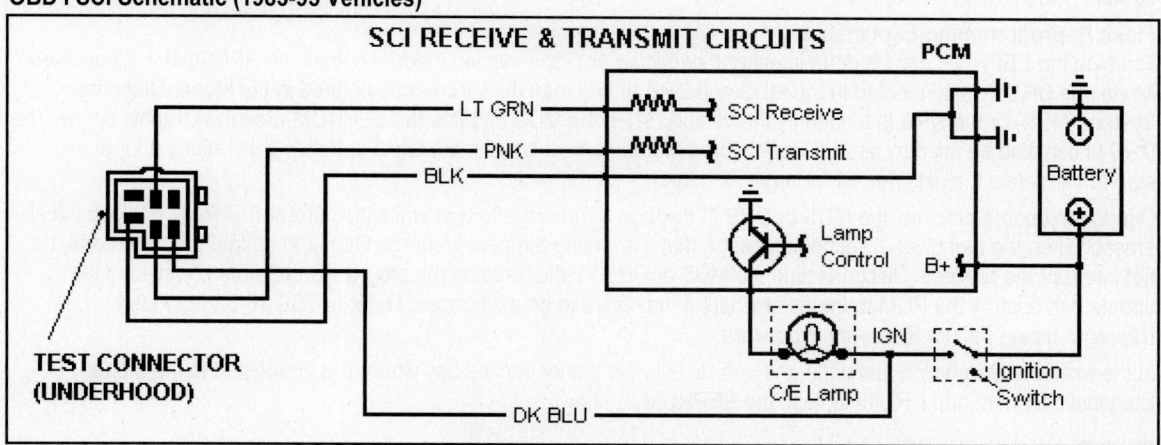

OBD II DLC Schematic (1999 LH Body)

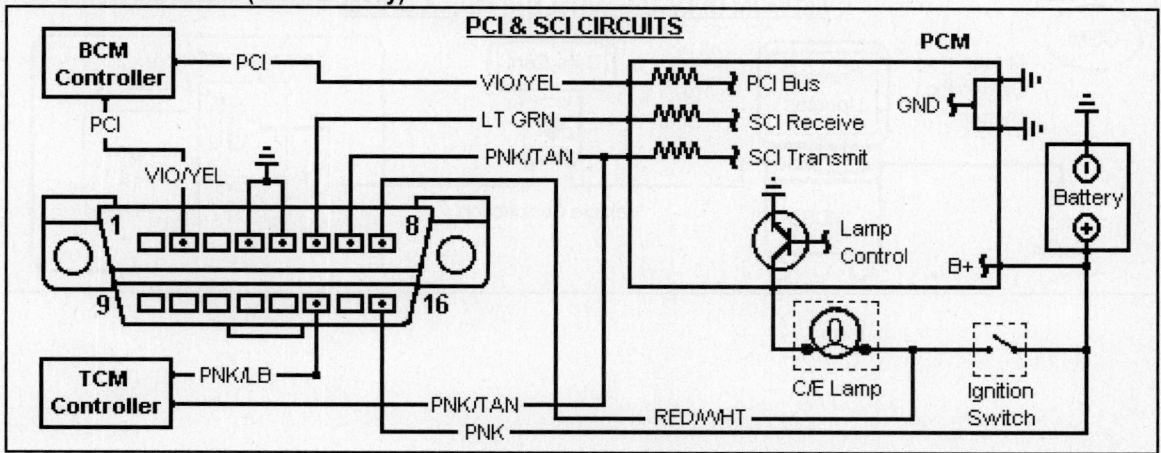

FLASH REPROGRAMMING

Introduction

Many factors influence vehicle emissions, fuel economy, and operating conditions. There are thousands of different combinations of engines, transmissions, accessories, load requirements, and climate considerations that must be included in vehicle controller programs. To account for all of these factors, Chrysler and Jeep vehicles use a PCM with a reprogrammable PROM with calibrations for various engine-operating parameters.

Vehicle operation is a compromise between maximum performance, fuel economy, and emissions control. The PROM is a critical link in the balancing of these systems. The amount of engine load at idle speed from the generator, air conditioning clutch, power steering pump, and other accessories can influence total performance. Tire size, gear ratios, and transmission shift points also influence vehicle operation. Service engineers for the vehicle manufacturers are constantly compiling and evaluating information on revised specifications and new diagnostic information in order to generate an updated program for maximum vehicle performance. Also, vehicle re-calibration may be necessary if a customer installs oversize tires or rims, or changes the gear ratios.

The reprogrammable PROM is called an Electronically Erasable PROM (EEPROM). The actual act of "flashing" the EEPROM requires the latest Chrysler or Jeep information and the use of either a Scan Tool or Mopar Diagnostic System (MDS) at the Dealership.

Flash Reprogramming Explanation

Flashing the EEPROM can be done in either *Connected* or *Disconnected* mode. To flash the EEPROM in *Connected Mode*, the DRB III is connected to the MDS and DRB III and then the VIN can be entered in the Mopar Diagnostic System (MDS) or the DRB III to identify the vehicle. Then the MDS updates the EEPROM. See the Graphic below. The DRB III can also be used by itself *(Disconnected Mode)*. It must contain the latest software calibrations prior to this step (it can retain the calibrations for up to 72 hours).

During the update process, the MDS or DRB III displays a message to wait while the current EEPROM is electronically erased. Then the tool gives a second message that it is writing the new program. Once the update process starts, do not interrupt the process. Disconnecting the MDS or DRB III after erasing the program and before completing the update can destroy the PCM to the point where it may have to be exchanged. Refer to TSB 18-32-98: Flash Reprogramming Failure Recovery Procedure.

In the example Graphic on this page, the vehicle is in the dealer service bay where it is connected to the Mopar Diagnostic System and DRB III to flash the EEPROM.

Flash Reprogramming Graphic

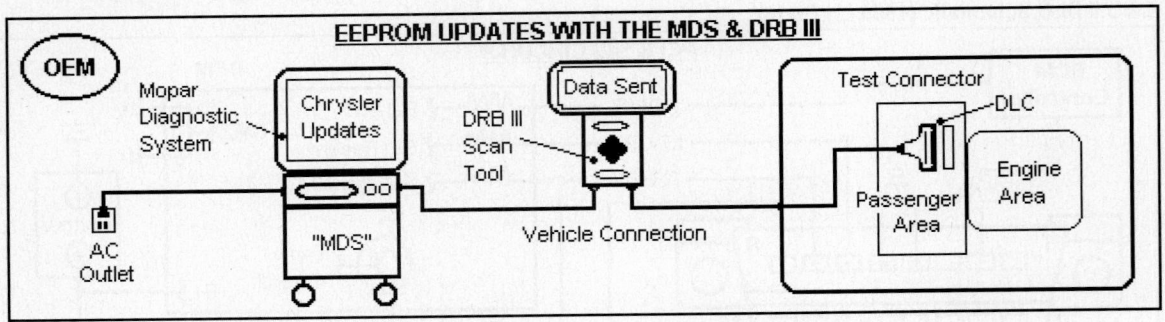

Aftermarket Reprogramming

There is currently only one Aftermarket Tool provider (SPX/OTC) that has the capability to perform Flash Reprogramming of Chrysler vehicles. The connection method is very simple and updates to the program are done through the OTC bulletin board service (BBS). This update program is then performed through the Diagnostic and Reprogram Tool (DART) provided by OTC to all service and repair facilities.

The OEM Scan Tool, the DRB III, is available to the aftermarket, along with reflashing capabilities.

Chrysler Alliance with SPX/OTC

An Aftermarket update program is used with the DART Tool in order to identify and reprogram the vehicle calibration or software. A calibration file is stored in programmable memory locations of the appropriate controller or module. In some cases, the EEPROM calibration is updated to provide a solution for emissions and drivability problems. The OTC Sales Phone Number is 800-533-6127. The Fax Phone Number is 800-283-8665.

DART Connection Graphic

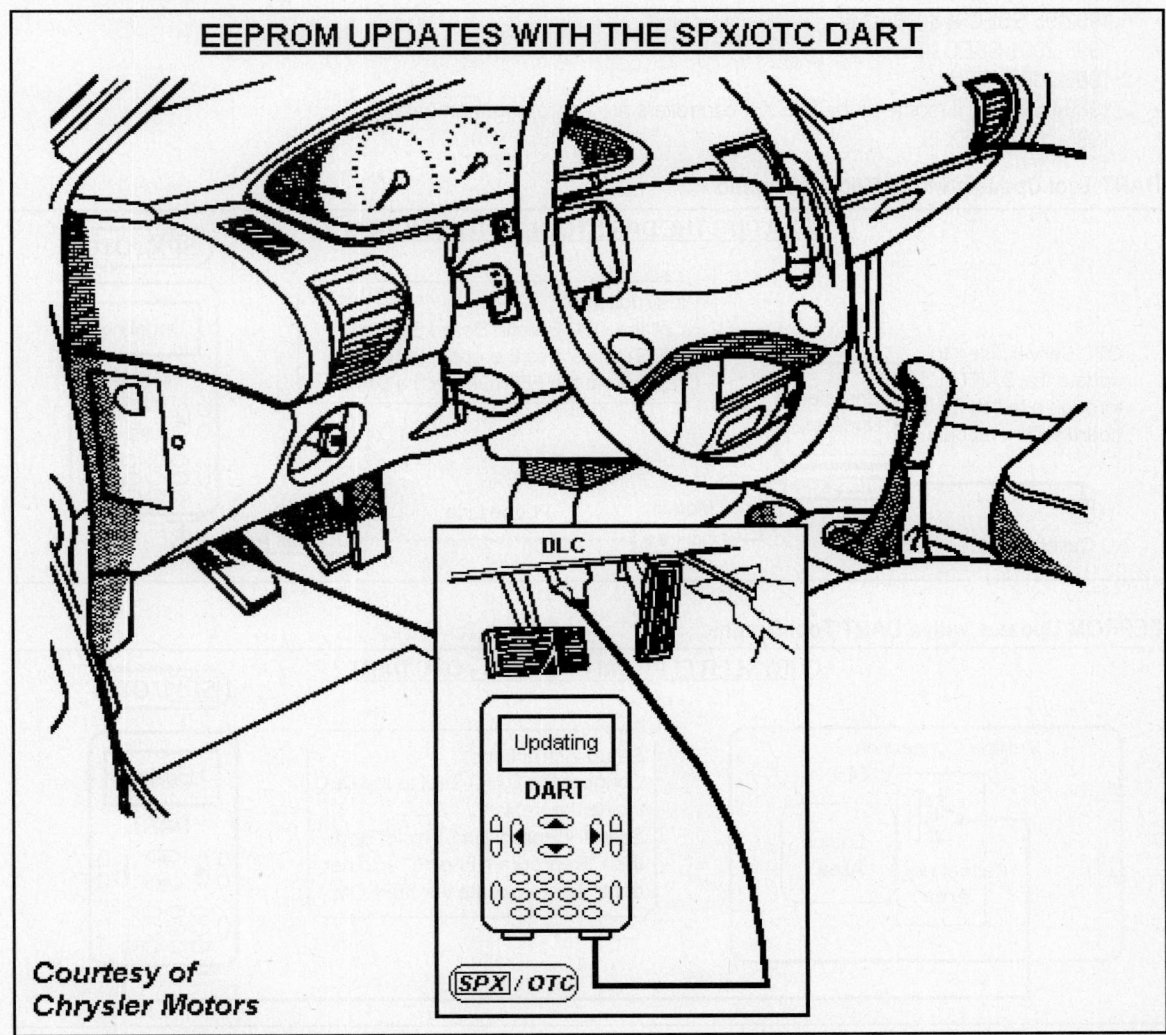

EEPROM UPDATES WITH THE SPX/OTC DART

Courtesy of Chrysler Motors

The DART Scan Tool

The DART Tool manufactured by SPX/OTC and approved by Chrysler can be used to reprogram controllers on 1989-2002 vehicles or to Read, Record & Clear codes.

According to SPX/OTC, this tool gives the user the ability to reprogram the vehicle computer in order to eliminate drivability problems (many of which cannot be repaired through conventional procedures). The reprogram function can be performed whenever a TSB is available. For example, by downloading the appropriate software update for the vehicle, you can eliminate various vehicle drivability problems (i.e., rough idle, stalling, hard shifting, or the MIL "on" when it should not be activated).

The user can update the DART Tool through the SPX BBS to add new DART features as well as to get updates for the vehicle computers (BCM, PCM or TCM).

Engine & Transmission Controller Coverage

*   1992-95 SBEC II
*   1995 FCC, 1995-2000 MMC
*   1992-95 SBEC II & SBEC III
*   1998-2001 SBEC IIIA
*   1996-2002 JTEC
*   1989-95 EATX II (some of the EATX II controllers are not reprogrammable)
*   1995-2002 EATX III

DART Tool Updates with a Modem Graphic

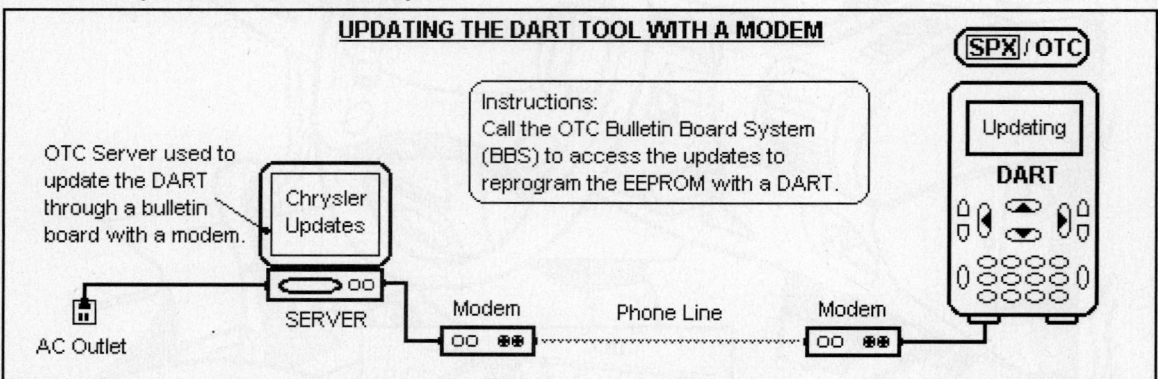

EEPROM Updates with a DART Tool Graphic

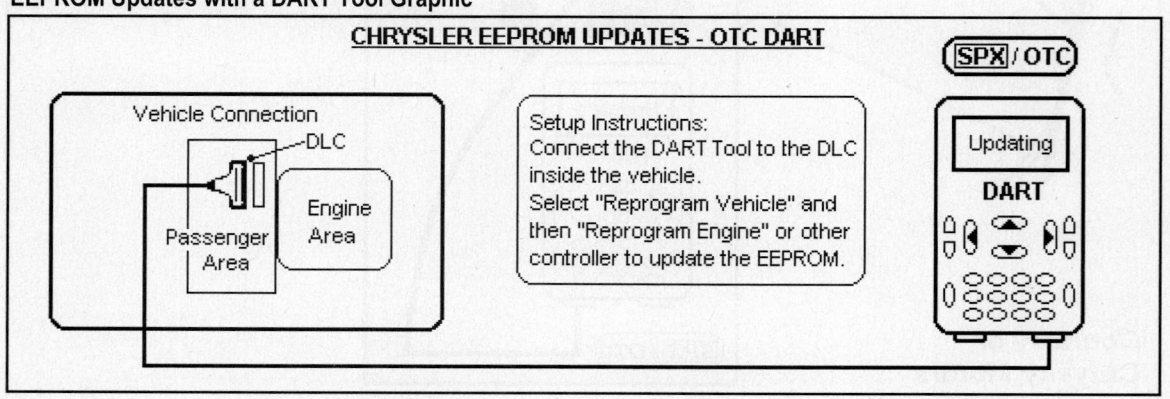

Powertrain Diagnostics

INTRODUCTION

The powertrain control module (PCM) used on Chrysler vehicles is a digital computer that contains a microprocessor. The PCM receives input signals from various sensors and switches that are referred to as PCM inputs. Based on these inputs, the PCM adjusts various engine and vehicle operations through devices that are referred to as PCM outputs. Examples of the input and output devices are shown in the graphic below.

Input & Output Device Graphic (1999 Neon)

PCM INPUTS & OUTPUTS
(OBD II)

Conditions Sensed

INPUT SENSORS
- Battery Temperature
- Camshaft Position
- Crankshaft Position
- Engine Coolant Temp.
- Fuel Level
- Intake Air Temperature
- Knock Detection
- Manifold Air Pressure
- Oxygen Content
- Throttle Position

INPUT SIGNALS
- A/C Select Switch
- Battery Voltage
- Brake Switch
- Leak Detection Pump Sw.
- Ignition Switch
- Park Neutral Switch
- Power Steering Press. Sw.
- SCI Receive
- Speed Control Switches

PCM

Systems Controlled

OUTPUT DEVICES
- A/C WOT Relay
- Auto Shutdown Relay
- Charge Indicator Lamp
- Data Link Connector
- EVAP Purge Solenoid
- EGR Solenoid
- Fuel Injectors
- Fuel Pump Relay
- Generator Field
- Idle Air Control Solenoid
- Ignition Coils
- Leak Detection Solenoid
- MIL or C/E Lamp
- Radiator Fan Relay
- SCI Transmit
- Speed Control Solenoids
- Tachometer
- TCC Solenoid

Powertrain Subsystems

A key to the diagnosis of the PCM and its subsystems is to determine which subsystems are on a vehicle. Examples of typical subsystems appear below:

- Cranking & Charging System
- Emission Control Systems
- Engine Cooling System
- Engine Air/Fuel Controls
- Exhaust System
- Ignition System
- Speed Control System
- Transaxle Controls

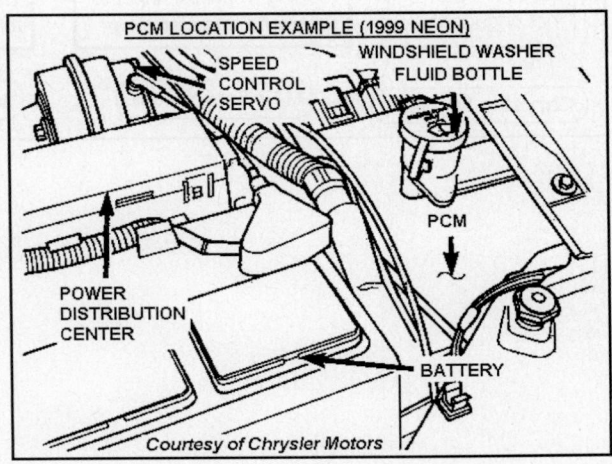

PCM LOCATION EXAMPLE (1999 NEON)

SPEED CONTROL SERVO

WINDSHIELD WASHER FLUID BOTTLE

PCM

POWER DISTRIBUTION CENTER

BATTERY

Courtesy of Chrysler Motors

PCM MODES OF OPERATION

Before you can get the full use out of "embedded" Powertrain Diagnostics built into a Daimler/Chrysler vehicle from 1996-2002, you must understand how the PCM operates. The various PCM Modes of Operation are discussed on the next few pages.

Note: ***This explanation of the PCM Modes of Operation refers to vehicles with OBD II systems. The operating modes of vehicles with an OBD I system are similar.***

As input signals to the PCM change, the PCM adjusts its response to output devices. For example, the PCM must calculate a different injector pulsewidth and ignition timing for idle speed operation than it does for wide open throttle operation. The two (2) basic modes of operation that determine how the PCM responds are Open and Closed Loop.

Open and Closed Loop Operation

During open loop mode, the PCM receives input signals and responds according to preset programming. Inputs from the front and rear oxygen sensors are not monitored during open loop mode, except for heated oxygen sensor test (this is a continuous test).

During closed loop mode, the PCM monitors the inputs from the front and rear heated oxygen sensors. The front heated oxygen sensor input tells the PCM if the calculated injector pulsewidth resulted in an ideal A/F ratio of 14.7 to1. By monitoring the exhaust oxygen content through the front heated oxygen sensor, the PCM can adjust the injector pulsewidth to achieve optimum fuel economy combined with low emissions.

The PCM will not enter closed loop operation until the following conditions occur:

1) The engine coolant temperature (ECT PID) must be more than 35ºF before the closed loop timer in the PCM begins its countdown.

- If the coolant is over 35ºF, the PCM will wait 44 seconds.
- If the coolant is over 50ºF, the PCM will wait 38 seconds.
- If the coolant is over 167ºF, the PCM will wait 11 seconds.
- For other coolant temperatures, the PCM will interpret the correct waiting time.

2) The O2 sensor signal must go over 0.74v and drop below 0.10v to enter closed loop.

Ignition On, Engine Off Mode (Open Loop)

These actions occur when the ignition switch activates the fuel injection system:

- The PCM determines the atmospheric air pressure from the MAP sensor input in order to determine a basic fuel control strategy.
- The PCM monitors the ECT and TP sensor inputs to modify its fuel control strategy.
- The ASD and fuel pump relays are turned off so that battery voltage is not supplied to the fuel pump, ignition coil, fuel injectors and heated oxygen sensors.

Ignition On, Engine Off Mode Graphic

OPERATING MODES - IGNITION ON, ENGINE OFF MODE

Ignition turned to "key on" position (starter motor not engaged)	**PCM Decisions ...** PCM determines BARO reading from MAP input to determine basic strategy	**PCM Decisions ...** PCM monitors the ECT and TP sensor inputs in order to modify this strategy	**PCM Outputs ...** The ASD and Fuel Pump relays are not energized at "key on"
Chrysler Example	MAP Sensor Table	MAP & TP Sensor Inputs	ASD/Fuel Pump Relay "off"

Startup Mode (Open Loop)

The following actions occur when the starter motor is engaged with the ignition switch:

- The ASD relay and fuel pump relays are energized. If the PCM does not receive CMP and CKP position sensor inputs within (1) second, the relays are de-energized.
- The PCM energizes the injectors until it detects the crankshaft position from the CMP and CKP sensor inputs within one engine revolution due to the configuration of the sensor signals. It energizes the injectors in sequence (it controls their ground paths).
- Once the engine idles within 64 rpm of its "target" idle speed, the PCM compares the current MAP sensor value with the value received from the "zero rpm" ignition on mode. A code is set if the PCM does not detect a minimum difference in the values.
- Once the ASD and fuel pump relays are energized, the PCM determines the injector pulsewidth based on the ECT, IAT, MAP and TP sensor inputs and engine speed.
- Once the engine is running, the PCM determines the amount of spark advance based on the ECT, CMP, CKP, IAT, MAP and TP sensor inputs.

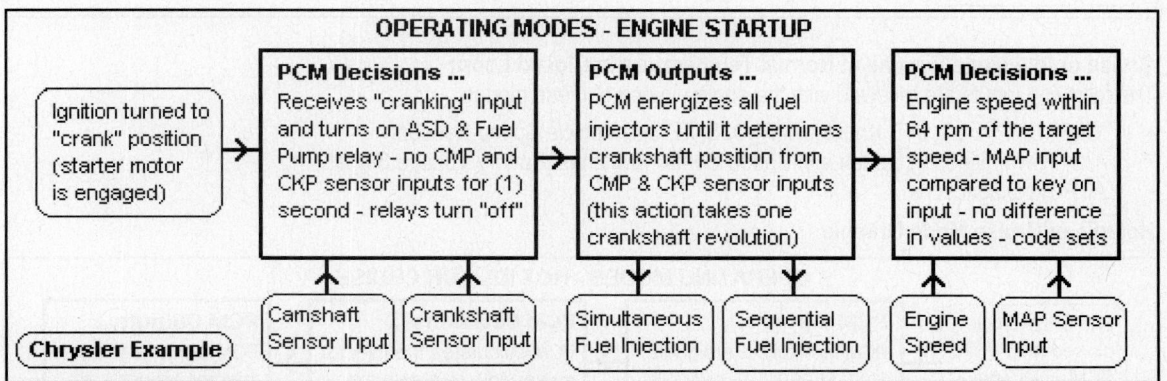

Engine Startup Mode Graphic

Warmup Mode (Open Loop)

The PCM adjusts the injector pulsewidth and controls the injector synchronization by controlling the various injector ground paths. The PCM adjusts the idle speed by controlling the IAC motor position and the amount of spark advance using input signals and responds according to preset PCM programming. The PCM receives inputs from the CKP, CMP, ECT, IAT, Knock, MAP, HO2S, TP and Vehicle Speed sensors during this period. It also receives A/C, PSP and Speed Control Switch and system voltage inputs.

Signals from the front and rear heated oxygen sensors are ignored. However, the PCM does run the O2 sensor Monitor Test during this mode (to check for a short circuit fault).

Engine Warmup Mode Graphic

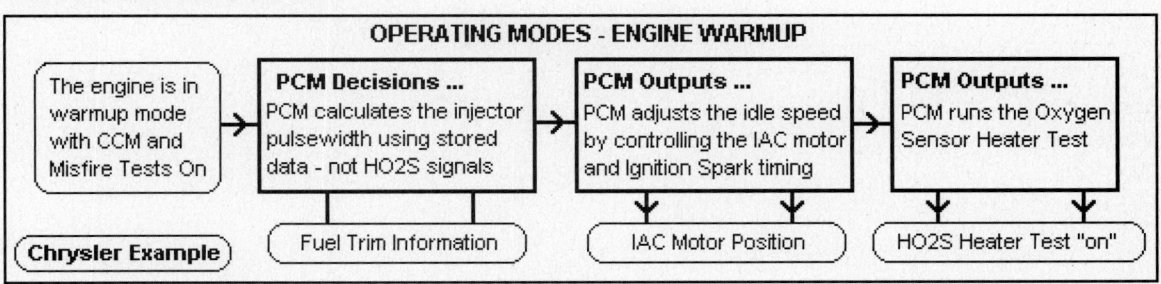

Closed Loop Mode

The PCM monitors the inputs from the front and rear heated oxygen sensors. The front heated oxygen sensor input tells the PCM if the calculated injector pulsewidth resulted in an ideal A/F ratio of 14.7:1. By monitoring the exhaust oxygen content of the front heated oxygen sensor, the PCM determine how to "fine tune" the injector pulsewidth so that it can achieve the optimum in fuel economy combined with low tailpipe emissions.

Closed Loop Mode Graphic

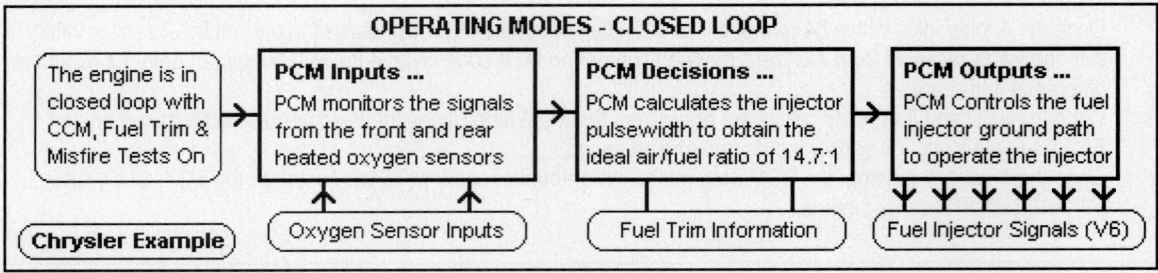

Cruise or Idle Mode (Engine at Normal Temperature in Closed Loop)

The following inputs are received with the engine in one of these modes:

- CKP, CMP, ECT, IAT, Knock, MAP, O2S, TP and Vehicle Speed Sensors
- A/C, Power Steering Switch and Speed Control Switches, Battery Voltage Signal
- All PCM Diagnostics

Hot Idle or Cruise Mode Graphic

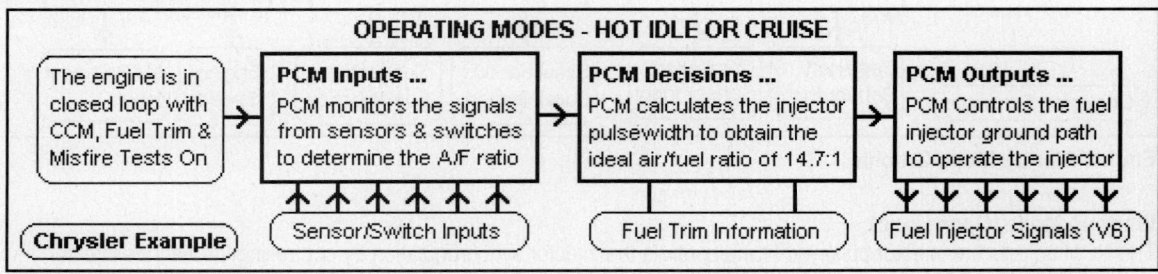

The PCM adjusts the injector pulsewidth and injector synchronization by turning each injector ground path "on" and "off" at the appropriate time. The PCM adjusts the engine idle speed and ignition timing. The PCM adjusts the A/F ratio according to the oxygen content in the exhaust gas (as measured by the oxygen sensors).

Diagnostics - The PCM monitors for engine misfire. During active misfire and depending on the severity, the PCM either continuously illuminates or flashes the MIL (C/E light on the Instrument panel). The PCM stores any CCM, Fuel Trim or Misfire codes in memory.

Acceleration Mode (Closed Loop)

The PCM recognizes an abrupt increase in the TP sensor or MAP sensor voltage as a demand for increased engine output and vehicle acceleration. The PCM increases injector pulsewidth in response to the increase in fuel demand.

During an acceleration condition, the PCM monitors the following inputs:

- MAP Sensor Signal
- TP Sensor Signal

Acceleration Mode Graphic

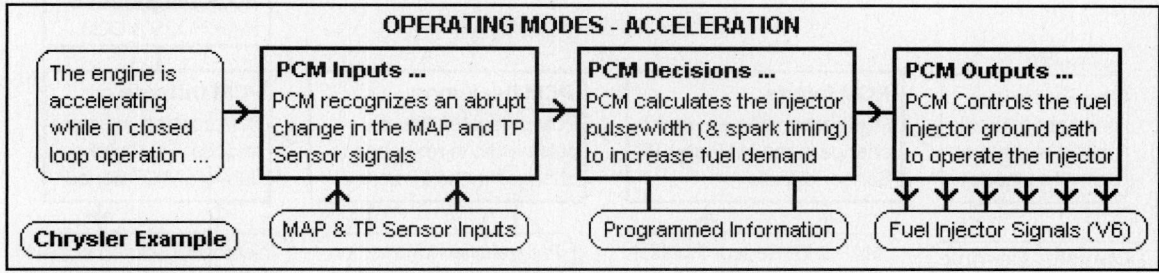

Deceleration Mode (Closed Loop)

During a deceleration condition, the PCM monitors the following inputs:

- A/C Pressure, A/C Sense, Battery Voltage and Crankshaft Position
- ECT, IAT, MAP, Knock, O2 and TP Sensor
- Power Steering Sense and the Vehicle Mileage
- IAC Motor control changes in response to MAP Sensor feedback
- If Decel Fuel Shutoff is detected, the PCM starts the rear Oxygen sensor diagnostics

Deceleration Mode Graphic

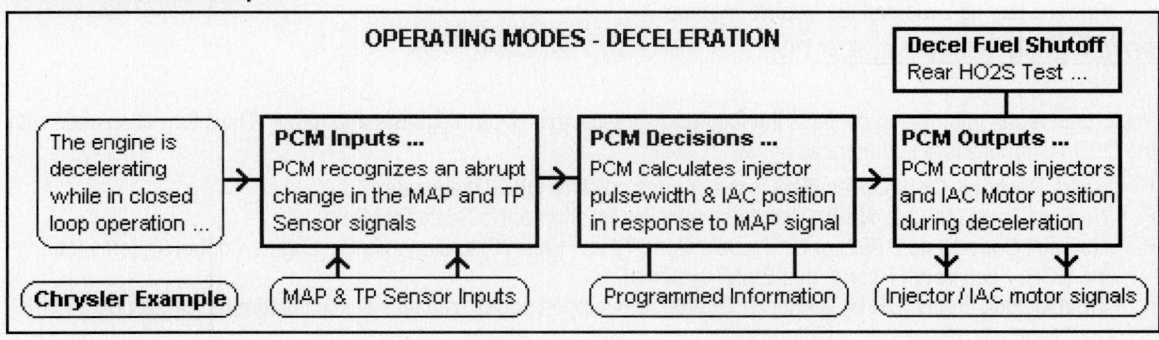

Wide Open Throttle Mode (Open Loop)

During wide-open throttle (WOT) conditions, these inputs are received by the PCM:

- CKP, ECT, IAT, MAP, Knock, and TP Sensors
- The PCM does not monitor the Oxygen sensor inputs during WOT except for the rear heated oxygen sensor, and the CCM test for a shorted Oxygen sensor circuit.

Wide Open Throttle Mode Graphic

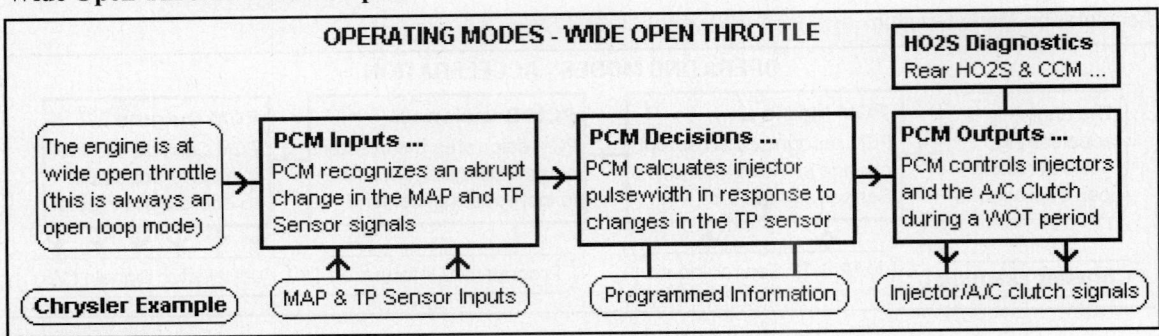

Diagnostic Modes

The PCM can operate in several diagnostic modes. These tests include checks of the EGR, EVAP, and Fuel Systems. They also include a diagnostic test called the Oxygen Sensor Monitor. These tests include monitoring all inputs for the proper voltage range.

During certain idle conditions, the PCM may enter a variable idle speed strategy. This is a strategy that allows the PCM to adjust engine speed based on the following inputs:

- A/C Sense, Battery Voltage and Battery Temperature
- Engine Coolant Temperature and Engine Run Time
- Power Steering Load and the Vehicle Mileage

POWERTRAIN REPAIR INFORMATION

Powertrain repair information contains the following subjects that cover diagnosis and repair. The information used with the OBD I diagnostics is summarized below:

- Read, Record & Clear Codes (This is the *starting point* for OBD I diagnostics).
- No Start Tests (NS-1 to NS-9). Use these tests for No Start or No Scan Tool Data.
- No Fault Code Tests (1992) or No Trouble Code Tests (1993-95) to diagnose the Engine Control systems (i.e., the EGR, Fuel, Spark Retard, and EVAP systems).
- Trouble code charts (DR or TC Repair Charts) with support pages containing circuit diagrams, circuit operation, and diagnostic repair support information.
- Component locations, wiring diagrams, PCM terminal definitions, and pin charts.

The information that is used with OBD II system diagnostics is summarized below:

- Chrysler DTC Check (The *starting point* for OBD II diagnostics)
- Charging, Drivability, Speed Control, Starting and Transmission Tests (these are numbered tests for OBD II system applications)
- Engine Control tests are included in the Drivability Tests found in repair manuals.
- Component locations, wiring diagrams, PCM terminal definitions, and pin charts.
- Trouble code tables or charts that include various forms of support information (i.e., schematics, circuit descriptions, enable criteria to run the test and why the code set).

Where To Begin

Diagnosis of engine performance or drivability problems on a vehicle with an onboard computer requires that you have a logical plan on how to approach the problem.

The Six Step Test Procedure provided by Chrysler is designed to provide a uniform approach to repair any problems that occur in one or more of the vehicle subsystems.

The diagnostic flow built into this test procedure has been field-tested for several years at Chrysler dealerships - *it is the starting point when a repair is required!*

It should be noted that a commonly overlooked part of the 'Problem Resolution' step is to check for any related Technical Service Bulletins.

Chrysler Test Procedure Graphic

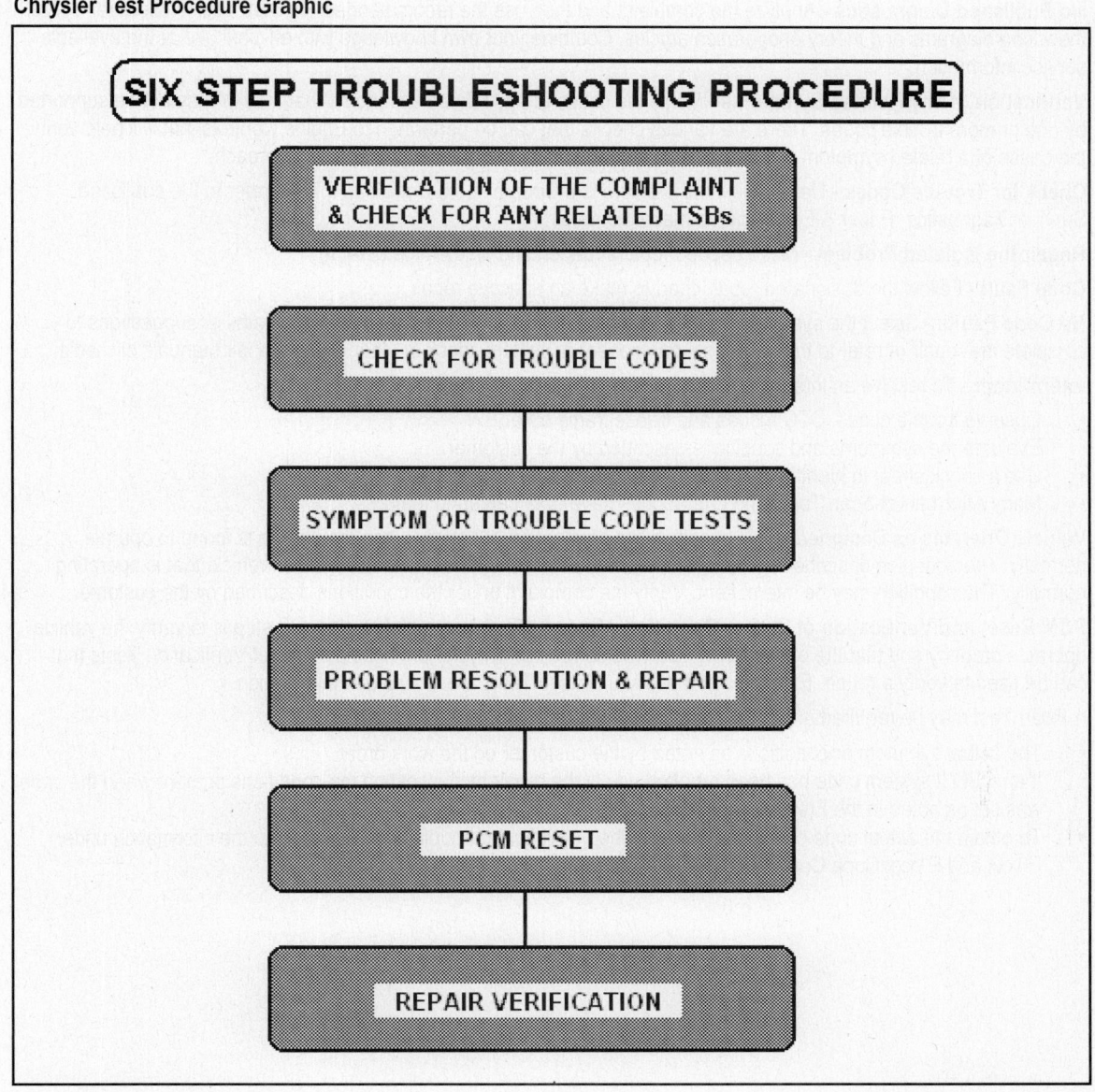

SIX STEP TROUBLESHOOTING PROCEDURE

VERIFICATION OF THE COMPLAINT & CHECK FOR ANY RELATED TSBs

CHECK FOR TROUBLE CODES

SYMPTOM OR TROUBLE CODE TESTS

PROBLEM RESOLUTION & REPAIR

PCM RESET

REPAIR VERIFICATION

SIX STEP TEST PROCEDURE

The steps outlined on this page were defined to help you determine how to perform a proper diagnosis. Refer to the flow chart that outlines the Six Step Test Procedure on the previous page as needed. The recommended steps include:

Verify the Complaint - To verify the customer complaint, the technician should understand the normal operation of the system.

Preliminary Checks - Conduct a thorough visual and operational inspection, review the service history, detect unusual sounds or odors, and gather diagnostic trouble code (DTC) information to achieve an effective repair.

Check Bulletins and Other Service Information - This check should include videos, newsletters, and any other information in the form of TSBs or Dealer Service Bulletins.

No Published Diagnostics - Analyze the complaint and then use the recommended Six Step Test Procedure. Utilize the wiring diagrams and theory of operation articles. Combine your own knowledge with efficient use of the available service information.

Verification of any Related Symptoms - Verify the cause of any related symptoms that may or may not be supported by one or more trouble codes. There are various checks that can be performed to Engine Controls that will help verify the cause of a related symptom. This step helps to lead you in an organized diagnostic approach.

Check for Trouble Codes - Determine if the problem is a Code or a No Code Fault. Then refer to the published Service Diagnostics (Paper & Electronic) to make the repair.

Repair the Isolated Problem - Make needed repairs (depending on the type of fault).

Code Fault - Follow the designated repair chart to make an effective repair.

No Code Faults - Select the symptom from the symptom tables and follow the diagnostic paths or suggestions to complete the repair or refer to the applicable component or system check in other Chilton repair manuals or media.

Intermittent - To resolve an intermittent fault, perform the following steps:

- Observe trouble codes, DTC modes and freeze frame data.
- Evaluate the symptoms and conditions described by the customer.
- Use a check sheet to identify the circuit or electrical system component.
- Many Aftermarket Scan Tools and Lab Scopes have data capturing features.

Vehicle Operates as Designed - This condition (no problem found) exists when the vehicle is found to operate normally. The condition described by the customer may be normal. Check against another vehicle that is operating normally. The condition may be intermittent. Verify the complaint under the conditions described by the customer.

PCM Reset and Verification of Proper Operation - Once a repair is completed, the next step is to verify the vehicle operates properly and that the original symptom was corrected. Chrysler provides a series of Verification Tests that can be used to verify a repair. Examples of these Verification Tests are included in this section.

A Road Test may be required under the conditions listed below to verify a complaint:

- The actual symptom or conditions as noted by the customer on the work order
- If an OBD II system code has been repaired, verify the repair by duplicating the conditions present when the code was set as noted in the Freeze Frame data.
- To obtain the actual code conditions present when a particular trouble code set, refer to the information under P0xxx and P1xxx Code Conditions later in this manual.

BASE ENGINE TESTS

To determine that an engine is mechanically sound, certain tests need to be performed to verify that the correct A/F mixture enters the engine, is compressed, ignited, burnt, and then discharged out of the Exhaust system. The tests in this article can be used to help determine the mechanical condition of the engine.

To diagnose an engine-related complaint, compare the results of the Compression, Cylinder Balance, Engine Cylinder Leakage (not included) and Engine Vacuum Tests.

Engine Compression Test

The Engine Compression Test is used to determine if each cylinder is contributing its equal share of power. The compression readings of all the cylinders are recorded and then compared to each other and to the manufacturer's specification (if available).

Cylinders that have low compression readings have lost their ability to seal. It this type of problem exists, the location of the compression leak must be identified. The leak can be in any of these areas: piston, head gasket, spark plugs, exhaust or intake valves.

The results of this test can be used to determine the overall condition of the engine and to identify any problem cylinders as well as the most likely cause of the problem.

✷✷ **CAUTION:** *Prior to starting this procedure, set the parking brake, place the gear selector in P/N and block the drive wheels for safety. The battery must be fully charged.*

Test Procedure (Compression Test)
- Allow the engine to run until it is fully warmed up.
- Remove the spark plugs and disable the Ignition system and the Fuel system for safety. Disconnecting the CKP sensor harness connector will disable both fuel and ignition (except on NGC vehicles – refer to Section 6 for NGC explanation).
- Carefully block the throttle to the wide-open position.
- Insert the compression gauge into the cylinder and tighten it firmly by hand.
- Use a remote starter switch or ignition key and crank the engine for 3-5 complete engine cycles. If the test is interrupted for any reason, release the gauge pressure and retest. Repeat this test procedure on all cylinders and record the readings.

The lowest cylinder compression reading should not be less than 70% of the highest cylinder compression reading and no cylinder should read less than 100 psi.

Evaluating the Test Results

To determine why an individual cylinder has a low compression reading, insert a small amount of engine oil (three squirts) into the suspect cylinder. Reinstall the compression gauge and retest the cylinder and record the reading. Review the explanations below.

Reading is higher - If the reading is higher at this point, oil inserted into the cylinder helped to seal the piston rings against the cylinder walls. Look for worn piston rings.

Reading did not change - If the reading didn't change, the most likely cause of the low cylinder compression reading is the head gasket or valves.

Low readings on companion cylinders - If low compression readings were recorded from cylinders located next to each other, the most likely cause is a blown head gasket.

Readings are higher than normal - If the compression readings are higher than normal, excessive carbon may have collected on the pistons and in the exhaust areas. One way to remove the carbon is with an approved brand of Top Engine Cleaner.

■ **Note:** *Always clean spark plug threads and seat with a spark plug thread chaser and seat cleaning tool prior to reinstallation. Use anti-seize compound on Aluminum heads.*

ENGINE VACUUM TESTS

An engine vacuum test can be used to determine if each cylinder is contributing an equal share of power. Engine vacuum, defined as any pressure lower than atmospheric pressure, is produced in each cylinder during the intake stroke. If each cylinder produces an equal amount of vacuum, the measured vacuum in the intake manifold will be even during engine cranking, at idle speed, and at off-idle speeds.

Engine vacuum is measured with a vacuum gauge calibrated to show the difference between engine vacuum (the lack of pressure in the intake manifold) and atmospheric pressure. Vacuum gauge measurements are usually shown in inches of Mercury (" Hg).

■ **Note:** *In the tests described in this article, connect the vacuum gauge to an intake manifold vacuum source at a point below the throttle plate on the throttle body.*

Engine Cranking Vacuum Test Procedure
The Engine Cranking Vacuum Test can be used to verify that low engine vacuum is not the cause of a No Start, Hard Start, Starts and Dies or Rough Idle condition (symptom).

The vacuum gauge needle fluctuations that occur during engine cranking are indications of individual cylinder problems. If a cylinder produces less than normal engine vacuum, the needle will respond by fluctuating between a steady high reading (from normal cylinders) and a lower reading (from the faulty cylinder). If more than one cylinder has a low vacuum reading, the needle will fluctuate very rapidly.

Prior to starting this test, set the parking brake, place the gearshift in P/N and block the drive wheels for safety. Then block the PCV valve and disable the idle air control device.

Disable the Fuel and/or Ignition system to prevent the vehicle from starting during the test (while it is cranking). Close the throttle plate and connect a vacuum gauge to an intake manifold vacuum source. Crank the engine for three seconds (do this step at least twice). The test results will vary due to engine design characteristics, the type of PCV valve and the position of the AIS or IAC motor and throttle plate. However, the engine vacuum should be steady between 1.0" to 4.0" of Hg during normal cranking.

Engine Running Vacuum Test Procedure
1) Allow the engine to run until fully warmed up. Connect a vacuum gauge to a clean intake manifold source. Connect a tachometer or Scan Tool to read engine speed.
2) Start the engine and let the idle speed stabilize. Raise the engine speed rapidly to just over 2000 rpm. Repeat the test (3) times. Compare the idle and cruise readings.

Evaluating the Test Results
If the engine wear is even, the gauge should read over 16" of vacuum and be steady. Test results can vary due to engine design and the altitude above or below sea level.

Engine Running Vacuum Test Graphic

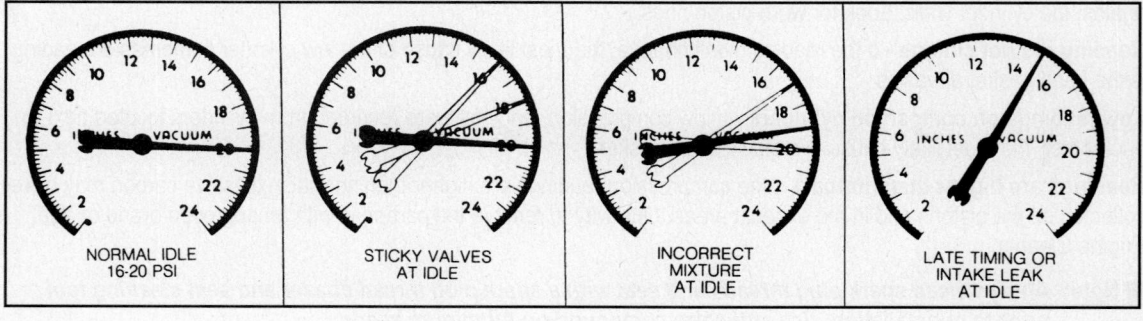

NORMAL IDLE 16-20 PSI   STICKY VALVES AT IDLE   INCORRECT MIXTURE AT IDLE   LATE TIMING OR INTAKE LEAK AT IDLE

IGNITION SYSTEM TESTS - DISTRIBUTOR

This article provides an overview of ignition tests with examples of Engine Analyzer patterns for the Distributor Ignition system used on Daimler/Chrysler vehicles.

Preliminary Inspection - Perform these checks prior to connecting the Engine Analyzer:
- Check the battery condition (verify that it can sustain a cranking voltage of 9.6v).
- Inspect the ignition coil for signs of damage or carbon tracking at the coil tower.
- Remove the coil wire and check for signs of corrosion on the wire or tower.
- Test the coil wire resistance with a DVOM (it should be less than 7 Kohm per foot).

Connect a *low* output spark tester to the coil wire and engine ground. Verify that the ignition coil can sustain adequate spark output while cranking for 3-6 seconds.

Ignition System Scope Patterns
Connect the Engine Analyzer to the Ignition System, and choose Parade display. Run the engine at 2000 RPM, and compare the display to the example below.

Ignition System Scope Patterns (4-Cylinder Engine)

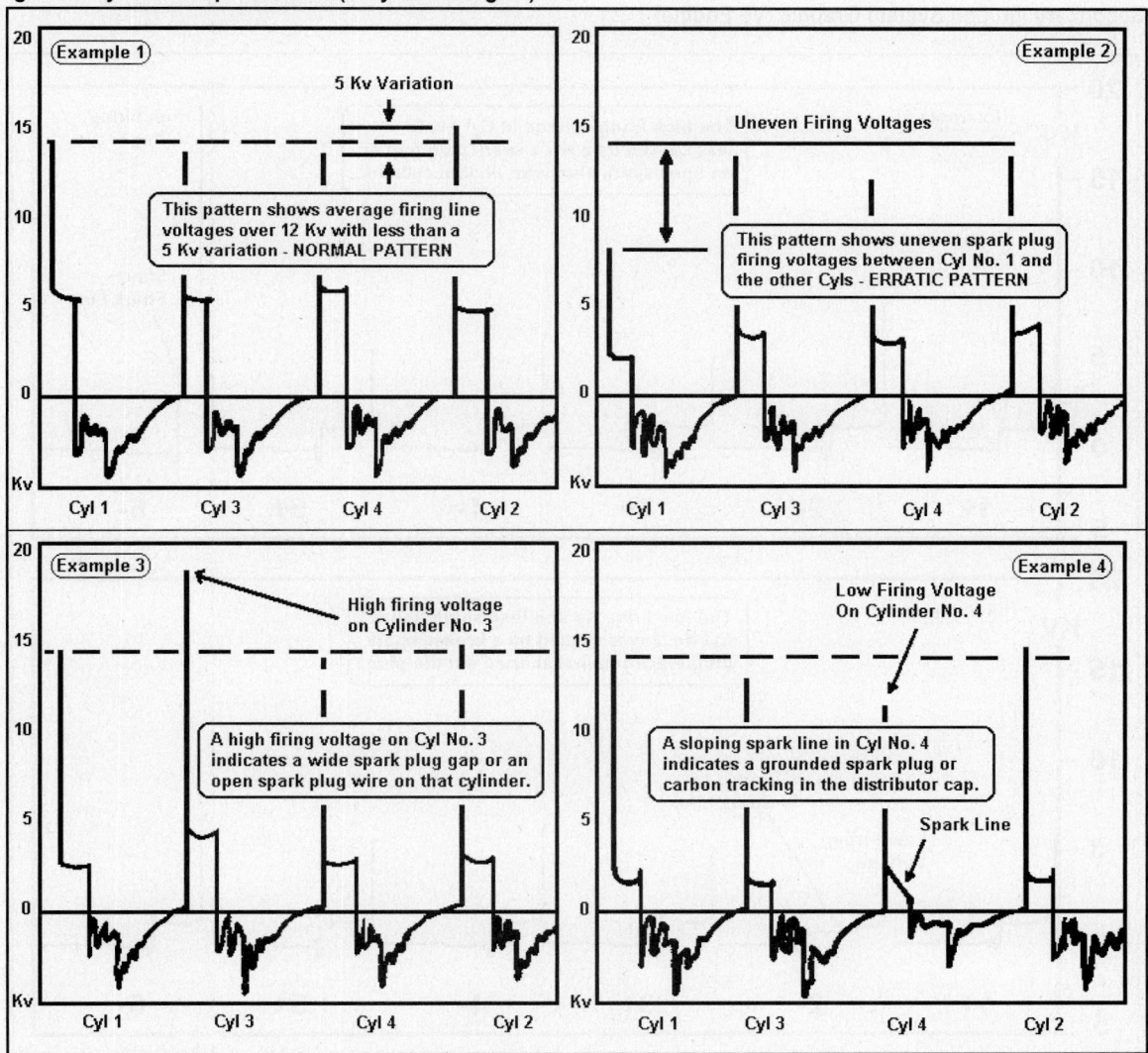

IGNITION SYSTEM TESTS - DISTRIBUTORLESS

This article provides an overview of ignition tests with examples of Engine Analyzer patterns for the Electronic Ignition (EI) system used on Daimler/Chrysler vehicles.

Preliminary Inspection - Perform these checks prior to connecting the Engine Analyzer:
- Check the battery condition (verify that it can sustain a cranking voltage of 9.6v).
- Inspect the ignition coils for signs of damage or carbon tracking at the coil towers.
- Remove the secondary ignition wires and check for signs of corrosion.
- Test the plug wire resistance with a DVOM (specification varies from 15-30 Kohm).

Connect a *low* output spark tester to a plug wire and to engine ground. Verify that the ignition coil can sustain adequate spark output for 3-6 seconds.

Secondary Ignition System Scope Patterns (V6 Engine)
Connect the Engine Analyzer to the Ignition system. Turn the scope selector to view the Parade Display of the ignition secondary. Start the engine in P/N and slowly increase the engine speed from idle to 2000 rpm. Compare actual display to the examples below.

Secondary Ignition System Graphic (V6 Engine)

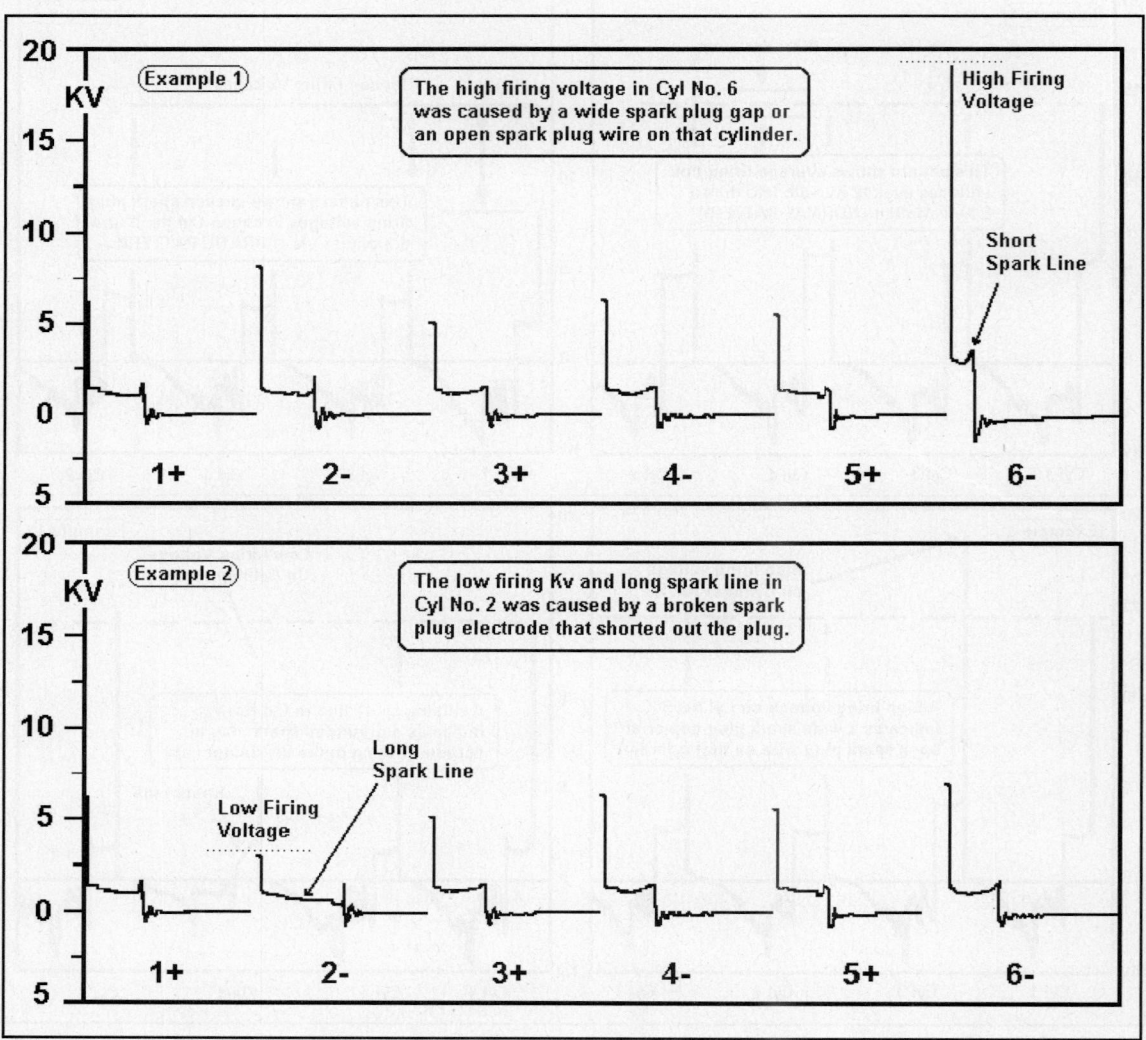

SYMPTOM DIAGNOSIS

Do not attempt to diagnose a Drivability Symptoms without having a logical plan to use to determine which Engine Control system is the cause of the symptom - this plan should include a way to determine which systems do not have a problem! *Drivability symptom diagnosis is a part of an organized approach to problem solving and repair.*

Symptom List

To use this list, locate the symptom that matches a particular problem and refer to the areas to test. The items listed under each symptom may not apply to all models, engines or vehicle systems. The repair steps indicate what vehicle component or system to test.

■ **Note:** *The Drivability Symptoms in this list are intended to be generic. While they apply to most Chrysler vehicles, some vehicles may not have all of the components listed. Refer to other Chilton repair manuals and electronic media for specific tests.*

Drivability Symptom Index Table

Symptom Description	Suggested Areas to Test
Test 1 - No Start, Hard Start Condition • No Crank • Hard Start, Long Crank, Erratic Crank • Stall After Start • No Start, Normal Crank • No Start, MIL is off (if the VREF shorts to ground)	- Check battery, battery circuits to starter - Check for a damaged flywheel, engine compression, base timing and minimum air rate - Check for a failed fuel pump relay - Check for distributor rotor "punch-through" - Check for a faulty ignition control module (ICM) - Check for a VREF circuit shorted to ground - Check SKIM (security system) with a Scan Tool
Test 2 - Rough Idle or Stalls Condition • Low or slow idle speed • Fast idle speed • Hunting or rolling idle speed • Slow return to idle speed • Stalls or almost stalls	- Check for engine vacuum leaks - Check the condition of the PCV valve and lines - Check for excessive carbon buildup - Check for a restricted exhaust (in Section 2) - Check base idle speed, check for low fuel pressure - Check the throttle linkage for sticking or binding
Test 3 - Runs Rough Condition • At idle speed • During acceleration • At cruise speed • During deceleration	- Check for engine vacuum leaks at intake manifold - Check condition of ignition secondary components - Check base timing and idle speed settings - Check for low or high fuel pressure - Check for dirty, leaking or shorted fuel injectors - Check for excessive carbon buildup on valves
Test 4 - Cuts-out, Misses Condition • At idle speed • During acceleration • At cruise speed • During deceleration	- Check for engine vacuum leaks at intake manifold - Check condition of ignition secondary components - Check that spark timing advance is available - Check for low or high fuel pressure - Check for dirty, leaking or shorted fuel injectors - Check for excessive carbon buildup on valves
Test 5 - Bucks, Jerks Condition • During acceleration • At cruise speed • During deceleration	- Check for engine vacuum leaks at intake manifold - Check condition of ignition secondary components - Check that spark timing advance is available - Check for low or high fuel pressure - Check for dirty, leaking or shorted fuel injectors - Check operation of the TCC solenoid, brake switch

SYMPTOM DIAGNOSIS - TEST 1

No Start, Hard Start Condition

■ **Note:** *If there is no spark output or fuel pressure available, check for a failed fuel pump relay, no power to the PCM, or loss of the ignition reference signal to the PCM.*

Preliminary Checks

Prior to starting this symptom test routine, inspect these underhood items:

• Check battery charge and condition, starter current draw.
• Verify the starter relay operation and that the engine cranks (turns over).
• Verify the check engine light (MIL) operation - if it does not activate, check the PCM power and ground circuits, and check for 5v supply at the MAP or TP sensor.
• Check Air Intake system for restrictions (inspect air inlet tubes, air filter for dirt, etc.).
• Check the status of the Smart Key Immobilizer System (SKIM) with the Scan Tool.

Test 1 - Step 1

Step	Action	Yes	No
1	Step Description: No Start Condition Only » Check battery cables, state of charge. » If the engine does not rotate, inspect for a locked engine (hydrostatic lockup condition). » Does the engine crank normally?	Go to Step 2.	Repair the fault in the battery, starter, or Base Engine. Retest for the symptom when all repairs are done.
2	Step Description: Check the Fuel System » Verify that the pump operates at key on. » Check the fuel pump relay operation. If the relay does not operate, check for blown fuse. » Inspect pump for a leak-down condition » Test fuel pressure, volume and quality. » Test the operation of the fuel regulator. » Are there any faults in the Fuel system?	Make needed repairs. Fuel Pressure Gauge / Fuel Rail Test Port	Go to Step 3.
3	Step Description: Check the Ignition System » Inspect ignition secondary components for damage (look for rotor "punch-through"). » Inspect the coils for signs of spark leakage at coil towers or primary connections. » Check the spark output with a spark tester. » Test Ignition system with an engine analyzer. » Are there any faults in the Ignition system?	Make repairs to the Ignition system. Then retest the symptom. CABLE / SPARK TESTER	Go to Step 4.
4	Step Description: Check the Exhaust System » Check Exhaust system for leaks or damage. » Check the Exhaust system for a restriction using the Vacuum or Pressure Gauge Test (e.g., exhaust backpressure reading should not exceed 1.5 psi at cruise speeds). » Are there any faults in the Exhaust system?	Make repairs to the Exhaust system. Then retest the symptom. **Inspect for Damage** 	Go to Step 5.
5	Step Description: Check the MAP Sensor » Disconnect the MAP sensor and attempt to start the engine. » Does the engine start and run normally?	Replace the MAP sensor. Retest for the symptom when repairs are completed.	Go to Step 6.

No Start, Hard Start Condition (Continued)

Test 1 - Step 6

Step	Action (Hard Start Only)	Yes	No
6	Step Description: Check for a Hot Engine » Check for signs of an engine overheating condition related to a Hard Start Symptom. » Does the engine appear to be overheated?	Make the repairs to correct the hot engine and then retest for the symptom when done.	Go to Step 7.
7	Step Description: Check ECT Sensor PID » Connect a Scan Tool and turn the key to on. » Read the ECT sensor (compare to chart). » Has the ECT sensor shifted out of range?	Replace the ECT sensor. Then retest for the symptom when all repairs are completed.	Go to Step 8.
8	Step Description: Check the PCV System » Inspect the PCV system components for broken parts or loose connections. » Test the operation of the PCV valve. » Are there any faults in the PCV system?	Repair the PCV system. Refer to the PCV system tests in this manual. Retest the symptom when all repairs are done.	Go to Step 9.
9	Step Description: Check the EVAP System » Inspect for damaged or disconnected EVAP system components. » Inspect for a fuel saturated charcoal canister. » Are there any faults in the EVAP system?	Refer to the EVAP system tests in this manual. Retest for the symptom when all repairs are completed.	Go to Step 10.
10	Step Description: Test the Base Engine » Check the engine compression. » Test valve timing and timing chain condition. » Check for a worn camshaft or valve train. » Check for any large intake manifold leaks. » Are there any faults in the Base Engine?	Repair the Base Engine. Refer to the Base Engine Tests in this manual. Retest symptom when done.	Return to Step 2 to repeat the test steps in this series to locate and repair the "No Start, Hard Start" condition.

Fuel Pump Schematic

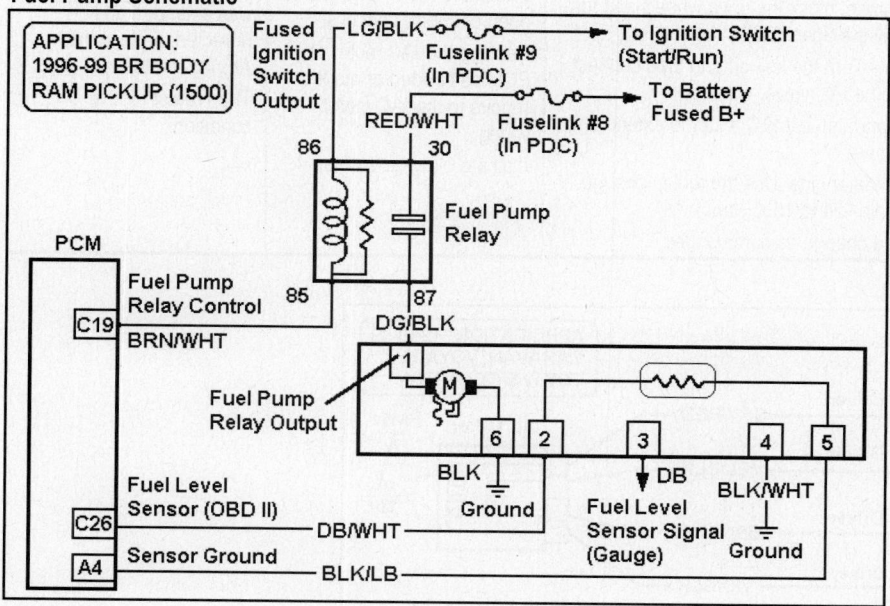

SYMPTOM DIAGNOSIS - TEST 2

Rough, Low or High Idle Speed Condition

■ Note: *If the vehicle has a rough idle and the base timing, idle speed and the IAC (or AIS) motor operates properly, check the engine for excessive carbon buildup.*

Preliminary Checks

Prior to starting this symptom test routine, inspect these underhood items:

- All related vacuum lines for proper routing and integrity.
- All related electrical connectors and wiring harnesses for faults (Wiggle Test).
- Check the throttle linkage for a sticking or binding condition.
- Air Intake system for restrictions (air inlet tubes, dirty air filter, etc.).
- Search for any technical service bulletins related to this symptom.
- Turn the key to off. Unplug the MAP sensor connection and restart the engine to recheck for the idle concern. If the condition is gone, replace the MAP sensor.

Test 2 - Step 1

Step	Action	Yes	No
1	Step Description: Verify the rough idle or stall » Does the engine have a warm engine rough idle, low idle or high idle condition in P or N?	Go to Step 2.	Fault is intermittent. Return to the Symptom List and select another fault.
2	Step Description: Verify idle speed & timing » Verify the base timing is within specifications » Verify that the base idle speed is set properly » Are the timing and idle speed set properly?	Go to Step 3.	Set the base idle speed and timing to the specifications and then retest for the symptom.
3	Step Description: Check AIS / IAC Operation » Check the AIS or IAC motor operation » Inspect the AIS/IAC housing in throttle body for restricted passages. Clean as needed. » Set the parking brake, block the drive wheels and turn the A/C off. Install the Scan Tool. » IAC Motor Tester - Turn the key off and then connect the IAC tester to the IAC valve. » Start the engine and use the IAC tester to extend and retract the IAC valve. » ATM Test - Start the engine. Use the tool to change the speed from min-idle to 1500 rpm. » Did the idle speed change as commanded?	Install an Aftermarket Noid light and check the operation of the PCM and AIS or IAC motor circuits. Check the motor for signs of open or shorted circuits. Replace the IAC motor or PCM as needed or make repairs to the IAC motor wiring. If all are okay, go to Step 4.	If the AIS/IAC motor passages are clean and engine speed did not change as described when the AIS/IAC motor was extended and retracted, replace the AIS/IAC motor. Then retest for the condition.

Idle Air Control Schematic

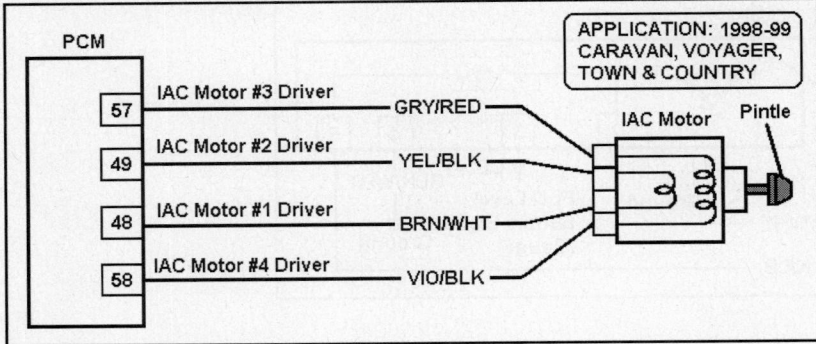

Rough, Low or High Idle Speed Condition (Continued)

Test 2 - Step 4

Step	Action	Yes	No
4	Step Description: Check/compare PID values » Connect Scan Tool & turn off all accessories. » Start the engine and allow it to fully warmup. » Monitor all related PIDs on the Scan Tool. » Verify the P/N switch input in gear and Park. » Check the O2S operation with a Lab Scope. » Are all PIDs within normal range?	Go to Step 5. Note: An IAC motor count of over 80 indicates the pintle is extended and an IAC count of (0) indicates the pintle is retracted.	One or more of the PIDs are out of range when compared to "known good" values. Make repairs to the system that is out of range, then retest for the symptom.
5	Step Description: Check the Ignition System » Inspect the coils for signs of spark leakage at coil towers or primary connections. » Check the spark output with a spark tester. » Test Ignition system with an engine analyzer. » Were any faults found in the Ignition system?	Make repairs as needed	Go to Step 6.
6	Step Description: Check the Fuel System » Inspect the Fuel delivery system for leaks. » Test the fuel pressure, quality and volume. » Test the operation of the pressure regulator. » Were any faults found in the Fuel system?	Make repairs as needed Fuel Pressure Gauge Fuel Rail Test Port	Go to Step 7.
7	Step Description: Check the Exhaust System » Check Exhaust system for leaks or damage. » Check the Exhaust system for a restriction using the Vacuum or Pressure Gauge Test (e.g., exhaust backpressure reading should not exceed 1.5 psi at cruise speeds). » Were any faults found in Exhaust System?	Make repairs to the Exhaust system. Then retest the symptom. Inspect for Damage	Go to Step 8.
8	Step Description: Check the PCV System » Inspect the PCV system components for broken parts or loose connections. » Test the operation of the PCV valve. » Were any faults found in the PCV system?	Make repairs to the PCV system. Refer to the PCV system tests in this manual. Then retest for the condition.	Go to Step 9.
9	Step Description: Check the EVAP System » Inspect for damaged or disconnected EVAP system components or a saturated canister. » Were any faults found in the EVAP system?	Make repairs to EVAP system (use the EVAP tests in this manual). Retest for the condition.	Go to Step 10.
10	Step Description: Check the Base Engine » Test the engine compression. » Test valve timing and timing chain condition. » Check for a worn camshaft or valve train. » Check for any large intake manifold leaks. » Were any faults found in the Base Engine?	Make repairs as needed to the Base Engine. Refer to the Base Engine tests in this manual. Then retest for the condition when repairs are completed.	Go to Step 2 and repeat the tests from the beginning to locate and repair the cause of the "Rough, Low or High Idle Speed" condition.

SYMPTOM DIAGNOSIS - TEST 3

Runs Rough Condition
Preliminary Checks

Prior to starting this symptom test routine, inspect these underhood items:

- All related vacuum lines for proper routing and integrity
- Air Intake system for restrictions (air inlet tubes, dirty air filter, etc.)
- Search for any technical service bulletins related to this symptom.

Test 3 - Step 1

Step	Action	Yes	No
1	Step Description: Verify engine runs rough » Start the engine and allow it to idle in P or N. » Does the engine run rough when warm in Park or Neutral position?	Check for any stored codes. If codes are set, repair codes and retest. If no codes are set, go to Step 3.	Go to Step 2.
2	Step Description: Condition does not exist! » Inspect various underhood items that could cause an intermittent Runs Rough condition (i.e., dirt in the throttle body, vacuum leaks, IAC motor connections, etc.). » Were any problems located in this step?	Correct the problems. Do a PCM reset and engine "idle relearn" procedure. Then verify the "runs rough" condition is repaired.	The problem is not present at this time. It may be an intermittent problem.
3	Step Description: Check/compare PID values » Connect a Scan Tool to the test connector. » Turn off all accessories. » Start the engine and allow it to fully warmup. » Monitor all related PIDs on the Scan Tool. » Were all PIDs within their normal range?	Go to Step 4. Note: The IAC motor should read from 5-50 counts. Check the LONGFT reading for a large shift into the negative range (due to a rich condition).	One or more of the PIDs are out of range when compared to "known good" values. Make repairs to the system that is out of range, then retest for the symptom.
4	Step Description: Check the Ignition System » Inspect the coils for signs of spark leakage at coil towers or primary connections. » Check the spark output with a spark tester. » Test Ignition system with an engine analyzer. » Were any faults found in the Ignition system?	Make repairs as needed 	Go to Step 5.
5	Step Description: Check the Fuel System » Inspect the Fuel delivery system for leaks. » Test the fuel pressure, quality and volume. » Test the operation of the pressure regulator. » Were any faults found in the Fuel system?	Make repairs as needed Fuel Pressure Gauge / Fuel Rail Test Port	Go to Step 6.
6	Step Description: Check the Exhaust System » Check Exhaust system for leaks or damage. » Check the Exhaust system for a restriction using the Vacuum or Pressure Gauge Test (e.g., exhaust backpressure reading should not exceed 1.5 psi at cruise speeds). » Were any faults found in Exhaust System?	Make repairs to the Exhaust system. Then retest the symptom. **Inspect for Damage** 	Go to Step 7.

Runs Rough Condition (Continued)

Test 3 - Step 7

Step	Action	Yes	No
7	Step Description: Check the PCV System » Inspect the PCV system components for broken parts or loose connections. » Test the operation of the PCV valve. » Were any faults found in the PCV system?	Make repairs to the PCV system. Refer to the PCV system tests in this manual. Then retest for the condition.	Go to Step 9.
8	Step Description: Check the EVAP System » Inspect for damaged or disconnected EVAP system components or a saturated canister. » Were any faults found in the EVAP system?	Make repairs to EVAP system (use the EVAP tests in this manual). Retest for the condition.	Go to Step 10.
9	Step Description: Check Engine Condition » Test the engine compression. » Test valve timing and timing chain condition. » Check for a worn camshaft or valve train. » Check for any large intake manifold leaks. » Were any faults found in the Base Engine?	Make repairs as needed to the Base Engine. Refer to the Base Engine tests in this manual. Then retest for the condition when repairs are completed.	Return to Step 2 and repeat the tests from the beginning to locate and repair the cause of the "Runs Rough" condition.

EVAP System Graphic

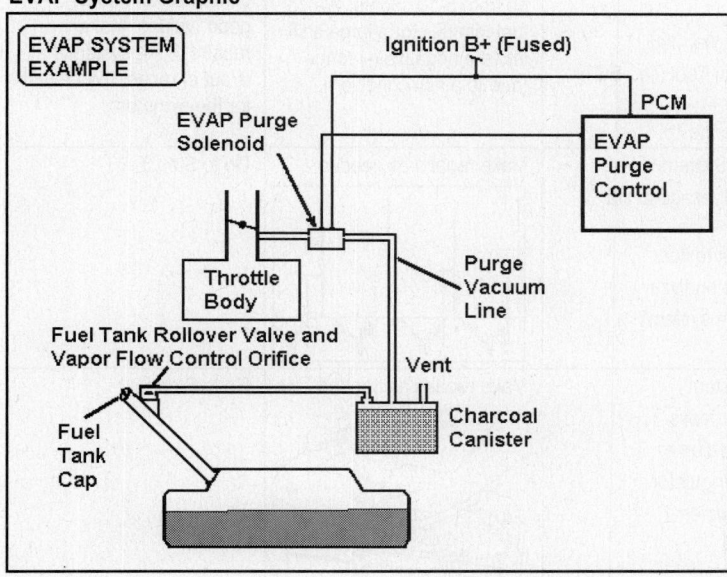

SYMPTOM DIAGNOSIS - TEST 4

Cuts-out or Misses Condition
Preliminary Checks

Prior to starting this symptom test routine, inspect these underhood items:

- All related vacuum lines for proper routing and integrity
- Search for any technical service bulletins related to this symptom.

Test 4 - Step 1

Step	Action	Yes	No
1	Step Description: Verify Cuts-out condition » Start the engine and attempt to verify the Cuts-out or misses condition. » Does the engine have a cuts-out condition?	Check for any stored codes. If codes are set, repair codes and retest. If no codes are set, go to Step 3.	Go to Step 2.
2	Step Description: Condition does not exist! » Inspect various underhood items that could cause an intermittent Cuts-out condition (i.e., EVAP, Fuel or Ignition system components). » Were any problems located in this step?	Correct the problems. Do a PCM reset and "Fuel Trim Relearn" procedure. Then verify condition is repaired.	The problem is not present at this time. It may be an intermittent problem.
3	Step Description: Check/compare PID values » Connect a Scan Tool to the test connector. » Turn off all accessories. » Start the engine and allow it to fully warmup. » Monitor all related PIDs on the Scan Tool (i.e., ECT IAC Counts and LONGFT at idle). » Were all PIDs within their normal range?	Go to Step 4. Note: The IAC motor should be from 5-50 counts. Watch fuel trim (%) for a large shift into the negative (-) range (due to a rich condition).	One or more of the PIDs are out of range when compared to "known good" values. Make repairs to the system that is out of range, then retest for the symptom.
4	Step Description: Check the Ignition System » Inspect the coils for signs of spark leakage at coil towers or primary connections. » Check the spark output with a spark tester. » Test Ignition system with an engine analyzer. » Were any faults found in the Ignition system?	Make repairs as needed 	Go to Step 5.
5	Step Description: Check the Fuel System » Inspect the Fuel delivery system for leaks. » Test the fuel pressure, quality and volume. » Test the operation of the pressure regulator. » Were any faults found in the Fuel system?	Make repairs as needed Fuel Pressure Gauge Fuel Rail Test Port	Go to Step 6.
6	Step Description: Check the Exhaust System » Check Exhaust system for leaks or damage. » Check the Exhaust system for a restriction using the Vacuum or Pressure Gauge Test (e.g., exhaust backpressure reading should not exceed 1.5 psi at cruise speeds). » Were any faults found in Exhaust System?	Make repairs to the Exhaust system. Then retest the symptom. Inspect for Damage 	Go to Step 7.

Cuts-out or Misses Condition (Continued)

Test 4 - Step 7

Step	Action	Yes	No
7	Step Description: Check the PCV System » Inspect the PCV system components for broken parts or loose connections. » Test the operation of the PCV valve. » Were any faults found in the PCV system?	Make repairs to the PCV system. Refer to the PCV system tests in this manual. Then retest for the condition.	Go to Step 8.
8	Step Description: Check the EVAP System » Inspect for damaged or disconnected EVAP system components » Check for a saturated EVAP canister. » Were any faults found in the EVAP system?	Make repairs to EVAP system (use the EVAP tests in this manual). Retest for the condition.	Go to Step 9.
9	Step Description: Check the AIR system » Inspect AIR system for broken parts, leaking valves or disconnected hoses (see graphic). » Test the operation of Secondary AIR system. » Were any faults found in the AIR system?	Make repairs as needed. Refer to the Secondary AIR system tests in this manual. Retest for the condition.	Go to Step 10.
10	Step Description: Check Engine Condition » Test the engine compression. » Test valve timing and timing chain condition. » Check for a worn camshaft or valve train. » Check for any large intake manifold leaks. » Were any faults found in the Base Engine?	Make repairs as needed to the Base Engine. Refer to the Base Engine tests in this manual. Then retest for the condition when repairs are completed.	Go to Step 2 and repeat the tests from the beginning to locate and repair the cause of the "Cuts Out or Misses" condition.

Secondary AIR System Graphic

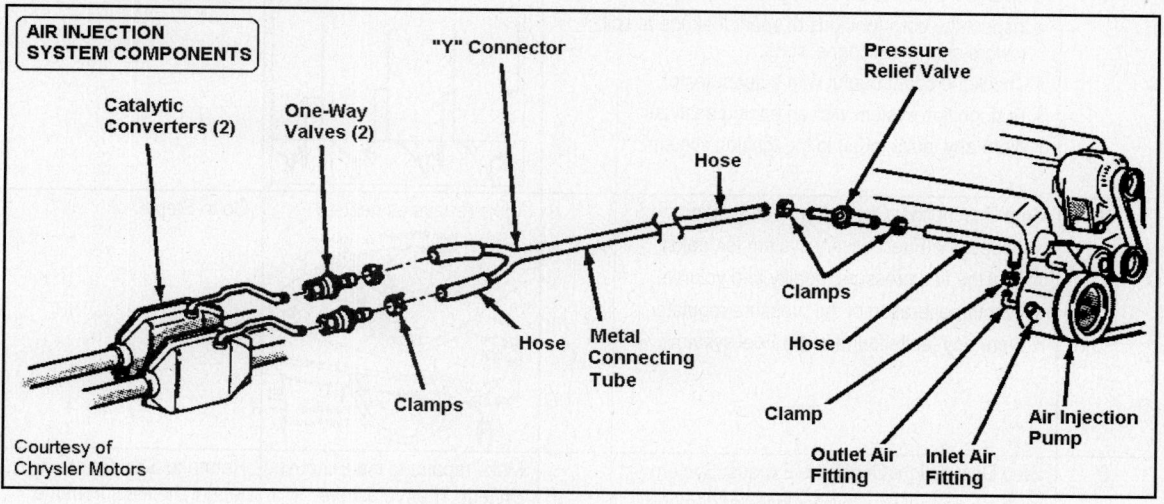

Courtesy of
Chrysler Motors

SYMPTOM DIAGNOSIS - TEST 5

Surge Condition

Preliminary Checks

- Discuss how the operation of the torque converter clutch (TCC) or air conditioning compressor can affect the "feel" of the vehicle during normal operation. Refer to the information in the Owner's Manual to explain how these devices normally operate.
- Search for any technical service bulletins related to this symptom.

Test 5 - Step 1

Step	Action	Yes	No
1	Step Description: Verify the surge condition » Drive the vehicle and attempt to verify that the vehicle surges at cruise speeds. » Does the engine have a surge condition?	Check for any stored codes. If codes are set, repair codes and retest. If no codes are set, go to Step 3.	Go to Step 2.
2	Step Description: Condition does not exist! » Inspect various underhood items that could cause an intermittent surge condition (check for leaks in the MAP sensor vacuum lines). » Were any problems located in this step?	Correct the problems. Do a PCM reset and "Fuel Trim Relearn" procedure. Then verify condition is repaired.	The problem is not present at this time. It may be an intermittent problem.
3	Step Description: Check/compare PID values » Connect a Scan Tool to the test connector. » Start the engine and allow it to fully warmup. » Monitor all related PIDs on Scan Tool (HO2S switching, LONGFT, and the TCC operation) » Compare VSS PID reading to speedometer. » Were all PIDs within their normal range?	Go to Step 4. Note: Verify that the front HO2S responds quickly to throttle changes. Check for silicon contamination on the front HO2S (this can cause a rich A/F signal).	One or more of the PIDs are out of range when compared to "known good" values. Make repairs to the system that is out of range, then retest for the symptom.
4	Step Description: Check the Ignition System » Inspect the coils for signs of spark leakage at coil towers or primary connections. » Check the spark output with a spark tester. » Test Ignition system with an engine analyzer. » Were any faults found in the Ignition system?	Make repairs as needed 	Go to Step 5.
5	Step Description: Check the Fuel System » Inspect the Fuel delivery system for leaks. » Test the fuel pressure, quality and volume. » Test the operation of the pressure regulator. » Were any faults found in the Fuel system?	Make repairs as needed Fuel Pressure Gauge / Fuel Rail Test Port	Go to Step 6.
6	Step Description: Check the Exhaust System » Check Exhaust system for leaks or damage. » Check the Exhaust system for a restriction using the Vacuum or Pressure Gauge Test (e.g., exhaust backpressure reading should not exceed 1.5 psi at cruise speeds). » Were any faults found in Exhaust System?	Make repairs to the Exhaust system. Then retest the symptom. **Inspect for Damage** 	Return to Step 2 and repeat the tests from the beginning to locate and repair the cause of the "Surge" condition.

INTERMITTENT TESTS

Many trouble code repair charts end with a result that reads "Fault Not Present at this Time." What this expression means is that the conditions that were present when a code set or drivability symptom occurred are no longer there or were not met. In effect, the problem was present at least once, but is not present at this time. However, it is likely to return in the future, so it should be diagnosed and repaired if at all possible.

One way to find an intermittent problem is to gather the information that was present when the problem occurred. In the case of a Code Fault, this can be done in two ways: by capturing the data in Snapshot or Movie mode or by driver observations.

The PCM has to detect the fault for a specific period of time before a trouble code will set. While intermittent problems may appear to be occasional in nature, they usually occur under specific conditions. Therefore, you should identify and duplicate these conditions. Since intermittent faults are difficult to duplicate, a logical routine (checklist) must be followed when attempting to find the faulty component, system or circuit. The tests on the next page can be used to help find the cause of an intermittent fault.

Some intermittent faults occur due to a loose connection, wiring problem or warped circuit board. An intermittent fault can also be caused by poor test techniques that cause damage to the male or female ends of a connector.

Tests for Loose Connectors
To test for a loose or damaged connection, take the male end of a connector from another wiring harness and carefully push it into the "suspect" female terminal to verify that the opening is tight. There should be some resistance felt as the male connector is inserted in the terminal connection.

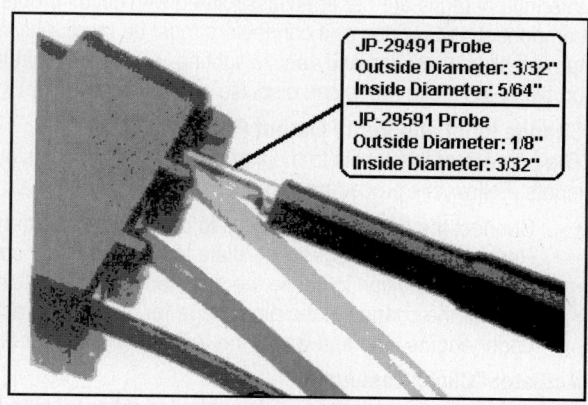

JP-29491 Probe
Outside Diameter: 3/32"
Inside Diameter: 5/64"

JP-29591 Probe
Outside Diameter: 1/8"
Inside Diameter: 3/32"

The Wiggle Test
A wiggle test can be used to locate the cause of some intermittent faults. The sensor, switch or the PCM wiring can be backprobed as shown while the test is done.

During testing, move or wiggle the suspect device, connector or wiring while watching for a change.
If the DVOM has a Min/Max record mode, use this mode during the test.

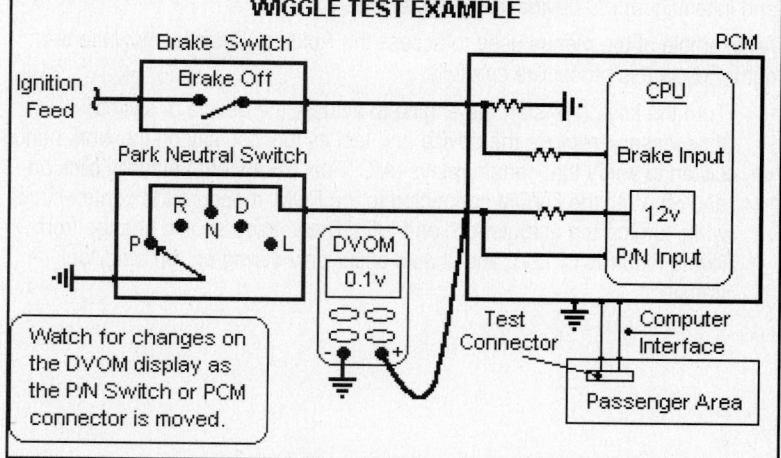

WIGGLE TEST EXAMPLE

Watch for changes on the DVOM display as the P/N Switch or PCM connector is moved.

INTERMITTENT TESTS - SPECIFIC TOOLS

The test procedure in this article contains instructions for isolating an intermittent fault while using a Breakout Box, Scan Tool, fuel pressure gauge, vacuum gauge or DVOM. Actual values from the vehicle under test can be compared to a typical set of values from the Reference Values in this manual, other Chilton repair manuals, or Chilton electronic media.

Preliminary Checks

Prior to testing for an intermittent fault, perform all of the preliminary checks listed below:

- Check all related electrical connections.
- Test for vacuum leaks in related components or mounting hardware.
- Check the fuel system - fuel level, quality (contamination), quantity, and pressure.
- Check the ignition wiring connections.
- Inspect the air intake system filters, tubes and gaskets.
- Check the base engine components (compression, valve and ignition timing, etc.).
- Look for any Aftermarket add-on devices.

Finding Intermittent Faults

Intermittent faults are generally associated with circuit problems. In order to pinpoint the fault area, a particular component or its wiring and connectors must be thoroughly inspected and tested. Prior to starting your test sequence, turn off all accessories and vehicle lighting. Also, verify that the battery and vehicle charging system are free of problems as these areas can disguise or mask a problem.

Change Input and Verify Output Response

The purpose of this test is to monitor how the PCM and its output devices respond to changes in sensor or switch inputs. Follow this procedure carefully:

- Connect the DVOM to the circuit to be tested and ground.
- Record any pin voltages that relate to an intermittent code or symptom.
- Create a condition to cause the selected input signal to change.
- Monitor the change in the pin voltage for a particular actuator signal on the DVOM (i.e., increase the throttle angle under engine load and watch the IAC and TP Sensor pin voltage change).

Actuator "Click" Testing

This test is used to monitor a PCM controlled relay or solenoid while watching and listening for the device to change states.

An example of the menus used to access the Actuator Tests is shown to the right. Follow this procedure carefully:

- Turn the key on or start the engine to actuate the device or switch
- If necessary, remove the device and test its functionality on the work bench
- Listen to verify that certain relays (A/C, Fuel Pump, etc.) actually click on and off. With the DVOM connected to the PCM, measure the control circuit while turning the outputs "on" and "off". The voltage should change from low to high (0v to 12v), and should occur only during an "on" and "off" transition.

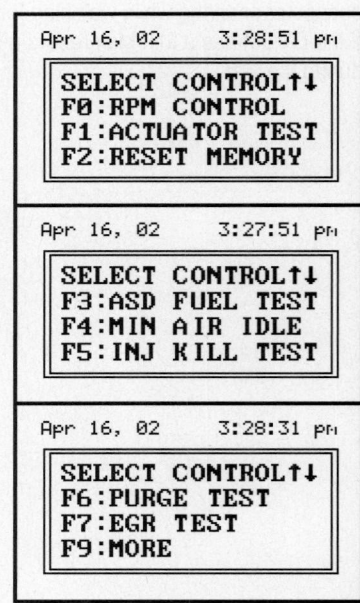

ACCESSING COMPONENTS & CIRCUITS

Every vehicle and every diagnostic situation is different. It is a good idea to first determine the best diagnostic path to follow using flow charts, wiring diagrams, TSBs, etc. Part of choosing steps is to determine how time-consuming and effective each step will be. It may be easy to access a component or circuit in one vehicle, but difficult in another. Many circuits are integrated into a large harness and are difficult to test. Many components are inaccessible without disassembly of unrelated systems.

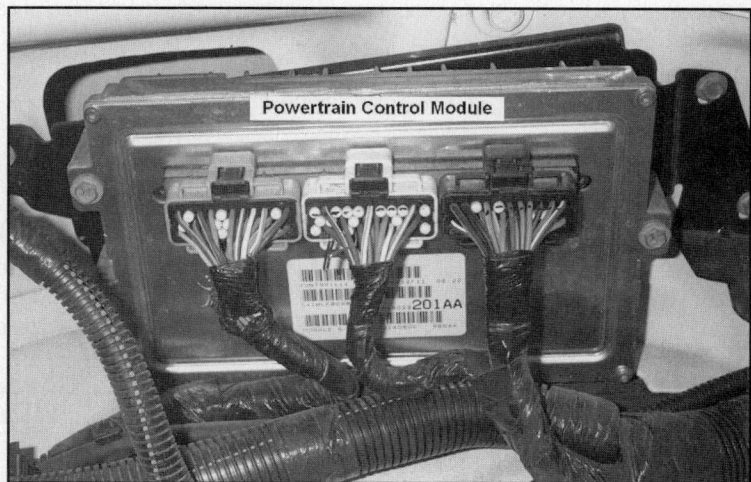

For example, when testing circuits on a 2002 Dodge Durango, a few minutes spent removing the air cleaner from the right fender provided the best access. See the Graphic to the right. Note that the protective covers have been removed from the PCM connectors, and any circuit can be easily identified and back probed. In other cases, PCM access is difficult, and it may be easier to access circuits at the component side of the harness.

Another important point to remember is that any circuit or component controlled by a relay or fused circuit can be monitored from the appropriate fuse box, called the Junction box or Power Distribution Center (PDC) on these Chrysler applications. See the 2002 Durango PDC in the Graphic to the right.

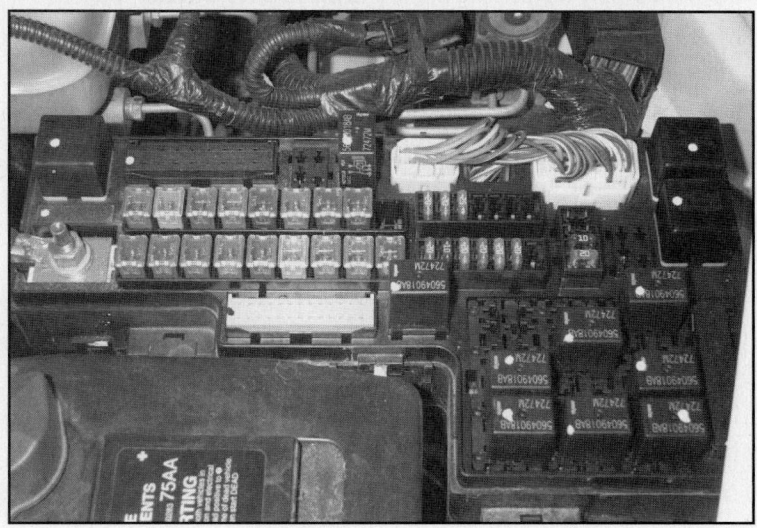

There is generally more than one of each type of relay or fuse. Therefore, swapping a suspect relay from another system may be more efficient than testing the relay itself. Relays and fuses may also be removed and replaced with fused jumper wires for testing circuits. Jumper wires can also provide a loop for inductive amperage tests.

Choosing the easiest way has its limitations, however. Remember that an appropriate signal on a PCM controlled circuit at an actuator means that the signal at the PCM is also good. However, a sensor signal at the sensor does not necessarily mean that the PCM is receiving the same signal. Think about the direction flow through a circuit, and not just what signal is appropriate, to save time without making costly assumptions.

<u>DTC TEST (READ TROUBLE CODES)</u>

The Chrysler DTC Test Procedure is one part of an organized approach to identifying a problem created by a fault in the engine control system. This DTC Test should be the starting point for diagnosis of a drivability problem.

The test steps in this procedure direct you to the next logical step during diagnosis of a customer complaint or vehicle condition. If you use this repair table you will reduce your diagnostic time. It can also help prevent you from replacing good parts!

Clear Codes Warning

Do not clear codes unless specifically directed to do so as part of a repair step. Important I/M Readiness Test and Freeze Frame data may be lost when clearing codes.

Wiring & Connector Inspection

Carefully inspect and check any wiring and connectors (including at the PCM) that could cause an intermittent fault.

Any circuit that is "suspected" of causing a fault should be thoroughly checked for backed-out terminals, improper mating of connectors and terminals, broken connector locks, improperly formed or damaged terminals, poor terminal to wiring connections, physical damage to any wiring harnesses, and corrosion on terminals or wiring.

DTC Test Table

Step	Action	Value	Yes	No
DTC1	*Attempt to start engine, read & record all codes* **Apply the parking brake and block drive wheels. Place the shift lever in (P) or (N) for a M/T.** **Verify that the battery is fully charged. Turn off all electrical loads.** **Try to start the engine (crank for 10 seconds).** **Connect a Scan Tool to DLC. Read and record all trouble codes. Are any codes displayed?** *Notes: If the Scan Tool screen displays No Response, go to Chrysler Test NS-6A. If the Scan Tool display is blank or if a self-test error message appears, go to the Scan Tool Instruction Manual.*	-	Go to the OBD II DTC Index in other Chilton repair manuals or electronic media and repair the code. Perform a PCM Reset. Then run VER-5A to verify the repair.	Go to Step 2.
DTC2	*Select a Symptom that matches the problem* **If no trouble codes are set, select a symptom. If the problem is related to Drivability, go to Drivability Symptom Index.** **If the problem is a No Start condition, go to the No Start Tests and start at step 1.** **If the problem is Speed Control related, go to the Speed Control System Tests.** **If the problem is Charging System related, go to the Charging System Tests.** **Is a repeatable symptom present?**	-	A repeatable drivability symptom is present. Refer to the Drivability Symptom Examples in this section.	The fault is intermittent. Refer to the Intermittent Tests Examples in this section.

DIAGNOSTIC FLOWCHART

■ NOTE: *The Diagnostic Flowchart is not meant to replace any other flowcharts found in other Chilton repair manuals or media.*

The Diagnostic Flowchart shown should be used as described next:

- Follow the flowchart through all of the linked steps.
- If the MIL is on and any trouble codes are set, refer to the Chrysler OBD II DTC Index and look up the code description and conditions.
- To repair a particular trouble code, look up the trouble code repair chart in other Chilton repair manuals or electronic media.
- Then follow all of the suggested trouble code repair steps to conclusion. When the repair is complete, refer to the appropriate repair verification table (at the end of this section) to look up the details that describe how to drive the vehicle to verify a repair.

Diagnostic Flowchart Graphic

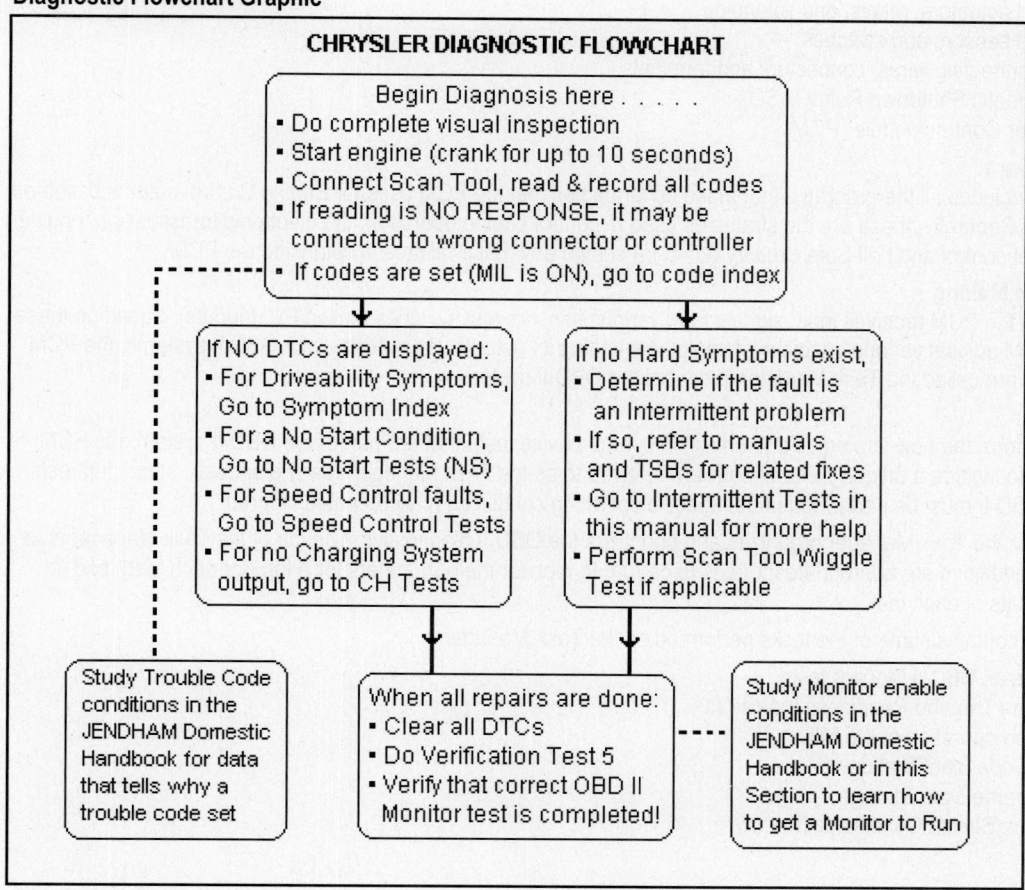

■ Note: *It is important to perform a Verification Test after all repairs are completed. This is true for Charging and Speed Control System as well as Non-OBD II related codes.*

Onboard Diagnostics

INTRODUCTION

The purpose of the Chrysler On-Board Diagnostic System is to provide optimum control of the engine and transmission while meeting the objectives of the OBD II regulations.

At the center of this system is the Powertrain Control Module (PCM) connected to various input and output devices through a wiring harness with two connectors with anywhere from 1 to 4 connectors, depending on controller type. The PCM receives input information from various sensors and switches, performs calculations based on data stored in long term memory, and controls output devices (actuators, relays, and solenoids).

PCM Hardware & Software

The PCM is divided into two main parts, the system hardware and system software. Hardware components include:

- All related actuators, relays, and solenoids
- All related sensors and switches
- All interconnecting wires, connectors and terminals
- The Automatic Shutdown Relay (ASD)
- The Power Control Module (PCM)

System Software

The software includes all the programs that make up strategies that the PCM uses for Engine Control outputs based on related inputs. Generally, these are the strategies used to control engine operation, the electronic transmission, engine idle speed, fuel control and Fail Safe circuitry (Limp-In) should any major failures occur inside the PCM.

PCM Decision Making

As in the past, the PCM receives input signals from various sensors and switches (called PCM inputs). Based on these inputs, the PCM adjusts various engine systems by controlling its outputs. On vehicles with OBD II systems, the PCM operates software called the Task Manager to control the OBD II system.

Task Manager

In order to perform the new strategies and emission control device tests that are part of the OBD II system, the PCM was changed to include a unique piece of software referred to as the Task Manager. Many diagnostic steps and tests required by OBD II must be performed under specific operating conditions (called enable criteria).

The software in the Task Manager organizes and prioritizes the OBD II diagnostics. The job of the Task Manager is to determine if conditions are appropriate for tests to be run, to monitor the parameters for a trip (for each test), and to record the results of each test.

The list below contains some of the tasks performed by the Task Manager:

- Sequence all OBD II Monitor tests
- Monitor the Trip and Readiness Indicators
- Control the operation of the MIL
- Trouble Code Identification
- Freeze Frame Data Storage
- Display the Similar Conditions Window

OBD II TERMINOLOGY

In order to diagnose Chrysler vehicles equipped with an OBD II System, it is important that you understand the terms related to these test procedures. Some of these terms and their definitions are discussed in the next few articles.

Two-Trip Detection

In many cases, an emission related system or component must fail a Monitor test more than once before it activates the MIL. The first time an OBD II Monitor detects a fault during a related trip, it sets a "pending code" in PCM memory. These codes appear when the Memory or Continuous codes are read. For a "pending code" to mature into a hard code (and illuminate the MIL), the original fault must occur for two consecutive trips (two-trip detection). However, a "pending code" can remain in the PCM for a long time before the conditions that caused the code to set <u>reappear</u>.

Fuel Trim and Misfire Detection trouble codes can cause the PCM to flash the MIL after <u>one</u> trip because faults in these systems can cause damage to the catalytic converter.

Pending Code

The term "pending code" is used to describe a fault that has been detected once and is stored in memory. This type of fault has not been detected on two consecutive trips (i.e., it has not matured into a hard code).

It is possible to access a "pending code" with a Generic Scan Tool (GST) on most Chrysler vehicles. Be aware that you may not be able to read a pending code with a Generic Scan Tool on some 1995 phase-in models.

Similar Conditions

If a "pending code" is set because of a Fuel System or Misfire Monitor detected fault, the vehicle must meet *similar conditions* for two consecutive trips before the code matures and the PCM activates the MIL and stores the code in memory. The meaning of *similar conditions* is important when you attempt to diagnose a fault detected by the Fuel System Monitor or Misfire Detection Monitor.

To achieve *similar conditions*, the vehicle must reach the following engine running conditions simultaneously (for the first failure recorded that set the code):

* Engine speed must be within 375 rpm
* Engine load must be within 10%
* Engine warmup state must match the previous state (cold or warm)

Summary - Similar conditions are defined as conditions that match those recorded when the fault was detected and a code was set.

Warmup Cycle

A warmup cycle is defined as vehicle operation (after a cool-down period) when the engine temperature increases by at least 40ºF and reaches at least 160ºF.

Most trouble codes are cleared from the PCM memory after 40 "warmup cycles" if the fault does not reappear.

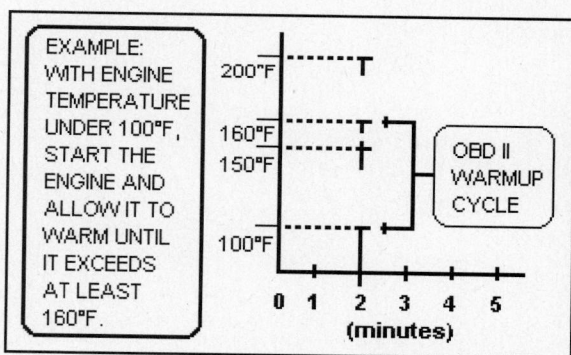

Cylinder Bank - A specific group of engine cylinders that share a common control sensor (e.g., Bank 1 identifies the location of cylinder 1 while Bank 2 identifies cylinders on the opposite bank). Refer to the Chrysler example in the Graphic.

Sensor - If sensors are numbered (Bank 1 Sensor 1), they follow the convention described above. If they are identified with letters ('A', 'B', 'C'), they are manufacturer defined. If only (1) sensor is used, the letter or number may be omitted.

Data Link Connector

Chrysler vehicles equipped with OBD II Systems use a standardized Data Link Connector (DLC). It is typically located between the left end of the instrument panel and 12 inches past vehicle centerline. The connector is mounted out of sight from the passengers, but should be easy to see from outside by a technician in a kneeling position (door open).

DLC Features

The DLC is rectangular in design. It can accept up to 16 terminals. The DLC in the graphic to the right shows many common pin designations, but does not represent any specific vehicle.

OBD II Chrysler vehicle will have power at pin 16, and ground at pins 4 and 5. All other pins vary by year and model. All applications use the SCI circuits, but the pin assignments vary. All vehicles will utilize either the PCI bus, or the CCD bus circuits.

Not all Chrysler vehicle DLCs are wired the same. Always use a wiring diagram when attempting to diagnose these circuits.

Both the DLC and Scan Tool have latching features that ensure that the Scan Tool will remain connected to the vehicle during operation.

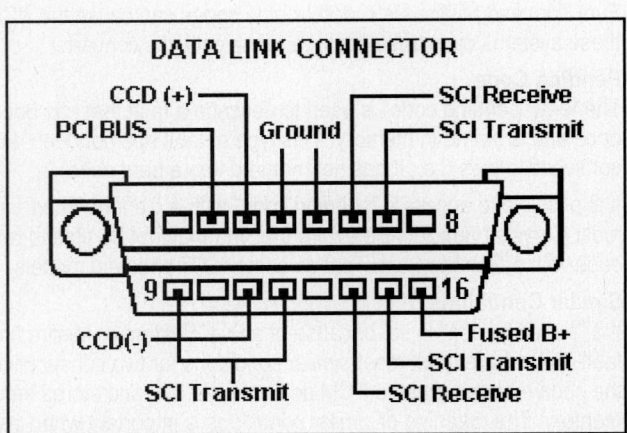

Common uses of the Scan Tool while connected to the DLC include:

- Display the results of the most current I/M Readiness Tests
- Read and clear any diagnostic trouble codes
- Read the Serial Data from the PCM
- Perform Enhanced Diagnostic Tests (System Tests or Actuator Tests)

PCM RESET STEP (CLEAR CODES)

The PCM Reset step allows the Scan Tool to command the PCM to clear all emission-related diagnostic information. Each time this step is done, the PCM will clear the status of the OBD II system Monitors or components that were tested to satisfy their particular "trip" (without any faults occurring). The PCM reset is done through the Generic OBD II Scan Tool software.

The following events occur when a PCM Reset is done:

* All emission-related trouble codes are cleared
* Clears the Freeze Frame data
* Clears the Diagnostic Monitoring results
* Resets the status of the OBD II system Monitors

An example of the warning displayed on a Vetronix Scan Tool prior to completing a PCM Reset step is shown in the example in the Graphic at the top of this page.

Freeze Frame Data

The term *Freeze Frame* is used to describe the engine conditions recorded in the PCM at the time an emission related fault is detected. These conditions include the fuel control state, spark timing, engine speed and load. Frame data is recorded during the first trip on a two-trip fault. It can be overwritten by codes with a higher priority. Some newer Chryslers can record from 2 to 5 unique Freeze Frames.

Importance of Resetting the PCM

It is important to reset the PCM to erase trouble codes and extinguish the Malfunction Indicator Light (MIL). Faults in memory sometimes cause certain engine management strategies to default to backup strategies.

Some faults are recognized by the PCM as possibly causing inaccurate OBD II monitor results, so those related monitors would be turned off until the fault is corrected.

Therefore, the only way to be certain the PCM is able to function normally in closed loop, and is able to run all necessary monitors, is to reset the PCM.

The PCM also adapts to operating conditions over time, and in many cases, adapts to the fault conditions. A PCM reset is the fastest way to reset all of the adaptive values to defaults. Otherwise, the PCM may compensate for a problem that has been repaired already, and a new symptom will be present.

When Not to Reset the PCM

There are some circumstances where a PCM reset is not desirable. For example, some states require that the I/M Readiness Monitors are complete before an emission test is performed. In this case, a trouble code that can be verified as repaired through other means, or was set accidentally during other procedures, should be cleared using the manufacturer specific Scan Tool software. This will not reset the I/M Readiness Monitors, so emission testing will not be delayed by the need to run the entire drive cycle.

```
      CLEAR INFO

     THIS OPERATION
   WILL CLEAR ALL DTC,
    FREEZE FRAME, AND
     READINESS TEST
         DATA.

    DO YOU WISH TO
```

```
   Scan Tool Menus
```

```
       DTC MENU

 F1:  READ DTCs
 F2:  FREEZE DATA
 F3:  PENDING DTCs
 F4:  CLEAR INFO
```

```
 DIAG. TROUBLE CODES
 ECU: $40 (Engine)
 Number of DTCs: 1
 *P0703 Torque
        Converter/Brake
        Switch B Circuit
        Malfunction

 ENTER = FREEZE FRAME
```

```
 DTC..............P0703
 ENGINE SPD....2208RPM
 ECT (°)..........203°F
 VEHICLE SPD.....31MPH
 ENGINE LOAD.....27.8%
 MAP (P).......19.8inHg
 FUEL STAT 1.........CL
 FUEL STAT 2....UNUSED
 ST FT 1............1.6%
 LT FT 1............1.6%
```

MALFUNCTION INDICATOR LAMP

The Malfunction Indicator Lamp (MIL) is located in the instrument cluster, and serves to warn the driver that an emissions management fault has been detected. If the MIL is on, then a Diagnostic Trouble Code (DTC) is set in the PCM memory.

The DTC can be retrieved using as Generic, Enhanced, or OEM Scan Tool. On many later models, the DTC can be displayed on the instrument cluster odometer by cycling the key On/Off/On/Off/On.

The MIL will illuminate for a four (4) seconds as a bulb check when the key is first turned on. If the MIL remains on, a mature DTC is set in the PCM.

Understanding MIL Conditions
The three (3) possible MIL conditions are explained next.

Condition 1: MIL Off
This condition indicates that the PCM has not detected any faults in an emission-related component or system, or that the MIL power or control circuit is not working properly. It is possible to have non-emissions codes set without the MIL illuminating.

Condition 2: MIL On Steady
This condition indicates a fault in an emission-related component or system that could cause tailpipe emissions to exceed the EPA standards.

Condition 3: MIL Flashing
This condition indicates either a misfire or fuel system related fault that could cause damage to a catalytic converter. If the fault is intermittent, the MIL will remain steadily illuminated.

Actions or Conditions to Turn On the MIL
The PCM will turn on the MIL any time a mature emission code is set in the PCM memory. A mature code is any 1-trip DTC that has failed once, or any 2-trip DTC that has failed during 2 consecutive trips.

Most of the codes set by the Comprehensive Component Monitor are 1-trip codes. Most of the codes set by the main monitors are 2-trip DTCs.

Actions or Conditions to Turn Off the MIL
The PCM will turn off the MIL if any of the following actions or conditions occurs:

- The codes are cleared with a Generic, Enhanced, or OEM Scan Tool
- Power to the PCM is removed (this is not recommended)
- The vehicle is driven on three consecutive trips (may include warmup cycles) and meets all of the particular code set conditions without the PCM detecting any faults

The PCM will set a code if a fault is detected that could cause tailpipe emissions to exceed 1.5 times the FTP Standard. However, the PCM will not de-activate the MIL until the vehicle has been driven on three consecutive trips with vehicle conditions similar to actual conditions present when the fault was detected. *This is not just three (3) vehicle startups and trips. It means three trips where certain engine operating conditions are met so that the OBD II Monitor that found the fault can "rerun" and pass that diagnostic test.*

Once the MIL is de-activated, the original code will remain in memory until forty (40) warmup cycles are completed without the fault reappearing. A warmup cycle is defined as a trip where with an engine temperature change of at least 40°F, and where the engine temperature reaches at least 160°F.

DIAGNOSTIC TROUBLE CODES

The OBD II system uses a Diagnostic Trouble Code (DTC) identification system that was established by SAE and the EPA. The first letter of a DTC is used to identify the type of vehicle computer system that failed. The types of systems are shown below:

- The letter 'P' indicates a Powertrain related device
- The letter 'C' indicates a Chassis related device
- The letter 'B' indicates a Body Control related device
- The letter 'U' indicates a Data Link or Network code.

The first number of a diagnostic trouble code (DTC) indicates either a generic (P0xxx) or a manufacturer specific (P1xxx) type of trouble code.

The number in the hundredth position indicates the specific vehicle system or subgroup that failed (i.e., P0300 for a Misfire code, P0400 for an emission system code, etc.).

Trouble Code Example

An example of how to navigate through the Vetronix Scan Tool menus to read a trouble code is shown step by step in the Graphic to the right. The Generic Scan Tool function was used in this example.

The vehicle application in this example was a 1999 Dodge Caravan with a 3.8L V6 (2v) engine. In this example, the cause of the problem was a Torque Converter Clutch (TCC) or brake switch circuit condition. The Comprehensive Component Monitor (CCM) identified the fault on the first failure and stored the code and the code conditions (engine operating conditions) in Freeze Frame. The MIL was illuminated and a hard trouble code was set (one-trip fault detection). Note the Freeze Frame code conditions shown in the Graphic.

1) Select from the three (3) choices on the screen. In this case, F1: SCAN TEST was selected.
2) Select the application from the choices on the screen. In this case, Global OBD II was selected.
3) Select the type of Test from the choices on the screen. In this case, F1: OBD II Functions.
4) Select the type of function from the choices. In this case, F1: READ DTCs was selected.
5) This screen shows an example of the DTC Menu and the Freeze Frame data for DTC P0703. Note the engine speed, vehicle speed and engine load values present when this trouble code was set.

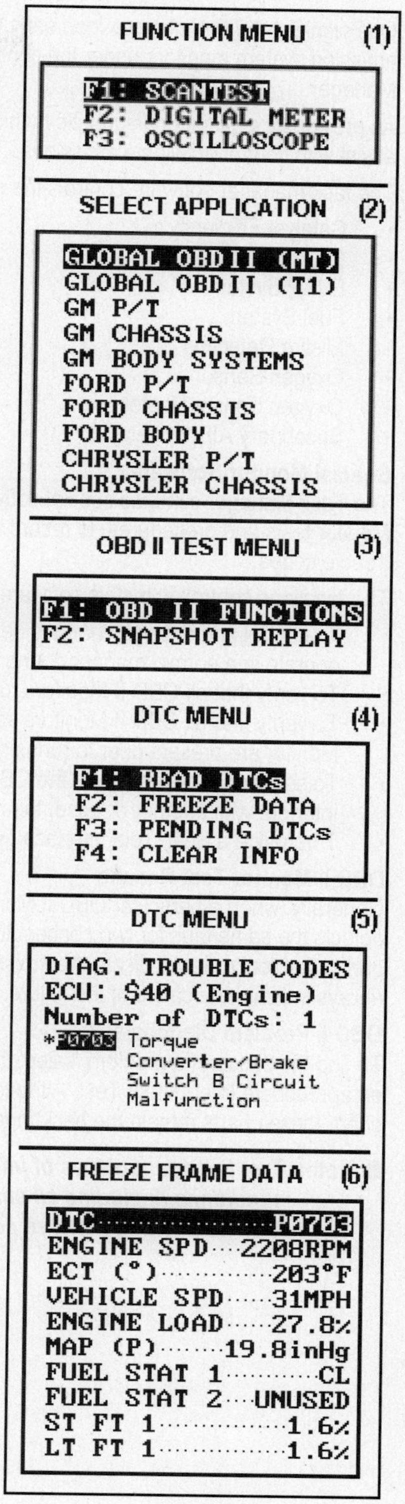

```
FUNCTION MENU        (1)

F1: SCANTEST
F2: DIGITAL METER
F3: OSCILLOSCOPE
```

```
SELECT APPLICATION   (2)

GLOBAL OBDII (MT)
GLOBAL OBDII (T1)
GM P/T
GM CHASSIS
GM BODY SYSTEMS
FORD P/T
FORD CHASSIS
FORD BODY
CHRYSLER P/T
CHRYSLER CHASSIS
```

```
OBD II TEST MENU     (3)

F1: OBD II FUNCTIONS
F2: SNAPSHOT REPLAY
```

```
DTC MENU             (4)

F1: READ DTCs
F2: FREEZE DATA
F3: PENDING DTCs
F4: CLEAR INFO
```

```
DTC MENU             (5)

DIAG. TROUBLE CODES
ECU: $40 (Engine)
Number of DTCs: 1
*P0703 Torque
       Converter/Brake
       Switch B Circuit
       Malfunction
```

```
FREEZE FRAME DATA    (6)

DTC............P0703
ENGINE SPD....2208RPM
ECT (°)..........203°F
VEHICLE SPD....31MPH
ENGINE LOAD....27.8%
MAP (P)......19.8inHg
FUEL STAT 1........CL
FUEL STAT 2..UNUSED
ST FT 1..........1.6%
LT FT 1..........1.6%
```

TASK MANAGER

One significant difference between early Chrysler OBD I and the OBD II version is the use of several dedicated emission system monitors under the control of a special piece of software inside the PCM referred to as the Task Manager.

As previously explained, diagnostic monitors are required to comply with regulations mandated by the EPA designed to assist with control of tailpipe emissions.

The task manager software controls the operation of these OBD II Monitors:

- Catalyst Efficiency
- EGR System
- EVAP System
- Fuel System
- Misfire Detection
- Oxygen Sensor
- Oxygen Sensor Heater
- Secondary AIR System

Special Monitor Software

The Task Manager contains special software designed to allow the PCM to organize and prioritize all of the Main Monitor tests and procedures, to record and display the test results (in Freeze Frame) and any related diagnostic trouble codes.

The functions controlled by this software include:

- To control and arrange changes between the "states" of the diagnostic system so that the vehicle will continue to operate in a normal manner during testing.
- To verify that all OBD II Monitors run during the first two (2) sample periods of the Federal Test Procedure (FTP).
- To verify that all OBD II Monitors and their related tests are sequenced so that the enable criteria for a particular Monitor are present prior to running that Monitor.
- To sequence the running of the OBD II Monitors to eliminate the chance that one of the Monitor tests might interfere with another or upset the normal vehicle operation.
- To provide a Scan Tool interface by coordinating the operation of special tests or data requests.

OBD II Monitor Test Results

Generally, when a particular OBD II Monitor is run and fails a test during a "trip", a "pending code" is set. If that Monitor detects the same fault for two consecutive trips, the MIL is activated and a hard code set in memory. The results of a particular Monitor test indicate the emission related system (and sometimes the component) that failed, but they do not always indicate the cause of the failure.

OBD II Problem Diagnosis

To find the cause of a problem, select the correct trouble code repair chart, a symptom from the Symptom List or select an appropriate Intermittent Test. Although it may not be necessary to do all of the test steps in a trouble code repair chart, these charts remain the backbone of any OBD II System diagnosis and repair.

■ **Note:** *Two important pieces of information that can help speed up the diagnosis are the DTC code conditions (including all enable criteria), and the parameter information (PIDs) stored in the Freeze Frame related to a stored code.*

COMPREHENSIVE COMPONENT MONITOR

OBD II regulations require that all emission related circuits and components controlled by the PCM that could affect emissions be monitored for circuit continuity and out-of-range faults. The comprehensive component monitor (CCM) consists of four different monitoring strategies: two for inputs and two for outputs. Some tests run continuously, some only after actuation. *The CCM Monitor is a 1- trip monitor for emission devices.*

Input Strategies
One input strategy is used to check devices with analog inputs for an open or shorted condition, or an input value that is out-of-range (i.e., IAT, ECT, MAP, TP and TR sensor).

Input Rationality Tests
The input signals to the PCM are constantly monitored for electrical circuit faults. As discussed, some input devices are also tested for *rationality*. In effect, the signal is compared against other inputs and information to see if it makes sense under current engine operating conditions.

Rationality Definition
Rationality is defined as a type of CCM test in which component input signals are compared against other component inputs to verify that the conditions match.

Output Strategies
An Output State Monitor in the PCM checks outputs for opens or shorts by watching the control voltage level of the related device. The control voltage should be low with it "on" and high with the device "off" (i.e., Injectors, Coils, TCC, Solenoids, and relay control circuits).

Output Functionality Tests
The output signals to the PCM are constantly monitored for electrical circuit faults. Some of the PCM outputs are tested for *functionality* in addition to testing for electrical circuit faults. The PCM can send a command to an output device and then monitor certain input signals related to that device for expected changes to verify the command was carried out (i.e., the PCM commands the IAC valve to a specific opening position, it expects to see a target idle speed). If it does not detect a change, the CCM fails and a code is set.

Functionality Definition
Functionality is defined as a type of comprehensive component test in which the PCM output commands are verified by monitoring specific input signals from other PCM components for an expected change.

OBD II Repair Verification "Trip" Graphic

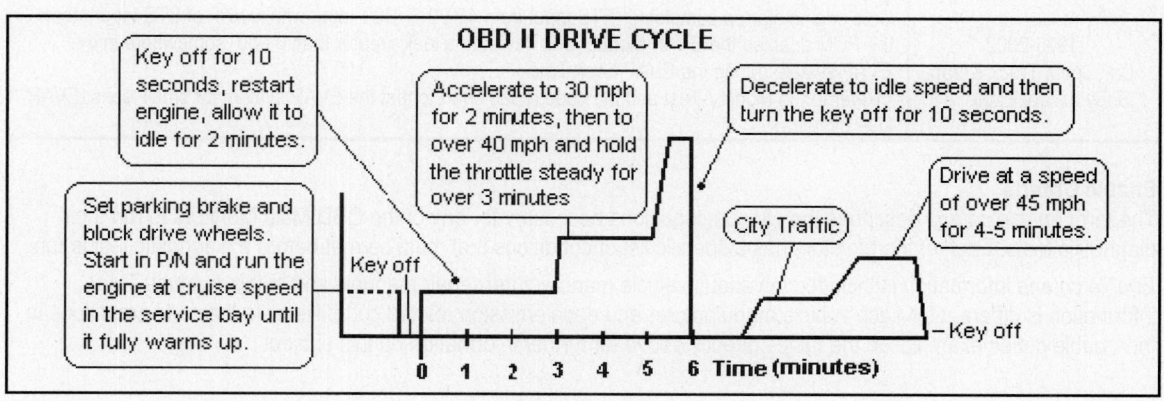

OBD II MAIN MONITORS

One of the key features of OBD Diagnostics is the use of several PCM controlled diagnostic monitors contained within the PCM software structure. These monitors are needed to meet CARB and U.S. EPA OBD II regulations. These monitors (Main Monitors) are advanced beyond OBD I tests. They include the following tests:

- Air Injection System (requires specific driving conditions)
- Catalytic Converter Efficiency (requires specific driving conditions)
- EGR System Monitor (requires idle, acceleration periods to complete)
- EVAP system function and flow monitoring.
- Fuel System Monitor (completes anytime during a trip)
- Misfire Detection (includes identification of the misfiring cylinder if possible)
- Oxygen Sensor Heater
- Oxygen Sensor Monitor (requires steady cruise speed to complete)

OBD II Trip Definition

The term OBD II "Trip" describes a method of driving the vehicle during which the following Main Monitors complete their tests: The actual "trip" requirements are different for each CCM and Main Monitor. Examples of these requirements are included below.

CCM "Trip" Requirements

The CCM "Trip" requirements for DTC P0121 are shown in the table below.

DTC	Trouble Code Title & Conditions
P0121 1T CCM 1995-2002 Car, Jeep, Truck & Van Body: All engines	TP Sensor Does Not Agree With MAP Conditions: No PCM or TCM codes set, and the PCM ran one of these tests: • High Input Test - Engine running, throttle closed and a high MAP input present, then the PCM detected the TP sensor input that was too high (it should be low). • Low Input Test - Engine running, VSS input over 25 mph, throttle open and a low MAP input present, then the PCM detected a TP sensor input that was too low. • Either condition met for 4 continuous seconds. • Refer to Mini-Test or code repair chart in other media.

Main Monitor "Trip" Requirements

The EVAP "Trip" requirements for DTC P0455 are shown in the table below.

DTC	Trouble Code Title & Conditions
P0455 2T EVAP 1996-2002 Car, Jeep, Truck & Van Body except Eagle: All	EVAP Large Leak Detected Conditions: Cold engine startup completed (BTS input from 40-90°F, ECT input within 10°F of BTS input), then the PCM enabled the EVAP Monitor and detected a leak greater than 0.080" somewhere in the EVAP system during the EVAP Leak Test. • Refer to the Monitor Test and the code repair chart to test the EVAP system for small leaks (EVAP LDP system).

Enable Criteria

The term *enable criteria* describes the various conditions necessary for any of the OBD Main Monitors to run their diagnostic tests. Each of the Monitors has a specific list of conditions that must be met before a diagnostic test is run.

Enable criteria information is included in various vehicle manufacturer repair manuals and in this manual. This information is different for each vehicle manufacturer and each emission related code. Refer to the code conditions in the trouble codes examples in the tables directly above for further information on this subject.

SCAN TOOL ACTUATOR TEST MODE

In the Actuator Test Mode (ATM), the Scan Tool is capable of commanding the PCM to cycle various relays and actuators. This function allows components and systems to be checked in the service bay under controlled conditions. The results can often prove whether or not the PCM has control over the component in question. If the individual ATM test results are as expected, the entire circuit (PCM driver, any relays involved, power, ground, and the component itself) has been tested and verified.

When this mode is selected on the Scan Tool, a menu of actuator choices appears. During the test, a command is sent from the tool to the PCM to cycle the actuators "on" and "off". The actuators are cycled according to a routine in the PCM, but can be terminated by the Scan Tool user. For example, ATM tests for relays often cycle "on" and "off" once per second for a 5-minute period. Others remain on for the duration of the test.

All the devices included in Actuator Test Mode are assigned two-digit numbers so that each particular device can be identified, though the numbers are not displayed on all Scan Tools.

Scan Tool Navigation

From the Snap-on Scan Tool main menu, select FUNCTIONAL TESTS, and then ACTUATOR TESTS. Scroll to the desired actuator tests and press the Y button.

From the Vetronix Mastertech SELECT MODE screen, select F4: OBD CONTROLS. From the SELECT CONTROLS screen, select F1: ACTUATOR TEST. Enter the 2-digit number from the desired ATM test on the list and press ENTER.

How to Use Actuator Tests

During each test, listen for the device to click "on" and "off". If a fault is suspected, test the control circuit to the relay or actuator with a DVOM to locate the fault.

It is also important to understand the circuit being actuated. Some are positive controlled circuits, and some are negative controlled. Know what results are desired before the test starts to avoid wasting time.

For example, in the case of the Actuator Test 54: O2 HTR. TEST, the heaters are not directly controlled by the PCM. The wiring diagram reveals that the heaters are on whenever the ASD relay is energized. All this test is doing is energizing the ASD relay with the key off and displaying the HO2S PIDs on the Scan Tool to allow the user to verify that the sensor output changes with the heaters on.

✳✳ CAUTIONS

On engine applications that use distributorless ignition, prior to performing ATM tests, disconnect all spark plug wires at the spark plugs and connect a spark tester between the spark plug and engine ground. Failure to follow these procedures could cause serious injury due to a backfire condition.

Some of the ATM tests (IGN1, IGN2 and IGN3) can cause the air/fuel mixture in the intake manifold to ignite and cause the engine to move or kick backwards or forwards.

Some of the ATM tests can cause the fuel pump to activate (run). Do not disconnect any fuel lines or components except when specifically directed to do so by a repair chart or test. After tests are done, reconnect all fuel lines and components (injectors) so that fuel leaks will not occur and cause injury.

ACTUATOR TEST MODE DESCRIPTIONS

All of the ATM tests are conducted with the key on, engine off. Refer to the Actuator Test Mode Tables for list of common tests. Not all tests are available on all models, and not all available tests are listed. It is best to enter the menu and scroll through the available list for each application, to make sure the capabilities of the PCM and Scan Tool are used to their fullest potential during diagnosis.

There are also many actuator tests available in other systems on the vehicle. When working on PCM, ABS, BCM, Transmission, Vehicle Theft or even Mechanical Instrument Cluster (MIC) problems, be sure to use the available tests.

In addition to these tests, some System Tests may be available, which allow systems and components to be activated with the engine running.

PCM Actuator Test Mode Table (Partial - Engine)

Mode	Relay & Solenoid Test Information	Mode	Relay & Solenoid Test Information
01	Ignition Coil #1	15	Speed Control Solenoids
02	Ignition Coil #2	16	Generator Field
03	Ignition Coil #3	19	EGR Solenoid
04	Injector # 1	34	Speed Control Power Relay
05	Injector # 2	47	Speed Control Vent Solenoid
06	Injector # 3	48	Speed Control Vacuum Solenoid
07	Injector # 4	49	Fuel Pump Relay
08	Injector # 5	54	HO2S Heater Test
09	Injector # 6	55	IAC Motor Open
10	AIS Motor (IAC)	56	IAC Motor Shut
11	Cooling Fan	57	Leak Detection Pump (LDP) Solenoid
12	A/C Clutch	62	Oxygen Sensor Bias
13	ASD Relay	99	All Solenoid and Relays
14	Canister Purge		

TCM Actuator Test Mode Table (Partial - Transmission)

Relay & Solenoid Test Information	Relay & Solenoid Test Information
Cycle Low-Reverse Solenoid	Cycle 4C Solenoid
Cycle 2C Solenoid	Cycle MS Solenoid
Cycle Underdrive Solenoid	Pressure Control Solenoid
Cycle Overdrive Solenoid	

BCM Actuator Test Mode Table (Partial - Body)

Relay & Solenoid Test Information	Relay & Solenoid Test Information
All External Lights	Fog Lamps
Chime	Headlamp Relay
Courtesy Lamp	Horn Relay
Decklid Release Solenoid	Park/Tail Lamp Relay
Door Lock Relay	Wiper Deicer Output
Door Unlock Relay	Wiper motor
Driver Door Unlock Relay	

SCAN TOOL SYSTEM TEST MODE

Unlike the ATM Tests, the System Tests are performed from the Scan Tool with the engine running. For the Powertrain Control Module (PCM), these tests include:

- RPM CONTROL
- PURGE TEST
- LDP SYSTEM TEST
- SET SYNCH TEST
- GENERATOR FIELD TEST
- EGR SYSTEM TEST
- MISFIRE COUNTERS
- INJECTOR KILL TEST

System Tests are designed to exercise these systems and components under more normal operating conditions. The result of these tests can be observed in the Scan Tool data lists (PIDs), or measured directly with a DVOM or Lab Scope.

For example, during the PURGE TEST, the Scan Tool PIDs can be observed. If the system is actually purging fuel vapors from the canister, the air/fuel mixture will be rich, and the short term fuel trim PIDs will respond by shifting negative (taking away fuel). By understanding how the system tests operate, and how the PCM uses sensor information to make decisions, complicated diagnostics become much simpler.

The Transmission Control Module (TCM) also has some available System Tests in addition to the engine-off ATM Tests. The following list contains examples of possible System Test choices, depending on year and model:

- CVI MONITOR
- RPM DISPLAY
- SET PINION FACTOR
- QUICK LEARN

TESTING EFFICIENTLY

Most of these PCM and TCM System Tests are discussed and demonstrated throughout Sections 3 through 6 of this manual. Detailed explanations of the tests are included along with examples of how to use the tests and tools to diagnose more efficiently.

Some of the tests available on each vehicle are not specifically covered in other Sections. However, the principles of logical testing and tool selection are universal. Learn to use the tools and tests that are available, as well as strategies to combine tools and tests to isolate faults more quickly. Most circuit and component tests can be logically eliminated before time-consuming test procedures are performed. Select a combination tools and tests that will result in the most information in the least amount of time.

For example, a suspect circuit should not be tested until the circuit state is observed on a Scan Tool. If the CKP sensor and circuit are to be tested, first observe the end result of CKP sensor activity – the RPM PID! If the Scan Tool shows that the RPM PID value is normal, then do not test the circuit.

Before testing any component or circuit, attempt to verify that a problem really exists. This will save diagnostic time and, more importantly, prevent the replacement of good components. Most flow charts are based on the assumption that a circuit or component fault is present when entering the chart. If the fault is not present at that time, the end result can be misleading.

OBD III DIAGNOSTICS

A third version of On-Board Diagnostics (OBD III) may be introduced in this next decade. This article includes comments about what this version of OBD might include and the role of the California Air Resources Board (CARB) in the design of this new system.

How will it work? While no final documents have been released about OBD III at this time, it now appears likely that several new technologies are being considered to detect and relay emissions faults as part of OBD III. The highlights of one program under study by Sierra Research (Sacramento, CA) for the Air Resources Board are presented next.

OBD III Highlights
- Allows wireless communication of the OBD system status to a manufacturer (e.g., GM "On Star" system) or to a state contractor responsible for emissions-related data. This could be similar to the role MCI has under the current Smog Check program.
- Vehicle owner decides whether the system is active or not active.
- Eliminates need for Smog Check (with system activated).
- Eliminates need for re-inspection of faulty vehicle after repair (with system activated).
- Life-cycle cost would be lower than the cost of the Smog Check program test fees.

OBD III System Graphic

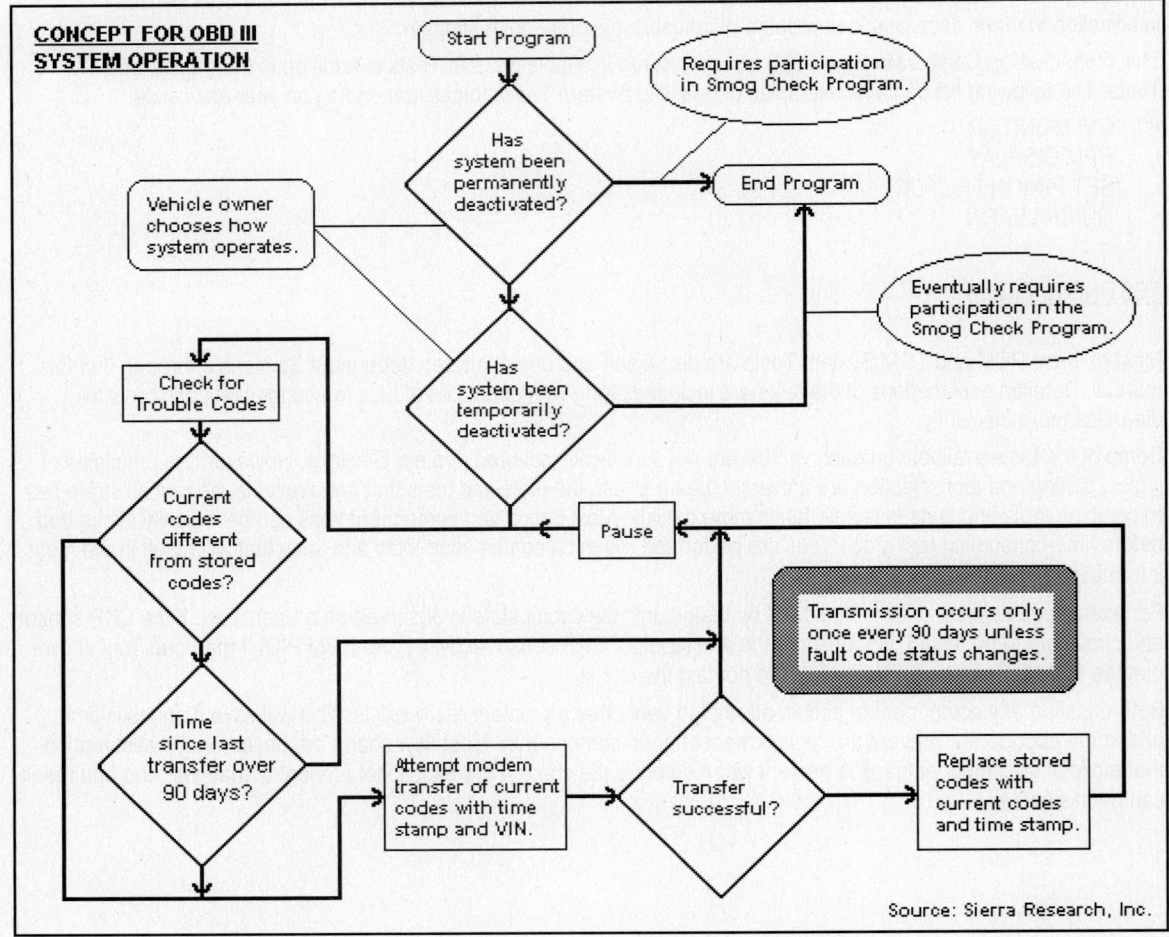

It also appears that the OBD II System diagnostics will continue to be the diagnostic "backbone" of the next version of On-Board Diagnostics. However, with OBD III, the vehicle owner may have the option of having a vehicle equipped with this system either activated or deactivated. Vehicles in this program may be equipped with a transmitter or transponder that would allow the vehicle to communicate with a monitoring service by wireless communication (i.e., a cell phone or a roadside reader).

As stated, the vehicle owner would have the choice of having the system "activated" or "deactivated". If the system were "on", they would not have to participate in the Smog Check program. Owners who decide not to activate the system would be required to be in the Smog Check program.

Cell Phone

A method of wireless communication that uses the vehicle cell phone has been studied. In this case, the phone would communicate with the regulatory agency when the vehicle has a trouble code set, and the agency would alert the owner of the emissions problem.

Roadside Reader

Another method of wireless communication that uses a roadside reader has also been studied (since 1994). This device can be used from either a fixed location "with a portable unit" or a mobile unit and is capable of reading eight lanes of bumper to bumper traffic at speeds of up to 100 mph.

In this system, the vehicle would have a transmitter or transponder capable of sending "stored" trouble code data to the reader unit. If the "roadside reader" detects a fault, it would send the vehicle identification number (VIN) plus the trouble code information to the regulatory agency. If a vehicle were out of compliance, the owner would receive an "out of compliance" notice from the regulatory agency to bring the vehicle in for testing.

How would OBD III be incorporated into the current Inspection and Maintenance (I/M) program? It could be used to generate an *out of cycle* inspection. Once a fault is detected, a notice would be mailed to the vehicle owner requiring an *out of cycle* inspection within a certain number of days, at the next registration date, or at resale.

OBD III Test Example Graphic

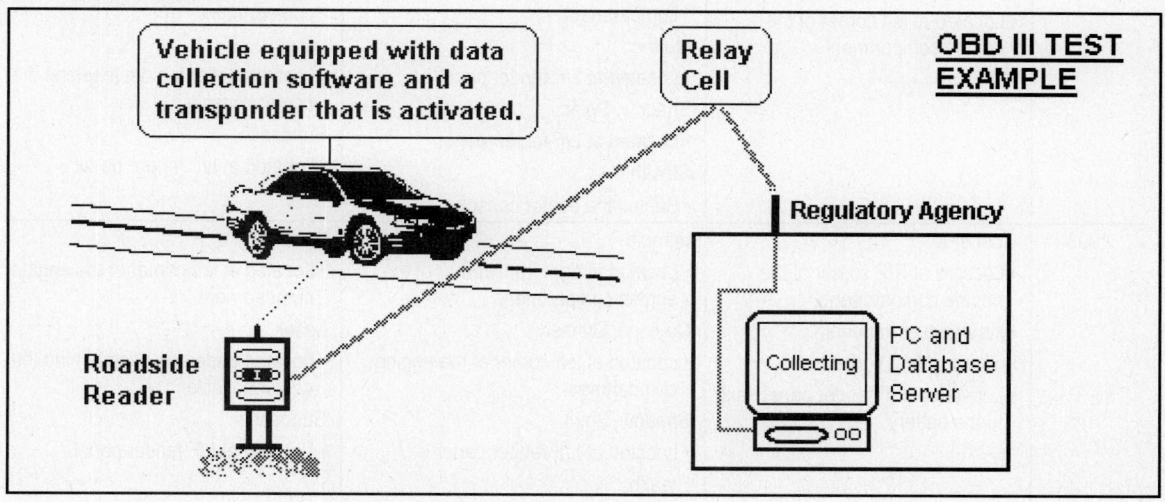

REFERENCE INFORMATION

This table can be used to identify the location of the Powertrain Control Module (PCM) for 1990-2002 Model Daimler/Chrysler vehicles.

PCM Location Table - Cars

Year	Chrysler	Dodge	Plymouth
1990	Fifth Avenue, Imperial, LeBaron, New Yorker • Located at left corner of the engine compartment	Daytona, Dynasty • Located at left corner of the engine compartment Monaco • Located at L/F fender panel Shadow, Spirit • Located at L/F fender panel	Acclaim • Located at left corner of the engine compartment Laser • Located under the dash behind the center console Sundance • Located at L/F fender panel
1991	Fifth Avenue, Imperial, LeBaron, New Yorker • Located at left corner of the engine compartment	Daytona, Dynasty • Located at left corner of the engine compartment Monaco • Located at L/F fender panel Shadow, Spirit • Located at L/F fender panel Stealth • Behind the center console	Acclaim • Located at left corner of the engine compartment Laser • Located under the dash behind the center console Sundance • Located at L/F fender panel
1992	Fifth Avenue, Imperial, LeBaron, New Yorker • Located at left corner of the engine compartment	Daytona, Dynasty • Located at left corner of the engine compartment Monaco • Located at L/F fender panel Shadow, Spirit • Located at L/F fender panel Stealth • Behind the center console	Acclaim • Located at left corner of the engine compartment Laser • Located under the dash behind the center console Sundance • Located at L/F fender panel
1993	Concorde • Located at R/F corner of the engine compartment Fifth Avenue, Imperial, LeBaron, New Yorker • Located at L/F fender panel next to the battery	Intrepid • Located at right front corner of the engine compartment Daytona, Dynasty • Located at left corner of the engine compartment Shadow, Spirit • Located at L/F fender panel Stealth • Behind the center console	Acclaim • Located at left corner of the engine compartment Laser • Located under the dash behind the center console Sundance • Located at L/F fender panel
1994	Concorde, LHS • Located at R/F corner of the engine compartment LeBaron • Located at L/F fender panel next to the battery New Yorker • Located at R/F corner of the engine compartment	Intrepid • Located at right front corner of the engine compartment Shadow, Spirit • Located at left front fender panel Stealth • Behind the center console	Acclaim • Located at left corner of the engine compartment Laser • Located under the dash behind the center console Sundance • Located at L/F fender panel

PCM Location Table - Cars (Continued)

Year	Chrysler	Dodge	Plymouth
1995	Cirrus • Located at left fender side shield Concorde, LHS • Located at right front corner of the engine compartment LeBaron, New Yorker • Located at left corner of the engine compartment Sebring • Located in front of battery at left side of engine compartment	Avenger • Located in front of battery at left side of engine compartment Intrepid • Located at right front corner of the engine compartment Neon, Spirit, Stratus • Located at left front inner fender Stealth • Behind the center console	Acclaim • Located at left corner of the engine compartment Neon • Located at the left front inner fender next to the washer fluid bottle.
1996	Cirrus • Located at left fender side shield Concorde, LHS, New Yorker • Located at right front corner of the engine compartment Sebring Convertible • Located at L/F inner fender Sebring Coupe • Located in front of battery at left side of engine compartment	Avenger • Located in front of battery at left side of engine compartment Intrepid • Located at right front corner of the engine compartment Neon, Stratus • Located at left fender side shield Stealth • Behind the center console	Breeze • Located at left fender side shield Neon • Located at the left front inner fender next to the washer fluid bottle.
1997	Cirrus • Located at left fender side shield Concorde, LHS • Located at right front corner of the engine compartment Sebring Convertible • Located at left front inner fender Sebring Coupe • Located in front of battery at left side of the engine compartment	Avenger • Located in front of battery at left side of engine compartment Intrepid • Located at right front corner of the engine compartment Neon • Located at left front inner fender Stratus • Located at left fender side shield	Breeze • Located at left fender side shield Neon • Located at the left front inner fender next to the washer fluid bottle.
1998	Cirrus • Located at left fender side shield at the power distribution center Concorde, LHS • Located at left rear of the engine compartment Sebring Convertible • Between the air cleaner and the power distribution center Sebring Coupe • Located in front of battery at left side of the engine compartment	Avenger • Located in front of battery at left side of engine compartment Intrepid • Located at left rear of the engine compartment Neon • Located at left front inner fender Stratus • Located at left fender side shield	Breeze • Located at left fender side shield at the power distribution center Neon • Located at the left front inner fender next to the washer fluid bottle.

PCM Location Table - Cars (Continued)

Year	Chrysler	Dodge	Plymouth
1999	300 M • Located at left rear engine area Cirrus • Located at left fender side shield at the power distribution center Concorde, LHS • Located at left rear of the engine compartment Sebring Convertible • Between the air cleaner and the power distribution center Sebring Coupe • Located in front of battery at left side of the engine compartment	Avenger • Located in front of battery at left side of engine compartment Intrepid • Located at left rear of the engine compartment Neon • Located at left front inner fender Stratus • Located at left fender side shield at the power distribution center	Breeze • Located at left fender side shield at the power distribution center Neon • Located at the left front inner fender next to the washer fluid bottle.
2000-02	Cirrus • Located at left fender side shield by the power distribution center 300 M, Concorde, LHS • Located at left rear of the engine area by power distribution center Sebring Convertible • Between the air cleaner and the power distribution center Sebring Coupe • Located in front of battery at left side of the engine compartment	Avenger • Located in front of battery at left side of the engine compartment Intrepid • Located at left rear of the engine area by power distribution center Neon • Located at left front of engine compartment below air cleaner Stratus • Located at left fender side shield at the power distribution center	Breeze • Located at left fender side shield at the power distribution center Neon • Located at left front of engine compartment below the air cleaner

PCM Location Table - Sports Utility Vehicles

Year	Chrysler	Dodge	Plymouth
1998	---	Durango • Located at the right front inner fender forward of windshield washer reservoir	---
1999	---	Durango • Located at the right front inner fender forward of windshield washer reservoir	---
2000-02	PT Cruiser • Located at the left side of engine compartment on firewall	Durango • Located at the right front inner fender forward of windshield washer reservoir	---

PCM Location Table - Ram Trucks & Ram Van/Wagons

Year	Chrysler	Dodge	Plymouth
1990	---	Dakota • Located at right inner fender panel in engine area Pickup, Ramcharger • Located at the left inner panel of engine area Ram-50 • Located behind right side of dash above kick panel Ram Van/Wagon • Located at the center of the firewall above engine	---
1991	---	Dakota • Located at right inner fender panel in engine area Pickup, Ramcharger • Located at the left inner panel of engine area Ram-50 • Located behind right side of dash above kick panel Ram Van/Wagon • Located at the center of the firewall above engine	---
1992	---	Dakota • Located at right inner fender panel in engine area Pickup, Ramcharger • Located at the left inner panel of engine area Ram-50 • Located behind right side of dash above kick panel Ram Van/Wagon • Located at the center of the firewall above engine	---
1993	---	Dakota • Located at right inner fender panel in engine area Pickup, Ramcharger • Located at the left inner panel of engine area Ram-50 • Located behind right side of dash above kick panel Ram Van/Wagon • Located at the center of the firewall above engine	---
1994	---	Dakota • Located at right inner fender panel in engine area Ram Pickup • Located at the right side of firewall in engine area Ram Van/Wagon • Located at the center of the firewall above engine	---
1995	---	Dakota • Located at right inner fender panel in engine area Ram Pickup • Located at the right side of firewall in engine area Ram Van/Wagon • Located at the center of the firewall above engine	---

PCM Location Table - Ram Trucks & Ram Van/Wagons

Year	Chrysler	Dodge	Plymouth
1996	---	Dakota • Located at right inner fender panel in engine area Pickup (R1500, R2500, R3500) • Located at the right side of firewall in engine area Ram Van (B1500, B2500, B3500) • Located at the center of the firewall above engine Ram Wagon (B1500, B2500, B3500) • Located at the center of the firewall above engine	---
1997	---	Dakota • Located at right inner fender panel in engine area Pickup (R1500, R2500, R3500) • Located at the right side of firewall in engine area Ram Van (B1500, B2500, B3500) • Located at the center of the firewall above engine Ram Wagon (B1500, B2500, B3500) • Located at the center of the firewall above engine	---
1998	---	Dakota • Located at right inner fender panel in engine area Pickup (R1500, R2500, R3500) • Located at the right side of firewall in engine area Ram Van (B1500, B2500, B3500) • Located at the center of the firewall above engine Ram Wagon (B1500, B2500, B3500) • Located at the center of the firewall above engine	---
1999	---	Dakota • Located at right inner fender panel in engine area Pickup (R1500, R2500, R3500) • Located at the right side of firewall in engine area Ram Van (B1500, B2500, B3500) • Located at the center of the firewall above engine Ram Wagon (B1500, B2500, B3500) • Located at the center of the firewall above engine	---
2000-02	---	Dakota • Located at right inner fender panel in engine area Pickup (R1500, R2500, R3500) • Located at the right side of firewall in engine area Ram Van (B1500, B2500, B3500) • Located at the center of the firewall above engine Ram Wagon (B1500, B2500, B3500) • Located at the center of the firewall above engine	---

PCM Location Table - Minivans

Year	Chrysler	Dodge	Plymouth
1990	Town & Country • Located at left inner fender panel in engine compartment	Caravan • Located at left inner fender panel in engine compartment	Voyager • Located at left inner fender panel in engine compartment
1991	Town & Country • Located at left inner fender panel in engine compartment	Caravan • Located at left inner fender panel in engine compartment	Voyager • Located at left inner fender panel in engine compartment
1992	Town & Country • Located at left inner fender panel in engine compartment	Caravan • Located at left inner fender panel in engine compartment	Voyager • Located at left inner fender panel in engine compartment
1993	Town & Country • Located at left inner fender panel in engine compartment	Caravan • Located at left inner fender panel in engine compartment	Voyager • Located at left inner fender panel in engine compartment
1994	Town & Country • Located at left inner fender panel in engine compartment	Caravan • Located at left inner fender panel in engine compartment	Voyager • Located at left inner fender panel in engine compartment
1995	Town & Country • Located at left inner fender panel in engine compartment	Caravan • Located at left inner fender panel in engine compartment	Voyager • Located at left inner fender panel in engine compartment
1996	Town & Country • Located next to the battery & PDC in engine compartment	Caravan • Located next to the battery & PDC in engine compartment	Voyager • Located next to the battery & PDC in engine compartment
1997	Town & Country • Located next to the battery & PDC in engine compartment	Caravan, Grand Caravan • Located next to the battery & PDC in engine compartment	Voyager, Grand Voyager • Located next to the battery & PDC in engine compartment
1998-1999	Town & Country • Located next to the battery & PDC in engine compartment	Caravan, Grand Caravan • Located next to the battery & PDC in engine compartment	Voyager, Grand Voyager • Located next to the battery & PDC in engine compartment
2000	Town & Country • Located at left side of engine compartment near the battery	Caravan, Grand Caravan • Located at left side of engine compartment near the battery	Voyager, Grand Voyager • Located at left side of engine compartment near the battery
2001-2002	Town & Country, Voyager • Located at left side of engine compartment near the battery	Caravan, Grand Caravan • Located at left side of engine compartment near the battery	

PCM Location Table - Jeep Applications

Year	Cherokee, Comanche	Grand Comanche	Wrangler
1991-95	Cherokee, Comanche • Located on left front apron behind air cleaner	Grand Comanche, Wagoneer • Located at right rear of engine area near A/C receiver Drier	Wrangler • Located on the firewall
1996-97	Cherokee • Located on the left fender side shield	Grand Cherokee • Located at right rear of engine area near A/C receiver Drier	Wrangler • Located in engine area on left side of firewall
1998-2002	Cherokee • Located on the left fender side shield	Grand Cherokee • Located at the right side of engine on the firewall	Wrangler • Located at right rear of engine on the firewall

REPAIR VERIFICATION TABLES

Carefully inspect the vehicle to verify that all engine components are connected and properly installed. Check the coolant level and oil level before continuing the procedure.

VER-5A Test Table

Verification Test 5A (OBD II Trouble Codes)
Step 1 - For ABS and Air Bag systems, enter the correct VIN and Mileage into the PCM. Then erase the codes in the ABS and Air Bag modules. Note: If the PCM has been changed and the correct VIN and Mileage are not programmed, a DTC will be set in the ABS and Air Bag modules.
Step 2 - If the vehicle has a SKIM theft-alarm system, connect a Scan Tool to the DLC and select Engine MISC and place the SKIM in secured access mode (use the appropriate PIN Code for the vehicle). Then select Update the Secret Key Data and the data will be transferred from the SKIM to the PCM.
Step 3 - Once all trouble codes have been repaired, return to the Scan Tool main menu and repeat the diagnostic test for the system (e.g., run the appropriate Monitor for a previously repair OBD trouble code).
Step 4 - Connect a Scan Tool to the DLC. Verify fuel tank is from 1/4 to 3/4 full and turn off all accessories.
Step 5 - The proper way to verify an OBD II code is repaired is to allow the PCM to run the appropriate Monitor(s) and increment a Global Good Trip. The Monitor running the test can be viewed on the tool. The DTC enable criteria must be met before the PCM will run the Monitor.
Step 6 - Monitor the pretest enable criteria on the tool until all conditions have been met. Once they have been met, watch the appropriate Monitor during a Road Test. If the trouble code resets (if it appears on the Monitor screen during the trip), the repair is not complete. If this is the case, check for any related TSB's or Flash Updates, and then return to the Scan Tool main menu. Retest for the trouble code or symptom.
Step 7 - If a new code sets go to the Code List and perform the tests specified for that particular code.
Step 8 - If the Monitor runs and the Good Trip Counter increments with no new trouble codes set, the repair was successful and is complete. Perform a PCM Reset to erase all diagnostic trouble codes.

VER-5A2 Test Table

Verification Test 5A2 (For Codes that use Similar Conditions Criteria)
Step 1 - Monitor the Similar Conditions to attempt to duplicate the conditions present when the code set. If the conditions can be duplicated, the Good Trip Counter will change to one (1) or higher.
Step 2 - If the conditions cannot be duplicated, clear the trouble codes. Then disconnect the Scan Tool.
Step 3 - If the repair code resets or the Monitor runs and fails, check for related TSB's or Flash Updates.
Step 4 - If a new code sets go to the Code List and perform the tests specified for that particular code.
Step 5 - If the Monitor ran (the Good Trip Counter incremented with no new codes) the repair is complete.
Step 6 - Erase the trouble codes and disconnect the Scan Tool.

VER-5A3 Test Table

Verification Test 5A3 (O2S Heater Failure & Slow Response, Catalyst Fault)
Step 1 - Monitor the pretest enable criteria on the tool until all conditions have been met. Once they have been met, watch the appropriate Monitor during a Road Test. If the trouble code resets (if it appears on the Monitor screen during the trip), the repair is not complete. If this is the case, check for any related TSB's or Flash Updates, and then return to the Scan Tool main menu. Retest for the trouble code or symptom.
Step 2 - If a new code sets go to the Code List and perform the tests specified for that particular code.
Step 3 - If the repair code resets or the Monitor runs and fails, check for related TSB's or Flash Updates.
Step 4 - If the Monitor ran (the Good Trip Counter incremented with no new codes) the repair is complete.
Step 5 - Erase the trouble codes and disconnect the Scan Tool.

VER-6A Test Table

Verification Test 6A (For EVAP Codes - use special LDP Dealer Test Mode)
Explanation - The LDP Dealer Test Mode has been added to the DRB III to verify repairs to the system. A DRB software program was written that causes the PCM to run the LDP Monitor as part of this test. Test failures are indicated through a DTC. The LDP Test Mode is used to run a total system performance test.

PCM OPERATING RANGE & RESISTANCE TABLES

ECT & IAT Sensor Conversion Table (Most Applications)

Degrees F	Voltage	Degrees F	Voltage
-20°F	4.70v	120°F	1.25v
-10°F	4.57v	130°F	3.77v
0°F	4.45v	140°F	3.60v
10°F	4.30v	150°F	3.40v
20°F	4.20v	160°F	3.20v
30°F	3.90v	170°F	3.02v
40°F	3.60v	180°F	2.80v
50°F	3.30v	190°F	2.60v
60°F	3.00v	200°F	2.40v
70°F	2.75v	210°F	2.20v
80°F	2.44v	220°F	2.00v
90°F	2.15v	230°F	1.80v
100°F	1.83v	240°F	1.62v
110°F	1.57v	250°F	1.45v

■ NOTE: *The PCM switches in a pullup resistor at 120°F to change to hot engine operation.*

MAP Sensor Table

Inches of Mercury Absolute	Inches of Mercury Vacuum	MAP Sensor Signal Voltage
31.0" Hg	0.50 psi	4.8v
29.92" Hg	0.00 psi	4.6v
27.00" Hg	2.92 psi	4.1v
25.00" Hg	4.92 psi	3.8v
23.00" Hg	6.92 psi	3.45v
20.00" Hg	9.92 psi	2.92v
15.00" Hg	14.92 psi	2.09v
10.00" Hg	19.92 psi	1.24v
5.00" Hg	24.92 psi	0.45v

Output Device Resistance Table (Most Applications)

Component	Static Ohm Value (at 68°F)
Air Injection Solenoid	33-39 ohms
EGR Solenoid	25-35 ohms
EGR Solenoid (Turbo)	33-39 ohms
EGR Transducer Solenoid	25-35 ohms
EGR Transducer Solenoid (Turbo)	36-44 ohms
EVAP Solenoid	25-70 ohms
EVAP Solenoid (Turbo)	33-39 ohms
Fuel Injectors	11-15 ohms
Fuel Pump Relay	35-75 ohms
Fuel Pressure Solenoid (Solenoid)	36-44 ohms
Fuel Injectors (Turbo)	2-3 ohms
MFI Relay	35-75 ohms
MFI Relay (Turbo)	90 ohms
IAC Motor Terminals 1 to 4 & 2 to 3	38-52 ohms
Wastegate Solenoid (Turbo)	36-44 ohms

CHRYSLER OBD II GLOSSARY

Glossary of Acronyms	
A/C - Air Conditioning	A/D - Analog to Digital Converter
A/F - Air Fuel Ratio	A/T - Automatic Transmission
AIR - Secondary Air Injection System	AIS - Automatic Idle Speed
ASD - Automatic Shutdown Relay	BARO - Barometric Pressure Sensor
BCM – Body Control Module	BTS - Battery Temperature Sensor
CARB - California Air Resources Board	CCP - Comprehensive Component Monitor
CCD - Chrysler Collision Detection	CNG - Certified Natural Gas
CNG - Certified Natural Gas	CO - Carbon Monoxide
CO2 - Carbon Dioxide	CYL - Cylinder
DLC - Data Link Connector	DTC - Diagnostic Trouble Code
DVOM - Digital Volt/Ohm Meter	ECT - Engine Coolant Temperature
EGR - Exhaust Gas Recirculation	EVAP - Evaporative Emissions System
EPA - Environmental Protection Agency	EEPROM - Electrically Erasable Programmable Read Only Memory
GPS - Governor Position Sensor	HC - Hydrocarbons
HO2S - Heated Oxygen Sensor	HO2S1 - Upstream HO2S
HO2S2 - Up or Downstream HO2S	HO2S3 - Downstream HO2S
Hz – Hertz	IAC - Idle Air Control Motor or Valve
ISO - International Standardization Organization	IAT - Intake Air Temperature
JTEC - Jeep/Truck Engine Control Module	KS - Knock Sensor
KSM - Knock Sensor Module	LDP - Leak Detection Pump
MAF - Mass Airflow Sensor	MAP - Manifold Air Pressure Sensor
MFI - Multiport Fuel Injection	MPD - Manifold Pressure Differential
MIL - Malfunction Indicator Lamp	M/T - Manual Transmission
MTV - Manifold Tuning Valve	mv - Millivolts
N.C. - Normally Closed Switch	NGC - New Generation Controller
N.O. - Normally Open Switch	NOx - Oxides of Nitrogen
NTC- Negative Temperature Coefficient	OBD I – On-Board Diagnostics, First Generation
OBD II – On-Board Diagnostics, Second Generation	OSS - Output Speed Sensor
PCI - Programmable Communications Interface	PCM - Powertrain Control Module
PDC - Power Distribution Center	PEP - Peripheral Expansion Port
P/N - Park/Neutral	PSS - Power Steering Pressure
PWM – Pulsewidth Modulated	REF, VREF - Voltage Reference
RWAL - Rear Wheel Antilock Brakes	RPM - Revolutions Per Minute
SBEC - Single Board Engine Controller	SCI - Standard Communication Interface
SRI - Service Reminder Indicator	SMEC - Single Module Engine Controller
TDC - Top Dead Center	TCC - Torque Converter Clutch
TCM - Transmission Control Module	TPS - Throttle Position Sensor
TRS - Transmission Range Sensor	TTS - Transmission Temp. Sensor
VSS - Vehicle Speed Sensor	WOT - Wide Open Throttle

OBD II TERMS

ABS - A two (or four wheel) Antilock Brake System designed to reduce the occurrence of wheel lockup during severe brake applications. It includes its own ABS controller.

A/F Ratio - The ratio (by weight) of air to gasoline entering the intake in a gasoline engine. The ideal ratio for complete combustion of air and fuel is 14.7 parts of gasoline to 1 part of fuel.

Backfire - The act of fuel igniting in either the intake system or the exhaust system.

Baud Rate - The rate at which a PCM is able to transfer and receive data. Baud rate is measured in bits per second.

Barometric Pressure (BARO) - Barometric pressure is the pressure created by atmosphere (this parameter changes as the altitude changes).

Big Slope - A term associated with Oxygen Sensor signals used to indicate the rate of HO2S input signal change over time.

Carbon Dioxide (CO2) - A relatively harmless gas that is a by-product of complete combustion. The chemical composition is CO_2.

Carbon Monoxide (CO) - A poisonous gas that is a result of incomplete combustion due to lack of oxygen. The chemical composition is CO.

Catalyst - This is a material that promotes a chemical reaction without being changed by the reaction. The noble metals Platinum, Palladium, and Rhodium are used as catalysts in automotive catalytic converters.

Closed Loop - A state in which the PCM controls and adjusts the A/R Mixture based on inputs from the upstream Heated Oxygen Sensor.

Comprehensive Component - Any component, other than a Main Monitor, that has any effect on vehicle emissions.

Detonation - A mild to severe ping, especially under loaded engine conditions.

Diagnostic Trouble Codes - Trouble codes associated with PCM fault messages that can be retrieved using a diagnostic Scan Tool.

Evaporative Emissions - Hydrocarbons emissions produced by evaporation of raw fuel.

Exhaust Gas Recirculation - The routing of exhaust gas into the intake manifold to dilute the A/F Mixture thus lowering combustion chamber temperature. The reduction of operating temperature reduces the emissions of NOx.

Federal Test Procedure - A transient-speed mass emissions test conducted on a loaded dynamometer. It is the test that, by law, vehicle manufacturers use to certify that new vehicles are in compliance with HC, CO and NOx emissions.

Flex Fuel - A term pertaining to a vehicle that is flexible in its fuel requirements (i.e., a vehicle that can use Natural Gas for its fuel).

Freeze Frame - A snapshot of engine operating conditions taken when a fault occurs. A Freeze Frame normally contains engine speed (RPM), Throttle Position, Engine Load, Vehicle Speed, Engine Temperature, and Open and Closed Loop Status information.

Fuel Metering - A method to control the A/F Mixture entering the combustion chamber.

Good Trip - A trip counter in which various OBD II Monitors passed a test under specific test conditions (i.e., it is the fulfillment of specific test parameters during a drive cycle). A Good Trip is counted and used for extinguishing the MIL and for erasing codes.

Half Cycle - When the HO2S voltage signal crosses over a predetermined threshold. The predetermined thresholds are the rich and lean switch points.

Heated Oxygen Sensor (HO2S) - An automotive oxygen sensor designed with a heater element built into the sensor body.

Hydrocarbons (HC) - A family of organic fuels containing only hydrogen and carbon. Gasoline consists almost entirely of a hydrocarbon mixture. Unburned hydrocarbons in the atmosphere are considered pollutants.

Long Term Adaptive - LONGFT (Adaptive) memory injector pulsewidth compensation stored by the PCM to maintain minimum emissions output. LONGFT (Adaptive) drives SHRTFT (Adaptive) to maximum operating efficiency.

Misfire - The lack of complete combustion in an engine cylinder.

Nitrogen (N) - A gas that makes up 78% of the atmosphere (air). Under conditions of high temperature and pressure in the combustion chamber, nitrogen can combine with oxygen to form harmful Oxides of Nitrogen (NOx). Oxides of nitrogen contribute to the formulation of ground level Ozone and Photochemical Smog.

Open Loop - An engine operating state in which the A/F Mixture is being controlled by the PCM according to a standard program in memory and not in response to signals from the HO2S. An engine operating state is normally encountered during the first few minutes of operation after a cold startup.

Oxides of Nitrogen (NOx) - These are harmful gases that form when Nitrogen from the air is combined with oxygen under conditions of high temperature and pressure in the combustion chamber. Oxides of Nitrogen contribute to the formulation of ground level Ozone and Photochemical Smog. The chemical composition is NOx.

PCI Bus - Various modules exchange information through a communications port called the Programmable Communications Interface (PCI) Bus. The PCI Bus connects to the ABS, ATC Display Head, BCM, IPC, Overhead Travel, PCM, SIR and TCM (modules).

Positive Thermal Coefficient Device (PTC) - An electrical component in which the molecular structure changes with temperature. Voltage through the PTC creates heat that changes the molecular structure of the device. The resistance of a PTC is proportional to the heat generated. These devices are used as self-resetting circuit breakers and component heaters.

Pulsewidth - The time duration of a voltage pulse activating a component (on-time).

Purge - The act of transferring fuel vapors from the vapor canister to the intake system by drawing fresh air into the canister.

Rationality - Rationality is defined as a type of comprehensive test in which component inputs are compared against other component inputs to verify that conditions match.

Similar Window - Pertaining to engine operation in which engine speed and load are within predetermined percentages. The enable criteria in which certain monitors must pass to extinguish the MIL or erase trouble codes.

Short Term Adaptive - SHRTFT (Adaptive) compensation controlled by the PCM to vary injector pulsewidth. Based upon HO2S input signals, the PCM changes pulsewidth by a percentage to maintain minimal emissions output.

Stoichiometric - A term used to describe the ideal A/F Ratio entering the intake manifold. This ratio helps to keep emissions at a minimum and catalyst conversion of emissions high. The stoichiometric A/F Ratio is 14.7:1 (measured in parts by weight).

Task Manager - The part of the PCM software designed to manage, regulate, and perform various OBD II Monitor diagnostic tests.

Trip - This is an ignition cycle or sequence in which the PCM runs the OBD II Monitors.

SECTION 2 CONTENTS

2002 Sebring –Powertrain

2002 Sebring - Body

2002 Sebring - Powertrain

HOW TO USE THIS SECTION

This section of the manual includes diagnostic and repair information for Chrysler Car applications. This information can be used to help you understand the Theory of Operation and Diagnostics of the electronic controls and devices on this vehicle. The articles in this section are separated into three sub categories:

- Body Control Module
- Powertrain Control Module
- *Transmission Control Module*

2002 Sebring 2.7L V6 VIN R (41TE Automatic Transaxle)

The articles listed for this vehicle include a variety of subjects that cover the key engine controls and vehicle diagnostics. Several sensor inputs and output devices are featured along with detailed descriptions of how they operate, what can go wrong with them, and most importantly, how to use the PCM, BCM, and TCM onboard diagnostics and common shop tools to determine if one or more devices or circuits has failed.

This vehicle uses an Integrated Coil-On-Plug (COP) Electronic Ignition system and Sequential Fuel Injection (SFI). These systems are primarily controlled by inputs from Hall effect camshaft and crankshaft position sensors. These devices and others (i.e., TP, VSS, MAP, Oxygen sensors, etc.) are covered along with how to test them.

Choose The Right Diagnostic Path

In most cases, the first step in any vehicle diagnosis is to follow the manufacturers recommended diagnostic path. In the case of Chrysler vehicle applications, the first step is to connect a Scan Tool (OBD II certified for this vehicle) and attempt to communicate with the engine controller or any other vehicle onboard controllers that apply. Diagnostic Trouble Codes (DTCs), data (PIDs), bi-directional controls, operating mode descriptions, and symptom charts can then be used to pinpoint the fault.

The Malfunction Indicator Lamp (MIL) can be a great help to you during diagnosis. You should note if the customer complaint included the fact that the MIL remained on during engine operation. And be sure to determine if the MIL comes on at key on and goes out after a 4 second bulb-check. There is more information on using the MIL as part of a diagnostic path in Section 1.

Diagnostic Points to Consider

If a problem is detected in a vehicle similar to this one (whether it is a trouble code or no code condition), you need to be able to determine which vehicle system (or systems) are involved before you proceed with any testing. While this may sound like a fairly routine way to approach problem solving, it is a key point to consider before you start testing.

Due to the complexity of these vehicles, a trouble code does not always point to a failed component - instead, it may point to a system that needs to be tested for some kind of problem. For example, if an Oxygen sensor trouble code is set, you need to consider what other conditions could cause an oxygen sensor to behave improperly.

Component Diagrams

You will find numerous "customized" component diagrams used throughout this section. Note the dashed lines in each Schematic, as they represent the circuits the control module uses to determine when a circuit has a fault.

Vehicle Identification

VECI DECAL

The Vehicle Emission Control Information (VECI) decal is located in the engine compartment. The information in this decal relates specifically to this vehicle and engine application. The specifications on this decal are critical to emissions system service.

2TWC/2HO2S (2)/SFI

These designators indicate this vehicle is equipped with 2 three-way catalysts (TWC), 4 Heated Oxygen Sensors (HO2S) and sequential fuel injection (SFI).

OBD II Certified

This designator indicates that this vehicle has been certified as OBD II compliant.

LEV

If this designator is used, the vehicle conforms to U.S. EPA and State of California Low Emissions Vehicle standards.

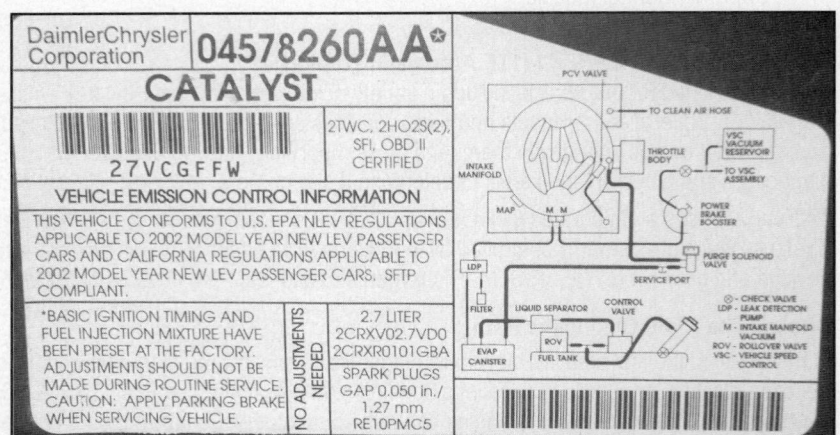

Underhood View Graphic

Electronic Engine Control System

OPERATING STRATEGIES

The Powertrain Control Module (PCM) in this application is a triple-microprocessor digital computer. The PCM, formerly known as the SBEC, provides optimum control of the engine and various other systems. The PCM uses several operating strategies to maintain good overall drivability and to meet the EPA mandated emission standards. These strategies are included in the tables on this page.

Based on various inputs it receives, the PCM adjusts the fuel injector pulsewidth, the idle speed, the amount of ignition advance or retard, the ignition coil dwell and operation of the canister purge valve, and battery charging rate.

The PCM also controls the Speed Control (S/C), A/C system and some transmission functions.

PCM Location
The PCM is located at the left side of the engine compartment, next to the air cleaner housing.

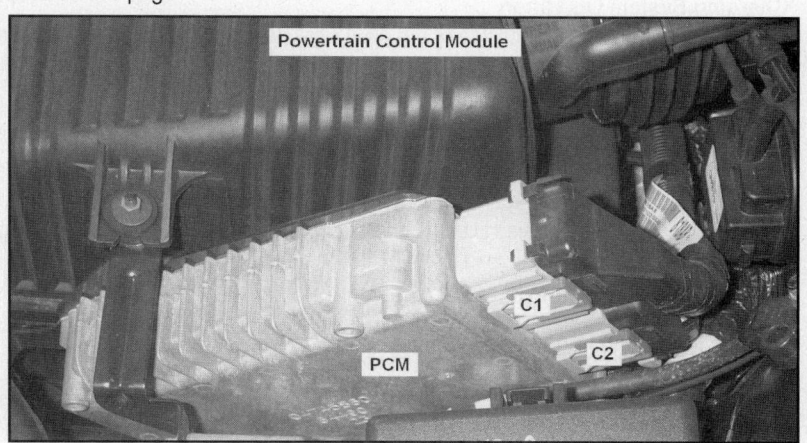

Operating Strategies

A/C Compressor Clutch Cycling Control	Closed Loop Fuel Control
Electric Fuel Pump Control	Charging Rate Control
Fuel Metering of Sequential Fuel Injectors	Speed Control (S/C)
Idle Speed Control	Ignition Timing Control
Vapor Canister Purge Control	

Operational Control of Components

A/C Compressor Clutch	Ignition Coils (Timing and Dwell)
Fuel Delivery (Injector Pulsewidth)	Fuel Pump Operation
Cooling Fan	Purge Solenoid/Leak Detection Pump
Charging System	Idle Air Control Motor
Malfunction Indicator Lamp	Integrated Speed Control

Items Provided (or Stored) By the Electronic Engine Control System

Shared PCI Data Items	Adaptive Values (Idle, Fuel Trim etc)
DTC Data for Non-Emission Faults	DTC Data for Emission Related Faults

Distinct Operating Modes

Key On	Cruise
Cranking	Acceleration (Enrichment)
Warm-up	Deceleration (Enleanment)
Idle	Wide Open Throttle (WOT)

Computer Controlled Charging System

INTRODUCTION

This application uses a 135-amp Denso generator. These generators do not contain voltage-regulating devices. The charging system voltage is regulated between 13.5-14.7 volts through an integrated circuit in the PCM, called the Electronic Voltage Regulator (EVR).

Charging System Operation

The PCM grounds the control side of the ASD relay, which closes the relay contacts, and provides power to many vehicle circuits, including the generator rotor. A magnetic field is produced any time the PCM grounds the field circuit, which provides a path for the battery voltage from the ASD relay.

The amount of DC current produced by the generator is controlled by the EVR circuitry in the PCM and engine speed.

Battery Temperature Sensor data (calculated) and system voltage are used by the PCM to vary the charging rate. The PCM accomplishes this by cycling the ground path to control the strength of the rotor magnetic field. Temperature dependent charging strategies like this one offer improved charging efficiency, slightly improved mileage and extended battery life.

Charging System Diagnosis

In addition to conventional voltage output and load testing, this charging system can be effectively diagnosed using a Scan Tool (PIDs, DTCs and bi-directional controls) and a Lab Scope or DVOM.

Many charging system faults will set DTCs. It is important to check for codes before beginning a conventional charging system diagnosis. The system is monitored for too high or too low voltage conditions. The PCM can also monitor it's own ability to switch the control circuit to ground to regulate charging rates.

Charging System Schematic

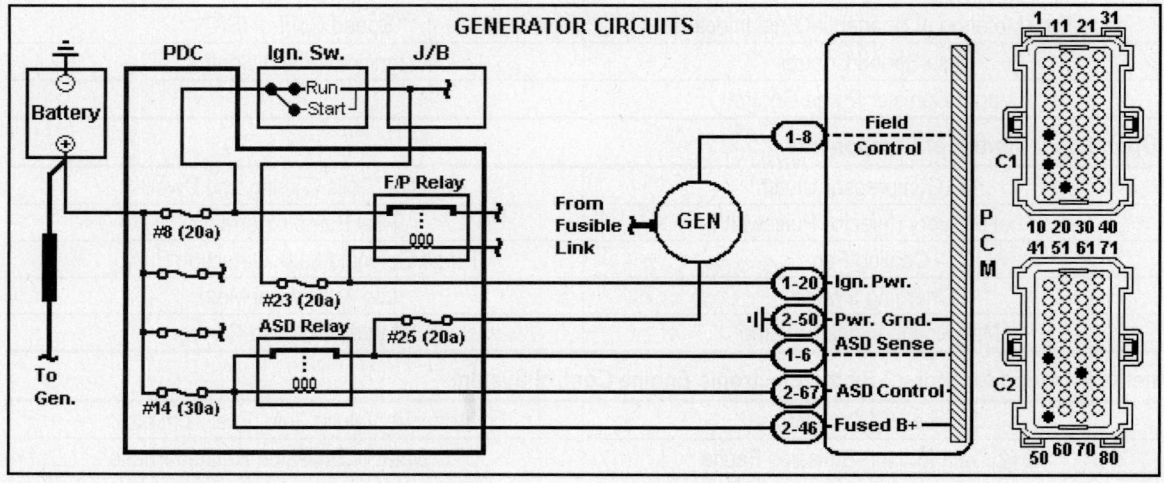

BI-DIRECTIONAL TEST (GENERATOR)

The PCM in this vehicle provides bi-directional testing capabilities, which can be accessed with a Scan Tool. If a system fault is suspected, it is much easier to determine the cause of the fault using these bi-directional controls (System Tests or Functional Tests). During the test, data values can be observed on the Scan Tool (PIDs) or measured directly with a DVOM or Lab Scope.

Scan Tool System Test Example
In this example, the Scan Tool System Test menu was accessed, and the GENERATOR FIELD TEST was chosen. See Section 1 for help navigating through Scan Tool menus.

While in the System Test, the "Y" button toggles the field command between NORM, FULL and OFF. The results can be seen as the BATTERY VOLTS PID value changes in response to the command. The FULL command actually "full fields" the generator, producing maximum current output.

```
+-------------------------------------------------+
|                                                 |
|        Snap-on Scan Tool Display                |
|       _____               |
|                                                 |
|  * Y TO SWITCH BETWEEN NORM, FULL & OFF         |
|  _____          |
|                                                 |
|  SYSTEM TESTS - GENERATOR FIELD TEST            |
|  CHRYSLER CAR                          A/T      |
|  2.7L V6 MPI                           A/C      |
|  CHRY ENG                                       |
|  NO CODES AVAILABLE FROM THIS MODE              |
|  GENERATR FIELD_NORM   ENGINE RPM_____640      |
|  DES CHARGE(V)__13.9   BATTERY VOLTS__14.0      |
|                                                 |
+-------------------------------------------------+
```

Using System Test Results
The System Tests available through the Scan Tool offer a very flexible range of testing options to the technician. The GENERATOR FIELD TEST in this example can "full field" the generator right from the Scan Tool. This option is much faster and more accurate than external load testing or a manual full field test.

Using Functional Test Results
If the System Test results are abnormal, use the GENERATOR FIELD test available in Functional Tests. These tests must be performed with the engine off. The field command is cycled at a fixed rate, but no current is produced, because the generator is not rotating. This Functional Test is used to verify that the PCM is capable of cycling the field command.

Using Scan Tool Data
Also, Scan Tool data (PIDs) can be observed to aid in diagnosis. BATT TEMP (V), BATT TEMP (°F), BATTERY VOLTS, GENERATOR LAMP, GENERATOR FIELD, and DESIRED CHARGE (V) PIDs should be observed on the Scan Tool to determine if inaccurate or missing data is the cause of the charging system fault.

Multiple Tool Diagnosis
Results from any of the above tests can be observed and verified using a DVOM or Lab Scope. A DVOM can be used to verify PID values. A current probe and a DVOM or Lab Scope can be used to measure current output while full-fielding with a Scan Tool. There are many other combinations available, depending on the trouble code or symptom. Start your diagnosis by thinking about which tests and tools will help to eliminate time-consuming or duplicate diagnostic steps and pinpoint the fault much faster.

LAB SCOPE TEST (GENERATOR)

The Lab Scope can be used to test the Computer Controlled Generator to determine if the generator is not capable of charging the battery, or if it is not being controlled by the PCM. Place the shift selector in Park and block the drive wheels for safety.

Scope Connections
Connect the Channel 'A' positive probe to the field control circuit (DK GR wire) at PCM Pin 1-8 and the negative probe to the battery negative post. Connect the low-amps probe to Channel 'B' and clamp probe tip around the large output wire. See the Schematic below.

Lab Scope Example (1)
In example (1), the trace shows the generator field command and current output (DC) with low accessory load. The generator is being commanded off more than on (voltage is high longer than it is low). When the circuit is switched to ground, the circuit is complete and current is produced. Therefore, the longer the trace is low, the more current is produced.

In the example, the current output varied from about 5-22 amps.

Lab Scope Example (2)
In example (2), Channel 'B' was changed to AC coupling, and 5 amps/division. This setting does not show absolute charging current, but does provide a better view of the *change* in output in response to the field command.

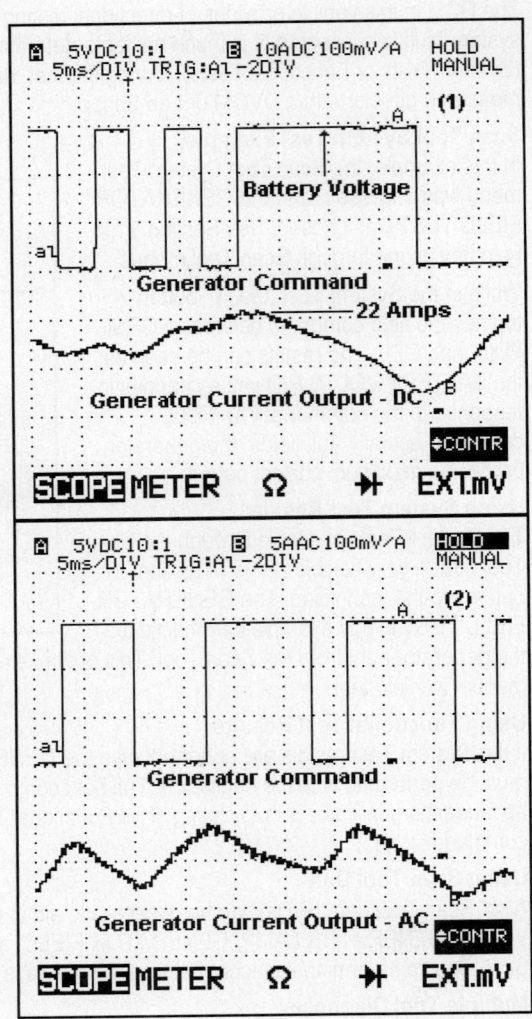

Charging System Test Schematic

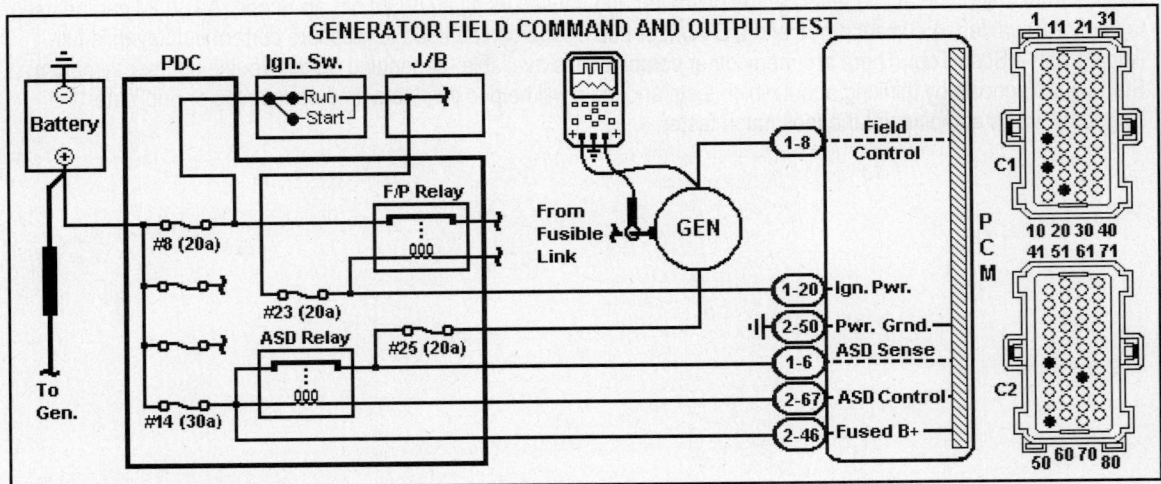

Computer Controlled Cooling Fans

INTRODUCTION

The two (2) electric radiator-cooling fans on this application are operated at low and high speeds through low-speed and high-speed fan relays located in the Power Distribution Center (PDC).

Both relays can be activated using the Scan Tool Functional Tests, and tested by observing the fan operation or measuring circuit activity with a DVOM or Lab Scope.

BI-DIRECTIONAL TEST (COOLING FANS)

The Lab Scope can be used to observe test results from the low or high-speed fan relay tests in the Scan Tool Functional Tests menu.

Lab Scope Connections
Connect the Channel 'A' positive probe to the high-speed fan power circuit (YL wire), and the negative probe to the battery negative post. Connect the low-amps probe to Channel 'B' and clamp the probe tip around the same wire. Refer to the Schematic below.

Lab Scope Example
In the example, the trace shows the voltage and current of the high-speed fan circuit during the HIGH-SPEED FAN RELAY test. Note that the circuit voltage does not drop below 3 volts during the test. During the test, the fans are cycled on for only 1 second at a time. The test duration is about 5 minutes.

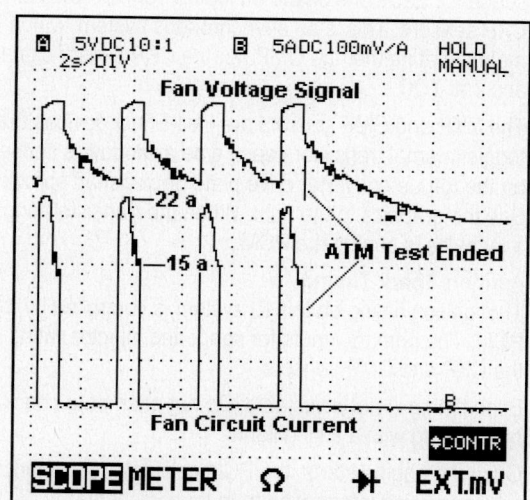

When electricity is used to perform a mechanical task (spin a motor), the current draw is the greatest when the circuit is initially completed. Cycling the fans over time is an effective test because the short-term effects of repeated loading can be seen as well as the long-term effects of continued circuit operation. In this example, the current exceeded 20 amps each time the fans were energized, and dropped to under 15 amps during the 1 second of on time. These results were constant for the duration of the test.

Cooling Fan Test Schematic

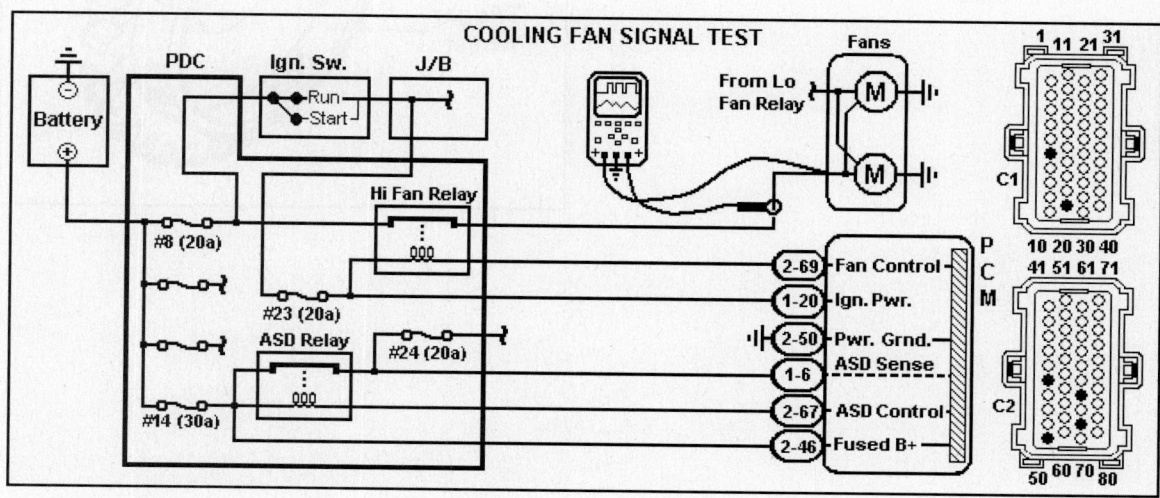

INTRODUCTION

The PCM controls the operation of the Integrated Electronic Ignition (EI) system on this vehicle application. Battery voltage is supplied to each coil through the ASD relay and fuse #24 (20a). These components are both in the Power Distribution Center (PDC).

The PCM controls the coils individually by grounding each coil primary circuit at the correct time. By switching the ground path for each coil "on" and "off" at the appropriate time, the PCM adjusts the ignition timing for each cylinder pair correctly to meet changing engine operating conditions.

The firing order on this application is 1-2-3-4-5-6.

EI System Sensor Operation

The PCM makes ignition and injector-timing and sequence decisions based on inputs from the CMP and CKP sensors. This is an asynchronous system, which means that neither the CMP nor the CKP sensor signals occur at TDC.

The CKP and CMP sensors are Hall-Effect devices that toggle internal transistors each time a window is sensed on the torque converter drive plate or camshaft sprocket. See the graphics to the right. When the transistor toggles, a 5v pulse is sent to the PCM.

Ignition Spark Timing

The ignition timing on this EI system is controlled by the PCM. The primary inputs for spark and injector timing are the CKP and CMP sensors.

Base timing is not adjustable. Do not attempt to check the base timing with a timing light.

Once the engine starts, the PCM calculates the spark advance using information from these sensors:

*   BARO Sensor (MAP signal at KOEO)
*   ECT Sensor
*   Engine Speed Signal (rpm)
*   MAP Sensor
*   Throttle Position

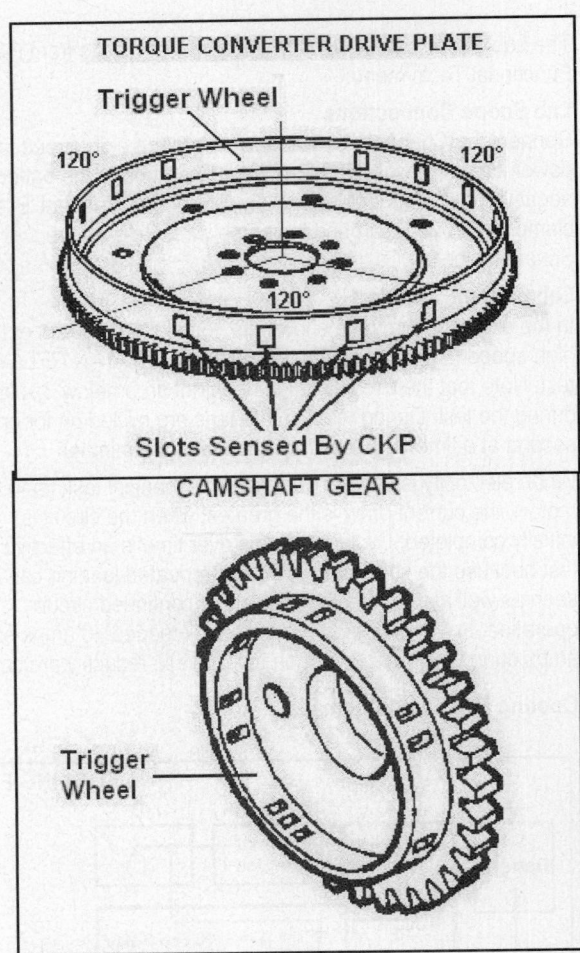

TORQUE CONVERTER DRIVE PLATE

Trigger Wheel

120°   120°

120°

Slots Sensed By CKP

CAMSHAFT GEAR

Trigger Wheel

IGNITION COIL OPERATION

This ignition system is referred to as a Coil On Plug (COP) system. It consists of six (6) individual coils. Each coil is mounted to one of the valve covers, and sits on top of one spark plug, eliminating the need for ignition wires.

All of the ignition module functions are integrated into the PCM. A separate driver in the PCM controls each coil.

The firing order on this engine is 1-2-3-4-5-6.

COP Circuits

All six (6) coils are powered by the ASD relay. Drivers in the PCM ground each coil circuit independently. Refer to the Ignition System Schematic below.

Trouble Code Help

Note the driver circuits that are shown with a dotted line in the schematic. The PCM monitors these circuits for peak current achieved during the dwell period.

If a preset current level is not achieved within 2.5 ms of dwell time, a DTC is set (DTC P0351 for Coil 1 through P0356 for Coil 6). The problem must occur continuously for 3 seconds during cranking, or 6 seconds with the engine running to set the trouble code.

If one of these trouble codes is set, compare the current between the suspect circuit and another coil. Even 1 Ohm of extra resistance in this circuit can set a trouble code.

Coil-On Plug - Bank 2

#2 #4 #6

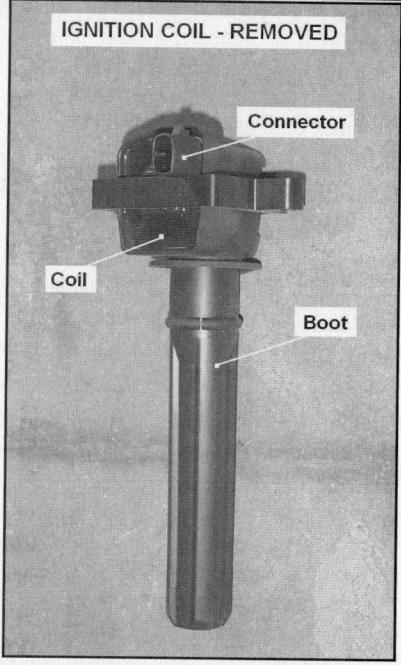

IGNITION COIL - REMOVED

Connector

Coil

Boot

Ignition System Schematic

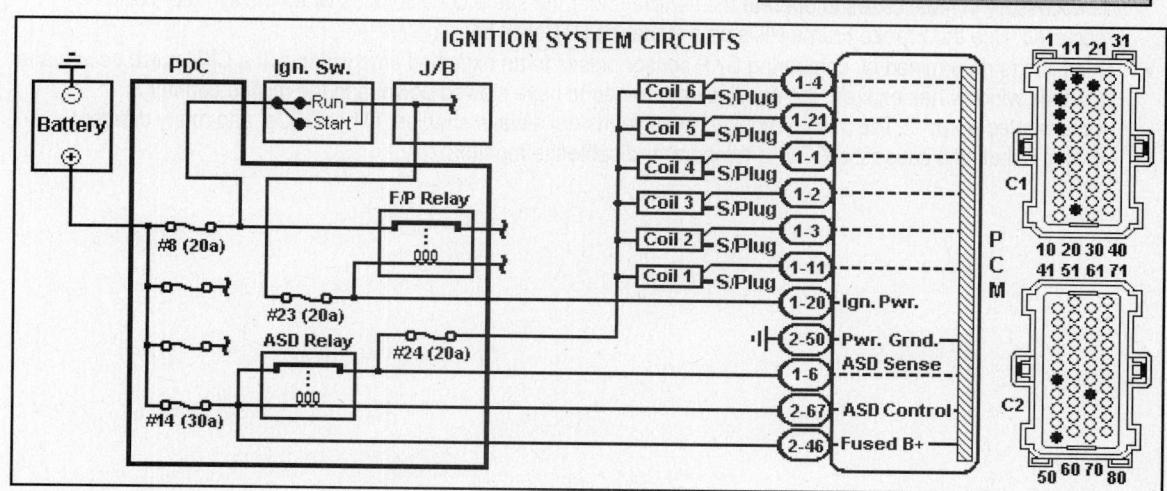

IGNITION SYSTEM CIRCUITS

SCAN TOOL TEST (MISFIRE)

The Scan Tool can be used to view PID values during misfire diagnosis. Also, a specific misfire PID list is available in the SYSTEM TESTS menu.

Misfires can occur for a number of reasons. Ignition system faults are certainly a common cause, but it is not safe to assume that a misfire DTC or symptom is caused by the ignition system.

It is important to view all available information, to determine if the problem is common to all cylinders, a bank, or an individual cylinder. Other sensor inputs, power and ground circuits, and fuel trim PIDs also offer valuable information.

```
        Snap-on Scan Tool Display
  ────────────────────────────────────────
  SYSTEM TESTS - MISFIRE
  CHRYSLER CAR                         A/T
  2.7L V6 MPI                          A/C
  CHRY ENG
  NO CODES AVAILABLE FROM THIS MODE
  CYL 1 MISFIRES_____0   CYL 2 MISFIRES_____0
  CYL 3 MISFIRES_____0   CYL 4 MISFIRES_____0
  CYL 5 MISFIRES_____0   CYL 6 MISFIRES_____0
  ADAP DONE LEARN_YES    MISFIRE ENABLED__NO
```

Scan Tool Example
In this example, the SYSTEM TESTS – MISFIRE screen is being monitored for misfire. Each cylinder counter value is 0, indicating no misfires.

The MISFIRE ENABLED PID value is NO, indicating that the PCM will not set a misfire DTC under the current conditions. Misfires will still be counted even if the PID value is NO, but no codes will be set. Under most positive load conditions, the value will be YES.

Using Scan Tool Test Results
It is important to know how to use the misfire PIDs, and what their limitations are, to avoid misdiagnosis. Some important points to remember when performing this Scan Tool test are:

- It is normal to occasionally see a few misfires. A PID with a low number of misfires does not necessarily indicate a problem. The PCM watches these values over time to determine if there is a persistent problem.
- The engine revolutions are counted in groups. This prevents the PCM from setting a DTC every time a cylinder misfires. There are 5 groups of 200 RPM in each 1000 RPM sampling block. There must be at least a 2.5% misfires rate in the 1000 RPM sample (25 misfires) in order to set a DTC. A DTC will also set if the misfire rate exceeds 10% in any of the 200 RPM blocks.
- The misfire counters reset every 1000 RPM, so these PIDs are only useful when a misfire is actually occurring. It is necessary in most cases to operate the vehicle under the same conditions as when the symptom or DTC occurred. Use the Freeze Frame PIDs as a guide if a DTC was set.
- A misfire is determined by comparing CKP sensor pulses to an expected time window. If a CKP pulse occurs after the time window has expired, the crank is determined to have slowed down, and the misfire counter is incremented up by 1. The misfire monitor, however, is not always enabled. Other DTCs, and many different driving conditions cause the PCM to temporarily disable the monitor.

LAB SCOPE TEST (COIL PRIMARY)

The ignition coil primary includes the switched ignition feed circuit, coil primary circuit and the coil driver control circuits at the PCM.

Lab Scope Connections
Connect the Channel 'A' positive probe to one of the six (6) coil control circuits at the PCM or coil, and the negative probe to the battery negative post.

Lab Scope Example (1)
In example (1), the trace shows the primary circuit at idle. The system voltage, dwell period, spark line, and coil oscillations can all be clearly seen in this example. Only the peak voltage is not seen with these settings.

Note the spark line in this example is only 24 volts. This capture was taken from a known good vehicle with 8,000 miles on the odometer.

Lab Scope Example (2)
In example (2), the time and voltage per division were changed. At 50 volts per division, the peak voltage can be seen. A Glitch Detect feature was used to ensure that the display was stable and accurate.

Compare the peak voltage to the other coils on the vehicle. If all are too low, check the ASD relay circuits and PCM power ground circuits for a voltage drop problem. If only one is too low, check the coil resistance, (only those circuits unique to that coil). Just one or two Ohms of added resistance in any of these circuits can significantly reduce the coil output.

Ignition Primary Test Schematic

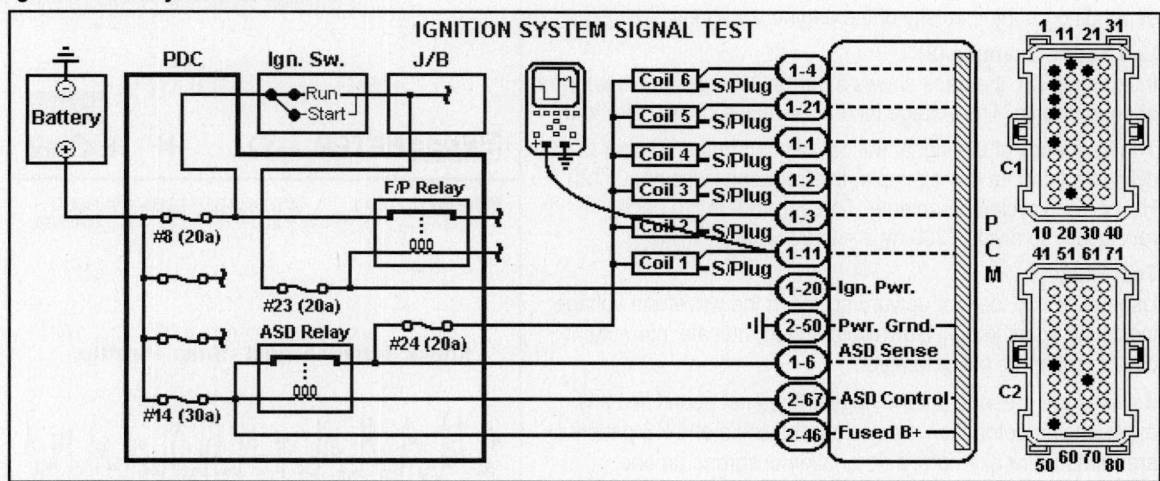

KNOCK SENSOR OPERATION

The Knock Sensor (KS) is mounted in the intake valley of the engine block, near cylinder #2. The sensor contains piezoelectric material which produces a voltage in direct relationship to engine vibration.

If the engine vibration is excessive, the PCM retards the ignition timing for all cylinders. Based on the severity of the detonation, the PCM will store retard values in either short-term or long-term memory.

Remember that the PCM cannot identify which cylinder is experiencing a knock. Therefore, when the KS value exceeds a preset limit, ignition timing is retarded equally for all cylinders. Also, the PCM ignores this input at idle.

LAB SCOPE TEST (KNOCK SENSOR)

Place the gear selector in Park and block the drive wheels for safety.

Lab Scope Connections
Connect the Channel 'A' positive probe to the Knock Sensor signal circuit (DB/LG wire) at PCM Pin 1-25, and connect the negative probe to the battery negative post.

Lab Scope Example (1)
In example (1), the trace shows a normal KS voltage waveform at idle. The voltage cycles from about 500 mv to almost a volt, depending on the intensity of the engine vibrations.

Lab Scope Example (2)
In example (2), the trace shows a normal KS voltage waveform at snap throttle. The voltage increased only slightly under load.

The frequency of the signal will correspond to the number of cylinders firing. In this example, the frequency was about 120 Hz, or 7200 cycles per minute. Three cylinders fire each revolution, so divide 7200 by 3 = 2400 RPM.

Test Results
Use this test to look for upward spikes in the waveform voltage, indicating spark knock. Downward spikes indicate intermittent circuit problems, not detonation.

If spikes are present, but the engine does not sound like it is experiencing detonation, make sure the spike and KS patterns are related. For example, a detonation occurring on one cylinder would recur every 6 cycles of the KS waveform. If the patterns do not match, it is not likely detonation. Test the alternator and fan motors for AC voltage output, and inspect the KS harness for damage. The KS circuit is very sensitive to Electro-Magnetic Interference (EMI) and Radio-Frequency Interference (RFI).

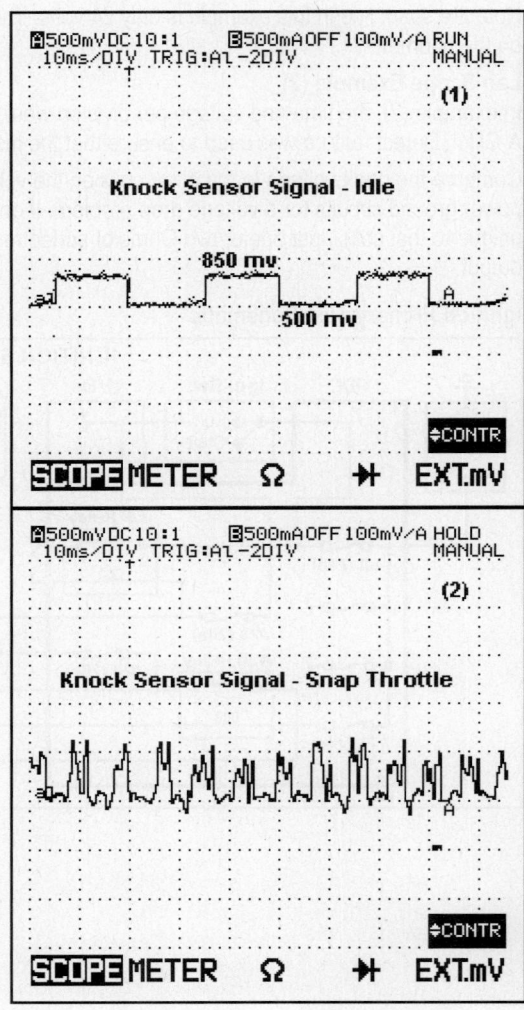

CHILTON DIAGNOSTIC SYSTEM - DTC P0352

The purpose of the Chilton Diagnostic System is to provide you with one or more tests that can be done with a DVOM, Lab Scope or Scan Tool *prior* to entering the complete trouble code repair procedure. The quick checks listed in the DVOM, Lab Scope or Scan Tool tests on the previous pages may help you quickly find the cause of the problem. If you cannot resolve the problem with these tests, a code repair chart is also included.

Code Description
DTC P0352 - The Comprehensive Component Monitor (CCM) detected a fault in the Coil 2 (or 'B') primary or secondary circuit with the engine running (1-trip code).

Code Conditions (Failure)
The PCM monitors the current in each of the six (6) coils during the dwell period. The monitoring conditions are: System voltage over 8v cranking or 12v engine running, and engine speed under 2016 RPM. If the coil peak current level is not achieved within 2.5 ms of dwell time, a DTC is set (DTC P0351 for Coil 1 through P0356 for Coil 6). The problem must occur continuously for 3 seconds during cranking, or 6 seconds with the engine running to set the DTC.

Quick Check Items
Inspect or physically check the following items listed below:

- ASD relay output circuit open
- Capacitor shorted to ground
- ASD relay output shorted to ground
- Ignition coil driver circuit open
- Ignition coil driver circuit shorted to ground
- *Ignition coil failure*

Drive Cycle Preparation
There are no drive cycle preparation steps required to run the Component Monitor.

Install the Scan Tool. Record any DTCs or Pending DTCs and any Freeze Frame data. Clear the codes (PCM reset).

Scan Tool Help
A Scan Tool can be used to view the trouble code ID and Freeze Frame Data (PIDs) for DTC P0352. It can also be used to clear stored trouble codes and Freeze Frame Data.

The Freeze Frame data indicates that the DTC set at idle shortly after startup. In this case, duplicating these conditions should be simple, and does not require a drive cycle. The elevated long-term fuel trim PID (LT FT 2) indicates that this circuit or component failure probably caused some degree of misfiring.

```
DIAG. TROUBLE CODES
ECU: $10 (Engine)
Number of DTCs: 1
* P0352  Ignition Coil B
         Primary/Secondary
         Circuit
         Malfunction




ENTER = FREEZE FRAME
```

```
DTC················P0352
ENGINE SPD······704RPM
ECT (°)············129°F
VEHICLE SPD·······0MPH
ENGINE LOAD·······5.8%
MAP (P)··········9.8inHg
FUEL STAT 1··········OL
FUEL STAT 2··········OL
ST FT 1·············0.0%
LT FT 1·············2.4%
ST FT 2·············0.0%
LT FT 2·············8.6%
```

Lab Scope testing

The Lab Scope and Low-Amp probe can be used to observe the ignition coil circuit activity. The captures on this page were taken from a 2002 Sebring with DTC P0352 set.

As seen in the Scan Tool DTC and Freeze Frame examples from the previous page, a circuit problem did exist which resulted in at least partial misfiring.

Lab Scope Example (1)

In example (1), the Glitch Detect feature was used with the coil primary signal to verify the coil output. Because a pattern was obtained, it is easy to rule out opens and shorts, as well as an ASD relay failure, as possible causes for the trouble code.

The peak firing voltage of 175 volts, however, is lower than that of a known-good coil, which was 225 volts. See Lab Scope Test (Coil Primary). Also, the duration of the spark is down to 1.5 ms, from 2 ms on the known good coil. Clearly the coil output is too low.

Lab Scope Examples (2) & (3)

To understand the source of the poor results in example (1), it is helpful to see the current flow through this circuit. In example (2), the traces show the coil primary voltage and current through coil 2. In example (3), the same settings were used to test the voltage and current in another coil on the same vehicle.

The coil dwell period in example (2) is longer than in example (3), and the current flow is much less (5 amps vs. 7 amps).

Understanding Test Results

Based on the low coil output in example (1), and the comparison of example (2) and (3), the next step was to test the coil 2 circuits and coil for high resistance. "High" resistance (1 ohm), in the driver circuit was found and repaired.

It is important to remember that these coils have less than 1 Ohm of resistance, so even a very small change in circuit or component resistance could cause these results.

As an alternative test procedure, you can use the flow chart on the next page.

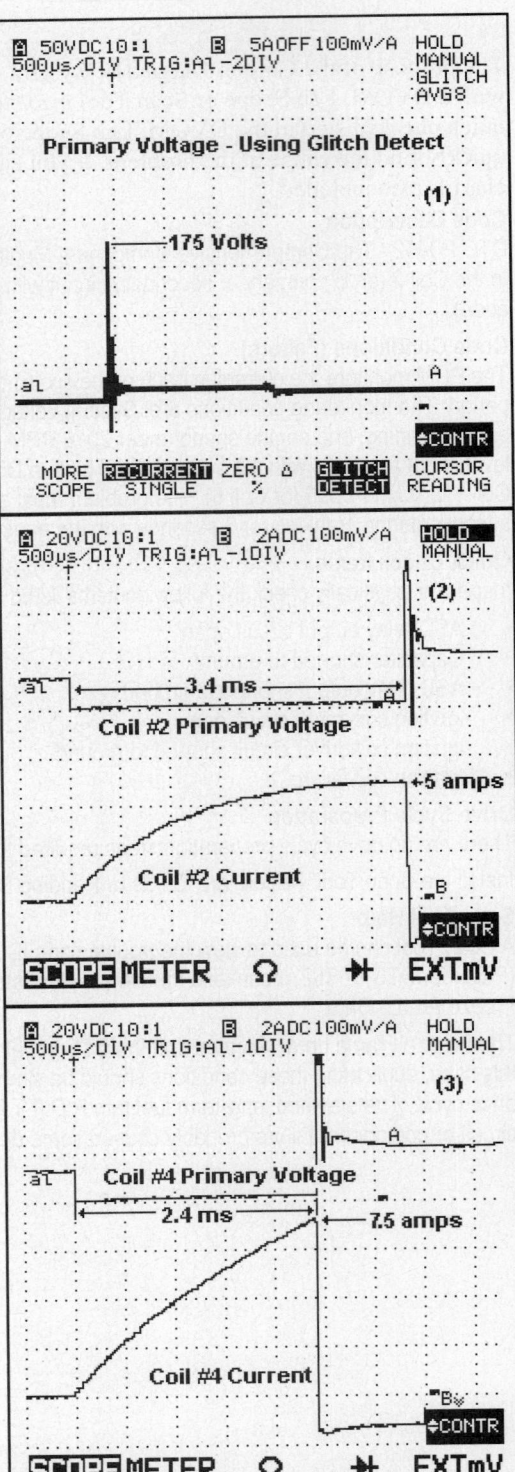

P0351-P0356 Repair Table

Step	Action	Yes	No
1	Record Freeze Frame. Turn ignition ON. Connect DRB-III. Is the GOOD TRIP display equal to zero?	Go To Step 2	Go To Step 8
2	Ignition OFF. Disconnect appropriate coil harness connector. Use Scan Tool to actuate ASD relay, and probe relay output circuit at coil connector. Did test light illuminate brightly?	Go To Step 3	Go To Step 6
3	Ignition OFF. Disconnect PCM harness connector. Measure resistance of driver circuit between PCM and coil harness connectors. Is resistance below 5.0 ohms?	Go To Step 4	Repair open coil driver circuit. Perform Test VER-5. (See Section 1).
4	Ignition OFF. Measure resistance between ground and the coil driver circuit at the PCM harness connector. Is the resistance below 100 ohms?	Repair short to ground in coil driver circuit. Perform Test VER-5. (See Section 1).	Go To Step 5
5	Using Scan Tool, erase DTCs. Ignition OFF. Install substitute coil in place of coil in question. Attempt to operate vehicle under Freeze Frame conditions and check for DTCs. Does the Scan Tool display the same DTC that was erased?	Replace PCM and reprogram. Perform Test VER-5. (See Section 1).	Test Complete
6	Ignition OFF, coil harness connector disconnected. Remove ASD relay from PDC. Measure resistance of the ASD relay output circuit between appropriate ASD relay cavity and coil harness connector. Is the resistance below 5 ohms?	Go To Step 7	Repair open in ASD output. Perform Test VER-5. (See Section 1).
7	Ignition OFF. Disconnect one capacitor harness connector. Install a good INJ/COIL fuse and actuate the ASD relay with the Scan Tool. Repeat for the other capacitor. Is fuse OK for both tests? (If the fuse is open during both tests, there is a short to ground in the ASD relay output circuit).	Replace capacitor(s). Perform Test VER-5. (See Section 1).	Repair ASD output short to ground. Perform Test VER-5. (See Section 1).
8	Condition not present at this time. While in Freeze Frame conditions, wiggle the harness and look for PID changes or DTCs. Refer to TSBs. Visually inspect harness and connectors for damage or corrosion. Were any of the above conditions present?	Repair As Necessary. Perform Test VER-5. (See Section 1).	Test Complete

Engine Position Sensors

CKP & CMP SENSOR OVERVIEW

The Crankshaft Position (CKP) and Camshaft Position (CMP) sensors used with this EI system provide engine position and speed data to the powertrain control module (PCM).

Both sensors are Hall effect devices. The CKP sensor is mounted on the transaxle, adjacent to a trigger wheel on the torque converter drive plate. The CMP sensor is mounted to the Bank 2 cylinder head, adjacent to a trigger wheel on the cam gear.

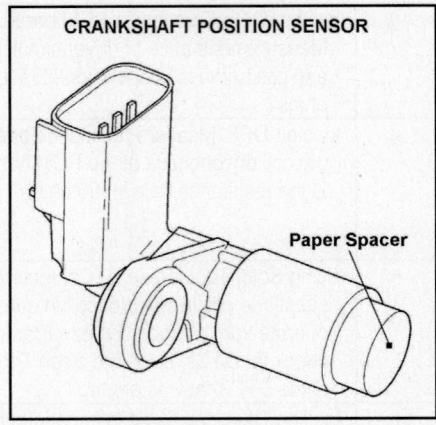

CRANKSHAFT POSITION SENSOR

Paper Spacer

Both trigger wheels have slots that the Hall effect circuit senses and then toggles a transistor. The transistors send 5v pulses to the PCM that represents the position and duration of the trigger wheel slot.

The signal to the PCM will remain high as long as the Hall effect circuitry 'senses' the slot in the trigger wheel. The PCM uses the pulsewidth from the CKP sensor to calculate engine speed. The Misfire Monitor also watches the pulsewidth to determine if crankshaft velocity has decreased more than allowed.

Sensor Signal Relationships

The CMP trigger wheel contains slots grouped in the following order: 1, 2, 3, 1, 3, and 2. Since each number is repeated, the PCM must keep track of the sequence over time in order to properly calculate cylinder position. Also, these pulses do not occur at TDC. They only signal the PCM which cylinder will be arriving at TDC next. The PCM then uses CKP sensor pulses to "count down" to TDC for the designated cylinder.

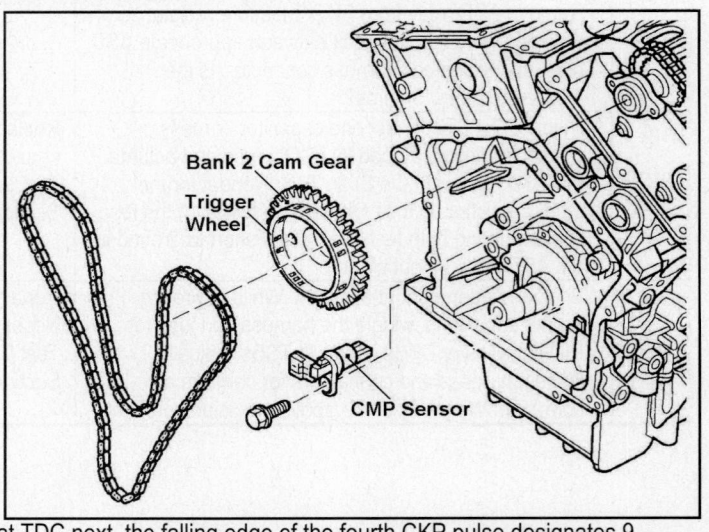

Bank 2 Cam Gear

Trigger Wheel

CMP Sensor

There are 13 slots in a ring on the CKP trigger wheel (in two groups of 4 and one group of 5 slots). The PCM determines engine position at the falling edge of the fourth pulse in each group.

The slots are located at 69°, 49°, 29°, and 9° before TDC for whichever pair of cylinders is approaching TDC. After the CMP pulse designates which cylinder is expected to arrive at TDC next, the falling edge of the fourth CKP pulse designates 9 before TDC.

The extra (thirteenth) slot occurs 11° after TDC, and helps the PCM to determine engine position at startup without having to wait for the complete CMP pulse sequence.

CKP & CMP Sensor Circuits

The CKP and CMP sensors are both connected to the PCM by 3 wires. The sensors share an 8v reference voltage from PCM Pin 2-44. They also share the sensor ground with most other engine sensors (PCM Pin 2-43). The 0-5 volt return signal is sent to PCM Pin 1-32. The CMP 0-5 volt return signal is sent to PCM Pin 1-33.

DVOM TEST (CKP SENSOR)

Place the shift selector in Park and block the drive wheels for safety.

DVOM Connections
Connect the DVOM positive lead to the CKP sensor signal circuit (GY/BK wire) at Pin 1-32 and the negative lead to the battery negative post.

DVOM Example
In this example, the display shows the CKP sensor signal at Hot Idle. The DVOM shows a CKP signal of only 0.48v DC because the instrument is averaging the signal voltage. The average value is low because the signal is low much longer than is high (note the 91.5 % duty cycle). Use the DVOM MIN/MAX feature, if available, to verify that the CKP signal is really switching from 0-5v.

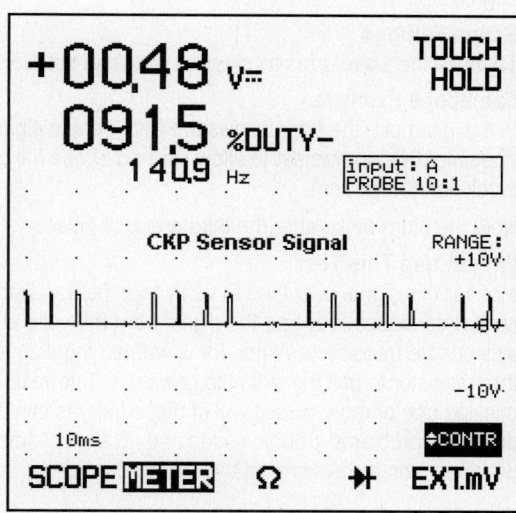

Another way to use the DVOM to test the CKP sensor signal is to observe the frequency of the signal. In this example, the display shows 140.9 Hz. (140.9 cycles per second X 60 seconds / 13 slots per revolution = 650.3 RPM). Using this formula for a no start, it is easy to calculate that at 200 RPM cranking speed, the DVOM should read about 43 Hz.

Why to Use This Test
Monitor the DC voltage and frequency (HZ) to look for circuit or sensor faults. If the transistor switches, but cannot pull the signal to ground, the average DC voltage displayed will jump significantly. Also, the transistor may fail and toggle rapidly, producing a frequency that is not appropriate for the engine RPM.

It is important to monitor these values over time, as many failures occur intermittently.

CKP Sensor Test Schematic

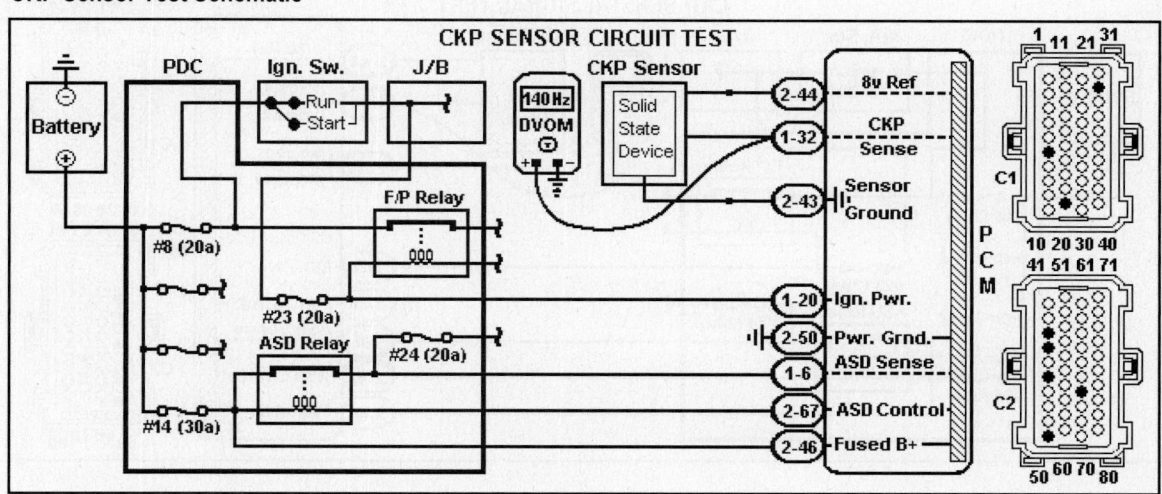

LAB SCOPE TEST (CKP SENSOR)

The Lab Scope can be used to test the CKP sensor as it provides a very accurate view of the sensor waveform and sensor relationships. Place the shift selector in Park and block the drive wheels for safety.

Scope Connections
Connect the positive lead to the CKP sensor signal circuit (GY/BK wire) at Pin 1-32 and the negative lead to the battery negative post.

Scope Settings
To make the waveforms as clear as possible, set the scope settings to match the examples.

Lab Scope Example
In this example, the trace shows the CKP sensor signal at idle. The time/division was set to allow viewing of one full crankshaft revolution (13 pulses).

Note the extra pulse after the third group of pulses.

Why to Use This Test
Monitor this signal over time to verify that the transistor in the sensor continues to toggle the signal from 0-5 volts at the appropriate frequency. Watch for unwanted toggling, failure of the transistor to pull the signal to ground, or intermittently missing one or more pulses. All of these failures can cause drivability problems, trouble codes and no start conditions, depending on the severity of the failure.

CKP Sensor Test Schematic

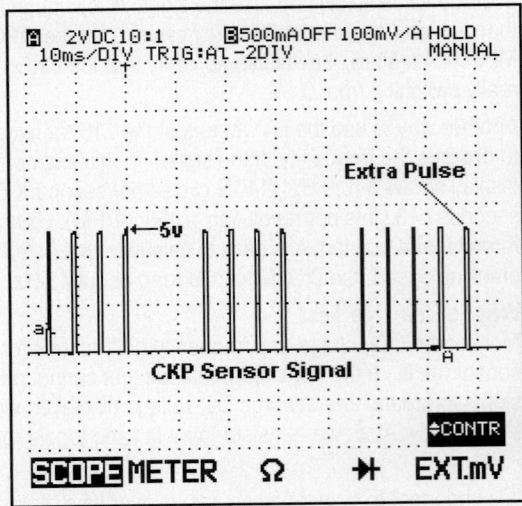

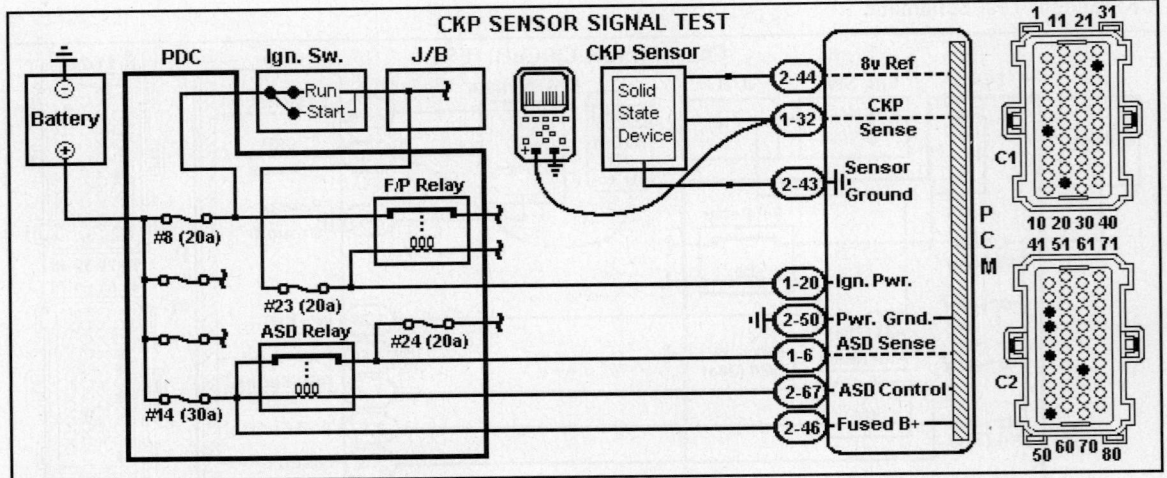

DVOM TEST (CMP SENSOR)

Place the shift selector in Park and block the drive wheels for safety.

DVOM Connections

Connect the DVOM positive lead to the CMP sensor signal circuit (TN/YL wire) at Pin 1-33 and the negative lead to the battery negative post.

DVOM Example

In this example, the display shows the CMP sensor signal at Hot Idle. The DVOM shows a CMP signal of only 0.46v DC because the instrument is averaging the signal voltage. The average value is low because the signal is low much longer than is high (note the 92.1 % duty cycle). Use the DVOM MIN/MAX feature, if available, to verify that the CMP signal is switching from 0-5v.

Note the HZ value in this example is OL. The frequency (Hz) measured by the DVOM could be a useful indicator of an intermittent signal. However, the frequency changes in relation to the camshaft position (number of windows) and can cause the DVOM to be unable to calculate a value. In this case, change the time base from 10 ms/division to 50 or 100 ms/division to have a better chance of seeing a stable and useful value.

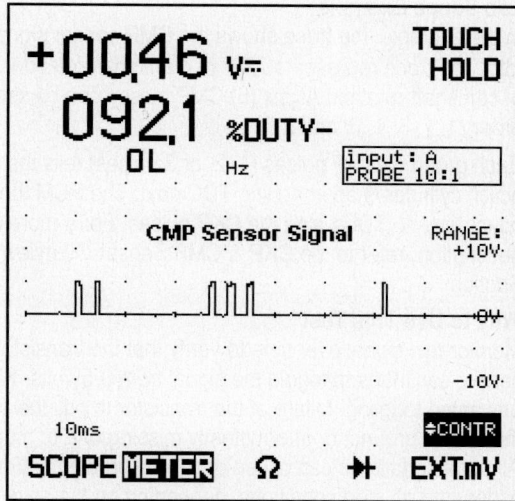

Why to Use This Test

Monitor the DC voltage to look for circuit or sensor faults. If the CMP sensor transistor switches, but cannot pull the signal to ground, the average DC voltage displayed will jump significantly. If the transistor stops switching, the average DC voltage will drop. Because the average voltage is already near 0v, this type of failure is more difficult to find with this test.

CMP Sensor Test Schematic

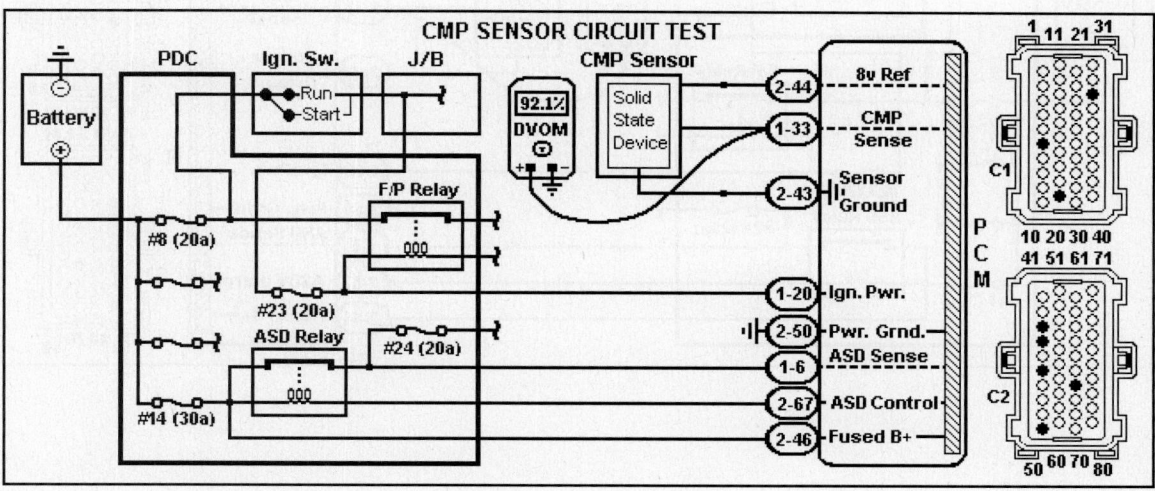

LAB SCOPE TEST (CMP SENSOR)

The Lab Scope can be used to test the CMP sensor circuit as it provides a very accurate view of circuit activity and any signal glitches. Place the shift selector in Park and block the drive wheels for safety.

Scope Connections
Connect the Channel 'A' positive probe to the CMP sensor circuit (TN/YL wire) at PCM Pin 1-33, and the negative probe to the battery negative post.

Lab Scope Example
In the example, the trace shows the CMP sensor signal at Hot Idle. The trace represents 720° of crankshaft rotation, or 360° of camshaft rotation. All six (6) CMP pulses can be seen, in order (1, 2, 3, 1, 3, and 2).

Each group of CMP pulses (1, 2, or 3 pulses) tells the PCM which cylinder is approaching TDC next. The PCM then starts a countdown to TDC using the CKP pulses. For a more complete description, refer to the CKP & CMP Sensor Overview in this Section.

Why to Use This Test
Monitor this signal over time to verify that the transistor in the sensor continues to toggle the signal from 0-5 volts. Watch for unwanted toggling, failure of the transistor to pull the signal all the way to ground, or intermittently missing one or more pulses. All of these failures can cause drivability problems, trouble codes and no start conditions, depending on the severity of the failure.

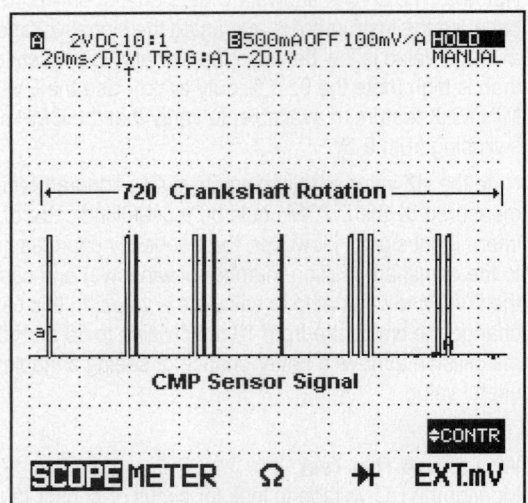

CMP Sensor Test Schematic

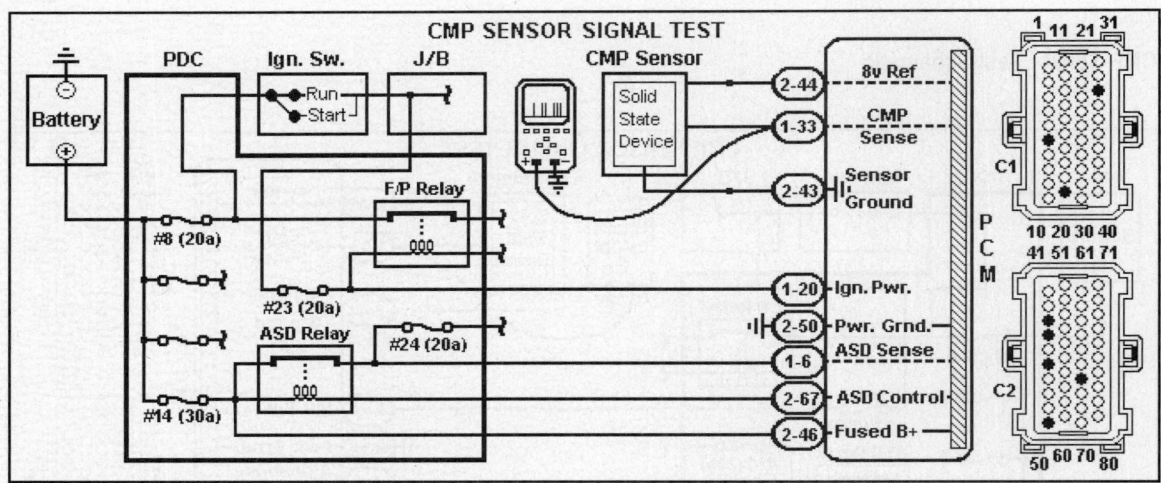

Enhanced Evaporative Emission System

INTRODUCTION

The Evaporative Emissions (EVAP) system is used to prevent the escape of fuel vapors to the atmosphere under hot soak, refueling and engine off conditions. Any fuel vapor pressure trapped in the fuel tank is vented through a vent hose, and stored in EVAP canister.

The most efficient way to dispose of fuel vapors without causing pollution is to burn them in the normal combustion process. The PCM cycles the purge valve on (open) during normal operation. With the valve open, manifold vacuum is applied to the canister and this allows it to draw fresh air and fuel vapors from the canister into the intake manifold. These vapors are drawn into each cylinder to be burned.

The PCM uses various sensor inputs to calculate the desired amount of EVAP purge flow. The PCM meters the purge flow by varying the duty cycle of the EVAP purge solenoid control signal. The EVAP purge solenoid will remain off during a cold start, and for a preprogrammed time after a hot start.

System Components:

The EVAP system contains many components for the management of fuel vapors and to test the integrity of the EVAP system. The EVAP system includes the following components:

- Fuel Tank
- Fuel Filler Cap
- Fuel Tank Level Sensor
- Fuel Vapor Vent
- Evaporative Charcoal Canister
- Leak Detection Pump
- EVAP Purge Solenoid
- PCM and Related Wiring
- *Fuel Vapor Lines with Test Port*

EVAP System Schematic

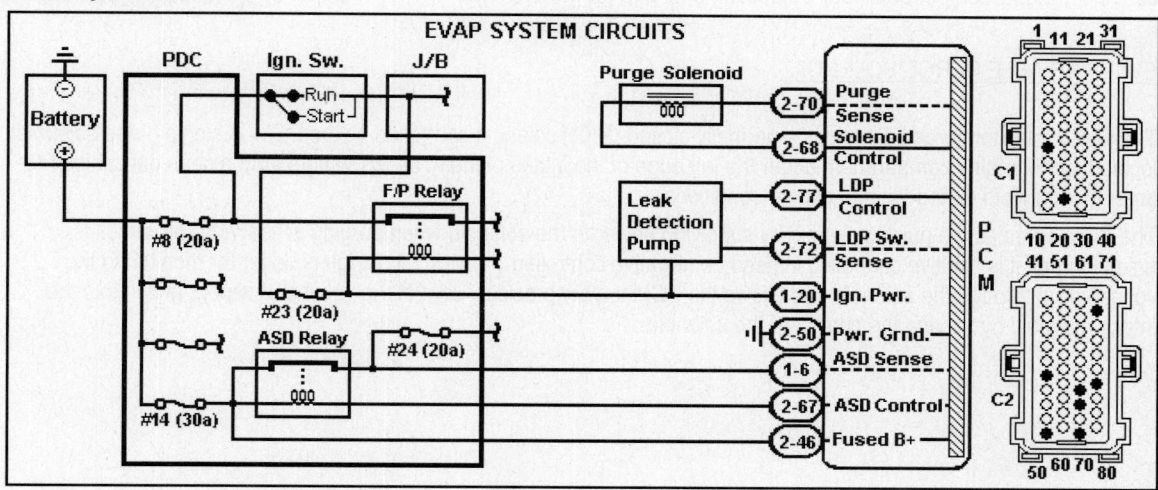

EVAPORATIVE CHARCOAL CANISTER

The EVAP canister is located on a bracket with the LDP on top of the fuel tank. Fuel vapors from the fuel tank are stored in the canister. The canister is connected in series between the fuel tank and the purge solenoid.

Although the Leak Detection Pump (LDP) pressurizes the entire EVAP system, the pressure enters the system through the canister.

Once the engine is in closed loop, fuel vapors stored in the charcoal canister are purged into the engine where they are burned during normal combustion.

The activated charcoal has the ability to purge, or release, any stored fuel vapors when it is exposed to fresh air.

EVAP SYSTEM TEST PORT

All OBD II vehicles must have an EVAP test port. This application incorporates the test port into the purge valve. There is specific test equipment designed for use with the test port. However, a smoke test analyzer with proper adaptors can be used to find even very small leaks. There are also Scan Tool functions that run the EVAP system monitor and control individual components (leak detection pump and purge solenoid).

EVAP VARIABLE PURGE SOLENOID

The EVAP canister purge solenoid is a normally closed (N.C.) solenoid controlled by the PCM. A single purge solenoid, located in the engine compartment under the left edge of the intake plenum, allows vapors from a charcoal canister to enter the manifold during various engine conditions.

The PCM operates the purge solenoid by supplying power to the solenoid when purging is desired. This circuit is unique in that it is positive controlled instead on negative controlled. This circuit design is better for monitoring the voltage drop through the solenoid and allows the PCM to very precisely control the volume of vapors drawn into the intake manifold by varying the current to the solenoid.

BI-DIRECTIONAL SYSTEM TESTS (PURGE)

The Scan Tool System Tests menu contains selections that can be used to trigger tests in the PCM. The System Tests are performed with the engine running to allow observation of the results with a Scan Tool (by viewing the PIDs), a DVOM, or a Lab Scope.

The controlled conditions of these tests remove many variables from the diagnostic process that would be present if the system were diagnosed under normal conditions. The results of normal purging could not easily be observed without the Purge Vapor Test. Under normal operation, the PCM continually anticipates its own purge requirements and compensates for the effect of the vapors on fuel trim. In this test, the Scan Tool controls the purge volume, so the PCM is not making any assumptions, and the results can be observed in the data (PIDs).

The Scan Tool can be used to force the purge solenoid to OFF, NORM, and FULL.

Test Example

In this example, the Scan Tool was used to command the solenoid to the NORM position, which means the PCM is cycling the solenoid command as it would under normal conditions.

The purpose of this test is to control the conditions so that the effects of purging or not purging can be seen in the HO2S signals (UPSTM O2S PIDs), and then in the fuel trim values (ST ADAP (%) PIDs).

In the example, "L" and "R" stand for left and right engine banks. The right bank is Bank 1, and the left bank is Bank 2.

```
Snap-on Scan Tool Display


SYSTEM TESTS - PURGE VAPOR TEST
CHRYSLER CAR                         A/T
2.7L V6 MPI                          A/C
CHRY ENG
NO CODES AVAILABLE FROM THIS MODE
PURGE STATUS___NORM   ENGINE RPM_____608
L UPSTM O2S(V)_0.51   R UPSTM O2S(V)_0.18
L DNSTM O2S(V)_0.86   R DNSTM O2S(V)_0.84
L ST ADAP(%)____-0.4  R ST ADAP(%)_____1.2
```

Using Test Results

When commanded to NORM, the HO2S and fuel trim values will not change significantly because the vehicle is normally purging at idle (as in this example).

Command the solenoid to FULL and watch the PID values change rapidly. When commanded to FULL, hydrocarbons flow from the canister, which drives the air-fuel ratio rich. The HO2S signals should rise and short-term fuel trim should shift negative. The amount of shift is not as important as the fact that a shift is seen.

Command the solenoid to OFF and watch the PID values again. When commanded to OFF, hydrocarbons stop flowing from the canister. Because the PCM normally commands some purge at idle, the air/fuel ratio is adjusted to compensate. When none of the anticipated vapors are flowing, the PCM calculation is wrong, which drives the air-fuel ratio lean. The HO2S signals should fall and short-term fuel trim should shift positive. Again, the amount of shift is not as important as the fact that a shift is seen.

If the solenoid is commanded to FULL and the PIDs reflect values that should be seen with an OFF command, then the EVAP system is likely purging outside air, not canister vapors. Look for a large leak or disconnected hose.

BI-DIRECTIONAL FUNCTIONAL TESTS (PURGE)

The Functional Test menu (on the Snap-on Scan Tool) is the same as the Actuator Test Mode (ATM) menu on the DRB III and other Scan Tools. The tests are used to cycle components with the engine off to allow testing of circuits under controlled conditions.

Most of the tests cycle components at a fixed rate for about 5 minutes. Repeatedly stressing the components and circuits over time also helps to pinpoint intermittent failures. The results can be measured with a DVOM or Lab Scope.

Lab Scope Connections
Connect the Channel 'A' positive probe to the purge solenoid control circuit (PK/BK wire) at PCM Pin 2-68, and connect the negative probe to the battery negative post.

Test Example
In this example, the actuator test has been turned ON with the Scan Tool. The purge solenoid is cycled at a fixed 50% duty cycle, and at a fixed frequency of 200 Hz.

The purge solenoid on this application is a positive-controlled actuator. The trace will look the same, but the solenoid is energized when the signal is high, not low.

Why To Use These Tests
The PCM makes purge decisions based on various criteria. It is much easier to simply command solenoid operation than to keep track of conditions. During this test, the desired operation is already known, which simplifies the test procedure.

The results of this actuator test will indicate the next test step. This actuator test helps to determine if the PCM is not capable of controlling the solenoid, or is just missing necessary sensor data to decide when to command the solenoid on.

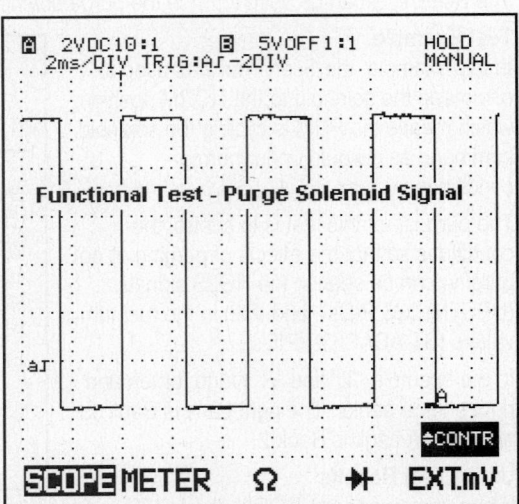

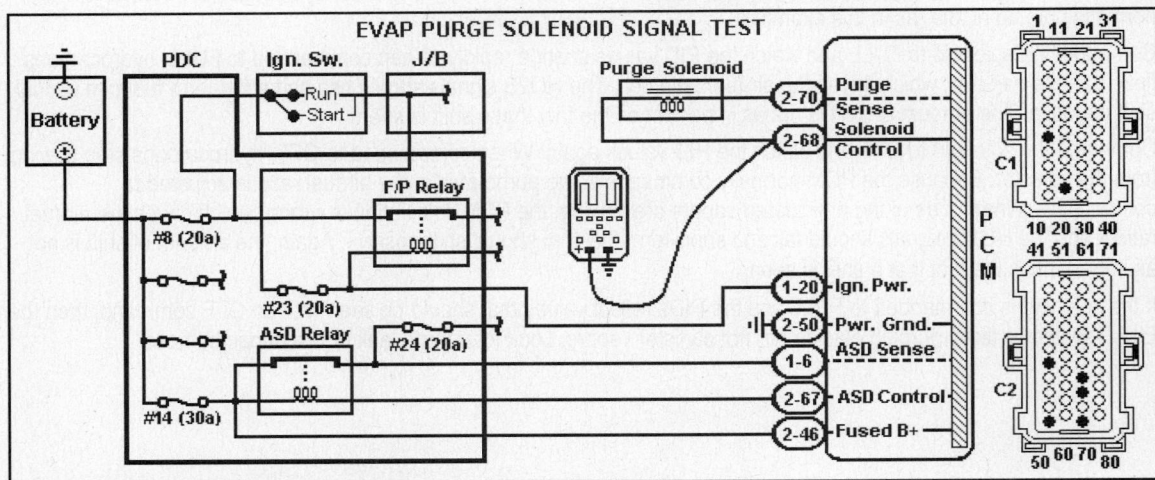

Purge Solenoid Signal Test Schematic

LAB SCOPE TEST (PURGE SENSE)

Place the gear selector in Park and block the drive wheels for safety

Lab Scope Connections
Connect the Channel 'A' positive probe to the purge sense circuit (WT/TN wire) at PCM Pin 2-70, and connect the negative probe to the battery negative post.

Lab Scope Example
In this example, the trace shows the purge sense signal during purging at Hot Idle. When the PCM energizes the purge solenoid, it supplies battery voltage to the coil inside the solenoid. Ground is supplied through the purge sense circuit.

In this example, when the purge solenoid is commanded on, the purge sense signal ranges from 300-500 mv. This value varies according to the duty cycle of the voltage command to the solenoid. At 0% duty cycle, the trace will show 0v (there is no voltage drop in an incomplete circuit).

Understanding Test Results
The purge solenoid is cycled at 200 Hz when energized, but this example shows a cycle every 3 seconds. That would be about 0.33 Hz. It is important to remember, however, that the Lab Scope cannot display 200 cycles per second when set to 1 second per division. At this setting, the trace is showing the *average* voltage drop in the circuit, measured in between the two loads in series (the solenoid and an internal PCM resistor).

The PCM monitors this drop and uses it to calculate the current and vapor flow. The current is then adjusted by varying the duty cycle of the voltage command. A higher duty cycle means the solenoid flows more current, which allows more vapors into the intake manifold. To see these PCM calculations, observe the following Scan Tool PIDs: PRGE FEDBCK (mA) and PRGE SOL FLW (%).

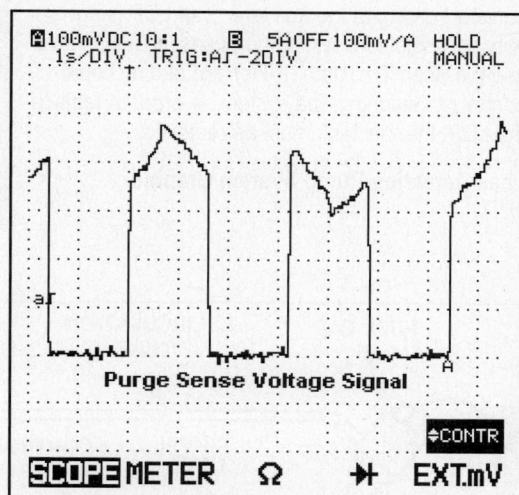

Purge Sense Voltage Signal

Purge Sense Signal Test Schematic

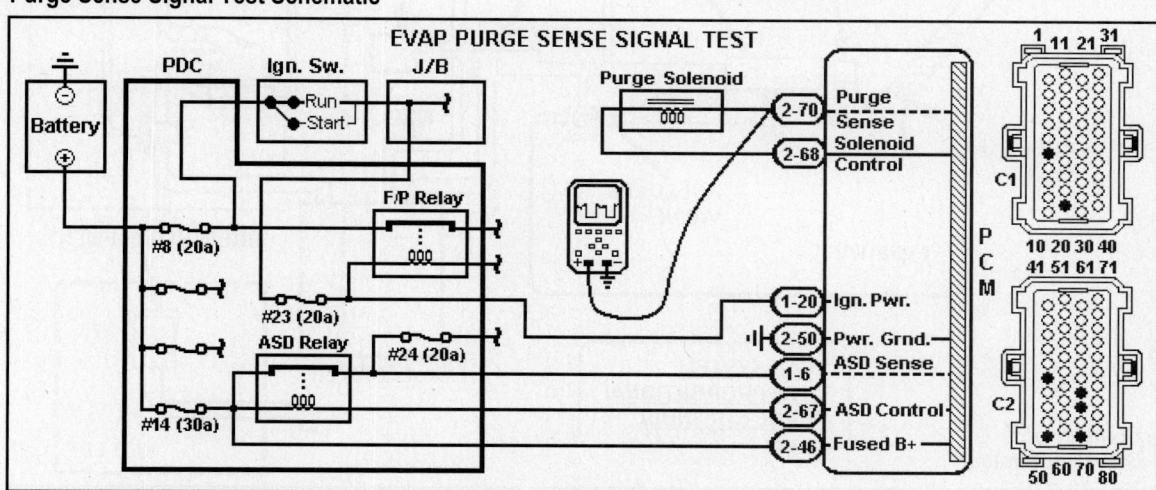

LEAK DETECTION PUMP (LDP)

The Leak Detection Pump (LDP) is a device used to pressurize the EVAP system during the EVAP System Monitor to check for leaks. The pump contains a 3-port solenoid, a pump that contains a calibrated pressure switch, a spring-loaded canister vent valve, 2 check valves and a spring/diaphragm assembly.

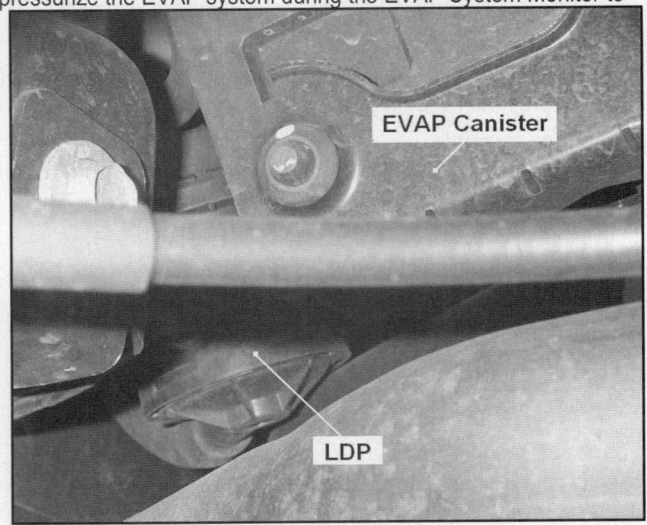

The LDP also seals the EVAP system from atmosphere so that the leak test can be performed by the PCM.

The LDP is located on a bracket (with the EVAP canister) on top of the fuel tank. The LDP is not very accessible for testing purposes, so it is best to use the Scan Tool data (PIDs) and bi-directional controls to diagnose the system. In order to replace the LDP, the fuel tank must be removed.

Leak Detection Pump System Graphic

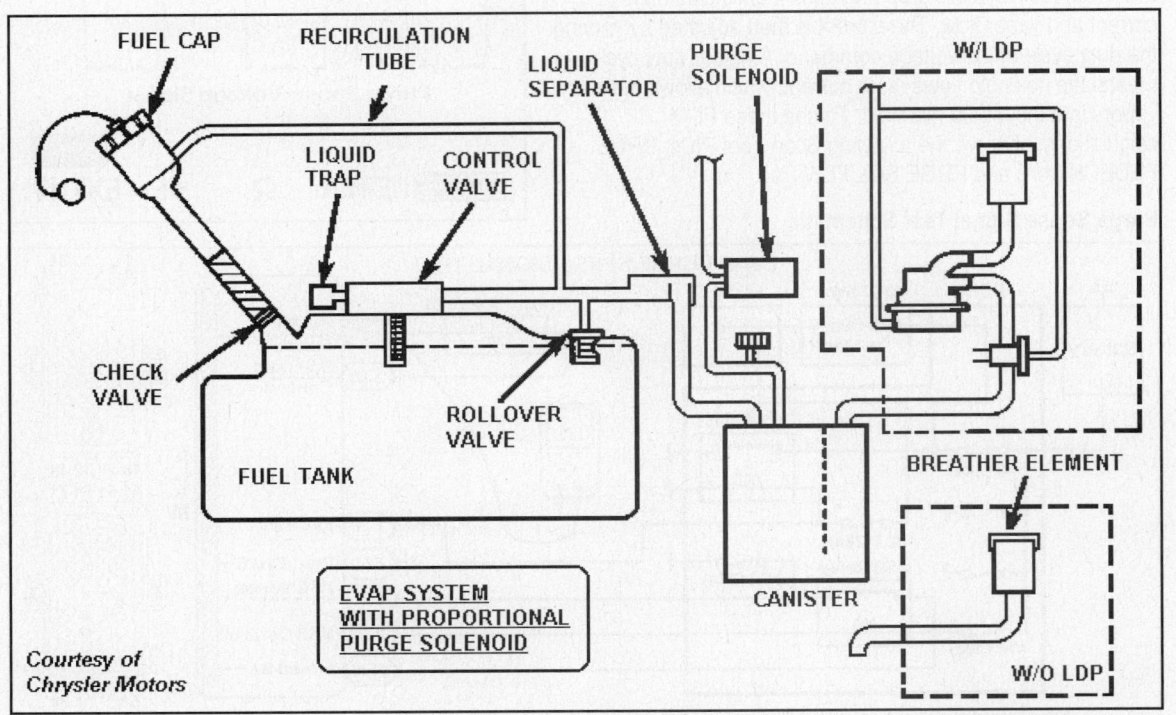

LDP MONITOR OPERATION

Leak Test

Immediately after a cold startup (a period established by predetermined temperature threshold limits), the three-port solenoid is briefly energized. This initializes the pump by drawing air into the pump cavity and also closes the vent seal.

During other non-test conditions, the pump diaphragm assembly holds the pump vent seal open (it pushes the vent seal open to the full travel position). The vent seal will remain closed while the pump is cycling because the reed switch triggers the three-port solenoid to prevent the diaphragm assembly from reaching full travel.

After the brief initialization period, the solenoid is de-energized allowing atmospheric pressure to enter the pump cavity, thus permitting the spring to drive the diaphragm that forces air out of the pump cavity and into the vent system. The solenoid is energized and de energized rapidly to create flow in typical diaphragm pump fashion.

The leak detection pump is controlled in two modes:

- Pump Mode - The pump is cycled at a fixed rate to achieve a rapid pressure build in order to shorten the overall test length.
- *Test Mode - The solenoid is energized with a fixed duration pulse. Subsequent fixed pulses occur when the diaphragm reaches the reed switch closure point.*

The spring in the pump is set so that the system will achieve an equalized pressure of about 7.5" H20. The cycle rate of pump strokes is quite rapid as the system begins to pump up to this pressure level. As the pressure increases, the cycle rate starts to drop off. If no leaks exist in the system, the pump will eventually stop pumping at the equalized pressure. If a leak is present in the system, the pump will continue to run at a rate that represents the flow characteristic of the size of the leak.

The PCM uses this information to determine if the leak is larger than the required detection limit. If a leak is revealed during the leak test portion of the test, the test is terminated at the end of the test mode and no further system checks will be performed.

After passing the leak detection phase of the test, system pressure is maintained by turning on the LDP solenoid until the purge system is activated. In effect purge activation creates a leak. The cycle rate is again interrogated and when it increases due to the flow through the purge system, the leak check portion of the diagnostic is complete.

The canister vent valve will unseal the system after completion of the test sequence as the pump diaphragm assembly moves to the full travel position.

Purge Test

The EVAP system functionality is verified by using the stricter EVAP Purge Flow monitor. At an appropriate warm idle period the LDP is energized to seal the canister vent. The purge flow will be ramped up from some small value in an attempt to see a shift in the Fuel Control (O2S) system. If fuel vapor, indicated by a shift in the fuel control is present, the test is passed. If not, it is assumed that the EVAP Purge system is not functioning in some respect. The LDP is again turned off and the test is complete.

BI-DIRECTIONAL TEST (LDP)

Place the shift selector in Park and block the drive wheels for safety.

Lab Scope Connections
Connect the Channel 'A' positive probe to the LDP control circuit (WT/DK GN wire) at PCM Pin 2-77, and connect the negative probe to the battery negative post.

Lab Scope Example (1)
In example (1), the FUNCTIONAL (actuator) TEST menu was selected on the Scan Tool. LDP SYSTEM TEST was selected. During this test, the PCM cycles the LDP on and off for 1.5 seconds (50% duty cycle).

This test pressurizes the EVAP system for leak testing, as well as for testing the LDP and pressure switch circuits

Lab Scope Example (2)
In example (2), the time base was changed to show the inductive kick each time the LDP is cycled off. Whenever a device that generates a magnetic field is turned off, the field collapses and a kick appears in the signal. The larger the field, the higher the voltage will be. However, Chrysler limits this kick to about 62 volts with a diode inside the PCM. This is also true for injectors and many other solenoids.

Observe the signal to verify that the trace has a flat top, indicating that the collapsing field was strong enough to reach the 62-volt limit.

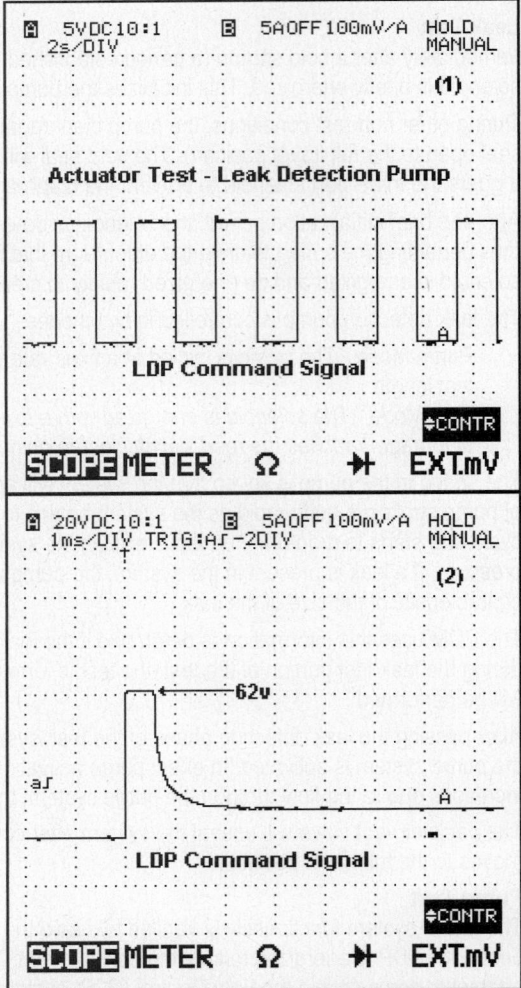

Leak Detection Pump Test Schematic

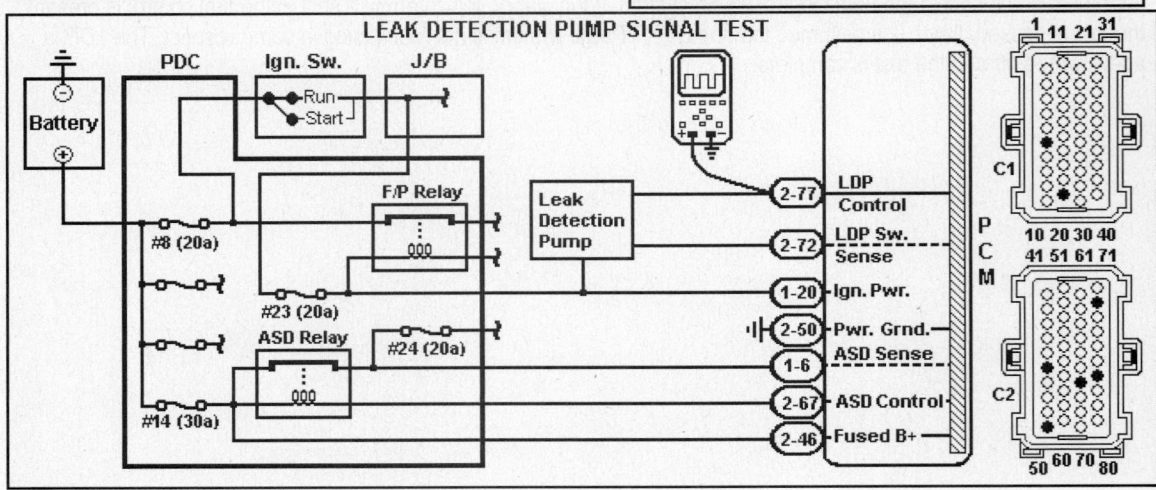

ONBOARD REFUELING VAPOR RECOVERY (ORVR)

A tremendous amount of hydrocarbons are released during the refueling process. When a tank is filled from empty, all of the fuel-enriched air in the tank is displaced. In addition to the problem of existing vapors, the refueling process generates a large volume of new vapors. Some states have programs with gas pump boots that recover the displaced and generated vapors. However, recent design changes allow the vehicle EVAP system to collect and store these vapors. The fuel vapors are then burned in the combustion process when the vehicle is started.

System Operation

The simple premise of the system is that the fuel tank filler tube can be designed in such a way that a venturi effect draws air into the tank instead of expelling it. The filler tube has a very small diameter (about 1").

The flow of fuel through this small tube causes fresh air to be drawn into the tank. The air in the tank is then forced out through a control valve in the tank to be stored in the EVAP canister. Hydrocarbons become trapped in the canister as the air passes through, and the 'cleaned' air is vented back to atmosphere. In effect, this process uses air as a vehicle to transport hydrocarbons to the desired location (canister) where they are stored.

As fuel enters the tank, a normally closed check valve is forced open. The check valve contains a float, so when the fuel level is high enough the valve floats up to seal the filler tube. This prevents overfilling by creating pressure in the filler tube that shuts off the pump nozzle. Also, the check valve prevents fuel from spraying out of the tank when the nozzle is shut off.

ORVR Problems

There are no electronic components used in this system, so diagnosis is limited to the physical condition of each component. The system operates by passively creating and using pressure and flow to accomplish the task of managing refueling vapors.

Therefore, most of the things that can go wrong with this system result in tank pressure buildup, which shuts off the pump nozzle. The complaint will be that the vehicle cannot be refueled, or sprays fuel when the nozzle is removed. Any restriction in the vapor lines from the tank to the EVAP canister can cause this condition. Also, any fuel saturation in the canister or the canister vent can restrict the system and cause pressure to build. Lastly, any mechanical defects in the fuel tank vapor control valve or the filler tube check valve can cause similar problems.

It is also possible to have refueling problems with nothing wrong. If the customer starts the car cold and drives to a nearby gas station, the EVAP Monitor may be interrupted when the key is shut off. When cold, the Monitor will enable the Leak Detection Pump (LDP), which *shuts off the canister vent* and pressurizes the system to check for leaks. See LDP Monitor Operation in this Section. If the system has residual pressure from this interrupted test, the filler tube check valve may stick closed. Even if it is not stuck, the vent valve may be sealed off from atmosphere, producing the same results. The best solution is to drive the car for a few minutes so the purge valve can bleed off pressures during normal operation.

A question to ask the customer is whether the engine was off during the refueling attempt. The system is designed only work properly with the engine off. This design prevents engine-on refueling by causing a buildup of pressure, and shuts off the pump.

Fuel Delivery System

INTRODUCTION

The Fuel system on this application is a mechanical-returnless system. The Fuel system delivers fuel at a controlled pressure and correct volume for efficient combustion. The PCM controls the Fuel system and injector timing and pulsewidth (duration).

On this system, there is no fuel return line from the fuel supply manifold back to the fuel tank. The fuel pump and fuel level module, mounted in the fuel tank, contains a regulator which returns extra fuel back through the pump assembly to the fuel tank. The pressure regulator is mechanical, but is not vacuum controlled. Therefore, fuel pressure should remain relatively constant under all engine-operating conditions.

Fuel Delivery

The vehicle is equipped with a sequential fuel injection (SFI) system that delivers the correct A/F mixture and injector timing under all operating conditions.

Air Induction

Air enters the system through the fresh air duct and flows through the air cleaner and into the intake plenum where it is mixed with fuel for combustion. The air volume is not directly measured on this application. The PCM uses the load (MAP sensor signal) and engine speed (CKP sensor signal) to calculate the amount of air entering the engine.

System Overview

A high-pressure (in-tank mounted) fuel pump delivers fuel to the fuel injection supply manifold. The fuel injection supply manifold incorporates electrically actuated fuel injectors mounted directly above each of the intake ports of the engine.

Power is supplied to the fuel injectors by the ASD relay, and to the fuel pump by the fuel pump relay. Drivers in the PCM ground the ASD relay control coil, fuel pump relay control coil, and individual injectors.

The PCM determines the amount of fuel the injectors spray under all engine-operating conditions. The PCM receives electrical signals from engine control sensors that monitor various factors such as manifold pressure (vacuum), battery temperature, engine coolant temperature, throttle position, and vehicle speed. The PCM evaluates the sensor information it receives and then signals the fuel injectors in order to control the fuel injector pulsewidth.

Fuel Pump Circuits Schematic

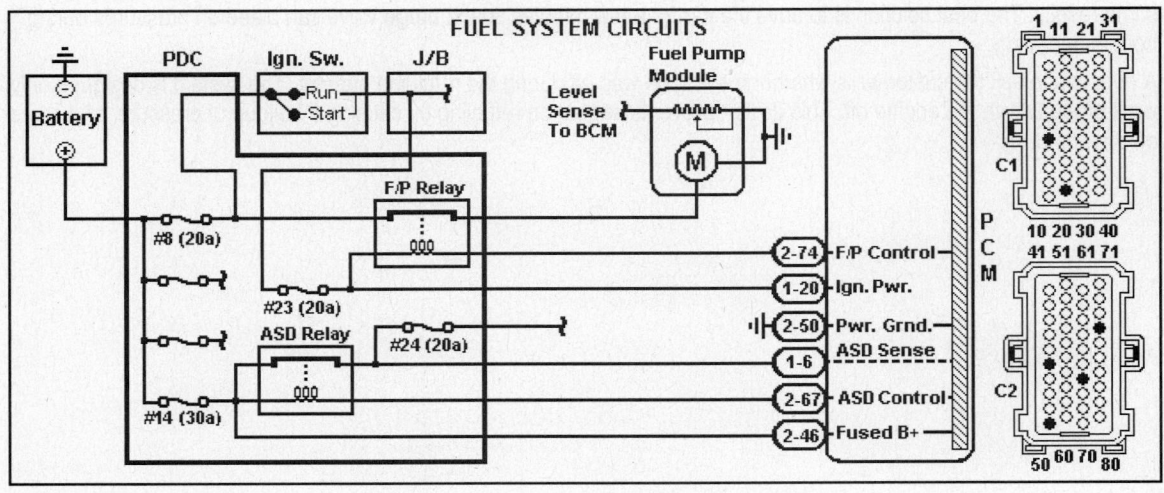

FUEL PUMP MODULE

The fuel pump is a positive-displacement electric motor. The fuel pump module is mounted in the top of the fuel tank. Fuel initially enters the module through a strainer in the bottom of the module, is pressurized and forced through the outlet to the fuel rail.

An integral regulator limits the pressure to the injectors to 58 psi. However, the pump has a maximum pressure output of 130 psi.

Fuel Pump Module Components

The fuel pump module includes an electric pump motor, reservoir, inlet strainer, filter, pressure regulator, fuel gauge sending unit, and a fuel line connection. Only the inlet strainer, level sending unit, and pressure regulator may be serviced separately.

Internal Valves

The pump module has two (2) internal check valves. One, located in the pump body, prevents excessive pressure from building inside the pump. The other, located in the outlet, allows the system pressure to be maintained with the ignition off for a short time. After a lengthy engine cold soak, it is normal for system pressure to drop to 0 psi.

FUEL FILTER & PRESSURE REGULATOR

The filter and regulator on this application are mounted to the top of the fuel pump module. The fuel filter is not normally replaced as a maintenance item. Replace this assembly when directed by a diagnostic procedure.

The regulator is calibrated to open a return port above 58 psi (+/- 5 psi). Excess fuel pump pressure is bled back into the tank.

The regulator traps fuel during engine shutdown. This eliminates the possibility of vapor formation in the fuel line, and provides quick restarts. However, residual pressure is not maintained, as the regulator only traps fuel to prevent it from draining back to the tank.

Filter Replacement

The fuel filter/pressure regulator assembly may be replaced separately without removing the fuel pump module. However, the fuel tank must be removed.

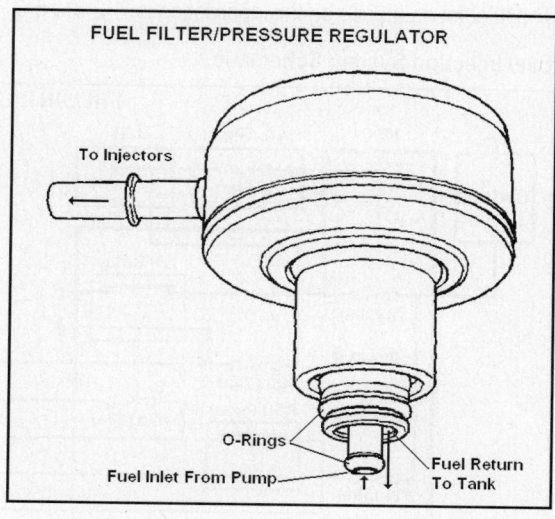

FUEL INJECTORS

Under normal operation, each injector is opened and closed at a specific time, once every other crankshaft revolution (once every engine cycle). The amount of fuel delivered is controlled by the length of time the injector is grounded (injector pulsewidth). The injectors are supplied voltage through the ASD relay.

Fuel Injection Timing

The PCM determines the fuel flow rate (pulsewidth) from the ECT, HO2S-11/21, TP and MAP sensors, but needs CKP and CMP signals to determine injector timing.

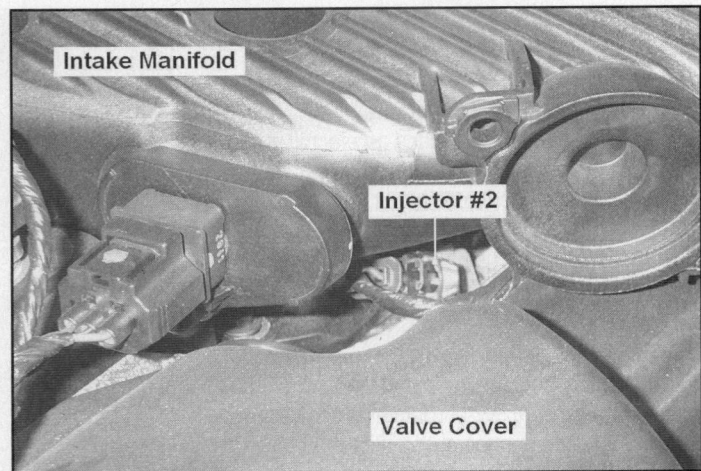

The PCM computes the desired injector pulsewidth and then grounds each fuel injector in sequence. Whenever an injector is grounded, the circuit through the injector is complete, and the current flow creates a magnetic field. This magnetic field pulls a pintle back against spring pressure and pressurized fuel escapes from the injector nozzle.

Circuit Description

The ASD relay connects the fuel injectors to direct battery power (B+) as shown in the schematic.

With the key in the "start or run" position, the ASD relay coil is connected to B+.

The ASD relay coil is grounded by the PCM, which creates a magnetic field that closes the relay contacts and supplies power to the fuel injectors.

The injector sequence on this application is: 1-2-3-4-5-6.

Fuel Injection System Schematic

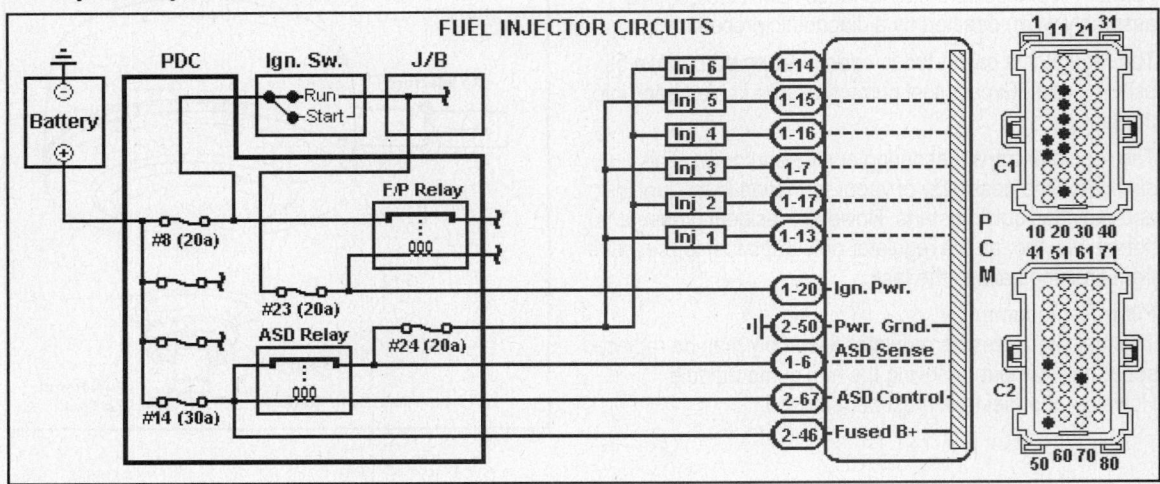

LAB SCOPE TEST (FUEL INJECTOR)

Place the gear selector in Park. Block the drive wheels for safety.

Scope Connections
Connect the Channel 'A' positive probe to the injector control wire at the injector or appropriate PCM Pin. Connect the negative probe to the battery negative post.

Lab Scope Example (1)
In example (1), the trace shows the injector voltage signal at idle speed. The 'on time' is the amount of time the PCM grounds the injector. In this example the on time is 3.3 ms.

The PCM clips the injector inductive kick at 62v. The PCM watches this voltage to determine if the circuit is operational. If the kick is less than 51 volts, a code will set for no inductive kick, (DTCs P0201 for injector 1, through P0206 for injector 6).

Lab Scope Example (2)
In example (2), a low-amps probe was added to the Channel 'B' to view the current flowing through the injector circuit.

The dip in the current trace is the point where the injector opens (left cursor). The hump in the voltage trace is the point where the injector closes (right cursor). Using voltage and current traces, it is easy to find differences in the injector mechanical operation ('squirt time') between cylinders.

Fuel Injector Test Schematic

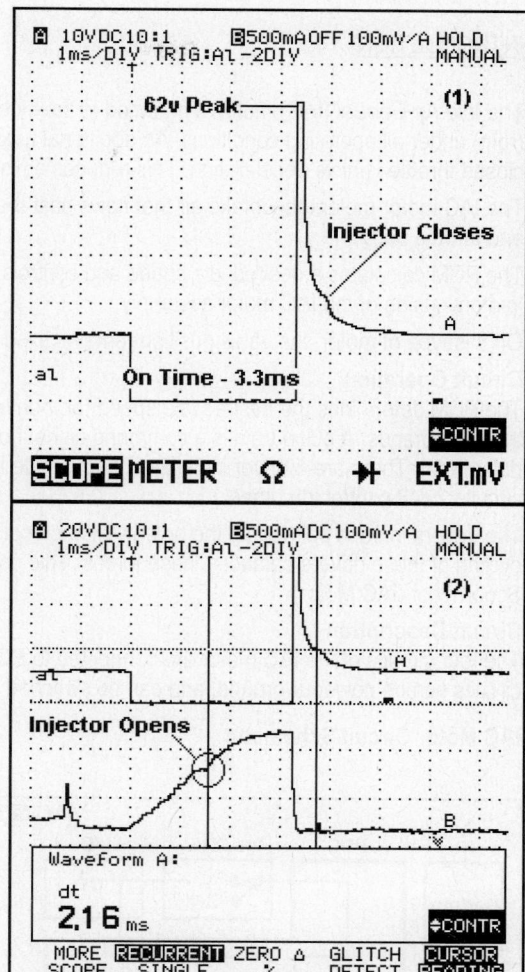

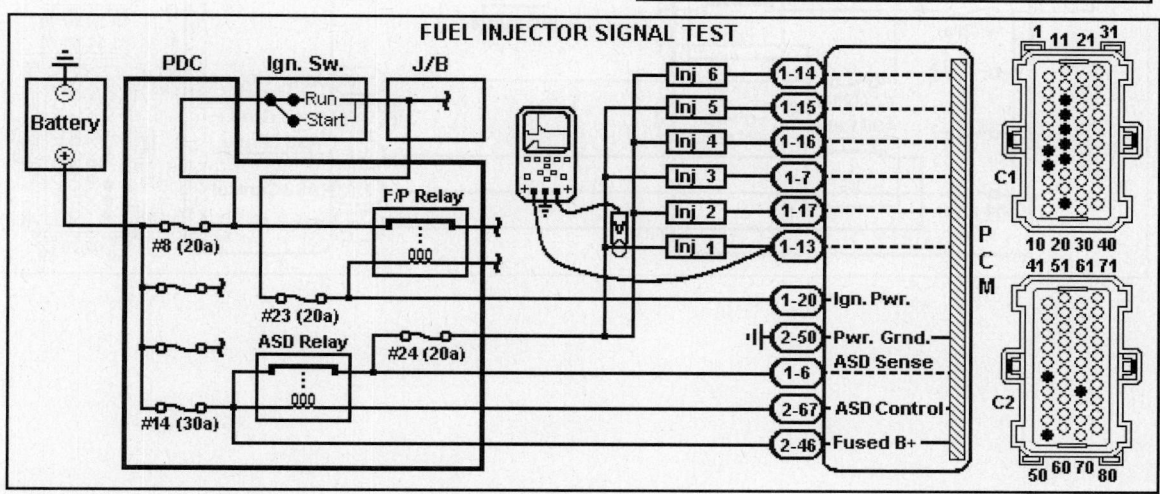

Idle Air Control Motor

INTRODUCTION

The Idle Air Control (IAC) motor is mounted to the side of the throttle body. It is designed to control engine idle speed (rpm) under all operating conditions. An additional function of the IAC motor is to control the engine deceleration during closed throttle vehicle deceleration. This reduces emissions and provides for smoother engine operation.

The IAC motor meters the intake air that flows past the throttle plate through a bypass area within the IAC assembly and throttle body.

The PCM calculates a desired idle speed and controls the IAC position through a pulse-train signal to the two (2) 'motors' inside of the IAC motor assembly.

On this type of motor, the valve position remains fixed if the harness is disconnected from the motor assembly.

Circuit Operation

The PCM determines the desired idle speed (amount of bypass air) and controls the IAC motor circuits through pulse train commands. A pulse train is a command signal from the PCM that continuously varies in both frequency and pulsewidth. There are 4 driver circuits connecting the IAC motor and the PCM. Circuits 1 & 4 control one 'motor' and circuits 2 & 3 control the other.

The drivers in each pair toggle the polarity of each coil to help push or pull the IAC motor pintle in its bore. The rapid cycling of these drivers balances these forces. This polarity switching can be observed on a Lab Scope. See Lab Scope Test (IAC Motor).

Circuit Description

The four circuits of the IAC motor are connected to PCM connector C2 at Pins 48, 49, 57, and 58. Each of these circuits can be power or ground, and can be switched rapidly according to current engine operating conditions.

IAC Motor Circuit Schematic

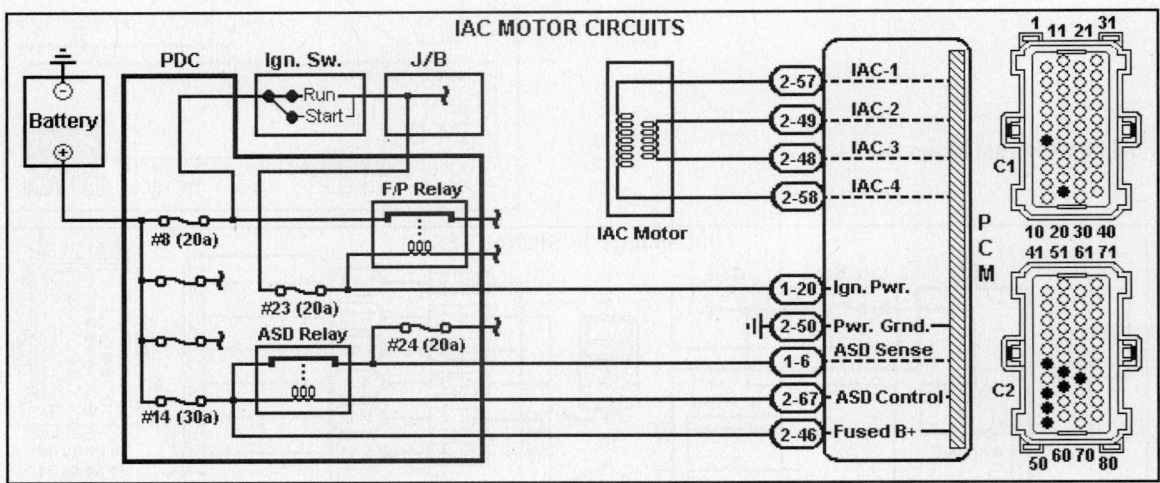

BI-DIRECTIONAL TEST (IAC MOTOR)

The Scan Tool bi-directional controls can be used to test the IAC motor and circuits. Test result can be monitored using a DVOM or Lab Scope. Place the gear selector in Park. Block the drive wheels for safety.

Lab Scope Connections

All four (4) drivers can be checked using 2 Lab Scope channels. The PCM controls one motor by alternating the polarity of drivers 1 & 4, and the other motor by alternating the polarity of drivers 2 & 3. Therefore, the Lab Scope positive probes should be connected to circuit 1 or 4, and circuit 2 or 3. A circuit problem in either pair will show up in this test.

Scan Tool Example

In this example, the Scan Tool was used to enter the IAC (AIS) MOTOR test in the FUNCTIONAL TESTS menu. This test, when activated, cycles the IAC motor back and forth in its bore. The pintle is commanded in each direction for about 1.5 seconds, and the pattern repeats for 5 minutes, or until commanded off using the Scan Tool.

Lab Scope Example

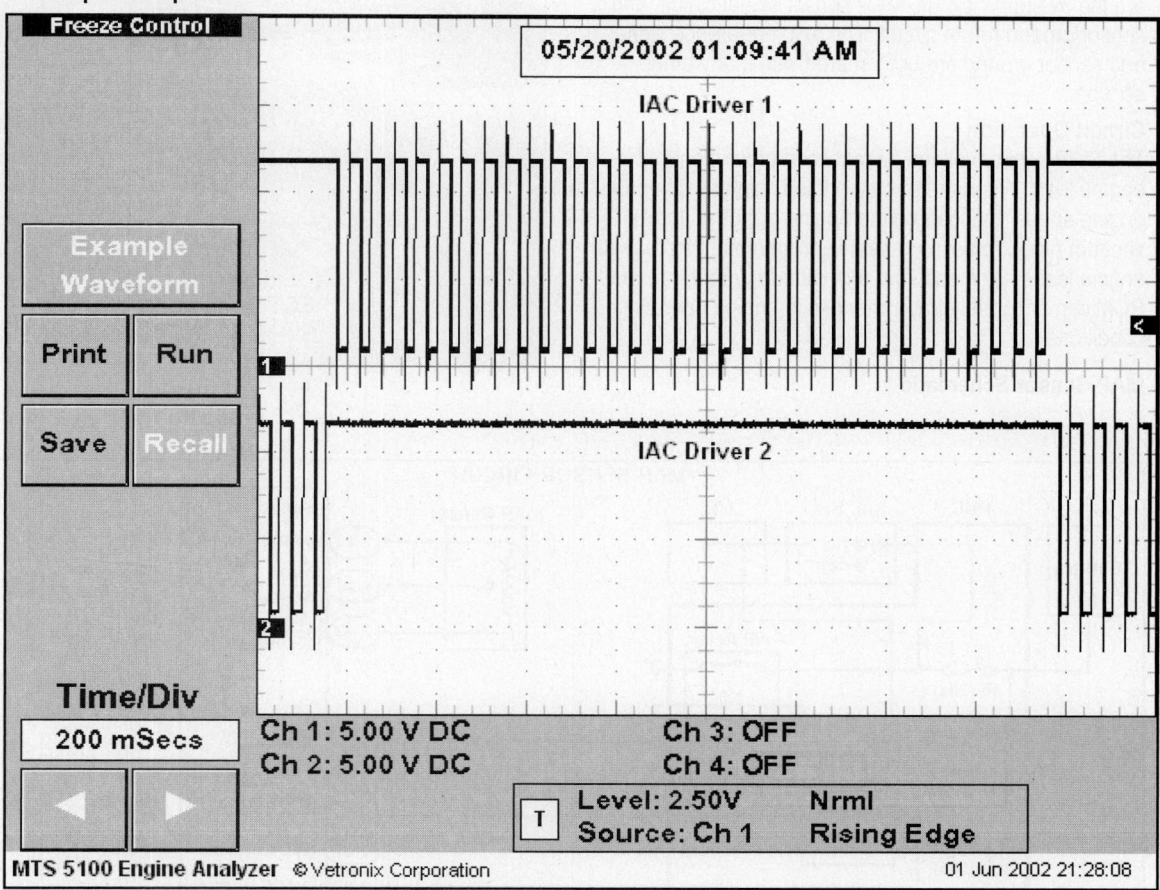

In this example, the traces show the PCM cycling the driver 1 and 2 command to exercise the motor from one limit to the other.

This test proves that the PCM is capable of controlling all four (4) IAC motor drivers, and that the IAC motor, circuit voltage, the ground circuit, and PCM drivers are all okay. This test does not verify that the IAC motor is in good mechanical condition.

Information Sensors

MANIFOLD ABSOLUTE PRESSURE SENSOR

The Manifold Absolute Pressure (MAP) sensor is located on the upper plenum of the intake manifold. The sensor is mounted with an o-ring directly to a port in the intake manifold (no vacuum hose is used). The MAP sensor input to the PCM is an analog voltage signal proportional to the intake manifold pressure.

This signal is compared to the TP sensor signal and other inputs to calculate engine conditions. The PCM uses theses calculations to determine the following outputs:

- Fuel Injector Pulsewidth
- Ignition Timing
- *Transmission Shift Points*

MAP Sensor Circuits

The three circuits that connect the MAP sensor to the PCM are the 5v supply circuit, MAP sensor signal circuit, and sensor ground return circuit. The 5-volt reference signal and sensor ground are both shared with many other sensors.

Circuit Operation

The MAP sensor voltage increases in proportion to the engine load. When the throttle angle is high as compared to engine speed, the load on the engine is higher. Intake vacuum drops (pressure rises) in proportion to increased engine loads. By measuring this vacuum (pressure) the PCM can accurately track changes in engine operating conditions.

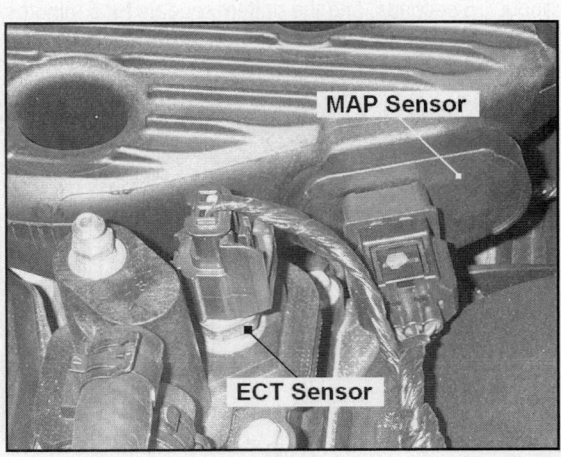

MAP Sensor Schematic

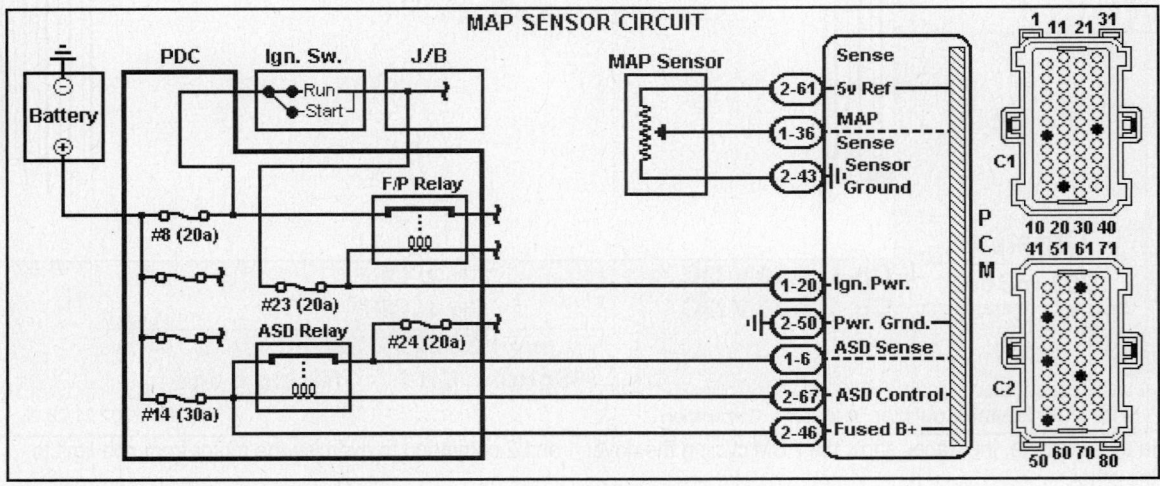

DVOM TEST (MAP SENSOR)

The DVOM is an excellent tool to test the condition of the MAP sensor circuits. The DVOM can be used to test the supply voltage, sensor ground voltage drop, and the MAP sensor signal for calibration or dropout.

The test results can be compared with the specifications in Map Sensor Table (located in Section 1).

DVOM Connections

Connect the DVOM positive probe to the MAP sensor signal circuit (DG/RD wire) at PCM Pin 1-36, or at the component harness connector. Connect the negative probe to the battery negative post.

DVOM Examples

In this example, the DVOM display shows the MAP sensor voltage at Hot Idle and key on, engine off (KOEO). The MAP sensor signal was 1.20v at Hot Idle.

The MAP sensor voltage increases in proportion to manifold pressure. In this example, the KOEO MAP sensor voltage was 4.58v at sea level. This signal is used by the PCM as a BARO reading for calculating base injector pulsewidth (duration).

The MAP sensor voltage supply should be 4.9-5.1v, and the ground voltage drop should not exceed 50 mv.

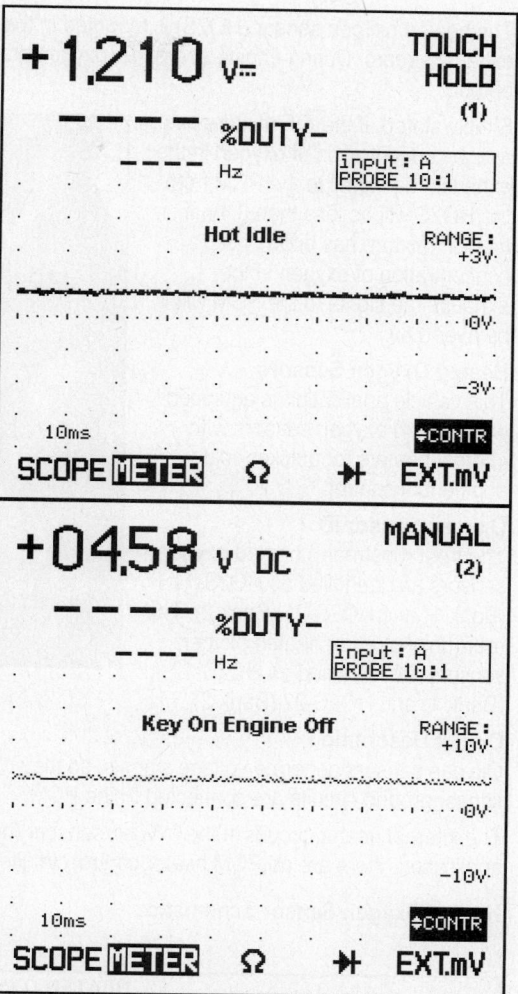

MAP Sensor Test Schematic

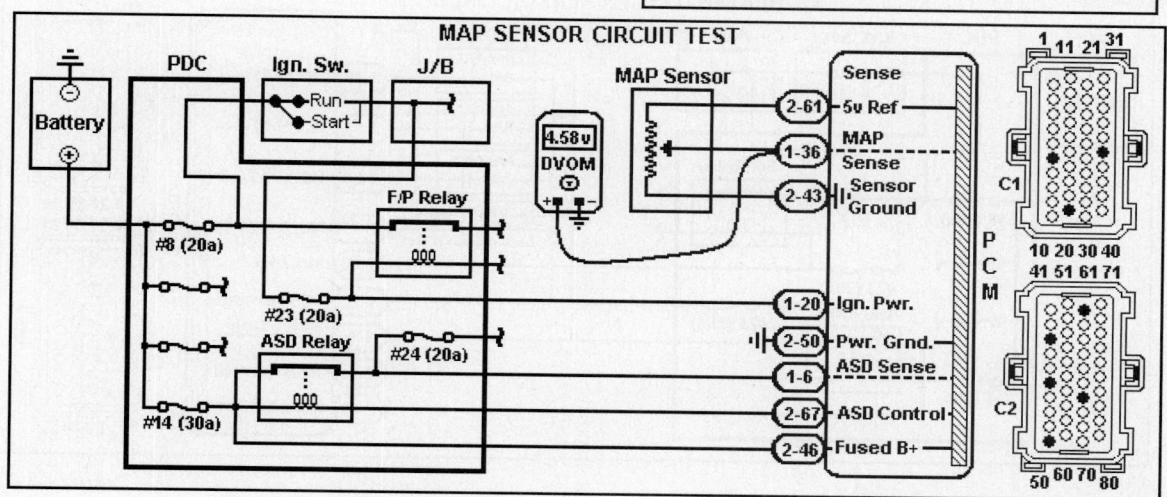

HEATED OXYGEN SENSORS

The heated oxygen sensor (HO2S) is mounted in the Exhaust system where it monitors the oxygen content of the exhaust stream. During engine operation, oxygen present in the exhaust reacts with the HO2S to produce a voltage output.

Simply stated, if the A/F mixture has a high concentration of oxygen in the exhaust, the signal to the PCM from the HO2S will be less than 0.4v. If the A/F mixture has a low concentration of oxygen in the exhaust, the signal to the PCM will be over 0.6v.

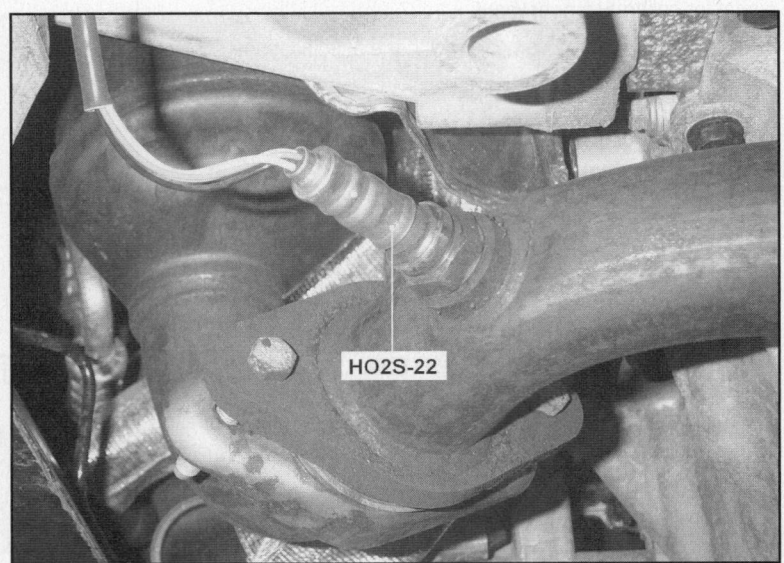

Heated Oxygen Sensors

This vehicle application is equipped with four (4) oxygen sensors with internal heaters for quicker cold engine fuel control.

Oxygen Sensor ID

The front (upstream) heated oxygen sensors are identified as HO2S-11 (Bank 1) and HO2S-21 (Bank 2). The rear (downstream) heated oxygen sensors are identified as HO2S-12 (Bank 1) and HO2S-22 (Bank 2).

Circuit Description

Oxygen sensors generate voltage signals, so they do not need any reference voltage. The HO2S signal circuits and sensor ground circuits are connected to the PCM.

The internal heater circuits in the oxygen sensors (both front and rear) receive power through the ASD relay. On this application, there are no PCM heater control circuits. The heaters circuits are connected directly to ground.

Heated Oxygen Sensor Schematic

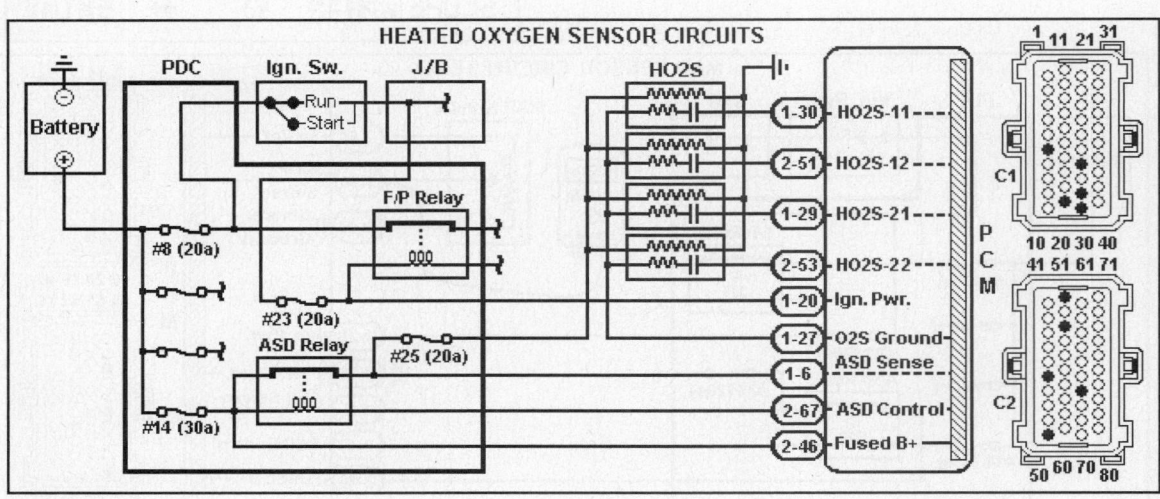

SCAN TOOL TEST (HO2S)

The Scan Tool is the most efficient way to compare HO2S values to other inputs to help determine the cause of a problem. It is also the best tool to use to get a quick idea of certain HO2S performance characteristics (i.e., time-to-activity from a cold start, general min/max values, etc.).

The Scan Tool has limitations, however, in that the data refresh rate is much slower than either a DVOM or Lab Scope. Therefore, much of the circuit activity is not viewable. It is important to determine before testing begins if the flexibility and ease of the Scan Tool is more important than the precision of a DVOM or Lab Scope.

Scan Tool Testing

Connect the Scan Tool to the DLC underdash connector. Navigate through the Scan Tool menus until you get to the Chrysler (OEM) Data List (a list of PID items).

Observe the HO2S PIDS and raise the engine speed to 2000 rpm for 3 minutes. Minimum and maximum voltage values, time-to-activity, and switch rates should all be evaluated.

Scan Tool Example

This example was captured at Hot Idle. L and R UPSTM O2S (V), and L and R DNSTM O2S (V) voltages were observed, as well as the RICH/CENTER/ LEAN values.

HO2S Heater Testing

It is possible to test the oxygen sensor heaters on this application using Scan Tool bi-directional controls.

There is no PID for heater status, as the heaters are not controlled by the PCM. They are on whenever the ASD relay is energized. The best way to know if the heaters are operational is to observe the effects of the heat on the sensor signal values. As the sensor temperature rises,

```
        Snap-on Scan Tool Display


CHRYSLER CAR                      A/T
2.7L V6 MPI                       A/C
CHRY ENG
NO CODES PRESENT
     RPM_____640   TPS(V)_____0.74
IGN CYCLES 1_____2   IGN CYCLES 2_____0
IGN CYCLES 3_____0   CLSD LOOP TMR__0:00
L-FUEL SY_CLSD LOOP     R-FUEL SY_CLSD LOOP
CRANK SENSOR____YES     CAM SENSOR_____YES
CURRENT SYNC_____OK    DIS SGNL__CAM&CRANK
MAP SNSR(V)_____1.3    MAN VAC(kPa)_____66
BARO PRES(kPa)__100     THROTTLE(%)_____0
TPS(V)_____0.74    MIN TPS(V)_____0.74
COOLANT(V)_____3.0    COOLANT(°F)_____165
IAT(V)_____2.25    IAT(°F)_____84
L BANK INJ(mS)__2.5     R BANK INJ(mS)__2.5
L UPSTM O2S(V)_0.33     R UPSTM O2S(V)_0.57
L BNK UP EXH___LEAN     R BNK UP EXH___LEAN
L DNSTM O2S(V)_0.88     R DNSTM O2S(V)_0.86
L ST ADAP(%)_____1.8    R ST ADAP(%)_____0.8
L LT ADAP(%)____10.7    R LT ADAP(%)_____7.2
TARGET IAC_____17    IAC (STEPS)_____17
ENGINE RPM_____640     DES IDLE RPM_____664
```

the resistance also rises, and signal voltage will drop. Observe the HO2S PID values on the scan tool or with a DVOM to verify that the signal voltages drop when the heater test is activated.

If the HO2S heaters are inoperative, the sensors will not become active quickly enough after a cold engine startup. This will delay closed loop cooperation. Also, even a warm sensor needs the heaters to stay at an efficient operating temperature during certain operating conditions, like extended idle.

LAB SCOPE TEST (COLD HO2S)

The Lab Scope can be used to observe the warm-up process of the HO2S and the catalytic converter. Place the gearshift selector in Park and block the drive wheels.

Lab Scope Connections

Connect the Channel 1 positive probe to the HO2S-11 signal circuit (BK/DG wire) at PCM Pin 2-51, and the negative probe to the battery negative post. Connect the Channel 2 probe to the HO2S-12 signal circuit (TN/WT Wire) at PCM Pin 1-30, the Channel 3 probe to the HO2S-21 signal circuit (LG/R wire) at PCM Pin 1-29, and the Channel 4 probe to the HO2S-22 signal circuit (PK/WT Wire) at PCM Pin 2-53.

Lab Scope Example

In this example, traces from the four (4) oxygen sensors were captured after a cold startup. The HO2S-11 and HO2S-12 signals were captured with the vehicle at idle from a cold start. Note that at 5 seconds per division, the HO2S-11 began switching (the vehicle entered closed loop) after about 45 seconds of engine operation.

In this example, the catalytic converters were warm and efficiently storing oxygen by the time the downstream sensors became active. It is not a good idea to test for converter efficiency in this way (a cold engine at idle). It is only important to realize that the downstream sensors may not be as active as the upstream sensors if the converters are functioning when the PCM enters closed loop.

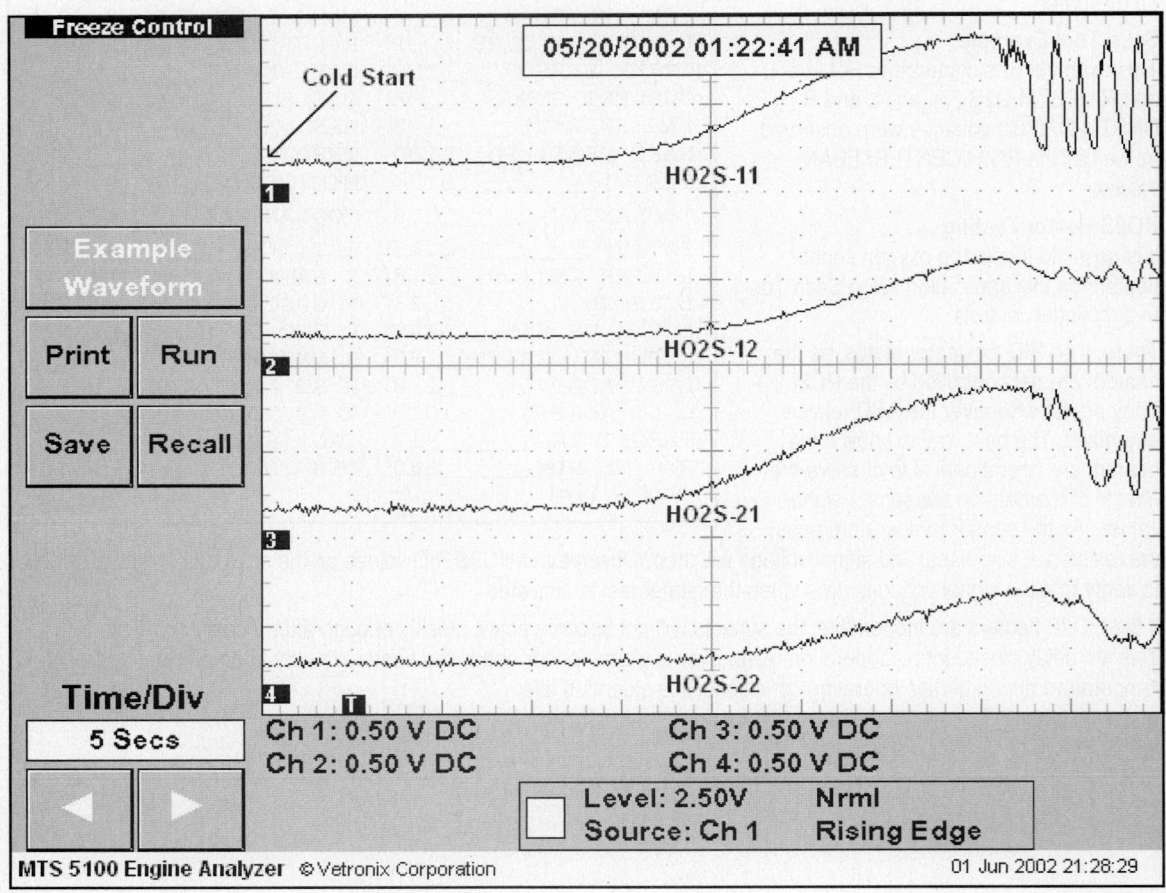

LAB SCOPE TEST (WARM HO2S)

Place the gearshift selector in Park and block the drive wheels for safety. This test is effective for monitoring converter efficiency and various oxygen sensor performance characteristics. For connections, see Lab Scope Connections on the previous page.

Lab Scope Example

In this example, traces from the four (4) oxygen sensors represent the 50 seconds of circuit activity just after PCM entered closed loop. Note that the downstream sensor signals have less amplitude change over time, but the frequency is similar to that of the upstream sensors. This is normal, as the vehicle is at idle and is just warming up. If these traces were obtained consistently at cruise, it could indicate that the catalysts were partially degraded.

Trouble Code Help

It is important to create the proper conditions when analyzing converter efficiency. The OBD II Catalyst Monitor runs only under the conditions where it is most likely to pass. The Monitor will run after 3 minutes from startup, and at steady speeds over 20 mph.

Under cruise conditions, the sensor and converter warming process is greatly accelerated. The PCM is attempting to verify that the switch rate of HO2S-12 as compared to that of HO2S-11 is less that 70% at steady cruise. If this ratio is exceeded, a DTC P0420 (converter efficiency) will set (2-trip code). This test can be used to recreate these conditions to verify converter failure.

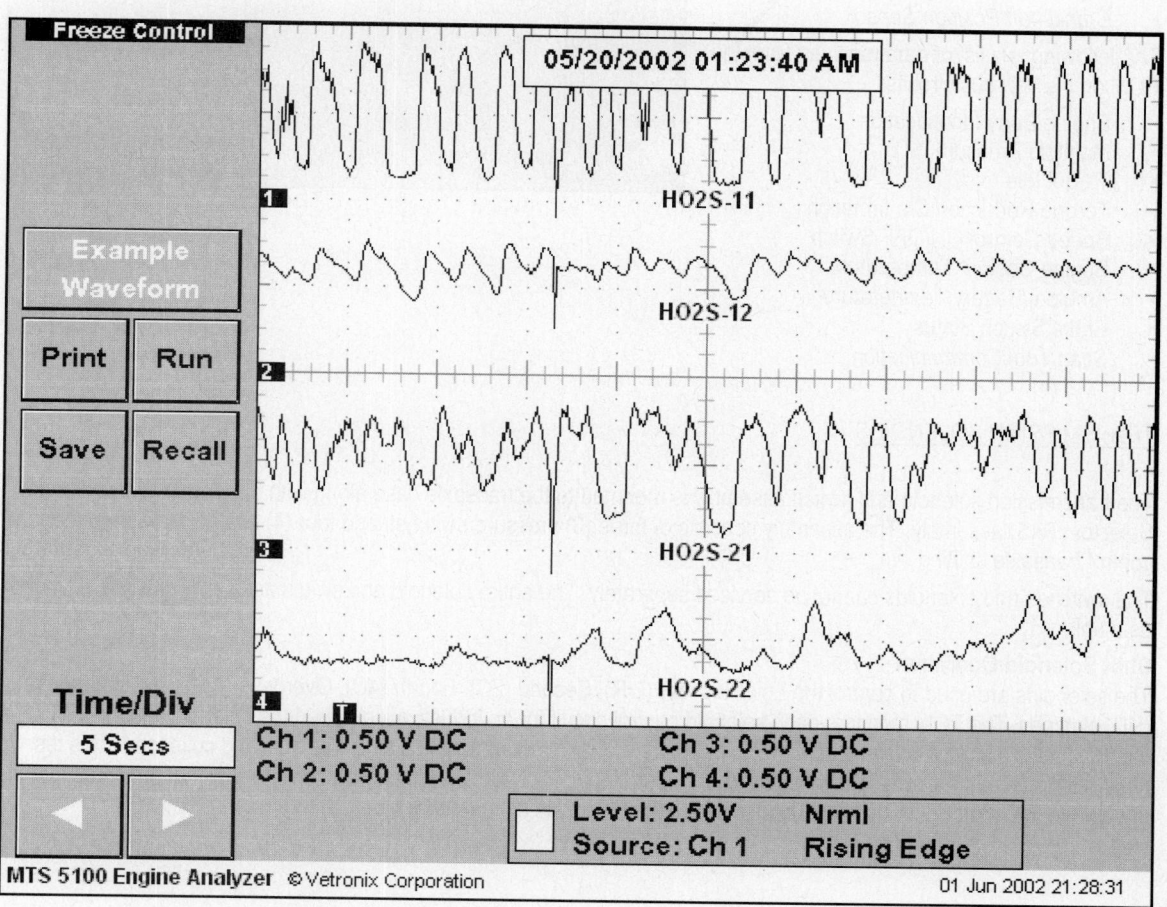

Transmission Controls

INTRODUCTION

This vehicle is equipped with a 41TE electronic automatic transaxle. This transmission features fully adaptive electronic control. The Transmission Control Module (TCM) is able to precisely control all shift schedules, line pressure, and torque converter clutch operation to deliver timely and quality shifts under all driving conditions.

The TCM is located in the left front corner of the engine compartment, next to the air cleaner housing and the PCM.

The TCM also performs many self-diagnostic functions, and provides for Scan Tool data (PID) access and bi-directional controls.

The direct TCM inputs include:

- Transmission Control Relay
- Pressure Switches
- Trans. Fluid Temp. Sensor
- Input Shaft Speed Sensor
- Output Shaft Speed Sensor
- Throttle Position Sensor
- Transmission Range Sensor
- *Crankshaft Position Sensor*

The following is a list of data received from the PCM on the SCI bus circuits:

- Engine/Body Identification
- Manifold Pressure
- Target Idle
- Torque Reduction Confirmation
- Speed Control ON/OFF Switch
- Engine Coolant Temperature
- Ambient/Battery Temperature
- Brake Switch Status
- *Scan Tool Communication*

TRANSMISSION SOLENOIDS

The transmission solenoid and switch assembly is mounted to the transaxle case along with the Transmission Range Selector (TRS) assembly. The assembly consists of three (3) pressure switches and four (4) solenoids that monitor and control transaxle shifting.

The switches and solenoids cannot be serviced separately. The entire solenoid and switch assembly must be replaced as a unit.

Shift Solenoid Operation

The solenoids are used to control the Low/Reverse (L/R), Second (2C), Fourth (4C), Overdrive (OD), and Underdrive (UD) clutches. The TCM monitors each solenoid control circuit for an inductive kick when the solenoid is turned off. If the TCM sees a kick, it knows that the circuit is complete, and that a magnetic field was built and collapsed. This test does not prove that the solenoid is mechanically operational. Pressure switches are used to verify that fluid pressure is diverted to the appropriate hydraulic circuit when the solenoid is commanded on.

TRANSMISSION PRESSURE SWITCHES

All of the pressure switches are calibrated to close at 23 psi, and open at 11 psi. The switch inputs are used by the TCM to verify normal solenoid operation.

Each pressure switch should open when the corresponding solenoid is off, and close when that solenoid is energized. The switch state should toggle rapidly as the corresponding solenoid command is cycled on and off.

Transmission Pressure Switch Position Table

Gear	L/R Switch	2/4 Switch	OD Switch
Park	Closed	Open	Open
Reverse	Open	Open	Open
Neutral	Closed	Open	Open
1	Closed	Open	Open
2	Open	Closed	Open
D	Open	Open	Closed
OD	Open	Closed	Closed

DIAGNOSTIC STRATEGY

The Transmission Clutch Application Table, shown below, indicates which clutches are in use for each Transaxle Range Selector position. Compare the information in the table to the Scan Tool data (PIDs) to help isolate system or circuit failures. The C1-C4 shift lever position switch, pressure switch, and Clutch Volume Index (CVI) PIDs are all useful in diagnosing electronic or hydraulic failures.

Other test procedures are covered in this article, and many more can be created using combinations of Scan Tool PIDs, bi-directional controls, a DVOM, and a Lab Scope. However, the test results all have to be compared to the expected results. Use the table below to determine which clutches will be used during each operating condition.

Transmission Clutch Application Table

TRS Pos.	UD - Underdrive	OD - Overdrive	Reverse	2/4	L/R – Low/Reverse
Park	OFF	OFF	OFF	OFF	ON
Reverse	OFF	OFF	ON	OFF	ON
Neutral	OFF	OFF	OFF	OFF	ON
OD-1	ON	OFF	OFF	OFF	ON
OD-2	ON	OFF	OFF	ON	OFF
OD-Direct	ON	ON	OFF	OFF	OFF
Overdrive	OFF	ON	OFF	ON	OFF
Drive-1	ON	OFF	OFF	OFF	ON
Drive-2	ON	OFF	OFF	ON	OFF
Drive-Direct	ON	ON	OFF	OFF	OFF
Low-1	ON	OFF	OFF	OFF	ON
Low-2	ON	OFF	OFF	ON	OFF
Low-Direct	ON	ON	OFF	OFF	OFF

TRANSMISSION RANGE SELECTOR (TRS)

The C1 through C4 PIDs are transmission range selection indicators. Each PID value means nothing on its own. The TCM looks at the combinations of switch positions in order to calculate the transmission range selector position.

The switch values are combined into possible expected combinations indicating valid gear positions or between gear conditions. If the switch value combination does not indicate the transaxle is in a valid gear or between gears, the TCM detects an impossible condition and sets a DTC. The TCM will continue to shift the transmission using only the pressure switches to determine shift lever position. This allows for close to normal transmission operation even with a TRS or circuit failure.

CLUTCH VOLUME INDEX (CVI)

An Important function of the TCM is to monitor the Clutch Volume Index (CVI). These indexes represent the volume of fluid needed to compress a clutch pack.

The TCM monitors gear ratio changes by watching the input and output speed sensor signals (ISS & OSS). By comparing the sensor signals, the TCM can determine the transmission gear ratio. This is important to the calculation because the TCM determines the CVI index by counting how long it takes for each gear change to occur. Once the ratio of the ISS and OSS has stabilized, the shift is considered complete.

The CVI values can be seen on the Scan Tool and compared to the Clutch Volume Index Table to help determine if a particular clutch pack is worn. Some other causes of failure include broken return springs, excessive clearance, and improper assembly.

Clutch Volume Index Table

Clutch	Updated During:	Throttle Angle	Fluid Temperature	Clutch Volume
L/R	2-1 or 3-1 Shift, coasting	< 5°	>70 °F	35-83
2/4	1-2 Shift	5-54°	>110°F	20-77
OD	2-3 Shift	5-54°	>110°F	48-150
UD	4-3 or 4-2 Shift	> 5°	>110°F	24-70

Diagnosing a Slipping Transmission

Using the logic described above, you can determine if the transmission is slipping and what clutch pack or packs have failed. The ISS and OSS signals generate the same number of pulses per revolution.

The gear ratio is determined by dividing the ISS frequency (Hz) by the OSS frequency (Hz). Measure both signals under the conditions where slippage seems to be occurring and then calculate the gear ratio. Compare the gear ratio to the Gear Ratio Table below to determine if they match. Refer to Scan Tool Test (RPM, ISS & OSS) in this section.

Gear Ratio Table

Gear	Ratio
First	2.84:1
Second	1.57:1
Third	1.00:1
Overdrive	0.69:1
Reverse	2.21:1

SCAN TOOL TEST (A/T PIDS)

The Scan Tool is an excellent tool to diagnose both electrical and mechanical problems in the transmission, TCM or circuits. While further testing may be required, the Scan Tool can help you determine the best diagnostic path to follow.

Scan Tool Example
In this example, the Scan Tool capture shows the PID values at Hot Idle with the transmission in neutral.

Line Pressure PIDs
Observe the following PIDs to check the pressure control solenoid, sensor and circuits: PCS DTY CYC (%), LINE PRES (V), LINE PRES (PSI), and DES PRES (PSI).

Pressure Switch PIDs
The solenoid commands cannot be seen as PIDs. However, the vehicle can be driven through the gears, and the following pressure switch PIDs can be observed: L-R PRESS SW, UD PRESS SW, 2/4 PRESS SW, and OD PRESS SW.

Range Selector PIDs
The C1 through C4 PIDs are transmission range selection indicators. The four switch values are used to calculate gear or between-gear position. Certain combinations indicate invalid gear positions, and will set a DTC.

```
+--------------------------------------------------------+
|                                                        |
|          Snap-on Scan Tool Display                     |
|                                                        |
|  --------------------------------------------------    |
|                                                        |
|  CHRYSLER CAR                         A/T              |
|  2.7L V6 MPI                          A/C              |
|  CHRY TRAN                                             |
|  NO CODES AVAILABLE FROM THIS MODE                     |
|      ENGINE RPM_2691     TPS(V)_____1.07         |
|  INPUT RPM_____2597   VERSION NO._____23         |
|  OUTPUT RPM_____3770    TPS ANGLE(°)_____9          |
|  MIN TPS(V)_____0.72    LR CLUTCH CVI_____42           |
|  2/4 CLUTCH CVI___57     OD CLUTCH CVI___118            |
|  UD CLUTCH CVI_____63    TRANS TEMP(°F)__136            |
|  TRANS TEMP(V)__1.04     LINE PRES(PSI)__114            |
|  LINE PRES(V)____2.12    DES PRES(PSI)___120            |
|  PCS DTY CYC(%)___34     IGNITION(V)_____12.0           |
|  SWITCH BATT(V)_14.1     L-R PRESS SW___OPEN            |
|  OD PRESS SW_____CLSD    UD PRESS SW_____OPEN           |
|  2/4 PRESS SW___CLSD     C1 SWITCH_____CLSD          |
|  C2 SWITCH_____OPEN     C3 SWITCH_____OPEN          |
|  C4 SWITCH_____OPEN     O/D OFF SW_____ON          |
|  GEAR_____O/D   PART NO._04896789AD           |
+--------------------------------------------------------+
```

SCAN TOOL TEST (RPM, ISS & OSS)

The Scan Tool can be used to test for transmission slippage. In addition to using the CVI to isolate worn clutch packs, the RPM PIDs can help to find worn and slipping components.

Observe the ENGINE RPM, INPUT RPM, and OUTPUT RPM. The PIDs in the example above were captured at 70 MPH in overdrive. The torque converter slip and any gear slip can be calculated from the PIDs. In this example, the torque converter slip is 94 RPM (ENGINE RPM minus INPUT RPM).

To calculate gear slippage, divide the INPUT RPM by the OUTPUT RPM. Compare the result to the Gear Ratio Table on the previous page. In this example, the ratio is 0.689:1, which is very close to the expected 0.69:1 in overdrive.

LAB SCOPE TEST (OSS & EATX)

The Lab Scope can be used to check the Input Speed Sensor (ISS) and Output Speed Sensor (OSS) signals. The processed EATX RPM signal and Vehicle Speed Sensor (VSS) signals from the TCM to the PCM can also be tested.

Circuit Description

The ISS and OSS sensors produce an AC signal with a frequency (Hz) proportional to the rotating speed of the appropriate shaft. This signal is sent to the TCM at Pin 52 for the ISS and Pin 14 for the OSS. The signals are then converted to a 0-5v digital signal by an A/D converter in the TCM and sent to the PCM. The ISS signal is sent as the EATX RPM signal to the PCM at Pin 2-71. The OSS signal is sent as the VSS signal to the PCM at Pin 2-66.

Lab Scope Connections

Connect the Channel 'A' positive probe to the ISS signal circuit (RD/BK wire) at TCM Pin 52, and connect the negative probe to the battery negative post. Connect the Channel 'B' probe to the OSS signal circuit (LG/WT Wire) at TCM Pin 14.

Lab Scope Example

In the example, the trace shows the ISS and OSS signals at about 60 MPH. The vehicle was in third gear (1.00:1 ratio) so the traces have the same frequency.

If slippage is suspected, use a DVOM to compare Hz readings on both sensors (use 2 DVOMs to compare values at the same point in time).

The TCM will detect the signal from either sensor as long as they produce at least 300 mv AC. The amplitude of these signals is about 3 volts AC, peak to peak. The amplitude will change very little at different speeds. Because of the way the circuits are constructed inside the TCM, normal amplitude variations are dampened.

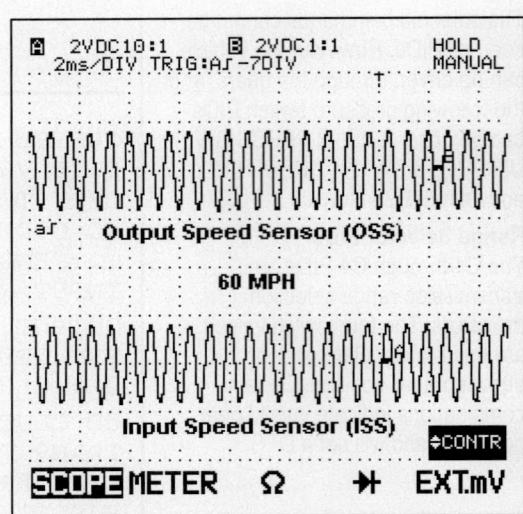

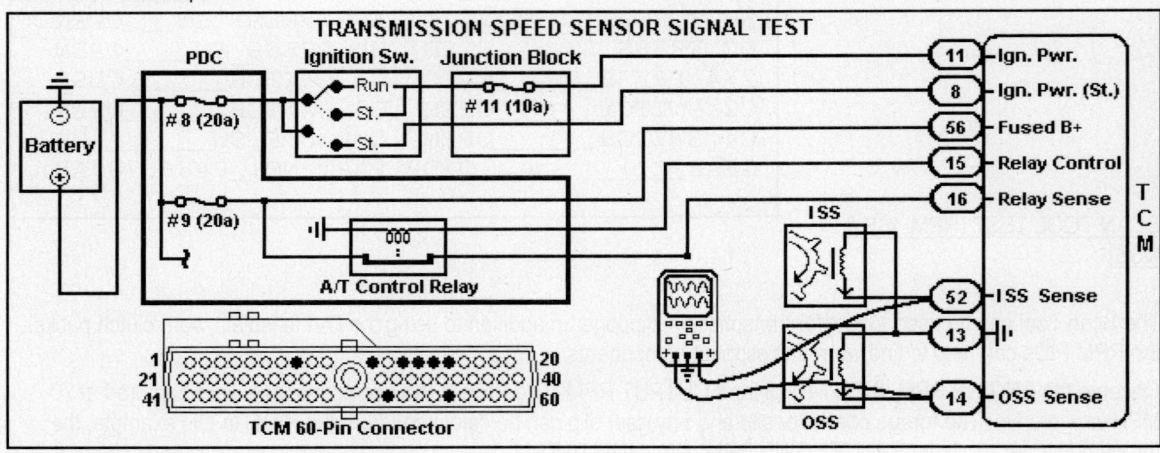

Transmission Speed Sensor Test Schematic

LAB SCOPE TEST (SHIFT SOLENOID)

The Lab Scope can be used to test the command signal from the TCM to any of the shift solenoids. Place the gear selector in Park and block the drive wheels for safety.

Lab Scope Connections
For this example, connect the Channel 'A' positive probe to the 2/4 shift solenoid signal circuit (WT Wire) at TCM Pin 19. Connect the negative probe to the battery negative post.

Lab Scope Example (1)
In example (1), the trace shows the TCM command to the 2/4 shift solenoid at idle. The waveform shows about a 35% duty cycle signal at about 550 Hz. The vehicle will not stall with this command at idle because the line pressure is low at idle speed.

Lab Scope Example (2)
In example (2), the trace shows the same signal as in example (1), but the volts/division setting has been changed to view the inductive kick from the solenoid.

When the TCM stops cycling the solenoid, the field collapses and an inductive kick is produced. The TCM uses an internal diode to clip this voltage to protect its circuits.

Shift Solenoid Diagnosis
This test does not prove that the shift solenoid is mechanically operational, only that the electrical circuit is good. Connect Channel 'B' to the 2/4-switch sense circuit to view the switch toggling along with the solenoid (not shown in these examples). This test confirms that the solenoid is mechanically operational.

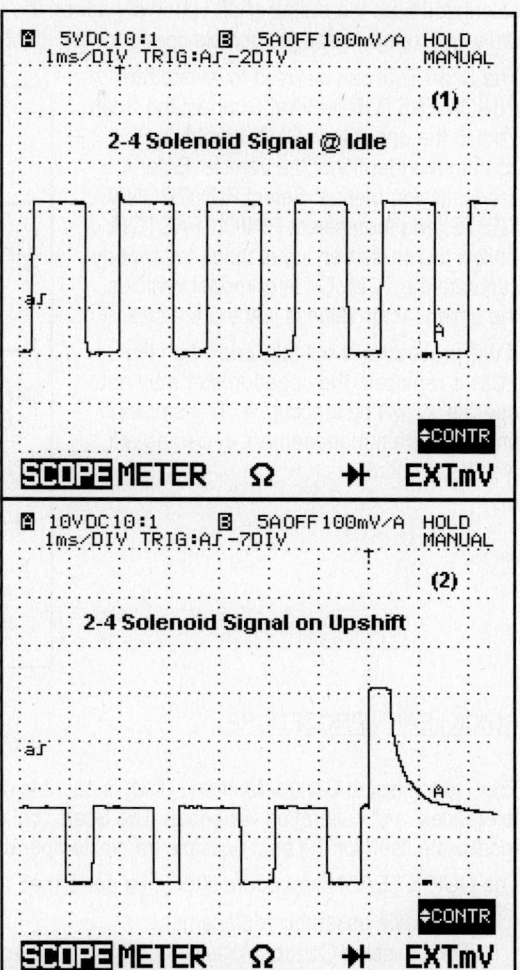

Transmission Solenoid Test Schematic

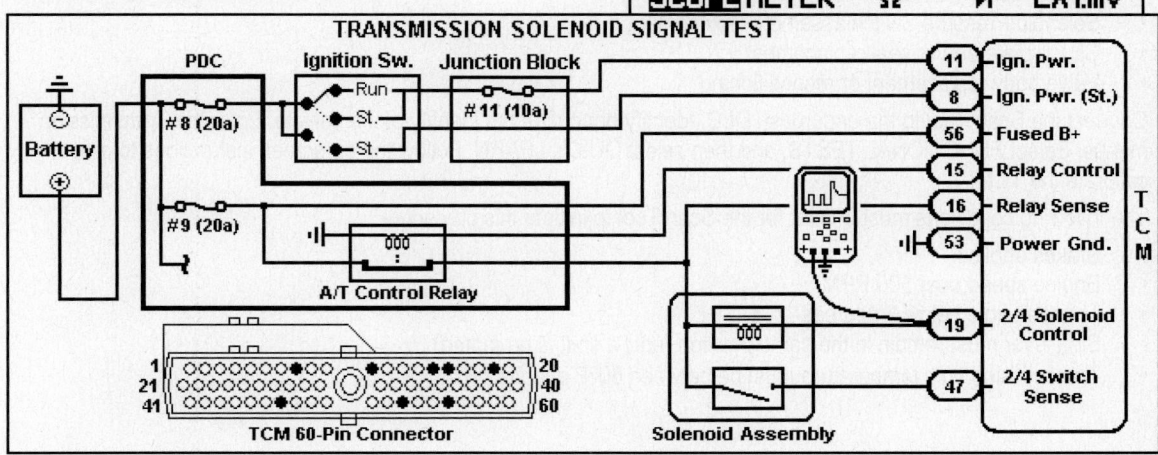

PINION FACTOR SETTING

When the Transmission Control Module (TCM) is replaced, the Pinion Factor must be set. The vehicle speed is calculated from the output shaft speed sensor on the transaxle. The TCM must know the final drive ratio and tire size in order to properly calculate vehicle speed for the PCM and the instrument cluster.

The Scan Tool can be used to initiate the PINION FACTOR feature. Connect the Scan Tool to the underdash DLC. Identify or confirm the identity of the vehicle. Enter the Transmission menus. Select FUNCTIONAL TESTS, and then select PINION FACTOR. Follow the on-screen instructions to properly calibrate the TCM. On later model vehicles, the pinion factor value is just the tire size.

If this procedure is not followed when the TCM is replaced, the speedometer may not operate, or will be inaccurate. Transmission and powertrain management decisions will also be affected.

```
    Snap-on Scan Tool Display
    ─────────────────────────────────────
    CHRYSLER CAR                      A/T
    2.7L V6 MPI                       A/C
    CHRY TRAN
    ─────────────────────────────────────

    MAIN MENU                OTHER SYSTEMS
    ~FUNCTIONAL TESTS        CODES & DATA MENU
     CUSTOM SETUP
     SYSTEM TESTS
    ─────────────────────────────────────

    FUNCTIONAL TESTS:
     ATM TESTS               PINION FACTOR
     BATTERY DISCONNECT      QUICK LEARN
```

QUICK LEARN PROCEDURE

The Transmission Control Module (TCM) is capable of adapting to changing conditions, mechanical wear, machining tolerances, and calibration variances. The Scan Tool controlled QUICK LEARN procedure allows the TCM to recalibrate itself for the best possible transaxle operation.

The QUICK LEARN procedure should be performed after any of the following events:

- Transaxle assembly replacement
- Transmission Control Module (TCM) replacement
- Solenoid/Pressure Switch assembly replacement
- Clutch plate and/or seal replacement
- Valve body replacement or reconditioning

Connect the Scan Tool to the underdash DLC. Identify or confirm the identity of the vehicle. Enter the Transmission menus. Select FUNCTIONAL TESTS, and then select QUICK LEARN. Follow the on-screen instructions to properly calibrate the TCM.

The following conditions must be met for the Scan Tool to initiate this procedure:

- Brakes applied
- Engine speed over 500 RPM
- Throttle angle below 3 degrees
- Shift lever must remain in the same position (until a shift is prompted)
- Transmission fluid temperature must be between 60°F and 200°F

Reference Information - PCM PID Examples

2002 CHRYSLER SEBRING 2.7L V6 DOHC SFI VIN R (A/T)

■ NOTE: *The following readings were obtained with the engine at idle speed, radiator hose hot, throttle closed, gear selector in Park and all accessories turned off.*

Acronym	Scan Tool Parameter	Range	Typical Value
A/C CLUTCH	A/C Clutch Relay Status	ON / OFF	OFF
A/C PRESS	A/C Pressure Sensor	0-5.1v	0.90v at 79 psi
ASD	Automatic Shutdown Relay Control	ON / OFF	ON
BARO	BARO (kPa)	10-110 kPa	100 kPa
BATT TEMP	Battery Temperature Sensor (volts)	0-5.1v	1.8v at 70°F
BRAKE SW	Brake Switch Position	ENGAGED/RELEASED	RELEASED
CL TIMER	Closed Loop Timer	0:00:0	0:20
CKP SENSOR	CKP Sensor Signal Present	NO / YES	YES
CMP SENSOR	CMP Sensor Signal Present	NO / YES	YES
CURR SYNC	Current CKP / CMP Sync	NOT OK / OK	OK
DES IDLE	Desired Idle Speed (rpm)	0-66635 rpm	664
ECT	ECT Sensor (°F)	-40 to 284°F	195°F
EGR %	EGR Solenoid Sensor	0-5.1v	4.98v (at 0% d/cycle)
ENG SPEED	Engine Speed (rpm)	0-66635 rpm	640
EVAP %	EVAP Purge Control (duty cycle)	0-100%	5%
FUEL LEVEL	Fuel Level Status (volts)	0-5.1v	1.17v
FUEL LEVEL	Fuel Level Sensor (gallons)	0-16 gallons	14 gallons
FUEL ALLOW	Fuel Allowed Status	NO / YES	YES
HO2S-11	HO2S-11 (B1 S1) Signal	0-1100 mv	570 mv
HO2S-12	HO2S-12 (B1 S2) Signal	0-1100 mv	860 mv
HO2S-21	HO2S-21 (B2 S1) Signal	0-1100 mv	330 mv
HO2S-22	HO2S-22 (B2 S2) Signal	0-1100 mv	880 mv
IGN CYCLES	Ignition Cycles 2	0-256	0
LOAD	Calculated Load Value (%)	0-100%	5%
IAT	IAT Sensor (°F)	-40 to 284°F	84°F
IAC POS	IAC Motor Position	0-256 steps	17
LB INJ P/W	Left Bank Injector Pulsewidth	0-999 ms	2.5 ms
LDP SOL	Leak Detection Pump Solenoid	OPEN / BLOCKED	BLOCKED
LDP SW	Leak Detection Pump Switch	OPEN / CLOSED	CLOSED
LB ST ADAPT	Left Bank Short Term Adaptive	0% (± 5%)	1.8%
MAN VAC	Manifold Vacuum (kPa)	10-110 kPa	66 kPa
MAP (V)	MAP Vacuum (volts)	0-5.1v	1.30v
MIN TPS	Minimum TP Sensor	0-5.1v	0.74v
MTV SOL	Manifold Tune Solenoid Valve	ON / OFF	OFF
MPH / KPH	Miles or Kilometers Per Hour	0-159	0 mph
P/N SW	Park Neutral Switch Position	P/N, DR, 2ND, 1ST	P/N
RB INJ P/W	Right Bank Injector Pulsewidth	0-999 ms	2.5 ms
RB ST ADAPT	Right Bank Short Term Adaptive	0% (± 5%)	0.8%
SPARK	Spark Advance (°)	-90° to +90°	11°
TARGET IAC	Target IAC Motor Position	0-256 steps	17
TP V	TP Sensor (volts)	0-5.1v	0.74v at 0%

Reference Information - Pin Voltage Table Examples

2002 SEBRING 2.7L V6 DOHC SFI VIN R (A/T) C1 CONNECTOR

Pin Number	Wire Color	Circuit Description (40-Pin)	Value at Hot Idle
1	TN/LG	COP 4 Driver Control	5°, 55 mph: 8° dwell
2	TN/OR	COP 3 Driver Control	5°, 55 mph: 8° dwell
3	TN/PK	COP 2 Driver Control	5°, 55 mph: 8° dwell
4	TN/LG	COP 6 Driver Control	5°, 55 mph: 8° dwell
5	YL/RD	S/C Power Supply	12-14v
6	DG/OR	ASD Relay Output	12-14v
7	YL/WT	Injector 3 Driver	1.0-4.0 ms
8	DG	Alternator Field Control	Digital Signals: 0-12-0v
9, 12	---	Not Used	---
10	BK/TN	Power Ground	<0.1v
11	TN/RD	COP 1 Driver Control	5°, 55 mph: 8° dwell
13	WT/DB	Injector 1 Driver	1.0-4.0 ms
14	BR/DB	Injector 6 Driver	1.0-4.0 ms
15	GY	Injector 5 Driver	1.0-4.0 ms
16	LB/BR	Injector 4 Driver	1.0-4.0 ms
17	TN	Injector 2 Driver	1.0-4.0 ms
18, 19	---	Not Used	---
20	DB/WT	Ignition Switch Output	12-14v
21	T/DG	COP 5 Driver Control	5°, 55 mph: 8° dwell
22-24, 28	---	Not Used	---
25	DB/LG	Knock Sensor Signal	0.080v AC
26	TN/BK	ECT Sensor Signal	At 180°F: 2.80v
27	DB/LG	HO2S Ground	<0.050v
29	LG/RD	HO2S-21 (B2 S1) Signal	0.1-1.1v
30	BK/DG	HO2S-11 (B1 S1) Signal	0.1-1.1v
31	TN	Starter Relay Control	KOEC: 9-11v
32	GY/BK	CKP Sensor Signal	Digital Signals: 0-5-0v
33	TN/YL	CMP Sensor Signal	Digital Signals: 0-5-0v
34	LG/PK	EGR Sensor Signal	0.6-0.8v
35	OR/DB	TP Sensor Signal	0.6-1.0v
36	DG/RD	MAP Sensor Signal	1.5-1.7v
37	BK/RD	IAT Sensor Signal	At 100°F: 1.83v
38	---	Not Used	---
39	VT/RD	Manifold Tune Solenoid Control	Solenoid Off: 12v, On: 1v
40	GY/Y	EGR Solenoid Control	12v, at 55 mph: 1v

2001 SEBRING 2.7L V6 DOHC SFI VIN R (A/T) C2 CONNECTOR

PCM Pin #	Wire Color	Circuit Description (40-Pin)	Value at Hot Idle
41	PK/LG	S/C Set Switch Signal	S/C & Set Sw. On: 3.8v
42	DB	A/C Pressure Switch Signal	A/C On: 0.45-4.85v
43	BK/LB	Sensor Ground	<0.050v
44	OR/WT	8-Volt Supply	7.9-8.1v
45, 54, 60	---	Not Used	---
46	RD/TN	Battery Power (Fused B+)	12-14v
47	BK	Power Ground	<0.1v
48	BR/WT	IAC 3 Driver	Pulse Signals
49	YL/BK	IAC 2 Driver	Pulse Signals
50	BK/TN	Power Ground	<0.1v
51	TN/WT	HO2S-12 (B1 S2) Signal	0.1-1.1v
52	VT/LG	Ambient Temperature Sensor	At 86°F: 1.96v
53	PK/WT	HO2S-22 (B2 S2) Signal	0.1-1.1v
55	DB/TN	Low Speed Fan Relay	Relay On: 1v, Off: 12v
56	TN/RD	S/C Vacuum Solenoid	Vacuum Increasing: 1v
57	GY/RD	IAC 1 Driver	Pulse Signals
58	VT/BK	IAC 4 Driver	Pulse Signals
59	OR	PCI Bus Signal	Digital Signals: 0-7.5v
61	VT/WT	5-Volt Supply	4.9-5.1v
62	WT/PK	Brake Switch Signal	Brake Off: 0v, On: 12v
63	YL/DG	Torque Management Request	Digital Signals
64	DB/OR	A/C Clutch Relay Control	Relay On: 1v, Off: 12v
65, 75	PK, LG	SCI Transmit, SCI Receive	0v
66	WT/OR	Vehicle Speed Signal	Digital Signals
67	DB/VT	ASD Relay Control	Relay On: 1v, Off: 12v
68	PK/BK	EVAP Purge Solenoid Control	PWM Signal: 0-12-0v
69	DB/PK	High Speed Fan Relay	Relay On: 1v, Off: 12v
70	WT/TN	EVAP Purge Solenoid Sense	12-14v
71	WT/RD	EATX RPM Signal	Digital Signals, 0-5v
72	OR	LDP Switch Sense Signal	Open: 12v, Closed: 0v
73, 78-79	---	Not Used	---
74	BR/LG	Fuel Pump Relay Control	Relay On: 1v, Off: 12v
76	BK/LB	TRS T41 Sense Signal	In P/N: 0v, Others: 5v
77	WT/DG	LDP Solenoid Control	PWM Signal: 0-12-0v
80	LG/RD	S/C Vent Solenoid	Vacuum Decreasing: 1v

STANDARD COLORS AND ABBREVIATIONS

Abbreviation	Color	Abbreviation	Color	Abbreviation	Color
BK	Black	GY	Gray	RD	Red
BL	Blue	GN	Green	TN	Tan
BR	Brown	LG	LT Green	VT	Purple
DB	Dark Blue	ON	Orange	WT	White
DG	DK Green	PK	Pink	YL	Yellow

Pin Connector Graphic

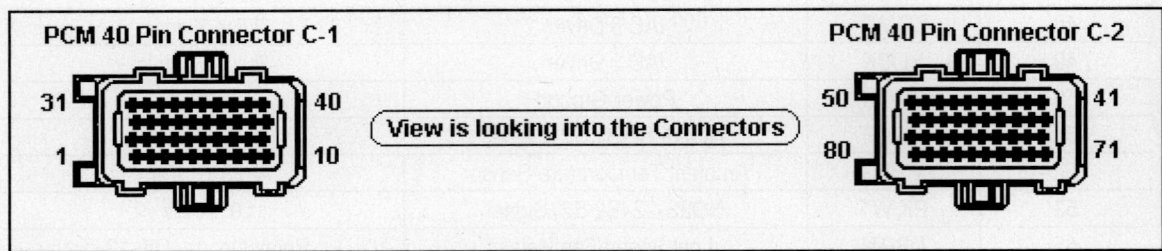

PCM Wiring Diagrams

FUEL & IGNITION SYSTEM DIAGRAMS (1 OF 14)

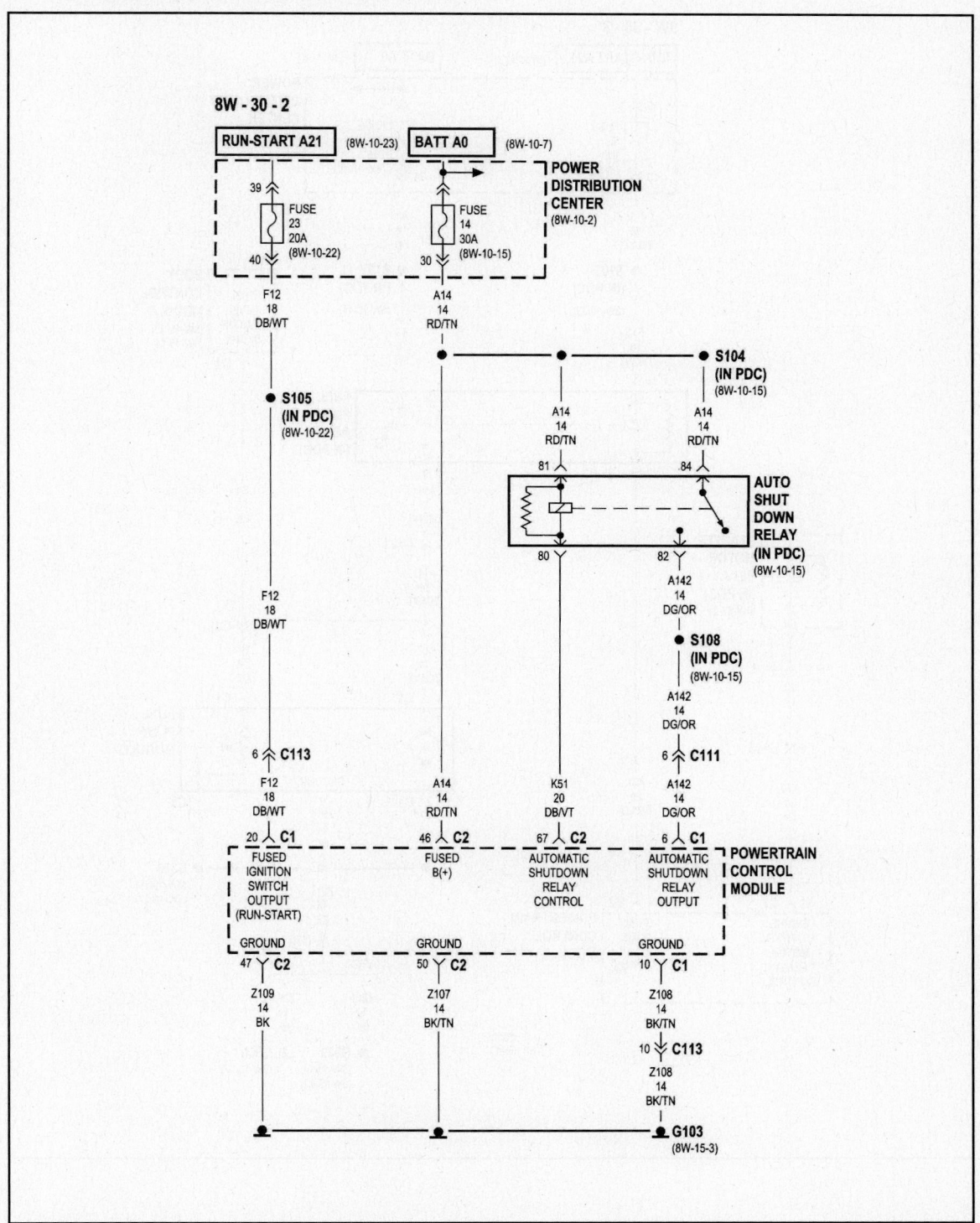

FUEL & IGNITION SYSTEM DIAGRAMS (2 OF 14)

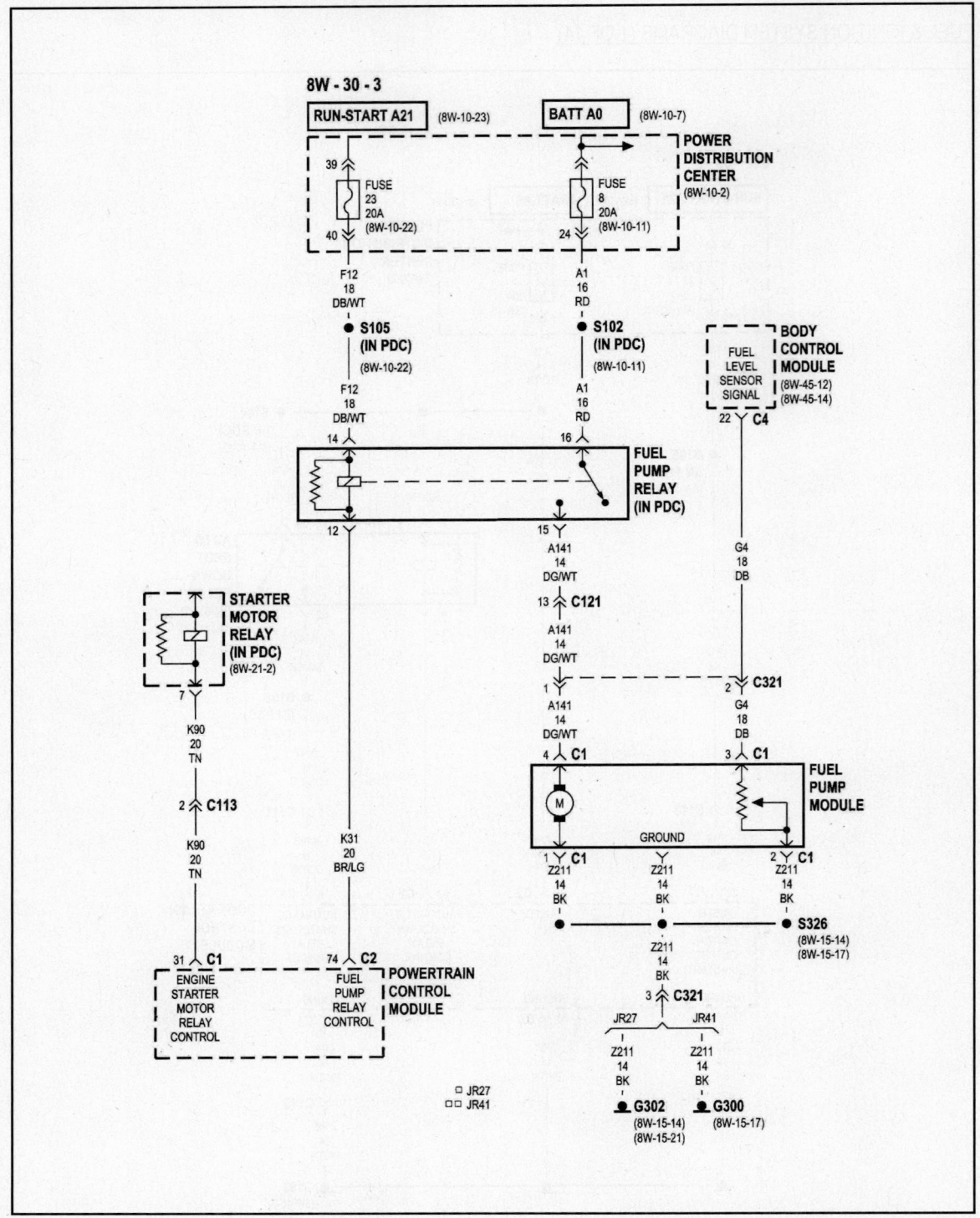

FUEL & IGNITION SYSTEM DIAGRAMS (3 OF 14)

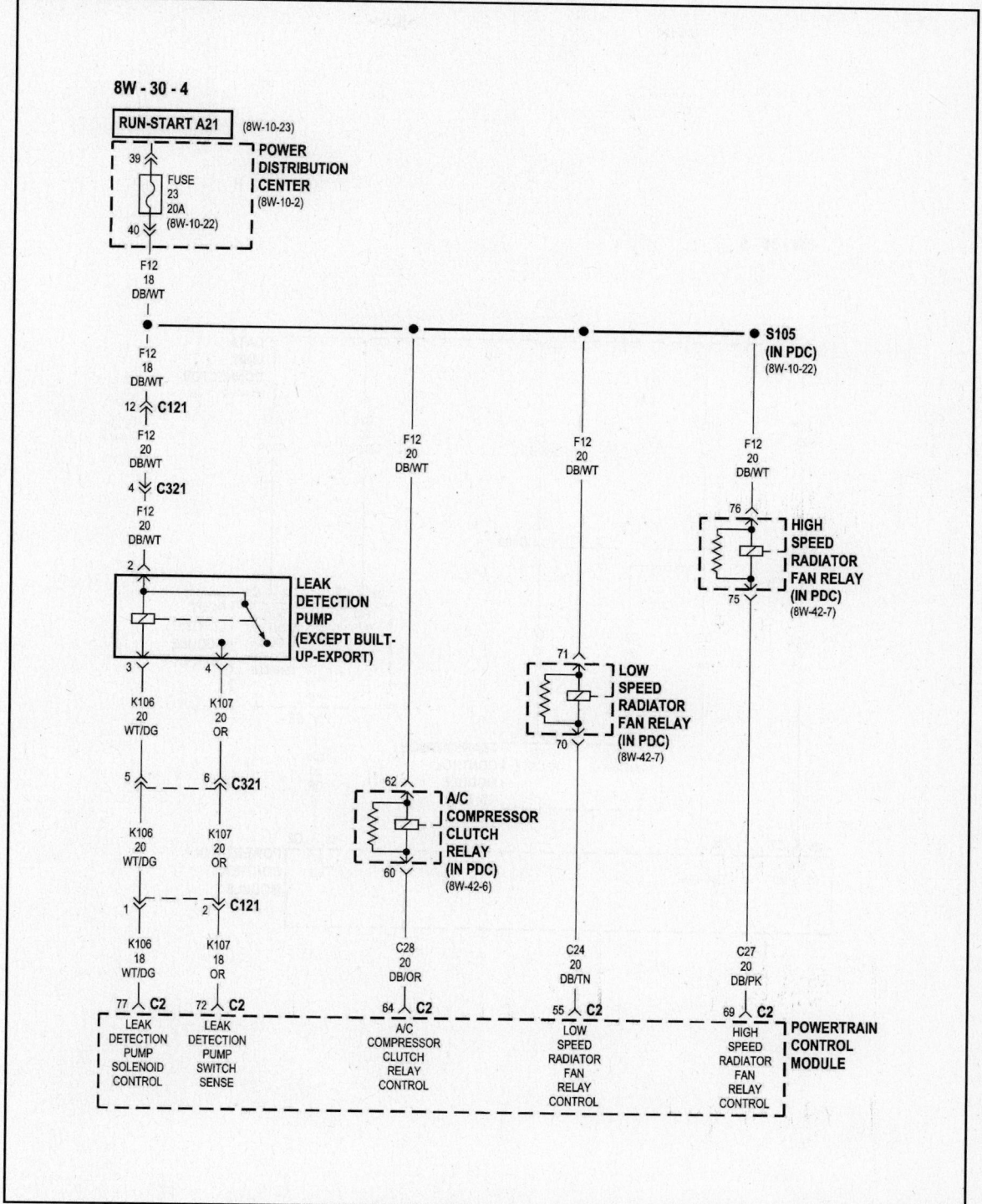

FUEL & IGNITION SYSTEM DIAGRAMS (4 OF 14)

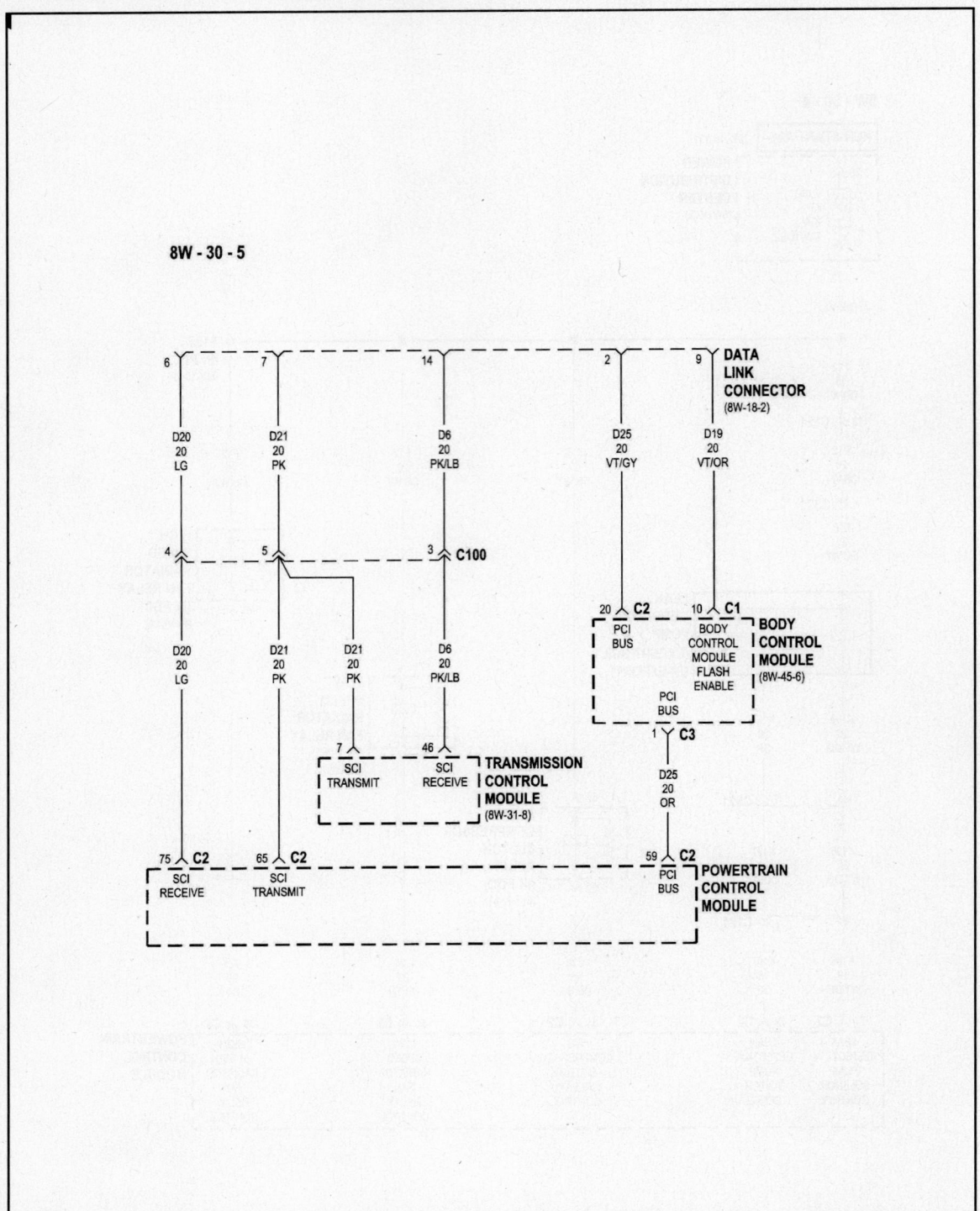

FUEL & IGNITION SYSTEM DIAGRAMS (5 OF 14)

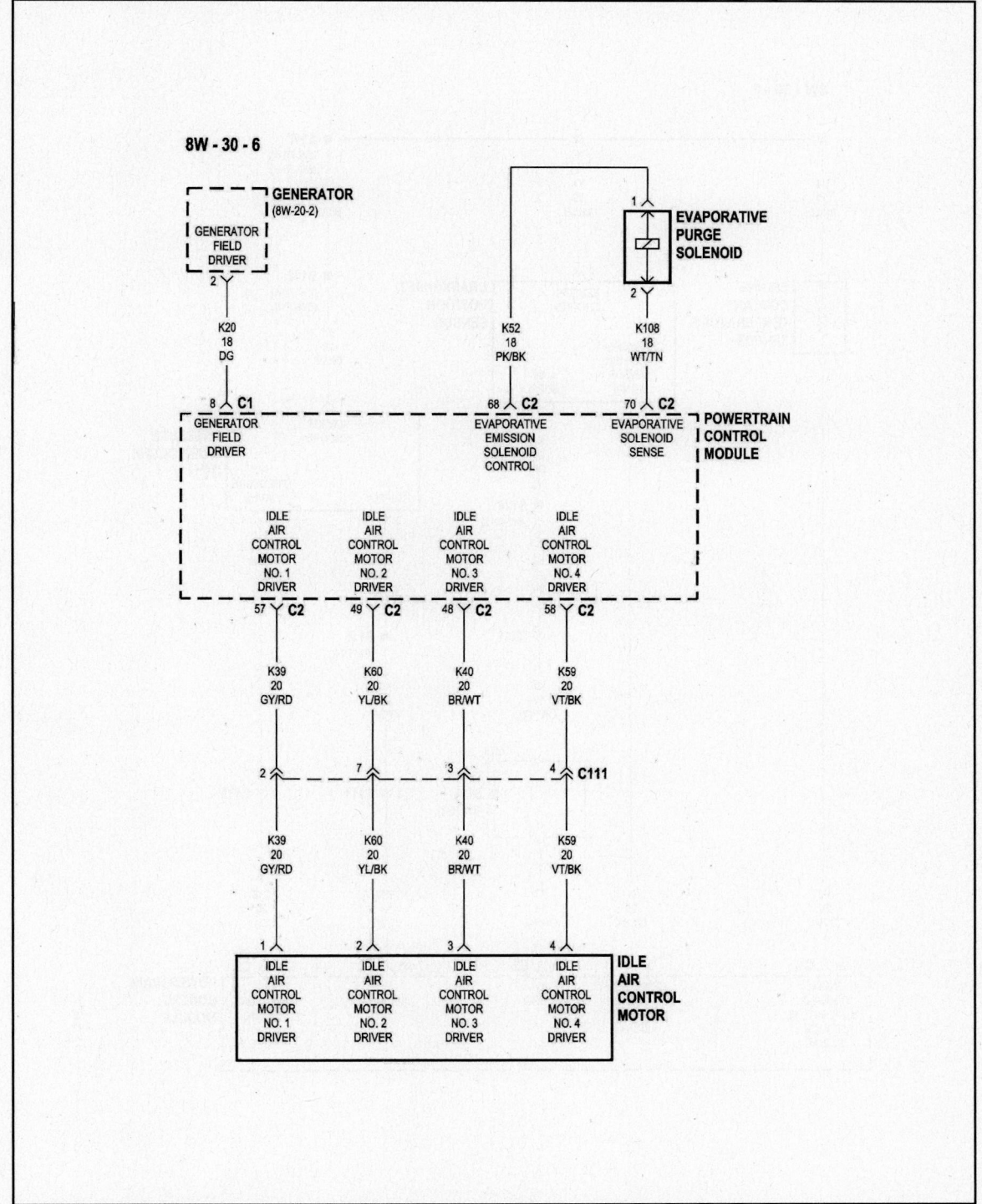

FUEL & IGNITION SYSTEM DIAGRAMS (6 OF 14)

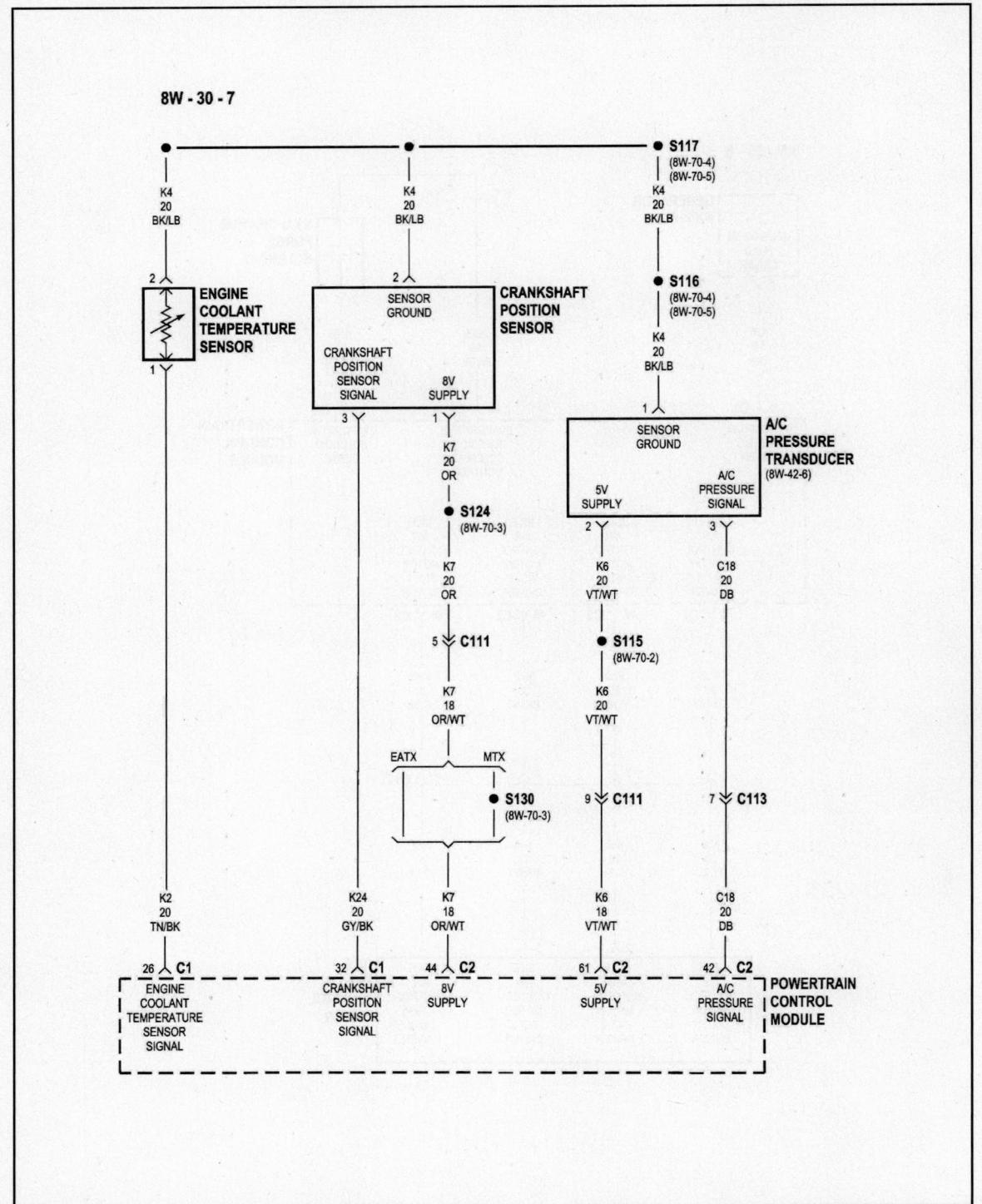

FUEL & IGNITION SYSTEM DIAGRAMS (7 OF 14)

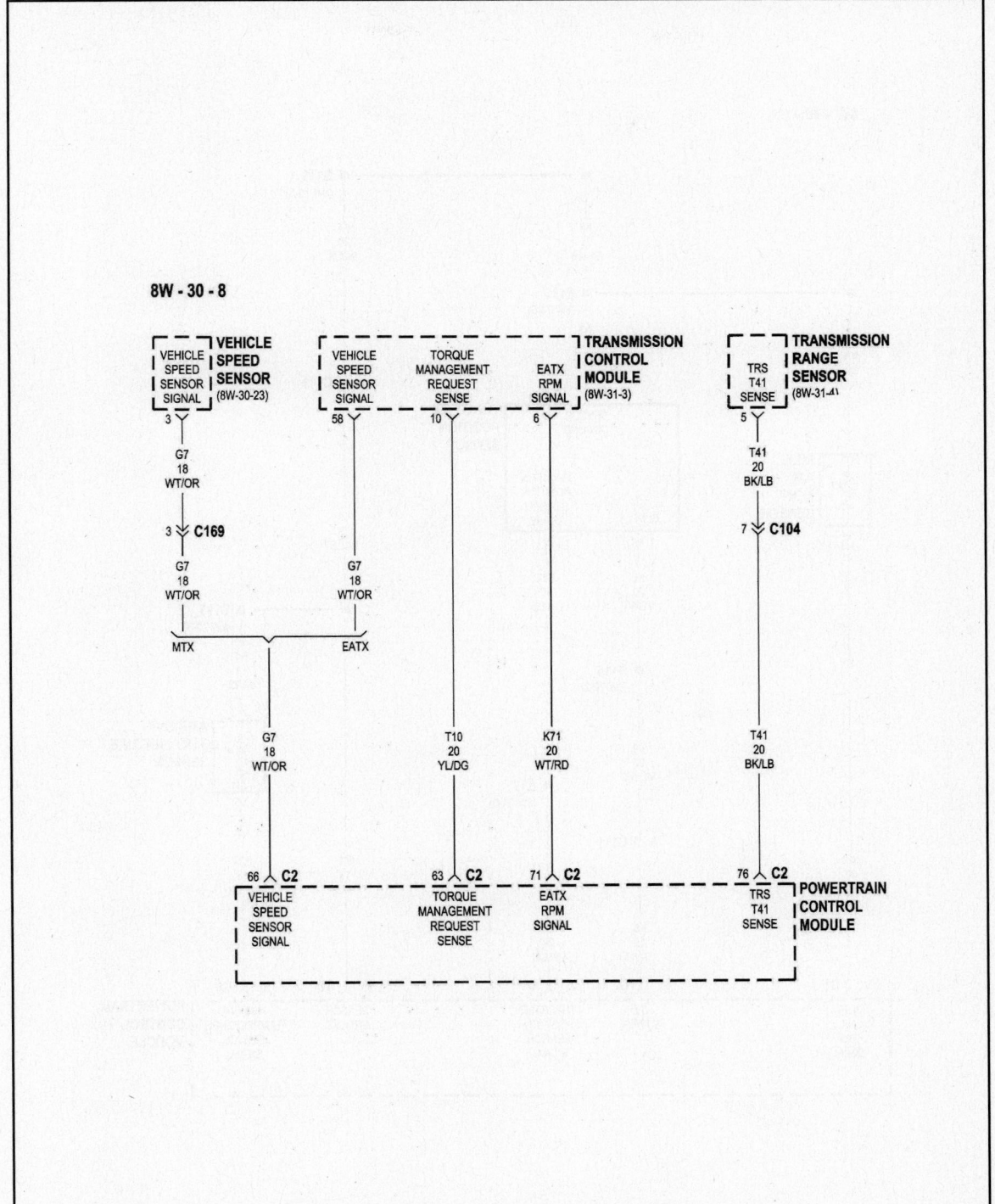

FUEL & IGNITION SYSTEM DIAGRAMS (8 OF 14)

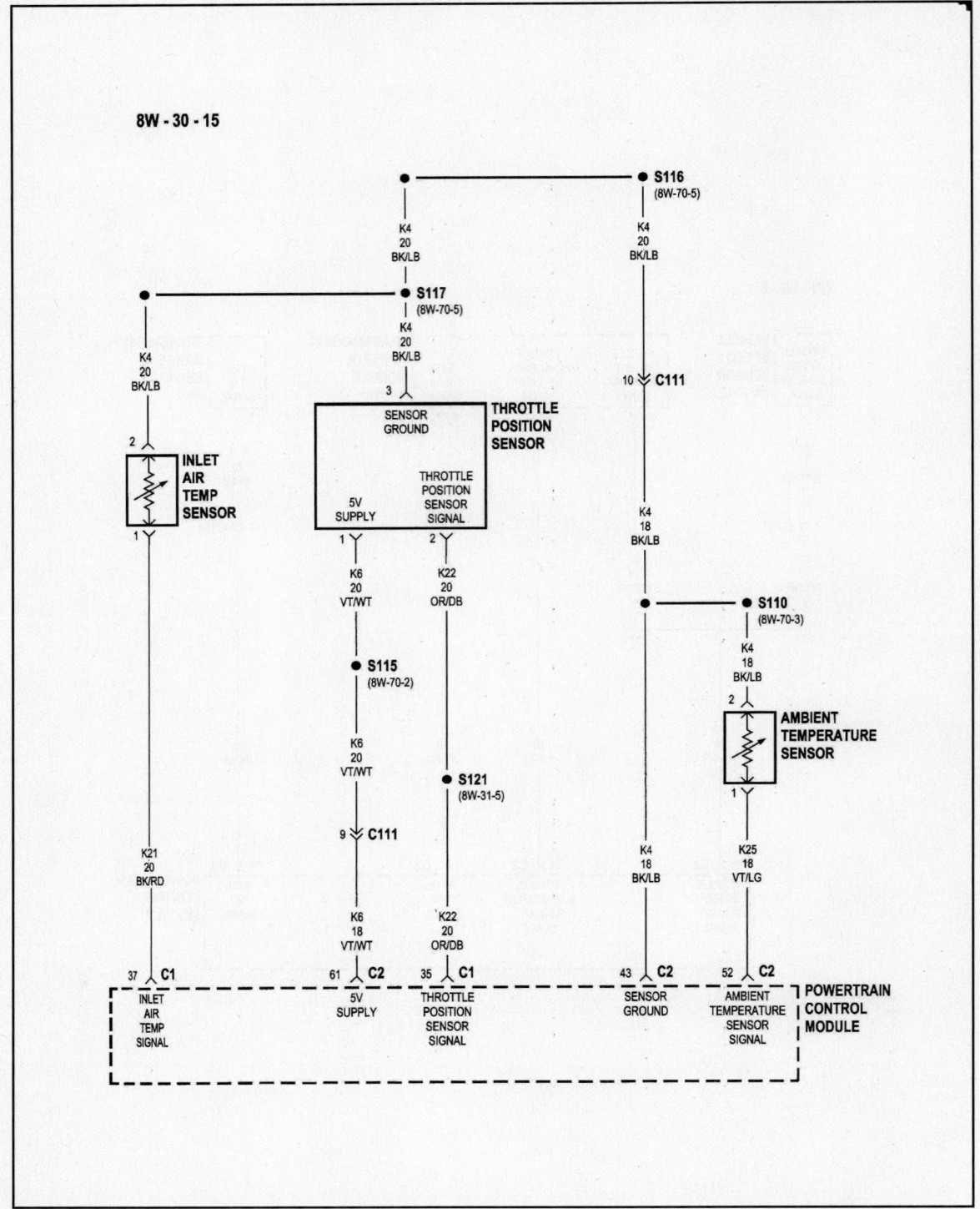

FUEL & IGNITION SYSTEM DIAGRAMS (9 OF 14)

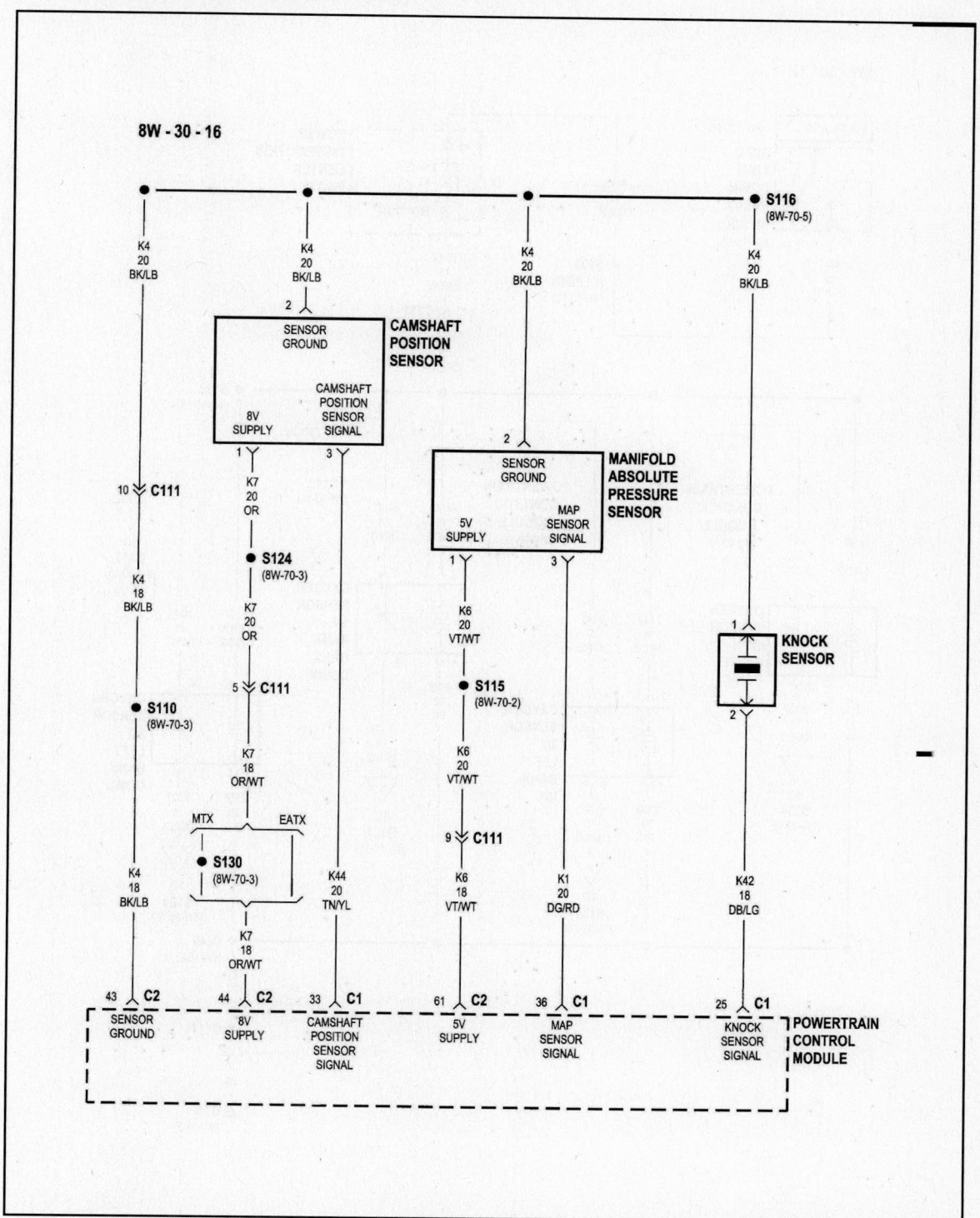

FUEL & IGNITION SYSTEM DIAGRAMS (10 OF 14)

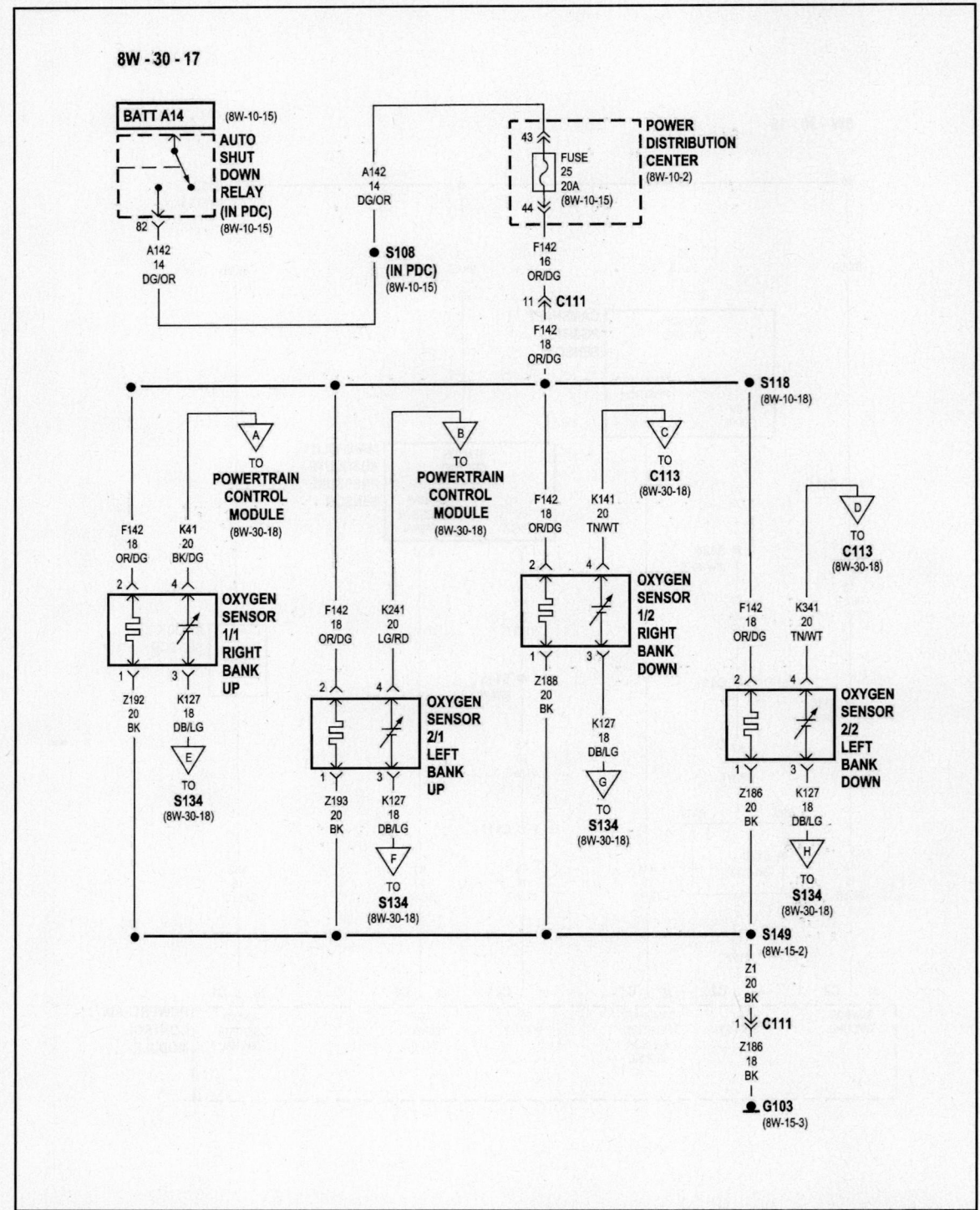

FUEL & IGNITION SYSTEM DIAGRAMS (11 OF 14)

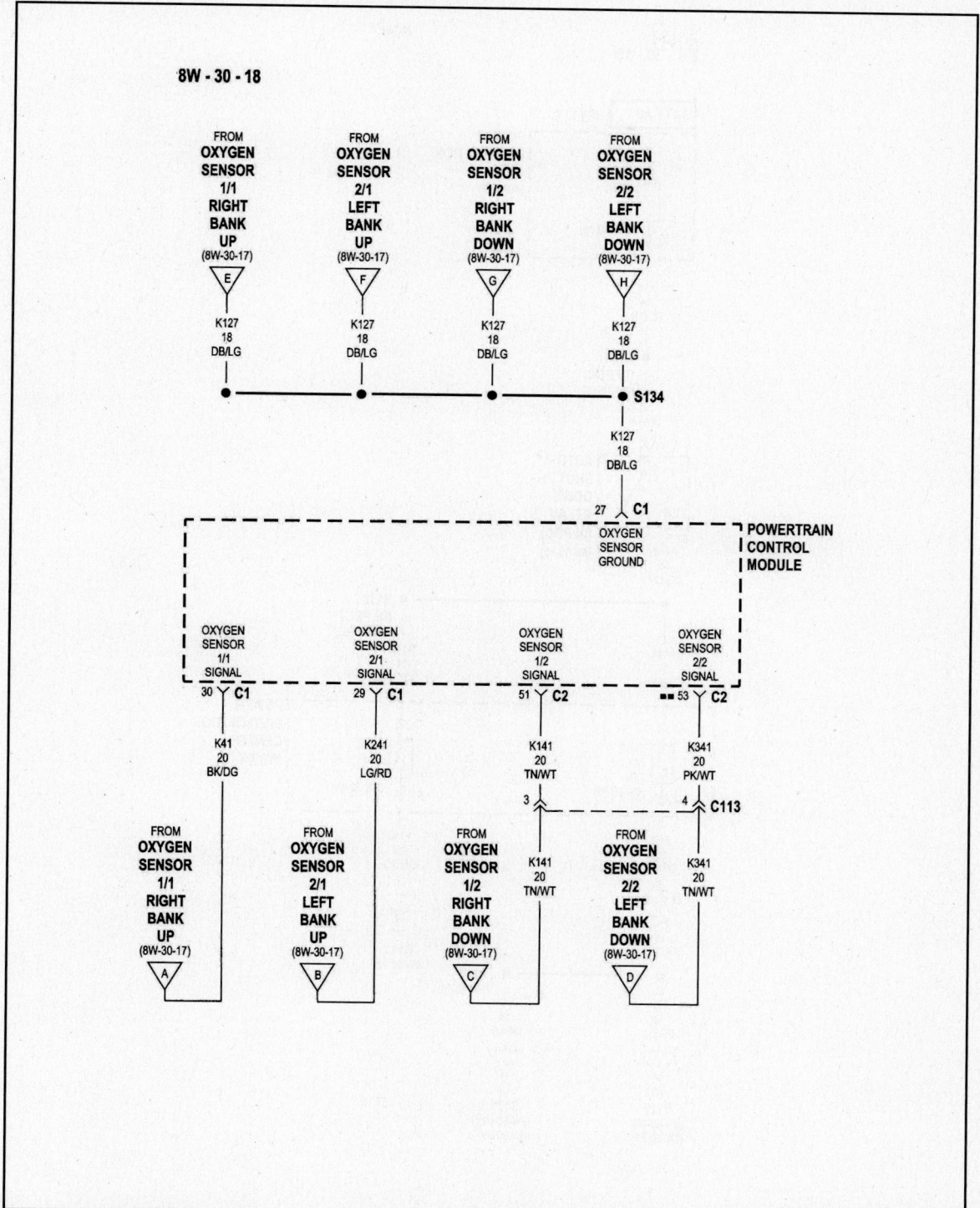

FUEL & IGNITION SYSTEM DIAGRAMS (12 OF 14)

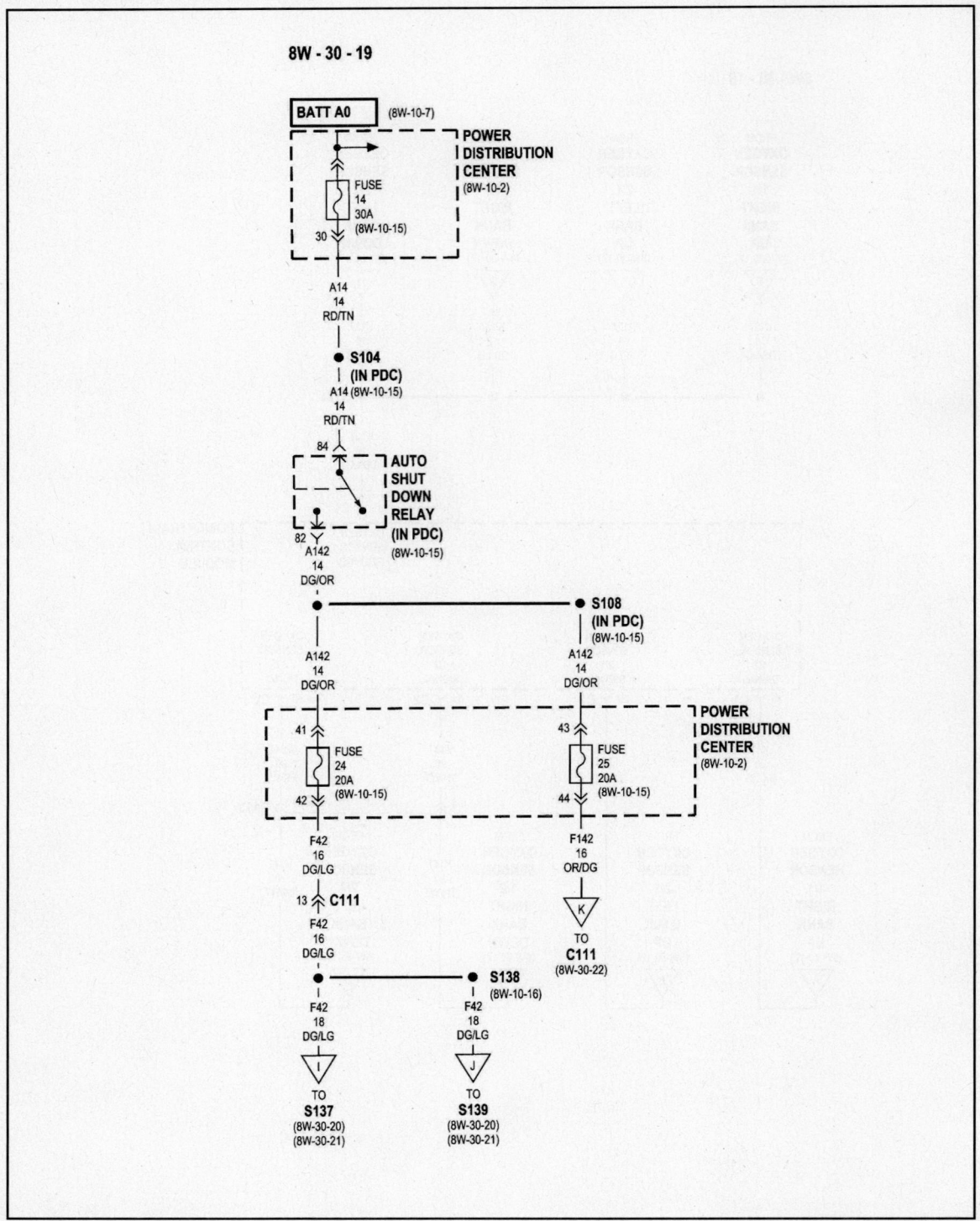

FUEL & IGNITION SYSTEM DIAGRAMS (13 OF 14)

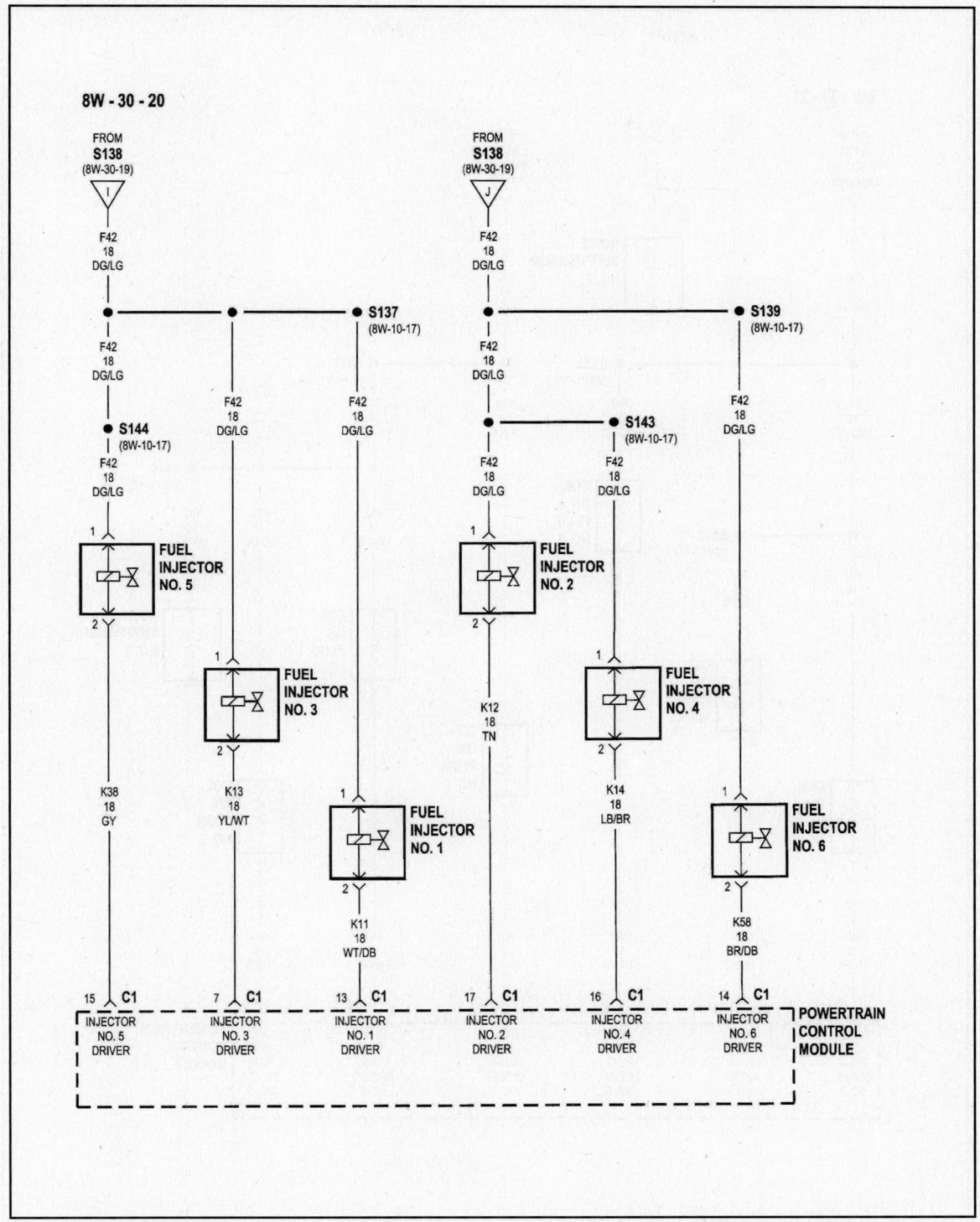

FUEL & IGNITION SYSTEM DIAGRAMS (14 OF 14)

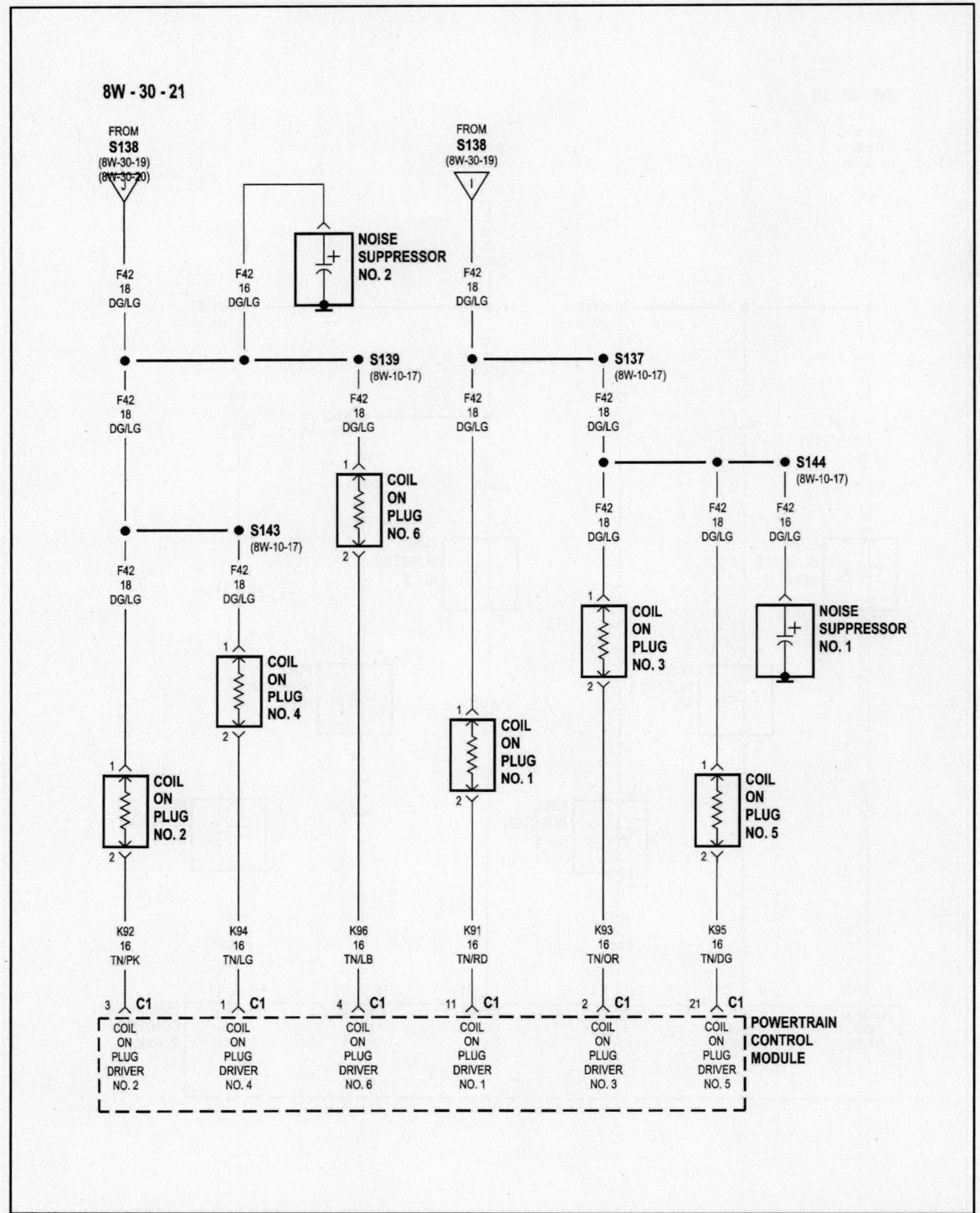

2002 SEBRING – BODY

BODY CONTROL MODULE

The Body Control Module (BCM) is concealed below the driver's side of the instrument panel, secured to the Junction Block (J/B). The BCM utilizes integrated circuitry and information carried on the Programmable Communications Interface (PCI) data bus along with many hard-wired inputs.

Sensor and switch data from throughout the vehicle is used to make decisions for those features the BCM is designated to control. Control of components is accomplished through the use of hard-wired circuits and the PCI data bus.

The BCM monitors its own internal circuits as well as many of its input and output circuits. The BCM has the capability to set and store trouble codes (DTCs) for any fault it detects. Specific Body System menus must be selected from the Scan Tool to access data (PIDs), bi-directional controls, and Diagnostic Trouble Codes (DTCs).

Controlled Systems

The following is a partial list of the body systems controlled by the BCM:

- Wipers, washers, and headlamp washers
- Power top up/down
- Electronic Odometer Support
- Power Locks and Windows
- All lighting (interior and exterior)
- Battery saver and lighting fade
- *Climate Control system support*

In addition to these systems, the BCM is involved in controlling other systems, and communicates with many more. For example, the Remote Keyless Entry (RKE) and Vehicle Theft Security System (VTSS) have their own control modules, but rely on the BCM for switch inputs and actuator outputs.

■ NOTE: *If the BCM must be replaced, the Vehicle Theft Security System (VTSS) must be enabled in the new BCM before the vehicle will start.*

BODY CONTROL DIAGNOSTICS

Scan Tool PIDs

The Scan Tool can be used to observe the data (PID) values for various BCM inputs and outputs. Diagnosis often involves observing a switch state change on the Scan Tool in response to an action (i.e., the doors can be opened and closed to observe the change in door switch state PIDs).

Scan Tool Bi-directional Controls

The Scan Tool can be used to control various lighting and body functions on the vehicle. These tests will confirm that the data bus, components, and circuits are all operational. In addition, these tests confirm that the BCM can process commands and toggle transistors to control devices.

The results of each test can be measured with a DVOM, Lab Scope, Scan Tool PIDs, or by simply observing the results visually or audibly.

Some of the components the Scan Tool can control are: All External Lights, Courtesy Lamp, Park/Tail Lamp Relay, Headlamp Relay, Decklid Release Solenoid, Door Lock Relay, Door Unlock Relay, Horn Relay, Wiper Motor and Wiper Deicer Output.

MODULE COMMUNICATIONS

The BCM communicates with other modules using the Programmable Communications Interface (PCI) bus. The following is a list of modules on the PCI network on this vehicle:

- Body Control Module (BCM)
- Powertrain Control Module (PCM)
- Transmission Control Module (TCM)
- Sentry Key Immobilizer Module (SKIM)
- Mechanical Instrument Cluster (MIC)
- Compass/Mini-Trip Computer (CMTC)
- Anti-Lock Brake System (ABS)
- Occupant Restraint Controller (ORC)
- CD/Radio

CD/Radio and BCM Communications

The data sent between modules varies according to the complex relationships between those modules. For example, the audio system receives only data for backlight dimming and volume equalization (for convertible models, the volume increases with the top down). The only data the audio system sends over the network is DTC status.

BCM and PCM Communications

The BCM uses certain PCM data, (i.e., RPM, ECT, VSS, injector on time, MIL status, VTSS arming status, engine model, and certain failure information). The BCM sends the PCM VTSS status and A/C switch status.

MIC and BCM Communications

The Mechanical Instrument Cluster (MIC) is a control module, but the only 'control' it has is in executing commands it receives from the BCM. It is important to understand how the MIC operates in the network in order to diagnose cluster problems. It is also helpful to understand these concepts because the MIC and BCM relationship is typical of how the BCM interacts with other systems on the network.

On this application, the Oil Pressure, Brake Warning, Turn Signal and Fog Lamp indicators are hardwired into the cluster. They are not part of the network or any self-diagnostics. All other gauges and indicators are controlled by the BCM via the PCI bus.

The PCM sends all gauge data to the BCM. The BCM sends this data and other indicator information to the MIC. The MIC translates the BCM messages and positions the gauges and indicators accordingly.

Whenever a cluster gauge or indicator appears to be malfunctioning, the first step should be to review service information or a wiring diagram to determine how that gauge or indicator is operated.

If the wiring diagram does not show a dedicated wire for that function, then the information is sent on the PCI bus. The way the MIC receives information determines which of the available diagnostic tools and methods should be used.

BI-DIRECTIONAL TESTS (DECKLID RELEASE SOLENOID)

The Scan Tool can be used to energize various BCM controlled components during diagnosis. The test results can be observed visually, audibly, or with a DVOM or Lab Scope.

Scan Tool ATM Test

Connect the Scan Tool to the DLC under the dash. Navigate through the menus to the BODY systems menu. Choose ATM TESTS, and then DECKLID RELEASE SOLENOID.

The message, TEST IS RUNNING indicates that the BCM will command the decklid release solenoid on every few seconds for about 5 minutes. This allows for repeated testing of the same event without manually commanding the solenoid on each time. The repetition also stresses the system, so intermittent failures have a good chance of occurring.

```
+---------------------------------------------------------+
|                                                         |
|         Snap-on Scan Tool Display                       |
|  ---------------------------------------------------    |
|                                                         |
|  MAIN MENU (CHRY BODY)     OTHER SYSTEMS                |
|   >ATM TESTS               CODES & DATA MENU            |
|    CUSTOM SETUP                                         |
|  ---------------------------------------------------    |
|  SCROLL TO SELECT A TEST                                |
|   >DECKLID RELEASE SOLENOID                             |
|    DOOR LOCK RELAY                                      |
|    DOOR UNLOCK RELAY                                    |
|  ---------------------------------------------------    |
|  DECKLID RELEASE SOLENOID                               |
|  TEST IS RUNNING                                        |
|  PRESS Y TO GO TO NEXT TEST                             |
|  PRESS N TO RETURN TO ATM MENU                          |
|                                                         |
+---------------------------------------------------------+
```

Lab Scope Test

The Lab Scope was used in this case to verify the circuit activity. This is one of many ways to use the Scan Tool ATM tests to verify this BCM controlled component and circuit, as well as many others.

Lab Scope Connections

Connecting the Lab Scope to the solenoid is much easier than accessing the BCM. Connect the Channel 'A' probe to the solenoid driver circuit (BK/WT Wire) at the solenoid harness connector, and connect the negative probe to the battery negative post.

Lab Scope Example

In this example, the trace shows the solenoid being commanded on for almost 500 ms. The circuit voltage rises to almost battery voltage as the solenoid is energized.

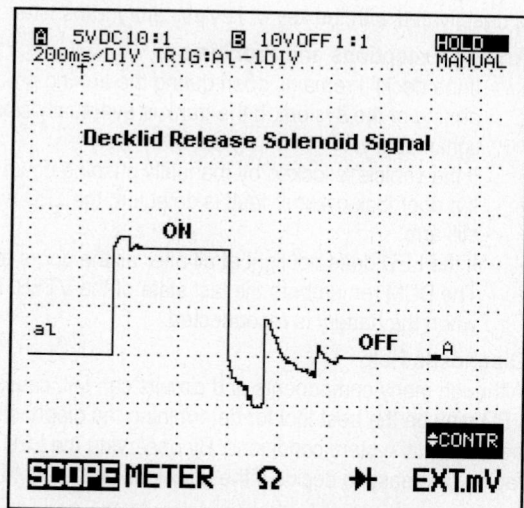

Note that the voltage oscillates when the solenoid is de-energized. This oscillation is proof that the circuit was completed and a magnetic field was built in the solenoid, and that the solenoid mechanically moved. The oscillations (coil ringing) are the end result of the field collapsing and the solenoid arm retracting. This trace does not prove that the solenoid moved enough to open the trunk, but it does eliminate many possible causes of the fault. If the ground circuit or the solenoid were open, the trace would look the same, except for the oscillations (coil ringing).

Vehicle Theft Security System

GENERAL DESCRIPTION

The Vehicle Theft Security System (VTSS) is a content theft deterrent system utilizing audible and visual alerts when activated. All of the VTSS electronic functions are contained in the BCM, although the system interacts with the Remote Keyless Entry (RKE) module as well. The VTSS functions as an alarm system only, and does not disable the vehicle. A Sentry Key Immobilizer System (SKIS) is available as an option, and provides starter and fuel pump interrupt. Refer to the Sentry Key Immobilizer System article on the next page.

VTSS Operation

Once enabled, the VTSS monitors the vehicle door switches, decklid switch, and ignition switch for unauthorized activity. When an unauthorized entry occurs, the horn will sound, accompanied by flashing park lamps, tail lamps, and the headlamps.

System Arming

The system is armed any time the doors are locked using the power door lock switches with the door ajar, the keyless entry transmitter, or the exterior door lock cylinders. If any door remains open, the system will not arm. Also, the system will not arm of the door lock switches are used to lock the vehicle when the doors are closed. In this case it is assumed that someone is in the vehicle or the windows are down.

After the vehicle is locked and the last door is closed, the VTSS LED in the instrument cluster will flash for 16 seconds. The system is not armed until the 16-second period has expired without any monitored switches closing (without any door or trunk opening). The LED will then light steadily.

The system is disarmed whenever the doors are unlocked using the key or the keyless entry transmitter. These inputs will also disarm the system if the alarm is activated.

If the system has been activated (an unauthorized entry was attempted), the horn will sound 3 times when the system is deactivated with the key or keyless entry transmitter.

Arming Exceptions and Problems

- If the decklid remains open during the arming process, the VTSS LED will flash slower indicating the conditional arming of the system. If the trunk is eventually closed, the system will fully arm in 16 seconds and the LED will light steadily.
- If the vehicle is locked by manually pushing down the locks, the system is bypassed and will not arm.
- If a door lock cylinder fault is detected, the LED will light steadily during the arming process, but the system will still arm.
- If the LED does not light at all after all the doors are closed, the system is not armed.
- The BCM remembers the last state of the VTSS. If the battery is disconnected with the system armed, it will rearm when the battery is reconnected.

Diagnostic Help

Although many components and circuits can fail, causing problems with the VTSS and other systems, monitoring the LED may be the best tool for determining the diagnostic path. The paragraphs above describe the LED behavior according to system conditions. By observing the LED (and instructing the customer to do the same for intermittent faults), it is easy to decide if the door-lock cylinders, trunk lock, or remote keyless entry may be the cause of the fault.

Once the LED status has been observed, the Scan Tool can be used to view any DTCs and observe the door and trunk switch PIDs.

Sentry Key Immobilizer System

GENERAL DESCRIPTION

The Sentry Key Immobilizer System (SKIS) is available as a factory installed option on this application. It provides passive protection against vehicle theft using starter and fuel pump interrupt. The system is "passive" because no action is required by the user for the system to work.

The SKIS uses Sentry Keys, a module with antenna, the BCM, the PCM, and an indicator light.

Fuel Pump and Starter Disable
The SKIS system will disable the fuel pump after 2 seconds of running whenever an invalid key is used to start the vehicle. The engine cannot be cranked again in that key cycle. The key must be cycled all the way off to enable the starter. After six (6) consecutive invalid key starts, the engine will not crank. A valid key must be used to re-enable the starter.

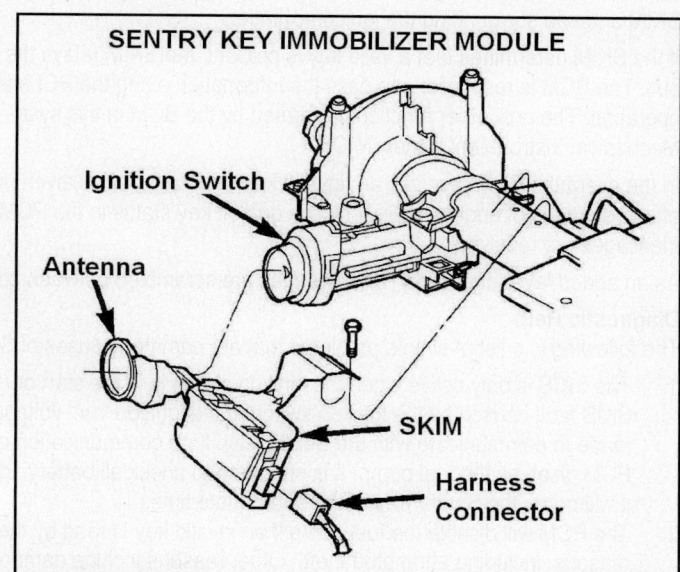

SENTRY KEY IMMOBILIZER MODULE

Ignition Switch
Antenna
SKIM
Harness Connector

Sentry Keys

The Sentry Keys are normal cut ignition keys with transponders molded into the head of each key. Each transponder communicates a unique identification code to the Sentry Key Immobilizer Module (SKIM). The SKIM knows and stores the unique identification code of each Sentry Key.

Each Sentry Key has to learn a "Secret Key" code programmed into the SKIM. For added security, the Secret Key code is also stored in the PCM. Forcing the PCM, SKIM and keys to agree on the Secret Key code prevents theft through the substitution of an alternate SKIM and Sentry Keys.

Once the Sentry Key has been learned by the SKIM, it is stored in memory. Also, a programmed key cannot be programmed to operate any other vehicle.

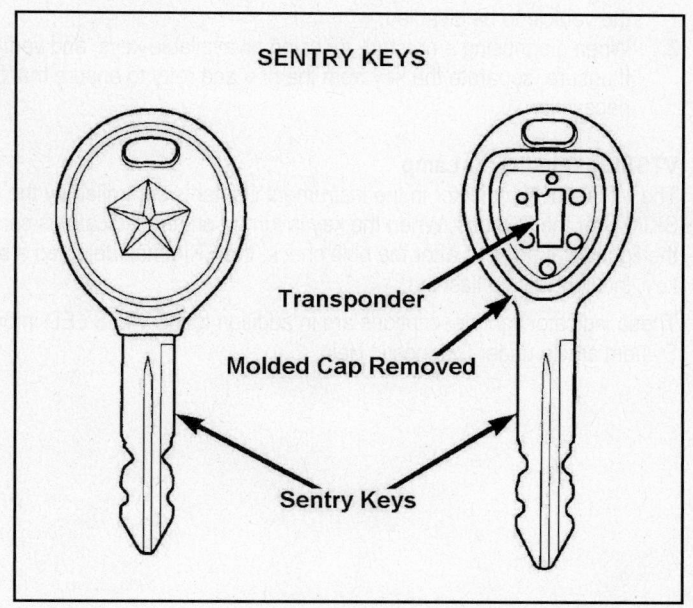

SENTRY KEYS

Transponder
Molded Cap Removed
Sentry Keys

IMMOBILIZER SYSTEM OPERATION

The Sentry Key Immobilizer System (SKIS) is only active when the ignition switch is turned to the start or run position. In the start or run position, the SKIM receives ignition switch voltage, signaling it to send an identification request to the transponder in the Sentry Key. The Sentry Key answers the query with the Secret Key code and its own unique code that has been learned by the SKIM.

The SKIM and Sentry key communicate using a radio frequency (RF) signal. The signals are sent and received by the SKIM antenna surrounding the ignition cylinder.

If the SKIM determines that a valid key is present, the SKIM relays the "valid key" message to the BCM via the PCI bus. The BCM is responsible to pass the information along the PCI bus to the PCM to command continued fuel pump operation. The only other function performed by the BCM in this system is the control of the VTSS indicator light in the Mechanical Instrument Cluster (MIC).

In the event the PCM receives an "invalid key" message, or receives no communication at all, the fuel pump is disabled after 2 seconds of engine operation. The default key status in the PCM is "invalid key" to prevent operation if no messages are received.

As an added layer of security, all messages are scrambled between components to prevent unauthorized access.

Diagnostic Help

The following is a list of simple problems that are common causes of SKIS failure:

1. The SKIS is only active when the ignition switch is in the start or run position. It is possible, therefore, to have a SKIS fault caused by the ignition switch. If the ignition 'run' voltage is not received by the SKIM, no attempt is made to communicate with the Sentry Key. If no communication occurs, no message is sent to the PCM and the PCM disables the fuel pump. It is important to check all battery, ignition, and ground circuits to the SKIM to ensure it will query the Sentry Key at the appropriate times.

2. The PCM will disable the fuel pump if an invalid key is read by the SKIM. A key can be invalid for a number of reasons, including attempted theft. Other reasons include damaged keys, multiple valid keys on the same ring, or other transponders responding to the SKIM request (i.e. gas station Speed Pass transponders). Anything within range of the antenna that contains a transponder can communicate incorrect information to the SKIM and cause the vehicle to be disabled.

3. When diagnosing a no-start, try using all available keys, and verify that no other transponders are on the key ring. If unsure, separate the key from the ring and retry to ensure that further diagnosis of the system is really necessary.

VTSS/SKIS Indicator Lamp

The VTSS/SKIS indicator in the instrument cluster is controlled by the BCM in response to messages received from the SKIM over the PCI bus. When the key is turned on, the indicator is commanded on for 4 seconds as a bulb check. If the light stays on solid after the bulb check, the SKIM has detected a system malfunction. If the SKIM detects an invalid key, the light will be flashing.

These indicator light descriptions are in addition to the VTSS LED information found in the Vehicle Theft Security System article under Diagnostic Help.

PROGRAMMING SENTRY KEYS

The SKIM is preprogrammed with Sentry Keys from the factory. Each programmed key is permanently remembered by the SKIM, and a total of eight keys can be programmed in the module. If the maximum number of keys has been programmed and are lost or damaged, the DRB III must be used to erase all keys and reprogram the existing keys and any new ones.

Programming New Keys With Less Than 2 Valid Keys

When one valid key or no keys are available, additional keys must be programmed using the DRB III Scan Tool. Customer and vehicle information is used to obtain a PIN from the Mopar Diagnostic System (MDS). The PIN must be used to allow the Scan Tool to access the SKIM to command the learning of new keys.

Programming New Keys With At Least 2 Valid Keys

If 2 valid keys are available, new keys can be programmed using the Customer Learn mode. Use the following procedure to add keys to the SKIM memory:

1. Cut the new key blank(s) to physically fit the ignition lock cylinder.
2. Insert one of the valid keys and turn the ignition switch to the run position for between 3 and 15 seconds. Turn the ignition switch to the off position and remove the key.
3. Within 15 seconds of step 2, insert the second valid key and turn to the run position.
4. After the second key has been in the run position for about 10 seconds, the VTSS/SKIM indicator should start to flash and a single chime will be heard. The chime and flashing indicator together signal that the SKIM has entered Customer Learn mode.
5. Within 60 seconds of entering Customer Learn mode, turn the ignition switch off, remove the second valid key, and insert one of the new keys to be programmed. Turn the ignition switch to the run position with the new key.
6. After about 10 seconds, a single chime will be heard and the indicator light will stop flashing. The indicator light will remain solid for 3 seconds, and then go out. This sequence indicates that the key has been successfully programmed.

As soon as step 6 is complete, the SKIM exits Customer Learn mode and the vehicle can be started using any of the 3 keys. The process must be repeated for each key to be programmed.

The programming process will abort and the Customer Learn modes will exit if any of the following conditions occur:

- If the SKIM sees a key that has already been programmed with another Secret Key code.
- If the SKIM already has eight (8) keys programmed.
- If the ignition switch is turned off for more than 50 seconds.

If the ignition switch is turned to the start position during the programming, the vehicle will see an invalid key and disable the fuel pump.

BODY CONTROLS SYSTEM DIAGRAMS (1 OF 12)

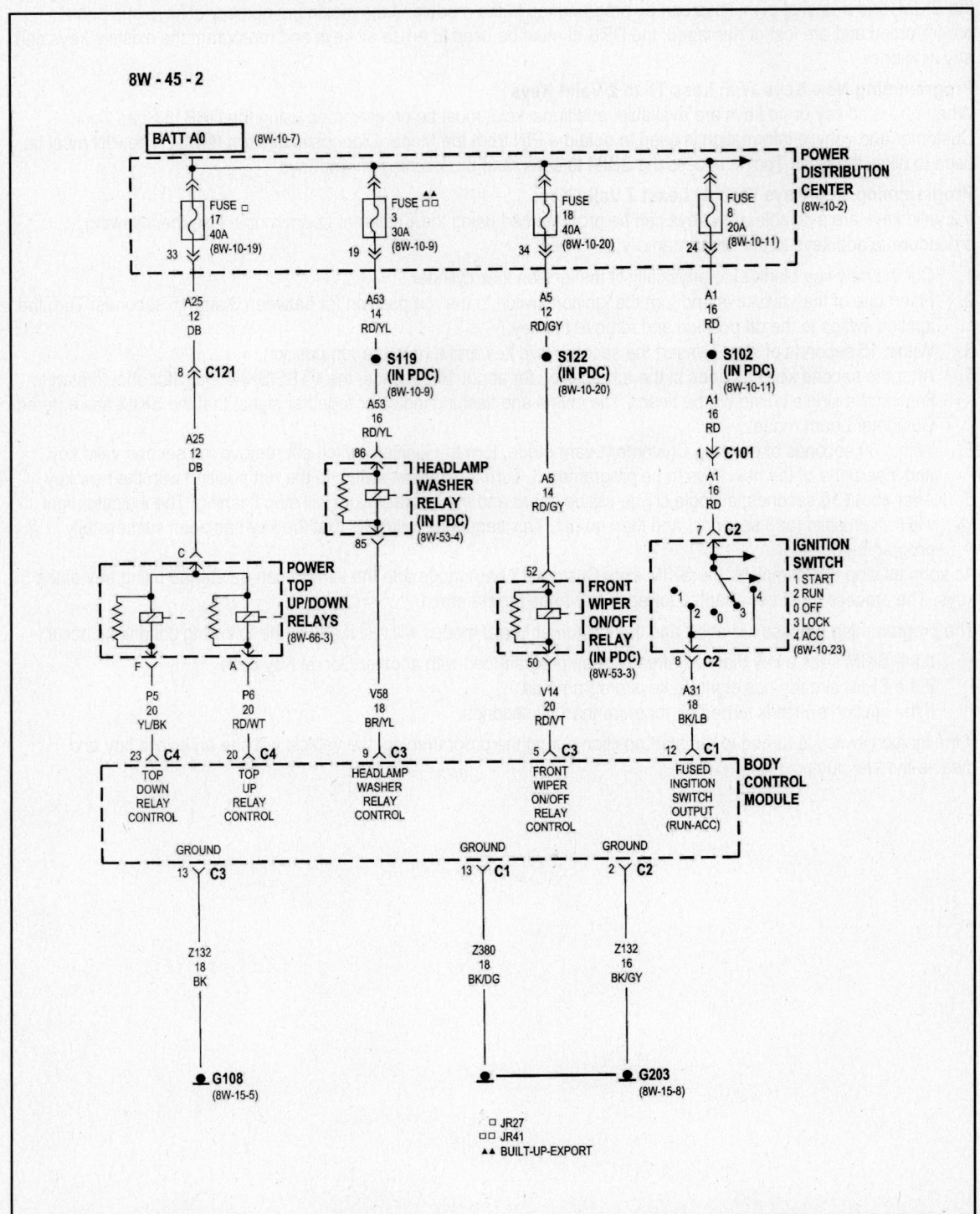

BODY CONTROLS SYSTEM DIAGRAMS (2 OF 12)

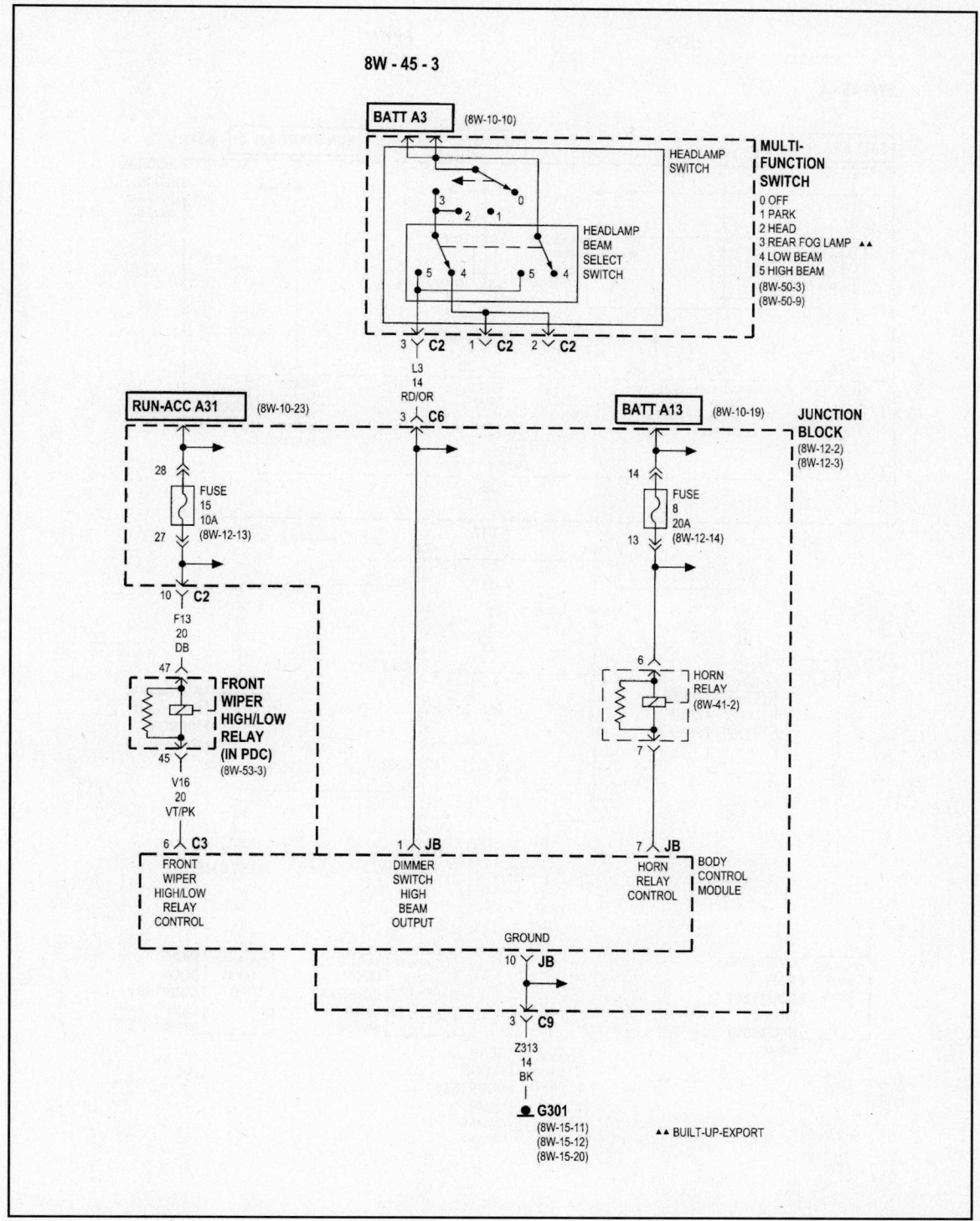

BODY CONTROLS SYSTEM DIAGRAMS (3 OF 12)

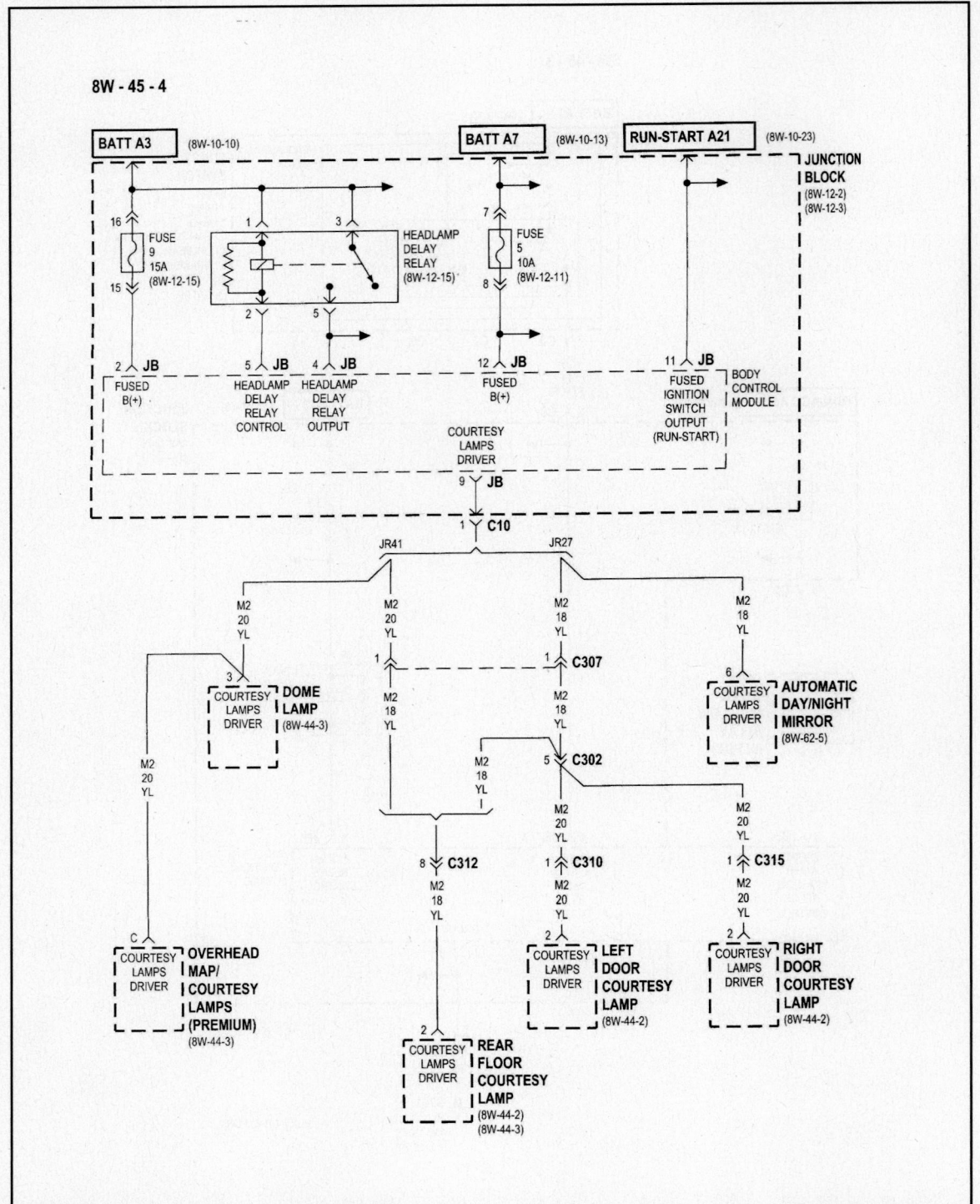

BODY CONTROLS SYSTEM DIAGRAMS (4 OF 12)

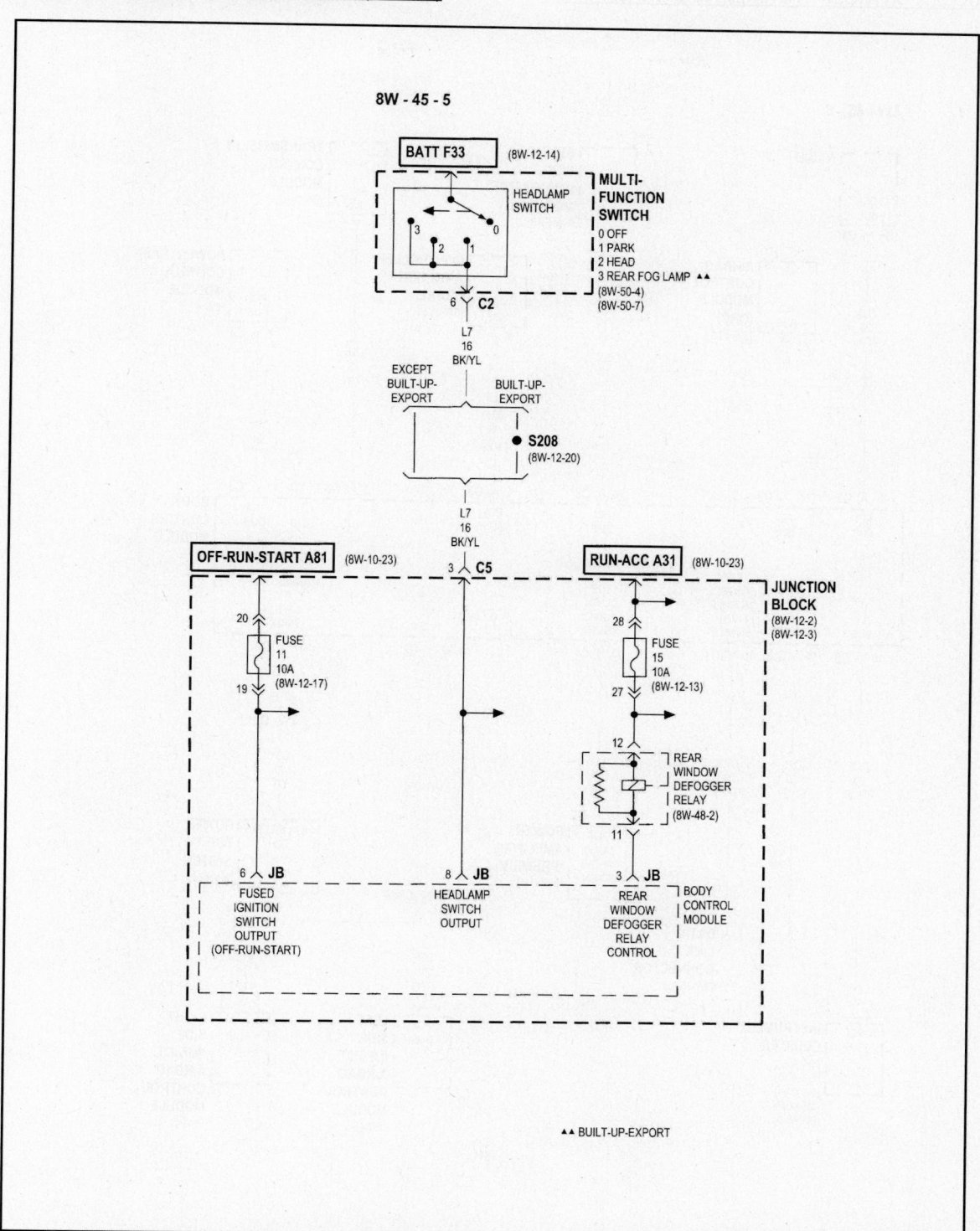

BODY CONTROLS SYSTEM DIAGRAMS (5 OF 12)

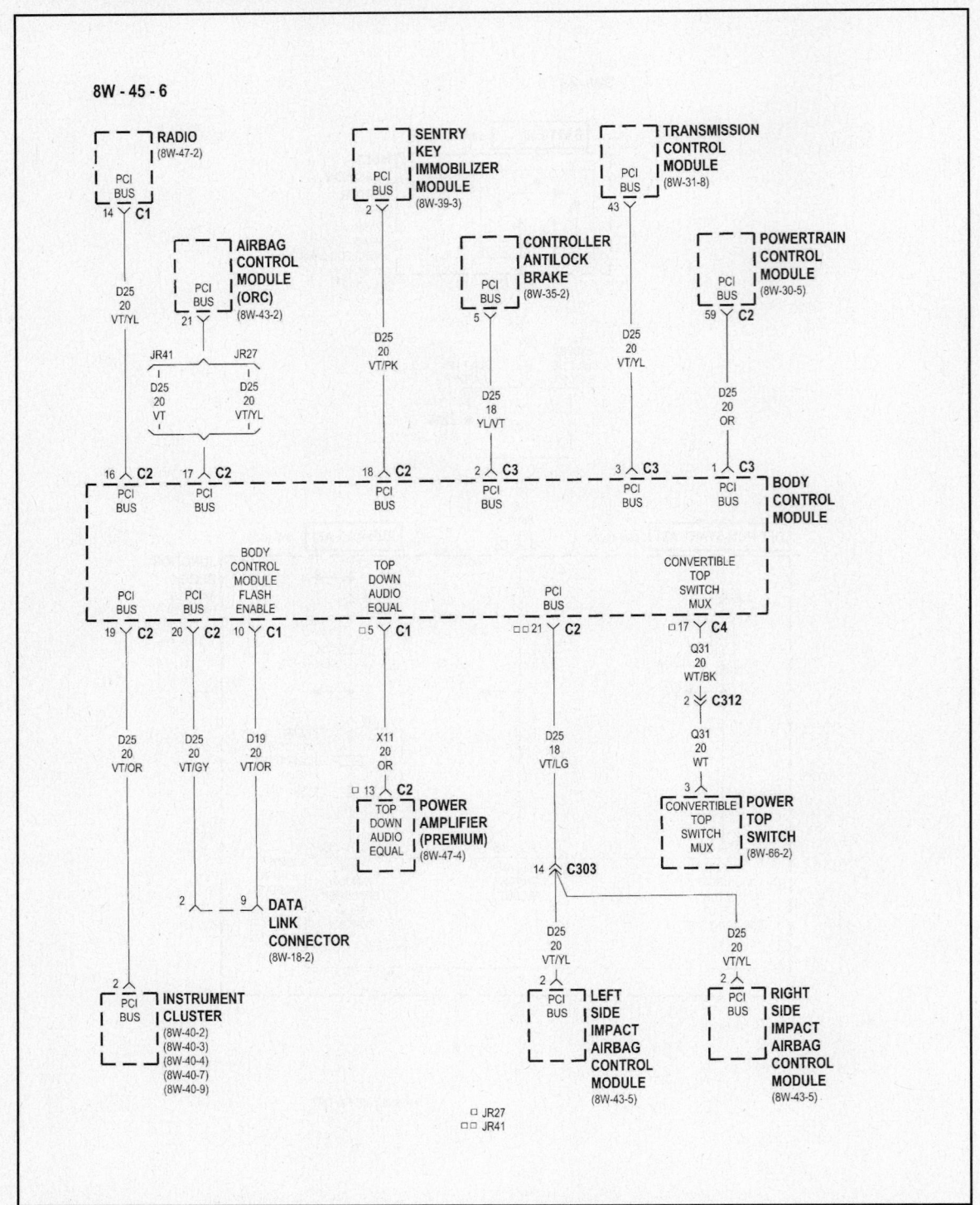

BODY CONTROLS SYSTEM DIAGRAMS (6 OF 12)

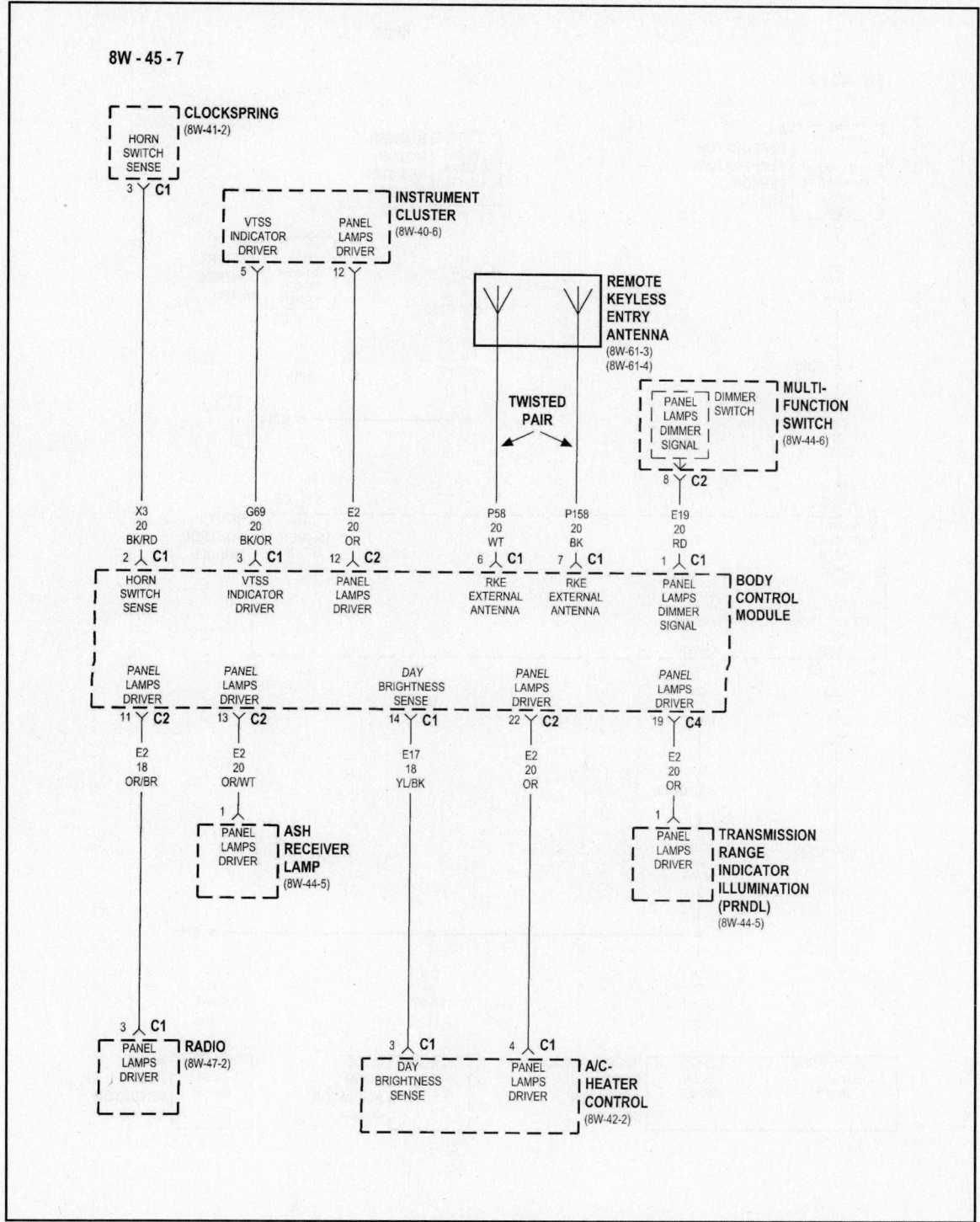

8W - 45 - 7

BODY CONTROLS SYSTEM DIAGRAMS (7 OF 12)

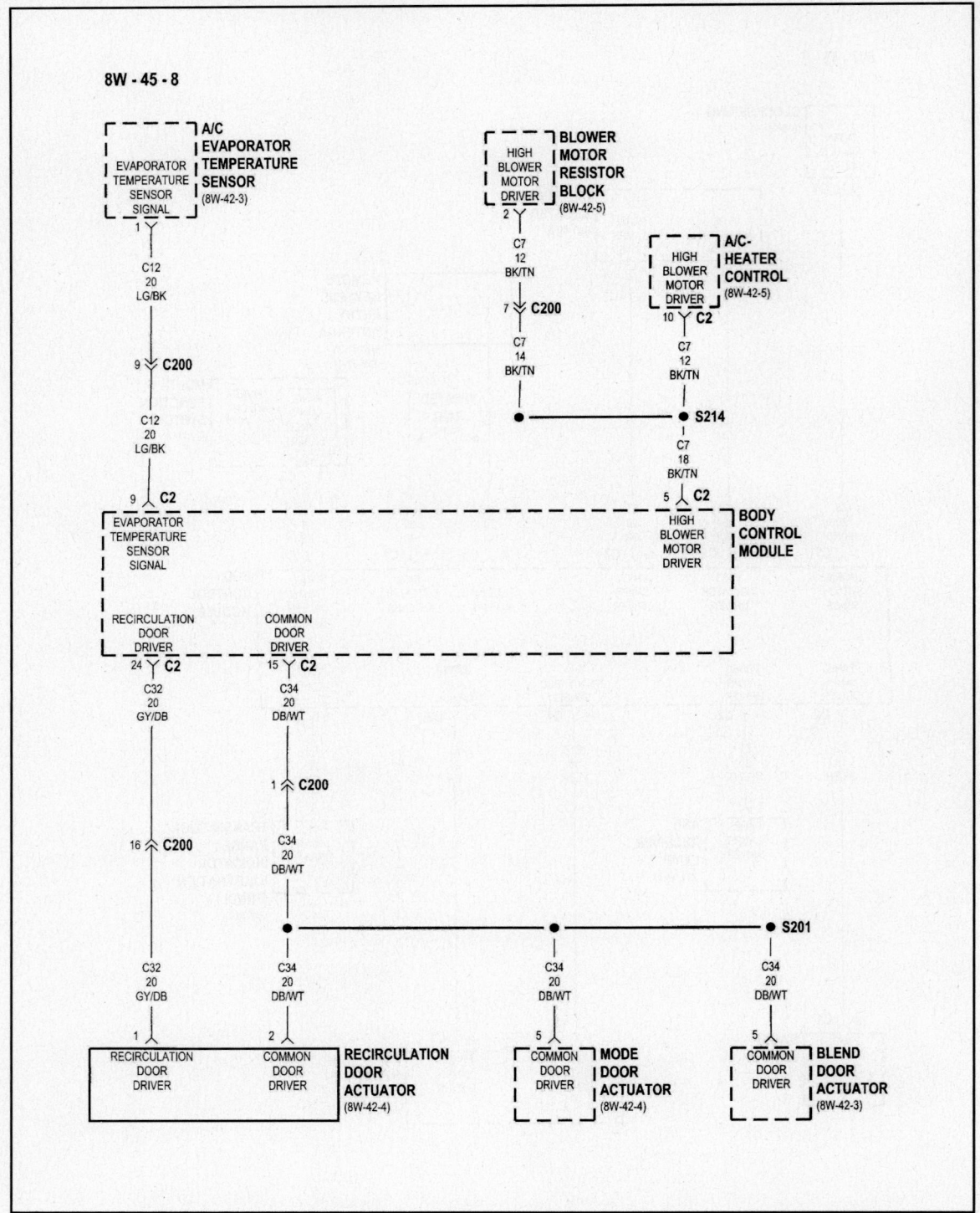

BODY CONTROLS SYSTEM DIAGRAMS (8 OF 12)

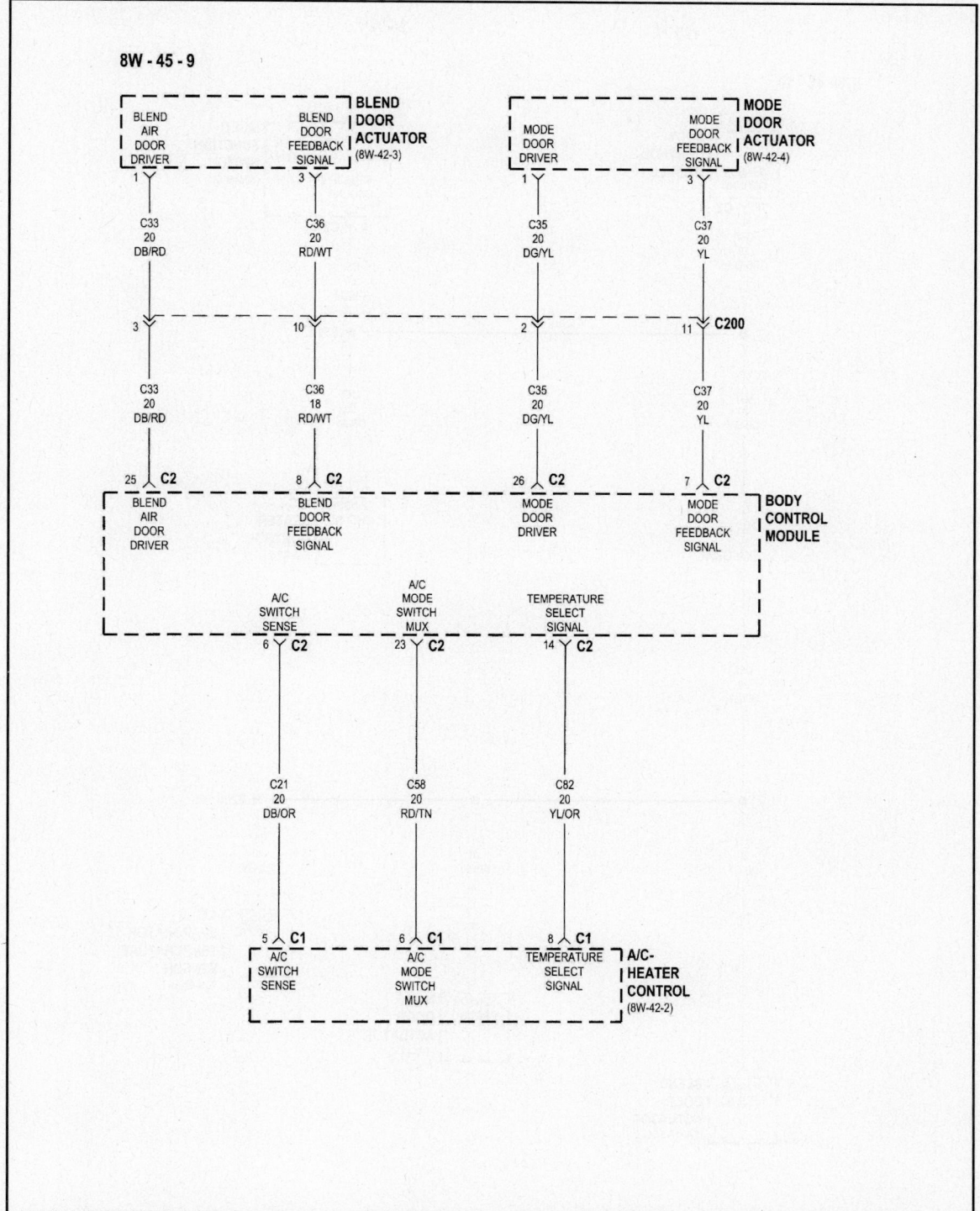

BODY CONTROLS SYSTEM DIAGRAMS (9 OF 12)

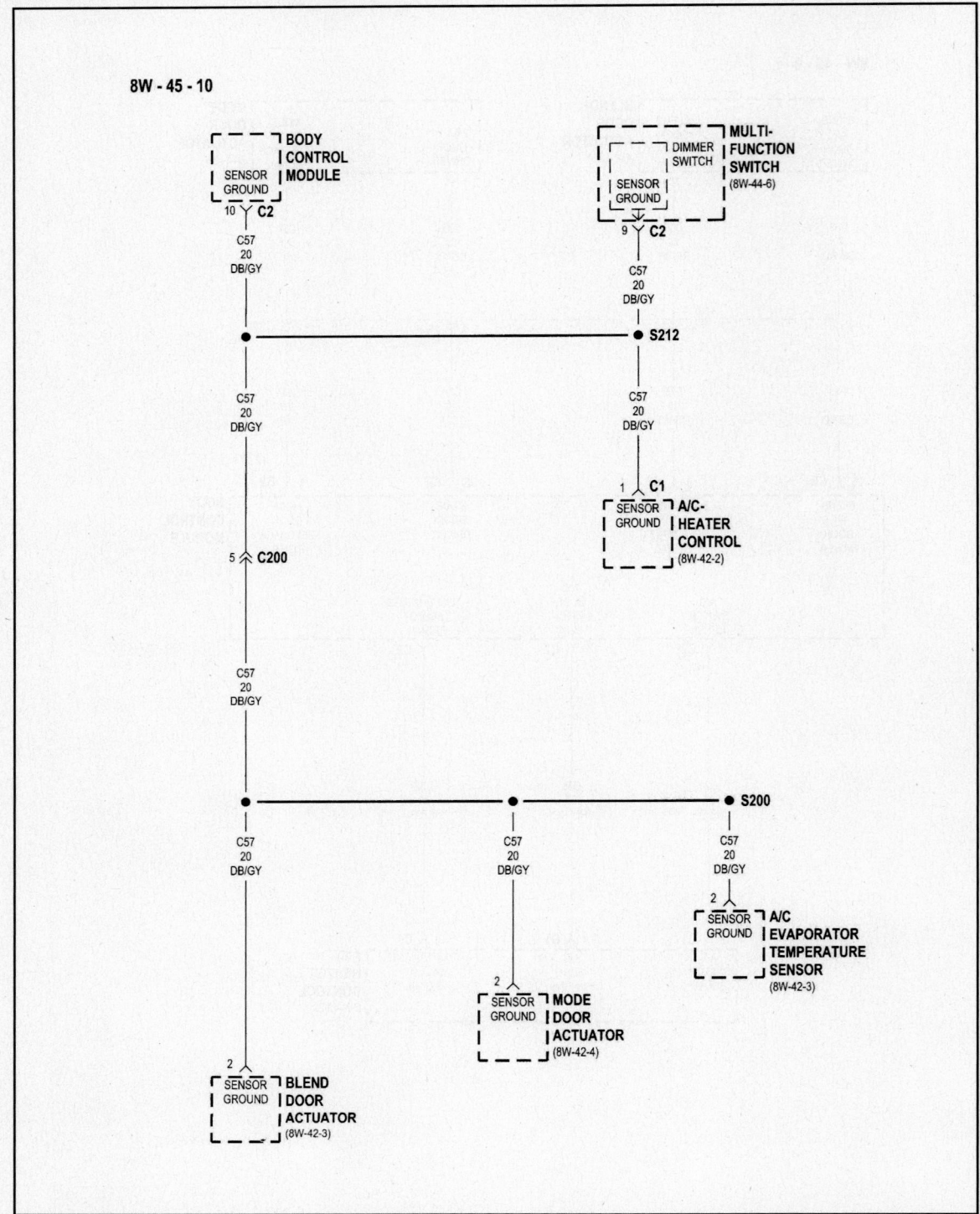

BODY CONTROLS SYSTEM DIAGRAMS (10 OF 12)

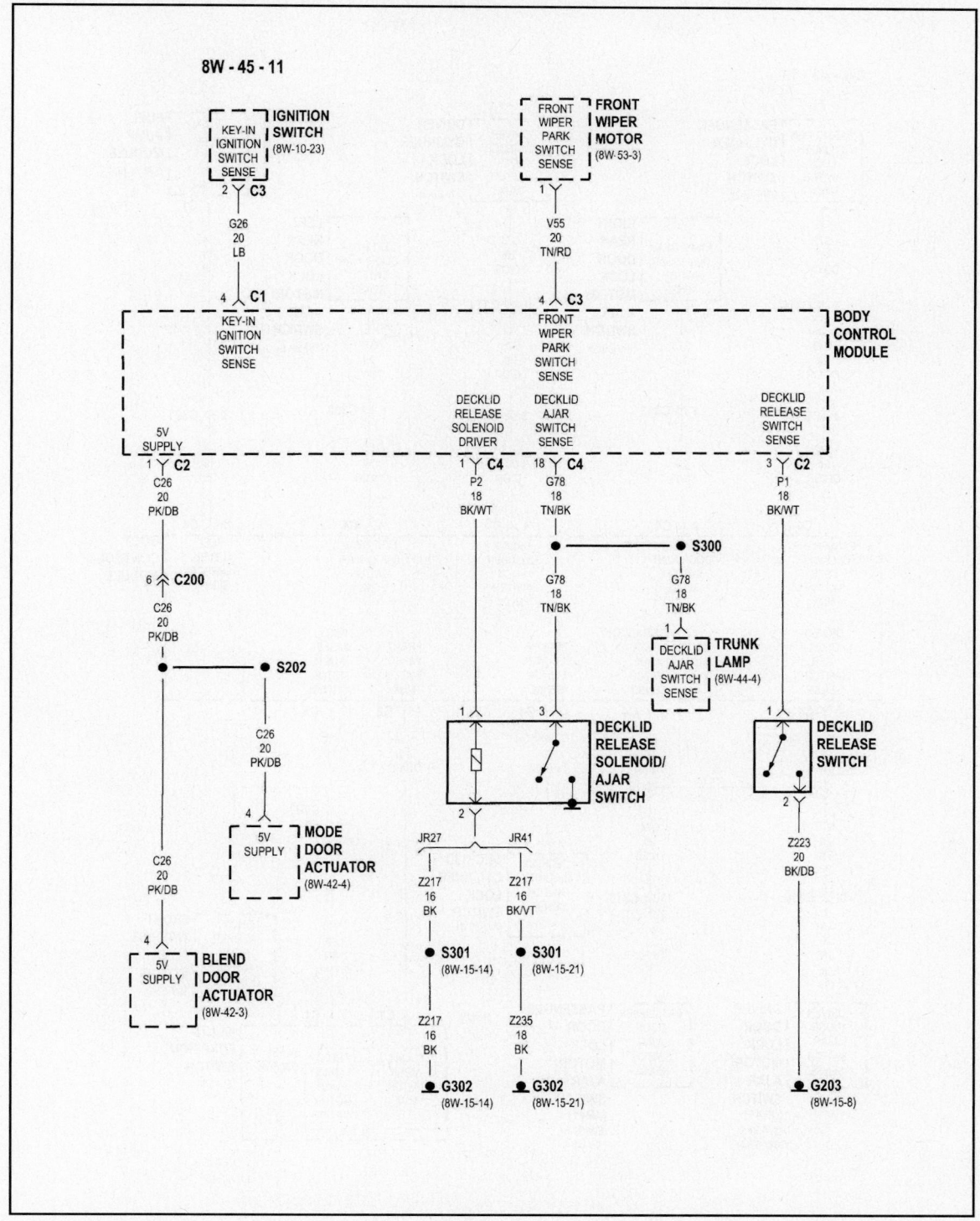

BODY CONTROLS SYSTEM DIAGRAMS (11 OF 12)

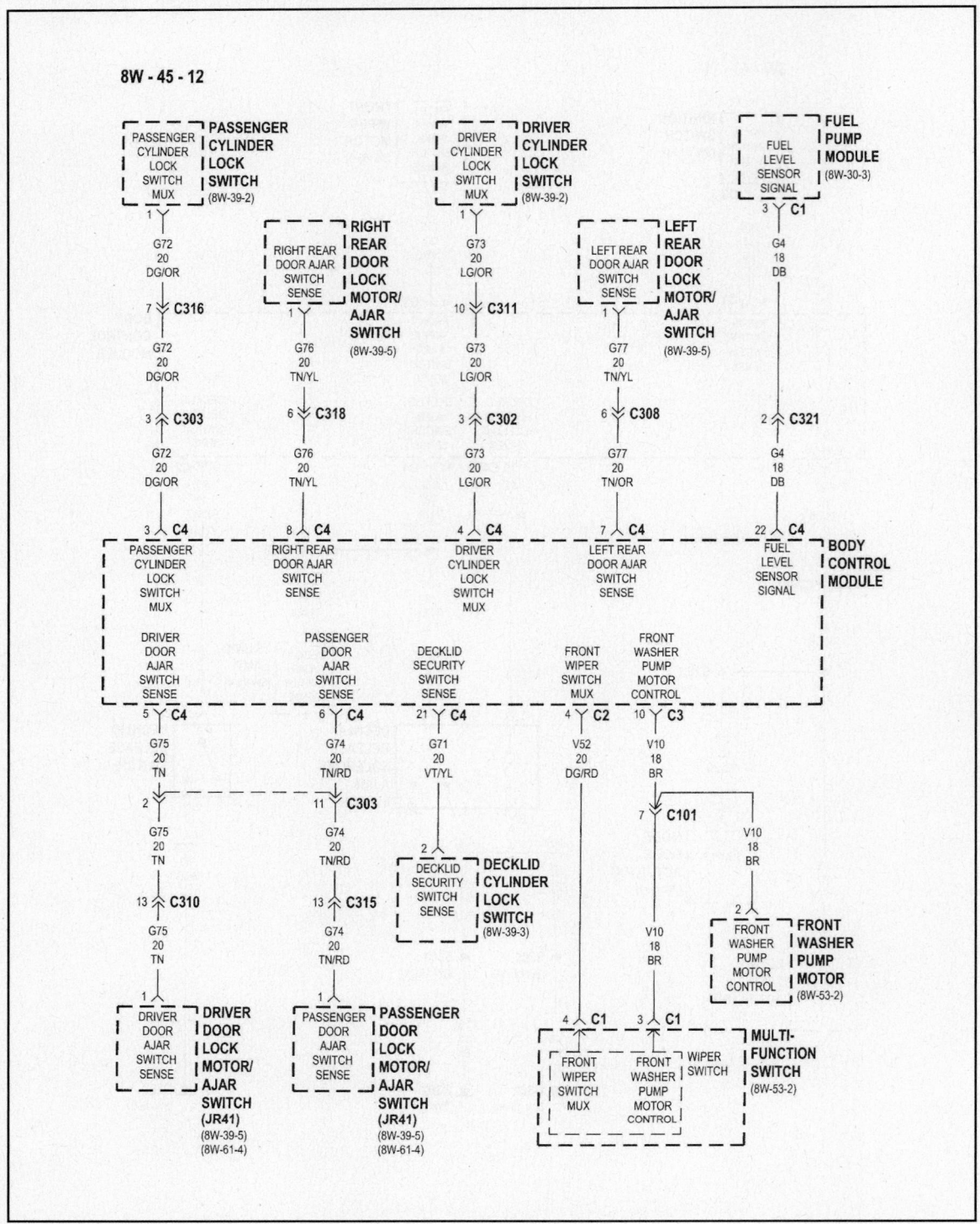

8W - 45 - 12

BODY CONTROLS SYSTEM DIAGRAMS (12 OF 12)

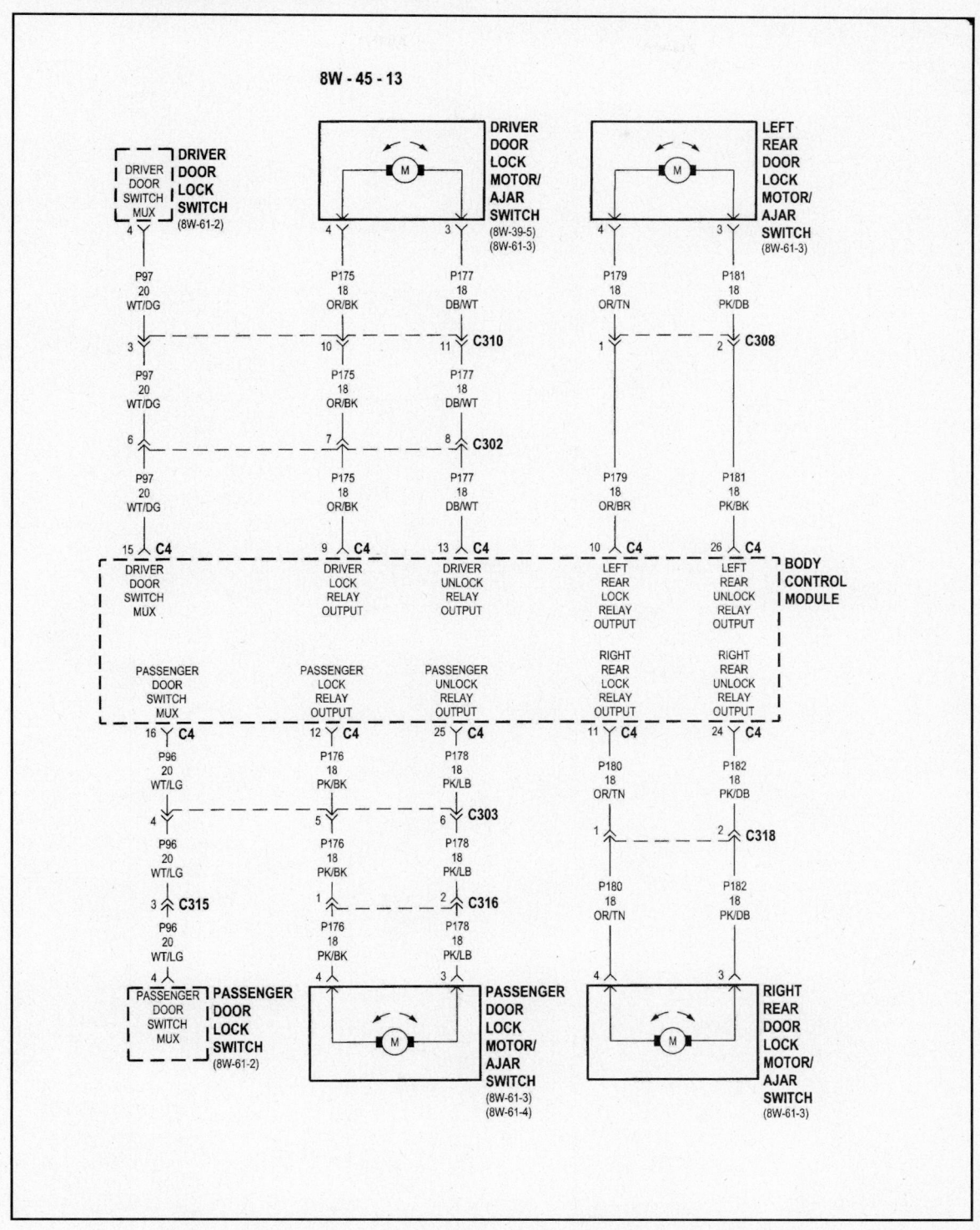

8W - 45 - 13

SECTION 3 CONTENTS

1998 RAM 1500 – POWERTRAIN

1998 RAM 1500 – POWERTRAIN

1998 RAM 1500 - POWERTRAIN

Introduction

HOW TO USE THIS SECTION

This section of the manual includes diagnostic and repair information for Chrysler Truck applications. This information can be used to help you understand the Theory of Operation and Diagnostics of the electronic controls and devices on this vehicle. The articles in this section are separated into two sub categories:

*   Powertrain Control Module
*   Transmission Controls – PCM Operated

1998 Dodge Ram 5.2L V8 OHV SFI VIN Y (42RE Electronic A/T)

The articles listed for this vehicle include a variety of subjects that cover the key engine controls and vehicle diagnostics. Several electronic control inputs and output devices are featured along with detailed descriptions of how they operate, what can go wrong with them, and most importantly, how to use the onboard diagnostics and common shop tools to determine if one or more devices or circuits have failed.

This vehicle uses a conventional coil and distributor Electronic Ignition system and Sequential Fuel Injection (SFI). These systems are primarily controlled by inputs from Hall effect camshaft and crankshaft position sensors. These devices and others (i.e., EVAP, IAC, MAP, TP, ECT, IAT, Oxygen sensors, etc.) are covered along with how to test them.

Choose The Right Diagnostic Path

In most cases, the first step in any vehicle diagnosis is to follow the manufacturers recommended diagnostic path. In the case of Chrysler vehicle applications, the first step is to connect a Scan Tool (OBD II certified for this vehicle) and attempt to communicate with the engine controller or any other vehicle onboard controllers that apply. Diagnostic Trouble Codes (DTCs), data (PIDs), bi-directional controls, operating mode descriptions, and symptom charts can then be used to pinpoint the fault.

The Malfunction Indicator Lamp (MIL) can be a great help to you during diagnosis. You should note if the customer complaint included the fact that the MIL remained on during engine operation. And be sure to determine if the MIL comes on at key on and goes out after a 4 second bulb-check. There is more information on using the MIL as part of the diagnostic path in Section 1.

Diagnostic Points to Consider

If a problem is detected in a vehicle similar to this one (whether it is a trouble code or no code condition), you need to be able to determine which vehicle system (or systems) are involved before you proceed with any testing. While this may sound like a fairly routine way to approach problem solving, it is a key point to consider before you start testing.

Due to the complexity of these vehicles, a trouble code does not always point to a failed component - instead, it may point to a system that needs to be tested for some kind of problem. For example, if an Oxygen sensor trouble code is set, you need to consider what other conditions could cause an oxygen sensor to behave improperly.

Component Diagrams

You will find numerous "customized" component diagrams used throughout this section. Note the dashed lines in each Schematic, as they represent the circuits the control module uses to determine when a circuit has a fault.

Vehicle Identification

VECI DECAL

The Vehicle Emission Control Information (VECI) decal is located in the engine compartment. The information in this decal relates specifically to this vehicle and engine application. The specifications on this decal are critical to emissions system service.

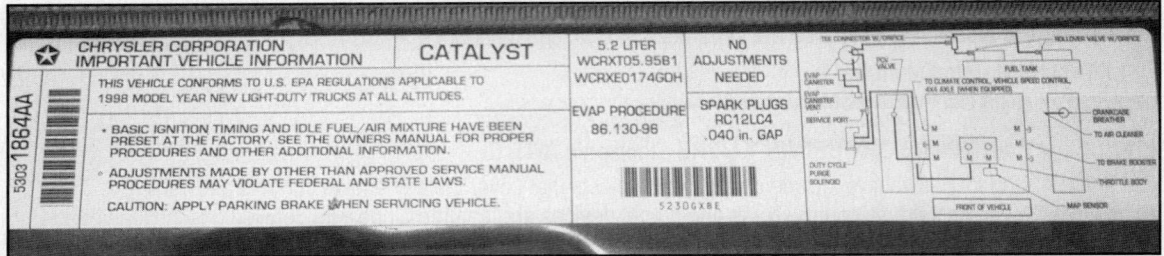

IMPORTANT VEHICLE INFORMATION

The idle speed, ignition timing and air/fuel mixture have all been preset at the factory. The idle speed can be reset, but this procedure requires a Scan Tool and a special calibrated air leak tool.

Vacuum Routing Diagram

Not all equipment is always written on the vehicle information decal. The vacuum routing diagram can be helpful in determining the required emission equipment. This vehicle does not have an EGR system, or an enhanced EVAP system.

LEV or ULEV

If these designators are used, the vehicle conforms to U.S. EPA and State of California Low or Ultra-Low Emissions Vehicle standards. This vehicle does not qualify as LEV or ULEV.

Underhood View Graphic

Electronic Engine Control System

OPERATING STRATEGIES

The Powertrain Control Module (PCM) in this application is a triple-microprocessor digital computer. The PCM, formerly known as the SBEC, provides optimum control of the engine and various other systems. The PCM uses several operating strategies to maintain good overall drivability and to meet the EPA mandated emission standards.

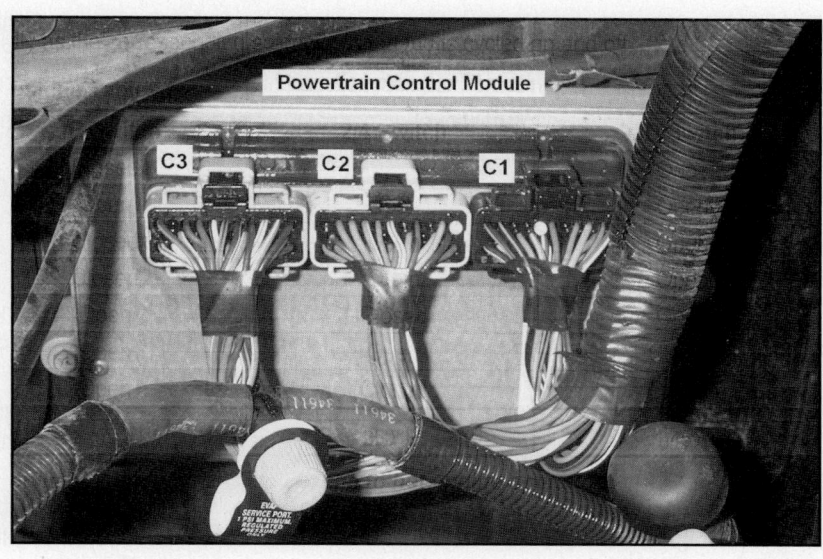

Based on various inputs it receives, the PCM adjusts the fuel injector pulsewidth, the idle speed, the amount of ignition advance or retard, the ignition coil dwell, and operation of the canister purge valve.

The PCM also controls the Speed Control (S/C), A/C, and the electronic transmission.

Operating Strategies

A/C Compressor Clutch Cycling Control	Closed Loop Fuel Control
Electric Fuel Pump Control	Limited Transmission Control
Fuel Metering of Sequential Fuel Injectors	Speed Control (S/C)
Idle Speed Control	Ignition Timing Control
Vapor Canister Purge Control	

Operational Control of Components

A/C Compressor Clutch	Ignition Coils (Timing and Dwell)
Fuel Delivery (Injector Pulsewidth)	Fuel Pump Operation
Purge Solenoid	Idle Air Control Motor
Integrated Speed Control	Malfunction Indicator Lamp

Items Provided (or Stored) By the Electronic Engine Control System

Shared CCD Data Items	Adaptive Values (Idle, Fuel Trim etc)
DTC Data for Non-Emission Faults	DTC Data for Emission Related Faults

Distinct Operating Modes

Key On	Cruise
Cranking	Acceleration (Enrichment)
Warm-up	Deceleration (Enleanment)
Idle	Wide Open Throttle (WOT)

Integrated Electronic Ignition System

INTRODUCTION

The PCM controls the operation of the integrated electronic ignition (DI) system on this vehicle application. The system consists of a conventional distributor, ignition coil, CKP sensor, CMP sensor, and a PCM. This is an integrated ignition system, meaning that the ignition control module is part of the PCM, and cannot be serviced separately.

The ignition coil, mounted to the right front corner of the engine, supplies spark to the distributor cap. The spark is distributed to the 8 cylinders through the spark plug wires.

Battery voltage is supplied to coil via the ASD relay in the Power Distribution Center (PDC). The PCM controls the ground circuit to ignition coil. By switching the ground path for the coil "on" and "off" at the appropriate time, the PCM adjusts the ignition timing to meet changing operating conditions. The firing order on this engine is 1-8-4-3-6-5-7-2.

Ignition System Sensor Operation
The PCM makes ignition-timing decisions based on inputs from the CKP sensor. Although the CMP sensor is mounted inside the distributor, the CMP sensor signal is not needed for the ignition system to operate. The CMP sensor signal is only necessary to calculate the proper fuel injector firing sequence.

The CKP and CMP sensors are Hall-Effect devices that toggle internal transistors each time a window is sensed on the crankshaft or the distributor pulse-ring. When the transistor toggles, a 5v pulse is sent to the PCM.

Ignition Spark Timing
The ignition timing on this ignition system is controlled by the PCM - base timing is not adjustable. The timing cover does have degree markings, but these are only to be used for reassembly. Rotating the distributor will have no effect on base timing. If the injector timing must be checked or adjusted, the SET SYNCH test must be performed using a Scan Tool.

Once the engine starts, the PCM calculates the spark advance using these signals:

• BARO Sensor (MAP signal at KOEO)
• ECT Sensor
• Engine Speed Signal (rpm)
• MAP Sensor
• Throttle Position

Ignition System Schematic

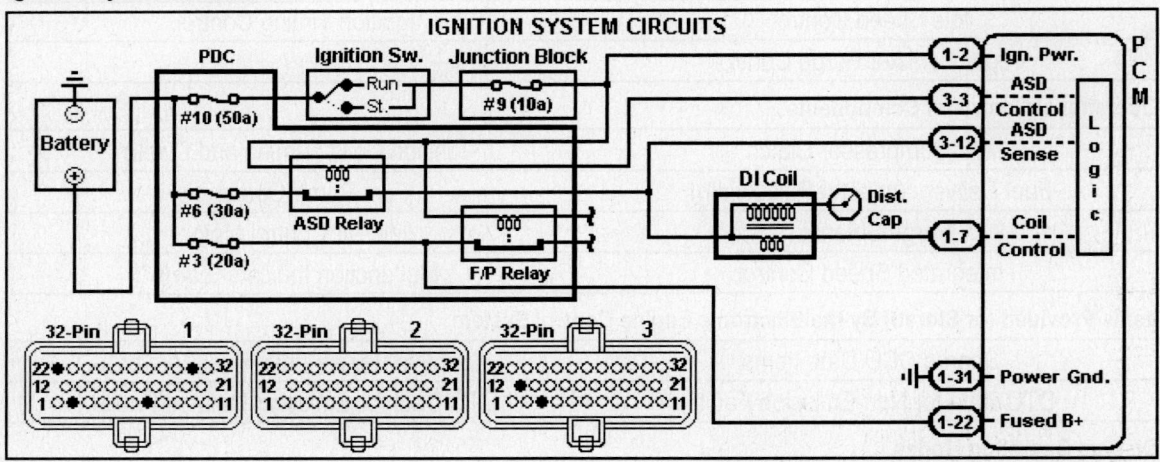

IGNITION COIL OPERATION

This ignition system uses a single ignition coil, mounted to a bracket on the right front corner of the engine.

Specifications

There are two types of ignition coils used on this application. The Diamond coil has a primary resistance of 0.97-1.18 Ohms, and a secondary resistance of 11,300-15,000 Ohms. The Toyodenso coil has a primary resistance of 0.95-1.20 Ohms, and a secondary resistance of 11,300-13,300 Ohms at room temperature.

The firing order on this engine is 1-8-4-3-6-5-7-2.

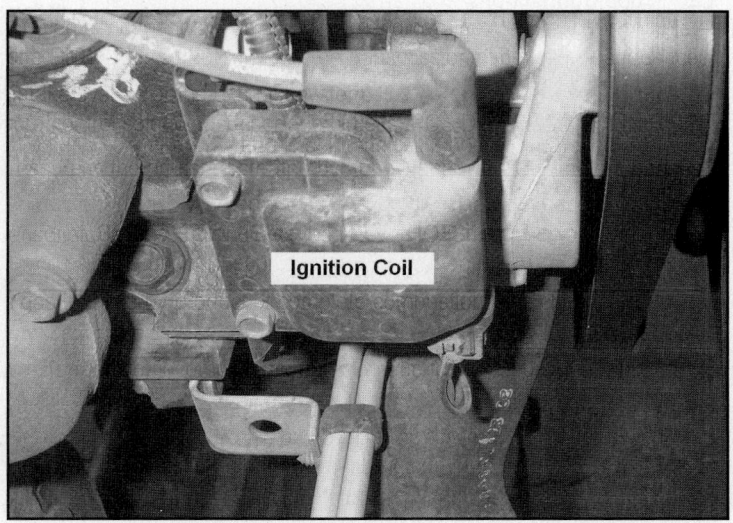

Coil Circuits

The ignition coil is powered by the ASD relay. A driver in the PCM controls the dwell (saturation time) and timing of the coil. Refer to the Ignition System Schematic below.

Drivability Help

Whenever a DTC or drivability problem exists, it is important to check for any related Technical Service Bulletins (TSBs). Because the ignition timing is not adjustable, and this vehicle is not equipped with an EGR system, diagnosing a complaint of detonation under load can be difficult.

TSB No. 18-48-98 addresses a common problem on many 1994 -1999 3.9L, 5.2L, and 5.9L engines. Ignition secondary-wire routing problems cause cross fire, resulting in a misfire. The misfire can occur on many combinations of cylinders, but cylinders 5 and 8 are the most common. The TSB includes detailed instructions on the installation of routing clips and a protective wire cover. Refer to TSB 09-05-00, concerning a leaking intake plenum gasket that produces similar symptoms.

Ignition System Schematic

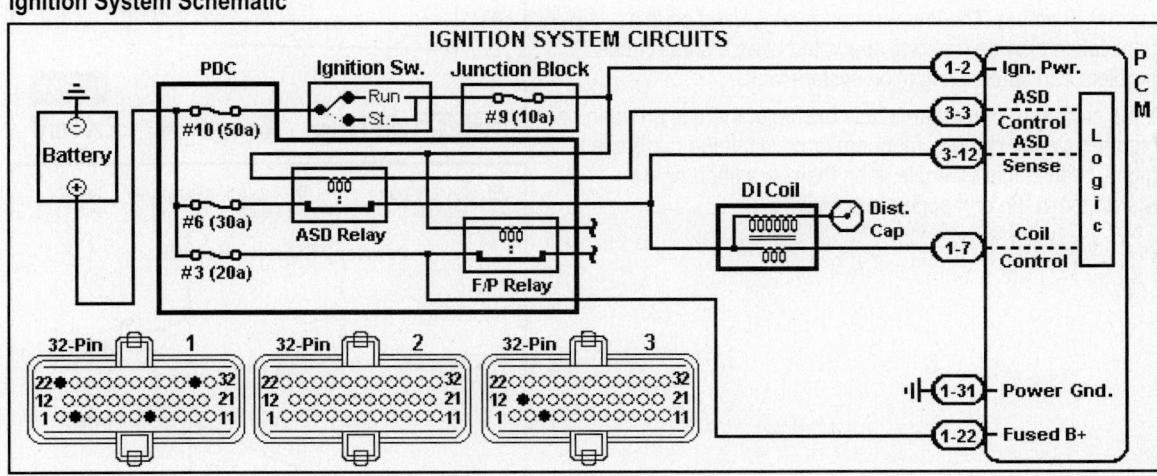

LAB SCOPE TEST (COIL PRIMARY)

Place the shift selector in Park and block the drive wheels for safety.

Lab Scope Connections
Connect the Channel 'A' positive probe to the coil control circuit (BK/GY wire) at PCM Pin 1-7 or at the coil connector. Connect the negative probe to the battery negative post. Connect the low-amps probe to channel 'B', and clamp the probe around the same wire, or around the coil power feed wire.

Lab Scope Example (1)
In example (1), the trace shows the coil driver (primary) circuit at idle. Use these Lab Scope settings to see system voltage, dwell (5.4 ms), spark line, and coil oscillations.

To see peak primary voltage, the settings need to be changed to 50 volts/division. On this application, the peak voltage was about 300v.

Lab Scope Example (2)
In example (2), the same settings were used, but the coil was fired using the Scan Tool bi-directional controls (ATM Test 01: COIL). This is a good test to use to isolate the cause of a no-spark, because this test is not dependent on PCM inputs.

The coil is fired every few seconds for 5 minutes. Because there is no compression or fuel, the firing voltage will be higher and the burn time will be shorter. Additionally, the rotor may not be in an appropriate position for spark distribution. Therefore, waveform analysis is not effective using this test. Use this test only to verify that the PCM, coil, and circuits are capable of generating and delivering a spark.

Lab Scope Example (3)
In example (3), the trace shows the coil primary voltage and current (6 amps). The low-resistance coils used on this application provide a strong spark, but even a small circuit problem can cause an ignition system fault.

If the current is lower than in this example, there is an open or high resistance in the component or circuit. If the current is higher than in this example, then there is a short or low resistance in the component or circuit.

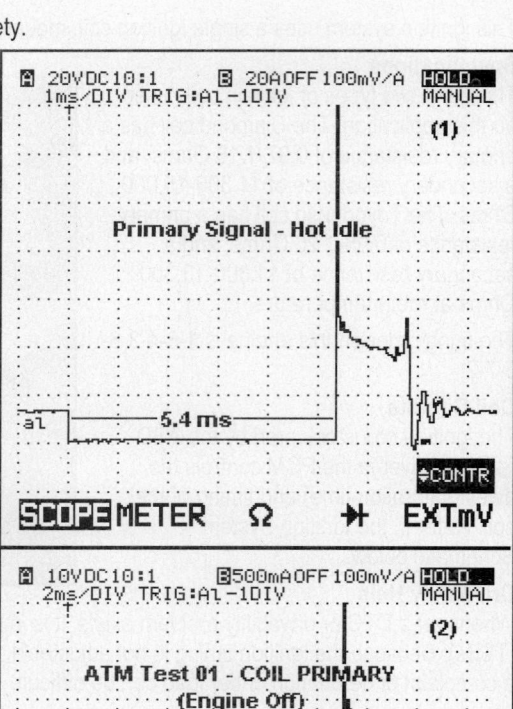

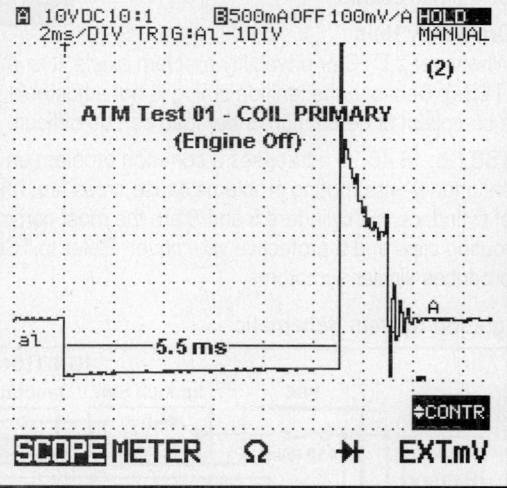

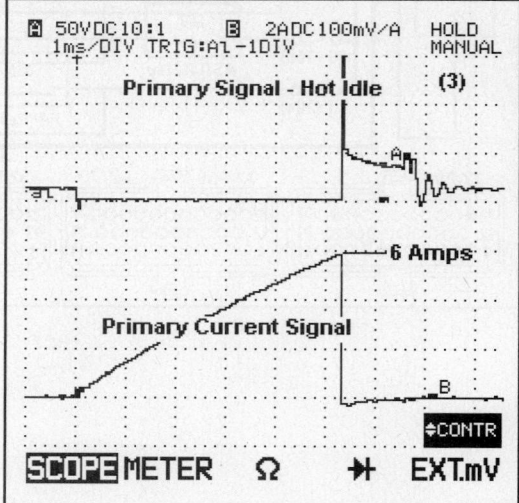

IGNITION SECONDARY TEST

An ignition analyzer can be used to view the secondary ignition pattern for all cylinders in order. Although extensive ignition diagnostics can be performed using a Lab Scope, an ignition analyzer offers the added benefit of seeing all cylinders at once. Many ignition analyzers offer the option of viewing primary or secondary patterns.

Analyzer Connections
Select the vehicle from the database, or manually enter the firing order. Connect the inductive secondary pickup around the coil wire, and the inductive synch probe around the ignition wire for cylinder 1. Different synch (trigger) cylinders can be selected if cylinder 1 is not easily accessible or if it is not firing. The synch cylinder must be firing to allow the analyzer a place to start its counting sequence.

Engine Analyzer Example
In this example, the trace shows the secondary waveform for all eight (8) cylinders in their correct order. Using a parade display, all of the firing events can be seen near each other for easy comparison. Using these settings, an intermittent ignition misfire can be seen.

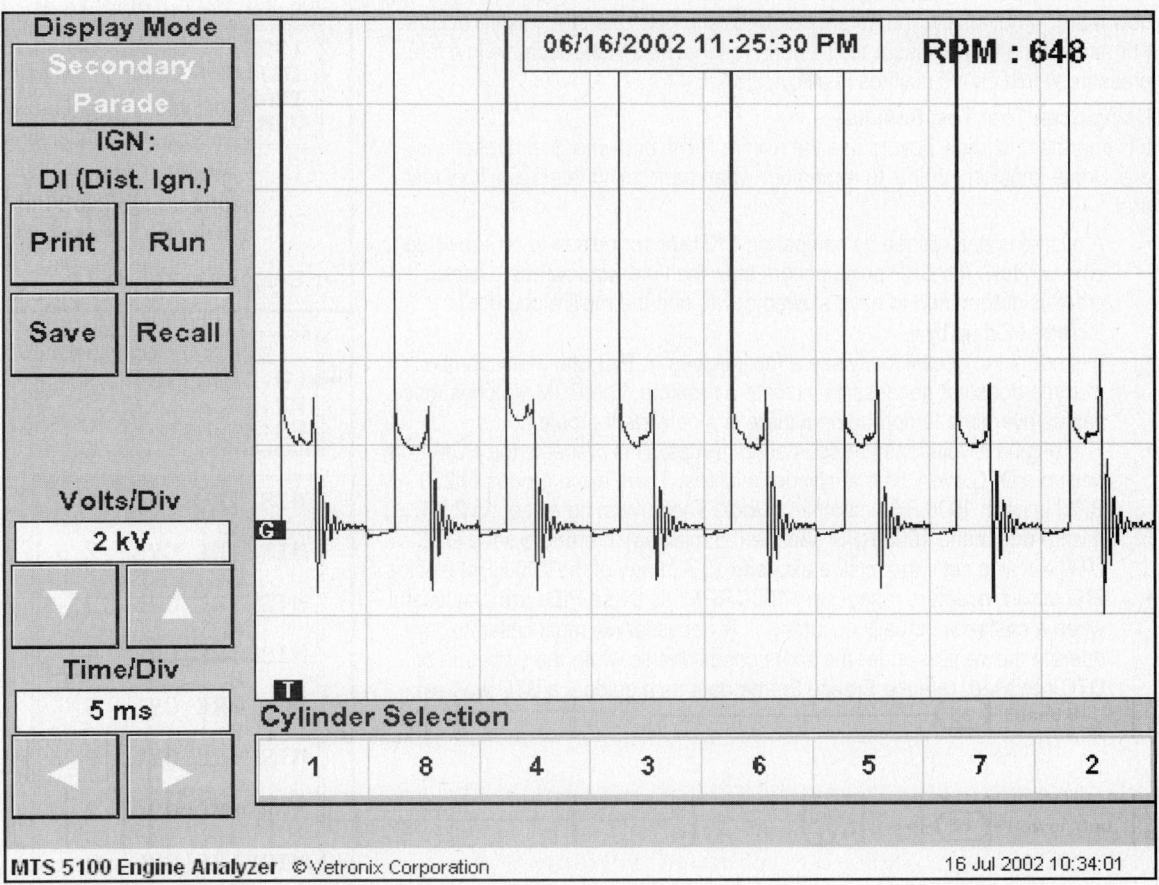

Alternate Views
The waveforms can also be arranged vertically (raster) to more precisely compare the burn time between cylinders. On many testers Firing Kv bar charts are also available to view peak firing voltage, as well as the minimum and average values for each cylinder.

SCAN TOOL TEST (MISFIRE PIDS)

The Scan Tool can be used to view PID values during misfire diagnosis. In the datalist, there is a MISFIRE PID for value each cylinder. Additionally, there may be a specific misfire PID list available in the SYSTEM TESTS menu.

Misfires can occur for any number of reasons. Ignition system faults are certainly a common cause, but do not assume that a misfire is caused by the ignition system.

It is important to view all available information, to determine if the problem is common to all cylinders, one bank, or an individual cylinder. Other sensor inputs, power and ground circuits, and fuel trim PIDs also provide clues when diagnosing an engine misfire.

Scan Tool Example

In this example, the MISFIRE PIDs are monitored for misfire. The counter value is 0 for most cylinders. The MSIFIRE CYL 1 and 2 PIDs indicate misfires. These misfires are significant and can be felt in the vehicle, but will not set a misfire DTC. See the bullet points below for an explanation of misfire DTCs. Even though no DTC is set, this msifire list is useful to isolate the cyclinder(s) involved in the fault.

Further diagnosis is necessary, but many questions can be resolved by knowing how many, if not all, cyclinders are involved, and how often the misfire occurs. For example, a fuel pressure test is not a rational next step, because low fuel pressure would cause misfires in all cylinders.

Using Scan Tool Test Results

It is important to know how to use the misfire PIDs, and what their limitations are. Some important points to remember when performing this Scan Tool test are:

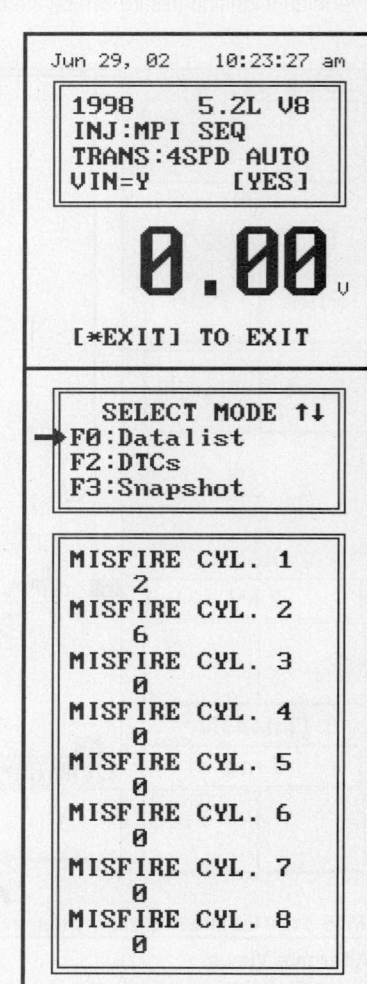

*   A misfire is determined by comparing CKP sensor pulses to an expected time window. If a CKP pulse occurs after the time window has expired, the crank is determined to have slowed down, and the misfire counter is incremented up by 1.
*   It is normal to occasionally see a few misfires. A PID with a low number of misfires does not necessarily indicate a problem. The PCM watches these values over time to determine if there is a persistent problem.
*   The engine revolutions are counted in groups. This prevents the PCM from setting a DTC every time a cylinder misfires. There are 5 groups of 200 RPM in each 1000 RPM sampling block. There must be at least a 2.5% misfire rate in the 1000 RPM sample (25 misfires) in order to set a DTC. A DTC will also set if the misfire exceeds 10% in any of the 200 RPM blocks.
*   The misfire counters reset every 1000 RPM, so these PIDs are only useful when a misfire is actually occurring. It is necessary in most cases to operate the vehicle under the same conditions as when the symptom or DTC occurred. Use the Freeze Frame data as a guide if a DTC was set.

CRANKSHAFT POSITION SENSOR (HALL EFFECT)

The CKP sensor is a Hall effect sensor mounted on the engine block, adjacent to notches on the flywheel. There are 8 notches evenly spaced around the flywheel. The CKP sensor Hall effect circuit senses each slot and toggles a transistor high. Each toggling sends a 5v pulse to the PCM for use in calculating engine position and speed.

By using all 8 slots, the PCM can track engine speed and variations in crankshaft rotation speed for misfire diagnosis.

All of the eight (8) notches are exactly the same size and spacing. Therefore, the CKP sensor signal is only used for relative position and crankshaft speed data. The CKP cannot identify the exact position for any one cylinder. To calculate which cylinder in on TDC on the compression stroke, the PCM requires the CMP sensor signal.

The CKP sensor is the most critical sensor on the vehicle. If this signal is missing, the vehicle will not start.

CKP Sensor Circuits
The CKP Sensor is connected to the PCM by 3 wires. The CKP and CMP sensors share a 5v supply voltage from PCM Pin 1-17. The 0-5v return signal is sent to PCM Pin 1-8. The shared sensor ground connects to the PCM at Pin 1-4.

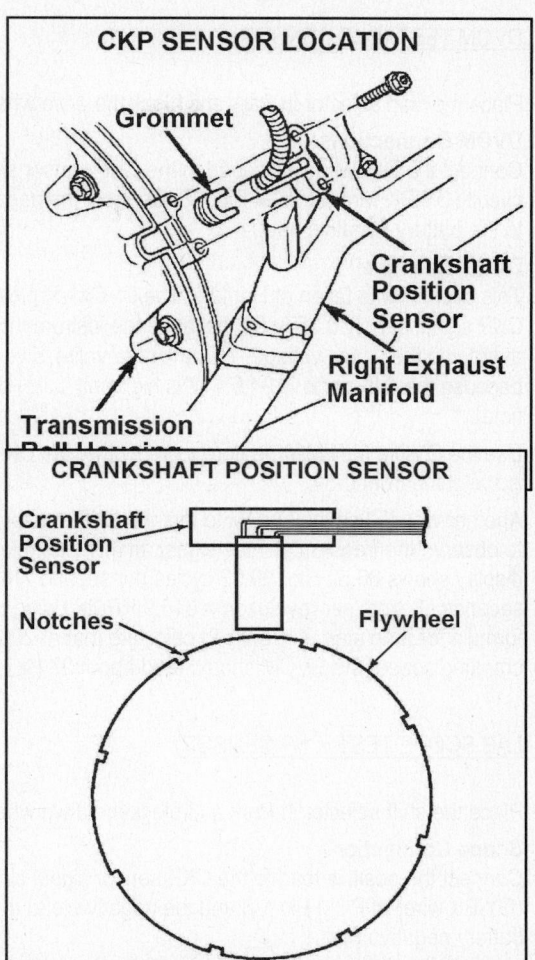

CKP SENSOR LOCATION

CRANKSHAFT POSITION SENSOR

CKP Sensor Schematic

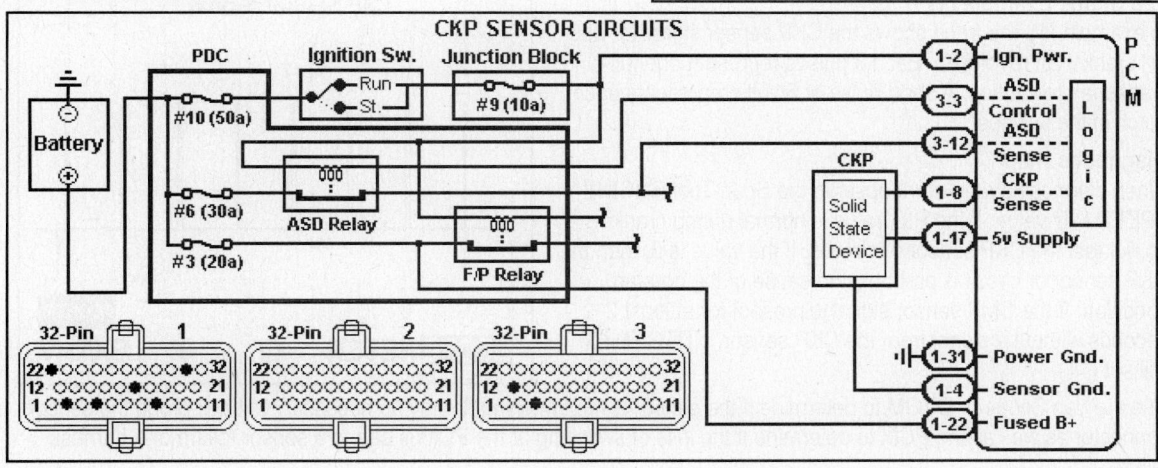

<u>DVOM TEST (CKP SENSOR)</u>

Place the shift selector in Park and block the drive wheels for safety.

DVOM Connections
Connect the DVOM positive lead to the CKP sensor signal circuit (GY/BK wire) at PCM Pin 1-8. Connect the negative lead to the battery negative post.

DVOM Example
This capture was taken at Hot Idle. The DVOM display shows a CKP signal of only 0.357v DC because the instrument is averaging the signal voltage. This average value is very low because the duty cycle is 94.6% (it is high only 3.4% of the time).

Use the DVOM MIN/MAX feature to verify that the CKP signal is switching from 0-5v.

Another way to use the DVOM to test the CKP sensor signal is to observe the frequency of the signal. In this example, the display shows 86.52 Hz. (86.52 cycles per second X 60 seconds / 8 slots per revolution = 648.9 RPM). Using this formula for a no start, it is easy to calculate that at 200-RPM cranking speed, the DVOM should read about 27 Hz.

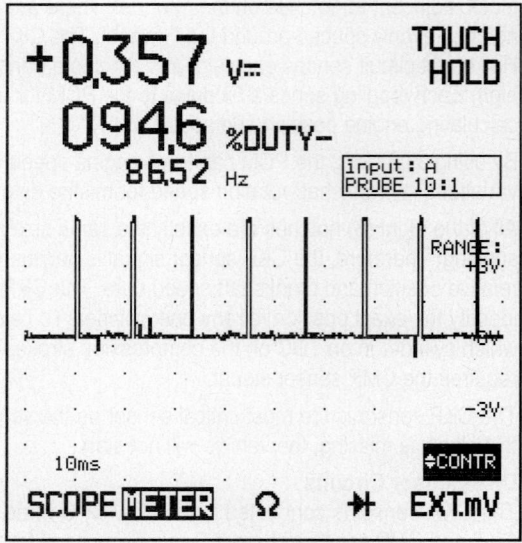

<u>LAB SCOPE TEST (CKP SENSOR)</u>

Place the shift selector in Park and block the drive wheels for safety.

Scope Connections
Connect the positive lead to the CKP sensor signal circuit (GY/BK wire) at PCM Pin 1-8 and the negative lead to the battery negative post.

Lab Scope Example (1)
In example (1), the trace shows the CKP sensor signal for one full crankshaft revolution. Each 8 pulses represent one full crankshaft revolution. A short pulse of 5 volts represents each notch in the flywheel.

Diagnostic Help
When diagnosing a no-start, observe the Scan Tool ENGINE SPEED PID value. If the PID value is normal during cranking, do not test the CKP sensor or circuits. If the value is 0, then the CKP sensor or circuit is probably the cause of the no-start condition. If the CMP sensor signal is present for at least 2 seconds without a signal from the CKP sensor, a DTC P0320 will set (1T).

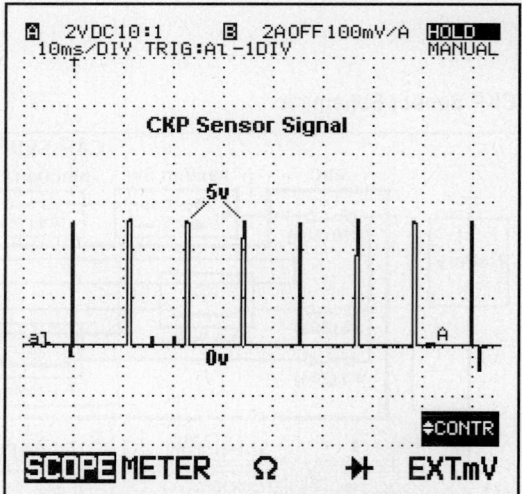

Use the Lab Scope or DVOM to determine if the sensor is not switching. It may be necessary to back probe the sensor connector as well as the PCM to determine if the loss of switching at the PCM is due to a sensor failure or a harness failure.

Evaporative Emission System

INTRODUCTION

The Evaporative Emissions (EVAP) system is used to prevent the escape of fuel vapors to the atmosphere under hot soak, refueling and engine off conditions. Fuel vapor pressure trapped in the fuel tank is vented through a vent hose, and stored in EVAP canister.

The most efficient way to dispose of fuel vapors without causing pollution is to burn them in the normal combustion process. The PCM cycles the purge valve on (open) during normal operation. With the valve open, manifold vacuum is applied to the canister and this allows it to draw fresh air and fuel vapors from the canister into the intake manifold. These vapors are drawn into each cylinder where they are burned.

The PCM uses various sensor inputs to calculate the desired amount of EVAP purge flow. The PCM meters the purge flow by varying the duty cycle of the EVAP purge solenoid control signal. The EVAP purge solenoid will remain off during a cold start, or for a preprogrammed time after a hot restart.

System Components:

The EVAP system contains many components for the management of fuel vapors and to test the integrity of the EVAP system. The EVAP system is constructed of the following components:

- Fuel Tank
- Fuel Filler Cap
- Fuel Tank Level Sensor
- Fuel Vapor Vent
- Evaporative Charcoal Canister
- EVAP Purge Solenoid
- PCM and related wiring
- Fuel vapor lines with Test Port

EVAP System Schematic

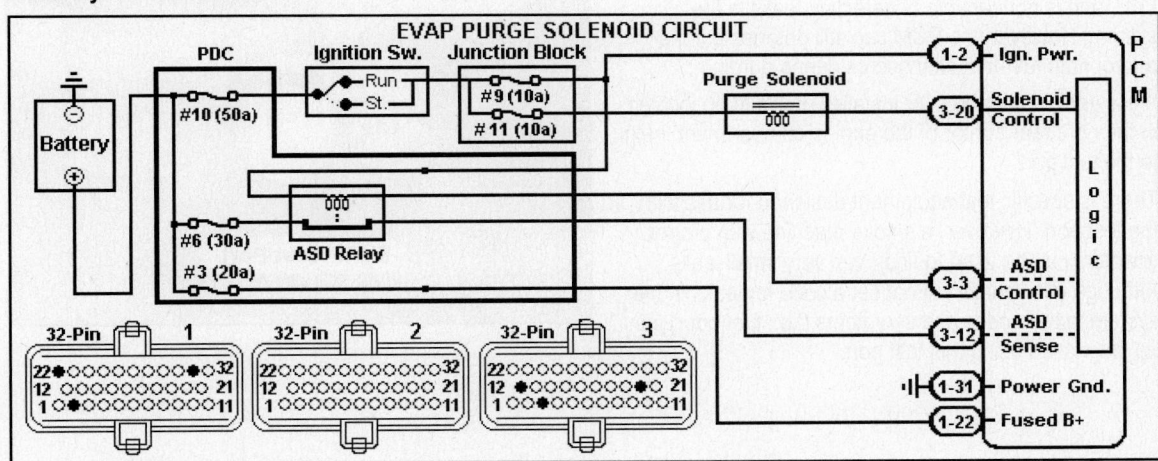

EVAPORATIVE CHARCOAL CANISTERS

The two (2) EVAP canisters are located on the outside of the right frame rail, near the rear axle. Fuel vapors from the fuel tank are routed through vapor lines to be stored in the canisters. The canisters are connected in series with each other and between the fuel tank and the purge solenoid.

Once the engine is in closed loop, fuel vapors stored in the charcoal canisters are purged into the engine where they are burned during normal combustion.

The activated charcoal in the canister has the ability to purge (release) any stored fuel vapors when it is exposed to fresh air.

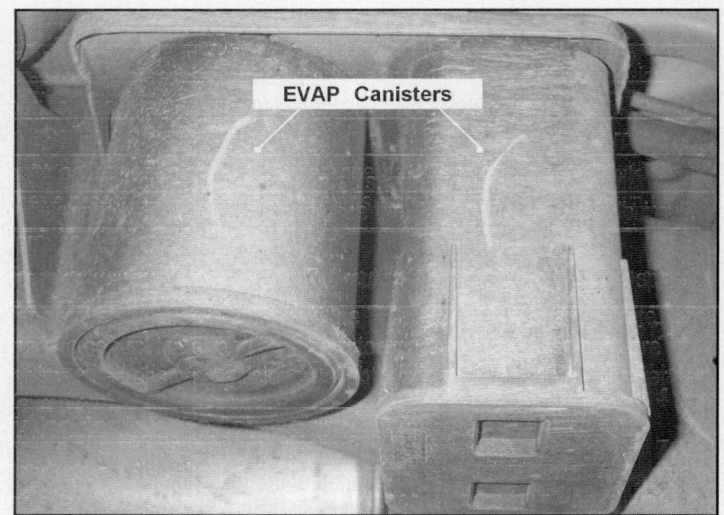

EVAP Canisters

EVAP SYSTEM TEST PORT

All OBD II vehicles must have an EVAP test port. The test port on this application is a separate component (it is not incorporated into the purge solenoid).

There are Scan Tool functions that run the EVAP system monitor and control individual components (leak detection pump and purge solenoid). However, this test is only available on applications equipped with an LDP.

The vehicle in this example was not equipped with an LPD, and is not capable of detecting leaks in the system. However, the PCM can still determine if the appropriate fuel-trim shift occurs during purging.

If this vehicle had the LDP installed, it would be located in the right rear corner of the engine compartment, next to the test port.

There is specific test equipment designed for use with the test port. However, a smoke machine with proper adaptors can be used to find even very small leaks. Although this vehicle will not set a code for leaks in the system, other codes and symptoms (i.e., fuel odor) can be diagnosed using this test port.

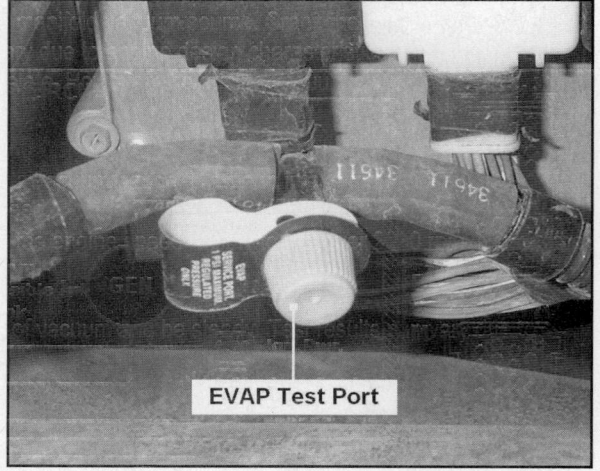

EVAP Test Port

EVAP PURGE SOLENOID

The EVAP canister purge solenoid is a normally closed (N.C.) solenoid controlled by the PCM. A single purge solenoid, located in the engine compartment on the right fender, allows vapors from the charcoal canisters to enter the manifold during various engine conditions.

The PCM controls purging by providing ground to the solenoid. By varying the percentage (%) of time the circuit is grounded and the purge solenoid is operating, the PCM can precisely control the volume of fuel vapors drawn into the intake manifold.

Purge Solenoid

DVOM TEST (PURGE SOLENOID)

Place the shift selector in Park and block the drive wheels for safety.

DVOM Connections
Connect the DVOM positive lead to the purge solenoid signal circuit (PK/WT wire) at PCM Pin 3-20 and the negative lead to the battery negative post.

DVOM Example
This capture was taken at Hot Idle. The DVOM display shows a 8.88v average purge solenoid signal. The Duty Cycle is 36.5%, but the engine vacuum is very high at idle, so this low duty cycle is still purging significant volumes of fuel vapors.

The frequency of this command is fixed at 5 Hz at idle, so it is not an important value to observe during diagnosis. The frequency switches to 10 Hz under all off-idle conditions, but the duty cycle is still variable.

This circuit was also captured during cruise conditions. The values at 2000 RPM were 4.25v and 72.3 % duty cycle at 9.989 Hz.

The higher duty cycle means that the purge solenoid is open and exposing the EVAP canisters to engine vacuum for more time during each of the 10 cycles per second. However, the engine vacuum is lower at cruise, so the volume of fuel vapors being purged is not necessarily directly related to duty cycle. The PCM determines the amount of engine vacuum from the MAP sensor, and calculates the purge duty cycle accordingly.

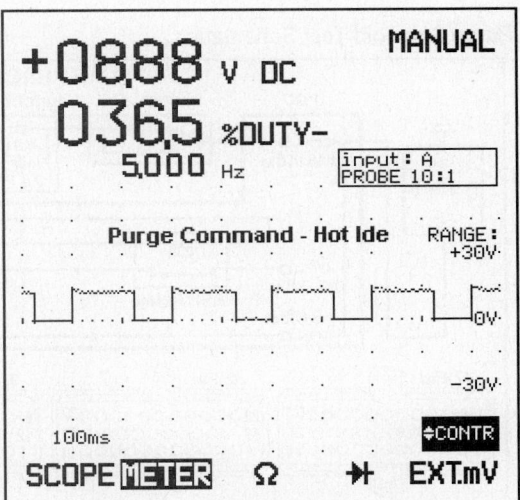

LAB SCOPE TEST (PURGE SOLENOID)

The Lab Scope is an excellent tool to view both the voltage and current waveforms for the purge solenoid and circuit. Place the shift selector in Park and block the drive wheels for safety.

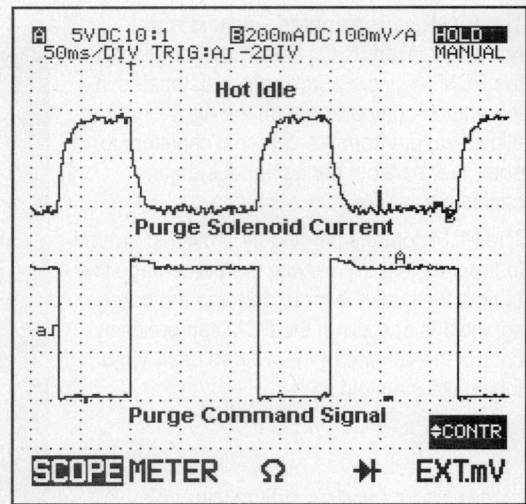

Scope Settings
To make the waveforms as clear as possible, set scope settings to match the examples.

Scope Connections
Connect the Channel 'A' positive lead to the Purge Solenoid control circuit (PK/WT wire) at PCM Pin 3-20 or at the component. Connect the negative lead to the battery negative post. Connect a low-amps probe to Channel 'B' and clamp the probe around either purge solenoid wire (PK/WH or DB/WH).

Lab Scope Example
In this example, the bottom trace shows the purge solenoid voltage command from the PCM. The top trace shows the current flow through the solenoid when the circuit is grounded.

How To Use This Test
Using these traces, the circuit voltage, circuit ground, and PCM control of the solenoid can all be verified. Observe both traces over time to look for the cause of intermittent circuit problems. Resistance problems in the solenoid cannot be seen easily in the voltage waveform, but will show up in the current waveform.

In this example, the peak current was almost 400 ma. If the current flow is higher all the time, look for a partial short to ground in the circuit. If the current flow is higher only when the solenoid is commanded on, check for a shorted solenoid.

Purge Solenoid Test Schematic

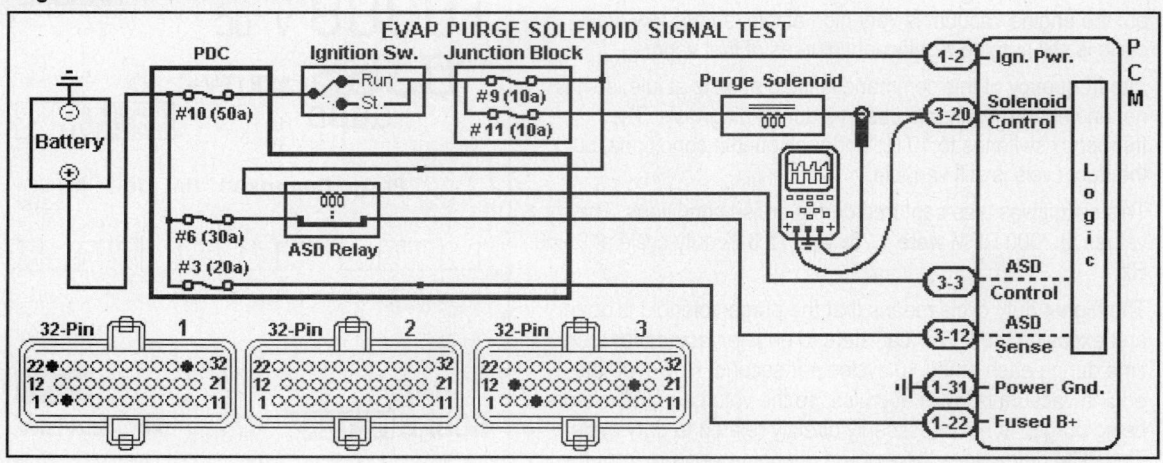

Vehicle Speed Control

INTRODUCTION

The Vehicle Speed Control (S/C) system on this application is controlled by the PCM. The PCM controls vacuum solenoids in the S/C servo to regulate servo movement. The servo acts through a cable to control throttle angle, and therefore vehicle speed.

Circuit Operation

The only 2 inputs required for the PCM to properly operate the system are the vehicle speed sensor signal and the steering wheel mounted S/C switches.

All of the steering wheel mounted switches are combined into a single input to the PCM. The PCM provides a 5v signal to the switch assembly. Each of the switches provides a path to ground with a different resistance. The PCM monitors the voltage drop in the circuit and can determine which of the switches was depressed.

The brake switch is not an actual input. When the brake pedal is applied, the power supply to the speed control servo is interrupted. The speed control servo contains three (3) solenoids powered by the PCM: the vent, vacuum and dump solenoids.

The PCM can only ground the vent and the vacuum solenoids to regulate vehicle speed. The dump solenoid is permanently grounded, and is therefore energized with the brake pedal up. When the brake pedal is depressed, the dump solenoid is de-energized, and the vacuum in the speed control servo is rapidly bled off to atmosphere.

Trouble Code Help

There are many trouble codes that can set due to various S/C component or circuit failures. In all cases, it is important to carefully read the code descriptions to determine the most efficient diagnostic path. For example, the definition for DTC P1597 is Speed Control Switch Always Low. The code description describes the circuit voltage as having been below 4.5v for more than 2 seconds.

Knowing these criteria, it is easy to measure the actual circuit voltage to see if the fault is present. However, it is also reasonable that this code could set due to operator error. If the Decel button is depressed for 2 seconds, this would not set a DTC (as this is an expected switch state during normal operation). The PCM expects certain voltage drops from the switch resistors, but if 2 or more buttons are held down for more than 2 seconds, an unexpected voltage drop would occur in the circuit, and this DTC would set.

Speed Control System Graphic

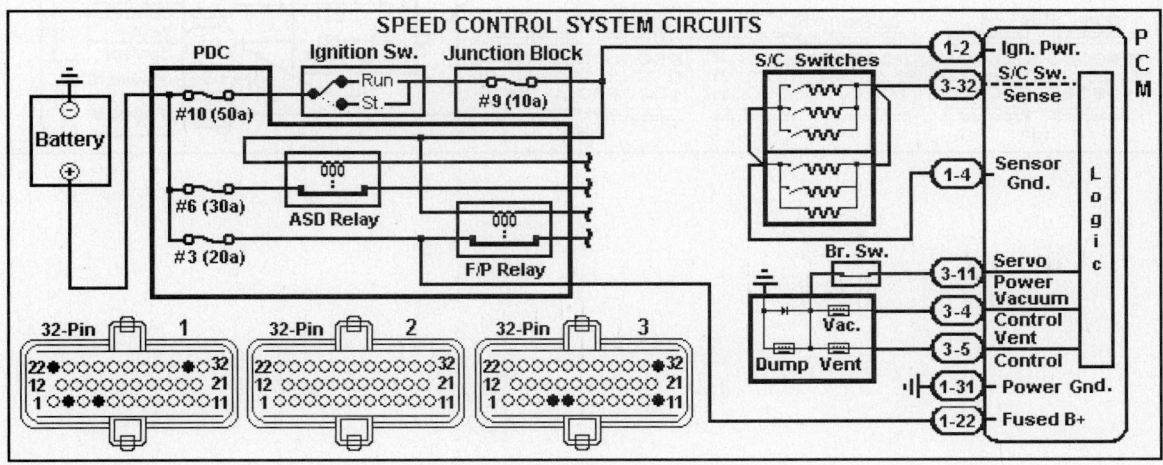

BI-DIRECTIONAL TEST (S/C SOLENOIDS)

The Lab Scope and Scan Tool can be used to diagnose suspected problems with the steering control mounted speed control switches, speed control servo, or circuits.

In this example, the Scan Tool was used to actuate the Vent Solenoid in the servo, and the results measured with the Lab Scope.

Place the gear selector in Park and block the drive wheels for safety.

Lab Scope Connections
Connect the Channel 'A' positive probe to the vent solenoid control circuit (LG/RD wire) at PCM Pin 3-5 and the negative probe to the battery negative post.

Lab Scope Example (1)
In example (1), the trace shows the vent solenoid command signal during the ATM Test 15: ALL S/C SOLENOIDS. The vent and vacuum solenoids were cycled, though only the vent signal is displayed in this example. The result of this test is that the alternating solenoid commands mechanically exercise the servo, which opens and closes the throttle plate.

The test cycles for 5 minutes, but is conducted with the engine off, so the vacuum reservoir is depleted after a few test cycles.

Lab Scope Example (2)
In example (2), the Scan Tool was used to perform ATM Test 47: S/C VENT. This test performs a similar function as in example (1), but only involves one solenoid. The vent solenoid will be cycled on for 1.5 seconds and off for 1.5 seconds for 5 minutes.

Speed Control System Test Graphic

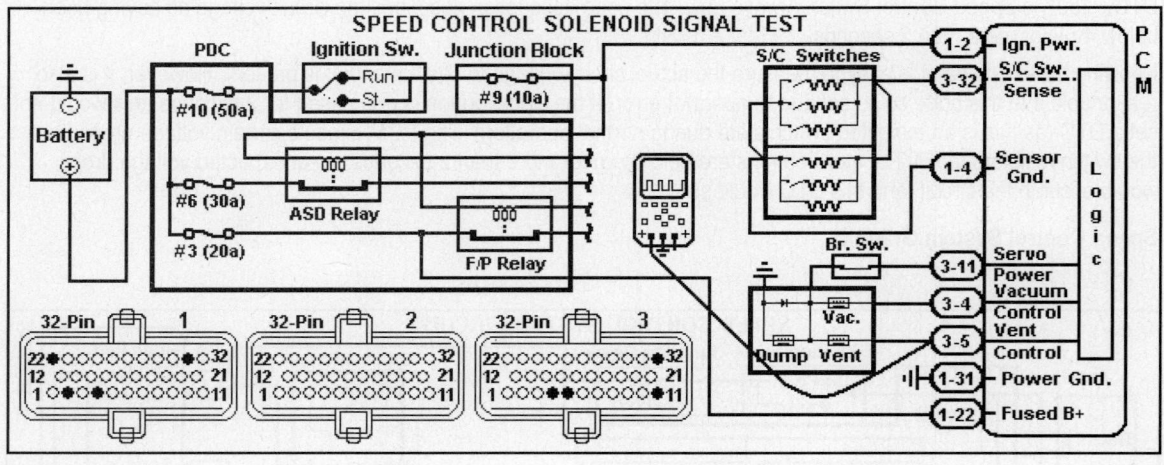

Fuel Delivery System

INTRODUCTION

The Fuel system delivers fuel at a controlled pressure and correct volume for efficient combustion. The PCM controls the Fuel system and provides the correct injector timing.

Fuel Delivery

This vehicle is equipped with a sequential fuel injection (SFI) system that delivers each cylinder the correct A/F mixture at the optimum time under all operating conditions.

Air Induction

Air enters the system through the fresh air duct and flows through the air cleaner, mounted on top of the throttle body. The unmetered air passes through another air duct and enters the throttle body. The incoming air passes through the throttle body and into the intake plenum to the intake manifold where it is mixed with fuel for combustion. The volume of intake air is not directly measured, but is inferred from manifold vacuum using a MAP sensor mounted on the front edge of the throttle body (Speed Density system).

System Overview

A high-pressure (in-tank mounted) fuel pump delivers fuel to the fuel injection supply manifold. The fuel injection supply manifold incorporates electrically actuated fuel injectors mounted directly above each of the intake ports of the engine.

Power is supplied to the fuel injectors by the ASD relay, and to the fuel pump by the fuel pump relay. Drivers in the PCM ground the injector circuits and the fuel pump relay control circuit to operate these devices.

This application uses a mechanical returnless fuel delivery system. There is no fuel return line from the fuel supply manifold. A fuel filter/pressure regulator assembly is mounted to the top of the in-tank fuel pump module. The regulator maintains a constant amount of fuel pressure in the system. The pressure regulator diverts extra fuel volume back through the fuel pump to the tank.

The PCM determines the amount of fuel the injectors spray under all engine-operating conditions. The PCM receives electrical signals from engine control sensors that monitor various factors such as manifold pressure, intake air temperature, engine coolant temperature, throttle position, and vehicle speed. The PCM evaluates the sensor information it receives and then signals the fuel injectors in order to control the fuel injector pulsewidth.

Fuel Delivery System Graphic

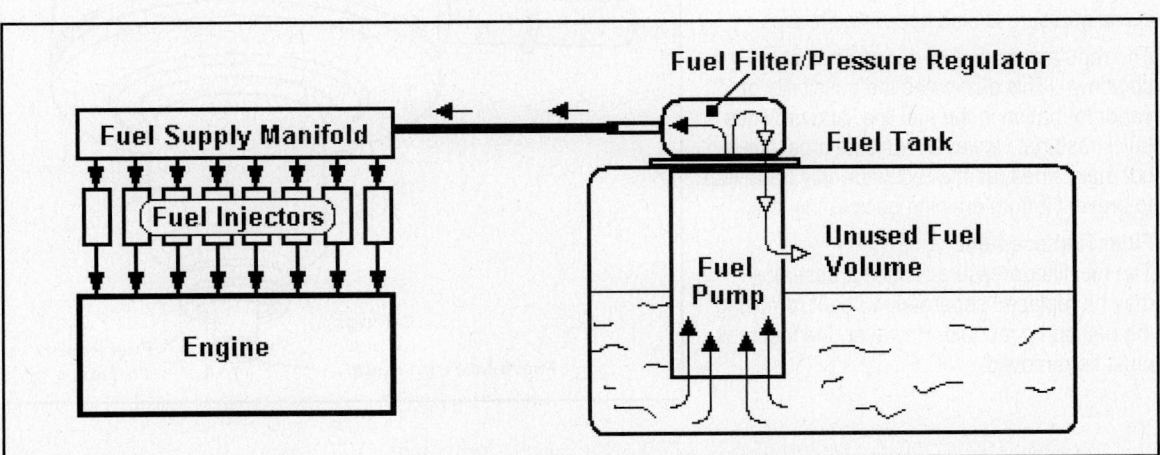

FUEL PUMP MODULE

The fuel pump module in this returnless fuel system is mounted inside the fuel tank.

The fuel pump produces more pressure and volume than necessary for proper system operation. When the pump pressure exceeds 49.2 psi (+/- 5 psi), excess fuel is bled back into the tank through the fuel filter and pressure regulator assembly mounted to the top of the fuel pump module. However, the pump has a maximum pressure output of 95 psi.

Fuel Pump Module Components
The fuel pump module includes an electric pump motor, inlet filter, fuel filter and pressure regulator assembly, fuel gauge sending unit, and a fuel line connection. Only the fuel filter and pressure regulator assembly may be serviced separately.

Internal Valve
The pump module has one internal check valve, located in the pump outlet, which prevents the system from draining back to the fuel tank with the ignition off. This valve also keeps the pump primed for quick pressurization of the system at startup, although residual fuel pressure of 0 psi on a cold engine is considered normal.

FUEL FILTER & PRESSURE REGULATOR

The filter and regulator on this application are mounted to the top of the fuel pump module. The fuel filter is not normally replaced as a maintenance item. Replace this assembly when directed to do so by a diagnostic procedure.

The regulator is calibrated to open a return port above 49.2 psi (+/- 5 psi). Excess fuel pump pressure is bled back into the tank.

The regulator traps fuel during engine shutdown. This eliminates the possibility of vapor formation in the fuel line, and provides quick restarts. However, residual pressure is not maintained, as the regulator only traps fuel to prevent it from draining back to the tank.

Filter Replacement
The fuel filter/pressure regulator assembly may be replaced separately without removing the fuel pump module. However, the fuel tank must be removed.

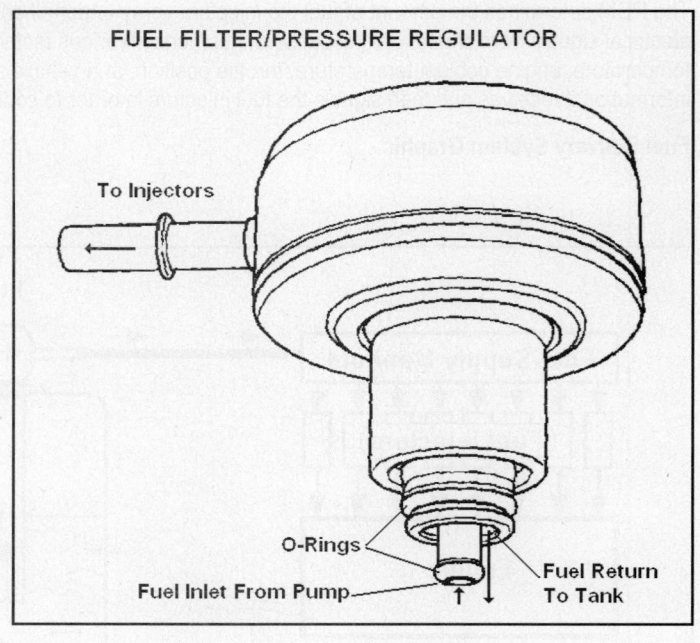

LAB SCOPE TEST (FUEL PUMP)

Testing fuel pump current draw with a Lab Scope is a very efficient method of determining fuel pump condition, since the pump can be tested without removing the fuel tank.

Scope Connections (Amp Probe)

Set the amp probe to 100 mv/amp, and zero the amp probe. Connect the probe around the fuel pump feed wire and connect the negative probe to chassis ground.

Scope Connection Tip

Accessing the fuel pump wiring harness for this test can be time consuming. In addition, a single wire must be isolated and completely surrounded by the bulky low amp probe tip.

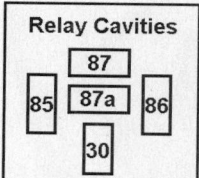

Remove the fuel pump relay from the PDC and install a fused jumper wire between relay cavities 30 and 87. The low amps probe can then be clipped around the jumper wire.

Lab Scope Examples

In example (1), the trace shows the fuel pump current (8 amps peak) at key on, engine off.

In example (2), the current waveform was captured when the key was turned ON. Note that the current peaked at 18 amps.

The fuel pump current specification for this application is under 10 amps during normal operating conditions.

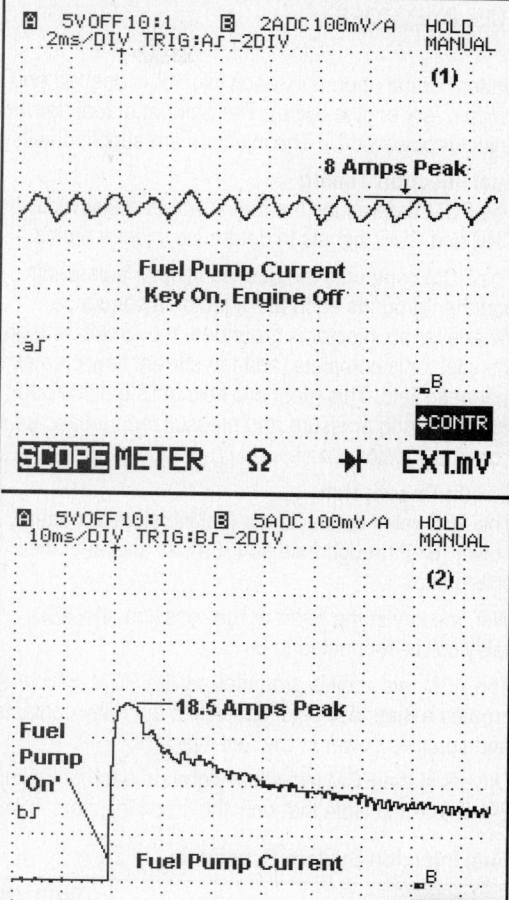

Fuel Pump Test Schematic

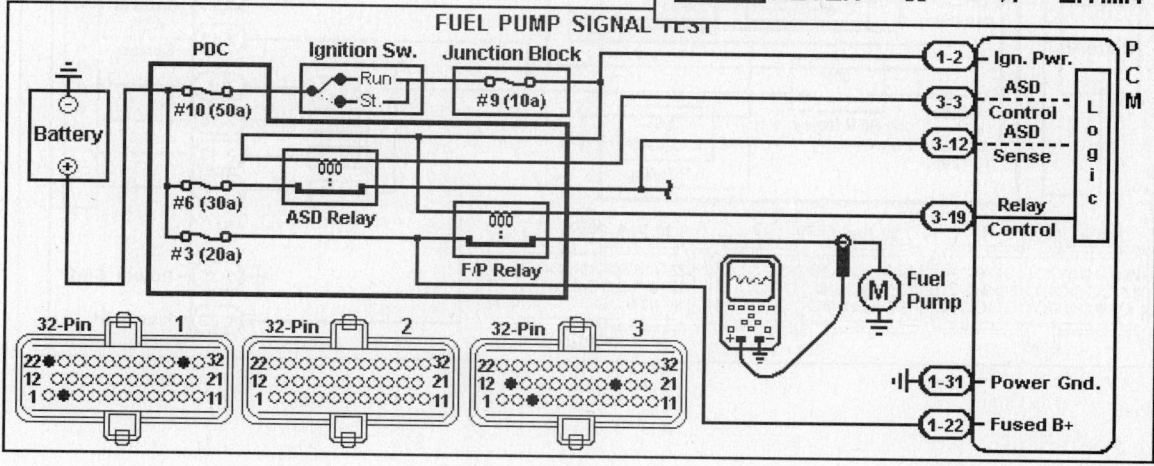

FUEL INJECTORS

Under normal operation, each injector is opened and closed at a specific time, once every other crankshaft revolution (once every engine cycle). The amount of fuel delivered is controlled by the length of time the injector is grounded (injector pulsewidth). The injectors are supplied voltage through the ASD relay.

Fuel Injection Timing

The PCM determines the fuel flow rate (pulsewidth) from the ECT, HO2S-11, and MAP sensor inputs, but needs the CKP and CMP signals to determine injector timing.

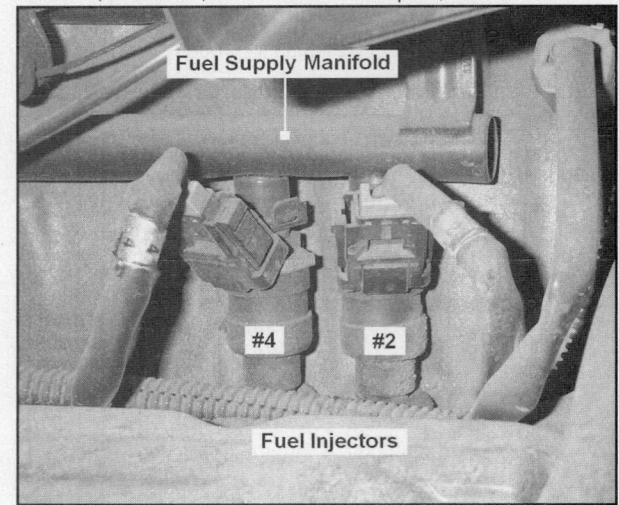

Fuel Supply Manifold

#4 #2

Fuel Injectors

The PCM computes the desired injector pulsewidth and then grounds each fuel injector in sequence. Whenever an injector is grounded, the circuit through the injector is complete, and the current flow creates a magnetic field. This magnetic field pulls a pintle back against spring pressure and pressurized fuel escapes from the injector nozzle.

Circuit Description

The ASD relay connects the fuel injectors to battery power (B+) through fuse #6 (30a), as shown in the schematic.

With the key in the "start or run" position, the ASD relay coil is connected to B+.

The ASD relay coil is grounded by the PCM, which creates a magnetic field that closes the relay contacts and supplies power to the fuel injectors.

Drivers in the PCM individually ground each of the injector circuits. The injector firing sequence on this application is: 1-8-4-3-6-5-7-2. Note that only the first 4 injectors in sequence are shown in the Schematic below.

Fuel Injection System Schematic

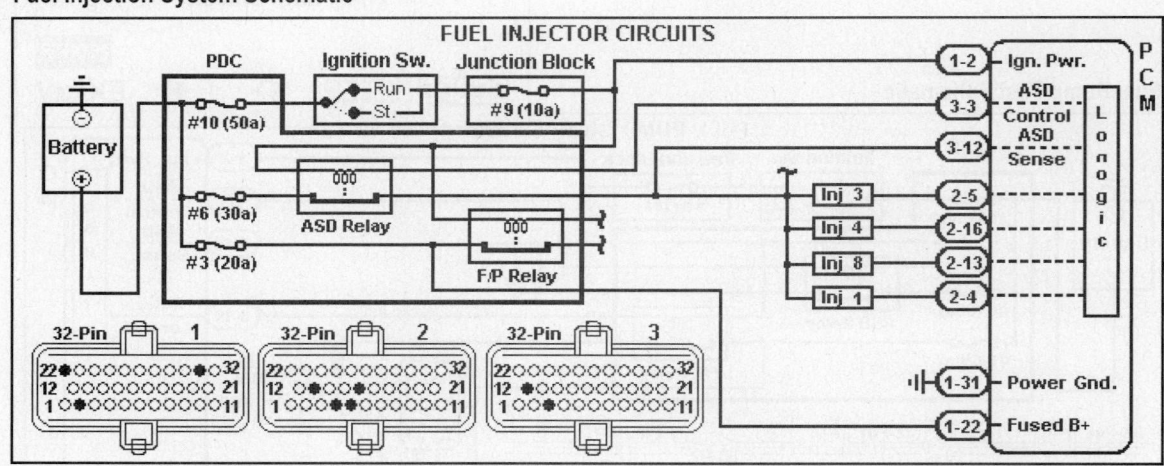

LAB SCOPE TEST (FUEL INJECTOR)

Place the gear selector in Park. Block the drive wheels for safety.

Scope Connections

Connect the Channel 'A' positive probe to an injector control wire at an injector or appropriate PCM Pin. Connect the negative probe to the battery negative post. Connect a low-amps probe to Channel 'B', and clamp the probe around the same wire, or around the injector power wire.

Lab Scope Example (1)

In example (1), the trace shows the injector voltage signal at Hot Idle.

In this example, the on time is 3.7 ms. By using these settings (1 ms/div. and 10v/div.), system voltage, on time, inductive kick (clipped), and injector pintle closure can all be clearly seen.

Diagnostic Help

Use these settings to verify that the inductive kick reaches 60v. The kick should be much higher, but is 'clipped' by a diode in the PCM. If the kick is less than 60v, the current flow through the circuit is too low. In this case look for extra resistance in the injector or circuit.

If the inductive kick is less than 51 volts, the PCM will set DTC P020X (X = the injector #).

Lab Scope Example (2)

In example (2), a low-amps probe was added to view the current in the circuit. The injectors on this application measure approximately 12 Ohms, so it can be expected that the maximum current flow through the circuit is about 1 Amp.

Diagnosing Injector Circuits

The low-amps probe and the Lab Scope are more effective than an ohmmeter in diagnosing circuit or component resistance problems. Resistance changes with temperature. Also, some circuits may measure within specifications when off, but go open when the circuit is loaded. This is similar to what occurs with a loose battery connection. An ohmmeter would show zero (0) Ohms of resistance, but when the vehicle starter is engaged, the circuit opens because the connection cannot flow the current required to operate the starter. Any injector circuit resistance problems, or problems that show up only when the circuit is stressed, can be found using this test.

The injector opening can also be seen in a current waveform, and injector closing can be seen in the voltage waveform. Also verify that the injector closes by observing the 'hump' in the falling edge of the voltage waveform. This hump is the result of the pintle being forced back to the seated position after the injector is turned off.

CAMSHAFT POSITION SENSOR (HALL EFFECT)

The camshaft position (CMP) sensor used with this EI system is a Hall effect sensor that sends a 0-5v pulse to the PCM corresponding to the position of a 180-degree pulse ring inside the distributor. The pulse ring rotates with the distributor shaft, and is sensed by the CMP sensor, mounted to a fixed plate in the distributor housing.

The CMP sensor provides a reference point for the PCM to calculate cylinder order and position. The PCM uses CMP data to determine which cylinder will be the next to reach TDC. The CKP sensor pulses are then counted to calculate actual cylinder position.

It is important to remember that the CMP input is only used to calculate which fuel injector to fire. The actual ignition and injector timing is determined by the CKP sensor input.

Cylinder Identification

The CMP sensor signal does not indicate specific engine position for each cylinder. The distributor rotates at one-half crankshaft speed, and the pulse ring covers one-half of the distributor rotation, so the pulse ring is in each position for one full crankshaft revolution. The signal is high (5 volts) for one full crankshaft revolution, and then low (0 volts) for one full crankshaft. These high and low signals are used as reference points for the PCM to start counting CKP pulses to calculate actual engine position.

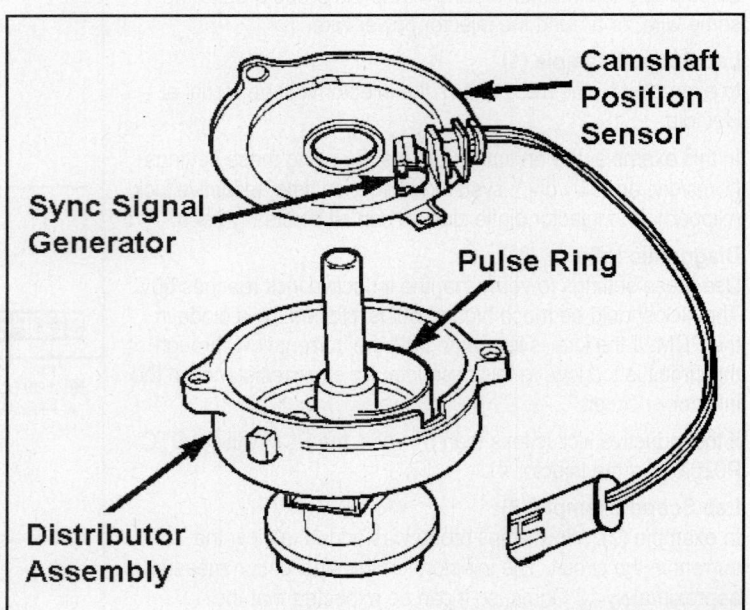

CMP Sensor Schematic

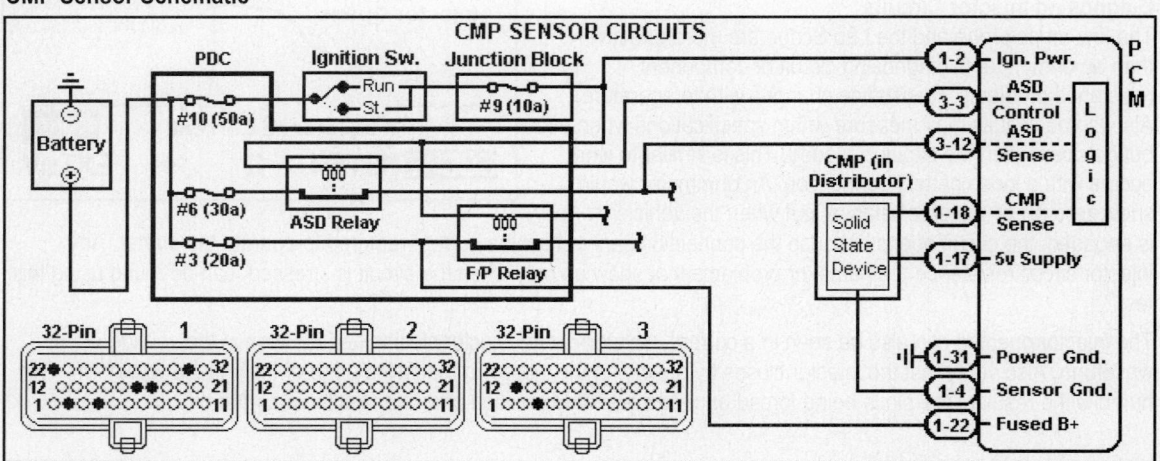

DVOM TEST (CMP SENSOR)

Place the shift selector in Park and block the drive wheels for safety.

DVOM Connections

Connect the DVOM positive lead to the CMP sensor signal circuit (TN/YL wire) at PCM Pin 1-18 and the negative lead to the battery negative post.

DVOM Example

This capture was taken at Hot Idle. The DVOM display shows a CMP signal of 2.62v DC because the instrument is averaging the signal voltage. This average value is about one-half of the 5-volt supply to the sensor because the duty cycle is very close to 50%.

Use the DVOM MIN/MAX feature, if available, to verify that the CMP signal is really switching from 0-5v.

How to Use This Test

Another way to use the DVOM to test the CMP sensor signal is to observe the frequency of the signal. In this example, the display shows 5.089 Hz. (5.089 cycles per second X 60 seconds / 0.5 notches per revolution = 610.7 RPM). An intermittent problem in the signal may be hard to see, but will usually cause a fluctuation in the average voltage or frequency that can be seen with the DVOM.

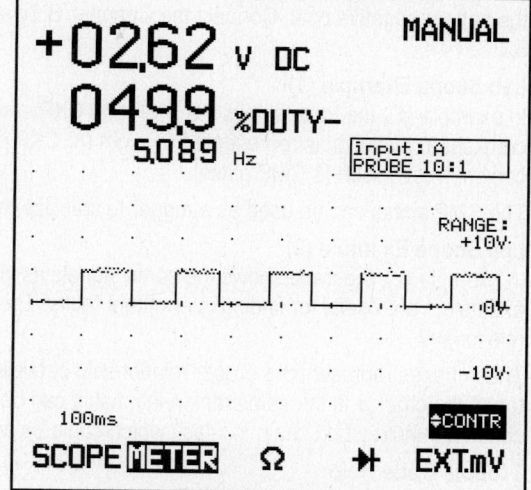

The CMP sensor signal is required for the PCM to correctly calculate sequential fuel injection. However, the PCM will default to a bank-fire injection strategy if this signal is missing. Therefore, the CMP sensor cannot cause a no-start. Do not test the CMP sensor or circuit for a no start condition.

CMP Sensor Test Schematic

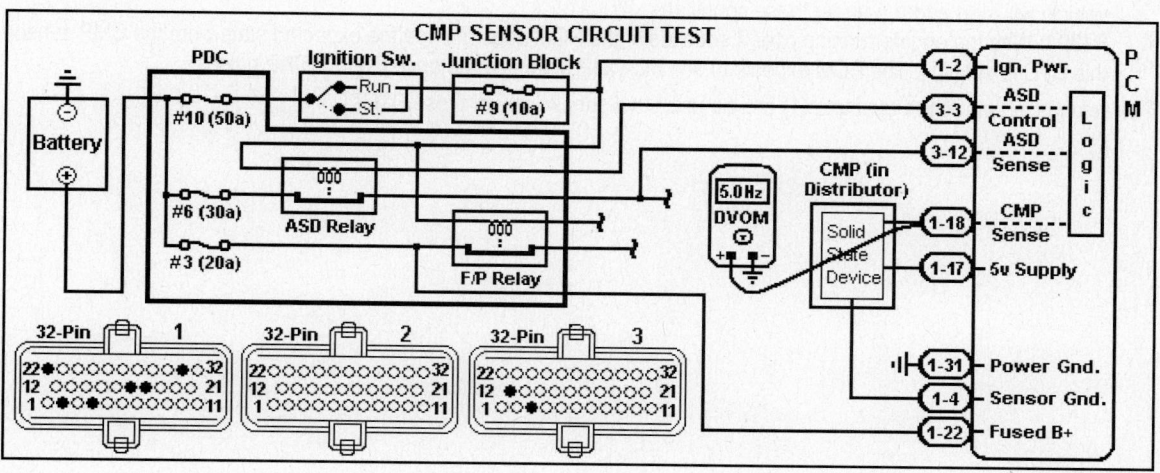

LAB SCOPE TEST (CMP & CKP SENSORS)

Place the gear selector in Park. Block the drive wheels for safety.

Scope Connections

Connect Channel 'A' positive probe to the CMP sensor circuit (TN/YL wire) at PCM Pin 1-18, and the negative probe to the battery negative post. Connect the Channel 'B' positive probe to the CKP sensor circuit (GY/BK wire) at PCM Pin 1-8.

Lab Scope Example (1)

In example (1), the trace shows the CMP and CKP sensor signals. Note the relationship between the traces. One crankshaft revolution is represented by eight (8) CKP pulses. Every 2 revolutions (16 pulses) represent one full camshaft revolution (1 CMP pulse).

The CMP signal can be used as a trigger to stabilize the CKP trace when looking for intermittent faults.

Lab Scope Example (2)

In example (2), the trace shows the same signals as in example (1), but using different time settings. The settings in example 1 are useful for finding intermittent faults. The settings in example (2) are useful for determining the cam/crank relationship.

These traces represent the proper relationship between the signals. If the timing chain is stretched or has jumped a tooth, the change in the cam/crank relationship can be clearly seen with these settings. Chain guide problems can cause an unstretched chain to 'slap', which could be seen as an erratic cam/crank relationship on the Lab Scope.

Trouble Code Help

The PCM does not monitor the CMP and CKP circuits for opens and shorts. The PCM watches both signals to determine sensor the state (high or low) expectations. There are three (3) codes that can be set as a result of problems with these sensors or circuits. Pay close attention to code definitions, as they help to determine the most efficient diagnostic path:

- P0320: If the vehicle is cranking and CMP signals are present, but no CKP signals are detected, this DTC is set (1T). Also, the vehicle will not start.
- P0340: If the PCM sees 32 CKP pulses, and then sees no CMP pulses for 3 seconds, this DTC is set (1T). The vehicle will start and run under these conditions.
- P1391: With the engine running over 7 seconds, if the PCM does not see the expected state from the CMP sensor this DTC is set (1T). The PCM expects to see the CMP state change once every 8 CKP pulses.

The settings in Lab Scope example (1) can be used to diagnose any of these DTCs.

Idle Air Control Motor

INTRODUCTION

The Idle Air Control (IAC) motor, mounted on the throttle body, is actually a dual stepper motor assembly. The IAC may

also be referred to as the Automatic Idle Speed (AIS) motor. This device is designed to control engine idle speed (rpm) under all engine-operating conditions and to provide a dashpot function (to control the rate of deceleration for emissions and idle quality).

The IAC motor meters the intake air that flows past the throttle plate by varying the position of a pintle in a bypass passage on the throttle body.

Circuit Operation

The PCM determines the desired idle speed (amount of bypass air) and controls the IAC motor circuits through pulse train commands. A pulse train is a command signal from the PCM that continuously varies in both frequency and pulsewidth. There are 4 driver circuits connecting the IAC motor and the PCM. Circuits 1 & 4 control one 'motor' and circuits 2 & 3 control the other. These 2 motors pull the pintle in opposite directions, and the rapid cycling of these drivers balances these forces. By slightly varying the strength of each field, the PCM can precisely control the pintle position.

The PCM can track the number of 'steps' the motor has been moved in either direction. This value is stored as a reference for startup idle control. If the PCM memory is cleared, the PCM will drive the IAC completely closed the first time the key is turned on. This sets the counter to '0' steps and allows the PCM to recalculate a reference point.

To keep the IAC motor in position when no idle correction is needed, the PCM will energize both windings at the same time. The opposing forces lock the pintle in place.

Idle Air Control Schematic

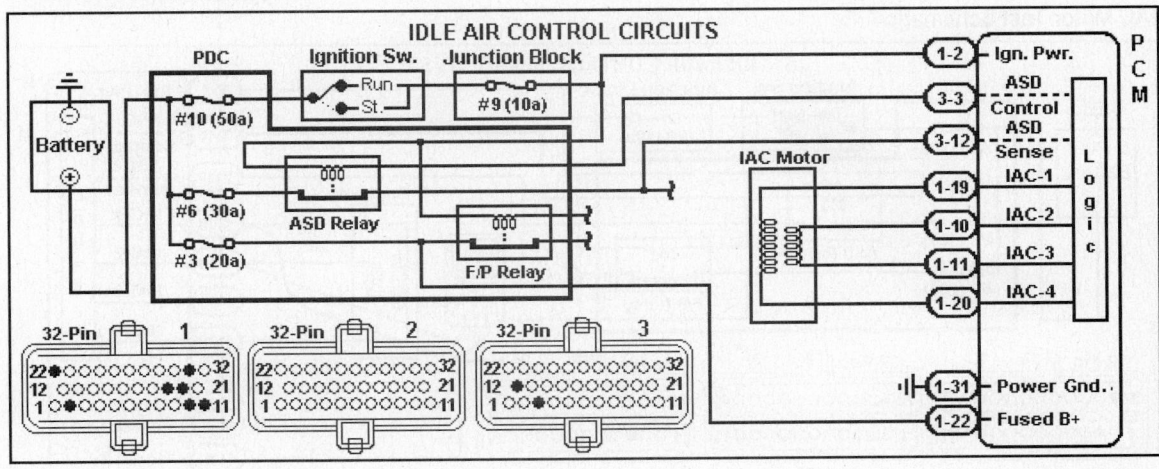

BI-DIRECTIONAL TEST (IAC MOTOR)

The Lab Scope is an excellent tool for testing the IAC motor circuit activity. Place the gear selector in Park. Block the drive wheels for safety.

Using the Scan Tool
Navigate through the Scan Tool menus to the OBD CONTROLS menu. Select ACTUATOR TEST, and then 10: AIS MOTOR. (AIS is another name for IAC). When the UP arrow is pressed, the IAC motor is cycled back and forth in its bore.

Observe the results with a Lab Scope, or the IAC POSITION and DESIRED IAC PIDs. IAC POSITION should stay within a few steps of DESIRED IAC. In this application, the value for each PID was 13 steps at idle, and 40 steps at cruise.

Lab Scope Connections
All 4 drivers can be checked using 2 channels. The PCM controls one motor by alternating the polarity of drivers 1 & 4, and the other motor by alternating the polarity of drivers 2 & 3.

Therefore, the Lab Scope positive probes should be connected to circuit 1 <u>or</u> 4, <u>and</u> circuit 2 <u>or</u> 3. A circuit problem in either pair will show up in this test.

Lab Scope Example
In this example, the traces show the two IAC motor windings cycling the pintle back and forth in it's bore in response to ATM test 10: AIS MOTOR. In this test, the PCM will cycle the commands to the IAC for 5 minutes.

Test Results
This test proves that the PCM is capable of electrically controlling all four (4) IAC motor drivers (i.e., the IAC motor, circuits, and PCM drivers are okay). Further diagnosis can be limited to PCM inputs, processing, or engine mechanical concerns (vacuum leaks).

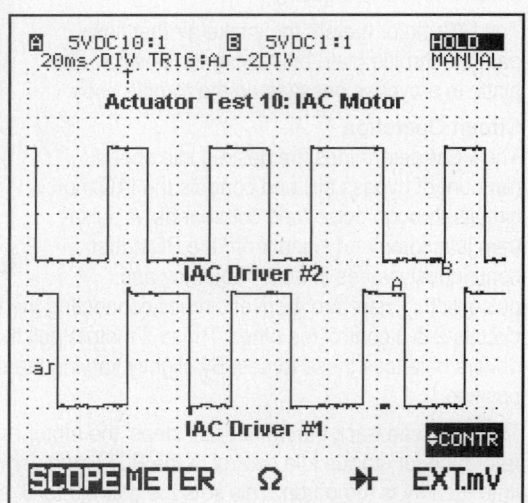

This test is limited in that it is performed with the engine off. To verify that the IAC motor is mechanically capable of controlling idle, use the Scan Tool RPM CONTROL test.

IAC Motor Test Schematic

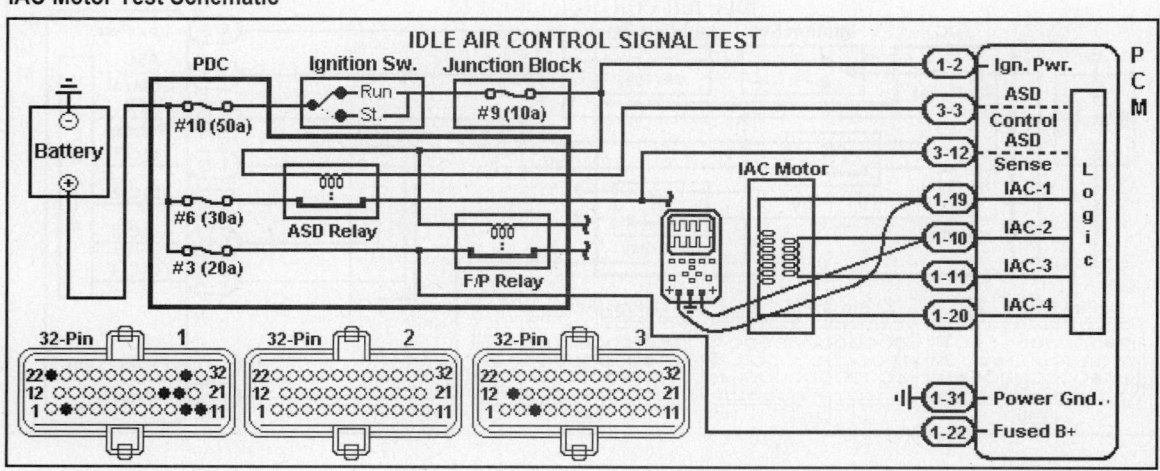

Information Sensors

MANIFOLD ABSOLUTE PRESSURE SENSOR

The Manifold Absolute Pressure (MAP) sensor is located on the front edge of the throttle body. The MAP sensor input to the PCM is an analog voltage signal proportional to the intake manifold pressure. The MAP sensor senses manifold pressure through a short 90° vacuum hose connected the throttle body just below the throttle plate.

This signal is compared to the TP sensor signal and other inputs to calculate engine conditions. The PCM uses theses calculations to determine fuel injector pulsewidth, ignition timing, and transmission shift points

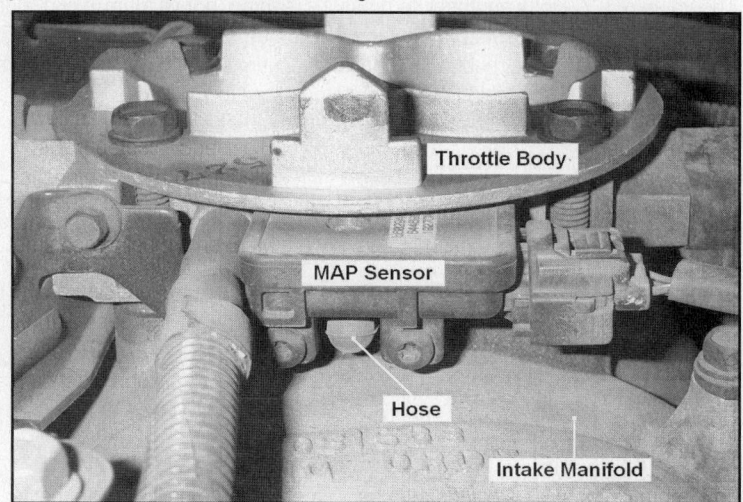

MAP Sensor Circuits

The three circuits that connect the MAP sensor to the PCM are the 5v supply circuit, sensor signal circuit, and sensor ground.

Circuit Operation

The MAP sensor voltage increases in proportion to the engine load. When the throttle angle is high as compared to engine RPM, the load on the engine is higher. Intake vacuum drops (pressure rises) in proportion to increased engine loads. By measuring this vacuum (pressure) the PCM has a very precise understanding of engine conditions.

Diagnostic Help

Any of the three (3) circuits connecting the MAP to the PCM can be the source of a fault resulting in a DTC or drivability problem. However, the 5-volt supply and sensor ground circuits are shared with other sensors. If the MAP sense circuit and MAP sensor are fine, check the operation of the other devices and compare the results to a wiring diagram to narrow down the possibilities. For example, the IAT, CMP, MAP, TP, CKP, Oil Pressure, Transmission Solenoid Assembly, Oxygen, and ECT sensors all share a common sensor ground circuit at the PCM.

MAP Sensor Schematic

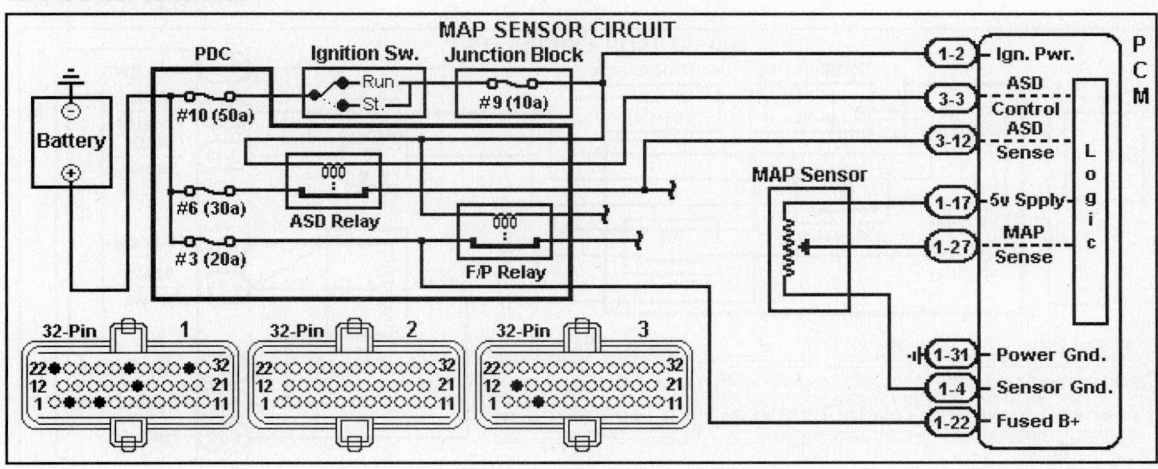

DVOM & SCAN TOOL TEST (MAP)

The DVOM is an excellent tool to test the condition of the MAP sensor circuits. The DVOM can be used to test the supply voltage, sensor ground voltage drop, and the MAP sensor signal for calibration or dropout. Place the gear selector in Park. Block the drive wheels for safety.

The test results can be compared with the specifications in Map Sensor Table (refer to Section 1).

The unit of measurement used for the MAP PID may cause some confusion. The value of 9.7 "Hg is an absolute pressure reading, not vacuum. This value is equal to the vacuum reading (on a vacuum gauge), subtracted from the BARO PID value. For example, subtracting an idle vacuum reading of 20.0 "Hg from a BARO PID of 29.5 "Hg yields the absolute pressure of 9.5 "hg.

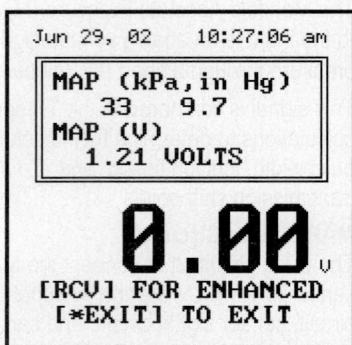

DVOM Test

On this vehicle, the MAP sensor signal at Hot Idle was 1.22v. The MAP sensor voltage increases in proportion to manifold pressure. In this example, the KOEO MAP sensor voltage was 4.68v at sea level, and was almost the same at snap throttle. The KOEO value is used by the PCM to calculate barometric pressure.

The voltage supply should be 4.9-5.1v, and the ground voltage drop should not exceed 50 mv.

In this example, the PID values were 1.21v and 9.7 in. Hg. Compare the DVOM readings and the Scan Tool PIDs to actual engine vacuum measurements.

If the Map sensor values are correct in this test, observe the snap-throttle response rate (in Park) using a Lab Scope to further test the MAP sensor performance.

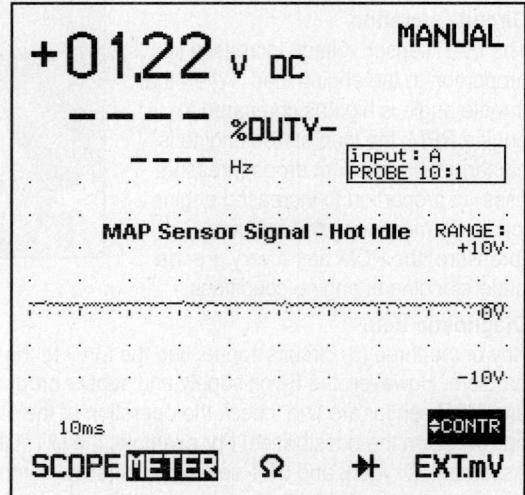

MAP Sensor Schematic

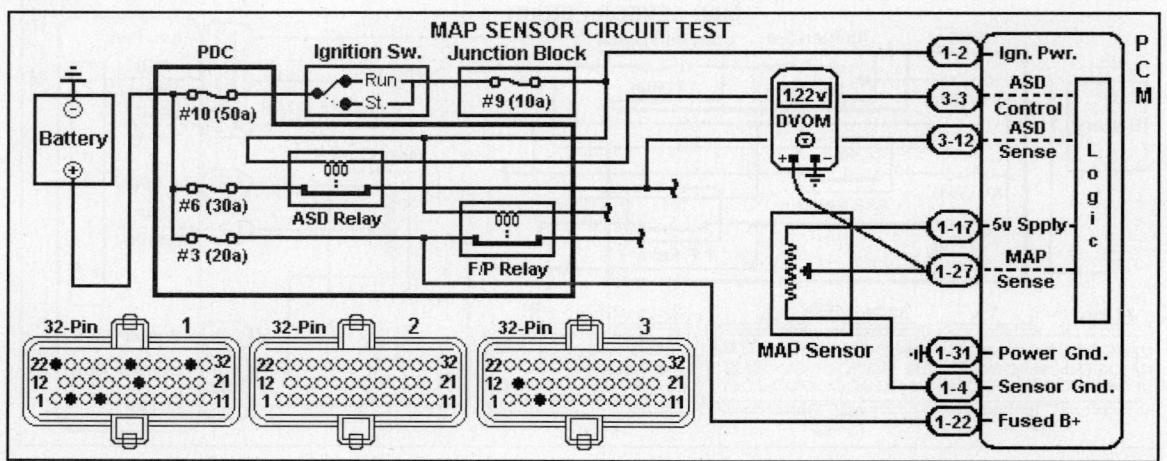

THROTTLE POSITION SENSOR

The Throttle Position (TP) sensor is mounted to the throttle body where it detects changes in the throttle valve angle.

The PCM uses the TP sensor analog voltage DC signal to detect the following vehicle driving conditions:

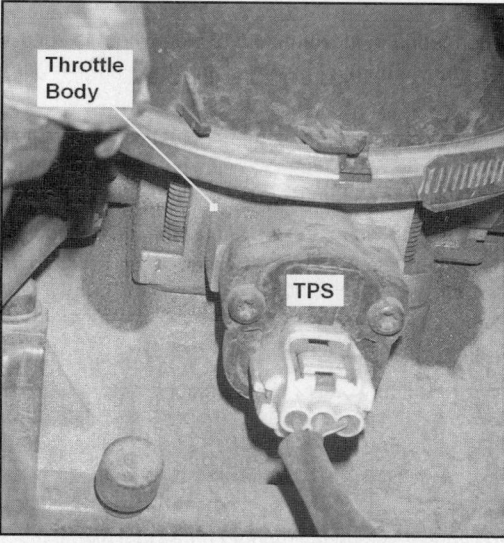

- Air Fuel Ratio Correction
- Fuel Cut Control
- Power Increase Correction

TP Sensor Circuits

The three circuits that connect the TP sensor to the PCM are listed below:

- The 5v supply circuit (shared)
- The TP sensor signal circuit
- The TP sensor ground circuit (shared)

Circuit Operation

With the throttle fully closed, the TP sensor signal to the PCM is from 0.6-1.0v at PCM Pin 1-23.

The TP sensor voltage increases in proportion to the throttle valve-opening angle. The voltage is used by the PCM to calculate throttle angle from 0% - 100%.

Trouble Code Help

Note the dashed line in the PCM for the TP sense circuit (Pin 1-23). In CHILTON diagrams, the dashed line indicates the circuit(s) the PCM uses to determine if a fault is present. The shared 5-volt supply circuit is also monitored, and will set a DTC P1496 if the circuit voltage falls below 3.5 volts.

Starting in 1999 model year trucks, the 5v supply circuit to the TP sensor is specifically monitored, and low-voltage faults will set trouble code P1295 (1-trip code).

Throttle Position Sensor Schematic

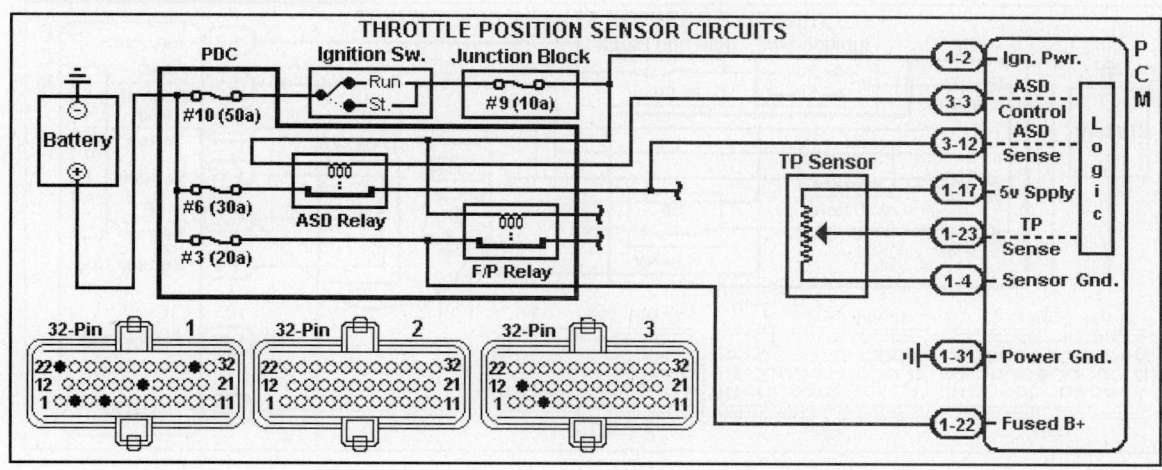

DVOM & SCAN TOOL TEST (TP)

The DVOM is an excellent tool to test the condition of the Throttle Position (TP) sensor circuits. The DVOM can be used to test the supply voltage, sensor ground voltage drop, and the TP sensor signal for calibration or dropout. Place the gear selector in Park. Block the drive wheels for safety.

DVOM Test

On this vehicle, the TP sensor signal at Hot Idle was 0.85v. The TP sensor voltage increases in proportion to throttle plate opening. To perform a sensor sweep test, slowly open and close the throttle with the engine off and watch for a smooth change in the voltage value.

The voltage supply should be 4.9-5.1v (the 5-volt supply actually measured 5.17v on this vehicle). The ground voltage drop should not exceed 50 mv.

Scan Tool Test

It is always a good idea to compare the DVOM readings to the Scan Tool PIDs. In this example, the PID values were 0.76v and 0%.

The Scan Tool values are rarely exactly the same as the DVOM values. Look for consistency, and verify that the values are in the same range. When performing the sensor sweep test under DVOM test, verify that the PID values match the DVOM. The comparison will help to determine if a signal fault is in the circuit or in the PCM.

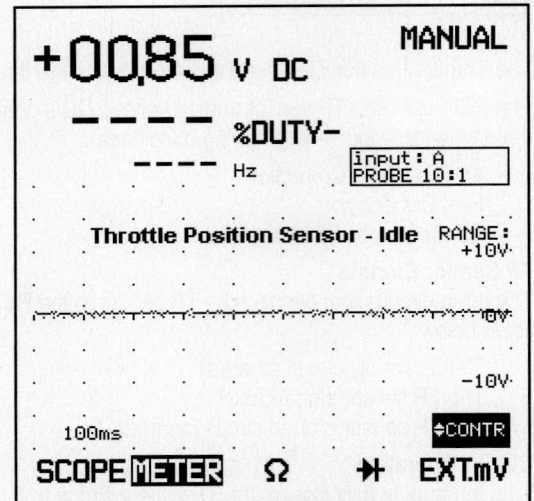

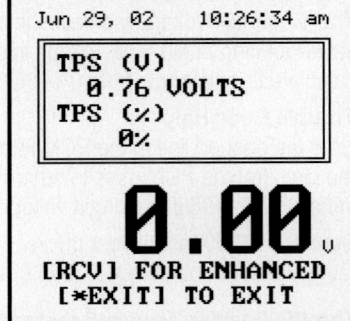

Throttle Position Sensor Test Schematic

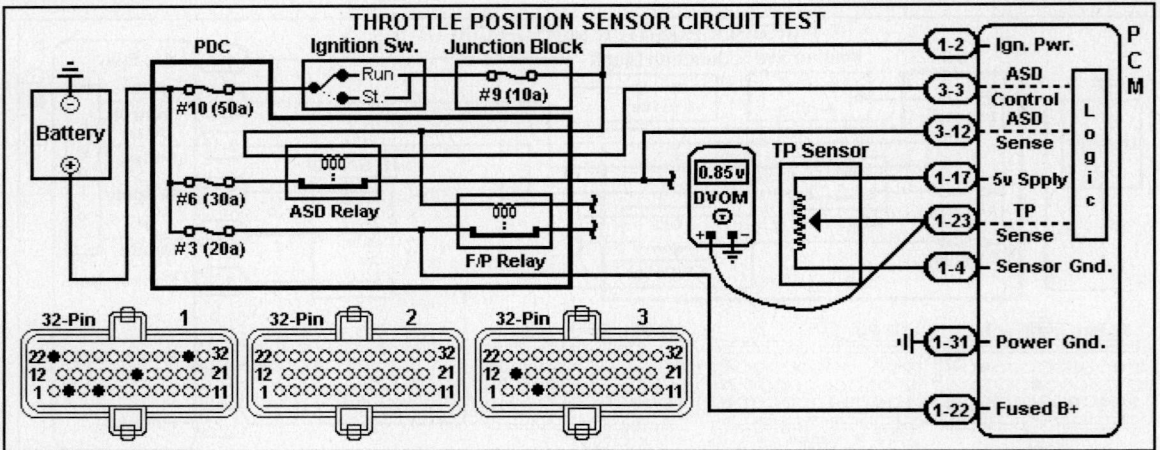

INTAKE AIR TEMPERATURE SENSOR

The Intake Air Temperature (IAT) Sensor is a negative temperature coefficient thermistor. This means that temperature and resistance are inversely related.

The sensor is located on the right front corner of the intake manifold.

DVOM & SCAN TOOL TEST (IAT)

Place the gear selector in Park and block the drive wheels for safety.

DVOM Connections

Connect the DVOM positive probe to the IAT sense circuit (BK/RD wire) at PCM Pin 1-15, or at the component harness connector. Connect the negative probe to the battery negative post.

DVOM Test

In this example, the sensor value is 1.58 volts with the engine running in closed loop. Compare the measurement to the Scan Tool PIDs to verify that the PCM receives the correct signal.

Scan Tool Test

Navigate through the Scan Tool menus, and select the DATA option to view the PIDs. Observe the IAT (°) and IAT (V) PIDs.

Make sure the PID values agree with the DVOM value. They should be very close, but will rarely be exactly the same.

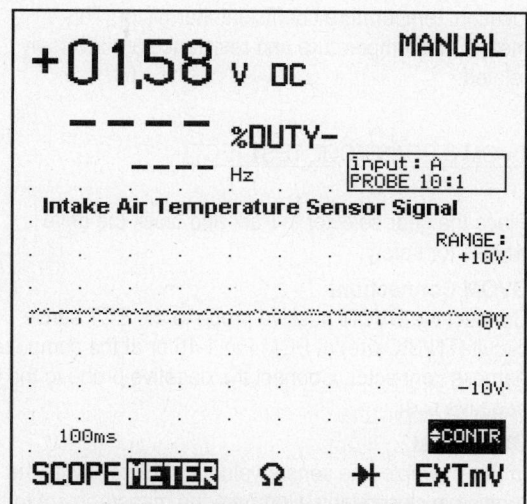

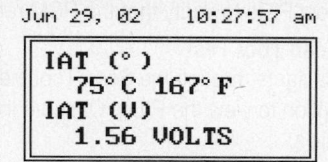

Intake Air Temperature Sensor Test Schematic

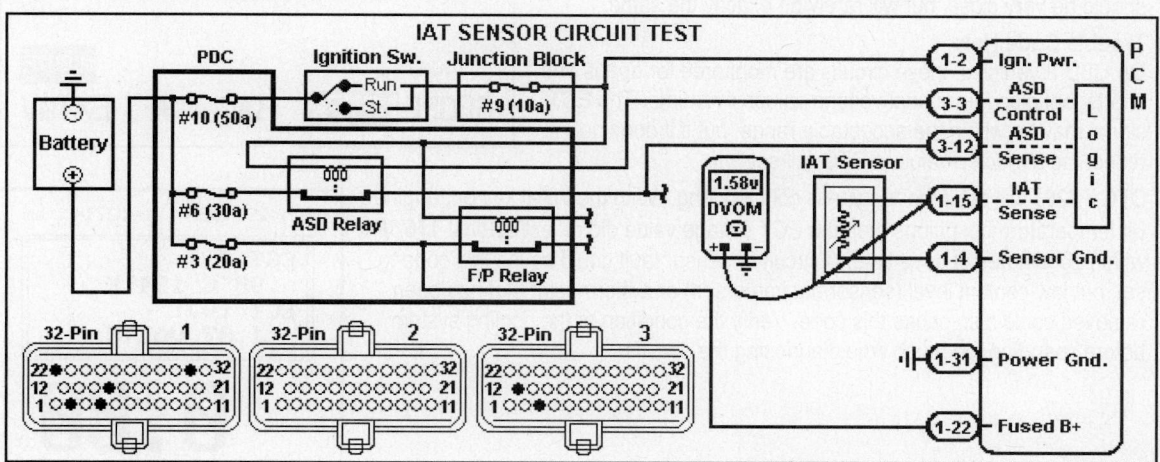

ENGINE COOLANT TEMPERATURE SENSOR

The Engine Coolant Temperature (ECT) Sensor is a negative

See the graphic to the right for the ECT sensor location. temperature coefficient thermistor. This means that temperature and resistance are inversely related.

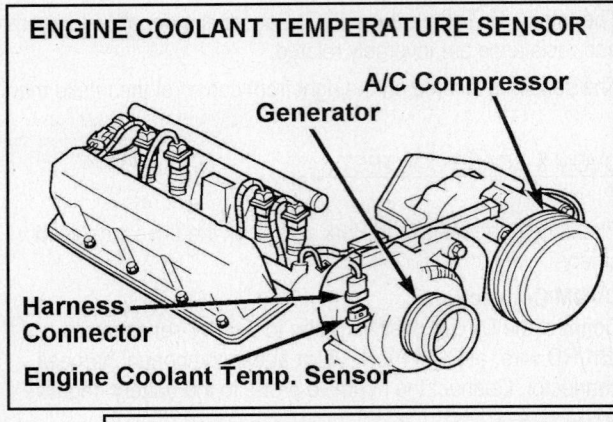

ENGINE COOLANT TEMPERATURE SENSOR

A/C Compressor
Generator
Harness Connector
Engine Coolant Temp. Sensor

DVOM & SCAN TOOL TEST (ECT)

Place the gear selector in Park and block the drive wheels for safety.

DVOM Connections
Connect the DVOM positive probe to the ECT sense circuit (TN/BK wire) at PCM Pin 1-16 or at the component harness connector. Connect the negative probe to the battery negative post.

DVOM Test
In this example, the sensor value is 1.19 volts with the engine running in closed loop. Compare the measurement to the Scan Tool PIDs to verify that the PCM receives the signal correctly.

Scan Tool Test
Navigate through the Scan Tool menus, and select the DATA option to view the PIDs. Observe the ECT (°) and ECT (V) PIDs.

Make sure the PID values agree with the DVOM value. They should be very close, but will rarely be exactly the same.

Trouble Code Help
On OBD II vehicles, these circuits are monitored for opens and shorts, but also for rational voltage values over time. The ECT signal may be within the acceptable range, but if it does not reflect the *expected* value, a DTC will set.

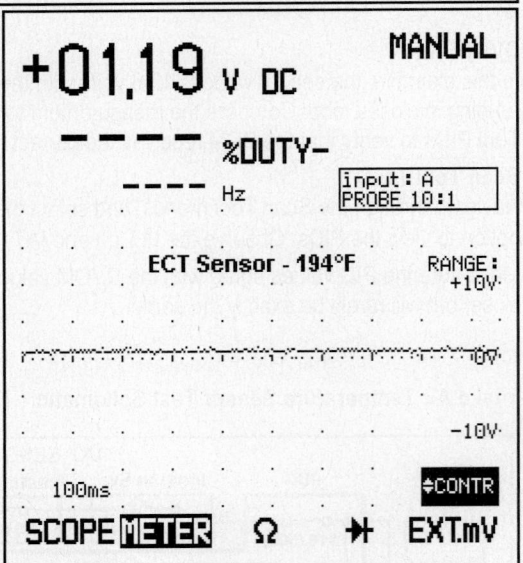

+01.19 v DC   MANUAL

---- %DUTY-
---- Hz   input: A
PROBE 10:1

ECT Sensor - 194°F   RANGE: +10V

-10V

100ms   ⬍CONTR
SCOPE METER   Ω   ✦   EXT.mV

DTC P1281 will set if the "engine is cold too long." With the initial key on, engine off temperature conditions met, the ECT voltage value did reflect at least 176 °F within 20 minutes of drive time. A circuit or sensor fault could cause this code to set, but low coolant level (sensor not immersed) or a thermostat that has been removed could also cause this code. Verify the condition of the cooling system before spending too much time diagnosing the circuits.

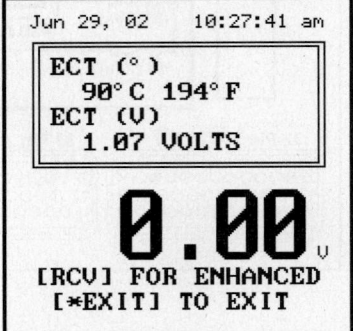

Jun 29, 02   10:27:41 am

ECT (°)
90°C 194°F
ECT (V)
1.07 VOLTS

0.00 v
[RCV] FOR ENHANCED
[*EXIT] TO EXIT

OIL PRESSURE SENSOR

The Oil Pressure Sensor (OPS) is mounted to the engine block where it taps into an oil galley. The PCM receives a voltage signal from the sensor and converts it to a pressure value. The value is monitored by the PCM, but is also sent through the data bus to the instrument cluster.

It is important to read circuit descriptions carefully to look for clues. In this case, the sensor is hardwired to the PCM, and the PCM communicates with the instrument cluster via the data bus circuits. Using a DVOM and Scan Tool, the circuit fault can be quickly isolated by determining at what point the value appears to be invalid.

DVOM & SCAN TOOL TEST (OPS)

The DVOM and Scan Tool can be used separate or together to help isolate faults in the oil pressure sensor or circuits. Place the gear selector in Park and block the drive wheels for safety.

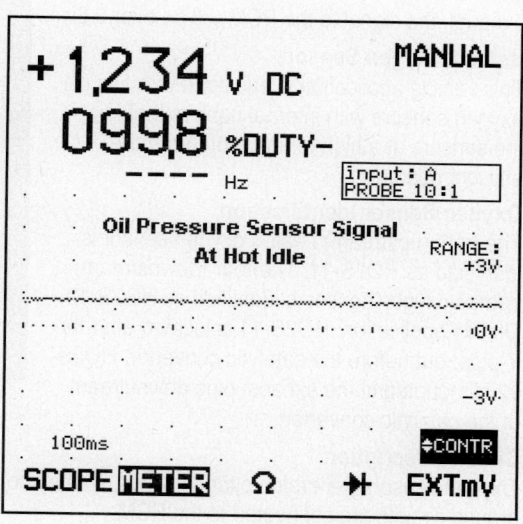

DVOM Connections
Connect the DVOM positive probe to the oil pressure sense circuit (GY/OR wire) at PCM Pin 2-23, or at the component harness connector. Connect the negative probe to the battery negative post.

DVOM Test
In this example, the sensor value is 1.234 volts with the engine running at Hot Idle. Compare the measurement to the Scan Tool PIDs and the instrument cluster oil pressure indicator to verify that the PCM receives the signal correctly, and that it sends the correct value to the instrument cluster over the CCD data bus.

Scan Tool Test
Navigate through the Scan Tool menus, and select the DATA option to view the PIDs. Observe the OIL PRESSURE and OIL PRESS SENSOR PIDs.

Make sure the PID values agree with the DVOM value. They should be very close, but will rarely be exactly the same.

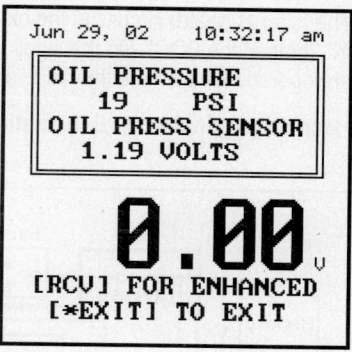

HEATED OXYGEN SENSORS

The heated oxygen sensors (HO2S) are mounted in the Exhaust system where they monitor the oxygen content of the exhaust stream. During engine operation, oxygen present in the exhaust reacts with the HO2S to produce a voltage output.

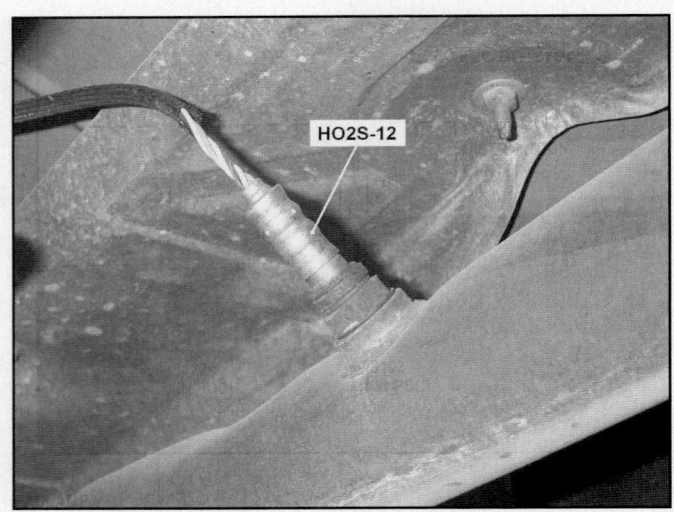

If the A/F mixture has a high concentration of oxygen in the exhaust, the signal to the PCM from the HO2S will be less than 0.4v. If the A/F mixture has a low concentration of oxygen in the exhaust, the signal to the PCM will be over 0.6v.

Heated Oxygen Sensors

This vehicle application is equipped with two (2) oxygen sensors with internal heaters that keep the sensors at a high temperature for more efficient operation.

Oxygen Sensor Identification

The front (upstream) heated oxygen sensor is identified as HO2S-11. The rear (downstream) heated oxygen sensor is identified as HO2S-12. On this application, HO2S-11 is located after the Y-pipe, but before the catalytic converter. HO2S-12 is mounted in the exhaust pipe downstream of the catalytic converter.

Circuit Description

Oxygen sensors generate voltage signals, so they do not need any reference voltage. The HO2S signal and sensor ground circuits are connected to the PCM.

The internal heater circuits in the oxygen sensors (both front and rear) receive power through the ASD relay and fuse 'K', (both in the PDC). On this application the sensor heaters are connected directly to ground. Therefore, the heaters are operational any time the ASD relay is energized.

Heated Oxygen Sensor Schematic

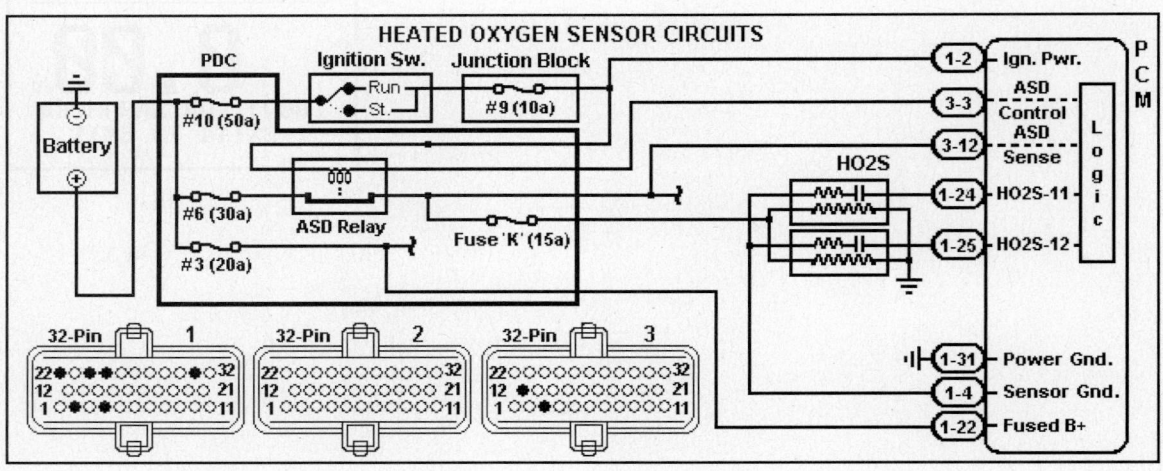

LAB SCOPE TEST (HO2S)

The Lab Scope can be used to observe the warm-up process for sensors and the catalytic converter. Place the gearshift selector in Park and block the drive wheels for safety.

Lab Scope Connections
Connect the Channel 'A' positive probe to the HO2S-11 sense circuit (TN/WT wire) at PCM Pin 1-24. Connect the negative probe to the battery negative post. Connect the Channel 'B' positive probe to the HO2S-12 sense circuit (OR/BK wire) at PCM Pin 1-25.

Lab Scope Example (1)
In example (2), traces from both oxygen sensors were captured for 10 seconds shortly after a cold start. Both of the sensors were active during this test, indicating that the vehicle was in closed loop, and that the downstream sensor (HO2S-12) became active before the catalytic converter heated to operating temperature.

Lab Scope Example (2)
In example (2), the traces represent 10 seconds of oxygen sensor signals after the vehicle was run at 2000 RPM for 1 minute from example (1).

Lab Scope Example (3)
In example (3), the traces show 10 seconds of oxygen sensor signals after 2 minutes from example (1). The catalytic converter has reached operating temperature, so the downstream (HO2S-12) signal has stabilized (is not cycling).

Using Test Results
These tests can be used to verify that the heated oxygen sensors are active, and that the catalytic converter is efficiently storing and releasing oxygen.

Observe the time it takes for the sensors to become active. In this example, these sensors (with 53,000 miles on them) became active within 20 seconds from a cold start.

If the catalytic converter becomes active before the downstream sensor, the trace may stay flat. If the sensor is suspect, drive the vehicle rich and then lean to force the signal high and then low. In this way, the sensor range can be checked even though the converter operation is causing a stable signal.

Transmission Controls

INTRODUCTION

This vehicle is equipped with a 42RE electronic automatic transmission. This transmission features electronic control of the torque converter clutch (TCC), the overdrive (OD) clutch, and governor pressure. This vehicle does not have a separate transmission control module. All transmission electronic functions are controlled by the PCM.

Circuit Description

The transmission pressure and speed sensors connect directly to the PCM. The PCM uses these inputs and others to calculate TCC operation, overdrive, and governor pressure control. The solenoids and sensors (except speed sensor) are contained in the transmission solenoid assembly. The assembly is powered by the transmission control relay in the Power Distribution Center (PDC). The solenoids are individually controlled by duty cycle ground commands from the PCM.

TORQUE CONVERTER CLUTCH

The torque converter is a hydraulic device that couples the engine to the transmission, and drives the transmission oil pump. For automatic transmission operation to be possible, various amounts of slippage must occur between the converter and the transmission. This allows smooth operation, and makes idling in gear possible.

Under steady cruise, however, a direct connection between the engine and transmission is desirable to reduce fluid temperatures and increase fuel mileage. The Torque Converter Clutch (TCC) solenoid performs this task by locking the normally slipping TCC inside the converter. The TCC is locked by hydraulic pressure that is routed to it when the solenoid is energized.

The TCC solenoid will be energized at steady speeds above 45 MPH in fourth gear with the overdrive on. The TCC solenoid will also be energized at steady speeds above 35 MPH in third gear with the overdrive off. The inputs for TCC operation are the ECT, TP, RPM, VSS, and MAP sensor signals.

GOVERNOR PRESSURE SOLENOID

The PCM controls the governor pressure inside the transmission through a duty cycle command to the governor pressure solenoid. The governor pressure sensor provides hydraulic pressure feedback. Both components are contained in the transmission solenoid assembly. The governor pressure solenoid may be referred to as a variable force solenoid in some repair information.

The governor pressure solenoid is an electro-hydraulic device used to regulate line pressure to produce governor pressure needed for upshifts and downshifts. The default solenoid position (0% duty cycle) allows full line pressure to pass through. When the duty cycle is increased until the circuit draws an average of 1 amp of current, the solenoid completely blocks off line pressure (the governor pressure is 0 psi).

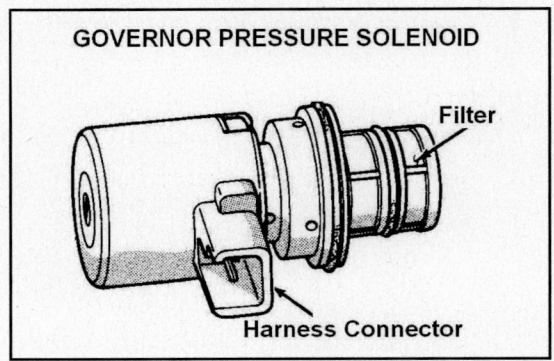

GOVERNOR PRESSURE SOLENOID

Filter

Harness Connector

SCAN TOOL TEST (A/T PIDS)

The Scan Tool can be used to view transmission data (PIDs) over time to help isolate the cause of a fault. Further testing is often required, but the Scan Tool will help to determine the best diagnostic path to follow.

Scan Tool Navigation

The PCM controls all transmission electronic functions on this application, so the engine data list will contain transmission data as well.

Connect the Scan Tool to the underdash DLC. Identify or confirm vehicle identity. Select F0: Datalist from the SELECT MODE screen. Scroll down to view the desired PIDs. Operate the vehicle under various conditions, focusing on ones where the fault occurs most frequently.

Scan Tool PID Example

In this example, the PID values for the transmission were captured at 30 MPH in closed loop.

PARK/NUETRAL and PRESENT GEAR

The PARK/NEUTRAL PID reflects whether or not the transmission is in Park or Neutral, not the actual gear. The PID value at 30 MPH is GEAR. If the gear selector is moved to the Park position, the PID should change to P/N.

The PRESENT GEAR PID represents the actual gear. The value at 30 MPH is 2 for second gear.

OUTPUT SPEED

The output speed PID is used to make transmission decisions, but is also used as a vehicle speed signal. The PID value is expressed as shaft RPM, from which the PCM calculates vehicle speed.

TRANS PWR RELAY

The transmission power relay is energized by the PCM, which supplies battery voltage to the solenoid assembly. This PID can be helpful when diagnosing a DTC P1765. This code will set when the PCM does not see 12v on the solenoid control circuits with the relays off. A DVOM may also be used, but knowing the commanded relay state solves one more variable and saves time.

Governor PIDs

The THEORY and ACTUAL governor pressure PIDs show what the PCM is trying to accomplish and if it is able to accomplish it. The GOV DUTY CYCLE PID shows the command, and the GOV. PRESS. SNSR PID gives pressure feedback.

There is not a direct relationship between duty cycle and governor pressure, because line pressure varies so much under different conditions. The PCM will vary the duty cycle according to its goals and pressure feedback, so line pressure variations are compensated for immediately.

```
Jun 29, 02    10:23:27 am

  1998      5.2L V8
  INJ:MPI SEQ
  TRANS:4SPD AUTO
  VIN=Y      [YES]


      0.00 v


 [*EXIT] TO EXIT
```

```
   SELECT MODE ↑↓
→ F0:Datalist
  F2:DTCs
  F3:Snapshot
```

```
PARK/NEUTRAL
  GEAR
TCC SOLENOID
  DISENGAGED    0
PRESENT GEAR
  2
OUTPUT SPEED
  1300 RPM      0
TRANS TEMP SNSR
  2.78 VOLTS
TRANS PWR RELAY
  ENERGIZED     0
OVERDRIVE SOL
  DE-ENRGZD.
OVERDRIVE LAMP
  OFF           0
THEORY GOV PRESS
  24    PSI
ACTUAL GOV PRESS
  24    PSI     0
GOV DUTY CYCLE
  29%
GOV. PRESS. SNSR
  0.60 VOLTS    0
TRANS PWR RELAY
  ENERGIZED
OD OVERRIDE SW
  OFF           0
```

LAB SCOPE TEST (GOVERNOR PRESSURE SOLENOID)

This vehicle is equipped with a 4-speed 42RE automatic transmission. The PCM controls governor pressure, overdrive operation, and TCC operations. The Lab Scope can be used to view any of the solenoids or pressure switches. Place the gear selector in Park and block the drive wheels for safety.

Lab Scope Connections

Connect the Channel 'A' positive probe to the governor pressure solenoid control circuit (P/WT Wire) at PCM Pin 2-8. The circuit may be labeled Variable Force Solenoid in some repair information. Connect the negative probe to the negative battery post. Connect the Channel 'B' positive probe to the governor pressure signal circuit (LG/WT Wire) at PCM Pin 2-29.

Lab Scope Example (1)

In example (1), the traces show the governor pressure solenoid command and feedback signals at idle. The solenoid command is cycling, but the line pressure is very low at idle, so the governor pressure is also very low. The pressure signal of about 650 mv reflects this low pressure.

Lab Scope Example (2)

In example (2), the vehicle was driven at approximately 12 MPH. The solenoid duty cycle has increased slightly, and there is a corresponding rise in the pressure sensor signal (about 800 mv).

Lab Scope Example (3)

In example (3), the vehicle was driven at approximately 20 MPH. The solenoid duty cycle decreased again from example (2), but the pressure sensor signal rose to about 1.25 volts. As discussed previously, as engine speed increases, the line pressure also increases, and the governor pressure solenoid requires lower duty cycles to produce the desired governor pressure.

These tests can be verified using a Scan Tool to observe the governor pressure and sensor PID values under the same conditions.

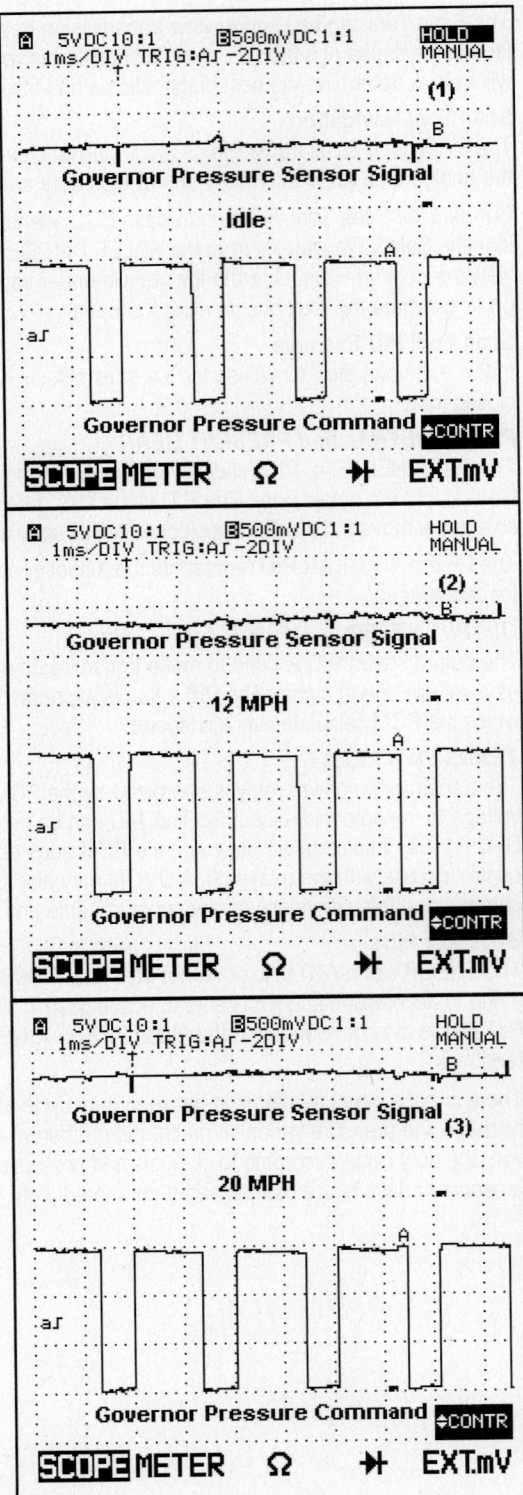

Reference Information - PCM PID Examples

<u>1998 DODGE RAM 5.2L V8 OHV SFI VIN Y (A/T)</u>

■ **NOTE:** *The following readings were obtained with the engine at idle speed, radiator hose hot, throttle closed, gear selector in Park and all accessories turned off.*

Acronym	Scan Tool Parameter	Range	Typical Value
A/C CLUTCH	A/C Clutch Relay Status	Energized/De-energized	De-energized
A/C REQ	A/C Request Signal	ON / OFF	OFF
ASD	Automatic Shutdown Relay Control	Energized/De-energized	Energized
BARO	BARO (kPa, In. Hg)	10-110 kPa, 0-30 "Hg	100 kPa, 29.5 "Hg
BRAKE SW	Brake Switch Position	ON / OFF	OFF
DES IAC	Desired Idle Air Control Position	0-255 steps	13
ECT	ECT Sensor (°F)	-40 to 284°F	194°F
ECT (V)	ECT Sensor (volts)	0-5.1v	1.07v
FUEL LEVEL	Fuel Level Sensor (volts)	0-5.1v	2.83v
FUEL STATUS	Fuel Status	CLOSED / OPEN	CLOSED
GOV D/C	Trans. Governor Duty Cycle	0-100%	38%
GOV P/S	Trans. Governor Press. Sensor	0-5.1v	0.58v
HO2S-11	HO2S-11 (B1 S1) Signal	0-1100 mv	84 mv
HO2S-11 ST	HO2S-11 S/T Fuel Trim Status	LEAN / RICH	RICH
HO2S-12	HO2S-12 (B1 S2) Signal	0-1100 mv	74 mv
HO2S-12 ST	HO2S-12 S/T Fuel Trim Status	LEAN / RICH	RICH
IAC	Idle Air Control Motor Position	0-255 steps	13
IAT	IAT Sensor (°F)	-40 to 284°F	167°F
IAT V	IAT Sensor (volts)	0-5.1v	1.56v
IGN SW	Ignition Switch Position	ON / OFF	ON
INJ P/W	Injector Pulsewidth (cranking)	0-999 ms	1.3 ms
INJ P/W	Injector Pulsewidth (running)	0-999 ms	3.6 ms
KEY CYCLES 1	Key On Cycles 1	0-255	61
KEY CYCLES 2	Key On Cycles 2	0-255	61
KEY CYCLES 3	Key On Cycles 2	0-255	61
LOAD	Engine Load (%)	0-100%	4%
LTFT ADAPT	Long Term Adaptive	0% (± 20%)	+1%
MAP (kPa)	MAP Vacuum (kPa, In. Hg)	10-110 kPa, 0-30 "Hg	33 kPa, 9.7 "Hg
MAP (V)	MAP Vacuum (volts)	0-5.1v	1.21v
MISF CYL1-6	Misfire Status Cylinder 1 to 6	CYL 1, 2 … (0-256)	CYL 1: 3, CYL 2: 3
OIL PRESS	Oil Pressure Sensor	0-5.1v	1.19v at 19 psi
P/N SW	Park Neutral Switch Position	P/N, DR, 2ND, 1ST	P/N
RPM	Engine Speed in RPM	0-66635 rpm	643
SHFT ADAPT	Short Term Adaptive	0% (± 20%)	-1%
SPARK	Spark Advance (°)	-90° to +90°	18°
T/IDLE	Target Idle Speed	0-66635 rpm	640
TCC SOL	TCC Solenoid Command	Energized/De-energized	De-energized
TFT	Transmission Temperature Sensor	0-5.1v	2.86v
TPS V	TP Sensor (volts)	0-5.1v	0.76v
T/RELAY	Transmission Power Relay	Energized/De-energized	Energized
VAC	Vacuum (kPa, In. Hg)	10-110 kPa, 0-30 "Hg	68 kPa, 20.1 "Hg

1998 DODGE RAM 5.2L V8 OHV SFI VIN Y (A/T) BLACK 32-PIN CONNECTOR 'A' OR 'C1'

PCM Pin #	Wire Color	Circuit Description (32-Pin)	Value at Hot Idle
A2	LG/BK	Ignition Switch Power (fused)	12-14v
A4	BLK/LG	Analog Sensor Ground	<0.050v
A6	BK/WT	PNP Switch Sense Signal	In P/N: 0v, Others: 5v
A7	BK/GY	Ignition Coil 1 Driver Control	Idle: 5º, 55 mph: 8º dwell
A8	GY/BK	CKP Sensor Signal	Digital Signals: 0-5-0-5v
A10	YL/BK	IAC 2 Driver Control	Pulse Signals
A11	BR/WT	IAC 3 Driver Control	Pulse Signals
A13	OR	Power Takeoff Switch Sense	Switch On: 0v, Off: 12v
A15	BK/RD	IAT Sensor Signal	At 100ºF: 1.83v
A16	TN/BK	ECT Sensor Signal	At 180ºF: 2.80v
A17	VT/WT	5-Volt Supply	4.9-5.1v
A18	TN/YL	CMP Sensor Signal	Digital Signals: 0-5-0-5v
A19	GY/RD	IAC 1 Driver Control	Pulse Signals
A20	VT/BK	IAC 4 Driver Control	Pulse Signals
A22	RD/WH	Fused Battery Power (B+)	12-14v
A23	OR/DB	TP Sensor Signal	Hot Idle: 0.6-1.0v
A24	TN/WT	HO2S-11 (B1 S1) Signal	0.1-1.1v
A25	OR/BK	HO2S-13 (B1 S3) Signal	0.1-1.1v
A26	LG/RD	HO2S-21 (B2 S1) Signal	0.1-1.1v
A27	DG/RD	MAP Sensor Signal	1.5-1.7v
A29	TN/WT	HO2S-12 (B1 S2) Signal	0.1-1.1v
A31-32	BK/TN	Power Ground	<0.1v

1998 DODGE RAM 5.2L V8 OHV SFI VIN Y (A/T) WHITE 32-PIN CONNECTOR 'B' OR 'C2'

PCM Pin #	Wire Color	Circuit Description (32-Pin)	Value at Hot Idle
B1	VT	Transmission Temperature Sensor	2.86v
B2	VT/TN	Injector 2 Driver	1-4 ms
B4	WT/DB	Injector 1 Driver	1-4 ms
B5	YL/WT	Injector 3 Driver	1-4 ms
B6	GY	Injector 5 Driver	1-4 ms
B8	VT/WT	Governor Pressure Solenoid	PWM Signal: 0-12-0v
B10	DG	Generator Field Control	Digital Signals: 0-12-0v
B11	OR/BK	TCC Solenoid Control	Solenoid Off: 12v, On: 1v
B12	LG/BK	Injector 6 Driver	1-4 ms
B13	GY/LB	Injector 8 Driver	1-4 ms
B15	TN	Injector 2 Driver	1-4 ms
B16	LB/BR	Injector 4 Driver	1-4 ms
B23	GY/YL	Engine Oil Pressure Sensor	1.6v at 23 psi
B25, B28	DB, LG	Output Shaft Speed Sensor N, P	AC pulse signals
B27	WT/OR	Vehicle Speed Signal	Digital Signals
B29	LG/WT	Governor Pressure Sensor Signal	0.58v
B30	PK	Transmission Relay Control	Relay Off: 12v, On: 1v
B31	OR	5-Volt Reference (Secondary)	4.9-5.1v

<u>1998 DODGE RAM 5.2L V8 OHV SFI VIN Y (A/T) GREY 32-PIN CONNECTOR 'C' OR 'C3'</u>

PCM Pin #	Wire Color	Circuit Description (32-Pin)	Value at Hot Idle
C1	DB/OR	A/C Clutch Relay Control	Relay Off: 12v, On: 1v
C2	---	Not Used	---
C3	DB/YL	Auto Shutdown Relay Control	Relay Off: 12v, On: 1v
C4	TN/RD	S/C Vacuum Solenoid Control	S/C on with vacuum increasing: 1v
C5	LG/RD	Speed Control Vent Solenoid	S/C on with vacuum decreasing: 1v
C6	LG/OR	Overdrive Off Lamp Driver	Lamp Off: 12v, On: 1v
C7-9	---	Not Used	---
C10	WT/DG	LDP Solenoid Control	Solenoid Off: 12v, On: 1v
C11	YL/RD	Speed Control Power	12-14v
C12	DG/OR	Auto Shutdown Relay Output	12-14v
C13	OR/WT	Overdrive Off Switch Sense	Switch Off: 12v, On: 1v
C14	---	Not Used	---
C15	OR	Battery Temperature Sensor	At 86°F: 1.96v
C16-18	---	Not Used	---
C19	BR/WT	Fuel Pump Relay Control	Relay On: 1v, Off: 12v
C20-21	---	Not Used	---
C22	BR	A/C Request Signal	A/C Off: 12v, On: 1v
C23	LG/WT	A/C Select Signal	A/C Off: 12v, On: 1v
C24	WT/PK	Brake Switch Sense Signal	Brake Off: 0v, On: 12v
C25	WT/OR	Generator Field Source	12-14v
C26	DB/WT	Fuel Level Sensor Signal	Full: 0.5v, 1/2 full: 2.5v
C27	PK/DB	SCI Transmit	0v
C28	WT/BK	CCD Bus (-)	<0.050v
C29	DG	SCI Receive	5v
C30	VT/BR	CCD Bus (+)	Digital Signals: 0-5-0v
C31	---	Not Used	---
C32	RD/LG	Speed Control Sw. Signal	S/C & Set Sw. On: 3.8v

Pin Connector Graphic

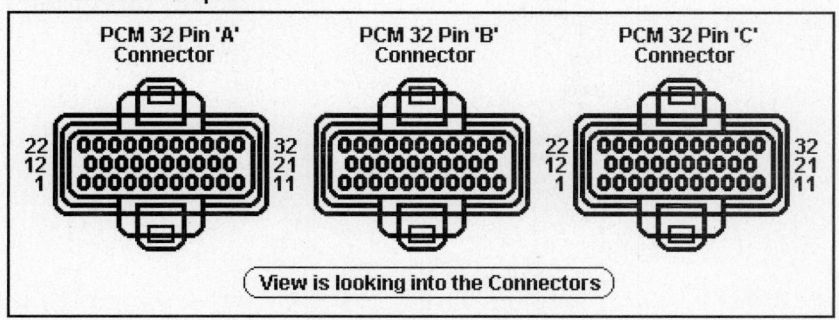

PCM Wiring Diagrams

FUEL & IGNITION SYSTEM DIAGRAMS (1 OF 14)

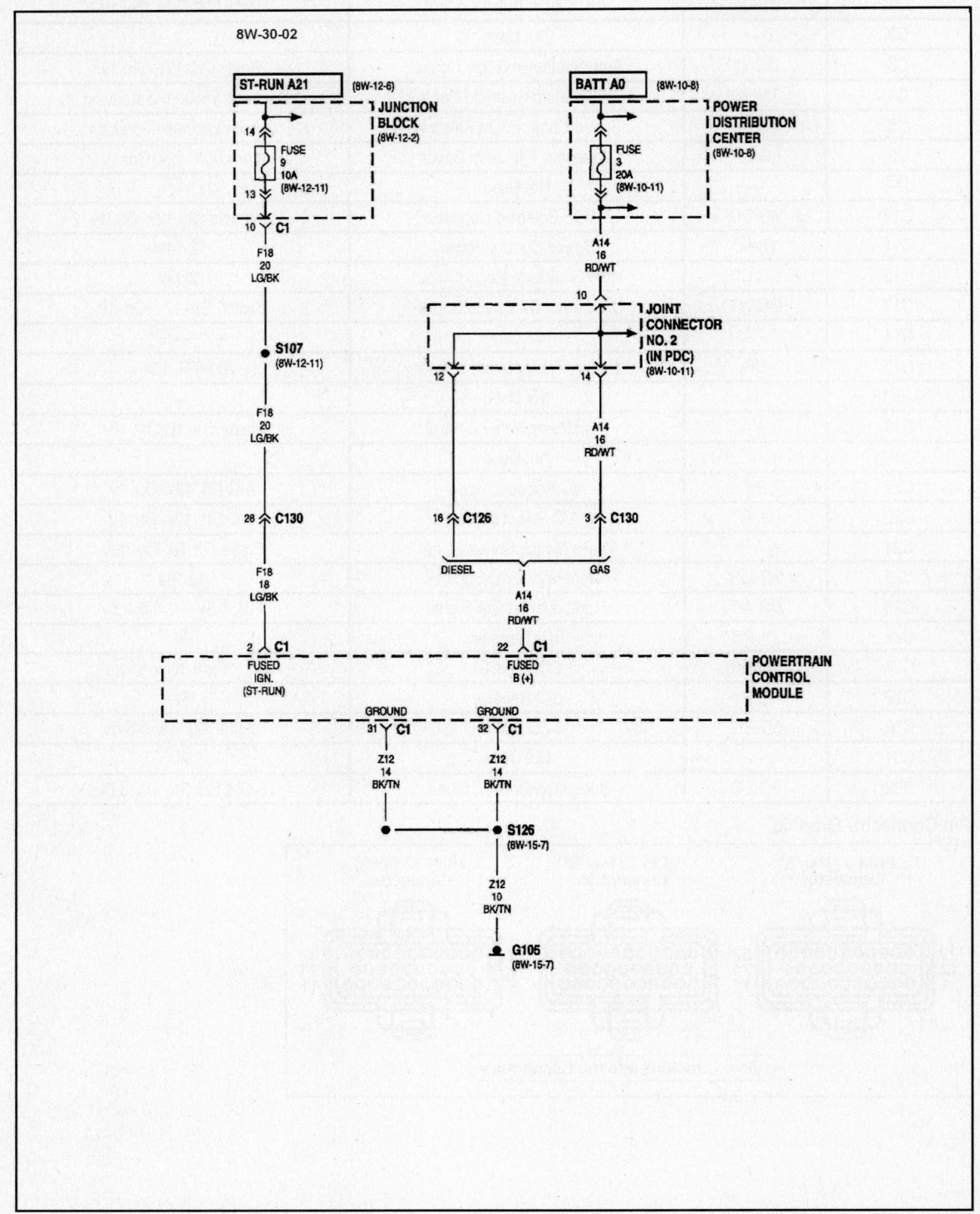

FUEL & IGNITION SYSTEM DIAGRAMS (2 OF 14)

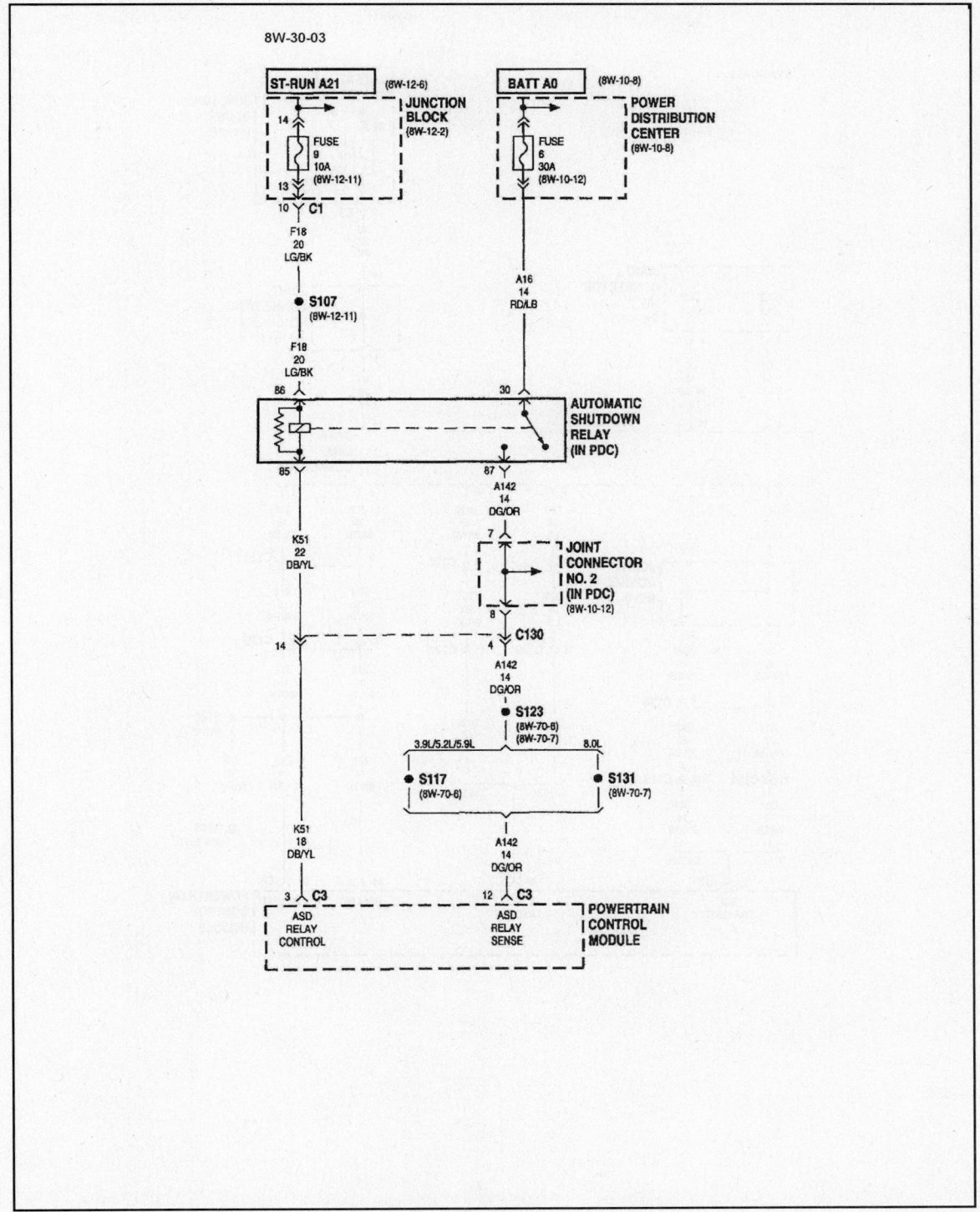

FUEL & IGNITION SYSTEM DIAGRAMS (3 OF 14)

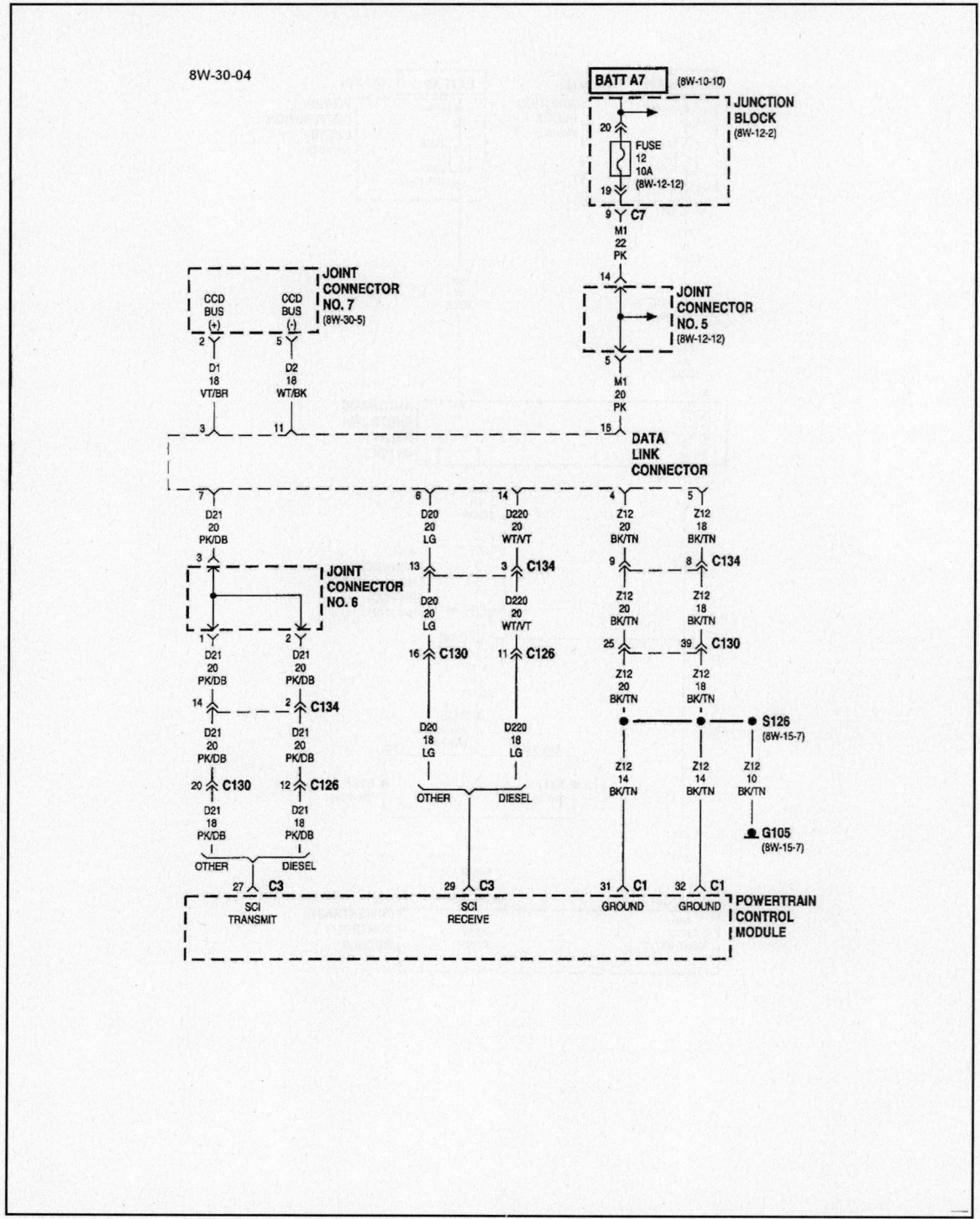

FUEL & IGNITION SYSTEM DIAGRAMS (4 OF 14)

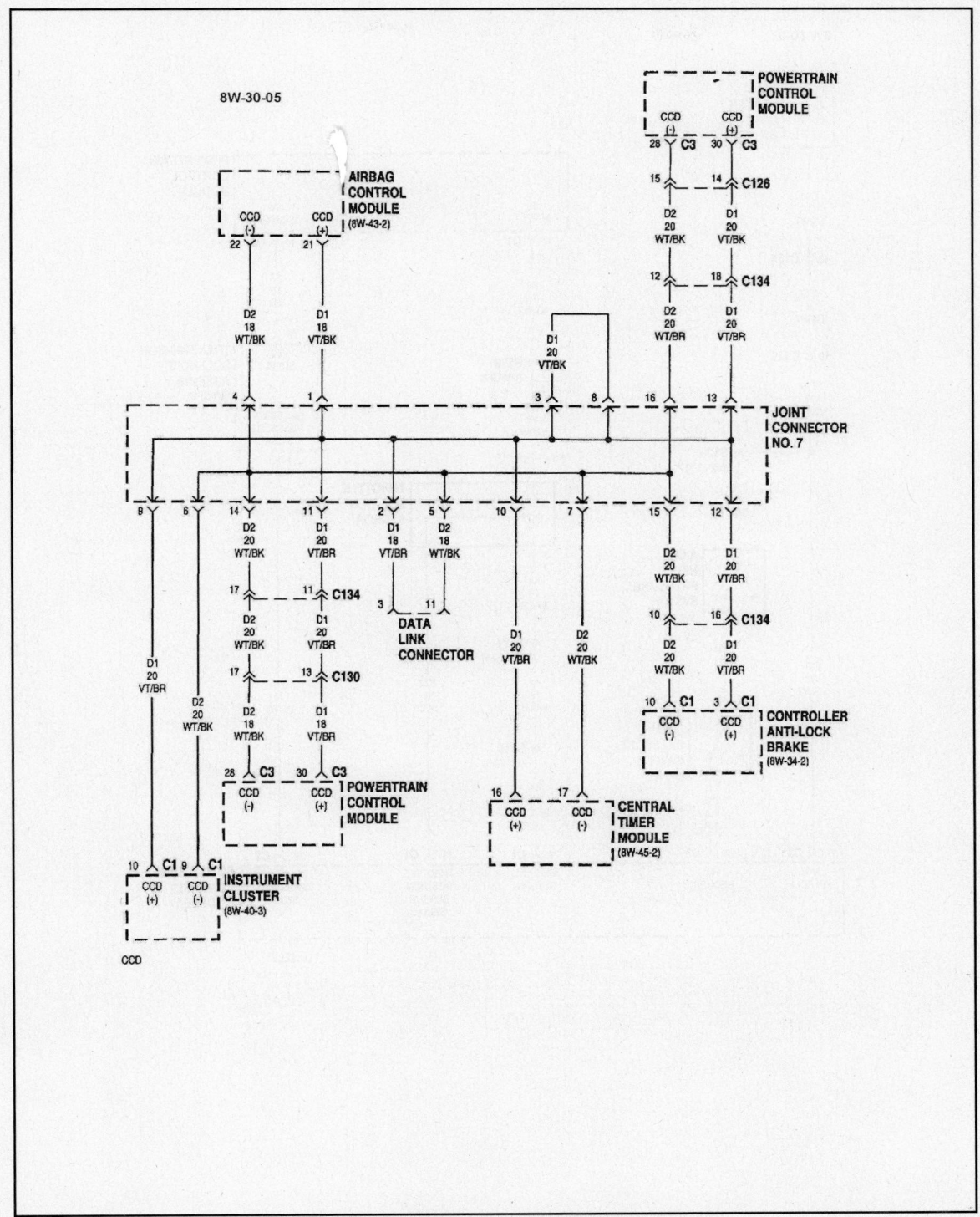

FUEL & IGNITION SYSTEM DIAGRAMS (5 OF 14)

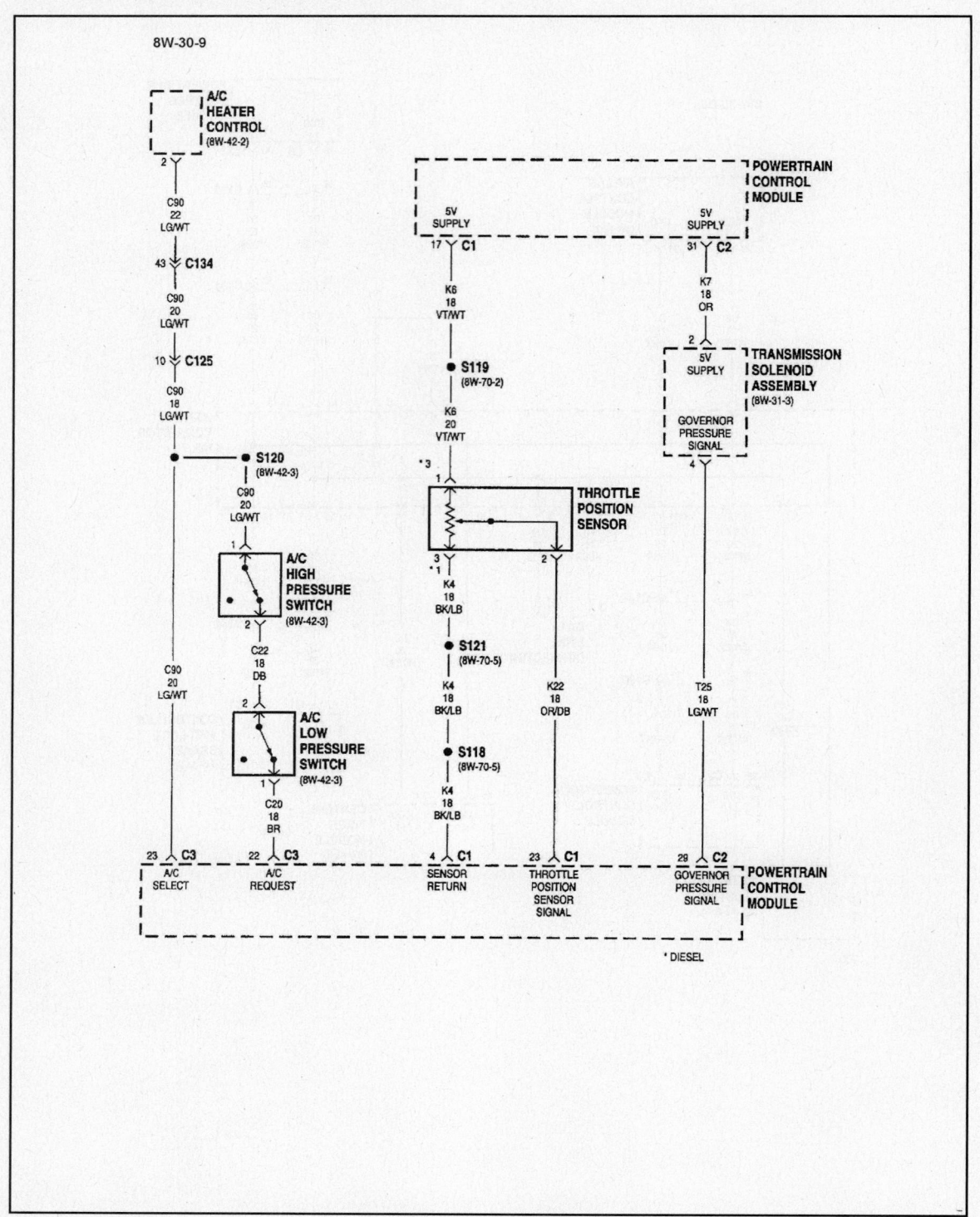

FUEL & IGNITION SYSTEM DIAGRAMS (6 OF 14)

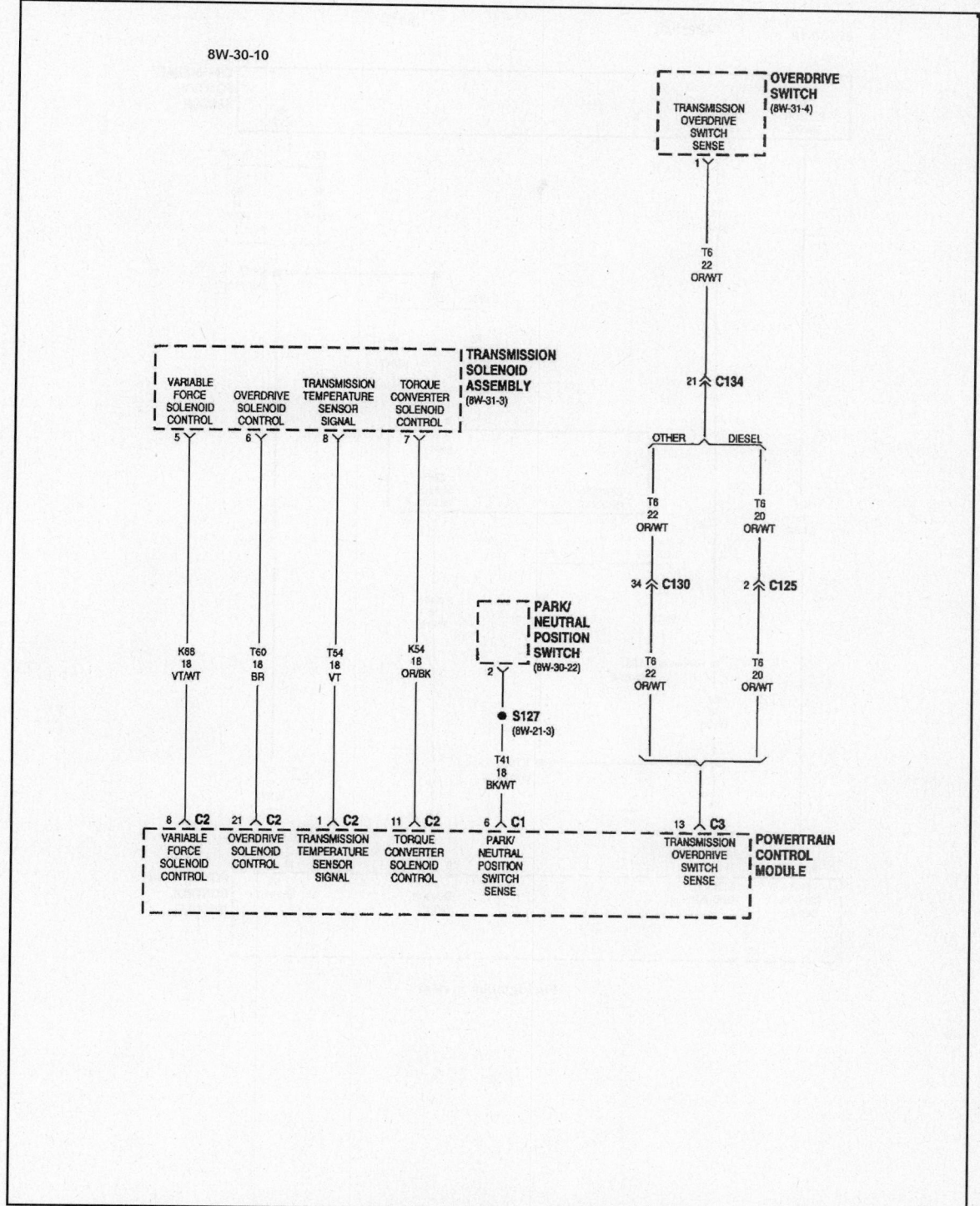

FUEL & IGNITION SYSTEM DIAGRAMS (7 OF 14)

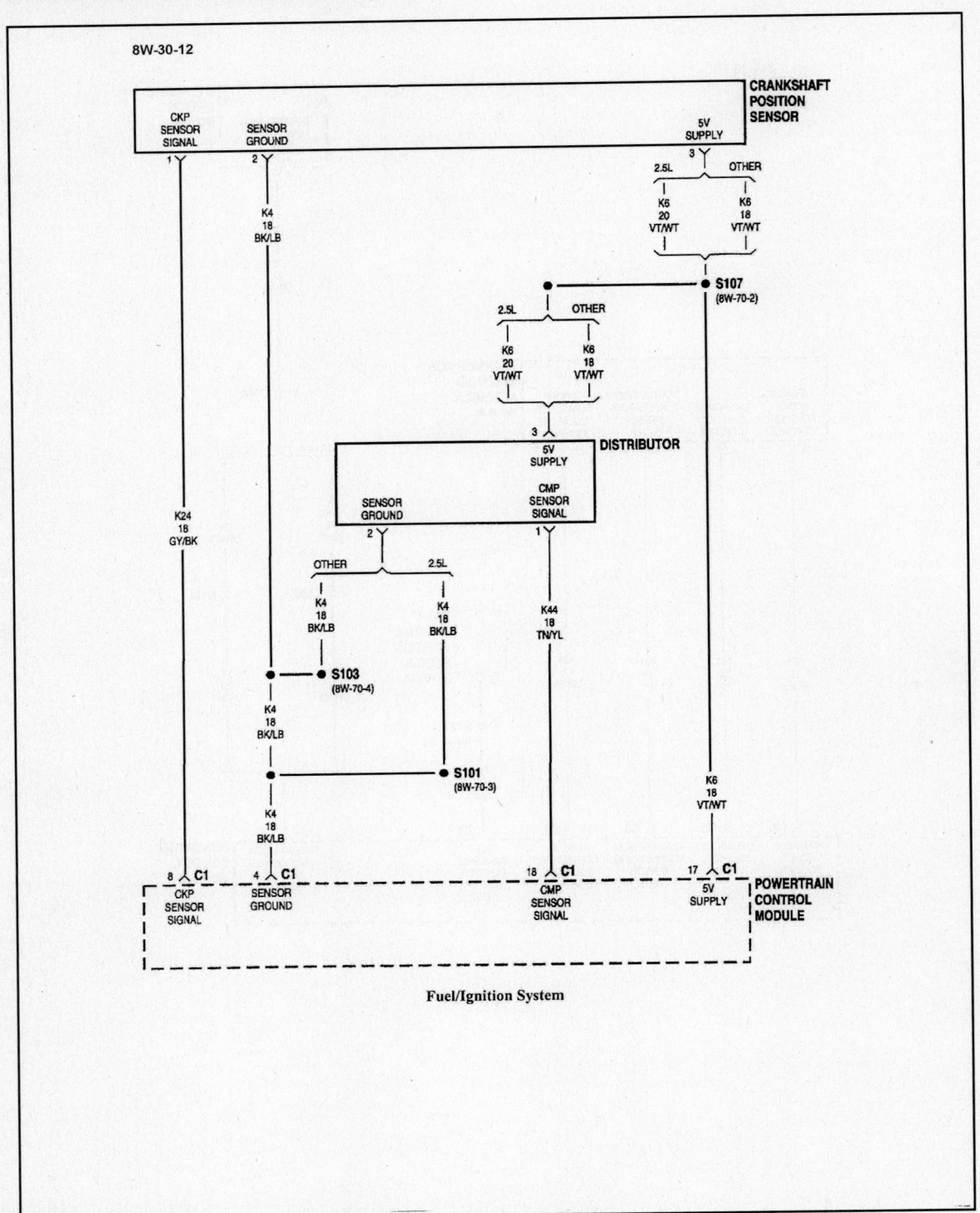

Fuel/Ignition System

FUEL & IGNITION SYSTEM DIAGRAMS (8 OF 14)

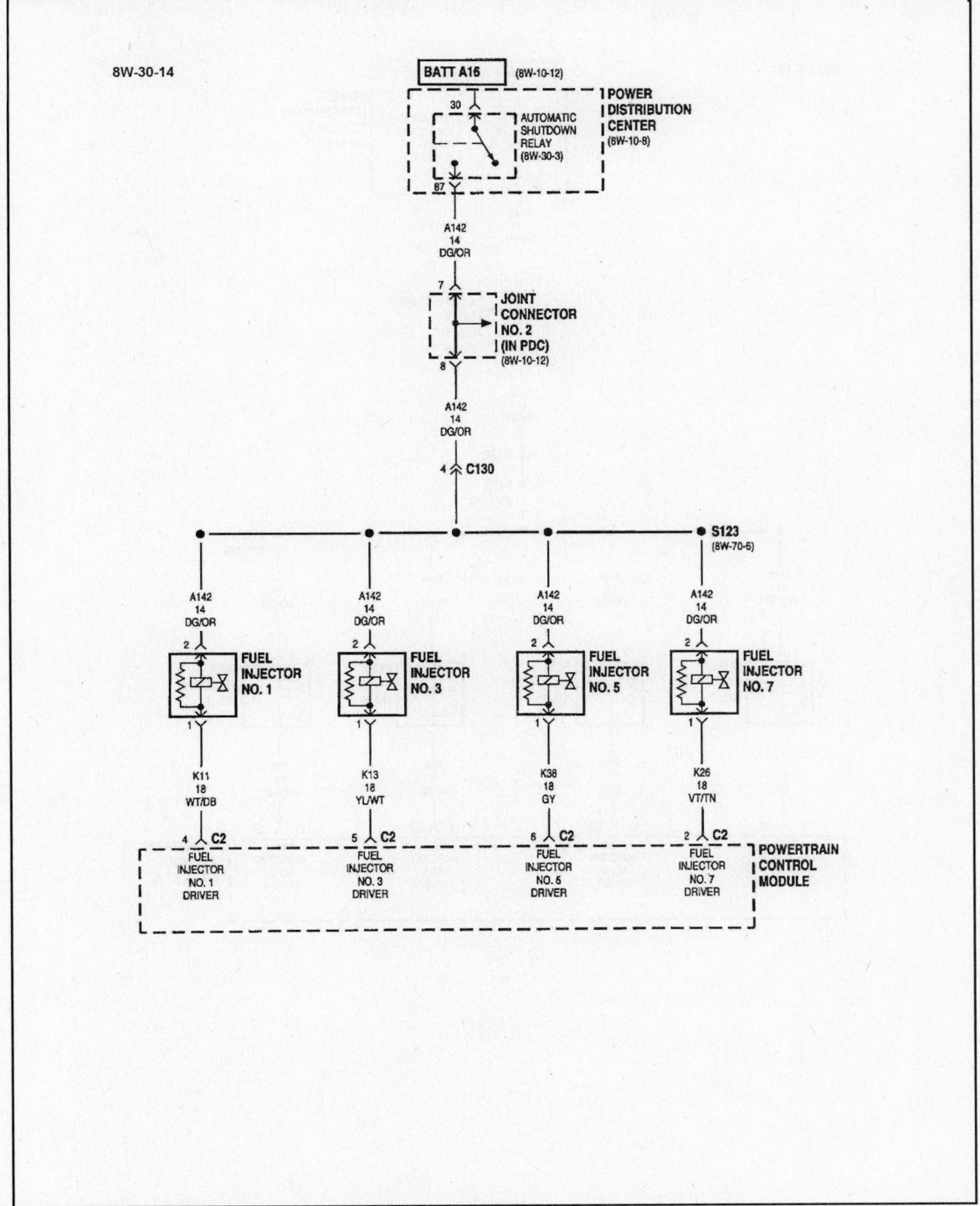

FUEL & IGNITION SYSTEM DIAGRAMS (9 OF 14)

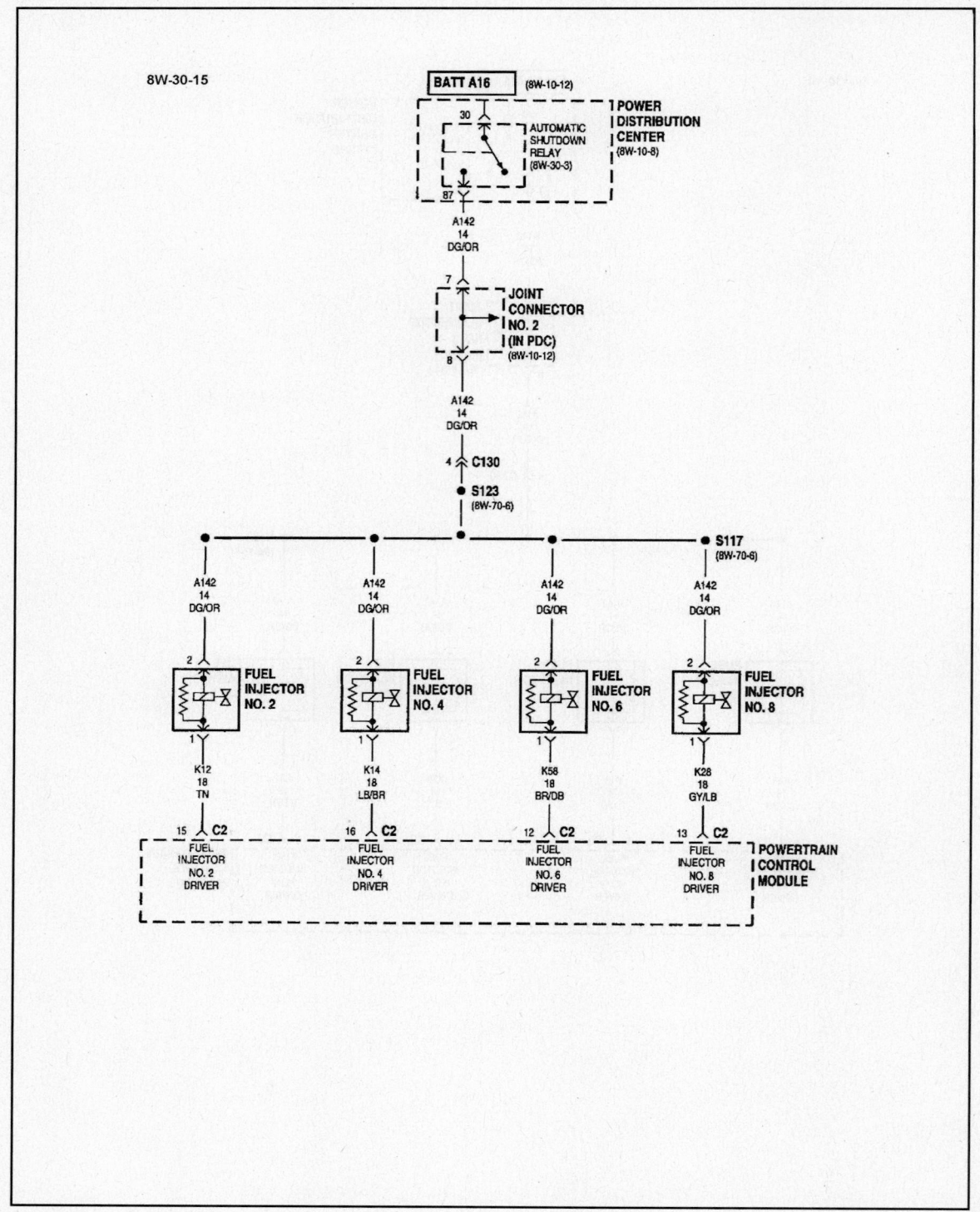

FUEL & IGNITION SYSTEM DIAGRAMS (10 OF 14)

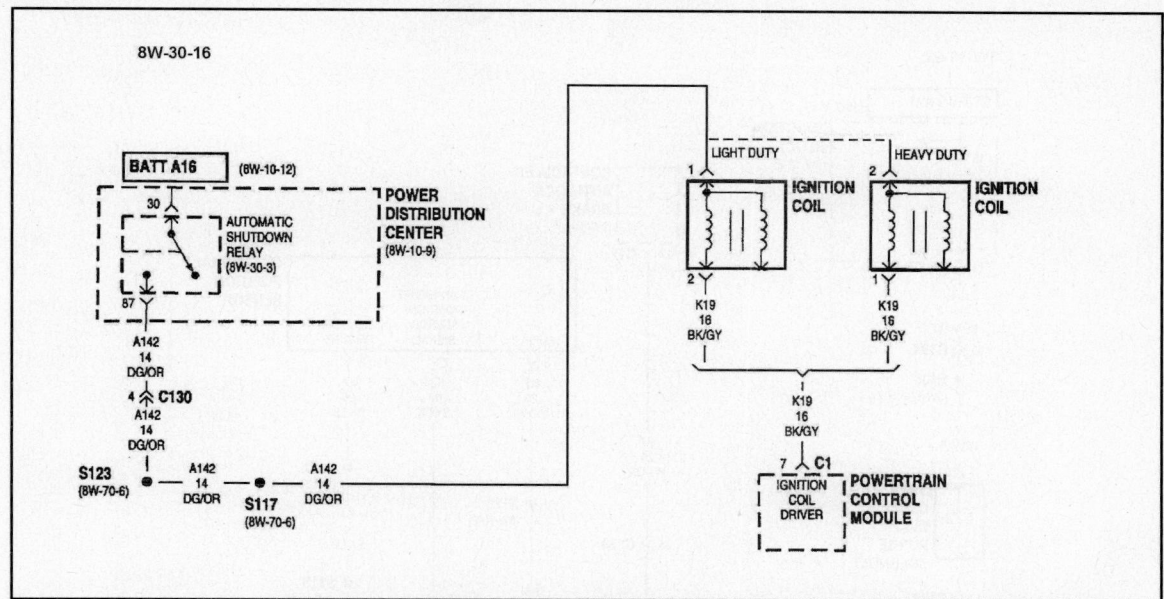

FUEL & IGNITION SYSTEM DIAGRAMS (11 OF 14)

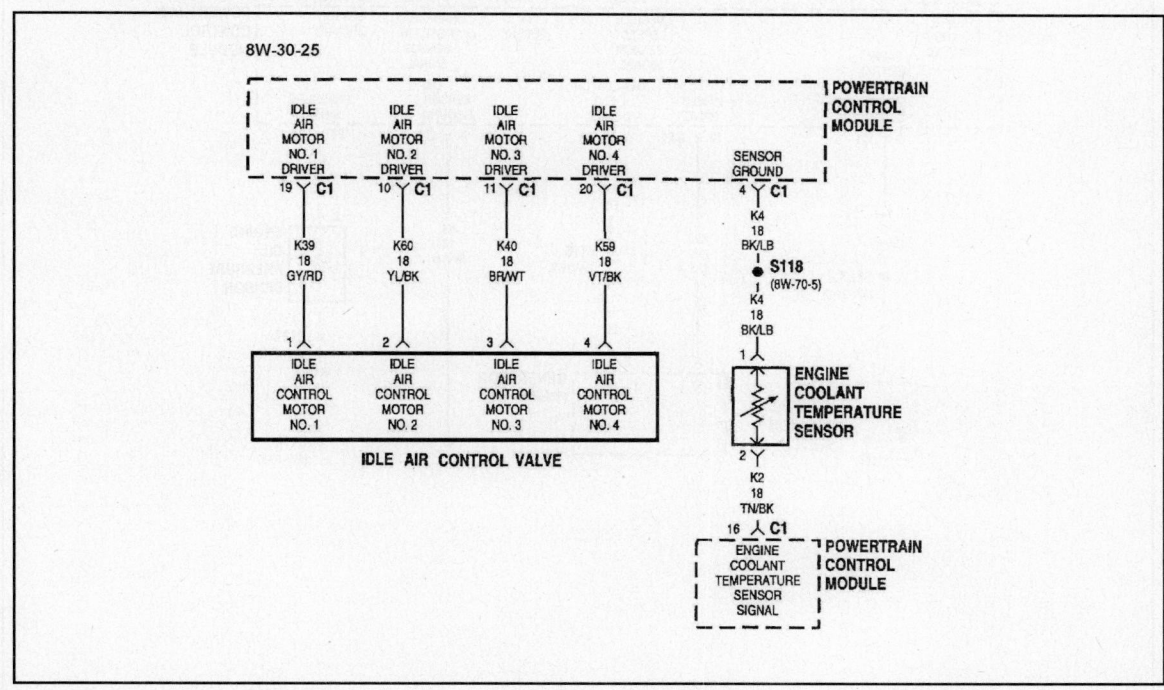

FUEL & IGNITION SYSTEM DIAGRAMS (12 OF 14)

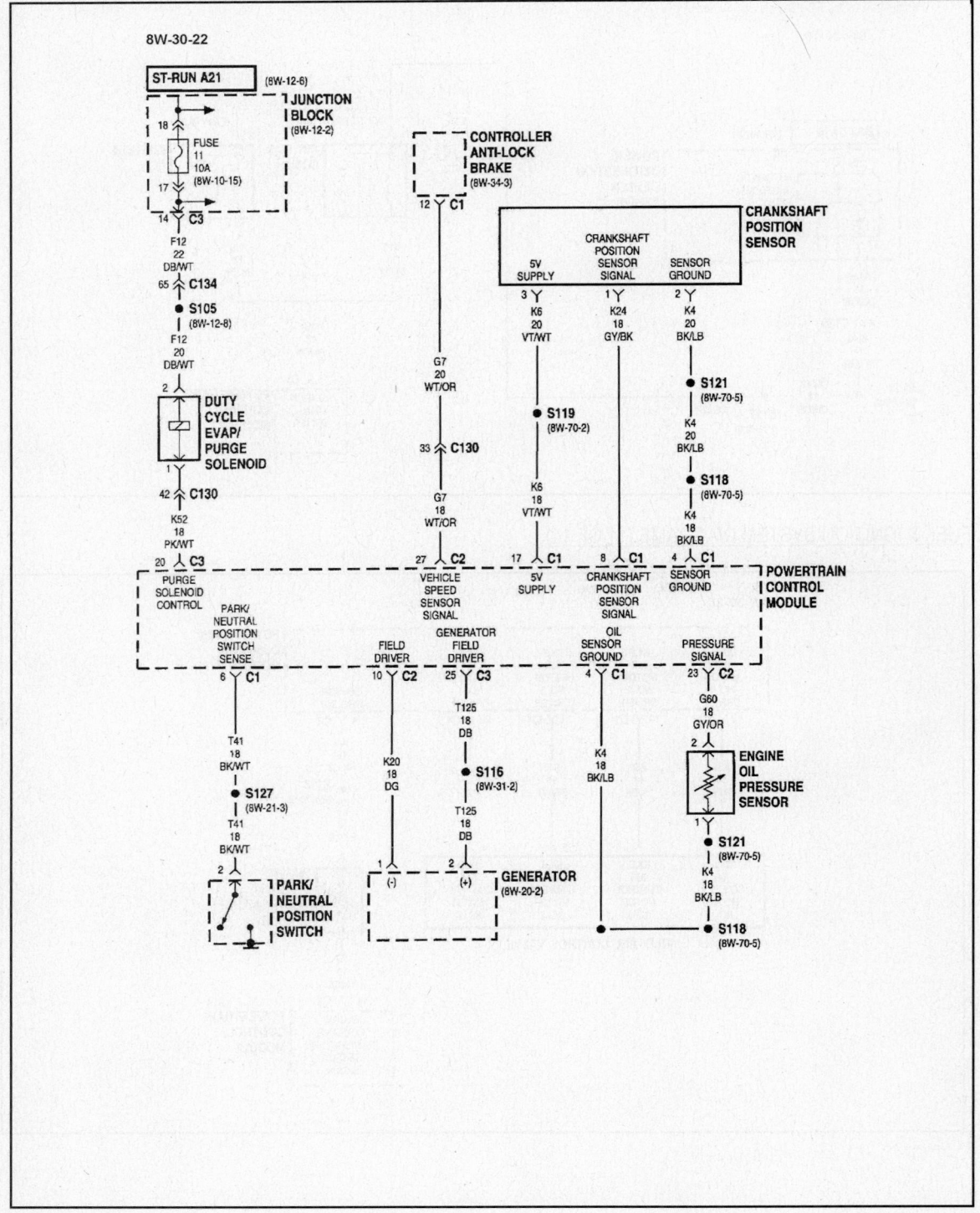

FUEL & IGNITION SYSTEM DIAGRAMS (13 OF 14)

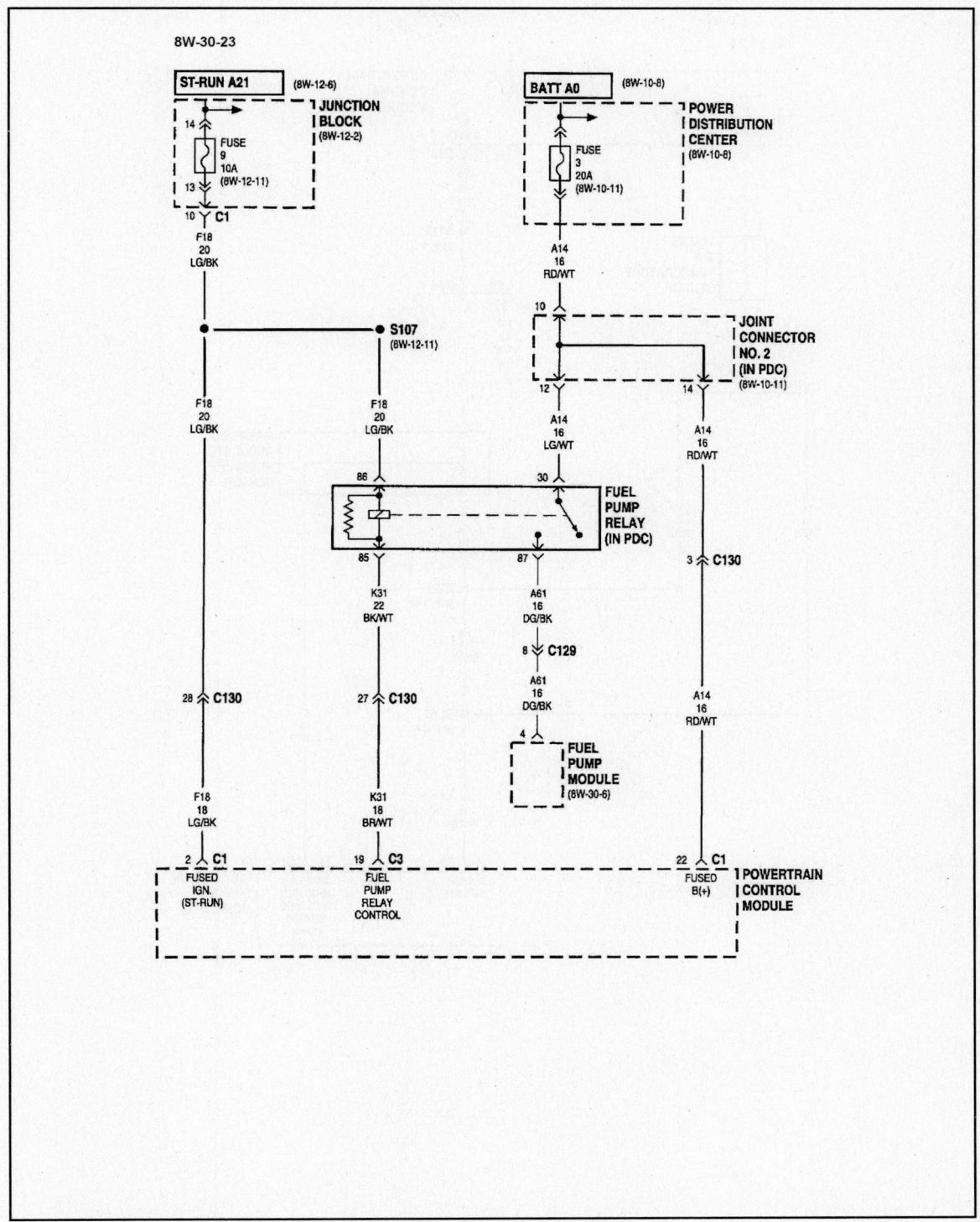

FUEL & IGNITION SYSTEM DIAGRAMS (14 OF 14)

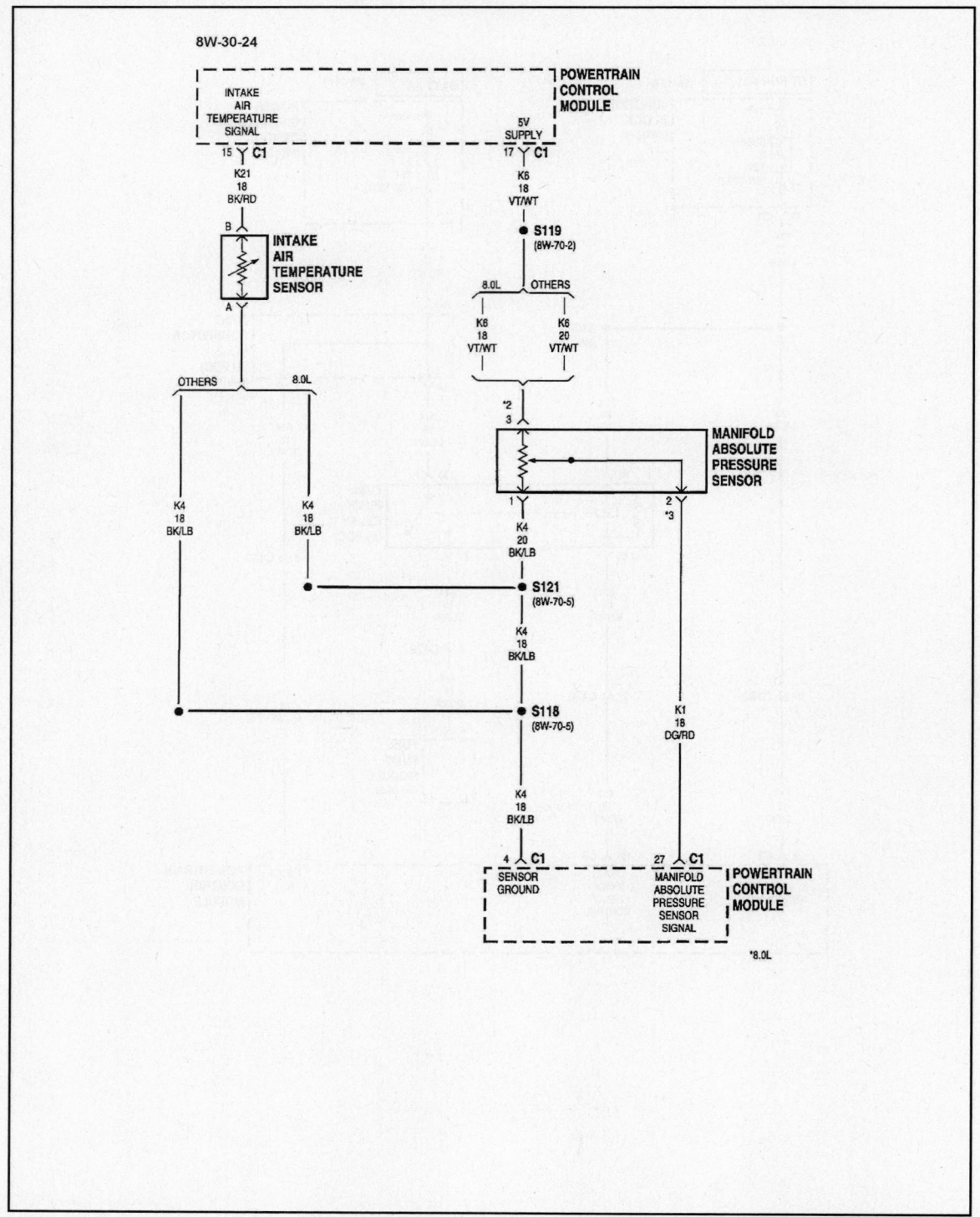

SECTION 4 CONTENTS

1999 Town & Country - Powertrain

1999 Town & Country - Body

1999 TOWN & COUNTRY – POWERTRAIN

Introduction

HOW TO USE THIS SECTION

This section of the manual includes diagnostic and repair information for Chrysler Minivan applications. This information can be used to help you understand the Theory of Operation and Diagnostics of the electronic controls and devices on this vehicle. The articles in this section are separated into three sub categories:

- Body Control Module
- Powertrain Control Module
- Transmission Control Module

1999 Town & Country 3.8L V6 VIN L (41TE Automatic Transaxle)

The articles listed for this vehicle include a variety of subjects that cover the key engine controls and vehicle diagnostics. Several sensor inputs and output devices are featured along with detailed descriptions of how they operate, what can go wrong with them, and most importantly, how to use the PCM, BCM, and TCM onboard diagnostics and common shop tools to determine if one or more devices or circuits has failed.

This vehicle uses an Integrated Waste-Spark Electronic Ignition system and Sequential Fuel Injection (SFI). These systems are primarily controlled by inputs from Hall effect camshaft and crankshaft position sensors. These devices and others (i.e., EVAP, EGR, IAC, MAP, Oxygen sensors, etc.) are covered along with how to test them.

Choose The Right Diagnostic Path

In most cases, the first step in any vehicle diagnosis is to follow the manufacturers recommended diagnostic path. In the case of Chrysler vehicle applications, the first step is to connect a Scan Tool (OBD II certified for this vehicle) and attempt to communicate with the engine controller or any other vehicle onboard controllers that apply. Diagnostic Trouble Codes (DTCs), data (PIDs), bi-directional controls, operating mode charts, and symptom charts can then be used to pinpoint the fault.

The Malfunction Indicator Lamp (MIL) can be a great help to you during diagnosis. You should note if the customer complaint included the fact that the MIL remained on during engine operation. And be sure to determine if the MIL comes on at key on and goes out after a 4 second bulb-check. There is more information on using the MIL as part of the diagnostic path in Section 1.

Diagnostic Points to Consider

If a problem is detected in a vehicle similar to this one (whether it is a trouble code or no code condition), you need to be able to determine which vehicle system (or systems) are involved before you proceed with any testing. While this may sound like a fairly routine way to approach problem solving, it is a key point to consider before you start testing.

Due to the complexity of these vehicles, a trouble code does not always point to a failed component - instead, it may point to a system that needs to be tested for some kind of problem. For example, if an Oxygen sensor trouble code is set, you need to consider what other conditions could cause an oxygen sensor to behave improperly.

Component Diagrams

You will find numerous "customized" component diagrams used throughout this section. Note the dashed lines in each Schematic, as they represent the circuits the control module uses to determine when a circuit has a fault.

Vehicle Identification

VECI DECAL

The Vehicle Emission Control Information (VECI) decal is located in the engine compartment. The information in this decal relates specifically to this vehicle and engine application. The specifications on this decal are critical to emissions system service.

EGR/TWC/HO2S (2)/SFI

These designators indicate this vehicle is equipped with an EGR, a three-way catalyst (TWC), 2 Heated Oxygen Sensors (HO2S) and sequential fuel injection (SFI).

OBD II Certified

This designator indicates that this vehicle has been certified as OBD II compliant.

LEV

If this designator is used, the vehicle conforms to U.S. EPA and State of California Low Emissions Vehicle standards.

Underhood View Graphic

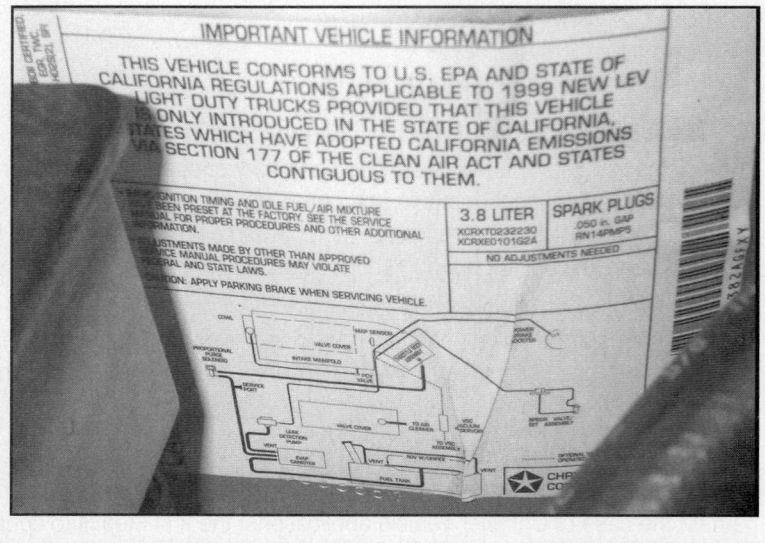

Electronic Engine Control System

OPERATING STRATEGIES

The Powertrain Control Module (PCM) in this application is a triple-microprocessor digital computer. The PCM, formerly known as the SBEC, provides optimum control of the engine and various other systems. The PCM uses several operating strategies to maintain good overall drivability and to meet the EPA mandated emission standards.

These strategies are included in the tables on this page.

Based on various inputs it receives, the PCM adjusts the fuel injector pulsewidth, the idle speed, the amount of ignition advance or retard, the ignition coil dwell and operation of the EVAP canister purge valve, EVAP leak detection (LDP) and EGR solenoid.

The PCM also controls the Speed Control (S/C), A/C and some transmission functions.

PCM Location

The PCM is located at the left side of the engine compartment.

Operating Strategies

A/C Compressor Clutch Cycling Control	Closed Loop Fuel Control
Electric Fuel Pump Control	Limited Transmission Control
Fuel Metering of Sequential Fuel Injectors	Speed Control (S/C)
Idle Speed Control	Ignition Timing Control for A/F Change
Vapor Canister Purge Control	EGR Control

Operational Control of Components

A/C Compressor Clutch	Ignition Coils (Timing and Dwell)
Fuel Delivery (Injector Pulsewidth)	Fuel Pump Operation
Cooling Fan	Purge Solenoid/Leak Detection Pump
EGR Solenoid	Idle Air Control Motor
Malfunction Indicator Lamp	Integrated Speed Control

Items Provided (or Stored) By the Electronic Engine Control System

Shared CCD Data Items	Adaptive Values (Idle, Fuel Trim, etc)
DTC Data for Non-Emission Faults	DTC Data for Emission Related Faults

Distinct Operating Modes

Key On	Cruise
Cranking	Acceleration (Enrichment)
Warm-up	Deceleration (Enleanment)
Idle	Wide Open Throttle (WOT)

Computer Controlled Charging System

INTRODUCTION

This application uses either a 90 or 120 amp Nippondenso generator. These generators do not contain voltage-regulating devices. The regulation of the charging system voltage is accomplished through an integrated circuit in the PCM, called the Electronic Voltage Regulator (EVR).

Charging System Operation

The Charging system is powered through the ignition switch. With the ignition in the "on" position, battery voltage is applied to the generator rotor, through one of the two field terminals, to produce a magnetic field.

The amount of DC current produced by the generator is controlled by the EVR circuitry in the PCM and engine speed. This circuitry is connected in series with the second rotor field terminal and ground.

Battery Temperature Sensor data (calculated) and system voltage are used by the PCM to vary the charging rate. The PCM accomplishes this by cycling the ground path to control the strength of the rotor magnetic field. Temperature dependant charging strategies like this one offer improved charging efficiency, slightly improved mileage and extended battery life.

Charging System Diagnosis

In addition to conventional voltage output and load testing, this charging system can be effectively diagnosed using a Scan Tool (PIDs, DTCs and bi-directional controls) and a Lab Scope or DVOM.

Many charging system faults will set codes. It is important to check for codes before beginning a conventional charging system diagnosis. The system is monitored for a too high or too low voltage condition. The PCM can also monitor it's own ability to switch the control circuit to ground to regulate the charging rates.

Charging System Schematic

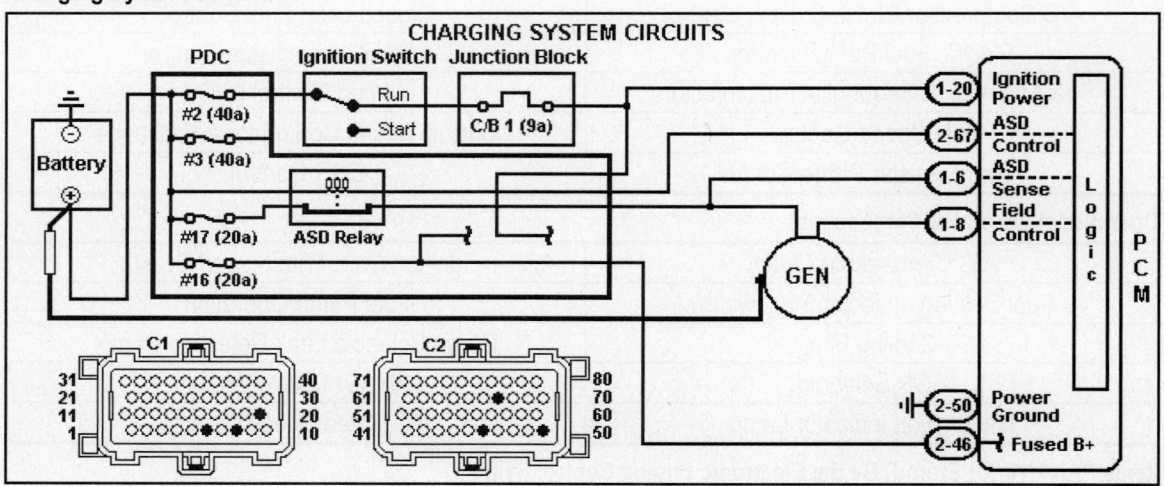

LAB SCOPE TEST (GENERATOR)

The Lab Scope can be used to test the Computer Controlled Generator to determine if the generator is not capable of charging the battery, or if it is not being controlled by the PCM. Place the shift selector in Park and block the drive wheels for safety.

Scope Connections
Connect the Channel 'A' positive probe to the field control circuit (DK GN wire) at PCM Pin 1-8 and the negative probe to the battery negative post.

Lab Scope Examples
In these examples, the traces show the generator field command from the EVR (in the PCM) to the generator at Hot Idle.

Lab Scope Example (1)
In example (1), the trace shows the generator command with no accessory load. The generator is being commanded "off" more than "on" (the voltage is high longer than it is low). When the circuit is switched to ground, the circuit is complete and current is produced. Therefore, the longer the trace is low, the higher the current output.

Lab Scope Example (2)
In example (2), multiple accessories were turned "on" to load the system. The PCM responded by increasing the amount of time the circuit was pulled to ground, allowing the generator to produce more current.

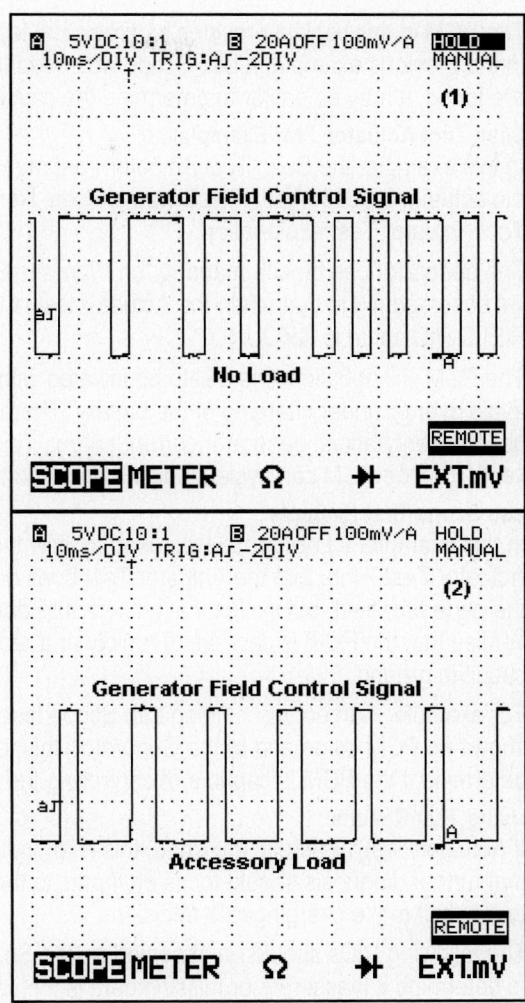

Charging System Test Schematic

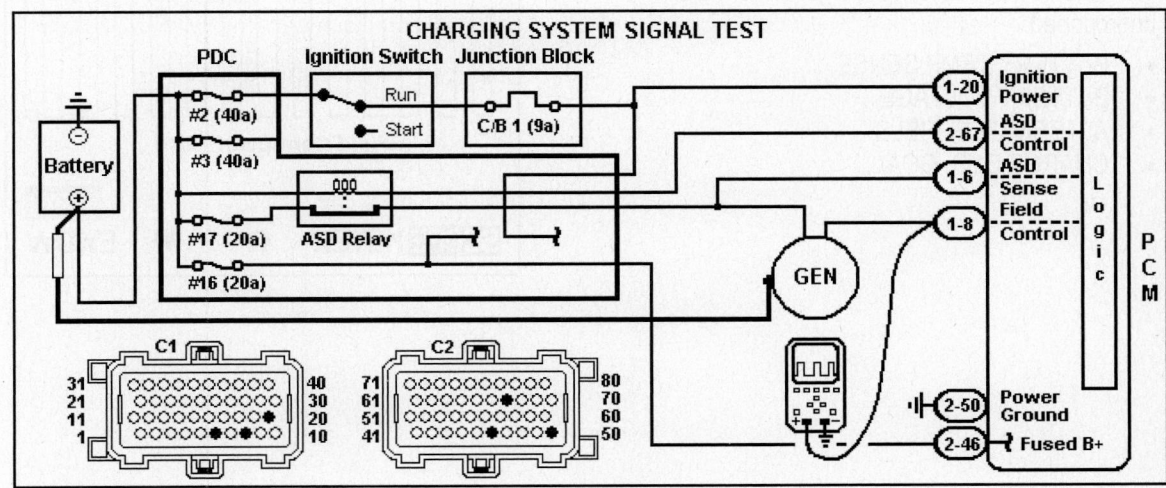

BI-DIRECTIONAL TEST (GENERATOR)

The PCM in this vehicle provides bi-directional testing capabilities, which can be accessed with a Scan Tool. In the previous example, a Lab Scope was used to view the PCM to generator command. If abnormalities are found, it may be easier to determine the cause of the fault using these bi-directional controls.

Scan Tool Actuator Test Example

In this example, the Scan Tool Actuator Test menu was accessed, and the actuator test 16: ALT. FIELD was chosen. Refer Section 1 for a Scan Tool Actuator Test explanation.

The generator can be commanded "on" from the Scan Tool, and the data list observed. Note that when the Actuator test is ON, the ALTERNATOR FIELD PID value is GROUND.

The PCM will not allow this test to be initiated with the engine running to avoid over or under charging of the battery. Therefore this test cannot be used to verify actual generator current output. This test is only used to verify that the PCM can cycle the field command to ground.

Lab Scope Test Example

In this example, a Lab Scope was used to view the results of the Actuator Test. Note that the Actuator Test does not constantly ground the generator field, but cycles it at a fixed 50% duty cycle. This test verifies that the PCM is capable of rapidly and repeatedly cycling this circuit to ground.

For example, if an engine running Lab Scope test (on the previous page) shows no PCM command to the Generator, this test should be used to determine if the PCM is capable of controlling the generator field.

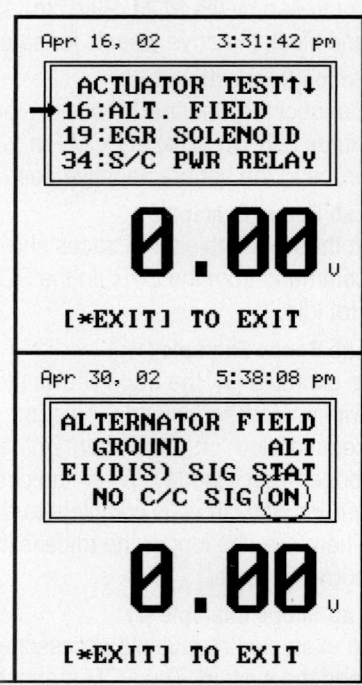

Using Test Results

If Actuator Test 16 results are good, this is an indication that further diagnosis should focus on inputs to the PCM that help it make charging rate decisions.

The following PIDs should be observed on the Scan Tool to determine if inaccurate or missing data is the cause of the PCM decision to leave the generator field ungrounded:

- BATTERY TEMP SENSE
- BATTERY VOLTAGE
- ALTERNATOR FIELD
- CHARGE SYS. GOAL.

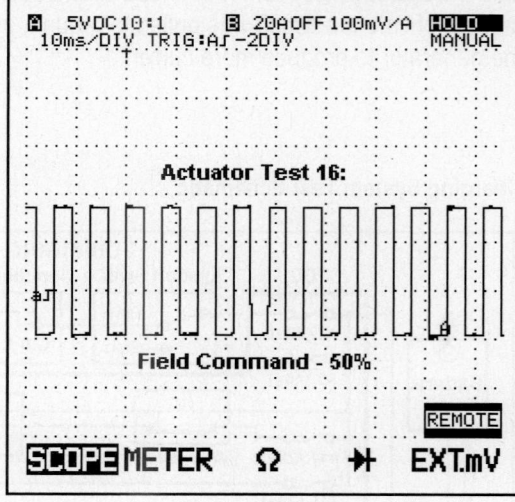

Integrated Electronic Ignition System

INTRODUCTION

The PCM controls the operation of the Integrated Electronic Ignition (EI) system on this vehicle application. Battery voltage is supplied to the coil pack (3 coils in one assembly) through the ASD relay and #17 fuse (20A) in the Power Distribution Center (PDC).

The PCM controls the individual ground circuits of the three (3) ignition coils. By switching the ground path for each coil "on" and "off" at the appropriate time, the PCM adjusts the ignition timing for each cylinder pair correctly to meet changing engine operating conditions. The coil pairs are identified as Coil packs 1, 2 and 3. Coil '1' fires cylinders 1 & 4; Coil '2' fires cylinders 2 & 5 and Coil '3' fires cylinders 3 & 6. The firing order on this engine is 1-2-3-4-5-6.

EI System Sensor Operation

The PCM makes ignition and injector-timing decisions based on inputs from the CMP and CKP sensors. This is an asynchronous system, which means that neither the CMP nor the CKP sensor signals occur at TDC. For ignition timing, the PCM needs both signals to calculate which cylinder pair is next in sequence, and when it is expected to arrive at TDC. Because this is a waste spark system, the PCM does not need to know which cylinder in the pair is at TDC on the compression stroke.

The CKP and CMP sensors are Hall-Effect devices that toggle internal transistors each time a window is sensed on the torque converter drive plate or camshaft sprocket. When the transistor toggles, a 5v pulse is sent to the PCM.

EI Ignition System Component Graphic

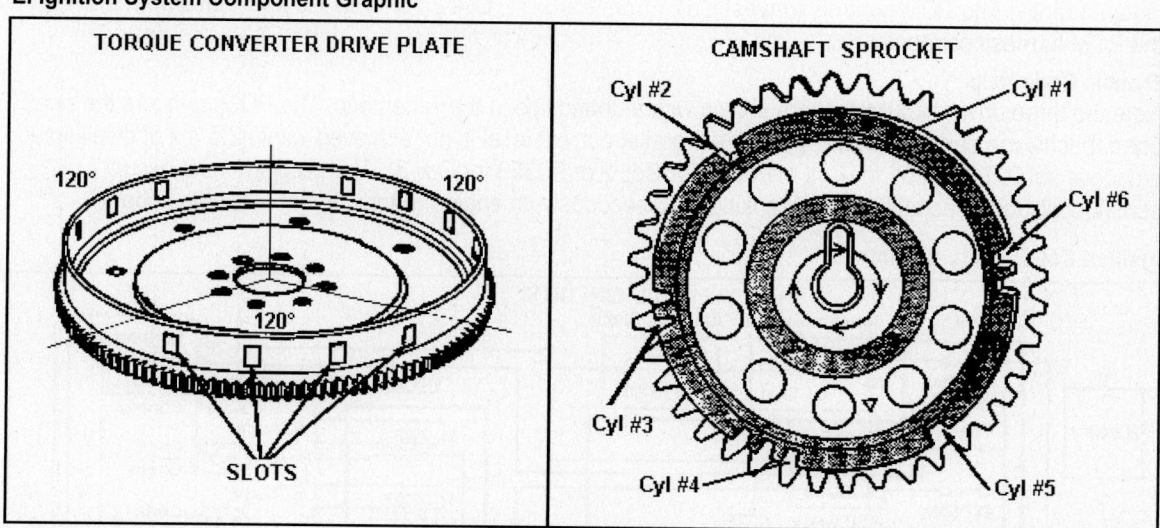

Ignition Spark Timing

The ignition timing on this EI system is entirely controlled by the PCM - base timing is not adjustable. Do not attempt to check the base timing - you will receive false readings.

Once the engine starts, the PCM calculates the spark advance using these inputs:

- BARO Sensor (MAP signal at KOEO)
- ECT Sensor
- Engine Speed Signal (rpm)
- MAP Sensor
- Throttle Position

IGNITION COIL OPERATION

The ignition coil pack consists of three (3) independent coils molded together. The coil assembly is mounted to the intake manifold. Spark plug cables rout to each cylinder from the coil assembly.

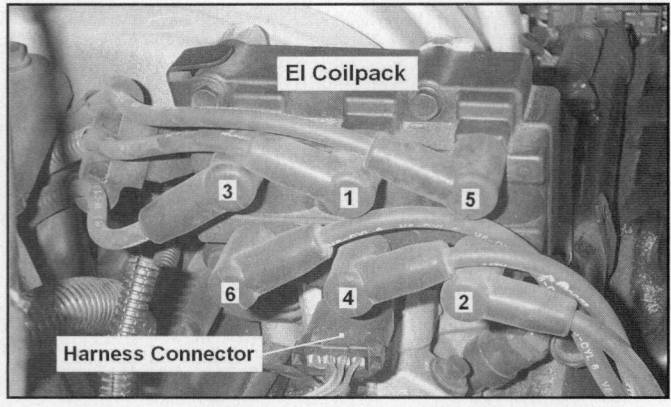

The coil fires two spark plugs before each power stroke. One cylinder fires on compression, the other cylinder fires on exhaust (waste spark method).

The PCM determines which coil to charge and fire. In this example, Coil '1' fires cylinders 1 & 4; Coil '2' fires cylinders 2 & 5 and Coil '3' fires cylinders 3 & 6. The firing order on this engine is 1-2-3-4-5-6.

The spark plug that fires the cylinder on the compression stroke uses the majority of the ignition coil energy and the spark plug that fires the cylinder on the exhaust stroke uses very little coil energy. The spark plugs are connected in series, so one spark plug firing voltage will be negative and the other spark plug will be positive with respect to ground.

Coil resistance is 0.45-0.65 Ohms primary, and 7K-15.8K Ohms secondary (70-80 °F).

El Coil pack Circuits

The coil pack assembly is powered by the ASD relay. However, drivers in the PCM control each coil independently. The PCM controls (drives) Coil 1 from Pin 1-11, Coil 2 from Pin 1-3 and Coil 3 from Pin 1-2 of the PCM harness connector.

Trouble Code Help

Note the three driver circuits that are shown with a dotted line in the schematic. The PCM monitors the peak current achieved during the dwell period. If a preset current level is not achieved within 2.5 ms of dwell time, a code is set (DTC P0351 for Coil 1, P0352 for Coil 2 or P0353 for Coil 3). The problem must occur continuously for 3 seconds during cranking, or 6 seconds with engine running to set a trouble code.

Ignition Coil Primary Schematic

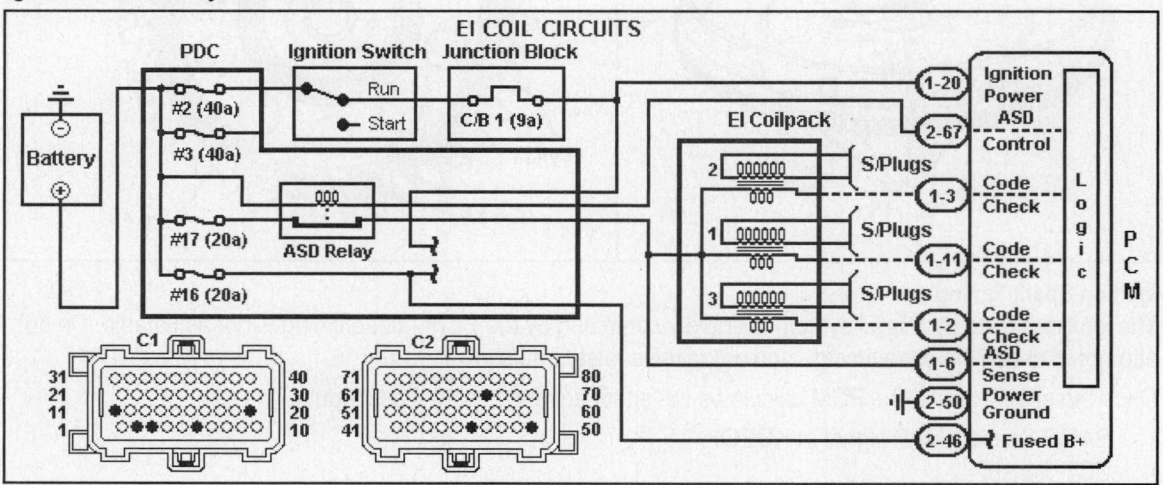

LAB SCOPE TEST (COIL PRIMARY)

The ignition coil primary includes the switched ignition feed circuit, coil primary circuit and the coil driver control circuits at the PCM.

Lab Scope Connections

Connect the Channel 'A' positive probe to one of the three (3) coil control circuits at the PCM or coil pack, and the negative probe to the battery negative post.

Scope Settings

To make the waveforms as clear as possible, set the scope settings to match the examples.

Lab Scope Example (1)

In example (1), the trace shows the coil driver (primary) circuit at idle. This is a known good waveform at Hot Idle. On DIS vehicles, primary ignition waveforms are generally more stable than secondary ignition waveforms.

Lab Scope Example (2)

In example (2), the time and voltage per division were changed. At only 2v per division, most of the waveform cannot be seen.

This setting can be used to check coil driver performance and ground voltage drop in the circuit under load. The normal rise of the waveform voltage above ground is referred to as 'counter-voltage.' It is more important to look for consistency among circuits than to apply an absolute limit to this voltage value.

Note that Channel 'B' was set to ground to help visualize the amount of counter-voltage.

Lab Scope Example (3)

In example (3), a low-amps probe was added to Channel "B" to check the current draw of the circuit. Clamp the probe around the same wire that was back probed for Channel 'A'. The plastic harness cover can be peeled back near the coil pack connector for easy probe connection.

This is a known good current value. Circuit resistance, opens and shorts affect the current waveform. Therefore, this test and is an excellent way to find these problems that are difficult to see in voltage waveforms.

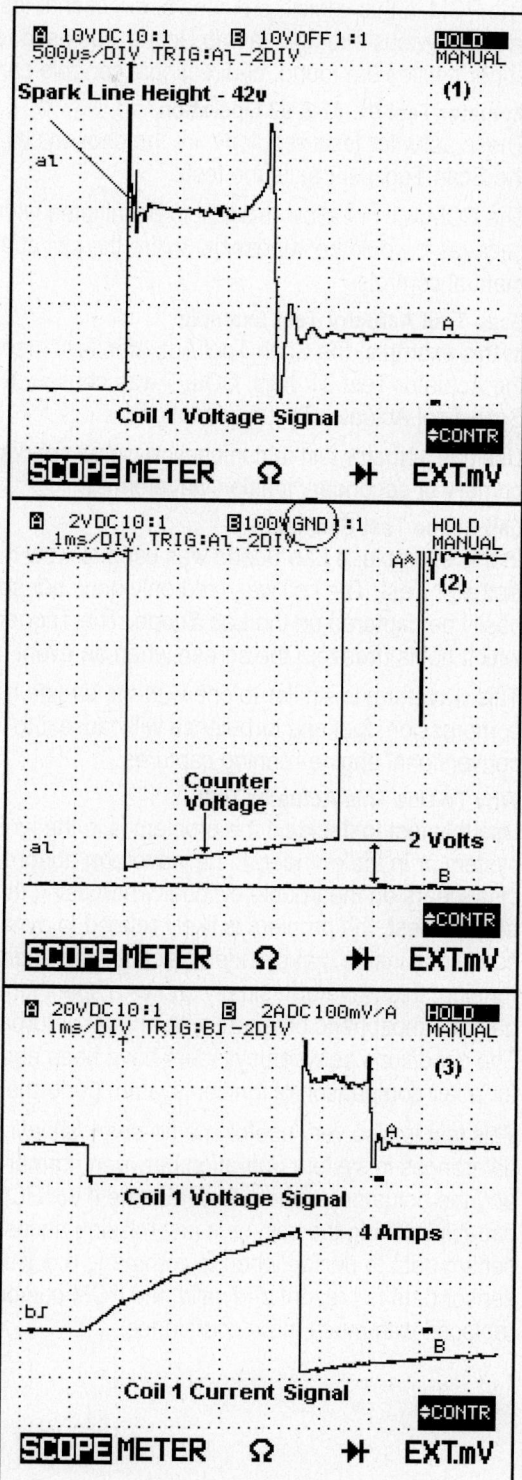

BI-DIRECTIONAL TEST (COIL PRIMARY)

The PCM in this vehicle provides bi-directional testing capabilities, which can be accessed with a Scan Tool. In the previous example, a Lab Scope was used to view the Ignition primary waveforms. If waveform abnormalities are found, bi-directional controls can help to isolate the problem.

Actuator Test 01, 02 & 03 Operation
These actuator tests regularly fire the chosen coil for 5 minutes, or until the Scan Tool user exits the tests.

The PCM will not allow this test to be initiated with the engine running because it would be impossible to fire the coil at the proper time using manual controls.

Scan Tool Actuator Test Example
In this example, the Scan Tool Actuator Test menu was accessed, and the Actuator Test 01: IGN. COIL 1 was chosen. Refer Section 1 for a Scan Tool Actuator Test explanation.

Use this actuator test in conjunction with a spark tester or Lab Scope (for primary or secondary ignition waveforms).

Lab Scope Test Example
In this example, a Lab Scope was used to view the results of the Actuator Test. The coil was fired only once per second and could not easily be captured on the Lab Scope. The Trigger was set to 'single', which holds (freezes) the screen when an event occurs.

This waveform is similar to one with the engine running. However, lack of compression, fuel and turbulence will cause it to look different than conventional engine-running captures.

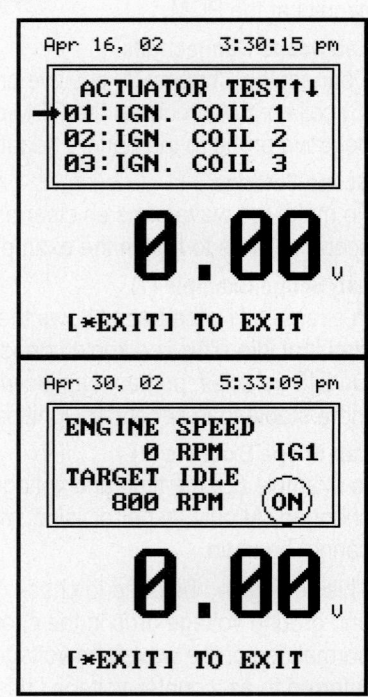

Why To Use This Actuator Test
Use this test to decide if the problem is in the ignition system or in the cylinder. If the waveform abnormality disappears on the trouble cylinder when using this actuator test, the problem is likely related to dynamic conditions inside that cylinder. Because the engine is not running, this waveform simply shows a spark jumping a gap at atmospheric pressure with no fuel or turbulence. The conditions between cylinders have been equalized for a fair comparison of ignition system performance

This test is also very useful in a no-spark situation. Any differences in system operation between normal cranking and the actuator test help to determine if the PCM is not capable of firing the coil, or is just missing necessary sensor data to decide when to command a coil on. If the sensor data is present and valid, the PCM can be replaced with much more confidence.

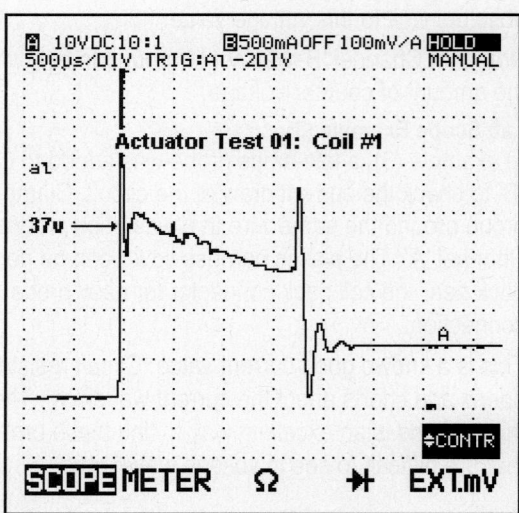

CRANKSHAFT POSITION SENSOR (HALL EFFECT)

The crankshaft position (CKP) sensor used with this EI system provides crankshaft position and speed data to the powertrain control module (PCM).

The CKP sensor is a Hall effect sensor mounted on the transaxle, adjacent to the torque converter drive plate. There are 12 slots in a ring on the drive plate (in three groups of 4 slots). The CKP sensor Hall effect circuit senses each window and toggles a transistor. Each toggling of the transistor sends a 5v pulse to the PCM that represents the drive plate slot.

By using all 12 slots, the PCM can track engine speed and variations in crankshaft rotation speed for misfire diagnosis.

Because the slots are evenly spaced around the

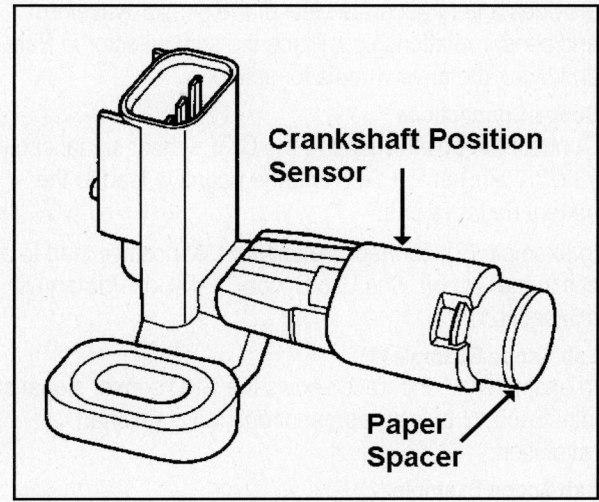

Crankshaft Position Sensor

Paper Spacer

ring, the PCM can also use these signals to keep track of general crankshaft position. However, the PCM needs the CMP sensor signal to calculate the exact crankshaft position.

CKP Sensor Circuits
The CKP Sensor is connected to the PCM by 3 wires. The CKP and CMP sensors share an 8v supply voltage from PCM Pin 2-44. The 0-5v CKP signal is sent to PCM Pin 1-32. The sensor ground connects to the PCM at Pin 2-43. The following sensors share this ground: CMP, CKP, ECT, MAP, KS, TP, TCM and A/C pressure transducer.

DVOM TEST (CKP SENSOR)

Place the shift selector in Park and block the drive wheels for safety.

DVOM Connections
Connect the DVOM positive lead to the CKP sensor signal circuit (GY/BK wire) at Pin 1-32 and the negative lead to the battery negative post.

Test Results
This capture was taken at Hot Idle. The DVOM display shows a CKP signal of only 0.44v DC because the instrument is averaging the signal voltage. The average value is low because the signal is low much longer than is high (note the 93.1 % duty cycle). Use the DVOM MIN/MAX feature, if available, to verify that the CKP signal is really switching from 0-5v.

Another way to use the DVOM to test the CKP sensor signal is to observe the frequency of the signal. In this

example, the display shows 156.7 Hz. (156.7 cycles per second X 60 seconds / 12 slots per revolution = 783.5 RPM). Using this formula for a no start, it is easy to calculate that at 200 RPM cranking speed, the DVOM should read about 40 Hz.

LAB SCOPE TEST (CKP SENSOR)

The Lab Scope can be used to test the CKP sensor as it provides a very accurate view of the sensor waveform and sensor relationships. Place the shift selector in Park and block the drive wheels for safety.

Scope Connections
Connect the positive lead to the CKP sensor signal circuit (GY/BK wire) at Pin 1-32 and the negative lead to the battery negative post.

In example (2), connect the channel 'A' positive lead to a coil trigger circuit. See Lab Scope Test (Coil Primary) in this section.

Lab Scope Example (1)
In example (1), the trace shows the CKP sensor signal at idle. Each 12 pulses represent one full crankshaft revolution.

Lab Scope Example (2)
In example (2), the trace shows the CKP and coil primary signals. Using another related signal, like coil primary, can help to stabilize the Lab Scope screen.

Some CKP pulses on this application are 20° apart and some 60° apart (between groups). Each time the display is refreshed the Lab Scope may trigger on a different pulse, and the trace will appear unstable. Because the coil fires at the same point every crankshaft revolution, primary is a much better trigger. This strategy will also make it much easier to see intermittent CKP signal failures.

CKP Sensor Test Schematic

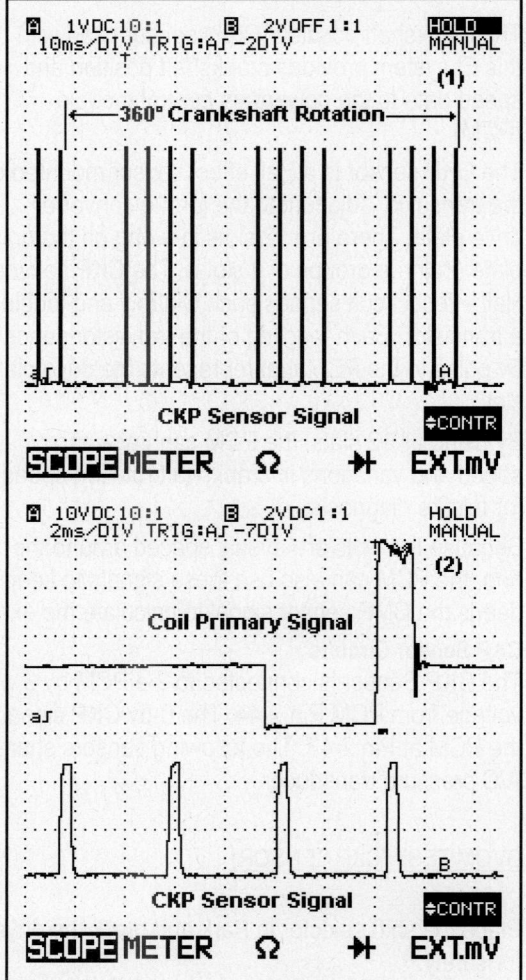

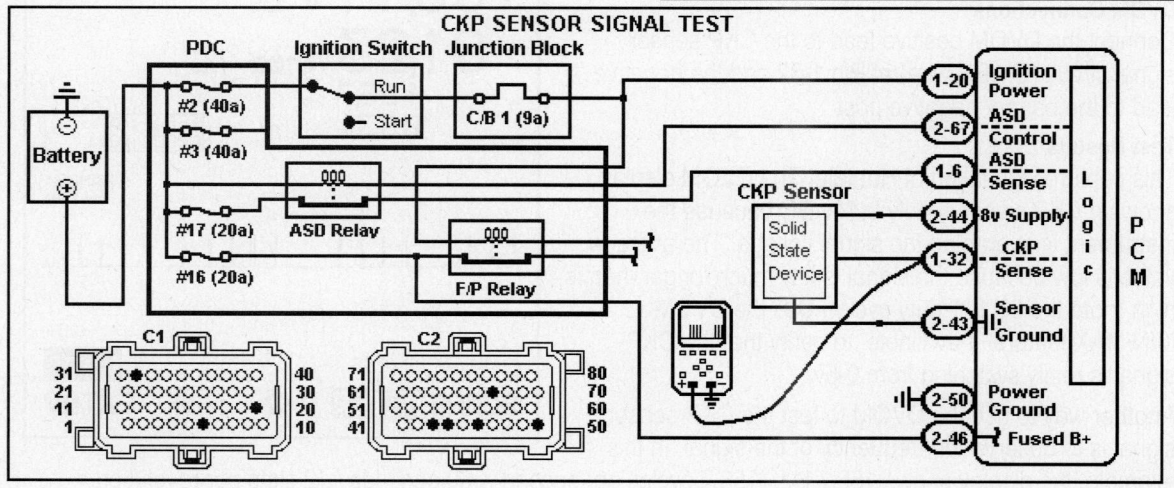

Enhanced Evaporative Emission System

INTRODUCTION

The Evaporative Emissions (EVAP) system is used to prevent the escape of fuel vapors to the atmosphere under hot soak, refueling and engine off conditions. Any fuel vapor pressure trapped in the fuel tank is vented through a vent hose, and stored in the EVAP canister.

The most efficient way to dispose of fuel vapors without causing pollution is to burn them in the normal combustion process. The PCM cycles the purge valve on (open) during normal operation. With the valve open, manifold vacuum is applied to the canister and this allows it to draw fresh air and fuel vapors from the canister into the intake manifold. These vapors are drawn into each cylinder to be burned in the engine during combustion.

The PCM uses various sensor inputs to calculate the desired amount of EVAP purge flow. The PCM meters the purge flow by varying the duty cycle of the EVAP purge solenoid control signal. The EVAP purge solenoid will remain off during a cold start, or for a preprogrammed time after a hot start.

This system uses a positive-control purge solenoid. The PCM senses voltage drop on the ground side of the solenoid circuit, to verify and control circuit operation.

System Components:

The EVAP system contains many components for the management of fuel vapors and to test the integrity of the EVAP system. The EVAP system is constructed of the following components:

- Fuel Tank
- Fuel Filler Cap
- Fuel Tank Level Sensor
- Fuel Vapor Vent
- Evaporative Charcoal Canister
- Leak Detection Pump
- EVAP Purge Solenoid
- PCM and related wiring
- Fuel vapor lines with Test Port

EVAP System Schematic

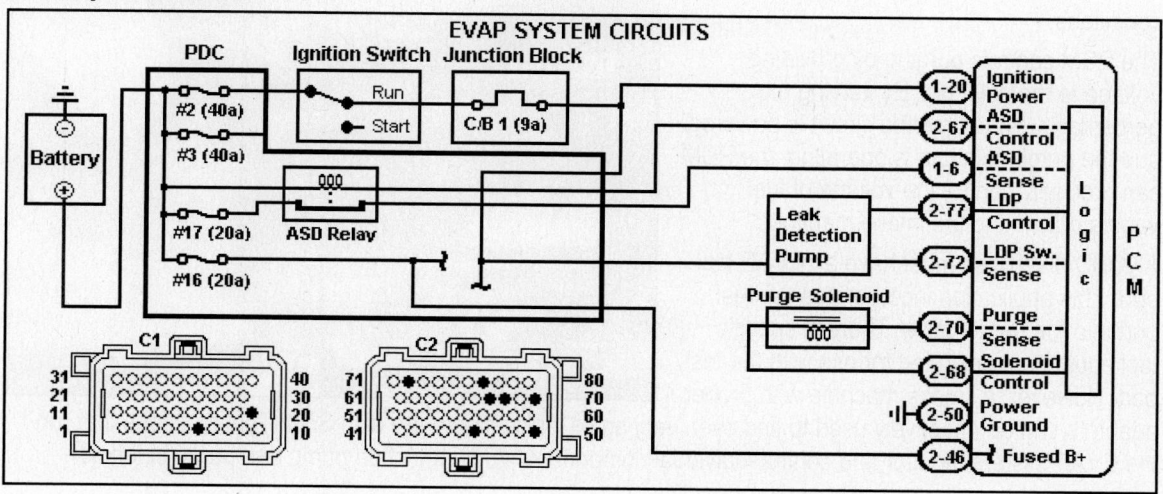

EVAPORATIVE CHARCOAL CANISTER

The EVAP canister is located on the frame rail under the driver's seat. Fuel vapors from the fuel tank are stored in the canister. The canister is connected in series between the fuel tank and the purge solenoid. Although the Leak Detection Pump (LDP) pressurizes the entire EVAP system, the pressure enters the system through the canister.

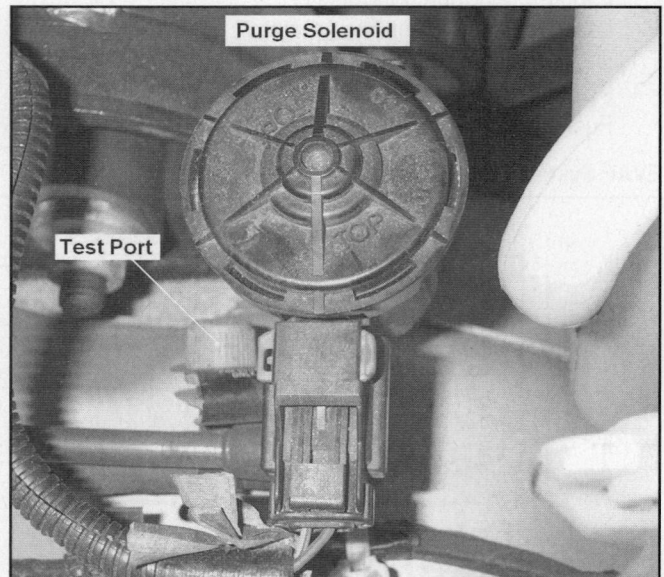

Once the engine is in closed loop, fuel vapors stored in the charcoal canister are purged into the engine where they are burned during normal combustion.

The activated charcoal has the ability to purge (or release) any stored fuel vapors when it is exposed to fresh air.

EVAP PURGE SOLENOID

The EVAP canister purge solenoid is a normally closed (N.C.) solenoid controlled by the PCM. A single purge solenoid, located by the upper motor mount on the right fender, allows vapors from a charcoal canister to enter the manifold during various engine conditions.

The PCM controls purging by providing voltage to the solenoid. By varying the percentage (%) of time the circuit is powered and the purge solenoid is operating, the PCM can precisely control the volume of fuel vapors drawn into the intake manifold.

All OBD II vehicles must have an EVAP test port. This application incorporates the test port into the purge valve. There is specific test equipment designed for use with the test port. However, a smoke machine with proper adaptors can be effectively used to find even very small leaks. There are also Scan Tool functions that run the EVAP system monitor and control individual components (leak detection pump and purge solenoid).

BI-DIRECTIONAL TEST (PURGE SOLENOID)

The PCM in this vehicle provides bi-directional testing capabilities, which can be accessed with a Scan Tool.

Actuator Test 14 Operation
This actuator test cycles the purge solenoid ground circuit at a fixed duty cycle and frequency for 5 minutes, or until the Scan Tool user exits the test.

Scan Tool Actuator Test Example
Connect the Scan Tool to the DLC under driver's side of I/P. In this example, the Scan Tool Actuator Test menu was accessed, and the actuator test 14: CANISTER PURG was chosen. Refer Section 1 for a Scan Tool Actuator Test explanation.

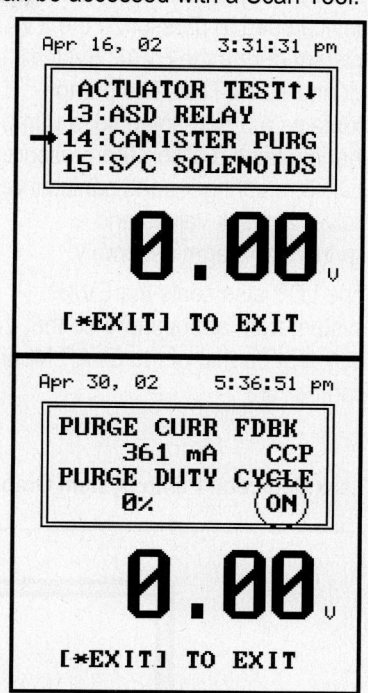

Observe the PURGE CURR FDBK PID to test the current flow through the solenoid windings. In this known good example, the PID value was 361 mA. Note that the PURGE DUTY CYCLE PID value is 0% during the test. These PID values may not always accurately reflect the actual circuit value during Actuator Tests. If in doubt, use a DVOM or Lab Scope to verify circuit activity.

Lab Scope Test Example
Connect the Channel 'A' positive probe to the solenoid control circuit (PK/BK wire), and connect the negative probe to the battery negative post.

Use this actuator test in conjunction with a Lab Scope to verify purge solenoid circuit activity, and that the PCM is capable of cycling the solenoid ground circuit.

In this example, the actuator test has been turned ON with the Scan Tool. The purge solenoid is cycled at a fixed 50% duty cycle, and at a fixed frequency of 200 Hz.

Why To Use This Actuator Test
The PCM makes purge decisions based on various criteria. It is much easier to simply command the solenoid to operate than to keep track of these various criteria.

During this test, the desired results are already known, which helps to keep the diagnosis focused. Use this test to decide if the problem is in the purge solenoid, wiring or PCM.

The results of this actuator test will indicate the next test step. This actuator test helps to determine if the PCM is not capable of controlling the solenoid, or is just missing necessary sensor data to decide when to cycle the command to the solenoid.

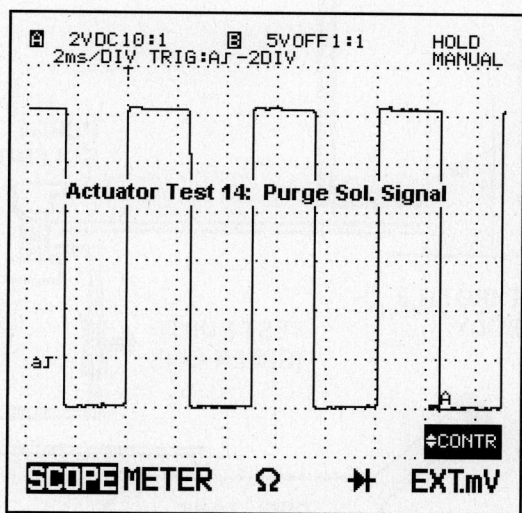

LEAK DETECTION PUMP (LDP)

The Leak Detection Pump (LDP) is a device used to pressurize the EVAP system during the EVAP System Monitor Leak Test. The pump contains a 3-port solenoid, a pump that contains a calibrated pressure switch, a spring-loaded canister vent valve, 2 check valves and a spring/diaphragm assembly.

The LDP also seals the EVAP system from atmosphere so that the leak test portion of the EVAP Monitor can be run.

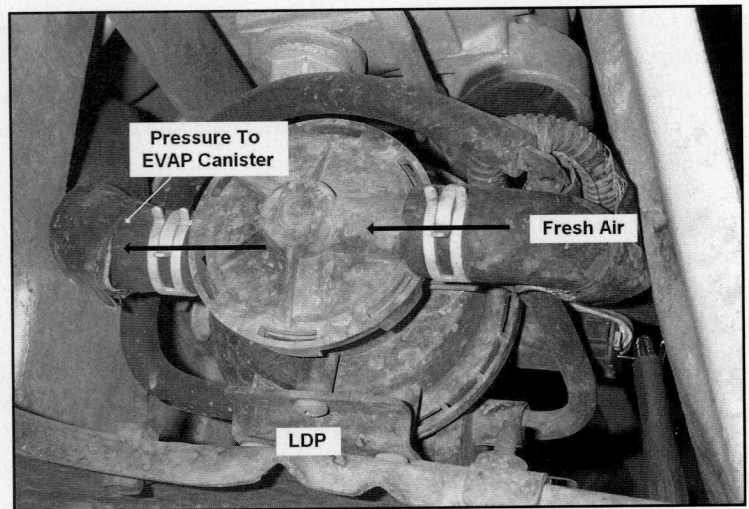

Leak Detection Pump System Graphic

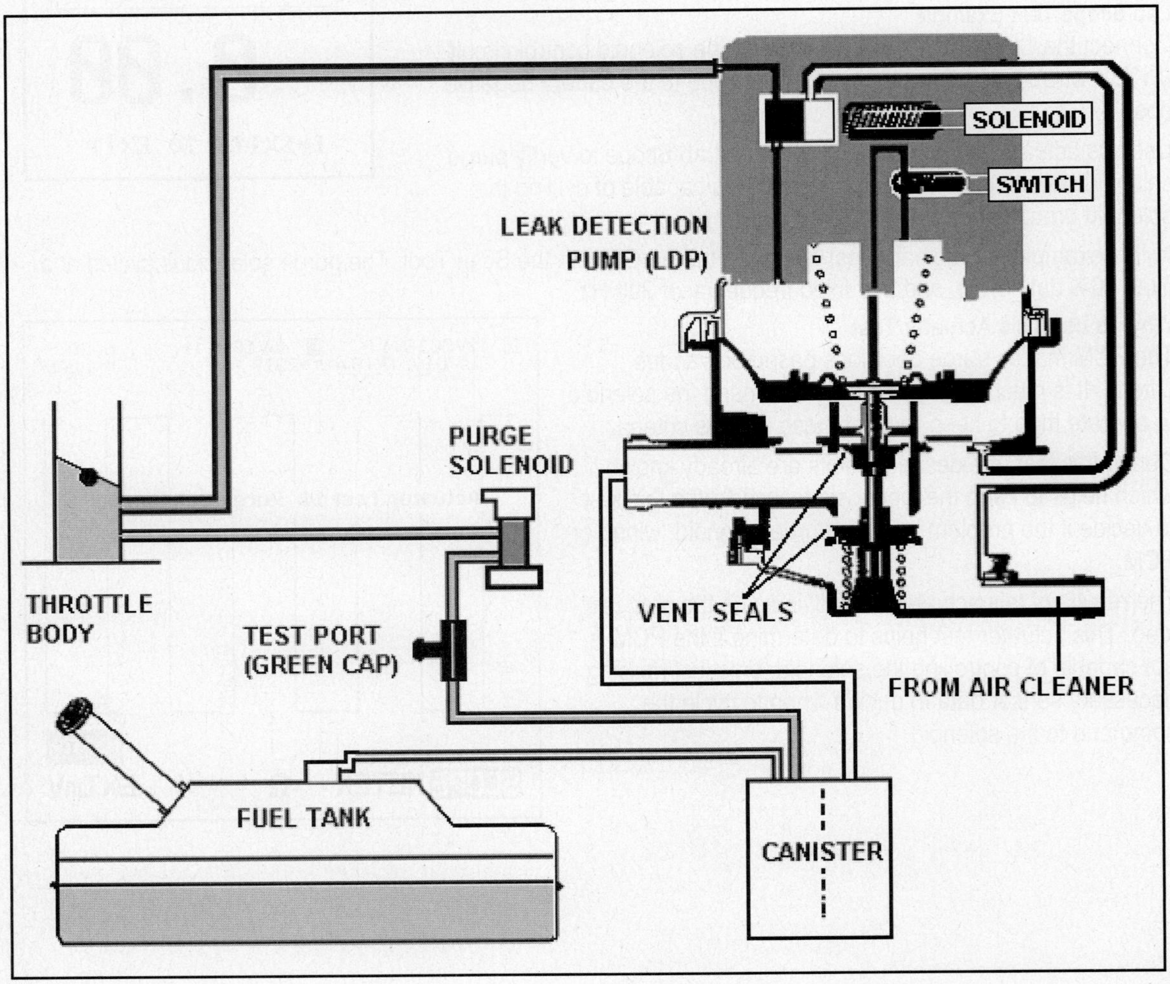

LDP MONITOR OPERATION

Leak Test

Immediately after a cold startup (a period established by predetermined temperature threshold limits), the three-port solenoid is briefly energized. This initializes the pump by drawing air into the pump cavity and also closes the vent seal.

During other non-test conditions, the pump diaphragm assembly holds the pump vent seal open (it pushes the vent seal open to the full travel position). The vent seal will remain closed while the pump is cycling because the reed switch triggers the three-port solenoid to prevent the diaphragm assembly from reaching full travel.

After the brief initialization period, the solenoid is de-energized allowing atmospheric pressure to enter the pump cavity, thus permitting the spring to drive the diaphragm that forces air out of the pump cavity and into the vent system. The solenoid is energized and de energized rapidly to create flow in typical diaphragm pump fashion.

The leak detection pump is controlled in two modes:

- Pump Mode - The pump is cycled at a fixed rate to achieve a rapid pressure build in order to shorten the overall test length.
- Test Mode - The solenoid is energized with a fixed duration pulse. Subsequent fixed pulses occur when the diaphragm reaches the reed switch closure point.

The spring in the pump is set so that the system will achieve an equalized pressure of about 7.5" H20. The cycle rate of pump strokes is quite rapid as the system begins to pump up to this pressure level. As the pressure increases, the cycle rate starts to drop off. If no leaks exist in the system, the pump will eventually stop pumping at the equalized pressure. If a leak is present in the system, the pump will continue to run at a rate that represents the flow characteristic of the size of the leak.

The PCM uses this information to determine if the leak is larger than the required detection limit. If a leak is revealed during the leak test portion of the test, the test is terminated at the end of the test mode and no further system checks will be performed.

After passing the leak detection phase of the test, system pressure is maintained by turning on the LDP solenoid until the purge system is activated. In effect purge activation creates a leak. The cycle rate is again interrogated and when it increases due to the flow through the purge system, the leak check portion of the diagnostic is complete.

The canister vent valve will unseal the system after completion of the test sequence as the pump diaphragm assembly moves to the full travel position.

Purge Test

The EVAP system functionality is verified by using the stricter EVAP Purge Flow monitor. At an appropriate warm idle period the LDP is energized to seal the canister vent. The purge flow will be ramped up from some small value in an attempt to see a shift in the Fuel Control (O2S) system. If fuel vapor, indicated by a shift in the fuel control is present, the test is passed. If not, it is assumed that the EVAP Purge system is not functioning in some respect. The LDP is again turned off and the test is complete.

SCAN TOOL TEST (LDP MONITOR)

Chrysler controllers (PCMs) include the ability to have certain components actuated from a Scan Tool. In addition to running LDP and purge solenoid tests from these menus, certain System Tests are available.

The LDP MON TEST forces the EVAP System Monitor to run as it would during normal operation. It is difficult to run some of the OBD II Monitor Tests under the required conditions because this criteria and service bay conditions are rarely the same. With the LDP MON TEST, you have a unique opportunity to observe an EVAP Monitor Leak Test in action.

Scan Tool System Test
Select Chrysler Powertrain and verify the vehicle identification. From the Scan Tool main menu, select SYSTEM TESTS and then LDP MON TEST.

A message appears explaining that this test is the actual LDP Monitor Test that the PCM normally runs on the system. Any leaks detected during this test will actually set codes and store Freeze Frame data. Be sure to record any codes or Freeze Frame data before this test is initiated.

The LDP is by definition a diagnostic tool, so this test can be used to find leaks or verify repairs to the system. The LDP can also require diagnosis. The LDP MON TEST can effectively isolate problems in the LDP when used in conjunction with a Lab Scope.

Interpreting Test Results
Unless the LDP pump itself is being tested, problems in the EVAP System will set codes during this test. Because this test forces the Monitor to run even if temperature and fuel level conditions are not correct, it is possible to receive incorrect results.

Although this is an effective method for finding EVAP system faults, it was designed to verify an EVAP system repair. Use this test even after routine repairs to verify that the vehicle was repaired properly.

After the test is complete, always check for pending trouble codes, even if the Test Failed value reads "NO".

```
Apr 16, 02      3:19:58 pm

    SELECT MODE ↑↓
  F4:OBD CONTROLS
→ F5:SYSTEM TESTS
  F8:INFORMATION

  [*EXIT] TO EXIT
```

```
    SELECT TEST
→ F2:LDP MON TEST
```

```
This test will
run the LDP
Monitor to
verify EVAP
Emissions System
repairs. Any LDP
leaks detected
during this
test will be
stored in DTCs
and Freeze Data.
The LDP Monitor
will be forced
to run
regardless of
low temperature
or fuel level.
The PCM will
change Engine
RPM during this
test. This test
takes 3-5
minutes.
    [ENTER]
```

```
Test in Progress
 Please Wait...
```

```
  Test Complete.

    [ENTER]
```

```
   THIS TRIP
Done/Stopped YES
Test Failed    NO
    [EXIT]
```

LAB SCOPE TEST (LDP MONITOR)

The LDP and LDP Monitor operation can be observed using a Lab Scope. Refer to Scan Tool System Test (LDP) to force the LDP Monitor to run in the service bay.

The entire duration of the test can be observed on the Lab Scope. These captures will be discussed separately, but they represent different time frames during the same LDP MON TEST.

Scope Connections
Connect the positive lead to the LDP control circuit (WT/DK GN wire) at Pin 2-77 or at the LDP connector at the component. Connect the negative lead to the battery negative post.

Scope Settings
To make the waveforms as clear as possible, set the scope settings to match the examples.

Lab Scope Example (1)
In example (1), the trace shows the LDP control circuit being pulled to ground to activate the pump. After 8 seconds the pump is cycled to begin pressurizing the EVAP system.

The cycle rate of the pump is actually much higher than shown in the capture. At two (2) seconds per division, the Lab Scope cannot display the actual frequency. It is not important in this case, as the purpose of this test is to observe the LDP Monitor progress, not the actual detail of the signal.

Lab Scope Example (2)
In example (2), the trace shows the pump reaching test pressure. The pump has stopped running, and is only commanded "on" again to replace any pressure that has leaked from the system during the test period.

Lab Scope Example (3)
In example (3), the trace shows a single pump pulse, indicating a small loss of pressure. If too many of these extra pulses are required to maintain test pressure, the LDP Monitor will fail, and a code will set.

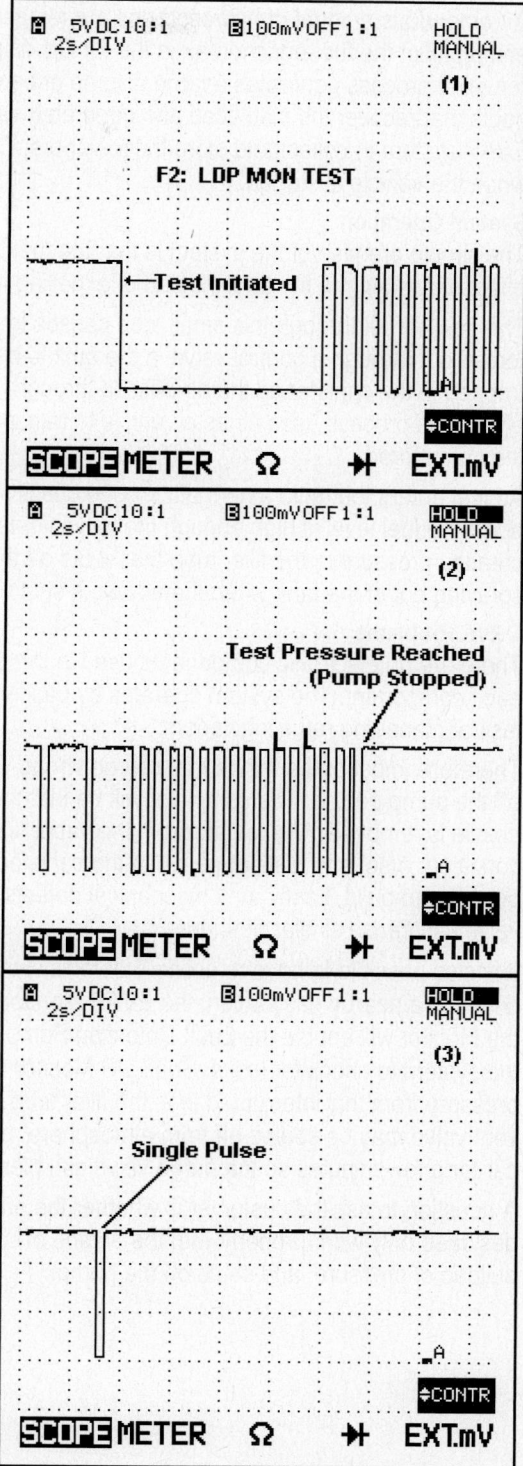

ONBOARD REFUELING VAPOR RECOVERY (ORVR)

A tremendous amount of hydrocarbons are released during the refueling process. When a tank is filled from empty, all of the fuel-enriched air in the tank is displaced. In addition to the problem of existing vapors, the refueling process generates a large volume of new vapors. Some states have programs with gas pump boots that recover the displaced and generated vapors. However, recent design changes allow the vehicle EVAP system to collect and store these vapors. The fuel vapors are then burned in the combustion process when the vehicle is started.

System Operation

The simple premise of the system is that the fuel tank filler tube can be designed in such a way that a venturi effect draws air into the tank instead of expelling it. The filler tube has a very small diameter (about 1").

The flow of fuel through this small tube causes fresh air to be drawn into the tank. The air in the tank is then forced out through a control valve in the tank to be stored in the EVAP canister. Hydrocarbons become trapped in the canister as the air passes through, and the 'cleaned' air is vented back to atmosphere. In effect, this process uses air as a vehicle to transport hydrocarbons to the desired location (canister) where they are stored.

As fuel enters the tank, a normally closed check valve is forced open. The check valve contains a float, so when the fuel level is high enough the valve floats up to seal the filler tube. This prevents overfilling by creating pressure in the filler tube that shuts off the pump nozzle. Also, the check valve prevents fuel from spraying out of the tank when the nozzle is shut off.

ORVR Problems

There are no electronic components used in this system, so diagnosis is limited to the physical condition of each component. The system operates by passively creating and using pressure and flow to accomplish the task of managing refueling vapors.

Therefore, most of the things that can go wrong with this system result in tank pressure buildup, which shuts off the pump nozzle. The complaint will be that the vehicle cannot be refueled, or sprays fuel when the nozzle is removed. Any restriction in the vapor lines from the tank to the EVAP canister can cause this condition. Also, any fuel saturation in the canister or the canister vent can restrict the system and cause pressure to build. Lastly, any mechanical defects in the fuel tank vapor control valve or the filler tube check valve can cause similar problems.

It is also possible to have refueling problems with nothing wrong. If the customer starts the car cold and drives to a nearby gas station, the EVAP Monitor may be interrupted when the key is shut off. When cold, the Monitor will enable the Leak Detection Pump (LDP), which *shuts off the canister vent* and pressurizes the system to check for leaks. See LDP Monitor Operation in this Section. If the system has residual pressure from this interrupted test, the filler tube check valve may stick closed. Even if it is not stuck, the vent valve may be sealed off from atmosphere, producing the same results. The best solution is to drive the car for a few minutes so the purge valve can bleed off pressures during normal operation.

A question to ask the customer is whether the engine was off during the refueling attempt. The system is designed only work properly with the engine off. This design prevents engine-on refueling by causing a buildup of pressure, and shuts off the pump.

Exhaust Gas Recirculation System

INTRODUCTION

The EGR system is designed to control the formation of NOx during combustion by controlling combustion temperatures. Exhaust gas has few oxygen or hydrocarbon molecules, the two main ingredients of combustion. The EGR system allows small amounts of exhaust gases to be recirculated back into the combustion chamber to mix with the A/F mixture. This reduces combustion temperatures and less NOx is formed.

The EGR system on this application contains the following components:

- EGR Valve
- EGR Tube
- Electronic Transducer (Includes EGR Solenoid)
- Connecting Hoses

Electronic Transducer

The term Electronic Transducer may be misleading. This assembly does contain an EGR solenoid, but the backpressure transducer portion is a mechanical device.

The backpressure transducer controls the strength of the vacuum signal to the EGR valve. As exhaust backpressure increases, the transducer diaphragm rises, restricting a vacuum bleed valve. When exhaust backpressure is high enough, the vacuum bleed valve is completely closed, and full vacuum is available to the EGR valve.

EGR SOLENOID

In this system, the PCM actually energizes the solenoid to cut off vacuum to the transducer, disabling the EGR system. The EGR valve only receives vacuum if the EGR solenoid is de-energized and the exhaust backpressure is sufficient to restrict or close the bleed valve in the transducer.

The EGR valve will not operate at idle speed or full throttle. In addition, the coolant temperature must be over 60°F, and the battery temperature must be over 7°F.

EGR System Graphic

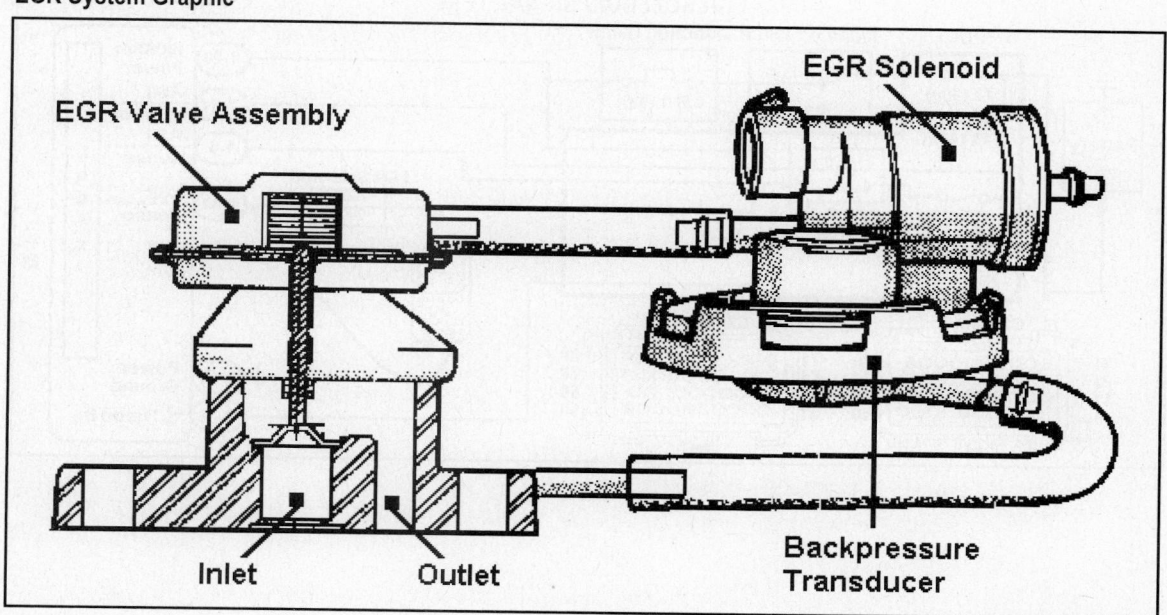

SCAN TOOL TEST (EGR SOLENOID)

Navigate through the Scan Tool menus to the PID Data List. Select the EGR SOLENOID from the Scan Tool data list.

Observe the PID status during normal vehicle operation, or during actuator test # 19. Note that when the PID value is "ENERGIZED', the EGR system is disabled. Vacuum is only available to the transducer when the PID value is 'DE-ENERGIZED', but backpressure is still required to get the vacuum to the EGR valve.

To test the EGR system, command the solenoid OFF using the actuator test while an assistant holds a rag over the tailpipe to build some backpressure. This should force the EGR valve to operate (open) in the service bay.

LAB SCOPE TEST (EGR SOLENOID)

The Lab Scope is an excellent tool to view the EGR solenoid command.

Lab Scope Connections
Connect the Channel 'A' positive probe the EGR solenoid control circuit (GY/YL wire) at PCM Pin 1-40 or at the solenoid connector. Connect the negative probe to the battery negative post.

Lab Scope Example
In this example, the trace shows the EGR solenoid circuit being de-energized during snap throttle. The PCM disables the EGR system under heavy acceleration by removing ground from the circuit.

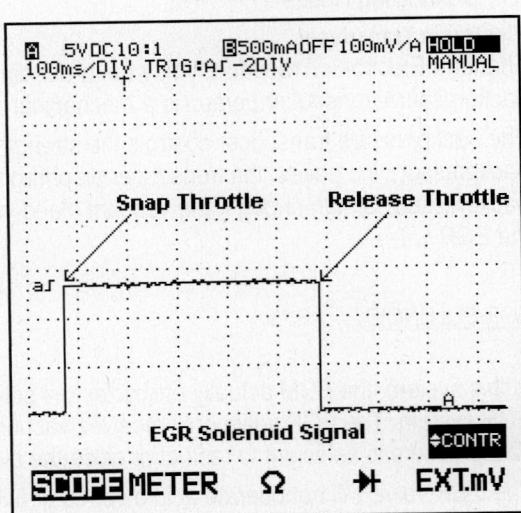

EGR Solenoid Test Schematic

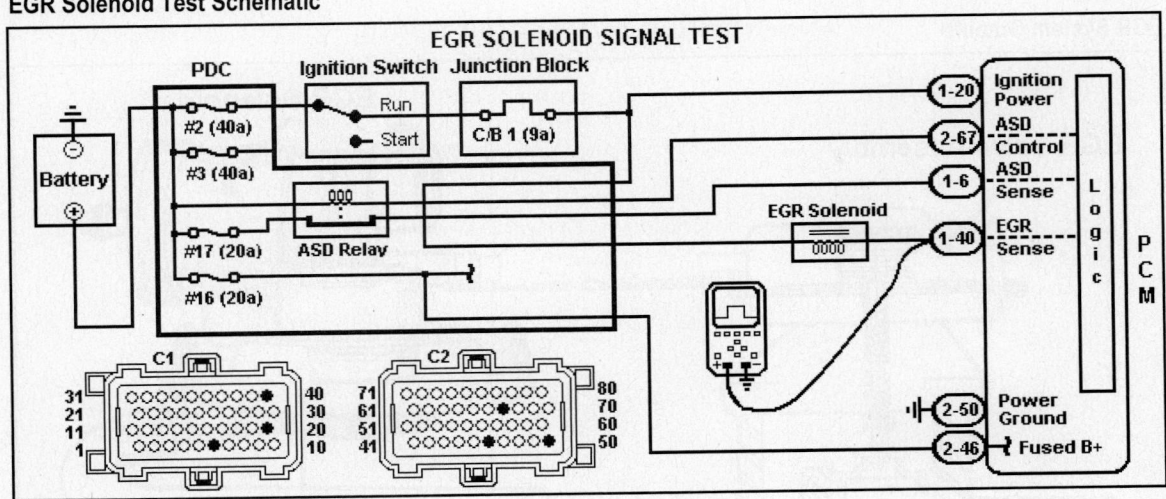

Vehicle Speed Control

INTRODUCTION

The Vehicle Speed Control (S/C) system on this application is controlled by the PCM. It controls vacuum solenoids in the S/C servo to regulate servo movement. The servo acts on a cable to control throttle angle, and therefore vehicle speed.

Circuit Operation

The only 2 inputs required for the PCM to properly operate the system are the vehicle speed sensor signal and the steering wheel mounted S/C switches.

All of the steering wheel mounted switches are combined into a single input to the PCM. The PCM provides a 5v signal to the switch assembly. Each of the switches provides a path to ground with a different resistance. The PCM monitors the voltage drop in the circuit in order to determine which of the switches was depressed.

The brake switch is not an actual input. When the brake pedal is applied, the power supply to the speed control servo is interrupted. The speed control servo contains 3 solenoids powered by the PCM: the vent, vacuum and dump solenoids.

The PCM can only ground the vent and the vacuum solenoids to regulate vehicle speed. The dump solenoid is permanently grounded, and is therefore energized with the brake pedal off. When the brake pedal is depressed, the dump solenoid is de-energized, and the vacuum in the speed control servo is rapidly bled off to atmosphere.

SCAN TOOL TEST (SPEED CONTROL PIDS)

Navigate through the Scan Tool menus to the PID Data List. Observe the S/C PIDs as the system is operated.

If necessary, Scan Tool actuator tests may be used to verify that the PCM has the ability to control the speed control solenoids. Available tests for this application are: 15: S/C SOLENOIDS, 47: S/C VENT SOL., and 47: S/C VAC. SOL. Results of the actuator tests can be verified with a DVOM, Lab Scope or vacuum gauge.

```
EGR SOLENOID...........ENRGZD
VEHICLE SPD................0MPH
S/C SET SPEED.............0MPH
S/C ON/OFF..................OFF
S/C SET......................OFF
S/C RESUME.................OFF
S/C CANCEL..................OFF
S/C PWR REL............DENRGZD
S/C STAT...........DIS:ON/OFF
S/C LC...................ON/OFF
S/C VENT SOL..........DENRGZD
S/C VAC SOL...........DENRGZD
MINIMUM TPS.............0.78V
ASD RELAY...............ENRGZD
BRAKE SWITCH................OFF
CHECK ENGINE.........DENRGZD
```

Speed Control System Graphic

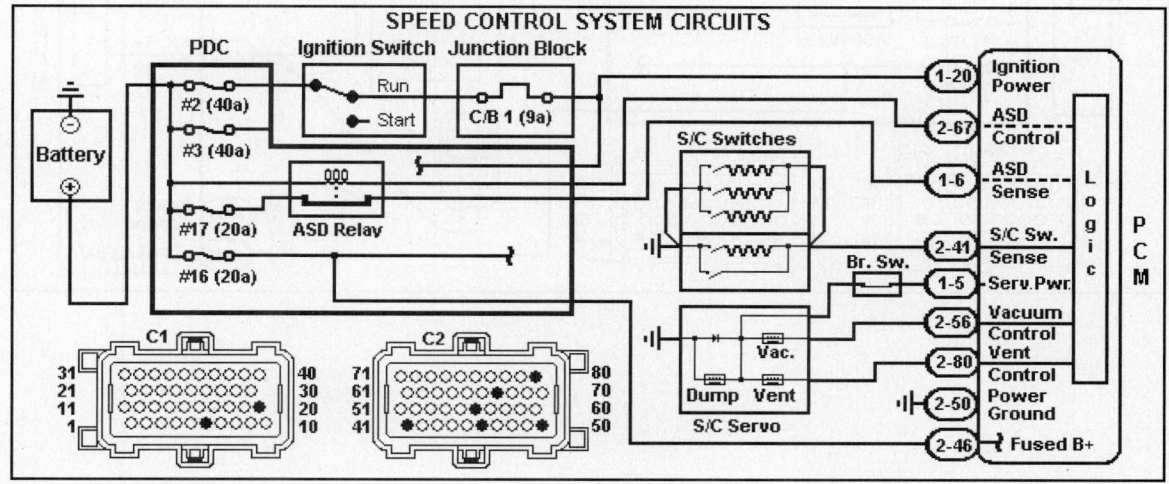

LAB SCOPE TEST (SPEED CONTROL SWITCHES)

The Lab Scope can be used to effectively diagnose problems with the steering control mounted S/C switches or circuits.

All of the steering wheel mounted switches are combined into a single input to the PCM. The PCM provides 5 volts to the switch assembly. Each of the switches provides a path to ground with a different resistance. The PCM monitors the voltage drop in the circuit in order to determine which of the switches was depressed.

Lab Scope Connections

Connect the Channel 'A' positive probe to the speed control switch circuit (RD/LG wire) at PCM Pin 2-41 and the negative probe to the battery negative post.

Lab Scope Settings

To make the waveforms as clear as possible, set the scope settings to match the examples.

Lab Scope Example

In this example, the trace shows the different voltage drops in the speed control switch circuit when each of the steering wheel mounted switches is depressed. A long time base was used (5 seconds per division) to capture all of the switches.

A much faster time base should be used to find any glitches in the circuit. This setting is useful for determining that the switches are functional, and that there is no extra resistance in the ground circuit. Note that pressing the ON button should cause the signal to drop to 0v with a good ground.

This is an efficient test, because all 5 switches, the 5v supplied by the PCM, and the circuit ground can be verified at the same time.

Speed Control System Graphic

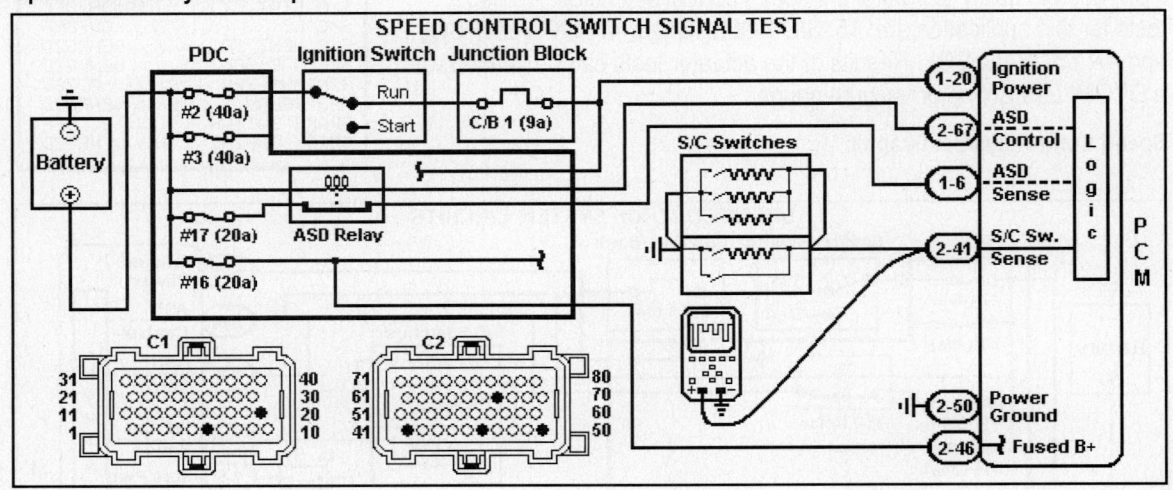

Fuel Delivery System

INTRODUCTION

The Fuel system on this application is a mechanical-returnless system. The Fuel system delivers fuel at a controlled pressure and correct volume for efficient combustion. The PCM controls the Fuel system and injector timing and duration.

On this returnless system, there is no fuel return line from the fuel supply manifold back to the fuel tank. The fuel pump and fuel level module, mounted in the fuel tank, contains a regulator which routs extra fuel back through the pump assembly to the fuel tank. The pressure regulator is mechanical, but is not vacuum controlled. Therefore, fuel pressure should remain relatively constant under all operating conditions.

Fuel Delivery

The vehicle is equipped with a sequential fuel injection (SFI) system that delivers the correct A/F mixture to the engine at the precise time throughout its entire speed range and under all operating conditions.

Air Induction

Air enters the system through the fresh air duct and flows through the air cleaner and into the intake plenum where it is mixed with fuel for combustion. The air volume is not directly measured, but is calculated by the PCM using the MAP sensor signal.

System Overview

A high-pressure (in-tank mounted) fuel pump delivers fuel to the fuel injection supply manifold. The fuel injection supply manifold incorporates electrically actuated fuel injectors mounted directly above each of the intake ports in the cylinder heads.

Power is supplied to the fuel injectors by the ASD relay, and to the fuel pump by the fuel pump relay. Drivers in the PCM ground the injector circuits and fuel pump relay control circuit to operate these devices.

The PCM determines the amount of fuel the injectors spray under all engine-operating conditions. The PCM receives electrical signals from engine control sensors that monitor various factors such as manifold pressure (vacuum), battery temperature, engine coolant temperature, throttle position, and vehicle speed. The PCM evaluates the sensor information it receives and then signals the fuel injectors in order to control the fuel injector pulsewidth.

Fuel Pump Circuits Schematic

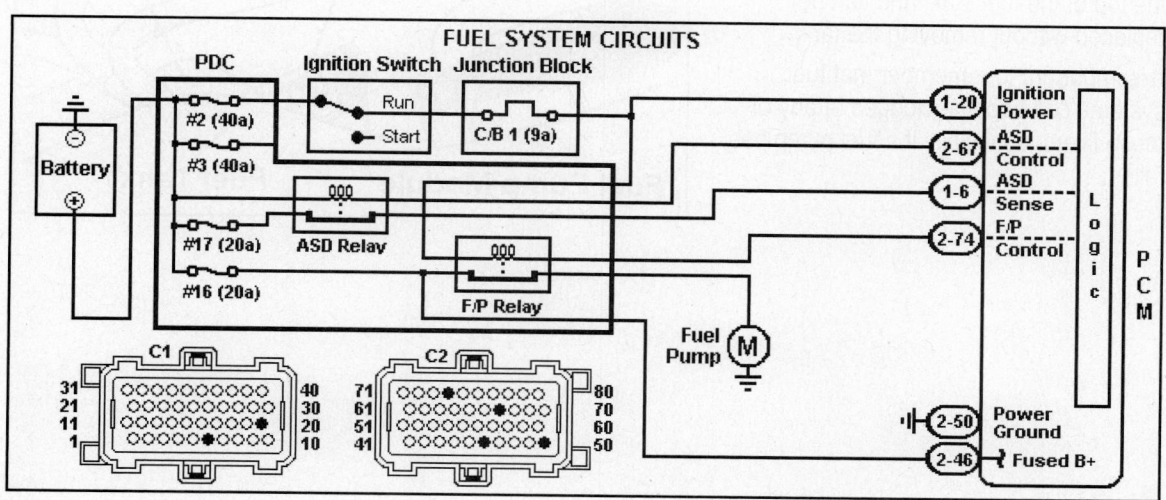

FUEL PUMP MODULE

The fuel pump module is mounted in the fuel tank. Fuel initially enters the module through a strainer in the bottom of the module, and primes the fuel pump.

An integral regulator limits the pressure to the injectors to 49 psi. However, the pump has a maximum pressure output of 95 psi.

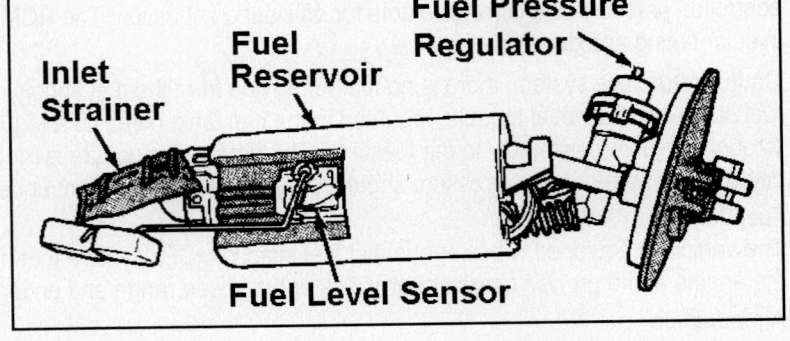

Fuel Pump Module Components
The fuel pump module includes an electric pump motor, reservoir, inlet strainer, pressure regulator, fuel gauge sending unit, and a fuel line connection. Only the inlet strainer and pressure regulator may be serviced separately.

Internal Valve
The pump module has one internal check valve, located in the pump outlet, which allows the system pressure to be maintained with the ignition off. This valve also keeps the pump primed for quick pressurization of the system at startup.

FUEL FILTER

A replaceable fuel filter is used to strain fuel particles through a screen or paper element. This filtration process reduces the possibility of obstruction in the fuel injector orifices.

The fuel filter is mounted to a bracket on the top of the fuel tank, and can be replaced without removing the tank.

It is important to remember that fuel systems can become clogged at any of these filters, especially if a fuel pump has failed.

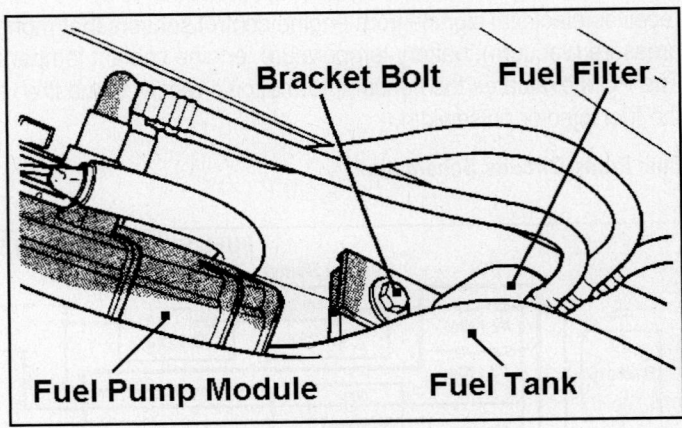

FUEL PRESSURE REGULATOR

The Fuel Pressure Regulator is a spring and diaphragm-operated relief valve attached to the fuel pump inside the fuel tank. Its function is to regulate the fuel pressure supplied to the injectors.

A constant fuel pressure of approximately 49 psi is supplied to the injectors. Any excess fuel pressure and volume are returned to the tank when pump pressure exceeds the calibrated spring tension in the regulator. Fuel is bypassed through the regulator and returned to the fuel tank.

The regulator is not controlled by the PCM or by engine vacuum. The regulator can be serviced separately, but requires fuel tank and pump removal.

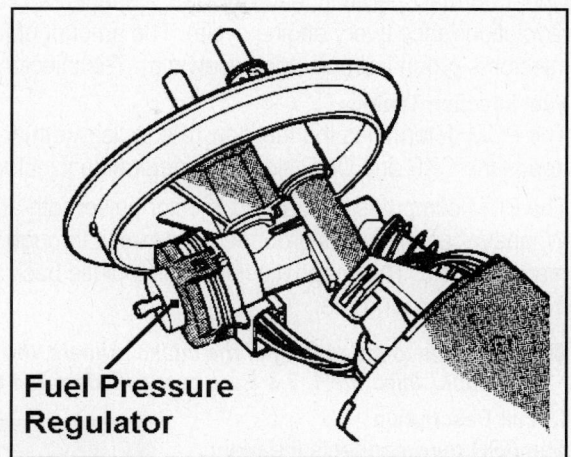

Fuel Pressure Regulator

LAB SCOPE TEST (FUEL PUMP)

A Lab Scope and low-amps probe can be used to test the fuel pump circuits and motor without having to remove the pump module from the tank.

The low-amps probe allows the Lab Scope to view the current flow through the entire fuel pump circuit.

Lab Scope Connections (Amp Probe)
Set the amp probe to 100 mv/amp, and zero the amp probe. Connect the probe around the fuel pump feed wire and connect the negative probe to chassis ground.

In this example, the Lab Scope was set so that each vertical division equals 5 amps.

Lab Scope Connection Tip
Accessing the fuel pump wiring harness for this test can be difficult because a single wire must be isolated and completely surrounded by the bulky low amp probe tip. As an alternative method, remove the fuel pump relay and install a fused jumper wire between relay cavities 30 and 87. The low amps probe can then be clipped around the jumper wire.

Lab Scope Example
In the example, the trace shows the fuel pump current when the ignition is initially turned "on". The peak current of 15 amps is typical. During normal pump operation, the current flow dropped to 7 amps.
If the pump motor is binding or shorted, the current will be higher. However, this type of pump tends to develop opens, which cause downward spikes in the current trace.

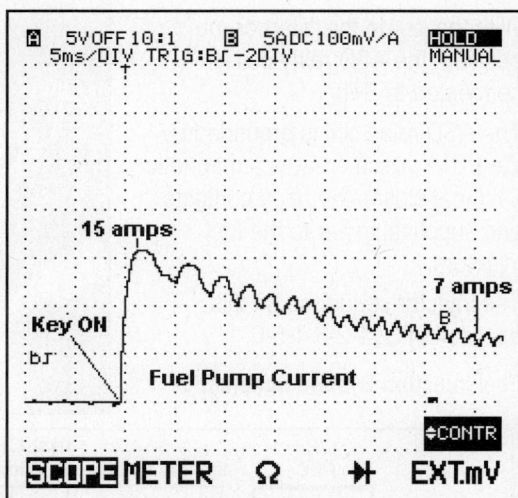

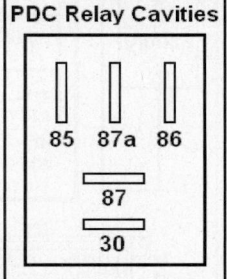

PDC Relay Cavities

FUEL INJECTORS

Under normal operation, each injector is opened and closed at a specific time, once every other crankshaft revolution (once every engine cycle). The amount of fuel delivered is controlled by the length of time the injector is grounded (injector pulsewidth). The injectors are supplied power through the ASD relay.

Fuel Injection Timing

The PCM determines the fuel flow rate (pulsewidth) from the ECT, HO2S-11, TP and MAP sensors, but needs the CKP and CMP signals to determine injector timing.

The PCM computes the desired injector pulsewidth and then grounds each fuel injector in sequence. Whenever an injector is grounded, the circuit through the injector is complete, and the current flow creates a magnetic field. This magnetic field pulls a pintle back against spring pressure and pressurized fuel escapes from the injector nozzle.

■ **NOTE:** *Due to the routing of the intake runners, the injector for each cylinder is located on the opposite bank. Injectors 1, 3 & 5 are near cylinders 2, 4 & 6. See the Graphic.*

Circuit Description

The ASD relay connects the fuel injectors to direct battery power (B+) as shown in the schematic.

With the key in the "start or run" position, the ASD relay coil is connected to B+.

The ASD relay coil is grounded by the PCM, which creates a magnetic field that closes the relay contacts and supplies power to the fuel injectors.

The injector sequence on this application is: 1-2-3-4-5-6.

Fuel Injection System Schematic

Injectors 1-3-5

Spark Plugs 2-4-6

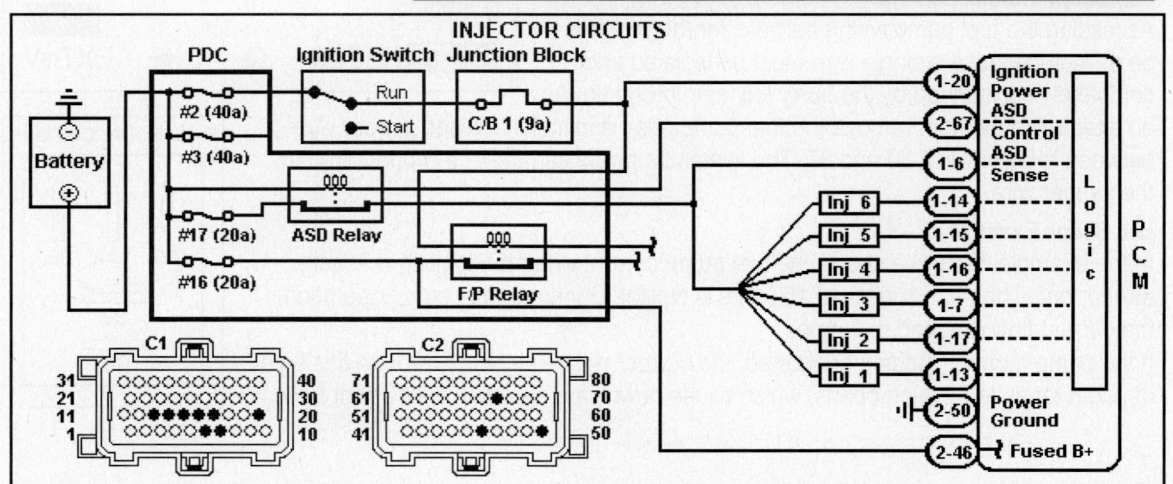

INJECTOR CIRCUITS

LAB SCOPE TEST (FUEL INJECTOR)

Place the gear selector in Park. Block the drive wheels for safety.

Scope Connections
Connect the Channel 'A' positive probe to the injector control wire at the injector or appropriate PCM Pin. Connect the negative probe to the battery negative post.

Scope Settings
To make the waveforms as clear as possible, set the scope settings to match the examples.

Lab Scope Example (1)
In example (1), the trace shows the injector voltage signal at idle speed. The 'on time' is the amount of time the PCM grounds the injector. In this example the on time is 3.5 ms.

Lab Scope Example (2)
In example (2), A low-amps probe was added the Channel 'B' to view the current flowing through the injector circuit.

Using both voltage and current traces, it is easy to find differences in the actual mechanical operation between cylinders. Refer to the example for details.

For example, even a few Ohms of resistance in an injector would lower the current flow through the injector. This would result in the injector opening dip in the current waveform occurring later, and less fuel would be injected into that cylinder.

Fuel Injector Test Schematic

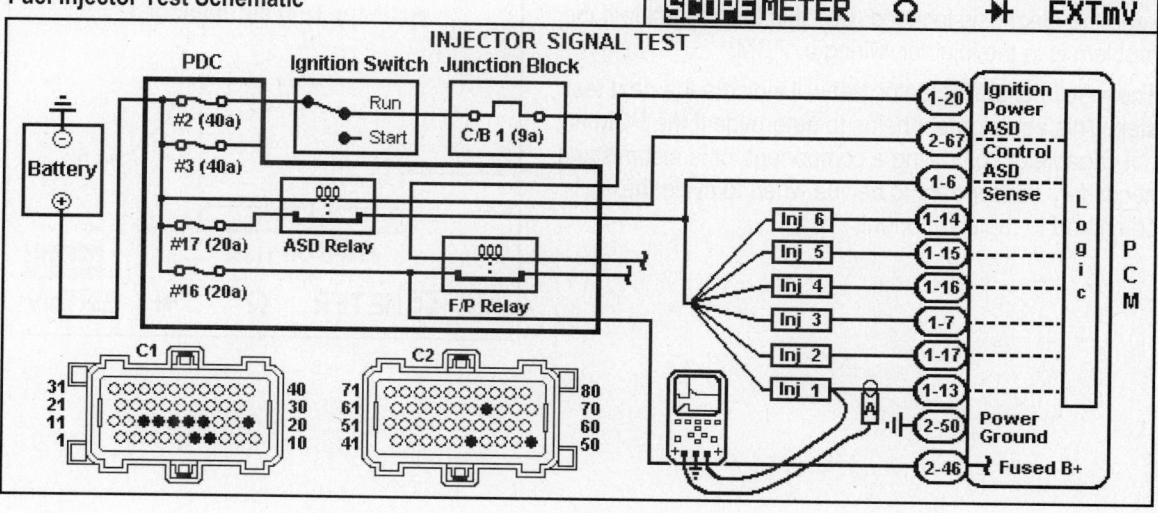

BI-DIRECTIONAL TEST (FUEL INJECTOR)

The PCM in this vehicle provides bi-directional testing capabilities, which can be accessed with a Scan Tool. The test is run with the key on, engine off.

Actuator Test 04-09 Operation
This actuator test grounds the chosen injector circuit at a fixed on time and frequency for 5 minutes, or until the Scan Tool user exits the tests.

Scan Tool Actuator Test Example
Connect the Scan Tool to the DLC under the driver's side of I/P. In this example, the Scan Tool Actuator Test menu was accessed, and the actuator test 04: INJECTOR (S) 1 was chosen. Refer to Section 1 for an explanation of Scan Tool Actuator Tests.

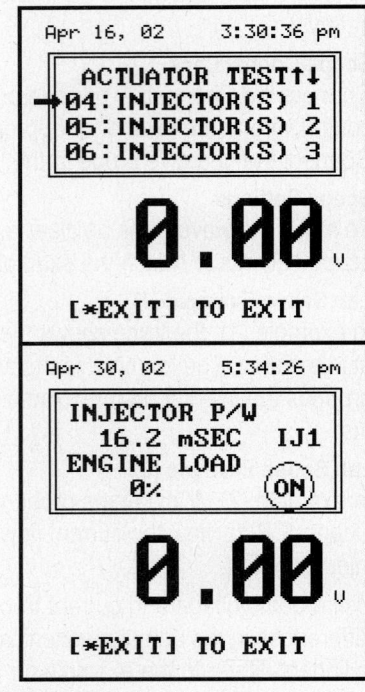

Note that, as in other actuator tests, the PIDs do not necessarily match measured values. The INJECTOR P/W PID value is 16.2ms with the actuator test ON or OFF (as shown on the Scan Tool). The INJECTOR P/W PID value represents the PCM command to the injector, but that command is being overridden by the actuator test.

Lab Scope Test Example
Use this actuator test in conjunction with a Lab Scope to verify injector circuit activity, and that the PCM is capable of cycling the ground.

In this example, the actuator test has been turned ON with the Scan Tool. The injector "on" time is fixed at 2 ms. With these Lab Scope settings, the on-time, inductive kick, circuit voltage, and pintle closing are all viewable.

This actuator test repeats, so it can be used to watch for inconsistency in the waveform over time.

Why To Use This Actuator Test
The PCM makes injector on time based on various criteria. It is much easier to command injector operation than to keep track of the various criteria. During this test, the desired results are already known, which helps to keep the diagnosis focused. Use this test to decide if the problem is in the injector, wiring or PCM.

The results of this actuator test will indicate the next test step. This actuator test helps to determine if the PCM is not capable of controlling a component, or is just missing necessary sensor data to decide when to cycle the command to the component.

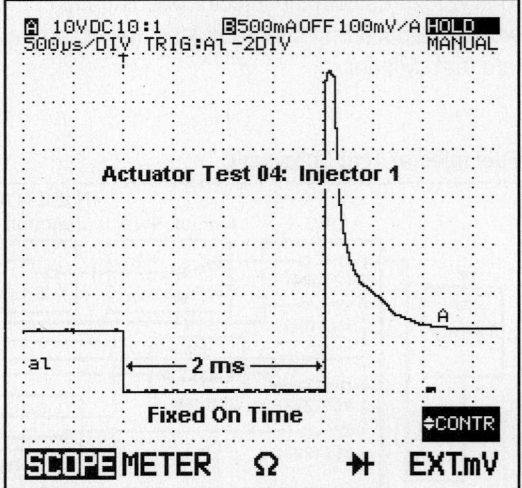

CAMSHAFT POSITION SENSOR (HALL EFFECT)

The camshaft position (CMP) sensor used with this EI system is a Hall effect sensor that sends a 0-5v pulse to the PCM corresponding to notches in the camshaft gear.

The CMP sensor provides a reference point for the PCM to calculate actual engine position. The PCM uses CMP data to determine which cylinder will be next to reach TDC. The CKP sensor pulses are then counted to calculate actual cylinder position.

Four (4) CKP pulses occur after each CMP pulse or group of CMP pulses. The CKP pulses are located 69°, 49°, 29° and 9° before TDC for the cylinder that was identified by the CMP sensor signal.

Note that cylinders #2 and #5 each have 1 notch and cylinders #3 and #6 each have 2 notches. Cylinder #4 has 3 notches, but Cylinder #1 does not have a notch.

With this arrangement, the PCM can determine which coil to fire next, as each coil fires a cylinder pair (1 & 4, 2 & 5, and 3 & 6).

However, for injection timing the PCM needs to determine which cylinder of each pair is on the intake stroke. The lack of notches for cylinder #1 provides a reference for the PCM to determine this information.

If the PCM sees 3 CMP pulses, it knows that the subsequent single CMP pulse indicates that cylinder #5 is coming up next. If the PCM sees no CMP pulses, it knows that the subsequent single CMP pulse indicates that cylinder #2 is coming up next.

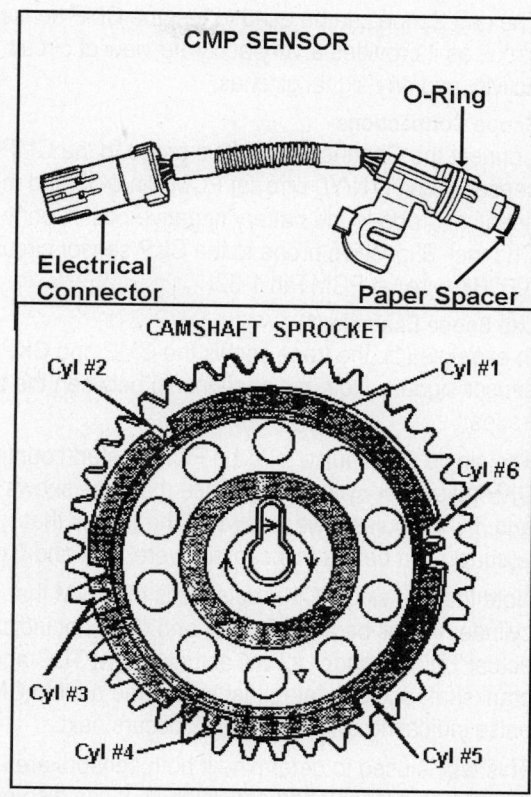

CMP Sensor Schematic

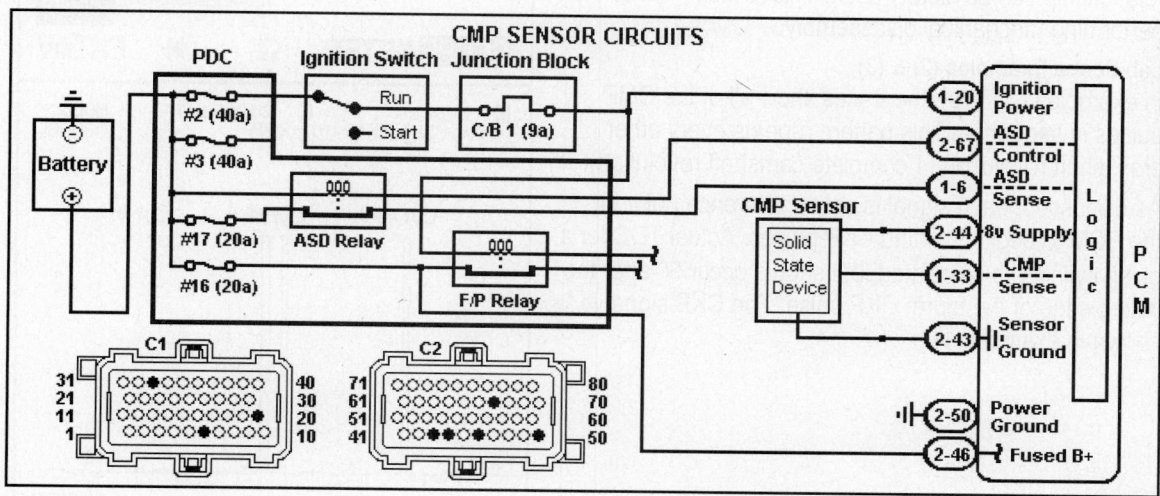

LAB SCOPE TEST (CMP SENSOR)

The Lab Scope can be used to test the CMP sensor circuit as it provides a very accurate view of circuit activity and any signal glitches.

Scope Connections
Connect the Channel 'A' positive probe to the CMP sensor circuit (TN/YL wire) at PCM Pin 1-33 and the negative probe to the battery negative post. Connect the Channel 'B' positive probe to the CKP sensor circuit (GY/BK wire) at PCM Pin 1-32.

Lab Scope Example (1)
In example (1), the trace shows the CMP and CKP sensor signals. Note the relationship between the two traces.

The single CMP pulse tells the PCM to start counting CKP pulses for cylinder #2. Since the trace shows 1, 2, and 3 CMP pulses, we know that the pulses that occurred just before this capture were 1, 2, and 0 pulses.

Note that the single CMP pulse tells the PCM that cylinder #2 will be at TDC next, and does not indicate actual TDC. Cylinder #2 will actually be at TDC about 90 crankshaft degrees later, just before the double CMP pulse indicating cylinder #3 TDC occurs next.

This test is used to determine if both sensors are functioning properly and consistently. It can also be used to verify the camshaft-to-crankshaft relationship. If the belt or chain jumps even 1 tooth, the change in sensor relationship can be clearly seen. This is much easier than performing mechanical disassembly.

Lab Scope Examples (2) & (3)
In examples (2) & (3), the traces show all of the CMP pulses in sequence. This pattern repeats every other crankshaft revolution (1 complete camshaft revolution).

As discussed, each signal is only a reference point for the PCM to begin counting CKP pulses. Actual TDC for a given cylinder is calculated, but should occur 9° after the falling edge of the fourth CKP pulse. The CKP signal is shown in example (1).

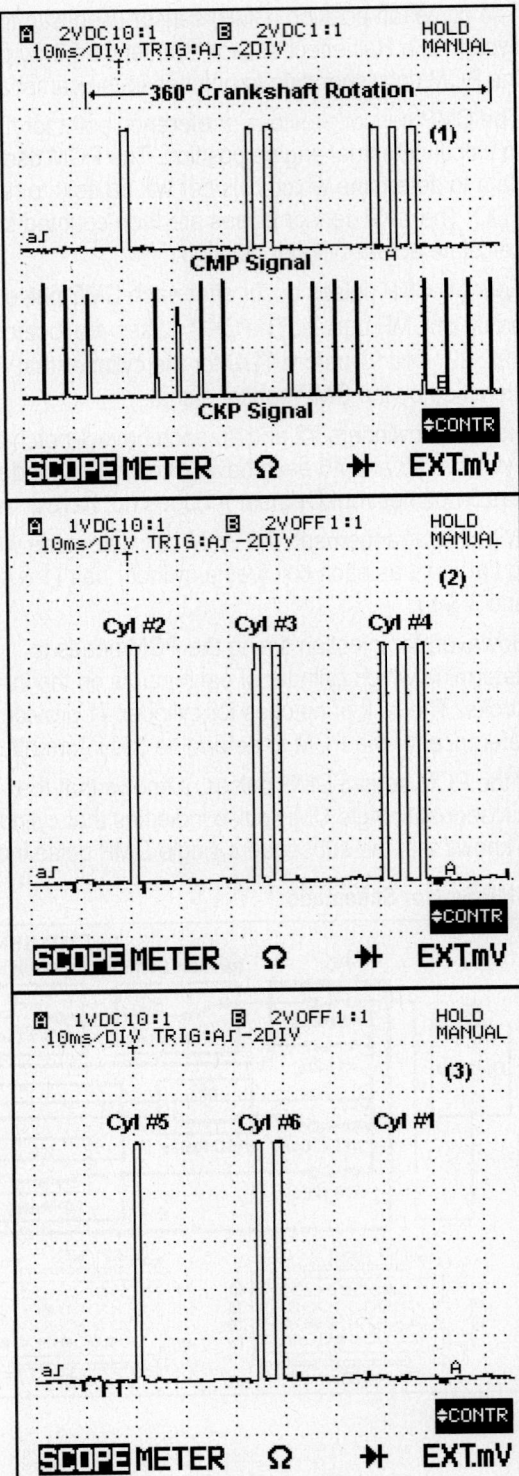

Idle Air Control Motor

INTRODUCTION

The Idle Air Control (IAC) motor, mounted on the throttle body, is actually a dual-motor assembly. The IAC may also be referred to as the Automatic Idle Speed (AIS) motor. This device is designed to control engine idle speed (rpm) under all engine-operating conditions and to provide a dashpot function (controls rate of deceleration for emissions and idle quality).

The IAC motor meters the intake air that flows past the throttle plate by varying the position of a pintle in the bypass passage on the throttle body.

Circuit Operation

The PCM determines the desired idle speed (amount of bypass air) and controls the IAC motor circuits through pulse train commands. A pulse train is a command signal from the PCM that continuously varies in both frequency and pulsewidth. There are 4 driver circuits connecting the IAC motor and the PCM. Circuits 1 & 4 control one 'motor' and circuits 2 & 3 control the other. These 2 motors pull the pintle in opposite directions, and the rapid cycling of these drivers balances these forces. By slightly varying the strength of each field, the PCM can precisely control the pintle position.

BI-DIRECTIONAL TEST (IAC MOTOR)

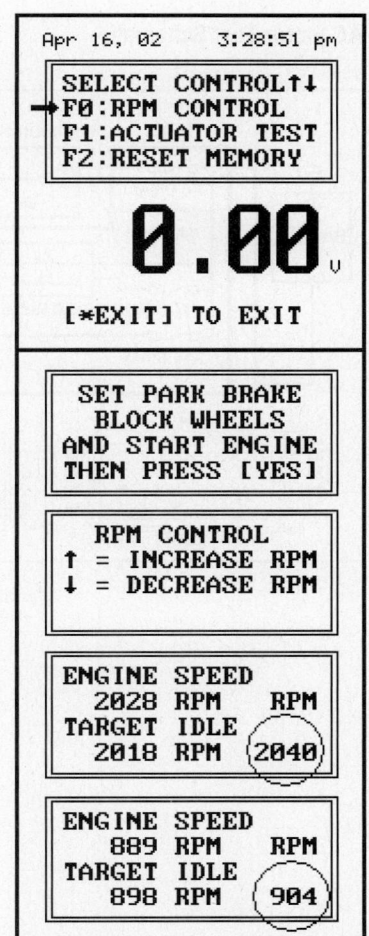

When testing the IAC motor circuits with a DVOM or Lab Scope, it is difficult to know exactly what to look for. The circuits are toggled from 0-12v in a pulse train, which means that both the frequency and the duty cycle of the signals are constantly changing. There are too many variables present to completely understand the changes in this signal.

By using a Scan Tool and the RPM CONTROL feature, it is possible to eliminate many of these variables to help- the diagnosis stay focused. With this control, the RPM can be varied from the Scan Tool between 1000 and 2000 RPM.

Connect the Scan Tool to the DLC underdash connector. Navigate through the Scan Tool menus and choose the OBD CONTROLS selection. From that menu, select RPM CONTROL.

Scan Tool Example

The Example shows the Scan Tool being used to command the idle up to 2040 RPM and then down to 904 RPM. The results of this test can be seen in the ENGINE SPEED and TARGET IDLE PIDs.

The Scan Tool can also be used with a DVOM or a Lab Scope to check the actual circuit activity at the PCM or IAC connector. Refer to Scan Tool & Lab Scope Test (IAC Motor) on the next page.

This is an excellent test to determine how the PCM reacts to changing engine conditions, and how the IAC reacts to changing commands.

BI-DIRECTIONAL TEST (IAC MOTOR – CONT.)

The examples on this page were captured using both the Scan Tool bi-directional controls and Lab Scope. For an explanation of Scan Tool RPM CONTROL, see the previous page.

Lab Scope Connections
All 4 drivers can be checked using 2 channels. The PCM controls one motor by alternating the polarity of drivers 1 & 4, and the other motor by alternating the polarity of drivers 2 & 3.

Therefore, the Lab Scope positive probes should be connected to circuit 1 _or_ 4, _and_ circuit 2 _or_ 3. A circuit problem in either pair will show up if you use this test method.

Lab Scope Example (1)
In example (1), the traces show the PCM increasing the RPM in response to the Scan Tool command. Circuit activity begins to slow down at the 2000-RPM target.

Lab Scope Example (2)
In example (2), the traces show the PCM decreasing the RPM in response to the Scan Tool exiting the RPM CONTROL test. Rapid circuit activity is seen as the test is ended, followed by about a 6-second return to idle.

Test Results
This test proves that the PCM is capable of controlling all 4 IAC motor drivers (the IAC motor, circuits, and PCM drivers are all good). Further diagnosis can be limited to PCM inputs, processing, or engine mechanical concerns (vacuum leaks).

IAC Motor Test Schematic

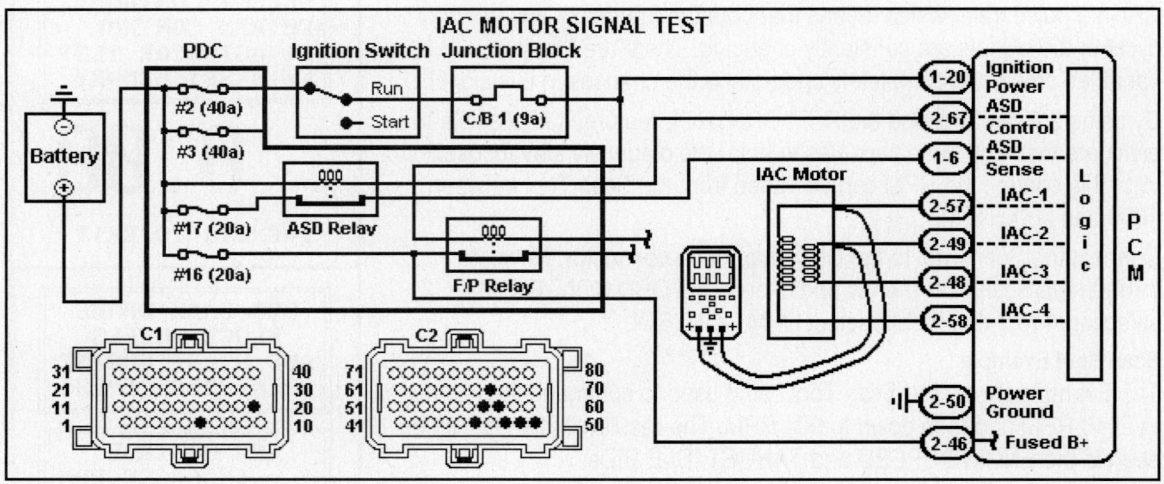

Information Sensors

MANIFOLD ABSOLUTE PRESSURE SENSOR

The Manifold Absolute Pressure (MAP) sensor is located on the intake manifold, near the throttle body. The sensor is mounted with an o-ring directly to a port in the intake manifold (no vacuum hose is used). The MAP sensor input to the PCM is an analog voltage signal proportional to the intake manifold pressure.

This signal is compared to the TP sensor signal and other inputs to calculate engine conditions. The PCM uses theses calculations to determine the following outputs:

- Fuel Injector Pulsewidth
- Ignition Timing
- Transmission Shift Points

MAP Sensor Circuits

The three circuits that connect the MAP sensor to the PCM are the 5v supply circuit, MAP sensor signal circuit, and sensor ground return circuit.

Circuit Operation

The MAP sensor voltage increases in proportion to the engine load. When the throttle angle is high as compared to engine speed, the load on the engine is higher. Intake vacuum drops (pressure rises) in proportion to increased engine loads. By measuring this vacuum (pressure) the PCM has a very precise understanding of engine conditions.

Trouble Code Help

It is important to remember that on OBD II vehicles, certain inputs are required to be within a specific range before many of the OBD II Monitors can run. The MAP sensor input is one such critical input. In addition, some tests compare inputs to make sure they make sense, even if they are within their normal operating range.

For example, P0121 will set if the TP sensor and the MAP sensor disagree about the current engine operating conditions. This is not necessarily a MAP sensor code, and both the MAP sensor and the TP sensor are operating within their normal range. If they were not, other codes would be set. Both sensors need to be checked under various operating conditions in order to verify the real cause of the code.

MAP Sensor Schematic

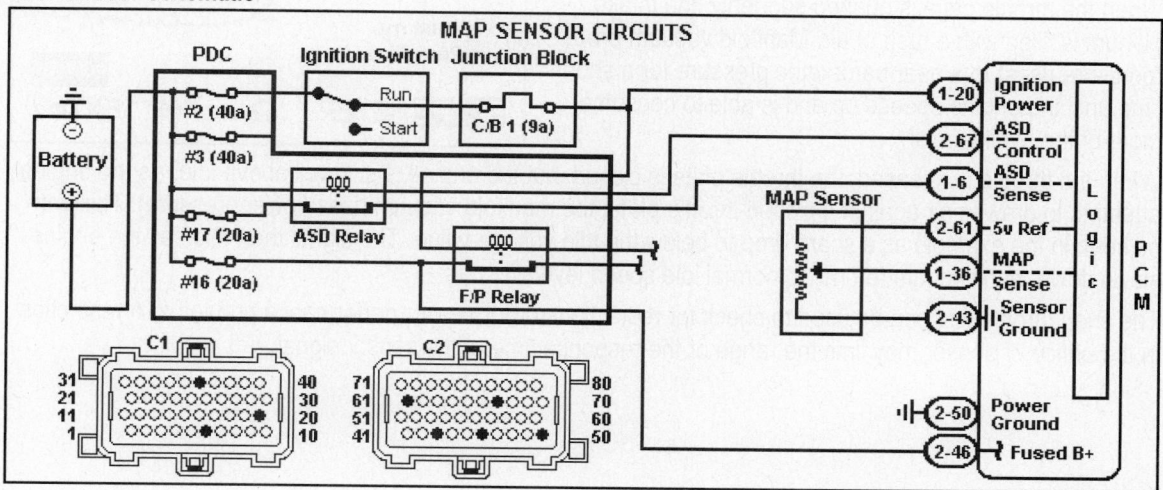

DVOM TEST (MAP SENSOR)

The DVOM is an excellent tool to test the condition of the MAP sensor <u>circuit</u>. The DVOM can be used to test the supply voltage, sensor ground voltage drop, and the MAP sensor signal for calibration or dropout.

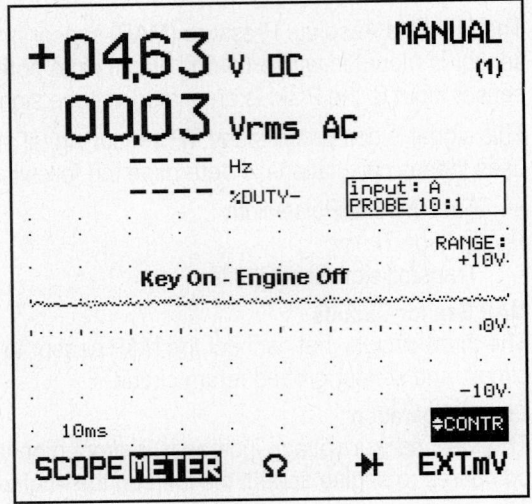

The test results can be compared with the specifications in Map Sensor Table (see Section 1). See the Lab Scope Test below for probe connections.

On this vehicle, the MAP sensor signal at Hot Idle was 1.20v.

The MAP sensor voltage increases in proportion to manifold pressure. In this example, the KOEO MAP sensor voltage was 4.63v at sea level, and was almost the same at snap throttle.

The voltage supply should be 4.9-5.1v, and the ground voltage drop should not exceed 50 mv.

LAB SCOPE TEST (MAP SENSOR)

The Lab Scope is an excellent tool to test the activity of the MAP sensor signal over time.

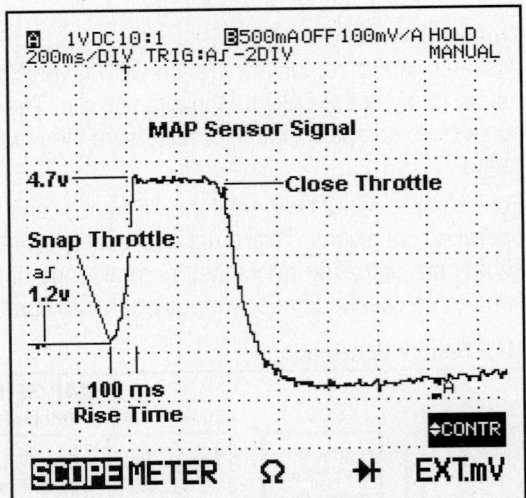

Lab Scope Connection
Connect the Channel 'A' positive probe to the MAP sensor signal circuit (DK GN/RD wire) at PCM Pin 1-36 or at the MAP sensor harness connector. Connect the negative probe to the battery negative post.

Lab Scope Example
In this example, the trace shows the MAP sensor signal during a snap throttle event starting from Hot Idle.

When the throttle plate is opened suddenly, the intake plenum is filled with a rush of air. Manifold vacuum drops (pressure rises) to a near barometric pressure for a short time until the engine speeds up and is able to consume more of the available air.

When the throttle is released, the throttle plate is closed, but the engine is still well above idle. As the engine attempts to draw in air against a closed throttle plate, the manifold vacuum rises (lower pressure). This can be seen in the example as a sharp drop to below the idle voltage value. The signal then rises as the engine slows down, and vacuum returns to normal idle speed levels.

The snap throttle test can be used to check for restrictions or for sensor performance problems. A restriction in the orifice or sensor may limit the range or the response time of the sensor signal.

HEATED OXYGEN SENSORS

The heated oxygen sensor (HO2S) is mounted in the Exhaust system where it monitors the oxygen content of the exhaust stream. During engine operation, oxygen present in the exhaust reacts with the HO2S to produce a voltage output.

Simply stated, if the A/F mixture has a high concentration of oxygen in the exhaust, the signal to the PCM from the HO2S will be less than 0.4v. If the A/F mixture has a low concentration of oxygen in the exhaust, the signal to the PCM will be over 0.6v.

Heated Oxygen Sensors

This vehicle application is equipped with two oxygen sensors with internal heaters that keep the sensors at a high temperature for more efficient operation.

Oxygen Sensor Identification

The front heated oxygen sensor is identified as Bank 1 HO2S-11. The rear heated oxygen sensor is identified as Bank 1 HO2S-12. HO2S-12 is mounted in the converter body on this application.

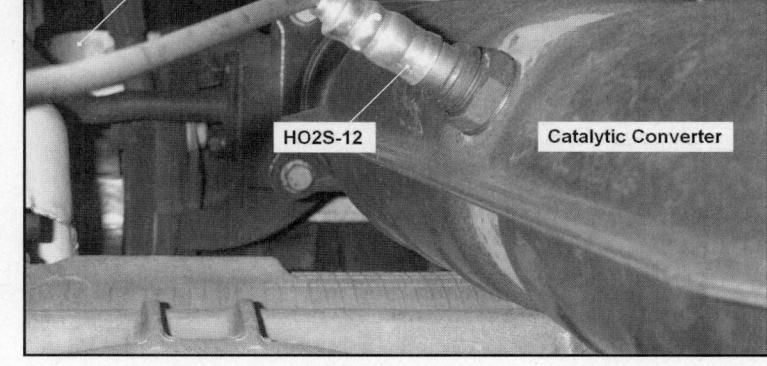

Circuit Description

Oxygen sensors generate voltage signals, so they do not need any reference voltage. The HO2S signal circuits and sensor ground circuits connect directly to the PCM.

The internal heater circuits in the oxygen sensors (both front and rear) receive power through the ASD relay. On this application, there are no PCM heater control circuits. The heaters circuits are always grounded, and operate whenever the ASD relay is energized.

Heated Oxygen Sensor Schematic

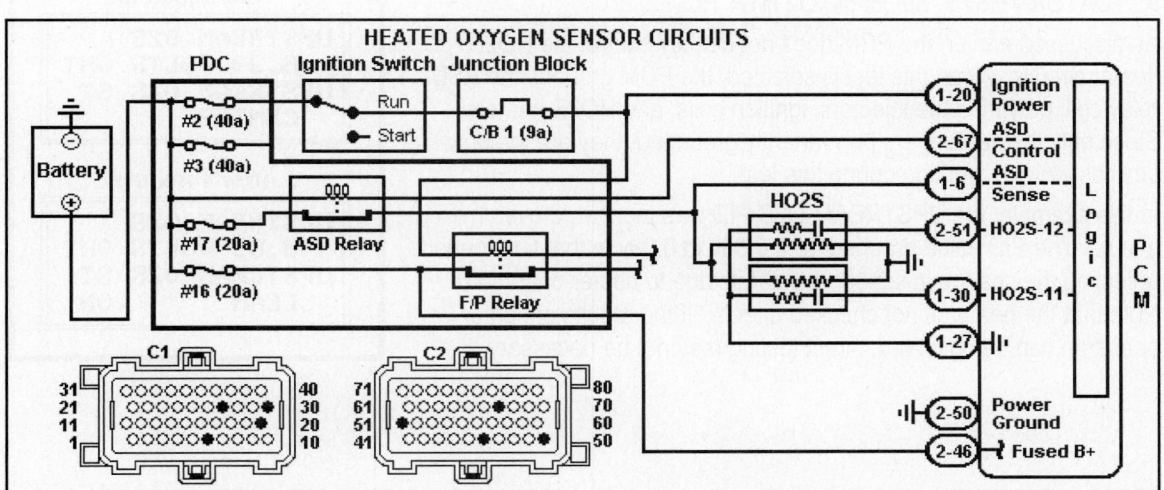

SCAN TOOL TEST (HO2S)

The Scan Tool is the most efficient way to compare HO2S values to other inputs to help determine the cause of a problem. It is also the best tool to use to get a quick idea of certain HO2S performance characteristics (time-to-activity from a cold start, general min/max values, etc.).

The Scan Tool is limited, however, in that the data refresh rate is much slower than either a DVOM or Lab Scope. Therefore, much of the subtle circuit activity is not viewable. It is important to determine before testing begins if the flexibility and ease of the Scan Tool is more important than the precision of a DVOM or Lab Scope.

It is also possible on this application to test the oxygen sensors using Scan Tool bi-directional controls. The HO2S heaters can be activated, which changes the resistance of the sensor, and signal voltage should drop.

Scan Tool Testing
Connect the Scan Tool to the DLC underdash connector. Navigate through the Scan Tool menus until you get to the Chrysler (OEM) Data List (a list of PID items).

Observe the HO2S PIDS and raise the engine speed to 2000 rpm for 3 minutes. Minimum and maximum voltage values, time-to-activity, and switch rates should all be evaluated under idle and off-idle conditions.

Scan Tool Example
This example was captured at Hot Idle. The upstream (HO2S-11) and downstream (HO2S-12) voltages were captured, as well as the RICH/CENTER/LEAN values.

There is no PID for heater status, as the heaters are not controlled by the PCM. They are on whenever the ASD relay is energized.

The HO2S PID values should be observed over a period of time. Individual captures such as these examples reveal very little about the condition of these circuits.

Actuator Test Example
Navigate through the Scan Tool menus to OBD CONTROLS, then ACTUATOR TESTS. Select 54: O2 HTR TEST.

As discussed earlier, the PCM does not directly control the HO2S heater circuits. When this test is selected, the PCM grounds the ASD relay coil, powering the injectors, ignition coils, and HO2S heaters. Since only the heaters are permanently grounded, they are the only devices actually working during this test.

In this example, the UPSTREAM O2S PID was monitored over 1 minute. The PID value dropped from 0.39v to 0.09v in that time period, reflecting the change in sensor resistance due to heater operation. Note that the heater is not checked directly. If the results of heater operation can be observed, direct testing may not be necessary.

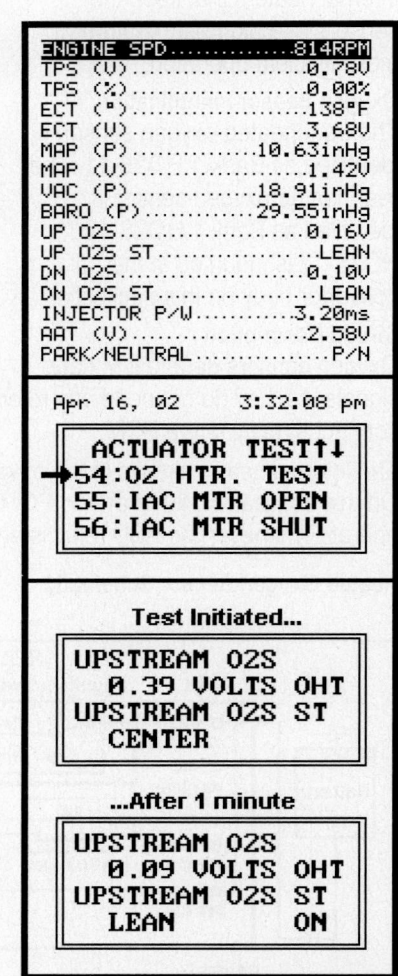

```
ENGINE SPD............814RPM
TPS (V)...............0.78V
TPS (%)...............0.00%
ECT (°)...............138°F
ECT (V)...............3.68V
MAP (P)...........10.63inHg
MAP (V)...............1.42V
VAC (P)...........18.91inHg
BARO (P)..........29.55inHg
UP O2S................0.16V
UP O2S ST.............LEAN
DN O2S................0.10V
DN O2S ST.............LEAN
INJECTOR P/W.........3.20ms
AAT (V)...............2.58V
PARK/NEUTRAL............P/N
```

```
Apr 16, 02      3:32:08 pm
  ACTUATOR TEST↑↓
→54:02 HTR. TEST
 55:IAC MTR OPEN
 56:IAC MTR SHUT
```

```
    Test Initiated...
UPSTREAM O2S
   0.39 VOLTS OHT
UPSTREAM O2S ST
   CENTER
```

```
    ...After 1 minute
UPSTREAM O2S
   0.09 VOLTS OHT
UPSTREAM O2S ST
   LEAN      ON
```

LAB SCOPE TEST (HO2S)

The Lab Scope can be used to observe the warm-up process for sensors and the catalytic converter. Place the gearshift selector in Park and block the drive wheels for safety.

Lab Scope Connections
Connect the Channel 'A' positive probe to the HO2S-12 sensor signal wire (TN/WT wire) at PCM Pin 2-51, and the negative probe to the battery negative post. Connect the Channel 'B' positive probe to the HO2S-11 sensor signal wire (BK/DK GN wire) at PCM Pin 1-30.

Lab Scope Example (1)
In example (1), traces from HO2S-11 and HO2S-12 were captured with the vehicle at idle after a cold start. Note that at 5 seconds per division, the HO2S-11 began switching after 40 seconds (vehicle entered closed loop).

HO2S-12 has not warmed up yet. Do not confuse a cold HO2S with a good converter. Both will result in a flat HO2S-12 trace.

Lab Scope Example (2)
In example (2), HO2S-12 has warmed enough to start switching after 3 minutes at idle, but the converter has not warmed enough to begin operating. The result is a trace that resembles HO2S-11, indicating that the converter is not storing and releasing oxygen at this time.

Lab Scope Example (3)
In example (3), HO2S-12 and the converter have warmed enough and are operating properly. Six minutes have elapsed at idle.

Trouble Code Help
It is important to create the proper conditions when analyzing converter efficiency. The OBD II Catalyst Monitor runs only under the conditions where it is most likely to pass. The Monitor will only run after 3 minutes after a cold start, at steady speeds over 20 mph.

Under cruise conditions, the sensor and converter warming process is greatly accelerated. The PCM is attempting to verify that the switch rate of HO2S-12 as compared to that of HO2S-11 is less that 70% at steady cruise. If this ratio is exceeded, a DTC P0420 (converter efficiency) will set (2-trip code). Use this test to recreate these conditions to verify converter failure.

VEHICLE SPEED SENSOR

The Vehicle Speed Sensor (VSS) on this application is actually a conditioned signal from the Transmission Control Module (TCM) based on the Output Speed Sensor (OSS) signal. The OSS is located on the transaxle housing. The TCM and PCM both use this input signal during certain calculations.

Circuit Description
The OSS produces an AC signal proportional to the rotating speed of the transaxle output shaft. This signal is sent to the TCM at Pin 14. The signal is then conditioned by an A/D converter in the TCM and sent to the PCM, at Pin 2-66, as a 0-5v digital signal.

DVOM TEST (VSS SIGNAL)

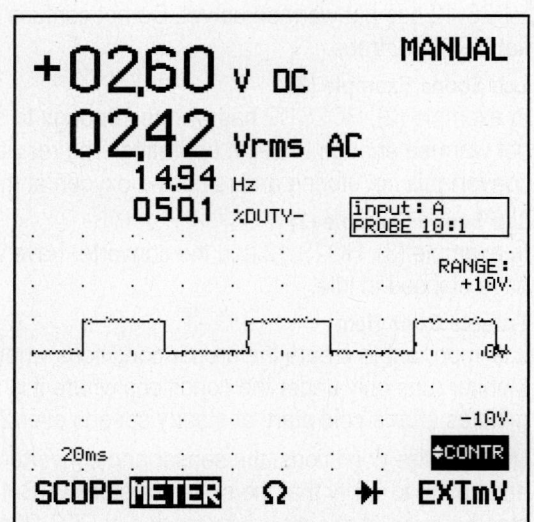

The OSS and VSS signals can be tested with a DVOM. The duty cycle of the VSS signal is fixed at 50%, so the DVOM should be set to measure signal voltage and/or frequency.

DVOM Connections
Connect the positive probe to the VSS signal circuit (WT/OR wire) at TCM Pin 58 or PCM Pin 2-66. Connect the negative probe to the battery negative post.

DVOM Example
In this example, the DVOM display shows the VSS signal at approximately 5 mph. The DC voltage value shown is the average of the 0-5v pulse from the TCM on this circuit. Use MIN/MAX settings on the DVOM to see the peak voltage of the signal, as well as the condition of the VSS ground circuit.

The frequency of the signal is 14.94 Hz in this example. Verify that this value changes in relation to vehicle speed.

Vehicle Speed Sensor Test Schematic

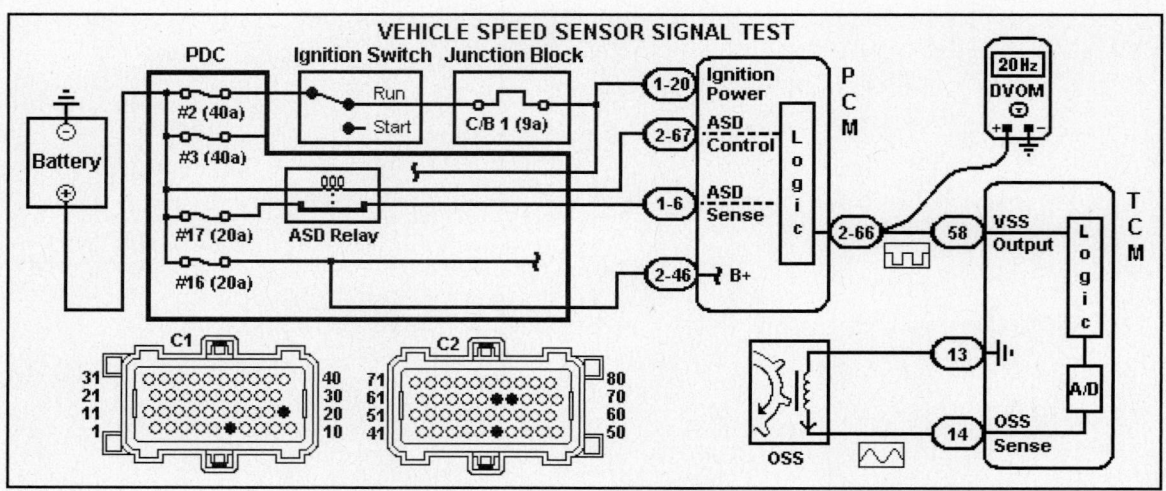

LAB SCOPE TEST (OSS & VSS)

The Lab Scope can be used to check the Output Speed Sensor (OSS) and the calculated Vehicle Speed Sensor (VSS) signal. The OSS is located on the transaxle housing. The Transmission Control Module (TCM) and PCM both need the vehicle speed sensor value to make certain calculations.

Circuit Description

The OSS produces an AC signal proportional to the rotating speed of the transaxle output shaft. This signal is sent to the TCM at Pin 14. The signal is then converted by an A/D converter in the TCM and sent to the PCM, at Pin 2-66, as a digital 0-5v signal.

Lab Scope Connections

Connect the positive probe to the appropriate sensor signal wire at the correct module, and the negative probe to the battery negative post.

The OSS signal wire is connected to the TCM at Pin 14. The VSS signal wire is connected to the TCM at Pin 58, and to the PCM at Pin 2-66.

Lab Scope Example (1)

In example (1), the trace shows the OSS and ISS. These sensors produce AC signals that increase in frequency and amplitude in proportion to the output shaft and input shaft speed. However, the TCM is only watching the change in frequency to determine the shaft speed.

This example was captured at about 3 mph.

Lab Scope Example (2)

In example (2), the trace shows VSS signal from the TCM to the PCM, representing vehicle speed. The OSS AC signal is processed by the TCM, converted into the VSS digital signal (0-5v), and sent to the PCM.

The waveform will toggle from about 200 mv to 5 volts.

This example was captured at about 8 mph.

Diagnostic Help

A problem with the OSS or its circuit will affect the VSS signal. However, the VSS signal may be missing with no OSS or circuit problems. Pay attention to symptoms and codes (and what modules set them) to determine the most efficient test procedure.

Generally, thinking through the circuit in this way can logically eliminate most potential tests procedures, which can save diagnostic time.

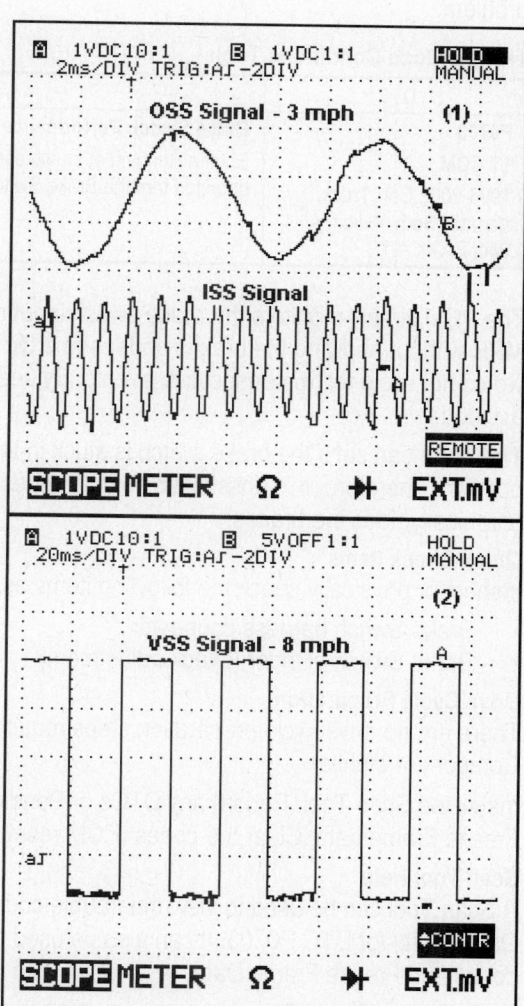

CHILTON DIAGNOSTIC SYSTEM – DTC P0703

The purpose of the Chilton Diagnostic System is to provide you with one or more tests that can be done with a DVOM, Lab Scope or Scan Tool *prior* to entering the complete trouble code repair procedure. The quick checks listed in the DVOM, Lab Scope or Scan Tool tests on the previous pages may help you quickly find the cause of the problem. If you cannot resolve the problem with these tests, a code repair chart is also included.

Code Description
DTC P0703 - The Comprehensive Component Monitor (CCM) detected a brake switch performance problem.

Trouble Code Conditions Table

DTC	Trouble Code Title & Code Conditions
P0703 1T CCM 1995-2002 Car, Truck, Van, LH Body Vision: All With EATX A/T	**Brake Switch Performance Conditions:** Engine Running at cruise speed followed by 16 Accel and Decel periods (30-0 mph), then the PCM detected that the Brake Switch signal did not cycle from low to high at all (it may be stuck).

The PCM simply monitors the brake switch input over a period of time. It is possible to decelerate from 30-0 MPH without applying the brakes under some circumstances, but it is highly unlikely the brake switch input would not cycle for 16 consecutive events. By keeping track of the circuit over time, false codes can be avoided.

This code can set if the brake switch is stuck in the retracted or extended position. Therefore, it is possible to set this code if the customer chronically rides the brakes.

Quick Check Items
Inspect or physically check the following items as described below:

- Brake Switch harness connector
- Brake Switch plunger (check adjustment)

Drive Cycle Preparation
There are no drive cycle preparation steps required to run the Component Monitor.

Install the Scan Tool. Record any DTCs or Pending DTCs and any Freeze Frame data. Clear the codes (PCM reset).

Scan Tool Help
A Scan Tool can be used to view the trouble code ID and Freeze Frame Data (PIDs) for DTC P0703. It can also be used to clear stored trouble codes and Freeze Frame Data.

```
DIAG. TROUBLE CODES
ECU: $40 (Engine)
Number of DTCs: 1
*P0703 Torque
       Converter/Brake
       Switch B Circuit
       Malfunction
```

```
DTC·················P0703
ENGINE SPD····2208RPM
ECT (°)·············203°F
VEHICLE SPD····31MPH
ENGINE LOAD····27.8%
MAP (P)·······19.8inHg
FUEL STAT 1··········CL
FUEL STAT 2···UNUSED
ST FT 1················1.6%
LT FT 1················1.6%
```

Brake Switch Circuit Description
The brake switch sense circuit is fed 12 volts by the PCM at Pin 2-62.
The PCM has an internal resistor that acts as a load when the circuit is completed, so the 12 volt signal can be pulled directly to ground. When the brake pedal is up (brakes off), the switch is closed and the 12-volt signal is in a low state (0v). When the pedal is depressed, the switch opens and the circuit is in a high state (12v).

CHILTON DIAGNOSTIC SYSTEM - DTC P0703 (CONT.)

The Scan Tool and Lab Scope can be used to observe the Brake Switch circuit activity.

As seen in the Scan Tool DTC and Freeze Frame examples from the previous page, a fault was detected in the Brake Switch circuit. The following procedures are an example of an efficient diagnostic path.

Scan Tool Test
If the PCM did not detect proper Brake Switch cycling during the test period, then the most efficient first step is to observe the Scan Tool PIDs. Regardless of the circuit conditions, the only critical question is whether or not the PCM is currently recognizing the switch status.

Connect the Scan Tool to the underdash DLC. Identify the vehicle or confirm identity. Navigate to the Datalist and observe the BRAKE SWITCH PID value while cycling the brake pedal.

In this example, the brake switch status consistently changed from OFF to ON when the pedal was depressed. Only when the brake pedal was released very slowly did the status remain ON. In many cases, the Freeze Frame will provide clues on how to operate the vehicle to duplicate the fault. In this case, however, the Freeze Frame did not help. It may be necessary to discuss the drivers' habits and try to duplicate them as close as possible. Once the vehicle was operated with a "lighter touch", the fault surfaced.

If the fault cannot be duplicated in the Scan Tool test, the problem may be intermittent. By using the Scan Tool first, the end result of the switch status was determined. Using this strategy eliminates unnecessary circuit testing. In this case, the fault was verified, and further circuit testing is justified.

Lab Scope Connections
Connect the Channel 'A' probe to the brake switch sense circuit (WT/PK wire) at PCM Pin 2-62, and connect the negative probe to the battery negative post.

Lab Scope Example
In this example, the Lab Scope was set to 2 seconds per division, to show circuit activity over a long period of time (20 sec. per screen).

This example shows the brake switch sense waveform while the brake pedal was depressed and released. When the pedal was released very slowly, the voltage remained high, indicating that the switch did not close.

Test Results
Through careful observation of the code definition, circuit description, and operating conditions (driver habits), the circuit fault was isolated. The Brake Switch, PCM, and circuits were all in good condition, but the switch plunger was not closing the switch contacts when the pedal was released very slowly. The brake switch was adjusted, and time-consuming diagnostic procedures were avoided.

Reference Information - PCM PID Examples

1999 TOWN & COUNTRY 3.8L V6 OHV MFI VIN L (A/T)

■ NOTE: *The following readings were obtained with the engine at idle speed, radiator hose hot, throttle closed, gear selector in Park and all accessories turned off.*

Acronym	Scan Tool Parameter	Range	Typical Value
A/C P	A/C Pressure Switch Signal	0-459 psi	79 psi
ASD	Automatic Shutdown Relay Control	ON / OFF	ON
ECT	Engine Coolant Temp. Sensor (°F)	-40 to 284°F	205°F
CAT-M	Catalyst Monitor	READY / NOT READY	READY
CCM	Component Monitor	READY / NOT READY	READY
EGR-M	EGR System Monitor	READY / NOT READY	READY
EVAP-M	EVAP System Monitor	READY / NOT READY	READY
F-SYS1	Fuel System Bank 1	CL/CLDRV/OL (Note 1)	CL
F-SYS2	Fuel System Bank 1	CL/CLDRV/OL (Note 1)	NOT USED
FP-M	Fuel System Monitor	READY / NOT READY	READY
FT-11	Fuel Trim B1 S1 (%)	0% (± 20%)	2.3%
FT-12	Fuel Trim B1 S2 (%)	0% (± 20%)	99.2% (Note 2)
HCAT-M	Heated Catalyst Monitor	READY / NOT READY	READY
HO2S-11	HO2S BK 1 S1	0-1100 mv	825 mv
HO2S-12	HO2S BK 1 S2	0-1100 mv	390 mv
HO2S11-M	Front Oxygen Sensor Monitor	READY / NOT READY	READY
HO2S-12-M	Rear Oxygen Sensor Monitor	READY / NOT READY	READY
LOAD	Engine Load Percentage (%)	0-100%	4.7%
LTFT B1	Long Term Fuel Trim Bank 1	0% (± 20%)	-1.1%
MAP	MAP Vacuum ("Hg)	0-30" Hg	16.9" Hg
MIL	Malfunction Indicator Lamp Status	ON / OFF	OFF
MISF-M	Misfire Monitor	READY / NOT READY	READY
RPM	Engine Speed in RPM	0-66635 rpm	735
SPARK	Ignition Spark Advance Degrees (°)	-90° to +90° BTDC	22°
STFT B1	Short Term Fuel Trim Bank 1	0% (± 20%)	-1.6%
TPS V	TP Sensor (volts)	0-5.1v	1.13v
VSS	Vehicle Speed	0-159 mph	0 mph

Note 1: This set of acronyms describes the various engine running loop conditions:

- CL: Closed Loop
- CLDRV: Closed Loop in Drive
- OL: Open Loop

Note 2: This reading (99.2%) indicates the PCM does not provide a fuel trim calculation from the rear heated oxygen sensor. This reading is not used to determine fuel trim.

1999 Town & Country 3.8L V6 OHV SFI VIN L (A/T) Black C1 40-Pin Connector

Pin Number	Wire Color	Circuit Description (40-Pin)	Value at Hot Idle
1	---	Not Used	---
2	RD/YL	Coil 3 Driver Control	5° dwell
3	DB/TN	Coil 2 Driver Control	5° dwell
4	DG	Generator Field Driver (+)	Digital Signals: 0-12-0-12v
5	YL/RD	S/C On/Off Switch Sense	0v or 6.7v
6	DG/OR	Auto Shutdown Relay Output	12-14v
7	YL/WT	Injector 3 Driver	1-4 ms
8	TN	Generator Field Control	Digital Signals: 0-12-0v
9, 12	---	Not Used	---
10	BK/TN	Power Ground	<0.1v
11	GY	Coil 1 Driver Control	5° dwell
13	WT/DB	Injector 1 Driver	1-4 ms
14	BR/DB	Injector 6 Driver	1-4 ms
15	GY	Injector 5 Driver	1-4 ms
16	LB/BR	Injector 4 Driver	1-4 ms
17	TN/WH	Injector 2 Driver	1-4 ms
18-19	---	Not Used	---
20	WT/BK	Ignition Switch Power (fused)	12-14v (start or run)
21	---	Not Used	---
22	BK/PK	MIL Driver Control (SES Lamp)	MIL Off: 12v, On: 1v
23-24	---	Not Used	---
25	DB/LG	Knock Sensor Signal	0.080v AC
26	TN/BK	ECT Sensor Signal	2.80v
27	BK/OR	Oxygen Sensor Ground	<0.050v
28-29	---	Not Used	---
30	BK/DG	HO2S-11 (B1 S1) Signal	0.1-1.1v
31	TN	SMART Start Relay Control	KOEC: 9-11v
32	GY/BK	CKP Sensor Signal	Digital Signals: 0-5-0v
33	TN/YL	CMP Sensor Signal	Digital Signals: 0-5-0v
34	---	Not Used	---
35	OR/DB	TP Sensor Signal	0.6-1.1v
36	DG/RD	MAP Sensor Signal	1.5-1.7v
37, 39	---	Not Used	---
38	DG/LB	A/C Switch Signal	A/C Off: 12v, On: 1v
40	GY/YL	EGR Solenoid Control	12v (Off-Idle: 1v)

Standard Colors and Abbreviations

Abbreviation	Color	Abbreviation	Color	Abbreviation	Color
BK	Black	GY	Gray	RD	Red
BL	Blue	GN	Green	TN	Tan
BR	Brown	LG	LT Green	VT	Purple
DB	Dark Blue	ON	Orange	WT	White
DG	DK Green	PK	Pink	YL	Yellow

1999 Town & Country 3.8L V6 OHV SFI VIN L (A/T) Gray C2 40-Pin Connector

PCM Pin #	Wire Color	Circuit Description (40-Pin)	Value at Hot Idle
41	RD/LG	S/C Switch Signal	S/C & Set Switch On: 3.8v
42	DB	A/C Pressure Signal	A/C On: 0.45 to 4.85v
43	BK/LB	Sensor Ground	<0.050v
44	OR	8-Volt Supply	7.9-8.1v
45, 47	---	Not Used	---
46	RD/WT	Fused Battery Power (B+)	12-14v
48	BR/WT	IAC 3 Driver Control	Pulse Signals
49	BL/BK	IAC 2 Driver Control	Pulse Signals
50	BK/TN	Power Ground	<0.1v
51	TN/WT	HO2S-12 (B1 S2) Signal	0.1-1.1v
52-55	---	Not Used	---
56	TN/RD	S/C Vacuum Solenoid	Vacuum Increasing: 1v (S/C On)
57	GY/RD	IAC 1 Driver Control	Pulse Signals
58	VT/BK	IAC 4 Driver Control	Pulse Signals
59	VT/BR	CCD Bus (+)	Digital Signals: 0-5-0v
60	WT/BK	CCD Bus (-)	<0.050v
61	VT/WT	5-Volt Supply	4.9-5.1v
62	WT/PK	Brake Switch Sense	Brake Off: 0v, On: 12v
63	YL/DG	Torque Mgmt. Request to TCM	Digital Signals
64	DB/OR	A/C Compressor Clutch Relay	Relay Off: 12v, On: 1v
65	PK	SCI Transmit	0v
66	WT/OR	Vehicle Speed Sensor Signal	Digital Signals
67	DB/YL	Auto Shutdown Relay Control	Relay Off: 12v, On: 0v
68	PK/BK	EVAP Solenoid Control	0-100% Duty Cycle
69, 71	---	Not Used	---
70	VT/RD	EVAP Solenoid Sense	12-14v
72	YL/BK	LDP Switch Sense	LDP Switch Closed: 0v, Open: 12v
73	LG/DG	Radiator Fan Relay Control	Relay Off: 12v, On: 0v
74	BR	Fuel Pump Relay Control	Relay Off: 12v, On: 0v
75	LG	SCI Receive	5v
76	BK/YL	TRS 41 Switch Sense	In P/N: 0v, Others: 5v
77	WT/DG	LDP Solenoid Control	Solenoid Off: 12v, On: 1v
78-79	---	Not Used	---
80	LG/RD	S/C Vent Solenoid Control	Vacuum Decreasing: 1v (S/C On)

Pin Connector Graphic

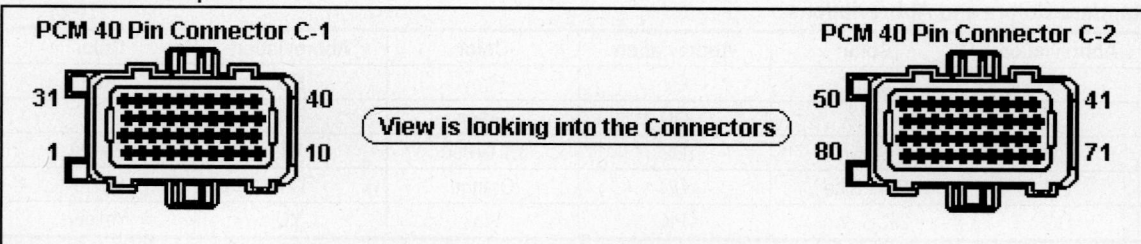

PCM Wiring Diagrams

FUEL & IGNITION SYSTEM DIAGRAMS (1 OF 12)

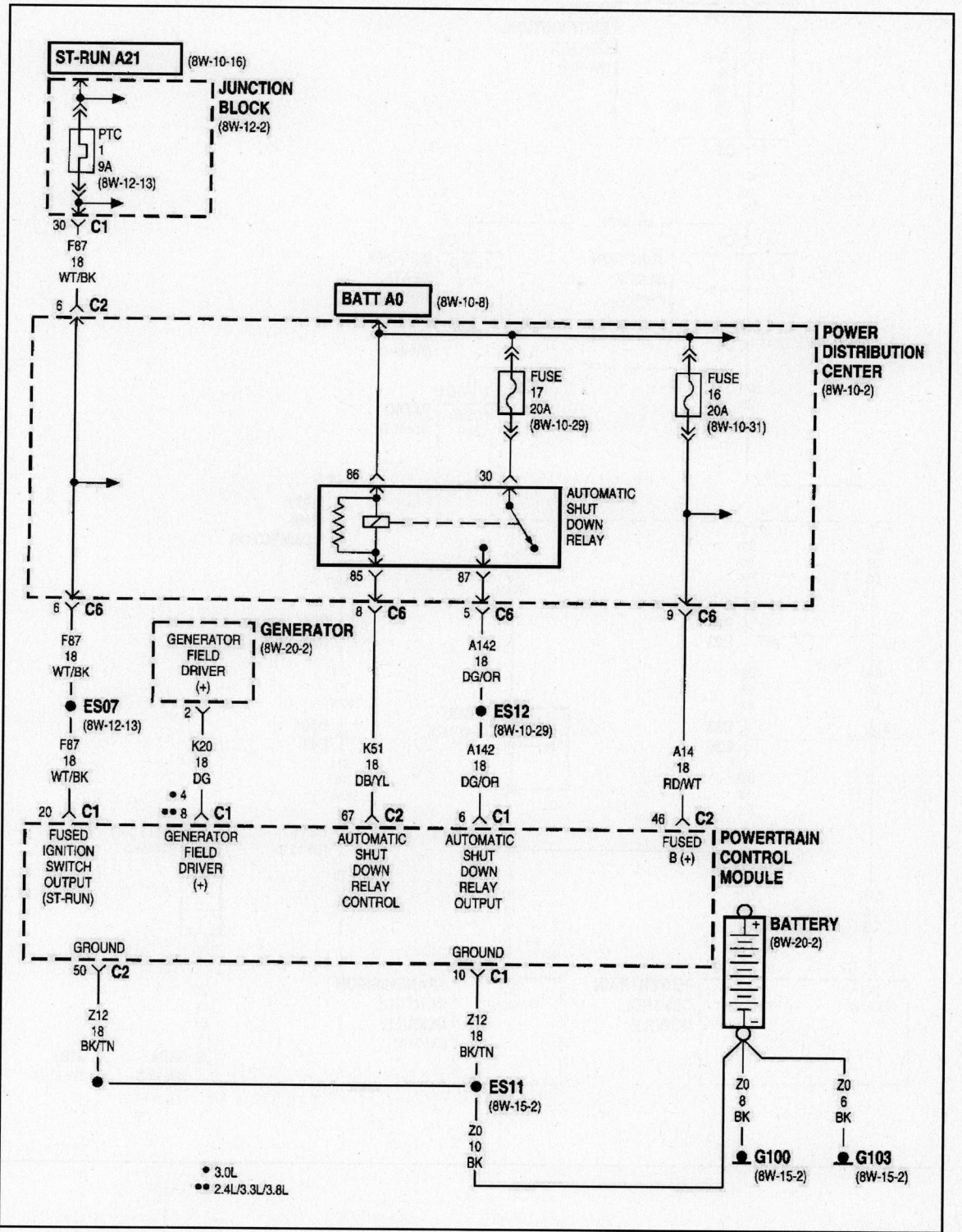

FUEL & IGNITION SYSTEM DIAGRAMS (2 OF 12)

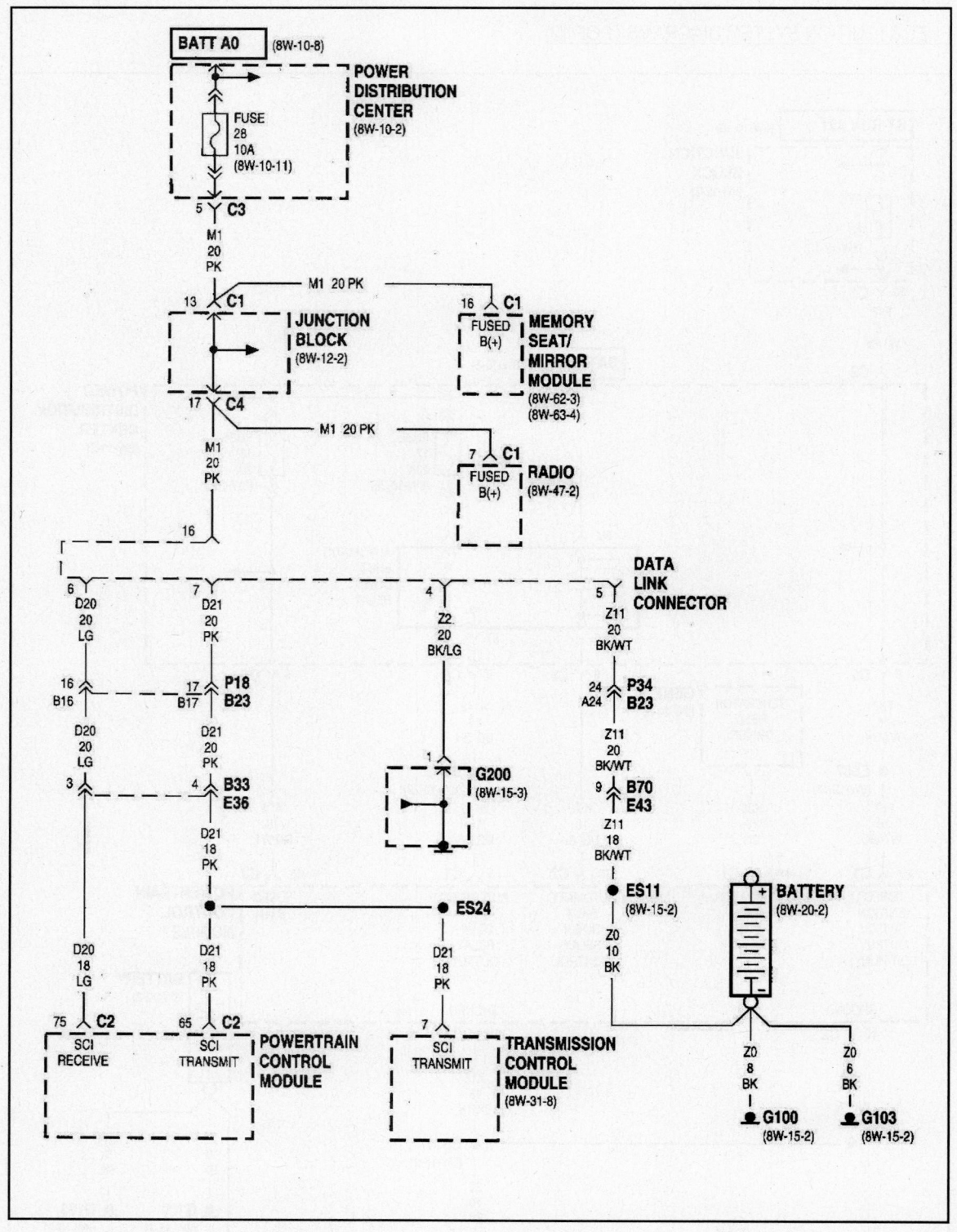

FUEL & IGNITION SYSTEM DIAGRAMS (3 OF 12)

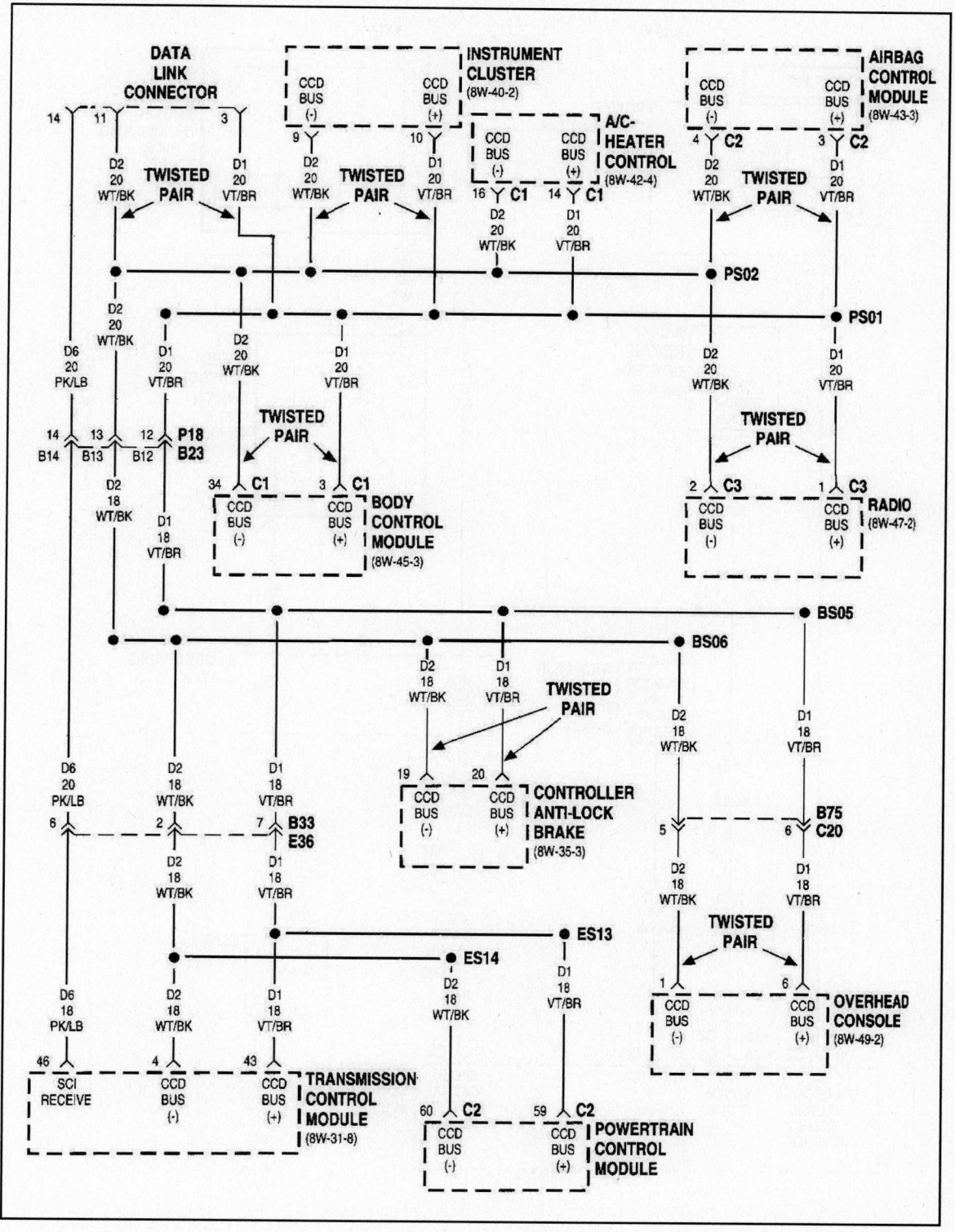

FUEL & IGNITION SYSTEM DIAGRAMS (4 OF 12)

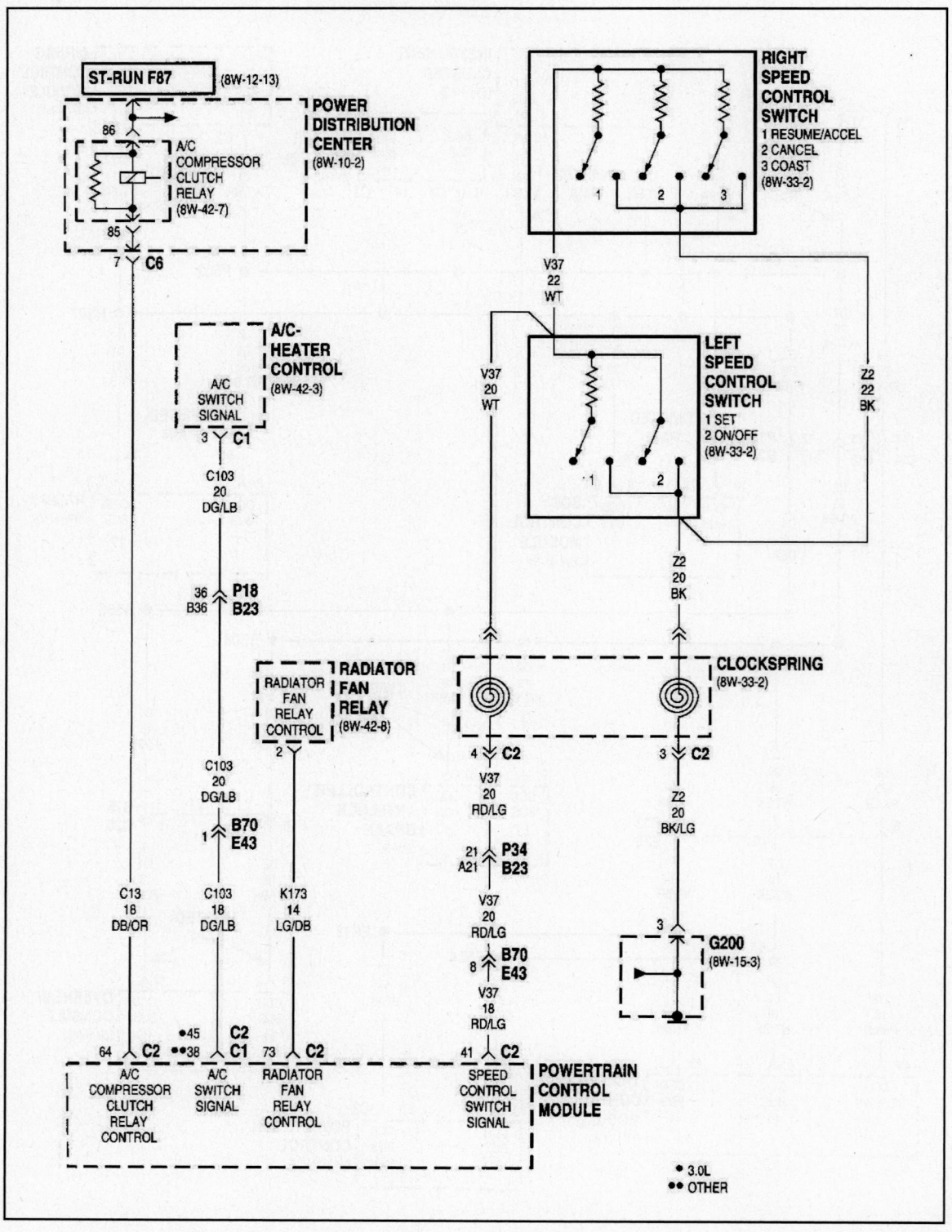

FUEL & IGNITION SYSTEM DIAGRAMS (5 OF 12)

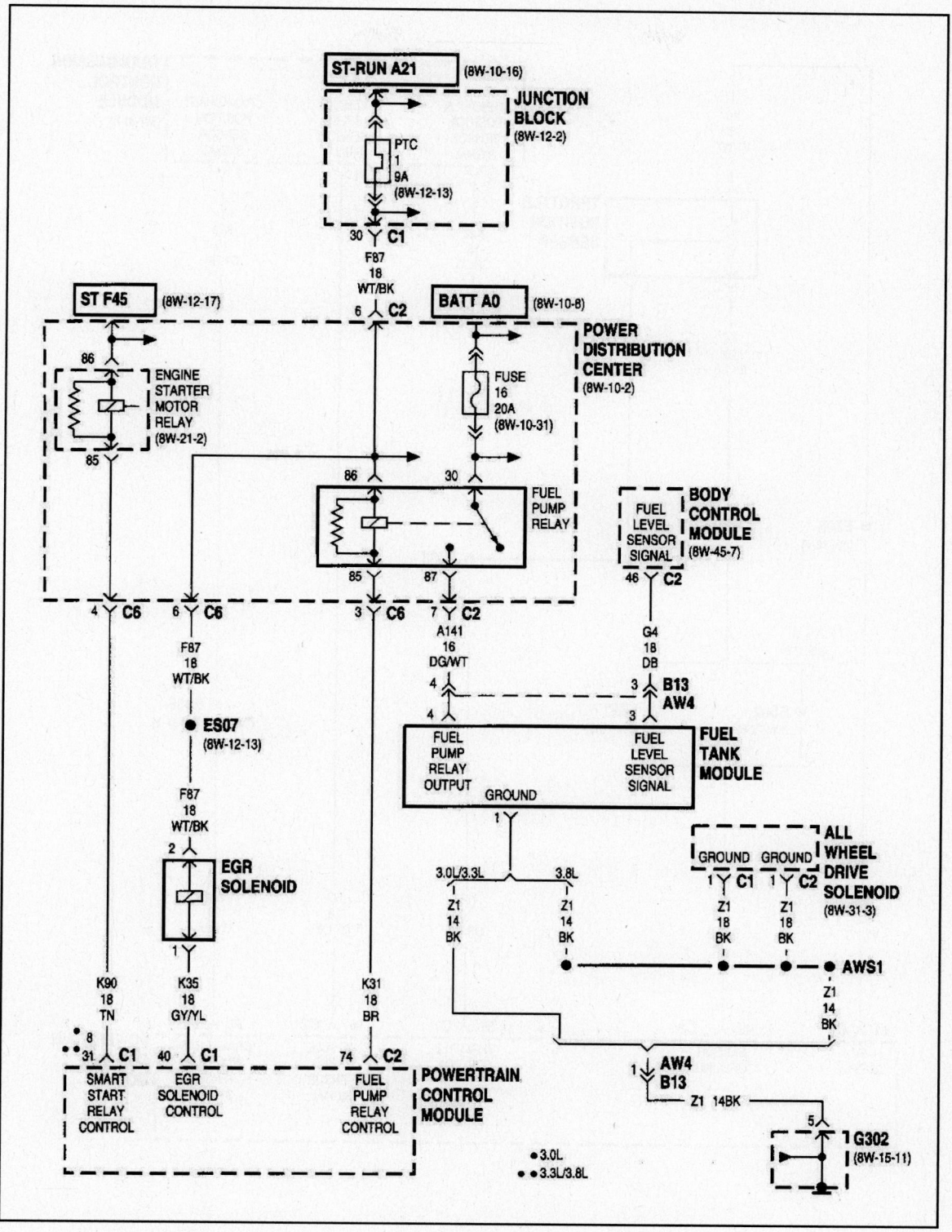

FUEL & IGNITION SYSTEM DIAGRAMS (6 OF 12)

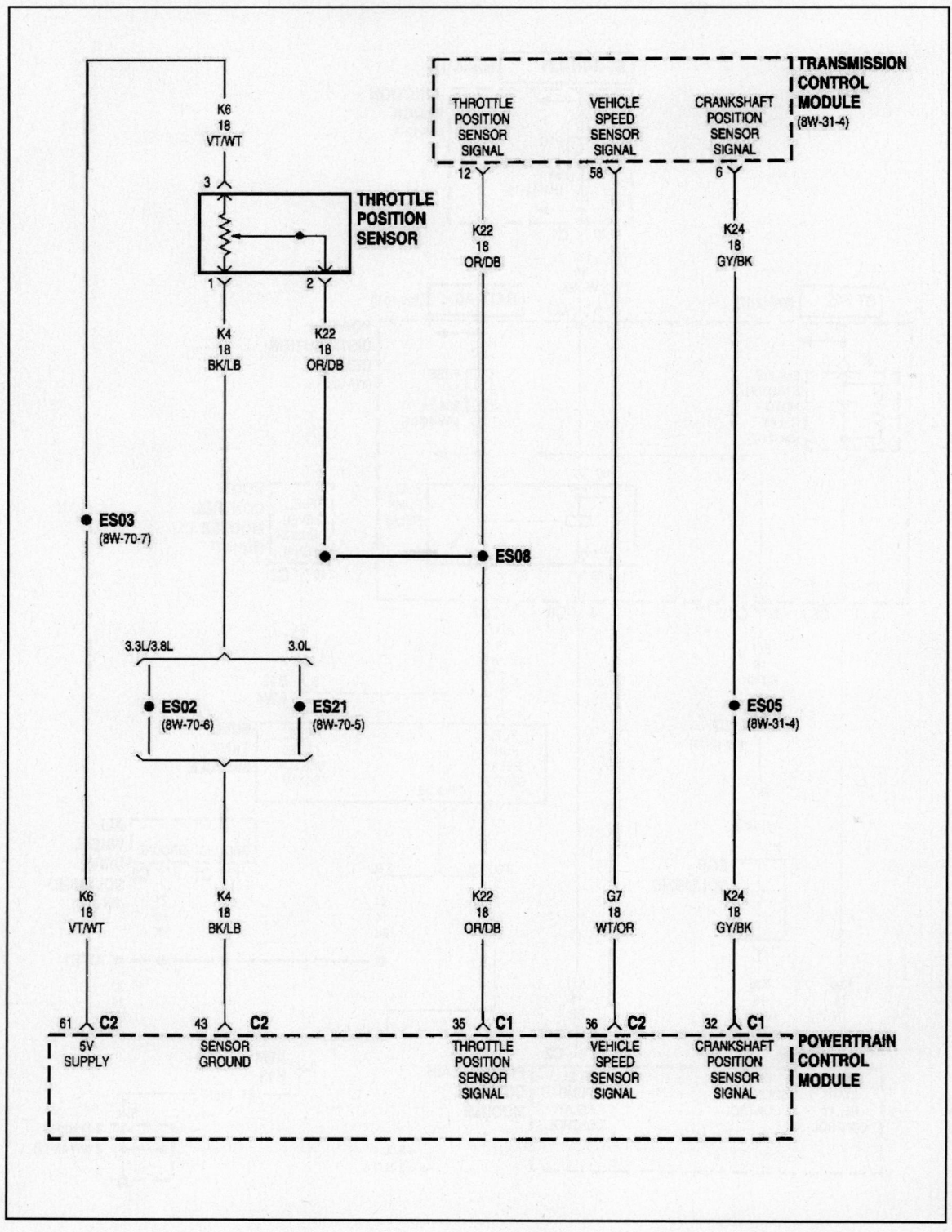

FUEL & IGNITION SYSTEM DIAGRAMS (7 OF 12)

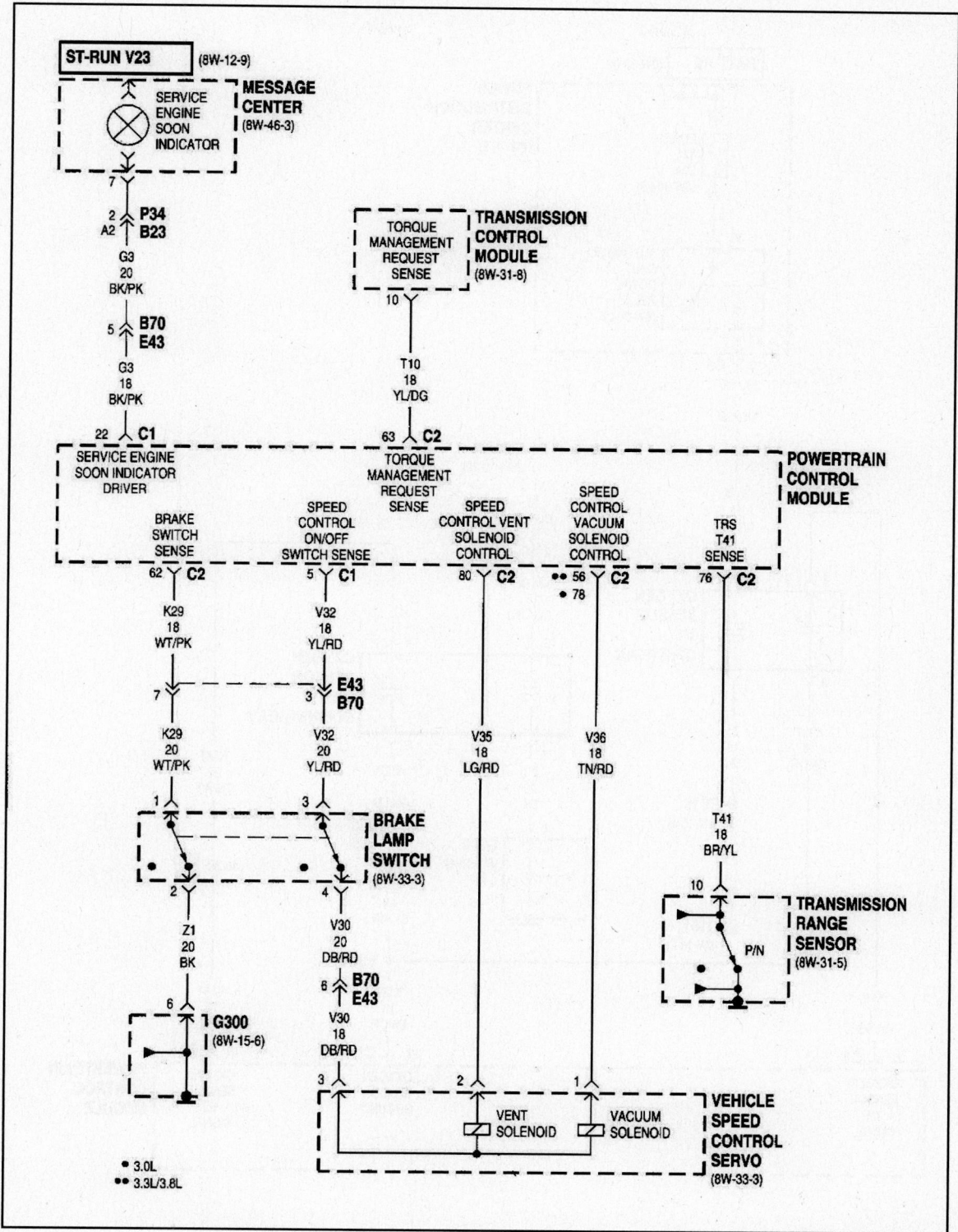

<u>FUEL & IGNITION SYSTEM DIAGRAMS (8 OF 12)</u>

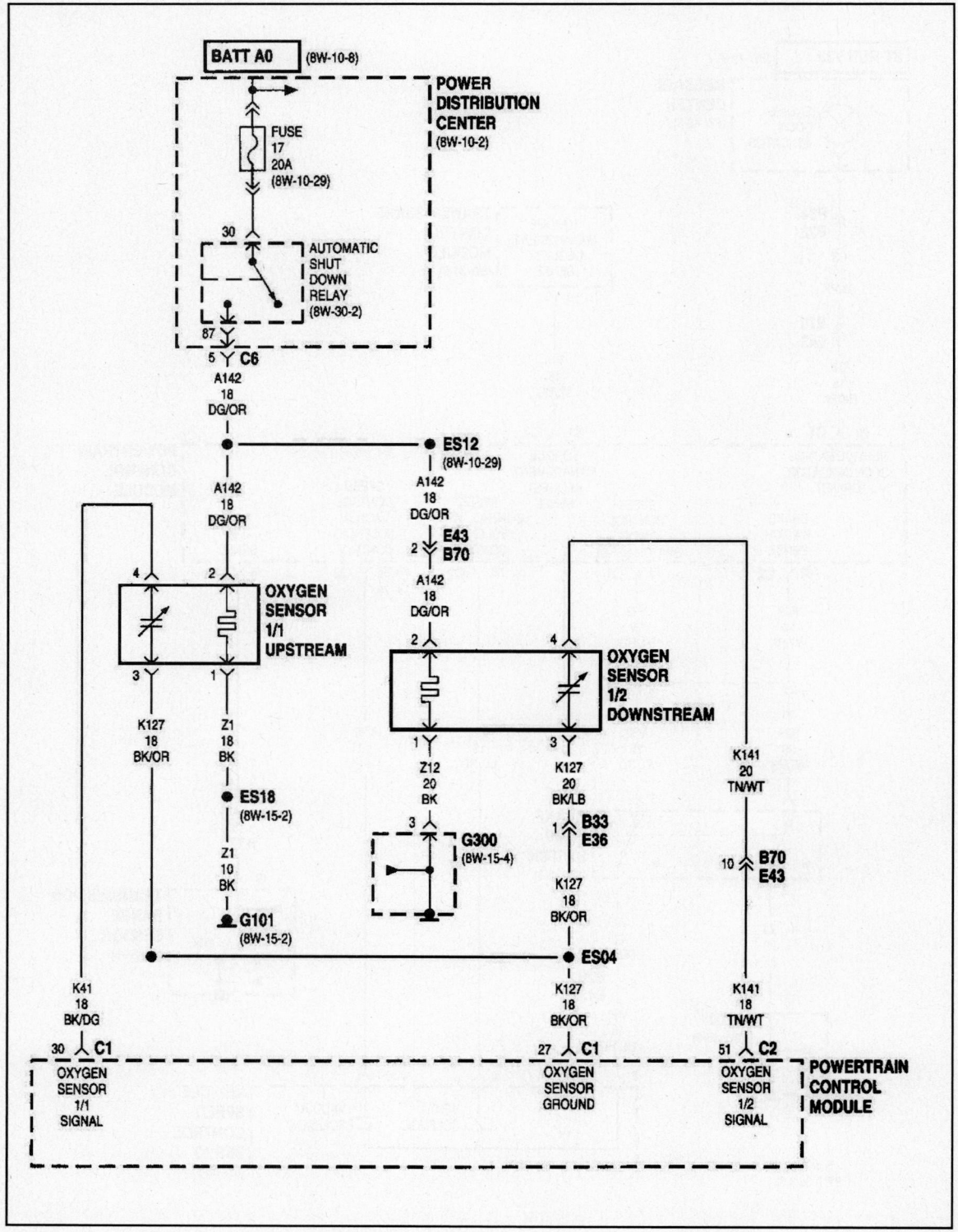

FUEL & IGNITION SYSTEM DIAGRAMS (9 OF 12)

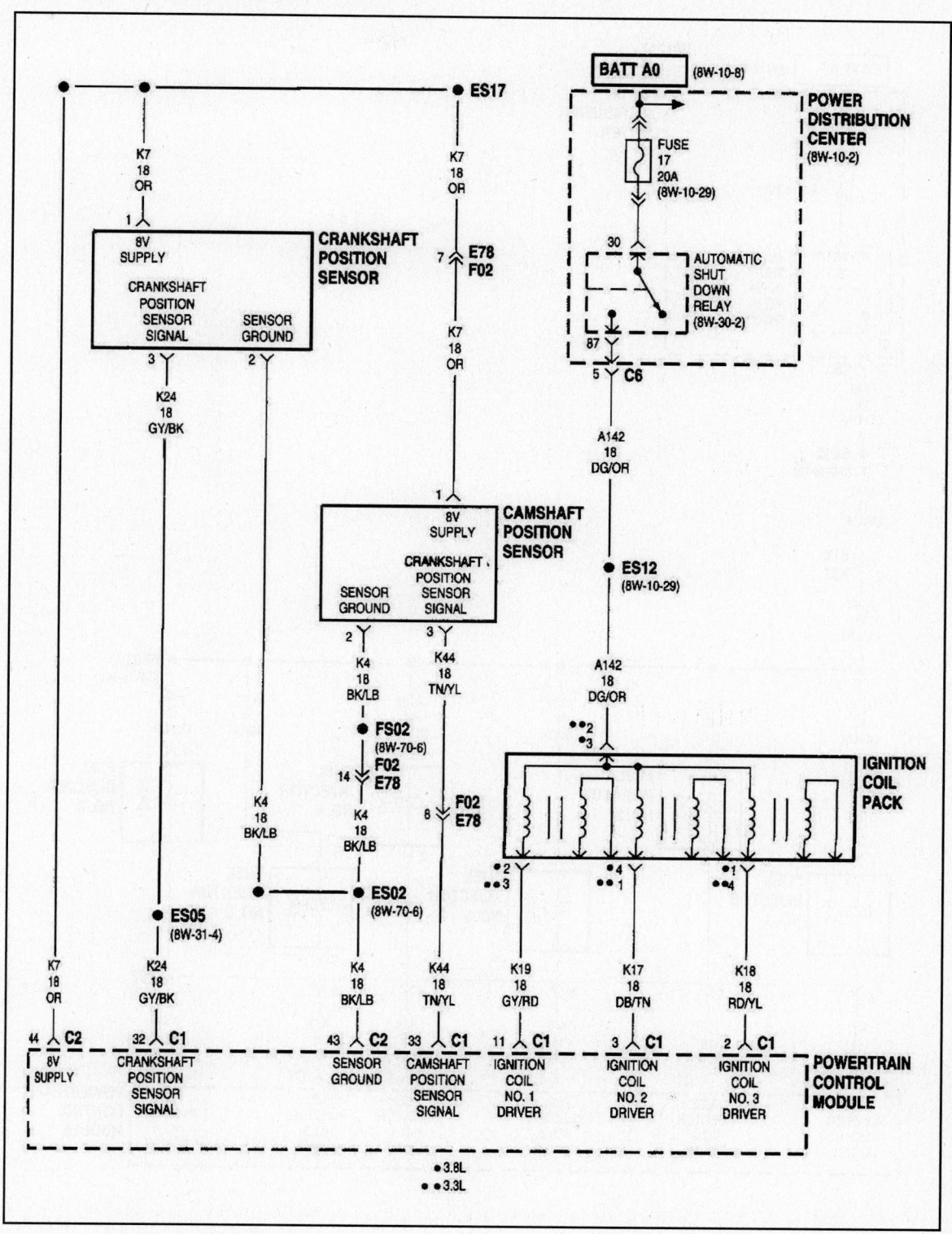

FUEL & IGNITION SYSTEM DIAGRAMS (10 OF 12)

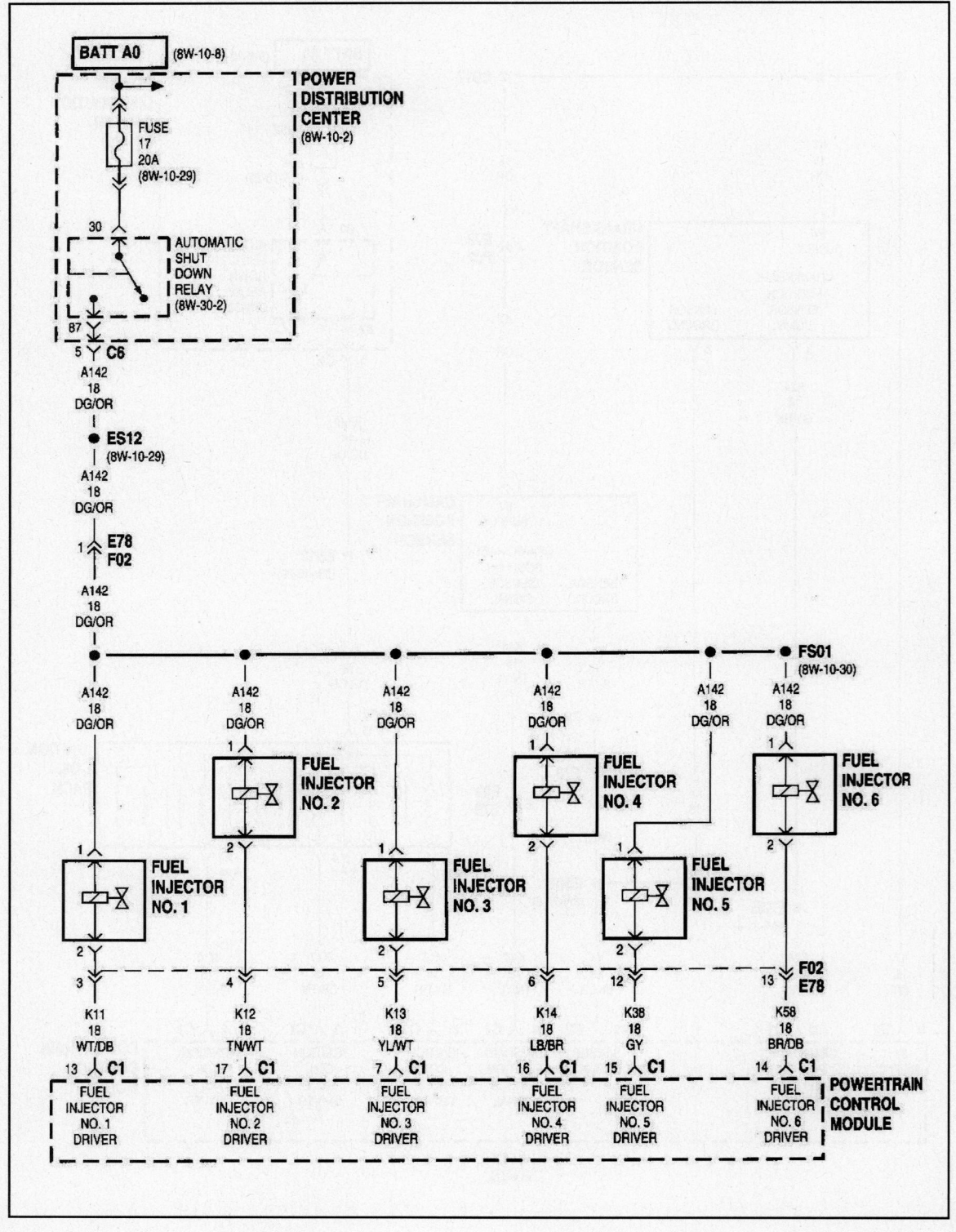

FUEL & IGNITION SYSTEM DIAGRAMS (11 OF 12)

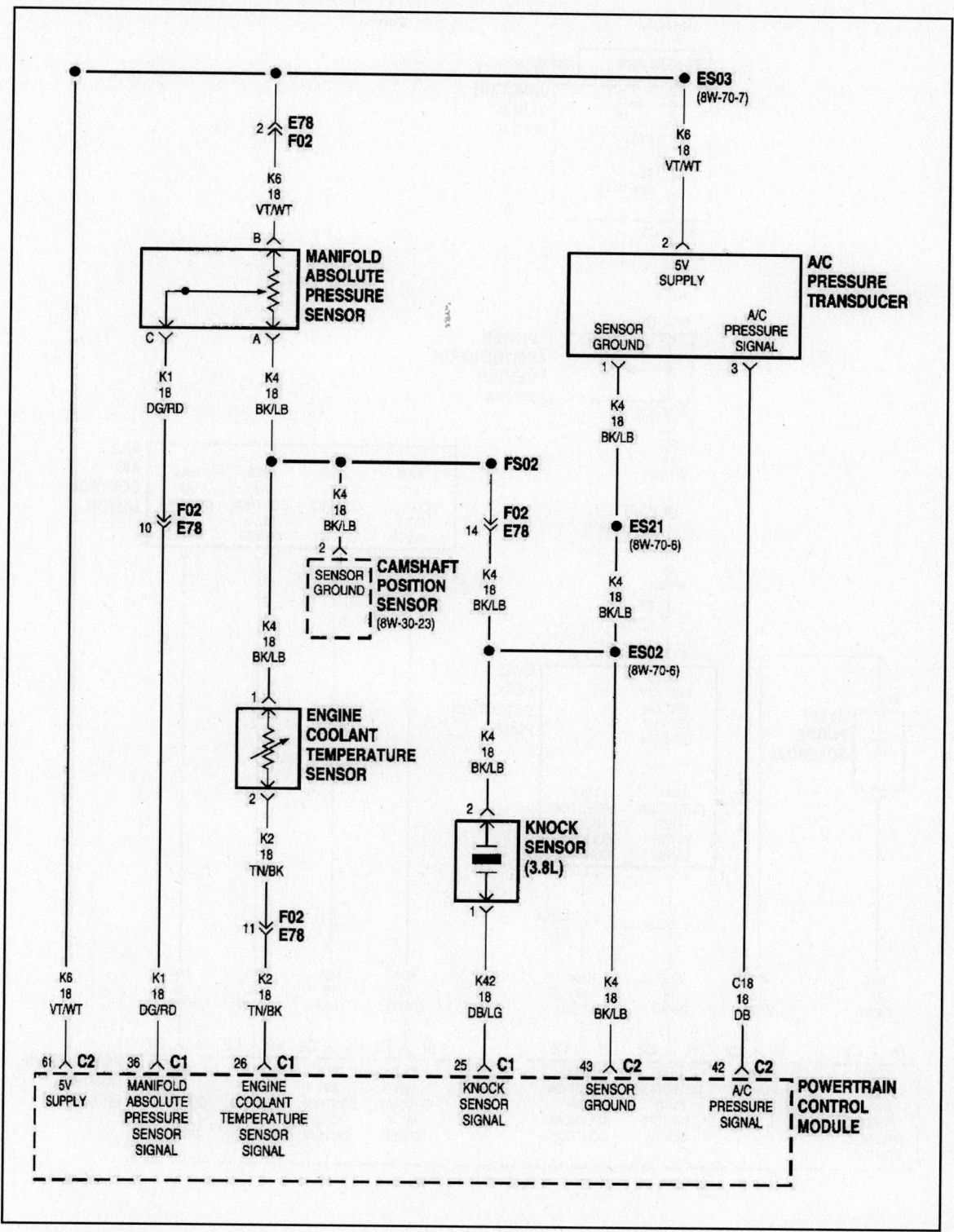

FUEL & IGNITION SYSTEM DIAGRAMS (12 OF 12)

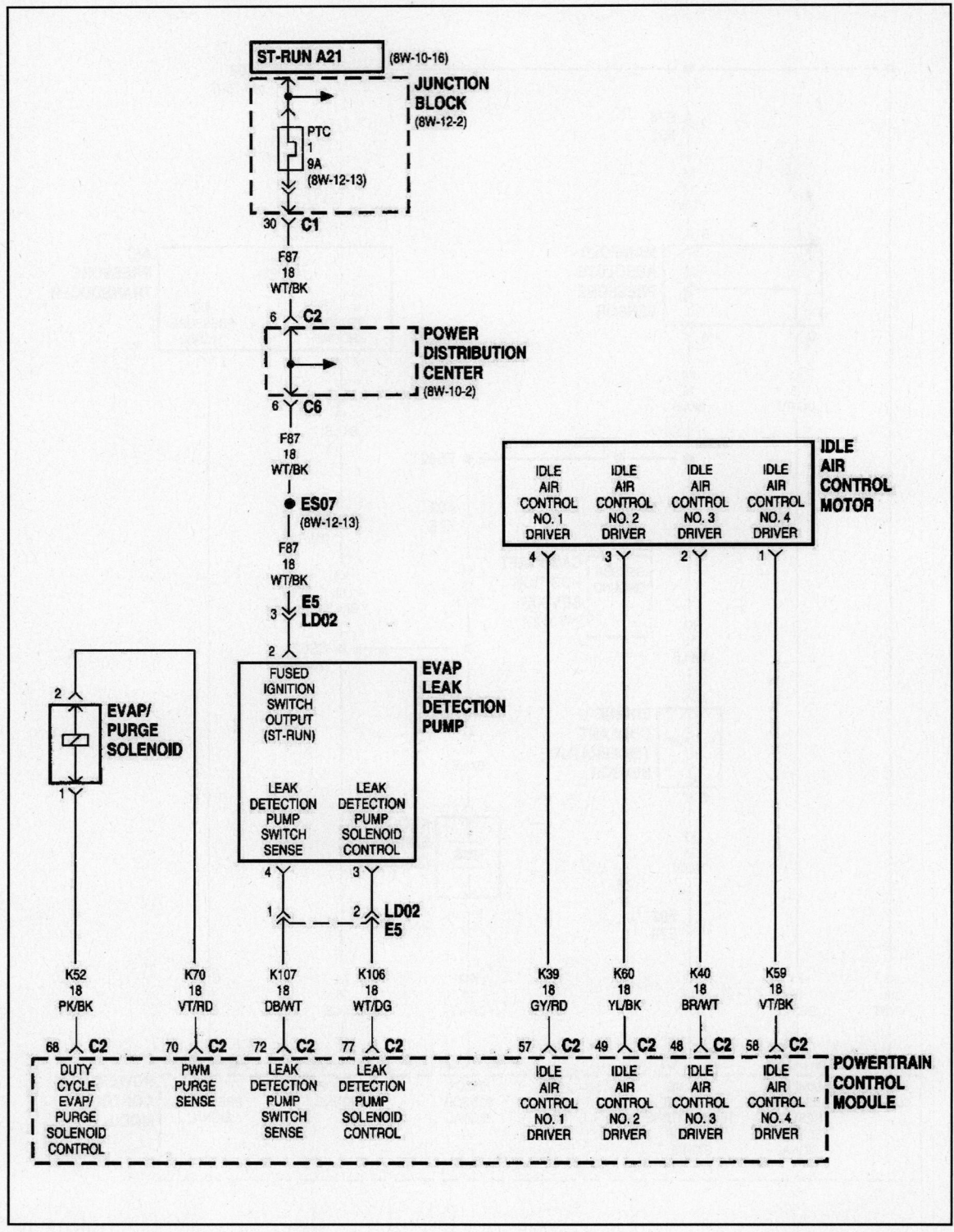

1999 TOWN & COUNTRY - BODY

Body Control Module

GENERAL DESCRIPTION

The Body Control Module (BCM) is concealed below the driver's side of the instrument panel, secured to the Junction Block (J/B). The BCM utilizes integrated circuitry and information carried on the Chrysler Collision Detection (CCD) data bus along with many hard-wired inputs.

Sensor and switch data from throughout the vehicle is used to make decisions for those features the BCM is designated to control. Control of components is accomplished through the use of hard-wired circuits and the CCD data bus.

The BCM monitors its own internal circuits as well as many of its input and output circuits. The BCM has the capability to set and store trouble codes (DTCs) for any fault it detects. Specific Body Systems menus must be selected from the Scan Tool to access data (PIDs), bi-directional controls, and codes (DTCs).

Controlled Systems

The following is a partial list of the body systems controlled by the BCM:

- Wipers and washers
- Headlamps, parking lamps, and turn signals
- Interior panel lamps, courtesy lamps, reading lamps, etc.
- Power Locks and Windows
- Automatic Day/Night Mirror
- Battery saver and lighting fade
- Message Center
- Instrument Cluster Support (many inputs are hard wired to the BCM)
- Horn

In addition to these systems, the BCM is involved in controlling many other systems, and communicates with many more. For example, the Remote Keyless Entry (RKE) and Vehicle Theft Security System (VTSS) have their own control modules, but rely on the BCM for switch inputs and actuator outputs.

■ NOTE: *If the BCM must be replaced, the Vehicle Theft Security System (VTSS) must be enabled in the new BCM before the vehicle will start.*

MODULE COMMUNICATIONS

The BCM communicates with other modules using the Chrysler Collision Detection (CCD) data bus. The following is a list of modules on the CCD data bus network on this vehicle:

- Body Control Module (BCM)
- Powertrain Control Module (PCM)
- Transmission Control Module (TCM)
- Overhead Console
- Mechanical Instrument Cluster (MIC)
- Controller Anti-Lock Brake (CAB)
- A/C-Heater Control
- Airbag Control Module
- CD/Radio

BODY CONTROL DIAGNOSTICS

Scan Tool PIDs

The Scan Tool can be used to observe the data (PID) values for various BCM inputs and outputs. Diagnosis often involves observing a switch state change on the Scan Tool in response to an action (i.e. the doors can be opened and closed to observe the change in door switch state PIDs).

Scan Tool Bi-directional Controls

The Scan Tool can be used to control various lighting and body functions on the vehicle. These tests will confirm that the data bus, components, and circuits are all operational. In addition, these tests confirm that the BCM can process commands and toggle transistors to control devices.

The results of each test can be measured with a DVOM or Lab Scope, or by simply observing the results visually or audibly.

Some of the components the Scan Tool can control are: Chime, Courtesy Lamp, Door Lock Relay, Door Unlock Relay, Driver Door Unlock Relay, Engine Temp Warning Lamp, Front Wiper Relay, Horn Relay, Rear Washer Relay, and Rear Wiper Motor Relay.

SCAN TOOL TEST (PIDS)

The Scan Tool can be used to monitors switch and sensor status to help diagnose body control faults.

Scan Tool Example

Connect the Scan Tool to the underdash DLC. Identify the vehicle or confirm identity. Select BODY, then CODES & DATA. The Scan Tool will communicate with the BCM through the CCD data bus.

In this example, the switch and sensor status PIDs were monitored. Although the BCM is a complex module, and communicates over a large network, most of the faults that occur are simple component or circuit faults.

Monitor a particular PID while cycling that component to verify the circuit. For example, the DOME LAMP SW PID should change from OFF to ON as the switch is cycled.

```
┌─────────────────────────────────────────────────┐
│                                                   │
│      Snap-on Scan Tool Display                    │
│                                                   │
│  ───────────────────────────────────────────     │
│                                                   │
│  MAIN MENU (CHRY BODY) OTHER SYSTEMS              │
│   ATM TESTS            ~CODES & DATA              │
│   CUSTOM SETUP                                    │
│                                                   │
│  ───────────────────────────────────────────     │
│                                                   │
│  PANEL(V)____3.8  FUEL LEVEL(V)___4.5             │
│  ** CODES & DATA.  OK TO DRIVE.  **               │
│  BATTERY(V)_____12.9      IGNITION(V)____12.9     │
│  KEY IN IGN SW___YES      SEAT BELT SW___OPEN     │
│  GATE AJAR SW___OPEN      DRVR DOOR AJAR_OPEN1     │
│  DOOR AJAR SW___OPEN      LR DOOR AJAR___OPEN      │
│  RR DOOR AJAR___OPEN      INTER WIPER(V)__0.0      │
│  HEADLAMP SW_____OFF      COURTESY LAMP___OFF      │
│  DOME LAMP SW____OFF      IGN CYCLES_____1      │
│  VERSION NO._____9.2                              │
│                                                   │
└─────────────────────────────────────────────────┘
```

CHILTON DIAGNOSTIC SYSTEM – WIPER PARK SWITCH

The purpose of the Chilton Diagnostic System is to provide you with one or more tests that can be done with a DVOM, Lab Scope or Scan Tool *prior* to entering the complete trouble code repair procedure.

Code Description

The Scan Tool displayed the error message "WIPER PARK SWITCH FAILURE". The BCM detected that the park switch state did not change as expected. The wiper park switch sense circuit at the BCM (Pin 2-41) should drop from 5v to 0v when the wipers park. The switch is closed in the park position, and the sense circuit voltage is pulled to ground.

When a park switch circuit fault is present, the wipers will continue to operate, but only at low speed. The wipers will not park, and will stop wherever they are when the wiper switch is turned off.

Quick Check Items

Verify that the wiper linkage functions properly, and that there is no mechanical interference.

Scan Tool Help

Use the Scan Tool to retrieve the error message, and to monitor the INTER WIPER(V) PID. The PID voltage value should be 0v when the wipers are in the park position.

In this example, the value is 2.9v with the wipers in the park position. This is an inappropriate voltage (should be 0v).

Diagnosis

A DVOM measurement should be made at the sense circuit at BCM Pin 2-41 to verify the value on the Scan Tool. Also, measure the voltage on the wiper motor side of the

---

Snap-on Scan Tool Display

MAIN MENU (CHRY BODY) OTHER SYSTEMS
ATM TESTS   ~CODES & DATA
CUSTOM SETUP

PANEL(V)____3.8 FUEL LEVEL(V)____4.5
** CODES & DATA. OK TO DRIVE. **
WIPER PARK SWITCH FAILURE
BATTERY(V)_____13.0   IGNITION(V)____13.0
KEY IN IGN SW___YES   SEAT BELT SW___OPEN
GATE AJAR SW___OPEN   DRVR DOOR AJAR_OPEN1
DOOR AJAR SW___OPEN   LR DOOR AJAR___OPEN
RR DOOR AJAR___OPEN   INTER WIPER(V)__2.9
HEADLAMP SW_____OFF   COURTESY LAMP___OFF
DOME LAMP SW___OFF   IGN CYCLES_____3
VERSION NO._____9.2

---

circuit with the wiper motor assembly harness disconnected. If the first measurement confirms the Scan Tool value, and the second measurement confirms that the open circuit voltage at the wiper motor connector is about 5 volts, then the park switch is defective.

The diagnosis could involve detailed circuit testing, but a look at the wiring diagram will rule out other possible causes. See the Front Wiper Wiring Diagram on the next page. The wiper motor and the park switch use a common ground. If the ground circuit was open or had high resistance, the wiper motor performance would also be affected.

BCM Wiring Diagrams

FRONT WIPER SYSTEM DIAGRAM (PARTIAL)

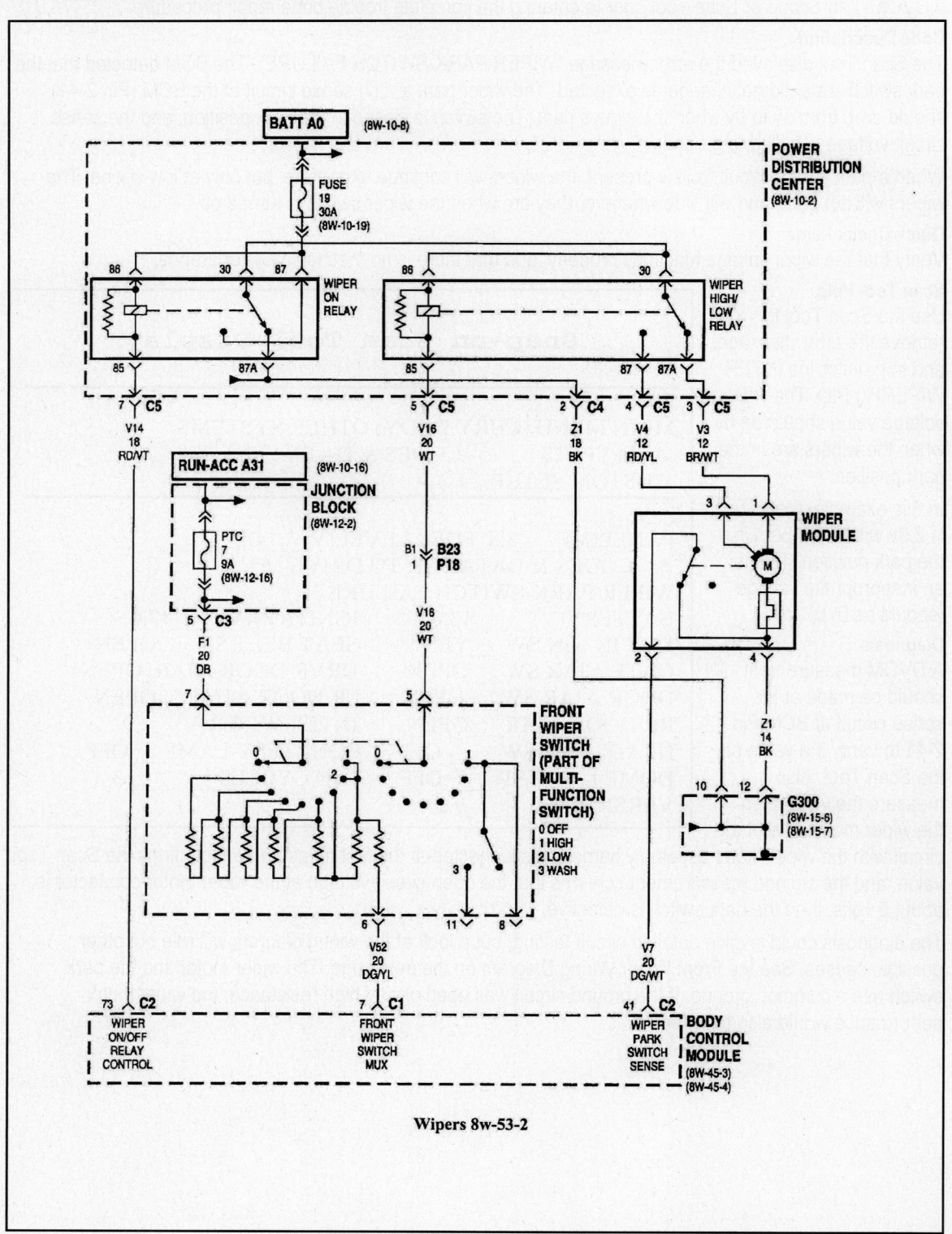

Wipers 8w-53-2

SPORT UTILITY VEHICLE CONTENTS

1999 CHEROKEE - POWERTRAIN

2002 ½ DURANGO - POWERTRAIN

2002 DURANGO - POWERTRAIN

1999 CHEROKEE - POWERTRAIN

Introduction

HOW TO USE THIS SECTION

This section of the manual includes diagnostic and repair information for Jeep and Chrysler SUV applications. This information can be used to help you understand the Theory of Operation and Diagnostics of the electronic controls on this vehicle.

The articles in this section are separated into two sub categories:

- Powertrain Control Module
- Transmission Controls

The articles listed for this vehicle include a variety of subjects that cover the key engine controls and vehicle diagnostics. Several sensor inputs and output devices are featured along with detailed descriptions of how they operate, what can go wrong with them, and most importantly, how to use the PCM and TCM onboard diagnostics and common shop tools to determine if one or more devices or circuits has failed.

1999 Cherokee 4.0L I6 VIN S (AW4 Automatic Transmission)

This vehicle uses a distributor Electronic Ignition system, along with camshaft and crankshaft position sensors (Hall Effect) to control the fuel injection timing. These devices and others (i.e. the Cooling Fan, EVAP, IAC, MAP, TP, etc.) are covered along with how to test them.

2002 ½ Durango 4.7L V8 VIN N (With New Generation Controller)

Starting in 2002 ½, this vehicle was equipped with the Next Generation Controller (NGC). The articles for this vehicle cover the major changes in engine management strategy and testing procedures. This includes the Natural Vacuum Leak Detection (NVLD) system, which replaces the LDP system. New cam and crank position sensing circuits are discussed, as well as a new heated oxygen sensor circuit design.

2002 Durango 4.7L V8 VIN N (45RFE Automatic Transmission)

This vehicle uses a Coil-On-Plug Electronic Ignition system, along with camshaft and crankshaft position sensors (Hall Effect) to control the ignition and fuel injection timing and sequence. These devices and others (i.e., the Generator, LDP, Speed Control, IAC, MAP, HO2S, etc.) are covered along with how to test them. Transmission electronic controls are also covered, along with various related test strategies.

Choose The Right Diagnostic Path

In most cases, the first step in any vehicle diagnosis is to follow the manufacturers recommended diagnostic path. In the case of Jeep or Chrysler vehicle applications, the first step is to connect a Scan Tool (OBD II certified for this vehicle) and attempt to communicate with the engine controller or any other applicable control modules.

The Malfunction Indicator Lamp (MIL) can be a great help to you during diagnosis. You should note if the customer complaint included the fact that the MIL remained on during engine operation. And be sure to determine if the MIL comes on at key on and goes out after a 4 second bulb-check. There is more information on using the MIL as part of a diagnostic path in Section 1.

Component Diagrams

You will find numerous "customized" component diagrams used throughout this section. These diagrams are really a mini-schematic of how a particular device (where an input or output device) is connected to the engine or transmission control module. Each diagram contains the relevant power, ground, signal and control circuits.

Vehicle Identification

VECI DECAL

The Vehicle Emission Control Information (VECI) decal is located on the underside of the hood. *This example is from a California 1999 Jeep Cherokee 4.0L I6 SOHC VIN S.*

Chrysler Corporation IMPORTANT VEHICLE INFORMATION
This vehicle is equipped with electronic engine control systems. Engine idle mixture, and ignition timing are not adjustable.

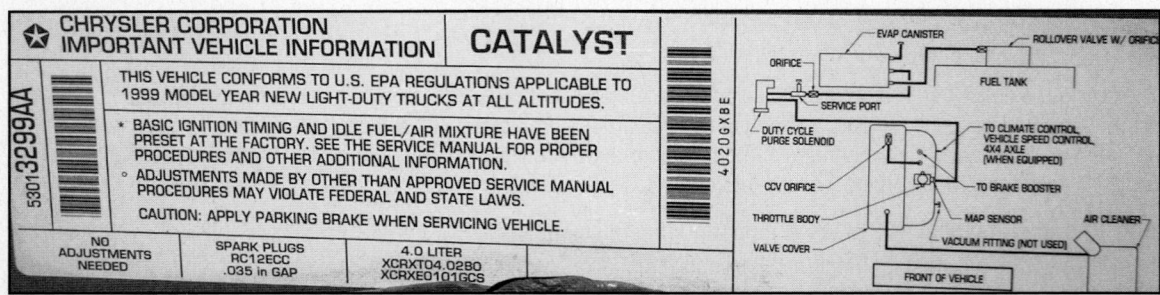

Vehicle Certification
This vehicle conforms to U.S. E.P.A regulations applicable to 1999 model year new light-duty trucks. It is OBD II Certified.

Other Identification Items
Engine: 4.0L – XCRXT04.O280/XCRXE0101GCS

Underhood View Graphic

Electronic Engine Control System

OPERATING STRATEGIES

The Powertrain Control Module (PCM) in this application is a triple-microprocessor digital computer. The PCM, formerly known as the SBEC, provides optimum control of the engine and various other systems. The PCM uses several operating strategies to maintain good overall drivability and to meet the EPA mandated emission standards. These strategies are included in the tables on this page.

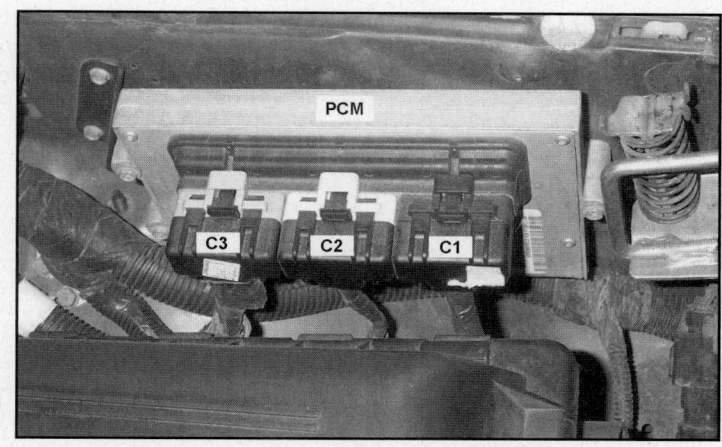

Based on various inputs it receives, the PCM adjusts the fuel injector pulsewidth, the idle speed, the amount of ignition advance or retard, the ignition coil dwell and operation of the EVAP canister purge valve.

The PCM also controls the Speed Control (S/C), air conditioning systems and some transmission functions.

PCM Location

The PCM is located at the left side of the engine compartment, next to the air cleaner.

Operating Strategies

A/C Compressor Clutch Cycling Control	Closed Loop Fuel Control
Electric Fuel Pump Control	Limited Transmission Control
Fuel Metering of Sequential Fuel Injectors	Speed Control (S/C)
Idle Speed Control	Ignition Timing Control for A/F Change
Vapor Canister Purge Control	

Operational Control of Components

A/C Compressor Clutch	Ignition Coil (Timing and Dwell)
Fuel Delivery (Injector Pulsewidth)	Fuel Pump Operation
Fuel Vapor Recovery System	Idle Speed
Malfunction Indicator Lamp	Integrated Speed Control

Items Provided (or Stored) By the Electronic Engine Control System

Shared CCD Data Items	Adaptive Values (Idle, Fuel Trim, etc)
DTC Data for Non-Emission Faults	DTC Data for Emission Related Faults

Distinct Operating Modes

Key On	Cruise
Cranking	Acceleration (Enrichment)
Warm-up	Deceleration (Enleanment)
Idle	Wide Open Throttle (WOT)

PCM Controlled Cooling Fan

GENERAL OVERVIEW

The engine-cooling fan on this application is an electric motor controlled by the radiator fan relay in the Power Distribution Center (PDC). The cooling fan relay pull-in coil is grounded by the PCM, which delivers battery power through PDC fuse #5 (40 Amp).

The fan circuit can be tested in a number of ways. By using the Scan Tool bi-directional controls to command fan operation, temperature requirements for fan operation can be ignored, and the fan can be tested under controlled conditions. The Lab Scope test, using a low-amps probe, is an efficient test because it looks at the end result of all components and circuits involved. If the current flow is normal, then the PCM, relay, fuses, fan motor, and circuits are all okay.

BI-DIRECTIONAL TEST (FAN)

Place the gear selector in Park and block the drive wheels for safety.

Lab Scope Connections
Connect a low-amps probe to the Lab Scope (Channel 'B' in this case). Clamp the probe tip around the fan power or ground wire at the harness connector on the fan shroud.

Scan Tool and Lab Scope Test
Using the Scan Tool, access the OBD CONTROLS menu, then ATM TESTS. Select the COOLING FAN control. When activated, the test cycles the fan "on" every few seconds for 5 minutes. This allows circuit testing without having to wait for the engine to heat up. Also, the repetition stresses the system, which helps to find intermittent faults.

Lab Scope Example (1)
In example (1), the trace shows the fan current waveform when the fan is initially energized. It takes much more current to start a motor than to keep it running. The peak current flow at startup is about 13.5 amps.

Lab Scope Example (2)
In example (2), the trace shows the fan current waveform after the fan speed has stabilized. Note the drop in the peak current flow (about 7 amps).

Using Test Results
High current flow indicates that the motor is binding due to worn bearings or interference. Low current flow would indicates low available voltage or ground problems. Also, some fan motor segments may be open or shorted, resulting in a current waveform with occasional sharp drops (open) or spikes (short).

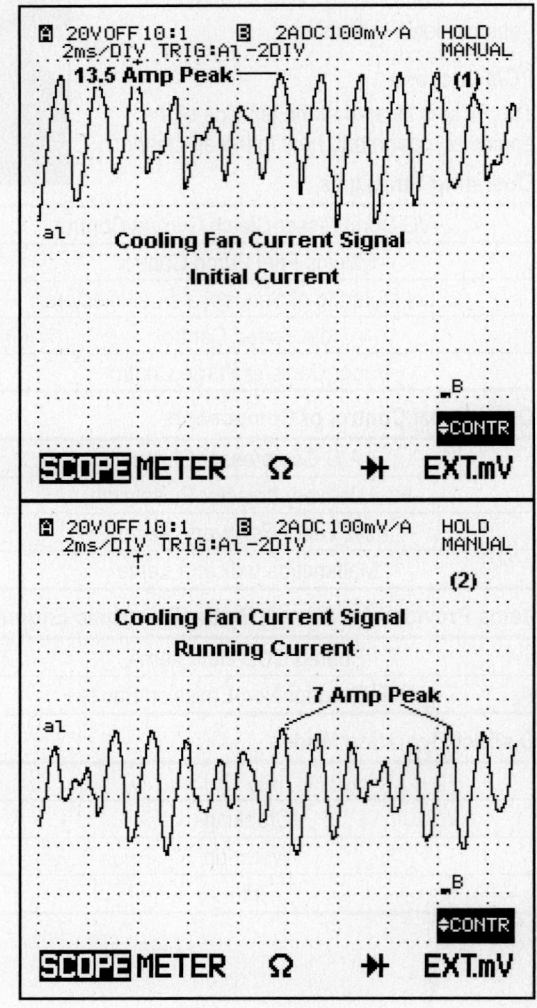

Integrated Electronic Ignition System

INTRODUCTION

The engine controller (PCM) controls the operation of the Distributor Electronic Ignition (DI) system on this vehicle. Battery voltage is supplied to the coil through the Auto Shut Down (ASD) relay and fuse #21, located in the Power Distribution Center (PDC).

This system is "integrated" because the ignition control module functions are incorporated into the PCM. The PCM controls the ground circuit of the ignition coil by switching the ground path for the coil "on" and "off" at the appropriate time. The PCM adjusts the ignition timing, advance and dwell to meet changes in the engine operating conditions.

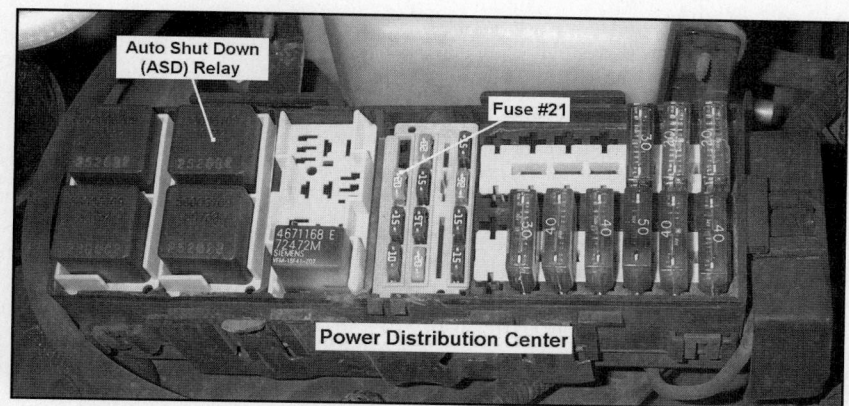

System Components

The main components of this Integrated DI system are the coil, distributor, crankshaft position sensor, ASD relay, PCM and the related wiring.

Ignition Spark Timing

The ignition timing on this DI system is entirely controlled by the PCM, as there are no vacuum or mechanical devices used to modify timing. Do not attempt to adjust base timing. Any rotation of the distributor housing will alter fuel injection timing, but will have no effect on ignition timing.

The PCM uses the Hall effect CKP sensor signal to indicate crankshaft position and speed by sensing a series of windows (3 sets of 4 windows) in the outer edge of the flywheel. From this signal, the PCM can determine that a pair of cylinders is approaching TDC. However, the CKP sensor does not indicate which cylinder is on compression stroke. Because one coil fires all 6 cylinders through a conventional distributor cap and rotor, it is not important for the PCM to identify specific cylinders for normal ignition system operation.

The PCM calculates cylinder identification using the CMP sensor, mounted inside the distributor. The CMP sensor is also a Hall effect device, and its signal is used as a reference point to help the PCM determine which cylinder that each set of CKP signals is related to. This signal is used only for sequential fuel injection timing and diagnostics, and has no effect on ignition timing.

After engine startup, the PCM calculates the ignition spark advance with these signals:

- MAP Sensor
- ECT Sensor
- Engine Speed Signal (rpm)
- Throttle Position Sensor
- Vehicle Speed Sensor
- Transmission Range Sensor
- Brake Lamp Switch

IGNITION SYSTEM OPERATION

The PCM controls the operation of the coil in order to provide ignition spark at the correct time in the combustion process.

DI Coil Control Circuit
The PCM grounds the Coil at Pin A7 of the PCM wiring harness (GY wire).

Coil Specifications
There are 2 different brands of ignition coil used in this application, Diamond and Toyodenso. The primary resistance specifications for the Diamond coil are 0.97-1.18 Ohms, and 0.95-1.20 Ohms for the Toyodenso. The coil specifications are at room temperature (68-72°F).

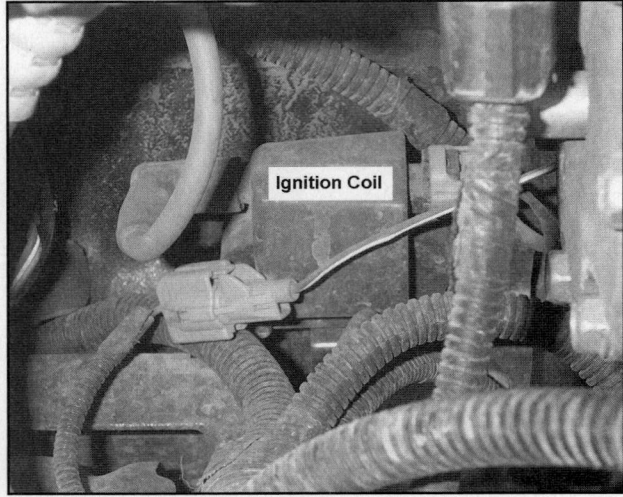

Trouble Code Help
The ignition coil is continuously monitored by the PCM for faults. However, the PCM cannot measure the condition of individual ignition circuits or components. Instead, the coil driver inside the PCM is monitored for current draw. If a predetermined peak current is not achieved for 3 seconds during cranking, or 6 seconds with the engine running, DTC P0351 is set (this is a 1 trip code).

It is important to understand why a particular code sets, so you can look at the same conditions the PCM used to determine a fault. The vehicle may run fine during these code conditions, and you may be fooled into determining that the failure is intermittent. By knowing what to look for (current draw), the system can be accurately diagnosed.

In this application, a very low resistance coil is used (about 1 Ohm). Because of this design, even a small amount of additional circuit resistance will significantly reduce the current consumed, and this code will set. This code may set as an indication that the coil or circuit is failing even though there are no drivability problems.

DI Coil Circuits Schematic

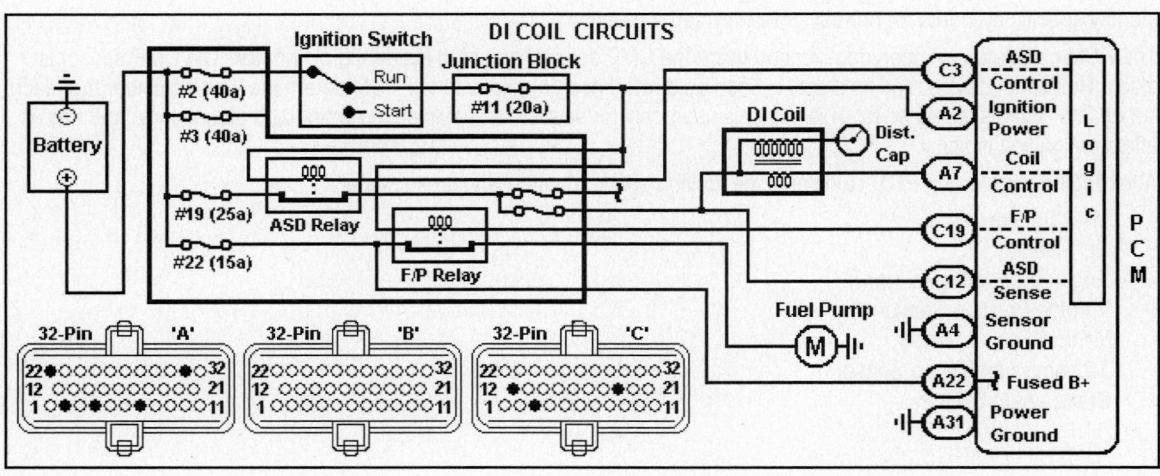

LAB SCOPE TEST (COIL PRIMARY)

The ignition coil primary circuit includes the Coil, Power Distribution Center (PDC), ASD relay, and coil driver control circuit in the PCM.

In this application, the ignition coil is powered by the ASD relay through fuse #21 (20a), and is grounded by the PCM at Pin A7.

Ignition primary waveforms can be used to diagnose ignition, mechanical or fuel concerns instead of secondary. Because no inductive probes are necessary, virtually any Lab Scope is capable of viewing ignition primary.

Lab Scope Connections
Connect the Channel 'A' positive probe to the coil primary at PCM connector 'A' Pin 7 (GY wire) or at the coil connector. Connect the negative probe to the battery negative post.

Lab Scope Example (1)
In example (1), the trace shows the coil primary circuit. The trace shows a spark line "intersect point" at about 45v (over 2 divisions at 20v per division). If this point is over 50v, it could indicate that the A/F mixture is lean, the cylinder compression is too high or the spark timing is retarded.

If this point is under 40v, it could indicate that the A/F mixture is too rich, the cylinder compression is too low or the spark timing is over advanced.

Lab Scope Example (2)
In example (2), the volt setting was changed to 100VDC to show the coil inductive kick (400v).

Lab Scope Example (3)
In example (3), the volt setting was changed to 200 mv to show the driver performance. Channel 'B' was set to 'ground coupled' to make it a ground reference point to help show the counter-voltage in the coil driver circuit. These settings are not shown in the capture, but are located in the menus on the Fluke 99 used for these examples.

In most cases, the voltage should be less than 750 mv. In this example, a counter-voltage of 150 mv proves that the driver and the PCM ground are in acceptable condition. Any further diagnosis of this circuit should not involve testing of the driver or PCM power ground.

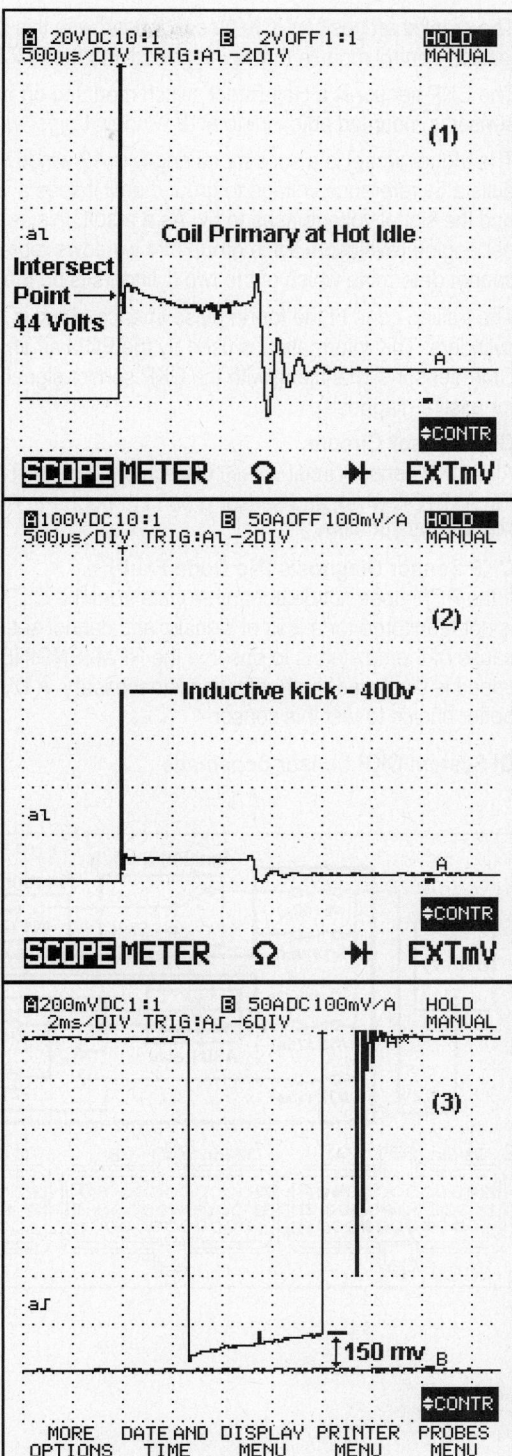

CRANKSHAFT POSITION SENSOR (HALL EFFECT)

The crankshaft position (CKP) sensor used with this DI system is the primary sensor for crankshaft position data to the ignition control module that is integrated into the PCM.

The CKP sensor is a Hall-Effect switch mounted on bell housing near the upper left corner of the engine block. The sensor is mounted adjacent to a 12-window trigger wheel that is part of the flywheel.

The trigger wheel is a solid metal ring with 12 windows arranged in 3 groups of 4 windows. A transistor in the sensor pulls a 5v reference voltage to ground until it sees a window. At the leading edge of each window, the transistor opens and the signal voltage rises to 5v. As a result, this sensor signal is 0v most of the time, and pulses to 5v twelve times per engine revolution. Each group of 4 windows represents the position of a pair of cylinders, but the CKP sensor cannot determine which of the two cylinders is on the compression stroke without an input from the CMP sensor.

The trailing edge of the fourth pulse in each group is located 4 degrees before TDC for the corresponding pair of cylinders. This information is used by the PCM as an engine speed input, and for ignition timing. The PCM uses the CMP sensor signal along with the CKP sensor signal to calculate cylinder identification for sequential fuel injection and for misfire diagnosis.

CKP Sensor Circuits

All 3 CKP sensor circuit wires connect to the PCM. The PCM supplies the sensor with a 5v reference voltage through Pin A17 (OR wire), and sensor ground through Pin A4 (BRN/YEL wire). The CKP sensor signal is sent to the PCM at Pin A8 (GY/BK wire).

CKP Sensor Diagnosis (No Code Fault)

If the PCM does not detect any signals from the CKP sensor with the engine cranking, it will not start. This sensor input is not monitored for a loss of signals, and cannot set a code. The quickest way to determine if the CKP sensor is the cause of the no start is to observe the RPM (ENGINE SPEED) PID on the Scan Tool. If '0', or intermittent engine speed is displayed, further testing is necessary. A DVOM can be used to check for "no output", but a Lab Scope is a better choice to test this sensor.

DI System CKP Sensor Schematic

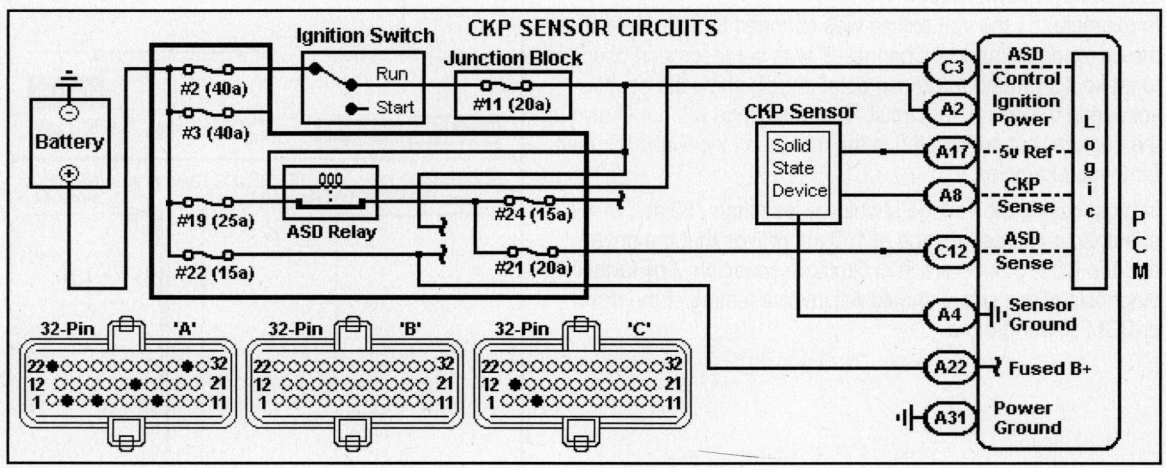

SCAN TOOL TEST (CKP SENSOR)

The Scan Tool provides an easy way to initially test a CKP sensor on any vehicle with serial data by observing the rpm (ENGINE SPEED) PID. However, Chrysler provides additional Scan Tool functions and PIDs to aid in diagnosis.

Scan Tool Example (1)
In example (1), the CAM/CRANK SYNC PID is an easy way to verify that the CMP and CKP sensors are synchronized. If the sensors are not synchronized, a separate Scan Tool control can be used to SET SYNCH. See CMP & CKP Sensor Synchronization in this Section.

Scan Tool Example (2)
In example (2), The CKP COUNTER and CMP COUNTER PIDs track the pulses of each sensor signal. There is no set value for these PIDs, as the numbers are rapidly incrementing while the engine is running.

These PIDs can be used to determine if one or both sensors are intermittently malfunctioning. In this case, one or both counters will stop incrementing while the problem is occurring.

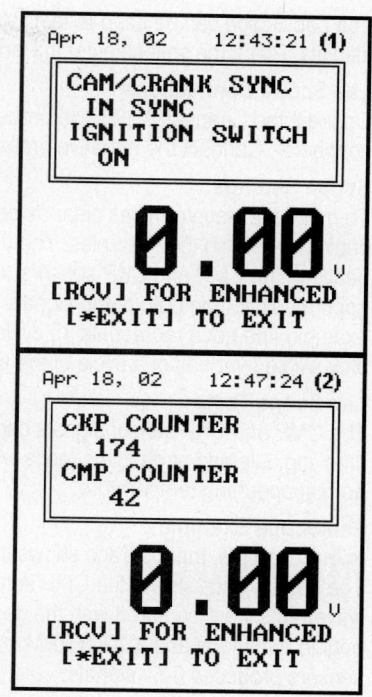

DVOM TEST (CKP SENSOR)

The DVOM is a useful tool to verify that there is a valid signal from the CKP sensor. The DC voltage, frequency (HZ) and duty cycle are all indicators that the circuit is active. For a precise view of the waveform, a Lab Scope is a better choice. Place the shift selector in Park and block the drive wheels for safety.

DVOM Connections
Connect the positive probe to the CKP sensor signal circuit (GY/BK wire) at PCM Pin A8, or at the CKP sensor harness connector. Connect the negative probe to the battery negative post.

DVOM Test Example
The DVOM display shows a CKP signal of 0.42v at a rate of 509.0 Hz and 93.3% (-) duty cycle. This capture was taken at approximately 2500 RPM. The values at idle will be similar, but the frequency should be about 150-160 Hz.

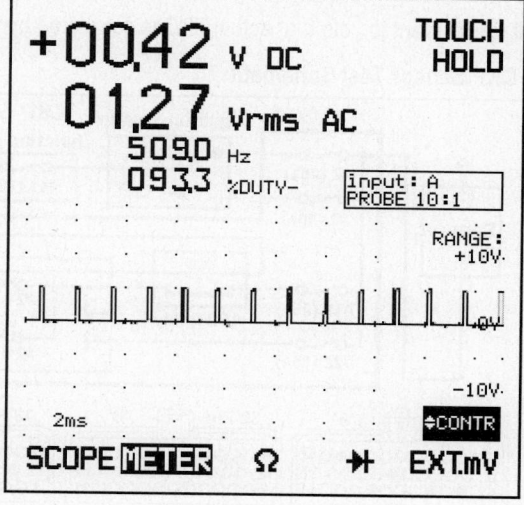

Use this test to verify that the frequency varies with rpm but remains steady at a stable engine speed. Any glitches would cause a fluctuation in the Hz signal. However, the voltage would be unaffected unless the signal was missing for an extended period of time.

Be aware that different DVOMs may display an accurate peak voltage, but may also display an average (which would be around 340 mv).

LAB SCOPE TEST (CKP SENSOR)

The Lab Scope can be used to test the CKP sensor as it provides a very accurate view of sensor waveform and of any glitches. Place the shift selector in Park (A/T) for safety.

Lab Scope Connections

Connect the Channel 'A' positive probe to the CKP signal wire at PCM connector Pin A8 (GY/BK wire), or at the CKP connector. Connect the negative probe to the battery negative post.

Scope Settings

To make the waveforms as clear as possible, set the scope settings to match the examples. The difference in frequency between the CMP and CKP sensors makes it difficult to see good resolution on both of the signals at the same time. In the example, the trace represents 15 cycles of the CKP signal, but only a little over half of a cycle of the CMP sensor signal.

Lab Scope Tests

The CMP and CKP sensor signals can be checked during cranking, idle and at off-idle speeds with the engine cold or at normal operating temperature.

Lab Scope Example

In this example, the top trace shows the CKP sensor signal. The bottom trace shows the CMP sensor signal. These waveforms were captured with the gear selector in Park, the engine running at about 2000 RPM in closed loop. Both sensors produced 0-5v signals.

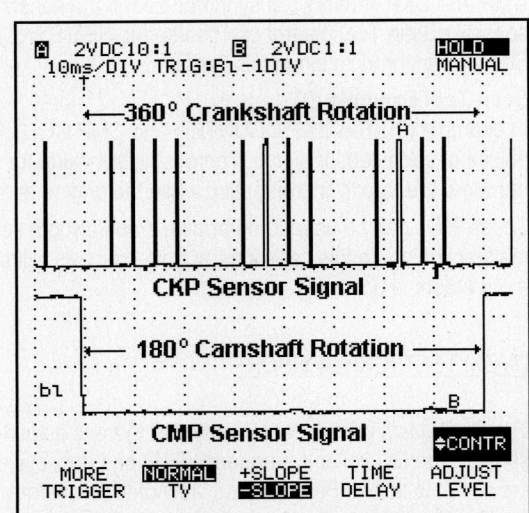

The top trace represents over 1 full engine revolution (3 sets of 4 pulses would equal 1 full revolution). The bottom trace represents over 180 degrees of camshaft revolution.

It is important to note that actual TDC is calculated by the PCM, as neither sensor switches at exactly TDC.

CKP Sensor Test Schematic

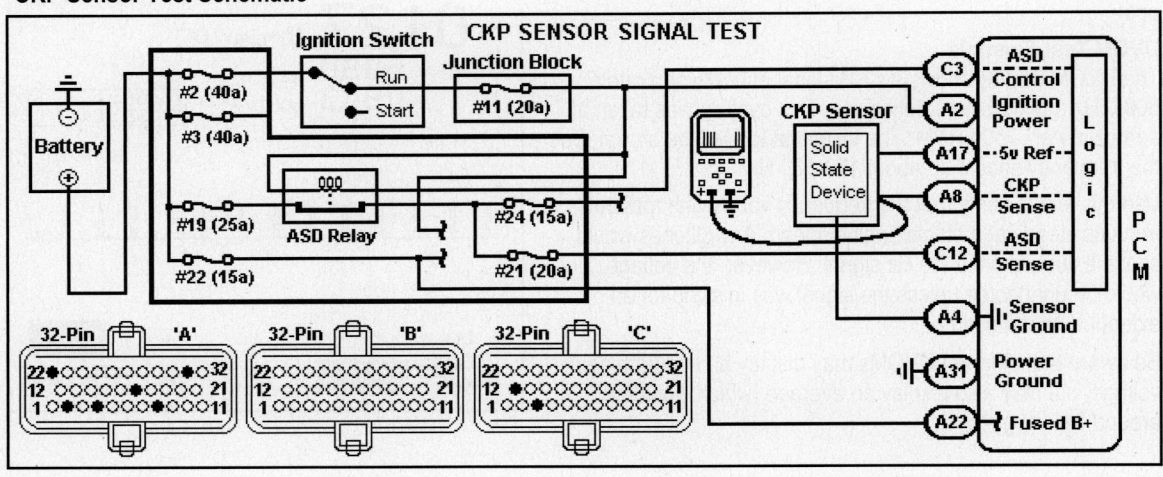

CMP & CKP SENSOR SYNCHRONIZATION

The Scan Tool can be used to verify that the CMP and CKP sensors are synchronized. Although this test will determine if the CMP and CKP sensor relationship is within specification, it may be more efficient to view the CAM/CRANK SYNC PID from the data list. See Scan Tool Test (CKP Sensor) in this section. If there is a problem with this PID, then it will be necessary to perform this procedure.

Scan Tool SET SYNCH Test

Connect the Scan Tool to the DLC underdash connector. From the Chrysler (OEM) menus, select SYSTEM TESTS, and then select SET SYNCH.

Follow the on-screen safety instructions, press YES, and wait for the Scan Tool to retrieve data from the PCM. A specification will appear (SET SYNCH SPEC) along with an actual value determined from the CMP and CKP relationship (DISTRIB).

Scan Tool Example

In this example, the specification is 0 degrees, and the actual value is 2 degrees. The test will conclude whether or not the value is within range. The message NO ADJUST REQ. is displayed.

The acceptable range is +/- 10°. It is not a problem to have such a wide range because this is not an actual timing specification. If the value is within this range, the PCM can properly calculate engine position. The value has no effect on engine performance as long as it is within this range.

Sensor Synchronization

If an adjustment is required in order to synchronize the sensors, the distributor hold-down bolt would need to be loosened and the distributor turned until the message changed to WITHIN SPEC.

There is very little adjustment available at the distributor. Generally, if the sensors are very far out of synchronization, look for a mechanical problem like a stretched or jumped timing chain.

When adjusting the distributor, set the idle above 1000 RPM and use the PIDs in the data list as a guide. When the DISTRIB PID reads 0°, tighten the hold-down bolt and verify the reading did not change.

It may be more efficient to use the PIDs than the SET SYNCH feature when making adjustments because the PIDs continue to update as changes are made.

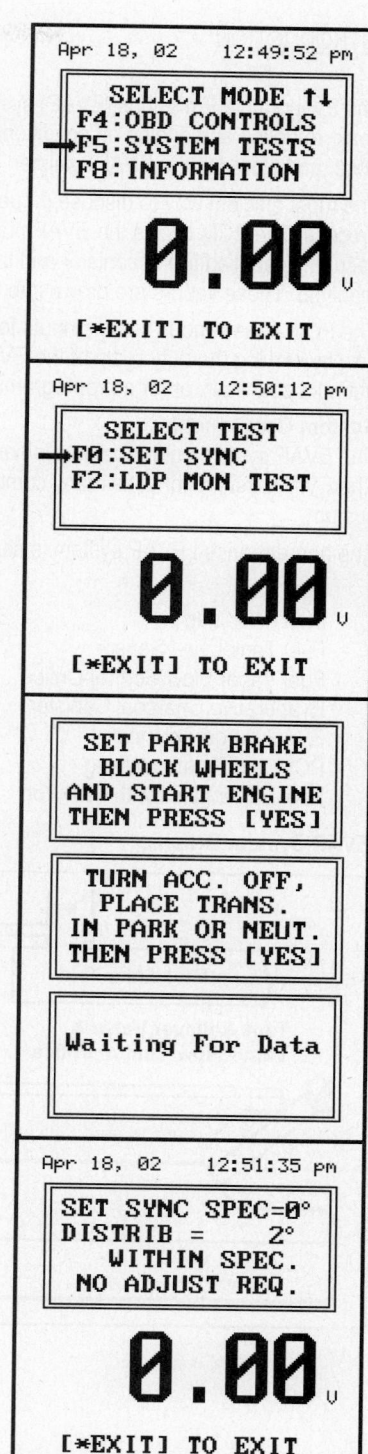

Evaporative Emission System

INTRODUCTION

The Evaporative Emissions (EVAP) system is used to prevent the escape of fuel vapors to the atmosphere under hot soak, refueling and engine off conditions. Any fuel vapor pressure trapped in the fuel tank is vented through a vent hose, and stored in the EVAP canister.

The most efficient way to dispose of fuel vapors without causing pollution is to burn them in the normal combustion process. The PCM cycles the EVAP purge valve "on" (open) during normal operation. With the valve open, manifold vacuum is applied to the canister and this allows it to draw fresh air and fuel vapors from the canister into the intake manifold. These vapors are drawn into each cylinder where they are burned.

The PCM uses various sensor inputs to calculate the desired amount of EVAP purge flow. The PCM meters the purge flow by varying the duty cycle of the EVAP purge solenoid control signal. The EVAP purge solenoid will remain off during a cold start, or for a preprogrammed time after a hot restart.

System Components:

The EVAP system on this vehicle is non-enhanced, which means the PCM cannot perform a leak check on the system. The EVAP system contains many components for the management of fuel vapors and to test the integrity of the EVAP system.

The non-enhanced EVAP system is constructed of the following components:

- Fuel Tank
- Fuel Filler Cap
- Fuel Tank Level Sensor
- Fuel Vapor Flow Control Orifice
- Evaporative Charcoal Canister
- EVAP Purge Solenoid
- PCM and Related Wiring
- Fuel Vapor Lines with Test Port

EVAP System Graphic

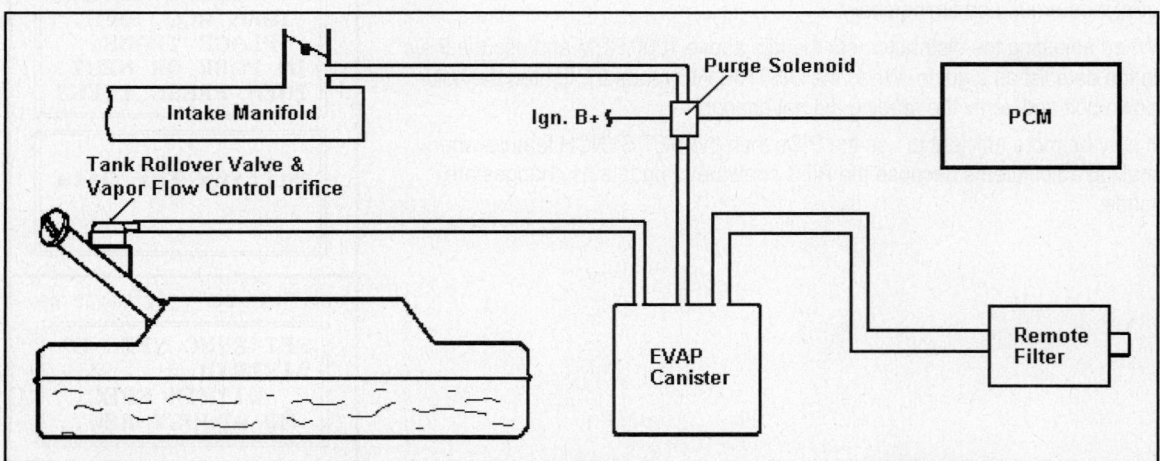

CANISTER PURGE SOLENOID

The EVAP canister purge solenoid is located underhood at the right rear of the engine area. Refer to the Graphic.

Solenoid Operation

The canister purge solenoid is located between the EVAP canister and the intake manifold. The function of this valve is to control the flow of vapors out of the canister, and to close off the flow of fuel vapors from the canister with the engine off.

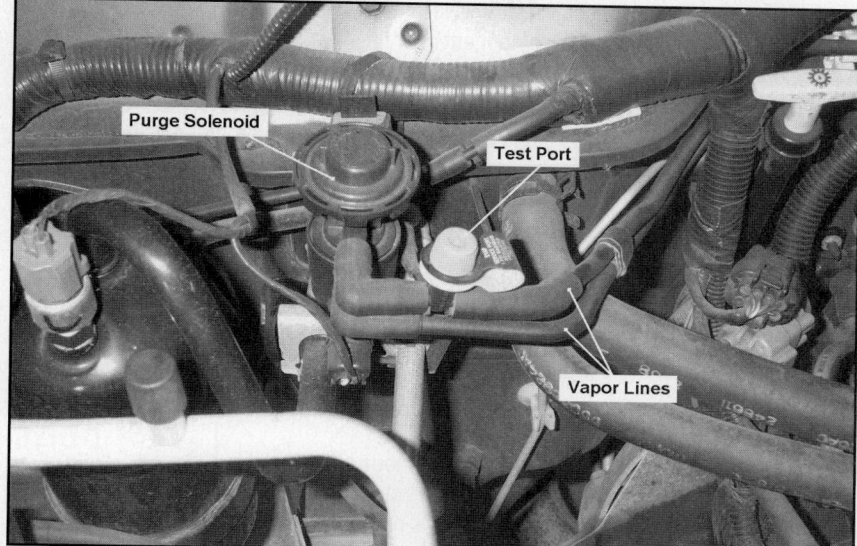

Once the engine is started, the PCM controls the operation of the solenoid The PCM opens and closes the solenoid to control the flow of fuel vapors from the charcoal canister during normal engine operation.

This device is a normally closed (N.C.) solenoid. The PCM controls the vapor flow rate from the EVAP canister to the intake manifold by completing the control circuit with a varying pulsewidth command. When the circuit is pulled to ground by a transistor in the PCM, the solenoid opens, and fuel vapor flows.

CHARCOAL CANISTER

The EVAP canister contains a mixture of activated carbon granules. This material has the ability to purge or release any stored fuel vapors when it is exposed to fresh air.

Fuel vapors from the fuel tank and the air cleaner are stored in the canister. This system has a remote air filter located at the rear of the vehicle to remove particles from air being drawn into the system.

Once the engine starts and reaches operating temperature, fuel vapors stored in the canister are purged into the engine to be burned during normal combustion.

The canister is located at the rear of the vehicle near the left frame rail.

BI-DIRECTIONAL TEST (PURGE SOLENOID)

The Scan Tool Actuator Test Mode (ATM) tests can be used to force the operation of a number of engine components and systems to aid in diagnosis. The ATM test 14: CANISTER PURG is used to cycle the purge solenoid at a fixed rate with the engine off to stress the component over time. The results can be measured with a DVOM or Lab Scope. This test is performed with the engine off.

Scan Tool Test Example
In this example, the actuator test has been turned ON with the Scan Tool. The solenoid duty cycle is fixed at 50%.

Lab Scope Test Example
In this example, the DVOM values reflect the circuit activity during ATM test 14.

The voltage value of 5.96 V DC represents the average circuit voltage as it is being cycled by the PCM. The purge solenoid is being cycled at a fixed 50% duty cycle, at a frequency of 5 Hz.

These values will remain constant for the duration of the test (about 5 minutes, unless the user exits the tests through the Scan Tool).

Why To Use These Tests
The PCM makes purge decisions based on various criteria. It is easier to simply command the solenoid to operate than to keep track of operating conditions. If the solenoid is not functioning during normal engine operation, it is hard to know if there are other conditions preventing the PCM from commanding it on. During this test, the desired operation is already known, which helps to simplify the diagnostic path.

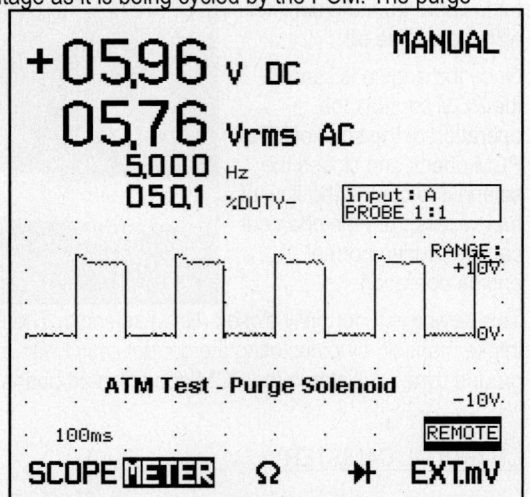

EVAP System Test Schematic

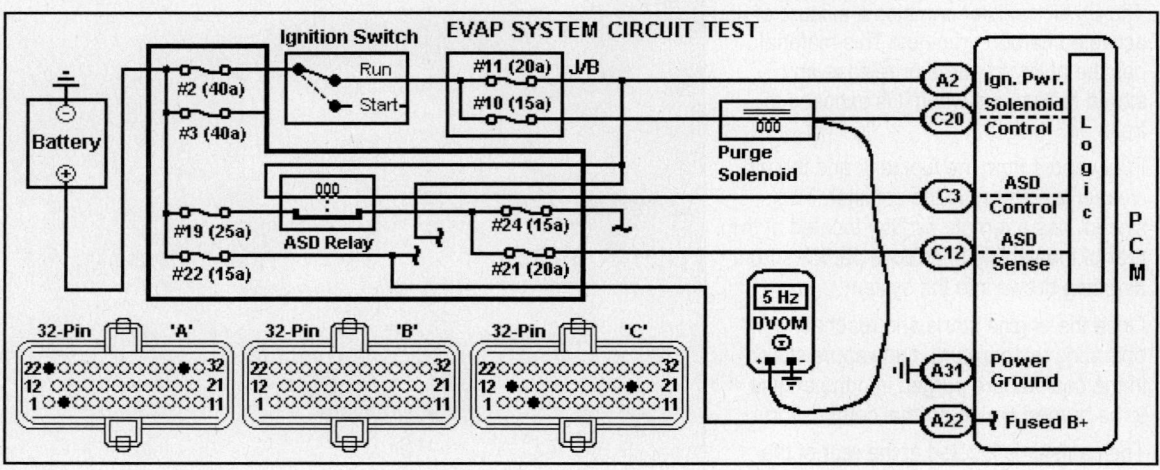

BI-DIRECTIONAL TEST (PURGE FLOW)

The Scan Tool can be used to force the purge solenoid command to BLOCK (off) or FLOW (on) positions with the engine running. This allows the user to observe the results of the test under real conditions.

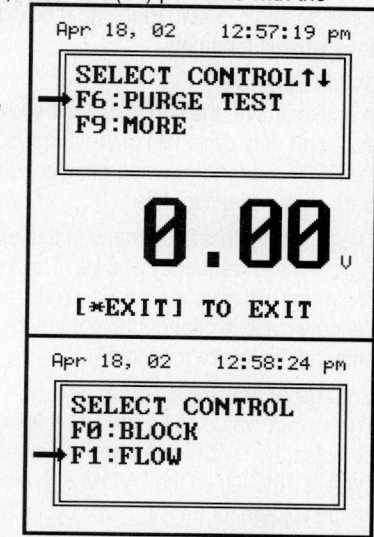

Test Example
In this example, the Scan Tool was used to command the solenoid to the FLOW position, which means that the PCM is commanding the purge solenoid to the "on" position.

The purpose of this test is to control the conditions so that the effects of purging or not purging can be seen in the HO2S signals (UPSTREAM O2S PIDs), and then in the fuel trim values (SHORT TERM ADAPT and LONG TERM ADAPT PIDs).

Using Test Results
When commanded to FLOW, the HO2S and fuel trim values should react quickly to changing air/fuel conditions.

Because purge vapors contain high concentrations of hydrocarbons, the oxygen sensors should almost immediately sense a rich condition (a rise in voltage). Next, short-term fuel trim will respond by going negative to achieve the proper air/fuel ratio. The more negative the value goes, the more fuel is being taken away to compensate.

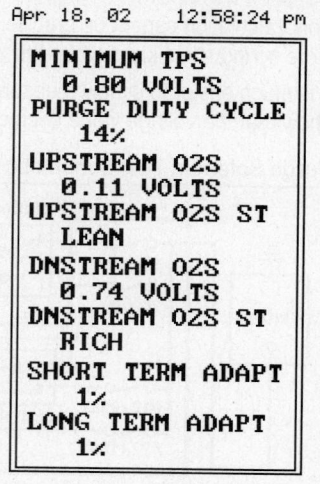

Long-term fuel trim will eventually go negative, forcing the short-term fuel trim value back to near 0% correction. This sequence confirms that the EVAP system is capable of purging vapors from the EVAP canister.

EVAP, O2S, and Fuel Trim PIDs
Observe the PIDs in the example to the right. These PIDs were captures under normal EVAP system operation. Monitor the values over time, as the O2S PIDs will cycle often and make it difficult to determine the systems response to the test from just one capture like this.

Expect the SHORT TERM ADAPT PID value to go negative by at least 2% from where it started. In other words, if the value was already +2%, the test may force the value back to 0%.

The LONG TERM ADAPT PID will react to a sustained SHORT TERM ADAPT value only if short term exceeds an *absolute* +/- 3%. For example, if the short term value was +2% before the test, and dropped to –2% in response to the purge flow, the long term value would not change, even though the effect of the test was a -4% fuel correction. The PID value change was –4%, but the absolute value never went below –3%.

DVOM TEST (PURGE SOLENOID)

The DVOM is an excellent tool to test the condition of the EVAP purge solenoid circuit. Place gear selector in Park and block the drive wheels for safety.

DVOM Example
In this example, the DVOM values show the voltage, frequency (Hz), and duty cycle (%) of the purge solenoid command from the PCM under normal operating conditions. These values were captured at Hot Idle.

The purge solenoid command in this example is at 16% duty cycle (-), at a frequency of 5 Hz. The frequency of the command will always be 5 Hz at idle, and 10 Hz off idle. Only the duty cycle is changed to control the volume of purge vapors entering the intake manifold.

Voltage Averaging Explanation
The voltage value of 11.06v DC is less than open circuit voltage because the DVOM averages the signal it receives. The duty cycle is 16% (-), so the DVOM is averaging a signal that is open circuit voltage for 84% of each cycle, and grounded for 16% of each cycle. Therefore 11.06v DC is 84% of open circuit voltage. It is easy to calculate that the open circuit voltage is 13.17 volts (11.06/0.84).

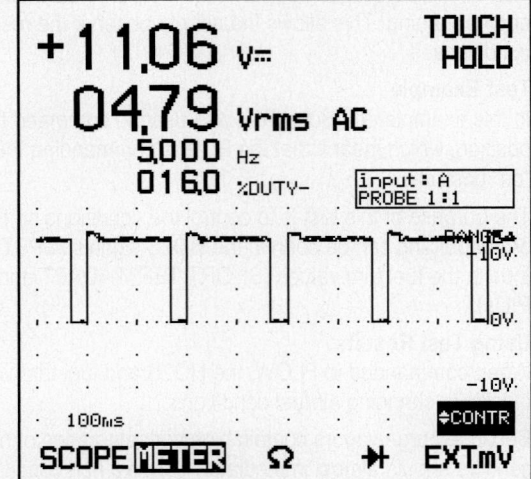

This calculation cannot be precise, because it assumes that the circuit was pulled down to exactly 0.0v, and that there were no inductive kicks when the circuit is cycled off. Note that a small kick can be seen in this example.

It is much easier to simply measure the voltage with the key on, engine off. The purpose of this calculation is just to show that the voltage value seen on the DVOM during this and many other tests does make sense.

Purge Solenoid Test Schematic

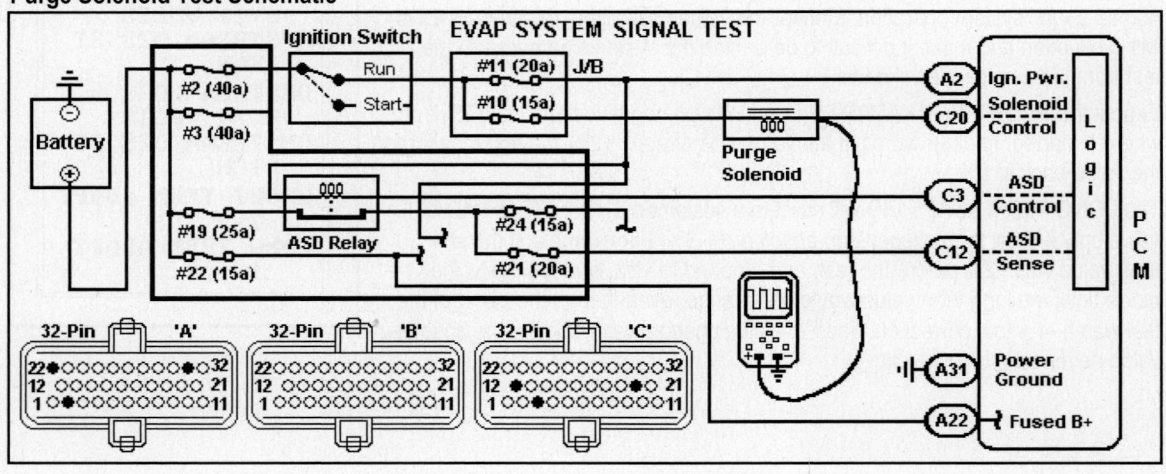

Fuel Delivery System

INTRODUCTION

The Fuel system delivers fuel at a controlled pressure and correct volume for efficient combustion. The PCM controls the Fuel system and provides the correct injector timing.

Fuel Delivery

This vehicle is equipped with a sequential fuel injection (SFI) system that delivers the correct A/F mixture to the engine at the precise time under all operating conditions.

Air Induction

Air enters the system through the fresh air duct and flows through the air cleaner. The unmetered air passes through another air duct and enters the throttle body. The incoming air passes through the throttle body and into the intake plenum to the intake manifold where it is mixed with fuel for combustion. The volume of intake air is not directly measured, but is inferred from manifold vacuum using a MAP sensor mounted to the intake manifold (Speed Density system).

Returnless Fuel Delivery System Operation

A high-pressure (in-tank mounted) fuel pump delivers fuel to the fuel injection supply manifold. The fuel injection supply manifold incorporates electrically actuated fuel injectors mounted directly above each of the intake ports of the engine.

This application uses a mechanical returnless fuel delivery system. There is no fuel return line from the fuel supply manifold. A fuel filter/pressure regulator assembly is mounted to the top of the in-tank fuel pump module. The regulator maintains a constant amount of fuel pressure in the system. The fuel pressure is constant while fuel demand is not, so the pressure regulator diverts extra fuel volume back through the fuel pump to the tank.

The PCM determines the amount of fuel the injectors spray under all engine-operating conditions. The PCM receives input signals from engine control sensors that monitor various factors such as manifold pressure, intake air temperature, engine coolant temperature, throttle position, and vehicle speed. The PCM evaluates the sensor information it receives and then signals the fuel injectors in order to control the fuel injector pulsewidth.

Fuel Delivery System Graphic

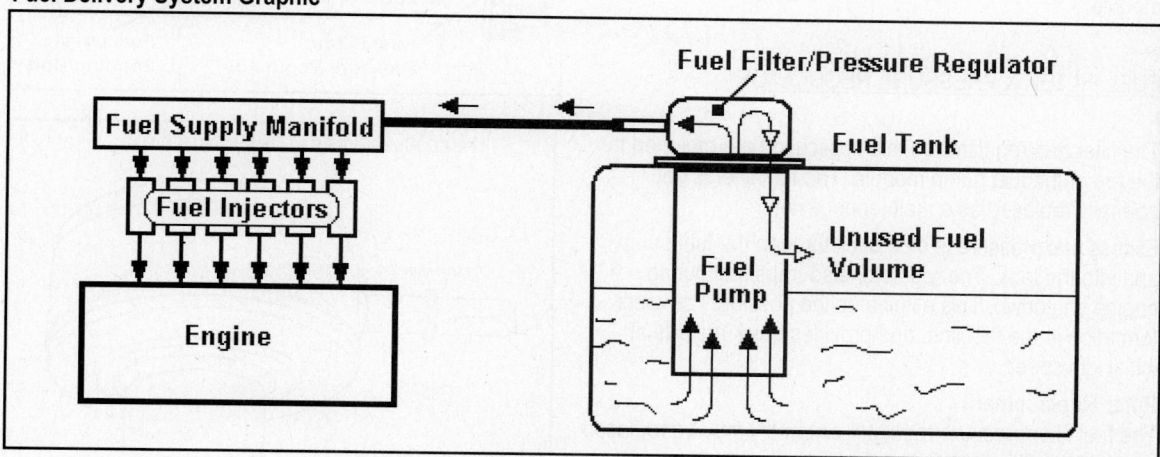

FUEL PUMP RELAY

The PCM controls the fuel pump through a fuel pump relay. Power is supplied from fuse #22 (15a) in the PDC, through the fuel pump relay, and to the fuel pump.

The relay pull-in coil is powered through the ignition switch in the Start or Run position, and is grounded by the PCM whenever fuel pump operation is required.

Component Location

The fuel pump relay is located in the engine area in the Power Distribution Center (PDC).

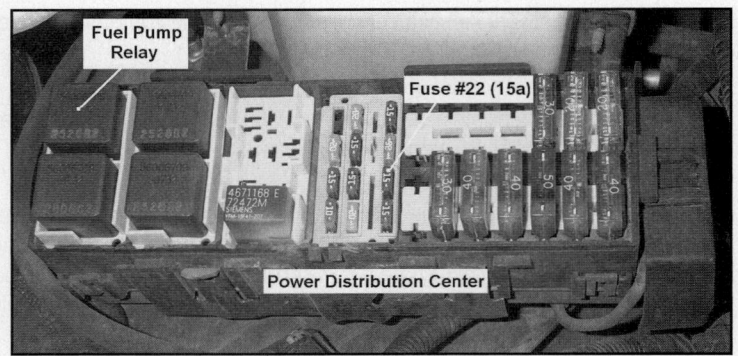

FUEL PUMP MODULE

The fuel pump module in this returnless fuel system is mounted inside the fuel tank.

The fuel level sensor is integrated into the fuel pump module. None of the module components can be serviced separately, except the fuel filter/pressure regulator, mounted on top of the fuel pump module.

The fuel pump produces more pressure and volume than necessary for proper system operation. The excess fuel is bled back into the tank through the fuel filter and pressure regulator assembly mounted to the top of the fuel pump module.

FUEL FILTER & PRESSURE REGULATOR

The filter and regulator on this application are mounted to the top of the fuel pump module. The fuel filter is not normally replaced as a maintenance item.

Excess fuel pressure is bled back through the fuel pump and into the tank. The regulator also traps fuel during engine shutdown. This eliminates the possibility of vapor formation in the fuel line, and provides quick restarts at initial idle speed.

Filter Replacement

The fuel filter/pressure regulator assembly may be replaced separately without removing the fuel pump module. However, the fuel tank must be removed.

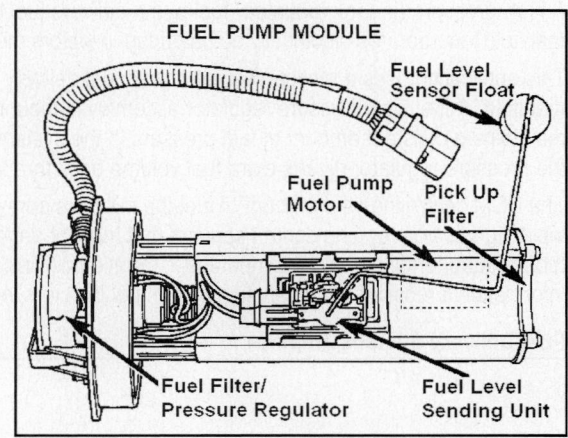

FUEL PUMP MODULE

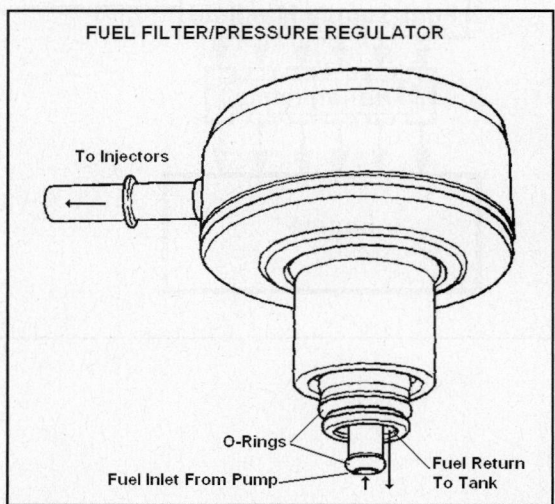

FUEL FILTER/PRESSURE REGULATOR

LAB SCOPE TEST (FUEL PUMP)

Testing fuel pump current draw with a Lab Scope is a very efficient way of determining fuel pump condition, since the pump can be tested without removing the fuel tank.

Scope Connections (Amp Probe)
Set the amp probe to 100 mv/amp, and zero the amp probe. Connect the probe around the fuel pump feed wire and connect the negative probe to chassis ground.

Scope Connection Tip
Accessing the fuel pump wiring harness for this test can be time consuming. In addition, a single wire must be isolated and completely surrounded by the bulky low amp probe tip. Remove the fuel pump relay from the PDC and install a fused jumper wire between relay cavities 30 and 87. The low amps probe can then be clipped around the jumper wire.

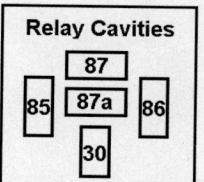

Lab Scope Examples
In example (1), the trace shows the fuel pump current (5 amps peak) at idle speed.

In example (2), the current waveform was captured when the key was turned "on". Note that the current peaked at 15 amps.

The OEM fuel pump current specification for this application is under 10 amps during normal operating conditions.

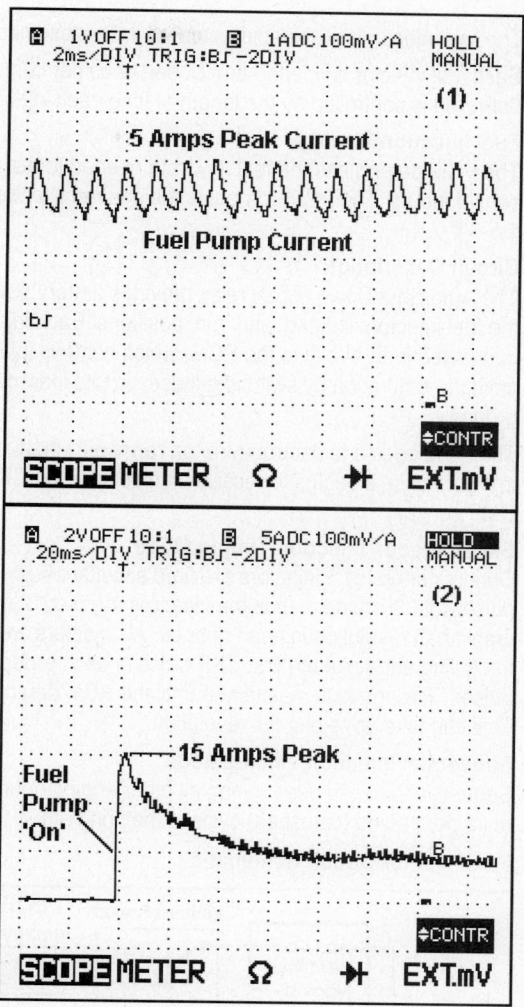

Fuel System Test Graphic

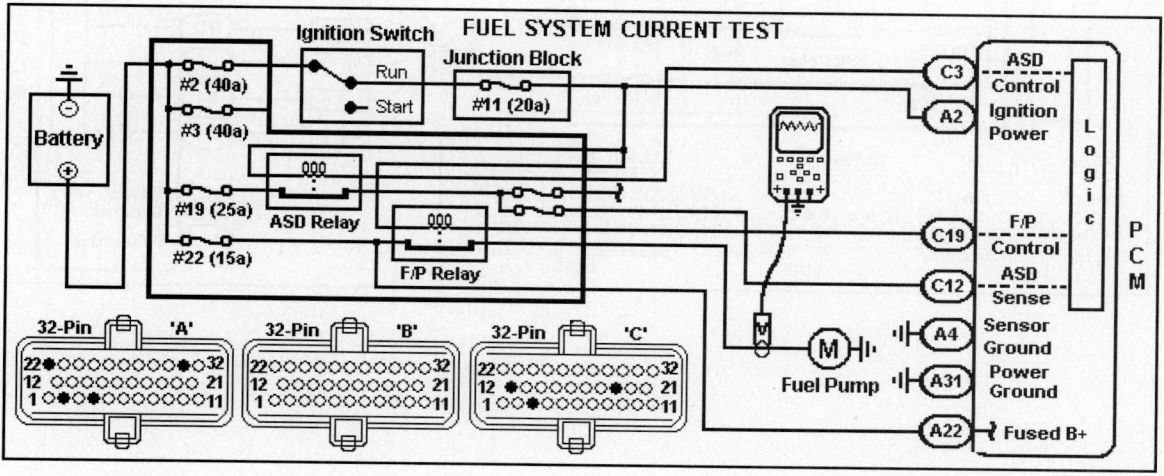

FUEL INJECTORS

The fuel injector is a solenoid-operated valve designed to meter the fuel flow to each combustion chamber (SFI).

Each fuel injector is opened and closed once per camshaft revolution during normal operation. The amount of fuel delivered is controlled by the length of time each injector circuit is grounded by the PCM (injector pulsewidth).

Fuel Injection Strategy
The PCM determines the fuel flow rate needed to maintain the correct A/F ratio from the ECT, IAT, HO2S-11, and MAP sensor inputs. The PCM computes the desired injector pulsewidth and then turns the injectors on/off in this sequence: 1-5-3-6-2-4.

Circuit Description
The Auto Shut Down (ASD) relay provides battery power to the fuel injectors in 'start' and 'run' positions. Each injector is connected individually to the PCM, which controls injector timing and duration by switching these circuits individually to ground.

The PCM begins to control (turn "on" and "off") the fuel injectors once it begins to receive signals from the CKP and CMP sensors.

Simultaneous Injection Timing Mode
During startup, all 6 injectors are fired simultaneously. The pulsewidth is shorter, since the injectors are fired 3 times per crankshaft revolution instead of once. All injectors are fired on the falling edge of each first CKP pulse in each group of four pulses. This mode is maintained until the PCM determines the exact engine position from the CKP and CMP sensors. This can take up to one full revolution.

Sequential Injection Timing Mode
Sequential mode is used during all other engine operating conditions. In this mode, fuel is injected into each cylinder once per engine (camshaft) cycle in the firing order.

Fuel Injector Circuit Schematic

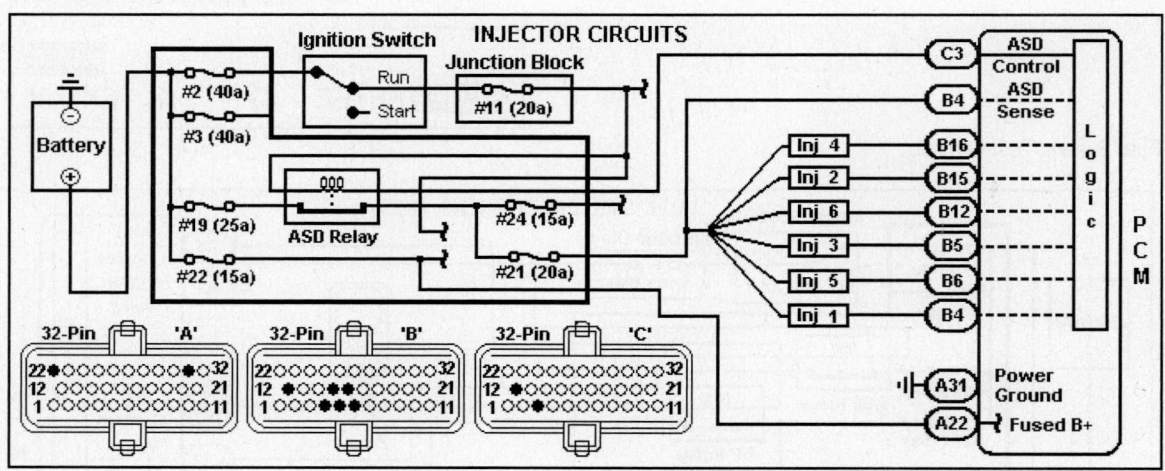

SCAN TOOL TEST (FUEL INJECTOR)

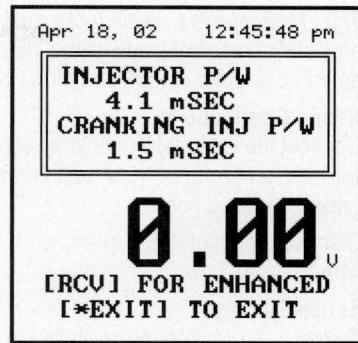

The Scan Tool can be used to quickly determine the fuel injector pulsewidth or duration. However, it is not the tool of choice for many fuel injector problems.

Connect the Scan Tool to the DLC connector. Navigate through the Scan Tool menus to the PID Data List.

Then select Injector Pulsewidth from the menu. The example to the right was captured with the vehicle in Park and running at Hot Idle speed. The values at 30 and 55 mph were 5.50 ms and 7.10 ms (not shown in this capture).

CRANKING INJ P/W Explanation
Normally, an injector pulsewidth of 1.5 ms would not deliver enough fuel to start the engine. It is important to remember that until the PCM calculates exact engine position from CKP and CMP inputs, the injectors are fired simultaneously 3 times per crankshaft revolution (6 times per camshaft revolution).

Therefore, the injector pulse width per camshaft revolution is closer to 9 ms (6 pulses X 1.5 ms = 9 ms). This is actually a very rich command, because much of this initial fuel is lost on the exhaust stroke, or absorbed by carbon deposits when fuel is squirted at the back of a closed intake valve.

SCAN TOOL TEST (FUEL TRIM)

Short Term fuel trim (SHORT TERM ADAPT) is an operating parameter that indicates the amount of short term fuel adjustment made by the PCM to compensate for variations from an ideal A/F ratio condition.

A positive short term fuel trim number (+15%) would mean the HO2S is indicating to the PCM that the A/F ratio is lean, and that the PCM is trying to enrich the A/F mixture. If A/F ratio conditions are ideal, the number will be close to 0% (+/- 5%).

Long Term fuel trim (LONG TERM ADAPT) is an engine operating parameter that indicates the amount of long term fuel adjustment made by the PCM to compensate for variations from an ideal A/F ratio condition.

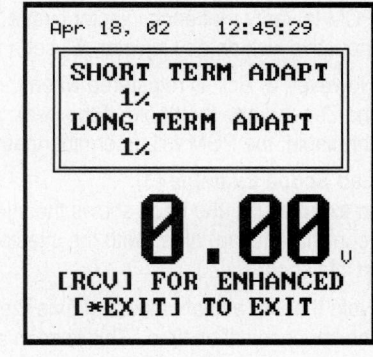

Scan Tool Example
In this Scan Tool example, the SHORT and LONG TERM ADAPT PIDs showed +1%. This vehicle is compensating very little for any unusual engine operating conditions.

Use this test to verify that the vehicle has no significant fuel pressure, vacuum, or engine mechanical concerns. If LONG TERM ADAPT exceeds +/- 5%, then further testing is required to determine the cause. Fuel pressure, vacuum, and engine mechanical concerns will affect Fuel Trim by changing the amount of fuel available.

It is important to remember that the PCM only uses the oxygen sensor signals to make adaptive fuel trim calculations. Any oxygen sensor calibration problems or sources of unwanted oxygen will cause the PCM to make incorrect air/fuel decisions. Unwanted oxygen sources include a defective air injection system and exhaust leaks upstream of the sensor.

LAB SCOPE TEST (FUEL INJECTOR)

The Lab Scope is a useful tool to test the fuel injectors as it provides an accurate view of the injector operation. Place the gearshift selector in Park and block the drive wheels.

Scope Connections

Connect the Channel 'A' positive probe to the injector control wire at the injector or PCM, and the negative probe to the battery negative post.

Clamp the amp probe in example (3) around the injector control wire.

Scope Settings

To make the waveforms as clear as possible, set the scope settings to match the examples.

Lab Scope Example (1)

In example (1), the trace shows the injector waveform with the sweep rate set to 1 ms in order to view injector circuit activity through an entire cycle.

Lab Scope Example (2)

In example (2), the trigger and time/div. were changed to show a more detailed view of the inductive kick and the closing of the injector.

Trouble Code Help

The inductive kick voltage from each injector circuit is 'clipped' by a Zener diode in the PCM to protect the circuits.

In this application, Zener diode operation is also used by the PCM to verify that each injector circuit 'kicks'. A loss of inductive kick on any injector will set a DTC P020x.

However, a 'kick' is registered when the voltage spike crosses the Zener diode threshold. If the peak voltage is lower than this threshold, the PCM will determine that the injector did not fire.

Lab Scope Example (3)

In example (3), the trace shows the injector current signal (current ramping) along with the injector volts over time signal at idle speed.

Note that the actual amount of time the injector sprayed fuel is the 'mechanical on time.' This is seen as the time between the pintle opening in the current waveform and the pintle closing in the voltage waveform.

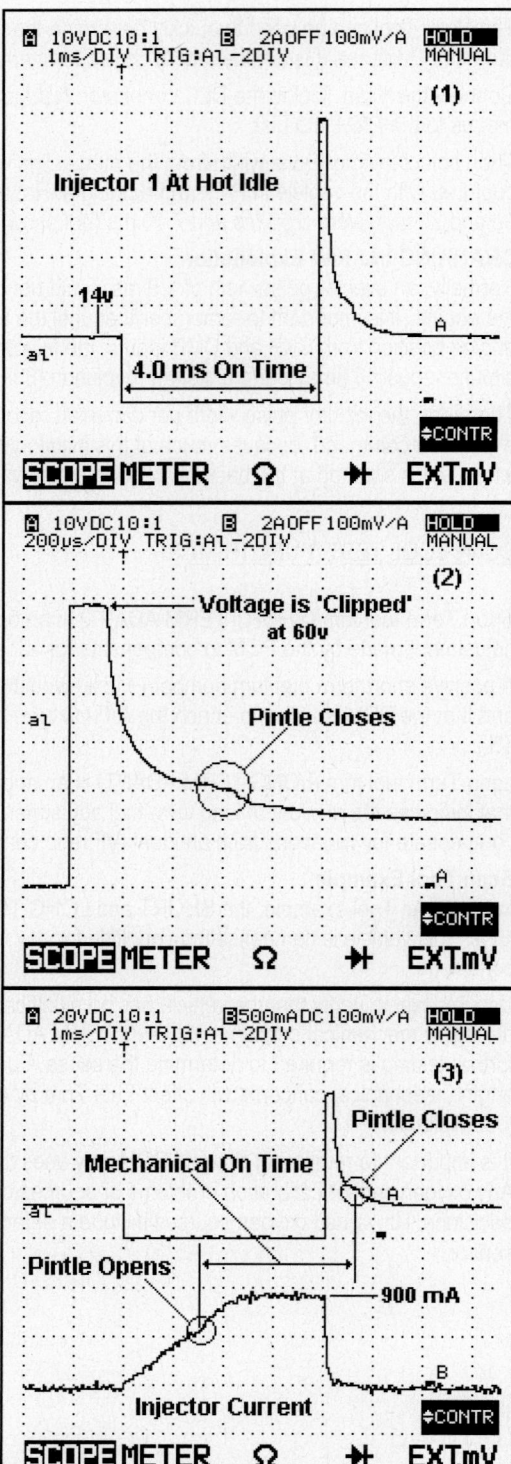

CAMSHAFT POSITION SENSOR (HALL EFFECT)

The camshaft position (CMP) sensor used with this application sends a 0-5v signal to the PCM that indicates the relative camshaft position. Camshaft position is actually calculated by the PCM based on the CKP sensor input. The CMP sensor is only used to determine if the camshaft is on the first or second half of its cycle.

The CMP sensor signal is high for one full rotation of the crankshaft, and then toggles low for another full rotation of the crankshaft. This signal is only used to calculate the injector sequence, and for Misfire diagnostics.

For the PCM to determine which cylinder pair is represented by the CKP signal, the CMP sensor is monitored for voltage switching (0v to 5v). The exact engine position is calculated based on these two signals. This calculation is used by the PCM in order to fire the six (6) fuel injectors in the correct sequence, and to keep track of each cylinder as part of the Misfire monitor.

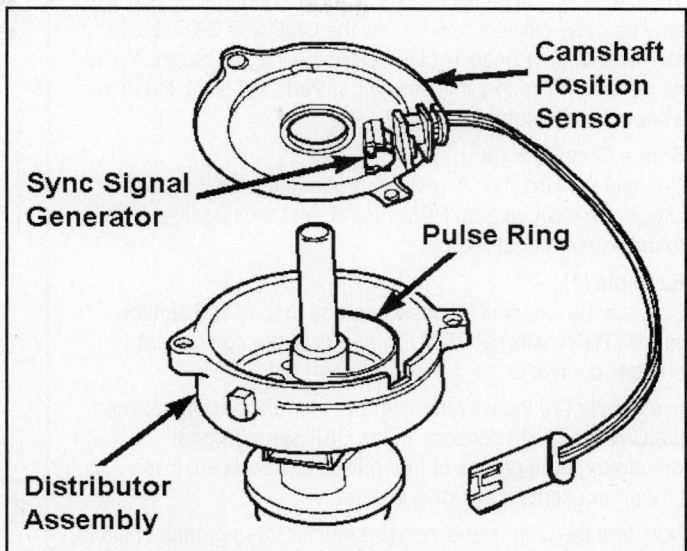

CMP Sensor Circuits
The CMP sensor receives a 5v reference voltage from PCM Pin A17, and sensor ground at Pin A4. Both the reference voltage and ground circuits are shared with other sensors. The 0-5v CMP signal is sent to PCM Pin A18.

Fuel System CMP Sensor Schematic

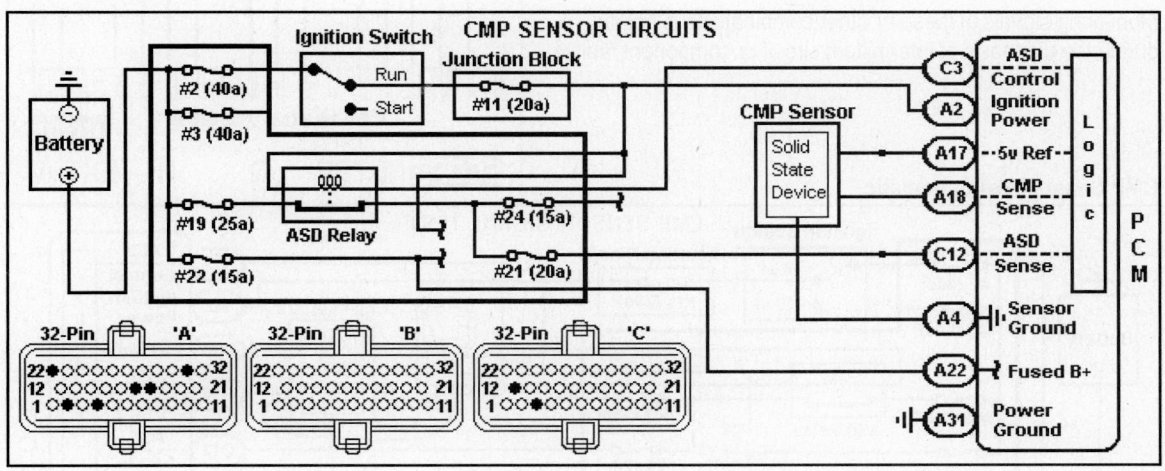

LAB SCOPE TEST (CMP SENSOR)

The Lab Scope is an excellent tool for viewing this sensor. It may be more efficient to first view the CMP and CKP sensor relationship on a Scan Tool to determine if a Lab Scope test is necessary. Place the shift selector in Park and block the drive wheels prior to starting this test.

Scope Connections
Connect the Channel 'A' positive probe to the CKP sensor circuit (GRY/BK wire) at PCM Pin A8, and the negative probe to the battery negative post.

Example (1)
Connect the Channel 'B' positive probe to the CMP sensor circuit (TN/YL wire) at PCM Pin A18 or at the component harness connector.

In example (1), the waveform shows the relationship between the CMP and CKP sensors. In the CKP sensor signal waveform, three groups of four pulses can be seen. Each group represents a pair of cylinders.

Note that the CMP signal remains low for three groups of CKP pulses.

Example (2)
Connect the Channel 'B' positive probe to the coil primary circuit (GY wire) at PCM Pin A7.

In example (2), the CMP signal is compared to coil primary. Note that for 720 degrees of crankshaft rotation, the coil fired six times.

Monitoring signals in these or other combinations is helpful during the diagnosis of intermittent circuit or component faults.

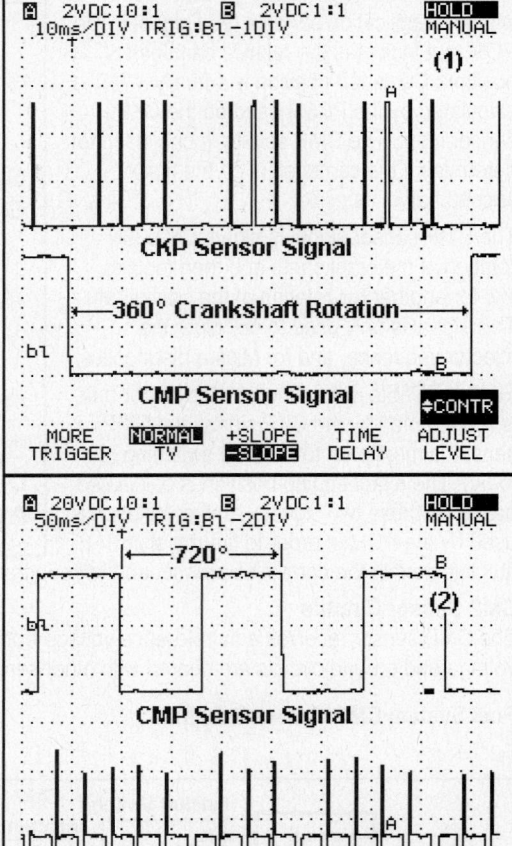

CMP Sensor Test Schematic

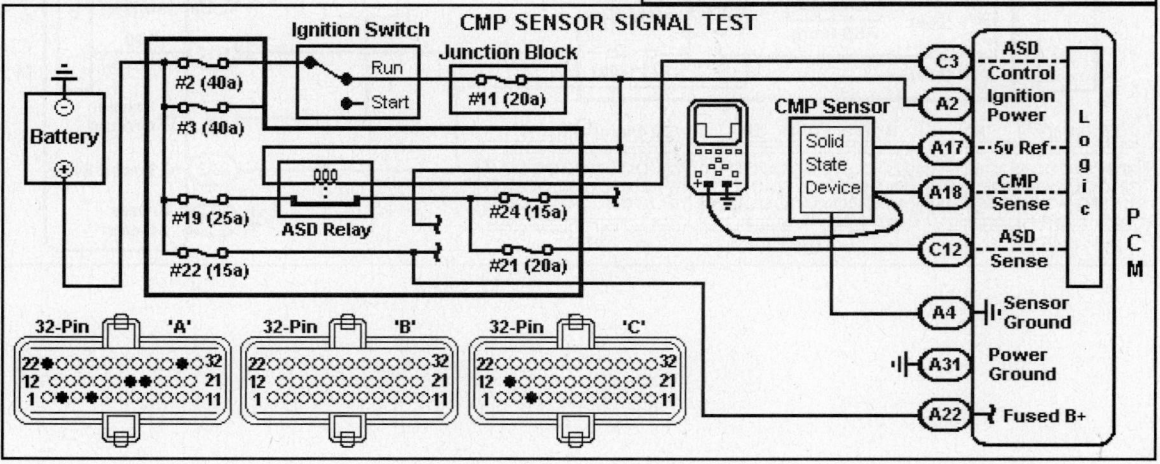

Idle Air Control Motor

GENERAL DESCRIPTION

The Idle Air Control (IAC) is mounted to the side of the throttle body. It is designed to control engine idle speed (rpm) under all operating conditions. An additional function of the IAC motor is to control the engine deceleration during closed throttle vehicle deceleration. This reduces emissions and provides for smoother engine operation.

The IAC motor meters the intake air that flows past the throttle plate through a bypass area within the IAC assembly and throttle body.

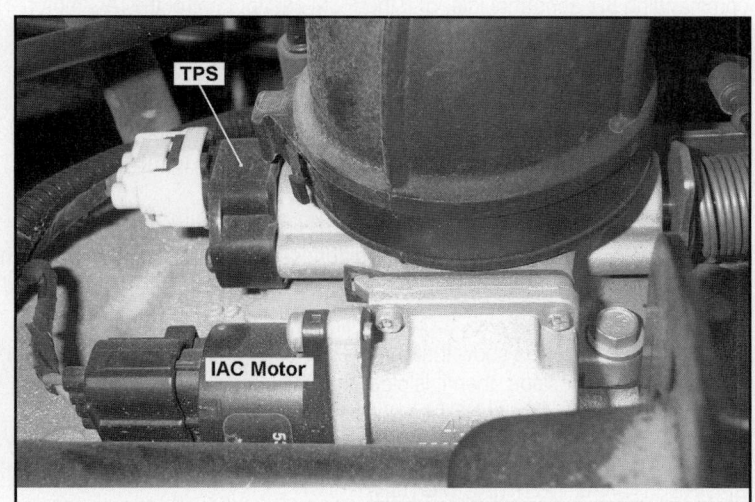

The PCM calculates a desired idle speed and controls the IAC position through a pulse-train signal to the two (2) 'motors' inside of the IAC motor assembly.

On these types of motors, the valve position remains fixed if the harness is disconnected from the motor.

Circuit Description

The four circuits of the IAC motor are connected to PCM connector 'A' at Pins 10, 11, 19, and 20. While the system has four drivers in the PCM, there are only 2 coils inside the IAC motor. Circuits 1 and 4 control one coil, and circuits 2 and 3 control the other.

The drivers in each pair toggle the polarity of each coil to help push or pull the IAC motor pintle in its bore. This polarity switching can be observed on a Lab Scope. See Lab Scope Test (IAC Motor).

IAC Motor Circuit Schematic

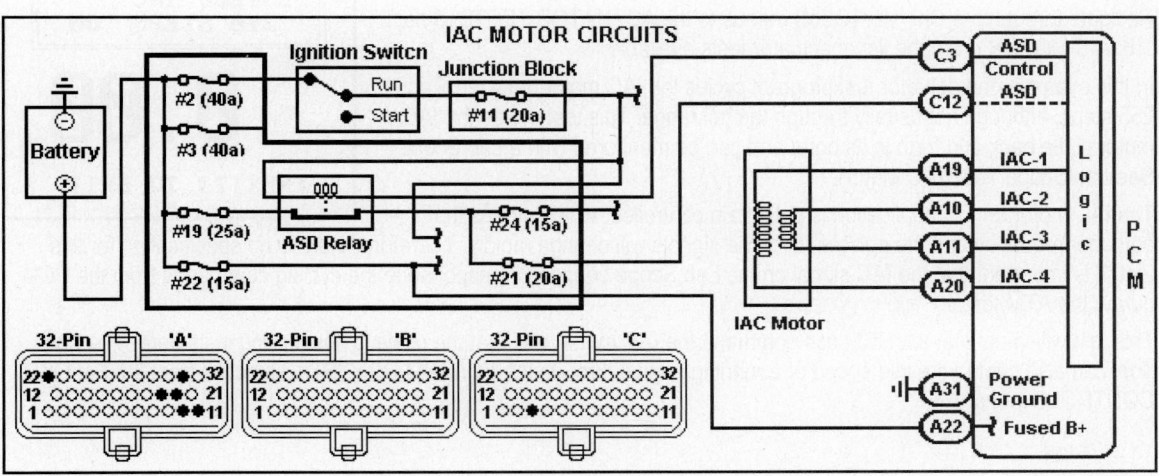

SCAN TOOL TEST (IAC MOTOR)

Connect the Scan Tool to the DLC connector. Navigate the Scan Tool menus and select the appropriate items from the PID (Data List) menu.

Scan Tool Example (1)
Examples of the ENGINE SPEED and TARGET IDLE PIDs for this vehicle at Hot Idle speed are shown.

When an idle concern is present, this is an excellent way to determine if PCM is commanding the abnormal idle, or if the PCM cannot control the IAC motor. The answer will help you determine the next testing step.

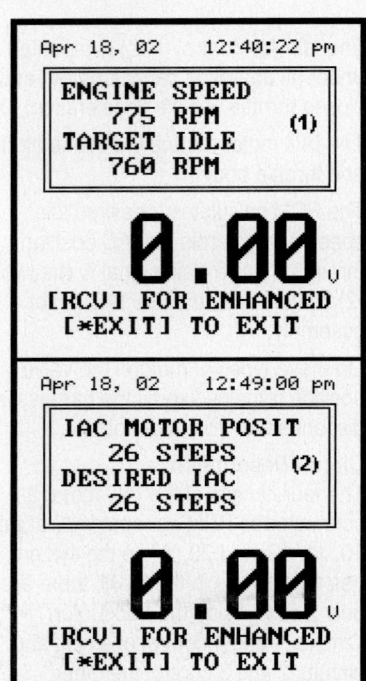

Scan Tool Example (2)
In Example (2), the IAC MOTOR POSIT and DESIRED IAC PIDs were captured at Hot Idle speed. Using these two PIDs, the command and feedback can be compared to quickly determine the ability of the PCM and related circuits to control the IAC motor.

Diagnostic Tip
Although a Scan Tool will almost never diagnose an idle concern without further testing, it is the most efficient way to determine what the next test step should be.

For example, if the PIDs show that the PCM is comanding the IAC motor to a valid and normal position, then it is clear that the inputs used for the PCM to make idle decisions should NOT be checked.

BI-DIRECTIONAL TEST (IAC MOTOR)

Scan Tool bi-directional testing allows the technician to force circuit operation while monitoring PIDs, or to observe the results with a Lab Scope. These controls are often the most efficient way to determine if the system is capable of responding to PCM commands.

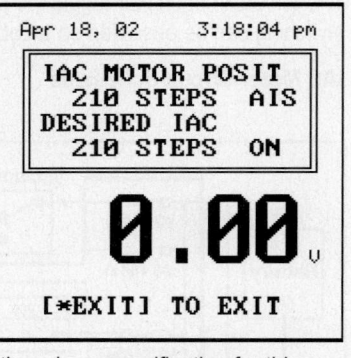

Navigate through the Chrysler (OEM) menus to the ACTUATOR TESTS. Select "10: AIS MOTOR" from the list of actuator tests available.

In this example, the actuator test program cycles the IAC motor between 0 and 255 steps, although not usually through the full range. This test cycles the IAC motor pintle back and forth in its bore, and can be monitored with a Lab Scope. See Lab Scope Test (IAC Motor).

The IAC motor on this application is pulsetrain controlled, which means that both the frequency and the pulsewidth of the signals will change rapidly. Therefore, there is no specification for this test. It is important that the IAC signal on the Lab Scope changes in response to the cycling commands from the PCM during the ATM test.

This test will verify that the PCM can command the IAC motor, and that the pintle is not sticking in its bore. The Scan Tool can also control the idle speed of a running engine through the FO: RPM CONTROL selection from the F4: OBD CONTROLS menu.

LAB SCOPE TEST (IAC MOTOR)

The Lab Scope can be used to monitor the IAC motor control signals under various engine load, speed and temperature conditions. Place the gearshift selector in Park and block the drive wheels prior to starting the test.

Scope Connections
Connect the Channel 'A' and 'B' positive probes to the appropriate IAC motor driver circuit at PCM Pin A10, A11, A19, or A20. Connect the negative probe to the battery negative post.

Scope Settings
To make the waveforms as clear as possible, set the scope settings to match the examples.

Lab Scope Example (1)
In this example, the trace shows the IAC motor signals on circuits 1 and 4. These two circuits are a pair, and work on the same coil inside the IAC motor. Circuits 2 and 3 are also paired. The polarity of these circuits will always be opposite. If one circuit is open or short, the other will be affected as well.

Therefore, it is more efficient to test the IAC motor signals on circuits 1 and 2, 1 and 3, 2 and 4, or 3 and 4.

Lab Scope Example (2)
In this example, the trace shows the IAC motor signals on circuits 3 and 4. By testing this pair of circuits, or any of those listed in example (1), any faults in the IAC circuits will be seen.

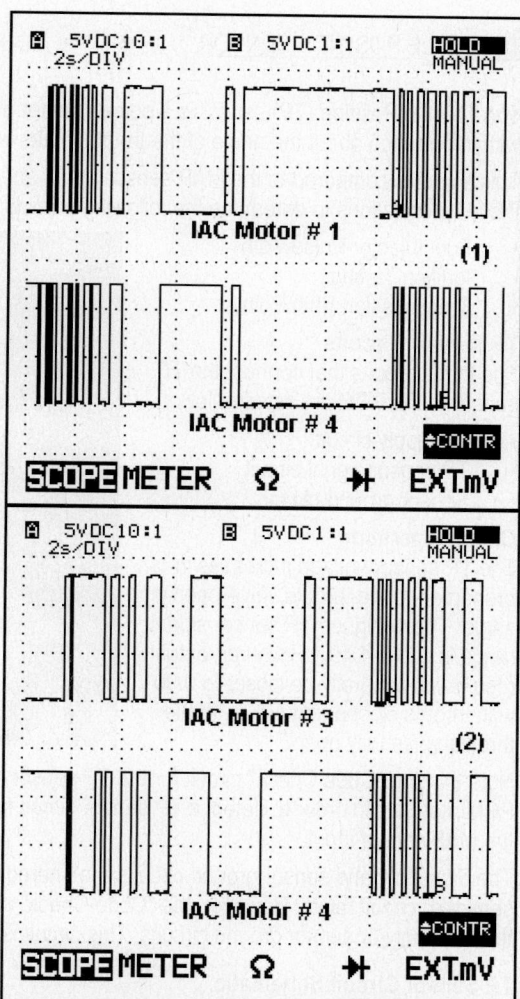

IAC Motor Test Schematic

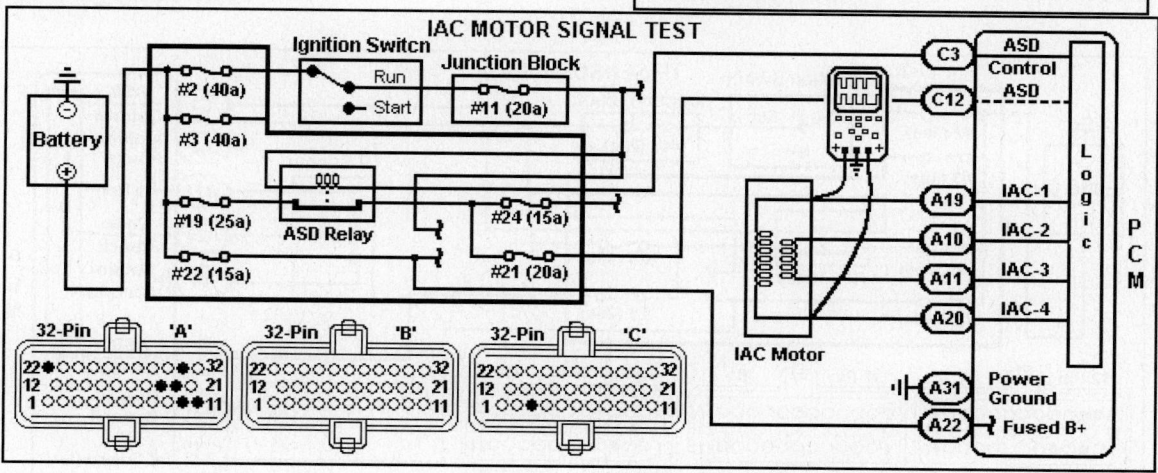

Information Sensors

THROTTLE POSITION SENSOR

The Throttle Position (TP) sensor is a potentiometer mounted to the throttle body. The TP sensor provides the PCM with information about the angle of the throttle plate, which is a function of driver demand.

This signal is compared to the MAP sensor signal and other inputs to calculate engine conditions. The PCM uses theses calculations to determine the following outputs:

- Fuel Injector Pulsewidth
- Ignition Timing
- Transmission Shift Points

TP Sensor Circuits
The three circuits that connect the TP sensor to the PCM are listed below:

- 5v supply circuit
- TP sensor signal circuit
- Sensor ground circuit

Circuit Operation
The TP sensor voltage increases in proportion to the throttle valve-opening angle. The designed TP sensor signal range is 0.26v-4.49v. However, actual results will generally be closer to 0.90 with throttle closed and 4.0v with the throttle valve fully open.

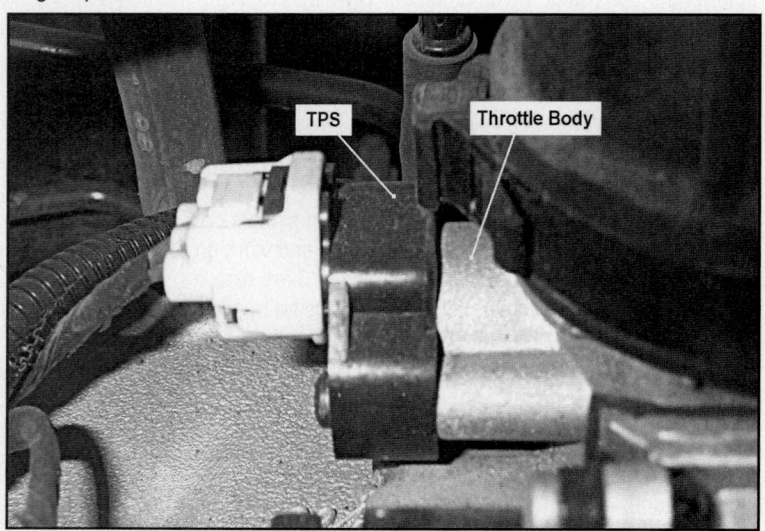

TPS   Throttle Body

Note that the "Code Check" circuit for the TP sensor connects to the PCM at Pin A23. This is the critical circuit that the PCM monitors in order to detect a TP sensor circuit fault or a fault in the relationship between the TP sensor input and the MAF sensor input.

The 5v supply and sensor ground circuits are shared with other sensors, and cannot be monitored to determine TP sensor or circuit faults. However, the "Code Check" circuit will monitor and set trouble codes for signal faults caused by the 5v supply or sensor ground circuits. This circuit is also a critical input for many OBD II monitor tests.

TP Sensor Circuit Schematic

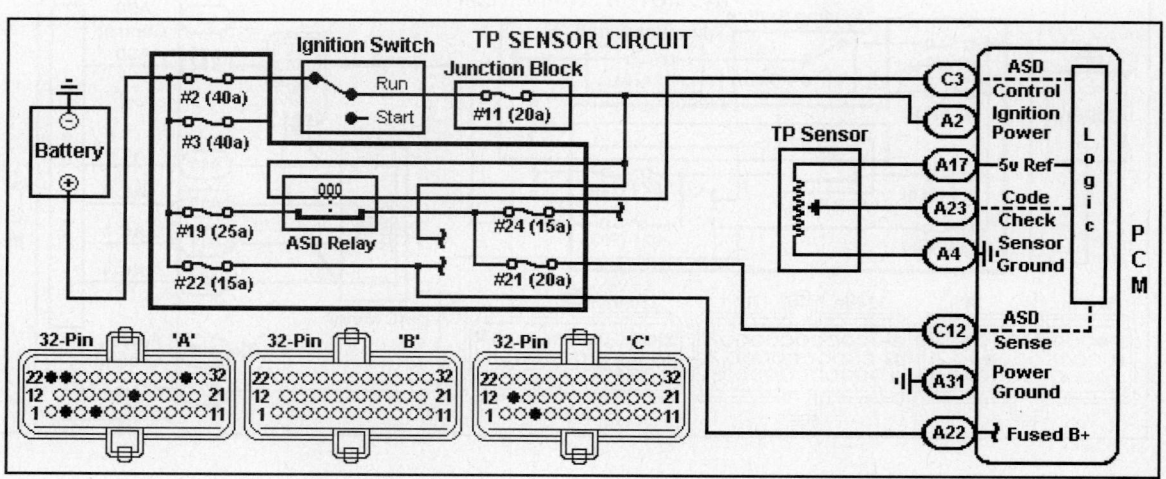

DVOM TEST (TP SENSOR)

The DVOM is an excellent tool to test the condition of the TP sensor circuit. The DVOM can be used to test the supply voltage, sensor ground voltage drop, and the TP sensor signal.

With the throttle fully closed, the TP sensor input to the PCM should be between 0.6 and 1.0v at PCM Pin A23. In this example, the closed throttle voltage was 0.87v.

The TP sensor voltage increases in proportion to the throttle valve-opening angle. The TP sensor signal should be under 4.99v at wide-open throttle (WOT). For this vehicle, the WOT TP signal was 4.0v.

The voltage supply should be 4.9-5.1v, and the ground voltage drop should not exceed 50 mv.

LAB SCOPE TEST (TP SENSOR)

The Lab Scope is an excellent tool to test the activity of the TP sensor circuit over time.

Lab Scope Connection
Connect the Channel 'A' positive probe to the TP sensor signal circuit (OR/DK BL wire) at PCM Pin A23 or at the component harness connector. Connect the negative probe to the battery negative post.

Lab Scope Example
In the KOEO mode, rotate the throttle plate slowly open and closed. Watch for any sudden dropouts (downward spikes) in the signal, indicating a 'dead spot' in the resistor.

TP Sensor Test Schematic

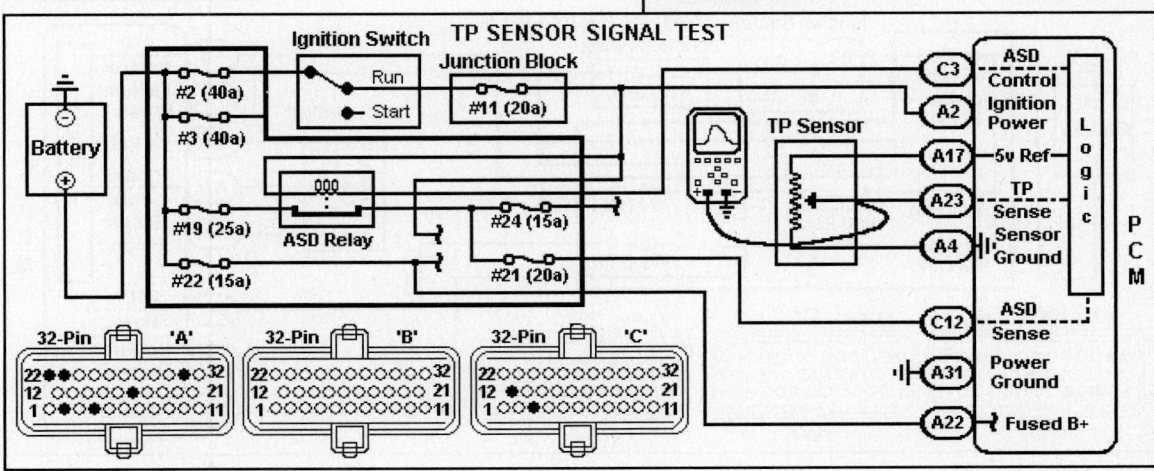

MANIFOLD ABSOLUTE PRESSURE SENSOR

The Manifold Absolute Pressure (MAP) sensor is located on the throttle body. The sensor is mounted with an o-ring directly to a port below the throttle plate (no vacuum hose is used). The MAP sensor input to the PCM is an analog voltage signal proportional to the intake manifold pressure.

The MAP sensor signal is compared to the TP sensor signal and other inputs to calculate engine conditions. The PCM uses theses calculations to determine the following outputs:

- Fuel Injector Pulsewidth
- Ignition Timing
- Transmission Shift Points

MAP Sensor Circuits

The three circuits that connect the MAP sensor to the PCM are the 5v supply circuit, MAP sensor signal circuit, and sensor ground return circuit.

Circuit Operation

The MAP sensor voltage increases in proportion to the engine load. When the throttle angle is high as compared to engine RPM, the load on the engine is higher. Intake vacuum drops (pressure rises) in proportion to increased engine loads. By measuring this vacuum (pressure) the PCM has a very precise understanding of engine conditions.

Note that the "Code Check" circuit for the MAP sensor connects to the PCM at Pin A27. This is the critical circuit that the PCM monitors in order to detect a MAP circuit fault or a fault in the relationship between the TP sensor input and the MAP sensor input.

The 5v supply and sensor ground circuits are shared with other sensors, and are not monitored to determine MAP sensor or circuit faults. However, the "Code Check" circuit will observe and set trouble codes for signal faults caused by the 5v supply or sensor ground circuits. This circuit is also a critical input for many OBD II monitor tests.

MAP Sensor Schematic

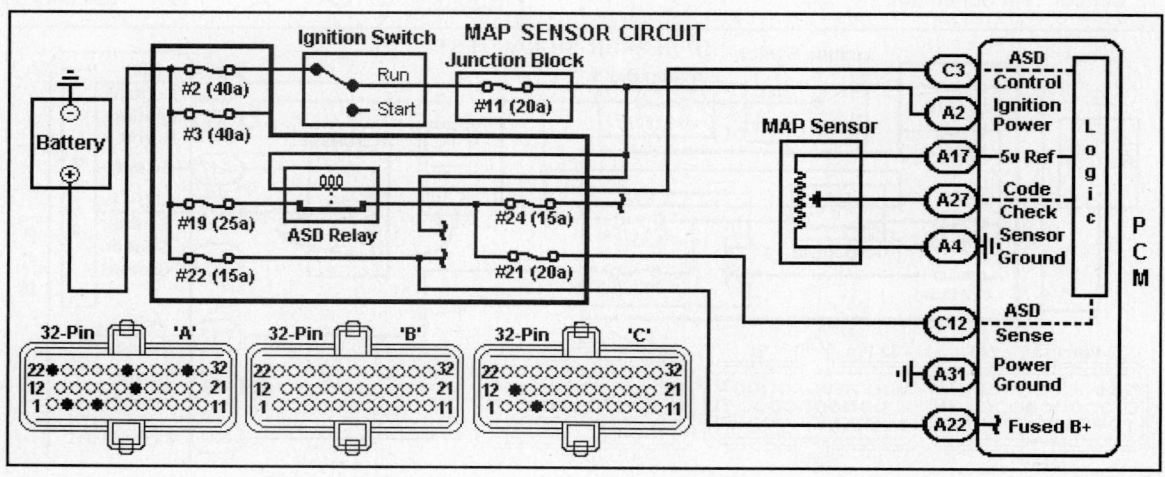

DVOM TEST (MAP SENSOR)

The DVOM is an excellent tool to test the condition of the MAP sensor circuit. The DVOM can be used to test the supply voltage, sensor ground voltage drop, and the MAP sensor signal for calibration or dropout.

The test results can be compared with the specifications in Map Sensor Table (Refer to Section 1).

With the engine at Hot Idle, the MAP sensor input to the PCM should be 1.5-1.7v at PCM Pin A27. In this example, the MAP signal voltage was 1.46v at Hot Idle.

The MAP sensor voltage increases in proportion to manifold pressure. On this vehicle, the KOEO MAP signal was 4.36v at sea level.

The voltage supply should be 4.9-5.1v, and the ground voltage drop should not exceed 50 mv.

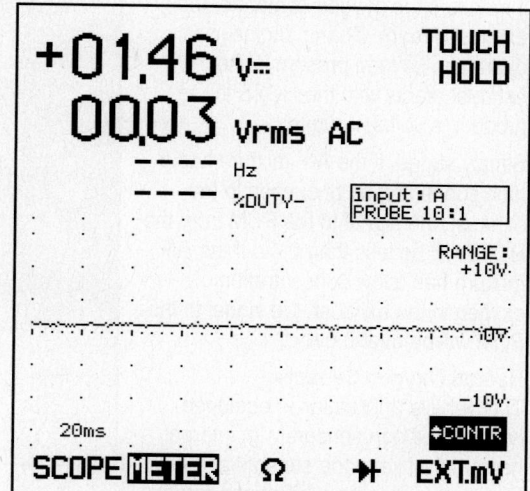

LAB SCOPE TEST (MAP SENSOR)

The Lab Scope is an excellent tool to test the activity of the MAP sensor signal over time.

Lab Scope Connection
Connect the Channel 'A' positive probe to the MAP sensor signal circuit (DK GR/RD wire) at PCM Pin A27 or at the MAP sensor harness connector. Connect the negative probe to the battery negative post.

Lab Scope Example
In this example, the trace shows the MAP sensor signal during a snap throttle event from Hot Idle.

When the throttle plate is opened suddenly, the intake plenum is filled with a rush of air. Manifold vacuum drops (pressure rises) to near barometric pressure for a short time until the engine speeds up and is able to use more of the available air. This can be seen in the example as the signal suddenly rises, then drops off slightly. The peak voltage in this test should be about 4v.

When the throttle is released, the throttle plate is closed, but the engine speed is still well above idle. As the engine attempts to draw in air against a closed throttle plate, the

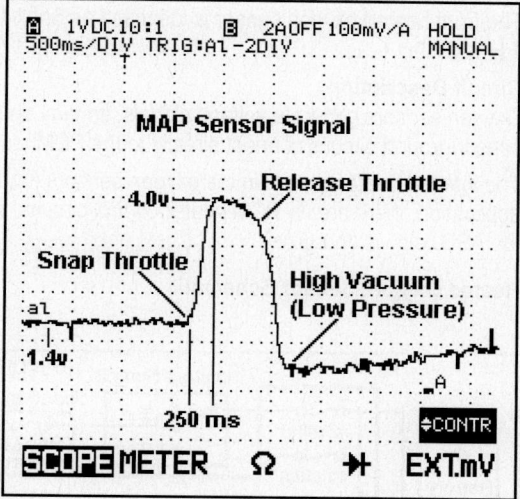

manifold vacuum rises (lower pressure). This can be seen in the example as a sharp drop to below the idle voltage value. The signal then rises as the engine slows down, and vacuum returns to normal idle levels.

The snap throttle test can be used to check for exhaust restrictions and for sensor performance problems. Also, a restriction in the orifice or sensor may limit the range or the response time of the MAP sensor signal.

HEATED OXYGEN SENSORS

The heated oxygen sensor (HO2S) is mounted in the Exhaust system where it monitors the oxygen content of the exhaust stream. During engine operation, oxygen present in the exhaust reacts with the HO2S to produce a voltage output.

Simply stated, if the A/F mixture has a high concentration of oxygen in the exhaust, the signal to the PCM from the HO2S will be less than 0.4v. If the A/F mixture has a low concentration of oxygen in the exhaust, the signal to the PCM will be over 0.6v.

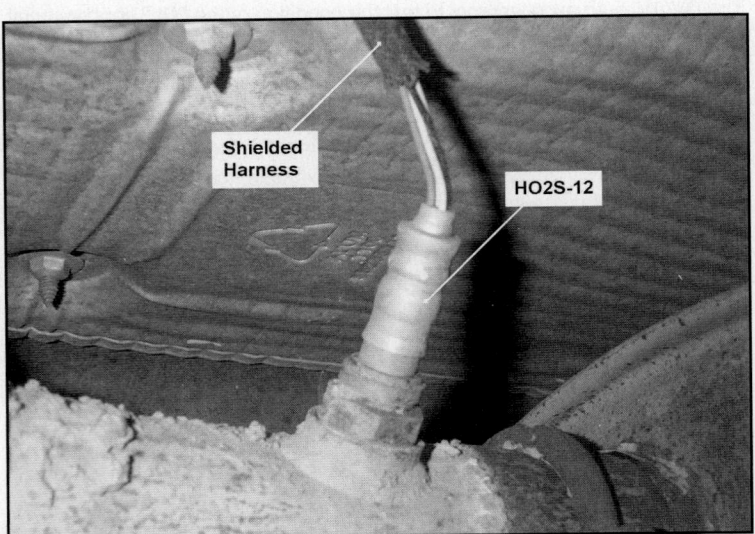

Heated Oxygen Sensors
This vehicle application is equipped with two oxygen sensors with internal heaters that keep the sensors at a high temperature for more efficient operation.

Oxygen Sensor Identification
The front heated oxygen sensor is identified as Bank 1 HO2S-11. The rear heated oxygen sensor is identified as Bank 1 HO2S-12.

Circuit Description
Oxygen sensors generate voltage signals, so they do not need any reference voltage. The HO2S signal circuits and sensor ground circuits connect directly to the PCM.

The internal heater circuits in the oxygen sensors (both front and rear) receive power through the ASD relay. On this application, there are no PCM heater control circuits. The heaters circuits are always grounded, and operate whenever the ASD relay is energized.

Heated Oxygen Sensor Schematic

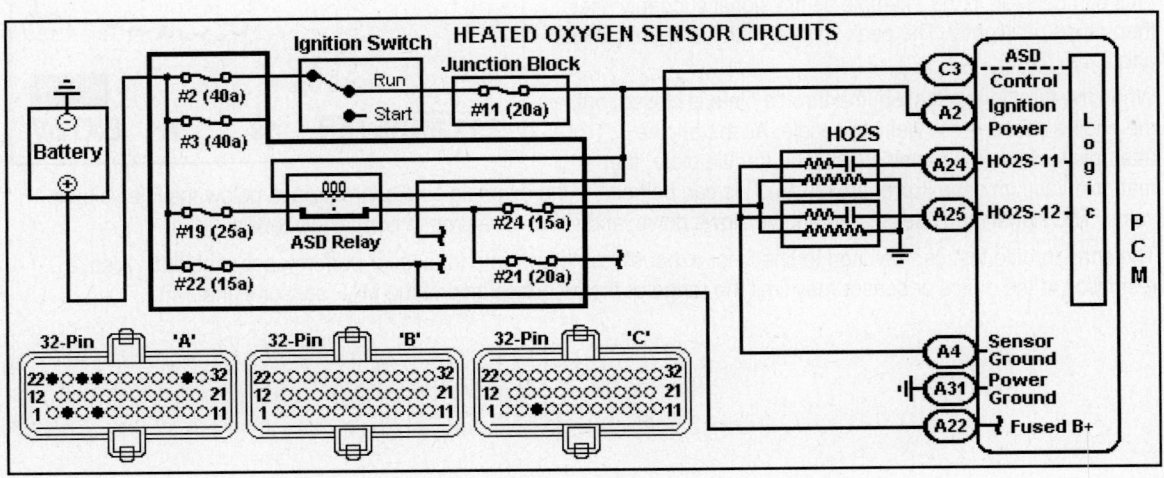

SCAN TOOL TEST (HO2S)

The Scan Tool is the most efficient way to compare HO2S values to other inputs to help determine the cause of a problem. It is also the best tool to use to get a quick idea of certain HO2S performance characteristics (time-to-activity from a cold start, general min/max values, etc.).

The Scan Tool is limited, however, in that the data refresh rate is much slower than either a DVOM or Lab Scope. Therefore, much of the subtle circuit activity is not viewable. It is important to determine before testing begins if the flexibility and ease of the Scan Tool is more important than the precision of a DVOM or Lab Scope.

Scan Tool Testing
Connect the Scan Tool to the DLC underdash connector. Navigate through the Scan Tool menus until you get to the Chrysler (OEM) Data List (a list of PID items). Raise the engine speed to 2000 rpm for 3 minutes to set up the test. These examples were captured at Hot Idle in Park.

Then select the type of oxygen sensor (O2S) information that you want to view from the various options in the menu.

Scan Tool Example (1)
This example shows the PID values for the upstream sensor (HO2S-11) signal and A/F ratio interpretation. In this application, the UPSTREAM O2S ST PID represents the PCMs interpretation of the A/F ratio. This PID value can be LEAN, RICH, or CENTER. The value changes at the same switching frequency as the UPSTREAM O2S PID.

The HO2S PID values should be observed over a period of time. Individual captures such as these examples reveal very little about the condition of these circuits.

Scan Tool Example (2)
This example shows the PID values for the downstream sensor (HO2S-12) signal and A/F ratio interpretation. These values are useful in diagnosing catalytic converter efficiency, but not A/F ratio, as the oxygen content has been modified by converter operation.

Scan Tool Example (3)
This example shows the Generic PIDs for this vehicle. Besides those required by the EPA, each manufacturer is allowed to include additional PIDs. In this application, HO2S-11 and HO2S-12 values are available as Generic PIDs, as well as the RICH, LEAN, and CENTER values.

Summary
There are numerous screens available on the Scan Tool that can help you diagnose the condition of the HO2S. You can vary the screens to obtain information from a particular sensor. On this application, there are no PIDs for the HO2S heaters because they are not controlled by the PCM.

```
Apr 18, 02      12:46:36 pm

UPSTREAM O2S      (1)
0.11 VOLTS
UPSTREAM O2S ST
LEAN


0.00 v

[RCV] FOR ENHANCED
[*EXIT] TO EXIT
```

```
Apr 18, 02      12:46:02 pm

DNSTREAM O2S      (2)
0.74 VOLTS
DNSTREAM O2S ST
RICH


0.00 v

[RCV] FOR ENHANCED
[*EXIT] TO EXIT
```

```
                         (3)
ENGINE SPD..............727RPM
TPS (V).................0.82V
TPS (%).................0.00%
ECT (°).................208°F
ECT (V).................0.92V
MAP (P)................11.50inHg
MAP (V).................1.58V
VAC (P)................18.00inHg
BARO (P)..............29.55inHg
UP O2S..................0.74V
UP O2S ST................RICH
DN O2S..................0.16V
DN O2S ST................LEAN
IAT (°).................135°F
IAT (V).................2.34V
INJECTOR P/W...........4.30ms
```

HO2S HEATER MONITOR

Neither the PCM nor the Scan Tool can directly monitor the HO2S heaters or circuits for performance faults. Instead, the PCM tests the heater circuits by watching the change in activity on the oxygen sensor circuits. These tests are the basis for the OBD II HO2S Heater Monitor, and can be monitored using a Scan Tool.

HO2S Heater Explanation
The voltages produced by the heated oxygen sensors are temperature sensitive. The values are not accurate below 300 °F. The sensors are heated by flowing current through a low-resistance heater element. This allows the engine to enter closed loop operation sooner after a cold start, and ensures optimum sensor performance during certain operating conditions (oxygen sensors tend to cool during extended idle periods).

HO2S Heater Monitor Operation
The heater element itself is not tested. During HO2S Heater Monitor operation, the PCM sends a 5-volt biased signal backwards through the sensor circuit and sensor ground, and then monitors the voltage drop across the sensor as it is heated. As the sensor is heated, the resistance decreases, and the drop in the bias voltage increases.

It is important to understand that the bias voltage is not used to heat the sensors. The heaters are energized in the normal manner through the ASD relay. The PCM is simply using the 5-volt signal to turn the oxygen sensors into negative-temperature coefficient thermistors for the duration of the test. Also, the test is performed in open loop, so the bias to the HO2S sensor signals has no effect on engine operation or fuel trim.

The sensor resistance varies greatly from about 4.5 mega Ohms when extremely cold to about 100 Ohms at 660 °F. When the test is initiated, the sensor resistance is so high that there is no measurable voltage drop, and the PCM will detect close to 5 volts.

When the HO2S Heater Monitor is enabled, the PCM monitors the circuit and measures the amount of time it takes to drop from 4 volts to 3 volts. If the voltage drop is not achieved during the test period, a DTC will set.

SCAN TOOL TEST (HO2S HEATERS)

Connect the Scan Tool and navigate to the data list (PIDs). Select UP O2S and DN O2S PIDS.

Scan Tool Test
Observe the PIDs from a cold engine startup. Both values should start at or near 5 volts, and drop steadily until the vehicle enters closed loop. In the examples to the right, graphing software was used to show the effects of the resistance dropping due to heater operation on a known-good engine.

Note that the downstream sensor takes longer to heat, due to its location (farther from the engine) and the cold catalytic converter.

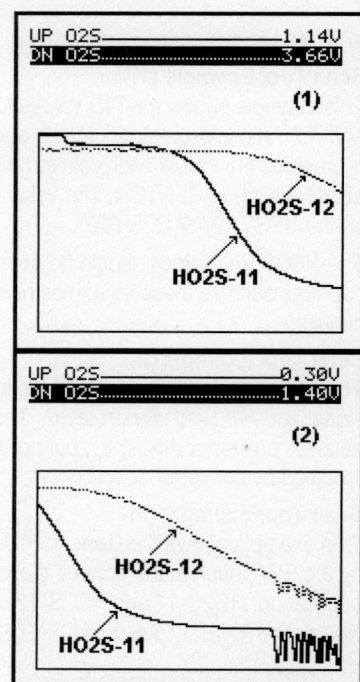

LAB SCOPE TEST (HO2S)

The Lab Scope is the "tool of choice" to test the oxygen sensor as it provides an accurate view of the sensor response and switch rate. Place the gearshift selector in Park and block the drive wheels for safety.

Lab Scope Connections
Connect the Channel 'A' positive probe to the HO2S-11 sensor signal wire (BR/YL wire) at PCM Pin A24, and the negative probe to the battery negative post. Connect the Channel 'B' positive probe to the HO2S-12 sensor signal wire (TN/WT Wire) at PCM Pin A25.

Lab Scope Example (1)
In example (1), traces from HO2S-11 and HO2S-12 were captured with the vehicle at 2000 rpm (warm engine). The HO2S-11 signal should switch at least 3-5 times per second. In this example, the sensor switched 22 times in 5 seconds (4.4 times per second).

Lab Scope Example (2)
In example (2), the upstream sensor (HO2S-11) trace was captured during a snap throttle event in order to view the response rate of the sensor. The screen was frozen (HOLD) an cursors were used to measure the rise time. The sensor should respond L-R or R-L in 100 ms or less (32 ms in this example). This vehicle had 79,000 miles on the odometer.

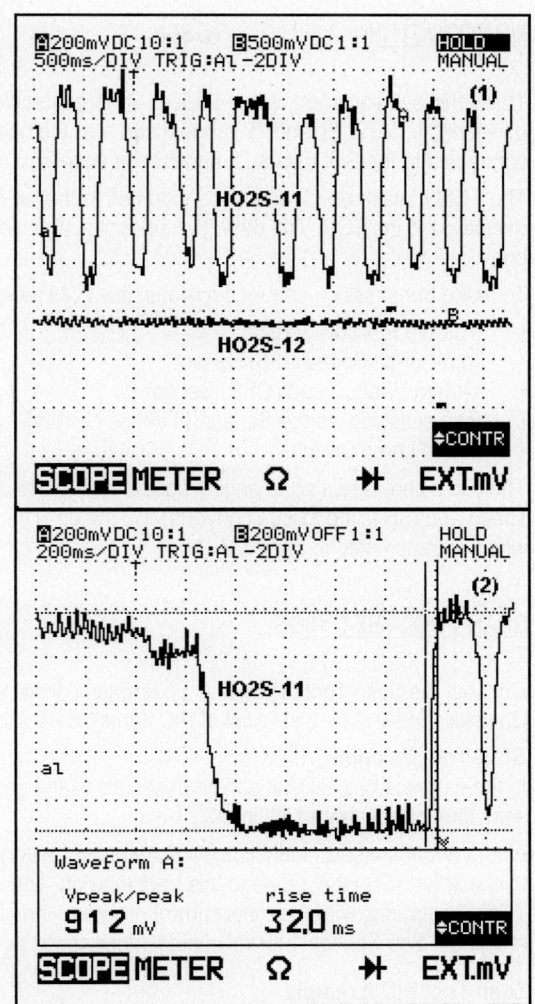

Heated Oxygen Sensor Schematic

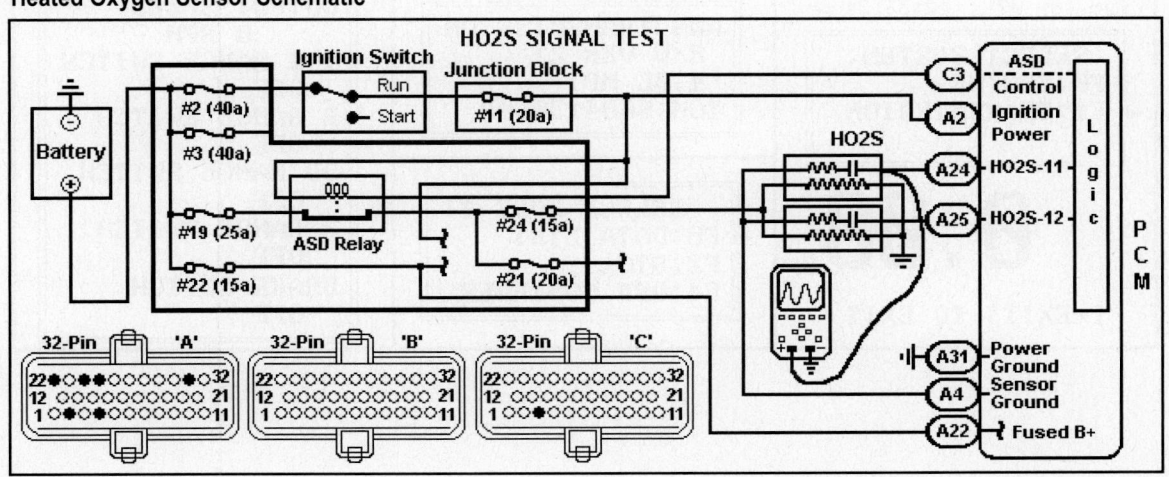

Transmission Controls

INTRODUCTION

This vehicle is equipped with a 4-speed AW4 electronic automatic transmission. The TCM controls the upshift, and downshift in '1', '2', '3' and 'D' ranges, and also the operation of the torque converter clutch (TCC). The TCM is connected to the SCI data bus, and can be accessed using a Scan Tool.

The TCM uses three (3) solenoids mounted to the valve body to control shift timing and TCC lockup. One solenoid is dedicated to the TCC. The other two solenoids direct hydraulic pressure to the shift valves in the transmission valve body.

To make transmission control decisions, the TCM relies on the following direct sensor inputs:

- Throttle Position Sensor
- Input Shaft Speed (ISS) Sensor
- Output Shaft Speed (OSS) Sensor
- Transmission Range Sensor (4 sense circuits)
- Brake Lamp Switch

The ISS signal is related to engine speed, but it is measured after the torque converter. The difference between engine speed and ISS is the torque converter slip speed. The OSS signal is used by the TCM to calculate the VSS value, which is transmitted to the PCM.

SCAN TOOL TEST (PIDS)

Connect the Scan Tool to the DLC underdash connector. Navigate through the Scan Tool menus until you get to the Chrysler (OEM) Data List (a list of PID items). Select the appropriate PIDs.

Scan Tool Example

In the example, the PID list can be observed as the gear selector is moved through its positions, to verify that the TCM sees the correct gear at all times.

Also, the transmission identification screen can be very useful. Use this screen to confirm the application. Remember to check for TSBs that relate to this transmission, and use the software version number (S/W VER 21) to verify that the TCM is operating with the most current program. If an updated software version is available, the TCM may need to be reflashed. See Section 1 for reflashing information.

Scan Tool PID Example

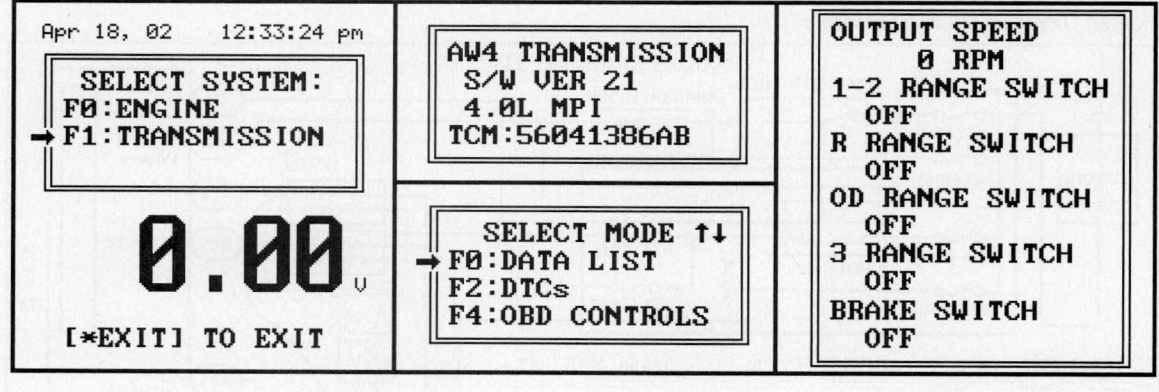

BI-DIRECTIONAL TESTS (SOLENOIDS)

The Scan Tool can control the three (3) solenoids in the valve body through the TCM. These bi-directional tests can only be used with the engine off. Solenoid and circuit operation can be checked, but the controls cannot be used to shift the transmission under normal operating conditions.

Scan Tool Bi-directional Example
Connect the Scan Tool to the DLC underdash connector. Navigate through the Scan Tool menus until you get to F4: OBD CONTROLS. Select F0: A/T OUTPUTS from the OBD CONTROLS menu. Then select the desired solenoid.

When the Scan Tool 'up' arrow is pressed, the TCM will cycle the designated solenoid at a fixed frequency and duty cycle for 5 minutes, or until the user exits the test from the Scan Tool.

This test cannot be used to check hydraulic circuit pressure in different gears, as the test only operates with the engine off, (when the line pressure is 0 psi).

Scan Tool Bi-directional Control Example

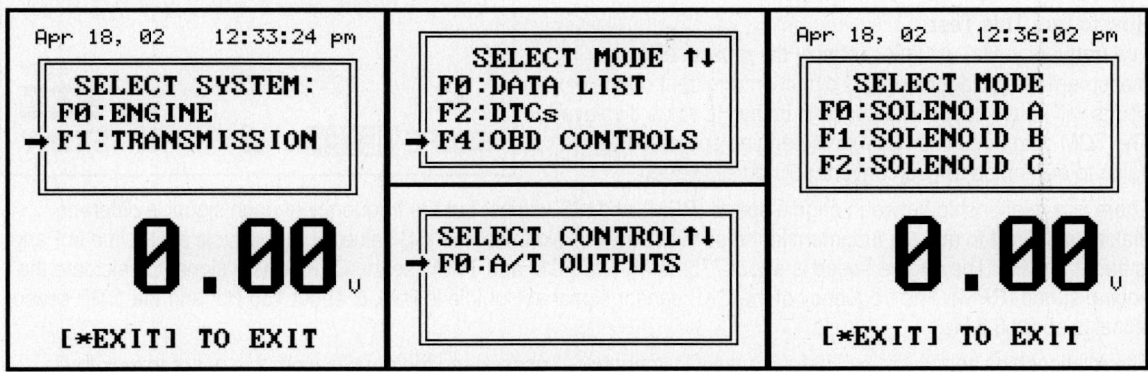

Why to Use This Test
Because these tests cannot be used with the engine running, the results cannot be measured by observing transmission performance. A DVOM or Lab Scope must be used to measure and display the circuit activity resulting from the use of these controls.

The TCM makes shift and TCC lockup decisions based on different operating criteria. If a solenoid is not operating or is suspected of causing a shifting or transmission performance fault, it can be difficult to determine if the circuit is incapable of directing fluid pressure properly, or if the TCM has simply made bad decisions based on missing or incorrect sensor data.

These tests can help to eliminate many assumptions, because the expected results are already known, and do not depend on outside conditions that need to be met.

While these tests may not pinpoint the fault, they are very useful in determining the best diagnostic path to follow.

DVOM TEST (INPUT SPEED SENSOR)

Place the gear selector in Park and block the drive wheels for safety.

DVOM Connections (ISS)
Connect the positive probe to the ISS circuit (RD/BK wire) at TCM Pin 2, and the negative probe to the battery negative post.

DVOM Example
In the example, the trace shows the ISS signal at cruise speed.

Observe the AC voltage value on the DVOM as the vehicle is driven. The voltage will be around 2.5v AC. Due to dampening circuitry inside the TCM, the voltage will not fluctuate like other AC sensor circuits. The signal should stay near 2.5v. A lower voltage will still be recognized by the TCM, as long as the signal voltage is greater than 300 mv.

How to Use This Test
The frequency (Hz) value is probably the most useful measurement in diagnosing this circuit. Intermittent opens or shorts will cause sudden fluctuations in the Hz value displayed. The TCM compares this value to the engine speed (RPM) value to determine torque converter clutch slip speed.

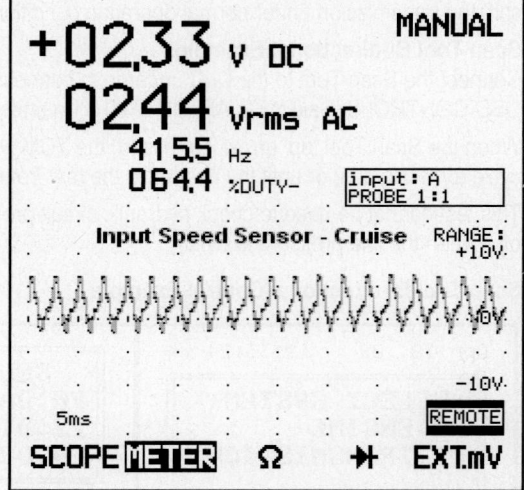

There is a relationship between engine speed (RPM) and ISS values, but the frequency of each signal is different, making it difficult to make a judgment in the service bay. For example, the ISS value for this vehicle at Hot Idle in Park is about 200 Hz. The engine speed is about 775 RPM. The PCM and TCM use the CKP sensor signal to calculate the engine speed (RPM). The frequency of the CKP sensor signal at Hot Idle in Park is about 155 Hz, and the CMP sensor signal is about 6.5 Hz.

The relationships above can be confusing, so it is important to understand their application. It is better to look for consistency than an absolute value. Multiple meters must be used to ensure valid results. With one meter or Lab Scope, observe the TCC circuit. When the TCC is engaged, compare the frequency of the CKP and ISS signals under very light load and moderate load. As long as the TCC does not disengage, the CKP and ISS frequency relationship should stay fairly constant.

Calculating the Results
The easiest way to calculate the results of this test is to divide the CKP frequency by the ISS frequency. This will yield a ratio that means nothing by itself. The ratio is only useful when compared under different circumstances. If the CKP value is 500 Hz @ 2500 RPM and the ISS value is 415 Hz at steady cruise, then the ratio is 1.20:1. If the vehicle is then lightly accelerated and the values are 520 Hz and 432 Hz, then the ratio is still 1.20:1, and the TCC is not slipping.

The same strategy can be used to test for slipping gears. Compare the frequency of the ISS and OSS signals under different loads and in different gears. Take into account the gear ratios. The ratios are available in other service information.

LAB SCOPE TEST (ISS & OSS)

The Lab Scope can be used to test the AC voltage output of the Input Speed Sensor (ISS) and Output Speed Sensor (OSS).

Place the gear selector in Park and block the drive wheels for safety.

Lab Scope Connections
Connect the Channel 'A' positive probe to the ISS circuit (RD/BK wire) at TCM Pin 2, and the negative probe to the battery negative post. Connect the Channel 'B' positive probe to the OSS circuit (LG/WT Wire) at TCM Pin 4.

Lab Scope Example (1)
In example (1), the trace shows the ISS and OSS signals with the key on, engine off. The sensors cannot generate voltage without rotation. The 2.5v is a diagnostic bias voltage used by the TCM to ensure the circuit is in good condition. The Lab Scope must be set to DC coupling (displays both AC and DC) to see this information.

Lab Scope Example (2)
In example (2), the trace shows the ISS and OSS signals at 25 MPH. Although the sensors generate an AC voltage, the signals are mostly positive because the bias voltage is present.

Observe the patterns over time to ensure the signals are stable, and that the frequency of the signals increases in relation to the engine and vehicle speed.

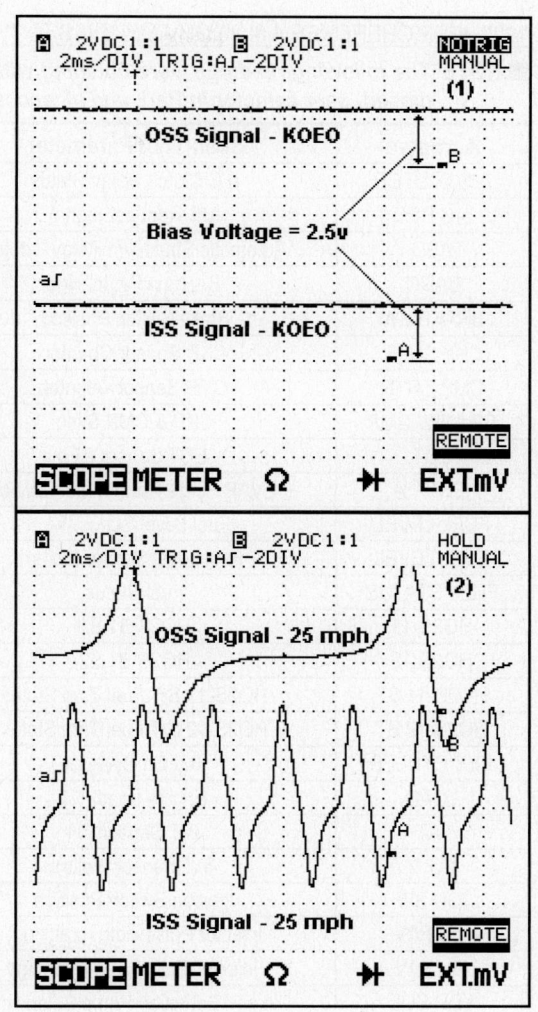

Speed Sensor Test Schematic

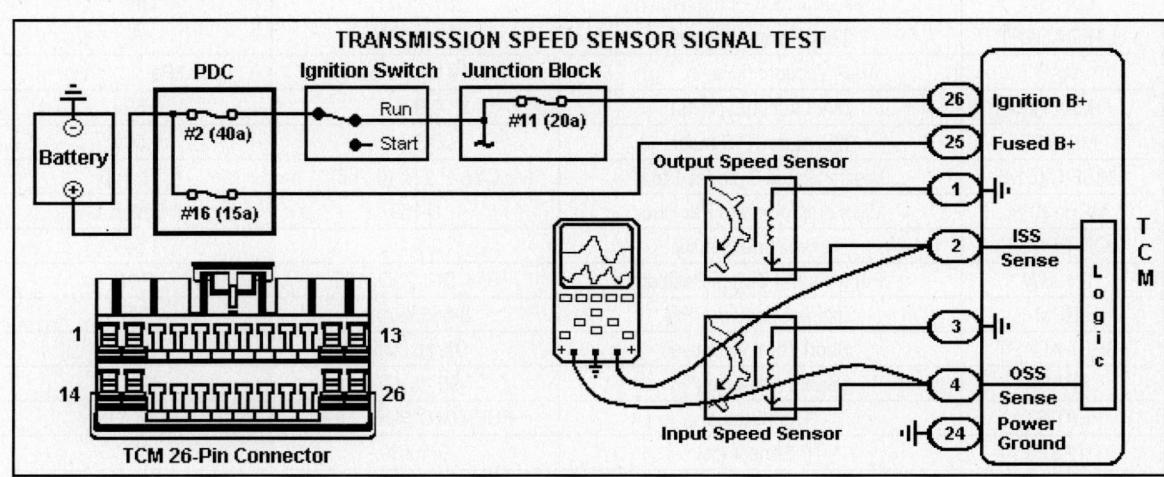

Reference Information - PCM PID Examples

1999 JEEP CHEROKEE 4.0L I6 OHV MFI VIN S (A/T)

■ NOTE: *The following readings were obtained with the engine at idle speed, radiator hose hot, throttle closed, gear selector in Park and all accessories turned off.*

Acronym	Scan Tool Parameter	Range	Typical Value
A/C CLUTCH	A/C Clutch Relay Status	ON / OFF	OFF
A/C REQ	A/C Request Signal	ON / OFF	OFF
ASD	Automatic Shutdown Relay Control	ON / OFF	ON
BARO	BARO (kPa, In. Hg)	10-110 kPa, 0-30 "Hg	101 kPa, 29.8 "Hg
BRAKE SW	Brake Switch Position	ON / OFF	OFF
CKP CNTR	CKP Sensor Counter	0-256	174
CMP CNTR	CMP Sensor Counter	0-256	42
CRANK / CMP	CKP & CMP Sync	In Sync / Not in Sync	In Sync
ECT	ECT Sensor (°F)	-40 to 284°F	216°F
EVAP %	EVAP Purge Control (duty cycle)	0-100%	14%
FUEL LEVEL	Fuel Level Status (%_	0-100%	53%
FUEL LEVEL	Fuel Level Sensor (volts)	0-5.1v	3.66v
FUEL STATUS	Fuel Status	CLOSED / OPEN	CLOSED
HO2S-11	HO2S B1 S1	0-1100 mv	110 mv
HO2S-12	HO2S B1 S2	0-1100 mv	74 mv
HO2S-11 ST	HO2S-11 S/T Fuel Trim Status	LEAN / RICH	LEAN
HO2S-12 ST	HO2S-12 S/T Fuel Trim Status	LEAN / RICH	RICH
KEY CYCLES	Key On Cycles 1	0-256	85
LOAD	Engine Load (%)	0-100%	5%
IAT	IAT Sensor (°F	-40 to 284°F	180°F
IAT V	IAT Sensor (volts)	0-5.1v	1.33v
IGN SW	Ignition Switch Position	ON / OFF	ON
INJ P/W	Injector Pulsewidth (cranking)	0-999 ms	1.5 ms
INJ P/W	Injector Pulsewidth (running)	0-999 ms	4.1 ms
LDP SOL	Leak Detection Pump Solenoid	ON / OFF	OFF
LDP SW	Leak Detection Pump Switch	ON / OFF	OFF
LTFT ADAPT	Long Term Adaptive	0% (± 20%)	1%
MAP	MAP Vacuum (kPa, In. Hg)	10-110 kPa, 0-30 "Hg	38 kPa, 11.2 "Hg
MAP (V)	MAP Vacuum (volts)	0-5.1v	1.52v
MIN TPS	Minimum TP Sensor	0-5.1v	0.00v
MISF CYL1-6	Misfire Status Cylinder 1 to 6	CYL 1, 2 ... (0-256)	CYL 1: 0
MPH / KPH	Miles or Kilometers Per Hour	0-159	0 mph
OIL PRESS	Oil Pressure Sensor	0-5.1v	1.60v
P/N SW	Park Neutral Switch Position	P/N, DR, 2ND, 1ST	P/N
RPM	Engine Speed in RPM	0-66635 rpm	775
SHFT ADAPT	Short Term Adaptive	0% (± 20%)	1%
SPARK	Spark Advance (°)	-90° to +90°	14°
THEFT STAT	Theft Status	FUEL ON / FUEL OFF	FUEL ON
TPS %	TP Sensor (%)	0-100%	0%
TPS V	TP Sensor (volts)	0-5.1v	0.88v
VAC	Vacuum (kPa, In. Hg)	10-110 kPa, 0-30 "Hg	63 kPa, 18.6 "Hg

1999 CHEROKEE 4.0L I6 OHV SFI VIN S BLACK 'A' 32-PIN CONNECTOR

PCM Pin #	Wire Color	Circuit Description (32-Pin)	Value at Hot Idle
1, 5, 9	---	Not Used	---
2	DB/WT	Ignition Switch Power (fused)	12-14v
4	BR/YL	A/T: Analog Sensor Ground	<0.050v
6	BK/WT	A/T: PNP Switch Sense Signal	In P/N: 0v, Others: 5v
7	GY	Coil Driver Control	Idle: 5°, 55 mph: 8° dwell
8	GY/BK	CKP Sensor Signal	Digital Signals: 0-5-0-5v
10	YL/BK	IAC 2 Driver Control	Pulse Signals
11	BR/WT	IAC 3 Driver Control	Pulse Signals
12	GY	Extended Idle Switch (Police)	Switch On: 1v, Off: 12v
13-14, 21	---	Not Used	---
15	BK/RD	IAT Sensor Signal	At 100°F: 1.83v
16	TN/BK	ECT Sensor Signal	At 180°F: 2.80v
17	OR	5-Volt Supply	4.9-5.1v
18	TN/YL	CMP Sensor Signal	Digital Signals: 0-5-0-5v
19	GY/RD	IAC 1 Driver Control	Pulse Signals
20	PK/BK	IAC 4 Driver Control	Pulse Signals
22	DG/BK	Fused Battery Power (B+)	12-14v
23	OR/DB	TP Sensor Signal	0.6-1.0v
24	BK/DG	HO2S-11 (B1 S1) Signal	0.1-1.1v
25	TN/YL	HO2S-12 (B1 S2) Signal	0.1-1.1v
26, 28-30	---	Not Used	---
27	DG/RD	MAP Sensor Signal	1.5-1.7v
31	BK/TN	Power Ground	<0.1v
32	BK/TN	Power Ground	<0.1v

1999 CHEROKEE 4.0L I6 OHV SFI VIN S WHITE 'B' 32-PIN CONNECTOR

PCM Pin #	Wire Color	Circuit Description (32-Pin)	Value at Hot Idle
1-3	---	Not Used	---
4	WT/DB	Injector 1 Driver	1-4 ms
5	YL/WT	Injector 3 Driver	1-4 ms
6	PK/BK	Injector 5 Driver	1-4 ms
7-9	---	Not Used	---
10	DG	Generator Field Control	Digital Signals: 0-12-0v
12	LG/BK	Injector 6 Driver	1-4 ms
11, 13-14	---	Not Used	---
15	TN	Injector 2 Driver	1-4 ms
16	LB/BR	Injector 4 Driver	1-4 ms
17-22	---	Not Used	---
23	GY/YL	Engine Oil Pressure Sensor	1.6v at 23 psi
24-26	---	Not Used	---
27	WT/OR	Vehicle Speed Signal	Digital Signals
28-30, 32	---	Not Used	---
31	PK/OR	5-Volt Supply	4.9-5.1v

<u>1999 CHEROKEE 4.0L I6 OHV SFI VIN S GRAY 'C' 32-PIN CONNECTOR</u>

PCM Pin #	Wire Color	Circuit Description (32-Pin)	Value at Hot Idle
1	DB/OR	A/C Clutch Relay Control	Relay On: 1v, Off: 12v
2	DB/PK	Radiator Fan Control Relay	Relay On: 1v, Off: 12v
3	DB/YL	Auto Shutdown Relay Control	Relay On: 1v, Off: 12v
4	TN/RD	S/C Vacuum Solenoid Control	Vacuum Increasing: 1v
5	LG/RD	Speed Control Vent Solenoid	Vacuum Decreasing: 1v
6-9	---	Not Used	---
10	WT/DG	LDP Solenoid Control	PWM Signal: 0-12-0v
11	YL/RD	Speed Control Power Supply	12-14v
12	DG/OR	Auto Shutdown Relay Output	12-14v
13	---	Not Used	---
14	OR	Battery Temperature Sensor	At 86ºF: 1.96v
15	PK/YL	LDP Switch Sense Signal	LDP Switch Closed: 0v
16-18	---	Not Used	---
19	BR	Fuel Pump Relay Control	Relay On: 1v, Off: 12v
20	PK/BK	EVAP Purge Solenoid Control	PWM Signal: 0-12-0v
21	---	Not Used	---
22	DB/WT	A/C Request Signal	A/C Off: 12v, On: 1v
23	LG	A/C Select Signal	A/C Off: 12v, On: 1v
24	WT/PK	Brake Switch Sense Signal	Brake Off: 0v, On: 12v
25	DG/OR	Generator Field Source	12-14v
26	DB/LG	Fuel Level Sensor Signal	Full: 0.5v, 1/2 full: 2.5v
27	PK	SCI Transmit	0v
28	WT/PK	CCD Bus (-)	<0.050v
29	LG/BK	SCI Receive	5v
30	PK/BR	CCD Bus (-), CCD Bus (+)	Digital Signals: 0-5-0v
31	---	Not Used	---
32	R/LG	Speed Control Sw. Signal	S/C & Set Sw. On: 3.8v

Pin Connector Graphic

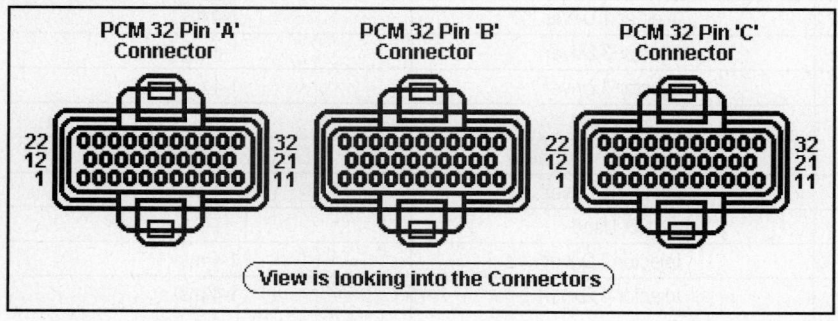

1999 CHEROKEE - POWERTRAIN

PCM Wiring Diagrams

FUEL & IGNITION SYSTEM DIAGRAMS (1 OF 14)

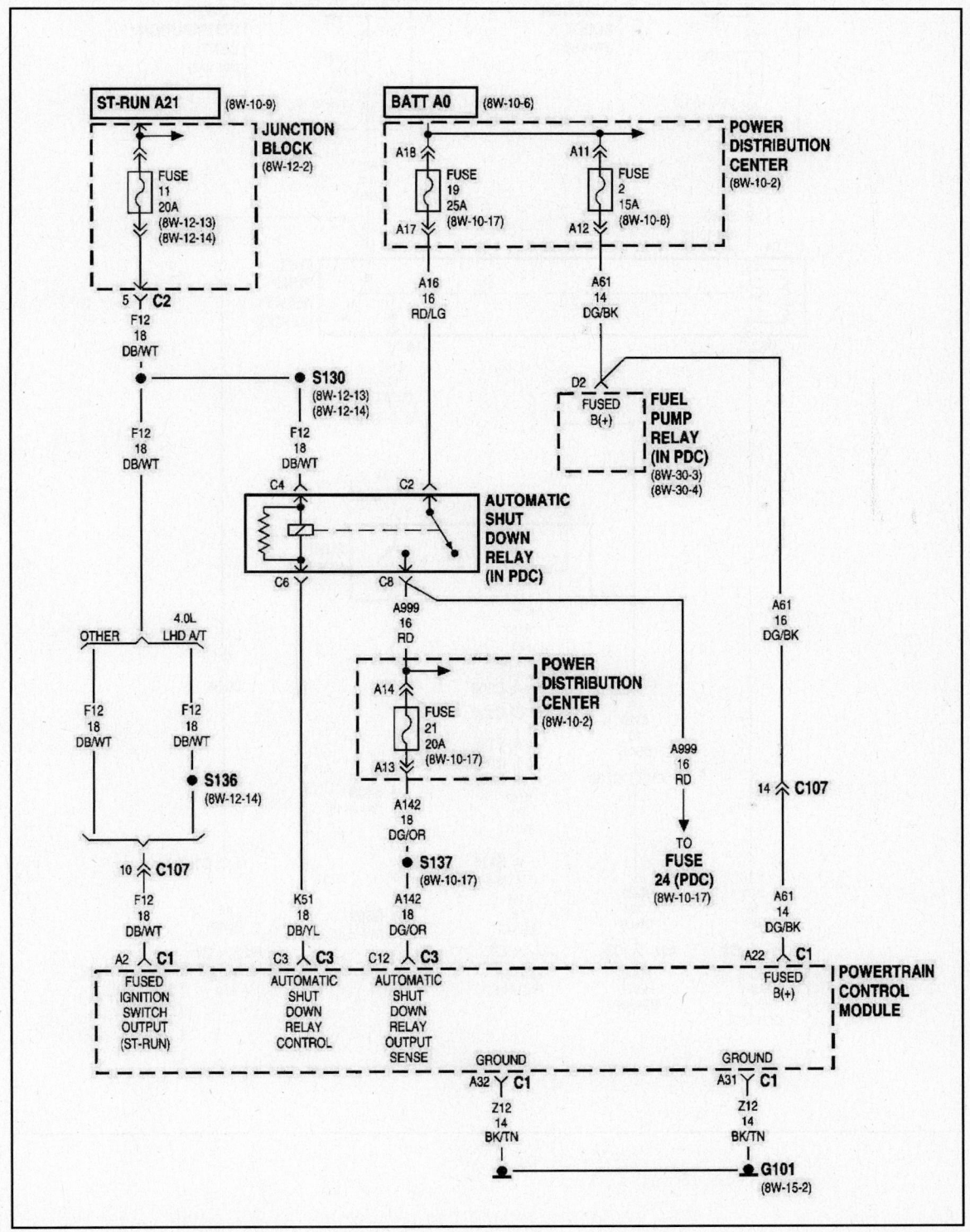

FUEL & IGNITION SYSTEM DIAGRAMS (2 OF 14)

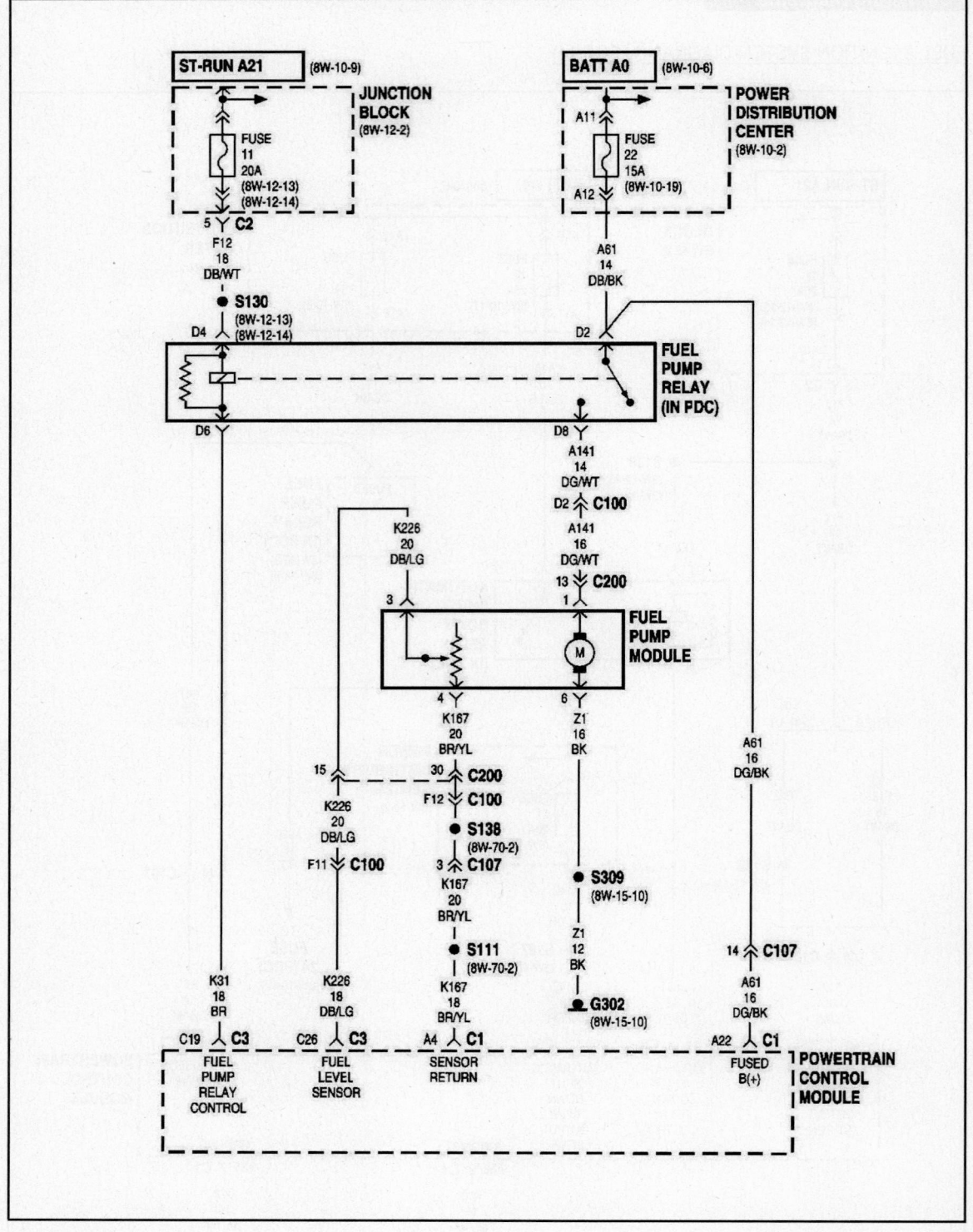

FUEL & IGNITION SYSTEM DIAGRAMS (3 OF 14)

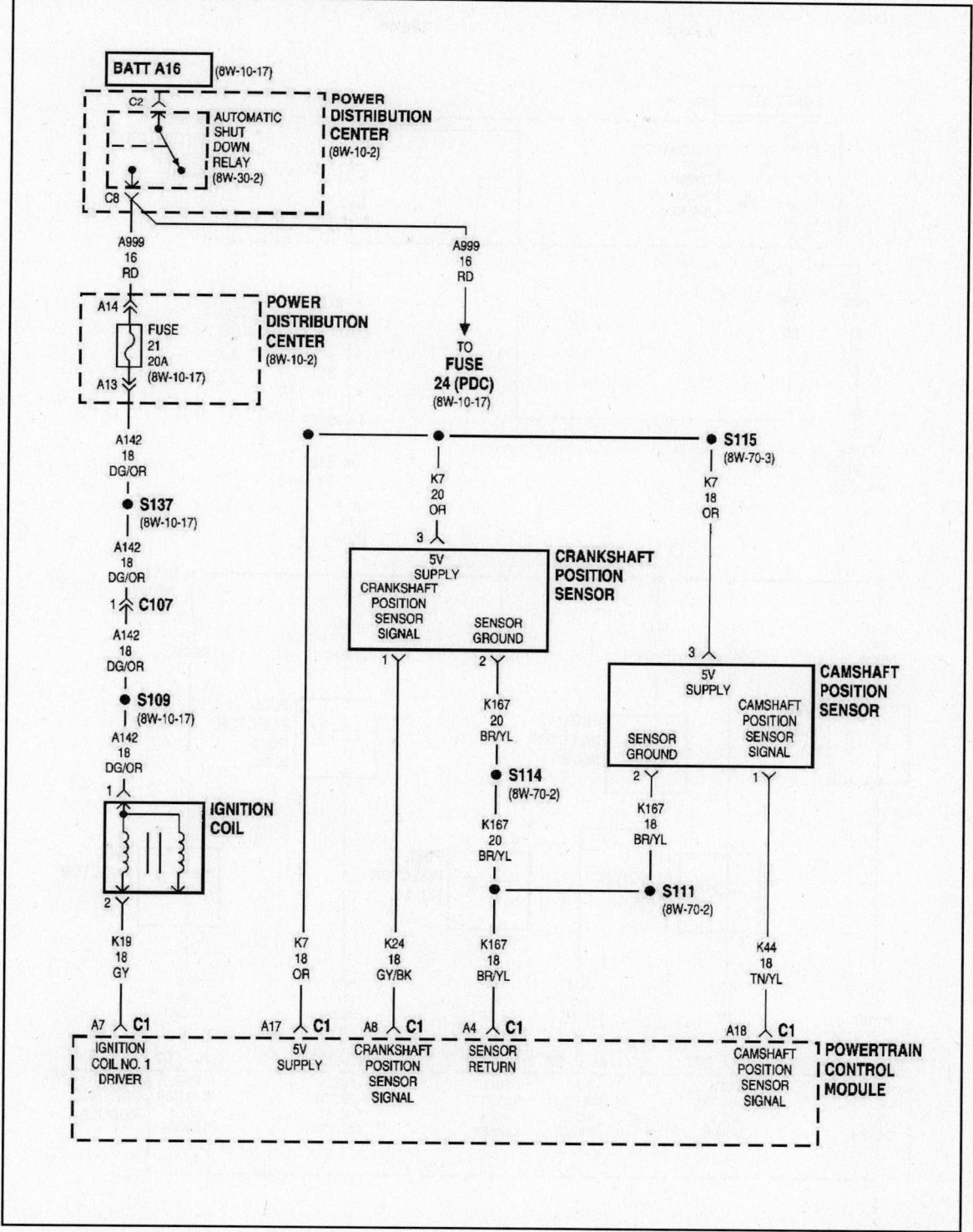

FUEL & IGNITION SYSTEM DIAGRAMS (4 OF 14)

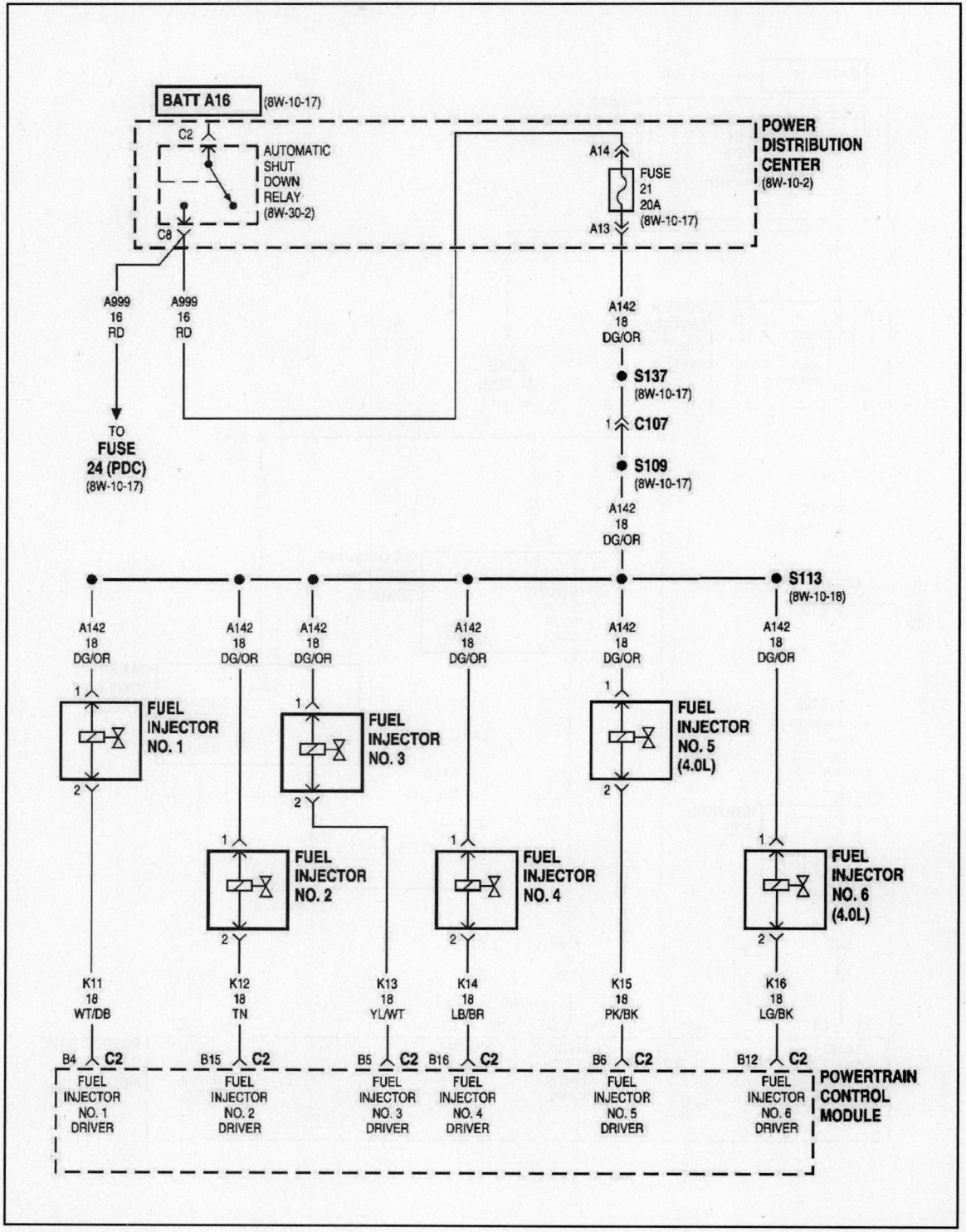

FUEL & IGNITION SYSTEM DIAGRAMS (5 OF 14)

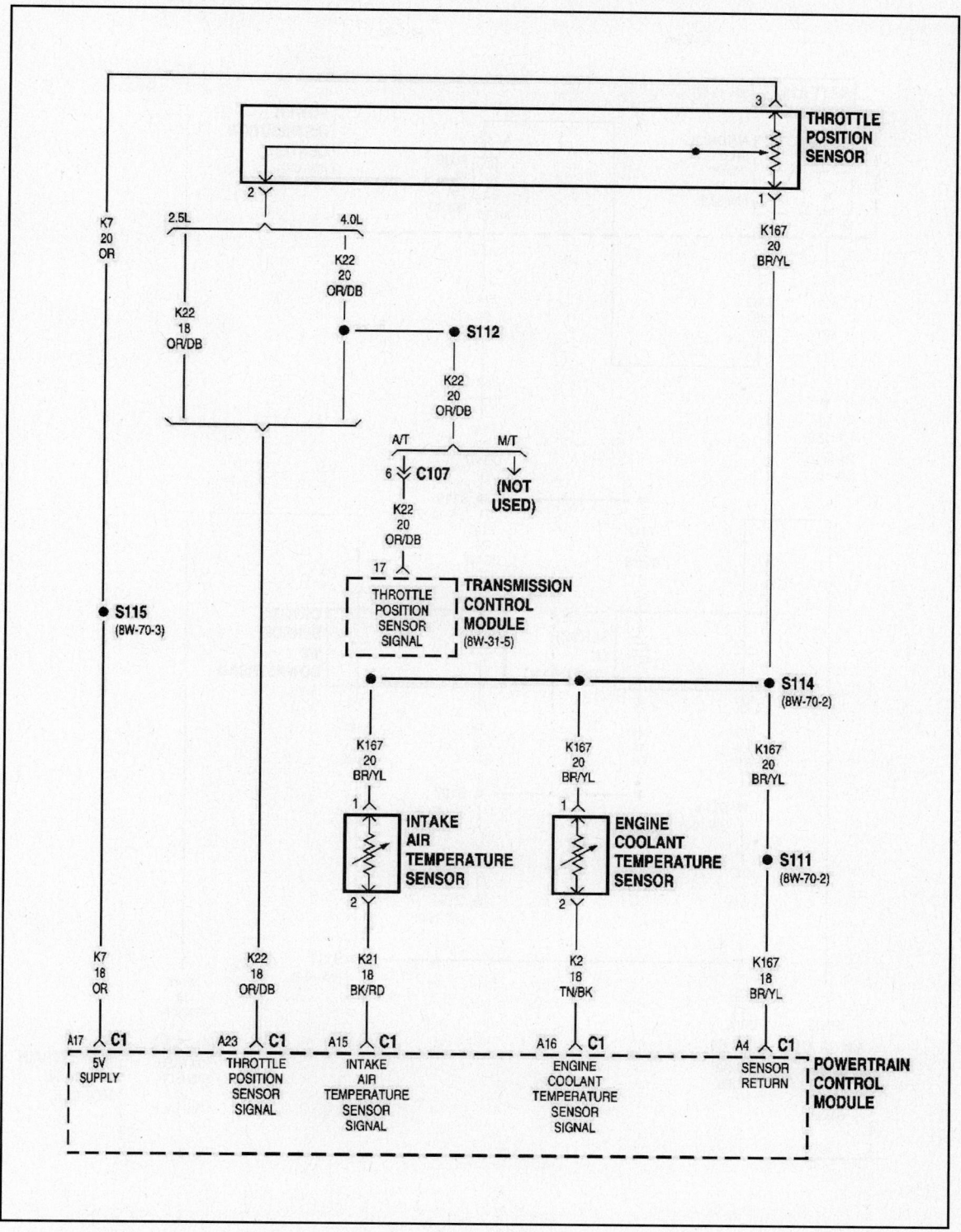

FUEL & IGNITION SYSTEM DIAGRAMS (6 OF 14)

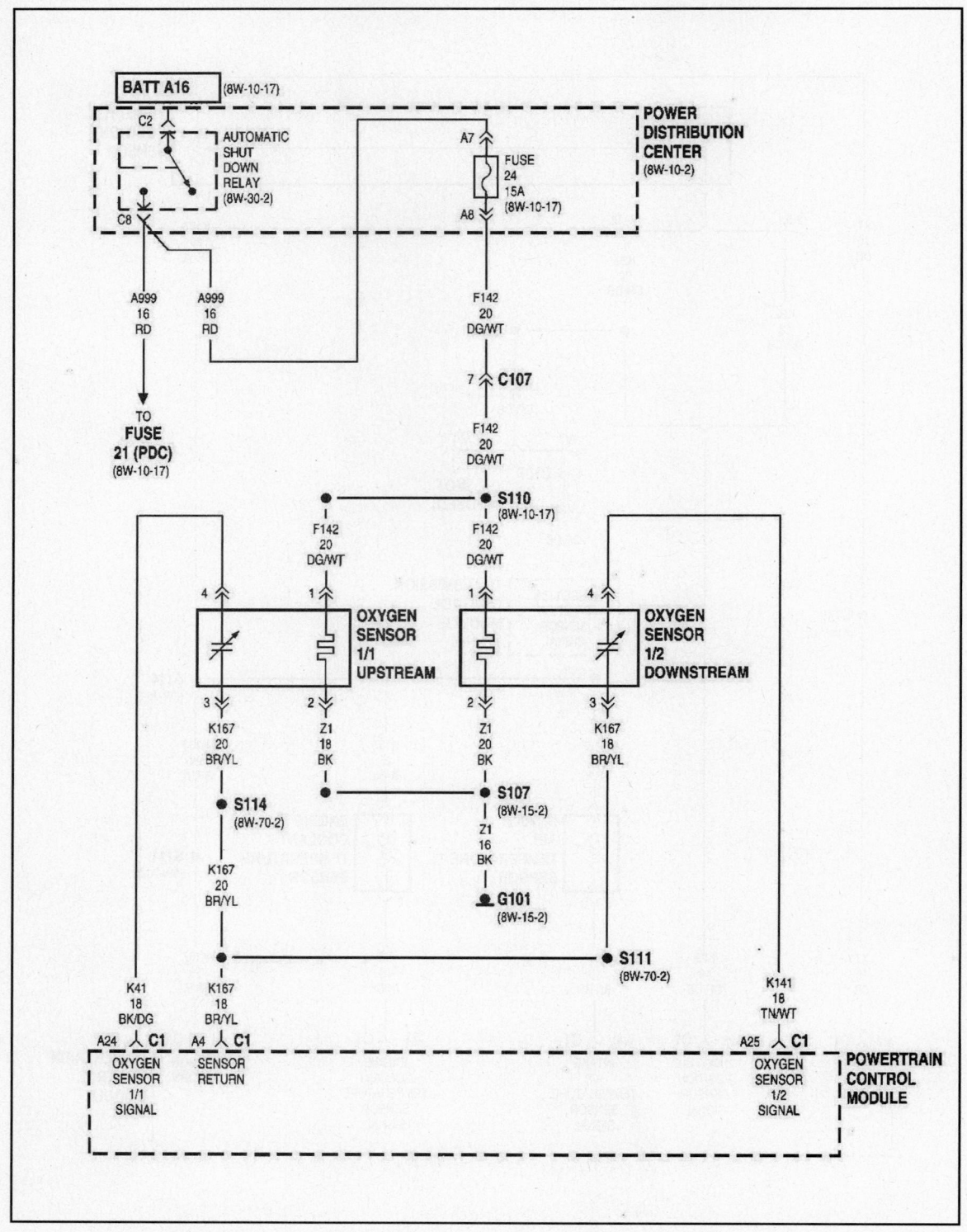

FUEL & IGNITION SYSTEM DIAGRAMS (7 OF 14)

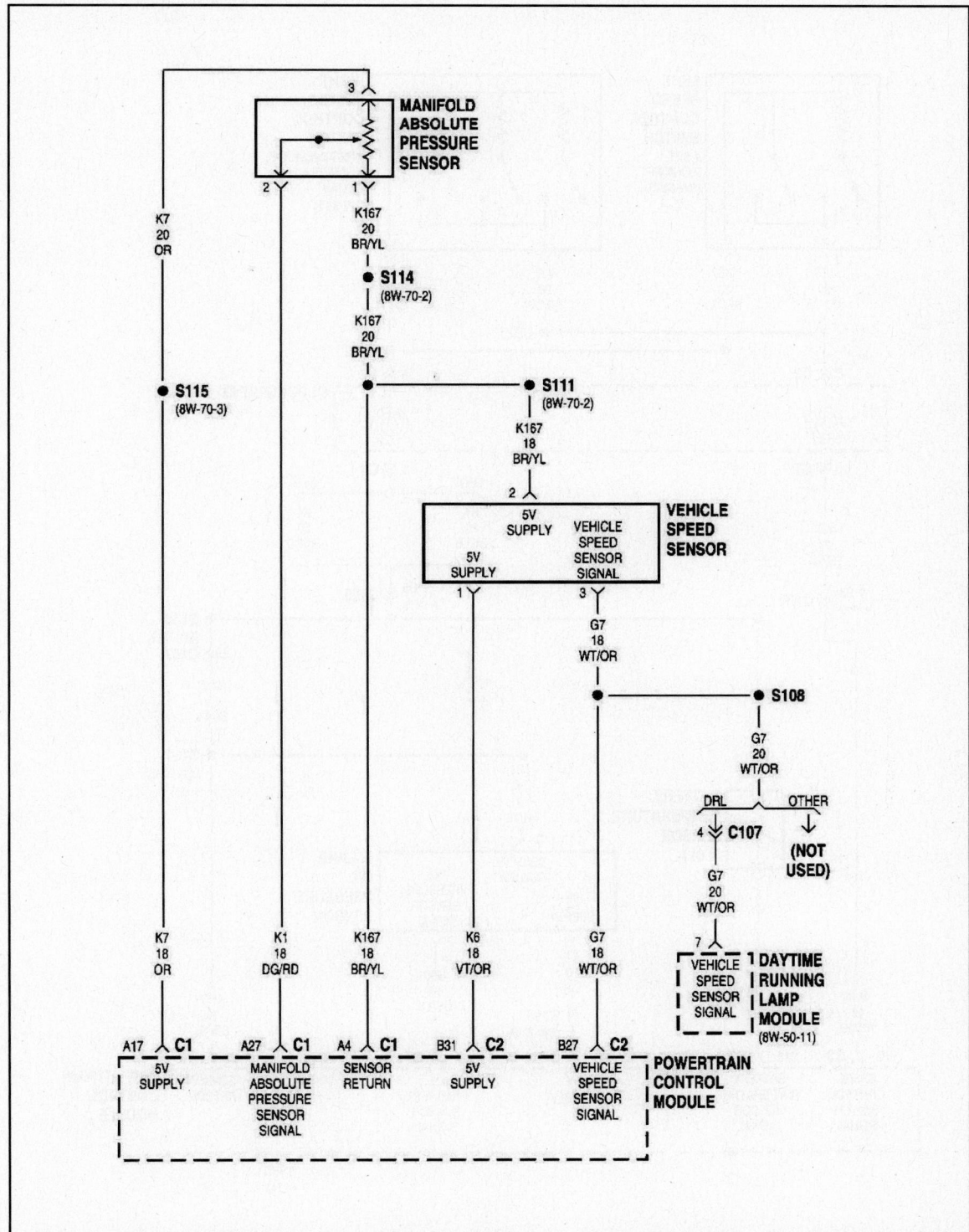

FUEL & IGNITION SYSTEM DIAGRAMS (8 OF 14)

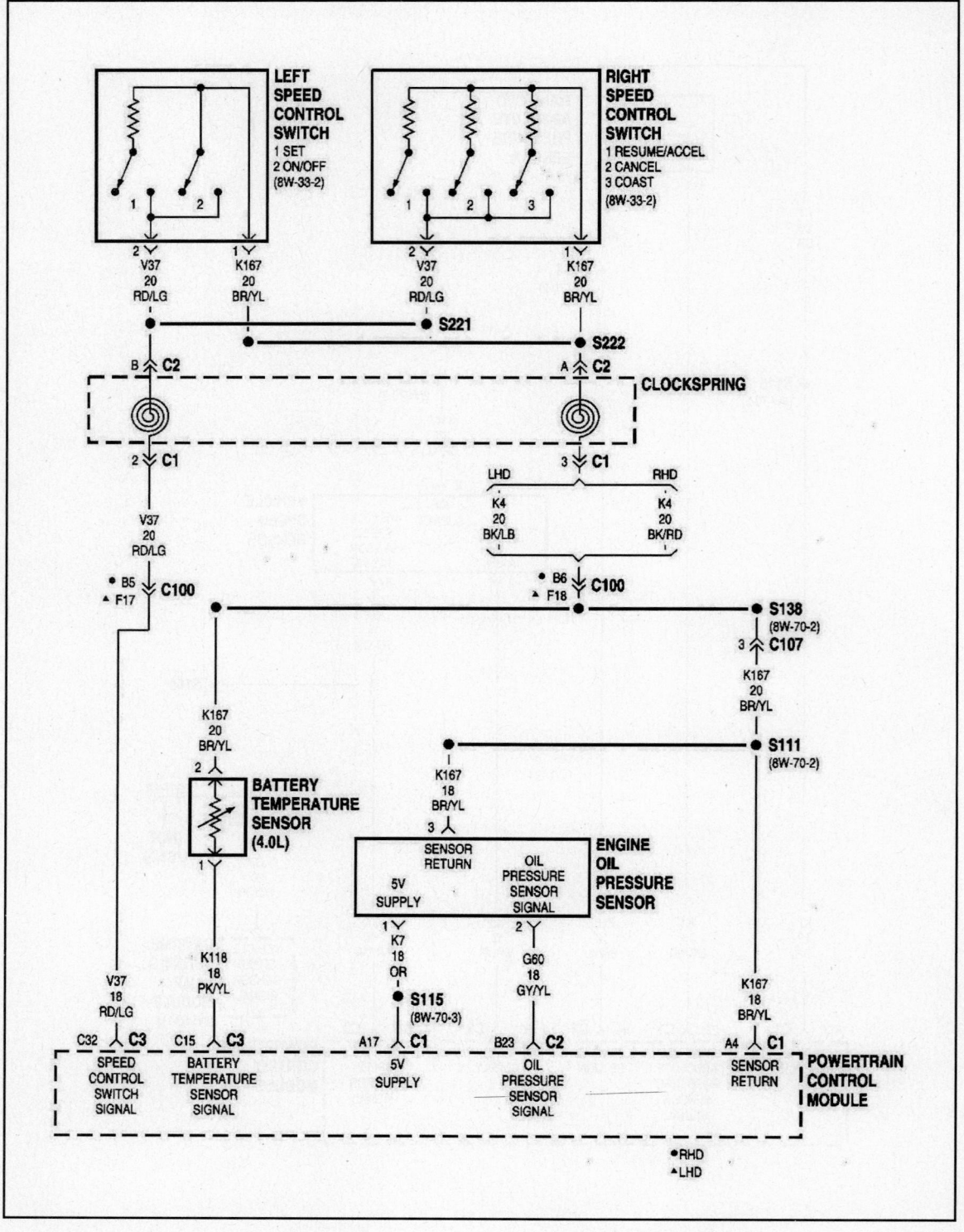

FUEL & IGNITION SYSTEM DIAGRAMS (9 OF 14)

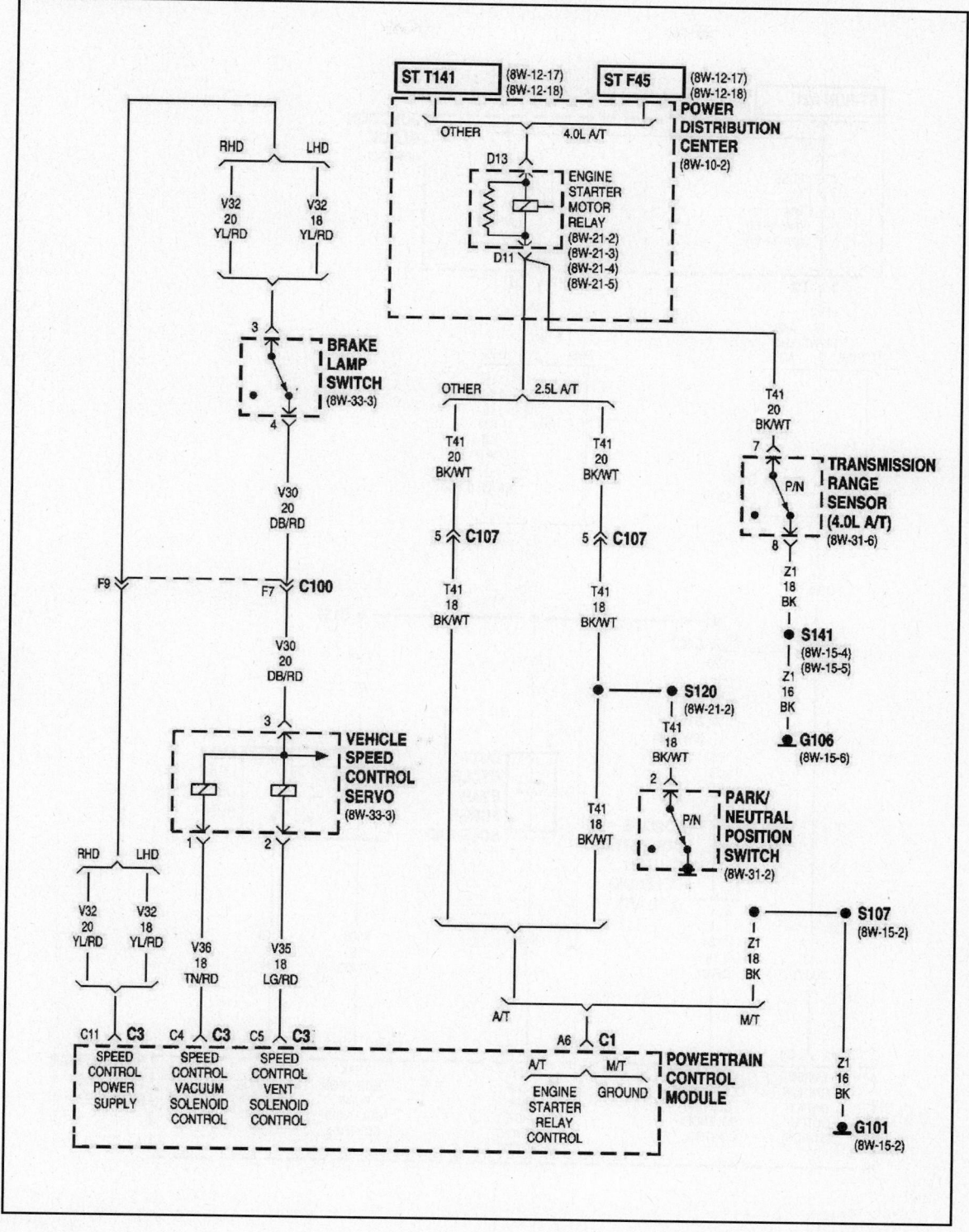

FUEL & IGNITION SYSTEM DIAGRAMS (10 OF 14)

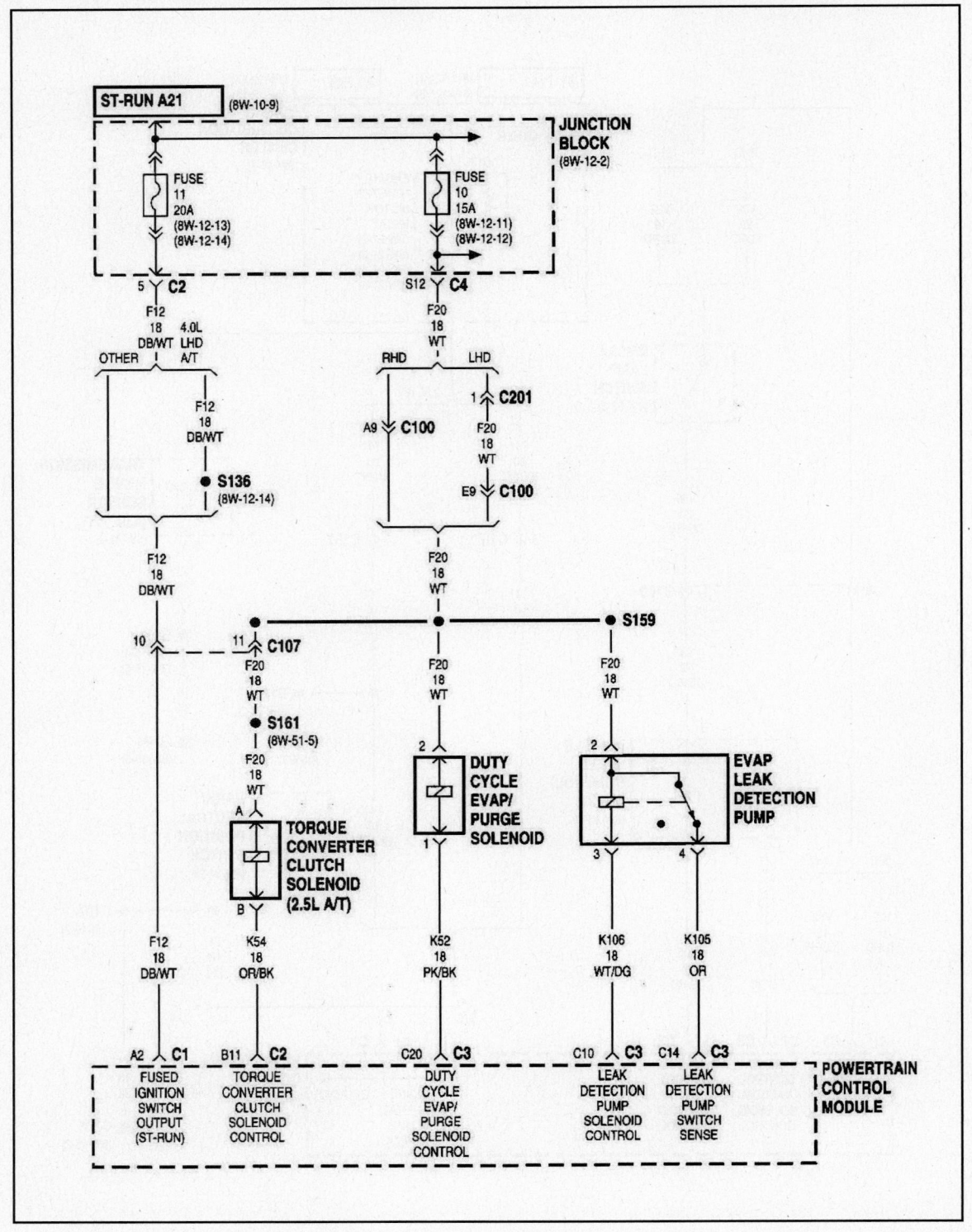

FUEL & IGNITION SYSTEM DIAGRAMS (11 OF 14)

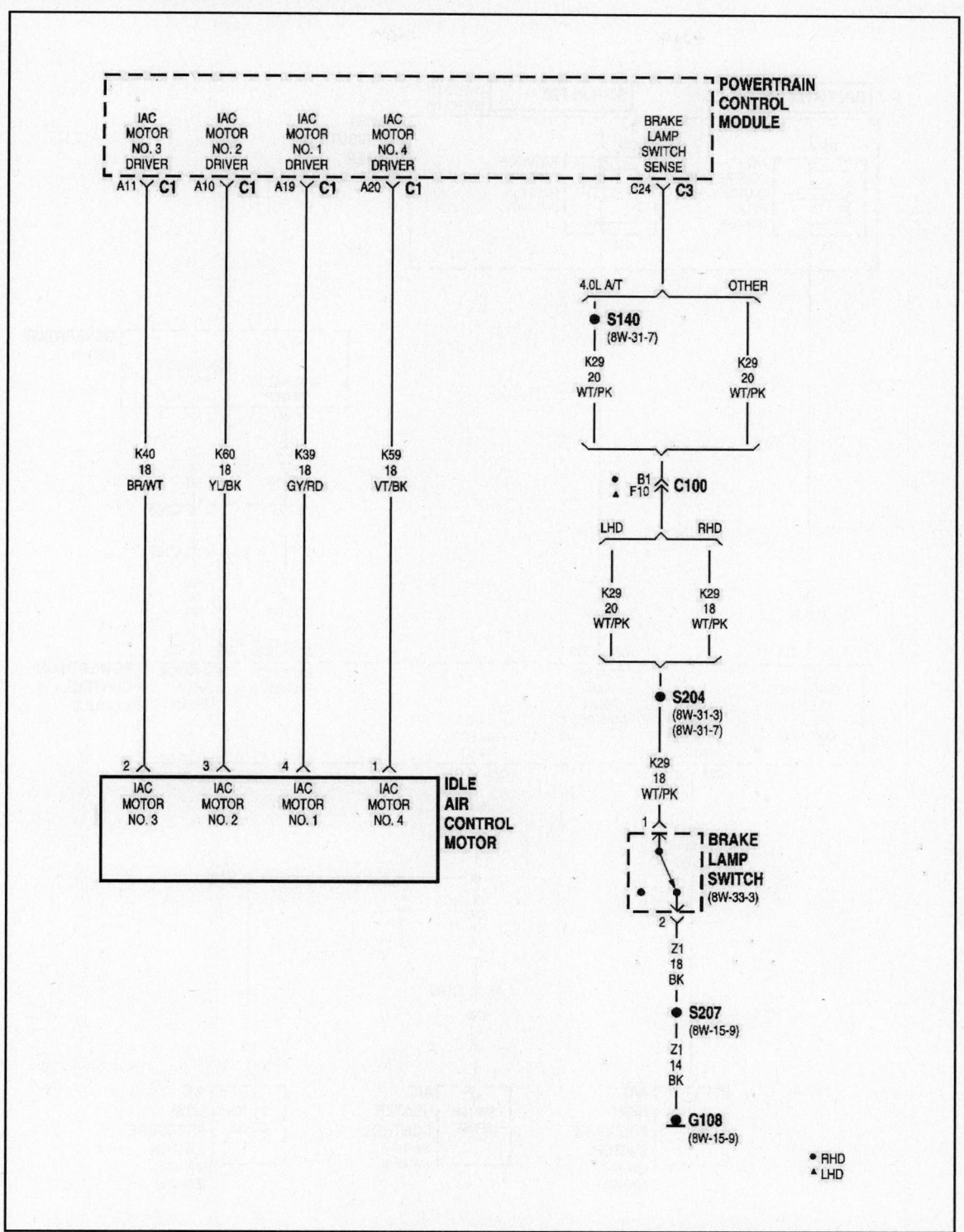

FUEL & IGNITION SYSTEM DIAGRAMS (12 OF 14)

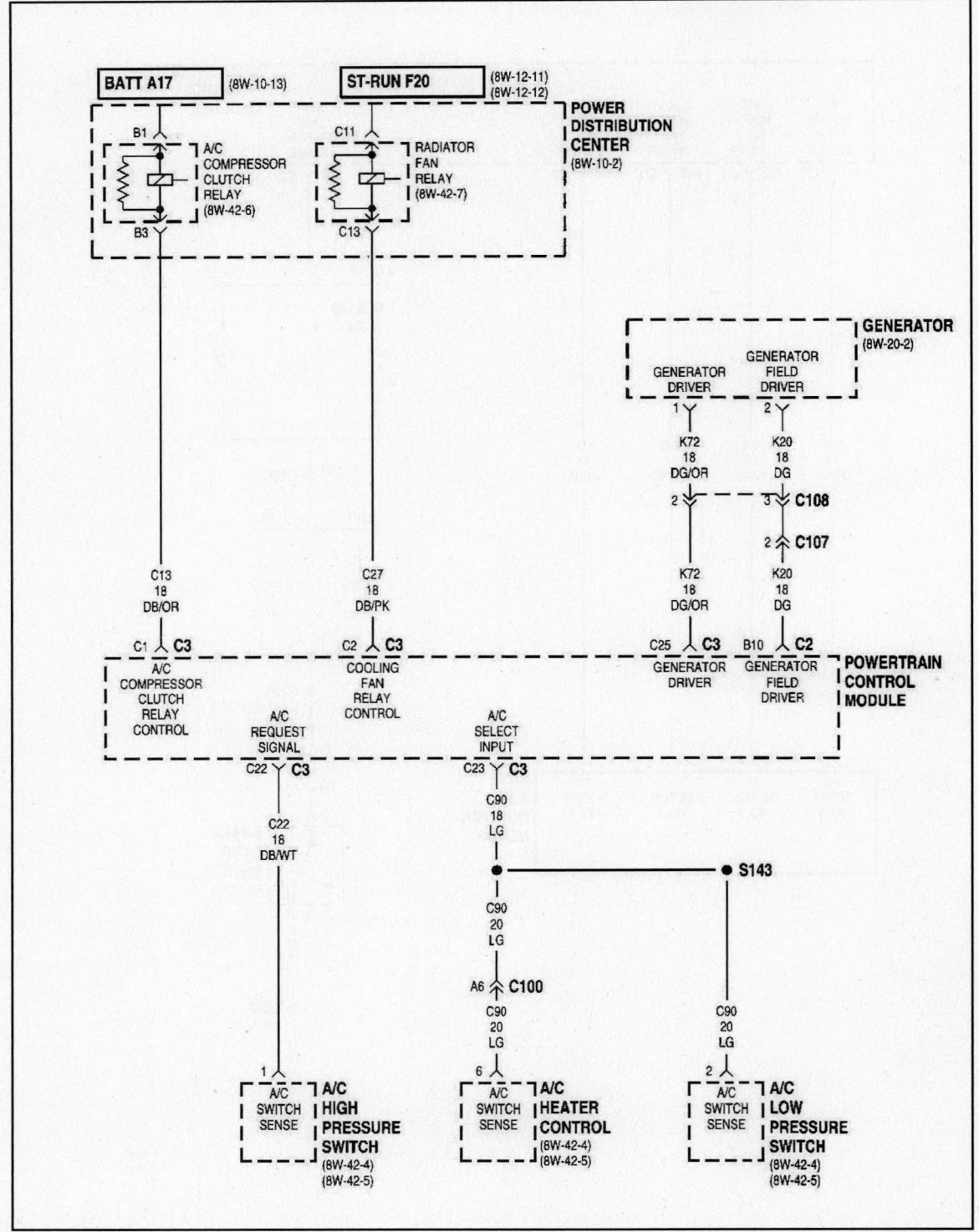

FUEL & IGNITION SYSTEM DIAGRAMS (13 OF 14)

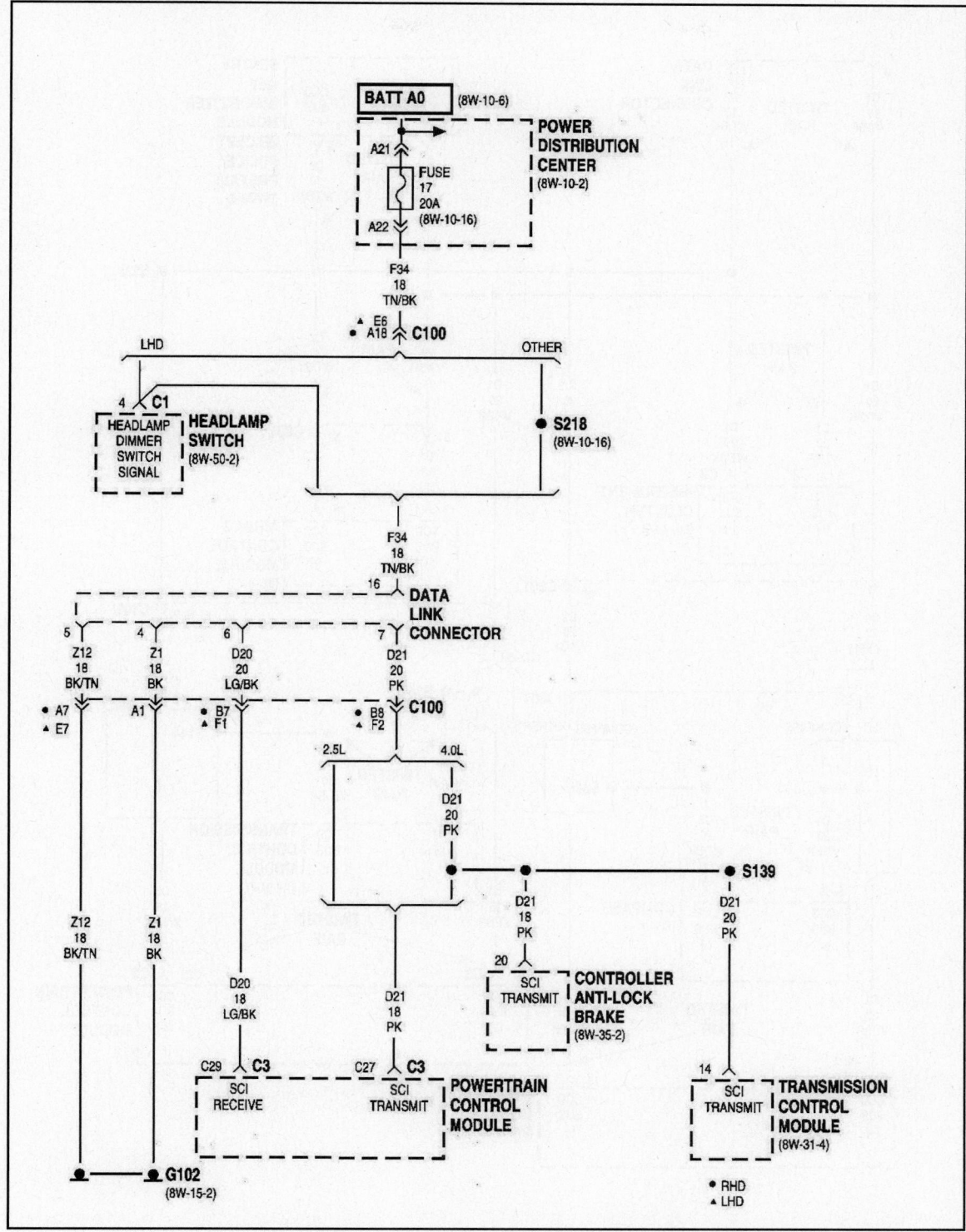

FUEL & IGNITION SYSTEM DIAGRAMS (14 OF 14)

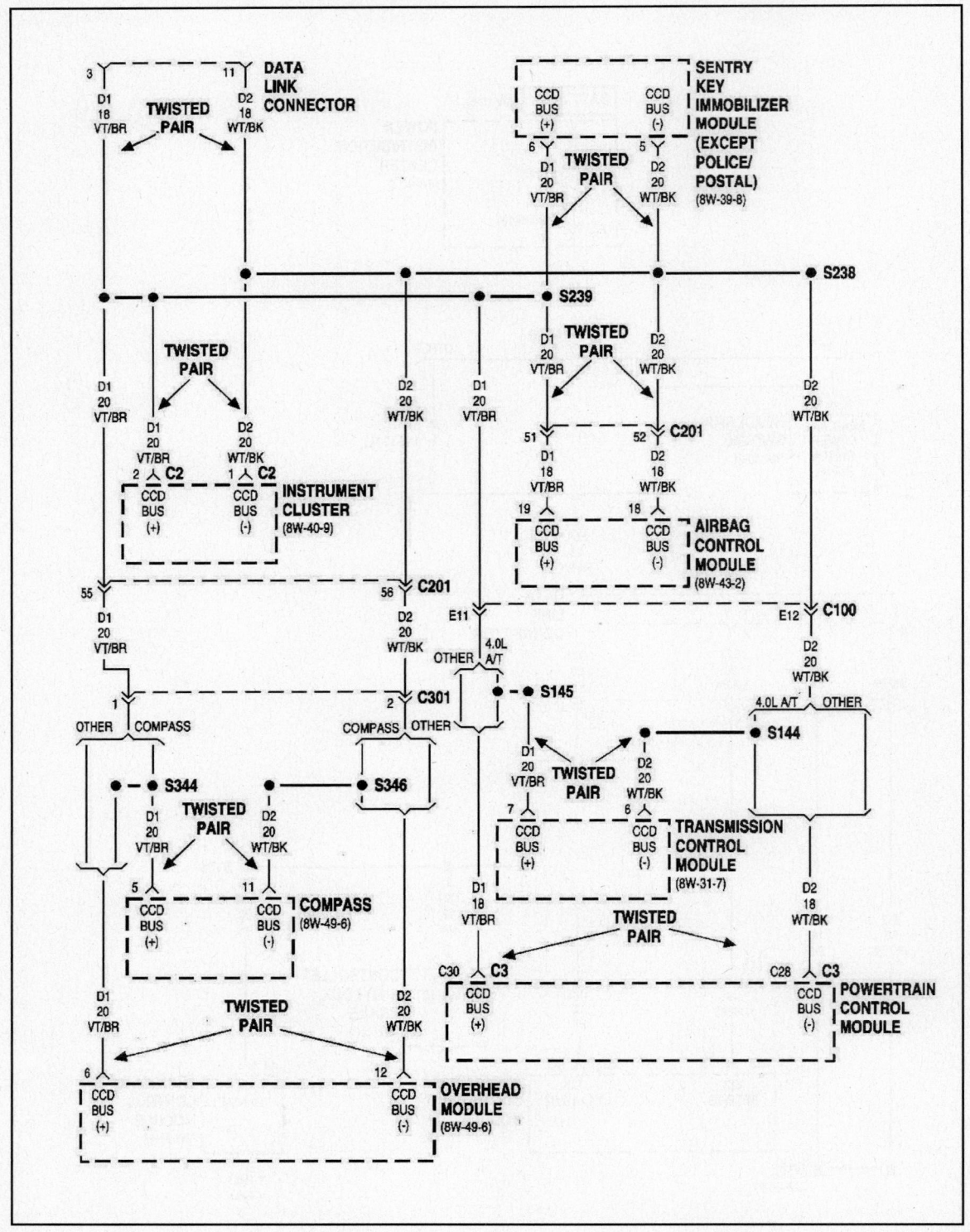

2002 ½ DURANGO - POWERTRAIN

Next Generation Controller (NGC)

GENERAL OVERVIEW (NGC)

The Next Generation Controller (NGC) is an advanced Powertrain Control Module (PCM) that will change some of the ways these vehicles are diagnosed. The NGC design controllers replace both the SBEC and JTEC style controllers. The NGC also includes the EATX Transmission Control Module (TCM) functions. This new style PCM is faster and capable of more accurately controlling vehicle emissions and drivability.

NGC Construction

The NGC uses a 32-bit, 32-MHz processor for the engine control functions, and a 16-bit, 16-MHz processor for the transmission control functions. Some of the advantages of combining the PCM and TCM include reduction in: weight, hardware cost, and wiring.

Although the PCM and TCM are contained inside the NGC, they are still separate processors and communicate on the PCI bus.

This module is easy to identify because of the four (4) PCM connectors. Each connector has 38 pins.

Vehicle Applications

The NGC is being used on the following vehicles:

- 2002 Concorde - all
- 2002 Intrepid - all
- 2002 300M - some engines
- 2002 1/2 Durango - 4.7L

All Chrysler vehicles will eventually move to this new style controller.

DIAGNOSTIC CHANGES

There are many significant changes to on-board diagnostics and diagnostic procedures.

Test Equipment

A new Break-Out Box (BOB) and terminal tool are available for the NGC 38-Pin connectors. The Miller tool description for the BOB is Essential Special Tool #8815. The Miller tool description for the terminal removal tool is Essential Special Tool # 8638.

PCM Circuit Construction and Protection

The NGC uses many positive-control drivers to control output devices. This allows for improved diagnostics, but can be confusing. Look carefully at wiring diagrams to determine output device polarity before diagnosing circuits.

Most of the driver and sensor circuits are protected against shorts to power or ground. In previous controllers, a short to ground would turn that PCM circuit off for the rest of the key cycle to protect the controller. In the NGC systems the 5-volt output is now self-recovering from intermittent faults. This means that the controller does not need a key cycle to turn the 5-volt output back on.

The following is a list of short circuit faults (and polarity) that may damage the controller:

- Fuel Injector Drivers (+)
- IAC Circuits (+)
- Purge Solenoid Sense (+)
- Sensor return (+)
- Other Grounds (+)
- Ignition Outputs (-)

CRANKSHAFT POSITION SENSOR

The Crankshaft Position Sensor (CKP) on this application is a Hall-effect device that senses slots in a trigger wheel. All NGC applications use the same trigger wheel mounted to the torque converter flex plate. The trigger wheel has 32 slots, 10 degrees apart, with 2 filled in and 2 connected 180° apart.

One advantage to this type of design is that the CKP scope pattern will look identical, no matter what engine is being tested. This feature can greatly simplify diagnostics. Using this arrangement, the NGC to detect crankshaft position within a tenth of a degree of crankshaft rotation.

NGC equipped vehicles will start with either the CKP or CMP signal missing.

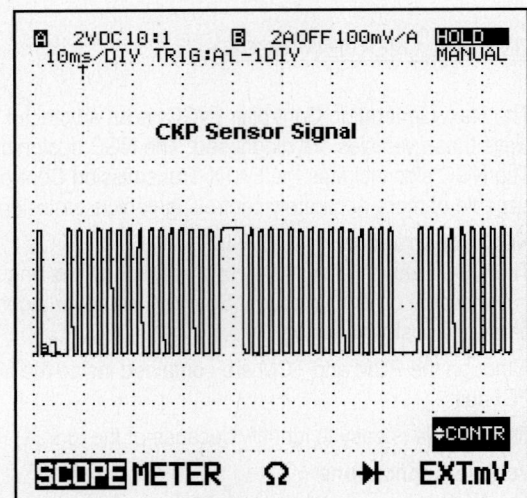

HEATED OXYGEN SENSORS

The heated oxygen sensor output, based on a comparison of oxygen content in outside air and in the exhaust, still varies from near 0v to under 1v. The difference in this system is not with the sensors, but with the controller.

Each HO2S signal is fed a bias voltage of 2.5v from the PCM. Therefore, the operating range of the sensors is from 2.5v to 3.5v. The presence of a bias voltage allows the PCM to differentiate between low sensor output and circuit faults, and prevents the sensor output from going below 0.0v if it becomes damaged.

Diagnostic Tip
The heated oxygen sensors on this application are the same as those on previous generation vehicles, but the bias voltage does change the way the circuits can be diagnosed.

If a Lab Scope is used, the volts per division setting should be changed to better view the waveform. At the usual setting of 100 mv or 200 mv per division, the waveform would have good resolution, but it would be off the top of the screen (15 to 25 divisions off the top of the screen!). At 500 mv per division, the entire waveform will be visible, but the actual sensor activity will be displayed very small.

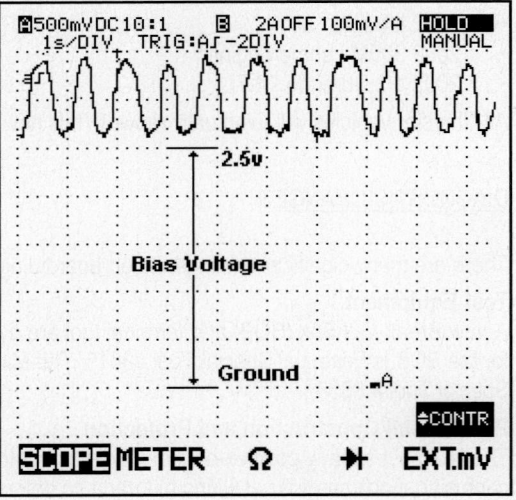

After using the 500 mv per division setting to establish the absolute voltage ranges of the sensor signal, switch to 100 mv per division and set the Lab Scope to AC coupling. The Lab Scope will then display only the change in voltage, so full resolution of the waveform can be seen. In effect, this setting will separate the sensor signal from the bias voltage. Absolute voltage cannot be viewed in AC coupling, but the peak-to-peak voltage will still accurately reflect the range of the sensor.

NATURAL VACUUM LEAK DETECTION (EVAP)

The EVAP system on this application no longer uses a Leak Detection Pump (LDP) for performing EVAP system leak checks as part of the EVAP Monitor. Because of the new engine-off diagnostic circuitry in the NGC type PCM, the system can be checked for leaks after the vehicle is driven.

The Natural Vacuum Leak Detection (NVLD) assembly has replaced the LDP in these vehicles. The System is called "Natural" because it uses vacuum that occurs naturally when the Fuel tank and EVAP system cool off. The system comes in two (2) different configurations. The in-line NVLD is found on the Durango, and the canister-mounted version is found on the car applications.

NVLD Construction

The NVLD assembly is smaller than the LDP, and contains a solenoid, poppet and seal, a diaphragm, and a switch. The 3-wire connector contains power, ground, and switch sense circuits.

The system still uses a conventional purge solenoid and other related components.

NVLD EVAP Monitor – Small Leak Detection

After the vehicle has been run to normal operating temperature and shut off, the PCM monitors the vacuum in the system through the switch in the NVLD assembly. As the temperature in the tank and EVAP system drops, the air in the system contracts and a vacuum is created. The PCM is not concerned about the quantity of vacuum, only that there is enough vacuum to trip the switch.

A temperature change of as little as 3.5°F will cause a 1" H2O rise in vacuum, which is enough to pull the diaphragm up and close the switch. If the switch closes within the test time frame, the small leak test passes. If the NVLD switch does not close within the time frame, a small leak is assumed (at least .020" in diameter) and a pending DTC is set.

NVLD EVAP Monitor – Medium or Large Leak Detection

If the small leak detection portion of the EVAP Monitor fails, the medium and large leak detection test is run. This portion of the test is referred to as the vacuum decay method.

During a cold start after the engine off period in which the small leak test failed, the PCM closes the vent solenoid and energizes the purge solenoid. This allows engine vacuum to pull a vacuum on the EVAP system. The NVLD switch closes, signaling the PCM that the preset test vacuum has been reached. The purge solenoid is de-energized, and the PCM monitors the rate of vacuum decay in the system.

The actual rate of vacuum decay is not directly monitored as it is in some other manufacturers vehicles. The PCM simply waits until the NVLD switch opens, and calculates the size of any vacuum leak by comparing the elapsed time to internal tables.

There are 3 possible outcomes of this test:

- The switch does not re-open during the test. There are no medium or large leaks. The small leak DTC remains in memory until matured or cleared by further testing.
- The switch reopens, and a medium or large leak DTC is set based on the amount of time it took to re-open.
- The switch never closes. The PCM will increase the purge duty cycle to try to force the switch closed. If the switch closes, a large leak DTC is set (2-trip code). If the switch remains open, a general EVAP leak DTC will set. This could indicate a switch or circuit failure. However, it is more likely there is loose gas cap or very large leak.

SERVICE BAY LEAK TESTING (EVAP)

As the EVAP monitors and hardware become more precise, finding the cause of the EVAP DTCs can be more difficult. If a thorough visual inspection does not locate the fault, use a smoke machine through the test port or a gas cap adapter to check for any leaks.

The recommended tool for this application is the Miller Essential Special Tool #8404. The tool includes special dyed smoke to help pinpoint even very small leaks.

When introducing smoke into the system, it is important to seal the system. Use rounded-jaw pliers to pinch off the vent hose. Never introduce more than 1 psi pressure into the system.

The NVLD is only serviced as an assembly. If any part of the unit fails, the entire assembly must be replaced. Be very careful to get the correct part number.

OTHER CONTROLLER CHANGES (NGC)

Speed Control Cutout Monitor
Another feature of the NGC type PCM is the ability to track Speed Control S/C) events to aid in the diagnosis of speed control performance issues. The PCM can record and display the last eight (8) speed control cutouts and the reason for their termination.

Cutout events include events like brake application and cancel switch input, so normal and abnormal cancellations are both recorded. At each cutout event, the following parameters are recorded:

- Cutout Reason (i.e., cancel switch input)
- Vacuum, Vent, and Dump Solenoid Desired Positions
- PCM Odometer
- Speed Control Set Speed
- Vehicle Speed (VSS)
- Speed Control Switch Voltage
- Engine RPM

The S/C inputs are also timed to differentiate between driver decisions and intermittent circuit faults.

PCM Freeze Frames
In the NGC type controllers, up to 5 Freeze Frames can be stored, with up to 27 PID values (on an OEM Scan Tool). In the event of multiple events of one failure (same DTC) Freeze Frames can be very helpful in defining the conditions under which the fault occurred. In the event of multiple unique codes, start by diagnosing the oldest fault code first.

Diagnostic Trouble Codes (DTCs)
P0xxx Generic trouble codes will remain the same, but many more P2xxx codes will be phased in with this new controller. Many of the current P1xxx trouble codes will no longer be used.

2002 DURANGO - POWERTRAIN

Vehicle Identification

VECI DECAL

The Vehicle Emission Control Information (VECI) decal is located in the engine compartment. The information in this decal relates specifically to this vehicle and engine application. The specifications on this decal are critical to emissions system service.

2HO2S (2)/2WUOC/SFI/TWC

These designators indicate this vehicle is equipped with 4 Heated Oxygen Sensors (HO2S), 2 Warm-up Oxidation Catalysts, Sequential Fuel Injection (SFI), and a Three-Way Catalyst (TWC).

OBD II Certified

This designator indicates that this vehicle has been certified as OBD II compliant.

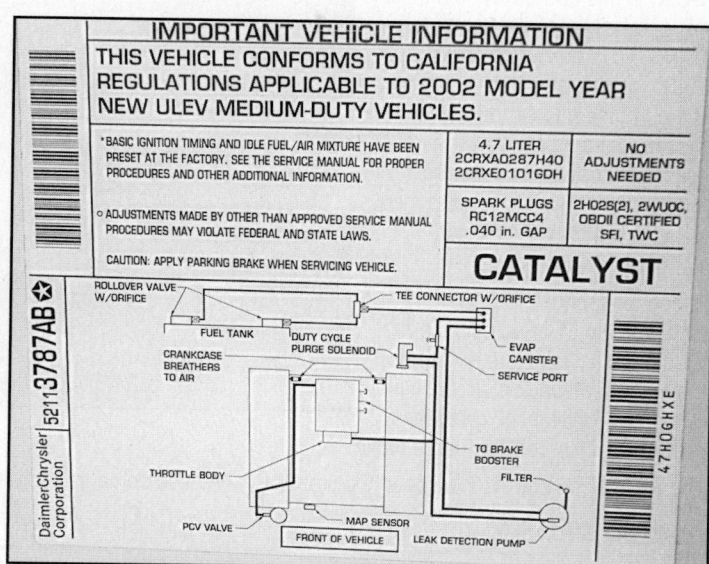

ULEV

If this designator is used, the vehicle conforms to U.S. EPA and State of California Ultra-Low Emissions Vehicle standards.

Underhood View Graphic

Electronic Engine Control System

OPERATING STRATEGIES

The Powertrain Control Module (PCM) in this application is a triple-microprocessor digital computer. The PCM, formerly known as the SBEC, provides optimum control of the engine and various other systems. The PCM uses several operating strategies to maintain good overall drivability and to meet the EPA mandated emission standards.

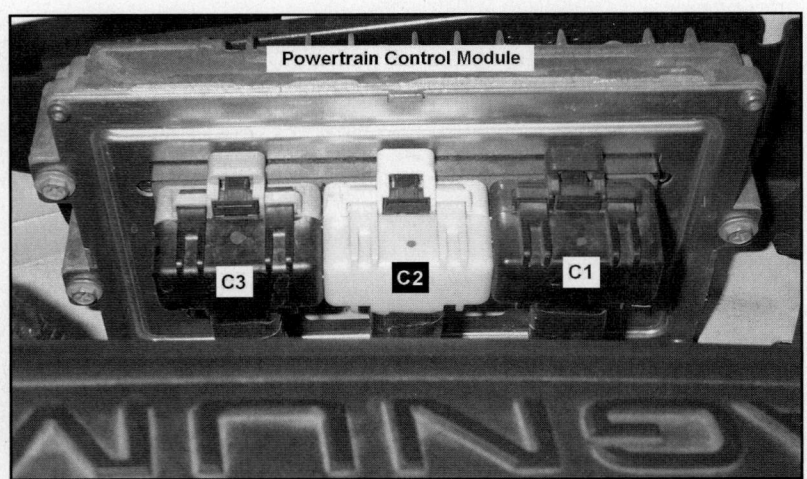

Based on various inputs it receives, the PCM adjusts the fuel injector pulsewidth, the idle speed, the amount of ignition advance or retard, the ignition coil dwell, and operation of the canister purge valve.

The PCM also controls the Speed Control (S/C), A/C and some transmission functions.

Operating Strategies

A/C Compressor Clutch Cycling Control	Closed Loop Fuel Control
Electric Fuel Pump Control	Limited Transmission Control
Fuel Metering of Sequential Fuel Injectors	Speed Control (S/C)
Idle Speed Control	Ignition Timing Control for A/F Change
Vapor Canister Purge Control	

Operational Control of Components

A/C Compressor Clutch	Ignition Coils (Timing and Dwell)
Fuel Delivery (Injector Pulsewidth)	Fuel Pump Operation
Purge Solenoid	EVAP Leak Detection Pump
Integrated Speed Control	Idle Air Control Motor
Malfunction Indicator Lamp	Cooling Fan

Items Provided (or Stored) By the Electronic Engine Control System

Shared CCD Data Items	Adaptive Values (Idle, Fuel Trim etc)
DTC Data for Non-Emission Faults	DTC Data for Emission Related Faults

Distinct Operating Modes

Key On	Cruise
Cranking	Acceleration (Enrichment)
Warm-up	Deceleration (Enleanment)
Idle	Wide Open Throttle (WOT)

Computer Controlled Charging System

INTRODUCTION

This application uses either a 136 or 160 amp Denso generator. These generators do not contain voltage-regulating devices. The regulation of charging system voltage is accomplished through the Electronic Voltage Regulator (EVR) inside the PCM.

Charging System Operation

The Charging system is turned on with the ignition switch. With the ignition in the ON position, battery voltage is applied to the generator rotor, through one of the two field terminals, to produce a magnetic field.

The amount of DC current produced by the generator is controlled by the EVR circuitry in the PCM and engine speed. This circuitry is connected in series with the second rotor field terminal and ground.

Battery Temperature Sensor data (calculated) and system voltage are inputs used by the PCM in order to vary the charging rate. The PCM accomplishes this by cycling the ground path to control the strength of the rotor magnetic field. Temperature dependant charging strategies like this one offer improved charging efficiency, slightly improved mileage and extended battery life.

Charging System Diagnosis

In addition to conventional voltage output and load testing, this charging system can be effectively diagnosed using a Scan Tool (PIDs, DTCs and bi-directional controls) and a Lab Scope (or DVOM).

Many charging system faults will set codes. It is important to check for codes before beginning a conventional charging system diagnosis. The charging system is monitored for too high or too low voltage conditions. The PCM can also monitor it's own ability to switch the field control circuit to ground to regulate charging rates.

Charging System Schematic

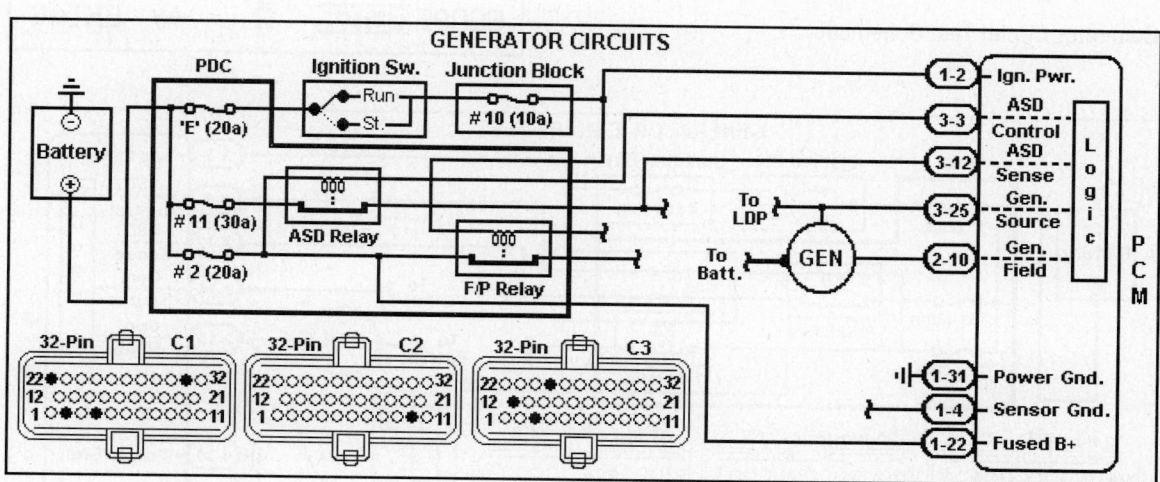

DVOM TEST (GENERATOR)

The DVOM can be used to test the Computer Controlled Generator. The PCM command in response to changing vehicle load (duty cycle) can be observed to verify that the PCM is in control of the charging system. Place the shift selector in Park and block the drive wheels.

DVOM Connections
Connect the positive probe to the field control circuit (DK GN wire) at PCM Pin 2-10 and the negative probe to the battery negative post.

DVOM Examples
In these examples, the traces show the generator field command from the EVR (in the PCM) to the generator at Hot Idle.

DVOM Example (1)
In example (1), the meter shows an average DC voltage of 8.95v and 39.2% duty cycle with minimum accessory electrical loads. The low duty cycle command means that the generator is commanded "on" for less time than it is commanded "off". This can also be seen in the waveform in example (1).

The frequency (Hz) of the signal changes constantly, and is not used during diagnosis.

DVOM Example (2)
In example (2), the meter shows an average DC voltage of 2.33v and 86.3% duty cycle with high accessory electrical loads. The high duty cycle command means that the generator is commanded on for more time than it is commanded off.

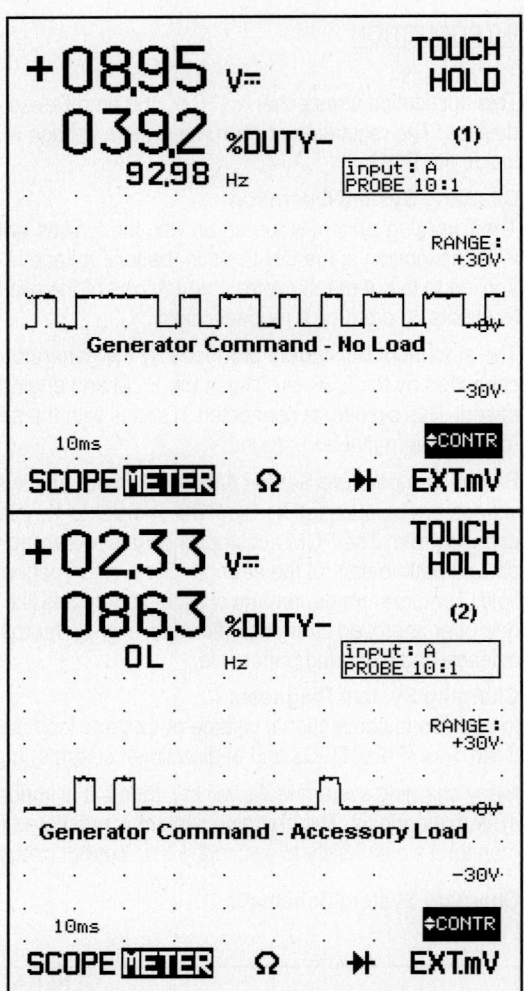

Generator Circuit Test Schematic

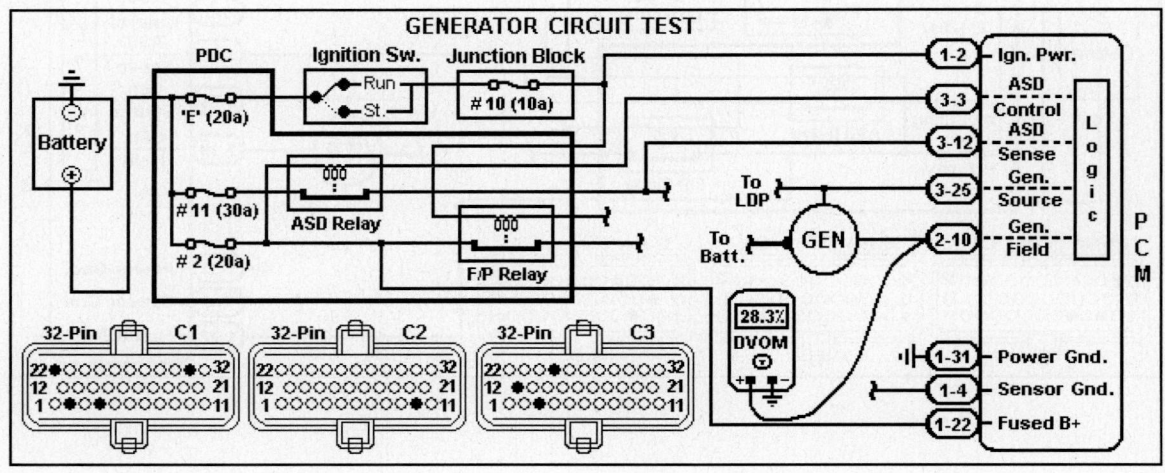

LAB SCOPE TEST (GENERATOR)

The Lab Scope can be used to test the Computer Controlled Generator to determine if the generator is not capable of charging the battery, or if it is not being commanded to by the PCM. Place the shift selector in Park and block the drive wheels for safety.

Scope Connections

Connect the Channel 'A' positive probe to the field control circuit (DK GR wire) at PCM Pin 2-10 and the negative probe to the battery negative post.

Clamp the low-amps probe around the large generator output wire.

Lab Scope Example

In this example, the trace shows the generator command and resulting generator current output. A moderate electrical load was placed on the system. The PCM commanded the generator "on" with approximately a 50% duty cycle signal.

When the EVR circuit is switched to ground, the generator produces current. Therefore, the longer the trace is low, the more current is produced. This example shows a direct relationship between the EVR command and the change in generator current output.

In the current trace, each vertical division is equal to 1 amp, so the generator output increased 3.5 amps during the 4 ms that the EVR circuit was grounded.

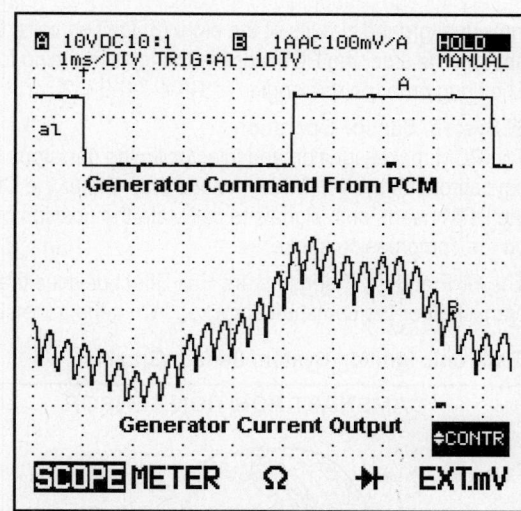

Note that Channel 'B' is set to AC coupling. At this setting, absolute amperage of the system cannot be observed. This setting was used because it allowed greater resolution of the *change* in current output. If DC amperage was used, and the generator output was 30 amps, the Lab Scope would be set to 10 ADC per division. At this resolution, the 3 amps change in output as a result of the PCM command would not be visible.

Charging System Test Schematic

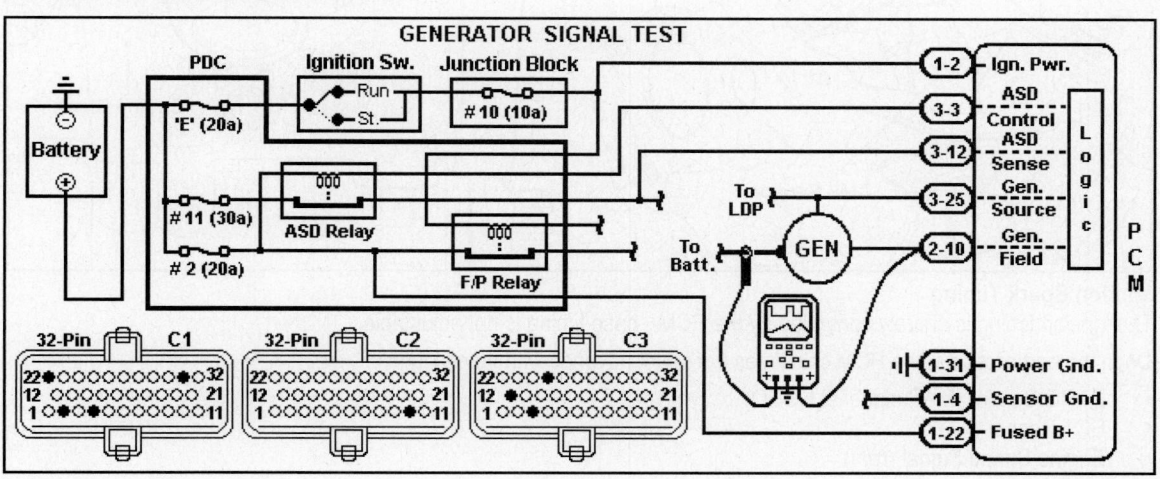

Integrated Electronic Ignition System

INTRODUCTION

The PCM controls the operation of the Coil-On-Plug Integrated Electronic Ignition (EI) system on this vehicle application. The system consists of a CKP sensor, CMP sensor, PCM, and eight (8) coils. There are no secondary ignition wires on this vehicle, as each coil is mounted on top of a spark plug.

Battery voltage is supplied to all coils via the ASD relay in the Power Distribution Center (PDC). The PCM controls the individual ground circuits of the eight (8) ignition coils. By switching the ground path for each coil "on" and "off" at the appropriate time, the PCM adjusts the ignition timing for each cylinder to meet changing engine operating conditions. The firing order on this engine is 1-8-4-3-6-5-7-2.

EI System Sensor Operation

The PCM makes ignition and injector-timing decisions based on inputs from the CMP and CKP sensors. This is an asynchronous system, which means that neither the CMP nor the CKP sensor signals occur at TDC. For ignition timing the PCM needs both signals to calculate the position of the engine, and to calculate which cylinder is approaching TDC on the compression stroke.

The CKP and CMP sensors are Hall-Effect devices that toggle internal transistors each time a window is sensed on the crankshaft or the camshaft sprocket. When the transistor toggles, a 5v pulse is sent to the PCM.

Electronic Ignition System Sensor Graphic

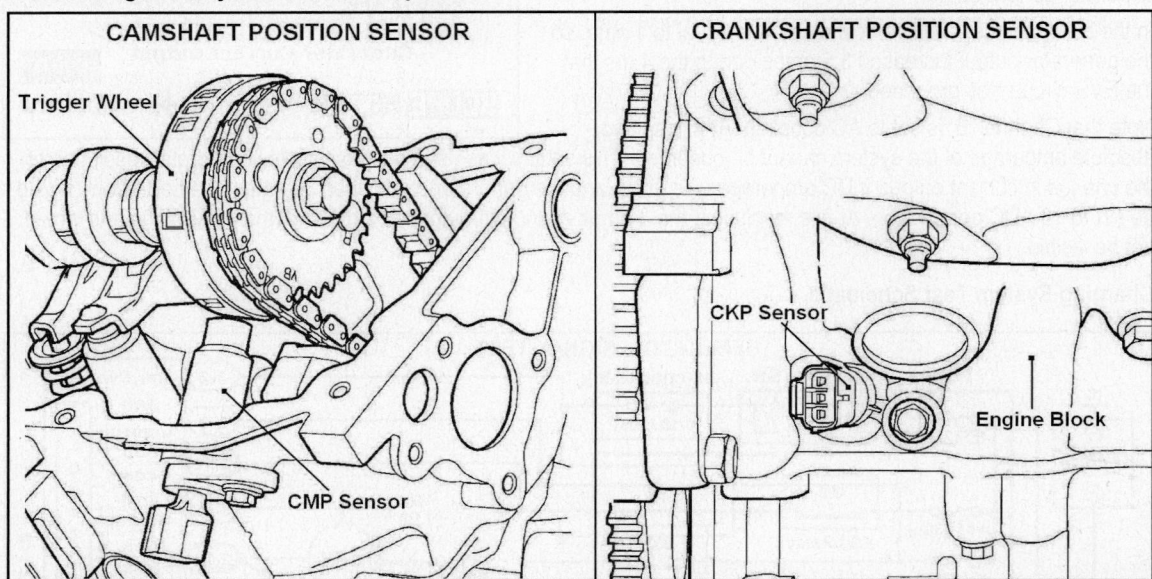

Ignition Spark Timing

The ignition timing is entirely controlled by the PCM - base timing is not adjustable.

Once the engine starts, the PCM calculates the spark advance with these signals:

- BARO Sensor (MAP signal at KOEO)
- ECT Sensor
- Engine Speed Signal (rpm)
- MAP Sensor
- Throttle Position

IGNITION COIL OPERATION

This ignition system is referred to as a Coil On Plug (COP) system. The ignition system consists of eight (8) individual coils. Each coil is mounted to the intake manifold, and sits on top of one spark plug, eliminating the need for ignition wires.

Ignition Coil #1

All of the ignition module functions are integrated into the PCM. A separate driver in the PCM controls each coil.

The firing order on this engine is 1-8-4-3-6-5-7-2.

COP Circuits

All eight (8) coils are powered by the ASD relay. However, drivers in the PCM control each coil independently. Refer to the Ignition System Schematic below. Note that only the first four coils in the firing order are shown in the Schematic.

Trouble Code Help

The PCM driver circuits are shown with a dotted line in the schematic. The dotted line in Chilton Schematics indicates that the circuit is monitored by the PCM in order to set codes. The PCM monitors the peak current achieved during the dwell period. If a preset current level is not achieved for 3 seconds during cranking, or 6 seconds with engine running, a code is set (DTC P0351 for Coil 1 through P0358 for Coil 8).

If a DTC P035x is set, compare the current between the suspect circuit and another coil. Even 1 or 2 Ohms of extra resistance in this circuit can set a code.

Ignition System Schematic

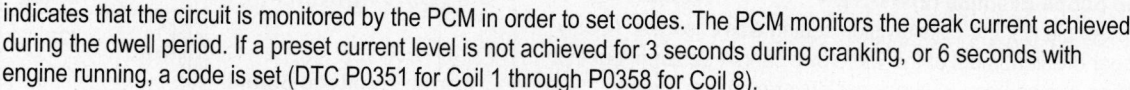

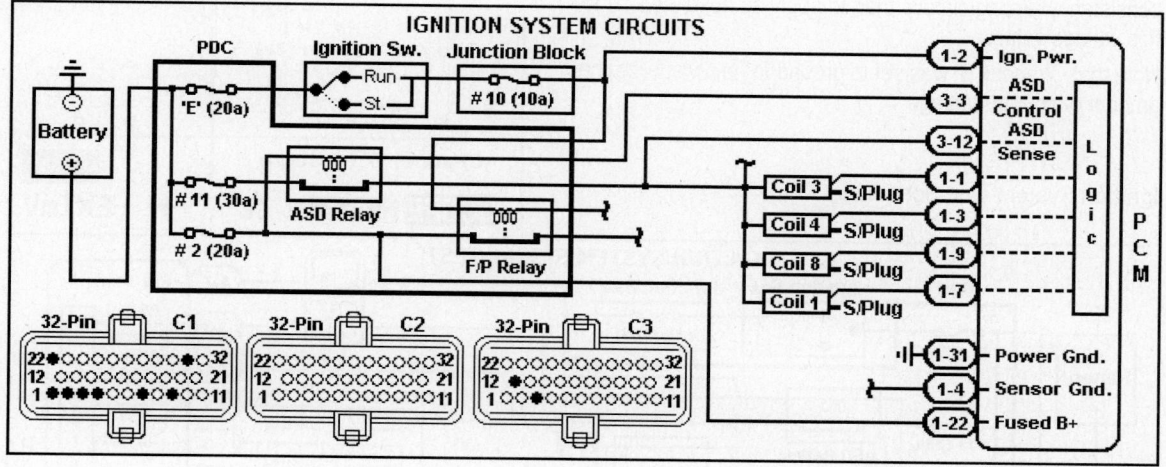

LAB SCOPE TEST (COIL PRIMARY)

Place the shift selector in Park and block the drive wheels for safety.

Lab Scope Connections
Connect the Channel 'A' positive probe to one of the eight (8) coil control circuits at the PCM or coil, and the negative probe to the battery negative post.

Scope Settings
To make the waveforms as clear as possible, set the scope settings to match the examples.

Lab Scope Example (1)
In example (1), the trace shows the coil driver (primary) circuit at idle. Use these Lab Scope settings to view the system voltage, spark line, and coil oscillations.

To view the peak primary voltage, the settings need to be changed to 50 volts/division. On this application, the peak voltage was 300v.

Lab Scope Example (2)
In example (2), the volts/division setting was changed. At only 2v per division, most of the waveform cannot be seen, but this setting can be used to check coil driver performance and ground voltage drop in the circuit under load.

The normal rise of the waveform voltage above ground is referred to as 'counter-voltage.' It is more important to look for consistency among circuits than to apply an absolute limit to this voltage value.

Note that Channel 'B' was set to ground to help visualize the amount of counter-voltage.

Ignition System Test Schematic

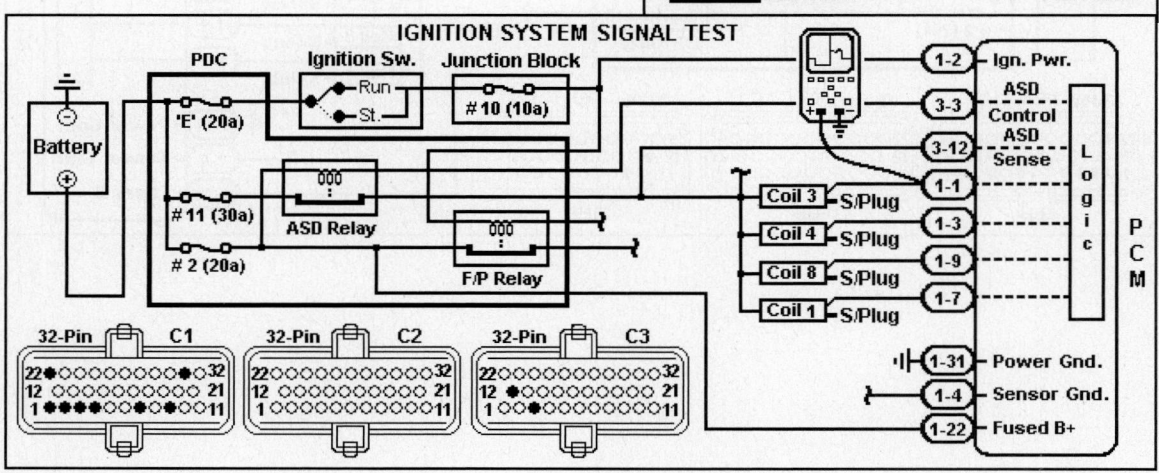

CRANKSHAFT POSITION SENSOR (HALL EFFECT)

The crankshaft position (CKP) sensor used with this EI system provides crankshaft position and engine speed data to the powertrain control module (PCM).

The CKP sensor is a Hall effect sensor mounted on the engine block, adjacent to a trigger wheel on the crankshaft. There are 16 slots on the trigger wheel. The CKP sensor circuit senses each slot and toggles a transistor. Each toggle or switch sends a 5v pulse to the PCM for use in calculating engine position and speed.

Using all 16 slots, the PCM can track engine speed and variations in crankshaft rotation speed for OBD II misfire diagnosis.

Every fourth slot on the crankshaft is wider than the other 3. Therefore, there are four (4) wider pulses sent to the PCM each crankshaft revolution to help the PCM calculate when a cylinder pair is approaching TDC. The PCM needs the CMP signal to determine exactly which cylinder is approaching TDC on the compression stroke.

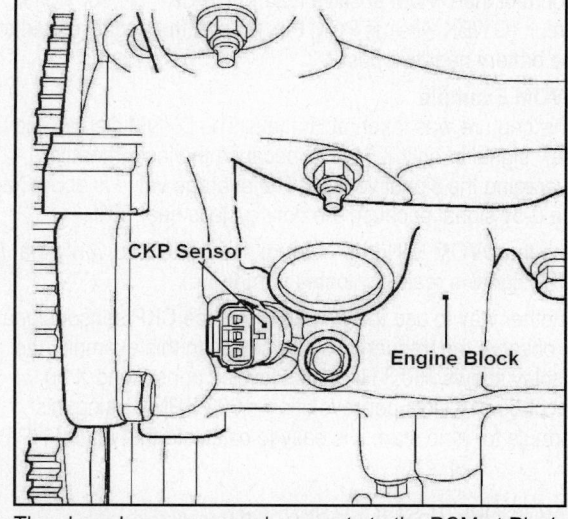

CRANKSHAFT POSITION SENSOR

CKP Sensor

Engine Block

CKP Sensor Circuits

The CKP Sensor is connected to the PCM by 3 wires. The CKP and CMP sensors share a 5v supply voltage from PCM Pin 1-17. The 0-5v CKP signal is sent to PCM Pin 1-8. The shared sensor ground connects to the PCM at Pin 1-4.

Diagnostic Help

The following sensors share the sensor ground: BTS, CMP, CKP, ECT, IAT, MAP, KS, O2, TP, TCM, fuel level, and A/C pressure transducer. Three (3) splices are used to connect all of these circuits, including a joint connector in the PDC. Determine which circuits or groups of circuits are not operating properly, and use a wiring diagram to logically isolate the problem area.

CKP Sensor Schematic

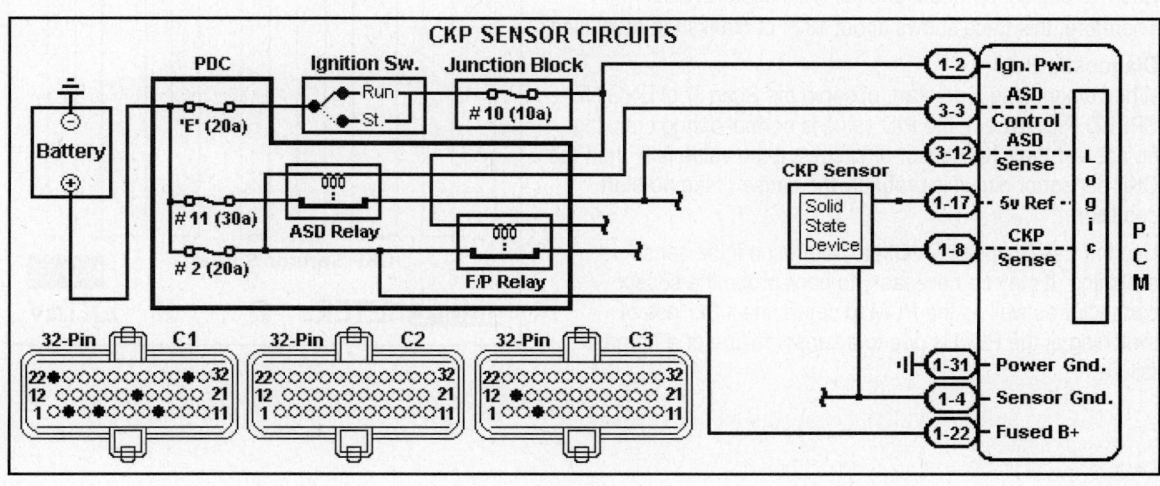

DVOM TEST (CKP SENSOR)

Place the shift selector in Park and block the drive wheels for safety.

DVOM Connections
Connect the DVOM positive lead to the CKP sensor signal circuit (GY/BK wire) at PCM Pin 1-8 and the negative lead to the battery negative post.

DVOM Example
This capture was taken at Hot Idle. The DVOM display shows a CKP signal of only 2.54v DC because the instrument is averaging the signal voltage. The average value is about half of the 0-5v signal because the duty cycle is near 50%.

Use the DVOM MIN/MAX feature, if available, to verify that the CKP signal is really switching from 0-5v.

Another way to use the DVOM to test the CKP sensor signal is to observe the frequency of the signal. In this example, the display shows 160.8 Hz. (160.8 cycles per second X 60 seconds / 16 slots per revolution = 603 RPM). Using this

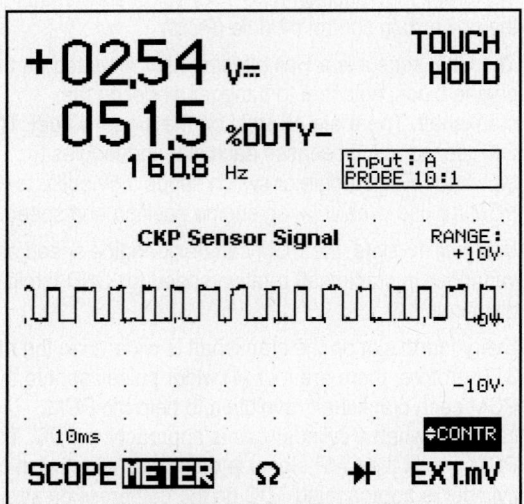

formula for a no start, it is easy to calculate that at 200-RPM cranking speed, the DVOM should read about 53 Hz.

LAB SCOPE TEST (CKP SENSOR)

Place the shift selector in Park and block the drive wheels for safety.

Scope Connections
Connect the positive lead to the CKP sensor signal circuit (GY/BK wire) at PCM Pin 1-8 and the negative lead to the battery negative post.

Lab Scope Example (1)
In example (1), the trace shows the CKP sensor signal at idle. Each 16 pulses represent one full crankshaft revolution. Therefore, this trace shows about 180° of crankshaft revolution.

Diagnostic Help
When diagnosing a no-start, observe the Scan Tool ENGINE SPEED PID value. If the PID value is normal during cranking, do not test the CKP sensor or circuits. If the value is 0, then the CKP sensor or circuit is probably the cause of the no-start condition.

Use the Lab Scope or DVOM to determine if the sensor is switching. It may be necessary to back probe the sensor connector as well as the PCM to determine if the loss of switching at the PCM is due to a sensor failure or a harness failure.

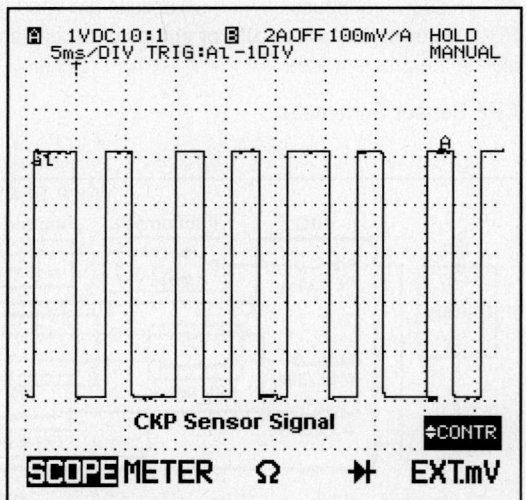

Enhanced Evaporative Emission System

INTRODUCTION

The Evaporative Emissions (EVAP) system is used to prevent the escape of fuel vapors to the atmosphere under hot soak, refueling and engine off conditions. Any fuel vapor pressure trapped in the fuel tank is vented through a vent hose, and stored in the EVAP canister.

The most efficient way to dispose of fuel vapors without causing pollution is to burn them in the normal combustion process. The PCM cycles the purge valve on (open) during normal operation. With the valve open, manifold vacuum is applied to the canister and this allows it to draw in fresh air and fuel vapors from the canister into the intake manifold. These vapors are drawn into each cylinder to be burned during normal combustion.

The PCM uses various sensor inputs to calculate the desired amount of EVAP purge flow. The PCM meters the purge flow by varying the duty cycle of the EVAP purge solenoid control signal. The EVAP purge solenoid will remain off during a cold start, or for a preprogrammed time after a hot engine restart.

System Components:
The EVAP system contains many components for the management of fuel vapors and to test the integrity of the EVAP system. The EVAP system is constructed of the following components:

- Fuel Tank
- Fuel Filler Cap
- Fuel Tank Level Sensor
- Fuel Vapor Vent
- Evaporative Charcoal Canister
- Leak Detection Pump
- EVAP Purge Solenoid
- PCM and related wiring
- Fuel vapor lines with Test Port

EVAP System Schematic

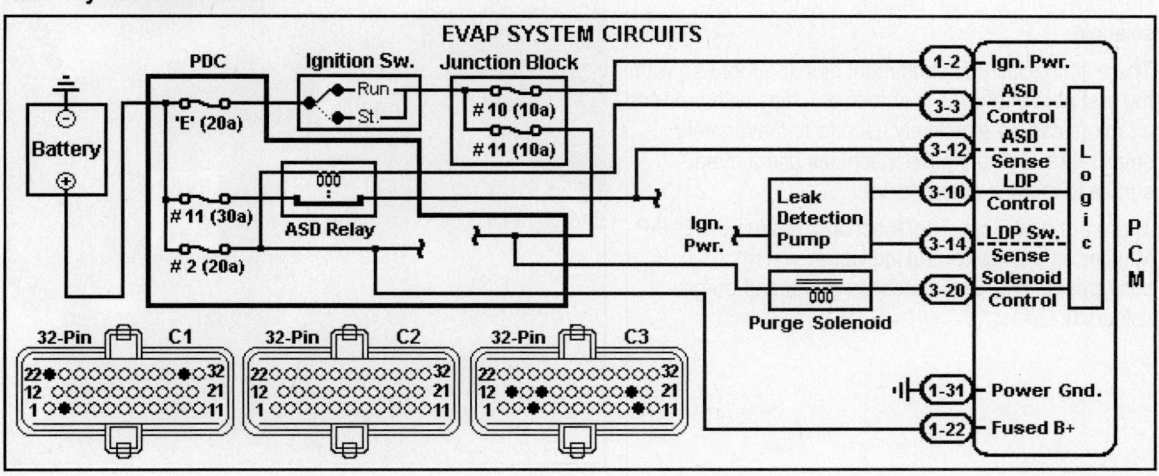

EVAPORATIVE CHARCOAL CANISTER

The EVAP canister is located on the frame rail near the front of the fuel tank. Fuel vapors from the fuel tank are stored in the canister. The canister is connected in series between the fuel tank and the purge solenoid. Although the Leak Detection Pump (LDP) pressurizes the entire EVAP system, the pressure enters the system through the canister.

Once the engine is in closed loop, fuel vapors stored in the charcoal canister are purged into the engine where they are burned during normal combustion.

The activated charcoal in the canister has the ability to purge, or release, any stored fuel vapors when it is exposed to fresh air.

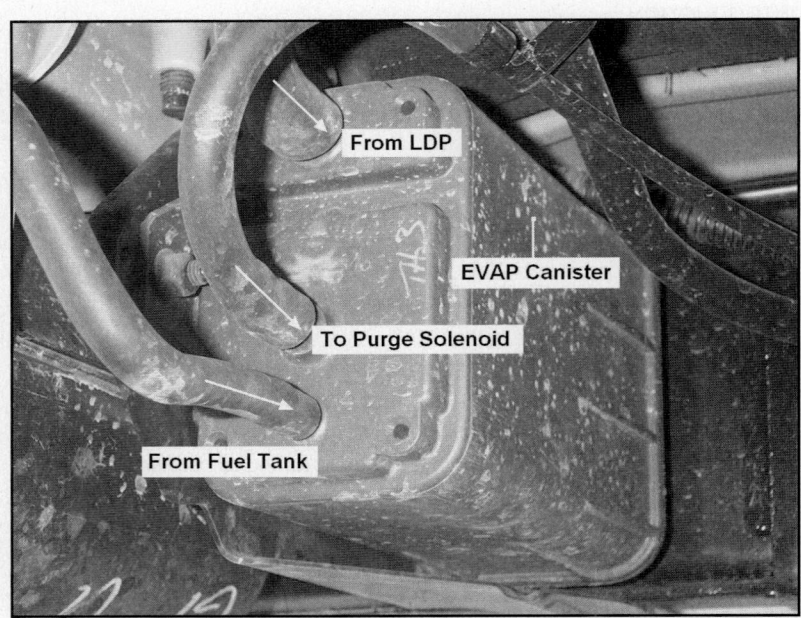

EVAP SYSTEM TEST PORT

All OBD II vehicles must have an EVAP test port. This application incorporates the test port into the purge line between the EVAP canister and the purge solenoid.

There is specific test equipment designed for use with the test port. However, a smoke machine with proper adaptors can be effectively used to find even very small leaks. Some smoke machines use a dyed smoke to make it easier to see.

There are also Scan Tool functions that run the EVAP Monitor as well as control individual system components (i.e., leak detection pump and purge solenoid).

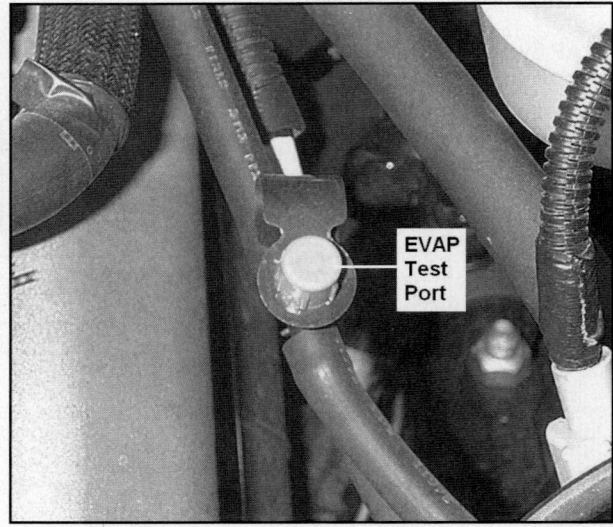

EVAP PURGE SOLENOID

The EVAP canister purge solenoid is a normally closed (N.C.) solenoid controlled by the PCM. A single purge solenoid, located next to the battery on the left fender, allows vapors from a charcoal canister to enter the manifold during various engine conditions.

The PCM controls purging by providing ground to the solenoid. By varying the percentage (%) of time the circuit is grounded and the purge solenoid is operating, the PCM can precisely control the volume of fuel vapors drawn into the intake manifold.

Purge Solenoid

DVOM TEST (PURGE SOLENOID)

Place the shift selector in Park and block the drive wheels for safety.

DVOM Connections
Connect the DVOM positive lead to the purge solenoid signal circuit (PK/BK wire) at PCM Pin 3-20 and the negative lead to the battery negative post.

DVOM Example
This capture was taken at Hot Idle. The DVOM display shows a 10.87v average purge solenoid signal. The duty cycle is 22.3%, which shows that the PCM is commanding very little purge under these conditions.

The frequency of this command is fixed, so it is not an important value to observe during diagnosis.

The Hz value in this example is OL, which means "Out of Limits." When this occurs, to change the time and voltage range until a value is shown.

In this case, it is easy to calculate that at 100ms/division, and 10 divisions per screen, that the width of the screen is equal to 1 second. Therefore, with 5 cycles of the signal on the screen, the frequency of the signal is 5 Hz.

The signal will remain at either 5 or 10 Hz even though the duty cycle command is varied according to desired purge flow. The frequency is switched between the 2 values based on current operating conditions.

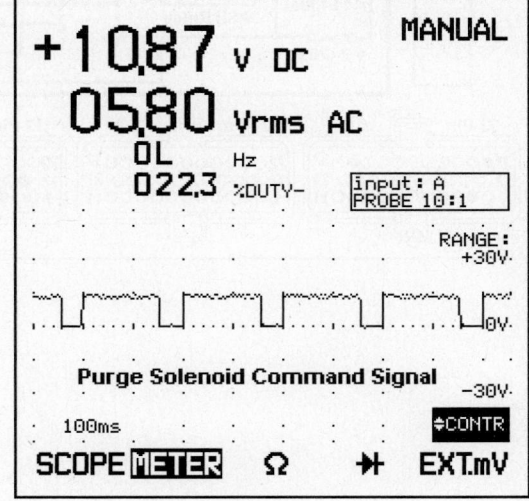

LAB SCOPE TEST (PURGE SOLENOID)

The Lab Scope is an excellent tool to view both the voltage and current waveforms of the purge solenoid and its circuit. Place the shift selector in Park and block the drive wheels for safety.

Scope Settings
To make the waveforms as clear as possible, set the scope settings to match the examples.

Scope Connections
Connect the Channel 'A' positive lead to the purge solenoid control circuit (PK/BK wire) at PCM Pin 3-20 or at the component. Connect the negative lead to the battery negative post. Connect a low-amps probe to Channel 'B' and clamp the probe around either purge solenoid wire (PK/BK or DB/WT).

Lab Scope Example
In this example, the top trace shows the purge solenoid command from the PCM. The bottom trace shows the current flow through the solenoid when the circuit is grounded.

Using these traces, the circuit voltage, ground circuit, and PCM control of the solenoid can all be verified. Observe both traces over time to look for the cause of intermittent circuit problems. Circuit or solenoid resistance problems that are difficult to see in the voltage waveform can be seen in the current waveform.

In this example, the peak current was about 400 ma. If the current flow is too high all the time, look for a partial short to ground in the circuit. If the current flow is too high only when the solenoid is commanded "on", check for a shorted solenoid.

Purge Solenoid Test Schematic

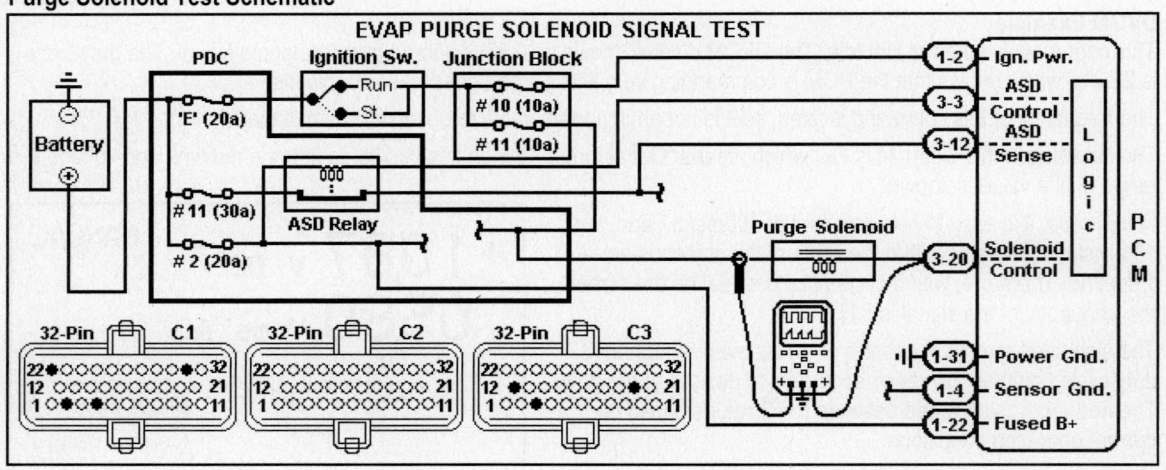

LEAK DETECTION PUMP (LDP)

The Leak Detection Pump (LDP) is a device used to pressurize the EVAP system during the EVAP Monitor to detect leaks and restrictions.

The pump contains a 3-port solenoid, a pump that contains a calibrated pressure switch, a spring-loaded canister vent valve, 2 check valves and a spring/diaphragm assembly.

The LDP also seals the EVAP system from atmosphere so that the OBD II EVAP Monitor Leak Test can be run.

Leak Detection Pump System Graphic

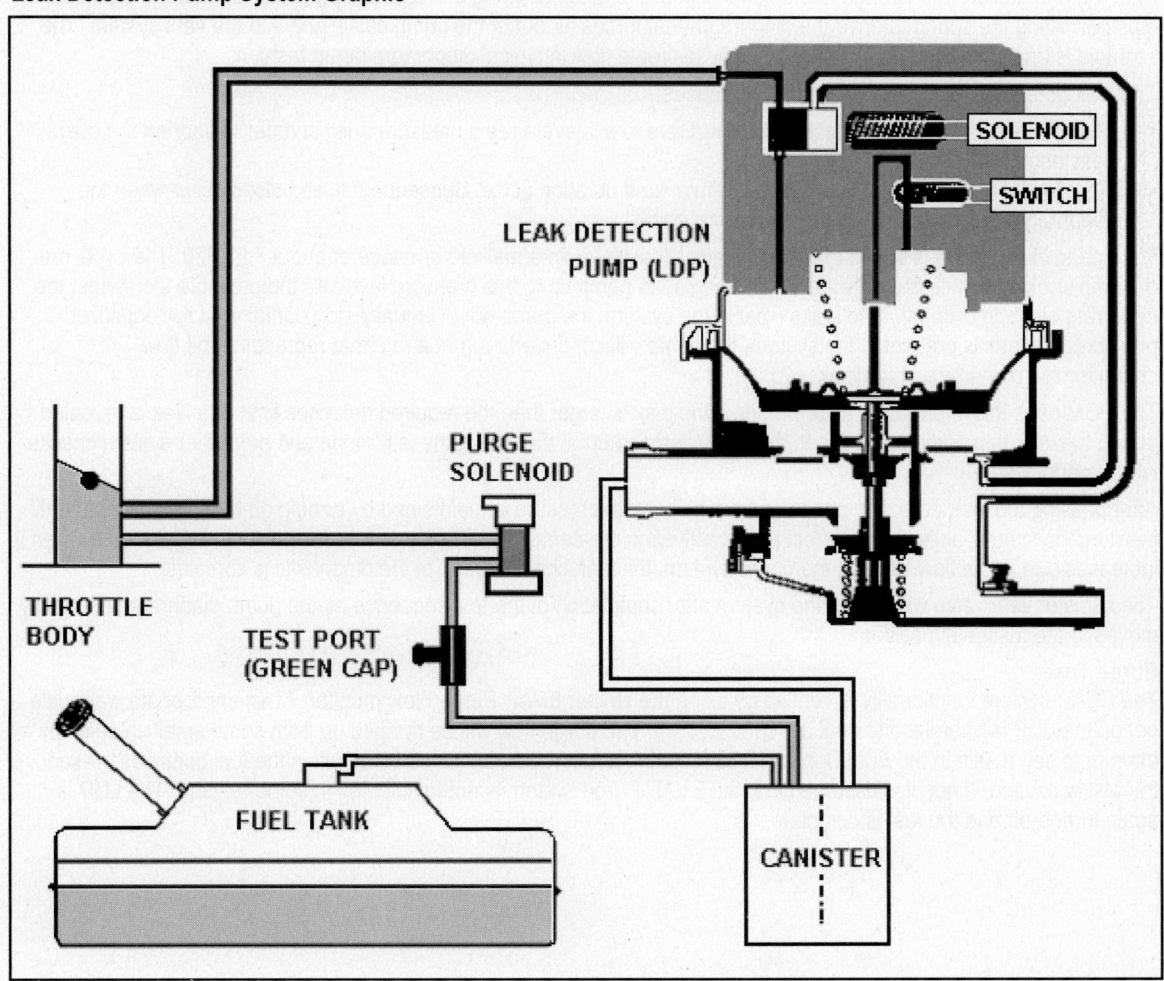

LDP MONITOR OPERATION

Leak Test

Immediately after a cold startup (a period established by predetermined temperature threshold limits), the three-port solenoid is briefly energized. This initializes the pump by drawing air into the pump cavity and also closes the vent seal.

During other non-test conditions, the pump diaphragm assembly holds the pump vent seal open (it pushes the vent seal open to the full travel position). The vent seal will remain closed while the pump is cycling because the reed switch triggers the three-port solenoid to prevent the diaphragm assembly from reaching full travel.

After the brief initialization period, the solenoid is de-energized allowing atmospheric pressure to enter the pump cavity, thus permitting the spring to drive the diaphragm that forces air out of the pump cavity and into the vent system. The solenoid is energized and de energized rapidly to create flow in typical diaphragm pump fashion.

The leak detection pump is controlled in two modes:

- Pump Mode - The pump is cycled at a fixed rate to achieve a rapid pressure build in order to shorten the overall test length.
- Test Mode - The solenoid is energized with a fixed duration pulse. Subsequent fixed pulses occur when the diaphragm reaches the reed switch closure point.

The spring in the pump is set so that the system will achieve an equalized pressure of about 7.5" H20. The cycle rate of pump strokes is quite rapid as the system begins to pump up to this pressure level. As the pressure increases, the cycle rate starts to drop off. If no leaks exist in the system, the pump will eventually stop pumping at the equalized pressure. If a leak is present in the system, the pump will continue to run at a rate that represents the flow characteristic of the size of the leak.

The PCM uses this information to determine if the leak is larger than the required detection limit. If a leak is revealed during the leak test portion of the test, the test is terminated at the end of the test mode and no further system checks will be performed.

After passing the leak detection phase of the test, system pressure is maintained by turning on the LDP solenoid until the purge system is activated. In effect purge activation creates a leak. The cycle rate is again interrogated and when it increases due to the flow through the purge system, the leak check portion of the diagnostic is complete.

The canister vent valve will unseal the system after completion of the test sequence as the pump diaphragm assembly moves to the full travel position.

Purge Test

The EVAP system functionality is verified by using the stricter EVAP Purge Flow monitor. At an appropriate warm idle period the LDP is energized to seal the canister vent. The purge flow will be ramped up from some small value in an attempt to see a shift in the Fuel Control (O2S) system. If fuel vapor, indicated by a shift in the fuel control is present, the test is passed. If not, it is assumed that the EVAP Purge system is not functioning in some respect. The LDP is again turned off and the test is complete.

ONBOARD REFUELING VAPOR RECOVERY (ORVR)

A tremendous amount of hydrocarbons are released during the refueling process. When a tank is filled from empty, all of the fuel-enriched air in the tank is displaced. In addition to the problem of existing vapors, the refueling process generates a large volume of new vapors. Some states have programs with gas pump boots that recover the displaced and generated vapors. However, recent design changes allow the vehicle EVAP system to collect and store these vapors. The fuel vapors are then burned in the combustion process when the vehicle is started.

System Operation

The simple premise of the system is that the fuel tank filler tube can be designed in such a way that a venturi effect draws air into the tank instead of expelling it. The filler tube has a very small diameter (about 1").

The flow of fuel through this small tube causes fresh air to be drawn into the tank. The air in the tank is then forced out through a control valve in the tank to be stored in the EVAP canister. Hydrocarbons become trapped in the canister as the air passes through, and the 'cleaned' air is vented back to atmosphere. In effect, this process uses air as a vehicle to transport hydrocarbons to the desired location (canister) where they are stored.

As fuel enters the tank, a normally closed check valve is forced open. The check valve contains a float, so when the fuel level is high enough the valve floats up to seal the filler tube. This prevents overfilling by creating pressure in the filler tube that shuts off the pump nozzle. Also, the check valve prevents fuel from spraying out of the tank when the nozzle is shut off.

ORVR Problems

There are no electronic components used in this system, so diagnosis is limited to the physical condition of each component. The system operates by passively creating and using pressure and flow to accomplish the task of managing refueling vapors.

Therefore, most of the things that can go wrong with this system result in tank pressure buildup, which shuts off the pump nozzle. The complaint will be that the vehicle cannot be refueled, or sprays fuel when the nozzle is removed. Any restriction in the vapor lines from the tank to the EVAP canister can cause this condition. Also, any fuel saturation in the canister or the canister vent can restrict the system and cause pressure to build. Lastly, any mechanical defects in the fuel tank vapor control valve or the filler tube check valve can cause similar problems.

It is also possible to have refueling problems with nothing wrong. If the customer starts the car cold and drives to a nearby gas station, the EVAP Monitor may be interrupted when the key is shut off. When cold, the Monitor will enable the Leak Detection Pump (LDP), which *shuts off the canister vent* and pressurizes the system to check for leaks. See LDP Monitor Operation in this Section. If the system has residual pressure from this interrupted test, the filler tube check valve may stick closed. Even if it is not stuck, the vent valve may be sealed off from atmosphere, producing the same results. The best solution is to drive the car for a few minutes so the purge valve can bleed off pressures during normal operation.

A question to ask the customer is whether the engine was off during the refueling attempt. The system is designed only work properly with the engine off. This design prevents engine-on refueling by causing a buildup of pressure, and shuts off the pump.

Vehicle Speed Control

INTRODUCTION

The Vehicle Speed Control system on this application is controlled by the PCM. The PCM controls vacuum solenoids in the speed control servo to regulate servo movement. The servo acts through a cable to control throttle angle, and therefore vehicle speed.

Circuit Operation

The only 2 inputs required for the PCM to properly operate the system are the vehicle speed sensor signal and the steering wheel mounted S/C switches.

All of the steering wheel mounted switches are combined into a single input to the PCM. The PCM provides a 5v signal to the switch assembly. Each of the switches provides a path to ground with a different resistance. The PCM monitors the voltage drop in the circuit and can determine which of the switches was depressed.

The brake switch is not an actual input. When the brake pedal is applied, the power supply to the speed control servo is interrupted. The S/C servo contains three (3) solenoids powered by the PCM: the vent, vacuum, and dump solenoids.

The PCM can only ground the vent and the vacuum solenoids to regulate vehicle speed. The dump solenoid is permanently grounded, and is therefore energized with the brake pedal off. When the brake pedal is depressed, the dump solenoid is de-energized, and the vacuum in the speed control servo is rapidly bled off to atmosphere.

Diagnostic Help

The PCM sends out 5 volts to the steering wheel switch assembly through a fixed resistor and monitors the voltage drop on the circuit to determine which switch was depressed.

There are only 5 buttons, but 6 resistors (refer to the Schematic below). The last resistor is not connected through a switch, so it provides a voltage drop even when none of the switches are depressed. This was done for diagnostic purposes. If the circuit is open anywhere past PCM Pin 3-32, the PCM will detect 5 volts and set a DTC P1596. If this code sets, there is an open in the circuit preventing voltage from dropping across the diagnostic resistor in the switch assembly.

Speed Control System Graphic

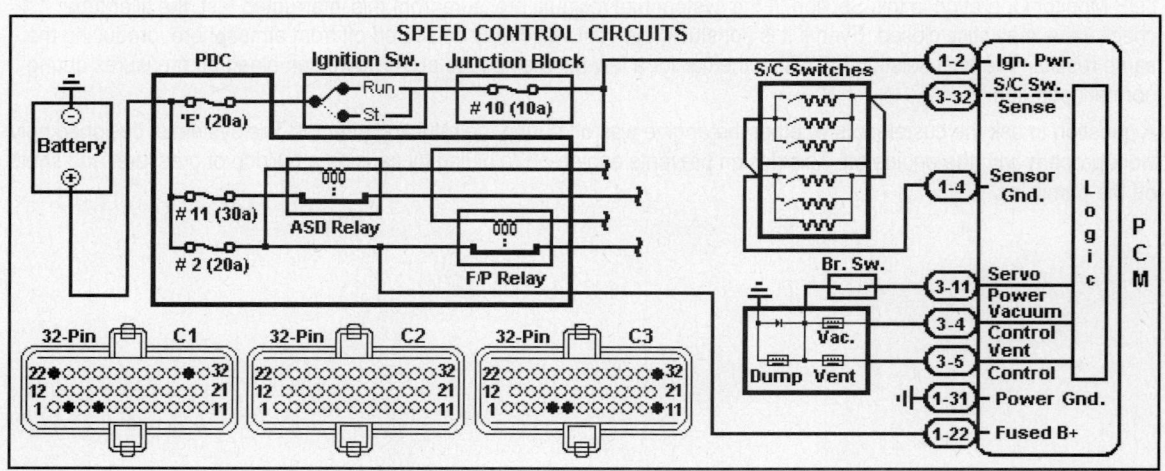

LAB SCOPE TEST (SPEED CONTROL SWITCHES)

The Lab Scope can be used to diagnose suspected problems with the steering control mounted Speed Control (S/C) switches or circuits.

All of the steering wheel mounted switches are combined into a single input to the PCM. The PCM provides 5 volts to the switch assembly. Each of the switches provides a path to ground with a different resistance. The PCM monitors the voltage drop in order to determine which of the S/C switches was depressed.

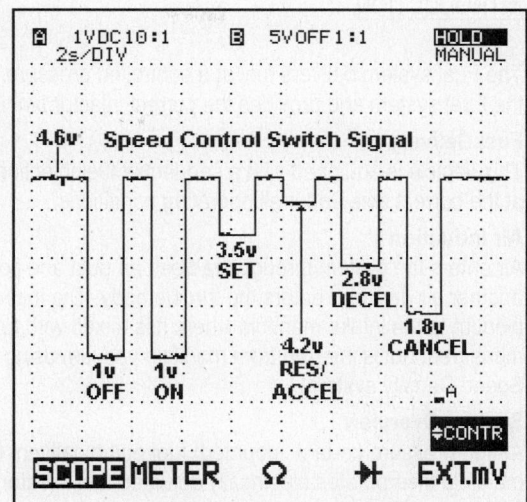

Lab Scope Connections
Connect the positive probe to the S/C switch circuit (RD/LG wire) at PCM Pin 3-32 and the negative probe to the battery negative post.

Lab Scope Example
In the example, the trace shows the different voltage drops in the S/C switch circuit when each of the steering wheel mounted switches is depressed. A long time base was used (2 seconds per division) in order to capture all of the switch conditions.

Note that the OFF/ON switch drops the same amount of voltage whether the system is being turned OFF or ON. The PCM keeps track of system status so one resistor can indicate 2 different conditions.

Diagnostic Resistor
The trace never rises to the 5v level because of the fixed diagnostic resistor in the switch assembly. Look for a steady 4.6v on this circuit at rest. If the circuit has an open or high resistance, even at rest, the voltage drop across the diagnostic resistor will be reduced or eliminated, causing the signal to rise above 4.6v and set a DTC. This diagnostic resistor is not in all Daimler-Chrysler vehicles, so test results may vary. Use a wiring diagram to determine if the S/C switch assembly has 5 or 6 resistors.

Speed Control System Test Graphic

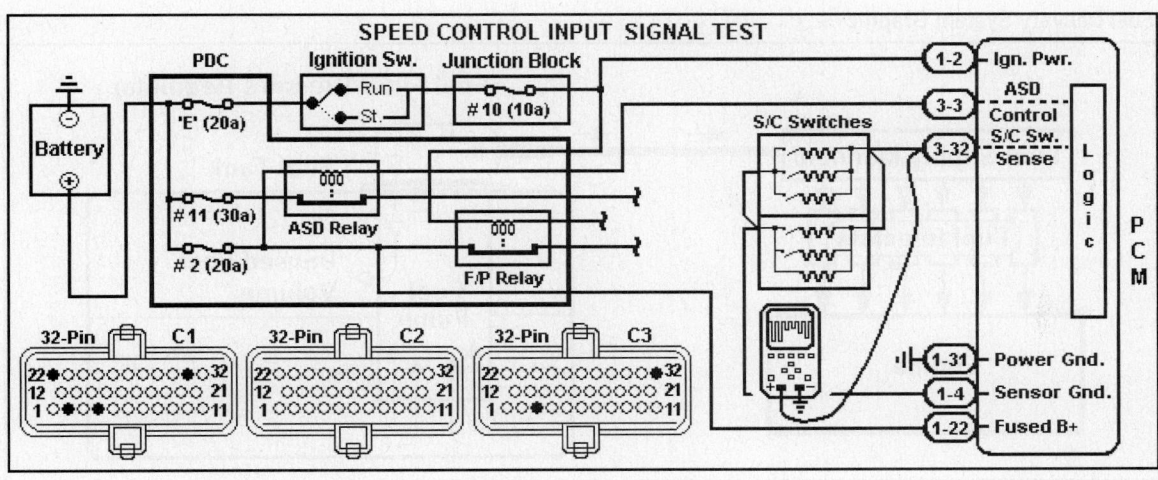

Fuel Delivery System

INTRODUCTION

The Fuel system delivers fuel at a controlled pressure and correct volume for efficient combustion. The PCM controls the Fuel system and provides the correct injector timing.

Fuel Delivery

This vehicle is equipped with a sequential fuel injection (SFI) system that delivers each cylinder the correct A/F mixture at the correct time under all operating conditions.

Air Induction

Air enters the system through the fresh air duct and flows through the air cleaner. The unmetered air passes through another air duct and enters the throttle body. The incoming air passes through the throttle body and into the intake plenum to the intake manifold where it is mixed with fuel for combustion. The volume of intake air is not directly measured, but is inferred from manifold vacuum using a MAP sensor mounted to the intake manifold (referred to as a Speed Density system).

System Overview

A high-pressure (in-tank mounted) fuel pump delivers fuel to the fuel injection supply manifold. The fuel injection supply manifold incorporates electrically actuated fuel injectors mounted directly above each of the intake ports in each cylinder head.

Power is supplied to the fuel injectors by the ASD relay, and to the fuel pump by the fuel pump relay. Drivers in the PCM ground the injector circuits and fuel pump relay control circuit to operate these devices.

This application uses a mechanical returnless fuel delivery system. There is no fuel return line from the fuel supply manifold. A fuel filter/pressure regulator assembly is mounted to the top of the in-tank fuel pump module. The regulator maintains a constant amount of fuel pressure in the system. The pressure regulator diverts extra fuel volume back through the fuel pump to the fuel tank.

The PCM determines the amount of fuel the injectors spray under all engine-operating conditions. The PCM receives input signals from engine control sensors that monitor various factors such as manifold pressure, intake air temperature, engine coolant temperature, throttle position, and vehicle speed. The PCM evaluates the sensor information it receives and then signals the fuel injectors in order to control the fuel injector pulsewidth.

Fuel Delivery System Graphic

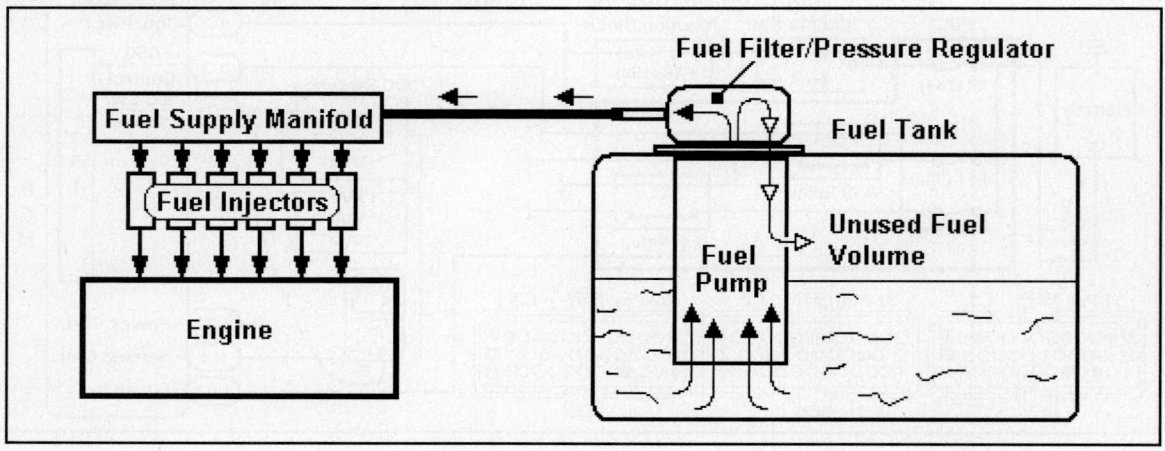

FUEL PUMP MODULE

The fuel pump module in this returnless fuel system is mounted inside the fuel tank.

The fuel pump produces more pressure and volume than necessary for proper system operation. When the pump pressure exceeds 49.2 psi, the excess is bled back into the tank through the fuel filter & pressure regulator assembly mounted to the top of the fuel pump module. However, the pump has a maximum pressure output of 95 psi.

Fuel Pump Module Components

The fuel pump module includes an electric pump motor, inlet filter, fuel filter and pressure regulator assembly, fuel gauge sending unit, and a fuel line connection. Only the fuel filter and pressure regulator assembly may be serviced separately.

Internal Valve

The pump module has one internal check valve, located in the pump outlet, which allows the system pressure to be maintained with the ignition off. This valve also keeps the pump primed for quick pressurization of the system at startup, although residual fuel pressure of 0 psi on a cold engine is considered normal.

FUEL PUMP MODULE

FUEL FILTER /PRESSURE REGULATOR — ELECTRICAL CONNECTOR — FUEL GAUGE FLOAT — ELECTRIC FUEL PUMP — MODULE LOCK TABS (3) — FUEL GAUGE SENDING UNIT — FUEL PUMP INLET FILTER

FUEL FILTER & PRESSURE REGULATOR

The filter and regulator on this application are mounted to the top of the fuel pump module. The fuel filter is not normally replaced as a maintenance item. Replace this assembly only when directed to do so by a diagnostic procedure.

The regulator is calibrated to open a return port above 49.2 psi. Excess fuel pump pressure is bled back into the fuel tank.

The regulator traps fuel during engine shutdown. This eliminates the possibility of vapor formation in the fuel line, and provides quick restarts. However, residual pressure is not maintained, as the regulator only traps fuel to prevent it from draining back to the tank.

Filter Replacement

The fuel filter/pressure regulator assembly may be replaced separately without removing the fuel pump module. However, the fuel tank must be removed.

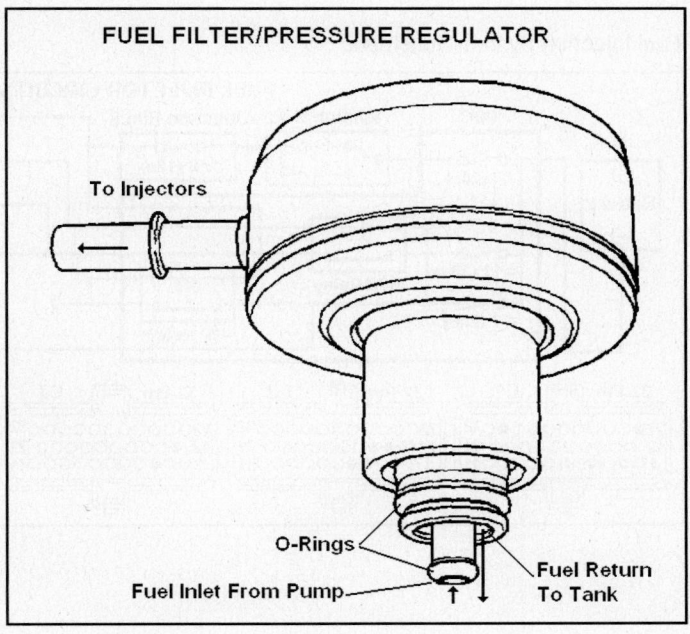

FUEL FILTER/PRESSURE REGULATOR

To Injectors

O-Rings

Fuel Inlet From Pump

Fuel Return To Tank

FUEL INJECTORS

Under normal operation, each injector is opened and closed at a specific time, once every other crankshaft revolution (once every engine cycle). The amount of fuel delivered is controlled by the length of time the injector is grounded (injector pulsewidth). The injectors are supplied voltage through the ASD relay.

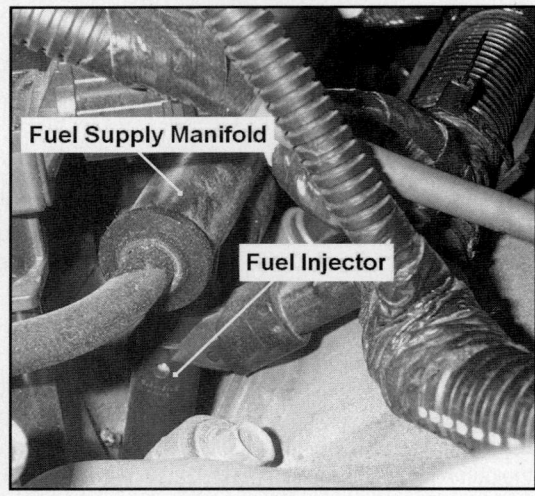

Fuel Supply Manifold

Fuel Injector

Fuel Injection Timing

The PCM determines the fuel flow rate (pulsewidth), from the ECT, HO2S-11, TP and MAP sensors, but needs the CKP and CMP signals to determine injector timing.

The PCM computes the desired injector pulsewidth and then grounds each fuel injector in sequence. Whenever an injector is grounded, the circuit through the injector is complete, and the current flow creates a magnetic field. This magnetic field pulls the pintle back against spring pressure and pressurized fuel escapes through the injector nozzle.

Circuit Description

The ASD relay connects the fuel injectors to battery power (B+) through fuse #11 (30a), as shown in the schematic.

With the key in the "start or run" position, the ASD relay coil is connected to B+.

The ASD relay coil is grounded by the PCM, which creates a magnetic field that closes relay contacts and supplies power to the fuel injectors.

Drivers in the PCM individually ground each injector circuits. The injector sequence on this application is: 1-8-4-3-6-5-7-2.

Fuel Injection System Schematic

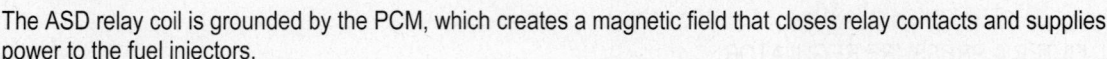

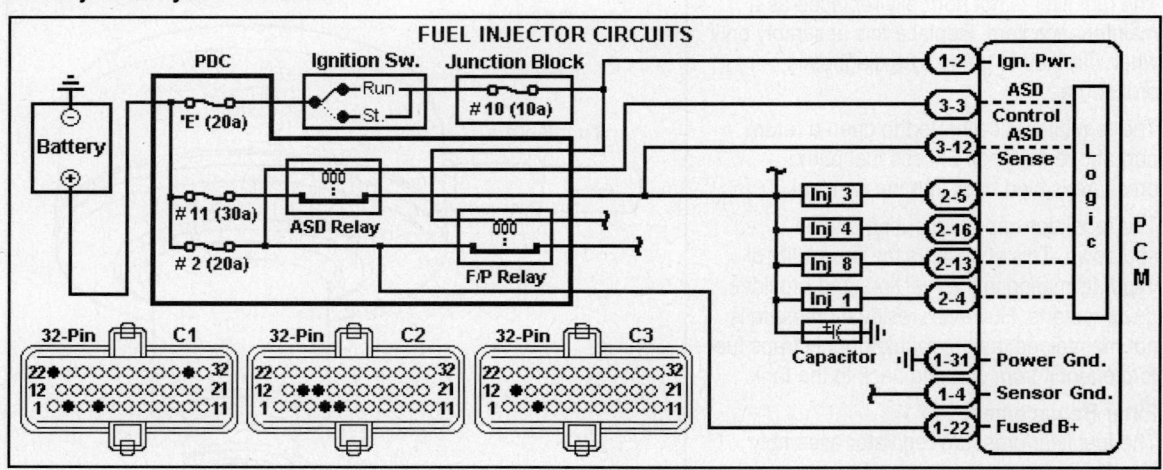

LAB SCOPE TEST (FUEL INJECTOR)

Place the gear selector in Park. Block the drive wheels for safety.

Scope Connections
Connect the Channel 'A' positive probe to an injector control wire at the injector or appropriate PCM Pin. Connect the negative probe to the battery negative post.

Scope Settings
To make the waveforms as clear as possible, set the scope settings to match the examples.

Lab Scope Example
In the example, the trace shows the injector voltage signal during snap throttle.

In this example the on time is 9 ms. Using these settings (2 ms/div. and 10v/div.), the system voltage, on time, inductive kick (clipped), and injector pintle closure can all be clearly seen.

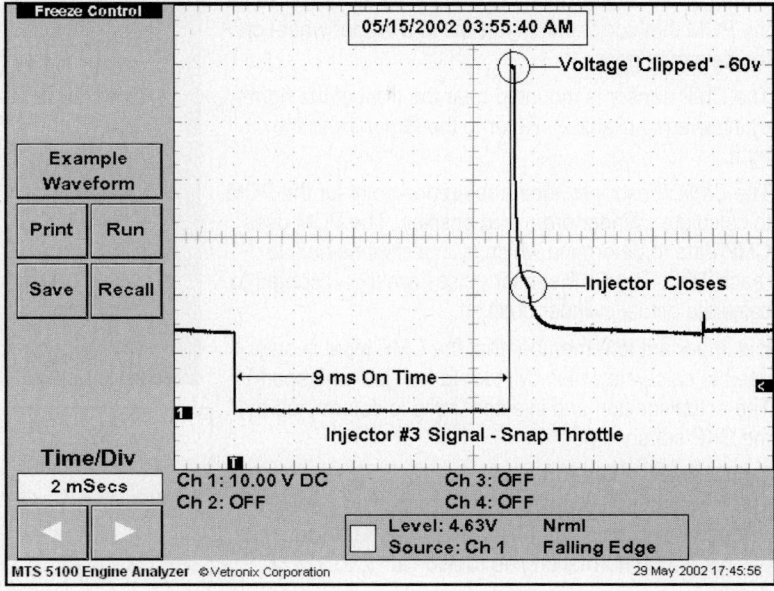

Use these settings to verify that the inductive kick reaches 60 volts. The kick should be much higher, but is 'clipped' by a diode in the PCM. If the kick is less than 60 volts, the current flow through the circuit is too low. Look for extra resistance in the injector or circuit. If the kick is less than 51 volts, the PCM will set a DTC for no inductive kick.

Also verify that the injector closes by observing the 'hump' in the falling edge of the waveform. This hump is the result of the pintle being forced back to the seated position after the injector is turned off. The injector opening cannot be seen in this waveform, but if it closes, it must have opened. The opening of the injector can be seen in a current waveform. Use a low-amps probe to view this event.

Fuel Injector Test Schematic

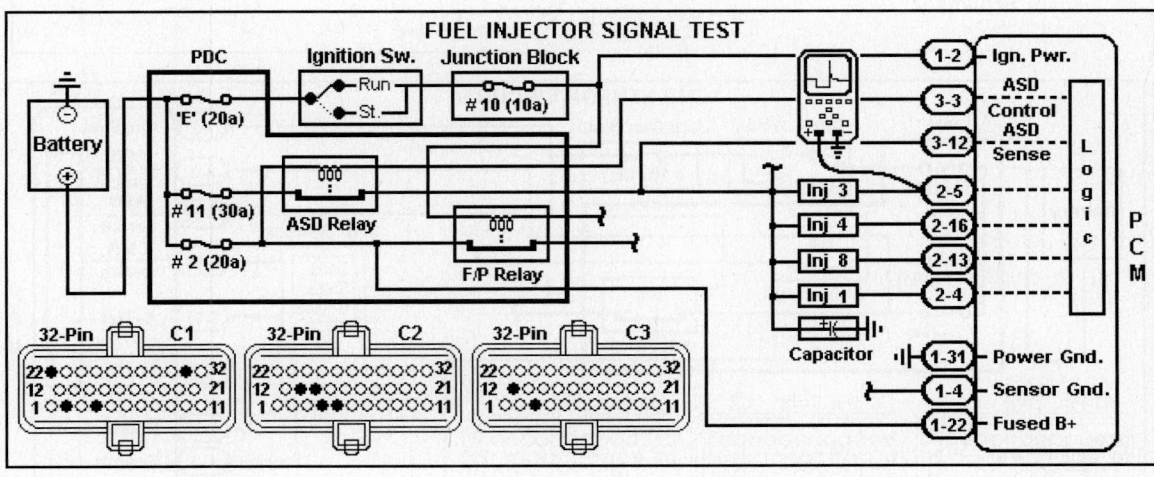

CAMSHAFT POSITION SENSOR (HALL EFFECT)

The camshaft position (CMP) sensor used with this EI system is a Hall effect sensor that sends a 0-5v pulse to the PCM that corresponds to slots in a trigger wheel on the camshaft gear.

The CMP sensor is mounted near the front of the right cylinder head (Bank 2). Refer to the Graphics to the right.

The CMP sensor provides a reference point for the PCM to calculate cylinder order and position. The PCM uses CMP data to determine which cylinder will be next to reach TDC. The CKP sensor pulses are then counted to calculate actual cylinder position.

It is important to remember that the CMP input is only used to calculate which cylinder to fire (fuel and spark). The actual ignition and injector timing is determined by the CKP sensor input.

Cylinder Identification

The trigger wheel contains groups of slots in a specific pattern. Refer to the Graphic to the right. The sequence of slots during normal engine rotation is: 1, 2, 3, 3, 2, 1, 3, and 1.

Each number of slots is repeated, so the PCM must keep track of the sequence. In other words, 3 CMP pulses in a row means nothing to the PCM unless it keeps track of the preceding number of pulses (1, 2 or 3).

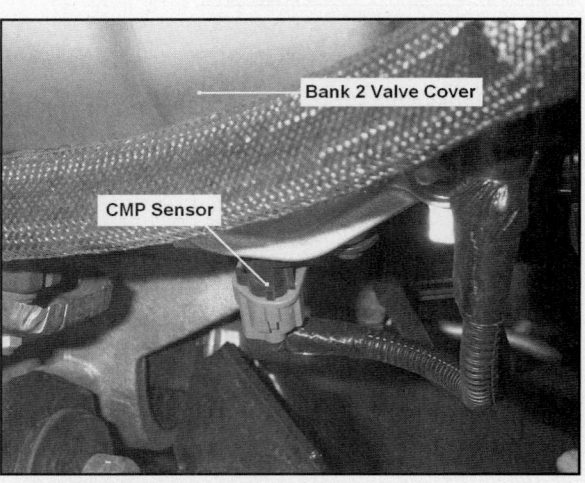

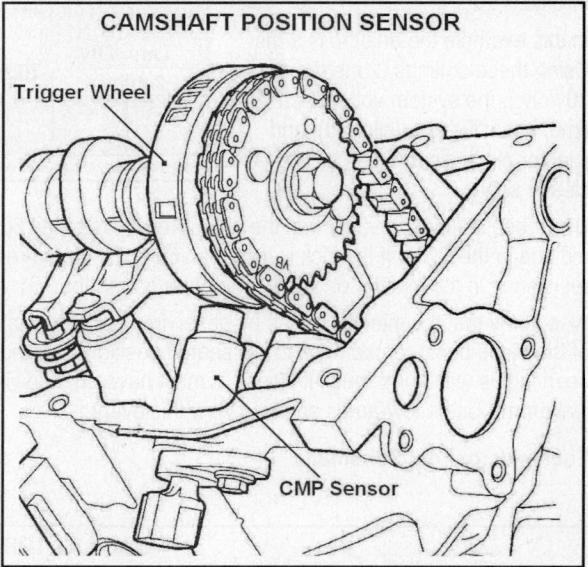

CAMSHAFT POSITION SENSOR

CMP Sensor Schematic

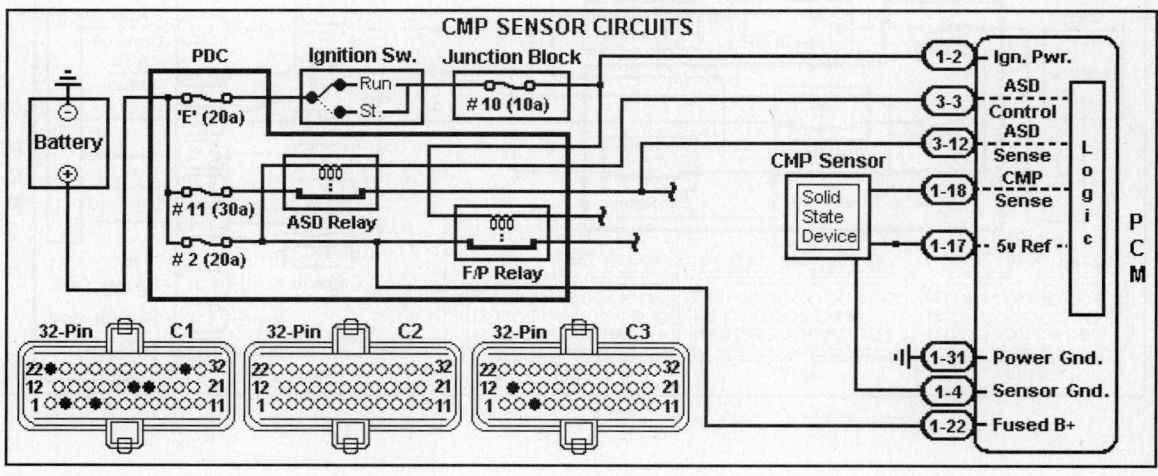

LAB SCOPE TEST (CMP & CKP SENSORS)

The Lab Scope can be used to test the CMP sensor circuit as it provides a very accurate view of circuit activity and any signal glitches. Place the gear selector in Park. Block the drive wheels for safety.

Scope Connections
In this example, the Channel '3' positive probe was connected to the CMP sensor circuit (TN/YL wire) at PCM Pin 1-18, and the negative probe was connected to the battery negative post. The Lab Scope was also connected the CKP sensor signal wire, and to the coil primary. See appropriate articles in this Section for those connections.

Lab Scope Example
In this example, the trace shows the coil primary signal, CMP and CKP sensor signals. Note the relationship between the three (3) traces.

The coil primary event occurs once every two (2) crankshaft revolutions (720°). The CKP trace shows 32 pulses in the same time period (16 pulses per revolution). There are eight (8) cylinder identification signals from the CMP sensor. Each cylinder identification signal consists of 1, 2 or 3 pulses. Note that the CMP signals occur in the proper sequence: 1, 2, 3, 3, 2, 1, 3, and 1.

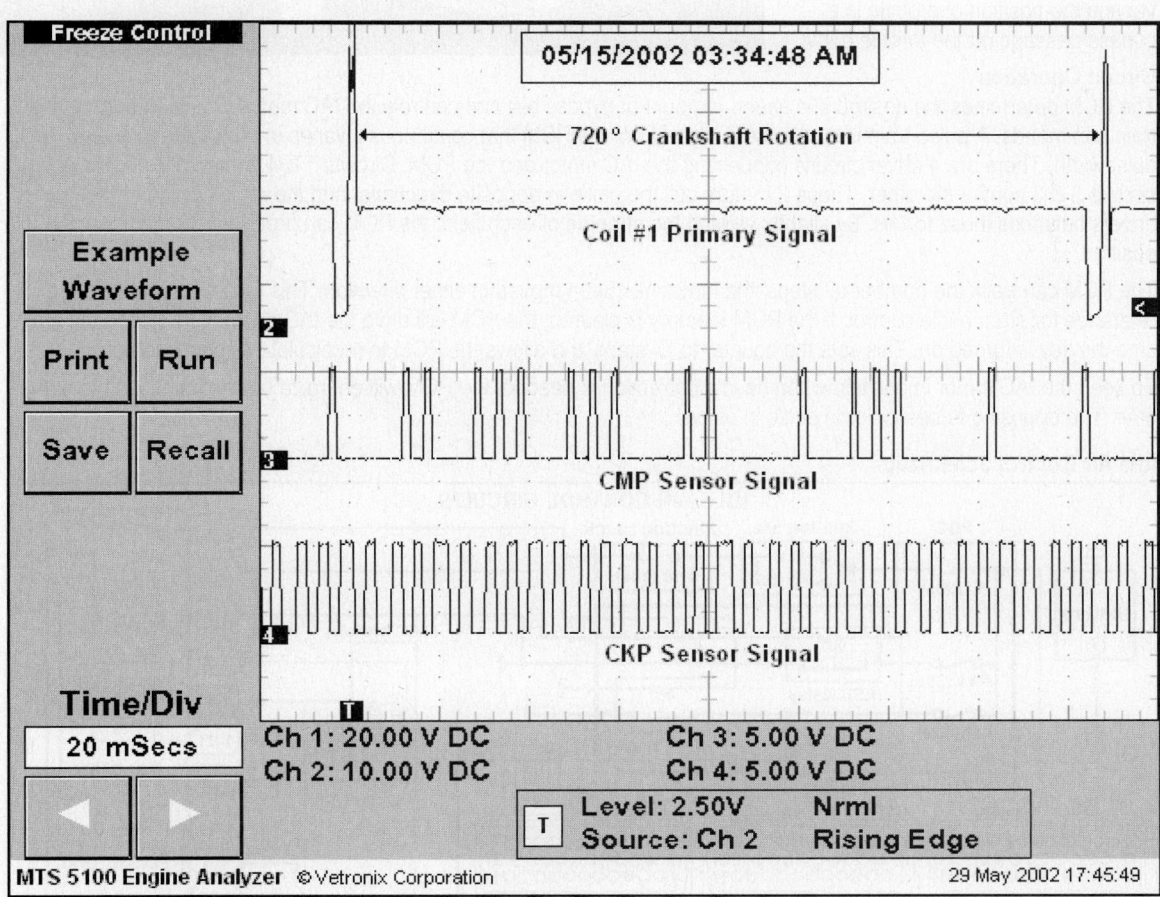

Idle Air Control Motor

INTRODUCTION

The Idle Air Control (IAC) motor, mounted on the throttle body, is actually a dual stepper motor assembly. The IAC may also be referred to as the Automatic Idle Speed (AIS) motor. This device is designed to control engine idle speed (rpm) under all engine-operating conditions and to provide a dashpot function (controls rate of deceleration for emissions and idle quality).

The IAC motor meters the intake air that flows past the throttle plate by varying the position of a pintle in a bypass passage on the throttle body.

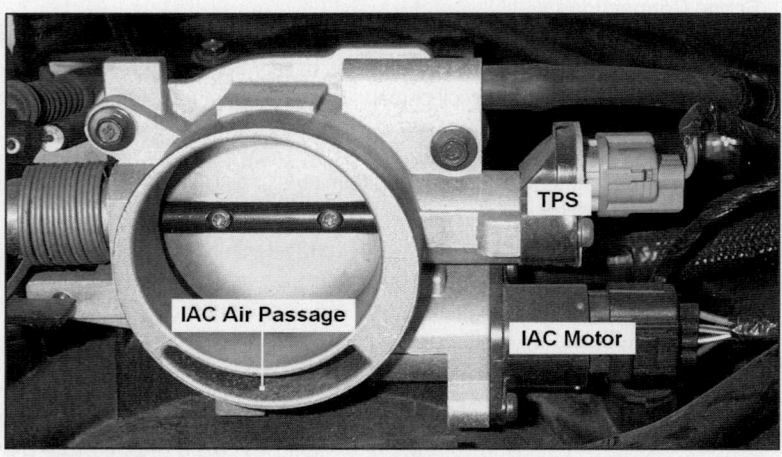

Circuit Operation

The PCM determines the desired idle speed (amount of bypass air) and controls the IAC motor circuits through pulse train commands. A pulse train is a command signal from the PCM that continuously varies in both frequency and pulsewidth. There are 4 driver circuits connecting the IAC motor and the PCM. Circuits 1 & 4 control one 'motor' and circuits 2 & 3 control the other. These 2 motors pull the pintle in opposite directions, and the rapid cycling of these drivers balances these forces. By slightly varying the strength of each field, the PCM can precisely control the pintle position.

The PCM can track the number of 'steps' the motor has been moved in either direction. This value is stored as a reference for startup idle control. If the PCM memory is cleared, the PCM will drive the IAC completely closed the first time the key is turned on. This sets the counter to '0' steps and allows the PCM to recalculate a reference point.

To keep the IAC motor in position when no idle correction is needed, the PCM will energize both windings at the same time. The opposing forces lock the pintle in place.

Idle Air Control Schematic

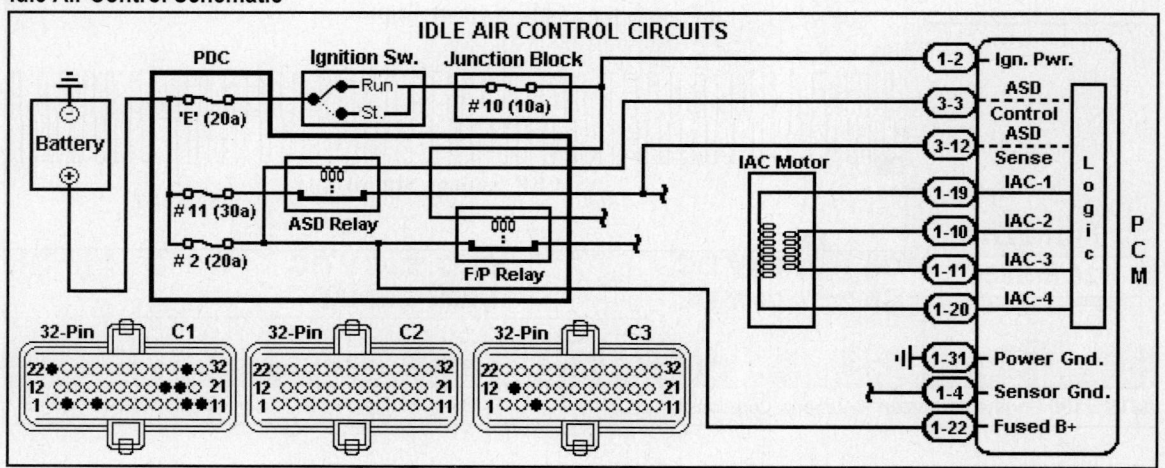

LAB SCOPE TEST (IAC MOTOR)

The Lab Scope is an excellent tool for testing the IAC motor circuit activity. Place the gear selector in Park. Block the drive wheels for safety.

Always check both motors when diagnosing these circuits. An IAC circuit or motor fault may not be seen unless you test both motors. A fault in only one circuit could cause the stepper motor to only move 1 step in either direction. Both motors are required for idle control.

Lab Scope Connections
All 4 drivers can be checked using 2 channels. The PCM controls one motor by alternating the polarity of drivers 1 & 4, and the other motor by alternating the polarity of drivers 2 & 3.

Therefore, the Lab Scope positive probes should be connected to circuit 1 or 4, and circuit 2 or 3. A circuit problem in either pair will show up with these connections.

Lab Scope Example
In this example, the traces show rapid circuit activity in response to light acceleration. Circuit activity can be seen slowing when the PCM finishes moving the IAC motor to its off-idle target.

The polarity of the circuits is toggled by the PCM to move the IAC motor. Therefore, connecting to one driver from each motor winding will reveal circuit problems in any of the four (4) circuits. In this case, drivers 1 & 2 are chosen. However, 1 & 3, 2 & 4, and 3 & 4 are also good testing combinations. If you connect to drivers 1 & 4 or 2 & 3, the pattern will only show the condition of one of the IAC motor windings.

Test Results
This test proves that the PCM is capable of electrically controlling all four (4) IAC motor drivers (the IAC motor, circuits, and PCM drivers are all good). Further diagnosis can be limited to PCM inputs, processing, or engine mechanical concerns (vacuum leaks).

IAC Motor Test Schematic

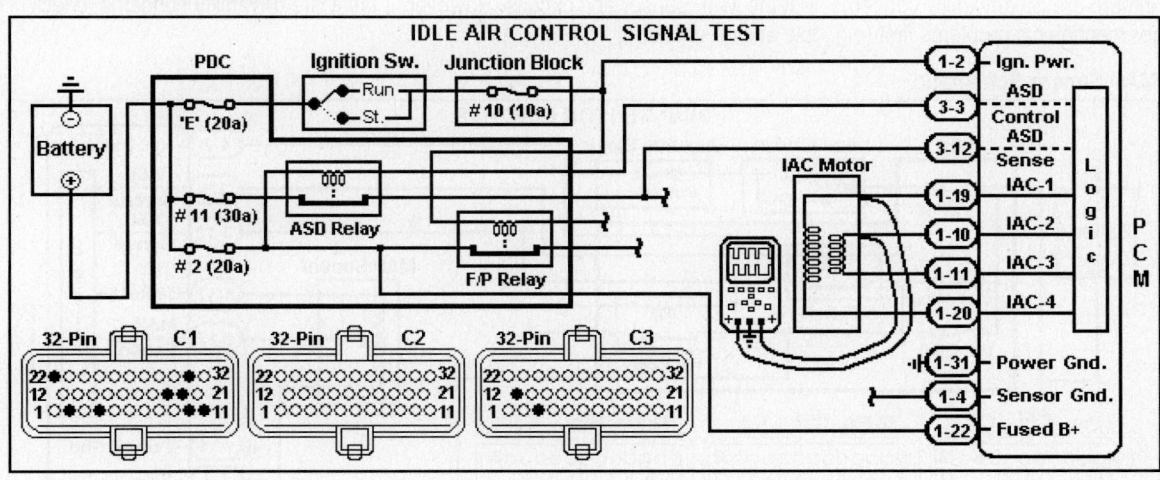

Information Sensors

MANIFOLD ABSOLUTE PRESSURE SENSOR

The Manifold Absolute Pressure (MAP) sensor is located on the front edge of the intake manifold, between the generator and A/C compressor. The sensor is mounted with an o-ring directly to a port in the intake manifold (no vacuum hose is used). The MAP sensor input to the PCM is an analog voltage signal proportional to the intake manifold pressure.

The MAP sensor signal is compared to the TP sensor signal and other inputs to calculate engine conditions. The PCM uses these calculations to determine fuel injector pulsewidth, ignition timing, and transmission shift points

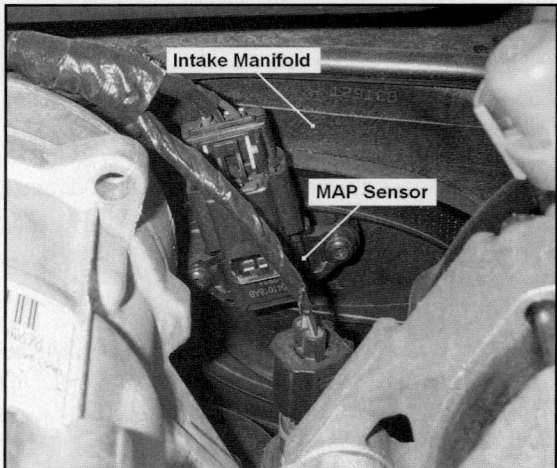

MAP Sensor Circuits
The three circuits that connect the MAP sensor to the PCM are the 5v supply circuit, sensor signal circuit, and sensor ground circuit.

Circuit Operation
The MAP sensor voltage increases in proportion to the engine load. When the throttle angle is high as compared to engine speed, the load on the engine is higher. Intake vacuum drops (pressure rises) in proportion to increased engine loads. By measuring this vacuum (pressure) the PCM has a very precise understanding of the current operating conditions.

Trouble Code Help
The PCM is capable of comparing values over time and under different conditions. This allows for very specific trouble codes to describe different conditions.

For example, the PCM takes a BARO reading from the MAP sensor at KOEO. If the engine is then started and there is not a large enough difference between the MAP values at KOEO and KOER, a DTC P1297 will set (1-trip DTC).

If there are no drivability concerns, test the MAP sensor and circuits. However, if there are drivability concerns, check for mechanical problems first (e.g., low engine vacuum).

MAP Sensor Schematic

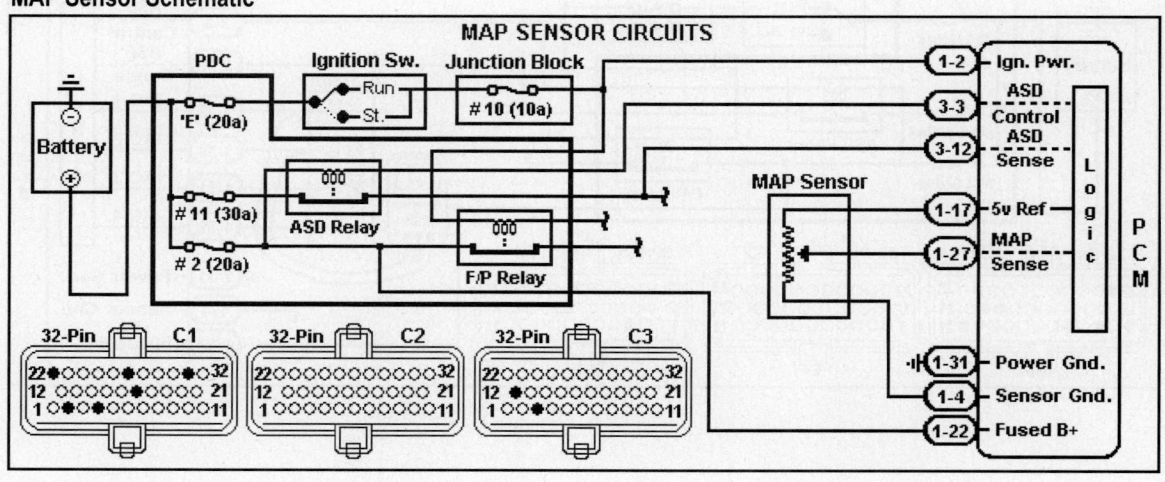

DVOM TEST (MAP SENSOR)

The DVOM is an excellent tool to test the condition of the MAP sensor circuit. The DVOM can be used to test the supply voltage, sensor ground voltage drop, and the MAP sensor signal for calibration or dropout. Place the gear selector in Park. Block the drive wheels for safety.

The test results can be compared with the specifications in Map Sensor Table (refer to Section 1). See the Lab Scope Test below for the probe connections.

DVOM Test

On this vehicle, the MAP sensor signal at Hot Idle was 1.49v. The MAP sensor voltage increases in proportion to manifold pressure. On this vehicle, the KOEO MAP sensor voltage was 4.78v at sea level, and was almost the same at snap throttle. The KOEO value is used by the PCM to calculate barometric pressure.

The voltage supply should be 4.9-5.1v. The ground circuit voltage drop should not exceed 50 mv.

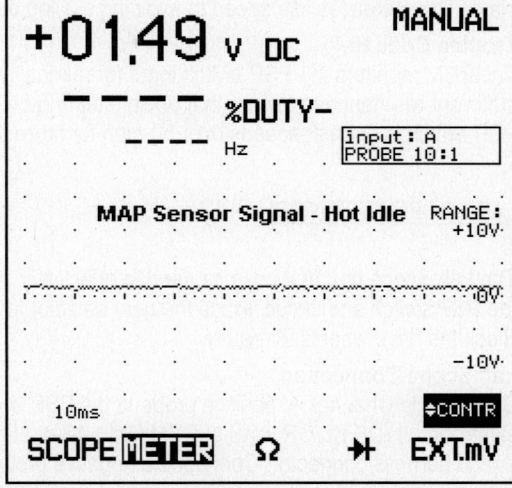

LAB SCOPE TEST (MAP SENSOR)

The Lab Scope can be used to test the activity of the MAP sensor signal. Place the gear selector in Park. Block the drive wheels for safety.

Lab Scope Connection

Connect the Channel 'A' positive probe to the MAP sensor signal circuit (DK GN/RD wire) at PCM Pin 1-27 or at the MAP sensor harness connector. Connect the negative probe to the battery negative post.

Scope Settings

To make the waveforms as clear as possible, set the scope settings to match the examples.

Lab Scope Example

In this example, the trace shows the MAP sensor signal during a snap throttle event.

The screen was frozen, and cursors were used to evaluate the sensor performance. A slow rise time usually indicates restrictions in the sensor orifice. The voltage change from idle to wide-open throttle should exceed 3 volts.

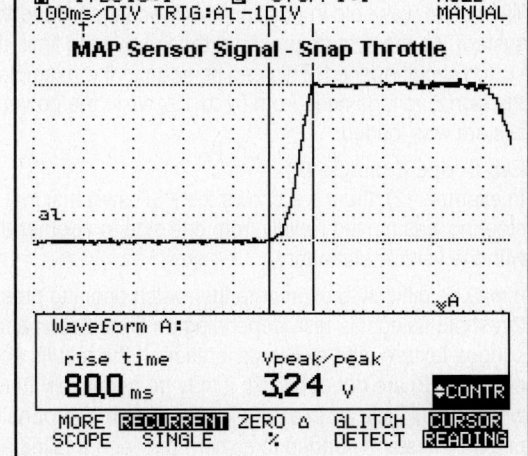

Although the snap throttle test can be used to check for restrictions or for sensor performance problems, it is important to remember that engine timing or mechanical problems can cause the MAP sensor signal to appear incorrect.

POWER STEERING PRESSURE SWITCH

The Power Steering Pressure (PSP) switch is a normally closed switch that provides an input to the PCM exclusively for idle control. The switch is mounted in the high-pressure power steering line. When system pressure exceeds 350 psi (+/- 125 psi) the switch opens. The PCM then commands the IAC motor to increase airflow around the throttle plate. This increases idle speed to avoiding stalling during low speed cornering.

Trouble Code Help
The PCM monitors the PSP switch input for rationality. At freeway speeds, it is not normally possible to experience sufficient resistance to turning that power steering pressure will cause the switch status to change. Therefore, if the PSP switch is open at speeds over 50 mph for more than 30 seconds, a DTC P0551 will set (1-trip).

LAB SCOPE TEST (PSP)

The Lab Scope or DVOM can be used to effectively diagnose the PSP switch and circuit. Place the gear selector in Park. Block the drive wheels for safety.

Lab Scope Connection
Connect the Channel 'A' positive probe to the PSP switch signal circuit (DK BL/OR wire) at PCM Pin 1-12 or at the PSP switch harness connector. Connect the negative probe to the battery negative post.

Lab Scope Example (1)
In example (1), the trace shows the PSP switch signal to the PCM when the steering wheel is turned all the way to one side.

The wheel was held in position for 4 seconds to load the system. As the system pressure rose above the limit, the PSP switch circuit opened. This can be seen in the example when the signal voltage rises from 0v to 13v while the power steering system was loaded.

Lab Scope Example (2)
In example (2), the trace shows the PSP switch signal as the steering was moved rapidly from one side to another at idle with the vehicle stationary.

It may be difficult to overcome the switch opening pressure threshold using this test, depending on tire inflation and size, surface texture, and switch calibration. If the results shown in example (2) are not obtained, it may be necessary to turn the wheel to full lock, as in example (1). Use an appropriate pressure tester if needed to discern between a failed sensor and poor power steering system performance.

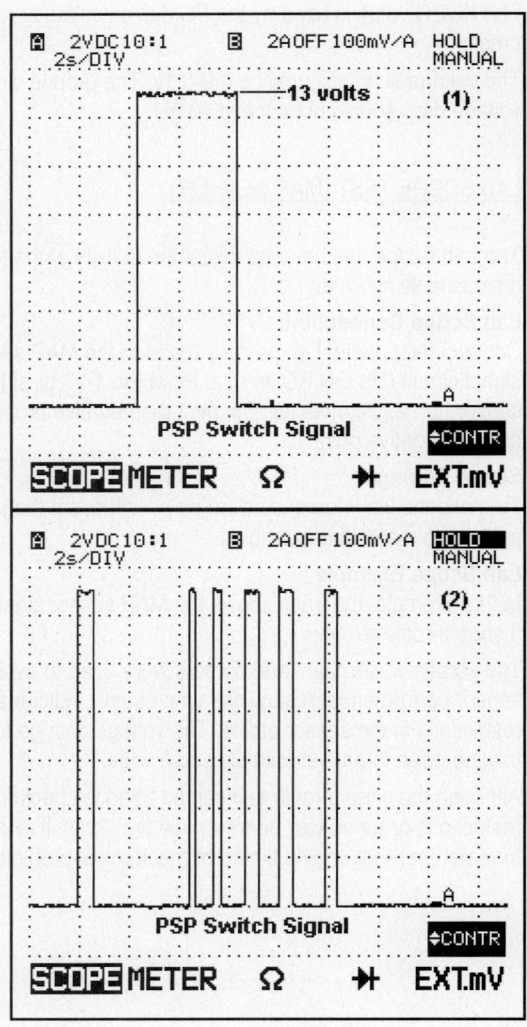

HEATED OXYGEN SENSORS

The heated oxygen sensors (HO2S) are mounted in the Exhaust system where they monitor the oxygen content of the exhaust stream. During engine operation, oxygen present in the exhaust reacts with the HO2S to produce a voltage signal.

If the A/F mixture has a high concentration of oxygen in the exhaust, the signal to the PCM from the HO2S will be less than 0.4v. If the A/F mixture has a low concentration of oxygen in the exhaust, the signal to the PCM will be over 0.6v.

Heated Oxygen Sensors

This vehicle application is equipped with four (4) oxygen sensors with internal heaters that keep the sensors at a high temperature for more efficient operation.

Oxygen Sensor Identification

The front heated oxygen sensor on the cylinder #1 side of the engine is identified as Bank 1 HO2S-11. The Bank 1 rear heated oxygen sensor is identified as HO2S-12. For Bank 2, the sensors are referred to as HO2S-21 and HO2S-22. On this application, HO2S-12 and HO2S-22 are mounted in the warm-up catalytic converter housings.

Circuit Description

Oxygen sensors generate voltage signals, so they do not need any reference voltage. The HO2S signal circuits and sensor ground circuits connect directly to the PCM.

The internal heater circuits in the oxygen sensors (both front and rear) receive power through the ASD relay. On this application, the rear sensor heaters are always on (permanently grounded). However, drivers in the PCM ground the front sensors whenever heater operation is desired to enhance oxygen sensor performance.

Heated Oxygen Sensor Schematic

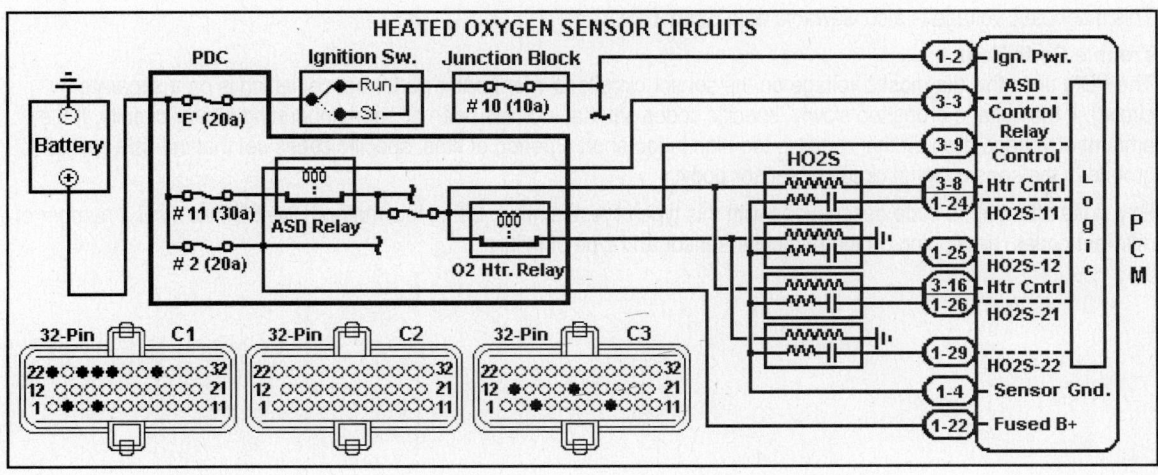

LAB SCOPE TEST (COLD HO2S)

The Lab Scope can be used to observe the warm-up process for the oxygen sensors and the catalytic converter. This test can also be used to observe the HO2S Heater Monitor on this application. Place the gearshift selector in Park and block the drive wheels for safety.

Lab Scope Connections
Connect the Channel '1' through '4' positive probes to the HO2S signal wires at the sensors or appropriate PCM Pin. Connect the negative probe to the battery negative post.

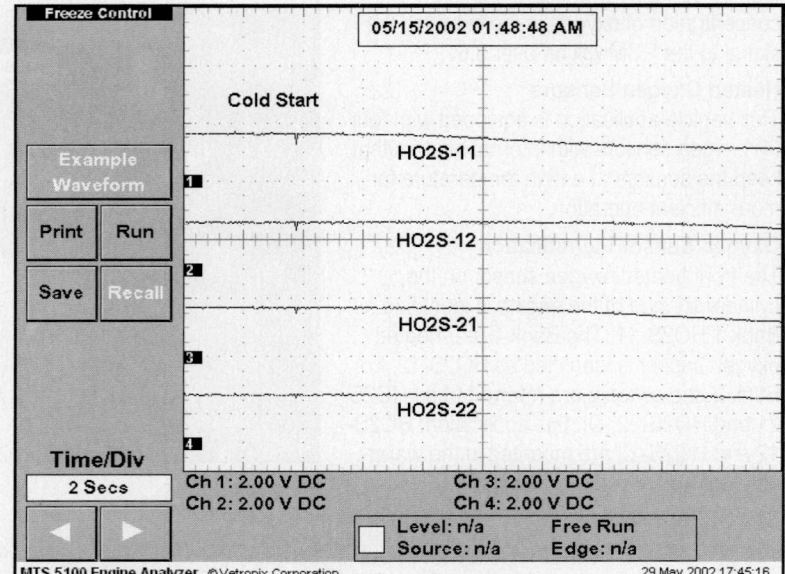

Lab Scope Example
In this example, traces from all four (4) oxygen sensors were captured after the first 20 seconds from a cold start. None of the sensors were active during this test, and the vehicle remained in open loop. Observe the waveforms over time to evaluate sensor performance under normal operating conditions.

HO2S Heater Monitor
Note that the Lab Scope was set to 2 volts per division for each channel. All four (4)-sensor waveforms were between 1.5 and 2 volts in this capture. This is not an actual sensor signal, as the normal sensor output range is from 0 to 1 volt.

During a cold start, the PCM applies 5 volts to this circuit and monitors the time it takes for the voltage to drop from 4 to 3 volts as each sensor is heated. The more effectively the heater circuits are working, the faster and farther the voltage will drop. This example represents heater performance on a known good vehicle with less than 10,000 miles on the odometer.

This diagnostic voltage is also viewable on a Scan Tool.

Trouble Code Help
The PCM uses this diagnostic voltage on the sensor circuits to infer heater performance (which is on a separate circuit). If the voltage drops too slowly, specific codes set that indicate open or poorly operating heater circuits. If the amount of voltage drop on this circuit is too high in too short a period of time, specific codes set that indicate a short to ground in the sensor signal circuit or sensor body.

Pay close attention to code descriptions with this type of test in mind. Using this method, the PCM can set a number of different codes, depending on the particular sensor and type of failure.

<u>LAB SCOPE TEST (WARM HO2S)</u>

The Lab Scope can be used to observe the warm-up process for sensors and the catalytic converter. Place the gearshift selector in Park and block the drive wheels for safety.

Lab Scope Connections
Connect the Channel '1' through '4' positive probes to the HO2S signal wires at the sensors or appropriate PCM Pin. Connect the negative probe to the battery negative post.

Lab Scope Example
In this example, the settings were changed to 0.50 volts per division to better view the normal sensor activity. The traces from all four (4) oxygen sensors were monitored as the vehicle entered closed loop.

This capture was taken after less than 1 minute had elapsed from a cold engine startup.

The downstream sensors (HO2S-12 and HO2S-22) remained flat from the cold start through this capture. In this case, the warm-up catalytic converters both became active (started storing and releasing oxygen) before the downstream oxygen

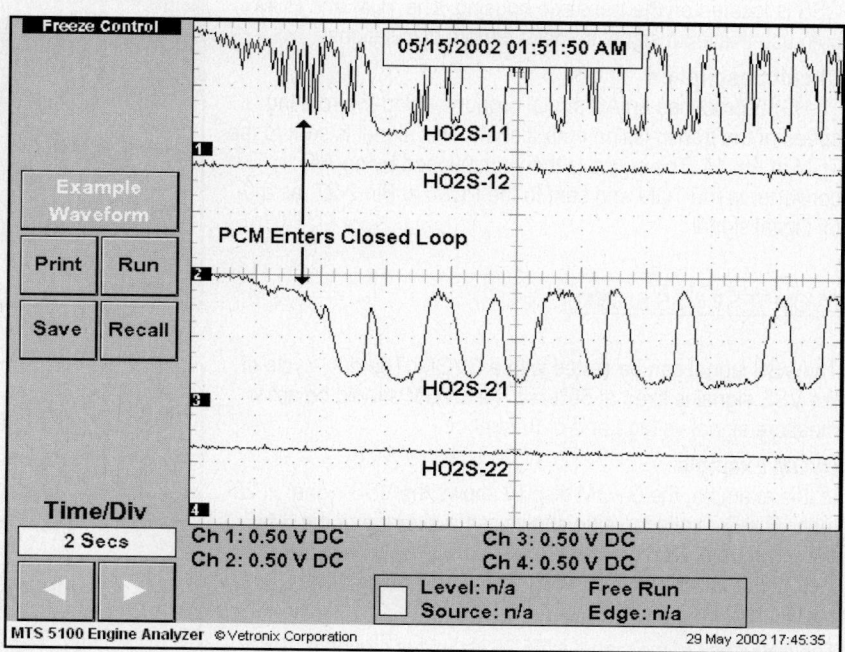

sensors became active. To verify that a flat downstream sensor trace indicates an efficient converter and not a failed or lazy sensor, snap the throttle a few times and look for a change in the downstream sensor value. Although a converter may be efficient, rapidly changing the oxygen content of the exhaust will result in a change in downstream sensor voltage.

Trouble Code Help
It is important to create the proper conditions when analyzing the efficiency of the catalytic converter. The OBD II Catalyst Monitor runs only under the conditions where it is most likely to pass. This Monitor will only run after 3 minutes from a start, at steady speeds over 20 mph.

Under cruise conditions, the sensor and converter warming process is greatly accelerated. The PCM is attempting to verify that the switch rate of HO2S-12 (or HO2S-22) as compared to that of HO2S-11 (or HO2S-21) is less that 70% at steady cruise. If this ratio is exceeded, a DTC P0420 (or P0430) will set (2-trip code). Use this test to recreate these conditions to verify that a catalytic converter has failed.

Vehicle Speed Sensor

INTRODUCTION (OSS & VSS)

The Vehicle Speed Sensor (VSS) on this application is actually a conditioned signal from the Transmission Control Module (TCM) based on the Output Speed Sensor (OSS) signal. The OSS is located on the transaxle housing. The TCM and PCM both use this input signal to make certain calculations.

Circuit Description
The OSS produces an AC signal proportional to the rotating speed of the transmission output shaft. This signal is sent to the TCM at Pin 14. The signal is then conditioned by an A/D converter in the TCM and sent to the PCM, at Pin 2-27, as a 0-5v digital signal.

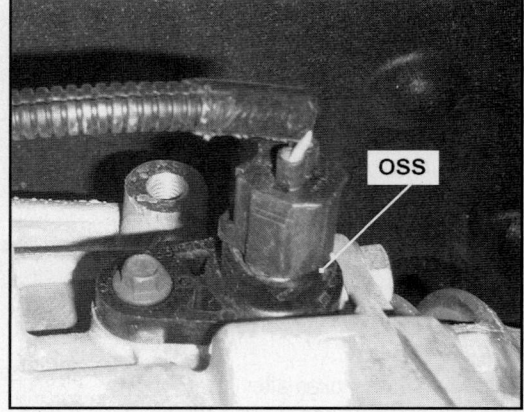

DVOM TEST (VSS SIGNAL)

The VSS signal can be tested with a DVOM. The duty cycle of the VSS signal is fixed at 50%, so the DVOM should be set to measure signal voltage and/or frequency.

DVOM Example
In this example, the DVOM display shows the VSS signal at 20 mph. The DC voltage value shown is the average of the 0-5v pulse from the TCM on this circuit. Use MIN/MAX settings on the DVOM to see the peak voltage of the signal and the condition of the ground circuit.

The frequency of the signal is 45.25 Hz in this example. Verify that this value changes as the vehicle speed changes.

Vehicle Speed Sensor Test Schematic

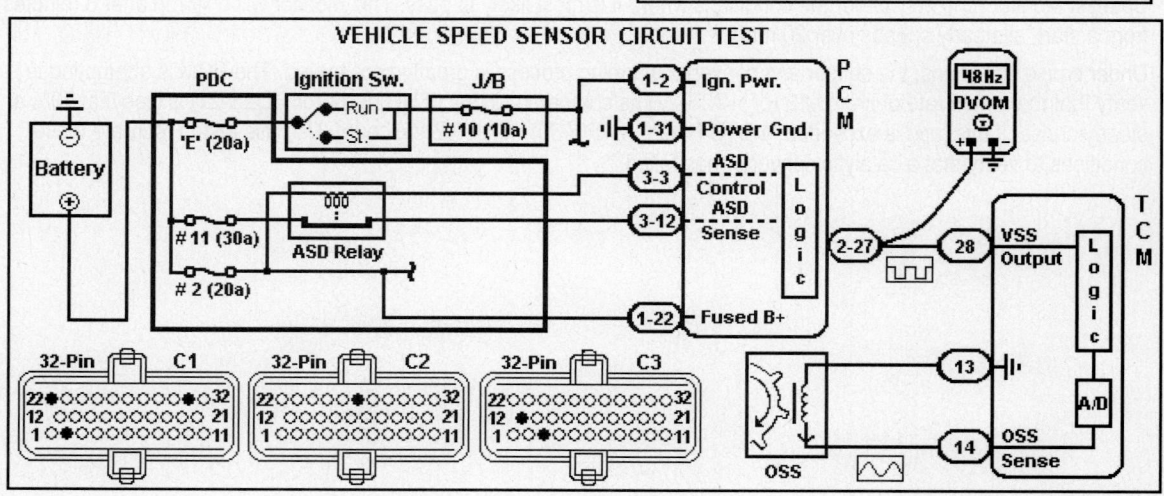

LAB SCOPE TEST (OSS & VSS)

The Lab Scope can be used to check the Output Speed Sensor (OSS) and the calculated Vehicle Speed Sensor (VSS) signal. The OSS is located on the transaxle housing. The Transmission Control Module (TCM) and PCM both need to know the vehicle speed to make certain calculations.

Circuit Description

The OSS produces an AC signal proportional to the rotating speed of the transmission output shaft. This signal is sent to the TCM at Pin 14. The signal is then converted by an A/D converter in the TCM and sent to the PCM, at Pin 2-27, as a 0-5v digital signal.

Lab Scope Connections

Connect the positive probe to the appropriate sensor signal wire at the correct module, and the negative probe to the battery negative post.

The OSS signal wire is connected to the TCM at Pin 14 (LG/WT wire). The VSS signal wire is connected to the TCM at Pin 28, and to the PCM at Pin 2-27 (WT/OR wire).

Lab Scope Example (1)

In example (1), the trace shows the OSS signal at 25 mph. This sensor produces an AC signal that increases in frequency and amplitude in proportion to output shaft speed. However, the TCM is only watching the change in frequency to determine shaft speed, as long as the amplitude is above 300 mv. In this example, the signal was 23 volts AC, peak to peak.

Lab Scope Example (2)

In example (2), the trace shows the 0-5v digital VSS signal from the TCM to the PCM. The VSS is calculated by the TCM from the OSS signal.

This example was captured at about 20 mph.

Diagnostic Help

A problem with the OSS or circuit will affect the VSS signal. However, the VSS signal may be missing with no OSS or circuit problems. Pay attention to symptoms and codes (and what modules set them) to determine the most efficient test procedure.

Generally, thinking through the circuit in this way can logically eliminate most potential test procedures, which can save

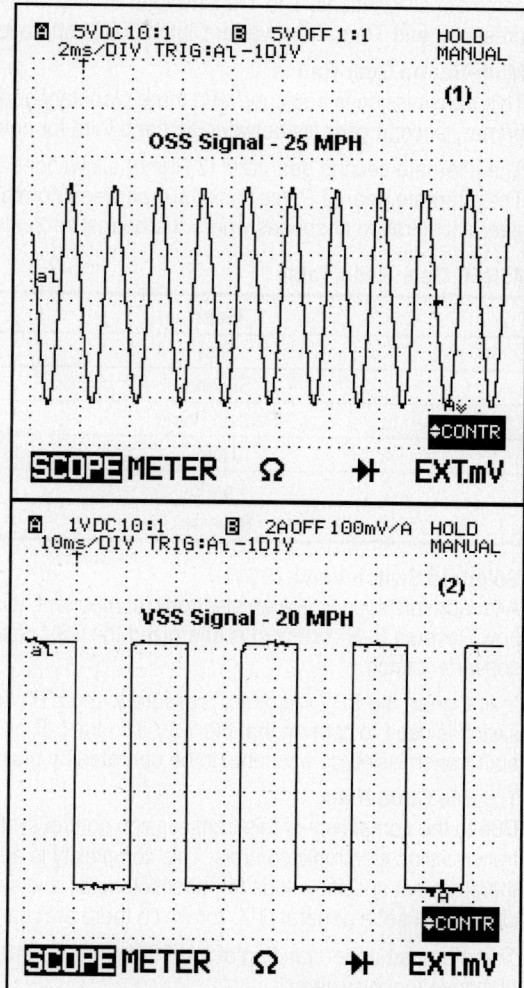

diagnostic time. For example, if the OSS signal is present at the TCM, but the VSS signal is not present at the PCM, then the fault can only be that the TCM is not processing the signal correctly, or the circuit between the TCM and PCM is damaged. By understanding the origin of the signal, you can isolate the problem to only two possibilities.

Transmission Controls

INTRODUCTION

This vehicle is equipped with a 45RFE electronic automatic transmission. This transmission features full electronic control of all functions. The Transmission Control Module (TCM) is able to precisely control all shift schedules, line pressure, and Torque Converter Clutch (TCC) operation to deliver both efficiency and performance.

Multi-Range Gear Ratios

This transmission is equipped with three planetary gear sets, which allow for a unique alternate second gear ratio. The primary second gear fits between first and third for normal sequential shifts on acceleration.

The alternate second gear ratio (2 Prime) allows for smoother 4-2 downshifts and better performance at high speeds. The alternate gear (2 Prime) ratio is in between normal second and third gears, so it offers much of the normal second gear acceleration at speeds where the normal second gear would be impractical.

45RFE Gear Ratio Table

Gear	Ratio
First	3.00:1
Second	1.67:1
Second Prime	1.50:1
Third	1.00:1
Fourth	0.75:1
Reverse	3.00:1

Solenoid Switch Valve (SSV)

Although this sounds like an electronic component, it is merely a mechanical piston in the valve body. When the Low/Reverse (L/R) solenoid is energized the SSV directs the pressurized fluid to the L/R clutch or to the torque converter clutch.

In first gear, the SSV will direct pressure to the L/R clutch, and to the TCC under all other conditions. The L/R pressure switch is used to confirm that the SSV is in the L/R position when the transmission is in first gear. This valve allows both transmission components to be operated by a single solenoid.

Trouble Code Help

Due to the complexity of this transmission control system, there are 17 different codes that are directly related to transmission electronic controls. This complexity is an advantage because the trouble codes are so specific. However, there are not specific codes for most solenoids and switches. A failure in most of the transmission control or sense circuits will set a general DTC for which there are many possible causes.

Therefore, an understanding of the TCM control strategy and function of components is necessary to effectively diagnose these systems.

Limp-In Operation

If a problem is detected that could damage the transmission or prevents the TCM from properly controlling transmission functions, a limp-in strategy is used. This strategy allows only second and third gear operation. Also, a code will be set in the TCM.

The Multi-Select (MS) solenoid, in the solenoid assembly, is used with the UD solenoid to enable only second and third gear operation during limp-in. The UD solenoid has other functions, but the MS solenoid is used exclusively for limp-in operation.

TRANSMISSION SOLENOIDS & SWITCHES

The transmission solenoid and switch assembly, mounted internally to the valve body, also includes the range selector assembly. The assembly consists of five (5) pressure switches and six (6) solenoids that control and monitor six (6) clutches.

Shift Solenoid Operation

The solenoids are used to control the Low/Reverse (L/R), Second (2C), Fourth (4C), Overdrive (OD), Underdrive (UD), and torque converter clutches. The L/R clutch and TCC are controlled by the same solenoid; so five (5) solenoids control six (6) clutches. The sixth solenoid (MS) controls limp-in operation.

The Transmission Solenoid Application Table shows which solenoids are energized under each operating condition. It is important to remember that these solenoids are duty cycle controlled. When they are "on" they are actually being cycled "on" and "off".

The TCM monitors each solenoid control circuit for an inductive kick when the solenoid is turned off. If the PCM sees a kick, it knows that the circuit is complete, and that a magnetic field was built and collapsed. This test does not prove that the solenoid is mechanically operational. Pressure switches are used to verify that fluid pressure is diverted to the appropriate hydraulic circuit when the solenoid is commanded on.

Transmission Solenoid Application Table

Gear	L/R Solenoid	2C Solenoid	4C Solenoid	UD Solenoid	OD Solenoid
Park	On	Off	Off	Off	Off
Reverse	On	Off	Off	Off	Off
Neutral	On	Off	Off	Off	Off
1	On	Off	Off	On	Off
2	Off	On	Off	On	Off
2 Prime	Off	Off	On	On	Off
3	Off	Off	Off	On	On
4	Off	Off	On	Off	On

Pressure Switch Operation

All of the pressure switches are calibrated to close at 23 psi, and reopen at 11 psi. The switch inputs are used by the TCM to verify correct solenoid operation.

Each pressure switch is open when the corresponding solenoid is off, and closes when the solenoid is energized. The switch state should toggle rapidly as the corresponding solenoid command is cycled.

Compare the transmission PIDs on the Scan Tool with the values in the table below.

Transmission Pressure Switch Position Table

Gear	L/R Switch	2C Switch	4C Switch	UD Switch	OD Switch
Park	Closed	Open	Open	Open	Open
Reverse	Open	Open	Open	Open	Open
Neutral	Closed	Open	Open	Open	Open
1	Closed	Open	Open	Closed	Open
2	Open	Closed	Open	Closed	Open
2 Prime	Open	Open	Closed	Closed	Open
3	Open	Open	Open	Closed	Closed
4	Open	Open	Closed	Open	Closed

CLUTCH VOLUME INDEX (CVI)

An Important function of the TCM is to monitor the Clutch Volume Index (CVI). This index represents the volume of fluid needed to compress a clutch pack.

The TCM monitors gear ratio changes by watching the input and output speed sensors (ISS & OSS). By comparing the speed sensors, the TCM can determine the actual transmission gear ratio. This is important to the calculation because the TCM determines the CVI index by counting how long it takes for each gear change to occur. Once the ratio of the ISS and OSS has stabilized, the shift is considered complete.

The TCM uses the CVI data to compensate for transmission wear by increasing the solenoid duty cycle to the affected clutch.

The CVI values can be seen on the Scan Tool and compared to the Clutch Volume Index Table to help determine if a particular clutch pack is worn. Some other causes of failure include broken return springs, excessive clearance, and improper assembly.

Clutch Volume Index Table

Clutch	Updated During:	Clutch Volume
L/R	2-1 or 3-1 Shift, coasting	45-134
2C	3-2 Kick down	25-85
OD	2-3 Shift	30-100
4C	3-4 Shift	30-85
UD	4-3 Kick down	30-100

SCAN TOOL TEST (CVI MONITOR)

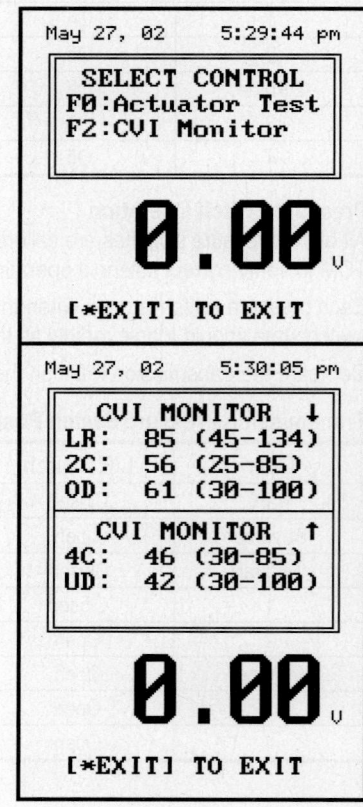

The Scan Tool can be used to observe the most recent TCM clutch volume calculations. The values are updated any time the specific shift occurs, as long as certain time and temperature criteria are satisfied.

Scan Tool Setup
Connect the Scan Tool to the underdash DLC. Identify or confirm vehicle identity. Choose Transmission, and then select F4: OBD CONTROLS. From the controls menu, select F2: CVI Monitor.

Scan Tool Test
The clutch volume calculations are displayed in this example. The vehicle has only a few thousand miles on the odometer, so it is expected that all of the calculations will be within the range. If any clutch pack was slipping or took too long to shift because of wear, the CVI calculation will be higher than the range. The range is shown in parentheses.

The Vetronix Mastertech, used in this example, displays the known-good range of CVI values. Not all Scan Tools display this information, but the acceptable values are readily available in other repair information. Be sure to look up the proper specifications, as the ranges are very different from one vehicle application to the next.

SCAN TOOL TEST (A/T PIDS)

The Scan Tool is an excellent tool to use to diagnose both electrical and mechanical problems in the transmission, TCM or related circuits. Further testing is often required, but the Scan Tool will help to determine the best diagnostic path to follow.

Scan Tool Setup
Connect the Scan Tool to the underdash DLC. Identify or confirm vehicle identity. Choose transmission, and confirm application (as shown in example). Select F0: Datalist from the SELECT MODE menu.

Scan Tool Example
There are many uses of the Scan Tool for transmission diagnosis. In this example, the capture shows the PID values at Hot Idle with the transmission in Neutral.

Line Pressure PIDs
Observe the following PIDs to check the Pressure Control Solenoid (PCS), line pressure sensor, and circuits: PCS Duty Cycle, Line Pressure (V), Line Pressure, and Desired Line Pressure.

The PCS Duty Cycle PID shows the amount of time the TCM grounds the Pressure Control Solenoid during each cycle. The function of the solenoid is actually to bleed off unwanted line pressure, so the PID value of 5% indicates the TCM is attempting to bleed off very little pressure

Pressure Switch PIDs
The solenoid commands cannot be seen as PIDs. However, the vehicle can be driven through the gears, and the pressure switch PIDs can be observed.

Range Selector Position PIDs
The C1 through C5 PIDs are transmission range selection indicators. Each PID means nothing on its own. The TCM looks at the combinations of switch positions to determine transmission range selector position.

The five (5) switch values are combined into 7 possible expected combinations indicating valid gear positions. The switch values can be combined to indicate a between-gear condition.

If the switch value combination does not indicate the transaxle is in a valid gear or between gears, the TCM sees an impossible condition and sets a code. The TCM will continue to shift the transmission using only the pressure switches to determine shift lever position. This allows for reasonably normal transmission operation even with a TRS or circuit failure.

```
May 27, 02          5:23:10 pm

 45RFE TRANS
S/W VER: 18
TCM:56028694AC
4.7L MPI    [YES]

      0.00 v

[*EXIT] TO EXIT
```

```
   SELECT MODE ↑↓
→F0:Datalist
 F2:DTCs
 F3:Snapshot
```

```
PCS Duty Cycle
     5%
Line Pressure(V)
   2.16 Volts
Line Pressure
   117  PSI
Des. Line Press.
   120  PSI
Applied Gear
   Neutral
Actual Gear
   Neutral
LR/CC Press. Sw.
   Closed
Limp-In Status
   Normal
2C Pressure Sw.
   Open
OD Pressure Sw.
   Open
4C Pressure Sw.
   Open
UD Pressure Sw.
   Open
C5
   Closed
O/D Lockout Sw.
   OD On
C3
   Open
C4
   Open
C1
   Closed
C2
   Closed
Trans Temp State
   Hot
Trans. Oil Temp
   167° F
Trans. Temp (V)
   3.06 Volts
```

LAB SCOPE TEST (SHIFT SOLENOIDS)

The solenoid commands are duty cycle controlled, so pressure is applied to the clutches in pulses. If the solenoid and its circuits are operating properly, the pressure switches should open and close in response to the solenoid commands.

Lab Scope Connections

Connect the Channel 'A' positive probe to the 2C Solenoid control circuit (WT/DB wire) at TCM Pin 19, and connect the negative probe to the battery negative post. Connect the Channel 'B' probe to the 2C Pressure Switch sense circuit (YL/BK wire) at TCM Pin 47.

Lab Scope Examples

In the examples, the traces show the 2C solenoid command cycling to apply hydraulic pressure to the 2C clutch, and the 2C pressure switch closing to provide the TCM with feedback about solenoid operation. These captures were taken while the transmission was shifting up to second gear.

Note that each time the solenoid is turned "off" (stops cycling) there is an inductive spike when the solenoid field collapses. The spike is limited to about 42 volts by a diode in the TCM.

Using Test Results

The TCM calculates the solenoid activity using the load, speed, temperature, throttle angle, and line pressure input signals. The volume of hydraulic fluid directed to the clutch is also calibrated by the TCM according to transmission wear. The Clutch Volume Index (CVI) is used by the TCM to pinpoint clutch wear.

Transmission 2C Solenoid and Switch Test Schematic

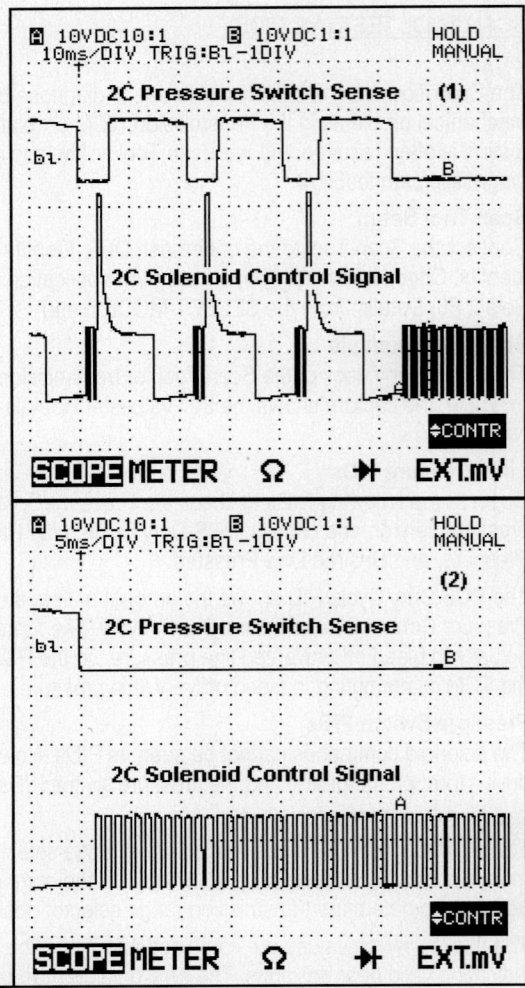

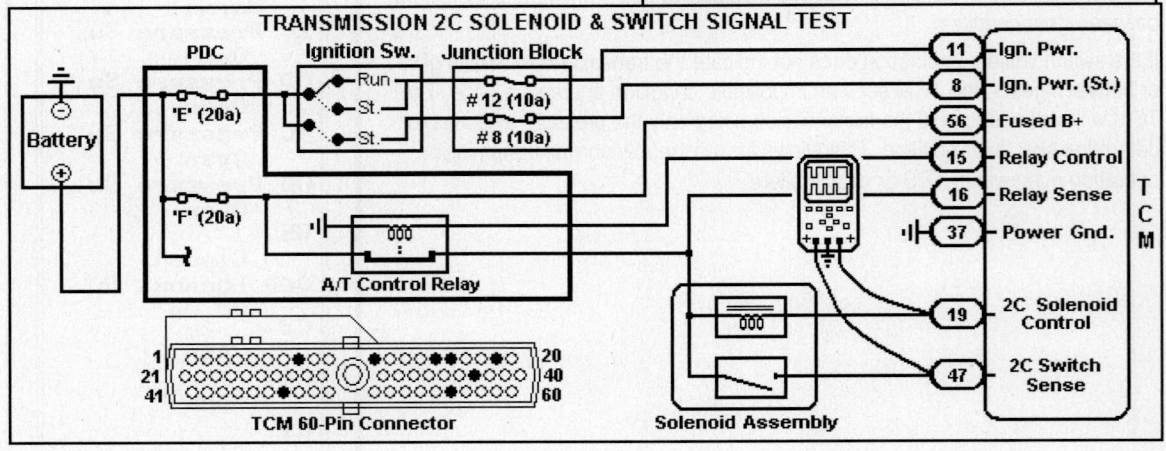

PRESSURE CONTROL SOLENOID AND SENSOR

The TCM is able to precisely control the transmission line pressure using a closed loop solenoid and sensor system. The Pressure Control Solenoid (or Line Pressure Solenoid) is a variable force style solenoid used to control hydraulic pressure. The solenoid is duty cycle controlled to bleed off unwanted line pressure. The default position is full line pressure, so a solenoid or circuit failure will not disable the transmission. The TCM makes duty cycle decisions using a Line Pressure Sensor. Both components are mounted in the transmission solenoid assembly.

LAB SCOPE TEST (PRESSURE CONTROL SOLENOID)

Place the gear selector in Park and block the drive wheels for safety.

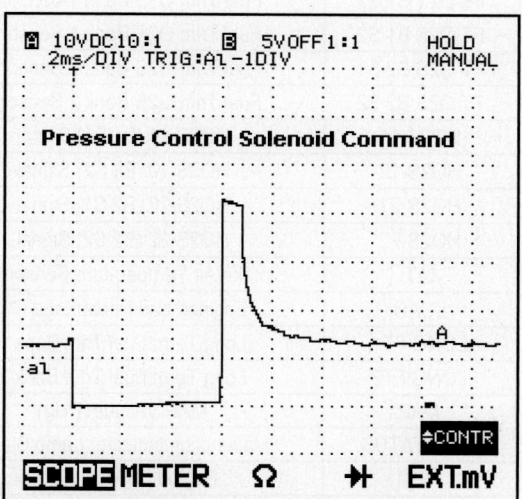

Lab Scope Connections
Connect the Channel 'A' positive probe to the pressure solenoid circuit (YL/DB wire) and TCM Pin 18, and the negative probe to the battery negative post.

Lab Scope Example
In this example, the trace shows the solenoid circuit being grounded for 6 ms to bleed off unwanted line pressure. When the solenoid is ungrounded, the field collapses and an inductive spike is produced. The spike is limited to about 42 volts by a diode in the TCM.

The pressure sensor signal can also be displayed on the Lab Scope if necessary.

The solenoid command trace will vary depending on the line pressure and driving conditions. However, the circuit voltage should always be pulled to near ground when the solenoid is commanded "on", and should always have an inductive spike when commanded "off".

Transmission Pressure Control Solenoid Test Schematic

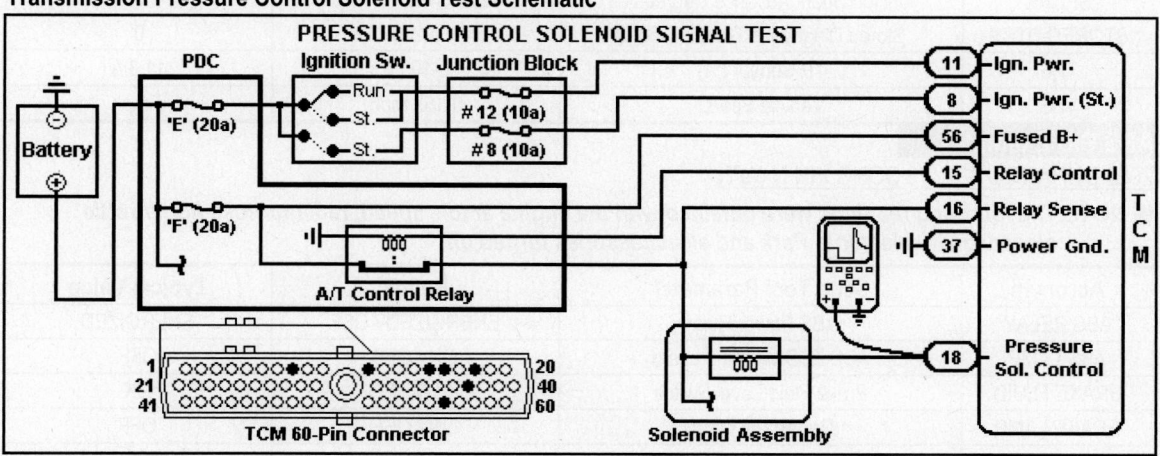

2002 DURANGO - POWERTRAIN

Reference Information - PCM PID Examples (Generic Scan Tool)

2002 DURANGO 4.7L V8 SOHC VIN N (A/T)

■ NOTE: *The following readings were obtained with the engine at idle speed, radiator hose hot, throttle closed, gear selector in Park and all accessories turned off.*

Acronym	Scan Tool Parameter	Range	Typical Value
ECT	Engine Coolant Temp. Sensor (°F)	-40 to 284°F	203°F
FUEL STAT1	Fuel Status Bank 1	CL / OPEN	CL
FUEL STAT2	Fuel Status Bank 2	CL / OPEN	CL
FT O2S B1 S1	Fuel Trim O2S Bank 1 Sensor 1	0% (± 20%)	0.0%
FT O2S B1 S2	Fuel Trim O2S Bank 1 Sensor 2	0% (± 20%)	Not Used
FT O2S B2 S1	Fuel Trim O2S Bank 2 Sensor 1	0% (± 20%)	Not Used
FT O2S B2 S2	Fuel Trim O2S Bank 2 Sensor 2	0% (± 20%)	Not Used
HO2S-11	HO2S-11 (B1 S1) Signal	0-1100 mv	100 mv
HO2S-12	HO2S-12 (B1 S2) Signal	0-1100 mv	740 mv
HO2S-21	HO2S-22 (B2 S1) Signal	0-1100 mv	760 mv
HO2S-22	HO2S-22 (B2 S2) Signal	0-1100 mv	680 mv
IAT	Intake Air Temperature Sensor (°F)	-40 to 284°F	147°F
LOAD	Engine Load Percentage (%)	0-100%	3.5%
LONG FT1	Long Term Fuel Trim Bank 1	0% (± 20%)	3.9%
LONG FT2	Long Term Fuel Trim Bank 2	0% (± 20%)	4.7%
MAP	MAP Vacuum ("Hg)	0-30" Hg	10.7" Hg
MIL STATUS	Malfunction Indicator Lamp Status	ON / OFF	OFF
RPM	Engine Speed in RPM	0-66635 rpm	608
SHRT FT1	Short Term Fuel Trim Bank 1	0% (± 20%)	2.3%
SHRT FT2	Short Term Fuel Trim Bank 2	0% (± 20%)	0.0%
SPARK	Ignition Spark Advance Degrees (°)	-90° to +90° BTDC	4°
STORED DTCS	Stored Diagnostic Trouble Codes	0-256	0
TPS	TP Sensor (%)	0-100%	11.3%
VSS	Vehicle Speed	0-159 mph	0 mph

ABS PID Examples

2002 DURANGO 4.7L V8 SOHC VIN N (A/T)

■ NOTE: *The following readings were obtained with the engine at idle speed, radiator hose hot, throttle closed, gear selector in Park and all accessories turned off.*

Acronym	Scan Tool Parameter	Range	Typical Value
ABS RELAY	ABS Power Relay	ENERGIZED / OFF	ENERGIZED
ABS LAMP	ABS EBD Warning Lamp	ON / OFF	OFF
BRAKE FLUID	Brake Fluid Level Switch	OK / NOT OK	OK
BRAKE LAMP	Red Brake Lamp	ON / OFF	OFF
BRAKE SW	Brake Light Switch Status	CLOSED / OPEN	OPEN
REAR DUMP	Rear Dump Valve Command	ON / OFF	OFF
REAR ISO VLV	Rear ISO Valve Command	ON / OFF	OFF
REAR WSS	Rear Wheel Speed Sensor	0-159 mph	0 mph

TCM PID Examples

2002 DURANGO 4.7L V8 SOHC VIN N (A/T)

■ **NOTE:** *The following readings were obtained with the engine at idle speed, radiator hose hot, throttle closed, gear selector in Park and all accessories turned off.*

Acronym	Scan Tool Parameter	Range	Typical Value
2C CVI	2C CVI	0-256	56
2C SW	2C Pressure Switch	CLOSED / OPEN	OPEN
4C CVI	4C CVI	0-256	46
4C SW	4C Pressure Switch	CLOSED / OPEN	OPEN
ACT GEAR	Actual Gear	NEUTRAL / DRIVE	NEUTRAL
APP GEAR	Applied Gear	NEUTRAL / DRIVE	NEUTRAL
C1 SW	C1 Switch	CLOSED / OPEN	CLOSED
C2 SW	C2 Switch	CLOSED / OPEN	CLOSED
C3 SW	C3 Switch	CLOSED / OPEN	OPEN
C4 SW	C4 Switch	CLOSED / OPEN	OPEN
C5 SW	C5 Switch	CLOSED / OPEN	CLOSED
ENG RPM	Engine Speed (rpm)	0-66635 rpm	595 rpm
IGN VOLTS	Ignition Feed (volts)	0-12.0v	12.0v
INPUT RPM	Input Shaft Speed (rpm)	0-10,000 rpm	594 rpm
LAST SHIFT	Last Shift Position	1st to Second, etc.	1ST to SECOND
LIMP IN	Limp In Status	Normal / Not Normal	NORMAL
LINE PRESS	Line Pressure (volts)	0-5.1v	2.6v
LINE PRESS	Line Pressure (psi)	0-1000 psi	120 psi
LR / CC	LR / CC Pressure Switch	CLOSED / OPEN	CLOSED
LR CVI	LR CVI	0-256	85
MIN TPS	Minimum TP Sensor (volts)	0-5.1v	0.58v
OD CVI	O/D CVI	0-256	61
OD LOCKOUT	O/D Lockout	OD ON / OD OFF	OD ON
OD SW	O/D Pressure Switch	CLOSED / OPEN	OPEN
OUTPUT RPM	Output Shaft Speed (rpm)	0-10,000 rpm	0 rpm
PCS	PCS Valve Duty Cycle	0-100%	5%
TP V	TP Sensor (volts)	0-5.1v	0.58v at 0°
UD CVI	U/D CVI	0-256	42
UD SW	UD Pressure Switch	CLOSED / OPEN	OPEN
SW BATTERY	Switched Battery (volts)	0-25.5v	13.3v
SHIFT LEVER	Shift Lever Position	PARK / DRIVE	PARK
TRANS TEMP	Transmission Temp. Status	COLD / HOT	HOT
TRANS TEMP	Transmission Oil Temp. (°F)	-40 to 284°F	167°F
TRANS TEMP	Transmission Oil Temp. (volts)	0-5.1v	3.06v

2002 DURANGO 4.7L V8 SOHC VIN N PCM BLACK 32-PIN CONNECTOR 'A' OR 'C1'

PCM Pin #	Wire Color	Circuit Description (32-Pin)	Value at Hot Idle
1	TN/OR	Coil 3 Driver Control	5°, 55 mph: 8° dwell
2	LG/BK	Ignition Switch Power (fused)	12-14v
3	TN/LG	Coil 4 Driver Control	5°, 55 mph: 8° dwell
4	BK/LB	Sensor Ground	<0.050v
5	TN/LB	Coil 6 Driver Control	5°, 55 mph: 8° dwell
6	BK/WT	PNP Switch Sense Signal	In P/N: 0v, Others: 5v
7	BK/GY	Coil 1 Driver Control	5°, 55 mph: 8° dwell
8	GY/BK	CKP Sensor Signal	Digital Signals: 0-5-0v
9	LB/RD	Coil 8 Driver Control	5°, 55 mph: 8° dwell
10	YL/BK	IAC 2 Driver Control	Pulse Signals
11	BR/WT	IAC 3 Driver Control	Pulse Signals
12	DB/OR	PSP Switch Sense Signal	Straight: 0v, Turning: 5v
13-14	---	Not Used	---
15	BK/RD	IAT Sensor Signal	At 100°F: 1.83v
16	TN/BK	ECT Sensor Signal	At 180°F: 2.80v
17	OR	5-Volt Supply	4.9-5.1v
18	TN/YL	CMP Sensor Signal	Digital Signals: 0-5-0v
19	GY/RD	IAC 1 Driver Control	Pulse Signals
20	VT/BK	IAC 4 Driver Control	Pulse Signals
21	TN/DG	Coil 5 Driver Control	5°, 55 mph: 8° dwell
22	RD/WT	Fused Battery Power (B+)	12-14v
23	OR/DB	TP Sensor Signal	0.6-1.0v
24	BK/DG	HO2S-11 (B1 S1) Signal	0.1-1.1v
25	TN/WT	HO2S-12 (B1 S2) Signal	0.1-1.1v
26	LG/RD	HO2S-21 (B2 S1) Signal (CAL)	0.1-1.1v
27	DG/RD	MAP Sensor Signal	1.5-1.7v
28	---	Not Used	---
29	TN/WT	HO2S-22 (B2 S2) Signal (CAL)	0.1-1.1v
30	---	Not Used	---
31	BK/TN	Power Ground	<0.1v
32	BK/TN	Power Ground	<0.1v

Pin Connector Graphic

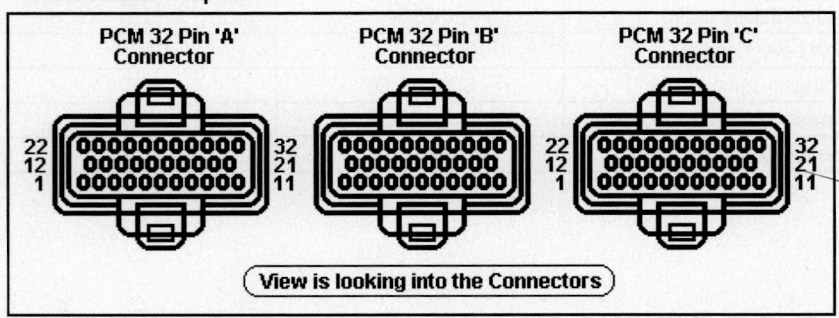

2002 DURANGO 4.7L V8 SOHC VIN N PCM WHITE 32-PIN CONNECTOR 'B' OR 'C2'

PCM Pin #	Wire Color	Circuit Description (32-Pin)	Value at Hot Idle
1	---	Not Used	---
2	PK	Injector 7 Driver	1-4 ms
3	---	Not Used	---
4	W/DB	Injector 1 Driver	1-4 ms
5	YL/WT	Injector 3 Driver	1-4 ms
6	GY	Injector 5 Driver	1-4 ms
7	DB/TN	Coil Driver 7 Control	5º, 55 mph: 8º dwell
8	---	Not Used	---
9	TN/PK	Coil Driver 2 Control	5º, 55 mph: 8º dwell
10	DG	Generator Field Control	Digital Signals: 0-12-0v
11	---	Not Used	---
12	BR/DB	Injector 6 Driver	1-4 ms
13	GY/LB	Injector 8 Driver	1-4 ms
14, 18	---	Not Used	---
15	TN	Injector 2 Driver	1-4 ms
16	LB/BR	Injector 4 Driver	1-4 ms
17	DB/PK	Radiator Fan Relay Control	Relay Off: 12v, On: 1v
19	DB	A/C Pressure Sensor Signal	0.90v at 79 psi
20-22	---	Not Used	---
23	GY/YL	Engine Oil Pressure Sensor	1.6v at 24 psi
24-26	---	Not Used	---
27	WT/OR	Vehicle Speed Sensor	Digital Signals
28-30	---	Not Used	---
31	VT/WT	5-Volt Supply	4.9-5.1v
32	---	Not Used	---

Pin Connector Graphic

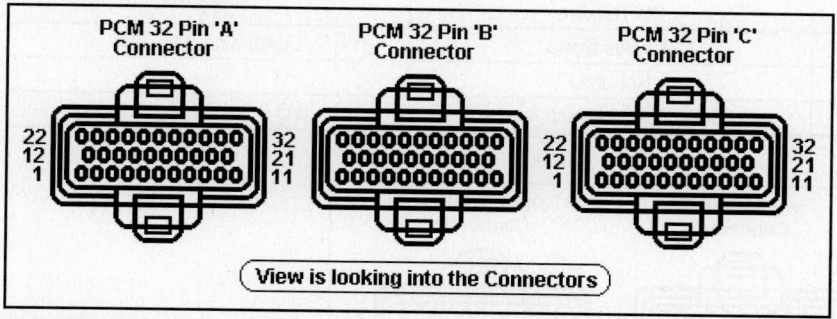

Standard Colors and Abbreviations

Abbreviation	Color	Abbreviation	Color	Abbreviation	Color
BK	Black	GY	Gray	RD	Red
BL	Blue	GN	Green	TN	Tan
BR	Brown	LG	LT Green	VT	Purple
DB	Dark Blue	ON	Orange	WT	White
DG	DK Green	PK	Pink	YL	Yellow

2002 DURANGO 4.7L V8 SOHC VIN N PCM GRAY 32-PIN CONNECTOR 'C' OR 'C3'

PCM Pin #	Wire Color	Circuit Description (32-Pin)	Value at Hot Idle
1	DB/OR	A/C Clutch Relay Control	Relay Off: 12v, On: 1v
2	---	Not Used	---
3	DB/YL	Auto Shutdown Relay Control	Relay Off: 12v, On: 1v
4	TN/RD	S/C Vacuum Solenoid Control	Vacuum Increasing: 1v (S/C On)
5	LG/RD	Speed Control Vent Solenoid	Vacuum Decreasing: 1v (S/C On)
6-7	---	Not Used	---
8	VT/WT	HO2S-11 HTR Control	Relay Off: 12v, On: 1v
9	DG/BK	HO2S-12 HTR Control	Relay Off: 12v, On: 1v
10	WT/DG	LDP Solenoid Control	Solenoid Off: 12v, On: 1v
11	YL/RD	Speed Control Power Supply	12-14v
12	DG/OR	Auto Shutdown Relay Output	12-14v
13	YL/DG	Torque Management Request	Digital Signals
14	OR	LDP Switch Sense Signal	Switch Closed: 0v, Open: 12v
15	PK/YL	Battery Temperature Sensor	At 86ºF: 1.96v
16 (CAL)	VT/OR	HO2S-21 Heater PWM Control	Heater On: duty cycle signals
16 (N/CAL)	VT/OR	HO2S-12 Heater PWM Control	Heater On: duty cycle signals
17-18	---	Not Used	---
19	BR	Fuel Pump Relay Control	Relay Off: 12v, On: 1v
20	PK/BK	EVAP Purge Solenoid Control	Duty cycle: 0-100%
21, 23	---	Not Used	---
22	BR	A/C Switch Signal	A/C Off: 12v, On: 1v
24	WT/PK	Brake Switch Sense Signal	Brake Off: 0v, On: 12v
25	WT/DB	Generator Field Source	12-14v
26	DB/WT	Fuel Pump Relay Control	Relay Off: 12v, On: 1v
27	PK	SCI Transmit	0v
28	---	Not Used	---
29	LG	SCI Receive	5v
30	VT/YL	PCI Bus Signal	Digital Signals: 0-7.5v
31	---	Not Used	---
32	RD/LG	Speed Control Switch Signal	S/C & Set Sw. On: 3.8v

Pin Connector Graphic

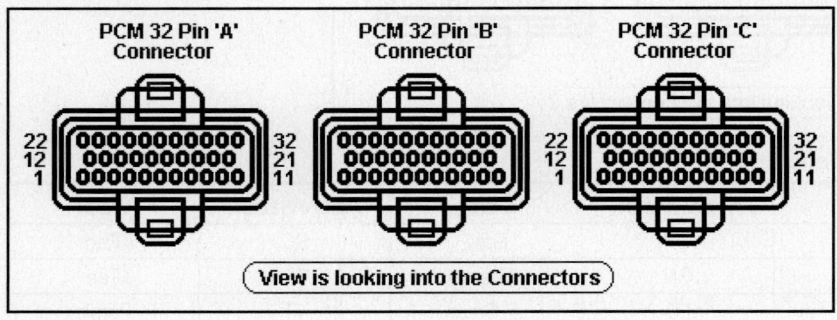

2002 DURANGO - POWERTRAIN

PCM Wiring Diagrams

FUEL & IGNITION SYSTEM DIAGRAMS (1 OF 15)

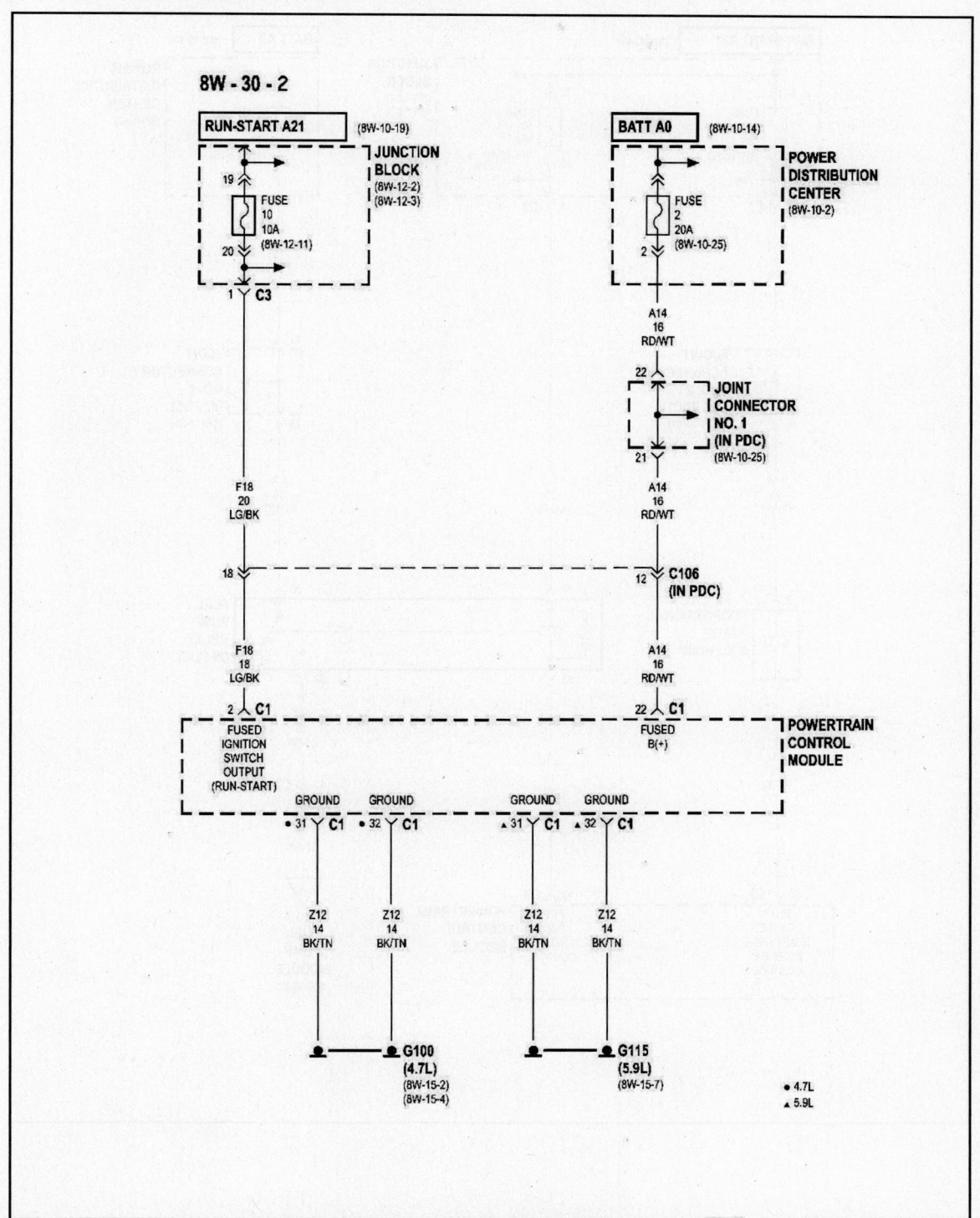

FUEL & IGNITION SYSTEM DIAGRAMS (2 OF 15)

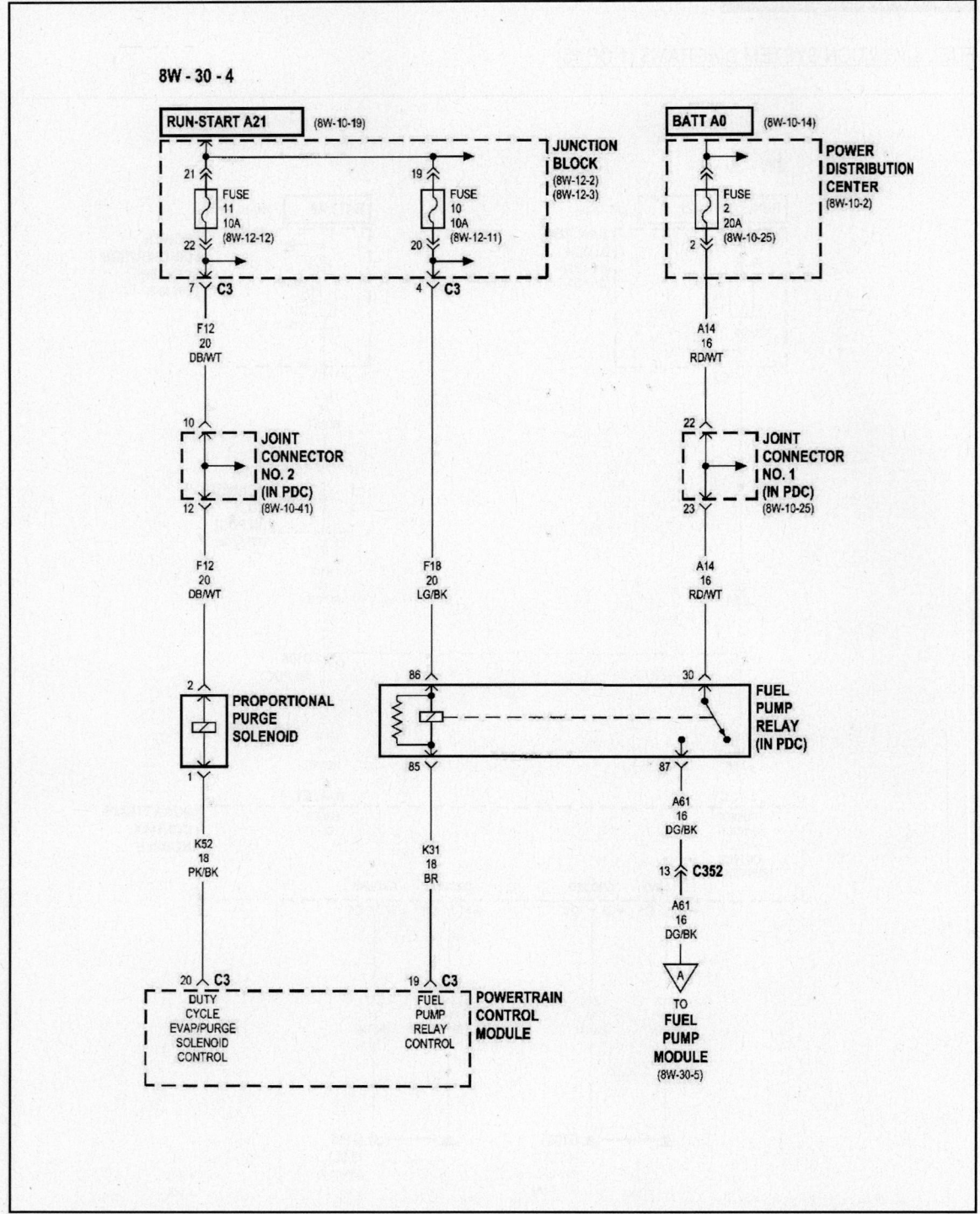

8W - 30 - 4

FUEL & IGNITION SYSTEM DIAGRAMS (3 OF 15)

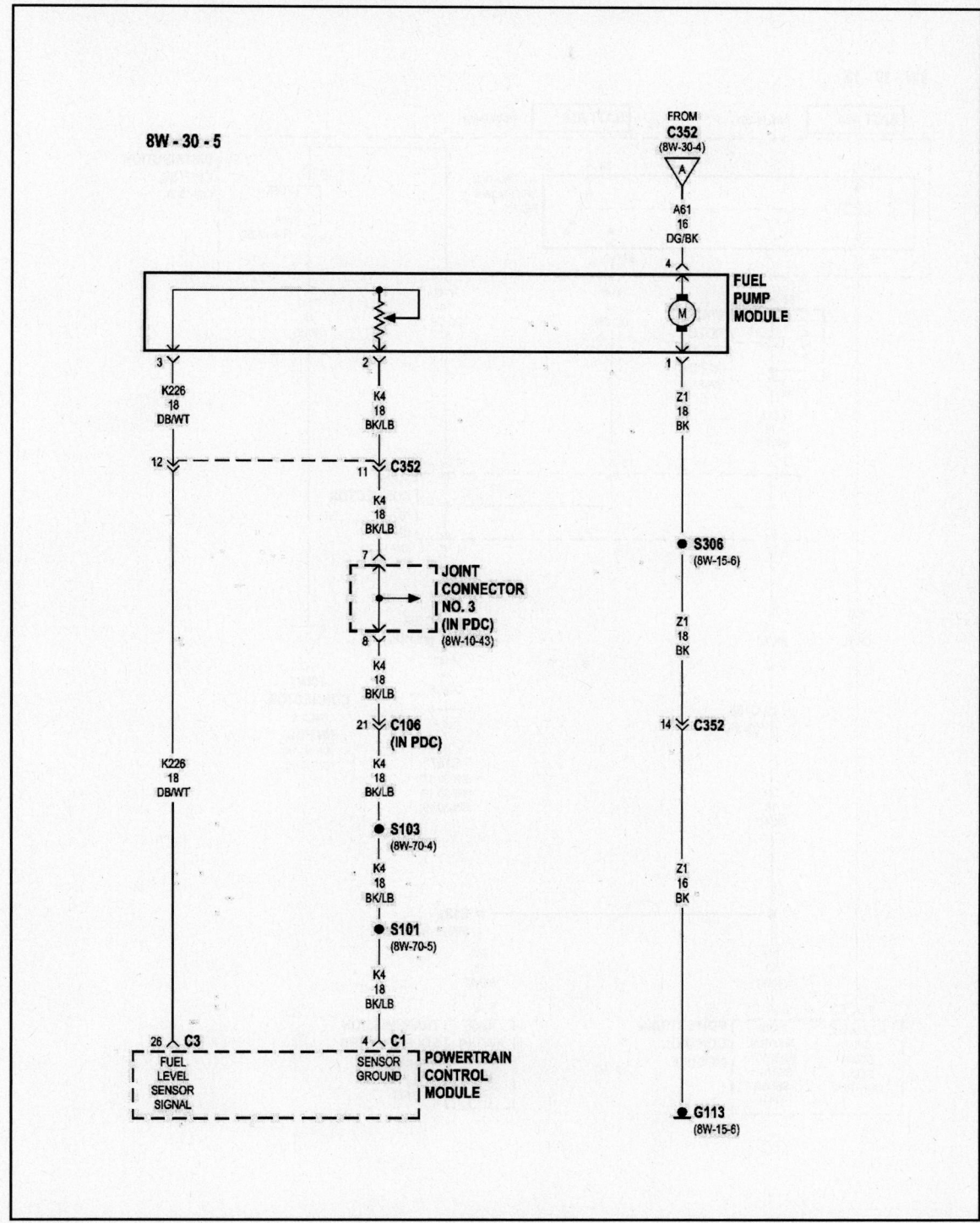

FUEL & IGNITION SYSTEM DIAGRAMS (4 OF 15)

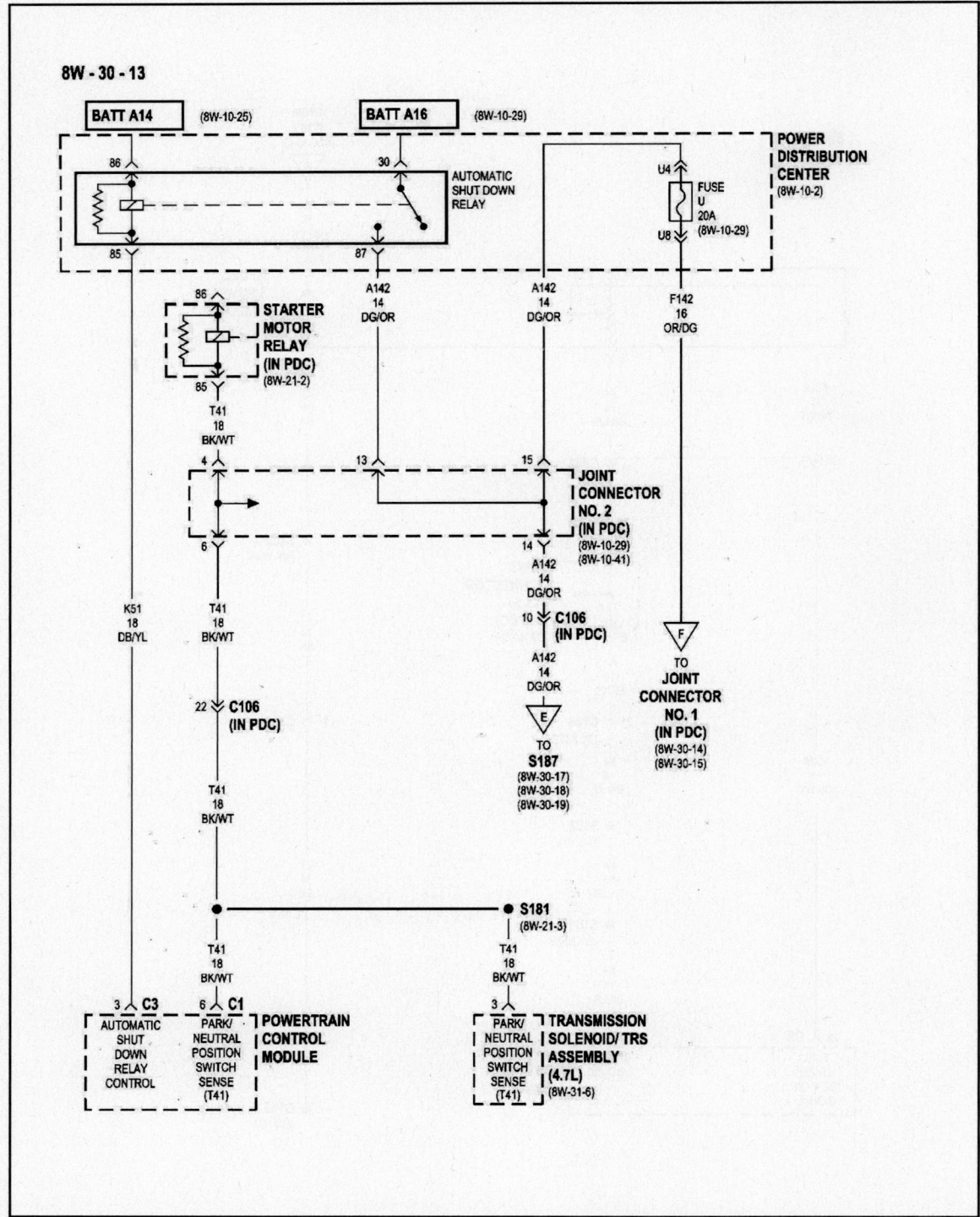

FUEL & IGNITION SYSTEM DIAGRAMS (5 OF 15)

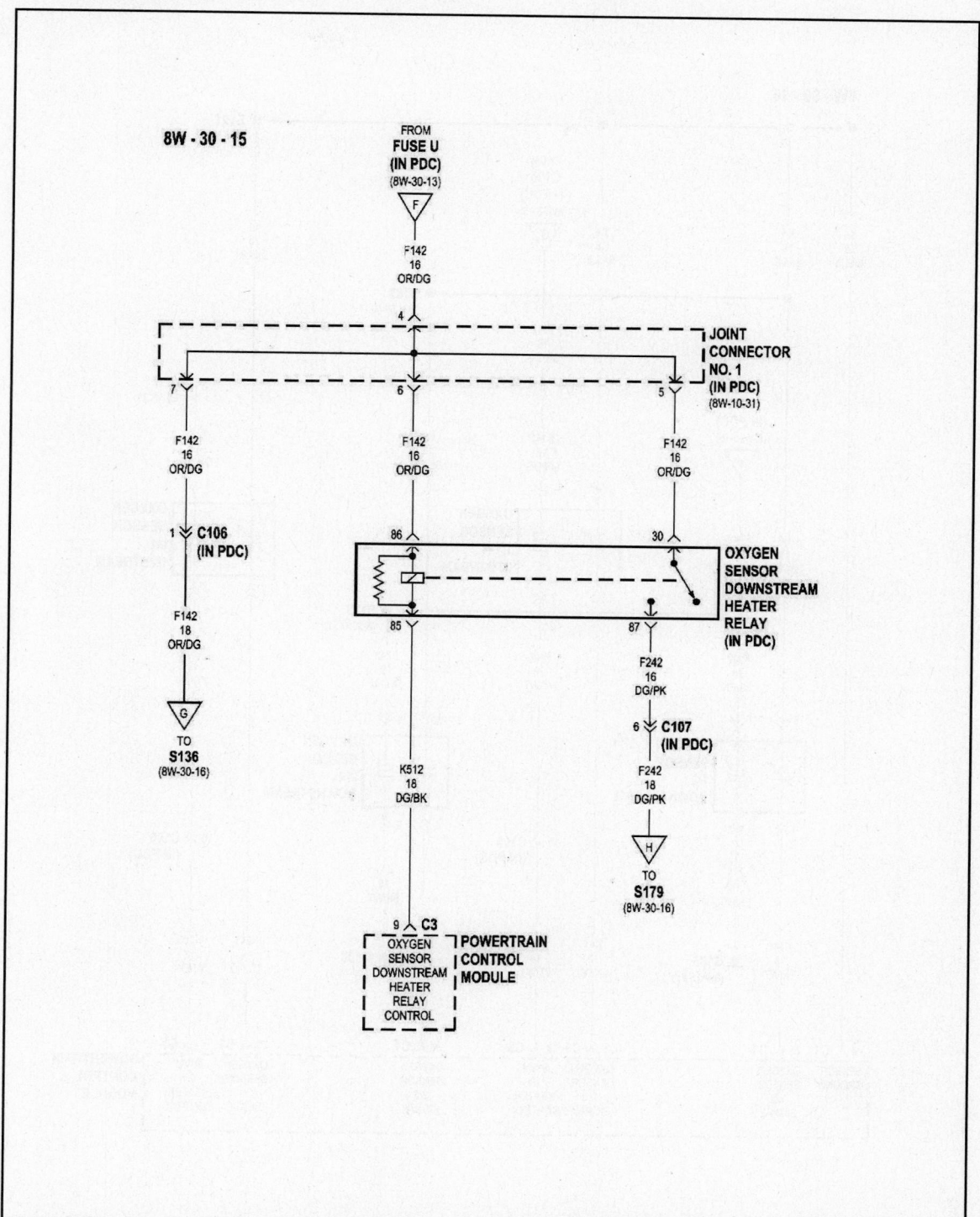

FUEL & IGNITION SYSTEM DIAGRAMS (6 OF 15)

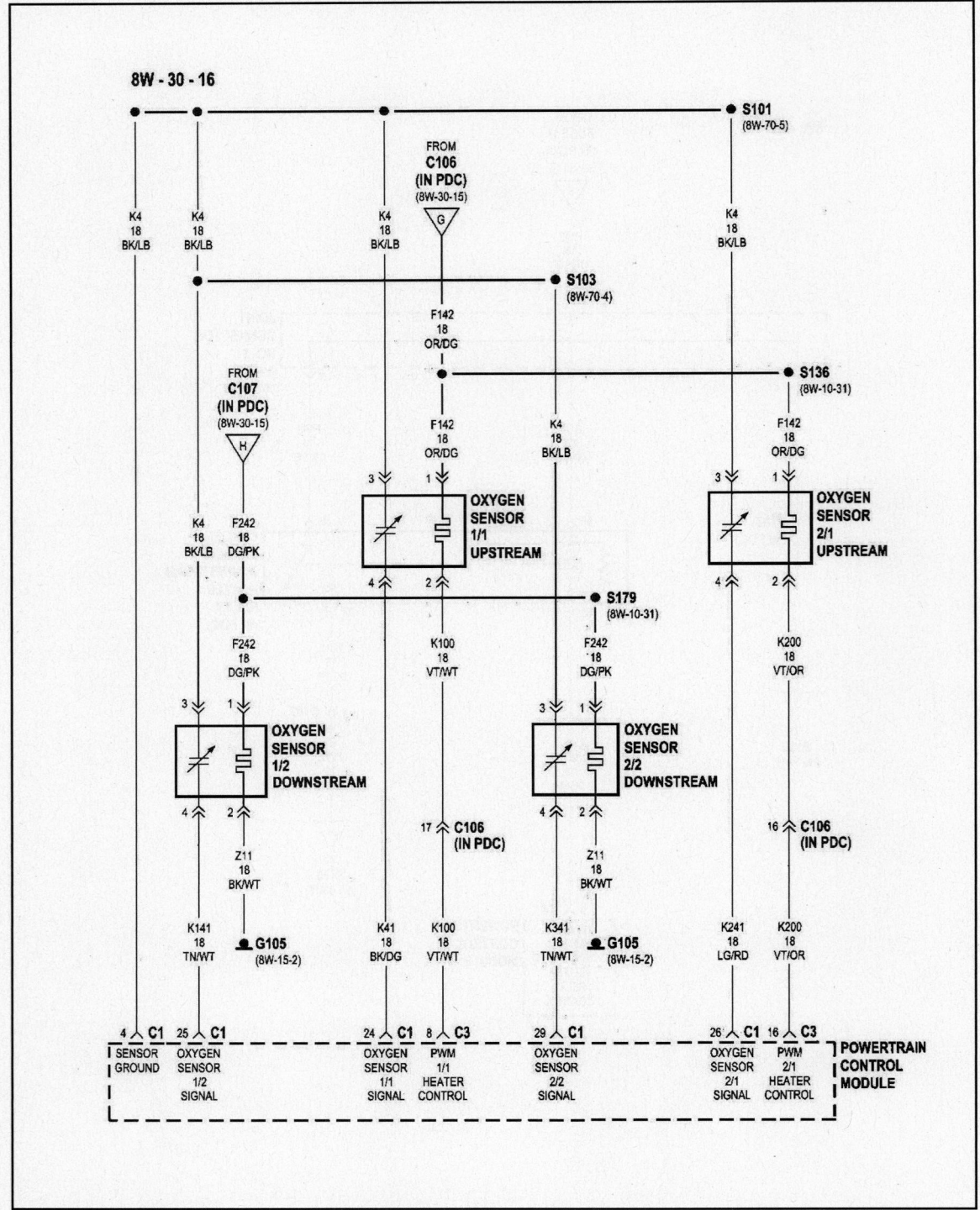

FUEL & IGNITION SYSTEM DIAGRAMS (7 OF 15)

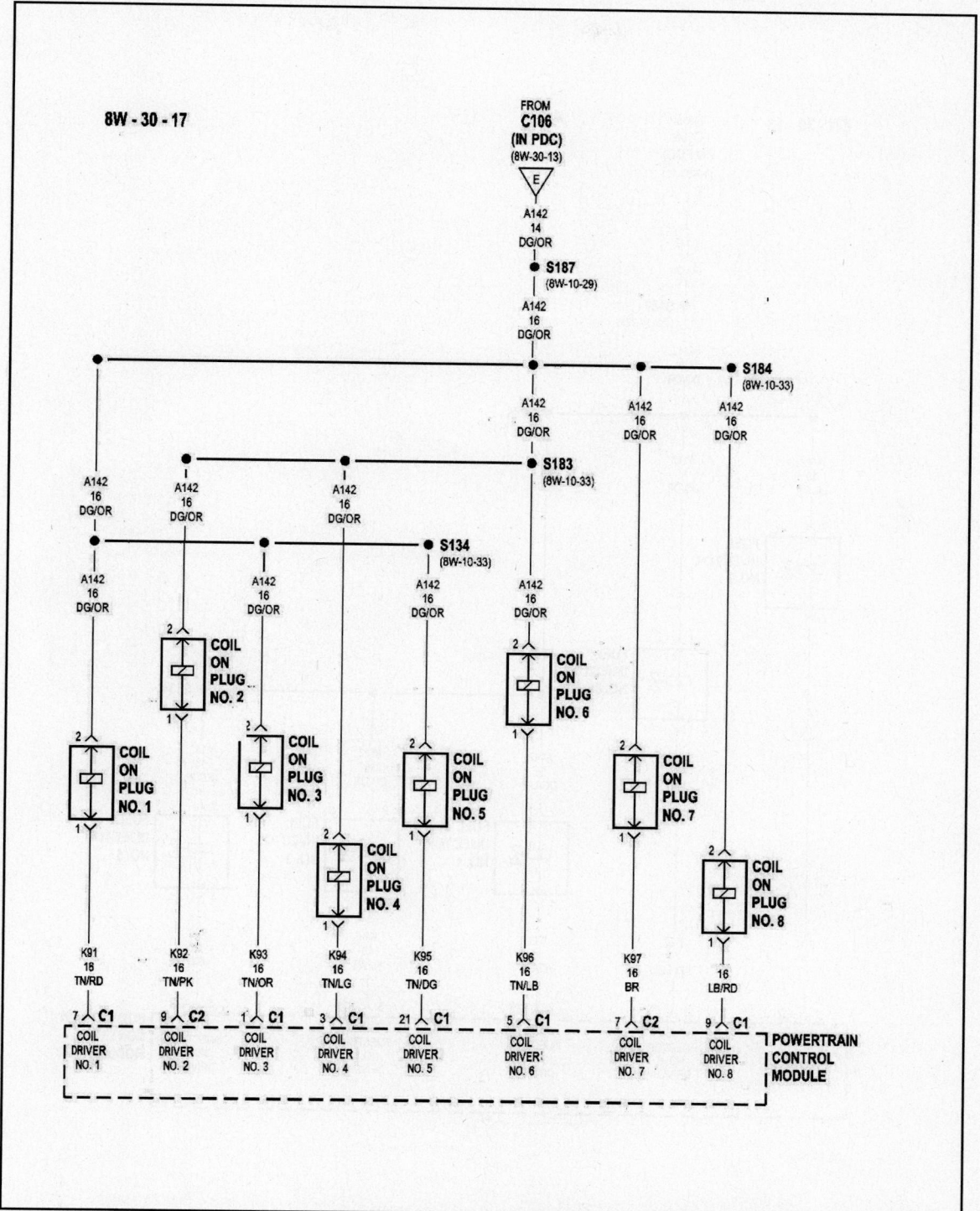

<u>FUEL & IGNITION SYSTEM DIAGRAMS (8 OF 15)</u>

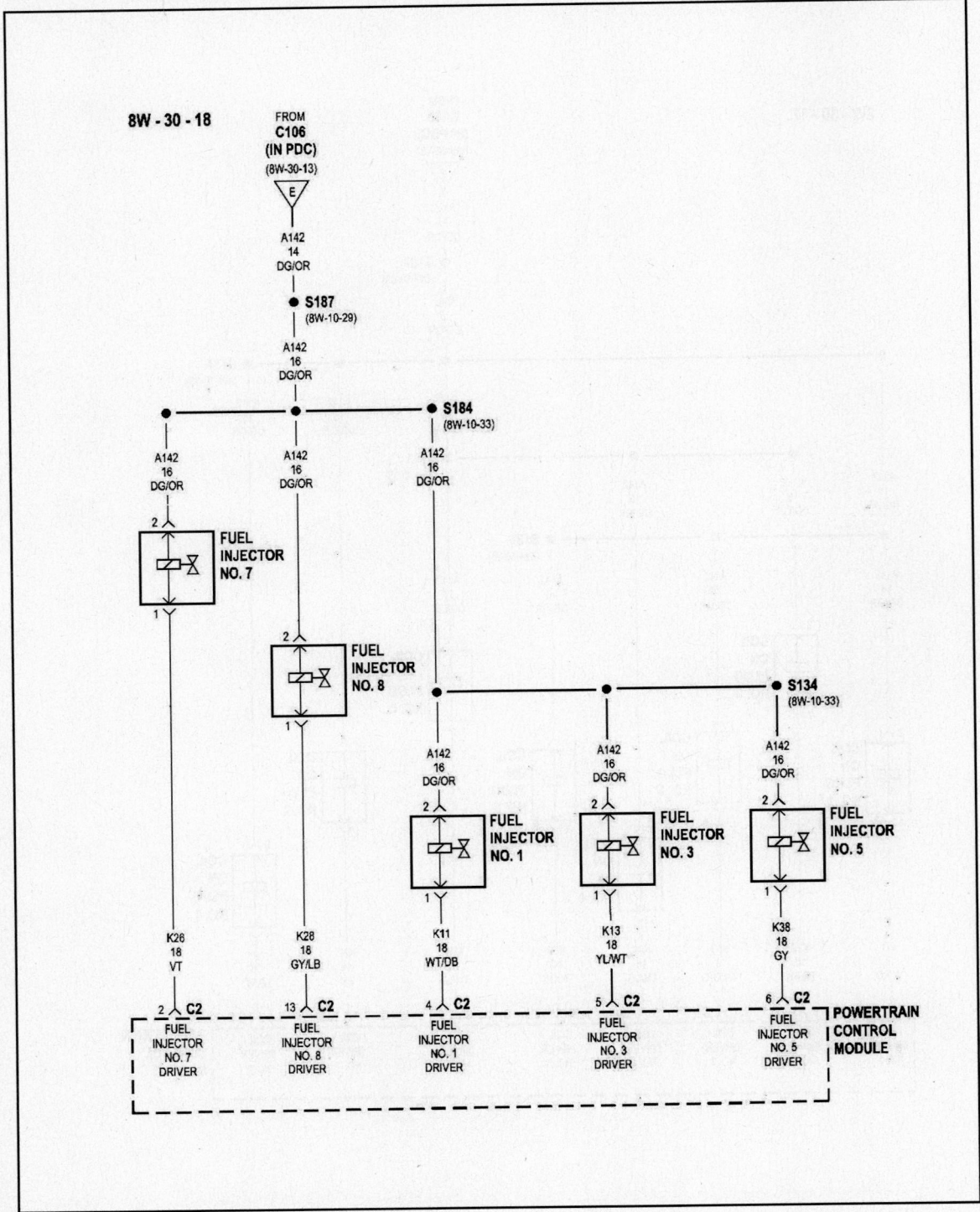

FUEL & IGNITION SYSTEM DIAGRAMS (9 OF 15)

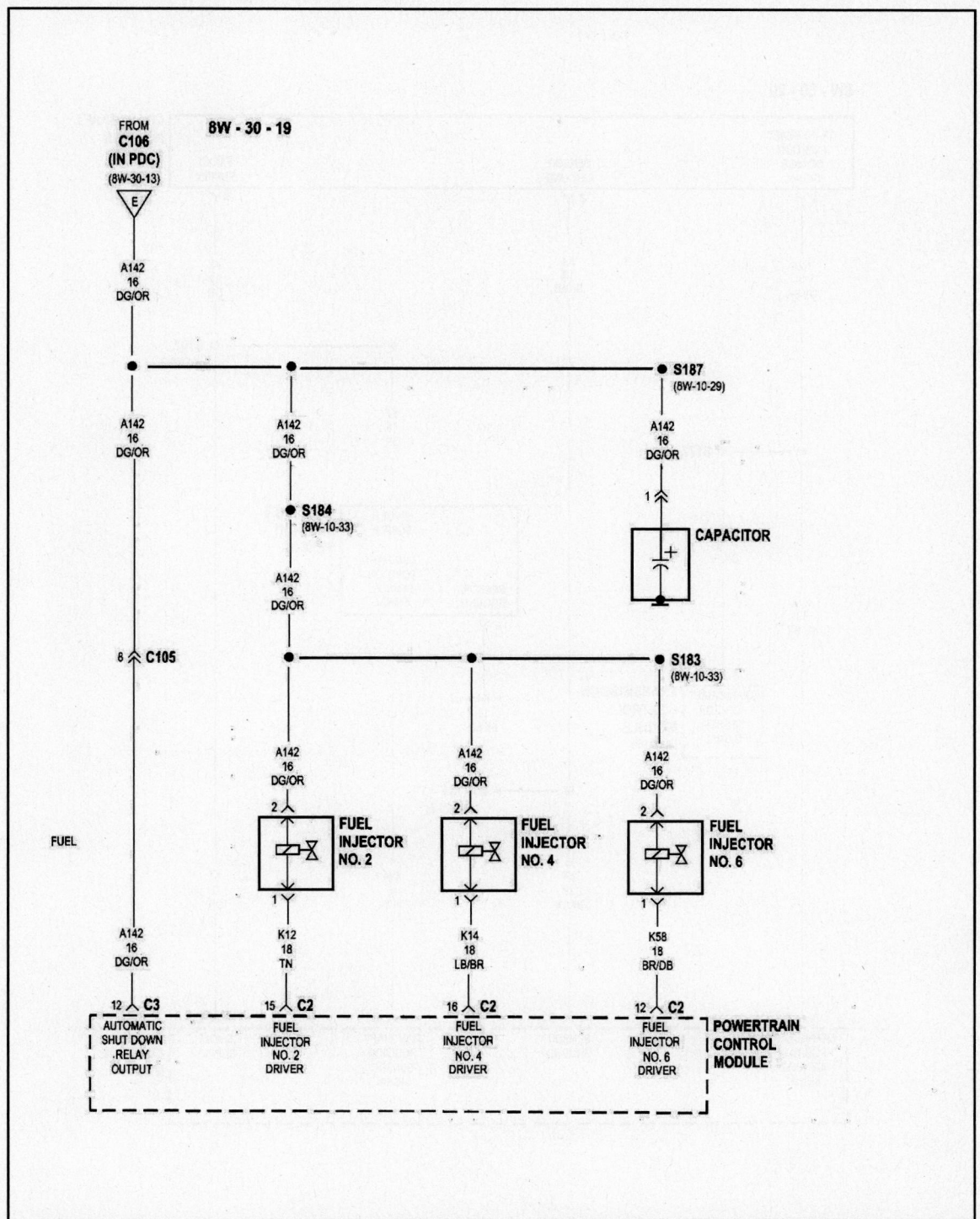

FUEL & IGNITION SYSTEM DIAGRAMS (10 OF 15)

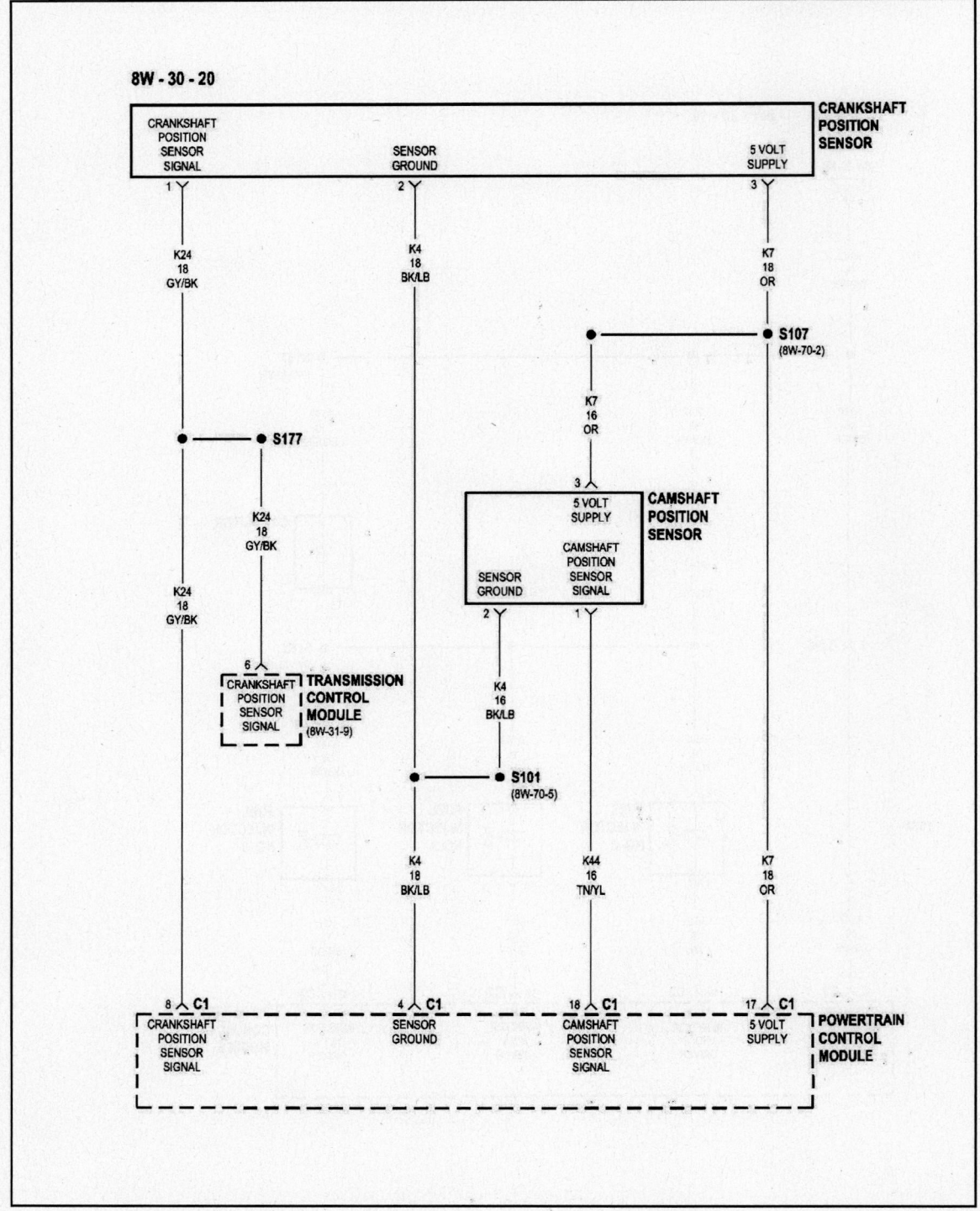

FUEL & IGNITION SYSTEM DIAGRAMS (11 OF 15)

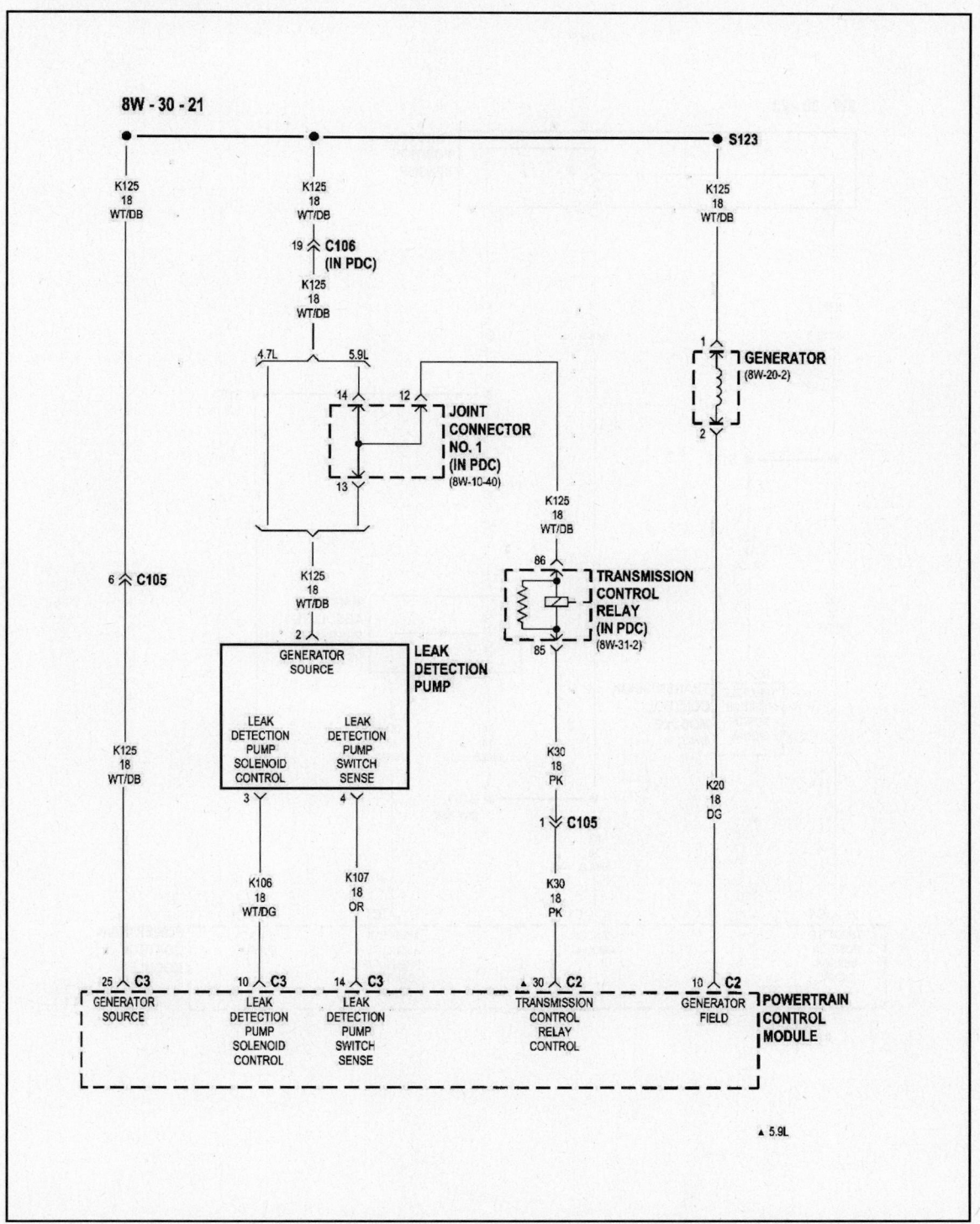

FUEL & IGNITION SYSTEM DIAGRAMS (12 OF 15)

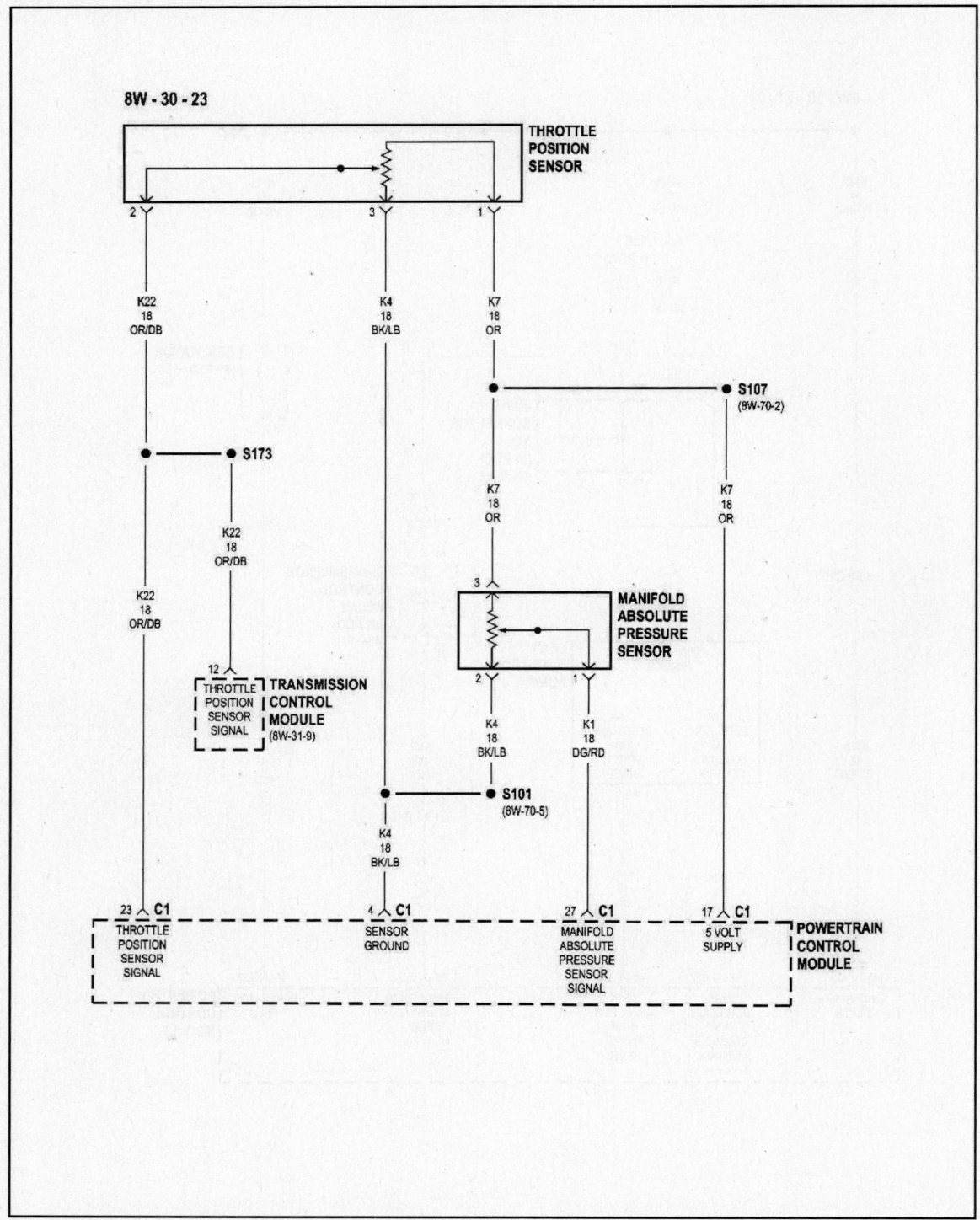

FUEL & IGNITION SYSTEM DIAGRAMS (13 OF 15)

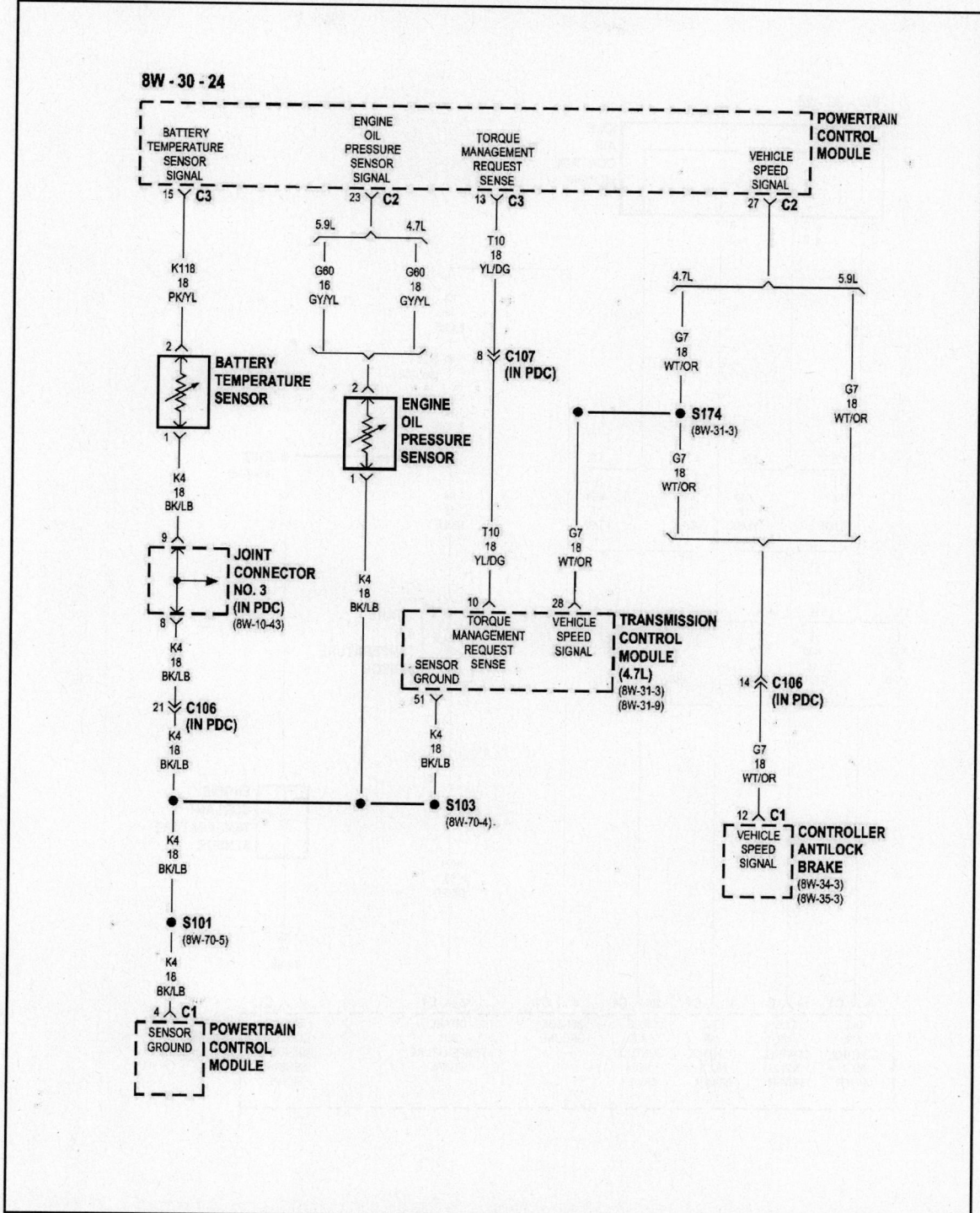

FUEL & IGNITION SYSTEM DIAGRAMS (14 OF 15)

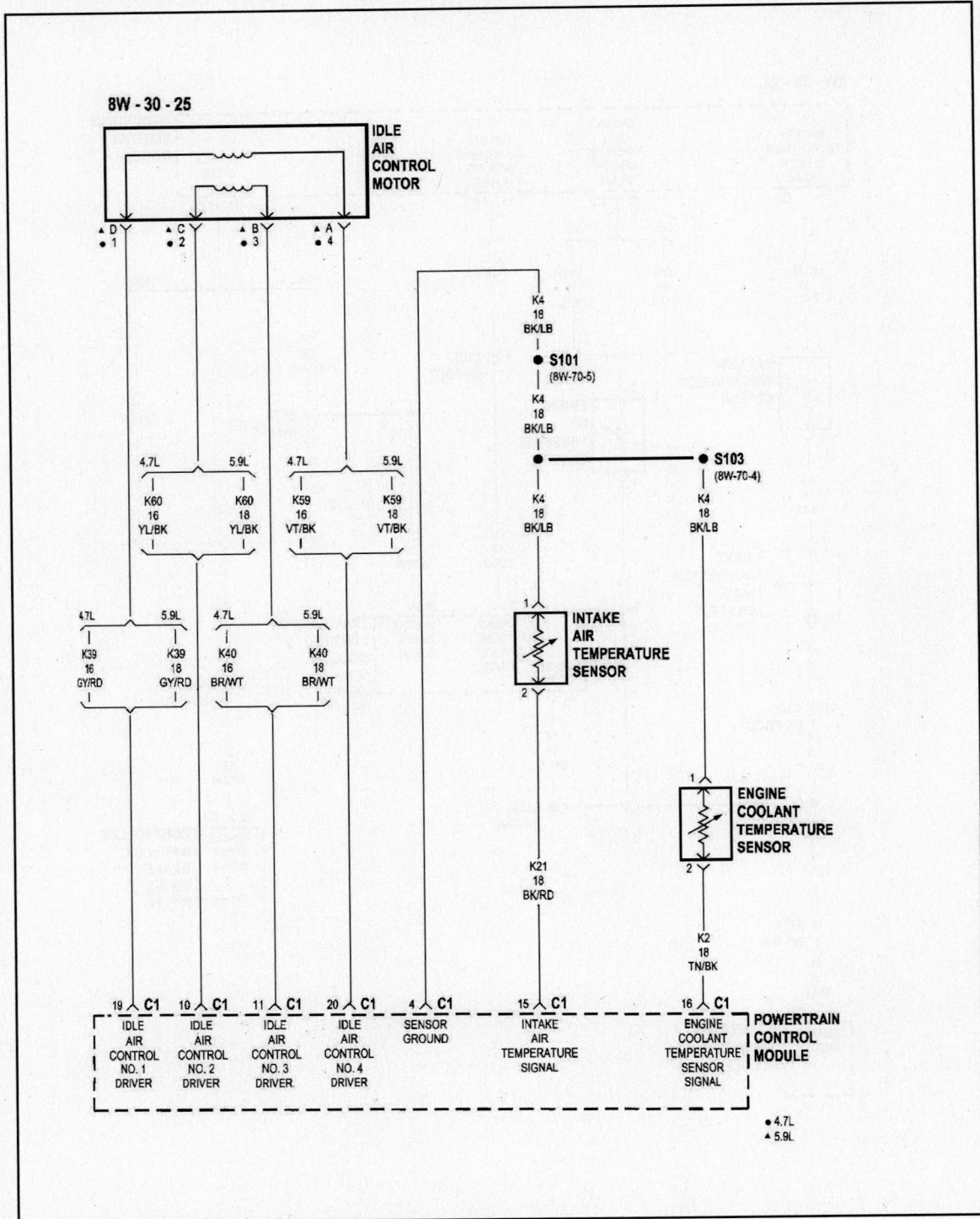

FUEL & IGNITION SYSTEM DIAGRAMS (15 OF 15)

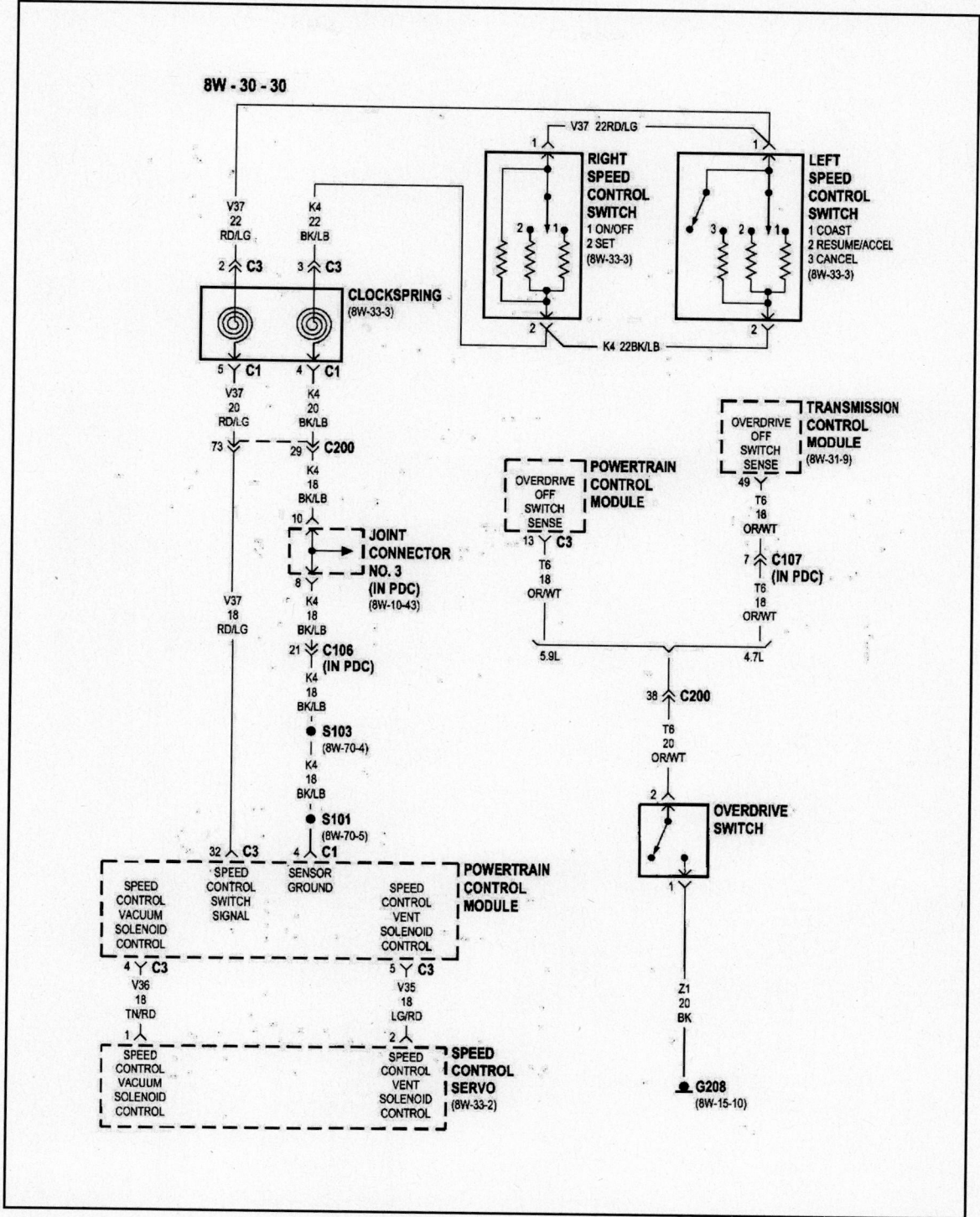

Chrysler OBD II System Contents

About This Section

This manual was developed to assist you during diagnosis and repair of problems related to the PCM and its subsystems.

HOW TO USE EACH SECTION

Section 6: OBD II Systems

• Chrysler OBD II Applications

Refer to this section to learn more about OBD II System diagnostics on Chrysler applications. If you want to clear codes or learn how to drive a particular vehicle to run the OBD II monitors, look in this section. If you want to know why a code set, or how to drive the vehicle to verify a code repair, look in this section.

Section 7: Pin Voltage Tables for Chrysler Applications

• 1990- 2003 Chrysler Car Applications

Refer to this section to identify PCM circuit descriptions, terminals, wire colors and pin voltage values for Chrysler Cars.

Section 8: Pin Voltage Tables for Dodge Applications

• 1990-2003 Dodge Car Applications

Refer to this section to identify PCM circuit descriptions, terminals, wire colors and pin voltage values for Dodge Cars.

Section 9: Pin Voltage Tables for Eagle Applications

• 1990-97 Eagle Car Applications

Refer to this section to identify PCM circuit descriptions, terminals, wire colors and pin voltage values for Eagle Cars.

Section 10: Pin Voltage Tables for Plymouth Applications

• 1990-2001 Plymouth Car Applications

Refer to this section to identify PCM circuit descriptions, terminals, wire colors and pin voltage values for Plymouth Cars.

Section 11: Pin Voltage Tables for Chrysler Van Applications

• 1998-2003 Chrysler Van Applications

Refer to this section to identify PCM circuit descriptions, terminals, wire colors and pin voltage values for Chrysler Vans.

Section 12: Pin Voltage Tables for Dodge Truck and Van Applications

• 1990-2003 Dodge Truck and V an Applications

Refer to this section to identify PCM circuit descriptions, terminals, wire colors and pin voltage values for Dodge Truck and Van Applications.

Section 13: Pin Voltage Tables for Chrysler SUV Applications

• 1990-2003 Chrysler SUV Applications

Refer to this section to identify PCM circuit descriptions, terminals, wire colors and pin voltage values for Dodge Durango and Jeep Applications.

DIAGNOSTIC HELP

The Onboard Diagnostics, PID information and Pin Voltage Tables in this manual contain *diagnostic help* in the form of "known good" voltage values and examples of what caused a trouble code to set. The examples in this manual were obtained with a Breakout Box (BOB), Digital Volt/Ohm Meter (DVOM) and a Scan Tool.

OBD II Vehicle Coverage

CAR APPLICATIONS

Avenger (FJ-Body Code) & Sebring (JX-Body Code)

1996-2000 ES, Sport 2D Convertible, 2D Coupe & 4D Sedan

Engine: 2.0L I4 4v DOHC MFI .... VIN C

Engine: 2.0L I4 4v DOHC MFI .... VIN Y

Engine: 2.4L I4 4v DOHC MFI .... VIN X

Engine: 2.5L V6 4v SOHC MFI ....VIN H, N

Eagle (FJ-Body Code)

1995-98 Talon ESi, TSi 2D Coupe

Engine: 2.0L I4 4v DOHC MFI .... VIN Y

Engine: 2.0L I4 4v DOHC Turbo MFI ....VIN F

Breeze, Cirrus & Stratus (JA-Body Code)

1996-2000 LX, LXi 4D Sedan

Engine: 2.0L I4 4v DOHC MFI .... VIN C

Engine: 2.4L I4 4v SOHC MFI .... VIN X

Engine: 2.5L V6 4v SOHC MFI .... VIN H

Sebring & Stratus (JR-Body Code)

2001-03 GT, JXi, Limited, LX, LXi 2D Convertible & 4D Sedan

Engine: 2.4L I4 4v DOHC MFI .... VIN S, X

Engine: 2.7L V6 4v DOHC MFI ....VIN R, U

Sebring & Stratus (ST-Body Code)

2001-03 JXi, Limited 2D Coupe

Engine: 2.4L I4 2v SOHC MFI .... VIN G

Engine: 3.0L V6 2v SOHC MFI .... VIN H

Concorde, Intrepid, LHS, New Yorker, Vision, 300M (LH Body)

1996-2003 ES, R/T & SE 4D Sedan

Engine: 2.7L V6 4v DOHC MFI .... VIN R, U, V

Engine: 3.2L V6 4v SOHC MFI ....VIN J

Engine: 3.3L V6 2v OHV MFI ....VIN T

Engine: 3.5L V6 4v SOHC MFI ....VIN F, G

Neon (PL-Body Code)

1995-2003 ACR, ES, R/T, SE & Sport 2D Coupe, 4D Sedan

Engine: 2.0L I4 2v SOHC MFI .... VIN C

Engine: 2.0L I4 2v SOHC MFI ....VIN F

Engine: 2.0L I4 4v DOHC MFI .... VIN Y

Prowler (PL-Body Code) & Viper (SR-Body Code)

1997-2001 2D Convertible (Prowler with A/T)

Engine: 3.5L V6 2v SOHC MFI ....VIN F

Engine: 3.5L V6 2v SOHC MFI ....VIN G

1996-2003 2D Coupe & 2D Convertible (Viper with M/T)

Engine: 8.0L V10 2v SOHC MFI .... VIN E

DODGE TRUCK & VAN APPLICATIONS
Ram Truck, Ram Van & Ram Wagon (B & R-Body Codes)
1996-2003 B/R 1500-3500 Series Trucks, Van Cargo, Van Passenger

Engine: 3.9L V6 2v OHV MFI .... VIN X
Engine: 5.2L V8 2v OHV MFI .... VIN 2, T & Y
Engine: 5.9L V8 2v OHV MFI LD .... VIN Z

DAKOTA (N), DURANGO (DN) & RAM TRUCK (R-BODY CODE)
1996-2003 R1500-3500 Series 2D & 4D Pickup, 2D & 4D Extra Cab

Engine: 3.7L V6 2v SOHC MFI .... VIN K
Engine: 4.7L V8 2v SOHC MFI .... VIN N
Engine: 5.2L V8 2v OHV MFI .... VIN T & Y
Engine: 5.9L V8 2v OHV MFI LD .... VIN 5 & Z
Engine: 5.9L I6 4v OHV Diesel .... VIN 6 & 7
Engine: 8.0L V10 2v OHV MFI .... VIN W

MINIVAN APPLICATIONS
Caravan, Town & Country Voyager (NS-Body Code)
1996-2003 ES, LE, SE, LS Van & LSi Van Passenger (AWD & FWD)

Engine: 2.4L I4 4v DOHC MFI .... VIN B, X
Engine: 3.0L V6 2v SOHC MFI .... VIN 3
Engine: 3.3L V6 2v OHV MFI .... VIN R
Engine: 3.3L V6 2v OHV MFI (Gas and Ethanol) .... VIN G
Engine: 3.8L V6 2v OHV MFI .... VIN L

SUV APPLICATIONS
Cherokee (XJ-Body Code) & Grand Cherokee (ZJ-Body Code)
1996-2003 Country, Limited, Sport & TSi 2D Utility, 4D Utility

Engine: 2.5L I4 2v OHV MFI .... VIN P
Engine: 4.0L I6 2v OHV MFI .... VIN S
Engine: 4.7L V8 2v SOHC MFI .... VIN N
Engine: 5.2L V8 2v OHV MFI .... VIN Y
Engine: 5.9L V8 2v SOHC MFI .... VIN Z

Liberty (KJ-Body Code)
2002-03 Limited & Sport 4D Utility

Engine: 2.4L I4 4v DOHC MFI .... VIN 1
Engine: 3.7L V6 2v SOHC MFI .... VIN K

PT Cruiser (PT-Body Code)
2001-03 4D Wagon & Limited 4D Wagon

Engine: 2.4L I4 4v DOHC MFI .... VIN B

Wrangler (TJ-Body Code)
1997-2003 Sahara, S, SE, Sport & X 2D Utility

Engine: 2.5L I4 2v OHV MFI .... VIN P
Engine: 4.0L I6 2v OHV MFI .... VIN S

ON-BOARD DIAGNOSTICS

Chrysler OBD II systems implement the usual test procedures built into the Powertrain Control Module (PCM) and Transmission Control Module (TCM). The first step in diagnosis of a problem on an OBD II system is to identify it as either a trouble code fault (Code Fault) or a driveability symptom (No Code Fault). The OBD II Drive Cycle is used to verify any repair to the system.

DIAGNOSTIC PROCEDURE

The OBD II Flowchart shown here should be used like this:

- Trouble Code Diagnosis - Refer to the Code List (in this section) or electronic media for a repair chart for a particular trouble code.
- Driveability Symptoms - Refer to the Driveability Symptom List in other manuals or other media.
- Intermittent Faults - Refer to the Intermittent Test Procedures.
- OBD II Drive Cycles - Refer to the CCM or the Main Monitor drive cycle articles that follow.

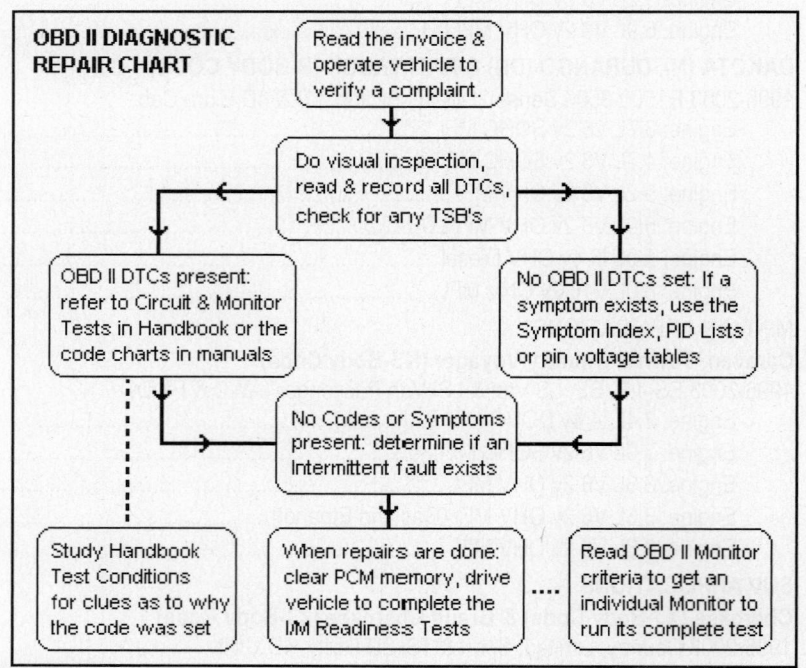

OBD II DIAGNOSTIC REPAIR CHART

Read the invoice & operate vehicle to verify a complaint.

Do visual inspection, read & record all DTCs, check for any TSB's

OBD II DTCs present: refer to Circuit & Monitor Tests in Handbook or the code charts in manuals

No OBD II DTCs set: If a symptom exists, use the Symptom Index, PID Lists or pin voltage tables

No Codes or Symptoms present: determine if an Intermittent fault exists

Study Handbook Test Conditions for clues as to why the code was set

When repairs are done: clear PCM memory, drive vehicle to complete the I/M Readiness Tests

Read OBD II Monitor criteria to get an individual Monitor to run its complete test

OBD II SYSTEM TERMINOLOGY

It is important to understand the terminology related to OBD II test procedures. Several essential OBD II terms and definitions are discussed on the following pages.

TWO-TRIP DETECTION

Frequently, an emission system or component must fail a Monitor test more than once before the MIL is activated. In these cases, the first time an OBD II Monitor detects a fault during any drive cycle it sets a "pending code" in the PCM memory.

A "pending code", which can be read by selecting trouble codes from the Scan Tool menu, appears when Pending Codes are selected. In order for a "pending code" to cause the MIL to activate, the original fault must be repeated on two consecutive trips. This is a critical issue to understand as a "pending code" could remain in the PCM for a long time before the conditions that caused the code to set reappear. This type of OBD II trouble code logic is frequently referred to as the "Two-Trip Detection Logic". Trouble codes that are related to a Misfire fault and Fuel Trim can cause the PCM to activate the MIL after <u>one</u> trip because they are related to critical emission systems that could cause emissions to exceed the federally mandated limits.

SIMILAR CONDITIONS

If a "pending code" is set because of a Misfire or Fuel System Monitor fault, the vehicle must meet *similar conditions* for a second trip before the code matures before the PCM activates the MIL and then stores the code in long-term memory. Refer to the Note above for exceptions to this rule. Understanding the meaning of *similar conditions* is important when repairing a fault detected by a Misfire or Fuel System Monitor.

To achieve *similar conditions*, the vehicle must reach these engine running conditions simultaneously:

- Engine speed must be within 375 rpm of the engine speed when the trouble code set
- Engine load must be within 10% of the engine load when the trouble code set
- Engine warmup state must match a previous cold or warm state

Summary - Similar conditions are defined as the engine operating conditions that match conditions recorded in Freeze Frame at the point the fault was detected and a trouble code was stored in the PCM memory.

OBD II WARMUP CYCLE

It is important to understand the meaning of the expression *"warmup cycle"*. Once the MIL is turned off and the fault that caused the code does not reappear, the OBD II trouble code that was stored in PCM memory will be erased after 40 warmup cycles.

The exceptions to this rule are codes related to the Fuel system and Misfire Detection tests.

These codes require that 80 warmup cycles occur without the fault reappearing before the code will be erased from the PCM memory.

■ **NOTE:** *A warmup cycle is defined as vehicle operation (after a cool-down period) where the engine temperature increases at least 40°F and reaches at least 160°F.*

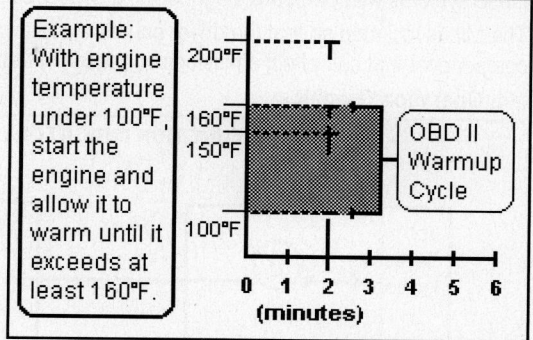

ELECTRONIC PINION FACTOR

The Vehicle Speed Sensor (VSS) provides distance pulses to the PCM. This information is used to calculate speed and mileage. The Pinion Factor is programmed into the TCM at the factory.

TCM PINION FACTOR RESET

If the TCM is replaced, the Pinion Factor must be set properly. If it is not set, or set incorrectly, speed-related accessories will not operate, or will operate inaccurately. Accessories include the speedometer, speed control, rolling door locks, and other devices that are operated by the PCM and Body Control Module (BCM).

FREEZE FRAME DATA

The term *Freeze Frame* is used to describe the engine conditions that are recorded in PCM memory at the time a Monitor detects an emissions related fault. These conditions include fuel control state, spark timing, engine speed and engine load. Freeze Frame data is recorded when a system fails the first time on a two-trip type fault. The Freeze Frame Data will only be overwritten by a different fault with a "higher emission priority."

<u>MALFUNCTION INDICATOR LAMP</u>

With the introduction of OBD II System diagnostics, the functions necessary to meet the requirements of the US-EPA and CARB OBD requirements are under the control of the PCM. A Malfunction Indicator Lamp (MIL) was added to the instrument panel of all 1994 and later passenger car, light duty truck, medium duty truck, van and SUV applications. On these systems with on-board diagnostics, a control circuit (driver) inside the PCM controls the MIL operation.

The MIL is included so that the driver can be made aware that a fault has occurred in one or more of the Powertrain components that can effect emissions or cause the vehicle to fail a smog inspection.

MIL Operation Graphic

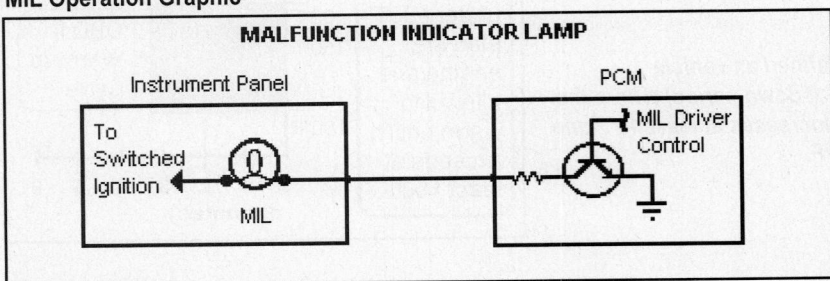

<u>MIL Operation</u>

The Malfunction Indicator Lamp (MIL) is illuminated in two ways. First, as a functional test at "key on" prior to Engine cranking; and second, whenever the PCM sets a trouble code that affects vehicle emissions. Some Monitors require that the test fail on two consecutive trips before the PCM will illuminate the MIL. The MIL is on continuously with the PCM in Limp-In Mode or when it identifies a failed emission component or system. When an active misfire condition is present, the MIL will either flash or remain on continuously.

<u>MIL Reset</u>

Most trouble codes related to a MIL are erased from KAM after 40 warmup periods if the same fault is not repeated. The MIL can be turned off after a repair by using the Scan Tool PCM Reset function.

If the MIL was activated due to a fault in one of the Main Monitors, the Diagnostic Task Manager can only turn the MIL off if the most recent fault passes its diagnostic test on three consecutive trips.

If either the Misfire or Fuel System Monitor detects a fault, the next time the same diagnostic test starts, the PCM requires that the engine return to the operating conditions at which the fault originated before the test will be repeated. These two Monitors will retest for the fault only if the engine is running within "similar operating conditions".

DIAGNOSTIC TROUBLE CODES

The OBD II system uses a *Diagnostic Trouble Code (DTC)* identification system established by the Society of Automotive Engineers (SAE) and the EPA. The first letter of the DTC is used to identify the type of computer system that has failed as shown below:

- The letter 'P' indicates a Powertrain related device.
- The letter 'C' indicates a Chassis related device.
- The letter 'B' indicates a Body related device.
- The letter 'U' indicates a Data Link or Network device code.

The first DTC number indicates a generic (P0xxx) or manufacturer (P1xxx) type code. A list of trouble codes is included in this section.

The number in the hundredth position indicates the specific vehicle system or subgroup that failed (i.e., P0300 for a Misfire code, P0400 for an emission system code, etc.).

STANDARD CORPORATE PROTOCOL

On vehicles equipped with OBD II, a Standard Corporate Protocol (SCP) communication language is used to exchange bi-directional messages between stand-alone modules and devices. With this type of system, two or more messages can be sent over one circuit.

OBD II MONITOR SOFTWARE

The Diagnostic Task Manager contains software designed to allow the PCM to organize and prioritize the Main Monitor tests and procedures, and to record and display test results and diagnostic trouble codes.

The functions controlled by this software include:

- To control the diagnostic system so that the vehicle will continue to operate in a normal manner during testing.

- To ensure that all of the OBD II Monitors run during the first two sample periods of the Federal Test Procedure.

- To ensure that all OBD II Monitors and their related tests are sequenced so that required inputs (enable criteria) for a particular Monitor are present prior to running that particular Monitor.

- To sequence the running of the Monitors to eliminate the possibility of different Monitor tests interfering with each other or upsetting normal vehicle operation.

- To provide a Scan Tool interface by coordinating the operation of special tests or data requests.

CYLINDER BANK IDENTIFICATION

Engine sensors are identified for each engine cylinder bank by SAE regulations as explained below.

BANK

A specific group of engine cylinders that share a common control sensor (e.g., Bank 1 identifies the location of Cylinder 1 while Bank 2 identifies the cylinders on the opposite bank).

An example of the cylinder bank configuration for a Dodge Caravan with FWD and a 3.0L V6 (VIN 3) transverse mounted engine is shown in the Graphic to the right.

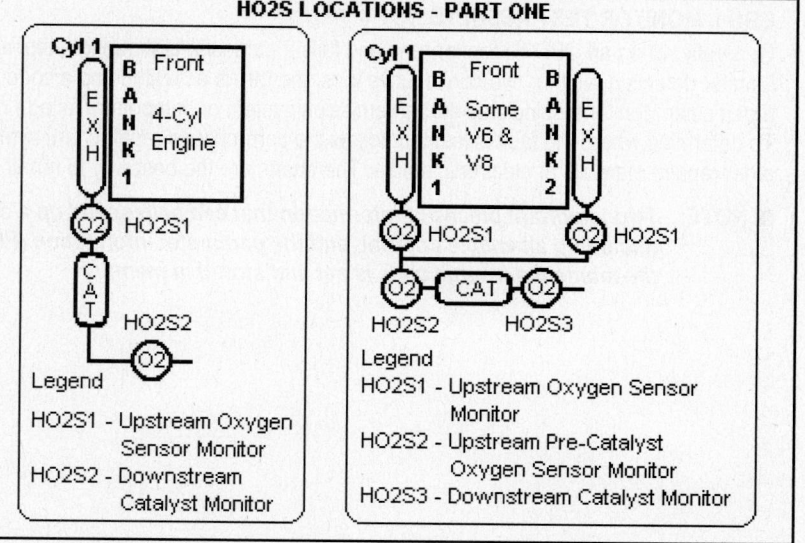

Right Bank
(Bank 1)
Firing Order:
1-2-3-4-5-6

Left Bank
(Bank 2)

HEATED OXYGEN SENSOR

The Heated Oxygen Sensor (HO2S) detects the presence of oxygen in the exhaust and produces a variable voltage according to the amount of oxygen detected. The HO2S generates a voltage between 0 and 1 volt. A value of less than 0.4v is an indication of a lean A/F ratio while a value over 0.6v indicates a rich A/F ratio.

Oxygen Sensor Identification

There are many references to oxygen sensors identified as the HO2S-11, HO2S-12 and HO2S-13. Oxygen sensors are identified to each engine cylinder bank as the front HO2S or rear HO2S. For example, HO2S-11 identifies the front oxygen sensor located in Bank 1 of the engine. HO2S-21 identifies the front oxygen sensor in Bank 2 of the engine, and so on. Note that Bank 1 always contains engine cylinder number 1 (**Cylinder 1**).

In the V6 and V8 picture in Graphic above, HO2S-11 refers to the front oxygen sensor, HO2S-12 refers to the middle pre-catalyst oxygen sensor and

HO2S LOCATIONS - PART ONE

Cyl 1 | EXH | BANK 1 | Front 4-Cyl Engine
O2 HO2S1
CAT
HO2S2
O2

Legend
HO2S1 - Upstream Oxygen Sensor Monitor
HO2S2 - Downstream Catalyst Monitor

Cyl 1 | EXH | BANK 1 | Front Some V6 & V8 | BANK 2 | EXH
O2 HO2S1   O2 HO2S1
O2 — CAT — O2
HO2S2   HO2S3

Legend
HO2S1 - Upstream Oxygen Sensor Monitor
HO2S2 - Upstream Pre-Catalyst Oxygen Sensor Monitor
HO2S3 - Downstream Catalyst Monitor

HO2S-13 refers to the rear oxygen sensor. The front HO2S-11 (or HO2S-12 shown above) signal is used with the HO2S Monitor test. The HO2S-12 (or HO2S-13) signal is used with the Catalyst Monitor.

PCM RESET INFORMATION

The PCM Reset function allows the technician to command the Scan Tool to clear all emissions related diagnostic information from the PCM. Emissions related codes are stored in the PCM until all the OBD II System monitors or components have been tested to satisfy a drive cycle without any other faults occurring. Refer to the OBD II Drive Cycle article in this section for additional information on this subject. The events listed below occur after the OBD II Drive Cycle is completed and the PCM Reset procedure is performed.

- The trouble code number total will be cleared
- All trouble codes will be cleared
- All Freeze Frame data will be cleared
- All Oxygen Sensor Test data will be cleared
- All OBD II System Monitors will have their status reset

PCM Reset Instructions

Refer to the Scan Tool Operating Manual for specific instructions on how to perform a PCM Reset. The list of instructions should include:

- Turn the key off
- Perform the necessary vehicle preparation and visual inspection
- Connect the Scan Tool to the DLC
- Turn the key on or start the vehicle
- Verify the Scan Tool is connected and communicating properly with the PCM
- Perform the PCM Reset step, then turn the key off and disconnect the Scan Tool

Reset Counter

The Reset Counter counts the number of times the vehicle has been started since a trouble codes was set, erased, or the battery was disconnected. The number of starts can be used to help determine when a trouble code actually set. This information is recorded by the PCM and displayed on the Scan Tool as the "RESET COUNTER." When there are no codes stored in memory, the Scan Tool will display "NO TROUBLE CODES FOUND" and the Reset Counter will display "RESET COUNT = XXX."

OBD II MONITOR TEST RESULTS

Generally, when an OBD II Monitor runs and fails a particular test during a trip, a pending code is set. If the same Monitor detects a fault for two consecutive trips, the MIL is activated and a code is set in PCM memory. The results of a particular Monitor test indicate that an emission system or component failed - not the circuit that failed!

To determine where the fault is located, follow the correct code repair chart, symptom diagnosis or intermittent test in other repairs manuals or electronic media. The charts are the best way to repair the system.

■ NOTE: *Two important pieces of information that can help speed up a diagnosis are code conditions (including all enable criteria), and the parameter information (PID) stored in the Freeze Frame at the moment a trouble code is set and stored in memory*

ADAPTIVE FUEL CONTROL STRATEGY

The PCM incorporates an Adaptive Fuel Control Strategy that includes an adaptive fuel control table stored in KAM to compensate for normal changes in fuel system devices due to age or engine wear.

During closed loop operation, the Fuel System Monitor has two methods of attempting to maintain an ideal A/F ratio of 14:7 to 1 (they are referred to as short term fuel trim and long term fuel trim).

■ NOTE: *If the fuel injector, fuel pressure regulator or oxygen sensor is replaced, the KAM in the PCM should be cleared by a PCM Reset step so that the PCM will not use a previously learned strategy.*

SHORT TERM FUEL TRIM

Short term fuel trim (SHRTFT) is an engine operating parameter that indicates the amount of short term fuel adjustment made by the PCM to compensate for operating conditions that vary from the ideal A/F ratio condition. A SHRTFT number that is negative (-15%) means that the HO2S is indicating a richer than normal condition to the PCM, and that the PCM is attempting to lean the A/F mixture. If the A/F ratio conditions are near ideal, the SHRTFT number will be close to 0%.

LONG TERM FUEL TRIM

Long term fuel trim (LONGFT) is an engine parameter that indicates the amount of long term fuel adjustment made by the PCM to correct for operating conditions that vary from ideal A/F ratios.

A LONGFT number that is positive (+15%) indicates that the HO2S is indicating a leaner than normal condition, and that it is attempting to add more fuel to the A/F mixture. If A/F ratio conditions are near ideal, the LONGFT number will be near 0%. The PCM adjusts the LONGFT in a range from -35 to +35%. The values are in (%) on the Scan Tool.

ENABLE CRITERIA

The terms *enable criteria* are used to the conditions required for an OBD II Monitor to run a certain test. Each Monitor has a specific list of conditions that must be met before it will run its diagnostic test.

Enable criteria information can be found in the Chrysler Manual and in this manual. Look under Diagnostics and then Trouble Code Conditions. This type of data can be different for each vehicle and engine type. Examples of trouble code conditions for DTC P0460 and P1281 are shown in the table below.

Code ID	Trouble Code Description and Code Conditions
P0460 N/MIL 1T CCM 1999-2003 All Models	Fuel Level Sending No Change Over Miles Conditions: Engine started; and PCM received (from the BCM) the low fuel sense input did not move enough over a certain number of miles. • Refer to the code chart in other manuals to repair this code.

The trouble code *enable criteria* information includes these examples:

- A/C, BARO, ECT, IAT, TFT, TP and Vehicle Speed sensors status
- Camshaft (CMP) and Crankshaft (CKP) sensors
- Canister Purge (duty cycle) and Ignition Control Module signals
- Short (SHRTFT) and Long Term (LONGFT) Fuel Trim values
- Transmission Shift Solenoid On/Off status

OBD II TRIP

The term *OBD II Trip* describes a method of driving the vehicle so that one or more of the following OBD II Monitors complete their tests:

- Comprehensive Component Monitor (completes anytime in a trip)
- Fuel System Monitor (completes anytime during a trip)
- EGR System Monitor (completes after accomplishing a specific idle and acceleration period)
- Oxygen Sensor Monitor (completes after accomplishing a specific steady state cruise speed for a certain amount of time)

OBD II DRIVE CYCLE

The term *drive cycle* has been used to describe a drive pattern used to verify that a trouble code, driveability symptom or intermittent fault had been fixed. With OBD II systems, this term is used to describe a vehicle drive pattern that would allow all the OBD II Monitors to initiate and run their diagnostic tests. For OBD II purposes, a minimum *drive cycle* includes an engine startup with continued vehicle operation that meets the amount of time required to enter closed loop fuel control.

The ambient or battery temperature must be from 40-90°F to initiate the OBD II drive cycle. Allow the engine to warm to 170°F prior to starting the test. The fuel level must be from 15-85% to start the test.

Connect the Scan Tool prior to beginning the drive cycle. Some tools are designed to emit a three-pulse beep when one or all of the OBD II Monitors complete their particular tests.

■ **NOTE:** *The IAT PID must be from 40-90°F to start the drive cycle. If it is less than 50°F at any time during the highway part of the drive cycle, the EVAP Monitor may not complete. The engine should reach 170°F before starting the trip (a cold engine startup of less than 100°F must occur prior to attempting to verify an EVAP system fault). For the EVAP LDP system, verify that the FLI PID is at 15-85%. Some Monitors require specific acceleration steps.*

DRIVE CYCLE PROCEDURE

The primary intention of the Chrysler OBD II drive cycle is to run all of the OBD II Monitors. The drive cycle can also be used to assist in identifying any OBD II concerns present through total Monitor testing. Perform all of the Vehicle Preparation steps. Then refer to the Drive Cycle Graphic below for details on how to run an OBD II Drive Cycle.

Connect a Scan Tool and have an assistant watch the Scan Tool I/M Readiness Status to determine when the Catalyst, EGR, EVAP, Fuel System, O2 Sensor, Secondary AIR and Misfire Monitors complete.

OBD II Drive Cycle Graphic

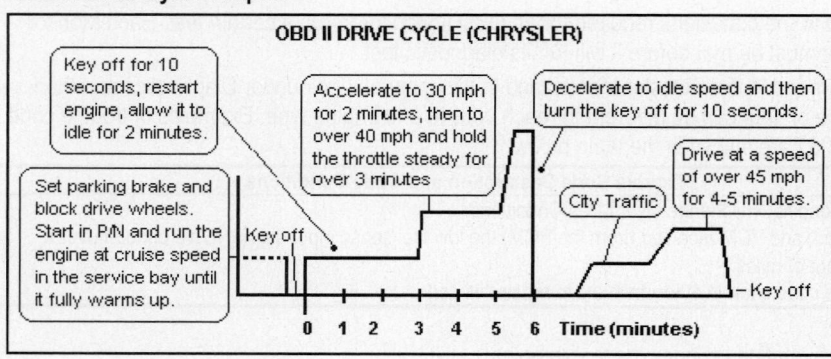

OBD II MONITORS

COMPREHENSIVE COMPONENT MONITOR

OBD II regulations require that all emission related circuits and components controlled by the PCM that could affect emissions are monitored for circuit continuity and out-of-range faults. The Comprehensive Component Monitor (CCM) consists of four different monitoring strategies: two for inputs and two for output signals. *Note that the CCM is a one-trip Monitor for emission faults on Chrysler vehicles.*

INPUT STRATEGIES

One input strategy is used to check devices with analog inputs for opens, shorts, or out-of-range values. The CCM accomplishes this task by monitoring A/D converter input voltages of various sensors.

A second input strategy is used to check devices with digital and frequency inputs by performing rationality checks. The PCM uses other sensor readings and calculations to determine if a sensor or switch reading is correct under existing conditions. Some tests of the CKP, CMP and VSS run continuously, some only after actuation.

OUTPUT STRATEGIES

An Output Device Monitor in the PCM checks for open and shorted circuits by observing the control voltage level of a particular device. The control voltage is low when the device is "on" and the voltage is high when the device is "off". Monitored outputs include the AC relay, ASD relay SS1, SS2, SS3, EGR and EVAP and HO2S heater control. In the Graphic below, the pair of eyes represents the point at which the PCM "watches" the circuit for the correct change in voltage.

Output Device Monitor Graphic

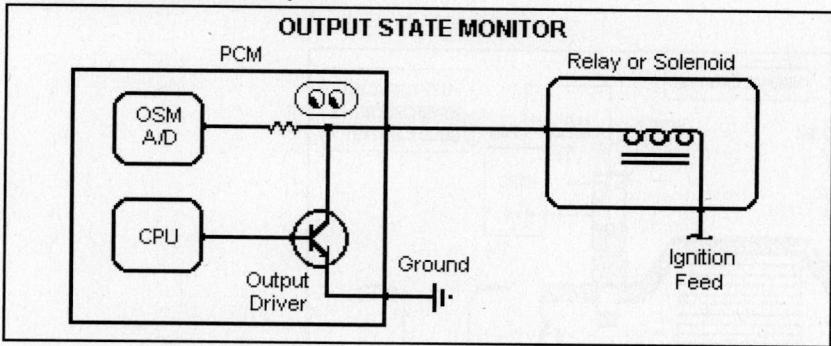

CRANKSHAFT RELEARN

Select Crank Relearn on the Scan Tool to allow the PCM to enable a fast CMP and CKP sensor relearn procedure. The PCM relearns the crank target window spacing during Decel Fuel Shutoff mode The PCM reset step requires three (3) closed throttle decelerations from 55 mph in order to "relearn" the Crank Target Spacing.

CATALYST MONITOR

OBD II regulations require that the functionality of the Catalyst system be monitored. If the Catalyst system deteriorates to a point where vehicle emissions increase by more than 1.5 times the Federal Test Procedure (FTP) standard, the MIL must be activated.

The oxygen content in a catalyst is important for efficient conversion of exhaust gases. When a lean A/F ratio is present for an extended period of time, the oxygen content in the catalyst can reach a maximum. Conversely, when a rich A/F ratio is present for too long a period, the oxygen content in the catalyst can become totally depleted.

Catalyst operation is dependent on its ability to store and release oxygen needed to complete the emissions-reducing chemical reactions. As a catalyst deteriorates, its ability to store oxygen is reduced. Since the catalyst's ability to store oxygen is somewhat related to proper operation, oxygen storage can be used as an indicator of catalyst performance.

FRONT AND REAR HEATED OXYGEN SENSORS

Two oxygen sensors are used to accomplish the task of monitoring the catalyst storage capability. One sensor is mounted in front of the catalyst (pre-catalyst HO2S) and another sensor is mounted in back of the catalyst (post-catalyst HO2S).

By utilizing an oxygen sensor in front of the catalytic converter, and a second sensor located behind the catalyst, the oxygen storage capability of the catalyst can be determined by comparing the voltage signals from the two sensors.

An example of the waveforms from the front and rear heated oxygen sensors on a "normal" catalyst is shown in the Graphic below.

Catalyst Monitor Graphic

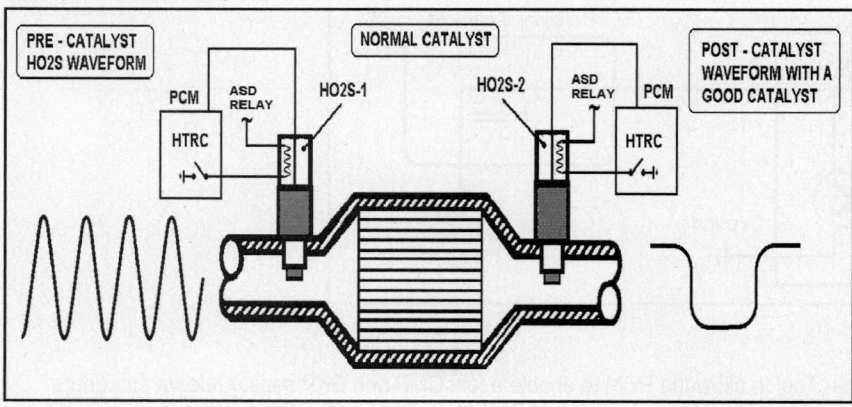

CATALYST MONITOR (CONTINUED)

Calculating the HO2S Switch Ratio

During the Catalyst Monitor test, the PCM expands the oxygen sensor lean and rich switch points to allow the A/F mixture to run leaner and then richer (in an attempt to overload the converter). And the PCM counts the number of HO2S switches. To count as an official switch, the HO2S signal must go from below the lean threshold to above the rich threshold. The switch ratio is calculated by dividing the number of HO2S-12 switches by HO2S-11 switches during a 20-second period.

As the catalyst degrades, the switch rate of the HO2S-12 approaches that of the HO2S-11. If the HO2S-12 switch rate reaches a preset value, a counter goes to (1) and the test continues. If the test fails (3) times, a pending code is set and Frame data is stored. If the counter reaches (3) on the next trip, the MIL is activated and a code is set. If the test passes the first time, no further testing is done on that trip.

Monitor Operation

To run this Monitor, warm the engine to normal temperature (ECT sensor input over 147°F) and then accelerate to 40-55 mph and maintain cruise speed for 3 minutes with the throttle open. The RPM should read 1200-1700 and the MAP PID should be 10.5-15.0" Hg.

Catalyst Monitor Trip Graphic

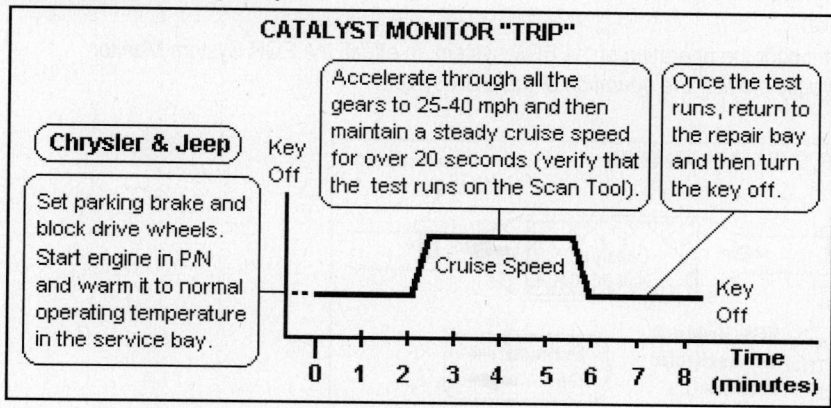

MIL Activation, How to Clear Memory Codes

The first time this Monitor fails, a "pending" code is set and the engine conditions are stored in Freeze Frame. If it fails on the very next trip, the fault matures, the MIL is activated and a code is set. The MIL will remain on for more than one trip, but will turn off if the test runs (3) consecutive trips without detecting a fault. If the MIL is off, a related code is erased after 40 warm-ups without a failure. A new rear HO2S installed with an aging front HO2S in Bank 1 can cause a code to set.

EGR SYSTEM MONITOR

Emissions of NOx increase proportionally with the temperature in the combustion chamber. The Exhaust Gas Recirculation (EGR) system is designed to circulate non-combustible exhaust gases into the manifold to dilute the A/F mixture. In this manner, the EGR system lowers the combustion temperature and reduces NOx emissions and pre-detonation (knocking).

OBD II regulations require that all vehicles equipped with an EGR system must test that system for high or low flow rates. If the EGR Monitor detects a fault in any of the system components, the MIL must be activated. If the flow rate exceeds the recommended flow rate limit at a level (too low or too high) that could cause NOx emissions to exceed 1.5 times the FTP standard on (2) consecutive trips, the MIL must be activated.

<u>Monitor Operation</u>

As discussed, this Monitor is designed to test the integrity and flow characteristics of the EGR system. To achieve this goal during the test, the PCM temporarily disables automatic fuel compensation, turns the EGR flow off (by controlling the EGR solenoid) and then monitors the front Oxygen sensor signals in order to determine the effects of the changes to fuel compensation.

If the EGR system is operating properly, turning off automatic fuel compensation shifts the A/F ratio in the lean direction and should cause a rise in HO2S voltage (due to an increase in oxygen in the exhaust). It should also cause Short Term fuel control to shift to rich.

The amount of the shift is used to monitor the operation of the EGR system. In effect, the EGR System Monitor watches the amount of shift to *indirectly monitor the operation of the EGR system*.

EGR System Monitor Graphic

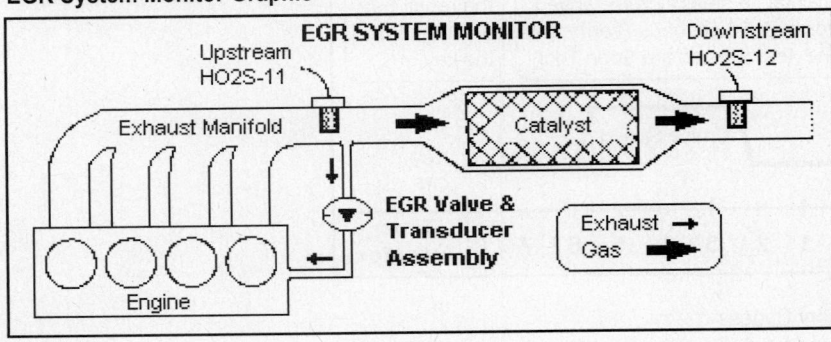

EGR SYSTEM MONITOR (CONTINUED)

EGR System Monitor Enable Criteria

The EGR System Monitor is run after these enable criteria are met:

- ECT input over 180ºF and engine run-time over three minutes
- Engine speed 2248-2688 rpm for A/T or 1952-2400 rpm for M/T
- MAP sensor input from 1.80 to 2.70v, BTS input more than 20ºF
- TP sensor input from 0.6-1.8v and VSS input more than 10 mph
- Short Term fuel trim adjusting pulsewidth < + 4.4% or > -8%

EGR System Monitor Suspend Conditions

If the Fuel Trim, Oxygen Sensor, Oxygen Sensor Heater or Misfire Monitor is running, the EGR System Monitor will be suspended.

EGR System Monitor Conflicting Conditions

If any of the "conflicting" conditions listed below are present, the Task Manager will not run the EGR System Monitor.

- Catalyst, EVAP or Fuel (rich intrusive test) Monitor in progress
- Fuel System Monitor 1-Trip "pending" rich or lean code is stored
- Misfire Monitor 1-Trip "pending" code is stored
- Oxygen Sensor or Heater Monitor 1-Trip "pending" code is stored

EGR System Monitor Trip Graphic

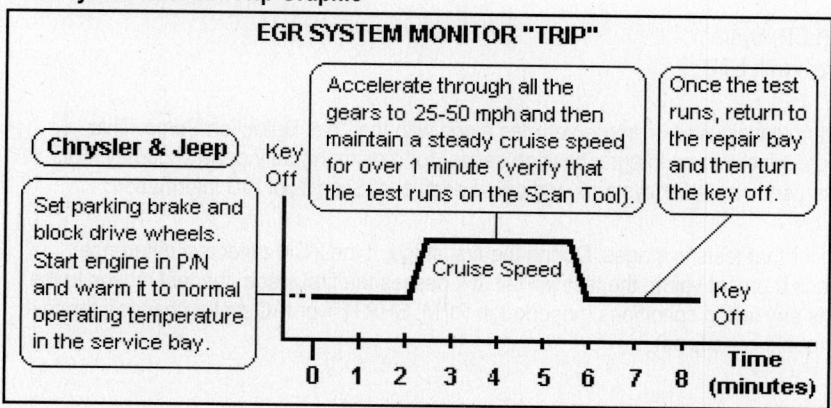

MIL Activation, Code Set Conditions

The EGR Monitor is run once each trip. If the measured change in A/F ratio in Short Term fuel trim is less than 7.4% or more than 20.5%, the EGR Monitor fails and a "pending" code is set. If the test fails for (2) consecutive trips, the MIL is activated and a code is stored. The PCM will turn off the MIL if the EGR Monitor passes for (3) consecutive trips. EGR system codes are erased after 40 official warmup cycles. Any type of restriction in the EGR tubes can cause a code to set.

EVAP SYSTEM MONITOR

OBD II regulations require that all vehicles equipped with an EVAP system be monitored for component integrity, system functionality and loss of hydrocarbons. The Enhanced EVAP system test is required to detect leaks as small as 0.040" in diameter. The test must detect leaks of 0.040-0.080" in diameter on fuel tanks larger than 25 gallons.

EVAP System Background

The EVAP system prevents fuel tank vapors from entering the atmosphere. Fuel evaporation emits Hydrocarbons (HC) directly into the atmosphere. As fuel evaporates in the fuel tank, vapors are routed into a charcoal canister. Through the use of an EVAP Purge solenoid, manifold vacuum draws these vapors into the combustion chamber.

Factory emissions tests have determined that an EVAP system with a leak as small as 0.020" can yield an average of about 1.35 grams of HC per vehicle driven mile. This is over 30 times the allowable exhaust emission standard. Because of this potential for unwanted release of HC into the atmosphere, OBD II regulations require that the EVAP System be monitored. The EVAP Monitor is designed to verify airflow through the system and monitor for leaks to prevent HC loss.

Currently, a 0.040" diameter leak is the smallest leak that can be detected on a regular basis. However, the vacuum tubing and its connections between the EVAP Purge valve and the intake manifold are excluded from these regulations (they require a visual inspection).

EVAP System Test Methods

Chrysler vehicles use (3) methods to monitor (test) the EVAP system:

- Stricter EVAP (Non-LDP)
- EVAP Leak Detection Pump (LDP) System
- California Stricter EVAP System with LDP

Linear DCP Solenoid

Starting in 1998, a Linear DCP solenoid was used on some vehicles along with the Leak Detection Pump. This solenoid is a pulsewidth-modulated device with an integral position sensor for monitoring duty cycle accuracy. This solenoid is more precise than those used on earlier systems and offers additional Scan Tool PID information.

Summary of Stricter Monitor Test

The EVAP Monitor is a 2-Trip Monitor that tests in stages. During the first stage, if the PCM detects a difference between the purge cells that exceeds a preset value, the test will fail. If it passes the first stage, the test moves to the second stage. If the Monitor detects any failing conditions (not enough RPM, SHRTFT or IAC motor change), the test fails and a related EVAP System trouble code is set.

EVAP SYSTEM MONITOR (CONTINUED)

EVAP Leak Detection Pump

A Leak Detection Pump (LDP) is used on Federal and California models equipped with the Stricter EVAP system. The LDP is designed to pressurize the EVAP System and to seal off the charcoal canister.

LDP System Components

The Leak Detection Pump contains these components:

- A 3-Port solenoid that activates the two-primary system functions
- A vacuum pump with a reed switch, check valves and diaphragm
- A canister vent valve that contains a spring loaded vent seal valve

Leak Detection Pump Graphic

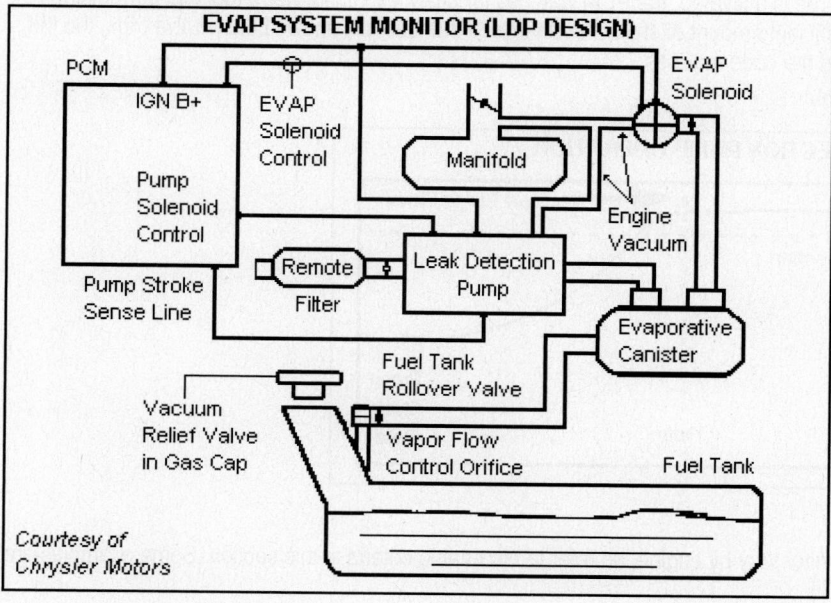

LDP Test Overview

Once outside air temperature is within preset parameters, the leak detection part of the EVAP Monitor test runs right after a cold engine startup. The 3-port solenoid is energized, allowing vacuum to pull the pump diaphragm upward (this draws air from the atmosphere into the pump). With the solenoid off, the pump is sealed and spring pressure drives the diaphragm down so that air is pumped into the system.

The solenoid and diaphragm are cycled to pressurize the EVAP system. The spring on the diaphragm is calibrated to 7.5 inches of water. If no leaks are present, the pressure in the system equalizes and the pump cycle rate falls to zero.

EVAP SYSTEM MONITOR (CONTINUED)

<u>LDP Monitor Operation</u>

The first part of the test determines whether or not a line is pinched. If the pump rate falls to zero rapidly, the PCM detects that there is not enough space to pressurize (a line may be pinched). If this part of the test passes, the LDP is stopped when the rate falls to zero.

If there is a leak present in the EVAP system, the cycle rate falls to a rate proportional to the size of the leak. By monitoring the cycle rate of the pump, the PCM can calculate the size of a leak. If the calculations reveal the presence of a hole that is 0.040" in diameter or more, a code is set and current engine conditions are stored in Freeze Frame.

When the leak test portions of the EVAP Monitor are completed and all test parameters are met, EVAP system operation is monitored during a warm idle condition. The pump is again actuated and cycled to a rate to achieve 7.5 inches of water pressure. Once purge is activated, the PCM watches for changes in short term fuel trim correction. If the EVAP Monitor detects an insufficient amount of shift, the test fails. If the test fails on two consecutive trips, the MIL is activated (two-trip detection) and the code matures.

Leak Detection Operation Graphic

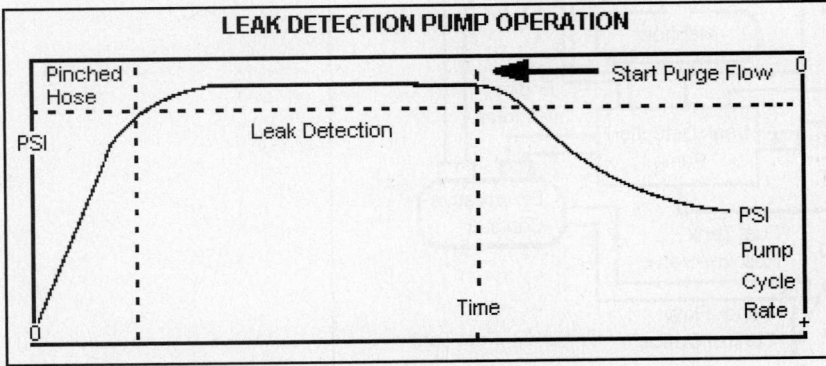

<u>LDP Monitor Enable Criteria</u>

The conditions to run the LDP Monitor vary by engine, so refer to the enable criteria in this section. Some examples are shown below:

- Test runs right after a cold startup (it can run for up to 3 minutes)
- Ambient Air Temperature Sensor from 40-90°F at engine startup
- BARO and MAP Sensor within preset limits
- ECT Sensor input within 10°F of the BTS input
- Engine run-time within limits, engine speed stable
- Fuel Level Sensor input more than 15% and less than 85%

EVAP SYSTEM MONITOR (CONTINUED)

<u>EVAP with Proportional Purge Solenoid</u>

The LDP assembly is designed to detect leaks in the EVAP system and to seal the system so that the required LDP Monitor can be run.

Some vehicles are equipped with a proportional purge solenoid that regulates the rate of vapor flow from the canister to the throttle body. The PCM controls the operation of the solenoid. During the cold start warmup periods and also during a hot start time delay period, the PCM leaves the solenoid off and fuel vapors are not purged.

The PCM control circuit senses current applied to the solenoid and adjusts the current to achieve the desired purge flow. The purge solenoid controls the purge rate of fuel vapors from the canister and fuel tank to the intake manifold. This solenoid operates at 200 Hz.

<u>System Components</u>

The leak detection pump assembly includes a 3-port leak detection solenoid valve, a pump unit with a spring loaded diaphragm, a reed switch which is used to monitor the pump diaphragm movement (position), two check valves, and a spring loaded vent seal valve.

EVAP System with Proportional Purge Solenoid Graphic

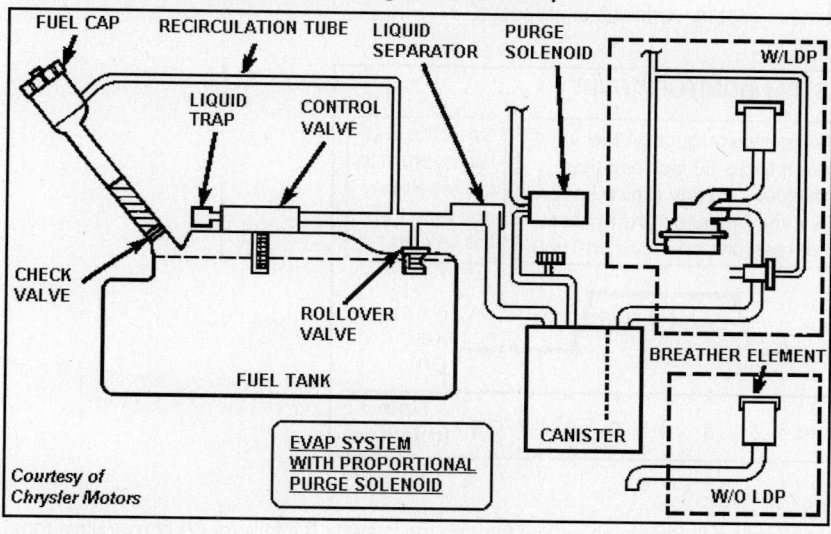

<u>LDP Pressure Switch</u>

The LDP pressure switch is monitored immediately after a cold engine startup. The PCM will set a "pending" code if the LDP switch is not in its expected state at key on (engine off) or if a change in the switch state is not detected by the PCM with the LDP solenoid energized to pull the pump diaphragm to its "up" position immediately after startup.

EVAP SYSTEM MONITOR (CONTINUED)

<u>Leak Detection Pump Modes</u>

Pump Mode - The pump is cycled at a fixed rate to achieve a rapid pressure build in order to shorten the overall test time.

Test Mode - The solenoid is energized with a fixed duration pulse. Subsequent fixed pulses then occur when the diaphragm reaches the switch closure point. The spring in the pump is set so that the system will achieve an equalized pressure of about 7.5 inches of water.

When the pump starts, the cycle rate is quite high. As the system becomes pressurized, the pump rate drops. If there is no leak, the pump will stop and the test is terminated at the end of the test mode. If there is no leak, the Purge Monitor is run. If the cycle rate increases due to the flow through the purge system, the test is passed and the test is complete. The canister vent valve will unseal the system after completion of the test sequence as the pump diaphragm assembly moves to the full travel position.

<u>EVAP Monitor Trip</u>

The EVAP Monitor trip in the Graphic below can be used to validate a repair of an EVAP System trouble code (i.e., DTC P0441, P0442, P0443, P0455 or P0456). It can also be used to "run" the EVAP Monitor to complete an Inspection/Maintenance (I/M) Readiness Test for certification purposes.

EVAP System Monitor Trip Graphic

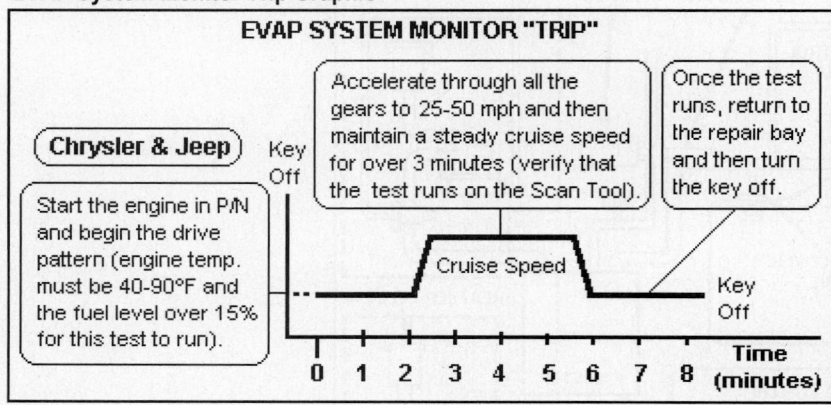

<u>MIL Activation, Code Set Conditions</u>

The EVAP Monitor will set a "pending" code if it detects no airflow through the system. If it fails for (2) consecutive trips, the MIL is activated and a code is set. The PCM will turn off the MIL if the EVAP Monitor passes for (3) consecutive trips. EVAP system codes are erased after 40 official warmup cycles. Any restrictions or disconnects in the EVAP system vacuum lines a faulty solenoid can set a code.

FUEL SYSTEM MONITOR

To comply with clean air regulations, Chrysler vehicles are equipped with catalytic converters that reduce the amount of HC, CO and NOx emissions released from the vehicle. The catalyst works best when the A/F ratio is at or near an optimum value of 14.7 to 1. The PCM is programmed to maintain the optimum A/F ratio. This is accomplished by making Short Term fuel trim corrections in injector pulsewidth based upon the oxygen sensor input. The programmed memory in the PCM acts as a self-calibration tool to compensate for variations in engine specifications, sensor tolerances and engine fatigue over the life span of the engine.

As injector pulsewidth increases, the amount of fuel delivered by the fuel injector is increased. A shorter pulsewidth decreases the amount of fuel delivered. The PCM monitors engine load, RPM, TP sensor, ECT sensor signals in order to make changes to injector pulsewidth.

SHORT TERM COMPENSATION

During open loop, the PCM changes pulsewidth without feedback from the oxygen sensors. Once the ECT sensor input is 30-35ºF, the PCM enters closed loop Short Term operation and utilizes feedback from the oxygen sensors. The PCM enters Long Term adaptive fuel control once the ECT sensor input reaches 170-190ºF. The PCM returns to open loop operation during wide open throttle condition.

To maintain the correct air fuel ratio, the PCM monitors the exhaust with an Oxygen sensor. Based on the oxygen content in the exhaust, the PCM can detect if the A/F mixture is rich or lean. As the PCM detects the percent of change required, it changes Short Term fuel compensation value in the pulsewidth equation.

LONG TERM COMPENSATION

If a Fuel system runs rich for an extended period of time, the O2S input will be high. The PCM will interpret this value as a rich mixture condition, and the SHRTFT value will be updated to a value of -17%. In effect, SHRTFT compensation is taking away fuel by decreasing the pulsewidth by 17%. At this point, the LONGFT adaptive then starts to decrease the amount of fuel delivered by changing pulsewidth in incremental steps. The PCM updates the LONGFT adaptive memory cells and the amount of SHRTFT compensation change required becomes less. In effect, the PCM used LONGFT adaptive to drive the SHRTFT compensation back to a value near 0%. This action provides SHRTFT compensation with a wider range of authority.

CATALYST MONITOR OPERATION

The PCM constantly monitors Short and Long Term adaptive. In effect, no pre-test is required.

Lean A/F Ratio - Any time that the engine is running lean and SHRTFT compensation multiplied by LONGFT adaptive exceeds a certain percentage for an extended period, the PCM will set a Fuel System Lean code for that trip and save the Freeze Frame data.

Rich A/F Ratio - Any time that the engine is running rich and SHRTFT compensation multiplied by LONGFT adaptive is less than a predetermined value, then the PCM will check the Purge Free Cells.

Purge Free Cells

Purge Free Cells are values placed in LONGFT adaptive memory cells when the Purge solenoid is off. There are 2, 3 or 4 Purge Free Cells. The PCM uses (2) purge cells to accomplish this task. One cell corresponds to an adaptive memory cell at idle and the other to a cell used for off-idle speed. A Purge Free Cell labeled PFC1 would hold the value for adaptive memory cell C1 under non-purge conditions.

If all of the Purge Free Cells are less than a certain value (%) and the adaptive memory factor is less than a certain (%), the PCM sets a Fuel System Rich code for that trip and saves the Freeze Frame data.

Monitor Test Conditions

Because the code conditions for the Fuel System Monitor vary by year and engine, look up each trouble code in this section. The code conditions are listed in the text area that follows the trouble code description.

MIL Operation, How to Clear Memory Codes

This Monitor is usually a two-trip monitor (it runs continuous in closed loop). The PCM sets a "pending" code and records engine data in Freeze Frame the first time it detects a fault. If a fault is detected for two consecutive trips, the PCM turns on the MIL and sets a code.

Before this Monitor will turn off the MIL, the engine must be within a similar value of the engine speed and load *recorded* when the code set before the PCM will record a Good Trip. If it passes (3) consecutive trips, the MIL is turned off. If it passes for 40 warmup cycles under these conditions, the PCM will clear the code. If the engine does not run under these "similar conditions", the code will be cleared after 80 warmup cycles.

MISFIRE MONITOR

OBD II regulations require that the OBD II system monitor the engine for misfire conditions and identify specific cylinders that experience a misfire. The PCM should also identify and set a different code to indicate if a multiple misfire condition exists. It should also be able to identify a specific cylinder that misfires in a multiple misfire condition.

This Monitor detects misfires related to engine mechanical, Ignition or Fuel system faults under positive load conditions (idle, cruise, etc.).

Misfire Explanation

Engine misfire is defined as the lack of combustion in a cylinder. When a misfire occurs, raw fuel and excess oxygen are released into the exhaust stream. The misfire causes unburned fuel in the exhaust continues to burn in the catalytic converter, the temperature of the catalyst to rise and the level of tailpipe emissions to increase. If the misfire condition is severe, damage to the catalyst could occur.

CRANKSHAFT SPEED FLUCTUATION METHOD

The Chrysler Misfire Monitor uses the Crankshaft Speed Fluctuation method to monitor for engine misfire. The CKP sensor is used to determine engine speed changes. The CKP sensor can detect slight variations in engine speed due to a misfire condition. Once the engine is running with all enable criteria met, the Misfire Monitor is activated. During this period of time, the PCM continuously monitors the speed of the crankshaft. When a misfire occurs, the PCM detects a decrease in crankshaft speed and begins to count the misfires.

Crankshaft Relearn

A Scan Tool can be used to reset (relearn) the crankshaft position. When the Crank Relearn function is selected on the Scan Tool, the PCM will enable a fast CMP and CKP sensor relearn procedure. To use this function on a Scan Tool refer the information that follows:

- OTC Scan Tool - select Special Tests and then Set Sync Mode
- Snap-On Scan Tool - select the Function Test menu
- Vetronix Scan Tool - select MISC Test and then Set Sync Mode

■ NOTE: *The engine must learn the spacing of the crank target windows while the engine is in Decel Fuel Shutoff. This reset step requires three closed throttle*

TWO TYPES OF MISFIRES

The OBD II regulations also require the use of two different tests to detect the severity of a misfire that is present in order to determine the type of failure criteria present at the time of the misfire. The two types of engine misfire are referred to as Type 'A' and Type B Misfires.

Type 'A' (One-Trip Monitor)

This type of Misfire (Type 'A') is set if the misfire condition is so excessive that damage to the catalyst will result at the current engine speed, engine load and coolant temperature. The PCM will flash the MIL once per second within 200 engine revolutions from the point where misfire was first detected with a Type 'A' misfire present. This action is necessary to warn the driver that if the vehicle continues to operate under these conditions that the catalytic converter will be damaged. If vehicle operating conditions change to conditions that will not cause converter damage, the MIL must stop flashing and remain "on" steady.

Type 'B' (Two-Trip Monitor)

This type of Misfire (Type 'B') is set if a misfire of more than 2% is detected that could cause tailpipe emissions to exceed 1.5 times the FTP Standard. The PCM will set a "pending" code and store engine data in Freeze Frame when a Type 'B' code is set. If the same fault is detected on (2) consecutive trips under similar conditions, the MIL is turned on and a code is set. The MIL is also turned on if a misfire is detected during (2) non-consecutive trips that are not more than 80 trips apart.

REVOLUTION COUNTER

The PCM contains a program called a Revolution Counter. In the logic portion of this counter, each 1000-revolution window contains five 200-revolution windows. The PCM counts the misfires for each 200-window and carries that value over to the 1000-revolution window.

MISFIRE MONITOR (CONTINUED)

<u>200-Revolution Counters</u>

With the engine running under positive load conditions, the PCM counts the number of misfires in every 200 revolutions of the crank. If, the misfire value (%) exceeds a preset during (5) of the 200 counters, a "pending" code is set and engine data is stored (the Freeze Frame data is recorded during the last 200 revolutions of the 1000-revolution period). If the same fault occurs on the next trip, a trouble code is set.

<u>1000-Revolution Counters</u>

The 1000-Revolution Counters are two-trip detection monitors. During 1000 revolution increments, this Monitor must be able to identify misfire values (%) that would cause a "durability demonstration vehicle" to fail an Inspection/Maintenance (I/M) test and vehicle manufacturers must determine a misfire value (%) that would cause the vehicle to fail a tailpipe test.

If an active misfire ends, the MIL will remain on continuously. The MIL will remain on for more than one trip, but will go out if the misfire conditions do not repeat during (3) consecutive trips with the vehicle operated within 375 rpm and 10% of the load condition stored in the Freeze Frame. The code is then erased from memory following the successful completion of 80 warmup cycles.

<u>Full Range Misfire Detection</u>

Chrysler began to phase in "Full Range Misfire Detection" in 1997. This term means that the PCM will detect misfires at any engine speed. Previously, misfire could only be detected at the lower end of the RPM range. 1997 Models with Full Range Misfire Detection:

* LH models with 3.3L and 3.5L engines (SBEC III+)
* NS models with 3.3L and 3.8L engines (SBEC III+)
* AN, BR and ZJ body models with a 5.2L engine (JTEC)

Full Range Misfire Detection Graphic

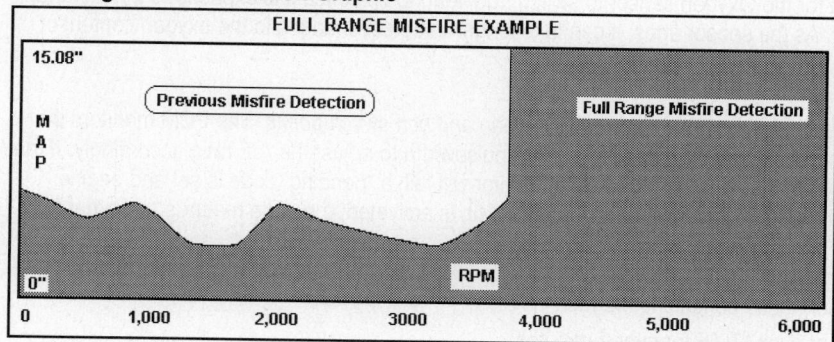

<u>Pending Faults - Comprehensive Component Monitor (MIL is on)</u>

If the CCM detects any of the faults below, the Misfire test will not run:

* CKP, CMP, ECT, EGR, IAT, MAP or Vehicle Speed Sensor (Electrical)
* ECT Sensor excessive time to enter closed loop (Rationality)
* Power Control Module Memory
* Upstream and Downstream HO2S (Electrical)

<u>Pending Faults - Main Monitors (MIL is on)</u>

If a fault is present in any of these below, the Misfire test will not run:

* EGR or EVAP System Monitor
* Fuel System or Oxygen Sensor Monitor

<u>MIL Activation, Code Set Conditions</u>

Once this Monitor is activated, the PCM continually tests for engine misfire as previously described. If the PCM detects more than a 2% misfire rate during two trips, or a 10-12% misfire rate during one trip, a "pending" code is set. The MIL is activated if the same fault is detected on (2) consecutive drive cycles. False misfire conditions can occur due to damp ignition components, running low on fuel, pulling heavy engine loads, low quality fuel or driving on a very rough road.

OXYGEN SENSOR MONITOR

The Oxygen Sensor Monitor is a test run once per trip designed to monitor the front Oxygen sensor (HO2S-11) and rear Oxygen sensor (HO2S-12) for faults or deterioration that could cause tailpipe emissions to exceed 1.5 times the FTP Standard.

Oxygen Sensor Monitor

The Oxygen Sensor Monitor is an on-board diagnostic designed to monitor the front and rear oxygen sensors for slow response, low and high voltage, a dynamic shift fault, an open circuit or short to voltage.

Oxygen Sensor Operation

The Oxygen Sensor Feedback system is used to control the exhaust emissions. The most important element of this system is the Oxygen sensor located in the exhaust path in front of the converter.

Once this sensor reaches an operating temperature of 572-662°F, it begins to generate a voltage that is inversely proportional to the amount of oxygen in the exhaust. The PCM uses information obtained from the Oxygen sensor to calculate the fuel injector pulsewidth. This action allows the PCM to maintain an ideal A/F ratio of 14:7 to 1. The catalyst works best at this A/F ratio to remove HC, CO and NOx from the exhaust. The O2 sensor also provides essential information to the PCM during operation of the EGR, Catalyst and Fuel System Monitor.

Oxygen Sensor Failure Conditions

The O2 sensor may fail in any of the following conditions:

- Due to a slow response rate
- Due to reduced voltage output
- Due to incorrect dynamic shift
- Due to a shorted or open circuit

Response Rate

Response rate is the time required for the Oxygen sensor to switch from lean to rich once it is exposed to a richer than optimum A/F mixture or vice versa. As the sensor ages, it can take longer to detect changes in the oxygen content of the exhaust gas.

Oxygen Sensor Switch Points

The O2 Sensor Switch Points (thresholds) are often referred to as *lean and rich switch points*. The PCM monitors the O2 sensor signal as it crosses a threshold and changes the injector pulsewidth to adjust the A/F ratio accordingly. If the O2 sensor voltage does not go beyond these switch points, the Monitor will fail, a "pending" code is set and engine data is stored in Freeze Frame. If it fails on (2) consecutive trips, the MIL is activated, the code matures and a hard code is set.

Oxygen Sensor Monitor Enable Criteria

The HO2S Monitor is activated when these conditions are met:

- Vehicle running over 24 mph at over 170°F for over 2 minutes
- Engine idling in Drive for A/T (Neutral for M/T) at 512-864 rpm
- High Pressure Power Steering switch is Off
- Air Conditioning is not cycling too rapidly (turn off for testing)
- Engine run-time over 3 minutes since startup

Oxygen Sensor Monitor Faults

The O2 Sensor Monitor will fail if it detects any of these conditions:

- A dynamic shift in the Oxygen sensor voltage
- Low response rate
- An open or shorted Oxygen sensor circuit condition
- Reduced signal voltage output

OXYGEN SENSOR MONITOR (CONTINUED)

Monitor Operation

The test conditions to run the Oxygen Sensor Monitor vary by engine application. Refer to the actual enable criteria in this section. The list below contains examples of this type of code condition information.

- Air Conditioning not cycling (turn the A/C Off during the test)
- Battery Voltage more than 10.5v
- BARO and Battery Temperature Sensor signals within limits
- Engine Coolant Temperature Sensor more than 170°F
- Engine runtime 3 minutes with throttle at off-idle position (operating in a Purge Memory Normal cell)
- Fuel system in closed loop with LONGFT (adaptive) enabled
- Power Steering switch in low PSI state (no steering load applied)
- Engine at idle speed and within the Desired Idle Speed range
- Fuel level from 15% to 85% (from Fuel Level sensor)
- Vehicle speed over 24 mph (A/T)

Pending Conditions - Main Monitors (MIL is on)

If a Misfire or Oxygen Sensor Heater Monitor "pending" code is set, the Oxygen Sensor Monitor will not run.

Suspend Conditions

If the Misfire or Oxygen Sensor Heater Monitor detects a fault during testing, the PCM will suspend the test.

MIL Operation, How to Clear Memory Codes

The Oxygen Sensor Monitor runs once per trip. It must pass for (3) consecutive trips before it will extinguish the MIL. Also, once the MIL is off, the code associated with the fault will be erased after 40 warmup periods without a failure. Any contamination in the O2 sensor or its connector will set a code.

OXYGEN SENSOR HEATER MONITOR

The Oxygen Heater Sensor Monitor is a PCM diagnostic designed to monitor the operation of the Oxygen Sensor Heater circuit. This Monitor runs at either "key on" or "key off" (depending on the vehicle and engine type). The Oxygen sensor voltage readings are very temperature sensitive and are not accurate below 300ºF.

Oxygen Sensor Heater Graphic

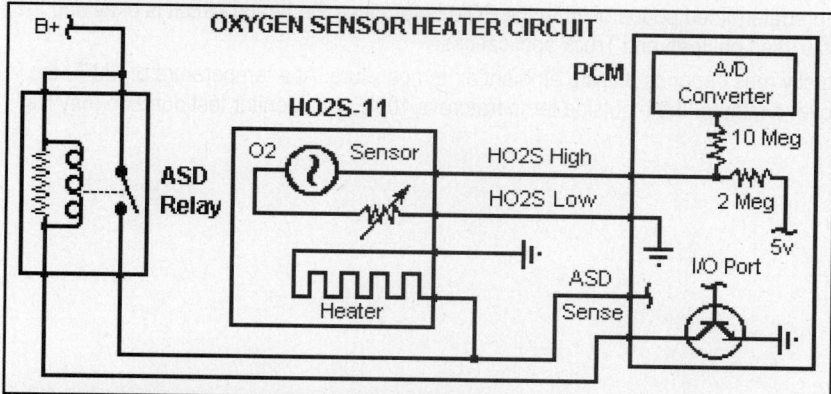

OXYGEN SENSOR HEATER EXPLANATION

The heater is used along with the sensor to allow the PCM to shift to closed loop control as soon as possible. The heating element in the O2 sensor must be tested to ensure that it is heating the sensor properly. The O2 sensor circuit is monitored for a drop in voltage. The sensor output is used to test the sensor by isolating the effect of the heater element on the O2 sensor output voltage *from other effects!*

The HO2S contains a heating element with a positive thermal coefficient device to control and maintain the correct sensor temperature. As current flows through the heater, its internal structure changes as temperature increases. The exhaust and heater combine to maintain the HO2S temperature at around 1200°F. Note that the PCM turns "on" the ASD relay in order to connect the heater to battery voltage with the key off.

OXYGEN SENSOR HEATER MONITOR (CONTINUED)

MIL Operation, How to Clear Memory Codes

This Monitor is a two-trip monitor. The first time a heater fault is detected, a "pending" code is set and engine data is stored in Freeze Frame. This Monitor must pass for (3) consecutive trips before it will turn off the MIL. Once the MIL is off, the memory code associated with the fault will be erased after 40 warmup periods without a failure.

SBEC III Oxygen Sensor Heater Test (Key Off)

This test is for Cars and Minivans. This Monitor does not test the heater element. Instead, it tests the resistance of the Oxygen Sensor signal circuit to determine if the heater is operating properly. This test is based on the fact that as the temperature of the sensor increases, the signal circuit resistance decreases. The resistance of a normal Oxygen sensor varies from 100 ohms to 4.5 Meg ohms.

This test is based on the principle that as the sensor temperature increases, the sensor resistance decreases, and the PCM will detect a lower voltage at the reference signal. Inversely, as temperature decreases, the resistance of the sensor increases, and the PCM should detect a higher voltage at the reference signal.

The enable criteria for this test include a key off power down period. Once the Oxygen sensor has had a chance to cool down after the key is off, the PCM sends a 5v bias voltage signal (35 ms long) every 1.6 seconds to the O2 sensor. After the voltage increases to a preset amount (more than the startup level), the PCM is ready to test the heater operation. The PCM turns on the ASD relay to apply voltage to the heater and continues to send the 5v bias signal to the sensor while testing the amount of voltage decrease. If it detects a decrease in the correct range for several 5v bias pulses, the test passes.

JTEC Key "On" Heater Monitor (Jeep & Truck Models)

This test is for Jeep and Truck models. The operation of this Monitor is similar to the test used for Cars and Minivans except that it is run with the key on or at engine startup. In this particular type of heater test, the PCM sends a 5v bias signal to the Oxygen sensor to measure the amount of voltage decrease as the sensor warms up.

JTEC III O2 Sensor Heater Test (Key On)

This test is run under cold startup conditions. If the PCM determines that the outside air temperature is within 10 degrees of the engine temperature, it determines that the engine has experienced a cold soak period. Once the enable criteria are met, the PCM tests the heater operation at startup.

During the test, the PCM monitors the decrease in O2 sensor signal voltage from a cold start. If the HO2S voltage signal drops to 3v or less during a predetermined period of time, the PCM determines the heater circuit is okay and the Monitor will pass the test. This test is used on Jeep and Truck applications.

The length of time that the test actually runs depends outside ambient air temperature. At a temperature of -20°F, the Monitor test duration may last for over a minute. If the outside temperature is 100°F, the Monitor test duration may last for only 40 seconds.

SECONDARY AIR MONITOR

This Monitor is a PCM diagnostic that monitors the Secondary AIR system for component integrity, system functionality and faults that could cause vehicle tailpipe levels to exceed 1.5 times the FTP Standard. The Secondary AIR System Monitor is run once per trip once the required enable criteria are met.

Secondary AIR System Explanation

Secondary Air Injection (AIR) is used when the PCM determines that the catalytic converter warmup time after engine startup will be too long. Outside air is injected through an electronically controlled aspirator upstream of the catalytic converter. Additional oxygen injected into the exhaust stream allows for a quick rise in converter temperature during warmup periods. On Eagle Talon models, Air Injection is required due to the relationship of the exhaust manifold location (before the exhaust) to the converter (under the floor pan).

Secondary AIR System Graphic

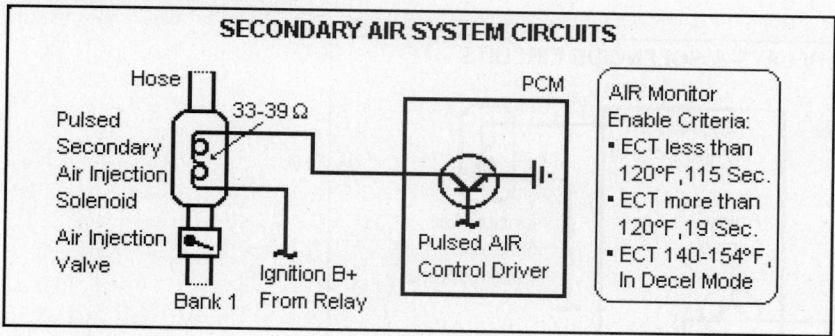

Secondary AIR Monitor Enable Criteria (all except for Eagle Talon)

The correct enable criteria must be met before the Secondary Air Monitor will begin its test. In addition, the Secondary AIR Monitor will not run if the Catalyst, EGR, Fuel System or Purge Monitor is activated, or if a Misfire, HO2S, HO2S Heater, or Fuel System Rich Monitor one trip "pending" code is present.

*   Engine at idle speed
*   Idle speed must not drop below 700 rpm during the test
*   ECT input is over 170°F

Monitor Operation

Once the enable criteria are met (engine runtime over 11 minutes since startup and engine at idle speed), the PCM temporarily disables automatic fuel control and enriches the injector pulsewidth by around 10%. The maximum time that this part of the test runs is 18 seconds.

When the O2 sensor voltage switches to a rich indication, the PCM energizes the AIR solenoid. If the Secondary AIR system is working properly, when the solenoid is energized, the O2 sensor voltage should shift in a lean direction. If the voltage does not switch to a rich indication (due to the additional fuel) and then back to a lean indication (due to the additional air) within 19 seconds, a "pending" code is set and engine data is stored in Freeze Frame. If the test fails for (2) consecutive trips, the MIL is activated and a code is stored.

Possible Causes of an AIR System Monitor Fault

If air is injected continuously into the exhaust, the catalytic converter could be damaged prematurely and cause a false indication of a lean A/F ratio. This action could cause the PCM to increase injector pulsewidth and the result would be an increase in tailpipe emissions. If the AIR System fails to operate, the catalyst would not warmup quickly enough. This could result in periods of time with high tailpipe emissions.

Pending Fault - Comprehensive Component Monitor (MIL is on)

If the CCM detects a fault in the Camshaft or Crankshaft Position Sensor, Secondary Air Component or the Vehicle Speed Sensor (Electrical faults), the Secondary Air Monitor will not run.

MIL Operation, How to Clear Memory Codes

The PCM will turn off the MIL if the Secondary AIR Monitor passes (3) consecutive trips without a fault detected. The code is erased by the PCM after 40 successful warmup cycles.

Onboard Diagnostics

CIRCUIT ACTUATION TEST MODE

When the Circuit Actuation Test Mode option is selected on the Scan Tool, you will have access to Actuator Test menus. The tests listed in the menu provide the capability to activate certain controlled output circuits to the PCM. The lists of tests available vary depending on engine and vehicle equipment. The tests may be used to find device faults that the PCM may not recognize.

During Actuator Tests, both electrical and mechanical activity can be checked and verified. Once a certain device is selected, the PCM activates the device that allows you to listen for an audible click or observe a visual indication of correct device operation.

The Scan Tool Actuator Test Mode can be used to diagnose many PCM controlled devices, and to help diagnose the cause of one or more trouble codes on Chrysler and Jeep vehicles:

Actuator Tests Graphic

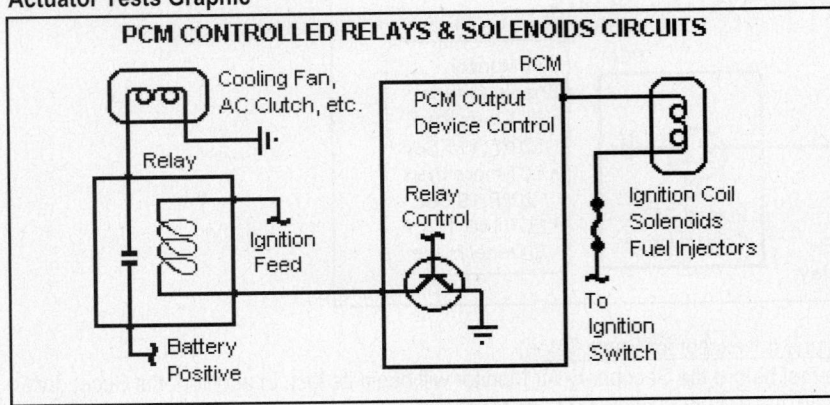

Testing Relays & Solenoids with a Scan Tool

Refer to the specific instructions in the Scan Tool Operating Manual to test a particular device or its circuit.

STATE DISPLAY SWITCH TEST MODE

The PCM switch inputs have two recognized states: they are either in a HIGH or LOW state. Because these inputs are either high or low, the PCM may not be able to detect the difference between when a switch is in a high or low state from when a switch has a circuit problem (i.e., when a circuit is open, grounded or shorted or even if the switch is defective).

If the Scan Tool State Display (meaning the Switch State) shows a change from HIGH to LOW; or from LOW to HIGH when the selected switch is activated, it can be assumed that the entire switch circuit to the PCM is functioning properly. The Scan Tool State Display Mode can be used to diagnose these codes:

- Brake Switch: P0703
- LDP Switch: P1494
- P/N Switch: P1899
- PSP Switch: P0551

TESTING SWITCHES WITH A SCAN TOOL

Refer to the specific instructions in the Scan Tool Operating Manual that cover how to test a particular switch or its related circuit.

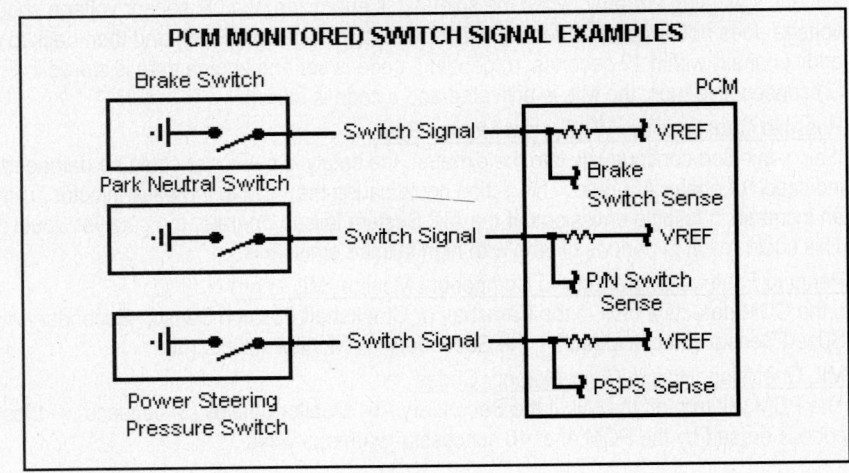

ONBOARD DIAGNOSTICS

NON-MONITORED SYSTEMS & CIRCUITS

The PCM and CCM cannot monitor all of the Base Engine systems and components for faults or conditions that might cause a driveability problem. This handicap can cause confusing diagnostic situations when a Base Engine problem occurs. If the fuel pressure is too low or high, a fuel pressure code will not be set, but a misfire or oxygen sensor code might be set due to a lean or rich A/F condition.

Fuel Pressure

On engines with fuel injection, the fuel pressure regulator controls fuel system pressure. The PCM cannot detect a restricted fuel pump inlet filter, a dirty in-line fuel filter, or a pinched fuel supply or return line. An O2 sensor or Fuel system code might set due to a lean A/F condition.

Ignition System Secondary

The PCM cannot detect a faulty ignition coil, fouled or worn out spark plugs, ignition wires that are cross firing, or an open spark plug wire. However, the Misfire Monitor would detect these faults during testing.

Engine Analyzer Testing the Ignition System

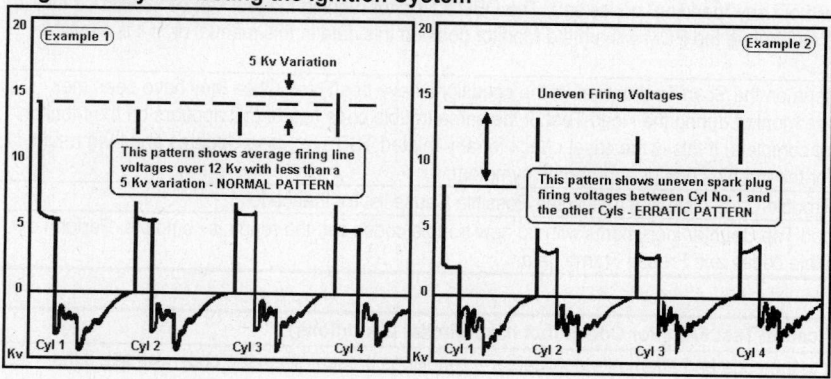

Engine Compression

The PCM cannot detect uneven, low, or high engine cylinder compression. However, a fault in one of these areas could cause the Oxygen Sensor or Misfire Monitor to fail during testing.

Exhaust System

The PCM cannot detect a restriction or leak in the Exhaust system. However, a fault in one of these areas could cause the EGR System, Fuel System, HO2S or Misfire Monitor to fail during testing.

Fuel Injector Mechanical Fault

The PCM cannot detect if a fuel injector is restricted, stuck open or closed. However, a fault in one of these areas could result in a rich or lean condition and cause the Fuel System or HO2S Monitor to fail.

ROAD TEST VERIFICATION - CODE FAULTS

Prior to proceeding with Road Test Verification for an OBD II system related trouble code, you should perform a complete visual inspection.

Visual Inspection

Verify that all engine components have been reconnected and installed properly. If this has not been done, reassemble or reconnect all components as needed. Check the oil level and coolant level.

If this Road Test Verification procedure is for an OBD II System code, refer to the OBD II Drive Cycle article in this section as needed. For previously read codes that have not been repaired, refer to the correct code repair chart and perform the repair test steps before continuing.

ROAD TEST VERIFICATION - CODE FAULTS (CONTINUED)

VER-5A Test Table

Verification Test 5A (OBD II Trouble Codes)
Step 1: For ABS and Air Bag systems, enter the correct VIN and Mileage into the PCM. Then erase the codes in the ABS and Air Bag modules. Note: If the PCM has been replaced and the correct VIN and Mileage are not programmed, a code will be set in the ABS and Air Bag modules.
Step 2: If the vehicle has a SKIM theft-alarm system, connect a Scan Tool to the DLC, select Engine MISC and place the SKIM in secured access mode (use the appropriate vehicle PIN Code). Select Update the Secret Key Data and the data will be transferred from the SKIM to the PCM.
Step 3: Once all of the codes are repaired, return to the Scan Tool main menu and repeat the diagnostic test for the system (e.g., run the appropriate OBD II Monitor for a previously repaired OBD II trouble code).
Step 4: Connect a Scan Tool. Verify that the fuel tank is from 1/4 to 3/4 full and turn off all accessories.
Step 5: The right way to verify an OBD II code has been repaired is to allow the PCM to run the related OBD II Monitor(s) and increment a Global Good Trip (a trip without any "pending" codes set). The OBD II Monitor running the test can be viewed on the tool. The code enable criteria must be met before the PCM will run the Monitor (look up this data in this manual or, if it is available, view it on the pretest screen).
Step 6: Monitor the pretest enable criteria on the Scan Tool until all of the conditions have been met. Once they have been met, have an assistant watch the appropriate Monitor during the Road Test. If the same trouble code resets (if it appears on the Monitor screen during the trip), the repair is not complete. If this is the case, check for any related TSB's or Flash Updates and then return to the Scan Tool main menu. Retest for the trouble code or appropriate symptom.
Step 7: If a new code sets, go to the Trouble Code Lists and review the Possible Cause list for that code.
Step 8: If the Monitor runs and the Good Trip Counter increments with no new trouble codes set, the repair is complete. Perform a PCM Reset to erase all diagnostic trouble codes and Freeze Frame data.

VER-5A2 TEST TABLE

Verification Test 5A2 (For Codes that have Similar Conditions)
Step 1: Monitor the conditions related to the code and attempt to duplicate the conditions present when the code was set. If the conditions can be duplicated, the Good Trip Counter will change to one (1) or higher.
Step 2: If the conditions cannot be duplicated, clear the trouble codes and then disconnect the Scan Tool.
Step 3: If the repair code resets or the Monitor fails, check for related TSB's or flash updates to the PCM.
Step 4: If a new code sets, go to the Trouble Code Lists and review the Possible Cause list for that code.
Step 5: If the OBD II Monitor runs and no new codes are set, then the repair is complete.
Step 6: Erase the trouble codes (PCM Reset) and remove the Scan Tool.

VER-5A3 TEST TABLE

Verification Test 5A3 (HO2S Slow Response, HO2S Heater or Catalyst)
Step 1: Monitor the pretest Enable Criteria on the Scan Tool until all of the code conditions have been met. Once they have been met, have an assistant watch the appropriate OBD II Monitor during a Road Test. If the trouble code resets (if it appears on the Monitor screen during the trip), the repair is not complete. If this is the case, check for any related TSB's or Flash Updates for the PCM, and then return to the Scan Tool main menu. Retest for the trouble code or any related symptoms.
Step 2: If a new code sets, go to the Trouble Code Lists and review the Possible Cause list for that code.
Step 3: If the code resets or the Monitor runs and fails check for related TSB's or updates to the PCM.
Step 4: If the OBD II Monitor runs and no new codes are set, then the repair is complete.
Step 5: Erase the trouble codes (PCM Reset) and remove the Scan Tool.

VER-6A TEST TABLE

Verification Test 6A (For EVAP Codes - LDP Dealer Test Mode)
Explanation: The LDP Dealer Test Mode was added to the DRB III to verify repairs to the system (i.e., a DRB software program was written that causes the PCM to run the LDP Monitor portion of this test). Test failures are indicated through a related trouble code. This Test Mode runs a total system performance test.

ONBOARD DIAGNOSTICS

ROAD TEST VERIFICATION - NO CODE FAULTS

Prior to proceeding with Road Test Verification for Non-OBD II codes, inspect the vehicle to ensure that all engine components are properly connected. Reassemble and connect all components as needed.

If the Road Test Verification procedure is for a No Code Fault, refer to the OBD II Drive Cycle article in this section. For trouble codes that are not OBD II related, continue with this Road Test Verification step. For previously read codes that have not been repaired, refer to the correct code repair test and perform the repair test steps.

For No Code Faults

To locate the cause of a "No Code Fault or a Driveability Symptom", do the following test steps:

- Check to see if the initial symptom is still present
- If the initial symptom or another symptom is present, the repair is not completed

Check all pertinent technical service bulletins for corrective actions that fit the symptom. Then refer to the Chrysler No Code Test Main Menu in other repair manuals or electronic media for a list of tests to run on non-monitored devices and their related circuits (i.e., secondary ignition, engine timing, etc.).

PCM Has Not Been Changed

If the PCM has not been changed, do the following steps:

- Connect a Scan Tool, read, record and erase all codes
- Use the Scan Tool to reset all values in the adaptive memory

Road Test Verification for A/C Relay Non-OBD II Codes

Drive the vehicle for at least 5 minutes with the A/C on. During the drive cycle, go at least 40 mph and at some point, stop the vehicle and turn off the engine for 10 seconds or more. Then restart the engine and drive the vehicle while shifting through all gears. Return to the service bay and verify that no codes are stored.

Road Test Verification for Charging System Non-OBD II Codes

If the PCM has been changed, and the vehicle is equipped with a factory theft alarm, start the vehicle 20 times to activate the alarm.

- Connect the Scan Tool, read, record and erase all trouble codes
- Start engine, raise and hold the engine speed at 2000 rpm for 30 seconds
- Return to idle speed, turn the engine off, then turn the key on and read the Charging System trouble codes

REFERENCE INFORMATION

OBD II TROUBLE CODE LISTS

The first step is to read and record all trouble codes and Freeze Frame data - this information will be used during diagnosis. *If the Reset function of the PCM is done first, all codes and related data will be lost!*

Look up the appropriate trouble code in the list on the following pages. This information includes the DTC number, Code Title and Conditions (the enable criteria that indicate conditions present when a code sets).

You can use this information to determine how to drive a vehicle to validate a repair procedure is complete. The enable criteria can be used to help diagnose a trouble code related to the AIR, Catalyst, CCM (component monitor), EGR or EVAP, Fuel and Misfire Monitors. The **(1T)** and **(2T)** designators in the first column indicate whether the code requires one or two "trips" to occur before the MIL is activated.

OBD II TROUBLE CODE LIST (P0XXX CODES)

DTC	Trouble Code Title, Conditions & Possible Causes
P0016 1T CCM 2000-03 300M, LHS, Caravan, Concorde, Intrepid; Neon, PT Cruiser, Town & Country, Sebring	Crankshaft/Camshaft Timing Misalignment Conditions: Engine cranking or running; and the PCM detected the camshaft was out of phase with the crankshaft during the CCM test period. Possible Causes • Base engine problem (i.e., the camshaft timing is not correct) • CKP sensor signal is erratic (intermittent loss of the signal • CMP sensor signal is erratic (intermittent loss of the signal) • Tone wheel or pulse wheel is damaged or contains debris • PCM has failed
P0030 1T CCM 2000-03 300M, LHS, Concorde, Intrepid; Grand Cherokee	HO2S-11 (B1 S1) Heater Relay Circuit Fault Conditions: Engine started; system voltage over 10.6v, and the PCM detected a fault in the HO2S-11 heater element feedback sense circuit. Possible Causes • HO2S assembly is damaged or it has failed • HO2S heater control circuit is open, shorted to ground or B+ • HO2S heater ground circuit is open • HO2S heater element is damaged or has failed • PCM has failed
P0031 1T CCM 2001-03 300M, LHS, Caravan, Concorde, Intrepid; Neon, PT Cruiser, Town & Country, Sebring, Viper & Jeep models; Truck & Van models	HO2S-11 (B1 S1) Heater Relay Circuit Low Conditions: Key on, system voltage over 10.6v, ASD relay on, HO2S-11 heater "on", and the PCM detected the Heater Relay circuit Actual state did not match the Desired state (low circuit). Possible Causes • HO2S assembly is damaged or it has failed • HO2S heater element is damaged or has failed • HO2S heater control circuit is shorted to ground • PCM has failed
P0032 1T CCM 2001-03 300M, LHS, Caravan, Concorde, Intrepid; Neon, PT Cruiser, Town & Country, Sebring, Viper & Jeep models; Truck & Van models	HO2S-11 (B1 S1) Heater Relay Circuit High Conditions: Key on, system voltage over 10.6v, ASD relay on, HO2S-11 heater "on", and the PCM detected the Heater Relay circuit Actual state did not match the Desired state (high circuit). Possible Causes • HO2S heater element is damaged or the heater has failed • HO2S heater control circuit is open or it is shorted to power • HO2S heater ground circuit is open • PCM has failed
P0033 1T CCM 2003 PT Cruiser & Sebring equipped with a 4-cylinder engine	Surge Valve Solenoid Circuit Fault Conditions: Key on or engine running; system voltage over 10.5v and the PCM detected the Actual state of the Surge Valve solenoid circuit did not match the Intended state during the CCM test. Possible Causes • Surge valve power supply is open (test power from ASD relay) • Surge valve solenoid circuit is open • Surge valve solenoid circuit is shorted to ground or power (B+) • Surge valve solenoid is damaged or it has failed • PCM has failed

OBD II TROUBLE CODE LIST (P0XXX CODES)

DTC	Trouble Code Title, Conditions & Possible Causes
P0036 1T CCM 2000-03 300M, LHS, Concorde, Intrepid; Grand Cherokee	HO2S-12 (B1 S2) Heater Relay Circuit Fault Conditions: Engine started; system voltage over 10.8v and the PCM detected a problem in the Heater Relay circuit. • HO2S assembly is damaged or it has failed • HO2S heater control circuit is open, shorted to ground or B+ • HO2S heater ground circuit is open • HO2S heater element is damaged or has failed • PCM has failed
P0037 1T CCM 2001-03 300M, LHS, Caravan, Concorde, Intrepid; Neon, PT Cruiser, Town & Country, Sebring, Viper & Jeep models; Truck & Van models	HO2S-12 (B1 S2) Heater Relay Circuit Low Conditions: Key on, system voltage over 10.8v, ASD relay on, HO2S-12 heater "on", and the PCM detected the Heater Relay circuit Actual state did not match the Desired state (the circuit status remained low). Possible Causes • HO2S assembly is damaged or it has failed • HO2S heater element is damaged or has failed • HO2S heater control circuit is shorted to ground • PCM has failed
P0038 1T CCM 2001-03 300M, LHS, Caravan, Concorde, Intrepid; Neon, PT Cruiser, Town & Country, Sebring, Viper & Jeep models; Truck & Van models	HO2S-12 (B1 S2) Heater Relay Circuit High Conditions: Key on, system voltage over 10.6v, ASD relay on, HO2S-12 heater "on", and the PCM detected the Heater Relay circuit Actual state did not match the Desired state (high circuit). Possible Causes • HO2S heater element is damaged or the heater has failed • HO2S heater control circuit is open or it is shorted to power • HO2S heater ground circuit is open • PCM has failed
P0043 1T CCM 2001-03 Truck & Van models equipped with a 3.9L V6, 5.9L V8 & 8.0L V10 engine	HO2S-13 (B1 S3) Heater Relay Circuit Low Conditions: Engine started; system voltage over 10.6v, ECT less within the test condition value, and the PCM detected the Post Catalyst HO2S-13 Heater Relay circuit voltage remained too high. Possible Causes • HO2S assembly is damaged or the HO2S heater element is damaged or has failed • HO2S heater control circuit is shorted to ground • PCM has failed
P0044 1T CCM 2001-03 Truck & Van models equipped with a 3.9L V6, 5.9L V8 & 8.0L V10 engine	HO2S-13 (B1 S3) Heater Relay Circuit High Conditions: Engine started; system voltage over 10.6v, ECT less within the test condition value, and the PCM detected the Post Catalyst HO2S-13 Heater Relay circuit voltage remained too high. Possible Causes • HO2S heater element is damaged, or the control circuit is open or shorted to power • HO2S heater ground circuit is open • PCM has failed
P0050 1T CCM 2002-03 300M, LHS, Concorde & Intrepid	HO2S-21 (B2 S1) Heater Relay Circuit Low Conditions: Key on, system voltage over 10.6v, ASD relay on, HO2S-21 heater "on", and the PCM detected the Heater Relay circuit Actual state did not match the Desired state (low circuit). Possible Causes • HO2S assembly is damaged, or the HO2S heater element is damaged or has failed • HO2S heater control circuit is shorted to ground • PCM has failed
P0051 1T CCM 2001-03 300M, LHS, Caravan, Concorde, Intrepid; Neon, PT Cruiser, Town & Country, Sebring, Viper & Jeep models; Truck & Van models	HO2S-21 (B2 S1) Heater Relay Circuit Low Conditions: Key on, system voltage over 10.6v, ASD relay on, HO2S-21 heater "on", and the PCM detected the Heater Relay circuit Actual state did not match the Desired state (low circuit). Possible Causes • HO2S assembly is damaged or it has failed • HO2S heater element is damaged or has failed • HO2S heater control circuit is shorted to ground • PCM has failed

OBD II TROUBLE CODE LIST (P0XXX CODES)

DTC	Trouble Code Title, Conditions & Possible Causes
P0052 1T CCM 2001-03 300M, LHS, Caravan, Concorde, Intrepid; Neon, PT Cruiser, Town & Country, Sebring, Viper & Jeep models; Truck & Van models	**HO2S-21 (B2 S1) Heater Relay Circuit High Conditions:** Key on, system voltage over 10.6v, ASD relay on, HO2S-21 heater "on", and the PCM detected the Heater Relay circuit Actual state did not match the Desired state (high circuit). Possible Causes • HO2S heater element is damaged or the heater has failed • HO2S heater control circuit is open or it is shorted to power • HO2S heater ground circuit is open • PCM has failed
P0056 1T CCM 2002-03 300M, LHS, Concorde & Intrepid	**HO2S-21 (B2 S1) Heater Relay Circuit Low Conditions:** Key on, system voltage over 10.6v, ASD relay on, HO2S-21 heater "on", and the PCM detected the Heater Relay circuit Actual state did not match the Desired state (low circuit). Possible Causes • HO2S assembly is damaged or it has failed • HO2S heater element is damaged or has failed • HO2S heater control circuit is shorted to ground • PCM has failed
P0057 1T CCM 2001-03 300M, LHS, Caravan, Concorde, Intrepid; Neon, PT Cruiser, Town & Country, Sebring, Viper & Jeep models; Truck & Van models	**HO2S-21 (B2 S1) Heater Relay Circuit Low Conditions:** Key on, system voltage over 10.6v, ECT input under test condition value, and the PCM detected the Heater Relay signal was too low. Possible Causes • HO2S assembly is damaged or it has failed • HO2S heater element is damaged or has failed • HO2S heater control circuit is shorted to ground • PCM has failed
P0058 1T CCM 2002-03 300M, LHS, Concorde & Intrepid	**HO2S-21 (B2 S1) Heater Relay Circuit High Conditions:** Key on, system voltage over 10.6v, ECT input under test condition value and the PCM detected the Heater Relay signal was too high. Possible Causes • HO2S heater element is damaged or the heater has failed • HO2S heater control circuit is open or it is shorted to power • HO2S heater ground circuit is open • PCM has failed
P0068 1T CCM 2003 300M, Caravan, Concorde, Intrepid; Neon, PT Cruiser, Town & Country, Sebring, Viper, Jeep & Dakota models	**MAP Sensor/TP Sensor Correction - High Flow/Vacuum Leak Conditions:** Engine started; engine speed over 2000 rpm, and the PCM detected the Manifold Air Pressure (MAP) value dropped to less than 1.5" Hg with the throttle closed during the test. Possible Causes • An engine vacuum leak present • High resistance in the MAP ground circuit, MAP sensor signal or VREF (5v) circuit • High resistance in the TP ground, TP circuit or the TP sensor VREF (5v) circuit • MAP sensor is damaged or it has failed • TP sensor is damaged or it has failed • PCM has failed
P0070 1T CCM 2001-03 300M, LHS, Caravan, Concorde, Intrepid; Neon, PT Cruiser, Town & Country, Sebring, Viper, Jeep & Dakota models	**Ambient Temperature Sensor Circuit Stuck Conditions:** Engine started 4 times, 4 warmup cycles completed, vehicle driven for 200 miles, and the PCM did not detect more than a 6°F change in the Air Temperature sensor signal in the test. Possible Causes • Air temperature sensor signal circuit shorted to power (VREF) • Air temperature sensor ground circuit is open • Air temperature sensor signal circuit is open • Air temperature sensor signal circuit is shorted to ground • PCM High or Low circuit is damaged or it has failed

OBD II TROUBLE CODE LIST (P0XXX CODES)

DTC	Trouble Code Title, Conditions & Possible Causes
P0071 1T CCM 2001-03 300M, LHS, Caravan, Concorde, Intrepid; Neon, PT Cruiser, Town & Country, Sebring, Viper, Jeep & Dakota & Prowler	Battery Temperature Sensor Performance Conditions: Engine "off" time over 8 hours, DTC P0072 and P0073 not set, engine started, ambient temperature more than 39ºF, and the PCM determined the Ambient Air Temperature sensor was not within 53ºF of the ECT and IAT sensor signals after a cool down period. Possible Causes • Ambient Temp. Sensor circuit open, shorted to ground or VREF • Ambient air temperature sensor ground circuit is open • Ambient air temperature sensor is damaged or it has failed • PCM High or Low circuit is damaged or it has failed
P0072 1T CCM 2002-03 300M, LHS, Caravan, Concorde, Intrepid; Neon, PT Cruiser, Town & Country, Sebring, Viper, Jeep & Dakota & Prowler	Ambient Temperature Sensor Circuit Low Input Conditions: DTC P0072 and P0073 not set, Key on or engine running; system voltage over 10.5v, at least 5 warmup cycles completed, odometer mileage change at least 196.6 miles, and the PCM detected the BTS signal change was less than 7.2ºF. Possible Causes • Sensor signal value less than 1.0v • Sensor signal or ground circuit has high resistance • PCM has failed
P0073 1T CCM 2002-03 300M, LHS, Caravan, Concorde, Intrepid; Neon, PT Cruiser, Town & Country, Sebring, Viper, Jeep & Dakota & Prowler	Battery Temperature Sensor Circuit High Input Conditions: Key on or engine running; system voltage over 10.5v and the PCM detected the Ambient Air Temperature sensor signal was more than 4.90v. Note that this code can be set due to an intermittent failure. Possible Causes • Ambient air temperature sensor signal shorted to VREF (5v) • Ambient air temperature sensor signal circuit is open • Ambient air temperature sensor is damaged (it may be open) • PCM has failed
P0100 1T CCM 1995-98 Eagle Talon equipped with a Turbo engine	Volume Airflow Sensor Circuit Conditions: Engine started; engine speed over 500 rpm, and the PCM detected the VAF sensor input was 3 Hz or less (the correct reading should be somewhere between 22-48 Hz). Possible Causes • Volume Airflow sensor signal shorted to power • Volume Airflow sensor signal shorted to ground • Volume Airflow sensor has failed • PCM has failed
P0105 1T CCM 1995-98 Eagle Talon equipped with a Turbo engine	Barometric Pressure Sensor Circuit Conditions: Key on for less than 60 seconds or right after startup, system voltage over 8v, and PCM detected a BARO input of more than 4.5v or less than 0.2v for 4 sec's (normal is 3.2-3.8v). Possible Causes • Loss of 5-volt supply from PCM (internal failure) • Sensor 5-volt supply circuit is shorted, open or grounded • Sensor signal circuit is open or shorted to ground • Sensor has failed • PCM has failed
P0106 1T CCM 1996-2003 All models	BARO Out-Of-Range at Key On / MAP Sensor Low Conditions: Key on for less than 350 ms; engine speed less than 255 rpm, and the PCM detected the MAP sensor input was less than 2.196v but more than 0.019v during a 300 ms period. Possible Causes • Loss of 5-volt supply from PCM (internal failure) • Sensor 5-volt supply circuit is shorted, open or grounded • Sensor signal circuit is open or shorted to ground • Sensor has failed • PCM has failed

DTC	Trouble Code Title, Conditions & Possible Causes
P0107 1T CCM 1995-98 Eagle Talon equipped with Non-Turbo engine	MAP Sensor Circuit Low Input Conditions: Engine speed from 400-1500 rpm; TP sensor input less than 1.3v, and PCM detected a MDP input less than 0.02v for 2 seconds (it should be between 0.8-2.5v). Possible Causes • Loss of 5-volt supply from PCM (internal failure) • Sensor 5-volt supply circuit is open or shorted to ground • Sensor signal circuit is shorted to ground • Sensor has failed • PCM has failed
P0107 1T CCM 1995-2003 Car & Mini Van models	MAP Sensor Circuit Low Input Conditions: Engine speed from 600-3500 rpm; TP sensor input less than 1.2v, system voltage over 10.6v, and the PCM detected the MAP input was below 0.0392v (conditions met for 1.76 seconds). Possible Causes • MAP sensor connector is damaged or shorted • MAP sensor 5v supply circuit is open or it is shorted to ground • MAP sensor signal circuit is shorted to ground • MAP sensor has failed • PCM has failed
P0107 1T CCM 2001-03 Sebring & Stratus models with a 2.4L & 3.0L engine	Barometric Pressure Sensor Circuit Low Input Conditions: Engine started, engine runtime over 2 seconds, system voltage over 10.5v and the PCM detected the BARO sensor was 1.95v or lower (a value equal to 50 kPa or 7.3" Hg or higher) for 10 seconds. Note: This value indicates the vehicle was at 15,000 feet above sea level. Possible Causes • BARO sensor connector is damaged or shorted • BARO sensor 5v supply circuit is open or it is shorted to ground • BARO sensor signal circuit is shorted to ground • BARO sensor has failed • PCM has failed
P0107 1T CCM 1996-2003 Jeep, Truck & Van Models	MAP Sensor Circuit Low Input Conditions: Engine speed from 416-1470 rpm; TP sensor input less than 1v, and the PCM detected the MAP sensor input was less than 2.35v at startup, or was less than 0.20v with the engine running (conditions met for 1.76 seconds). Possible Causes • Loss of 5-volt supply from PCM (internal failure) • Sensor 5-volt supply circuit is open or shorted to ground • Sensor signal circuit is shorted to ground • Sensor has failed • PCM has failed
P0108 1T CCM 1995-98 Eagle Talon with a Non-Turbo engine	MAP Sensor Circuit High Input Conditions: Engine speed from 600-3500 rpm; TP input less than 1.3v; and the PCM detected the MAP input was more than 4.7v for 2 seconds. Possible Causes • MAP sensor signal circuit open, or the ground circuit is open • MAP sensor signal circuit is shorted to VREF (5v) • MAP sensor is damaged or it has failed • PCM has failed
P0108 1T CCM 1995-2003 Car & Mini Van models	MAP Sensor Circuit High Input Conditions: Engine speed from 600-3500 rpm, TP sensor input less than 1.2v, system voltage over 10.6v, and the PCM detected the MAP input was over 4.60v (conditions met for 1.76 seconds). Possible Causes • MAP sensor signal circuit is open • MAP sensor ground circuit is open • MAP sensor signal circuit shorted to VREF (5v) • MAP sensor has failed or it has failed (possible open circuit) • PCM has failed

OBD II TROUBLE CODE LIST (P0XXX CODES)

DTC	Trouble Code Title, Conditions & Possible Causes
P0108 1T CCM 2001-03 Sebring & Stratus models with a 2.4L & 3.0L engine	Barometric Pressure Sensor Circuit High Input Conditions: Engine started; engine runtime over 2 seconds, system voltage over 8.0v, and the PCM detected the BARO sensor was 4.45v or higher (a value equal to 114 kPa or 16.5" psi or higher) for 10 seconds. Note: This value indicates the vehicle is 4,000 feet below sea level. Possible Causes • MAP sensor signal circuit open, or the ground circuit is open • MAP sensor signal circuit is shorted to VREF (5v) • MAP sensor is damaged or it has failed • PCM has failed
P0108 1T CCM 1996-2003 Jeep, Truck & Van models	MAP Sensor Circuit High Input Conditions: Engine speed from 416-1470 rpm with the throttle closed; and the PCM detected a MAP sensor input of more than 4.6v (conditions met for 1 second). Possible Causes • MAP sensor signal circuit is open • MAP sensor ground circuit is open • MAP sensor signal circuit shorted to VREF (5v) • MAP sensor has failed or it has failed (possible open circuit) • PCM has failed
P0110 1T CCM 1995-98 Eagle Talon models	IAT Sensor Circuit High or Low Input Conditions: Key on for 60 seconds or right after startup; and the PCM detected an IAT sensor input of 4.6v or less than 0.2v for 4 seconds. Possible Causes • IAT sensor signal circuit is open, shorted to ground or to VREF • IAT sensor signal circuit has a high resistance condition • IAT sensor is damaged or it has failed • PCM has failed
P0110 1T CCM 2002-03 300M, Concorde, Intrepid, Dakota, Neon, Sebring, Caravan, Town & Country, PT Cruiser	IAT Sensor Circuit High or Low Input Conditions: Key on for 60 seconds or right after startup; and the PCM detected an IAT sensor input of 4.6v or less than 0.2v for 4 seconds. Possible Causes • IAT sensor signal circuit is open, shorted to ground or to VREF • IAT sensor signal circuit has a high resistance condition • IAT sensor is damaged or it has failed • PCM has failed
P0111 1T CCM 2001-03 All Car, Jeep, Truck & Van models	IAT Sensor Performance Conditions: DTC P0112 and P0113 not set, key on, ECT sensor input more than 160°F at startup, at least 5 warmup cycles have occurred with a vehicle mileage change of more than 196.6 miles, and the PCM detected the IAT input changed less than 5.4°F during this period. Possible Causes • IAT sensor signal circuit is open, shorted to ground or VREF • IAT sensor ground circuit is open • IAT sensor is damaged or it has failed • PCM High or Low circuit is damaged or it has failed
P0112 1T CCM 1998-2003 Trucks equipped with a 5.9L Diesel engine	IAT Sensor Circuit Low Input Conditions: Engine started; system voltage over 10.5v and the PCM detected that the IAT sensor indicated less than 0.098v for 2 seconds. Possible Causes • IAT sensor signal circuit is shorted to chassis ground • IAT sensor signal circuit is shorted to sensor ground • IAT sensor is damaged or it has failed (an internal short circuit) • PCM has failed

OBD II TROUBLE CODE LIST (P0XXX CODES)

DTC	Trouble Code Title, Conditions & Possible Causes
P0112 1T CCM 1995-98 Eagle Talon models	IAT Sensor Circuit Low Input Conditions: Engine started; system voltage over 10.5v and the PCM detected the IAT sensor was less than 0.157v for 2 seconds. Possible Causes • IAT sensor signal circuit is shorted to chassis ground • IAT sensor signal circuit is shorted to sensor ground • IAT sensor is damaged or it has failed (an internal short circuit) • PCM has failed
P0112 1T CCM 1995-2003 Car, Jeep, Truck & Van models	IAT Sensor Circuit Low Input Conditions: Engine started and the PCM detected an IAT sensor input of less than 0.157v for 3 seconds. Possible Causes • IAT sensor signal circuit is shorted to chassis ground • IAT sensor signal circuit is shorted to sensor ground • IAT sensor is damaged or it has failed (an internal short circuit) • PCM has failed
P0113 1T CCM 1995-98 Eagle Talon models	IAT Sensor Circuit High Input Conditions: Engine started and the PCM detected an IAT sensor input was more than 4.96v for 3 seconds. • IAT sensor signal circuit is shorted to VREF (5v) • IAT sensor signal circuit is open, or the ground circuit is open • IAT sensor is damaged or it has failed (an internal open circuit) • PCM has failed
P0113 1T CCM 1995-2003 Car, Jeep, Truck & Van models	IAT Sensor Circuit High Input Conditions: Engine started; and the PCM detected the IAT sensor input was over 4.90v for 3 seconds. Possible Causes • IAT sensor signal circuit shorted to VREF (5v) • IAT sensor signal circuit is open, or the ground circuit is open • IAT sensor is damaged or it has failed (an internal open circuit) • PCM has failed
P0113 1T CCM 1998-2003 Trucks equipped with a 5.9L Diesel engine	IAT Sensor Circuit High Input Conditions: Engine started; system voltage over 10.5v and the PCM detected the IAT sensor input was more than 4.90v for 2 seconds. Possible Causes • IAT sensor signal circuit shorted to VREF (5v) • IAT sensor signal circuit is open, or the ground circuit is open • IAT sensor is damaged or it has failed (an internal open circuit) • PCM has failed
P0115 1T CCM 1995-98 Eagle Talon models	ECT Sensor Circuit High or Low Input Conditions: Key on for 60 seconds or right after startup, and the PCM detected the ECT sensor was more than 4.6v or less than 0.2v for 4 seconds, or with engine running, the signal went from under 1.6v to over 1.6v and remained over 1.6v for 5 minutes, or with ECT and IAT inputs over 68ºF at startup, the ECT sensor took 5 minutes to reach 122ºF. Possible Causes • ECT sensor signal circuit is open or shorted to ground • ECT sensor signal circuit is shorted to VREF (5v) • ECT sensor signal circuit has a high resistance condition • ECT sensor is damaged or it has failed • PCM has failed.
P0116 1T CCM 2001-03 Car, Jeep, Truck & Van models	ECT Sensor Circuit Performance Conditions: Engine started; engine runtime over 10 minutes, and the PCM detected the ECT sensor did not reach a calibrated level during the test period (i.e., it failed the CCM rationality test). Possible Causes • ECT sensor signal circuit is open or it is shorted to ground • ECT sensor signal circuit is shorted to VREF (5v) • ECT sensor ground circuit is open • ECT sensor is damaged or it has failed • PCM High or Low circuit is damaged or it has failed

OBD II TROUBLE CODE LIST (P0XXX CODES)

DTC	Trouble Code Title, Conditions & Possible Causes
P0117 1T CCM 1995-2003 All Models	ECT Sensor Circuit Low Input Conditions: Engine started and the PCM detected the ECT sensor input was below 0.51v for 3 seconds. Possible Causes • ECT sensor signal circuit is shorted to chassis ground • ECT sensor signal circuit is shorted to sensor ground • ECT sensor is damaged or it has failed (it may be shorted) • PCM has failed
P0118 1T CCM 1995-2003 All Models	ECT Sensor Circuit High Input Conditions: Engine started and the PCM detected the ECT sensor input was over 4.9v for 3 seconds. Possible Causes • ECT sensor signal circuit is shorted to VREF (5v) • ECT sensor signal circuit is open • ECT sensor ground circuit is open • ECT sensor is damaged or it has failed (possible open circuit) • PCM has failed
P0120 1T CCM 1995-98 Eagle Talon models	Throttle Position Sensor Circuit Conditions: Key on for 60 seconds or right after startup, closed throttle switch on; and the PCM detected a TP sensor input of 2v or more or a TP sensor input of 0.2v or less for 4 seconds. Possible Causes • TP sensor 5-volt supply circuit is shorted, open or grounded • TP sensor signal circuit is open or shorted to ground • TP sensor is damaged or it has failed • PCM has failed
P0120 1T CCM 2001-03 PT Cruiser models	Throttle Position Sensor Circuit Conditions: Engine started; and the TCM detected an unexpected change in the throttle angle, or that the throttle angle went out-of-range abruptly. Possible Causes • TP sensor signal circuit is open or shorted to ground • TP sensor ground circuit is open • TP sensor signal circuit to TCM is open or shorted to ground • TP sensor is damaged or it has failed • PCM has failed
P0121 1T CCM 1995-2003 Car, Jeep, Truck & Van models	TP Sensor Does Not Agree With MAP Conditions: DTC P0107, P0108, P0122 and P0123 not set, engine running with throttle plate closed (at high vacuum), and the PCM detected the TP sensor input was high when it should have been low (High Input Test), or with the engine running and the VSS input over 25 mph, throttle plate open (at low vacuum), the PCM detected the TP sensor input was low when it should have been high, conditions met for 4 seconds. Possible Causes • MAP sensor and/or TP sensor VREF (5v) not present • MAP or TP sensor ground circuit has high resistance • MAP or TP sensor signal circuit has high resistance • TP sensor is damaged or has failed • PCM has failed
P0121 1T CCM 1998-2003 Trucks equipped with a 5.9L Diesel engine	Accelerator Position Sensor Signal Does Not Agree With IVS: Engine running at idle speed (throttle indicating less than 15% open) for 1 second or running at off-idle with the VSS input indicating over 10 mph for 1 second, and the PCM detected the IVS indicated idle when the APPS signal indicated off-idle, or the IVS indicated off-idle when the APPS indicated the engine was at idle position. Possible Causes • APP sensor signal circuit open • APP sensor signal circuit shorted to voltage (IVS 2 circuit) • APP sensor short to vehicle power • APP sensor is damaged or it has failed • PCM has failed

OBD II TROUBLE CODE LIST (P0XXX CODES)

DTC	Trouble Code Title, Conditions & Possible Causes
P0122 1T CCM 1995-98 Eagle Talon models	TP Sensor Circuit Low Input Conditions: Key on, and the PCM detected the TP sensor was less than 0.20v, or with engine speed over 1500 rpm, VSS over 20 mph, engine vacuum below 2" Hg, it was less than 0.5v for 704 ms. Possible Causes • Sensor 5-volt supply circuit open or shorted to ground • Sensor signal circuit is shorted to ground • TP sensor is damaged or it has failed • PCM has failed
P0122 1T CCM 1995-2003 Car & Mini Van models	TP Sensor Circuit Low Input Conditions: Key on or engine running; system voltage over 10.5v and the PCM detected the TP sensor indicated less than 0.0978v for 700 ms. Possible Causes • TP sensor VREF (5v) circuit is open or shorted to ground • TP sensor 5v supply circuit is shorted to ground • TP sensor signal circuit is shorted to ground • TP sensor is damaged or it has failed • PCM has failed
P0122 1T CCM 1995-2003 Jeep, Truck & Van models	TP Sensor Circuit Low Input Conditions: Key on or engine running; system voltage over 10.5v and the PCM detected the TP sensor was less than 0.10v, condition met for 1.3 seconds. Possible Causes • Sensor 5-volt supply circuit open or shorted to ground • Sensor signal circuit is shorted to ground • TP sensor is damaged or it has failed • PCM has failed
P0122 1T CCM 2001-03 Trucks equipped with a 5.9L Diesel engine	Accelerator Position Sensor Circuit Low Conditions: Engine started; and the PCM detected the APPS signal was to low, or the APPS signal between the ECM and the Powertrain Control Module (PCM) was too low at any time. Possible Causes • APP sensor 5-volt supply circuit open or shorted to ground • APP sensor signal shorted to ground • APP sensor is damaged or it has failed • PCM has failed
P0123 1T CCM 1995-98 Eagle Talon models	TP Sensor Circuit High Input Conditions: Key on or engine running; and the PCM detected the TP sensor input was more than 4.7v (all conditions met for 1 second). • Refer to Mini-Test (Section 1) or code repair chart in other media.
P0123 1T CCM 1995-2003 Car, Jeep, Mini Van, Truck & Van models	TP Sensor Circuit High Input Conditions: Key on or engine running; system voltage over 10.5v and the PCM detected the TP sensor indicated more than 4.50v for 700 ms. Possible Causes • TP sensor ground circuit is open • TP sensor signal circuit is open • TP sensor signal circuit shorted to VREF (5v) • TP sensor is damaged or has failed (an internal open circuit) • PCM has failed
P0123 1T CCM 2001-03 Trucks equipped with a 5.9L Diesel engine	Accelerator Position Sensor Circuit High Conditions: Engine started; and the PCM detected the APPS signal was too high, or the APPS signal between the Engine Control Module (ECM) and the Powertrain Control Module (PCM) was too high at any time. Possible Causes • APP sensor circuit is open, or sensor ground circuit is open • APP sensor signal shorted to power • APP sensor has failed or the PCM has failed • PCM has failed

OBD II TROUBLE CODE LIST (P0XXX CODES)

DTC	Trouble Code Title, Conditions & Possible Causes
P0125 2T CCM 1995-98 Eagle Talon models	**Closed Loop Temperature Not Reached Conditions:** DTC P0117 and P0118 not set, engine speed from 2400-4000 rpm, IAT sensor from 32-122°F, and the PCM detected the ECT sensor value did not exceed 140°F after 10 minutes. **Possible Causes** • Check the operation of the thermostat (it may be stuck open) • Inspect for low coolant level or for an incorrect coolant mixture • ECT sensor is damaged or it has failed.
P0125 2T CCM 1995-2003 Car & Mini Van models	**Closed Loop Temperature Not Reached Conditions:** DTC P0117 and P0118 not set, ECT Sensor between -20°F and +21.2°F at startup, engine runtime over 10 minutes, vehicle speed more than 28 mph, and the PCM detected the ECT sensor did not reach 174°F after 20 minutes of sustained engine operation. **Possible Causes** • Check the operation of the thermostat (it may be stuck open) • ECT sensor is damaged or it is out-of-calibration • Inspect for low coolant level or for an incorrect coolant mixture
P0125 2T CCM 1996-2003 Jeep, Truck & Van models	**Closed Loop Temperature Not Reached Conditions:** DTC P0117 and P0118 not set, engine runtime over 10 minutes, system voltage over 10.5v and the PCM detected the ECT sensor did not reach 60°F after 10 minutes of operation. **Possible Causes** • Check the operation of the thermostat (it may be stuck open) • ECT sensor is damaged or it is out-of-calibration • Inspect for low coolant level or for an incorrect coolant mixture
P0125 2T CCM 1998-2003 Trucks equipped with a 5.9L Diesel engine	**Engine Does Not Reach Closed Loop Conditions:** DTC P0112, P0113, P0117, P0118, P1492 and P1493 not set, key on, IAT sensor signal less than -80°F, system voltage over 10.4v, and the PCM detected the ECT sensor was more than 140°F, or with the ECT sensor between -40°F and 150°F at startup, or it detected the ECT sensor did not change more than 6°F after 10 minutes. **Possible Causes** • Check the operation of the thermostat (it may be stuck open) • ECT sensor is damaged or it is out-of-calibration • Inspect for low coolant level or for an incorrect coolant mixture
P0128 1T CCM 2001-03 Car & Mini Van models	**Thermostat Rationality Test Conditions:** DTC P0117, P0118, P1492 and P1493 not set, ECT sensor between 20°F and 130°F at startup, and the PCM detected the ECT sensor did not exceed 170°F after 10-32 minutes of sustained engine operation (the actual time depends on the ECT sensor at startup). **Possible Causes** • Check the operation of the thermostat (it may be stuck open) • ECT sensor is contaminated, damaged or it has failed • Inspect for low coolant level or for an incorrect coolant mixture
P0128 1T CCM 2001-03 Sebring & Stratus models equipped with a 2.4L & 3.0L engine	**Thermostat Rationality Test Conditions:** ECT sensor from 14°F and 82°F at startup with the ECT and IAT sensor inputs within 48°F and the IAT input less than 36°F, volume airflow from 50-100 Hz for under 300 seconds, and the PCM detected the ECT sensor took too long to reach 180°F. The actual time vary with the ECT sensor at startup (e.g., from 11-23 minutes to reach 180°F if the ECT sensor is over 68°F at startup, or 20-54 minutes to reach 180°F if the ECT sensor is less than 68F at startup). **Possible Causes** • Check the operation of the thermostat (it may be stuck open) • Inspect for low coolant level or for an incorrect coolant mixture
P0128 1T CCM 2001-03 Trucks equipped with a 5.9L Diesel engine	**Thermostat Rationality Test Conditions:** DTC P0117, P0118, P1492 and P1493 not set, ECT sensor between 20°F and 130°F at startup, and the PCM detected the ECT sensor did not exceed 170°F after 10-32 minutes of sustained engine operation (the actual time depends on the ECT sensor at startup). **Possible Causes** • Check the operation of the thermostat (it may be stuck open) • ECT sensor is damaged or it is out-of-calibration • Inspect for low coolant level or for an incorrect coolant mixture

OBD II TROUBLE CODE LIST (P0XXX CODES)

DTC	Trouble Code Title, Conditions & Possible Causes
P0129 1T CCM 2003 Caravan, Neon, PT Cruiser, Town & Country, Sebring	**Barometric Pressure Out-Of-Range Low Conditions:** Engine cranking at less than 250 rpm, no CKP or CMP sensor signals for 75 ms, and the PCM detected the BARO sensor signal range was 0.0392-2.196v for 300 ms during testing. Possible Causes • IAC motor control low or control high circuit has failed • MAP sensor VREF (5v) circuit is open or shorted to ground • MAP sensor signal circuit is open or shorted to ground • MAP sensor is damaged or it has failed • PCM has failed
P0130 1T CCM 1995-98 Eagle Talon models	**HO2S-11 (B1 S1) Circuit Fault Conditions:** Engine running in closed loop, ECT sensor over 176ºF, IAT sensor signal over 14ºF, BARO signal more than 75 kPa, engine speed above 1200 rpm, volumetric efficiency over 25%, then after the PCM commanded the fuel injector pulsewidth lean and rich, it detected the HO2S response time was too slow, or it detected the HO2S signal was less than 0.10v for 3 minutes. Possible Causes • HO2S signal circuit is open or shorted to ground • HO2S ground circuit is open • HO2S may be contaminated or it has failed • PCM has failed
P0130 1T CCM 2001 Sebring & Stratus models equipped with a 2.4L & 3.0L engine	**HO2S-11 (B1 S1) Circuit Fault Conditions:** Engine runtime over 3 minutes, engine speed more than 1200 rpm, volumetric efficiency more than 25%, ECT sensor more than 180ºF, and the PCM detected the HO2S-11 signal was less than 0.2v during the 7 second test period, and the PCM detected the HO2S signal remained at more than 4.5v with 5v applied to the circuit during the HO2S Test period. Possible Causes • HO2S signal circuit is open or shorted to ground • HO2S ground circuit is open • HO2S may be contaminated or it has failed • PCM has failed
P0130 1T CCM 1999-2000 Dakota, Durango, Jeep & Truck models with a 4.7L engine	**HO2S-11 (B1 S1) Circuit Fault Conditions:** Key on or engine running; system voltage over 10.5v and the PCM detected the state of the HO2S relay coil circuit (between the PCM and the relay) did not match the expected state. Possible Causes • Fused ignition feed (power) circuit open • Heater relay control circuit open • Heater relay has failed (an internal winding is open) • PCM has failed
P0131 1T CCM 1995-98 Eagle Talon models	**HO2S-11 (B1 S1) Short to Ground Conditions:** ECT sensor 170ºF or more at previous key on, ECT sensor below 90ºF and BTS signal within 27ºF of the ECT, and the PCM detected the HO2S input was below 156 mv for 28 seconds. Possible Causes • HO2S signal circuit is shorted to chassis or sensor ground • HO2S signal circuit is open (circuit reads 450 mv when open) • HO2S ground circuit is open (circuit reads 170 mv when open) • HO2S may be contaminated or it has failed • PCM has failed
P0131 1T CCM 1995-2003 Concorde, Intrepid, LHS, Avenger, Neon, Dakota, Durango, PT Cruiser, Sebring, Stratus, Prowler, Viper	**HO2S-11 (B1 S1) Short to Ground Conditions:** ECT sensor over 170ºF on previous key on, ECT sensor under 98ºF and BTS within ±27ºF of ECT sensor, and the PCM detected the HO2S signal was below 156 mv for 28 seconds. Possible Causes • HO2S signal circuit is shorted to chassis or sensor ground • HO2S ground circuit is open (circuit reads 170 mv when open) • HO2S may be contaminated or it has failed • PCM has failed

OBD II TROUBLE CODE LIST (P0XXX CODES)

DTC	Trouble Code Title, Conditions & Possible Causes
P0131 1T CCM 1996-2003 Jeep, Van, Truck & Van models	HO2S-11 (B1 S1) Circuit Short to Ground Conditions: Engine running with ECT sensor over 146°F, followed by an engine off period in which the HO2S Heater Test ran and passed, then after an engine cool-down period, and a cold engine startup (ECT sensor under 100°F and the BTS signal within 44°F of the ECT sensor), the PCM detected the HO2S signal was less than 78 mv for 5 seconds. Possible Causes • HO2S signal circuit is shorted to chassis or sensor ground • HO2S ground circuit is open (circuit reads 170 mv when open) • HO2S may be contaminated or it has failed • PCM has failed
P0132 1T CCM 1995-98 Eagle Talon models	HO2S-11 (B1 S1) Circuit Short to Power Conditions: Engine started; engine runtime over 2 minutes, ECT sensor more than 176°F, and the PCM detected the HO2S signal was over 1.2v. Possible Causes • HO2S signal circuit shorted to heater B+ circuit (inspect the connector for oil or moisture inside the terminal area) • HO2S signal circuit is open • PCM has failed
P0132 1T CCM 1995-2003 Concorde, Intrepid, LHS, Avenger, Neon, PT Cruiser, Sebring, Stratus, Dakota, Durango, PT Cruiser, Prowler, Viper models	HO2S-11 (B1 S1) Circuit Short to Voltage Conditions: Engine started; engine runtime over 119 seconds, system voltage over 10.5v, ECT sensor more than 176°F, and the PCM detected the HO2S signal was more than 1.29v for 3 seconds. Possible Causes • HO2S signal circuit shorted to heater B+ circuit (inspect the connector for oil or moisture inside the terminal area) • HO2S signal circuit is open • PCM has failed
P0132 1T CCM 1996-2003 Jeep, Mini Van, Truck & Van models	HO2S-11 (B1 S1) Circuit Short to Voltage Conditions: Engine started; engine runtime over 119 seconds, system voltage over 10.99v, HO2S Heater temperature over 1085°F and the PCM detected the HO2S input was over 3.70v for 1 minute. Possible Causes • HO2S signal tracking (wet/oily) in connector causing a short between the signal and heater power circuits • HO2S ground circuit is open • HO2S signal circuit is open • PCM has failed
P0133 2T HO2S 1995-98 Eagle Talon models	HO2S-11 (B1 S1) Slow Response Conditions: Engine started; engine runtime over 3 minutes, ECT sensor more than 170°F, vehicle driven to a speed over 24 mph at an engine speed of 518-864 rpm (near idle speed), A/C and PSP switches indicating off, the PCM detected the HO2S signal did not cross 670 mv or the HO2S switch rate was too low during the HO2S Test. Possible Causes • Exhaust leak present in the exhaust manifold or exhaust pipes • HO2S element is fuel contaminated • HO2S element is deteriorated or it has failed
P0133 2T HO2S 1996-2003 300M, Concorde, LHS, Intrepid, Jeep, Mini Van, Truck & Van models	HO2S-11 (B1 S1) Slow Response Conditions: Engine runtime over 2 minutes, ECT sensor more than 147°F, VSS over 10 mph, A/C and PSPS indicating off, then at idle speed in Drive (A/T) or Neutral (M/T), and the PCM detected the HO2S signal did not switch enough times from 270-620 mv in the test period. Possible Causes • Exhaust leak present in the exhaust manifold or exhaust pipes • HO2S element is fuel contaminated • HO2S element is deteriorated or it has failed

OBD II TROUBLE CODE LIST (P0XXX CODES)

DTC	Trouble Code Title, Conditions & Possible Causes
P0133 2T HO2S 1995-2003 Avenger, Sebring, Stratus, PT Cruiser, Neon, Dakota, Durango, PT Cruiser, Prowler, Viper models	HO2S-11 (B1 S1) Slow Response Conditions: Engine at idle speed, system voltage over 10.5v, then engine speed from 1216-1984 rpm, VSS input from 19-46 mph, BTS signal more than 44°F, BARO signal more than 22.16" Hg, MAP sensor signal from 11.9-18.15" Hg, then back to idle speed in Drive, and the PCM detected the HO2S signal did not switch enough times from below 330 mv to above 610 mv, condition met for 60 seconds. Possible Causes • Exhaust leak present in the exhaust manifold or exhaust pipes • HO2S element is fuel contaminated • HO2S element is deteriorated or it has failed
P0134 2T HO2S 1995-98 Eagle Talon models	HO2S-11 (B1 S1) Remains At Center Conditions: Engine runtime over 2 minutes, system voltage over 10.5v, ECT sensor more than 170°F, and the PCM detected the HO2S signal remained fixed between 350-550 mv for 1.5 minutes. Possible Causes • Exhaust leak present in exhaust manifold or exhaust pipes • HO2S element is fuel contaminated or it is deteriorated • PCM has failed
P0134 2T HO2S 1995-2003 Concorde, Intrepid, LHS, Avenger, Neon, PT Cruiser, Sebring, Stratus, Dakota, Durango, PT Cruiser, Prowler, Viper models	HO2S-11 (B1 S1) Remains At Center Conditions: Engine runtime more than 121 seconds, system voltage over 10.5v, ECT sensor more than 150.8°F, fuel control system in closed loop mode, and the PCM detected the HO2S signal remained fixed in a range between 350-580 mv, condition met for 60 seconds. Possible Causes • Exhaust leak present in exhaust manifold or exhaust pipes • HO2S element is fuel contaminated or has deteriorated • O2S signal circuit or ground circuit has high resistance • PCM has failed
P0134 2T HO2S 1996-2003 Jeep, Truck & Van models	HO2S-11 (B1 S1) Remains At Center Conditions: Engine runtime more than 121 seconds, system voltage over 10.5v, ECT sensor more than 150.8°F, fuel control system in closed loop mode, and the PCM detected the HO2S signal remained fixed in a range between 350-580 mv, condition met for 60 seconds. Possible Causes • Exhaust leak present in exhaust manifold or exhaust pipes • HO2S element is fuel contaminated or has deteriorated • O2S signal circuit or ground circuit has high resistance • PCM has failed
P0135 1T CCM 1995-98 Eagle Talon models	HO2S-11 (B1 S1) Heater Circuit Conditions: ECT sensor less than 147°F, BTS signal within ±27°F of ECT sensor, system voltage over 10.5v, running at idle for 12 seconds, and the PCM detected the HO2S signal was more than 3.0v for 30-90 seconds. Possible Causes • ASD relay output (power) circuit to the heater is open • HO2S heater ground circuit is open or has high resistance • HO2S heater element is damaged or it has failed • PCM has failed
P0135 1T CCM 2001-03 Sebring & Stratus models equipped with a 2.4L & 3.0L engine	HO2S-11 (B1 S1) Heater Circuit Conditions: Engine started; system voltage from 11-16v, ECT sensor over 68°F, and the PCM detected the HO2S-11 Heater current was less than 0.16 amps or more than 7.5 amps for 4 seconds. Possible Causes • HO2S heater ground circuit open or HO2S signal circuit is open • HO2S heater element has high resistance • HO2S heater element has failed (open or shorted) • MFI relay output (power) circuit to the heater is open • PCM has failed

OBD II TROUBLE CODE LIST (P0XXX CODES)

DTC	Trouble Code Title, Conditions & Possible Causes
P0135 1T CCM 1995-2003 Concorde, Intrepid, LHS, Avenger, Neon, PT Cruiser, Sebring, Stratus, Dakota, Durango, PT Cruiser, Prowler, Viper models	HO2S-11 (B1 S1) Heater Circuit Conditions: ECT sensor less than 147ºF and BTS signal within 27ºF of the ECT sensor (cold engine), engine started, engine running at idle speed for over 12 seconds, system voltage over 10.5v and the PCM detected the HO2S signal was more than 3.0v for 30-90 seconds. Possible Causes • ASD relay output (power) circuit to the heater is open • HO2S heater ground circuit is open • HO2S heater element is damaged or has high resistance • PCM has failed
P0135 1T CCM 1996-2003 Jeep, Truck & Van models	HO2S-11 (B1 S1) Heater Circuit Conditions: Key off after a warm engine drive cycle, engine cool-down finished (at least 5 seconds after the key is turned off), system voltage over 10.5v, then with the ASD relay energized, the PCM detected the HO2S signal rose to 0.49v or more within a 144 second period, and the initial rise of the oxygen sensor signal was less than 1.57v. Possible Causes • HO2S heater ground circuit open or HO2S signal circuit is open • HO2S heater element has high resistance • HO2S heater element has failed (open or shorted) • PCM has failed
P0136 1T CCM 1995-98 Eagle Talon models	HO2S-12 (B1 S2) Circuit Conditions: Engine started; engine speed above 1200 rpm, ECT sensor over 176ºF, IAT sensor over 14ºF, BARO signal over 75 kPa, volumetric efficiency over 25%, Fuel Injector pulsewidth commanded lean and rich, and the PCM detected the HO2S response time was too slow, or it detected the HO2S signal was less than 0.10v for 3 minutes. Possible Causes • HO2S signal circuit is open or shorted to ground • HO2S ground circuit is open • HO2S may be contaminated or it has failed • PCM has failed
P0136 1T CCM 2000-03 Dakota, Durango, Jeep & Truck models	HO2S-12 (Bank 1 Sensor 2) Heater Circuit Low Input Conditions Key on, cold engine conditions, system voltage over 10.5v, ASD relay on, and the PCM detected the Actual state of the Heater relay circuit did not match the Intended state (i.e., the circuit status remained in a low state). Possible Causes • Fused ignition feed (power) circuit to the relay is open • Heater relay control circuit is open • Heater relay is damaged or it has failed • PCM has failed
P0137 1T CCM 1995-2003 300M, Concorde, Intrepid, LHS, Avenger, Neon, PT Cruiser, Sebring, Stratus, Dakota, Durango, PT Cruiser, Prowler, Viper models	HO2S-12 (B1 S2) Circuit Short to Ground Conditions: Engine started; engine runtime over 119 seconds, system voltage over 10.99v, HO2S Heater temperature over 1085ºF and the PCM detected the HO2S input was over 3.70v for 1 minute. Possible Causes • HO2S signal tracking in connector due to wet/oil conditions • HO2S signal circuit is open, or the ground circuit is open • HO2S heater supply circuit is open • PCM has failed
P0137 1T CCM 1996-2003 Jeep, Truck & Van models	HO2S-12 (B1 S2) Circuit Short to Ground Conditions: ECT sensor over 170ºF at previous key off, HO2S Heater Test completed, engine cool down period finished, engine started, ECT sensor less than 98ºF and BTS signal within 27ºF of the ECT sensor, and PCM detected the HO2S signal was below 156 mv for 28 seconds. Possible Causes • HO2S signal circuit is shorted to chassis or sensor ground • HO2S ground circuit is open (circuit reads 170 mv when open) • HO2S may be contaminated or it has failed • PCM has failed

OBD II TROUBLE CODE LIST (P0XXX CODES)

DTC	Trouble Code Title, Conditions & Possible Causes
P0138 1T CCM 1996-2003 Jeep, Truck & Van models	**HO2S-12 (B1 S2) Circuit Short to Voltage Conditions:** Engine started; engine runtime more than 4 minutes, system voltage over 10.5v, ECT sensor more than 180ºF, and the PCM detected the HO2S signal was more than 1.50v for 3 seconds. Possible Causes • HO2S signal tracking (wet/oily) in connector due to short from signal to power circuit • HO2S signal circuit is open • HO2S ground circuit is open • HO2S is damaged or it has failed • PCM has failed
P0138 1T CCM 1995-2003 300M, Concorde, Intrepid, LHS, Avenger, Neon, PT Cruiser, Sebring, Stratus, Dakota, Durango, PT Cruiser, Prowler, Viper models	**HO2S-12 (B1 S2) Circuit Short to Voltage Conditions:** Engine runtime over 2 minutes, system voltage over 10.5v, ECT sensor more than 170ºF, and the PCM detected the HO2S signal was more than 1.20v for over 3 seconds. Possible Causes • HO2S signal tracking (wet/oily) causing a short to heater power • HO2S signal circuit open, or ground circuit open • PCM has failed
P0139 1T CCM 1996-2003 300M, Concorde, Intrepid, LHS, Avenger, Neon, PT Cruiser, Sebring, Stratus, Dakota, Durango, PT Cruiser, Prowler, Viper models	**HO2S-12 (B1 S2) Slow Response Conditions:** Engine started; vehicle driven at 20-55 mph with the throttle open for 2 minutes, ECT sensor more than 158ºF, Converter temperature over 1112ºF, EVAP purge "on", and the PCM detected the HO2S signal switched from rich-to-lean less than 11 times in 20 seconds. Possible Causes • Exhaust leak present in the exhaust manifold or exhaust pipes • HO2S element is contaminated, deteriorated or it has failed • HO2S ground or HO2S signal circuit has high resistance
P0139 2T HO2S 2002-03 300M, Concorde, Intrepid, LHS, Jeep	**HO2S-12 (B1 S2) Slow Response Conditions:** Engine started; vehicle driven at 20-55 mph with the throttle open for 2 minutes, ECT sensor more than 158ºF, Converter Temperature over 1112ºF, EVAP purge on, and the PCM detected the HO2S signal switched from rich-to-lean less than 11 times in 20 seconds. Possible Causes • Exhaust leak present in the exhaust manifold or exhaust pipes • HO2S element is contaminated, deteriorated or it has failed • HO2S ground or HO2S signal circuit has high resistance
P0139 2T HO2S 2002-03 Sebring & Stratus with a 2.4L & 3.0L engine	**HO2S-12 (B1 S2) Slow Response Conditions:** Engine started; ECT sensor more than 169ºF, front HO2S-11 active, volumetric airflow sensor more than 4000 Hz, then vehicle speed over 18.7 mph at over 1500 rpm with the volumetric efficiency over 40%, then vehicle speed below 0.9 mph, Fuel Shutoff active, and the PCM detected the HO2S signal was less than 0.78v for 38 seconds. Possible Causes • Exhaust leak present in the exhaust manifold or exhaust pipes • HO2S element is fuel contaminated or it has failed • HO2S ground or HO2S signal circuit has high resistance
P0140 2T HO2S 1996-2003 Car, Jeep, Truck & Van models	**HO2S-12 (B1 S2) Remains At Center Conditions:** Engine started; system voltage over 10.5v, ECT sensor over 150.8ºF, engine running in closed loop, and the PCM detected the HO2S signal was fixed at 350-580 mv for 60 seconds. Possible Causes • Exhaust leak present in exhaust manifold or exhaust pipes • HO2S element is fuel contaminated or has deteriorated • HO2S signal circuit or ground circuit has high resistance • PCM has failed

OBD II TROUBLE CODE LIST (P0XXX CODES)

DTC	Trouble Code Title, Conditions & Possible Causes
P0141 1T CCM 1995-98 Eagle Talon models	HO2S-12 (B1 S2) Heater Circuit Conditions: ECT sensor less than 147ºF and BTS signal within 27ºF of the ECT sensor (cold engine), engine running at idle speed for over 12 seconds, system voltage over 10.5v and the PCM detected the HO2S signal was more than 3.0v for 30-90 seconds. Possible Causes • ASD relay power supply circuit to the heater is open • HO2S heater ground circuit or the HO2S signal circuit is open • HO2S heater element has high resistance or it has failed • PCM has failed
P0141 1T CCM 2001-03 Sebring & Stratus with a 2.4L & 3.0L engine	HO2S-11 (B1 S1) Heater Circuit Conditions: Engine started; system voltage from 11-16v, ECT sensor over 68ºF, and the PCM detected the HO2S-12 Heater current was less than 0.16 amps or more than 7.5 amps for 4 seconds. Possible Causes • HO2S heater ground circuit open or HO2S signal circuit is open • HO2S heater element has high resistance • HO2S heater element has failed (open or shorted) • MFI relay output (power) circuit to the heater is open • PCM has failed
P0141 1T CCM 1996-2003 300M, Concorde, Intrepid, LHS & Mini Vans	HO2S-12 (B1 S2) Heater Circuit Conditions: Key off after the vehicle has been driven for at least 10 miles with the throttle open for 3 minutes, system voltage over 11v, then the HO2S Heater Test is enabled (i.e., the PCM energizes the ASD relay to provide power to the heater, and the PCM monitors the HO2S signal voltage. If it continues to increase (instead of decreasing), the O2S Monitor Test fails. Possible Causes • ASD relay output (power) circuit to the heater is open • HO2S heater ground circuit is open or has high resistance • HO2S heater element has failed (open or shorted) • PCM has failed
P0141 1T CCM 1995-2003 Avenger, Neon, PT Cruiser, Sebring, Stratus, Dakota, Durango, PT Cruiser, Prowler, Viper, Jeep, Truck & Van models	HO2S-12 (B1 S2) Heater Circuit Conditions: ECT sensor less than 147ºF, BTS signal within ±27ºF of ECT sensor (cold engine); engine running at idle speed for over 12 seconds, system voltage over 10.5v and the PCM detected the HO2S signal was more than 3.0v for 30-90 seconds. Possible Causes • ASD relay output (power) circuit to the heater is open • HO2S heater ground circuit open or HO2S signal circuit is open • HO2S heater element has high resistance • HO2S heater element has failed (open or shorted) • PCM has failed
P0143 1T CCM 1996-2002 Truck models with a 5.9L & 8.0L engine (California models)	HO2S-13 (B1 S3) Short to Ground Conditions: ECT sensor more than 170ºF on previous key on, ECT sensor less than 98ºF and the BTS signal within ±59ºF of the ECT sensor at startup (cold engine), and the PCM detected the HO2S signal was less than 156 mv for 28 seconds. Possible Causes • HO2S signal circuit is shorted to chassis or sensor ground • HO2S signal circuit is open (circuit reads 450 mv when open) • HO2S ground circuit is open (circuit reads 170 mv when open) • HO2S may be contaminated or it has failed
P0144 1T CCM 1996-2002 Truck models with a 5.9L & 8.0L engine (California models)	HO2S-13 (B1 S3) Short to Voltage Conditions: Engine runtime over 2 minutes, ECT sensor more than 180ºF, and the PCM detected the HO2S-13 signal was more than 1.50v. Possible Causes • HO2S signal tracking in connector due to a short between the signal and power circuits • HO2S signal circuit shorted to system power • HO2S signal circuit open, or ground circuit open • HO2S heater supply circuit is open • PCM has failed

OBD II TROUBLE CODE LIST (P0XXX CODES)

DTC	Trouble Code Title, Conditions & Possible Causes
P0145 2T HO2S 1996-2002 Truck models with a 5.9L & 8.0L engine (California models)	HO2S-13 (B1 S3) Slow Response Conditions: Engine started; ECT sensor more than 170ºF, VSS input over 10 mph with throttle open for 2 minutes, then back to idle speed in Drive (A/T) or Neutral (M/T), and the PCM detected the HO2S signal did not switch enough times from 270-620 mv during the HO2S Test. Possible Causes • Exhaust leak present in the exhaust manifold or exhaust pipes • HO2S element is fuel contaminated or the HO2S element is deteriorated • PCM has failed
P0147 1T CCM 1996-2002 Truck models with a 5.9L & 8.0L engine (California models)	HO2S-13 (B1 S3) Heater Fault Conditions: ECT sensor less than 147ºF, BTS signal within ±27ºF of ECT sensor (cold engine), engine running at idle speed for over 12 seconds, system voltage over 10.5v and the PCM detected the HO2S signal was more than 3.0v for 30-90 seconds. Possible Causes • ASD relay output (power) circuit to the heater is open • HO2S heater ground circuit open or HO2S signal circuit is open • HO2S heater element has high resistance • HO2S heater element has failed (open or shorted) • PCM has failed
P0151 1T CCM 1996-2003 Avenger, Sebring & Stratus with V6 engine	HO2S-21 (B2 S1) Circuit Short to Ground Conditions: Engine runtime less than 3 seconds, ECT sensor less than 120ºF at engine startup, and the PCM detected the HO2S signal indicated less than 160 mv right after engine startup. Possible Causes • HO2S signal circuit is shorted to chassis or sensor ground • HO2S signal circuit is open (circuit reads 450 mv when open) • HO2S ground circuit is open (circuit reads 170 mv when open) • HO2S may be contaminated or it has failed • PCM has failed
P0151 1T CCM 1996-2003 300M, Concorde, Intrepid & LHS	HO2S-21 (B2 S1) Circuit Short to Ground Conditions: Engine runtime less than 3 seconds, ECT sensor less than 120ºF at engine startup, and the PCM detected the HO2S signal indicated less than 160 mv right after engine startup. Possible Causes • HO2S signal circuit is shorted to chassis or sensor ground • HO2S signal circuit is open (circuit reads 450 mv when open) • HO2S ground circuit is open (circuit reads 170 mv when open) • HO2S may be contaminated or it has failed • PCM has failed
P0151 1T CCM 2002-03 300M, Concorde, Intrepid & LHS	HO2S-21 (B2 S1) Circuit Short to Ground Conditions: Engine runtime under 20 seconds, system voltage over 10.9v, HO2S Heater temperature below 484ºF, and the PCM detected the HO2S signal was below 2.5196 volts for 3 seconds. Possible Causes • HO2S signal circuit is shorted to chassis or sensor ground • HO2S signal circuit is open (circuit reads 450 mv when open) • HO2S ground circuit is open (circuit reads 170 mv when open) • HO2S may be contaminated or it has failed • PCM has failed
P0151 1T CCM 1996-2003 Dakota, Durango, Prowler, Viper, Jeep, Truck & Van models	HO2S-21 (B2 S1) Circuit Short to Ground Conditions: ECT sensor more than 170ºF on previous key on, ECT sensor less than 98ºF and the BTS signal within ±27ºF of the ECT sensor at startup (cold engine), and the PCM detected the HO2S signal was less than 156 mv for 28 seconds. Possible Causes • HO2S signal circuit is shorted to chassis or sensor ground • HO2S signal circuit is open (circuit reads 450 mv when open) • HO2S ground circuit is open (circuit reads 170 mv when open) • HO2S may be contaminated or it has failed • PCM has failed

OBD II TROUBLE CODE LIST (P0XXX CODES)

DTC	Trouble Code Title, Conditions & Possible Causes
P0152 1T CCM 1996-2003 Avenger, Sebring & Stratus with V6 engine	HO2S-21 (B2 S1) Circuit Shorted to Voltage Conditions: Engine runtime more than 2 minutes, ECT sensor more than 176ºF, and the PCM detected the HO2S signal was more than 1.2v, condition met for 3 seconds. Possible Causes • HO2S signal tracking due to wet/oily condition in the connector • HO2S signal circuit shorted to system power • HO2S signal circuit open, or ground circuit open • PCM has failed
P0152 1T CCM 1996-2001 300M, Concorde, Intrepid & LHS	HO2S-21 (B2 S1) Circuit Shorted to Power Conditions: Engine runtime more than 2 minutes, system voltage over 10.5v, ECT sensor more than 176ºF, and the PCM detected the HO2S signal was more than 1.2v for 3 seconds. Possible Causes • HO2S signal tracking in the connector due to wet/oily condition • HO2S signal circuit is open, or the ground circuit open • HO2S heater supply circuit is open • PCM has failed
P0152 1T CCM 2002-03 300M, Concorde, Intrepid & LHS	HO2S-21 (B2 S1) Circuit Shorted to Power Conditions: Engine runtime over 119 seconds, system voltage over 10.99v, HO2S Heater temperature more than 1085ºF, and the PCM detected the HO2S signal was more than 3.70v for 1 minute. Possible Causes • HO2S signal tracking (wet/oily) in connector causing a short between the signal and heater power circuits • HO2S ground circuit is open • HO2S signal circuit is open • PCM has failed
P0152 1T CCM 1996-2003 Dakota, Durango, Prowler, Viper, Jeep, Truck & Van models	HO2S-21 (B2 S1) Short to Voltage Conditions: Engine runtime over 4 minutes, system voltage over 10.5v, ECT sensor more than 180ºF, and the PCM detected the HO2S signal was more than 1.50v for 3 seconds during the CCM test. Possible Causes • HO2S signal tracking in connector due to wet/oily conditions • HO2S ground circuit is open • HO2S heater supply circuit is open • HO2S signal circuit is open • HO2S element is damaged or it has failed • PCM has failed
P0153 1T CCM 1996-2003 Avenger, Sebring & Stratus with V6 engine	HO2S-21 (B2 S1) Slow Response Conditions: Engine runtime over 3 minutes, ECT sensor over 170ºF, VSS more than 24 mph for 75 seconds, A/C and PSPS both indicating off, then back to idle in Drive or Neutral (M/T), and the PCM detected the HO2S signal did not reach 670 mv, or the HO2S switched from 350-550 mv too few times, condition met for 6 seconds. Possible Causes • Exhaust leak present in the exhaust manifold or exhaust pipes • HO2S signal or ground circuit has a high resistance condition • HO2S element is fuel contaminated • HO2S element is deteriorated or it has failed
P0153 1T CCM 1996-2001 300M, Concorde, Intrepid & LHS	HO2S-21 (B2 S1) Slow Response Conditions: Engine runtime over 2 minutes, ECT sensor more than 147ºF, VSS indicating over 10 mph, then back to idle speed in Drive (A/T) or Neutral (M/T), and the PCM detected the HO2S signal did not switch enough times from 270-620 mv in the HO2S Monitor Test. Possible Causes • Exhaust leak present in the exhaust manifold or exhaust pipes • HO2S signal or ground circuit has a high resistance condition • HO2S element is fuel contaminated • HO2S element is deteriorated or it has failed

OBD II TROUBLE CODE LIST (P0XXX CODES)

DTC	Trouble Code Title, Conditions & Possible Causes
P0153 1T CCM 2002-03 300M, Concorde, Intrepid & LHS	HO2S-21 (B2 S1) Slow Response Conditions: Engine started; vehicle driven at a steady speed of 20-55 mph with the throttle open for at least 2 minutes, ECT sensor more than 158ºF, Catalytic Converter temperature more than 1115ºF, and the PCM detected the HO2S signal switched from rich-to-lean less than 12 times within a 60 second period. Possible Causes • Exhaust leak present in the exhaust manifold or exhaust pipes • HO2S signal or ground circuit has a high resistance condition • HO2S element is fuel contaminated • HO2S element is deteriorated or it has failed
P0153 1T CCM 1996-2003 Dakota, Durango, Prowler, Viper, Jeep, Truck & Van models	HO2S-21 (B2 S1) Slow Response Conditions: Engine started; ECT sensor more than 147ºF, VSS input over 10 mph with throttle open for 2 minutes, then back to idle speed in Drive (A/T) or Neutral (M/T), and the PCM detected the HO2S signal did not switch enough from 270-620 mv during the HO2S Test. Possible Causes • Exhaust leak present in the exhaust manifold or exhaust pipes • HO2S signal or ground circuit has a high resistance condition • HO2S element is fuel contaminated • HO2S element is deteriorated or it has failed
P0154 1T CCM 1996-2003 Avenger, Sebring & Stratus with V6 engine	HO2S-21 (B2 S1) Remains At Center Conditions: Engine runtime over 2 minutes, system voltage over 10.5v, ECT sensor more than 170ºF, and the PCM detected the HO2S signal remained fixed between 350-550 mv for 1.5 minutes. Possible Causes • Exhaust leak present in exhaust manifold or exhaust pipes • HO2S element is fuel contaminated or has deteriorated • O2S signal circuit or ground circuit has high resistance • PCM has failed
P0154 1T CCM 1996-2001 300M, Concorde, Intrepid, LHS	HO2S-21 (B2 S1) Remains At Center Conditions: Engine runtime over 2 minutes, system voltage over 10.5v, ECT sensor more than 170ºF, and the PCM detected the HO2S signal remained fixed between 350-550 mv for 1.5 minutes. Possible Causes • Exhaust leak present in exhaust manifold or exhaust pipes • HO2S element is fuel contaminated or has deteriorated • O2S signal circuit or ground circuit has high resistance • PCM has failed
P0154 1T CCM 2002-03 300M, Concorde, Intrepid & LHS	HO2S-21 (B2 S1) Remains At Center Conditions: Engine runtime more than 121 seconds, system voltage over 10.5v, ECT sensor more than 150.8ºF, fuel control system in closed loop mode, and the PCM detected the HO2S signal remained fixed in a range between 350-580 mv, condition met for 60 seconds. Possible Causes • Exhaust leak present in exhaust manifold or exhaust pipes • HO2S element is fuel contaminated or has deteriorated • O2S signal circuit or ground circuit has high resistance • PCM has failed
P0154 1T CCM 1996-2003 Dakota, Durango, Prowler, Viper, Jeep, Truck & Van models	HO2S-21 (B2 S1) Remains At Center Conditions: Engine started; ECT sensor more than 147ºF, VSS input over 10 mph with throttle open for 2 minutes, then back to idle speed in Drive (A/T) or Neutral (M/T), and the PCM detected the HO2S signal did not switch enough from 270-620 mv during the HO2S Test. Possible Causes • Exhaust leak present in the exhaust manifold or exhaust pipes • HO2S signal or ground circuit has a high resistance condition • HO2S element is fuel contaminated • HO2S element is deteriorated or it has failed

OBD II TROUBLE CODE LIST (P0XXX CODES)

DTC	Trouble Code Title, Conditions & Possible Causes
P0155 1T CCM 1996-2003 Avenger, Sebring & Stratus with V6 engine	**HO2S-21 (B2 S1) Heater Circuit Conditions:** Key off after a warm engine drive cycle, engine cool-down finished (at least 5 seconds after the key is turned off), system voltage over 10.5v, then with the ASD relay energized, the PCM detected the HO2S signal rose to 0.49v or more within a 144 second period, and the initial rise of the oxygen sensor signal was less than 1.57v. Possible Causes • ASD relay power supply circuit to the heater is open • HO2S heater ground circuit open or HO2S signal circuit is open • HO2S heater element has high resistance • HO2S heater element has failed (open or shorted) • PCM has failed
P0155 1T CCM 1996-2001 300M, Concorde, Intrepid, LHS	**HO2S-21 (B2 S1) Heater Circuit Conditions:** Key off after a warm engine drive cycle, engine cool-down finished (at least 5 seconds after the key is turned off), system voltage over 10.5v, then with the ASD relay energized, the PCM detected the HO2S signal rose to 0.49v or more within a 144 second period, and the initial rise of the oxygen sensor signal was less than 1.57v. Possible Causes • ASD relay power supply circuit to the heater is open • HO2S heater ground circuit open or HO2S signal circuit is open • HO2S heater element has high resistance • HO2S heater element has failed (open or shorted) • PCM has failed
P0155 1T CCM 2002-03 300M, Concorde, Intrepid & LHS	**HO2S-21 (B2 S1) Heater Circuit Conditions:** Engine started; HO2S Heater duty cycle from 1-99%, and the PCM detected the Heater temperature did not reach 959°F in 90 seconds. Possible Causes • HO2S heater control circuit is open • HO2S heater ground circuit is open • HO2S heater element has high resistance • HO2S heater is damaged or it has failed (open or shorted) • PCM has failed
P0155 1T CCM 1996-2003 Dakota, Durango, Prowler, Viper, Jeep, Truck & Van models	**HO2S-21 (B2 S1) Heater Circuit Conditions:** ECT sensor less than 147°F, BTS signal within ±27°F of ECT sensor (cold engine), engine running at idle speed for over 12 seconds, system voltage over 10.5v and the PCM detected the HO2S signal was more than 3.0v for 30-90 seconds. Possible Causes • ASD relay output (power) circuit to the heater is open • HO2S heater ground circuit open or HO2S signal circuit is open • HO2S heater element has high resistance • HO2S heater element has failed (open or shorted) • PCM has failed
P0157 1T CCM 1998-2003 Avenger, Sebring & Stratus with V6 engine	**HO2S-22 (B2 S2) Short to Ground Conditions:** Engine runtime less than 3 seconds, ECT sensor 120°F or less at engine startup, and the PCM detected the HO2S signal indicated less than 160 mv during the CCM test period. Possible Causes • HO2S signal circuit is shorted to chassis or sensor ground • HO2S signal circuit is open (circuit reads 450 mv when open) • HO2S ground circuit is open (circuit reads 170 mv when open) • HO2S may be contaminated or it has failed • PCM has failed

OBD II TROUBLE CODE LIST (P0XXX CODES)

DTC	Trouble Code Title, Conditions & Possible Causes
P0157 1T CCM 1996-2001 300M, Concorde, Intrepid, LHS	HO2S-22 (B2 S2) Circuit Shorted to Ground Conditions: ECT sensor more than 170ºF on previous key on, ECT sensor less than 98ºF and the BTS signal within ±27ºF of ECT sensor at startup (cold engine), and the PCM detected the HO2S signal was less than 156 mv for 28 seconds. Possible Causes • HO2S signal circuit is shorted to chassis or sensor ground • HO2S may be contaminated or it has failed • PCM has failed
P0157 1T CCM 2002-03 300M, Concorde, Intrepid & LHS	HO2S-22 (B2 S2) Circuit Shorted to Ground Conditions: Engine runtime under 20 seconds, system voltage over 10.99v, HO2S Heater temperature below 705ºF, and the PCM detected the HO2S signal was less than 1.50 volts for 3 seconds. Possible Causes • HO2S signal circuit is shorted to chassis or sensor ground • HO2S is damaged or it has failed • PCM has failed
P0157 1T CCM 1996-2003 Dakota, Durango, Prowler, Viper, Jeep, Truck & Van models	HO2S-22 (B2 S2) Circuit Shorted to Ground Conditions: ECT sensor more than 170ºF at previous key off, HO2S Heater Test completed after shutdown, engine cool-down period completed, then engine started, ECT sensor less than 98ºF and the BTS signal within 27ºF of the ECT sensor (cold engine), and the PCM detected the HO2S signal was less than 156 mv for 28 seconds during the test. Possible Causes • HO2S signal circuit is shorted to chassis or sensor ground • HO2S may be contaminated or it has failed • PCM return or signal circuit is damaged or has failed
P0158 1T CCM 1998-2003 Avenger, Sebring & Stratus with V6 engine	HO2S-22 (B2 S2) Circuit Shorted to Power Conditions: Engine runtime over 2 minutes, ECT sensor over 176ºF, and the PCM detected the HO2S signal was more than 1.2v for 3 seconds. Possible Causes • HO2S signal tracking in connector due to wet/oily condition • HO2S signal circuit is open or shorted to system power • PCM has failed
P0158 1T CCM 1996-2001 300M, Concorde, Intrepid, LHS	HO2S-22 (B2 S2) Circuit Shorted to Power Conditions: Engine runtime over 2 minutes, system voltage over 10.5v, ECT sensor more than 176ºF, and the PCM detected the HO2S signal was more than 1.21v, condition met for 3 seconds. Possible Causes • HO2S signal tracking due to wet/oily condition in the connector • HO2S signal circuit is open or shorted to system power • PCM has failed
P0158 1T CCM 2002-03 300M, Concorde, Intrepid & LHS	HO2S-22 (B2 S2) Circuit Shorted to Power Conditions: Engine runtime over 119 seconds, system voltage over 10.99v, HO2S Heater temperature more than 1085ºF, and the PCM detected the HO2S signal was more than 3.70v for 1 minute. Possible Causes • HO2S signal tracking due to wet/oily condition in the connector • HO2S ground circuit or the signal circuit is open • HO2S is damaged or it has failed • PCM has failed
P0158 1T CCM 1996-2003 Dakota, Durango, Prowler, Viper, Jeep, Truck & Van models	HO2S-22 (B2 S2) Circuit Shorted to Ground Conditions: Engine runtime over 4 minutes, system voltage over 10.5v, ECT sensor more than 180ºF, and the PCM detected the HO2S signal was more than 1.50v, condition met for 3 seconds. Possible Causes • HO2S signal tracking due to wet/oily condition in the connector • HO2S signal circuit is open or shorted to system power • HO2S is damaged or is has failed • PCM has failed

OBD II TROUBLE CODE LIST (P0XXX CODES)

DTC	Trouble Code Title, Conditions & Possible Causes
P0159 2T HO2S 1998-2003 Avenger, Sebring & Stratus with V6 engine	HO2S-22 (B2 S2) Slow Response Conditions: Engine runtime over 3 minutes, ECT sensor over 170ºF, VSS over 24 mph for 75 seconds, A/C and PSPS off, then back to idle speed in Drive or Neutral, and the PCM detected the HO2S signal did not reach 670 mv, or switched from 350-550 mv too few times in 6 seconds. Possible Causes • Exhaust leak present in the exhaust manifold or exhaust pipes • HO2S element is fuel contaminated • HO2S element is deteriorated or it has failed
P0159 2T HO2S 1996-2001 300M, Concorde, Intrepid, LHS	HO2S-22 (B2 S2) Slow Response Conditions: Engine runtime over 2 minutes, ECT sensor more than 147ºF, VSS indicating over 10 mph, then back to idle speed in Drive (A/T) or Neutral (M/T), and the PCM detected the HO2S signal did not switch enough times from 270-620 mv in the HO2S Monitor Test. Possible Causes • Exhaust leak present in the exhaust manifold or exhaust pipes • HO2S element is fuel contaminated • HO2S element is deteriorated or it has failed
P0159 2T HO2S 2002-03 300M, Concorde, Intrepid & LHS	HO2S-22 (B2 S2) Slow Response Conditions: Engine speed from 1200-2000 rpm at 20-60 mph with throttle open for 2 minutes, MAP sensor from 28-65 kPa, ECT sensor over 158ºF, Converter temperature over 1115ºF, and the PCM detected the HO2S signal switched from rich-to-lean less than 12 times in 60 seconds. Possible Causes • Exhaust leak present in the exhaust manifold or exhaust pipes • HO2S element is fuel contaminated • HO2S element is deteriorated or it has failed
P0159 2T HO2S 1996-2003 Dakota, Durango, Prowler, Viper, Jeep, Truck & Van models	HO2S-22 (B2 S2) Slow Response Conditions: Engine started; ECT sensor more than 170ºF, VSS input over 10 mph with throttle open for 2 minutes, then back to idle speed in Drive (A/T) or Neutral (M/T), and the PCM detected the HO2S signal did not switch enough times from 270-620 mv during the HO2S Test. Possible Causes • Exhaust leak present in the exhaust manifold or exhaust pipes • HO2S element is fuel contaminated • HO2S element is deteriorated or it has failed
P0160 2T HO2S 1998-2003 Avenger, Sebring & Stratus with V6 engine	HO2S-22 (B2 S2) Slow Response Conditions: Engine runtime over 2 minutes, system voltage over 10.5v, ECT sensor more than 170ºF, and the PCM detected the HO2S signal remained fixed between 350-550 mv for 1.5 minutes. Possible Causes • Exhaust leak present in exhaust manifold or exhaust pipes • HO2S element is fuel contaminated or has deteriorated • HO2S signal circuit or ground circuit has high resistance • PCM has failed
P0160 2T HO2S 1996-2001 300M, Concorde, Intrepid, LHS	HO2S-22 (B2 S2) Remains at Center Conditions: Engine runtime over 2 minutes, system voltage over 10.5v, ECT sensor more than 170ºF, and the PCM detected the HO2S signal remained fixed between 350-550 mv for 1.5 minutes. Possible Causes • Exhaust leak present in exhaust manifold or exhaust pipes • HO2S element is fuel contaminated or has deteriorated • HO2S signal circuit or ground circuit has high resistance • PCM has failed
P0160 2T HO2S 2002-03 300M, Concorde, Intrepid & LHS	HO2S-22 (B2 S2) Remains at Center Conditions: Engine runtime over 121 seconds, system voltage over 10.5v, ECT sensor more than 170ºF, and the PCM detected the HO2S signal remained fixed between 350-550 mv for 1.5 minutes. Possible Causes • Exhaust leak present in exhaust manifold or exhaust pipes • HO2S element is fuel contaminated or has deteriorated • HO2S signal circuit or ground circuit has high resistance • PCM has failed

OBD II TROUBLE CODE LIST (P0XXX CODES)

DTC	Trouble Code Title, Conditions & Possible Causes
P0161 1T CCM 1998-2003 Avenger, Sebring & Stratus with V6 engine	HO2S-22 (B2 S2) Heater Circuit Conditions: Key off after a warm engine drive cycle, engine cool-down finished (at least 5 seconds after the key is turned off), system voltage over 10.5v, then with the ASD relay energized, the PCM detected the HO2S signal rose to 0.49v or more within a 144 second period, and the initial rise of the oxygen sensor signal was less than 1.57v. Possible Causes • ASD relay output (power) circuit to the heater is open • HO2S heater ground circuit open or HO2S signal circuit is open • HO2S heater element has high resistance or it has failed • PCM has failed
P0161 1T CCM 1996-2001 300M, Concorde, Intrepid, LHS	HO2S-22 (B2 S2) Heater Circuit Conditions: Key off after a warm engine drive cycle, engine cool-down finished (at least 5 seconds after the key is turned off), system voltage over 10.5v, then with the ASD relay energized, the PCM detected the HO2S signal rose to 0.49v or more within a 144 second period, and the initial rise of the oxygen sensor signal was less than 1.57v. Possible Causes • ASD relay output (power) circuit to the heater is open • O2S heater ground circuit is open or O2S signal circuit is open • O2S heater element has high resistance or it has failed • PCM has failed
P0161 1T CCM 2002-03 300M, Concorde, Intrepid & LHS	HO2S-22 (B2 S2) Heater Circuit Conditions: Engine runtime over 121 seconds, system voltage over 10.5v, ECT sensor more than 170°F, and the PCM detected the HO2S signal remained fixed between 350-550 mv for 1.5 minutes. Possible Causes • Exhaust leak present in exhaust manifold or exhaust pipes • HO2S element is fuel contaminated or has deteriorated • HO2S signal circuit or ground circuit has high resistance • PCM has failed
P0161 1T CCM 1996-2003 Dakota, Durango, Prowler, Viper, Jeep, Truck & Van models	HO2S-22 (B2 S2) Slow Response Conditions: ECT sensor less than 147°F and BTS signal within 27°F of the ECT sensor (cold engine), engine started, system voltage over 10.5v, engine running at idle speed for over 12 seconds, and the PCM detected the HO2S signal was more than 3.0v for 30-90 seconds. Possible Causes • ASD relay output (power) circuit to the heater is open • HO2S heater ground circuit is open or HO2S signal circuit open • HO2S heater element has high resistance, is open or shorted • PCM has failed
P0165 1T CCM 2001-03 All models	Starter Relay Circuit Conditions: Engine cranking and the PCM did not detect the correct voltage signal from the Starter relay control circuit. Possible Causes • Starter Relay control circuit is open • Starter Relay Control circuit is grounded • PCM has failed
P0168 1T CCM 2001-03 Trucks equipped with a 5.9L Diesel engine	Decreased Performance (High Injection Pump Fuel Temperature) Conditions: Engine started; and the PCM detected decreased performance from the Injector Pump due to a high injection fuel pump temperature. Possible Causes • Fuel Injection pump module has failed • Overflow Valve Test failed

OBD II TROUBLE CODE LIST (P0XXX CODES)

DTC	Trouble Code Title, Conditions & Possible Causes
P0170 2T Fuel 1995-98 Eagle Talon models	Fuel Trim Too Lean or Too Rich (Bank 1) Conditions: Engine running in closed loop, IAT sensor signal more than 14°F, BARO sensor signal more than 75 kPa, and the PCM detected the Long Term fuel trim value was less than -12.5% (a rich A/F ratio) or was more than +12.5% (a lean A/F ratio), or the Short Term fuel trim value was +10% or higher, or it was -10% or lower for 10 seconds. Possible Causes • Air leaks present in the exhaust manifold or exhaust pipes • Air is being drawn in from leaks in gaskets or other seals • Fuel control sensor is out calibration (BARO, ECT, IAT or VAF) • Fuel pressure too high or low, leaking or restricted fuel injector • HO2S element is contaminated, deteriorated or it has failed
P0171 2T Fuel 1996-2000 Avenger, Breeze, Cirrus, Eagle Talon, Eagle Vision, Sebring & Stratus	Fuel System Lean (Bank 1) Conditions: Engine running in closed loop, IAT sensor signal over 20°F, altitude less than 8,000 feet, and the PCM detected too large an amount of Fuel Trim correction due to a lean A/F condition. Possible Causes • Air leaks in intake manifold, exhaust pipes or exhaust manifold • Base engine mechanical problem causing a lean A/F condition • Fuel control sensor is out of calibration (e.g., ECT, IAT or MAP) • Fuel delivery component fault (clogged filter, low fuel pressure) • HO2S element is contaminated, deteriorated or it has failed • Vacuum hose is disconnected, broken, leaking or loose
P0171 2T Fuel 1996-2003 300M, Concorde, Intrepid, LHS, Neon, PT Cruiser, Caravan, Town & Country & Voyager	Fuel System Lean (Bank 1) Conditions: Engine running in closed loop, IAT sensor signal over 20°F, altitude less than 8,000 feet, and the PCM detected too large an amount of Fuel Trim correction due to a lean A/F condition. Possible Causes • Air leaks in intake manifold, exhaust pipes or exhaust manifold • Base engine mechanical problem causing a lean A/F condition • Fuel control sensor is out of calibration (e.g., ECT, IAT or MAP) • Fuel delivery component fault (clogged filter, low fuel pressure) • HO2S element is contaminated, deteriorated or it has failed • Vacuum hose is disconnected, broken, leaking or loose
P0171 2T Fuel 2001-03 Sebring & Stratus Coupe with a 2.4L I4 or 3.0L V6 engine	Fuel System Lean (Bank 1) Conditions: ECT sensor between 140°F to 212°F, IAT sensor less than 140°F at startup, engine running in closed loop, volume airflow sensor less than 88 Hz, and the PCM detected the Long Term fuel trim exceed +12.5% or the Short Term fuel trim exceeded +25% for 5 seconds. Possible Causes • Air leaks in intake manifold, exhaust pipes or exhaust manifold • Base engine mechanical problem causing a lean A/F condition • Fuel control sensor is out of calibration (e.g., BARO or VAF) • Fuel delivery component fault (clogged filter, low fuel pressure) • HO2S element is contaminated, deteriorated or it has failed • Vacuum hose is disconnected, broken, leaking or loose
P0171 2T Fuel 1996-2003 Dakota, Durango, Jeep, Prowler, Viper, Truck & Van Models	Fuel System Lean (Bank 1) Conditions: Engine running in closed loop, IAT sensor signal over 20°F, altitude less than 8,000 feet, and the PCM detected too large an amount of Fuel Trim correction due to a lean A/F condition. Possible Causes • Air leaks in intake manifold, exhaust pipes or exhaust manifold • Base engine mechanical problem causing a lean A/F condition • Fuel control sensor is out of calibration (e.g., ECT, IAT or MAP) • Fuel delivery component fault (clogged filter, low fuel pressure) • HO2S element is contaminated, deteriorated or it has failed • Vacuum hose is disconnected, broken, leaking or loose

OBD II TROUBLE CODE LIST (P0XXX CODES)

DTC	Trouble Code Title, Conditions & Possible Causes
P0172 2T Fuel 1996-2000 Avenger, Breeze, Cirrus, Eagle Talon, Eagle Vision, Sebring & Stratus	Fuel System Rich (Bank 1) Conditions: Engine running in closed loop, IAT sensor signal over 20°F, altitude less than 8,000 feet, and the PCM detected too large an amount of Fuel Trim correction due to a rich A/F condition. Possible Causes • Base engine fault (i.e., cam timing incorrect, oil level too high) • EVAP vapor recovery system failure (e.g., canister full of fuel) • Fuel control sensor is out of calibration (e.g., ECT, IAT or MAP) • Fuel delivery component fault (injector leak, high fuel pressure) • HO2S element is contaminated, deteriorated or it has failed • HO2S heater is damaged or it has failed
P0172 2T Fuel 1996-2003 300M, Concorde, Intrepid, LHS, Neon, PT Cruiser, Caravan, Town & Country & Voyager	Fuel System Rich (Bank 1) Conditions: Engine running in closed loop, IAT sensor signal over 20°F, altitude less than 8,000 feet, and the PCM detected too large an amount of Fuel Trim correction due to a rich A/F condition. Possible Causes • Base engine fault (i.e., cam timing incorrect, oil level too high) • EVAP vapor recovery system failure (e.g., canister full of fuel) • Fuel control sensor is out of calibration (e.g., ECT, IAT or MAP) • Fuel delivery component fault (injector leak, high fuel pressure) • HO2S element is contaminated, deteriorated or it has failed • HO2S heater is damaged or it has failed
P0172 2T Fuel 2001-03 Sebring & Stratus Coupe with a 2.4L I4 or 3.0L V6 engine	Fuel System Rich (Bank 1) Conditions: ECT sensor from 140-212°F, IAT sensor less than 140°F at startup, engine running in closed loop, volume airflow sensor less than 88 Hz, and the PCM detected the Long Term fuel trim exceeded -12.5% or the Short Term fuel trim exceeded -25% for 5 seconds. Possible Causes • Base engine fault (i.e., cam timing incorrect, oil level too high) • EVAP vapor recovery system has failed (canister full of fuel) • High fuel pressure (fuel pressure regulator is sticking or failed) • One or more injectors leaking or pressure regulator is leaking • HO2S element is contaminated, deteriorated or it has failed
P0172 2T Fuel 1996-2003 Dakota, Durango, Jeep, Prowler, Viper, Truck & Van Models	Fuel System Rich (Bank 1) Conditions: Engine running in closed loop, IAT sensor signal over 20°F, altitude less than 8,000 feet, and the PCM detected too large an amount of Fuel Trim correction due to a rich A/F condition. Possible Causes • Base engine fault (i.e., cam timing incorrect, oil level too high) • EVAP vapor recovery system has failed (canister full of fuel) • Fuel injector (one or more) leaking or pressure regulator leaking • HO2S element is deteriorated or has failed
P0174 2T Fuel 1996-2000 Avenger, Eagle Vision, Sebring & Stratus with a V6 engine	Fuel System Lean (Bank 2) Conditions: Engine running in closed loop, IAT sensor signal over 20°F, altitude less than 8,000 feet, and the PCM detected too large an amount of Fuel Trim correction due to a lean A/F condition. Possible Causes • Air leaks in intake manifold, exhaust pipes or exhaust manifold • Base engine mechanical problem causing a lean A/F condition • Fuel control sensor is out of calibration (e.g., ECT, IAT or MAP) • Fuel delivery component fault (clogged filter, low fuel pressure) • HO2S element is contaminated, deteriorated or it has failed • Vacuum hose is disconnected, broken, leaking or loose

OBD II TROUBLE CODE LIST (P0XXX CODES)

DTC	Trouble Code Title, Conditions & Possible Causes
P0174 2T Fuel 1996-2003 300M, Concorde, Intrepid, LHS, Neon, PT Cruiser, Caravan, Town & Country & Voyager	Fuel System Lean (Bank 2) Conditions: Engine running in closed loop, IAT sensor signal over 20ºF, altitude less than 8,000 feet, and the PCM detected too large an amount of Fuel Trim correction due to a lean A/F condition. Possible Causes • Air leaks in intake manifold, exhaust pipes or exhaust manifold • Base engine mechanical problem causing a lean A/F condition • Fuel control sensor is out of calibration (e.g., ECT, IAT or MAP) • Fuel delivery component fault (clogged filter, low fuel pressure) • HO2S element is contaminated, deteriorated or it has failed • Vacuum hose is disconnected, broken, leaking or loose
P0174 2T Fuel 2001-03 Sebring & Stratus Coupe with a 3.0L V6 engine	Fuel System Lean (Bank 2) Conditions: ECT sensor between 140ºF to 212ºF, IAT sensor less than 140ºF at startup, engine running in closed loop, volume airflow sensor less than 88 Hz, and the PCM detected the Long Term fuel trim exceed +12.5% or the Short Term fuel trim exceeded +25% for 5 seconds. Possible Causes • Air leaks in intake manifold, exhaust pipes or exhaust manifold • Base engine mechanical problem causing a lean A/F condition • Fuel control sensor is out of calibration (e.g., BARO or VAF) • Fuel delivery component fault (clogged filter, low fuel pressure) • HO2S element is contaminated, deteriorated or it has failed • Vacuum hose is disconnected, broken, leaking or loose
P0174 2T Fuel 1996-2003 Dakota, Durango, Jeep, Prowler, Viper, Truck & Van Models	Fuel System Lean (Bank 2) Conditions: Engine running in closed loop, IAT sensor signal over 20ºF, altitude less than 8,000 feet, and the PCM detected too large an amount of Fuel Trim correction due to a lean A/F condition. Possible Causes • Air leaks in intake manifold, exhaust pipes or exhaust manifold • Base engine mechanical problem causing a lean A/F condition • Fuel control sensor is out of calibration (e.g., ECT, IAT or MAP) • Fuel delivery component fault (clogged filter, low fuel pressure) • HO2S element is contaminated, deteriorated or it has failed • Vacuum hose is disconnected, broken, leaking or loose
P0175 2T Fuel 1996-2000 Avenger, Eagle Vision, Sebring & Stratus with a V6 engine	Fuel System Rich (Bank 2) Conditions: Engine running in closed loop, IAT sensor signal over 20ºF, altitude less than 8,000 feet, and the PCM detected too large an amount of Fuel Trim correction due to a rich A/F condition. Possible Causes • Base engine fault (i.e., cam timing incorrect, oil level too high) • EVAP vapor recovery system failure (e.g., canister full of fuel) • Fuel control sensor is out of calibration (e.g., ECT, IAT or MAP) • Fuel delivery component fault (injector leak, high fuel pressure) • HO2S element is contaminated, deteriorated or it has failed • HO2S heater is damaged or it has failed
P0175 2T Fuel 1996-2003 300M, Concorde, Intrepid, LHS, Neon, PT Cruiser, Caravan, Town & Country & Voyager	Fuel System Rich (Bank 2) Conditions: Engine running in closed loop, IAT sensor signal over 20ºF, altitude less than 8,000 feet, and the PCM detected too large an amount of Fuel Trim correction due to a rich A/F condition. Possible Causes • Base engine fault (i.e., cam timing incorrect, oil level too high) • EVAP vapor recovery system failure (e.g., canister full of fuel) • Fuel control sensor is out of calibration (e.g., ECT, IAT or MAP) • Fuel delivery component fault (injector leak, high fuel pressure) • HO2S element is contaminated, deteriorated or it has failed • HO2S heater is damaged or it has failed

OBD II TROUBLE CODE LIST (P0XXX CODES)

DTC	Trouble Code Title, Conditions & Possible Causes
P0175 2T Fuel 2001-03 Sebring & Stratus Coupe with a 3.0L V6 engine	Fuel System Rich (Bank 2) Conditions: ECT sensor from 140-212ºF, IAT sensor less than 140ºF at startup, engine running in closed loop, volume airflow sensor less than 88 Hz, and the PCM detected the Long Term fuel trim exceeded -12.5% or the Short Term fuel trim exceeded -25% for 5 seconds. Possible Causes • Base engine fault (i.e., cam timing incorrect, oil level too high) • EVAP vapor recovery system has failed (canister full of fuel) • High fuel pressure (fuel pressure regulator is sticking or failed) • One or more injectors leaking or pressure regulator is leaking • HO2S element is contaminated, deteriorated or it has failed
P0175 2T Fuel 1996-2003 Dakota, Durango, Jeep, Prowler, Viper, Truck & Van Models	Fuel System Rich (Bank 2) Conditions: Engine running in closed loop, IAT sensor signal over 20ºF, altitude less than 8,000 feet, and the PCM detected too large an amount of Fuel Trim correction due to a rich A/F condition. Possible Causes • Base engine fault (i.e., cam timing incorrect, oil level too high) • EVAP vapor recovery system has failed (canister full of fuel) • Fuel injector (one or more) leaking or pressure regulator leaking • HO2S element is deteriorated or has failed
P0176 1T CCM 1998-2002 Caravan, Town & Country & Voyager	Loss Of Flexible Fuel Calibration Signal Conditions: Key on, and the PCM detected the Flexible Fuel (FF) sensor signal was too low or too high. Possible Causes • Flexible Fuel sensor signal circuit shorted to sensor ground • Flexible Fuel sensor signal circuit is open • Flexible Fuel sensor has failed • PCM has failed
P0177 1T CCM 1998-2003 Trucks equipped with a 5.9L Diesel engine	Water In Fuel Signal Circuit Fault Conditions: Key on, and the PCM detected the Water In Fuel (WIF) sensor signal was out of range low. Possible Causes • Water In Fuel sensor has detected water in the fuel supply (drain the water from the fuel filter and then retest for the code) • Water In Fuel sensor signal circuit shorted to sensor ground • Water In Fuel sensor signal circuit shorted to chassis ground • PCM has failed
P0178 1T CCM 1998-2002 Caravan, Town & Country & Voyager	Flexible Fuel Sensor Voltage Too Low Conditions: Key on, and the PCM detected the Flexible Fuel (FF) sensor signal was less than 0.51v. Possible Causes • Flexible Fuel sensor signal circuit shorted to sensor ground • Flexible Fuel sensor signal circuit is shorted to chassis ground • Flexible Fuel sensor has failed • PCM has failed
P0178 1T CCM 1998-2003 Trucks equipped with a 5.9L Diesel engine	Water in Fuel Sensor Low Input Conditions: Key on, and the PCM detected the Water In Fuel (WIF) sensor signal was less than 0.51v. Possible Causes • Water In Fuel sensor signal circuit shorted to sensor ground • Water In Fuel sensor signal circuit shorted to chassis ground • Water In Fuel sensor has failed • PCM has failed
P0179 1T CCM 1998-2002 Caravan, Town & Country & Voyager	Flexible Fuel Sensor Voltage Too High Conditions: Key on or engine running; and the PCM detected the Flexible Fuel (FF) sensor signal was more than 4.96v during the test period. Possible Causes • Flexible Fuel sensor signal circuit open or ground circuit is open • Flexible Fuel sensor signal circuit shorted to vehicle power • Flexible Fuel sensor has failed • PCM has failed

OBD II TROUBLE CODE LIST (P0XXX CODES)

DTC	Trouble Code Title, Conditions & Possible Causes
P0180 1T CCM 1996-2003 Trucks equipped with a 5.9L Diesel engine	Fuel Injection Pump Temperature Out-Of-Range Conditions: Engine started; and the PCM detected the temperature of the fuel was out of its normal operating range during the test period. Possible Causes • Fuel injection pump has failed (this code is normally caused by an internal failure of the fuel injection pump) • PCM has failed
P0181 1T CCM 2001-03 Sebring & Stratus Coupe with a 2.4L I4 or 3.0L V6 engine	Fuel Temperature Sensor Circuit Malfunction Conditions: ECT sensor from 14-97ºF and IAT sensor within 5ºF of the ECT at startup, ECT sensor over 140ºF during testing, engine running with the VSS less than 17 mph, and the PCM detected the difference between the Fuel Temperature and ECT sensor signals was more than 27ºF. Possible Causes • Fuel Temperature sensor connector is damaged or loose • Fuel Temperature sensor is damaged or it has failed • PCM has failed
P0181 1T CCM 1996-2003 Trucks equipped with a 5.9L Diesel engine	Fuel Injection Pump Failure Conditions: Key on or engine running; and the PCM detected the signal from the Fuel Temperature sensor signal was too high or too low in the test. Note: The sensor is located inside the Bosch VP44 Pump Controller. Possible Causes • Fuel Temperature sensor has failed (this sensor is not serviceable without replacing the VP44 pump controller unit). • Engine power is "de-rated" when this trouble code is set • PCM has failed
P0182 1T CCM 2001-03 Sebring & Stratus Coupe with a 2.4L I4 or 3.0L V6 engine	Fuel Temperature Sensor Circuit Low Input Conditions: Engine started; engine runtime more than 2 seconds and the PCM detected the Fuel Temperature sensor indicated less than 0.10v. Possible Causes • Fuel temperature sensor connector is damaged or shorted • Fuel temperature sensor circuit is shorted to ground • Fuel temperature sensor is damaged or it has failed • PCM has failed
P0182 1T CCM 1996-2003 Trucks & Vans equipped with a 5.2L VIN T CNG engine	CNG Temperature Sensor Low Input Conditions: Key on or engine running; and the PCM detected the CNG temperature sensor signal was less than 0.51v during the test. Possible Causes • CNG sensor signal circuit is shorted to sensor ground • CNG sensor signal circuit is shorted to chassis ground • CNG sensor has failed (it may be shorted internally) • PCM has failed
P0183 1T CCM 2001-03 Sebring & Stratus Coupe with a 2.4L I4 or 3.0L V6 engine	Fuel Temperature Sensor Circuit High Input Conditions: Engine started; engine runtime more than 2 seconds and the PCM detected the Fuel Temperature sensor indicated less than 4.60v. Possible Causes • Fuel temperature sensor connector is damaged or shorted • Fuel temperature sensor signal circuit is shorted to VREF (5v) • Fuel temperature sensor is damaged or it has failed • PCM has failed
P0183 1T CCM 1996-2003 Trucks & Vans equipped with a 5.2L VIN T CNG engine	CNG Temperature Sensor High Input Conditions: Key on or engine running; and the PCM detected the CNG temperature sensor signal was more than 4.96v during the test. Possible Causes • CNG sensor signal circuit is open • CNG sensor signal circuit is shorted to vehicle power • CNG sensor has failed (it may be open internally) • PCM has failed

OBD II TROUBLE CODE LIST (P0XXX CODES)

DTC	Trouble Code Title, Conditions & Possible Causes
P0201-P0204 1T CCM 1995-98 Eagle Talon models	Fuel Injector 1, 2, 3 or 4 Control Conditions: Engine speed below 1000 rpm, system voltage over 12v, Actuator Test off, TP sensor signal under 1.16v, and the PCM did not detect any injector coil surge voltage for 3 ms after Injector 1-4 was turned "off" (a surge voltage of 2v higher than system voltage is expected). Possible Causes • Fuel injector 1-4 control circuit is open or grounded • Fuel injector 1-4 power circuit from the ASD relay is open • Fuel injector 1 has failed or PCM injector 1 driver has failed
P0201-P0210 1T CCM 1995-2003 Car, Mini-Van, Pickup, Truck and Van models	Injector 1, 2, 3, 4, 5, 6, 7, 8, 9 or 10 Control Conditions: ASD relay "on", engine speed under 3000 rpm, injector pulsewidth under 10 ms, system voltage over 12v, and the PCM did not detect any inductive spike from the injector for 0.18 ms after it is turned off. Note: This code takes 0.64-10 seconds to set once the injector is off. Possible Causes • Fuel injector 1-10 control circuit is open or grounded • Fuel injector 1-10 power circuit from the ASD relay is open • Fuel injector 1-10 has failed • PCM injector 1-10 driver has failed
P0215 1T CCM 1998-2003 Trucks equipped with a 5.9L Diesel engine	Fuel Injection Pump Control Circuit Conditions: Key on, and the PCM did not detect any power present on the Fuel Injection Pump control circuit (VP44) during the CCM test period. Possible Causes • Fuel injection pump relay control circuit open, shorted to ground • Fuel injection pump relay ground circuit is open • Fuel injection pump relay has failed • Engine Control Module has failed
P0216 1T CCM 1998-2003 Trucks equipped with a 5.9L Diesel engine	Fuel Injection Pump Timing Failure Conditions: No other fuel injection pump trouble codes set, engine speed more than 300 rpm, fuel command 5 ml/stroke, and the PCM detected the pump timing command was out of range. Possible Causes • Fuel injection pump gear not aligned properly • Fuel filter plugged or restricted • Fuel injection pump is damaged • Transfer pump inlet line restricted
P0217 1T CCM 1998-2003 Trucks equipped with a 5.9L Diesel engine	Decreased Performance (Engine Overheat) Conditions: Engine started; and the PCM detected that the signal from the ECT sensor indicated an engine overheat condition existed. Note: Measure coolant temperature (with thermal couple) and compare to the Scan Tool ECT PID (they should be within 10°F of each other). Possible Causes • ECT sensor circuit has high resistance • ECT sensor is damaged or has deteriorated • PCM has failed
P0218 1T CCM 2002-03 Trucks & Jeep Liberty with a 3.7L V6 engine	A/T High Temperature Operation Activated Conditions: Engine started; vehicle driven in gear, and the TCM indicated the Overheat shift schedule was activated (i.e., the TCM had detected a transmission oil temperature of more than 240°F). Possible Causes • Engine cooling system malfunction present • High temperature operations activated • Transmission oil pump flow is too low or it is restricted
P0219 1T CCM 1998-2003 Trucks equipped with a 5.9L Diesel engine	Crankshaft Position Sensor Overspeed Signal Conditions: Engine started; and the PCM detected the engine had exceeded its Overspeed value. Possible Causes • CKP sensor or its tone wheel is damaged (incorrect engine speed reading) • Induction of an alternate fuel source (i.e., ether or propane) • Improper operating conditions (i.e., motoring downhill) • Turbocharger seals are leaking

OBD II TROUBLE CODE LIST (P0XXX CODES)

DTC	Trouble Code Title, Conditions & Possible Causes
P0222 1T CCM 1998-2003 Trucks equipped with a 5.9L Diesel engine	Idle Validation Signals Both Low Conditions: Key on, and the PCM detected that the Idle Validation Signal #1 and Idle Validation Signal #2 both indicated "no" voltage during the test. Possible Causes • APP sensor is damaged • IVS #1 or IVS #2 circuit(s) is shorted to sensor ground • IVS #1 or IVS #2 circuit(s) is shorted to chassis ground • IVS #1 or IVS #2 is damaged or has failed • Engine Control Module has failed
P0223 1T CCM 1998-2003 Trucks equipped with a 5.9L Diesel engine	Idle Validation Signals Both High Conditions: Key on, and the PCM detected that the Idle Validation Signal #1 and Idle Validation Signal #2 both indicated high voltage during the test. Possible Causes • APP sensor is damaged, or the sensor ground circuit is open • IVS #1 or IVS #2 circuit(s) is open or shorted to system power • IVS #1 or IVS #2 is damaged or has failed • Engine Control Module has failed
P0224 1T CCM 1996-2003 Trucks equipped with a 5.9L Diesel engine	Turbocharger Boost Limited Exceeded Conditions: No Intake Air Pressure Sensor codes set, engine started, engine running at a speed of more than 2200 rpm, and the PCM detected that the Turbocharger Boost Limit had been exceeded. Possible Causes • Wastegate has a mechanical problem or it has failed • Wastegate stuck open
P0230 1T CCM 1998-2002 Trucks equipped with a 5.9L Diesel engine	Transfer Pump (Lift Pump) Circuit Out-Of-Range Conditions: Key on, and the PCM detected the Transfer Pump signal was too high or low during testing. Possible Causes • Transfer Pump power feed circuit open, or shorted to ground • Transfer Pump ground circuit is open • Transfer Pump (internal) resistance out of range • Engine Control Module has failed
P0232 1T CCM 1998-2002 Trucks equipped with a 5.9L Diesel engine	Fuel Shutoff Signal Voltage To High Conditions: Key on, and the PCM detected that the Fuel Shutoff Signal was too high due a problem. Possible Causes • Transfer Pump may be damaged • Engine Control Module has failed
P0234 1T CCM 1998-2003 Trucks equipped with a 5.9L Diesel engine	Turbo Boost Limit Exceeded Conditions: No Intake Air Pressure sensor codes set, engine speed over 2200 rpm, and the PCM detected the Turbo Boost exceeded its limit due to a problem in the Turbocharger Wastegate Unit. Possible Causes • Wastegate is stuck • Wastegate has failed (due to a mechanical failure)
P0236 1T CCM 1998-2003 Trucks equipped with a 5.9L Diesel engine	MAP Sensor Voltage Too High To Long Conditions: DTC P0106, P0107 and P0108 not set, engine started and the PCM detected the Pressure Test failed because the MAP Sensor signal was too high. This occurs if the Boost Pressure sensor indicates high when other related signals (load and speed signals) indicate low. Possible Causes • MAP sensor signal circuit open, shorted to ground or to power • MAP sensor has failed • Engine Control Module has failed
P0237 1T CCM 1998-2003 Trucks equipped with a 5.9L Diesel engine	MAP Sensor Voltage Too Low Conditions: Engine started, engine speed from 416-3520 rpm, system voltage over 10.5v and the PCM detected the MAP sensor was under 0.10v. Possible Causes • Boost Pressure Test did not pass, or a MAP VREF DTC is present • MAP sensor signal circuit shorted to chassis or sensor ground • MAP sensor has failed

OBD II TROUBLE CODE LIST (P0XXX CODES)

DTC	Trouble Code Title, Conditions & Possible Causes
P0238 1T CCM 1998-2003 Trucks equipped with a 5.9L Diesel engine	MAP Sensor Voltage Too High Conditions: Engine speed more than 416 rpm but less than 3520 rpm, system voltage over 10.5v and the PCM detected the MAP sensor signal was more than 4.88v, condition met for 2 seconds. Possible Causes • Boost Pressure Test did not pass, ECT or IAT DTC present • MAP sensor signal circuit open or shorted to vehicle power • MAP sensor has failed • Engine Control Module has failed
P0243 1T CCM 2003 Sebring Sedan 2.4L I4	Wastegate Solenoid Circuit Malfunction Conditions: Key on or engine running; system voltage over 10.5v and the PCM detected an unexpected voltage condition on the Wastegate Solenoid control circuit during the CCM test period. Possible Causes • Wastegate solenoid control circuit is open or shorted to ground • Wastegate solenoid control circuit is shorted to system power • Wastegate solenoid is damaged or it has failed • PCM has failed
P0251 1T CCM 1998-2003 Trucks equipped with a 5.9L Diesel engine	Injector Pump Mechanical / Fuel Valve Feedback Circuit Fault Conditions: DTC P1284 not set, key on, system voltage over 12.0v, fuel command 20 ml/stroke, Fuel Delivery valve enabled, and the PCM detected the Fuel Valve feedback signal was over 9.0v. Possible Causes • Fuel injection pump relay control circuit is open or grounded • Fused battery power (B+) circuit or relay output circuit is open • Fuel injection pump relay or fuel injection pump is damaged
P0252 1T CCM 1998-2003 Trucks equipped with a 5.9L Diesel engine	Fuel Valve Signal Missing Conditions: Engine started; system voltage over 10.5v, Fuel Valve current feedback test completed, and the PCM detected a signal from the Fuel Metering valve that indicated the valve had moved. Possible Causes • Fuel injection pump relay control circuit is open or grounded • Fused battery power (B+) circuit or relay output circuit is open • Fuel injection pump relay or fuel injection pump is damaged
P0253 1T CCM 1998-2003 Trucks equipped with a 5.9L Diesel engine	Injector Pump Fuel Valve Current Too Low Conditions: Engine started; engine speed more than 100 rpm and the PCM detected a condition of low current (or no current) at the Fuel Metering Valve during the test period (there may be an open circuit). • Refer to code repair chart in other manuals to test this code.
P0254 1T CCM 1998-2003 Trucks equipped with a 5.9L Diesel engine	Injector Pump Fuel Valve Current Too High Conditions: Key "off", and the PCM detected system voltage (power) at the Fuel Metering Valve VP44 circuit during the test period. Possible Causes • Fuel injection pump is damaged or it has failed
P0300 1T Catalyst 2T Emission 1995-2000 Eagle Talon models	Multiple Cylinder Misfire Detected Conditions: Engine started; adaptive numerator updated, engine speed less than 3000 rpm, and the PCM detected a misfire condition (1-2%) within 1000 revolutions (High Emissions) or a 2-10% misfire condition within 200 revolutions (Catalyst Damaging) in two or more cylinders. Note: If the misfire is severe, the MIL will flash on/off on the 1st trip! Possible Causes • Air leak in the intake manifold, or in the EGR or EVAP system • Base engine mechanical fault that affects one or more cylinders • Erratic or interrupted CKP or CMP sensor signals • Fuel delivery component fault that affects one or more cylinders (i.e., a contaminated, dirty or sticking fuel injector) • Ignition system problem (coil or plug) in one or more cylinders • Vehicle driven with low fuel pressure or while very low on fuel

OBD II TROUBLE CODE LIST (P0XXX CODES)

DTC	Trouble Code Title, Conditions & Possible Causes
P0300 1T Catalyst 2T Emission 1995-2003 Cars, Mini-Vans & PT Cruiser, Prowler, Viper	Multiple Cylinder Misfire Detected Conditions: Engine speed less than 3000 rpm, adaptive numerator updated, and the PCM detected a misfire rate of 1-2% (High Emissions 2T), or a misfire rate of 6-30% (Catalyst Damaging 1T) in 2 or more cylinders. Note: If the misfire is severe, the MIL will flash on/off on the 1st trip! Possible Causes • Air leak in the intake manifold, or in the EGR or EVAP system • Base engine mechanical fault that affects two or more cylinders • Erratic or interrupted CKP or CMP sensor signals • Fuel delivery component fault that affects two or more cylinders (i.e., a contaminated, dirty or sticking fuel injector) • Ignition system problem (coil or plug) in two or more cylinders • Vehicle driven with low fuel pressure or while very low on fuel
P0300 1T Catalyst 2T Emission 2001-03 Sebring & Stratus Coupe with a 2.4L I4 or 3.0L V6 engine	Multiple Cylinder Misfire Detected Conditions: Engine speed from 500-6000 rpm at a steady throttle, ECT and IAT sensors over 14ºF, BARO sensor over 76 kPa (11 psi), and the PCM detected a misfire rate of 1.5% (High Emissions 2T) or a high misfire rate with Converter temperature over 1742ºF in 2 or more cylinders. Note: If the misfire is severe, the MIL will flash on/off on the 1st trip! Possible Causes • Air leak in the intake manifold, or in the EGR or EVAP systems • Base engine mechanical fault (e.g., a skipped timing belt) • Erratic or interrupted CKP or CMP sensor signals • Fuel delivery component fault that affects two or more cylinders (i.e., a contaminated, dirty or sticking fuel injector) • Ignition system problem (coil or plug) in two or more cylinders • Vehicle driven with low fuel pressure or while very low on fuel • TSB 18-028-02 contains a repair procedure for this code
P0300 1T Catalyst 2T Emission 1996-2003 Dakota, Durango, Jeep, Truck and Van models	Multiple Cylinder Misfire Detected Conditions: Engine speed less than 3000 rpm, adaptive numerator updated, and PCM detected a misfire rate of 1% (High Emissions 2T), or a misfire rate of 6-30% (Catalyst Damaging 1T) in two or more cylinders. Note: If the misfire is severe, the MIL will flash on/off on the 1st trip! Possible Causes • Air leak in the intake manifold, or in the EGR or EVAP system • Base engine mechanical fault that affects two or more cylinders • Erratic or interrupted CKP or CMP sensor signals • Fuel delivery component fault that affects two or more cylinders (i.e., a contaminated, dirty or sticking fuel injector) • Ignition system problem (coil or plug) in two or more cylinders Vehicle driven with low fuel pressure or while very low on fuel
P0300 1T Catalyst 2T Emission 1998-2003 Trucks equipped with a 5.9L Diesel engine	Multiple Cylinder Misfire Conditions: Engine speed less than 880 rpm, ECT sensor more than 140ºF, VSS at 0 mph, PTO Inactive, and the PCM detected the time for the crankshaft to turn 120 degrees during the firing event of one or more cylinders under test compared to the time for the crank to turn 120 degrees for the previous cylinder exceeded 240 microseconds. Note: If the misfire is severe, the MIL will flash on/off on the 1st trip! Possible Causes • Air leak (manifold, PCV) that affects more than 1 cylinder • Base engine problem that affects more than 1 cylinder • CMP sensor, sensor wiring harness or tone wheel is damaged • Ignition system problem that affects more than one cylinder • Fuel metering problem (fuel injectors, fuel pressure) that affects more than one cylinder • EGR, or EVAP Purge problem that affects more than 1 cylinder

OBD II TROUBLE CODE LIST (P0XXX CODES)

DTC	Trouble Code Title, Conditions & Possible Causes
P0301-P0304 1T Catalyst 2T Emission 1995-98 Eagle Talon models	Cylinder 1, 2, 3 or 4 Misfire Detected Conditions: Engine speed less than 3000 rpm, adaptive numerator updated, and the PCM detected a misfire rate of 1-2% (High Emissions 2T), or a misfire rate of 6-30% (Catalyst Damaging 1T) in a single cylinder. Note: If the misfire is severe, the MIL will flash on/off on the 1st trip! Possible Causes • Air leak in the intake manifold, or in the EGR or PCV system • Base engine mechanical fault that affects only one cylinder • Fuel delivery component fault that affects only one cylinder (i.e., a contaminated, dirty or sticking fuel injector) • Ignition system problem (coil or plug) that affects one cylinder
P0301-P0310 1T Catalyst 2T Emission 1995-2003 Cars, Mini-Vans & PT Cruiser, Prowler, Viper	Cylinder 1-10 Misfire Detected Conditions: Engine speed less than 3000 rpm, adaptive numerator updated, and the PCM detected a misfire rate of 1-2% (High Emissions 2T), or a misfire rate of 6-30% (Catalyst Damaging 1T) in only one cylinder. Note: If the misfire is severe, the MIL will flash on/off on the 1st trip! Possible Causes • Air leak in the intake manifold, or in the EGR or EVAP system • Base engine mechanical fault that affects only one cylinder • Fuel delivery component fault that affects only one cylinder (e.g., a dirty fuel injector) • Ignition system problem (coil or plug) in only one cylinder
P0301-P0306 1T Catalyst 2T Emission 2001-03 Sebring & Stratus Coupe with a 2.4L I4 or 3.0L V6 engine	Cylinder 1-6 Misfire Detected Conditions: Engine speed from 500-6000 rpm at a steady throttle, ECT and IAT sensors over 14ºF, BARO sensor over 76 kPa (11 psi), and the PCM detected a misfire rate of 1.5% (High Emissions 2T) or a high misfire rate with Converter temperature over 1742ºF in only one cylinder. Note: If the misfire is severe, the MIL will flash on/off on the 1st trip! Possible Causes • Air leak in the intake manifold, or in the EGR or EVAP systems • Base engine mechanical fault (e.g., a skipped timing belt) • Fuel delivery component fault that affects only one cylinder (e.g., a dirty fuel injector) • Ignition system problem (coil or plug) in only one cylinder • TSB 18-028-02 contains a repair procedure for this code
P0301-P0308 1T Catalyst 2T Emission 1996-2003 Dakota, Durango, Jeep, Truck and Van models	Cylinder 1-8 Misfire Detected Conditions: Engine speed less than 3000 rpm, adaptive numerator updated, and PCM detected a misfire rate of 1% (High Emissions 2T), or a misfire rate of 6-30% (Catalyst Damaging 1T) in a single cylinder. Note: If the misfire is severe, the MIL will flash on/off on the 1st trip! Possible Causes • Air leak in the intake manifold, or in the EGR or EVAP system • Base engine mechanical fault that affects only one cylinder • Erratic or interrupted CKP or CMP sensor signals • Fuel delivery component fault that affects only one cylinder (e.g., a dirty fuel injector) • Ignition system problem (coil or plug) that affects only one cylinder
P0301-P0306 1T Catalyst 2T Emission 1998-2003 Trucks equipped with a 5.9L Diesel engine	Cylinder 1-6 Misfire Detected Conditions: Engine started; engine speed below 860 rpm, ECT sensor over 140ºF, VSS at 0 mph, PTO "off", and the PCM detected the time for the crankshaft to turn 120 degrees while firing one cylinder when compared to the time for the crankshaft to turn 120 degrees for the previous cylinder exceeded 40 ms. Note: If the misfire is severe, the MIL will flash on/off on 1st trip! Possible Causes • Base engine mechanical fault that affects only one cylinder • CMP sensor, sensor wiring harness or tone wheel is damaged • Fuel delivery component fault that affects only one cylinder (e.g., a dirty fuel injector)

OBD II TROUBLE CODE LIST (P0XXX CODES)

DTC	Trouble Code Title, Conditions & Possible Causes
P0315 N/MIL 1T CCM 2001-03 All vehicle applications	No Crankshaft Position Sensor Learned Conditions: Engine started; engine runtime more than 50 seconds under closed throttle operating conditions, ECT sensor more than 167ºF, and the PCM detected that one of the CKP sensor windows had too much variance (e.g., over 2.86%) from its calibrated reference point. Possible Causes • Crankshaft tone wheel flex plate is damaged • Erratic CKP sensor signals • PCM has failed
P0320 N/MIL 1T CCM 1996-2003 All vehicle applications except the 5.9L diesel	No Crank Reference Signal to PCM Conditions: Engine cranking, vacuum signals present; and the PCM detected from 3-8 CMP sensor signals without detecting a CKP sensor signal. Possible Causes • CKP sensor signal circuit is open or grounded • CKP sensor VREF circuit is open or grounded • Tone wheel or flex plate damage or erratic CKP sensor signals • CKP sensor is damaged or has failed • PCM has failed
P0320 N/MIL 1T CCM 1998-2002 Trucks equipped with a 5.9L Diesel engine	No Crank Reference Signal to PCM Conditions: Engine started; system voltage from 10-15v, and the PCM detected the engine speed signal indicated under 800 rpm while the engine speed signal at the ECM indicated over 1024 rpm. Possible Causes • CKP sensor signal circuit is open or grounded • CKP sensor VREF circuit is open or grounded • Tone wheel or flex plate damage or erratic CKP sensor signals • CKP sensor is damaged or has failed • PCM has failed
P0325 N/MIL 1T CCM 1995-98 Eagle Talon models	Knock Sensor 1 Circuit Conditions: Engine running at idle or in deceleration mode, and the PCM detected the Knock sensor signal was below a minimum value (value depends on engine speed), or was more than 5.0v Possible Causes • Knock sensor signal circuit open or grounded • Knock sensor not tightened properly • Knock sensor damaged or has failed (it may be open internally) • PCM has failed
P0325 N/MIL 1T CCM 1995-2003 Car, SUV, Truck & Van models	Knock Sensor 1 Circuit Conditions: Engine running at idle or in deceleration mode, and the PCM detected the Knock sensor signal was below a minimum value (value depends on engine speed), or was more than 5.0v Possible Causes • Knock sensor connector is damaged or shorted • Knock sensor signal circuit open or grounded • Knock sensor not tightened properly • Knock sensor damaged or has failed (it may be open internally) • PCM has failed
P0330 1T CCM 2002-03 Jeep Liberty & Ram Pickup models with a 2.4L I4 & 3.7L V6 engine	Knock Sensor 2 Circuit Conditions: Engine running at idle or in deceleration mode, and the PCM detected the Knock Sensor 2 signal was below a minimum value (value depends on engine speed), or was more than 5.0v. Possible Causes • Knock sensor signal circuit open or grounded • Knock sensor not tightened properly • Knock sensor damaged or has failed (it may be open internally) • PCM has failed

OBD II TROUBLE CODE LIST (P0XXX CODES)

DTC	Trouble Code Title, Conditions & Possible Causes
P0335 1T CCM 1995-2000 Avenger, Eagle Talon & Sebring models	Crankshaft Position Sensor Circuit Conditions: Engine started and the PCM did not detect any change in the CKP sensor input for 2 seconds. Possible Causes • CKP sensor signal circuit is open or shorted to ground • CKP sensor VREF circuit is open or shorted to ground • CKP sensor is damaged or has failed • PCM has failed
P0335 N/MIL 1T CCM 2001-03 Car models	Crankshaft Position Sensor Circuit Conditions: Engine cranking with at least 8 CMP sensor signals detected, and the PCM did not detect any CKP sensor signals for 2 seconds. Possible Causes • CKP sensor signal circuit is open or it is shorted to ground • CKP sensor VREF circuit is open or shorted to ground • CKP sensor ground circuit is open • CKP sensor is damaged or has failed • PCM has failed
P0336 1T CCM 1998-2003 Trucks equipped with a 5.9L Diesel engine	Crankshaft Position Sensor Signal Conditions: Engine started; CMP sensor signals detected, and the PCM did not detect any change in the CKP sensor signal during the test period. Possible Causes • Check the tone or pulse ring for damage or debris collection • CKP sensor signal circuit is open or shorted to ground • CKP sensor VREF circuit is open or shorted to ground • CKP sensor is damaged or has failed • PCM has failed
P0339 N/MIL 1T CCM 2000-03 Car, SUV, Truck & Van models	Crankshaft Position Sensor Circuit Intermittent Conditions: Engine started; CMP sensor signals detected, and the PCM detected an intermittent loss of the CKP sensor signal. The Failure counter must reach 20 before this code will set. Possible Causes • Check the tone wheel/pulse ring for damage or debris collection • CKP sensor signal circuit is open or shorted to ground • CKP sensor VREF circuit is open or shorted to ground • CKP sensor is damaged or it has failed • PCM has failed
P0340 1T CCM 1995-98 Eagle Talon models	No Camshaft Synch Signal To PCM Conditions: Engine cranking, CKP sensor pulses detected, and the PCM did not detect any CMP sensor pulses, condition met for 2 seconds. Possible Causes • CMP sensor signal circuit is open or shorted to ground • CMP sensor VREF circuit is open or shorted to ground • CMP sensor is damaged or has failed • PCM has failed
P0340 1T CCM 1995-20003 Car, SUV, Truck & Van models	No Camshaft Fuel (Sync) Signal To PCM Conditions: Engine cranking, system voltage over 10.0v, and the PCM detected CKP pulses without detecting any CMP sensor pulses for 2 seconds. Possible Causes • CMP sensor connector is damaged, open or it is shorted • CMP sensor signal circuit is open or shorted to ground • CMP sensor VREF circuit is open or shorted to ground • CMP sensor is damaged or has failed • PCM has failed

OBD II TROUBLE CODE LIST (P0XXX CODES)

DTC	Trouble Code Title, Conditions & Possible Causes
P0341 1T CCM 1998-2003 Trucks equipped with a 5.9L Diesel engine	Camshaft Position Sensor Signal Conditions: Engine started; system voltage from 8.0-15v, and the PCM detected an incorrect tone wheel tooth signal occurred 25 times in 8 seconds, or the PCM did not detect any CMP sensor signals for 2 seconds. Possible Causes • CMP sensor signal circuit is open or shorted to ground • CMP sensor VREF circuit is open or shorted to ground • CMP sensor is damaged or has failed • PCM has failed
P0342 1T CCM 1998-2003 Trucks equipped with a 5.9L Diesel engine	CMP Sensor Supply Voltage Too Low Conditions: Engine started; system voltage from 8.0-15v, and the PCM detected the CMP sensor VREF signal was less than 4.2v for 2 seconds. Possible Causes • CMP sensor connector is damaged, open or it is shorted • CMP sensor VREF circuit is open • CMP sensor VREF circuit is shorted to ground • CMP sensor is damaged or has failed • PCM has failed
P0343 1T CCM 1998-2003 Trucks equipped with a 5.9L Diesel engine	CMP Sensor Supply Voltage Too Low Conditions: Engine cranking or engine running, system voltage over 10.5v and the PCM detected an intermittent loss of the CMP sensor signal during the period of 2.5 complete engine revolutions. The failure counter must reach 20 before this code matures and a code is set. Possible Causes • CMP sensor signal circuit is open (intermittent fault) • CMP sensor signal circuit shorted to ground (intermittent fault) • CMP sensor ground circuit is open (an intermittent fault) • PCM has failed
P0344 1T CCM 2000-03 Car, SUV, Truck & Van models	Camshaft Position Sensor Circuit Intermittent Conditions: Engine cranking or engine running, system voltage over 10.5v and the PCM detected an intermittent loss of the CMP sensor signal during the period of 2.5 complete engine revolutions. The failure counter must reach 20 before this code matures and a code is set. Possible Causes • CMP sensor signal circuit is open (intermittent fault) • CMP sensor signal circuit shorted to ground (intermittent fault) • CMP sensor ground circuit is open (an intermittent fault) • PCM has failed

OBD II TROUBLE CODE LIST (P0XXX CODES)

DTC	Trouble Code Title, Conditions & Possible Causes
P0350 1T CCM 2001-2002 Trucks equipped with a 5.9L Diesel engine	Ignition Coil Current Too High Conditions: Engine started; and the PCM detected the Ignition Coil current level was too high. Possible Causes • Ignition Coil 1, 2 or 3 control circuit (the coil windings) shorted internally • Ignition Coil 1, 2 or 3 control circuit is shorted to ground • PCM has failed
P0351, P0152, P0353 1T CCM 1995-98 Eagle Talon models	Ignition Coil 1, 2 or 3 Primary Circuit Conditions: Engine speed less than 4000 rpm, MFI relay on, system voltage over 13v, and the PCM detected the Coil 1, 2 or 3 primary circuit did not achieve peak current (dwell) for 3 seconds. Possible Causes • Ignition coil 1, 2 or 3 primary "driver" circuit is open or it is grounded • Ignition coil 1, 2 or 3 is damaged or it has failed • MFI relay power circuit to ignition coil is open • PCM has failed
P0351, P0152, P0153 1T CCM 1996-2003 300M, Avenger, Caravan, Concorde, Intrepid, Jeep, LHS, Neon, Prowler, PT Cruiser, Stratus, Sebring, Viper, Town/Country, Vision, & Voyager models	Ignition Coil 1, 2 or 3 Primary Circuit Conditions: Engine cranking or running; ASD relay on, system voltage over 10v, Coil 1, 2 or 3 was not in dwell period when it was checked, and the PCM detected that the peak coil current was not reached in 2.5 ms of dwell for 3-6 seconds. The coil primary resistance is less than 2 ohms. Possible Causes • ASD relay power circuit to the ignition coil is open • Ignition Coil 1, 2 or 3 primary "driver" circuit open or grounded • Ignition Coil 1, 2 or 3 is damaged or it has failed • PCM has failed
P0351-P0358 N/MIL Codes 1T CCM 1999-2003 Dakota, Durango, Truck & Van models	Ignition Coil 1-8 Primary Circuit Conditions: Engine cranking or speed less than 2012 rpm, ASD relay on, system voltage 8v or 12v, coils not in dwell period, and the PCM detected peak coil current was not reached in 2.5 ms of dwell for 3-6 seconds. The coil primary resistance is less than 2 ohms. Possible Causes • ASD relay power circuit to the ignition coil is open • Ignition coil primary "driver" circuit open or grounded • Ignition coil is damaged or it has failed • PCM has failed
P0370 1T CCM 1998-2003 Trucks equipped with a 5.9L Diesel engine	Fuel Injection Pump Speed Sensor Signal Lost Conditions: Engine started; and the PCM detected that it lost the Fuel Injection Pump Speed/Position Signal (due to a fault in the internal Fuel injection Pump (the pump may have failed). Possible Causes • ECM cannot control the engine speed through circuit WP44 • Fuel Injection pump has failed due to an internal problem
P0380 1T CCM 1998-2003 Trucks equipped with a 5.9L Diesel engine	Intake Air Heater Relay #1 Control Circuit Conditions: Key on for 1 second, and the PCM detected the Intake Air Heater Relay No. 1 circuit was not activated for more than one second. This problem is not due to a fault in the heater element. Possible Causes • Intake Air Heater relay control circuit open or shorted to ground • Intake Air Heater relay ground circuit is open • Intake Air Heater relay is damaged or has failed • PCM has failed
P0381 1T CCM 1998-2003 Trucks equipped with a 5.9L Diesel engine	Wait To Start Lamp Inoperative Conditions: Key on for 2 seconds, and the PCM detected the Wait To Start control circuit was not activated for more than two seconds. The PCM expects to detect a voltage drop of 0.5v. Possible Causes • Wait To Start lamp driver circuit shorted to ground (the lamp will be on continuously) • Wait To Start lamp driver circuit open (lamp will not come on) • Wait To Start bulb has failed • Wait To Start ignition feed circuit is open, or the Cluster is open • PCM has failed

OBD II TROUBLE CODE LIST (P0XXX CODES)

DTC	Trouble Code Title, Conditions & Possible Causes
P0382 1T CCM 1998-2003 Trucks equipped with a 5.9L Diesel engine	Intake Air Heater Relay #2 Control Circuit Conditions: Key on for one second; and the PCM detected the Intake Air Heater Relay 2 Signal circuit was not activated for more than one second. Possible Causes • Intake air heater relay control circuit is open or shorted to ground • Intake air heater relay power circuit is open • Intake air heater relay is damaged or has failed • PCM has failed
P0387 1T CCM 1998-2000 Trucks equipped with a 5.9L Diesel engine	Crankshaft Position Sensor Supply Voltage Low Conditions: Engine started and the PCM did not detect any CKP sensor position or engine speed signals. Possible Causes • CKP sensor signal circuit is open or shorted to ground • CKP sensor is damaged or has failed • PCM has failed
P0388 1T CCM 1998-2000 Trucks equipped with a 5.9L Diesel engine	Crankshaft Position Sensor Supply Voltage High Conditions: Engine started and the PCM did not detect any CKP sensor position or engine speed signals. Possible Causes • CKP sensor ground circuit is open • CKP sensor signal circuit is shorted to VREF or to system power • CKP sensor is damaged or has failed • PCM has failed
P0400 2T EGR 1995-98 Eagle Talon equipped with a VIN F (turbo)	EGR System Fault Conditions: Engine runtime over 3 minutes in closed loop, ECT input over 170ºF, engine speed at 1952-2400 rpm, MAP input at 1.80-2.70v, TP input from 0.6-1.8v, VSS over 3 mph, EGR Test "on", and the PCM detected too little change in EGR flow with the valve cycled. Possible Causes • EGR valve source vacuum supply line open or blocked • EGR exhaust transfer tubes blocked in the exhaust manifold • EGR transducer tube to the transducer is blocked or restricted • EGR valve or solenoid transducer is damaged or has failed • PCM has failed
P0401 2T EGR 1995-2000 Avenger, Breeze, Cirrus, Eagle Talon, Sebring, Stratus models	EGR System Malfunction Conditions: Engine running with throttle open under steady load in closed loop, ECT sensor more than 180ºF, BTS signal more than 40ºF, fuel trim not operating near its limits, then the EGR valve was cycled off to on and the PCM detected very little change in the HO2S signal. Note: This test is repeated up to three (3) times for each vehicle trip. Possible Causes • EGR valve source vacuum supply line open or blocked • EGR exhaust transfer tubes blocked in the exhaust manifold • EGR transducer tube to the transducer is blocked or restricted • EGR valve or solenoid transducer is damaged or has failed • PCM has failed
P0401 2T EGR 1995-2003 300M, Concorde, LHS, Intrepid, Neon, PT Cruiser, Prowler, Viper, Caravan, Town & Country, Voyager models	EGR System Fault Conditions: Engine running with throttle open under steady load in closed loop, ECT sensor more than 180ºF, BTS signal more than 0ºF, system voltage over 10.5v, fuel trim not operating near its limits, then the EGR valve was cycled off to on and the PCM detected very little change in the HO2S signal. Note: This test is repeated up to three (3) times for each vehicle trip. Possible Causes • EGR valve control circuit open, grounded or shorted to power • EGR exhaust transfer tubes blocked in the exhaust manifold • ASD relay power circuit open to the EGR solenoid • EGR valve is damaged or has failed • PCM has failed • TSB 18-33-98 contains a repair procedure for this code

OBD II TROUBLE CODE LIST (P0XXX CODES)

DTC	Trouble Code Title, Conditions & Possible Causes
P0403 1T CCM 1995-98 Eagle Talon models	EGR Solenoid Circuit Conditions: Engine started; system voltage over 10.5v and the PCM detected the EGR Solenoid control circuit was not in its expected state when requested to operate by the PCM. Note: The EGR solenoid resistance range is 25-35 ohms at 68°F. Possible Causes • EGR solenoid control circuit is open or shorted to ground • EGR solenoid power circuit is open • EGR solenoid is damaged or has failed • PCM has failed
P0403 1T CCM 300M, Concorde, LHS, Intrepid, Neon, PT Cruiser, Prowler, Viper, Caravan, Town & Country, Voyager models	EGR Solenoid Circuit Conditions: Engine started; system voltage over 10.5v and the PCM detected the EGR Solenoid control circuit was not in its expected state when requested to operate by the PCM. Note: The EGR solenoid resistance range is 25-35 ohms at 68°F. Possible Causes • EGR solenoid control circuit is open or shorted to ground • EGR solenoid power circuit is open • EGR solenoid is damaged or has failed • PCM has failed
P0404 1T CCM 1995-2003 300M, Concorde, LHS, Intrepid, Prowler, Sebring & Vision models	EGR Position Sensor Signal Performance Conditions: Engine started; system voltage over 10.5v and the PCM detected that the EGR flow (or valve movement) was not what was expected during the test period. The normal range is 0.8-4.3v. Possible Causes • EGR sensor signal circuit is open or shorted to ground • EGR sensor 5v supply circuit is open • EGR sensor ground circuit is open • EGR valve actuator loose, sticking or blocked • EGR sensor is damaged or has failed • PCM has failed
P0405 1T CCM 1995-2003 300M, Concorde, LHS, Intrepid, Prowler, Sebring models	EGR Position Sensor Circuit Low Input Conditions: Key on or engine running; system voltage over 10.5v and the PCM detected that the EGR sensor signal indicated less than 0.1026v. Possible Causes • EGR sensor signal circuit is shorted to ground • EGR sensor VREF (5v) circuit is open or shorted to ground • EGR sensor is damaged (shorted internally) or it has failed • PCM has failed
P0406 1T CCM 1995-2003 300M, Concorde, LHS, Intrepid, Prowler, Sebring models	EGR Position Sensor Circuit High Input Conditions: Key on or engine running; system voltage over 10.5v and the PCM detected the EGR sensor indicated more than 4.89v for 6 seconds. Possible Causes • EGR sensor signal is shorted to VREF (5v) supply circuit • EGR sensor ground circuit is open • EGR sensor is damaged (it may have an internal open circuit) • PCM has failed
P0411 2T AIR 1995-96 Eagle Talon equipped with a VIN Y engine	Too Little or Too Much Secondary Air Conditions: Engine runtime over 11 minutes, engine speed over 700 rpm, then after the PCM shifted the A/F mixture rich by 10% for 18 seconds, it detected the HO2S signal did not switch to rich, or with the AIR solenoid commanded on, the HO2S-11 did not indicate a lean ratio. Possible Causes • AIR solenoid air injection valve is damaged or has failed • AIR solenoid source vacuum hoses loose or disconnected • AIR solenoid air injection tube is restricted or clogged • PCM has failed

OBD II TROUBLE CODE LIST (P0XXX CODES)

DTC	Trouble Code Title, Conditions & Possible Causes
P0412 1T CCM 1995-96 Eagle Talon equipped with a VIN Y engine	Secondary Air Solenoid Circuit Conditions: Engine started; system voltage over 10.5v and the PCM detected an unexpected voltage state after the AIR solenoid was enabled. The AIR solenoid resistance is from 33-39 ohms at 68°F. Possible Causes • AIR solenoid control circuit is open or shorted to ground • AIR solenoid power circuit is open • AIR solenoid is damaged or has failed • PCM has failed
P0420 2T Catalyst 1995-2003 300M, Avenger, Breeze, Eagle Talon, Cirrus, LHS, Concorde, PT Cruiser, Intrepid, Sebring, Caravan, Voyager, Town & Country, Viper, Prowler models	Catalyst Efficiency Below Normal (Bank 1) Conditions: Engine speed at 1200-1700 rpm in closed loop with the throttle open for over 2 minutes, ECT sensor more than 147°F, MAP sensor signal from 15.0-21.0" Hg, and the PCM detected the switch rate of the rear HO2S reached 70% of the switch rate of the front HO2S. Possible Causes • Air leaks in at the exhaust manifold or exhaust pipes • Base engine problems (high coolant or engine oil consumption) • Catalytic converter damaged or has failed • Front HO2S older (aged) than the rear HO2S (HO2S-12 is lazy)
P0420 2T Catalyst 1996-2003 Jeep, Truck, Dakota, Durango & Van models	Catalyst Efficiency Below Normal (Bank 1) Conditions: Engine started; then vehicle driven at 1200-1700 rpm at over 20 mph with the throttle open and steady for 3 minutes, ECT sensor over 147°F, MAP sensor at 15.0-21.0" Hg, and the PCM detected the switch rate of the rear HO2S approached the switch rate of the front HO2S-11 during the Catalyst Monitor test period. Possible Causes • Air leaks in at the exhaust manifold or exhaust pipes • Base engine problems (high coolant or engine oil consumption) • Catalytic converter damaged or has failed • Front HO2S older (aged) than the rear HO2S (HO2S-12 is lazy)
P0421 2T Catalyst 2001-03 Sebring & Stratus Coupe with a 2.4L I4 or 3.0L V6 engine	Catalyst Efficiency Fault (Bank 1) Conditions: Engine started, engine speed below 3000 rpm in closed loop with throttle open at more than 1 mph for 84 seconds, IAT sensor over 14°F, VAF sensor over 4000 Hz, BARO sensor over 75 kPa, volumetric efficiency from 63-169 Hz, and the PCM detected the frequency of the front and rear HO2S exceeded 0.8. Possible Causes • Air leaks in at the exhaust manifold or exhaust pipes • Base engine problems (high coolant or engine oil consumption) • Catalytic converter is damaged or it has failed • Front HO2S older (aged) than the rear HO2S (HO2S-12 is lazy)
P0421 2T Catalyst 2001-03 Sebring & Stratus Coupe with a 3.0L V6 engine	Catalyst Efficiency Fault (Bank 1) Conditions: Engine speed below 3000 rpm with throttle open at over 1 mph for 84 seconds, IAT sensor over 14°F, VAF sensor over 4000 Hz, volumetric efficiency from 63-169 Hz, BARO sensor over 75 kPa, and the PCM detected the frequency of the front and rear HO2S exceeded 0.8. Possible Causes • Air leaks in at the exhaust manifold or exhaust pipes • Base engine problems (high coolant or engine oil consumption) • Catalytic converter is damaged or it has failed • Front HO2S older (aged) than the rear HO2S (HO2S-12 is lazy)
P0422 2T Catalyst 1995-98 Eagle Talon models	Catalyst Efficiency Below Normal (Bank 1) Conditions: Engine speed from 1248-2400 in closed loop with the throttle open for over 2 minutes, ECT sensor more than 170°F, MAP sensor signal from 1.50-2.60v, and the PCM detected the switch rate of the rear HO2S reached 70% of the switch rate of the front HO2S. Possible Causes • Air leaks in at the exhaust manifold or exhaust pipes • Base engine problems (high coolant or engine oil consumption) • Catalytic converter damaged or has failed • Front HO2S older (aged) than the rear HO2S (HO2S-12 is lazy)

OBD II TROUBLE CODE LIST (P0XXX CODES)

DTC	Trouble Code Title, Conditions & Possible Causes
P0432 2T Catalyst 1995-2003 300M, LHS, Concorde, Intrepid, Sebring, Caravan, Voyager, Town & Country, Viper, Prowler models	Catalyst Efficiency Below Normal (Bank 2) Conditions: Engine speed at 1200-1700 rpm in closed loop with the throttle open for over 2 minutes, ECT sensor more than 147ºF, MAP sensor signal from 15.0-21.0" Hg, and the PCM detected the switch rate of the rear HO2S reached 70% of the switch rate of the front HO2S. Possible Causes • Air leaks in at the exhaust manifold or exhaust pipes • Base engine problems (high coolant or engine oil consumption) • Catalytic converter damaged or has failed • Front HO2S older (aged) than the rear HO2S (HO2S-12 is lazy)
P0432 2T Catalyst 1996-2003 Jeep, Truck, Dakota, Durango & Van models	Catalyst Efficiency Fault (Bank 2) Conditions: Engine started; then vehicle driven at 1200-1700 rpm at over 20 mph with the throttle open and steady for 3 minutes, ECT sensor over 147ºF, MAP sensor at 15.0-21.0" Hg, and the PCM detected the switch rate of the rear HO2S approached the switch rate of the front HO2S-11 during the Catalyst Monitor test period. Possible Causes • Air leaks in at the exhaust manifold or exhaust pipes • Base engine problems (high coolant or engine oil consumption) • Catalytic converter damaged or has failed • Front HO2S older (aged) than the rear HO2S (HO2S-12 is lazy)
P0440 2T EVAP 1995-98 Eagle Talon with a 2.0L Turbo engine	EVAP Purge System Fault Conditions: ECT sensor from 40-90ºF and IAT sensor within 10ºF of ECT at startup (cold engine), engine speed under 2048 rpm, Purge solenoid commanded on, EVAP Stricter test completed and passed, and the PCM detected too little change in the Fuel Control system. Possible Causes • EVAP system vacuum hoses plugged, loose or disconnected • EVAP purge control solenoid is damaged or has failed • EVAP purge control valve is damaged
P0440 2T EVAP 2002-03 300M, Concorde, LHS, Intrepid, Sebring, Neon, PT Cruiser, Caravan, Town & Country, Dakota & Durango models	EVAP Purge System Fault Conditions: Ambient Air Temperature from 39-89ºF at startup, engine running, Fuel Level over 12%, and the PCM detected that the NVLD switch did not close. Once this event occurs, the PCM will increase the amount of vacuum in the system that flows past the purge valve. If the NVLD switch does not close under these conditions, the PCM will set this code. Possible Causes • EVAP purge valve vacuum supply is leaking or clogged • EVAP purge valve is stuck closed • NVLD assembly (leak detection) is damaged or has failed • NVLD switch circuit is open or the NVLD switch has failed • PCM has failed
P0441 2T EVAP 1995-2003 Car, Jeep, SUV, Truck & Van models	EVAP Purge Flow Monitor Fault Conditions: Engine at idle speed in closed loop for 200 seconds, BARO sensor signal less than 8,000 feet, ECT sensor more than 160ºF, no Low Fuel, MAP sensor signal less than 23.7" Hg, and the PCM did not detect any purge flow through the EVAP system during this test. Possible Causes • EVAP purge solenoid vacuum line loose, leaking or restricted • EVAP purge solenoid stuck leaking, stuck open or stuck closed • EVAP purge vacuum line to canister leaking or disconnected • EVAP canister leaking, damaged or has failed
P0442 2T EVAP 1996-2003 Car, SUV, Truck & Van models	EVAP System Small Leak Detected Conditions: ECT sensor from 40-90ºF and within 10ºF of the BTS input at startup (cold engine), fuel level more than 1/2 full, engine started, EVAP leak test enabled, and the PCM detected a leak greater than 0.040" but less than 0.080" somewhere in the EVAP system. Possible Causes • EVAP fuel tank or canister vapor hoses leaking or damaged • EVAP system component leaking, damaged or has failed • Fuel tank cap is loose, or the cap release pressure is incorrect • Leak detection pump damaged or leaking

OBD II TROUBLE CODE LIST (P0XXX CODES)

DTC	Trouble Code Title, Conditions & Possible Causes
P0442 2T EVAP 2001-03 Sebring & Stratus Coupe with a 2.4L I4 or 3.0L V6 engine	EVAP System Small Leak Detected Conditions: ECT and IAT sensors less than 86ºF at startup, engine runtime less than 12 minutes, BARO sensor over 76 kPa, engine speed over 1600 rpm, volumetric efficiency from 20-80%, ECT sensor more than 140ºF, PSP switch indicating off. The test stops when the FTP sensor indicates 451 Pa (0.065 psi) and fluctuates less than 647 Pa (0.094 psi) with a FTP input of 1-4.0v, an ECT input over 140ºF and an IAT input under 41ºF with the Purge and Vent solenoid closed. The test fails if the PCM detects that the FTP signal changes more than 785 Pa (0.114 psi) within 20 seconds. Possible Causes • EVAP canister vent solenoid is damaged or has failed • EVAP emission canister seal is damaged or leaking • Fuel tank cap is loose, or the cap release pressure is incorrect • Fuel tank, vapor line or vacuum seal is damaged or leaking
P0443 1T CCM 1995-98 Eagle Talon models	EVAP Purge Solenoid Circuit Fault Conditions: Engine started; system voltage from 10-16v, and after the EVAP purge solenoid was commanded on and off, the PCM did not detect any solenoid surge voltage (system voltage +2v) for 2 seconds. Note: The solenoid resistance is from 25-35 ohms at 68ºF. Possible Causes • EVAP purge solenoid control circuit open or shorted to ground • EVAP purge solenoid power circuit open • EVAP purge solenoid damaged or has failed • PCM has failed
P0443 1T CCM 1995-2003 Car, Jeep, SUV, Truck & Van models	EVAP Purge Solenoid Circuit Fault Conditions: Engine started; system voltage over 10v, EVAP Purge solenoid commanded on and off, and the PCM detected an unexpected voltage condition at the EVAP purge solenoid for 3 seconds. Note: The solenoid resistance at 68ºF is from 25-35 ohms. Possible Causes • EVAP purge solenoid control circuit open or shorted to ground • EVAP purge solenoid power circuit open • EVAP purge solenoid is damaged or it has failed • PCM has failed
P0443 1T CCM 2001-03 Sebring & Stratus Coupe with a 2.4L I4 or 3.0L V6 engine	EVAP Purge Solenoid Circuit Fault Conditions: Engine started; system voltage from 10-16v, and after the EVAP purge solenoid was commanded on and off, the PCM did not detect enough solenoid surge voltage (system voltage +2v) for 200 ms. Note: The solenoid resistance at 68ºF is from 25-35 ohms. Possible Causes • EVAP purge solenoid control circuit open or shorted to ground • EVAP purge solenoid power circuit open • EVAP purge solenoid is damaged or it has failed • PCM has failed
P0446 1T CCM 2001-03 Sebring & Stratus Coupe with a 2.4L I4 or 3.0L V6 engine	EVAP Canister Vent Solenoid Circuit Fault Conditions: Engine started; system voltage from 10-16v, EVAP canister vent solenoid commanded on and off, and the PCM did not detect a solenoid surge voltage (system voltage +2v) for 30 ms. Possible Causes • EVAP vent solenoid circuit open, shorted to ground or power • EVAP canister vent solenoid has failed (solenoid resistance is 25-35 ohms at 68ºF) • PCM has failed
P0450 1T CCM 1998 Eagle Talon models	EVAP Fuel Tank Pressure Sensor Range/Performance Conditions: Engine speed 1600 rpm or higher, volumetric efficiency at 20-80%, Purge solenoid "on", and the PCM detected the EVAP pressure sensor input was over 4.50v, or with the solenoid "off", the sensor signal was 0.5v, or the sensor signal varied 0.2v over 20 times with throttle open. Possible Causes • EVAP pressure sensor signal circuit open or shorted to ground • EVAP pressure sensor is damaged or has failed • EVAP fuel vent valve or fuel vapor line clogged or restricted • PCM has failed

OBD II TROUBLE CODE LIST (P0XXX CODES)

DTC	Trouble Code Title, Conditions & Possible Causes
P0450 1T CCM 2001-03 Sebring & Stratus Coupe with a 2.4L I4 or 3.0L V6 engine	EVAP Fuel Tank Pressure Sensor Range/Performance Conditions: Engine speed 1600 rpm or higher, volumetric efficiency at 20-80%, IAT sensor signal more than 41°F, and with the purge solenoid on (100% duty cycle), the PCM detected the EVAP pressure sensor signal was more than 4.50v, or with the purge solenoid off (0%), the pressure sensor signal was 0.5v, or the sensor signal varied 0.2v at least 20 times with either the throttle open, or at idle with it closed. Possible Causes • EVAP pressure sensor signal circuit open or shorted to ground • EVAP pressure sensor is damaged or has failed • EVAP fuel vent valve or fuel vapor line clogged or restricted • PCM has failed
P0451 1T CCM 2001-03 Sebring & Stratus Coupe with a 2.4L I4 or 3.0L V6 engine	EVAP Fuel Tank Pressure Sensor Performance Conditions: Engine started; engine at idle speed, throttle closed (throttle switch is "on"), and the PCM detected the fuel tank pressure differential sensor value fluctuated over 0.20v at least 20 times, or with engine speed over 2500 rpm at over 9.5 mph with the volumetric efficiency over 55%, the sensor value changed over 0.20v at least 20 times. Possible Causes • Fuel tank pressure differential sensor connector is damaged • Fuel tank pressure differential sensor is damaged or has failed • PCM has failed
P0452 1T CCM 2001-03 Sebring & Stratus Coupe with a 2.4L I4 or 3.0L V6 engine	Fuel Tank Differential Pressure Sensor Circuit Low Input Conditions: Engine started; IAT sensor more than 41°F, engine speed over 1600 rpm, volumetric efficiency from 20-80%, and the PCM detected the Fuel Tank pressure differential sensor indicated less than 0.1v. Possible Causes • Fuel tank pressure differential sensor connector is shorted • Fuel tank pressure differential sensor circuit shorted to ground • Fuel tank pressure differential sensor is damaged or has failed • PCM has failed
P0452 1T CCM 2002-03 300M, Concorde, LHS, Intrepid, Sebring, Neon, PT Cruiser, Caravan, Town & Country, Dakota & Durango models	NVLD Pressure Switch Sense Circuit Low Input Conditions: Engine started; and immediately after the engine is running, the PCM activates the NVLD solenoid to test the NVLD switch circuit. If the switch is not open, the PCM sets this code. Possible Causes • EVAP purge solenoid control circuit is shorted to ground • EVAP purge solenoid is leaking or it is stuck in open position • NVLD assembly or NVLD switch is damaged or it has failed • NVLD switch signal circuit is shorted to ground • PCM has failed
P0453 1T CCM 2001-03 Sebring & Stratus Coupe with a 2.4L I4 or 3.0L V6 engine	Fuel Tank Differential Pressure Sensor Circuit High Input Conditions: Engine speed over 1600 rpm, IAT sensor from 41-113°F, volumetric efficiency from 20-80%, and the PCM detected the Fuel Tank pressure differential sensor indicated less than 4.0v. Possible Causes • Fuel tank pressure differential sensor connector is open • Fuel tank pressure differential sensor circuit is open • Fuel tank pressure differential sensor is damaged or has failed • PCM has failed
P0453 1T CCM 2002-03 300M, Concorde, LHS, Intrepid, Sebring, Neon, PT Cruiser, Caravan, Town & Country, Dakota & Durango models	NVLD Pressure Switch Sense Circuit High Input Conditions: Engine started; and immediately after the engine is running, the PCM activates the NVLD solenoid to test the NVLD switch circuit. If the switch does not close under these conditions, this code is set. Possible Causes • NVLD assembly ground circuit is open • NVLD switch signal circuit is open or shorted to power (B+) • NVLD switch assembly is damaged or it has failed • PCM has failed

OBD II TROUBLE CODE LIST (P0XXX CODES)

DTC	Trouble Code Title, Conditions & Possible Causes
P0455 2T EVAP 1998 Eagle Talon models	EVAP Large Leak (0.80") Detected Conditions: ECT sensor from 40-90ºF and within 10ºF of BTS input at startup, Fuel Level over 50%, engine running, EVAP leak test "on", and the PCM detected a leak greater than 0.080" somewhere in the EVAP system. Possible Causes • EVAP fuel tank or canister vapor hoses leaking or damaged • EVAP system component leaking, fuel cap loose or missing • Leak detection pump damaged or leaking
P0455 2T EVAP 2002-03 300M, Concorde, LHS, Intrepid, Sebring, Neon, PT Cruiser, Caravan, Town & Country, Dakota & Durango models	EVAP Large Leak (0.80") Detected Conditions: Ambient Air Temperature from 39-89ºF at engine startup, engine running under closed loop conditions, Fuel Level over 12%, then with the EVAP purge solenoid enabled (to pull vacuum into the system to close the NVLD switch) and the EVAP "small leak" test maturing, the PCM turns "off" the EVAP purge solenoid once the NVLD switch closes. If the NVLD switch reopens before a calibrated amount of time expires, a "medium" leak in the system is detected. Possible Causes • EVAP purge solenoid is damaged or it has failed • Fuel tank cap is damaged, missing or the wrong part number • NVLD switch is damaged or it has failed
P0455 2T EVAP 2001-03 Sebring & Stratus Coupe with a 2.4L I4 or 3.0L V6 engine	EVAP System Large Leak (0.080") Detected Conditions: ECT and IAT sensor signals less than 86ºF at startup, engine runtime under 12 minutes, BARO sensor over 76 kPa, engine speed over 1600 rpm, volumetric efficiency 20-80%, ECT sensor more than 140ºF, PSP switch indicating off. The test stops when the FTP sensor indicates 451 Pa (0.065 psi) and fluctuates less than 647 Pa (0.094 psi) with a FTP input of 1-4.0v, an ECT input over 140ºF and an IAT input under 41ºF with the purge and vent solenoid closed. The test fails if the PCM detects that the FTP input changes less than 324 Pa (0.047 psi) within 20 seconds. Test takes from 75-125 seconds (depends on fuel level, etc.). Possible Causes • EVAP emission canister seal is damaged or leaking • Fuel tank, vapor line or vacuum seal is damaged or leaking • EVAP canister vent solenoid is damaged or has failed • Fuel tank cap is loose, or the cap release pressure is incorrect
P0455 2T EVAP 1996-2003 Car, Jeep, SUV, Truck & Van models	EVAP Large Leak (0.80") Detected Conditions: ECT sensor from 40-90ºF and within 10ºF of the BTS input at startup (cold startup), fuel level more than 1/2 full, engine started, EVAP leak test enabled, and the PCM detected a leak greater than 0.080" existed somewhere in the EVAP system. Possible Causes • EVAP fuel tank or canister vapor hoses leaking or damaged • EVAP system component leaking, fuel cap loose or missing • Leak detection pump damaged or leaking
P0456 2T EVAP 1998-2003 Car, Jeep, SUV, Truck & Van models	EVAP Small Leak (0.020") Detected Conditions: ECT sensor from 40-90ºF and within 10ºF of BTS sensor at startup, Fuel Level from 15-85%, engine started, EVAP leak test enabled, and the PCM detected a leak greater than 0.020" but less than 0.040" somewhere in the EVAP system. Possible Causes • EVAP fuel tank or canister vapor hoses leaking or damaged • EVAP system component leaking, fuel cap loose or missing • Leak detection pump damaged or leaking
P0456 2T EVAP 2002-03 300M, Concorde, LHS, Intrepid, Sebring, Neon, PT Cruiser, Caravan, Town & Country, Dakota & Durango models	EVAP System Small Leak (0.020") Detected Conditions: Ambient Air Temperature from 39-109ºF at engine startup, engine running under closed loop conditions, Fuel Level below 88%, then with the EVAP system sealed, the PCM monitors the NVLD switch. If the NVLD switch does not close within a calibrated amount of time expires, a "small" leak in the EVAP system was detected. Possible Causes • Fuel tank cap is damaged, loose or the wrong part number • Small leak present somewhere in the EVAP system

OBD II Trouble Code List (P0xxx Codes)

DTC	Trouble Code Title, Conditions & Possible Causes
P0460 1T CCM 1998-2003 Car, Jeep, SUV, Truck & Van models	Fuel Level Sending No Change Over Miles Conditions: Engine started; fuel level less than 15% or more than 85%, and the PCM detected a Low Fuel condition that the fuel level was less than 15% for 120 miles, or that the fuel level remained at more than 85% and did not change by at least 10% after traveling over 100 miles. Possible Causes • Fuel level sending unit signal circuit open or shorted to ground • Fuel level sending unit ground circuit is open • Fuel level sensing unit is damaged or the fuel tank is damaged
P0460 1T CCM 2001-03 Sebring & Stratus Coupe with a 2.4L I4 or 3.0L V6 engine	Fuel Gauge Unit Circuit Malfunction Conditions: Engine started; then after the vehicle was driven enough miles for the fuel calculation from the fuel injector usage to reach 20 liters, the PCM detected the diversity of the amount of fuel in the fuel tank calculated from the fuel level sensor indicated less than 2 liters. Possible Causes • Fuel gauge unit connector is damaged, open or shorted • Fuel gauge unit signal circuit is open or shorted to ground • Fuel gauge unit is damaged or the fuel tank is damaged
P0462 N/MIL 1T CCM 1998-2003 Car, SUV, Truck & Van models	Fuel Level Sensing Unit Low Input Conditions: Key on or engine running; system voltage over 10.5v and the PCM detected the fuel level sensing unit signal indicated less than 0.98v, condition met for 200 seconds. Possible Causes • Fuel level sending unit signal circuit shorted to sensor ground • Fuel level sending unit signal circuit shorted to chassis ground • Fuel level sensing unit is damaged or the fuel tank is damaged • BCM or PCM has failed
P0463 N/MIL 1T CCM 1998-2003 Car, Jeep, SUV, Truck & Van models	Fuel Level Sending Unit High Input Conditions: Key on or engine running; system voltage over 10.5v and the PCM detected the fuel level sensing unit signal indicated more than 4.90v, condition met for 200 seconds. Possible Causes • Fuel level sending unit signal circuit shorted to sensor or chassis ground • Fuel level sensing unit is damaged or the fuel tank is damaged • BCM or PCM has failed
P0480 N/MIL 1T CCM 2001-03 Car models	Low Speed Fan Relay Control Circuit Conditions: Key on or engine running; and the PCM detected an unexpected low or high voltage condition on the Low Speed Fan Relay control circuit. Possible Causes • LFAN relay power circuit is open from the relay to fused power • LFAN relay control circuit is open or shorted to chassis ground • LFAN relay is damaged or has failed • PCM has failed
P0481 N/MIL 1T CCM 2001-03 Car models	High Speed Fan Relay Control Circuit Conditions: Key on or engine running; and the PCM detected an unexpected low or high voltage condition on the High Speed Fan Relay circuit. Possible Causes • HFAN relay power circuit is open from the relay to fused power • HFAN relay control circuit is open or shorted to chassis ground • HFAN relay is damaged or has failed • PCM has failed.
P0498 N/MIL 1T CCM 2001-03 300M, Concorde, LHS, Intrepid, Sebring, Neon, PT Cruiser, Caravan, Town & Country, Dakota & Durango models	NVLD Canister Vent Solenoid Circuit Low Conditions: Key on or engine running; and the PCM detected an unexpected low voltage condition on the Natural Vacuum Leak Detection (NVLD) control circuit during the CCM test period. Possible Causes • NVLD canister vent solenoid control circuit is shorted to ground • NVLD canister vent solenoid is damaged or it has failed • PCM has failed

OBD II Trouble Code List (P0xxx Codes)

DTC	Trouble Code Title, Conditions & Possible Causes
P0499 N/MIL 1T CCM 2001-03 300M, Concorde, LHS, Intrepid, Sebring, Neon, PT Cruiser, Caravan, Town & Country, Dakota & Durango models	**NVLD Canister Vent Solenoid Circuit High Conditions:** Key on or engine running; and the PCM detected an unexpected high voltage condition on the Natural Vacuum Leak Detection circuit. Possible Causes • NVLD canister vent solenoid control circuit is shorted to power • NVLD canister vent solenoid ground circuit is open • NVLD canister vent solenoid is damaged or it has failed • PCM has failed
P0500 1T CCM 1995-98 Eagle Talon models	**Vehicle Speed Sensor Circuit Fault Conditions:** Engine started; engine runtime over 31 seconds, ECT sensor more than 120ºF, transaxle not in Park or Neutral, brakes not applied, engine speed over 1800 rpm, MAP sensor less than 11" Hg, and the PCM did not detect a VSS signal from the TCM for over 11 seconds. Possible Causes • OSS signal circuit to the TCM is open or shorted to ground • OSS is damaged or has failed • VSS circuit between the PCM and TCM is open or shorted • TCM has failed or speedometer pinion factor not programmed
P0500 1T CCM 1995-2003 Car, Jeep, SUV, Truck & Van models	**Vehicle Speed Sensor Circuit Fault Conditions:** Engine started; engine runtime over 31 seconds, ECT sensor more than 180ºF, transaxle not in Park or Neutral, brakes not applied, engine speed over 1800 rpm, MAP sensor less than 11" Hg, and the PCM did not detect a VSS signal from the TCM for over 11 seconds. Possible Causes • VSS or OSS signal circuit to the PCM is open or shorted to ground • VSS or OSS power circuit (8v) is open or shorted to ground • VSS or OSS ground circuit to the PCM is open • Speedometer pinion or the VSS is damaged or has failed
P0500 2T CCM 1998-2002 Trucks equipped with a 5.9L Diesel engine	**Vehicle Speed Sensor Circuit Fault Conditions:** Engine started; engine speed from 1024-2784 rpm, engine boost more than 7 psi, and the PCM did not detect a VSS signal from the CAB module for over 15 seconds during testing. Possible Causes • CAB circuit from the CAB to PCM is open or it is shorted • OSS signal circuit to the TCM is open or shorted to ground • OSS is damaged or has failed • TCM or PCM has failed
P0501 2T CCM 2001-03 300M, Concorde, LHS, Intrepid, Sebring, Neon, PT Cruiser, Caravan, Town & Country, Dakota & Durango models	**Vehicle Speed Sensor Performance Conditions:** Engine started; vehicle driven at over 1500 rpm for 10 seconds, gear selector not in Park or Neutral, brakes not applied, and the PCM did not receive any VSS signals the ABS/RWAL Module or the TCM. Possible Causes • Check for any ABS/RWAL or TCM codes related to the VSS • VSS connector is damaged, open or it is shorted • VSS signal is open, shorted to ground or shorted to power • ABS/RWAL controller, TCM or the PCM has failed
P0501 1T CCM 2001-03 Trucks equipped with a 5.9L Diesel engine	**Vehicle Speed Sensor Signal Range/Performance Conditions:** Engine started; vehicle driven to a speed over 20 mph for 2 seconds (as indicated by the CCD bus signal to the PCM), ECM vehicle speed signal less than 10 mph, and the PCM did not receive a CAB vehicle speed signal during the CCM Rationality test. Possible Causes • Check for a CAB controller code related to the VSS signal • CAB (ABS) controller has failed • VSS signal is open, shorted to ground or shorted to power • ECM or PCM has failed

OBD II Trouble Code List (P0xxx Codes)

DTC	Trouble Code Title, Conditions & Possible Causes
P0505 2T IAC 1995-98 Eagle Talon models	Idle Air Control Motor System Fault Conditions: IAT sensor less than 131°F during last key cycle, engine started, system voltage over 10v, engine running at hot idle speed, ECT sensor over 176°F, IAT sensor over 14°F, Short Term fuel trim from -8% to +8%, BARO sensor over 76 kPa, and the PCM detected the Actual idle speed was more than 200 rpm above or 100 rpm below the Target idle speed for 10 seconds. Possible Causes • Stepper motor Coil 1, 2, 3 or 4 circuit open or shorted to ground • Stepper motor coil circuit(s) shorted to system power (B+) • Stepper motor is damaged or has failed • PCM has failed
P0505 2T IAC 1995-2003 Car, Jeep, SUV, Truck & Van models	Idle Air Control Motor Circuit Conditions: Engine started; system voltage over 11.5v and the PCM detected an unexpected voltage condition on one or more of the IAC motor circuits for 2.75 seconds with the motor active. Possible Causes • Stepper motor Coil 1, 2, 3 or 4 circuit open or shorted to ground • Stepper motor coil circuit(s) shorted to system power (B+) • Stepper motor is damaged or has failed • PCM has failed
P0506 1T CCM 2002-03 Car, Jeep, SUV, Truck & Van models	Idle Speed Low Performance Conditions: Engine running at idle speed in closed loop, and the PCM detected the Actual idle speed was not within the expected low idle speed limit when compared to the Target idle speed limit. Possible Causes • Air induction system restrictions (clogged air filter, etc.) • Idle air control passage is clogged or dirty (clean and retest) • Throttle body or linkage is binding, damaged or sticking
P0506 1T CCM 2001-03 Sebring & Stratus Coupe with a 2.4L I4 or 3.0L V6 engine	Idle Control System RPM Lower Than Expected Conditions: Engine started; system voltage over 10v, ECT sensor over 171°F, volumetric efficiency less than 40%, BARO sensor over 76 kPa, IAT sensor over 14°F, PSPS is off, 25 seconds since last test, Target IAC position over 100 steps, and the PCM detected the Actual idle speed was 100 rpm less than the Target idle speed for 12 seconds. Possible Causes • Stepper motor Coil A1, A2, B1 and B2 circuit is open • Stepper motor Coil A1, A2, B1 and B2 circuit shorted to ground • Stepper motor Coil A1, A2, B1 and B2 circuit shorted to B+ • Stepper motor is damaged or it has failed • PCM has failed
P0507 1T CCM 2002-03 Car, Jeep, SUV, Truck & Van models	Idle Speed High Performance Conditions: No CKP, ECT, MAF, MAP or VSS codes set, engine started, engine running at idle speed, BTS sensor from 0-19.4°F, ECT sensor from 158-266°F, MAF sensor under 250 mg, and the PCM detected the Actual speed was 200 rpm above the Target speed for 7 seconds. Possible Causes • Idle air control passage is clogged or dirty (clean and retest) • Throttle body or linkage is binding, damaged or sticking • Vacuum leaks in the engine or PCV system components
P0507 1T CCM 2001-03 Sebring & Stratus Coupe with a 2.4L I4 or 3.0L V6 engine	Idle Control System RPM Higher Than Expected Conditions: Engine running in closed loop, ECT sensor over 171°F, system voltage over 10v, volumetric efficiency less than 40%, BARO sensor over 76 kPa, IAT sensor over 14°F, PSPS signal "off", 25 seconds have elapsed since last test, Target IAC position at (0) steps, and the PCM detected the Actual idle speed was 200 rpm more than the Target idle speed for 12 seconds. Possible Causes • Stepper motor Coil A1 or A2 circuit is open or shorted to ground • Stepper motor Coil B1 or B2 circuit is open or shorted to ground • Stepper motor coil circuit(s) is shorted to system power (B+) • Stepper motor power circuit is open (test power from MFI relay) • Stepper motor is damaged or it has failed • PCM has failed

OBD II Trouble Code List (P0xxx Codes)

DTC	Trouble Code Title, Conditions & Possible Causes
P0508 2T CCM 2001-03 300M, Concorde, LHS, Intrepid, Sebring, Neon, PT Cruiser, Caravan, Town & Country, Dakota & Durango models	Idle Air Control Motor Sense Circuit Low Input Conditions: Engine started; system voltage over 10.5v and the PCM detected the IAC Motor Sense circuit current was less than 175 mA during the CCM test period. Possible Causes • IAC motor driver circuit is open or shorted to ground • IAC motor sense circuit is open or shorted to ground • IAC motor is damaged or it has failed • PCM has failed
P0509 2T CCM 2001-03 300M, Concorde, LHS, Intrepid, Sebring, Neon, PT Cruiser, Caravan, Town & Country, Dakota & Durango models	Idle Air Control Motor Circuit High Conditions: Engine started; system voltage over 10.5v, IAC motor activated, and the PCM detected a high voltage on one or more of the IAC motor circuits during the CCM test period. Possible Causes • IAC motor driver circuit is shorted to power • IAC motor sense circuit is shorted to power • IAC motor is damaged or it has failed • PCM has failed
P0513 1T PCM 2001-03 300M, Concorde, LHS, Intrepid, Sebring, Neon, PT Cruiser, Caravan, Town & Country, Dakota & Durango models	Invalid SKIM Key Detected Conditions: Key on, and the PCM detected an invalid Sentry Key Immobilizer key had been inserted into the ignition key assembly. Possible Causes • Incorrect VIN in the PCM • No communication between the PCM and the SKIM • SKIM trouble codes present (check for any SKIM codes) • Valid SKIM key not present • VIN not programmed into the PCM • PCM has failed
P0516 1T CCM 2003 Neon, PT Cruiser, Caravan, Town & Country, Dakota & Durango models	Battery Temperature Sensor Circuit Low Input Conditions: Key on or engine running; and the PCM detected a Battery Temperature sensor signal that indicated less than 0.10v. Possible Causes • BTS signal circuit is shorted to sensor or chassis ground • BTS assembly is damaged or it has failed • PCM has failed
P0517 1T CCM 2003 Neon, PT Cruiser, Caravan, Town & Country, Dakota & Durango models	Battery Temperature Sensor Circuit High Input Conditions: Key on or engine running; and the PCM detected a Battery Temperature sensor signal that indicated more than 4.90v. Possible Causes • BTS signal circuit is shorted to VREF (5v) • BTS signal circuit is open or the BTS ground circuit is open • BTS assembly is damaged or it has failed • PCM has failed
P0519 N/MIL 2T IAC 2002-03 300M, Concorde, LHS, Intrepid, Neon, PT Cruiser, Caravan, Town & Country, Dakota & Durango models	Idle Air Performance Conditions: DTC P0106, P0107, P0108, P0121, P0122 and P0123 not set, engine started, engine running with the gear selector indicating Drive position, and the PCM detected the engine idle speed was not within 200 rpm of the high idle limit or within 100 rpm of the low idle limit when compared to the Target Idle Speed limit for 40 seconds. Possible Causes • Idle air control passage is clogged or dirty (clean and retest) • Throttle body or linkage is binding, damaged or sticking • Vacuum leaks in the engine or PCV system components

OBD II TROUBLE CODE LIST (P0XXX CODES)

DTC	Trouble Code Title, Conditions & Possible Causes
P0522 N/MIL 1T CCM 1999-2003 300M, Concorde, LHS, Intrepid, Jeep, Neon, PT Cruiser, Caravan, Town & Country, Voyager, Dakota & Durango models	Engine Oil Pressure Sensor Rationality Conditions: Key on (engine not started), engine speed (rpm) indicating 0 rpm, and the PCM detected an engine oil pressure reading. Possible Causes • Oil pressure sensor signal circuit is open • Oil pressure sensor signal circuit is shorted to ground • Oil pressure sensor is damaged or has failed • PCM has failed
P0522 N/MIL 1T CCM 2002-03 Trucks equipped with a 5.9L Diesel engine	Engine Oil Pressure Sensor Circuit Low Input Conditions: Key on or engine running; system voltage more than 10.4v, and the PCM detected the Oil Pressure sensor signal was less than 0.10v. Possible Causes • Check for multiple trouble codes set related to this circuit • Oil pressure sensor signal circuit is open or shorted to ground • Oil pressure sensor VREF circuit is open • Oil pressure sensor is damaged or has failed • PCM has failed
P0523 N/MIL 1T CCM 1999-2003 Jeep, Neon, PT Cruiser, Caravan, Town & Country, Voyager, Dakota & Durango models	Engine Oil Pressure Sensor Rationality Conditions: Key on (engine not started), engine speed (rpm) indicating 0 rpm, and the PCM detected an engine oil pressure reading. Possible Causes • Oil pressure sensor signal circuit is open • Oil pressure sensor signal circuit is shorted to ground • Oil pressure sensor is damaged or has failed • PCM has failed
P0523 N/MIL 1T CCM 1998-2003 R Body Truck: All	Engine Oil Pressure Sensor Circuit High Input Conditions: Key on or engine running; system voltage more than 10.4v, and the PCM detected the Oil Pressure sensor signal indicated more than 4.90v during the CCM test. Possible Causes • Oil pressure sensor signal circuit is shorted to VREF • Oil pressure sensor ground circuit is open • Oil pressure sensor is damaged or has failed • PCM has failed
P0524 N/MIL 1T CCM 1998-2003 Trucks equipped with a 5.9L Diesel engine	Engine Oil Pressure Sensor Signal Too Low Conditions: Key on or engine running; system voltage more than 10.4v, and the PCM detected a signal from the Oil Pressure sensor that indicated the engine oil pressure was too low. Possible Causes • Engine oil level is very low • Oil pressure sensor is damaged or has failed • PCM has failed
P0532 1T CCM 2002-03 300M, Concorde, LHS, Intrepid, Jeep, Neon, PT Cruiser, Caravan, Town & Country, Voyager, Dakota & Durango models	Air Conditioning Pressure Sensor Circuit Low Input Conditions: Engine running with the A/C relay energized, and the PCM detected the signal from the A/C Pressure sensor indicated less than 0.58v for over 2.6 seconds during the CCM test period. Possible Causes • A/C pressure sensor power supply (VREF) circuit is open • A/C pressure sensor power (VREF) circuit shorted to ground • A/C pressure sensor signal circuit is shorted to ground • A/C pressure sensor is damaged or it has failed • PCM has failed

OBD II TROUBLE CODE LIST (P0XXX CODES)

DTC	Trouble Code Title, Conditions & Possible Causes
P0533 N/MIL 1T CCM 2002-03 300M, Concorde, LHS, Intrepid, Jeep, Neon, PT Cruiser, Caravan, Town & Country, Voyager, Dakota & Durango models	Air Conditioning Pressure Sensor Circuit High Input Conditions: Engine running with the A/C relay energized, and the PCM detected the signal from the A/C Pressure sensor indicated less than 4.92v for over 2.6 seconds during the CCM test period. Possible Causes • A/C pressure sensor signal circuit is shorted to VREF power • A/C pressure sensor signal circuit is open • A/C pressure sensor ground circuit is open • A/C pressure sensor is damaged or it has failed • PCM has failed
P0545 1T CCM 1996-2002 Trucks equipped with a V8 engine	Air Conditioning Clutch Circuit Failure Conditions: Key on or engine running; and the PCM detected an unexpected voltage condition on the A/C clutch control circuit in the CCM test. Possible Causes • A/C clutch control circuit is open between the clutch and PCM • A/C clutch control circuit is shorted to ground • A/C clutch power circuit is open • A/C clutch is damaged or has failed • PCM has failed
P0551 1T CCM 1995-2003 Car, Jeep, SUV, Truck & Van models	Power Steering Pressure Switch Circuit Failure Conditions: Vehicle driven at more than 56 mph for more than 30 seconds and the PCM detected a high voltage input on the PSPS circuit (pressure exceeds 500 psi). This is a normally open switch. Possible Causes • PSPS sense circuit is shorted to ground • PSPS is damaged or has failed • PCM has failed
P0551 1T CCM 2001-03 Sebring & Stratus Coupe with a 2.4L I4 or 3.0L V6 engine	Power Steering Pressure Switch Circuit Failure Conditions: Engine started; vehicle driven at more than 31 mph for 4 seconds, BARO sensor over 75 kPa, ECT sensor more than 86°F, followed by a deceleration period to a stop with VSS at 0.93 mph, and the PCM detected the PSPS switch signal indicated "on" (test fails 10 times). Possible Causes • PSPS sense circuit is open between the switch and the PCM • PSPS ground circuit is open between the switch and ground • PSPS is damaged or it has failed • PCM has failed
P0562 1T CCM 1996-2003 Cars, Dakota, Durango, Truck & Van models	Battery Sense Circuit Low Input Conditions: Engine running at a speed over 380 rpm, and the PCM detected the Battery Sense circuit voltage indicated one volt less than the Charging system "goal" (11.50v) for 13.47 seconds. Possible Causes • Battery sense circuit has a high resistance condition • Generator ground circuit has a high resistance condition • Generator field ground circuit is open • Generator field control circuit is open or shorted to ground • Generator is damaged or it has failed • PCM has failed
P0562 N/MIL 1T CCM 1998-2003 Trucks equipped with a 5.9L Diesel engine	Charging System Voltage Low Input Conditions: Engine started; engine speed over 1000 rpm during testing, and the PCM detected the system voltage was less than 6.0v at any time. Possible Causes • Battery connections corroded (high resistance) or loose • Ignition system voltage circuit is open at the PCM terminals • Generator is damaged or has failed (output is too low) • PCM has failed

OBD II TROUBLE CODE LIST (P0XXX CODES)

DTC	Trouble Code Title, Conditions & Possible Causes
P0563 1T CCM 1996-2003 Cars, Dakota, Durango, Truck & Van models	Battery Sense Circuit High Input Conditions: Engine running at a speed over 380 rpm, and the PCM detected the Battery Sense circuit voltage indicated one volt higher than the Charging system "goal" during the CCM test. Possible Causes • Generator field control circuit is shorted to system power (B+) • Generator is damaged or it has failed • PCM has failed
P0563 N/MIL 1T CCM 1998-2003 Trucks equipped with a 5.9L Diesel engine	Charging System Voltage High Input Conditions: Engine started; engine speed over 1000 rpm during the test period, and the PCM detected the system voltage was more than 17.0v at any time during the CCM test. Possible Causes • Generator is damaged or has failed (output is too high) • PCM has failed
P0572 N/MIL 1T CCM 1998-2003 Trucks equipped with a 5.9L Diesel engine	Brake Switch Signal No. 1 Circuit Failure Conditions: Vehicle driven at over 55 mph for 1 minute, then returned to idle speed, test performed at least 10 times, and the PCM did not detect any change in the Brake Switch No. 1 signal. Possible Causes • Brake switch signal circuit to open or shorted to ground • Brake switch power circuit is open • Brake switch is damaged or has failed • PCM has failed
P0573 N/MIL 1T CCM 1998-2003 Trucks equipped with a 5.9L Diesel engine	Brake Switch Signal No. 2 Circuit Malfunction Conditions: Engine started; engine running and the PCM did not detect any Brake Switch No. 2 signals on the CCD bus circuit during the test. Possible Causes • Brake switch signal to the PCM is missing (circuit problem) • CCD bus circuit is open or shorted to ground • CCD bus circuit is shorted to system power • PCM has failed
P0575 N/MIL 1T CCM 2001-03 Trucks equipped with a 5.9L Diesel engine	Cruise Control Switch Signal Low Input Conditions: Engine started; engine running and the PCM detected a continuous "low" voltage condition on the Cruise Control switch circuit. Possible Causes • C/C switch signal shorted to ground between switch and PCM • C/C clockspring is shorted to ground • C/C switch is damaged or has failed • PCM has failed
P0576 N/MIL 1T CCM 2001-03 Trucks equipped with a 5.9L Diesel engine	Cruise Control Switch Signal High Input Conditions: Engine started; engine running and the PCM detected a continuous "high" voltage condition on the Cruise Control switch circuit. Possible Causes • C/C switch circuit is open or shorted to VREF or system power • C/C switch or clockspring ground circuit is open • C/C switch is damaged or has failed • PCM has failed
P0577 N/MIL 1T CCM 2001-03 Trucks equipped with a 5.9L Diesel engine	Cruise Control Switch Signal High Input Conditions: Engine started; engine running and the PCM detected a continuous "high" voltage condition on the Cruise Control switch circuit. Possible Causes • C/C switch circuit is open or shorted to VREF or system power • C/C switch or clockspring ground circuit is open • C/C switch is damaged or has failed • PCM has failed

OBD II TROUBLE CODE LIST (P0XXX CODES)

DTC	Trouble Code Title, Conditions & Possible Causes
P0579 1T CCM 2002-03 Cars, Dakota, Durango, Truck & Van models	Speed Control Switch No. 1 Performance Conditions: Engine started; and the PCM detected a continuous invalid voltage condition on the Speed Control Switch No. 1 signal. Possible Causes • S/C switch signal circuit is shorted to chassis or sensor ground • S/C switch signal circuit is open or shorted to system power • S/C switch is damaged or it has failed • Sensor ground circuit is open • PCM has failed
P0580 1T CCM 2002-03 Cars, Dakota, Durango, Truck & Van models	Speed Control Switch No. 1 Circuit Low Input Conditions: Engine started; system voltage over 10.5v and the PCM detected the Speed Control Switch No. 1 signal indicated less than 0.43v. Possible Causes • S/C switch signal circuit is shorted to chassis or sensor ground • S/C On/Off switch is damaged or it has failed • S/C Resume/Accelerator switch is damaged or it has failed • PCM has failed.
P0581 1T CCM 2002-03 Cars, Dakota, Durango, Truck & Van models	Speed Control Switch No. 1 Circuit High Input Conditions: Engine started; system voltage over 10.5v and the PCM detected the Speed Control Switch No. 1 signal indicated less than 0.43v. Possible Causes • S/C switch No.1 signal circuit is shorted to system power (B+) • S/C switch ground circuit is open • S/C switch (one or more) is damaged or has failed • PCM has failed
P0582 1T CCM 2002-03 Cars, Dakota, Durango, Truck & Van models	Speed Control Vacuum Solenoid Circuit Malfunction Conditions: Engine started; Speed Control (S/C) system activated, and the PCM detected an unexpected voltage condition on the S/C Vacuum solenoid circuit during the CCM test period. Possible Causes • S/C vacuum solenoid control circuit is open • S/C vacuum solenoid control circuit is shorted to ground • S/C vacuum solenoid is damaged or has failed • PCM has failed
P0586 1T CCM 2002-03 Cars, Dakota, Durango, Truck & Van models	Speed Control Vent Solenoid Circuit Malfunction Conditions: Engine started; Speed Control (S/C) system activated, and the PCM detected an unexpected voltage condition on the S/C Vent solenoid circuit during the CCM test period. Possible Causes • S/C vent solenoid control circuit is open • S/C vent solenoid control circuit is shorted to ground • S/C vent solenoid is damaged or has failed • PCM has failed
P0594 N/MIL 1T CCM 2002-03 Cars, Dakota, Durango, Truck & Van models	Speed Control Servo Power Circuit Malfunction Conditions: Engine started; Speed Control (S/C) system activated, and the PCM detected an unexpected voltage condition on the S/C Vent solenoid circuit during the CCM test period. Possible Causes • Brake switch is damaged or it has failed • S/C brake switch circuit is open or it is shorted to ground • S/C power circuit is open or it is shorted to ground • PCM has failed

OBD II TROUBLE CODE LIST (P0XXX CODES)

DTC	Trouble Code Title, Conditions & Possible Causes
P0600 N/MIL 1T PCM 1995-2003 Car, Jeep, SUV, Truck & Van models	PCM Internal Failure, No SPI Communications Conditions: Key on, system voltage over 10.5v and the PCM detected the initial serial data communications attempt failed 8 times in succession. Possible Causes • Turn the key off. Remove the PCM connector(s) and inspect the wiring harness connector pins and PCM pins for damaged, bent or missing pins. If any problems are located, make the repair and then perform the PCM Reset function. If this code resets after the repairs are made, the PCM has failed. • PCM needs to be replaced, and then reprogrammed
P0601 N/MIL 1T PCM 1995-2003 Car, Jeep, SUV, Truck & Van models	PCM Random Access Memory Failure, Self-Test Failed Conditions: Key on, system voltage over 10.5v and the PCM detected the Random Access Memory (RAM) test failed in the initial Self-Test. Possible Causes • PCM needs to be replaced, and then reprogrammed
P0602 N/MIL 1T PCM 2001-03 Trucks equipped with a 5.9L Diesel engine	PCM Fueling Calibration Conditions: Key on or engine starting, and the PCM detected one or more parameters was out-of-range for over one (1) second during testing. Possible Causes • PCM needs to be replaced, and then reprogrammed
P0604 N/MIL 1T PCM 1999-2003 Truck & Van models	TCM Random Access Memory Failure, Self-Test Failed Conditions: Key on, system voltage over 10.5v and the TCM detected the Random Access Memory (RAM) test failed during the initial Self-Test. Possible Causes • TCM needs to be replaced, and then reprogrammed
P0605 N/MIL 2T PCM 1995-98 Eagle Talon models	PCM Fault, SPI Communications Conditions: Key on or engine running; and the PCM detected that the serial communications inside the controller failed 8 times. Possible Causes • PCM needs to be replaced, and then reprogrammed
P0605 N/MIL 1T PCM 1999-2003 Truck & Van models	TCM Random Access Memory Failure, Self-Test Failed Conditions: Key on, system voltage over 10.5v and the TCM detected the Read Only Memory (ROM) test failed during the initial Self-Test. Possible Causes • TCM needs to be replaced, and then reprogrammed
P0606 1T PCM 2002-03 Trucks equipped with a 5.9L Diesel engine	Powertrain Control Module Malfunction Conditions: Key on or engine running; and the PCM detected an internal malfunction had occurred during the initialization step. Possible Causes • PCM needs to be replaced, and then reprogrammed
P0613 N/MIL 1T PCM 2002-03 Truck & Van models	TCM Internal Failure Conditions: Key on, system voltage over 10.5v and the TCM detected an internal failure (possible open condition in the TCM ground circuit). Possible Causes • Turn the key off. Remove the TCM connector(s) and inspect the wiring harness connector pins and TCM pins for damaged, bent or missing pins. Make repairs as needed. Then perform the TCM Reset function. If the code resets, the PCM has failed.
P0622 (GEN) 1T CCM 1999-2003 300M, Concorde, LHS, Intrepid, Jeep, Neon, PT Cruiser, Caravan, Town & Country, Voyager, Dakota & Durango models	Generator Field Control Circuit Malfunction Conditions: Engine started and the PCM detected the Generator Field control circuit had malfunctioned. Possible Causes • Generator field control circuit is open or is shorted to ground • Generator field control circuit is shorted to system power (B+) • Generator field ground circuit is open • Generator is damaged or the PCM has failed

OBD II TROUBLE CODE LIST (P0XXX CODES)

DTC	Trouble Code Title, Conditions & Possible Causes
P0627 1T CCM 2002-03 300M, Concorde, LHS, Intrepid, Jeep, Neon, PT Cruiser, Caravan, Town & Country, & Voyager models	Fuel Pump Relay Control Circuit Malfunction Conditions: Engine started; system voltage over 10.5v and the PCM detected an unexpected voltage condition on the Fuel Pump relay control circuit during the CCM test period. Possible Causes • Fuel pump relay control circuit is open or is shorted to ground • Fuel pump relay control circuit is shorted to system power (B+) • Fuel pump relay power circuit (fused ignition) circuit is open • Fuel pump relay is damaged or it has failed • PCM has failed
P0628 1T CCM 2002-03 300M, Concorde, LHS, & Intrepid models	Fuel Pump Relay Control Circuit Malfunction Conditions: Engine started; system voltage over 10.5v and the PCM detected an unexpected voltage condition on the Fuel Pump relay control circuit during the CCM test period. Possible Causes • Fuel pump relay circuit is open, shorted to ground or to power • Fuel pump relay power circuit (fused ignition) circuit is open • Fuel pump relay is damaged or it has failed • PCM has failed
P0630 1T CCM 2002-03 300M, Concorde, LHS, Intrepid, Jeep, Neon, PT Cruiser, Caravan, Town & Country, Voyager, Dakota & Durango models	VIN Not Programmed Into The PCM Conditions: Key on, and the PCM determined that the vehicle identification number (VIN) had not been programmed into its memory. Possible Causes • Reprogram the correct VIN into the PCM (use a Scan Tool) and then retest for this trouble code. If the same code resets, the PCM may be damaged and need replacement.
P0632 1T CCM 2002-03 300M, Concorde, LHS, Intrepid, Jeep, Neon, PT Cruiser, Caravan, Town & Country, Voyager, Dakota & Durango models	Mileage Not Programmed Into The PCM Conditions: Key on, and the PCM detected the vehicle mileage had not been programmed into memory. Possible Causes • Reprogram the correct mileage into the PCM (use a Scan Tool) and then retest for this trouble code. If the same code resets, the PCM may be damaged and need replacement.
P0633 1T CCM 2002-03 300M, Concorde, LHS, Intrepid, Jeep, Neon, PT Cruiser, Caravan, Town & Country, Voyager, Dakota & Durango models	SKIM Key Not Programmed Into The PCM Conditions: Key on, and the PCM determined that the Security Key Immobilizer (SKIM) information had not been programmed into its memory. Possible Causes • Reprogram the correct mileage into the PCM (use a Scan Tool) and then retest for this trouble code. If the same code resets, the PCM may be damaged and need replacement.
P0645 N/MIL 1T CCM 2002-03 300M, Concorde, LHS, Intrepid, Jeep, Neon, PT Cruiser, Caravan, Town & Country, Voyager, Dakota, Jeep, Viper, Durango, Truck & Van models	A/C Clutch Relay Circuit Malfunction Conditions: Engine started; system voltage over 10.0v, A/C switch "on", and the PCM detected an unexpected voltage condition on the A/C Clutch relay control circuit during the CCM test. Possible Causes • A/C relay clutch control circuit is open or it is shorted to ground • A/C relay clutch power supply (fused ignition) circuit is open • A/C relay is damaged or it has failed PCM has failed

OBD II TROUBLE CODE LIST (P0XXX CODES)

DTC	Trouble Code Title, Conditions & Possible Causes
P0660 1T CCM 2002-03 300M, Concorde, LHS, Intrepid, Neon models	Manifold Tune Valve Solenoid Circuit Malfunction Conditions: Engine started; ASD relay "on", system voltage over 10.0v, and the PCM detected an unexpected voltage condition on the Manifold Tune Valve solenoid control circuit. Possible Causes • Manifold tune solenoid circuit is open or it is shorted to ground • Manifold tune solenoid circuit is shorted to power • Manifold tune solenoid ground circuit is open • Manifold tune valve solenoid is damaged or it has failed • PCM has failed
P0685 1T CCM 2002-03 300M, Concorde, LHS, Intrepid, Jeep, Neon, PT Cruiser, Caravan, Town & Country, Voyager, Dakota & Durango models	ASD Relay Control Circuit Malfunction Conditions: Key on, system voltage over 10.0v, and the PCM detected an unexpected voltage condition on the Automatic Shutdown (ASD) relay control circuit during the CCM test period. Possible Causes • ASD relay connector is damaged, loose or shorted • ASD relay control circuit is open or it is shorted to ground • ASD power supply (fused B+) circuit is open • ASD relay is damaged or it has failed • PCM has failed.
P0688 1T CCM 2002-03 300M, Concorde, LHS, Intrepid, Jeep, Neon, PT Cruiser, Caravan, Town & Country, Voyager, Dakota & Durango models	ASD Relay Sense Circuit Low Input Conditions: Key on, ASD relay energized, system voltage over 10.0v, and the PCM did not detect any voltage on the Automatic Shutdown (ASD) Sense circuit during the CCM test period. Possible Causes • ASD relay output circuit is open • ASD power supply (fused B+) circuit is open • ASD relay is damaged or it has failed • PCM has failed
P0700 2T TCM 1995-98 Eagle Talon models	Transaxle Control System Fault Conditions: Engine started and the PCM received a signal from the TCM that it had detected a fault. Possible Causes • The presence of this code means the TCM detected a problem • TCM related sensor has solenoid is damaged or has failed • This code is for information only - check for other TCM codes • TCM or PCM has failed
P0700 2T TCM 1995-2003 Car, Jeep, SUV and Mini-Van models	Automatic Transmission System Malfunction Detected Conditions: Engine started; and the PCM received a message over the CCD bus from the Transmission Control Module (TCM) that it had detected a problem and set a trouble code in memory. Possible Causes • The presence of this code means the TCM detected a problem • TCM related sensor has solenoid is damaged or has failed • This code is for information only - check for other TCM codes • TCM or PCM has failed
P0700 2T TCM 1998-2003 Truck & Van models	Automatic Transmission System Malfunction Detected Conditions: Engine started; and the PCM received a message over the CCD bus from the Transmission Control Module (TCM) that it had detected a problem and set a trouble code in memory. Possible Causes • The presence of this code means the TCM detected a problem • TCM related sensor has solenoid is damaged or has failed • This code is for information only - check for other TCM codes • TCM or PCM has failed

OBD II TROUBLE CODE LIST (P0XXX CODES)

DTC	Trouble Code Title, Conditions & Possible Causes
P0703 1T CCM 1996-2003 300M, Concorde, LHS, Intrepid, Jeep, Neon, PT Cruiser, Caravan, Town & Country, Voyager, Dakota & Durango models	A/T Brake Switch Sense Circuit Malfunction Conditions: Engine started; vehicle driven to over 20 mph with the TP sensor over 0.02v for 6 seconds, followed by a deceleration period to 0 mph at least 16 times, and the PCM detected the Brake switch signal status was not correct during the acceleration/deceleration periods. Possible Causes • Brake switch signal circuit or the switch ground circuit is open • Brake switch is damaged, out of adjustment or has failed • PCM has failed
P0705 1T CCM 1995-98 Eagle Talon models	A/T Transmission Range Switch Circuit Malfunction Conditions: Key on or engine running; and the PCM detected an invalid PRNDL switch signal occurred (i.e., an invalid PRNDL switch signal occurred three times for 100 ms) during the CCM test. Possible Causes • Manual Lever (Rooster Comb) is worn out (check the contacts) • TR switch signal circuit is open, shorted to ground or to power • TR switch is damaged or has failed • TCM or the PCM has failed
P0705 1T CCM 1996-2003 300M, Concorde, LHS, Intrepid, Jeep, Neon, PT Cruiser, Caravan, Town & Country, Van, Voyager, Dakota, Jeep, Viper, Durango, Truck	A/T Check Shifter Signal Circuit Malfunction Conditions: Key on or engine running; and the PCM detected at least three occurrences of an invalid TR sensor PRNDL signal for over 100 ms. Possible Causes • TR1, T3, T41 or T42 sense circuit is open • TR1, T3, T41 or T42 sense circuit is shorted to ground • TR1, T3, T41 or T42 sensor circuit is shorted to system power • Transmission Range sensor is damaged or has failed • PCM has failed
P0710 1T CCM 2001-03 Sebring & Stratus Coupe with a 2.4L I4 or 3.0L V6 engine	TCM Transmission Fluid Temperature Sensor Circuit Malfunction Conditions: Key on or engine running; and the PCM detected an unexpected low or high voltage condition on the TFT sensor circuit in the CCM test. Possible Causes • TFT sensor signal circuit is open between the sensor and PCM • TFT sensor signal circuit is shorted to ground • TFT sensor is damaged or has failed (open or shorted) • PCM has failed
P0711 1T CCM 1996-2003 Dakota, Durango Jeep, SUV, Truck and Van models with an A/T	A/T Transmission Fluid Temperature Sensor Signal - No Rise After Startup Conditions: Engine started; engine runtime over 20 minutes, and the PCM detected the TFT sensor signal did not increase at least 16°F; or the PCM detected the TFT sensor indicated more than 260°F with the ECT sensor signal indicating less than 100°F during the CCM test. Possible Causes • TFT sensor signal has an intermittent high resistance condition • TFT sensor is damaged, skewed or has drifted out of range • PCM has failed
P0712 1T CCM 1996-2003 Dakota, Durango Jeep, SUV, Truck and Van models with an A/T	A/T Transmission Fluid Temperature Sensor Low Input Conditions: Engine started and the PCM detected the TFT sensor signal was under 1.55v for 2.2 seconds. Possible Causes • TFT sensor signal circuit is shorted to ground • TFT sensor is damaged or has failed (it may be shorted) • PCM has failed
P0713 1T CCM 1996-2003 Dakota, Durango Jeep, SUV, Truck and Van models with an A/T	A/T Transmission Fluid Temperature Sensor High Input Conditions: Engine started and the PCM detected the TFT sensor signal was over 3.76v for 2.2 seconds. Possible Causes • TFT sensor signal circuit is open between the sensor and PCM • TFT sensor ground circuit is open between sensor and PCM • TFT sensor is damaged or has failed (it may be open internally) • PCM has failed

OBD II TROUBLE CODE LIST (P0XXX CODES)

DTC	Trouble Code Title, Conditions & Possible Causes
P0715 1T CCM 1996-2003 Car, Jeep, SUV, Truck & Van models (A/T)	**A/T Input Speed Sensor Circuit Malfunction Conditions:** Vehicle driven in 1st, 2nd or 3rd gear with Output Speed Sensor (OSS) signals present, and the PCM did not detect any ISS signals less for 1 second with the vehicle in 2nd gear. **Possible Causes** • ISS signal (+) circuit is open, shorted to ground or to power • ISS signal (-) circuit is open, shorted to ground or to power • ISS is damaged or has failed (it may be open internally) • Overdrive clutch drum lugs are damaged or missing • PCM has failed
P0715 1T CCM 2001-03 Sebring & Stratus Coupe with a 2.4L I4 or 3.0L V6 engine	**TCM Input Shaft Speed Sensor Circuit Malfunction Conditions:** Engine started; and the PCM detected a signal from the TCM indicating it had detected an unexpected voltage condition on the Input Speed Sensor (ISS) signal circuit during the test. **Possible Causes** • ISS positive (+) circuit is open, shorted to ground or to power • ISS positive (-) circuit is open, shorted to ground or to power • ISS sensor is damaged or it has failed
P0720 1T CCM 1996-2003 Car, SUV, Truck & Van models (A/T)	**A/T Output Speed Sensor - Low Output Above 15 MPH Conditions:** Engine started; vehicle driven to a speed over 15 mph, and the PCM detected the OSS signal was less than 60 RPM for 2.6 seconds. **Possible Causes** • OSS signal (+) circuit is open, shorted to ground or to power • OSS signal (-) circuit is open, shorted to ground or to power • OSS is damaged or has failed (it may be open internally) • Parking pawl lugs damaged or missing • PCM has failed
P0720 1T CCM 2001-03 Sebring & Stratus Coupe with a 2.4L I4 or 3.0L V6 engine	**TCM Output Shaft Speed Sensor Circuit Malfunction Conditions:** Engine started; and the PCM detected a signal from the TCM indicating it had detected an unexpected voltage condition on the Output Speed Sensor (OSS) circuit during the test. **Possible Causes** • OSS positive (+) circuit is open, shorted to ground or to power • OSS positive (-) circuit is open, shorted to ground or to power • OSS sensor is damaged or it has failed • PCM has failed
P0720 N/MIL 1T CCM 1996-2003 Jeep: All with an A/T	**A/T Output Speed Sensor - Low Output Above 15 MPH Conditions:** Engine started; vehicle driven to a speed over 15 mph (determined by the VSS inputs), and the PCM detected the OSS input was less than 60 RPM for 2.6 seconds during the CCM test. **Possible Causes** • OSS signal (+) circuit is open, shorted to ground or to power • OSS signal (-) circuit is open, shorted to ground or to power • OSS is damaged or has failed (it may be open internally) • Parking pawl lugs damaged or missing • PCM has failed
P0725 1T CCM 2001-03 Caravan, Jeep, Dakota, Durango, PT Cruiser, Neon, Town & Country, Voyager and Truck models	**A/T Engine Speed Sensor Circuit Malfunction Conditions:** Engine started; and the PCM detected an Engine Speed sensor reading of less than 390 rpm or more than 800 rpm occurred for 2 seconds during the CCM test. **Possible Causes** • Check for trouble codes related to the CKP sensor • CKP sensor signal circuit open, shorted to ground or to power • CKP sensor is damaged or has failed (open or shorted)
P0725 1T CCM 2001-03 Sebring & Stratus Coupe with a 2.4L I4 or 3.0L V6 engine	**TCM Engine Speed Sensor Circuit Malfunction Conditions:** Engine started; and the PCM detected a signal from the TCM indicating it had detected an unexpected voltage condition on the CKP sensor signal circuit during the CCM test. **Possible Causes** • CKP sensor signal circuit open, shorted to ground or to power • CKP sensor is damaged or has failed (open or shorted) • TCM engine speed signal circuit from the PCM has failed

OBD II TROUBLE CODE LIST (P0XXX CODES)

DTC	Trouble Code Title, Conditions & Possible Causes
P0731 1T CCM 1999-2003 Car, Jeep, SUV, Truck & Van models (A/T)	A/T Additional Gear Ratio Error In First Gear Conditions: Vehicle driven to a speed of 4-37 mph at more than 500 rpm, gear selector in 1 Gear in Drive, and the PCM detected too much difference in the engine speed and output shaft speed. Possible Causes • U/D and 2-4 pressures less than 105 psi • U/D or L/R clutch volume indexes out of specification • L/R accumulator seals are damaged or L/R switch valve stuck • L/R switch valve stuck in first gear position • Transmission has an internal leak (U/D or other seals leaking)
P0732 1T CCM 1999-2003 Car, Jeep, SUV, Truck & Van models (A/T)	A/T Additional Gear Ratio Error In Second Gear Conditions: Vehicle driven to a speed of 4-37 mph at more than 500 rpm, gear selector in 2nd Gear in Drive, and the PCM detected too much difference in the engine speed and output shaft speed. Possible Causes • U/D and L/R pressures less than 95 psi at 2000 rpm • U/D or 2-4 clutch volume indexes out of specification • L/R accumulator seals are damaged or L/R switch valve stuck • Transmission has an internal leak (U/D or 2-4 seals leaking) • L/R switch valve stuck in first gear position
P0733 1T CCM 1999-2003 Car, Jeep, SUV, Truck & Van models (A/T)	A/T Additional Gear Ratio Error In Third Gear Conditions: Vehicle driven to a speed of 4-37 mph at more than 500 rpm, gear selector in 3rd Gear in Drive, and the PCM detected too much difference in the engine speed and output shaft speed. Possible Causes • O/D and U/D pressures less than 75 psi • O/D or U/D clutch volume indexes out of specification • O/D accumulator seals are damaged • L/R switch valve stuck in first gear position • Transmission has an internal leak (U/D seals leaking)
P0734 1T CCM 1999-2003 Car, Jeep, SUV, Truck & Van models (A/T)	A/T Additional Gear Ratio Error In Fourth Gear Conditions: Vehicle driven to a speed of 4-37 mph at more than 500 rpm, gear selector in 4th Gear in Drive, and the PCM detected too much difference in the engine speed and output shaft speed. Possible Causes • Customer may complain of intermittent 4th gear operation • O/D and 2-4 pressures less than 75 psi • O/D or 2-4 clutch volume indexes out of specification • O/D accumulator seals are damaged • Transmission enters "limp in" mode running in 4th gear
P0735 1T CCM 1999-2003 Car, Jeep, SUV, Truck & Van models (A/T)	A/T Gear Ratio Error Fourth Prime Conditions: Vehicle driven any forward Gear, and the TCM detected the ratio of the Input speed to the Output Speed did not match the current Gear Ratio (this test can take up to 5 minutes). Possible Causes • Related Gear Ratio trouble codes may be stored (note that some of these Gear Ratio trouble codes may be intermittent) • Transmission has internal problems or damage present
P0736 1T CCM 1999-2003 Car, Jeep, SUV, Truck & Van models (A/T)	A/T Additional Gear Ratio Error In Reverse Gear Conditions: Vehicle driven in Reverse Gear at more than 500 rpm, and the PCM detected the ratio of the Input speed to the Output Speed did not match the current Gear Ratio during the test. Possible Causes • Customer may complain of intermittent Reverse gear operation • Related Gear Ratio trouble codes may be stored • Transmission has internal problems or damage present

OBD II TROUBLE CODE LIST (P0XXX CODES)

DTC	Trouble Code Title, Conditions & Possible Causes
P0740 1T CCM 1995-2003 Car, Jeep, SUV, Truck & Van models (A/T)	A/T Torque Converter Clutch - No RPM Drop At Lockup Conditions: Engine started; vehicle driven at over 1750 rpm at 50 mph for 20 seconds, TCC command at 100%, and the PCM detected the engine speed was not within 60 rpm of ISS speed with the TP angle less than 30 degrees. The test must fail 3 times on one trip to set a code. Possible Causes • Internal transmission component fault (Input or Reaction shaft) • Internal transmission leakage (valve body leaking or seal rings) • Converter clutch valve or clutch timing valve is sticking • TCC assembly has failed
P0740 1T CCM 2001-03 Sebring & Stratus Coupe with a 2.4L I4 or 3.0L V6 engine	TCM Torque Converter Clutch System Failure Conditions: Engine started; and the TCM indicating it had detected a malfunction in the TCC system. Possible Causes • TCC solenoid circuit open, shorted to ground or power • TCC solenoid is damaged or it has failed • TCC System is damaged (possible mechanical fault present) • PCM has failed
P0743 1T CCM 1995-2003 Car, Jeep, SUV, Truck & Van models (A/T)	A/T Torque Converter Clutch Solenoid Circuit Failure Conditions: Key on or engine started, and the PCM detected an unexpected voltage condition on the TCC solenoid control circuit during the test. Possible Causes • TCC power circuit is open between the solenoid and Fused B+ • TCC solenoid control circuit is open or shorted to ground • TCC solenoid control circuit is shorted to system power • TCC solenoid is damaged or has failed • PCM has failed
P0748 1T CCM 1996-2003 Dakota, Durango, Jeep, Truck & Van models with an A/T	A/T Governor Pressure Solenoid/Transmission Relay Fault Conditions: Engine started, and the PCM detected an unexpected voltage on the GPS control circuit. This solenoid is used to control governor pressure so the transmission can determine shift points. Possible Causes • A/T transmission relay power circuit is open (test power to B+) • A/T transmission relay is damaged or has failed • Governor pressure solenoid circuit is open or shorted to ground • Governor pressure solenoid is damaged or has failed • PCM has failed
P0750 1T CCM 1998-2003 Car, Jeep, SUV, Truck & Van models	A/T Low/Reverse Solenoid Circuit Failure Conditions: Key on, and the PCM detected an unexpected voltage condition on the L/R Solenoid control circuit after the solenoid was enabled and disabled (the PCM checks for an inductive spike). Possible Causes • L/R solenoid control circuit is open or shorted to ground • L/R solenoid control circuit is shorted to system power • L/R Solenoid is damaged or has failed • TCM or PCM has failed
P0750 1T CCM 2001-03 Sebring & Stratus Coupe with a 2.4L I4 or 3.0L V6 engine	TCM Shift Solenoid 'A' Control Circuit Failure Conditions: Engine started; and the TCM detected a problem in the Low/Reverse or Shift Solenoid 'A' Control circuit during the CCM continuous test. Possible Causes • L/R solenoid control circuit open, shorted to ground or power • L/R solenoid power circuit is open (test power to Fused B+) • L/R control solenoid is damaged or has failed • TCM or the PCM has failed

OBD II TROUBLE CODE LIST (P0XXX CODES)

DTC	Trouble Code Title, Conditions & Possible Causes
P0751 1T CCM 1996-2003 Car, Jeep, SUV, Truck & Van models	A/T Overdrive Switch Pressed Low For Over 5 Minutes Conditions: Engine started; engine runtime over 10 seconds and the PCM detected the Overdrive Off Switch indicated "low" for over 5 minutes. Possible Causes • Overdrive Switch Off" circuit is shorted to ground • Overdrive Switch is damaged or has failed • PCM has failed
P0753 1T CCM 1996-2003 Jeep, SUV, Truck & Van models with A/T	A/T 3-4 Solenoid/Transmission Relay Circuit Failure Conditions: Key on or engine started, and the PCM detected an unexpected voltage condition on the 3-4 Solenoid control circuit during the test. Possible Causes • 3-4 Solenoid control circuit is open or shorted to ground • 3-4 Solenoid is damaged or has failed • Transmission relay power circuit is open (check the fused B+) • Transmission relay is damaged or has failed • PCM has failed
P0755 1T CCM 1996-2003 Car, Jeep, SUV, Truck & Van models (A/T)	A/T 2-4 Solenoid Circuit Failure Conditions: Key on or engine running; and the PCM detected an unexpected voltage condition on the2-4 Solenoid control circuit immediately the solenoid was energized or de-energized. The PCM checks for an inductive spike when the solenoid is turned "on" and then "off". Possible Causes • 2-4 Solenoid control circuit is open or shorted to ground • 2-4 Solenoid control circuit is shorted to system power • 2-4 Solenoid is damaged or has failed • TCM or PCM has failed
P0755 1T CCM 2001-03 Sebring & Stratus Coupe with a 2.4L I4 or 3.0L V6 engine	TCM Shift Solenoid 'B' Control Circuit Failure Conditions: Engine started; and the PCM detected a signal from the TCM indicating it had detected a problem in the Underdrive or Shift Solenoid 'B' Control circuit during the CCM test. Possible Causes • U/D solenoid control circuit open, shorted to ground or power • U/D solenoid power circuit is open (test power to Fused B+) • U/D control solenoid is damaged or has failed • TCM or the PCM has failed
P0756 1T CCM 1999-2003 Jeep models with A/T	A/T Shift Solenoid 'B' Mechanical Failure Conditions: DTC P1746 and P1747 not set, engine started, vehicle driven at cruise speed under heavy load conditions, and the PCM detected a problem while operating the Shift Solenoid 'B'. Possible Causes • ATF level is low, or the ATF fluid is contaminated • A/T SSB is damaged (mechanical fault) or has failed • Automatic transmission has internal damage or has failed
P0760 1T CCM 1999-2003 Car, Jeep, SUV, Truck & Van models (A/T)	A/T Overdrive Solenoid Circuit Failure Conditions: Engine started, and the PCM detected an unexpected voltage condition on the Overdrive solenoid control circuit. The solenoids are tested at power-up and then ever 10 seconds. Possible Causes • O/D Solenoid control circuit is open, shorted to ground or power • O/D Solenoid is damaged or has failed • TCM relay power circuit to O/D solenoid is open (loss of B+) • TCM or PCM has failed
P0760 1T CCM 2001-03 Sebring & Stratus Coupe with a 2.4L I4 or 3.0L V6 engine	TCM Shift Solenoid 'C' Control Circuit Failure Conditions: Engine started; and the PCM detected a signal from the TCM indicating it had detected a problem in the 2nd Solenoid or Shift Solenoid 'C' control circuit during the CCM test. Possible Causes • 2nd solenoid control circuit open, shorted to ground or power • 2nd solenoid power circuit is open (test power to Fused B+) • 2nd control solenoid is damaged or has failed • TCM or the PCM has failed

OBD II TROUBLE CODE LIST (P0XXX CODES)

DTC	Trouble Code Title, Conditions & Possible Causes
P0765 1T CCM 1999-2003 Car, Jeep, SUV, Truck & Van models (A/T)	A/T Underdrive Solenoid Circuit Malfunction Conditions: Key on or engine running; and the PCM detected an unexpected voltage condition on the U/D Solenoid control circuit immediately after a Gear Ratio or Pressure Switch error was detected (this code can set due to a single or multiple circuit faults). Possible Causes • U/D Solenoid control circuit is open or shorted to ground • U/D Solenoid control circuit is shorted to system power • U/D Solenoid is damaged or has failed • TCM or PCM has failed
P0783 N/MIL 1T CCM 1996-2003 Car, Jeep, SUV, Truck & Van models (A/T	A/T 3-4 Shift Solenoid - No RPM Drop At Lockup Conditions: Engine started; vehicle driven to over 30 mph, and the PCM did not detect the correct amount of rpm drop after a gear change occurred. Possible Causes • ATF fluid level is burnt, contaminated or low • 3-4 Solenoid control circuit is open or shorted (intermittent fault) • 3-4 Solenoid is damaged or has failed • Transmission oil pan has excessive debris
P0801 1T CCM 1996-2003 Viper models with A/T	Reverse Gear Lockout Control Circuit Malfunction Conditions: Vehicle driven at over 5 mph at an engine speed over 608 rpm, and the PCM detected an unexpected "low" or high voltage condition on the Reverse Gear Lockout solenoid circuit. Possible Causes • Reverse gear lockout solenoid circuit open or shorted to ground • Reverse gear lockout solenoid is damaged or has failed • PCM has failed
P0830 1T CCM 1999-2003 Truck models equipped with a Manual Transmission (M/T)	Clutch Pedal Depressed Circuit Malfunction Conditions: Engine cranking or vehicle driven to over 15 mph at an engine speed of 1550-2880 rpm with the delta throttle at over 1.1v for 4 seconds, and the PCM detected an unexpected voltage condition on the Clutch Pedal switch circuit (test must fail 5 times during one trip). Possible Causes • CPP switch signal circuit is open or shorted to ground • CPP switch power circuit is open (test the power at the PDC) • Clutch pedal is damaged or has failed • PCM has failed
P0833 N/MIL 1T CCM 1999-2003 Truck models equipped with a Manual Transmission (M/T	Clutch Pedal Released Circuit Malfunction Conditions: Engine cranking or vehicle driven to over 15 mph at an engine speed of 1550-2880 rpm with the delta throttle at over 1.1v for 4 seconds, and the PCM detected an unexpected voltage condition on the Clutch Pedal switch circuit (test must fail 5 times during one trip). Possible Causes • CPP switch signal circuit is open or shorted to ground • CPP switch power circuit is open (test the power at the PDC) • Clutch pedal is damaged or has failed • PCM has failed
P0833 1T CCM 2001-03 PT Cruise & Neon models with a M/T	Clutch Pedal Position Switch Circuit Malfunction Conditions: Engine cranking, or vehicle speed over 25 mph at an engine speed from 1550-2880 rpm with the delta throttle over 1.1v for 4 seconds, and the PCM detected an unexpected voltage condition on the Clutch Pedal switch circuit (test must fail 5 times during one trip). Possible Causes • CPP switch signal circuit is open or shorted to ground • CPP switch power circuit is open (test the power at the PDC) • Clutch pedal is damaged or has failed • PCM has failed

OBD II TROUBLE CODE LIST (P0XXX CODES)

DTC	Trouble Code Title, Conditions & Possible Causes
P0841 1T CCM 2001-03 300M, Concorde, LHS, Intrepid, Jeep, Neon, PT Cruiser, Sebring, Caravan, Town & Country, Voyager, Dakota & Durango	Low/Reverse Pressure Switch Sense Circuit Malfunction Conditions" Engine started; vehicle driven in a forward gear, and the PCM detected an unexpected voltage condition on the Low/Reverse Pressure Switch Sense circuit during the CCM test. Possible Causes • L/R pressure switch sense circuit is open or shorted to ground • L/R pressure switch sense circuit is shorted to power (B+) • L/R pressure switch is damaged or it has failed • TCM relay power circuit to L/R switch is open (loss of B+) • TCM has failed
P0845 1T CCM 2001-03 300M, Concorde, LHS, Intrepid, Jeep, Neon, PT Cruiser, Sebring, Caravan, Town & Country, Voyager, Dakota & Durango	A/T 2-4 Hydraulic Pressure Test Malfunction Conditions: Engine speed over 1000 rpm, then immediately after a shift event, the PCM detected a failure in one or more of the Pressure Switch circuits (i.e., it tests switches that are not operating). Possible Causes • 2-4 pressure is incorrect, or internal transmission faults exist • 2-4 pressure switch circuit is open, shorted to ground or power • 2-4 pressure switch is damaged or it has failed • TCM relay power circuit to 2-4 switch is open (loss of B+) • TCM has failed
P0846 1T CCM 2001-03 300M, Concorde, LHS, Intrepid, Jeep, Neon, PT Cruiser, Sebring, Caravan, Town & Country, Voyager, Dakota & Durango	A/T 2-4 Pressure Switch Circuit Malfunction Conditions: Engine started; vehicle driven in a forward gear, and the PCM detected that the 2-4 Pressure Switch circuit indicated open or closed at the wrong time. Possible Causes • 2-4 pressure is incorrect, or internal transmission faults exist • 2-4 pressure switch circuit is open, shorted to ground or power • 2-4 pressure switch is damaged or it has failed • TCM relay power circuit to L/R switch is open (loss of B+) • TCM has failed
P0850 1T CCM 2001-03 Cars, Dakota, Durango and Mini-Van models	Park Neutral Switch Performance Conditions: Engine running with the gearshift selector in Park, Neutral or Drive position (not in Limp mode), and the PCM detected an invalid Park Neutral switch state during vehicle operation. Possible Causes • Check for any TCM related codes stored in the TCM controller (use the Scan Tool to access any TCM codes in memory) • Good Trip equals zero (0) • PCM has failed
P0867 1T CCM 2002-03 Jeep & Truck models	A/T Line Pressure Malfunction Conditions: Engine started; vehicle driven, and the PCM detected the Actual Line Pressure was not within 10 psi of the Desired Line pressure. Possible Causes • Check for related trouble codes • Check TCM connectors for loose or damaged terminals • A/T line pressure is out of range • TCM line pressure is out of range
P0868 1T CCM 2002-03 Jeep & Truck models	A/T Line Pressure Low Conditions: Engine started; vehicle driven, and the PCM detected the Actual Line Pressure indicated 10 psi less than the Desired Line pressure. Possible Causes • Check for related trouble codes • Check TCM connectors for loose or damaged terminals • A/T line pressure is out of range • TCM line pressure is out of range

OBD II TROUBLE CODE LIST (P0XXX CODES)

DTC	Trouble Code Title, Conditions & Possible Causes
P0869 1T CCM 2002-03 Jeep & Truck models	A/T Line Pressure High Conditions: Engine started; vehicle driven, and the PCM detected the Actual Line Pressure indicated 10 psi more than the Desired Line pressure. Possible Causes • Check for related trouble codes • Check TCM connectors for loose or damaged terminals • A/T line pressure is out of range • TCM line pressure is out of range
P0870 1T CCM 2002-03 Car, Jeep, SUV, Truck & Van models (A/T)	A/T Hydraulic Pressure Line Malfunction Conditions: Engine started; vehicle driven at over 1000 rpm, then immediately after a shift, the PCM detected a malfunction in one or more of the Pressure Switch circuits (it detected the switch did not close twice). Possible Causes • Check for related line pressure trouble codes • Check for related speed ratio and pressure switch codes • Excessive debris in the oil pan • Line pressure sensor connector is loose or damaged • Oil pressure switch is damaged or has failed • TCM 5v supply circuit is open
P0871 1T CCM 2002-03 Car, Jeep, SUV, Truck & Van models (A/T)	A/T O/D Pressure Switch Circuit Malfunction Conditions: Engine started; vehicle driven in a forward gear, and the PCM detected that the O/D Pressure Switch circuit indicated open or closed at the wrong time. Possible Causes • O/D pressure is incorrect, or internal transmission faults exist • O/D pressure switch circuit is open, shorted to ground or power • O/D pressure switch is damaged or it has failed • TCM relay power circuit to O/D switch is open (loss of B+) • TCM has failed
P0875 1T CCM 2002-03 Jeep & Truck models	A/T U/D Hydraulic Pressure Test Malfunction Conditions: Engine speed over 1000 rpm, then immediately after a shift event, the PCM detected a fault in one or more of the Pressure Switch circuits (it detected the switch did not close twice). Possible Causes • Check for related line pressure trouble codes • Check for related speed ratio and pressure switch codes • Excessive debris in the oil pan • Line pressure sensor connector is loose or damaged • U/D Oil pressure switch is damaged or it has failed • TCM 5v supply circuit is open
P0876 1T CCM 2002-03 Jeep & Truck models	A/T U/D Pressure Switch Sense Malfunction Conditions: Engine started; vehicle driven in a forward gear, and the PCM detected that the U/D Pressure Switch Sense circuit indicated open or closed at the wrong time during the CCM test period. Possible Causes • U/D pressure is incorrect, or internal transmission faults exist • U/D pressure switch circuit is open, shorted to ground or power • U/D pressure switch is damaged or it has failed • TCM relay power circuit to L/R switch is open (loss of B+) • TCM has failed

OBD II TROUBLE CODE LIST (P0XXX CODES)

DTC	Trouble Code Title, Conditions & Possible Causes
P0884 1T CCM 2002-03 Jeep & Truck models	A/T U/D Pressure Switch Sense Malfunction Conditions: Engine started; vehicle driven in a forward gear, and the PCM detected that the U/D Pressure Switch Sense circuit indicated open or closed at the wrong time during the CCM test period. Possible Causes • U/D pressure is incorrect, or internal transmission faults exist • U/D pressure switch circuit is open, shorted to ground or power • U/D pressure switch is damaged or it has failed • TCM relay power circuit to L/R switch is open (loss of B+) • TCM has failed
P0884 1T CCM 2002-03 Jeep & Truck models	Power-Up Automatic Transmission Speed Malfunction Conditions: Engine started, TCM relay enabled; and the TCM detected a valid forward gear PNDRL signal with the Output Speed more than 800 rpm indicating a vehicle speed of over 20 mph. Possible Causes • TCM power supply circuit to direct battery is open • TCM power supply circuit to the ignition switch is open • TCM power ground circuit is open or the connector is loose • TCM has failed
P0890 1T CCM 2001-03 Car, Jeep, SUV, Truck & Van models (A/T)	A/T TCM Switched Battery Circuit Malfunction Conditions: Key on or engine cranking; TCM relay "not" energized, and the TCM detected voltage present at any of the Pressure switch input circuits. Possible Causes • 2-4 switch circuit is shorted to system power (B+) • L/R switch circuit is shorted to system power (B+) • O/D switch circuit is shorted to system power (B+) • TCM switched battery circuit is damaged • TCM has failed
P0891 1T CCM 2001-03 Car, Jeep, SUV, Truck & Van models (A/T)	A/T TCM Relay Always On Conditions: Key on or engine cranking; TCM relay "not" energized, and the TCM detected voltage present at the TCM output circuit during the test. Possible Causes • TCM relay output circuit is shorted to system power (B+) • TCM relay control circuit is shorted to system power (B+) • TCM relay is damaged or it has failed (it may be stuck closed) • TCM has failed
P0897 1T CCM 2001-03 Car, Dakota & Durango models (A/T)	A/T Transmission Fluid Burnt Or Worn Out Conditions: Engine started; vehicle driven, and immediately after a transition from full TCC lockup to partial TCC engagement (for A/C bump prevention), the TCM detected vehicle shutter during engagement. Possible Causes • Automatic transmission fluid is burnt or contaminated • Automatic transmission fluid is worn out
P0932 1T CCM 2002-03 Jeep & Truck models	A/T Line Pressure Malfunction Conditions: Engine started; vehicle driven in any forward gear, and the TCM detected the Line Pressure Sensor signal was less than 0.20v or more than 4.75v during the CCM test period. Possible Causes • Line pressure sensor connector is damaged or loose • Line pressure sensor circuit is open or shorted to ground • Line pressure sensor circuit is shorted to VREF (5v) • Line pressure sensor ground circuit is open • Line pressure sensor supply circuit (5v) is open or missing • TCM Line Pressure Low or High circuit is open

OBD II TROUBLE CODE LIST (P0XXX CODES)

DTC	Trouble Code Title, Conditions & Possible Causes
P0944 1T CCM 2002-03 Jeep & Truck models	**A/T Loss Of Prime Pressure Conditions:** Engine started; vehicle driven, and immediately after a slipping condition is detected with the pressure switches "not" indicating pressure, the PCM detected a loss of prime pressure. In effect, the TCM turns "on" available elements to detect if prime pressure exists. Possible Causes • A/T pressure switch connector is damaged, loose or shorted • Automatic transmission fluid level is too low • Shift lever position error • Transmission oil filter is clogged or severely restricted • Transmission oil pump is damaged or weak
P0951 1T CCM 2001-03 Car, Dakota, Durango & Mini-Van models with an A/T	**A/T Autostick Sensor Circuit Malfunction Conditions:** Engine started; vehicle driven, transmission not in Autostick position, and the TCM that either the Upshift or Downshift switch was closed, or if the Upshift and Downshift switches are closed at the same time. Possible Causes • Autostick assembly is damaged or has failed • Autostick connector is damaged, loose or shorted • Downshift switch circuit is shorted too ground • Upshift switch circuit is shorted to ground
P0987 1T CCM 2002-03 Jeep & Truck models	**A/T 4C Hydraulic Pressure Test Malfunction Conditions:** Engine started; vehicle driven at over 1000 rpm, then immediately after a shift, the PCM detected a malfunction in one or more of the Pressure Switch circuits (it detected the switch did not close twice). Possible Causes • Check for related line pressure trouble codes • Check for related speed ratio and pressure switch codes • Excessive debris in the oil pan • Line pressure sensor connector is loose or damaged • 4C Oil pressure switch is damaged or it has failed • TCM 5v supply circuit is open
P0988 1T CCM 2002-03 Jeep & Truck models	**A/T 4C Pressure Switch Sense Circuit Malfunction Conditions:** Engine started; vehicle driven in any forward gear, and the PCM detected that the 4C Pressure Switch circuit indicated open or closed at the wrong time during the CCM test. Possible Causes • 4C pressure is incorrect, or internal transmission faults exist • 4C pressure switch circuit is open, shorted to ground or power • 4C pressure switch is damaged or it has failed • TCM relay power circuit to L/R switch is open (loss of B+) • TCM has failed
P0992 1T CCM 2001-03 Car, Dakota, Durango & Mini-Van models with an A/T	**A/T 2-4 & O/D Hydraulic Pressure Test Malfunction Conditions:** Engine started; vehicle driven at over 1000 rpm, then immediately after a shift, the PCM detected a malfunction in one or more of the Pressure Switch circuits (it tests the switches that are not operating). Possible Causes • 2-4 pressure switch circuit is open, shorted to ground or power • 2-4 pressure switch is damaged or it has failed • O/D pressure switch circuit is open, shorted to ground or power • O/D pressure switch is damaged or it has failed • Internal transmission faults exist • TCM relay power circuit to 2-4 or O/D switch open (loss of B+) • TCM has failed

OBD II TROUBLE CODE LIST (P1XXX CODES)

DTC	Trouble Code Title, Conditions & Possible Causes
P1103 1T CCM 1995-98 Eagle Talon models with a Turbo engine	Turbocharger Wastegate Actuator Malfunction Conditions: Engine started, ECT sensor signal 176°F, and the PCM detected the volumetric efficiency was more than 200% for over 1.5 seconds during the CCM Rationality test. Possible Causes • Charging pressure control system has failed • Turbocharger Wastegate actuator is damaged or has failed • Vacuum hose routing is incorrect • PCM has failed
P1104 1T CCM 1995-98 Eagle Talon models with a Turbo engine	Turbocharger Wastegate Actuator Circuit Malfunction Conditions: Engine started, system voltage over 10.5v, and the PCM did not detect any surge voltage (system voltage +2V) on the Wastegate actuator control circuit when the solenoid was cycled "on" to "off". Note: The Scan Tool Actuator test can be used to cycle the device. Possible Causes • Wastegate solenoid control circuit is open or shorted to ground • Wastegate solenoid power circuit open (power from MPI relay) • Wastegate solenoid is damaged or has failed • PCM has failed
P1105 1T CCM 1995-98 Eagle Talon models with a Turbo engine	Fuel Pressure Control Solenoid Circuit Malfunction Conditions: Engine started, system voltage over 10.5v, and the PCM did not detect any surge voltage (system voltage +2V) on the Fuel Pressure Solenoid control circuit when the solenoid was cycled "on" to "off". Note: The Scan Tool Actuator test can be used to cycle the device. Possible Causes • Fuel pressure solenoid circuit is open or shorted to ground • Fuel pressure solenoid power circuit open (power from relay) • Fuel pressure control solenoid is damaged or has failed • PCM has failed
P1105 2T CCM 2003 PT Cruiser with a 2.4L I4 VIN G Turbo engine	Throttle Inlet Pressure Sensor Solenoid Circuit Malfunction Conditions: Engine started, system voltage over 10.5v, Turbo Boost mode enabled, and the PCM detected the Actual and Intended state of the Throttle Inlet Pressure Sensor solenoid did not match. Possible Causes • ASD output circuit to the TIP solenoid is open • Throttle inlet pressure sensor solenoid is damaged or has failed • TIP solenoid control circuit is open, shorted to ground or power • PCM has failed
P1106 2T CCM 2003 PT Cruiser with a 2.4L I4 VIN G Turbo engine	Throttle Inlet Pressure Sensor Solenoid Circuit Malfunction Conditions: Engine started, system voltage over 10.5v, Turbo Boost mode enabled, and the PCM did not detect enough difference between the BARO sensor and TIP sensor values while under boost. Possible Causes • Check the vacuum supply to the turbo surge solenoid unit • Inspect the hoses and tubing to the turbo charger assembly • Review results of Solenoid Tests (Test 1, 2, 3 and 4 results) • Turbocharger assembly is damaged or it has failed • Wastegate actuator has failed (due to a mechanical failure)
P1110 N/MIL 1T CCM 1998-99 Trucks equipped with a 5.9L Diesel engine	Decreased Engine Performance Due To High Intake Air Temperature Conditions: Engine started, engine running, and the PCM detected the Intake Air Temperature exceeded its normal operating range during the test. Possible Causes • IAT sensor is contaminated, dirty or skewed • Base engine conditions causing the high operating temperature • ECM has failed

OBD II TROUBLE CODE LIST (P1XXX CODES)

DTC	Trouble Code Title, Conditions & Possible Causes
P1115 1T CCM 2002-03 300M, Concorde, Intrepid, Dakota & Durango models	General Temperature Sensor Performance Conditions: Engine "off" more than 8 hours, then engine started, ambient temperature above -10ºF; and after a calibrated amount of cool-down time, the PCM compares the values from the Ambient Air Temperature, ECT and IAT sensors. If the PCM detects that the value of any combination of these sensors (AAT-IAT, AAT-ECT or ECT-IAT) is less than a calibrated value, it will set this trouble code. Possible Causes • Sensor signal circuit is open or shorted to ground • Sensor ground circuit is open or shorted to VREF (5v) • One or more of the identified sensors is out-of-calibration • Ambient air temperature sensor is damaged or it has failed • PCM High or Low circuit is damaged or it has failed
P1188 2T CCM 2003 PT Cruiser with a 2.4L I4 VIN G Turbo engine	Throttle Inlet Pressure Sensor Signal Range/Performance Conditions: Engine started, engine running in Turbo Boost or Non-Boost mode, and the PCM detected a significant difference between the BARO sensor and TIP sensor signals (i.e., the TIP sensor cannot read the signal correctly). Possible Causes • ASD output circuit to the TIP solenoid is open • Throttle inlet pressure sensor solenoid is damaged or has failed • TIP solenoid control circuit is open, shorted to ground or power • PCM has failed
P1189 2T CCM 2003 PT Cruiser with a 2.4L I4 VIN G Turbo engine	Throttle Inlet Pressure Sensor Circuit Low Input Conditions: Engine started, TP sensor less than 1.2v, system voltage over 10.5v, and the PCM detected the Throttle Inlet Pressure (TIP) sensor was less than 0.0782v for a period of 1-7 seconds. Possible Causes • TIP sensor VREF circuit is open • TIP sensor signal circuit is open • TIP sensor signal circuit is shorted to chassis or sensor ground • TIP sensor is damaged or it has failed • PCM has failed
P1190 2T CCM 2003 PT Cruiser with a 2.4L I4 VIN G Turbo engine	Throttle Inlet Pressure Sensor Circuit High Input Conditions: Engine started, TP sensor less than 1.2v, system voltage over 10.5v, and the PCM detected the Throttle Inlet Pressure (TIP) sensor was more than 4.92v for a period of 1-7 seconds. Possible Causes • TIP sensor VREF circuit is open • TIP sensor signal circuit is open • TIP sensor signal circuit is shorted to chassis or sensor ground • TIP sensor is damaged or it has failed • PCM has failed
P1192 1T CCM 2001-03 Neon, Sebring, Stratus, PT Cruiser, Caravan, Town & Country & Voyager models	Intake Air Temperature Sensor Circuit Low Input Conditions: Engine started; system voltage over 10.5v and the PCM detected the IAT sensor signal was less than 0.80v during the CCM test period. Possible Causes • IAT sensor signal circuit is shorted to ground • IAT sensor is damaged or has failed • PCM has failed
P1193 1T CCM 2001-03 Neon, Sebring, Stratus, PT Cruiser, Caravan, Town & Country & Voyager models	Intake Air Temperature Sensor Circuit High Input Conditions: Engine started; system voltage over 10.5v and the PCM detected the IAT sensor signal was more than 4.90v during the CCM test. Possible Causes • IAT sensor connector is damaged or it is open • IAT sensor signal circuit is open between the sensor and PCM • IAT sensor ground circuit is open between the sensor and PCM • IAT sensor damaged or has failed • PCM has failed

OBD II TROUBLE CODE LIST (P1XXX CODES)

DTC	Trouble Code Title, Conditions & Possible Causes
P1194 1T CCM 2001-02 PT Cruiser	HO2S-11 (Bank 1 Sensor 1) Heater Performance Conditions: Key off after a warm engine drive cycle, engine cool-down finished (at least 5 seconds after the key is turned off), system voltage over 10.5v, then with the ASD relay energized, the PCM detected the HO2S signal rose to 0.49v or more within a 144 second period, and the initial rise of the oxygen sensor signal was less than 1.57v. Note: This test is done at key off. Possible Causes • ASD relay output (power) circuit to the heater is open • HO2S heater ground circuit open or HO2S signal circuit is open • HO2S heater element has high resistance • HO2S heater element has failed (open or shorted) • PCM has failed
P1195 1T CCM 1998-2003 Car, Jeep, SUV, Truck & Van models	HO2S-11 (Bank 1 Sensor 1) Circuit Insufficient Activity Conditions: Engine started, vehicle driven with the throttle open at a speed over 18 mph at light engine load for over 5 minutes, ECT sensor more than 170ºF, and the PCM detected the HO2S signal switched from 0.39v to 0.60v too few times during the Oxygen Sensor Monitor test. Possible Causes • Base engine mechanical fault affecting more than one cylinder • Exhaust leak present in exhaust manifold or exhaust pipes • HO2S element fuel contamination or has deteriorated • HO2S signal circuit or ground circuit has high resistance • PCM has failed
P1196 1T CCM 1998-2003 Car, Jeep, SUV, Truck & Van models	HO2S-21 (Bank 2 Sensor 1) Circuit Insufficient Activity Conditions: Engine started, vehicle driven with the throttle open at a speed over 18-55 mph at light engine load for over 5 minutes, ECT sensor more than 170ºF, and the PCM detected the HO2S signal switched from 0.39v to 0.60v too few times in the Oxygen Sensor Monitor test. Possible Causes • Base engine mechanical fault affecting more than one cylinder • Exhaust leak present in exhaust manifold or exhaust pipes • HO2S element fuel contamination or has deteriorated • HO2S signal circuit or ground circuit has high resistance
P1197 1T CCM 1998-2003 SUV, Truck & Van models with V8, V10	HO2S-12 (Bank 1 Sensor 2) Circuit Insufficient Activity Conditions: Engine started, vehicle driven with the throttle open at a speed over 18-55 mph at light engine load for over 5 minutes, ECT sensor more than 170ºF, and the PCM detected the HO2S signal switched from 0.39v to 0.60v too few times in the Oxygen Sensor Monitor test. Possible Causes • Base engine mechanical fault affecting more than one cylinder • Exhaust leak present in exhaust manifold or exhaust pipes • HO2S element fuel contamination or has deteriorated • HO2S signal circuit or ground circuit has high resistance
P1281 2T ECT 1996-2003 Car, Jeep, SUV, Truck & Van models	Engine Is Cold Too Long Conditions: Engine started, engine runtime more than 20 minutes, and the PCM detected the engine temperature did not exceed 176ºF in the period. Possible Causes • Check the operation of the thermostat (it may be stuck open) • ECT sensor signal circuit has high resistance • ECT sensor is damaged or it has failed • Inspect for low coolant level or an incorrect coolant mixture
P1282 1T CCM 1996-2003 Car, Jeep, SUV, Truck & Van models	Fuel Pump Relay Control Circuit Malfunction Conditions: Key on or engine started, system voltage over 10.5v, and the PCM detected an unexpected voltage condition on the Fuel Pump Relay control circuit during the CCM test period. Possible Causes • Fuel pump relay control circuit is open or shorted to ground • Fuel pump relay power circuit is open (test power from Ignition) • Fuel pump relay is damaged or has failed • PCM has failed

OBD II TROUBLE CODE LIST (P1XXX CODES)

DTC	Trouble Code Title, Conditions & Possible Causes
P1283 1T CCM 1998-2002 Trucks equipped with a 5.9L Diesel engine	Idle Select Signal Invalid Conditions: Key on and the Fuel Pump Control Module (VP44) detected an invalid Low Idle Select signal from the PCM (ECM controller). Possible Causes • Low idle select signal circuit is open or shorted to ground • Low idle select signal circuit is shorted to VREF or to power • ECM internal circuit is shorted to ground • ECM internal regulator output is more than 6.0v • Fuel injection pump is damaged or has failed
P1284 1T CCM 1998-2002 Trucks equipped with a 5.9L Diesel engine	Fuel Injection Pump Battery Voltage Out-Of-Range Conditions: Key on and the Fuel Pump Control Module (VP44) detected an invalid Low Idle Select signal from the PCM (ECM controller). Possible Causes • Fuel injection pump is damaged or has failed
P1285 1T CCM 1998-2002 Trucks equipped with a 5.9L Diesel engine	Fuel Injection Pump Controller Always On Conditions: Engine started, engine running, and the PCM detected the Fuel Injection Pump Controller was in an "always on" condition. Possible Causes • Fuel injection pump relay driver circuit shorted to system power • Fuel injection pump relay output circuit shorted to power • Fuel injection pump relay is damaged or has failed • ECM has failed
P1286 1T CCM 1998-2002 Trucks equipped with a 5.9L Diesel engine	Accelerator Position Sensor Supply Voltage High Input Conditions: Key on or engine running; and the PCM detected the supply voltage circuit to the Accelerator Position (APP) sensor was too high. Possible Causes • APP sensor supply voltage circuit is shorted to system power • APP sensor signal circuit is shorted to power • APP sensor ground circuit is open • ECM has failed
P1287 1T CCM 1998-2002 Trucks equipped with a 5.9L Diesel engine	Fuel Injection Pump Battery Voltage Out-Of-Range Conditions: Engine started and the PCM detected the Fuel Injection Pump battery voltage was too low. Possible Causes • Fuel injection pump ground circuit is open • Generator voltage is less than 12.0v (the Generator has failed) • Battery voltage is less than 8.0v (the battery is defective) • PCM has failed
P1288 1T CCM 1998-2003 Concorde, Intrepid, Sebring, Stratus & Prowler with a 2.7L V6	Short Runner Valve Control Circuit Malfunction Conditions: Key on or engine cranking; and the PCM detected an unexpected voltage condition on the Short Runner Solenoid (SRV) Control circuit during the CCM test. Possible Causes • SRV control circuit is open between the solenoid and PCM • SRV control circuit is shorted to sensor or chassis ground • SRV is damaged or has failed • PCM has failed
P1289 1T CCM 1998-2002 300M, Concorde, LHS, Intrepid, & Prowler with a 3.2L or 3.5L V6	Manifold Tuning Valve Control Circuit Malfunction Conditions: Key on or engine cranking; and the PCM detected an unexpected voltage condition on the Manifold Tuning Valve (MTV) control circuit. Possible Causes • MTV control circuit is open between the solenoid and PCM • MTV control circuit is shorted to sensor or chassis ground • MTV is damaged or has failed • PCM has failed

OBD II TROUBLE CODE LIST (P1XXX CODES)

DTC	Trouble Code Title, Conditions & Possible Causes
P1290 N/MIL 1T CCM 1996-2002 Ram Van, Van Wagon with a CNG VIN T engine	Certified Natural Gas Fuel System Pressure Too High Conditions: Engine started, and the PCM detected the CNG Fuel System was operating outside of its normal operating range during the CCM test. Possible Causes • CNG pressure sensor signal is skewed • CNG fuel system component is damaged or has failed • PCM has failed
P1290 N/MIL 1T CCM 1998-2002 Caravan, Town & Country, Voyager with a 3.3L CNG engine	CNG Pressure Sensor Circuit High Input Conditions: Key on or engine running; and the PCM detected the Certified Natural Gas (CNG) sensor was more than 4.96v during the test. Possible Causes • CNG sensor signal circuit open between the sensor and PCM • CNG sensor ground circuit open between the sensor and PCM • CNG sensor is damaged or has failed (it may be open) • PCM has failed
P1291 1T CCM 1998-2002 Caravan, Town & Country, Voyager with a 3.3L CNG engine	CNG Pressure Sensor Circuit Low Input Conditions: Engine started and the PCM detected the Certified Natural Gas (CNG) sensor was less than 0.49v. Possible Causes • CNG sensor signal circuit is shorted to chassis or sensor ground between the sensor and PCM • CNG sensor is damaged or has failed (it may be shorted) • PCM has failed
P1291 1T CCM 1998-2002 Trucks with 5.9L Diesel engine	No Temperature Rise Detected From The Intake Heaters Conditions: No IAT or IAH Relay codes set, preheat function completed before startup, post-heat function active, Engine cranking for less than 5 seconds, engine runtime over 15 seconds, IAT sensor from 0-66°F, BTS and IAT sensors with 10°F of each other, time between engine preheat period and engine run state is less than 30 seconds, and the PCM did not detect a temperature increase at the Intake Heaters. Possible Causes • Battery cable from No. 1 Relay to No. 1 Heater is open • Battery cable from No. 2 Relay to No. 1 Heater is open • Battery cable to the No. 1 Relay is open or has high resistance • Battery cable to the No. 2 Relay is open or has high resistance • Intake Air Heater Relay is damaged or has failed • PCM has failed
P1292 1T CCM 1996-2002 Ram Van, Van Wagon with a CNG VIN T	Certified Natural Gas Pressure Sensor Circuit Low Input Conditions: Engine started and the PCM detected the CNG Pressure sensor indicated less than 0.49v. Possible Causes • CNG pressure sensor signal circuit is shorted to ground • CNG pressure sensor is damaged or has failed • PCM has failed
P1294 1T CCM 1995-98 Eagle Talon models	Target Idle Speed Not Reached Conditions: DTC P0106, P0107, P0108, P0121, P0122 and P0123 not set, engine started, engine running at idle in Drive or Neutral, and the PCM detected the Actual idle speed was more than 200 rpm over or more than 100 rpm less than the Target speed for over 14 seconds. Possible Causes • Engine vacuum leak in a hose, Brake Booster or in the engine • IAC motor control circuits open or grounded in the wire harness • Throttle body dirty or restricted (trying cleaning it and retesting) • Throttle linkage or throttle plate not in the correct position • PCM has failed

OBD II TROUBLE CODE LIST (P1XXX CODES)

DTC	Trouble Code Title, Conditions & Possible Causes
P1294 1T CCM 1996-2003 Car, Mini-Van & SUV models	Target Idle Speed Not Reached Conditions: DTC P0106, P0107, P0108, P0121, P0122 and P0123 not set, engine started, running at idle in Drive or Neutral, and the PCM detected the Actual idle speed was more than 200 rpm over or more than 100 rpm less than the Target speed for over 14 seconds. Possible Causes • Engine vacuum leak in a hose, brake booster or in the engine • IAC motor control circuits open or grounded in the wire harness • Throttle body dirty or restricted (trying cleaning it and retesting) • Throttle linkage or throttle plate not in the correct position • PCM has failed
P1294 1T CCM 1996-2003 Jeep, Truck & Van models	Target Idle Speed Not Reached Conditions: DTC P0106, P0107, P0108, P0121, P0122 and P0123 not set, engine started, engine running at idle in Drive or Neutral, and the PCM detected the Actual idle speed was more than 200 rpm over or more than 100 rpm less than the Target speed for over 14 seconds. Possible Causes • Engine vacuum leak in a hose, Brake Booster or in the engine • IAC motor control circuits open or grounded in the wire harness • Throttle body dirty or restricted (trying cleaning it and retesting) • Throttle linkage or throttle plate not in the correct position • PCM has failed
P1295 N/MIL 1T CCM 1996-2000 Breeze, Cirrus, Eagle Talon, Sebring & Stratus models	5-Volt VREF Missing To Position Sensor Conditions: Engine started, vehicle driven to over 20 mph at more than 1500 rpm with MAP sensor less than 13 kPa, and the PCM detected the TP sensor signal was less than a specified value during the test. Possible Causes • TP sensor VREF circuit is open between the sensor and PCM • TP sensor ground circuit is open between the sensor and PCM • TP sensor is damaged or has failed • PCM has failed
P1295 N/MIL 1T CCM 1998-2002 Trucks with 5.9L Diesel engine	5-Volt VREF Missing To APP Sensor Conditions: Key on or engine running; and the PCM detected the supply voltage circuit to the Accelerator Position (APP) sensor was too low. Possible Causes • APP sensor supply voltage circuit is open • ACCEL pedal position sensor is damaged or has failed • APP sensor supply circuit shorted to chassis or sensor ground • PCM has failed
P1296 N/MIL 1T CCM 1999-2003 Car, Jeep, SUV, Truck & Van models	5-Volt VREF Supply Not Present Conditions: Key on, altitude indicating zero feet above seal level, then the PCM detected the MAP sensor was near 101 kPa; or with altitude at 1200 feet above sea level, the MAP sensor was near 88 kPa. Possible Causes • MAP sensor VREF circuit open between the sensor and PCM • MAP sensor ground circuit open between the sensor and PCM • MAP sensor is damaged or has failed • PCM has failed
P1297 1T CCM 1996-2000 Breeze, Cirrus, Eagle Talon, Sebring & Stratus models	No Change In MAP Signal From Start To Run Transition Conditions: Engine started, engine speed between 400 and 1200 rpm, and the PCM detected too small a difference between the BARO signal at key "on" and the engine running MAP sensor input for 1.76 seconds. Possible Causes • Engine vacuum port to MAP sensor clogged, dirty or restricted • MAP sensor signal is skewed or the sensor is out-of-calibration • MAP sensor VREF circuit open or grounded (intermittent fault) • PCM has failed

OBD II TROUBLE CODE LIST (P1XXX CODES)

DTC	Trouble Code Title, Conditions & Possible Causes
P1297 1T CCM 1996-2003 Car, Mini-Van & SUV models	No Change In MAP Signal From Start To Run Transition Conditions: Engine started, and with the engine speed within ±64 rpm of the Target idle speed, the PCM detected too small a difference between the BARO and MAP sensor signals for 8.80 seconds. Possible Causes • Engine vacuum port to MAP sensor clogged, dirty or restricted • MAP sensor signal is skewed or the sensor is out-of-calibration • MAP sensor VREF circuit open or grounded (intermittent fault) • PCM has failed
P1297 1T CCM 1996-2003 Jeep, Truck & Van models	No Change In MAP Signal From Start To Run Transition Conditions: Engine started, and with the engine speed within ±64 rpm of the Target idle speed, the PCM detected too small a difference between the BARO and MAP sensor signals for 8.80 seconds. Possible Causes • Engine vacuum port to MAP sensor clogged, dirty or restricted • MAP sensor signal is skewed or the sensor is out-of-calibration • MAP sensor VREF circuit open or grounded (intermittent fault) • PCM has failed
P1299 2T IAC 1996-2003 Car, Jeep, Mini-Van, SUV, Truck & Van models	Vacuum Leak Present With IAC Valve Fully Seated Conditions: Engine running at idle speed in closed loop, and the PCM detected the MAP sensor signal did not correlate to the TP sensor signal under these operating conditions during the test. Possible Causes • Engine vacuum leak in a hose, Brake Booster or in the engine MAP sensor is out-of-calibration or skewed • TP sensor is damaged or has failed (perform a sweep test) • PCM has failed
P1388 1T CCM 1996-2003 Car, Jeep, Mini-Van, SUV, Truck & Van models	Auto Shutdown Relay Control Circuit Malfunction Conditions: Key on or engine cranking; and the PCM detected an unexpected voltage condition on the ASD Relay Control circuit. The ASD Relay coil resistance is 95-105 ohms at 68ºF. Possible Causes • ASD relay control circuit is open between the relay and PCM • ASD relay control circuit is shorted to ground • ASD relay power circuit is open (test power from Fused B+) • ASD relay is damaged or has failed • PCM has failed
P1388 N/MIL 1T CCM 1998-2002 Truck applications with a 5.9L Diesel engine	Auto Shutdown Relay Control Circuit Malfunction Conditions: Key on or engine cranking; and the PCM detected an unexpected voltage condition on the ASD Relay Control circuit. The ASD Relay coil resistance is 95-105 ohms at 68ºF. Possible Causes • ASD relay control circuit is open between the relay and PCM • ASD relay control circuit is shorted to ground • ASD relay power circuit is open (test power from Fused B+) • ASD relay is damaged or has failed • PCM has failed
P1389 1T CCM 1996-2003 Car, Jeep, Mini-Van, SUV, Truck & Van models	No Auto Shutdown Relay Output Voltage To PCM Conditions: Engine cranking; and the PCM did not detect any voltage on the ASD Relay Output circuit to the PCM during the CCM test. Possible Causes • ASD relay connector is damaged, loose or shorted • ASD relay output circuit is open between the relay and PCM • ASD relay power circuit is open (test power from Fused B+) • ASD relay is damaged or has failed • PCM has failed

OBD II TROUBLE CODE LIST (P1XXX CODES)

DTC	Trouble Code Title, Conditions & Possible Causes
P1390 1T CCM 1995-2003 Car & PT Cruiser models with a 4-Cyl engine	Timing Belt Skipped One Tooth Or More Conditions: Engine started, engine running, then with the Inhibit Test not active, the PCM checked the CKP and CMP sensor alignment. If the PCM detects the CMP sensor is offset from the CKP sensor signal by 1 tooth, this trouble code is set. The PCM performs the Inhibit Test whenever the engine is cold, if the engine speed is outside of a given window, or if there is a large change in the MAP sensor signal. Possible Causes • Camshaft timing is out of specifications • Valve timing is out of specifications
P1391 1T CCM 1995-2003 Car, Jeep, Mini-Van, SUV, Truck & Van models	CKP Or CMP Sensor Signal Intermittent Conditions: Engine started, engine running, and after every 69-degree CKP sensor leading edge and trailing signal edge is determined, the PCM updates this data and compares it to the true CMP sensor port level. If the PCM detects a disagreement between these two values 20 times in succession, this trouble code is set. Possible Causes • Camshaft sensor is not installed properly • Engine valve timing is not within specifications • Perform a CKP and CMP Sensor relearn with the Scan Tool • Tone wheel or pulse ring is damaged
P1398 1T CCM 1995-2003 Car, Jeep, Mini-Van, SUV, Truck & Van models	Misfire Adaptive Numerator At Limit Conditions: Engine started, ECT sensor less than 75ºF, engine runtime over 50 seconds, A/C "off", vehicle driven to over 36 mph in 1st gear, or to over 65 mph in high gear, followed by a closed throttle deceleration period. This code sets if the PCM detects one of the CKP sensor target windows varies more than 2.86% from the reference window. Background - The PCM needs to learn any variation in the engine machining to detect when a misfire is present. The CKP sensor has two 40-degree windows that are 180 degrees apart. The window for Cylinders 1 and 4 is the reference window. It is checked against the window for Cylinders 2 and 3. The PCM checks for any variation to make engine speed adjustments. Possible Causes • Base engine problem (i.e., low cylinder compression) • CKP sensor crankshaft target variation too large • CKP sensor improperly installed or the CKP sensor has failed • CKP sensor signal circuit open or shorted (intermittent fault) • Tone wheel or pulse ring is damaged
P1400 1T CCM 1995-98 Eagle Talon models with 2.0L Turbo engine	Manifold Differential Pressure Sensor Circuit Malfunction Conditions: Engine running at low to medium load, ECT sensor over 65.4ºF, and the PCM detected the Manifold Differential Pressure (MDP) sensor was over 4.50v or under 0.20v for 4 seconds. Possible Causes • MDP sensor signal circuit is open or shorted to ground • MDP sensor power (VREF) circuit is open • MDP sensor is damaged or has failed • PCM has failed
P1400 1T CCM 2001-03 Sebring & Stratus Coupe with a 2.4L I4 or 3.0L V6 engine	Manifold Differential Pressure Sensor Circuit Malfunction Conditions: Engine runtime over 8 minutes if the ECT sensor is less than 32ºF at startup, ECT more than 113ºF during testing, volumetric efficiency from 30-45%, IAT sensor over 14ºF and the PCM detected the Manifold Differential Pressure (MDP) sensor was more than 4.60v (Scan Tool reads over 108 kPa) or less than 0.10v (Scan Tool reads under 2.4 kPa) for 2 seconds. If the volumetric efficiency is less than 30%, P1400 sets if the MDP sensor is over 4.20v (Scan Tool reads over 108 kPa) for 2 seconds). If the volumetric efficiency is more than 70%, P1400 sets if the MDP sensor is less than 1.80v (San Tool read 46 kPa) for 2 seconds. Possible Causes • MDP sensor signal circuit is open or shorted to ground • MDP sensor ground circuit is open (fault may be intermittent) • MDP power supply (VREF) circuit is open • MDP sensor is damaged, skewed or it has failed • PCM has failed

OBD II TROUBLE CODE LIST (P1XXX CODES)

DTC	Trouble Code Title, Conditions & Possible Causes
P1475 N/MIL 1T CCM 1998-2002 Truck applications with a 5.9L Diesel engine	Auxiliary 5-Volt Supply Circuit High Input Conditions: Key on or engine running; and the PCM detected an unexpected "high" voltage condition on the Auxiliary 5-volt power circuit. Possible Causes • Auxiliary 5v supply circuit shorted to system power • MAP sensor VREF circuit is shorted to ground • MAP sensor is open internally • PCM has failed
P1478 1T CCM 2002-03 Caravan, Town & Country, Voyager models	Battery Temperature Sensor Circuit Out-Of-Limits Conditions: Key on or engine running; and the PCM detected the Battery Temperature sensor was under 0.1v or over 4.90v for 3.2 seconds. Possible Causes • Battery temperature sensor is damaged or it has failed. • Clear the codes and retest for this trouble code. If the same code resets, the PCM will have to be replaced, as the Battery Temperature sensor is located inside the controller.
P1479 N/MIL 1T CCM 1997-2002 Prowler models	A/T Fan Relay Circuit Malfunction Conditions: Engine started, engine running, and the PCM detected an unexpected voltage condition on the A/T Transmission Relay circuit. Possible Causes • A/T fan relay control circuit is open or shorted to ground • A/T fan relay power circuit is open (test power to Fused IGN) • A/T fan relay is damaged or has failed • PCM has failed
P1480 1T CCM 2001-03 Truck & Van models	Positive Crankcase Ventilation Solenoid Circuit Failure Conditions: Engine started and the PCM detected an unexpected voltage on the PCV Solenoid circuit. Possible Causes • PCV solenoid control circuit is open or shorted to ground • PCV solenoid power circuit is open to the fuse in the PDC • PCV solenoid is damaged or has failed • PCM has failed
P1481 1T CCM 2001-03 Truck & Van models	EVAP Leak Detection Monitor Pinched Hose Detected Conditions: BTS from 40-96°F, ECT sensor within 10°F of the BTS signal at startup (cold engine); engine started, and after the EVAP Leak test started, the PCM detected a no flow condition. Possible Causes • EVAP vapor hose blocked between the fuel tank and the LDP (check rollover valve) • EVAP ventilation solenoid is damaged or has failed • Purge line is loose, damaged or incorrectly routed
P1486 2T EVAP 1996-2003 Cars, Jeep, Mini-Vans, SUV, Truck & Van models	EVAP Leak Detection Monitor Pinched Hose Detected Conditions: BTS from 40-96°F and ECT sensor within 20°F of the BTS signal at startup (cold engine), engine started, and after the EVAP Leak Detection test was enabled, the PCM detected the LDP switch did not reach 3 closures (i.e., a "no flow" condition was present). Possible Causes • EVAP vapor hose blocked between the fuel tank and the LDP (i.e., in the OLFV, rollover or vapor hose) • EVAP canister is clogged or full of dirt or moisture • EVAP ventilation solenoid is damaged or has failed • Purge line is loose, damaged or incorrectly routed • PCM has failed
P1487 1T CCM 1995-98 Eagle Talon, Sebring & Stratus models	High Speed Radiator Fan Relay Circuit Failure Conditions: Key on or engine running; system voltage over 10.5v, and the PCM detected an unexpected voltage condition on the High Speed Radiator Fan Relay control circuit during the CCM test. Possible Causes • HFAN radiator relay control circuit is open or shorted to ground • HFAN radiator relay power circuit is open (test power to IGN) • HFAN radiator fan relay is damaged or has failed • PCM has failed

OBD II TROUBLE CODE LIST (P1XXX CODES)

DTC	Trouble Code Title, Conditions & Possible Causes
P1488 1T CCM 1996-2002 Trucks with a 5.9L Diesel engine	Auxiliary 5-Volt Supply Circuit Low Input Conditions: Key on or engine running; and the PCM detected an unexpected "low" voltage condition on the Auxiliary 5-volt power circuit. Possible Causes • Auxiliary 5v supply circuit shorted to sensor or chassis ground • Camshaft position sensor VREF circuit is shorted to ground • MAP sensor VREF circuit is shorted to ground • Oil Pressure sensor VREF circuit is shorted to ground • PCM has failed
P1489 1T CCM 1995-2003 Car & PT Cruiser models	High Speed Radiator Fan Relay Circuit Malfunction Conditions: Key on or engine running; system voltage over 10.5v, and the PCM detected an unexpected voltage condition on the High Speed Radiator Fan Relay control circuit during the CCM test. Possible Causes • HFAN radiator relay control circuit is open or shorted to ground • HFAN radiator relay power circuit is open (test power to IGN) • HFAN radiator fan relay is damaged or has failed • PCM has failed
P1490 1T CCM 1995-2003 Car & PT Cruiser models	Low Speed Radiator Fan Relay Circuit Malfunction Conditions: Key on or engine running; system voltage over 10.5v, and the PCM detected an unexpected voltage condition on the Low Speed Radiator Fan Relay control circuit during the CCM test. Possible Causes • LFAN radiator relay control circuit is open or shorted to ground • LFAN radiator relay power circuit is open (test power to IGN) • LFAN radiator fan relay is damaged or has failed • PCM has failed
P1491 1T CCM 1995-2003 Car, Jeep, Mini-Van, SUV, Truck & Van models	Radiator Fan Control Relay Circuit Malfunction Conditions: Key on or engine running; system voltage over 10.5v, and the PCM detected an unexpected voltage condition on the Radiator Fan Control Relay circuit during the CCM test period. Possible Causes • Radiator fan control relay circuit is open or shorted to ground • Radiator fan control relay circuit is open (test power to IGN) • Radiator fan control relay is damaged or has failed • PCM has failed
P1492 1T CCM 1995-2003 Car, Jeep, Mini-Van, SUV, Truck & Van models	Battery Temperature Sensor Circuit High Input Conditions: Key on or engine running; and the PCM detected the BTS signal indicated more than 4.90v for 3 seconds during the CCM test. Possible Causes • BTS signal circuit is open between the sensor and the PCM • BTS ground circuit is open between the sensor and the PCM • BTS (sensor) is damaged or the PCM has failed
P1493 1T CCM 1995-2003 Car, Jeep, Mini-Van, SUV, Truck & Van models	Battery Temperature Sensor Circuit Low Input Conditions: Key on or engine running; and the PCM detected the BTS signal indicated less than 0.30v for 3 seconds during the CCM test. Possible Causes • BTS circuit is shorted to ground between sensor and the PCM • BTS (sensor) is damaged or has failed • PCM has failed
P1494 1T CCM 1996-2003 Car, Jeep, Mini-Van, SUV, Truck & Van models	EVAP Leak Detection Pump Switch Or Mechanical Fault Conditions: BTS from 40-96ºF and ECT sensor within 10ºF of the BTS signal at startup (cold engine), engine started, and the PCM detected the LDP switch was not in its expected state at key "on" or engine running. Possible Causes • LDP switch signal circuit is open or shorted to ground • LDP switch power circuit is open (test power to Fused Ignition) • LDP vacuum hose is clogged, loose or restricted • LDP assembly is damaged or has failed (the switch has failed)

OBD II TROUBLE CODE LIST (P1XXX CODES)

DTC	Trouble Code Title, Conditions & Possible Causes
P1495 1T CCM 1996-2003 Car, Jeep, Mini-Van, SUV, Truck & Van models	Leak Detection Pump Solenoid Circuit Malfunction Conditions: Engine started, ECT sensor from 40-90ºF and within 10ºF of the Battery Temperature sensor signal, engine running, and the PCM detected the Actual state of the Leak Detection Pump solenoid did not match the Intended state of the solenoid during the test period. Possible Causes • LDP power supply circuit from the ignition switch is open • LDP solenoid control circuit is open or shorted to ground • LDP assembly is damaged or it has failed • PCM has failed
P1496 1T CCM 1996-2003 Car, Jeep, Mini-Van, SUV, Truck & Van models	5-Volt VREF Supply Voltage Too Low Conditions: Key on or engine running; and the PCM detected the 5-volt VREF supply was less than 3.5v for 4 seconds during the CCM test. Possible Causes • A/C pressure sensor has failed (a short to ground condition) • MAP sensor has failed (a short to ground condition) • TP sensor has failed (a short to ground condition) • PCM has failed
P1497 1T CCM 1996-97 Concorde, Intrepid, LHS & Prowler models	PCM Failure (SRI Mileage Not Stored) Conditions: Key on, and the PCM detected an unsuccessful attempt to "write" the Service Reminder Indicator (SRI) or Emission Mileage Request (EMR) mileage to an EEPROM located occurred during initialization. Possible Causes • Clear the trouble codes and retest for the same trouble code. If DTC P1697 resets, replace the PCM and then reprogram it.
P1498 1T CCM 1996-2002 Avenger, Eagle Talon, Sebring, 300M, Concorde, Intrepid, LHS & Prowler models	No CCD Messages Received From The TCM Conditions: Key on or engine running; and the PCM detected a failure to communicate with the TCM over the CCD data bus circuit. Possible Causes • CCD data bus circuit is open, shorted to ground or to power • TCM or the PCM has failed
P1499 1T CCM 1996-2003 Jeep models	Radiator (Hydraulic) Fan Solenoid Circuit Failure Conditions: Key on or engine running; and the PCM detected an unexpected voltage condition on the Radiator Fan Solenoid Control circuit. Possible Causes • Radiator fan solenoid control circuit is open • Radiator fan solenoid ground circuit is open • Radiator fan solenoid power circuit is open • Radiator fan solenoid is damaged or has failed • PCM has failed
P1500 1T CCM 1995-98 Eagle Talon models	Generator 'FR' Terminal Circuit Failure Conditions: Engine started, and the PCM detected the Generator 'FR' terminal remained at more than 4.5v for over 20 seconds during the test. Possible Causes • Generator 'FR' terminal circuit is open between the Generator and the PCM terminal • Generator is damaged or has failed • PCM has failed
P1500 1T CCM 2001-03 Sebring & Stratus Coupe with a 2.4L I4 or 3.0L V6 engine	Generator 'FR' Terminal Circuit Failure Conditions: Engine started, engine running and the PCM detected the Generator 'FR' terminal input signal indicated more than 4.50v for 20 seconds. Possible Causes • Generator 'FR' circuit is open between the Generator and PCM • Generator is damaged or it has failed • PCM has failed

OBD II TROUBLE CODE LIST (P1XXX CODES)

DTC	Trouble Code Title, Conditions & Possible Causes
P1594 1T CCM 1996-97 Concorde, Intrepid, LHS & Prowler models	Charging System Voltage Too High Conditions: Engine started, and the PCM detected the Charging System voltage was too high even after it tried to lower the output. Note: The Generator illuminates when this code sets. Possible Causes • Battery temperature sensor is damaged or has failed (skewed) • Generator field driver circuit is shorted to ground • Generator has an internal short circuit condition • PCM has failed
P1594 1T CCM 1996-2003 Jeep models	Charging System Voltage Too High Conditions: Engine started, engine running, and the PCM detected the Charging System voltage was too high even after it tried to lower the generator output by controlling the Field control circuit (Generator Lamp is on). Possible Causes • Battery temperature sensor is damaged or has failed (skewed) • Generator field driver circuit is shorted to ground • Generator has an internal short circuit condition • PCM has failed
P1595 N/MIL 1T CCM 1996-2003 Car, Jeep, Mini-Van, SUV, Truck & Van models	Speed Control Solenoid Circuit Failure Conditions: Engine started, vehicle driven at over 35 mph, S/C enabled with the Set switch "on", and the PCM detected it could not control the operation of the vacuum and vent control solenoids. Possible Causes • S/C power supply circuit is open (test power from Brake switch) • S/C vacuum solenoid control circuit open or shorted to ground • S/C vent solenoid control circuit is open or shorted to ground • S/C vacuum or vent solenoid is damaged or has failed • PCM has failed
P1596 N/MIL 1T CCM 1996-2003 Car, Jeep, Mini-Van, SUV, Truck & Van models	Speed Control Switch Continuous High Input Conditions: Key on or engine running; and the PCM detected the S/C switch was in a continuous high voltage state (over 4.70v) during the CCM test. Possible Causes • S/C On/Off switch is open • S/C switch (MUX switch) is open • S/C switch (MUX switch) is shorted to VREF or system power • S/C switch (MUX switch) is damaged or has failed • PCM has failed
P1597 N/MIL 1T CCM 1996-2003 Car, Jeep, Mini-Van, SUV, Truck & Van models	Speed Control Switch Continuous Low Input Conditions: Key on or engine running; and the PCM detected the S/C switch was in a continuous low voltage state (below 4.50v) during the CCM test. Possible Causes • S/C On/Off switch is shorted to ground • S/C switch (MUX switch) is shorted to ground • S/C switch (MUX switch) is damaged or has failed • PCM has failed
P1598 N/MIL 1T CCM 1996-2003 Car, Jeep, Mini-Van, SUV, Truck & Van models	A/C Pressure Sensor Circuit High Input Conditions: Engine started, engine running, A/C Relay is "on", and the PCM detected the A/C Pressure sensor indicated more than 4.90v. Possible Causes • A/C pressure sensor circuit is open or shorted to VREF (5v) • A/C pressure sensor ground circuit is open • A/C pressure sensor is damaged or has failed • PCM has failed

OBD II TROUBLE CODE LIST (P1XXX CODES)

DTC	Trouble Code Title, Conditions & Possible Causes
P1599 N/MIL 1T CCM 1996-2003 Car, Jeep, Mini-Van, SUV, Truck & Van models	A/C Pressure Sensor Circuit Low Input Conditions: Engine started, engine running, A/C Relay is "on", and the PCM detected the A/C Pressure sensor indicated less than 0.70v. Possible Causes • A/C pressure sensor circuit is shorted to ground • A/C pressure sensor power circuit is open • A/C pressure sensor is damaged or has failed • PCM has failed
P1602 1T PCM 1995-98 Eagle Talon models	A/T Serial Communication Link Circuit Malfunction Conditions: Key on or engine running; and the PCM detected an unexpected voltage condition on the serial communication link used to communicate between it and the TCM. Possible Causes • CCD data bus (+) circuit is open or shorted to ground • CCD data bus (-) circuit is open • TCM has failed • PCM has failed
P1602 1T PCM 2001-03 Car, Mini-Van & PT Cruiser models	PCM Not Programmed Conditions: Key on; and the PCM detected that it had not been programmed. Possible Causes • Program the PCM and then retest for this same trouble code
P1603 1T PCM 2001-03 Car, Mini-Van & PT Cruiser models	Powertrain Control Module Internal Dual-Port Ram Communication Conditions: Key on; and the PCM detected an error message that indicated that it had not been programmed or that it was programmed properly. Possible Causes • Fused ignition switch output is missing (off-start-run circuit) • PCM is damaged or it has an internal failure
P1603 1T PCM 2001-03 Sebring & Stratus with a 2.4L I4 and 3.0L V6 engine	Powertrain Control Module Internal Dual-Port Ram Communication Conditions: Key on; and the PCM detected an error message that indicated that it had not been programmed or that it was programmed properly. Possible Causes • Fused ignition switch output is missing (off-start-run circuit) • PCM is damaged or it has an internal failure
P1604 N/MIL 1T PCM 2001-03 Car, Mini-Van & PT Cruiser models	PCM Internal Dual-Port Ram Read/Write Integrity Failure Conditions: Key on; and the PCM detected an error message that indicated it had not been programmed, or it was not programmed properly. Possible Causes • Fused ignition switch output is missing (off-start-run circuit) • PCM is damaged or it has an internal failure
P1607 N/MIL 1T PCM 2001-03 Car, PT Cruiser, Dakota & Durango models	Powertrain Control Module Internal Shutdown Timer Rationality Conditions: Cold engine startup, and after the PCM compared the coolant temperature to the shutdown time, it detected a rationality fault. Possible Causes • Fused ignition switch output is missing (off-start-run circuit) • PCM is damaged or it has an internal failure
P1610 1T PCM 2001-03 Sebring & Stratus with a 2.4L I4 and 3.0L V6 engine	PCM Signal Line To Immobilizer Circuit Malfunction Conditions: Key on, and the PCM detected an unexpected voltage condition on the communication line between the Immobilizer ECU and the PCM. Possible Causes • Immobilizer communication line to the PCM is open • Immobilizer communication line to the PCM is shored to ground • Immobilizer communication line to the PCM is shorted to power • Immobilizer ECU is damaged or it has failed • PCM has failed

OBD II TROUBLE CODE LIST (P1XXX CODES)

DTC	Trouble Code Title, Conditions & Possible Causes
P1652 1T CCM 2001-03 Car, Mini-Van & PT Cruiser models	Serial Communication Link Malfunction Conditions: Engine started; and after the TCM did not detect any signals on the Serial Communication Line for more than 20 seconds. Possible Causes • TCM cannot communicate with the Instrument Cluster (MIC) • TCM cannot communicate with the Powertrain Control Module • TCM is damaged or it has an internal failure
P1652 1T CCM 2001-03 Car, Mini-Van & PT Cruiser models	Serial Communication Link Malfunction Conditions: Engine started; and after the TCM did not detect any signals on the Serial Communication Line for more than 20 seconds. Possible Causes • TCM cannot communicate with the Instrument Cluster (MIC) • TCM cannot communicate with the Powertrain Control Module • TCM is damaged or it has an internal failure
P1681 1T PCM 2003 Car, PT Cruiser & Truck models	No Fuel Level Bus Messages Conditions: Key on, and the PCM determined that it did not receive any Fuel Level messages over the Data Bus line for 20 seconds. Possible Causes • Data bus circuit from BCM to PCM is damaged or it has failed • Fuel level Bus message to the PCM is invalid • BCM is damaged or has failed • PCM unable to communicate with the Body Control Module
P1682 1T CCM 1996-2003 Car, Jeep, Mini-Van, SUV, Truck & Van models	Charging System Voltage Too Low Conditions: Engine started; engine speed over 1152 rpm, and the PCM detected the Battery Sense circuit was 1.0v less than the Charging System circuit for 25 seconds during the CCM test (Generator Lamp is "on"). Possible Causes • Battery positive or Fused Ignition circuit has high resistance • Generator drive belt out-of-adjustment or worn out • Generator field circuit has a high resistance condition • PCM has failed
P1683 1T CCM 1996-2003 Car, Jeep, Mini-Van, SUV, Truck & Van models	Speed Control Relay Or Driver Circuit Malfunction Conditions: Engine started, engine running with the S/C switch "on"; and the PCM detected an unusual voltage condition on the S/C Relay control circuit during the CCM test. Possible Causes • S/C power supply circuit is open or shorted to ground • S/C dump solenoid (servo) is damaged or has failed • PCM has failed
P1684 1T PCM 2001-03 Car, Mini-Van, PT Cruiser models	Battery Has Been Disconnected Conditions: Key on, and the TCM detected that it had been disconnected from the Battery Direct (B+) circuit or its Power Ground circuit. Possible Causes • Quick Learn procedure was performed with a Scan Tool • TCM battery direct (B+) circuit is open or disconnected • TCM power ground circuit is open • TCM was disconnected or it has been replaced
P1685 1T PCM 1996-2003 Car, Jeep, Mini-Van, SUV, Truck & Van models	Smart Key Immobilizer Module Invalid Key Conditions: Key on, and the PCM received a message from the Smart Key Immobilizer Module (SKIM) that an invalid key had been inserted. Possible Causes • A theft attempt may have occurred. • Obtain the correct key and attempt to start the vehicle • Do a PCM Reset function to clear the code after engine startup

OBD II TROUBLE CODE LIST (P1XXX CODES)

DTC	Trouble Code Title, Conditions & Possible Causes
P1686 1T PCM 1999-2003 Car, Jeep, Mini-Van, SUV, Truck & Van models	No SKIM Bus Messages Received Conditions: Engine started; and the PCM did not detect any MIC (I/P Cluster) messages over the Data Bus circuit for 20 seconds. Possible Causes • Data bus circuit from MIC to the PCM is damaged or it is open • Instrument Cluster is damaged or has failed • PCM unable to communicate with the Instrument Cluster • PCM has failed
P1687 1T PCM 1999-2003 Car, Jeep, Mini-Van, SUV, Truck & Van models	No Cluster Bus Messages Conditions: Key on or engine running; and the PCM determined that it did not receive any Security Key Bus Messages over the Data Bus line for 20 seconds. This malfunction may be an intermittent problem. Possible Causes • Data bus circuit from SKIM to PCM is damaged or it is open • PCM unable to communicate with the Body Control Module • PCM has failed, or the SKIM is damaged or has failed
P1688 N/MIL 1T CCM 1998-2002 Trucks equipped with a 5.9L Diesel engine	Fuel Injection Pump Internal Malfunction Conditions: Key on or engine running; and the PCM detected an unexpected "low" voltage condition on the Auxiliary 5-volt power circuit. Possible Causes • Fuel injection pump DTC counter malfunction • Fuel injection pump Good Trip counter malfunction • Fuel injection pump is damaged or has failed • ECM has failed
P1689 N/MIL 1T PCM 1998-2002 Trucks equipped with a 5.9L Diesel engine	No Communication Between ECM And Injection Pump Module Conditions: Key on, and the PCM detected the time between the CAN messages received from the Instrument Panel (I/P) module was more than 3 seconds, or it detected no messages arrived. Possible Causes • Fuel injection pump ground circuit open or has high resistance • Fuel injection module wiring harness is damaged or has failed • CAN data bus (+) circuit is open, shorted to ground or to power • ECM has failed
P1690 N/MIL 1T CCM 1998-2002 Trucks equipped with a 5.9L Diesel engine	Injection Pump CKP Signal Different Than The CKP Signal Conditions: No CKP or CMP sensor codes set, engine started, and the PCM detected that the CKP signal received by the Instrument Panel (I/P) module was not within its normal operating range. Possible Causes • DTC counter did not change to (0) • Fuel injection module wiring harness is damaged or has failed • Fuel injection static timing is incorrect • Fuel "sync" circuit is open or shorted to ground • ECM has failed
P1691 N/MIL 1T PCM 1998-2002 Trucks equipped with a 5.9L Diesel engine	Fuel Injection Pump Calibration Error Conditions: Key on or engine running; and the PCM detected a calibration error related to the Fuel Injection pump operation during the CCM test. Possible Causes • Fuel Injection pump error at key "on" • Fuel injection pump error at startup • Fuel injection pump is damaged or has failed • ECM has failed
P1692 N/MIL 1T PCM 1998-2002 Trucks equipped with a 5.9L Diesel engine	Diagnostic Trouble Code Set In The Companion Module Conditions: Key on and the PCM detected a diagnostic trouble code was set in the Companion Module. Possible Causes • Companion Module detected a problem and set a trouble code in memory • ECM has failed

OBD II TROUBLE CODE LIST (P1XXX CODES)

DTC	Trouble Code Title, Conditions & Possible Causes
P1693 N/MIL 1T PCM 1998-2002 Trucks equipped with a 5.9L Diesel engine	Diagnostic Trouble Code Set In The ECM Conditions: Key on or engine running; and the PCM detected a diagnostic trouble code was set in the Electronic Control Module (ECM). Possible Causes • PCM has detected a problem and set a trouble code in memory • PCM has failed
P1694 N/MIL 1999-2003 Car, Jeep, Mini-Van & PT Cruiser models	No PCM Bus Messages Conditions: Engine started; system voltage over 10.5v and the PCM determined that it did not receive any Bus messages for 10 seconds. Possible Causes • Data bus circuit connector is damaged, open or it is shorted • Data bus circuit to the PCM is damaged or it is open • PCM unable to communicate with the body control module • PCM has failed
P1694 N/MIL 1998-2002 Trucks equipped with a 5.9L Diesel engine	No PCM Bus Messages Conditions: Engine started; system voltage over 10.5v and the PCM determined that it did not receive any Bus messages for 10 seconds. Possible Causes • Data bus circuit connector is damaged, open or it is shorted • Data bus circuit to the PCM is damaged or it is open • PCM unable to communicate with the body control module • PCM has failed
P1695 N/MIL 1T PCM 1998-2003 Car & Mini-Van models	No PCM Bus Messages Conditions: Engine started; system voltage over 10.5v and the TCM determined that it did not receive any PCM messages for 20 seconds. Possible Causes • Data bus circuit from TCM to the PCM is damaged or it is open • Powertrain control module is damaged or has failed • TCM unable to communicate with the TCM • TCM has failed
P1696 1T PCM 1998-2003 Car, Jeep, Mini-Van, SUV, Truck & Van models	PCM EEPROM Write Operation Denied Conditions: Engine started; and the PCM detected an unsuccessful attempt to "write" to an EEPROM location occurred at initialization or shutdown. Possible Causes • DRB or Scan Tool displays a "write" failure occurred • DRB or Scan Tool displays "write" refused a second time • DRB or Scan Tool displays SRI mileage invalid (compare the SRI mileage reading to the reading on the odometer) • Clear the trouble codes and retest for the same trouble code. If DTC P1696 resets, replace the PCM and then reprogram it.
P1696 1T PCM 1998-2002 Trucks equipped with a 5.9L Diesel engine	PCM Failure (EEPROM Write Operation Denied) Conditions: Key on, and the PCM detected an unsuccessful attempt to "write" to an EEPROM location occurred during initialization. Possible Causes • Clear the trouble codes and retest for the same trouble code. If DTC P1696 resets, replace the PCM and then reprogram it.
P1697 1T PCM 1996-2003 Car, Jeep & Mini-Van models	PCM Failure (SRI Mileage Not Stored) Conditions: Key on, and the PCM detected an unsuccessful attempt to "write" the Service Reminder Indicator (SRI) or Emission Mileage Request (EMR) mileage to an EEPROM located occurred during initialization. Possible Causes • Clear the trouble codes and retest for the same trouble code. If DTC P1697 resets, replace the PCM and then reprogram it.

OBD II TROUBLE CODE LIST (P1XXX CODES)

DTC	Trouble Code Title, Conditions & Possible Causes
P1697 1T PCM 1996-2003 Dakota, Durango, Truck & Van models	EMR Or SRI Mileage Not Stored Conditions: Engine started; and the PCM detected an unsuccessful attempt to "write" the Service Reminder Indicator (SRI) or Emission Mileage Request (EMR) mileage to an EEPROM located occurred during initialization or at engine shutdown. Possible Causes • DRB or Scan Tool displays a "write" failure occurred • DRB or Scan Tool displays "write" refused a second time • DRB or Scan Tool displays SRI mileage invalid (compare the SRI mileage reading to the reading on the odometer) • Clear the trouble codes and retest for the same trouble code. If DTC P1696 resets, replace the PCM and then reprogram it.
P1698 1T CCM 1995-98 Eagle Talon models	No CCD Messages Received From The TCM Conditions: Key on or engine running; and the PCM detected a failure to communicate with the TCM over the CCD data bus circuit. Possible Causes • CCD data bus circuit is open, shorted to ground or to power • TCM or the PCM has failed
P1698 1T PCM 1995-2003 Car, Jeep, Mini-Van Truck & Van models	No CCD Messages Received From The TCM Conditions: Key on or engine running; and the PCM detected a failure to communicate with the TCM over the CCD data bus circuit. Possible Causes • CCD data bus circuit is open, shorted to ground or to power • TCM or the PCM has failed
P1714 1T CCM 1999-2003 Breeze, Cirrus, Stratus, Sebring & Prowler models	A/T Transmission Control Relay Low Battery Voltage Conditions: Engine started; and the PCM detected a "low" voltage condition on the Transmission Control Relay output circuit after the relay was energized during the CCM test. Possible Causes • TCM relay output circuit(s) to the TCM have high resistance • TCM relay control circuit is open or shorted to ground • TCM relay control power circuit is open (check the fused B+) • TCM relay ground circuit is open • TCM control relay is damaged or has failed • TCM has failed
P1715 1T CCM 1995-98 Eagle Talon with a 2.0L Turbo engine	A/T Pulse Generator Assembly Circuit Malfunction Conditions: Engine started; vehicle driven to over 10 mph and the PCM detected an unexpected voltage condition on the Pulse Generator circuit. Possible Causes • Pulse generator positive (+) circuit is open or shorted to ground • Pulse generator negative (-) circuit is open or shorted to ground • Pulse generator is damaged or has failed
P1715 1T CCM 2002-03 Jeep & Truck models	A/T Pulse Generator Assembly Circuit Malfunction Conditions: Engine started; vehicle driven to over 10 mph and the PCM detected an unexpected voltage condition on the Pulse Generator circuit. Possible Causes • Pulse generator positive (+) circuit is open or shorted to ground • Pulse generator negative (-) circuit is open or shorted to ground • Pulse generator is damaged or has failed
P1716 1T CCM 1999-2002 Car & Mini-Van models	Bus Communication Failure With The PCM Conditions: Key on, and the PCM detected it could not communicate with the TCM for over 10 seconds. Possible Causes • CCD data bus circuit is open, shorted to ground or to power • Extremely low battery (system) voltage • TCM or the PCM has failed

OBD II TROUBLE CODE LIST (P1XXX CODES)

DTC	Trouble Code Title, Conditions & Possible Causes
P1717 1T CCM 2002-03 Breeze, Cirrus, Sebring, Stratus, 300M, Concorde & LHS models	Bus Communication Failure With The MIC Conditions: Key on or engine started, and the PCM detected it could not communicate with the Mechanical Instrument Cluster (MIC) for over 25 seconds during the test. Possible Causes • CCD data bus circuit is open, shorted to ground or to power • Extremely low battery (system) voltage • MIC or the TCM has failed
P1718 1T CCM 1999-2003 Jeep Models	TCM Internal Malfunction Conditions: Key on or engine started, and the PCM received a signal from the TCM over the data bus circuit that the TCM had an internal problem. Possible Causes • TCM has failed. • Replace the TCM • Perform Transmission Verification Test VER-1A
P1719 N/MIL 1T CCM 1996-2003 Sebring & Stratus Coupe, Viper models	A/T Skip Shift Solenoid Control Circuit Malfunction Conditions: Engine started; vehicle driven to a speed of 12-18 mph in 1st Gear at light to moderate engine load at an engine speed over 608 rpm, and the PCM detected an unexpected "low" or high voltage condition on the Reverse Gear Lockout solenoid circuit. Possible Causes • Skip Shift solenoid control circuit is open • Skip Shift solenoid control circuit shorted to ground • Skip Shift solenoid is damaged or has failed • PCM has failed
P1736 1T CCM 2002-03 Jeep & Truck models with a 3.7L V6 engine	A/T Gear Ratio Error In Second Prime Conditions: Engine started; vehicle driven in any forward gear, and the PCM detected that the ratio of the Input speed (rpm) to the Output speed did not match the current gear (up to 5 minutes to set). Possible Causes • Check for related trouble codes, and it can set due to an intermittent fault condition • Transmission internal gear ratio 2nd prime error or malfunction
P1738 1T CCM 1999-2003 Car, Dakota, Durango & Mini-Van models	A/T High Temperature Operation Activated Conditions: Engine started and the PCM detected the TFT sensor was more than 240ºF during the test. Possible Causes • Engine Cooling Fan System is not operating properly • Engine cooling fan has failed • ATF oil level is too high (overfilled) • Transmission oil cooler capacity too low, or cooler is plugged • TCM or the PCM has failed
P1739 1T CCM 1999-2002 Car, Dakota, Durango & Mini-Van models	A/T Power Up Circuit At Speed Conditions: Key on or engine running; and the PCM detected an unexpected "low" voltage on the TCM power up circuit (battery direct circuit). Possible Causes • Fused Ignition power circuit is open between TCM and PCM • TCM Power UP circuit is open (test power from PDC or Fuse) • TCM Power UP circuit is grounded (test power from PDC or • TCM or the PCM has failed
P1740 N/MIL 2T PCM 1996-2003 Dakota, Durango, Jeep, Truck & Van models	TCM Internal Malfunction Conditions: Engine started; vehicle driven at over 30 mph in 3rd gear with TCC and Overdrive Clutch engaged, and PCM detected an invalid engine speed to output shaft ratio with the TCC or O/D Clutch engaged. Possible Causes • ATF fluid is burnt, contaminated or contains excessive debris • A/T cooler flow restriction or problems in the oil pump shaft • Internal transmission problem (O/D clutch seals or valve body) • Overdrive clutch or TCC clutch is damaged or has failed • Transmission valve body is leaking, damaged or has failed

OBD II TROUBLE CODE LIST (P1XXX CODES)

DTC	Trouble Code Title, Conditions & Possible Causes
P1742 1T CCM 1999-2003 Jeep Models	TCM Internal Malfunction Conditions: Key on or engine started, and the PCM received a signal from the TCM over the data bus circuit that the TCM had an internal problem. Possible Causes • TCM has failed. • Replace the TCM • Perform Transmission Verification Test VER-1A
P1743 1T CCM 1998-2002 Jeep Models	TCM Internal Malfunction Conditions: Key on or engine started, and the PCM received a signal from the TCM over the data bus circuit that the TCM had an internal problem. Possible Causes • TCM has failed. • Replace the TCM • Perform Transmission Verification Test VER-1A
P1744 1T CCM 1998-2002 Jeep Models	A/T Shift Solenoid 'A' Control Circuit Low Input Conditions: Engine started; vehicle driven in 1st or 2nd gear position, and the PCM detected an unexpected "low" voltage condition on the SSA control circuit during the CCM test. Possible Causes • A/T SSA control circuit is shorted to ground • A/T SSA is damaged or has failed (it may be shorted) • PCM has failed
P1745 1T CCM 1998-2002 Jeep Models	A/T Shift Solenoid 'A' Control Circuit High Input Conditions: Engine started; vehicle driven in 1st or 4th gear position, and the PCM detected an unexpected "high" voltage condition on the SSA control circuit during the CCM test. Possible Causes • A/T SSA control circuit is open between the solenoid and PCM • A/T SSA control circuit is shorted to system power (B+) • A/T SSA is damaged or has failed (it may be shorted) • PCM has failed
P1746 1T CCM 1998-2002 Jeep Models	A/T Shift Solenoid 'B' Control Circuit Low Input Conditions: Engine started; vehicle driven in 1st or 2nd gear position, and the PCM detected an unexpected "low" voltage condition on the SSB control circuit during the CCM test. Possible Causes • A/T SSB control circuit is shorted to ground • A/T SSB is damaged or has failed (it may be shorted) • PCM has failed
P1747 1T CCM 1998-2002 Jeep Models	A/T Shift Solenoid 'B' Control Circuit High Input Conditions: Engine started; vehicle driven in 3rd or 4th gear position, and the PCM detected an unexpected "high" voltage condition on the SSB control circuit during the CCM test. Possible Causes • A/T SSB control circuit is open between the solenoid and PCM • A/T SSB control circuit is shorted to system power (B+) • A/T SSB is damaged or has failed (it may be shorted) • PCM has failed
P1748 1T CCM 1998-2002 Jeep Models	A/T TCC Solenoid 'C' Control Circuit Low Input Conditions: Engine started; vehicle driven at cruise with the TCC engaged, and the PCM detected an unexpected "low" voltage condition for 12.5 seconds on the TCC control circuit in the test. Possible Causes • A/T TCC control circuit is shorted to ground • A/T TCC is damaged or has failed (it may be shorted) • PCM has failed

OBD II TROUBLE CODE LIST (P1XXX CODES)

DTC	Trouble Code Title, Conditions & Possible Causes
P1749 1T CCM 1998-2002 Jeep Models	A/T Shift Solenoid 'C' Control Circuit High Input Conditions: Engine started; vehicle driven at cruise with the TCC engaged, and the PCM detected an unexpected "high" voltage condition on the SSC control circuit during the CCM test. Possible Causes • A/T SSC control circuit is open between the solenoid and PCM • A/T SSC control circuit is shorted to system power (B+) • A/T SSC is damaged or has failed (it may be shorted) • PCM has failed
P1750 1T CCM 1995-98 Eagle Talon with a 2.0L VIN F models	A/T Solenoid Assembly Control Circuit Malfunction Conditions: Key on or engine running; and the PCM detected an unexpected voltage condition on the A/T Solenoid Assembly Control circuit. Possible Causes • A/T converter clutch solenoid circuit open or shorted to ground • A/T pressure solenoid circuit is open or shorted to ground • A/T shift solenoid control circuit is open or shorted to ground • A/T solenoid assembly power circuit is open (power from relay)
P1751 1T CCM 2001 Sebring & Stratus models with a 2.4L I4 & 3.0L V6 engines	A/T Control Relay Circuit Malfunction Conditions: Engine started; and the TCM detected a malfunction in the A/T Control Relay circuit, and then sent a signal to the PCM indicating that the malfunction had occurred. Possible Causes • A/T control relay circuit is open, shorted to ground or to power • A/T control relay power circuit is open (test power to Fused B+) • A/T control relay is damaged or has failed • PCM has failed
P1756 1T CCM 1996-2003 Dakota, Durango, Jeep, Truck & Van models	A/T Governor Pressure Not Equal To Target At 15-20 PSI Conditions: Engine started; vehicle driven to a speed over 30 mph, and the PCM detected the Governor Pressure sensor indicated less than 15 psi or more than 30 psi when the requested pressure was 20-25 psi for 2.2 seconds. This fault must occur 5 times on one trip to set this code. Possible Causes • Check for the presence of other A/T related trouble codes • Governor pressure sensor is damaged or has failed • Governor pressure sensor VREF (5v) circuit is open or shorted • Transmission valve body is leaking, damaged or has failed • PCM has failed
P1757 1T CCM 1996-2003 Dakota, Durango, Jeep, Truck & Van models	A/T Governor Pressure Above 3 PSI In Gear At 0 MPH Conditions: Engine started; vehicle driven to a speed over 30 mph, and the PCM detected the Governor Pressure sensor was more than 3 psi with the requested pressure at 0 psi (95% duty cycle command) for 2.65 seconds. The fault must occur twice on one trip to set the code. Possible Causes • Check for the presence of other A/T related trouble codes • Governor pressure sensor is damaged or has failed • Governor pressure sensor VREF (5v) circuit is open or shorted • Transmission valve body is leaking, damaged or has failed • PCM has failed
P1762 1T CCM 1996-2003 Dakota, Durango, Jeep, Truck & Van models	A/T Governor Pressure Sensor Offset Volts Too High/Low Conditions: Key on or engine running; gear selector in Park or Neutral, and the PCM detected an out-of-range Governor Pressure sensor signal for 1.3 seconds. The fault must occur 3 times on one trip to set a code. Possible Causes • Governor pressure sensor VREF (5v) circuit is open • Governor pressure sensor ground circuit is open • P/N switch is damaged or has failed (not operating properly) • Transmission fluid has excessive debris or it is contaminated • PCM has failed

OBD II TROUBLE CODE LIST (P1XXX CODES)

DTC	Trouble Code Title, Conditions & Possible Causes
P1763 N/MIL 1T CCM 1996-2003 Dakota, Durango, Jeep, Truck & Van models	A/T Governor Pressure Sensor Signal Too High Conditions: Key on or engine running; and the PCM detected the Governor Pressure sensor signal indicated more than 4.89v for 8.5 seconds. Possible Causes • Governor pressure sensor signal circuit is open • Governor pressure sensor ground circuit is open • Governor pressure sensor signal circuit is shorted to VREF • Governor pressure sensor is damaged or has failed • Transmission harness solenoid circuit has a problem • PCM has failed
P1764 N/MIL 1T CCM 1996-2003 Dakota, Durango, Jeep, Truck & Van models	A/T Governor Pressure Sensor Signal Too Low Conditions: Key on or engine running; and the PCM detected the Governor Pressure sensor signal indicated less than 0.10v for 8.5 seconds. Possible Causes • Governor pressure sensor signal circuit is shorted to ground • Governor pressure sensor VREF (5v) circuit is open • Governor pressure sensor is damaged or has failed • Transmission harness solenoid circuit has a problem • PCM has failed
P1765 1T CCM 1996-2003 Car, SUV, Jeep, Truck & Van models	A/T Transmission Relay Circuit Malfunction Conditions: Key on or engine started, and the PCM detected an unexpected voltage condition on the TCC solenoid control circuit during the test. Possible Causes • Generator power source circuit to relay control circuit is open • Transmission relay power circuit is open (check fuse in PCC) • Transmission control circuit is open or shorted to ground • Transmission relay is damaged or has failed • PCM has failed
P1767 1T CCM 1996-2003 Car, SUV, Jeep, Truck & Van models	A/T Transmission Relay Circuit Malfunction Conditions: Key on or engine started, and the PCM detected an unexpected voltage condition on the TCC solenoid control circuit during the test. Possible Causes • Generator power source circuit to relay control circuit is open • Transmission relay power circuit is open (check fuse in PCC) • Transmission control circuit is open or shorted to ground • Transmission relay is damaged or has failed • PCM has failed
P1768 1T CCM 1996-2003 Car, SUV, & Mini-Van models	A/T Transmission Relay Output Always Off Conditions: Engine started; and the TCM did not detect voltage on the Transmission Control Relay output circuit with the relay "on". Possible Causes • 2-4 or L/R Solenoid control circuit is open • O/D or U/D Solenoid control circuit is open • Transmission control relay power circuit is open to Fused B+ • Transmission control relay is damaged (contacts have failed) • TCM has failed
P1775 1T CCM 1999-2003 Car, Jeep, Mini-Van, Truck & Van models	A/T Solenoid Switch Latched In TCC Position Conditions: Engine started; vehicle driven to over 15 mph and the TCM detected the Transmission did not shift into 1st Gear (test must fail 3 times). Possible Causes • Extremely low battery (system) voltage • L/R Solenoid pressure switch circuit is open or switch has failed • Transmission solenoid pack is damaged or has failed • Transmission control relay circuit is shorted to L/R solenoid • Valve body engine idle too high • Valve body solenoid switch stuck in "lockup" position

OBD II TROUBLE CODE LIST (P1XXX CODES)

DTC	Trouble Code Title, Conditions & Possible Causes
P1776 1T CCM 1999-2003 Car, Jeep, Mini-Van, Truck & Van models	A/T Solenoid Switch Latched In Low/Reverse Position Conditions: Engine started; vehicle driven to over 30 mph and the TCM detected the L/R switch was closed while performing PEMCC or FEMCC. Possible Causes • Extremely low battery (system) voltage • Transmission pan has debris caused by valve body damage • Transmission internal problems or valve body damage
P1781 1T CCM 1999-2003 Car, Jeep, Mini-Van, Truck & Van models	A/T Overdrive Pressure Switch Circuit Malfunction Conditions: Engine started; engine runtime over 2 seconds, engine speed over 500 rpm, no "loss of prime" test in progress, no pressure switch mismatch detected, and the TCM detected the Overdrive pressure switch was open or closed at the wrong time in any gear position. Possible Causes • O/D pressure switch circuit is open or shorted to ground • O/D pressure switch circuit is shorted to the TCR solenoid pack • O/D pressure switch is damaged or has failed • O/D solenoid is damaged or has failed • Transmission seals are damaged or have failed
P1782 1T CCM 1999-2003 Car, Jeep, Mini-Van, Truck & Van models	A/T Overdrive Pressure Switch Circuit Malfunction Conditions: Engine started; and the TCM detected an unexpected voltage condition on the Overdrive Pressure Switch while driving in any gear positions during the CCM test. Possible Causes • 2-4 line pressure is too high • 2-4 pressure switch is open or shorted to ground • Solenoid pack is damaged or has failed • Transmission has internal problems or has failed • Valve body torque is out of specification
P1784 1T CCM 1999-2003 Car, Jeep, Mini-Van, Truck & Van models	A/T Additional L/R Pressure Switch Circuit Malfunction Conditions: Engine started; vehicle driven to a speed over 30 mph in Drive, and the PCM detected an invalid Low/Reverse Pressure switch value. Possible Causes • L/R pressure switch circuit open, shorted to ground or to power • L/R solenoid switch valve is stuck in "lockup" position • L/R solenoid is damaged or has failed • Solenoid pack is damaged or has failed (causing this code) • Valve body bolt torque is out of specification
P1787 1T CCM 1999-2002 Breeze, Cirrus, Stratus & Sebring models	A/T Additional O/D Pressure Switch Circuit Malfunction Conditions: Engine started; vehicle driven to a speed over 30 mph in Drive, and the PCM detected an invalid Overdrive Pressure Switch pressure. Possible Causes • O/D pressure switch circuit is open or shorted to ground • O/D pressure switch circuit is shorted to TCR output circuit • O/D solenoid is damaged or has failed
P1788 1T CCM 1999-2003 Car, Jeep, Mini-Van, Truck & Van models	A/T Additional 2-4 Pressure Switch Circuit Malfunction Conditions: Engine started; vehicle driven to a speed of 4-37 mph in Drive, and the PCM detected an invalid 2-4 Pressure Switch circuit pressure. Possible Causes • 2-4 pressure switch circuit is open or shorted to ground • 2-4 pressure switch circuit is shorted to TCR output circuit • 2-4 solenoid is damaged or has failed
P1789 1T CCM 1999-2003 Car & Mini-Van models	A/T Overdrive/2-4 Pressure Switch Circuit Malfunction Conditions: Engine started; vehicle driven in 1st, 2nd or 3rd Gear at an engine speed over 1000 rpm, and the TCM detected the O/D or 2-4 Pressure switch did not close (test fails twice to set a code). Possible Causes • Extremely low battery (system) voltage • 2-4 pressure switch circuit is open or shorted to ground • O/D pressure switch circuit is open or shorted to ground

OBD II TROUBLE CODE LIST (P1XXX CODES)

DTC	Trouble Code Title, Conditions & Possible Causes
P1790 1T CCM 1999-2003 Car, SUV, Jeep, Mini-Van, Truck & Van	A/T Malfunction Immediately After Shift Event Conditions: Engine started; vehicle driven to a speed over 10 mph in Drive, and the TCM detected a Speed Ratio Code within 1.3 seconds of a shift. Possible Causes • Transmission has an internal mechanical problem
P1791 1T CCM 1995-98 Eagle Talon models	A/T Engine Coolant Temperature Circuit Malfunction Conditions: Key on or engine running; and the PCM detected an unexpected voltage condition on the Engine Coolant Temperature (ECT) sensor circuit shared by the PCM and TCM in the test. Possible Causes • ECT sensor signal circuit is open or shorted to ground • ECT sensor ground circuit is open between sensor and PCM • ECT sensor is damaged o has failed (It is open or shorted)
P1791 1T CCM 1999-2003 Car, SUV, Jeep, Mini-Van, Truck & Van models	A/T Loss Of Prime Malfunction Conditions: Engine started; vehicle driven to a speed of 4-37 mph in Drive, and the vehicle exhibited a "no drive" condition during the CCM test. Possible Causes • ATF cooler lines damaged or leaking • ATF fluid level incorrect or filled with debris • Customer complains of "no drive" condition • Oil pump pressure drop occurs with "no drive" condition • Transmission filter clogged or restricted
P1792 1T CCM 1999-2003 Car, SUV, Jeep, Mini-Van, Truck & Van models	Battery Was Disconnected Conditions: Key on and the TCM detected an interruption of the Battery Direct Power (B+) circuit. Possible Causes • Battery direct power (B+) circuit is open (test power to PDC) • Battery direct power (B+) circuit is shorted to ground (test fuse) • Battery was discharged to a low voltage state or disconnected • Scan Tool "quick learn" for quick disconnect function performed
P1793 1T CCM 1999-2003 Car, SUV, Jeep, Mini-Van, Truck & Van models	Torque Management Request Circuit Malfunction Conditions: Key on and the TCM did not detect any signals on the Torque Management Request circuit. Possible Causes • Extremely low battery (system) voltage • Torque management request circuit is open, shorted to ground or shorted to system power (i.e., the TRD circuit has failed) or the TCM has failed
P1794 1T CCM 1999-2003 Car, SUV, Jeep, Mini-Van, Truck & Van models	A/T Speed Sensor Ground Circuit Malfunction Conditions: Engine started; gear selector position indicating Neutral, and the PCM an error in the Output Speed Sensor signal during the test. Possible Causes • Extremely low battery (system) voltage • TCM "reset" function has just been performed
P1795 1T CCM 1999-2003 Car, SUV, Jeep, Mini-Van, Truck & Van models	Transmission Control Module Internal Malfunction Conditions: Engine started; gear position is Neutral, and the PCM detected the engine model stored in RAM was different than the EEPROM value. Possible Causes • Extremely low battery (system) voltage • Engine "starts" since set counter less than three (3)
P1796 1T CCM 1999-2003 Car & Mini-Van models	A/T Additional Autostick Input Circuit Malfunction Conditions: Engine started; vehicle driven to a speed of 4-37 mph in Drive, and the PCM detected a problem on the Autostick signal circuit. Possible Causes • Autostick switch ground circuit is open • Autostick downshift switch is stuck in open or closed position • Autostick downshift sensor circuit is open or shorted to ground • Autostick switch power circuit is open (test power at Fused B+) • Autostick Upshift switch is stuck in open or closed position • Autostick Upshift switch sensor is open or shorted to ground

OBD II TROUBLE CODE LIST (P1XXX CODES)

DTC	Trouble Code Title, Conditions & Possible Causes
P1797 1T CCM 1999-2003 Car & Mini-Van models	A/T Manual Shift Overheat Malfunction Conditions: Engine started; vehicle driven to a speed over 30 mph in Drive, and the TCM detected A Manual Shift Overheat condition was present. Possible Causes • ATF fluid level too high (transmission may be overfilled) • Engine Cooling System or engine cooling fan malfunction • Excessive drive time in low gear, or aggressive drive patterns • Transmission oil cooler is clogged or restricted
P1798 1T CCM 1999-2003 Car & Mini-Van models	A/T Transmission Fluid Is Burnt Or Contaminated Conditions: Engine started; vehicle driven to a speed over 30 mph in Drive, and the TCM detected a vehicle "shutter" condition during TCC lockup. Possible Causes • ATF is burnt, contaminated or very dirty • Extremely low battery (system) voltage • Transmission pan has debris • Transmission has internal damage causing the vehicle shudder
P1799 1T CCM 1999-2003 Car, Jeep, SUV, Mini-Van, Truck & Van models	A/T Calculated Oil Temperature In Use Malfunction Conditions: Engine started; vehicle driven to a speed over 30 mph in Drive, and the TCM detected A Manual Shift Overheat condition was present. Possible Causes • TFT sensor signal circuit is open • TFT sensor ground circuit is open • TFT sensor signal circuit is shorted to ground • TFT sensor signal circuit is shorted to VREF (5v) • TFT sensor signal not correct for the ATF fluid temperature • Speed sensor signal circuit is open or shorted to ground • Speed sensor signal circuit is shorted to system power • TCM oil temperature sensor high or low circuit has failed
P1830 1T CCM 2002-03 Jeep with a 2.4L VIN 1	Clutch Override Relay Control Circuit Malfunction Conditions: Key on or engine running; and the PCM detected an unexpected voltage condition on the Clutch Override Relay Control circuit. Possible Causes • Clutch override relay control circuit is open • Clutch override relay control circuit is shorted to ground • Clutch override relay is damaged or it has failed • Fused ignition power circuit to the clutch override relay is open • PCM has failed
P1854 1T CCM 2003 PT Cruiser models	Throttle Inlet Pressure BARO Reading Out Of Range Conditions: Engine started; and the PCM detected the BARO sensor indicated an incorrect reading. Possible Causes • Inspect the hoses and tubing to the turbo charger assembly • Review results of Solenoid Tests (Test 1, 2, 3 and 4 results) • TIP signal circuit is open (may be an intermittent fault) • TIP ground circuit is open (may be an intermittent fault) • PCM has failed
P1899 1T CCM 1995-2003 Car, Jeep, Mini-Van, SUV, Truck & Van models	A/T Park Neutral Switch Stuck in Park Position or In Gear Conditions: Key on or engine running (while not in Limp-In mode), gear selector in Park, Neutral or Drive position, and the PCM detected an invalid P/N position switch state for a given mode of vehicle operation. Possible Causes • P/N switch signal circuit is open between the switch and PCM • P/N switch signal circuit is shorted to sensor or chassis ground • P/N switch ground circuit is open (at the switch mounting point) • P/N switch is stuck in P/N position, or stuck in Drive position • P/N switch is damaged or has failed • PCM has failed

OBD II TROUBLE CODE LIST (P2XXX CODES)

DTC	Trouble Code Title, Conditions & Possible Causes
P2008 1T CCM 2002-03 300M, Concorde & Intrepid models	Short Runner Solenoid Circuit Malfunction Conditions: Engine started; ASD relay energized, and the PCM detected the Short Runner solenoid circuit was not in its expected voltage state. Possible Causes • S/R solenoid control circuit is open • S/R solenoid control circuit is shorted to ground or power (B+) • S/R solenoid power supply circuit is open to the ASD relay • Short runner solenoid is damaged or it has failed • PCM has failed
P2302 1T CCM 2002-03 300M, Concorde & Intrepid, Neon, PT Cruiser, Caravan models	Ignition Coil 1 Secondary Circuit Insufficient Ionization Conditions: Engine started; and the PCM detected the Ignition Coil No. 1 secondary "burn time" was insufficient, or it was missing. Possible Causes • Cylinder No. 1 spark plug is damaged or it has failed • Ignition Coil No. 1 is damaged or it has failed • PCM has failed
P2305 1T CCM 2002-03 300M, Concorde & Intrepid, Caravan models	Ignition Coil 2 Secondary Circuit Insufficient Ionization Conditions: Engine started; and the PCM detected the Ignition Coil No. 2 secondary "burn time" was insufficient, or it was missing. Possible Causes • Cylinder No. 2 spark plug is damaged or it has failed • Ignition Coil No. 2 is damaged or it has failed • PCM has failed
P2308 1T CCM 2002-03 300M, Concorde & Intrepid models	Ignition Coil 3 Secondary Circuit Insufficient Ionization Conditions: Engine started; and the PCM detected the Ignition Coil No. 3 secondary "burn time" was insufficient, or it was missing. Possible Causes • Cylinder No. 3 spark plug is damaged or it has failed • Ignition Coil No. 3 is damaged or it has failed • PCM has failed
P2311 1T CCM 2002-03 300M, Concorde & Intrepid models	Ignition Coil 4 Secondary Circuit Insufficient Ionization Conditions: Engine started; and the PCM detected the Ignition Coil No. 4 secondary "burn time" was insufficient, or it was missing. Possible Causes • Cylinder No. 4 spark plug is damaged or it has failed • Ignition Coil No. 4 is damaged or it has failed • PCM has failed
P2314 1T CCM 2002-03 300M, Concorde & Intrepid models	Ignition Coil 5 Secondary Circuit Insufficient Ionization Conditions: Engine started; and the PCM detected the Ignition Coil No. 5 secondary "burn time" was insufficient, or it was missing. Possible Causes • Cylinder No. 5 spark plug is damaged or it has failed • Ignition Coil No. 5 is damaged or it has failed • PCM has failed
P2317 1T CCM 2002-03 300M, Concorde & Intrepid models	Ignition Coil 6 Secondary Circuit Insufficient Ionization Conditions: Engine started; and the PCM detected the Ignition Coil No. 6 secondary "burn time" was insufficient, or it was missing. Possible Causes • Cylinder No. 6 spark plug is damaged or it has failed • Ignition Coil No. 6 is damaged or it has failed • PCM has failed

OBD II TROUBLE CODE LIST (P2XXX CODES)

DTC	Trouble Code Title, Conditions & Possible Causes
P2503 1T CCM 2002-03 300M, Concorde & Intrepid, Neon, PT Cruiser, Caravan models	Charging System Voltage Low Conditions: Engine started; engine speed over 1157 rpm, and the PCM detected the Battery Sense voltage was 1v less than the Charging system voltage "goal" for 13.47 seconds during the CCM test. Possible Causes • Battery sense circuit has a high resistance condition • Generator ground circuit has a high resistance condition • Generator field ground circuit is open • Generator field control circuit is open or shorted to ground • Generator is damaged or it has failed
P2700 1T CCM 2002-03 Dakota, Durango, Jeep Liberty & Truck with a 3.7L & 4.7L V6 engine	A/T L/R Inadequate Element Volume Detected Conditions: Engine started; transmission fluid temperature more than 110ºF, vehicle driven, and the PCM updated the L/R volume (during a 3-1 or 2-1 Manual downshift) with the throttle angle less than 5 degrees, and it detected that the L/R volume fell below 16 during the test. Possible Causes • L/R volume clutch index is too low • TCM L/R volume clutch circuit is damaged or has failed
P2701 1T CCM 2002-03 Dakota, Durango, Jeep Liberty & Truck with a 3.7L & 4.7L V6 engine	A/T 2C Inadequate Element Volume Detected Conditions: Engine started; transmission fluid temperature more than 110ºF, vehicle driven, then after the PCM updated the 2C volume (during a 3-2 kickdown event) with the throttle angle from 10-54 degrees, the PCM detected that the 2C volume fell below 5 during the CCM test. Possible Causes • 2C volume clutch index is too low • TCM 2C volume clutch circuit is damaged or has failed
P2702 1T CCM 2002-03 Dakota, Durango, Jeep Liberty & Truck with a 3.7L & 4.7L V6 engine	A/T O/D Inadequate Element Volume Detected Conditions: Engine started; transmission fluid temperature more than 110ºF, vehicle driven, then after he PCM updated the O/D volume (during a 2-3 Upshift event) with the throttle angle from 10-54 degrees, the PCM detected that the O/D volume fell below 5 during the CCM test. Possible Causes • O/D volume clutch index is too low • TCM O/D volume clutch circuit is damaged or has failed
P2703 1T CCM 2002-03 Dakota, Durango, Jeep Liberty & Truck with a 3.7L & 4.7L V6 engine	A/T U/D Inadequate Element Volume Detected Conditions: Engine started; transmission fluid temperature more than 110ºF, vehicle driven, and the TCM updated the U/D volume (during a 4-3 kickdown) with the throttle angle from 10-54 degrees, and it detected that the U/D volume fell below 11 during the test. Possible Causes • U/D volume clutch index is too low • TCM U/D volume clutch circuit is damaged or has failed
P2704 1T CCM 2002-03 Dakota, Durango, Jeep Liberty & Truck with a 3.7L & 4.7L V6 engine	A/T 4C Inadequate Element Volume Detected Conditions: Engine started; transmission fluid temperature more than 110ºF, vehicle driven, then after the TCM updated the 4C volume (during a 3-4 Upshift event) with the throttle angle from 10-54 degrees, the PCM detected that the 4C volume fell below 5 during the CCM test. Possible Causes • 4C volume clutch index is too low • TCM 4C volume clutch circuit is damaged or has failed
P2706 1T CCM 2002-03 Dakota, Durango, Jeep Liberty & Truck with a 3.7L & 4.7L V6 engine	A/T MS Solenoid Circuit Malfunction Conditions: Engine started; vehicle driven in a forward gear, and immediately after a gear ratio or pressure switch change, the TCM detected a detected a MS solenoid error. The PCM sets this code when it detects three consecutive solenoid continuity test faults; or 1 failure if the test is run in response to a gear ratio of pressure switch fault. Possible Causes • Check for a loose connector to the MS solenoid (intermittent) • MS solenoid control circuit is open or shorted to ground • MS solenoid control circuit is shorted to system power (B+) • MS solenoid is damaged or it has failed • Transmission control relay output supply circuit is open • TCM MS solenoid circuit is damaged or it has failed

1998 Dodge Caravan 3.8L V6 MFI VIN L Engine

PCM PID Acronym	Parameter Identification	Parameter Range	PID Value (Hot Idle)	PID Value (21 mph)	PID Value (55 mph)
A/C 1	A/C Cycling Clutch	ON/OFF	ON (if on)	OFF	OFF
A/C SE	A/C Select Switch	ON/OFF	ON (if on)	OFF	OFF
A/C P	A/C Pressure Value	0-500 psi		72.6 psi	
ADV	Spark Advance	-90° - 90°		20°	
ASD S	ASD Sense	HIGH/LO		LOW	
ASD 1	Desired ASD Relay	ON / OFF		OFF	
BARO	BARO Pressure	0-30"Hg		29.5Hg	
BRAKE	Brake Switch	REL/Apply	RELEASE	RELEASE	RELEASE
BATT	Battery (volts)	0-25.5v	14.1v	14.1v	14.1v
BTS	Battery Temp. SEN	-40-284°F		71°F	
CELL	Fuel Cell	1-20		12	
COUNT	Newest DTC Count	0-255	0	0	0
Dwell 1-3	Coil Dwell 1 to 3	0-999 ms		0.00 ms	
KILC1	CUR Fuel Shutoff 1	Decel		Decel	
KILC2	CUR Fuel Shutoff 2	Torque		Torque	
KILC3	CUR Fuel Shutoff 3	Rev Limit		Rev Limit	
KILC4	CUR Fuel Shutoff 4	>112mph		>112mph	
ECT F	ECT Sensor (°F)	-40-284°F		201°F	
ECT V	ECT Sensor (v)	0-5v		2.35v	
EGR 1	Desired EGR SOL.	Allow/BLK	ALLOW	ALLOW	ALLOW
EVAP %	Purge Duty Cycle	0-100%	0	36	95
FAN %	Fan Duty Cycle (%)	0-100%		69.8	
U-O2S	HO2S BK1 SEN 1	0-1100mv		0.51mv	
D-O2S	HO2S BK1 SEN 2	0-1100mv		0.06mv	
IAC	IAC Motor Steps	0-255		19	21
INJ PW	Injector Pulsewidth	0-999 ms		2.476	
KNK V	Knock Sensor	0-5v	0.28	0.29	0.30
LDP 1	Desired LDP SOL.	VAC/Apply	VAC	VAC	VAC
LDP SW	LDP Switch	OP/ Down	OPEN	OPEN	OPEN
LIMP 1-4	Limp-In Reason 1-4	ECT Fault		ECT Fault	
LOOP	Open/Closed Loop	OL / CL	CL	CL	CL
MAP	MAP Vacuum	0-30"Hg		20.1"Hg	
MAP V	MAP Sensor	0-5v		1.25v	
M TPS	Minimum TPS	0-5v		0.80v	
P/N P	PNP Switch	P/N-D-R	P/N	D-R	D
READY	Closed Loop Timer	Min/Sec		0.633	
RPM	Engine RPM	0-65535		743	
S-AD P	Short term Adaptive	0-100%	0	-1	-1
L-AD P	Long term Adaptive	0-100%	1	1	1
TIME	Time From start/run	Min/Sec		11.42m/s	
T IDL	Desired Idle Speed	0-3150rpm		752rpm	
TPS V	TP Sensor	0-5v		0.80v	

2000 Dodge Caravan 3.8L V6 MFI VIN L Engine

PCM PID Acronym	Parameter Identification	Parameter Range	PID Value (Hot Idle)	PID Value (30 mph)	PID Value (55 mph)
A/C P	A/C Pressure (psi)	0-459psi	79	79	79
ASD	ASD Relay Control	ON/OFF	ON	ON	ON
ECT	ECT Sensor (°F)	-40-284°F	207	205	205
CAT-M	Catalyst Monitor	Ready/Not	READY	READY	READY
CCM	Component Monitor	Ready/Not	READY	READY	READY
EGR-M	EGR SYS Monitor	Ready/Not	READY	READY	READY
EVAP	EVAP SYS Monitor	Ready/Not	READY	READY	READY
FP-M	Fuel SYS Monitor	Ready/Not	READY	READY	READY
F-SYS1	Fuel System Bank 1 Loop Status	CL/CL DRV/OL	CL	CL DRV	CL DRV
FT-11	Fuel Trim B1S1 (%)	0-100%	2.3	-1.1	1.0
F-SYS2	Fuel System Bank 2 Loop Status	CL/CL DRV/OL	Not Used	Not Used	Not Used
FT-12	Fuel Trim B1S2 (%)	0-100%	99.2	99.2	99.2
Load	Engine Load (%)	0-100%	4.7	24	21
LT B1	Long Term Fuel Trim Bank 1 (%)	0% (±20)	-6.1	-5.9	-5.8
MAP	MAP Vacuum	0-30"Hg	16.9	12.7	22.2
MISF	Misfire Monitor	Ready/Not	READY	READY	READY
HCAT-M	Heated Cat Monitor	Ready/Not	READY	READY	READY
HO2S-11	HO2S-11 (Bank 1 Sensor 1) Monitor	Ready/Not	READY	READY	READY
HO2S-11	HO2S-11 (Bank 1 Sensor 1) Signal	0-1100 mv	0.825	0.215	0.815
HO2S-12	HO2S-12 (Bank 1 Sensor 2) Monitor	Ready/Not	READY	READY	READY
HO2S-12	HO2S (Bank 1 Sensor 2) Signal	0-1100 mv	0.390	0.645	0.655
MIL	MIL Status	ON/OFF	OFF	OFF	OFF
RPM	Engine RPM	0-6000rpm	735rpm	2293rpm	1533rpm
SPARK	Ignition Advance (°)	-90° - +90°	22	20	18
ST B1	Short Term Fuel Trim Bank 1 (%)	0% (±20)	-1.6	-1.6	-1.2
TP V	TP Sensor (V)	0-5.1v	1.13	1.39	2.95
VSS	Vehicle Speed	0-159 mph	0	30	55

2001 Dodge Neon 2.0L I4 SOHC VIN C Engine

PCM PID Acronym	Parameter Identification	Parameter Range	PID Value (Hot Idle)	PID Value (30 mph)	PID Value (55 mph)
L-AD P	Adaptive L/T	0-100%	-4	-4	-4
L-AD P	Adaptive L/T	0-100%	-0	-4	-0
BARO	BARO Sensor	0-30" Hg	29.6	29.4	30.2
BTS (°F)	Battery Temp. (°F)	-40-284°F	66	64	55
BTS (V)	Battery Temp. (V)	0-5.1v	1.88	1.86	1.53
BATT	Battery (V)	0-25.5v	13.8	14.0	14.0
CHG S	Charging System	0-25.5v	13.7	14	14
CL ENAB	CL Enable (sec's)	0-65855	0	0	0
ECT (°F)	Coolant Temp. (°F)	-40-284°F	186	190	188
ECT (V)	Coolant Temp. (V)	0-5.1v	1.88	1.86	1.53
DES IDL	Desired Idle (rpm)	0-10,000	976	840	840
RPM	Engine Speed (rpm)	0-10,000	976	2080	2144
EVAP	Purge Duty Cycle	0-100%	9	13	78
FLVL	Fuel Level (V)	0-5.1v	0.56 (full)	0.56 (full)	0.56 (full)
IAC DES	IAC Desired Counts	0-255	28	37	37
IAC MTR	IAC Motor Counts	0-255	28	37	37
IAT (°F)	IAT Sensor (°F)	-40-284°F	116	111	107
IAT (V)	IAT Sensor (V)	0-5.1v	1.41	1.53	1.61
INJ PW	Injector P/Width 1	0-99 ms	2.11	5.91	8.33
RETARD	Knock Retard (°)	-90 to 90°	0	0	0
KNOCK	Knock Sensor (V)	0-5.1v	0.56	0.52	0.52
MAP (Hg)	MAP Sensor (" Hg)	0-30" Hg	9.11	22.01	24.91
MAP (V)	MAP Sensor (V)	0-5.1v	1.13	3.28	3.83
MVAC	Manifold Vacuum	0-30" Hg	20.50	7.41	5.34
HO2S-11	HO2S-11 (B1 S1)	0-1100 mv	210	170	330
HO2S-12	HO2S-12 (B1 S2)	0-1100 mv	090	780	820
S/C ACT	S/C Active	Active/Not	Not Active	Not Active	Not Active
S/C HIST	S/C History	KEY	KEY	CRSW	CRSW
S/C SET	S/C Set Speed	0-256 mph	0	30	55
S/C STS	S/C Status	CRSW	CRSW	CRSW	CRSW
S/C SW	S/C Switch (V)	0-5.1v	5.01	0.01	0.01
SPARK	Spark Advance	-90 to 90°	9	31	22
T/ANGLE	Throttle Angle (%)	0-100%	0	10	37
TP MIN	TPS Minimum (V)	0-5.1v	0.70	0.74	0.74
TP V	TP Sensor (V)	0.5.1v	0.70	1.01	1.69
VSS	Vehicle Speed	0-159 mph	0	30	55

2001 Sebring 2.7L V6 MFI VIN R Engine (A/T)

PCM PID Acronym	Parameter Identification	Parameter Range	PID Value (Hot Idle)	PID Value (30 mph)	PID Value (55 mph)
A/C CLU	A/C Clutch Relay	ON/OFF	OFF (if off)	ON (if on)	ON (if on)
A/C PSI	A/C Pressure (psi)	0-459 psi	79	79	79
A/C PSI	A/C Pressure (V)	0-5.1v	0.90	0.90	0.90
ASD	ASD Relay	ON / OFF	ON	ON	ON
BARO	BARO Pressure	10-110 kPa	100	100	100
BTS	Battery Temp. (°F)	-40-284°F	70	70	70
BTS V	Battery Temp. (V)	0-5.1v	1.8	1.8	1.8
BRAKE	Brake Switch	ON / OFF	ON (if on)	OFF	OFF
CL Timer	Closed Loop Timer	NO / YES	YES	YES	YES
CKP	CKP sensor signal	NO / YES	YES	YES	YES
CMP	CMP sensor signal	NO / YES	YES	YES	YES
DES IDL	Desired Idle Speed	0-6000 rpm	664	---	---
ECT F	ECT Sensor (°F)	-40-284°F	195	198	199
EGR %	EGR Sensor (%)	0-100%	0 (4.98v)	29	45
EVAP %	EVAP Purge D/C	0-100%	5	55	95
F/LEVEL	Fuel Level Gallons	0-16 (Gal)	14	14	14
F/LEVEL	Fuel Level (V)	0-5.1v	1.17	1.17	1.17
F/ALLOW	Fuel Allowed	NO / YES	YES	YES	YES
HO2S-11	HO2S-11 Signal	0-1100 mv	570	790	210
HO2S-12	HO2S-12 Signal	0-1100 mv	630	220	810
HO2S-21	HO2S-21 Signal	0-1100 mv	470	660	720
HO2S-22	HO2S-22 Signal	0-1100 mv	110	890	340
IAC	IAC Motor (steps)	0-255	17	43	44
IAT F	IAT Sensor (°F)	-40-284°F	84	86	88
IGN CY1	Ignition Cycles 1	0-256	0	0	0
LBINJPW	Left Bank Injector	0-999 ms	2.5	4.8	3.3
LBSTAD	L/B FT Adaptive	0% (±20)	1.8	1.6	1.4
LDP SOL	LDP Solenoid	OPEN/BLK	BLOCKED	BLOCKED	BLOCKED
LDP SW	LDP Switch Status	OPEN/CLD	CLOSED	CLOSED	CLOSED
MTV SOL	Manifold Solenoid	ON / OFF	OFF	OFF	OFF
MAN VAC	Manifold Vacuum	10-110 kPa	66	88	90
MAP V	MAP Sensor (V)	0-5v	1.25	2.70	2.10
M TPS	Minimum TPS	0-5.1v	0.74	0.74	0.74
MPH	Miles Per Hour	0-159 mph	0	30	55
P/N SW	Park/Neutral Switch	PN-Dr-2nd	P-N	DR	DR
RBINJPW	Left Bank Injector	0-999 ms	0.8	0.9	1.1
RBSTAD	R/B FT Adaptive	0% (±20)	1.8	1.6	1.4
RPM	Engine Speed	0-6000 rpm	640	1350	1655
SPARK	Spark Advance	-90° - 90°	11	18	26
TARG IAC	Target Idle Speed	0-256 step	17	---	---
TP (%)	Throttle (%)	0-100%	0	11	19
TP (V)	TP Sensor (V)	0-5.1v	0.74	1.20	1.26

Glossary of Terms & Acronyms

(<) - Indicates less than the value	(>) - Indicates more than the value
A/C - Air Conditioning System	A/D - Analog to Digital Converter
A/F - Air Fuel Ratio	A/T - Automatic Transmission
ABS - Antilock Brake System	ACM - Airbag control Module
AIR - Secondary Air Injection System	AIS - Automatic Idle Speed
ASD - Automatic Shutdown Relay	AWD - All Wheel Drive
B+ - Battery Voltage	BARO - Barometric Pressure Sensor
BCM - Body Control Module	BOB - Breakout Box
CANP - EVAP Canister Purge Solenoid	CARB - California Air Resources Board
CCD - Chrysler Collision Detection Serial Data Bus	CCP - Comprehensive Component Monitor
CCS - Coast Clutch Solenoid	CKP - Crankshaft Position
CKT - Circuit	CMP - Camshaft Position
CNG - Certified Natural Gas	CO - Carbon Monoxide
CO2 - Carbon Dioxide	CTRL - Control
CYL - Cylinder	DI - Distributor Ignition
DIS - Direct Ignition System	DLC - Data Link Connector
DOHC - Double Overhead Cam Engine	DTC - Diagnostic Trouble Code
DRL - Daytime Running Lights	DVOM - Digital Volt/Ohm Meter
EATX - Electronic Automatic Transaxle	EBCM - Electronic Brake Control Module
EBTCM - Electronic Brake T/C Module	ECT - Engine Coolant Temperature
EEPROM - Electronic Erasable Programmable Read Only Memory	EFI - Electronic Fuel Injection
EGR - Exhaust Gas Recirculation	EGR Monitor - OBD II EGR Test
EI - Electronic Ignition System	EMCC - Electronically Modulated Converter Clutch
EPA - Environmental Protection Agency	EVAP - Evaporative Emission System
FAN - Cooling Fan (Low or High Speed)	FF - Flexible Fuel Vehicle
FTP - Fuel Tank Pressure	FWD - Front Wheel Drive
GEM - Generic Electronic Module	GND - Electrical ground connection
GPS - Governor Position Sensor	GVW - Gross Vehicle Weight
HC - Hydrocarbons	HMSL - High Mounted Stop Lamp
HO2S-11 (Bank 1 Sensor 1) Signal	HO2S-12 (Bank 1 Sensor 2) Signal
HO2S-21 (Bank 2 Sensor 1) Signal	HO2S-22 (Bank 2 Sensor 2) Signal
HO2S - Heated Oxygen Sensor	Hz - Hertz
IAC - Idle Air Control Sensor	IAT - Intake Air Temperature Sensor
ICM - Ignition Control Module	IGN GND - Ignition Ground
ISO - International Standards Organization	JTEC - Jeep/Truck Engine Control Module
KAM - Keep Alive Memory	KAPWR - Direct Battery Power
Kg/cm² - Kilograms/Cubic Centimeters	KOEC - Key On, Engine Cranking
KOEO - Key On, Engine Off	KOER - Key On, Engine Running
KS - Knock Sensor	LDP - Leak Detection Pump
LED - Light Emitting Diode	LONGFT - Long Term Fuel Trim
LOOP - Engine Operating Loop Status	LPG - Liquid Petroleum Gas

Glossary of Terms & Acronyms

MAF - Mass Airflow (sensor)	MAP - Manifold Air Pressure
MFI - Multiport Fuel Injection	MIL - Malfunction Indicator Lamp
MPD - Manifold Pressure Differential	MPH - Miles Per Hour
MPI - Multiport Injection (relay)	M/T - Manual Transmission
MTV - Manifold Tuning Valve	Ms - Milliseconds
Mv - Millivolt	NGV - Natural Gas Vehicles
NOx - Oxides of Nitrogen	NTC - Negative Temperature Coefficient
O2S-11 (Bank 1 Sensor 1) Signal	O2S-21 (Bank 2 Sensor 1) Signal
OBD I - On Board Diagnostics Version I	OBD II - On Board Diagnostics Version II
ORD - Overdrive Running Clutch	OSS - Output Speed Shaft
PCI - Programmable Communications Interface	PCM - Powertrain Control Module
PCV - Positive Crankcase Ventilation	PDC - Power Distribution Center
PFI - Port Fuel Injection	PID - Parameter Identification Location
PNP - Park Neutral Position (switch)	PSP - Power Steering Pressure (switch)
PWR GND - Power Ground for PCM	PWM - Pulsewidth Modulated (signal)
RAM - Random Access Memory	ROM - Read Only Memory
RPM - Revolutions Per Minute	RWD - Rear Wheel Drive
S/C - Speed Control	SBEC - Single Board Engine Controller
SCI - Serial Communication Interface	SFI - Sequential Fuel Injection
SIL - Shift Indicator Lamp	SKIM: Sentry Key Immobilizer Module
SMEC- Single Module Engine Controller	SOHC - Single Overhead Cam Engine
SRI - Service Reminder Indicator	SRS - Supplemental Restraint System
TAC - Thermostatic Air Cleaner	TACH - Tachometer (signal)
TBI - Throttle Body Injection	TCC - Torque Converter Clutch
TCCS - Torque Converter Clutch Solenoid	TCM - Transmission Control Module
TCS - Traction Control Switch	TDC - Top Dead Center
TPS - Throttle Position Sensor	TRS - Transmission Range Switch
TSB - Technical Service Bulletin	TTS - Transmission Temp. Sensor
Turbo - Turbo Charged	TWC - Three Way Catalyst
VAC - Vacuum	VAF - Volume Airflow (sensor)
VECI - Vehicle Emission Control Information (Label)	VREF - Reference Voltage (from PCM)
VSS - Vehicle Speed Sensor	WOT - Wide Open Throttle

CHRYSLER CAR CONTENTS

CHRYSLER CAR CONTENTS

About This Section

Introduction

This section of the manual contains Pin Tables for Chrysler vehicles from 1990-2003. It can be used to help you repair Trouble Code and No Code problems related to the PCM.

Vehicle Coverage

The following vehicle applications are covered in this section:

- 1999-2003 300M
- 1995-2000 Cirrus
- 1993-2003 Concorde, LHS
- 1990-93 Imperial
- 1990-93 LeBaron, LeBaron GTS
- 1990-96 New Yorker, New Yorker Fifth Avenue
- 1995-2003 Sebring

How to Use This Section

This section of the manual can be used to look up the location of a particular pin, a wire color or to find a "known good" value of a circuit. To locate the PCM information for a particular vehicle, find the model, correct engine size (with VIN Code) and finally the year of the vehicle.

For example, to look up the PCM terminals for a 1999 Cirrus 2.5L 4v V6 VIN H, go to Contents Page 1 to find the text string shown below.

Then turn to Page 7-16 to find the following PCM related information.

1998-2000 Cirrus 2.5L 4v V6 MFI VIN H (A/T) 'C1' Connector

PCM Pin #	Wire Color	Circuit Description (40-Pin)	Value at Hot Idle
4	BK/GY	**Coil 1 Driver Control**	5°, at 55 mph: 8° dwell
6	DG/OR	**ASD Relay Output (B+)**	12-14v
7	YL/WT	**Injector 3 Driver**	1.0-4.0 ms
8	DG	**Generator Field Driver**	Digital Signal: 0-12-0v

In this example, the Coil Driver circuit is connected to Pin 4 of the 40-Pin connector with a BK/GY wire. The Hot Idle value is 5° while the 55 mph value is 8°. Note the change in dwell as the mph changed.

The ASD relay output signal is connected to Pin 6 of the 40-Pin connector (DG/OR wire). This signal indicates the voltage output of the ASD relay during vehicle operation. This signal should always read near battery voltage with the engine running.

The acronym A/T that appears in the title for the table indicates the information in this table is for a vehicle with an automatic transaxle.

CHRYSLER PIN TABLES

1999 300M 3.5L V6 SOHC 24v MFI VIN G 'C1' Connector

PCM Pin #	Wire Color	Circuit Description (40-Pin)	Value at Hot Idle
1	TN/LG	Coil 4 Driver	5°, at 55 mph: 8° dwell
2	TN/OR	Coil 3 Driver	5°, at 55 mph: 8° dwell
3	TN/PK	Coil 2 Driver	5°, at 55 mph: 8° dwell
4	TN/LB	Coil 6 Driver	5°, at 55 mph: 8° dwell
5	YL/RD	S/C Power Supply	12-14v
6	DG/OR	ASD Relay Output	12-14v
7	YL/WT	Injector 3 Driver	1.0-4.0 ms
8	DG	Generator Field Driver	Digital Signal: 0-12-0v
9	---	Not Used	---
10	BK/TN	Power Ground	<0.1v
11	TN/RD	Coil 1 Driver	5°, at 55 mph: 8° dwell
12	---	Not Used	---
13	WT/DB	Injector 1 Driver	1.0-4.0 ms
14	BR/DB	Injector 6 Driver	1.0-4.0 ms
15	GY	Injector 5 Driver	1.0-4.0 ms
16	LB/BR	Injector 4 Driver	1.0-4.0 ms
17	TN	Injector 2 Driver	1.0-4.0 ms
18	GY/PK	Short Runner Valve Solenoid	Valve On: 1v, Off: 12v
19	---	Not Used	---
20	DB/WT	Ignition Switch Output	12-14v
21	TN/DG	Coil 5 Driver	5°, at 55 mph: 8° dwell
22-23	---	Not Used	---
24	DB/LG	Knock Sensor Signal	0.080v AC
25	BK/PK	Knock Sensor Ground	<0.050v
26	TN/BK	ECT Sensor Signal	At 180°F: 2.80v
27	BK/OR	Oxygen Sensor Ground	<0.050v
28	---	Not Used	---
29	LG/RD	HO2S-21 (B2 S1) Signal	0.1-1.1v
30	BK/DG	HO2S-11 (B1 S1) Signal	0.1-1.1v
31	TN/WT	Starter Relay Control	KOEC: 9-11v
32	GY/BK	CKP Sensor Signal	Digital Signal: 0-5-0v
33	TN/YL	CMP Sensor Signal	Digital Signal: 0-5-0v
34	LG/PK	EGR Sensor Signal	0.6-0.8v
35	OR/DB	TP Sensor Signal	0.6-1.0v
36	DG/RD	MAP Sensor Signal	1.5-1.7v
37	BK/RD	IAT Sensor Signal	At 100°F: 1.83v
38	---	Not Used	---
39	PK/RD	Manifold Tuning Valve Control	Valve Off: 12v, On: 1v
40	GY/YL	EGR Solenoid Control	12v, at 55 mph: 1v

Standard Colors and Abbreviations

Abbreviation	Color	Abbreviation	Color	Abbreviation	Color
BK	Black	GY	Gray	RD	Red
BL	Blue	GN	Green	TN	Tan
BR	Brown	LG	Light Green	VT	Violet
DB	Dark Blue	OR	Orange	WT	White
DG	Dark Green	PK	Pink	YL	Yellow

1999 300M 3.5L V6 SOHC 24v MFI VIN G 'C2' Connector

PCM Pin #	Wire Color	Circuit Description (40-Pin)	Value at Hot Idle
41	RD/LG	S/C Set Switch Signal	S/C & Set Switch On: 3.8v
42	DB	A/C Pressure Switch Signal	A/C On: 0.45-4.85v
43	BK/LB	Sensor Ground	<0.050v
44	OR	8-Volt Supply	7.9-8.1v
45	DB/LG	PSP Switch Signal	Straight: 0v, Turning: 5v
46	RD/WT	Battery Power (Fused B+)	12-14v
47	BK/LB	Sensor Ground	<0.050v
48	BR/WT	IAC 3 Driver	DC pulse signals: 0.8-11v
49	YL/BK	IAC 2 Driver	DC pulse signals: 0.8-11v
50	BK/TN	Power Ground	<0.1v
51	TN/WT	HO2S-12 (B1 S2) Signal	0.1-1.1v
53	PK/WT	HO2S-22 (B2 S2) Signal	0.1-1.1v
55	DB/PK	Low Speed Fan Relay	Relay Off: 12v, On: 1v
56	TN/RD	S/C Vacuum Solenoid	Vacuum Increasing: 1v
57	GY/RD	IAC 1 Driver	DC pulse signals: 0.8-11v
58	PK/BK	IAC 4 Driver	DC pulse signals: 0.8-11v
59	YL/PK	PCI Data Bus (J1850)	Digital Signals: 0-7-0v
61	PK/WT	5-Volt Supply	4.9-5.1v
62	WT/PK	Brake Switch Signal	Brake Off: 0v, On: 12v
63	YL/DG	Torque Management Request	Digital Signals
64	DB/OR	A/C Clutch Relay Control	Relay Off: 12v, On: 1v
65	PK	SCI Transmit	0v
66	WT/OR	Vehicle Speed Signal	Digital Signal
67	DB/YL	ASD Relay Control	Relay Off: 12v, On: 1v
68	PK/BK	EVAP Purge Solenoid Control	PWM Signal: 0-12-0v
69	DB/LG	High Speed Fan Relay	Relay Off: 12v, On: 1v
70	DG/LG	EVAP Purge Solenoid Sense	0-1v
71	WT/RD	EATX RPM Signal	Digital Signals
72	OR/RD	LDP Switch Sense	Open: 12v, Closed: 0v
73, 78-79	---	Not Used	---
74	BR	Fuel Pump Relay Control	Relay Off: 12v, On: 1v
75	LG	SCI Receive	0v
76	BK/PK	PNP Switch Signal	In P/N: 0v, Others: 5v
77	WT/DG	LDP Solenoid Control	PWM Signal: 0-12-0v
80	LG/RD	S/C Vent Solenoid	Vacuum Decreasing: 1v

Pin Connector Graphic

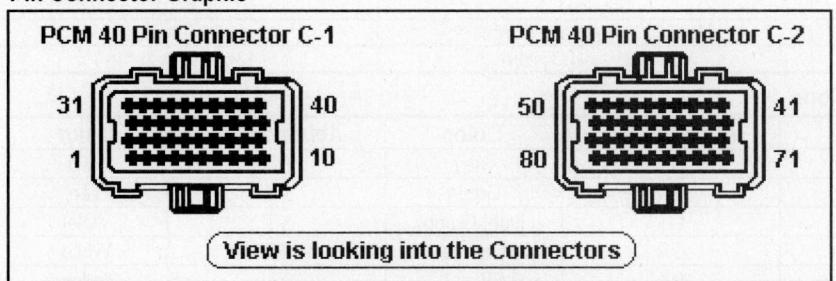

PCM 40 Pin Connector C-1

PCM 40 Pin Connector C-2

31   40
1   10

50   41
80   71

View is looking into the Connectors

2000-01 300M 3.5L V6 SOHC 24v MFI VIN G 'C1' Connector

PCM Pin #	Wire Color	Circuit Description (40-Pin)	Value at Hot Idle
1	TN/LG	Coil 4 Driver	5°, at 55 mph: 8° dwell
2	TN/OR	Coil 3 Driver	5°, at 55 mph: 8° dwell
3	TN/PK	Coil 2 Driver	5°, at 55 mph: 8° dwell
4	TN/LB	Coil 6 Driver	5°, at 55 mph: 8° dwell
5	YL/RD	S/C Power Supply	12-14v
6	DG/OR	ASD Relay Output	12-14v
7	YL/WT	Injector 3 Driver	1.0-4.0 ms
8	DG	Generator Field Driver	Digital Signal: 0-12-0v
9	---	Not Used	---
10	BK/TN	Power Ground	<0.1v
11	TN/RD	Coil 1 Driver	5°, at 55 mph: 8° dwell
12	---	Not Used	---
13	WT/DB	Injector 1 Driver	1.0-4.0 ms
14	BR/DB	Injector 6 Driver	1.0-4.0 ms
15	GY	Injector 5 Driver	1.0-4.0 ms
16	LB/BR	Injector 4 Driver	1.0-4.0 ms
17	TN	Injector 2 Driver	1.0-4.0 ms
18	GY/PK	Short Runner Valve Solenoid	Valve On: 1v, Off: 12v
19	---	Not Used	---
20	DB/WT	Ignition Switch Output	12-14v
21	TN/DG	Coil 5 Driver	5°, at 55 mph: 8° dwell
22-23	---	Not Used	---
24	DB/LG	Knock Sensor Signal	0.080v AC
25	BK/VT	Knock Sensor Return	<0.050v
26	TN/BK	ECT Sensor Signal	At 180°F: 2.80v
27	BK/OR	Oxygen Sensor Ground	<0.050v
28	---	Not Used	---
29	LG/RD	HO2S-21 (B2 S1) Signal	0.1-1.1v
30	BK/DG	HO2S-11 (B1 S1) Signal	0.1-1.1v
31	TN/WT	Starter Relay Control	KOEC: 9-11v
32	GY/BK	CKP Sensor Signal	Digital Signal: 0-5-0v
33	TN/YL	CMP Sensor Signal	Digital Signal: 0-5-0v
34	LG/PK	EGR Sensor Signal	0.6-0.8v
35	OR/DB	TP Sensor Signal	0.6-1.0v
36	DG/RD	MAP Sensor Signal	1.5-1.7v
37	BK/RD	IAT Sensor Signal	At 100°F: 1.83v
38	---	Not Used	---
39	VT/RD	Manifold Tuning Valve Control	Valve Off: 12v, On: 1v
40	GY/YL	EGR Solenoid Control	12v, at 55 mph: 1v

Standard Colors and Abbreviations

Abbreviation	Color	Abbreviation	Color	Abbreviation	Color
BK	Black	GY	Gray	RD	Red
BL	Blue	GN	Green	TN	Tan
BR	Brown	LG	Light Green	VT	Violet
DB	Dark Blue	OR	Orange	WT	White
DG	Dark Green	PK	Pink	YL	Yellow

2000-01 300M 3.5L V6 SOHC 24V MFI VIN G 'C2' CONNECTOR

PCM Pin #	Wire Color	Circuit Description (40-Pin)	Value at Hot Idle
41	RD/LG	S/C Set Switch Signal	S/C & Set Switch On: 3.8v
42	DB	A/C Pressure Switch Signal	A/C On: 0.45-4.85v
43	BK/LB	Sensor Ground	<0.050v
44	OR	8-Volt Supply	7.9-8.1v
45	DB/LG	PSP Switch Signal	Straight: 0v, Turning: 5v
46	RD/WT	Battery Power (Fused B+)	12-14v
47	BK/LB	Sensor Ground	<0.050v
48	BR/WT	IAC 3 Driver	DC pulse signals: 0.8-11v
49	YL/BK	IAC 2 Driver	DC pulse signals: 0.8-11v
50	BK/TN	Power Ground	<0.1v
51	TN/WT	HO2S-12 (B1 S2) Signal	0.1-1.1v
53	PK/WT	HO2S-22 (B2 S2) Signal	0.1-1.1v
55	DB/PK	Low Speed Fan Relay	Relay Off: 12v, On: 1v
56	TN/RD	S/C Vacuum Solenoid	Vacuum Increasing: 1v
57	GY/RD	IAC 1 Driver	DC pulse signals: 0.8-11v
58	PK/BK	IAC 4 Driver	DC pulse signals: 0.8-11v
59	YL/PK	PCI Data Bus (J1850)	Digital Signals: 0-7-0v
61	PK/WT	5-Volt Supply	4.9-5.1v
62	WT/PK	Brake Switch Signal	Brake Off: 0v, On: 12v
63	YL/DG	Torque Management Request	Digital Signals
64	DB/OR	A/C Clutch Relay Control	Relay Off: 12v, On: 1v
65	LG	SCI Transmit	0v
66	WT/OR	Vehicle Speed Signal	Digital Signal
67	DB/YL	ASD Relay Control	Relay Off: 12v, On: 1v
68	PK/BK	EVAP Purge Solenoid Control	PWM Signal: 0-12-0v
69	DB/LG	High Speed Fan Relay	Relay Off: 12v, On: 1v
70	DG/LG	EVAP Purge Solenoid Sense	0-1v
71	WT/RD	EATX RPM	Digital Signals
72	OR/RD	LDP Switch Sense	Open: 12v, Closed: 0v
73, 78-79	---	Not Used	---
74	BR	Fuel Pump Relay Control	Relay Off: 12v, On: 1v
75	LG	SCI Receive	5v
76	BK/PK	PNP Switch Signal	In P/N: 0v, Others: 5v
77	WT/DG	LDP Solenoid Control	PWM Signal: 0-12-0v
80	LG/RD	S/C Vent Solenoid	Vacuum Decreasing: 1v

Pin Connector Graphic

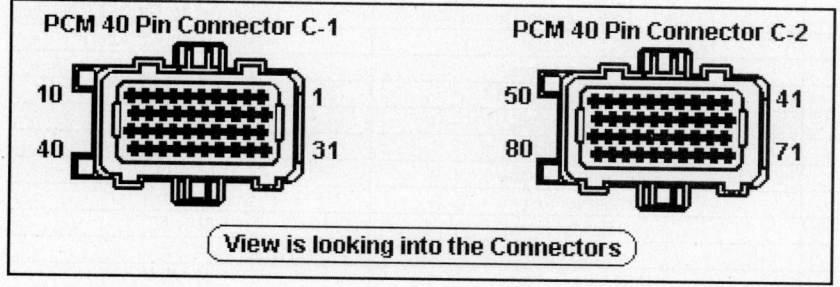

PCM 40 Pin Connector C-1

PCM 40 Pin Connector C-2

View is looking into the Connectors

2002-03 300 M 3.5L V6 SOHC 24v MFI VIN G, K 'C1' Black Connector

PCM Pin #	Wire Color	Circuit Description (38-Pin)	Value at Hot Idle
8.	---	Not Used	---
9	TN	Power Ground	<0.1v
11	DB/WT	Ignition Switch Output (Run-Start)	12-14v
12	RD/WT	Ignition Switch Output (Off-Run-Start)	12-14v
13-15	---	Not Used	---
16	GY/PK	Short Runner Valve Solenoid	Valve On: 1v, Off: 12v
17, 19-20	---	Not Used	---
18	BK/TN	Power Ground	<0.1v
21	DB	A/C Pressure Switch Signal	A/C On: 0.45-4.85v
22-24	---	Not Used	---
25	LG	SCI Receive (PCM)	5v
26	VT/OR	SCI Receive (TCM)	5v
27-28	---	Not Used	---
29	RD	Battery Power (Fused B+)	12-14v
30	YL/BK	Ignition Switch Output (tart)	8-11v (cranking)
31	TN/WT	HO2S-12 (B1 S2) Signal	0.1-1.1v
32	DB/LG	Oxygen Sensor Return (down)	<0.050v
33	PK/WT	HO2S-22 (B2 S2) Signal	0.1-1.1v
34-35	---	Not Used	---
36	PK/TN	SCI Transmit (PCM)	0v
37	WT/DG	SCI Transmit (TCM)	0v
38	VT/YL	PCI Data Bus (J1850)	Digital Signals: 0-7-0v

2002-03 300M 3.5L V6 SOHC 24v MFI VIN G, K 'C2' Gray Connector

PCM Pin #	Wire Color	Circuit Description (38-Pin)	Value at Hot Idle
1	TN/LB	Coil 6 Driver	5°, at 55 mph: 8° dwell
2	TN/DG	Coil 5 Driver	5°, at 55 mph: 8° dwell
3	TN/LG	Coil 4 Driver	5°, at 55 mph: 8° dwell
4	BR/DB	Injector 6 Driver	1.0-4.0 ms
5	GY	Injector 5 Driver	1.0-4.0 ms
6, 15, 26	---	Not Used	---
7	TN/OR	Coil 3 Driver	5°, at 55 mph: 8° dwell
8	GY/YL	EGR Solenoid Control	12v, at 55 mph: 1v
9	TN/PK	Coil 2 Driver	5°, at 55 mph: 8° dwell
10	TN/RD	Coil 1 Driver	5°, at 55 mph: 8° dwell
11	LB/BR	Injector 4 Driver	1.0-4.0 ms
12	YL/WT	Injector 3 Driver	1.0-4.0 ms
13	TN	Injector 2 Driver	1.0-4.0 ms
14	WT/DB	Injector 1 Driver	1.0-4.0 ms
16	VT/RD	Manifold Tuning Solenoid Control	Valve Off: 12v, On: 1v
17	BR/WT	HO2S-21 (B2 S1) Heater Control	Heater On: 1v, Off: 12v
18	BR/OR	HO2S-11 (B1 S1) Heater Control	Heater On: 1v, Off: 12v
19	DG	Generator Field Driver	Digital Signal: 0-12-0v
20	TN/BK	ECT Sensor Signal	At 180°F: 2.80v
21	OR/DB	TP Sensor Signal	0.6-1.0v
22	LG/PK	EGR Sensor Signal	0.6-0.8v
23	DG/RD	MAP Sensor Signal	1.5-1.7v
24	BK/VT	Knock Sensor Return	<0.050v
25	DB/LG	Knock Sensor Signal	0.080v AC
27	BK/LB	Sensor Ground	<0.1v
28	YL/BK	IAC Motor Sense	12-14v
29	VT/WT	5-Volt Supply	4.9-5.1v
30	BK/RD	IAT Sensor Signal	At 100°F: 1.83v
31	BK/DG	HO2S-11 (B1 S1) Signal	0.1-1.1v
32	BR/DG	HO2S-11 (B1 S1) Ground (Up)	<0.1v
33	LG/RD	HO2S-21 (B2 S1) Signal	0.1-1.1v
34	TN/YL	CMP Sensor Signal	Digital Signal: 0-5-0v
35	GY/BK	CKP Sensor Signal	Digital Signal: 0-5-0v
36-37	---	Not Used	---
38	GY/RD	IAC Motor Driver	DC pulse signals: 0.8-11v

2002-03 300M 3.5L V6 SOHC 24V MFI VIN G, K 'C3' WHITE CONNECTOR

PCM Pin #	Wire Color	Circuit Description (38-Pin)	Value at Hot Idle
1-2, 13-17	---	Not Used	---
3	DB/YL	ASD Relay Control	Relay Off: 12v, On: 1v
4	DB/PK	High Speed Radiator Fan Relay	Relay Off: 12v, On: 1v
5	LG/RD	S/C Vent Solenoid	Vacuum Decreasing: 1v
6	DB/PK	Low Speed Radiator Fan Relay	Relay Off: 12v, On: 1v
7	YL/RD	S/C Power Supply	12-14v
8	WT/DG	Natural Vacuum Leak Detection Solenoid	Solenoid Off: 12v, On: 1v
9	BR/VT	HO2S-21 (B2 S1) Heater Control	Heater On: 1v, Off: 12v
10	BR/GY	HO2S-11 (B1 S1) Heater Control	Heater On: 1v, Off: 12v
11	DB/OR	A/C Clutch Relay Control	Relay Off: 12v, On: 1v
12	TN/RD	S/C Vacuum Solenoid	Vacuum Increasing: 1v
18, 19	OR/DG	Automatic Shutdown Relay Output	12-14v
20	PK/BK	EVAP Purge Solenoid Control	PWM Signal: 0-12-0v
21-22, 24-25	---	Not Used	---
23	WT/PK	Brake Switch Sense	Brake Off: 0v, On: 12v
26	YL	Autostick Downshift Switch	Digital Signal: 0v or 12v
27	LG/RD	Autostick Upshift Switch	Digital Signal: 0v or 12v
28	OR/DG	Automatic Shutdown Relay Output	12-14v
29	DG/LG	EVAP Purge Solenoid Sense	0-1v
30-31, 33, 36	---	Not Used	---
32	VT/LG	Ambient Air Temperature Sensor	At 100ºF: 1.83v
34	RD/LG	S/C Set Switch Signal	S/C & Set Switch On: 3.8v
35	OR/RD	Natural Vacuum Leak Detection Switch Sense	0.1v
37	BR	Fuel Pump Relay Control	Relay Off: 12v, On: 1v
38	TN	Starter Relay Control	KOEC: 9-11v

2002-03 300M 3.5L V6 SOHC 24v MFI VIN G, K 'C4' Green Connector

PCM Pin #	Wire Color	Circuit Description (38-Pin)	Value at Hot Idle
1, 3-5	---	Not Used	---
2	BR	Overdrive Solenoid Control	Solenoid Off: 12v, On: 1v
3	PK	Underdrive Solenoid Control	Solenoid Off: 12v, On: 1v
6	WT	2-4 Solenoid Control	Solenoid Off: 12v, On: 1v
7-9, 11	---	Not Used	---
10	LB	Low/Reverse Solenoid Control	Solenoid Off: 12v, On: 1v
12	BK/YL	Power Ground	<0.050v
13, 14	BK/RD	Power Ground	<0.050v
15	LG/BK	TRS T1 Sense	<0.050v
16	VT	TRS T3 Sense	<0.050v
17, 20-21	---	Not Used	---
18, 19	LG, RD	Transmission Control Relay Output	Relay Off: 12v, On: 1v
19	RD	Transmission Control Relay Output	Relay Off: 12v, On: 1v
23-26, 31, 36	---	Not Used	---
27	BK/WT	TRS T41 Sense	<0.050v
28, 38	RD	Transmission Control Relay Output	Relay Off: 12v, On: 1v
29	LG	Low/Reverse Pressure Switch Sense	12-14v
30	YL/BK	2-4 Pressure Switch Sense	In Low/Reverse: 2-4v
32	LG/WT	Output Speed Sensor Signal	In 2-4 Position: 2-4v
33	RD/BK	Input Speed Sensor Signal	Moving: AC voltage
34	DB/BK	Speed Sensor Ground	Moving: AC voltage
35	VT/PK	Transmission Temperature Sensor Signal	<0.050v
37	VT/WT	TRS T42 Sense	In PRNL: 0v, Others 5v

Pin Connector Graphic

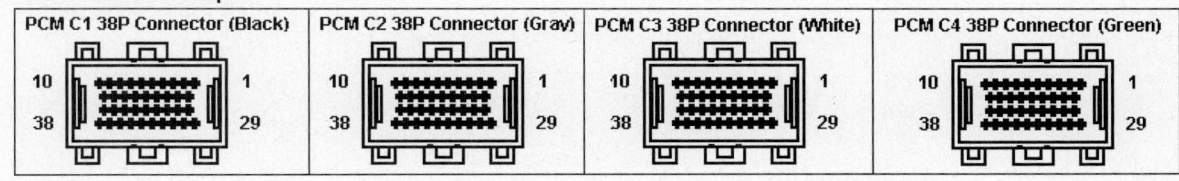

| PCM C1 38P Connector (Black) | PCM C2 38P Connector (Gray) | PCM C3 38P Connector (White) | PCM C4 38P Connector (Green) |

1995-97 Cirrus 2.4L I4 DOHC MFI VIN X (A/T) 'C1' Connector

PCM Pin #	Wire Color	Circuit Description (40-Pin)	Value at Hot Idle
1, 8-9	---	Not Used	---
2	BK/GY	Coil 1 Driver	5º, at 55 mph: 8º dwell
3	DB/DG	Coil 2 Driver	5º, at 55 mph: 8º dwell
4	DG	Generator Field Driver	Digital Signal: 0-12-0v
5	YL/PK	S/C Power Supply	12-14v
6	DG/OR	ASD Relay Output	12-14v
7	YL/WT	Injector 3 Driver	1.0-4.0 ms
8-9	---	Not Used	---
10	BK/TN	Power Ground	<0.1v
11-12	---	Not Used	---
13	WT/LB	Injector 1 Driver	1.0-4.0 ms
14-15	---	Not Used	---
16	LB/BR	Injector 4 Driver	1.0-4.0 ms
17	TN	Injector 2 Driver	1.0-4.0 ms
18-19	---	Not Used	---
20	DB/WT	Ignition Switch Output	12-14v
21-23	---	Not Used	---
24 ('95)	BK/LG	Knock Sensor Signal	0.080v AC
24 ('96-'97)	GY/BK	Knock Sensor Signal	0.080v AC
25	---	Not Used	---
26	TN/BK	ECT Sensor Signal	At 180ºF: 2.80v
27-29	---	Not Used	---
30	BK/DG	HO2S-11 (B1 S1) Signal	0.1-1.1v
31	---	Not Used	---
32	GY/BK	CKP Sensor Signal	Digital Signal: 0-5-0v
33	TN/YL	CMP Sensor Signal	Digital Signal: 0-5-0v
34	---	Not Used	---
35	OR/LB	TP Sensor Signal	0.6-1.0v
36	DG/RD	MAP Sensor Signal	1.5-1.7v
37	BK/RD	IAT Sensor Signal	At 100ºF: 1.83v
38-39	---	Not Used	---
40	GY/YL	EGR Solenoid Control	12v, at 55 mph: 1v

Pin Connector Graphic

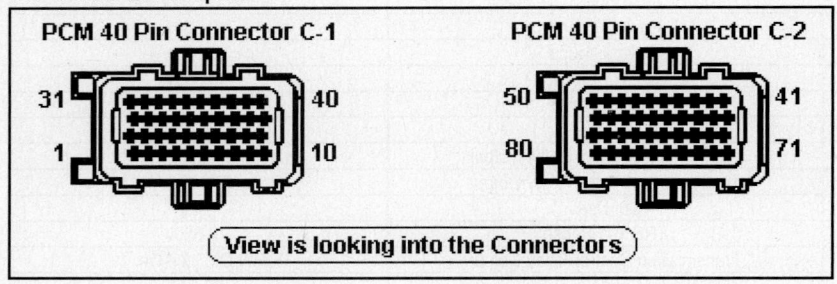

PCM 40 Pin Connector C-1

31   40
1   10

PCM 40 Pin Connector C-2

50   41
80   71

View is looking into the Connectors

1995-97 Cirrus 2.4L I4 DOHC MFI VIN X (A/T) 'C2' Connector

PCM Pin #	Wire Color	Circuit Description (40-Pin)	Value at Hot Idle
41	PK/LG	S/C Set Switch Signal	S/C & Set Switch On: 3.8v
42	DB/YL	A/C Pressure Switch Signal	A/C On: 0.45-4.85v
43	BK/LB	Sensor Ground	<0.050v
44	OR/WT	8-Volt Supply	7.9-8.1v
45	DB/LG	PSP Switch Signal	Straight: 0v, Turning: 5v
46 ('95)	RD/WT	Battery Power (Fused B+)	12-14v
46 ('96-'97)	RD/TN	Battery Power (Fused B+)	12-14v
47	BK	Power Ground	<0.1v
48	BR/GY	IAC 3 Driver	DC pulse signals: 0.8-11v
49	YL/BK	IAC 2 Driver	DC pulse signals: 0.8-11v
50	BK/TN	Power Ground	<0.1v
51 ('96-'97)	TN/WT	HO2S-12 (B1 S2) Signal	0.1-1.1v
52	PK/LG	Battery Temperature Sensor	At 86°F: 1.96v
55	DB/TN	Low Speed Fan Relay	Relay Off: 12v, On: 1v
57	GY/RD	IAC 1 Driver	DC pulse signals: 0.8-11v
58	PK/GY	IAC 4 Driver	DC pulse signals: 0.8-11v
59 ('95)	YL/PK	CCD Bus (+)	Digital Signal: 0-5-0v
59 ('96-'97)	PK/BR	CCD Bus (+)	Digital Signal: 0-5-0v
60	WT/BK	CCD Bus (-)	<0.050v
61	PK/WT	5-Volt Supply	4.9-5.1v
62	WT/RD	Brake Switch Signal	Brake Off: 0v, On: 12v
63 ('96-'97)	YL/DG	Torque Management Request	Digital Signals
64	DB/OR	A/C Clutch Relay Control	Relay Off: 12v, On: 1v
65	PK/LB	SCI Transmit	0v
66	WT/OR	Vehicle Speed Signal	Digital: 0-8-0-8v
67	DB/PK	ASD Relay Control	Relay Off: 12v, On: 1v
68 ('95)	DG/LG	EVAP Purge Solenoid Control	PWM Signal: 0-12-0v
68 ('96-'97)	PK/GY	EVAP Purge Solenoid Control	PWM Signal: 0-12-0v
69	DB/PK	High Speed Fan Relay	Relay Off: 12v, On: 1v
72 ('96-'97)	OR/DG	LDP Switch Sense	Open: 12v, Closed: 0v
74	BR/LG	Fuel Pump Relay Control	Relay Off: 12v, On: 1v
75	LG/WT	SCI Receive	0v
76 ('96-'97)	BK/PK	PNP Switch Signal	In P/N: 0v, Others: 5v
77 ('96-'97)	WT/DG	LDP Solenoid Control	PWM Signal: 0-12-0v
78	WT/PK	S/C Vacuum Solenoid	Vacuum Increasing: 1v
80	LG/RD	S/C Vent Solenoid	Vacuum Decreasing: 1v

Standard Colors and Abbreviations

Abbreviation	Color	Abbreviation	Color	Abbreviation	Color
BK	Black	GY	Gray	RD	Red
BL	Blue	GN	Green	TN	Tan
BR	Brown	LG	Light Green	VT	Violet
DB	Dark Blue	OR	Orange	WT	White
DG	Dark Green	PK	Pink	YL	Yellow

1998-2000 Cirrus 2.4L I4 DOHC MFI VIN X (A/T) 'C1' Connector

PCM Pin #	Wire Color	Circuit Description (40-Pin)	Value at Hot Idle
1-2	---	Not Used	---
3	DB/TN	Coil 2 Driver	5°, at 55 mph: 8° dwell
4	---	Not Used	---
5	YL/RD	S/C On/Off Switch Signal	0v or 6.7v
6	DG/OR	ASD Relay Output	12-14v
7	YL/WT	Injector 3 Driver	1.0-4.0 ms
8	DG	Generator Field Driver	Digital Signal: 0-12-0v
9	---	Not Used	---
10	BK/TN	Power Ground	<0.1v
11	BK/GY	Coil 2 Driver	5°, at 55 mph: 8° dwell
12	---	Not Used	---
13	WT/LB	Injector 1 Driver	1.0-4.0 ms
14-15	---	Not Used	---
16	LB/BR	Injector 4 Driver	1.0-4.0 ms
17	TN	Injector 2 Driver	1.0-4.0 ms
18-19	---	Not Used	---
20	DB/WT	Ignition Switch Output	12-14v
21-24	---	Not Used	---
25	DB/LG	Knock Sensor Signal	0.080v AC
26	TN/BK	ECT Sensor Signal	At 180°F: 2.80v
27	BK/OR	HO2S-11 (B1 S1) Ground	<0.050v
28-29	---	Not Used	---
30	BK/DG	HO2S-11 (B1 S1) Signal	0.1-1.1v
31	---	Not Used	---
32	GY/BK	CKP Sensor Signal	Digital Signal: 0-5-0v
33	TN/YL	CMP Sensor Signal	Digital Signal: 0-5-0v
34	---	Not Used	---
35	OL/LB	TP Sensor Signal	0.6-1.0v
36	DG/RD	MAP Sensor Signal	1.5-1.7v
37	BK/RD	IAT Sensor Signal	At 100°F: 1.83v
38-39	---	Not Used	---
40	GY/YL	EGR Solenoid Control	12v, at 55 mph: 1v

Pin Connector Graphic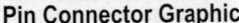

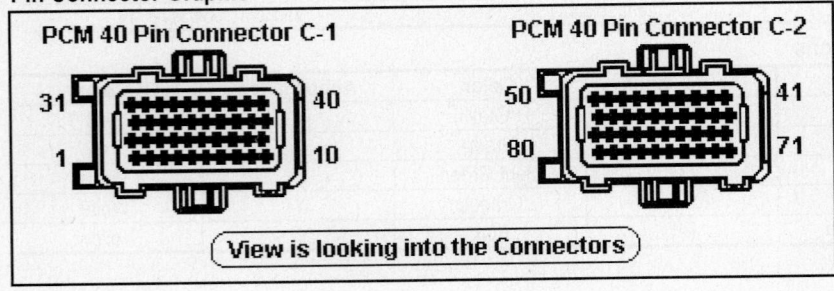

PCM 40 Pin Connector C-1   PCM 40 Pin Connector C-2

31   40   50   41
1   10   80   71

View is looking into the Connectors

1998-2000 Cirrus 2.4L I4 DOHC MFI VIN X (A/T) 'C2' Connector

PCM Pin #	Wire Color	Circuit Description (40-Pin)	Value at Hot Idle
41	RD/LG	S/C Set Switch Signal	S/C & Set Switch On: 3.8v
42	DB/YL	A/C Pressure Switch Signal	A/C On: 0.45-4.85v
43	BK/LB	Sensor Ground	<0.050v
44	OR/WT	8-Volt Supply	7.9-8.1v
45	DB/LG	PSP Switch Signal	Straight: 0v, Turning: 5v
46	RD/TN	Battery Power (Fused B+)	12-14v
47	BK	Power Ground	<0.1v
48	BR/WT	IAC 3 Driver	DC pulse signals: 0.8-11v
49	YL/BK	IAC 2 Driver	DC pulse signals: 0.8-11v
50	BK/TN	Power Ground	<0.1v
51	TN/WT	HO2S-12 (B1 S2) Signal	0.1-1.1v
52	VT/LG	Battery Temperature Sensor	At 86°F: 1.96v
53-54	---	Not Used	---
55	DB/TN	Low Speed Fan Relay	Relay Off: 12v, On: 1v
56	WT/VT	S/C Vacuum Solenoid	Vacuum Increasing: 1v
57	GY/RD	IAC 1 Driver	DC pulse signals: 0.8-11v
58	VT/BK	IAC 4 Driver	DC pulse signals: 0.8-11v
59	VT/BR	CCD Bus (+)	Digital Signal: 0-5-0v
60	WT/BK	CCD Bus (-)	<0.050v
61	VT/WT	5-Volt Supply	4.9-5.1v
62	WT/RD	Brake Switch Signal	Brake Off: 0v, On: 12v
63	YL/DG	Torque Management Request	Digital Signals
64	DB/OR	A/C Clutch Relay Control	Relay Off: 12v, On: 1v
65	PK/LB	SCI Transmit	0v
66	WT/OR	Vehicle Speed Signal	Digital: 0-8-0-8v
67	DB/VT	ASD Relay Control	Relay Off: 12v, On: 1v
68	PK/GY	EVAP Purge Solenoid Control	PWM Signal: 0-12-0v
69	DB/PK	High Speed Fan Relay	Relay Off: 12v, On: 1v
70	WT/TN	EVAP Purge Solenoid Sense	0-1v
71, 73	---	Not Used	---
72	OR/DG	LDP Switch Sense	Open: 12v, Closed: 0v
74	BR/LG	Fuel Pump Relay Control	Relay Off: 12v, On: 1v
75	LG	SCI Receive	0v
76	BK/WT	PNP Switch Signal	In P/N: 0v, Others: 5v
77	WT/DG	LDP Solenoid Control	PWM Signal: 0-12-0v
78-79	---	Not Used	---
80	LG/RD	S/C Vent Solenoid	Vacuum Decreasing: 1v

Standard Colors and Abbreviations

Abbreviation	Color	Abbreviation	Color	Abbreviation	Color
BK	Black	GY	Gray	RD	Red
BL	Blue	GN	Green	TN	Tan
BR	Brown	LG	Light Green	VT	Violet
DB	Dark Blue	OR	Orange	WT	White
DG	Dark Green	PK	Pink	YL	Yellow

1995-97 Cirrus 2.5L V6 SOHC MFI VIN H (A/T) 'C1' Connector

PCM Pin #	Wire Color	Circuit Description (40-Pin)	Value at Hot Idle
1-3	---	Not Used	---
4	DG	Generator Field Driver	Digital Signal: 0-12-0v
5	YL/PK	S/C Power Supply	12-14v
6	DG/OR	ASD Relay Output	12-14v
7	YL/WT	Injector 3 Driver	1.0-4.0 ms
8-9	---	Not Used	---
10	BK/TN	Power Ground	<0.1v
11	BK/GY	Coil Driver	5º, at 55 mph: 8º dwell
12	---	Not Used	---
13	WT/LB	Injector 1 Driver	1.0-4.0 ms
14	BR/DG	Injector 6 Driver	1.0-4.0 ms
15	GY	Injector 5 Driver	1.0-4.0 ms
16	LB/BR	Injector 4 Driver	1.0-4.0 ms
17	TN	Injector 2 Driver	1.0-4.0 ms
18-19	---	Not Used	---
20	DB/WT	Ignition Switch Output	12-14v
21-25	---	Not Used	---
26	TN/BK	ECT Sensor Signal	At 180ºF: 2.80v
27-28	---	Not Used	---
29 ('95)	BK/DG	HO2S-11 (B1 S1) Signal	0.1-1.1v
30	BK/DG	HO2S-11 (B1 S1) Signal	0.1-1.1v
30 ('95)	TN/WT	HO2S-12 (B1 S2) Signal	0.1-1.1v
31	---	Not Used	---
32	GY/BK	CKP Sensor Signal	Digital Signal: 0-5-0v
33	TN/YL	CMP Sensor Signal	Digital Signal: 0-5-0v
34	---	Not Used	---
35	OR/LB	TP Sensor Signal	0.6-1.0v
36	DG/RD	MAP Sensor Signal	1.5-1.7v
37	BK/RD	IAT Sensor Signal	At 100ºF: 1.83v
38-39	---	Not Used	---
40	GY/YL	EGR Solenoid Control	12v, at 55 mph: 1v

Standard Colors and Abbreviations

Abbreviation	Color	Abbreviation	Color	Abbreviation	Color
BK	Black	GY	Gray	RD	Red
BL	Blue	GN	Green	TN	Tan
BR	Brown	LG	Light Green	VT	Violet
DB	Dark Blue	OR	Orange	WT	White
DG	Dark Green	PK	Pink	YL	Yellow

1995-97 Cirrus 2.5L V6 SOHC MFI VIN H (A/T) 'C2' Connector

PCM Pin #	Wire Color	Circuit Description (40-Pin)	Value at Hot Idle
41	PK/LG	S/C Set Switch Signal	S/C & Set Switch On: 3.8v
42	DB/YL	A/C Pressure Switch Signal	A/C On: 0.45-4.85v
43	BK/LB	Sensor Ground	<0.050v
44	OR/WT	8-Volt Supply	7.9-8.1v
45	DB/LG	PSP Switch Signal	Straight: 0v, Turning: 5v
46	RD/TN	Battery Power (Fused B+)	12-14v
47, 53-54	---	Not Used	---
48	BR/GY	IAC 1 Driver	DC pulse signals: 0.8-11v
49	YL/BK	IAC 2 Driver	DC pulse signals: 0.8-11v
50	BK/TN	Power Ground	<0.1v
51 ('96-'97)	TN/WT	HO2S-12 (B1 S2) Signal	0.1-1.1v
52	PK/LG	Battery Temperature Sensor	At 86°F: 1.96v
55	DB/TN	Low Speed Fan Relay	Relay Off: 12v, On: 1v
57	GY/RD	IAC 3 Driver	DC pulse signals: 0.8-11v
58	PK/GY	IAC 4 Driver	DC pulse signals: 0.8-11v
59	PK/BR	CCD Bus (+)	Digital Signal: 0-5-0v
60	WT/BK	CCD Bus (-)	<0.050v
61	PK/WT	5-Volt Supply	4.9-5.1v
62	WT/RD	Brake Switch Signal	Brake Off: 0v, On: 12v
63	YL/DG	Torque Management Request	Digital Signals
64	DB/OR	A/C Clutch Relay Control	Relay Off: 12v, On: 1v
65, 75	PK, LG	SCI Transmit, SCI Receive	0v
66	WT/OR	Vehicle Speed Signal	Digital: 0-8-0-8v
67	DB/PK	ASD Relay Control	Relay Off: 12v, On: 1v
68	PK/GY	EVAP Purge Solenoid Control	PWM Signal: 0-12-0v
69	DB/PK	High Speed Fan Relay	Relay Off: 12v, On: 1v
70-71, 73	---	Not Used	---
72 ('96-'97)	OR/DG	LDP Switch Sense	Open: 12v, Closed: 0v
74	BR/LG	Fuel Pump Relay Control	Relay Off: 12v, On: 1v
76	BK/WT	PNP Switch Signal	In P/N: 0v, Others: 5v
77 ('96-'97)	WT/DG	LDP Solenoid Control	PWM Signal: 0-12-0v
78	WT/PK	S/C Vacuum Solenoid	Vacuum Increasing: 1v
79	---	Not Used	---
80	LG/RD	S/C Vent Solenoid	Vacuum Decreasing: 1v

Pin Connector Graphic

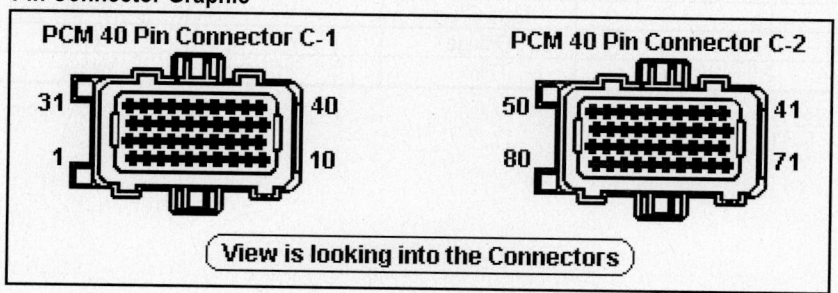

PCM 40 Pin Connector C-1
31 40
1 10

PCM 40 Pin Connector C-2
50 41
80 71

(View is looking into the Connectors)

1998-2000 Cirrus 2.5L V6 SOHC MFI VIN H (A/T) 'C1' Connector

PCM Pin #	Wire Color	Circuit Description (40-Pin)	Value at Hot Idle
1-3	---	Not Used	---
4	BK/GY	Coil 1 Driver	5°, at 55 mph: 8° dwell
5	YL/RD	S/C Power Supply	12-14v
6	DG/OR	ASD Relay Output	12-14v
7	YL/WT	Injector 3 Driver	1.0-4.0 ms
8	DG	Generator Field Driver	Digital Signal: 0-12-0v
9	---	Not Used	---
10	BK/TN	Power Ground	<0.1v
11-12	---	Not Used	---
13	WT/DB	Injector 1 Driver	1.0-4.0 ms
14	BR/DG	Injector 6 Driver	1.0-4.0 ms
15	GY	Injector 5 Driver	1.0-4.0 ms
16	LB/BR	Injector 4 Driver	1.0-4.0 ms
17	TN	Injector 2 Driver	1.0-4.0 ms
18-19	---	Not Used	---
20	DB/WT	Ignition Switch Output	12-14v
21-25	---	Not Used	---
26	TN/BK	ECT Sensor Signal	At 180°F: 2.80v
27	BK/OR	Oxygen Sensor Ground	<0.050v
28-29	---	Not Used	---
30	BK/DG	HO2S-11 (B1 S1) Signal	0.1-1.1v
31	---	Not Used	---
32	GY/BK	CKP Sensor Signal	Digital Signal: 0-5-0v
33	TN/YL	CMP Sensor Signal	Digital Signal: 0-5-0v
34	---	Not Used	---
35	OR/DB	TP Sensor Signal	0.6-1.0v
36	DG/RD	MAP Sensor Signal	1.5-1.7v
37	BK/RD	IAT Sensor Signal	At 100°F: 1.83v
38-39	---	Not Used	---
40	GY/YL	EGR Solenoid Control	12v, at 55 mph: 1v

Standard Colors and Abbreviations

Abbreviation	Color	Abbreviation	Color	Abbreviation	Color
BK	Black	GY	Gray	RD	Red
BL	Blue	GN	Green	TN	Tan
BR	Brown	LG	Light Green	VT	Violet
DB	Dark Blue	OR	Orange	WT	White
DG	Dark Green	PK	Pink	YL	Yellow

1998-2000 Cirrus 2.5L V6 SOHC MFI VIN H (A/T) 'C2' Connector

PCM Pin #	Wire Color	Circuit Description (40-Pin)	Value at Hot Idle
41	RD/LG	S/C Set Switch Signal	S/C & Set Switch On: 3.8v
42	DB/YL	A/C Pressure Switch Signal	A/C On: 0.45-4.85v
43	BK/LB	Sensor Ground	<0.050v
44	OR/WT	8-Volt Supply	7.9-8.1v
45	DB/LG	PSP Switch Signal	Straight: 0v, Turning: 5v
46	RD/TN	Battery Power (Fused B+)	12-14v
47, 50	BK/TN	Power Ground	<0.1v
48	BR/WT	IAC 3 Driver	DC pulse signals: 0.8-11v
49	YL/BK	IAC 2 Driver	DC pulse signals: 0.8-11v
51	TN/WT	HO2S-12 (B1 S2) Signal	0.1-1.1v
52	VT/LG	Battery Temperature Sensor	At 86°F: 1.96v
53-54	---	Not Used	---
55	DB/TN	Low Speed Fan Relay	Relay Off: 12v, On: 1v
56	WT/PK	S/C Vacuum Solenoid	Vacuum Increasing: 1v
57	GY/RD	IAC 1 Driver	DC pulse signals: 0.8-11v
58	VT/BK	IAC 4 Driver	DC pulse signals: 0.8-11v
59	VT/BR	CCD Bus (+)	Digital Signal: 0-5-0v
60	WT/BK	CCD Bus (-)	<0.050v
61	VT/WH	5-Volt Supply	4.9-5.1v
62	WT/RD	Brake Switch Signal	Brake Off: 0v, On: 12v
63	YL/DG	Torque Management Request	Digital Signals
64	DB/OR	A/C Clutch Relay Control	Relay Off: 12v, On: 1v
65, 75	PK, LG	SCI Transmit, SCI Receive	0v
66	WT/OR	Vehicle Speed Signal	Digital: 0-8-0-8v
67	DB/VT	ASD Relay Control	Relay Off: 12v, On: 1v
68	PK/GY	EVAP Purge Solenoid Control	PWM Signal: 0-12-0v
69	DB/PK	High Speed Fan Relay	Relay Off: 12v, On: 1v
70	WT/TN	EVAP Purge Solenoid Sense	0-1v
71, 73	---	Not Used	---
72	OR/DG	LDP Switch Sense	Open: 12v, Closed: 0v
74	BR/LG	Fuel Pump Relay Control	Relay Off: 12v, On: 1v
76	BK/WT	PNP Switch Signal	In P/N: 0v, Others: 5v
77	WT/DG	LDP Solenoid Control	PWM Signal: 0-12-0v
78-79	---	Not Used	---
80	LG/RD	S/C Vent Solenoid	Vacuum Decreasing: 1v

Pin Connector Graphic

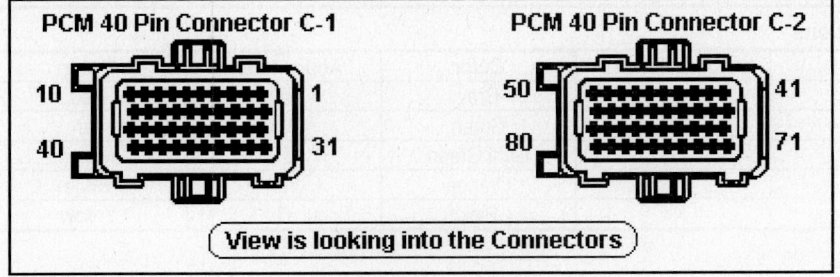

1998-99 Concorde 2.7L V6 DOHC MFI VIN R (A/T) 'C1' Connector

PCM Pin #	Wire Color	Circuit Description (40-Pin)	Value at Hot Idle
1	TN/LG	Coil 4 Driver	5°, at 55 mph: 8° dwell
2	TN/OR	Coil 3 Driver	5°, at 55 mph: 8° dwell
3	TN/PK	Coil 2 Driver	5°, at 55 mph: 8° dwell
4	TN/LB	Coil 6 Driver	5°, at 55 mph: 8° dwell
5	YL/RD	S/C Power Supply	12-14v
6	DG/OR	ASD Relay Output	12-14v
7	YL/WT	Injector 3 Driver	1.0-4.0 ms
8	DG	Generator Field Driver	Digital Signal: 0-12-0v
9	---	Not Used	---
10	BK/TN	Power Ground	<0.1v
11	TN/RD	Coil 1 Driver	5°, at 55 mph: 8° dwell
12	---	Not Used	---
13	WT/DB	Injector 1 Driver	1.0-4.0 ms
14	BR/DB	Injector 6 Driver	1.0-4.0 ms
15	GY	Injector 5 Driver	1.0-4.0 ms
16	LB/BR	Injector 4 Driver	1.0-4.0 ms
17	TN/WT	Injector 2 Driver	1.0-4.0 ms
18-19	---	Not Used	---
20	DB/WT	Ignition Switch Output	12-14v
21	TN/DG	Coil 5 Driver	5°, at 55 mph: 8° dwell
22-24	---	Not Used	---
25	BK/PK	Knock Sensor Signal	0.080v AC
26	TN/BK	ECT Sensor Signal	At 180°F: 2.80v
27	BK/OR	Oxygen Sensor Ground	<0.050v
28	---	Not Used	---
29	LG/RD	HO2S-21 (B2 S1) Signal	0.1-1.1v
30	BK/DG	HO2S-11 (B1 S1) Signal	0.1-1.1v
31	TN	Starter Relay Control	KOEC: 9-11v
32	GY/BK	CKP Sensor Signal	Digital Signal: 0-5-0v
33	TN/YL	CMP Sensor Signal	Digital Signal: 0-5-0v
34	LG/PK	EGR Sensor Signal	0.6-0.8v
35	OR/DB	TP Sensor Signal	0.6-1.0v
36	DG/RD	MAP Sensor Signal	1.5-1.7v
37	BK/RD	IAT Sensor Signal	At 100°F: 1.83v
38	---	Not Used	---
39	VT/RD	Manifold Tuning Valve Control	Valve Off: 12v, On: 1v
40	GY/YL	EGR Solenoid Control	12v, at 55 mph: 1v

Standard Colors and Abbreviations

Abbreviation	Color	Abbreviation	Color	Abbreviation	Color
BK	Black	GY	Gray	RD	Red
BL	Blue	GN	Green	TN	Tan
BR	Brown	LG	Light Green	VT	Violet
DB	Dark Blue	OR	Orange	WT	White
DG	Dark Green	PK	Pink	YL	Yellow

1998-99 Concorde 2.7L V6 DOHC MFI VIN R (A/T) 'C2' Connector

PCM Pin #	Wire Color	Circuit Description (40-Pin)	Value at Hot Idle
41	RD/LG	S/C Set Switch Signal	S/C & Set Switch On: 3.8v
42	DB	A/C Pressure Switch Signal	A/C On: 0.45-4.85v
43, 47	BK/LB	Sensor Ground	<0.050v
44	OR	8-Volt Supply	7.9-8.1v
45	DB/LG	PSP Switch Signal	Straight: 0v, Turning: 5v
46	RD/WT	Battery Power (Fused B+)	12-14v
48	BR/WT	IAC 3 Driver	DC pulse signals: 0.8-11v
49	YL/BK	IAC 2 Driver	DC pulse signals: 0.8-11v
50	BK/TN	Power Ground	<0.1v
51	TN/WT	HO2S-12 (B1 S2) Signal	0.1-1.1v
53	PK/WT	HO2S-22 (B2 S2) Signal	0.1-1.1v
52, 54, 60	---	Not Used	---
55	DB/PK	Low Speed Fan Relay	Relay Off: 12v, On: 1v
56	TN/RD	S/C Vacuum Solenoid	Vacuum Increasing: 1v
57	GY/RD	IAC 1 Driver	DC pulse signals: 0.8-11v
58	PK/BK	IAC 4 Driver	DC pulse signals: 0.8-11v
59	YL/PK	PCI Data Bus (J1850)	Digital Signals: 0-7-0v
61	PK/WT	5-Volt Supply	4.9-5.1v
62	WT/PK	Brake Switch Signal	Brake Off: 0v, On: 12v
63	YL/DG	Torque Management Request	Digital Signals
64	DB/OR	A/C Clutch Relay Control	Relay Off: 12v, On: 1v
65, 75	PK, LG	SCI Transmit, SCI Receive	0v
66	WT/OR	Vehicle Speed Signal	Digital Signal
67	DB/YL	ASD Relay Control	Relay Off: 12v, On: 1v
68	PK/BK	EVAP Purge Solenoid Control	PWM Signal: 0-12-0v
69	DB/LG	High Speed Fan Relay	Relay Off: 12v, On: 1v
70	DG/LG	EVAP Purge Solenoid Sense	0-1v
71	WT/RD	EATX RPM	Digital Signals
72	OR/RD	LDP Switch Sense	Open: 12v, Closed: 0v
73, 78-79	---	Not Used	---
74	BR	Fuel Pump Relay Control	Relay Off: 12v, On: 1v
76	BK/PK	PNP Switch Signal	In P/N: 0v, Others: 5v
77	WT/DG	LDP Solenoid Control	PWM Signal: 0-12-0v
80	LG/RD	S/C Vent Solenoid	Vacuum Decreasing: 1v

Pin Connector Graphic

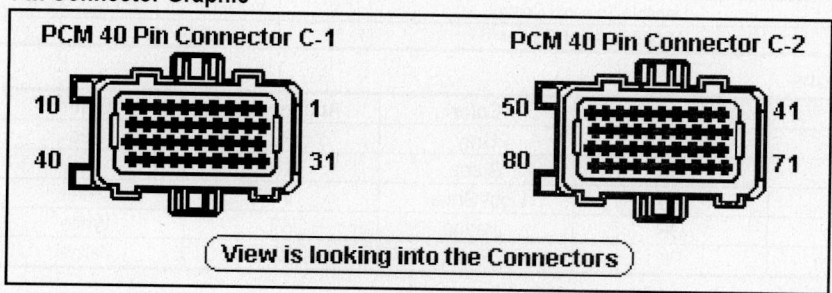

PCM 40 Pin Connector C-1
10 1
40 31

PCM 40 Pin Connector C-2
50 41
80 71

View is looking into the Connectors

2000-01 Concorde 2.7L V6 DOHC MFI VIN R, U & V 'C1' Connector

PCM Pin #	Wire Color	Circuit Description (40-Pin)	Value at Hot Idle
1	TN/LG	Coil 4 Driver	5°, at 55 mph: 8° dwell
2	TN/OR	Coil 3 Driver	5°, at 55 mph: 8° dwell
3	TN/PK	Coil 2 Driver	5°, at 55 mph: 8° dwell
4	TN/LB	Coil 6 Driver	5°, at 55 mph: 8° dwell
5	YL/RD	S/C Power Supply	12-14v
6	DG/OR	Auto Shutdown Relay Sense	12-14v
7	YL/WT	Injector 3 Driver	1.0-4.0 ms
8	DG	Generator Field Driver	Digital Signal: 0-12-0v
9	---	Not Used	---
10	BK/TN	Power Ground	<0.1v
11	TN/RD	Coil 1 Driver	5°, at 55 mph: 8° dwell
12	---	Not Used	---
13	WT/DB	Injector 1 Driver	1.0-4.0 ms
14	BR/DB	Injector 6 Driver	1.0-4.0 ms
15	GY	Injector 5 Driver	1.0-4.0 ms
16	LB/BR	Injector 4 Driver	1.0-4.0 ms
17	TN/WT	Injector 2 Driver	1.0-4.0 ms
18-19	---	Not Used	---
20	DB/WT	Ignition Switch Output	12-14v
21	TN/DG	Coil 5 Driver	5°, at 55 mph: 8° dwell
22-24	---	Not Used	---
25	BK/VT	Knock Sensor Signal	0.080v AC
26	TN/BK	ECT Sensor Signal	At 180°F: 2.80v
27	BK/OR	Oxygen Sensor Ground	<0.050v
28	---	Not Used	---
29	LG/RD	HO2S-21 (B2 S1) Signal	0.1-1.1v
30	BK/DG	HO2S-11 (B1 S1) Signal	0.1-1.1v
31	TN	Starter Relay Control	KOEC: 9-11v
32	GY/BK	CKP Sensor Signal	Digital Signal: 0-5-0v
33	TN/YL	CMP Sensor Signal	Digital Signal: 0-5-0v
34	LG/PK	EGR Sensor Signal	0.6-0.8v
35	OR/DB	TP Sensor Signal	0.6-1.0v
36	DG/RD	MAP Sensor Signal	1.5-1.7v
37	BK/RD	IAT Sensor Signal	At 100°F: 1.83v
38	---	Not Used	---
39	VT/RD	Manifold Solenoid Control	Valve On: 1v, Off: 12v
40	GY/YL	EGR Solenoid Control	12v, at 55 mph: 1v

Standard Colors and Abbreviations

Abbreviation	Color	Abbreviation	Color	Abbreviation	Color
BK	Black	GY	Gray	RD	Red
BL	Blue	GN	Green	TN	Tan
BR	Brown	LG	Light Green	VT	Violet
DB	Dark Blue	OR	Orange	WT	White
DG	Dark Green	PK	Pink	YL	Yellow

2000-01 Concorde 2.7L V6 DOHC MFI VIN R, U & V 'C2' Connector

PCM Pin #	Wire Color	Circuit Description (40-Pin)	Value at Hot Idle
41	RD/LG	S/C Set Switch Signal	S/C & Set Switch On: 3.8v
42	DB	A/C Pressure Switch Signal	A/C On: 0.45-4.85v
43, 47	BK/LB	Sensor Ground	<0.050v
44	OR	8-Volt Supply	7.9-8.1v
45	DB/LG	PSP Switch Signal	Straight: 0v, Turning: 5v
46	RD/WT	Battery Power (Fused B+)	12-14v
48	BR/WT	IAC 3 Driver	DC pulse signals: 0.8-11v
49	YL/BK	IAC 2 Driver	DC pulse signals: 0.8-11v
50	BK/TN	Power Ground	<0.1v
51	TN/WT	HO2S-12 (B1 S2) Signal	0.1-1.1v
52 ('01)	VT/LG	Battery Temperature Sensor	At 86°F: 1.96v
53	PK/WT	HO2S-22 (B2 S2) Signal	0.1-1.1v
54, 60	---	Not Used	---
55	DB/PK	Low Speed Fan Relay	Relay Off: 12v, On: 1v
56	TN/RD	S/C Vacuum Solenoid	Vacuum Increasing: 1v
57	GY/RD	IAC 1 Driver	DC pulse signals: 0.8-11v
58	PK/BK	IAC 4 Driver	DC pulse signals: 0.8-11v
59	YL/PK	PCI Data Bus (J1850)	Digital Signals: 0-7-0v
61	PK/WT	5-Volt Supply	4.9-5.1v
62	WT/PK	Brake Switch Signal	Brake Off: 0v, On: 12v
63	YL/DG	Torque Management Request	Digital Signals
64	DB/OR	A/C Clutch Relay Control	Relay Off: 12v, On: 1v
65, 75	PK, LG	SCI Transmit, SCI Receive	0v
66	WT/OR	Vehicle Speed Signal	Digital Signal
67	DB/YL	ASD Relay Control	Relay Off: 12v, On: 1v
68	PK/BK	EVAP Purge Solenoid Control	PWM Signal: 0-12-0v
69	DB/LG	High Speed Fan Relay	Relay Off: 12v, On: 1v
70	DG/LG	EVAP Purge Solenoid Sense	0-1v
71	WT/RD	EATX RPM	Digital Signals
72	OR/RD	LDP Switch Sense	Open: 12v, Closed: 0v
73, 78-79	---	Not Used	---
74	BR	Fuel Pump Relay Control	Relay Off: 12v, On: 1v
76	BK/PK	PNP Switch Signal	In P/N: 0v, Others: 5v
77	WT/DG	LDP Solenoid Control	PWM Signal: 0-12-0v
80	LG/RD	S/C Vent Solenoid	Vacuum Decreasing: 1v

Pin Connector Graphic

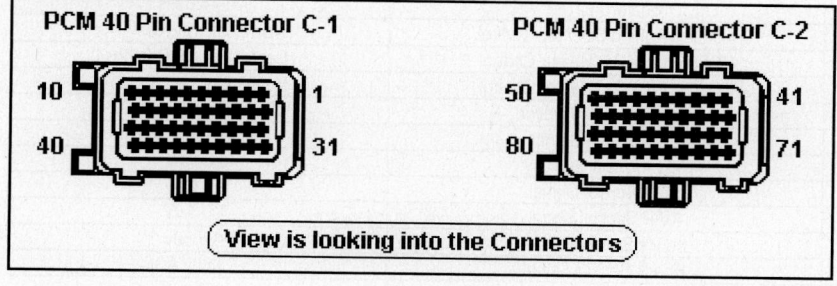

PCM 40 Pin Connector C-1

10   1
40   31

PCM 40 Pin Connector C-2

50   41
80   71

View is looking into the Connectors

2002-03 Concorde 2.7L V6 SOHC 24v MFI VIN R 'C1' Black Connector

PCM Pin #	Wire Color	Circuit Description (38-Pin)	Value at Hot Idle
1-8, 10	---	Not Used	---
9	BK/TN	Power Ground	<0.1v
11	DB/WT	Ignition Switch Output (Run-Start)	12-14v
12	RD/WT	Ignition Switch Output (Off-Run-Start)	12-14v
13-16	---	Not Used	---
17	---	Not Used	---
18	BK/TN	Power Ground	<0.1v
19-20	---	Not Used	---
21	DB	A/C Pressure Switch Signal	A/C On: 0.45-4.85v
22-24	---	Not Used	---
25	LG	SCI Receive (PCM)	5v
26	VT/OR	SCI Receive (TCM)	5v
27-28	---	Not Used	---
29	RD	Battery Power (Fused B+)	12-14v
30	YL/BK	Ignition Switch Output (tart)	8-11v (cranking)
31	TN/WT	HO2S-12 (B1 S2) Signal	0.1-1.1v
32	DB/LG	Oxygen Sensor Return (down)	<0.050v
33	PK/WT	HO2S-22 (B2 S2) Signal	0.1-1.1v
34-35	---	Not Used	---
36	PK/TN	SCI Transmit (PCM)	0v
37	WT/DG	SCI Transmit (TCM)	0v
38	VT/YL	PCI Data Bus (J1850)	Digital Signals: 0-7-0v

2002-03 Concorde 2.7L V6 SOHC 24v MFI VIN R 'C2' Gray Connector

PCM Pin #	Wire Color	Circuit Description (38-Pin)	Value at Hot Idle
1	TN/LB	Coil 6 Driver	5°, at 55 mph: 8° dwell
2	TN/DG	Coil 5 Driver	5°, at 55 mph: 8° dwell
3	TN/LG	Coil 4 Driver	5°, at 55 mph: 8° dwell
4	BR/DB	Injector 6 Driver	1.0-4.0 ms
5	GY	Injector 5 Driver	1.0-4.0 ms
6, 15, 26	---	Not Used	---
7	TN/OR	Coil 3 Driver	5°, at 55 mph: 8° dwell
8	GY/YL	EGR Solenoid Control	12v, at 55 mph: 1v
9	TN/PK	Coil 2 Driver	5°, at 55 mph: 8° dwell
10	TN/RD	Coil 1 Driver	5°, at 55 mph: 8° dwell
11	LB/BR	Injector 4 Driver	1.0-4.0 ms
12	YL/WT	Injector 3 Driver	1.0-4.0 ms
13	TN	Injector 2 Driver	1.0-4.0 ms
14	WT/DB	Injector 1 Driver	1.0-4.0 ms
16	VT/RD	Manifold Tuning Solenoid Control	Valve Off: 12v, On: 1v
17	BR/WT	HO2S-21 (B2 S1) Heater Control	Heater On: 1v, Off: 12v
18	BR/OR	HO2S-11 (B1 S1) Heater Control	Heater On: 1v, Off: 12v
19	DG	Generator Field Driver	Digital Signal: 0-12-0v
20	TN/BK	ECT Sensor Signal	At 180°F: 2.80v
21	OR/DB	TP Sensor Signal	0.6-1.0v
22	LG/PK	EGR Sensor Signal	0.6-0.8v
23	DG/RD	MAP Sensor Signal	1.5-1.7v
24	BK/VT	Knock Sensor Return	<0.050v
25	DB/LG	Knock Sensor Signal	0.080v AC
27	BK/LB	Sensor Ground	<0.1v
28	YL/BK	IAC Motor Sense	12-14v
29	VT/WT	5-Volt Supply	4.9-5.1v
30	BK/RD	IAT Sensor Signal	At 100°F: 1.83v
31	BK/DG	HO2S-11 (B1 S1) Signal	0.1-1.1v
32	BR/DG	HO2S-11 (B1 S1) Ground (Up)	<0.1v
33	LG/RD	HO2S-21 (B2 S1) Signal	0.1-1.1v
34	TN/YL	CMP Sensor Signal	Digital Signal: 0-5-0v
35	GY/BK	CKP Sensor Signal	Digital Signal: 0-5-0v
36-37	---	Not Used	---
38	GY/RD	IAC Motor Driver	DC pulse signals: 0.8-11v

2002-03 Concorde 2.7L V6 SOHC 24v MFI VIN R 'C3' White Connector

PCM Pin #	Wire Color	Circuit Description (38-Pin)	Value at Hot Idle
1-2, 13-17	---	Not Used	---
3	DB/YL	ASD Relay Control	Relay Off: 12v, On: 1v
4	DB/PK	High Speed Radiator Fan Relay	Relay Off: 12v, On: 1v
5	LG/RD	S/C Vent Solenoid	Vacuum Decreasing: 1v
6	DB/PK	Low Speed Radiator Fan Relay	Relay Off: 12v, On: 1v
7	YL/RD	S/C Power Supply	12-14v
8	WT/DG	Natural Vacuum Leak Detection Solenoid	Solenoid Off: 12v, On: 1v
9	BR/VT	HO2S-21 (B2 S1) Heater Control	Heater On: 1v, Off: 12v
10	BR/GY	HO2S-11 (B1 S1) Heater Control	Heater On: 1v, Off: 12v
11	DB/OR	A/C Clutch Relay Control	Relay Off: 12v, On: 1v
12	TN/RD	S/C Vacuum Solenoid	Vacuum Increasing: 1v
18, 19	OR/DG	Automatic Shutdown Relay Output	12-14v
20	PK/BK	EVAP Purge Solenoid Control	PWM Signal: 0-12-0v
21-22, 24-25	---	Not Used	---
23	WT/PK	Brake Switch Sense	Brake Off: 0v, On: 12v
26	YL	Autostick Downshift Switch	Digital Signal: 0v or 12v
27	LG/RD	Autostick Upshift Switch	Digital Signal: 0v or 12v
28	OR/DG	Automatic Shutdown Relay Output	12-14v
29	DG/LG	EVAP Purge Solenoid Sense	0-1v
30-31, 33, 36	---	Not Used	---
32	VT/LG	Ambient Air Temperature Sensor	At 100°F: 1.83v
34	RD/LG	S/C Set Switch Signal	S/C & Set Switch On: 3.8v
35	OR/RD	Natural Vacuum Leak Detection Switch Sense	0.1v
37	BR	Fuel Pump Relay Control	Relay Off: 12v, On: 1v
38	TN	Starter Relay Control	KOEC: 9-11v

2002-03 Concorde 2.7L V6 SOHC 24v MFI VIN R 'C4' Green Connector

PCM Pin #	Wire Color	Circuit Description (38-Pin)	Value at Hot Idle
1, 3-5	---	Not Used	---
2	BR	Overdrive Solenoid Control	Solenoid Off: 12v, On: 1v
3	PK	Underdrive Solenoid Control	Solenoid Off: 12v, On: 1v
6	WT	2-4 Solenoid Control	Solenoid Off: 12v, On: 1v
7-9, 11	---	Not Used	---
10	LB	Low/Reverse Solenoid Control	Solenoid Off: 12v, On: 1v
12	BK/YL	Power Ground	<0.050v
13, 14	BK/RD	Power Ground	<0.050v
15	LG/BK	TRS T1 Sense	<0.050v
16	VT	TRS T3 Sense	<0.050v
17, 20-21	---	Not Used	---
18, 19	LG, RD	Transmission Control Relay Output	Relay Off: 12v, On: 1v
19	RD	Transmission Control Relay Output	Relay Off: 12v, On: 1v
23-26, 31, 36	---	Not Used	---
27	BK/WT	TRS T41 Sense	<0.050v
28, 38	RD	Transmission Control Relay Output	Relay Off: 12v, On: 1v
29	LG	Low/Reverse Pressure Switch Sense	12-14v
30	YL/BK	2-4 Pressure Switch Sense	In Low/Reverse: 2-4v
32	LG/WT	Output Speed Sensor Signal	In 2-4 Position: 2-4v
33	RD/BK	Input Speed Sensor Signal	Moving: AC voltage
34	DB/BK	Speed Sensor Ground	Moving: AC voltage
35	VT/PK	Transmission Temperature Sensor Signal	<0.050v
37	VT/WT	TRS T42 Sense	In PRNL: 0v, Others 5v

Pin Connector Graphic

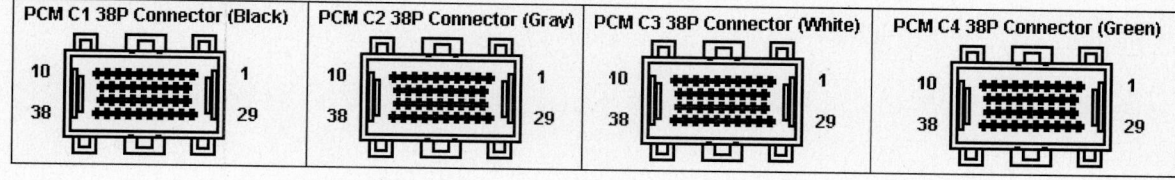

| PCM C1 38P Connector (Black) | PCM C2 38P Connector (Gray) | PCM C3 38P Connector (White) | PCM C4 38P Connector (Green) |

1998-99 Concorde 3.2L V6 SOHC MFI VIN J (A/T) 'C1' Connector

PCM Pin #	Wire Color	Circuit Description (40-Pin)	Value at Hot Idle
1	TN/LG	Coil 4 Driver	5°, at 55 mph: 8° dwell
2	TN/OR	Coil 3 Driver	5°, at 55 mph: 8° dwell
3	TN/PK	Coil 2 Driver	5°, at 55 mph: 8° dwell
4	TN/LB	Coil 6 Driver	5°, at 55 mph: 8° dwell
5	YL/RD	S/C Power Supply	12-14v
6	DG/OR	ASD Relay Output	12-14v
7	YL/WT	Injector 3 Driver	1.0-4.0 ms
8	DG	Generator Field Driver	Digital Signal: 0-12-0v
9	---	Not Used	---
10	BK/TN	Power Ground	<0.1v
11	TN/RD	Coil 1 Driver	5°, at 55 mph: 8° dwell
12	---	Not Used	---
13	WT/DB	Injector 1 Driver	1.0-4.0 ms
14	BR/DB	Injector 6 Driver	1.0-4.0 ms
15	GY	Injector 5 Driver	1.0-4.0 ms
16	LB/BR	Injector 4 Driver	1.0-4.0 ms
17	TN	Injector 2 Driver	1.0-4.0 ms
18	GY/PK	Short Runner Valve Control	Valve Off: 12v, On: 1v
19	---	Not Used	---
20	DB/WT	Ignition Switch Output	12-14v
21	TN/DG	Coil 5 Driver	5°, at 55 mph: 8° dwell
22-23	---	Not Used	---
24	DB/LG	Knock Sensor Signal	0.080v AC
25	BK/VT	Knock Sensor Ground	<0.050v
26	TN/BK	ECT Sensor Signal	At 180°F: 2.80v
27	BK/OR	Oxygen Sensor Ground	<0.050v
28	---	Not Used	---
29	LG/RD	HO2S-21 (B2 S1) Signal	0.1-1.1v
30	BK/DG	HO2S-11 (B1 S1) Signal	0.1-1.1v
31	TN/WT	Starter Relay Control	KOEC: 9-11v
32	GY/BK	CKP Sensor Signal	Digital Signal: 0-5-0v
33	TN/YL	CMP Sensor Signal	Digital Signal: 0-5-0v
34	LG/PK	EGR Sensor Signal	0.6-0.8v
35	OR/DB	TP Sensor Signal	0.6-1.0v
36	DG/RD	MAP Sensor Signal	1.5-1.7v
37	BK/RD	IAT Sensor Signal	At 100°F: 1.83v
38	---	Not Used	---
39	VT/RD	Manifold Tuning Valve Control	Valve Off: 12v, On: 1v
40	GY/YL	EGR Solenoid Control	12v, at 55 mph: 1v

Standard Colors and Abbreviations

Abbreviation	Color	Abbreviation	Color	Abbreviation	Color
BK	Black	GY	Gray	RD	Red
BL	Blue	GN	Green	TN	Tan
BR	Brown	LG	Light Green	VT	Violet
DB	Dark Blue	OR	Orange	WT	White
DG	Dark Green	PK	Pink	YL	Yellow

1998-99 Concorde 3.2L V6 SOHC MFI VIN J (A/T) 'C2' Connector

PCM Pin #	Wire Color	Circuit Description (40-Pin)	Value at Hot Idle
41	RD/LG	S/C Set Switch Signal	S/C & Set Switch On: 3.8v
42	DB	A/C Pressure Switch Signal	A/C On: 0.45-4.85v
43	BK/LB	Sensor Ground	<0.050v
44	OR	8-Volt Supply	7.9-8.1v
45	DB/LG	PSP Switch Signal	Straight: 0v, Turning: 5v
46	RD/WT	Battery Power (Fused B+)	12-14v
47, 50	BK/WT	Power Ground	<0.1v
48	BR/WT	IAC 3 Driver	DC pulse signals: 0.8-11v
49	YL/BK	IAC 2 Driver	DC pulse signals: 0.8-11v
51	TN/WT	HO2S-12 (B1 S2) Signal	0.1-1.1v
52, 54, 60	---	Not Used	---
53	PK/WT	HO2S-22 (B2 S2) Signal	0.1-1.1v
55	DB/PK	Low Speed Fan Relay	Relay Off: 12v, On: 1v
56	TN/RD	S/C Vacuum Solenoid	Vacuum Increasing: 1v
57	GY/RD	IAC 1 Driver	DC pulse signals: 0.8-11v
58	VT/BK	IAC 4 Driver	DC pulse signals: 0.8-11v
59	YL/VT	PCI Data Bus (J1850)	Digital Signals: 0-7-0v
61	VT/WT	5-Volt Supply	4.9-5.1v
62	WT/PK	Brake Switch Signal	Brake Off: 0v, On: 12v
63	YL/DG	Torque Management Request	Digital Signals
64	DB/OR	A/C Clutch Relay Control	Relay Off: 12v, On: 1v
65, 75	PK, LG	SCI Transmit, SCI Receive	0v
66	WT/OR	Vehicle Speed Signal	Digital Signal
67	DB/YL	ASD Relay Control	Relay Off: 12v, On: 1v
68	PK/BK	EVAP Purge Solenoid Control	PWM Signal: 0-12-0v
69	DB/LG	High Speed Fan Relay	Relay Off: 12v, On: 1v
70	DG/LG	EVAP Purge Solenoid Sense	0-1v
71	WT/RD	EATX RPM	Digital Signals
72	OR/RD	LDP Switch Sense	Open: 12v, Closed: 0v
73, 78-79	---	Not Used	---
74	BR	Fuel Pump Relay Control	Relay Off: 12v, On: 1v
76	BK/PK	PNP Switch Signal	In P/N: 0v, Others: 5v
77	WT/DG	LDP Solenoid Control	PWM Signal: 0-12-0v
80	LG/RD	S/C Vent Solenoid	Vacuum Decreasing: 1v

Pin Connector Graphic

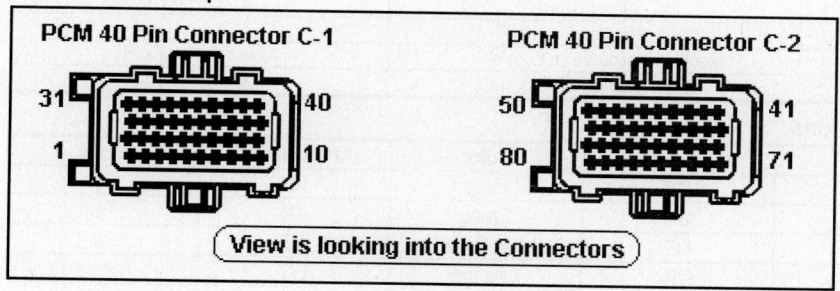

PCM 40 Pin Connector C-1

31 40
1 10

PCM 40 Pin Connector C-2

50 41
80 71

View is looking into the Connectors

2000-01 Concorde 3.2L V6 SOHC MFI VIN J (A/T) 'C1' Connector

PCM Pin #	Wire Color	Circuit Description (40-Pin)	Value at Hot Idle
1	TN/LG	Coil 4 Driver	5°, at 55 mph: 8° dwell
2	TN/OR	Coil 3 Driver	5°, at 55 mph: 8° dwell
3	TN/PK	Coil 2 Driver	5°, at 55 mph: 8° dwell
4	TN/LB	Coil 6 Driver	5°, at 55 mph: 8° dwell
5	YL/RD	S/C Power Supply	12-14v
6	DG/OR	ASD Relay Output	12-14v
7	YL/WT	Injector 3 Driver	1.0-4.0 ms
8	DG	Generator Field Driver	Digital Signal: 0-12-0v
9	---	Not Used	---
10	BK/TN	Power Ground	<0.1v
11	TN/RD	Coil 1 Driver	5°, at 55 mph: 8° dwell
12	---	Not Used	---
13	WT/DB	Injector 1 Driver	1.0-4.0 ms
14	BR/DB	Injector 6 Driver	1.0-4.0 ms
15	GY	Injector 5 Driver	1.0-4.0 ms
16	LB/BR	Injector 4 Driver	1.0-4.0 ms
17	TN	Injector 2 Driver	1.0-4.0 ms
18	GY/PK	Short Runner Valve Control	Valve Off: 12v, On: 1v
19	---	Not Used	---
20	DB/WT	Ignition Switch Output	12-14v
21	TN/DG	Coil 5 Driver	5°, at 55 mph: 8° dwell
22-23	---	Not Used	---
24	DB/LG	Knock Sensor Signal	0.080v AC
25	BK/VT	Knock Sensor Ground	<0.050v
26	TN/BK	ECT Sensor Signal	At 180°F: 2.80v
27	BK/OR	Oxygen Sensor Ground	<0.050v
28	---	Not Used	---
29	LG/RD	HO2S-21 (B2 S1) Signal	0.1-1.1v
30	BK/DG	HO2S-11 (B1 S1) Signal	0.1-1.1v
31	TN/WH	Starter Relay Control	KOEC: 9-11v
32	GY/BK	CKP Sensor Signal	Digital Signal: 0-5-0v
33	TN/YL	CMP Sensor Signal	Digital Signal: 0-5-0v
34	LG/PK	EGR Sensor Signal	0.6-0.8v
35	OR/DB	TP Sensor Signal	0.6-1.0v
36	DG/RD	MAP Sensor Signal	1.5-1.7v
37	BK/RD	IAT Sensor Signal	At 100°F: 1.83v
38	---	Not Used	---
39	VT/RD	Manifold Tuning Valve Control	Valve Off: 12v, On: 1v
40	GY/YL	EGR Solenoid Control	12v, at 55 mph: 1v

Standard Colors and Abbreviations

Abbreviation	Color	Abbreviation	Color	Abbreviation	Color
BK	Black	GY	Gray	RD	Red
BL	Blue	GN	Green	TN	Tan
BR	Brown	LG	Light Green	VT	Violet
DB	Dark Blue	OR	Orange	WT	White
DG	Dark Green	PK	Pink	YL	Yellow

2000-01 Concorde 3.2L V6 SOHC MFI VIN J (A/T) 'C2' Connector

PCM Pin #	Wire Color	Circuit Description (40-Pin)	Value at Hot Idle
41	RD/LG	S/C Set Switch Signal	S/C & Set Switch On: 3.8v
42	DB	A/C Pressure Switch Signal	A/C On: 0.45-4.85v
43	BK/LB	Sensor Ground	<0.050v
44	OR	8-Volt Supply	7.9-8.1v
45	DB/LG	PSP Switch Signal	Straight: 0v, Turning: 5v
46	RD/WT	Battery Power (Fused B+)	12-14v
47, 50	BK/WT	Power Ground	<0.1v
48	BR/WT	IAC 3 Driver	DC pulse signals: 0.8-11v
49	YL/BK	IAC 2 Driver	DC pulse signals: 0.8-11v
51	TN/WT	HO2S-12 (B1 S2) Signal	0.1-1.1v
52 ('01)	VT/LG	Battery Temperature Sensor	At 86°F: 1.96v
53	PK/WT	HO2S-22 (B2 S2) Signal	0.1-1.1v
54, 60	---	Not Used	---
55	DB/PK	Low Speed Fan Relay	Relay Off: 12v, On: 1v
56	TN/RD	S/C Vacuum Solenoid	Vacuum Increasing: 1v
57	GY/RD	IAC 1 Driver	DC pulse signals: 0.8-11v
58	VT/BK	IAC 4 Driver	DC pulse signals: 0.8-11v
59	VT/YL	PCI Data Bus (J1850)	Digital Signals: 0-7-0v
61	VT/WT	5-Volt Supply	4.9-5.1v
62	WT/PK	Brake Switch Signal	Brake Off: 0v, On: 12v
63	YL/DG	Torque Management Request	Digital Signals
64	DB/OR	A/C Clutch Relay Control	Relay Off: 12v, On: 1v
65, 75	PK, LG	SCI Transmit, SCI Receive	0v
66	WT/OR	Vehicle Speed Signal	Digital Signal
67	DB/YL	ASD Relay Control	Relay Off: 12v, On: 1v
68	PK/BK	EVAP Purge Solenoid Control	PWM Signal: 0-12-0v
69	DB/LG	High Speed Fan Relay	Relay Off: 12v, On: 1v
70	DG/LG	EVAP Purge Solenoid Sense	0-1v
71	WT/RD	EATX RPM	Digital Signals
72	OR/RD	LDP Switch Sense	Open: 12v, Closed: 0v
73, 78-79	---	Not Used	---
74	BR	Fuel Pump Relay Control	Relay Off: 12v, On: 1v
76	BR/YL	TRS T41 Sense Signal	In P/N: 0v, Others: 5v
77	WT/DG	LDP Solenoid Control	PWM Signal: 0-12-0v
80	LG/RD	S/C Vent Solenoid	Vacuum Decreasing: 1v

Pin Connector Graphic

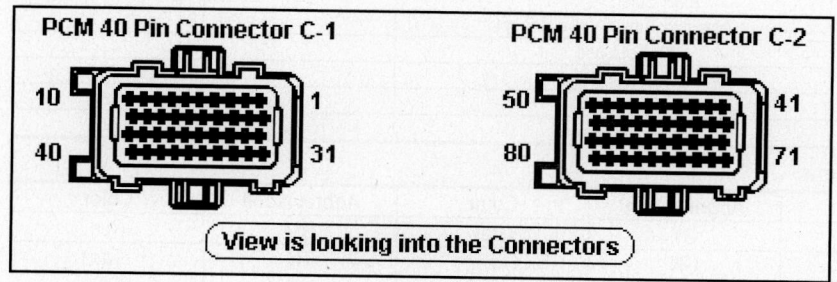

1993-95 Concorde 3.3L V6 MFI VIN T & F/F VIN U (A/T)

PCM Pin #	Wire Color	Circuit Description (60 Pin)	Value at Hot Idle
1	DG/RD	MAP Sensor Signal	1.5-1.7v
2	TN/BK	ECT Sensor Signal	At 180°F: 2.80v
3	RD/WT	Battery Power (Fused B+)	12-14v
4	BK/LB	Sensor Ground	<0.050v
5	BK/WT	Sensor Ground	<0.050v
6	PK/WT	5-Volt Supply	4.9-5.1v
7	OR	8-Volt Supply	7.9-8.1v
8	YL/DG	Torque Management Request	Digital Signals
9	DB/WT	Ignition Switch Output	12-14v
10	GY/BK	Knock Sensor 2 Signal	0.080v AC
11	BK/TN	Power Ground	<0.1v
12	BK/TN	Power Ground	<0.1v
13	LB/BR	Injector 4 Driver	1.0-4.0 ms
14	YL/WT	Injector 3 Driver	1.0-4.0 ms
15	TN	Injector 2 Driver	1.0-4.0 ms
16	WT/DB	Injector 1 Driver	1.0-4.0 ms
17 ('93)	DB/YL	Coil 2 Driver	5°, at 55 mph: 8° dwell
17 ('94-'95)	WT	Coil 2 Driver	5°, at 55 mph: 8° dwell
18 ('93)	RD/YL	Coil 3 Driver	5°, at 55 mph: 8° dwell
18 ('94-'95)	RD	Coil 3 Driver	5°, at 55 mph: 8° dwell
19 ('93)	GY	Coil 1 Driver	5°, at 55 mph: 8° dwell
19 ('94-'95)	BK	Coil 1 Driver	5°, at 55 mph: 8° dwell
20	DG	Generator Field Driver	Digital Signal: 0-12-0v
21	BK/RD	IAT Sensor Signal	At 100°F: 1.83v
22	OR/DB	TP Sensor Signal	0.6-1.0v
23	RD/LG	S/C Set Switch Signal	S/C & Set Switch On: 3.8v
24	LB/DB	CKP Sensor Signal	Digital Signal: 0-5-0v
25	PK	SCI Transmit	0v
26	PK/BR	CCD Bus (+)	Digital Signal: 0-5-0v
27-28	---	Not Used	---
29	WT/PK	Brake Switch Signal	Brake Off: 0v, On: 12v
30	BK/LG	PNP Switch Signal	In P/N: 0v, Others: 5v
31	DB/PK	High Speed Fan Relay	Relay Off: 12v, On: 1v
32	WT	Low Speed Fan Relay	Relay Off: 12v, On: 1v
33	TN/RD	S/C Vacuum Solenoid	Vacuum Increasing: 1v
34	DB/OR	A/C Clutch Relay Control	Relay Off: 12v, On: 1v
35	GY/YL	EGR Solenoid Control	12v, at 55 mph: 1v
36 ('93)	PK	Manifold Tuning Valve Control	Valve Off: 12v, On: 1v
37	---	Not Used	---
38	BR/RD	Injector 5 Driver	1.0-4.0 ms
39	GY/RD	IAC 1 Driver	DC pulse signals: 0.8-11v
40	BR/WT	IAC 3 Driver	DC pulse signals: 0.8-11v

Standard Colors and Abbreviations

Abbreviation	Color	Abbreviation	Color	Abbreviation	Color
BK	Black	GY	Gray	RD	Red
BL	Blue	GN	Green	TN	Tan
BR	Brown	LG	Light Green	VT	Violet
DB	Dark Blue	OR	Orange	WT	White
DG	Dark Green	PK	Pink	YL	Yellow

1993-95 Concorde 3.3L V6 MFI VIN T & F/F VIN U (A/T)

PCM Pin #	Wire Color	Circuit Description (60 Pin)	Value at Hot Idle
41	BK/DG	HO2S-11 (B1 S1) Signal	0.1-1.1v
42	BK/LG	Knock Sensor 1 Signal	0.080v AC
43, 54, 56	---	---	---
44	TN/YL	CMP Sensor Signal	Digital Signal: 0-5-0v
45	LG	SCI Receive	0v
46	WT/BK	CCD Bus (-)	<0.050v
47	WT/OR	Vehicle Speed Signal	Digital Signal
48	DB	A/C Pressure Switch Signal	A/C On: 0.45-4.85v
49	TN/WT	HO2S-12 (B1 S2) Signal	0.1-1.1v
50	YL/WT	Flexible Fuel Sensor Signal	0.5-4.5v
51	DB/YL	ASD Relay Output	12-14v
52	PK/BK	EVAP Purge Solenoid Control	Solenoid Off: 12v, On: 1v
53	LG/RD	S/C Vent Solenoid	Vacuum Decreasing: 1v
55	TN/RD	S/C Power Supply	12-14v
57	DG/OR	ASD Relay Control	Relay Off: 12v, On: 1v
58	BR/BK	Injector 6 Driver	1.0-4.0 ms
59	PK/BK	IAC 4 Driver	DC pulse signals: 0.8-11v
60	YL/BK	IAC 2 Driver	DC pulse signals: 0.8-11v

Pin Connector Graphic

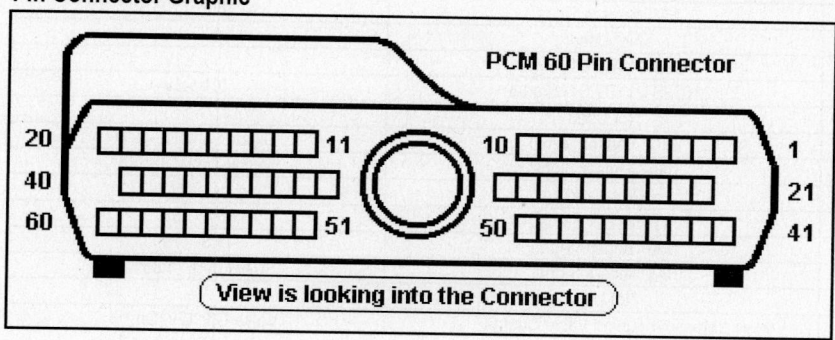

1996 Concorde 3.3L V6 MFI VIN T (A/T) 'C1' Connector

PCM Pin #	Wire Color	Circuit Description (40-Pin)	Value at Hot Idle
1	LG/RD	HO2S-12 (B1 S2) Signal	0.1-1.1v
2	RD	Coil 3 Driver	5°, at 55 mph: 8° dwell
3	WT	Coil 2 Driver	5°, at 55 mph: 8° dwell
4	DG	Generator Field Driver	Digital Signal: 0-12-0v
5	YL/RD	S/C Power Supply	12-14v
6	DG/OR	ASD Relay Output	12-14v
7	YL/WT	Injector 3 Driver	1.0-4.0 ms
8-9	---	Not Used	---
10	BK/TN	Power Ground	<0.1v
11	BK	Coil 1 Driver	5°, at 55 mph: 8° dwell
12	---	Not Used	---
13	WT/LB	Injector 1 Driver	1.0-4.0 ms
14	BR/BK	Injector 6 Driver	1.0-4.0 ms
15	BR/RD	Injector 5 Driver	1.0-4.0 ms
16	LB/BR	Injector 4 Driver	1.0-4.0 ms
17	TN	Injector 2 Driver	1.0-4.0 ms
18-19	---	Not Used	---
20	DB/WT	Ignition Switch Output	12-14v
21-25	---	Not Used	---
26	TN/BK	ECT Sensor Signal	At 180°F: 2.80v
27-28	---	Not Used	---
29	TN/WT	HO2S-11 (B1 S1) Signal	0.1-1.1v
30	BK/DG	HO2S-21 (B2 S1) Signal	0.1-1.1v
31	---	Not Used	---
32	LB/DB	CKP Sensor Signal	Digital Signal: 0-5-0v
33	TN/YL	CMP Sensor Signal	Digital Signal: 0-5-0v
34	---	Not Used	---
35	OR/DB	TP Sensor Signal	0.6-1.0v
36	DG/RD	MAP Sensor Signal	1.5-1.7v
37	BK/RD	IAT Sensor Signal	At 100°F: 1.83v
38	---	Not Used	---
39	PK/RD	Manifold Tuning Valve Control	Valve Off: 12v, On: 1v
40	GY/YL	EGR Solenoid Control	12v, at 55 mph: 1v

Standard Colors and Abbreviations

Abbreviation	Color	Abbreviation	Color	Abbreviation	Color
BK	Black	GY	Gray	RD	Red
BL	Blue	GN	Green	TN	Tan
BR	Brown	LG	Light Green	VT	Violet
DB	Dark Blue	OR	Orange	WT	White
DG	Dark Green	PK	Pink	YL	Yellow

1996 Concorde 3.3L V6 MFI VIN T (A/T)

PCM Pin #	Wire Color	Circuit Description (40-Pin)	Value at Hot Idle
41	RD/LG	S/C Set Switch Signal	S/C & Set Switch On: 3.8v
42	DB	A/C Pressure Switch Signal	A/C On: 0.45-4.85v
43	BK/LB	Sensor Ground	<0.050v
44	OR	8-Volt Supply	7.9-8.1v
45	---	Not Used	---
46	RD/WT	Battery Power (Fused B+)	12-14v
47	BK/WT	Power Ground	<0.1v
48	BR/WT	IAC 1 Driver	DC pulse signals: 0.8-11v
49	YL/BK	IAC 2 Driver	DC pulse signals: 0.8-11v
50	BK/TN	Power Ground	<0.1v
51	TN/WT	HO2S-22 (B2 S2) Signal	0.1-1.1v
52-54	---	Not Used	---
55	WT	Low Speed Fan Relay	Relay Off: 12v, On: 1v
56	---	Not Used	---
57	GY/RD	IAC 3 Driver	DC pulse signals: 0.8-11v
58	PK/GY	IAC 4 Driver	DC pulse signals: 0.8-11v
59	PK/BR	CCD Bus (+)	Digital Signal: 0-5-0v
60	WT/BK	CCD Bus (-)	<0.050v
61	PK/WT	5-Volt Supply	4.9-5.1v
62	WT/PK	Brake Switch Signal	Brake Off: 0v, On: 12v
63	YL/DG	Torque Management Request	Digital Signals
64	DB/OR	A/C Clutch Relay Control	Relay Off: 12v, On: 1v
65	PK	SCI Transmit	0v
66	WT/OR	Vehicle Speed Signal	Digital: 0-8-0-8v
67	DB/YL	ASD Relay Control	Relay Off: 12v, On: 1v
68	PK/BK	EVAP Purge Solenoid Control	PWM Signal: 0-12-0v
69	DB/PK	High Speed Fan Relay	Relay Off: 12v, On: 1v
70-73	---	Not Used	---
74	BR	Fuel Pump Relay Control	Relay Off: 12v, On: 1v
75	LG	SCI Receive	0v
76	BK/DG	PNP Switch Signal	In P/N: 0v, Others: 5v
77	---	Not Used	---
78	TN/RD	S/C Vacuum Solenoid	Vacuum Increasing: 1v
79	---	Not Used	---
80	LG/RD	S/C Vent Solenoid	Vacuum Decreasing: 1v

Pin Connector Graphic

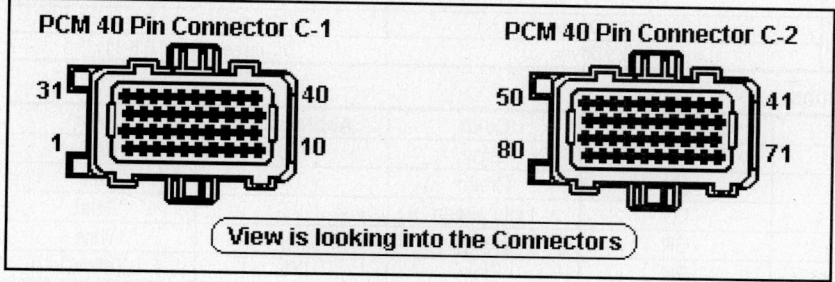

PCM 40 Pin Connector C-1   PCM 40 Pin Connector C-2

View is looking into the Connectors

1993-95 Concorde 3.5L V6 SOHC MFI VIN F (A/T)

PCM Pin #	Wire Color	Circuit Description (60 Pin)	Value at Hot Idle
1	DG/RD	MAP Sensor Signal	1.5-1.7v
2	TN/BK	ECT Sensor Signal	At 180°F: 2.80v
3	RD/WT	Battery Power (Fused B+)	12-14v
4	BK/LB	Sensor Ground	<0.050v
5	BK/WT	Sensor Ground	<0.050v
6	PK/WT	5-Volt Supply	4.9-5.1v
7	OR	8-Volt Supply	7.9-8.1v
8	YL/DG	Torque Management Request	Digital Signals
9	DB/WT	Ignition Switch Output	12-14v
10	GY/BK	Knock Sensor 2 Signal	0.080v AC
11	BK/TN	Power Ground	<0.1v
12	BK/TN	Power Ground	<0.1v
13	LB/BR	Injector 4 Driver	1.0-4.0 ms
14	YL/WT	Injector 3 Driver	1.0-4.0 ms
15	TN	Injector 2 Driver	1.0-4.0 ms
16	WT/DB	Injector 1 Driver	1.0-4.0 ms
17	DB/YL	Coil 2 Driver	5°, at 55 mph: 8° dwell
18	RD/YL	Coil 3 Driver	5°, at 55 mph: 8° dwell
19	GY	Coil 1 Driver	5°, at 55 mph: 8° dwell
20	DG	Generator Field Driver	Digital Signal: 0-12-0v
21	BK/RD	IAT Sensor Signal	At 100°F: 1.83v
22	OR/DB	TP Sensor Signal	0.6-1.0v
23	RD/LG	S/C Set Switch Signal	S/C & Set Switch On: 3.8v
24	LB/DB	CKP Sensor Signal	Digital Signal: 0-5-0v
25	PK	SCI Transmit	0v
26	PK/BR	CCD Bus (+)	Digital Signal: 0-5-0v
27-28	---	Not Used	---
29	WT/PK	Brake Switch Signal	Brake Off: 0v, On: 12v
30	BK/LG	PNP Switch Signal	In P/N: 0v, Others: 5v
31	DB/PK	High Speed Fan Relay	Relay Off: 12v, On: 1v
32	WT	Low Speed Fan Relay	Relay Off: 12v, On: 1v
33	TN/RD	S/C Vacuum Solenoid	Vacuum Increasing: 1v
34	DB/OR	A/C Clutch Relay Control	Relay Off: 12v, On: 1v
35	GY/YL	EGR Solenoid Control	12v, at 55 mph: 1v
36	PK	Manifold Tuning Valve Control	Valve Off: 12v, On: 1v
37	---	Not Used	---
38	BR/RD	Injector 5 Driver	1.0-4.0 ms
39	GY/RD	IAC 1 Driver	DC pulse signals: 0.8-11v
40	BR/WT	IAC 3 Driver	DC pulse signals: 0.8-11v

Standard Colors and Abbreviations

Abbreviation	Color	Abbreviation	Color	Abbreviation	Color
BK	Black	GY	Gray	RD	Red
BL	Blue	GN	Green	TN	Tan
BR	Brown	LG	Light Green	VT	Violet
DB	Dark Blue	OR	Orange	WT	White
DG	Dark Green	PK	Pink	YL	Yellow

1993-95 Concorde 3.5L V6 SOHC MFI VIN F (A/T)

PCM Pin #	Wire Color	Circuit Description (60 Pin)	Value at Hot Idle
41	BK/DG	HO2S-11 (B1 S1) Signal	0.1-1.1v
42	BK/LG	Knock Sensor 1 Signal	0.080v AC
43	---	Not Used	---
44	TN/YL	CMP Sensor Signal	Digital Signal: 0-5-0v
45	LG	SCI Receive	0v
46	WT/BK	CCD Bus (-)	<0.050v
47	WT/OR	Vehicle Speed Signal	Digital Signal
48	DB	A/C Pressure Switch Signal	A/C On: 0.45-4.85v
49	TN/WT	HO2S-12 (B1 S2) Signal	0.1-1.1v
50	YL/WT	Flexible Fuel Sensor Signal	0.5-4.5v
51	DB/YL	ASD Relay Output	12-14v
52	PK/BK	EVAP Purge Solenoid Control	Solenoid Off: 12v, On: 1v
53	LG/RD	S/C Vent Solenoid	Vacuum Decreasing: 1v
54	---	Not Used	---
55	TN/RD	S/C Power Supply	12-14v
56	---	Not Used	---
57	DG/OR	ASD Relay Control	Relay Off: 12v, On: 1v
58	BR/BK	Injector 6 Driver	1.0-4.0 ms
59	PK/BK	IAC 4 Driver	DC pulse signals: 0.8-11v
60	YL/BK	IAC 2 Driver	DC pulse signals: 0.8-11v

Pin Connector Graphic

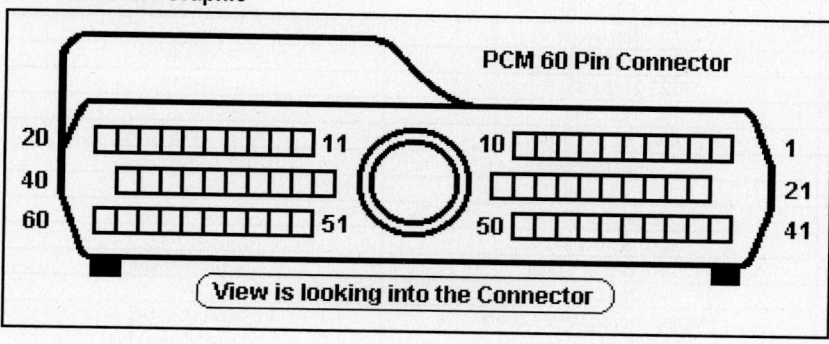

PCM 60 Pin Connector

20   11   10   1
40
60   51   50   41
21

View is looking into the Connector

1996-97 Concorde 3.5L V6 SOHC MFI VIN F (A/T) 'C1' Connector

PCM Pin #	Wire Color	Circuit Description (40-Pin)	Value at Hot Idle
1	LG/RD	HO2S-12 (B1 S2) Signal	0.1-1.1v
2	RD	Coil 3 Driver	5°, at 55 mph: 8° dwell
3	WT	Coil 2 Driver	5°, at 55 mph: 8° dwell
4	DG	Generator Field Driver	Digital Signal: 0-12-0v
5	YL/RD	S/C Power Supply	12-14v
6	DG/OR	ASD Relay Output	12-14v
7	YL/WT	Injector 3 Driver	1.0-4.0 ms
8-9	---	Not Used	---
10	BK/TN	Power Ground	<0.1v
11	BK	Coil 1 Driver	5°, at 55 mph: 8° dwell
12	---	Not Used	---
13	WT/LB	Injector 1 Driver	1.0-4.0 ms
14	BR/BK	Injector 6 Driver	1.0-4.0 ms
15	BR/RD	Injector 5 Driver	1.0-4.0 ms
16	LB/BR	Injector 4 Driver	1.0-4.0 ms
17	TN	Injector 2 Driver	1.0-4.0 ms
18-19	---	Not Used	---
20	DB/WT	Ignition Switch Output	12-14v
21-23	---	Not Used	---
24	BK/LG	Knock Sensor 1 Signal	0.080v AC
25	GY/BK	Knock Sensor 2 Signal	0.080v AC
26	TN/BK	ECT Sensor Signal	At 180°F: 2.80v
27-28	---	Not Used	---
29	TN/WT	HO2S-11 (B1 S1) Signal	0.1-1.1v
30	BK/DG	HO2S-21 (B2 S1) Signal	0.1-1.1v
31	---	Not Used	---
32	LB/DB	CKP Sensor Signal	Digital Signal: 0-5-0v
33	TN/YL	CMP Sensor Signal	Digital Signal: 0-5-0v
34	---	Not Used	---
35	OR/DB	TP Sensor Signal	0.6-1.0v
36	DG/RD	MAP Sensor Signal	1.5-1.7v
37	BK/RD	IAT Sensor Signal	At 100°F: 1.83v
38	---	Not Used	---
39	PK/RD	Manifold Tuning Valve Control	Valve Off: 12v, On: 1v
40	GY/YL	EGR Solenoid Control	12v, at 55 mph: 1v

Standard Colors and Abbreviations

Abbreviation	Color	Abbreviation	Color	Abbreviation	Color
BK	Black	GY	Gray	RD	Red
BL	Blue	GN	Green	TN	Tan
BR	Brown	LG	Light Green	VT	Violet
DB	Dark Blue	OR	Orange	WT	White
DG	Dark Green	PK	Pink	YL	Yellow

1996-97 Concorde 3.5L V6 SOHC MFI VIN F (A/T) 'C2' Connector

PCM Pin #	Wire Color	Circuit Description (40-Pin)	Value at Hot Idle
41	RD/LG	S/C Set Switch Signal	S/C & Set Switch On: 3.8v
42	DB	A/C Pressure Switch Signal	A/C On: 0.45-4.85v
43	BK/LB	Sensor Ground	<0.050v
44	OR	8-Volt Supply	7.9-8.1v
45	---	Not Used	---
46	RD/WT	Battery Power (Fused B+)	12-14v
47	BK/WT	Power Ground	<0.1v
48	BR/WT	IAC 1 Driver	DC pulse signals: 0.8-11v
49	YL/BK	IAC 2 Driver	DC pulse signals: 0.8-11v
50	BK/TN	Power Ground	<0.1v
51	TN/WT	HO2S-22 (B2 S2) Signal	0.1-1.1v
52-54	---	Not Used	---
55	WT	Low Speed Fan Relay	Relay Off: 12v, On: 1v
56	---	Not Used	---
57	GY/RD	IAC 3 Driver	DC pulse signals: 0.8-11v
58	PK/GY	IAC 4 Driver	DC pulse signals: 0.8-11v
59	PK/BR	CCD Bus (+)	Digital Signal: 0-5-0v
60	WT/BK	CCD Bus (-)	<0.050v
61	PK/WT	5-Volt Supply	4.9-5.1v
62	WT/PK	Brake Switch Signal	Brake Off: 0v, On: 12v
63	YL/DG	Torque Management Request	Digital Signals
64	DB/OR	A/C Clutch Relay Control	Relay Off: 12v, On: 1v
65	PK	SCI Transmit	0v
66	WT/OR	Vehicle Speed Signal	Digital: 0-8-0-8v
67	DB/YL	ASD Relay Control	Relay Off: 12v, On: 1v
68	PK/BK	EVAP Purge Solenoid Control	PWM Signal: 0-12-0v
69	DB/PK	High Speed Fan Relay	Relay Off: 12v, On: 1v
70-73	---	Not Used	---
74	BR	Fuel Pump Relay Control	Relay Off: 12v, On: 1v
75	LG	SCI Receive	0v
76	BK/DG	PNP Switch Signal	In P/N: 0v, Others: 5v
77, 79	---	Not Used	---
78	TN/RD	S/C Vacuum Solenoid	Vacuum Increasing: 1v
79	---	Not Used	---
80	LG/RD	S/C Vent Solenoid	Vacuum Decreasing: 1v

Pin Connector Graphic

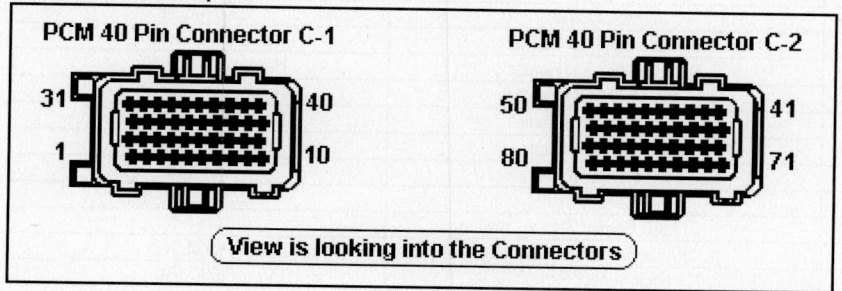

View is looking into the Connectors

2002-3 Concorde 3.5L V6 SOHC MFI VIN G, M (A/T) 'C1' Black Connector

PCM Pin #	Wire Color	Circuit Description (38-Pin)	Value at Hot Idle
1-8, 10	---	Not Used	---
9	BK/TN	Power Ground	<0.1v
11	DB/WT	Fused Ignition Power (B+)	12-14v
12	RD/WT	Fused Ignition Power (B+)	12-14v
13-15	---	Not Used	---
16	GY/PK	Short Runner Valve Solenoid	Valve On: 1v, Off: 12v
17, 19-20	---	Not Used	---
18	BK/TN	Power Ground	<0.1v
21	DB	A/C Pressure Sensor	A/C On: 0.45-4.85v
22-24	---	Not Used	---
25	LG	SCI Receive	5v
26	VT/OR	SCI Transmit	0v
27-28	---	Not Used	---
29	RD	Fused Power (B+)	12-14v
30	YL/BK	Fused Ignition Output (Start)	10-12v
31	TN/WT	HO2S-12 (B1 S2) Signal	0.1-1.1v
32	DB/DG	HO2S-12 (B1 S2) Ground	<0.050v
33	PK/WT	HO2S-22 (B2 S2) Signal	0.1-1.1v
34-35	---	Not Used	---
36	PK/TN	SCI Transmit (PCM)	0v
37	WT/DG	SCI Transmit (PCM)	0v
38	VT/YL	PCI Data Bus (J1850)	Digital Signals: 0-7-0v

2002-3 Concorde 3.5L V6 SOHC MFI VIN G, M (A/T) 'C2' Gray Connector

PCM Pin #	Wire Color	Circuit Description (38-Pin)	Value at Hot Idle
1	TN/LB	Coil On Plug 6 Driver	5°, at 55 mph: 8° dwell
2	TN/DG	Coil On Plug 5 Driver	5°, at 55 mph: 8° dwell
3	TN/LG	Coil On Plug 4 Driver	5°, at 55 mph: 8° dwell
4	BR/DB	Injector 6 Driver	1.0-4.0 ms
5	GY	Injector 5 Driver	1.0-4.0 ms
6, 15, 26	---	Not Used	---
7	TN/OR	Coil On Plug 3 Driver	5°, at 55 mph: 8° dwell
8	GY/YL	EGR Solenoid Control	12v, at 55 mph: 1v
9	TN/PK	Coil On Plug 2 Driver	5°, at 55 mph: 8° dwell
10	TN/RD	Coil On Plug 1 Driver	5°, at 55 mph: 8° dwell
11	LB/BR	Injector 4 Driver	1.0-4.0 ms
12	YL/WT	Injector 3 Driver	1.0-4.0 ms
13	TN/WT	Injector 2 Driver	1.0-4.0 ms
14	WT/DB	Injector 1 Driver	1.0-4.0 ms
16	VT/RD	Manifold Solenoid Control	Valve Off: 12v, On: 1v
17	BR/WT	HO2S-21 (B2 S1) Heater	Heater Off: 12v, On: 1v
18	BR/OR	HO2S-11 (B1 S1) Heater	Heater Off: 12v, On: 1v
19	DG	Generator Field Driver	Digital Signals: 0-12-0v
20	TN/BK	ECT Sensor Signal	At 180°F: 2.80v
21	DB	TP Sensor Signal	0.6-1.0v
22	LG/PK	EGR Sensor Signal	0.6-0.8v
23	DG/RD	MAP Sensor Signal	1.5-1.7v
24	BK/VT	Knock Sensor Ground	<0.050v
25	DB/LG	Knock Sensor Signal	0.080v AC
27	BK/LB	Sensor Ground	<0.050v
28	YL/BK	IAC Motor Sense	12-14v
29	VT/WT	5-Volt Supply	4.9-5.1v
30	BK/RD	IAT Sensor Signal	At 100°F: 1.83v
31	BK/DG	HO2S-11 (B1 S1) Signal	0.1-1.1v
32	BR/DG	HO2S-11 (B1 S2) Ground	<0.050v
33	LG/RD	HO2S-21 (B2 S1) Signal	0.1-1.1v
34	TN/YL	CMP Sensor Signal	Digital Signal: 0-5-0v
35	GY/BK	CKP Sensor Signal	Digital Signal: 0-5-0v
36-37	---	Not Used	---
38	GY/RD	IAC Motor Driver Control	DC pulses: 0.8-11v

2002-3 Concorde 3.5L V6 SOHC MFI VIN G, M (A/T) 'C3' White Connector

PCM Pin #	Wire Color	Circuit Description (38-Pin)	Value at Hot Idle
1-2, 13-17	---	Not Used	---
3	DB/YL	ASD Relay Control	Relay Off: 12v, On: 1v
4	DB/PK	High Speed Fan Relay	Relay Off: 12v, On: 1v
5	DB/PK	S/C Vent Solenoid	Vacuum Decreasing: 1v
6	DB/PK	Low Speed Fan Relay	Relay Off: 12v, On: 1v
7	YL/RD	S/C Power Supply (B+)	12-14v
8	WT/DG	Natural Vacuum Leak Detection Solenoid	Solenoid Off: 12v, On: 1v
9	BR/VT	HO2S-12 (B1 S2) Heater	Heater Off: 12v, On: 1v
10	BR/GY	HO2S-22 (B2 S2) Heater	Heater Off: 12v, On: 1v
11	DB/OR	A/C Clutch Relay Control	Relay Off: 12v, On: 1v
12	TN/RD	S/C Vacuum Solenoid	Vacuum Increasing: 1v
18, 19	OR/DG	Fused ASD Relay Power (B+)	12-14v
20	PK/BK	EVAP Purge Solenoid Control	PWM Signal: 0-12-0v
21-22, 24-25	---	Not Used	---
23	WT/PK	Brake Switch Sense	Brake Off: 0v, On: 12v
26	YL	Autostick Downshift Switch	Digital Signal: 0v or 12v
27	LG/RD	Autostick Upshift Switch	Digital Signal: 0v or 12v
28	OR/DG	Fused ASD Relay Power (B+)	12-14v
29	DG/LG	EVAP Purge Solenoid Sense	<0.1v
30-31, 33, 36	---	Not Used	---
32	VT/LG	Ambient Temperature Sensor	At 86ºF: 1.96v
34	RD/LG	S/C Set Switch Signal	S/C & Set Switch On: 3.8v
35	BR	Natural Vacuum Leak Detection Switch	Open: 12v, Closed: 0v
37	BR	Fuel Pump Relay Control	Relay Off: 12v, On: 1v
38	TN	Starter Relay Control	Relay Off: 12v, On: 1v

2002 Concorde 3.5L V6 SOHC MFI VIN G, J (A/T) 'C4' Green Connector

1, 3-5	---	Not Used	---
2	BR	Overdrive Solenoid Control	Solenoid Off: 12v, On: 1v
3	PK	Underdrive Solenoid Control	Solenoid Off: 12v, On: 1v
6	WT	A/T: 2-4 Solenoid Control	Solenoid Off: 12v, On: 1v
7-9, 11	---	Not Used	---
10	LB	Low/Reverse Solenoid control	Solenoid Off: 12v, On: 1v
12	BK/YL	Power Ground	<0.1v
13, 14	BK/RD	Power Ground	<0.1v
15	LG/BK	A/T: TRS T1 Sense	In NOL: 0v, Others: 5v
16	VT	A/T: TRS T3 Sense	In P3L: 0v, Others: 5v
17, 20-21	---	Not Used	---
18, 19	LG, RD	Transmission Control Relay Control	Relay Off: 12v, On: 1v
22	OR/BK	Overdrive Pressure Switch	In Overdrive: 2-4v
23-26, 31, 36	---	Not Used	---
27	BK/WT	A/T: TRS T41 Sense	In P/N: 0v, Others: 5v
28, 38	RD	Trans. Control Relay Output	12-14v
29	DG	Low/Reverse Pressure Switch	In Low/Reverse: 2-4v
30	YL/BK	A/T: 2-4 Pressure Switch	In 2-4 Position: 2-4v
32	LG/WT	A/T: Output Speed Sensor	Moving: AC voltage
33	RD/BK	A/T: Input Speed Sensor	Moving: AC voltage
34	DB/BK	A/T: Speed Sensor Ground	<0.050v
35	VT/PK	Trans. Temperature Sensor	3.2-3.4v at 104ºF
37	VT/WT	A/T: TRS T42 Sense	In PRNL: 0v, Others 5v

Pin Connector Graphic

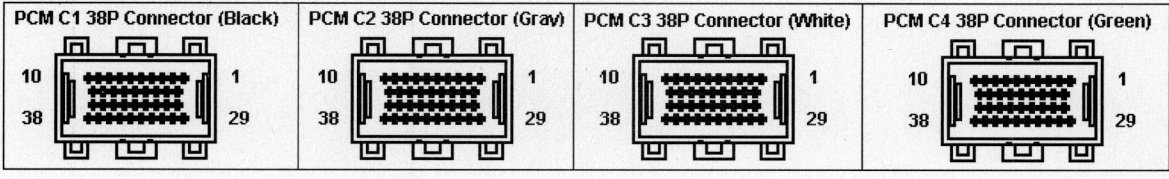

PCM C1 38P Connector (Black)	PCM C2 38P Connector (Gray)	PCM C3 38P Connector (White)	PCM C4 38P Connector (Green)

1990-91 Imperial 3.3L V6 MFI VIN R (A/T) 60 Pin Connector

PCM Pin #	Wire Color	Circuit Description (60 Pin)	Value at Hot Idle
1	DG/RD	MAP Sensor Signal	1.5-1.7v
2	TN/WT	ECT Sensor Signal	At 180°F: 2.80v
3	RD	Battery Power (Fused B+)	12-14v
4	BK/LB	Sensor Ground	<0.050v
5	---	Not Used	---
6	PK/WT	5-Volt Supply	4.9-5.1v
7	OR	8-Volt Supply	7.9-8.1v
8	DG/BK	Fuel Pump ASD Relay	Relay Off: 12v, On: 1v
9	DB	Ignition Switch Output	12-14v
10	---	Not Used	---
11-12	LB/RD	Power Ground	<0.1v
13	---	Not Used	---
14	YL/WT	Injector 3 Driver	1.0-4.0 ms
15	TN	Injector 2 Driver	1.0-4.0 ms
16	WT/DB	Injector 1 Driver	1.0-4.0 ms
17	DB/BK	Coil 2 Driver	5°, at 55 mph: 8° dwell
18	DB/GY	Coil 3 Driver	5°, at 55 mph: 8° dwell
19	BK/GY	Coil 1 Driver	5°, at 55 mph: 8° dwell
20	DG	Voltage Regulator	Digital Signal: 0-12-0v
21	BK/RD	IAT Sensor Signal	At 100°F: 1.83v
22	OR/DB	TP Sensor Signal	0.6-1.0v
24	GY/BK	CKP Sensor Signal	Digital Signal: 0-5-0v
25	PK	SCI Transmit	0v
26	BK	CCD Bus (+)	Digital Signal: 0-5-0v
27	BR	A/C Damped Pressure Switch	Relay Off: 12v, On: 1v
29	WT/PK	Brake Switch Signal	Brake Off: 0v, On: 12v
30	BR/YL	PNP Switch Signal	In P/N: 0v, Others: 5v
31	DB/PK	Radiator Fan Relay Control	Relay Off: 12v, On: 1v
32	BK/PK	Amber MIL Control	MIL On: 1v, MIL Off: 12v
33	TN/RD	S/C Vacuum Solenoid	Vacuum Increasing: 1v
34	DB/OR	A/C Clutch Relay Control	Relay Off: 12v, On: 1v
35	GY/YL	EGR Solenoid Control (Cal)	12v, at 55 mph: 1v
36-38	---	Not Used	---
39	GY/RD	IAC 3 Driver	DC pulse signals: 0.8-11v
40	BR/WT	IAC 1 Driver	DC pulse signals: 0.8-11v

Standard Colors and Abbreviations

Abbreviation	Color	Abbreviation	Color	Abbreviation	Color
BK	Black	GY	Gray	RD	Red
BL	Blue	GN	Green	TN	Tan
BR	Brown	LG	Light Green	VT	Violet
DB	Dark Blue	OR	Orange	WT	White
DG	Dark Green	PK	Pink	YL	Yellow

1990-91 Imperial 3.3L V6 MFI VIN R (A/T) 60 Pin Connector

PCM Pin #	Wire Color	Circuit Description (60 Pin)	Value at Hot Idle
41	BK/DG	Heated Oxygen Sensor	0.1-1.1v
42	BK/LG	Knock Sensor Signal	0.080v AC
43	---	Not Used	---
44	TN/YL	Distributor Reference Signal	Digital Signal: 0-5-0v
45	LG	SCI Receive	5v
46	WT/BK	CCD Bus (-)	<0.050v
47	WT/OR	Vehicle Speed Signal	Digital Signal
48	BR/RD	Speed Control Set Switch	S/C & Set Switch On: 3.8v
49	YL/RD	Speed Control Power Switch	12-14v
50	WT/LG	S/C Resume Switch	S/C On: 1v, Off: 12v
51	DB/YL	ASD Relay Output	12-14v
52	PK	EVAP Purge Solenoid Control	Solenoid Off: 12v, On: 1v
53	LG/RD	S/C Vent Solenoid	Vacuum Decreasing: 1v
54	OR/BK	Lockup Torque Converter	At Cruise w/TCC On: <1v
56-58	---	Not Used	---
59	PK/BK	IAC 4 Driver	DC pulse signals: 0.8-11v
60	YL/BK	IAC 2 Driver	DC pulse signals: 0.8-11v

Pin Connector Graphic

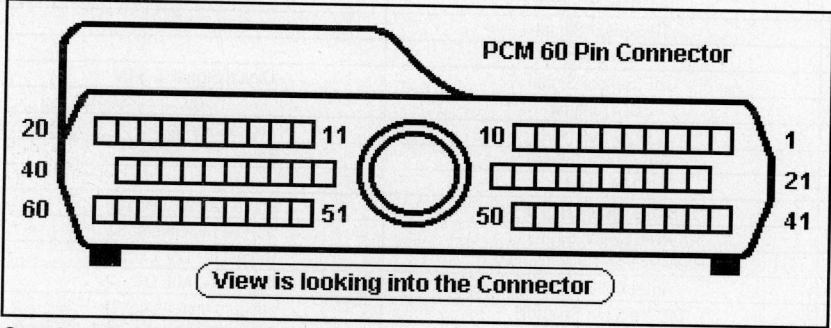

Standard Colors and Abbreviations

Abbreviation	Color	Abbreviation	Color	Abbreviation	Color
BK	Black	GY	Gray	RD	Red
BL	Blue	GN	Green	TN	Tan
BR	Brown	LG	Light Green	VT	Violet
DB	Dark Blue	OR	Orange	WT	White
DG	Dark Green	PK	Pink	YL	Yellow

1991 Imperial 3.8L V6 MFI VIN L (A/T) 60 Pin Connector

PCM Pin #	Wire Color	Circuit Description (60 Pin)	Value at Hot Idle
1	DG/RD	MAP Sensor Signal	1.5-1.7v
2	TN/WT	ECT Sensor Signal	At 180°F: 2.80v
3	RD/WT	Battery Power (Fused B+)	12-14v
4-5	BK/LB	Sensor Ground	<0.050v
6	PK/WT	5-Volt Supply	4.9-5.1v
7	OR	9-Volt Supply	8.9-9.1v
8	DG/BK	Fuel Pump ASD Relay	Relay Off: 12v, On: 1v
9	DB	Ignition Switch Output	12-14v
10	---	Not Used	---
11-12	LB/RD	Power Ground	<0.1v
13	---	Not Used	---
14	YL/WT	Injector 3 Driver	1.0-4.0 ms
15	TN	Injector 2 Driver	1.0-4.0 ms
16	WT/DB	Injector 1 Driver	1.0-4.0 ms
17	DB/BK	Coil 2 Driver	5°, at 55 mph: 8° dwell
18	DB/GY	Coil 3 Driver	5°, at 55 mph: 8° dwell
19	BK/GY	Coil 1 Driver	5°, at 55 mph: 8° dwell
20	DG	Voltage Regulator	Digital Signal: 0-12-0v
21	BK/RD	IAT Sensor Signal	At 100°F: 1.83v
22	OR/DB	TP Sensor Signal	0.6-1.0v
23	---	Not Used	---
24	GY/BK	CKP Sensor Signal	Digital Signal: 0-5-0v
25	PK	SCI Transmit	0v
26	BK/PK	CCD Bus (+)	Digital Signal: 0-5-0v
27	BR	A/C Clutch Sense	Relay Off: 12v, On: 1v
28	---	Not Used	---
29	WT/PK	Brake Switch Signal	Brake Off: 0v, On: 12v
30	BR/YL	PNP Switch Signal	In P/N: 0v, Others: 5v
31	DB/PK	Radiator Fan Relay Control	Relay Off: 12v, On: 1v
32	BK/PK	Amber MIL Control	MIL On: 1v, MIL Off: 12v
33	TN/RD	S/C Vacuum Solenoid	Vacuum Increasing: 1v
34	DB/OR	A/C Clutch Relay Control	Relay Off: 12v, On: 1v
35	GY/YL	EGR Solenoid Control (Cal)	12v, at 55 mph: 1v
36-38	---	Not Used	---
39	GY/RD	IAC 1 Driver	DC pulse signals: 0.8-11v
40	BR/WT	IAC 3 Driver	DC pulse signals: 0.8-11v

Standard Colors and Abbreviations

Abbreviation	Color	Abbreviation	Color	Abbreviation	Color
BK	Black	GY	Gray	RD	Red
BL	Blue	GN	Green	TN	Tan
BR	Brown	LG	Light Green	VT	Violet
DB	Dark Blue	OR	Orange	WT	White
DG	Dark Green	PK	Pink	YL	Yellow

1991 Imperial 3.8L V6 MFI VIN L (A/T) 60 Pin Connector

PCM Pin #	Wire Color	Circuit Description (60 Pin)	Value at Hot Idle
41	BK/DG	Heated Oxygen Sensor	0.1-1.1v
42	BK/LG	Knock Sensor Signal	0.080v AC
43	---	Not Used	---
44	TN/YL	CMP Sensor Signal	Digital Signal: 0-5-0v
45	LG	SCI Receive	5v
46	WT/BK	CCD Bus (-)	<0.050v
47	WT/OR	Vehicle Speed Signal	Digital Signal
48	BR/RD	Speed Control Set Switch	S/C & Set Switch On: 3.8v
49	YL/RD	Speed Control Power Switch	12-14v
50	WT/LG	S/C Resume Switch	S/C On: 1v, Off: 12v
51	DB/YL	ASD Relay Output	12-14v
52	PK	EVAP Purge Solenoid Control	Solenoid Off: 12v, On: 1v
53	LG/RD	S/C Vent Solenoid	Vacuum Decreasing: 1v
54-58	---	Not Used	---
59	PK/BK	IAC 4 Driver	DC pulse signals: 0.8-11v
60	YL/BK	IAC 2 Driver	DC pulse signals: 0.8-11v

Pin Connector Graphic

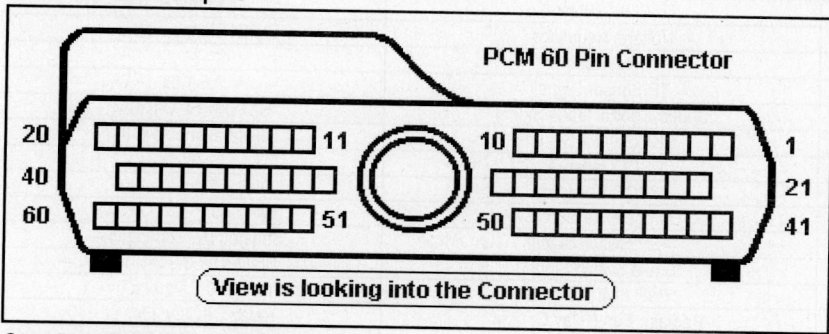

Standard Colors and Abbreviations

Abbreviation	Color	Abbreviation	Color	Abbreviation	Color
BK	Black	GY	Gray	RD	Red
BL	Blue	GN	Green	TN	Tan
BR	Brown	LG	Light Green	VT	Violet
DB	Dark Blue	OR	Orange	WT	White
DG	Dark Green	PK	Pink	YL	Yellow

1992-93 Imperial 3.8L V6 MFI VIN L (A/T) 60 Pin Connector

PCM Pin #	Wire Color	Circuit Description (60 Pin)	Value at Hot Idle
1	DG/RD	MAP Sensor Signal	1.5-1.7v
2	TN/BK	ECT Sensor Signal	At 180ºF: 2.80v
3	RD/WT	Fuel Pump ASD Relay	Relay Off: 12v, On: 1v
4	BK/LB	Sensor Ground	<0.050v
5	BK/WT	Sensor Ground	<0.050v
6	PK/WT	5-Volt Supply	4.9-5.1v
7	OR	8-Volt Supply	7.9-8.1v
8	---	Not Used	---
9	DB	Ignition Switch Output	12-14v
10	---	Not Used	---
11-12	BK/TN	Power Ground	<0.1v
13	LB/BR	Injector 4 Driver	1.0-4.0 ms
14	YL/WT	Injector 3 Driver	1.0-4.0 ms
15	TN	Injector 2 Driver	1.0-4.0 ms
16	WT/DB	Injector 1 Driver	1.0-4.0 ms
17	DB/TN	Coil 2 Driver	5º, at 55 mph: 8º dwell
18	DB/GY	Coil 3 Driver	5º, at 55 mph: 8º dwell
19	BK/GY	Coil 1 Driver	5º, at 55 mph: 8º dwell
20	DG	Voltage Regulator	Digital Signal: 0-12-0v
21	---	Not Used	---
22	OR/DB	TP Sensor Signal	0.6-1.0v
23	RD/LG	Speed Control Mode Signal	S/C On: 1v, Off: 12v
24	GY/BK	CKP Sensor Signal	Digital Signal: 0-5-0v
25	PK	SCI Transmit	0v
26	PK/BR	CCD Bus (+)	Digital Signal: 0-5-0v
27	BR	A/C Damped Pressure Switch	Switch Off: 12v, On: 1v
28 ('93)	DB/OR	PSP Switch Signal	Straight: 0v, Turning: 5v
29	WT/PK	Brake Switch Signal	Brake Off: 0v, On: 12v
30	BR/YL	PNP Switch Signal	In P/N: 0v, Others: 5v
31	DB/PK	Radiator Fan Relay Control	Relay Off: 12v, On: 1v
32	BK/PK	Amber MIL Control	MIL On: 1v, MIL Off: 12v
33	TN/RD	S/C Vacuum Solenoid	Vacuum Increasing: 1v
34	DB/OR	A/C Clutch Relay Control	Relay Off: 12v, On: 1v
35	GY/YL	EGR Solenoid Control (Cal)	12v, at 55 mph: 1v
36-37	---	---	---
38	GY	Injector 5 Driver	1.0-4.0 ms
39	GY/RD	IAC Open Driver	DC pulse signals: 0.8-11v
40	BR/WT	IAC Close Driver	DC pulse signals: 0.8-11v

Standard Colors and Abbreviations

Abbreviation	Color	Abbreviation	Color	Abbreviation	Color
BK	Black	GY	Gray	RD	Red
BL	Blue	GN	Green	TN	Tan
BR	Brown	LG	Light Green	VT	Violet
DB	Dark Blue	OR	Orange	WT	White
DG	Dark Green	PK	Pink	YL	Yellow

1992-93 Imperial 3.8L V6 MFI VIN L (A/T) 60 Pin Connector

PCM Pin #	Wire Color	Circuit Description (60 Pin)	Value at Hot Idle
41	BK/DG	Heated Oxygen Sensor	0.1-1.1v
42	BK/LG	Knock Sensor Signal	0.080v AC
43	---	---	---
44	TN/YL	CMP Sensor Signal	Digital Signal: 0-5-0v
45	LG	SCI Receive	5v
46	WT/BK	CCD Bus (-)	<0.050v
47	WT/OR	Vehicle Speed Signal	Digital Signal
48-50	---	---	---
51	DB/YL	ASD Relay Output	12-14v
52	PK/BK	EVAP Purge Solenoid Control	Solenoid Off: 12v, On: 1v
53	LG/RD	S/C Vent Solenoid	Vacuum Decreasing: 1v
54-56	TN/YL	CMP Sensor Signal	Digital Signal: 0-5-0v
57	DG/OR	Injector Feed Signal	12-14v
58	BR/DB	Injector 6 Driver	1.0-4.0 ms
59	PK/BK	IAC Close Driver	DC pulse signals: 0.8-11v
60	YL/BK	IAC Close Driver	DC pulse signals: 0.8-11v

Pin Connector Graphic

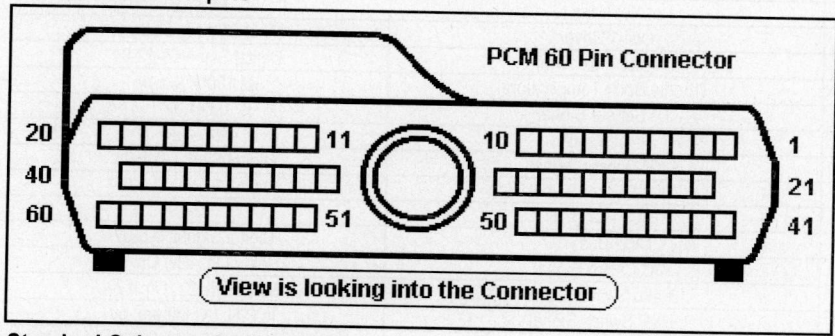

PCM 60 Pin Connector

20   11   10   1
40
60   51   50   21
          41

View is looking into the Connector

Standard Colors and Abbreviations

Abbreviation	Color	Abbreviation	Color	Abbreviation	Color
BK	Black	GY	Gray	RD	Red
BL	Blue	GN	Green	TN	Tan
BR	Brown	LG	Light Green	VT	Violet
DB	Dark Blue	OR	Orange	WT	White
DG	Dark Green	PK	Pink	YL	Yellow

1990-92 LeBaron 2.5L I4 SOHC TBI VIN K 60 Pin Connector

PCM Pin #	Wire Color	Circuit Description (60 Pin)	Value at Hot Idle
1	DG/RD	MAP Sensor Signal	1.5-1.7v
2	TN/WT	ECT Sensor Signal	At 180°F: 2.80v
2 ('91-'92)	TN/BK	ECT Sensor Signal	At 180°F: 2.80v
3	RD/WT	Battery Power (Fused B+)	12-14v
3 ('91)	RD	Battery Power (Fused B+)	12-14v
4	BK/LB	Sensor Ground	<0.050v
5	BK/WT	Sensor Ground	<0.050v
6	PK/WT	5-Volt Supply	4.9-5.1v
6 ('91)	PK	5-Volt Supply	4.9-5.1v
7	OR	9-Volt Supply	8.9-9.1v
8	WT	Auto Shutdown Relay Input	12-14v
9	DB	Ignition Switch Output	12-14v
10	---	Not Used	---
11	LB/RD	Power Ground	<0.1v
12	LB/RD	Power Ground	<0.1v
13-15	---	Not Used	---
16	WT/DB	Injector 1 Driver	1.0-4.0 ms
17-18	---	Not Used	---
19	BK/GY	Coil 1 Driver	5°, at 55 mph: 8° dwell
20	DG	Alternator Field Control	Digital Signal: 0-12-0v
21	BK/RD	Throttle Body Temperature	At 100°F: 2.51v
22	OR/DB	TP Sensor Signal	0.6-1.0v
23, 28	---	Not Used	---
24	GY/BK	CKP Sensor Signal	Digital Signal: 0-5-0v
24 ('91)	GY	CKP Sensor Signal	Digital Signal: 0-5-0v
25	PK	SCI Transmit	0v
26	PK/BR	CCD Bus (+)	Digital Signal: 0-5-0v
27	BR	A/C Clutch Signal	Relay Off: 12v, On: 1v
29	WT/PK	Brake Switch Signal	Brake Off: 0v, On: 12v
30	BR/YL	PNP Switch Signal	In P/N: 0v, Others: 5v
31	DB/PK	Radiator Fan Relay Control	Relay Off: 12v, On: 1v
32	BK/PK	Amber MIL Control	MIL On: 1v, MIL Off: 12v
33	TN/RD	S/C Vacuum Solenoid	Vacuum Increasing: 1v
34	DB/OR	A/C WOT Relay Control	Relay Off: 12v, On: 1v
35	GY/YL	EGR Solenoid Control (Cal)	12v, at 55 mph: 1v
36-38	---	Not Used	---
39	GY/RD	AIS 3 Motor Control	DC pulse signals: 0.8-11v
40	BR/WT	AIS 1 Motor Control	DC pulse signals: 0.8-11v
40 ('91)	BR	AIS 1 Motor Control	DC pulse signals: 0.8-11v

1990-92 LeBaron 2.5L I4 SOHC TBI VIN K 60 Pin Connector

PCM Pin #	Wire Color	Circuit Description (60 Pin)	Value at Hot Idle
41	BK/DG	Heated Oxygen Sensor	0.1-1.1v
42	---	Not Used	
43	GY/LB	Tachometer Signal	Pulse Signals
44	---	Not Used	---
45	LG	SCI Receive	0v
46	WT/BK	CCD Bus (-)	<0.050v
46 ('91)	WT	CCD Bus (-)	<0.050v
47	WT/OR	Vehicle Speed Signal	Digital Signal
48	BR/RD	Speed Control Set Switch	S/C & Set Switch On: 3.8v
49	YL/RD	Speed Control Power Switch	12-14v
50	WT/LG	S/C Resume Switch	S/C On: 1v, Off: 12v
51	DB/YL	ASD Relay Output	12-14v
52	PK/BK	EVAP Purge Solenoid Control	Solenoid Off: 12v, On: 1v
53	LG/RD	S/C Vent Solenoid	Vacuum Decreasing: 1v
54	OR/BK	Part Throttle Unlock Solenoid	At Cruise w/TCC On: <1v
55-58	---	Not Used	
59	PK/BK	AIS 4 Motor Control	DC pulse signals: 0.8-11v
60	YL/BK	AIS 2 Motor Control	DC pulse signals: 0.8-11v

Pin Connector Graphic

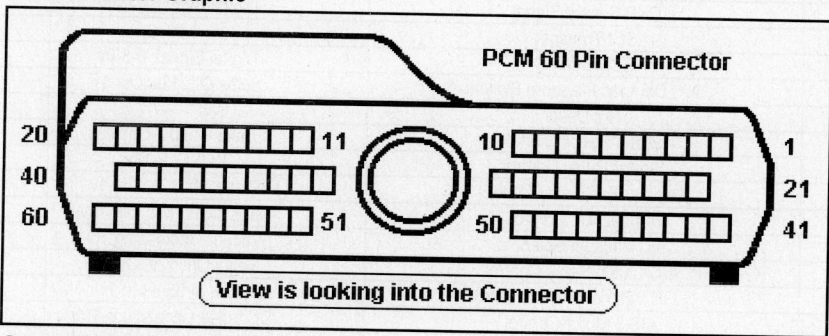

Standard Colors and Abbreviations

Abbreviation	Color	Abbreviation	Color	Abbreviation	Color
BK	Black	GY	Gray	RD	Red
BL	Blue	GN	Green	TN	Tan
BR	Brown	LG	Light Green	VT	Violet
DB	Dark Blue	OR	Orange	WT	White
DG	Dark Green	PK	Pink	YL	Yellow

1993 LeBaron 2.5L I4 SOHC TBI VIN K (All) 60 Pin Connector

PCM Pin #	Wire Color	Circuit Description (60 Pin)	Value at Hot Idle
1	DG/RD	MAP Sensor Signal	1.5-1.7v
2	TN/BK	ECT Sensor Signal	At 180°F: 2.80v
3	RD/WT	Battery Power (Fused B+)	12-14v
4	BK/LB	Sensor Ground	<0.050v
5	BK/WT	Sensor Ground	<0.050v
6	PK/WT	5-Volt Supply	4.9-5.1v
7	OR	8-Volt Supply	7.9-8.1v
8	WT	Auto Shutdown Relay Input	12-14v
9	DB	Ignition Switch Output	12-14v
10	---	Not Used	---
11-12	BK/TN	Power Ground	<0.1v
13-15	---	Not Used	---
16	WT/DB	Injector 1 Driver	1.0-4.0 ms
17-18	---	Not Used	---
19	BK/GY	Coil 1 Driver	5°, at 55 mph: 8° dwell
20	DG	Alternator Field Control	Digital Signal: 0-12-0v
21	---	Not Used	---
22	OR/DB	TP Sensor Signal	0.6-1.0v
23	RD/LG	S/C Set Switch Signal	S/C & Set Switch On: 3.8v
24	GY/BK	CKP Sensor Signal	Digital Signal: 0-5-0v
25	PK	SCI Transmit	0v
26	PK/BR	CCD Bus (+)	Digital Signal: 0-5-0v
27	BR	A/C Damped Pressure Switch	Relay Off: 12v, On: 1v
28, 36-38	---	Not Used	---
29	WT/PK	Brake Switch Signal	Brake Off: 0v, On: 12v
30	BR/LB	Starter Relay Control	KOEC: 9-11v
31	DB/PK	Radiator Fan Relay Control	Relay Off: 12v, On: 1v
32	BK/PK	Amber MIL Control	MIL On: 1v, MIL Off: 12v
33	TN/RD	S/C Vacuum Solenoid	Vacuum Increasing: 1v
34	DB/OR	A/C Clutch Relay Control	Relay Off: 12v, On: 1v
35	GY/YL	EGR Solenoid Control (Cal)	12v, at 55 mph: 1v
39	GY/RD	AIS 3 Motor Control	DC pulse signals: 0.8-11v
40	BR/WT	AIS 1 Motor Control	DC pulse signals: 0.8-11v

Standard Colors and Abbreviations

Abbreviation	Color	Abbreviation	Color	Abbreviation	Color
BK	Black	GY	Gray	RD	Red
BL	Blue	GN	Green	TN	Tan
BR	Brown	LG	Light Green	VT	Violet
DB	Dark Blue	OR	Orange	WT	White
DG	Dark Green	PK	Pink	YL	Yellow

1993 LeBaron 2.5L I4 SOHC TBI VIN K (All) 60 Pin Connector

PCM Pin #	Wire Color	Circuit Description (60 Pin)	Value at Hot Idle
41	BK/DG	Heated Oxygen Sensor	0.1-1.1v
42	---	Not Used	---
43	GY/LB	Tachometer Signal	Pulse Signals
44	---	Not Used	---
45	LG	SCI Receive	0v
46	WT/BK	CCD Bus (-)	<0.050v
47	WT/OR	Vehicle Speed Signal	Digital Signal
48-49	---	Not Used	---
51	DB/YL	ASD Relay Output	12-14v
52	PK/BK	EVAP Purge Solenoid Control	Solenoid Off: 12v, On: 1v
53	LG/RD	S/C Vent Solenoid	Vacuum Decreasing: 1v
54	LG/WT	EMCC Solenoid Control	At Cruise w/TCC On: <1v
55-56	---	Not Used	---
57	DG/OR	ASD Relay Control	Relay Off: 12v, On: 1v
58	---	Not Used	---
59	PK/BK	AIS 4 Motor Control	DC pulse signals: 0.8-11v
60	YL/BK	AIS 2 Motor Control	DC pulse signals: 0.8-11v

Pin Connector Graphic

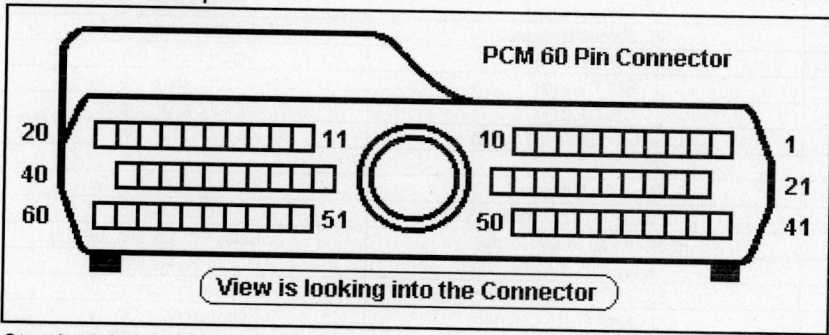

Standard Colors and Abbreviations

Abbreviation	Color	Abbreviation	Color	Abbreviation	Color
BK	Black	GY	Gray	RD	Red
BL	Blue	GN	Green	TN	Tan
BR	Brown	LG	Light Green	VT	Violet
DB	Dark Blue	OR	Orange	WT	White
DG	Dark Green	PK	Pink	YL	Yellow

1993 LeBaron 2.5L I4 SOHC MFI VIN B (All) 60 Pin Connector

PCM Pin #	Wire Color	Circuit Description (60 Pin)	Value at Hot Idle
1	DG/RD	MAP Sensor Signal	1.5-1.7v
2	TN/BK	ECT Sensor Signal	At 180°F: 2.80v
3	RD/WT	Battery Power (Fused B+)	12-14v
4	BK/LB	Sensor Ground	<0.050v
5	BK/WT	Sensor Ground	<0.050v
6	PK/WT	5-Volt Supply	4.9-5.1v
7	OR	8-Volt Supply	7.9-8.1v
8	WT	Auto Shutdown Relay Input	12-14v
9	DB	Ignition Switch Output	12-14v
10	---	Not Used	---
11	BK/TN	Power Ground	<0.1v
12	BK/TN	Power Ground	<0.1v
13-15	---	Not Used	---
16	WT/DB	Injector 1 Driver	1.0-4.0 ms
17-18	---	Not Used	---
19	BK/GY	Coil 1 Driver	5°, at 55 mph: 8° dwell
20	DG	Alternator Field Control	Digital Signal: 0-12-0v
21	---	Not Used	---
22	OR/DB	TP Sensor Signal	0.6-1.0v
23	RD/LG	S/C Set Switch Signal	S/C & Set Switch On: 3.8v
24	GY/BK	CKP Sensor Signal	Digital Signal: 0-5-0v
25	PK	SCI Transmit	0v
26	PK/BR	CCD Bus (+)	Digital Signal: 0-5-0v
27	BR	A/C Damped Pressure Switch	Relay Off: 12v, On: 1v
28	---	Not Used	---
29	WT/PK	Brake Switch Signal	Brake Off: 0v, On: 12v
30	BR/LB	Starter Relay Control	KOEC: 9-11v
31	DB/PK	Radiator Fan Relay Control	Relay Off: 12v, On: 1v
32	BK/PK	Amber MIL Control	MIL On: 1v, MIL Off: 12v
33	TN/RD	S/C Vacuum Solenoid	Vacuum Increasing: 1v
34	DB/OR	A/C Clutch Relay Control	Relay Off: 12v, On: 1v
35	GY/YL	EGR Solenoid Control (Cal)	12v, at 55 mph: 1v
36-38	---	Not Used	---
39	GY/RD	AIS 3 Motor Control	DC pulse signals: 0.8-11v
40	BR/WT	AIS 1 Motor Control	DC pulse signals: 0.8-11v

Standard Colors and Abbreviations

Abbreviation	Color	Abbreviation	Color	Abbreviation	Color
BK	Black	GY	Gray	RD	Red
BL	Blue	GN	Green	TN	Tan
BR	Brown	LG	Light Green	VT	Violet
DB	Dark Blue	OR	Orange	WT	White
DG	Dark Green	PK	Pink	YL	Yellow

1993 LeBaron 2.5L I4 SOHC MFI VIN B (All) 60 Pin Connector

PCM Pin #	Wire Color	Circuit Description (60 Pin)	Value at Hot Idle
41	BK/DG	Heated Oxygen Sensor	0.1-1.1v
42	---	Not Used	---
43	GY/LB	Tachometer Signal	Pulse Signals
44	TN/YL	CMP Sensor Signal	Digital Signal: 0-5-0v
45	LG	SCI Receive	0v
46	WT/BK	CCD Bus (-)	<0.050v
47	WT/OR	Vehicle Speed Signal	Digital Signal
48-50	---	Not Used	---
51	DB/YL	ASD Relay Output	12-14v
52	PK/BK	EVAP Purge Solenoid Control	Solenoid Off: 12v, On: 1v
53	LG/RD	S/C Vent Solenoid	Vacuum Decreasing: 1v
54	LG/WT	EMCC Solenoid Control	At Cruise w/TCC On: <1v
55-56	---	Not Used	---
57	DG/OR	ASD Relay Control	Relay Off: 12v, On: 1v
58	---	Not Used	---
59	PK/BK	AIS 4 Motor Control	DC pulse signals: 0.8-11v
60	YL/BK	AIS 2 Motor Control	DC pulse signals: 0.8-11v

Pin Connector Graphic

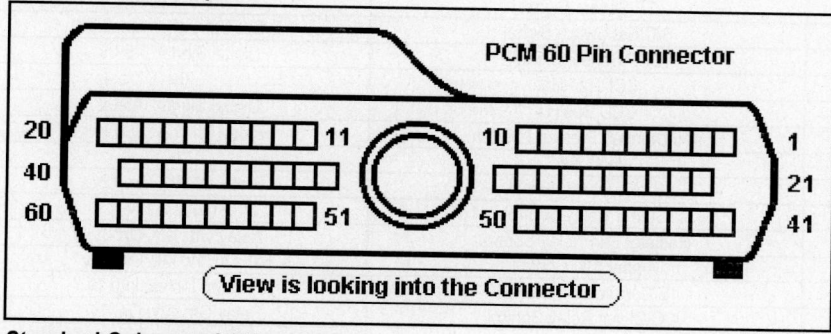

Standard Colors and Abbreviations

Abbreviation	Color	Abbreviation	Color	Abbreviation	Color
BK	Black	GY	Gray	RD	Red
BL	Blue	GN	Green	TN	Tan
BR	Brown	LG	Light Green	VT	Violet
DB	Dark Blue	OR	Orange	WT	White
DG	Dark Green	PK	Pink	YL	Yellow

1993 LeBaron 2.5L I4 Flexible Fuel VIN V (All) 60 Pin Connector

PCM Pin #	Wire Color	Circuit Description (60 Pin)	Value at Hot Idle
1	DG/RD	MAP Sensor Signal	1.5-1.7v
2	TN/BK	ECT Sensor Signal	At 180°F: 2.80v
3	RD/WT	Battery Power (Fused B+)	12-14v
4	BK/LB	Sensor Ground	<0.050v
5	BK/WT	Sensor Ground	<0.050v
6	PK/WT	5-Volt Supply	4.9-5.1v
7	OR	8-Volt Supply	7.9-8.1v
8	WT	Auto Shutdown Relay Input	12-14v
9	DB	Ignition Switch Output	12-14v
10	---	Not Used	---
11	BK/TN	Power Ground	<0.1v
12	BK/TN	Power Ground	<0.1v
13-15	---	Not Used	---
16	WT/DB	Injector 1 Driver	1.0-4.0 ms
17-18	---	Not Used	---
19	BK/GY	Coil 1 Driver	5°, at 55 mph: 8° dwell
20	DG	Alternator Field Control	Digital Signal: 0-12-0v
21	BK/RD	Flexible Fuel Sensor Signal	0.5-4.5v
22	OR/DB	TP Sensor Signal	0.6-1.0v
23	RD/LG	S/C Set Switch Signal	S/C & Set Switch On: 3.8v
24	GY/BK	CKP Sensor Signal	Digital Signal: 0-5-0v
25	PK	SCI Transmit	0v
26	PK/BR	CCD Bus (+)	Digital Signal: 0-5-0v
27	BR	A/C Damped Pressure Switch	Relay Off: 12v, On: 1v
28	---	Not Used	---
29	WT/PK	Brake Switch Signal	Brake Off: 0v, On: 12v
30	BR/LB	Starter Relay Control	KOEC: 9-11v
31	DB/PK	Radiator Fan Relay Control	Relay Off: 12v, On: 1v
32	BK/PK	Amber MIL Control	MIL On: 1v, MIL Off: 12v
33	TN/RD	S/C Vacuum Solenoid	Vacuum Increasing: 1v
34	DB/OR	A/C Clutch Relay Control	Relay Off: 12v, On: 1v
35	GY/YL	EGR Solenoid Control (Cal)	12v, at 55 mph: 1v
36-38	---	Not Used	---
39	GY/RD	AIS 3 Motor Control	DC pulse signals: 0.8-11v
40	BR/WT	AIS 1 Motor Control	DC pulse signals: 0.8-11v

Standard Colors and Abbreviations

Abbreviation	Color	Abbreviation	Color	Abbreviation	Color
BK	Black	GY	Gray	RD	Red
BL	Blue	GN	Green	TN	Tan
BR	Brown	LG	Light Green	VT	Violet
DB	Dark Blue	OR	Orange	WT	White
DG	Dark Green	PK	Pink	YL	Yellow

1993 LeBaron 2.5L I4 Flexible Fuel VIN V (All) 60 Pin Connector

PCM Pin #	Wire Color	Circuit Description (60 Pin)	Value at Hot Idle
41	BK/DG	Heated Oxygen Sensor	0.1-1.1v
42	---	Not Used	---
43	GY/LB	Tachometer Signal	Pulse Signals
44	TN/YL	CMP Sensor Signal	Digital Signal: 0-5-0v
45	LG	SCI Receive	0v
46	WT/BK	CCD Bus (-)	<0.050v
47	WT/OR	Vehicle Speed Signal	Digital Signal
48-50	---	Not Used	---
51	DB/YL	ASD Relay Output	12-14v
52	PK/BK	EVAP Purge Solenoid Control	Solenoid Off: 12v, On: 1v
53	LG/RD	S/C Vent Solenoid	Vacuum Decreasing: 1v
54	LG/WT	EMCC Solenoid Control	At Cruise w/TCC On: <1v
55-56	---	Not Used	---
57	DG/OR	ASD Relay Control	Relay Off: 12v, On: 1v
58	---	Not Used	---
59	PK/BK	AIS 4 Motor Control	DC pulse signals: 0.8-11v
60	YL/BK	AIS 2 Motor Control	DC pulse signals: 0.8-11v

Pin Connector Graphic

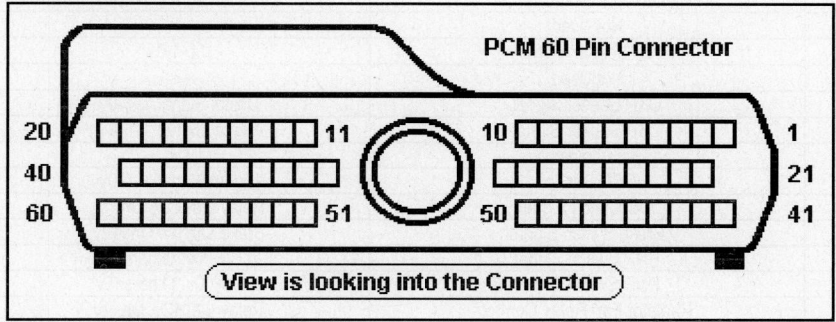

Standard Colors and Abbreviations

Abbreviation	Color	Abbreviation	Color	Abbreviation	Color
BK	Black	GY	Gray	RD	Red
BL	Blue	GN	Green	TN	Tan
BR	Brown	LG	Light Green	VT	Violet
DB	Dark Blue	OR	Orange	WT	White
DG	Dark Green	PK	Pink	YL	Yellow

1990-92 LeBaron 3.0L V6 SOHC MFI VIN 3 (All) 60 Pin Connector

PCM Pin #	Wire Color	Circuit Description (60 Pin)	Value at Hot Idle
1	DG/RD	MAP Sensor Signal	1.5-1.7v
2	TN/WT	ECT Sensor Signal	At 180°F: 2.80v
3	RD/WT	Battery Power (Fused B+)	12-14v
4	BK/LB	Sensor Ground	<0.050v
5	BK/WT	Sensor Ground	<0.050v
6	PK/WT	5-Volt Supply	4.9-5.1v
7	OR	9-Volt Supply	8.9-9.1v
8	DG/BK	Auto Shutdown Relay Input	12-14v
9	DB	Ignition Switch Output	12-14v
10	---	Not Used	---
11	BK/TN	Power Ground	<0.1v
12	BK/TN	Power Ground	<0.1v
13	---	Not Used	---
14	YL/WT	Injector 3 Driver	1.0-4.0 ms
15	TN	Injector 2 Driver	1.0-4.0 ms
16	WT/DB	Injector 1 Driver	1.0-4.0 ms
17-18	---	Not Used	---
19	BK/GY	Coil 1 Driver	5°, at 55 mph: 8° dwell
20	DG	Alternator Field Control	Digital Signal: 0-12-0v
21	---	Not Used	---
22	OR/DB	TP Sensor Signal	0.6-1.0v
23, 28	---	Not Used	---
24	GY/BK	CKP Sensor Signal	Digital Signal: 0-5-0v
24 ('91)	GY/WT	CKP Sensor Signal	Digital Signal: 0-5-0v
25	PK	SCI Transmit	0v
26	PK/BR	CCD Bus (+)	Digital Signal: 0-5-0v
27	BR	A/C Clutch Signal	Relay Off: 12v, On: 1v
29	WT/PK	Brake Switch Signal	Brake Off: 0v, On: 12v
30	BR/YL	PNP Switch Signal	In P/N: 0v, Others: 5v
30 ('91)	BR/LB	PNP Switch Signal	In P/N: 0v, Others: 5v
31	DB/PK	Radiator Fan Relay Control	Relay Off: 12v, On: 1v
32	BK/PK	Amber MIL Control	MIL On: 1v, MIL Off: 12v
33	TN/RD	S/C Vacuum Solenoid	Vacuum Increasing: 1v
34	DB/OR	A/C WOT Relay Control	Relay Off: 12v, On: 1v
35	GY/YL	EGR Solenoid Control (Cal)	12v, at 55 mph: 1v
36-38	---	Not Used	---
39	GY/RD	AIS 3 Motor Control	DC pulse signals: 0.8-11v
40	BR/WT	AIS 1 Motor Control	DC pulse signals: 0.8-11v

Standard Colors and Abbreviations

Abbreviation	Color	Abbreviation	Color	Abbreviation	Color
BK	Black	GY	Gray	RD	Red
BL	Blue	GN	Green	TN	Tan
BR	Brown	LG	Light Green	VT	Violet
DB	Dark Blue	OR	Orange	WT	White
DG	Dark Green	PK	Pink	YL	Yellow

1990-92 LeBaron 3.0L V6 SOHC MFI VIN 3 (All) 60 Pin Connector

PCM Pin #	Wire Color	Circuit Description (60 Pin)	Value at Hot Idle
41	BK/DG	Heated Oxygen Sensor	0.1-1.1v
42	---	Not Used	---
43	GY/LB	Tachometer Signal	Pulse Signals
44	TN/YL	CMP Sensor Signal	Digital Signal: 0-5-0v
45	LG	SCI Receive	0v
46	WT/BK	CCD Bus (-)	<0.050v
46 ('91)	WT	CCD Bus (-)	<0.050v
47	WT/OR	Vehicle Speed Signal	Digital Signal
48	BR/RD	Speed Control Set Switch	S/C & Set Switch On: 3.8v
49	YL/RD	S/C On/Off Switch Signal	S/C On: 1v, Off: 12v
50	WT/LG	S/C Resume Switch	S/C On: 1v, Off: 12v
51	DB/YL	ASD Relay Output	12-14v
52	PK/BK	EVAP Purge Solenoid Control	Solenoid Off: 12v, On: 1v
52 ('91)	PK/WT	EVAP Purge Solenoid Control	Solenoid Off: 12v, On: 1v
53	LG/RD	S/C Vent Solenoid	Vacuum Decreasing: 1v
54-58	---	Not Used	---
59	PK/BK	AIS 4 Motor Control	DC pulse signals: 0.8-11v
60	YL/BK	AIS 2 Motor Control	DC pulse signals: 0.8-11v

Pin Connector Graphic

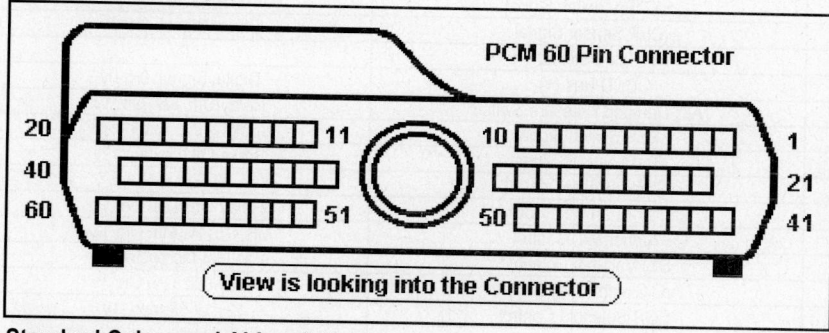

Standard Colors and Abbreviations

Abbreviation	Color	Abbreviation	Color	Abbreviation	Color
BK	Black	GY	Gray	RD	Red
BL	Blue	GN	Green	TN	Tan
BR	Brown	LG	Light Green	VT	Violet
DB	Dark Blue	OR	Orange	WT	White
DG	Dark Green	PK	Pink	YL	Yellow

1993-95 LeBaron 3.0L V6 SOHC MFI VIN 3 (All) 60 Pin Connector

PCM Pin #	Wire Color	Circuit Description (60 Pin)	Value at Hot Idle
1	DG/RD	MAP Sensor Signal	1.5-1.7v
2	TN/BK	ECT Sensor Signal	At 180°F: 2.80v
3	RD/WT	Battery Power (Fused B+)	12-14v
4-5	BK/LB	Sensor Ground	<0.050v
6	PK/WT	5-Volt Supply	4.9-5.1v
7	OR	8-Volt Supply	7.9-8.1v
8	---	Not Used	---
9	DB	Ignition Switch Output	12-14v
10	---	Not Used	---
11-12	BK/TN	Power Ground	<0.1v
13	LB/BR	Injector 4 Driver	1.0-4.0 ms
14	YL/WT	Injector 3 Driver	1.0-4.0 ms
15	TN	Injector 2 Driver	1.0-4.0 ms
16	WT/DB	Injector 1 Driver	1.0-4.0 ms
17-18	---	Not Used	---
19	BK/GY	Coil 1 Driver	5°, at 55 mph: 8° dwell
20	DG	Alternator Field Control	Digital Signal: 0-12-0v
21	---	Not Used	---
22	OR/DB	TP Sensor Signal	0.6-1.0v
23	RD/LG	S/C Set Switch Signal	S/C & Set Switch On: 3.8v
24	GY/BK	CKP Sensor Signal	Digital Signal: 0-5-0v
25	PK	SCI Transmit	0v
26	PK/BR	CCD Bus (+)	Digital Signal: 0-5-0v
27	BR	A/C Damped Pressure Switch	Relay Off: 12v, On: 1v
28	DB/OR	PSP Switch Signal	Straight: 0v, Turning: 5v
29	WT/PK	Brake Switch Signal	Brake Off: 0v, On: 12v
30	BR/LB	Starter Relay Control	KOEC: 9-11v
31	DB/PK	Radiator Fan Relay Control	Relay Off: 12v, On: 1v
32	BK/PK	Amber MIL Control	MIL On: 1v, MIL Off: 12v
33	TN/RD	S/C Vacuum Solenoid	Vacuum Increasing: 1v
34	DB/OR	A/C Clutch Relay Control	Relay Off: 12v, On: 1v
35	GY/YL	EGR Solenoid Control	12v, at 55 mph: 1v
36-37	---	Not Used	---
38	GY	Injector 5 Driver	1.0-4.0 ms
39	GY/RD	IAC 3 Driver	DC pulse signals: 0.8-11v
40	BR/WT	IAC 1 Driver	DC pulse signals: 0.8-11v

Standard Colors and Abbreviations

Abbreviation	Color	Abbreviation	Color	Abbreviation	Color
BK	Black	GY	Gray	RD	Red
BL	Blue	GN	Green	TN	Tan
BR	Brown	LG	Light Green	VT	Violet
DB	Dark Blue	OR	Orange	WT	White
DG	Dark Green	PK	Pink	YL	Yellow

1993-95 LeBaron 3.0L V6 SOHC MFI VIN 3 (All) 60 Pin Connector

PCM Pin #	Wire Color	Circuit Description (60 Pin)	Value at Hot Idle
41	BK/DG	Heated Oxygen Sensor	0.1-1.1v
42	---	Not Used	---
43	GY/LB	Tachometer Signal	Pulse Signals
44	TN/YL	CMP Sensor Signal	Digital Signal: 0-5-0v
45	LG	SCI Receive	0v
46	WT/BK	CCD Bus (-)	<0.050v
47	WT/OR	Vehicle Speed Signal	Digital Signal
48-50	---	Not Used	---
51	DB/YL	ASD Relay Output	12-14v
52	PK/BK	EVAP Purge Solenoid Control	Solenoid Off: 12v, On: 1v
53	LG/RD	S/C Vent Solenoid	Vacuum Decreasing: 1v
54	LG/WT	EMCC Solenoid Control	At Cruise w/TCC On: <1v
55-56	---	Not Used	---
57	DG/OR	ASD Relay Control	Relay Off: 12v, On: 1v
58	BR/DB	Injector 6 Driver	1.0-4.0 ms
59	PK/BK	IAC 4 Driver	DC pulse signals: 0.8-11v
60	YL/BK	IAC 2 Driver	DC pulse signals: 0.8-11v

Pin Connector Graphic

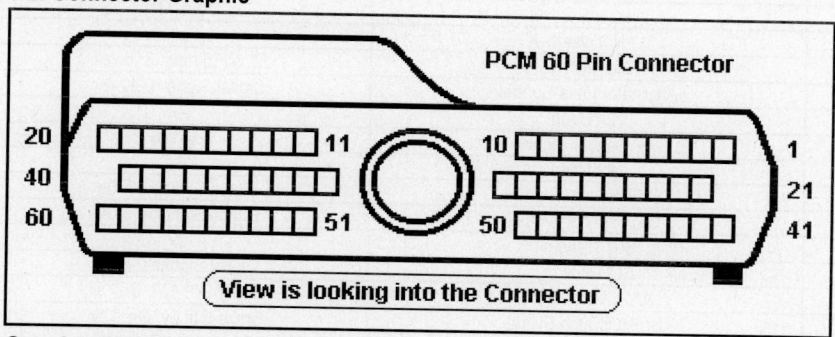

Standard Colors and Abbreviations

Abbreviation	Color	Abbreviation	Color	Abbreviation	Color
BK	Black	GY	Gray	RD	Red
BL	Blue	GN	Green	TN	Tan
BR	Brown	LG	Light Green	VT	Violet
DB	Dark Blue	OR	Orange	WT	White
DG	Dark Green	PK	Pink	YL	Yellow

1990-92 LeBaron GTS 2.5L I4 SOHC Turbo MFI VIN J (A/T)

PCM Pin #	Wire Color	Circuit Description (60 Pin)	Value at Hot Idle
1	DG/RD	MAP Sensor Signal	1.5-1.7v
2	TN/WT	ECT Sensor Signal	At 180°F: 2.80v
2 ('91)	TN/BK	ECT Sensor Signal	At 180°F: 2.80v
3	RD/WT	Battery Power (Fused B+)	12-14v
3 ('91)	RD	Battery Power (Fused B+)	12-14v
4	BK/LB	Sensor Ground	<0.050v
5	BK/WT	Sensor Ground	<0.050v
5 ('91)	BK	Sensor Ground	<0.050v
6	PK/WT	5-Volt Supply	4.9-5.1v
7	OR	9-Volt Supply	8.9-9.1v
8	DG/BK	Auto Shutdown Relay Input	12-14v
9	DB	Ignition Switch Output	12-14v
10	---	Not Used	---
11	BK/TN	Power Ground	<0.1v
12	BK/TN	Power Ground	<0.1v
13	LB/BR	Injector 4 Driver	1.0-4.0 ms
14	YL/WT	Injector 3 Driver	1.0-4.0 ms
15	TN	Injector 2 Driver	1.0-4.0 ms
16	WT/DB	Injector 1 Driver	1.0-4.0 ms
17-18	---	Not Used	---
19	BK/GY	Coil 1 Driver	5°, at 55 mph: 8° dwell
20	DG	Alternator Field Control	Digital Signal: 0-12-0v
21	---	Not Used	---
22	OR/DB	TP Sensor Signal	0.6-1.0v
23, 28	---	Not Used	---
24	GY/BK	CKP Sensor Signal	Digital Signal: 0-5-0v
24 ('91)	GY	CKP Sensor Signal	Digital Signal: 0-5-0v
25	PK	SCI Transmit	0v
26	PK/BR	CCD Bus (+)	Digital Signal: 0-5-0v
27	BR	A/C Clutch Signal	Relay Off: 12v, On: 1v
29	WT/PK	Brake Switch Signal	Brake Off: 0v, On: 12v
30	BR/LB	PNP Switch Signal	In P/N: 0v, Others: 5v
31	DB/PK	Radiator Fan Relay Control	Relay Off: 12v, On: 1v
32	BK/PK	Amber MIL Control	MIL On: 1v, MIL Off: 12v
33	TN/RD	S/C Vacuum Solenoid	Vacuum Increasing: 1v
34	DB/OR	A/C WOT Relay Control	Relay Off: 12v, On: 1v
35	GY/YL	EGR Solenoid Control (Cal)	12v, at 55 mph: 1v
36	LG/BK	Wastegate Solenoid Control	Valve Off: 12v, On: 1v
37-38	---	Not Used	---
39	GY/RD	AIS 3 Motor Control	DC pulse signals: 0.8-11v
40	BR/WT	AIS 1 Motor Control	DC pulse signals: 0.8-11v
40 ('91)	BR	AIS 1 Motor Control	DC pulse signals: 0.8-11v

1990-92 LeBaron GTS 2.5L I4 SOHC Turbo MFI VIN J (A/T)

PCM Pin #	Wire Color	Circuit Description (60 Pin)	Value at Hot Idle
41	BK/DG	Heated Oxygen Sensor	0.1-1.1v
42	BK/LG	Knock Sensor Signal	0.080v AC
43	GY/LB	Tachometer Signal	Pulse Signals
44	TN/YL	Fuel Monitor Output Signal	Relay Off: 12v, On: 1v
45	LG	SCI Receive	0v
46	WT/BK	CCD Bus (-)	<0.050v
46 ('91)	WT	CCD Bus (-)	<0.050v
47	WT/OR	Vehicle Speed Signal	Digital Signal
48	BR/RD	Speed Control Set Switch	S/C & Set Switch On: 3.8v
49	YL/RD	S/C On/Off Switch Signal	S/C On: 1v, Off: 12v
50	WT/LG	S/C Resume Switch	S/C On: 1v, Off: 12v
51	DB/YL	ASD Relay Output	12-14v
52	PK/BK	EVAP Purge Solenoid Control	Solenoid Off: 12v, On: 1v
52 ('91)	PK	EVAP Purge Solenoid Control	Solenoid Off: 12v, On: 1v
53	LG/RD	S/C Vent Solenoid	Vacuum Decreasing: 1v
54	---	Not Used	---
55	LB	BARO Read Solenoid	Valve Off: 12v, On: 1v
56-58	---	Not Used	---
59	PK/BK	AIS 4 Motor Control	DC pulse signals: 0.8-11v
60	YL/BK	AIS 2 Motor Control	DC pulse signals: 0.8-11v
60 ('91)	YL	AIS 2 Motor Control	DC pulse signals: 0.8-11v

Pin Connector Graphic

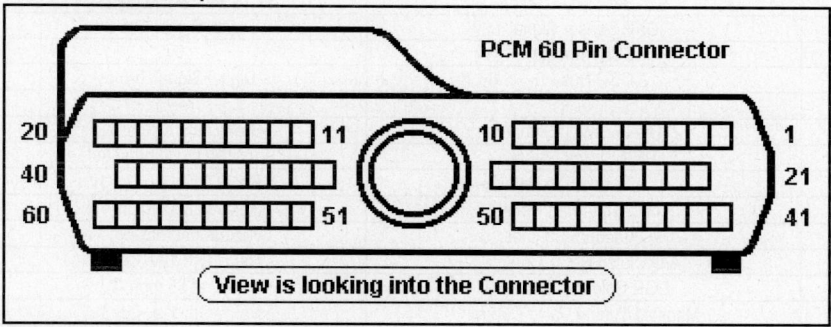

Standard Colors and Abbreviations

Abbreviation	Color	Abbreviation	Color	Abbreviation	Color
BK	Black	GY	Gray	RD	Red
BL	Blue	GN	Green	TN	Tan
BR	Brown	LG	Light Green	VT	Violet
DB	Dark Blue	OR	Orange	WT	White
DG	Dark Green	PK	Pink	YL	Yellow

1994-95 LHS 3.5L V6 SOHC MFI VIN F (A/T) 60 Pin Connector

PCM Pin #	Wire Color	Circuit Description (60 Pin)	Value at Hot Idle
1	DG/RD	MAP Sensor Signal	1.5-1.7v
2	TN/BK	ECT Sensor Signal	At 180°F: 2.80v
3	RD/WT	Battery Power (Fused B+)	12-14v
4	BK/LB	Sensor Ground	<0.050v
5	BK/WT	Sensor Ground	<0.050v
6	PK/WT	5-Volt Supply	4.9-5.1v
7	OR	8-Volt Supply	7.9-8.1v
8	YL/DG	Torque Management Request	Digital Signals
9	DB/WT	Ignition Switch Output	12-14v
10	GY/BK	Knock Sensor 2 Signal	0.080v AC
11	BK/TN	Power Ground	<0.1v
12	BK/TN	Power Ground	<0.1v
13	LB/BR	Injector 4 Driver	1.0-4.0 ms
14	YL/WT	Injector 3 Driver	1.0-4.0 ms
15	TN	Injector 2 Driver	1.0-4.0 ms
16	WT/DB	Injector 1 Driver	1.0-4.0 ms
17	WT	Coil 2 Driver	5°, at 55 mph: 8° dwell
18	RD	Coil 3 Driver	5°, at 55 mph: 8° dwell
19	BK	Coil 1 Driver	5°, at 55 mph: 8° dwell
20	DG	Generator Field Driver	Digital Signal: 0-12-0v
21	BK/RD	IAT Sensor Signal	At 100°F: 1.83v
22	OR/DB	TP Sensor Signal	0.6-1.0v
23	RD/LG	S/C Set Switch Signal	S/C & Set Switch On: 3.8v
24	LB/DB	CKP Sensor Signal	Digital Signal: 0-5-0v
25	PK	SCI Transmit	0v
26	PK/BR	CCD Bus (+)	Digital Signal: 0-5-0v
27-28	---	---	---
29	WT/PK	Brake Switch Signal	Brake Off: 0v, On: 12v
30	BK/LG	PNP Switch Signal	In P/N: 0v, Others: 5v
31	DB/PK	High Speed Fan Relay	Relay Off: 12v, On: 1v
32	WT	Low Speed Fan Relay	Relay Off: 12v, On: 1v
33	TN/RD	S/C Vacuum Solenoid	Vacuum Increasing: 1v
34	DB/OR	A/C Clutch Relay Control	Relay Off: 12v, On: 1v
35	GY/YL	EGR Solenoid Control	12v, at 55 mph: 1v
36	PK	Manifold Tuning Valve Control	Valve Off: 12v, On: 1v
37	---	---	---
38	BR/RD	Injector 5 Driver	1.0-4.0 ms
39	GY/RD	IAC 1 Driver	DC pulse signals: 0.8-11v
40	BR/WT	IAC 3 Driver	DC pulse signals: 0.8-11v

1994-95 LHS 3.5L V6 SOHC MFI VIN F (A/T) 60 Pin Connector

PCM Pin #	Wire Color	Circuit Description (60 Pin)	Value at Hot Idle
41	BK/DG	HO2S-11 (B1 S1) Signal	0.1-1.1v
42	BK/LG	Knock Sensor 1 Signal	0.080v AC
43	---	---	---
44	TN/YL	CMP Sensor Signal	Digital Signal: 0-5-0v
45	LG	SCI Receive	0v
46	WT/BK	CCD Bus (-)	<0.050v
47	WT/OR	Vehicle Speed Signal	Digital Signal
48	DB	A/C Pressure Switch Signal	A/C On: 0.45-4.85v
49	TN/WT	HO2S-12 (B1 S2) Signal	0.1-1.1v
50, 54, 56	---	---	---
51	DB/YL	ASD Relay Output	12-14v
52	PK/BK	EVAP Purge Solenoid Control	Solenoid Off: 12, On: 1v
53	LG/RD	S/C Vent Solenoid	Vacuum Decreasing: 1v
55	TN/RD	S/C Power Supply	12-14v
57	DG/OR	ASD Relay Control	Relay Off: 12v, On: 1v
58	BR/BK	Injector 6 Driver	1.0-4.0 ms
59	PK/BK	IAC 4 Driver	DC pulse signals: 0.8-11v
60	YL/BK	IAC 2 Driver	DC pulse signals: 0.8-11v

Pin Connector Graphic

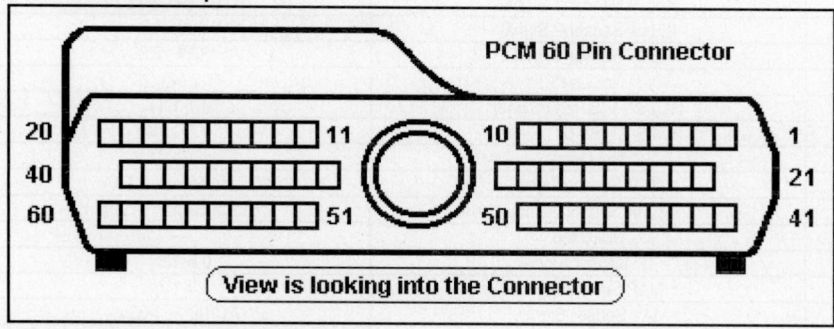

PCM 60 Pin Connector

View is looking into the Connector

Standard Colors and Abbreviations

Abbreviation	Color	Abbreviation	Color	Abbreviation	Color
BK	Black	GY	Gray	RD	Red
BL	Blue	GN	Green	TN	Tan
BR	Brown	LG	Light Green	VT	Violet
DB	Dark Blue	OR	Orange	WT	White
DG	Dark Green	PK	Pink	YL	Yellow

1996-97 LHS 3.5L V6 SOHC MFI VIN F (A/T) 'C1' Connector

PCM Pin #	Wire Color	Circuit Description (40-Pin)	Value at Hot Idle
1	LG/RD	HO2S-12 (B1 S2) Signal	0.1-1.1v
2	RD	Coil 3 Driver	5°, at 55 mph: 8° dwell
3	WT	Coil 2 Driver	5°, at 55 mph: 8° dwell
4	DG	Generator Field Driver	Digital Signal: 0-12-0v
5	YL/RD	S/C Power Supply	12-14v
6	DG/OR	ASD Relay Output	12-14v
7	YL/WT	Injector 3 Driver	1.0-4.0 ms
8-9	---	Not Used	---
10	BK/TN	Power Ground	<0.1v
11	BK	Coil 1 Driver	5°, at 55 mph: 8° dwell
12	---	Not Used	---
13	WT/LB	Injector 1 Driver	1.0-4.0 ms
14	BR/BK	Injector 6 Driver	1.0-4.0 ms
15	BR/RD	Injector 5 Driver	1.0-4.0 ms
16	LB/BR	Injector 4 Driver	1.0-4.0 ms
17	TN	Injector 2 Driver	1.0-4.0 ms
18-19	---	Not Used	---
20	DB/WT	Ignition Switch Output	12-14v
21-23	---	Not Used	---
24	BK/LG	Knock Sensor 1 Signal	0.080v AC
25	GY/BK	Knock Sensor 2 Signal	0.080v AC
26	TN/BK	ECT Sensor Signal	At 180°F: 2.80v
27-28	---	Not Used	---
29	TN/WT	HO2S-11 (B1 S1) Signal	0.1-1.1v
30	BK/DG	HO2S-21 (B2 S1) Signal	0.1-1.1v
31	---	Not Used	---
32	LB/DB	CKP Sensor Signal	Digital Signal: 0-5-0v
33	TN/YL	CMP Sensor Signal	Digital Signal: 0-5-0v
34	---	Not Used	---
35	OR/DB	TP Sensor Signal	0.6-1.0v
36	DG/RD	MAP Sensor Signal	1.5-1.7v
37	BK/RD	IAT Sensor Signal	At 100°F: 1.83v
38	---	Not Used	---
39	PK/RD	Manifold Tuning Valve Control	Valve Off: 12v, On: 1v
40	GY/YL	EGR Solenoid Control	12v, at 55 mph: 1v

Standard Colors and Abbreviations

Abbreviation	Color	Abbreviation	Color	Abbreviation	Color
BK	Black	GY	Gray	RD	Red
BL	Blue	GN	Green	TN	Tan
BR	Brown	LG	Light Green	VT	Violet
DB	Dark Blue	OR	Orange	WT	White
DG	Dark Green	PK	Pink	YL	Yellow

1996-97 LHS 3.5L V6 SOHC MFI VIN F (A/T) 'C2' Connector

PCM Pin #	Wire Color	Circuit Description (40-Pin)	Value at Hot Idle
41	RD/LG	S/C Set Switch Signal	S/C & Set Switch On: 3.8v
42	DB	A/C Pressure Switch Signal	A/C On: 0.45-4.85v
43	BK/LB	Sensor Ground	<0.050v
44	OR	8-Volt Supply	7.9-8.1v
45	---	Not Used	---
46	RD/WT	Battery Power (Fused B+)	12-14v
47	BK/WT	Power Ground	<0.1v
48	BR/WT	IAC 1 Driver	DC pulse signals: 0.8-11v
49	YL/BK	IAC 2 Driver	DC pulse signals: 0.8-11v
50	BK/TN	Power Ground	<0.1v
51	TN/WT	HO2S-22 (B2 S2) Signal	0.1-1.1v
52-54	---	Not Used	---
55	WT	Low Speed Fan Relay	Relay Off: 12v, On: 1v
56	---	Not Used	---
57	GY/RD	IAC 3 Driver	DC pulse signals: 0.8-11v
58	PK/GY	IAC 4 Driver	DC pulse signals: 0.8-11v
59	PK/BR	CCD Bus (+)	Digital Signal: 0-5-0v
60	WT/BK	CCD Bus (-)	<0.050v
61	PK/WT	5-Volt Supply	4.9-5.1v
62	WT/PK	Brake Switch Signal	Brake Off: 0v, On: 12v
63	YL/DG	Torque Management Request	Digital Signals
64	DB/OR	A/C Clutch Relay Control	Relay Off: 12v, On: 1v
65	PK	SCI Transmit	0v
66	WT/OR	Vehicle Speed Signal	Digital: 0-8-0-8v
67	DB/YL	ASD Relay Control	Relay Off: 12v, On: 1v
68	PK/BK	EVAP Purge Solenoid Control	PWM Signal: 0-12-0v
69	DB/PK	High Speed Fan Relay	Relay Off: 12v, On: 1v
70-73	---	Not Used	---
74	BR	Fuel Pump Relay Control	Relay Off: 12v, On: 1v
75	LG	SCI Receive	0v
76	BK/DG	PNP Switch Signal	In P/N: 0v, Others: 5v
77	---	Not Used	---
78	TN/RD	S/C Vacuum Solenoid	Vacuum Increasing: 1v
79	---	Not Used	---
80	LG/RD	S/C Vent Solenoid	Vacuum Decreasing: 1v

Pin Connector Graphic

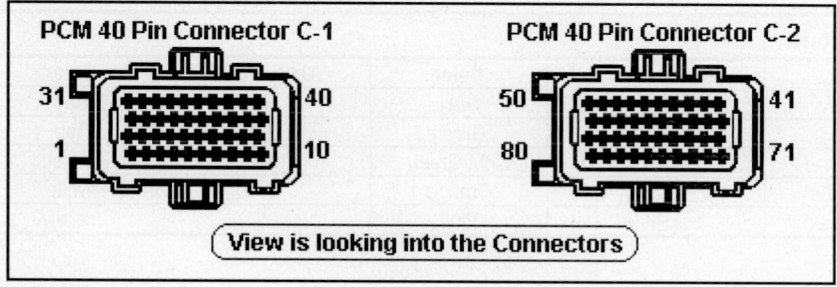

PCM 40 Pin Connector C-1

PCM 40 Pin Connector C-2

View is looking into the Connectors

1999 LHS 3.5L V6 SOHC 24v MFI VIN G (A/T) 'C1' Connector

PCM Pin #	Wire Color	Circuit Description (40-Pin)	Value at Hot Idle
1	TN/LG	Coil 4 Driver	5°, at 55 mph: 8° dwell
2	TN/OR	Coil 3 Driver	5°, at 55 mph: 8° dwell
3	TN/PK	Coil 2 Driver	5°, at 55 mph: 8° dwell
4	TN/LB	Coil 6 Driver	5°, at 55 mph: 8° dwell
5	YL/RD	S/C Power Supply	12-14v
6	DG/OR	ASD Relay Output	12-14v
7	YL/WT	Injector 3 Driver	1.0-4.0 ms
8	DG	Generator Field Driver	Digital Signal: 0-12-0v
9, 12	---	Not Used	---
10	BK/TN	Power Ground	<0.1v
11	TN/RD	Coil 1 Driver	5°, at 55 mph: 8° dwell
13	WT/DB	Injector 1 Driver	1.0-4.0 ms
14	BR/DB	Injector 6 Driver	1.0-4.0 ms
15	GY	Injector 5 Driver	1.0-4.0 ms
16	LB/BR	Injector 4 Driver	1.0-4.0 ms
17	TN	Injector 2 Driver	1.0-4.0 ms
18	GY/PK	Short Runner Valve Solenoid	Valve On: 1v, Off: 12v
19	---	Not Used	---
20	DB/WT	Ignition Switch Output	12-14v
21	TN/DG	Coil 5 Driver	5°, at 55 mph: 8° dwell
22-23	---	Not Used	---
24	DB/LG	Knock Sensor Signal	0.080v AC
25	BK/PK	Knock Sensor Ground	<0.050v
26	TN/BK	ECT Sensor Signal	At 180°F: 2.80v
27	BK/OR	Oxygen Sensor Ground	<0.050v
28, 38	---	Not Used	---
29	LG/RD	HO2S-21 (B2 S1) Signal	0.1-1.1v
30	BK/DG	HO2S-11 (B1 S1) Signal	0.1-1.1v
31	TN/WT	Starter Relay Control	KOEC: 9-11v
32	GY/BK	CKP Sensor Signal	Digital Signal: 0-5-0v
33	TN/YL	CMP Sensor Signal	Digital Signal: 0-5-0v
34	LG/PK	EGR Sensor Signal	0.6-0.8v
35	OR/DB	TP Sensor Signal	0.6-1.0v
36	DG/RD	MAP Sensor Signal	1.5-1.7v
37	BK/RD	IAT Sensor Signal	At 100°F: 1.83v
39	PK/RD	Manifold Tuning Valve Control	Valve Off: 12v, On: 1v
40	GY/YL	EGR Solenoid Control	12v, at 55 mph: 1v

Standard Colors and Abbreviations

Abbreviation	Color	Abbreviation	Color	Abbreviation	Color
BK	Black	GY	Gray	RD	Red
BL	Blue	GN	Green	TN	Tan
BR	Brown	LG	Light Green	VT	Violet
DB	Dark Blue	OR	Orange	WT	White
DG	Dark Green	PK	Pink	YL	Yellow

1999 LHS 3.5L V6 SOHC 24v MFI VIN G (A/T) 'C2' Connector

PCM Pin #	Wire Color	Circuit Description (40-Pin)	Value at Hot Idle
41	RD/LG	S/C Set Switch Signal	S/C & Set Switch On: 3.8v
42	DB	A/C Pressure Switch Signal	A/C On: 0.45-4.85v
43	BK/LB	Sensor Ground	<0.050v
44	OR	8-Volt Supply	7.9-8.1v
45	DB/LG	PSP Switch Signal	Straight: 0v, Turning: 5v
46	RD/WT	Battery Power (Fused B+)	12-14v
47	BK/LB	Power Ground	<0.1v
48	BR/WT	IAC 3 Driver	DC pulse signals: 0.8-11v
49	YL/BK	IAC 2 Driver	DC pulse signals: 0.8-11v
50	BK/TN	Power Ground	<0.1v
51	TN/WT	HO2S-12 (B1 S2) Signal	0.1-1.1v
53	PK/WT	HO2S-22 (B2 S2) Signal	0.1-1.1v
55	DB/PK	Low Speed Fan Relay	Relay Off: 12v, On: 1v
56	TN/RD	S/C Vacuum Solenoid	Vacuum Increasing: 1v
57	GY/RD	IAC 1 Driver	DC pulse signals: 0.8-11v
58	PK/BK	IAC 4 Driver	DC pulse signals: 0.8-11v
59	YL/PK	PCI Data Bus (J1850)	Digital Signals: 0-7-0v
61	PK/WT	5-Volt Supply	4.9-5.1v
62	WT/PK	Brake Switch Signal	Brake Off: 0v, On: 12v
63	YL/DG	Torque Management Request	Digital Signals
64	DB/OR	A/C Clutch Relay Control	Relay Off: 12v, On: 1v
65, 75	PK, LG	SCI Transmit, SCI Receive	0v
66	WT/OR	Vehicle Speed Signal	Digital Signal
67	DB/YL	ASD Relay Control	Relay Off: 12v, On: 1v
68	PK/BK	EVAP Purge Solenoid Control	PWM Signal: 0-12-0v
69	DB/LG	High Speed Fan Relay	Relay Off: 12v, On: 1v
70	DG/LG	EVAP Purge Solenoid Sense	0-1v
71	WT/RD	EATX RPM	Digital Signals
72	OR/RD	LDP Switch Sense	Open: 12v, Closed: 0v
73	---	Not Used	---
74	BR	Fuel Pump Relay Control	Relay Off: 12v, On: 1v
76	BK/PK	PNP Switch Signal	In P/N: 0v, Others: 5v
77	WT/DG	LDP Solenoid Control	PWM Signal: 0-12-0v
78-79	---	Not Used	---
80	LG/RD	S/C Vent Solenoid	Vacuum Decreasing: 1v

Pin Connector Graphic

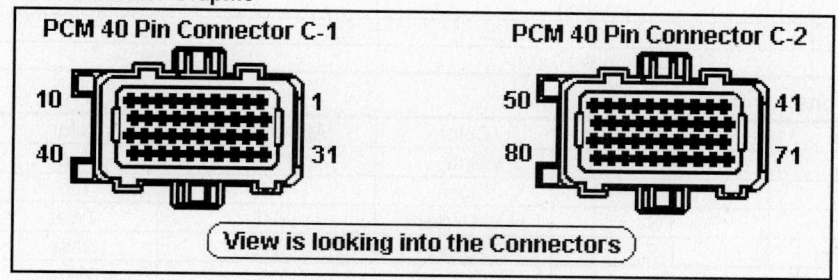

PCM 40 Pin Connector C-1

PCM 40 Pin Connector C-2

View is looking into the Connectors

2000-01 LHS 3.5L V6 SOHC 24v MFI VIN G (A/T) 'C1' Connector

PCM Pin #	Wire Color	Circuit Description (40-Pin)	Value at Hot Idle
1	TN/LG	Coil 4 Driver	5°, at 55 mph: 8° dwell
2	TN/OR	Coil 3 Driver	5°, at 55 mph: 8° dwell
3	TN/PK	Coil 2 Driver	5°, at 55 mph: 8° dwell
4	TN/LB	Coil 6 Driver	5°, at 55 mph: 8° dwell
5	YL/RD	S/C Power Supply	12-14v
6	DG/OR	ASD Relay Output	12-14v
7	YL/WT	Injector 3 Driver	1.0-4.0 ms
8	DG	Generator Field Driver	Digital Signal: 0-12-0v
9	---	Not Used	---
10	BK/TN	Power Ground	<0.1v
11	TN/RD	Coil 1 Driver	5°, at 55 mph: 8° dwell
12	---	Not Used	---
13	WT/DB	Injector 1 Driver	1.0-4.0 ms
14	BR/DB	Injector 6 Driver	1.0-4.0 ms
15	GY	Injector 5 Driver	1.0-4.0 ms
16	LB/BR	Injector 4 Driver	1.0-4.0 ms
17	TN	Injector 2 Driver	1.0-4.0 ms
18	GY/PK	Short Runner Valve Solenoid	Valve On: 1v, Off: 12v
19	---	Not Used	---
20	DB/WT	Ignition Switch Output	12-14v
21	TN/DG	Coil 5 Driver	5°, at 55 mph: 8° dwell
22-23	---	Not Used	---
24	DB/LG	Knock Sensor Signal	0.080v AC
25	BK/VT	Knock Sensor Ground	<0.050v
26	TN/BK	ECT Sensor Signal	At 180°F: 2.80v
27	BK/OR	Oxygen Sensor Ground	<0.050v
28	---	Not Used	---
29	LG/RD	HO2S-21 (B2 S1) Signal	0.1-1.1v
30	BK/DG	HO2S-11 (B1 S1) Signal	0.1-1.1v
31	TN/WT	Starter Relay Control	KOEC: 9-11v
32	GY/BK	CKP Sensor Signal	Digital Signal: 0-5-0v
33	TN/YL	CMP Sensor Signal	Digital Signal: 0-5-0v
34	LG/PK	EGR Sensor Signal	0.6-0.8v
35	OR/DB	TP Sensor Signal	0.6-1.0v
36	DG/RD	MAP Sensor Signal	1.5-1.7v
37	BK/RD	IAT Sensor Signal	At 100°F: 1.83v
38	---	Not Used	---
39	VT/RD	Manifold Tuning Valve Control	Valve Off: 12v, On: 1v
40	GY/YL	EGR Solenoid Control	12v, at 55 mph: 1v

Standard Colors and Abbreviations

Abbreviation	Color	Abbreviation	Color	Abbreviation	Color
BK	Black	GY	Gray	RD	Red
BL	Blue	GN	Green	TN	Tan
BR	Brown	LG	Light Green	VT	Violet
DB	Dark Blue	OR	Orange	WT	White
DG	Dark Green	PK	Pink	YL	Yellow

2000-01 LHS 3.5L V6 SOHC 24v MFI VIN G (A/T) 'C2' Connector

PCM Pin #	Wire Color	Circuit Description (40-Pin)	Value at Hot Idle
41	RD/LG	S/C Set Switch Signal	S/C & Set Switch On: 3.8v
42	DB	A/C Pressure Switch Signal	A/C On: 0.45-4.85v
43	BK/LB	Sensor Ground	<0.050v
44	OR	8-Volt Supply	7.9-8.1v
45	DB/LG	PSP Switch Signal	Straight: 0v, Turning: 5v
46	RD/WT	Battery Power (Fused B+)	12-14v
47	BK	Power Ground	<0.1v
48	BR/WT	IAC 3 Driver	DC pulse signals: 0.8-11v
49	YL/BK	IAC 2 Driver	DC pulse signals: 0.8-11v
50	BK	Power Ground	<0.1v
51	TN/WT	HO2S-12 (B1 S2) Signal	0.1-1.1v
52 ('01)	VT/LG	Battery Temperature Sensor	At 86°F: 1.96v
53	PK/WT	HO2S-22 (B2 S2) Signal	0.1-1.1v
54	---	Not Used	---
55	DB/PK	Low Speed Fan Relay	Relay Off: 12v, On: 1v
56	TN/RD	S/C Vacuum Solenoid	Vacuum Increasing: 1v
57	GY/RD	IAC 1 Driver	DC pulse signals: 0.8-11v
58	VT/BK	IAC 4 Driver	DC pulse signals: 0.8-11v
59	VT/YL	PCI Data Bus (J1850)	Digital Signals: 0-7-0v
60	---	Not Used	---
61	VT/WT	5-Volt Supply	4.9-5.1v
62	WT/PK	Brake Switch Signal	Brake Off: 0v, On: 12v
63	YL/DG	Torque Management Request	Digital Signals
64	DB/OR	A/C Clutch Relay Control	Relay Off: 12v, On: 1v
65	PK	SCI Transmit	0v
66	WT/OR	Vehicle Speed Signal	Digital Signal
67	DB/YL	ASD Relay Control	Relay Off: 12v, On: 1v
68	PK/BK	EVAP Purge Solenoid Control	PWM Signal: 0-12-0v
69	DB/LG	High Speed Fan Relay	Relay Off: 12v, On: 1v
70	DG/LG	EVAP Purge Solenoid Sense	0-1v
71	WT/RD	EATX RPM	Digital Signals
72	OR/RD	LDP Switch Sense	Open: 12v, Closed: 0v
73	---	Not Used	---
74	BR	Fuel Pump Relay Control	Relay Off: 12v, On: 1v
75	LG	SCI Receive	0v
76	LG	TRS T41 Sense	In P/N: 0v, Others: 5v
77	WT/DG	LDP Solenoid Control	PWM Signal: 0-12-0v
78-79	---	Not Used	---
80	LG/RD	S/C Vent Solenoid	Vacuum Decreasing: 1v

Pin Connector Graphic

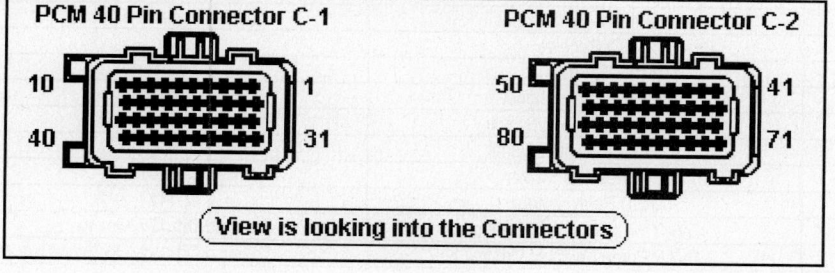

1990 New Yorker 3.0L V6 SOHC MFI VIN 3 (A/T) 60 Pin Connector

PCM Pin #	Wire Color	Circuit Description (60 Pin)	Value at Hot Idle
1	DG/RD	MAP Sensor Signal	1.5-1.7v
2	TN/WT	ECT Sensor Signal	At 180ºF: 2.80v
3	RD	Battery Power (Fused B+)	12-14v
3	RD/WT	Battery Power (Fused B+)	12-14v
4	BK/LB	Sensor Ground	<0.050v
5	BK/WT	Sensor Ground	<0.050v
6	PK/WT	5-Volt Supply	4.9-5.1v
7	OR	9-Volt Supply	8.9-9.1v
8	DG/BK	Auto Shutdown Relay Input	12-14v
9	DB	Ignition Switch Output	12-14v
10	---	Not Used	---
11	BK/TN	Power Ground	<0.1v
12	BK/TN	Power Ground	<0.1v
13	---	Not Used	---
14	YL/WT	Injector 3 Driver	1.0-4.0 ms
15	TN	Injector 2 Driver	1.0-4.0 ms
16	WT/DB	Injector 1 Driver	1.0-4.0 ms
17-18	---	Not Used	---
19	BK/GY	Coil 1 Driver	5º, at 55 mph: 8º dwell
20	DG	Alternator Field Control	Digital Signal: 0-12-0v
21	---	Not Used	---
22	OR/DB	TP Sensor Signal	0.6-1.0v
23	---	Not Used	---
24	GY/BK	Distributor Reference Signal	Digital Signal: 0-5-0v
25	PK	SCI Transmit	0v
26	VT	CCD Bus (+)	Digital Signal: 0-5-0v
27	BR	A/C Clutch Signal	Relay Off: 12v, On: 1v
28	---	Not Used	---
29	WT/PK	Brake Switch Signal	Brake Off: 0v, On: 12v
30	BR/YL	PNP Switch Signal	In P/N: 0v, Others: 5v
31	DB/PK	Radiator Fan Relay Control	Relay Off: 12v, On: 1v
32	BK/PK	Amber MIL Control	MIL On: 1v, MIL Off: 12v
33	TN/RD	S/C Vacuum Solenoid	Vacuum Increasing: 1v
34	DB/OR	A/C WOT Relay Control	Relay Off: 12v, On: 1v
35	GY/YL	EGR Solenoid Control (Cal)	12v, at 55 mph: 1v
36-38	---	Not Used	---
39	GY/RD	AIS 3 Motor Control	DC pulse signals: 0.8-11v
40	BR/WT	AIS 1 Motor Control	DC pulse signals: 0.8-11v
41	BK/DG	Heated Oxygen Sensor	0.1-1.1v
42	---	Not Used	---
43	GY/LB	Tachometer Signal	Pulse Signals
44	TN/YL	Distributor Sync Signal	Digital Signal: 0-5-0v
45	LG	SCI Receive	5v
46	WT	CCD Bus (-)	<0.1v
47	WT/OR	Vehicle Speed Signal	Digital Signal
48	BR/RD	Speed Control Set Switch	S/C & Set Switch On: 3.8v
49	YL/RD	S/C On/Off Switch Signal	S/C On: 1v, Off: 12v
50	WT/LG	S/C Resume Switch	S/C On: 1v, Off: 12v
51	DB/YL	ASD Relay Output	12-14v
52	PK	EVAP Purge Solenoid Control	Solenoid Off: 12v, On: 1v
52	PK/BK	EVAP Purge Solenoid Control	Solenoid Off: 12v, On: 1v
53	LG/RD	S/C Vent Solenoid	Vacuum Decreasing: 1v
54-58	---	Not Used	---
59	PK/BK	AIS 4 Motor Control	DC pulse signals: 0.8-11v
60	YL/BK	AIS 2 Motor Control	DC pulse signals: 0.8-11v

1991 New Yorker 3.3L V6 MFI VIN R (A/T) 60 Pin Connector

PCM Pin #	Wire Color	Circuit Description (60 Pin)	Value at Hot Idle
1	DG/RD	MAP Sensor Signal	1.5-1.7v
2	TN/WT	ECT Sensor Signal	At 180°F: 2.80v
3	RD/WT	Battery Power (Fused B+)	12-14v
4-5	BK/LB	Sensor Ground	<0.050v
6	PK/WT	5-Volt Supply	4.9-5.1v
7	OR	9-Volt Supply	8.9-9.1v
8	DG/BK	Auto Shutdown Relay Input	12-14v
9	DB	Ignition Switch Output	12-14v
10	---	Not Used	---
11	LB/RD	Power Ground	<0.1v
12	LB/RD	Power Ground	<0.1v
13	---	Not Used	---
14	YL/WT	Injector 3 Driver	1.0-4.0 ms
15	TN	Injector 2 Driver	1.0-4.0 ms
16	WT/DB	Injector 1 Driver	1.0-4.0 ms
17	DB/BK	Coil 2 Driver	5°, at 55 mph: 8° dwell
18	DB/GY	Coil 3 Driver	5°, at 55 mph: 8° dwell
19	BK/GY	Coil 1 Driver	5°, at 55 mph: 8° dwell
20	DG	Alternator Field Control	Digital Signal: 0-12-0v
21	BK/RD	Throttle Body Temperature	At 100°F: 2.51v
22	OR/DB	TP Sensor Signal	0.6-1.0v
24	GY/BK	CKP Sensor Signal	Digital Signal: 0-5-0v
25	PK	SCI Transmit	0v
26	BK	CCD Bus (+)	Digital Signal: 0-5-0v
27	BR	A/C Clutch Signal	Relay Off: 12v, On: 1v
28	---	Not Used	---
29	WT/PK	Brake Switch Signal	Brake Off: 0v, On: 12v
30	BR/YL	PNP Switch Signal	In P/N: 0v, Others: 5v
31	DB/PK	Radiator Fan Relay Control	Relay Off: 12v, On: 1v
32	BK/PK	Amber MIL Control	MIL On: 1v, MIL Off: 12v
33	TN/RD	S/C Vacuum Solenoid	Vacuum Increasing: 1v
34	DB/OR	A/C WOT Relay Control	Relay Off: 12v, On: 1v
35	GY/YL	EGR Solenoid Control (Cal)	12v, at 55 mph: 1v
37	---	Not Used	---
38	---	Not Used	---
39	GY/RD	AIS 3 Motor Control	DC pulse signals: 0.8-11v
40	BR/WT	AIS 1 Motor Control	DC pulse signals: 0.8-11v
41	BK/DG	Heated Oxygen Sensor	0.1-1.1v
42	BK/LG	Knock Sensor Signal	0.080v AC
43	---	Not Used	---
44	TN/YL	CMP Sensor Signal	Digital Signal: 0-5-0v
45	LG	SCI Receive	0v
46	WT	CCD Bus (-)	<0.1v
47	WT/OR	Vehicle Speed Signal	Digital Signal
48	BR/RD	Speed Control Set Switch	S/C & Set Switch On: 3.8v
49	YL/RD	S/C On/Off Switch Signal	S/C On: 1v, Off: 12v
50	WT/LG	S/C Resume Switch	S/C On: 1v, Off: 12v
51	DB/YL	ASD Relay Output	12-14v
52	PK/BK	EVAP Purge Solenoid Control	Solenoid Off: 12v, On: 1v
53	LG/RD	S/C Vent Solenoid	Vacuum Decreasing: 1v
54-58	---	Not Used	---
59	PK/BK	AIS 4 Motor Control	DC pulse signals: 0.8-11v
60	YL/BK	AIS 2 Motor Control	DC pulse signals: 0.8-11v

1992-93 New Yorker 3.3L V6 MFI VIN R (A/T) 60 Pin Connector

PCM Pin #	Wire Color	Circuit Description (60 Pin)	Value at Hot Idle
1	DG/RD	MAP Sensor Signal	1.5-1.7v
2	TN/BK	ECT Sensor Signal	At 180°F: 2.80v
3	RD/WT	Fuel Pump ASD Relay	Relay Off: 12v, On: 1v
4	BK/LB	Sensor Ground	<0.050v
5	BK/WT	Sensor Ground	<0.050v
6	PK/WT	5-Volt Supply	4.9-5.1v
7	OR	8-Volt Supply	7.9-8.1v
8	---	Not Used	---
9	DB	Ignition Switch Output	12-14v
10	---	Not Used	---
11	BK/TN	Power Ground	<0.1v
12	BK/TN	Power Ground	<0.1v
13	LB/BR	Injector 4 Driver	1.0-4.0 ms
14	YL/WT	Injector 3 Driver	1.0-4.0 ms
15	TN	Injector 2 Driver	1.0-4.0 ms
16	WT/DB	Injector 1 Driver	1.0-4.0 ms
17	DB/BK	Coil 2 Driver	5°, at 55 mph: 8° dwell
18	DB/GY	Coil 3 Driver	5°, at 55 mph: 8° dwell
19	BK/GY	Coil 1 Driver	5°, at 55 mph: 8° dwell
20	DG	Alternator Field Control	Digital Signal: 0-12-0v
21	---	Not Used	---
22	OR/DB	TP Sensor Signal	0.6-1.0v
23	RD/LG	S/C Set Switch Signal	S/C & Set Switch On: 3.8v
24	GY/BK	CKP Sensor Signal	Digital Signal: 0-5-0v
25	PK	SCI Transmit	0v
26	PK/BK	CCD Bus (+)	Digital Signal: 0-5-0v
27	BR	A/C Damped Pressure Switch	Relay Off: 12v, On: 1v
28	DB/OR	PSP Switch Signal	Straight: 0v, Turning: 5v
29	WT/PK	Brake Switch Signal	Brake Off: 0v, On: 12v
30	BR/YL	PNP Switch Signal	In P/N: 0v, Others: 5v
31	DB/PK	Radiator Fan Relay Control	Relay Off: 12v, On: 1v
32	BK/PK	Amber MIL Control	MIL On: 1v, MIL Off: 12v
33	TN/RD	S/C Vacuum Solenoid	Vacuum Increasing: 1v
34	DB/OR	A/C WOT Relay Control	Relay Off: 12v, On: 1v
35	GY/YL	EGR Solenoid Control (Cal)	12v, at 55 mph: 1v
36-37	---	---	---
38	GY	Injector 5 Driver	1.0-4.0 ms
39	GY/RD	AIS Motor Signal (open)	DC pulse signals: 0.8-11v
40	BR/WT	AIS Motor Signal (close)	DC pulse signals: 0.8-11v

1992-93 New Yorker 3.3L V6 MFI VIN R (A/T) 60 Pin Connector

PCM Pin #	Wire Color	Circuit Description (60 Pin)	Value at Hot Idle
41	BK/DG	Heated Oxygen Sensor	0.1-1.1v
42	BK/LG	Knock Sensor Signal	0.080v AC
43	---	Not Used	---
44	TN/YL	CMP Sensor Signal	Digital Signal: 0-5-0v
45	LG	SCI Receive	0v
46	WT/BK	CCD Bus (-)	<0.050v
47	WT/OR	Vehicle Speed Signal	Digital Signal
48-50	---	Not Used	---
51	DB/YL	ASD Relay Output	12-14v
52	PK/BK	EVAP Purge Solenoid Control	Solenoid Off: 12v, On: 1v
53	LG/RD	S/C Vent Solenoid	Vacuum Decreasing: 1v
54-56	---	Not Used	---
57	DG/OR	ASD Relay Control	Relay Off: 12v, On: 1v
58	BR/DB	Injector 6 Driver	1.0-4.0 ms
59	PK/BK	AIS Motor Signal (close)	DC pulse signals: 0.8-11v
60	YL/BK	AIS Motor Signal (close)	DC pulse signals: 0.8-11v

Pin Connector Graphic

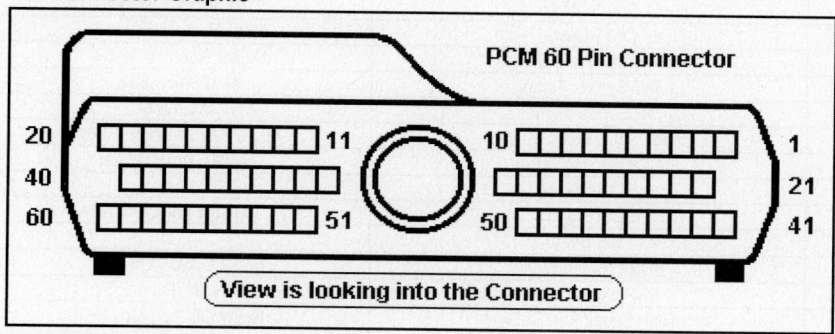

Standard Colors and Abbreviations

Abbreviation	Color	Abbreviation	Color	Abbreviation	Color
BK	Black	GY	Gray	RD	Red
BL	Blue	GN	Green	TN	Tan
BR	Brown	LG	Light Green	VT	Violet
DB	Dark Blue	OR	Orange	WT	White
DG	Dark Green	PK	Pink	YL	Yellow

1994-95 New Yorker 3.5L V6 SOHC MFI VIN F 60 Pin Connector

PCM Pin #	Wire Color	Circuit Description (60 Pin)	Value at Hot Idle
1	DG/RD	MAP Sensor Signal	1.5-1.7v
2	TN/BK	ECT Sensor Signal	At 180°F: 2.80v
3	RD/WT	Battery Power (Fused B+)	12-14v
4	BK/LB	Sensor Ground	<0.050v
5	BK/WT	Sensor Ground	<0.050v
6	PK/WT	5-Volt Supply	4.9-5.1v
7	OR	8-Volt Supply	7.9-8.1v
8	YL/DG	Torque Management Request	Digital Signals
9	DB/WT	Ignition Switch Output	12-14v
10	GY/BK	Knock Sensor 2 Signal	0.080v AC
11	BK/TN	Power Ground	<0.1v
12	BK/TN	Power Ground	<0.1v
13	LB/BR	Injector 4 Driver	1.0-4.0 ms
14	YL/WT	Injector 3 Driver	1.0-4.0 ms
15	TN	Injector 2 Driver	1.0-4.0 ms
16	WT/DB	Injector 1 Driver	1.0-4.0 ms
17	WT	Coil 2 Driver	5°, at 55 mph: 8° dwell
18	RD	Coil 3 Driver	5°, at 55 mph: 8° dwell
19	BK	Coil 1 Driver	5°, at 55 mph: 8° dwell
20	DG	Generator Field Driver	Digital Signal: 0-12-0v
21	BK/RD	IAT Sensor Signal	At 100°F: 1.83v
22	OR/DB	TP Sensor Signal	0.6-1.0v
23	RD/LG	S/C Set Switch Signal	S/C & Set Switch On: 3.8v
24	LB/DB	CKP Sensor Signal	Digital Signal: 0-5-0v
25	PK	SCI Transmit	0v
26	PK/BR	CCD Bus (+)	Digital Signal: 0-5-0v
27-28	---	Not Used	---
29	WT/PK	Brake Switch Signal	Brake Off: 0v, On: 12v
30	BK/LG	PNP Switch Signal	In P/N: 0v, Others: 5v
31	DB/PK	High Speed Fan Relay	Relay Off: 12v, On: 1v
32	WT	Low Speed Fan Relay	Relay Off: 12v, On: 1v
33	TN/RD	S/C Vacuum Solenoid	Vacuum Increasing: 1v
34	DB/OR	A/C Clutch Relay Control	Relay Off: 12v, On: 1v
35	GY/YL	EGR Solenoid Control	12v, at 55 mph: 1v
36	PK	Manifold Tuning Valve Control	Valve Off: 12v, On: 1v
37	---	Not Used	---
38	BR/RD	Injector 5 Driver	1.0-4.0 ms
39	GY/RD	IAC 1 Driver	DC pulse signals: 0.8-11v
40	BR/WT	IAC 3 Driver	DC pulse signals: 0.8-11v

1994-95 New Yorker 3.5L V6 SOHC MFI VIN F 60 Pin Connector

PCM Pin #	Wire Color	Circuit Description (60 Pin)	Value at Hot Idle
41	BK/DG	HO2S-11 (B1 S1) Signal	0.1-1.1v
42	BK/LG	Knock Sensor 1 Signal	0.080v AC
43	---	Not Used	---
44	TN/YL	CMP Sensor Signal	Digital Signal: 0-5-0v
45	LG	SCI Receive	0v
46	WT/BK	CCD Bus (-)	<0.050v
47	WT/OR	Vehicle Speed Signal	Digital Signal
48	DB	A/C Pressure Switch Signal	A/C On: 0.45-4.85v
49	TN/WT	HO2S-12 (B1 S2) Signal	0.1-1.1v
50	---	Not Used	---
51	DB/YL	ASD Relay Output	12-14v
52	PK/BK	EVAP Purge Solenoid Control	Solenoid Off: 12v, On: 1v
53	LG/RD	S/C Vent Solenoid	Vacuum Decreasing: 1v
54	---	Not Used	---
55	TN/RD	S/C Power Supply	12-14v
56	---	Not Used	---
57	DG/OR	ASD Relay Control	Relay Off: 12v, On: 1v
58	BR/BK	Injector 6 Driver	1.0-4.0 ms
59	PK/BK	IAC 4 Driver	DC pulse signals: 0.8-11v
60	YL/BK	IAC 2 Driver	DC pulse signals: 0.8-11v

Pin Connector Graphic

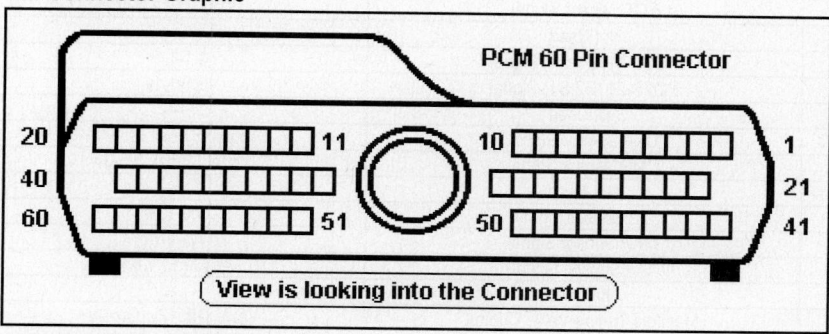

PCM 60 Pin Connector

View is looking into the Connector

Standard Colors and Abbreviations

Abbreviation	Color	Abbreviation	Color	Abbreviation	Color
BK	Black	GY	Gray	RD	Red
BL	Blue	GN	Green	TN	Tan
BR	Brown	LG	Light Green	VT	Violet
DB	Dark Blue	OR	Orange	WT	White
DG	Dark Green	PK	Pink	YL	Yellow

1996 New Yorker 3.5L V6 SOHC MFI VIN F (A/T) 'C1' Connector

PCM Pin #	Wire Color	Circuit Description (40-Pin)	Value at Hot Idle
1	LG/RD	HO2S-12 (B1 S2) Signal	0.1-1.1v
2	RD	Coil 3 Driver	5°, at 55 mph: 8° dwell
3	WT	Coil 2 Driver	5°, at 55 mph: 8° dwell
4	DG	Generator Field Driver	Digital Signal: 0-12-0v
5	YL/RD	S/C Power Supply	12-14v
6	DG/OR	ASD Relay Output	12-14v
7	YL/WT	Injector 3 Driver	1.0-4.0 ms
8-9	---	Not Used	---
10	BK/TN	Power Ground	<0.1v
11	BK	Coil 1 Driver	5°, at 55 mph: 8° dwell
12	---	Not Used	---
13	WT/LB	Injector 1 Driver	1.0-4.0 ms
14	BR/BK	Injector 6 Driver	1.0-4.0 ms
15	BR/RD	Injector 5 Driver	1.0-4.0 ms
16	LB/BR	Injector 4 Driver	1.0-4.0 ms
17	TN	Injector 2 Driver	1.0-4.0 ms
18-19	---	Not Used	---
20	DB/WT	Ignition Switch Output	12-14v
21-23	---	Not Used	---
24	BK/LG	Knock Sensor 1 Signal	0.080v AC
25	GY/BK	Knock Sensor 2 Signal	0.080v AC
26	TN/BK	ECT Sensor Signal	At 180°F: 2.80v
27-28	---	Not Used	---
29	TN/WT	HO2S-11 (B1 S1) Signal	0.1-1.1v
30	BK/DG	HO2S-21 (B2 S1) Signal	0.1-1.1v
31	---	Not Used	---
32	LB/DB	CKP Sensor Signal	Digital Signal: 0-5-0v
33	TN/YL	CMP Sensor Signal	Digital Signal: 0-5-0v
34	---	Not Used	---
35	OR/DB	TP Sensor Signal	0.6-1.0v
36	DG/RD	MAP Sensor Signal	1.5-1.7v
37	BK/RD	IAT Sensor Signal	At 100°F: 1.83v
38	---	Not Used	---
39	PK/RD	Manifold Tuning Valve Control	Valve Off: 12v, On: 1v
40	GY/YL	EGR Solenoid Control	12v, at 55 mph: 1v

Standard Colors and Abbreviations

Abbreviation	Color	Abbreviation	Color	Abbreviation	Color
BK	Black	GY	Gray	RD	Red
BL	Blue	GN	Green	TN	Tan
BR	Brown	LG	Light Green	VT	Violet
DB	Dark Blue	OR	Orange	WT	White
DG	Dark Green	PK	Pink	YL	Yellow

1996 New Yorker 3.5L V6 SOHC MFI VIN F (A/T) 'C2' Connector

PCM Pin #	Wire Color	Circuit Description (40-Pin)	Value at Hot Idle
41	RD/LG	S/C Set Switch Signal	S/C & Set Switch On: 3.8v
42	DB	A/C Pressure Switch Signal	A/C On: 0.45-4.85v
43	BK/LB	Sensor Ground	<0.050v
44	OR	8-Volt Supply	7.9-8.1v
45	---	Not Used	---
46	RD/WT	Battery Power (Fused B+)	12-14v
47	BK/WT	Power Ground	<0.1v
48	BR/WT	IAC 1 Driver	DC pulse signals: 0.8-11v
49	YL/BK	IAC 2 Driver	DC pulse signals: 0.8-11v
50	BK/TN	Power Ground	<0.1v
51	TN/WT	HO2S-22 (B2 S2) Signal	0.1-1.1v
52-54	---	Not Used	---
55	WT	Low Speed Fan Relay	Relay Off: 12v, On: 1v
56	---	Not Used	---
57	GY/RD	IAC 3 Driver	DC pulse signals: 0.8-11v
58	PK/GY	IAC 4 Driver	DC pulse signals: 0.8-11v
59	PK/BR	CCD Bus (+)	Digital Signal: 0-5-0v
60	WT/BK	CCD Bus (-)	<0.050v
61	PK/WT	5-Volt Supply	4.9-5.1v
62	WT/PK	Brake Switch Signal	Brake Off: 0v, On: 12v
63	YL/DG	Torque Management Request	Digital Signals
64	DB/OR	A/C Clutch Relay Control	Relay Off: 12v, On: 1v
65	PK	SCI Transmit	0v
66	WT/OR	Vehicle Speed Signal	Digital: 0-8-0-8v
67	DB/YL	ASD Relay Control	Relay Off: 12v, On: 1v
68	PK/BK	EVAP Purge Solenoid Control	PWM Signal: 0-12-0v
69	DB/PK	High Speed Fan Relay	Relay Off: 12v, On: 1v
70-73	---	Not Used	---
74	BR	Fuel Pump Relay Control	Relay Off: 12v, On: 1v
75	LG	SCI Receive	0v
76	BK/DG	PNP Switch Signal	In P/N: 0v, Others: 5v
77	---	Not Used	---
78	TN/RD	S/C Vacuum Solenoid	Vacuum Increasing: 1v
79	---	Not Used	---
80	LG/RD	S/C Vent Solenoid	Vacuum Decreasing: 1v

Pin Connector Graphic

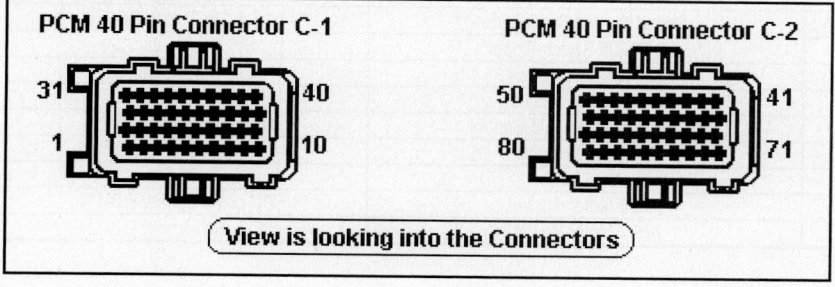

PCM 40 Pin Connector C-1

PCM 40 Pin Connector C-2

View is looking into the Connectors

1990-91 New Yorker Fifth Avenue 3.3L V6 MFI VIN R (A/T)

PCM Pin #	Wire Color	Circuit Description (60 Pin)	Value at Hot Idle
1	DG/RD	MAP Sensor Signal	1.5-1.7v
2	TN/WT	ECT Sensor Signal	At 180°F: 2.80v
3	RD/WT	Battery Power (Fused B+)	12-14v
4-5	BK/LB	Sensor Ground	<0.050v
6, 7	PK/WT, OR	5-Volt Supply, 8-Volt Supply	4.9-5.1v, 7.9-8.1v
8	DG/BK	Auto Shutdown Relay Input	12-14v
9	DB	Ignition Switch Output	12-14v
10, 13	---	Not Used	---
11-12	LB/RD	Power Ground	<0.1v
14	YL/WT	Injector 3 Driver (Cyl 4 & 5)	1.0-4.0 ms
15	TN	Injector 2 Driver (Cyl 2 & 3)	1.0-4.0 ms
16	WT/DB	Injector 1 Driver (Cyl 1 & 6)	1.0-4.0 ms
17	DB/BK	Coil 2 Driver	5°, at 55 mph: 8° dwell
18	DB/GY	Coil 3 Driver	5°, at 55 mph: 8° dwell
19	BK/GY	Coil 1 Driver	5°, at 55 mph: 8° dwell
20	DG	Alternator Field Control	Digital Signal: 0-12-0v
21	BK/RD	IAT Sensor Signal	At 100°F: 1.83v
22	OR/DB	TP Sensor Signal	0.6-1.0v
23, 28	---	Not Used	---
24	GY/BK	CKP Sensor Signal	Digital Signal: 0-5-0v
25, 45	PK, LG	SCI Transmit, SCI Receive	0v, 5v
26, 46	BK, WT/BK	CCD Bus (+), CCD Bus (-)	Digital Signal: 0-5-0v, <0.1v
27	BR	A/C Clutch Signal	Relay Off: 12v, On: 1v
29	WT/PK	Brake Switch Signal	Brake Off: 0v, On: 12v
30	BR/YL	PNP Switch Signal	In P/N: 0v, Others: 5v
31	DB/PK	Radiator Fan Relay Control	Relay Off: 12v, On: 1v
32	BK/PK	Amber MIL Control	MIL On: 1v, MIL Off: 12v
33	TN/RD	S/C Vacuum Solenoid	Vacuum Increasing: 1v
34	DB/OR	A/C WOT Relay Control	Relay Off: 12v, On: 1v
35	GY/YL	EGR Solenoid Control (California)	12v, at 55 mph: 1v
36-38, 43	---	Not Used	---
39	GY/RD	AIS 3 Motor Control	DC pulse signals: 0.8-11v
40	BR/WT	AIS 1 Motor Control	DC pulse signals: 0.8-11v
41	BK/DG	Heated Oxygen Sensor	0.1-1.1v
42	BK/LG	Knock Sensor Signal	0.080v AC
44	TN/YL	CMP Sensor Signal	Digital Signal: 0-5-0v
47	WT/OR	Vehicle Speed Signal	Digital Signal
48	BR/RD	Speed Control Set Switch	S/C & Set Switch On: 3.8v
49	YL/RD	S/C On/Off Switch Signal	S/C On: 1v, Off: 12v
50	WT/LG	S/C Resume Switch	S/C On: 1v, Off: 12v
51, 57	DB/YL	ASD Relay Output	12-14v
52	PK	EVAP Purge Solenoid Control	Solenoid Off: 12v, On: 1v
53	LG/RD	S/C Vent Solenoid	Vacuum Decreasing: 1v
54-58	---	Not Used	---
59	PK/BK	AIS 4 Motor Control	DC pulse signals: 0.8-11v
60	YL/BK	AIS 2 Motor Control	DC pulse signals: 0.8-11v

Pin Connector Graphic

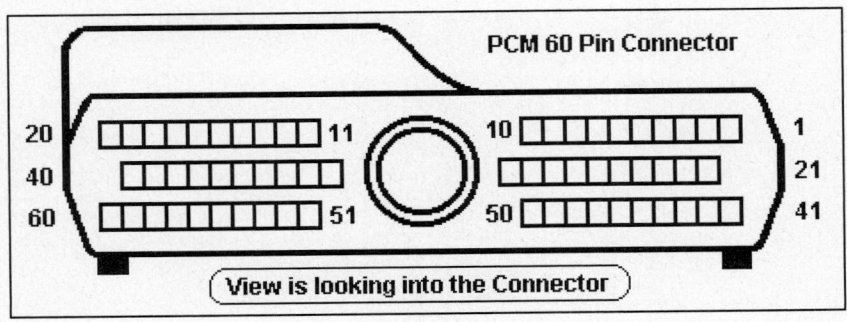

PCM 60 Pin Connector

View is looking into the Connector

1992-93 New Yorker Fifth Avenue 3.3L V6 MFI VIN R (A/T)

PCM Pin #	Wire Color	Circuit Description (60 Pin)	Value at Hot Idle
1	DG/RD	MAP Sensor Signal	1.5-1.7v
2	TN/BK	ECT Sensor Signal	At 180°F: 2.80v
3	RD/WT	Fuel Pump Relay Control	Relay Off: 12v, On: 1v
4, 5	BK/LB, BK/WT	Sensor Ground	<0.050v
6, 7	PK/WT, OR	5-Volt Supply, 8-Volt Supply	4.9-5.1v, 7.9-8.1v
8	DG/BK	Fuel Pump Ignition Feed	12-14v
9	DB	Ignition Switch Output	12-14v
10, 36-37, 43	---	Not Used	---
11, 12	BK/TN	Power Ground	<0.1v
13	LB/BR	Injector 4 Driver	1.0-4.0 ms
14	YL/WT	Injector 3 Driver	1.0-4.0 ms
15	TN	Injector 2 Driver	1.0-4.0 ms
16	WT/DB	Injector 1 Driver	1.0-4.0 ms
17	DB/BK	Coil 2 Driver	5°, at 55 mph: 8° dwell
18	DB/GY	Coil 3 Driver	5°, at 55 mph: 8° dwell
19	BK/GY	Coil 1 Driver	5°, at 55 mph: 8° dwell
20	DG	Alternator Field Control	Digital Signal: 0-12-0v
21	BK/RD	IAT Sensor Signal	At 100°F: 1.83v
22	OR/DB	TP Sensor Signal	0.6-1.0v
23	RD/LG	S/C Set Switch Signal	S/C & Set Switch On: 3.8v
24	GY/BK	CKP Sensor Signal	Digital Signal: 0-5-0v
25, 45	PK, LG	SCI Transmit, SCI Receive	0v, 5v
26, 46	PK/BK, WT/BK	CCD Bus (+), CCD Bus (-)	Digital Signal: 0-5-0v, <0.1v
27	BR	A/C Damped Pressure Switch	Relay Off: 12v, On: 1v
28	DB/OR	PSP Switch Signal	Straight: 0v, Turning: 5v
29	WT/PK	Brake Switch Signal	Brake Off: 0v, On: 12v
30	BR/YL	PNP Switch Signal	In P/N: 0v, Others: 5v
31	DB/PK	Radiator Fan Relay Control	Relay Off: 12v, On: 1v
32	BK/PK	Amber MIL Control	MIL On: 1v, MIL Off: 12v
33	TN/RD	S/C Vacuum Solenoid	Vacuum Increasing: 1v
34	DB/OR	A/C WOT Relay Control	Relay Off: 12v, On: 1v
35	GY/YL	EGR Solenoid Control (California)	12v, at 55 mph: 1v
38	GY	Injector 5 Driver	1.0-4.0 ms
39	GY/RD	AIS 3 Motor Control	DC pulse signals: 0.8-11v
40	BR/WT	AIS 1 Motor Control	DC pulse signals: 0.8-11v
41	BK/DG	Heated Oxygen Sensor	0.1-1.1v
42	BK/LG	Knock Sensor Signal	0.080v AC
44	TN/YL	CMP Sensor Signal	Digital Signal: 0-5-0v
47	WT/OR	Vehicle Speed Signal	Digital Signal
48-50, 54-56	---	Not Used	---
51, 57	DB/YL, DG/OR	ASD Relay Output	12-14v
52	PK	EVAP Purge Solenoid Control	Solenoid Off: 12v, On: 1v
53	LG/RD	S/C Vent Solenoid	Vacuum Decreasing: 1v
58	BR/DB	Injector 6 Driver	1.0-4.0 ms
59	PK/BK	AIS 4 Motor Control	DC pulse signals: 0.8-11v
60	YL/BK	AIS 2 Motor Control	DC pulse signals: 0.8-11v

Pin Connector Graphic

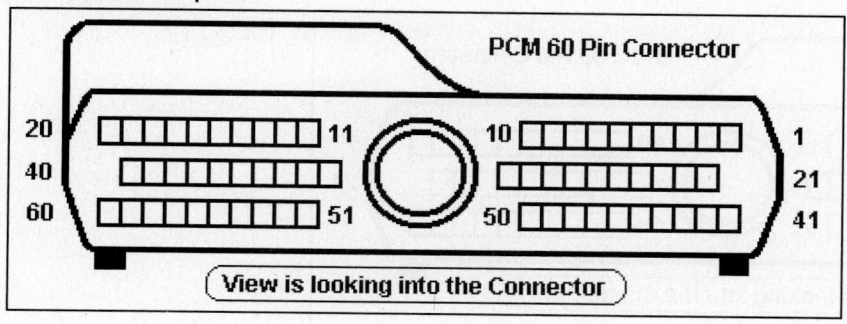

1991 New Yorker Fifth Avenue 3.8L V6 MFI VIN L (A/T)

PCM Pin #	Wire Color	Circuit Description (60 Pin)	Value at Hot Idle
1	DG/RD	MAP Sensor Signal	1.5-1.7v
2	TN/BK	ECT Sensor Signal	At 180ºF: 2.80v
3	RD	Fuel Pump ASD Relay	Relay Off: 12v, On: 1v
4-5	BK/LB	Sensor Ground	<0.050v
6	PK/WT	5-Volt Supply	4.9-5.1v
7	OR	8-Volt Supply	7.9-8.1v
8	DG/BK	Fuel Pump Ignition Feed	12-14v
9	DB	Ignition Switch Output	12-14v
10, 13	---	Not Used	---
11, 12, 23, 28	BK/TN	Power Ground	<0.1v
14	YL/WT	Injector 3 Driver	1.0-4.0 ms
15	TN	Injector 2 Driver	1.0-4.0 ms
16	WT/DB	Injector 1 Driver	1.0-4.0 ms
17	DB/BK	Coil 2 Driver	5º, at 55 mph: 8º dwell
18	DB/GY	Coil 3 Driver	5º, at 55 mph: 8º dwell
19	BK/GY	Coil 1 Driver	5º, at 55 mph: 8º dwell
20	DG	Voltage Regulator	Digital Signal: 0-12-0v
21	BK/RD	IAT Sensor Signal	At 100ºF: 1.83v
22	OR/DB	TP Sensor Signal	0.6-1.0v
24	GY/BK	CKP Sensor Signal	Digital Signal: 0-5-0v
25, 45	PK, LG	SCI Transmit, SCI Receive	0v, 5v
26, 46	PK, WT/BK	CCD Bus (+), CCD Bus (-)	Digital Signal: 0-5-0v, <0.1v
27	BR	A/C Clutch Sense	Relay Off: 12v, On: 1v
36-38, 54, 58	---	Not Used	---
29	WT/PK	Brake Switch Signal	Brake Off: 0v, On: 12v
30	BR/YL	PNP Switch Signal	In P/N: 0v, Others: 5v
31	DB/PK	Radiator Fan Relay Control	Relay Off: 12v, On: 1v
32	BK/PK	Amber MIL Control	MIL On: 1v, MIL Off: 12v
33	TN/RD	S/C Vacuum Solenoid	Vacuum Increasing: 1v
34	DB/OR	A/C Clutch Relay Control	Relay Off: 12v, On: 1v
35	GY/YL	EGR Solenoid Control (Cal)	12v, at 55 mph: 1v
39	GY/RD	IAC 1 Driver	DC pulse signals: 0.8-11v
40	BR/WT	IAC 3 Driver	DC pulse signals: 0.8-11v
41	BK/DG	Heated Oxygen Sensor	0.1-1.1v
42	BK/LG	Knock Sensor Signal	0.080v AC
44	TN/YL	CMP Sensor Signal	Digital Signal: 0-5-0v
47	WT/OR	Vehicle Speed Signal	Digital Signal
48	BR/RD	Speed Control Set Switch	S/C & Set Switch On: 3.8v
49	YL/RD	S/C On/Off Switch Signal	S/C On: 1v, Off: 12v
50	WT/LG	S/C Resume Switch	S/C On: 1v, Off: 12v
51	DB/YL	ASD Relay Output	12-14v
52	PK	EVAP Purge Solenoid Control	Solenoid Off: 12v, On: 1v
53	LG/RD	S/C Vent Solenoid	Vacuum Decreasing: 1v
59	PK/BK	IAC 4 Driver	DC pulse signals: 0.8-11v
60	YL/BK	IAC 2 Driver	DC pulse signals: 0.8-11v

Pin Connector Graphic

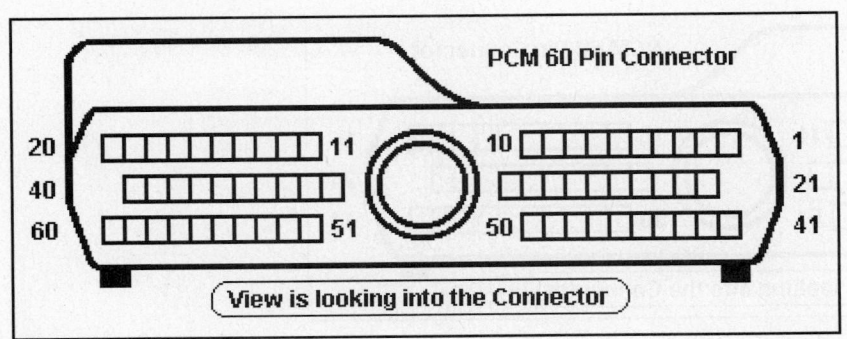

1992-93 New Yorker Fifth Avenue 3.8L V6 VIN L 60 Pin Connector

PCM Pin #	Wire Color	Circuit Description (60 Pin)	Value at Hot Idle
1	DG/RD	MAP Sensor Signal	1.5-1.7v
2	TN/BK	ECT Sensor Signal	At 180ºF: 2.80v
3	RD/WT	Fuel Pump Relay Control	Relay Off: 12v, On: 1v
4, 5	BK/LB	Sensor Ground	<0.050v
6, 7	PK/WT, OR	5-Volt Supply, 8-Volt Supply	4.9-5.1v, 7.9-8.1v
8, 10, 21	---	Not Used	---
9	DB	Ignition Switch Output	12-14v
11, 12	BK/TN	Power Ground	<0.1v
13	LB/BR	Injector 4 Driver	1.0-4.0 ms
14	YL/WT	Injector 3 Driver	1.0-4.0 ms
15	TN	Injector 2 Driver	1.0-4.0 ms
16	WT/DB	Injector 1 Driver	1.0-4.0 ms
17	DB/TN	Coil 2 Driver	5º, at 55 mph: 8º dwell
18	DB/GY	Coil 3 Driver	5º, at 55 mph: 8º dwell
19	BK/GY	Coil 1 Driver	5º, at 55 mph: 8º dwell
20	DG	Voltage Regulator	Digital Signal: 0-12-0v
22	OR/DB	TP Sensor Signal	0.6-1.0v
23	RD/LG	Speed Control Mode Signal	S/C & Set Switch On: 3.8v
24	GY/BK	CKP Sensor Signal	Digital Signal: 0-5-0v
25, 45	PK, LG	SCI Transmit, SCI Receive	0v, 5v
26, 46	PK, WT	CCD Bus (+), CCD Bus (-)	Digital Signal: 0-5-0v
27	BR	A/C Damped Pressure Switch	Relay Off: 12v, On: 1v
28 ('93)	DB/OR	PSP Switch Signal	Straight: 0v, Turning: 5v
29	WT/PK	Brake Switch Signal	Brake Off: 0v, On: 12v
30	BR/YL	PNP Switch Signal	In P/N: 0v, Others: 5v
31	DB/PK	Radiator Fan Relay Control	Relay Off: 12v, On: 1v
32	BK/PK	Amber MIL Control	MIL On: 1v, MIL Off: 12v
33	TN/RD	S/C Vacuum Solenoid	Vacuum Increasing: 1v
34	DB/OR	A/C Clutch Relay Control	Relay Off: 12v, On: 1v
35	GY/YL	EGR Solenoid Control (California)	12v, at 55 mph: 1v
36-37, 43	---	Not Used	---
38	GY	Injector 5 Driver	1.0-4.0 ms
39	GY/RD	IAC Open Driver	DC pulse signals: 0.8-11v
40	BR/WT	IAC Close Driver	DC pulse signals: 0.8-11v
41	BK/DG	Heated Oxygen Sensor	0.1-1.1v
42	BK/LG	Knock Sensor Signal	0.080v AC
44	TN/YL	CMP Sensor Signal	Digital Signal: 0-5-0v
47	WT/OR	Vehicle Speed Signal	Digital Signal
48-50, 54-56	---	Not Used	---
51, 57	DB/YL, DG/OR	ASD Relay Output	12-14v
52	PK/BK	EVAP Purge Solenoid Control	Solenoid Off: 12v, On: 1v
53	LG/RD	S/C Vent Solenoid	Vacuum Decreasing: 1v
58	BR/DB	Injector 6 Driver	1.0-4.0 ms
59	PK/BK	IAC Close Driver	DC pulse signals: 0.8-11v
60	YL/BK	IAC Close Driver	DC pulse signals: 0.8-11v

Pin Connector Graphic

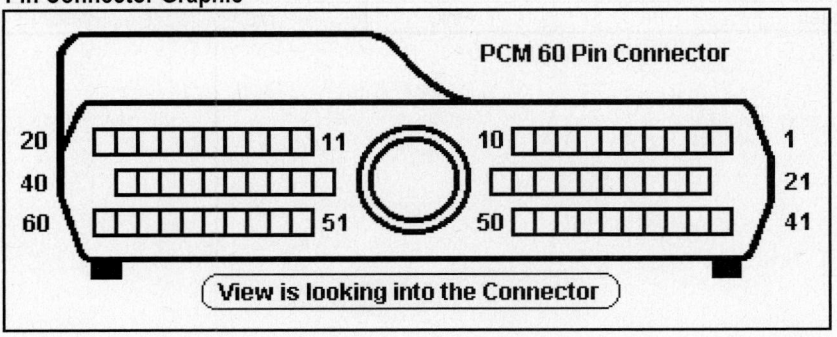

2001-02 Prowler 3.5L V6 24v SOHC MFI VIN G (A/T) 'C1' Connector

PCM Pin #	Wire Color	Circuit Description (40-Pin)	Value at Hot Idle
1	TN/LG	COP 4 Driver	6°, 55 mph: 9° dwell
2	RD/YL	COP 3 Driver	6°, 55 mph: 9° dwell
3	DB/TN	COP 2 Driver	6°, 55 mph: 9° dwell
4	TN/VT	COP 6 Driver	6°, 55 mph: 9° dwell
5	YL/RD	S/C On/Off Switch Power	12-14v
6	DG/OR	ASD Relay Output	12-14v
7	YL/WT	Injector 3 Driver	1-4 ms
8	DG	Generator Field Driver	Digital Signal: 0-12-0v
9	---	Not Used	---
10	BK/TN	Power Ground	<0.1v
11	GY/OR	COP 1 Driver	6°, 55 mph: 9° dwell
12	---	Not Used	---
13	WT/DB	Injector 1 Driver	1-4 ms
14	BR/DB	Injector 6 Driver	1-4 ms
15	GY	Injector 5 Driver	1-4 ms
16	LB/BR	Injector 4 Driver	1-4 ms
17	TN	Injector 2 Driver	1-4 ms
18	DB/BK	Short Runner Valve Solenoid	Valve On: 1v, Off: 12v
19	---	Not Used	---
20	DB/YL	Ignition Switch Output	12-14v
21	TN/PK	COP 5 Driver	6°, 55 mph: 9° dwell
22-23	---	Not Used	---
24	DB/LG	Knock Sensor Signal	0.080v AC
25	GY/LG	Knock Sensor Return	<0.050v
26	TN/BK	ECT Sensor Signal	At 180°F: 2.80v
27	BK/YL	HO2S Ground	<0.050v
28	BK/LG	Transmission Fan Relay	Relay Off: 12v, On: 1v
29	LG/RD	HO2S-21 (B2 S1) Signal	0.1-1.1v
30	BK/DG	HO2S-11 (B1 S1) Signal	0.1-1.1v
31	TN/DG	Starter Relay Control	KOEC: 9-11v
32	GY/BK	CKP Sensor Signal	Digital Signal: 0-5-0v
33	TN/YL	CMP Sensor Signal	Digital Signal: 0-5-0v
34	LG/PK	EGR Sensor Signal	0.6-0.8v
35	OR/DB	TP Sensor Signal	0.6-1.0v
36	DG/RD	MAP Sensor Signal	1.5-1.7v
37	BK/RD	IAT Sensor Signal	At 100°F: 1.83v
38	---	Not Used	---
39	VT/RD	Manifold Tuning Valve Control	Solenoid Off: 12v, On: 1v
40	GY/YL	EGR Solenoid Control	12v, 55 mph: 1v

Standard Colors and Abbreviations

Abbreviation	Color	Abbreviation	Color	Abbreviation	Color
BK	Black	GY	Gray	RD	Red
BL	Blue	GN	Green	TN	Tan
BR	Brown	LG	Light Green	VT	Violet
DB	Dark Blue	OR	Orange	WT	White
DG	Dark Green	PK	Pink	YL	Yellow

2001-02 Prowler 3.5L V6 24v SOHC MFI VIN G (A/T) 'C2' Connector

PCM Pin #	Wire Color	Circuit Description (40-Pin)	Value at Hot Idle
41	RD/LG	S/C Set Switch Signal	S/C & Set Switch On: 3.8v
42	DB	A/C Pressure Switch Signal	A/C On: 0.45-4.85v
43, 47	BK/LB	Sensor & Power Ground	<0.1v
44	OR	8-Volt Supply	7.9-8.1v
45, 54	---	Not Used	---
46	RD/WT	Battery Power (Fused B+)	12-14v
48	BR/WT	IAC 3 Driver	DC pulse signals: 0.8-11v
49	YL/BK	IAC 2 Driver	DC pulse signals: 0.8-11v
50	BK/TN	Power Ground	<0.1v
51	TN/WT	HO2S-12 (B1 S2) Signal	0.1-1.1v
52	PK/YL	Battery Temperature Sensor	At 86°F: 1.96v
53	BK/OR	HO2S-22 (B2 S2) Signal	0.1-1.1v
55	DB/RD	Low Speed Fan Relay	Relay Off: 12v, On: 1v
56	TN/RD	S/C Vacuum Solenoid	Vacuum Increasing: 1v
57	GY/RD	IAC 1 Driver	DC pulse signals: 0.8-11v
58	VT/BK	IAC 4 Driver	DC pulse signals: 0.8-11v
59, 60	VT, WT	CCD Bus (+), CCD Bus (-)	Digital Signal: 0-5-0v
61	VT/WT	5-Volt Supply	4.9-5.1v
62	WT/PK	Brake Switch Signal	Brake Off: 0v, On: 12v
63	YL/DG	Torque Management Request	Digital Signals
64	DB/OR	A/C Clutch Relay Control	Relay Off: 12v, On: 1v
65, 75	PK, LG	SCI Transmit, SCI Receive	0v
66	WT/OR	Vehicle Speed Signal	Digital Signal
67	DB/WT	ASD Relay Control	Relay Off: 12v, On: 1v
68	PK/BK	EVAP Purge Solenoid Control	PWM Signal: 0-12-0v
69	DB/LG	High Speed Fan Control	Relay Off: 12v, On: 1v
70	PK/LB	EVAP Purge Solenoid Sense	0-1v
71	WT/RD	EATX RPM Signal	Digital Signals
72	LB	LDP Switch Signal	Closed: 0v, Open: 12v
73	GY/LB	Tachometer Signals	Pulse Signals
74	BR	Fuel Pump Relay Control	Relay Off: 12v, On: 1v
76	BK/VT	TRS T41 Sense Signal	In P/N: 0v, Others: 5v
77	WT/DG	LDP Solenoid Control	PWM Signal: 0-12-0v
78-79	---	Not Used	---
80	LG/WT	S/C Vent Solenoid	Vacuum Decreasing: 1v

Pin Connector Graphic

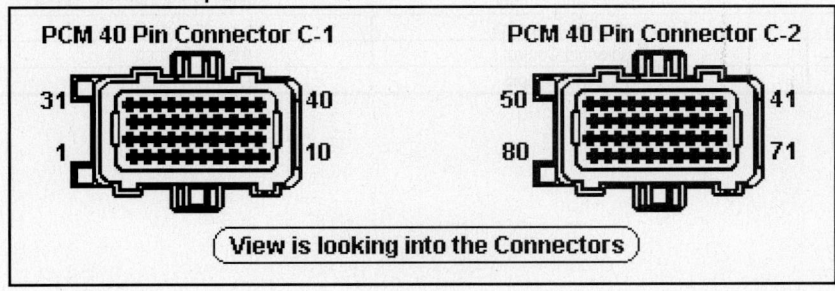

1995 Sebring 2.0L I4 DOHC MFI VIN Y (All) 60 Pin Connector

PCM Pin #	Wire Color	Circuit Description (60 Pin)	Value at Hot Idle
1	BK/GY	Coil 2 Driver	5°, at 55 mph: 8° dwell
2	BK	Power Ground	<0.1v
3	YL/DG	Injector 3 Driver	1.0-4.0 ms
4	LG/BK	Injector 1 Driver	1.0-4.0 ms
5	YL/WT	Vehicle Speed Signal	Digital Signal
6	BR/DB	IAT Sensor Signal	At 100°F: 1.83v
7	WT/DG	HO2S-12 (B1 S2) Signal	0.1-1.1v
8	WT/BK	HO2S-11 (B1 S1) Signal	0.1-1.1v
9	DB/BK	SCI Receive	0v
10	BR/RD	TP Sensor Signal	0.6-1.0v
11	RD/BK	Battery Power (Fused B+)	12-14v
12	---	Not Used	---
13	YL/DB	Fuel Level Sensor Signal	Digital Signal
14	OR	IAC 2 Driver	DC pulse signals: 0.8-11v
15	GY	IAC 3 Driver	DC pulse signals: 0.8-11v
16	LG/BK	EVAP Purge Solenoid Control	PWM Signal: 0-12-0v
17	LG	Speed Control Lamp Driver	S/C On: 1v, S/C Off: 12v
18	RD/WT	ASD Relay Control	Relay Off: 12v, On: 1v
19	DG/BK	Low Speed Fan Relay	Relay Off: 12v, On: 1v
20	RD/YL	Secondary Air Solenoid	Valve Off: 12v, On: 1v
21	BK/DB	Coil 1 Driver	5°, at 55 mph: 8° dwell
22	BK	Power Ground	<0.1v
23	YL/RD	Injector 2 Driver	1.0-4.0 ms
24	LG/RD	Injector 4 Driver	1.0-4.0 ms
25	DB/WT	CKP Sensor Signal	Digital Signal: 0-5-0v
26	DB/RD	CMP Sensor Signal	Digital Signal: 0-5-0v
27	WT/YL	Knock Sensor Signal	0.080v AC
28	DG/WT	ECT Sensor Signal	At 180°F: 2.80v
29	YL/BK	MAP Sensor Signal	1.5-1.7v
30	PK	SCI Transmit	0v
31	RD	S/C Set Switch Signal	S/C & Set Switch On: 3.8v
32	BR/WT	Brake Switch Signal	Brake Off: 0v, On: 12v
33	DG/RD	A/C Clutch Switch Sense	Relay Off: 12v, On: 1v
34	YL/DB	IAC 4 Driver	DC pulse signals: 0.8-11v
35	GY/DB	IAC 2 Driver	DC pulse signals: 0.8-11v
36	DG/RD	Amber MIL Control	MIL On: 1v, MIL Off: 12v
37	DB	Generator Lamp Control	Lamp On: 1v, Off: 12v
38	WT/RD	Fuel Pump Relay Control	Relay Off: 12v, On: 1v
39	RD/DB	EGR Solenoid Control	12v, at 55 mph: 1v
40	LG/WT	S/C Vacuum Solenoid	Vacuum Increasing: 1v

1995 Sebring 2.0L I4 DOHC MFI VIN Y (All) 60 Pin Connector

PCM Pin #	Wire Color	Circuit Description (60 Pin)	Value at Hot Idle
41	DB	Alternator Field Control	Digital Signal: 0-12-0v
42	BK/RD	ASD Relay Output	12-14v
43	DG/YL	5-Volt Supply	4.9-5.1v
44	YL	8-Volt Supply	7.9-8.1v
45	WT/DB	CCD Bus (-)	<0.050v
46	BK/DB	CCD Bus (+)	Digital Signal: 0-5-0v
47	---	Not Used	---
48	WT	Tachometer Signal	Pulse Signals
49	---	Not Used	---
50	BK/YL	PNP Switch Signal	In P/N: 0v, Others: 5v
51	BK/DG	Sensor Ground	<0.050v
52	BK	Power Ground	<0.1v
53	---	Not Used	---
54	BK/WT	Ignition Switch Output	12-14v
55	RD/BK	Speed Control Mode Signal	S/C On: 1v, Off: 12v
56	DB/YL	PSP Switch Signal	Straight: 0v, Turning: 5v
57	DG/OR	High Speed Fan Relay	Relay Off: 12v, On: 1v
58	DB/YL	Bulb (lamp) Check Control	B/C On: 1v, B/C Off: 12v
59	DG	A/C Clutch Relay Control	Relay Off: 12v, On: 1v
60	BK/YL	S/C Vent Solenoid	Vacuum Decreasing: 1v

Pin Connector Graphic

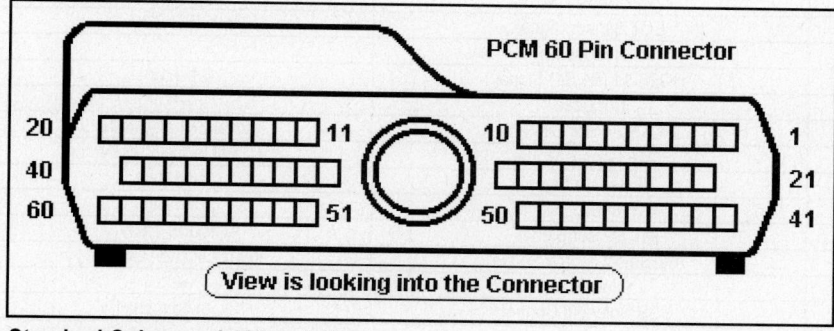

PCM 60 Pin Connector

20 11 10 1
40
60 51 50 41

View is looking into the Connector

Standard Colors and Abbreviations

Abbreviation	Color	Abbreviation	Color	Abbreviation	Color
BK	Black	GY	Gray	RD	Red
BL	Blue	GN	Green	TN	Tan
BR	Brown	LG	Light Green	VT	Violet
DB	Dark Blue	OR	Orange	WT	White
DG	Dark Green	PK	Pink	YL	Yellow

1996-97 Sebring 2.0L I4 DOHC MFI VIN Y (All) 'C1' Connector

PCM Pin #	Wire Color	Circuit Description (40-Pin)	Value at Hot Idle
1, 5	---	Not Used	---
2	BK/DB	Coil 1 Driver	5°, at 55 mph: 8° dwell
3	BR	Coil 2 Driver	5°, at 55 mph: 8° dwell
4	DB	Alternator Field Control	Digital Signal: 0-12-0v
6	BK/RD	ASD Relay Output	12-14v
7	YL/DG	Injector 3 Driver	1.0-4.0 ms
8	DG/RD	Amber MIL Control	MIL On: 1v, MIL Off: 12v
9	DB/YL	Bulb (lamp) Check Control	B/C On: 1v, B/C Off: 12v
10	BK	Power Ground	<0.1v
11	---	Not Used	---
12	RD/BK	S/C On Switch Signal	S/C On: 1v, Off: 12v
13	LG/BK	Injector 1 Driver	1.0-4.0 ms
14-15	---	Not Used	---
16	LG/RD	Injector 4 Driver	1.0-4.0 ms
17	YL/RD	Injector 2 Driver	1.0-4.0 ms
18	---	Not Used	---
19	DG/OR	High Speed Fan Relay	Relay Off: 12v, On: 1v
20	BK/WT	Ignition Switch Output	12-14v
21, 25	---	---	---
22	LG	Speed Control Lamp Driver	S/C On: 1v, S/C Off: 12v
23	YL/DB	Fuel Level Sensor Signal	70 ohms (±20) w/full tank
24	WT/YL	Knock Sensor Signal	0.080v AC
26	DG/WT	ECT Sensor Signal	At 180°F: 2.80v
27-29	---	Not Used	---
30	WT/BK	HO2S-11 (B1 S1) Signal	0.1-1.1v
31, 34	---	Not Used	---
32	DB/WT	CKP Sensor Signal	Digital Signal: 0-5-0v
33	DB/RD	CMP Sensor Signal	Digital Signal: 0-5-0v
35	BR/RD	TP Sensor Signal	0.6-1.0v
36	YL/BK	MAP Sensor Signal	1.5-1.7v
37	BR/DB	IAT Sensor Signal	At 100°F: 1.83v
38	DG/RD	A/C Select Switch Sense	Relay Off: 12v, On: 1v
39	---	Not Used	---
40	RD/DB	EGR Solenoid Control	12v, at 55 mph: 1v

Pin Connector Graphic

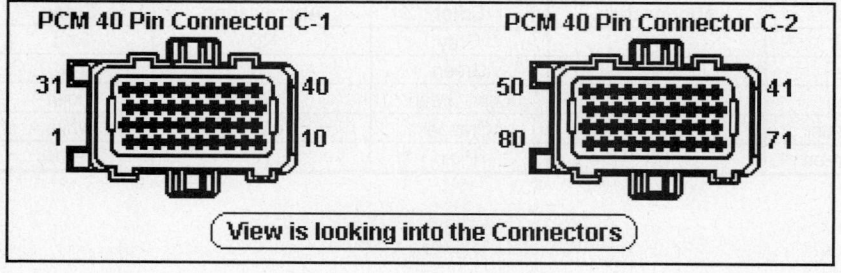

PCM 40 Pin Connector C-1

31   40

1   10

PCM 40 Pin Connector C-2

50   41

80   71

View is looking into the Connectors

1996-97 Sebring 2.0L I4 DOHC MFI VIN Y 'C2' Connector

PCM Pin #	Wire Color	Circuit Description (40-Pin)	Value at Hot Idle
41	RD	S/C Set Switch Signal	S/C & Set Switch On: 3.8v
42	---	Not Used	---
43	BK/DG	Sensor Ground	<0.050v
44	YL	8-Volt Supply	7.9-8.1v
45	DB/YL	PSP Switch Signal	Straight: 0v, Turning: 5v
46	RD/BK	Battery Power (Fused B+)	12-14v
47	BK/DG	Sensor Ground	<0.050v
48	GY/DB	IAC 3 Driver	DC pulse signals: 0.8-11v
49	OR	IAC 2 Driver	DC pulse signals: 0.8-11v
50	BK	Power Ground	<0.1v
51	WT/DG	HO2S-12 (B1 S2) Signal	0.1-1.1v
52-54	---	Not Used	---
55	DG/BK	Low Speed Fan Relay	Relay Off: 12v, On: 1v
56	DB	Generator Lamp Control	Lamp On: 1v, Off: 12v
57	GY	IAC 1 Driver	DC pulse signals: 0.8-11v
58	YL/DB	IAC 4 Driver	DC pulse signals: 0.8-11v
59	BK/DB	CCD Bus (+)	Digital Signal: 0-5-0v
60	WT/DB	CCD Bus (-)	<0.050v
61	DG/YL	5-Volt Supply	4.9-5.1v
62	BR/WT	Brake Switch Signal	Brake Off: 0v, On: 12v
63	OR/BK	Torque Management Request	Digital Signals
64	DG	A/C Clutch Relay Control	Relay Off: 12v, On: 1v
65	PK	SCI Transmit	0v
66	YL/WT	Vehicle Speed Signal	Digital Signal
67	RD/WT	ASD Relay Control	Relay Off: 12v, On: 1v
68	LG/BK	EVAP Purge Solenoid Control	PWM Signal: 0-12-0v
69	DG/OR	High Speed Fan Relay	Relay Off: 12v, On: 1v
70-71	---	Not Used	---
72	RD/DB	LDP Switch Sense	Open: 12v, Closed: 0v
73	WT	Tachometer Signal	Pulse Signals
74	BK/DB	Fuel Pump Relay Control	Relay Off: 12v, On: 1v
75	DB/BK	SCI Receive	0v
76	BK/YL	PNP Switch Signal	In P/N: 0v, Others: 5v
77	RD/YL	LDP Solenoid Control	PWM Signal: 0-12-0v
78	LG/WT	S/C Vacuum Solenoid	Vacuum Increasing: 1v
79	RD/YL	M/T: Aspirator Solenoid	Valve Off: 12v, On: 1v
80	BK/YL	S/C Vent Solenoid	Vacuum Decreasing: 1v

Standard Colors and Abbreviations

Abbreviation	Color	Abbreviation	Color	Abbreviation	Color
BK	Black	GY	Gray	RD	Red
BL	Blue	GN	Green	TN	Tan
BR	Brown	LG	Light Green	VT	Violet
DB	Dark Blue	OR	Orange	WT	White
DG	Dark Green	PK	Pink	YL	Yellow

1998-99 Sebring 2.0L I4 DOHC MFI VIN Y 'C1' Connector

PCM Pin #	Wire Color	Circuit Description (40-Pin)	Value at Hot Idle
1-2	---	Not Used	---
3	BR	Coil 2 Driver	5°, at 55 mph: 8° dwell
4-5	---	Not Used	---
6	BK/RD	ASD Relay Output	12-14v
7	YL/DG	Injector 3 Driver	1.0-4.0 ms
8	DB	Alternator Field Control	Digital Signal: 0-12-0v
9	DB/YL	Bulb (lamp) Check Control	B/C On: 1v, B/C Off: 12v
10	BK	Power Ground	<0.1v
11	BK/DB	Coil 1 Driver	5°, at 55 mph: 8° dwell
12	RD/BK	S/C On Switch Signal	S/C On: 1v, Off: 12v
13	LG/BK	Injector 1 Driver	1.0-4.0 ms
14-15	---	Not Used	---
16	LG/RD	Injector 4 Driver	1.0-4.0 ms
17	YL/RD	Injector 2 Driver	1.0-4.0 ms
18	LG	Speed Control Lamp Driver	S/C On: 1v, S/C Off: 12v
19	DG/OR	High Speed Fan Relay	Relay Off: 12v, On: 1v
20	BK/WT	Ignition Switch Output	12-14v
21	---	Not Used	---
22	DG/RD	Amber MIL Control	MIL On: 1v, MIL Off: 12v
23	YL/DB	Fuel Level Sensor Signal	70 ohms (±20) w/full tank
24	---	Not Used	---
25	WT/YL	Knock Sensor Signal	0.080v AC
26	DG/WT	ECT Sensor Signal	At 180°F: 2.80v
27-29	---	Not Used	---
30	WT/BK	HO2S-11 (B1 S1) Signal	0.1-1.1v
31	---	Not Used	---
32	DB/WT	CKP Sensor Signal	Digital Signal: 0-5-0v
33	DB/RD	CMP Sensor Signal	Digital Signal: 0-5-0v
34	---	Not Used	---
35	BR/RD	TP Sensor Signal	0.6-1.0v
36	YL/BK	MAP Sensor Signal	1.5-1.7v
37	BR/DB	IAT Sensor Signal	At 100°F: 1.83v
38	DG/RD	A/C Select Switch Sense	Relay Off: 12v, On: 1v
39	DB	Generator Lamp Control	Lamp On: 1v, Off: 12v
40	RD/DB	EGR Solenoid Control	12v, at 55 mph: 1v

Standard Colors and Abbreviations

Abbreviation	Color	Abbreviation	Color	Abbreviation	Color
BK	Black	GY	Gray	RD	Red
BL	Blue	GN	Green	TN	Tan
BR	Brown	LG	Light Green	VT	Violet
DB	Dark Blue	OR	Orange	WT	White
DG	Dark Green	PK	Pink	YL	Yellow

1998-99 Sebring 2.0L I4 DOHC MFI VIN Y (All) 'C2' Connector

PCM Pin #	Wire Color	Circuit Description (40-Pin)	Value at Hot Idle
41	RD	S/C Set Switch Signal	S/C & Set Switch On: 3.8v
42-43	---	Not Used	---
44	YL	8-Volt Supply	7.9-8.1v
45	DB/YL	PSP Switch Signal	Straight: 0v, Turning: 5v
46	RD/BK	Battery Power (Fused B+)	12-14v
47	BK	Sensor Ground	<0.050v
48	GY/DB	IAC 3 Driver	DC pulse signals: 0.8-11v
49	OR	IAC 2 Driver	DC pulse signals: 0.8-11v
50	BK	Power Ground	<0.1v
51	WT/DG	HO2S-12 (B1 S2) Signal	0.1-1.1v
52-54	---	Not Used	---
55	DG/BK	Low Speed Fan Relay	Relay Off: 12v, On: 1v
56	LG/WT	S/C Vacuum Solenoid	Vacuum Increasing: 1v
57	GY	IAC 1 Driver	DC pulse signals: 0.8-11v
58	YL/DB	IAC 4 Driver	DC pulse signals: 0.8-11v
59	BK/DB	CCD Bus (+)	Digital Signal: 0-5-0v
60	WT/DB	CCD Bus (-)	<0.050v
61	DG/YL	5-Volt Supply	4.9-5.1v
62	BR/WT	Brake Switch Signal	Brake Off: 0v, On: 12v
63	OR/BK	Torque Management Request	Digital Signals
64	DG	A/C Clutch Relay Control	Relay Off: 12v, On: 1v
65	PK	SCI Transmit	0v
66	YL/WT	Vehicle Speed Signal	Digital Signal
67	RD/WT	ASD Relay Control	Relay Off: 12v, On: 1v
68	LG/BK	EVAP Purge Solenoid Control	PWM Signal: 0-12-0v
69	DG/OR	High Speed Fan Relay	Relay Off: 12v, On: 1v
70	BK/WT	EVAP Purge Solenoid Sense	0-1v
72	RD/DB	LDP Switch Sense	Open: 12v, Closed: 0v
73	WT	Tachometer Signal	Pulse Signals
74	BK/DB	Fuel Pump Relay Control	Relay Off: 12v, On: 1v
75	DB/BK	SCI Receive	0v
76	BK/YL	PNP Switch Signal	In P/N: 0v, Others: 5v
77	RD/YL	LDP Solenoid Control	PWM Signal: 0-12-0v
78-79	---	Not Used	---
80	BK/YL	S/C Vent Solenoid	Vacuum Decreasing: 1v

Pin Connector Graphic

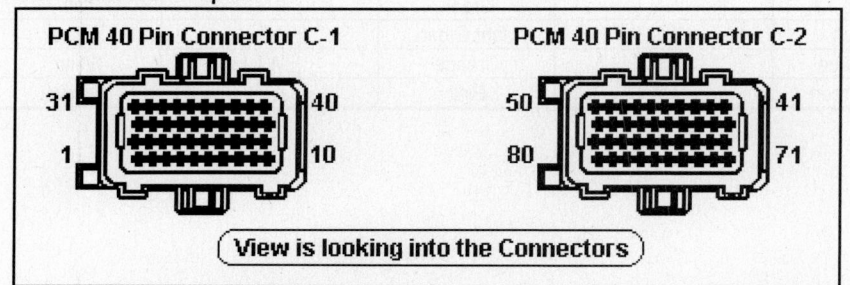

PCM 40 Pin Connector C-1

PCM 40 Pin Connector C-2

31   40   50   41

1   10   80   71

View is looking into the Connectors

1996-97 Sebring 2.4L I4 DOHC MFI VIN X (A/T) 'C1' Connector

PCM Pin #	Wire Color	Circuit Description (40-Pin)	Value at Hot Idle
1	---	Not Used	---
2	BK/GY	Coil 1 Driver	5°, at 55 mph: 8° dwell
3	DB/DG	Coil 2 Driver	5°, at 55 mph: 8° dwell
4	DB	Alternator Field Control	Digital Signal: 0-12-0v
4 ('97)	DG	Alternator Field Control	Digital Signal: 0-12-0v
5	YL/PK	S/C Power Supply	12-14v
6	DG/OR	ASD Relay Output	12-14v
7	YL/WT	Injector 3 Driver	1.0-4.0 ms
8-9	---	Not Used	---
10	BK/TN	Power Ground	<0.1v
11-12	---	Not Used	---
13	WT/LB	Injector 1 Driver	1.0-4.0 ms
14-15	---	Not Used	---
16	LB/BR	Injector 4 Driver	1.0-4.0 ms
17	TN	Injector 2 Driver	1.0-4.0 ms
18-19	---	Not Used	---
20	DB/WT	Ignition Switch Output	12-14v
21-23	---	Not Used	---
24	GY/BK	Knock Sensor Signal	0.080v AC
25	---	Not Used	---
26	TN/BK	ECT Sensor Signal	At 180°F: 2.80v
27-29	---	Not Used	---
30	BK/DG	HO2S-11 (B1 S1) Signal	0.1-1.1v
31	---	Not Used	---
32	GY/BK	CKP Sensor Signal	Digital Signal: 0-5-0v
33	TN/YL	CMP Sensor Signal	Digital Signal: 0-5-0v
34	---	Not Used	---
35	OR/LB	TP Sensor Signal	0.6-1.0v
36	DG/RD	MAP Sensor Signal	1.5-1.7v
37	BK/RD	IAT Sensor Signal	At 100°F: 1.83v
38-39	---	Not Used	---
40	GY/YL	EGR Solenoid Control	12v, at 55 mph: 1v

Standard Colors and Abbreviations

Abbreviation	Color	Abbreviation	Color	Abbreviation	Color
BK	Black	GY	Gray	RD	Red
BL	Blue	GN	Green	TN	Tan
BR	Brown	LG	Light Green	VT	Violet
DB	Dark Blue	OR	Orange	WT	White
DG	Dark Green	PK	Pink	YL	Yellow

1996-97 Sebring 2.4L I4 DOHC MFI VIN X (A/T) 'C2' Connector

PCM Pin #	Wire Color	Circuit Description (40-Pin)	Value at Hot Idle
41	PK/LG	S/C Set Switch Signal	S/C & Set Switch On: 3.8v
42	DB/YL	A/C Pressure Switch Signal	A/C On: 0.45-4.85v
43	BK/LB	Sensor Ground	<0.050v
44	OR/WT	8-Volt Supply	7.9-8.1v
45	DB/LG	PSP Switch Signal	Straight: 0v, Turning: 5v
46	RD/TN	Battery Power (Fused B+)	12-14v
47, 56	---	Not Used	---
48	BR/GY	IAC 1 Driver	DC pulse signals: 0.8-11v
49	YL/BK	IAC 2 Driver	DC pulse signals: 0.8-11v
50	BK/TN	Power Ground	<0.1v
51	TN/WT	HO2S-12 (B1 S2) Signal	0.1-1.1v
52	PK/LG	Battery Temperature Sensor	At 86°F: 1.96v
55	DB/TN	Low Speed Fan Relay	Relay Off: 12v, On: 1v
57	GY/RD	IAC 3 Driver	DC pulse signals: 0.8-11v
58	PK/GY	IAC 4 Driver	DC pulse signals: 0.8-11v
59	PK/BR	CCD Bus (+)	Digital Signal: 0-5-0v
60	WT/BK	CCD Bus (-)	<0.050v
61	PK/WT	5-Volt Supply	4.9-5.1v
62	WT/RD	Brake Switch Signal	Brake Off: 0v, On: 12v
63	YL/DG	Torque Management Request	Digital Signals
64	DB/OR	A/C Clutch Relay Control	Relay Off: 12v, On: 1v
65	PK/LB	SCI Transmit	0v
66	WT/OR	Vehicle Speed Signal	Digital Signal
67	DB/PK	ASD Relay Control	Relay Off: 12v, On: 1v
68	PK/GY	EVAP Purge Solenoid Control	PWM Signal: 0-12-0v
69	DB/PK	High Speed Fan Relay	Relay Off: 12v, On: 1v
70-71, 79	---	Not Used	---
72	OR/DG	LDP Switch Sense	Open: 12v, Closed: 0v
74	BR/LG	Fuel Pump Relay Control	Relay Off: 12v, On: 1v
75	LG/WT	SCI Receive	0v
76	BK/WT	PNP Switch Signal	In P/N: 0v, Others: 5v
77	WT/DG	LDP Solenoid Control	PWM Signal: 0-12-0v
78	WT/PK	S/C Vacuum Solenoid	Vacuum Increasing: 1v
80	LG/RD	S/C Vent Solenoid	Vacuum Decreasing: 1v

Pin Connector Graphic

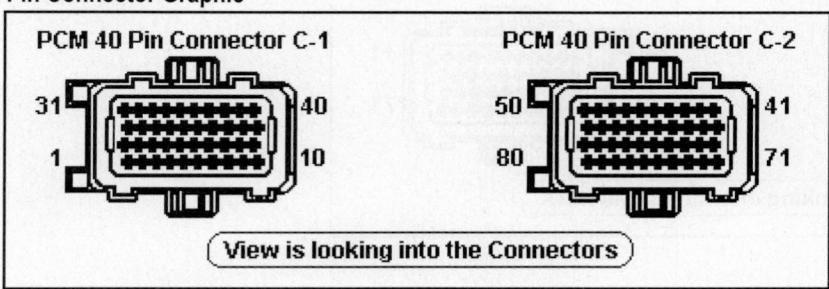

1998-99 Sebring 2.4L I4 DOHC MFI VIN X (A/T) 'C1' Connector

PCM Pin #	Wire Color	Circuit Description (40-Pin)	Value at Hot Idle
1-2	---	Not Used	---
3	DB/TN	Coil 2 Driver	5°, at 55 mph: 8° dwell
4	---	Not Used	---
5	YL/RD	S/C Power Supply	12-14v
6	DG/OR	ASD Relay Output	12-14v
7	YL/WT	Injector 3 Driver	1.0-4.0 ms
8	DG	Alternator Field Control	Digital Signal: 0-12-0v
9	---	Not Used	---
10	BK/TN	Power Ground	<0.1v
11	BK/GY	Coil 1 Driver	5°, at 55 mph: 8° dwell
12	---	Not Used	---
13	WT/DB	Injector 1 Driver	1.0-4.0 ms
14-15	---	Not Used	---
16	LB/BR	Injector 4 Driver	1.0-4.0 ms
17	TN	Injector 2 Driver	1.0-4.0 ms
18-19	---	Not Used	---
20	DB/WT	Ignition Switch Output	12-14v
21-24	---	Not Used	---
25	DB/LG	Knock Sensor Signal	0.080v AC
26	TN/BK	ECT Sensor Signal	At 180°F: 2.80v
27	BK/OR	Oxygen Sensor Ground	<0.050v
28-29	---	Not Used	---
30	BK/DG	HO2S-11 (B1 S1) Signal	0.1-1.1v
31	TN	Starter Relay Control	KOEC: 9-11v
32	GY/BK	CKP Sensor Signal	Digital Signal: 0-5-0v
33	TN/YL	CMP Sensor Signal	Digital Signal: 0-5-0v
34	---	Not Used	---
35	OR/DB	TP Sensor Signal	0.6-1.0v
36	DG/RD	MAP Sensor Signal	1.5-1.7v
37	BK/RD	IAT Sensor Signal	At 100°F: 1.83v
38-39	---	Not Used	---
40	GY/YL	EGR Solenoid Control	12v, at 55 mph: 1v

Pin Connector Graphic

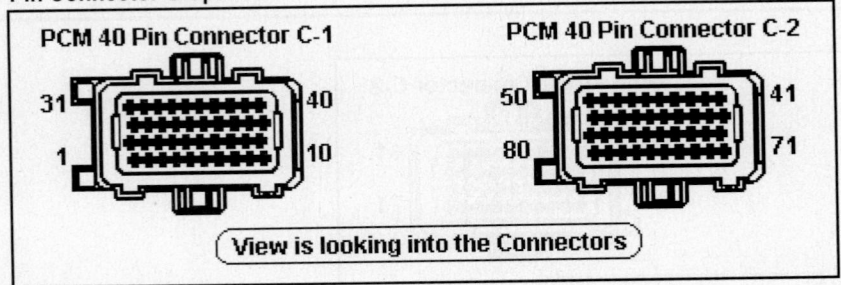

PCM 40 Pin Connector C-1   PCM 40 Pin Connector C-2

31   40   50   41
1   10   80   71

View is looking into the Connectors

1998-99 Sebring 2.4L I4 DOHC MFI VIN X (A/T) 'C2' Connector

PCM Pin #	Wire Color	Circuit Description (40-Pin)	Value at Hot Idle
41	RD/LG	S/C Set Switch Signal	S/C & Set Switch On: 3.8v
42	DB/YL	A/C Pressure Switch Signal	A/C On: 0.45-4.85v
43	BK/LB	Sensor Ground	<0.050v
44	OR/WT	8-Volt Supply	7.9-8.1v
45	DB/LG	PSP Switch Signal	Straight: 0v, Turning: 5v
46	RD/TN	Battery Power (Fused B+)	12-14v
47	BK/WT	Sensor Ground	<0.050v
48	BR/WT	IAC 3 Driver	DC pulse signals: 0.8-11v
49	YL/BK	IAC 2 Driver	DC pulse signals: 0.8-11v
50	BK/TN	Power Ground	<0.1v
51	TN/WT	HO2S-12 (B1 S2) Signal	0.1-1.1v
52	PK/YL	Battery Temperature Sensor	At 86°F: 1.96v
53-54	---	Not Used	---
55	DB/TN	Low Speed Fan Relay	Relay Off: 12v, On: 1v
56	WT/PK	S/C Vacuum Solenoid	Vacuum Increasing: 1v
57	GY/RD	IAC 1 Driver	DC pulse signals: 0.8-11v
58	PK/BK	IAC 4 Driver	DC pulse signals: 0.8-11v
59	PK/BR	CCD Bus (+)	Digital Signal: 0-5-0v
60	WT/BK	CCD Bus (-)	<0.050v
61	PK/WT	5-Volt Supply	4.9-5.1v
62	WT/RD	Brake Switch Signal	Brake Off: 0v, On: 12v
63	YL/DG	Torque Management Request	Digital Signals
64	DB/OR	A/C Clutch Relay Control	Relay Off: 12v, On: 1v
65	PK/LB	SCI Transmit	0v
66	WT/OR	Vehicle Speed Signal	Digital Signal
67	DB/PK	ASD Relay Control	Relay Off: 12v, On: 1v
68	PK/GY	EVAP Purge Solenoid Control	PWM Signal: 0-12-0v
69	DB/PK	High Speed Fan Relay	Relay Off: 12v, On: 1v
70	WT/TN	EVAP Purge Solenoid Sense	0-1v
71-73	---	Not Used	---
74	BR/LG	Fuel Pump Relay Control	Relay Off: 12v, On: 1v
75	LG	SCI Receive	0v
76	BK/LB	PNP Switch Signal	In P/N: 0v, Others: 5v
77-79	---	Not Used	---
80	LG/RD	S/C Vent Solenoid	Vacuum Decreasing: 1v

Standard Colors and Abbreviations

Abbreviation	Color	Abbreviation	Color	Abbreviation	Color
BK	Black	GY	Gray	RD	Red
BL	Blue	GN	Green	TN	Tan
BR	Brown	LG	Light Green	VT	Violet
DB	Dark Blue	OR	Orange	WT	White
DG	Dark Green	PK	Pink	YL	Yellow

2001 Sebring Sedan 2.4L I4 DOHC MFI VIN X (A/T) 'C1' Connector

PCM Pin #	Wire Color	Circuit Description (40-Pin)	Value at Hot Idle
1-2	---	Not Used	---
3	DB/TN	Coil 2 Driver	5°, at 55 mph: 8° dwell
4	---	Not Used	---
5	YL/RD	S/C Power Supply	12-14v
6	DG/OR	ASD Relay Output	12-14v
7	YL/WT	Injector 3 Driver	1.0-4.0 ms
8	DG	Alternator Field Control	Digital Signal: 0-12-0v
9	---	Not Used	---
10	BK/TN	Power Ground	<0.1v
11	BK/GY	Coil 1 Driver	5°, at 55 mph: 8° dwell
12	---	Not Used	---
13	WT/DB	Injector 1 Driver	1.0-4.0 ms
14-15	---	Not Used	---
16	LB/BR	Injector 4 Driver	1.0-4.0 ms
17	TN	Injector 2 Driver	1.0-4.0 ms
18	OR/RD	HO2S-11Heater Control	Heater On: 1v, Off: 12v
19	---	Not Used	---
20	DB/WT	Ignition Switch Output	12-14v
21-24	---	Not Used	---
25	DB/LG	Knock Sensor Signal	0.080v AC
26	TN/BK	ECT Sensor Signal	At 180°F: 2.80v
27	DB/OR	Oxygen Sensor Ground	<0.050v
28-29	---	Not Used	---
30	BK/DG	HO2S-11 (B1 S1) Signal	0.1-1.1v
31	TN	Starter Relay Control	KOEC: 9-11v
32	GY/BK	CKP Sensor Signal	Digital Signal: 0-5-0v
33	TN/YL	CMP Sensor Signal	Digital Signal: 0-5-0v
34	---	Not Used	---
35	OR/DB	TP Sensor Signal	0.6-1.0v
36	DG/RD	MAP Sensor Signal	1.5-1.7v
37	BK/RD	IAT Sensor Signal	At 100°F: 1.83v
38-39	---	Not Used	---
40	GY/YL	EGR Solenoid Control	12v, at 55 mph: 1v

Pin Connector Graphic

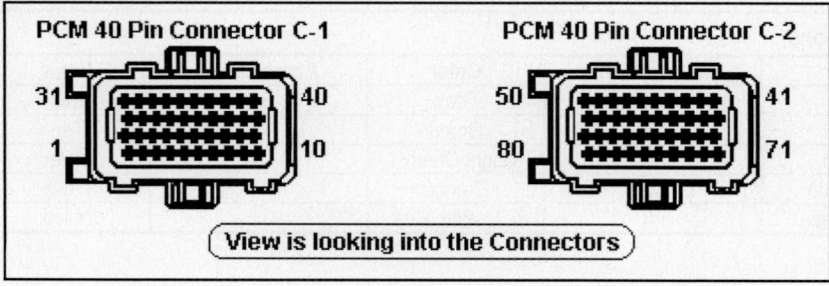

PCM 40 Pin Connector C-1

PCM 40 Pin Connector C-2

View is looking into the Connectors

2001 Sebring Sedan 2.4L I4 DOHC MFI VIN X (A/T) 'C2' Connector

PCM Pin #	Wire Color	Circuit Description (40-Pin)	Value at Hot Idle
41	PK/LG	S/C Set Switch Signal	S/C & Set Switch On: 3.8v
42	DB	A/C Pressure Switch Signal	A/C On: 0.45-4.85v
43	BK/LB	Sensor Ground	<0.050v
44	OR/WT	8-Volt Supply	7.9-8.1v
45	DB/OR	PSP Switch Signal	Straight: 0v, Turning: 5v
46	RD/TN	Battery Power (Fused B+)	12-14v
47	BK	Power Ground	<0.1v
48	BR/WT	IAC 3 Driver	DC pulse signals: 0.8-11v
49	YL/BK	IAC 2 Driver	DC pulse signals: 0.8-11v
50	BK/TN	Power Ground	<0.1v
51	TN/WT	HO2S-12 (B1 S2) Signal	0.1-1.1v
52	VT/LG	Ambient Temperature Sensor	At 86°F: 1.96v
53-54	---	Not Used	---
55	DB/TN	Low Speed Fan Relay	Relay Off: 12v, On: 1v
56	TN/RD	S/C Vacuum Solenoid	Vacuum Increasing: 1v
57	GY/RD	IAC 1 Driver	DC pulse signals: 0.8-11v
58	VT/BK	IAC 4 Driver	DC pulse signals: 0.8-11v
59	OR	PCI Data Bus (J1850)	Digital Signals: 0-7-0v
60	---	Not Used	---
61	VT/WT	5-Volt Supply	4.9-5.1v
62	WT/PK	Brake Switch Signal	Brake Off: 0v, On: 12v
63	YL/DG	Torque Management Request	Digital Signals
64	DB/OR	A/C Clutch Relay Control	Relay Off: 12v, On: 1v
65	PK	SCI Transmit	0v
66	WT/OR	Vehicle Speed Signal	Digital Signal
67	DB/VT	ASD Relay Control	Relay Off: 12v, On: 1v
68	PK/BK	EVAP Purge Solenoid Control	PWM Signal: 0-12-0v
69	DB/PK	High Speed Fan Relay	Relay Off: 12v, On: 1v
70	WT/TN	EVAP Purge Solenoid Sense	0-1v
71	WT/RD	EATX RPM Signal	Digital Signals
72	OR	LDP Switch Sense	Open: 12v, Closed: 0v
73	---	Not Used	---
74	BR/LG	Fuel Pump Relay Control	Relay Off: 12v, On: 1v
75	LG	SCI Receive	0v
76	BK/LB	TRS T41 Sense	In P/N: 0v, Others: 5v
77	WT/DG	LDP Solenoid Control	PWM Signal: 0-12-0v
78-79	---	Not Used	---
80	LG/RD	S/C Vent Solenoid	Vacuum Decreasing: 1v

Standard Colors and Abbreviations

Abbreviation	Color	Abbreviation	Color	Abbreviation	Color
BK	Black	GY	Gray	RD	Red
BL	Blue	GN	Green	TN	Tan
BR	Brown	LG	Light Green	VT	Violet
DB	Dark Blue	OR	Orange	WT	White
DG	Dark Green	PK	Pink	YL	Yellow

2001-03 Sebring Coupe 2.4L SOHC VIN G (A/T) 'C110' Connector

PCM Pin #	Wire Color	Circuit Description (35-Pin)	Value at Hot Idle
1	YL/BL	Injector 1 Driver	1.0-4.0 ms
2	GN/YL	Injector 4 Driver	1.0-4.0 ms
3	BR/WT	HO2S-11 Heater Control	Digital Signal: 0-12-0v
4-5	---	Not Used	---
6	BL/RD	EGR Solenoid Control	Solenoid Off: 12v, On: 1v
7	---	Not Used	---
8	BK/RD	Generator Signal (lights on)	0.2-3.5v
9	YL/RD	Injector 2 Driver	1.0-4.0 ms
10	---	Not Used	---
11	BL/YL	Ignition Coil 1 Control	Digital Signal: 0-12-0v
12	WT/GN	Ignition Coil 2 Control	Digital Signal: 0-12-0v
12-13	---	Not Used	---
14	GN/BK	IAC Stepper Motor 'A' Signal	Pulse Signals
15	GN/WT	IAC Stepper Motor 'B' Signal	Pulse Signals
16-17	---	Not Used	---
18	BL/OR	Radiator Fan Relay Control	Fan Off: 0.1v, On: 0.7v
19	GN/BL	Volume Airflow Sensor Reset	1-3v, 3000 rpm: 6-9v
20	GN	A/C Clutch Relay Control	Relay Off: 12v, On: 1v
21	BK	Fuel Pump Relay Control	Pump Off: 12v, On: 1v
22	RD/YL	MIL (lamp) Control	Lamp Off: 12v, On: 1v
23	---	Not Used	---
24	YL/BK	Injector 3 Driver	1.0-4.0 ms
25	---	Not Used	---
26	BL/WT	HO2S-12 Heater Control	Digital Signal: 0-12-0v
27	---	Not Used	---
28	GN/RD	IAC Stepper Motor 'A' Signal	Pulse Signals
29	BK/YL	IAC Stepper Motor 'B' Signal	Pulse Signals
30-33	---	Not Used	---
34	BL	EVAP Purge Solenoid Control	PWM Signal: 0-12-0v
35	YL	EVAP Vent Solenoid Control	Solenoid Off: 12v, On: 1v

PCM C110 Wire Harness Connector Graphic

C110 Connector

```
| 1| 2|   | 3| 4|            | 5| 6|   | 7| 8|
| 9|10|11|12|13|14|15|16|17|18|19|20|21|22|23|
|24|25|   |26|27|28|29|   |30|31|32|33|   |34|35|
```

View in into Front of Wire Harness (Connector Removed)

2001-03 Sebring Coupe 2.4L SOHC VIN G (A/T) 'C114' Connector

PCM Pin #	Wire Color	Circuit Description (26-Pin)	Value at Hot Idle
41	RD	Ignition Power (MPI Relay)	12-14v
42	BK	Power Ground	<0.1v
43	WT/RD	Tachometer Signal	Pulse Signals
44	YL/GN	ECT Sensor Signal	0.3-0.9v at 176°F
45	GN/RD	Distributor CKP Sensor Signal	Digital Signal: 0-5-0v
46	GN/YL	Sensor Voltage Reference	4.9-5.1v
47	RD	Ignition Power (MPI Relay)	12-14v
48	BK	Power Ground	<0.1v
49	WT/VT	MPI (Power) Relay Control	Relay Off: 12v, On: 1v
50	WT/BL	A/T Control Relay	Relay Off: 12v, Off: 1v
51	---	Not Used	---
52	YL	PSP Switch Signal	Straight: 5v, Turning: 0v
53	---	Not Used	---
54	YL/BK	Generator 'FR' Terminal	0.5-4.5v
55	GN/WT	BARO Sensor Signal	3.7-4.3 at Sea Level
56	BL/YL	CMP Sensor Signal	Digital Signal: 0-5-0v
57	BK	Sensor Ground	<0.050v
58	BK/RD	Park Neutral Switch	In P/N: 12v, Others: 0v
59	BK/RD	Starter (Cranking) Signal	9-11v (cranking)
60	---	Not Used	---
61	BL/WT	A/C Switch 2 Signal (Hi Blow)	12v
62-63	---	Not Used	---
64	RD/BL	IAT Sensor Signal	1.5-2.1v at 104°F
65	WT/GN	Volume Airflow Sensor	2.2-3.2v
66	OR/BL	Keep Alive Power	12-14v

PCM C114 Wire Harness Connector Graphic

```
              C114 Connector
    41 42 43          44 45 46
    47 48 49 50 51 52 53 54 55 56 57
    58 59    60 61 62 63    64 65 66
  View in into Front of Wire Harness (Connector Removed)
```

Standard Colors and Abbreviations

Abbreviation	Color	Abbreviation	Color	Abbreviation	Color
BK	Black	GY	Gray	RD	Red
BL	Blue	GN	Green	TN	Tan
BR	Brown	LG	Light Green	VT	Violet
DB	Dark Blue	OR	Orange	WT	White
DG	Dark Green	PK	Pink	YL	Yellow

2001-03 Sebring Coupe 2.4L SOHC VIN G (A/T) 'C117' Connector

PCM Pin #	Wire Color	Circuit Description (28-Pin)	Value at Hot Idle
71	WT	HO2S-11 (B1 S1) Signal	0.1-1.1v
72	---	Not Used	---
73	GN	HO2S-12 (B1 S2) Signal	0.1-1.1v
74	---	Not Used	---
75	GY/BL	Cruise Control Switch Signal	12v or 0v
76	BK	Power Ground	<0.1v
77	RD/BL	A/T Control Relay Output	12-14v
78	YL	TP Sensor Signal	0.53-0.73v
79	YL/RD	Idle Position Switch Signal	0v, Switch Open: 4-5v
80	WT/BL	Vehicle Speed Signal	Digital Signal
81-82	---	Not Used	---
83	GN/RD	A/C Switch On/Off Signal	A/C Off: 0v, On: 12v
84	GY/RD	DLC Diagnosis Control (#1)	0v
85	RD/WT	DLC ISO 9141 Bus (#7)	12v
86-87	---	Not Used	---
88	BK	Power Ground	<0.1v
89	RD/BL	A/T Control Relay Output	12-14v
90	WT	Knock Sensor Signal	0.080v AC
91	BL/RD	MAP Sensor Signal	0.8-1.1v
92	BR/WT	FTP Sensor Signal	2.5v (fuel cap off)
93-96	---	Not Used	---
97	BK	Power Ground	<0.1v
98	BK/WT	Ignition Switch Power	12-14v

PCM C117 Wire Harness Connector Graphic

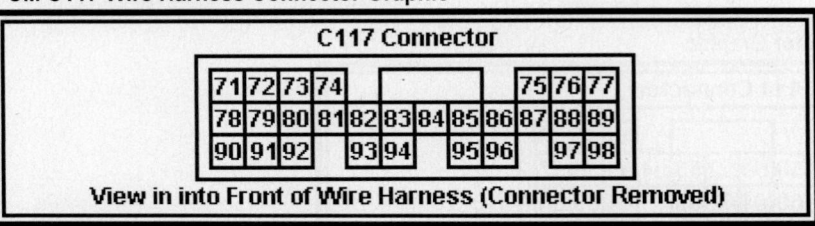

C117 Connector

View in into Front of Wire Harness (Connector Removed)

Standard Colors and Abbreviations

Abbreviation	Color	Abbreviation	Color	Abbreviation	Color
BK	Black	GY	Gray	RD	Red
BL	Blue	GN	Green	TN	Tan
BR	Brown	LG	Light Green	VT	Violet
DB	Dark Blue	OR	Orange	WT	White
DG	Dark Green	PK	Pink	YL	Yellow

2001-03 Sebring Coupe 2.4L SOHC VIN G (A/T) 'C121' Connector

PCM Pin #	Wire Color	Circuit Description (30-Pin)	Value at Hot Idle
101	BK/BL	PNP Switch 'P' Signal	In Park: 12v, Or 0v
102	YL	PNP Switch 'D' Signal	In Drive: 12v, Or 0v
103	WT	Input Shaft Speed Sensor	Moving: 0-12-0v
104	GN/YL	Output Shaft Speed Sensor	Moving: 0-12-0v
105	BR/RD	A/T Shift Mode 1st Indicator	0v
106	RD/YL	A/T Second Solenoid	12-14v
107	YL/RD	A/T TCC Solenoid	12-14v
108	RD/BL	PNP Switch 'R' Signal	In Reverse: 12v, Or: 0v
109	WT	PNP Switch 'D3' Signal	In Drive 3: 12v, Or 0v
110	GY	PNP Switch 'Low' Signal	In Low: 12v, Or 0v
111	WT/GN	Immobilizer System Signal	Digital Signals
112	---	Not Used	---
113	WT/VT	DLC No. 2 Signal	0v
114	WT/RD	DLC No. 2 Signal	0v
115	---	Not Used	---
116	---	Not Used	---
117	WT/BL	A/T Shift Mode 3rd Indicator	0v
118	YL/BL	A/T Shift Mode 2nd Indicator	0v
120	RD	A/T Underdrive Solenoid	12-14v
121	BR	PNP Switch 'N' Signal	In Neutral: 12v, Or 0v
122	YL/BL	PNP Switch 'D2' Signal	In Drive 2: 12v, Or 0v
123	GN/OR	Stop Light Switch Signal	Brake Off: 0v, On: 12v
124	BK/PK	TFT Sensor Signal	3.2-3.4v at 104°F
128	YL/BL	A/T Shift Mode 4th Indicator	0v
129	RD/WT	A/T Low/Reverse Solenoid	12-14v
130	BL	A/T Overdrive Solenoid	12-14v

PCM C121 Wire Harness Connector Graphic

Standard Colors and Abbreviations

Abbreviation	Color	Abbreviation	Color	Abbreviation	Color
BK	Black	GY	Gray	RD	Red
BL	Blue	GN	Green	TN	Tan
BR	Brown	LG	Light Green	VT	Violet
DB	Dark Blue	OR	Orange	WT	White
DG	Dark Green	PK	Pink	YL	Yellow

1996-97 Sebring 2.5L V6 24v SOHC MFI VIN H (A/T) 'C1' Connector

PCM Pin #	Wire Color	Circuit Description (40-Pin)	Value at Hot Idle
1-3	---	Not Used	---
4	DG	Alternator Field Control	Digital Signal: 0-12-0v
5	YL/PK	S/C On Switch Signal	S/C On: 1v, Off: 12v
6	DG/OR	ASD Relay Output	12-14v
7	YL/WT	Injector 3 Driver	1.0-4.0 ms
8-9	---	Not Used	---
10	BK/TN	Power Ground	<0.1v
11	BK/GY	Coil 1 Driver	5°, at 55 mph: 8° dwell
12	---	Not Used	---
13	WT/LB	Injector 1 Driver	1.0-4.0 ms
14	BR/DG	Injector 6 Driver	1.0-4.0 ms
15	GY	Injector 5 Driver	1.0-4.0 ms
16	LB/BR	Injector 4 Driver	1.0-4.0 ms
17	TN	Injector 2 Driver	1.0-4.0 ms
18-19	---	Not Used	---
20	DB/WT	Ignition Switch Output	12-14v
21-25	---	Not Used	---
26	TN/BK	ECT Sensor Signal	At 180°F: 2.80v
27-29	---	Not Used	---
30	BK/DG	HO2S-11 (B1 S1) Signal	0.1-1.1v
31	---	Not Used	---
32	GY/BK	CKP Sensor Signal	Digital Signal: 0-5-0v
33	TN/YL	CMP Sensor Signal	Digital Signal: 0-5-0v
34	---	Not Used	---
35	OR/LB	TP Sensor Signal	0.6-1.0v
36	DG/RD	MAP Sensor Signal	1.5-1.7v
37	BK/RD	IAT Sensor Signal	At 100°F: 1.83v
38-39	---	Not Used	---
40	GY/YL	EGR Solenoid Control	12v, at 55 mph: 1v

Standard Colors and Abbreviations

Abbreviation	Color	Abbreviation	Color	Abbreviation	Color
BK	Black	GY	Gray	RD	Red
BL	Blue	GN	Green	TN	Tan
BR	Brown	LG	Light Green	VT	Violet
DB	Dark Blue	OR	Orange	WT	White
DG	Dark Green	PK	Pink	YL	Yellow

1996-97 Sebring 2.5L V6 24v SOHC MFI VIN H (A/T) 'C2' Connector

PCM Pin #	Wire Color	Circuit Description (40-Pin)	Value at Hot Idle
41	RD/LG	S/C Set Switch Signal	S/C & Set Switch On: 3.8v
42	DB/YL	A/C Pressure Switch Signal	A/C On: 0.45-4.85v
43	BK/LB	Sensor Ground	<0.050v
44	OR/WT	8-Volt Supply	7.9-8.1v
45	DB/LG	PSP Switch Signal	Straight: 0v, Turning: 5v
46	RD/TN	Battery Power (Fused B+)	12-14v
47, 53-54	---	Not Used	---
48	BR/GY	IAC 3 Driver	DC pulse signals: 0.8-11v
49	YL/BK	IAC 2 Driver	DC pulse signals: 0.8-11v
50	BK/TN	Power Ground	<0.1v
51	WT/DG	HO2S-12 (B1 S2) Signal	0.1-1.1v
52	PK/YL	Battery Temperature Sensor	At 86°F: 1.96v
55	DB/TN	Low Speed Fan Relay	Relay Off: 12v, On: 1v
57	GY/RD	IAC 1 Driver	DC pulse signals: 0.8-11v
58	PK/GY	IAC 4 Driver	DC pulse signals: 0.8-11v
59	PK/BR	CCD Bus (+)	Digital Signal: 0-5-0v
60	WT/BK	CCD Bus (-)	<0.050v
61	PK/WT	5-Volt Supply	4.9-5.1v
62	WT/RD	Brake Switch Signal	Brake Off: 0v, On: 12v
63	YL/DG	Torque Management Request	Digital Signals
64	DB/OR	A/C Clutch Relay Control	Relay Off: 12v, On: 1v
65	PK/LG	SCI Transmit	0v
66	WT/OR	Vehicle Speed Signal	Digital Signal
67	DB/PK	ASD Relay Control	Relay Off: 12v, On: 1v
68	PK/GY	EVAP Purge Solenoid Control	PWM Signal: 0-12-0v
69	DB/PK	High Speed Fan Relay	Relay Off: 12v, On: 1v
70-71	---	Not Used	---
72	OR	LDP Switch Sense	Open: 12v, Closed: 0v
74	BR/LG	Fuel Pump Relay Control	Relay Off: 12v, On: 1v
75	LG	SCI Receive	0v
76	BK/WT	PNP Switch Signal	In P/N: 0v, Others: 5v
77	WT/DG	LDP Solenoid Control	PWM Signal: 0-12-0v
78	WT/PK	S/C Vacuum Solenoid	Vacuum Increasing: 1v
79	---	Not Used	---
80	BK/YL	S/C Vent Solenoid	Vacuum Decreasing: 1v

Pin Connector Graphic

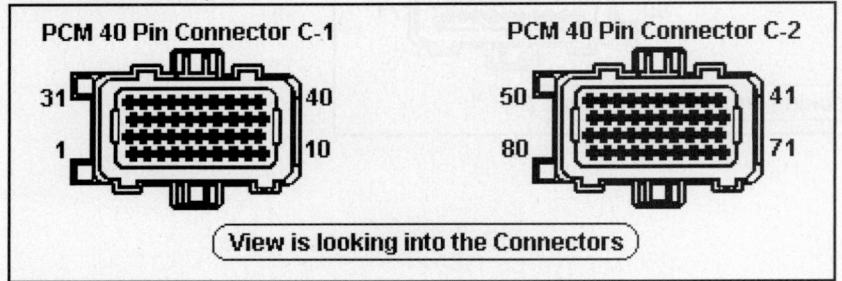

PCM 40 Pin Connector C-1   PCM 40 Pin Connector C-2

31 ... 40   50 ... 41
1 ... 10   80 ... 71

(View is looking into the Connectors)

1998 Sebring 2.5L V6 24v SOHC MFI VIN H (A/T) 'C1' Connector

PCM Pin #	Wire Color	Circuit Description (40-Pin)	Value at Hot Idle
1-3	---	Not Used	---
4	BK/GY	Coil 1 Driver	5°, at 55 mph: 8° dwell
5	YL/RD	S/C On Switch Signal	S/C On: 1v, Off: 12v
6	DG/OR	ASD Relay Output	12-14v
7	YL/WT	Injector 3 Driver	1.0-4.0 ms
8	DG	Alternator Field Control	Digital Signal: 0-12-0v
9	---	Not Used	---
10	BK/TN	Power Ground	<0.1v
11-12	---	Not Used	---
13	WT/DB	Injector 1 Driver	1.0-4.0 ms
14	BR/DB	Injector 6 Driver	1.0-4.0 ms
15	GY	Injector 5 Driver	1.0-4.0 ms
16	LB/BR	Injector 4 Driver	1.0-4.0 ms
17	TN	Injector 2 Driver	1.0-4.0 ms
18-19	---	Not Used	---
20	DB/WT	Ignition Switch Output	12-14v
21-25	---	Not Used	---
26	TN/BK	ECT Sensor Signal	At 180°F: 2.80v
27	BK/OR	Oxygen Sensor Ground	<0.050v
28-29	---	Not Used	---
30	BK/DG	HO2S-11 (B1 S1) Signal	0.1-1.1v
31	TN	Starter Relay Control	KOEC: 9-11v
32	GY/BK	CKP Sensor Signal	Digital Signal: 0-5-0v
33	TN/YL	CMP Sensor Signal	Digital Signal: 0-5-0v
34	---	Not Used	---
35	OR/DB	TP Sensor Signal	0.6-1.0v
36	DG/RD	MAP Sensor Signal	1.5-1.7v
37	BK/RD	IAT Sensor Signal	At 100°F: 1.83v
38-39	---	Not Used	---
40	GY/YL	EGR Solenoid Control	12v, at 55 mph: 1v

Pin Connector Graphic

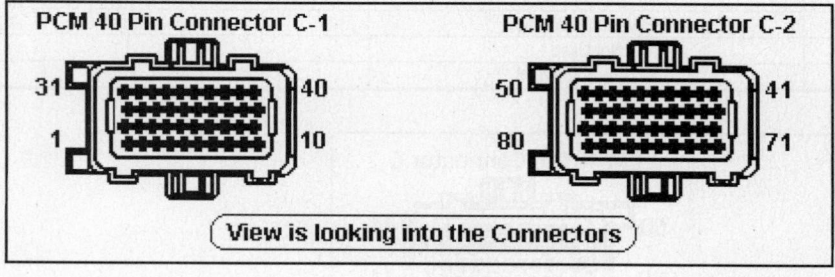

PCM 40 Pin Connector C-1

31   40
1   10

PCM 40 Pin Connector C-2

50   41
80   71

View is looking into the Connectors

1998 Sebring 2.5L V6 24v SOHC MFI VIN H (A/T) 'C2' Connector

PCM Pin #	Wire Color	Circuit Description (40-Pin)	Value at Hot Idle
41	RD/LG	S/C Set Switch Signal	S/C & Set Switch On: 3.8v
42	DB/YL	A/C Pressure Switch Signal	A/C On: 0.45-4.85v
43	BK/LB	Sensor Ground	<0.050v
44	OR/WT	8-Volt Supply	7.9-8.1v
45	DB/LG	PSP Switch Signal	Straight: 0v, Turning: 5v
46	RD/TN	Battery Power (Fused B+)	12-14v
47	BK/WT	Sensor Ground	<0.050v
48	BR/WT	IAC 1 Driver	DC pulse signals: 0.8-11v
49	YL/BK	IAC 2 Driver	DC pulse signals: 0.8-11v
50	BK/TN	Power Ground	<0.1v
51	TN/WT	HO2S-12 (B1 S2) Signal	0.1-1.1v
52	PK/YL	Battery Temperature Sensor	At 86°F: 1.96v
53-54	---	Not Used	---
55	DB/TN	Low Speed Fan Relay	Relay Off: 12v, On: 1v
56	WT/VT	S/C Vacuum Solenoid	Vacuum Increasing: 1v
57	GY/RD	IAC 3 Driver	DC pulse signals: 0.8-11v
58	VT/GY	IAC 4 Driver	DC pulse signals: 0.8-11v
59	VT/BR	CCD Bus (+)	Digital Signal: 0-5-0v
60	WT/BK	CCD Bus (-)	<0.050v
61	VT/WT	5-Volt Supply	4.9-5.1v
62	WT/RD	Brake Switch Signal	Brake Off: 0v, On: 12v
63	YL/DG	Torque Management Request	Digital Signals
64	DB/OR	A/C Clutch Relay Control	Relay Off: 12v, On: 1v
65	PK/LB	SCI Transmit	0v
66	WT/OR	Vehicle Speed Signal	Digital Signal
67	DB/VT	ASD Relay Control	Relay Off: 12v, On: 1v
68	PK/GY	EVAP Purge Solenoid Control	PWM Signal: 0-12-0v
69	DB/PK	High Speed Fan Relay	Relay Off: 12v, On: 1v
70	WT/TN	EVAP Purge Solenoid Sense	0-1v
71, 73	---	Not Used	---
72	OR	LDP Switch Sense	Open: 12v, Closed: 0v
74	BR/LG	Fuel Pump Relay Control	Relay Off: 12v, On: 1v
75	LG	SCI Receive	0v
76	BK/LB	PNP Switch Signal	In P/N: 0v, Others: 5v
77	WT/DG	LDP Solenoid Control	PWM Signal: 0-12-0v
78-79	---	Not Used	---
80	LG/RD	S/C Vent Solenoid	Vacuum Decreasing: 1v

Standard Colors and Abbreviations

Abbreviation	Color	Abbreviation	Color	Abbreviation	Color
BK	Black	GY	Gray	RD	Red
BL	Blue	GN	Green	TN	Tan
BR	Brown	LG	Light Green	VT	Violet
DB	Dark Blue	OR	Orange	WT	White
DG	Dark Green	PK	Pink	YL	Yellow

1999-2000 Sebring 2.5L V6 SOHC MFI VIN H (A/T) 'C1' Connector

PCM Pin #	Wire Color	Circuit Description (40-Pin)	Value at Hot Idle
1-3	---	Not Used	---
4	BK/GY	Coil 1 Driver	5°, at 55 mph: 8° dwell
5	YL/RD	S/C On Switch Signal	S/C On: 1v, Off: 12v
6	DG/OR	ASD Relay Output	12-14v
7	YL/WT	Injector 3 Driver	1.0-4.0 ms
8	DG	Alternator Field Control	Digital Signal: 0-12-0v
9	---	Not Used	---
10	BK/TN	Power Ground	<0.1v
11-12	---	Not Used	---
13	WT/DB	Injector 1 Driver	1.0-4.0 ms
14	BR/DB	Injector 6 Driver	1.0-4.0 ms
15	GY	Injector 5 Driver	1.0-4.0 ms
16	LB/BR	Injector 4 Driver	1.0-4.0 ms
17	TN	Injector 2 Driver	1.0-4.0 ms
18-19	---	Not Used	---
20	DB/WT	Ignition Switch Output	12-14v
21-25	---	Not Used	---
26	TN/BK	ECT Sensor Signal	At 180°F: 2.80v
27	BK/OR	Oxygen Sensor Ground	<0.050v
28-29	---	Not Used	---
30	BK/DG	HO2S-11 (B1 S1) Signal	0.1-1.1v
31	TN	Starter Relay Control	KOEC: 9-11v
32	GY/BK	CKP Sensor Signal	Digital Signal: 0-5-0v
33	TN/YL	CMP Sensor Signal	Digital Signal: 0-5-0v
34	---	Not Used	---
35	OR/LB	TP Sensor Signal	0.6-1.0v
36	DG/RD	MAP Sensor Signal	1.5-1.7v
37	BK/RD	IAT Sensor Signal	At 100°F: 1.83v
38-39	---	Not Used	---
40	GY/YL	EGR Solenoid Control	12v, at 55 mph: 1v

Standard Colors and Abbreviations

Abbreviation	Color	Abbreviation	Color	Abbreviation	Color
BK	Black	GY	Gray	RD	Red
BL	Blue	GN	Green	TN	Tan
BR	Brown	LG	Light Green	VT	Violet
DB	Dark Blue	OR	Orange	WT	White
DG	Dark Green	PK	Pink	YL	Yellow

1999-2000 Sebring 2.5L V6 SOHC MFI VIN H (A/T) 'C2' Connector

PCM Pin #	Wire Color	Circuit Description (40-Pin)	Value at Hot Idle
41	RD/LG	S/C Set Switch Signal	S/C & Set Switch On: 3.8v
42	DB/YL	A/C Pressure Switch Signal	A/C On: 0.45-4.85v
43	BK/LB	Sensor Ground	<0.050v
44	OR/WT	8-Volt Supply	7.9-8.1v
45	DB/LG	PSP Switch Signal	Straight: 0v, Turning: 5v
46	RD/TN	Battery Power (Fused B+)	12-14v
47	BK/WT	Sensor Ground	<0.050v
48	BR/GY	IAC 1 Driver	DC pulse signals: 0.8-11v
49	YL/BK	IAC 2 Driver	DC pulse signals: 0.8-11v
50	BK/TN	Power Ground	<0.1v
51	TN/WT	HO2S-12 (B1 S2) Signal	0.1-1.1v
52	PK/YL	Battery Temperature Sensor	At 86°F: 1.96v
53-54	---	Not Used	---
55	DB/TN	Low Speed Fan Relay	Relay Off: 12v, On: 1v
56	WT/VT	S/C Vacuum Solenoid	Vacuum Increasing: 1v
57	GY/RD	IAC 3 Driver	DC pulse signals: 0.8-11v
58	VT/BK	IAC 4 Driver	DC pulse signals: 0.8-11v
59	VT/BR	CCD Bus (+)	Digital Signal: 0-5-0v
60	WT/BK	CCD Bus (-)	<0.050v
61	VT/WT	5-Volt Supply	4.9-5.1v
62	WT/RD	Brake Switch Signal	Brake Off: 0v, On: 12v
63	YL/DG	Torque Management Request	Digital Signals
64	DB/OR	A/C Clutch Relay Control	Relay Off: 12v, On: 1v
65	PK/LG	SCI Transmit	0v
66	WT/OR	Vehicle Speed Signal	Digital Signal
67	DB/VT	ASD Relay Control	Relay Off: 12v, On: 1v
68	PK/GY	EVAP Purge Solenoid Control	PWM Signal: 0-12-0v
69	DB/PK	High Speed Fan Relay	Relay Off: 12v, On: 1v
70	WT/TN	EVAP Purge Solenoid Sense	0-1v
71-73	---	Not Used	---
74	BR/LG	Fuel Pump Relay Control	Relay Off: 12v, On: 1v
75	LG	SCI Receive	0v
76	BK/LB	TRS T41 Sense	In P/N: 0v, Others: 5v
77-79	---	Not Used	---
80	LG/RD	S/C Vent Solenoid	Vacuum Decreasing: 1v

Pin Connector Graphic

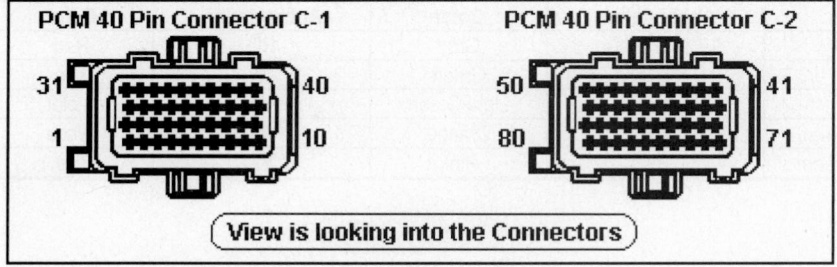

1995 Sebring 2.5L V6 24v SOHC MFI VIN N (A/T) 'C1' Connector

PCM Pin #	Wire Color	Circuit Description (40-Pin)	Value at Hot Idle
1-3	---	Not Used	---
4	DB	Alternator Field Control	Digital Signal: 0-12-0v
5	---	Not Used	---
6	BK/RD	ASD Relay Output	12-14v
7	YL/DG	Injector 3 Driver	1.0-4.0 ms
8	DG/RD	Amber MIL Control	MIL On: 1v, MIL Off: 12v
9	DB/YL	Bulb Check Control	Lamp On: 1v, Off: 12v
10	BK	Power Ground	<0.1v
11	RD/DG	Coil Driver	5°, at 55 mph: 8° dwell
12	RD/BK	S/C On Switch Signal	S/C On: 1v, Off: 12v
13	LG/BK	Injector 1 Driver	1.0-4.0 ms
14	BR/RD	Injector 6 Driver	1.0-4.0 ms
15	RD/WT	Injector 5 Driver	1.0-4.0 ms
16	LG/RD	Injector 4 Driver	1.0-4.0 ms
17	YL/RD	Injector 2 Driver	1.0-4.0 ms
18	---	Not Used	---
19	DG/OR	High Speed Fan Relay	Relay Off: 12v, On: 1v
20	BK/WT	Ignition Switch Output	12-14v
21	---	Not Used	---
22	LG	Speed Control Lamp Driver	Lamp On: 1v, Off: 12v
23-25	---	Not Used	---
26	DG/WT	ECT Sensor Signal	At 180°F: 2.80v
27-28	---	Not Used	---
29	DG/BK	HO2S-12 (B1 S2) Signal	0.1-1.1v
30	WT/BK	HO2S-11 (B1 S1) Signal	0.1-1.1v
31	---	Not Used	---
32	LB/WT	CKP Sensor Signal	Digital Signal: 0-5-0v
33	BR	CMP Sensor Signal	Digital Signal: 0-5-0v
34	---	Not Used	---
35	BR/RD	TP Sensor Signal	0.6-1.0v
36	YL/BK	MAP Sensor Signal	1.5-1.7v
37	BR/LB	IAT Sensor Signal	At 100°F: 1.83v
38	DG/RD	A/C Clutch Switch Sense	Relay Off: 12v, On: 1v
39	---	Not Used	---
40	RD/LB	EGR Solenoid Control	12v, at 55 mph: 1v

Standard Colors and Abbreviations

Abbreviation	Color	Abbreviation	Color	Abbreviation	Color
BK	Black	GY	Gray	RD	Red
BL	Blue	GN	Green	TN	Tan
BR	Brown	LG	Light Green	VT	Violet
DB	Dark Blue	OR	Orange	WT	White
DG	Dark Green	PK	Pink	YL	Yellow

1995 Sebring 2.5L V6 24v SOHC MFI VIN N (A/T) 'C2' Connector

PCM Pin #	Wire Color	Circuit Description (40-Pin)	Value at Hot Idle
41	RD	S/C Set Switch Signal	S/C & Set Switch On: 3.8v
42	---	Not Used	---
43	BK/DG	Sensor Ground	<0.050v
44	YL	8-Volt Supply	7.9-8.1v
45	LB/YL	PSP Switch Signal	Straight: 0v, Turning: 5v
46	RD/BK	Battery Power (Fused B+)	12-14v
47	BK	Sensor Ground	<0.050v
48	DG/LB	IAC 1 Driver	DC pulse signals: 0.8-11v
49	OR	IAC 2 Driver	DC pulse signals: 0.8-11v
50	BK	Power Ground	<0.1v
51-54	---	Not Used	---
55	DG/BK	Low Speed Fan Relay	Relay Off: 12v, On: 1v
56	LB	Generator Lamp Control	Lamp On: 1v, Off: 12v
57	DG	IAC 3 Driver	DC pulse signals: 0.8-11v
58	YL/LB	IAC 4 Driver	DC pulse signals: 0.8-11v
59	BK/DB	CCD Bus (+)	Digital Signal: 0-5-0v
60	WT/LB	CCD Bus (-)	<0.050v
61	DG/YL	5-Volt Supply	4.9-5.1v
62	BR/WT	Brake Switch Signal	Brake Off: 0v, On: 12v
63	OR/BK	Torque Management Request	Digital Signals
64	DG	A/C Clutch Relay Control	Relay Off: 12v, On: 1v
65	PK	SCI Transmit	0v
66	YL/WT	Vehicle Speed Signal	Digital Signal
67	RD/WT	ASD Relay Control	Relay Off: 12v, On: 1v
68	LG/BK	EVAP Purge Solenoid Control	PWM Signal: 0-12-0v
69	DG/WT	High Speed Fan Relay	Relay Off: 12v, On: 1v
70-72	---	Not Used	---
73	WT	Tachometer Signal	Pulse Signals
74	WT/RD	Fuel Pump Relay Control	Relay Off: 12v, On: 1v
75	DB/BK	SCI Receive	0v
76	BK/YL	PNP Switch Signal	In P/N: 0v, Others: 5v
77, 79	---	Not Used	---
78	LG/WT	S/C Vacuum Solenoid	Vacuum Increasing: 1v
80	BK/YL	S/C Vent Solenoid	Vacuum Decreasing: 1v

Pin Connector Graphic

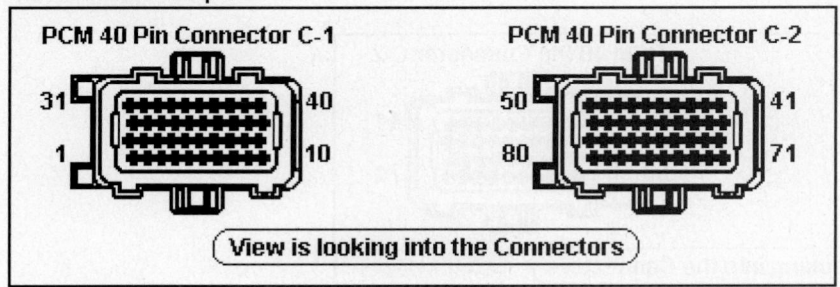

PCM 40 Pin Connector C-1

31   40
1   10

PCM 40 Pin Connector C-2

50   41
80   71

View is looking into the Connectors

1996-97 Sebring 2.5L V6 24v SOHC MFI VIN N (A/T) 'C1' Connector

PCM Pin #	Wire Color	Circuit Description (40-Pin)	Value at Hot Idle
1	WT/RD	HO2S-12 (B1 S2) Signal	0.1-1.1v
2-3, 5	---	Not Used	---
4	DB	Alternator Field Control	Digital Signal: 0-12-0v
6	BK/RD	ASD Relay Output	12-14v
7	YL/DG	Injector 3 Driver	1.0-4.0 ms
8	DG/RD	Amber MIL Control	MIL On: 1v, MIL Off: 12v
9	DB/YL	Bulb (lamp) Check Control	B/C On: 1v, B/C Off: 12v
10	BK	Power Ground	<0.1v
11	RD/DG	Coil Driver	5°, at 55 mph: 8° dwell
12	RD/BK	S/C On Switch Signal	S/C On: 1v, Off: 12v
13	LG/BK	Injector 1 Driver	1.0-4.0 ms
14	BR/RD	Injector 6 Driver	1.0-4.0 ms
15	RD/WT	Injector 5 Driver	1.0-4.0 ms
16	LG/RD	Injector 4 Driver	1.0-4.0 ms
17	YL/RD	Injector 2 Driver	1.0-4.0 ms
19	DG/OR	High Speed Fan 2 Control	Relay Off: 12v, On: 1v
20	BK/WT	Ignition Switch Output	12-14v
21	---	Not Used	---
22	LG	Speed Control Lamp Driver	S/C On: 1v, S/C Off: 12v
23	YL/DB	Fuel Level Sensor Signal	70 ohms (±20) w/full tank
24-25	---	Not Used	---
26	DG/WT	ECT Sensor Signal	At 180°F: 2.80v
27	---	Not Used	---
28	DG/YL	High Speed Fan 1 Control	Relay Off: 12v, On: 1v
29	DG/BK	HO2S-11 (B1 S1) Signal	0.1-1.1v
30	WT/BK	HO2S-21 (B2 S1) Signal	0.1-1.1v
32	LB/WT	CKP Sensor Signal	Digital Signal: 0-5-0v
33	BR	CMP Sensor Signal	Digital Signal: 0-5-0v
34	---	Not Used	---
35	BR/RD	TP Sensor Signal	0.6-1.0v
36	YL/BK	MAP Sensor Signal	1.5-1.7v
37	BR/DB	IAT Sensor Signal	At 100°F: 1.83v
38	DG/RD	A/C Clutch Switch Sense	Relay Off: 12v, On: 1v
39	---	Not Used	---
40	RD/DB	EGR Solenoid Control	12v, at 55 mph: 1v

Pin Connector Graphic

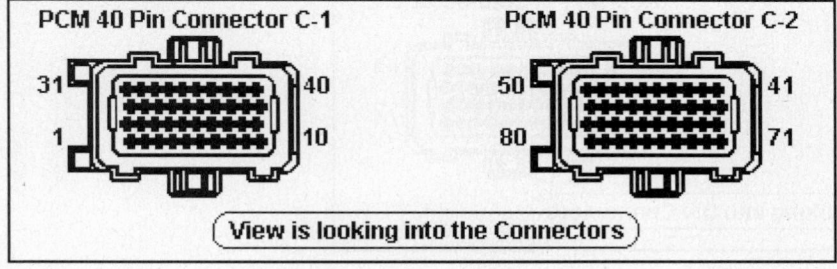

1996-97 Sebring 2.5L V6 24v SOHC MFI VIN N (A/T) 'C2' Connector

PCM Pin #	Wire Color	Circuit Description (40-Pin)	Value at Hot Idle
41	RD	S/C Set Switch Signal	S/C & Set Switch On: 3.8v
42	---	Not Used	---
43	BK/DG	Sensor Ground	<0.050v
44	YL	8-Volt Supply	7.9-8.1v
45	DB/YL	PSP Switch Signal	Straight: 0v, Turning: 5v
46	RD/BK	Battery Power (Fused B+)	12-14v
47	BK	Sensor Ground	<0.050v
48	DG/DB	IAC 3 Driver	DC pulse signals: 0.8-11v
49	OR	IAC 2 Driver	DC pulse signals: 0.8-11v
50	BK	Power Ground	<0.1v
51	WT/DG	HO2S-22 (B2 S2) Signal	0.1-1.1v
52-54	---	Not Used	---
55	DG/BK	Low Speed Fan Relay	Relay Off: 12v, On: 1v
56	DB	Generator Lamp Control	Lamp On: 1v, Off: 12v
57	DG	IAC 1 Driver	DC pulse signals: 0.8-11v
58	YL/DB	IAC 4 Driver	DC pulse signals: 0.8-11v
59	BK/DB	CCD Bus (+)	Digital Signal: 0-5-0v
60	WT/DB	CCD Bus (-)	<0.050v
61	DG/YL	5-Volt Supply	4.9-5.1v
62	BR/WT	Brake Switch Signal	Brake Off: 0v, On: 12v
63	OR/BK	Torque Management Request	Digital Signals
64	DG	A/C Clutch Relay Control	Relay Off: 12v, On: 1v
65	PK	SCI Transmit	0v
66	YL/WT	Vehicle Speed Signal	Digital Signal
67	RD/WT	ASD Relay Control	Relay Off: 12v, On: 1v
68	LG/BK	EVAP Purge Solenoid Control	PWM Signal: 0-12-0v
69	DG/WT	High Condenser Fan Relay	Relay Off: 12v, On: 1v
70-71, 79	---	Not Used	---
72	RD/DB	LDP Switch Sense	Open: 12v, Closed: 0v
73	WT	Tachometer Signal	Pulse Signals
74	WT/RD	Fuel Pump Relay Control	Relay Off: 12v, On: 1v
75	DB/BK	SCI Receive	0v
76	BK/YL	PNP Switch Signal	In P/N: 0v, Others: 5v
77	RD/YL	LDP Solenoid Control	PWM Signal: 0-12-0v
78	LG/WT	S/C Vacuum Solenoid	Vacuum Increasing: 1v
80	BK/YL	S/C Vent Solenoid	Vacuum Decreasing: 1v

Standard Colors and Abbreviations

Abbreviation	Color	Abbreviation	Color	Abbreviation	Color
BK	Black	GY	Gray	RD	Red
BL	Blue	GN	Green	TN	Tan
BR	Brown	LG	Light Green	VT	Violet
DB	Dark Blue	OR	Orange	WT	White
DG	Dark Green	PK	Pink	YL	Yellow

1998-2000 Sebring 2.5L V6 SOHC MFI VIN N (A/T) 'C1' Connector

PCM Pin #	Wire Color	Circuit Description (40-Pin)	Value at Hot Idle
1-3	---	Not Used	---
4	RD/DG	Coil Driver	5°, at 55 mph: 8° dwell
5	---	Not Used	---
6	BK/RD	ASD Relay Output	12-14v
7	YL/DG	Injector 3 Driver	1.0-4.0 ms
8	DB	Alternator Field Control	Digital Signal: 0-12-0v
9	DB/YL	Bulb (lamp) Check Control	B/C On: 1v, B/C Off: 12v
10	BK	Power Ground	<0.1v
11	---	Not Used	---
12	RD/BK	S/C On Switch Signal	S/C On: 1v, Off: 12v
13	LG/BK	Injector 1 Driver	1.0-4.0 ms
14	BR/RD	Injector 6 Driver	1.0-4.0 ms
15	RD/WT	Injector 5 Driver	1.0-4.0 ms
16	LG/RD	Injector 4 Driver	1.0-4.0 ms
17	YL/RD	Injector 2 Driver	1.0-4.0 ms
18	LG	Speed Control Lamp Driver	S/C On: 1v, S/C Off: 12v
19	DG/OR	High Speed Fan 2 Control	Relay Off: 12v, On: 1v
20	BK/WT	Ignition Switch Output	12-14v
21	---	Not Used	---
22	DG/RD	Amber MIL Control	MIL On: 1v, MIL Off: 12v
23	YL/DB	Fuel Level Sensor Signal	70 ohms (±20) w/full tank
24-25	---	Not Used	---
26	DG/WT	ECT Sensor Signal	At 180°F: 2.80v
27	---	Not Used	---
28	DG/YL	High Speed Fan 1 Control	Relay Off: 12v, On: 1v
29	DG/BK	HO2S-11 (B1 S1) Signal	0.1-1.1v
30	WT/BK	HO2S-12 (B1 S2) Signal	0.1-1.1v
31	---	Not Used	---
32	DB/WT	CKP Sensor Signal	Digital Signal: 0-5-0v
33	BR	CMP Sensor Signal	Digital Signal: 0-5-0v
34	---	Not Used	---
35	BR/RD	TP Sensor Signal	0.6-1.0v
36	YL/BK	MAP Sensor Signal	1.5-1.7v
37	BR/DB	IAT Sensor Signal	At 100°F: 1.83v
38	DG/RD	A/C Clutch Switch Sense	Relay Off: 12v, On: 1v
39	DB	Generator Lamp Control	Lamp On: 1v, Off: 12v
40	RD/DB	EGR Solenoid Control	12v, at 55 mph: 1v

Pin Connector Graphic

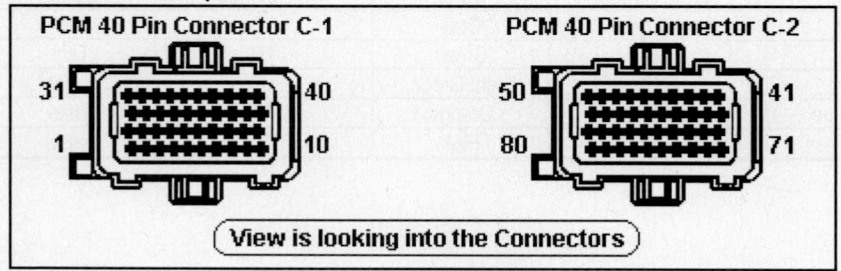

1998-2000 Sebring 2.5L V6 SOHC MFI VIN N (A/T) 'C2' Connector

PCM Pin #	Wire Color	Circuit Description (40-Pin)	Value at Hot Idle
41	RD	S/C Set Switch Signal	S/C & Set Switch On: 3.8v
42-43	---	Not Used	---
44	YL	8-Volt Supply	7.9-8.1v
45	DB/YL	PSP Switch Signal	Straight: 0v, Turning: 5v
46	RD/BK	Battery Power (Fused B+)	12-14v
47	BK	Sensor Ground	<0.050v
48	GY/DB	IAC 3 Driver	DC pulse signals: 0.8-11v
49	OR	IAC 2 Driver	DC pulse signals: 0.8-11v
50	BK	Power Ground	<0.1v
51	WT/DG	HO2S-22 (B2 S2) Signal	0.1-1.1v
52, 54	---	Not Used	---
53	WT/RD	HO2S-21 (B2 S1) Signal	0.1-1.1v
55	DG/BK	Low Speed Fan Relay	Relay Off: 12v, On: 1v
56	LG/WT	S/C Vacuum Solenoid	Vacuum Increasing: 1v
57	GY	IAC 1 Driver	DC pulse signals: 0.8-11v
58	YL/DB	IAC 4 Driver	DC pulse signals: 0.8-11v
59	BK/DB	CCD Bus (+)	Digital Signal: 0-5-0v
60	WT/DB	CCD Bus (-)	<0.050v
61	DG/YL	5-Volt Supply	4.9-5.1v
62	BR/WT	Brake Switch Signal	Brake Off: 0v, On: 12v
63	OR/BK	Torque Management Request	Digital Signals
64	DG	A/C Clutch Relay Control	Relay Off: 12v, On: 1v
65	PK	SCI Transmit	0v
66	YL/WT	Vehicle Speed Signal	Digital Signal
67	RD/WT	ASD Relay Control	Relay Off: 12v, On: 1v
68	LG/BK	EVAP Purge Solenoid Control	PWM Signal: 0-12-0v
69	DG/WT	High Condenser Fan Relay	Relay Off: 12v, On: 1v
70	B/WT	EVAP Purge Solenoid Sense	0-1v
71	---	Not Used	---
72	RD/DB	LDP Switch Sense	Open: 12v, Closed: 0v
73	WT	Tachometer Signal	Pulse Signals
74	WT/RD	Fuel Pump Relay Control	Relay Off: 12v, On: 1v
75	DB/BK	SCI Receive	0v
76	BK/YL	PNP Switch Signal	PWM Signal: 0-12-0v
77	RD/YL	LDP Solenoid Control	PWM Signal: 0-12-0v
78-79	---	Not Used	---
80	BK/YL	S/C Vent Solenoid	Vacuum Decreasing: 1v

Standard Colors and Abbreviations

Abbreviation	Color	Abbreviation	Color	Abbreviation	Color
BK	Black	GY	Gray	RD	Red
BL	Blue	GN	Green	TN	Tan
BR	Brown	LG	Light Green	VT	Violet
DB	Dark Blue	OR	Orange	WT	White
DG	Dark Green	PK	Pink	YL	Yellow

2001 Sebring 2.7L V6 DOHC MFI VIN R (A/T) 'C1' Connector

PCM Pin #	Wire Color	Circuit Description (40-Pin)	Value at Hot Idle
1	TN/LG	COP 4 Driver Control	5°, at 55 mph: 8° dwell
2	TN/OR	COP 3 Driver Control	5°, at 55 mph: 8° dwell
3	TN/PK	COP 2 Driver Control	5°, at 55 mph: 8° dwell
4	TN/LG	COP 6 Driver Control	5°, at 55 mph: 8° dwell
5	YL/RD	S/C Power Supply	12-14v
6	DG/OR	ASD Relay Output	12-14v
7	YL/WT	Injector 3 Driver	1.0-4.0 ms
8	DG	Alternator Field Control	Digital Signal: 0-12-0v
9	---	Not Used	---
10	BK/TN	Power Ground	<0.1v
11	TN/RD	COP 1 Driver Control	5°, at 55 mph: 8° dwell
12	---	Not Used	---
13	WT/DB	Injector 1 Driver	1.0-4.0 ms
14	BR/DB	Injector 6 Driver	1.0-4.0 ms
15	GY	Injector 5 Driver	1.0-4.0 ms
16	LB/BR	Injector 4 Driver	1.0-4.0 ms
17	TN	Injector 2 Driver	1.0-4.0 ms
18-19	---	Not Used	---
20	DB/WT	Ignition Switch Output	12-14v
21	TN/DG	COP 5 Driver Control	5°, at 55 mph: 8° dwell
22-24	---	Not Used	---
25	DB/LG	Knock Sensor Signal	0.080v AC
26	TN/BK	ECT Sensor Signal	At 180°F: 2.80v
27	DB/LG	HO2S Sensor Ground	<0.050v
28	---	Not Used	---
29	LG/RD	HO2S-21 (B2 S1) Signal	0.1-1.1v
30	BK/DG	HO2S-11 (B1 S1) Signal	0.1-1.1v
31	TN	Starter Relay Control	KOEC: 9-11v
32	GY/BK	CKP Sensor Signal	Digital Signal: 0-5-0v
33	TN/YL	CMP Sensor Signal	Digital Signal: 0-5-0v
34	LG/PK	EGR Sensor Signal	0.6-0.8v
35	OR/DB	TP Sensor Signal	0.6-1.0v
36	DG/RD	MAP Sensor Signal	1.5-1.7v
37	BK/RD	IAT Sensor Signal	At 100°F: 1.83v
38	---	Not Used	---
39	VT/RD	Manifold Solenoid Control	Solenoid Off: 12v, On: 1v
40	GY/YL	EGR Solenoid Control	12v, at 55 mph: 1v

Standard Colors and Abbreviations

Abbreviation	Color	Abbreviation	Color	Abbreviation	Color
BK	Black	GY	Gray	RD	Red
BL	Blue	GN	Green	TN	Tan
BR	Brown	LG	Light Green	VT	Violet
DB	Dark Blue	OR	Orange	WT	White
DG	Dark Green	PK	Pink	YL	Yellow

2001 Sebring 2.7L V6 DOHC MFI VIN R (A/T) 'C2' Connector

PCM Pin #	Wire Color	Circuit Description (40-Pin)	Value at Hot Idle
41	PK/LG	S/C Set Switch Signal	S/C & Set Switch On: 3.8v
42	DB	A/C Pressure Switch Signal	A/C On: 0.45-4.85v
43	BK/LB	Sensor Ground	<0.050v
44	OR/WT	8-Volt Supply	7.9-8.1v
45	---	Not Used	---
46	RD/TN	Battery Power (Fused B+)	12-14v
47	BK	Power Ground	<0.1v
48	BR/WT	IAC 3 Driver	DC pulse signals: 0.8-11v
49	YL/BK	IAC 2 Driver	DC pulse signals: 0.8-11v
50	BK/TN	Power Ground	<0.1v
51	TN/WT	HO2S-12 (B1 S2) Signal	0.1-1.1v
52	VT/LG	Battery Temperature Sensor	At 86°F: 1.96v
53	PK/WT	HO2S-22 (B2 S2) Signal	0.1-1.1v
54	---	Not Used	---
55	DB/TN	Low Speed Fan Relay	Relay Off: 12v, On: 1v
56	TN/RD	S/C Vacuum Solenoid	Vacuum Increasing: 1v
57	GY/RD	IAC 1 Driver	DC pulse signals: 0.8-11v
58	VT/BK	IAC 4 Driver	DC pulse signals: 0.8-11v
59	OR	PCI Data Bus (J1850)	Digital Signals: 0-7-0v
60	---	Not Used	---
61	VT/WT	5-Volt Supply	4.9-5.1v
62	WT/PK	Brake Switch Signal	Brake Off: 0v, On: 12v
63	YL/DG	Torque Management Request	Digital Signals
64	DB/OR	A/C Clutch Relay Control	Relay Off: 12v, On: 1v
65	PK	SCI Transmit	Digital Signal: 0-5-0v
66	WT/OR	Vehicle Speed Signal	Digital Signal
67	DB/VT	ASD Relay Control	Relay Off: 12v, On: 1v
68	PK/BK	EVAP Purge Solenoid Control	PWM Signal: 0-12-0v
69	DB/PK	High Speed Fan Relay	Relay Off: 12v, On: 1v
70	WT/TN	EVAP Purge Solenoid Sense	0-1v
71	WT/RD	EATX RPM Signal	Digital Signals
72	OR	LDP Switch Sense	Open: 12v, Closed: 0v
73	---	Not Used	---
74	BR/LG	Fuel Pump Relay Control	Relay Off: 12v, On: 1v
75	LG	SCI Receive	0v
76	BK/LB	TRS T41 Sense	In P/N: 0v, Others: 5v
77	WT/DG	LDP Solenoid Control	PWM Signal: 0-12-0v
78-79	---	Not Used	---
80	LG/RD	S/C Vent Solenoid	Vacuum Decreasing: 1v

Pin Connector Graphic

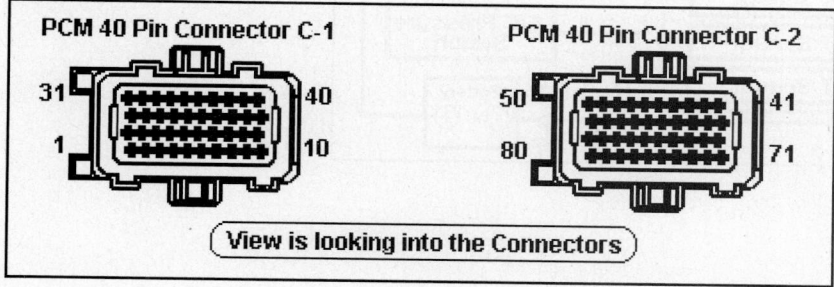

PCM 40 Pin Connector C-1   PCM 40 Pin Connector C-2

31   40   50   41

1   10   80   71

View is looking into the Connectors

2001 Sebring Coupe 2.7L V6 VIN R Wiring Diagram (Part 1)

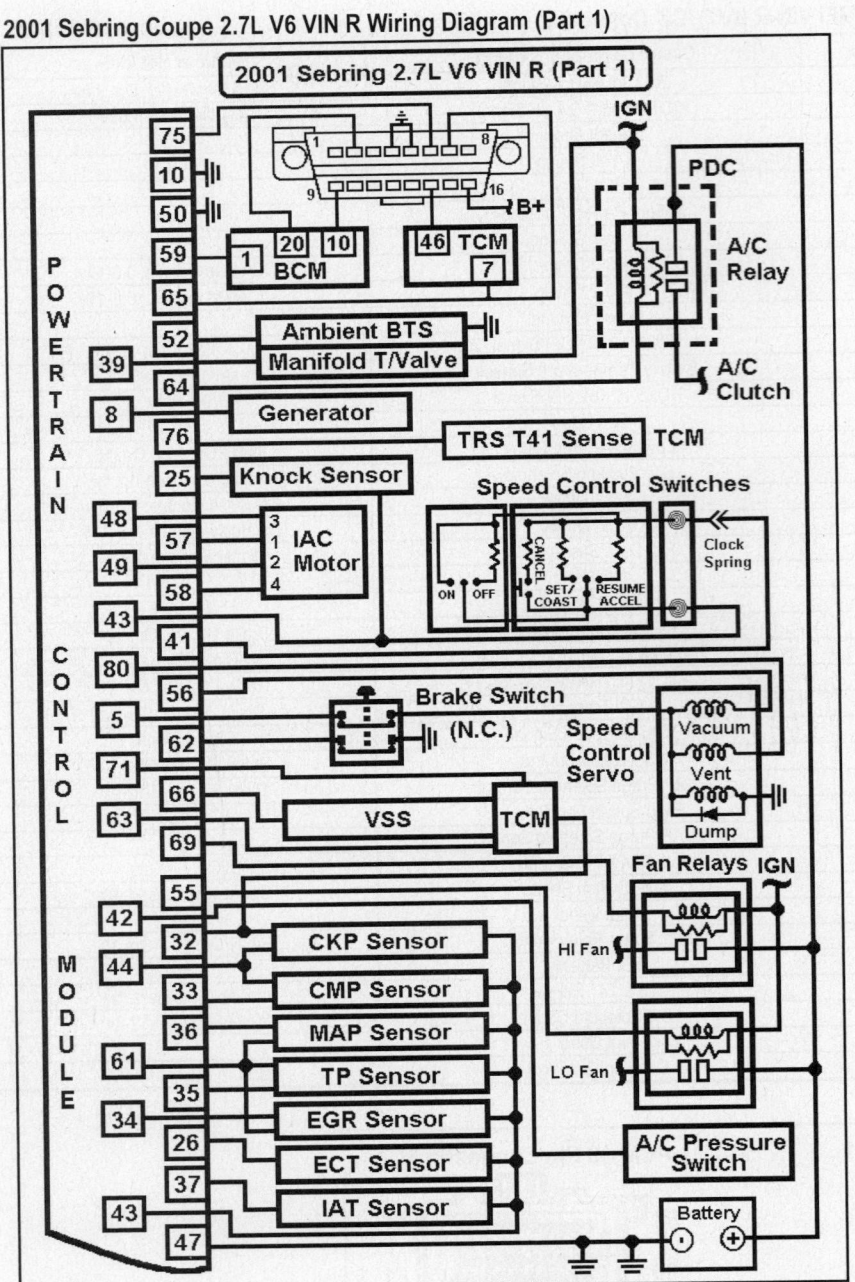

2001 Sebring Coupe 2.7L V6 VIN R Wiring Diagram (Part 2)

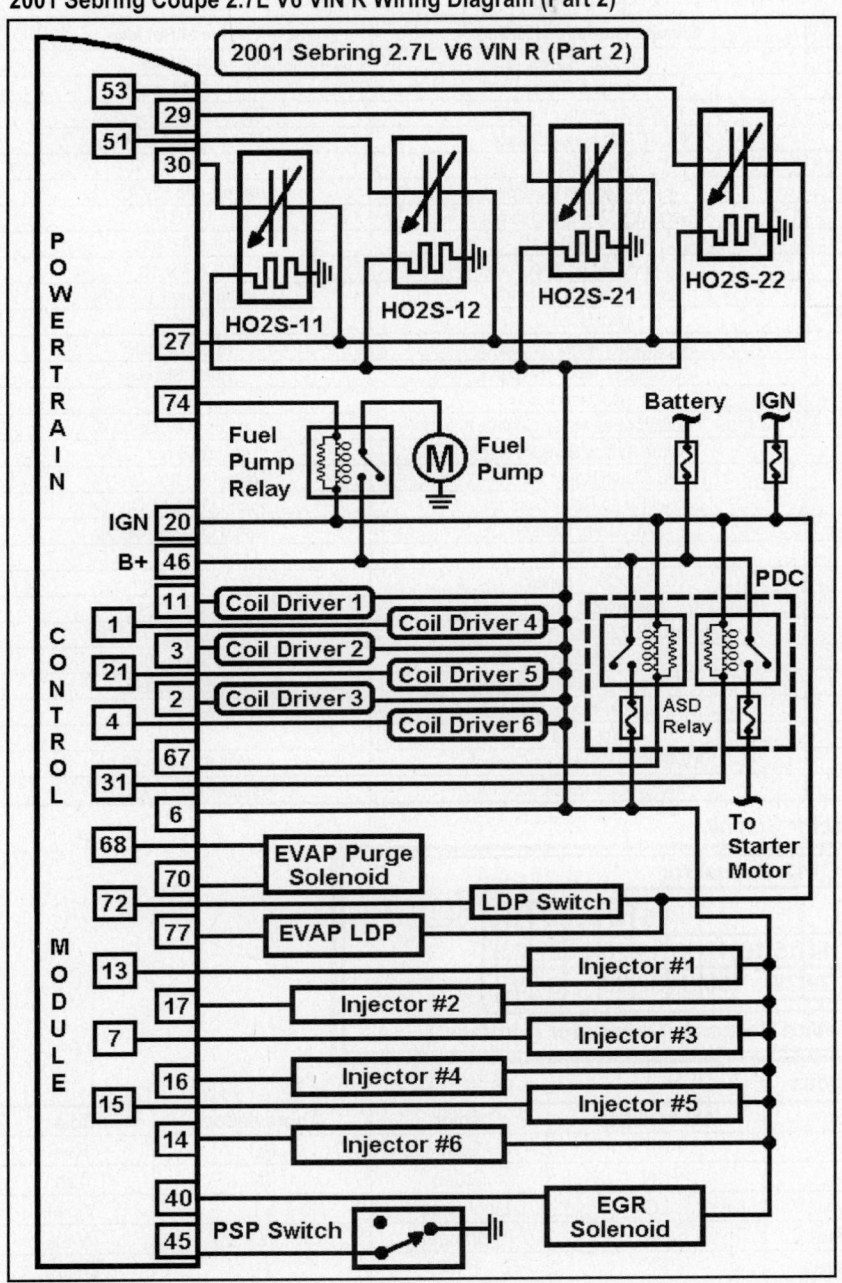

2001-03 Sebring Coupe 3.0L V6 VIN H (A/T) 'C112' Connector

PCM Pin #	Wire Color	Circuit Description (35-Pin)	Value at Hot Idle
1	YL/BL	Injector 1 Driver	1.0-4.0 ms
2	GN/YL	Injector 4 Driver	1.0-4.0 ms
3	BR/WT	HO2S-21 Heater Control	Digital Signal: 0-12-0v
4	BL/WT	HO2S-11 Heater Control	Digital Signal: 0-12-0v
5, 7	---	Not Used	---
6	BL/RD	EGR Solenoid Control	Solenoid Off: 12v, On: 1v
8	BK/RD	Generator Signal (lights on)	0.2-3.5v
9	YL/RD	Injector 2 Driver	1.0-4.0 ms
10	RD/BL	Injector 5 Driver	1.0-4.0 ms
11	BK/BL	Power Transistor Control	Digital Signal: 0-12-0v
12-13	---	Not Used	---
14	GN/BK	IAC Stepper Motor 'A' Signal	Pulse Signals
15	GN/WT	IAC Stepper Motor 'B' Signal	Pulse Signals
16-17	---	Not Used	---
18	BL/OR	Radiator Fan Relay Control	Fan Off: 0.1v, On: 0.7v
19	GN/BL	Volume Airflow Sensor Reset	1-3v, 3000 rpm: 6-9v
20	GN	A/C Clutch Relay Control	Relay Off: 12v, On: 1v
21	BK	Fuel Pump Relay Control	Pump Off: 12v, On: 1v
22	RD/YL	MIL (lamp) Control	Lamp Off: 12v, On: 1v
23	---	Not Used	---
24	YL/BK	Injector 3 Driver	1.0-4.0 ms
25	LG	Injector 6 Driver	1.0-4.0 ms
26	BL/WT	HO2S-22 Heater Control	Digital Signal: 0-12-0v
27	BR	HO2S-12 Heater Control	Digital Signal: 0-12-0v
28	GN/RD	IAC Stepper Motor 'A' Signal	Pulse Signals
29	BL/YL	IAC Stepper Motor 'B' Signal	Pulse Signals
30-33	---	Not Used	---
34	BL	EVAP Purge Solenoid Control	PWM Signal: 0-12-0v
35	YL	EVAP Vent Solenoid Control	Solenoid Off: 12v, On: 1v

PCM C112 Wire Harness Connector Graphic

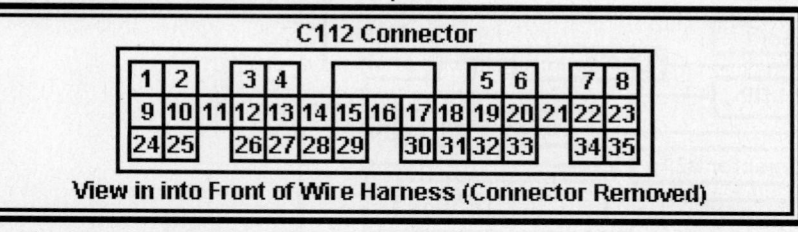

Standard Colors and Abbreviations

Abbreviation	Color	Abbreviation	Color	Abbreviation	Color
BK	Black	GY	Gray	RD	Red
BL	Blue	GN	Green	TN	Tan
BR	Brown	LG	Light Green	VT	Violet
DB	Dark Blue	OR	Orange	WT	White
DG	Dark Green	PK	Pink	YL	Yellow

Coupe 3.0L V6 VIN H (A/T) 'C115' Connector

2001-0

Wire Color	Circuit Description (26-Pin)	Value at Hot Idle
RD	Ignition Power (MPI Relay)	12-14v
BK	Power Ground	<0.1v
WT	Tachometer Signal	Pulse Signals
YL/GN	ECT Sensor Signal	0.3-0.9v at 176°F
GN/RD	Distributor CKP Sensor Signal	Digital Signal: 0-5-0v
GN	Sensor Voltage Reference	4.9-5.1v
RD	Ignition Power (MPI Relay)	12-14v
BK	Power Ground	<0.1v
WT/VT	MPI (Power) Relay Control	Relay Off: 12v, On: 1v
WT/BL	A/T Control Relay	Relay Off: 12v, On: 1v
---	Not Used	---
YL	PSP Switch Signal	Straight: 5v, Turning: 0v
---	Not Used	---
YL/BK	Generator 'FR' Terminal	0.5-4.5v
GN/WT	BARO Sensor Signal	3.7-4.3 at Sea Level
RD/WT	CMP (TDC) Sensor Signal	Digital Signal: 0-5-0v
BK	Sensor Ground	<0.050v
BR/RD	Starter (Cranking) Signal	9-11v (cranking)
---	Not Used	---
BL/WT	A/C Switch 2 Signal (Hi Blow)	12v
---	Not Used	---
64	RD/BL	IAT Sensor Signal	1.5-2.1v at 104°F
65	WT/GN	Volume Airflow Sensor	2.2-3.2v
66	OR/BL	Keep Alive Power	12-14v

PCM C115 Wire Harness Connector Graphic

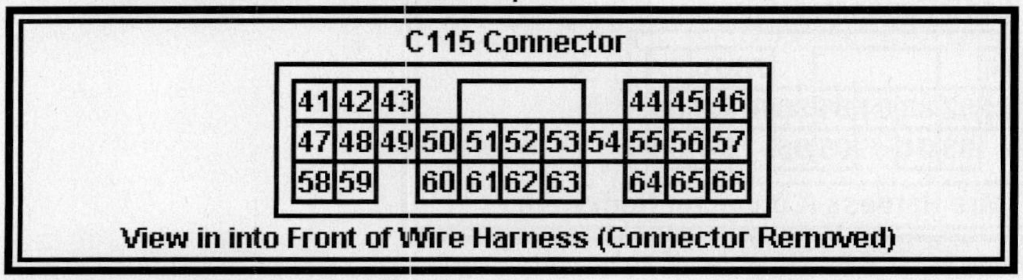

C115 Connector

```
41 42 43        44 45 46
47 48 49 50 51 52 53 54 55 56 57
58 59    60 61 62 63    64 65 66
```

View in into Front of Wire Harness (Connector Removed)

Standard Colors and Abbreviations

Abbreviation	Color	Abbreviation	Color	Abbreviation	Color
BK	Black	GY	Gray	RD	Red
BL	Blue	GN	Green	TN	Tan
BR	Brown	LG	Light Green	VT	Violet
DB	Dark Blue	OR	Orange	WT	White
DG	Dark Green	PK	Pink	YL	Yellow

2001-03 Sebring Coupe 3.0L V6 VIN H (A/T) 'C119' Connector

PCM Pin #	Wire Color	Circuit Description (28-Pin)	Value at Hot Idle
71	WT	HO2S-21 (B2 S1) Signal	0.1-1.1v
72	BL	HO2S-11 (B1 S1) Signal	0.1-1.1v
73	GN	HO2S-22 (B2 S2) Signal	0.1-1.1v
74	BR	HO2S-12 (B1 S2) Signal	0.1-1.1v
75	GY/BL	Cruise Control Switch Signal	N/A
76	BK	Power Ground	<0.1v
77	RD/BL	A/T Control Relay Output	12-14v
78	YL	TP Sensor Signal	0.53-0.73v
79	YL/RD	Idle Position Switch Signal	0v, Switch Open: 4-5v
80	WT/YL	Vehicle Speed Signal	Digital Signal
81-82	---	Not Used	---
83	GN/RD	A/C Switch On/Off Signal	A/C Off: 0v, On: 12v
84	GY/RD	Diagnosis Control (DLC #1)	0v
85	RD/WT	ISO 9141 Bus (DLC #7)	12v
86-87	---	Not Used	---
88	BK	Power Ground	<0.1v
89	RD/BL	A/T Control Relay Output	12-14v
90	WT	Knock Sensor Signal	0.080v AC
91	BL/RD	MAP Sensor Signal	0.8-1.1v
92	BR/WT	FTP Sensor Signal	2.5v (fuel cap off)
93-96	---	Not Used	---
97	BK	Power Ground	<0.1v
98	BK/WT	Ignition Switch Power	12-14v

PCM C119 Wire Harness Connector Graphic

```
            C119 Connector
      71 72 73 74        75 76 77
      78 79 80 81 82 83 84 85 86 87 88 89
      90 91 92    93 94    95 96    97 98
```

View in into Front of Wire Harness (Connector Removed)

Standard Colors and Abbreviations

Abbreviation	Color	Abbreviation	Color	Abbreviation	Color
BK	Black	GY	Gray	RD	Red
BL	Blue	GN	Green	TN	Tan
BR	Brown	LG	Light Green	VT	Violet
DB	Dark Blue	OR	Orange	WT	White
DG	Dark Green	PK	Pink	YL	Yellow

2001-03 Sebring Coupe 3.0L V6 VIN H (A/T) 'C123' Connector

PCM Pin #	Wire Color	Circuit Description (30-Pin)	Value at Hot Idle
101	BK/BL	PNP Switch 'P' Signal	In Park: 12v, Or 0v
102	YL	PNP Switch 'D' Signal	In Drive: 12v, Or 0v
103	WT	Input Shaft Speed Sensor	Moving: 0-12-0v
104	GN/YL	Output Shaft Speed Sensor	Moving: 0-12-0v
105	BR/RD	A/T Shift Mode 1st Indicator	0v
106	RD/YL	A/T Second Solenoid	12-14v
107	YL/RD	A/T TCC Solenoid	12-14v
108	RD/BL	PNP Switch 'R' Signal	In Reverse: 12v, Or: 0v
109	WT	PNP Switch 'D3' Signal	In Drive 3: 12v, Or 0v
110	GN	PNP Switch 'Low' Signal	In Low: 12v, Or 0v
111	WT/GN	Immobilizer System Signal	Digital Signals
112	---	Not Used	---
113	WT/BL	DLC No. 2 Signal	N/A
114-116	---	Not Used	---
117	WT/BL	A/T Shift Mode 3rd Indicator	0v
118	YL/BL	A/T Shift Mode 2nd Indicator	0v
120	RD	A/T Underdrive Solenoid	12-14v
121	BR	PNP Switch 'N' Signal	In Neutral: 12v, Or 0v
122	YL/BL	PNP Switch 'D2' Signal	In Drive 2: 12v, Or 0v
123	GN/OR	Stop Light Switch Signal	Brake Off: 0v, On: 12v
124	BL/PK	TFT Sensor Signal	3.2-3.4v at 104°F
125-127	---	Not Used	---
128	YL/BL	A/T Shift Mode 4th Indicator	0v
129	RD/WT	A/T Low/Reverse Solenoid	12-14v
130	BL	A/T Overdrive Solenoid	12-14v

PCM C123 Wire Harness Connector Graphic

```
                    C123 Connector
 101 102    103 104              105 106 107
 108 109 110 111 112 113 114 115 116 117 118 119 120
 121 122 123    124 125    126 127 128    129 130
        View in into Front of Wire Harness (Connector Removed)
```

Standard Colors and Abbreviations

Abbreviation	Color	Abbreviation	Color	Abbreviation	Color
BK	Black	GY	Gray	RD	Red
BL	Blue	GN	Green	TN	Tan
BR	Brown	LG	Light Green	VT	Violet
DB	Dark Blue	OR	Orange	WT	White
DG	Dark Green	PK	Pink	YL	Yellow

2001-03 Sebring Coupe 3.0L V6 VIN H (M/T) 'C111' Connector

PCM Pin #	Wire Color	Circuit Description (35-Pin)	Value at Hot Idle
1	YL/BL	Injector 1 Driver	1.0-4.0 ms
2	GN/YL	Injector 4 Driver	1.0-4.0 ms
3	BR/WT	HO2S-21 Heater Control	Digital Signal: 0-12-0v
4	BL/WT	HO2S-11 Heater Control	Digital Signal: 0-12-0v
5, 7	---	Not Used	---
6	BL/RD	EGR Solenoid Control	Solenoid Off: 12v, On: 1v
8	BR/RD	Generator Signal (Lights "on")	0.2-3.5v
9	YL/RD	Injector 2 Driver	1.0-4.0 ms
10	GN/RD	Injector 5 Driver	1.0-4.0 ms
11	BK/BL	Power Transistor Control	Digital Signal: 0-12-0v
12-13	---	Not Used	---
14	GN/BK	IAC Stepper Motor 'A' Signal	Pulse Signals
15	GN/WT	IAC Stepper Motor 'B' Signal	Pulse Signals
16	BL	EVAP Purge Solenoid Control	PWM Signal: 0-12-0v
17	---	Not Used	
18	BL/OR	Radiator Cooling Fan Control	Fan Off: 0.1v, On: 0.7v
19	GN/BL	Volume Airflow Sensor Reset	1-3v, 3000 rpm: 6-9v
20	GN	A/C Clutch Relay Control	Relay Off: 12v, On: 1v
21	BK	Fuel Pump Relay Control	Pump Off: 12v, On: 1v
22	RD/YL	MIL (lamp) Control	Lamp Off: 12v, On: 1v
23	---	Not Used	---
24	YL/BL	Injector 3 Driver	1.0-4.0 ms
25	LG	Injector 6 Driver	1.0-4.0 ms
26	BL/WT	HO2S-22 Heater Control	Digital Signal: 0-12-0v
27	BR	HO2S-12 Heater Control	Digital Signal: 0-12-0v
28	GN/RD	IAC Stepper Motor 'A' Signal	Pulse Signals
29	BK/YL	IAC Stepper Motor 'B' Signal	Pulse Signals
30-34	---	Not Used	---
35	YL	EVAP Vent Solenoid Control	Solenoid Off: 12v, On: 1v

PCM C111 Wire Harness Connector Graphic

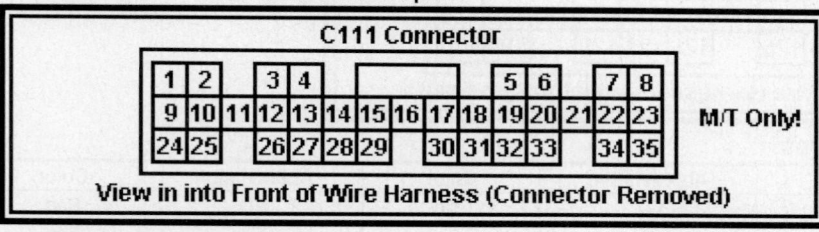

Standard Colors and Abbreviations

Abbreviation	Color	Abbreviation	Color	Abbreviation	Color
BK	Black	GY	Gray	RD	Red
BL	Blue	GN	Green	TN	Tan
BR	Brown	LG	Light Green	VT	Violet
DB	Dark Blue	OR	Orange	WT	White
DG	Dark Green	PK	Pink	YL	Yellow

2001-03 Sebring Coupe 3.0L V6 VIN H (M/T) 'C118' Connector

PCM Pin #	Wire Color	Circuit Description (28-Pin)	Value at Hot Idle
41	---	Not Used	---
42	GN/YL	Sensor Voltage Reference	4.9-5.1v
43	GN/RD	Distributor CKP Sensor Signal	Digital Signal: 0-5-0v
44	YL/GN	ECT Sensor Signal	0.3-0.9v at 176°F
45	WT	Tachometer Signals	Pulse Signals
46	BK	Power Ground	<0.1v
47	RD	Ignition Power (MPI Relay)	12-14v
48	---	Not Used	---
49	BK	Sensor Ground	<0.050v
50	RD/WT	CMP (TDC) Sensor Signal	Digital Signal: 0-5-0v
51	GN/WT	BARO Sensor Signal	3.7-4.3 at Sea Level
52	YL/BK	Generator 'FR' Terminal	0.5-4.5v
53	---	Not Used	---
54	YL	PSP Switch Signal	Straight: 5v, Turning: 0v
55-56	---	Not Used	---
57	WT/VT	MPI (Power) Relay Control	Relay Off: 12v, On: 1v
58	BK	Power Ground	<0.1v
59	RD	Ignition Power (MPI Relay)	12-14v
60	OR/BK	Keep Alive Power	12-14v
61	WT/GN	Volume Airflow Sensor	2.2-3.2v
62	RD/BL	IAT Sensor Signal	1.5-2.1v at 104°F
63-64	---	Not Used	---
65	BL/WT	A/C Switch 2 Signal (Hi Blow)	12v
66-67	---	Not Used	---
68	BR/RD	Starter (Cranking) Signal	9-11v (cranking)

PCM C118 Wire Harness Connector Graphic

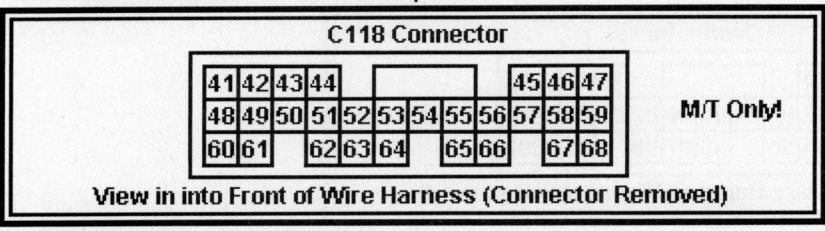

Standard Colors and Abbreviations

Abbreviation	Color	Abbreviation	Color	Abbreviation	Color
BK	Black	GY	Gray	RD	Red
BL	Blue	GN	Green	TN	Tan
BR	Brown	LG	Light Green	VT	Violet
DB	Dark Blue	OR	Orange	WT	White
DG	Dark Green	PK	Pink	YL	Yellow

2001-03 Sebring Coupe 3.0L V6 VIN H (M/T) 'C122' Connector

PCM Pin #	Wire Color	Circuit Description (30-Pin)	Value at Hot Idle
71	WT	HO2S-21 (B2 S1) Signal	0.1-1.1v
72	BL	HO2S-11 (B1 S1) Signal	0.1-1.1v
73	GN	HO2S-22 (B2 S2) Signal	0.1-1.1v
74	YL	HO2S-12 (B1 S2) Signal	0.1-1.1v
75	---	Not Used	---
76	BK	Power Ground	<0.1v
77	---	Not Used	---
78	YL	TP Sensor Signal	0.53-0.73v
79	YL/RD	Idle Position Switch Signal	0v, Off-Idle: 4v
80	WT/BL	Vehicle Speed Signal	Digital Signal
81-82	---	Not Used	---
83	GN/RD	A/C Switch On/Off Signal	A/C Off: 0v, On: 12v
84	GY/RD	Diagnosis Control (DLC #1)	0v
85	RD/WT	ISO 9141 Bus (DLC #7)	12v
86-87	---	Not Used	---
88	BK	Power Ground	<0.1v
89-90	---	Not Used	---
91	WT	Knock Sensor Signal	0.080v AC
92	BL/RD	MAP Sensor Signal	0.8-1.1v
93	BR/WT	FTP Sensor Signal	2.5v (fuel cap off)
94-96	---	Not Used	---
97	BK	Power Ground	<0.1v
98	WT/GN	Immobilizer System Signal	Digital Signals
99	BK/WT	Ignition Switch Power	12-14v
100	WT/VT	DLC No. 2 Signal	0v

PCM C122 Wire Harness Connector Graphic

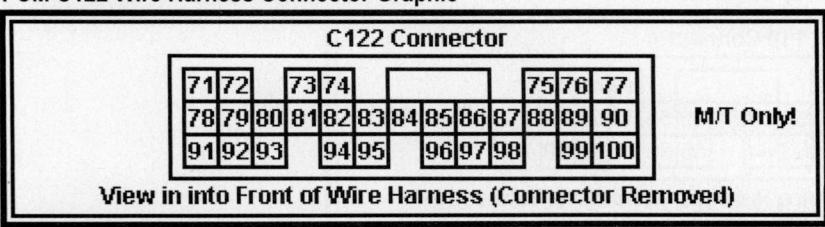

Standard Colors and Abbreviations

Abbreviation	Color	Abbreviation	Color	Abbreviation	Color
BK	Black	GY	Gray	RD	Red
BL	Blue	GN	Green	TN	Tan
BR	Brown	LG	Light Green	VT	Violet
DB	Dark Blue	OR	Orange	WT	White
DG	Dark Green	PK	Pink	YL	Yellow

DODGE CAR CONTENTS

About This Section

Introduction

This section of the manual contains Pin Tables for Dodge vehicles from 1990-2001. It can be used to help you repair Trouble Code and No Code problems related to the PCM.

Vehicle Coverage

The following vehicle applications are covered in this section:

- 1995-2000 Avenger
- 1990-93 Daytona, Dynasty & Monaco
- 1994-2003 Intrepid
- 1995-2003 Neon
- 1990-95 Shadow & Spirit
- 1995-2003 Stratus
- 1992-2003 Viper

How to Use This Section

This section of the manual can be used to look up the location of a particular pin, a wire color or to find a "known good" value of a circuit. To locate the PCM information for a particular vehicle, find the model, correct engine size (with VIN Code) and finally the year of the vehicle.

For example, to look up the PCM terminals for a 2001 Intrepid equipped with a 2.7L V6 DOHC VIN R go to Page 1 of the contents to find the text string below.

Then turn to Page 8-44 to find the following PCM related information.

2000-02 Intrepid 2.7L V6 DOHC VIN R, U & V (A/T) 'C1' Black Connector

PCM Pin #	Wire Color	Circuit Description (40 Pin)	Value at Hot Idle
4	TN/LB	Coil 6 Driver	5°, at 55 mph: 8° dwell
5	YL/RD	S/C Power Supply (B+)	12-14v
6	DG/OR	ASD Relay Output	12-14v
7	YL/WT	Injector 3 Driver	1-4 ms
8	DG	Generator Field Driver	Digital Signal: 0-12-0v

In this example, the Coil Driver control circuit is connected to Pin 4 of the 'C1' Black Connector with a TN/LB wire. The Hot Idle value is 5° while the 55 mph value is 8°. Note the change in dwell as the mph changed.

The ASD relay output signal is connected to Pin 6 of the C1 40 Pin connector with a DG/OR wire. This signal indicates the voltage output of the ASD relay during vehicle operation. This signal should always read near battery voltage with the engine running.

The acronym (A/T) that appears in the title for the table indicates the information in this table is for a vehicle with an automatic transaxle.

DODGE Pin Tables

1995 Avenger 2.0L I4 DOHC MFI VIN Y (All) 60 Pin Connector

PCM Pin #	Wire Color	Circuit Description (60 Pin)	Value at Hot Idle
1	BR	Coil 2 Driver	5°, at 55 mph: 8° dwell
2	BK	Power Ground	<0.1v
3	YL/DG	Injector 3 Driver	1-4 ms
4	LG/BK	Injector 1 Driver	1-4 ms
5	YL/WT	Vehicle Speed Signal	Digital Signal
6	BR/DB	IAT Sensor Signal	At 100°F: 1.83v
7	WT/DG	HO2S-12 (B1 S2) Signal	0.1-1.1v
8	WT/BK	HO2S-11 (B1 S1) Signal	0.1-1.1v
9	DB/BK	SCI Receive	0v
10	BR/RD	TP Sensor Signal	0.6-1.0v
11	RD/BK	Battery Power (Fused B+)	12-14v
12	---	Not Used	---
13	YL/DB	Fuel Level Sensor Signal	70 ohms (±20) with full tank
14	OR	IAC 2 Driver	DC pulse signals: 0.8-11v
15	GY	IAC 3 Driver	DC pulse signals: 0.8-11v
16	LG/BK	EVAP Purge Solenoid Control	Solenoid Off: 12v, On: 1v
17	LG	S/C Indicator Driver Control	Lamp On: 1v, Off: 12v
18	RD/WT	ASD Relay Control	Relay Off: 12v, On: 1v
19	DG/BK	Low Speed Fan Relay	Relay Off: 12v, On: 1v
20	RD/YL	Secondary Air Solenoid	Solenoid Off: 12v, On: 1v
21	BK/DB	Coil 1 Driver	5°, at 55 mph: 8° dwell
22	BK	Power Ground	<0.1v
23	YL/RD	Injector 2 Driver	1-4 ms
24	LG/RD	Injector 4 Driver	1-4 ms
25	DB/WT	CKP Sensor Signal	Digital Signal: 0-5-0v
26	DB/RD	CMP Sensor Signal	Digital Signal: 0-5-0v
27	WT/YL	Knock Sensor Signal	0.080v AC
28	DG/WT	ECT Sensor Signal	At 180°F: 2.80v
29	YL/BK	MAP Sensor Signal	1.5-1.7v
30	PK	SCI Transmit	0v
31	RD	S/C Set Switch Signal	S/C & Set Switch On: 3.8v
32	BR/WT	Brake Switch Signal	Brake Off: 0v, On: 12v
33	DG/RD	A/C Select Switch Sense	A/C On: 1v, Off: 12v
34	YL/DB	IAC 4 Driver	DC pulse signals: 0.8-11v
35	GY/DB	IAC 1 Driver	DC pulse signals: 0.8-11v
36	DG/RD	MIL (lamp) Control	Lamp On: 1v, Off: 12v
37	DB	Generator Lamp Control	Lamp On: 1v, Off: 12v
38	WT/RD	Fuel Pump Relay Control	Relay Off: 12v, On: 1v
39	RD/DB	EGR Solenoid Control	12v, 55 mph: 1v
40	LG/WT	S/C Vacuum Solenoid	Vacuum Increasing: 1v

1995 Avenger 2.0L I4 DOHC MFI VIN Y (All) 60 Pin Connector

PCM Pin #	Wire Color	Circuit Description (60 Pin)	Value at Hot Idle
41	DB	Generator Field Driver	Digital Signal: 0-12-0v
42	BK/RD	ASD Relay Output	12-14v
43	DG/YL	5-Volt Supply	4.9-5.1v
44	YL	8-Volt Supply	7.9-8.1v
45	WT/DB	CCD Bus (-)	<0.050v
46	BK/DB	CCD Bus (+)	Digital Signal: 0-5-0v
47	---	Not Used	---
48	WT	Tachometer Signal	Pulse Signals
49	---	Not Used	---
50	BK/YL	PNP Switch Signal	In P/N: 0v, Others: 5v
51	BK/DG	Sensor Ground	<0.050v
52	BK	Power Ground	<0.1v
53	---	Not Used	---
54	BK/WT	Ignition Switch Output	12-14v
55	RD/BK	S/C On/Off Sense Signal	S/C On: 1v, Off: 12v
56	DB/YL	PSP Switch Signal	Straight: 0v, Turning: 5v
57	DG/OR	High Speed Fan Relay	Relay Off: 12v, On: 1v
58	DB/YL	Bulb Check Driver Control	Lamp On: 1v, Off: 12v
59	DG	A/C Clutch Relay Control	Relay Off: 12v, On: 1v
60	BK/YL	S/C Vent Solenoid	Vacuum Decreasing: 1v

Pin Connector Graphic

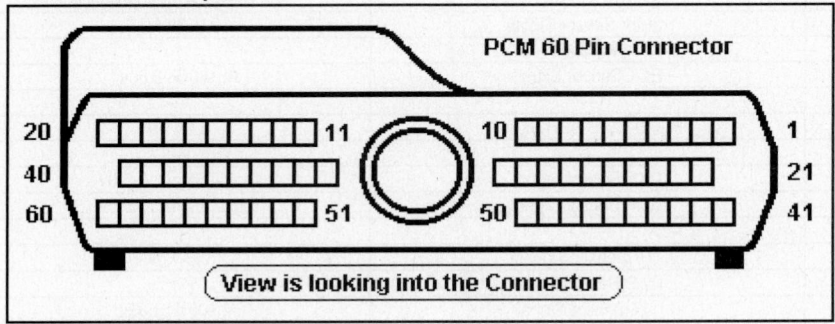

Standard Colors and Abbreviations

Abbreviation	Color	Abbreviation	Color	Abbreviation	Color
BK	Black	GY	Gray	RD	Red
BL	Blue	GN	Green	TN	Tan
BR	Brown	LG	Light Green	VT	Violet
DB	Dark Blue	OR	Orange	WT	White
DG	Dark Green	PK	Pink	YL	Yellow

1996-97 Avenger 2.0L I4 DOHC MFI VIN Y (All) 'C1' Black Connector

PCM Pin #	Wire Color	Circuit Description (40 Pin)	Value at Hot Idle
1	---	Not Used	---
2	BK/DB	Coil 1 Driver	5°, at 55 mph: 8° dwell
3	BR	Coil 2 Driver	5°, at 55 mph: 8° dwell
4	DB	Generator Field Driver	Digital Signal: 0-12-0v
5	---	Not Used	---
6	BK/RD	ASD Relay Output	12-14v
7	YL/DG	Injector 3 Driver	1-4 ms
8	DG/RD	MIL (lamp) Control	Lamp On: 1v, Off: 12v
9	DB/YL	Bulb Check Driver Control	Lamp On: 1v, Off: 12v
10	BK	Power Ground	<0.1v
11	---	Not Used	---
12	RD/BK	S/C On/Off Switch Signal	S/C On: 1v, Off: 12v
13	LG/BK	Injector 1 Driver	1-4 ms
14-15	---	Not Used	---
16	LG/RD	Injector 4 Driver	1-4 ms
17	YL/RD	Injector 2 Driver	1-4 ms
18	---	Not Used	---
19	DG/OR	High Speed Fan Relay	Relay Off: 12v, On: 1v
20	BK/WT	Ignition Switch Output	12-14v
21	---	Not Used	---
22	LG	S/C Indicator Driver Control	Lamp On: 1v, Off: 12v
23	YL/DB	Fuel Level Sensor Signal	70 ohms (±20) with full tank
24	WT/YL	Knock Sensor Signal	0.080v AC
25	---	Not Used	---
26	DG/WT	ECT Sensor Signal	At 180°F: 2.80v
27-29	---	---	---
30	WT/BK	HO2S-11 (B1 S1) Signal	0.1-1.1v
31	---	---	---
32	DB/WT	CKP Sensor Signal	Digital Signal: 0-5-0v
33	DB/RD	CMP Sensor Signal	Digital Signal: 0-5-0v
34	---	---	---
35	BR/RD	TP Sensor Signal	0.6-1.0v
36	YL/BK	MAP Sensor Signal	1.5-1.7v
37	BR/DB	IAT Sensor Signal	At 100°F: 1.83v
38	DG/RD	A/C Select Switch Sense	A/C On: 1v, Off: 12v
39	---	Not Used	---
40	RD/DB	EGR Solenoid Control	12v, 55 mph: 1v

Standard Colors and Abbreviations

Abbreviation	Color	Abbreviation	Color	Abbreviation	Color
BK	Black	GY	Gray	RD	Red
BL	Blue	GN	Green	TN	Tan
BR	Brown	LG	Light Green	VT	Violet
DB	Dark Blue	OR	Orange	WT	White
DG	Dark Green	PK	Pink	YL	Yellow

1996-97 Avenger 2.0L I4 DOHC MFI VIN Y (All) 'C2' White Connector

PCM Pin #	Wire Color	Circuit Description (40 Pin)	Value at Hot Idle
41	RD	S/C Set Switch Signal	S/C & Set Switch On: 3.8v
42	---	Not Used	---
43	BK/DG	Sensor Ground	<0.050v
44	YL	8-Volt Supply	7.9-8.1v
45	DB/YL	PSP Switch Signal	Straight: 0v, Turning: 5v
46	RD/BK	Battery Power (Fused B+)	12-14v
47	BK/DG	Sensor Ground	<0.050v
48	DG/DB	IAC 3 Driver	DC pulse signals: 0.8-11v
49	OR	IAC 2 Driver	DC pulse signals: 0.8-11v
50	BK	Power Ground	<0.1v
51	WT/DG	HO2S-12 (B1 S2) Signal	0.1-1.1v
52-54	---	Not Used	---
55	DG/BK	Low Speed Fan Relay	Relay Off: 12v, On: 1v
56	DB	Generator Lamp Control	Lamp On: 1v, Off: 12v
57	DG, GY	IAC 1 Driver	DC pulse signals: 0.8-11v
58	YL/DB	IAC 4 Driver	DC pulse signals: 0.8-11v
59	BK/DB	CCD Bus (+)	Digital Signal: 0-5-0v
60	WT/DB	CCD Bus (-)	<0.050v
61	DG/YL	5-Volt Supply	4.9-5.1v
62	BR/WT	Brake Switch Signal	Brake Off: 0v, On: 12v
63	OR/BK	Torque Management Request	Digital Signals
64	DG	A/C Clutch Relay Control	Relay Off: 12v, On: 1v
65	PK	SCI Transmit	0v
66	YL/WT	Vehicle Speed Signal	Digital Signal
67	RD/WT	ASD Relay Control	Relay Off: 12v, On: 1v
68	LG/BK	EVAP Purge Solenoid Control	PWM signal: 0-12-0v
69	DG/OR	High Speed Fan Relay	Relay Off: 12v, On: 1v
70-71	---	Not Used	---
72	RD/DB	LDP Switch Signal	Switch Closed: 0v, open: 12v
73	WT	Tachometer Signal	Pulse Signals
74 ('96)	BK/DB	Fuel Pump Relay Control	Relay Off: 12v, On: 1v
74 ('97)	WT/RD	Fuel Pump Relay Control	Relay Off: 12v, On: 1v
75	DB	SCI Receive	5v
77	RD/YL	LDP Solenoid Control	PWM signal: 0-12-0v
78	LG/WT	S/C Vacuum Solenoid	Vacuum Increasing: 1v
79	RD/YL	M/T: Aspirator Solenoid	Solenoid Off: 12v, On: 1v
80	BK/YL	S/C Vent Solenoid	Vacuum Decreasing: 1v

Pin Connector Graphic

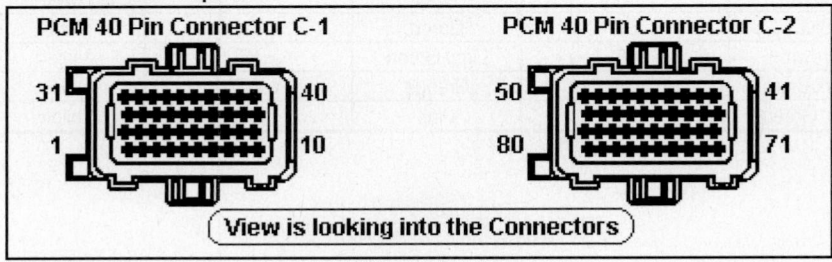

1998-99 Avenger 2.0L I4 DOHC MFI VIN Y (All) 'C1' Black Connector

PCM Pin #	Wire Color	Circuit Description (40 Pin)	Value at Hot Idle
1-2	---	Not Used	---
3	BR	Coil 2 Driver	5°, at 55 mph: 8° dwell
4-5	---	Not Used	---
6	BK/RD	ASD Relay Output	12-14v
7	YL/DG	Injector 3 Driver	1-4 ms
8	DB	Generator Field Driver	Digital Signal: 0-12-0v
9	DB/YL	Bulb Check Driver Control	Lamp On: 1v, Off: 12v
10	BK	Power Ground	<0.1v
11	BK/DB	Coil 1 Driver	5°, at 55 mph: 8° dwell
12	RD/BK	S/C On/Off Switch Signal	S/C On: 1v, Off: 12v
13	LG/BK	Injector 1 Driver	1-4 ms
14-15	---	Not Used	---
16	LG/RD	Injector 4 Driver	1-4 ms
17	YL/RD	Injector 2 Driver	1-4 ms
18	LG	S/C Indicator Driver Control	Lamp On: 1v, Off: 12v
19	DG/OR	High Speed Fan Relay	Relay Off: 12v, On: 1v
20	BK/WT	Ignition Switch Output	12-14v
21	---	Not Used	---
22	DG/RD	MIL (lamp) Control	Lamp On: 1v, Off: 12v
23	YL/DB	Fuel Level Sensor Signal	70 ohms (±20) with full tank
24	---	Not Used	---
25	WT/YL	Knock Sensor Signal	0.080v AC
26	DG/WT	ECT Sensor Signal	At 180°F: 2.80v
27-29	---	Not Used	---
30	WT/BK	HO2S-11 (B1 S1) Signal	0.1-1.1v
31	---	Not Used	---
32	DB/WT	CKP Sensor Signal	Digital Signal: 0-5-0v
33	DB/RD	CMP Sensor Signal	Digital Signal: 0-5-0v
34	---	Not Used	---
35	BR/RD	TP Sensor Signal	0.6-1.0v
36	YL/BK	MAP Sensor Signal	1.5-1.7v
37	BR/DB	IAT Sensor Signal	At 100°F: 1.83v
38	DG/RD	A/C Select Switch Sense	A/C On: 1v, Off: 12v
39	DB	Generator Lamp Control	Lamp On: 1v, Off: 12v
40	RD/DB	EGR Solenoid Control	12v, 55 mph: 1v

Standard Colors and Abbreviations

Abbreviation	Color	Abbreviation	Color	Abbreviation	Color
BK	Black	GY	Gray	RD	Red
BL	Blue	GN	Green	TN	Tan
BR	Brown	LG	Light Green	VT	Violet
DB	Dark Blue	OR	Orange	WT	White
DG	Dark Green	PK	Pink	YL	Yellow

1998-99 Avenger 2.0L I4 DOHC MFI VIN Y (All) 'C2' White Connector

PCM Pin #	Wire Color	Circuit Description (40 Pin)	Value at Hot Idle
41	RD	S/C Set Switch Signal	S/C & Set Switch On: 3.8v
42-43	---	Not Used	---
44	YL	8-Volt Supply	7.9-8.1v
45	DB/YL	PSP Switch Signal	Straight: 0v, Turning: 5v
46	RD/BK	Battery Power (Fused B+)	12-14v
47	BK	Sensor Ground	<0.050v
48	GY/DB	IAC 3 Driver	DC pulse signals: 0.8-11v
49	OR	IAC 2 Driver	DC pulse signals: 0.8-11v
50	BK	Power Ground	<0.1v
51	WT/DG	HO2S-12 (B1 S2) Signal	0.1-1.1v
52-54	---	Not Used	---
55	DG/BK	Low Speed Fan Relay	Relay Off: 12v, On: 1v
56	LG/WT	S/C Vacuum Solenoid	Vacuum Increasing: 1v
57	GY	IAC 1 Driver	DC pulse signals: 0.8-11v
58	YL/DB	IAC 4 Driver	DC pulse signals: 0.8-11v
59	BK/DB	CCD Bus (+)	Digital Signal: 0-5-0v
60	WT/DB	CCD Bus (-)	<0.050v
61	DG/YL	5-Volt Supply	4.9-5.1v
62	BR/WT	Brake Switch Signal	Brake Off: 0v, On: 12v
63	OR/BK	Torque Management Request	Digital Signals
64	DG	A/C Clutch Relay Control	Relay Off: 12v, On: 1v
65	PK	SCI Transmit	0v
66	YL/WT	Vehicle Speed Signal	Digital Signal
67	RD/WT	ASD Relay Control	Relay Off: 12v, On: 1v
68	LG/BK	EVAP Purge Solenoid Control	0-100%
69	DG/OR	High Speed Fan Relay	Relay Off: 12v, On: 1v
70	BK/WT	EVAP Purge Sense Signal	12-14v
71	---	Not Used	---
72	RD/DB	LDP Switch Signal	Switch Closed: 0v, open: 12v
73	WT	Tachometer Signal	Pulse Signals
74	BK/DB	Fuel Pump Relay Control	Relay Off: 12v, On: 1v
75	DB/BK	SCI Receive	5v
76	BK/YL	PNP Switch Signal	In P/N: 0v, Others: 5v
77	RD/YL	LDP Solenoid Control	PWM signal: 0-12-0v
78-79	---	Not Used	---
80	BK/YL	S/C Vent Solenoid	Vacuum Decreasing: 1v

Pin Connector Graphic

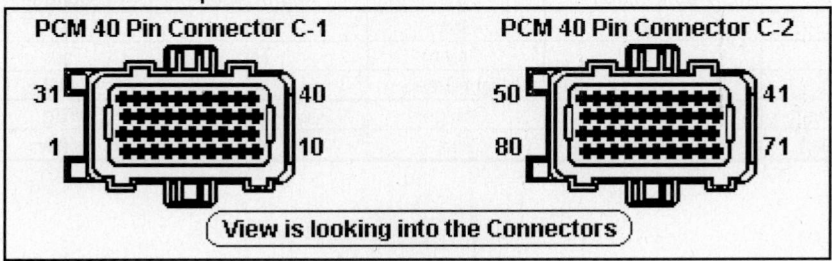

1995 Avenger 2.5L V6 24v SOHC MFI VIN N (A/T) 'C1' Black Connector

PCM Pin #	Wire Color	Circuit Description (40 Pin)	Value at Hot Idle
1-3	---	Not Used	---
4	DB	Generator Field Driver	Digital Signal: 0-12-0v
5	---	Not Used	---
6	BK/RD	ASD Relay Output	12-14v
7	YL/DG	Injector 3 Driver	1-4 ms
8	DG/RD	MIL (lamp) Control	Lamp On: 1v, Off: 12v
9	DB/YL	Lamp Check Driver	Lamp On: 1v, Off: 12v
10	BK	Power Ground	<0.1v
11	RD/DG	Coil Driver	5°, at 55 mph: 8° dwell
12	RD/BK	S/C On/Off Switch Signal	S/C On: 1v, Off: 12v
13	LG/BK	Injector 1 Driver	1-4 ms
14	BR/RD	Injector 6 Driver	1-4 ms
15	RD/WT	Injector 5 Driver	1-4 ms
16	LG/RD	Injector 4 Driver	1-4 ms
17	YL/RD	Injector 2 Driver	1-4 ms
18	---	Not Used	---
19	DG/OR	High Speed Fan Relay	Relay Off: 12v, On: 1v
20	BK/WT	Ignition Switch Output	12-14v
21	---	Not Used	---
22	DG	S/C Indicator Driver Control	Lamp On: 1v, Off: 12v
23-25	---	Not Used	---
26	DG/WT	ECT Sensor Signal	At 180°F: 2.80v
27-28	---	Not Used	---
29	DG/BK	HO2S-12 (B1 S2) Signal	0.1-1.1v
30	WT/BK	HO2S-11 (B1 S1) Signal	0.1-1.1v
31	---	Not Used	---
32	LB/WT	CKP Sensor Signal	Digital Signal: 0-5-0v
33	BR	CMP Sensor Signal	Digital Signal: 0-5-0v
34	---	Not Used	---
35	BR/RD	TP Sensor Signal	0.6-1.0v
36	YL/BK	MAP Sensor Signal	1.5-1.7v
37	BR/LB	IAT Sensor Signal	At 100°F: 1.83v
38	DG/RD	A/C Clutch Switch Sense	A/C On: 1v, Off: 12v
39	---	Not Used	---
40	RD/LB	EGR Solenoid Control	12v, 55 mph: 1v

Standard Colors and Abbreviations

Abbreviation	Color	Abbreviation	Color	Abbreviation	Color
BK	Black	GY	Gray	RD	Red
BL	Blue	GN	Green	TN	Tan
BR	Brown	LG	Light Green	VT	Violet
DB	Dark Blue	OR	Orange	WT	White
DG	Dark Green	PK	Pink	YL	Yellow

1995 Avenger 2.5L V6 24v SOHC MFI VIN N (A/T) 'C2' White Connector

PCM Pin #	Wire Color	Circuit Description (40 Pin)	Value at Hot Idle
41	RD	S/C Set Switch Signal	S/C & Set Switch On: 3.8v
42	---	Not Used	---
43	BK/DG	Sensor Ground	<0.050v
44	YL	8-Volt Supply	7.9-8.1v
45	LB/YL	PSP Switch Signal	Straight: 0v, Turning: 5v
46	RD/BK	Battery Power (Fused B+)	12-14v
47	BK	Sensor Ground	<0.050v
48	DG/LB	IAC 1 Driver	DC pulse signals: 0.8-11v
49	OR	IAC 2 Driver	DC pulse signals: 0.8-11v
50	BK	Power Ground	<0.1v
51-54	---	Not Used	---
55	DG/BK	Low Speed Fan Relay	Relay Off: 12v, On: 1v
56	LB	Generator Lamp Control	Lamp On: 1v, Off: 12v
57	DG	IAC 3 Driver	DC pulse signals: 0.8-11v
58	YL/LB	IAC 4 Driver	DC pulse signals: 0.8-11v
59	BK/DB	CCD Bus (+)	Digital Signal: 0-5-0v
60	WT/LB	CCD Bus (-)	<0.050v
61	DG/YL	5-Volt Supply	4.9-5.1v
62	BR/WT	Brake Switch Signal	Brake Off: 0v, On: 12v
63	OR/BK	Torque Management Request	Digital Signals
64	DG	A/C Clutch Relay Control	Relay Off: 12v, On: 1v
65	PK	SCI Transmit	0v
66	YL/WT	Vehicle Speed Signal	Digital Signal
67	RD/WT	ASD Relay Control	Relay Off: 12v, On: 1v
68	LG/BK	EVAP Purge Solenoid Control	PWM signal: 0-12-0v
69	DG/WT	High Speed Fan Relay	Relay Off: 12v, On: 1v
70-72	---	Not Used	---
73	WT	Tachometer Signal	Pulse Signals
74	WT/RD	Fuel Pump Relay Control	Relay Off: 12v, On: 1v
75	DB/BK	SCI Receive	0v
76	BK/YL	PNP Switch Signal	In P/N: 0v, Others: 5v
77	---	Not Used	---
78	LG/WT	S/C Vacuum Solenoid	Vacuum Increasing: 1v
79	---	Not Used	---
80	BK/YL	S/C Vent Solenoid	Vacuum Decreasing: 1v

Pin Connector Graphic

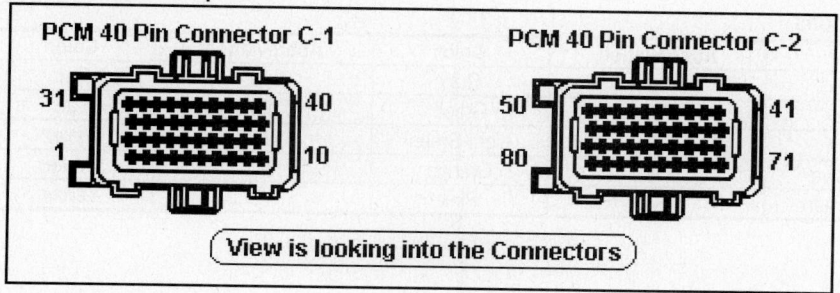

PCM 40 Pin Connector C-1   PCM 40 Pin Connector C-2

View is looking into the Connectors

1996-97 Avenger 2.5L V6 SOHC MFI VIN N (A/T) 'C1' Black Connector

PCM Pin #	Wire Color	Circuit Description (40 Pin)	Value at Hot Idle
1	WT/RD	HO2S-12 (B1 S2) Signal	0.1-1.1v
2-3	---	Not Used	---
4	DB	Generator Field Driver	Digital Signal: 0-12-0v
5	---	Not Used	---
6	BK/RD	ASD Relay Output	12-14v
7	YL/DG	Injector 3 Driver	1-4 ms
8	DG/RD	MIL (lamp) Control	Lamp On: 1v, Off: 12v
9	DB/YL	Lamp Check Driver	Lamp On: 1v, Off: 12v
10	BK	Power Ground	<0.1v
11	RD/DG	Coil Driver	5°, at 55 mph: 8° dwell
12	RD/BK	S/C On/Off Switch Signal	S/C On: 1v, Off: 12v
13	LG/BK	Injector 1 Driver	1-4 ms
14	BR/RD	Injector 6 Driver	1-4 ms
15	RD/WT	Injector 5 Driver	1-4 ms
16	LG/RD	Injector 4 Driver	1-4 ms
17	YL/RD	Injector 2 Driver	1-4 ms
19	DG/OR	High Speed Fan 2 Control	Relay Off: 12v, On: 1v
20	BK/WT	Ignition Switch Output	12-14v
21	---	Not Used	---
22	LG	S/C Indicator Driver Control	Lamp On: 1v, Off: 12v
23	YL/DB	Fuel Level Sensor Signal	70 ohms (±20) with full tank
24-25	---	Not Used	---
26	DG/WT	ECT Sensor Signal	At 180°F: 2.80v
27	---	Not Used	---
28	DG/YL	High Speed Fan 1 Control	Relay Off: 12v, On: 1v
29	DG/BK	HO2S-11 (B1 S1) Signal	0.1-1.1v
30	WT/BK	HO2S-21 (B2 S1) Signal	0.1-1.1v
32	DB/WT	CKP Sensor Signal	Digital Signal: 0-5-0v
33	BR	CMP Sensor Signal	Digital Signal: 0-5-0v
34	---	Not Used	---
35	BR/RD	TP Sensor Signal	0.6-1.0v
36	YL/BK	MAP Sensor Signal	1.5-1.7v
37	BR/DB	IAT Sensor Signal	At 100°F: 1.83v
38	DG/RD	A/C Clutch Switch Sense	A/C On: 1v, Off: 12v
39	---	Not Used	---
40	RD/DB	EGR Solenoid Control	12v, 55 mph: 1v

Standard Colors and Abbreviations

Abbreviation	Color	Abbreviation	Color	Abbreviation	Color
BK	Black	GY	Gray	RD	Red
BL	Blue	GN	Green	TN	Tan
BR	Brown	LG	Light Green	VT	Violet
DB	Dark Blue	OR	Orange	WT	White
DG	Dark Green	PK	Pink	YL	Yellow

1996-97 Avenger 2.5L V6 SOHC MFI VIN N (A/T) 'C2' White Connector

PCM Pin #	Wire Color	Circuit Description (40 Pin)	Value at Hot Idle
41	RD	S/C Set Switch Signal	S/C & Set Switch On: 3.8v
42	---	Not Used	---
43	BK/DG	Sensor Ground	<0.050v
44	YL	8-Volt Supply	7.9-8.1v
45	DB/YL	PSP Switch Signal	Straight: 0v, Turning: 5v
46	RD/BK	Battery Power (Fused B+)	12-14v
47	BK	Sensor Ground	<0.050v
48	GY/DB	IAC 3 Driver	DC pulse signals: 0.8-11v
49	OR	IAC 2 Driver	DC pulse signals: 0.8-11v
50	BK	Power Ground	<0.1v
51	WT/DG	HO2S-22 (B2 S2) Signal	0.1-1.1v
55	DG/BK	Low Speed Fan Relay	Relay Off: 12v, On: 1v
56	DB	Generator Lamp Control	Lamp On: 1v, Off: 12v
57	GY	IAC 1 Driver	DC pulse signals: 0.8-11v
58	YL/DB	IAC 4 Driver	DC pulse signals: 0.8-11v
59	BK/DB	CCD Bus (+)	Digital Signal: 0-5-0v
60	WT/DB	CCD Bus (-)	<0.050v
61	DG/YL	5-Volt Supply	4.9-5.1v
62	BR/WT	Brake Switch Signal	Brake Off: 0v, On: 12v
63	OR/BK	Torque Management Request	Digital Signals
64	DG	A/C Clutch Relay Control	Relay Off: 12v, On: 1v
65	PK	SCI Transmit	0v
66	YL/WT	Vehicle Speed Signal	Digital Signal
67	RD/WT	ASD Relay Control	Relay Off: 12v, On: 1v
68	LG/BK	EVAP Purge Solenoid Control	PWM signal: 0-12-0v
69	DG/WT	High Speed Condenser Fan	Relay Off: 12v, On: 1v
72	RD/DB	LDP Switch Signal	Switch Closed: 0v, open: 12v
73	WT	Tachometer Signal	Pulse Signals
74	WT/RD	Fuel Pump Relay Control	Relay Off: 12v, On: 1v
75	DB/BK	SCI Receive	0v
76	BK/YL	PNP Switch Signal	In P/N: 0v, Others: 5v
77	RD/YL	LDP Solenoid Control	PWM signal: 0-12-0v
78	LG/WT	S/C Vacuum Solenoid	Vacuum Increasing: 1v
80	BK/YL	S/C Vent Solenoid	Vacuum Decreasing: 1v

Pin Connector Graphic

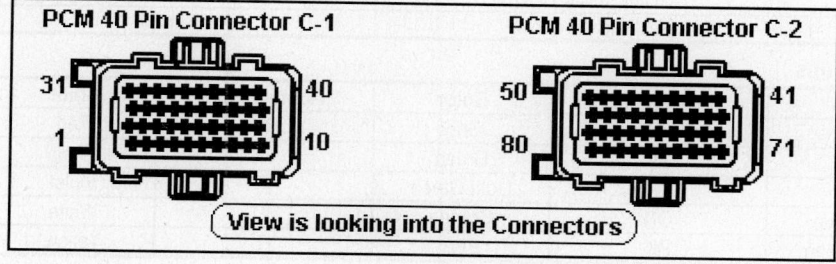

PCM 40 Pin Connector C-1

31 · 40
1 · 10

PCM 40 Pin Connector C-2

50 · 41
80 · 71

View is looking into the Connectors

1998-2000 Avenger 2.5L V6 SOHC MFI VIN N (A/T) 'C1' Black Connector

PCM Pin #	Wire Color	Circuit Description (40 Pin)	Value at Hot Idle
1-3	---	Not Used	---
4	RD/DG	Coil Driver Control	5°, at 55 mph: 8° dwell
5	---	Not Used	---
6	BK/RD	ASD Relay Output	12-14v
7	YL/DG	Injector 3 Driver	1-4 ms
8	DB	Generator Field Driver	Digital Signal: 0-12-0v
9	DB/YL	Lamp Check Driver	Lamp On: 1v, Off: 12v
10	BK	Power Ground	<0.1v
11	---	Not Used	---
12	RD/BK	S/C On Switch Sense	12-14v
13	LG/BK	Injector 1 Driver	1-4 ms
14	BR/RD	Injector 6 Driver	1-4 ms
15	RD/WT	Injector 5 Driver	1-4 ms
16	LG/RD	Injector 4 Driver	1-4 ms
17	YL/RD	Injector 2 Driver	1-4 ms
18	LG	S/C Lamp Driver	Lamp On: 1v, Off: 12v
19	DG/OR	High Speed Fan 2 Control	Relay Off: 12v, On: 1v
20	BK/WT	Ignition Switch Output	12-14v
21	---	Not Used	---
22	RD/DG	MIL (lamp) Control	Lamp On: 1v, Off: 12v
23	YL/DB	Fuel Level Sensor Signal	70 ohms (±20) with full tank
24	---	Not Used	---
26	DG/WT	ECT Sensor Signal	At 180°F: 2.80v
27	---	Not Used	---
28	DG/YL	High Speed Fan 1 Control	Relay Off: 12v, On: 1v
29	DG/BK	HO2S-11 (B1 S1) Signal	0.1-1.1v
30	WT/BK	HO2S-12 (B1 S2) Signal	0.1-1.1v
31	---	Not Used	---
32	DB/WT	CKP Sensor Signal	Digital Signal: 0-5-0v
33	BR	CMP Sensor Signal	Digital Signal: 0-5-0v
34	---	Not Used	---
35	BR/RD	TP Sensor Signal	0.6-1.0v
36	YL/BK	MAP Sensor Signal	1.5-1.7v
37	BR/DB	IAT Sensor Signal	At 100°F: 1.83v
38	DG/RD	A/C Clutch Switch Sense	A/C On: 1v, Off: 12v
39	DB	Generator Lamp Control	Lamp On: 1v, Off: 12v
40	RD/DB	EGR Solenoid Control	12v, 55 mph: 1v

Standard Colors and Abbreviations

Abbreviation	Color	Abbreviation	Color	Abbreviation	Color
BK	Black	GY	Gray	RD	Red
BL	Blue	GN	Green	TN	Tan
BR	Brown	LG	Light Green	VT	Violet
DB	Dark Blue	OR	Orange	WT	White
DG	Dark Green	PK	Pink	YL	Yellow

1998-2000 Avenger 2.5L V6 SOHC MFI VIN N (A/T) 'C2' White Connector

PCM Pin #	Wire Color	Circuit Description (40 Pin)	Value at Hot Idle
41	RD	S/C Set Switch Signal	S/C & Set Switch On: 3.8v
42-43	---	Not Used	---
44	YL	8-Volt Supply	7.9-8.1v
45	DB/YL	PSP Switch Signal	Straight: 0v, Turning: 5v
46	RD/BK	Battery Power (Fused B+)	12-14v
47	BK	Sensor Ground	<0.050v
48	GY/DB	IAC 3 Driver	DC pulse signals: 0.8-11v
49	OR	IAC 2 Driver	DC pulse signals: 0.8-11v
50	BK	Power Ground	<0.1v
51	WT/DG	HO2S-22 (B2 S2) Signal	0.1-1.1v
53	WT/RD	HO2S-21 (B2 S1) Signal	0.1-1.1v
55	DG/BK	Low Speed Fan Relay	Relay Off: 12v, On: 1v
56	LG/WT	S/C Vacuum Solenoid	Vacuum Increasing: 1v
57	GY	IAC 1 Driver	DC pulse signals: 0.8-11v
58	YL/DB	IAC 4 Driver	DC pulse signals: 0.8-11v
59	BK/DB	CCD Bus (+)	Digital Signal: 0-5-0v
60	WT/DB	CCD Bus (-)	<0.050v
61	DG/YL	5-Volt Supply	4.9-5.1v
62	BR/WT	Brake Switch Signal	Brake Off: 0v, On: 12v
63	OR/BK	Torque Management Request	Digital Signals
64	DG	A/C Clutch Relay Control	Relay Off: 12v, On: 1v
65	PK	SCI Transmit	0v
66	YL/WT	Vehicle Speed Signal	Digital Signal
67	RD/WT	ASD Relay Control	Relay Off: 12v, On: 1v
68	LG/BK	EVAP Purge Solenoid Control	PWM signal: 0-12-0v
69	DG/WT	High Speed Condenser Fan	Relay Off: 12v, On: 1v
70	B/WT	EVAP Purge Solenoid Sense	0-1v
72	RD/DB	LDP Switch Signal	Switch Closed: 0v, open: 12v
73	WT	Tachometer Signal	Pulse Signals
74	WT/RD	Fuel Pump Relay Control	Relay Off: 12v, On: 1v
75	DB/BK	SCI Receive	0v
76	BK/YL	PNP Switch Signal	In P/N: 0v, Others: 5v
77	RD/YL	LDP Solenoid Control	PWM signal: 0-12-0v
80	BK/YL	S/C Vent Solenoid	Vacuum Decreasing: 1v

Pin Connector Graphic

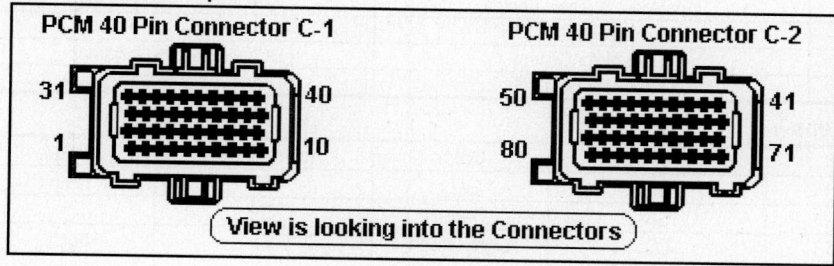

PCM 40 Pin Connector C-1

PCM 40 Pin Connector C-2

31   40   50   41

1   10   80   71

View is looking into the Connectors

1992-93 Daytona 2.2L I4 DOHC Turbo VIN A (M/T) 60 Pin Connector

PCM Pin #	Wire Color	Circuit Description (60 Pin)	Value at Hot Idle
1	DG/RD	MAP Sensor Signal	1.5-1.7v
2	TN/BK	ECT Sensor Signal	At 180°F: 2.80v
3	RD/WT	Battery Power (Fused B+)	12-14v
4	BK/LB	Sensor Ground	<0.050v
5	BK/LB	Sensor Ground	<0.050v
6	PK/WT	5-Volt Supply	4.9-5.1v
7	OR	8-Volt Supply	7.9-8.1v
8	---	Not Used	---
9	DB	Ignition Switch Start Signal	12-14v
10	---	Not Used	---
11	BK/TN	Power Ground	<0.1v
12	BK/TN	Power Ground	<0.1v
13	LB/BR	Injector 4 Driver	1-4 ms
14	YL/WT	Injector 3 Driver	1-4 ms
15	TN	Injector 2 Driver	1-4 ms
16	WT/DB	Injector 1 Driver	1-4 ms
17	DB/YL	Coil 1 Driver	5°, at 55 mph: 8° dwell
18	---	Not Used	---
19	BK/GY	Coil 2 Driver	5°, at 55 mph: 8° dwell
20	DG	Generator Field Driver	Digital Signal: 0-12-0v
21	BK/RD	Air Temperature Sensor	At 100°F: 2.51v
22	OR/DB	TP Sensor Signal	0.6-1.0v
23	RD/LG	S/C Set Switch Signal	S/C & Set Switch On: 3.8v
24	GY/BK	CKP Sensor Signal	Digital Signal: 0-5-0v
25	PK	SCI Transmit	0v
26	PK	CCD Bus (+)	Digital Signal: 0-5-0v
27	BR	A/C Clutch Signal	A/C On: 1v, Off: 12v
28	---	Not Used	---
29	WT/PK	Brake Switch Signal	Brake Off: 0v, On: 12v
30	BR/YL	PNP Switch Signal	In P/N: 0v, Others: 5v
31	DB/PK	Radiator Fan Relay Control	Relay Off: 12v, On: 1v
32	BK/PK	MIL (lamp) Control	Lamp On: 1v, Off: 12v
33	TN/RD	S/C Vacuum Solenoid	Vacuum Decreasing: 1v
34	DB/OR	A/C Clutch Relay Control	Relay Off: 12v, On: 1v
35	---	Not Used	---
36	LG/BK	Wastegate Solenoid Control	Solenoid Off: 12v, On: 1v
37-38	---	Not Used	---
39	GY/RD	AIS 3 Motor Control	DC pulse signals: 0.8-11v
40	BR/WT	AIS 1 Motor Control	DC pulse signals: 0.8-11v

Standard Colors and Abbreviations

Abbreviation	Color	Abbreviation	Color	Abbreviation	Color
BK	Black	GY	Gray	RD	Red
BL	Blue	GN	Green	TN	Tan
BR	Brown	LG	Light Green	VT	Violet
DB	Dark Blue	OR	Orange	WT	White
DG	Dark Green	PK	Pink	YL	Yellow

1992-93 Daytona 2.2L I4 DOHC Turbo VIN A (M/T) 60 Pin Connector

PCM Pin #	Wire Color	Circuit Description (60 Pin)	Value at Hot Idle
41	BK/DG	HO2S-11 (B1 S1) Signal	0.1-1.1v
42	BK/LG	Knock Sensor Signal	0.080v AC
43	GY/LB	Tachometer Signal	Pulse Signals
44	TN/YL	CMP Sensor Signal	Digital Signal: 0-5-0v
45	LG	SCI Receive	5v
46	PK	CCD Bus (-)	<0.050v
48-50	---	Not Used	---
51	DB	ASD Relay Control	Relay Off: 12v, On: 1v
52	PK/BK	EVAP Purge Solenoid Control	Solenoid Off: 12v, On: 1v
53	LG/RD	S/C Vent Solenoid	Vacuum Decreasing: 1v
54	---	Not Used	---
55	LB	BARO Read Solenoid	Solenoid Off: 12v, On: 1v
56	---	Not Used	---
57	DG	ASD Relay Control	Relay Off: 12v, On: 1v
58	---	Not Used	---
59	PK/BK	AIS 4 Motor Control	DC pulse signals: 0.8-11v
60	YL/BK	AIS 2 Motor Control	DC pulse signals: 0.8-11v

Pin Connector Graphic

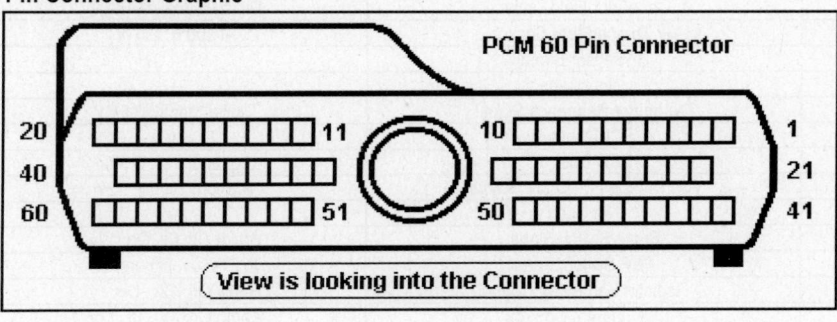

1990 Daytona 2.2L I4 SOHC Turbo VIN C (M/T) 60 Pin Connector

PCM Pin #	Wire Color	Circuit Description (60 Pin)	Value at Hot Idle
1	DG/RD	MAP Sensor Signal	1.5-1.7v
2	TN/BK	ECT Sensor Signal	At 180ºF: 2.80v
3	RD/WT	Battery Power (Fused B+)	12-14v
4	BK/LB	Sensor Ground	<0.050v
5	BK/WT	Sensor Ground	<0.050v
6	PK/WT	5-Volt Supply	4.9-5.1v
7	OR	8-Volt Supply	7.9-8.1v
8	DG/BK	ASD Relay Output	12-14v
9	DB	Ignition Switch Start Signal	12-14v
10	---	Not Used	---
11	BK/TN	Power Ground	<0.1v
12	BK/TN	Power Ground	<0.1v
13	LB/BR	Injector 4 Driver	1-4 ms
14	YL/WT	Injector 3 Driver	1-4 ms
15	TN	Injector 2 Driver	1-4 ms
16	WT/DB	Injector 1 Driver	1-4 ms
17-18	---	Not Used	---
19	BK/GY	Coil 1 Driver	5º, at 55 mph: 8º dwell
20	DG	Generator Field Driver	Digital Signal: 0-12-0v
21	BK/RD	Air Temperature Sensor	At 100ºF: 2.51v
22	OR/DB	TP Sensor Signal	0.6-1.0v
23	---	Not Used	---
24	GY/BK	Distributor Reference Signal	Digital Signal: 0-5-0v
25	PK	SCI Transmit	0v
26	PK/BR	CCD Bus (+)	Digital Signal: 0-5-0v
27	BR	A/C Damped Pressure Switch	A/C On: 1v, Off: 12v
28	---	Not Used	---
29	WT/PK	Brake Switch Signal	Brake Off: 0v, On: 12v
30	BR/YL	PNP Switch Signal	In P/N: 0v, Others: 5v
31	DB/PK	Radiator Fan Relay Control	Relay Off: 12v, On: 1v
32	BK/PK	MIL (lamp) Control	Lamp On: 1v, Off: 12v
33	TN/RD	S/C Vacuum Solenoid	Vacuum Increasing: 1v
34	DB/OR	A/C Clutch Relay Control	Relay Off: 12v, On: 1v
35	GY/YL	EGR Solenoid Control (Calif.)	12v, 55 mph: 1v
36	LG/BK	Wastegate Solenoid Control	Solenoid Off: 12v, On: 1v
37-38	---	Not Used	---
39	GY/RD	AIS 3 Motor Control	DC pulse signals: 0.8-11v
40	BR/WT	AIS 1 Motor Control	DC pulse signals: 0.8-11v

Standard Colors and Abbreviations

Abbreviation	Color	Abbreviation	Color	Abbreviation	Color
BK	Black	GY	Gray	RD	Red
BL	Blue	GN	Green	TN	Tan
BR	Brown	LG	Light Green	VT	Violet
DB	Dark Blue	OR	Orange	WT	White
DG	Dark Green	PK	Pink	YL	Yellow

1990 Daytona 2.2L I4 SOHC Turbo VIN C (M/T) 60 Pin Connector

PCM Pin #	Wire Color	Circuit Description (60 Pin)	Value at Hot Idle
41	BK/DG	HO2S-11 (B1 S1) Signal	0.1-1.1v
42	BK/LG	Knock Sensor Signal	0.080v AC
43	GY/LB	Tachometer Signal	Pulse Signals
44	TN/YL	Distributor Sync Signal	Digital Signal: 0-5-0v
45	LG	SCI Receive	5v
46	WT/BK	CCD Bus (-)	<0.050v
47	WT/OR	Vehicle Speed Signal	Digital Signal
48	BR/RD	S/C Set Switch Signal	S/C On: 1v, Off: 12v
49	YL/RD	S/C On/Off Switch Signal	S/C On: 1v, Off: 12v
50	WT/LG	S/C Resume Switch	S/C On: 1v, Off: 12v
51	DB/YL	ASD Relay Control	Relay Off: 12v, On: 1v
52	PK/BK	EVAP Purge Solenoid Control	Valve On: 1v, Off: 12v
53	LG/RD	S/C Vent Solenoid	Vacuum Decreasing: 1v
54	LG/WT	Turbo Solenoid No. 1 Control	Solenoid Off: 12v, On: 1v
55	LB	BARO Read Solenoid	Solenoid Off: 12v, On: 1v
56	LG/OR	Turbo Solenoid No. 2 Control	Solenoid Off: 12v, On: 1v
57-58	---	Not Used	---
59	PK/BK	AIS 4 Motor Control	DC pulse signals: 0.8-11v
60	YL/BK	AIS 2 Motor Control	DC pulse signals: 0.8-11v

Pin Connector Graphic

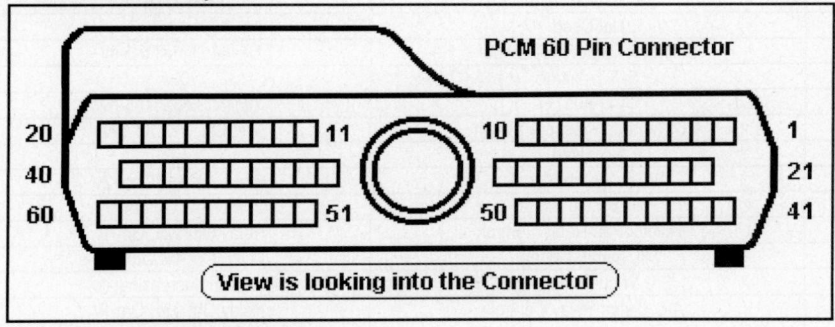

Standard Colors and Abbreviations

Abbreviation	Color	Abbreviation	Color	Abbreviation	Color
BK	Black	GY	Gray	RD	Red
BL	Blue	GN	Green	TN	Tan
BR	Brown	LG	Light Green	VT	Violet
DB	Dark Blue	OR	Orange	WT	White
DG	Dark Green	PK	Pink	YL	Yellow

1990-92 Daytona 2.5L I4 SOHC Turbo VIN J (All) 60 Pin Connector

PCM Pin #	Wire Color	Circuit Description (60 Pin)	Value at Hot Idle
1	DG/RD	MAP Sensor Signal	1.5-1.7v
2	TN/BK	ECT Sensor Signal	At 180°F: 2.80v
3	RD/WT	Battery Power (Fused B+)	12-14v
4	BK/LB	Sensor Ground	<0.050v
5	BK/WT	Sensor Ground	<0.050v
6	PK/WT	5-Volt Supply	4.9-5.1v
7	OR	8-Volt Supply	7.9-8.1v
8	DG/BK	Auto Shutdown Relay Input	12-14v
9	DB	Ignition Switch Output	12-14v
10	---	Not Used	---
11	BK/TN	Power Ground	<0.1v
12	BK/TN	Power Ground	<0.1v
13	LB/BR	Injector 4 Driver	1-4 ms
14	YL/WT	Injector 3 Driver	1-4 ms
15	TN	Injector 2 Driver	1-4 ms
16	WT/DB	Injector 1 Driver	1-4 ms
17-18	---	Not Used	---
19	BK/GY	Coil 1 Driver	5°, at 55 mph: 8° dwell
20	DG	Generator Field Driver	Digital Signal: 0-12-0v
21	---	Not Used	---
22	OR/DB	TP Sensor Signal	0.6-1.0v
23	---	Not Used	---
24	GY/BK	Distributor Reference Signal	Digital Signal: 0-5-0v
25	PK	SCI Transmit	0v
26	PK/BR	CCD Bus (+)	Digital Signal: 0-5-0v
27	BR	A/C Damped Pressure Switch	A/C On: 1v, Off: 12v
28	---	Not Used	---
29	WT/PK	Brake Switch Signal	Brake Off: 0v, On: 12v
30	BR/LB	PNP Switch Signal	In P/N: 0v, Others: 5v
31	DB/PK	Radiator Fan Relay Control	Relay Off: 12v, On: 1v
32	BK/PK	MIL (lamp) Control	Lamp On: 1v, Off: 12v
33	TN/RD	S/C Vacuum Solenoid	Vacuum Increasing: 1v
34	DB/OR	A/C Clutch Relay Control	Relay Off: 12v, On: 1v
35	GY/YL	EGR Solenoid Control (Calif.)	12v, 55 mph: 1v
36	LG/BK	Wastegate Solenoid Control	Solenoid Off: 12v, On: 1v
37-38	---	Not Used	---
39	GY/RD	AIS 3 Motor Control	DC pulse signals: 0.8-11v
40	BR/WT	AIS 1 Motor Control	DC pulse signals: 0.8-11v

Standard Colors and Abbreviations

Abbreviation	Color	Abbreviation	Color	Abbreviation	Color
BK	Black	GY	Gray	RD	Red
BL	Blue	GN	Green	TN	Tan
BR	Brown	LG	Light Green	VT	Violet
DB	Dark Blue	OR	Orange	WT	White
DG	Dark Green	PK	Pink	YL	Yellow

1990-92 Daytona 2.5L I4 SOHC Turbo MFI VIN J 60 Pin Connector

PCM Pin #	Wire Color	Circuit Description (60 Pin)	Value at Hot Idle
41	BK/DG	HO2S-11 (B1 S1) Signal	0.1-1.1v
42	BK/LG	Knock Sensor Signal	0.080v AC
43	GY/LB	Tachometer Signal	Pulse Signals
44	TN/YL	Distributor Sync Signal	Digital Signal: 0-5-0v
45	LG	SCI Receive	0v
46	WT/BK	CCD Bus (-)	<0.050v
47	WT/OR	Vehicle Speed Signal	Digital Signal
48	BR/RD	S/C Set Switch Signal	S/C & Set Switch On: 3.8v
49	YL/RD	S/C On/Off Switch Signal	S/C On: 1v, Off: 12v
50	WT/LG	S/C Resume Switch	S/C On: 1v, Off: 12v
51	DB/YL	ASD Relay Control	Relay Off: 12v, On: 1v
52	PK/BK	EVAP Purge Solenoid Control	Solenoid Off: 12v, On: 1v
53	LG/RD	S/C Vent Solenoid	Vacuum Decreasing: 1v
54	LG/WT	Turbo Solenoid No. 1 Control	Solenoid Off: 12v, On: 1v
55	LB	BARO Read Solenoid	Solenoid Off: 12v, On: 1v
56	GY/PK	EVAP Amber (lamp) Control	Lamp On: 1v, Off: 12v
57	DG/OR	ASD Relay Output	12-14v
58	---	Not Used	---
59	PK/BK	AIS 4 Motor Control	DC pulse signals: 0.8-11v
60	YL/BK	AIS 2 Motor Control	DC pulse signals: 0.8-11v

Pin Connector Graphic

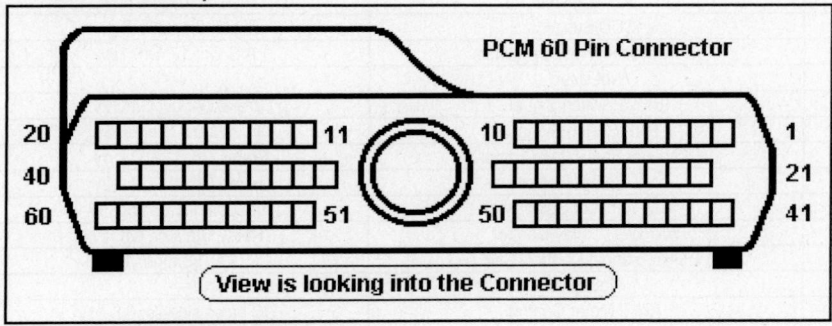

Standard Colors and Abbreviations

Abbreviation	Color	Abbreviation	Color	Abbreviation	Color
BK	Black	GY	Gray	RD	Red
BL	Blue	GN	Green	TN	Tan
BR	Brown	LG	Light Green	VT	Violet
DB	Dark Blue	OR	Orange	WT	White
DG	Dark Green	PK	Pink	YL	Yellow

1990-91 Daytona 2.5L I4 SOHC TBI VIN K (All) 60 Pin Connector

PCM Pin #	Wire Color	Circuit Description (60 Pin)	Value at Hot Idle
1	DG/RD	MAP Sensor Signal	1.5-1.7v
2	TN/BK	ECT Sensor Signal	At 180°F: 2.80v
3	RD/WT	Battery Power (Fused B+)	12-14v
4	BK/LB	Sensor Ground	<0.050v
5	BK/WT	Sensor Ground	<0.050v
6	PK/WT	5-Volt Supply	4.9-5.1v
7	OR	8-Volt Supply	7.9-8.1v
8	WT	Ignition Switch Output	12-14v
9	DB	Ignition Switch Run Output	12-14v
10	---	Not Used	---
11	BK/TN	Power Ground	<0.1v
12	BK/TN	Power Ground	<0.1v
13-15	---	Not Used	---
16	WT/DB	Injector Control Driver	1-4 ms
17-18	---	Not Used	---
19	BK/GY	Coil 1 Driver	5°, at 55 mph: 8° dwell
20	DG	Generator Field Driver	Digital Signal: 0-12-0v
17-18	---	Not Used	---
22	OR/DB	TP Sensor Signal	0.6-1.0v
23	---	Not Used	---
24	GY/BK	Distributor Reference Signal	Digital Signal: 0-5-0v
25	PK	SCI Transmit	0v
26	PK/BR	CCD Bus (+)	Digital Signal: 0-5-0v
27	BR	A/C Damped Pressure Switch	A/C On: 1v, Off: 12v
28	---	Not Used	---
29	WT/PK	Brake Switch Signal	Brake Off: 0v, On: 12v
30	BR/LB	PNP Switch Signal	In P/N: 0v, Others: 5v
31	DB/PK	Radiator Fan Relay Control	Relay Off: 12v, On: 1v
32	BK/PK	MIL (lamp) Control	Lamp On: 1v, Off: 12v
33	TN/RD	S/C Vacuum Solenoid	Vacuum Increasing: 1v
34	DB/OR	A/C WOT Relay Control	Relay Off: 12v, On: 1v
35	GY/YL	EGR Solenoid Control (Calif.)	12v, 55 mph: 1v
36-38	---	Not Used	---
39	GY/RD	AIS 3 Motor Control	DC pulse signals: 0.8-11v
40	BR/WT	AIS 1 Motor Control	DC pulse signals: 0.8-11v

Standard Colors and Abbreviations

Abbreviation	Color	Abbreviation	Color	Abbreviation	Color
BK	Black	GY	Gray	RD	Red
BL	Blue	GN	Green	TN	Tan
BR	Brown	LG	Light Green	VT	Violet
DB	Dark Blue	OR	Orange	WT	White
DG	Dark Green	PK	Pink	YL	Yellow

1990-91 Daytona 2.5L I4 SOHC TBI VIN K (All) 60 Pin Connector

PCM Pin #	Wire Color	Circuit Description (60 Pin)	Value at Hot Idle
41	BK/DG	HO2S-11 (B1 S1) Signal	0.1-1.1v
42	---	Not Used	---
43	GY/LB	Tachometer Signal	Pulse Signals
44	---	Not Used	---
45	LG	SCI Receive	0v
46	WT/BK	CCD Bus (-)	<0.050v
47	WT/OR	Vehicle Speed Signal	Digital Signal
48	BR/RD	S/C Set Switch Signal	S/C & Set Switch On: 3.8v
49	YL/RD	S/C On/Off Switch Signal	S/C On: 1v, Off: 12v
50	WT/LG	S/C Resume Switch	S/C On: 1v, Off: 12v
51	DB/YL	ASD Relay Control	Relay Off: 12v, On: 1v
52	PK/BK	EVAP Purge Solenoid Control	Solenoid Off: 12v, On: 1v
53	LG/RD	S/C Vent Solenoid	Vacuum Decreasing: 1v
54	LG/WT	Part Throttle Unlock Solenoid	Solenoid Off: 12v, On: 1v
55-58	---	Not Used	---
59	PK/BK	AIS 4 Motor Control	DC pulse signals: 0.8-11v
60	YL/BK	AIS 2 Motor Control	DC pulse signals: 0.8-11v

Pin Connector Graphic

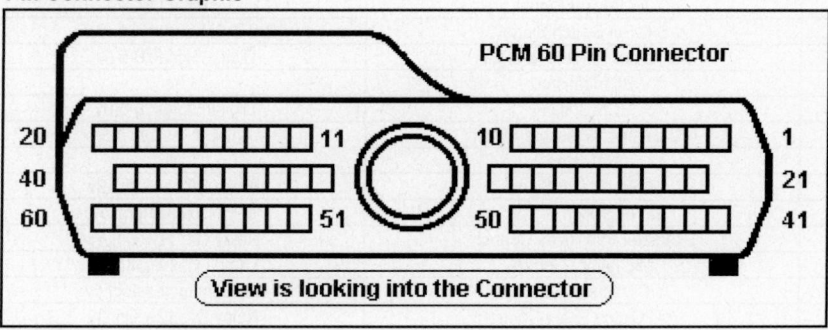

1992-93 Daytona 2.5L I4 SOHC TBI VIN K (All) 60 Pin Connector

PCM Pin #	Wire Color	Circuit Description (60 Pin)	Value at Hot Idle
1	DG/RD	MAP Sensor Signal	1.5-1.7v
2	TN/BK	ECT Sensor Signal	At 180°F: 2.80v
3	RD/WT	Battery Power (Fused B+)	12-14v
4	BK/LB	Sensor Ground	<0.050v
5	BK/WT	Sensor Ground	<0.050v
6	PK/WT	5-Volt Supply	4.9-5.1v
7	OR	8-Volt Supply	7.9-8.1v
8	WT	Ignition Switch Output	12-14v
9	DB	Ignition Switch Run Output	12-14v
10	---	Not Used	---
11	BK/TN	Power Ground	<0.1v
12	BK/TN	Power Ground	<0.1v
13-15	---	Not Used	---
16	WT/DB	Injector Control Driver	1-4 ms
17-18	---	Not Used	---
19	BK/GY	Coil 1 Driver	5°, at 55 mph: 8° dwell
20	DG	Generator Field Driver	Digital Signal: 0-12-0v
21	---	Not Used	---
22	OR/DB	TP Sensor Signal	0.6-1.0v
23	RD/LG	S/C Set Switch Signal	S/C & Set Switch On: 3.8v
24	GY/BK	Distributor Reference Signal	Digital Signal: 0-5-0v
25	PK	SCI Transmit	0v
26	PK/BR	CCD Bus (+)	Digital Signal: 0-5-0v
27	BR	A/C Damped Pressure Switch	A/C On: 1v, Off: 12v
28	---	Not Used	---
29	WT/PK	Brake Switch Signal	Brake Off: 0v, On: 12v
30	BR/LB	PNP Switch Signal	In P/N: 0v, Others: 5v
31	DB/PK	Radiator Fan Relay Control	Relay Off: 12v, On: 1v
32	BK/PK	MIL (lamp) Control	Lamp On: 1v, Off: 12v
33	TN/RD	S/C Vacuum Solenoid	Vacuum Increasing: 1v
34	DB/OR	A/C Clutch Relay Control	Relay Off: 12v, On: 1v
35	GY/YL	EGR Solenoid Control (Calif.)	12v, 55 mph: 1v
36-38	---	Not Used	---
39	GY/RD	AIS 3 Motor Control	DC pulse signals: 0.8-11v
40	BR/WT	AIS 1 Motor Control	DC pulse signals: 0.8-11v

1992-93 Daytona 2.5L I4 SOHC TBI VIN K (All)

PCM Pin #	Wire Color	Circuit Description (60 Pin)	Value at Hot Idle
41	BK/DG	HO2S-11 (B1 S1) Signal	0.1-1.1v
42	---	Not Used	---
43	GY/LB	Tachometer Signal	Pulse Signals
44	---	Not Used	---
45	LG	SCI Receive	0v
46	WT/BK	CCD Bus (-)	<0.050v
47	WT/OR	Vehicle Speed Signal	Digital Signal
48-50	---	Not Used	---
51	DB/YL	ASD Relay Control	Relay Off: 12v, On: 1v
52	PK/BK	EVAP Purge Solenoid Control	Solenoid Off: 12v, On: 1v
53	LG/RD	S/C Vent Solenoid	Vacuum Decreasing: 1v
54	LG/WT	Part Throttle Unlock Solenoid	Solenoid Off: 12v, On: 1v
55-56	---	Not Used	---
57	DG/OR	ASD Relay Output	12-14v
58	---	Not Used	---
59	PK/BK	AIS 4 Motor Control	DC pulse signals: 0.8-11v
60	YL/BK	AIS 2 Motor Control	DC pulse signals: 0.8-11v

Pin Connector Graphic

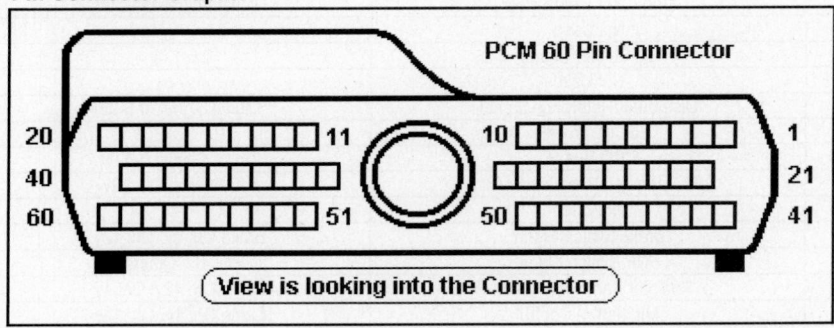

Standard Colors and Abbreviations

Abbreviation	Color	Abbreviation	Color	Abbreviation	Color
BK	Black	GY	Gray	RD	Red
BL	Blue	GN	Green	TN	Tan
BR	Brown	LG	Light Green	VT	Violet
DB	Dark Blue	OR	Orange	WT	White
DG	Dark Green	PK	Pink	YL	Yellow

1990-91 Daytona 3.0L V6 SOHC MFI VIN 3 (All) 60 Pin Connector

PCM Pin #	Wire Color	Circuit Description (60 Pin)	Value at Hot Idle
1	DG/RD	MAP Sensor Signal	1.5-1.7v
2	TN/BK	ECT Sensor Signal	At 180°F: 2.80v
3	RD	Battery Power (Fused B+)	12-14v
4	BK/LB	Sensor Ground	<0.050v
5	BK/WT	Sensor Ground	<0.050v
6	PK/WT	5-Volt Supply	4.9-5.1v
7	OR	8-Volt Supply	7.9-8.1v
8	DG/BK	Ignition Switch Start Signal	12-14v
9	DB	Ignition Switch Output	12-14v
10	---	Not Used	---
11	BK/TN	Power Ground	<0.1v
12	BK/TN	Power Ground	<0.1v
13	---	Not Used	---
14	YL/WT	Injector Control 3 & 4 Driver	1-4 ms
15	TN	Injector Control 2 & 5 Driver	1-4 ms
16	WT/DB	Injector Control 1 & 3 Driver	1-4 ms
17-18	---	Not Used	---
19	BK/GY	Coil Driver	5°, at 55 mph: 8° dwell
20	DG	Generator Field Driver	Digital Signal: 0-12-0v
21	---	Not Used	---
22	OR/DB	TP Sensor Signal	0.6-1.0v
23	---	Not Used	---
24	GY/BK	Distributor Reference Signal	Digital Signal: 0-5-0v
25	PK	SCI Transmit	0v
26	PK/BR	CCD Bus (+)	Digital Signal: 0-5-0v
27	BR	A/C Damped Pressure Switch	A/C On: 1v, Off: 12v
28	---	Not Used	---
29	WT/PK	Brake Switch Signal	Brake Off: 0v, On: 12v
30	BR/LB	PNP Switch Signal	In P/N: 0v, Others: 5v
31	DB/PK	Radiator Fan Relay Control	Relay Off: 12v, On: 1v
32	BK/PK	MIL (lamp) Control	Lamp On: 1v, Off: 12v
33	TN/RD	S/C Vacuum Solenoid	Vacuum Increasing: 1v
34	DB/OR	A/C Clutch Relay Control	Relay Off: 12v, On: 1v
35	GY/YL	EGR Solenoid Control (Calif.)	12v, 55 mph: 1v
36-38	---	Not Used	---
39	GY/RD	AIS 3 Motor Control	DC pulse signals: 0.8-11v
40	OR/WT	AIS 1 Motor Control	DC pulse signals: 0.8-11v

Standard Colors and Abbreviations

Abbreviation	Color	Abbreviation	Color	Abbreviation	Color
BK	Black	GY	Gray	RD	Red
BL	Blue	GN	Green	TN	Tan
BR	Brown	LG	Light Green	VT	Violet
DB	Dark Blue	OR	Orange	WT	White
DG	Dark Green	PK	Pink	YL	Yellow

1990-91 Daytona 3.0L V6 SOHC MFI VIN 3 (All) 60 Pin Connector

PCM Pin #	Wire Color	Circuit Description (60 Pin)	Value at Hot Idle
41	BK/DG	HO2S-11 (B1 S1) Signal	0.1-1.1v
42	---	Not Used	---
43	GY/LB	Tachometer Signal	Pulse Signals
44	TN/YL	Distributor Sync Signal	Digital Signal: 0-5-0v
45	LG	SCI Receive	0v
46	WT	CCD Bus (-)	<0.050v
47	WT/OR	Vehicle Speed Signal	Digital Signal
48	BR/RD	S/C Set Switch Signal	S/C & Set Switch On: 3.8v
49	YL/RD	S/C On/Off Switch Signal	S/C On: 1v, Off: 12v
50	WT/LG	S/C Resume Switch	S/C On: 1v, Off: 12v
51	DB/YL	ASD Relay Control	Relay Off: 12v, On: 1v
52	PK	EVAP Purge Solenoid Control	Solenoid Off: 12v, On: 1v
53	LG/RD	S/C Vent Solenoid	Vacuum Decreasing: 1v
54-58	---	Not Used	---
59	PK/BK	AIS 4 Motor Control	DC pulse signals: 0.8-11v
60	YL/BK	AIS 2 Motor Control	DC pulse signals: 0.8-11v

Pin Connector Graphic

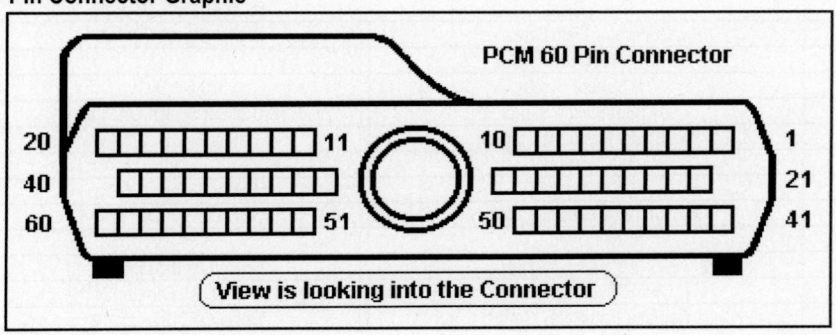

PCM 60 Pin Connector

View is looking into the Connector

1992-93 Daytona 3.0L V6 SOHC MFI VIN 3 (A/T) 60 Pin Connector

PCM Pin #	Wire Color	Circuit Description (60 Pin)	Value at Hot Idle
1	DG/RD	MAP Sensor Signal	1.5-1.7v
2	TN/BK	ECT Sensor Signal	At 180°F: 2.80v
3	RD/WT	Battery Power (Fused B+)	12-14v
4	BK/LB	Sensor Ground	<0.050v
5	BK/WT	Sensor Ground	<0.050v
6	PK/WT	5-Volt Supply	4.9-5.1v
7	OR	8-Volt Supply	7.9-8.1v
8	---	Not Used	---
9	DB	Ignition Switch Output	12-14v
10	---	Not Used	---
11	BK/TN	Power Ground	<0.1v
12	BK/TN	Power Ground	<0.1v
13	LB/BR	Injector 4 Driver	1-4 ms
14	YL/WT	Injector 3 Driver	1-4 ms
15	TN	Injector 2 Driver	1-4 ms
16	WT/DB	Injector 1 Driver	1-4 ms
17-18	---	Not Used	---
19	BK/GY	Coil 1 Driver	5°, at 55 mph: 8° dwell
20	DG	Generator Field Driver	Digital Signal: 0-12-0v
21	---	Not Used	---
22	OR/DB	TP Sensor Signal	0.6-1.0v
23	RD/LG	S/C Set Switch Signal	S/C & Set Switch On: 3.8v
24	GY/BK	Distributor Reference Signal	Digital Signal: 0-5-0v
25	PK	SCI Transmit	0v
26	PK/BR	CCD Bus (+)	Digital Signal: 0-5-0v
27	BR	A/C Damped Pressure Switch	A/C On: 1v, Off: 12v
28	---	Not Used	---
29	WT/PK	Brake Switch Signal	Brake Off: 0v, On: 12v
30	BR/LB	PNP Switch Signal	In P/N: 0v, Others: 5v
31	DB/PK	Radiator Fan Relay Control	Relay Off: 12v, On: 1v
32	BK/PK	MIL (lamp) Control	Lamp On: 1v, Off: 12v
33	TN/RD	S/C Vacuum Solenoid	Vacuum Increasing: 1v
34	DB/OR	A/C Clutch Relay Control	Relay Off: 12v, On: 1v
35	GY/YL	EGR Solenoid Control (Calif.)	12v, 55 mph: 1v
36-37	---	Not Used	---
38	GY	Injector 5 Driver	1-4 ms
39	GY/RD	AIS 3 Motor Control	DC pulse signals: 0.8-11v
40	BR/WT	AIS 1 Motor Control	DC pulse signals: 0.8-11v

Standard Colors and Abbreviations

Abbreviation	Color	Abbreviation	Color	Abbreviation	Color
BK	Black	GY	Gray	RD	Red
BL	Blue	GN	Green	TN	Tan
BR	Brown	LG	Light Green	VT	Violet
DB	Dark Blue	OR	Orange	WT	White
DG	Dark Green	PK	Pink	YL	Yellow

1992-93 Daytona 3.0L V6 SOHC MFI VIN 3 (A/T) 60 Pin Connector

PCM Pin #	Wire Color	Circuit Description (60 Pin)	Value at Hot Idle
41	BK/DG	HO2S-11 (B1 S1) Signal	0.1-1.1v
42	---	Not Used	
43	GY/LB	Tachometer Signal	Pulse Signals
44	TN/YL	Distributor Sync Signal	Digital Signal: 0-5-0v
45	LG	SCI Receive	0v
46	WT/BK	CCD Bus (-)	<0.050v
47	WT/OR	Vehicle Speed Signal	Digital Signal
48-50	---	Not Used	---
51	DB/YL	ASD Relay Control	Relay Off: 12v, On: 1v
52	PK/BK	EVAP Purge Solenoid Control	Solenoid Off: 12v, On: 1v
53	LG/RD	S/C Vent Solenoid	Vacuum Decreasing: 1v
54-56	---	Not Used	---
57	DG/OR	ASD Relay Output	12-14v
58	BR/DB	Injector 6 Driver	1-4 ms
59	PK/BK	AIS 4 Motor Control	DC pulse signals: 0.8-11v
60	YL/BK	AIS 2 Motor Control	DC pulse signals: 0.8-11v

Pin Connector Graphic

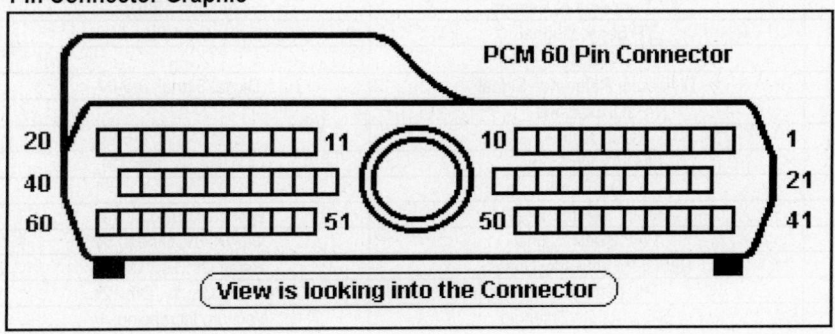

1990-91 Dynasty 2.5L I4 SOHC TBI VIN K (A/T) 60 Pin Connector

PCM Pin #	Wire Color	Circuit Description (60 Pin)	Value at Hot Idle
1	DG/RD	MAP Sensor Signal	1.5-1.7v
2	TN/WT	ECT Sensor Signal	At 180°F: 2.80v
3	RD	Battery Power (Fused B+)	12-14v
4	BK/LB	Sensor Ground	<0.050v
5	BK	Sensor Ground	<0.050v
6	PK/WT	5-Volt Supply	4.9-5.1v
7	OR	8-Volt Supply	7.9-8.1v
8	WT	Ignition Switch Output	12-14v
9	DB	Ignition Switch Run Output	12-14v
10	---	Not Used	---
11	BK/TN	Power Ground	<0.1v
12	BK/TN	Power Ground	<0.1v
13-15	---	Not Used	---
16	WT/DB	Injector Control Driver	1-4 ms
17-18	---	Not Used	---
19	BK/GY	Coil 1 Driver	5°, at 55 mph: 8° dwell
20	DG	Generator Field Driver	Digital Signal: 0-12-0v
21	BK/RD	Air Temperature Sensor	At 100°F: 2.51v
22	OR/DB	TP Sensor Signal	0.6-1.0v
23	---	Not Used	---
24	GY/BK	Distributor Reference Signal	Digital Signal: 0-5-0v
25	PK	SCI Transmit	0v
26	BK	CCD Bus (+)	Digital Signal: 0-5-0v
27	BR	A/C Clutch Signal	A/C On: 1v, Off: 12v
28	---	Not Used	---
29	WT/PK	Brake Switch Signal	Brake Off: 0v, On: 12v
30	BR/YL	PNP Switch Signal	In P/N: 0v, Others: 5v
31	DB/PK	Radiator Fan Relay Control	Relay Off: 12v, On: 1v
32	BK/PK	MIL (lamp) Control	Lamp On: 1v, Off: 12v
33	TN/RD	S/C Vacuum Solenoid	Vacuum Increasing: 1v
34	DB/OR	A/C Clutch Relay Control	Relay Off: 12v, On: 1v
35	GY/YL	EGR Solenoid Control (Calif.)	12v, 55 mph: 1v
36-38	---	Not Used	---
39	GY/RD	AIS 3 Motor Control	DC pulse signals: 0.8-11v
40	BR/WT	AIS 1 Motor Control	DC pulse signals: 0.8-11v

1990-91 Dynasty 2.5L I4 SOHC TBI VIN K (A/T) 60 Pin Connector

PCM Pin #	Wire Color	Circuit Description (60 Pin)	Value at Hot Idle
41	BK/DG	HO2S-11 (B1 S1) Signal	0.1-1.1v
42-44	---	Not Used	---
45	LG	SCI Receive	0v
46	WT/BK	CCD Bus (-)	<0.050v
47	WT/OR	Vehicle Speed Signal	Digital Signal
48	BR/RD	S/C Set Switch Signal	S/C & Set Switch On: 3.8v
49	YL/RD	S/C On/Off Switch Signal	S/C On: 1v, Off: 12v
50	WT/LG	S/C Resume Switch	S/C On: 1v, Off: 12v
51	DB/YL	ASD Relay Control	Relay Off: 12v, On: 1v
52	PK	EVAP Purge Solenoid Control	Solenoid Off: 12v, On: 1v
53	LG/RD	S/C Vent Solenoid	Vacuum Decreasing: 1v
54	OR/BK	Part Throttle Unlock Solenoid	Solenoid Off: 12v, On: 1v
55-58	---	Not Used	---
59	PK/BK	AIS 4 Motor Control	DC pulse signals: 0.8-11v
60	YL/BK	AIS 2 Motor Control	DC pulse signals: 0.8-11v

Pin Connector Graphic

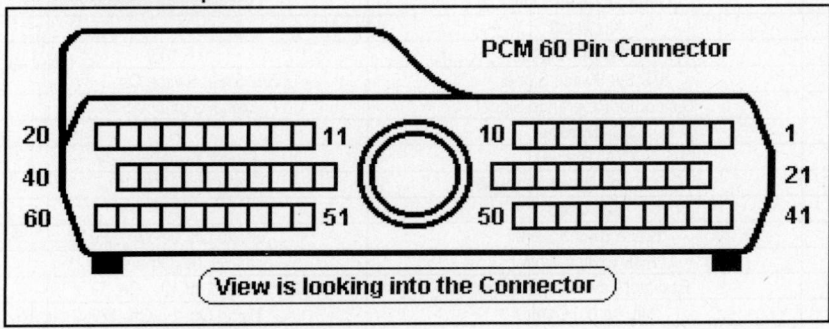

Standard Colors and Abbreviations

Abbreviation	Color	Abbreviation	Color	Abbreviation	Color
BK	Black	GY	Gray	RD	Red
BL	Blue	GN	Green	TN	Tan
BR	Brown	LG	Light Green	VT	Violet
DB	Dark Blue	OR	Orange	WT	White
DG	Dark Green	PK	Pink	YL	Yellow

1992-93 Dynasty 2.5L I4 SOHC TBI VIN K (A/T) 60 Pin Connector

PCM Pin #	Wire Color	Circuit Description (60 Pin)	Value at Hot Idle
1	DG/RD	MAP Sensor Signal	1.5-1.7v
2	TN/BK	ECT Sensor Signal	At 180°F: 2.80v
3	RD/WT	Battery Power (Fused B+)	12-14v
4	BK/LB	Sensor Ground	<0.050v
5	BK/WT	Sensor Ground	<0.050v
6	PK/WT	5-Volt Supply	4.9-5.1v
7	OR	8-Volt Supply	7.9-8.1v
8	WT	Ignition Switch Output	12-14v
9	DB	Ignition Switch Run Output	12-14v
10	---	Not Used	---
11	BK/TN	Power Ground	<0.1v
12	BK/TN	Power Ground	<0.1v
13-15	---	Not Used	---
16	WT/DB	Injector Control Driver	1-4 ms
17-18	---	Not Used	---
19	BK/GY	Coil 1 Driver	5°, at 55 mph: 8° dwell
20	DG	Generator Field Driver	Digital Signal: 0-12-0v
21	---	Not Used	---
22	OR/DB	TP Sensor Signal	0.6-1.0v
23	RD/LG	S/C Set Switch Signal	S/C & Set Switch On: 3.8v
24	GY/BK	Distributor Reference Signal	Digital Signal: 0-5-0v
25	PK	SCI Transmit	0v
26	BK	CCD Bus (+)	Digital Signal: 0-5-0v
27	BR	A/C Damped Pressure Switch	A/C On: 1v, Off: 12v
28	---	Not Used	---
29	WT/PK	Brake Switch Signal	Brake Off: 0v, On: 12v
30	BR/YL	PNP Switch Signal	In P/N: 0v, Others: 5v
31	DB/PK	Radiator Fan Relay Control	Relay Off: 12v, On: 1v
32	BK/PK	MIL (lamp) Control	Lamp On: 1v, Off: 12v
33	TN/RD	S/C Vacuum Solenoid	Vacuum Increasing: 1v
34	DB/OR	A/C Clutch Relay Control	Relay Off: 12v, On: 1v
35	GY/YL	EGR Solenoid Control (Calif.)	12v, 55 mph: 1v
36-38	---	Not Used	---
39	GY/RD	AIS 3 Motor Control	DC pulse signals: 0.8-11v
40	BR/WT	AIS 1 Motor Control	DC pulse signals: 0.8-11v

1992-93 Dynasty 2.5L I4 SOHC TBI VIN K (A/T) 60 Pin Connector

PCM Pin #	Wire Color	Circuit Description (60 Pin)	Value at Hot Idle
41	BK/DG	HO2S-11 (B1 S1) Signal	0.1-1.1v
42-44	---	Not Used	---
45	LG	SCI Receive	0v
46	WT/BK	CCD Bus (-)	<0.050v
47	WT/OR	Vehicle Speed Signal	Digital Signal
48-50	---	Not Used	---
51	DB/YL	ASD Relay Control	Relay Off: 12v, On: 1v
52	PK	EVAP Purge Solenoid Control	Solenoid Off: 12v, On: 1v
53	LG/RD	S/C Vent Solenoid	Vacuum Decreasing: 1v
54	OR/BK	Part Throttle Unlock Solenoid	Solenoid Off: 12v, On: 1v
55-58	---	Not Used	---
59	PK/BK	AIS 4 Motor Control	DC pulse signals: 0.8-11v
60	YL/BK	AIS 2 Motor Control	DC pulse signals: 0.8-11v

Pin Connector Graphic

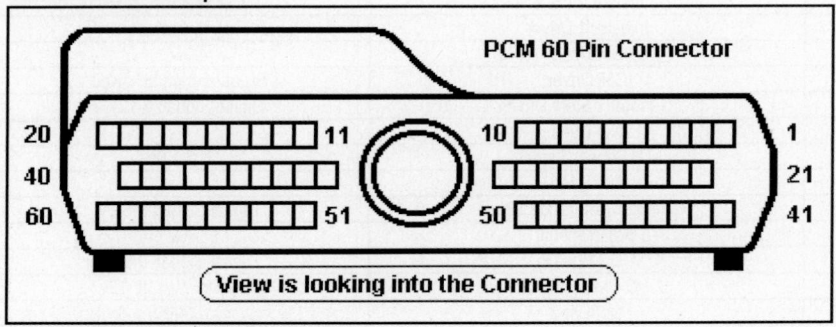

Standard Colors and Abbreviations

Abbreviation	Color	Abbreviation	Color	Abbreviation	Color
BK	Black	GY	Gray	RD	Red
BL	Blue	GN	Green	TN	Tan
BR	Brown	LG	Light Green	VT	Violet
DB	Dark Blue	OR	Orange	WT	White
DG	Dark Green	PK	Pink	YL	Yellow

1990-91 Dynasty 3.0L V6 SOHC MFI VIN 3 (A/T) 60 Pin Connector

PCM Pin #	Wire Color	Circuit Description (60 Pin)	Value at Hot Idle
1	DG/RD	MAP Sensor Signal	1.5-1.7v
2	TN/WT	ECT Sensor Signal	At 180ºF: 2.80v
3	RD	Battery Power (Fused B+)	12-14v
4	BK/LB	Sensor Ground	<0.050v
5	BK/WT	Sensor Ground	<0.050v
6	PK/WT	5-Volt Supply	4.9-5.1v
7	OR	8-Volt Supply	7.9-8.1v
8	DG/BK	Ignition Switch Start Signal	12-14v
9	DB	Ignition Switch Output	12-14v
10	---	Not Used	---
11	LB/RD	Power Ground	<0.1v
12	LB/RD	Power Ground	<0.1v
13	---	Not Used	---
14	YL/WT	Injector 3 Driver	1-4 ms
15	TN	Injector 2 Driver	1-4 ms
16	WT/DB	Injector 1 Driver	1-4 ms
17-18	---	Not Used	---
19	BK/GY	Coil Driver	5º, at 55 mph: 8º dwell
20	DG	Generator Field Driver	Digital Signal: 0-12-0v
21	---	Not Used	---
22	OR/DB	TP Sensor Signal	0.6-1.0v
23	---	Not Used	---
24	GY/BK	Distributor Reference Signal	Digital Signal: 0-5-0v
25	PK	SCI Transmit	0v
26	BK/PK	CCD Bus (+)	Digital Signal: 0-5-0v
27	BR	A/C Clutch Signal	A/C On: 1v, Off: 12v
28	---	Not Used	---
29	WT/PK	Brake Switch Signal	Brake Off: 0v, On: 12v
30	BR/YL	PNP Switch Signal	In P/N: 0v, Others: 5v
31	DB/PK	Radiator Fan Relay Control	Relay Off: 12v, On: 1v
32	BK/PK	MIL (lamp) Control	Lamp On: 1v, Off: 12v
33	TN/RD	S/C Vacuum Solenoid	Vacuum Increasing: 1v
34	DB/OR	A/C Clutch Relay Control	Relay Off: 12v, On: 1v
35	GY/YL	EGR Solenoid Control (Calif.)	12v, 55 mph: 1v
36-38	---	Not Used	---
39	GY/RD	AIS 3 Motor Control	DC pulse signals: 0.8-11v
40	OR/WT	AIS 1 Motor Control	DC pulse signals: 0.8-11v

1990-91 Dynasty 3.0L V6 SOHC MFI VIN 3 (A/T) 60 Pin Connector

PCM Pin #	Wire Color	Circuit Description (60 Pin)	Value at Hot Idle
41	BK/DG	HO2S-11 (B1 S1) Signal	0.1-1.1v
42	---	Not Used	---
43	GY/LB	Tachometer Signal	Pulse Signals
44	TN/YL	Distributor Sync Signal	Digital Signal: 0-5-0v
45	LG	SCI Receive	0v
46	WT/BK	CCD Bus (-)	<0.050v
47	WT/OR	Vehicle Speed Signal	Digital Signal
48	BR/RD	S/C Set Switch Signal	S/C & Set Switch On: 3.8v
49	YL/RD	S/C On/Off Switch Signal	S/C On: 1v, Off: 12v
50	WT/LG	S/C Resume Switch	S/C On: 1v, Off: 12v
51	DB/YL	ASD Relay Control	Relay Off: 12v, On: 1v
52	PK/BK	EVAP Purge Solenoid Control	Solenoid Off: 12v, On: 1v
53	LG/RD	S/C Vent Solenoid	Vacuum Decreasing: 1v
54-58	---	Not Used	---
59	PK/BK	AIS 4 Motor Control	DC pulse signals: 0.8-11v
60	YL/BK	AIS 2 Motor Control	DC pulse signals: 0.8-11v

Pin Connector Graphic

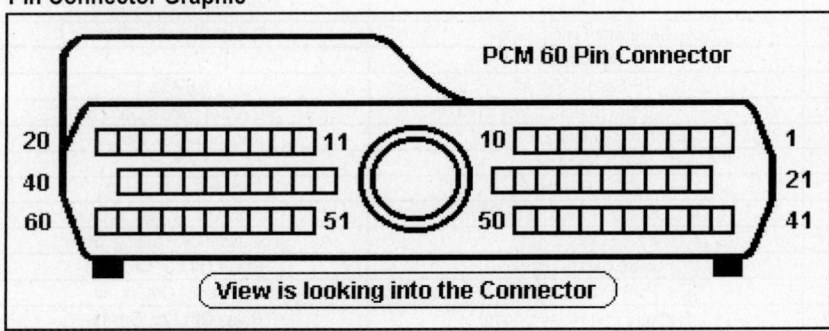

Standard Colors and Abbreviations

Abbreviation	Color	Abbreviation	Color	Abbreviation	Color
BK	Black	GY	Gray	RD	Red
BL	Blue	GN	Green	TN	Tan
BR	Brown	LG	Light Green	VT	Violet
DB	Dark Blue	OR	Orange	WT	White
DG	Dark Green	PK	Pink	YL	Yellow

1992-93 Dynasty 3.0L V6 SOHC MFI VIN 3 (A/T) 60 Pin Connector

PCM Pin #	Wire Color	Circuit Description (60 Pin)	Value at Hot Idle
1	DG/RD	MAP Sensor Signal	1.5-1.7v
2	TN/BK	ECT Sensor Signal	At 180°F: 2.80v
3	RD/BK	Battery Power (Fused B+)	12-14v
4	BK/LB	Sensor Ground	<0.050v
5	BK/WT	Sensor Ground	<0.050v
6	PK/WT	5-Volt Supply	4.9-5.1v
7	OR	8-Volt Supply	7.9-8.1v
8	---	Not Used	---
9	DB	Ignition Switch Output	12-14v
10	---	Not Used	---
11	BK/TN	Power Ground	<0.1v
12	BK/TN	Power Ground	<0.1v
13	LB/BR	Injector 4 Driver	1-4 ms
14	YL/WT	Injector 3 Driver	1-4 ms
15	TN	Injector 2 Driver	1-4 ms
16	WT/DB	Injector 1 Driver	1-4 ms
17-18	---	Not Used	---
19	BK/GY	Coil 1 Driver	5°, at 55 mph: 8° dwell
20	DG	Generator Field Driver	Digital Signal: 0-12-0v
21	---	Not Used	---
22	OR/DB	TP Sensor Signal	0.6-1.0v
23	RD/LG	S/C Set Switch Signal	S/C & Set Switch On: 3.8V
24	GY/BK	Distributor Reference Signal	Digital Signal: 0-5-0v
25	PK	SCI Transmit	0v
26	PK/BR	CCD Bus (+)	Digital Signal: 0-5-0v
27	BR	A/C Damped Pressure Switch	A/C On: 1v, Off: 12v
28 ('93)	DB/YL	PSP Switch Signal	Straight: 0v, Turning: 5v
29	WT/PK	Brake Switch Signal	Brake Off: 0v, On: 12v
30	BR/YL	PNP Switch Signal	In P/N: 0v, Others: 5v
31	DB/PK	Radiator Fan Relay Control	Relay Off: 12v, On: 1v
32	BK/PK	MIL (lamp) Control	Lamp On: 1v, Off: 12v
33	TN/RD	S/C Vacuum Solenoid	Vacuum Increasing: 1v
34	DB/OR	A/C Clutch Relay Control	Relay Off: 12v, On: 1v
35	GY/YL	EGR Solenoid Control (Calif.)	12v, 55 mph: 1v
36-37	---	Not Used	---
38	GY	Injector 5 Driver	1-4 ms
39	GY/RD	AIS Motor Control (open)	DC pulse signals: 0.8-11v
40	BR/WT	AIS Motor Control (close)	DC pulse signals: 0.8-11v

1992-93 Dyna...

...ty 3.0L V6 SOHC MFI VIN 3 (A/T) 60 Pin Connector

Wire Color	Circuit Description (60 Pin)	Value at Hot Idle
BK/DG	HO2S-11 (B1 S1) Signal	0.1-1.1v
---	Not Used	---
TN/YL	Distributor Sync Signal	Digital Signal: 0-5-0v
LG	SCI Receive	0v
WT/BK	CCD Bus (-)	<0.050v
WT/OR	Vehicle Speed Signal	Digital Signal
---	Not Used	---
DB/YL	ASD Relay Control	Relay Off: 12v, On: 1v
PK/BK	EVAP Purge Solenoid Control	Solenoid Off: 12v, On: 1v
LG/RD	S/C Vent Solenoid	Vacuum Decreasing: 1v
---	Not Used	---
G/OR	ASD Relay Output	12-14v
?/DB	Injector 6 Driver	1-4 ms
?/BK	AIS Motor Control (close)	DC pulse signals: 0.8-11v
YL/BK	AIS Motor Control (close)	DC pulse signals: 0.8-11v

...aphic

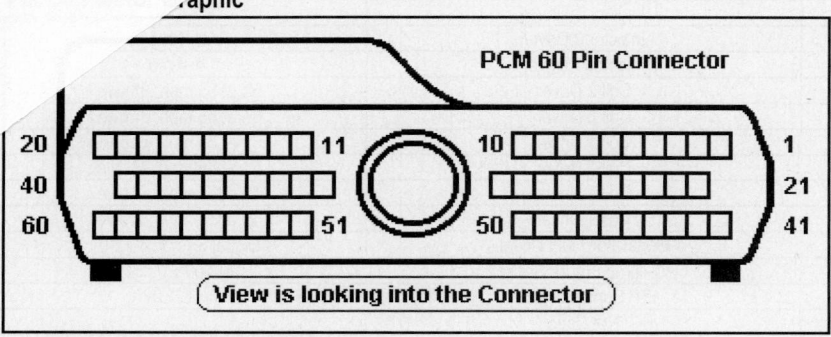

PCM 60 Pin Connector

View is looking into the Connector

Standard Colors and Abbreviations

Abbreviation	Color	Abbreviation	Color	Abbreviation	Color
BK	Black	GY	Gray	RD	Red
BL	Blue	GN	Green	TN	Tan
BR	Brown	LG	Light Green	VT	Violet
DB	Dark Blue	OR	Orange	WT	White
DG	Dark Green	PK	Pink	YL	Yellow

1990-91 Dynasty 3.3L V6 MFI VIN R (A/T) 60 Pin Connector

PCM Pin #	Wire Color	Circuit Description (60 Pin)	Value at Hot
1	DG/RD	MAP Sensor Signal	1.5-1.7v
2	TN/BK	ECT Sensor Signal	At 180°F: 2.80v
2 ('90)	TN/WT	ECT Sensor Signal	At 180°F: 2.80v
3	RD	Battery Power (Fused B+)	12-14v
4	BK/LB	Sensor Ground	<0.050v
5	BK/TN	Sensor Ground	<0.050v
6	PK/WT	5-Volt Supply	4.9-5.1v
7	OR	8-Volt Supply	7.9-8.1v
8	DG/BK	Ignition Switch Start Signal	12-14v
9	DB	Ignition Switch Output	12-14v
10	---	Not Used	---
11	BK/TN	Power Ground	<0.1v
12	BK/TN	Power Ground	<0.1v
11 ('90)	LB/RD	Power Ground	<0.1v
12 ('90)	LB/RD	Power Ground	<0.1v
13	---	Not Used	---
14	YL/WT	Injector 3 Driver	1-4 ms
15	TN	Injector 2 Driver	1-4 ms
16	WT/DB	Injector 1 Driver	1-4 ms
17	DB/TN	Coil 2 Driver	5°, at 55 mph: 8° dwell
18	DB/GY	Coil 3 Driver	5°, at 55 mph: 8° dwell
19	GY	Coil 1 Driver	5°, at 55 mph: 8° dwell
17 ('90)	DB/BK	Coil 2 Driver	5°, at 55 mph: 8° dwell
18 ('90)	DB/GY	Coil 3 Driver	5°, at 55 mph: 8° dwell
19 ('90)	BK/GY	Coil 1 Driver	5°, at 55 mph: 8° dwell
20	DG	Generator Field Driver	Digital Signal: 0-12-0v
21	BK/RD	Air Temperature Sensor	At 100°F: 2.51v
22	OR/DB	TP Sensor Signal	0.6-1.0v
23	---	Not Used	---
24	GY/BK	Distributor Reference Signal	Digital Signal: 0-5-0v
25	PK	SCI Transmit	0v
26	PK/BR	CCD Bus (+)	Digital Signal: 0-5-0v
26 ('90)	BK	CCD Bus (+)	Digital Signal: 0-5-0v
27	BR	A/C Clutch Signal	A/C On: 1v, Off: 12v
28	---	Not Used	---
29	WT/PK	Brake Switch Signal	Brake Off: 0v, On: 12v
30	BR/YL	PNP Switch Signal	In P/N: 0v, Others: 5v
31	DB/PK	Radiator Fan Relay Control	Relay Off: 12v, On: 1v
32	BK/PK	MIL (lamp) Control	Lamp On: 1v, Off: 12v
33	TN/RD	S/C Vacuum Solenoid	Vacuum Increasing: 1v
34	DB/OR	A/C Clutch Relay Control	Relay Off: 12v, On: 1v
35	GY/YL	EGR Solenoid Control (Calif.)	12v, 55 mph: 1v
36-38	---	Not Used	---
39	GY/RD	AIS 3 Motor Control	DC pulse signals: 0.8-11v
40	BR/WT	AIS 1 Motor Control	DC pulse signals: 0.8-11v

Standard Colors and Abbreviations

Abbreviation	Color	Abbreviation	Color	Abbreviation	Color
BK	Black	GY	Gray	RD	Red
BL	Blue	GN	Green	TN	Tan
BR	Brown	LG	Light Green	VT	Violet
DB	Dark Blue	OR	Orange	WT	White
DG	Dark Green	PK	Pink	YL	Yellow

1990-91 Dynasty 3.3L V6 MFI VIN R (A/T) 60 Pin Connector

PCM Pin #	Wire Color	Circuit Description (60 Pin)	Value at Hot Idle
41	BK/DG	HO2S-11 (B1 S1) Signal	0.1-1.1v
42	BK/LG	Knock Sensor Signal	0.080v AC
43	---	Not Used	---
44	TN/YL	Distributor Sync Signal	Digital Signal: 0-5-0v
45	LG	SCI Receive	0v
46	WT/BK	CCD Bus (-)	<0.050v
47	WT/OR	Vehicle Speed Signal	Digital Signal
48	BR/RD	S/C Set Switch Signal	S/C & Set Switch On: 3.8v
49	YL/RD	S/C On/Off Switch Signal	S/C On: 1v, Off: 12v
50	WT/LG	S/C Resume Switch	S/C On: 1v, Off: 12v
51	DB/YL	ASD Relay Control	Relay Off: 12v, On: 1v
52	PK	EVAP Purge Solenoid Control	Solenoid Off: 12v, On: 1v
53	LG/RD	S/C Vent Solenoid	Vacuum Decreasing: 1v
54-58	---	Not Used	---
59	PK/BK	AIS 4 Motor Control	DC pulse signals: 0.8-11v
60	YL/BK	AIS 2 Motor Control	DC pulse signals: 0.8-11v

Pin Connector Graphic

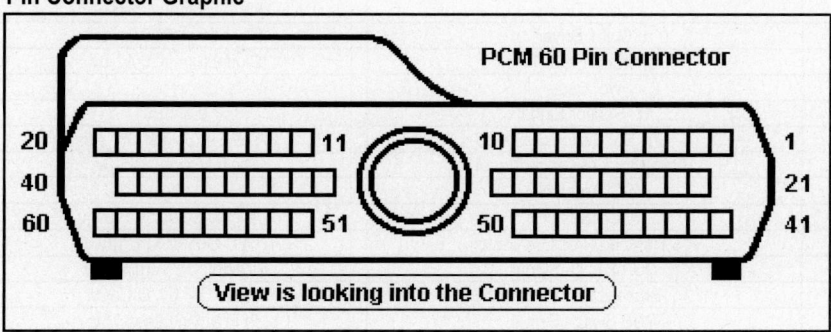

1992-93 Dynasty 3.3L V6 MFI VIN R (A/T) 60 Pin Connector

PCM Pin #	Wire Color	Circuit Description (60 Pin)	Value at Hot Idle
1	DG/RD	MAP Sensor Signal	1.5-1.7v
2	TN/BK	ECT Sensor Signal	At 180°F: 2.80v
3	RD/WT	Battery Power (Fused B+)	12-14v
4	BK/LB	Sensor Ground	<0.050v
5	BK/WT	Sensor Ground	<0.050v
6	PK/WT	5-Volt Supply	4.9-5.1v
7	OR	8-Volt Supply	7.9-8.1v
8	---	Not Used	---
9	DB	Ignition Switch Output	12-14v
10	---	Not Used	---
11	BK/TN	Power Ground	<0.1v
12	BK/TN	Power Ground	<0.1v
13	LB/BR	Injector 4 Driver	1-4 ms
14	YL/WT	Injector 3 Driver	1-4 ms
15	TN	Injector 2 Driver	1-4 ms
16	WT/DB	Injector 1 Driver	1-4 ms
17	DB/TN	Coil 2 Driver	5°, at 55 mph: 8° dwell
18	DB/GY	Coil 3 Driver	5°, at 55 mph: 8° dwell
19	BK/GY	Coil 1 Driver	5°, at 55 mph: 8° dwell
20	DG	Generator Field Driver	Digital Signal: 0-12-0v
21	---	Not Used	---
22	OR/DB	TP Sensor Signal	0.6-1.0v
23	RD/LG	S/C Set Switch Signal	S/C & Set Switch On: 3.8v
24	GY/BK	Distributor Reference Signal	Digital Signal: 0-5-0v
25	PK	SCI Transmit	0v
26	PK/BR	CCD Bus (+)	Digital Signal: 0-5-0v
27	BR	A/C Damped Pressure Switch	A/C On: 1v, Off: 12v
28 ('93)	DB/OR	PSP Switch Signal	Straight: 0v, Turning: 5v
29	WT/PK	Brake Switch Signal	Brake Off: 0v, On: 12v
30	BR/YL	PNP Switch Signal	In P/N: 0v, Others: 5v
31	DB/PK	Radiator Fan Relay Control	Relay Off: 12v, On: 1v
32	BK/PK	MIL (lamp) Control	Lamp On: 1v, Off: 12v
33	TN/RD	S/C Vacuum Solenoid	Vacuum Increasing: 1v
34	DB/OR	A/C Clutch Relay Control	Relay Off: 12v, On: 1v
35	GY/YL	EGR Solenoid Control (Calif.)	12v, 55 mph: 1v
36-37	---	Not Used	---
38	GY	Injector 5 Driver	1-4 ms
39	GY/RD	AIS Motor Control (open)	DC pulse signals: 0.8-11v
40	BR/WT	AIS Motor Control (close)	DC pulse signals: 0.8-11v

Standard Colors and Abbreviations

Abbreviation	Color	Abbreviation	Color	Abbreviation	Color
BK	Black	GY	Gray	RD	Red
BL	Blue	GN	Green	TN	Tan
BR	Brown	LG	Light Green	VT	Violet
DB	Dark Blue	OR	Orange	WT	White
DG	Dark Green	PK	Pink	YL	Yellow

L V6 MFI VIN R (A/T) 60 Pin Connector

Wire Color	Circuit Description (60 Pin)	Value at Hot Idle
BK/DG	HO2S-11 (B1 S1) Signal	0.1-1.1v
BK/LG	Knock Sensor Signal	0.080v AC
—	Not Used	---
TN/YL	Distributor Sync Signal	Digital Signal: 0-5-0v
LG	SCI Receive	0v
WT/BK	CCD Bus (-)	<0.050v
WT/OR	Vehicle Speed Signal	Digital Signal
—	Not Used	—
DB/YL	ASD Relay Control	Relay Off: 12v, On: 1v
PK/BK	EVAP Purge Solenoid Control	Solenoid Off: 12v, On: 1v
LG/RD	S/C Vent Solenoid	Vacuum Decreasing: 1v
—	Not Used	—
DG/OR	ASD Relay Output	12-14v
BR/DB	Injector 6 Driver	1-4 ms
PK/BK	AIS Motor Control (close)	DC pulse signals: 0.8-11v
YL/BK	AIS Motor Control (close)	DC pulse signals: 0.8-11v

Pin Connector Graphic

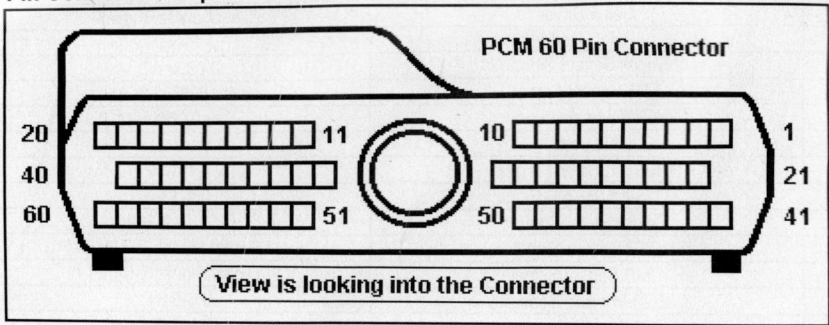

1998-99 Intrepid 2.7L V6 DOHC MFI VIN R (A/T) 'C1' Black Connector

PCM Pin #	Wire Color	Circuit Description (40 Pin)	Value
1	TN/LG	Coil 4 Driver	5°, at 55
2	TN/OR	Coil 3 Driver	5°, at 55 mph
3	TN/PK	Coil 2 Driver	5°, at 55 mph:
4	TN/LB	Coil 6 Driver	5°, at 55 mph: 8°
5	YL/RD	S/C Power Supply	12-14v
6	DG/OR	ASD Relay Output	12-14v
7	YL/WT	Injector 3 Driver	1-4 ms
8	DG	Generator Field Driver	Digital Signal: 0-12-0v
9	---	Not Used	---
10	BK/TN	Power Ground	<0.1v
11	TN/RD	Coil 1 Driver	5°, at 55 mph: 8° dwell
12	---	Not Used	---
13	WT/DB	Injector 1 Driver	1-4 ms
14	BR/DB	Injector 6 Driver	1-4 ms
15	GY	Injector 5 Driver	1-4 ms
16	LB/BR	Injector 4 Driver	1-4 ms
17	TN/WT	Injector 2 Driver	1-4 ms
18-19	---	Not Used	
20	DB/WT	Ignition Switch Output	12-14v
21	TN/DG	Coil 5 Driver	5°, at 55 mph: 8° dwell
22-24	---	Not Used	
25	BK/PK	Knock Sensor Signal	0.080v AC
26	TN/BK	ECT Sensor Signal	At 180°F: 2.80v
27	BK/OR	HO2S Signal Ground	<0.050v
28	---	Not Used	---
29	LG/RD	HO2S-21 (B2 S1) Signal	0.1-1.1v
30	BK/DG	HO2S-11 (B1 S1) Signal	0.1-1.1v
31	TN	Starter Relay Control	KOEC: 9-11v
32	GY/BK	CKP Sensor Signal	Digital Signal: 0-5-0v
33	TN/YL	CMP Sensor Signal	Digital Signal: 0-5-0v
34	LG/PK	EGR Position Sensor	0.6-0.8v
35	OR/DB	TP Sensor Signal	0.6-1.0v
36	DG/RD	MAP Sensor Signal	1.5-1.7v
37	BK/RD	IAT Sensor Signal	At 100°F: 1.83v
38-39	---	Not Used	---
40	GY/YL	EGR Solenoid Control	12v, 55 mph: 1v

Standard Colors and Abbreviations

Abbreviation	Color	Abbreviation	Color	Abbreviation	Color
BK	Black	GY	Gray	RD	Red
BL	Blue	GN	Green	TN	Tan
BR	Brown	LG	Light Green	VT	Violet
DB	Dark Blue	OR	Orange	WT	White
DG	Dark Green	PK	Pink	YL	Yellow

1998-99 Intrepid 2.7L V6 DOHC MFI VIN R (A/T) 'C2' White Connector

PCM Pin #	Wire Color	Circuit Description (40 Pin)	Value at Hot Idle
41	RD/LG	S/C Set Switch Signal	S/C & Set Switch On: 3.8v
42	DB	A/C Pressure Switch Signal	A/C On: 0.45-4.85v
43, 47	BK/LB	Sensor Ground	<0.050v
44	OR	8-Volt Supply	7.9-8.1v
45	DB/LG	PSP Switch Signal	Straight: 0v, Turning: 5v
46	RD/WT	Battery Power (Fused B+)	12-14v
48	BR/WT	IAC 3 Driver	DC pulse signals: 0.8-11v
49	YL/BK	IAC 2 Driver	DC pulse signals: 0.8-11v
50	BK/TN	Power Ground	<0.1v
51	TN/WT	HO2S-12 (B1 S2) Signal	0.1-1.1v
52, 60	---	Not Used	---
53	PK/WT	HO2S-22 (B2 S2) Signal	0.1-1.1v
55	DB/PK	Low Speed Fan Relay	Relay Off: 12v, On: 1v
56	TN/RD	S/C Vacuum Solenoid	Vacuum Increasing: 1v
57	GY/RD	IAC 1 Driver	DC pulse signals: 0.8-11v
58	PK/BK	IAC 4 Driver	DC pulse signals: 0.8-11v
59	YL/PK	CCD Bus (+)	Digital Signal: 0-5-0v
61	PK/WT	5-Volt Supply	4.9-5.1v
62	WT/PK	Brake Switch Signal	Brake Off: 0v, On: 12v
63	YL/DG	Torque Management Request	Digital Signals
64	DB/OR	A/C Clutch Relay Control	Relay Off: 12v, On: 1v
65	PK	SCI Transmit	0v
66	WT/OR	Vehicle Speed Signal	Digital Signal
67	DB/YL	ASD Relay Control	Relay Off: 12v, On: 1v
68	PK/BK	EVAP Purge Solenoid Control	PWM signal: 0-12-0v
69	DB/LG	High Speed Fan Relay	Relay Off: 12v, On: 1v
70	DG/LG	EVAP Purge Solenoid Sense	0-1v
71	WT/RD	EATX RPM Signal	Digital Signals
72	OR/RD	LDP Switch Signal	Switch Closed: 0v, open: 12v
74	BR	Fuel Pump Relay Control	Relay Off: 12v, On: 1v
75	LG	SCI Receive	0v
76	BK/PK	PNP Switch Sense (EATX)	In P/N: 0v, Others: 5v
77	WT/DG	LDP Solenoid Control	PWM signal: 0-12-0v
78-89	---	Not Used	---
80	LG/RD	S/C Vent Solenoid	Vacuum Decreasing: 1v

Pin Connector Graphic

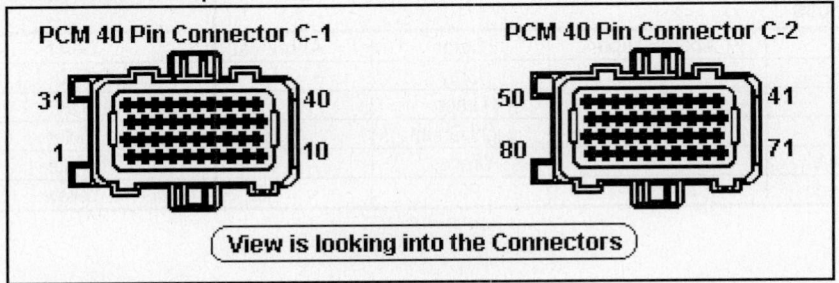

PCM 40 Pin Connector C-1

PCM 40 Pin Connector C-2

View is looking into the Connectors

2000-01 Intrepid 2.7L V6 24v DOHC VIN R, & U & V 'C1' Black Connector

PCM Pin #	Wire Color	Circuit Description (40 Pin)	Value at Hot Idle
1	TN/LG	Coil 4 Driver	5°, at 55 mph: 8° dwell
2	TN/OR	Coil 3 Driver	5°, at 55 mph: 8° dwell
3	TN/PK	Coil 2 Driver	5°, at 55 mph: 8° dwell
4	TN/LB	Coil 6 Driver	5°, at 55 mph: 8° dwell
5	YL/RD	S/C Power Supply	12-14v
6	DG/OR	ASD Relay Output	12-14v
7	YL/WT	Injector 3 Driver	1-4 ms
8	DG	Generator Field Driver	Digital Signal: 0-12-0v
9	---	Not Used	---
10	BK/TN	Power Ground	<0.1v
11	TN/RD	Coil 1 Driver	5°, at 55 mph: 8° dwell
12	---	Not Used	---
13	WT/DB	Injector 1 Driver	1-4 ms
14	BR/DB	Injector 6 Driver	1-4 ms
15	GY	Injector 5 Driver	1-4 ms
16	LB/BR	Injector 4 Driver	1-4 ms
17	TN/WT	Injector 2 Driver	1-4 ms
18-19	---	Not Used	---
20	DB/WT	Ignition Switch Output	12-14v
21	TN/DG	Coil 5 Driver	5°, at 55 mph: 8° dwell
22-24	---	Not Used	---
25	BK/VT	Knock Sensor Signal	0.080v AC
26	TN/BK	ECT Sensor Signal	At 180°F: 2.80v
27	BK/OR	HO2S Signal Ground	<0.050v
28	---	Not Used	---
29	LG/RD	HO2S-21 (B2 S1) Signal	0.1-1.1v
30	BK/DG	HO2S-11 (B1 S1) Signal	0.1-1.1v
31	TN	Starter Relay Control	KOEC: 9-11v
32	GY/BK	CKP Sensor Signal	Digital Signal: 0-5-0v
33	TN/YL	CMP Sensor Signal	Digital Signal: 0-5-0v
34	LG/PK	EGR Position Sensor	0.6-0.8v
35	OR/DB	TP Sensor Signal	0.6-1.0v
36	DG/RD	MAP Sensor Signal	1.5-1.7v
37	BK/RD	IAT Sensor Signal	At 100°F: 1.83v
38-39	---	Not Used	---
40	GY/YL	EGR Solenoid Control	12v, 55 mph: 1v

Standard Colors and Abbreviations

Abbreviation	Color	Abbreviation	Color	Abbreviation	Color
BK	Black	GY	Gray	RD	Red
BL	Blue	GN	Green	TN	Tan
BR	Brown	LG	Light Green	VT	Violet
DB	Dark Blue	OR	Orange	WT	White
DG	Dark Green	PK	Pink	YL	Yellow

2000-01 Intrepid 2.7L V6 24v DOHC VIN R, & U & V 'C2' White Connector

PCM Pin #	Wire Color	Circuit Description (40 Pin)	Value at Hot Idle
41	RD/LG	S/C Set Switch Signal	S/C & Set Switch On: 3.8v
42	DB	A/C Pressure Switch Signal	A/C On: 0.45-4.85v
43, 47	BK/LB	Sensor Ground	<0.050v
44	OR	8-Volt Supply	7.9-8.1v
45	DB/LG	PSP Switch Signal	Straight: 0v, Turning: 5v
46	RD/WT	Battery Power (Fused B+)	12-14v
48	BR/WT	IAC 3 Driver	DC pulse signals: 0.8-11v
49	YL/BK	IAC 2 Driver	DC pulse signals: 0.8-11v
50	BK/TN	Power Ground	<0.1v
51	TN/WT	HO2S-12 (B1 S2) Signal	0.1-1.1v
52 ('01-'02)	VT/LG	Battery Temperature Sensor	At 86°F: 1.96v
53	PK/WT	HO2S-22 (B2 S2) Signal	0.1-1.1v
54, 60, 73	---	Not Used	
55	DB/PK	Low Speed Fan Relay	Relay Off: 12v, On: 1v
56	TN/RD	S/C Vacuum Solenoid	Vacuum Increasing: 1v
57	GY/RD	IAC 1 Driver	DC pulse signals: 0.8-11v
58	VT/BK	IAC 4 Driver	DC pulse signals: 0.8-11v
59	VT/YL	CCD Bus (+)	Digital Signal: 0-5-0v
61	V/TN/WT	5-Volt Supply	4.9-5.1v
62	WT/PK	Brake Switch Signal	Brake Off: 0v, On: 12v
63	YL/DG	Torque Management Request	Digital Signals
64	DB/OR	A/C Clutch Relay Control	Relay Off: 12v, On: 1v
65, 75	PK, LG	SCI Transmit, SCI Receive	0v
66	WT/OR	Vehicle Speed Signal	Digital Signal
67	DB/YL	ASD Relay Control	Relay Off: 12v, On: 1v
68	PK/BK	EVAP Purge Solenoid Control	PWM signal: 0-12-0v
69	DB/LG	High Speed Fan Relay	Relay Off: 12v, On: 1v
70	DG/LG	EVAP Purge Solenoid Sense	0-1v
71	WT/RD	EATX RPM Signal	Digital Signals
72	OR/RD	LDP Switch Signal	Switch Closed: 0v, open: 12v
74	BR	Fuel Pump Relay Control	Relay Off: 12v, On: 1v
76	BK/PK	TRS T41 Sense	In P/N: 0v, Others: 5v
77	WT/DG	LDP Solenoid Control	PWM signal: 0-12-0v
78-79	---	Not Used	---
80	LG/RD	S/C Vent Solenoid	Vacuum Decreasing: 1v

Pin Connector Graphic

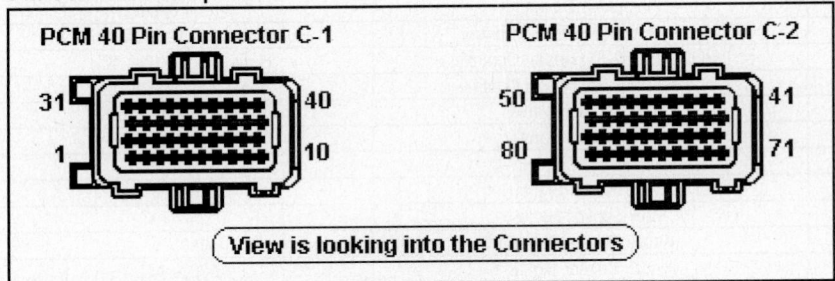

PCM 40 Pin Connector C-1

PCM 40 Pin Connector C-2

31   40

1   10

50   41

80   71

View is looking into the Connectors

2002-03 Intrepid 2.7L V6 SOHC 24v MFI VIN R 'C1' Black Connector

PCM Pin #	Wire Color	Circuit Description (38 Pin)	Value at Hot Idle
1-8, 10	---	Not Used	---
9	BK/TN	Power Ground	<0.1v
11	DB/WT	Ignition Switch Output (Run-Start)	12-14v
12	RD/WT	Ignition Switch Output (Off-Run-Start)	12-14v
13-16	---	Not Used	---
17	---	Not Used	---
18	BK/TN	Power Ground	<0.1v
19-20	---	Not Used	---
21	DB	A/C Pressure Switch Signal	A/C On: 0.45-4.85v
22-24	---	Not Used	---
25	LG	SCI Receive (PCM)	5v
26	VT/OR	SCI Receive (TCM)	5v
27-28	---	Not Used	---
29	RD	Battery Power (Fused B+)	12-14v
30	YL/BK	Ignition Switch Output (tart)	8-11v (cranking)
31	TN/WT	HO2S-12 (B1 S2) Signal	0.1-1.1v
32	DB/LG	Oxygen Sensor Return (down)	<0.050v
33	PK/WT	HO2S-22 (B2 S2) Signal	0.1-1.1v
34-35	---	Not Used	---
36	PK/TN	SCI Transmit (PCM)	0v
37	WT/DG	SCI Transmit (TCM)	0v
38	VT/YL	PCI Data Bus (J1850)	Digital Signals: 0-7-0v

2002-03 Intrepid 2.7L V6 SOHC 24v MFI VIN R 'C2' Gray Connector

PCM Pin #	Wire Color	Circuit Description (38 Pin)	Value at Hot Idle
1	TN/LB	Coil 6 Driver	5º, at 55 mph: 8º dwell
2	TN/DG	Coil 5 Driver	5º, at 55 mph: 8º dwell
3	TN/LG	Coil 4 Driver	5º, at 55 mph: 8º dwell
4	BR/DB	Injector 6 Driver	1.0-4.0 ms
5	GY	Injector 5 Driver	1.0-4.0 ms
6, 15, 26	---	Not Used	---
7	TN/OR	Coil 3 Driver	5º, at 55 mph: 8º dwell
8	GY/YL	EGR Solenoid Control	12v, at 55 mph: 1v
9	TN/PK	Coil 2 Driver	5º, at 55 mph: 8º dwell
10	TN/RD	Coil 1 Driver	5º, at 55 mph: 8º dwell
11	LB/BR	Injector 4 Driver	1.0-4.0 ms
12	YL/WT	Injector 3 Driver	1.0-4.0 ms
13	TN	Injector 2 Driver	1.0-4.0 ms
14	WT/DB	Injector 1 Driver	1.0-4.0 ms
16	VT/RD	Manifold Tuning Solenoid Control	Valve Off: 12v, On: 1v
17	BR/WT	HO2S-21 (B2 S1) Heater Control	Heater On: 1v, Off: 12v
18	BR/OR	HO2S-11 (B1 S1) Heater Control	Heater On: 1v, Off: 12v
19	DG	Generator Field Driver	Digital Signal: 0-12-0v
20	TN/BK	ECT Sensor Signal	At 180ºF: 2.80v
21	OR/DB	TP Sensor Signal	0.6-1.0v
22	LG/PK	EGR Sensor Signal	0.6-0.8v
23	DG/RD	MAP Sensor Signal	1.5-1.7v
24	BK/VT	Knock Sensor Return	<0.050v
25	DB/LG	Knock Sensor Signal	0.080v AC
27	BK/LB	Sensor Ground	<0.1v
28	YL/BK	IAC Motor Sense	12-14v
29	VT/WT	5-Volt Supply	4.9-5.1v
30	BK/RD	IAT Sensor Signal	At 100ºF: 1.83v
31	BK/DG	HO2S-11 (B1 S1) Signal	0.1-1.1v
32	BR/DG	HO2S-11 (B1 S1) Ground (Up)	<0.1v
33	LG/RD	HO2S-21 (B2 S1) Signal	0.1-1.1v
34	TN/YL	CMP Sensor Signal	Digital Signal: 0-5-0v
35	GY/BK	CKP Sensor Signal	Digital Signal: 0-5-0v
36-37	---	Not Used	---
38	GY/RD	IAC Motor Driver	DC pulse signals: 0.8-11v

2002-03 Intrepid 2.7L V6 SOHC 24v MFI VIN R 'C3' White Connector

PCM Pin #	Wire Color	Circuit Description (38 Pin)	Value at Hot Idle
1-2, 13-17	---	Not Used	---
3	DB/YL	ASD Relay Control	Relay Off: 12v, On: 1v
4	DB/PK	High Speed Radiator Fan Relay	Relay Off: 12v, On: 1v
5	LG/RD	S/C Vent Solenoid	Vacuum Decreasing: 1v
6	DB/PK	Low Speed Radiator Fan Relay	Relay Off: 12v, On: 1v
7	YL/RD	S/C Power Supply	12-14v
8	WT/DG	Natural Vacuum Leak Detection Solenoid	Solenoid Off: 12v, On: 1v
9	BR/VT	HO2S-21 (B2 S1) Heater Control	Heater On: 1v, Off: 12v
10	BR/GY	HO2S-11 (B1 S1) Heater Control	Heater On: 1v, Off: 12v
11	DB/OR	A/C Clutch Relay Control	Relay Off: 12v, On: 1v
12	TN/RD	S/C Vacuum Solenoid	Vacuum Increasing: 1v
18, 19	OR/DG	Automatic Shutdown Relay Output	12-14v
20	PK/BK	EVAP Purge Solenoid Control	PWM Signal: 0-12-0v
21-22, 24-25	---	Not Used	---
23	WT/PK	Brake Switch Sense	Brake Off: 0v, On: 12v
26	YL	Autostick Downshift Switch	Digital Signal: 0v or 12v
27	LG/RD	Autostick Upshift Switch	Digital Signal: 0v or 12v
28	OR/DG	Automatic Shutdown Relay Output	12-14v
29	DG/LG	EVAP Purge Solenoid Sense	0-1v
30-31, 33, 36	---	Not Used	---
32	VT/LG	Ambient Air Temperature Sensor	At 100°F: 1.83v
34	RD/LG	S/C Set Switch Signal	S/C & Set Switch On: 3.8v
35	OR/RD	Natural Vacuum Leak Detection Switch Sense	0.1v
37	BR	Fuel Pump Relay Control	Relay Off: 12v, On: 1v
38	TN	Starter Relay Control	KOEC: 9-11v

2002-03 Intrepid 2.7L V6 SOHC 24v MFI VIN R 'C4' Green Connector

PCM Pin #	Wire Color	Circuit Description (38 Pin)	Value at Hot Idle
1, 3-5	---	Not Used	---
2	BR	Overdrive Solenoid Control	Solenoid Off: 12v, On: 1v
3	PK	Underdrive Solenoid Control	Solenoid Off: 12v, On: 1v
6	WT	2-4 Solenoid Control	Solenoid Off: 12v, On: 1v
7-9, 11	---	Not Used	---
10	LB	Low/Reverse Solenoid Control	Solenoid Off: 12v, On: 1v
12	BK/YL	Power Ground	<0.050v
13, 14	BK/RD	Power Ground	<0.050v
15	LG/BK	TRS T1 Sense	<0.050v
16	VT	TRS T3 Sense	<0.050v
17, 20-21	---	Not Used	---
18, 19	LG, RD	Transmission Control Relay Output	Relay Off: 12v, On: 1v
19	RD	Transmission Control Relay Output	Relay Off: 12v, On: 1v
23-26, 31, 36	---	Not Used	---
27	BK/WT	TRS T41 Sense	<0.050v
28, 38	RD	Transmission Control Relay Output	Relay Off: 12v, On: 1v
29	LG	Low/Reverse Pressure Switch Sense	12-14v
30	YL/BK	2-4 Pressure Switch Sense	In Low/Reverse: 2-4v
32	LG/WT	Output Speed Sensor Signal	In 2-4 Position: 2-4v
33	RD/BK	Input Speed Sensor Signal	Moving: AC voltage
34	DB/BK	Speed Sensor Ground	Moving: AC voltage
35	VT/PK	Transmission Temperature Sensor Signal	<0.050v
37	VT/WT	TRS T42 Sense	In PRNL: 0v, Others 5v

Pin Connector Graphic

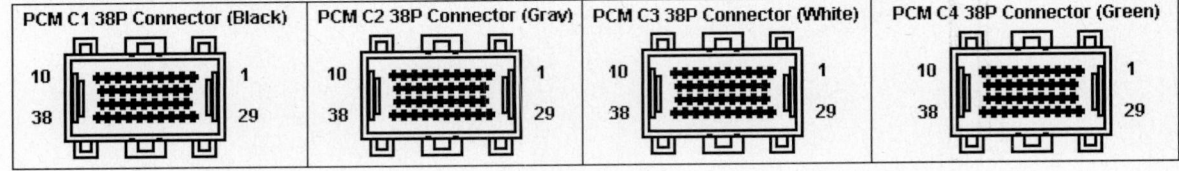

1998-99 Intrepid 3.2L V6 SOHC MFI VIN J (A/T) 'C1' Black Connector

PCM Pin #	Wire Color	Circuit Description (40 Pin)	Value at Hot Idle
1	TN/LG	Coil 4 Driver	5°, at 55 mph: 8° dwell
2	TN/OR	Coil 3 Driver	5°, at 55 mph: 8° dwell
3	TN/PK	Coil 2 Driver	5°, at 55 mph: 8° dwell
4	TN/LB	Coil 6 Driver	5°, at 55 mph: 8° dwell
5	YL/RD	S/C Power Supply	12-14v
6	DG/OR	ASD Relay Output	12-14v
7	YL/WT	Injector 3 Driver	1-4 ms
8	DG	Generator Field Driver	Digital Signal: 0-12-0v
9	---	Not Used	---
10	BK/TN	Power Ground	<0.1v
11	TN/RD	Coil 1 Driver	5°, at 55 mph: 8° dwell
12	---	Not Used	---
13	WT/DB	Injector 1 Driver	1-4 ms
14	BR/DB	Injector 6 Driver	1-4 ms
15	GY	Injector 5 Driver	1-4 ms
16	LB/BR	Injector 4 Driver	1-4 ms
17	TN	Injector 2 Driver	1-4 ms
18	GY/PK	Short Runner Valve	Valve Off: 12, On: 1v
19	---	Not Used	---
20	DB/WT	Ignition Switch Output	12-14v
21	TN/DG	Coil 5 Driver	5°, at 55 mph: 8° dwell
22-23	---	Not Used	---
24	DB/LG	Knock Sensor Signal	0.080v AC
25	BK/PK	Knock Sensor Ground	<0.050v
26	TN/BK	ECT Sensor Signal	At 180°F: 2.80v
27	BK/OR	HO2S Signal Ground	<0.050v
28	---	Not Used	---
29	LG/RD	HO2S-21 (B2 S1) Signal	0.1-1.1v
30	BK/DG	HO2S-11 (B1 S1) Signal	0.1-1.1v
31	TN/WT	Starter Relay Control	KOEC: 9-11v
32	GY/BK	CKP Sensor Signal	Digital Signal: 0-5-0v
33	TN/YL	CMP Sensor Signal	Digital Signal: 0-5-0v
34	LG/PK	EGR Position Sensor	0.6-0.8v
35	OR/DB	TP Sensor Signal	0.6-1.0v
36	DG/RD	MAP Sensor Signal	1.5-1.7v
37	BK/RD	IAT Sensor Signal	At 100°F: 1.83v
38	---	Not Used	---
39	PK/RD	Manifold Solenoid Valve	Solenoid Off: 12v, On: 1v
40	GY/YL	EGR Solenoid Control	12v, 55 mph: 1v

Standard Colors and Abbreviations

Abbreviation	Color	Abbreviation	Color	Abbreviation	Color
BK	Black	GY	Gray	RD	Red
BL	Blue	GN	Green	TN	Tan
BR	Brown	LG	Light Green	VT	Violet
DB	Dark Blue	OR	Orange	WT	White
DG	Dark Green	PK	Pink	YL	Yellow

1998-99 Intrepid 3.2L V6 SOHC MFI VIN J (A/T) 'C2' White Connector

PCM Pin #	Wire Color	Circuit Description (40 Pin)	Value at Hot Idle
41	RD/LG	S/C Set Switch Signal	S/C & Set Switch On: 3.8v
42	DB	A/C Pressure Switch Signal	A/C On: 0.45-4.85v
43	BK/LB	Sensor Ground	<0.050v
44	OR	8-Volt Supply	7.9-8.1v
45	DB/LG	PSP Switch Signal	Straight: 0v, Turning: 5v
46	RD/WT	Battery Power (Fused B+)	12-14v
47, 50	BK/WT	Power Ground	<0.1v
48	BR/WT	IAC 3 Driver	DC pulse signals: 0.8-11v
49	YL/BK	IAC 2 Driver	DC pulse signals: 0.8-11v
51	TN/WT	HO2S-12 (B1 S2) Signal	0.1-1.1v
52, 54	---	Not Used	---
53	PK/WT	HO2S-22 (B2 S2) Signal	0.1-1.1v
55	DB/PK	Low Speed Fan Relay	Relay Off: 12v, On: 1v
56	TN/RD	S/C Vacuum Solenoid	Vacuum Increasing: 1v
57	GY/RD	IAC 1 Driver	DC pulse signals: 0.8-11v
58	PK/BK	IAC 4 Driver	DC pulse signals: 0.8-11v
59	YL/PK	CCD Bus (+)	Digital Signal: 0-5-0v
60, 73	---	Not Used	---
61	PK/WT	5-Volt Supply	4.9-5.1v
62	WT/PK	Brake Switch Signal	Brake Off: 0v, On: 12v
63	YL/DG	Torque Management Request	Digital Signals
64	DB/OR	A/C Clutch Relay Control	Relay Off: 12v, On: 1v
65	PK	SCI Transmit	0v
66	WT/OR	Vehicle Speed Signal	Digital Signal
67	DB/YL	ASD Relay Control	Relay Off: 12v, On: 1v
68	PK/BK	EVAP Purge Solenoid Control	PWM signal: 0-12-0v
69	DB/PK	High Speed Fan Relay	Relay Off: 12v, On: 1v
70	DG/LG	EVAP Purge Solenoid Sense	0-1v
71	WT/RD	EATX RPM Signal	Digital Signals
72	OR	LDP Switch Signal	Switch Closed: 0v, open: 12v
74	BR	Fuel Pump Relay Control	Relay Off: 12v, On: 1v
75	LG	SCI Receive	5v
76	BK/WT	PNP Switch Sense (EATX)	In P/N: 0v, Others: 5v
77	WT/DG	LDP Solenoid Control	PWM signal: 0-12-0v
78-79	---	Not Used	---
80	LG/RD	S/C Vent Solenoid	Vacuum Decreasing: 1v

Pin Connector Graphic

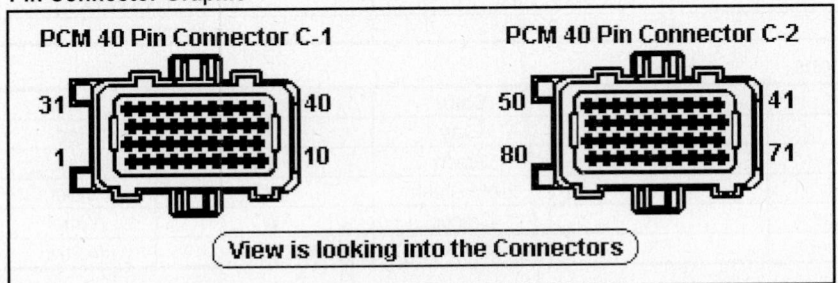

2000-01 Intrepid 3.2L V6 SOHC MFI VIN J (A/T) 'C1' Black Connector

PCM Pin #	Wire Color	Circuit Description (40 Pin)	Value at Hot Idle
1	TN/LG	Coil 4 Driver	5°, at 55 mph: 8° dwell
2	TN/OR	Coil 3 Driver	5°, at 55 mph: 8° dwell
3	TN/PK	Coil 2 Driver	5°, at 55 mph: 8° dwell
4	TN/LB	Coil 6 Driver	5°, at 55 mph: 8° dwell
5	YL/RD	S/C Power Supply	12-14v
6	DG/OR	ASD Relay Output	12-14v
7	YL/WT	Injector 3 Driver	1-4 ms
8	DG	Generator Field Driver	Digital Signal: 0-12-0v
9	---	Not Used	---
10	BK/TN	Power Ground	<0.1v
11	TN/RD	Coil 1 Driver	5°, at 55 mph: 8° dwell
12	---	Not Used	---
13	WT/DB	Injector 1 Driver	1-4 ms
14	BR/DB	Injector 6 Driver	1-4 ms
15	GY	Injector 5 Driver	1-4 ms
16	LB/BR	Injector 4 Driver	1-4 ms
17	TN	Injector 2 Driver	1-4 ms
18	GY/PK	Short Runner Valve Control	Valve Off: 12, On: 1v
19	---	Not Used	---
20	DB/WT	Ignition Switch Output	12-14v
21	TN/DG	Coil 5 Driver	5°, at 55 mph: 8° dwell
22-23	---	Not Used	---
24	DB/LG	Knock Sensor Signal	0.080v AC
25	BK/VT	Knock Sensor Ground	<0.050v
26	TN/BK	ECT Sensor Signal	At 180°F: 2.80v
27	BK/OR	HO2S Signal Ground	<0.050v
28	---	Not Used	---
29	LG/RD	HO2S-21 (B2 S1) Signal	0.1-1.1v
30	BK/DG	HO2S-11 (B1 S1) Signal	0.1-1.1v
31	TN/WT	Starter Relay Control	KOEC: 9-11v
32	GY/BK	CKP Sensor Signal	Digital Signal: 0-5-0v
33	TN/YL	CMP Sensor Signal	Digital Signal: 0-5-0v
34	LG/PK	EGR Position Sensor	0.6-0.8v
35	OR/DB	TP Sensor Signal	0.6-1.0v
36	DG/RD	MAP Sensor Signal	1.5-1.7v
37	BK/RD	IAT Sensor Signal	At 100°F: 1.83v
38	---	Not Used	---
39	VT/RD	Manifold Solenoid Valve	Solenoid Off: 12v, On: 1v
40	GY/YL	EGR Solenoid Control	12v, 55 mph: 1v

Standard Colors and Abbreviations

Abbreviation	Color	Abbreviation	Color	Abbreviation	Color
BK	Black	GY	Gray	RD	Red
BL	Blue	GN	Green	TN	Tan
BR	Brown	LG	Light Green	VT	Violet
DB	Dark Blue	OR	Orange	WT	White
DG	Dark Green	PK	Pink	YL	Yellow

2000-01 Intrepid 3.2L V6 SOHC MFI VIN J (A/T) 'C2' White Connector

PCM Pin #	Wire Color	Circuit Description (40 Pin)	Value at Hot Idle
41	RD/LG	S/C Set Switch Signal	S/C & Set Switch On: 3.8v
42	DB	A/C Pressure Switch Signal	A/C On: 0.45-4.85v
43	BK/LB	Sensor Ground	<0.050v
44	OR	8-Volt Supply	7.9-8.1v
45	DB/LG	PSP Switch Signal	Straight: 0v, Turning: 5v
46	RD/WT	Battery Power (Fused B+)	12-14v
47, 50	BK/WT	Power Ground	<0.1v
48	BR/WT	IAC 3 Driver	DC pulse signals: 0.8-11v
49	YL/BK	IAC 2 Driver	DC pulse signals: 0.8-11v
51	TN/WT	HO2S-12 (B1 S2) Signal	0.1-1.1v
52 ('01)	VT/LG	Battery Temperature Sensor	At 86°F: 1.96v
53	PK/WT	HO2S-22 (B2 S2) Signal	0.1-1.1v
54, 60, 73	---	Not Used	---
55	DB/PK	Low Speed Fan Relay	Relay Off: 12v, On: 1v
56	TN/RD	S/C Vacuum Solenoid	Vacuum Increasing: 1v
57	GY/RD	IAC 1 Driver	DC pulse signals: 0.8-11v
58	VT/BK	IAC 4 Driver	DC pulse signals: 0.8-11v
59	VT/YL	CCD Bus (+)	Digital Signal: 0-5-0v
61	VT/WT	5-Volt Supply	4.9-5.1v
62	WT/PK	Brake Switch Signal	Brake Off: 0v, On: 12v
63	YL/DG	Torque Management Request	Digital Signals
64	DB/OR	A/C Clutch Relay Control	Relay Off: 12v, On: 1v
65	PK	SCI Transmit	0v
66	WT/OR	Vehicle Speed Signal	Digital Signal
67	DB/YL	ASD Relay Control	Relay Off: 12v, On: 1v
68	PK/BK	EVAP Purge Solenoid Control	PWM signal: 0-12-0v
69	DB/LG	High Speed Fan Relay	Relay Off: 12v, On: 1v
70	DG/LG	EVAP Purge Solenoid Sense	0-1v
71	WT/RD	EATX RPM Signal	Digital Signals
72	OR/RD	LDP Switch Signal	Switch Closed: 0v, open: 12v
74	BR	Fuel Pump Relay Control	Relay Off: 12v, On: 1v
75	LG	SCI Receive	5v
76	BK/YL	TRS T41 Sense	In P/N: 0v, Others: 5v
77	WT/DG	LDP Solenoid Control	PWM signal: 0-12-0v
78-79	---	Not Used	---
80	LG/RD	S/C Vent Solenoid	Vacuum Decreasing: 1v

Pin Connector Graphic

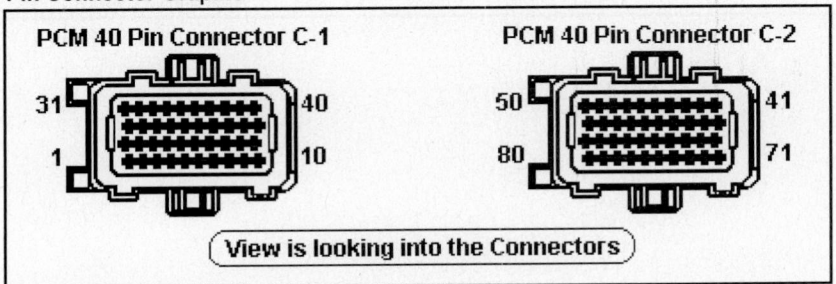

PCM 40 Pin Connector C-1

31   40
1   10

PCM 40 Pin Connector C-2

50   41
80   71

(View is looking into the Connectors)

1993-1995 Intrepid 3.3L V6 MFI VIN T (A/T) 60 Pin Connector

PCM Pin #	Wire Color	Circuit Description (60 Pin)	Value at Hot Idle
1	DG/RD	MAP Sensor Signal	1.5-1.7v
2	TN/BK	ECT Sensor Signal	At 180°F: 2.80v
3	RD/WT	Battery Power (Fused B+)	12-14v
4	BK/LB	Sensor Ground	<0.050v
5	BK/WT	Sensor Ground	<0.050v
6	PK/WT	5-Volt Supply	4.9-5.1v
7	OR	9-Volt Supply	8.9-9.1v
8	YL/DG	Torque Management Request	Digital Signals
9	DB/WT	Ignition Switch Output	12-14v
10	GY/BK	Knock Sensor 2 Signal	0.080v AC
11	BK/TN	Power Ground	<0.1v
12	BK/TN	Power Ground	<0.1v
13	LB/BR	Injector 4 Driver	1-4 ms
14	YL/WT	Injector 3 Driver	1-4 ms
15	TN	Injector 2 Driver	1-4 ms
16	WT/DB	Injector 1 Driver	1-4 ms
17	DB/YL	Coil 2 Driver	5°, at 55 mph: 8° dwell
18	RD/YL	Coil 3 Driver	5°, at 55 mph: 8° dwell
19	GY	Coil 1 Driver	5°, at 55 mph: 8° dwell
20	DG	Generator Field Control	Digital Signal: 0-12-0v
21	BK/RD	Air Temperature Sensor	At 100°F: 2.51v
22	OR/DB	TP Sensor Signal	0.6-1.0v
23	RD/LG	S/C Switch Signal	S/C & Set Switch On: 3.8V
24	LB/DB	CKP Sensor Signal	Digital Signal: 0-5-0v
25	PK	SCI Transmit	0v
26	PK/BR	CCD Bus (+)	Digital Signal: 0-5-0v
27-28	---	Not Used	---
29	WT/PK	Brake Switch Signal	Brake Off: 0v, On: 12v
30	BK/LG	PNP Switch Signal	In P/N: 0v, Others: 5v
31	DB/PK	High Speed Fan Relay	Relay Off: 12v, On: 1v
32	WT	Low Speed Fan Relay	Relay Off: 12v, On: 1v
33	TN/RD	S/C Vacuum Solenoid	Vacuum Increasing: 1v
34	DB/OR	A/C Clutch Relay Control	Relay Off: 12v, On: 1v
35	GY/YL	EGR Solenoid Control	Solenoid Off: 12v, On: 1v
36-37	---	Not Used	---
38	BR/RD	Injector 5 Driver	1-4 ms
39	GY/RD	AIS 1 Motor Control	DC pulse signals: 0.8-11v
40	BR/WT	AIS 3 Motor Control	DC pulse signals: 0.8-11v

1993-1995 Intrepid 3.3L V6 MFI VIN T (A/T) 60 Pin Connector

PCM Pin #	Wire Color	Circuit Description (60 Pin)	Value at Hot Idle
41	BK/DG	HO2S-11 (B1 S1) Signal	0.1-1.1v
42	BK/LG	Knock Sensor 1 Signal	0.080v AC
43	---	Not Used	---
44	TN/YL	CMP Sensor Signal	Digital Signal: 0-5-0v
45	LG	SCI Receive	0v
46	WT/BK	CCD Bus (-)	<0.050v
47	WT/OR	Vehicle Speed Signal	Digital Signal
48	DB	A/C Pressure Switch Signal	A/C On: 0.45-4.85v
49	TN/WT	HO2S-21 (B2 S1) Signal	0.1-1.1v
50	---	Not Used	---
51	DB/YL	ASD Relay Control	Relay Off: 12v, On: 1v
52	PK/BK	EVAP Purge Solenoid Control	Solenoid Off: 12v, On: 1v
53	LG/RD	S/C Vent Solenoid	Vacuum Decreasing: 1v
54	---	Not Used	---
55	TN/RD	S/C Power Supply	12-14v
56	---	Not Used	---
57	DG/OR	ASD Relay Output	12-14v
58	BR/BK	Injector 6 Driver	1-4 ms
59	PK/BK	AIS 4 Motor Control	DC pulse signals: 0.8-11v
60	YL/BK	AIS 2 Motor Control	DC pulse signals: 0.8-11v

Pin Connector Graphic

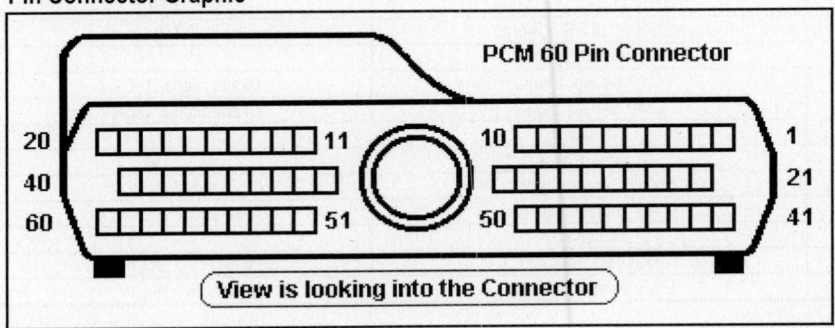

Standard Colors and Abbreviations

Abbreviation	Color	Abbreviation	Color	Abbreviation	Color
BK	Black	GY	Gray	RD	Red
BL	Blue	GN	Green	TN	Tan
BR	Brown	LG	Light Green	VT	Violet
DB	Dark Blue	OR	Orange	WT	White
DG	Dark Green	PK	Pink	YL	Yellow

1996-97 Intrepid 3.3L V6 MFI VIN T (A/T) 'C1' Black Connector

PCM Pin #	Wire Color	Circuit Description (40 Pin)	Value at Hot Idle
1	LG/RD	HO2S-12 (B1 S2) Signal	0.1-1.1v
2	RD	Coil 3 Driver	5°, at 55 mph: 8° dwell
3	WT	Coil 2 Driver	5°, at 55 mph: 8° dwell
4	DG	Generator Field Driver	Digital Signal: 0-12-0v
5	YL/RD	S/C Power Supply	12-14v
6	DG/OR	ASD Relay Output	12-14v
7	YL/WT	Injector 3 Driver	1-4 ms
8-9	---	Not Used	---
10	BK/TN	Power Ground	<0.1v
11	BK	Coil 1 Driver	5°, at 55 mph: 8° dwell
12	---	Not Used	
13	WT/LB	Injector 1 Driver	1-4 ms
14	BR/BK	Injector 6 Driver	1-4 ms
15	BR/RD	Injector 5 Driver	1-4 ms
16	LB/BR	Injector 4 Driver	1-4 ms
17	TN	Injector 2 Driver	1-4 ms
18-19	---	Not Used	---
20	DB/WT	Ignition Switch Output	12-14v
21-25	---	Not Used	---
26	TN/BK	ECT Sensor Signal	At 180°F: 2.80v
27-28	---	Not Used	---
29	TN/WT	HO2S-11 (B1 S1) Signal	0.1-1.1v
30	BK/DG	HO2S-21 (B2 S1) Signal	0.1-1.1v
31	---	Not Used	
32	LB/DB	CKP Sensor Signal	Digital Signal: 0-5-0v
33	TN/YL	CMP Sensor Signal	Digital Signal: 0-5-0v
34	---	Not Used	
35	OR/DB	TP Sensor Signal	0.6-1.0v
36	DG/RD	MAP Sensor Signal	1.5-1.7v
37	BK/RD	IAT Sensor Signal	At 100°F: 1.83v
38	---	Not Used	---
39	PK/RD	Manifold Solenoid	Solenoid On: 1v: Off: 12v
40	GY/YL	EGR Solenoid Control	12v, 55 mph: 1v

Standard Colors and Abbreviations

Abbreviation	Color	Abbreviation	Color	Abbreviation	Color
BK	Black	GY	Gray	RD	Red
BL	Blue	GN	Green	TN	Tan
BR	Brown	LG	Light Green	VT	Violet
DB	Dark Blue	OR	Orange	WT	White
DG	Dark Green	PK	Pink	YL	Yellow

1996-97 Intrepid 3.3L V6 MFI VIN T (A/T) 'C2' White Connector

PCM Pin #	Wire Color	Circuit Description (40 Pin)	Value at Hot Idle
41	RD/LG	S/C Set Switch Signal	S/C & Set Switch On: 3.8v
42	DB	A/C Pressure Switch Signal	A/C On: 0.45-4.85v
43	BK/LB	Sensor Ground	<0.050v
44	OR	8-Volt Supply	7.9-8.1v
45	---	Not Used	---
46	RD/WT	Battery Power (Fused B+)	12-14v
47	BK/LB	Sensor Ground	<0.050v
48	BR/WT	IAC 1 Driver	DC pulse signals: 0.8-11v
49	YL/BK	IAC 2 Driver	DC pulse signals: 0.8-11v
50	BK/TN	Power Ground	<0.1v
51	TN/WT	HO2S-22 (B2 S2) Signal	0.1-1.1v
52-54	---	Not Used	---
55	WT	Low Speed Fan Relay	Relay Off: 12v, On: 1v
56	---	Not Used	---
57	GY/RD	IAC 3 Driver	DC pulse signals: 0.8-11v
58	PK/BK	IAC 4 Driver	DC pulse signals: 0.8-11v
59	PK/BR	CCD Bus (+)	Digital Signal: 0-5-0v
60	WT/BK	CCD Bus (-)	<0.050v
61	PK/WT	5-Volt Supply	4.9-5.1v
62	WT/PK	Brake Switch Signal	Brake Off: 0v, On: 12v
63	YL/DG	Torque Management Request	Digital Signals
64	DB/OR	A/C Clutch Relay Control	Relay Off: 12v, On: 1v
65	PK	SCI Transmit	0v
66	WT/OR	Vehicle Speed Signal	Digital Signal
67	DB/YL	ASD Relay Control	Relay Off: 12v, On: 1v
68	PK/BK	EVAP Purge Solenoid Control	PWM signal: 0-12-0v
69	DB/PK	High Speed Fan Relay	Relay Off: 12v, On: 1v
70-73	---	Not Used	---
74	BR	Fuel Pump Relay Control	Relay Off: 12v, On: 1v
75	LG	SCI Receive	5v
76	BK/DG	PNP Switch Sense (EATX)	In P/N: 0v, Others: 5v
77	---	Not Used	---
78	TN/RD	S/C Vacuum Solenoid	Vacuum Increasing: 1v
79	---	Not Used	---
80	LG/RD	S/C Vent Solenoid	Vacuum Decreasing: 1v

Pin Connector Graphic

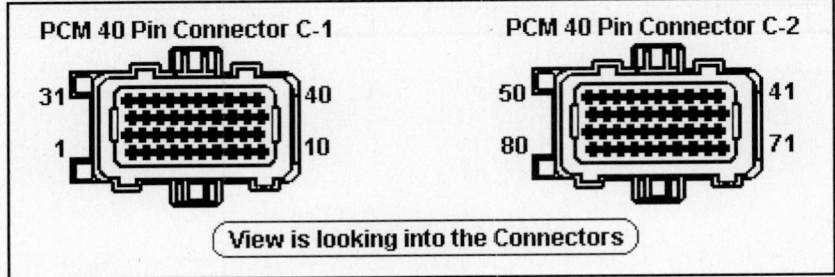

1994-95 Intrepid 3.3L V6 Flexible Fuel VIN U 60 Pin Connector

PCM Pin #	Wire Color	Circuit Description (60 Pin)	Value at Hot Idle
1	DG/RD	MAP Sensor Signal	1.5-1.7v
2	TN/BK	ECT Sensor Signal	At 180ºF: 2.80v
3	RD/WT	Battery Power (Fused B+)	12-14v
4	BK/LB	Sensor Ground	<0.050v
5	BK/WT	Sensor Ground	<0.050v
6	PK/WT	5-Volt Supply	4.9-5.1v
7	OR	9-Volt Supply	8.9-9.1v
8	YL/DG	Torque Management Request	Digital Signals
9	DB/WT	Ignition Switch Output	12-14v
10	GY/BK	Knock Sensor 2 Signal	0.080v AC
11	BK/TN	Power Ground	<0.1v
12	BK/TN	Power Ground	<0.1v
13	LB/BR	Injector 4 Driver	1-4 ms
14	YL/WT	Injector 3 Driver	1-4 ms
15	TN	Injector 2 Driver	1-4 ms
16	WT/DB	Injector 1 Driver	1-4 ms
17	DB/YL	Coil 2 Driver	5º, at 55 mph: 8º dwell
18	RD/YL	Coil 3 Driver	5º, at 55 mph: 8º dwell
19	GY	Coil 1 Driver	5º, at 55 mph: 8º dwell
20	DG	Generator Field Control	Digital Signal: 0-12-0v
21	BK/RD	Air Temperature Sensor	At 100ºF: 2.51v
22	OR/DB	TP Sensor Signal	0.6-1.0v
23	RD/LG	S/C Switch Signal	S/C & Set Switch On: 3.8V
24	LB/DB	CKP Sensor Signal	Digital Signal: 0-5-0v
25	PK	SCI Transmit	0v
26	PK/BR	CCD Bus (+)	Digital Signal: 0-5-0v
27-28	---	Not Used	---
29	WT/PK	Brake Switch Signal	Brake Off: 0v, On: 12v
30	BK/LG	PNP Switch Signal	In P/N: 0v, Others: 5v
31	DB/PK	High Speed Fan Relay	Relay Off: 12v, On: 1v
32	WT	Low Speed Fan Relay	Relay Off: 12v, On: 1v
33	TN/RD	S/C Vacuum Solenoid	Vacuum Increasing: 1v
34	DB/OR	A/C Clutch Relay Control	Relay Off: 12v, On: 1v
35	GY/YL	EGR Solenoid Control	Solenoid Off: 12v, On: 1v
36-37	---	Not Used	---
38	BR/RD	Injector 5 Driver	1-4 ms
39	GY/RD	AIS 1 Motor Control	DC pulse signals: 0.8-11v
40	BR/WT	AIS 3 Motor Control	DC pulse signals: 0.8-11v

1994-95 Intrepid 3.3L V6 Flexible Fuel VIN U 60 Pin Connector

PCM Pin #	Wire Color	Circuit Description (60 Pin)	Value at Hot Idle
41	BK/DG	HO2S-11 (B1 S1) Signal	0.1-1.1v
42	BK/LG	Knock Sensor 1 Signal	0.080v AC
43	---	Not Used	---
44	TN/YL	CMP Sensor Signal	Digital Signal: 0-5-0v
45	LG	SCI Receive	0v
46	WT/BK	CCD Bus (-)	<0.050v
47	WT/OR	Vehicle Speed Signal	Digital Signal
48	DB	A/C Pressure Switch Signal	A/C On: 0.45-4.85v
49	TN/WT	HO2S-21 (B2 S1) Signal	0.1-1.1v
50	YL/WT	Flexible Fuel Sensor Signal	0.5-4.5v
51	DB/YL	ASD Relay Control	Relay Off: 12v, On: 1v
52	PK/BK	EVAP Purge Solenoid Control	Solenoid Off: 12v, On: 1v
53	LG/RD	S/C Vent Solenoid	Vacuum Decreasing: 1v
54	---	Not Used	---
55	TN/RD	S/C Power Supply	12-14v
56	---	Not Used	---
57	DG/OR	ASD Relay Output	12-14v
58	BR/BK	Injector 6 Driver	1-4 ms
59	PK/BK	AIS 4 Motor Control	DC pulse signals: 0.8-11v
60	YL/BK	AIS 2 Motor Control	DC pulse signals: 0.8-11v

Pin Connector Graphic

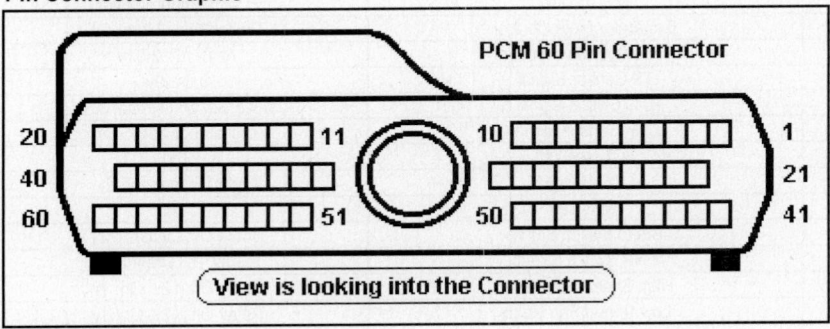

PCM 60 Pin Connector

View is looking into the Connector

Standard Colors and Abbreviations

Abbreviation	Color	Abbreviation	Color	Abbreviation	Color
BK	Black	GY	Gray	RD	Red
BL	Blue	GN	Green	TN	Tan
BR	Brown	LG	Light Green	VT	Violet
DB	Dark Blue	OR	Orange	WT	White
DG	Dark Green	PK	Pink	YL	Yellow

1993-95 Intrepid 3.5L V6 SOHC MFI VIN F (A/T) 60 Pin Connector

PCM Pin #	Wire Color	Circuit Description (60 Pin)	Value at Hot Idle
1	DG/RD	MAP Sensor Signal	1.5-1.7v
2	TN/BK	ECT Sensor Signal	At 180°F: 2.80v
3	RD/WT	Battery Power (Fused B+)	12-14v
4	BK/LB	Sensor Ground	<0.050v
5	BK/WT	Sensor Ground	<0.050v
6	PK/WT	5-Volt Supply	4.9-5.1v
7	OR	8-Volt Supply	7.9-8.1v
8	YL/DG	Torque Management Request	Digital Signals
9	DB/WT	Ignition Switch Output	12-14v
10	GY/BK	Knock Sensor 2 Signal	0.080v AC
11	BK/TN	Power Ground	<0.1v
12	BK/TN	Power Ground	<0.1v
13	LB/BR	Injector 4 Driver	1-4 ms
14	YL/WT	Injector 3 Driver	1-4 ms
15	TN	Injector 2 Driver	1-4 ms
16	WT/DB	Injector 1 Driver	1-4 ms
17	DB/YL	Coil 2 Driver	5°, at 55 mph: 8° dwell
17 ('95)	WT	Coil 2 Driver	5°, at 55 mph: 8° dwell
18	RD/YL	Coil 3 Driver	5°, at 55 mph: 8° dwell
18 ('95)	RD	Coil 3 Driver	5°, at 55 mph: 8° dwell
19	GY	Coil 1 Driver	5°, at 55 mph: 8° dwell
19 ('95)	BK	Coil 1 Driver	5°, at 55 mph: 8° dwell
20	DG	Generator Field Control	Digital Signal: 0-12-0v
21	BK/RD	Air Temperature Sensor	At 100°F: 2.51v
22	OR/DB	TP Sensor Signal	0.6-1.0v
23	RD/LG	S/C Switch Signal	S/C & Set Switch On: 3.8v
24	LB/DB	CKP Sensor Signal	Digital Signal: 0-5-0v
25	PK	SCI Transmit	0v
26	PK/BR	CCD Bus (+)	Digital Signal: 0-5-0v
27-28	---	Not Used	---
29	WT/PK	Brake Switch Signal	Brake Off: 0v, On: 12v
30	BK/LG	PNP Switch Signal	In P/N: 0v, Others: 5v
31	DB/PK	High Speed Fan Relay	Relay Off: 12v, On: 1v
32	WT	Low Speed Fan Relay	Relay Off: 12v, On: 1v
33	TN/RD	S/C Vacuum Solenoid	Vacuum Increasing: 1v
34	DB/OR	A/C Clutch Relay Control	Relay Off: 12v, On: 1v
35	GY/YL	EGR Solenoid Control	12v, 55 mph: 1v
36	PK	Manifold Solenoid Control	Solenoid On: 1v: Off: 12v
37	---	Not used	---
38	BR/RD	Injector 5 Driver	1-4 ms
39	GY/RD	AIS 1 Motor Control	DC pulse signals: 0.8-11v
40	BR/WT	AIS 3 Motor Control	DC pulse signals: 0.8-11v

1993-95 Intrepid 3.5L V6 SOHC MFI VIN F (A/T) 60 Pin Connector

PCM Pin #	Wire Color	Circuit Description (60 Pin)	Value at Hot Idle
41	BK/DG	HO2S-11 (B1 S1) Signal	0.1-1.1v
42	BK/LG	Knock Sensor 1 Signal	0.080v AC
43	---	Not used	---
44	TN/YL	CMP Sensor Signal	Digital Signal: 0-5-0v
45	LG	SCI Receive	0v
46	WT/BK	CCD Bus (-)	<0.050v
47	WT/OR	Vehicle Speed Signal	Digital Signal
48	DB	A/C Pressure Switch Signal	A/C On: 0.45-4.85v
49	TN/WT	HO2S-21 (B2 S1) Signal	0.1-1.1v
50	---	Not Used	---
51	DB/YL	ASD Relay Control	Relay Off: 12v, On: 1v
52	PK/BK	EVAP Purge Solenoid Control	0-12-0v
53	LG/RD	S/C Vent Solenoid	Vacuum Decreasing: 1v
54	---	Not Used	---
55	TN/RD	S/C Power Supply	12-14v
56	---	Not Used	---
57	DG/OR	ASD Relay Output	12-14v
58	BR/BK	Injector 6 Driver	1-4 ms
59	PK/BK	AIS 4 Motor Control	DC pulse signals: 0.8-11v
60	YL/BK	AIS 2 Motor Control	DC pulse signals: 0.8-11v

Pin Connector Graphic

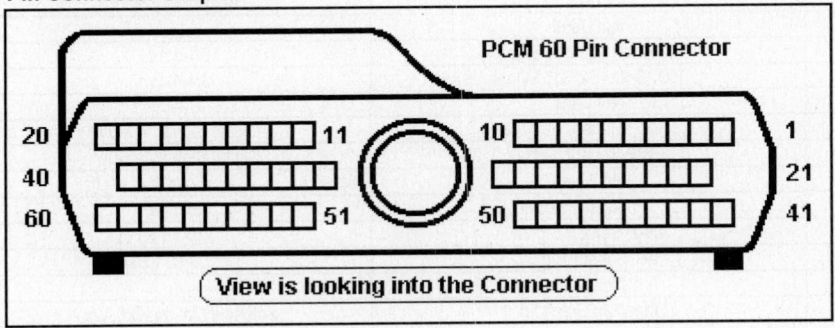

PCM 60 Pin Connector

View is looking into the Connector

Standard Colors and Abbreviations

Abbreviation	Color	Abbreviation	Color	Abbreviation	Color
BK	Black	GY	Gray	RD	Red
BL	Blue	GN	Green	TN	Tan
BR	Brown	LG	Light Green	VT	Violet
DB	Dark Blue	OR	Orange	WT	White
DG	Dark Green	PK	Pink	YL	Yellow

1996-97 Intrepid 3.5L V6 MFI VIN F (A/T) 'C1' Black Connector

PCM Pin #	Wire Color	Circuit Description (40 Pin)	Value at Hot Idle
1	LG/RD	HO2S-12 (B1 S2) Signal	0.1-1.1v
2	RD	Coil 3 Driver	5°, at 55 mph: 8° dwell
3	WT	Coil 2 Driver	5°, at 55 mph: 8° dwell
4	DG	Generator Field Driver	Digital Signal: 0-12-0v
5	YL/RD	S/C Power Supply	12-14v
6	DG/OR	ASD Relay Output	12-14v
7	YL/WT	Injector 3 Driver	1-4 ms
8-9	---	Not Used	---
10	BK/TN	Power Ground	<0.1v
11	BK	Coil 1 Driver	5°, at 55 mph: 8° dwell
12	---	Not Used	---
13	WT/LB	Injector 1 Driver	1-4 ms
14	BR/BK	Injector 6 Driver	1-4 ms
15	BR/RD	Injector 5 Driver	1-4 ms
16	LB/BR	Injector 4 Driver	1-4 ms
17	TN	Injector 2 Driver	1-4 ms
18-19	---	Not Used	---
20	DB/WT	Ignition Switch Output	12-14v
21-23	---	Not Used	---
24	BK/LG	Knock Sensor 1 Signal	0.080v AC
25	GY/BK	Knock Sensor 2 Signal	0.080v AC
26	TN/BK	ECT Sensor Signal	At 180°F: 2.80v
27-28	---	Not Used	---
29	TN/WT	HO2S-11 (B1 S1) Signal	0.1-1.1v
30	BK/DG	HO2S-21 (B2 S1) Signal	0.1-1.1v
31	---	Not Used	---
32	LB/DB	CKP Sensor Signal	Digital Signal: 0-5-0v
33	TN/YL	CMP Sensor Signal	Digital Signal: 0-5-0v
34	---	Not Used	---
35	OR/DB	TP Sensor Signal	0.6-1.0v
36	DG/RD	MAP Sensor Signal	1.5-1.7v
37	BK/RD	IAT Sensor Signal	At 100°F: 1.83v
38	---	Not Used	---
39	PK/RD	Manifold Solenoid	Solenoid On: 1v: Off: 12v
40	GY/YL	EGR Solenoid Control	12v, 55 mph: 1v

Standard Colors and Abbreviations

Abbreviation	Color	Abbreviation	Color	Abbreviation	Color
BK	Black	GY	Gray	RD	Red
BL	Blue	GN	Green	TN	Tan
BR	Brown	LG	Light Green	VT	Violet
DB	Dark Blue	OR	Orange	WT	White
DG	Dark Green	PK	Pink	YL	Yellow

1996-97 Intrepid 3.5L V6 MFI VIN F (A/T) 'C2' White Connector

PCM Pin #	Wire Color	Circuit Description (40 Pin)	Value at Hot Idle
41	RD/LG	S/C Set Switch Signal	S/C & Set Switch On: 3.8v
42	DB	A/C Pressure Switch Signal	A/C On: 0.45-4.85v
43	BK/LB	Sensor Ground	<0.050v
44	OR	8-Volt Supply	7.9-8.1v
45	---	Not Used	---
46	RD/WT	Battery Power (Fused B+)	12-14v
47	BK/WT	Power Ground	<0.1v
48	BR/WT	IAC 1 Driver	DC pulse signals: 0.8-11v
49	YL/BK	IAC 2 Driver	DC pulse signals: 0.8-11v
50	BK/TN	Power Ground	<0.1v
51	TN/WT	HO2S-22 (B2 S2) Signal	0.1-1.1v
52-54	---	Not Used	---
55	WT	Low Speed Fan Relay	Relay Off: 12v, On: 1v
56	---	Not Used	---
57	GY/RD	IAC 3 Driver	DC pulse signals: 0.8-11v
58	PK/BK	IAC 4 Driver	DC pulse signals: 0.8-11v
59	PK/BR	CCD Bus (+)	Digital Signal: 0-5-0v
60	WT/BK	CCD Bus (-)	<0.050v
61	PK/WT	5-Volt Supply	4.9-5.1v
62	WT/PK	Brake Switch Signal	Brake Off: 0v, On: 12v
63	YL/DG	Torque Management Request	Digital Signals
64	DB/OR	A/C Clutch Relay Control	Relay Off: 12v, On: 1v
65	PK	SCI Transmit	0v
66	WT/OR	Vehicle Speed Signal	Digital Signal
67	DB/YL	ASD Relay Control	Relay Off: 12v, On: 1v
68	PK/BK	EVAP Purge Solenoid Control	Solenoid Off: 12v, On: 1v
69	DB/PK	High Speed Fan Relay	Relay Off: 12v, On: 1v
70-73	---	Not Used	---
74	BR	Fuel Pump Relay Control	Relay Off: 12v, On: 1v
75	LG	SCI Receive	5v
76	BK/DG	PNP Switch Sense (EATX)	In P/N: 0v, Others: 5v
77	---	Not Used	---
78	TN/RD	S/C Vacuum Solenoid	Vacuum Increasing: 1v
79	---	Not Used	---
80	LG/RD	S/C Vent Solenoid	Vacuum Decreasing: 1v

Pin Connector Graphic

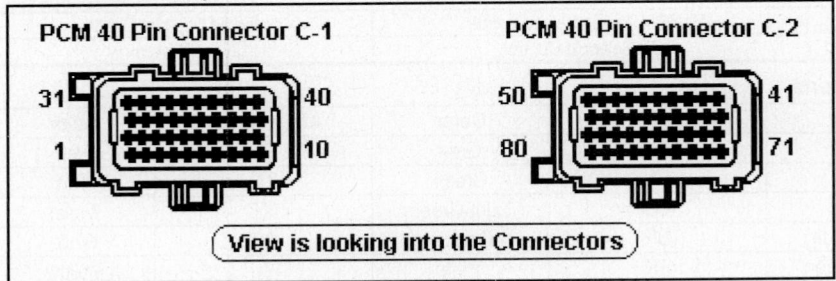

PCM 40 Pin Connector C-1   PCM 40 Pin Connector C-2

31   40   50   41

1   10   80   71

(View is looking into the Connectors)

2000-01 Intrepid 3.5L V6 SOHC 24v MFI VIN V 'C1' Black Connector

PCM Pin #	Wire Color	Circuit Description (40 Pin)	Value at Hot Idle
1	TN/LG	Coil 4 Driver Control	5°, at 55 mph: 8° dwell
2	TN/OR	Coil 3 Driver Control	5°, at 55 mph: 8° dwell
3	TN/PK	Coil 2 Driver Control	5°, at 55 mph: 8° dwell
4	TN/LB	Coil 6 Driver Control	5°, at 55 mph: 8° dwell
5	YL/RD	S/C Power Supply	12-14v
6	DG/OR	ASD Relay Output	12-14v
7	YL/WT	Injector 3 Driver	1.0-4.0 ms
8	DG	Generator Field Driver	Digital Signal: 0-12-0v
9	---	Not Used	---
10	BK/TN	Power Ground	<0.1v
11	TN/RD	Coil 1 Driver Control	5°, at 55 mph: 8° dwell
12	---	Not Used	---
13	WT/DB	Injector 1 Driver	1.0-4.0 ms
14	BR/DB	Injector 6 Driver	1.0-4.0 ms
15	GY	Injector 5 Driver	1.0-4.0 ms
16	LB/BR	Injector 4 Driver	1.0-4.0 ms
17	TN	Injector 2 Driver	1.0-4.0 ms
18	GY/PK	Short Runner Valve Solenoid	Valve On: 1v, Off: 12v
19	---	Not Used	---
20	DB/WT	Ignition Switch Output	12-14v
21	TN/DG	Coil 5 Driver Control	5°, at 55 mph: 8° dwell
22-23	---	Not Used	---
24	DB/LG	Knock Sensor Signal	0.080v AC
25	BK/VT	Knock Sensor Ground	<0.050v
26	TN/BK	ECT Sensor Signal	At 180°F: 2.80v
27	BK/OR	HO2S Signal Ground	<0.050v
28	---	Not Used	---
29	LG/RD	HO2S-21 (B2 S1) Signal	0.1-1.1v
30	BK/DG	HO2S-11 (B1 S1) Signal	0.1-1.1v
31	TN/WT	Starter Relay Control	KOEC: 9-11v
32	GY/BK	CKP Sensor Signal	Digital Signal: 0-5-0v
33	TN/YL	CMP Sensor Signal	Digital Signal: 0-5-0v
34	LG/PK	EGR Sensor Signal	0.6-0.8v
35	OR/DB	TP Sensor Signal	0.6-1.0v
36	DG/RD	MAP Sensor Signal	1.5-1.7v
37	BK/RD	IAT Sensor Signal	At 100°F: 1.83v
38	---	Not Used	---
39	VT/RD	Manifold Tuning Valve Control	Solenoid Off: 12v, On: 1v
40	GY/YL	EGR Solenoid Control	12v, at 55 mph: 1v

Standard Colors and Abbreviations

Abbreviation	Color	Abbreviation	Color	Abbreviation	Color
BK	Black	GY	Gray	RD	Red
BL	Blue	GN	Green	TN	Tan
BR	Brown	LG	Light Green	VT	Violet
DB	Dark Blue	OR	Orange	WT	White
DG	Dark Green	PK	Pink	YL	Yellow

2000-01 Intrepid 3.5L V6 SOHC 24v MFI VIN V 'C2' White Connector

PCM Pin #	Wire Color	Circuit Description (40 Pin)	Value at Hot Idle
41	RD/LG	S/C Set Switch Signal	S/C & Set Switch On: 3.8v
42	DB	A/C Pressure Switch Signal	A/C On: 0.45-4.85v
43	BK/LB	Sensor Ground	<0.050v
44	OR	8-Volt Supply	7.9-8.1v
45	DB/LG	PSP Switch Signal	Straight: 0v, Turning: 5v
46	RD/WT	Battery Power (Fused B+)	12-14v
47, 50	BK/WT	Power Ground	<0.1v
48	BR/WT	IAC 3 Driver Control	DC pulse signals: 0.8-11v
49	YL/BK	IAC 2 Driver Control	DC pulse signals: 0.8-11v
51	TN/WT	HO2S-12 (B1 S2) Signal	0.1-1.1v
52 ('01-'02)	VT/LG	Battery Temperature Sensor	At 86ºF: 1.96v
53	PK/WT	HO2S-22 (B2 S2) Signal	0.1-1.1v
54, 60, 73	---	Not Used	---
55	DB/PK	Low Speed Fan Relay	Relay Off: 12v, On: 1v
56	TN/RD	S/C Vacuum Solenoid	Vacuum Increasing: 1v
57	GY/RD	IAC 1 Driver Control	DC pulse signals: 0.8-11v
58	VT/BK	IAC 4 Driver Control	DC pulse signals: 0.8-11v
59	VT/YL	PCI Data Bus (J1850)	Digital Signals: 0-7-0v
61	VT/WT	5-Volt Supply	4.9-5.1v
62	WT/PK	Brake Switch Signal	Brake Off: 0v, On: 12v
63	YL/DG	Torque Management Request	Digital Signals
64	DB/OR	A/C Clutch Relay Control	Relay Off: 12v, On: 1v
65	LG	SCI Transmit	0v
66	WT/OR	Vehicle Speed Signal	Digital Signal
67	DB/YL	ASD Relay Control	Relay Off: 12v, On: 1v
68	PK/BK	EVAP Purge Solenoid Control	PWM signal: 0-12-0v
69	DB/LG	High Speed Fan Relay	Relay Off: 12v, On: 1v
70	DG/LG	EVAP Purge Solenoid Sense	0-1v
71	WT/RD	EATX RPM	Digital Signals
72	OR/RD	LDP Switch Sense	Open: 12v, Closed: 0v
74	BR	Fuel Pump Relay Control	Relay Off: 12v, On: 1v
75	LG	SCI Receive	5v
76	BR/YL	TRS T41 Sense	In P/N: 0v, Others: 5v
77	WT/DG	LDP Solenoid Control	PWM signal: 0-12-0v
78-79	---	Not Used	---
80	LG/RD	S/C Vent Solenoid	Vacuum Decreasing: 1v

Pin Connector Graphic

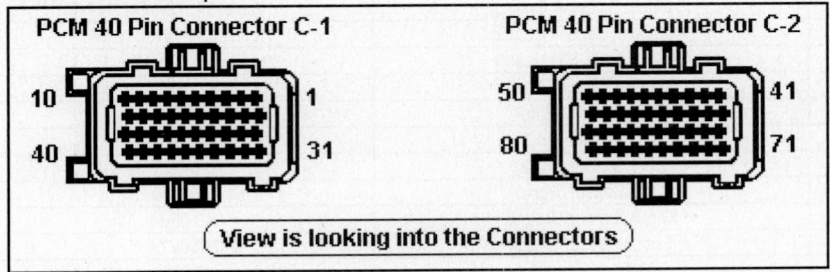

2002-3 Intrepid 3.5L V6 SOHC MFI VIN G, M & V (A/T) 'C1' Black Connector

PCM Pin #	Wire Color	Circuit Description (38 Pin)	Value at Hot Idle
1-8, 10	---	Not Used	---
9	BK/TN	Power Ground	<0.1v
11	DB/WT	Fused Ignition Power (B+)	12-14v
12	RD/WT	Fused Ignition Power (B+)	12-14v
13-15	---	Not Used	---
16	GY/PK	Short Runner Valve Solenoid	Valve On: 1v, Off: 12v
17, 19-20	---	Not Used	---
18	BK/TN	Power Ground	<0.1v
21	DB	A/C Pressure Sensor	A/C On: 0.45-4.85v
22-24	---	Not Used	---
25	LG	SCI Receive	5v
26	VT/OR	SCI Transmit	0v
27-28	---	Not Used	---
29	RD	Fused Power (B+)	12-14v
30	YL/BK	Fused Ignition Output (Start)	10-12v
31	TN/WT	HO2S-12 (B1 S2) Signal	0.1-1.1v
32	DB/DG	HO2S-12 (B1 S2) Ground	<0.050v
33	PK/WT	HO2S-22 (B2 S2) Signal	0.1-1.1v
34-35	---	Not Used	---
36	PK/TN	SCI Transmit (PCM)	0v
37	WT/DG	SCI Transmit (PCM)	0v
38	VT/YL	PCI Data Bus (J1850)	Digital Signals: 0-7-0v

2002-3 Intrepid 3.5L V6 SOHC MFI VIN G, M & V (A/T) 'C2' Gray Connector

PCM Pin #	Wire Color	Circuit Description (38 Pin)	Value at Hot Idle
1	TN/LB	Coil On Plug 6 Driver	5°, at 55 mph: 8° dwell
2	TN/DG	Coil On Plug 5 Driver	5°, at 55 mph: 8° dwell
3	TN/LG	Coil On Plug 4 Driver	5°, at 55 mph: 8° dwell
4	BR/DB	Injector 6 Driver	1.0-4.0 ms
5	GY	Injector 5 Driver	1.0-4.0 ms
6, 15, 26	---	Not Used	---
7	TN/OR	Coil On Plug 3 Driver	5°, at 55 mph: 8° dwell
8	GY/YL	EGR Solenoid Control	12v, at 55 mph: 1v
9	TN/PK	Coil On Plug 2 Driver	5°, at 55 mph: 8° dwell
10	TN/RD	Coil On Plug 1 Driver	5°, at 55 mph: 8° dwell
11	LB/BR	Injector 4 Driver	1.0-4.0 ms
12	YL/WT	Injector 3 Driver	1.0-4.0 ms
13	TN/WT	Injector 2 Driver	1.0-4.0 ms
14	WT/DB	Injector 1 Driver	1.0-4.0 ms
16	VT/RD	Manifold Solenoid Control	Valve Off: 12v, On: 1v
17	BR/WT	HO2S-21 (B2 S1) Heater	Heater Off: 12v, On: 1v
18	BR/OR	HO2S-11 (B1 S1) Heater	Heater Off: 12v, On: 1v
19	DG	Generator Field Driver	Digital Signals: 0-12-0v
20	TN/BK	ECT Sensor Signal	At 180°F: 2.80v
21	DB	TP Sensor Signal	0.6-1.0v
22	LG/PK	EGR Sensor Signal	0.6-0.8v
23	DG/RD	MAP Sensor Signal	1.5-1.7v
24	BK/VT	Knock Sensor Ground	<0.050v
25	DB/LG	Knock Sensor Signal	0.080v AC
27	BK/LB	Sensor Ground	<0.050v
28	YL/BK	IAC Motor Sense	12-14v
29	VT/WT	5-Volt Supply	4.9-5.1v
30	BK/RD	IAT Sensor Signal	At 100°F: 1.83v
31	BK/DG	HO2S-11 (B1 S1) Signal	0.1-1.1v
32	BR/DG	HO2S-11 (B1 S2) Ground	<0.050v
33	LG/RD	HO2S-21 (B2 S1) Signal	0.1-1.1v
34	TN/YL	CMP Sensor Signal	Digital Signal: 0-5-0v
35	GY/BK	CKP Sensor Signal	Digital Signal: 0-5-0v
36-37	---	Not Used	---
38	GY/RD	IAC Motor Driver Control	DC pulses: 0.8-11v

2002-3 Intrepid 3.5L V6 SOHC MFI VIN G, M & V (A/T) 'C3' White Connector

PCM Pin #	Wire Color	Circuit Description (38 Pin)	Value at Hot Idle
1-2, 13-17	---	Not Used	---
3	DB/YL	ASD Relay Control	Relay Off: 12v, On: 1v
4	DB/PK	High Speed Fan Relay	Relay Off: 12v, On: 1v
5	DB/PK	S/C Vent Solenoid	Vacuum Decreasing: 1v
6	DB/PK	Low Speed Fan Relay	Relay Off: 12v, On: 1v
7	YL/RD	S/C Power Supply (B+)	12-14v
8	WT/DG	Natural Vacuum Leak Detection Solenoid	Solenoid Off: 12v, On: 1v
9	BR/VT	HO2S-12 (B1 S2) Heater	Heater Off: 12v, On: 1v
10	BR/GY	HO2S-22 (B2 S2) Heater	Heater Off: 12v, On: 1v
11	DB/OR	A/C Clutch Relay Control	Relay Off: 12v, On: 1v
12	TN/RD	S/C Vacuum Solenoid	Vacuum Increasing: 1v
18, 19	OR/DG	Fused ASD Relay Power (B+)	12-14v
20	PK/BK	EVAP Purge Solenoid Control	PWM Signal: 0-12-0v
21-22, 24-25	---	Not Used	---
23	WT/PK	Brake Switch Sense	Brake Off: 0v, On: 12v
26	YL	Autostick Downshift Switch	Digital Signal: 0v or 12v
27	LG/RD	Autostick Upshift Switch	Digital Signal: 0v or 12v
28	OR/DG	Fused ASD Relay Power (B+)	12-14v
29	DG/LG	EVAP Purge Solenoid Sense	<0.1v
30-31, 33, 36	---	Not Used	---
32	VT/LG	Ambient Temperature Sensor	At 86°F: 1.96v
34	RD/LG	S/C Set Switch Signal	S/C & Set Switch On: 3.8v
35	BR	Natural Vacuum Leak Detection Switch	Open: 12v, Closed: 0v
37	BR	Fuel Pump Relay Control	Relay Off: 12v, On: 1v
38	TN	Starter Relay Control	Relay Off: 12v, On: 1v

2002 Intrepid 3.5L V6 SOHC MFI VIN G, M & V (A/T) 'C4' Green Connector

1, 3-5	---	Not Used	---
2	BR	Overdrive Solenoid Control	Solenoid Off: 12v, On: 1v
3	PK	Underdrive Solenoid Control	Solenoid Off: 12v, On: 1v
6	WT	A/T: 2-4 Solenoid Control	Solenoid Off: 12v, On: 1v
7-9, 11	---	Not Used	---
10	LB	Low/Reverse Solenoid control	Solenoid Off: 12v, On: 1v
12	BK/YL	Power Ground	<0.1v
13, 14	BK/RD	Power Ground	<0.1v
15	LG/BK	A/T: TRS T1 Sense	In NOL: 0v, Others: 5v
16	VT	A/T: TRS T3 Sense	In P3L: 0v, Others: 5v
17, 20-21	---	Not Used	---
18, 19	LG, RD	Transmission Control Relay Control	Relay Off: 12v, On: 1v
22	OR/BK	Overdrive Pressure Switch	In Overdrive: 2-4v
23-26, 31, 36	---	Not Used	---
27	BK/WT	A/T: TRS T41 Sense	In P/N: 0v, Others: 5v
28, 38	RD	Trans. Control Relay Output	12-14v
29	DG	Low/Reverse Pressure Switch	In Low/Reverse: 2-4v
30	YL/BK	A/T: 2-4 Pressure Switch	In 2-4 Position: 2-4v
32	LG/WT	A/T: Output Speed Sensor	Moving: AC voltage
33	RD/BK	A/T: Input Speed Sensor	Moving: AC voltage
34	DB/BK	A/T: Speed Sensor Ground	<0.050v
35	VT/PK	Trans. Temperature Sensor	3.2-3.4v at 104°F
37	VT/WT	A/T: TRS T42 Sense	In PRNL: 0v, Others 5v

Pin Connector Graphic

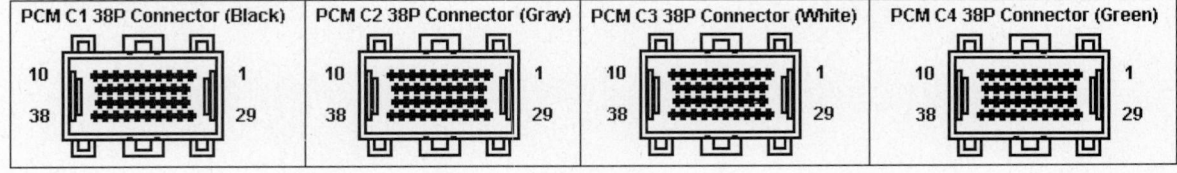

1990-1991 Monaco 3.0L V6 MFI VIN 3 (A/T) Row A 24P Connector

PCM Pin #	Wire Color	Circuit Description (24-Pin)	Value at Hot Idle
1-4	---.	Not Used	---
5	OR/LG	Fuel Pump Relay Control	Relay Off: 12v, On: 1v
6-8	---	Not Used	---
9	BK/LG	B+ Latch Relay Control	Relay Off: 12v, On: 1v
10	GY/YL	EGR Solenoid Control (Calif.)	12v, 55 mph: 1v
11	---	Not Used	---
12	YL/BK	A/C WOT Relay Control	Relay Off: 12v, On: 1v

1990-1991 Monaco 3.0L V6 MFI VIN 3 (A/T) Row B 24P Connector

PCM Pin #	Wire Color	Circuit Description (24-Pin)	Value at Hot Idle
1	PK/BK	Injector Control Driver	1-4 ms
2	LG/BK	Injector Control Driver	1-4 ms
3	BR	Idle Speed Control Motor	DC pulse signals: 0.8-11v
4	LB/RD	Diagnostic 2 Input	0v
5-6	---	Not Used	---
7	RD	Battery Power (B+)	12-14v
8	PK/WT	Ignition Switch Run Output	12-14v
9	LG/RD	TDC Signal To TCM	Digital Signal: 0-5-0v
10	PK	B+ Latch Relay Output	Relay Off: 12v, On: 1v
11	BK	Power Ground	<0.1v
12	BK	Power Ground	<0.1v

Pin Connector Graphic

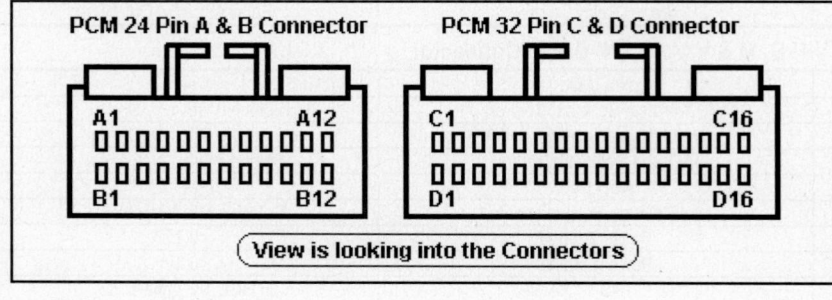

1990-1991 Monaco 3.0L V6 MFI VIN 3 (A/T) Row C 32P Connector

PCM Pin #	Wire Color	Circuit Description (32-Pin)	Value at Hot Idle
1	WT/BK	CKP Sensor (+)	Pulse Signals
2	LB	A/C Relay Control	Relay Off: 12v, On: 1v
3	BR	Starter Motor Relay	KOEC: 9-11v
4	BK/WT	PNP Switch Signal	In P/N: 0v, Others: 5v
5	---	Not Used	---
6	PK	MAP Sensor Signal	1.5-1.7v
7	OR/DB	TP Sensor Signal	0.6-1.0v
8	BK/RD	Air Temperature Sensor	At 100ºF: 2.51v
9	---	Not Used	---
10	TN/BK	ECT Sensor Signal	At 180ºF: 2.80v
11	DG/BK	Fuel Pump Relay Output	Relay Off: 12v, On: 1v
12	BK/PK	Diagnostic 2 Output	0v
13	---	Not Used	---
14	PK/RD	5-Volt Supply	4.9-5.1v
15	PK/WT	5-Volt Supply	4.9-5.1v
16	---	Not Used	---

1990-1991 Monaco 3.0L V6 MFI VIN 3 (A/T) Row D 32P Connector

PCM Pin #	Wire Color	Circuit Description (32-Pin)	Value at Hot Idle
1	RD/LG	CKP Sensor (-)	Pulse Signals
2	LG	A/C Select	A/C On: 1v, Off: 12v
3	BK/LB	Sensor Ground	<0.050v
4-5	---	Not Used	---
6	BK/RD	Power Ground	<0.1v
7	---	Not Used	---
8	YL/BK	Knock Sensor Ground	<0.050v
9	BK/DG	HO2S-11 (B1 S1) Signal	0.1-1.1v
10	DG/BK	Fuel Pump Relay Output	Relay Off: 12v, On: 1v
11	BK/WT	Diagnostic 2	0v
12	LB/BK	Fuel Flow Output	Relay Off: 12v, On: 1v
13	YL	Ignition Dwell	5º, at 55 mph: 8º dwell
14	---	Not Used	---
15	WT/OR	Vehicle Speed Signal	Digital Signal
16	YL/LG	Knock Sensor Signal	0.080v AC

Standard Colors and Abbreviations

Abbreviation	Color	Abbreviation	Color	Abbreviation	Color
BK	Black	GY	Gray	RD	Red
BL	Blue	GN	Green	TN	Tan
BR	Brown	LG	Light Green	VT	Violet
DB	Dark Blue	OR	Orange	WT	White
DG	Dark Green	PK	Pink	YL	Yellow

1991-1992 Monaco 3.0L V6 SOHC MFI VIN U 60 Pin Connector

PCM Pin #	Wire Color	Circuit Description (60 Pin)	Value at Hot Idle
1	DG/RD	MAP Sensor Signal	1.5-1.7v
2	TN/BK	ECT Sensor Signal	At 180°F: 2.80v
3	RD	Battery Power (Fused B+)	12-14v
4	BK/LB	Sensor Ground	<0.050v
5	BK/WT	Sensor Ground	<0.050v
6	PK/WT	5-Volt Supply	4.9-5.1v
7	OR	8-Volt Supply	7.9-8.1v
8	LG	A/C Select Signal	A/C On: 1v, Off: 12v
9	DB	Ignition Switch Output	12-14v
10	---	Not Used	---
11	BK/TN	Power Ground	<0.1v
12	BK/TN	Power Ground	<0.1v
13	---	Not Used	---
14	YL/WT	Injector 3 Driver	1-4 ms
15	TN	Injector 2 Driver	1-4 ms
16	WT/DB	Injector 1 Driver	1-4 ms
17	DB/YL	Coil 2 Driver	5°, at 55 mph: 8° dwell
18	RD/YL	Coil 3 Driver	5°, at 55 mph: 8° dwell
19	GY	Coil 1 Driver	5°, at 55 mph: 8° dwell
20	---	Not Used	---
21	BK/RD	Air Temperature Sensor	At 100°F: 2.51v
22	OR/DB	TP Sensor Signal	0.6-1.0v
23	RD/LG	S/C Set Switch Signal	S/C & Set Switch On: 3.8v
24	GY/BK	CKP Sensor Signal	Digital Signal: 0-5-0v
25	PK	SCI Transmit	0v
26	---	Not Used	---
27	LB	A/C Request Signal	A/C On: 1v, Off: 12v
28	---	Not Used	---
29	WT/PK	Brake Switch Signal	Brake Off: 0v, On: 12v
30	BR/YL	PNP Switch Signal	In P/N: 0v, Others: 5v
31	DB/PK	Radiator Fan Relay Control	Relay Off: 12v, On: 1v
32	BK/PK	MIL (lamp) Control	Lamp On: 1v, Off: 12v
33	TN/RD	S/C Vacuum Solenoid	Vacuum Increasing: 1v
34	DB/OR	A/C Clutch Relay Control	Relay Off: 12v, On: 1v
35	GY/YL	EGR Solenoid Control (Calif.)	12v, 55 mph: 1v
36-38	---	Not Used	---
39	GY/RD	AIS 3 Motor Control	DC pulse signals: 0.8-11v
40	BR/WT	AIS 1 Motor Control	DC pulse signals: 0.8-11v

1991-1992 Monaco 3.0L V6 SOHC MFI VIN U 60 Pin Connector

PCM Pin #	Wire Color	Circuit Description (60 Pin)	Value at Hot Idle
41	BK/DG	HO2S-11 (B1 S1) Signal	0.1-1.1v
42	---	Not Used	---
43	GY/LB	Tachometer Signal	Pulse Signals
44	TN/YL	CMP Sensor Signal	Digital Signal: 0-5-0v
45	LG	SCI Receive	0v
46	---	Not Used	---
47	DB/WT	Vehicle Speed Signal	Digital Signal
48-50	---	Not Used	---
51	DG/YL	ASD Relay Control	Relay Off: 12v, On: 1v
52	LB/BK	Fuel Monitor Output Signal	Relay Off: 12v, On: 1v
53	LG/RD	S/C Vent Solenoid	Vacuum Decreasing: 1v
54	---	Not Used	---
55	BK/LB	Speed Control Relay Control	Relay Off: 12v, On: 1v
56	PK	S/C Indicator Driver Control	Lamp On: 1v, Off: 12v
57	DG/RD	ASD Relay Output	Relay Off: 12v, On: 1v
58	---	Not Used	---
59	PK/BK	AIS 4 Motor Control	DC pulse signals: 0.8-11v
60	YL/BK	AIS 2 Motor Control	DC pulse signals: 0.8-11v

Pin Connector Graphic

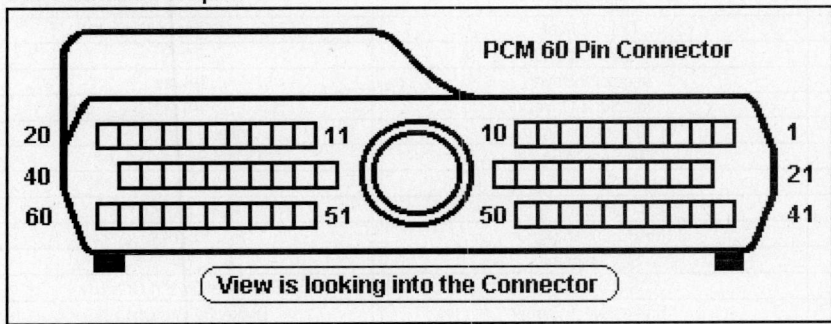

PCM 60 Pin Connector

20 11 10 1
40
60 51 50 41

View is looking into the Connector

Standard Colors and Abbreviations

Abbreviation	Color	Abbreviation	Color	Abbreviation	Color
BK	Black	GY	Gray	RD	Red
BL	Blue	GN	Green	TN	Tan
BR	Brown	LG	Light Green	VT	Violet
DB	Dark Blue	OR	Orange	WT	White
DG	Dark Green	PK	Pink	YL	Yellow

1995 Neon 2.0L I4 SOHC MFI VIN C (All) 60 Pin Connector

PCM Pin #	Wire Color	Circuit Description (60 Pin)	Value at Hot Idle
1	BK/GY	Coil 2 Driver	5°, at 55 mph: 8° dwell
2	BK/TN	Power Ground	<0.1v
3	YL/WT	Injector 3 Driver	1-4 ms
4	WT/DB	Injector 1 Driver	1-4 ms
5	WT/OR	Vehicle Speed Signal	Digital Signal
6	BK/RD	IAT Sensor Signal	At 100°F: 1.83v
7	TN/WT	HO2S-12 (B1 S2) Signal	0.1-1.1v
8	BK/DG	HO2S-11 (B1 S1) Signal	0.1-1.1v
9	LG	SCI Receive	0v
10	OR/DB	TP Sensor Signal	0.6-1.0v
11	RD/WT	Battery Power (Fused B+)	12-14v
12	---	Not Used	---
13	DB	Fuel Level Sensor Signal	70 ohms (±20) with full tank
14	YL/BK	IAC 2 Driver	DC pulse signals: 0.8-11v
15	GY/RD	IAC 3 Driver	DC pulse signals: 0.8-11v
16	PK/BK	EVAP Purge Solenoid Control	Digital Signal: 0-12-0v
17	OR/BK	TCC Solenoid Control	At Cruise w/TCC On: <1v
18	DB/YL	ASD Relay Control	Relay Off: 12v, On: 1v
19	DB/PK	Radiator Fan Relay Control	Relay Off: 12v, On: 1v
20	---	Not Used	---
21	DB/TN	Coil 1 Driver	5°, at 55 mph: 8° dwell
22	BK/TN	Power Ground	<0.1v
23	TN	Injector 2 Driver	1-4 ms
24	LB/BR	Injector 4 Driver	1-4 ms
25	GY/BK	CKP Sensor Signal	Digital Signal: 0-5-0v
26	TN/YL	CMP Sensor Signal	Digital Signal: 0-5-0v
27	DB/LG	Knock Sensor Signal	0.080v AC
28	TN/DB	ECT Sensor Signal	At 180°F: 2.80v
29	DG/RD	MAP Sensor Signal	1.5-1.7v
30	PK	SCI Transmit	0v
31	RD/LG	S/C Set Switch Signal	S/C & Set Switch On: 3.8v
32	WT/PK	Brake Switch Signal	Brake Off: 0v, On: 12v
33	BR/OR	A/C Select Switch Sense	A/C On: 1v, Off: 12v
34	PK/BK	IAC 4 Driver	DC pulse signals: 0.8-11v
35	BR/WT	IAC 1 Driver	DC pulse signals: 0.8-11v
36	BK/PK	MIL (lamp) Control	Lamp On: 1v, Off: 12v
37	TN/BK	Generator Lamp Control	Lamp On: 1v, Off: 12v
38	BR	Fuel Pump Relay Control	Relay Off: 12v, On: 1v
39	GY/YL	EGR Solenoid Control	12v, 55 mph: 1v
40	TN/RD	S/C Vacuum Solenoid	Vacuum Increasing: 1v

1995 Neon 2.0L I4 SOHC MFI VIN C (All) 60 Pin Connector

PCM Pin #	Wire Color	Circuit Description (60 Pin)	Value at Hot Idle
41	DG	Generator Field Control	Digital Signal: 0-12-0v
42	DG/OR	ASD Relay Output	12-14v
43	PK/WT	5-Volt Supply	4.9-5.1v
44	OR	8-Volt Supply	7.9-8.1v
45-46	---	Not Used	---
47	YL/RD	S/C Power Supply	12-14v
48	---	Not Used	---
49	PK/LG	Battery Temperature Sensor	At 86°F: 1.96v
50	BR/YL	PNP Switch Signal	In P/N: 0v, Others: 5v
51	BK/LB	Sensor Ground	<0.050v
52	BK/WT	Power Ground	<0.1v
53	---	Not Used	---
54	LG/BK	Ignition Switch Output	12-14v
55	PK/BK	Reverse Indicator Control	Lamp On: 1v, Off: 12v
56	WT	PSP Switch Signal	Straight: 0v, Turning: 5v
57-58	---	Not Used	---
59	DB/OR	A/C Clutch Relay Control	Relay Off: 12v, On: 1v
60	LG/RD	S/C Vent Solenoid	Vacuum Increasing: 1v

Pin Connector Graphic

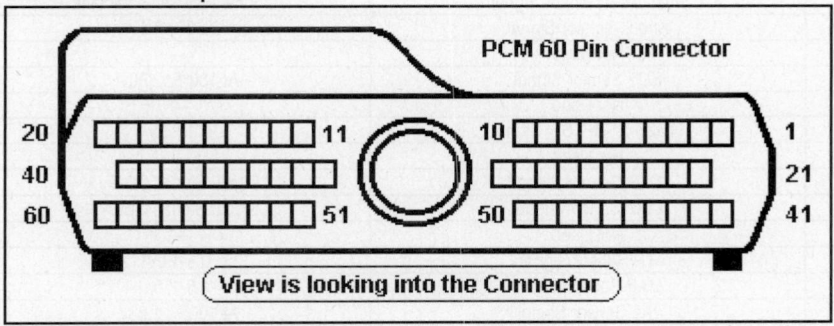

PCM 60 Pin Connector

View is looking into the Connector

Standard Colors and Abbreviations

Abbreviation	Color	Abbreviation	Color	Abbreviation	Color
BK	Black	GY	Gray	RD	Red
BL	Blue	GN	Green	TN	Tan
BR	Brown	LG	Light Green	VT	Violet
DB	Dark Blue	OR	Orange	WT	White
DG	Dark Green	PK	Pink	YL	Yellow

1996-99 Neon 2.0L I4 SOHC MFI VIN C (All) 'C1' Black Connector

PCM Pin #	Wire Color	Circuit Description (40 Pin)	Value at Hot Idle
1	---	Not Used	---
2	BK/GY	Coil 1 Driver	5°, at 55 mph: 8° dwell
3	DB/TN	Coil 2 Driver	5°, at 55 mph: 8° dwell
4	DG	Generator Field Control	Digital Signal: 0-12-0v
5	YL/RD	S/C Power Supply	12-14v
6	DG/OR	ASD Relay Output	12-14v
7	YL/WT	Injector 3 Driver	1-4 ms
8	BK/PK	MIL (lamp) Control	Lamp On: 1v, Off: 12v
9	---	Not Used	---
10	BK/TN	Power Ground	<0.1v
11-12	---	Not Used	---
13	WT/LB	Injector 1 Driver	1-4 ms
14-15	---	Not Used	---
16	LB/BR	Injector 4 Driver	1-4 ms
17	TN	Injector 2 Driver	1-4 ms
18	LG	Radiator Fan Relay Control	Relay Off: 12v, On: 1v
19	---	Not Used	---
20	DB/WT	Ignition Switch Output	12-14v
21-22	---	Not Used	---
23	DB	Fuel Level Sensor Signal	70 ohms (±20) with full tank
24	DB/LG	Knock Sensor Signal	0.080v AC
25	---	Not Used	---
26	TN/DB	ECT Sensor Signal	At 180°F: 2.80v
27-29	---	Not Used	---
30	BK/DG	HO2S-11 (B1 S1) Signal	0.1-1.1v
31	---	Not Used	---
32	GY/BK	CKP Sensor Signal	Digital Signal: 0-5-0v
33	TN/YL	CMP Sensor Signal	Digital Signal: 0-5-0v
34	---	Not Used	---
35	OR/LB	TP Sensor Signal	0.6-1.0v
36	DG/RD	MAP Sensor Signal	1.5-1.7v
37	BK/RD	IAT Sensor Signal	At 100°F: 1.83v
38	BR/OR	A/C Select Switch Sense	A/C On: 1v, Off: 12v
39	---	Not Used	---
40	GY/YL	EGR Solenoid Control	12v, 55 mph: 1v

Standard Colors and Abbreviations

Abbreviation	Color	Abbreviation	Color	Abbreviation	Color
BK	Black	GY	Gray	RD	Red
BL	Blue	GN	Green	TN	Tan
BR	Brown	LG	Light Green	VT	Violet
DB	Dark Blue	OR	Orange	WT	White
DG	Dark Green	PK	Pink	YL	Yellow

1996-99 Neon 2.0L I4 SOHC MFI VIN C (All) 'C2' White Connector

PCM Pin #	Wire Color	Circuit Description (40 Pin)	Value at Hot Idle
41	RD/LG	S/C Set Switch Signal	S/C & Set Switch On: 3.8v
42	---	Not Used	---
43	BK/LB	Sensor Ground	<0.050v
44	OR	8-Volt Supply	7.9-8.1v
45	WT	PSP Switch Signal	Straight: 0v, Turning: 5v
46	RD/WT	Battery Power (Fused B+)	12-14v
47	BK/WT	Sensor Ground	<0.050v
48	BR/WT	IAC 3 Driver	DC pulse signals: 0.8-11v
49	YL/BK	IAC 2 Driver	DC pulse signals: 0.8-11v
50	BK/TN	Power Ground	<0.1v
51	TN/WT	HO2S-12 (B1 S2) Signal	0.1-1.1v
52	PK/LG	Battery Temperature Sensor	At 86°F: 1.96v
53-55	---	Not Used	
56	TN/BK	Generator Lamp Control	Lamp On: 1v, Off: 12v
57	GY/RD	IAC 1 Driver	DC pulse signals: 0.8-11v
58	PK	IAC 4 Driver	DC pulse signals: 0.8-11v
59-60	---	Not Used	---
61	PK/WT	5-Volt Supply	4.9-5.1v
62	WT/PK	Brake Switch Signal	Brake Off: 0v, On: 12v
63	---	Not Used	
64	DB/OR	A/C Clutch Relay Control	Relay Off: 12v, On: 1v
65	PK	SCI Transmit	0v
66	WT/OR	Vehicle Speed Signal	Digital Signal
67	DB/YL	ASD Relay Control	Relay Off: 12v, On: 1v
68	PK/BK	EVAP Purge Solenoid Control	PWM signal: 0-12-0v
69-71	---	Not Used	---
72	OR	LDP Switch Signal	Switch Closed: 0v, open: 12v
73	GY/LB	Tachometer Signal	Pulse Signals
74	BR	Fuel Pump Relay Control	Relay Off: 12v, On: 1v
75	LG	SCI Receive	5v
76	BR/YL	PNP Switch Signal	In P/N: 0v, Others: 5v
77	WT/LG	LDP Solenoid Control	PWM signal: 0-12-0v
78	TN/RD	S/C Vacuum Solenoid	Vacuum Increasing: 1v
79	OR/BK	TCC Solenoid Control	At Cruise w/TCC On: <1v
80	LG/RD	S/C Vent Solenoid	Vacuum Decreasing: 1v

Pin Connector Graphic

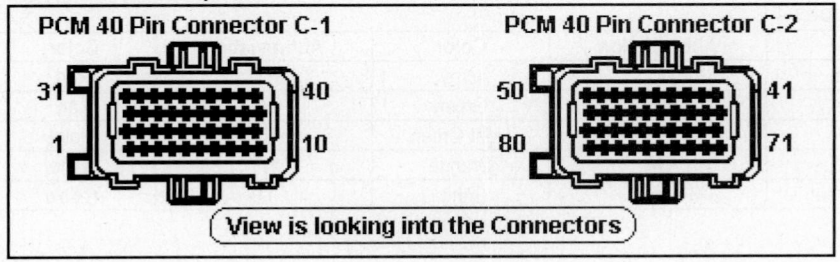

2000-03 Neon 2.0L I4 SOHC MFI VIN C (All) 'C1' Black Connector

PCM Pin #	Wire Color	Circuit Description (40 Pin)	Value at Hot Idle
1-2	---	Not Used	---
3	DB/TN	Coil 2 Driver	5°, at 55 mph: 8° dwell
4	---	Not Used	---
5	YL/RD	S/C Power Supply	12-14v
6	DG/OR	ASD Relay Output	12-14v
7	YL/WT	Injector 3 Driver	1-4 ms
8	DG	Generator Field Control	Digital Signal: 0-12-0v
9	---	Not Used	---
10	BK/TN	Power Ground	<0.1v
11	BK/GY	Coil 1 Driver	5°, at 55 mph: 8° dwell
12	GY	Engine Oil Pressure Sense	Switch open: 12v, closed: 0v
13	WT/LB	Injector 1 Driver	1-4 ms
14-15	---	Not Used	---
16	LB/BR	Injector 4 Driver	1-4 ms
17	TN	Injector 2 Driver	1-4 ms
18	OR/RD	HO2S-11 (B1 S1) Heater	Heater On: 1v, Off: 12v
19	---	Not Used	---
20	DB/WT	Ignition Switch Output	12-14v
21-22	---	Not Used	---
23	LG/BK	Clutch Switch Signal	Clutch Out: 5v, In: 0v
24	---	Not Used	---
25	DB/LG	Knock Sensor Signal	0.080v AC
26	TN/DB	ECT Sensor Signal	At 180°F: 2.80v
27	BK/OR	HO2S-11 Ground	<0.050v
28-29	---	Not Used	---
30	BK/DG	HO2S-11 (B1 S1) Signal	0.1-1.1v
31	TN	Starter Relay Control	KOEC: 9-11v
32	GY/BK	CKP Sensor Signal	Digital Signal: 0-5-0v
33	TN/YL	CMP Sensor Signal	Digital Signal: 0-5-0v
34	---	Not Used	---
35	OR/DB	TP Sensor Signal	0.6-1.0v
36	DG/RD	MAP Sensor Signal	1.5-1.7v
37	---	Not Used	---
38	BR/OR	A/C Select Switch Sense	A/C On: 1v, Off: 12v
39	BR/WT	Manifold Tuning Valve Relay	Relay Off: 12v, On: 1v
40	---	Not Used	---

Standard Colors and Abbreviations

Abbreviation	Color	Abbreviation	Color	Abbreviation	Color
BK	Black	GY	Gray	RD	Red
BL	Blue	GN	Green	TN	Tan
BR	Brown	LG	Light Green	VT	Violet
DB	Dark Blue	OR	Orange	WT	White
DG	Dark Green	PK	Pink	YL	Yellow

2000-03 Neon 2.0L I4 SOHC MFI VIN C (All) 'C2' White Connector

PCM Pin #	Wire Color	Circuit Description (40 Pin)	Value at Hot Idle
41	RD/LG	S/C Set Switch Signal	S/C & Set Switch On: 3.8v
42, 48	---	Not Used	---
43	BK/LB	Sensor Ground	<0.050v
44	OR	8-Volt Supply	7.9-8.1v
45	WT	PSP Switch Signal	Straight: 0v, Turning: 5v
46	RD/WT	Battery Power (Fused B+)	12-14v
47	BK/WT	Power Ground	<0.1v
49	YL/BK	IAC 2 Driver	DC pulse signals: 0.8-11v
50	BK/TN	Power Ground	<0.1v
51	TN/WT	HO2S-12 (B1 S2) Signal	0.1-1.1v
52	TN/LG	Inlet Air Temperature Sensor	At 86°F: 1.96v
53-54	---	Not Used	---
55	DB/PK	Radiator Fan Relay Control	Relay Off: 12v, On: 1v
56	TN/RD	S/C Vacuum Solenoid	Vacuum Increasing: 1v
57	GY/RD	IAC 1 Motor Sense (A/T)	12-14v
57	BR/WT	IAC 1 Motor Sense (M/T)	12-14v
58, 60	---	Not Used	---
59	VT/YL	PCM Data Bus (J1950)	Digital Signal: 0-5-0v
61	VT/WT	5-Volt Supply	4.9-5.1v
62	WT/PK	Brake Switch Signal	Brake Off: 0v, On: 12v
63	YL/DG	Torque Management Request	Digital Signals
64	DB/OR	A/C Clutch Relay Control	Relay Off: 12v, On: 1v
65	PK	SCI Transmit	0v
66	WT/OR	Vehicle Speed Signal	Digital Signal
67	DB/YL	ASD Relay Control	Relay Off: 12v, On: 1v
68	PK/BK	EVAP Purge Solenoid Control	PWM signal: 0-12-0v
69	---	Not Used	---
70	DB	EVAP Purge Solenoid Sense	0-1v
71	GY/RD	Brake Fluid Level Switch	Switch open: 12v, closed; 0v
72	YL	LDP Switch Signal	Switch Closed: 0v, open: 12v
73, 79	---	Not Used	---
74	BR	Fuel Pump Relay Control	Relay Off: 12v, On: 1v
75	LG	SCI Receive	5v
76	BR/YL	TRS TR1 Sense / PNP Signal	In P/N: 0v, Others: 5v
77	WT/DB	LDP Solenoid Control	PWM signal: 0-12-0v
78	OR/BK	TCC Solenoid Control	Digital Signal: 0-12-0v
80	LG/RD	S/C Vent Solenoid	Vacuum Decreasing: 1v

Pin Connector Graphic

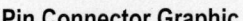

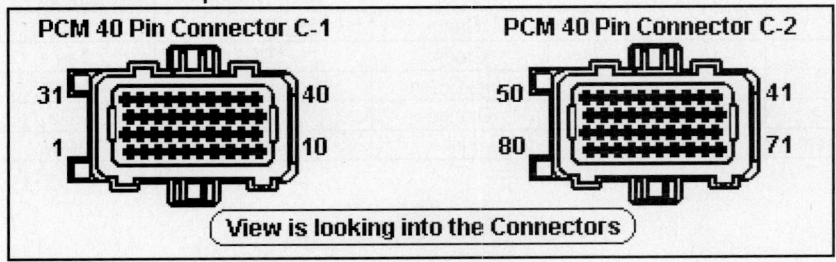

2001-03 Neon ACR & R/T 2.0L I4 SOHC MFI VIN F 'C1' Black Connector

PCM Pin #	Wire Color	Circuit Description (40 Pin)	Value at Hot Idle
1-2	---	Not Used	---
3	DB/TN	Coil 2 Driver	5°, at 55 mph: 8° dwell
4	---	Not Used	
5	YL/RD	S/C Power Supply	12-14v
6	DG/OR	ASD Relay Output	12-14v
7	YL/WT	Injector 3 Driver	1-4 ms
8	DG	Generator Field Control	Digital Signal: 0-12-0v
9	---	Not Used	---
10	BK/TN	Power Ground	<0.1v
11	BK/GY	Coil 1 Driver	5°, at 55 mph: 8° dwell
12	GY	Engine Oil Pressure Sense	Switch open: 12v, closed: 0v
13	WT/DB	Injector 1 Driver	1-4 ms
14-15	---	Not Used	---
16	LB/BR	Injector 4 Driver	1-4 ms
17	TN	Injector 2 Driver	1-4 ms
18	OR/RD	HO2S-11 (B1 S1) Heater	Heater On: 1v, Off: 12v
19	---	Not Used	---
20	DB/WT	Ignition Switch Output	12-14v
21-22	---	Not Used	---
23	LG/BK	Clutch Switch Signal	Clutch Out: 5v, In: 0v
24	---	Not Used	---
25	DB/LG	Knock Sensor Signal	0.080v AC
26	TN/BK	ECT Sensor Signal	At 180°F: 2.80v
27	BK/OR	HO2S-11 Ground	<0.050v
28-29	---	Not Used	---
30	BK/DG	HO2S-11 (B1 S1) Signal	0.1-1.1v
31	TN	Starter Relay Control	KOEC: 9-11v
32	GY/BK	CKP Sensor Signal	Digital Signal: 0-5-0v
33	TN/YL	CMP Sensor Signal	Digital Signal: 0-5-0v
34	---	Not Used	---
35	OR/DB	TP Sensor Signal	0.6-1.0v
36	DG/RD	MAP Sensor Signal	1.5-1.7v
37	---	Not Used	---
38	BR/OR	A/C Select Switch Sense	A/C On: 1v, Off: 12v
39	BR/WT	Manifold Tuning Valve Relay	Relay Off: 12v, On: 1v
40	---	Not Used	---

Standard Colors and Abbreviations

Abbreviation	Color	Abbreviation	Color	Abbreviation	Color
BK	Black	GY	Gray	RD	Red
BL	Blue	GN	Green	TN	Tan
BR	Brown	LG	Light Green	VT	Violet
DB	Dark Blue	OR	Orange	WT	White
DG	Dark Green	PK	Pink	YL	Yellow

2001-03 Neon ACR & R/T 2.0L I4 SOHC MFI VIN F 'C2' White Connector

PCM Pin #	Wire Color	Circuit Description (40 Pin)	Value at Hot Idle
41	RD/LG	S/C Set Switch Signal	S/C & Set Switch On: 3.8v
42, 48	---	Not Used	---
43	BK/LB	Sensor Ground	<0.050v
44	OR	8-Volt Supply	7.9-8.1v
45	WT	PSP Switch Signal	Straight: 0v, Turning: 5v
46	RD/WT	Battery Power (Fused B+)	12-14v
47	BK/WT	Power Ground	<0.1v
49	YL/BK	IAC 2 Driver	DC pulse signals: 0.8-11v
50	BK/TN	Power Ground	<0.1v
51	TN/WT	HO2S-12 (B1 S2) Signal	0.1-1.1v
52	VT/LG	Inlet Air Temperature Sensor	At 86°F: 1.96v
53-54	---	Not Used	---
55	DB/PK	Radiator Fan Relay Control	Relay Off: 12v, On: 1v
56	TN/RD	S/C Vacuum Solenoid	Vacuum Increasing: 1v
57	GY/RD	IAC Motor Sense (w/ EATX)	12-14v
57	BR/WT	IAC Motor Sense (w/o EATX)	12-14v
58, 60	---	Not Used	---
59	VT/YL	PCM Data Bus (J1950)	Digital Signal: 0-5-0v
61	VT/WT	5-Volt Supply	4.9-5.1v
62	WT/PK	Brake Switch Signal	Brake Off: 0v, On: 12v
63	YL/DG	Torque Management Request	Digital Signals
64	DB/OR	A/C Clutch Relay Control	Relay Off: 12v, On: 1v
65	PK	SCI Transmit	0v
66	WT/OR	Vehicle Speed Signal	Digital Signal
67	DB/YL	ASD Relay Control	Relay Off: 12v, On: 1v
68	PK/BK	EVAP Purge Solenoid Control	PWM signal: 0-12-0v
69, 73, 79	---	Not Used	---
70	DB	EVAP Purge Solenoid Sense	0-1v
71	GY/RD	Brake Fluid Level Sense	Switch open: 12v, closed; 0v
72	YL	LDP Switch Signal	Switch Closed: 0v, open: 12v
74	BR	Fuel Pump Relay Control	Relay Off: 12v, On: 1v
75	LG	SCI Receive	0v
76	BR/YL	TRS TR1 Sense / PNP Signal	In P/N: 0v, Others: 5v
76	YL/RD	Clutch Interlock Switch Sense	Clutch Out: 12v, In: 0v
77	WT/DB	LDP Solenoid Control	PWM signal: 0-12-0v
78	OR/BK	TCC Solenoid Control	Digital Signal: 0-12-0v
80	LG/RD	S/C Vent Solenoid	Vacuum Decreasing: 1v

Pin Connector Graphic

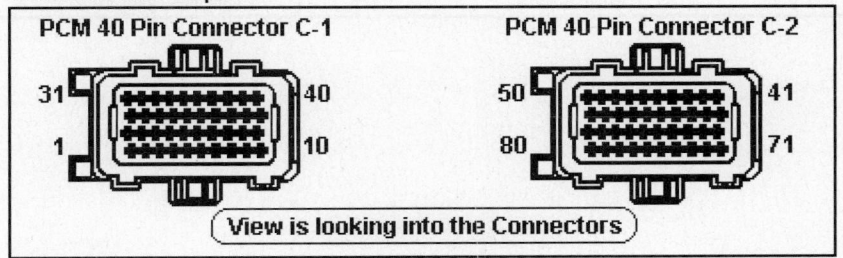

PCM 40 Pin Connector C-1

PCM 40 Pin Connector C-2

View is looking into the Connectors

1995 Neon 2.0L I4 DOHC MFI VIN Y (All) 60 Pin Connector

PCM Pin #	Wire Color	Circuit Description (60 Pin)	Value at Hot Idle
1	BK/GY	Coil 2 Driver	5°, at 55 mph: 8° dwell
2	BK/TN	Power Ground	<0.1v
3	YL/WT	Injector 3 Driver	1-4 ms
4	WT/DB	Injector 1 Driver	1-4 ms
5	WT/OR	Vehicle Speed Signal	Digital Signal
6	BK/RD	IAT Sensor Signal	At 100°F: 1.83v
7	DG/BK	HO2S-12 (B1 S2) Signal	0.1-1.1v
8	BK/DG	HO2S-11 (B1 S1) Signal	0.1-1.1v
9	LG	SCI Receive	0v
10	OR/DB	TP Sensor Signal	0.6-1.0v
11	RD/WT	Battery Power (Fused B+)	12-14v
12-13	---	Not Used	---
14	YL/BK	IAC 2 Driver	DC pulse signals: 0.8-11v
15	GY/RD	IAC 3 Driver	DC pulse signals: 0.8-11v
16	PK/BK	EVAP Purge Solenoid Control	PWM signal: 0-12-0v
17	OR/BK	TCC Solenoid Control	At Cruise w/TCC On: <1v
18	DB/YL	ASD Relay Control	Relay Off: 12v, On: 1v
19	DB/PK	Radiator Fan Relay Control	Relay Off: 12v, On: 1v
20	---	Not Used	---
21	DB/YL	Coil 1 Driver	5°, at 55 mph: 8° dwell
22	BK/TN	Power Ground	<0.1v
23	TN	Injector 2 Driver	1-4 ms
24	LB/BR	Injector 4 Driver	1-4 ms
25	GY/BK	CKP Sensor Signal	Digital Signal: 0-5-0v
26	TN/YL	CMP Sensor Signal	Digital Signal: 0-5-0v
27	BK/LG	Knock Sensor Signal	0.080v AC
28	TN/BK	ECT Sensor Signal	At 180°F: 2.80v
29	DG/RD	MAP Sensor Signal	1.5-1.7v
30	PK	SCI Transmit	0v
31	RD/LG	S/C Set Switch Signal	S/C & Set Switch On: 3.8v
32	WT/PK	Brake Switch Signal	Brake Off: 0v, On: 12v
33	DG/RD	A/C Damped Pressure Switch	A/C On: 1v, Off: 12v
34	PK/BK	IAC 4 Driver	DC pulse signals: 0.8-11v
35	BR/WT	IAC 1 Driver	DC pulse signals: 0.8-11v
36	BK/PK	MIL (lamp) Control	Lamp On: 1v, Off: 12v
37	TN/BK	Generator Lamp Control	Lamp On: 1v, Off: 12v
38	BR	Fuel Pump Relay Control	Relay Off: 12v, On: 1v
39	GY/YL	EGR Solenoid Control	12v, 55 mph: 1v
40	TN/RD	S/C Vacuum Solenoid	Vacuum Increasing: 1v

1995 Neon 2.0L I4 DOHC MFI VIN Y (All) 60 Pin Connector

PCM Pin #	Wire Color	Circuit Description (60 Pin)	Value at Hot Idle
41	DG	Generator Field Control	Digital Signal: 0-12-0v
42	DG/OR	ASD Relay Output	12-14v
43	PK/WT	5-Volt Supply	4.9-5.1v
44	OR	8-Volt Supply	7.9-8.1v
45-46	---	Not Used	---
47	YL/RD	S/C Power Supply	12-14v
48	GY/LB	Tachometer Signal	Pulse Signals
49	PK/LG	Battery Temperature Sensor	At 86°F: 1.96v
50	BR/YL	PNP Switch Signal	In P/N: 0v, Others: 5v
51	BK/LB	Sensor Ground	<0.050v
52	BK/WT	Sensor Ground	<0.050v
53	---	Not Used	---
54	LG/BK	Ignition Switch Output	12-14v
55	---	Not Used	---
56	WT	PSP Switch Signal	Straight: 0v, Turning: 5v
57-58	---	Not Used	---
59	DB/OR	A/C Clutch Relay Control	Relay Off: 12v, On: 1v
60	LG/RD	S/C Vent Solenoid	Vacuum Decreasing: 1v

Pin Connector Graphic

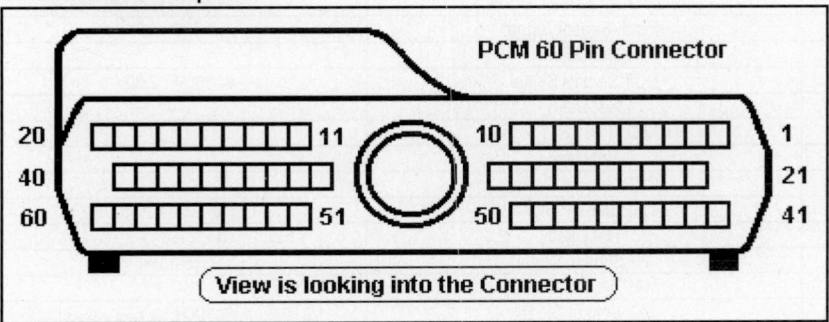

Standard Colors and Abbreviations

Abbreviation	Color	Abbreviation	Color	Abbreviation	Color
BK	Black	GY	Gray	RD	Red
BL	Blue	GN	Green	TN	Tan
BR	Brown	LG	Light Green	VT	Violet
DB	Dark Blue	OR	Orange	WT	White
DG	Dark Green	PK	Pink	YL	Yellow

1996-99 Neon 2.0L I4 DOHC MFI VIN Y (All) 'C1' Black Connector

PCM Pin #	Wire Color	Circuit Description (40 Pin)	Value at Hot Idle
1	---	Not Used	---
2	BK/GY	Coil 1 Driver	5°, at 55 mph: 8° dwell
3	DB/TN	Coil 2 Driver	5°, at 55 mph: 8° dwell
4	DG	Generator Field Control	Digital Signal: 0-12-0v
5	YL/RD	S/C Power Supply	12-14v
6	DG/OR	ASD Relay Output	12-14v
7	YL/WT	Injector 3 Driver	1-4 ms
8	BK/PK	MIL (lamp) Control	Lamp On: 1v, Off: 12v
9	---	Not Used	---
10	BK/TN	Power Ground	<0.1v
11-12	---	Not Used	---
13	WT/LB	Injector 1 Driver	1-4 ms
14-15	---	Not Used	---
16	LB/BR	Injector 4 Driver	1-4 ms
17	TN	Injector 2 Driver	1-4 ms
18	LG	Radiator Fan Relay Control	Relay Off: 12v, On: 1v
19	---	Not Used	---
20	DB/WT	Ignition Switch Output	12-14v
21-22	---	Not Used	---
23	DB	Fuel Level Sensor Signal	70 ohms (±20) with full tank
24	BK/LG	Knock Sensor Signal	0.080v AC
25	---	Not Used	---
26	TN/BK	ECT Sensor Signal	At 180°F: 2.80v
27-29	---	Not Used	---
30	BK/DG	HO2S-11 (B1 S1) Signal	0.1-1.1v
31	---	Not Used	---
32	GY/BK	CKP Sensor Signal	Digital Signal: 0-5-0v
33	TN/YL	CMP Sensor Signal	Digital Signal: 0-5-0v
34	---	Not Used	---
35	OR/LB	TP Sensor Signal	0.6-1.0v
36	DG/RD	MAP Sensor Signal	1.5-1.7v
37	BK/RD	IAT Sensor Signal	At 100°F: 1.83v
38	BR/OR	A/C Select Switch Sense	A/C On: 1v, Off: 12v
39	---	Not Used	---
40	GY/YL	EGR Solenoid Control	12v, 55 mph: 1v

Standard Colors and Abbreviations

Abbreviation	Color	Abbreviation	Color	Abbreviation	Color
BK	Black	GY	Gray	RD	Red
BL	Blue	GN	Green	TN	Tan
BR	Brown	LG	Light Green	VT	Violet
DB	Dark Blue	OR	Orange	WT	White
DG	Dark Green	PK	Pink	YL	Yellow

1996-99 Neon 2.0L I4 DOHC MFI VIN Y (All) 'C2' White Connector

PCM Pin #	Wire Color	Circuit Description (40 Pin)	Value at Hot Idle
41	RD/LG	S/C Set Switch Signal	S/C & Set Switch On: 3.8v
42	---	Not Used	---
43	BK/LB	Sensor Ground	<0.050v
44	OR	8-Volt Supply	7.9-8.1v
45	WT	PSP Switch Signal	Straight: 0v, Turning: 5v
46	RD/WT	Battery Power (Fused B+)	12-14v
47	BK/WT	Sensor Ground	<0.050v
48	BR/WT	IAC 3 Driver	DC pulse signals: 0.8-11v
49	YL/BK	IAC 2 Driver	DC pulse signals: 0.8-11v
50	BK/TN	Power Ground	<0.1v
51	TN/WT	HO2S-12 (B1 S2) Signal	0.1-1.1v
52	PK/LG	Battery Temperature Sensor	At 86°F: 1.96v
53-55	---	Not Used	---
56	TN/BK	Generator Lamp Control	Lamp On: 1v, Off: 12v
57	GY/RD	IAC 1 Driver	DC pulse signals: 0.8-11v
58	PK	IAC 4 Driver	DC pulse signals: 0.8-11v
59-60	---	Not Used	---
61	PK/WT	5-Volt Supply	4.9-5.1v
62	WT/PK	Brake Switch Signal	Brake Off: 0v, On: 12v
63	---	Not Used	---
64	DB/OR	A/C Clutch Relay Control	Relay Off: 12v, On: 1v
65	PK	SCI Transmit	0v
66	WT/OR	Vehicle Speed Signal	Digital Signal
67	DB/YL	ASD Relay Control	Relay Off: 12v, On: 1v
68	PK/BK	EVAP Purge Solenoid Control	PWM signal: 0-12-0v
69-71	---	Not Used	---
72	OR	LDP Switch Signal	Switch Closed: 0v, open: 12v
73	GY/LB	Tachometer Signal	Pulse Signals
74	BR	Fuel Pump Relay Control	Relay Off: 12v, On: 1v
75	LG	SCI Receive	5v
76	BR/YL	PNP Switch Signal	In P/N: 0v, Others: 5v
77	WT/LG	LDP Solenoid Control	PWM signal: 0-12-0v
78	TN/RD	S/C Vacuum Solenoid	Vacuum Increasing: 1v
79	OR/BK	TCC Solenoid Control	At Cruise w/TCC On: <1v
80	LG/RD	S/C Vent Solenoid	Vacuum Decreasing: 1v

Pin Connector Graphic

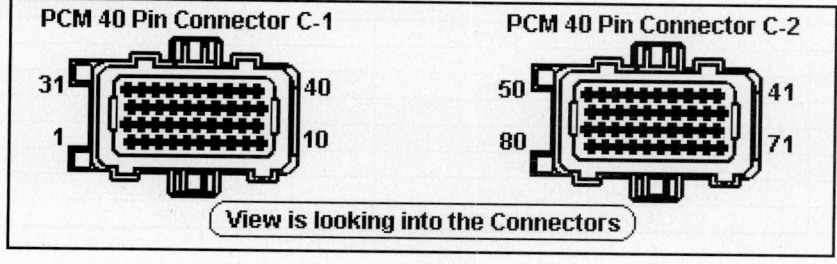

PCM 40 Pin Connector C-1

PCM 40 Pin Connector C-2

31 40
1 10

50 41
80 71

View is looking into the Connectors

1990 Shadow 2.2L I4 SOHC Turbo VIN C (M/T) 60 Pin Connector

PCM Pin #	Wire Color	Circuit Description (60 Pin)	Value at Hot Idle
1	DG/RD	MAP Sensor Signal	1.5-1.7v
2	TN/BK	ECT Sensor Signal	At 180°F: 2.80v
3	RD	Battery Power (7B+)	12-14v
4, 5	BK/LB, BK	Sensor Ground	<0.050v
6	PK/WT	5-Volt Supply	4.9-5.1v
7	OR	8-Volt Supply	7.9-8.1v
8	DG/BK	ASD Relay Output	12-14v
9	DB	Ignition Switch Start Signal	12-14v
10, 17-18	---	Not Used	---
11, 12	BK/TN	Power Ground	<0.1v
13	LB/BR	Injector 4 Driver	1-4 ms
14	YL/WT	Injector 3 Driver	1-4 ms
15	TN	Injector 2 Driver	1-4 ms
16	WT/DB	Injector 1 Driver	1-4 ms
19	BK/GY	Coil 1 Driver	5°, at 55 mph: 8° dwell
20	DG	Generator Field Driver	Digital Signal: 0-12-0v
21	BK/RD	Air Temperature Sensor	At 100°F: 2.51v
22	OR/DB	TP Sensor Signal	0.6-1.0v
23, 26, 28	---	Not Used	---
24	GY	Distributor Reference Signal	Digital Signal: 0-5-0v
25, 45	PK, LG	SCI Transmit, SCI Receive	0v, 5v
27	BR	A/C Damped Pressure Switch	A/C On: 1v, Off: 12v
29	WT/PK	Brake Switch Signal	Brake Off: 0v, On: 12v
30	BR	PNP Switch Signal	In P/N: 0v, Others: 5v
31	DB/PK	Radiator Fan Relay Control	Relay Off: 12v, On: 1v
32	BK/PK	MIL (lamp) Control	Lamp On: 1v, Off: 12v
33	TN/RD	Speed Control Vacuum Solenoid	Vacuum Increasing: 1v
34	DB/OR	A/C Clutch Relay Control	Relay Off: 12v, On: 1v
35	GY/YL	EGR Solenoid Control (California)	12v, 55 mph: 1v
36	LG/BK	Wastegate Solenoid Control	Solenoid Off: 12v, On: 1v
37-38, 46, 57-58	---	Not Used	---
39, 40	GY/RD, BR	AIS 3 Motor, AIS 1 Motor Control	DC pulse signals: 0.8-11v
41	BK/DG	HO2S-11 (B1 S1) Signal	0.1-1.1v
42	BK/LG	Knock Sensor Signal	0.080v AC
43	GY/LB	Tachometer Signal	Pulse Signals
44	TN/YL	Distributor Sync Signal	Digital Signal: 0-5-0v
47	WT/OR	Vehicle Speed Signal	Digital Signal
48	BR/RD	Speed Control Set Switch Signal	S/C & Set Switch On: 3.8v
49	YL/RD	Speed Control Switch Signal	S/C On: 1v, Off: 12v
50	WT/LG	Speed Control Resume Switch	S/C On: 1v, Off: 12v
51	DB/YL	ASD Relay Control	Relay Off: 12v, On: 1v
52	PK	EVAP Purge Solenoid Control	Solenoid Off: 12v, On: 1v
53	LG/RD	S/C Vent Solenoid	Vacuum Decreasing: 1v
54	LG/WT	Turbo Solenoid No. 1 Control	Solenoid Off: 12v, On: 1v
55	LB	BARO Read Solenoid	Solenoid Off: 12v, On: 1v
56	LG/OR	Turbo Solenoid No. 2 Control	Solenoid Off: 12v, On: 1v
59, 60	PK, YL/BK	AIS 4 Motor, AIS Motor Control	DC pulse signals: 0.8-11v

Pin Connector Graphic

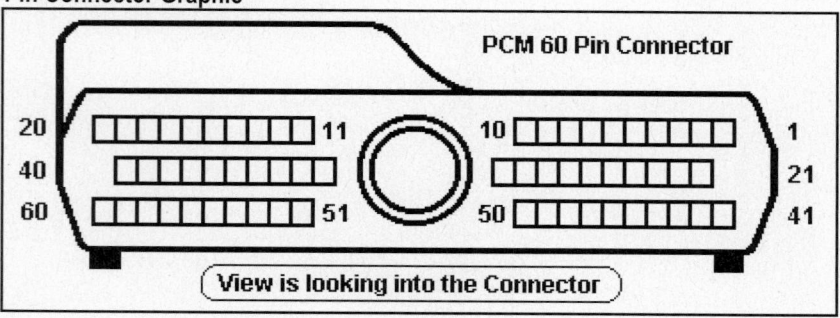

1990-92 Shadow 2.2L I4 SOHC TBI VIN D (All) 60 Pin Connector

PCM Pin #	Wire Color	Circuit Description (60 Pin)	Value at Hot Idle
1	DG/RD	MAP Sensor Signal	1.5-1.7v
2 ('90)	TN/WT	ECT Sensor Signal	At 180°F: 2.80v
2	TN/BK	ECT Sensor Signal	At 180°F: 2.80v
3 ('90)	RD	Battery Power (Fused B+)	12-14v
3	RD/WT	Battery Power (Fused B+)	12-14v
4	BK/LB	Sensor Ground	<0.050v
5 ('90)	BK	Sensor Ground	<0.050v
5	BK/WT	Sensor Ground	<0.050v
6	PK/WT	5-Volt Supply (MAP Sensor)	4.9-5.1v
7	OR	8-Volt Supply	7.9-8.1v
8	WT	Ignition Switch Start Signal	12-14v
9	DB	Ignition Switch Output	12-14v
10	---	Not Used	---
11 ('90)	LB/RD	Power Ground	<0.1v
12 ('90)	LB/RD	Power Ground	<0.1v
11	BK/TN	Power Ground	<0.1v
12	BK/TN	Power Ground	<0.1v
13-15	---	Not Used	---
16	WT/DB	Injector 1 Driver	1-4 ms
17-18	---	Not Used	---
19 ('90)	YL	Coil Driver	5°, at 55 mph: 8° dwell
19	BK/GY	Coil Driver	5°, at 55 mph: 8° dwell
20	DG	Generator Field Driver	Digital Signal: 0-12-0v
21 ('90)	BK/RD	Throttle Body Temperature	At 100°F: 2.51v
22	OR/DB	TP Sensor Signal	0.6-1.0v
23	---	Not Used	---
24	GY/BK	Distributor Reference Signal	Digital Signal: 0-5-0v
25	PK	SCI Transmit	0v
26	---	Not Used	---
27	BR	A/C Clutch Signal	A/C On: 1v, Off: 12v
28	---	Not Used	---
29	WT/PK	Brake Switch Signal	Brake Off: 0v, On: 12v
30 ('90)	BR	PNP Switch Signal	In P/N: 0v, Others: 5v
30	BR/YL	PNP Switch Signal	In P/N: 0v, Others: 5v
31	DB/PK	Radiator Fan Relay Control	Relay Off: 12v, On: 1v
32	BK/PK	MIL (lamp) Control	Lamp On: 1v, Off: 12v
33	TN/RD	S/C Vacuum Solenoid	Vacuum Increasing: 1v
34	DB/OR	A/C WOT Relay Control	Relay Off: 12v, On: 1v
35	GY/YL	EGR Solenoid Control (Calif.)	12v, 55 mph: 1v
36-38	---	Not Used	---
39	GY/RD	AIS Motor Control	DC pulse signals: 0.8-11v
40	BR/WT	AIS Motor Control	DC pulse signals: 0.8-11v

1990-92 Shadow 2.2L I4 SOHC TBI VIN D (All) 60 Pin Connector

PCM Pin #	Wire Color	Circuit Description (60 Pin)	Value at Hot Idle
41	BK/DG	HO2S-11 (B1 S1) Signal	0.1-1.1v
42	---	Not Used	---
43	GY/LB	Tachometer Signal	Pulse Signals
44	---	Not Used	---
45	LG	SCI Receive	0v
46	---	Not Used	---
47	WT/OR	Vehicle Speed Signal	Digital Signal
48	BR/RD	S/C Set Switch Signal	S/C & Set Switch On: 3.8v
49	YL/RD	S/C On/Off Switch Signal	S/C On: 1v, Off: 12v
50	WT/LG	S/C Resume Switch	S/C On: 1v, Off: 12v
51	DB/YL	ASD Relay Control	Relay Off: 12v, On: 1v
52 ('90)	PK	EVAP Purge Solenoid Control	Solenoid Off: 12v, On: 1v
52	PK/BK	EVAP Purge Solenoid Control	Solenoid Off: 12v, On: 1v
53	LG/RD	S/C Vent Solenoid	Vacuum Decreasing: 1v
54	OR/BK	Part Throttle Unlock Solenoid	Solenoid Off: 12v, On: 1v
55-58	---	Not Used	---
59	PK/BK	AIS Motor Control	DC pulse signals: 0.8-11v
60	YL/BK	AIS Motor Control	DC pulse signals: 0.8-11v

Pin Connector Graphic

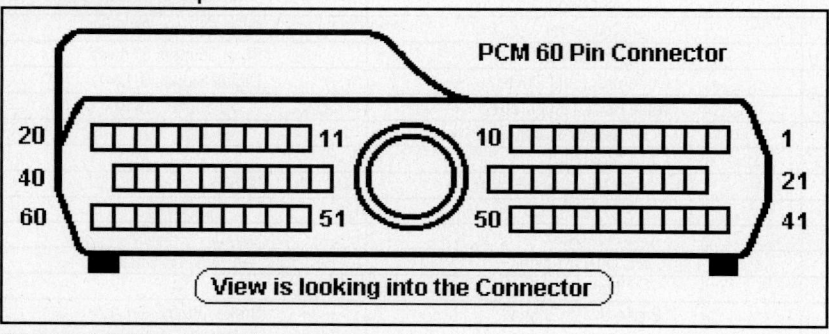

Standard Colors and Abbreviations

Abbreviation	Color	Abbreviation	Color	Abbreviation	Color
BK	Black	GY	Gray	RD	Red
BL	Blue	GN	Green	TN	Tan
BR	Brown	LG	Light Green	VT	Violet
DB	Dark Blue	OR	Orange	WT	White
DG	Dark Green	PK	Pink	YL	Yellow

1993-94 Shadow 2.2L I4 SOHC TBI VIN D (All) 60 Pin Connector

PCM Pin #	Wire Color	Circuit Description (60 Pin)	Value at Hot Idle
1	DG/RD	MAP Sensor Signal	1.5-1.7v
2	TN/BK	ECT Sensor Signal	At 180°F: 2.80v
3	RD/WT	Battery Power (Fused B+)	12-14v
4	BK/LB	Sensor Ground	<0.050v
5	BK/WT	Sensor Ground	<0.050v
6	PK/WT	5-Volt Supply (MAP Sensor)	4.9-5.1v
7	OR	8-Volt Supply	7.9-8.1v
8	WT	Ignition Switch Start Signal	12-14v
9	DB	Ignition Switch Output	12-14v
10	---	Not Used	---
11	BK/TN	Power Ground	<0.1v
12	BK/TN	Power Ground	<0.1v
13-15	---	Not Used	---
16	WT/DB	Injector 1 Driver	1-4 ms
17-18	---	Not Used	---
19	BK/GY	Coil Driver	5°, at 55 mph: 8° dwell
20	DG	Generator Field Driver	Digital Signal: 0-12-0v
21	---	Not Used	---
22	OR/DB	TP Sensor Signal	0.6-1.0v
23	RD/LG	S/C Set Switch Signal	S/C & Set Switch On: 3.8v
24	GY/BK	Distributor Reference Signal	Digital Signal: 0-5-0v
25	PK	SCI Transmit	0v
26	---	Not Used	---
27	BR	A/C Damped Pressure Switch	A/C On: 1v, Off: 12v
28	---	Not Used	---
29	WT/PK	Brake Switch Signal	Brake Off: 0v, On: 12v
30	BR/YL	PNP Switch Signal	In P/N: 0v, Others: 5v
31	DB/PK	Radiator Fan Relay Control	Relay Off: 12v, On: 1v
32	BK/PK	MIL (lamp) Control	Lamp On: 1v, Off: 12v
33	TN/RD	S/C Vacuum Solenoid	Vacuum Increasing: 1v
34	DB/OR	A/C Clutch Relay Control	Relay Off: 12v, On: 1v
35	GY/YL	EGR Solenoid Control (Calif.)	12v, 55 mph: 1v
36-38	---	Not Used	---
39	GY/RD	AIS Motor Control	DC pulse signals: 0.8-11v
40	BR/WT	AIS Motor Control	DC pulse signals: 0.8-11v

Standard Colors and Abbreviations

Abbreviation	Color	Abbreviation	Color	Abbreviation	Color
BK	Black	GY	Gray	RD	Red
BL	Blue	GN	Green	TN	Tan
BR	Brown	LG	Light Green	VT	Violet
DB	Dark Blue	OR	Orange	WT	White
DG	Dark Green	PK	Pink	YL	Yellow

1993-94 Shadow 2.2L I4 SOHC TBI VIN D (All)

PCM Pin #	Wire Color	Circuit Description (60 Pin)	Value at Hot Idle
41	BK/DG	HO2S-11 (B1 S1) Signal	0.1-1.1v
42	---	Not Used	---
43	GY/LB	Tachometer Signal	Pulse Signals
44	---	Not Used	---
45	LG	SCI Receive	0v
46	---	Not Used	---
47	WT/OR	Vehicle Speed Signal	Digital Signal
48-50	---	Not Used	---
51	DB/YL	ASD Relay Control	Relay Off: 12v, On: 1v
52	PK/BK	EVAP Purge Solenoid Control	Solenoid Off: 12v, On: 1v
53	LG/RD	S/C Vent Solenoid	Vacuum Decreasing: 1v
54	OR/BK	EMCC Solenoid	Solenoid Off: 12v, On: 1v
55-56, 58	---	Not Used	---
57	DG/OR	Generator Field Source	Digital Signal: 0-12-0v
59	PK/BK	AIS Motor Control	DC pulse signals: 0.8-11v
60	YL/BK	AIS Motor Control	DC pulse signals: 0.8-11v

Pin Connector Graphic

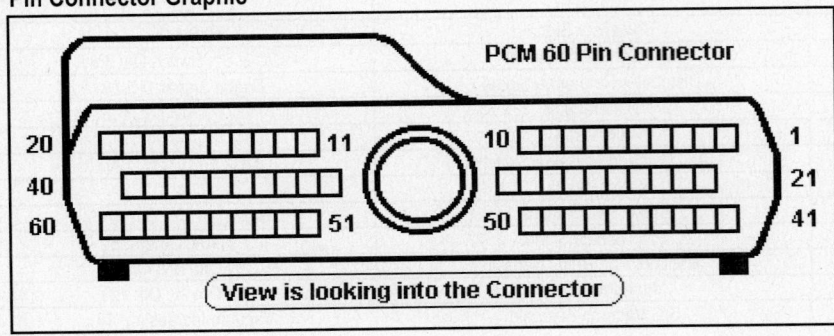

PCM 60 Pin Connector

View is looking into the Connector

1990-92 Shadow 2.5L I4 SOHC Turbo VIN J (All) 60 Pin Connector

PCM Pin #	Wire Color	Circuit Description (60 Pin)	Value at Hot Idle
1	DG/RD	MAP Sensor Signal	1.5-1.7v
2	TN/BK	ECT Sensor Signal	At 180°F: 2.80v
3	RD/WT	Battery Power (Fused B+)	12-14v
4	BK/LB	Sensor Ground	<0.050v
5 ('90)	BK	Sensor Ground	<0.050v
5	BK/WT	Sensor Ground	<0.050v
6	PK/WT	5-Volt Supply	4.9-5.1v
7	OR	8-Volt Supply	7.9-8.1v
8	DG/BK	Auto Shutdown Relay Input	12-14v
9	DB	Ignition Switch Output	12-14v
10	---	Not Used	---
11	BK/TN	Power Ground	<0.1v
12	BK/TN	Power Ground	<0.1v
13	LB/BR	Injector 4 Driver	1-4 ms
14	YL/WT	Injector 3 Driver	1-4 ms
15	TN	Injector 2 Driver	1-4 ms
16	WT/DB	Injector 1 Driver	1-4 ms
17-18	---	Not Used	---
19	BK/GY	Coil 1 Driver	5°, at 55 mph: 8° dwell
20	DG	Generator Field Driver	Digital Signal: 0-12-0v
21	BK/RD	Throttle Body Temperature	At 100°F: 2.51v
22	OR/DB	TP Sensor Signal	0.6-1.0v
23	---	Not Used	---
24	GY/BK	Distributor Reference Signal	Digital Signal: 0-5-0v
25	PK	SCI Transmit	0v
26	---	Not Used	---
27	BR	A/C Damped Pressure Switch	A/C On: 1v, Off: 12v
28	---	Not Used	---
29	WT/PK	Brake Switch Signal	Brake Off: 0v, On: 12v
30	BR/YL	PNP Switch Signal	In P/N: 0v, Others: 5v
31	DB/PK	Radiator Fan Relay Control	Relay Off: 12v, On: 1v
32	BK/PK	MIL (lamp) Control	Lamp On: 1v, Off: 12v
33	TN/RD	S/C Vacuum Solenoid	Vacuum Increasing: 1v
34	DB/OR	A/C Clutch Relay Control	Relay Off: 12v, On: 1v
35	GY/YL	EGR Solenoid Control (Calif.)	12v, 55 mph: 1v
36	LG/BK	Wastegate Solenoid Control	Solenoid Off: 12v, On: 1v
37-38	---	Not Used	---
39	GY/RD	AIS Motor Control	DC pulse signals: 0.8-11v
40	BR/WT	AIS Motor Control	DC pulse signals: 0.8-11v

1990-92 Shadow 2.5L I4 SOHC Turbo VIN J (All) 60 Pin Connector

PCM Pin #	Wire Color	Circuit Description (60 Pin)	Value at Hot Idle
41	BK/DG	HO2S-11 (B1 S1) Signal	0.1-1.1v
42	BK/LG	Knock Sensor Signal	0.080v AC
43	GY/LB	Tachometer Signal	Pulse Signals
44	TN/YL	Distributor Sync Signal	Digital Signal: 0-5-0v
45	LG	SCI Receive	0v
46	---	Not Used	---
47	WT/OR	Vehicle Speed Signal	Digital Signal
48	BR/RD	S/C Set Switch Signal	S/C On: 1v, Off: 12v
49	YL/RD	S/C On/Off Switch Signal	S/C On: 1v, Off: 12v
50	WT/LG	S/C Resume Switch	S/C On: 1v, Off: 12v
51	DB/YL	ASD Relay Control	Relay Off: 12v, On: 1v
52 ('90)	PK	EVAP Purge Solenoid Control	Solenoid Off: 12v, On: 1v
52	PK/BK	EVAP Purge Solenoid Control	Solenoid Off: 12v, On: 1v
53	LG/RD	S/C Vent Solenoid	Vacuum Decreasing: 1v
54	LG/WT	Turbo Solenoid No. 1 Control	Solenoid Off: 12v, On: 1v
55	LB	BARO Read Solenoid	Solenoid Off: 12v, On: 1v
56-58	---	Not Used	---
59	PK/BK	AIS Motor Control	DC pulse signals: 0.8-11v
60	YL/BK	AIS Motor Control	DC pulse signals: 0.8-11v

Pin Connector Graphic

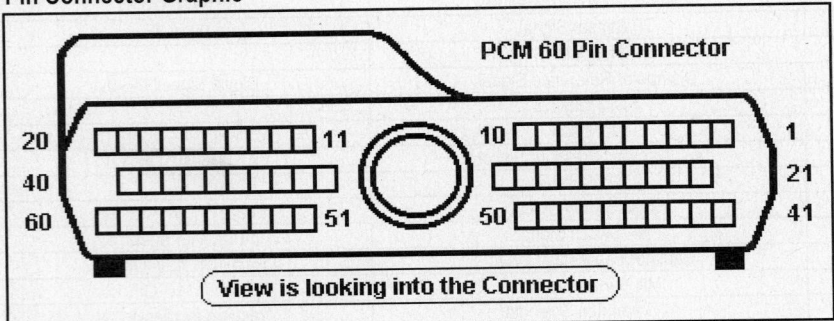

Standard Colors and Abbreviations

Abbreviation	Color	Abbreviation	Color	Abbreviation	Color
BK	Black	GY	Gray	RD	Red
BL	Blue	GN	Green	TN	Tan
BR	Brown	LG	Light Green	VT	Violet
DB	Dark Blue	OR	Orange	WT	White
DG	Dark Green	PK	Pink	YL	Yellow

1990-92 Shadow 2.5L I4 SOHC TBI VIN K (All) 60 Pin Connector

PCM Pin #	Wire Color	Circuit Description (60 Pin)	Value at Hot Idle
1	DG/RD	MAP Sensor Signal	1.5-1.7v
2 ('90)	TN/WT	ECT Sensor Signal	At 180°F: 2.80v
2	TN/BK	ECT Sensor Signal	At 180°F: 2.80v
3 ('90)	RD	Battery Power (Fused B+)	12-14v
3	RD/WT	Battery Power (Fused B+)	12-14v
4	BK/LB	Sensor Ground	<0.050v
5 ('90)	BK	Sensor Ground	<0.050v
5	BK/WT	Sensor Ground	<0.050v
6	PK/WT	5-Volt Supply (MAP Sensor)	4.9-5.1v
7	OR	8-Volt Supply	7.9-8.1v
8	WT	Ignition Switch Start Signal	12-14v
9	DB	Ignition Switch Output	12-14v
10	---	Not Used	---
11 ('90)	LB/RD	Power Ground	<0.1v
12 ('90)	LB/RD	Power Ground	<0.1v
11	BK/TN	Power Ground	<0.1v
12	BK/TN	Power Ground	<0.1v
13-15	---	Not Used	---
16	WT/DB	Injector 1 Driver	1-4 ms
17-18	---	Not Used	---
19 ('90)	YL	Coil Driver	5°, at 55 mph: 8° dwell
19	BK/GY	Coil Driver	5°, at 55 mph: 8° dwell
20	DG	Generator Field Driver	Digital Signal: 0-12-0v
21 ('90)	BK/RD	Throttle Body Temperature	At 100°F: 2.51v
22	OR/DB	TP Sensor Signal	0.6-1.0v
23	---	Not Used	---
24	GY/BK	Distributor Reference Signal	Digital Signal: 0-5-0v
25	PK	SCI Transmit	0v
26	---	Not Used	---
27	BR	A/C Clutch Signal	A/C On: 1v, Off: 12v
28	---	Not Used	---
29	WT/PK	Brake Switch Signal	Brake Off: 0v, On: 12v
30 ('90)	BR	PNP Switch Signal	In P/N: 0v, Others: 5v
30	BR/YL	PNP Switch Signal	In P/N: 0v, Others: 5v
31	DB/PK	Radiator Fan Relay Control	Relay Off: 12v, On: 1v
32	BK/PK	MIL (lamp) Control	Lamp On: 1v, Off: 12v
33	TN/RD	S/C Vacuum Solenoid	Vacuum Increasing: 1v
34	DB/OR	A/C WOT Relay Control	Relay Off: 12v, On: 1v
35	GY/YL	EGR Solenoid Control (Calif.)	12v, 55 mph: 1v
36-38	---	Not Used	---
39	GY/RD	AIS Motor Control	DC pulse signals: 0.8-11v
40	BR/WT	AIS Motor Control	DC pulse signals: 0.8-11v

1990-92 Shadow 2.5L I4 SOHC TBI VIN K (All) 60 Pin Connector

PCM Pin #	Wire Color	Circuit Description (60 Pin)	Value at Hot Idle
41	BK/DG	HO2S-11 (B1 S1) Signal	0.1-1.1v
42	---	Not Used	---
43	GY/LB	Tachometer Signal	Pulse Signals
44	---	Not Used	---
45	LG	SCI Receive	5v
46	---	Not Used	---
47	WT/OR	Vehicle Speed Signal	Digital Signal
48	BR/RD	S/C Set Switch Signal	S/C On: 1v, Off: 12v
49	YL/RD	S/C On/Off Switch Signal	S/C On: 1v, Off: 12v
50	WT/LG	S/C Resume Switch	S/C On: 1v, Off: 12v
51	DB/YL	ASD Relay Control	Relay Off: 12v, On: 1v
52 ('90)	PK	EVAP Purge Solenoid Control	Solenoid Off: 12v, On: 1v
52	PK/BK	EVAP Purge Solenoid Control	Solenoid Off: 12v, On: 1v
53	LG/RD	S/C Vent Solenoid	Vacuum Decreasing: 1v
54	OR/BK	Part Throttle Unlock Solenoid	Solenoid Off: 12v, On: 1v
55-58	---	Not Used	---
59	PK/BK	AIS Motor Control	DC pulse signals: 0.8-11v
60	YL/BK	AIS Motor Control	DC pulse signals: 0.8-11v

Pin Connector Graphic

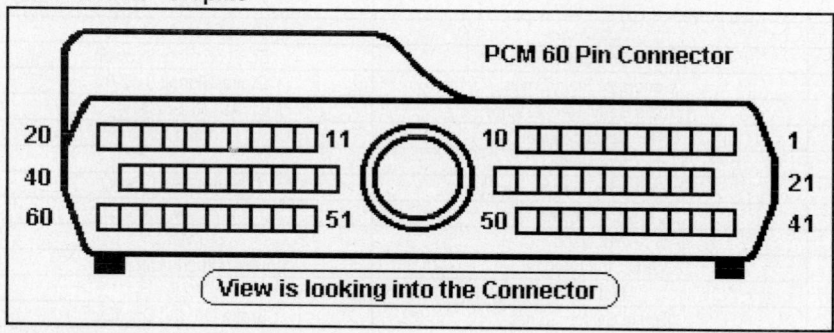

Standard Colors and Abbreviations

Abbreviation	Color	Abbreviation	Color	Abbreviation	Color
BK	Black	GY	Gray	RD	Red
BL	Blue	GN	Green	TN	Tan
BR	Brown	LG	Light Green	VT	Violet
DB	Dark Blue	OR	Orange	WT	White
DG	Dark Green	PK	Pink	YL	Yellow

1993-94 Shadow 2.5L I4 SOHC TBI VIN K (All) 60 Pin Connector

PCM Pin #	Wire Color	Circuit Description (60 Pin)	Value at Hot Idle
1	DG/RD	MAP Sensor Signal	1.5-1.7v
2	TN/BK	ECT Sensor Signal	At 180°F: 2.80v
3	RD/WT	Battery Power (Fused B+)	12-14v
4	BK/LB	Sensor Ground	<0.050v
5	BK/WT	Sensor Ground	<0.050v
6	PK/WT	5-Volt Supply (MAP Sensor)	4.9-5.1v
7	OR	8-Volt Supply	7.9-8.1v
8	WT	Ignition Switch Start Signal	12-14v
9	DB	Ignition Switch Output	12-14v
10	---	Not Used	---
11	BK/TN	Power Ground	<0.1v
12	BK/TN	Power Ground	<0.1v
13-15	---	Not Used	---
16	WT/DB	Injector 1 Driver	1-4 ms
17-18	---	Not Used	---
19	BK/GY	Coil Driver	5°, at 55 mph: 8° dwell
20	DG	Generator Field Driver	Digital Signal: 0-12-0v
21	---	Not Used	---
22	OR/DB	TP Sensor Signal	0.6-1.0v
23	RD/LG	S/C Set Switch Signal	S/C & Set Switch On: 3.8V
24	GY/BK	Distributor Reference Signal	Digital Signal: 0-5-0v
25	PK	SCI Transmit	0v
26	---	Not Used	---
27	BR	A/C Damped Pressure Switch	A/C On: 1v, Off: 12v
28	---	Not Used	---
29	WT/PK	Brake Switch Signal	Brake Off: 0v, On: 12v
30	BR/YL	PNP Switch Signal	In P/N: 0v, Others: 5v
31	DB/PK	Radiator Fan Relay Control	Relay Off: 12v, On: 1v
32	BK/PK	MIL (lamp) Control	Lamp On: 1v, Off: 12v
33	TN/RD	S/C Vacuum Solenoid	Vacuum Increasing: 1v
34	DB/OR	A/C Clutch Relay Control	Relay Off: 12v, On: 1v
35	GY/YL	EGR Solenoid Control (Calif.)	12v, 55 mph: 1v
36-38	---	Not Used	---
39	GY/RD	AIS Motor Control	DC pulse signals: 0.8-11v
40	BR/WT	AIS Motor Control	DC pulse signals: 0.8-11v

1993-94 Shadow 2.5L I4 SOHC TBI VIN K (All) 60 Pin Connector

PCM Pin #	Wire Color	Circuit Description (60 Pin)	Value at Hot Idle
41	BK/DG	HO2S-11 (B1 S1) Signal	0.1-1.1v
42	---	Not Used	---
43	GY/LB	Tachometer Signal	Pulse Signals
44	---	Not Used	---
45	LG	SCI Receive	0v
46	---	Not Used	---
47	WT/OR	Vehicle Speed Signal	Digital Signal
48-50	---	Not Used	---
51	DB/YL	ASD Relay Control	Relay Off: 12v, On: 1v
52	PK/BK	EVAP Purge Solenoid Control	Solenoid Off: 12v, On: 1v
53	LG/RD	S/C Vent Solenoid	Vacuum Decreasing: 1v
54	OR/BK	EMCC Solenoid	Solenoid Off: 12v, On: 1v
55-56	---	Not Used	---
57	DG/OR	Generator Field Source	12-14v
58	---	Not Used	---
59	PK/BK	AIS Motor Control	DC pulse signals: 0.8-11v
60	YL/BK	AIS Motor Control	DC pulse signals: 0.8-11v

Pin Connector Graphic

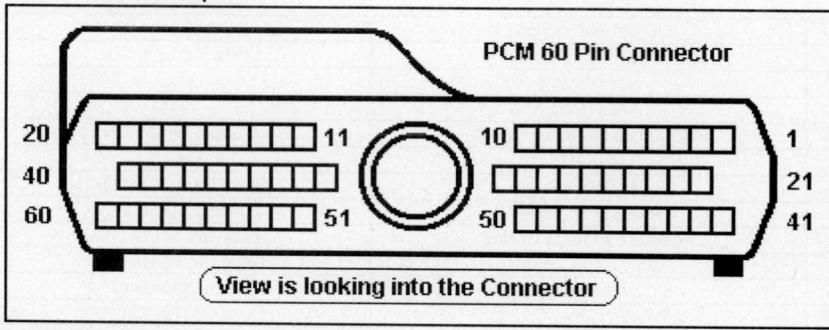

Standard Colors and Abbreviations

Abbreviation	Color	Abbreviation	Color	Abbreviation	Color
BK	Black	GY	Gray	RD	Red
BL	Blue	GN	Green	TN	Tan
BR	Brown	LG	Light Green	VT	Violet
DB	Dark Blue	OR	Orange	WT	White
DG	Dark Green	PK	Pink	YL	Yellow

1992-94 Shadow 3.0L V6 SOHC MFI VIN 3 (A/T) 60 Pin Connector

PCM Pin #	Wire Color	Circuit Description (60 Pin)	Value at Hot Idle
1	DG/RD	MAP Sensor Signal	1.5-1.7v
2	TN/BK	ECT Sensor Signal	At 180ºF: 2.80v
3	OR	Battery Power (Fused B+)	12-14v
4	BK/LB	Sensor Ground	<0.050v
5	BK/WT	Sensor Ground	<0.050v
6	PK/WT	5-Volt Supply	4.9-5.1v
7	OR	9-Volt Supply	8.9-9.1v
8	---	Not Used	---
9	DB	Ignition Switch Output	12-14v
10	---	Not Used	---
11	BK/TN	Power Ground	<0.1v
12	BK/TN	Power Ground	<0.1v
13	LB/BR	Injector 4 Driver	1-4 ms
14	YL/WT	Injector 3 Driver	1-4 ms
15	TN	Injector 2 Driver	1-4 ms
16	WT/DB	Injector 1 Driver	1-4 ms
17-18	---	Not Used	---
19	BK/GY	Coil Driver	5º, at 55 mph: 8º dwell
20	DG	Generator Field Driver	Digital Signal: 0-12-0v
21	---	Not Used	---
22	OR/DB	TP Sensor Signal	0.6-1.0v
23	RD/LG	S/C Set Switch Signal	S/C & Set Switch On: 3.8V
24	GY/BK	Distributor Reference Signal	Digital Signal: 0-5-0v
25	PK	SCI Transmit	0v
26	PK/BR	CCD Bus (+)	Digital Signal: 0-5-0v
27	BR	A/C Damped Pressure Switch	A/C On: 1v, Off: 12v
28	DB/OR	PSP Switch Signal	Straight: 0v, Turning: 5v
29	WT/PK	Brake Switch Signal	Brake Off: 0v, On: 12v
30	BR/YL	PNP Switch Signal	In P/N: 0v, Others: 5v
31	DB/PK	Radiator Fan Relay Control	Relay Off: 12v, On: 1v
32	BK/PK	MIL (lamp) Control	Lamp On: 1v, Off: 12v
33	TN/RD	S/C Vacuum Solenoid	Vacuum Increasing: 1v
34	DB/OR	A/C Clutch Relay Control	Relay Off: 12v, On: 1v
35	GY/YL	EGR Solenoid Control (Calif.)	12v, 55 mph: 1v
36-37	---	Not Used	---
38	GY	Injector 5 Driver	1-4 ms
39	GY/RD	AIS Motor Control (open)	DC pulse signals: 0.8-11v
40	BR/WT	AIS Motor Control (close)	DC pulse signals: 0.8-11v

1992-94 Shadow 3.0L V6 SOHC MFI VIN 3 (A/T) 60 Pin Connector

PCM Pin #	Wire Color	Circuit Description (60 Pin)	Value at Hot Idle
41	BK/DG	HO2S-11 (B1 S1) Signal	0.1-1.1v
42	---	Not Used	---
43	GY/LB	Tachometer Signal	Pulse Signals
44	TN/YL	Distributor Sync Signal	Digital Signal: 0-5-0v
45	LG	SCI Receive	0v
46	WT/BK	CCD Bus (-)	<0.050v
47	WT/OR	Vehicle Speed Signal	Digital Signal
48-50	---	Not Used	---
51	DB/YL	ASD Relay Control	Relay Off: 12v, On: 1v
52	PK/BK	EVAP Purge Solenoid Control	Solenoid Off: 12v, On: 1v
53	LG/RD	S/C Vent Solenoid	Vacuum Decreasing: 1v
54-56	---	Not Used	---
57	DG/OR	ASD Relay Output	12-14v
58	BR/DB	Injector 6 Driver	1-4 ms
59	PK/BK	AIS Motor Control (close)	DC pulse signals: 0.8-11v
60	YL/BK	AIS Motor Control (close)	DC pulse signals: 0.8-11v

Pin Connector Graphic

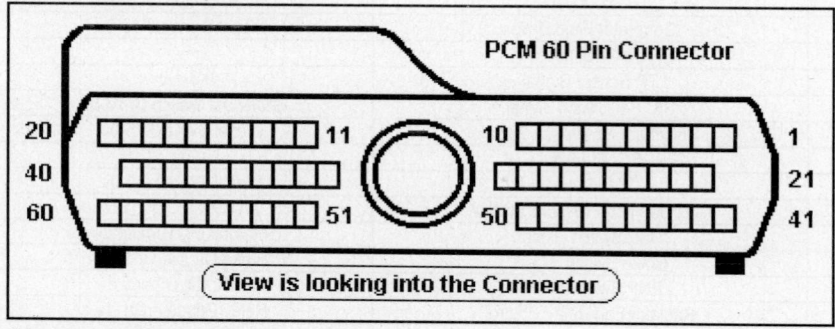

Standard Colors and Abbreviations

Abbreviation	Color	Abbreviation	Color	Abbreviation	Color
BK	Black	GY	Gray	RD	Red
BL	Blue	GN	Green	TN	Tan
BR	Brown	LG	Light Green	VT	Violet
DB	Dark Blue	OR	Orange	WT	White
DG	Dark Green	PK	Pink	YL	Yellow

1991 Spirit 2.2L I4 DOHC Turbo VIN A (M/T) 60 Pin Connector

PCM Pin #	Wire Color	Circuit Description (60 Pin)	Value at Hot Idle
1	DG/RD	MAP Sensor Signal	1.5-1.7v
2	TN/BK	ECT Sensor Signal	At 180°F: 2.80v
3	RD	Battery Power (Fused B+)	12-14v
4	BK/LB	Sensor Ground	<0.050v
5	BK/WT	Sensor Ground	<0.050v
6	PK/WT	5-Volt Supply	4.9-5.1v
7	OR	8-Volt Supply	7.9-8.1v
8	DG/BK	Auto Shutdown Relay Input	12-14v
9	DB	Ignition Switch Start Signal	12-14v
10	---	Not Used	---
11	BK/TN	Power Ground	<0.1v
12	BK/TN	Power Ground	<0.1v
13	LB/BR	Injector 4 Driver	1-4 ms
14	YL/WT	Injector 3 Driver	1-4 ms
15	TN	Injector 2 Driver	1-4 ms
16	WT/DB	Injector 1 Driver	1-4 ms
17	---	Not Used	---
18	RD/YL	Coil 1 Driver	5°, at 55 mph: 8° dwell
19	BK/GY	Coil 2 Driver	5°, at 55 mph: 8° dwell
20	DG	Generator Field Driver	Digital Signal: 0-12-0v
21	BK/RD	Air Temperature Sensor	At 100°F: 2.51v
22	OR/DB	TP Sensor Signal	0.6-1.0v
23	---	Not Used	---
24	GY/BK	CKP Sensor Signal	Digital Signal: 0-5-0v
25	PK	SCI Transmit	0v
26	PK/BR	CCD Bus (+)	Digital Signal: 0-5-0v
27	BR	A/C Clutch Signal	A/C On: 1v, Off: 12v
28	---	Not Used	---
29	WT/PK	Brake Switch Signal	Brake Off: 0v, On: 12v
30	BR/LB	PNP Switch Signal	In P/N: 0v, Others: 5v
31	DB/PK	Radiator Fan Relay Control	Relay Off: 12v, On: 1v
32	BK/PK	MIL (lamp) Control	Lamp On: 1v, Off: 12v
33	TN/RD	S/C Vacuum Solenoid	Vacuum Increasing: 1v
34	DB/OR	A/C Clutch Relay Control	Relay Off: 12v, On: 1v
35	---	Not Used	---
36	LG	Wastegate Solenoid Control	Solenoid Off: 12v, On: 1v
37-38	---	Not Used	---
39	GY/RD	AIS 3 Motor Control	DC pulse signals: 0.8-11v
40	BR	AIS 1 Motor Control	DC pulse signals: 0.8-11v

Standard Colors and Abbreviations

Abbreviation	Color	Abbreviation	Color	Abbreviation	Color
BK	Black	GY	Gray	RD	Red
BL	Blue	GN	Green	TN	Tan
BR	Brown	LG	Light Green	VT	Violet
DB	Dark Blue	OR	Orange	WT	White
DG	Dark Green	PK	Pink	YL	Yellow

1991 Spirit 2.2L I4 DOHC Turbo VIN A (M/T) 60 Pin Connector

PCM Pin #	Wire Color	Circuit Description (60 Pin)	Value at Hot Idle
41	BK/DG	HO2S-11 (B1 S1) Signal	0.1-1.1v
42	BK/LG	Knock Sensor Signal	0.080v AC
43	GY/LB	Tachometer Signal	Pulse Signals
44	TN/YL	CMP Sensor Signal	Digital Signal: 0-5-0v
45	LG	SCI Receive	0v
46	WT/BK	CCD Bus (-)	<0.050v
47	WT/OR	Vehicle Speed Signal	Digital Signal
48	BR/RD	S/C Set Switch Signal	S/C & Set Switch On: 3.8v
49	YL/RD	S/C On/Off Switch Signal	S/C On: 1v, Off: 12v
50	WT/LG	S/C Resume Switch	S/C On: 1v, Off: 12v
51	DB/YL	ASD Relay Control	Relay Off: 12v, On: 1v
52	PK	EVAP Purge Solenoid Control	Solenoid Off: 12v, On: 1v
53	LG/RD	S/C Vent Solenoid	Vacuum Decreasing: 1v
54	---	Not Used	---
55	LB	BARO Read Solenoid	Solenoid Off: 12v, On: 1v
56-58	---	Not Used	
59	PK/BK	AIS 4 Motor Control	DC pulse signals: 0.8-11v
60	YL/BK	AIS 2 Motor Control	DC pulse signals: 0.8-11v

Pin Connector Graphic

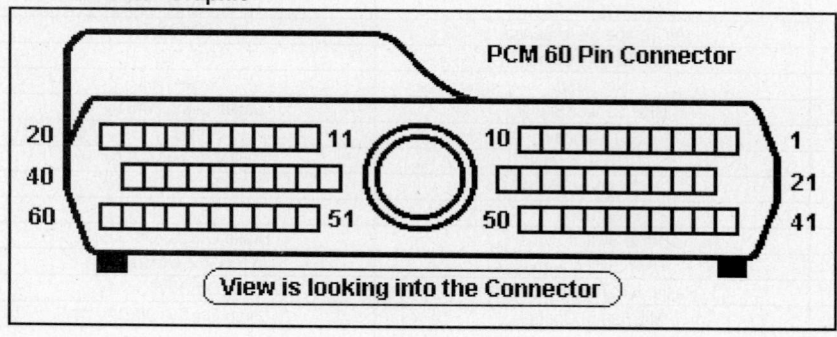

1992 Spirit 2.2L I4 DOHC Turbo VIN A (M/T) 60 Pin Connector

PCM Pin #	Wire Color	Circuit Description (60 Pin)	Value at Hot Idle
1	DG/RD	MAP Sensor Signal	1.5-1.7v
2	TN/BK	ECT Sensor Signal	At 180°F: 2.80v
3	RD/WT	Battery Power (Fused B+)	12-14v
4	BK/LB	Sensor Ground	<0.050v
5	BK/WT	Sensor Ground	<0.050v
6	PK/WT	5-Volt Supply	4.9-5.1v
7	OR	8-Volt Supply	7.9-8.1v
8	---	Not Used	---
9	DB	Ignition Switch Start Signal	12-14v
10	---	Not Used	---
11	BK/TN	Power Ground	<0.1v
12	BK/TN	Power Ground	<0.1v
13	LB/BR	Injector 4 Driver	1-4 ms
14	YL/WT	Injector 3 Driver	1-4 ms
15	TN	Injector 2 Driver	1-4 ms
16	WT/DB	Injector 1 Driver	1-4 ms
17	DB/YL	Coil 1 Driver	5°, at 55 mph: 8° dwell
18	---	Not Used	---
19	BK/GY	Coil 2 Driver	5°, at 55 mph: 8° dwell
20	DG	Generator Field Driver	Digital Signal: 0-12-0v
21	BK/RD	Air Temperature Sensor	At 100°F: 2.51v
22	OR/DB	TP Sensor Signal	0.6-1.0v
23	RD/LG	S/C Set Switch Signal	S/C & Set Switch On: 3.8V
24	GY/BK	CKP Sensor Signal	Digital Signal: 0-5-0v
25	PK	SCI Transmit	0v
26	PK/BR	CCD Bus (+)	Digital Signal: 0-5-0v
27	BR	A/C Clutch Signal	A/C On: 1v, Off: 12v
28	---	Not Used	---
29	WT/PK	Brake Switch Signal	Brake Off: 0v, On: 12v
30	BR/YL	PNP Switch Signal	In P/N: 0v, Others: 5v
31	DB/PK	Radiator Fan Relay Control	Relay Off: 12v, On: 1v
32	BK/PK	MIL (lamp) Control	Lamp On: 1v, Off: 12v
33	TN/RD	S/C Vacuum Solenoid	Vacuum Increasing: 1v
34	DB/OR	A/C Clutch Relay Control	Relay Off: 12v, On: 1v
35	---	Not Used	---
36	LG	Wastegate Solenoid Control	Solenoid Off: 12v, On: 1v
37-38	---	Not Used	---
39	GY/RD	AIS 3 Motor Control	DC pulse signals: 0.8-11v
40	BR/WT	AIS 1 Motor Control	DC pulse signals: 0.8-11v

Standard Colors and Abbreviations

Abbreviation	Color	Abbreviation	Color	Abbreviation	Color
BK	Black	GY	Gray	RD	Red
BL	Blue	GN	Green	TN	Tan
BR	Brown	LG	Light Green	VT	Violet
DB	Dark Blue	OR	Orange	WT	White
DG	Dark Green	PK	Pink	YL	Yellow

1992 Spirit 2.2L I4 DOHC Turbo VIN A (M/T) 60 Pin Connector

PCM Pin #	Wire Color	Circuit Description (60 Pin)	Value at Hot Idle
41	BK/DG	HO2S-11 (B1 S1) Signal	0.1-1.1v
42	BK/LG	Knock Sensor Signal	0.080v AC
43	GY/LB	Tachometer Signal	Pulse Signals
44	TN/YL	CMP Sensor Signal	Digital Signal: 0-5-0v
45	LG	SCI Receive	5v
46	WT/BK	CCD Bus (-)	<0.050v
47	WT/OR	Vehicle Speed Signal	Digital Signal
48-50	---	Not Used	---
51	DB/YL	ASD Relay Control	Relay Off: 12v, On: 1v
52	PK/BK	EVAP Purge Solenoid Control	Solenoid Off: 12v, On: 1v
53	LG/RD	S/C Vent Solenoid	Vacuum Decreasing: 1v
54	---	Not Used	---
55	LB	BARO Read Solenoid	Solenoid Off: 12v, On: 1v
56	---	Not Used	---
57	DG/OR	Generator Field Source	Digital Signal: 0-12-0v
58	---	Not Used	---
59	PK/BK	AIS 4 Motor Control	DC pulse signals: 0.8-11v
60	YL/BK	AIS 2 Motor Control	DC pulse signals: 0.8-11v

Pin Connector Graphic

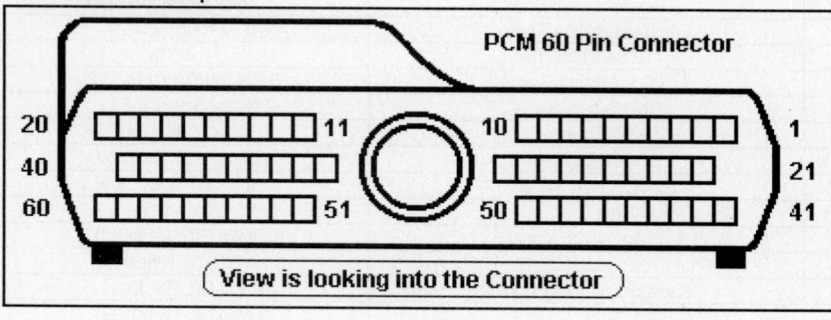

PCM 60 Pin Connector

View is looking into the Connector

1990-92 Spirit 2.5L I4 SOHC Turbo VIN J (All) 60 Pin Connector

PCM Pin #	Wire Color	Circuit Description (60 Pin)	Value at Hot Idle
1	DG/RD	MAP Sensor Signal	1.5-1.7v
2	TN/BK	ECT Sensor Signal	At 180°F: 2.80v
3	RD	Battery Power (Fused B+)	12-14v
4	BK/LB	Sensor Ground	<0.050v
5	BK/WT	Sensor Ground	<0.050v
6	PK/WT	5-Volt Supply	4.9-5.1v
7	OR	8-Volt Supply	7.9-8.1v
8	WT or DG/BK	Voltage Sense	12-14v
9	DB	Ignition Switch Output	12-14v
10	---	Not Used	---
11	BK/TN	Power Ground	<0.1v
12	BK/TN	Power Ground	<0.1v
13	LB/BR	Injector 4 Driver	1-4 ms
14	YL/WT	Injector 3 Driver	1-4 ms
15	TN	Injector 2 Driver	1-4 ms
16	WT/DB	Injector 1 Driver	1-4 ms
17-18	---	Not Used	---
19	BK/GY	Coil 1 Driver	5°, at 55 mph: 8° dwell
20	DG	Generator Field Driver	Digital Signal: 0-12-0v
21	---	Not Used	---
22	OR/DB	TP Sensor Signal	0.6-1.0v
23	RD/LG	S/C Set Switch Signal	S/C & Set Switch On: 3.8V
24	GY/BK	Distributor Reference Signal	Digital Signal: 0-5-0v
25	PK	SCI Transmit	0v
26	PK/BR	CCD Bus (+)	Digital Signal: 0-5-0v
27	BR	A/C Damped Pressure Switch	A/C On: 1v, Off: 12v
28	---	Not Used	---
29	WT/PK	Brake Switch Signal	Brake Off: 0v, On: 12v
30	BR/LB or BR/YL	PNP Switch Signal	In P/N: 0v, Others: 5v
31	DB/PK	Radiator Fan Relay Control	Relay Off: 12v, On: 1v
32	BK/PK	MIL (lamp) Control	Lamp On: 1v, Off: 12v
33	TN/RD	S/C Vacuum Solenoid	Vacuum Increasing: 1v
34	DB/OR	A/C Clutch Relay Control	Relay Off: 12v, On: 1v
35	GY/YL	EGR Solenoid Control (California)	12v, 55 mph: 1v
36	LG/BK	Wastegate Solenoid Control	Solenoid Off: 12v, On: 1v
37-38	---	Not Used	---
39	GY/RD	AIS Motor Control	DC pulse signals: 0.8-11v
40	BR/WT	AIS Motor Control	DC pulse signals: 0.8-11v

Standard Colors and Abbreviations

Abbreviation	Color	Abbreviation	Color	Abbreviation	Color
BK	Black	GY	Gray	RD	Red
BL	Blue	GN	Green	TN	Tan
BR	Brown	LG	Light Green	VT	Violet
DB	Dark Blue	OR	Orange	WT	White
DG	Dark Green	PK	Pink	YL	Yellow

1990-92 Spirit 2.5L I4 SOHC Turbo VIN J (All) 60 Pin Connector

PCM Pin #	Wire Color	Circuit Description (60 Pin)	Value at Hot Idle
41	BK/DG	HO2S-11 (B1 S1) Signal	0.1-1.1v
42	BK/LG	Knock Sensor Signal	0.080v AC
43	GY/LB	Tachometer Signal	Pulse Signals
44	TN/YL	Distributor Sync Signal	Digital Signal: 0-5-0v
45	PK	SCI Receive	0v
46	WT/BK	CCD Bus (-)	<0.050v
47	WT/OR	Vehicle Speed Signal	Digital Signal
48 ('90)	BR/RD	S/C Set Switch Signal	S/C & Set Switch On: 3.8v
49 ('90)	YL/RD	S/C On/Off Switch Signal	S/C On: 1v, Off: 12v
50 ('90)	WT/LG	S/C Resume Switch	S/C On: 1v, Off: 12v
51	DB/YL	ASD Relay Control	Relay Off: 12v, On: 1v
52	PK/BK	EVAP Purge Solenoid Control	Solenoid Off: 12v, On: 1v
53	LG/RD	S/C Vent Solenoid	Vacuum Increasing: 1v
54	LG/WT	Turbo Solenoid Control	Solenoid Off: 12v, On: 1v
55	LB	BARO Read Solenoid	Solenoid Off: 12v, On: 1v
56	---	Not Used	---
57 ('91-'92)	DG/OR	Generator Field Source	12-14v
58	---	Not Used	---
59	PK/BK	AIS Motor Control	DC pulse signals: 0.8-11v
60	YL/BK	AIS Motor Control	DC pulse signals: 0.8-11v

Pin Connector Graphic

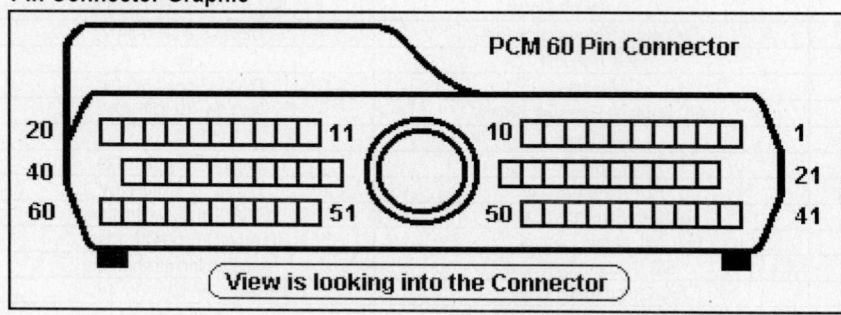

1990-92 Spirit 2.5L I4 SOHC TBI VIN K (All) 60 Pin Connector

PCM Pin #	Wire Color	Circuit Description (60 Pin)	Value at Hot Idle
1	DG/RD	MAP Sensor Signal	1.5-1.7v
2	TN/BK	ECT Sensor Signal	At 180°F: 2.80v
3 ('90)	RD/WT	Battery Power (Fused B+)	12-14v
3	RD	Battery Power (Fused B+)	12-14v
4	BK/LB	Sensor Ground	<0.050v
5 ('91)	BK	Sensor Ground	<0.050v
5	BK/WT	Sensor Ground	<0.050v
6 ('91)	PK	5-Volt Supply (MAP Sensor)	4.9-5.1v
6	PK/WT	5-Volt Supply (MAP Sensor)	4.9-5.1v
7	OR	8-Volt Supply	7.9-8.1v
8	WT	Ignition Switch Start Signal	12-14v
9	DB	Ignition Switch Output	12-14v
10, 13-15	---	Not Used	---
11	BK/TN	Power Ground	<0.1v
12	BK/TN	Power Ground	<0.1v
16	WT/DB	Injector 1 Driver	1-4 ms
17-18	---	Not Used	---
19	BK/GY	Coil Driver	5°, at 55 mph: 8° dwell
20	DG	Generator Field Driver	Digital Signal: 0-12-0v
21 ('90)	BK/RD	Throttle Body Temperature	At 100°F: 2.51v
22	OR/DB	TP Sensor Signal	0.6-1.0v
23 ('91-'92)	RD/LG	S/C Set Switch Signal	S/C & Set Switch On: 3.8V
24	GY/BK	Distributor Reference Signal	Digital Signal: 0-5-0v
25	PK	SCI Transmit	0v
26	PK/BR	CCD Bus (+)	Digital Signal: 0-5-0v
27	BR	A/C Clutch Signal	A/C On: 1v, Off: 12v
28	---	Not Used	---
29	WT/PK	Brake Switch Signal	Brake Off: 0v, On: 12v
30 ('90)	BR/LB	PNP Switch Signal	In P/N: 0v, Others: 5v
30	BR/YL	PNP Switch Signal	In P/N: 0v, Others: 5v
31	DB/PK	Radiator Fan Relay Control	Relay Off: 12v, On: 1v
32	BK/PK	MIL (lamp) Control	Lamp On: 1v, Off: 12v
33	TN/RD	S/C Vacuum Solenoid	Vacuum Increasing: 1v
34	DB/OR	A/C WOT Relay Control	Relay Off: 12v, On: 1v
35	GY/YL	EGR Solenoid Control (Calif.)	12v, 55 mph: 1v
36-38	---	Not Used	---
39	GY/RD	AIS Motor Control	DC pulse signals: 0.8-11v
40	BR	AIS Motor Control	DC pulse signals: 0.8-11v
40 ('91-92)	BR/WT	AIS Motor Control	DC pulse signals: 0.8-11v

Standard Colors and Abbreviations

Abbreviation	Color	Abbreviation	Color	Abbreviation	Color
BK	Black	GY	Gray	RD	Red
BL	Blue	GN	Green	TN	Tan
BR	Brown	LG	Light Green	VT	Violet
DB	Dark Blue	OR	Orange	WT	White
DG	Dark Green	PK	Pink	YL	Yellow

1990-92 Spirit 2.5L I4 SOHC TBI VIN K (All) 60 Pin Connector

PCM Pin #	Wire Color	Circuit Description (60 Pin)	Value at Hot Idle
41	BK/DG	HO2S-11 (B1 S1) Signal	0.1-1.1v
42	---	Not Used	---
43	GY/LB	Tachometer Signal	Pulse Signals
44	---	Not Used	---
45	LG	SCI Receive	0v
46	WT/BK	CCD Bus (-)	<0.050v
47	WT/OR	Vehicle Speed Signal	Digital Signal
48 ('90)	BR/RD	S/C Set Switch Signal	S/C On: 1v, Off: 12v
49 ('90)	YL/RD	S/C On/Off Switch Signal	S/C On: 1v, Off: 12v
50 ('90)	WT/LG	S/C Resume Switch	S/C On: 1v, Off: 12v
51	DB/YL	ASD Relay Control	Relay Off: 12v, On: 1v
52	PK	EVAP Purge Solenoid Control	Solenoid Off: 12v, On: 1v
52 ('90)	PK/BK	EVAP Purge Solenoid Control	Solenoid Off: 12v, On: 1v
53	LG/RD	S/C Vent Solenoid	Vacuum Decreasing: 1v
54	OR/BK	Part Throttle Unlock Solenoid	Solenoid Off: 12v, On: 1v
55-56	---	Not Used	---
57 ('91-'92)	DG/OR	Generator Field Source	12-14v
58	---	Not Used	---
59	PK/BK	AIS Motor Control	DC pulse signals: 0.8-11v
60	YL/BK	AIS Motor Control	DC pulse signals: 0.8-11v

Pin Connector Graphic

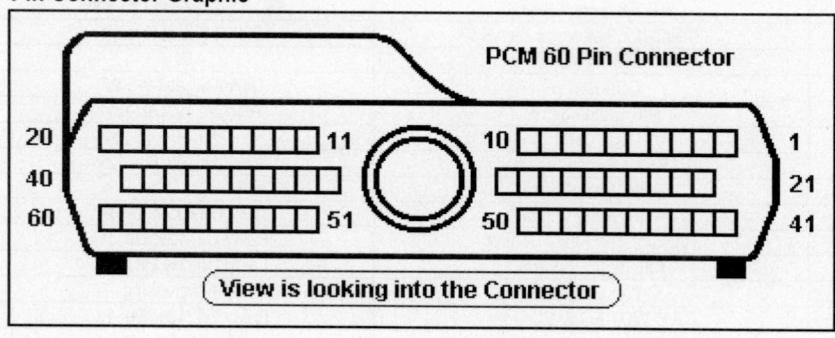

PCM 60 Pin Connector

View is looking into the Connector

1993-95 Spirit 2.5L I4 SOHC TBI VIN K (All) 60 Pin Connector

PCM Pin #	Wire Color	Circuit Description (60 Pin)	Value at Hot Idle
1	DG/RD	MAP Sensor Signal	1.5-1.7v
2	TN/BK	ECT Sensor Signal	At 180°F: 2.80v
3	RD/WT	Battery Power (Fused B+)	12-14v
4	BK/LB	Sensor Ground	<0.050v
5	BK/WT	Sensor Ground	<0.050v
6	PK/WT	5-Volt Supply (MAP Sensor)	4.9-5.1v
7	OR	8-Volt Supply	7.9-8.1v
8	WT	Ignition Switch Start Signal	12-14v
9	DB	Ignition Switch Output	12-14v
10	---	Not Used	---
11	BK/TN	Power Ground	<0.1v
12	BK/TN	Power Ground	<0.1v
13-15	---	Not Used	---
16	WT/DB	Injector 1 Driver	1-4 ms
17-18	---	Not Used	---
19	BK/GY	Coil Driver	5°, at 55 mph: 8° dwell
20	DG	Generator Field Driver	Digital Signal: 0-12-0v
21	---	Not Used	---
22	OR/DB	TP Sensor Signal	0.6-1.0v
23	RD/LG	S/C Set Switch Signal	S/C & Set Switch On: 3.8V
24	GY/BK	Distributor Reference Signal	Digital Signal: 0-5-0v
25	PK	SCI Transmit	0v
26	PK/BR	CCD Bus (+)	Digital Signal: 0-5-0v
27	BR	A/C Damped Pressure Switch	A/C On: 1v, Off: 12v
28	---	Not Used	---
29	WT/PK	Brake Switch Signal	Brake Off: 0v, On: 12v
30	BR/YL	PNP Switch Signal	In P/N: 0v, Others: 5v
31	DB/PK	Radiator Fan Relay Control	Relay Off: 12v, On: 1v
32	BK/PK	MIL (lamp) Control	Lamp On: 1v, Off: 12v
33	TN/RD	S/C Vacuum Solenoid	Vacuum Increasing: 1v
34	DB/OR	A/C Clutch Relay Control	Relay Off: 12v, On: 1v
35	GY/YL	EGR Solenoid Control (Calif.)	12v, 55 mph: 1v
36-38	---	Not Used	---
39	GY/RD	AIS Motor Control	DC pulse signals: 0.8-11v
40	BR/WT	AIS Motor Control	DC pulse signals: 0.8-11v

Standard Colors and Abbreviations

Abbreviation	Color	Abbreviation	Color	Abbreviation	Color
BK	Black	GY	Gray	RD	Red
BL	Blue	GN	Green	TN	Tan
BR	Brown	LG	Light Green	VT	Violet
DB	Dark Blue	OR	Orange	WT	White
DG	Dark Green	PK	Pink	YL	Yellow

1993-95 Spirit 2.5L I4 SOHC TBI VIN K (All) 60 Pin Connector

PCM Pin #	Wire Color	Circuit Description (60 Pin)	Value at Hot Idle
41	BK/DG	HO2S-11 (B1 S1) Signal	0.1-1.1v
42	---	Not Used	---
43	GY/LB	Tachometer Signal	Pulse Signals
44	---	Not Used	---
45	LG	SCI Receive	0v
46	WT/BK	CCD Bus (-)	<0.050v
47	WT/OR	Vehicle Speed Signal	Digital Signal
48-50	---	Not Used	---
51	DB/YL	ASD Relay Control	Relay Off: 12v, On: 1v
52	PK/BK	EVAP Purge Solenoid Control	Solenoid Off: 12v, On: 1v
53	LG/RD	S/C Vent Solenoid	Vacuum Decreasing: 1v
54	OR/BK	EMCC Solenoid	Solenoid Off: 12v, On: 1v
55	---	Not Used	---
57	DG/OR	Generator Field Source	12-14v
58	---	Not Used	---
59	PK/BK	AIS Motor Control	DC pulse signals: 0.8-11v
60	YL/BK	AIS Motor Control	DC pulse signals: 0.8-11v

Pin Connector Graphic

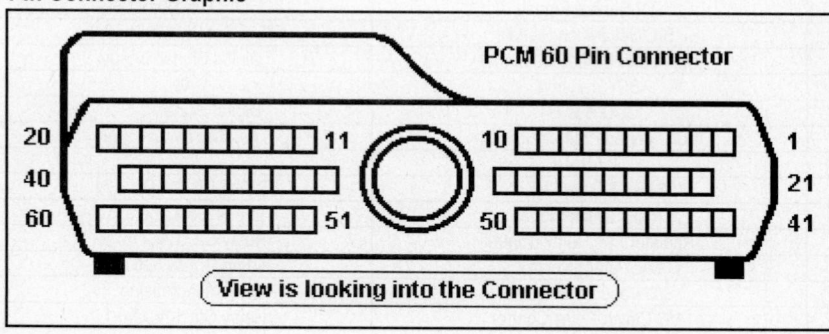

1994-95 Spirit 2.5L I4 Flexible Fuel VIN V (All) 60 Pin Connector

PCM Pin #	Wire Color	Circuit Description (60 Pin)	Value at Hot Idle
1	DG/RD	MAP Sensor Signal	1.5-1.7v
2	TN/BK	ECT Sensor Signal	At 180°F: 2.80v
3	RD/WT	Battery Power (Fused B+)	12-14v
4	BK/LB	Sensor Ground	<0.050v
5	BK/LB	Sensor Ground	<0.050v
6	PK/WT	5-Volt Supply (MAP Sensor)	4.9-5.1v
7	OR	8-Volt Supply	7.9-8.1v
8	WT	Ignition Switch Start Signal	12-14v
9	DB	Ignition Switch Output	12-14v
10	---	Not Used	---
11	BK/TN	Power Ground	<0.1v
12	BK/TN	Power Ground	<0.1v
13	LB/BR	Injector 4 Driver	1-4 ms
14	YL/WT	Injector 3 Driver	1-4 ms
15	TN	Injector 2 Driver	1-4 ms
16	WT/DB	Injector 1 Driver	1-4 ms
17-18	---	Not Used	---
19	BK/GY	Coil Driver	5°, at 55 mph: 8° dwell
20	DG	Generator Field Driver	Digital Signal: 0-12-0v
21	BK/RD	Flexible Fuel Sensor Signal	0.5-4.5v
22	OR/DB	TP Sensor Signal	0.6-1.0v
23	RD/LG	S/C Set Switch Signal	S/C & Set Switch On: 3.8V
24	GY/BK	Distributor Reference Signal	Digital Signal: 0-5-0v
25	PK	SCI Transmit	0v
26	PK/BR	CCD Bus (+)	Digital Signal: 0-5-0v
27	BR	A/C Damped Pressure Switch	A/C On: 1v, Off: 12v
28	---	Not Used	---
29	WT/PK	Brake Switch Signal	Brake Off: 0v, On: 12v
30	BR/YL	PNP Switch Signal	In P/N: 0v, Others: 5v
31	DB/PK	Radiator Fan Relay Control	Relay Off: 12v, On: 1v
32	BK/PK	MIL (lamp) Control	Lamp On: 1v, Off: 12v
33	TN/RD	S/C Vacuum Solenoid	Vacuum Increasing: 1v
34	DB/OR	A/C Clutch Relay Control	Relay Off: 12v, On: 1v
35	GY/YL	EGR Solenoid Control (Calif.)	12v, 55 mph: 1v
36-38	---	Not Used	---
39	GY/RD	AIS Motor Control	DC pulse signals: 0.8-11v
40	BR/WT	AIS Motor Control	DC pulse signals: 0.8-11v

Standard Colors and Abbreviations

Abbreviation	Color	Abbreviation	Color	Abbreviation	Color
BK	Black	GY	Gray	RD	Red
BL	Blue	GN	Green	TN	Tan
BR	Brown	LG	Light Green	VT	Violet
DB	Dark Blue	OR	Orange	WT	White
DG	Dark Green	PK	Pink	YL	Yellow

1994-95 Spirit 2.5L I4 Flexible Fuel VIN V (All) 60 Pin Connector

PCM Pin #	Wire Color	Circuit Description (60 Pin)	Value at Hot Idle
41	BK/DG	HO2S-11 (B1 S1) Signal	0.1-1.1v
42	BK/RD	Knock Sensor Signal	0.080v AC
43	GY/LB	Tachometer Signal	Pulse Signals
44	TN/YL	CMP Sensor Signal	Digital Signal: 0-5-0v
45	LG	SCI Receive	0v
46	WT/BK	CCD Bus (-)	<0.050v
47	WT/OR	Vehicle Speed Signal	Digital Signal
48-50	---	Not Used	---
51	DB/YL	ASD Relay Control	Relay Off: 12v, On: 1v
52	PK/BK	EVAP Purge Solenoid Control	Solenoid Off: 12v, On: 1v
53	LG/RD	S/C Vent Solenoid	Vacuum Decreasing: 1v
54	OR/BK	EMCC Solenoid	Solenoid Off: 12v, On: 1v
56	---	Not Used	---
57	DG/OR	Generator Field Source	12-14v
58	---	Not Used	---
59	PK/BK	AIS Motor Control	DC pulse signals: 0.8-11v
60	YL/BK	AIS Motor Control	DC pulse signals: 0.8-11v

Pin Connector Graphic

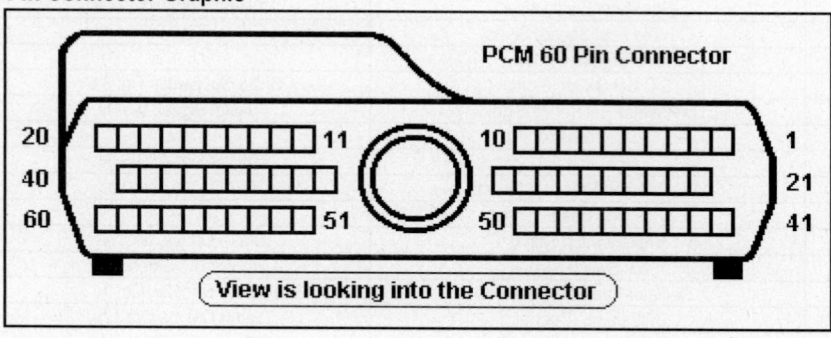

PCM 60 Pin Connector

View is looking into the Connector

1990-91 Spirit 3.0L V6 SOHC MFI VIN 3 (A/T) 60 Pin Connector

PCM Pin #	Wire Color	Circuit Description (60 Pin)	Value at Hot Idle
1	DG/RD	MAP Sensor Signal	1.5-1.7v
2	TN	ECT Sensor Signal	At 180°F: 2.80v
2 ('90-'91)	TN/BK	ECT Sensor Signal	At 180°F: 2.80v
3	RD	Battery Power (Fused B+)	12-14v
3 ('90-'91)	RD/WT	Battery Power (Fused B+)	12-14v
4	BK/LB	Sensor Ground	<0.050v
5	BK/WT	Sensor Ground	<0.050v
6	PK/WT	5-Volt Supply	4.9-5.1v
7	OR	8-Volt Supply	7.9-8.1v
8	DG/BK	ASD Relay Output	12-14v
9	DB	Ignition Switch Output	12-14v
10	---	Not Used	---
11	LB/RD	Power Ground	<0.1v
12	LB/RD	Power Ground	<0.1v
11 ('90-'91)	BK/TN	Power Ground	<0.1v
12 ('90-'91)	BK/TN	Power Ground	<0.1v
13	---	Not Used	---
14	YL/WT	Injector 3 Driver	1-4 ms
15	TN	Injector 2 Driver	1-4 ms
16	WT/DB	Injector 1 Driver	1-4 ms
17-18	---	Not Used	---
19	BK/GY	Coil Driver	5°, at 55 mph: 8° dwell
20	DG	Generator Field Driver	Digital Signal: 0-12-0v
21	---	Not Used	---
22	OR/DB	TP Sensor Signal	0.6-1.0v
23	---	Not Used	---
24	GY/BK	Distributor Reference Signal	Digital Signal: 0-5-0v
24 ('91)	GY/WT	Distributor Reference Signal	Digital Signal: 0-5-0v
25	PK	SCI Transmit	0v
26	BK/PK	CCD Bus (+)	Digital Signal: 0-5-0v
26 ('90-'91)	PK/BR	CCD Bus (+)	Digital Signal: 0-5-0v
27	BR	A/C Clutch Signal	A/C On: 1v, Off: 12v
28	---	Not Used	---
29	WT/PK	Brake Switch Signal	Brake Off: 0v, On: 12v
30	BR/YL	PNP Switch Signal	In P/N: 0v, Others: 5v
30 ('91)	BR/YL	PNP Switch Signal	In P/N: 0v, Others: 5v
31	DB/PK	Radiator Fan Relay Control	Relay Off: 12v, On: 1v
32	BK/PK	MIL (lamp) Control	Lamp On: 1v, Off: 12v
33	TN/RD	S/C Vacuum Solenoid	Vacuum Increasing: 1v
34	DB/OR	A/C WOT Relay Control	Relay Off: 12v, On: 1v
35	GY/YL	EGR Solenoid Control (Calif.)	12v, 55 mph: 1v
36-38	---	Not Used	---
39	GY/RD	AIS Motor Control	DC pulse signals: 0.8-11v
40	BR/WT	AIS Motor Control	DC pulse signals: 0.8-11v

1990-91 Spirit 3.0L V6 SOHC MFI VIN 3 (A/T) 60 Pin Connector

PCM Pin #	Wire Color	Circuit Description (60 Pin)	Value at Hot Idle
41	BK/DG	HO2S-11 (B1 S1) Signal	0.1-1.1v
42	---	Not Used	---
43	GY/LB	Tachometer Signal	Pulse Signals
44	TN/YL	Distributor Sync Signal	Digital Signal: 0-5-0v
45	LG	SCI Receive	0v
46	WT/BK	CCD Bus (-)	<0.050v
47	WT/OR	Vehicle Speed Signal	Digital Signal
48	BR/RD	S/C Set Switch Signal	S/C & Set Switch On: 3.8v
49	YL/RD	S/C On/Off Switch Signal	S/C On: 1v, Off: 12v
50	WT/LG	S/C Resume Switch	S/C On: 1v, Off: 12v
51	DB/YL	ASD Relay Control	Relay Off: 12v, On: 1v
52	PK	EVAP Purge Solenoid Control	Solenoid Off: 12v, On: 1v
52 ('91)	PK/WT	EVAP Purge Solenoid Control	Solenoid Off: 12v, On: 1v
53	LG/RD	S/C Vent Solenoid	Vacuum Decreasing: 1v
54-58	---	Not Used	
59	PK/BK	AIS Motor Control	DC pulse signals: 0.8-11v
60	YL/BK	AIS Motor Control	DC pulse signals: 0.8-11v

Pin Connector Graphic

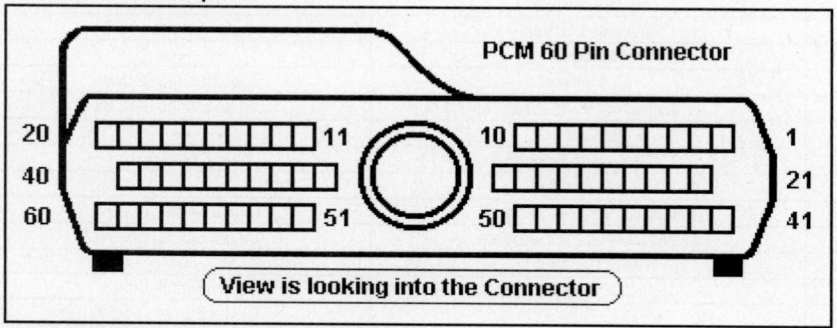

PCM 60 Pin Connector

View is looking into the Connector

Standard Colors and Abbreviations

Abbreviation	Color	Abbreviation	Color	Abbreviation	Color
BK	Black	GY	Gray	RD	Red
BL	Blue	GN	Green	TN	Tan
BR	Brown	LG	Light Green	VT	Violet
DB	Dark Blue	OR	Orange	WT	White
DG	Dark Green	PK	Pink	YL	Yellow

1992-95 Spirit 3.0L V6 SOHC MFI VIN 3 (A/T) 60 Pin Connector

PCM Pin #	Wire Color	Circuit Description (60 Pin)	Value at Hot Idle
1	DG/RD	MAP Sensor Signal	1.5-1.7v
2	TN/BK	ECT Sensor Signal	At 180°F: 2.80v
3	RD/WT	Battery Power (Fused B+)	12-14v
4	BK/LB	Sensor Ground	<0.050v
5	BK/LB	Sensor Ground	<0.050v
6	PK/WT	5-Volt Supply	4.9-5.1v
7	OR	8-Volt Supply	7.9-8.1v
8	---	Not Used	---
9	DB	ASD Relay Output	12-14v
10	---	Not Used	---
11-12	BK/TN	Power Ground	<0.1v
13	LB/BR	Injector 4 Driver	1-4 ms
14	YL/WT	Injector 3 Driver	1-4 ms
15	TN	Injector 2 Driver	1-4 ms
16	WT/DB	Injector 1 Driver	1-4 ms
17-18	---	Not Used	---
19	BK/GY	Coil 1 Driver	5°, at 55 mph: 8° dwell
20	DG	Generator Field Control	Digital Signal: 0-12-0v
21	---	Not Used	---
22	OR/DB	TP Sensor Signal	0.6-1.0v
23	RD/LG	S/C Set Switch Signal	S/C & Set Switch On: 3.8V
24	GY/WT	Distributor Reference Signal	Digital Signal: 0-5-0v
25	PK	SCI Transmit	0v
26	PK/BR	CCD Bus (+)	Digital Signal: 0-5-0v
27	BR	A/C Damped Pressure Switch	A/C On: 1v, Off: 12v
28	DB/OR	PSP Switch Signal	Straight: 0v, Turning: 5v
29	WT/PK	Brake Switch Signal	Brake Off: 0v, On: 12v
30	BR/YL	PNP Switch Signal	In P/N: 0v, Others: 5v
31	DB/PK	Radiator Fan Relay Control	Relay Off: 12v, On: 1v
32	BK/PK	MIL (lamp) Control	Lamp On: 1v, Off: 12v
33	TN/RD	S/C Vacuum Solenoid	Vacuum Increasing: 1v
34	DB/OR	A/C Clutch Relay Control	Relay Off: 12v, On: 1v
35	GY/YL	EGR Solenoid Control (Calif.)	12v, 55 mph: 1v
36-37	---	Not Used	---
38	GY	Injector 5 Driver	1-4 ms
39	GY/RD	AIS 3 Motor Control	DC pulse signals: 0.8-11v
40	BR/WT	AIS 1 Motor Control	DC pulse signals: 0.8-11v

1992-95 Spirit 3.0L V6 SOHC MFI VIN 3 (A/T) 60 Pin Connector

PCM Pin #	Wire Color	Circuit Description (60 Pin)	Value at Hot Idle
41	BK/DG	HO2S-11 (B1 S1) Signal	0.1-1.1v
42	---	Not Used	---
43	GY/LB	Tachometer Signal	Pulse Signals
44	TN/YL	Distributor Sync Signal	Digital Signal: 0-5-0v
45	LG	SCI Receive	0v
46	WT/BK	CCD Bus (-)	<0.050v
47	WT/OR	Vehicle Speed Signal	Digital Signal
51	DB/YL	ASD Relay Control	Relay Off: 12v, On: 1v
52	PK/BK	EVAP Purge Solenoid Control	Solenoid Off: 12v, On: 1v
53	LG/RD	S/C Vent Solenoid	Vacuum Decreasing: 1v
54	OR/BK	EMCC Solenoid	Solenoid Off: 12v, On: 1v
55-56	---	Not Used	---
57	DG/OR	Generator Field Source	12-14v
58	BR/DB	Injector 5 Driver	1-4 ms
59	PK/BK	AIS 4 Motor Control	DC pulse signals: 0.8-11v
60	YL/BK	AIS 2 Motor Control	DC pulse signals: 0.8-11v

Pin Connector Graphic

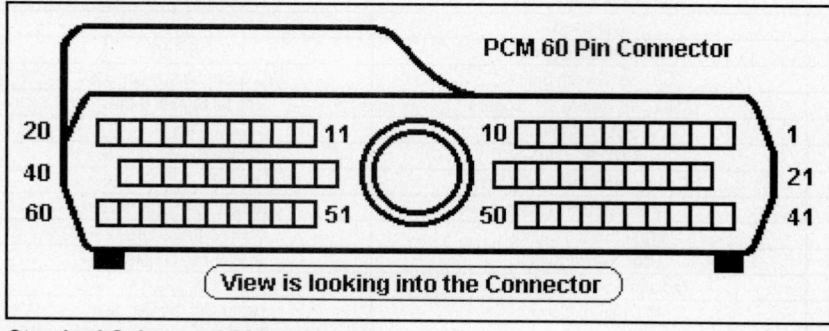

Standard Colors and Abbreviations

Abbreviation	Color	Abbreviation	Color	Abbreviation	Color
BK	Black	GY	Gray	RD	Red
BL	Blue	GN	Green	TN	Tan
BR	Brown	LG	Light Green	VT	Violet
DB	Dark Blue	OR	Orange	WT	White
DG	Dark Green	PK	Pink	YL	Yellow

1995-97 Stratus 2.0L I4 SOHC MFI VIN C (M/T) 'C1' Black Connector

PCM Pin #	Wire Color	Circuit Description (40 Pin)	Value at Hot Idle
1	---	Not Used	---
2	BK/GY	Coil 1 Driver	5°, at 55 mph: 8° dwell
3	DB/DG	Coil 2 Driver	5°, at 55 mph: 8° dwell
4	DG	Generator Field Control	Digital Signal: 0-12-0v
5	YL/PK	S/C Power Supply	12-14v
6	DG/OR	ASD Relay Output	12-14v
7	YL/WT	Injector 3 Driver	1-4 ms
8-9	---	Not Used	---
10	BK/TN	Power Ground	<0.1v
10 ('95)	BK/WT	Power Ground	<0.1v
11-12	---	Not Used	---
13	WT/LB	Injector 1 Driver	1-4 ms
14-15	---	Not Used	---
16	LB/BR	Injector 4 Driver	1-4 ms
17	TN	Injector 2 Driver	1-4 ms
18-19	---	Not Used	---
20	DB/WT	Ignition Switch Output	12-14v
21-23	---	Not Used	---
24	GY/BK	Knock Sensor Signal	0.080v AC
24 ('95)	DB/LG	Knock Sensor Signal	0.080v AC
25	---	Not Used	---
26	TN/BK	ECT Sensor Signal	At 180°F: 2.80v
27-29	---	Not Used	---
30	BK/DG	HO2S-11 (B1 S1) Signal	0.1-1.1v
31	---	Not Used	---
32	GY/BK	CKP Sensor Signal	Digital Signal: 0-5-0v
33	TN/YL	CMP Sensor Signal	Digital Signal: 0-5-0v
34	---	Not Used	---
35	OR/LB	TP Sensor Signal	0.6-1.0v
36	DG/RD	MAP Sensor Signal	1.5-1.7v
37	BK/RD	IAT Sensor Signal	At 100°F: 1.83v
38-39	---	Not Used	---
40	GY/YL	EGR Solenoid Control	12v, 55 mph: 1v

Standard Colors and Abbreviations

Abbreviation	Color	Abbreviation	Color	Abbreviation	Color
BK	Black	GY	Gray	RD	Red
BL	Blue	GN	Green	TN	Tan
BR	Brown	LG	Light Green	VT	Violet
DB	Dark Blue	OR	Orange	WT	White
DG	Dark Green	PK	Pink	YL	Yellow

1995-97 Stratus 2.0L I4 SOHC MFI VIN C (M/T) 'C2' White Connector

PCM Pin #	Wire Color	Circuit Description (40 Pin)	Value at Hot Idle
41	PK/LG	S/C Set Switch Signal	S/C & Set Switch On: 3.8V
42	DB/YL	A/C Damped Pressure Switch	A/C On: 1v, Off: 12v
43	BK/LB	Sensor Ground	<0.050v
44	OR/WT	8-Volt Supply	7.9-8.1v
45	DB/LG	PSP Switch Signal	Straight: 0v, Turning: 5v
46	RD/TN	Battery Power (Fused B+)	12-14v
47	---	Not Used	---
48	BR/GY	IAC 3 Driver	DC pulse signals: 0.8-11v
49	YL/BK	IAC 2 Driver	DC pulse signals: 0.8-11v
50	BK/TN	Power Ground	<0.1v
51	TN/WT	HO2S-12 (B1 S2) Signal	0.1-1.1v
52	PK/LG	Battery Temperature Sensor	At 86°F: 1.96v
53-54	---	Not Used	---
55	DB/TN	Low Speed Fan Relay	Relay Off: 12v, On: 1v
56	---	Not Used	---
57	GY/RD	IAC 1 Driver	DC pulse signals: 0.8-11v
58	PK/GY	IAC 4 Driver	DC pulse signals: 0.8-11v
59	PK/BR	CCD Bus (+)	Digital Signal: 0-5-0v
60	WT/BK	CCD Bus (-)	<0.050v
61	PK/WT	5-Volt Supply	4.9-5.1v
62	WT/RD	Brake Switch Signal	Brake Off: 0v, On: 12v
63	YL/DG	Torque Management Request	Digital Signals
64	DB/OR	A/C Clutch Relay Control	Relay Off: 12v, On: 1v
65	PK/LB	SCI Transmit	0v
66	WT/OR	Vehicle Speed Signal	Digital Signal
67	DB/PK	ASD Relay Control	Relay Off: 12v, On: 1v
68	PK/GY	EVAP Purge Solenoid Control	PWM signal: 0-12-0v
69	DB/PK	High Speed Fan Relay	Relay Off: 12v, On: 1v
72 ('96-'97)	OR/DG	LDP Switch Signal	Switch Closed: 0v, open: 12v
73	---	Not Used	---
74	BR/LG	Fuel Pump Relay Control	Relay Off: 12v, On: 1v
75	LG/WT	SCI Receive	0v
76	BR/WT	PNP Switch Signal	In P/N: 0v, Others: 5v
77 ('96-'97)	WT/DG	LDP Solenoid Control	PWM signal: 0-12-0v
78	WT/PK	S/C Vacuum Solenoid	Vacuum Increasing: 1v
80	LG/RD	S/C Vent Solenoid	Vacuum Decreasing: 1v

Pin Connector Graphic

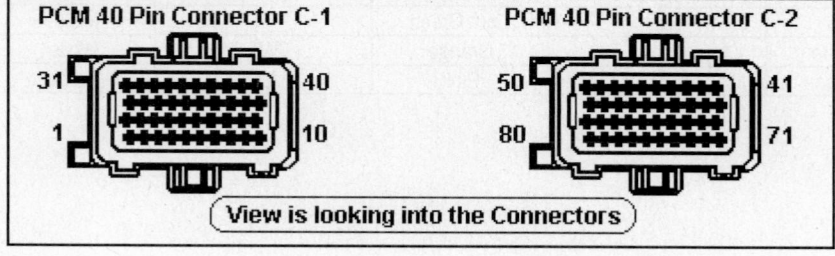

PCM 40 Pin Connector C-1

31 — 40
1 — 10

PCM 40 Pin Connector C-2

50 — 41
80 — 71

(View is looking into the Connectors)

1998-2000 Stratus 2.0L I4 SOHC MFI VIN C (M/T) 'C1' Black Connector

PCM Pin #	Wire Color	Circuit Description (40 Pin)	Value at Hot Idle
1-2	---	Not Used	---
3	DB/TN	Coil 2 Driver	5°, at 55 mph: 8° dwell
4	---	Not Used	---
5	YL/RD	S/C Power Supply	12-14v
6	DG/OR	ASD Relay Output	12-14v
7	YL/WT	Injector 3 Driver	1-4 ms
8	DG	Generator Field Control	Digital Signal: 0-12-0v
9	---	Not Used	---
10	BK/TN	Power Ground	<0.1v
11	BK/GY	Coil 1 Driver	5°, at 55 mph: 8° dwell
12	---	Not Used	---
13	WT/DB	Injector 1 Driver	1-4 ms
14-15	---	Not Used	---
16	LB/BR	Injector 4 Driver	1-4 ms
17	TN	Injector 2 Driver	1-4 ms
18-19	---	Not Used	---
20	DB/WT	Ignition Switch Output	12-14v
21-24	---	Not Used	---
25	DB/LG	Knock Sensor Signal	0.080v AC
26	TN/BK	ECT Sensor Signal	At 180°F: 2.80v
27	BK/OR	HO2S Signal Ground	<0.050v
28-29	---	Not Used	---
30	BK/DG	HO2S-11 (B1 S1) Signal	0.1-1.1v
31	TN	Starter Relay Control	KOEC: 9-11v
32	GY/BK	CKP Sensor Signal	Digital Signal: 0-5-0v
33	TN/YL	CMP Sensor Signal	Digital Signal: 0-5-0v
34	---	Not Used	---
35	OR/LB	TP Sensor Signal	0.6-1.0v
36	DG/RD	MAP Sensor Signal	1.5-1.7v
37	BK/RD	IAT Sensor Signal	At 100°F: 1.83v
38-39	---	Not Used	---
40	GY/YL	EGR Solenoid Control	12v, 55 mph: 1v

Standard Colors and Abbreviations

Abbreviation	Color	Abbreviation	Color	Abbreviation	Color
BK	Black	GY	Gray	RD	Red
BL	Blue	GN	Green	TN	Tan
BR	Brown	LG	Light Green	VT	Violet
DB	Dark Blue	OR	Orange	WT	White
DG	Dark Green	PK	Pink	YL	Yellow

1998-2000 Stratus 2.0L I4 SOHC MFI VIN C (M/T) 'C2' White Connector

PCM Pin #	Wire Color	Circuit Description (40 Pin)	Value at Hot Idle
41	RD/LG	S/C Set Switch Signal	S/C & Set Switch On: 3.8V
42	DB/YL	A/C Damped Pressure Switch	A/C On: 1v, Off: 12v
43	BK/LB	Sensor Ground	<0.050v
44	OR/WT	8-Volt Supply	7.9-8.1v
45	DB/LG	PSP Switch Signal	Straight: 0v, Turning: 5v
46	RD/TN	Battery Power (Fused B+)	12-14v
47	BK	Power Ground	<0.1v
48	BR/WT	IAC 3 Driver	DC pulse signals: 0.8-11v
49	YL/BK	IAC 2 Driver	DC pulse signals: 0.8-11v
50	BK/TN	Power Ground	<0.1v
51	TN/WT	HO2S-12 (B1 S2) Signal	0.1-1.1v
52	VT/LG	Battery Temperature Sensor	At 86°F: 1.96v
53-54, 71	---	Not Used	---
55	DB/TN	Low Speed Fan Relay	Relay Off: 12v, On: 1v
56	WT/VT	S/C Vacuum Solenoid	Vacuum Increasing: 1v
57	GY/RD	IAC 1 Driver	DC pulse signals: 0.8-11v
58	VT/BK	IAC 4 Driver	DC pulse signals: 0.8-11v
59	VT/BR	CCD Bus (+)	Digital Signal: 0-5-0v
60	WT/BK	CCD Bus (-)	<0.050v
61	VT/WT	5-Volt Supply	4.9-5.1v
62	WT/RD	Brake Switch Signal	Brake Off: 0v, On: 12v
63	YL/DG	Torque Management Request	Digital Signals
64	DB/OR	A/C Clutch Relay Control	Relay Off: 12v, On: 1v
65, 75	PK, LG	SCI Transmit, SCI Receive	0v
66	WT/OR	Vehicle Speed Signal	Digital Signal
67	DB/VT	ASD Relay Control	Relay Off: 12v, On: 1v
68	PK/GY	EVAP Purge Solenoid Control	PWM signal: 0-12-0v
69	DB/PK	High Speed Fan Relay	Relay Off: 12v, On: 1v
70	WT/TN	EVAP Purge Solenoid Sense	0-1v
72	OR/DG	LDP Switch Signal	Switch Closed: 0v, open: 12v
73, 78-79	---	Not Used	---
74	BR/LG	Fuel Pump Relay Control	Relay Off: 12v, On: 1v
76	BK/WT	PNP Switch Signal	In P/N: 0v, Others: 5v
77	WT/DG	LDP Solenoid Control	PWM signal: 0-12-0v
80	LG/RD	S/C Vent Solenoid	Vacuum Decreasing: 1v

Pin Connector Graphic

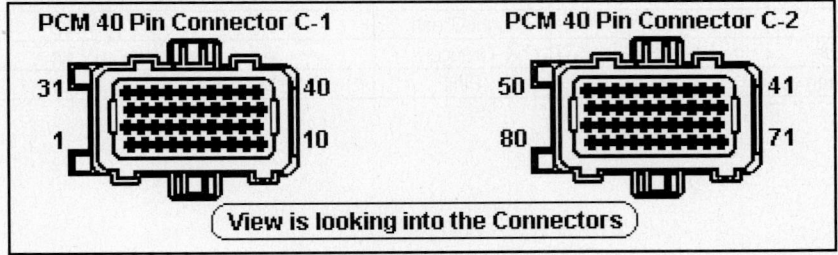

PCM 40 Pin Connector C-1

31   40

1   10

PCM 40 Pin Connector C-2

50   41

80   71

View is looking into the Connectors

1995-97 Stratus 2.4L I4 DOHC MFI VIN X (A/T) 'C1' Black Connector

PCM Pin #	Wire Color	Circuit Description (40 Pin)	Value at Hot Idle
1	---	Not Used	---
2	BK/GY	Coil 1 Driver	5º, at 55 mph: 8º dwell
3	DB/DG	Coil 2 Driver	5º, at 55 mph: 8º dwell
4	DG	Generator Field Control	Digital Signal: 0-12-0v
5	YL/PK	S/C Power Supply	12-14v
6	DG/OR	ASD Relay Output	12-14v
7	YL/WT	Injector 3 Driver	1-4 ms
8-9	---	Not Used	---
10	BK/TN	Power Ground	<0.1v
11-12	---	Not Used	---
13	WT/LB	Injector 1 Driver	1-4 ms
14-15	---	Not Used	---
16	LB/BR	Injector 4 Driver	1-4 ms
17	TN	Injector 2 Driver	1-4 ms
18-19	---	Not Used	---
20	DB/WT	Ignition Switch Output	12-14v
21-23	---	Not Used	---
24	GY/BK	Knock Sensor Signal	0.080v AC
24 ('95)	BK/LG	Knock Sensor Signal	0.080v AC
25	---	Not Used	---
26	TN/BK	ECT Sensor Signal	At 180ºF: 2.80v
27-29	---	Not Used	---
30	BK/DG	HO2S-11 (B1 S1) Signal	0.1-1.1v
31	---	Not Used	---
32	GY/BK	CKP Sensor Signal	Digital Signal: 0-5-0v
33	TN/YL	CMP Sensor Signal	Digital Signal: 0-5-0v
34	---	Not Used	---
35	OR/LB	TP Sensor Signal	0.6-1.0v
36	DG/RD	MAP Sensor Signal	1.5-1.7v
37	BK/RD	IAT Sensor Signal	At 100ºF: 1.83v
38-39	---	Not Used	---
40	GY/YL	EGR Solenoid Control	12v, 55 mph: 1v

Standard Colors and Abbreviations

Abbreviation	Color	Abbreviation	Color	Abbreviation	Color
BK	Black	GY	Gray	RD	Red
BL	Blue	GN	Green	TN	Tan
BR	Brown	LG	Light Green	VT	Violet
DB	Dark Blue	OR	Orange	WT	White
DG	Dark Green	PK	Pink	YL	Yellow

1995-97 Stratus 2.4L I4 DOHC MFI VIN X (A/T) 'C2' White Connector

PCM Pin #	Wire Color	Circuit Description (40 Pin)	Value at Hot Idle
41	PK/LG	S/C Set Switch Signal	S/C & Set Switch On: 3.8V
42	DB/YL	A/C Damped Pressure Switch	A/C On: 1v, Off: 12v
43	BK/LB	Sensor Ground	<0.050v
44	OR/WT	8-Volt Supply	7.9-8.1v
45	DB/LG	PSP Switch Signal	Straight: 0v, Turning: 5v
46	RD/TN	Battery Power (Fused B+)	12-14v
47	---	Not Used	---
48	BR/GY	IAC 3 Driver	DC pulse signals: 0.8-11v
49	YL/BK	IAC 2 Driver	DC pulse signals: 0.8-11v
50	BK/TN	Power Ground	<0.1v
51	TN/WT	HO2S-12 (B1 S2) Signal	0.1-1.1v
52	PK/LG	Battery Temperature Sensor	At 86°F: 1.96v
53-54	---	Not Used	---
55	DB/TN	Low Speed Fan Relay	Relay Off: 12v, On: 1v
57	GY/RD	IAC 1 Driver	DC pulse signals: 0.8-11v
58	PK/GY	IAC 4 Driver	DC pulse signals: 0.8-11v
59	PK/BR	CCD Bus (+)	Digital Signal: 0-5-0v
60	WT/BK	CCD Bus (-)	<0.050v
61	PK/WT	5-Volt Supply	4.9-5.1v
62	WT/RD	Brake Switch Signal	Brake Off: 0v, On: 12v
63 ('96-'97)	YL/DG	Torque Management Request	Digital Signals
64	DB/OR	A/C Clutch Relay Control	Relay Off: 12v, On: 1v
65	PK/LB	SCI Transmit	0v
66	WT/OR	Vehicle Speed Signal	Digital Signal
67	DB/PK	ASD Relay Control	Relay Off: 12v, On: 1v
68	PK/GY	EVAP Purge Solenoid Control	PWM signal: 0-12-0v
69	DB/PK	High Speed Fan Relay	Relay Off: 12v, On: 1v
70-71, 79	---	Not Used	---
72 ('96-'97)	OR/DG	LDP Switch Signal	Switch Closed: 0v, open: 12v
74	BR/LG	Fuel Pump Relay Control	Relay Off: 12v, On: 1v
75	LG/WT	SCI Receive	0v
76	BR/WT	PNP Switch Signal	In P/N: 0v, Others: 5v
77 ('96-'97)	WT/DG	LDP Solenoid Control	PWM signal: 0-12-0v
78	WT/PK	S/C Vacuum Solenoid	Vacuum Increasing: 1v
80	LG/RD	S/C Vent Solenoid	Vacuum Decreasing: 1v

Pin Connector Graphic

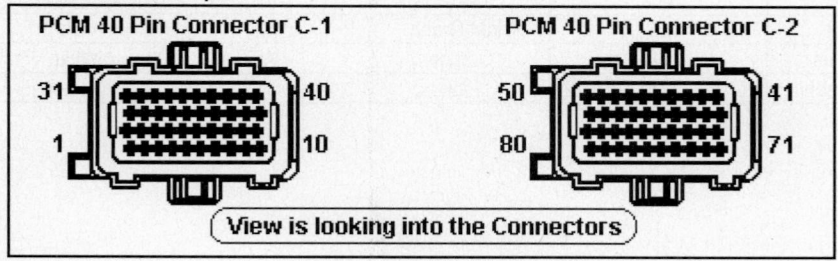

PCM 40 Pin Connector C-1   PCM 40 Pin Connector C-2

31 [...] 40   50 [...] 41
1 [...] 10   80 [...] 71

View is looking into the Connectors

1998-2000 Stratus 2.4L I4 DOHC MFI VIN X (A/T) 'C1' Black Connector

PCM Pin #	Wire Color	Circuit Description (40 Pin)	Value at Hot Idle
1-2	---	Not Used	---
3	DB/TN	Coil 2 Driver	5°, at 55 mph: 8° dwell
4	---	Not Used	---
5	YL/RD	S/C Power Supply	12-14v
6	DG/OR	ASD Relay Output	12-14v
7	YL/WT	Injector 3 Driver	1-4 ms
8	DG	Generator Field Control	Digital Signal: 0-12-0v
9	---	Not Used	---
10	BK/TN	Power Ground	<0.1v
11	BK/GY	Coil 1 Driver	5°, at 55 mph: 8° dwell
12	---	Not Used	---
13	WT/DB	Injector 1 Driver	1-4 ms
14-15	---	Not Used	---
16	LB/BR	Injector 4 Driver	1-4 ms
17	TN	Injector 2 Driver	1-4 ms
18-19	---	Not Used	---
20	DB/WT	Ignition Switch Output	12-14v
21-24	---	Not Used	---
25	DB/LG	Knock Sensor Signal	0.080v AC
26	TN/BK	ECT Sensor Signal	At 180°F: 2.80v
27	BK/OR	HO2S Signal Ground	<0.050v
28-29	---	Not Used	---
30	BK/DG	HO2S-11 (B1 S1) Signal	0.1-1.1v
31	---	Not Used	---
32	GY/BK	CKP Sensor Signal	Digital Signal: 0-5-0v
33	TN/YL	CMP Sensor Signal	Digital Signal: 0-5-0v
34	---	Not Used	---
35	OR/DB	TP Sensor Signal	0.6-1.0v
36	DG/RD	MAP Sensor Signal	1.5-1.7v
37	BK/RD	IAT Sensor Signal	At 100°F: 1.83v
38-39	---	Not Used	---
40	GY/YL	EGR Solenoid Control	12v, 55 mph: 1v

Standard Colors and Abbreviations

Abbreviation	Color	Abbreviation	Color	Abbreviation	Color
BK	Black	GY	Gray	RD	Red
BL	Blue	GN	Green	TN	Tan
BR	Brown	LG	Light Green	VT	Violet
DB	Dark Blue	OR	Orange	WT	White
DG	Dark Green	PK	Pink	YL	Yellow

1998-2000 Stratus 2.4L I4 DOHC MFI VIN X (A/T) 'C2' White Connector

PCM Pin #	Wire Color	Circuit Description (40 Pin)	Value at Hot Idle
41	RD/LG	S/C Set Switch Signal	S/C & Set Switch On: 3.8V
42	DB/YL	A/C Damped Pressure Switch	A/C On: 1v, Off: 12v
43	BK/LB	Sensor Ground	<0.050v
44	OR/WT	8-Volt Supply	7.9-8.1v
45	DB/LG	PSP Switch Signal	Straight: 0v, Turning: 5v
46	RD/TN	Battery Power (Fused B+)	12-14v
47	BK	Power Ground	<0.1v
48	BR/GY	IAC 3 Driver	DC pulse signals: 0.8-11v
49	YL/BK	IAC 2 Driver	DC pulse signals: 0.8-11v
50	BK/TN	Power Ground	<0.1v
51	TN/WT	HO2S-12 (B1 S2) Signal	0.1-1.1v
52	VT/LG	Battery Temperature Sensor	At 86°F: 1.96v
53-54	---	Not Used	---
55	DB/TN	Low Speed Fan Relay	Relay Off: 12v, On: 1v
56	WT/VT	S/C Vacuum Solenoid	Vacuum Increasing: 1v
57	GY/RD	IAC 1 Driver	DC pulse signals: 0.8-11v
58	VT/BK	IAC 4 Driver	DC pulse signals: 0.8-11v
59	VT/BR	CCD Bus (+)	Digital Signal: 0-5-0v
60	WT/BK	CCD Bus (-)	<0.050v
61	VT/WT	5-Volt Supply	4.9-5.1v
62	WT/RD	Brake Switch Signal	Brake Off: 0v, On: 12v
63	YL/DG	Torque Management Request	Digital Signals
64	DB/OR	A/C Clutch Relay Control	Relay Off: 12v, On: 1v
65	PK	SCI Transmit	0v
66	WT/OR	Vehicle Speed Signal	Digital Signal
67	DB/VT	ASD Relay Control	Relay Off: 12v, On: 1v
68	PK/GY	EVAP Purge Solenoid Control	PWM signal: 0-12-0v
69	DB/PK	High Speed Fan Relay	Relay Off: 12v, On: 1v
70	WT/TN	EVAP Purge Solenoid Sense	0-1v
71, 73	---	Not Used	---
72	OR/DG	LDP Switch Signal	Switch Closed: 0v, open: 12v
74	BR/LG	Fuel Pump Relay Control	Relay Off: 12v, On: 1v
75	LG	SCI Receive	5v
76	BK/WT	PNP Switch Signal	In P/N: 0v, Others: 5v
77	WT/LG	LDP Solenoid Control	PWM signal: 0-12-0v
78-79	---	Not Used	---
80	LG/RD	S/C Vent Solenoid	Vacuum Decreasing: 1v

Pin Connector Graphic

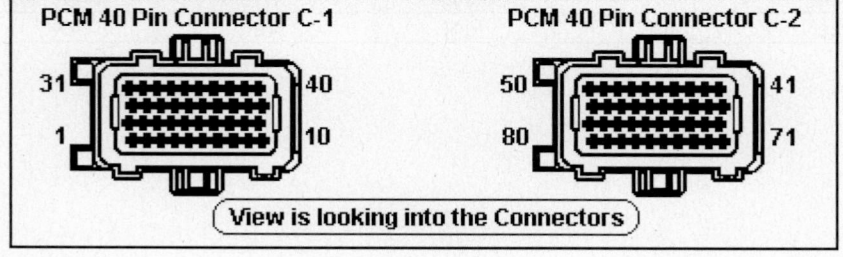

2001-02 Stratus Sedan 2.4L I4 DOHC MFI VIN X (A/T) 'C1' Black Connector

PCM Pin #	Wire Color	Circuit Description (40 Pin)	Value at Hot Idle
1-2	---	Not Used	---
3	DB/TN	Coil 2 Driver	5°, at 55 mph: 8° dwell
4	---	Not Used	---
5	YL/RD	S/C Power Supply	12-14v
6	DG/OR	ASD Relay Output	12-14v
7	YL/WT	Injector 3 Driver	1.0-4.0 ms
8	DG	Alternator Field Control	Digital Signal: 0-12-0v
9	---	Not Used	---
10	BK/TN	Power Ground	<0.1v
11	BK/GY	Coil 1 Driver	5°, at 55 mph: 8° dwell
12	---	Not Used	---
13	WT/DB	Injector 1 Driver	1.0-4.0 ms
14-15	---	Not Used	---
16	LB/BR	Injector 4 Driver	1.0-4.0 ms
17	TN	Injector 2 Driver	1.0-4.0 ms
18	OR/RD	HO2S-11 Heater Control	Heater On: 1v, Off: 12v
19	---	Not Used	---
20	DB/WT	Ignition Switch Output	12-14v
21-24	---	Not Used	---
25	DB/LG	Knock Sensor Signal	0.080v AC
26	TN/BK	ECT Sensor Signal	At 180°F: 2.80v
27	DB/OR	HO2S Ground	<0.050v
28-29	---	Not Used	---
30	BK/DG	HO2S-11 (B1 S1) Signal	0.1-1.1v
31	TN	Starter Relay Control	KOEC: 9-11v
32	GY/BK	CKP Sensor Signal	Digital Signal: 0-5-0v
33	TN/YL	CMP Sensor Signal	Digital Signal: 0-5-0v
34	---	Not Used	---
35	OR/DB	TP Sensor Signal	0.6-1.0v
36	DG/RD	MAP Sensor Signal	1.5-1.7v
37	BK/RD	IAT Sensor Signal	At 100°F: 1.83v
38-39	---	Not Used	---
40	GY/YL	EGR Solenoid Control	12v, at 55 mph: 1v

Standard Colors and Abbreviations

Abbreviation	Color	Abbreviation	Color	Abbreviation	Color
BK	Black	GY	Gray	RD	Red
BL	Blue	GN	Green	TN	Tan
BR	Brown	LG	Light Green	VT	Violet
DB	Dark Blue	OR	Orange	WT	White
DG	Dark Green	PK	Pink	YL	Yellow

2001-02 Stratus Sedan 2.4L I4 DOHC MFI VIN X (A/T) 'C2' White Connector

PCM Pin #	Wire Color	Circuit Description (40 Pin)	Value at Hot Idle
41	PK/LG	S/C Set Switch Signal	S/C & Set Switch On: 3.8v
42	DB	A/C Pressure Switch Signal	A/C On: 0.45-4.85v
43	BK/LB	Sensor Ground	<0.050v
44	OR/WT	8-Volt Supply	7.9-8.1v
45	DB/OR	PSP Switch Signal	Straight: 0v, Turning: 5v
46	RD/TN	Battery Power (Fused B+)	12-14v
47, 50	BK	Power Ground	<0.1v
48	BR/WT	IAC 3 Driver	DC pulse signals: 0.8-11v
49	YL/BK	IAC 2 Driver	DC pulse signals: 0.8-11v
51	TN/WT	HO2S-12 (B1 S2) Signal	0.1-1.1v
52	VT/LG	Ambient Temperature Sensor	At 86ºF: 1.96v
53-54, 60	---	Not Used	
55	DB/TN	Low Speed Fan Relay	Relay Off: 12v, On: 1v
56	TN/RD	S/C Vacuum Solenoid	Vacuum Increasing: 1v
57	GY/RD	IAC 1 Driver	DC pulse signals: 0.8-11v
58	VT/BK	IAC 4 Driver	DC pulse signals: 0.8-11v
59	OR	PCI Data Bus (J1850)	Digital Signals: 0-7-0v
61	VT/WT	5-Volt Supply	4.9-5.1v
62	WT/PK	Brake Switch Signal	Brake Off: 0v, On: 12v
63	YL/DG	Torque Management Request	Digital Signals
64	DB/OR	A/C Clutch Relay Control	Relay Off: 12v, On: 1v
65	PK	SCI Transmit	0v
66	WT/OR	Vehicle Speed Signal	Digital Signal
67	DB/VT	ASD Relay Control	Relay Off: 12v, On: 1v
68	PK/BK	EVAP Purge Solenoid Control	PWM signal: 0-12-0v
69	DB/PK	High Speed Fan Relay	Relay Off: 12v, On: 1v
70	WT/TN	EVAP Purge Solenoid Sense	0-1v
71	WT/RD	EATX RPM Signal	Digital Signals
72	OR	LDP Switch Sense	Open: 12v, Closed: 0v
73	---	Not Used	---
74	BR/LG	Fuel Pump Relay Control	Relay Off: 12v, On: 1v
75	LG	SCI Receive	5v
76	BK/LB	TRS T41 Sense	In P/N: 0v, Others: 5v
77	WT/DG	LDP Solenoid Control	PWM signal: 0-12-0v
78-79	---	Not Used	---
80	LG/RD	S/C Vent Solenoid	Vacuum Decreasing: 1v

Pin Connector Graphic

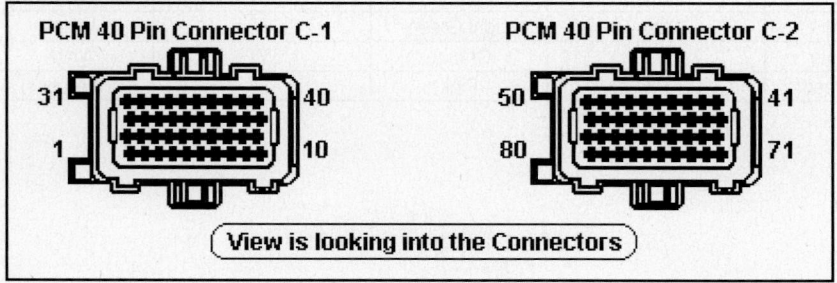

2003 Stratus 2.4L I4 DOHC SFI NGC VIN X 'C1' Black Connector

PCM Pin #	Wire Color	Circuit Description (38 Pin)	Value at Hot Idle
1-8, 10	---	Not Used	---
9	BK/TN	Power Ground	<0.1v
11	DB/WT	Ignition Switch Output (Run-Start)	12-14v
12	RD/WT	Ignition Switch Output (Off-Run-Start)	12-14v
13	WT/OR	Vehicle Speed Signal	Digital Signal
14-17	---	Not Used	---
18	BK/TN	Power Ground	<0.1v
19-20	---	Not Used	---
21	DB	A/C Pressure Switch Signal	A/C On: 0.45-4.85v
22-24	---	Not Used	---
25	LG	SCI Receive (PCM)	5v
26	PK/LB	SCI Receive (TCM)	5v
27-28	---	Not Used	---
29	RD/TN	Battery Power (Fused B+)	12-14v
30	YL	Ignition Switch Output (tart)	8-11v (cranking)
31	TN/WT	HO2S-12 (B1 S2) Signal	0.1-1.1v
32	DB/LG	Oxygen Sensor Return (down)	<0.050v
33-35	---	Not Used	---
36	PK	SCI Transmit (PCM)	0v
37	WT/DG	SCI Transmit (TCM)	0v
38	YL/VT	PCI Data Bus (J1850)	Digital Signals: 0-7-0v

2003 Stratus 2.4L I4 DOHC SFI NGC VIN X 'C2' Gray Connector

PCM Pin #	Wire Color	Circuit Description (38 Pin)	Value at Hot Idle
1-8	---	Not Used	---
9	DB/TN	Coil 2 Driver	5°, at 55 mph: 8° dwell
10	BK/GY	Coil 1 Driver	5°, at 55 mph: 8° dwell
11	LB/BR	Injector 4 Driver	1.0-4.0 ms
12	YL/WT	Injector 3 Driver	1.0-4.0 ms
13	TN	Injector 2 Driver	1.0-4.0 ms
14	WT/DB	Injector 1 Driver	1.0-4.0 ms
15-16	---	Not Used	---
17	BR/VT	HO2S-21 (B2 S1) Heater Control	Heater On: 1v, Off: 12v
18	BR/OR	HO2S-11 (B1 S1) Heater Control	Heater On: 1v, Off: 12v
19	DG	Generator Field Driver	Digital Signal: 0-12-0v
20	TN/BK	ECT Sensor Signal	At 180°F: 2.80v
21	OR/DB	TP Sensor Signal	0.6-1.0v
22, 25	---	Not Used	---
23	OR/RD	MAP Sensor Signal	1.5-1.7v
24	BK/VT	Knock Sensor Return	<0.050v
27	BK/LB	Sensor Ground	<0.1v
28	YL/BK	IAC Motor Sense	12-14v
29	VT/WT	5-Volt Supply	4.9-5.1v
30	BK/RD	IAT Sensor Signal	At 100°F: 1.83v
31	BK/DG	HO2S-11 (B1 S1) Signal	0.1-1.1v
32	BR/DG	HO2S-11 (B1 S1) Ground (Up)	<0.1v
33	TN/WT	HO2S-12 (B1 S2) Signal	0.1-1.1v
34	TN/YL	CMP Sensor Signal	Digital Signal: 0-5-0v
35	GY/BK	CKP Sensor Signal	Digital Signal: 0-5-0v
36-37	---	Not Used	---
38	GY/RD	IAC Motor Driver	DC pulse signals: 0.8-11v

Standard Colors and Abbreviations

Abbreviation	Color	Abbreviation	Color	Abbreviation	Color
BK	Black	GY	Gray	RD	Red
BL	Blue	GN	Green	TN	Tan
BR	Brown	LG	Light Green	VT	Violet
DB	Dark Blue	OR	Orange	WT	White
DG	Dark Green	PK	Pink	YL	Yellow

2003 Stratus 2.4L I4 DOHC SFI NGC VIN X 'C3' White Connector

PCM Pin #	Wire Color	Circuit Description (38 Pin)	Value at Hot Idle
1-2, 10	---	Not Used	---
3	DB/VT	ASD Relay Control	Relay Off: 12v, On: 1v
4	DB/PK	High Speed Radiator Fan Relay	Relay Off: 12v, On: 1v
5	LG/RD	S/C Vent Solenoid	Vacuum Decreasing: 1v
6	DB/TN	Low Speed Radiator Fan Relay	Relay Off: 12v, On: 1v
7	YL/RD	S/C Power Supply	12-14v
8	WT/DG	Natural Vacuum Leak Detection Solenoid	Solenoid Off: 12v, On: 1v
9	BR/VT	HO2S-12 (B1 S2) Heater Control	Heater On: 1v, Off: 12v
11	DB/OR	A/C Clutch Relay Control	Relay Off: 12v, On: 1v
12	TN/RD	S/C Vacuum Solenoid	Vacuum Increasing: 1v
13-16, 18	---	Not Used	---
17	LB	Sensor Ground No. 2	<0.050v
19	OR/DG	Automatic Shutdown Relay Output	12-14v
20	PK/BK	EVAP Purge Solenoid Control	PWM Signal: 0-12-0v
21-22, 24-25	---	Not Used	---
23	WT/PK	Brake Switch Sense	Brake Off: 0v, On: 12v
26	YL	Autostick Downshift Switch	Digital Signal: 0v or 12v
27	LG	Autostick Upshift Switch	Digital Signal: 0v or 12v
28	OR/DG	Automatic Shutdown Relay Output	12-14v
29	WT/TN	EVAP Purge Solenoid Return	0-1v
30	DB/LG	Power Steering Pressure Switch Signal	Straight: 0v, Turning: 5v
31, 33, 36	---	Not Used	---
32	VT/LG	Ambient Air Temperature Sensor	At 100°F: 1.83v
34	PK/LG	Speed Control Set Switch Signal	S/C & Set Switch On: 3.8v
35	OR	Natural Vacuum Leak Detection Switch Sense	0.1v
37	BR/LG	Fuel Pump Relay Control	Relay Off: 12v, On: 1v
38	TN	Starter Relay Control	KOEC: 9-11v

2003 Stratus 2.4L I4 DOHC SFI NGC VIN X 'C4' Green Connector

PCM Pin #	Wire Color	Circuit Description (38 Pin)	Value at Hot Idle
1	BR	Overdrive Solenoid Control	Solenoid Off: 12v, On: 1v
2	PK	Underdrive Solenoid Control	Solenoid Off: 12v, On: 1v
3-5, 7-9, 11	---	Not Used	---
6	WT	2-4 Solenoid Control	Solenoid Off: 12v, On: 1v
10	LB	Low/Reverse Solenoid Control	Solenoid Off: 12v, On: 1v
12	BK/YL	Power Ground	<0.050v
13, 14	BK/RD	Power Ground	<0.050v
15	LG/BK	TRS T1 Sense	<0.050v
16	VT	TRS T3 Sense	<0.050v
17, 20-21	---	Not Used	---
18, 19	LG, RD	Transmission Control Relay Output	Relay Off: 12v, On: 1v
23-26	---	Not Used	---
27	BK/WT	TRS T41 Sense	<0.050v
28, 38	RD	Transmission Control Relay Output	Relay Off: 12v, On: 1v
29	DG	Low/Reverse Pressure Switch Sense	12-14v
30	YL/WT	2-4 Pressure Switch Sense	In Low/Reverse: 2-4v
31, 36	---	Not Used	---
32	LG/WT	Output Speed Sensor Signal	In 2-4 Position: 2-4v
33	RD/BK	Input Speed Sensor Signal	Moving: AC voltage
34	DB/BK	Speed Sensor Ground	Moving: AC voltage
35	VT/YL	Transmission Temperature Sensor Signal	<0.050v
37	VT/WT	TRS T42 Sense	In PRNL: 0v, Others 5v

Pin Connector Graphic

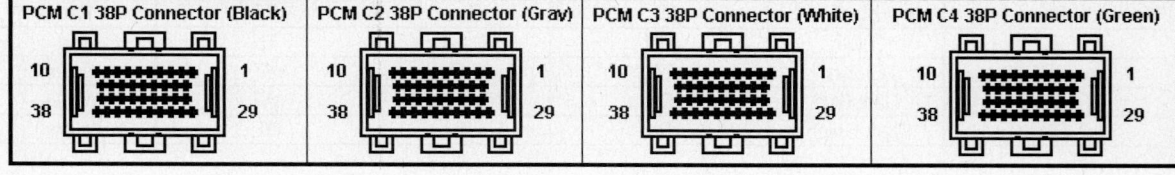

PCM C1 38P Connector (Black) PCM C2 38P Connector (Gray) PCM C3 38P Connector (White) PCM C4 38P Connector (Green)

2001-03 Stratus Coupe 2.4L SOHC VIN G (A/T) C110 Connector

PCM Pin #	Wire Color	Circuit Description (35 Pin)	Value at Hot Idle
1	YL/BL	Injector 1 Driver	1.0-4.0 ms
2	GN/YL	Injector 4 Driver	1.0-4.0 ms
3	BR/WT	HO2S-11 Heater Control	Digital Signal: 0-12-0v
4-5, 7	---	Not Used	---
6	BL/RD	EGR Solenoid Control	Solenoid Off: 12v, On: 1v
8	BK/RD	Generator Signal (lights on)	0.2-3.5v
9	YL/RD	Injector 2 Driver	1.0-4.0 ms
10	---	Not Used	---
11	BL/YL	Ignition Coil 1 Control	Digital Signal: 0-12-0v
12	WT/GN	Ignition Coil 2 Control	Digital Signal: 0-12-0v
12-13	---	Not Used	---
14	GN/BK	IAC Stepper Motor 'A' Signal	Pulse Signals
15	GN/WT	IAC Stepper Motor 'B' Signal	Pulse Signals
16-17	---	Not Used	---
18	BL/OR	Radiator Fan Control	Fan Off: 0.1v, On: 0.7v
19	GN/BL	Volume Airflow Sensor Reset	1-3v, 3000 rpm: 6-9v
20	GN	A/C Clutch Relay Control	Relay Off: 12v, On: 1v
21	BK	Fuel Pump Relay Control	Pump Off: 12v, On: 1v
22	RD/YL	MIL (lamp) Control	Lamp Off: 12v, On: 1v
23	---	Not Used	---
24	YL/BK	Injector 3 Driver	1.0-4.0 ms
25	---	Not Used	---
26	BL/WT	HO2S-12 Heater Control	Digital Signal: 0-12-0v
27	---	Not Used	---
28	GN/RD	IAC Stepper Motor 'A' Signal	Pulse Signals
29	BK/YL	IAC Stepper Motor 'B' Signal	Pulse Signals
30-33	---	Not Used	---
34	BL	EVAP Purge Solenoid Control	Solenoid Off: 12v, On: 1v
35	YL	EVAP Vent Solenoid Control	Solenoid Off: 12v, On: 1v

PCM C110 Wire Harness Connector Graphic

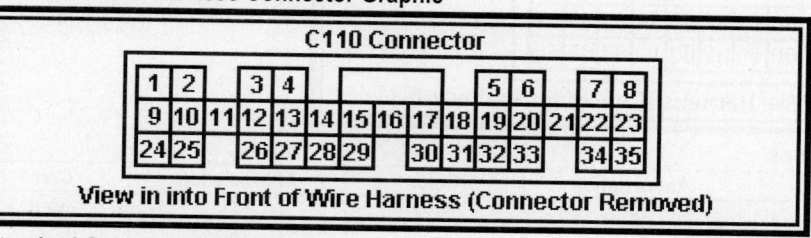

Standard Colors and Abbreviations

Abbreviation	Color	Abbreviation	Color	Abbreviation	Color
BK	Black	GY	Gray	RD	Red
BL	Blue	GN	Green	TN	Tan
BR	Brown	LG	Light Green	VT	Violet
DB	Dark Blue	OR	Orange	WT	White
DG	Dark Green	PK	Pink	YL	Yellow

2001-03 Stratus Coupe 2.4L I4 SOHC VIN G (A/T) C114 Connector

PCM Pin #	Wire Color	Circuit Description (26 Pin)	Value at Hot Idle
41	RD	Ignition Power (MPI Relay)	12-14v
42	BK	Power Ground	<0.1v
43	WT/RD	Tachometer Signal	Pulse Signals
44	YL/GN	ECT Sensor Signal	0.3-0.9v at 176°F
45	GN/RD	Distributor CKP Sensor Signal	Digital Signal: 0-5-0v
46	GN/YL	Sensor Voltage Reference	4.9-5.1v
47	RD	Ignition Power (MPI Relay)	12-14v
48	BK	Power Ground	<0.1v
49	WT/VT	MPI (Power) Relay Control	Relay Off: 12v, On: 1v
50	WT/BL	A/T Control Relay	Relay Off: 12v, Off: 1v
51	---	Not Used	---
52	YL	PSP Switch Signal	Straight: 5v, Turning: 0v
53	---	Not Used	---
54	YL/BK	Generator 'FR' Terminal	0.5-4.5v
55	GN/WT	BARO Sensor Signal	3.7-4.3 at Sea Level
56	BL/YL	CMP Sensor Signal	Digital Signal: 0-5-0v
57	BK	Sensor Ground	<0.050v
58	BK/RD	Park Neutral Switch	In P/N: 12v, Others: 0v
59	BR/RD	Starter (Cranking) Signal	9-11v (cranking)
60	---	Not Used	---
61	BL/WT	A/C Switch 2 Signal (Hi Blow)	12v
62-63	---	Not Used	---
64	RD/BL	IAT Sensor Signal	1.5-2.1v at 104°F
65	WT/GN	Volume Airflow Sensor	2.2-3.2v
66	OR/BL	Keep Alive Power	12-14v

PCM C114 Wire Harness Connector Graphic

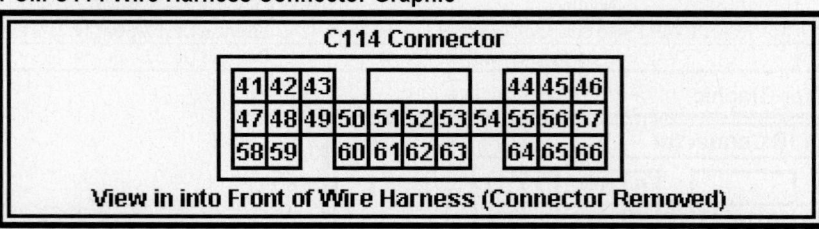

Standard Colors and Abbreviations

Abbreviation	Color	Abbreviation	Color	Abbreviation	Color
BK	Black	GY	Gray	RD	Red
BL	Blue	GN	Green	TN	Tan
BR	Brown	LG	Light Green	VT	Violet
DB	Dark Blue	OR	Orange	WT	White
DG	Dark Green	PK	Pink	YL	Yellow

2001-03 Stratus Coupe 2.4L I4 SOHC VIN G (A/T) C117 Connector

PCM Pin #	Wire Color	Circuit Description (28 Pin)	Value at Hot Idle
71	WT	HO2S-11 (B1 S1) Signal	0.1-1.1v
72	---	Not Used	---
73	GN	HO2S-12 (B1 S2) Signal	0.1-1.1v
74	---	Not Used	---
75	GY/BL	Cruise Control Switch Signal	N/A
76	BK	Power Ground	<0.1v
77	RD/BL	A/T: Control Relay Output	12-14v
78	YL	TP Sensor Signal	0.53-0.73v
79	YL/RD	Idle Position Switch Signal	0-1v, Switch Open: 4-5v
80	WT/BL	Vehicle Speed Signal	Digital Signal
81-82	---	Not Used	---
83	GN/RD	A/C Switch On/Off Signal	A/C Off: 0v, On: 12v
84	GY/RD	DLC Diagnosis Control (#1)	0v
85	RD/WT	DLC ISO 9141 Bus (#7)	12v (no Scan Tool)
86-87	---	Not Used	---
88	BK	Power Ground	<0.1v
89	RD/BL	A/T: Control Relay Output	12-14v
90	WT	Knock Sensor Signal	0.080v AC
91	BL/RD	MAP Sensor Signal	0.8-1.1v
92	BN/WT	FTP Sensor Signal	2.5v (fuel cap off)
93-96	---	Not Used	---
97	BK	Power Ground	<0.1v
98	BK/WT	Ignition Switch Power	12-14v

PCM C117 Wire Harness Connector Graphic

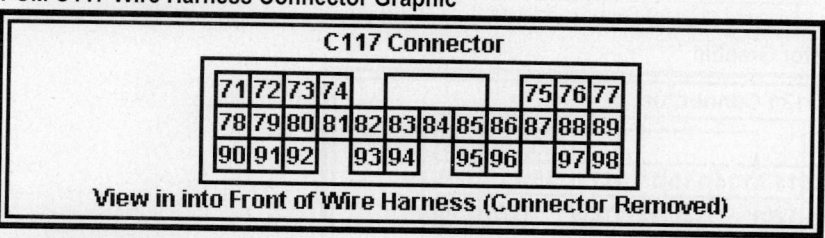

Standard Colors and Abbreviations

Abbreviation	Color	Abbreviation	Color	Abbreviation	Color
BK	Black	GY	Gray	RD	Red
BL	Blue	GN	Green	TN	Tan
BR	Brown	LG	Light Green	VT	Violet
DB	Dark Blue	OR	Orange	WT	White
DG	Dark Green	PK	Pink	YL	Yellow

2001-03 Stratus Coupe 2.4L I4 SOHC VIN G (A/T) C121 Connector

PCM Pin #	Wire Color	Circuit Description (30 Pin)	Value at Hot Idle
101	BK/BL	PNP Switch 'P' Signal	In Park: 12v, Or 0v
102	YL	PNP Switch 'D' Signal	In Drive: 12v, Or 0v
103	WT	Input Shaft Speed Sensor	Moving: 0-12-0v
104	GN/YL	Output Shaft Speed Sensor	Moving: 0-12-0v
105	BR/RD	A/T Shift Mode 1st Indicator	0v
106	RD/YL	A/T Second Solenoid	12-14v
107	YL/RD	A/T TCC Solenoid	12-14v
108	RD/BL	PNP Switch 'R' Signal	In Reverse: 12v, Or: 0v
109	WT	PNP Switch 'D3' Signal	In Drive 3: 12v, Or 0v
110	GY	PNP Switch 'Low' Signal	In Low: 12v, Or 0v
111	WT/GN	Immobilizer System Signal	Digital Signals
112	---	Not Used	---
113	WT/VT	DLC No. 2 Signal	N/A
114	WT/RD	DLC No. 2 Signal	N/A
115-116	---	Not Used	---
117	WT/BL	A/T Shift Mode 3rd Indicator	0v
118	YL/BL	A/T Shift Mode 2nd Indicator	0v
120	RD	A/T Underdrive Solenoid	12-14v
121	BR	PNP Switch 'N' Signal	In Neutral: 12v, Or 0v
122	YL/BL	PNP Switch 'D2' Signal	In Drive 2: 12v, Or 0v
123	GN/OR	Stop Light Switch Signal	Brake Off: 0v, On: 12v
124	BK/PK	TFT Sensor Signal	3.2-3.4v at 104°F
128	YL/BL	A/T Shift Mode 4th Indicator	0v
129	RD/WT	A/T Low/Reverse Solenoid	12-14v
130	BL	A/T Overdrive Solenoid	12-14v

PCM C121 Wire Harness Connector Graphic

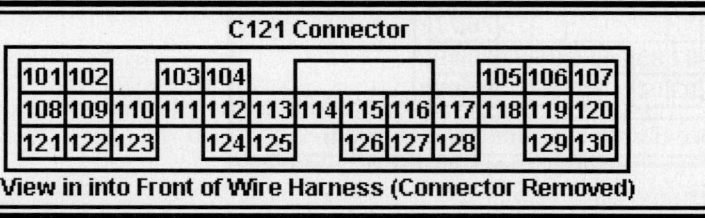

Standard Colors and Abbreviations

Abbreviation	Color	Abbreviation	Color	Abbreviation	Color
BK	Black	GY	Gray	RD	Red
BL	Blue	GN	Green	TN	Tan
BR	Brown	LG	Light Green	VT	Violet
DB	Dark Blue	OR	Orange	WT	White
DG	Dark Green	PK	Pink	YL	Yellow

2001-03 Stratus Coupe 2.4L I4 SOHC VIN G (M/T) C109 Connector

PCM Pin #	Wire Color	Circuit Description (26 Pin)	Value at Hot Idle
1	YL/BL	Injector 1 Driver	1.0-4.0 ms
2	YL/BK	Injector 3 Driver	1.0-4.0 ms
3	---	Not Used	---
4	GN/BK	IAC Stepper Motor 'A' Signal	Pulse Signals
5	GN/WT	IAC Stepper Motor 'B' Signal	Pulse Signals
6	BL/RD	EGR Solenoid Control	Solenoid Off: 12v, On: 1v
7	---	Not Used	---
8	GN	A/C Clutch Relay Control	Relay Off: 12v, On: 1v
9	BL	EVAP Purge Solenoid Control	Solenoid Off: 12v, On: 1v
10	BK/BL	Ignition Coil 1 Control	Digital Signal: 0-12-0v
11	---	Not Used	---
12	RD	Ignition Power (MPI Relay)	12-14v
13	BK	Power Ground	<0.1v
14	YL/BL	Injector 2 Driver	1.0-4.0 ms
15	GN/YL	Injector 4 Driver	1.0-4.0 ms
16	---	Not Used	---
17	GN/RD	IAC Stepper Motor 'A' Signal	Pulse Signals
18	BK/YL	IAC Stepper Motor 'B' Signal	Pulse Signals
19	GN/BL	Volume Airflow Sensor Reset	1-3v, 3000 rpm: 6-9v
20	---	Not Used	---
21	BL/OR	Radiator Cooling Fan Control	Fan Off: 0.1v, On: 0.7v
22	BK	Fuel Pump Relay Control	Pump Off: 12v, On: 1v
23	WT/GN	Ignition Coil 2 Control	Digital Signal: 0-12-0v
24	---	Not Used	---
25	RD	Ignition Power (MPI Relay)	12-14v
26	BK	Power Ground	<0.1v

PCM C109 Wire Harness Connector Graphic

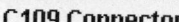

C109 Connector

1	2	3	4	5	6	7	8	9	10	11	12	13
14	15	16	17	18	19	20	21	22	23	24	25	26

M/T Only!

View in into Front of Wire Harness (Connector Removed)

Standard Colors and Abbreviations

Abbreviation	Color	Abbreviation	Color	Abbreviation	Color
BK	Black	GY	Gray	RD	Red
BL	Blue	GN	Green	TN	Tan
BR	Brown	LG	Light Green	VT	Violet
DB	Dark Blue	OR	Orange	WT	White
DG	Dark Green	PK	Pink	YL	Yellow

2001-03 Stratus Coupe 2.4L I4 SOHC VIN G (M/T) C113 Connector

PCM Pin #	Wire Color	Circuit Description (16 Pin)	Value at Hot Idle
31-32	---	Not Used	---
33	BR/RD	Generator Signal (Lights "on")	0.2-3.5v
34-35	---	Not Used	---
36	RD/YL	MIL (lamp) Control	Lamp Off: 12v, On: 1v
37	YL	PSP Switch Signal	Straight: 5v, Turning: 0v
38	WT/VT	MPI (Power) Relay Control	Relay Off: 12v, On: 1v
39-41	---	Not Used	---
41	YL/BK	Generator 'FR' Terminal	0.5-4.5v
42-44	---	Not Used	---
45	GN/RD	A/C Switch On/Off Signal	A/C Off: 0v, On: 12v
46	---	Not Used	---

2001-03 Stratus Coupe 2.4L I4 SOHC VIN G (M/T) C116 Connector

PCM Pin #	Wire Color	Circuit Description (12 Pin)	Value at Hot Idle
51	WT/BL	Immobilizer System Signal	Digital Signals
52-53	---	Not Used	---
54	BL/WT	HO2S-12 Heater Control	Digital Signal: 0-12-0v
55	YL	EVAP Vent Solenoid Control	Solenoid Off: 12v, On: 1v
56	GY/RD	Diagnosis Control (DLC #1)	0v
57	---	Not Used	---
58	WT/RD	Tachometer Signals	Pulse Signals
59	---	Not Used	---
60	BR/WT	HO2S-11 Heater Control	Digital Signal: 0-12-0v
61	BR/WT	FTP Sensor Signal	2.5v (fuel cap off)
62	RD/WT	ISO 9141 Bus (DLC #7)	12v (no Scan Tool)

PCM C113 & C116 Wire Harness Connector Graphic

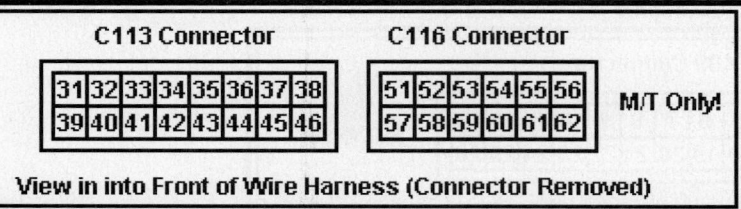

Standard Colors and Abbreviations

Abbreviation	Color	Abbreviation	Color	Abbreviation	Color
BK	Black	GY	Gray	RD	Red
BL	Blue	GN	Green	TN	Tan
BR	Brown	LG	Light Green	VT	Violet
DB	Dark Blue	OR	Orange	WT	White
DG	Dark Green	PK	Pink	YL	Yellow

2001-03 Stratus Coupe 2.4L I4 SOHC VIN G (M/T) C120 Connector

PCM Pin #	Wire Color	Circuit Description (22 Pin)	Value at Hot Idle
71	BR/RD	Starter (Cranking) Signal	9-11v (cranking)
72	RD/BL	IAT Sensor Signal	1.5-2.1v at 104°F
73	BL/RD	MAP Sensor Signal	0.8-1.1v
75	GN	HO2S-12 (B1 S2) Signal	0.1-1.1v
76	WT	HO2S-11 (B1 S1) Signal	0.1-1.1v
78	WT	Knock Sensor Signal	0.080v AC
79	WT/VT	DLC No. 2 Signal	N/A
80	OR/BK	Keep Alive Power	12-14v
81	GN/YL	Sensor Voltage Reference	4.9-5.1v
82	BK/WT	Ignition Switch Power	12-14v
83	YL/GN	ECT Sensor Signal	0.3-0.9v at 176°F
84	YL	TP Sensor Signal	0.53-0.73v
85	GN/WT	BARO Sensor Signal	3.7-4.3 at Sea Level
86	WT/BL	Vehicle Speed Signal	Digital Signal
87	YL/RD	Idle Position Switch Signal	0v, Off-Idle: 4v
88	BL/YL	CMP (TDC) Sensor Signal	Digital Signal: 0-5-0v
89	GN/RD	CKP Sensor Signal	Digital Signal: 0-5-0v
90	WT/GN	Volume Airflow Sensor	2.2-3.2v
91	BK	Power Ground	<0.1v
92	BK	Sensor Ground	<0.050v

PCM C120 Wire Harness Connector Graphic

C120 Connector

```
71 72 73 74 75 76 77 78 79 80 81      M/T Only!
82 83 84 85 86 87 88 89 90 91 92
```

View in into Front of Wire Harness (Connector Removed)

Standard Colors and Abbreviations

Abbreviation	Color	Abbreviation	Color	Abbreviation	Color
BK	Black	GY	Gray	RD	Red
BL	Blue	GN	Green	TN	Tan
BR	Brown	LG	Light Green	VT	Violet
DB	Dark Blue	OR	Orange	WT	White
DG	Dark Green	PK	Pink	YL	Yellow

1995-97 Stratus 2.5L V6 SOHC 24v MFI VIN H (A/T) 'C1' Black Connector

PCM Pin #	Wire Color	Circuit Description (40 Pin)	Value at Hot Idle
1-3	---	Not Used	---
4	DG	Generator Field Control	Digital Signal: 0-12-0v
5	YL/PK	S/C Power Supply	12-14v
6	DG/OR	ASD Relay Output	12-14v
7	YL/WT	Injector 3 Driver	1-4 ms
8-9	---	Not Used	---
10	BK/TN	Power Ground	<0.1v
11	BK/GY	Coil 1 Driver	5°, at 55 mph: 8° dwell
12	---	Not Used	---
13	WT/LB	Injector 1 Driver	1-4 ms
14	BR/DG	Injector 6 Driver	1-4 ms
15	GY	Injector 5 Driver	1-4 ms
16	LB/BR	Injector 4 Driver	1-4 ms
17	TN	Injector 2 Driver	1-4 ms
18-19	---	Not Used	---
20	DB/WT	Ignition Switch Output	12-14v
21-23	---	Not Used	---
24 ('96-'97)	GY/BK	Knock Sensor Signal	0.080v AC
25	---	Not Used	---
26	TN/BK	ECT Sensor Signal	At 180°F: 2.80v
27-28	---	Not Used	---
29 ('95)	TN/WT	HO2S-12 (B1 S2) Signal	0.1-1.1v
30	BK/DG	HO2S-11 (B1 S1) Signal	0.1-1.1v
31	---	Not Used	---
32	GY/BK	CKP Sensor Signal	Digital Signal: 0-5-0v
33	TN/YL	CMP Sensor Signal	Digital Signal: 0-5-0v
34	---	Not Used	---
35	OR/LB	TP Sensor Signal	0.6-1.0v
36	DG/RD	MAP Sensor Signal	1.5-1.7v
37	BK/RD	IAT Sensor Signal	At 100°F: 1.83v
38-39	---	Not Used	---
40	GY/YL	EGR Solenoid Control	12v, 55 mph: 1v

Standard Colors and Abbreviations

Abbreviation	Color	Abbreviation	Color	Abbreviation	Color
BK	Black	GY	Gray	RD	Red
BL	Blue	GN	Green	TN	Tan
BR	Brown	LG	Light Green	VT	Violet
DB	Dark Blue	OR	Orange	WT	White
DG	Dark Green	PK	Pink	YL	Yellow

1995-97 Stratus 2.5L V6 SOHC 24v MFI VIN H (A/T) 'C2' White Connector

PCM Pin #	Wire Color	Circuit Description (40 Pin)	Value at Hot Idle
41	PK/LG	S/C Set Switch Signal	S/C & Set Switch On: 3.8V
42	DB/YL	A/C Damped Pressure Switch	A/C On: 1v, Off: 12v
43	BK/LB	Sensor Ground	<0.050v
44	OR/WT	8-Volt Supply	7.9-8.1v
45	DB/LG	PSP Switch Signal	Straight: 0v, Turning: 5v
47, 53-54	---	Not Used	---
46	RD/TN	Battery Power (Fused B+)	12-14v
48	BR/GY	IAC 3 Driver	DC pulse signals: 0.8-11v
49	YL/BK	IAC 2 Driver	DC pulse signals: 0.8-11v
50	BK/TN	Power Ground	<0.1v
51 ('96-'97)	TN/WT	HO2S-12 (B1 S2) Signal	0.1-1.1v
52	PK/LG	Battery Temperature Sensor	At 86°F: 1.96v
55	DB/TN	Low Speed Fan Relay	Relay Off: 12v, On: 1v
57	GY/RD	IAC 1 Driver	DC pulse signals: 0.8-11v
58	PK/GY	IAC 4 Driver	DC pulse signals: 0.8-11v
59	PK/BR	CCD Bus (+)	Digital Signal: 0-5-0v
60	WT/BK	CCD Bus (-)	<0.050v
61	PK/WT	5-Volt Supply	4.9-5.1v
62	WT/RD	Brake Switch Signal	Brake Off: 0v, On: 12v
63	YL/DG	Torque Management Request	Digital Signals
64	DB/OR	A/C Clutch Relay Control	Relay Off: 12v, On: 1v
65	PK/LB	SCI Transmit	0v
66	WT/OR	Vehicle Speed Signal	Digital Signal
67	DB/PK	ASD Relay Control	Relay Off: 12v, On: 1v
68	PK/GY	EVAP Purge Solenoid Control	Digital Signal: 0-12-0v
69	DB/PK	High Speed Fan Relay	Relay Off: 12v, On: 1v
71, 73	---	Not Used	---
72 ('96-'97)	OR/DG	LDP Switch Signal	Switch Closed: 0v, open: 12v
74	BR/LG	Fuel Pump Relay Control	Relay Off: 12v, On: 1v
75	LG/WT	SCI Receive	0v
76	BR/WT	PNP Switch Signal	In P/N: 0v, Others: 5v
77 ('96-'97)	WT/DG	LDP Solenoid Control	PWM signal: 0-12-0v
78	WT/PK	S/C Vacuum Solenoid	Vacuum Increasing: 1v
79	---	Not Used	---
80	LG/RD	S/C Vent Solenoid	Vacuum Decreasing: 1v

Pin Connector Graphic

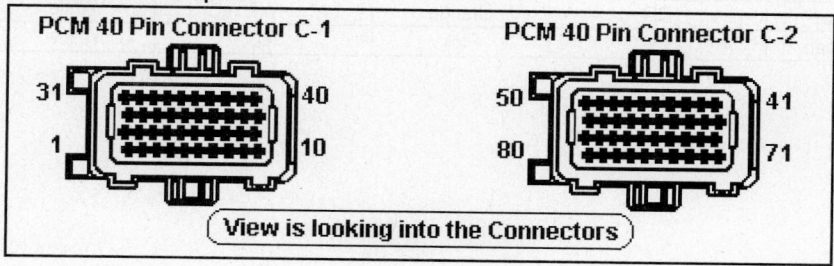

PCM 40 Pin Connector C-1
31   40
1   10

PCM 40 Pin Connector C-2
50   41
80   71

View is looking into the Connectors

1998-2000 Stratus 2.5L V6 SOHC 24v VIN H (A/T) 'C1' Black Connector

PCM Pin #	Wire Color	Circuit Description (40 Pin)	Value at Hot Idle
1-3	---	Not Used	---
4	BK/GY	Coil 1 Driver	5°, at 55 mph: 8° dwell
5	YL/RD	S/C Power Supply	12-14v
6	DG/OR	ASD Relay Output	12-14v
7	YL/WT	Injector 3 Driver	1-4 ms
8	DG	Generator Field Control	Digital Signal: 0-12-0v
9	---	Not Used	---
10	BK/TN	Power Ground	<0.1v
11-12	---	Not Used	---
13	WT/DB	Injector 1 Driver	1-4 ms
14	BR/DB	Injector 6 Driver	1-4 ms
15	GY	Injector 5 Driver	1-4 ms
16	LB/BR	Injector 4 Driver	1-4 ms
17	TN	Injector 2 Driver	1-4 ms
18-19	---	Not Used	---
20	DB/WT	Ignition Switch Output	12-14v
21-25	---	Not Used	---
26	TN/BK	ECT Sensor Signal	At 180°F: 2.80v
27	BK/OR	HO2S Signal Ground	<0.050v
28-29	---	Not Used	---
30	BK/DG	HO2S-11 (B1 S1) Signal	0.1-1.1v
31	---	Not Used	---
32	GY/BK	CKP Sensor Signal	Digital Signal: 0-5-0v
33	TN/YL	CMP Sensor Signal	Digital Signal: 0-5-0v
34	---	Not Used	---
35	OR/DB	TP Sensor Signal	0.6-1.0v
36	DG/RD	MAP Sensor Signal	1.5-1.7v
37	BK/RD	IAT Sensor Signal	At 100°F: 1.83v
38-39	---	Not Used	---
40	GY/YL	EGR Solenoid Control	12v, 55 mph: 1v

Standard Colors and Abbreviations

Abbreviation	Color	Abbreviation	Color	Abbreviation	Color
BK	Black	GY	Gray	RD	Red
BL	Blue	GN	Green	TN	Tan
BR	Brown	LG	Light Green	VT	Violet
DB	Dark Blue	OR	Orange	WT	White
DG	Dark Green	PK	Pink	YL	Yellow

1998-2000 Stratus 2.5L V6 SOHC 24v VIN H (A/T) 'C2' White Connector

PCM Pin #	Wire Color	Circuit Description (40 Pin)	Value at Hot Idle
41	RD/LG	S/C Set Switch Signal	S/C & Set Switch On: 3.8V
42	DB/YL	A/C Damped Pressure Switch	A/C On: 1v, Off: 12v
43	BK/LB	Sensor Ground	<0.050v
44	OR/WT	8-Volt Supply	7.9-8.1v
45	DB/LG	PSP Switch Signal	Straight: 0v, Turning: 5v
46	RD/TN	Battery Power (Fused B+)	12-14v
47	BK	Power Ground	<0.1v
48	BR/WT	IAC 3 Driver	DC pulse signals: 0.8-11v
49	YL/BK	IAC 2 Driver	DC pulse signals: 0.8-11v
50	BK/TN	Power Ground	<0.1v
51	TN/WT	HO2S-12 (B1 S2) Signal	0.1-1.1v
52	VT/LG	Battery Temperature Sensor	At 86°F: 1.96v
53-54	---	Not Used	---
55	DB/TN	Low Speed Fan Relay	Relay Off: 12v, On: 1v
56	WT/VT	S/C Vacuum Solenoid	Vacuum Increasing: 1v
57	GY/RD	IAC 1 Driver	DC pulse signals: 0.8-11v
58	VT/BK	IAC 4 Driver	DC pulse signals: 0.8-11v
59	VT/BR	CCD Bus (+)	Digital Signal: 0-5-0v
60	WT/BK	CCD Bus (-)	<0.050v
61	VT/WT	5-Volt Supply	4.9-5.1v
62	WT/RD	Brake Switch Signal	Brake Off: 0v, On: 12v
63	YL/DG	Torque Management Request	Digital Signals
64	DB/OR	A/C Clutch Relay Control	Relay Off: 12v, On: 1v
65, 75	PK, LG	SCI Transmit, SCI Receive	0v
66	WT/OR	Vehicle Speed Signal	Digital Signal
67	DB/VT	ASD Relay Control	Relay Off: 12v, On: 1v
68	PK/GY	EVAP Purge Solenoid Control	PWM signal: 0-12-0v
69	DB/PK	High Speed Fan Relay	Relay Off: 12v, On: 1v
70	WT/TN	EVAP Purge Solenoid Sense	0-1v
71, 73	---	Not Used	---
72	OR/DG	LDP Switch Signal	Switch Closed: 0v, open: 12v
74	BR/LG	Fuel Pump Relay Control	Relay Off: 12v, On: 1v
76	BK/WT	PNP Switch Signal	In P/N: 0v, Others: 5v
77	WT/DG	LDP Solenoid Control	PWM signal: 0-12-0v
80	LG/RD	S/C Vent Solenoid	Vacuum Decreasing: 1v

Pin Connector Graphic

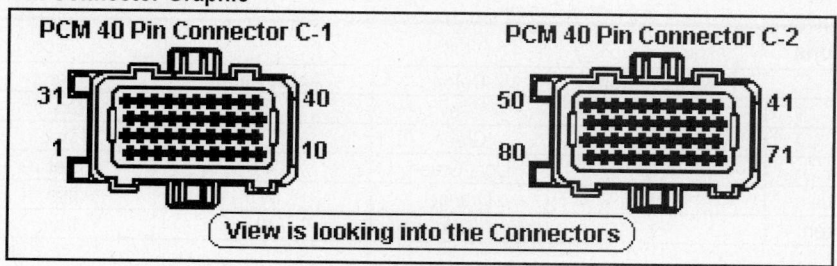

PCM 40 Pin Connector C-1

PCM 40 Pin Connector C-2

31   40
1   10

50   41
80   71

View is looking into the Connectors

2001-03 Stratus 2.7L V6 DOHC 24v VIN R (A/T) 'C1' Black Connector

PCM Pin #	Wire Color	Circuit Description (40 Pin)	Value at Hot Idle
1	TN/LG	COP 4 Driver Control	5°, at 55 mph: 8° dwell
2	TN/OR	COP 3 Driver Control	5°, at 55 mph: 8° dwell
3	TN/PK	COP 2 Driver Control	5°, at 55 mph: 8° dwell
4	TN/LB	COP 6 Driver Control	5°, at 55 mph: 8° dwell
5	YL/RD	S/C Power Supply	12-14v
6	DG/OR	ASD Relay Output	12-14v
7	YL/WT	Injector 3 Driver	1.0-4.0 ms
8	DG	Alternator Field Control	Digital Signal: 0-12-0v
9	---	Not Used	---
10	BK/TN	Power Ground	<0.1v
11	TN/RD	COP 1 Driver Control	5°, at 55 mph: 8° dwell
12	---	Not Used	---
13	WT/DB	Injector 1 Driver	1.0-4.0 ms
14	BR/DB	Injector 6 Driver	1.0-4.0 ms
15	GY	Injector 5 Driver	1.0-4.0 ms
16	LB/BR	Injector 4 Driver	1.0-4.0 ms
17	TN	Injector 2 Driver	1.0-4.0 ms
18-19	---	Not Used	---
20	DB/WT	Ignition Switch Output	12-14v
21	TN/DG	COP 5 Driver Control	5°, at 55 mph: 8° dwell
22-24	---	Not Used	---
25	DB/LG	Knock Sensor Signal	0.080v AC
26	TN/BK	ECT Sensor Signal	At 180°F: 2.80v
27	DB/LG	HO2S Signal Ground	<0.050v
28	---	Not Used	---
29	LG/RD	HO2S-21 (B2 S1) Signal	0.1-1.1v
30	BK/DG	HO2S-11 (B1 S1) Signal	0.1-1.1v
31	TN	Starter Relay Control	KOEC: 9-11v
32	GY/BK	CKP Sensor Signal	Digital Signal: 0-5-0v
33	TN/YL	CMP Sensor Signal	Digital Signal: 0-5-0v
34	LG/PK	EGR Sensor Signal	0.6-0.8v
35	OR/DB	TP Sensor Signal	0.6-1.0v
36	DG/RD	MAP Sensor Signal	1.5-1.7v
37	BK/RD	IAT Sensor Signal	At 100°F: 1.83v
38	---	Not Used	---
39	VT/RD	Manifold Solenoid Control	Solenoid Off: 12v, On: 1v
40	GY/YL	EGR Solenoid Control	12v, at 55 mph: 1v

Standard Colors and Abbreviations

Abbreviation	Color	Abbreviation	Color	Abbreviation	Color
BK	Black	GY	Gray	RD	Red
BL	Blue	GN	Green	TN	Tan
BR	Brown	LG	Light Green	VT	Violet
DB	Dark Blue	OR	Orange	WT	White
DG	Dark Green	PK	Pink	YL	Yellow

V6 DOHC 24v VIN R (A/T) 'C2' White Connector

Wire Color	Circuit Description (40 Pin)	Value at Hot Idle
PK/LG	S/C Set Switch Signal	S/C & Set Switch On: 3.8v
DB	A/C Pressure Switch Signal	A/C On: 0.45-4.85v
BK/LB	Sensor Ground	<0.050v
OR/WT	8-Volt Supply	7.9-8.1v
---	Not Used	---
RD/TN	Battery Power (Fused B+)	12-14v
BK	Power Ground	<0.1v
BR/WT	IAC 3 Driver	DC pulse signals: 0.8-11v
YL/BK	IAC 2 Driver	DC pulse signals: 0.8-11v
BK	Power Ground	<0.1v
TN/WT	HO2S-12 (B1 S2) Signal	0.1-1.1v
VT/LG	Battery Temperature Sensor	At 86°F: 1.96v
PK/WT	HO2S-22 (B2 S2) Signal	0.1-1.1v
---	Not Used	---
DB/TN	Low Speed Fan Relay	Relay Off: 12v, On: 1v
TN/RD	S/C Vacuum Solenoid	Vacuum Increasing: 1v
GY/RD	IAC 1 Driver	DC pulse signals: 0.8-11v
VT/BK	IAC 4 Driver	DC pulse signals: 0.8-11v
OR	PCI Data Bus (J1850)	Digital Signals: 0-7-0v
---	Not Used	---
VT/WT	5-Volt Supply	4.9-5.1v
WT/PK	Brake Switch Signal	Brake Off: 0v, On: 12v
YL/DG	Torque Management Request	Digital Signals
DB/OR	A/C Clutch Relay Control	Relay Off: 12v, On: 1v
PK	SCI Transmit	0v
WT/OR	Vehicle Speed Signal	Digital Signal
DB/VT	ASD Relay Control	Relay Off: 12v, On: 1v
PK/BK	EVAP Purge Solenoid Control	PWM signal: 0-12-0v
DB/PK	High Speed Fan Relay	Relay Off: 12v, On: 1v
WT/TN	EVAP Purge Solenoid Sense	0-1v
WT/RD	EATX RPM Signal	Digital Signals
OR	LDP Switch Sense	Open: 12v, Closed: 0v
---	Not Used	---
BR/LG	Fuel Pump Relay Control	Relay Off: 12v, On: 1v
LG	SCI Receive	5v
BK/LB	TRS T41 Sense	In P/N: 0v, Others: 5v
WT/DG	LDP Solenoid Control	PWM signal: 0-12-0v
---	Not Used	---
LG/RD	S/C Vent Solenoid	Vacuum Decreasing: 1v

Connector Graphic

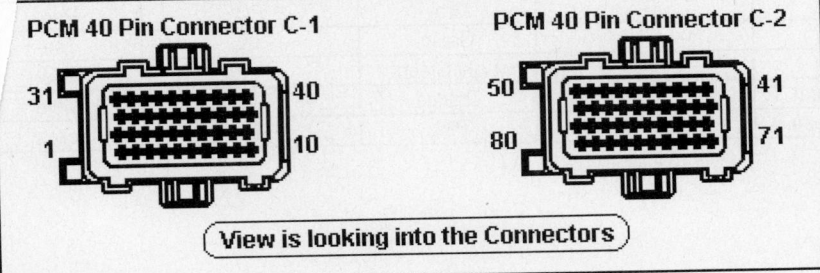

2001-03 Stratus Coupe 3.0L SOHC VIN H (A/T) C112 Connector

PCM Pin #	Wire Color	Circuit Description (35 Pin)	Value
1	YL/BL	Injector 1 Driver	1
2	GN/YL	Injector 4 Driver	1.0
3	BN/WT	HO2S-21 Heater Control	Digital Signal
4	BL/WT	HO2S-11 Heater Control	Digital Signal:
5, 7	---	Not Used	---
6	BL/RD	EGR Solenoid Control	Solenoid Off: 12v, O
8	BK/RD	Generator Signal (lights on)	0.2-3.5v
9	YL/RD	Injector 2 Driver	1.0-4.0 ms
10	RD/BL	Injector 5 Driver	1.0-4.0 ms
11	BK/BL	Power Transistor Control	Digital Signal: 0-12-0
12-13	---	Not Used	---
14	GN/BK	IAC Stepper Motor 'A' Signal	Pulse Signals
15	GN/WT	IAC Stepper Motor 'B' Signal	Pulse Signals
16-17, 23	---	Not Used	---
18	BL/OR	Radiator Fan Control	Fan Off: 0.1v, On: 0.7v
19	GN/BL	Volume Airflow Sensor Reset	1-3v, 3000 rpm: 6-9v
20	GN	A/C Clutch Relay Control	Relay Off: 12v, On: 1v
21	BK	Fuel Pump Relay Control	Pump Off: 12v, On: 1v
22	RD/L	MIL (lamp) Control	Lamp Off: 12v, On: 1v
24	YL/BK	Injector 3 Driver	1.0-4.0 ms
25	LG	Injector 6 Driver	1.0-4.0 ms
26	BL/WT	HO2S-22 Heater Control	Digital Signal: 0-12-0v
27	BR	HO2S-12 Heater Control	Digital Signal: 0-12-0v
28	GN/RD	IAC Stepper Motor 'A' Signal	Pulse Signals
29	BL/YL	IAC Stepper Motor 'B' Signal	Pulse Signals
30-33	---	Not Used	---
34	BL	EVAP Purge Solenoid Control	Solenoid Off: 12v, On: 1v
35	YL	EVAP Vent Solenoid Control	Solenoid Off: 12v, On: 1v

PCM C112 Wire Harness Connector Graphic

```
+------------------------------------------------------+
|                  C112 Connector                       |
|                                                       |
|   [1][2]  [3][4]          [5][6]    [7][8]            |
|   [9][10][11][12][13][14][15][16][17][18][19][20][21][22][23] |
|   [24][25]  [26][27][28][29]   [30][31][32][33]  [34][35]     |
|                                                       |
|      View in into Front of Wire Harness (Connector Removed) |
+------------------------------------------------------+
```

Standard Colors and Abbreviations

Abbreviation	Color	Abbreviation	Color	Abbreviation	Color
BK	Black	GY	Gray	RD	Red
BL	Blue	GN	Green	TN	Tan
BR	Brown	LG	Light Green	VT	Violet
DB	Dark Blue	OR	Orange	WT	White
DG	Dark Green	PK	Pink	YL	Yellow

s Coupe 3.0L SOHC VIN H (A/T) C115 Connector

Wire Color	Circuit Description (26 Pin)	Value at Hot Idle
RD	Ignition Power (MPI Relay)	12-14v
BK	Power Ground	<0.1v
WT	Tachometer Signal	Pulse Signals
YL/GN	ECT Sensor Signal	0.3-0.9v at 176°F
GN/RD	Distributor CKP Sensor Signal	Digital Signal: 0-5-0v
GN	Sensor Voltage Reference	4.9-5.1v
RD	Ignition Power (MPI Relay)	12-14v
BK	Power Ground	<0.1v
WT/VT	MPI (Power) Relay Control	Relay Off: 12v, On: 1v
WT/BL	A/T Control Relay	Relay Off: 12v, On: 1v
---	Not Used	---
YL	PSP Switch Signal	Straight: 5v, Turning: 0v
---	Not Used	---
YL/BK	Generator 'FR' Terminal	0.5-4.5v
GN/WT	BARO Sensor Signal	3.7-4.3 at Sea Level
RD/WT	CMP (TDC) Sensor Signal	Digital Signal: 0-5-0v
BK	Sensor Ground	<0.050v
BR/RD	Starter (Cranking) Signal	9-11v (cranking)
---	Not Used	---
BL/WT	A/C Switch 2 Signal (Hi Blow)	12v
---	Not Used	---
RD/BL	IAT Sensor Signal	1.5-2.1v at 104°F
WT/GN	Volume Airflow Sensor	2.2-3.2v
OR/BL	Keep Alive Power	12-14v

rness Connector Graphic

```
        C115 Connector
   ┌───────────────────────────────┐
   │ 41 42 43       44 45 46        │
   │ 47 48 49 50 51 52 53 54 55 56 57│
   │ 58 59     60 61 62 63  64 65 66 │
   └───────────────────────────────┘
```

to Front of Wire Harness (Connector Removed)

d Abbreviations

Color	Abbreviation	Color	Abbreviation	Color
Black	GY	Gray	RD	Red
Blue	GN	Green	TN	Tan
Brown	LG	Light Green	VT	Violet
Dark Blue	OR	Orange	WT	White
Dark Green	PK	Pink	YL	Yellow

2001-03 Stratus Coupe 3.0L SOHC VIN H (A/T) C119 Connector

PCM Pin #	Wire Color	Circuit Description (28 Pin)	Value at Hot Idle
71	WT	HO2S-21 (B2 S1) Signal	0.1-1.1v
72	BL	HO2S-11 (B1 S1) Signal	0.1-1.1v
73	GN	HO2S-22 (B2 S2) Signal	0.1-1.1v
74	BN	HO2S-12 (B1 S2) Signal	0.1-1.1v
75	GY/BL	Cruise Control Switch Signal	N/A
76	BK	Power Ground	<0.1v
77	RD/BL	A/T: Control Relay Output	12-14v
78	YL	TP Sensor Signal	0.53-0.73v
79	YL/RD	Idle Position Switch Signal	0-1v, Switch Open: 4-5v
80	WT/YL	Vehicle Speed Signal	Digital Signal
81-82	---	Not Used	---
83	GN/RD	A/C Switch On/Off Signal	A/C Off: 0v, On: 12v
84	GY/RD	Diagnosis Control (DLC #1)	0v
85	RD/WT	ISO 9141 Bus (DLC #7)	12v (no Scan Tool)
86-87	---	Not Used	---
88	BK	Power Ground	<0.1v
89	RD/BL	A/T: Control Relay Output	12-14v
90	WT	Knock Sensor Signal	0.080v AC
91	BL/RD	MAP Sensor Signal	0.8-1.1v
92	BN/WT	FTP Sensor Signal	2.5v (fuel cap off)
93-96	---	Not Used	---
97	BK	Power Ground	<0.1v
98	BK/WT	Ignition Switch Power	12-14v

PCM C119 Wire Harness Connector Graphic

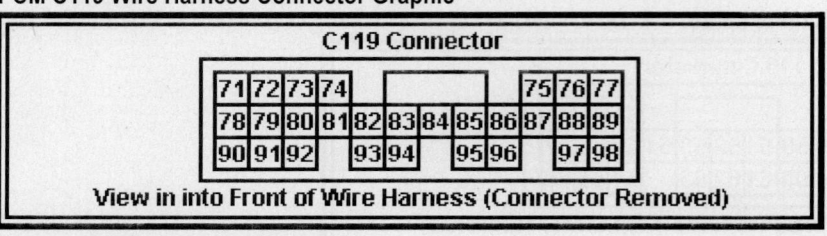

Standard Colors and Abbreviations

Abbreviation	Color	Abbreviation	Color	Abbreviation	Color
BK	Black	GY	Gray	RD	Red
BL	Blue	GN	Green	TN	Tan
BR	Brown	LG	Light Green	VT	Violet
DB	Dark Blue	OR	Orange	WT	White
DG	Dark Green	PK	Pink	YL	Yellow

2001-03 Stratus Coupe 3.0L SOHC VIN H (A/T) C123 Connector

PCM Pin #	Wire Color	Circuit Description (30 Pin)	Value at Hot Idle
101	BK/BL	PNP Switch 'P' Signal	In Park: 12v, Or 0v
102	YL	PNP Switch 'D' Signal	In Drive: 12v, Or 0v
103	WT	Input Shaft Speed Sensor	Moving: 0-12-0v
104	GN/YL	Output Shaft Speed Sensor	Moving: 0-12-0v
105	BR/RD	A/T Shift Mode 1st Indicator	0v
106	RD/YL	A/T Second Solenoid	12-14v
107	YL/RD	A/T TCC Solenoid	12-14v
108	RD/BL	PNP Switch 'R' Signal	In Reverse: 12v, Or: 0v
109	WT	PNP Switch 'D3' Signal	In Drive 3: 12v, Or 0v
110	GN	PNP Switch 'Low' Signal	In Low: 12v, Or 0v
111	WT/GN	Immobilizer System Signal	Digital Signals
112	---	Not Used	---
113	WT/BL	DLC No. 2 Signal	N/A
114-116	---	Not Used	---
117	WT/BL	A/T Shift Mode 3rd Indicator	0v
118	YL/BL	A/T Shift Mode 2nd Indicator	0v
120	RD	A/T Underdrive Solenoid	12-14v
121	BN	PNP Switch 'N' Signal	In Neutral: 12v, Or 0v
122	YL/BL	PNP Switch 'D2' Signal	In Drive 2: 12v, Or 0v
123	GN/OR	Stop Light Switch Signal	Brake Off: 0v, On: 12v
124	BL/PK	TFT Sensor Signal	3.2-3.4v at 104ºF
128	YL/BL	A/T Shift Mode 4th Indicator	0v
129	RD/WT	A/T Low/Reverse Solenoid	12-14v
130	BL	A/T Overdrive Solenoid	12-14v

PCM C123 Wire Harness Connector Graphic

```
                    C123 Connector
 ┌───┬───┐   ┌───┬───┐           ┌───┬───┬───┐
 │101│102│   │103│104│           │105│106│107│
 ├───┼───┼───┼───┼───┼───┼───┼───┼───┼───┼───┤
 │108│109│110│111│112│113│114│115│116│117│118│119│120│
 ├───┼───┼───┼───┼───┼───┼───┼───┼───┼───┼───┤
 │121│122│123│   │124│125│   │126│127│128│   │129│130│
 └───┴───┴───┘   └───┴───┘   └───┴───┴───┘   └───┴───┘
```

View in into Front of Wire Harness (Connector Removed)

Standard Colors and Abbreviations

Abbreviation	Color	Abbreviation	Color	Abbreviation	Color
BK	Black	GY	Gray	RD	Red
BL	Blue	GN	Green	TN	Tan
BR	Brown	LG	Light Green	VT	Violet
DB	Dark Blue	OR	Orange	WT	White
DG	Dark Green	PK	Pink	YL	Yellow

2001-03 Stratus Coupe 3.0L SOHC VIN H (M/T) C111 Connector

PCM Pin #	Wire Color	Circuit Description (35 Pin)	Value at Hot Idle
1	YL/BL	Injector 1 Driver	1.0-4.0 ms
2	GN/YL	Injector 4 Driver	1.0-4.0 ms
3	BN/WT	HO2S-21 Heater Control	Digital Signal: 0-12-0v
4	BL/WT	HO2S-11 Heater Control	Digital Signal: 0-12-0v
5, 7	---	Not Used	---
6	BL/RD	EGR Solenoid Control	Solenoid Off: 12v, On: 1v
8	BR/RD	Generator Signal (Lights "on")	0.2-3.5v
9	YL/RD	Injector 2 Driver	1.0-4.0 ms
10	GN/RD	Injector 5 Driver	1.0-4.0 ms
11	BK/BL	Power Transistor Control	Digital Signal: 0-12-0v
12-13	---	Not Used	---
14	GN/BK	IAC Stepper Motor 'A' Signal	Pulse Signals
15	GN/WT	IAC Stepper Motor 'B' Signal	Pulse Signals
16	BL	EVAP Purge Solenoid Control	Solenoid Off: 12v, On: 1v
17, 23	---	Not Used	---
18	BL/OR	Radiator Cooling Fan Control	Fan Off: 0.1v, On: 0.7v
19	GN/BL	Volume Airflow Sensor Reset	1-3v, 3000 rpm: 6-9v
20	GN	A/C Clutch Relay Control	Relay Off: 12v, On: 1v
21	BK	Fuel Pump Relay Control	Pump Off: 12v, On: 1v
22	RD/YL	MIL (lamp) Control	Lamp Off: 12v, On: 1v
24	YL/BL	Injector 3 Driver	1.0-4.0 ms
25	LG	Injector 6 Driver	1.0-4.0 ms
26	BL/WT	HO2S-22 Heater Control	Digital Signal: 0-12-0v
27	BN	HO2S-12 Heater Control	Digital Signal: 0-12-0v
28	GN/RD	IAC Stepper Motor 'A' Signal	Pulse Signals
29	BK/YL	IAC Stepper Motor 'B' Signal	Pulse Signals
30-34	---	Not Used	---
35	YL	EVAP Vent Solenoid Control	Solenoid Off: 12v, On: 1v

PCM C111 Wire Harness Connector Graphic

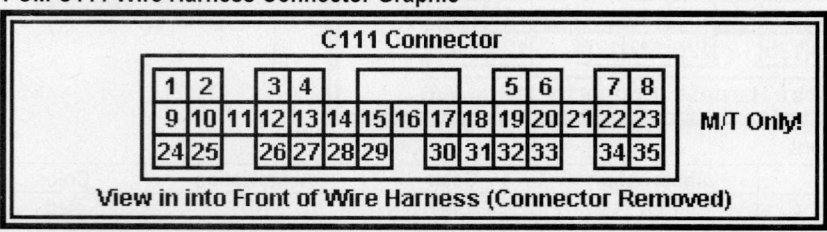

Standard Colors and Abbreviations

Abbreviation	Color	Abbreviation	Color	Abbreviation	Color
BK	Black	GY	Gray	RD	Red
BL	Blue	GN	Green	TN	Tan
BR	Brown	LG	Light Green	VT	Violet
DB	Dark Blue	OR	Orange	WT	White
DG	Dark Green	PK	Pink	YL	Yellow

2001-03 Stratus Coupe 3.0L SOHC VIN H (M/T) C118 Connector

PCM Pin #	Wire Color	Circuit Description (28 Pin)	Value at Hot Idle
41	---	Not Used	---
42	GN/YL	Sensor Voltage Reference	4.9-5.1v
43	GN/RD	Distributor CKP Sensor Signal	Digital Signal: 0-5-0v
44	YL/GN	ECT Sensor Signal	0.3-0.9v at 176°F
45	WT	Tachometer Signals	Pulse Signals
46	BK	Power Ground	<0.1v
47	RD	Ignition Power (MPI Relay)	12-14v
48	---	Not Used	---
49	BK	Sensor Ground	<0.050v
50	RD/WT	CMP (TDC) Sensor Signal	Digital Signal: 0-5-0v
51	GN/WT	BARO Sensor Signal	3.7-4.3 at Sea Level
52	YL/BK	Generator 'FR' Terminal	0.5-4.5v
53	---	Not Used	---
54	YL	PSP Switch Signal	Straight: 5v, Turning: 0v
55-56	---	Not Used	---
57	WT/VT	MPI (Power) Relay Control	Relay Off: 12v, On: 1v
58	BK	Power Ground	<0.1v
59	RD	Ignition Power (MPI Relay)	12-14v
60	OR/BK	Keep Alive Power	12-14v
61	WT/GN	Volume Airflow Sensor	2.2-3.2v
62	RD/BL	IAT Sensor Signal	1.5-2.1v at 104°F
63-64	---	Not Used	---
65	BL/WT	A/C Switch 2 Signal (Hi Blow)	12v
66-67	---	Not Used	---
68	BR/RD	Starter (Cranking) Signal	9-11v (cranking)

PCM C118 Wire Harness Connector Graphic

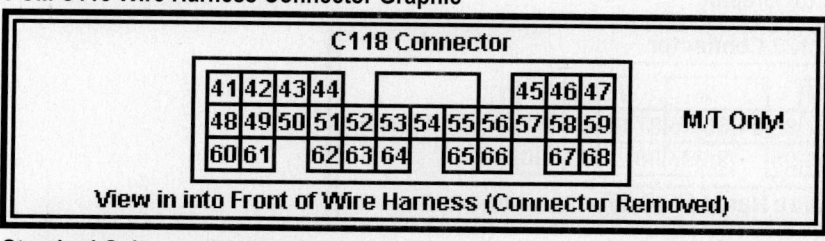

C118 Connector

| 41 | 42 | 43 | 44 | | | | 45 | 46 | 47 |

| 48 | 49 | 50 | 51 | 52 | 53 | 54 | 55 | 56 | 57 | 58 | 59 |

| 60 | 61 | | 62 | 63 | 64 | | 65 | 66 | | 67 | 68 |

M/T Only!

View in into Front of Wire Harness (Connector Removed)

Standard Colors and Abbreviations

Abbreviation	Color	Abbreviation	Color	Abbreviation	Color
BK	Black	GY	Gray	RD	Red
BL	Blue	GN	Green	TN	Tan
BR	Brown	LG	Light Green	VT	Violet
DB	Dark Blue	OR	Orange	WT	White
DG	Dark Green	PK	Pink	YL	Yellow

2001-03 Stratus Coupe 3.0L SOHC VIN H (M/T) C122 Connector

PCM Pin #	Wire Color	Circuit Description (30 Pin)	Value at Hot Idle
71	WT	HO2S-21 (B2 S1) Signal	0.1-1.1v
72	BL	HO2S-11 (B1 S1) Signal	0.1-1.1v
73	GN	HO2S-22 (B2 S2) Signal	0.1-1.1v
74	YL	HO2S-12 (B1 S2) Signal	0.1-1.1v
75	---	Not Used	---
76	BK	Power Ground	<0.1v
77	---	Not Used	---
78	YL	TP Sensor Signal	0.53-0.73v
79	YL/RD	Idle Position Switch Signal	0v, Off-Idle: 4v
80	WT/BL	Vehicle Speed Signal	Digital Signal
81-82	---	Not Used	---
83	GN/RD	A/C Switch On/Off Signal	A/C Off: 0v, On: 12v
84	GY/RD	Diagnosis Control (DLC #1)	0v
85	RD/WT	ISO 9141 Bus (DLC #7)	12v (no Scan Tool)
86-87	---	Not Used	---
88	BK	Power Ground	<0.1v
89-90	---	Not Used	---
91	WT	Knock Sensor Signal	0.080v AC
92	BL/RD	MAP Sensor Signal	0.8-1.1v
93	BN/WT	FTP Sensor Signal	2.5v (fuel cap off)
94-96	---	Not Used	---
97	BK	Power Ground	<0.1v
98	WT/GN	Immobilizer System Signal	Digital Signals
99	BK/WT	Ignition Switch Power	12-14v
100	WT/VT	DLC No. 2 Signal	N/A

PCM C122 Wire Harness Connector Graphic

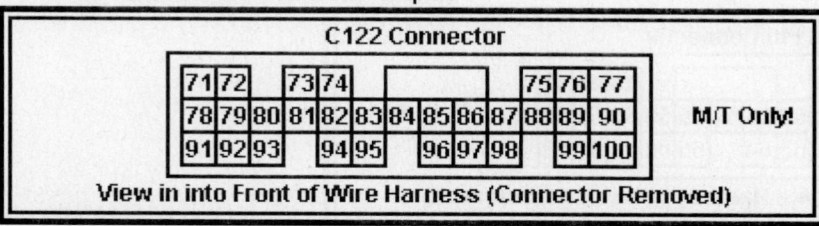

C122 Connector

71 72 73 74 75 76 77
78 79 80 81 82 83 84 85 86 87 88 89 90
91 92 93 94 95 96 97 98 99 100

M/T Only!

View in into Front of Wire Harness (Connector Removed)

Standard Colors and Abbreviations

Abbreviation	Color	Abbreviation	Color	Abbreviation	Color
BK	Black	GY	Gray	RD	Red
BL	Blue	GN	Green	TN	Tan
BR	Brown	LG	Light Green	VT	Violet
DB	Dark Blue	OR	Orange	WT	White
DG	Dark Green	PK	Pink	YL	Yellow

1992-95 Viper 8.0L V10 MFI VIN E (M/T) 60 Pin Connector

PCM Pin #	Wire Color	Circuit Description (60 Pin)	Value at Hot Idle
1	DG/RD	MAP Sensor Signal	1.5-1.7v
2	TN/BK	ECT Sensor Signal	At 180°F: 2.80v
3	RD/WT	ASD Relay Output (B+)	12-14v
4	BK/LB	Sensor Ground	<0.050v
5	BK/WT	Sensor Ground	<0.050v
6	VT/WT	5-Volt Supply	4.9-5.1v
7	OR	8-Volt Supply	7.9-8.1v
8	---	Not Used	---
9	DB	Ignition Switch Output	12-14v
10	---	Not Used	---
11	BK/TN	Power Ground	<0.1v
12	BK/TN	Power Ground	<0.1v
13	BR/BK	Injector 6 Driver	1-4 ms
14	YL/WT	Injector 3 Driver	1-4 ms
15	WT/LG	Injector 9 Driver	1-4 ms
16	TN	Injector 2 Driver	1-4 ms
17	BK/LG	Injector 10 Driver	1-4 ms
18	LB/BR	Injector 4 Driver	1-4 ms
19-20	---	Not Used	---
21	BK/RD	Air Temperature Sensor	At 100°F: 1.83v
22	OR/DB	TP Sensor Signal	0.6-1.0v
23	LG/RD	HO2S-21 (B2 S1) Signal	0.1-1.1v
24	GY/BK	CKP Sensor Signal	Digital Signal: 0-5-0v
25	PK	SCI Transmit	0v
26	VT/BR	CCD Bus (+)	Digital Signal: 0-5-0v
27	BR	A/C Damped Pressure Switch	A/C On: 1v, Off: 12v
28	---	Not Used	---
29	WT/PK	Brake Switch Signal	Brake Off: 0v, On: 12v
30	---	Not Used	---
31	DB/PK	Radiator Fan Relay Control	Relay Off: 12v, On: 1v
32	BK/PK	MIL (lamp) Control	Lamp On: 1v, Off: 12v
33	---	Not Used	---
34	DB/OR	A/C Clutch Relay Control	Relay Off: 12v, On: 1v
35 ('94-'95)	YL	Low / High Speed Fan Relay	Relay Off: 12v, On: 1v
36 ('94-'95)	LB/BK	Reverse Lockout Solenoid	Solenoid Off: 12v, On: 1v
37	PK/VT	Serial COMM (multiplex)	Digital Signals
38	GY/LG	Injector 8 Driver	1-4 ms
39	GY/RD	IAC 1 Driver (open)	DC pulse signals: 0.8-11v
40	BR/WT	IAC 3 Driver (open)	DC pulse signals: 0.8-11v

Standard Colors and Abbreviations

Abbreviation	Color	Abbreviation	Color	Abbreviation	Color
BK	Black	GY	Gray	RD	Red
BL	Blue	GN	Green	TN	Tan
BR	Brown	LG	Light Green	VT	Violet
DB	Dark Blue	OR	Orange	WT	White
DG	Dark Green	PK	Pink	YL	Yellow

1992-95 Viper 8.0L V10 MFI VIN E (M/T) 60 Pin Connector

PCM Pin #	Wire Color	Circuit Description (60 Pin)	Value at Hot Idle
41	BK/DG	HO2S-11 (B1 S1) Signal	0.1-1.1v
42	---	Not Used	---
43	GY/LB	Tachometer Signal	Pulse Signals
44	TN/YL	CMP Sensor Signal	Digital Signal: 0-5-0v
45	LG	SCI Receive	0v
46	WT/BK	CCD Bus (-)	<0.050v
47	WT/TN	Vehicle Speed Signal	Digital Signal
48-50	---	Not Used	---
51	DB/YL	ASD Relay Control	Relay Off: 12v, On: 1v
52	PK/BK	EVAP Purge Solenoid Control	Solenoid Off: 12v, On: 1v
53	---	Not Used	---
54	OR/BK	Shift Indicator Lamp Control	Lamp On: 1v, Off: 12v
55	GY/OR	2-3 Skip Shift Solenoid	Solenoid Off: 12v, On: 1v
56	---	Not Used	---
57	DG/OR	Voltage Sensor to PCM	12-14v
58	VT	Injector 7 Driver	1-4 ms
59	VT/BK	IAC 4 Driver (Close)	DC pulse signals: 0.8-11v
60	YL/BK	IAC 2 Driver (Close)	DC pulse signals: 0.8-11v

Pin Connector Graphic

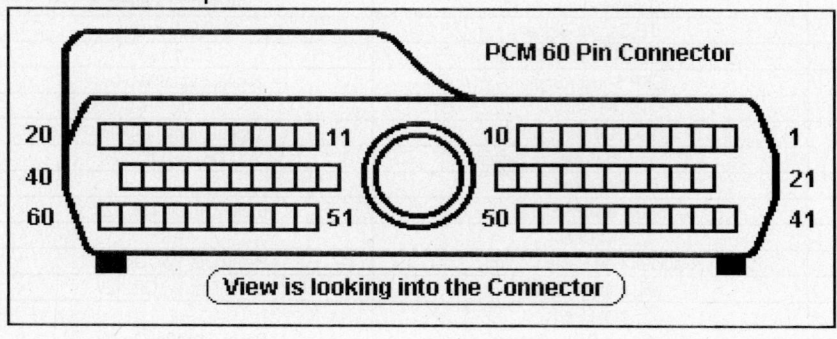

1996-97 Viper 8.0L V10 MFI VIN E (M/T) Black 'A' Connector

PCM Pin #	Wire Color	Circuit Description (32-Pin)	Value at Hot Idle
1	YL/GY	Coil 4 Driver	5°, at 55 mph: 8° dwell
2	DB	Ignition Switch Output	12-14v
3	RD/YL	Coil 3 Driver	5°, at 55 mph: 8° dwell
4	BK/LB	Sensor Ground	<0.050v
5	DG/GY	Coil 5 Driver	5°, at 55 mph: 8° dwell
6	---	Not Used	---
7	BK/GY	Coil 1 Driver	5°, at 55 mph: 8° dwell
8	GY/BK	CKP Sensor Signal	Digital Signal: 0-5-0v
9	DB/WT	Coil 2 Driver	5°, at 55 mph: 8° dwell
10	YL/BK	IAC 2 Driver	DC pulse signals: 0.8-11v
11	BR/WT	IAC 3 Driver	DC pulse signals: 0.8-11v
12-14	---	Not Used	---
15	BK/RD	IAT Sensor Signal	At 100°F: 1.83v
16	TN/BK	ECT Sensor Signal	At 180°F: 2.80v
17	OR	5-Volt Supply	4.9-5.1v
18	TN/YL	CMP Sensor Signal	Digital Signal: 0-5-0v
19	GY/RD	IAC 1 Driver	DC pulse signals: 0.8-11v
20	PK/BK	IAC 4 Driver	DC pulse signals: 0.8-11v
21	---	Not Used	---
22	RD/WT	Battery Power (Fused B+)	12-14v
23	OR/DB	TP Sensor Signal	0.6-1.0v
24	LG/RD	HO2S-11 (B1 S1) Signal	0.1-1.1v
25	PK/WT	HO2S-12 (B1 S2) Signal	0.1-1.1v
26	BK/DG	HO2S-21 (B2 S1) Signal	0.1-1.1v
27	DG/RD	MAP Sensor Signal	1.5-1.7v
28	---	Not Used	---
29	TN/WT	HO2S-22 (B2 S2) Signal	0.1-1.1v
30	---	Not Used	---
31	DB	Power Ground	<0.1v
32	DB	Power Ground	<0.1v

Pin Connector Graphic

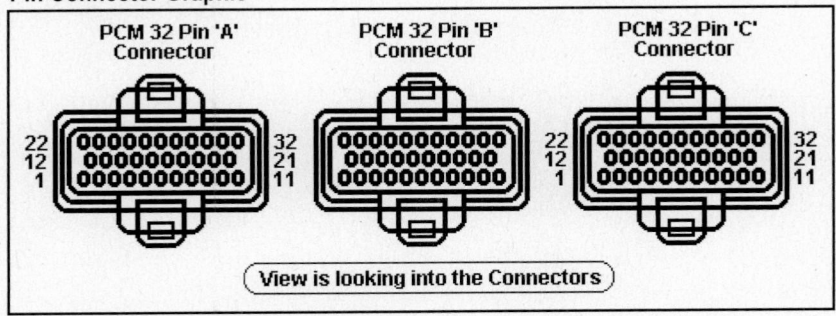

1996-97 Viper 8.0L V10 MFI VIN E (M/T) White 'B' Connector

PCM Pin #	Wire Color	Circuit Description (32-Pin)	Value at Hot Idle
1	---	Not Used	---
2	PK	Injector 7 Driver	1-4 ms
3	WT/LG	Injector 9 Driver	1-4 ms
4	WT/DB	Injector 1 Driver	1-4 ms
5	YL/WT	Injector 3 Driver	1-4 ms
6	GY/OR	Injector 5 Driver	1-4 ms
7-10	---	Not Used	---
11	OR/BK	Shift Indicator Lamp Control	Lamp On: 1v, Off: 12v
12	BR/DB	Injector 6 Driver	1-4 ms
13	GY/LB	Injector 8 Driver	1-4 ms
14	BK/LG	Injector 10 Driver	1-4 ms
15	TN	Injector 2 Driver	1-4 ms
16	LB/BR	Injector 4 Driver	1-4 ms
17-20	---	Not Used	---
21	LB/BK	Reverse Lockout Solenoid	Solenoid Off: 12v, On: 1v
22-26	---	Not Used	---
27	WT/TN	OSS Sensor Signal (+)	Moving: AC Pulse Signals
28	WT/OR	Vehicle Speed Signal	Digital Signal
29-30	---	Not Used	---
31	PK/WT	5-Volt Supply	4.9-5.1v
32	---	Not Used	---

Standard Colors and Abbreviations

Abbreviation	Color	Abbreviation	Color	Abbreviation	Color
BK	Black	GY	Gray	RD	Red
BL	Blue	GN	Green	TN	Tan
BR	Brown	LG	Light Green	VT	Violet
DB	Dark Blue	OR	Orange	WT	White
DG	Dark Green	PK	Pink	YL	Yellow

1996-97 Viper 8.0L V10 MFI VIN E (M/T) Grey 'C' Connector

PCM Pin #	Wire Color	Circuit Description (32-Pin)	Value at Hot Idle
1	DB/OR	A/C Clutch Relay Control	Relay Off: 12v, On: 1v
2	DB/PK	Low Speed Fan Relay	Relay Off: 12v, On: 1v
3	DB/YL	ASD Relay Control	Relay Off: 12v, On: 1v
4-5	---	Not Used	---
6	GY/OR	2-3 Skip Shift Solenoid	Solenoid Off: 12v, On: 1v
7-11	---	Not Used	---
12	DG/OR	ASD Relay Output	12-14v
13-14	---	Not Used	---
15	PK/LG	Battery Temperature Sensor	At 86°F: 1.96v
16	---	Not Used	---
17	BK/PK	MIL (lamp) Control	Lamp On: 1v, Off: 12v
18	---	Not Used	---
19	BR/PK	Fuel Pump Relay Control	Relay Off: 12v, On: 1v
20	PK/BK	EVAP Purge Solenoid Control	PWM signal: 0-12-0v
21	YL	High Speed Fan Relay	Relay Off: 12v, On: 1v
22	BR	A/C Request Signal	A/C On: 1v, Off: 12v
23	LG	A/C Select Signal	A/C On: 1v, Off: 12v
24	WT/PK	Brake Switch Signal	Brake Off: 0v, On: 12v
25	---	Not Used	---
26	DB/WT	Fuel Level Sensor Signal	70 ohms (±20) with full tank
27	PK	SCI Transmit	0v
28	WT/BK	CCD Bus (-)	<0.050v
29	LG	SCI Receive	0v
30	PK/BR	CCD Bus (+)	Digital Signal: 0-5-0v
31	GY/LB	Tachometer Signal	Pulse Signals
32	---	Not Used	---

Pin Connector Graphic

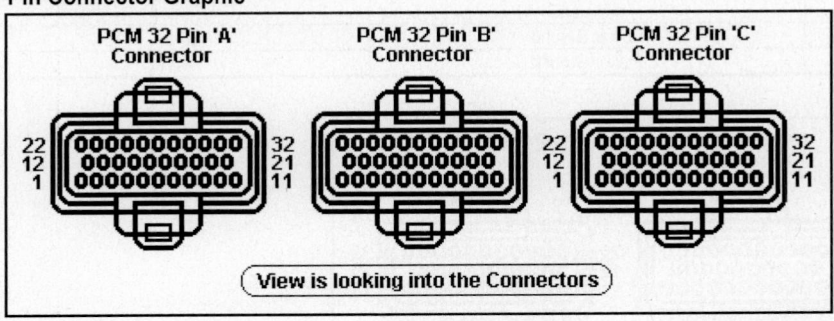

1998-99 Viper 8.0L V10 MFI VIN E (M/T) Black 'A' Connector

PCM Pin #	Wire Color	Circuit Description (32-Pin)	Value at Hot Idle
1	YL/GY	Coil 4 Driver	6°, 55 mph: 9° dwell
2	DB/BK	Ignition Switch Output	12-14v
3	RD/YL	Coil 3 Driver	6°, 55 mph: 9° dwell
4	BK/LB	Sensor Ground	<0.050v
5	DG/GY	Coil 5 Driver	6°, 55 mph: 9° dwell
6	---	Not Used	---
7	BK/GY	Coil 1 Driver	6°, 55 mph: 9° dwell
8	GY/BK	CKP Sensor Signal	Digital Signal: 0-5-0v
9	DB/TN	Coil 2 Driver	6°, 55 mph: 9° dwell
10	YL/BK	IAC 2 Driver	DC pulse signals: 0.8-11v
11	BR/WT	IAC 3 Driver	DC pulse signals: 0.8-11v
12-14	---	Not Used	---
15	BK/RD	IAT Sensor Signal	At 100°F: 1.83v
16	TN/BK	ECT Sensor Signal	At 180°F: 2.80v
17	OR	5-Volt Supply	4.9-5.1v
18	TN/YL	CMP Sensor Signal	Digital Signal: 0-5-0v
19	GY/RD	IAC 1 Driver	DC pulse signals: 0.8-11v
20	PK/BK	IAC 4 Driver	DC pulse signals: 0.8-11v
21	---	Not Used	---
22	RD/WT	Battery Power (Fused B+)	12-14v
23	OR/DB	TP Sensor Signal	0.6-1.0v
24	LG/RD	HO2S-11 (B1 S1) Signal	0.1-1.1v
25	PK/WT	HO2S-12 (B1 S2) Signal	0.1-1.1v
26	BK/DG	HO2S-21 (B2 S1) Signal	0.1-1.1v
27	DG/RD	MAP Sensor Signal	1.5-1.7v
28	---	Not Used	---
29	TN/WT	HO2S-22 (B2 S2) Signal	0.1-1.1v
30	---	Not Used	---
31	BK/TN	Power Ground	<0.1v
32	BK/TN	Power Ground	<0.1v

Pin Connector Graphic

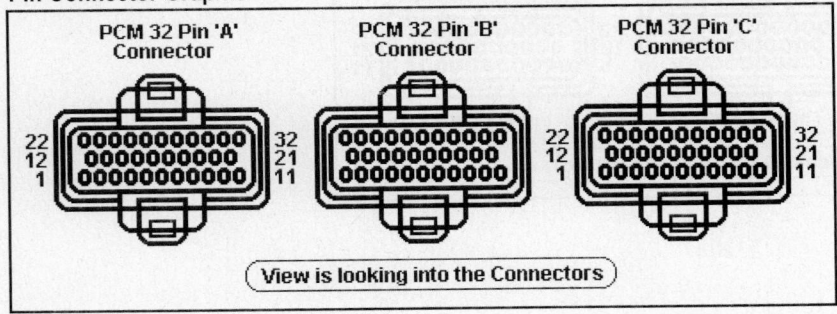

PCM 32 Pin 'A' Connector   PCM 32 Pin 'B' Connector   PCM 32 Pin 'C' Connector

View is looking into the Connectors

1998-99 Viper 8.0L V10 MFI VIN E (M/T) White 'B' Connector

PCM Pin #	Wire Color	Circuit Description (32-Pin)	Value at Hot Idle
1	---	Not Used	---
2	PK	Injector 7 Driver	1-4 ms
3	WT/LG	Injector 9 Driver	1-4 ms
4	WT/DB	Injector 1 Driver	1-4 ms
5	YL/WT	Injector 3 Driver	1-4 ms
6	GY	Injector 5 Driver	1-4 ms
7-9	---	Not Used	---
10	DG	Generator Field Control	Digital Signal: 0-12-0v
11	OR/BK	Shift Indicator Lamp Control	Lamp On: 1v, Off: 12v
12	BR/BK	Injector 6 Driver	1-4 ms
13	GY/LG	Injector 8 Driver	1-4 ms
14	BK/LG	Injector 10 Driver	1-4 ms
15	TN	Injector 2 Driver	1-4 ms
16	LB/BR	Injector 4 Driver	1-4 ms
17-20	---	Not Used	---
21	LB/BK	Reverse Lockout Solenoid	Solenoid Off: 12v, On: 1v
22-26	---	Not Used	---
27	WT/TN	OSS Sensor Signal (+)	Moving: AC Pulse Signals
28	WT/OR	Vehicle Speed Signal	Digital Signal
29-30	---	Not Used	---
31	PK/WT	5-Volt Supply	4.9-5.1v
32	---	Not Used	---

Standard Colors and Abbreviations

Abbreviation	Color	Abbreviation	Color	Abbreviation	Color
BK	Black	GY	Gray	RD	Red
BL	Blue	GN	Green	TN	Tan
BR	Brown	LG	Light Green	VT	Violet
DB	Dark Blue	OR	Orange	WT	White
DG	Dark Green	PK	Pink	YL	Yellow

1998-99 Viper 8.0L V10 MFI VIN E (M/T) Grey 'C' Connector

PCM Pin #	Wire Color	Circuit Description (32-Pin)	Value at Hot Idle
1	DB/OR	A/C Clutch Relay Control	Relay Off: 12v, On: 1v
2	DB/PK	Low Speed Fan Relay	Relay Off: 12v, On: 1v
3	DB/YL	ASD Relay Control	Relay Off: 12v, On: 1v
4-5	---	Not Used	---
6	GY/OR	2-3 Skip Shift Solenoid	Solenoid Off: 12v, On: 1v
7-9	---	Not Used	---
10	WT/DG	LDP Solenoid Control	PWM signal: 0-12-0v
11	---	Not Used	---
12	DG/OR	ASD Relay Output	12-14v
13	---	Not Used	---
14	OR	LDP Switch Signal	Switch Closed: 0v, open: 12v
15	PK/LG	Battery Temperature Sensor	At 86°F: 1.96v
16	DG/YL	Generator Lamp Control	Lamp On: 1v, Off: 12v
17	BK/PK	MIL (lamp) Control	Lamp On: 1v, Off: 12v
18	---	Not Used	---
19	BR/PK	Fuel Pump Relay Control	Relay Off: 12v, On: 1v
20	PK/BK	EVAP Purge Solenoid Control	Digital Signal: 0-12-0v
21	YL	High Speed Fan Relay	Relay Off: 12v, On: 1v
22	BR	A/C Request Signal	A/C On: 1v, Off: 12v
23	BR	A/C Select Signal	A/C On: 1v, Off: 12v
24	WT/PK	Brake Switch Signal	Brake Off: 0v, On: 12v
25	---	Not Used	---
26	DB	Fuel Level Sensor Signal	70 ohms (±20) with full tank
27	PK	SCI Transmit	0v
28	WT/BK	CCD Bus (-)	<0.050v
29	LG	SCI Receive	0v
30	PK/BR	CCD Bus (+)	Digital Signal: 0-5-0v
31	GY/LB	Tachometer Signal	Pulse Signals
32	---	Not Used	---

Pin Connector Graphic

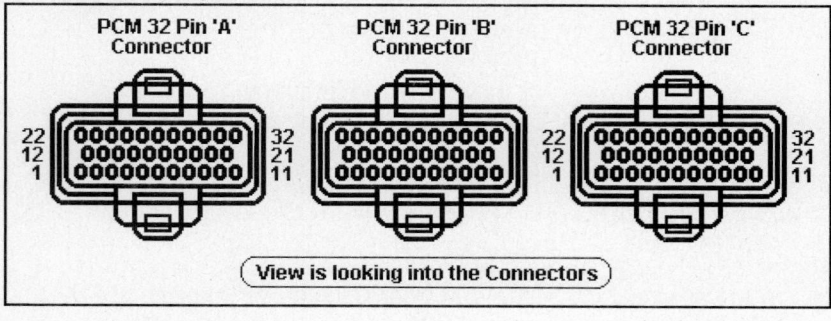

2000-02 Viper 8.0L V10 MFI VIN E (M/T) 32P C1 Black Connector

PCM Pin #	Wire Color	Circuit Description (32-Pin)	Value at Hot Idle
1	YL/GY	Coil 4 Driver	6°, 55 mph: 9° dwell
2	DB/GY	Ignition Switch Output	12-14v
3	RD/YL	Coil 3 Driver	6°, 55 mph: 9° dwell
4	BK/LB	Sensor Ground	<0.050v
5	DG/GY	Coil 5 Driver	6°, 55 mph: 9° dwell
6	---	Not Used	---
7	BK/GY	Coil 1 Driver	6°, 55 mph: 9° dwell
8	GY/BK	CKP Sensor Signal	Digital Signal: 0-5-0v
9	DB/TN	Coil 2 Driver	6°, 55 mph: 9° dwell
10	YL/BK	IAC 2 Driver	DC pulse signals: 0.8-11v
11	BR/WT	IAC 3 Driver	DC pulse signals: 0.8-11v
12-14	---	Not Used	---
15	BK/RD	IAT Sensor Signal	At 100°F: 1.83v
16	TN/BK	ECT Sensor Signal	At 180°F: 2.80v
17	OR	5-Volt Supply	4.9-5.1v
18	TN/YL	CMP Sensor Signal	Digital Signal: 0-5-0v
19	GY/RD	IAC 1 Driver	DC pulse signals: 0.8-11v
20	VT/BK	IAC 4 Driver	DC pulse signals: 0.8-11v
21	---	Not Used	---
22	PK/WT	Battery Power (Fused B+)	12-14v
23	OR/DB	TP Sensor Signal	0.6-1.0v
24	LG/RD	HO2S-11 (B1 S1) Signal	0.1-1.1v
25	PK/WT	HO2S-12 (B1 S2) Signal	0.1-1.1v
26	BK/DG	HO2S-21 (B2 S1) Signal	0.1-1.1v
27	DG/RD	MAP Sensor Signal	1.5-1.7v
28	---	Not Used	---
29	TN/WT	HO2S-22 (B2 S2) Signal	0.1-1.1v
30	---	Not Used	---
31	BK/TN	Power Ground	<0.1v
32	BK/TN	Power Ground	<0.1v

Pin Connector Graphic

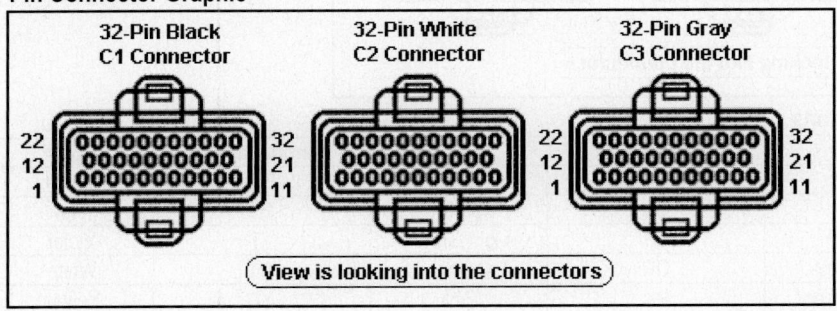

32-Pin Black C1 Connector   32-Pin White C2 Connector   32-Pin Gray C3 Connector

View is looking into the connectors

2000-02 Viper 8.0L V10 MFI VIN E (M/T) 32P C2 White Connector

PCM Pin #	Wire Color	Circuit Description (32-Pin)	Value at Hot Idle
1	---	Not Used	---
2	VT	Injector 7 Driver	1-4 ms
3	WT/LG	Injector 9 Driver	1-4 ms
4	WT/DB	Injector 1 Driver	1-4 ms
5	YL/WT	Injector 3 Driver	1-4 ms
6	GY	Injector 5 Driver	1-4 ms
7-9	---	Not Used	---
10	DG	Generator Field Control	Digital Signal: 0-12-0v
11	OR/BK	Shift Indicator Lamp Control	Lamp On: 1v, Off: 12v
12	BR/BK	Injector 6 Driver	1-4 ms
13	GY/LG	Injector 8 Driver	1-4 ms
14	BK/LG	Injector 10 Driver	1-4 ms
15	TN	Injector 2 Driver	1-4 ms
16	LB/BR	Injector 4 Driver	1-4 ms
17-20	---	Not Used	---
21	LB/BK	Reverse Lockout Solenoid	Solenoid Off: 12v, On: 1v
22-26	---	Not Used	---
27	WT/TN	OSS Sensor Signal (+)	Moving: AC Pulse Signals
28	WT/OR	Vehicle Speed Signal	Digital Signal
29-30	---	Not Used	---
31	VT/WT	5-Volt Supply	4.9-5.1v
32	---	Not Used	---

Pin Connector Graphic

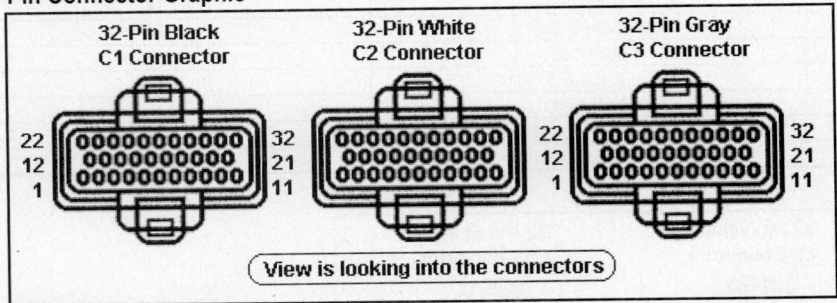

Standard Colors and Abbreviations

Abbreviation	Color	Abbreviation	Color	Abbreviation	Color
BK	Black	GY	Gray	RD	Red
BL	Blue	GN	Green	TN	Tan
BR	Brown	LG	Light Green	VT	Violet
DB	Dark Blue	OR	Orange	WT	White
DG	Dark Green	PK	Pink	YL	Yellow

2000-02 Viper 8.0L V10 MFI VIN E (M/T) 32P C3 Gray Connector

PCM Pin #	Wire Color	Circuit Description (32-Pin)	Value at Hot Idle
1	DB/OR	A/C Clutch Relay Control	Relay Off: 12v, On: 1v
2	DB/PK	Low Speed Fan Relay	Relay Off: 12v, On: 1v
3	DB/YL	ASD Relay Control	Relay Off: 12v, On: 1v
4-5	---	Not Used	---
6	GY/OR	2-3 Skip Shift Solenoid	Solenoid Off: 12v, On: 1v
7-9	---	Not Used	---
10	WT/DG	LDP Solenoid Control	PWM signal: 0-12-0v
11	---	Not Used	---
12	DG/OR	ASD Relay Output	12-14v
13	---	Not Used	---
14	OR	LDP Switch Signal	Switch Closed: 0v, open: 12v
15	VT/LG	Battery Temperature Sensor	At 86°F: 1.96v
16	DG/YL	Generator Lamp Control	Lamp On: 1v, Off: 12v
17	BK/PK	MIL (lamp) Control	Lamp On: 1v, Off: 12v
18	---	Not Used	---
19	BR/VT	Fuel Pump Relay Control	Relay Off: 12v, On: 1v
20	PK/BK	EVAP Purge Solenoid Control	PWM signal: 0-12-0v
21	YL	High Speed Fan Relay	Relay Off: 12v, On: 1v
22	BR	A/C Switch Sense	A/C On: 1v, Off: 12v
23	BR	A/C Switch Sense	A/C On: 1v, Off: 12v
24	WT/PK	Brake Switch Signal	Brake Off: 0v, On: 12v
25	---	Not Used	---
26	DB	Fuel Level Sensor Signal	70 ohms (±20) with full tank
27	PK	SCI Transmit	0v
28	WT/BK	CCD Bus (-)	<0.050v
29	LG	SCI Receive	0v
30	VT/BR	CCD Bus (+)	Digital Signal: 0-5-0v
31	GY/LB	Tachometer Signal	Pulse Signals
32	---	Not Used	---

Pin Connector Graphic

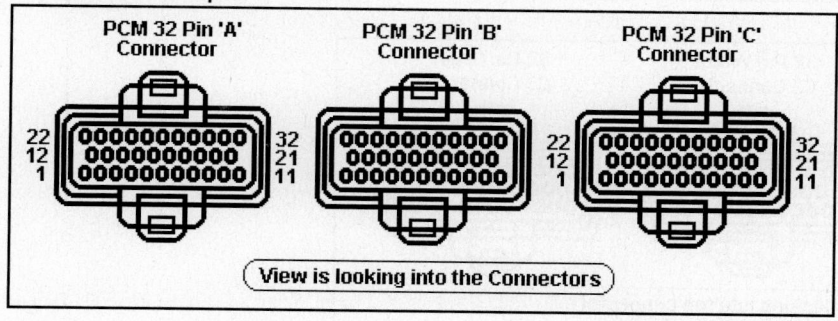

2003 Viper 8.3L V10 MFI VIN Z (M/T) 32P C1 Black Connector

PCM Pin #	Wire Color	Circuit Description (32-Pin)	Value at Hot Idle
1	YL/GY	Coil 4 Driver	6°, 55 mph: 9° dwell
2	DB/GY	Ignition Switch Output	12-14v
3	RD/YL	Coil 3 Driver	6°, 55 mph: 9° dwell
4	BK/LB	Sensor Ground	<0.050v
5	DG/GY	Coil 5 Driver	6°, 55 mph: 9° dwell
6	---	Not Used	---
7	BK/GY	Coil 1 Driver	6°, 55 mph: 9° dwell
8	GY/BK	CKP Sensor Signal	Digital Signal: 0-5-0v
9	DB/TN	Coil 2 Driver	6°, 55 mph: 9° dwell
10	YL/BK	IAC 2 Driver	DC pulse signals: 0.8-11v
11	BR/WT	IAC 3 Driver	DC pulse signals: 0.8-11v
12-14	---	Not Used	---
15	BK/RD	IAT Sensor Signal	At 100°F: 1.83v
16	TN/BK	ECT Sensor Signal	At 180°F: 2.80v
17	OR	5-Volt Supply	4.9-5.1v
18	TN/YL	CMP Sensor Signal	Digital Signal: 0-5-0v
19	GY/RD	IAC 1 Driver	DC pulse signals: 0.8-11v
20	VT/BK	IAC 4 Driver	DC pulse signals: 0.8-11v
21	---	Not Used	---
22	PK/WT	Battery Power (Fused B+)	12-14v
23	OR/DB	TP Sensor Signal	0.6-1.0v
24	LG/RD	HO2S-11 (B1 S1) Signal	0.1-1.1v
25	PK/WT	HO2S-12 (B1 S2) Signal	0.1-1.1v
26	BK/DG	HO2S-21 (B2 S1) Signal	0.1-1.1v
27	DG/RD	MAP Sensor Signal	1.5-1.7v
28	---	Not Used	---
29	TN/WT	HO2S-22 (B2 S2) Signal	0.1-1.1v
30	---	Not Used	---
31	BK/TN	Power Ground	<0.1v
32	BK/TN	Power Ground	<0.1v

Pin Connector Graphic

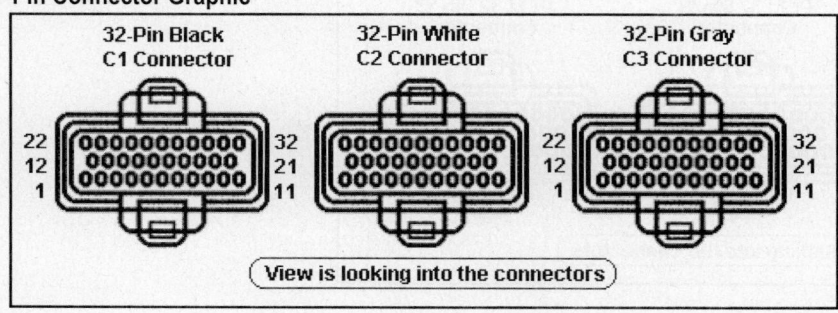

2003 Viper 8.3L V10 MFI VIN Z (M/T) 32P C2 White Connector

PCM Pin #	Wire Color	Circuit Description (32-Pin)	Value at Hot Idle
1	---	Not Used	---
2	VT	Injector 7 Driver	1-4 ms
3	WT/LG	Injector 9 Driver	1-4 ms
4	WT/DB	Injector 1 Driver	1-4 ms
5	YL/WT	Injector 3 Driver	1-4 ms
6	GY	Injector 5 Driver	1-4 ms
7-9	---	Not Used	---
10	DG	Generator Field Control	Digital Signal: 0-12-0v
11	OR/BK	Shift Indicator Lamp Control	Lamp On: 1v, Off: 12v
12	BR/BK	Injector 6 Driver	1-4 ms
13	GY/LG	Injector 8 Driver	1-4 ms
14	BK/LG	Injector 10 Driver	1-4 ms
15	TN	Injector 2 Driver	1-4 ms
16	LB/BR	Injector 4 Driver	1-4 ms
17-20	---	Not Used	---
21	LB/BK	Reverse Lockout Solenoid	Solenoid Off: 12v, On: 1v
22-26	---	Not Used	---
27	WT/TN	OSS Sensor Signal (+)	Moving: AC Pulse Signals
28	WT/OR	Vehicle Speed Signal	Digital Signal
29-30	---	Not Used	---
31	VT/WT	5-Volt Supply	4.9-5.1v
32	---	Not Used	---

Pin Connector Graphic

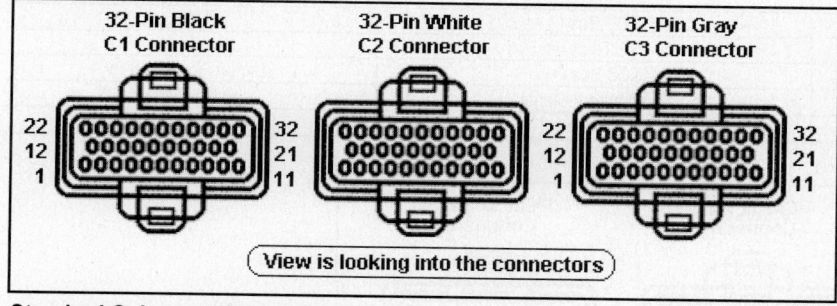

Standard Colors and Abbreviations

Abbreviation	Color	Abbreviation	Color	Abbreviation	Color
BK	Black	GY	Gray	RD	Red
BL	Blue	GN	Green	TN	Tan
BR	Brown	LG	Light Green	VT	Violet
DB	Dark Blue	OR	Orange	WT	White
DG	Dark Green	PK	Pink	YL	Yellow

2003 Viper 8.3L V10 MFI VIN Z (M/T) 32P C3 Gray Connector

PCM Pin #	Wire Color	Circuit Description (32-Pin)	Value at Hot Idle
1	DB/OR	A/C Clutch Relay Control	Relay Off: 12v, On: 1v
2	DB/PK	Low Speed Fan Relay	Relay Off: 12v, On: 1v
3	DB/YL	ASD Relay Control	Relay Off: 12v, On: 1v
4-5	---	Not Used	---
6	GY/OR	2-3 Skip Shift Solenoid	Solenoid Off: 12v, On: 1v
7-9	---	Not Used	---
10	WT/DG	LDP Solenoid Control	PWM signal: 0-12-0v
11	---	Not Used	---
12	DG/OR	ASD Relay Output	12-14v
13	---	Not Used	---
14	OR	LDP Switch Signal	Switch Closed: 0v, open: 12v
15	VT/LG	Battery Temperature Sensor	At 86°F: 1.96v
16	DG/YL	Generator Lamp Control	Lamp On: 1v, Off: 12v
17	BK/PK	MIL (lamp) Control	Lamp On: 1v, Off: 12v
18	---	Not Used	---
19	BR/VT	Fuel Pump Relay Control	Relay Off: 12v, On: 1v
20	PK/BK	EVAP Purge Solenoid Control	PWM signal: 0-12-0v
21	YL	High Speed Fan Relay	Relay Off: 12v, On: 1v
22	BR	A/C Switch Sense	A/C On: 1v, Off: 12v
23	BR	A/C Switch Sense	A/C On: 1v, Off: 12v
24	WT/PK	Brake Switch Signal	Brake Off: 0v, On: 12v
25	---	Not Used	---
26	DB	Fuel Level Sensor Signal	70 ohms (±20) with full tank
27	PK	SCI Transmit	0v
28	WT/BK	CCD Bus (-)	<0.050v
29	LG	SCI Receive	0v
30	VT/BR	CCD Bus (+)	Digital Signal: 0-5-0v
31	GY/LB	Tachometer Signal	Pulse Signals
32	---	Not Used	---

Pin Connector Graphic

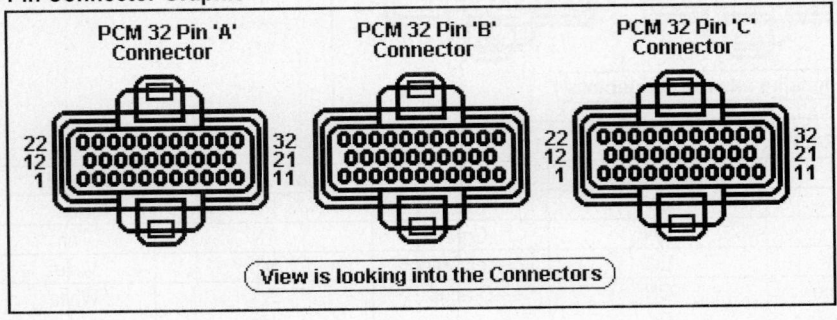

PCM 32 Pin 'A' Connector   PCM 32 Pin 'B' Connector   PCM 32 Pin 'C' Connector

View is looking into the Connectors

EAGLE CAR CONTENTS

About This Section

Introduction

This sectin contains Pin Tables for Eagle vehicles from 1990-1997. It can be used to help you repair Trouble Code and No Code problems related to the PCM.

<u>Vehicle Coverage</u>

The following vehicle applications are covered:

- 1990-92 Premier 3.0L V6 Engine
- 1993-97 Vision 3.3L V6 Engine
- 1993-97 Vision 3.5L V6 Engine

How to Use This Section

This section can be used to look up the location of a particular pin, a wire color or to find a "known good" value of a circuit. To locate the PCM information for a particular vehicle, find the model, correct engine size (with VIN Code) and finally the year of the vehicle.

For example, to look up the PCM terminals for a 1997 Vision 3.3L V6 VIN F, go to Page 1 of the contents to find the text string below.

Then turn to Page 9-15 to find the following PCM related information.

1996-97 Vision 3.3L V6 SOHC VIN F (A/T) 'C1' Black Connector

PCM Pin #	Wire Color	Circuit Description (40 Pin)	Value at Hot Idle
1	LG/RD	HO2S-22 (B2 S2) Signal	0.1-1.1v
2	RD	Coil 3 Driver Control	5°, at 55 mph: 8° dwell
3	WT	Coil 2 Driver Control	5°, at 55 mph: 8° dwell
4	DG	Generator Field Driver	Digital Signal: 0-12-0v
5	YL/RD	S/C Power Supply (B+)	12-14v

In this example, the HO2S-22 (B2 S2) circuit is connected to Pin 1 of the C1 40 Pin connector with an LG/RD wire. The value at Hot Idle shown here indicates this signal varies from 0.1-1.1v in closed loop.

The Generator Field Driver that connects to the DG wire at Pin 4 of the 40 Pin 'C1' Black Connector is used by the PCM to control the output of the Generator. The PCM accomplishes this task by varying the duty cycle of the control signal (0-12v) to the generator field circuit. The PCM varies the duty cycle to meet changes in the vehicle's charging system (the electrical load) whenever the engine is running.

The acronym (A/T) that appears in the title for the table indicates the information in this table is for a vehicle with an automatic transaxle.

EAGLE PIN TABLES

1990-91 Premier Early 3.0L V6 VIN U (A/T) 24P Row 'A' Connector

PCM Pin #	Wire Color	Circuit Description (24 Pin)	Value at Hot Idle
A1-4	---	Not Used	---
A5	DB/YL	Fuel Pump Relay Control	Relay Off: 12v, On: 1v
A6-8	---	Not Used	---
A9	BK/LG	B+ Latch Relay Control	Relay Off: 12v, On: 1v
A10	GY/YL	EGR Solenoid Control	12v, at 55 mph: 1v
A11	---	Not Used	---
A12	YL/BK	A/C WOT Relay Control	Relay Off: 12v, On: 1v

1990-91 Premier Early 3.0L V6 VIN U (A/T) 24P Row 'B' Connector

PCM Pin #	Wire Color	Circuit Description (24 Pin)	Value at Hot Idle
B1	PK/BK	Injector Driver Control	1-4 ms
B2	LG/BK	Injector Driver Control	1-4 ms
B3	BR	ISC Solenoid Control	DC pulse signals
B4	BR/RD	SCI Receive	0v
B5-6	---	Not Used	---
B7	RD	Fused Battery Power (B+)	12-14v
B8	PK/WT	Ignition Switch Power (B+)	12-14v
B9	LG/RD	Engine Speed (PCM to TCM)	DC Signals: 0.2-2.0v
B10	PK	Latch Relay Output (B+)	Relay Off: 12v, On: 1v
B11	BK	Power Ground	<0.1v
B12	BK	Power Ground	<0.1v

Pin Connector Graphic

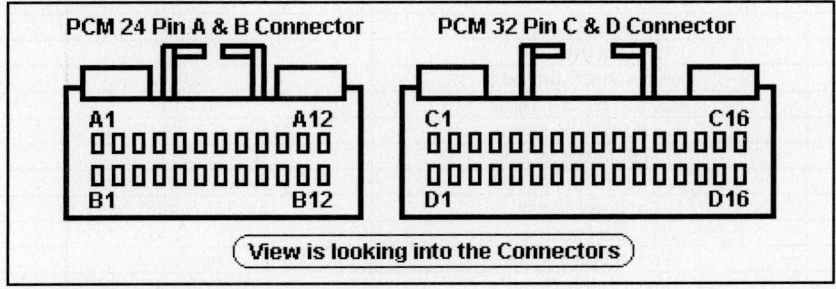

Standard Colors and Abbreviations

Abbreviation	Color	Abbreviation	Color	Abbreviation	Color
BK	Black	GY	Gray	RD	Red
BL	Blue	GN	Green	TN	Tan
BR	Brown	LG	Light Green	VT	Violet
DB	Dark Blue	OR	Orange	WT	White
DG	Dark Green	PK	Pink	YL	Yellow

1990-91 Premier Early 3.0L V6 VIN U (A/T) 32P Row 'C' Connector

PCM Pin #	Wire Color	Circuit Description (32 Pin)	Value at Hot Idle
C1	WT/BK	CKP Sensor (+) Signal	AC pulse signals
C2	LB	A/C Clutch Relay Control	A/C Off: 12v, On: 1v
C3	BR	Starter Input Signal	KOEC: 9-11v
C4	BK/WT	PNP Switch Signal	In P/N: 0v, Others: 5v
C5	---	Not Used	---
C6	VT	MAP Sensor Signal	1.5-1.7v
C7	OR/DB	TP Sensor Signal	0.6-1.0v
C8	BK/RD	Air Temperature Sensor	At 100ºF: 2.51v
C9	---	Not Used	---
C10	TN/BK	ECT Sensor Signal	At 180ºF: 2.80v
C11	DG/BK	Fuel Pump Relay Output	12-14v
C12	BK/PK	SCI Transmit	0v
C13	---	Not Used	---
C14	VT/RD	MAP Sensor VREF	4.9-5.1v
C15	VT/WT	TP Sensor VREF	4.9-5.1v
C16	---	Not Used	---

1990-91 Premier Early 3.0L V6 VIN U (A/T) 32P Row 'D' Connector

PCM Pin #	Wire Color	Circuit Description (32 Pin)	Value at Hot Idle
D1	RD/GN	CKP Sensor (-) Signal	AC pulse signals
D2	LG	A/C Request Switch Sense	A/C Off: 12v, On: 1v
D3	BK/LB	Sensor Ground	<0.050v
D4-5	---	Not Used	---
D6	BK/RD	Power Ground	<0.1v
D7	---	Not Used	---
D8	YL/BK	Knock Sensor Ground	<0.050v
D9	BK/DG	HO2S-11 (B1 S1) Signal	0.1-1.1v
D10	DG/BK	Fuel Pump Relay Control	Relay Off: 12v, On: 1v
D11	BK/WT	SCI Receive	5v
D12	LB/BK	Fuel Pump Relay Control	Relay Off: 12v, On: 1v
D13	YL	Coil Driver Control	5º, at 55 mph: 8º dwell
D14	---	Not Used	---
D15	WT/OR	Vehicle Speed Signal	Digital Signal
D16	YL/BK	Knock Sensor Signal	0.080v AC

Standard Colors and Abbreviations

Abbreviation	Color	Abbreviation	Color	Abbreviation	Color
BK	Black	GY	Gray	RD	Red
BL	Blue	GN	Green	TN	Tan
BR	Brown	LG	Light Green	VT	Violet
DB	Dark Blue	OR	Orange	WT	White
DG	Dark Green	PK	Pink	YL	Yellow

1991-92 Premier 3.0L V6 MFI VIN U (A/T) 60 Pin Connector

PCM Pin #	Wire Color	Circuit Description (60 Pin)	Value at Hot Idle
1	DG/RD	MAP Sensor Signal	1.5-1.7v
2	TN/BK	ECT Sensor Signal	At 180ºF: 2.80v
3	RD/WT	Fused Battery Power (B+)	12-14v
4	BK/LB	Sensor Ground	<0.050v
5	BK/WT	Sensor Ground	<0.050v
6	PK/WT	5-Volt Supply	4.9-5.1v
7	OR	9-Volt Supply	8.9-9.1v
8	LG	A/C Select Switch Sense	A/C Off: 12v, On: 1v
9	DB	Ignition Switch Power (B+)	12-14v
10	---	Not Used	---
11	BK/TN	Power Ground	<0.1v
12	BK/TN	Power Ground	<0.1v
13	---	Not Used	---
14	YL/WT	Injector 3 Driver	1-4 ms
15	TN	Injector 2 Driver	1-4 ms
16	WT/DB	Injector 1 Driver	1-4 ms
17	DB/YL	Coil 2 Driver	5º, at 55 mph: 8º dwell
18	RD/YL	Coil 3 Driver	5º, at 55 mph: 8º dwell
19	GY	Coil 1 Driver	5º, at 55 mph: 8º dwell
20	---	Not Used	---
21	BK	Air Temperature Sensor	At 100ºF: 2.51v
22	OR/DB	TP Sensor Signal	0.6-1.0v
23	RD/LG	S/C Set Switch Signal	S/C & Set Switch On: 3.8v
24	GY/BK	CKP Sensor Signal	Digital Signal: 0-5-0v
25	PK	SCI Transmit	0v
26	---	Not Used	---
27	LB	A/C Request Signal	A/C Off: 12v, On: 1v
28	---	Not Used	---
29	WT/PK	Brake Pedal Position Switch	Brake Off: 0v, On: 12v
30	BR/YL	PNP Switch Signal	In P/N: 0v, Others: 5v
31	DB/PK	Radiator Fan Relay Control	Relay Off: 12v, On: 1v
32	BK/PK	MIL (lamp) Control	MIL On: 1v, Off: 12v
33	TN/RD	S/C Vacuum Solenoid Control	Vacuum Increasing: 1v
34	DB/OR	A/C Clutch Relay Control	Relay Off: 12v, On: 1v
35	GY/YL	EGR Solenoid Control	12v, at 55 mph: 1v
36-38	---	Not Used	---
39	GY/RD	AIS 3 Motor Control	DC pulse signals
40	BR/WT	AIS 1 Motor Control	DC pulse signals

Standard Colors and Abbreviations

Abbreviation	Color	Abbreviation	Color	Abbreviation	Color
BK	Black	GY	Gray	RD	Red
BL	Blue	GN	Green	TN	Tan
BR	Brown	LG	Light Green	VT	Violet
DB	Dark Blue	OR	Orange	WT	White
DG	Dark Green	PK	Pink	YL	Yellow

1991-92 Premier 3.0L V6 MFI VIN U (A/T) 60 Pin Connector

PCM Pin #	Wire Color	Circuit Description (60 Pin)	Value at Hot Idle
41	BK/DG	HO2S-11 (B1 S1) Signal	0.1-1.1v
42	---	Not Used	---
43	GY/LB	Tachometer Signals	DC pulse signals
44	TN/YL	CMP Sensor Signal	Digital Signal: 0-5-0v
45	LG	SCI Receive	5v
46	---	Not Used	---
47	DB/WT	Vehicle Speed Signal	Digital Signal
48-50	---	Not Used	---
51	DB/YL	ASD Relay Output	12-14v
52	LB/BK	Fuel Pump Relay Control	Relay Off: 12v, On: 1v
53	LG/RD	S/C Vent Solenoid Control	Vacuum Decreasing: 1v
54	---	Not Used	---
55	PK/WT	S/C Relay Output	12-14v
56	PK	S/C Indicator Control	Lamp Off: 12v, On: 1v
57	DG/RD	ASD Relay Control	Relay Off: 12v, On: 1v
58	---	Not Used	---
59	PK/BK	AIS 4 Motor Control	DC pulse signals
60	YL/BK	AIS 2 Motor Control	DC pulse signals

Pin Connector Graphic

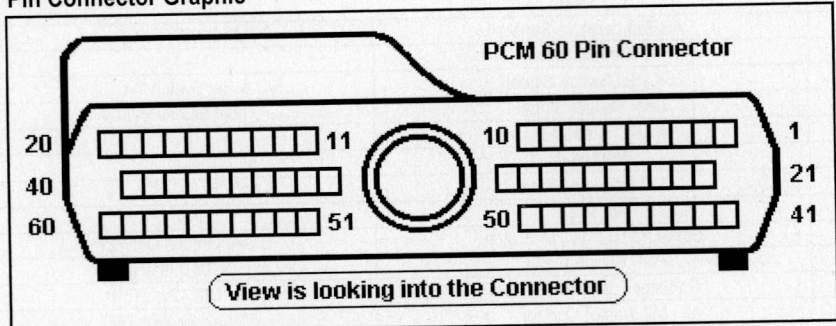

PCM 60 Pin Connector

20   11   10   1
40   21
60   51   50   41

View is looking into the Connector

Standard Colors and Abbreviations

Abbreviation	Color	Abbreviation	Color	Abbreviation	Color
BK	Black	GY	Gray	RD	Red
BL	Blue	GN	Green	TN	Tan
BR	Brown	LG	Light Green	VT	Violet
DB	Dark Blue	OR	Orange	WT	White
DG	Dark Green	PK	Pink	YL	Yellow

1993-95 Vision 3.3L V6 MFI VIN T (A/T) 60 Pin Connector

PCM Pin #	Wire Color	Circuit Description (60 Pin)	Value at Hot Idle
1	DG/RD	MAP Sensor Signal	1.5-1.7v
2	TN/BK	ECT Sensor Signal	At 180ºF: 2.80v
3	RD/WT	Fused Battery Power (B+)	12-14v
4	BK/LB	Sensor Ground	<0.050v
5	BK/WT	Sensor Ground	<0.050v
6	PK/WT	5-Volt Supply	4.9-5.1v
7	OR	8-Volt Supply	7.9-8.1v
8	YL/DG	Torque Management Request	Digital Signals
9	DB/WT	Ignition Switch Power (B+)	12-14v
10	GY/BK	Knock Sensor 2 Signal	0.080v AC
11	BK/TN	Power Ground	<0.1v
12	BK/TN	Power Ground	<0.1v
13	LB/BR	Injector 4 Driver	1-4 ms
14	YL/WT	Injector 3 Driver	1-4 ms
15	TN	Injector 2 Driver	1-4 ms
16	WT/DB	Injector 1 Driver	1-4 ms
17	DB/YL	Coil 2 Driver	5º, at 55 mph: 8º dwell
17 ('94-'95)	WT	Coil 2 Driver	5º, at 55 mph: 8º dwell
18	RD/YL	Coil 3 Driver	5º, at 55 mph: 8º dwell
18 ('94-'95)	RD	Coil 3 Driver	5º, at 55 mph: 8º dwell
19	GY	Coil 1 Driver	5º, at 55 mph: 8º dwell
19 ('94-'95)	BK	Coil 1 Driver	5º, at 55 mph: 8º dwell
20	DG	Alternator Field Control	Digital Signal: 0-12-0v
21	BK/RD	IAT Sensor Signal	At 100ºF: 1.83v
22	OR/DB	TP Sensor Signal	0.6-1.0v
23	RD/LG	S/C Set Switch Signal	S/C & Set Switch On: 3.8v
24	LB/DB	CKP Sensor Signal	Digital Signal: 0-5-0v
25	PK	SCI Transmit	0v
26	VT/BR	CCD Bus (+)	Digital Signal: 0-5-0v
27-28	---	Not Used	---
29	WT/PK	Brake Pedal Position Switch	Brake Off: 0v, On: 12v
30	BK/LG	PNP Switch Signal	In P/N: 0v, Others: 5v
31	DB/PK	High Speed Fan Control	Relay Off: 12v, On: 1v
32	WT	Low Speed Fan Relay	Relay Off: 12v, On: 1v
33	TN/RD	S/C Vacuum Solenoid Control	Vacuum Increasing: 1v
34	DB/OR	A/C Clutch Relay Control	Relay Off: 12v, On: 1v
35	GY/YL	EGR Solenoid Control (Cal)	12v, at 55 mph: 1v
36	VT	Manifold Tuning Valve Control	Solenoid Off: 12v, On: 1v
37	---	Not Used	---
38	BR/RD	Injector 5 Driver	1-4 ms
39	GY/RD	AIS 1 Motor Control	DC pulse signals
40	BR/WT	AIS 3 Motor Control	DC pulse signals

Standard Colors and Abbreviations

Abbreviation	Color	Abbreviation	Color	Abbreviation	Color
BK	Black	GY	Gray	RD	Red
BL	Blue	GN	Green	TN	Tan
BR	Brown	LG	Light Green	VT	Violet
DB	Dark Blue	OR	Orange	WT	White
DG	Dark Green	PK	Pink	YL	Yellow

1993-95 Vision 3.3L V6 MFI VIN T (A/T) 60 Pin Connector

PCM Pin #	Wire Color	Circuit Description (60 Pin)	Value at Hot Idle
41	BK/DG	HO2S-11 (B1 S1) Signal	0.1-1.1v
42	BK/LG	Knock Sensor 1 Signal	0.080v AC
43	---	Not Used	---
44	TN/YL	CMP Sensor Signal	Digital Signal: 0-5-0v
45	LG	SCI Receive	5v
46	WT/BK	CCD Bus (-)	<0.050v
47	WT/OR	Vehicle Speed Signal	Digital Signal
48	DB	A/C Pressure Switch Signal	A/C On: 0.45-4.85v
49	TN/WT	HO2S-21 (B2 S1) Signal	0.1-1.1v
50	---	Not Used	---
51	DB/YL	ASD Relay Control	Relay Off: 12v, On: 1v
52	PK/BK	EVAP Purge Solenoid Control	Solenoid Off: 12v, On: 1v
53	LG/RD	S/C Vent Solenoid Control	Vacuum Decreasing: 1v
54	---	Not Used	---
55	TN/RD	S/C Relay Control	Relay Off: 12v, On: 1v
56	---	Not Used	---
57	DG/OR	ASD Relay Output	12-14v
58	BR/BK	Injector 6 Driver	1-4 ms
59	PK/BK	AIS 4 Motor Control	DC pulse signals
60	YL/BK	AIS 2 Motor Control	DC pulse signals

Pin Connector Graphic

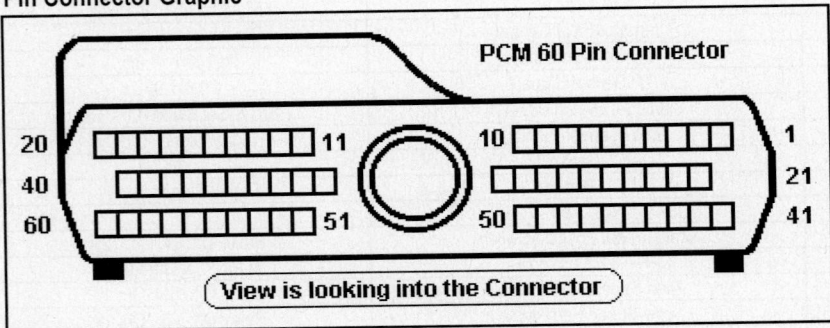

1996 Vision 3.3L V6 MFI VIN T (A/T) 'C1' Black Connector

PCM Pin #	Wire Color	Circuit Description (40 Pin)	Value at Hot Idle
1	LG/RD	HO2S-22 (B2 S2) Signal	0.1-1.1v
2	RD	Coil 3 Driver	5°, at 55 mph: 8° dwell
3	WT	Coil 2 Driver	5°, at 55 mph: 8° dwell
4	DG	Generator Field Driver	Digital Signal: 0-12-0v
5	YL/RD	S/C Power Supply (B+)	12-14v
6	DG/OR	ASD Relay Output	12-14v
7	YL/WT	Injector 3 Driver	1-4 ms
8-9	---	Not Used	---
10	BK/TN	Power Ground	<0.1v
11	BK	Coil 1 Driver	5°, at 55 mph: 8° dwell
13	WT/LB	Injector 1 Driver	1-4 ms
14	BR/BK	Injector 6 Driver	1-4 ms
15	BR/RD	Injector 5 Driver	1-4 ms
16	LB/BR	Injector 4 Driver	1-4 ms
17	TN	Injector 2 Driver	1-4 ms
18-19	---	Not Used	---
20	DB/WT	Ignition Switch Power (B+)	12-14v
21-25	---	Not Used	---
26	TN/BK	ECT Sensor Signal	At 180°F: 2.80v
27-28	---	Not Used	---
29	TN/WT	HO2S-21 (B2 S1) Signal	0.1-1.1v
30	BK/DG	HO2S-11 (B1 S1) Signal	0.1-1.1v
31	---	Not Used	---
32	LB/DB	CKP Sensor Signal	Digital Signal: 0-5-0v
33	TN/YL	CMP Sensor Signal	Digital Signal: 0-5-0v
34	---	Not Used	---
35	OR/DB	TP Sensor Signal	0.6-1.0v
36	DG/RD	MAP Sensor Signal	1.5-1.7v
37	BK/RD	IAT Sensor Signal	At 100°F: 1.83v
38	---	Not Used	---
39	PK/RD	Manifold Tuning Valve Control	Valve On: 1v, Off: 12v
40	GY/YL	EGR Solenoid Control	12v, at 55 mph: 1v

Standard Colors and Abbreviations

Abbreviation	Color	Abbreviation	Color	Abbreviation	Color
BK	Black	GY	Gray	RD	Red
BL	Blue	GN	Green	TN	Tan
BR	Brown	LG	Light Green	VT	Violet
DB	Dark Blue	OR	Orange	WT	White
DG	Dark Green	PK	Pink	YL	Yellow

1996 Vision 3.3L V6 MFI VIN T (A/T) 'C2' White Connector

PCM Pin #	Wire Color	Circuit Description (40 Pin)	Value at Hot Idle
41	RD/LG	S/C Set Switch Signal	S/C & Set Switch On: 3.8v
42	DB	A/C Pressure Sensor	A/C On: 0.45-4.85v
43	BK/LB	Sensor Ground	<0.050v
44	OR	8-Volt Supply	7.9-8.1v
45	---	Not Used	---
46	RD/WT	Fused Battery Power (B+)	12-14v
47	BK/WT	Power Ground	<0.1v
48	BR/WT	IAC 1 Driver	DC pulse signals
49	YL/BK	IAC 2 Driver	DC pulse signals
50	BK/TN	Power Ground	<0.1v
51	TN/WT	HO2S-12 (B1 S2) Signal	0.1-1.1v
52-54	---	Not Used	---
55	WT	Low Speed Fan Relay	Relay Off: 12v, On: 1v
56	---	Not Used	---
57	GY/RD	IAC 3 Driver	DC pulse signals
58	PK/BK	IAC 4 Driver	DC pulse signals
59	PK/BR	CCD Bus (+)	Digital signals: 0-5-0-5v
60	WT/BK	CCD Bus (-)	<0.050v
61	PK/WT	5-Volt Supply	4.9-5.1v
62	WT/PK	Brake Pedal Position Switch	Brake Off: 0v, On: 12v
63	YL/DG	Torque Management Request	Digital Signals
64	DB/OR	A/C Clutch Relay Control	Relay Off: 12v, On: 1v
65	PK	SCI Transmit	0v
66	WT/OR	Vehicle Speed Signal	Digital Signal
67	DB/YL	ASD Relay Control	Relay Off: 12v, On: 1v
68	PK/BK	EVAP Purge Solenoid Control	PWM Signal: 0-12-0v
69	DB/PK	High Speed Fan Control	Relay Off: 12v, On: 1v
70-73	---	Not Used	---
74	BR	Fuel Pump Relay Control	Relay Off: 12v, On: 1v
75	LG	SCI Receive	5v
76	BK/DG	PNP Switch Signal	In P/N: 0v, Others: 5v
77	---	Not Used	---
78	TN/RD	S/C Vacuum Solenoid Control	Vacuum Increasing: 1v
79	---	Not Used	---
80	LG/RD	S/C Vent Solenoid Control	Vacuum Decreasing: 1v

Pin Connector Graphic

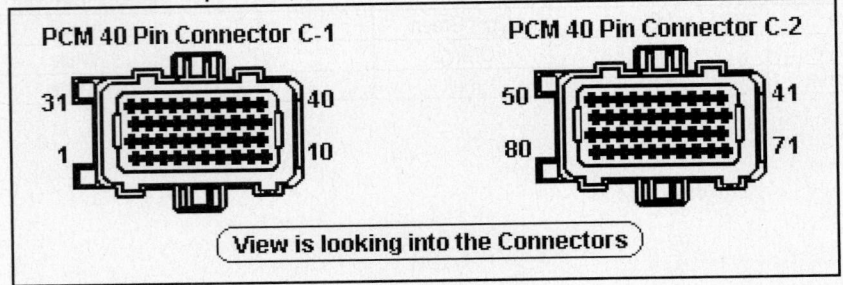

PCM 40 Pin Connector C-1

31   40

1   10

PCM 40 Pin Connector C-2

50   41

80   71

View is looking into the Connectors

1994 Vision 3.3L V6 Flexible Fuel VIN U (A/T) 60 Pin Connector

PCM Pin #	Wire Color	Circuit Description (60 Pin)	Value at Hot Idle
1	DG/RD	MAP Sensor Signal	1.5-1.7v
2	TN/BK	ECT Sensor Signal	At 180°F: 2.80v
3	RD/WT	Fused Battery Power (B+)	12-14v
4	BK/LB	Sensor Ground	<0.050v
5	BK/WT	Sensor Ground	<0.050v
6	PK/WT	5-Volt Supply	4.9-5.1v
7	OR	8-Volt Supply	7.9-8.1v
8	YL/DG	Torque Management Request	Digital Signals
9	DB/WT	Ignition Switch Power (B+)	12-14v
10	---	Not Used	---
11	BK/TN	Power Ground	<0.1v
12	BK/TN	Power Ground	<0.1v
13	LB/BR	Injector 4 Driver	1-4 ms
14	YL/WT	Injector 3 Driver	1-4 ms
15	TN	Injector 2 Driver	1-4 ms
16	WT/DB	Injector 1 Driver	1-4 ms
17	WT	Coil 2 Driver	5°, at 55 mph: 8° dwell
18	RD/YL	Coil 3 Driver	5°, at 55 mph: 8° dwell
19	BK	Coil 1 Driver	5°, at 55 mph: 8° dwell
20	DG	Alternator Field Control	Digital Signal: 0-12-0v
21	BK/RD	IAT Sensor Signal	At 100°F: 1.83v
22	OR/DB	TP Sensor Signal	0.6-1.0v
23	RD/LG	S/C Set Switch Signal	S/C & Set Switch On: 3.8v
24	GY/BK	CKP Sensor Signal	Digital Signal: 0-5-0v
25	PK	SCI Transmit	0v
26	PK/BR	CCD Bus (+)	Digital Signal: 0-5-0v
27-28	---	Not Used	---
29	WT/PK	Brake Pedal Position Switch	Brake Off: 0v, On: 12v
30	BK/LG	PNP Switch Signal	In P/N: 0v, Others: 5v
31	DB/PK	High Speed Fan Control	Relay Off: 12v, On: 1v
32	WT	Low Speed Fan Relay	Relay Off: 12v, On: 1v
33	TN/RD	S/C Vacuum Solenoid Control	Vacuum Increasing: 1v
34	DB/OR	A/C Clutch Relay Control	Relay Off: 12v, On: 1v
35	GY/YL	EGR Solenoid Control (Cal)	12v, at 55 mph: 1v
36	PK/RD	Manifold Tuning Valve Control	Solenoid Off: 12v, On: 1v
37	---	Not Used	---
38	BR/RD	Injector 5 Driver	1-4 ms
39	GY/RD	AIS 1 Motor Control	DC pulse signals
40	BR/WT	AIS 3 Motor Control	DC pulse signals

Standard Colors and Abbreviations

Abbreviation	Color	Abbreviation	Color	Abbreviation	Color
BK	Black	GY	Gray	RD	Red
BL	Blue	GN	Green	TN	Tan
BR	Brown	LG	Light Green	VT	Violet
DB	Dark Blue	OR	Orange	WT	White
DG	Dark Green	PK	Pink	YL	Yellow

1994 Vision 3.3L V6 Flexible Fuel VIN U (A/T) 60 Pin Connector

PCM Pin #	Wire Color	Circuit Description (60 Pin)	Value at Hot Idle
41	BK/DG	HO2S-11 (B1 S1) Signal	0.1-1.1v
42	BK/LG	Knock Sensor Signal	0.080v AC
43	---	Not Used	---
44	TN/YL	CMP Sensor Signal	Digital Signal: 0-5-0v
45	LG	SCI Receive	5v
46	WT/BK	CCD Bus (-)	<0.050v
47	WT/OR	Vehicle Speed Signal	Digital Signal
48	DB	A/C Pressure Switch Signal	A/C On: 0.45-4.85v
49	TN/WT	HO2S-21 (B2 S1) Signal	0.1-1.1v
50	YL/WT	Flexible Fuel Sensor Signal	0.5-4.5v
51	DB/YL	ASD Relay Control	Relay Off: 12v, On: 1v
52	PK/BK	EVAP Purge Solenoid Control	Solenoid Off: 12v, On: 1v
53	LG/RD	S/C Vent Solenoid Control	Vacuum Decreasing: 1v
54	---	Not Used	---
55	TN/RD	S/C Relay Output	12-14v
56	---	Not Used	---
57	DG/OR	ASD Relay Output	12-14v
58	BR/BK	Injector 6 Driver	1-4 ms
59	PK/BK	AIS 4 Motor Control	DC pulse signals
60	YL/BK	AIS 2 Motor Control	DC pulse signals

Pin Connector Graphic

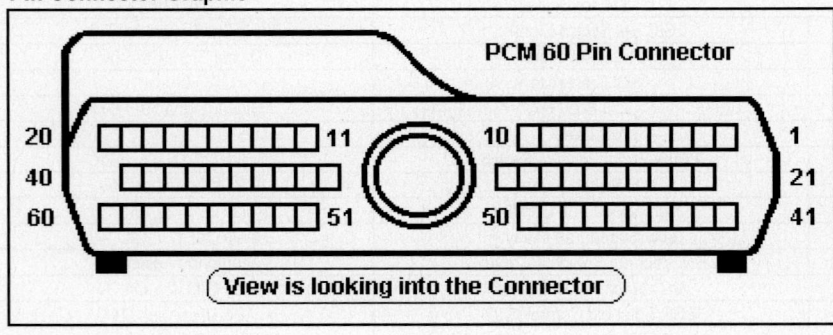

1993-95 Vision 3.5L V6 SOHC MFI VIN F (A/T) 60 Pin Connector

PCM Pin #	Wire Color	Circuit Description (60 Pin)	Value at Hot Idle
1	DG/RD	MAP Sensor Signal	1.5-1.7v
2	TN/BK	ECT Sensor Signal	At 180°F: 2.80v
3	RD/WT	Fused Battery Power (B+)	12-14v
4	BK/LB	Sensor Ground	<0.050v
5	BK/WT	Sensor Ground	<0.050v
6	PK/WT	5-Volt Supply	4.9-5.1v
7	OR	8-Volt Supply	7.9-8.1v
8	YL/DG	Torque Management Request	Digital Signals
9	DB/WT	Ignition Switch Power (B+)	12-14v
10	GY/BK	Knock Sensor 2 Signal	0.080v AC
11	BK/TN	Power Ground	<0.1v
12	BK/TN	Power Ground	<0.1v
13	LB/BR	Injector 4 Driver	1-4 ms
14	YL/WT	Injector 3 Driver	1-4 ms
15	TN	Injector 2 Driver	1-4 ms
16	WT/DB	Injector 1 Driver	1-4 ms
17	DB/YL	Coil 2 Driver	5°, at 55 mph: 8° dwell
17 ('94-'95)	WT	Coil 2 Driver	5°, at 55 mph: 8° dwell
18	RD/YL	Coil 3 Driver	5°, at 55 mph: 8° dwell
18 ('94-'95)	RD	Coil 3 Driver	5°, at 55 mph: 8° dwell
19	GY	Coil 1 Driver	5°, at 55 mph: 8° dwell
19 ('94-'95)	BK	Coil 1 Driver	5°, at 55 mph: 8° dwell
20	DG	Generator Field Driver	Digital Signal: 0-12-0v
21	BK/RD	IAT Sensor Signal	At 100°F: 1.83v
22	OR/DB	TP Sensor Signal	0.6-1.0v
23	RD/LG	S/C Set Switch Signal	S/C & Set Switch On: 3.8v
24	LB/DB	CKP Sensor Signal	Digital Signal: 0-5-0v
25	PK	SCI Transmit	0v
26	PK/BR	CCD Bus (+)	Digital Signal: 0-5-0v
27-28	---	Not Used	---
29	WT/PK	Brake Pedal Position Switch	Brake Off: 0v, On: 12v
30	BK/LG	PNP Switch Signal	In P/N: 0v, Others: 5v
31	DB/PK	High Speed Fan Control	Relay Off: 12v, On: 1v
32	WT	Low Speed Fan Relay	Relay Off: 12v, On: 1v
33	TN/RD	S/C Vacuum Solenoid Control	Vacuum Increasing: 1v
34	DB/OR	A/C Clutch Relay Control	Relay Off: 12v, On: 1v
35	GY/YL	EGR Solenoid Control	12v, at 55 mph: 1v
36	PK	Manifold Tuning Valve Control	Solenoid Off: 12v, On: 1v
37	---	Not Used	---
38	BR/RD	Injector 5 Driver	1-4 ms
39	GY/RD	IAC 1 Driver	DC pulse signals
40	BR/WT	IAC 3 Driver	DC pulse signals

Standard Colors and Abbreviations

Abbreviation	Color	Abbreviation	Color	Abbreviation	Color
BK	Black	GY	Gray	RD	Red
BL	Blue	GN	Green	TN	Tan
BR	Brown	LG	Light Green	VT	Violet
DB	Dark Blue	OR	Orange	WT	White
DG	Dark Green	PK	Pink	YL	Yellow

1993-95 Vision 3.5L V6 SOHC MFI VIN F (A/T) 60 Pin Connector

PCM Pin #	Wire Color	Circuit Description (60 Pin)	Value at Hot Idle
41	BK/DG	HO2S-11 (B1 S1) Signal	0.1-1.1v
42	BK/LG	Knock Sensor 1 Signal	0.080v AC
43	---	Not Used	---
44	TN/YL	CMP Sensor Signal	Digital Signal: 0-5-0v
45	LG	SCI Receive	5v
46	WT/BK	CCD Bus (-)	<0.050v
47	WT/OR	Vehicle Speed Signal	Digital Signal
48	DB	A/C Pressure Switch Signal	A/C On: 0.45-4.85v
49	TN/WT	HO2S-21 (B2 S1) Signal	0.1-1.1v
50	---	Not Used	---
51	DB/YL	ASD Relay Control	Relay Off: 12v, On: 1v
52	PK/BK	EVAP Purge Solenoid Control	Solenoid Off: 12v, On: 1v
53	LG/RD	S/C Vent Solenoid Control	Vacuum Decreasing: 1v
54	---	Not Used	---
55	TN/RD	S/C Relay Power Output	Relay Off: 12v, On: 1v
56	---	Not Used	---
57	DG/OR	ASD Relay Output	12-14v
58	BR/BK	Injector 6 Driver	1-4 ms
59	PK/BK	IAC 4 Driver	DC pulse signals
60	YL/BK	IAC 2 Driver	DC pulse signals

Pin Connector Graphic

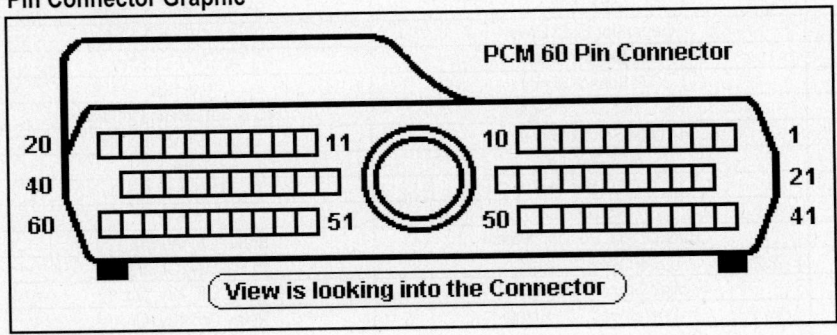

1996-97 Vision 3.5L V6 SOHC MFI VIN F (A/T) 'C1' Black Connector

PCM Pin #	Wire Color	Circuit Description (40 Pin)	Value at Hot Idle
1	LG/RD	HO2S-22 (B2 S2) Signal	0.1-1.1v
2	RD	Coil 3 Driver	5°, at 55 mph: 8° dwell
3	WT	Coil 2 Driver	5°, at 55 mph: 8° dwell
4	DG	Generator Field Driver	Digital Signal: 0-12-0v
5	YL/RD	S/C Power Supply (B+)	12-14v
6	DG/OR	ASD Relay Output	12-14v
7	YL/WT	Injector 3 Driver	1-4 ms
8-9	---	Not Used	---
10	BK/TN	Power Ground	<0.1v
11	BK	Coil 1 Driver	5°, at 55 mph: 8° dwell
12	---	Not Used	---
13	WT/LB	Injector 1 Driver	1-4 ms
14	BR/BK	Injector 6 Driver	1-4 ms
15	BR/RD	Injector 5 Driver	1-4 ms
16	LB/BR	Injector 4 Driver	1-4 ms
17	TN	Injector 2 Driver	1-4 ms
18-19	---	Not Used	---
20	DB/WT	Ignition Switch Power (B+)	12-14v
21-23	---	Not Used	---
24	BK/LG	Knock Sensor 1 Signal	0.080v AC
25	GY/BK	Knock Sensor 2 Signal	0.080v AC
26	TN/BK	ECT Sensor Signal	At 180°F: 2.80v
27-28	---	Not Used	---
29	TN/WT	HO2S-21 (B2 S1) Signal	0.1-1.1v
30	BK/DG	HO2S-11 (B1 S1) Signal	0.1-1.1v
31	---	Not Used	---
32	LB/DB	CKP Sensor Signal	Digital Signal: 0-5-0v
33	TN/YL	CMP Sensor Signal	Digital Signal: 0-5-0v
34	---	Not Used	---
35	OR/DB	TP Sensor Signal	0.6-1.0v
36	DG/RD	MAP Sensor Signal	1.5-1.7v
37	BK/RD	IAT Sensor Signal	At 100°F: 1.83v
38	---	Not Used	---
39	PK/RD	Manifold Tuning Valve Control	Solenoid Off: 12v, On: 1v
40	GY/YL	EGR Solenoid Control	12v, 55 mph: 1v

Standard Colors and Abbreviations

Abbreviation	Color	Abbreviation	Color	Abbreviation	Color
BK	Black	GY	Gray	RD	Red
BL	Blue	GN	Green	TN	Tan
BR	Brown	LG	Light Green	VT	Violet
DB	Dark Blue	OR	Orange	WT	White
DG	Dark Green	PK	Pink	YL	Yellow

1996-97 Vision 3.5L V6 SOHC MFI VIN F (A/T) 'C2' White Connector

PCM Pin #	Wire Color	Circuit Description (40 Pin)	Value at Hot Idle
41	RD/LG	S/C Set Switch Signal	S/C & Set Switch On: 3.8v
42	DB	A/C Pressure Switch Signal	A/C On: 0.45-4.85v
43	BK/LB	Sensor Ground	<0.050v
44	OR	8-Volt Supply	7.9-8.1v
45	---	Not Used	---
46	RD/WT	Fused Battery Power (B+)	12-14v
47	BK/WT	Power Ground	<0.1v
48	BR/WT	IAC 1 Driver	DC pulse signals
49	YL/BK	IAC 2 Driver	DC pulse signals
50	BK/TN	Power Ground	<0.1v
51	TN/WT	HO2S-12 (B1 S2) Signal	0.1-1.1v
52-54	---	Not Used	---
55	WT	Low Speed Fan Relay	Relay Off: 12v, On: 1v
56	---	Not Used	---
57	GY/RD	IAC 3 Driver	DC pulse signals
58	PK/BK	IAC 4 Driver	DC pulse signals
59	PK/BR	CCD Bus (+)	Digital Signal: 0-5-0v
60	WT/BK	CCD Bus (-)	<0.050v
61	PK/WT	5-Volt Supply	4.9-5.1v
62	WT/PK	Brake Pedal Position Switch	Brake Off: 0v, On: 12v
63	YL/DG	Torque Management Request	Digital Signals
64	DB/OR	A/C Clutch Relay Control	A/C Off: 12v, On: 1v
65	PK	SCI Transmit	0v
66	WT/OR	Vehicle Speed Signal	Digital Signal
67	DB/YL	ASD Relay Control	Relay Off: 12v, On: 1v
68	PK/BK	EVAP Purge Solenoid Control	PWM Signal: 0-12-0v
69	DB/PK	High Speed Fan Control	Relay Off: 12v, On: 1v
70-73	---	Not Used	---
74	BR	Fuel Pump Relay Control	Relay Off: 12v, On: 1v
75	LG	SCI Receive	5v
76	BK/DG	PNP Switch Signal	In P/N: 0v, Others: 5v
77	---	Not Used	---
78	TN/RD	S/C Vacuum Solenoid Control	Vacuum Increasing: 1v
77	---	Not Used	---
80	LG/RD	S/C Vent Solenoid Control	Vacuum Decreasing: 1v

Pin Connector Graphic

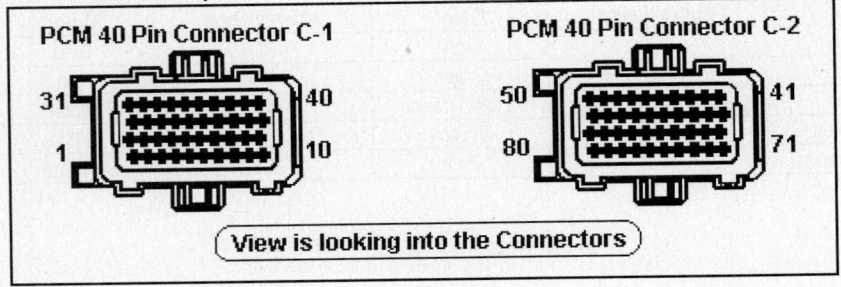

PLYMOUTH CAR CONTENTS

About This Section

This section of the manual contains Pin Tables for Plymouth vehicles from 1990-2001. It can be used to help you repair Trouble Code and No Code problems related to the PCM.

Vehicle Coverage

The following vehicle applications are covered:

- 1990-95 Acclaim
- 1996-2000 Breeze
- 1990-94 Laser
- 1995-2001 Neon
- 1997-2001 Prowler
- 1990-94 Sundance

How to Use This Section

This section of the manual can be used to look up the location of a particular pin, a wire color or to find a "known good" value of a circuit. To locate the PCM information for a particular vehicle, find the model, correct engine size (with VIN Code) and finally the year of the vehicle.

For example, to look up the PCM terminals for a 1999 Breeze 2.4L I4 VIN X, go to Contents Page 1 and find the text string shown below:

2.4L I4 4v MFI VIN X (A/T) - 104 Pin **(1998-2000)** ....Page 10-21

Then turn to Page 10-22 to find the following PCM related information.

1998-2000 Breeze 2.4L I4 DOHC VIN X (A/T) 'C1' Black Connector

PCM Pin #	Wire Color	Circuit Description (40-Pin)	Value at Hot Idle
26	TN/BK	ECT Sensor Signal	At 180°F: 2.80v
27	BK/OR	HO2S Ground	<0.050v
30	BK/DG	HO2S-11 (B1 S1) Signal	0.1-1.1v
32	GY/BK	CKP Sensor Signal	Digital Signal: 0-5-0v
33	TN/YL	CMP Sensor Signal	Digital Signal: 0-5-0v

In this example, the ECT sensor circuit is connected to Pin 26 of the C1 40-pin connector with a TN/BK wire. The value at Hot Idle shown here indicates a voltage of 2.80v at an engine temperature of 180°F.

The CKP sensor, which connects to Pin 32 of the 'C1' Black Connector with a GY/BK wire, provides crankshaft position and speed information.

The CMP sensor connects to Pin 33 with a TN/YL wire. Note that the CKP and CMP Sensor signals are digital signals that vary from 0-5-0v with the engine running. At key on, engine off, these circuits read either 0v or 5v (depending upon the position of the Hall-Effect sensor).

The acronym (A/T) that appears in the title for the table indicates the information in this table is for a vehicle with an automatic transaxle.

PLYMOUTH PIN TABLES

1990 Acclaim 2.5L I4 MFI Turbo VIN J (All) 60 Pin Connector

PCM Pin #	Wire Color	Circuit Description (60 Pin)	Value at Hot Idle
1	DG/RD	MAP Sensor Signal	1.5-1.7v
2	TN/BK	ECT Sensor Signal	At 180°F: 2.80v
3	RD/WT	Battery Power (Fused B+)	12-14v
4	BK/LB	Sensor Ground	<0.050v
5	BK/WT	Sensor Ground	<0.050v
6	PK/WT	5-Volt Supply	4.9-5.1v
7	OR	9-Volt Supply	8.9-9.1v
8	DG/BK	ASD Relay Output	12-14v
9	DB	Ignition Switch Output	12-14v
10	---	Not Used	---
11	BK/TN	Power Ground	<0.1v
12	BK/TN	Power Ground	<0.1v
13	LB/BR	Injector 4 Driver	1-4 ms
14	YL/WT	Injector 3 Driver	1-4 ms
15	TN	Injector 2 Driver	1-4 ms
16	WT/DB	Injector 1 Driver	1-4 ms
17-18	---	Not Used	---
19	BK/GY	Coil 1 Driver	5°, at 55 mph: 8° dwell
20	DG	Alternator Field Control	Digital Signal: 0-12-0v
21	---	Not Used	---
22	OR/DB	TP Sensor Signal	0.6-1.0v
23	---	Not Used	---
24	GY/BK	Distributor Reference Signal	Digital Signal: 0-5-0v
25	PK	SCI Transmit	0v
26	PK/BR	CCD Bus (+)	Digital Signal: 0-5-0v
27	BR	A/C Clutch Signal	A/C On: 1v, Off: 12v
28	---	Not Used	---
29	WT/PK	Brake Switch Signal	Brake Off: 0v, On: 12v
30	BR/LB	PNP Switch Signal	In P/N: 0v, Others: 5v
31	DB/PK	Radiator Fan Relay Control	Relay Off: 12v, On: 1v
32	BK/PK	MIL (lamp) Control	Lamp On: 1v, Off: 12v
33	TN/RD	S/C Vacuum Solenoid	Vacuum Increasing: 1v
34	DB/OR	A/C WOT Relay Control	Relay Off: 12v, On: 1v
35	GY/YL	EGR Solenoid Control (California)	12v, 55 mph: 1v
36	LG/BK	Wastegate Solenoid Control	Solenoid Off: 12v, On: 1v
37-38	---	Not Used	---
39	GY/RD	AIS 3 Motor Control	DC pulse signals: 0.8-11v
40	BR/WT	AIS 1 Motor Control	DC pulse signals: 0.8-11v

Standard Colors and Abbreviations

Abbreviation	Color	Abbreviation	Color	Abbreviation	Color
BK	Black	GY	Gray	RD	Red
BL	Blue	GN	Green	TN	Tan
BR	Brown	LG	Light Green	VT	Violet
DB	Dark Blue	OR	Orange	WT	White
DG	Dark Green	PK	Pink	YL	Yellow

1990 Acclaim 2.5L I4 MFI Turbo VIN J (All) 60 Pin Connector

PCM Pin #	Wire Color	Circuit Description (60 Pin)	Value at Hot Idle
41	BK/DG	HO2S-11 (B1 S1) Signal	0.1-1.1v
42	BK/LG	Knock Sensor Signal	0.080v AC
43	GY/LB	Tachometer Signals	DC pulse signals
44	TN/YL	Distributor Sync Signals	Digital Signal: 0-5-0v
45	LG	SCI Receive	0v
46	WT/BK	CCD Bus (-)	<0.050v
47	WT/OR	Vehicle Speed Signal	Digital Signal
48	BR/RD	Speed Control Set Switch Signal	S/C & Set Switch On: 3.8v
49	YL/RD	S/C On & Off Switch	S/C On: 1v, Off: 12v
50	WT/LG	S/C Resume Switch	S/C On: 1v, Off: 12v
51	DB/YL	ASD Relay Output	12-14v
52	PK/BK	EVAP Purge Solenoid Control	Solenoid Off: 12v, On: 1v
53	LG/RD	S/C Vent Solenoid	Vacuum Decreasing: 1v
54	---	Not Used	---
55	LB	BARO Read Solenoid	Solenoid Off: 12v, On: 1v
56-58	---	Not Used	---
59	PK/BK	AIS 4 Motor Control	DC pulse signals: 0.8-11v
60	YL/BK	AIS 2 Motor Control	DC pulse signals: 0.8-11v

Pin Connector Graphic

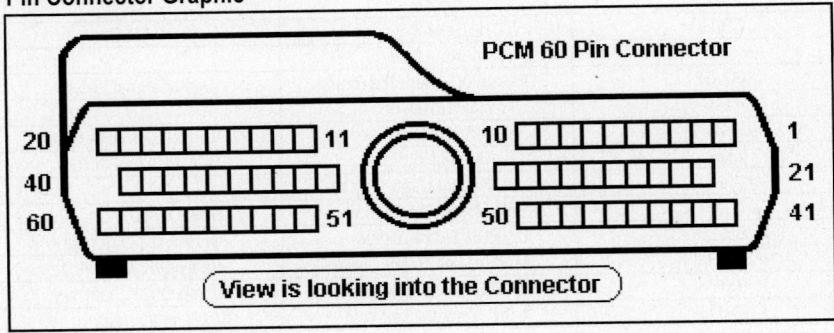

PCM 60 Pin Connector

20   11   10   1
40   21
60   51   50   41

View is looking into the Connector

1990-91 Acclaim 2.5L I4 SOHC TBI VIN K (All) 60 Pin Connector

PCM Pin #	Wire Color	Circuit Description (60 Pin)	Value at Hot Idle
1	DG/RD	MAP Sensor Signal	1.5-1.7v
2	TN/BK	ECT Sensor Signal	At 180°F: 2.80v
3	RD/WT	Battery Power (Fused B+)	12-14v
4	BK/LB	Sensor Ground	<0.050v
5	BK/WT	Sensor Ground	<0.050v
6	PK/WT	5-Volt Supply	4.9-5.1v
7	OR	9-Volt Supply	8.9-9.1v
8	WT	ASD Relay Output	12-14v
9	DB	Ignition Switch Output	12-14v
10	---	Not Used	---
11	BK/TN	Power Ground	<0.1v
12	BK/TN	Power Ground	<0.1v
13-15	---	Not Used	---
16	WT/DB	Injector 1 Driver	1-4 ms
17-18	---	Not Used	---
19	BK/GY	Coil 1 Driver	5°, at 55 mph: 8° dwell
20	DG	Alternator Field Control	Digital Signal: 0-12-0v
21 ('90)	BK/RD	Throttle Body Temperature	At 100°F: 2.51v
22	OR/DB	TP Sensor Signal	0.6-1.0v
23	---	Not Used	---
24	GY/BK	Distributor Reference Signal	Digital Signal: 0-5-0v
25	PK	SCI Transmit	0v
26	PK/BR	CCD Bus (+)	Digital Signal: 0-5-0v
27	BR	A/C Clutch Signal	A/C On: 1v, Off: 12v
28	---	Not Used	---
29	WT/PK	Brake Switch Signal	Brake Off: 0v, On: 12v
30	BR/YL	PNP Switch Signal	In P/N: 0v, Others: 5v
31	DB/PK	Radiator Fan Relay Control	Relay Off: 12v, On: 1v
32	BK/PK	MIL (lamp) Control	Lamp On: 1v, Off: 12v
33	TN/RD	S/C Vacuum Solenoid	Vacuum Increasing: 1v
34	DB/OR	A/C WOT Relay Control	Relay Off: 12v, On: 1v
35	GY/YL	EGR Solenoid Control (California)	12v, 55 mph: 1v
36-38	---	Not Used	---
39	GY/RD	AIS 3 Motor Control	DC pulse signals: 0.8-11v
40	BR/WT	AIS 1 Motor Control	DC pulse signals: 0.8-11v

Standard Colors and Abbreviations

Abbreviation	Color	Abbreviation	Color	Abbreviation	Color
BK	Black	GY	Gray	RD	Red
BL	Blue	GN	Green	TN	Tan
BR	Brown	LG	Light Green	VT	Violet
DB	Dark Blue	OR	Orange	WT	White
DG	Dark Green	PK	Pink	YL	Yellow

1990-91 Acclaim 2.5L I4 SOHC TBI VIN K (All) 60 Pin Connector

PCM Pin #	Wire Color	Circuit Description (60 Pin)	Value at Hot Idle
41	BK/DG	HO2S-11 (B1 S1) Signal	0.1-1.1v
42	---	Not Used	---
43	GY/LB	Tachometer Signals	Pulse Signals
44	---	Not Used	---
45	LG	SCI Receive	0v
46	WT/BK	CCD Bus (-)	<0.050v
47	WT/OR	Vehicle Speed Signal	Digital Signal
48	BR/RD	Speed Control Set Switch Signal	S/C & Set Switch On: 3.8v
49	YL/RD	S/C On & Off Switch	S/C On: 1v, Off: 12v
50	WT/LG	S/C Resume Switch	S/C On: 1v, Off: 12v
51	DB/YL	ASD Relay Output	12-14v
52	PK/BK	EVAP Purge Solenoid Control	Solenoid Off: 12v, On: 1v
53	LG/RD	S/C Vent Solenoid	Vacuum Decreasing: 1v
54	OR/BK	Part Throttle Unlock Solenoid	At Cruise w/TCC On: 1v
55	---	Not Used	---
56 ('90)	GY/PK	Emissions Lamp Control	Lamp On: 1v, Off: 12v
57-58	---	Not Used	---
59	PK/BK	AIS 4 Motor Control	DC pulse signals: 0.8-11v
60	YL/BK	AIS 2 Motor Control	DC pulse signals: 0.8-11v

Pin Connector Graphic

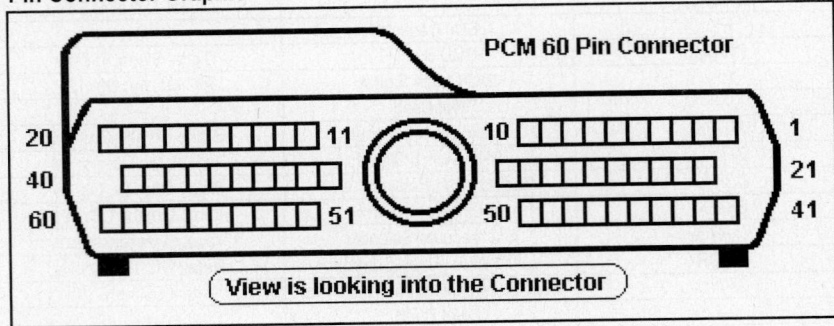

PCM 60 Pin Connector

20 11   10   1
40
60 51   50   41

View is looking into the Connector

1992-95 Acclaim 2.5L I4 SOHC TBI VIN K (All) 60 Pin Connector

PCM Pin #	Wire Color	Circuit Description (60 Pin)	Value at Hot Idle
1	DG/RD	MAP Sensor Signal	1.5-1.7v
2	TN/BK	ECT Sensor Signal	At 180°F: 2.80v
3	RD/WT	Battery Power (Fused B+)	12-14v
4	BK/LB	Sensor Ground	<0.050v
5	BK/WT	Sensor Ground	<0.050v
6	PK/WT	5-Volt Supply	4.9-5.1v
7	OR	8-Volt Supply	7.9-8.1v
8	WT	ASD Relay Output	12-14v
9	DB	Ignition Switch Output	12-14v
10	---	Not Used	---
11	BK/TN	Power Ground	<0.1v
12	BK/TN	Power Ground	<0.1v
13-15	---	Not Used	---
16	WT/DB	Injector 1 Driver	1-4 ms
17-18	---	Not Used	---
19	BK/GY	Coil 1 Driver	5°, at 55 mph: 8° dwell
20	DG	Alternator Field Control	Digital Signal: 0-12-0v
21	---	Not Used	---
22	OR/DB	TP Sensor Signal	0.6-1.0v
23	RD/LG	Speed Control Set Switch Signal	S/C & Set Switch On: 3.8v
24	GY/BK	CKP Sensor Signal	Digital Signal: 0-5-0v
25	PK	SCI Transmit	0v
26	PK/BR	CCD Bus (+)	Digital Signal: 0-5-0v
27	BR	A/C Damped Pressure Switch	A/C On: 1v, Off: 12v
28	---	Not Used	---
29	WT/PK	Brake Switch Signal	Brake Off: 0v, On: 12v
30	BR/YL	Starter Relay Control	KOEC: 9-11v
31	DB/PK	Radiator Fan Relay Control	Relay Off: 12v, On: 1v
32	BK/PK	MIL (lamp) Control	Lamp On: 1v, Off: 12v
33	TN/RD	S/C Vacuum Solenoid	Vacuum Increasing: 1v
34	DB/OR	A/C Clutch Relay Control	Relay Off: 12v, On: 1v
35	GY/YL	EGR Solenoid Control (California)	12v, 55 mph: 1v
36-38	---	Not Used	---
39	GY/RD	AIS 3 Motor Control	DC pulse signals: 0.8-11v
40	BR/WT	AIS 1 Motor Control	DC pulse signals: 0.8-11v

Standard Colors and Abbreviations

Abbreviation	Color	Abbreviation	Color	Abbreviation	Color
BK	Black	GY	Gray	RD	Red
BL	Blue	GN	Green	TN	Tan
BR	Brown	LG	Light Green	VT	Violet
DB	Dark Blue	OR	Orange	WT	White
DG	Dark Green	PK	Pink	YL	Yellow

1992-95 Acclaim 2.5L I4 SOHC TBI VIN K (All) 60 Pin Connector

PCM Pin #	Wire Color	Circuit Description (60 Pin)	Value at Hot Idle
41	BK/DG	HO2S-11 (B1 S1) Signal	0.1-1.1v
42	---	Not Used	---
43	GY/LB	Tachometer Signals	Pulse Signals
44	---	Not Used	---
45	LG	SCI Receive	0v
46	WT/BK	CCD Bus (-)	<0.050v
47	WT/OR	Vehicle Speed Signal	Digital Signal
48-50	---	Not Used	---
51	DB/YL	ASD Relay Output	12-14v
52	PK/BK	EVAP Purge Solenoid Control	Solenoid Off: 12v, On: 1v
53	LG/RD	S/C Vent Solenoid	Vacuum Decreasing: 1v
54	OR/BK	Part Throttle Unlock Solenoid	At Cruise w/TCC On: 1v
55-56	---	Not Used	---
57	DG/OR	Generator Field Source	12-14v
58	---	Not Used	---
59	PK/BK	AIS 4 Motor Control	DC pulse signals: 0.8-11v
60	YL/BK	AIS 2 Motor Control	DC pulse signals: 0.8-11v

Pin Connector Graphic

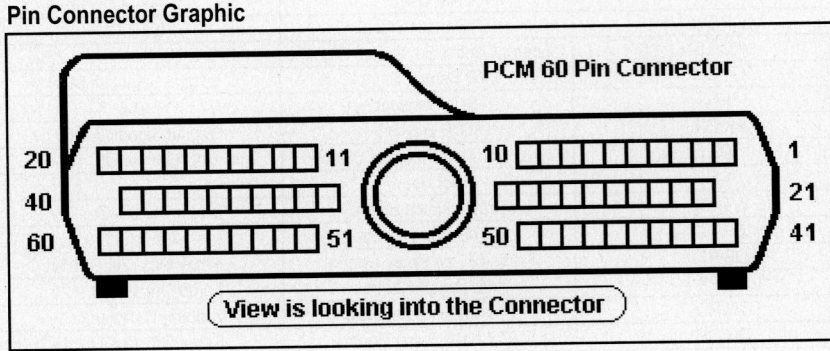

1993 Acclaim 2.5L I4 Flexible Fuel VIN V (M/T) 60 Pin Connector

PCM Pin #	Wire Color	Circuit Description (60 Pin)	Value at Hot Idle
1	DG/RD	MAP Sensor Signal	1.5-1.7v
2	TN/BK	ECT Sensor Signal	At 180°F: 2.80v
3	RD/WT	Battery Power (Fused B+)	12-14v
4	BK/LB	Sensor Ground	<0.050v
5	BK/WT	Sensor Ground	<0.050v
6	PK/WT	5-Volt Supply	4.9-5.1v
7	OR	8-Volt Supply	7.9-8.1v
8	WT	ASD Relay Output	12-14v
9	DB	Ignition Switch Output	12-14v
10	---	Not Used	---
11	BK/TN	Power Ground	<0.1v
12	BK/TN	Power Ground	<0.1v
13	LB/BR	Injector 4 Driver	1-4 ms
14	YL/WT	Injector 3 Driver	1-4 ms
15	TN	Injector 2 Driver	1-4 ms
16	WT/DB	Injector 1 Driver	1-4 ms
17-18	---	Not Used	---
19	BK/GY	Coil 1 Driver	5°, at 55 mph: 8° dwell
20	DG	Alternator Field Control	Digital Signal: 0-12-0v
21	BK/RD	Flexible Fuel Sensor Signal	0.5-4.5v
22	OR/DB	TP Sensor Signal	0.6-1.0v
23	RD/LG	Speed Control Set Switch Signal	S/C & Set Switch On: 3.8v
24	GY/BK	CKP Sensor Signal	Digital Signal: 0-5-0v
25	PK	SCI Transmit	0v
26	PK/BR	CCD Bus (+)	Digital Signal: 0-5-0v
27	BR	A/C Damped Pressure Switch	A/C On: 1v, Off: 12v
28	---	Not Used	---
29	WT/PK	Brake Switch Signal	Brake Off: 0v, On: 12v
30	BR/YL	Starter Relay Control	KOEC: 9-11v
31	DB/PK	Radiator Fan Relay Control	Relay Off: 12v, On: 1v
32	BK/PK	MIL (lamp) Control	Lamp On: 1v, Off: 12v
33	TN/RD	S/C Vacuum Solenoid	Vacuum Increasing: 1v
34	DB/OR	A/C Clutch Relay Control	Relay Off: 12v, On: 1v
35-38	---	Not Used	---
39	GY/RD	AIS 3 Motor Control	DC pulse signals: 0.8-11v
40	BR/WT	AIS 1 Motor Control	DC pulse signals: 0.8-11v

1993 Acclaim 2.5L I4 Flexible Fuel VIN V (M/T) 60 Pin Connector

PCM Pin #	Wire Color	Circuit Description (60 Pin)	Value at Hot Idle
41	BK/DG	HO2S-11 (B1 S1) Signal	0.1-1.1v
42	---	Not Used	---
43	GY/LB	Tachometer Signals	Pulse Signals
44	TN/YL	CMP Sensor Signal	Digital Signal: 0-5-0v
45	LG	SCI Receive	5v
46	WT/BK	CCD Bus (-)	<0.050v
47	WT/OR	Vehicle Speed Signal	Digital Signal
48-50	---	Not Used	---
51	DB/YL	ASD Relay Output	12-14v
52	PK/BK	EVAP Purge Solenoid Control	Solenoid Off: 12v, On: 1v
53	LG/RD	S/C Vent Solenoid	Vacuum Decreasing: 1v
54	OR/BK	EMCC Solenoid Control	At Cruise w/TCC On: 1v
55-56	---	Not Used	---
57	DG/OR	ASD Relay Control	Relay Off: 12v, On: 1v
58	---	Not Used	---
59	PK/BK	AIS 4 Motor Control	DC pulse signals: 0.8-11v
60	YL/BK	AIS 2 Motor Control	DC pulse signals: 0.8-11v

Pin Connector Graphic

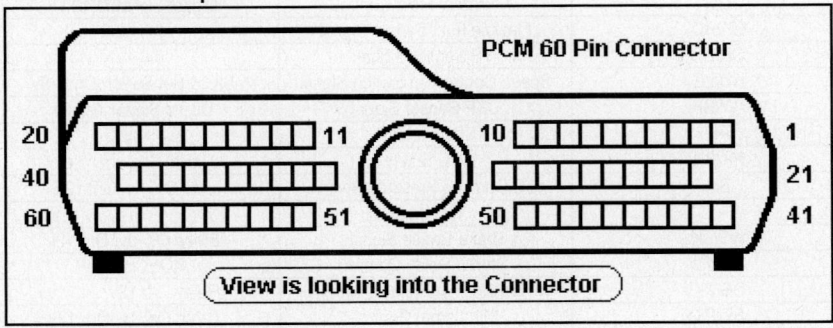

PCM 60 Pin Connector

View is looking into the Connector

Standard Colors and Abbreviations

Abbreviation	Color	Abbreviation	Color	Abbreviation	Color
BK	Black	GY	Gray	RD	Red
BL	Blue	GN	Green	TN	Tan
BR	Brown	LG	Light Green	VT	Violet
DB	Dark Blue	OR	Orange	WT	White
DG	Dark Green	PK	Pink	YL	Yellow

1990-92 Acclaim 3.0L V6 SOHC MFI VIN 3 (A/T) 60 Pin Connector

PCM Pin #	Wire Color	Circuit Description (60 Pin)	Value at Hot Idle
1	DG/RD	MAP Sensor Signal	1.5-1.7v
2	TN/WT	ECT Sensor Signal	At 180°F: 2.80v
2 ('90-'91)	TN/BK	ECT Sensor Signal	At 180°F: 2.80v
3	RD	Battery Power (Fused B+)	12-14v
3 ('90-'91)	RD/WT	Battery Power (Fused B+)	12-14v
4	BK/LB	Sensor Ground	<0.050v
5	BK/WT	Sensor Ground	<0.050v
6	PK/WT	5-Volt Supply	4.9-5.1v
7	OR	8-Volt Supply	7.9-8.1v
8	DG/BK	ASD Relay Output	12-14v
9	DB	Fused Ignition Switch Input	12-14v
10	---	Not Used	---
11	BK/TN	Power Ground	<0.1v
12	BK/TN	Power Ground	<0.1v
13	---	Not Used	---
14	YL/WT	Injector 3 Driver	1-4 ms
15	TN	Injector 2 Driver	1-4 ms
16	WT/DB	Injector 1 Driver	1-4 ms
17-18	---	Not Used	---
19	BK/GY	Coil 1 Driver	5°, at 55 mph: 8° dwell
20	DG	Alternator Field Control	Digital Signal: 0-12-0v
21	---	Not Used	---
22	OR/DB	TP Sensor Signal	0.6-1.0v
23	---	Not Used	---
24	GY/BK	Distributor Reference Signal	Digital Signal: 0-5-0v
25	PK	SCI Transmit	0v
26	BK/PK	CCD Bus (+)	Digital Signal: 0-5-0v
26 ('90-'91)	PK/BR	CCD Bus (+)	Digital Signal: 0-5-0v
27	BR	A/C Clutch Signal	A/C On: 1v, Off: 12v
28	---	Not Used	---
29	WT/PK	Brake Switch Signal	Brake Off: 0v, On: 12v
30	BR/YL	PNP Switch Signal	In P/N: 0v, Others: 5v
31	DB/PK	Radiator Fan Relay Control	Relay Off: 12v, On: 1v
32	BK/PK	MIL (lamp) Control	Lamp On: 1v, Off: 12v
33	TN/RD	S/C Vacuum Solenoid	Vacuum Increasing: 1v
34	DB/OR	A/C WOT Relay Control	Relay Off: 12v, On: 1v
35	GY/YL	EGR Solenoid Control (California)	12v, 55 mph: 1v
36-38	---	Not Used	---
39	GY/RD	AIS 3 Motor Control	DC pulse signals: 0.8-11v
40	BR/WT	AIS 1 Motor Control	DC pulse signals: 0.8-11v

1990-92 Acclaim 3.0L V6 SOHC MFI VIN 3 (A/T) 60 Pin Connector

PCM Pin #	Wire Color	Circuit Description (60 Pin)	Value at Hot Idle
41	BK/DG	HO2S-11 (B1 S1) Signal	0.1-1.1v
42	---	Not Used	---
43	LB	Tachometer Signals	Pulse Signals
43 ('91)	GY/LB	Tachometer Signals	Pulse Signals
44	TN/YL	Distributor Sync Signals	Digital Signal: 0-5-0v
45	LG	SCI Receive	5v
46	WT/BK	CCD Bus (-)	<0.050v
47	WT/OR	Vehicle Speed Signal	Digital Signal
48	BR/RD	Speed Control Set Switch Signal	S/C & Set Switch On: 3.8v
49	YL/RD	S/C On & Off Switch	S/C On: 1v, Off: 12v
50	WT/LG	S/C Resume Switch	S/C On: 1v, Off: 12v
51	DB/YL	ASD Relay Output	12-14v
52	PK	EVAP Purge Solenoid Control	Solenoid Off: 12v, On: 1v
52 ('90-'91)	PK/BK	EVAP Purge Solenoid Control	Solenoid Off: 12v, On: 1v
53	LG/RD	S/C Vent Solenoid	Vacuum Decreasing: 1v
54-56	---	Not Used	---
57	DG/OR	ASD Relay Control	Relay Off: 12v, On: 1v
58	---	Not Used	---
59	PK/BK	AIS 4 Motor Control	DC pulse signals: 0.8-11v
60	YL/BK	AIS 2 Motor Control	DC pulse signals: 0.8-11v

Pin Connector Graphic

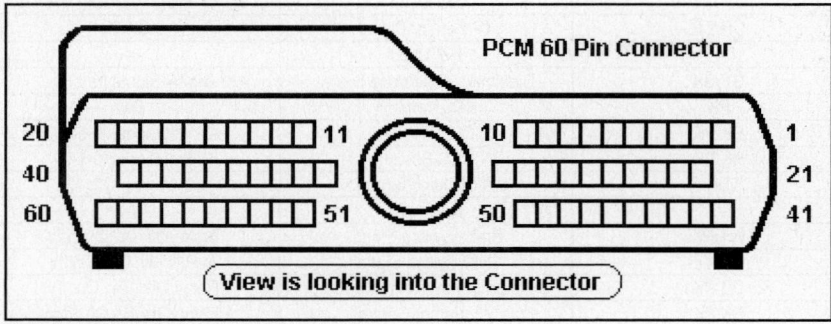

PCM 60 Pin Connector

20 | 11   10 | 1
40 | 21
60 | 51   50 | 41

View is looking into the Connector

Standard Colors and Abbreviations

Abbreviation	Color	Abbreviation	Color	Abbreviation	Color
BK	Black	GY	Gray	RD	Red
BL	Blue	GN	Green	TN	Tan
BR	Brown	LG	Light Green	VT	Violet
DB	Dark Blue	OR	Orange	WT	White
DG	Dark Green	PK	Pink	YL	Yellow

1993-95 Acclaim 3.0L V6 SOHC MFI VIN 3 (A/T) 60 Pin Connector

PCM Pin #	Wire Color	Circuit Description (60 Pin)	Value at Hot Idle
1	DG/RD	MAP Sensor Signal	1.5-1.7v
2	TN/BK	ECT Sensor Signal	At 180°F: 2.80v
3	RD/WT	Battery Power (Fused B+)	12-14v
4	BK/LB	Sensor Ground	<0.050v
5	BK/WT	Sensor Ground	<0.050v
6	PK/WT	5-Volt Supply	4.9-5.1v
7	OR	8-Volt Supply	7.9-8.1v
8	---	Not Used	---
9	DB	ASD Relay Output	12-14v
10	---	Not Used	---
11	BK/TN	Power Ground	<0.1v
12	BK/TN	Power Ground	<0.1v
13	LB/BR	Injector 4 Driver	1-4 ms
14	YL/WT	Injector 3 Driver	1-4 ms
15	TN	Injector 2 Driver	1-4 ms
16	WT/DB	Injector 1 Driver	1-4 ms
17-18	---	Not Used	---
19	BK/GY	Coil 1 Driver	5°, at 55 mph: 8° dwell
20	DG	Alternator Field Control	Digital Signal: 0-12-0v
21	---	Not Used	
22	OR/DB	TP Sensor Signal	0.6-1.0v
23	RD/LG	Speed Control Set Switch Signal	S/C & Set Switch On: 3.8v
24	GY/BK	CKP Sensor Signal	Digital Signal: 0-5-0v
25	PK	SCI Transmit	0v
26	PK/BR	CCD Bus (+)	Digital Signal: 0-5-0v
27	BR	A/C Damped Pressure Switch	A/C On: 1v, Off: 12v
28	DB/OR	PSP Switch Signal	Straight: 0v, Turning: 5v
29	WT/PK	Brake Switch Signal	Brake Off: 0v, On: 12v
30	BR/YL	PNP Switch Signal	In P/N: 0v, Others: 5v
31	DB/PK	Radiator Fan Relay Control	Relay Off: 12v, On: 1v
32	BK/PK	MIL (lamp) Control	Lamp On: 1v, Off: 12v
33	TN/RD	S/C Vacuum Solenoid	Vacuum Increasing: 1v
34	DB/OR	A/C Clutch Relay Control	Relay Off: 12v, On: 1v
35	GY/YL	EGR Solenoid Control (California)	12v, 55 mph: 1v
36-37	---	Not Used	---
38	GY	Injector 5 Driver	1-4 ms
39	GY/RD	IAC 3 Driver	DC pulse signals: 0.8-11v
40	BR/WT	IAC 1 Driver	DC pulse signals: 0.8-11v

Standard Colors and Abbreviations

Abbreviation	Color	Abbreviation	Color	Abbreviation	Color
BK	Black	GY	Gray	RD	Red
BL	Blue	GN	Green	TN	Tan
BR	Brown	LG	Light Green	VT	Violet
DB	Dark Blue	OR	Orange	WT	White
DG	Dark Green	PK	Pink	YL	Yellow

1993-95 Acclaim 3.0L V6 SOHC MFI VIN 3 (A/T) 60 Pin Connector

PCM Pin #	Wire Color	Circuit Description (60 Pin)	Value at Hot Idle
41	BK/DG	HO2S-11 (B1 S1) Signal	0.1-1.1v
42	---	Not Used	---
43	GY/LB	Tachometer Signals	Pulse Signals
44	TN/YL	CMP Sensor Signal	Digital Signal: 0-5-0v
45	LG	SCI Receive	5v
46	WT/BK	CCD Bus (-)	<0.050v
47	WT/OR	Vehicle Speed Signal	Digital Signal
48-50	---	Not Used	---
51	DB/YL	ASD Relay Output	12-14v
52	PK/BK	EVAP Purge Solenoid Control	Solenoid Off: 12v, On: 1v
53	LG/RD	S/C Vent Solenoid	Vacuum Decreasing: 1v
54	OR/BK	EMCC Solenoid Control	At Cruise w/TCC On: 1v
55-56	---	Not Used	---
57	DG/OR	ASD Relay Control	Relay Off: 12v, On: 1v
58	BR/DB	Injector 6 Driver	1-4 ms
59	PK/BK	IAC 4 Driver	DC pulse signals: 0.8-11v
60	YL/BK	IAC 2 Driver	DC pulse signals: 0.8-11v

Pin Connector Graphic

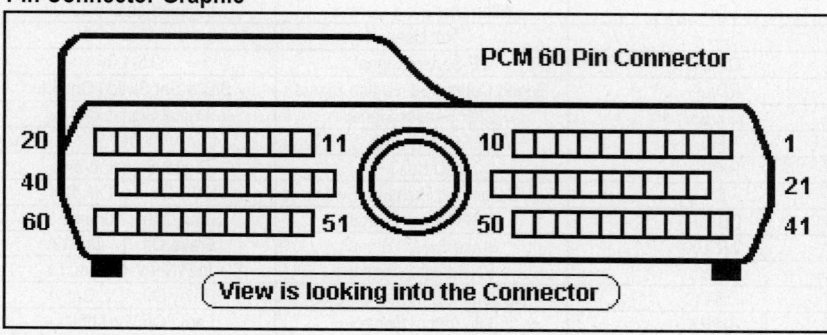

1996-97 Breeze 2.0L I4 SOHC MFI VIN C (All) 'C1' Black Connector

PCM Pin #	Wire Color	Circuit Description (40-Pin)	Value at Hot Idle
1	---	Not Used	---
2	BK/GY	Coil 1 Driver	5°, at 55 mph: 8° dwell
3	DB/DG	Coil 2 Driver	5°, at 55 mph: 8° dwell
4	DG	Generator Field Control	Digital Signal: 0-12-0v
5	YL/PK	S/C Power Supply	12-14v
6	DG/OR	ASD Relay Output	12-14v
7	YL/WT	Injector 3 Driver	1-4 ms
8-9	---	Not Used	---
10	BK/TN	Power Ground	<0.1v
11-12	---	Not Used	---
13	WT/LB	Injector 1 Driver	1-4 ms
14-15	---	Not Used	---
16	LB/BR	Injector 4 Driver	1-4 ms
17	TN	Injector 2 Driver	1-4 ms
18-19	---	Not Used	---
20	DB/WT	Ignition Switch Output	12-14v
21-23	---	Not Used	---
24	GY/BK	Knock Sensor Signal	0.080v AC
25	---	Not Used	---
26	TN/BK	ECT Sensor Signal	At 180°F: 2.80v
27-29	---	Not Used	---
30	BK/DG	HO2S-11 (B1 S1) Signal	0.1-1.1v
31	---	Not Used	---
32	GY/BK	CKP Sensor Signal	Digital Signal: 0-5-0v
33	TN/YL	CMP Sensor Signal	Digital Signal: 0-5-0v
34	---	Not Used	---
35	OR/LB	TP Sensor Signal	0.6-1.0v
36	DG/RD	MAP Sensor Signal	1.5-1.7v
37	BK/RD	IAT Sensor Signal	At 100°F: 1.83v
38-39	---	Not Used	---
40	GY/YL	EGR Solenoid Control	12v, 55 mph: 1v

Standard Colors and Abbreviations

Abbreviation	Color	Abbreviation	Color	Abbreviation	Color
BK	Black	GY	Gray	RD	Red
BL	Blue	GN	Green	TN	Tan
BR	Brown	LG	Light Green	VT	Violet
DB	Dark Blue	OR	Orange	WT	White
DG	Dark Green	PK	Pink	YL	Yellow

1996-97 Breeze 2.0L I4 SOHC MFI VIN C (All) 'C2' White Connector

PCM Pin #	Wire Color	Circuit Description (40-Pin)	Value at Hot Idle
41	RD/LG	Speed Control Set Switch Signal	S/C & Set Switch On: 3.8v
42	DB/YL	A/C Pressure Switch Signal	A/C On: 0.45-4.85v
43	BK/LB	Sensor Ground	<0.050v
44	OR/WT	8-Volt Supply	7.9-8.1v
45	DB/LG	PSP Switch Signal	Straight: 0v, Turning: 5v
46	RD/TN	Battery Power (Fused B+)	12-14v
47	---	Not Used	---
48	BR/GY	IAC 3 Driver	DC pulse signals: 0.8-11v
49	YL/BK	IAC 2 Driver	DC pulse signals: 0.8-11v
50	BK/TN	Power Ground	<0.1v
51	TN/WT	HO2S-12 (B1 S2) Signal	0.1-1.1v
52	PK/LG	Battery Temperature Sensor	At 86°F: 1.96v
53-54	---	Not Used	---
55	DB/TN	Low Speed Fan Relay	Relay Off: 12v, On: 1v
56	---	Not Used	---
57	GY/RD	IAC 1 Driver	DC pulse signals: 0.8-11v
58	VT/GY	IAC 4 Driver	DC pulse signals: 0.8-11v
59	VT/BR	CCD Bus (+)	Digital Signal: 0-5-0v
60	WT/BK	CCD Bus (-)	<0.050v
61	PK/WT	5-Volt Supply	4.9-5.1v
62	WT/RD	Brake Switch Signal	Brake Off: 0v, On: 12v
63	YL/DG	Torque Management Request	Digital Signals
64	DB/OR	A/C Clutch Relay Control	Relay Off: 12v, On: 1v
65	PK/LB	SCI Transmit	0v
66	WT/OR	Vehicle Speed Signal	Digital Signal
67	DB/PK	ASD Relay Control	Relay Off: 12v, On: 1v
68	PK/GY	EVAP Purge Solenoid Control	PWM Signal: 0-12-0v
69	DB/PK	High Speed Fan Control	Relay Off: 12v, On: 1v
71, 73	---	Not Used	---
72	OR/DG	LDP Switch Signal	Closed: 0v, Open: 12v
74	BR/LG	Fuel Pump Relay Control	Relay Off: 12v, On: 1v
75	LG/WT	SCI Receive	0v
76	BK/WT	PNP Switch Signal	In P/N: 0v, Others: 5v
77	WT/DG	LDP Solenoid Control	PWM Signal: 0-12-0v
78	TN/RD	S/C Vacuum Solenoid	Vacuum Increasing: 1v
79	---	Not Used	---
80	LG/RD	S/C Vent Solenoid	Vacuum Decreasing: 1v

Pin Connector Graphic

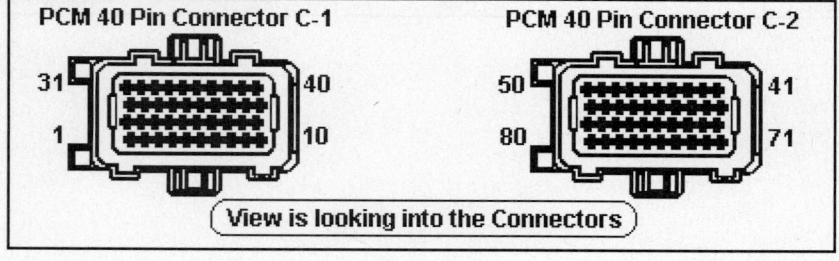

1998-2000 Breeze 2.0L I4 SOHC MFI VIN C (All) 'C1' Black Connector

PCM Pin #	Wire Color	Circuit Description (40-Pin)	Value at Hot Idle
1-2	---	Not Used	---
3	DB/TN	Coil 2 Driver	5°, at 55 mph: 8° dwell
4	---	Not Used	---
5	YL/RD	S/C Power Supply	12-14v
6	DG/OR	ASD Relay Output	12-14v
7	YL/WT	Injector 3 Driver	1-4 ms
8	DG	Generator Field Control	Digital Signal: 0-12-0v
9	---	Not Used	---
10	BK/TN	Power Ground	<0.1v
11	BK/GY	Coil 1 Driver	5°, at 55 mph: 8° dwell
12	---	Not Used	---
13	WT/DB	Injector 1 Driver	1-4 ms
14-15	---	Not Used	---
16	LB/BR	Injector 4 Driver	1-4 ms
17	TN	Injector 2 Driver	1-4 ms
18-19	---	Not Used	---
20	DB/WT	Ignition Switch Output	12-14v
21-24	---	Not Used	---
25	DB/LG	Knock Sensor Signal	0.080v AC
26	TN/BK	ECT Sensor Signal	At 180°F: 2.80v
27	BK/OR	HO2S-11 (B1 S1) Ground	<0.050v
28-29	---	Not Used	---
30	BK/DG	HO2S-11 (B1 S1) Signal	0.1-1.1v
31	TN	Starter Relay Control	KOEC: 9-11v
32	GY/BK	CKP Sensor Signal	Digital Signal: 0-5-0v
33	TN/YL	CMP Sensor Signal	Digital Signal: 0-5-0v
34	---	Not Used	---
35	OR/LB	TP Sensor Signal	0.6-1.0v
36	DG/RD	MAP Sensor Signal	1.5-1.7v
37	BK/RD	IAT Sensor Signal	At 100°F: 1.83v
38-39	---	Not Used	---
40	GY/YL	EGR Solenoid Control	12v, 55 mph: 1v

Standard Colors and Abbreviations

Abbreviation	Color	Abbreviation	Color	Abbreviation	Color
BK	Black	GY	Gray	RD	Red
BL	Blue	GN	Green	TN	Tan
BR	Brown	LG	Light Green	VT	Violet
DB	Dark Blue	OR	Orange	WT	White
DG	Dark Green	PK	Pink	YL	Yellow

1998-2000 Breeze 2.0L I4 SOHC MFI VIN C (All) 'C2' White Connector

PCM Pin #	Wire Color	Circuit Description (40-Pin)	Value at Hot Idle
41	RD/LG	Speed Control Set Switch Signal	S/C & Set Switch On: 3.8v
42	DB/YL	A/C Pressure Switch Signal	A/C On: 0.45-4.85v
43	BK/LB	Sensor Ground	<0.050v
44	OR/WT	8-Volt Supply	7.9-8.1v
45	DB/LG	PSP Switch Signal	Straight: 0v, Turning: 5v
46	RD/TN	Battery Power (Fused B+)	12-14v
47	BK	Power Ground	<0.1v
48	BR/WT	IAC 3 Driver	DC pulse signals: 0.8-11v
49	YL/BK	IAC 2 Driver	DC pulse signals: 0.8-11v
50	BK/TN	Power Ground	<0.1v
51	TN/WT	HO2S-12 (B1 S2) Signal	0.1-1.1v
52	VT/LG	Battery Temperature Sensor	At 86°F: 1.96v
53-54	---	Not Used	---
55	DB/TN	Low Speed Fan Relay	Relay Off: 12v, On: 1v
56	WT/VT	S/C Vacuum Solenoid	Vacuum Increasing: 1v
57	GY/RD	IAC 1 Driver	DC pulse signals: 0.8-11v
58	VT/BK	IAC 4 Driver	DC pulse signals: 0.8-11v
59	VT/BR	CCD Bus (+)	Digital Signal: 0-5-0v
60	WT/BK	CCD Bus (-)	<0.050v
61	VT/WT	5-Volt Supply	4.9-5.1v
62	WT/RD	Brake Switch Signal	Brake Off: 0v, On: 12v
63	YL/DG	Torque Management Request	Digital Signals
64	DB/OR	A/C Clutch Relay Control	Relay Off: 12v, On: 1v
65	PK/LB	SCI Transmit	0v
66	WT/OR	Vehicle Speed Signal	Digital Signal
67	DB/VT	ASD Relay Control	Relay Off: 12v, On: 1v
68	PK/GY	EVAP Purge Solenoid Control	PWM Signal: 0-12-0v
69	DB/PK	High Speed Fan Control	Relay Off: 12v, On: 1v
70	WT/TN	EVAP Purge Solenoid Sense	0-1v
71, 73	---	Not Used	---
72	OR/DG	LDP Switch Signal	Closed: 0v, Open: 12v
74	BR/LG	Fuel Pump Relay Control	Relay Off: 12v, On: 1v
75	LG/WT	SCI Receive	0v
76	BK/WT	PNP Switch Signal	In P/N: 0v, Others: 5v
77	WT/DG	LDP Solenoid Control	PWM Signal: 0-12-0v
78-79	---	Not Used	---
80	LG/RD	S/C Vent Solenoid	Vacuum Decreasing: 1v

Pin Connector Graphic

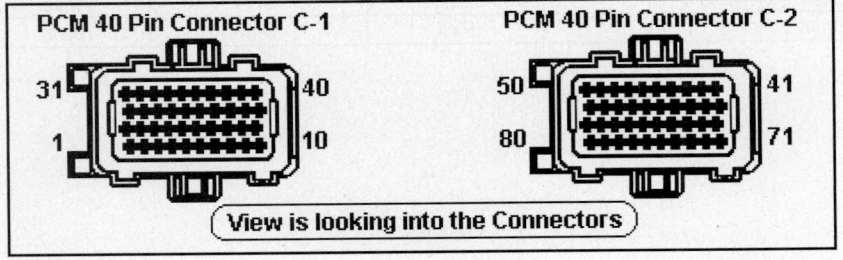

1997 Breeze 2.4L I4 DOHC MFI VIN X (A/T) 'C1' Black Connector

PCM Pin #	Wire Color	Circuit Description (40-Pin)	Value at Hot Idle
1	---	Not Used	---
2	BK/GY	Coil 1 Driver	5°, at 55 mph: 8° dwell
3	DB/DG	Coil 2 Driver	5°, at 55 mph: 8° dwell
4	DG	Alternator Field Control	Digital Signal: 0-12-0v
5	YL/PK	S/C Power Supply	12-14v
6	DG/OR	ASD Relay Output	12-14v
7	YL/WT	Injector 3 Driver	1-4 ms
8-9	---	Not Used	---
10	BK/TN	Power Ground	<0.1v
11-12	---	Not Used	---
13	WT/LB	Injector 1 Driver	1-4 ms
14-15	---	Not Used	---
16	LB/BR	Injector 4 Driver	1-4 ms
17	TN	Injector 2 Driver	1-4 ms
18-19	---	Not Used	---
20	DB/WT	Ignition Switch Output	12-14v
21-23	---	Not Used	---
24	GY/BK	Knock Sensor Signal	0.080v AC
25	---	Not Used	---
26	TN/BK	ECT Sensor Signal	At 180°F: 2.80v
27-29	---	Not Used	---
30	BK/DG	HO2S-11 (B1 S1) Signal	0.1-1.1v
31	---	Not Used	---
32	GY/BK	CKP Sensor Signal	Digital Signal: 0-5-0v
33	TN/YL	CMP Sensor Signal	Digital Signal: 0-5-0v
34	---	Not Used	---
35	OR/LB	TP Sensor Signal	0.6-1.0v
36	DG/RD	MAP Sensor Signal	1.5-1.7v
37	BK/RD	IAT Sensor Signal	At 100°F: 1.83v
38-39	---	Not Used	---
40	GY/YL	EGR Solenoid Control	12v, 55 mph: 1v

Standard Colors and Abbreviations

Abbreviation	Color	Abbreviation	Color	Abbreviation	Color
BK	Black	GY	Gray	RD	Red
BL	Blue	GN	Green	TN	Tan
BR	Brown	LG	Light Green	VT	Violet
DB	Dark Blue	OR	Orange	WT	White
DG	Dark Green	PK	Pink	YL	Yellow

1997 Breeze 2.4L I4 DOHC MFI VIN X (A/T) 'C2' White Connector

PCM Pin #	Wire Color	Circuit Description (40-Pin)	Value at Hot Idle
41	PK/LG	Speed Control Set Switch Signal	S/C & Set Switch On: 3.8v
42	DB/YL	A/C Pressure Switch Signal	A/C On: 0.45-4.85v
43	BK/LB	Sensor Ground	<0.050v
44	OR/WT	8-Volt Supply	7.9-8.1v
45	DB/LG	PSP Switch Signal	Straight: 0v, Turning: 5v
46	RD/TN	Battery Power (Fused B+)	12-14v
47	---	Not Used	---
48	BR/GY	IAC 1 Driver	DC pulse signals: 0.8-11v
49	YL/BK	IAC 2 Driver	DC pulse signals: 0.8-11v
50	BK/TN	Power Ground	<0.1v
51	TN/WT	HO2S-12 (B1 S2) Signal	0.1-1.1v
52	PK/LG	Battery Temperature Sensor	At 86°F: 1.96v
53-54	---	Not Used	---
55	DB/TN	Low Speed Fan Relay	Relay Off: 12v, On: 1v
56	---	Not Used	---
57	GY/RD	IAC 3 Driver	DC pulse signals: 0.8-11v
58	PK/GY	IAC 4 Driver	DC pulse signals: 0.8-11v
59	PK/BR	CCD Bus (+)	Digital Signal: 0-5-0v
60	WT/BK	CCD Bus (-)	<0.050v
61	PK/WT	5-Volt Supply	4.9-5.1v
62	WT/RD	Brake Switch Signal	Brake Off: 0v, On: 12v
63	YL/DG	Torque Management Request	Digital Signals
64	DB/OR	A/C Clutch Relay Control	Relay Off: 12v, On: 1v
65	PK/LB	SCI Transmit	0v
66	WT/OR	Vehicle Speed Signal	Digital Signal
67	DB/PK	ASD Relay Control	Relay Off: 12v, On: 1v
68	PK/GY	EVAP Purge Solenoid Control	PWM Signal: 0-12-0v
69	DB/PK	High Speed Fan Control	Relay Off: 12v, On: 1v
71	---	Not Used	---
72	OR/DG	LDP Switch Signal	Closed: 0v, Open: 12v
73	---	Not Used	---
74	BR/LG	Fuel Pump Relay Control	Relay Off: 12v, On: 1v
75	LG/WT	SCI Receive	5v
76	BK/WT	PNP Switch Signal	In P/N: 0v, Others: 5v
77	WT/DG	LDP Solenoid Control	PWM Signal: 0-12-0v
78	WT/PK	S/C Vacuum Solenoid	Vacuum Increasing: 1v
79	---	Not Used	---
80	LG/RD	S/C Vent Solenoid	Vacuum Decreasing: 1v

Pin Connector Graphic

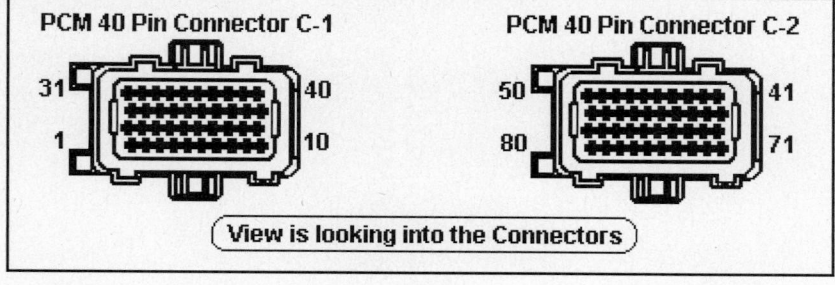

1998-2000 Breeze 2.4L I4 DOHC MFI VIN X (A/T) 'C1' Black Connector

PCM Pin #	Wire Color	Circuit Description (40-Pin)	Value at Hot Idle
1-2	---	Not Used	---
3	DB/TN	Coil 2 Driver	5°, at 55 mph: 8° dwell
4	---	Not Used	---
5	YL/RD	S/C Power Supply	12-14v
6	DG/OR	ASD Relay Output	12-14v
7	YL/WT	Injector 3 Driver	1-4 ms
8	DG	Alternator Field Control	Digital Signal: 0-12-0v
9	---	Not Used	---
10	BK/TN	Power Ground	<0.1v
11	BK/GY	Coil 1 Driver	5°, at 55 mph: 8° dwell
12	---	Not Used	---
13	WT/DB	Injector 1 Driver	1-4 ms
14-15	---	Not Used	---
16	LB/BR	Injector 4 Driver	1-4 ms
17	TN	Injector 2 Driver	1-4 ms
18-19	---	Not Used	---
20	DB/WT	Ignition Switch Output	12-14v
21-24	---	Not Used	---
25	DB/LG	Knock Sensor Signal	0.080v AC
26	TN/BK	ECT Sensor Signal	At 180°F: 2.80v
27	BK/OR	HO2S Ground	<0.050v
28-29	---	Not Used	---
30	BK/DG	HO2S-11 (B1 S1) Signal	0.1-1.1v
31	---	Not Used	---
32	GY/BK	CKP Sensor Signal	Digital Signal: 0-5-0v
33	TN/YL	CMP Sensor Signal	Digital Signal: 0-5-0v
34	---	Not Used	---
35	OR/LB	TP Sensor Signal	0.6-1.0v
36	DG/RD	MAP Sensor Signal	1.5-1.7v
37	BK/RD	IAT Sensor Signal	At 100°F: 1.83v
38-39	---	Not Used	---
40	GY/YL	EGR Solenoid Control	12v, 55 mph: 1v

Standard Colors and Abbreviations

Abbreviation	Color	Abbreviation	Color	Abbreviation	Color
BK	Black	GY	Gray	RD	Red
BL	Blue	GN	Green	TN	Tan
BR	Brown	LG	Light Green	VT	Violet
DB	Dark Blue	OR	Orange	WT	White
DG	Dark Green	PK	Pink	YL	Yellow

1998-2000 Breeze 2.4L I4 DOHC MFI VIN X (A/T) 'C2' White Connector

PCM Pin #	Wire Color	Circuit Description (40-Pin)	Value at Hot Idle
41	RD/LG	Speed Control Set Switch Signal	S/C & Set Switch On: 3.8v
42	DB/YL	A/C Pressure Switch Signal	A/C On: 0.45-4.85v
43	BK/LB	Sensor Ground	<0.050v
44	OR/WT	8-Volt Supply	7.9-8.1v
45	DB/LG	PSP Switch Signal	Straight: 0v, Turning: 5v
46	RD/TN	Battery Power (Fused B+)	12-14v
47	BK/TN	Power Ground	<0.1v
48	BR/WT	IAC 3 Driver	DC pulse signals: 0.8-11v
49	YL/BK	IAC 2 Driver	DC pulse signals: 0.8-11v
50	BK/TN	Power Ground	<0.1v
51	TN/WT	HO2S-12 (B1 S2) Signal	0.1-1.1v
52	VT/LG	Battery Temperature Sensor	At 86°F: 1.96v
53-54	---	Not Used	---
55	DB/TN	Low Speed Fan Relay	Relay Off: 12v, On: 1v
56	WT/VT	S/C Vacuum Solenoid	Vacuum Increasing: 1v
57	GY/RD	IAC 1 Driver	DC pulse signals: 0.8-11v
58	VT/BK	IAC 4 Driver	DC pulse signals: 0.8-11v
59	VT/BR	CCD Bus (+)	Digital Signal: 0-5-0v
60	WT/BK	CCD Bus (-)	<0.050v
61	VT/WT	5-Volt Supply	4.9-5.1v
62	WT/RD	Brake Switch Signal	Brake Off: 0v, On: 12v
63	YL/DG	Torque Management Request	Digital Signals
64	DB/OR	A/C Clutch Relay Control	Relay Off: 12v, On: 1v
65	PK	SCI Transmit	0v
66	WT/OR	Vehicle Speed Signal	Digital Signal
67	DB/VT	ASD Relay Control	Relay Off: 12v, On: 1v
68	PK/GY	EVAP Purge Solenoid Control	PWM Signal: 0-12-0v
69	DB/PK	High Speed Fan Control	Relay Off: 12v, On: 1v
70	WT/TN	EVAP Purge Solenoid Sense	0-1v
71	---	Not Used	---
72	OR/DG	LDP Switch Signal	Closed: 0v, Open: 12v
73	---	Not Used	---
74	BR/LG	Fuel Pump Relay Control	Relay Off: 12v, On: 1v
75	LG	SCI Receive	5v
76	BK/WT	PNP Switch Signal	In P/N: 0v, Others: 5v
77	WT/DG	LDP Solenoid Control	PWM Signal: 0-12-0v
78-79	---	Not Used	---
80	LG/RD	S/C Vent Solenoid	Vacuum Decreasing: 1v

Pin Connector Graphic

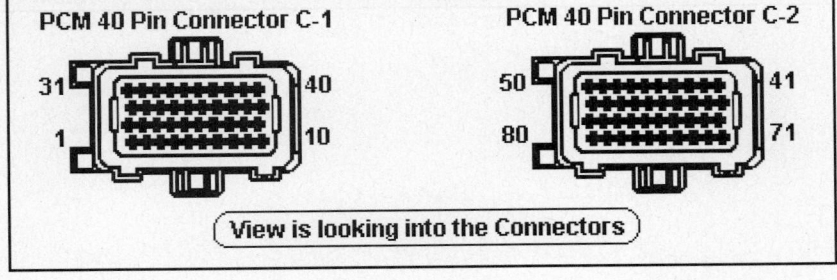

PCM 40 Pin Connector C-1

PCM 40 Pin Connector C-2

View is looking into the Connectors

1990 Horizon 2.2L I4 SOHC TBI VIN D (All) 60 Pin Connector

PCM Pin #	Wire Color	Circuit Description (60 Pin)	Value at Hot Idle
1	DG/RD	MAP Sensor Signal	1.5-1.7v
2	TN/WT	ECT Sensor Signal	At 180°F: 2.80v
3	RD/WT	Battery Power (Fused B+)	12-14v
4	BK/LB	Sensor Ground	<0.050v
5	BK/WT	Sensor Ground	<0.050v
6	PK/WT	5-Volt Supply	4.9-5.1v
7	OR	8-Volt Supply	7.9-8.1v
8	WT	Ignition Switch Output	12-14v
9	DB	Ignition Switch Output	12-14v
10	---	Not Used	---
11	LB/RD	Power Ground	<0.1v
12	LB/RD	Power Ground	<0.1v
13-15	---	Not Used	---
16	WT/DB	Injector 1 Driver	1-4 ms
17-18	---	Not Used	---
19	BK/GY	Coil Driver	5°, at 55 mph: 8° dwell
20	DG	Alternator Field Control	Digital Signal: 0-12-0v
21	BK/RD	Throttle Body Temperature	At 100°F: 2.51v
22	OR/DB	TP Sensor Signal	0.6-1.0v
23	---	Not Used	---
24	GY/BK	Distributor Reference Signal	Digital Signal: 0-5-0v
25	PK	SCI Transmit	0v
26	---	Not Used	---
27	BR	A/C Clutch Signal	A/C On: 1v, Off: 12v
28	---	Not Used	---
29	WT/PK	Brake Switch Signal	Brake Off: 0v, On: 12v
30	BR/YL	PNP Switch Signal	In P/N: 0v, Others: 5v
31	DB/PK	Radiator Fan Relay Control	Relay Off: 12v, On: 1v
32	BK/PK	MIL (lamp) Control	MIL On: 1v, Off: 12v
33	---	---	---
34	DB/OR	A/C WOT Relay Control	Relay Off: 12v, On: 1v
35	GY/YL	EGR Solenoid Control (California)	12v, 55 mph: 1v
36-38	---	Not Used	---
39	GY/RD	Auto Idle Speed Motor	DC pulse signals: 0.8-11v
40	BR	Auto Idle Speed Motor	DC pulse signals: 0.8-11v

Standard Colors and Abbreviations

Abbreviation	Color	Abbreviation	Color	Abbreviation	Color
BK	Black	GY	Gray	RD	Red
BL	Blue	GN	Green	TN	Tan
BR	Brown	LG	Light Green	VT	Violet
DB	Dark Blue	OR	Orange	WT	White
DG	Dark Green	PK	Pink	YL	Yellow

1990 Horizon 2.2L I4 SOHC TBI VIN D (All) 60 Pin Connector

PCM Pin #	Wire Color	Circuit Description (60 Pin)	Value at Hot Idle
41	BK/DG	HO2S-11 (B1 S1) Signal	0.1-1.0v
42	---	Not Used	---
43	GY/LB	Tachometer Signals	Pulse Signals
44	---	Not Used	---
45	LG	SCI Receive	5v
46	---	Not Used	---
47	WT/BK	VSS Signal	Digital Signal: 0-5-0v
48-50	---	Not Used	
51	DB/YL	ASD Relay Control	Relay Off: 12v, On: 1v
52	PK	EVAP Purge Solenoid Control	Solenoid Off: 12v, On: 1v
53	---	Not Used	---
54	OR/BK	A/T: Lockup Torque Converter	At Cruise w/TCC On: 1v
54	OR/BK	M/T: Shift Indicator Control	Lamp On: 1v, Off: 12v
55-58	---	Not Used	---
59	PK/BK	AIS Motor Control	DC pulse signals: 0.8-11v
60	YL/BK	AIS Motor Control	DC pulse signals: 0.8-11v

Pin Connector Graphic

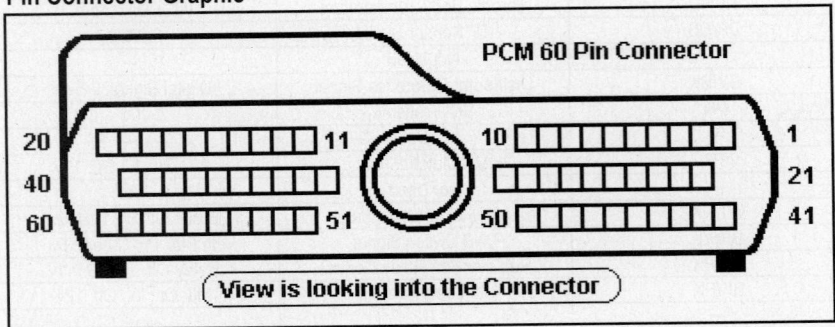

1990-92 Laser 1.8L I4 SOHC MFI VIN T (All) C22 Connector

PCM Pin #	Wire Color	Circuit Description (10 Pin)	Value at Hot Idle
101	BK	Power Ground	<0.1v
102	RD	Ignition Switch Output	12-14v
103	RD/BK	Battery Power (Fused B+)	12-14v
104	BK/YL	A/T: PNP Switch Signal	In P/N: 0v, Others: 5v
104	BK	M/T: Ground	<0.1v
105	---	Not Used	---
106	BK	Power Ground	<0.1v
107	RD	Ignition Switch Output	12-14v
108	BK/YL	Starter Relay Control	KOEC: 9-11v
109	BK/WT	Fuel Pump Relay Control	Relay Off: 12v, On: 1v
110	---	Not Used	---

1990-92 Laser 1.8L I4 SOHC MFI VIN T (All) C23 Connector

PCM Pin #	Wire Color	Circuit Description (18 Pin)	Value at Hot Idle
51	YL/BL	Injector 1 Driver	1-4 ms
52	YL/BK	Injector 2 Driver	1-4 ms
53	BR/WT	EGR Solenoid Control (California)	12v, 55 mph: 1v
54	GN/YL	Coil Driver	5°, at 55 mph: 8° dwell
55	---	Not Used	---
56	WT/RD	MPI Relay Control	Relay Off: 12v, On: 1v
57	---	Not Used	---
58	BL/WT	Idle Speed Motor (-) Retract	DC pulse signals: 0.8-11v
59	GN/BK	Idle Speed Motor (+) Extend	DC pulse signals: 0.8-11v
60	LG	Injector 3 Driver	1-4 ms
61	LG/WT	Injector 4 Driver	1-4 ms
62	BK/WT	EVAP Purge Solenoid Control	Solenoid Off: 12v, On: 1v
63	---	Not Used	---
64	RD/G	Emissions Lamp Control	Lamp On: 1v, Off: 12v
65	RD/BK	A/C Clutch Relay Control	A/C On: 1v, Off: 12v
66-68	---	Not Used	---

Standard Colors and Abbreviations

Abbreviation	Color	Abbreviation	Color	Abbreviation	Color
BK	Black	GY	Gray	RD	Red
BL	Blue	GN	Green	TN	Tan
BR	Brown	LG	Light Green	VT	Violet
DB	Dark Blue	OR	Orange	WT	White
DG	Dark Green	PK	Pink	YL	Yellow

1990-92 Laser 1.8L I4 SOHC MFI VIN T (All) C24 Connector

PCM Pin #	Wire Color	Circuit Description (24 Pin)	Value at Hot Idle
1	YL	Data Link Connector	0v
2	WT	Data Link Connector	0v
3	---	Not Used	---
4	WT	HO2S-11 (B1 S1) Signal	0.1-1.1v
5	YL/BK	PSP Switch Signal	Straight: 0v, Turning: 5v
6	GN	Idle Speed Control Switch	0-6v, at WOT: 12v
7	BK/GN	A/C Select Switch Sense	A/C On: 1v, Off: 12v
8	GN/OR	Air Temperature Sensor	At 100°F: 2.51v
9	---	Not Used	---
10	GN/BL	Airflow Sensor	2.2-3.2v
11	---	Not Used	---
12	YL/RD	Ignition Timing Adjustment	4.0-5.5v
13	GN/RD	5-Volt Supply	4.9-5.1v
14	GN/BK	Sensor Ground	<0.050v
15	BL/YL	EGR Solenoid Control (California)	12v, 55 mph: 1v
16	PK	BARO Read Sensor	Sea level: 3.5-4.2v
16 ('91-'92)	GN/YL	BARO Read Sensor	Sea level: 3.5-4.2v
17	BR	ISC Motor Position Sensor	0.8-1.2v
18	YL/WT	Vehicle Speed Signal	Digital Signal
19	GN/WT	TP Sensor Signal	0.6-1.0v
20	YL/GN	ECT Sensor Signal	At 180°F: 2.80v
21	BR/YL	CKP Sensor Signal	Digital Signal: 0-5-0v
22	BK/BL	TDC Signal Output	Pulse Signals: 0.2-2.0v
23	GN/RD	5-Volt Supply	4.9-5.1v
24	BK	Sensor Ground	<0.050v
24 ('91-'92)	GN/BK	Sensor Ground	<0.050v

Pin Connector Graphic

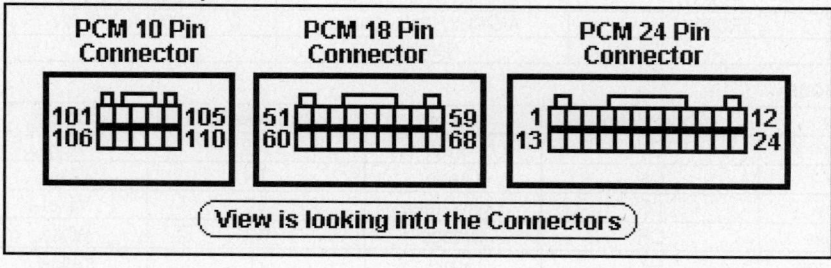

PCM 10 Pin Connector · PCM 18 Pin Connector · PCM 24 Pin Connector

View is looking into the Connectors

1993-94 Laser 1.8L I4 SOHC MFI VIN B (All) C22 Connector

PCM Pin #	Wire Color	Circuit Description (10 Pin)	Value at Hot Idle
101	BK	Power Ground	<0.1v
102	RD	Ignition Switch Output	12-14v
103	RD/BK	Battery Power (Fused B+)	12-14v
104	BK/YL	A/T: PNP Switch Signal	In P/N: 0v, Others: 5v
104	BK	M/T: Ground	<0.1v
105	---	Not Used	---
106	BK	Power Ground	<0.1v
107	RD	Ignition Switch Output	12-14v
108	BK/YL	Starter Relay Control	KOEC: 9-11v
109	BK/WT	Fuel Pump Relay Control	Relay Off: 12v, On: 1v
110	---	Not Used	---

1993-94 Laser 1.8L I4 SOHC MFI VIN B (All) C23 Connector

PCM Pin #	Wire Color	Circuit Description (18 Pin)	Value at Hot Idle
51	YL/BL	Injector 1 Driver	1-4 ms
52	YL/BK	Injector 2 Driver	1-4 ms
53	BR/WT	EGR Solenoid Control (California)	12v, 55 mph: 1v
54	GN/YL	Coil Driver	5°, at 55 mph: 8° dwell
55	---	Not Used	---
56	WT/RD	MPI Relay Control	Relay Off: 12v, On: 1v
57	---	Not Used	---
58	BL/WT	Idle Speed Motor (-) Retract	DC pulse signals: 0.8-11v
59	GN/BK	Idle Speed Motor (+) Extend	DC pulse signals: 0.8-11v
60	LG	Injector 3 Driver	1-4 ms
61	LG/WT	Injector 4 Driver	1-4 ms
62	BK/WT	EVAP Purge Solenoid Control	Solenoid Off: 12v, On: 1v
63	---	Not Used	---
64	RD/G	Emissions Lamp Control	Lamp On: 1v, Off: 12v
65	RD/BK	A/C Clutch Relay Control	A/C On: 1v, Off: 12v
66-68	---	Not Used	---

Standard Colors and Abbreviations

Abbreviation	Color	Abbreviation	Color	Abbreviation	Color
BK	Black	GY	Gray	RD	Red
BL	Blue	GN	Green	TN	Tan
BR	Brown	LG	Light Green	VT	Violet
DB	Dark Blue	OR	Orange	WT	White
DG	Dark Green	PK	Pink	YL	Yellow

1993-94 Laser 1.8L I4 SOHC MFI VIN B (All) C24 Connector

PCM Pin #	Wire Color	Circuit Description (24 Pin)	Value at Hot Idle
1	YL	Data Link Connector	0v
2	WT	Data Link Connector	0v
3	---	Not Used	---
4	WT	HO2S-11 (B1 S1) Signal	0.1-1.1v
5	YL/BK	PSP Switch Signal	Straight: 0v, Turning: 5v
6	GN	Idle Speed Control Switch	0-6v, at WOT: 12v
7	BK/GN	A/C Select Switch Sense	A/C On: 1v, Off: 12v
8	GN/OR	Air Temperature Sensor	At 100°F: 2.51v
9	---	Not Used	---
10	GN/BL	Airflow Sensor	2.2-3.2v
11	---	Not Used	---
12	YL/RD	Ignition Timing Adjustment	4.0-5.5v
13	GN/RD	5-Volt Supply	4.9-5.1v
14	GN/BK	Sensor Ground	<0.050v
15	BL/YL	EGR Solenoid Control (California)	12v, 55 mph: 1v
16	GN/YL	BARO Read Sensor	Sea level: 3.5-4.2v
17	BR	ISC Motor Position Sensor	0.8-1.2v
18	YL/WT	Vehicle Speed Signal	Digital Signal
19	GN/WT	TP Sensor Signal	0.6-1.0v
20	YL/GN	ECT Sensor Signal	At 180°F: 2.80v
21	BR/YL	CKP Sensor Signal	Digital Signal: 0-5-0v
22	BK/BL	TDC Signal Output	Pulse Signals: 0.2-2.0v
23	GN/RD	5-Volt Supply	4.9-5.1v
24	GN/BK	Sensor Ground	<0.050v

Pin Connector Graphic

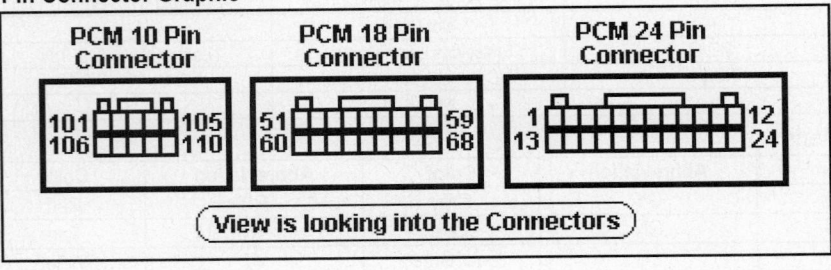

PCM 10 Pin Connector

101 105
106 110

PCM 18 Pin Connector

51 59
60 68

PCM 24 Pin Connector

1 12
13 24

(View is looking into the Connectors)

1990-92 Laser 2.0L I4 DOHC MFI VIN R (All) C22 Connector

PCM Pin #	Wire Color	Circuit Description (10 Pin)	Value at Hot Idle
101	BK	Power Ground	<0.1v
102	RD	Ignition Switch Output	12-14v
103	RD/BK	Battery Power (Fused B+)	12-14v
104	BK/YL	A/T: PNP Switch Signal	In P/N: 0v, Others: 5v
104	BK	M/T: Ground	<0.1v
105	---	Not Used	---
106	BK	Power Ground	<0.1v
107	RD	Ignition Switch Output	12-14v
108	BK/YL	Starter Relay Control	KOEC: 9-11v
109	WT	Fuel Pump Relay Control	Relay Off: 12v, On: 1v
110	BK/WT	MPI Relay Control	Relay Off: 12v, On: 1v

1990-92 Laser 2.0L I4 DOHC MFI VIN R (All) C23 Connector

PCM Pin #	Wire Color	Circuit Description (18 Pin)	Value at Hot Idle
51	YL/BL	Injector 1 Driver	1-4 ms
52	YL/BK	Injector 2 Driver	1-4 ms
53	BR/WT	EGR Solenoid Control (California)	12v, 55 mph: 1v
54	YL	Coil Driver	5°, at 55 mph: 8° dwell
55	YL/RD	Coil Driver	5°, at 55 mph: 8° dwell
56	WT/RD	MPI Relay Control	Relay Off: 12v, On: 1v
57	---	Not Used	---
58	BL	ISC Motor Control	DC pulse signals: 0.8-11v
59	YL	ISC Motor Control	DC pulse signals: 0.8-11v
60	LG	Injector 3 Driver	1-4 ms
61	LG/WT	Injector 4 Driver	1-4 ms
62	BK/WT	EVAP Purge Solenoid Control	Solenoid Off: 12v, On: 1v
63	BK/BL	MPI Relay Control	Relay Off: 12v, On: 1v
64	RD/G	Emissions Lamp Control	Lamp On: 1v, Off: 12v
65	RD/BK	A/C Clutch Relay Control	A/C On: 1v, Off: 12v
66	BK/BL	MPI Relay Control	Relay Off: 12v, On: 1v
67	WT	ISC Motor Control	DC pulse signals: 0.8-11v
68	BK	ISC Motor Control	DC pulse signals: 0.8-11v

Standard Colors and Abbreviations

Abbreviation	Color	Abbreviation	Color	Abbreviation	Color
BK	Black	GY	Gray	RD	Red
BL	Blue	GN	Green	TN	Tan
BR	Brown	LG	Light Green	VT	Violet
DB	Dark Blue	OR	Orange	WT	White
DG	Dark Green	PK	Pink	YL	Yellow

1990-92 Laser 2.0L I4 DOHC MFI VIN R (All) C24 Connector

PCM Pin #	Wire Color	Circuit Description (24 Pin)	Value at Hot Idle
1	YL	Data Link Connector	0v
2	WT	Data Link Connector	0v
3	---	Not Used	---
4	WT	HO2S-11 (B1 S1) Signal	0.1-1.1v
5	YL/BK	PSP Switch Signal	Straight: 0v, Turning: 5v
6	---	Not Used	---
7	BK/GN	A/C Select Switch Sense	A/C On: 1v, Off: 12v
8	GN/OR	Air Temperature Sensor	At 100°F: 2.51v
9	---	Not Used	---
10	GN/BL	Airflow Sensor	2.2-3.2v
11	---	Not Used	---
12	YL/RD	Ignition Timing Adjustment	4.0-5.5v
13	BK/GN	5-Volt Supply	4.9-5.1v
14	GN	Idle Speed Control Switch	0-6v, at WOT: 12v
15	BL/YL	EGR Solenoid Control (California)	12v, 55 mph: 1v
16	GN/YL	BARO Read Sensor	Sea level: 3.5-4.2v
17	GN/BK	Sensor Ground	<0.050v
18	YL/WT	Vehicle Speed Signal	Digital Signal
19	GN/WT	TP Sensor Signal	0.6-1.0v
20	YL/GN	ECT Sensor Signal	At 180°F: 2.80v
21	BK	CKP Sensor Signal	Digital Signal: 0-5-0v
22	WT	TDC Signal Output	Pulse Signals: 0.2-2.0v
23	GN/RD	5-Volt Supply	4.9-5.1v
24	GN/BK	Sensor Ground	<0.050v

Pin Connector Graphic

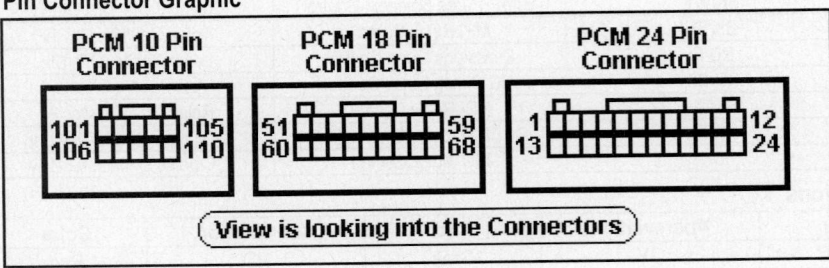

View is looking into the Connectors

1993-94 Laser 2.0L I4 DOHC MFI VIN E (All) C22 Connector

PCM Pin #	Wire Color	Circuit Description (10 Pin)	Value at Hot Idle
101	BK	Power Ground	<0.1v
102	RD	Ignition Switch Output	12-14v
103	RD/BK	Battery Power (Fused B+)	12-14v
104	BK/YL	A/T: PNP Switch Signal	In P/N: 0v, Others: 5v
104	BK	M/T: Ground	<0.1v
105	---	Not Used	---
106	BK	Power Ground	<0.1v
107	RD	Ignition Switch Output	12-14v
108	BK/YL	Starter Relay Control	KOEC: 9-11v
109	WT	Fuel Pump Relay Control	Relay Off: 12v, On: 1v
110	BK/WT	MPI Relay Control	Relay Off: 12v, On: 1v

1993-94 Laser 2.0L I4 DOHC MFI VIN E (All) C23 Connector

PCM Pin #	Wire Color	Circuit Description (18 Pin)	Value at Hot Idle
51	YL/BL	Injector 1 Driver	1-4 ms
52	YL/BK	Injector 2 Driver	1-4 ms
53	BR/WT	EGR Solenoid Control (California)	12v, 55 mph: 1v
54	YL	Coil Driver	5°, at 55 mph: 8° dwell
55	YL/RD	Coil Driver	5°, at 55 mph: 8° dwell
56	WT/RD	MPI Relay Control	Relay Off: 12v, On: 1v
57	---	Not Used	---
58	BL	ISC Motor Control	DC pulse signals: 0.8-11v
59	YL	ISC Motor Control	DC pulse signals: 0.8-11v
60	LG	Injector 3 Driver	1-4 ms
61	LG/WT	Injector 4 Driver	1-4 ms
62	BK/WT	EVAP Purge Solenoid Control	Solenoid Off: 12v, On: 1v
63	BK/BL	MPI Relay Control	Relay Off: 12v, On: 1v
64	RD/G	Emissions Lamp Control	Lamp On: 1v, Off: 12v
65	RD/BK	A/C Clutch Relay Control	A/C On: 1v, Off: 12v
66	BK/BL	MPI Relay Control	Relay Off: 12v, On: 1v
67	WT	ISC Motor Control	DC pulse signals: 0.8-11v
68	BK	ISC Motor Control	DC pulse signals: 0.8-11v

Standard Colors and Abbreviations

Abbreviation	Color	Abbreviation	Color	Abbreviation	Color
BK	Black	GY	Gray	RD	Red
BL	Blue	GN	Green	TN	Tan
BR	Brown	LG	Light Green	VT	Violet
DB	Dark Blue	OR	Orange	WT	White
DG	Dark Green	PK	Pink	YL	Yellow

1993-94 Laser 2.0L I4 DOHC MFI VIN E (All) C24 Connector

PCM Pin #	Wire Color	Circuit Description (24 Pin)	Value at Hot Idle
1	YL	Data Link Connector	0v
2	WT	Data Link Connector	0v
3	---	Not Used	---
4	WT	HO2S-11 (B1 S1) Signal	0.1-1.1v
5	YL/BK	PSP Switch Signal	Straight: 0v, Turning: 5v
6	---	Not Used	---
7	BK/GN	A/C Select Switch Sense	A/C On: 1v, Off: 12v
8	GN/OR	Air Temperature Sensor	At 100°F: 2.51v
9	---	Not Used	---
10	GN/BL	Airflow Sensor	2.2-3.2v
11	---	Not Used	---
12	YL/RD	Ignition Timing Adjustment	4.0-5.5v
13	BK/GN	5-Volt Supply	4.9-5.1v
14	GN	Idle Speed Control Switch	0-6v, at WOT: 12v
15	BL/YL	EGR Solenoid Control (California)	12v, 55 mph: 1v
16	GN/YL	BARO Read Sensor	Sea level: 3.5-4.2v
17	GN/BK	Sensor Ground	<0.050v
18	YL/WT	Vehicle Speed Signal	Digital Signal
19	GN/WT	TP Sensor Signal	0.6-1.0v
20	YL/GN	ECT Sensor Signal	At 180°F: 2.80v
21	BK	CKP Sensor Signal	Digital Signal: 0-5-0v
22	WT	TDC Signal Output	Pulse Signals: 0.2-2.0v
23	GN/RD	5-Volt Supply	4.9-5.1v
24	GN/BK	Sensor Ground	<0.050v

Pin Connector Graphic

PCM 10 Pin Connector

101 105
106 110

PCM 18 Pin Connector

51 59
60 68

PCM 24 Pin Connector

1 12
13 24

View is looking into the Connectors

1990-92 Laser 2.0L I4 MFI Turbo VIN U (All) C22 Connector

PCM Pin #	Wire Color	Circuit Description (10 Pin)	Value at Hot Idle
101	BK	Power Ground	<0.1v
102	RD	Ignition Switch Output	12-14v
103	RD/BK	Battery Power (Fused B+)	12-14v
104	BK/YL	A/T: PNP Switch Signal	In P/N: 0v, Others: 5v
104	BK	M/T: Ground	<0.1v
105	OR	Wastegate Solenoid Control	Solenoid Off: 12v, On: 1v
106	BK	Power Ground	<0.1v
107	RD	Ignition Switch Output	12-14v
108	BK/YL	Starter Relay Control	KOEC: 9-11v
109	WT	Fuel Pump Relay Control	Relay Off: 12v, On: 1v
110	BK/WT	MPI Relay Control	Relay Off: 12v, On: 1v

1990-92 Laser 2.0L I4 MFI Turbo VIN U (All) C23 Connector

PCM Pin #	Wire Color	Circuit Description (18 Pin)	Value at Hot Idle
51	YL/BL	Injector 1 Driver	1-4 ms
52	YL/BK	Injector 2 Driver	1-4 ms
53	BK/YL	EGR Solenoid Control (California)	12v, 55 mph: 1v
54	YL	Coil Driver	5°, at 55 mph: 8° dwell
55	YL/RD	Coil Driver	5°, at 55 mph: 8° dwell
56	WT/RD	MPI Relay Control	Relay Off: 12v, On: 1v
57	WT	Fuel Pressure Solenoid	Solenoid Off: 12v, On: 1v
58	BL	ISC Motor Control	DC pulse signals: 0.8-11v
59	YL	ISC Motor Control	DC pulse signals: 0.8-11v
60	LG	Injector 3 Driver	1-4 ms
61	LG/WT	Injector 4 Driver	1-4 ms
62	BK/WT	EVAP Purge Solenoid Control	Solenoid Off: 12v, On: 1v
63	BK/BL	MPI Relay Control	Relay Off: 12v, On: 1v
64	RD/G	Emissions Lamp Control	Lamp On: 1v, Off: 12v
65	RD/BK	A/C Clutch Relay Control	A/C On: 1v, Off: 12v
66	BK/BL	MPI Relay Control	Relay Off: 12v, On: 1v
67	WT	ISC Motor Control	DC pulse signals: 0.8-11v
68	BK	ISC Motor Control	DC pulse signals: 0.8-11v

Standard Colors and Abbreviations

Abbreviation	Color	Abbreviation	Color	Abbreviation	Color
BK	Black	GY	Gray	RD	Red
BL	Blue	GN	Green	TN	Tan
BR	Brown	LG	Light Green	VT	Violet
DB	Dark Blue	OR	Orange	WT	White
DG	Dark Green	PK	Pink	YL	Yellow

1990-92 Laser 2.0L I4 MFI Turbo VIN U (All) C24 Connector

PCM Pin #	Wire Color	Circuit Description (24 Pin)	Value at Hot Idle
1	YL	Data Link Connector	0v
2	WT	Data Link Connector	0v
3	BK/RD	Boost Pressure Gauge	0-5.1v
4	WT	HO2S-11 (B1 S1) Signal	0.1-1.1v
5	YL/BK	PSP Switch Signal	Straight: 0v, Turning: 5v
6	GN	Idle Speed Control Switch	0-6v, at WOT: 12v
7	BK/GN	A/C Select Switch Sense	A/C On: 1v, Off: 12v
8	GN/OR	Air Temperature Sensor	At 100°F: 2.51v
9	WT	Knock Sensor Signal	0.080v AC
10	GN/BL	Airflow Sensor	2.2-3.2v
11	---	Not Used	---
12	YL/RD	Ignition Timing Adjustment	4.0-5.5v
13	BK/WT	5-Volt Supply	4.9-5.1v
14	GN/WT	Airflow Sensor VREF	4.9-5.1v
15	BL/YL	EGR Solenoid Control (California)	12v, 55 mph: 1v
16	GN/YL	BARO Read Sensor	Sea level: 3.5-4.2v
17	GN/BK	Sensor Ground	<0.050v
18	YL/WT	Vehicle Speed Signal	Digital Signal
19	GN/WT	TP Sensor Signal	0.6-1.0v
20	YL/GN	ECT Sensor Signal	At 180°F: 2.80v
21	BK	CKP Sensor Signal	Digital Signal: 0-5-0v
22	WT	TDC Signal Output	Pulse Signals: 0.2-2.0v
23	GN/RD	5-Volt Supply	4.9-5.1v
24	GN/BK	Sensor Ground	<0.050v

Pin Connector Graphic

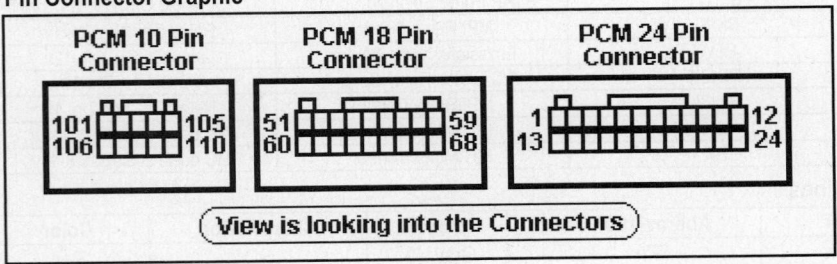

PCM 10 Pin Connector
101 105
106 110

PCM 18 Pin Connector
51 59
60 68

PCM 24 Pin Connector
1 12
13 24

View is looking into the Connectors

Turbo VIN F (All) C22 Connector

Wire Color	Circuit Description (10 Pin)	Value at Hot Idle
BK	Power Ground	<0.1v
RD	Ignition Switch Output	12-14v
RD/BK	Battery Power (Fused B+)	12-14v
BK/YL	A/T: PNP Switch Signal	In P/N: 0v, Others: 5v
BK	M/T: Ground	<0.1v
OR	Wastegate Solenoid Control	Solenoid Off: 12v, On: 1v
BK	Power Ground	<0.1v
RD	Ignition Switch Output	12-14v
BK/YL	Starter Relay Control	KOEC: 9-11v
WT	Fuel Pump Relay Control	Relay Off: 12v, On: 1v
BK/WT	MPI Relay Control	Relay Off: 12v, On: 1v

.0L I4 MFI Turbo VIN F (All) C23 Connector

Pin #	Wire Color	Circuit Description (18 Pin)	Value at Hot Idle
1	YL/BL	Injector 1 Driver	1-4 ms
52	YL/BK	Injector 2 Driver	1-4 ms
53	BK/YL	EGR Solenoid Control (California)	12v, 55 mph: 1v
54	YL	Coil Driver	5°, at 55 mph: 8° dwell
55	YL/RD	Coil Driver	5°, at 55 mph: 8° dwell
56	WT/RD	MPI Relay Control	Relay Off: 12v, On: 1v
57	WT	Fuel Pressure Solenoid	Solenoid Off: 12v, On: 1v
58	BL	ISC Motor Control	DC pulse signals: 0.8-11v
59	YL	ISC Motor Control	DC pulse signals: 0.8-11v
60	LG	Injector 3 Driver	1-4 ms
61	LG/WT	Injector 4 Driver	1-4 ms
62	BK/WT	EVAP Purge Solenoid Control	Solenoid Off: 12v, On: 1v
63	BK/BL	MPI Relay Control	Relay Off: 12v, On: 1v
64	RD/G	Emissions Lamp Control	Lamp On: 1v, Off: 12v
65	RD/BK	A/C Clutch Relay Control	A/C On: 1v, Off: 12v
66	BK/BL	MPI Relay Control	Relay Off: 12v, On: 1v
67	WT	ISC Motor Control	DC pulse signals: 0.8-11v
68	BK	ISC Motor Control	DC pulse signals: 0.8-11v

Standard Colors and Abbreviations

Abbreviation	Color	Abbreviation	Color	Abbreviation	Color
BK	Black	GY	Gray	RD	Red
BL	Blue	GN	Green	TN	Tan
BR	Brown	LG	Light Green	VT	Violet
DB	Dark Blue	OR	Orange	WT	White
DG	Dark Green	PK	Pink	YL	Yellow

1993-94 Laser
PC

1993-94 Laser 2.0L I4 MFI Turbo VIN F (All) C24 Connector

PCM Pin #	Wire Color	Circuit Description (24 Pin)	
1	YL	Data Link Connector	
2	WT	Data Link Connector	
3	BK/RD	Boost Pressure Gauge	
4	WT	HO2S-11 (B1 S1) Signal	0.
5	YL/BK	PSP Switch Signal	Straight: 0v,
6	GN/WT	Airflow Sensor	2.2-3.
7	BK/GN	A/C Select Switch Sense	A/C On: 1v, O
8	GN/OR	Air Temperature Sensor	At 100°F: 2.5
9	WT	Knock Sensor Signal	0.080v AC
10	GN/BL	Airflow Sensor VREF	2.2-3.2v
11	---	Not Used	---
12	YL/RD	Ignition Timing Adjustment	4.0-5.5v
13	BK/GN	5-Volt Supply	4.9-5.1v
14	GN	Idle Speed Control Switch	0-6v, at WOT: 12v
15	BL/YL	EGR Solenoid Control (California)	12v, 55 mph: 1v
16	GN/YL	BARO Read Sensor	Sea level: 3.5-4.2v
17	GN/BK	Sensor Ground	<0.050v
18	YL/WT	Vehicle Speed Signal	Digital Signal
19	GN/WT	TP Sensor Signal	0.6-1.0v
20	YL/GN	ECT Sensor Signal	At 180°F: 2.80v
21	BK	CKP Sensor Signal	Digital Signal: 0-5-0v
22	WT	TDC Signal Output	Pulse Signals: 0.2-2.0v
23	GN/RD	5-Volt Supply	4.9-5.1v
24	GN/BK	Sensor Ground	<0.050v

Pin Connector Graphic

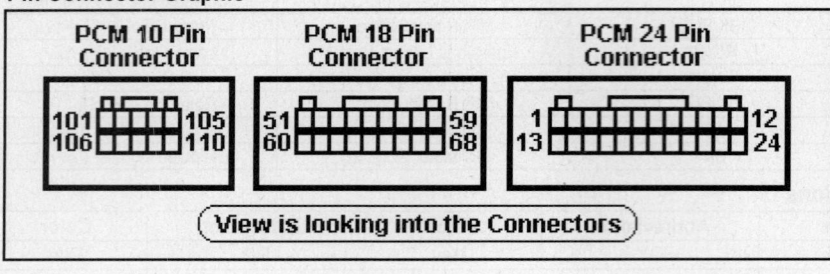

PCM 10 Pin Connector
101 105
106 110

PCM 18 Pin Connector
51 59
60 68

PCM 24 Pin Connector
1 12
13 24

View is looking into the Connectors

1995 Neon 2.0L I4 SOHC MFI VIN C (All) 60 Pin Connector

PCM Pin #	Wire Color	Circuit Description (60 Pin)	Value at Hot Idle
1	BK/GY	Coil 2 Driver	5°, at 55 mph: 8° dwell
2	BK/TN	Power Ground	<0.1v
3	YL/WT	Injector 3 Driver	1-4 ms
4	WT/DB	Injector 1 Driver	1-4 ms
5	WT/OR	Vehicle Speed Signal	Digital Signal
6	BK/RD	IAT Sensor Signal	At 100°F: 1.83v
7	TN/WT	HO2S-12 (B1 S2) Signal	0.1-1.1v
8	BK/DG	HO2S-11 (B1 S1) Signal	0.1-1.1v
9	LG	SCI Receive	0v
10	OR/DB	TP Sensor Signal	0.6-1.0v
11	RD/WT	Battery Power (Fused B+)	12-14v
12	---	Not Used	---
13	DB	Fuel Level Sensor Signal	70 ohms (±20) w/full tank
14	YL/BK	IAC 2 Driver	DC pulse signals: 0.8-11v
15	GY/RD	IAC 3 Driver	DC pulse signals: 0.8-11v
16	PK/BK	EVAP Purge Solenoid Control	PWM Signal: 0-12-0v
17	OR/BK	TCC Solenoid Control	At Cruise w/TCC On: 1v
18	DB/YL	ASD Relay Control	Relay Off: 12v, On: 1v
19	DB/PK	Radiator Fan Relay Control	Relay Off: 12v, On: 1v
20	---	Not Used	---
21	DB/TN	Coil 1 Driver	5°, at 55 mph: 8° dwell
22	BK/TN	Power Ground	<0.1v
23	TN	Injector 2 Driver	1-4 ms
24	LB/BR	Injector 4 Driver	1-4 ms
25	GY/BK	CKP Sensor Signal	Digital Signal: 0-5-0v
26	TN/YL	CMP Sensor Signal	Digital Signal: 0-5-0v
27	DB/LG	Knock Sensor Signal	0.080v AC
28	TN/DB	ECT Sensor Signal	At 180°F: 2.80v
29	DG/RD	MAP Sensor Signal	1.5-1.7v
30	PK	SCI Transmit	0v
31	RD/LG	Speed Control Set Switch Signal	S/C & Set Switch On: 3.8v
32	WT/PK	Brake Switch Signal	Brake Off: 0v, On: 12v
33	BR/OR	A/C Select Switch Sense	A/C On: 1v, Off: 12v
34	PK/BK	IAC 4 Driver	DC pulse signals: 0.8-11v
35	BR/WT	IAC 1 Driver	DC pulse signals: 0.8-11v
36	BK/PK	MIL (lamp) Control	Lamp On: 1v, Off: 12v
37	TN/BK	Generator Indicator Control	Lamp On: 1v, Off: 12v
38	BR	Fuel Pump Relay Control	Relay Off: 12v, On: 1v
39	GY/YL	EGR Solenoid Control	12v, 55 mph: 1v
40	TN/RD	S/C Vacuum Solenoid	Vacuum Increasing: 1v

Standard Colors and Abbreviations

Abbreviation	Color	Abbreviation	Color	Abbreviation	Color
BK	Black	GY	Gray	RD	Red
BL	Blue	GN	Green	TN	Tan
BR	Brown	LG	Light Green	VT	Violet
DB	Dark Blue	OR	Orange	WT	White
DG	Dark Green	PK	Pink	YL	Yellow

1995 Neon 2.0L I4 SOHC MFI VIN C (All) 60 Pin Connector

PCM Pin #	Wire Color	Circuit Description (60 Pin)	Value at Hot Idle
41	DG	Generator Field Control	Digital Signal: 0-12-0v
42	DG/OR	ASD Relay Output	12-14v
43	PK/WT	5-Volt Supply	4.9-5.1v
44	OR	8-Volt Supply	7.9-8.1v
45-46	---	Not Used	---
47	YL/RD	S/C Power Supply	12-14v
48	---	Not Used	---
49	PK/LG	Battery Temperature Sensor	At 86°F: 1.96v
50	BR/YL	PNP Switch Signal	In P/N: 0v, Others: 5v
51	BK/LB	Sensor Ground	<0.050v
52	BK/WT	Power Ground	<0.1v
53	---	Not Used	---
54	LG/BK	Ignition Switch Output	12-14v
55	PK/BK	Reverse Indicator Control	Lamp On: 1v, Off: 12v
56	WT	PSP Switch Signal	Straight: 0v, Turning: 5v
57-58	---	Not Used	---
59	DB/OR	A/C Clutch Relay Control	Relay Off: 12v, On: 1v
60	LG/RD	S/C Vent Solenoid	Vacuum Decreasing: 1v

Pin Connector Graphic

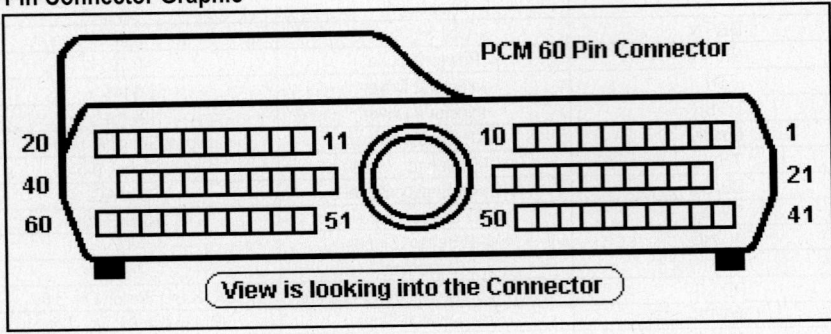

1996-98 Neon 2.0L I4 SOHC MFI VIN C (All) 'C1' Black Connector

PCM Pin #	Wire Color	Circuit Description (40-Pin)	Value at Hot Idle
1	---	Not Used	---
2	BK/GY	Coil 1 Driver	5°, at 55 mph: 8° dwell
3	DB/TN	Coil 2 Driver	5°, at 55 mph: 8° dwell
4	DG	Generator Field Control	Digital Signal: 0-12-0v
5	YL/RD	S/C Power Supply	12-14v
6	DG/OR	ASD Relay Output	12-14v
7	YL/WT	Injector 3 Driver	1-4 ms
8	BK/PK	MIL (lamp) Control	Lamp On: 1v, Off: 12v
9	---	Not Used	---
10	BK/TN	Power Ground	<0.1v
11-12	---	Not Used	---
13	WT/LB	Injector 1 Driver	1-4 ms
14-15	---	Not Used	---
16	LB/BR	Injector 4 Driver	1-4 ms
17	TN	Injector 2 Driver	1-4 ms
18	LG	Radiator Fan Relay Control	Relay Off: 12v, On: 1v
19	---	Not Used	---
20	DB/WT	Ignition Switch Output	12-14v
21-22	---	Not Used	---
23	DB	Fuel Level Sensor Signal	70 ohms (±20) w/full tank
24	DB/LG	Knock Sensor Signal	0.080v AC
25	---	Not Used	---
26	TN/DB	ECT Sensor Signal	At 180°F: 2.80v
27-29	---	Not Used	---
30	BK/DG	HO2S-11 (B1 S1) Signal	0.1-1.1v
31	---	Not Used	---
32	GY/BK	CKP Sensor Signal	Digital Signal: 0-5-0v
33	TN/YL	CMP Sensor Signal	Digital Signal: 0-5-0v
34	---	Not Used	---
35	OR/LB	TP Sensor Signal	0.6-1.0v
36	DG/RD	MAP Sensor Signal	1.5-1.7v
37	BK/RD	IAT Sensor Signal	At 100°F: 1.83v
38	BR/OR	A/C Select Switch Sense	A/C On: 1v, Off: 12v
39	---	Not Used	---
40	GY/YL	EGR Solenoid Control	12v, 55 mph: 1v

Standard Colors and Abbreviations

Abbreviation	Color	Abbreviation	Color	Abbreviation	Color
BK	Black	GY	Gray	RD	Red
BL	Blue	GN	Green	TN	Tan
BR	Brown	LG	Light Green	VT	Violet
DB	Dark Blue	OR	Orange	WT	White
DG	Dark Green	PK	Pink	YL	Yellow

1996-98 Neon 2.0L I4 SOHC MFI VIN C (All) 'C2' White Connector

PCM Pin #	Wire Color	Circuit Description (40-Pin)	Value at Hot Idle
41	RD/LG	Speed Control Set Switch Signal	S/C & Set Switch On: 3.8v
42	---	Not Used	---
43	BK/LB	Sensor Ground	<0.050v
44	OR	8-Volt Supply	7.9-8.1v
45	WT	PSP Switch Signal	Straight: 0v, Turning: 5v
46	RD/WT	Battery Power (Fused B+)	12-14v
47	BK/WT	Sensor Ground	<0.050v
48	BR/WT	IAC 3 Driver	DC pulse signals: 0.8-11v
49	YL/BK	IAC 2 Driver	DC pulse signals: 0.8-11v
50	BK/TN	Power Ground	<0.1v
51	TN/WT	HO2S-12 (B1 S2) Signal	0.1-1.1v
52	PK/LG	Battery Temperature Sensor	At 86°F: 1.96v
53-54	---	Not Used	---
55 ('96)	DB/PK	Radiator Fan Relay Control	Relay Off: 12v, On: 1v
56	TN/BK	Generator Indicator Control	Lamp On: 1v, Off: 12v
57	GY/RD	IAC 1 Driver	DC pulse signals: 0.8-11v
58	VT	IAC 4 Driver	DC pulse signals: 0.8-11v
59-60	---	Not Used	---
61	VT/WT	5-Volt Supply	4.9-5.1v
62	WT/PK	Brake Switch Signal	Brake Off: 0v, On: 12v
63	---	Not Used	---
64	DB/OR	A/C Clutch Relay Control	Relay Off: 12v, On: 1v
65	PK	SCI Transmit	0v
66	WT/OR	Vehicle Speed Signal	Digital Signal
67	DB/YL	ASD Relay Control	Relay Off: 12v, On: 1v
68	PK/BK	EVAP Purge Solenoid Control	PWM Signal: 0-12-0v
70-71	---	Not Used	---
72	OR	LDP Switch Signal	Closed: 0v, Open: 12v
73	GY/LB	Tachometer Signals	Pulse Signals
74	BR	Fuel Pump Relay Control	Relay Off: 12v, On: 1v
75	LG	SCI Receive	5v
76	BR/YL	PNP Switch Signal	In P/N: 0v, Others: 5v
77	WT/LG	LDP Solenoid Control	PWM Signal: 0-12-0v
78	TN/RD	S/C Vacuum Solenoid	Vacuum Increasing: 1v
79	OR/BK	TCC Solenoid Control	At Cruise w/TCC On: 1v
80	LG/RD	S/C Vent Solenoid	Vacuum Decreasing: 1v

Pin Connector Graphic

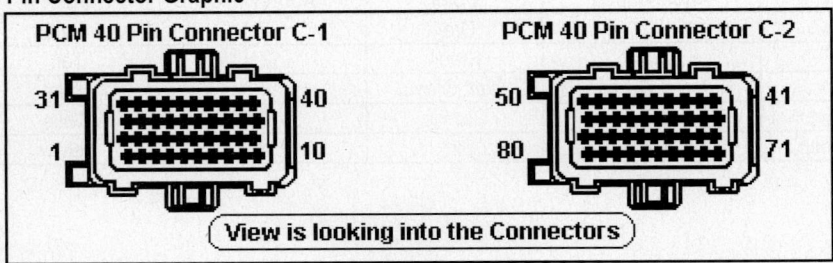

1999 Neon 2.0L I4 SOHC MFI VIN C (All) 'C1' Black Connector

PCM Pin #	Wire Color	Circuit Description (40-Pin)	Value at Hot Idle
1	---	Not Used	---
2	BK/GY	Coil 1 Driver	5°, at 55 mph: 8° dwell
3	DB/TN	Coil 2 Driver	5°, at 55 mph: 8° dwell
4	DG	Generator Field Control	Digital Signal: 0-12-0v
5	YL/RD	S/C Power Supply	12-14v
6	DG/OR	ASD Relay Output	12-14v
7	YL/WT	Injector 3 Driver	1-4 ms
8	BK/PK	MIL (lamp) Control	Lamp On: 1v, Off: 12v
9	---	Not Used	---
10	BK/TN	Power Ground	<0.1v
11-12	---	Not Used	---
13	WT/DB	Injector 1 Driver	1-4 ms
14-15	---	Not Used	---
16	LB/BR	Injector 4 Driver	1-4 ms
17	TN	Injector 2 Driver	1-4 ms
18	LG	Radiator Fan Relay Control	Relay Off: 12v, On: 1v
19	---	Not Used	---
20	DB/WT	Ignition Switch Output	12-14v
21-22	---	Not Used	---
23	DB	Fuel Level Sensor Signal	70 ohms (±20) w/full tank
24-25	---	Not Used	---
26	TN/DB	ECT Sensor Signal	At 180°F: 2.80v
27-29	---	Not Used	---
30	BK/DG	HO2S-11 (B1 S1) Signal	0.1-1.1v
31	---	Not Used	---
32	GY/BK	CKP Sensor Signal	Digital Signal: 0-5-0v
33	TN/YL	CMP Sensor Signal	Digital Signal: 0-5-0v
34	---	Not Used	---
35	OR/LB	TP Sensor Signal	0.6-1.0v
36	DG/RD	MAP Sensor Signal	1.5-1.7v
37	BK/RD	IAT Sensor Signal	At 100°F: 1.83v
38	BR/OR	A/C Select Switch Sense	A/C On: 1v, Off: 12v
39	---	Not Used	---
40	GY/YL	EGR Solenoid Control	12v, 55 mph: 1v

Standard Colors and Abbreviations

Abbreviation	Color	Abbreviation	Color	Abbreviation	Color
BK	Black	GY	Gray	RD	Red
BL	Blue	GN	Green	TN	Tan
BR	Brown	LG	Light Green	VT	Violet
DB	Dark Blue	OR	Orange	WT	White
DG	Dark Green	PK	Pink	YL	Yellow

1999 Neon 2.0L I4 SOHC MFI VIN C (All) 'C2' White Connector

PCM Pin #	Wire Color	Circuit Description (40-Pin)	Value at Hot Idle
41	RD/LG	Speed Control Set Switch Signal	S/C & Set Switch On: 3.8v
42	---	Not Used	---
43	BK/LB	Sensor Ground	<0.050v
44	OR	8-Volt Supply	7.9-8.1v
45	WT	PSP Switch Signal	Straight: 0v, Turning: 5v
46	RD/WT	Battery Power (Fused B+)	12-14v
47	BK/WT	Sensor Ground	<0.050v
48	BR/WT	IAC 3 Driver	DC pulse signals: 0.8-11v
49	YL/BK	IAC 2 Driver	DC pulse signals: 0.8-11v
50	BK/TN	Power Ground	<0.1v
51	TN/WT	HO2S-12 (B1 S2) Signal	0.1-1.1v
52	VT/LG	Battery Temperature Sensor	At 86°F: 1.96v
53-55	---	Not Used	---
56	TN/BK	Generator Indicator Control	Lamp On: 1v, Off: 12v
57	GY/RD	IAC 1 Driver	DC pulse signals: 0.8-11v
58	VT	IAC 4 Driver	DC pulse signals: 0.8-11v
59-60	---	Not Used	---
61	VT/WT	5-Volt Supply	4.9-5.1v
62	WT/PK	Brake Switch Signal	Brake Off: 0v, On: 12v
63	---	Not Used	---
64	DB/OR	A/C Clutch Relay Control	Relay Off: 12v, On: 1v
65	PK	SCI Transmit	0v
66	WT/OR	Vehicle Speed Signal	Digital Signal
67	PK/BK	EVAP Purge Solenoid Control	PWM Signal: 0-12-0v
68-71	---	Not Used	---
72	OR	LDP Switch Signal	Closed: 0v, Open: 12v
73	GY/LB	Tachometer Signals	Pulse Signals
74	BR	Fuel Pump Relay Control	Relay Off: 12v, On: 1v
75	LG	SCI Receive	5v
76	BR/YL	PNP Switch Signal	In P/N: 0v, Others: 5v
77	WT/LG	LDP Solenoid Control	PWM Signal: 0-12-0v
78	TN/RD	S/C Vacuum Solenoid	Vacuum Increasing: 1v
79	OR/BK	TCC Solenoid Control	At Cruise w/TCC On: 1v
80	LG/RD	S/C Vent Solenoid	Vacuum Decreasing: 1v

Pin Connector Graphic

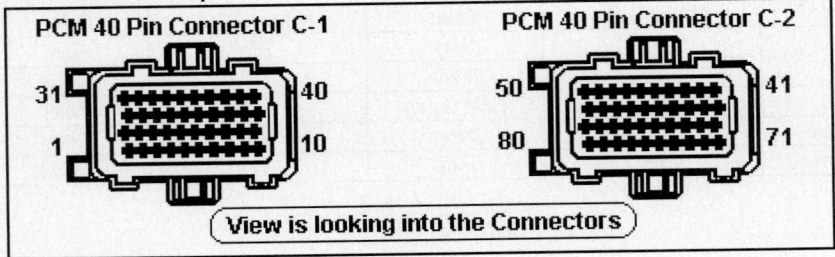

PCM 40 Pin Connector C-1   PCM 40 Pin Connector C-2

31   40   50   41
1   10   80   71

View is looking into the Connectors

2000-01 Neon 2.0L I4 SOHC MFI VIN C (All) 'C1' Black Connector

PCM Pin #	Wire Color	Circuit Description (40-Pin)	Value at Hot Idle
1-2	---	Not Used	---
3	DB/TN	Coil 2 Driver	5°, at 55 mph: 8° dwell
4	---	Not Used	---
5	YL/RD	S/C Power Supply	12-14v
6	DG/OR	ASD Relay Output	12-14v
7	YL/WT	Injector 3 Driver	1-4 ms
8	DG	Generator Field Control	Digital Signal: 0-12-0v
9	---	Not Used	---
10	BK/TN	Power Ground	<0.1v
11	BK/GY	Coil 1 Driver	5°, at 55 mph: 8° dwell
12	GY	Engine Oil Pressure Signal	Switch open: 12v, closed: 0v
13	WT/DB	Injector 1 Driver	1-4 ms
14-15	---	Not Used	---
16	LB/BR	Injector 4 Driver	1-4 ms
17	TN	Injector 2 Driver	1-4 ms
18	OR/RD	HO2S-11 (B1 S1) Signal	0.1-1.1v
19	---	Not Used	---
20	DB/WT	Ignition Switch Output	12-14v
21-22	---	Not Used	---
23	LG/BK	Clutch Switch Signal	Clutch Out: 5v, In: 0v
24	---	Not Used	---
25	DB/LG	Knock Sensor Signal	0.080v AC
26	TN/BK	ECT Sensor Signal	At 180°F: 2.80v
27	BK/OR	HO2S-11 Ground	<0.050v
28-29	---	Not Used	---
30	BK/DG	HO2S-11 (B1 S1) Signal	0.1-1.1v
31	TN	Starter Relay Control	KOEC: 9-11v
32	GY/BK	CKP Sensor Signal	Digital Signal: 0-5-0v
33	TN/YL	CMP Sensor Signal	Digital Signal: 0-5-0v
34	---	Not Used	---
35	OR/LB	TP Sensor Signal	0.6-1.0v
36	DG/RD	MAP Sensor Signal	1.5-1.7v
37	---	Not Used	---
38	BR/OR	A/C Select Switch Sense	A/C On: 1v, Off: 12v
39	BR/WT	Manifold Tuning Valve Relay	Relay Off: 12v, On: 1v
40	---	Not Used	---

Standard Colors and Abbreviations

Abbreviation	Color	Abbreviation	Color	Abbreviation	Color
BK	Black	GY	Gray	RD	Red
BL	Blue	GN	Green	TN	Tan
BR	Brown	LG	Light Green	VT	Violet
DB	Dark Blue	OR	Orange	WT	White
DG	Dark Green	PK	Pink	YL	Yellow

2000-01 Neon 2.0L I4 SOHC MFI VIN C (All) 'C2' White Connector

PCM Pin #	Wire Color	Circuit Description (40-Pin)	Value at Hot Idle
41	RD/LG	Speed Control Set Switch Signal	S/C & Set Switch On: 3.8v
42	---	Not Used	---
43	BK/LB	Sensor Ground	<0.050v
44	OR	8-Volt Supply	7.9-8.1v
45	WT	PSP Switch Signal	Straight: 0v, Turning: 5v
46	RD/WT	Battery Power (Fused B+)	12-14v
47	BK/WT	Sensor Ground	<0.050v
48	---	Not Used	---
49	YL/BK	IAC Driver	DC pulse signals: 0.8-11v
50	BK/TN	Power Ground	<0.1v
51	TN/WT	HO2S-12 (B1 S2) Signal	0.1-1.1v
52	VT/LG	Inlet Air Temperature Sensor	At 86°F: 1.96v
53-54	---	Not Used	---
55	DB/PK	Radiator Fan Relay Control	Relay Off: 12v, On: 1v
56	TN/RD	S/C Vacuum Solenoid	Vacuum Increasing: 1v
57	BR/WT	IAC Motor Sense	12-14v
58, 60	---	Not Used	---
59	VT/YL	PCI Data Bus (J1950)	Digital Signals: 0-7-0v
61	VT/WT	5-Volt Supply	4.9-5.1v
62	WT/PK	Brake Switch Signal	Brake Off: 0v, On: 12v
63	YL/DG	Torque Mgmt. Request Signal	Digital Signals
64	DB/OR	A/C Clutch Relay Control	Relay Off: 12v, On: 1v
65	PK	SCI Transmit	0v
66	WT/OR	Vehicle Speed Signal	Digital Signal
67	DB/YL	ASD Relay Control	Relay Off: 12v, On: 1v
68	PK/BK	EVAP Purge Solenoid Control	PWM Signal: 0-12-0v
69, 73	---	Not Used	---
71	GY/RD	Brake Fluid Level Switch	Switch open: 12v, closed: 0v
72	YL	LDP Switch Signal	Closed: 0v, Open: 12v
73, 79		Not Used	---
74	BR	Fuel Pump Relay Control	Relay Off: 12v, On: 1v
75	LG	SCI Receive	0v
76	BR/YL	PNP Switch or TRS Signal	In P/N: 0v, Others: 5v
77	WT/DB	LDP Solenoid Control	PWM Signal: 0-12-0v
78	OR/BK	TCC Solenoid Control	At Cruise w/TCC On: 1v
80	LG/RD	S/C Vent Solenoid	Vacuum Decreasing: 1v

Pin Connector Graphic

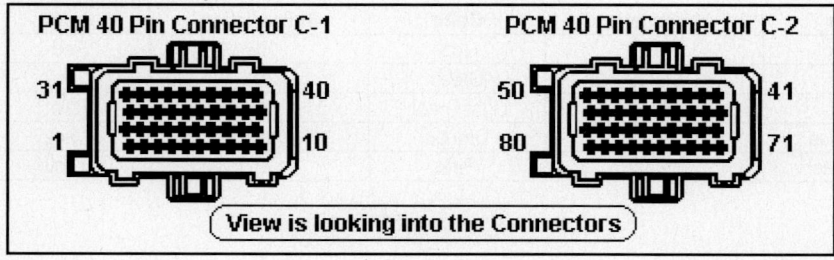

PCM 40 Pin Connector C-1   PCM 40 Pin Connector C-2

31 ... 40   50 ... 41
1 ... 10   80 ... 71

View is looking into the Connectors

1995 Neon 2.0L I4 DOHC MFI VIN Y (All) 60 Pin Connector

PCM Pin #	Wire Color	Circuit Description (60 Pin)	Value at Hot Idle
1	BK/GY	Coil 2 Driver	5°, at 55 mph: 8° dwell
2	BK/TN	Power Ground	<0.1v
3	YL/WT	Injector 3 Driver	1-4 ms
4	WT/DB	Injector 1 Driver	1-4 ms
5	WT/OR	Vehicle Speed Signal	Digital Signal
6	BK/RD	IAT Sensor Signal	At 100°F: 1.83v
7	TN/WT	HO2S-12 (B1 S2) Signal	0.1-1.1v
8	BK/DG	HO2S-11 (B1 S1) Signal	0.1-1.1v
9	LG	SCI Receive	0v
10	OR/DB	TP Sensor Signal	0.6-1.0v
11	RD/WT	Battery Power (Fused B+)	12-14v
12	---	Not Used	---
13	DB	Fuel Level Sensor Signal	70 ohms (±20) w/full tank
14	YL/BK	IAC 2 Driver	DC pulse signals: 0.8-11v
15	GY/RD	IAC 3 Driver	DC pulse signals: 0.8-11v
16	PK/BK	EVAP Purge Solenoid Control	PWM Signal: 0-12-0v
17	OR/BK	TCC Solenoid Control	At Cruise w/TCC On: 1v
18	DB/YL	ASD Relay Control	Relay Off: 12v, On: 1v
19	DB/PK	Radiator Fan Relay Control	Relay Off: 12v, On: 1v
20	---	Not Used	---
21	DB/TN	Coil 1 Driver	5°, at 55 mph: 8° dwell
22	BK/TN	Power Ground	<0.1v
23	TN	Injector 2 Driver	1-4 ms
24	LB/BR	Injector 4 Driver	1-4 ms
25	GY/BK	CKP Sensor Signal	Digital Signal: 0-5-0v
26	TN/YL	CMP Sensor Signal	Digital Signal: 0-5-0v
27	DB/LG	Knock Sensor Signal	0.080v AC
28	TN/DB	ECT Sensor Signal	At 180°F: 2.80v
29	DG/RD	MAP Sensor Signal	1.5-1.7v
30	PK	SCI Transmit	0v
31	RD/LG	Speed Control Set Switch Signal	S/C & Set Switch On: 3.8v
32	WT/PK	Brake Switch Signal	Brake Off: 0v, On: 12v
33	BR/OR	A/C Select Switch Sense	A/C On: 1v, Off: 12v
34	PK/BK	IAC 4 Driver	DC pulse signals: 0.8-11v
35	BR/WT	IAC 1 Driver	DC pulse signals: 0.8-11v
36	BK/PK	MIL (lamp) Control	Lamp On: 1v, Off: 12v
37	TN/BK	Generator Indicator Control	Lamp On: 1v, Off: 12v
38	BR	Fuel Pump Relay Control	Relay Off: 12v, On: 1v
39	GY/YL	EGR Solenoid Control	12v, 55 mph: 1v
40	TN/RD	S/C Vacuum Solenoid	Vacuum Increasing: 1v

1995 Neon 2.0L I4 DOHC MFI VIN Y (All) 60 Pin Connector

PCM Pin #	Wire Color	Circuit Description (60 Pin)	Value at Hot Idle
41	DG	Generator Field Control	Digital Signal: 0-12-0v
42	DG/OR	ASD Relay Output	12-14v
43	PK/WT	5-Volt Supply	4.9-5.1v
44	OR	8-Volt Supply	7.9-8.1v
45-46	---	Not Used	---
47	YL/RD	S/C Power Supply	12-14v
48	---	Not Used	---
49	PK/LG	Battery Temperature Sensor	At 86°F: 1.96v
50	BR/YL	PNP Switch Signal	In P/N: 0v, Others: 5v
51	BK/LB	Sensor Ground	<0.050v
52	BK/WT	Sensor Ground	<0.050v
53	---	Not Used	---
54	LG/BK	Ignition Switch Output	12-14v
55	PK/BK	Reverse Indicator Control	Lamp On: 1v, Off: 12v
56	WT	PSP Switch Signal	Straight: 0v, Turning: 5v
57-58	---	Not Used	---
59	DB/OR	A/C Clutch Relay Control	Relay Off: 12v, On: 1v
60	LG/RD	S/C Vent Solenoid	Vacuum Decreasing: 1v

Pin Connector Graphic

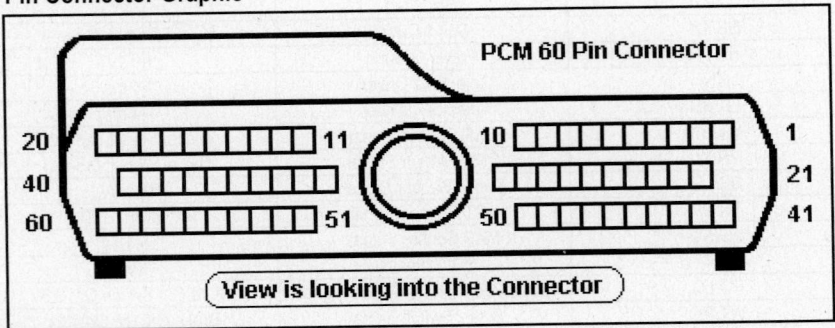

Standard Colors and Abbreviations

Abbreviation	Color	Abbreviation	Color	Abbreviation	Color
BK	Black	GY	Gray	RD	Red
BL	Blue	GN	Green	TN	Tan
BR	Brown	LG	Light Green	VT	Violet
DB	Dark Blue	OR	Orange	WT	White
DG	Dark Green	PK	Pink	YL	Yellow

1996-98 Neon 2.0L I4 DOHC MFI VIN Y (All) 'C1' Black Connector

PCM Pin #	Wire Color	Circuit Description (40-Pin)	Value at Hot Idle
1	---	Not Used	---
2	BK/GY	Coil 1 Driver	5º, at 55 mph: 8º dwell
3	DB/TN	Coil 2 Driver	5º, at 55 mph: 8º dwell
4	DG	Generator Field Control	Digital Signal: 0-12-0v
5	YL/RD	S/C Power Supply	12-14v
6	DG/OR	ASD Relay Output	12-14v
7	YL/WT	Injector 3 Driver	1-4 ms
8	BK/PK	MIL (lamp) Control	Lamp On: 1v, Off: 12v
9	---	Not Used	---
10	BK/TN	Power Ground	<0.1v
11-12	---	Not Used	---
13	WT/LB	Injector 1 Driver	1-4 ms
14-15	---	Not Used	---
16	LB/BR	Injector 4 Driver	1-4 ms
17	TN	Injector 2 Driver	1-4 ms
18	LG	Radiator Fan Relay Control	Relay Off: 12v, On: 1v
19	---	Not Used	---
20	DB/WT	Ignition Switch Output	12-14v
21-22	---	Not Used	---
23	DB	Fuel Level Sensor Signal	70 ohms (±20) w/full tank
24	BK/LG	Knock Sensor Signal	0.080v AC
25	---	Not Used	---
26	TN/BK	ECT Sensor Signal	At 180ºF: 2.80v
27-29	---	Not Used	---
30	BK/DG	HO2S-11 (B1 S1) Signal	0.1-1.1v
31	---	Not Used	---
32	GY/BK	CKP Sensor Signal	Digital Signal: 0-5-0v
33	TN/YL	CMP Sensor Signal	Digital Signal: 0-5-0v
34	---	Not Used	---
35	OR/LB	TP Sensor Signal	0.6-1.0v
36	DG/RD	MAP Sensor Signal	1.5-1.7v
37	BK/RD	IAT Sensor Signal	At 100ºF: 1.83v
38	BR/OR	A/C Select Switch Sense	A/C On: 1v, Off: 12v
39	---	Not Used	---
40	GY/YL	EGR Solenoid Control	12v, 55 mph: 1v

Standard Colors and Abbreviations

Abbreviation	Color	Abbreviation	Color	Abbreviation	Color
BK	Black	GY	Gray	RD	Red
BL	Blue	GN	Green	TN	Tan
BR	Brown	LG	Light Green	VT	Violet
DB	Dark Blue	OR	Orange	WT	White
DG	Dark Green	PK	Pink	YL	Yellow

1996-98 Neon 2.0L I4 DOHC MFI VIN Y (All) 'C2' White Connector

PCM Pin #	Wire Color	Circuit Description (40-Pin)	Value at Hot Idle
41	RD/LG	Speed Control Set Switch Signal	S/C & Set Switch On: 3.8v
42	---	Not Used	---
43	BK/LB	Sensor Ground	<0.050v
44	OR	8-Volt Supply	7.9-8.1v
45	WT	PSP Switch Signal	Straight: 0v, Turning: 5v
46	RD/WT	Battery Power (Fused B+)	12-14v
47	BK/WT	Sensor Ground	<0.050v
48	BR/WT	IAC 3 Driver	DC pulse signals: 0.8-11v
49	YL/BK	IAC 2 Driver	DC pulse signals: 0.8-11v
50	BK/TN	Power Ground	<0.1v
51	TN/WT	HO2S-12 (B1 S2) Signal	0.1-1.1v
52	PK/LG	Battery Temperature Sensor	At 86°F: 1.96v
53-55	---	Not Used	---
55 ('96)	DB/PK	Radiator Fan Relay Control	Relay Off: 12v, On: 1v
56	TN/BK	Generator Indicator Control	Lamp On: 1v, Off: 12v
57	GY/RD	IAC 1 Driver	DC pulse signals: 0.8-11v
58	PK	IAC 4 Driver	DC pulse signals: 0.8-11v
59-60	---	Not Used	---
61	PK/WT	5-Volt Supply	4.9-5.1v
62	WT/PK	Brake Switch Signal	Brake Off: 0v, On: 12v
63	---	Not Used	---
64	DB/OR	A/C Clutch Relay Control	Relay Off: 12v, On: 1v
65	PK	SCI Transmit	0v
66	WT/OR	Vehicle Speed Signal	Digital Signal
67	DB/YL	ASD Relay Control	Relay Off: 12v, On: 1v
68	PK/BK	EVAP Purge Solenoid Control	PWM Signal: 0-12-0v
70-71	---	Not Used	---
72	OR	LDP Switch Signal	Closed: 0v, Open: 12v
73	GY/LB	Tachometer Signals	Pulse Signals
74	BR	Fuel Pump Relay Control	Relay Off: 12v, On: 1v
75	LG	SCI Receive	5v
76	BR/YL	PNP Switch Signal	In P/N: 0v, Others: 5v
77	WT/LG	LDP Solenoid Control	PWM Signal: 0-12-0v
78	TN/RD	S/C Vacuum Solenoid	Vacuum Increasing: 1v
79	OR/BK	TCC Solenoid Control	At Cruise w/TCC On: 1v
80	LG/RD	S/C Vent Solenoid	Vacuum Decreasing: 1v

Pin Connector Graphic

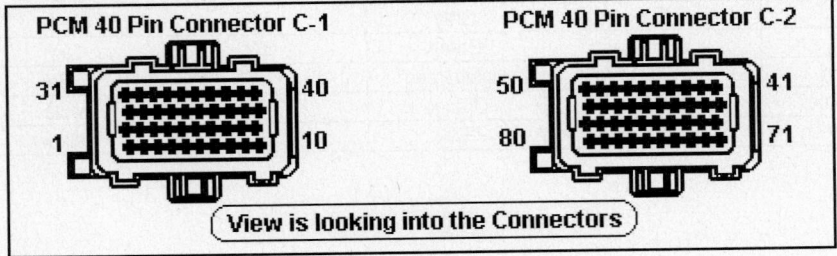

PCM 40 Pin Connector C-1

PCM 40 Pin Connector C-2

31 40
1 10

50 41
80 71

View is looking into the Connectors

1999 Neon 2.0L I4 DOHC MFI VIN Y (All) 'C1' Black Connector

PCM Pin #	Wire Color	Circuit Description (40-Pin)	Value at Hot Idle
1	---	Not Used	---
2	BK/GY	Coil 1 Driver	5°, at 55 mph: 8° dwell
3	DB/TN	Coil 2 Driver	5°, at 55 mph: 8° dwell
4	DG	Generator Field Control	Digital Signal: 0-12-0v
5	YL/RD	S/C Power Supply	12-14v
6	DG/OR	ASD Relay Output	12-14v
7	YL/WT	Injector 3 Driver	1-4 ms
8	BK/PK	MIL (lamp) Control	Lamp On: 1v, Off: 12v
9	---	Not Used	---
10	BK/TN	Power Ground	<0.1v
11-12	---	Not Used	---
13	WT/DB	Injector 1 Driver	1-4 ms
14-15	---	Not Used	---
16	LB/BR	Injector 4 Driver	1-4 ms
17	TN	Injector 2 Driver	1-4 ms
18	LG	Radiator Fan Relay Control	Relay Off: 12v, On: 1v
19	---	Not Used	---
20	DB/WT	Ignition Switch Output	12-14v
21-22	---	Not Used	---
23	DB	Fuel Level Sensor Signal	70 ohms (±20) w/full tank
24	BK/LG	Knock Sensor Signal	0.080v AC
25	---	Not Used	---
26	TN/BK	ECT Sensor Signal	At 180°F: 2.80v
27-29	---	Not Used	---
30	BK/DG	HO2S-11 (B1 S1) Signal	0.1-1.1v
31	---	Not Used	---
32	GY/BK	CKP Sensor Signal	Digital Signal: 0-5-0v
33	TN/YL	CMP Sensor Signal	Digital Signal: 0-5-0v
34	---	Not Used	---
35	OR/DB	TP Sensor Signal	0.6-1.0v
36	DG/RD	MAP Sensor Signal	1.5-1.7v
37	BK/RD	IAT Sensor Signal	At 100°F: 1.83v
38	BR/OR	A/C Select Switch Sense	A/C On: 1v, Off: 12v
39	---	Not Used	---
40	GY/YL	EGR Solenoid Control	12v, 55 mph: 1v

Standard Colors and Abbreviations

Abbreviation	Color	Abbreviation	Color	Abbreviation	Color
BK	Black	GY	Gray	RD	Red
BL	Blue	GN	Green	TN	Tan
BR	Brown	LG	Light Green	VT	Violet
DB	Dark Blue	OR	Orange	WT	White
DG	Dark Green	PK	Pink	YL	Yellow

1999 Neon 2.0L I4 DOHC MFI VIN Y (All) 'C2' White Connector

PCM Pin #	Wire Color	Circuit Description (40-Pin)	Value at Hot Idle
41	RD/LG	Speed Control Set Switch Signal	S/C & Set Switch On: 3.8v
42	---	Not Used	---
43	BK/LB	Sensor Ground	<0.050v
44	OR	8-Volt Supply	7.9-8.1v
45	WT	PSP Switch Signal	Straight: 0v, Turning: 5v
46	RD/WT	Battery Power (Fused B+)	12-14v
47	BK/WT	Sensor Ground	<0.050v
48	BR/WT	IAC 3 Driver	DC pulse signals: 0.8-11v
49	YL/BK	IAC 2 Driver	DC pulse signals: 0.8-11v
50	BK/TN	Power Ground	<0.1v
51	TN/WT	HO2S-12 (B1 S2) Signal	0.1-1.1v
52	VT/LG	Battery Temperature Sensor	At 86°F: 1.96v
53-55	---	Not Used	---
56	TN/BK	Generator Indicator Control	Lamp On: 1v, Off: 12v
57	GY/RD	IAC 1 Driver	DC pulse signals: 0.8-11v
58	VT	IAC 4 Driver	DC pulse signals: 0.8-11v
59-60	---	Not Used	---
61	VT/WT	5-Volt Supply	4.9-5.1v
62	WT/PK	Brake Switch Signal	Brake Off: 0v, On: 12v
63	---	Not Used	---
64	DB/OR	A/C Clutch Relay Control	Relay Off: 12v, On: 1v
65	PK	SCI Transmit	0v
66	WT/OR	Vehicle Speed Signal	Digital Signal
67	PK/BK	EVAP Purge Solenoid Control	PWM Signal: 0-12-0v
68-71	---	Not Used	---
72	OR	LDP Switch Signal	Closed: 0v, Open: 12v
73	GY/LB	Tachometer Signals	Pulse Signals
74	BR	Fuel Pump Relay Control	Relay Off: 12v, On: 1v
75	LG	SCI Receive	5v
76	BR/YL	PNP Switch Signal	In P/N: 0v, Others: 5v
77	WT/LG	LDP Solenoid Control	PWM Signal: 0-12-0v
78	TN/RD	S/C Vacuum Solenoid	Vacuum Increasing: 1v
79	OR/BK	TCC Solenoid Control	At Cruise w/TCC On: 1v
80	LG/RD	S/C Vent Solenoid	Vacuum Decreasing: 1v

Pin Connector Graphic

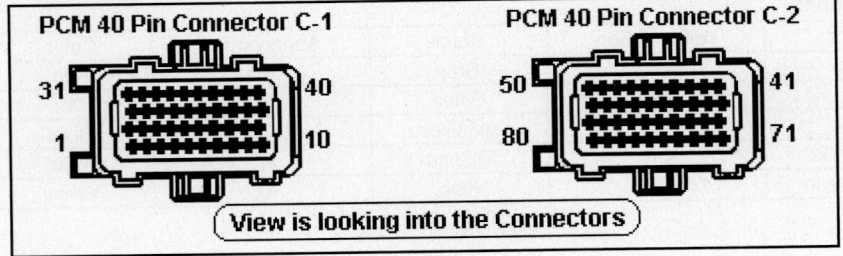

PCM 40 Pin Connector C-1   PCM 40 Pin Connector C-2

31   40   50   41

1   10   80   71

View is looking into the Connectors

1997 Prowler 3.5L V6 24v SOHC MFI VIN F (A/T) 'C1' Black Connector

PCM Pin #	W/Color	Circuit Description (40-Pin)	Value at Hot Idle
1	BK/OR	HO2S-22 (B2 S2) Signal	0.1-1.1v
2	DB/YL	Coil 3 Driver	6°, 55 mph: 9° dwell
3	DB/TN	Coil 2 Driver	6°, 55 mph: 9° dwell
4	DG	Generator Field Driver	Digital Signal: 0-12-0v
5	YL/RD	S/C Power Supply	12-14v
6	DG/OR	ASD Relay Output	12-14v
7	YL/WT	Injector 3 Driver	1-4 ms
8	TN/DG	Starter Relay Control	KOEC: 9-11v
9	---	Not Used	---
10	BK/TN	Power Ground	<0.1v
11	GY/OR	Coil 1 Driver	6°, 55 mph: 9° dwell
12	---	Not Used	---
13	WT/DB	Injector 1 Driver	1-4 ms
14	BR/DB	Injector 6 Driver	1-4 ms
15	GY	Injector 5 Driver	1-4 ms
16	LB/BR	Injector 4 Driver	1-4 ms
17	TN	Injector 2 Driver	1-4 ms
18-19	---	Not Used	---
20	DB/YL	Ignition Switch Output	12-14v
21-23	---	Not Used	---
24	DB/LG	Knock Sensor 1 Signal	0.080v AC
25	GY/LG	Knock Sensor 2 Signal	0.080v AC
26	TN/BK	ECT Sensor Signal	At 180°F: 2.80v
27-28	---	Not Used	---
29	LG/RD	HO2S-21 (B2 S1) Signal	0.1-1.1v
30	BK/DG	HO2S-11 (B1 S1) Signal	0.1-1.1v
31	---	Not Used	---
32	GY/BK	CKP Sensor Signal	Digital Signal: 0-5-0v
33	TN/YL	CMP Sensor Signal	Digital Signal: 0-5-0v
34	---	Not Used	---
35	OR/DB	TP Sensor Signal	0.6-1.0v
36	DG/RD	MAP Sensor Signal	1.5-1.7v
37	BK/RD	IAT Sensor Signal	At 100°F: 1.83v
38	---	Not Used	---
39	VT/RD	Manifold Tuning Valve Control	Solenoid Off: 12v, On: 1v
40	GY/YL	EGR Solenoid Control	Idle: 12v, 55 mph: 1v

Standard Colors and Abbreviations

Abbreviation	Color	Abbreviation	Color	Abbreviation	Color
BK	Black	GY	Gray	RD	Red
BL	Blue	GN	Green	TN	Tan
BR	Brown	LG	Light Green	VT	Violet
DB	Dark Blue	OR	Orange	WT	White
DG	Dark Green	PK	Pink	YL	Yellow

1997 Prowler 3.5L V6 24v SOHC MFI VIN F (A/T) 'C2' White Connector

PCM Pin #	Wire Color	Circuit Description (40-Pin)	Value at Hot Idle
41	RD/LG	Speed Control Set Switch Signal	S/C & Set Switch On: 3.8v
42	DB	A/C Pressure Switch Signal	A/C On: 0.45-4.85v
43	BK/LB	Sensor Ground	<0.050v
44	OR	8-Volt Supply	7.9-8.1v
46	RD/DB	Battery Power (Fused B+)	12-14v
47	BK/WT	Power Ground	<0.1v
48	BR/WT	IAC 1 Driver	DC pulse signals: 0.8-11v
49	YL/BK	IAC 2 Driver	DC pulse signals: 0.8-11v
50, 71	BK/TN	Power Ground	<0.1v
51	TN/WT	HO2S-12 (B1 S2) Signal	0.1-1.1v
52	PK/YL	Battery Temperature Sensor	At 86°F: 1.96v
53-54	---	Not Used	---
55	DB/RD	Low Speed Fan Relay	Relay Off: 12v, On: 1v
56	---	Not Used	---
57	GY/RD	IAC 3 Driver	DC pulse signals: 0.8-11v
58	VT/BK	IAC 4 Driver	DC pulse signals: 0.8-11v
59	VT/BR	CCD Bus (+)	Digital Signal: 0-5-0v
60	WT/BK	CCD Bus (-)	<0.050v
61	VT/WT	5-Volt Supply	4.9-5.1v
62	WT/PK	Brake Switch Signal	Brake Off: 0v, On: 12v
63	YL/DG	Torque Management Request	Digital Signals
64	DB/OR	A/C Clutch Relay Control	A/C On: 1v, Off: 12v
65	PK/DB	SCI Transmit	0v
66	WT/OR	Vehicle Speed Signal	Digital Signal
67	DB/WT	ASD Relay Control	Relay Off: 12v, On: 1v
68	PK/BK	EVAP Purge Solenoid Control	PWM Signal: 0-12-0v
69	DB/PK	High Speed Fan Control	Relay Off: 12v, On: 1v
73	GY/LB	Tachometer Signals	Pulse Signals
74	BR	Fuel Pump Relay Control	Relay Off: 12v, On: 1v
75	LG/TN	SCI Receive	5v
76	BK/VT	PNP Switch Signal	In P/N: 0v, Others: 5v
78	TN/RD	S/C Vacuum Solenoid	Vacuum Increasing: 1v
80	LG/WT	S/C Vent Solenoid	Vacuum Decreasing: 1v

Pin Connector Graphic

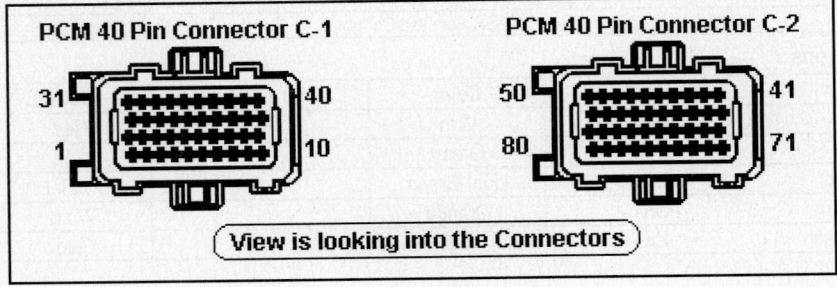

1999 Prowler 3.5L V6 24v SOHC MFI VIN G (A/T) 'C1' Black Connector

PCM Pin #	Wire Color	Circuit Description (40-Pin)	Value at Hot Idle
1	TN/LG	Coil 4 Driver	6°, 55 mph: 9° dwell
2	RD/YL	Coil 3 Driver	6°, 55 mph: 9° dwell
3	DB/TN	Coil 2 Driver	6°, 55 mph: 9° dwell
4	TN/VT	Coil 6 Driver	6°, 55 mph: 9° dwell
5	YL/RD	S/C Power Supply	12-14v
6	DG/OR	ASD Relay Output	12-14v
7	YL/WT	Injector 3 Driver	1-4 ms
8	DG	Generator Field Driver	Digital Signal: 0-12-0v
9	---	Not Used	---
10	BK/TN	Power Ground	<0.1v
11	GY/OR	Coil 1 Driver	6°, 55 mph: 9° dwell
12	---	Not Used	---
13	WT/DB	Injector 1 Driver	1-4 ms
14	BR/DB	Injector 6 Driver	1-4 ms
15	GY	Injector 5 Driver	1-4 ms
16	LB/BR	Injector 4 Driver	1-4 ms
17	TN	Injector 2 Driver	1-4 ms
18	DB/BK	Short Runner Valve Solenoid	Valve On: 1v, Off: 12v
19	---	Not Used	---
20	DB/YL	Ignition Switch Output	12-14v
21	TN/PK	Coil 5 Driver	6°, 55 mph: 9° dwell
22-23	---	Not Used	---
24	DB/LG	Knock Sensor Signal	0.080v AC
25	GY/LG	Knock Sensor Return	<0.050v
26	TN/BK	ECT Sensor Signal	At 180°F: 2.80v
27	---	Not Used	---
28	BK/LG	Transmission Fan Relay	Relay Off: 12v, On: 1v
29	LG/RD	HO2S-21 (B2 S1) Signal	0.1-1.1v
30	BK/DG	HO2S-11 (B1 S1) Signal	0.1-1.1v
31	TN/DG	Starter Relay Control	KOEC: 9-11v
32	GY/BK	CKP Sensor Signal	Digital Signal: 0-5-0v
33	TN/YL	CMP Sensor Signal	Digital Signal: 0-5-0v
34	LG/PK	EGR Sensor Signal	0.6-0.8v
35	OR/DB	TP Sensor Signal	0.6-1.0v
36	DG/RD	MAP Sensor Signal	1.5-1.7v
37	BK/RD	IAT Sensor Signal	At 100°F: 1.83v
38	---	Not Used	---
39	VT/RD	Manifold Tuning Valve Control	Solenoid Off: 12v, On: 1v
40	GY/YL	EGR Solenoid Control	12v, 55 mph: 1v

Standard Colors and Abbreviations

Abbreviation	Color	Abbreviation	Color	Abbreviation	Color
BK	Black	GY	Gray	RD	Red
BL	Blue	GN	Green	TN	Tan
BR	Brown	LG	Light Green	VT	Violet
DB	Dark Blue	OR	Orange	WT	White
DG	Dark Green	PK	Pink	YL	Yellow

1999 Prowler 3.5L V6 24v SOHC MFI VIN G (A/T) 'C2' White Connector

PCM Pin #	Wire Color	Circuit Description (40-Pin)	Value at Hot Idle
41	RD/LG	Speed Control Set Switch Signal	S/C & Set Switch On: 3.8v
42	DB	A/C Pressure Switch Signal	A/C On: 0.45-4.85v
43	BK/LB	Sensor Ground	<0.050v
44	OR	8-Volt Supply	7.9-8.1v
45	---	Not Used	---
46	RD/DB	Battery Power (Fused B+)	12-14v
47	BK/WT	Power Ground	<0.1v
48	BR/WT	IAC 3 Driver	DC pulse signals: 0.8-11v
49	YL/BK	IAC 2 Driver	DC pulse signals: 0.8-11v
50	BK/TN	Power Ground	<0.1v
51	TN/WT	HO2S-12 (B1 S2) Signal	0.1-1.1v
52	---	Not Used	---
53	BK/OR	HO2S-22 (B2 S2) Signal	0.1-1.1v
54	---	Not Used	---
55	DB/RD	Low Speed Fan Relay	Relay Off: 12v, On: 1v
56	TN/RD	S/C Vacuum Solenoid	Vacuum Increasing: 1v
57	GY/RD	IAC 1 Driver	DC pulse signals: 0.8-11v
58	VT/BK	IAC 4 Driver	DC pulse signals: 0.8-11v
59	VT/BR	CCD Bus (+)	Digital Signal: 0-5-0v
60	WT/BK	CCD Bus (-)	<0.050v
61	VT/WT	5-Volt Supply	4.9-5.1v
62	WT/PK	Brake Switch Signal	Brake Off: 0v, On: 12v
63	YL/DG	Torque Management Request	Digital Signals
64	DB/OR	A/C Clutch Relay Control	Relay Off: 12v, On: 1v
65	PK	SCI Transmit	5v
66	WT/OR	Vehicle Speed Signal	Digital Signal
67	DB/WT	ASD Relay Control	Relay Off: 12v, On: 1v
68	PK/BK	EVAP Purge Solenoid Control	PWM Signal: 0-12-0v
69	DB/PK	High Speed Fan Control	Relay Off: 12v, On: 1v
70	PK/LB	EVAP Purge Solenoid Sense	0-1v
71	WT/RD	EATX RPM Signal	Digital Signals
72	WT/OR or LB	LDP Switch Signal	Closed: 0v, Open: 12v
74	BR	Fuel Pump Relay Control	Relay Off: 12v, On: 1v
75	LG	SCI Receive	0v
76	BK/VT	PNP Switch Signal	In P/N: 0v, Others: 5v
77	WT/DG	LDP Solenoid Control	PWM Signal: 0-12-0v
78-79	---	Not Used	---
80	LG/WT	S/C Vent Solenoid	Vacuum Decreasing: 1v

Pin Connector Graphic

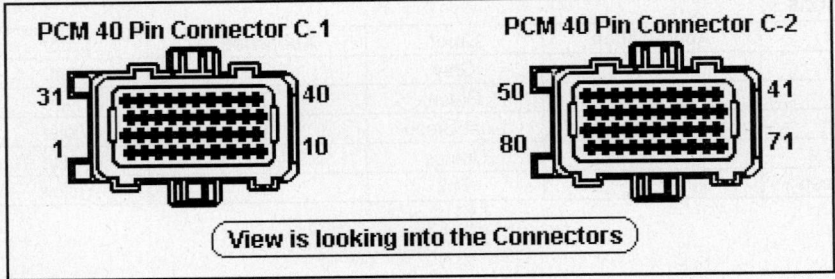

PCM 40 Pin Connector C-1   PCM 40 Pin Connector C-2

31 40   50 41
1   10   80 71

View is looking into the Connectors

2000-01 Prowler 3.5L V6 24v SOHC MFI VIN G (A/T) 'C1' Black Connector

PCM Pin #	Wire Color	Circuit Description (40-Pin)	Value at Hot Idle
1	TN/LG	COP 4 Driver	6°, 55 mph: 9° dwell
2	RD/YL	COP 3 Driver	6°, 55 mph: 9° dwell
3	DB/TN	COP 2 Driver	6°, 55 mph: 9° dwell
4	TN/VT	COP 6 Driver	6°, 55 mph: 9° dwell
5	YL/RD	S/C On/Off Switch Power	12-14v
6	DG/OR	ASD Relay Output	12-14v
7	YL/WT	Injector 3 Driver	1-4 ms
8	DG	Generator Field Driver	Digital Signal: 0-12-0v
9	---	Not Used	---
10	BK/TN	Power Ground	<0.1v
11	GY/OR	COP 1 Driver	6°, 55 mph: 9° dwell
12	---	Not Used	---
13	WT/DB	Injector 1 Driver	1-4 ms
14	BR/DB	Injector 6 Driver	1-4 ms
15	GY	Injector 5 Driver	1-4 ms
16	LB/BR	Injector 4 Driver	1-4 ms
17	TN	Injector 2 Driver	1-4 ms
18	DB/BK	Short Runner Valve Solenoid	Valve On: 1v, Off: 12v
19	---	Not Used	---
20	DB/YL	Ignition Switch Output	12-14v
21	TN/PK	COP 5 Driver	6°, 55 mph: 9° dwell
22-23	---	Not Used	---
24	DB/LG	Knock Sensor Signal	0.080v AC
25	GY/LG	Knock Sensor Return	<0.050v
26	TN/BK	ECT Sensor Signal	At 180°F: 2.80v
27	BK/YL	HO2S Ground	<0.050v
28	BK/LG	Transmission Fan Relay	Relay Off: 12v, On: 1v
29	LG/RD	HO2S-21 (B2 S1) Signal	0.1-1.1v
30	BK/DG	HO2S-11 (B1 S1) Signal	0.1-1.1v
31	TN/DG	Starter Relay Control	KOEC: 9-11v
32	GY/BK	CKP Sensor Signal	Digital Signal: 0-5-0v
33	TN/YL	CMP Sensor Signal	Digital Signal: 0-5-0v
34	LG/PK	EGR Sensor Signal	0.6-0.8v
35	OR/DB	TP Sensor Signal	0.6-1.0v
36	DG/RD	MAP Sensor Signal	1.5-1.7v
37	BK/RD	IAT Sensor Signal	At 100°F: 1.83v
38	---	Not Used	---
39	VT/RD	Manifold Tuning Valve Control	Solenoid Off: 12v, On: 1v
40	GY/YL	EGR Solenoid Control	12v, 55 mph: 1v

Standard Colors and Abbreviations

Abbreviation	Color	Abbreviation	Color	Abbreviation	Color
BK	Black	GY	Gray	RD	Red
BL	Blue	GN	Green	TN	Tan
BR	Brown	LG	Light Green	VT	Violet
DB	Dark Blue	OR	Orange	WT	White
DG	Dark Green	PK	Pink	YL	Yellow

2000-01 Prowler 3.5L V6 24v SOHC MFI VIN G (A/T) 'C2' White Connector

PCM Pin #	Wire Color	Circuit Description (40-Pin)	Value at Hot Idle
41	RD/LG	Speed Control Set Switch Signal	S/C & Set Switch On: 3.8v
42	DB	A/C Pressure Switch Signal	A/C On: 0.45-4.85v
43	BK/LB	Sensor Ground	<0.1v
44	OR	8-Volt Supply	7.9-8.1v
45	---	Not Used	---
46	RD/WT	Battery Power (Fused B+)	12-14v
47	BK/LB	Power Ground	<0.1v
48	BR/WT	IAC 3 Driver	DC pulse signals: 0.8-11v
49	YL/BK	IAC 2 Driver	DC pulse signals: 0.8-11v
50	BK/TN	Power Ground	<0.1v
51	TN/WT	HO2S-12 (B1 S2) Signal	0.1-1.1v
52	PK/YL	Battery Temperature Sensor	At 86°F: 1.96v
53	BK/OR	HO2S-22 (B2 S2) Signal	0.1-1.1v
54	---	Not Used	---
55	DB/RD	Low Speed Fan Relay	Relay Off: 12v, On: 1v
56	TN/RD	S/C Vacuum Solenoid	Vacuum Increasing: 1v
57	GY/RD	IAC 1 Driver	DC pulse signals: 0.8-11v
58	VT/BK	IAC 4 Driver	DC pulse signals: 0.8-11v
59	VT	CCD Bus (+)	Digital Signal: 0-5-0v
60	WT	CCD Bus (-)	<0.050v
61	VT/WT	5-Volt Supply	4.9-5.1v
62	WT/PK	Brake Switch Signal	Brake Off: 0v, On: 12v
63	YL/DG	Torque Management Request	Digital Signals
64	DB/OR	A/C Clutch Relay Control	Relay Off: 12v, On: 1v
65	PK	SCI Transmit	5v
66	WT/OR	Vehicle Speed Signal	Digital Signal
67	DB/WT	ASD Relay Control	Relay Off: 12v, On: 1v
68	PK/BK	EVAP Purge Solenoid Control	PWM Signal: 0-12-0v
69	DB/LG	High Speed Fan Control	Relay Off: 12v, On: 1v
70	PK/LB	EVAP Purge Solenoid Sense	0-1v
71	WT/RD	EATX RPM Signal	Digital Signals
72	LB	LDP Switch Signal	Closed: 0v, Open: 12v
73	GY/LB	Tachometer Signals	Pulse Signals
74	BR	Fuel Pump Relay Control	Relay Off: 12v, On: 1v
75	LG	SCI Receive	0v
76	BK/VT	TRS T41 Sense Signal	In P/N: 0v, Others: 5v
77	WT/DG	LDP Solenoid Control	PWM Signal: 0-12-0v
78-79	---	Not Used	---
80	LG/WT	S/C Vent Solenoid	Vacuum Decreasing: 1v

Pin Connector Graphic

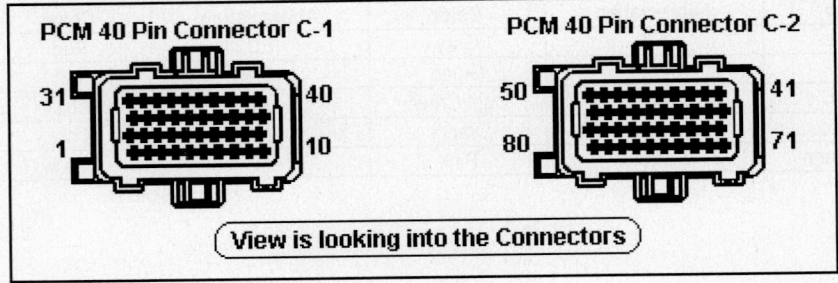

PCM 40 Pin Connector C-1

PCM 40 Pin Connector C-2

31 40
1   10

50 41
80 71

View is looking into the Connectors

1990-91 Sundance 2.2L I4 SOHC TBI VIN D (All) 60 Pin Connector

PCM Pin #	Wire Color	Circuit Description (60 Pin)	Value at Hot Idle
1	DG/RD	MAP Sensor Signal	1.5-1.7v
2	TN/WT	ECT Sensor Signal	At 180°F: 2.80v
2 ('91)	TN/BK	ECT Sensor Signal	At 180°F: 2.80v
3	RD	Battery Power (Fused B+)	12-14v
4	BK/LB	Sensor Ground	<0.050v
5	BK	Sensor Ground	<0.050v
6	PK/WT	5-Volt Supply	4.9-5.1v
6 ('91)	PK	5-Volt Supply	4.9-5.1v
7	OR	8-Volt Supply	7.9-8.1v
8	WT	ASD Relay Output	12-14v
8 ('91)	DG/BK	ASD Relay Output	12-14v
9	DB	Ignition Switch Output	12-14v
10	---	Not Used	---
11	LB/RD	Power Ground	<0.1v
12	LB/RD	Power Ground	<0.1v
11 ('91)	BK/TN	Power Ground	<0.1v
12 ('91)	BK/TN	Power Ground	<0.1v
13-15	---	Not Used	---
16	WT/DB	Injector 1 Driver	1-4 ms
17-18	---	Not Used	---
19	BK/GY	Coil 1 Driver	5°, at 55 mph: 8° dwell
20	DG	Alternator Field Control	Digital Signal: 0-12-0v
21	BK/RD	Throttle Body Temperature	At 100°F: 2.51v
22	OR/DB	TP Sensor Signal	0.6-1.0v
23	---	Not Used	---
24	GY/BK	Distributor Reference Signal	Digital Signal: 0-5-0v
24 ('91)	GY	Distributor Reference Signal	Digital Signal: 0-5-0v
25	PK	SCI Transmit	0v
26	---	Not Used	---
27	BR	A/C Switch Signal	A/C On: 1v, Off: 12v
28	---	Not Used	---
29	WT/PK	Brake Switch Signal	Brake Off: 0v, On: 12v
30	BR	PNP Switch Signal	In P/N: 0v, Others: 5v
30 ('91)	BR/YL	PNP Switch Signal	In P/N: 0v, Others: 5v
31	DB/PK	Radiator Fan Relay Control	Relay Off: 12v, On: 1v
32	BK/PK	MIL (lamp) Control	Lamp On: 1v, Off: 12v
33	TN/RD	S/C Vacuum Solenoid	Vacuum Increasing: 1v
34	DB/OR	A/C WOT Relay Control	Relay Off: 12v, On: 1v
35	GY/YL	EGR Solenoid Control (California)	12v, 55 mph: 1v
36-38	---	Not Used	---
39	GY/RD	AIS 3 Motor Control	DC pulse signals: 0.8-11v
40	BR/WT	AIS 1 Motor Control	DC pulse signals: 0.8-11v
40 ('91)	BR	AIS 1 Motor Control	DC pulse signals: 0.8-11v

Standard Colors and Abbreviations

Abbreviation	Color	Abbreviation	Color	Abbreviation	Color
BK	Black	GY	Gray	RD	Red
BL	Blue	GN	Green	TN	Tan
BR	Brown	LG	Light Green	VT	Violet
DB	Dark Blue	OR	Orange	WT	White
DG	Dark Green	PK	Pink	YL	Yellow

1990-91 Sundance 2.2L I4 SOHC TBI VIN D (All) 60 Pin Connector

PCM Pin #	Wire Color	Circuit Description (60 Pin)	Value at Hot Idle
41	BK/DG	HO2S-11 (B1 S1) Signal	0.1-1.1v
42	---	Not Used	---
43	GY/LB	Tachometer Signals	Pulse Signals
44	TN/YL	Distributor Sync Signals	Digital Signal: 0-5-0v
45	LG	SCI Receive	5v
46	---	Not Used	---
47	WT/OR	Vehicle Speed Signal	Digital Signal
48	BR/RD	Speed Control Set Switch Signal	S/C & Set Switch On: 3.8v
49	YL/RD	S/C On & Off Switch	S/C On: 1v, Off: 12v
50	WT/LG	S/C Resume Switch	S/C On: 1v, Off: 12v
51	DB/YL	ASD Relay Output	12-14v
52	PK	EVAP Purge Solenoid Control	Solenoid Off: 12v, On: 1v
53	LB/RD	S/C Vent Solenoid	Vacuum Decreasing: 1v
54 ('91)	LG/WT	A/T: Lockup Torque Converter	At Cruise w/TCC On: 1v
54	OR/BK	A/T: Lockup Torque Converter	At Cruise w/TCC On: 1v
54	OR/BK	M/T: Shift Indicator Control	Lamp On: 1v, Off: 12v
55-57	---	Not Used	
59	PK/BK	AIS 4 Motor Control	DC pulse signals: 0.8-11v
60	YL/BK	AIS 2 Motor Control	DC pulse signals: 0.8-11v

Pin Connector Graphic

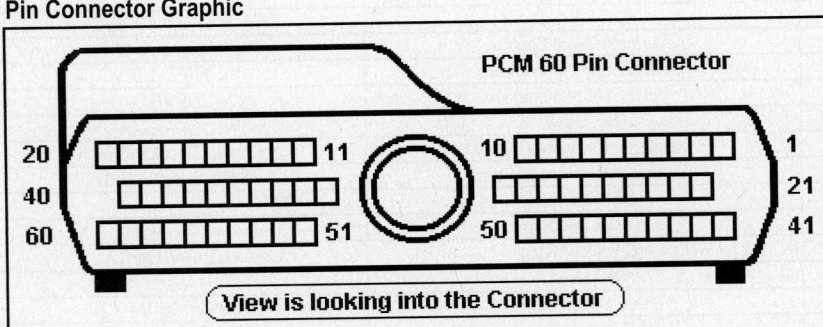

1992-94 Sundance 2.2L I4 SOHC TBI VIN D (All) 60 Pin Connector

PCM Pin #	Wire Color	Circuit Description (60 Pin)	Value at Hot Idle
1	DG/RD	MAP Sensor Signal	1.5-1.7v
2	TN/BK	ECT Sensor Signal	At 180°F: 2.80v
3	RD/WT	Battery Power (Fused B+)	12-14v
4	BK/LB	Sensor Ground	<0.050v
5	BK/WT	Sensor Ground	<0.050v
6	PK/WT	5-Volt Supply	4.9-5.1v
7	OR	8-Volt Supply	7.9-8.1v
8	WT	ASD Relay Output	12-14v
9	DB	Ignition Switch Output	12-14v
10	---	Not Used	---
11	BK/TN	Power Ground	<0.1v
12	BK/TN	Power Ground	<0.1v
13-15	---	Not Used	---
16	WT/DB	Injector 1 Driver	1-4 ms
17-18	---	Not Used	---
19	BK/GY	Coil 1 Driver	5°, at 55 mph: 8° dwell
20	DG	Alternator Field Control	Digital Signal: 0-12-0v
21	---	Not Used	---
22	OR/DB	TP Sensor Signal	0.6-1.0v
23	RD/LG	Speed Control Set Switch Signal	S/C & Set Switch On: 3.8v
24	GY/BK	Distributor Reference Signal	Digital Signal: 0-5-0v
25	PK	SCI Transmit	0v
26	---	Not Used	---
27	BR	A/C Switch Signal	A/C On: 1v, Off: 12v
28	---	Not Used	---
29	WT/PK	Brake Switch Signal	Brake Off: 0v, On: 12v
30	BR/YL	PNP Switch Signal	In P/N: 0v, Others: 5v
31	DB/PK	Radiator Fan Relay Control	Relay Off: 12v, On: 1v
32	BK/PK	MIL (lamp) Control	Lamp On: 1v, Off: 12v
33	TN/RD	S/C Vacuum Solenoid	Vacuum Increasing: 1v
34	DB/OR	A/C Clutch Relay Control	Relay Off: 12v, On: 1v
35	GY/YL	EGR Solenoid Control (California)	12v, 55 mph: 1v
36-38	---	Not Used	---
39	GY/RD	AIS 3 Motor Control	DC pulse signals: 0.8-11v
40	BR/WT	AIS 1 Motor Control	DC pulse signals: 0.8-11v

Standard Colors and Abbreviations

Abbreviation	Color	Abbreviation	Color	Abbreviation	Color
BK	Black	GY	Gray	RD	Red
BL	Blue	GN	Green	TN	Tan
BR	Brown	LG	Light Green	VT	Violet
DB	Dark Blue	OR	Orange	WT	White
DG	Dark Green	PK	Pink	YL	Yellow

1992-94 Sundance 2.2L I4 SOHC TBI VIN D (All) 60 Pin Connector

PCM Pin #	W/Color	Circuit Description (60 Pin)	Value at Hot Idle
41	BK/DG	HO2S-11 (B1 S1) Signal	0.1-1.1v
42	---	Not Used	---
43	GY/LB	Tachometer Signals	Pulse Signals
44	---	Not Used	---
45	LG	SCI Receive	5v
46	---	Not Used	---
47	WT/OR	Vehicle Speed Signal	Digital Signal
48-50	---	Not Used	---
51	DB/YL	ASD Relay Output	12-14v
52	PK/BK	EVAP Purge Solenoid Control	Solenoid Off: 12v, On: 1v
53	LG/RD	S/C Vent Solenoid	Vacuum Decreasing: 1v
54	LG/WT	A/T: Lockup Torque Converter	At Cruise w/TCC On: 1v
54	OR/BK	A/T: Lockup Torque Converter	At Cruise w/TCC On: 1v
55-56	---	Not Used	---
57	DG/OR	ASD Relay Control	Relay Off: 12v, On: 1v
58	---	Not Used	---
59	PK/BK	AIS 4 Motor Control	DC pulse signals: 0.8-11v
60	YL/BK	AIS 2 Motor Control	DC pulse signals: 0.8-11v

Pin Connector Graphic

PCM 60 Pin Connector

20 11   10   1
40   21
60 51   50   41

View is looking into the Connector

1990-91 Sundance 2.5L I4 Turbo MFI VIN J (All) 60 Pin Connector

PCM Pin #	Wire Color	Circuit Description (60 Pin)	Value at Hot Idle
1	DG/RD	MAP Sensor Signal	1.5-1.7v
2	TN/BK	ECT Sensor Signal	At 180°F: 2.80v
3	RD/WT	Battery Power (Fused B+)	12-14v
4	BK/LB	Sensor Ground	<0.050v
5	BK/WT	Sensor Ground	<0.050v
6	PK/WT	5-Volt Supply	4.9-5.1v
7	OR	9-Volt Supply	8.9-9.1v
8	DG/BK	ASD Relay Output	12-14v
9	DB	Ignition Switch Output	12-14v
10	---	Not Used	---
11 ('90)	LB/RD	Power Ground	<0.1v
12 ('90)	LB/RD	Power Ground	<0.1v
11	BK/TN	Power Ground	<0.1v
12	BK/TN	Power Ground	<0.1v
13	LB/BR	Injector 4 Driver	1-4 ms
14	YL/WT	Injector 3 Driver	1-4 ms
15	TN	Injector 2 Driver	1-4 ms
16	WT/DB	Injector 1 Driver	1-4 ms
17-18	---	Not Used	---
19	BK/GY	Coil 1 Driver	5°, at 55 mph: 8° dwell
20	DG	Alternator Field Control	Digital Signal: 0-12-0v
21	---	Not Used	---
22	OR/DB	TP Sensor Signal	0.6-1.0v
23, 28	---	Not Used	---
24	GY/BK	Distributor Reference Signal	Digital Signal: 0-5-0v
25	PK	SCI Transmit	0v
26 ('90)	BK/PK	CCD Bus (+)	Digital Signal: 0-5-0v
26	PK/BR	CCD Bus (+)	Digital Signal: 0-5-0v
27 ('90)	BR	A/C Clutch Signal	A/C On: 1v, Off: 12v
27	BR	A/C Switch Signal	A/C On: 1v, Off: 12v
29	WT/PK	Brake Switch Signal	Brake Off: 0v, On: 12v
30 ('90)	BR/YL	PNP Switch Signal	In P/N: 0v, Others: 5v
30	BR/LB	PNP Switch Signal	In P/N: 0v, Others: 5v
31	DB/PK	Radiator Fan Relay Control	Relay Off: 12v, On: 1v
32	BK/PK	MIL (lamp) Control	Lamp On: 1v, Off: 12v
33	TN/RD	S/C Vacuum Solenoid	Vacuum Increasing: 1v
34	DB/OR	A/C WOT Relay Control	Relay Off: 12v, On: 1v
35	GY/YL	EGR Solenoid Control (California)	12v, 55 mph: 1v
36	LG/BK	Wastegate Solenoid Control	Solenoid Off: 12v, On: 1v
37-38	---	Not Used	---
39	GY/RD	AIS 3 Motor Control	DC pulse signals: 0.8-11v
40	BR/WT	AIS 1 Motor Control	DC pulse signals: 0.8-11v

Standard Colors and Abbreviations

Abbreviation	Color	Abbreviation	Color	Abbreviation	Color
BK	Black	GY	Gray	RD	Red
BL	Blue	GN	Green	TN	Tan
BR	Brown	LG	Light Green	VT	Violet
DB	Dark Blue	OR	Orange	WT	White
DG	Dark Green	PK	Pink	YL	Yellow

1990-91 Sundance 2.5L I4 Turbo MFI VIN J (All) 60 Pin Connector

PCM Pin #	Wire Color	Circuit Description (60 Pin)	Value at Hot Idle
41	BK/DG	HO2S-11 (B1 S1) Signal	0.1-1.1v
42	BK/LG	Knock Sensor Signal	0.080v AC
43	GY/LB	Tachometer Signals	Pulse Signals
44	TN/YL	Distributor Sync Signals	Relay Off: 12v, On: 1v
45	LG	SCI Receive	5v
46	WT/BK	CCD Bus (-)	<0.050v
47	WT/OR	Vehicle Speed Signal	Digital Signal
48	BR/RD	Speed Control Set Switch Signal	S/C & Set Switch On: 3.8v
49	YL/RD	S/C On & Off Switch	S/C On: 1v, Off: 12v
50	WT/LG	S/C Resume Switch	S/C On: 1v, Off: 12v
51	DB/YL	ASD Relay Output	12-14v
52	PK/BK	EVAP Purge Solenoid Control	Solenoid Off: 12v, On: 1v
53	LG/RD	S/C Vent Solenoid	Vacuum Decreasing: 1v
54	LG/WT	A/T: Lockup Torque Converter	At Cruise w/TCC On: 1v
55	LB	BARO Read Solenoid	Solenoid Off: 12v, On: 1v
56-58	---	Not Used	---
59	PK/BK	AIS 4 Motor Control	DC pulse signals: 0.8-11v
60	YL/BK	AIS 2 Motor Control	DC pulse signals: 0.8-11v

Pin Connector Graphic

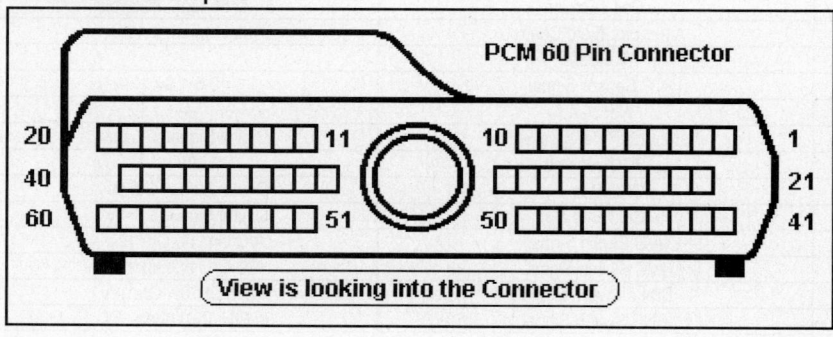

1990-91 Sundance 2.5L I4 SOHC TBI VIN K (All) 60 Pin Connector

PCM Pin #	Wire Color	Circuit Description (60 Pin)	Value at Hot Idle
1	DG/RD	MAP Sensor Signal	1.5-1.7v
2	TN/WT	ECT Sensor Signal	At 180°F: 2.80v
2 ('91)	TN/BK	ECT Sensor Signal	At 180°F: 2.80v
3	RD/WT	Battery Power (Fused B+)	12-14v
4	BK/LB	Sensor Ground	<0.050v
5	BK	Sensor Ground	<0.050v
5 ('91)	BK/WT	Sensor Ground	<0.050v
6	PK/WT	5-Volt Supply	4.9-5.1v
7	OR	9-Volt Supply	8.9-9.1v
8	WT	ASD Relay Output	12-14v
9	DB	Ignition Switch Output	12-14v
10	---	Not Used	---
11	LB/RD	Power Ground	<0.1v
11 ('91)	BK/TN	Power Ground	<0.1v
12	LB/RD	Power Ground	<0.1v
12 ('91)	BK/TN	Power Ground	<0.1v
13-15	---	Not Used	---
16	WT/DB	Injector 1 Driver	1-4 ms
17-18	---	Not Used	---
19	BK/GY	Coil 1 Driver	5°, at 55 mph: 8° dwell
20	DG	Alternator Field Control	Digital Signal: 0-12-0v
21 ('90)	BK/RD	Throttle Body Temperature	At 100°F: 2.51v
22	OR/DB	TP Sensor Signal	0.6-1.0v
23, 28	---	Not Used	---
24	GY/BK	Distributor Reference Signal	Digital Signal: 0-5-0v
25	PK	SCI Transmit	0v
26 ('91)	BK/BR	CCD Bus (+)	Digital Signal: 0-5-0v
27	BR	A/C Switch Signal	A/C On: 1v, Off: 12v
29	WT/PK	Brake Switch Signal	Brake Off: 0v, On: 12v
30	BR	PNP Switch Signal	In P/N: 0v, Others: 5v
30 ('91)	BR/YL	PNP Switch Signal	In P/N: 0v, Others: 5v
31	DB/PK	Radiator Fan Relay Control	Relay Off: 12v, On: 1v
32	BK/PK	MIL (lamp) Control	Lamp On: 1v, Off: 12v
33	TN/RD	S/C Vacuum Solenoid	Vacuum Increasing: 1v
34	DB/OR	A/C Clutch Relay Control	Relay Off: 12v, On: 1v
35	GY/YL	EGR Solenoid Control (California)	12v, 55 mph: 1v
36-38	---	Not Used	---
39	GY/RD	AIS 3 Motor Control	DC pulse signals: 0.8-11v
40	BR/WT	AIS 1 Motor Control	DC pulse signals: 0.8-11v

Standard Colors and Abbreviations

Abbreviation	Color	Abbreviation	Color	Abbreviation	Color
BK	Black	GY	Gray	RD	Red
BL	Blue	GN	Green	TN	Tan
BR	Brown	LG	Light Green	VT	Violet
DB	Dark Blue	OR	Orange	WT	White
DG	Dark Green	PK	Pink	YL	Yellow

1990-91 Sundance 2.5L I4 SOHC TBI VIN K (All) 60 Pin Connector

PCM Pin #	Wire Color	Circuit Description (60 Pin)	Value at Hot Idle
41	BK/DG	HO2S-11 (B1 S1) Signal	0.1-1.1v
42	---	Not Used	---
43	GY/LB	Tachometer Signals	Pulse Signals
45	LG	SCI Receive	5v
46 ('91)	WT/BK	CCD Bus (-)	<0.050v
47	WT/OR	Vehicle Speed Signal	Digital Signal
48	BR/RD	Speed Control Set Switch Signal	S/C & Set Switch On: 3.8v
49	YL/RD	S/C On & Off Switch	S/C On: 1v, Off: 12v
50	WT/LG	S/C Resume Switch	S/C On: 1v, Off: 12v
51	DG/YL	ASD Relay Output	12-14v
52	PK	EVAP Purge Solenoid Control	Solenoid Off: 12v, On: 1v
52 ('91)	PK/BK	EVAP Purge Solenoid Control	Solenoid Off: 12v, On: 1v
53	LB/RD	S/C Vent Solenoid	Vacuum Decreasing: 1v
54	OR/BK	A/T: Lockup Torque Converter	At Cruise w/TCC On: 1v
54	OR/BK	M/T: Shift Indicator Control	Lamp On: 1v, Off: 12v
55	---	Not Used	---
56 ('91)	GY/PK	Emissions Lamp Control	Lamp On: 1v, Off: 12v
57-58	---	Not Used	---
59	PK/BK	AIS 4 Motor Control	DC pulse signals: 0.8-11v
60	YL/BK	AIS 2 Motor Control	DC pulse signals: 0.8-11v

Pin Connector Graphic

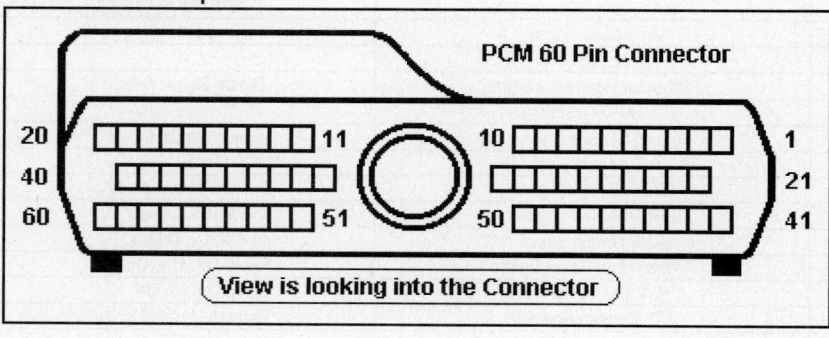

PCM 60 Pin Connector

View is looking into the Connector

1992-94 Sundance 2.5L I4 SOHC TBI VIN K (All) 60 Pin Connector

PCM Pin #	Wire Color	Circuit Description (60 Pin)	Value at Hot Idle
1	DG/RD	MAP Sensor Signal	1.5-1.7v
2	TN/BK	ECT Sensor Signal	At 180°F: 2.80v
3	RD/WT	Battery Power (Fused B+)	12-14v
4	BK/LB	Sensor Ground	<0.050v
5	BK/WT	Sensor Ground	<0.050v
6	PK/WT	5-Volt Supply	4.9-5.1v
7	OR	8-Volt Supply	7.9-8.1v
8	WT	ASD Relay Output	12-14v
9	DB	Ignition Switch Output	12-14v
10	---	Not Used	---
11-12	BK/TN	Power Ground	<0.1v
13-15	---	Not Used	---
16	WT/DB	Injector 1 Driver	1-4 ms
17-18	---	Not Used	---
19	BK/GY	Coil 1 Driver	5°, at 55 mph: 8° dwell
20	DG	Alternator Field Control	Digital Signal: 0-12-0v
21	---	Not Used	---
22	OR/DB	TP Sensor Signal	0.6-1.0v
23	RD/LG	Speed Control Set Switch Signal	S/C & Set Switch On: 3.8v
24	GY/BK	Distributor Reference Signal	Digital Signal: 0-5-0v
25	PK	SCI Transmit	0v
26	---	Not Used	---
27	BR	A/C Switch Signal	A/C On: 1v, Off: 12v
28	---	Not Used	---
29	WT/PK	Brake Switch Signal	Brake Off: 0v, On: 12v
30	BR/YL	PNP Switch Signal	In P/N: 0v, Others: 5v
31	DB/PK	Radiator Fan Relay Control	Relay Off: 12v, On: 1v
32	BK/PK	MIL (lamp) Control	Lamp On: 1v, Off: 12v
33	TN/RD	S/C Vacuum Solenoid	Vacuum Increasing: 1v
34	DB/OR	A/C Clutch Relay Control	Relay Off: 12v, On: 1v
35	GY/YL	EGR Solenoid Control (California)	12v, 55 mph: 1v
36-38	---	Not Used	---
39	GY/RD	AIS 3 Motor Control	DC pulse signals: 0.8-11v
40	BR/WT	AIS 1 Motor Control	DC pulse signals: 0.8-11v

Standard Colors and Abbreviations

Abbreviation	Color	Abbreviation	Color	Abbreviation	Color
BK	Black	GY	Gray	RD	Red
BL	Blue	GN	Green	TN	Tan
BR	Brown	LG	Light Green	VT	Violet
DB	Dark Blue	OR	Orange	WT	White
DG	Dark Green	PK	Pink	YL	Yellow

1992-94 Sundance 2.5L I4 SOHC TBI VIN K (All) 60 Pin Connector

PCM Pin #	Wire Color	Circuit Description (60 Pin)	Value at Hot Idle
41	BK/DG	HO2S-11 (B1 S1) Signal	0.1-1.1v
42	---	Not Used	---
43	GY/LB	Tachometer Signals	Pulse Signals
44	---	Not Used	---
45	LG	SCI Receive	5v
46	---	Not Used	---
47	WT/OR	Vehicle Speed Signal	Digital Signal
48-50	---	Not Used	---
51	DB/YL	ASD Relay Output	12-14v
52	PK/BK	EVAP Purge Solenoid Control	Solenoid Off: 12v, On: 1v
53	LG/RD	S/C Vent Solenoid	Vacuum Decreasing: 1v
54	LG/WT	A/T: Lockup Torque Converter	At Cruise w/TCC On: 1v
55-56	---	Not Used	---
57	DG/OR	ASD Relay Control	Relay Off: 12v, On: 1v
58	---	Not Used	---
59	PK/BK	AIS 4 Motor Control	DC pulse signals: 0.8-11v
60	YL/BK	AIS 2 Motor Control	DC pulse signals: 0.8-11v

Pin Connector Graphic

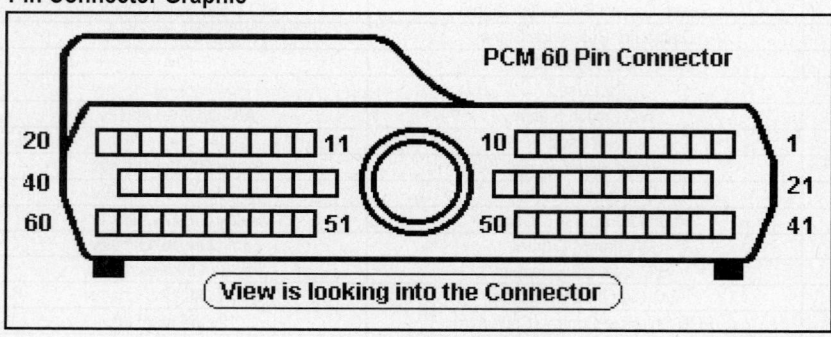

1992-94 Sundance 3.0L V6 MFI VIN 3 (All) 60 Pin Connector

PCM Pin #	Wire Color	Circuit Description (60 Pin)	Value at Hot Idle
1	DG/RD	MAP Sensor Signal	1.5-1.7v
2	TN/BK	ECT Sensor Signal	At 180°F: 2.80v
3	RD	Battery Power (Fused B+)	12-14v
3	OR	Battery Power (Fused B+)	12-14v
4	BK/LB	Sensor Ground	<0.050v
5	BK/WT	Sensor Ground	<0.050v
6	PK/WT	5-Volt Supply	4.9-5.1v
7	OR	8-Volt Supply	7.9-8.1v
8	---	Not Used	---
9	DB	ASD Relay Output	12-14v
10	---	Not Used	---
11	BK/TN	Power Ground	<0.1v
12	BK/TN	Power Ground	<0.1v
13	LB/BR	Injector 4 Driver	1-4 ms
14	YL/WT	Injector 3 Driver	1-4 ms
15	TN	Injector 2 Driver	1-4 ms
16	WT/DB	Injector 1 Driver	1-4 ms
17-18	---	Not Used	---
19	BK/GY	Coil 1 Driver	5°, at 55 mph: 8° dwell
20	DG	Alternator Field Control	Digital Signal: 0-12-0v
21	---	Not Used	---
22	OR/DB	TP Sensor Signal	0.6-1.0v
23	RD/LG	Speed Control Set Switch Signal	S/C & Set Switch On: 3.8v
24	GY/BK	Distributor Reference Signal	Digital Signal: 0-5-0v
25	PK	SCI Transmit	0v
26	PK/BK	CCD Bus (+)	Digital Signal: 0-5-0v
26 ('93-94)	PK/BR	CCD Bus (+)	Digital Signal: 0-5-0v
27	BR	A/C Damped Pressure Switch	A/C On: 1v, Off: 12v
28 ('93-'94)	DB/OR	PSP Switch Signal	Straight: 0v, Turning: 5v
29	WT/PK	Brake Switch Signal	Brake Off: 0v, On: 12v
30	BR/YL	PNP Switch Signal	In P/N: 0v, Others: 5v
31	DB/PK	Radiator Fan Relay Control	Relay Off: 12v, On: 1v
32	BK/PK	MIL (lamp) Control	Lamp On: 1v, Off: 12v
33	TN/RD	S/C Vacuum Solenoid	Vacuum Increasing: 1v
34	DB/OR	A/C Clutch Relay Control	Relay Off: 12v, On: 1v
35	GY/YL	EGR Solenoid Control (California)	12v, 55 mph: 1v
36-37	---	Not Used	---
38	GY	Injector 5 Driver	1-4 ms
39	GY/RD	IAC 3 Driver	DC pulse signals: 0.8-11v
40	BR/WT	IAC 1 Driver	DC pulse signals: 0.8-11v

Standard Colors and Abbreviations

Abbreviation	Color	Abbreviation	Color	Abbreviation	Color
BK	Black	GY	Gray	RD	Red
BL	Blue	GN	Green	TN	Tan
BR	Brown	LG	Light Green	VT	Violet
DB	Dark Blue	OR	Orange	WT	White
DG	Dark Green	PK	Pink	YL	Yellow

1992-94 Sundance 3.0L V6 MFI VIN 3 (All) 60 Pin Connector

PCM Pin #	Wire Color	Circuit Description (60 Pin)	Value at Hot Idle
41	BK/DG	HO2S-11 (B1 S1) Signal	0.1-1.1v
42	---	Not Used	---
43	GY/LB	Tachometer Signals	Pulse Signals
44	TN/YL	Distributor Sync Signals	Digital Signal: 0-5-0v
45	LG	SCI Receive	5v
46	WT/BK	CCD Bus (-)	<0.050v
47	WT/OR	Vehicle Speed Signal	Digital Signal
48-50	---	Not Used	---
51	DB/YL	ASD Relay Output	12-14v
52	PK/BK	EVAP Purge Solenoid Control	Solenoid Off: 12v, On: 1v
53	LG/RD	S/C Vent Solenoid	Vacuum Decreasing: 1v
54-56	---	Not Used	---
57	DG/OR	ASD Relay Control	Relay Off: 12v, On: 1v
58	BR/DB	Injector 6 Driver	1-4 ms
59	PK/BK	IAC 4 Driver	DC pulse signals: 0.8-11v
60	YL/BK	IAC 2 Driver	DC pulse signals: 0.8-11v

Pin Connector Graphic

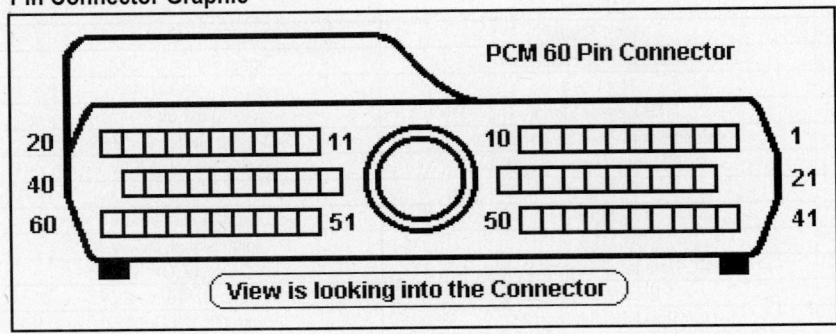

CHRYSLER VAN CONTENTS

CHRYSLER VAN PIN TABLES

Introduction
This section contains Pin Tables for Chrysler Vans from 1990-2003. It can be used to assist in the repair of Trouble Code and No Code faults related to the PCM.

Vehicle Coverage
- 2001-03 PT Cruiser Van Applications
- 1990-2003 Town & Country Van Applications

How to Use This Section
This section can be used to look up the location of a particular pin, a wire color or a "known good" value for a PCM circuit. To locate this information for a particular vehicle, find the model, correct engine size, VIN Code and model year on the Contents page.

To view the PCM terminals of a 1996 Town & Country Van with a 3.3L VIN R, go to Contents Page 1 and find the text string shown below:

3.3L V6 MFI VIN R A/T 80 Pin **(1996)** ....Page 11-15

Then turn to Page 11-15 to find the following PCM related information.

1996 Town & Country 3.3L V6 MFI A/T VIN R 'C1' Connector

PCM Pin #	Wire Color	Circuit Description (40 Pin)	Value at Hot Idle
1	---	Not Used	---
2	RD/YL	Coil Driver 3 (Cyl 3 & 6)	5°, 55 mph: 8° dwell
3	DB/TN	Coil Driver 2 (Cyl 2 & 5)	5°, 55 mph: 8° dwell
4	DG	Generator Field Driver	Digital Signal: 0-12-0v
5	YL/RD	Speed Control Power Supply	12-14v
6	DG/OR	ASD Relay Output (B+)	12-14v
7	YL/WT	Injector 3 Driver Control	1-4 ms
8	TN	Smart Start Relay Control	KOEC: 9-11v
9	---	Not Used	---
10	BK/TN	Power Ground	<0.1v
11	GY/RD	Coil Driver 1 (Cyl 1 & 4)	5°, 55 mph: 8° dwell
12	---	Not Used	---
13	WT/DB	Injector 1 Driver Control	1-4 ms
14	BR/DB	Injector 6 Driver Control	1-4 ms
15	GY	Injector 5 Driver Control	1-4 ms
16	LB/BR	Injector 4 Driver Control	1-4 ms
17	TN	Injector 2 Driver Control	1-4 ms

In this example, the Injector 3 Driver control circuit is connected to Pin 7 of the "C1" 40 Pin connector (i.e., to the YL/WT wire). The value at Hot Idle shown here is for this circuit is 1-4 milliseconds (use Lab Scope).

The acronym A/T in the Title at the top of the table indicates that information in this table is for vehicles with an automatic transmission.

CHRYSLER PIN TABLES

1998-99 Town & Country 3.3L V6 FLEXIBLE FUEL VIN G 'C1' Connector

PCM Pin #	Wire Color	Circuit Description (40 Pin)	Value at Hot Idle
1	---	Not Used	---
2	RD/YL	Coil Driver 3 (Cyl 3 & 6)	5°, 55 mph: 8° dwell
3	DB/TN	Coil Driver 2 (Cyl 2 & 5)	5°, 55 mph: 8° dwell
4	---	Not Used	---
5	YL/RD	Speed Control Power Supply	12-14v
6	DG/OR	ASD Relay Output (B+)	12-14v
7	WT/DB	Injector 3 Driver Control	1-4 ms
8	DG	Generator Field Driver	Digital Signal: 0-12-0v
9	---	Not Used	---
10	BK/TN	Power Ground	<0.1v
11	GY/RD	Coil Driver 1 (Cyl 1 & 4)	5°, 55 mph: 8° dwell
12	---	Not Used	---
13	YL/WT	Injector 1 Driver Control	1-4 ms
14	BR/DB	Injector 6 Driver Control	1-4 ms
15	GY	Injector 5 Driver Control	1-4 ms
16	LB/BR	Injector 4 Driver Control	1-4 ms
17	TN	Injector 2 Driver Control	1-4 ms
18-19	---	Not Used	---
20	WT/BK	Ignition Switch Output (B+)	12-14v
21	---	Not Used	---
22	BK/PK	Malfunction Indicator Lamp	MIL Off: 12v, On: 1v
23-24	---	Not Used	---
25	DB/LG	Knock Sensor Signal	0.080v AC
26	TN/BK	ECT Sensor Signal	At 180°F: 2.80v
27	BK/OR	Oxygen Sensor Ground	<0.050v
28-29	---	Not Used	---
30	BK/DG	HO2S-11 (B1 S1) Signal	0.1-1.1v
31	TN	Smart Start Relay Control	KOEC: 9-11v
32	GY/BK	CKP Sensor Signal	Digital Signal: 0-5-0v
33	TN/YL	CMP Sensor Signal	Digital Signal: 0-5-0v
34	---	Not Used	---
35	OR/DB	TP Sensor Signal	0.6-1.0v
36	DG/RD	MAP Sensor Signal	1.5-1.7v
37	---	Not Used	---
38	DG	A/C On/Off Switch Signal	A/C Off: 12v, On: 1v
39	---	Not Used	---
40	GY/YL	EGR Solenoid Control	12v, 55 mph: 1v

Pin Connector Graphic

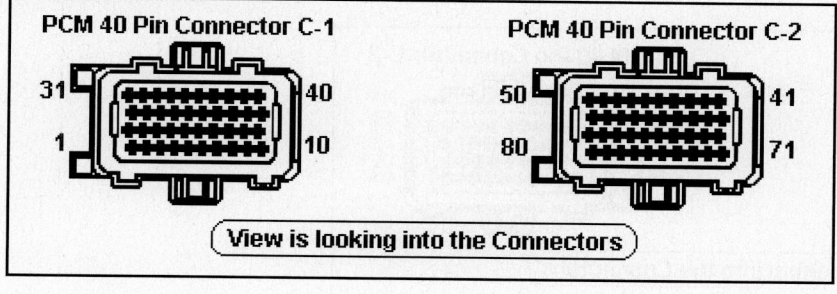

PCM 40 Pin Connector C-1   PCM 40 Pin Connector C-2

View is looking into the Connectors

1998-99 Town & Country 3.3L V6 FLEXIBLE FUEL VIN G 'C2' Connector

PCM Pin #	Wire Color	Circuit Description (40 Pin)	Value at Hot Idle
41	RD/LG	Speed Control Set Switch Signal	S/C & Set Switch On: 3.8v
42	DB	A/C Pressure Switch Signal	A/C On: 0.451-4.850v
43	BK/LB	Sensor Ground	<0.050v
44	OR	8-Volt Supply	7.9-8.1v
45	DG	A/C Switch Sense	A/C Off: 12v, On: 1v
46	RD/WT	Fused Battery Power (B+)	12-14v
47	---	Not Used	---
48	BR/WT	IAC 3 Driver Control	Pulse Signal: 0.8-11v
49	YL/BK	IAC 2 Driver Control	Pulse Signal: 0.8-11v
50	BK/TN	Power Ground	<0.1v
51	TN/WT	HO2S-12 (B1 S2) Signal	0.1-1.1v
52-53	---	Not Used	---
54	DG/LB	Flexible Fuel Sensor	Digital Signal
55-56	---	Not Used	---
57	GY/RD	IAC 1 Driver Control	Pulse Signal: 0.8-11v
58	VT/BK	IAC 4 Driver Control	Pulse Signal: 0.8-11v
59	VT/BR	CCD Bus (+)	Digital Signal: 0-5-0v
60	WT/BK	CCD Bus (-)	<0.050v
61	VT/WT	5-Volt Supply	4.9-5.1v
62	WT/PK	Brake Switch Sense Signal	Brake Off: 0v, On: 12v
63	YL/DG	Torque Management Request	Digital Signals
64	DB/OR	A/C Clutch Relay Control	Relay Off: 12v, On: 1v
65	PK	SCI Transmit	0v
66	WT/OR	Vehicle Speed Signal	Digital Signal
67	DB/YL	Auto Shutdown Relay Control	Relay Off: 12v, On: 1v
68	PK/BK	EVAP Purge Solenoid Control	PWM Signal: 0-12-0v
69	---	Not Used	---
70	DG/LG	EVAP Purge Solenoid Sense	0-1v
71	---	Not Used	---
72	DB/WT	LDP Switch Sense Signal	LDP Switch Closed: 0v
73	LG/DB	Radiator Fan Control Relay	Relay Off: 12v, On: 1v
74	BR	Fuel Pump Relay Control	Relay Off: 12v, On: 1v
75	LG	SCI Receive	5v
76	BK/WT	PNP Switch Sense Signal	In P/N: 0v, Others: 5v
77	WT/DG	LDP Solenoid Control	PWM Signal: 0-12-0v
78	TN/RD	Speed Control Vacuum Solenoid Control	Vacuum Increasing: 1v
79	---	Not Used	---
80	LG/RD	Speed Control Vent Solenoid Control	Vacuum Decreasing: 1v

Pin Connector Graphic

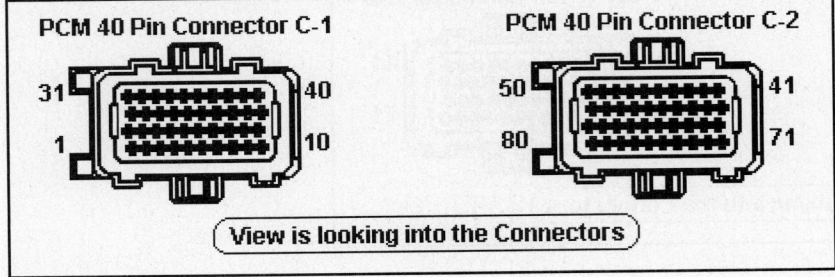

PCM 40 Pin Connector C-1

PCM 40 Pin Connector C-2

View is looking into the Connectors

2000 Town & Country 3.3L V6 FLEXIBLE FUEL VIN G 'C1' Connector

PCM Pin #	Wire Color	Circuit Description (40 Pin)	Value at Hot Idle
1	---	Not Used	---
2	RD/YL	Coil Driver 3 (Cyl 3 & 6)	5°, 55 mph: 8° dwell
3	DB/TN	Coil Driver 2 (Cyl 2 & 5)	5°, 55 mph: 8° dwell
4	---	Not Used	---
5	YL/RD	Speed Control Power Supply	12-14v
6	DG/OR	ASD Relay Output (B+)	12-14v
7	YL/WT	Injector 3 Driver Control	1-4 ms
8	DG	Generator Field Driver	Digital Signal: 0-12-0v
9	---	Not Used	---
10	BK/TN	Power Ground	<0.1v
11	GY/RD	Coil Driver 1 (Cyl 1 & 4)	5°, 55 mph: 8° dwell
12	---	Not Used	---
13	WT/DB	Injector 1 Driver Control	1-4 ms
14	BR/DB	Injector 6 Driver Control	1-4 ms
15	GY	Injector 5 Driver Control	1-4 ms
16	LB/BR	Injector 4 Driver Control	1-4 ms
17	TN/WT	Injector 2 Driver Control	1-4 ms
18-19	---	Not Used	---
20	WT/BK	Ignition Switch Output (B+)	12-14v
21	---	Not Used	---
22	BK/PK	Malfunction Indicator Lamp	MIL Off: 12v, On: 1v
23-24	---	Not Used	---
25	DB/LG	Knock Sensor Signal	0.080v AC
26	TN/BK	ECT Sensor Signal	At 180°F: 2.80v
27	BK/OR	Oxygen Sensor Ground	<0.050v
28-29	---	Not Used	---
30	BK/DG	HO2S-11 (B1 S1) Signal	0.1-1.1v
31	TN	Smart Start Relay Control	KOEC: 9-11v
32	GY/BK	CKP Sensor Signal	Digital Signal: 0-5-0v
33	TN/YL	CMP Sensor Signal	Digital Signal: 0-5-0v
34	---	Not Used	---
35	OR/BK	TP Sensor Signal	0.6-1.0v
36	DG/RD	MAP Sensor Signal	1.5-1.7v
37	---	Not Used	---
38	DG/LB	A/C On/Off Switch Signal	A/C Off: 12v, On: 1v
39	---	Not Used	---
40	GY/YL	EGR Solenoid Control	12, 55 mph: 1v

Pin Connector Graphic

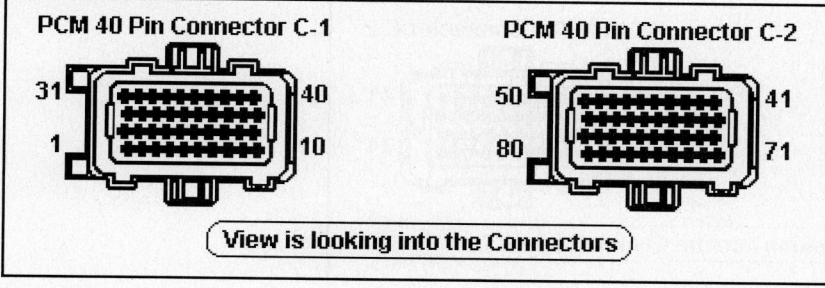

PCM 40 Pin Connector C-1

PCM 40 Pin Connector C-2

31   40

1   10

50   41

80   71

View is looking into the Connectors

2000 Town & Country 3.3L V6 FLEXIBLE FUEL VIN G 'C2' Connector

PCM Pin #	Wire Color	Circuit Description (40 Pin)	Value at Hot Idle
41	RD/LG	Speed Control Set Switch Signal	S/C & Set Switch On: 3.8v
42	DB	A/C Pressure Switch Signal	A/C On: 0.451-4.850v
43	BK/LB	Sensor Ground	<0.050v
44	OR	8-Volt Supply	7.9-8.1v
45, 47	---	Not Used	---
46	RD/WT	Fused Battery Power (B+)	12-14v
48	BR/WT	IAC 3 Driver Control	Pulse Signal: 0.8-11v
49	YL/BK	IAC 2 Driver Control	Pulse Signal: 0.8-11v
50	BK/TN	Power Ground	<0.1v
51	TN/WT	HO2S-12 (B1 S2) Signal	0.1-1.1v
52-53	---	Not Used	---
54	YL/WT	Flexible Fuel Sensor	Digital Signal
55	---	Not Used	---
56	TN/RD	Speed Control Vacuum Solenoid Control	Vacuum Increasing: 1v
57	GY/RD	IAC 1 Driver Control	Pulse Signal: 0.8-11v
58	PK/BK	IAC 4 Driver Control	Pulse Signal: 0.8-11v
59	PK/BR	CCD Bus (+)	Digital Signal: 0-5-0v
60	WT/BK	CCD Bus (-)	<0.050v
61	PK/WT	5-Volt Supply	4.9-5.1v
62	WT/PK	Brake Switch Sense Signal	Brake Off: 0v, On: 12v
63	YL/DG	Torque Management Request	Digital Signals
64	DB/OR	A/C Clutch Relay Control	Relay Off: 12v, On: 1v
65	PK	SCI Transmit	0v
66	WT/OR	Vehicle Speed Signal	Digital Signal
67	DB/YL	Auto Shutdown Relay Control	Relay Off: 12v, On: 1v
68	PK/BK	EVAP Purge Solenoid Control	PWM Signal: 0-12-0v
69	---	Not Used	---
70	PK/RD	EVAP Purge Solenoid Sense	0-1v
71	---	Not Used	---
72	DB/WT	LDP Switch Sense Signal	LDP Switch Closed: 0v
73	LG/DB	Radiator Fan Control Relay	Relay Off: 12v, On: 1v
74	BR	Fuel Pump Relay Control	Relay Off: 12v, On: 1v
75	LG	SCI Receive	5V
76	BR/YL	PNP Switch Sense Signal	In P/N: 0v, Others: 5v
77	WT/DG	LDP Solenoid Control	PWM Signal: 0-12-0v
78-79	---	Not Used	---
80	LG/RD	Speed Control Vent Solenoid Control	Vacuum Decreasing: 1v

Pin Connector Graphic

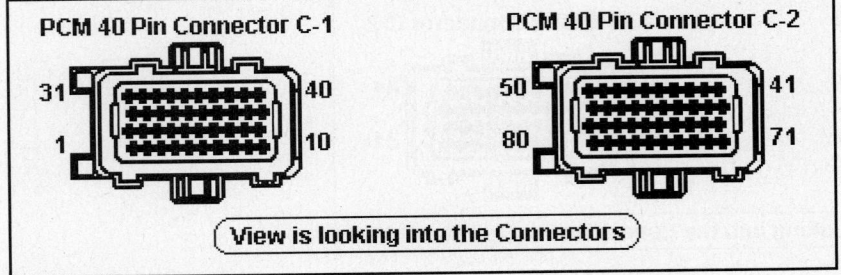

PCM 40 Pin Connector C-1   PCM 40 Pin Connector C-2

31   40   50   41

1   10   80   71

(View is looking into the Connectors)

2001-02 Town & Country 3.3L V6 FLEXIBLE FUEL VIN G A/T 'C1' Connector

PCM Pin #	Wire Color	Circuit Description (40 Pin)	Value at Hot Idle
1	---	Not Used	---
2	DB/OR	Coil 3 Driver Control	5°, 55 mph: 8° dwell
3	DB/TN	Coil 2 Driver Control	5°, 55 mph: 8° dwell
4	---	Not Used	---
5	VT/YL	Speed Control On/Off Switch Sense	12-14v
6	BR/WT	ASD Relay Power (B+)	12-14v
7	BR/LB	Injector 3 Driver	1-4 ms
8	BR/GY	Generator Field Driver	Digital Signal: 0-12-0v
9	---	Not Used	---
10	BK/TN	Power Ground	<0.1v
11	DB/DG	Coil 1 Driver Control	5°, 55 mph: 8° dwell
12	VT/GY	Engine Oil Pressure Sensor	1.6v at 24 psi
13	BR/YL	Injector 1 Driver Control	1-4 ms
14	BR/VT	Injector 6 Driver Control	1-4 ms
15	BR/OR	Injector 5 Driver Control	1-4 ms
16	BR/TN	Injector 4 Driver Control	1-4 ms
17	BR/DB	Injector 2 Driver Control	1-4 ms
18	BR/LG	HO2S-11 (B1 S1) Heater	PWM Signal: 0-12-0v
19	---	Not Used	---
20	PK/GY	Ignition Switch Power (B+)	12-14v
21-24	---	Not Used	---
25	DB/YL	Knock Sensor Signal	0.080v AC
26	VT/OR	ECT Sensor Signal	At 180°F: 2.80v
27	BR/DG	Oxygen Sensor Ground	<0.050v
28-29	---	Not Used	---
30	DB/LB	HO2S-11 (B1 S1) Signal	0.1-1.1v
31	DG/OR	Double Start Override Signal	KOEC: 9-11v
32	BR/LB	CKP Sensor Signal	Digital Signal: 0-5-0v
33	DB/GY	CMP Sensor Signal	Digital Signal: 0-5-0v
34	---	Not Used	---
35	BR/OR	TP Sensor Signal	0.6-1.0v
36	VT/BR	MAP Sensor Signal	1.5-1.6v
37	DB/LG	Intake Air Temp. Signal	At 100°F: 1.83v
38-39	---	Not Used	---
40	DB/VT	EGR Solenoid Control	12v, at 55 mph: 1v

Pin Connector Graphic

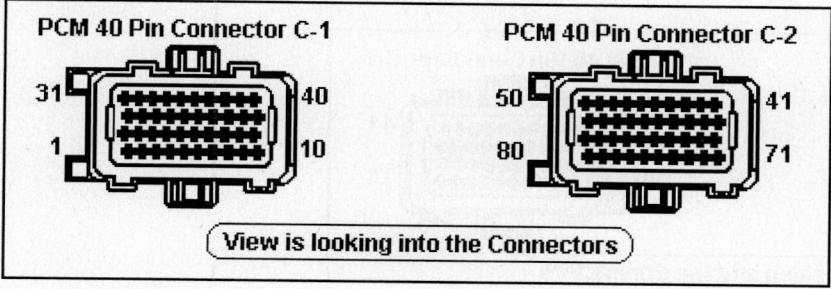

2001-02 Town & Country 3.3L V6 FLEXIBLE FUEL VIN G A/T 'C2' Connector

PCM Pin #	Wire Color	Circuit Description (40 Pin)	Value at Hot Idle
41	VT	Speed Control Set Switch Signal	S/C & Set Switch On: 3.8v
42	LB/BR	A/C Pressure Switch Signal	A/C On: 0.451-4.850v
43	DB/DG	Sensor Ground	<0.050v
44	BR/PK	8-Volt Supply	7.9-8.1v
45	---	Not Used	---
46	OR/RD	Keep Alive Power (B+)	12-14v
47-48	---	Not Used	---
49	VT/DG	IAC 1 Driver Control	Pulse Signals
50	BK/DG	Power Ground	<0.1v
51	DB/YL	HO2S-12 (B1 S2) Signal	0.1-1.1v
52-53	---	Not Used	---
54	YL/WT	Flexible Fuel Sensor	Digital Signals
56	VT/YL	Speed Control Vacuum Solenoid	Vacuum Increasing: 1v
57	VT/LG	IAC 2 Driver Control	Pulse Signals
58	---	Not Used	---
59	WT/VT	PCI Data Bus (J1850)	Digital Signals: 0-7-0v
60	---	Not Used	---
61	PK/YL	5-Volt Supply	4.9-5.1v
62	DG/WT	Brake Switch Sense Signal	Brake Off: 0v, On: 12v
63	DG/LG	Torque Management Request	Digital Signals
64	LB/OR	A/C Clutch Relay Control	Relay Off: 12v, On: 1v
65	WT/BR	SCI Transmit Signal	0v
66	DB/OR	Vehicle Speed Signal	Digital Signal
67	BR/WT	ASD Relay Control	Relay Off: 12v, On: 1v
68	DB/WT	EVAP Purge Solenoid Control	PWM Signal: 0-12-0v
69	---	Not Used	---
70	DB/BR	EVAP Purge Solenoid Sense	0-1v
71	---	Not Used	---
72	VT/WT	LDP Switch Sense Signal	LDP Switch Closed: 0v
73	BR/VT	Radiator Fan Control Relay	Relay Off: 12v, On: 1v
74	BR	Fuel Pump Relay Control	Relay Off: 12v, On: 1v
75	WT/LG	SCI Receive Signal	5v
76	YL/LB	TRS T41 Signal	In P/N: 0v, Others: 5v
77	VT/LB	LDP Solenoid Control	PWM Signal: 0-12-0v
78-79	---	Not Used	---
80	LG/RD	Speed Control Vent Solenoid Control	Vacuum Decreasing: 1v

Pin Connector Graphic

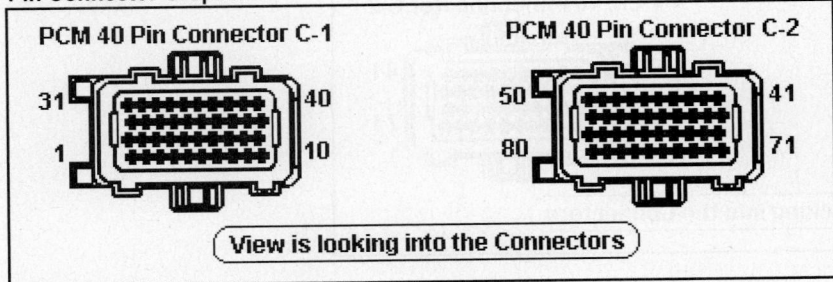

PCM 40 Pin Connector C-1   PCM 40 Pin Connector C-2

View is looking into the Connectors

1990-91 Town & Country 3.3L V6 SOHC VIN R 60 Pin Connector

PCM Pin #	Wire Color	Circuit Description (60 Pin)	Value at Hot Idle
1	DG/RD	MAP Sensor Signal	1.5-1.7v
2	TN/WT	ECT Sensor Signal	At 180°F: 2.80v
3	RD/WT	Fused Battery Power (B+)	12-14v
4	BK/LB	Sensor Ground	<0.050v
5	BK/WT	Sensor Ground	<0.050v
6	PK/WT	5-Volt Supply	4.9-5.1v
7	OR	9-Volt Supply	8.9-9.1v
8	DG/BK	Auto Shutdown Relay Input	12-14v
9	DB	Ignition Switch Output (B+)	12-14v
10	---	Not Used	---
11	BK/TN	Power Ground	<0.1v
12	BK/TN	Power Ground	<0.1v
14	YL/WT	Injector Control 3 Driver	1-4 ms
15	TN	Injector Control 2 Driver	1-4 ms
16	WT/DB	Injector Control 1 Driver	1-4 ms
17-18	---	Not Used	---
19	BK/GY	Coil Driver 1 Control	5°, 55 mph: 8° dwell
20	DG	Alternator Field Control	Digital Signal: 0-12-0v
21	---	Not Used	---
22	OR/DB	TP Sensor Signal	0.6-1.0v
24	GY/BK	Distributor Reference Pick-up	Digital Signal: 0-5-0v
25	PK	SCI Transmit	0v
26	PK/BR	CCD Bus (+)	Digital Signal: 0-5-0v
27	BR	A/C Clutch Signal	A/C Off: 12v, On: 1v
28	---	Not Used	---
29	WT/PK	Brake Switch Sense Signal	Brake Off: 0v, On: 12v
30	BR/YL	PNP Switch Sense Signal	In P/N: 0v, Others: 5v
31	DB/PK	Radiator Fan Control Relay	Relay Off: 12v, On: 1v
32	BK/PK	Malfunction Indicator Lamp	MIL Off: 12v, On: 1v
33	TN/RD	Speed Control Vacuum Solenoid Control	Vacuum Increasing: 1v
34	DB/OR	A/C WOT Relay Control	Relay Off: 12v, On: 1v
35	GY/YL	EGR Solenoid Control (Cal)	12v, 55 mph: 1v
36-40	---	Not Used	---

Standard Colors and Abbreviations

Abbreviation	Color	Abbreviation	Color	Abbreviation	Color
BK	Black	GY	Gray	RD	Red
BL	Blue	GN	Green	TN	Tan
BR	Brown	LG	Light Green	VT	Violet
DB	Dark Blue	OR	Orange	WT	White
DG	Dark Green	PK	Pink	YL	Yellow

1990-91 Town & Country 3.3L V6 SOHC VIN R 60 Pin Connector

PCM Pin #	Wire Color	Circuit Description (60 Pin)	Value at Hot Idle
41	BK/DG	HO2S-11 (B1 S1) Signal	0.1-1.1v
42	BK/LG	Knock Sensor Signal	0.080v AC
43	GY/LB	Tachometer Signal	Pulse Signals
44	TN/YL	Distributor Sync Signal	Digital Signal: 0-5-0v
45	LG	SCI Receive	5v
46	WT/BK	CCD Bus (-)	<0.050v
47	WT/OR	Vehicle Speed Signal	Digital Signal
48	BR/RD	Speed Control Set Switch Signal	S/C & Set Switch On: 3.8v
49	YL/RD	S/C On/Off Switch Signal	S/C Off: 12v, On: 1v
50	W/LG	S/C Resume Switch Signal	S/C Off: 12v, On: 1v
51	DB/YL	ASD Relay Output (B+)	12-14v
52	PK/BK	EVAP Purge Solenoid Control	Solenoid Off: 12v, On: 1v
53	LG/RD	Speed Control Vent Solenoid Control	Vacuum Decreasing: 1v
54-55	---	Not Used	---
56	GY/PK	EMR Indicator (lamp) Control	Lamp Off: 12v, On: 1v
57-58	---	Not Used	---
59	PK/BK	AIS Motor 4 Control	Pulse Signal: 0.8-11v
60	YL/BK	AIS Motor 2 Control	Pulse Signal: 0.8-11v

Pin Connector Graphic

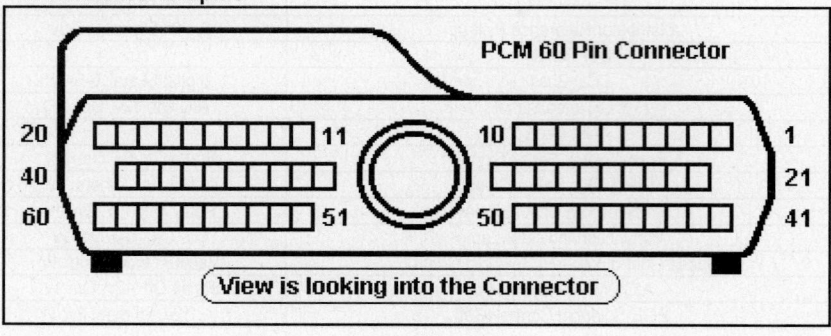

1992 Town & Country 3.3L V6 MFI VIN R A/T 60 Pin Connector

PCM Pin #	Wire Color	Circuit Description (60 Pin)	Value at Hot Idle
1	DG/RD	MAP Sensor Signal	1.5-1.7v
2	TN/BK	ECT Sensor Signal	At 180ºF: 2.80v
3	RD/WT	Fused Battery Power (B+)	12-14v
4	BK/LB	Sensor Ground	<0.050v
5	BK/WT	Sensor Ground	<0.050v
6	PK/WT	5-Volt Supply	4.9-5.1v
7	OR	9-Volt Supply	8.9-9.1v
9	DB	Ignition Switch Output (B+)	12-14v
10	---	Not Used	---
11	BK/TN	Power Ground	<0.1v
12	BK/TN	Power Ground	<0.1v
13	LB/BR	Injector Control 4 Driver	1-4 ms
14	YL/WT	Injector Control 3 Driver	1-4 ms
15	TN	Injector Control 2 Driver	1-4 ms
16	WT/DB	Injector Control 1 Driver	1-4 ms
17	DB/YL	Coil Driver 2 Control	5º, 55 mph: 8º dwell
18	RD/YL	Coil Driver 3 Control	5º, 55 mph: 8º dwell
19	GY	Coil Driver 1 Control	5º, 55 mph: 8º dwell
20	DG	Alternator Field Control	Digital Signal: 0-12-0v
21	BK/RD	Air Temperature Sensor	At 100ºF: 2.51v
22	OR/DB	TP Sensor Signal	0.6-1.0v
23	RD/LG	Speed Control Set Switch Signal	S/C & Set Switch On: 3.8v
24	GY/BK	CKP Sensor Signal	Digital Signal: 0-5-0v
25	PK	SCI Transmit	0v
26	PK/BR	CCD Bus (+)	Digital Signal: 0-5-0v
27	BR	A/C Damped Pressure Switch	A/C Off: 12v, On: 1v
28	---	Not Used	---
29	WT/PK	Brake Switch Sense Signal	Brake Off: 0v, On: 12v
30	BR/YL	PNP Switch Sense Signal	In P/N: 0v, Others: 5v
31	DB/PK	Radiator Fan Control Relay	Relay Off: 12v, On: 1v
32	---	Not Used	---
33	TN/RD	Speed Control Vacuum Solenoid Control	Vacuum Increasing: 1v
34	DB/OR	A/C WOT Relay Control	Relay Off: 12v, On: 1v
35-37	---	Not Used	---
38	GY	Injector Control 5 Driver	1-4 ms
39	GY/RD	AIS Motor 3 Control	Pulse Signal: 0.8-11v
40	BR/WT	AIS Motor 1 Control	Pulse Signal: 0.8-11v

Standard Colors and Abbreviations

Abbreviation	Color	Abbreviation	Color	Abbreviation	Color
BK	Black	GY	Gray	RD	Red
BL	Blue	GN	Green	TN	Tan
BR	Brown	LG	Light Green	VT	Violet
DB	Dark Blue	OR	Orange	WT	White
DG	Dark Green	PK	Pink	YL	Yellow

1992 Town & Country 3.3L V6 MFI VIN R A/T 60 Pin Connector

PCM Pin #	Wire Color	Circuit Description (60 Pin)	Value at Hot Idle
41	BK/DG	HO2S-11 (B1 S1) Signal	0.1-1.1v
42-43	---	Not Used	---
44	TN/YL	CMP Sensor Signal	Digital Signal: 0-5-0v
45	LG	SCI Receive	5v
46	WT/BK	CCD Bus (-)	<0.050v
47	WT/OR	Vehicle Speed Signal	Digital Signal
48-50	---	Not Used	---
51	DB/YL	ASD Relay Output (B+)	12-14v
52	PK/BK	EVAP Purge Solenoid Control	Solenoid Off: 12v, On: 1v
53	LG/RD	Speed Control Vent Solenoid Control	Vacuum Decreasing: 1v
54-55	---	Not Used	---
56	GY/PK	EMR Indicator (lamp) Control	Lamp On: <1v, Off: 12v
57	DG/OR	Fuel Injector Power Supply	12-14v
58	BR/DB	Injector 6 Driver Control	1-4 ms
59	PK/BK	AIS Motor 4 Control	Pulse Signal: 0.8-11v
60	YL/BK	AIS Motor 2 Control	Pulse Signal: 0.8-11v

Pin Connector Graphic

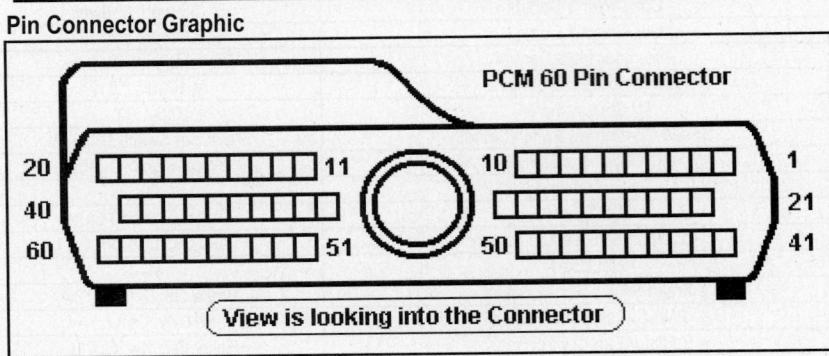

1993 Town & Country 3.3L V6 MFI VIN R A/T 60 Pin Connector

PCM Pin #	Wire Color	Circuit Description (60 Pin)	Value at Hot Idle
1	DG/RD	MAP Sensor Signal	1.5-1.7v
2	TN/BK	ECT Sensor Signal	At 180°F: 2.80v
3	RD/WT	Fused Battery Power (B+)	12-14v
4	BK/LB	Sensor Ground	<0.050v
5	BK/WT	Sensor Ground	<0.050v
6	PK/WT	5-Volt Supply	4.9-5.1v
7	OR	9-Volt Supply	8.9-9.1v
8	---	Not Used	---
9	DB	Ignition Switch Output (B+)	12-14v
10	---	Not Used	---
11	BK/TN	Power Ground	<0.1v
12	BK/TN	Power Ground	<0.1v
13	LB/BR	Injector Control 4 Driver	1-4 ms
14	YL/WT	Injector Control 3 Driver	1-4 ms
15	TN	Injector Control 2 Driver	1-4 ms
16	WT/DB	Injector Control 1 Driver	1-4 ms
17	DB/YL	Coil Driver 2 Control	5°, 55 mph: 8° dwell
18	RD/YL	Coil Driver 3 Control	5°, 55 mph: 8° dwell
19	BK/GY	Coil Driver 1 Control	5°, 55 mph: 8° dwell
20	DG	Alternator Field Control	Digital Signal: 0-12-0v
21	---	Not Used	---
22	OR/DB	TP Sensor Signal	0.6-1.0v
23	RD/LG	Speed Control Set Switch Signal	S/C & Set Switch On: 3.8v
24	GY/BK	CKP Sensor Signal	Digital Signal: 0-5-0v
25	PK	SCI Transmit	0v
26	PK/BR	CCD Bus (+)	Digital Signal: 0-5-0v
27	BR	A/C Damped Pressure Switch	A/C Off: 12v, On: 1v
28	---	Not Used	---
29	WT/PK	Brake Switch Sense Signal	Brake Off: 0v, On: 12v
30	BR/YL	PNP Switch Sense Signal	In P/N: 0v, Others: 5v
31	DB/PK	Low Speed Fan Control	Relay Off: 12v, On: 1v
32	---	Not Used	---
33	TN/RD	Speed Control Vacuum Solenoid Control	Vacuum Increasing: 1v
34	DB/OR	A/C Clutch Relay Control	Relay Off: 12v, On: 1v
35	GY/YL	EGR Solenoid Control (Cal)	12v, 55 mph: 1v
36-37	---	Not Used	---
38	GY	Injector Control 5 Driver	1-4 ms
39	GY/RD	AIS Motor 3 Control	Pulse Signal: 0.8-11v
40	BR/WT	AIS Motor 1 Control	Pulse Signal: 0.8-11v

Standard Colors and Abbreviations

Abbreviation	Color	Abbreviation	Color	Abbreviation	Color
BK	Black	GY	Gray	RD	Red
BL	Blue	GN	Green	TN	Tan
BR	Brown	LG	Light Green	VT	Violet
DB	Dark Blue	OR	Orange	WT	White
DG	Dark Green	PK	Pink	YL	Yellow

1993 Town & Country 3.3L V6 MFI VIN R A/T 60 Pin Connector

PCM Pin #	Wire Color	Circuit Description (60 Pin)	Value at Hot Idle
41	BK/DG	HO2S-11 (B1 S1) Signal	0.1-1.1v
42-43	---	Not Used	---
44	TN/YL	CMP Sensor Signal	Digital Signal: 0-5-0v
45	LG	SCI Receive	5v
46	WT/BK	CCD Bus (-)	<0.050v
47	WT/OR	Vehicle Speed Signal	Digital Signal
48-50	---	Not Used	---
51	DB/YL	ASD Relay Output (B+)	12-14v
52	PK/BK	EVAP Purge Solenoid Control	Solenoid Off: 12v, On: 1v
53	LG/RD	Speed Control Vent Solenoid Control	Vacuum Decreasing: 1v
54	---	Not Used	---
55	YL	High Speed Fan Control	Relay Off: 12v, On: 1v
56	---	Not Used	---
57	DG/OR	Fuel Injector Power Supply	12-14v
58	BR/DB	Injector 6 Driver Control	1-4 ms
59	PK/BK	AIS Motor 4 Control	Pulse Signal: 0.8-11v
60	YL/BK	AIS Motor 2 Control	Pulse Signal: 0.8-11v

Pin Connector Graphic

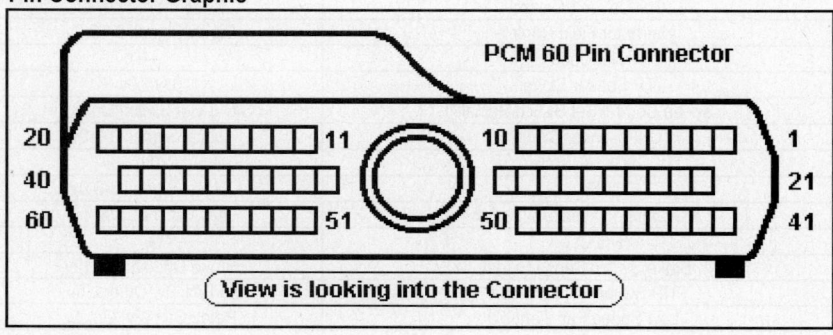

1996 Town & Country 3.3L V6 MFI VIN R 'C1' Connector

PCM Pin #	Wire Color	Circuit Description (40 Pin)	Value at Hot Idle
1	---	Not Used	---
2	RD/YL	Coil Driver 3 (Cyl 3 & 6)	5°, 55 mph: 8° dwell
3	DB/TN	Coil Driver 2 (Cyl 2 & 5)	5°, 55 mph: 8° dwell
4	DG	Generator Field Driver	Digital Signal: 0-12-0v
5	YL/RD	Speed Control Power Supply	12-14v
6	DG/OR	ASD Relay Output (B+)	12-14v
7	YL/WT	Injector 3 Driver Control	1-4 ms
8	TN	Smart Start Relay Control	KOEC: 9-11v
9	---	Not Used	---
10	BK/TN	Power Ground	<0.1v
11	GY/RD	Coil Driver 1 (Cyl 1 & 4)	5°, 55 mph: 8° dwell
12	---	Not Used	---
13	WT/DB	Injector 1 Driver Control	1-4 ms
14	BR/DB	Injector 6 Driver Control	1-4 ms
15	GY	Injector 5 Driver Control	1-4 ms
16	LB/BR	Injector 4 Driver Control	1-4 ms
17	TN	Injector 2 Driver Control	1-4 ms
18-19	---	Not Used	---
20	WT/BK	Ignition Switch Output (B+)	12-14v
21-25	---	Not Used	---
26	TN/BK	ECT Sensor Signal	At 180°F: 2.80v
27-29	---	Not Used	---
30	BK/DG	HO2S-11 (B1 S1) Signal	0.1-1.1v
31	---	Not Used	---
32	GY/BK	CKP Sensor Signal	Digital Signal: 0-5-0v
33	TN/YL	CMP Sensor Signal	Digital Signal: 0-5-0v
34	---	Not Used	---
35	OR/DB	TP Sensor Signal	0.6-1.0v
36	DG/RD	MAP Sensor Signal	1.5-1.7v
37-39	---	Not Used	---
40	GY/YL	EGR Solenoid Control	12v, 55 mph: 1v

Pin Connector Graphic

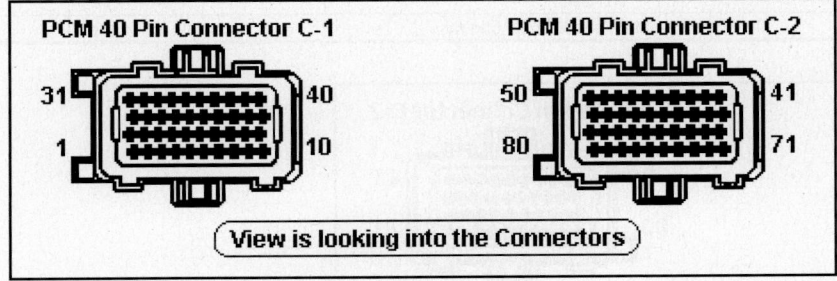

1996 Town & Country 3.3L V6 MFI VIN R 'C2' Connector

PCM Pin #	Wire Color	Circuit Description (40 Pin)	Value at Hot Idle
41	RD/LG	Speed Control Set Switch Signal	S/C & Set Switch On: 3.8v
42	DB	A/C Pressure Switch Signal	A/C On: 0.451-4.850v
43	BK/LB	Sensor Ground	<0.050v
44	OR	8-Volt Supply	7.9-8.1v
45	DB	A/C Switch Sense Signal	A/C Off: 12v, On: 1v
46	RD/WT	Fused Battery Power (B+)	12-14v
47	---	Not Used	---
48	BR/WT	IAC 1 Driver Control	Pulse Signal: 0.8-11v
49	YL/BK	IAC 2 Driver Control	Pulse Signal: 0.8-11v
50	BK/TN	Power Ground	<0.1v
52-56	---	Not Used	---
51	TN/WT	HO2S-12 (B1 S2) Signal	0.1-1.1v
57	GY/RD	IAC 3 Driver Control	Pulse Signal: 0.8-11v
58	PK/BK	IAC 4 Driver Control	Pulse Signal: 0.8-11v
59	PK/BR	CCD Bus (+)	Digital Signal: 0-5-0v
60	WT/BK	CCD Bus (-)	<0.050v
61	PK/WT	5-Volt Supply	4.9-5.1v
62	WT/PK	Brake Switch Sense Signal	Brake Off: 0v, On: 12v
63	YL/DG	Torque Management Request	Digital Signals
64	DB/OR	A/C Clutch Relay Control	Relay Off: 12v, On: 1v
65	PK	SCI Transmit	0v
66	WT/OR	Vehicle Speed Signal	Digital Signal
67	DB/YL	Auto Shutdown Relay Control	Relay Off: 12v, On: 1v
68	PK/BK	EVAP Purge Solenoid Control	PWM Signal: 0-12-0v
69-71	---	Not Used	---
72	YL/BK	LDP Switch Sense Signal	LDP Switch Closed: 0v
73	LG/DB	Radiator Fan Control Relay	Relay Off: 12v, On: 1v
74	BR	Fuel Pump Relay Control	Relay Off: 12v, On: 1v
75	LG	SCI Receive	5v
76	BK/WT	PNP Switch Sense Signal	In P/N: 0v, Others: 5v
77	WT/DG	LDP Solenoid Control	PWM Signal: 0-12-0v
78	TN/RD	Speed Control Vacuum Solenoid Control	Vacuum Increasing: 1v
79	---	Not Used	---
80	LG/RD	Speed Control Vent Solenoid Control	Vacuum Decreasing: 1v

Pin Connector Graphic
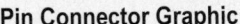

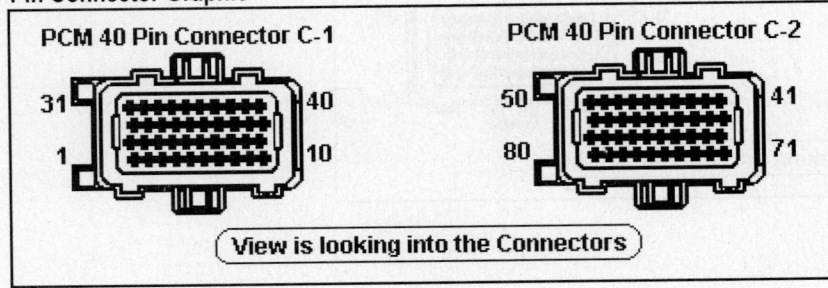

PCM 40 Pin Connector C-1

PCM 40 Pin Connector C-2

31   40
1   10

50   41
80   71

View is looking into the Connectors

1997-98 Town & Country 3.3L V6 MFI VIN R 'C1' Connector

PCM Pin #	Wire Color	Circuit Description (40 Pin)	Value at Hot Idle
1	---	Not Used	---
2	RD/YL	Coil Driver 3 (Cyl 3 & 6)	5°, 55 mph: 8° dwell
3	DB/TN	Coil Driver 2 (Cyl 2 & 5)	5°, 55 mph: 8° dwell
4	DG	Generator Field Driver	Digital Signal: 0-12-0v
5	YL/RD	Speed Control Power Supply	12-14v
6	DG/OR	ASD Relay Output (B+)	12-14v
7	YL/WT	Injector 3 Driver Control	1-4 ms
8	TN	Smart Start Relay Control	KOEC: 9-11v
9	---	Not Used	---
10	BK/TN	Power Ground	<0.1v
11	GY/RD	Coil Driver 1 (Cyl 1 & 4)	5°, 55 mph: 8° dwell
12	---	Not Used	---
13	WT/DB	Injector 1 Driver Control	1-4 ms
14	BR/DB	Injector 6 Driver Control	1-4 ms
15	GY	Injector 5 Driver Control	1-4 ms
16	LB/BR	Injector 4 Driver Control	1-4 ms
17	TN	Injector 2 Driver Control	1-4 ms
18-19	---	Not Used	---
20	WT/BK	Ignition Switch Output (B+)	12-14v
21, 23	---	Not Used	---
22	BK/PK	Malfunction Indicator Lamp	MIL Off: 12v, On: 1v
24	DB/LG	Knock Sensor Signal	No knock: 2.5v DC
25	---	Not Used	---
26	TN/BK	ECT Sensor Signal	At 180°F: 2.80v
27-29	---	Not Used	---
30	BK/DG	HO2S-11 (B1 S1) Signal	0.1-1.1v
31	---	Not Used	---
32	GY/BK	CKP Sensor Signal	Digital Signal: 0-5-0v
33	TN/YL	CMP Sensor Signal	Digital Signal: 0-5-0v
34	---	Not Used	---
35	OR/DB	TP Sensor Signal	0.6-1.0v
36	DG/RD	MAP Sensor Signal	1.5-1.7v
37-39	---	Not Used	---
40	GY/YL	EGR Solenoid Control	12v, 55 mph: 1v

Pin Connector Graphic

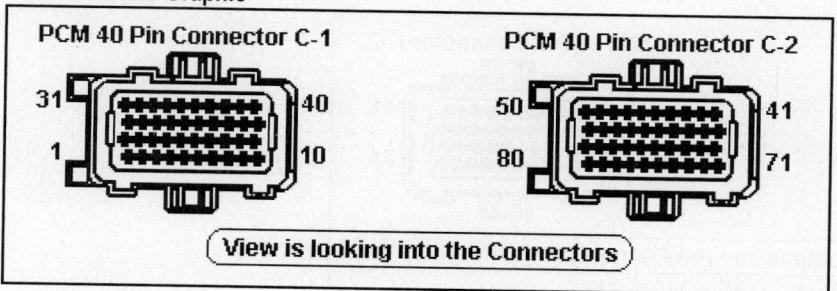

PCM 40 Pin Connector C-1

31   40
1   10

PCM 40 Pin Connector C-2

50   41
80   71

View is looking into the Connectors

1997-98 Town & Country 3.3L V6 MFI VIN R 'C2' Connector

PCM Pin #	Wire Color	Circuit Description (40 Pin)	Value at Hot Idle
41	RD/LG	Speed Control Set Switch Signal	S/C & Set Switch On: 3.8v
42	DB	A/C Pressure Switch Signal	A/C On: 0.451-4.850v
43	BK/LB	Sensor Ground	<0.050v
44	OR	8-Volt Supply	7.9-8.1v
45	DG	A/C Switch Sense Signal	A/C Off: 12v, On: 1v
46	RD/WT	Fused Battery Power (B+)	12-14v
47	---	Not Used	---
48	BR/WT	IAC 3 Driver Control	Pulse Signal: 0.8-11v
49	YL/BK	IAC 2 Driver Control	Pulse Signal: 0.8-11v
50	BK/TN	Power Ground	<0.1v
51	TN/WT	HO2S-12 (B1 S2) Signal	0.1-1.1v
52-56	---	Not Used	---
57	GY/RD	IAC 1 Driver Control	Pulse Signal: 0.8-11v
58	PK/BK	IAC 4 Driver Control	Pulse Signal: 0.8-11v
59	PK/BR	CCD Bus (+)	Digital Signal: 0-5-0v
60	WT/BK	CCD Bus (-)	<0.050v
61	PK/WT	5-Volt Supply	4.9-5.1v
62	WT/PK	Brake Switch Sense Signal	Brake Off: 0v, On: 12v
63	YL/DG	Torque Management Request	Digital Signals
64	DB/OR	A/C Clutch Relay Control	Relay Off: 12v, On: 1v
65	PK	SCI Transmit	0v
66	WT/OR	Vehicle Speed Signal	Digital Signal
67	DB/YL	Auto Shutdown Relay Control	Relay Off: 12v, On: 1v
68	PK/BK	EVAP Purge Solenoid Control	PWM Signal: 0-12-0v
69-71	---	Not Used	---
72	DB/WT	LDP Switch Sense Signal	LDP Switch Closed: 0v
73	LG/DG	Radiator Fan Control Relay	Relay Off: 12v, On: 1v
74	BR	Fuel Pump Relay Control	Relay Off: 12v, On: 1v
75	LG	SCI Receive	5v
76	BK/Y	PNP Switch Sense Signal	In P/N: 0v, Others: 5v
77	WT/DG	LDP Solenoid Control	PWM Signal: 0-12-0v
78	TN/RD	Speed Control Vacuum Solenoid Control	Vacuum Increasing: 1v
79	---	Not Used	---
80	LG/RD	Speed Control Vent Solenoid Control	Vacuum Decreasing: 1v

Pin Connector Graphic

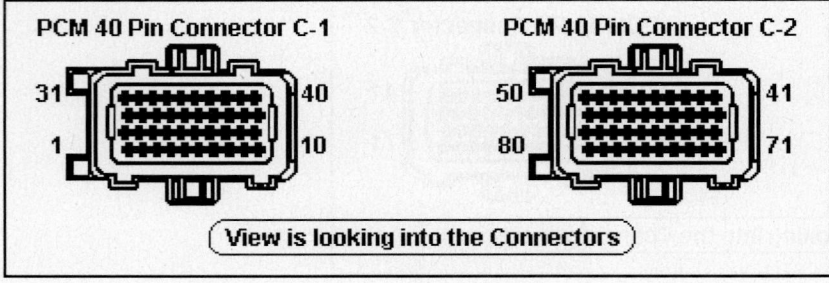

PCM 40 Pin Connector C-1

31 · · · 40
1 · · · 10

PCM 40 Pin Connector C-2

50 · · · 41
80 · · · 71

View is looking into the Connectors

1999-2000 Town & Country 3.3L MFI VIN R 'C1' Connector

PCM Pin #	Wire Color	Circuit Description (40 Pin)	Value at Hot Idle
1, 4	---	Not Used	---
2	RD/YL	Coil Driver 3 (Cyl 3 & 6)	5°, 55 mph: 8° dwell
3	DB/TN	Coil Driver 2 (Cyl 2 & 5)	5°, 55 mph: 8° dwell
5	YL/RD	Speed Control Power Supply	12-14v
6	DG/OR	ASD Relay Output (B+)	12-14v
7	YL/WT	Injector 3 Driver Control	1-4 ms
8	DG	Generator Field Driver	Digital Signal: 0-12-0v
9	---	Not Used	---
10	BK/TN	Power Ground	<0.1v
11	GY/RD	Coil Driver 1 (Cyl 1 & 4)	5°, 55 mph: 8° dwell
12	---	Not Used	---
13	WT/DB	Injector 1 Driver Control	1-4 ms
14	BR/DB	Injector 6 Driver Control	1-4 ms
15	GY	Injector 5 Driver Control	1-4 ms
16	LB/BR	Injector 4 Driver Control	1-4 ms
17	TN/WT	Injector 2 Driver Control	1-4 ms
18-19	---	Not Used	---
20	WT/BK	Ignition Switch Output (B+)	12-14v
21, 23	---	Not Used	---
22	BK/PK	Malfunction Indicator Lamp	MIL Off: 12v, On: 1v
24	---	Not Used	---
25	DB/LG	Knock Sensor Signal	No knock: 2.5v DC
26	TN/BK	ECT Sensor Signal	At 180°F: 2.80v
27	BK/OR	HO2S-11 Ground	<0.1v
28-29	---	Not Used	---
30	BK/DG	HO2S-11 (B1 S1) Signal	0.1-1.1v
31	TN	Smart Start Relay Control	KOEC: 9-11v
32	GY/BK	CKP Sensor Signal	Digital Signal: 0-5-0v
33	TN/YL	CMP Sensor Signal	Digital Signal: 0-5-0v
34	---	Not Used	---
35	OR/DB	TP Sensor Signal	0.6-1.0v
36	DG/RD	MAP Sensor Signal	1.5-1.7v
37, 39	---	Not Used	---
38	DG/LB	A/C Switch Sense Signal	A/C Off: 12v, On: 1v
40	GY/YL	EGR Solenoid Control	12v, 55 mph: 1v

Pin Connector Graphic

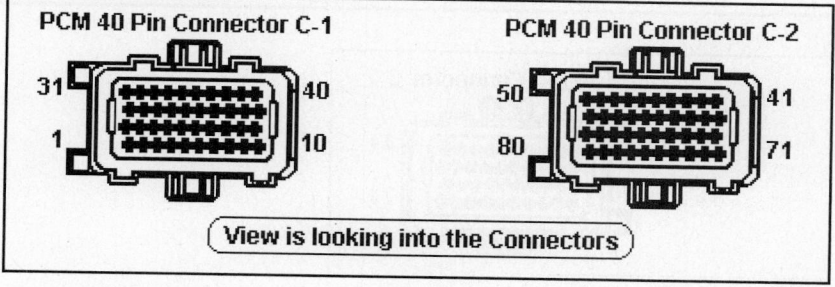

PCM 40 Pin Connector C-1

31   40
1   10

PCM 40 Pin Connector C-2

50   41
80   71

View is looking into the Connectors

1999-2000 Town & Country 3.3L MFI VIN R 'C2' Connector

PCM Pin #	Wire Color	Circuit Description (40 Pin)	Value at Hot Idle
41	RD/LG	Speed Control Set Switch Signal	S/C & Set Switch On: 3.8v
42	DB	A/C Pressure Switch Signal	A/C On: 0.451-4.850v
43	BK/LB	Sensor Ground	<0.050v
44	OR	8-Volt Supply	7.9-8.1v
45	---	Not Used	---
46	RD/WT	Fused Battery Power (B+)	12-14v
47	---	Not Used	---
48	BR/WT	IAC 3 Driver Control	Pulse Signal: 0.8-11v
49	YL/BK	IAC 2 Driver Control	Pulse Signal: 0.8-11v
50	BK/TN	Power Ground	<0.1v
51	TN/WT	HO2S-12 (B1 S2) Signal	0.1-1.1v
52-55	---	Not Used	---
56	TN/RD	Speed Control Vacuum Solenoid Control	Vacuum Increasing: 1v
57	GY/RD	IAC 1 Driver Control	Pulse Signal: 0.8-11v
58	PK/BK	IAC 4 Driver Control	Pulse Signal: 0.8-11v
59	PK/BR	CCD Bus (+)	Digital Signal: 0-5-0v
60	WT/BK	CCD Bus (-)	<0.050v
61	PK/WT	5-Volt Supply	4.9-5.1v
62	WT/PK	Brake Switch Sense Signal	Brake Off: 0v, On: 12v
63	YL/DG	Torque Management Request	Digital Signals
64	DB/OR	A/C Clutch Relay Control	Relay Off: 12v, On: 1v
65	PK	SCI Transmit Signal	0v
66	WT/OR	Vehicle Speed Signal	Digital Signal
67	DB/YL	Auto Shutdown Relay Control	Relay Off: 12v, On: 1v
68	PK/BK	EVAP Purge Solenoid Control	PWM Signal: 0-12-0v
69	---	Not Used	---
70 ('99)	DG/LG	EVAP Purge Solenoid Sense	0-1v
70	PK/RD	EVAP Purge Solenoid Sense	0-1v
71	---	Not Used	---
72	DB/WT	LDP Switch Sense Signal	LDP Switch Closed: 0v
73	LG/DG	Radiator Fan Control Relay	Relay Off: 12v, On: 1v
74	BR	Fuel Pump Relay Control	Relay Off: 12v, On: 1v
75	LG	SCI Receive Signal	5v
76	BK/Y	PNP Switch Sense Signal	In P/N: 0v, Others: 5v
77	WT/DG	LDP Solenoid Control	PWM Signal: 0-12-0v
78-79	---	Not used	---
80	LG/RD	Speed Control Vent Solenoid Control	Vacuum Decreasing: 1v

Pin Connector Graphic

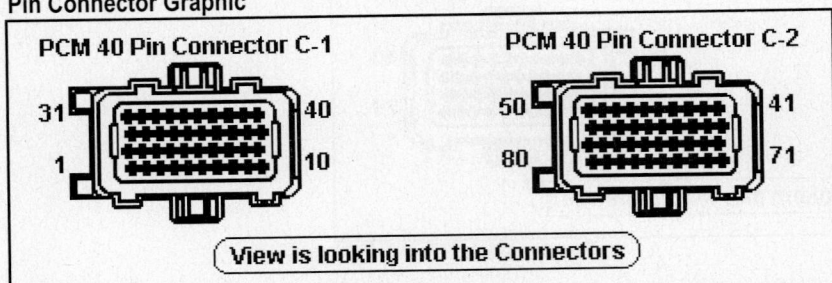

2002-03 Town and Country 3.3L V6 MFI VIN 3 NGC A/T 'C1' Black Connector

PCM Pin #	Wire Color	Circuit Description (40 Pin)	Value at Hot Idle
1-8	---	Not Used	---
9	BK/BR	Power Ground	<0.1v
10	---	Not Used	---
11	PK/GY	Ignition Switch Output (Start-Run)	12-14v
12	PK/WT	Ignition Switch Output (Run/Start)	12-14v
13-17	---	Not Used	---
18	BK/DG	Power Ground	<0.1v
19, 21	---	Not Used	---
20	VT/GY	Engine Oil Pressure Sensor	1.6v at 24 psi
22	VT/LG	Ambient Air Temperature Sensor	At 100°F: 1.83v
23-24	---	Not Used	---
25	WT/LG	SCI Receive Signal	5v
26	WT/BK	Flash Program Enable	---
27-28	---	Not Used	---
29	OR/RD	Fused B+	12-14v
30	YL	Ignition Switch Output (Start)	12-14v
31	DB/YL	HO2S-12 (B1 S2) Signal	0.1-1.1v
32	DB/DG	Oxygen Sensor Ground	<0.050v
33-35	---	Not Used	---
36	WT/BR	SCI Transmit (PCM)	0v
37	DG/YL	SCI Transmit (TCM)	0v
38	WT/VT	PCI Data Bus (J1850)	Digital Signals: 0-7-0v

2002-03 Town and Country 3.3L V6 MFI VIN 3 NGC A/T 'C2' Black Connector

PCM Pin #	Wire Color	Circuit Description (40 Pin)	Value at Hot Idle
1-8	---	Not Used	---
9	DB/TN	Coil 2 Driver	5°, 55 mph: 8° dwell
10	DB/DG	Coil 1 Driver	5°, 55 mph: 8° dwell
11	BR/TN	Injector 4 Driver Control	1-4 ms
12	BR/LB	Injector 3 Driver Control	1-4 ms
13	BR/DB	Injector 2 Driver Control	1-4 ms
14	BR/YL	Injector 1 Driver Control	1-4 ms
15-17	---	Not Used	---
18	BR/LG	HO2S-11 (B1 S1) Heater Control	PWM Signal: 0-12-0v
19	BR/GY	Generator Field Driver	Digital Signal: 0-12-0v
20	VT/OR	ECT Sensor Signal	At 180°F: 2.80v
21	BR/OR	TP Sensor Signal	0.6-1.0v
22, 26	---	Not Used	---
23	VT/BR	MAP Sensor Signal	1.5-1.6v
24	BR/LG	Knock Sensor Return	<0.050v
25	DB/YL	Knock Sensor Signal	0.080v AC
27	DB/DG	Sensor Ground	<0.1v
28	BR/VT	IAC Motor Return	<0.050v
29	PK/YL	5-Volt Supply	4.9-5.1v
30	DB/LG	IAT Sensor Signal	At 100°F: 1.83v
31	DB/LG	HO2S-11 (B1 S1) Sensor	0.1-1.1v
32	BR/DG	HO2S-11 (B1 S1) Ground (Up)	<0.050v
33, 36-37	---	Not Used	---
34	BR/LB	CKP Sensor Signal	Digital Signal: 0-5-0v
35	DB/GY	CMP Sensor Signal	Digital Signal: 0-5-0v
38	VT/GY	IAC 1 Motor Control	Pulse Signals

Pin Connector Graphic

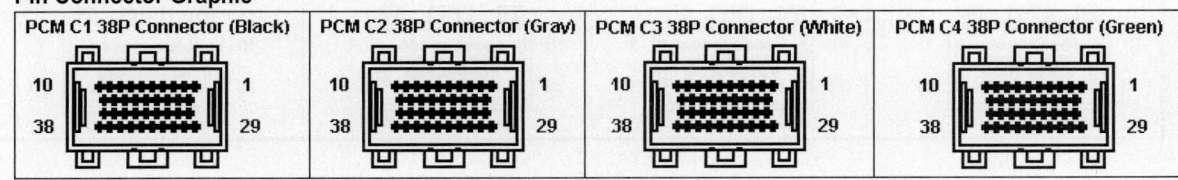

PCM C1 38P Connector (Black)	PCM C2 38P Connector (Gray)	PCM C3 38P Connector (White)	PCM C4 38P Connector (Green)

2002-03 Town and Country 3.3L V6 MFI VIN 3 NGC A/T 'C3' Black Connector

PCM Pin #	Wire Color	Circuit Description (38 Pin)	Value at Hot Idle
1-2	---	Not Used	---
3	BR/WT	ASD Relay Control	Relay Off: 12v, On: 1v
4	---	Not Used	---
5	VT/OR	Speed Control Vent Solenoid	Vacuum Decreasing: 1v
6	BR/VT	Radiator Fan Relay Control	Relay Off: 12v, On: 1v
7	VT/YL	Speed Control Power Supply	12-14v
8	WT/DG	Natural Vacuum Leak Detection Solenoid	Solenoid Off: 12v, On: 1v
9	BR/WT	HO2S-12 (Bank 1 Sensor 2) Heater Control	---
11	LB/OR	A/C Clutch Relay Control	Relay Off: 12v, On: 1v
12	VT/YL	Speed Control Vacuum Solenoid	Vacuum Increasing: 1v
13-18	---	Not Used	---
19	BR/WT	Automatic Shutdown Relay Output	12-14v
20	DB/WT	EVAP Purge Solenoid Control	PWM Signal: 0-12-0v
21	---	Not Used	---
23	DG/WT	Brake Switch Sense	Brake Off: 0v, On: 12v
24-27	---	Not Used	---
28	BR/WT	Automatic Shutdown Relay Output	12-14v
29	DB/BR	EVAP Purge Solenoid Sense	0-1v
30	---	Not Used	---
31	LB/BR	A/C Pressure Signal	A/C On: 0.451-4.850v
32	DB/YL	Battery Temperature Sensor	At 100ºF: 1.83v
33, 36	---	Not Used	---
34	VT	Speed Control Set Switch Signal	S/C & Set Switch On: 3.8v
35	VT/WT	Natural Vacuum Leak Detection Switch Sense	0.1v
37	BR	Fuel Pump Relay Control	Relay Off: 12v, On: 1v
38	DG/OR	Starter Relay Control	KOEC: 9-11v

2003 Town and Country 2.4L I4 DOHC MFI VIN B All 'C4' Green Connector

PCM Pin #	Wire Color	Circuit Description (38 Pin)	Value at Hot Idle
1	YL/LG	Overdrive Solenoid Control	Solenoid Off: 12v, On: 1v
2	YL/LB	Underdrive Solenoid Control	Solenoid Off: 12v, On: 1v
3-5, 7-9, 11, 13	---	Not Used	---
6	YL/DB	2-4 Solenoid Control	Solenoid Off: 12v, On: 1v
10	DG/WT	Low/Reverse Solenoid Control	Solenoid Off: 12v, On: 1v
12, 13	BK/LG	Power Ground	<0.050v
15	DG/LG	TRS T1 Sense	<0.050v
16	DG/DB	TRS T3 Sense	<0.050v
17, 20-21, 23-26	---	Not Used	---
18	YL/BR	Transmission Control Relay Control	Relay Off: 12v, On: 1v
19, 38	YL/OR	Transmission Control Relay Output	12-14v
23-26	---	Not Used	---
27	DG/GY	TRS T41 Sense	<0.050v
28	YL/OR	Transmission Control Relay Output	12-14v
29	YL/TN	Low/Reverse Pressure Switch Sense	12-14v
30	YL/DG	2-4 Pressure Switch Sense	In Low/Reverse: 2-4v
31, 36	---	Not Used	---
32	DG/BR	Output Speed Sensor Signal	In 2-4 Position: 2-4v
33	DG/WT	Input Speed Sensor Signal	Moving: AC voltage
34	DG/VT	Speed Sensor Ground	Moving: AC voltage
35	DG/OR	Transmission Temperature Sensor Signal	3.2-3.4v at 104ºF
37	DG/YL	TRS T42 Sense	In PRNL: 0v, Others 5v

Pin Connector Graphic

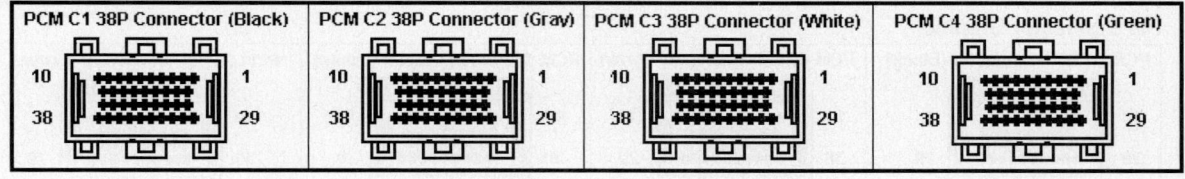

PCM C1 38P Connector (Black)   PCM C2 38P Connector (Gray)   PCM C3 38P Connector (White)   PCM C4 38P Connector (Green)

1994-95 Town & Country 3.8L V6 MFI VIN L 60 Pin Connector

PCM Pin #	Wire Color	Circuit Description (60 Pin)	Value at Hot Idle
1	DG/RD	MAP Sensor Signal	1.5-1.7v
2	TN/BK	ECT Sensor Signal	At 180°F: 2.80v
3	RD/WT	Fused Battery Power (B+)	12-14v
4	BK/LB	Sensor Ground	<0.050v
5	BK/WT	Sensor Ground	<0.050v
6	PK/WT	5-Volt Supply	4.9-5.1v
7	OR	8-Volt Supply	7.9-8.1v
9	DB/WT	Ignition Switch Output (B+)	12-14v
11	BK/TN	Power Ground	<0.1v
12	BK/TN	Power Ground	<0.1v
13	LB/BR	Injector Control 4 Driver	1-4 ms
14	YL/WT	Injector Control 3 Driver	1-4 ms
15	TN	Injector Control 2 Driver	1-4 ms
16	LB	Injector Control 1 Driver	1-4 ms
17	DB/YL	Coil Driver 2 Control	5°, 55 mph: 8° dwell
18	DB/GY	Coil Driver 3 Control	5°, 55 mph: 8° dwell
19	BK/GY	Coil Driver 1 Control	5°, 55 mph: 8° dwell
20	DG	Alternator Field Control	Digital Signal: 0-12-0v
22	DB	TP Sensor Signal	0.6-1.0v
23	RD	Speed Control Set Switch Signal	S/C & Set Switch On: 3.8v
24	GY/PK	CKP Sensor Signal	Digital Signal: 0-5-0v
25	PK	SCI Transmit	0v
26	PK/BR	CCD Bus (+)	Digital Signal: 0-5-0v
27	BR	A/C Damped Pressure Switch	A/C Off: 12v, On: 1v
29	WT/PK	Brake Switch Sense Signal	Brake Off: 0v, On: 12v
30	BK/OR	PNP Switch Sense Signal	In P/N: 0v, Others: 5v
31	DB/PK	Low Speed Fan Control	Relay Off: 12v, On: 1v
32	---	Not Used	---
33	TN/RD	Speed Control Vacuum Solenoid Control	Vacuum Increasing: 1v
34	DB/OR	A/C Clutch Relay Control	Relay Off: 12v, On: 1v
35	GY/YL	EGR Solenoid Control	12v, 55 mph: 1v
36-37	---	Not Used	---
38	GY	Injector Control 5 Driver	1-4 ms
39	GY/RD	AIS Motor 3 Control	Pulse Signal: 0.8-11v
40	BR/WT	AIS Motor 1 Control	Pulse Signal: 0.8-11v

Standard Colors and Abbreviations

Abbreviation	Color	Abbreviation	Color	Abbreviation	Color
BK	Black	GY	Gray	RD	Red
BL	Blue	GN	Green	TN	Tan
BR	Brown	LG	Light Green	VT	Violet
DB	Dark Blue	OR	Orange	WT	White
DG	Dark Green	PK	Pink	YL	Yellow

1994-95 Town & Country 3.8L V6 MFI VIN L 60 Pin Connector

PCM Pin #	Wire Color	Circuit Description (60 Pin)	Value at Hot Idle
41	BK/DG	HO2S-11 (B1 S1) Signal	0.1-1.1v
42-43	---	Not Used	---
44	TN/YL	CMP Sensor Signal	Digital Signal: 0-5-0v
45	LG	SCI Receive	5v
46	WT/BK	CCD Bus (-)	<0.050v
47	WT/OR	Vehicle Speed Signal	Digital Signal
50	---	Not Used	---
51	DB/YL	ASD Relay Output (B+)	12-14v
52	PK/BK	EVAP Purge Solenoid Control	Solenoid Off: 12v, On: 1v
53	LG/RD	Speed Control Vent Solenoid Control	Vacuum Decreasing: 1v
55	Y	High Speed Fan Control	Relay Off: 12v, On: 1v
56	---	Not Used	---
57	DG/OR	Fuel Injector Power Supply	12-14v
58	BR/DB	Injector 6 Driver Control	1-4 ms
59	PK/BK	AIS Motor 4 Control	Pulse Signal: 0.8-11v
60	YL/BK	AIS Motor 2 Control	Pulse Signal: 0.8-11v

Pin Connector Graphic

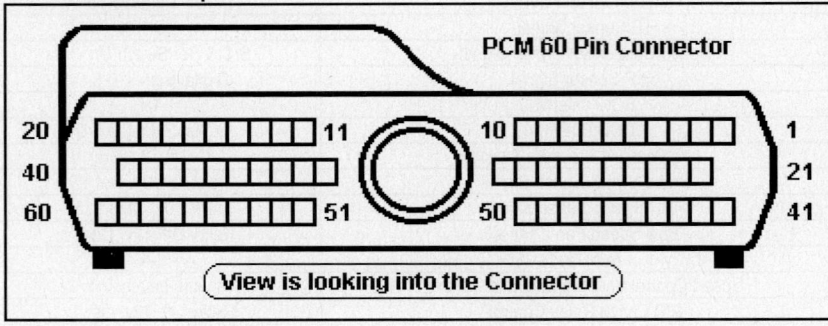

PCM 60 Pin Connector

View is looking into the Connector

1996 Town & Country 3.8L V6 MFI VIN L 'C1' Connector

PCM Pin #	Wire Color	Circuit Description (40 Pin)	Value at Hot Idle
1	---	Not Used	---
2	RD/YL	Coil Driver 3 (Cyl 3 & 6)	5°, 55 mph: 8° dwell
3	DB/TN	Coil Driver 2 (Cyl 2 & 5)	5°, 55 mph: 8° dwell
4	DG	Generator Field Driver	Digital Signal: 0-12-0v
5	YL/RD	Speed Control Power Supply	12-14v
6	DG/OR	ASD Relay Output (B+)	12-14v
7	YL/WT	Injector 3 Driver Control	1-4 ms
8	TN	Smart Start Relay Control	KOEC: 9-11v
9	---	Not Used	---
10	BK/TN	Power Ground	<0.1v
11	GY	Coil Driver 1 (Cyl 1 & 4)	5°, 55 mph: 8° dwell
12	---	Not Used	---
13	WT/DB	Injector 1 Driver Control	1-4 ms
14	BR/DB	Injector 6 Driver Control	1-4 ms
15	GY	Injector 5 Driver Control	1-4 ms
16	LB/BR	Injector 4 Driver Control	1-4 ms
17	TN	Injector 2 Driver Control	1-4 ms
18-19	---	Not Used	---
20	WT/BK	Ignition Switch Output (B+)	12-14v
21-23	---	Not Used	---
24	DB/LG	Knock Sensor Signal	No knock: 2.5v DC
25	---	Not Used	---
26	TN/BK	ECT Sensor Signal	At 180°F: 2.80v
27-29	---	Not Used	---
30	BK/DG	HO2S-11 (B1 S1) Signal	0.1-1.1v
31	---	Not Used	---
32	GY/BK	CKP Sensor Signal	Digital Signal: 0-5-0v
33	TN/YL	CMP Sensor Signal	Digital Signal: 0-5-0v
34	---	Not Used	---
35	OR/DB	TP Sensor Signal	0.6-1.0v
36	DG/RD	MAP Sensor Signal	1.5-1.7v
37-39	---	Not Used	---
40	GY/YL	EGR Solenoid Control	12v, 55 mph: 1v

Pin Connector Graphic

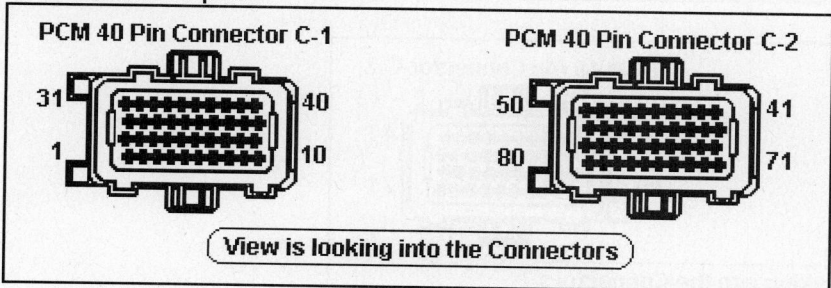

1996 Town & Country 3.8L V6 MFI VIN L 'C2' Connector

PCM Pin #	Wire Color	Circuit Description (40 Pin)	Value at Hot Idle
41	RD/LG	Speed Control Set Switch Signal	S/C & Set Switch On: 3.8v
42	DB	A/C Pressure Switch Signal	A/C On: 0.451-4.850v
43	BK/LB	Sensor Ground	<0.050v
44	OR	8-Volt Supply	7.9-8.1v
45	DB	A/C Switch Sense Signal	A/C Off: 12v, On: 1v
46	RD/WT	Fused Battery Power (B+)	12-14v
47	---	Not Used	---
48	BR/WT	IAC 1 Driver Control	Pulse Signal: 0.8-11v
49	YL/BK	IAC 2 Driver Control	Pulse Signal: 0.8-11v
50	BK/TN	Power Ground	<0.1v
51	TN/WT	HO2S-12 (B1 S2) Signal	0.1-1.1v
52-56	---	Not Used	---
57	GY/RD	IAC 3 Driver Control	Pulse Signal: 0.8-11v
58	PK/BK	IAC 4 Driver Control	Pulse Signal: 0.8-11v
59	PK/BR	CCD Bus (+)	Digital Signal: 0-5-0v
60	WT/BK	CCD Bus (-)	<0.050v
61	PK/WT	5-Volt Supply	4.9-5.1v
62	WT/PK	Brake Switch Sense Signal	Brake Off: 0v, On: 12v
63	YL/DG	Torque Management Request	Digital Signals
64	DB/OR	A/C Clutch Relay Control	Relay Off: 12v, On: 1v
65	PK	SCI Transmit	0v
66	WT/OR	Vehicle Speed Signal	Digital Signal
67	DB/YL	Auto Shutdown Relay Control	Relay Off: 12v, On: 1v
68	PK/BK	EVAP Purge Solenoid Control	PWM Signal: 0-12-0v
69-71	---	Not Used	---
72	YL/BK	LDP Switch Sense Signal	LDP Switch Closed: 0v
73	LG/DB	Radiator Fan Control Relay	Relay Off: 12v, On: 1v
74	BR	Fuel Pump Relay Control	Relay Off: 12v, On: 1v
75	LG	SCI Receive	5v
76	BK/WT	PNP Switch Sense Signal	In P/N: 0v, Others: 5v
77	WT/DG	LDP Solenoid Control	PWM Signal: 0-12-0v
78	TN/RD	Speed Control Vacuum Solenoid Control	Vacuum Increasing: 1v
79	---	Not Used	---
80	LG/RD	Speed Control Vent Solenoid Control	Vacuum Decreasing: 1v

Pin Connector Graphic

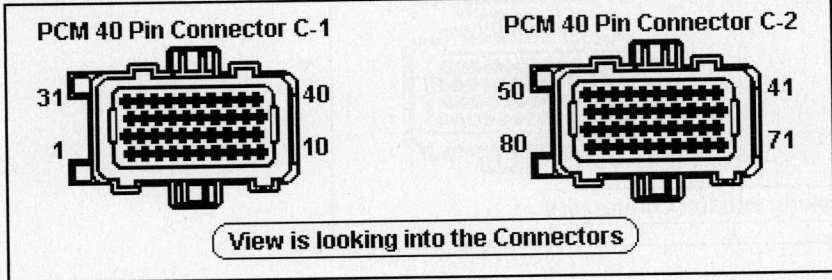

PCM 40 Pin Connector C-1

PCM 40 Pin Connector C-2

31   40   50   41

1   10   80   71

(View is looking into the Connectors)

1997-98 Town & Country 3.8L V6 MFI VIN L 'C1' Connector

PCM Pin #	Wire Color	Circuit Description (40 Pin)	Value at Hot Idle
1	---	Not Used	---
2	RD/YL	Coil Driver 3 (Cyl 3 & 6)	5°, 55 mph: 8° dwell
3	DB/TN	Coil Driver 2 (Cyl 2 & 5)	5°, 55 mph: 8° dwell
4	DG	Generator Field Driver	Digital Signal: 0-12-0v
5	YL/RD	Speed Control Power Supply	12-14v
6	DG/OR	ASD Relay Output (B+)	12-14v
7	YL/WT	Injector 3 Driver Control	1-4 ms
8	TN	Smart Start Relay Control	KOEC: 9-11v
9	---	Not Used	---
10	BK/TN	Power Ground	<0.1v
11	GY/RD	Coil Driver 1 (Cyl 1 & 4)	5°, 55 mph: 8° dwell
12	---	---	---
13	WT/DB	Injector 1 Driver Control	1-4 ms
14	BR/DB	Injector 6 Driver Control	1-4 ms
15	GY	Injector 5 Driver Control	1-4 ms
16	LB/BR	Injector 4 Driver Control	1-4 ms
17	TN	Injector 2 Driver Control	1-4 ms
18-19	---	Not Used	---
20	WT/BK	Ignition Switch Output (B+)	12-14v
21, 23	---	Not Used	---
22	BK/PK	Malfunction Indicator Lamp	MIL Off: 12v, On: 1v
24	DB/LG	Knock Sensor Signal	No knock: 2.5v DC
25	---	Not Used	---
26	TN/BK	ECT Sensor Signal	At 180°F: 2.80v
27-29	---	Not Used	---
30	BK/DG	HO2S-11 (B1 S1) Signal	0.1-1.1v
31	---	Not Used	---
32	GY/BK	CKP Sensor Signal	Digital Signal: 0-5-0v
33	TN/YL	CMP Sensor Signal	Digital Signal: 0-5-0v
34	---	Not Used	---
35	OR/DB	TP Sensor Signal	0.6-1.0v
36	DG/RD	MAP Sensor Signal	1.5-1.7v
37-39	---	Not Used	---
40	GY/YL	EGR Solenoid Control	12v, 55 mph: 1v

Pin Connector Graphic

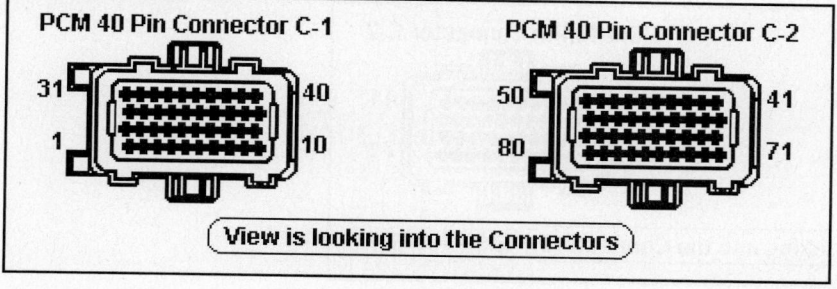

PCM 40 Pin Connector C-1

PCM 40 Pin Connector C-2

View is looking into the Connectors

1997-98 Town & Country 3.8L V6 MFI VIN L 'C2' Connector

PCM Pin #	Wire Color	Circuit Description (40 Pin)	Value at Hot Idle
41	RD/LG	Speed Control Set Switch Signal	S/C & Set Switch On: 3.8v
42	DB	A/C Pressure Switch Signal	A/C On: 0.451-4.850v
43	BK/LB	Sensor Ground	<0.050v
44	OR	8-Volt Supply	7.9-8.1v
45	DG	A/C Switch Sense Signal	A/C Off: 12v, On: 1v
46	RD/WT	Fused Battery Power (B+)	12-14v
47	---	Not Used	---
48	BR/WT	IAC 3 Driver Control	Pulse Signal: 0.8-11v
49	YL/BK	IAC 2 Driver Control	Pulse Signal: 0.8-11v
50	BK/TN	Power Ground	<0.1v
51	TN/WT	HO2S-12 (B1 S2) Signal	0.1-1.1v
52-56	---	Not Used	---
57	GY/RD	IAC 1 Driver Control	Pulse Signal: 0.8-11v
58	PK/BK	IAC 4 Driver Control	Pulse Signal: 0.8-11v
59	PK/BR	CCD Bus (+)	Digital Signal: 0-5-0v
60	WT/BK	CCD Bus (-)	<0.050v
61	PK/WT	5-Volt Supply	4.9-5.1v
62	WT/PK	Brake Switch Sense Signal	Brake Off: 0v, On: 12v
63	YL/DG	Torque Management Request	Digital Signals
64	DB/OR	A/C Clutch Relay Control	Relay Off: 12v, On: 1v
65	PK	SCI Transmit	0v
66	WT/OR	Vehicle Speed Signal	Digital Signal
67	DB/YL	Auto Shutdown Relay Control	Relay Off: 12v, On: 1v
68	PK/BK	EVAP Purge Solenoid Control	PWM Signal: 0-12-0v
69-71	---	Not Used	---
72	DB/WT	LDP Switch Sense Signal	LDP Switch Closed: 0v
73	LG/DG	Radiator Fan Control Relay	Relay Off: 12v, On: 1v
74	BR	Fuel Pump Relay Control	Relay Off: 12v, On: 1v
75	LG	SCI Receive	5v
76	BR/YL	PNP Switch Sense Signal	In P/N: 0v, Others: 5v
77	WT/DG	LDP Solenoid Control	PWM Signal: 0-12-0v
78	TN/RD	Speed Control Vacuum Solenoid Control	Vacuum Increasing: 1v
79	---	Not Used	---
80	LG/RD	Speed Control Vent Solenoid Control	Vacuum Decreasing: 1v

Pin Connector Graphic

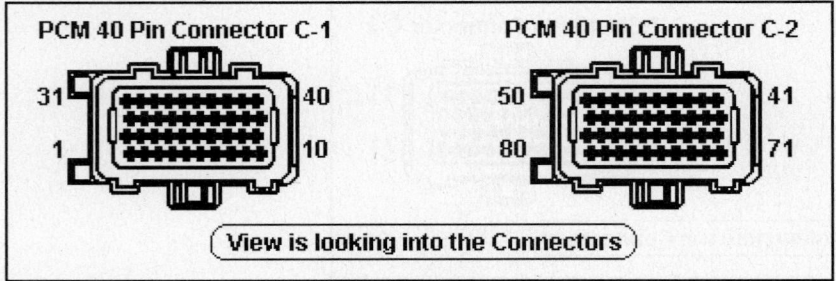

1999-2000 Town & Country 3.8L MFI VIN L 'C1' Connector

PCM Pin #	Wire Color	Circuit Description (40 Pin)	Value at Hot Idle
1	---	Not Used	---
2	RD/YL	Coil Driver 3 (Cyl 3 & 6)	5°, 55 mph: 8° dwell
3	DB/TN	Coil Driver 2 (Cyl 2 & 5)	5°, 55 mph: 8° dwell
4	---	Not Used	
5	YL/RD	Speed Control Power Supply	12-14v
6	DG/OR	ASD Relay Output (B+)	12-14v
7	YL/WT	Injector 3 Driver Control	1-4 ms
8	DG	Generator Field Driver	Digital Signal: 0-12-0v
9	---	Not Used	---
10	BK/TN	Power Ground	<0.1v
11	GY/RD	Coil Driver 1 (Cyl 1 & 4)	5°, 55 mph: 8° dwell
12	---	Not Used	---
13	WT/DB	Injector 1 Driver Control	1-4 ms
14	BR/DB	Injector 6 Driver Control	1-4 ms
15	GY	Injector 5 Driver Control	1-4 ms
16	LB/BR	Injector 4 Driver Control	1-4 ms
17	TN/WT	Injector 2 Driver Control	1-4 ms
18-19	---	Not Used	---
20	WT/BK	Ignition Switch Output (B+)	12-14v
21	---	Not Used	---
22	BK/PK	Malfunction Indicator Lamp	MIL Off: 12v, On: 1v
23-24	---	Not Used	---
25	DB/LG	Knock Sensor Signal	No knock: 2.5v DC
26	TN/BK	ECT Sensor Signal	At 180°F: 2.80v
27	BK/OR	HO2S-11 & HO2S-12 Ground	<0.1v
28-29	---	Not Used	---
30	BK/DG	HO2S-11 (B1 S1) Signal	0.1-1.1v
31	TN	Smart Start Relay Control	KOEC: 9-11v
32	GY/BK	CKP Sensor Signal	Digital Signal: 0-5-0v
33	TN/YL	CMP Sensor Signal	Digital Signal: 0-5-0v
34	---	Not Used	---
35	OR/DB	TP Sensor Signal	0.6-1.0v
36	DG/RD	MAP Sensor Signal	1.5-1.7v
37	---	Not Used	---
38	DG/LB	A/C Switch Sense Signal	A/C Off: 12v, On: 1v
39	---	Not Used	---
40	GY/YL	EGR Solenoid Control	12v, 55 mph: 1v

Pin Connector Graphic

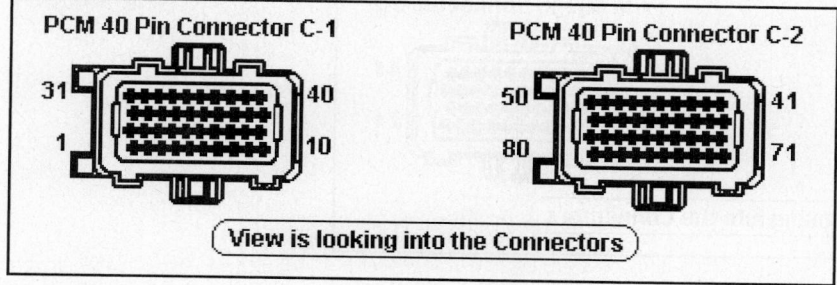

1999-2000 Town & Country 3.8L MFI VIN L 'C2' Connector

PCM Pin #	Wire Color	Circuit Description (40 Pin)	Value at Hot Idle
41	RD/LG	Speed Control Set Switch Signal	S/C & Set Switch On: 3.8v
42	DB	A/C Pressure Switch Signal	A/C On: 0.451-4.850v
43	BK/LB	Sensor Ground	<0.050v
44	OR	8-Volt Supply	7.9-8.1v
45	---	Not Used	---
46	RD/WT	Fused Battery Power (B+)	12-14v
47	---	Not Used	---
48	BR/WT	IAC 3 Driver Control	Pulse Signal: 0.8-11v
49	YL/BK	IAC 2 Driver Control	Pulse Signal: 0.8-11v
50	BK/TN	Power Ground	<0.1v
51	TN/WT	HO2S-12 (B1 S2) Signal	0.1-1.1v
52-55	---	Not Used	---
56	TN/RD	Speed Control Vacuum Solenoid Control	Vacuum Increasing: 1v
57	GY/RD	IAC 1 Driver Control	Pulse Signal: 0.8-11v
58	PK/BK	IAC 4 Driver Control	Pulse Signal: 0.8-11v
59	PK/BR	CCD Bus (+)	Digital Signal: 0-5-0v
60	WT/BK	CCD Bus (-)	<0.050v
61	PK/WT	5-Volt Supply	4.9-5.1v
62	WT/PK	Brake Switch Sense Signal	Brake Off: 0v, On: 12v
63	YL/DG	Torque Management Request	Digital Signals
64	DB/OR	A/C Clutch Relay Control	Relay Off: 12v, On: 1v
65	PK	SCI Transmit	0v
66	WT/OR	Vehicle Speed Signal	Digital Signal
67	DB/YL	Auto Shutdown Relay Control	Relay Off: 12v, On: 1v
68	PK/BK	EVAP Purge Solenoid Control	PWM Signal: 0-12-0v
69	---	Not Used	---
70	PK/RD	EVAP Purge Solenoid Sense	0-1v
71	---	Not Used	---
72	DB/WT	LDP Switch Sense Signal	LDP Switch Closed: 0v
73	LG/DG	Radiator Fan Control Relay	Relay Off: 12v, On: 1v
74	BR	Fuel Pump Relay Control	Relay Off: 12v, On: 1v
75	LG	SCI Receive	5v
76	BR/YL	PNP Switch Sense Signal	In P/N: 0v, Others: 5v
77	WT/DG	LDP Solenoid Control	PWM Signal: 0-12-0v
78-79	---	Not Used	---
80	LG/RD	Speed Control Vent Solenoid Control	Vacuum Decreasing: 1v

Pin Connector Graphic

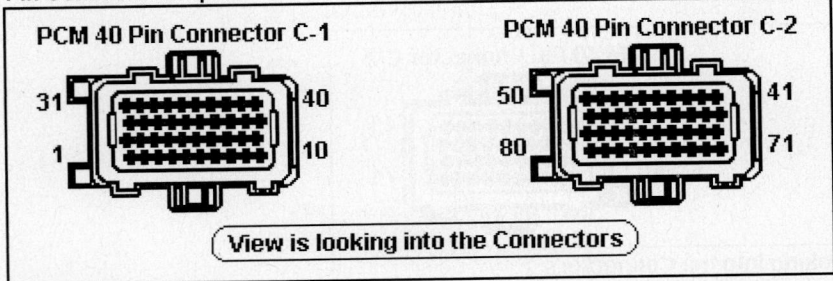

PCM 40 Pin Connector C-1

31   40
1   10

PCM 40 Pin Connector C-2

50   41
80   71

(View is looking into the Connectors)

2001-02 Town & Country 3.8L V6 MFI VIN L A/T 'C1' Connector

PCM Pin #	Wire Color	Circuit Description (40 Pin)	Value at Hot Idle
1	---	Not Used	---
2	DB/OR	Coil 3 Driver Control	5°, 55 mph: 8° dwell
3	DB/TN	Coil 2 Driver Control	5°, 55 mph: 8° dwell
4	---	Not Used	---
5	VT/YL	Speed Control On/Off Switch Sense	12-14v
6	BR/WT	ASD Relay Power (B+)	12-14v
7	BR/LB	Injector 3 Driver	1-4 ms
8	BR/GY	Generator Field Driver	Digital Signal: 0-12-0v
9	---	Not Used	---
10	BK/BR	Power Ground	<0.1v
11	DB/GY	Coil 1 Driver Control	5°, 55 mph: 8° dwell
12	VT/GY	Engine Oil Pressure Sensor	1.6v at 24 psi
13	BR/YL	Injector 1 Driver Control	1-4 ms
14	BR/VT	Injector 6 Driver Control	1-4 ms
15	BR/OR	Injector 5 Driver Control	1-4 ms
16	BR/TN	Injector 4 Driver Control	1-4 ms
17	BR/DB	Injector 2 Driver Control	1-4 ms
18	BR/LG	HO2S-11 (B1 S1) Heater	PWM Signal: 0-12-0v
19	---	Not Used	---
20	PK/GY	Ignition Switch Power (B+)	12-14v
21-24	---	Not Used	---
25	DB/YL	Knock Sensor Signal	0.080v AC
26	VT/OR	ECT Sensor Signal	At 180°F: 2.80v
27	BR/LG	HO2S Ground	<0.050v
28-29	---	Not Used	---
30	DB/LB	HO2S-11 (B1 S1) Signal	0.1-1.1v
31	DG/OR	Double Start Override Signal	KOEC: 9-11v
32	BR/LB	CKP Sensor Signal	Digital Signal: 0-5-0v
33	DB/GY	CMP Sensor Signal	Digital Signal: 0-5-0v
34	---	Not Used	---
35	DB/OR	TP Sensor Signal	0.6-1.0v
36	VT/BR	MAP Sensor Signal	1.5-1.6v
37	DB/LG	IAT Sensor Signal	At 100°F: 1.83v
38-39	---	Not Used	---
40	DB/VT	EGR Solenoid Control	12v, at 55 mph: 1v

Pin Connector Graphic

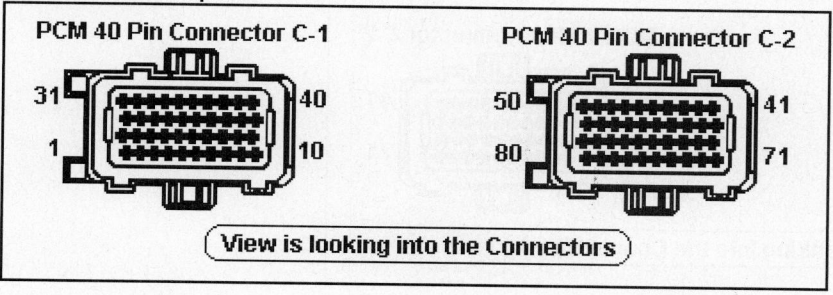

2001 Town & Country 3.8L V6 MFI VIN L A/T 'C2' Connector

PCM Pin #	Wire Color	Circuit Description (40 Pin)	Value at Hot Idle
41	VT	Speed Control Set Switch Signal	S/C & Set Switch On: 3.8v
42	LB/BR	A/C Pressure Switch Signal	A/C On: 0.451-4.850v
43	DB/DG	Sensor Ground	<0.050v
44	BR/PK	8-Volt Supply	7.9-8.1v
45	---	Not Used	---
46	OR/RD	Keep Alive Power (B+)	12-14v
47-48	---	Not Used	---
49	VT/DG	IAC 1 Driver Control	Pulse Signals
50	BK/DG	Power Ground	<0.1v
51	DB/YL	HO2S-12 (B1 S2) Signal	0.1-1.1v
52-55	---	Not Used	---
56	VT/YL	Speed Control Vacuum Solenoid	Vacuum Increasing: 1v
57	VT/LG	IAC 2 Driver Control	Pulse Signals
58	---	Not Used	---
59	WT/VT	PCI Data Bus (J1850)	Digital Signals: 0-7-0v
60	---	Not Used	---
61	PK/YL	5-Volt Supply	4.9-5.1v
62	DG/WT	Brake Switch Sense Signal	Brake Off: 0v, On: 12v
63	DG/LG	Torque Management Request	Digital Signals
64	LB/OR	A/C Clutch Relay Control	Relay Off: 12v, On: 1v
65	WT/BR	SCI Transmit Signal	0v
66	DB/OR	Vehicle Speed Signal	Digital Signal
67	BR/WT	ASD Relay Control	Relay Off: 12v, On: 1v
68	DB/WT	EVAP Purge Solenoid Control	PWM Signal: 0-12-0v
69	---	Not Used	---
70	DB/BR	EVAP Purge Solenoid Sense	0-1v
71	---	Not Used	---
72	VT/WT	LDP Switch Sense Signal	LDP Switch Closed: 0v
73	BR/VT	Radiator Fan Control Relay	Relay Off: 12v, On: 1v
74	BR	Fuel Pump Relay Control	Relay Off: 12v, On: 1v
75	WT/LG	SCI Receive Signal	5v
76	YL/DB	TRS TR1 Signal	In P/N: 0v, Others: 5v
77	VT/LB	LDP Solenoid Control	PWM Signal: 0-12-0v
78-79	---	Not Used	---
80	VT/OR	Speed Control Vent Solenoid	Vacuum Decreasing: 1v

Pin Connector Graphic

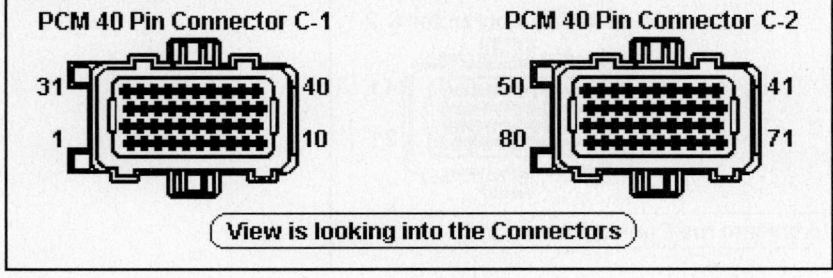

PCM 40 Pin Connector C-1

31   40
1   10

PCM 40 Pin Connector C-2

50   41
80   71

(View is looking into the Connectors)

2003 Town and Country 3.3L V6 MFI VIN L SBEC A/T 'C1' Black Connector

PCM Pin #	Wire Color	Circuit Description (40 Pin)	Value at Hot Idle
1-8	---	Not Used	---
9	BK/BR	Power Ground	<0.1v
10	---	Not Used	---
11	PK/GY	Ignition Switch Output (Start-Run)	12-14v
12	PK/WT	Ignition Switch Output (Run/Start)	12-14v
13-17	---	Not Used	---
18	BK/DG	Power Ground	<0.1v
19, 21	---	Not Used	---
20	VT/GY	Engine Oil Pressure Sensor	1.6v at 24 psi
22	VT/GY	Ambient Air Temperature Sensor	At 100°F: 1.83v
23-24	---	Not Used	---
25	WT/LG	SCI Receive Signal - DLC Pin 12	5v
26	WT/BR	Flash Program Enable	0v
27-28	---	Not Used	---
29	OR/RD	Fused B+	12-14v
30	YL	Ignition Switch Output (Start)	12-14v
31	DB/YL	HO2S-12 (B1 S2) Signal	0.1-1.1v
32	DB/DG	Oxygen Sensor Return (Downstream)	<0.050v
33-35	---	Not Used	---
36	WT/BR	SCI Transmit (PCM) - DLC Pin 7	0v
37	DG/YL	SCI Transmit (TCM)	0v
38	WT/VT	PCI Data Bus (J1850) - DLC Pin 2	Digital Signals: 0-7-0v

2002-03 Town and Country 3.3L V6 MFI VIN 3 NGC A/T 'C2' Gray Connector

PCM Pin #	Wire Color	Circuit Description (40 Pin)	Value at Hot Idle
1-8	---	Not Used	---
9	DB/TN	Coil 2 Driver	5°, 55 mph: 8° dwell
10	DB/DG	Coil 1 Driver	5°, 55 mph: 8° dwell
11	BR/TN	Injector 4 Driver Control	1-4 ms
12	BR/LB	Injector 3 Driver Control	1-4 ms
13	BR/DB	Injector 2 Driver Control	1-4 ms
14	BR/YL	Injector 1 Driver Control	1-4 ms
15-17	---	Not Used	---
18	BR/LG	HO2S-11 (B1 S1) Heater Control	PWM Signal: 0-12-0v
19	BR/GY	Generator Field Driver	Digital Signal: 0-12-0v
20	VT/OR	ECT Sensor Signal	At 180°F: 2.80v
21	BR/OR	TP Sensor Signal	0.6-1.0v
22, 26	---	Not Used	---
23	VT/BR	MAP Sensor Signal	1.5-1.6v
24	BR/LG	Knock Sensor Return	<0.050v
25	DB/YL	Knock Sensor Signal	0.080v AC
27	DB/DG	Sensor Ground	<0.1v
28	BR/VT	IAC Motor Return	<0.050v
29	PK/YL	5-Volt Supply	4.9-5.1v
30	DB/LG	IAT Sensor Signal	At 100°F: 1.83v
31	DB/LB	HO2S-11 (B1 S1) Sensor	0.1-1.1v
32	BR/DG	Oxygen Sensor Return (Upstream)	<0.050v
33	---	Not Used	---
34	DB/GY	CKP Sensor Signal	Digital Signal: 0-5-0v
35	BR/L	CMP Sensor Signal	Digital Signal: 0-5-0v
36-37	---	Not Used	---
38	VT/GY	IAC 1 Motor Control	Pulse Signals

Pin Connector Graphic

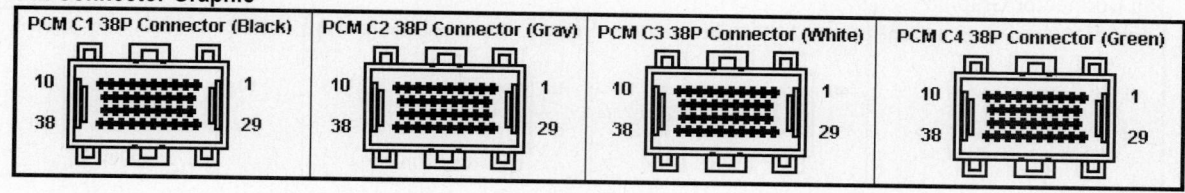

2002-03 Town and Country 3.3L V6 MFI VIN 3 NGC A/T 'C3' White Connector

PCM Pin #	Wire Color	Circuit Description (38 Pin)	Value at Hot Idle
1-2, 4	---	Not Used	---
3	BR/WT	ASD Relay Control	Relay Off: 12v, On: 1v
5	VT/OR	Speed Control Vent Solenoid	Vacuum Decreasing: 1v
6	BR/VT	Radiator Fan Relay Control	Relay Off: 12v, On: 1v
7	VT/YL	Speed Control Power Supply	12-14v
8	WT/DG	Natural Vacuum Leak Detection Solenoid	Solenoid Off: 12v, On: 1v
9	BR/WT	HO2S-12 (Bank 1 Sensor 2) Heater Control	---
11	LB/OR	A/C Clutch Relay Control	Relay Off: 12v, On: 1v
12	VT/YL	Speed Control Vacuum Solenoid	Vacuum Increasing: 1v
13-18	---	Not Used	---
19	BR/WT	Automatic Shutdown Relay Output	12-14v
20	DB/WT	EVAP Purge Solenoid Control	PWM Signal: 0-12-0v
21	YL	Ignition Switch Output (Start) on MTX	Cranking: 9-11v
22, 24-27	---	Not Used	---
23	DG/WT	Brake Switch Sense	Brake Off: 0v, On: 12v
28	BR/WT	Automatic Shutdown Relay Output	12-14v
29	DB/BR	EVAP Purge Solenoid Sense	0-1v
30	---	Not Used	---
31	LB/BR	A/C Pressure Signal	A/C On: 0.451-4.850v
32	DB/YL	Battery Temperature Sensor	At 100ºF: 1.83v
33, 36	---	Not Used	---
34	VT	Speed Control Set Switch Signal	S/C & Set Switch On: 3.8v
35	VT/WT	Natural Vacuum Leak Detection Switch Sense	0.1v
37	BR	Fuel Pump Relay Control	Relay Off: 12v, On: 1v
38	DG/OR	Starter Relay Control	KOEC: 9-11v

2003 Town and Country 2.4L I4 DOHC MFI VIN B All 'C4' Green Connector

PCM Pin #	Wire Color	Circuit Description (38 Pin)	Value at Hot Idle
1	YL/LG	Overdrive Solenoid Control	Solenoid Off: 12v, On: 1v
2	YL/LB	Underdrive Solenoid Control	Solenoid Off: 12v, On: 1v
3-5, 7-9	---	Not Used	---
6	YL/DB	2-4 Solenoid Control	Solenoid Off: 12v, On: 1v
10	DG/WT	Low/Reverse Solenoid Control	Solenoid Off: 12v, On: 1v
11, 13, 17	---	Not Used	---
12, 13	BK/LG	Power Ground	<0.050v
15	DG/LB	TRS T1 Sense	<0.050v
16	DG/DB	TRS T3 Sense	<0.050v
18	YL/BR	Transmission Control Relay Control	Relay Off: 12v, On: 1v
19, 28	YL/OR	Transmission Control Relay Output	12-14v
20-21	---	Not Used	---
22	DG/TN	Overdrive Pressure Switch Sense	12-14v
23-26	---	Not Used	---
27	DG/GY	TRS T41 Sense	<0.050v
29	YL/TN	Low/Reverse Pressure Switch Sense	12-14v
30	YL/DG	2-4 Pressure Switch Sense	In Low/Reverse: 2-4v
31, 36	---	Not Used	---
32	DG/BR	Output Speed Sensor Signal	In 2-4 Position: 2-4v
33	DG/WT	Input Speed Sensor Signal	Moving: AC voltage
34	DG/VT	Speed Sensor Ground	Moving: AC voltage
35	DG/OR	Transmission Temperature Sensor Signal	3.2-3.4v at 104ºF
37	DG/YL	TRS T42 Sense	In PRNL: 0v, Others 5v
38	YL/OR	Transmission Control Relay Output	12-14v

Pin Connector Graphic

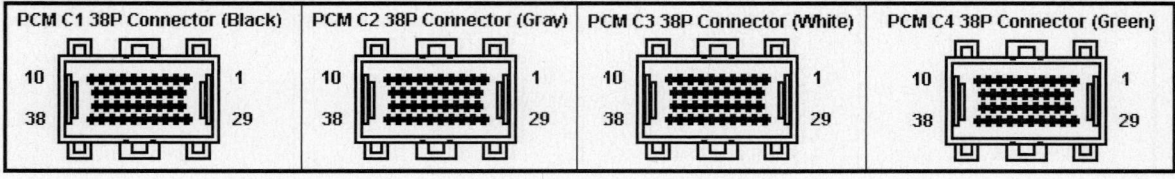

| PCM C1 38P Connector (Black) | PCM C2 38P Connector (Gray) | PCM C3 38P Connector (White) | PCM C4 38P Connector (Green) |

2001-02 PT Cruiser 2.4L I4 DOHC MFI VIN B A/T 'C1' Black Connector

PCM Pin #	Wire Color	Circuit Description (40 Pin)	Value at Hot Idle
1-2	---	Not Used	---
3	DB/TN	Coil 2 Driver	5°, 55 mph: 8° dwell
4	---	Not Used	---
5	YL/RD	Speed Control Power Supply	0v or 6.7v
6	DG/OR	ASD Relay Output	12-14v
7	YL/WT	Injector 3 Driver	1.0-4.0 ms
8	DG	Generator Field Driver	Digital Signal: 0-12-0v
9	---	Not Used	---
10	BK/TN	Power Ground	<0.1v
11	BK/GY	Coil 2 Driver	5°, 55 mph: 8° dwell
12	GY	Engine Oil Pressure Sensor	1.6v at 24 psi
13	WT/DB	Injector 1 Driver	1.0-4.0 ms
14-15	---	Not Used	---
16	LB/BR	Injector 4 Driver	1.0-4.0 ms
17	TN	Injector 2 Driver	1.0-4.0 ms
18	OR/RD	HO2S-11 (B1 S1) Heater	PWM Signal: 0-12-0v
19	---	Not Used	---
20	DB/WT	Ignition Switch Output	12-14v
21-22	---	Not Used	---
23	LG/BK	Clutch Upstop Switch Signal	Pedal Up: 0v, Down: 1v
24	---	Not Used	---
25	DB/LG	Knock Sensor Signal	0.080v AC
26	TN/BK	ECT Sensor Signal	At 180°F: 2.80v
27	BK/OR	Oxygen Sensor Ground	<0.050v
28-29	---	Not Used	---
30	BK/DG	HO2S-11 (B1 S1) Signal	0.1-1.1v
31	TN	Engine Starter Motor Relay	KOEC: 9-11v
32	GY/BK	CKP Sensor Signal	Digital Signal: 0-5-0v
33	TN/YL	CMP Sensor Signal	Digital Signal: 0-5-0v
34	---	Not Used	---
35	OR/DB	TP Sensor Signal	0.6-1.0v
36	DG/RD	MAP Sensor Signal	1.5-1.7v
37	---	Not Used	---
38	BR	A/C Switch Sense	A/C Off: 12v, On: 1v
37	---	Not Used	---
40	GY/YL	EGR Solenoid Control	12v, at 55 mph: 1v

Standard Colors and Abbreviations

Abbreviation	Color	Abbreviation	Color	Abbreviation	Color
BK	Black	GY	Gray	RD	Red
BL	Blue	GN	Green	TN	Tan
BR	Brown	LG	Light Green	VT	Violet
DB	Dark Blue	OR	Orange	WT	White
DG	Dark Green	PK	Pink	YL	Yellow

2001-02 PT Cruiser 2.4L I4 DOHC MFI VIN B A/T 'C2' Gray Connector

PCM Pin #	Wire Color	Circuit Description (40 Pin)	Value at Hot Idle
41	OR/DG	Speed Control Set Switch Signal	S/C & Set Switch On: 3.8v
42	---	Not Used	---
43	BK/LB	Sensor Ground	<0.050v
44	OR	8-Volt Supply	7.9-8.1v
45	DB/OR	PSP Switch Signal	Straight: 0v, Turning: 5v
46	RD/WT	Battery Power (Fused B+)	12-14v
47	BK/WT	Power Ground	<0.1v
48	BR/WT	IAC 3 Driver	Pulse Signals: 0.8-11v
49	YL/BK	IAC 2 Driver	Pulse Signals: 0.8-11v
50	BK/TN	Power Ground	<0.1v
51	TN/WT	HO2S-12 (B1 S2) Signal	0.1-1.1v
52	BK/RD	Inlet Air Temperature Sensor	At 86°F: 1.96v
53-54	---	Not Used	---
55	YL/RD	Low Speed Fan Relay	Relay Off: 12v, On: 1v
56	TN/RD	Speed Control Vacuum Solenoid	Vacuum Increasing: 1v
57	GY/RD	IAC 1 Driver	Pulse Signals: 0.8-11v
58	VT/BK	IAC 4 Driver	Pulse Signals: 0.8-11v
59	VT/YL	PCI Data Bus (J1850)	Digital Signal: 0-7-0v
60-61	---	Not Used	---
61	WT/PK	5-Volt Supply	4.9-5.1v
62	WT/PK	Brake Switch Signal	Brake Off: 0v, On: 12v
63	YL/DG	Torque Management Request	Digital Signals
64	DB/OR	A/C Clutch Relay Control	Relay Off: 12v, On: 1v
65	PK	SCI Transmit	0v
66	WT/OR	Vehicle Speed Signal	Digital Signal
67	DB/YL	ASD Relay Control	Relay Off: 12v, On: 1v
68	PK/BK	EVAP Purge Solenoid Control	PWM Signal: 0-12-0v
69	DB/PK	High Speed Fan Relay	Relay Off: 12v, On: 1v
70	VT/RD	EVAP Purge Solenoid Sense	0-1v
71	GY/BK	Red Brake Warning Indicator	Lamp Off: 12v, On: 1v
72	OR/YL	LDP Switch Sense	Open: 12v, Closed: 0v
73	---	Not Used	---
74	BR	Fuel Pump Relay Control	Relay Off: 12v, On: 1v
75	LG	SCI Receive	5v
76	BK/WT	TRS T41 Signal	In P/N: 0v, Others: 5v
77	WT/DG	LDP Solenoid Control	PWM Signal: 0-12-0v
78	OR/BK	TCC Solenoid Control	Solenoid Off: 12v, On: 1v
79	---	Not Used	---
80	LG/RD	Speed Control Vent Solenoid	Vacuum Decreasing: 1v

Pin Connector Graphic

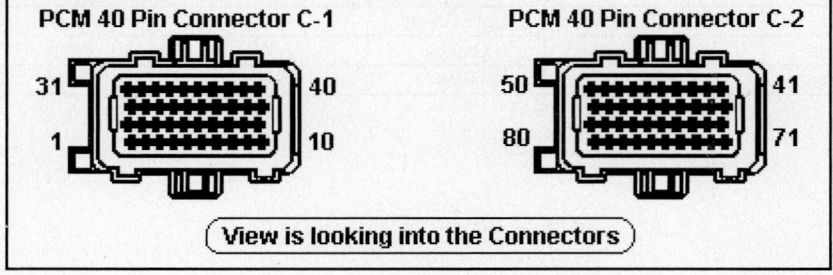

2003 PT Cruiser 2.4L I4 DOHC MFI VIN B All 'C1' Black Connector

PCM Pin #	Wire Color	Circuit Description (38 Pin)	Value at Hot Idle
1-8	---	Not Used	---
9	BK/TN	Power Ground	<0.1v
10	---	Not Used	---
11	DB/WT	Ignition Switch Output (Run-Start)	12-14v
12	RD/WT	Ignition Switch Output (Run-Start)	12-14v
12	RD/WT	Ignition Output - Autostick (Unlock-Run-Start)	12-14v
13	WT/OR	Vehicle Speed Signal	Digital Signal
14	GY/BK	Brake Fluid Level Switch Signal	Switch Open: 12v, Closed: 0v
15-16	---	Not Used	---
17	BR/YL	Sensor 2 Ground	<0.050v
18	BK/TN	Power Ground	<0.1v
19	---	Not Used	---
20	GY	Oil Pressure Signal	1.6v at 24 psi
21	DB	A/C Pressure Switch Signal	A/C On: 0.45-4.85v
22	BR/OR	Ambient Air Temperature Sensor	At 100ºF: 1.83v
23-24	---	Not Used	---
25	LG	SCI Receive (PCM) - DLC Pin 12	5v
26	PK/LB	SCI Receive (TCM)	5v
27-28	---	Not Used	---
29	RD/WT	Battery Power (Fused B+)	12-14v
30	YL	Ignition Switch Output (Start)	8-11v (cranking)
31-35	---	Not Used	---
36	PK	SCI Transmit (PCM) - DLC Pin 7	0v
37	WT/DG	SCI Transmit (TCM)	0v
38	VT/YL	PCI Data Bus (J1850) - DLC Pin 2	Digital Signals: 0-7-0v

2003 PT Cruiser 2.4L I4 DOHC MFI VIN B All 'C2' Gray Connector

PCM Pin #	Wire Color	Circuit Description (38 Pin)	Value at Hot Idle
1-8	---	Not Used	---
9	DB/TN	Coil 2 Driver	5º, at 55 mph: 8º dwell
10	BK/GY	Coil 1 Driver	5º, at 55 mph: 8º dwell
11	LB/BR	Injector 4 Driver	1.0-4.0 ms
12	YL/WT	Injector 3 Driver	1.0-4.0 ms
13	TN	Injector 2 Driver	1.0-4.0 ms
14	WT/DB	Injector 1 Driver	1.0-4.0 ms
15-16	---	Not Used	---
17	BR/VT	HO2S-12 (B1 S2) Heater Control	Heater Off: 12v, On: 1v
18	BR/OR	HO2S-11 (B1 S1) Heater Control	Heater Off: 12v, On: 1v
19	DG	Generator Field Driver	Digital Signal: 0-12-0v
20	TN/BK	ECT Sensor Signal	At 180ºF: 2.80v
21	OR/DB	TP Sensor Signal	0.6-1.0v
22, 26	---	Not Used	---
23	OR/RD	MAP Sensor Signal	1.5-1.7v
24	BK/VT	Knock Sensor Return	<0.050v
25	DB/LG	Knock Sensor Signal	0.080v AC
27	BK/LB	Sensor 1 Ground	<0.050v
28	BR/WT	IAC Motor Return	12-14v
29	OR	5-Volt Supply	4.9-5.1v
30	BK/RD	IAT Sensor Signal	At 100ºF: 1.83v
31	TN/VT	HO2S-12 (B1 S2) Signal	0.1-1.1v
32	DB/DG	Oxygen Sensor Ground	<0.1v
33	TN/WT	HO2S-12 (B1 S2) Signal	0.1-1.1v
34	TN/YL	CMP Sensor Signal	Digital Signal: 0-5-0v
35	GY/BK	CKP Sensor Signal	Digital Signal: 0-5-0v
36-37	---	Not Used	---
38	VT/GY	IAC Motor Driver	DC pulse signals: 0.8-11v

2003 PT Cruiser 2.4L I4 DOHC MFI VIN B All 'C3' White Connector

PCM Pin #	Wire Color	Circuit Description (38 Pin)	Value at Hot Idle
1-2, 9-10	---	Not Used	---
3	DB/YL	ASD Relay Control	Relay Off: 12v, On: 1v
4	DB/PK	High Speed Radiator Fan Relay	Relay Off: 12v, On: 1v
5	LG/RD	Speed Control Vent Solenoid	Vacuum Decreasing: 1v
6	DB/TN	Low Speed Radiator Fan Relay	Relay Off: 12v, On: 1v
7	YL/RD	Speed Control Power Supply	12-14v
8	WT/DG	Natural Vacuum Leak Detection Solenoid	Solenoid Off: 12v, On: 1v
13-16, 18	---	Not Used	---
11	DB/OR	A/C Clutch Relay Control	Relay Off: 12v, On: 1v
12	TN/RD	Speed Control Vacuum Solenoid	Vacuum Increasing: 1v
17	BR/YL	Sensor Ground No. 2	<0.050v
19	LG/RD	Automatic Shutdown Relay Output	12-14v
20	PK/BK	EVAP Purge Solenoid Control	PWM Signal: 0-12-0v
21	YL/RD	Clutch Interlock Switch Signal	Clutch Pedal Out: 12v, In: 0v
22, 25	---	Not Used	---
23	WT/PK	Brake Switch Sense	Brake Off: 0v, On: 12v
24	BR	A/C High Pressure Signal	A/C On: 0.45-4.85v
26	YL/LB	Autostick Downshift Switch	Digital Signal: 0v or 12v
27	LG/LB	Autostick Upshift Switch	Digital Signal: 0v or 12v
28	OR	Automatic Shutdown Relay Output	12-14v
29	WT/TN	EVAP Purge Solenoid Return	0-1v
30	DB/OR	Power Steering Pressure Switch Signal	Straight: 0v, Turning: 5v
31, 33, 36	---	Not Used	---
32	PK/YL	Battery Temperature Sensor	At 100°F: 1.83v
34	PK/LG	Speed Control Set Switch Signal	S/C & Set Switch On: 3.8v
35	OR/YL	Natural Vacuum Leak Detection Switch Sense	0.1v
37	BR	Fuel Pump Relay Control	Relay Off: 12v, On: 1v
38	TN	Starter Relay Control	KOEC: 9-11v

2003 PT Cruiser 2.4L I4 DOHC MFI VIN B All 'C4' Green Connector

PCM Pin #	Wire Color	Circuit Description (38 Pin)	Value at Hot Idle
1	BR	Overdrive Solenoid Control	Solenoid Off: 12v, On: 1v
2	PK	Underdrive Solenoid Control	Solenoid Off: 12v, On: 1v
3-5, 7-9, 11, 13	---	Not Used	---
6	WT	2-4 Solenoid Control	Solenoid Off: 12v, On: 1v
10	LB	Low/Reverse Solenoid Control	Solenoid Off: 12v, On: 1v
12	BK/YL	Power Ground	<0.050v
14	BK/RD	Power Ground	<0.050v
15	LG/BK	TRS T1 Sense	<0.050v
16	VT	TRS T3 Sense	<0.050v
17, 20-21, 23-26	---	Not Used	---
18	LG, RD	Transmission Control Relay Output	Relay Off: 12v, On: 1v
19, 28	RD	Transmission Control Relay Output	Relay Off: 12v, On: 1v
27	BK/WT	TRS T41 Sense	<0.050v
29	DG	Low/Reverse Pressure Switch Sense	12-14v
30	YL/WT	2-4 Pressure Switch Sense	In Low/Reverse: 2-4v
31, 36, 38	---	Not Used	---
32	LG/WT	Output Speed Sensor Signal	In 2-4 Position: 2-4v
33	RD/BK	Input Speed Sensor Signal	Moving: AC voltage
34	DB/BK	Speed Sensor Ground	Moving: AC voltage
35	VT/YL	Transmission Temperature Sensor Signal	<0.050v
37	VT/WT	TRS T42 Sense	In PRNL: 0v, Others 5v

Pin Connector Graphic

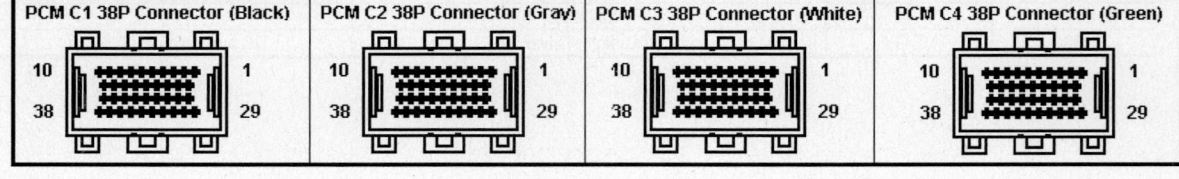

PCM C1 38P Connector (Black)	PCM C2 38P Connector (Gray)	PCM C3 38P Connector (White)	PCM C4 38P Connector (Green)

2003 PT Cruiser 2.4L I4 DOHC Turbo VIN G All 'C1' Black Connector

PCM Pin #	Wire Color	Circuit Description (38 Pin)	Value at Hot Idle
1-8, 10	---	Not Used	---
9	BK/TN	Power Ground	<0.1v
11	DB/WT	Ignition Switch Output (Run-Start)	12-14v
12	RD/WT	Ignition Switch Output (Unlock-Run-Start)	12-14v
13	WT/OR	Vehicle Speed Signal	Digital Signal
14	GY/BK	Brake Fluid Level Switch Signal	Switch Open: 12v, Closed: 0v
15	LB	TIP Solenoid Control	Solenoid Off: 12v, On: <1v
16, 19	---	Not Used	---
17	DB/YL	Surge Solenoid Control	Solenoid Off: 12v, On: <1v
18	BK/TN	Power Ground	<0.1v
20	GY	Oil Pressure Signal	1.6v at 24 psi
21	DB	A/C Pressure Sensor Signal	A/C On: 0.45-4.85v
22	BR/OR	Ambient Air Temperature Sensor	At 100ºF: 1.83v
22, 24	---	Not Used	---
23	DB/LG	TIP Signal	---
25	LG	SCI Receive (PCM) - DLC Pin 12	5v
26	PK/LB	SCI Receive (TCM)	5v
27	VT/WT	5-Volt Supply	4.9-5.1v
28	DB/GY	Wastegate Solenoid Control	Solenoid Off: 12v, On: <1v
29	RD/WT	Battery Power (Fused B+)	12-14v
30	YL	Ignition Switch Output (Start)	8-11v (cranking)
31-35	---	Not Used	---
36	PK	SCI Transmit (PCM) - DLC Pin 7	0v
37	WT/DG	SCI Transmit (TCM)	0v
38	VT/YL	PCI Data Bus (J1850) - DLC Pin 2	Digital Signals: 0-7-0v

2003 PT Cruiser 2.4L I4 DOHC Turbo VIN B All 'C2' Gray Connector

PCM Pin #	Wire Color	Circuit Description (38 Pin)	Value at Hot Idle
1-8	---	Not Used	---
9	DB/TN	Coil 2 Driver	5º, at 55 mph: 8º dwell
10	BK/GY	Coil 1 Driver	5º, at 55 mph: 8º dwell
11	LB/BR	Injector 4 Driver	1.0-4.0 ms
12	YL/WT	Injector 3 Driver	1.0-4.0 ms
13	TN	Injector 2 Driver	1.0-4.0 ms
14	WT/DB	Injector 1 Driver	1.0-4.0 ms
15-16	---	Not Used	---
17	BR/VT	HO2S-12 (B1 S2) Heater Control	Heater On: 1v, Off: 12v
18	BR/OR	HO2S-11 (B1 S1) Heater Control	Heater On: 1v, Off: 12v
19	DG	Generator Field Driver	Digital Signal: 0-12-0v
20	TN/BK	ECT Sensor Signal	At 180ºF: 2.80v
21	OR/DB	TP Sensor Signal	0.6-1.0v
22, 26	---	Not Used	---
23	OR/RD	MAP Sensor Signal	1.5-1.7v
24	BK/VT	Knock Sensor Return	<0.050v
25	DB/LG	Knock Sensor Signal	0.080v AC
27	BK/LB	Sensor Ground	<0.1v
28	BR/WT	IAC Motor Return	12-14v
29	OR	5-Volt Supply	4.9-5.1v
30	BK/RD	IAT Sensor Signal	At 100ºF: 1.83v
31	TN/VT	HO2S-12 (B1 S2) Signal	0.1-1.1v
32	DB/DG	Oxygen Sensor Return (both sensors)	<0.1v
33	TN/WT	HO2S-12 (B1 S2) Signal	0.1-1.1v
34	TN/YL	CMP Sensor Signal	Digital Signal: 0-5-0v
35	GY/BK	CKP Sensor Signal	Digital Signal: 0-5-0v
36-37	---	Not Used	---
38	VT/GY	IAC Motor Driver	DC pulse signals: 0.8-11v

2003 PT Cruiser 2.4L I4 DOHC Turbo VIN G All 'C3' White Connector

PCM Pin #	Wire Color	Circuit Description (38 Pin)	Value at Hot Idle
1-2, 9-10	---	Not Used	---
3	DB/YL	ASD Relay Control	Relay Off: 12v, On: 1v
5	LG/RD	Speed Control Vent Solenoid	Vacuum Decreasing: 1v
6	DB/DG	Radiator Fan Motor Relay	Relay Off: 12v, On: 1v
7	YL/RD	Speed Control Power Supply	12-14v
8	WT/DG	Natural Vacuum Leak Detection Solenoid	Solenoid Off: 12v, On: 1v
11	DB/OR	A/C Clutch Relay Control	Relay Off: 12v, On: 1v
12	TN/RD	Speed Control Vacuum Solenoid	Vacuum Increasing: 1v
13-16, 18	---	Not Used	---
17	BR/YL	Sensor Ground No. 2	<0.050v
19	LG/RD	Automatic Shutdown Relay Output	12-14v
20	PK/BK	EVAP Purge Solenoid Control	PWM Signal: 0-12-0v
21	YL/RD	Clutch Interlock Switch Signal	Clutch Pedal Out: 12v, In: 0v
22, 25	---	Not Used	---
23	WT/PK	Brake Switch Sense	Brake Off: 0v, On: 12v
24	DB/WT	A/C Low Pressure Switch	Switch Open: 0v, Closed: 12v
26	YL	Autostick Downshift Switch	Digital Signal: 0v or 12v
27	LG	Autostick Upshift Switch	Digital Signal: 0v or 12v
28	OR	Automatic Shutdown Relay Output	12-14v
29	WT/TN	EVAP Purge Solenoid Return	0-1v
30	DB/OR	Power Steering Pressure Switch Signal	Straight: 0v, Turning: 5v
31, 33, 36	---	Not Used	---
32	PK/YL	Battery Temperature Sensor	At 100ºF: 1.83v
34	PK/LG	Speed Control Set Switch Signal	S/C & Set Switch On: 3.8v
35	OR/YL	Natural Vacuum Leak Detection Switch Sense	0.1v
37	BR	Fuel Pump Relay Control	Relay Off: 12v, On: 1v
38	TN	Starter Relay Control	KOEC: 9-11v

2003 PT Cruiser 2.4L I4 DOHC Turbo VIN G All 'C4' Green Connector

PCM Pin #	Wire Color	Circuit Description (38 Pin)	Value at Hot Idle
1	BR	Overdrive Solenoid Control	Solenoid Off: 12v, On: 1v
2	PK/BK	Underdrive Solenoid Control	Solenoid Off: 12v, On: 1v
3-5, 7-9, 11-13	---	Not Used	---
6	WT	2-4 Solenoid Control	Solenoid Off: 12v, On: 1v
10	LB	Low/Reverse Solenoid Control	Solenoid Off: 12v, On: 1v
12	BK/YL	Power Ground	<0.050v
14	BK/RD	Power Ground	<0.050v
15	LG/BK	TRS T1 Sense	<0.050v
16	VT	TRS T3 Sense	<0.050v
17, 20-21	---	Not Used	---
18	LG, RD	Transmission Control Relay Output	Relay Off: 12v, On: 1v
19, 28	RD	Transmission Control Relay Output	Relay Off: 12v, On: 1v
23-26	---	Not Used	---
27	BK/WT	TRS T41 Sense	<0.050v
29	DG	Low/Reverse Pressure Switch Sense	12-14v
30	YL/WT	2-4 Pressure Switch Sense	In Low/Reverse: 2-4v
31, 36, 38	---	Not Used	---
32	LG/WT	Output Speed Sensor Signal	In 2-4 Position: 2-4v
33	RD/BK	Input Speed Sensor Signal	Moving: AC voltage
34	DB/BK	Speed Sensor Ground	Moving: AC voltage
35	VT/YL	Transmission Temperature Sensor Signal	<0.050v
37	VT/WT	TRS T42 Sense	In PRNL: 0v, Others 5v

Pin Connector Graphic

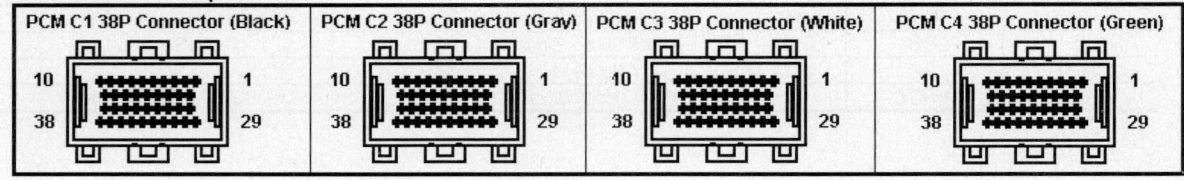

DODGE TRUCK & VAN CONTENTS

DODGE TRUCK & VAN CONTENTS

DODGE TRUCK & VAN CONTENTS

DODGE TRUCK & VAN CONTENTS

DODGE Pin Tables

Introduction

This section contains Pin Tables for Dodge vehicles from 1990-2003. It can be used to test and repair the cause of both Code and No Code faults related to the PCM.

Vehicle Coverage

- 1990-2003 Dakota Applications
- 1990-2003 Pickup Applications
- 1990-93 Ram/Power Ram 50 Truck Applications
- 1990-93 Ramcharger Truck Applications
- 1990-2003 Caravan Van Applications
- 1990-2003 Grand Caravan Van Applications
- 1990-2003 Ram Van Cargo/Passenger Applications

How to Use This Section

This Section can be used to look up the location of a particular pin, a wire color or a "known good" value of a PCM circuit. To locate the PCM information for a particular vehicle, find the model, correct engine size (with VIN Code) and finally the year of the vehicle.

For example, to look up the PCM terminals for a 1998 Pickup 5.2L VIN Y, go to Contents Page 1 and find the text string shown below:

5.2L V8 MFI VIN Y (All) 96 Pin **(1998-2001)** ....Page 12-75

Then turn to Page 12-75 to find the following PCM related information.

1998-2001 Pickup 5.2L V8 MFI (All) VIN Y 'C1' Black Connector

PCM Pin #	Wire Color	Circuit Description (32 Pin)	Value at Hot Idle
22	RD/WT	Keep Alive Power (B+)	12-14v
23	OR/DB	TP Sensor Signal	0.6-1.0v
24	TN/WT	HO2S-11 (B1 S1) Signal	0.1-1.1v
25	OR/BK	HO2S-12 (B1 S2) Signal	0.1-1.1v
26	LG/RD	HO2S-21 (B2 S1) Signal (Cal)	0.1-1.1v
27	DG/RD	MAP Sensor Signal	1.5-1.6v

In this example, the HO2S-11 signal is connected to Pin 24 of the 'C1' Black Connector with a TN/WT wire. The value at Hot Idle shown here indicates that the signal varies between 0.1 and 1.1v.

The acronym "All" in the Title at the top of the table indicates that information in this table is for 1998-2001 Dodge Pickup models with an automatic transmission or a manual transmission.

DODGE PIN TABLES

1990 Dakota 2.5L I4 TBI VIN G (M/T) 60 Pin Connector

PCM Pin #	Wire Color	Circuit Description (60 Pin)	Value at Hot Idle
1	DG/RD	MAP Sensor Signal	1.5-1.6v
2	---	Not Used	---
3	TN/WT	ECT Sensor Signal	At 180°F: 2.80v
4, 5	BK/LB	Sensor Ground	<0.050v
6	---	Not Used	---
7	WT/LG	Speed Control Resume Switch	S/C On: 1v, Off: 12v
8	YL/RD	Speed Control On/Off Switch	S/C On: 1v, Off: 12v
9	BR/RD	Speed Control Set Switch Signal	S/C & Set Switch On: 3.8v
10	WT	Ignition Switch (Start) Output	12-14v
11	---	Not Used	---
12	DB/WT	Ignition Switch Power (B+)	12-14v
13	PK	5-Volt Supply	4.9-5.1v
14	DG/OR	Alternator Field Control	Digital Signals: 0-12-0v
15, 16	LB/RD	Power Ground	<0.1v
17	BR/WT	AIS Motor 1 Control	Pulse Signals
18	YL/BK	AIS Motor 2 Control	Pulse Signals
19	GY/RD	AIS Motor 3 Control	Pulse Signals
20	WT/BK	AIS Motor 4 Control	Pulse Signals
21	BK/RD	Air Temperature Sensor	At 100°F: 2.51v
22	OR/DB	TP Sensor Signal	0.6-1.0v
23	BK/DG	HO2S-11 (B1 S1) Signal	0.1-1.1v
24-28	---	Not Used	---
29	WT/PK	Brake Switch Sense Signal	Brake Off: 0v, On: 12v
30	BR/YL	PNP Switch Sense Signal	In P/N: 0v, Others: 5v
31	LG	SCI Receive Signal	5v
32	---	Not Used	---
33	PK/YL	Injector 1 Driver Control	1-4 ms
34	YL	Coil Driver Control	5°, 55 mph: 8° dwell
35-36	---	Not Used	---
37	GY/PK	Maintenance Indicator Lamp	Lamp Off: 12v, On: 1v
38	OR/WT	Overdrive Lockout Switch	O/D On: 1v, Off: 12v
39	---	Not Used	---
40	GY/YL	EGR Solenoid Control (Cal)	12v, at 55 mph: 1v

Standard Colors and Abbreviations

Abbreviation	Color	Abbreviation	Color	Abbreviation	Color
BK	Black	GY	Gray	RD	Red
BL	Blue	GN	Green	TN	Tan
BR	Brown	LG	Light Green	VT	Violet
DB	Dark Blue	OR	Orange	WT	White
DG	Dark Green	PK	Pink	YL	Yellow

1990 Dakota 2.5L I4 TBI VIN G (M/T) 60 Pin Connector

PCM Pin #	Wire Color	Circuit Description (60 Pin)	Value at Hot Idle
41	RD	Keep Alive Power (B+)	12-14v
42-44	---	Not Used	---
45	BR	A/C Clutch Signal	A/C Off: 12v, On: 1v
46	---	Not Used	---
47	GY/BK	Distributor Reference Signal	Digital Signals: 0-5-0v
48	WT/OR	Vehicle Speed Signal	Digital Signal
49	---	Not Used	---
50	GY/LB	Tachometer Signal	Pulse Signals
51	PK	SCI Transmit Signal	0v
52	OR	8-Volt Supply	7.9-8.1v
53	TN/RD	Speed Control Vacuum Solenoid	Vacuum Increasing: 1v
54	PK/BK	EVAP Purge Solenoid Control	Solenoid Off: 12v, On: 1v
55	OR/BK	Lockup Torque Converter	At Cruise w/TCC On: 1v
56	DB/OR	A/C WOT Relay Control	Relay Off: 12v, On: 1v
57	DB/PK	Radiator Fan Control Relay	Relay Off: 12v, On: 1v
58	DB/YL	ASD Relay Control	Relay Off: 12v, On: 1v
59	BK/PK	MIL (lamp) Control	MIL Off: 12v, On: 1v
60	LG/RD	Speed Control Vent Solenoid	Vacuum Decreasing: 1v

Pin Connector Graphic

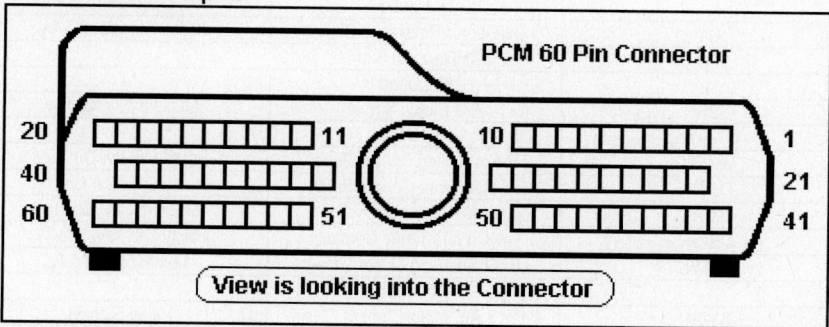

Standard Colors and Abbreviations

Abbreviation	Color	Abbreviation	Color	Abbreviation	Color
BK	Black	GY	Gray	RD	Red
BL	Blue	GN	Green	TN	Tan
BR	Brown	LG	Light Green	VT	Violet
DB	Dark Blue	OR	Orange	WT	White
DG	Dark Green	PK	Pink	YL	Yellow

1991-92 Dakota 2.5L I4 TBI SOHC VIN G (M/T) 60 Pin Connector

PCM Pin #	Wire Color	Circuit Description (60 Pin)	Value at Hot Idle
1	DG/RD	MAP Sensor Signal	1.5-1.6v
2	TN	ECT Sensor Signal	At 180°F: 2.80v
3	RD	Keep Alive Power (B+)	12-14v
4	BK/LB	Sensor Ground	<0.050v
5	BK/WT	Sensor Ground	<0.050v
6	PK/WT	5-Volt Supply	4.9-5.1v
7	OR	8-Volt Supply	7.9-8.1v
8	WT	Ignition Switch Power (B+)	12-14v
9	DB	Ignition Switch Power (B+)	12-14v
10	---	Not Used	---
11	BK/TN	Power Ground	<0.1v
12	BK/TN	Not Used	---
13-15	---	Not Used	---
16	WT/DB	Injector 1 Driver Control	1-4 ms
17-18	---	Not Used	---
19	BK/GY	Coil Driver Control	5°, 55 mph: 8° dwell
20	DG	Alternator Field Control	Digital Signals: 0-12-0v
21	BK/RD	Throttle Body Temperature	At 100°F: 2.51v
22	OR/DB	TP Sensor Signal	0.6-1.0v
23	---	Not Used	---
24	GY	Distributor Reference Signal	Digital Signals: 0-5-0v
25	LG	SCI Transmit Signal	0v
26	---	Not Used	---
27	BR	A/C Clutch Signal	A/C Off: 12v, On: 1v
28, 30, 33	---	Not Used	---
29	WT/PK	Brake Switch Sense Signal	Brake Off: 0v, On: 12v
31	DB/PK	Radiator Fan Control Relay	Relay Off: 12v, On: 1v
32	BK/PK	MIL (lamp) Control	MIL Off: 12v, On: 1v
34	DB/OR	A/C WOT Relay Control	Relay Off: 12v, On: 1v
35	GY/YL	EGR Solenoid Control (Cal)	12v, at 55 mph: 1v
36-38	---	Not Used	---
39	GY/RD	AIS Motor Control	Pulse Signals
40	BR	AIS Motor Control	Pulse Signals
40 ('92)	BR/WT	AIS Motor Control	Pulse Signals

Standard Colors and Abbreviations

Abbreviation	Color	Abbreviation	Color	Abbreviation	Color
BK	Black	GY	Gray	RD	Red
BL	Blue	GN	Green	TN	Tan
BR	Brown	LG	Light Green	VT	Violet
DB	Dark Blue	OR	Orange	WT	White
DG	Dark Green	PK	Pink	YL	Yellow

1991-92 Dakota 2.5L I4 TBI SOHC VIN G (M/T) 60 Pin Connector

PCM Pin #	Wire Color	Circuit Description (60 Pin)	Value at Hot Idle
41	BK/DG	HO2S-11 (B1 S1) Signal	0.1-1.1v
42-45	---	Not Used	---
45	LG	SCI Receive Signal	5v
46	---	Not Used	---
47	WT/OR	Vehicle Speed Signal	Digital Signal
48-50	---	Not Used	---
51	DB/YL	ASD Relay Control	Relay Off: 12v, On: 1v
52	PK	EVAP Purge Solenoid Control	Solenoid Off: 12v, On: 1v
53-55	---	Not Used	---
56	GY/PK	Maintenance Indicator Lamp	Lamp Off: 12v, On: 1v
57-58	---	Not Used	---
59	BR/WT	AIS Motor Control	Pulse Signals
60	YL/BK	AIS Motor Control	Pulse Signals

Pin Connector Graphic

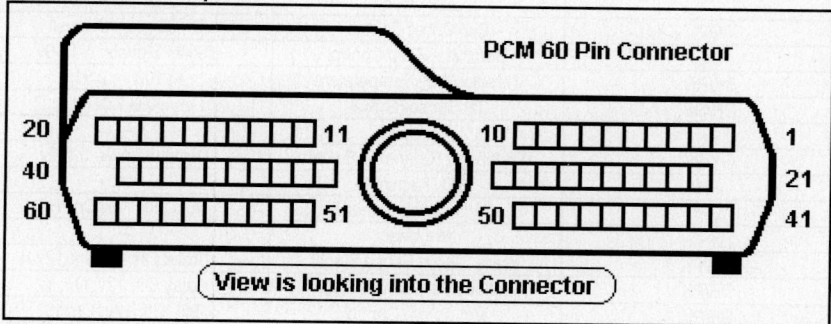

1993-95 Dakota 2.5L I4 TBI SOHC VIN G (M/T) Connector

PCM Pin #	Wire Color	Circuit Description (60 Pin)	Value at Hot Idle
1	DG/RD	MAP Sensor Signal	1.5-1.6v
2	TN/BK	ECT Sensor Signal	At 180°F: 2.80v
3	RD/WT	Keep Alive Power (B+)	12-14v
4	BK/LB	Sensor Ground	<0.050v
5	BK/LB	Sensor Ground	<0.050v
6	PK/WT	5-Volt Supply	4.9-5.1v
7	OR	8-Volt Supply	7.9-8.1v
8	WT	Ignition Switch Power (B+)	12-14v
9	DB	Ignition Switch Power (B+)	12-14v
10	---	Not Used	---
11	BK/TN	Power Ground	<0.1v
12	BK/TN	Power Ground	<0.1v
13-15	---	Not Used	---
16	WT/DB	Injector 1 Driver Control	1-4 ms
17-18	---	Not Used	---
19	GY	Coil Driver Control	5°, 55 mph: 8° dwell
20	DG	Generator Field Control	Digital Signals: 0-12-0v
21	BK/RD	Throttle Body Temperature	At 100°F: 2.51v
22	OR/DB	TP Sensor Signal	0.6-1.0v
23	---	Not Used	---
24	GY/BK	Distributor Reference Signal	Digital Signals: 0-5-0v
25	PK	SCI Transmit	0v
27	BR	A/C Clutch Signal	A/C Off: 12v, On: 1v
28, 30	---	Not Used	---
29	WT/PK	Brake Switch Sense Signal	Brake Off: 0v, On: 12v
31	DB/PK	Radiator Fan Control Relay	Relay Off: 12v, On: 1v
32	BK/PK	MIL (lamp) Control	MIL Off: 12v, On: 1v
33	---	Not Used	---
34	DB/OR	A/C WOT Relay Control	Relay Off: 12v, On: 1v
35	GY/YL	EGR Solenoid Control	Solenoid Off: 12v, On: 1v
36-38	---	Not Used	---
39	GY/RD	AIS Motor Control	Pulse Signals
40	BR/WT	AIS Motor Control	Pulse Signals

Standard Colors and Abbreviations

Abbreviation	Color	Abbreviation	Color	Abbreviation	Color
BK	Black	GY	Gray	RD	Red
BL	Blue	GN	Green	TN	Tan
BR	Brown	LG	Light Green	VT	Violet
DB	Dark Blue	OR	Orange	WT	White
DG	Dark Green	PK	Pink	YL	Yellow

1993-95 Dakota 2.5L I4 TBI SOHC VIN G & K (M/T) Connector

PCM Pin #	Wire Color	Circuit Description (60 Pin)	Value at Hot Idle
41	BK/DG	HO2S-11 (B1 S1) Signal	0.1-1.1v
42-45	---	Not Used	---
45	LG	SCI Receive	5v
46	---	Not Used	---
47	WT/OR	Vehicle Speed Signal	Digital Signal
48-49	---	Not Used	---
51	DB/YL	ASD Relay Control	Relay Off: 12v, On: 1v
52	PK/BK	EVAP Purge Solenoid Control	Solenoid Off: 12v, On: 1v
53-55	---	Not Used	---
56 ('93)	GY/PK	Maintenance Indicator Lamp	Lamp Off: 12v, On: 1v
57-58	---	Not Used	---
59	PK/BK	AIS Motor Control	Pulse Signals
60	YL/BK	AIS Motor Control	Pulse Signals

Pin Connector Graphic

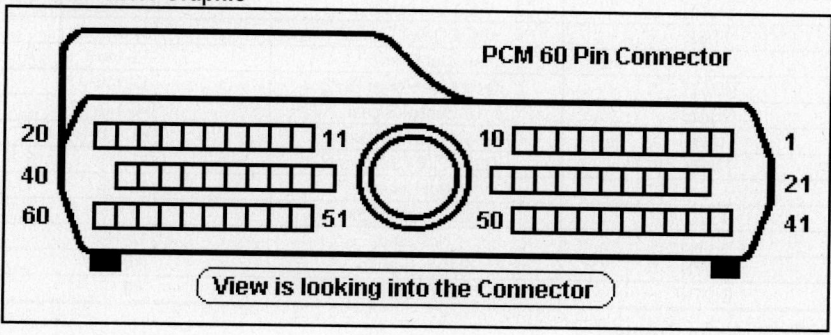

PCM 60 Pin Connector

View is looking into the Connector

1996 Dakota 2.5L I4 MFI VIN P (M/T) 'A' Black Connector

PCM Pin #	Wire Color	Circuit Description (32 Pin)	Value at Hot Idle
1, 3	---	Not Used	12-14v
2	DB	Ignition Switch Power (B+)	12-14v
4	BK/LB	Sensor Ground	<0.050v
5-6	---	Not Used	12-14v
7	GY	Coil 1 Driver Control	5°, 55 mph: 8° dwell
8	GY/BK	CKP Sensor Signal	Digital Signals: 0-5-0v
10	YL/BK	IAC 2 Driver Control	Pulse Signals
11	BR/WT	IAC 3 Driver Control	Pulse Signals
12	OR	PSP Pressure Switch Sense	Straight: 0v, Turning: 5v
13-14	---	Not Used	12-14v
15	BK/RD	IAT Sensor Signal	At 100°F: 1.83v
16	TN/BK	ECT Sensor Signal	At 180°F: 2.80v
17	VT/WT	5-Volt Supply	4.9-5.1v
18	TN/YL	CMP Sensor Signal	Digital Signals: 0-5-0v
19	GY/RD	IAC 1 Driver Control	Pulse Signals
20	VT/BK	IAC 4 Driver Control	Pulse Signals
21	---	Not Used	12-14v
22	RD/WT	Keep Alive Power (B+)	12-14v
23	OR/DB	TP Sensor Signal	0.6-1.0v
24	TN/WT	HO2S-11 (B1 S1) Signal	0.1-1.1v
25	TN/PK	HO2S-12 (B1 S2) Signal	0.1-1.1v
26	---	Not Used	12-14v
27	DG/RD	MAP Sensor Signal	1.5-1.6v
28-30	---	Not Used	12-14v
31	BK/TN	Power Ground	<0.1v
32	BK/TN	Power Ground	<0.1v

1996 Dakota 2.5L I4 MFI VIN P (M/T) 'B' White Connector

PCM Pin #	Wire Color	Circuit Description (32 Pin)	Value at Hot Idle
1-3	---	Not Used	---
4	WT/DB	Injector 1 Driver	1-4 ms
5	YL/WT	Injector 3 Driver	1-4 ms
6-9	---	Not Used	---
10	DG	Generator Field Control	Digital Signals: 0-12-0v
11	OR/BK	Upshift Lamp Driver	Lamp Off: 12v, On: 1v
12-14	---	Not Used	---
15	TN	Injector 2 Driver	1-4 ms
16	LB/BR	Injector 4 Driver	1-4 ms
17-26	---	Not Used	---
27	WT/OR	Vehicle Speed Signal	Digital Signal
28-30	---	Not Used	---
31	RD/YL	5-Volt Supply	4.9-5.1v
32	---	Not Used	---

Standard Colors and Abbreviations

Abbreviation	Color	Abbreviation	Color	Abbreviation	Color
BK	Black	GY	Gray	RD	Red
BL	Blue	GN	Green	TN	Tan
BR	Brown	LG	Light Green	VT	Violet
DB	Dark Blue	OR	Orange	WT	White
DG	Dark Green	PK	Pink	YL	Yellow

1996 Dakota 2.5L I4 MFI VIN P (M/T) 'C' Gray Connector

PCM Pin #	Wire Color	Circuit Description (32 Pin)	Value at Hot Idle
1	DB/OR	A/C Clutch Relay Control	Relay Off: 12v, On: 1v
2	DB/PK	Radiator Fan Control Relay	Relay Off: 12v, On: 1v
3	DB/YL	ASD Relay Control	Relay Off: 12v, On: 1v
4	TN/RD	Speed Control Vacuum Solenoid	Vacuum Increasing: 1v
5	LG/RD	Speed Control Vent Solenoid	Vacuum Decreasing: 1v
6	LG/OR	Overdrive Off Lamp Control	Lamp Off: 12v, On: 1v
7	PK/BK	Transmission Temperature Lamp Control	Lamp Off: 12v, On: 1v
8-10	---	Not Used	---
11	YL/RD	Speed Control Power Supply	12-14v
12	DG/OR	ASD Relay Power (B+)	12-14v
13	OR/WT	Overdrive Off Switch Sense	Switch Off: 12v, On: 1v
14	---	Not Used	---
15	PK/YL	Battery Temperature Sensor	At 86°F: 1.96v
16	DG/YL	Generator Lamp Control	Lamp Off: 12v, On: 1v
17	BK/PK	MIL (lamp) Control	MIL Off: 12v, On: 1v
18	---	Not Used	---
19	BR/YL	Fuel Pump Relay Control	Relay Off: 12v, On: 1v
20	PK/BK	EVAP Purge Solenoid Control	PWM Signal: 0-12-0v
21	---	Not Used	---
22	BR	A/C Request Signal	A/C Off: 12v, On: 1v
23	LG/BK	A/C Select Signal	A/C Off: 12v, On: 1v
24	WT/PK	Brake Switch Sense Signal	Brake Off: 0v, On: 12v
25	---	Not Used	---
26	LB/BK	Fuel Level Sensor Signal	Full: 0.56v, 1/2 full: 2.5v
27	PK	SCI Transmit	0v
28	WT/BK	CCD Bus (-)	<0.050v
29	LG	SCI Receive	5v
30	VT/BR	CCD Bus (+)	Digital Signals: 0-5-0v
31	GY/LB	Tachometer Signal	Pulse Signals
32	WT/LG	Speed Control Set Switch Signal	S/C & Set Switch On: 3.8v

Pin Connector Graphic

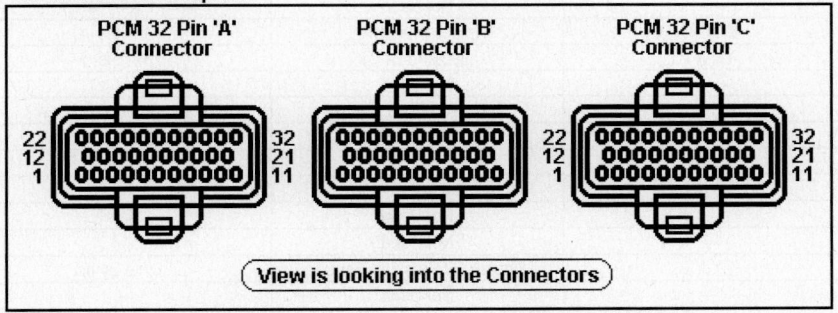

1997 Dakota 2.5L I4 MFI VIN P (M/T) 'A' Black Connector

PCM Pin #	Wire Color	Circuit Description (32 Pin)	Value at Hot Idle
1	---	Not Used	---
2	LG/BK	Ignition Switch Power (B+)	12-14v
3	---	Not Used	---
4	BK/LB	Sensor Ground	<0.050v
5	---	Not Used	---
6	BK/WT	PNP Switch Sense Signal	In P/N: 0v, Others: 5v
7	BK/GY	Coil 1 Driver Control	5°, 55 mph: 8° dwell
8	GY/BK	CKP Sensor Signal	Digital Signals: 0-5-0v
9	---	Not Used	---
10	YL/BK	IAC 2 Driver Control	Pulse Signals
11	BR/WT	IAC 3 Driver Control	Pulse Signals
12	GY/WT	PSP Switch Sense Signal	Straight: 0v, Turning: 5v
13-14	---	Not Used	---
15	BK/RD	IAT Sensor Signal	At 100°F: 1.83v
16	TN/BK	ECT Sensor Signal	At 180°F: 2.80v
17	VT/WT	5-Volt Supply	4.9-5.1v
18	TN/YL	CMP Sensor Signal	Digital Signals: 0-5-0v
19	GY/RD	IAC 1 Driver Control	Pulse Signals
20	VT/BK	IAC 4 Driver Control	Pulse Signals
21	---	Not Used	---
22	RD/WT	Keep Alive Power (B+)	12-14v
23	OR/DB	TP Sensor Signal	0.6-1.0v
24	TN/WT	HO2S-11 (B1 S1) Signal	0.1-1.1v
25	TN/PK	HO2S-12 (B1 S2) Signal	0.1-1.1v
26	---	Not Used	---
27	DG/RD	MAP Sensor Signal	1.5-1.6v
28-30	---	Not Used	---
31	BK	Power Ground	<0.1v
32	BK	Power Ground	<0.1v

1997 Dakota 2.5L I4 MFI VIN P (M/T) 'B' White Connector

PCM Pin #	Wire Color	Circuit Description (32 Pin)	Value at Hot Idle
1-3	---	Not Used	---
4	WT/DB	Injector 1 Driver	1-4 ms
5	YL/WT	Injector 3 Driver	1-4 ms
6-8	---	Not Used	---
10	DG	Generator Field Control	Digital Signals: 0-12-0v
11-14	---	Not Used	---
15	TN	Injector 2 Driver	1-4 ms
16	LB/BR	Injector 4 Driver	1-4 ms
17-21	---	Not Used	---
22	---	Not Used	---
23	GY/YL	Engine Oil Pressure Sensor	1.6v at 24 psi
24-26	---	Not Used	---
27	WT/OR	Vehicle Speed Signal	Digital Signal
28-30	---	Not Used	---
31	OR	5-Volt Supply	4.9-5.1v
32	---	Not Used	---

Standard Colors and Abbreviations

Abbreviation	Color	Abbreviation	Color	Abbreviation	Color
BK	Black	GY	Gray	RD	Red
BL	Blue	GN	Green	TN	Tan
BR	Brown	LG	Light Green	VT	Violet
DB	Dark Blue	OR	Orange	WT	White
DG	Dark Green	PK	Pink	YL	Yellow

1997 Dakota 2.5L I4 MFI VIN P (M/T) 'C' Gray Connector

PCM Pin #	Wire Color	Circuit Description (32 Pin)	Value at Hot Idle
1	DB/OR	A/C Clutch Relay Control	Relay Off: 12v, On: 1v
2	DB/PK	Radiator Fan Control Relay	Relay Off: 12v, On: 1v
3	DB/YL	ASD Relay Control	Relay Off: 12v, On: 1v
4	TN/RD	Speed Control Vacuum Solenoid	Vacuum Increasing: 1v
5	LG/RD	Speed Control Vent Solenoid	Vacuum Decreasing: 1v
6	LG/OR	Overdrive 'Off' Lamp Control	Lamp Off: 12v, On: 1v
7-10	---	Not Used	---
11	YL/RD	Speed Control Power Supply	12-14v
12	DG/OR	ASD Relay Power (B+)	12-14v
13	OR/WT	Overdrive Off Switch Sense	Switch Off: 12v, On: 1v
14	---	Not Used	---
15	PK/YL	Battery Temperature Sensor	At 86°F: 1.96v
16-18	---	Not Used	---
19	LB/OR	Fuel Pump Relay Control	Relay Off: 12v, On: 1v
20	PK/BK	EVAP Purge Solenoid Control	PWM Signal: 0-12-0v
21	---	Not Used	---
22	BR	A/C Request Signal	A/C Off: 12v, On: 1v
23	LG/WT	A/C Select Signal	A/C Off: 12v, On: 1v
24	WT/PK	Brake Switch Sense Signal	Brake Off: 0v, On: 12v
25	DG/BK	Generator Field Source	12-14v
26	DB	Fuel Level Sensor Signal	Full: 0.56v, 1/2 full: 2.5v
27	PK	SCI Transmit	0v
28	WT/PK	CCD Bus (-)	<0.050v
29	LG	SCI Receive	5v
30	VT/BR	CCD Bus (+)	Digital Signals: 0-5-0v
31	---	Not Used	---
32	GY/LB	Speed Control Set Switch Signal	S/C & Set Switch On: 3.8v

Pin Connector Graphic

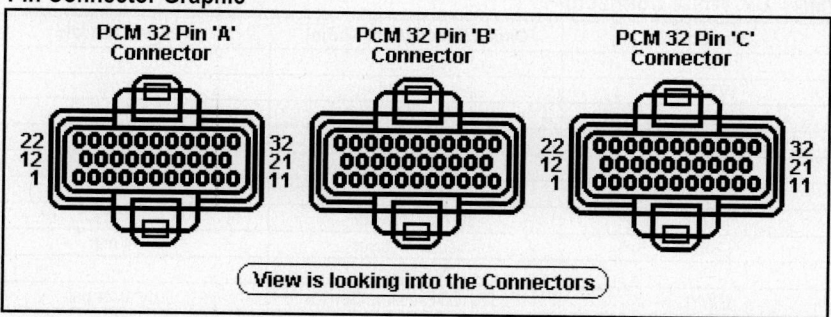

PCM 32 Pin 'A' Connector

PCM 32 Pin 'B' Connector

PCM 32 Pin 'C' Connector

View is looking into the Connectors

1998-99 Dakota 2.5L MFI VIN P (M/T) 'C1' Black Connector

PCM Pin #	Wire Color	Circuit Description (32 Pin)	Value at Hot Idle
1	---	Not Used	---
2	LG/BK	Ignition Switch Power (B+)	12-14v
3	---	Not Used	---
4	BK/LB	Sensor Ground	<0.050v
5-6	---	Not Used	---
7	BK/GY	Coil 1 Driver Control	5°, 55 mph: 8° dwell
8	GY/BK	CKP Sensor Signal	Digital Signals: 0-5-0v
9	---	Not Used	---
10	YL/BK	IAC 2 Driver Control	Pulse Signals
11	BR/WT	IAC 3 Driver Control	Pulse Signals
12	GY/WT	PSP Switch Sense Signal	Straight: 0v, Turning: 5v
13-14	---	Not Used	---
15	BK/RD	IAT Sensor Signal	At 100°F: 1.83v
16	TN/BK	ECT Sensor Signal	At 180°F: 2.80v
17	PK/WT	5-Volt Supply	4.9-5.1v
18	TN/YL	CMP Sensor Signal	Digital Signals: 0-5-0v
19	GY/RD	IAC 1 Driver Control	Pulse Signals
20	VT/BK	IAC 4 Driver Control	Pulse Signals
21	---	Not Used	---
22	RD/WT	Keep Alive Power (B+)	12-14v
23	OR/DB	TP Sensor Signal	0.6-1.0v
24	TN/WT	HO2S-11 (B1 S1) Signal	0.1-1.1v
25	OR/BK	HO2S-12 (B1 S2) Signal	0.1-1.1v
26	---	Not Used	---
27	DG/RD	MAP Sensor Signal	1.5-1.6v
28-30	---	Not Used	---
31-32	BK/TN	Power Ground	<0.1v

1998-99 Dakota 2.5L MFI VIN P (M/T) 'C2' White Connector

PCM Pin #	Wire Color	Circuit Description (32 Pin)	Value at Hot Idle
1-3	---	Not Used	---
4	WT/DB	Injector 1 Driver	1-4 ms
5	YL/WT	Injector 3 Driver	1-4 ms
6-9	---	Not Used	---
10	DG	Generator Field Control	Digital Signals: 0-12-0v
11-14	---	Not Used	---
15	TN	Injector 2 Driver	1-4 ms
16	LB/BR	Injector 4 Driver	1-4 ms
17-23	---	Not Used	---
23	GY/YL	Engine Oil Pressure Sensor	1.6v at 24 psi
24-26	---	Not Used	---
27	WT/OR	Vehicle Speed Signal	Digital Signal
28-32	---	Not Used	---

Standard Colors and Abbreviations

Abbreviation	Color	Abbreviation	Color	Abbreviation	Color
BK	Black	GY	Gray	RD	Red
BL	Blue	GN	Green	TN	Tan
BR	Brown	LG	Light Green	VT	Violet
DB	Dark Blue	OR	Orange	WT	White
DG	Dark Green	PK	Pink	YL	Yellow

1998-99 Dakota 2.5L MFI VIN P (M/T) 'C3' Gray Connector

PCM Pin #	Wire Color	Circuit Description (32 Pin)	Value at Hot Idle
1	DB/OR	A/C Clutch Relay Control	Relay Off: 12v, On: 1v
2	DB/PK	Radiator Fan Control Relay	Relay Off: 12v, On: 1v
3	DB/YL	ASD Relay Control	Relay Off: 12v, On: 1v
4	TN/RD	Speed Control Vacuum Solenoid	Vacuum Increasing: 1v
5	LG/RD	Speed Control Vent Solenoid	Vacuum Decreasing: 1v
6	LG/OR	Overdrive 'Off' Lamp Control	Lamp Off: 12v, On: 1v
8-9	---	Not Used	---
10	YL/DG	LDP Solenoid Control	PWM Signal: 0-12-0v
11	---	Not Used	---
12	DG/OR	ASD Relay Power (B+)	12-14v
13	---	Not Used	---
14	YL/WT	LDP Switch Sense Signal	LDP Switch Closed: 0v
15	PK/YL	Battery Temperature Sensor	At 86°F: 1.96v
16-18	---	Not Used	---
19	LB/OR	Fuel Pump Relay Control	Relay Off: 12v, On: 1v
20	PK/BK	EVAP Purge Solenoid Control	PWM Signal: 0-12-0v
21	---	Not Used	---
22	BR	A/C Switch Sense	A/C Off: 12v, On: 1v
23	LG/WT	A/C Select Signal	A/C Off: 12v, On: 1v
24	WT/PK	Brake Switch Sense Signal	Brake Off: 0v, On: 12v
25	DG/BK	Generator Field Source	12-14v
26	DB	Fuel Level Sensor Signal	Full: 0.56v, 1/2 full: 2.5v
27	PK	SCI Transmit	0v
28	WT/PK	CCD Bus (-)	<0.050v
29	LG	SCI Receive	5v
30	VT/BR	CCD Bus (+)	Digital Signals: 0-5-0v
31	---	Not Used	---
32	GY/LB	Speed Control Set Switch Signal	S/C & Set Switch On: 3.8v

Pin Connector Graphic

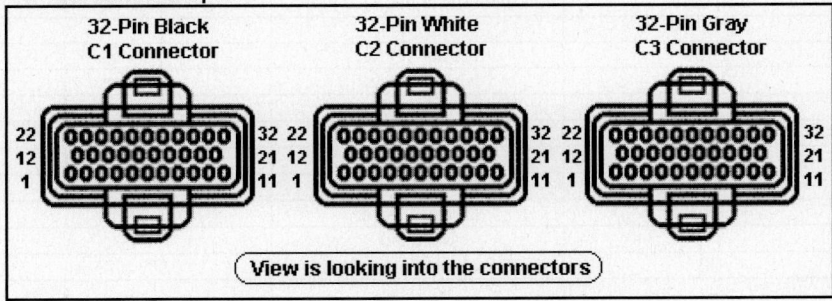

2000-01 Dakota 2.5L MFI VIN P (All) 'C1' Black Connector

PCM Pin #	Wire Color	Circuit Description (32 Pin)	Value at Hot Idle
1	---	Not Used	---
2	LG/BK	Ignition Switch Power (B+)	12-14v
3	---	Not Used	---
4	BK/LB	Sensor Ground	<0.050v
5-6	---	Not Used	---
7	BK/GY	Coil 1 Driver Control	5°, 55 mph: 8° dwell
8	GY/BK	CKP Sensor Signal	Digital Signals: 0-5-0v
9	---	Not Used	---
10	YL/BK	IAC 2 Driver Control	Pulse Signals
11	BR/WT	IAC 3 Driver Control	Pulse Signals
12	DB/OR	PSP Switch Sense Signal	Straight: 0v, Turning: 5v
13-14	---	Not Used	---
15	BK/RD	IAT Sensor Signal	At 100°F: 1.83v
16	TN/BK	ECT Sensor Signal	At 180°F: 2.80v
17	OR	5-Volt Supply	4.9-5.1v
18	TN/YL	CMP Sensor Signal	Digital Signals: 0-5-0v
19	GY/RD	IAC 1 Driver Control	Pulse Signals
20	VT/BK	IAC 4 Driver Control	Pulse Signals
21	---	Not Used	---
22	RD/WT	Keep Alive Power (B+)	12-14v
23	OR/DB	TP Sensor Signal	0.6-1.0v
24	TN/WT	HO2S-11 (B1 S1) Signal	0.1-1.1v
25	OR/BK	HO2S-12 (B1 S2) Signal	0.1-1.1v
26	---	Not Used	---
27	DG/RD	MAP Sensor Signal	1.5-1.6v
28-30	---	Not Used	---
31-32	BK/TN	Power Ground	<0.1v

2000-01 Dakota 2.5L MFI VIN P (All) 'C2' White Connector

PCM Pin #	Wire Color	Circuit Description (32 Pin)	Value at Hot Idle
1-3	---	Not Used	---
4	WT/DB	Injector 1 Driver	1-4 ms
5	YL/WT	Injector 3 Driver	1-4 ms
6-9	---	Not Used	---
10	DG	Generator Field Control	Digital Signals: 0-12-0v
11-14	---	Not Used	---
15	TN	Injector 2 Driver	1-4 ms
16	LB/BR	Injector 4 Driver	1-4 ms
17-18	---	Not Used	---
19	DB	AC Pressure Sensor Signal	0.90v at 79 psi
20-22	---	Not Used	---
23	GY/YL	Engine Oil Pressure Sensor	1.6v at 24 psi
24-26	---	Not Used	---
27	WT/OR	Vehicle Speed Signal	Digital Signal
28-32	---	Not Used	---

Standard Colors and Abbreviations

Abbreviation	Color	Abbreviation	Color	Abbreviation	Color
BK	Black	GY	Gray	RD	Red
BL	Blue	GN	Green	TN	Tan
BR	Brown	LG	Light Green	VT	Violet
DB	Dark Blue	OR	Orange	WT	White
DG	Dark Green	PK	Pink	YL	Yellow

2000-01 Dakota 2.5L MFI VIN P (All) 'C3' Gray Connector

PCM Pin #	Wire Color	Circuit Description (32 Pin)	Value at Hot Idle
1	DB/OR	A/C Clutch Relay Control	Relay Off: 12v, On: 1v
2	DB/PK	Radiator Fan Control Relay	Relay Off: 12v, On: 1v
3	DB/YL	ASD Relay Control	Relay Off: 12v, On: 1v
4	TN/RD	Speed Control Vacuum Solenoid	Vacuum Increasing: 1v
5	LG/RD	Speed Control Vent Solenoid	Vacuum Decreasing: 1v
6-9	---	Not Used	---
10	WT/DG	LDP Solenoid Control	PWM Signal: 0-12-0v
11	YL/RD	Speed Control Power Supply	12-14v
12	DG/OR	ASD Relay Power (B+)	12-14v
13	YL/DG	Torque Mgmt. Request Sense	Digital Signals
14	OR	LDP Switch Sense Signal	LDP Switch Closed: 0v
15	PK/YL	Battery Temperature Sensor	At 86°F: 1.96v
16-18	---	Not Used	---
19 ('00)	LB/OR	Fuel Pump Relay Control	Relay Off: 12v, On: 1v
19 ('01)	BR	Fuel Pump Relay Control	Relay Off: 12v, On: 1v
20	PK/BK	EVAP Purge Solenoid Control	PWM Signal: 0-12-0v
21	---	Not Used	---
22	BR	A/C Switch Sense	A/C Off: 12v, On: 1v
23	LG/WT	A/C Select Signal	A/C Off: 12v, On: 1v
24	WT/PK	Brake Switch Sense Signal	Brake Off: 0v, On: 12v
25	WT/DB	Generator Field Source	12-14v
26	DB	Fuel Level Sensor Signal	Full: 0.56v, 1/2 full: 2.5v
27	PK	SCI Transmit	0v
27 ('00)	PK/DB	SCI Transmit	0v
28 ('00)	WT/BK	CCD Bus (-)	<0.050v
28 ('01)	---	Not Used	---
29	LG	SCI Receive	5v
30 ('00)	VT/BR	CCD Bus (+)	Digital Signals: 0-5-0v
30 ('01)	VT/BR	PCI Bus Signal (J1850)	Digital Signals: 0-5-0v
31	---	Not Used	---
32	RD/LG	Speed Control Set Switch Signal	S/C & Set Switch On: 3.8v

Pin Connector Graphic

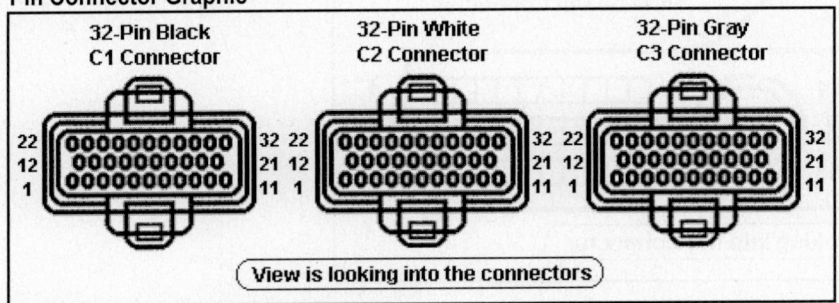

1990-91 Dakota 3.9L V6 TBI VIN X (All) 60 Pin Connector

PCM Pin #	Wire Color	Circuit Description (60 Pin)	Value at Hot Idle
1	DG/RD	MAP Sensor Signal	1.5-1.6v
2	TN/WT	ECT Sensor Signal	At 180°F: 2.80v
3	RD	Keep Alive Power (B+)	12-14v
4	BK/LB	Sensor Ground	<0.050v
5	BK	Sensor Ground	<0.050v
6	PK	5-Volt Supply	4.9-5.1v
7	OR	8-Volt Supply	7.9-8.1v
8	DG/BK	ASD Relay Power (B+)	12-14v
9	DB	Ignition Switch Power (B+)	12-14v
11	LB/RD	Power Ground	<0.1v
12	LB/RD	Power Ground	<0.1v
15	TN	Injector 2 Driver	1-4 ms
16	WT/DB	Injector 1 Driver	1-4 ms
19	BK/GY	Coil Driver Control	5°, 55 mph: 8° dwell
20	DG	Alternator Field Control	Digital Signals: 0-12-0v
22	OR/DB	TP Sensor Signal	0.6-1.0v
24	GY	Distributor Reference Signal	Digital Signals: 0-5-0v
25	PK	SCI Transmit	0v
27	BR	A/C Clutch Signal	A/C Off: 12v, On: 1v
28	PK	Closed Throttle Switch	Switch Off: 12v, On: 1v
29	WT/PK	Brake Switch Sense Signal	Brake Off: 0v, On: 12v
30	BR/YL	PNP Switch Sense Signal	In P/N: 0v, Others: 5v
32	PK/BK	MIL (lamp) Control	MIL Off: 12v, On: 1v
33	TN/RD	Speed Control Vacuum Solenoid	Vacuum Increasing: 1v
34	DB/OR	A/C WOT Relay Control	Relay Off: 12v, On: 1v
35	GY/YL	EGR Solenoid Control (Cal)	12v, at 55 mph: 1v
36	BK/OR	Air Switching Solenoid	Solenoid Off: 12v, On: 1v
37-39	---	Not Used	---
40	BR	AIS Motor Control	Pulse Signals

Pin Connector Graphic

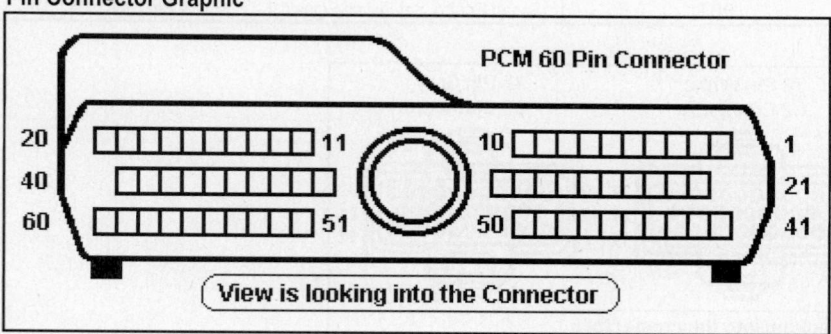

1990-91 Dakota 3.9L V6 TBI VIN X (All) 60 Pin Connector

PCM Pin #	Wire Color	Circuit Description (60 Pin)	Value at Hot Idle
41	BK/DG	HO2S-11 (B1 S1) Signal	0.1-1.1v
43	GY/LB	Tachometer Signal	Pulse Signals
44	OR/WT	Overdrive Lockout Switch	Switch Off: 12v, On: 1v
45	LG	SCI Receive	5v
47	WT/OR	Vehicle Speed Signal	Digital Signal
48	BR/RD	Speed Control Set Switch Signal	S/C & Set Switch On: 3.8v
49	YL/RD	Speed Control On/Off Switch	S/C On: 1v, Off: 12v
50	WT/LG	Speed Control Resume Switch	S/C On: 1v, Off: 12v
51	DB/YL	ASD Relay Control	Relay Off: 12v, On: 1v
52	PK	EVAP Purge Solenoid Control	Solenoid Off: 12v, On: 1v
53	LG/RD	Speed Control Vent Solenoid	Vacuum Decreasing: 1v
54	OR/BK	Part Throttle Unlock Solenoid	Solenoid Off: 12v, On: 1v
55	OR/LG	Overdrive Lockout Solenoid	Solenoid Off: 12v, On: 1v
56	GY/PK	Maintenance Indicator Lamp	Lamp Off: 12v, On: 1v
60	GY/RD	AIS Motor Control	Pulse Signals

Pin Connector Graphic

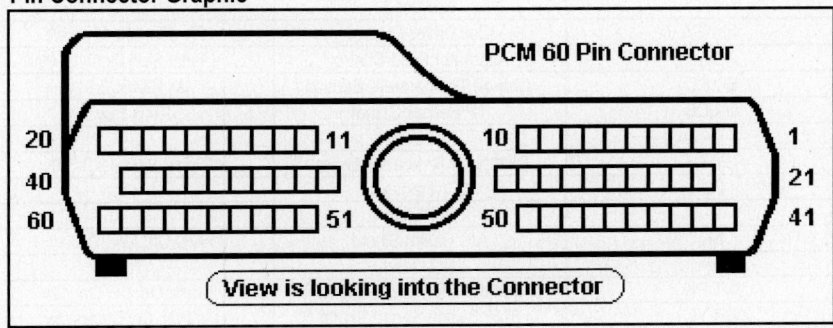

1992-95 Dakota 3.9L V6 MFI VIN X (All) 60 Pin Connector

PCM Pin #	Wire Color	Circuit Description (60 Pin)	Value at Hot Idle
1	DG/RD	MAP Sensor Signal	1.5-1.6v
2	TN/BK	ECT Sensor Signal	At 180°F: 2.80v
3	RD/WT	Keep Alive Power (B+)	12-14v
4	BK/LB	Sensor Ground	<0.050v
5 ('92-'93)	BK/WT	Sensor Ground	<0.050v
6	PK/WT	5-Volt Supply	4.9-5.1v
7	OR	8-Volt Supply	7.9-8.1v
8	---	Not Used	---
9	DB	Ignition Switch Power (B+)	12-14v
10	OR/WT	Overdrive Override Switch	Switch Off: 12v, On: 1v
11	BK/TN	Power Ground	<0.1v
12	BK/TN	Power Ground	<0.1v
13	LB/BR	Injector 4 Driver Control	1-4 ms
14	YL/WT	Injector 3 Driver Control	1-4 ms
15	TN	Injector 2 Driver Control	1-4 ms
16	WT/DG	Injector 1 Driver Control	1-4 ms
17-18	---	Not Used	---
19	BK/GY	Coil Driver Control	5°, 55 mph: 8° dwell
20	DG	Alternator Field Control	Digital Signals: 0-12-0v
21	BK/RD	Charge Temperature Sensor	At 100°F: 2.51v
22	OR/DB	TP Sensor Signal	0.6-1.0v
23	---	Not Used	---
24	GY/BK	Distributor Reference Signal	Digital Signals: 0-5-0v
25	PK	SCI Transmit	0v
26	---	Not Used	---
27	BR	A/C Clutch Signal	A/C Off: 12v, On: 1v
28	---	Not Used	---
29	WT/PK	Brake Switch Sense Signal	Brake Off: 0v, On: 12v
30	BR/YL	PNP Switch Sense Signal	In P/N: 0v, Others: 5v
31	---	Not Used	---
32	PK/BK	MIL (lamp) Control	MIL Off: 12v, On: 1v
33	TN/RD	Speed Control Vacuum Solenoid	Vacuum Increasing: 1v
34	DB/OR	A/C WOT Relay Control	Relay Off: 12v, On: 1v
35	GY/YL	EGR Solenoid Control	12v, at 55 mph: 1v

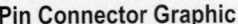

Pin Connector Graphic

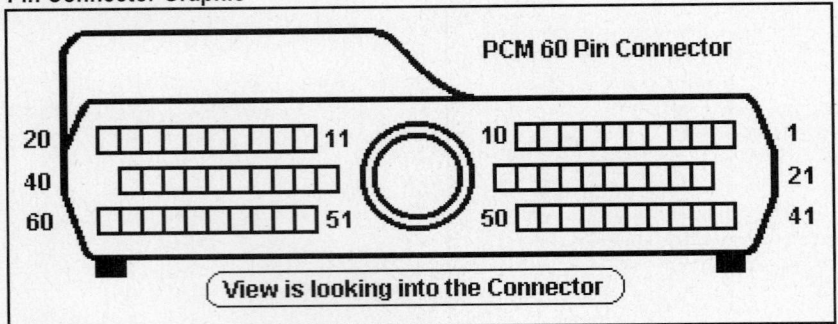

1992-95 Dakota 3.9L V6 MFI VIN X (All) 60 Pin Connector

PCM Pin #	Wire Color	Circuit Description (60 Pin)	Value at Hot Idle
36	---	Not Used	---
37	BK/OR	Overdrive Lamp Control	Lamp Off: 12v, On: 1v
38	PK/BK	Injector 5 Driver Control	1-4 ms
39	GY/RD	AIS Motor 4 Control	Pulse Signals
40	BR/WT	AIS Motor 2 Control	Pulse Signals
41	BK/DG	HO2S-11 (B1 S1) Signal	0.1-1.1v
42	---	Not Used	---
43	GY/LB	Tachometer Signal	Pulse Signals
44	TN/YL	Distributor Sync Signal	Digital Signals: 0-5-0v
45	LG	SCI Receive	5v
46	---	Not Used	---
47	WT/OR	Vehicle Speed Signal	Digital Signal
48	BR/RD	Speed Control Set Switch Signal	S/C & Set Switch On: 3.8v
49	YL/RD	Speed Control On/Off Switch	S/C On: 1v, Off: 12v
50	WT/LG	Speed Control Resume Switch	S/C On: 1v, Off: 12v
51	DB/YL	ASD Relay Control	Relay Off: 12v, On: 1v
52	PK/BK	EVAP Purge Solenoid Control	Solenoid Off: 12v, On: 1v
53	LG/RD	Speed Control Vent Solenoid	Vacuum Decreasing: 1v
54	OR/BK	Part Throttle Unlock Solenoid	Solenoid Off: 12v, On: 1v
55	OR/WT	Overdrive Lockout Solenoid	Solenoid Off: 12v, On: 1v
56 ('92-'93)	GY/PK	Maintenance Indicator Lamp	Lamp Off: 12v, On: 1v
57	DG/BK	ASD Relay Power (B+)	12-14v
58	LG/BK	Injector 6 Driver Control	1-4 ms
59	PK/BK	AIS Motor 1 Control	Pulse Signals
60	GY/RD	AIS Motor 3 Control	Pulse Signals

Pin Connector Graphic

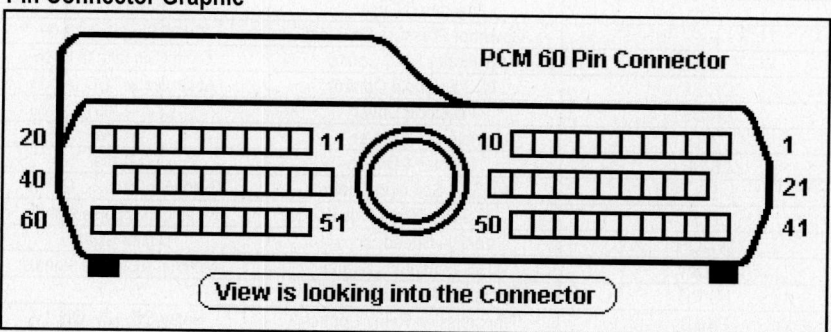

PCM 60 Pin Connector

20 11   10 1
40
60 51   50 41

View is looking into the Connector

Standard Colors and Abbreviations

Abbreviation	Color	Abbreviation	Color	Abbreviation	Color
BK	Black	GY	Gray	RD	Red
BL	Blue	GN	Green	TN	Tan
BR	Brown	LG	Light Green	VT	Violet
DB	Dark Blue	OR	Orange	WT	White
DG	Dark Green	PK	Pink	YL	Yellow

1996 Dakota 3.9L V6 MFI VIN X (All) 'A' Black Connector

PCM Pin #	Wire Color	Circuit Description (32 Pin)	Value at Hot Idle
2	DB	Ignition Switch Power (B+)	12-14v
4	BK/LB	Sensor Ground	<0.050v
6	BR/YL	PNP Switch Sense Signal	In P/N: 0v, Others: 5v
7	GY	Coil 1 Driver Control	5°, 55 mph: 8° dwell
8	GY/BK	CKP Sensor Signal	Digital Signals: 0-5-0v
10	YL/BK	IAC 2 Driver Control	Pulse Signals
11	BR/WT	IAC 3 Driver Control	Pulse Signals
15	BK/RD	IAT Sensor Signal	At 100°F: 1.83v
16	TN/BK	ECT Sensor Signal	At 180°F: 2.80v
17	PK/WT	5-Volt Supply	4.9-5.1v
18	TN/YL	CMP Sensor Signal	Digital Signals: 0-5-0v
19	GY/RD	IAC 1 Driver Control	Pulse Signals
20	PK/BK	IAC 4 Driver Control	Pulse Signals
22	RD/WT	Keep Alive Power (B+)	12-14v
23	OR/DB	TP Sensor Signal	0.6-1.0v
24	TN/WT	HO2S-11 (B1 S1) Signal	0.1-1.1v
25	TN/PK	HO2S-12 (B1 S2) Signal	0.1-1.1v
27	DG/RD	MAP Sensor Signal	1.5-1.6v
31-32	BK/TN	Power Ground	<0.1v

1996 Dakota 3.9L V6 MFI VIN X (All) 'B' White Connector

PCM Pin #	Wire Color	Circuit Description (32 Pin)	Value at Hot Idle
1	PK	Transmission Temperature Sensor Signal	At 200°F: 2.40v
4	WT/DB	Injector 1 Driver	1-4 ms
5	YL/WT	Injector 3 Driver	1-4 ms
6	GY	Injector 5 Driver	1-4 ms
8	PK	Governor Pressure Solenoid	PWM Signal: 0-12-0v
10	DG	Generator Field Control	Digital Signals: 0-12-0v
11	OR/BK	TCC Solenoid Control	At Cruise w/TCC On: 1v
12	BR/DB	Injector 6 Driver	1-4 ms
15	TN	Injector 2 Driver	1-4 ms
16	LB/BR	Injector 4 Driver	1-4 ms
21	BR	3-4 Shift Solenoid Control	Solenoid Off: 12v, On: 1v
25	DB/BK	OSS Sensor (-) Signal	Moving: AC pulse signals
27	WT/OR	Vehicle Speed Signal	Digital Signal
28	LG/WT	OSS Sensor (+) Signal	Moving: AC pulse signals
29	LG/RD	Governor Pressure Sensor	0.58v
30	PK/LB	Transmission Relay Control	Relay Off: 12v, On: 1v
31	RD/YL	5-Volt Supply	4.9-5.1v

Standard Colors and Abbreviations

Abbreviation	Color	Abbreviation	Color	Abbreviation	Color
BK	Black	GY	Gray	RD	Red
BL	Blue	GN	Green	TN	Tan
BR	Brown	LG	Light Green	VT	Violet
DB	Dark Blue	OR	Orange	WT	White
DG	Dark Green	PK	Pink	YL	Yellow

1996 Dakota 3.9L V6 MFI VIN X (All) 'C' Gray Connector

PCM Pin #	Wire Color	Circuit Description (32 Pin)	Value at Hot Idle
1	DB/OR	A/C Clutch Relay Control	Relay Off: 12v, On: 1v
2	DB/PK	Radiator Fan Control Relay	Relay Off: 12v, On: 1v
3	DB/YL	ASD Relay Control	Relay Off: 12v, On: 1v
4	TN/RD	Speed Control Vacuum Solenoid	Vacuum Increasing: 1v
5	LG/RD	Speed Control Vent Solenoid	Vacuum Decreasing: 1v
6	LG/OR	Overdrive 'Off' Lamp Control	Lamp Off: 12v, On: 1v
7	PK/BK	Transmission Temperature Lamp Control	Lamp Off: 12v, On: 1v
8-10	---	Not Used	---
11	YL/RD	Speed Control Power Supply	12-14v
12	DG/OR	ASD Relay Power (B+)	12-14v
13	OR/WT	Overdrive Off Switch Sense	Switch Off: 12v, On: 1v
14	---	Not Used	---
15	PK/YL	Battery Temperature Sensor	At 86°F: 1.96v
16	DG/Y	Generator Lamp Control	Lamp Off: 12v, On: 1v
17	BK/PK	MIL (lamp) Control	MIL Off: 12v, On: 1v
18	---	Not Used	---
19	BR	Fuel Pump Relay Control	Relay Off: 12v, On: 1v
20	PK/BK	EVAP Purge Solenoid Control	PWM Signal: 0-12-0v
21	---	Not Used	---
22	BR	A/C Request Signal	A/C Off: 12v, On: 1v
23	LG	A/C Select Signal	A/C Off: 12v, On: 1v
24	WT/PK	Brake Switch Sense Signal	Brake Off: 0v, On: 12v
25	---	Not Used	---
26	LB/BK	Fuel Level Sensor Signal	Full: 0.56v, 1/2 full: 2.5v
27	PK	SCI Transmit	0v
28	WT/PK	CCD Bus (-)	<0.050v
29	LG	SCI Receive	5v
30	PK/BR	CCD Bus (+)	Digital Signals: 0-5-0v
31	GY/LB	Tachometer Signal	Pulse Signals
32	WT/LG	Speed Control Set Switch Signal	S/C & Set Switch On: 3.8v

Pin Connector Graphic

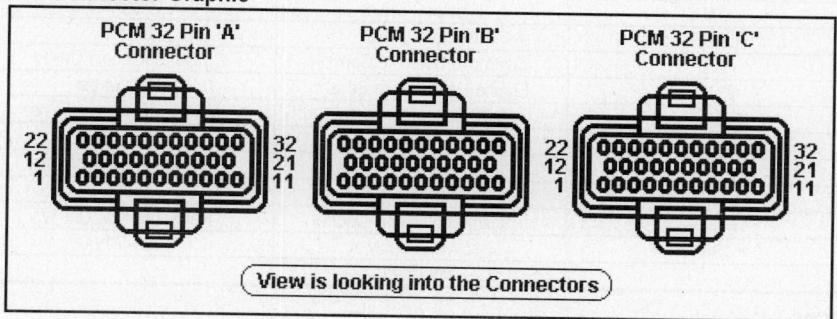

PCM 32 Pin 'A' Connector

PCM 32 Pin 'B' Connector

PCM 32 Pin 'C' Connector

View is looking into the Connectors

1997 Dakota 3.9L V6 MFI VIN X (All) 'A' Black Connector

PCM Pin #	Wire Color	Circuit Description (32 Pin)	Value at Hot Idle
2	LG/BK	Ignition Switch Power (B+)	12-14v
4	BK/LB	Sensor Ground	<0.050v
6	BK/WT	PNP Switch Sense Signal	In P/N: 0v, Others: 5v
7	BK/GY	Coil 1 Driver Control	5°, 55 mph: 8° dwell
8	GY/BK	CKP Sensor Signal	Digital Signals: 0-5-0v
10	YL/BK	IAC 2 Driver Control	Pulse Signals
11	BR/WT	IAC 3 Driver Control	Pulse Signals
12	GY/WT	PSP Switch Sense Signal	Straight: 0v, Turning: 5v
15	BK/RD	IAT Sensor Signal	At 100°F: 1.83v
16	TN/BK	ECT Sensor Signal	At 180°F: 2.80v
17	PK/WT	5-Volt Supply	4.9-5.1v
18	TN/YL	CMP Sensor Signal	Digital Signals: 0-5-0v
19	GY/RD	IAC 1 Driver Control	Pulse Signals
20	PK/BK	IAC 4 Driver Control	Pulse Signals
22	RD/WT	Keep Alive Power (B+)	12-14v
23	OR/DB	TP Sensor Signal	0.6-1.0v
24	OR/TN	HO2S-11 (B1 S1) Signal	0.1-1.1v
25	OR/BK	HO2S-12 (B1 S2) Signal	0.1-1.1v
27	DG/RD	MAP Sensor Signal	1.5-1.6v
31-32	BK/TN	Power Ground	<0.1v

1997 Dakota 3.9L V6 MFI VIN X (All) 'B' White Connector

PCM Pin #	Wire Color	Circuit Description (32 Pin)	Value at Hot Idle
1	PK	Transmission Temperature Sensor Signal	At 200°F: 2.40v
4	WT/DB	Injector 1 Driver	1-4 ms
5	YL/WT	Injector 3 Driver	1-4 ms
6	GY	Injector 5 Driver	1-4 ms
8	PK	Governor Pressure Solenoid	PWM Signal: 0-12-0v
10	DG	Generator Field Control	Digital Signals: 0-12-0v
11	OR/BK	TCC Solenoid Control	At Cruise w/TCC On: 1v
12	BR/DB	Injector 6 Driver	1-4 ms
15	TN	Injector 2 Driver	1-4 ms
16	LB/BR	Injector 4 Driver	1-4 ms
21	BR	Overdrive Solenoid Control	Cruise w/solenoid On: 1v
23	GY/YL	Engine Oil Pressure Sensor	1.6v at 24 psi
25	DB/BK	OSS Sensor (-) Signal	Moving: AC pulse signals
27	WT/OR	Vehicle Speed Signal	Digital Signal
28	LG/WT	OSS Sensor (+) Signal	Moving: AC pulse signals
29	LG/RD	Governor Pressure Sensor	0.58v
30	PK/BK	Transmission Relay Control	Relay Off: 12v, On: 1v
31	OR	5-Volt Supply	4.9-5.1v
32	---	Not Used	---

Standard Colors and Abbreviations

Abbreviation	Color	Abbreviation	Color	Abbreviation	Color
BK	Black	GY	Gray	RD	Red
BL	Blue	GN	Green	TN	Tan
BR	Brown	LG	Light Green	VT	Violet
DB	Dark Blue	OR	Orange	WT	White
DG	Dark Green	PK	Pink	YL	Yellow

1997 Dakota 3.9L V6 MFI VIN X (All) 'C' Gray Connector

PCM Pin #	Wire Color	Circuit Description (32 Pin)	Value at Hot Idle
1	DB/OR	A/C Clutch Relay Control	Relay Off: 12v, On: 1v
2	DB/PK	Radiator Fan Control Relay	Relay Off: 12v, On: 1v
3	DB/YL	ASD Relay Control	Relay Off: 12v, On: 1v
4	TN/RD	Speed Control Vacuum Solenoid	Vacuum Increasing: 1v
5	LG/RD	Speed Control Vent Solenoid	Vacuum Decreasing: 1v
6	LG/OR	Overdrive 'Off' Lamp Control	Lamp Off: 12v, On: 1v
7-10	---	Not Used	---
11	YL/RD	Speed Control Power Supply	12-14v
12	DG/OR	ASD Relay Power (B+)	12-14v
13	OR/WT	Overdrive Off Switch Sense	Switch Off: 12v, On: 1v
14	---	Not Used	---
15	PK/YL	Battery Temperature Sensor	At 86°F: 1.96v
16-18	---	Not Used	---
19	LB/OR	Fuel Pump Relay Control	Relay Off: 12v, On: 1v
20	PK/BK	EVAP Purge Solenoid Control	PWM Signal: 0-12-0v
21	---	Not Used	---
22	BR	A/C Request Signal	A/C Off: 12v, On: 1v
23	LG/WT	A/C Select Signal	A/C Off: 12v, On: 1v
24	WT/PK	Brake Switch Sense Signal	Brake Off: 0v, On: 12v
25	DG/BK	Generator Field Source	12-14v
26	DB	Fuel Level Sensor Signal	Full: 0.56v, 1/2 full: 2.5v
27	PK	SCI Transmit	0v
28	WT/BK	CCD Bus (-)	<0.050v
29	LG	SCI Receive	5v
30	PK/BR	CCD Bus (+)	Digital Signals: 0-5-0v
31	---	Not Used	---
32	GY/LB	Speed Control Set Switch Signal	S/C & Set Switch On: 3.8v

Pin Connector Graphic

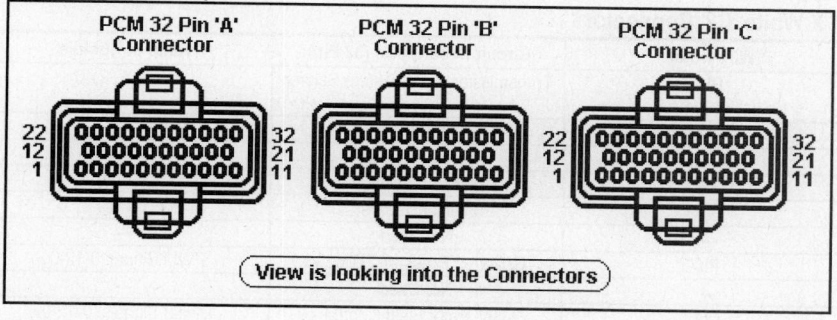

PCM 32 Pin 'A' Connector   PCM 32 Pin 'B' Connector   PCM 32 Pin 'C' Connector

View is looking into the Connectors

1998-99 Dakota 3.9L V6 MFI VIN X 'C1' Black Connector

PCM Pin #	Wire Color	Circuit Description (32 Pin)	Value at Hot Idle
1	---	Not Used	---
2	LG/BK	Ignition Switch Power (B+)	12-14v
3	---	Not Used	---
4	BK/LB	Sensor Ground	<0.050v
5	---	Not Used	---
6	BK/WT	PNP Switch Sense Signal	In P/N: 0v, Others: 5v
7	BK/GY	Coil 1 Driver Control	5°, 55 mph: 8° dwell
8	GY/BK	CKP Sensor Signal	Digital Signals: 0-5-0v
9	---	Not Used	---
10	YL/BK	IAC 2 Driver Control	Pulse Signals
11	BR/WT	IAC 3 Driver Control	Pulse Signals
12-14	---	Not Used	---
15	BK/RD	IAT Sensor Signal	At 100°F: 1.83v
16	TN/BK	ECT Sensor Signal	At 180°F: 2.80v
17	PK/WT	5-Volt Supply	4.9-5.1v
18	TN/YL	CMP Sensor Signal	Digital Signals: 0-5-0v
19	GY/RD	IAC 1 Driver Control	Pulse Signals
20	PK/BK	IAC 4 Driver Control	Pulse Signals
21	---	Not Used	---
22	RD/WT	Keep Alive Power (B+)	12-14v
23	OR/DB	TP Sensor Signal	0.6-1.0v
24	OR/TN	HO2S-11 (B1 S1) Signal	0.1-1.1v
25	OR/BK	HO2S-12 (B1 S2) Signal	0.1-1.1v
26	---	Not Used	---
27	DG/RD	MAP Sensor Signal	1.5-1.6v
28-30	---	Not Used	---
31	BK/TN	Power Ground	<0.1v
32	BK/TN	Power Ground	<0.1v

1998-99 Dakota 3.9L V6 MFI VIN X White 'C2' Connector

PCM Pin #	Wire Color	Circuit Description (32 Pin)	Value at Hot Idle
1	PK	Transmission Temperature Sensor Signal	At 200°F: 2.40v
3	---	Not Used	---
4	WT/DB	Injector 1 Driver	1-4 ms
5	YL/WT	Injector 3 Driver	1-4 ms
6	GY	Injector 5 Driver	1-4 ms
7	---	Not Used	---
8	PK	Governor Pressure Solenoid	PWM Signal: 0-12-0v
9	---	Not Used	---
10	DG	Generator Field Control	Digital Signals: 0-12-0v
11	OR/BK	TCC Solenoid Control	At Cruise w/TCC On: 1v
12	BR/DB	Injector 6 Driver	1-4 ms
13-14	---	Not Used	---
15	TN	Injector 2 Driver	1-4 ms
16	LB/BR	Injector 4 Driver	1-4 ms
17-20	---	Not Used	---
21	BR	Overdrive Solenoid Control	Cruise w/solenoid On: 1v
22	---	Not Used	---
23	GY/YL	Engine Oil Pressure Sensor	1.6v at 24 psi
24	---	Not Used	---
25	DB/BK	OSS Sensor (-) Signal	Moving: AC pulse signals
26	---	Not Used	---
27	WT/OR	Vehicle Speed Signal	Digital Signal
28	LG/WT	OSS Sensor (+) Signal	Moving: AC pulse signals
29	LG/RD	Governor Pressure Sensor	0.58v
30	PK/BK	Transmission Relay Control	Relay Off: 12v, On: 1v
31	OR	5-Volt Supply	4.9-5.1v
32	---	Not Used	---

1998-99 Dakota 3.9L V6 MFI VIN X 'C3' Gray Connector

PCM Pin #	Wire Color	Circuit Description (32 Pin)	Value at Hot Idle
1	DB/OR	A/C Clutch Relay Control	Relay Off: 12v, On: 1v
2	DB/PK	Radiator Fan Control Relay	Relay Off: 12v, On: 1v
3	DB/YL	ASD Relay Control	Relay Off: 12v, On: 1v
4	TN/RD	Speed Control Vacuum Solenoid	Vacuum Increasing: 1v
5	LG/RD	Speed Control Vent Solenoid	Vacuum Decreasing: 1v
6-9	---	Not Used	---
10	YL/DG	LDP Solenoid Control	PWM Signal: 0-12-0v
11	YL/RD	Speed Control Power Supply	12-14v
12	DG/OR	ASD Relay Power (B+)	12-14v
13	OR/WT	Overdrive Off Switch Sense	Switch Off: 12v, On: 1v
14	YL/WT	LDP Switch Sense Signal	LDP Switch Closed: 0v
15	PK/YL	Battery Temperature Sensor	At 86°F: 1.96v
16-18	---	Not Used	---
19	LB/OR	Fuel Pump Relay Control	Relay Off: 12v, On: 1v
20	PK/BK	EVAP Purge Solenoid Control	PWM Signal: 0-12-0v
21	---	Not Used	
22	BR	A/C Switch Signal	A/C Off: 12v, On: 1v
23	LG/WT	A/C Select Signal	A/C Off: 12v, On: 1v
24	WT/PK	Brake Switch Sense Signal	Brake Off: 0v, On: 12v
25	DG/BK	Generator Field Source	12-14v
26	DB	Fuel Level Sensor Signal	Full: 0.56v, 1/2 full: 2.5v
27	PK	SCI Transmit	0v
28	WT/BK	CCD Bus (-)	<0.050v
29	LG	SCI Receive	5v
30	PK/BR	CCD Bus (+)	Digital Signals: 0-5-0v
31	---	Not Used	---
32	RD/LG	Speed Control Set Switch Signal	S/C & Set Switch On: 3.8v

Pin Connector Graphic

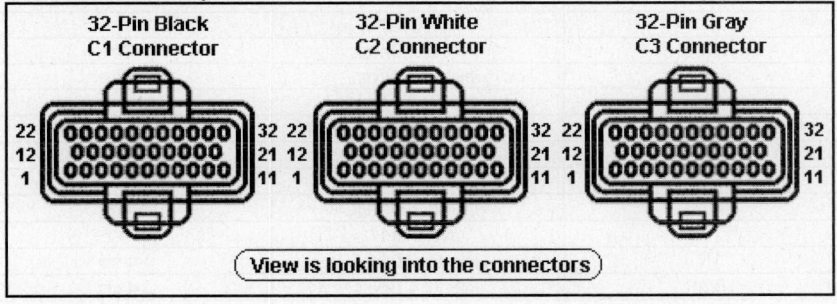

2000 Dakota 3.9L V6 MFI VIN X (All) 'C1' Black Connector

PCM Pin #	Wire Color	Circuit Description (32 Pin)	Value at Hot Idle
1, 3, 5, 9	---	Not Used	---
2	LG/BK	Ignition Switch Power (B+)	12-14v
4	BK/LB	Sensor Ground	<0.050v
6	BK/WT	PNP Switch Sense Signal	In P/N: 0v, Others: 5v
7	BK/GY	Coil 1 Driver Control	5º, 55 mph: 8º dwell
8	GY/BK	CKP Sensor Signal	Digital Signals: 0-5-0v
10	YL/BK	IAC 2 Driver Control	Pulse Signals
11	BR/WT	IAC 3 Driver Control	Pulse Signals
12-14	---	Not Used	---
15	BK/RD	IAT Sensor Signal	At 100ºF: 1.83v
16	TN/BK	ECT Sensor Signal	At 180ºF: 2.80v
17	OR	5-Volt Supply	4.9-5.1v
18	TN/YL	CMP Sensor Signal	Digital Signals: 0-5-0v
19	GY/RD	IAC 1 Driver Control	Pulse Signals
20	VT/BK	IAC 4 Driver Control	Pulse Signals
21	---	Not Used	---
22	RD/WT	Keep Alive Power (B+)	12-14v
23	OR/DB	TP Sensor Signal	0.6-1.0v
24	TN/WT	HO2S-11 (B1 S1) Signal	0.1-1.1v
25	OR/BK	HO2S-12 (B1 S2) Signal	0.1-1.1v
26, 28	---	Not Used	---
27	DG/RD	MAP Sensor Signal	1.5-1.6v
29-30	---	Not Used	---
31, 32	BK/TN	Power Ground	<0.1v

2000 Dakota 3.9L V6 MFI VIN X (All) White 'C2' Connector

PCM Pin #	Wire Color	Circuit Description (32 Pin)	Value at Hot Idle
1	PK	Transmission Temperature Sensor Signal	At 200ºF: 2.40v
2-3, 7, 9	---	Not Used	---
4	WT/DB	Injector 1 Driver	1-4 ms
5	YL/WT	Injector 3 Driver	1-4 ms
6	GY	Injector 5 Driver	1-4 ms
8	PK	Governor Pressure Solenoid	PWM Signal: 0-12-0v
10	DG	Generator Field Control	Digital Signals: 0-12-0v
11	OR/BK	TCC Solenoid Control	At Cruise w/TCC On: 1v
12	BR/DB	Injector 6 Driver	1-4 ms
13-14	---	Not Used	---
15	TN	Injector 2 Driver	1-4 ms
16	LB/BR	Injector 4 Driver	1-4 ms
17	DB/PK	Radiator Fan Control Relay	Relay Off: 12v, On: 1v
18-20, 22	---	Not Used	---
21	BR	Overdrive Solenoid Control	Cruise w/solenoid On: 1v
23	GY/YL	Engine Oil Pressure Sensor	1.6v at 24 psi
24	---	Not Used	---
25	DB/BK	OSS Sensor (-) Signal	Moving: AC pulse signals
26, 32	---	Not Used	---
27	WT/OR	Vehicle Speed Signal	Digital Signal
28	LG/WT	OSS Sensor (+) Signal	Moving: AC pulse signals
29	LG/RD	Governor Pressure Sensor	0.58v
30	PK/BK	Transmission Relay Control	Relay Off: 12v, On: 1v
31	PK/WT	5-Volt Supply	4.9-5.1v

Standard Colors and Abbreviations

Abbreviation	Color	Abbreviation	Color	Abbreviation	Color
BK	Black	GY	Gray	RD	Red
BL	Blue	GN	Green	TN	Tan
BR	Brown	LG	Light Green	VT	Violet
DB	Dark Blue	OR	Orange	WT	White
DG	Dark Green	PK	Pink	YL	Yellow

2000 Dakota 3.9L V6 MFI VIN X (All) 'C3' Gray Connector

PCM Pin #	Wire Color	Circuit Description (32 Pin)	Value at Hot Idle
1	DB/OR	A/C Clutch Relay Control	Relay Off: 12v, On: 1v
2	---	Not Used	---
3	DB/YL	ASD Relay Control	Relay Off: 12v, On: 1v
4	TN/RD	Speed Control Vacuum Solenoid	Vacuum Increasing: 1v
5	LG/RD	Speed Control Vent Solenoid	Vacuum Decreasing: 1v
6-9	---	Not Used	---
10	WT/DG	LDP Solenoid Control	PWM Signal: 0-12-0v
11	YL/RD	Speed Control Power Supply	12-14v
12	DG/OR	ASD Relay Power (B+)	12-14v
13	OR/WT	Overdrive Off Switch Sense	Switch Off: 12v, On: 1v
14	OR	LDP Switch Sense Signal	LDP Switch Closed: 0v
15	PK/YL	Battery Temperature Sensor	At 86°F: 1.96v
16-18	---	Not Used	---
19	LB/OR	Fuel Pump Relay Control	Relay Off: 12v, On: 1v
20	PK/BK	EVAP Purge Solenoid Control	PWM Signal: 0-12-0v
21	---	Not Used	---
22	BR	A/C Switch Signal	A/C Off: 12v, On: 1v
23	LG/WT	A/C Select Signal	A/C Off: 12v, On: 1v
24	WT/PK	Brake Switch Sense Signal	Brake Off: 0v, On: 12v
25	WT/DB	Generator Field Source	12-14v
26	DB	Fuel Level Sensor Signal	Full: 0.56v, 1/2 full: 2.5v
27	PK/DB	SCI Transmit	0v
28	WT/BK	CCD Bus (-)	<0.050v
29	LG	SCI Receive	5v
30	PK/BR	CCD Bus (+)	Digital Signals: 0-5-0v
31	---	Not Used	---
32	RD/LG	Speed Control Set Switch Signal	S/C & Set Switch On: 3.8v

Pin Connector Graphic

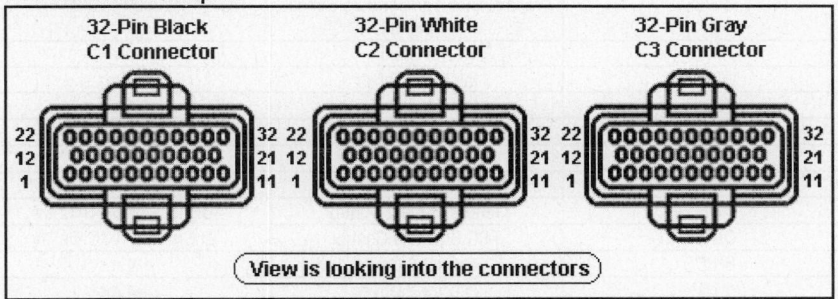

2001 Dakota 3.9L V6 MFI VIN X (All) 'C1' Black Connector

PCM Pin #	Wire Color	Circuit Description (32 Pin)	Value at Hot Idle
1, 3, 5, 9	---	Not Used	---
2	LG/BK	Ignition Switch Power (B+)	12-14v
4	BK/LB	Sensor Ground	<0.050v
6	BK/WT	PNP Switch Sense Signal	In P/N: 0v, Others: 5v
7	BK/GY	Coil 1 Driver Control	5°, 55 mph: 8° dwell
8	GY/BK	CKP Sensor Signal	Digital Signals: 0-5-0v
10	YL/BK	IAC 2 Driver Control	Pulse Signals
11	BR/WT	IAC 3 Driver Control	Pulse Signals
12-14	---	Not Used	---
15	BK/RD	IAT Sensor Signal	At 100°F: 1.83v
16	TN/BK	ECT Sensor Signal	At 180°F: 2.80v
17	OR	5-Volt Supply	4.9-5.1v
18	TN/YL	CMP Sensor Signal	Digital Signals: 0-5-0v
19	GY/RD	IAC 1 Driver Control	Pulse Signals
20	VT/BK	IAC 4 Driver Control	Pulse Signals
21	---	Not Used	---
22	RD/WT	Keep Alive Power (B+)	12-14v
23	OR/DB	TP Sensor Signal	0.6-1.0v
24	BK/DG	HO2S-11 (B1 S1) Signal	0.1-1.1v
25	TN/WT	HO2S-12 (B1 S2) Signal	0.1-1.1v
26	LG/RD	HO2S-21 (B2 S1) Signal	0.1-1.1v
27	DG/RD	MAP Sensor Signal	1.5-1.6v
28, 30	---	Not Used	---
29	TN/WT	HO2S-22 (B2 S2) Signal	0.1-1.1v
31, 32	BK/TN	Power Ground	<0.1v

2001 Dakota 3.9L V6 MFI VIN X (All) White 'C2' Connector

PCM Pin #	Wire Color	Circuit Description (32 Pin)	Value at Hot Idle
1	GY/BK	Transmission Temperature Sensor Signal	At 200°F: 2.40v
2-3	---	Not Used	---
4	WT/DB	Injector 1 Driver	1-4 ms
5	YL/WT	Injector 3 Driver	1-4 ms
6	GY	Injector 5 Driver	1-4 ms
7, 9	---	Not Used	---
8	VT/WT	Variable Force Solenoid	PWM Signal: 0-12-0v
10	DB	Generator Field Control	Digital Signals: 0-12-0v
11	OR/BK	TCC Solenoid Control	At Cruise w/TCC On: 1v
12	BR/DB	Injector 6 Driver	1-4 ms
15	TN	Injector 2 Driver	1-4 ms
16	LB/BR	Injector 4 Driver	1-4 ms
17	DB/PK	Radiator Fan Control Relay	Relay Off: 12v, On: 1v
19	DB	AC Pressure Sensor Signal	0.90v at 79 psi
21	BR	3-4 Shift Solenoid Control	Cruise w/solenoid on: 1v
23	GY/YL	Engine Oil Pressure Sensor	1.6v at 24 psi
25	DB/BK	OSS Sensor (-) Signal	Moving: AC pulse signals
26, 32	---	Not Used	---
27	DB	Vehicle Speed Signal	Digital Signal
28	LG/WT	OSS Sensor (+) Signal	Moving: AC pulse signals
29	LG/RD	Governor Pressure Sensor	0.58v
30	PK	Transmission Relay Control	Relay Off: 12v, On: 1v
31	VT/WT	5-Volt Supply	4.9-5.1v

Standard Colors and Abbreviations

Abbreviation	Color	Abbreviation	Color	Abbreviation	Color
BK	Black	GY	Gray	RD	Red
BL	Blue	GN	Green	TN	Tan
BR	Brown	LG	Light Green	VT	Violet
DB	Dark Blue	OR	Orange	WT	White
DG	Dark Green	PK	Pink	YL	Yellow

2001 Dakota 3.9L V6 MFI VIN X (All) 'C3' Gray Connector

PCM Pin #	Wire Color	Circuit Description (32 Pin)	Value at Hot Idle
1	DB/OR	A/C Clutch Relay Control	Relay Off: 12v, On: 1v
2	---	Not Used	---
3	DB/YL	ASD Relay Control	Relay Off: 12v, On: 1v
4	TN/RD	Speed Control Vacuum Solenoid	Vacuum Increasing: 1v
5	LG/RD	Speed Control Vent Solenoid	Vacuum Decreasing: 1v
6-7	---	Not Used	---
8	VT/WT	HO2S-11 (B1 S1) Heater	PWM Signal: 0-12-0v
9	---	Not Used	---
10	WT/DG	LDP Solenoid Control	PWM Signal: 0-12-0v
11	YL/RD	Speed Control Power Supply	12-14v
12	DG/OR	ASD Relay Power (B+)	12-14v
13	OR/WT	Overdrive Off Switch Sense	Switch Off: 12v, On: 1v
14	OR	LDP Switch Sense Signal	LDP Switch Closed: 0v
15	PK/YL	Battery Temperature Sensor	At 86°F: 1.96v
16	VT/OR	HO2S-21 (B2 S1) Heater	PWM Signal: 0-12-0v
17-18	---	Not Used	---
19	BR	Fuel Pump Relay Control	Relay Off: 12v, On: 1v
20	PK/BK	EVAP Purge Solenoid Control	PWM Signal: 0-12-0v
21	---	Not Used	---
22	BR	A/C Switch Signal	A/C Off: 12v, On: 1v
23	---	Not Used	---
24	WT/PK	Brake Switch Sense Signal	Brake Off: 0v, On: 12v
25	WT/DB	Generator Field Source	12-14v
26	DB/WT	Fuel Level Sensor Signal	Full: 0.56v, 1/2 full: 2.5v
27	PK	SCI Transmit	0v
28	---	Not Used	---
29	LG	SCI Receive	5v
30	VT/YL	PCI Bus Signal (J1850)	Digital Signals: 0-5-0v
31	---	Not Used	---
32	RD/LG	Speed Control Set Switch Signal	S/C & Set Switch On: 3.8v

Pin Connector Graphic

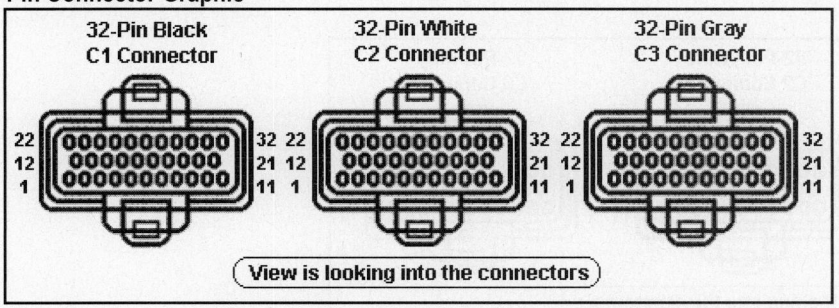

2000 Dakota 4.7L V8 MFI VIN N (All) 'C1' Black Connector

PCM Pin #	Wire Color	Circuit Description (32 Pin)	Value at Hot Idle
1	TN/OR	Coil 3 Driver Control	5°, 55 mph: 8° dwell
2	LG/BK	Ignition Switch Power (B+)	12-14v
3	TN/LG	Coil 4 Driver Control	5°, 55 mph: 8° dwell
4	BK/LB	Sensor Ground	<0.050v
5	TN/LB	Coil 6 Driver Control	5°, 55 mph: 8° dwell
6	BK/WT	PNP Switch Sense Signal	In P/N: 0v, Others: 5v
7	BK/GY	Coil 1 Driver Control	5°, 55 mph: 8° dwell
8	GY/BK	CKP Sensor Signal	Digital Signals: 0-5-0v
9	DB/GY	Coil 8 Driver Control	5°, 55 mph: 8° dwell
10	YL/BK	IAC 2 Driver Control	Pulse Signals
11	BR/WT	IAC 3 Driver Control	Pulse Signals
12	DB/OR	PSP Switch Sense Signal	Straight: 0v, Turning: 5v
13-14	---	Not Used	---
15	BK/RD	IAT Sensor Signal	At 100°F: 1.83v
16	TN/BK	ECT Sensor Signal	At 180°F: 2.80v
17	OR	5-Volt Supply	4.9-5.1v
18	TN/YL	CMP Sensor Signal	Digital Signals: 0-5-0v
19	GY/RD	IAC 1 Driver Control	Pulse Signals
20	VT/BK	IAC 4 Driver Control	Pulse Signals
21	TN/DG	Coil 5 Driver Control	5°, 55 mph: 8° dwell
22	RD/WT	Keep Alive Power (B+)	12-14v
23	OR/DB	TP Sensor Signal	0.6-1.0v
24	LG/RD	HO2S-11 (B1 S1) Signal (Cal)	0.1-1.1v
25	TN/WT	HO2S-12 (B1 S2) Signal (Cal)	0.1-1.1v
26	OR/TN	HO2S-21 (B2 S1) Signal (Cal)	0.1-1.1v
27	DG/RD	MAP Sensor Signal	1.5-1.6v
28	---	Not Used	---
29	PK/WT	HO2S-22 (B2 S2) Signal (Cal)	0.1-1.1v
30	---	Not Used	---
31	BK/TN	Power Ground	<0.1v
32	BK/TN	Power Ground	<0.1v

Pin Connector Graphic

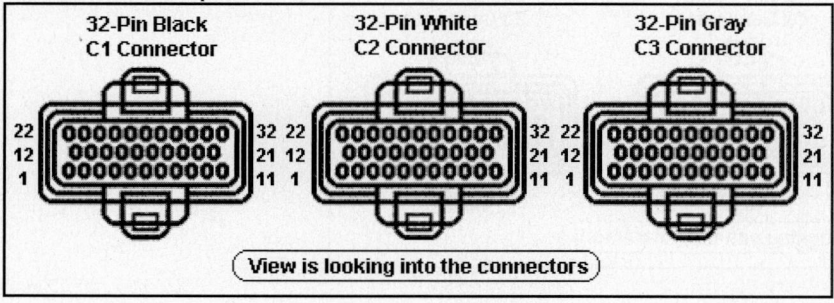

2000 Dakota 4.7L V8 MFI VIN N (All) Gray 'C2' Connector

PCM Pin #	Wire Color	Circuit Description (32 Pin)	Value at Hot Idle
1	---	Not Used	---
2	VT	Injector 7 Driver	1-4 ms
3	---	Not Used	---
4	WT/DB	Injector 1 Driver	1-4 ms
5	YL/WT	Injector 3 Driver	1-4 ms
6	GY	Injector 5 Driver	1-4 ms
7	DB/TN	Coil 7 Driver Control	5°, 55 mph: 8° dwell
8	---	Not Used	---
9	TN/PK	Coil 2 Driver Control	5°, 55 mph: 8° dwell
10	DG	Generator Field Control	Digital Signals: 0-12-0v
11	---	Not Used	---
12	BR/DB	Injector 6 Driver	1-4 ms
13	GY/LB	Injector 8 Driver	1-4 ms
14	---	Not Used	---
15	TN	Injector 2 Driver	1-4 ms
16	LB/BR	Injector 4 Driver	1-4 ms
17	DB/PK	Radiator Fan Relay Control	Relay Off: 12v, On: 1v
18-22	---	Not Used	---
23	GY/YL	Engine Oil Pressure Sensor	1.6v at 24 psi
24-26	---	Not Used	---
27	WT/OR	Vehicle Speed Signal	Digital Signal
28-32	---	Not Used	---

2000 Dakota 4.7L V8 MFI VIN N (All) Black 'C3' Connector

PCM Pin #	Wire Color	Circuit Description (32 Pin)	Value at Hot Idle
1	DB/OR	A/C Clutch Relay Control	Relay Off: 12v, On: 1v
2	---	Not Used	---
3	DB/YL	ASD Relay Control	Relay Off: 12v, On: 1v
4	TN/RD	Speed Control Vacuum Solenoid	Vacuum Increasing: 1v
5	LG/RD	Speed Control Vent Solenoid	Vacuum Decreasing: 1v
6-7	---	Not Used	---
8	DG/WT	HO2S-11 HTR Control (Cal)	Relay Off: 12v, On: 1v
9	DG/BK	HO2S-12 HTR Control (Cal)	Relay Off: 12v, On: 1v
10	WT/DG	LDP Solenoid Control	PWM Signal: 0-12-0v
11	YL/RD	Speed Control Power Supply	12-14v
12	DG/OR	ASD Relay Power (B+)	12-14v
13	YL/DG	Torque Management Relay	Relay Off: 12v, On: 1v
14	OR	LDP Switch Sense Signal	LDP Switch Closed: 0v
15	PK/YL	Battery Temperature Sensor	At 86°F: 1.96v
16-18	---	Not Used	---
19	LB/OR	Fuel Pump Relay Control	Relay Off: 12v, On: 1v
20	PK/BK	EVAP Purge Solenoid Control	PWM Signal: 0-12-0v
21	---	Not Used	---
22	BR	A/C Switch Signal	A/C Off: 12v, On: 1v
23	LG/WT	A/C Select Signal	A/C Off: 12v, On: 1v
24	WT/PK	Brake Switch Sense Signal	Brake Off: 0v, On: 12v
25	WT/DB	Generator Field Source	12-14v
26	DB	Fuel Level Sensor Signal	Full: 0.56v, 1/2 full: 2.5v
27	PK/DB	SCI Transmit	0v
28	WT/BK	CCD Bus (-)	<0.050v
29	LG	SCI Receive	5v
30	PK/BR	CCD Bus (+)	Digital Signals: 0-5-0v
31	---	Not Used	---
32	RD/LG	Speed Control Set Switch Signal	S/C & Set Switch On: 3.8v

DAKOTA PIN VOLTAGE TABLES

2000 Dakota 4.7L V8 MFI VIN N Wiring Schematic Part 1

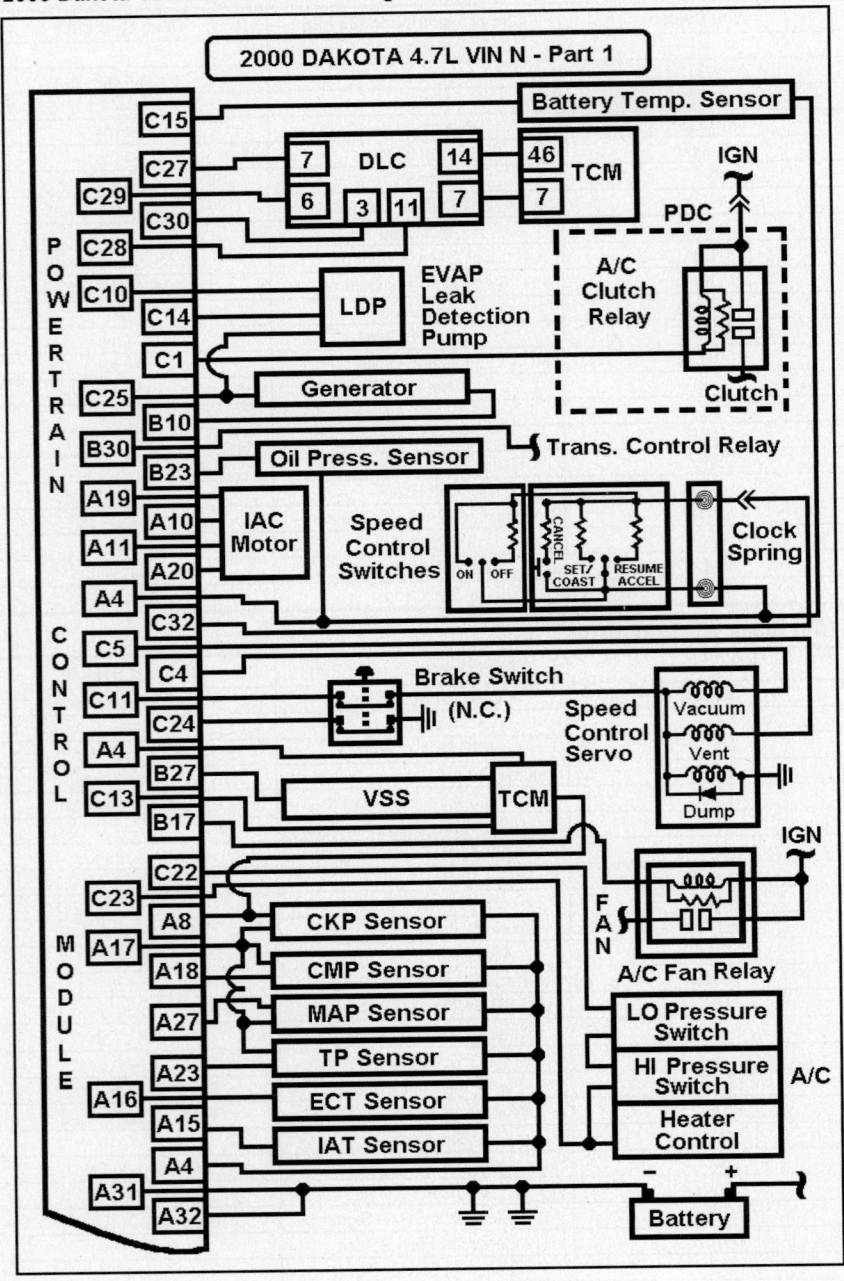

DAKOTA PIN VOLTAGE TABLES

2000 Dakota 4.7L V8 MFI VIN N Wiring Schematic Part 2

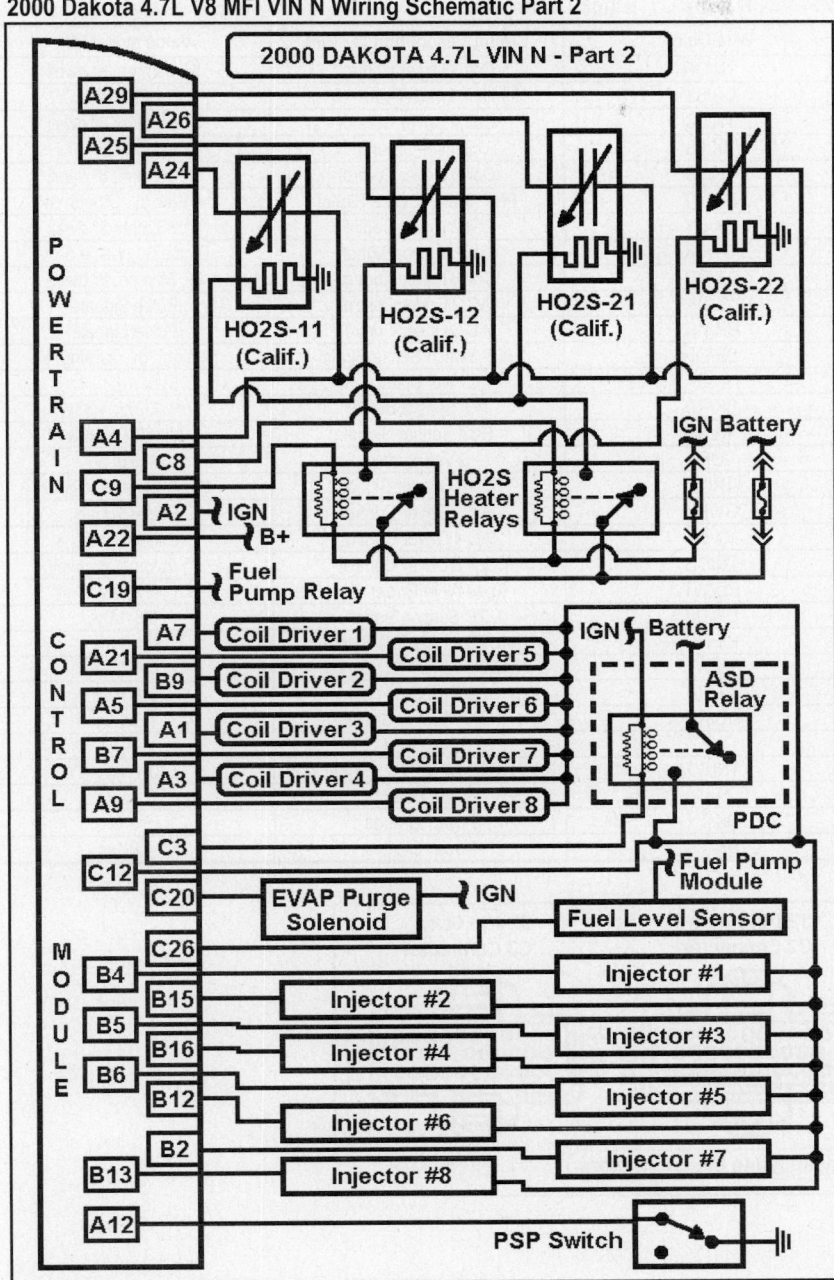

2001-02 Dakota 4.7L V8 SOHC MFI VIN N (All) 'C1' Black Connector

PCM Pin #	Wire Color	Circuit Description (32 Pin)	Value at Hot Idle
1	TN/OR	Coil 3 Driver Control	5°, 55 mph: 8° dwell
2	LG/BK	Ignition Switch Power (B+)	12-14v
3	TN/LG	Coil 4 Driver Control	5°, 55 mph: 8° dwell
4	BK/LB	Sensor Ground	<0.050v
5	TN/LB	Coil 6 Driver Control	5°, 55 mph: 8° dwell
6	BK/WT	PNP Switch Sense Signal	In P/N: 0v, Others: 5v
7	BK/GY	Coil 1 Driver Control	5°, 55 mph: 8° dwell
8	GY/BK	CKP Sensor Signal	Digital Signals: 0-5-0v
9	LB/RD	Coil 8 Driver Control	5°, 55 mph: 8° dwell
10	YL/BK	IAC 2 Driver Control	Pulse Signals
11	BR/WT	IAC 3 Driver Control	Pulse Signals
12	DB/OR	PSP Switch Sense Signal	Straight: 0v, Turning: 5v
13-14	---	Not Used	---
15	BK/RD	IAT Sensor Signal	At 100°F: 1.83v
16	TN/BK	ECT Sensor Signal	At 180°F: 2.80v
17	OR	5-Volt Supply	4.9-5.1v
18	TN/YL	CMP Sensor Signal	Digital Signals: 0-5-0v
19	GY/RD	IAC 1 Driver Control	Pulse Signals
20	VT/BK	IAC 4 Driver Control	Pulse Signals
21	TN/DG	Coil 5 Driver Control	5°, 55 mph: 8° dwell
22	RD/WT	Keep Alive Power (B+)	12-14v
23	OR/DB	TP Sensor Signal	0.6-1.0v
24	BK/DG	HO2S-11 (B1 S1) Signal	0.1-1.1v
25	TN/WT	HO2S-12 (B1 S2) Signal	0.1-1.1v
26	LG/RD	HO2S-21 (B2 S1) Signal	0.1-1.1v
27	DG/RD	MAP Sensor Signal	1.5-1.6v
28	---	Not Used	---
29	TN/WT	HO2S-22 (B2 S2) Signal	0.1-1.1v
30	---	Not Used	---
31	BK/TN	Power Ground	<0.1v
32	BK/TN	Power Ground	<0.1v

Pin Connector Graphic

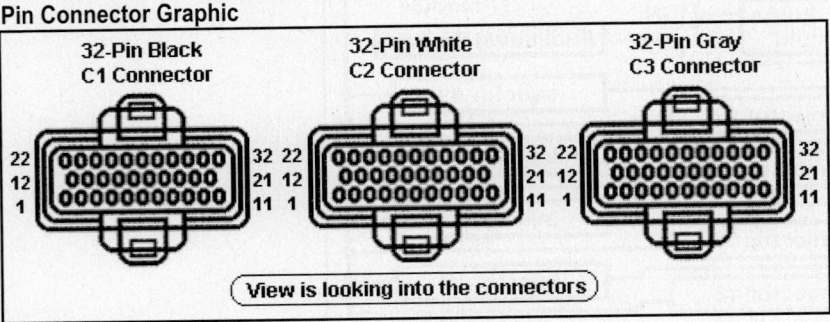

32-Pin Black C1 Connector   32-Pin White C2 Connector   32-Pin Gray C3 Connector

View is looking into the connectors

2001-02 Dakota 4.7L V8 SOHC MFI VIN N (All) Gray 'C2' Connector

PCM Pin #	Wire Color	Circuit Description (32 Pin)	Value at Hot Idle
1, 3	---	Not Used	---
2	VT	Injector 7 Driver	1-4 ms
4	WT/DB	Injector 1 Driver	1-4 ms
5	YL/WT	Injector 3 Driver	1-4 ms
6	GY	Injector 5 Driver	1-4 ms
7	DB/TN	Coil 7 Driver Control	5°, 55 mph: 8° dwell
8, 11	---	Not Used	---
9	TN/PK	Coil 2 Driver Control	5°, 55 mph: 8° dwell
10	DG	Generator Field Control	Digital Signals: 0-12-0v
12	BR/DB	Injector 6 Driver	1-4 ms
13	GY/LB	Injector 8 Driver	1-4 ms
14	---	Not Used	---
15	TN	Injector 2 Driver	1-4 ms
16	LB/BR	Injector 4 Driver	1-4 ms
17	DB/PK	Radiator Fan Relay Control	Relay Off: 12v, On: 1v
18	---	Not Used	---
19	DB	AC Pressure Sensor Signal	0.90v at 79 psi
20-22	---	Not Used	---
23	GY/YL	Engine Oil Pressure Sensor	1.6v at 24 psi
24-26	---	Not Used	---
27	WT/OR	Vehicle Speed Signal	Digital Signal
28-30	---	Not Used	---
31	VT/WT	5-Volt Supply	4.9-5.1v
32	---	Not Used	---

2001-02 Dakota 4.7L V8 SOHC MFI VIN N (All) Black 'C3' Connector

PCM Pin #	Wire Color	Circuit Description (32 Pin)	Value at Hot Idle
1	DB/OR	A/C Clutch Relay Control	Relay Off: 12v, On: 1v
3	DB/YL	ASD Relay Control	Relay Off: 12v, On: 1v
4	TN/RD	Speed Control Vacuum Solenoid	Vacuum Increasing: 1v
5	LG/RD	Speed Control Vent Solenoid	Vacuum Decreasing: 1v
8	DG/WT	HO2S-11 HTR Control (Cal)	Relay Off: 12v, On: 1v
9	DG/BK	HO2S-12 HTR Control (Cal)	Relay Off: 12v, On: 1v
10	WT/DG	LDP Solenoid Control	PWM Signal: 0-12-0v
11	YL/RD	Speed Control Power Supply	12-14v
12	DG/OR	ASD Relay Power (B+)	12-14v
13	YL/DG	Torque Management Relay	Relay Off: 12v, On: 1v
14	OR	LDP Switch Sense Signal	LDP Switch Closed: 0v
15	PK/YL	Battery Temperature Sensor	At 86°F: 1.96v
19	LB/OR	Fuel Pump Relay Control	Relay Off: 12v, On: 1v
20	PK/BK	EVAP Purge Solenoid Control	PWM Signal: 0-12-0v
22	BR	A/C Switch Signal	A/C Off: 12v, On: 1v
23	LG/WT	A/C Select Signal	A/C Off: 12v, On: 1v
24	WT/PK	Brake Switch Sense Signal	Brake Off: 0v, On: 12v
25	WT/DB	Generator Field Source	12-14v
26	DB	Fuel Level Sensor Signal	Full: 0.56v, 1/2 full: 2.5v
27	PK/DB	SCI Transmit	0v
28	WT/BK	CCD Bus (-)	<0.050v
29	LG	SCI Receive	5v
30	PK/BR	CCD Bus (+)	Digital Signals: 0-5-0v
32	RD/LG	Speed Control Set Switch Signal	S/C & Set Switch On: 3.8v

Standard Colors and Abbreviations

Abbreviation	Color	Abbreviation	Color	Abbreviation	Color
BK	Black	GY	Gray	RD	Red
BL	Blue	GN	Green	TN	Tan
BR	Brown	LG	Light Green	VT	Violet
DB	Dark Blue	OR	Orange	WT	White
DG	Dark Green	PK	Pink	YL	Yellow

2003 Dakota 4.7L V8 SOHC MFI VIN N (All) 'C1' 38 Pin Black Connector

PCM Pin #	Wire Color	Circuit Description (38 Pin)	Value at Hot Idle
1	LB/RD	Coil On Plug 8 Driver	5°, at 55 mph: 8° dwell
2, 6-8, 10	---	Not Used	---
3	BR	Coil On Plug 7 Driver	5°, at 55 mph: 8° dwell
4	GY/LG	Injector 8 Driver	1.0-4.0 ms
5	VT	Injector 7 Driver	1.0-4.0 ms
9, 18	BK/TN	Power Ground	<0.050v
11	LG/BK	Ignition Switch Output (Run-Start)	12-14v
12	RD/WT	Ignition Switch Output (Off-Run-Start)	12-14v
13	WT/OR	Vehicle Speed Signal	Digital Signals
14-17, 19	---	Not Used	---
20	GY/YL	Oil Pressure Sensor	1.6v at 24 psi
21	DB	A/C Pressure Sensor	A/C On: 0.45-4.85v
22	VT/LG	Ambient Temperature Sensor	At 86°F: 1.96v
23-24, 28	---	Not Used	---
25	LG	SCI Receive (PCM)	5v
26	PK/LB	SCI Receive (PCM)	5v
27	VT/WT	5-Volt Supply	4.9-5.1v
29	RD/WT	Fused Power (B+)	12-14v
30	RD/YL	Fused Ignition Output (Start)	9-11v
31	TN/WT	HO2S-12 (B1 S2) Signal	0.1-1.1v
32	DB/DG	HO2S Return (Downstream)	<0.050v
33	TN/WT	HO2S-22 (B1 S2) Signal	0.1-1.1v
34-35	---	Not Used	---
36	PK	SCI Transmit (PCM)	0v
37	WT/DG	SCI Transmit (PCM)	0v
38	WT/VT	PCI Data Bus (J1850)	Digital Signals: 0-7-0v

2003 Dakota 4.7L V8 SOHC MFI VIN N (All) 'C2' 38 Pin Gray Connector

PCM Pin #	Wire Color	Circuit Description (38 Pin)	Value at Hot Idle
1	TN/LB	Coil On Plug 6 Driver	5°, at 55 mph: 8° dwell
2	TN/DG	Coil On Plug 5 Driver	5°, at 55 mph: 8° dwell
3	TN/LG	Coil On Plug 4 Driver	5°, at 55 mph: 8° dwell
4	BR/DB	Injector 6 Driver	1.0-4.0 ms
5	GY	Injector 5 Driver	1.0-4.0 ms
6, 8, 15-16	---	Not Used	---
7	TN/OR	Coil On Plug 3 Driver	5°, at 55 mph: 8° dwell
9	TN/PK	Coil On Plug 2 Driver	5°, at 55 mph: 8° dwell
10	TN/RD	Coil On Plug 1 Driver	5°, at 55 mph: 8° dwell
11	LB/BR	Injector 4 Driver	1.0-4.0 ms
12	YL/WT	Injector 3 Driver	1.0-4.0 ms
13	TN	Injector 2 Driver	1.0-4.0 ms
14	WT/DG	Injector 1 Driver	1.0-4.0 ms
17	BR/VT	HO2S-21 (B2 S1) Heater	Heater Off: 12v, On: 1v
18	BR/OR	HO2S-11 (B1 S1) Heater	Heater Off: 12v, On: 1v
19	DG	Generator Field Driver	Digital Signals: 0-12-0v
20	VT/OR	ECT Sensor Signal	At 180°F: 2.80v
21	BR/OR	TP Sensor 1 Signal	0.6-1.0v
22, 24-26	---	Not Used	---
23	VT/BR	MAP Sensor Signal	1.5-1.7v
27	BK/LB	Sensor Ground	<0.050v
28	YL/BK	IAC Motor Return	12-14v
29	OR	5-Volt Supply	4.9-5.1v
30	BK/RD	IAT Sensor Signal	At 100°F: 1.83v
31	BK/DG	HO2S-11 (B1 S1) Signal	0.1-1.1v
32	DB/DG	HO2S Return (Upstream)	<0.050v
33	LG/RD	HO2S-21 (B2 S1) Signal	0.1-1.1v
34	DB/GY	CMP Sensor Signal	Digital Signal: 0-5-0v
35	DB/WT	CKP Sensor Signal	Digital Signal: 0-5-0v
36-37	---	Not Used	---
38	GY/RD	IAC Motor Driver Control	DC pulses: 0.8-11v

2003 Dakota 4.7L V8 SOHC MFI VIN N (All) 'C3' 38 Pin White Connector

PCM Pin #	Wire Color	Circuit Description (38 Pin)	Value at Hot Idle
1-2, 4, 13-18	---	Not Used	---
3	DB/YL	ASD Relay Control	Relay Off: 12v, On: 1v
5	LG/RD	S/C Vent Solenoid	Vacuum Decreasing: 1v
6	DB/PK	Low Speed Fan Relay	Relay Off: 12v, On: 1v
7	VT/YL	Speed Control Power Supply	12-14v
8	WT/DG	Natural Vacuum Leak Detection Solenoid	Solenoid Off: 12v, On: 1v
9	BR/WT	HO2S-12 (B1 S2) Heater	Heater Off: 12v, On: 1v
10	BR/GY	HO2S-22 (B2 S2) Heater	Heater Off: 12v, On: 1v
11	LB/OR	A/C Clutch Relay Control	Relay Off: 12v, On: 1v
12	VT/DG	Speed Control Vacuum Solenoid	Vacuum Increasing: 1v
19	DG/OR	ASD Relay Output	12-14v
20	PK/BK	EVAP Purge Solenoid Control	PWM Signal: 0-12-0v
21-22, 24-27	---	Not Used	---
23	WT/PK	Brake Switch Signal	Brake Off: 0v, On: 12v
28	DG/OR	ASD Relay Output	12-14v
29	DB/BR	EVAP Purge Solenoid Sense	<0.1v
30	DB/OR	Power Steering Pressure Switch	Straight: 0v, Turning: 5v
31, 36	---	Not Used	---
33	DB/WT	Fuel Level Signal	Full: 0.56v, 1/2 full: 2.5v
34	VT/TN	Speed Control Set Switch Signal	S/C & Set Switch On: 3.8v
35	OR	Natural Vacuum Leak Detection Switch	Open: 12v, Closed: 0v
37	BR	Fuel Pump Relay Control	Relay Off: 12v, On: 1v
38	DG/OR	Starter Relay Control	Relay Off: 12v, On: 1v

2003 Dakota 4.7L V8 SOHC MFI VIN N (A/T Only) 'C4' 38 Pin Green Connector

1	YL/GY	Overdrive Solenoid Control	Solenoid Off: 12v, On: 1v
2	DG/WT	4C Solenoid Control	Solenoid Off: 12v, On: 1v
3, 5, 7, 9	---	Not Used	---
4	YL/DB	MS Solenoid Control	Solenoid Off: 12v, On: 1v
6	WT/DB	2C Solenoid Control	Solenoid Off: 12v, On: 1v
8	YL/LB	Underdrive Solenoid Control	Solenoid Off: 12v, On: 1v
10	LG	Low/Reverse Solenoid control	Solenoid Off: 12v, On: 1v
11	VT/LG	Pressure Control Solenoid	PWM Signals (0-12-0v)
12-14	BK/RD	Power Ground	<0.1v
15	DG/LB	TRS T1 Sense	In NOL: 0v, Others: 5v
16	DG/DB	TRS T3 Sense	In P3L: 0v, Others: 5v
17	DB/WT	Overdrive Off Switch Sense	In Overdrive: 2-4v
18	YL/BR	Transmission Control Relay Control	Relay Off: 12v, On: 1v
19, 28, 38	RD	Transmission Control Relay Output	12-14v
20	DB	4C Pressure Switch Sense	In 4th Position: 2-4v
21	GY	Underdrive Pressure Switch Sense	In Underdrive Position: 2-4v
22	YL/LG	Overdrive Pressure Switch Sense	In Overdrive Position: 2-4v
23-25, 27, 36	---	Not Used	
26	PK/BK	TRS T2 Sense	In P/N: 0v, Others: 5v
29	DG	Low/Reverse Pressure Switch	In Low/Reverse: 2-4v
30	GY/DG	2C Pressure Switch	In 2-4 Position: 2-4v
31	VT/TN	Line Pressure Switch	In 2-4 Position: 2-4v
32	LG/WT	Output Speed Sensor	Moving: AC voltage
33	RD/BK	Input Speed Sensor	Moving: AC voltage
34	DB/BK	Speed Sensor Ground	<0.050v
35	DG/OR	Transmission Temperature Sensor	3.2-3.4v at 104°F
37	DG/YL	TRS T42 Sense	In PRNL: 0v, Others 5v

Pin Connector Graphic

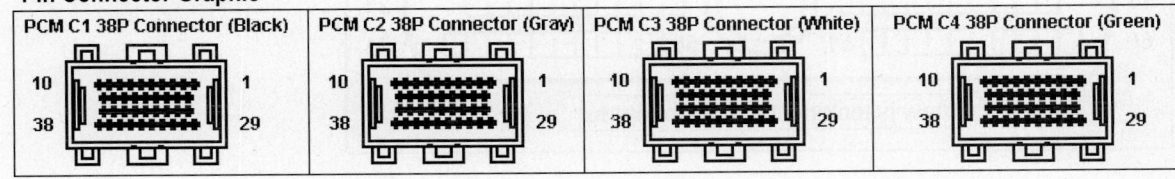

1990-91 Dakota 5.2L V8 TBI VIN Y (A/T) 60 Pin Connector

PCM Pin #	Wire Color	Circuit Description (60 Pin)	Value at Hot Idle
1	DG/RD	MAP Sensor Signal	1.5-1.6v
2	TN/WT	ECT Sensor Signal	At 180°F: 2.80v
3	RD	Keep Alive Power (B+)	12-14v
4	BK/LB	Sensor Ground	<0.050v
5	BK	Sensor Ground	<0.050v
6	PK	5-Volt Supply	4.9-5.1v
7	OR	8-Volt Supply	7.9-8.1v
8	DG/BK	ASD Relay Power (B+)	12-14v
9	DB	Ignition Switch Power (B+)	12-14v
11	LB/RD	Power Ground	<0.1v
12	LB/RD	Power Ground	<0.1v
15	TL	Injector 2 Driver	1-4 ms
16	WT/DG	Injector 1 Driver	1-4 ms
19	BK/GY	Coil Driver Control	5°, 55 mph: 8° dwell
20	DG	Alternator Field Control	Digital Signals: 0-12-0v
21	BK/RD	Throttle Body Temperature	At 100°F: 2.51v
22	OR/DB	TP Sensor Signal	0.6-1.0v
24	GY	Distributor Reference Signal	Digital Signals: 0-5-0v
25	PK	SCI Transmit	0v
27	BR	A/C Clutch Signal	A/C Off: 12v, On: 1v
28	PK	Closed Throttle Switch	Switch Off: 12v, On: 1v
29	WT/PK	Brake Switch Sense Signal	Brake Off: 0v, On: 12v
30	BR/YL	PNP Switch Sense Signal	In P/N: 0v, Others: 5v
32	PK/BK	MIL (lamp) Control	MIL Off: 12v, On: 1v
33	TN/RD	Speed Control Vacuum Solenoid	Vacuum Increasing: 1v
34	DB/OR	A/C WOT Relay Control	Relay Off: 12v, On: 1v
35	GY/YL	EGR Solenoid Control (Cal)	12v, at 55 mph: 1v
36	BK/OR	Air Switching Solenoid	Solenoid Off: 12v, On: 1v
37-39	---	Not Used	---
40	BR	AIS Motor Control	Pulse Signals
41	BK/DG	HO2S-11 (B1 S1) Signal	0.1-1.1v
43	GY/LB	Tachometer Signal	Pulse Signals
44	OR/WT	Overdrive Lockout Switch	Switch Off: 12v, On: 1v
45	LG	SCI Receive	5v
47	WT/OR	Vehicle Speed Signal	Digital Signal
48	BR/RD	Speed Control Set Switch Signal	S/C & Set Switch On: 3.8v
49	YL/RD	Speed Control On/Off Switch	S/C On: 1v, Off: 12v
50	WT/LG	Speed Control Resume Switch	S/C On: 1v, Off: 12v
51	DB/YL	ASD Relay Control	Relay Off: 12v, On: 1v
52	PK	EVAP Purge Solenoid Control	Solenoid Off: 12v, On: 1v
53	LG/RD	Speed Control Vent Solenoid	Vacuum Decreasing: 1v
54	OR/BK	Part Throttle Unlock Solenoid	Solenoid Off: 12v, On: 1v
55	OR/WT	Overdrive Lockout Solenoid	Solenoid Off: 12v, On: 1v
56	GY/PK	Maintenance Indicator Lamp	Lamp Off: 12v, On: 1v
60	GY/RD	AIS Motor Control	Pulse Signals

Pin Connector Graphic

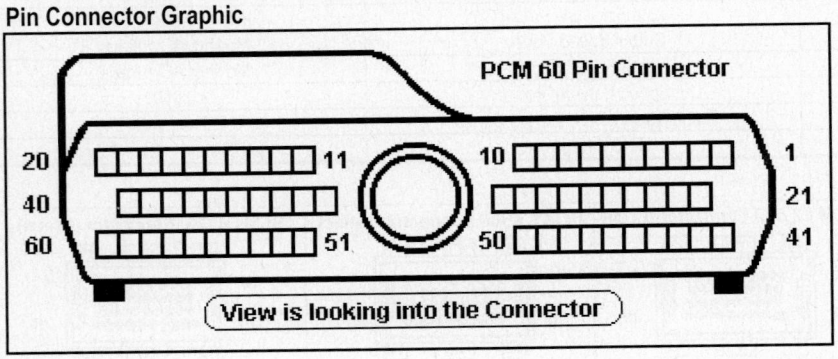

1992-95 Dakota 5.2L V8 MFI VIN Y (A/T) 60 Pin Connector

PCM Pin #	Wire Color	Circuit Description (60 Pin)	Value at Hot Idle
1	DG/RD	MAP Sensor Signal	1.5-1.6v
2	TN/BK	ECT Sensor Signal	At 180°F: 2.80v
3	RD/WT	Keep Alive Power (B+)	12-14v
4	BK/LB	Sensor Ground	<0.050v
5 ('92-'93)	BK/WT	Sensor Ground	<0.050v
6	PK/WT	5-Volt Supply	4.9-5.1v
7	OR	8-Volt Supply	7.9-8.1v
8	---	Not Used	---
9	DB	Ignition Switch Power (B+)	12-14v
10	OR/WT	Overdrive Override Switch	Switch Off: 12v, On: 1v
11	BK/TN	Power Ground	<0.1v
12	BK/TN	Power Ground	<0.1v
13	LB/BR	Injector 4 Driver	1-4 ms
14	YL/WT	Injector 3 Driver	1-4 ms
15	TN	Injector 2 Driver	1-4 ms
16	WT/DG	Injector 1 Driver	1-4 ms
17	DB/TN	Injector 7 Driver	1-4 ms
18	DB/GY	Injector 8 Driver	1-4 ms
19	BK/GY	Coil Driver Control	5°, 55 mph: 8° dwell
20	DG	Alternator Field Control	Digital Signals: 0-12-0v
21	BK/RD	Charge Temperature Sensor	At 100°F: 2.51v
22	OR/DB	TP Sensor Signal	0.6-1.0v
23	---	Not Used	---
24	GY/BK	Distributor Reference Signal	Digital Signals: 0-5-0v
25	PK	SCI Transmit	0v
26, 28	---	Not Used	---
27	BR	A/C Clutch Signal	A/C Off: 12v, On: 1v
29	WT/PK	Brake Switch Sense Signal	Brake Off: 0v, On: 12v
30	BR/YL	PNP Switch Sense Signal	In P/N: 0v, Others: 5v
31	---	Not Used	---
32	PK/BK	MIL (lamp) Control	MIL Off: 12v, On: 1v
33	TN/RD	Speed Control Vacuum Solenoid	Vacuum Increasing: 1v
34	DB/OR	A/C WOT Relay Control	Relay Off: 12v, On: 1v
35 ('92)	GY/YL	EGR Solenoid Control (Cal)	12v, at 55 mph: 1v
35	GY/YL	EGR Solenoid Control	12v, at 55 mph: 1v
36	---	Not Used	---
37	BK/OR	Overdrive Lamp Control	Lamp Off: 12v, On: 1v
38	PK/BK	Injector 5 Driver	1-4 ms
39	GY/RD	AIS Motor 4 Control	Pulse Signals
40	BR/WT	AIS Motor 2 Control	Pulse Signals
41	BK/DG	HO2S-11 (B1 S1) Signal	0.1-1.1v
42	---	Not Used	---
43	GY/LB	Tachometer Signal	Pulse Signals
44	TN/YL	Distributor Sync Signal	Digital Signals: 0-5-0v
45	LG	SCI Receive	5v
46	---	Not Used	---
47	WT/OR	Vehicle Speed Signal	Digital Signal
48	BR/RD	Speed Control Set Switch Signal	S/C & Set Switch On: 3.8v
49	YL/RD	Speed Control On/Off Switch	S/C On: 1v, Off: 12v
50	WT/LG	Speed Control Resume Switch	S/C On: 1v, Off: 12v
51	DB/YL	ASD Relay Control	Relay Off: 12v, On: 1v
52	PK/BK	EVAP Purge Solenoid Control	Solenoid Off: 12v, On: 1v
53	LG/RD	Speed Control Vent Solenoid	Vacuum Decreasing: 1v
54	OR/BK	Part Throttle Unlock Solenoid	Solenoid Off: 12v, On: 1v
55	OR/WT	Overdrive Lockout Solenoid	Solenoid Off: 12v, On: 1v
56 ('92-'93)	GY/PK	Maintenance Indicator Lamp	Lamp Off: 12v, On: 1v
57	DG/BK	ASD Relay Power (B+)	12-14v
58	LG/BK	Injector 6 Driver	1-4 ms
59	PK/BK	AIS Motor 1 Control	Pulse Signals
60	GY/RD	AIS Motor 3 Control	Pulse Signals

1996 Dakota 5.2L V8 MFI VIN Y (All) 'A' Black Connector

PCM Pin #	Wire Color	Circuit Description (32 Pin)	Value at Hot Idle
1, 3, 5	---	Not Used	---
2	DB	Ignition Switch Power (B+)	12-14v
4	BK/LB	Sensor Ground	<0.050v
6	BR/YL	PNP Switch Sense Signal	In P/N: 0v, Others: 5v
7	GY	Coil 1 Driver Control	5°, 55 mph: 8° dwell
8	GY/BK	CKP Sensor Signal	Digital Signals: 0-5-0v
9	---	Not Used	---
10	YL/BK	IAC 2 Driver Control	Pulse Signals
11	BR/WT	IAC 3 Driver Control	Pulse Signals
12-14	---	Not Used	---
15	BK/RD	IAT Sensor Signal	At 100°F: 1.83v
16	TN/BK	ECT Sensor Signal	At 180°F: 2.80v
17	PK/WT	5-Volt Supply	4.9-5.1v
18	TN/YL	CMP Sensor Signal	Digital Signals: 0-5-0v
19	GY/RD	IAC 1 Driver Control	Pulse Signals
20	PK/BK	IAC 4 Driver Control	Pulse Signals
21	---	Not Used	---
22	RD/WT	Keep Alive Power (B+)	12-14v
23	OR/DB	TP Sensor Signal	0.6-1.0v
24	TN/WT	HO2S-11 (B1 S1) Signal	0.1-1.1v
25	TN/PK	HO2S-12 (B1 S2) Signal	0.1-1.1v
26	---	Not Used	---
27	DG/RD	MAP Sensor Signal	1.5-1.6v
28-30	---	Not Used	---
31	BK/TN	Power Ground	<0.1v
32	BK/TN	Power Ground	<0.1v

Pin Connector Graphic

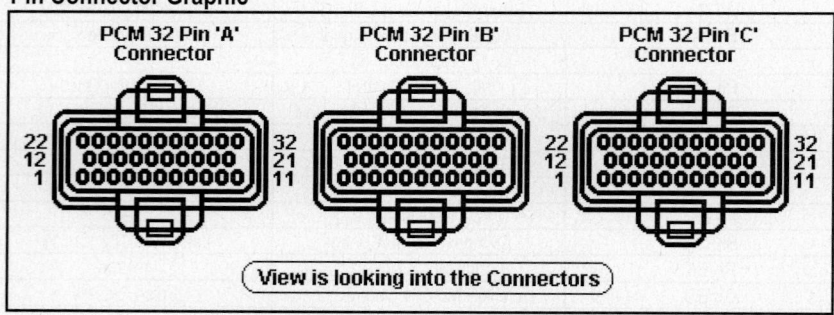

1996 Dakota 5.2L V8 MFI VIN Y (All) 'B' White Connector

PCM Pin #	Wire Color	Circuit Description (32 Pin)	Value at Hot Idle
1	PK	Transmission Temperature Sensor Signal	At 200°F: 2.40v
2	DB	Injector 7 Driver	1-4 ms
3, 7, 9, 14	---	Not Used	---
4	WT/DB	Injector 1 Driver	1-4 ms
5	YL/WT	Injector 3 Driver	1-4 ms
6	GY	Injector 5 Driver	1-4 ms
8	PK	Governor Pressure Solenoid	PWM Signal: 0-12-0v
10	DG	Generator Field Control	Digital Signals: 0-12-0v
11	OR/BK	TCC Solenoid Control	At Cruise w/TCC On: 1v
12	BR/DB	Injector 6 Driver	1-4 ms
13	GY/LB	Injector 8 Driver	1-4 ms
15	TN	Injector 2 Driver	1-4 ms
16	LB/BR	Injector 4 Driver	1-4 ms
17-20, 22-24	---	Not Used	---
21	BR	3-4 Shift Solenoid Control	Solenoid Off: 12v, On: 1v
25, 28	DB, LG	OSS Sensor (-) Signal, (+)	Moving: AC pulse signals
27	WT/OR	Vehicle Speed Signal	Digital Signal
29	LG/RD	Governor Pressure Sensor	0.58v
30	PK/LB	Transmission Relay Control	Relay Off: 12v, On: 1v
31	RD/YL	5-Volt Supply	4.9-5.1v

1996 Dakota 5.2L V8 MFI VIN Y (All) 'C' Gray Connector

PCM Pin #	Wire Color	Circuit Description (32 Pin)	Value at Hot Idle
1	DB/OR	A/C Clutch Relay Control	Relay Off: 12v, On: 1v
2	DB/PK	Radiator Fan Control Relay	Relay Off: 12v, On: 1v
3	DB/YL	ASD Relay Control	Relay Off: 12v, On: 1v
4	TN/RD	Speed Control Vacuum Solenoid	Vacuum Increasing: 1v
5	LG/RD	Speed Control Vent Solenoid	Vacuum Decreasing: 1v
6	LG/OR	Overdrive 'Off' Lamp Control	Lamp Off: 12v, On: 1v
7	PK/BK	Transmission Temperature Lamp Control	Lamp Off: 12v, On: 1v
8-10	---	Not Used	---
11	YL/RD	Speed Control Power Supply	12-14v
12	DG/OR	ASD Relay Power (B+)	12-14v
13	OR/WT	Overdrive Off Switch Sense	Switch Off: 12v, On: 1v
14	---	Not Used	---
15	PK/YL	Battery Temperature Sensor	At 86°F: 1.96v
16	DG/Y	Generator Lamp Control	Lamp Off: 12v, On: 1v
17	BK/PK	MIL (lamp) Control	MIL Off: 12v, On: 1v
18	---	Not Used	---
19	BR	Fuel Pump Relay Control	Relay Off: 12v, On: 1v
20	PK/BK	EVAP Purge Solenoid Control	PWM Signal: 0-12-0v
22	BR	A/C Request Signal	A/C Off: 12v, On: 1v
23	LG	A/C Select Signal	A/C Off: 12v, On: 1v
24	WT/PK	Brake Switch Sense Signal	Brake Off: 0v, On: 12v
26	LB/BK	Fuel Level Sensor Signal	Full: 0.56v, 1/2 full: 2.5v
27	PK	SCI Transmit	0v
28	WT	CCD Bus (-) Signal	<0.050v
29	LG	SCI Receive	5v
30	PK/BR	CCD Bus (+) Signal	Digital Signals: 0-5-0v
31	GY/LB	Tachometer Signal	Pulse Signals
32	WT/LG	Speed Control Set Switch Signal	S/C & Set Switch On: 3.8v

Standard Colors and Abbreviations

Abbreviation	Color	Abbreviation	Color	Abbreviation	Color
BK	Black	GY	Gray	RD	Red
BL	Blue	GN	Green	TN	Tan
BR	Brown	LG	Light Green	VT	Violet
DB	Dark Blue	OR	Orange	WT	White
DG	Dark Green	PK	Pink	YL	Yellow

1997 Dakota 5.2L V8 MFI VIN Y (All) 'A' Black Connector

PCM Pin #	Wire Color	Circuit Description (32 Pin)	Value at Hot Idle
1	---	Not Used	---
2	LG/BK	Ignition Switch Power (B+)	12-14v
3	---	Not Used	---
4	BK/LB	Sensor Ground	<0.050v
5	---	Not Used	---
6	BK/WT	PNP Switch Sense Signal	In P/N: 0v, Others: 5v
7	BK/GY	Coil 1 Driver Control	5°, 55 mph: 8° dwell
8	GY/BK	CKP Sensor Signal	Digital Signals: 0-5-0v
9	---	Not Used	---
10	YL/BK	IAC 2 Driver Control	Pulse Signals
11	BR/WT	IAC 3 Driver Control	Pulse Signals
12	GY/WT	PSP Switch Sense Signal	Straight: 0v, Turning: 5v
13-14	---	Not Used	---
15	BK/RD	IAT Sensor Signal	At 100°F: 1.83v
16	TN/BK	ECT Sensor Signal	At 180°F: 2.80v
17	PK/WT	5-Volt Supply	4.9-5.1v
18	TN/YL	CMP Sensor Signal	Digital Signals: 0-5-0v
19	GY/RD	IAC 1 Driver Control	Pulse Signals
20	PK/BK	IAC 4 Driver Control	Pulse Signals
21	---	Not Used	---
22	RD/WT	Keep Alive Power (B+)	12-14v
23	OR/DB	TP Sensor Signal	0.6-1.0v
24	OR/TN	HO2S-11 (B1 S1) Signal	0.1-1.1v
25	OR/BK	HO2S-12 (B1 S2) Signal	0.1-1.1v
26	---	Not Used	---
27	DG/RD	MAP Sensor Signal	1.5-1.6v
28-30	---	Not Used	---
31	BK/TN	Power Ground	<0.1v
32	BK/TN	Power Ground	<0.1v

Pin Connector Graphic

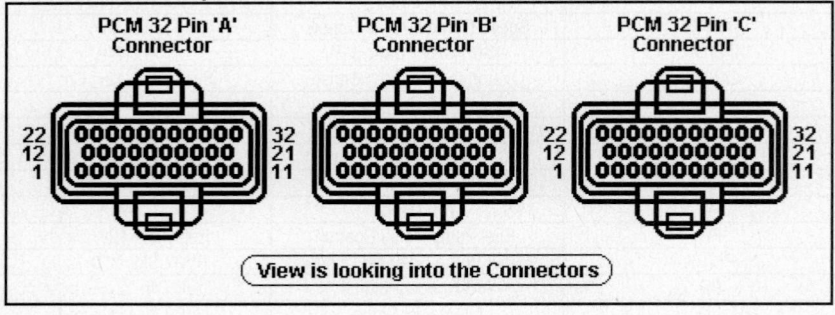

1997 Dakota 5.2L V8 MFI VIN Y (All) 'B' White Connector

PCM Pin #	Wire Color	Circuit Description (32 Pin)	Value at Hot Idle
1	PK	Transmission Temperature Sensor Signal	At 200°F: 2.40v
2	PK/WT	Injector 7 Driver	1-4 ms
3, 7	---	Not Used	---
4	WT/DB	Injector 1 Driver	1-4 ms
5	YL/WT	Injector 3 Driver	1-4 ms
6	GY	Injector 5 Driver	1-4 ms
8	PK	Governor Pressure Solenoid	PWM Signal: 0-12-0v
9, 14	---	Not Used	---
10	DG	Generator Field Control	Digital Signals: 0-12-0v
11	OR/BK	TCC Solenoid Control	At Cruise w/TCC On: 1v
12	BR/DB	Injector 6 Driver	1-4 ms
13	GY/LB	Injector 8 Driver	1-4 ms
15	TN	Injector 2 Driver	1-4 ms
16	LB/BR	Injector 4 Driver	1-4 ms
21	BR	Overdrive Solenoid Control	Cruise w/solenoid On: 1v
23	GY/YL	Engine Oil Pressure Sensor	1.6v at 24 psi
25	DB/BK	OSS Sensor (-) Signal	Moving: AC pulse signals
27	WT/OR	Vehicle Speed Signal	Digital Signal
28	LG/WT	OSS Sensor (+) Signal	Moving: AC pulse signals
29	LG/RD	Governor Pressure Sensor	0.58v
30	PK/BK	Transmission Relay Control	Relay Off: 12v, On: 1v
31	OR	5-Volt Supply	4.9-5.1v

1997 Dakota 5.2L V8 MFI VIN Y (All) 'C' Gray Connector

PCM Pin #	Wire Color	Circuit Description (32 Pin)	Value at Hot Idle
1	DB/OR	A/C Clutch Relay Control	Relay Off: 12v, On: 1v
2	DB/PK	Radiator Fan Control Relay	Relay Off: 12v, On: 1v
3	DB/YL	ASD Relay Control	Relay Off: 12v, On: 1v
4	TN/RD	Speed Control Vacuum Solenoid	Vacuum Increasing: 1v
5	LG/RD	Speed Control Vent Solenoid	Vacuum Decreasing: 1v
6	LG/OR	Overdrive Off Lamp Control	Lamp Off: 12v, On: 1v
7-10	---	Not Used	---
11	YL/RD	Speed Control Power Supply	12-14v
12	DG/OR	ASD Relay Power (B+)	12-14v
13	OR/WT	Overdrive Off Switch Sense	Switch Off: 12v, On: 1v
14	---	Not Used	---
15	PK/YL	Battery Temperature Sensor	At 86°F: 1.96v
16-18, 21	---	Not Used	---
19	LB/OR	Fuel Pump Relay Control	Relay Off: 12v, On: 1v
20	PK/BK	EVAP Purge Solenoid Control	PWM Signal: 0-12-0v
22	BR	A/C Request Signal	A/C Off: 12v, On: 1v
23	LG/WT	A/C Select Signal	A/C Off: 12v, On: 1v
24	WT/PK	Brake Switch Sense Signal	Brake Off: 0v, On: 12v
25	DG/BK	Generator Field Source	12-14v
26	DB	Fuel Level Sensor Signal	Full: 0.56v, 1/2 full: 2.5v
27	PK	SCI Transmit Signal	0v
28	WT/BK	CCD Bus (-)	<0.050v
29	LG	SCI Receive Signal	5v
30	PK/BR	CCD Bus (+)	Digital Signals: 0-5-0v
31	---	Not Used	---
32	GY/LB	Speed Control Set Switch Signal	S/C & Set Switch On: 3.8v

Standard Colors and Abbreviations

Abbreviation	Color	Abbreviation	Color	Abbreviation	Color
BK	Black	GY	Gray	RD	Red
BL	Blue	GN	Green	TN	Tan
BR	Brown	LG	Light Green	VT	Violet
DB	Dark Blue	OR	Orange	WT	White
DG	Dark Green	PK	Pink	YL	Yellow

1998-99 Dakota 5.2L V8 MFI VIN Y 'A' Black Connector

PCM Pin #	Wire Color	Circuit Description (32 Pin)	Value at Hot Idle
1	---	Not Used	---
2	LG/BK	Ignition Switch Power (B+)	12-14v
3	---	Not Used	---
4	BK/LB	Sensor Ground	<0.050v
5	---	Not Used	---
6	BK/WT	PNP Switch Sense Signal	In P/N: 0v, Others: 5v
7	BK/GY	Coil 1 Driver Control	5°, 55 mph: 8° dwell
8	GY/BK	CKP Sensor Signal	Digital Signals: 0-5-0v
9	---	Not Used	---
10	YL/BK	IAC 2 Driver Control	Pulse Signals
11	BR/WT	IAC 3 Driver Control	Pulse Signals
12-14	---	Not Used	---
15	BK/RD	IAT Sensor Signal	At 100°F: 1.83v
16	TN/BK	ECT Sensor Signal	At 180°F: 2.80v
17	PK/WT	5-Volt Supply	4.9-5.1v
18	TN/YL	CMP Sensor Signal	Digital Signals: 0-5-0v
19	GY/RD	IAC 1 Driver Control	Pulse Signals
20	PK/BK	IAC 4 Driver Control	Pulse Signals
21	---	Not Used	---
22	RD/WT	Keep Alive Power (B+)	12-14v
23	OR/DB	TP Sensor Signal	0.6-1.0v
24	OR/TN	HO2S-11 (B1 S1) Signal	0.1-1.1v
25	OR/BK	HO2S-12 (B1 S2) Signal	0.1-1.1v
26	---	Not Used	---
27	DG/RD	MAP Sensor Signal	1.5-1.6v
28-30	---	Not Used	---
31	BK/TN	Power Ground	<0.1v
32	BK/TN	Power Ground	<0.1v

Pin Connector Graphic

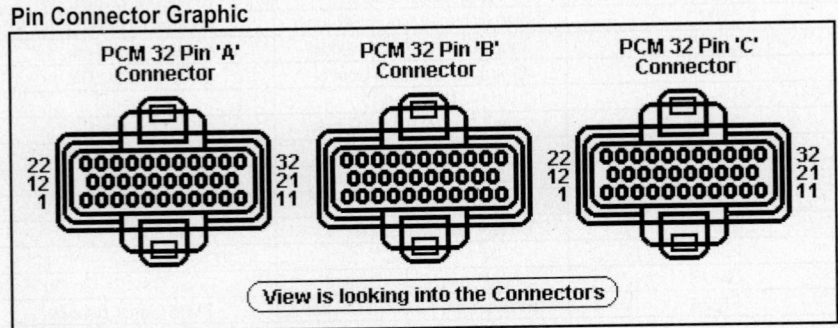

PCM 32 Pin 'A' Connector   PCM 32 Pin 'B' Connector   PCM 32 Pin 'C' Connector

View is looking into the Connectors

1998-99 Dakota 5.2L V8 MFI VIN Y 'B' White Connector

PCM Pin #	Wire Color	Circuit Description (32 Pin)	Value at Hot Idle
1	GY/BK	Transmission Temperature Sensor Signal	At 200°F: 2.40v
2	PK	Injector 7 Driver	1-4 ms
3	---	Not Used	---
4	WT/DB	Injector 1 Driver	1-4 ms
5	YL/WT	Injector 3 Driver	1-4 ms
6	GY	Injector 5 Driver	1-4 ms
7	---	Not Used	---
8	PK/WT	Variable Force Solenoid	PWM Signal: 0-12-0v
9	---	Not Used	---
10	DG	Generator Field Control	Digital Signals: 0-12-0v
11	OR/BK	TCC Solenoid Control	At Cruise w/TCC On: 1v
12	BR/DB	Injector 6 Driver	1-4 ms
13	GY/LB	Injector 8 Driver	1-4 ms
14	---	Not Used	---
15	TN	Injector 2 Driver	1-4 ms
16	LB/BR	Injector 4 Driver	1-4 ms
17 ('00)	DB/P	Radiator Fan Relay Control	Relay Off: 12v, On: 1v
21	BR	Overdrive Solenoid Control	Cruise w/solenoid On: 1v
23	GY/YL	Engine Oil Pressure Sensor	1.6v at 24 psi
25	DB/BK	OSS Sensor (-) Signal	Moving: AC pulse signals
27	WT/OR	Vehicle Speed Signal	Digital Signal
28	LG/WT	OSS Sensor (+) Signal	Moving: AC pulse signals
29	LG/RD	Governor Pressure Sensor	0.58v
30	PK/BK	Transmission Relay Control	Relay Off: 12v, On: 1v
31	OR	5-Volt Supply	4.9-5.1v

Pin Connector Graphic

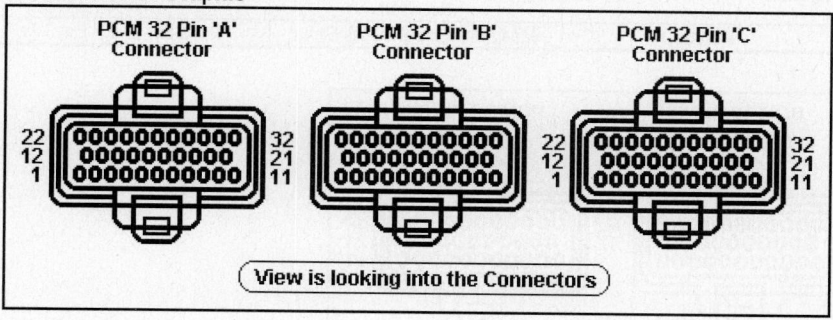

PCM 32 Pin 'A' Connector   PCM 32 Pin 'B' Connector   PCM 32 Pin 'C' Connector

View is looking into the Connectors

1998-99 Dakota 5.2L V8 MFI VIN Y 'C' Gray Connector

PCM Pin #	Wire Color	Circuit Description (32 Pin)	Value at Hot Idle
1	DB/OR	A/C Clutch Relay Control	Relay Off: 12v, On: 1v
2 ('98-'99)	DB/PK	Radiator Fan Control Relay	Relay Off: 12v, On: 1v
3	DB/YL	ASD Relay Control	Relay Off: 12v, On: 1v
4	TN/RD	Speed Control Vacuum Solenoid	Vacuum Increasing: 1v
5	LG/RD	Speed Control Vent Solenoid	Vacuum Decreasing: 1v
6-7	---	Not Used	---
8	DG/BK	HO2S-11 HTR Control (California)	Relay Off: 12v, On: 1v
9	OR/RD	HO2S-12 HTR Control (California)	Relay Off: 12v, On: 1v
10	YL/DG	LDP Solenoid Control	PWM Signal: 0-12-0v
11	YL/RD	Speed Control Power Supply	12-14v
12	DG/OR	ASD Relay Power (B+)	12-14v
13	OR/WT	Overdrive Off Switch Sense	Switch Off: 12v, On: 1v
14	YL/WT	LDP Switch Sense Signal	LDP Switch Closed: 0v
15	PK/YL	Battery Temperature Sensor	At 86°F: 1.96v
16-18	---	Not Used	---
19	LB/OR	Fuel Pump Relay Control	Relay Off: 12v, On: 1v
20	PK/BK	EVAP Purge Solenoid Control	PWM Signal: 0-12-0v
21	---	Not Used	---
22	BR	A/C Request Signal	A/C Off: 12v, On: 1v
23	LG/WT	A/C Select Signal	A/C Off: 12v, On: 1v
24	WT/PK	Brake Switch Sense Signal	Brake Off: 0v, On: 12v
25	DG/BK	Generator Field Source	12-14v
26	DB	Fuel Level Sensor Signal	Full: 0.56v, 1/2 full: 2.5v
27	PK	SCI Transmit Signal	0v
28	WT	CCD Bus (-) Signal	<0.050v
29	LG	SCI Receive Signal	5v
30	PK/BR	CCD Bus (+) Signal	Digital Signals: 0-5-0v
31	---	Not Used	---
32	RD/LG	Speed Control Set Switch Signal	S/C & Set Switch On: 3.8v

Pin Connector Graphic

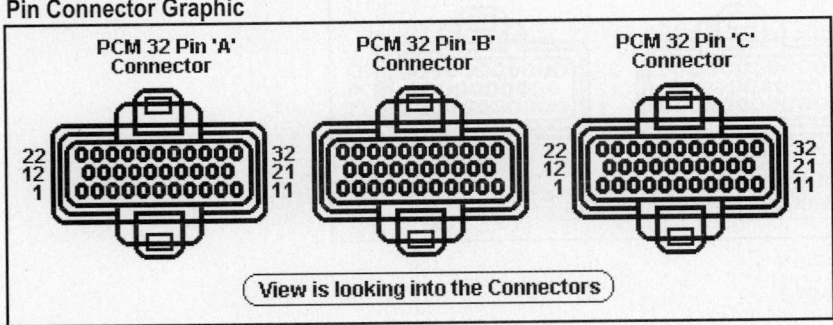

PCM 32 Pin 'A' Connector   PCM 32 Pin 'B' Connector   PCM 32 Pin 'C' Connector

View is looking into the Connectors

1998-99 Dakota 5.9L V8 MFI VIN Z (A/T) 'C1' Black Connector

PCM Pin #	Wire Color	Circuit Description (32 Pin)	Value at Hot Idle
1	---	Not Used	---
2	LG/BK	Ignition Switch Power (B+)	12-14v
3	---	Not Used	---
4	BK/LB	Sensor Ground	<0.050v
5	---	Not Used	---
6	BK/WT	PNP Switch Sense Signal	In P/N: 0v, Others: 5v
7	BK/GY	Coil 1 Driver Control	5°, 55 mph: 8° dwell
8	GY/BK	CKP Sensor Signal	Digital Signals: 0-5-0v
9	---	Not Used	---
10	YL/BK	IAC 2 Driver Control	Pulse Signals
11	BR/WT	IAC 3 Driver Control	Pulse Signals
12-14	---	Not Used	---
15	BK/RD	IAT Sensor Signal	At 100°F: 1.83v
16	TN/BK	ECT Sensor Signal	At 180°F: 2.80v
17	PK/WT	5-Volt Supply	4.9-5.1v
18	TN/YL	CMP Sensor Signal	Digital Signals: 0-5-0v
19	GY/RD	IAC 1 Driver Control	Pulse Signals
20	PK/BK	IAC 4 Driver Control	Pulse Signals
21	---	Not Used	---
22	RD/WT	Keep Alive Power (B+)	12-14v
23	OR/DB	TP Sensor Signal	0.6-1.0v
24	TN/WT	HO2S-11 (B1 S1) Signal	0.1-1.1v
25	OR/BK	HO2S-12 (B1 S2) Signal	0.1-1.1v
26	---	Not Used	---
27	DG/RD	MAP Sensor Signal	1.5-1.6v
28-29	---	Not Used	---
30	---	Not Used	---
31	BK/TN	Power Ground	<0.1v
32	BK/TN	Power Ground	<0.1v

Pin Connector Graphic

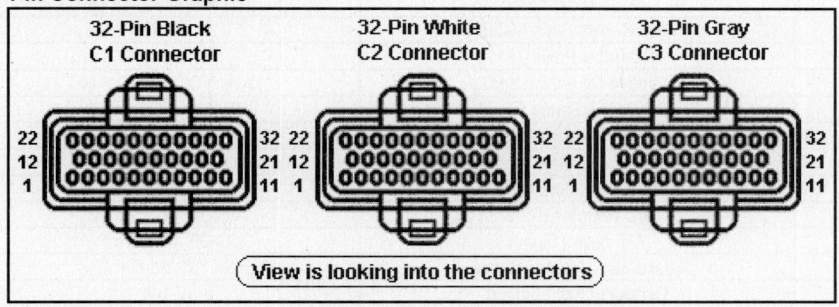

1998-99 Dakota 5.9L V8 MFI VIN Z (A/T) 'C2' White Connector

PCM Pin #	Wire Color	Circuit Description (32 Pin)	Value at Hot Idle
1	PK	Transmission Temperature Sensor Signal	At 200°F: 2.40v
2	VT	Injector 7 Driver	1-4 ms
4	WT/DB	Injector 1 Driver	1-4 ms
5	YL/WT	Injector 3 Driver	1-4 ms
6	GY	Injector 5 Driver	1-4 ms
8	PK	Governor Pressure Solenoid	PWM Signal: 0-12-0v
10	DG	Generator Field Control	Digital Signals: 0-12-0v
11	OR/BK	TCC Solenoid Control	At Cruise w/TCC On: 1v
12	BR/DB	Injector 6 Driver	1-4 ms
13	GY/LB	Injector 8 Driver	1-4 ms
15	TN	Injector 2 Driver	1-4 ms
16	LB/BR	Injector 4 Driver	1-4 ms
17-20	---	Not Used	---
21	BR	Overdrive Solenoid Control	Cruise w/solenoid On: 1v
22, 32	---	Not Used	---
23	GY/YL	Engine Oil Pressure Sensor	1.6v at 24 psi
25	DB/BK	OSS Sensor (-) Signal	Moving: AC pulse signals
27	WT/OR	Vehicle Speed Signal	Digital Signal
28	LG/WT	OSS Sensor (+) Signal	Moving: AC pulse signals
29	LG/RD	Governor Pressure Sensor	0.58v
30	PK/BK	Transmission Relay Control	Relay Off: 12v, On: 1v
31	VT/WT	5-Volt Supply	4.9-5.1v

1998-99 Dakota 5.9L V8 MFI VIN Z (A/T) 'C3' Gray Connector

PCM Pin #	Wire Color	Circuit Description (32 Pin)	Value at Hot Idle
1	DB/OR	A/C Clutch Relay Control	Relay Off: 12v, On: 1v
3	DB/YL	ASD Relay Control	Relay Off: 12v, On: 1v
4	TN/RD	Speed Control Vacuum Solenoid	Vacuum Increasing: 1v
5	LG/RD	Speed Control Vent Solenoid	Vacuum Decreasing: 1v
6-9	---	Not Used	---
10	YL/DG	LDP Solenoid Control	PWM Signal: 0-12-0v
11	YL/RD	Speed Control Power Supply	12-14v
12	DG/OR	ASD Relay Power (B+)	12-14v
13	OR/WT	Overdrive Off Switch Sense	Switch Off: 12v, On: 1v
14	YL/WT	LDP Switch Sense Signal	LDP Switch Closed: 0v
15	PK/YL	Battery Temperature Sensor	At 86°F: 1.96v
19	LB/OR	Fuel Pump Relay Control	Relay Off: 12v, On: 1v
20	PK/BK	EVAP Purge Solenoid Control	PWM Signal: 0-12-0v
22	BR	A/C Request Signal	A/C Off: 12v, On: 1v
23	LG/WT	A/C Select Signal	A/C Off: 12v, On: 1v
24	WT/PK	Brake Switch Sense Signal	Brake Off: 0v, On: 12v
25	DG, W	Generator Field Source	12-14v
26	DB	Fuel Level Sensor Signal	Full: 0.56v, 1/2 full: 2.5v
27	PK/DB	SCI Transmit Signal	0v
28	WT/BK	CCD Bus (-) Signal	<0.050v
29	LG	SCI Receive Signal	5v
30	PK/BR	CCD Bus (+) Signal	Digital Signals: 0-5-0v
32	RD/LG	Speed Control Set Switch Signal	S/C & Set Switch On: 3.8v

Standard Colors and Abbreviations

Abbreviation	Color	Abbreviation	Color	Abbreviation	Color
BK	Black	GY	Gray	RD	Red
BL	Blue	GN	Green	TN	Tan
BR	Brown	LG	Light Green	VT	Violet
DB	Dark Blue	OR	Orange	WT	White
DG	Dark Green	PK	Pink	YL	Yellow

2000 Dakota 5.9L V8 MFI VIN Z (A/T) 'C1' Black Connector

PCM Pin #	Wire Color	Circuit Description (32 Pin)	Value at Hot Idle
1	---	Not Used	---
2	LG/BK	Ignition Switch Power (B+)	12-14v
3	---	Not Used	---
4	BK/LB	Sensor Ground	<0.050v
5	---	Not Used	---
6	BK/WT	PNP Switch Sense Signal	In P/N: 0v, Others: 5v
7	BK/GY	Coil 1 Driver Control	5°, 55 mph: 8° dwell
8	GY/BK	CKP Sensor Signal	Digital Signals: 0-5-0v
9	---	Not Used	---
10	YL/BK	IAC 2 Driver Control	Pulse Signals
11	BR/WT	IAC 3 Driver Control	Pulse Signals
12-14	---	Not Used	---
15	BK/RD	IAT Sensor Signal	At 100°F: 1.83v
16	TN/BK	ECT Sensor Signal	At 180°F: 2.80v
17	VT/WT	5-Volt Supply	4.9-5.1v
18	TN/YL	CMP Sensor Signal	Digital Signals: 0-5-0v
19	GY/RD	IAC 1 Driver Control	Pulse Signals
20	VT/BK	IAC 4 Driver Control	Pulse Signals
21	---	Not Used	---
22	RD/WT	Keep Alive Power (B+)	12-14v
23	OR/DB	TP Sensor Signal	0.6-1.0v
24	TN/WT	HO2S-11 (B1 S1) Signal	0.1-1.1v
25	OR/BK	HO2S-12 (B1 S2) Signal	0.1-1.1v
26	---	Not Used	---
27	DG/RD	MAP Sensor Signal	1.5-1.6v
28-30	---	Not Used	---
31	BK/TN	Power Ground	<0.1v
32	BK/TN	Power Ground	<0.1v

Pin Connector Graphic

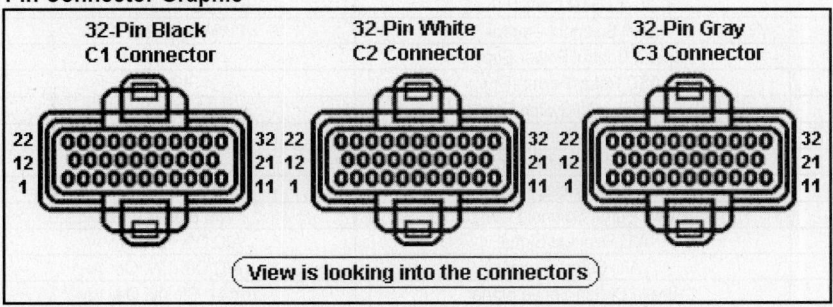

2000 Dakota 5.9L V8 MFI VIN Z (A/T) White 'C2' Connector

PCM Pin #	Wire Color	Circuit Description (32 Pin)	Value at Hot Idle
1	PK	Transmission Temperature Sensor Signal	At 200°F: 2.40v
2	VT	Injector 7 Driver	1-4 ms
4	WT/DB	Injector 1 Driver	1-4 ms
5	YL/WT	Injector 3 Driver	1-4 ms
6	GY	Injector 5 Driver	1-4 ms
8	PK	Governor Pressure Solenoid	PWM Signal: 0-12-0v
10	DG	Generator Field Control	Digital Signals: 0-12-0v
11	OR/BK	TCC Solenoid Control	At Cruise w/TCC On: 1v
12	BR/DB	Injector 6 Driver	1-4 ms
13	GY/LB	Injector 8 Driver	1-4 ms
15	TN	Injector 2 Driver	1-4 ms
16	LB/BR	Injector 4 Driver	1-4 ms
17	DB/PK	Radiator Fan Control Relay	Relay Off: 12v, On: 1v
21	BR	Overdrive Solenoid Control	Cruise w/solenoid On: 1v
23	GY/YL	Engine Oil Pressure Sensor	1.6v at 24 psi
25	DB	OSS Sensor (-) Signal	Moving: AC pulse signals
27	WT/OR	Vehicle Speed Signal	Digital Signal
28	LG	OSS Sensor (+) Signal	Moving: AC pulse signals
29	LG/RD	Governor Pressure Sensor	0.58v
30	PK	Transmission Relay Control	Relay Off: 12v, On: 1v
31	VT/WT	5-Volt Supply	4.9-5.1v

2000 Dakota 5.9L V8 MFI VIN Z (A/T) 'C3' Gray Connector

PCM Pin #	Wire Color	Circuit Description (32 Pin)	Value at Hot Idle
1	DB/OR	A/C Clutch Relay Control	Relay Off: 12v, On: 1v
3	DB/YL	ASD Relay Control	Relay Off: 12v, On: 1v
4	TN/RD	Speed Control Vacuum Solenoid	Vacuum Increasing: 1v
5	LG/RD	Speed Control Vent Solenoid	Vacuum Decreasing: 1v
8	DG/BK	HO2S-11 HTR Control (Cal)	Relay Off: 12v, On: 1v
9	OR/RD	HO2S-12 HTR Control (Cal)	Relay Off: 12v, On: 1v
10	YL/DG	LDP Solenoid Control	PWM Signal: 0-12-0v
11	YL/RD	Speed Control Power Supply	12-14v
12	DG/OR	ASD Relay Power (B+)	12-14v
13	OR/WT	Overdrive Off Switch Sense	Switch Off: 12v, On: 1v
14	YL/WT	LDP Switch Sense Signal	LDP Switch Closed: 0v
15	PK/YL	Battery Temperature Sensor	At 86°F: 1.96v
19	LB/OR	Fuel Pump Relay Control	Relay Off: 12v, On: 1v
20	PK/BK	EVAP Purge Solenoid Control	PWM Signal: 0-12-0v
22	BR	A/C Request Signal	A/C Off: 12v, On: 1v
23	LG/WT	A/C Select Signal	A/C Off: 12v, On: 1v
24	WT/PK	Brake Switch Sense Signal	Brake Off: 0v, On: 12v
25	DG, W	Generator Field Source	12-14v
26	DB	Fuel Level Sensor Signal	Full: 0.56v, 1/2 full: 2.5v
27	PK/DB	SCI Transmit Signal	0v
28	WT/BK	CCD Bus (-) Signal	<0.050v
29	LG	SCI Receive Signal	5v
30	PK/BR	CCD Bus (+) Signal	Digital Signals: 0-5-0v
32	RD/LG	Speed Control Set Switch Signal	S/C & Set Switch On: 3.8v

Standard Colors and Abbreviations

Abbreviation	Color	Abbreviation	Color	Abbreviation	Color
BK	Black	GY	Gray	RD	Red
BL	Blue	GN	Green	TN	Tan
BR	Brown	LG	Light Green	VT	Violet
DB	Dark Blue	OR	Orange	WT	White
DG	Dark Green	PK	Pink	YL	Yellow

2001-03 Dakota 5.9L V8 MFI VIN Z (A/T) 'C1' Black Connector

PCM Pin #	Wire Color	Circuit Description (32 Pin)	Value at Hot Idle
1	---	Not Used	---
2	LG/BK	Ignition Switch Power (B+)	12-14v
3	---	Not Used	---
4	BK/LB	Sensor Ground	<0.050v
5	---	Not Used	---
6	BK/WT	PNP Switch Sense Signal	In P/N: 0v, Others: 5v
7	BK/GY	Coil 1 Driver Control	5°, 55 mph: 8° dwell
8	GY/BK	CKP Sensor Signal	Digital Signals: 0-5-0v
9	---	Not Used	---
10	YL/BK	IAC 2 Driver Control	Pulse Signals
11	BR/WT	IAC 3 Driver Control	Pulse Signals
12-14	---	Not Used	---
15	BK/RD	IAT Sensor Signal	At 100°F: 1.83v
16	TN/BK	ECT Sensor Signal	At 180°F: 2.80v
17	OR	5-Volt Supply	4.9-5.1v
18	TN/YL	CMP Sensor Signal	Digital Signals: 0-5-0v
19	GY/RD	IAC 1 Driver Control	Pulse Signals
20	VT/BK	IAC 4 Driver Control	Pulse Signals
21	---	Not Used	---
22	RD/WT	Keep Alive Power (B+)	12-14v
23	OR/DB	TP Sensor Signal	0.6-1.0v
24	BK/DG	HO2S-11 (B1 S1) Signal	0.1-1.1v
25	TN/WT	HO2S-12 (B1 S2) Signal	0.1-1.1v
26	LG/RD	HO2S-21 (B2 S1) Signal	0.1-1.1v
27	DG/RD	MAP Sensor Signal	1.5-1.6v
28	---	Not Used	---
29	TN/WT	HO2S-22 (B2 S2) Signal	0.1-1.1v
30	---	Not Used	---
31	BK/TN	Power Ground	<0.1v
32	BK/TN	Power Ground	<0.1v

Pin Connector Graphic

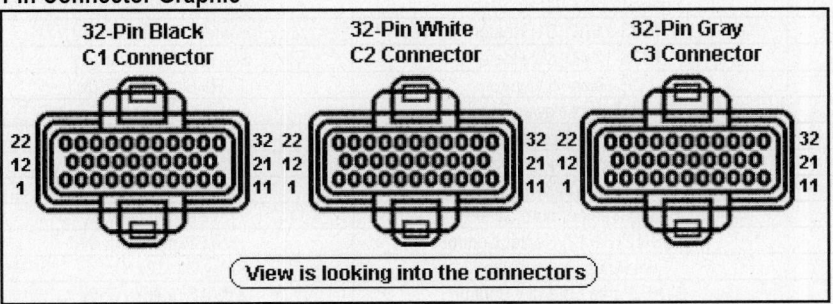

2001-03 Dakota 5.9L V8 MFI VIN Z (A/T) White 'C2' Connector

PCM Pin #	Wire Color	Circuit Description (32 Pin)	Value at Hot Idle
1	GY/BK	Transmission Temperature Sensor Signal	At 200°F: 2.40v
2	VT	Injector 7 Driver	1-4 ms
3, 7, 9, 14, 18	---	Not Used	---
4	WT/DB	Injector 1 Driver	1-4 ms
5	YL/WT	Injector 3 Driver	1-4 ms
6	GY	Injector 5 Driver	1-4 ms
8	VT/WT	Variable Force Solenoid	PWM Signal: 0-12-0v
10	DG	Generator Field Control	Digital Signals: 0-12-0v
11	OR/BK	TCC Solenoid Control	At Cruise w/TCC On: 1v
12	BR/DB	Injector 6 Driver	1-4 ms
13	GY/LB	Injector 8 Driver	1-4 ms
15	TN	Injector 2 Driver	1-4 ms
16	LB/BR	Injector 4 Driver	1-4 ms
17	DB/PK	Radiator Fan Control Relay	Relay Off: 12v, On: 1v
19	DB	AC Pressure Sensor Signal	0.90v at 79 psi
20, 22, 26, 32	---	Not Used	---
21	BR	3-4 Shift Solenoid Control	Cruise w/solenoid on: 1v
23	GY/YL	Engine Oil Pressure Sensor	1.6v at 24 psi
25	DB/BK	OSS Sensor (-) Signal	Moving: AC pulse signals
27	WT/OR	Vehicle Speed Signal	Digital Signals
28	LG/WT	OSS Sensor (+) Signal	Moving: AC pulse signals
29	LG/RD	Governor Pressure Sensor	0.58v
30	PK	Transmission Relay Control	Relay Off: 12v, On: 1v
31	VT/WT	5-Volt Supply	4.9-5.1v

2001-03 Dakota 5.9L V8 MFI VIN Z (A/T) 'C3' Gray Connector

PCM Pin #	Wire Color	Circuit Description (32 Pin)	Value at Hot Idle
1	DB/OR	A/C Clutch Relay Control	Relay Off: 12v, On: 1v
2, 6-7, 17-18	---	Not Used	---
3	DB/YL	ASD Relay Control	Relay Off: 12v, On: 1v
4	TN/RD	Speed Control Vacuum Solenoid	Vacuum Increasing: 1v
5	LG/RD	Speed Control Vent Solenoid	Vacuum Decreasing: 1v
8	VT/WT	HO2S-11 (B1 S1) Heater	Heater Off: 12v, On: 1v
9	DG/BK	HO2S-12 (B1 S2) Heater	Heater Off: 12v, On: 1v
10	WT/DG	LDP Solenoid Control	PWM Signal: 0-12-0v
11	YL/RD	Speed Control Power Supply	12-14v
12	DG/OR	ASD Relay Power (B+)	12-14v
13	OR/WT	Overdrive Off Switch Sense	Switch Off: 12v, On: 1v
14	OR	LDP Switch Sense Signal	LDP Switch Closed: 0v
15	PK/YL	Battery Temperature Sensor	At 86°F: 1.96v
16	VT/OR	HO2S-12 (B1 S2) Heater Control	PWM Signal: 0-12-0v
19	BR	Fuel Pump Relay Control	Relay Off: 12v, On: 1v
20	PK/BK	EVAP Purge Solenoid Control	PWM Signal: 0-12-0v
21, 23, 28, 31	---	Not Used	---
22	BR	A/C Request Signal	A/C Off: 12v, On: 1v
24	WT/PK	Brake Switch Sense	Brake Off: 0v, On: 12v
25	WT/DB	Generator Field Source	12-14v
26	DB/WT	Fuel Level Sensor Signal	Full: 0.56v, 1/2 full: 2.5v
27	PK, LG	SCI Transmit, SCI Receive	0v, 5v
29	LG	SCI Receive	5v
32	RD/LG	Speed Control Switch Signal	S/C & Set Switch On: 3.8v

Standard Colors and Abbreviations

Abbreviation	Color	Abbreviation	Color	Abbreviation	Color
BK	Black	GY	Gray	RD	Red
BL	Blue	GN	Green	TN	Tan
BR	Brown	LG	Light Green	VT	Violet
DB	Dark Blue	OR	Orange	WT	White
DG	Dark Green	PK	Pink	YL	Yellow

2002-03 Pickup 3.7L V6 MFI SOHC VIN K (All) 'C1' Black Connector

PCM Pin #	Wire Color	Circuit Description (32 Pin)	Value at Hot Idle
1	TN/OR	Coil 3 Driver Control	5°, 55 mph: 8° dwell
2	LG/BK	Ignition Switch Power (Start-Run)	12-14v
3	TN/LG	Coil 4 Driver Control	5°, 55 mph: 8° dwell
4	BK/LB	Sensor Ground	<0.050v
5	TN/LB	Coil 6 Driver Control	5°, 55 mph: 8° dwell
6	BR/YL	PNP Switch Sense Signal	In P/N: 0v, Others: 5v
7	TN/RD	Coil 1 Driver Control	5°, 55 mph: 8° dwell
8	GY/BK	CKP Sensor Signal	Digital Signals: 0-5-0v
9	---	Not Used	---
10	GY/RD	IAC 3 Driver Control	Pulse Signals
11	YL/BK	IAC 2 Driver Control	Pulse Signals
12	YL/BK	Power Steering Pressure Switch	Straight: 0v, Turning: 5v
13	YL	Ignition Switch Power (Start-Run)	12-14v
14	---	Not Used	---
15	BK/RD	IAT Sensor Signal	At 100°F: 1.83v
16	TN/BK	ECT Sensor Signal	At 180°F: 2.80v
17	OR	5-Volt Supply	4.9-5.1v
18	TN/YL	CMP Sensor Signal	Digital Signals: 0-5-0v
19	VT/BK	IAC 4 Driver Control	Pulse Signals
20	BR/WT	IAC 1 Driver Control	Pulse Signals
21	TN/DG	Coil 3 Driver Control	5°, 55 mph: 8° dwell
22	RD/WT	Keep Alive Power (B+)	12-14v
23	OR/DB	TP Sensor Signal	0.6-1.0v
24	BK/DG	HO2S-11 (B1 S1) Signal	0.1-1.1v
25	TN/WT	HO2S-12 (B1 S2) Signal	0.1-1.1v
26	---	Not Used	---
27	OR/RD	MAP Sensor Signal	1.5-1.6v
28-30	---	Not Used	---
31	BK/TN	Power Ground	<0.1v
32	BK/TN	Power Ground	<0.1v

Pin Connector Graphic

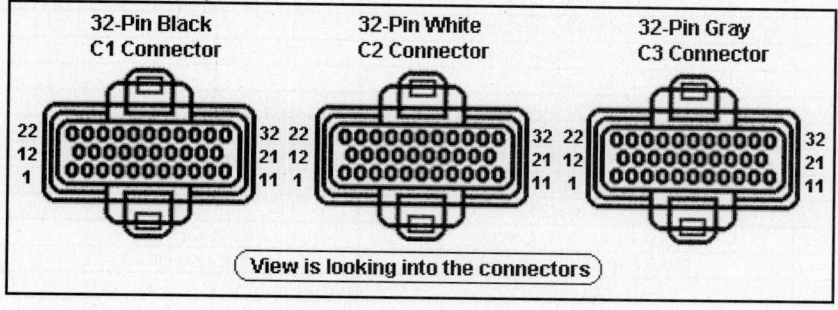

32-Pin Black C1 Connector 32-Pin White C2 Connector 32-Pin Gray C3 Connector

View is looking into the connectors

2001-03 Dakota 5.9L V8 MFI VIN Z (A/T) White 'C2' Connector

PCM Pin #	Wire Color	Circuit Description (32 Pin)	Value at Hot Idle
1-3	---	Not Used	---
4	WT/DB	Injector 1 Driver	1-4 ms
5	YL/WT	Injector 3 Driver	1-4 ms
6	GY	Injector 5 Driver	1-4 ms
7-8	---	Not Used	---
9	TN/PK	Coil 1 Driver Control	5º, 55 mph: 8º dwell
10	DG	Generator Field Control	Digital Signal: 0-12-0v
11-13	---	Not Used	---
15	TN	Injector 2 Driver	1-4 ms
16	LB/BR	Injector 4 Driver	1-4 ms
17	DB/YL	Condenser Fan Control Relay	Relay Off: 12v, On: 1v
18	---	Not Used	---
19	DB	AC Pressure Sensor Signal	0.90v at 79 psi
20-22	---	Not Used	---
23	GY/YL	Engine Oil Pressure Sensor	1.6v at 24 psi
24-26	---	Not Used	---
27	WT/OR	Vehicle Speed Signal	Digital Signals
28-30	---	Not Used	---
31	VT/WT	5-Volt Supply	4.9-5.1v
32	---	Not Used	---

2001-03 Dakota 5.9L V8 MFI VIN Z (A/T) 'C3' Gray Connector

PCM Pin #	Wire Color	Circuit Description (32 Pin)	Value at Hot Idle
1	DB/OR	A/C Clutch Relay Control	Relay Off: 12v, On: 1v
2, 6	---	Not Used	---
3	DB/YL	ASD Relay Control	Relay Off: 12v, On: 1v
4	TN/RD	Speed Control Vacuum Solenoid	Vacuum Increasing: 1v
5	LG/RD	Speed Control Vent Solenoid	Vacuum Decreasing: 1v
7	YL/BK	Knock Sensor 1 Signal	0.080v AC
8	BR/VT	HO2S-11 (B1 S1) Heater Control	Heater Off: 12v, On: 1v
9	DB/OR	Heater Relay Control (Downstream)	Relay Off: 12v, On: 1v
10	WT/DG	LDP Switch Sense	LDP Switch Closed: 0v
11	YL/RD	Speed Control Power Supply	12-14v
12	DG/OR	ASD Relay Power (B+)	12-14v
13	---	Not Used	---
14	OR	LDP Solenoid Control	PWM Signal: 0-12-0v
15	PK/YL	Battery Temperature Sensor	At 86ºF: 1.96v
16	BR/WT	HO2S-12 (B1 S2) Heater Control	Heater Off: 12v, On: 1v
17	---	Not Used	---
18	BR/VT	Knock Sensor 2 Signal	0.080v AC
19	BR	Fuel Pump Relay Control	Relay Off: 12v, On: 1v
20	PK/BK	EVAP Purge Solenoid Control	PWM Signal: 0-12-0v
21-23	---	Not Used	---
24	WT/PK	Brake Switch Sense	Brake Off: 0v, On: 12v
25	WT/DB	Generator Field Source	12-14v
26	LB/BK	Fuel Level Sensor Signal	Full: 0.56v, 1/2 full: 2.5v
27	PK	SCI Transmit	0v
28, 31	---	Not Used	---
29	LG	SCI Receive	5v
30	WT/BR	PCI Data Bus (J1850)	Digital Signals: 0-7-0v
32	RD/LG	Speed Control Switch Signal	S/C & Set Switch On: 3.8v

Standard Colors and Abbreviations

Abbreviation	Color	Abbreviation	Color	Abbreviation	Color
BK	Black	GY	Gray	RD	Red
BL	Blue	GN	Green	TN	Tan
BR	Brown	LG	Light Green	VT	Violet
DB	Dark Blue	OR	Orange	WT	White
DG	Dark Green	PK	Pink	YL	Yellow

1990-91 D & W Series Pickup 3.9L V6 TBI VIN X 60 Pin Connector

PCM Pin #	Wire Color	Circuit Description (60 Pin)	Value at Hot Idle
1	DG/RD	MAP Sensor Signal	1.5-1.6v
2	TN	ECT Sensor Signal	At 180°F: 2.80v
3	RD	Keep Alive Power (B+)	12-14v
4	BK/LB	Sensor Ground	<0.050v
5	BK/WT	Sensor Ground	<0.050v
6	PK/WT	5-Volt Supply	4.9-5.1v
7	OR	8-Volt Supply	7.9-8.1v
8	DG/BK	ASD Relay Power (B+)	12-14v
9	DB	Ignition Switch Power (B+)	12-14v
10	---	Not Used	---
11	LB/RD	Power Ground	<0.1v
12	LB/RD	Power Ground	<0.1v
13-14	---	Not Used	---
15	TN	Injector 2 Driver	1-4 ms
16	WT/DB	Injector 1 Driver	1-4 ms
17-18	---	Not Used	---
19	BK/GY	Coil Driver Control	5°, 55 mph: 8° dwell
20	DG	Alternator Field Control	Digital Signals: 0-12-0v
21	BK/RD	Throttle Body Temperature	At 100°F: 2.51v
22	OR/DB	TP Sensor Signal	0.6-1.0v
23, 26	---	Not Used	---
24	GY	Distributor Reference Signal	Digital Signals: 0-5-0v
25	PK	SCI Transmit	0v
27	BR	A/C Clutch Signal	A/C Off: 12v, On: 1v
28	PK	Closed Throttle Switch	Switch Off: 12v, On: 1v
29	WT/PK	Brake Switch Sense Signal	Brake Off: 0v, On: 12v
30	BR/YL	PNP Switch Sense Signal	In P/N: 0v, Others: 5v
31, 37-39	---	Not Used	---
32	BK/PK	MIL (lamp) Control	MIL Off: 12v, On: 1v
33	TN/RD	Speed Control Vacuum Solenoid	Vacuum Increasing: 1v
34	DB/OR	A/C WOT Relay Control	Relay Off: 12v, On: 1v
35	GY/YL	EGR Solenoid Control	12v, at 55 mph: 1v
36	BK/OR	Air Switching Solenoid	Solenoid Off: 12v, On: 1v
40	BR	AIS Motor Control	Pulse Signals

Standard Colors and Abbreviations

Abbreviation	Color	Abbreviation	Color	Abbreviation	Color
BK	Black	GY	Gray	RD	Red
BL	Blue	GN	Green	TN	Tan
BR	Brown	LG	Light Green	VT	Violet
DB	Dark Blue	OR	Orange	WT	White
DG	Dark Green	PK	Pink	YL	Yellow

1990-91 D & W Series Pickup 3.9L V6 TBI VIN X 60 Pin Connector

PCM Pin #	Wire Color	Circuit Description (60 Pin)	Value at Hot Idle
41	BK/DG	HO2S-11 (B1 S1) Signal	0.1-1.1v
42-43	---	Not Used	---
44	OR/WT	Overdrive Lockout Switch	Switch Off: 12v, On: 1v
45	LG	SCI Receive	5v
46	---	Not Used	---
47	WT/OR	Vehicle Speed Signal	Digital Signal
48	BR/RD	Speed Control Set Switch Signal	S/C & Set Switch On: 3.8v
49	YL/RD	Speed Control On/Off Switch	S/C On: 1v, Off: 12v
50	WT/LG	Speed Control Resume Switch	S/C On: 1v, Off: 12v
51	DB/YL	ASD Relay Control	Relay Off: 12v, On: 1v
52	PK/BK	EVAP Purge Solenoid Control	Solenoid Off: 12v, On: 1v
53	LG/RD	Speed Control Vent Solenoid	Vacuum Decreasing: 1v
54	OR/BK	Part Throttle Unlock Solenoid	Solenoid Off: 12v, On: 1v
55	OR/LG	Overdrive Lockout Solenoid	Solenoid Off: 12v, On: 1v
56	GY/PK	Maintenance Indicator Lamp	Lamp Off: 12v, On: 1v
57-59	---	Not Used	---
60	GY/RD	AIS Motor Control	Pulse Signals

Pin Connector Graphic

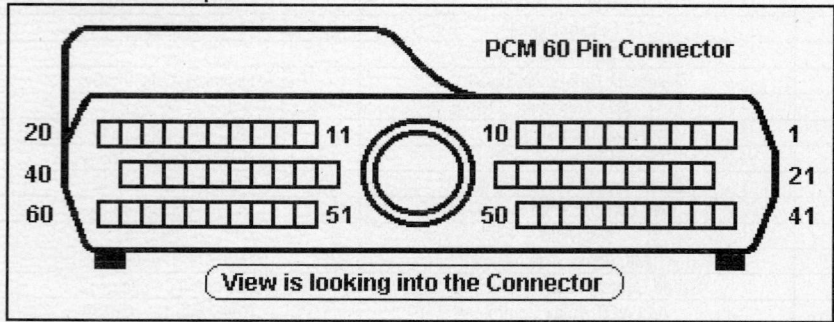

Standard Colors and Abbreviations

Abbreviation	Color	Abbreviation	Color	Abbreviation	Color
BK	Black	GY	Gray	RD	Red
BL	Blue	GN	Green	TN	Tan
BR	Brown	LG	Light Green	VT	Violet
DB	Dark Blue	OR	Orange	WT	White
DG	Dark Green	PK	Pink	YL	Yellow

1992-93 D, R & W Series Pickup 3.9L MFI VIN X 60 Pin Connector

PCM Pin #	Wire Color	Circuit Description (60 Pin)	Value at Hot Idle
1	DG/RD	MAP Sensor Signal	1.5-1.6v
2	TN/BK	ECT Sensor Signal	At 180°F: 2.80v
3	RD	Keep Alive Power (B+)	12-14v
4	BK/LB	Sensor Ground	<0.050v
5	BK/WT	Sensor Ground	<0.050v
6	PK/WT	5-Volt Supply	4.9-5.1v
7	OR	8-Volt Supply	7.9-8.1v
8	---	Not Used	---
9	DB	Ignition Switch Power (B+)	12-14v
9	LG/BK	Ignition Switch Power (B+)	12-14v
10	OR/WT	Overdrive Override Switch	Switch Off: 12v, On: 1v
11	BK/TN	Power Ground	<0.1v
12	BK/TN	Power Ground	<0.1v
13	LB/BR	Injector 4 Driver	1-4 ms
14	YL/WT	Injector 3 Driver	1-4 ms
15	TN	Injector 2 Driver	1-4 ms
16	WT/DB	Injector 1 Driver	1-4 ms
17-18	---	Not Used	---
19	BK/GY	Coil Driver Control	5°, 55 mph: 8° dwell
20	DG	Alternator Field Control	Digital Signals: 0-12-0v
21	BK/RD	Throttle Body Temperature	At 100°F: 2.51v
22	OR/DB	TP Sensor Signal	0.6-1.0v
23	---	Not Used	---
24	GY/BK	Distributor Reference Signal	Digital Signals: 0-5-0v
25	PK	SCI Transmit Signal	0v
26	---	Not Used	---
27	BR	A.C Clutch Signal	A/C Off: 12v, On: 1v
28, 31	---	Not Used	---
29	WT/PK	Brake Switch Sense Signal	Brake Off: 0v, On: 12v
30	BK/WT	PNP Switch Sense Signal	In P/N: 0v, Others: 5v
32	BK/PK	MIL (lamp) Control	MIL Off: 12v, On: 1v
33	TN/RD	Speed Control Vacuum Solenoid	Vacuum Increasing: 1v
34	DB/OR	A/C WOT Relay Control	Relay Off: 12v, On: 1v
35	GY/YL	EGR Solenoid Control	12v, at 55 mph: 1v
36	---	Not Used	---
37	BK/OR	Overdrive Lamp Control	Lamp Off: 12v, On: 1v
38	PK/BK	Injector 5 Driver	1-4 ms
39	GY/RD	AIS Motor 4 Control	Pulse Signals
40	BR/WT	AIS Motor 2 Control	Pulse Signals
41	BK/DG	HO2S-11 (B1 S1) Signal	0.1-1.1v
42, 43	---	Not Used	---
44	GY	Distributor Sync Signal	Digital Signals: 0-5-0v
45	LG	SCI Receive Signal	5v
46	---	Not Used	---
47	WT/OR	Vehicle Speed Signal	Digital Signal
48	BR/RD	Speed Control Set Switch Signal	S/C & Set Switch On: 3.8v
49	YL/RD	Speed Control On/Off Switch	S/C On: 1v, Off: 12v
50	WT/LG	Speed Control Resume Switch	S/C On: 1v, Off: 12v
51	DB/YL	ASD Relay Control	Relay Off: 12v, On: 1v
52	PK/BK	EVAP Purge Solenoid Control	Solenoid Off: 12v, On: 1v
53	LG/RD	Speed Control Vent Solenoid	Vacuum Decreasing: 1v
54	OR/BK	A/T: Overdrive Solenoid	Solenoid Off: 12v, On: 1v
54	OR/BK	M/T: SIL (lamp) Control	Lamp Off: 12v, On: 1v
55	OR/WT	Overdrive Lockout Solenoid	Solenoid Off: 12v, On: 1v
56	GY/PK	Maintenance Indicator Lamp	Lamp Off: 12v, On: 1v
57	DG/OR	ASD Relay Power (B+)	12-14v
58	LG/BK	Injector 6 Driver	1-4 ms
58	BR/DB	Injector 6 Driver	1-4 ms
59	PK/BK	AIS Motor 1 Control	Pulse Signals
60	YL/BK	AIS Motor 3 Control	Pulse Signals

1994-95 D, R & W Series Pickup 3.9L MFI VIN X 60 Pin Connector

PCM Pin #	Wire Color	Circuit Description (60 Pin)	Value at Hot Idle
1	DG/RD	MAP Sensor Signal	1.5-1.6v
2	TN/BK	ECT Sensor Signal	At 180ºF: 2.80v
3	RD/WT	Keep Alive Power (B+)	12-14v
4	BK/LB	Sensor Ground	<0.050v
5	---	Not Used	---
6	PK/WT	5-Volt Supply	4.9-5.1v
7	OR	8-Volt Supply	7.9-8.1v
8	---	Not Used	---
9	LG/BK	Ignition Switch Power (B+)	12-14v
10	OR/WT	Overdrive Override Switch	Switch Off: 12v, On: 1v
11	BK/TN	Power Ground	<0.1v
12	BK/TN	Power Ground	<0.1v
13	LB/BR	Injector 4 Driver	1-4 ms
14	YL/WT	Injector 3 Driver	1-4 ms
15	TN	Injector 2 Driver	1-4 ms
16	WT/DB	Injector 1 Driver	1-4 ms
17-18	---	Not Used	---
19	BK/GY	Coil Driver Control	5º, 55 mph: 8º dwell
20	DG	Alternator Field Control	Digital Signals: 0-12-0v
21	BK/RD	Throttle Body Temperature	At 100ºF: 2.51v
22	OR/DB	TP Sensor Signal	0.6-1.0v
23	---	Not Used	---
24	GY/BK	Distributor Reference Signal	Digital Signals: 0-5-0v
25	PK	SCI Transmit Signal	0v
26, 28	---	Not Used	---
27	BR	A/C Damped Pressure Switch	A/C Off: 12v, On: 1v
29	WT/PK	Brake Switch Sense Signal	Brake Off: 0v, On: 12v
30	BK/WT	PNP Switch Sense Signal	In P/N: 0v, Others: 5v
31	---	Not Used	---
32	BK/PK	MIL (lamp) Control	MIL Off: 12v, On: 1v
33	TN/RD	Speed Control Vacuum Solenoid	Vacuum Increasing: 1v
34	DB/OR	A/C Clutch Relay Control	Relay Off: 12v, On: 1v
35	GY/YL	EGR Solenoid Control	12v, at 55 mph: 1v
36	---	Not Used	---
37	LG/OR	Overdrive Lamp Control	Lamp Off: 12v, On: 1v
38	GY	Injector 5 Driver	1-4 ms
39	GY/RD	AIS Motor 4 Control	Pulse Signals
40	BR/WT	AIS Motor 2 Control	Pulse Signals
41	BK/DG	HO2S-11 (B1 S1) Signal	0.1-1.1v
42	PK	Transmission Temperature Sensor Signal	At 200ºF: 2.40v
43	GY/LB	Tachometer Signal	Pulse Signals
44	TN/YL	Distributor Sync Signal	Digital Signals: 0-5-0v
45	LG	SCI Receive Signal	5v
46	---	Not Used	---
47	WT/OR	Vehicle Speed Signal	Digital Signal
48	BR/RD	Speed Control Set Switch Signal	S/C & Set Switch On: 3.8v
49	YL/RD	Speed Control On/Off Switch	S/C On: 1v, Off: 12v
50	WT/LG	Speed Control Resume Switch	S/C On: 1v, Off: 12v
51	DB/YL	ASD Relay Control	Relay Off: 12v, On: 1v
52	PK/BK	EVAP Purge Solenoid Control	Solenoid Off: 12v, On: 1v
53	LG/RD	Speed Control Vent Solenoid	Vacuum Decreasing: 1v
54	OR/BK	A/T: Overdrive Solenoid	Solenoid Off: 12v, On: 1v
54	OR/BK	M/T: SIL (lamp) Control	Lamp Off: 12v, On: 1v
55	OR/WT	Overdrive Lockout Solenoid	Solenoid Off: 12v, On: 1v
56	---	Not Used	---
57	DG/OR	ASD Relay Power (B+)	12-14v
58	BR/DB	Injector 6 Driver	1-4 ms
59	PK/BK	AIS Motor 1 Control	Pulse Signals
60	YL/BK	AIS Motor 3 Control	Pulse Signals

1996-97 Pickup 3.9L V6 MFI VIN X 'A' Black Connector

PCM Pin #	Wire Color	Circuit Description (32 Pin)	Value at Hot Idle
2	LG/BK	Ignition Switch Power (B+)	12-14v
4	BK/LB	Sensor Ground	<0.050v
6	BK/WT	PNP Switch Sense Signal	In P/N: 0v, Others: 5v
7	BK/GY	Coil 1 Driver Control	5°, 55 mph: 8° dwell
8	GY/BK	CKP Sensor Signal	Digital Signals: 0-5-0v
10	YL/BK	IAC 2 Driver Control	Pulse Signals
11	BR/WT	IAC 3 Driver Control	Pulse Signals
15	BK/RD	IAT Sensor Signal	At 100°F: 1.83v
16	TN/BK	ECT Sensor Signal	At 180°F: 2.80v
17	PK/WT	5-Volt Supply	4.9-5.1v
18	TN/YL	CMP Sensor Signal	Digital Signals: 0-5-0v
19	GY/RD	IAC 1 Driver Control	Pulse Signals
20	PK/BK	IAC 4 Driver Control	Pulse Signals
22	RD/WT	Keep Alive Power (B+)	12-14v
23	OR/DB	TP Sensor Signal	0.6-1.0v
24	TN/WT	HO2S-11 (B1 S1) Signal	0.1-1.1v
25	OR/BK	HO2S-12 (B1 S2) Signal	0.1-1.1v
27	DG/RD	MAP Sensor Signal	1.5-1.6v
31	BK/TN	Power Ground	<0.1v
32	BK/TN	Power Ground	<0.1v

1996-97 Pickup 3.9L V6 MFI VIN X 'B' White Connector

PCM Pin #	Wire Color	Circuit Description (32 Pin)	Value at Hot Idle
1	PK	Transmission Temperature Sensor Signal	At 200°F: 2.40v
4	WT/DB	Injector 1 Driver	1-4 ms
5	YL/WT	Injector 3 Driver	1-4 ms
6	GY	Injector 5 Driver	1-4 ms
8	PK/W	Governor Pressure Solenoid	PWM Signal: 0-12-0v
10	DG	Generator Field Control	Digital Signals: 0-12-0v
11	OR/BK	TCC Solenoid Control	At Cruise w/TCC On: 1v
12	BR/DB	Injector 6 Driver	1-4 ms
15	TN	Injector 2 Driver	1-4 ms
16	LB/BR	Injector 4 Driver	1-4 ms
21	BR	3-4 Shift Solenoid Control	Solenoid Off: 12v, On: 1v
25	DB/BK	OSS Sensor (-) Signal	Moving: AC pulse signals
27	WT/OR	Vehicle Speed Signal	Digital Signal
28	LG/BK	OSS Sensor (+) Signal	Moving: AC pulse signals
29	LG/WT	Governor Pressure Sensor	0.58v
30	PK	Transmission Relay Control	Relay Off: 12v, On: 1v
31	OR	5-Volt Supply	4.9-5.1v
32	GY/YL	EGR Solenoid Control	12v, at 55 mph: 1v

Standard Colors and Abbreviations

Abbreviation	Color	Abbreviation	Color	Abbreviation	Color
BK	Black	GY	Gray	RD	Red
BL	Blue	GN	Green	TN	Tan
BR	Brown	LG	Light Green	VT	Violet
DB	Dark Blue	OR	Orange	WT	White
DG	Dark Green	PK	Pink	YL	Yellow

1996-97 Pickup 3.9L V6 MFI VIN X 'C' Gray Connector

PCM Pin #	Wire Color	Circuit Description (32 Pin)	Value at Hot Idle
1	DB/OR	A/C Clutch Relay Control	Relay Off: 12v, On: 1v
2	---	Not Used	---
3	DB/YL	ASD Relay Control	Relay Off: 12v, On: 1v
4	TN/RD	Speed Control Vacuum Solenoid	Vacuum Increasing: 1v
5	LG/RD	Speed Control Vent Solenoid	Vacuum Decreasing: 1v
6	LG/OR	Overdrive Off Lamp Control	Lamp Off: 12v, On: 1v
7	PK/BK	Transmission Temperature Lamp Control	Lamp Off: 12v, On: 1v
8-10	---	Not Used	---
11	YL/RD	Speed Control Power Supply	12-14v
12	DG/OR	ASD Relay Power (B+)	12-14v
13	OR/WT	Overdrive Off Switch Sense	Switch Off: 12v, On: 1v
14	---	Not Used	---
15	PK/YL	Battery Temperature Sensor	At 86°F: 1.96v
16	TN/YL	Generator Lamp Control	Lamp Off: 12v, On: 1v
17	BK/PK	MIL (lamp) Control	MIL Off: 12v, On: 1v
18	GY/PK	Maintenance Indicator Lamp	Lamp Off: 12v, On: 1v
19	BR/WT	Fuel Pump Relay Control	Relay Off: 12v, On: 1v
20	PK	EVAP Purge Solenoid Control	PWM Signal: 0-12-0v
21	---	Not Used	---
22	BR	A/C Request Signal	A/C Off: 12v, On: 1v
23	LG	A/C Select Signal	A/C Off: 12v, On: 1v
24	WT/PK	Brake Switch Sense Signal	Brake Off: 0v, On: 12v
25	---	Not Used	---
26	DB/WT	Fuel Level Sensor Signal	Full: 0.56v, 1/2 full: 2.5v
27	PK/DB	SCI Transmit	0v
28	---	Not Used	---
29	DG	SCI Receive	5v
30	---	Not Used	---
31	GY/LB	Tachometer Signal	Pulse Signals
32	RD/LG	Speed Control Set Switch Signal	S/C & Set Switch On: 3.8v

Pin Connector Graphic

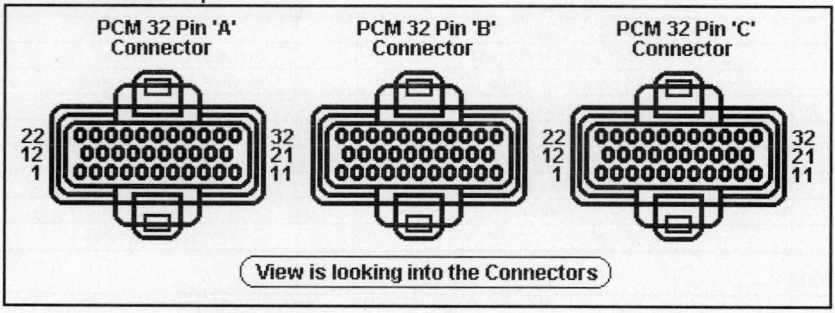

PCM 32 Pin 'A' Connector   PCM 32 Pin 'B' Connector   PCM 32 Pin 'C' Connector

View is looking into the Connectors

1998-99 Pickup 3.9L V6 MFI VIN X (All) 'C1' Black Connector

PCM Pin #	Wire Color	Circuit Description (32 Pin)	Value at Hot Idle
1, 3	---	Not Used	---
2	LG/BK	Ignition Switch Power (B+)	12-14v
4	BK/LB	Sensor Ground	<0.050v
5	---	Not Used	---
6	BK/WT	PNP Switch Sense Signal	In P/N: 0v, Others: 5v
7	BK/GY	Coil 1 Driver Control	5°, 55 mph: 8° dwell
8	GY/BK	CKP Sensor Signal	Digital Signals: 0-5-0v
9	---	Not Used	---
10	YL/BK	IAC 2 Driver Control	Pulse Signals
11	BR/WT	IAC 3 Driver Control	Pulse Signals
12	---	Not Used	---
13	OR	Power Takeoff Switch Sense	Switch Off: 12v, On: 1v
14	---	Not Used	---
15	BK/RD	IAT Sensor Signal	At 100°F: 1.83v
16	TN/BK	ECT Sensor Signal	At 180°F: 2.80v
17	PK/WT	5-Volt Supply	4.9-5.1v
18	TN/YL	CMP Sensor Signal	Digital Signals: 0-5-0v
19	GY/RD	IAC 1 Driver Control	Pulse Signals
20	PK/BK	IAC 4 Driver Control	Pulse Signals
21	---	Not Used	---
22	RD/WT	Keep Alive Power (B+)	12-14v
23	OR/DB	TP Sensor Signal	0.6-1.0v
24	TN/WT	HO2S-11 (B1 S1) Signal	0.1-1.1v
25	OR/BK	HO2S-12 (B1 S2) Signal	0.1-1.1v
26	---	Not Used	---
27	DG/RD	MAP Sensor Signal	1.5-1.6v
28-30	---	Not Used	---
31	BK/TN	Power Ground	<0.1v
32	BK/TN	Power Ground	<0.1v

Pin Connector Graphic

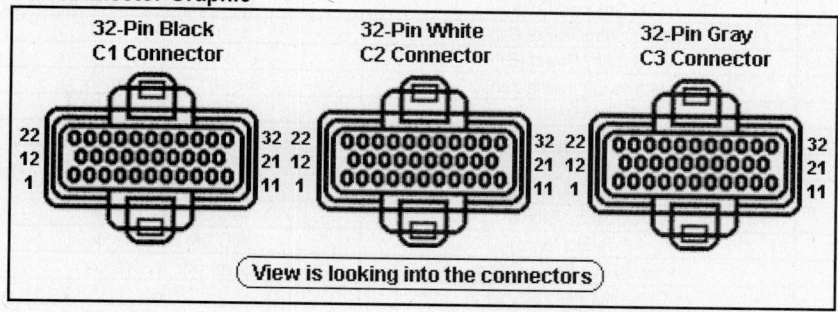

32-Pin Black C1 Connector   32-Pin White C2 Connector   32-Pin Gray C3 Connector

View is looking into the connectors

1998-99 Pickup 3.9L V6 MFI VIN X (All) 'C2' White Connector

PCM Pin #	Wire Color	Circuit Description (32 Pin)	Value at Hot Idle
1	PK	Transmission Temperature Sensor Signal	At 200°F: 2.40v
2-3	---	Not Used	
4	WT/DB	Injector 1 Driver	1-4 ms
5	YL/WT	Injector 3 Driver	1-4 ms
6	GY	Injector 5 Driver	1-4 ms
7, 9	---	Not Used	---
8	PK/W	Governor Pressure Solenoid	PWM Signal: 0-12-0v
10	DG	Generator Field Control	Digital Signals: 0-12-0v
11	OR/BK	TCC Solenoid Control	At Cruise w/TCC On: 1v
12	BR/DB	Injector 6 Driver	1-4 ms
15	TN	Injector 2 Driver	1-4 ms
16	LB/BR	Injector 4 Driver	1-4 ms
21	BR	Overdrive Solenoid Control	Cruise w/solenoid On: 1v
23	GY/O	Engine Oil Pressure Sensor	1.6v at 24 psi
25	DB/BK	OSS Sensor (-) Signal	Moving: AC pulse signals
27	WT/OR	Vehicle Speed Signal	Digital Signal
28	LG/BK	OSS Sensor (+) Signal	Moving: AC pulse signals
29	LG/WT	Governor Pressure Sensor	0.58v
30	P	Transmission Relay Control	Relay Off: 12v, On: 1v
31	O	5-Volt Supply	4.9-5.1v

1998-99 Pickup 3.9L V6 MFI VIN X (All) 'C3' Gray Connector

PCM Pin #	Wire Color	Circuit Description (32 Pin)	Value at Hot Idle
1	DB/OR	A/C Clutch Relay Control	Relay Off: 12v, On: 1v
3	DB/YL	ASD Relay Control	Relay Off: 12v, On: 1v
4	TN/RD	Speed Control Vacuum Solenoid	Vacuum Increasing: 1v
5	LG/RD	Speed Control Vent Solenoid	Vacuum Decreasing: 1v
6 ('98)	LG/OR	Overdrive Off Lamp Control	Lamp Off: 12v, On: 1v
8	BR/VT	HO2S-11 (B1 S1) Heater	Heater Off: 12v, On: 1v
9	DG/PK	HO2S-12 (B1 S2) Heater	Heater Off: 12v, On: 1v
10	WT/DG	LDP Solenoid Control	PWM Signal: 0-12-0v
11	YL/RD	Speed Control Power Supply	12-14v
12	DG/OR	ASD Relay Power (B+)	12-14v
13	OR/WT	Overdrive Off Switch Sense	Switch Off: 12v, On: 1v
14	OR	LDP Switch Sense Signal	LDP Switch Closed: 0v
15	PK/YL	Battery Temperature Sensor	At 86°F: 1.96v
19	BR/WT	Fuel Pump Relay Control	Relay Off: 12v, On: 1v
20	PK/WT	EVAP Purge Solenoid Control	PWM Signal: 0-12-0v
22	BR	A/C Switch Signal	A/C Off: 12v, On: 1v
23	LG/WT	A/C Select Signal	A/C Off: 12v, On: 1v
24	WT/PK	Brake Switch Sense Signal	Brake Off: 0v, On: 12v
25	DB	Generator Field Source	12-14v
26	DB/WT	Fuel Level Sensor Signal	Full: 0.56v, 1/2 full: 2.5v
27	PK/DB	SCI Transmit Signal	0v
28	WT/BK	CCD Bus (-) Signal	<0.050v
29	DG	SCI Receive Signal	5v
30	VT/BR	CCD Bus (+) Signal	Digital Signals: 0-5-0v
32	RD/LG	Speed Control Set Switch Signal	S/C & Set Switch On: 3.8v

Standard Colors and Abbreviations

Abbreviation	Color	Abbreviation	Color	Abbreviation	Color
BK	Black	GY	Gray	RD	Red
BL	Blue	GN	Green	TN	Tan
BR	Brown	LG	Light Green	VT	Violet
DB	Dark Blue	OR	Orange	WT	White
DG	Dark Green	PK	Pink	YL	Yellow

2000-01 Pickup 3.9L V6 MFI VIN X (All) 'C1' Black Connector

PCM Pin #	Wire Color	Circuit Description (32 Pin)	Value at Hot Idle
1	---	Not Used	---
2	LG/BK	Ignition Switch Power (B+)	12-14v
3	---	Not Used	---
4	BK/LB	Sensor Ground	<0.050v
5	---	Not Used	---
6	BK/WT	PNP Switch Sense Signal	In P/N: 0v, Others: 5v
7	BK/GY	Coil 1 Driver Control	5°, 55 mph: 8° dwell
8	GY/BK	CKP Sensor Signal	Digital Signals: 0-5-0v
9	---	Not Used	---
10	YL/BK	IAC 2 Driver Control	Pulse Signals
11	BR/WT	IAC 3 Driver Control	Pulse Signals
12	---	Not Used	---
13	OR	Power Takeoff Switch Sense	Switch Off: 12v, On: 1v
14	---	Not Used	---
15	BK/RD	IAT Sensor Signal	At 100°F: 1.83v
16	TN/BK	ECT Sensor Signal	At 180°F: 2.80v
17	VT/WT	5-Volt Supply	4.9-5.1v
18	TN/YL	CMP Sensor Signal	Digital Signals: 0-5-0v
19	GY/RD	IAC 1 Driver Control	Pulse Signals
20	VT/BK	IAC 4 Driver Control	Pulse Signals
21	---	Not Used	---
22	RD/WT	Keep Alive Power (B+)	12-14v
23	OR/DB	TP Sensor Signal	0.6-1.0v
24	TN/WT	HO2S-11 (B1 S1) Signal	0.1-1.1v
25	OR/BK	HO2S-12 (B1 S2) Signal	0.1-1.1v
26	LG/RD	HO2S-21 (B2 S1) Signal (Cal)	0.1-1.1v
27	DG/RD	MAP Sensor Signal	1.5-1.6v
28	---	Not Used	---
29	OR/TN	HO2S-22 (B2 S2) Signal (Cal)	0.1-1.1v
30	---	Not Used	---
31	BK/TN	Power Ground	<0.1v
32	BK/TN	Power Ground	<0.1v

Pin Connector Graphic

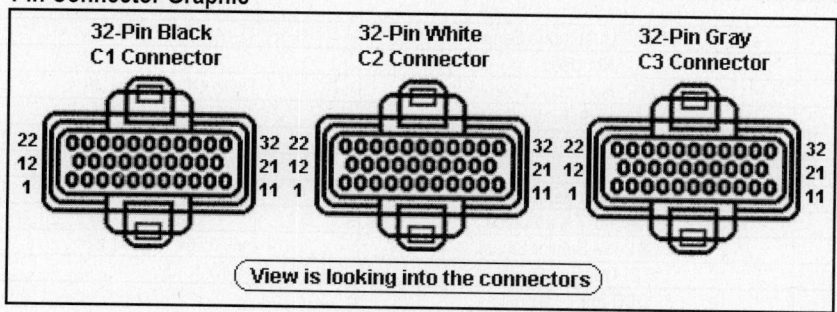

View is looking into the connectors

2000-01 Pickup 3.9L V6 MFI VIN X (All) 'C2' White Connector

PCM Pin #	Wire Color	Circuit Description (32 Pin)	Value at Hot Idle
1	VT	Transmission Temperature Sensor Signal	At 200°F: 2.40v
2-3, 7, 9	---	Not Used	---
4	WT/DB	Injector 1 Driver	1-4 ms
5	YL/WT	Injector 3 Driver	1-4 ms
6	GY	Injector 5 Driver	1-4 ms
8	VT/WT	Governor Pressure Solenoid	PWM Signal: 0-12-0v
10	DG	Generator Field Control	Digital Signals: 0-12-0v
11	OR/BK	TCC Solenoid Control	At Cruise w/TCC On: 1v
12	BR/DB	Injector 6 Driver	1-4 ms
15	TN	Injector 2 Driver	1-4 ms
16	LB/BR	Injector 4 Driver	1-4 ms
21	BR	3-4 Shift Solenoid Control	Cruise w/solenoid On: 1v
23	GY/OR	Engine Oil Pressure Sensor	1.6v at 24 psi
25	DB/BK	OSS Sensor (-) Signal	Moving: AC pulse signals
27	WT/OR	Vehicle Speed Signal	Digital Signal
28	LG/BK	OSS Sensor (+) Signal	Moving: AC pulse signals
29	LG/WT	Governor Pressure Sensor	0.58v
30	PK	Transmission Relay Control	Relay Off: 12v, On: 1v
31	OR	5-Volt Supply	4.9-5.1v

2000-01 Pickup 3.9L V6 MFI VIN X (All) 'C3' Gray Connector

PCM Pin #	Wire Color	Circuit Description (32 Pin)	Value at Hot Idle
1	DB/OR	A/C Clutch Relay Control	Relay Off: 12v, On: 1v
3	DB/YL	ASD Relay Control	Relay Off: 12v, On: 1v
4	TN/RD	Speed Control Vacuum Solenoid	Vacuum Increasing: 1v
5	LG/RD	Speed Control Vent Solenoid	Vacuum Decreasing: 1v
6-7	---	Not Used	---
8	BR/VT	HO2S-11 Heater Control (Cal)	PWM Signal: 0-12-0v
9	DG/PK	HO2S-12 Heater Relay (Cal)	Relay Off: 12v, On: 1v
10	WT/DG	LDP Solenoid Control	PWM Signal: 0-12-0v
11	YL/RD	Speed Control Power Supply	12-14v
12	DG/OR	ASD Relay Power (B+)	12-14v
13	OR/WT	Overdrive Off Switch Sense	Switch Off: 12v, On: 1v
14	OR	LDP Switch Sense Signal	LDP Switch Closed: 0v
15	PK/YL	Battery Temperature Sensor	At 86°F: 1.96v
16-18, 21	---	Not Used	---
19	BR/WT	Fuel Pump Relay Control	Relay Off: 12v, On: 1v
20	PK/WT	EVAP Purge Solenoid Control	PWM Signal: 0-12-0v
22	BR	A/C Switch Signal	A/C Off: 12v, On: 1v
23	LG/WT	A/C Select Signal	A/C Off: 12v, On: 1v
24	WT/PK	Brake Switch Sense Signal	Brake Off: 0v, On: 12v
25	DB	Generator Field Source	12-14v
26	DB/WT	Fuel Level Sensor Signal	Full: 0.56v, 1/2 full: 2.5v
27	PK/DB	SCI Transmit Signal	0v
28	WT/BK	CCD Bus (-) Signal	<0.050v
29	DG	SCI Receive Signal	5v
30	VT/BR	CCD Bus (+) Signal	Digital Signals: 0-5-0v
32	RD/LG	Speed Control Set Switch Signal	S/C & Set Switch On: 3.8v

Standard Colors and Abbreviations

Abbreviation	Color	Abbreviation	Color	Abbreviation	Color
BK	Black	GY	Gray	RD	Red
BL	Blue	GN	Green	TN	Tan
BR	Brown	LG	Light Green	VT	Violet
DB	Dark Blue	OR	Orange	WT	White
DG	Dark Green	PK	Pink	YL	Yellow

1996-97 Pickup 5.2L V8 CNG VIN T (All) 'A' Black Connector

PCM Pin #	Wire Color	Circuit Description (32 Pin)	Value at Hot Idle
1	---	Not Used	---
2	LG/BK	Ignition Switch Power (B+)	12-14v
3	---	Not Used	---
4	BK/LB	Sensor Ground	<0.050v
5	---	Not Used	---
6	BK/WT	PNP Switch Sense Signal	In P/N: 0v, Others: 5v
7	BK/GY	Coil 1 Driver Control	5°, 55 mph: 8° dwell
8	GY/BK	CKP Sensor Signal	Digital Signals: 0-5-0v
9	---	Not Used	---
10	YL/BK	IAC 2 Driver Control	Pulse Signals
11	BR/WT	IAC 3 Driver Control	Pulse Signals
12-14	---	Not Used	---
15	BK/RD	IAT Sensor Signal	At 100°F: 1.83v
16	TN/BK	ECT Sensor Signal	At 180°F: 2.80v
17	PK/WT	5-Volt Supply	4.9-5.1v
18	TN/YL	CMP Sensor Signal	Digital Signals: 0-5-0v
19	GY/RD	IAC 1 Driver Control	Pulse Signals
20	PK/BK	IAC 4 Driver Control	Pulse Signals
21	---	Not Used	---
22	RD/WT	Keep Alive Power (B+)	12-14v
23	OR/DB	TP Sensor Signal	0.6-1.0v
24	TN/WT	HO2S-11 (B1 S1) Signal	0.1-1.1v
25	OR/BK	HO2S-12 (B1 S2) Signal	0.1-1.1v
26	---	Not Used	---
27	DG/RD	MAP Sensor Signal	1.5-1.6v
28	DG/OR	Fuel Pressure Sensor	0-255 psi
29-30	---	Not Used	---
31	BK/TN	Power Ground	<0.1v
32	BK/TN	Power Ground	<0.1v

Pin Connector Graphic

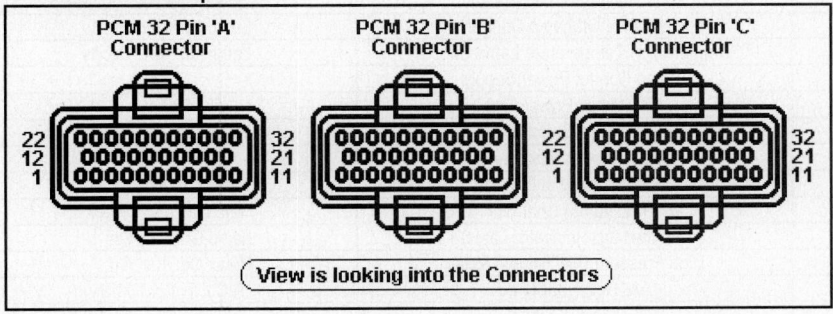

PCM 32 Pin 'A' Connector   PCM 32 Pin 'B' Connector   PCM 32 Pin 'C' Connector

View is looking into the Connectors

1996-97 Pickup 5.2L V8 CNG VIN T (All) 'B' White Connector

PCM Pin #	Wire Color	Circuit Description (32 Pin)	Value at Hot Idle
1	PK	Transmission Temperature Sensor Signal	At 200°F: 2.40v
2	PK/TN	Injector 7 Driver	1-4 ms
3, 7, 9, 14	---	Not Used	---
4	WT/DB	Injector 1 Driver	1-4 ms
5	YL/WT	Injector 3 Driver	1-4 ms
6	GY	Injector 5 Driver	1-4 ms
8	PK/WT	Governor Pressure Solenoid	PWM Signal: 0-12-0v
10	DG	Generator Field Control	Digital Signals: 0-12-0v
11	OR/BK	TCC Solenoid Control	At Cruise w/TCC On: 1v
12	BR/DB	Injector 6 Driver	1-4 ms
13	GY/LB	Injector 8 Driver	1-4 ms
15	TN	Injector 2 Driver	1-4 ms
16	LB/BR	Injector 4 Driver	1-4 ms
17-20	---	Not Used	---
21	BR	3-4 Shift Solenoid Control	Solenoid Off: 12v, On: 1v
22	TN/PK	Fuel Temperature Sensor	0.5-4.5v
23-24	---	Not Used	---
25	DB	OSS Sensor (+	Moving: AC pulse signals
27	WT/OR	Vehicle Speed Signal	Digital Signal
28	LG	OSS Sensor (- Signals	Moving: AC pulse signals
29	LG/WT	Governor Pressure Sensor	0.58v
30	PK	Transmission Relay Control	Relay Off: 12v, On: 1v
31	OR	5-Volt Supply	4.9-5.1v
32	GY/YL	EGR Solenoid Control	12v, at 55 mph: 1v

1996-97 Pickup 5.2L V8 CNG VIN T (All) 'C' Gray Connector

PCM Pin #	Wire Color	Circuit Description (32 Pin)	Value at Hot Idle
1	DB/OR	A/C Clutch Relay Control	Relay Off: 12v, On: 1v
3	DB/YL	ASD Relay Control	Relay Off: 12v, On: 1v
4	TN/RD	Speed Control Vacuum Solenoid	Vacuum Increasing: 1v
5	LG/RD	Speed Control Vent Solenoid	Vacuum Decreasing: 1v
6	LG/OR	Overdrive Off Lamp Control	Lamp Off: 12v, On: 1v
7	PK/BK	Transmission Temperature Lamp Control	Lamp Off: 12v, On: 1v
11	YL/RD	Speed Control Power Supply	12-14v
12	DG/OR	ASD Relay Power (B+)	12-14v
13	OR/WT	Overdrive Off Switch Sense	Switch Off: 12v, On: 1v
15	PK/YL	Battery Temperature Sensor	At 86°F: 1.96v
16	TN/YL	Generator Lamp Control	Lamp Off: 12v, On: 1v
17	BK/PK	MIL (lamp) Control	MIL Off: 12v, On: 1v
18	GY/PK	Maintenance Indicator Lamp	Lamp Off: 12v, On: 1v
19	BR/WT	High Pressure Fuel Shutoff Relay	Relay Off: 12v, On: 1v
22	BR	A/C Request Signal	A/C Off: 12v, On: 1v
23	LG	A/C Select Signal	A/C Off: 12v, On: 1v
24	WT/PK	Brake Switch Sense Signal	Brake Off: 0v, On: 12v
26	DB/WT	Fuel Level Sensor Signal	Full: 0.56v, 1/2 full: 2.5v
27	PK	SCI Transmit	0v
28	---	Not Used	---
29	DG	SCI Receive	5v
30	---	Not Used	---
31	GY/LB	Tachometer Signal	Pulse Signals
32	RD/LG	Speed Control Set Switch Signal	S/C & Set Switch On: 3.8v

Standard Colors and Abbreviations

Abbreviation	Color	Abbreviation	Color	Abbreviation	Color
BK	Black	GY	Gray	RD	Red
BL	Blue	GN	Green	TN	Tan
BR	Brown	LG	Light Green	VT	Violet
DB	Dark Blue	OR	Orange	WT	White
DG	Dark Green	PK	Pink	YL	Yellow

1990-91 Pickup 5.2L V8 TBI VIN Y (All) 60 Pin Connector

PCM Pin #	Wire Color	Circuit Description (60 Pin)	Value at Hot Idle
1	DG/RD	MAP Sensor Signal	1.5-1.6v
2	TN	ECT Sensor Signal	At 180°F: 2.80v
3	RD	Keep Alive Power (B+)	12-14v
4	BK/LB	Sensor Ground	<0.050v
5	BK/WT	Sensor Ground	<0.050v
6	PK/WT	5-Volt Supply	4.9-5.1v
7	OR	8-Volt Supply	7.9-8.1v
8	DG/BK	ASD Relay Power (B+)	12-14v
9	DB	Ignition Switch Power (B+)	12-14v
10, 13-14, 17-18	---	Not Used	---
11, 12	LB/RD	Power Ground	<0.1v
15	TN	Injector 2 Driver	1-4 ms
16	WT/DB	Injector 1 Driver	1-4 ms
19	BK/GY	Coil Driver Control	5°, 55 mph: 8° dwell
20	DG	Alternator Field Control	Digital Signals: 0-12-0v
21	BK/RD	Throttle Body Temperature	At 100°F: 2.51v
22	OR/DB	TP Sensor Signal	0.6-1.0v
23, 26, 31	---	Not Used	---
24	GY	Distributor Reference Signal	Digital Signals: 0-5-0v
25	PK	SCI Transmit	0v
27	BR	A/C Clutch Signal	A/C Off: 12v, On: 1v
28	PK	Closed Throttle Switch	Switch Off: 12v, On: 1v
29	WT/PK	Brake Switch Sense Signal	Brake Off: 0v, On: 12v
30	BR/YL	PNP Switch Sense Signal	In P/N: 0v, Others: 5v
32	PK/BK	MIL (lamp) Control	MIL Off: 12v, On: 1v
33	TN/RD	Speed Control Vacuum Solenoid	Vacuum Increasing: 1v
34	DB/OR	A/C WOT Relay Control	Relay Off: 12v, On: 1v
35	GY/YL	EGR Solenoid Control	12v, at 55 mph: 1v
36	BK/OR	Air Switching Solenoid	Solenoid Off: 12v, On: 1v
37-39, 42-43	---	Not used	---
40	BR	AIS Motor Control	Pulse Signals
41	BK/DG	HO2S-11 (B1 S1) Signal	0.1-1.1v
44	OR/WT	Overdrive Lockout Switch	Switch Off: 12v, On: 1v
45	LG	SCI Receive	5v
46, 57-59	---	Not Used	---
47	WT/OR	Vehicle Speed Signal	Digital Signal
48	BR/RD	Speed Control Set Switch Signal	S/C & Set Switch On: 3.8v
49	YL/RD	Speed Control On/Off Switch	S/C On: 1v, Off: 12v
50	WT/LG	Speed Control Resume Switch	S/C On: 1v, Off: 12v
51	DB/YL	ASD Relay Control	Relay Off: 12v, On: 1v
52	PK/BK	EVAP Purge Solenoid Control	Solenoid Off: 12v, On: 1v
53	LG/RD	Speed Control Vent Solenoid	Vacuum Decreasing: 1v
54	OR/BK	Part Throttle Unlock Solenoid	Solenoid Off: 12v, On: 1v
55	OR/LG	Overdrive Lockout Solenoid	Solenoid Off: 12v, On: 1v
56	GY/PK	Maintenance Indicator Lamp	Lamp Off: 12v, On: 1v
60	GY/RD	AIS Motor Control	Pulse Signals

Pin Connector Graphic

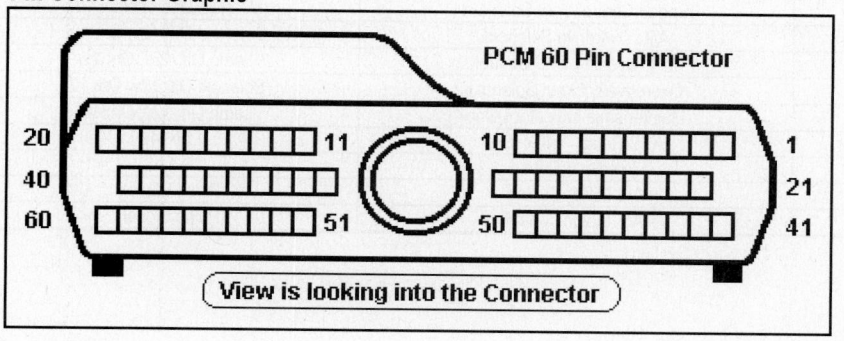

1992-93 Pickup 5.2L V8 MFI VIN Y (All) 60 Pin Connector

PCM Pin #	Wire Color	Circuit Description (60 Pin)	Value at Hot Idle
1	DG/RD	MAP Sensor Signal	1.5-1.6v
2	TN/BK	ECT Sensor Signal	At 180ºF: 2.80v
3	RD	Keep Alive Power (B+)	12-14v
4	BK/LB	Sensor Ground	<0.050v
5	BK/WT	Sensor Ground	<0.050v
6	PK/WT	5-Volt Supply	4.9-5.1v
7	OR	8-Volt Supply	7.9-8.1v
9	DB	Ignition Switch Power (B+)	12-14v
10	OR/WT	Overdrive Override Switch	Switch Off: 12v, On: 1v
11	BK/TN	Power Ground	<0.1v
12	BK/TN	Power Ground	<0.1v
13	LB/BR	Injector 4 Driver	1-4 ms
14	YL/WT	Injector 3 Driver	1-4 ms
15	TN	Injector 2 Driver	1-4 ms
16	WT/DB	Injector 1 Driver	1-4 ms
17	DB/TN	Injector 7 Driver	1-4 ms
18	RD/YL	Injector 8 Driver	1-4 ms
19	BK/GY	Coil Driver Control	5º, 55 mph: 8º dwell
20	DG	Alternator Field Control	Digital Signals: 0-12-0v
21	BK/RD	IAT Sensor Signal	At 100ºF: 1.83v
22	OR/DB	TP Sensor Signal	0.6-1.0v
23, 26, 28	---	Not Used	---
24	GY/BK	Distributor Reference Signal	Digital Signals: 0-5-0v
25	PK	SCI Transmit	0v
27	BR	A/C Clutch Signal	A/C Off: 12v, On: 1v
29	WT/PK	Brake Switch Sense Signal	Brake Off: 0v, On: 12v
30	BR/YL	PNP Switch Sense Signal	In P/N: 0v, Others: 5v
31	---	Not Used	---
32	BK/PK	MIL (lamp) Control	MIL Off: 12v, On: 1v
33	TN/RD	Speed Control Vacuum Solenoid	Vacuum Increasing: 1v
34	DB/OR	A/C WOT Relay Control	Relay Off: 12v, On: 1v
35	GY/YL	EGR Solenoid Control	12v, at 55 mph: 1v
37	BK/OR	Overdrive Lamp Control	Lamp Off: 12v, On: 1v
38	PK/BK	Injector 5 Driver	1-4 ms
39	GY/RD	AIS Motor Control	Pulse Signals
40	BR/WT	AIS Motor Control	Pulse Signals
41	BK/DG	HO2S-11 (B1 S1) Signal	0.1-1.1v
42-43	---	Not Used	---
44	GY	Distributor Sync Signal	Digital Signals: 0-5-0v
45	LG	SCI Receive	5v
46	---	Not Used	---
47	WT/OR	Vehicle Speed Signal	Digital Signal
48	BR/RD	Speed Control Set Switch Signal	S/C & Set Switch On: 3.8v
49	YL/RD	Speed Control On/Off Switch	S/C On: 1v, Off: 12v
50	WT/LG	Speed Control Resume Switch	S/C On: 1v, Off: 12v
51	DB/YL	ASD Relay Control	Relay Off: 12v, On: 1v
52	PK/BK	EVAP Purge Solenoid Control	Solenoid Off: 12v, On: 1v
53	LG/RD	Speed Control Vent Solenoid	Vacuum Decreasing: 1v
54	OR/BK	A/T: Overdrive Solenoid	Solenoid Off: 12v, On: 1v
54	OR/BK	M/T: Upshift Lamp Control	Lamp Off: 12v, On: 1v
55	OR/LG	Overdrive Lockout Solenoid	Solenoid Off: 12v, On: 1v
56	GY/PK	Maintenance Indicator Lamp	Lamp Off: 12v, On: 1v
57	DG/OR	ASD Relay Power (B+)	12-14v
58	LG/BK	Injector 6 Driver	1-4 ms
59	PK/BK	AIS Motor Control	Pulse Signals
60	YL/BK	AIS Motor Control	Pulse Signals

1994-95 Pickup 5.2L V8 MFI VIN Y (All) 60 Pin Connector

PCM Pin #	Wire Color	Circuit Description (60 Pin)	Value at Hot Idle
1	DG/RD	MAP Sensor Signal	1.5-1.6v
2	TN/BK	ECT Sensor Signal	At 180°F: 2.80v
3	RD/WT	Keep Alive Power (B+)	12-14v
4	BK/LB	Sensor Ground	<0.050v
5	---	Not Used	---
6	PK/WT	5-Volt Supply	4.9-5.1v
7	OR	8-Volt Supply	7.9-8.1v
8	---	Not Used	---
9	DB	Ignition Switch Power (B+)	12-14v
10	OR/WT	Overdrive Override Switch	Switch Off: 12v, On: 1v
11	BK/TN	Power Ground	<0.1v
12	BK/TN	Power Ground	<0.1v
13	LB/BR	Injector 4 Driver	1-4 ms
14	YL/WT	Injector 3 Driver	1-4 ms
15	TN	Injector 2 Driver	1-4 ms
16	WT/DB	Injector 1 Driver	1-4 ms
17	DB/TN	Injector 7 Driver	1-4 ms
18	RD/YL	Injector 8 Driver	1-4 ms
19	BK/GY	Coil Driver Control	5°, 55 mph: 8° dwell
20	DG	Alternator Field Control	Digital Signals: 0-12-0v
21	BK/RD	IAT Sensor Signal	At 100°F: 1.83v
22	OR/DB	TP Sensor Signal	0.6-1.0v
23	TN/WT	Left HO2S-11 (B1 S1) Signal	0.1-1.1v
24	GY/BK	Distributor Reference Signal	Digital Signals: 0-5-0v
25	PK	SCI Transmit	0v
26	---	Not Used	---
27	BR	A/C Damped Pressure Switch	A/C Off: 12v, On: 1v
29	WT/PK	Brake Switch Sense Signal	Brake Off: 0v, On: 12v
30	BK/WT	PNP Switch Sense Signal	In P/N: 0v, Others: 5v
31	PK/BK	Transmission Temperature Lamp Control	Lamp Off: 12v, On: 1v
32	BK/PK	MIL (lamp) Control	MIL Off: 12v, On: 1v
33	TN/RD	Speed Control Vacuum Solenoid	Vacuum Increasing: 1v
34	DB/OR	A/C WOT Relay Control	Relay Off: 12v, On: 1v
35	GY/YL	EGR Solenoid Control	12v, at 55 mph: 1v
36	---	Not Used	---
37	LG/OR	Overdrive Lamp Control	Lamp Off: 12v, On: 1v
38	GY	Injector 5 Driver	1-4 ms
39	GY/RD	AIS Motor Control	Pulse Signals
40	BR/WT	AIS Motor Control	Pulse Signals
41	BK/DG	HO2S-11 (B1 S1) Signal	0.1-1.1v
42	PK	Transmission Temperature Sensor Signal	At 200°F: 2.40v
43	GY/LB	Tachometer Signal	Pulse Signals
44	TN/YL	Distributor Sync Signal	Digital Signals: 0-5-0v
45	LG	SCI Receive	5v
47	WT/OR	Vehicle Speed Signal	Digital Signal
48	BR/RD	Speed Control Set Switch Signal	S/C & Set Switch On: 3.8v
49	YL/RD	Speed Control On/Off Switch	S/C On: 1v, Off: 12v
50	WT/LG	Speed Control Resume Switch	S/C On: 1v, Off: 12v
51	DB/YL	ASD Relay Control	Relay Off: 12v, On: 1v
52	PK/BK	EVAP Purge Solenoid Control	Solenoid Off: 12v, On: 1v
53	LG/RD	Speed Control Vent Solenoid	Vacuum Decreasing: 1v
54	OR/BK	A/T: Overdrive Solenoid	Solenoid Off: 12v, On: 1v
54	OR/BK	M/T: Upshift Lamp Control	Lamp Off: 12v, On: 1v
55	BR	Overdrive Lockout Solenoid	Solenoid Off: 12v, On: 1v
56	GY/PK	Maintenance Indicator Lamp	Lamp Off: 12v, On: 1v
57	DG/OR	ASD Relay Power (B+)	12-14v
58	BR/DB	Injector 6 Driver	1-4 ms
59	PK/BK	AIS Motor Control	Pulse Signals
60	YL/BK	AIS Motor Control	Pulse Signals

1996-97 Pickup 5.2L V8 MFI VIN Y 'A' Black Connector

PCM Pin #	Wire Color	Circuit Description (32 Pin)	Value at Hot Idle
1	---	Not Used	---
2	LG/BK	Ignition Switch Power (B+)	12-14v
3	---	Not Used	---
4	BK/LB	Sensor Ground	<0.050v
5	---	Not Used	---
6	BK/WT	PNP Switch Sense Signal	In P/N: 0v, Others: 5v
7	BK/GY	Coil 1 Driver Control	5°, 55 mph: 8° dwell
8	GY/BK	CKP Sensor Signal	Digital Signals: 0-5-0v
9	---	Not Used	---
10	YL/BK	IAC 2 Driver Control	Pulse Signals
11	BR/WT	IAC 3 Driver Control	Pulse Signals
12-14	---	Not Used	---
15	BK/RD	IAT Sensor Signal	At 100°F: 1.83v
16	TN/BK	ECT Sensor Signal	At 180°F: 2.80v
17	PK/WT	5-Volt Supply	4.9-5.1v
18	TN/YL	CMP Sensor Signal	Digital Signals: 0-5-0v
19	GY/RD	IAC 1 Driver Control	Pulse Signals
20	OR/BK	IAC 4 Driver Control	Pulse Signals
21	---	Not Used	---
22	RD/WT	Keep Alive Power (B+)	12-14v
23	OR/DB	TP Sensor Signal	0.6-1.0v
24	TN/WT	HO2S-11 (B1 S1) Signal	0.1-1.1v
25	OR/BK	HO2S-12 (B1 S2) Signal	0.1-1.1v
26	---	Not Used	---
27	DG/RD	MAP Sensor Signal	1.5-1.6v
28-30	---	Not Used	---
31	BK/TN	Power Ground	<0.1v
32	BK/TN	Power Ground	<0.1v

Pin Connector Graphic

1996-97 Pickup 5.2L V8 MFI VIN Y 'B' White Connector

PCM Pin #	Wire Color	Circuit Description (32 Pin)	Value at Hot Idle
1	VT	Transmission Temperature Sensor Signal	At 200ºF: 2.40v
2	PK/TN	Injector 7 Driver	1-4 ms
3, 7, 9	---	Not Used	---
4	WT/DB	Injector 1 Driver	1-4 ms
5	YL/WT	Injector 3 Driver	1-4 ms
6	GY	Injector 5 Driver	1-4 ms
8	PK/WT	Governor Pressure Solenoid	PWM Signal: 0-12-0v
10	DG	Generator Field Control	Digital Signals: 0-12-0v
11	OR/BK	TCC Solenoid Control	At Cruise w/TCC On: 1v
12	BR/DB	Injector 6 Driver	1-4 ms
13	GY/LB	Injector 8 Driver	1-4 ms
14	---	Not Used	---
15	TN	Injector 2 Driver	1-4 ms
16	LB/BR	Injector 4 Driver	1-4 ms
17-20	---	Not Used	
21	BR	3-4 Shift Solenoid Control	Solenoid Off: 12v, On: 1v
22-24	---	Not Used	
25, 28	DB, LG	OSS Sensor (-), (+) Signals	Moving: AC pulse signals
26	---	Not Used	---
27	WT/OR	Vehicle Speed Signal	Digital Signal
29	LG/WT	Governor Pressure Sensor	0.58v
30	PK	Transmission Relay Control	Relay Off: 12v, On: 1v
31	OR	5-Volt Supply	4.9-5.1v
32	GY/YL	EGR Solenoid Control	12v, at 55 mph: 1v

1996-97 Pickup 5.2L V8 MFI VIN Y 'C' Gray Connector

PCM Pin #	Wire Color	Circuit Description (32 Pin)	Value at Hot Idle
1	DB/OR	A/C Clutch Relay Control	Relay Off: 12v, On: 1v
3	DB/YL	ASD Relay Control	Relay Off: 12v, On: 1v
4	TN/RD	Speed Control Vacuum Solenoid	Vacuum Increasing: 1v
5	LG/RD	Speed Control Vent Solenoid	Vacuum Decreasing: 1v
6	LG/OR	Overdrive Off Lamp Control	Lamp Off: 12v, On: 1v
7	PK/BK	Transmission Temperature Lamp Control	Lamp Off: 12v, On: 1v
11	YL/RD	Speed Control Power Supply	12-14v
12	DG/OR	ASD Relay Power (B+)	12-14v
13	OR/WT	Overdrive Off Switch Sense	Switch Off: 12v, On: 1v
15	PK/YL	Battery Temperature Sensor	At 86ºF: 1.96v
16	TN/YL	Generator Lamp Control	Lamp Off: 12v, On: 1v
17	BK/PK	MIL (lamp) Control	MIL Off: 12v, On: 1v
18	GY/PK	Maintenance Indicator Lamp	Lamp Off: 12v, On: 1v
19	BR/WT	Fuel Pump Relay Control	Relay Off: 12v, On: 1v
20	PK	EVAP Purge Solenoid Control	PWM Signal: 0-12-0v
22	BR	A/C Request Signal	A/C Off: 12v, On: 1v
23	LG	A/C Select Signal	A/C Off: 12v, On: 1v
24	WT/PK	Brake Switch Sense Signal	Brake Off: 0v, On: 12v
26	DB/WT	Fuel Level Sensor Signal	Full: 0.56v, 1/2 full: 2.5v
27	PK	SCI Transmit Signal	0v
29	LG	SCI Receive Signal	5v
31	GY/LB	Tachometer Signal	Pulse Signals
32	RD/LG	Speed Control Set Switch Signal	S/C & Set Switch On: 3.8v

Standard Colors and Abbreviations

Abbreviation	Color	Abbreviation	Color	Abbreviation	Color
BK	Black	GY	Gray	RD	Red
BL	Blue	GN	Green	TN	Tan
BR	Brown	LG	Light Green	VT	Violet
DB	Dark Blue	OR	Orange	WT	White
DG	Dark Green	PK	Pink	YL	Yellow

1998-2001 Pickup 5.2L V8 MFI (All) VIN Y 'C1' Black Connector

PCM Pin #	Wire Color	Circuit Description (32 Pin)	Value at Hot Idle
1	---	Not Used	---
2	LG/BK	Ignition Switch Power (B+)	12-14v
3	---	Not Used	---
4	BK/LB	Sensor Ground	<0.050v
5	---	Not Used	---
6	BK/WT	PNP Switch Sense Signal	In P/N: 0v, Others: 5v
7	BK/GY	Coil 1 Driver Control	5°, 55 mph: 8° dwell
8	GY/BK	CKP Sensor Signal	Digital Signals: 0-5-0v
9	---	Not Used	---
10	YL/BK	IAC 2 Driver Control	Pulse Signals
11	BR/WT	IAC 3 Driver Control	Pulse Signals
12	---	Not Used	---
13	OR	Power Takeoff Switch Sense	Switch Off: 12v, On: 1v
14	---	Not Used	---
15	BK/RD	IAT Sensor Signal	At 100°F: 1.83v
16	TN/BK	ECT Sensor Signal	At 180°F: 2.80v
17	VT/WT	5-Volt Supply	4.9-5.1v
18	TN/YL	CMP Sensor Signal	Digital Signals: 0-5-0v
19	GY/RD	IAC 1 Driver Control	Pulse Signals
20	PK/BK	IAC 4 Driver Control	Pulse Signals
21	---	Not Used	---
22	RD/WT	Keep Alive Power (B+)	12-14v
23	OR/DB	TP Sensor Signal	0.6-1.0v
24	TN/WT	HO2S-11 (B1 S1) Signal	0.1-1.1v
25	OR/BK	HO2S-12 (B1 S2) Signal	0.1-1.1v
26	LG/RD	HO2S-21 (B2 S1) Signal (Cal)	0.1-1.1v
27	DG/RD	MAP Sensor Signal	1.5-1.6v
28	---	Not Used	---
29	OR/TN	HO2S-22 (B2 S2) Signal (Cal)	0.1-1.1v
30	---	Not Used	---
31	BK/TN	Power Ground	<0.1v
32	BK/TN	Power Ground	<0.1v

Pin Connector Graphic

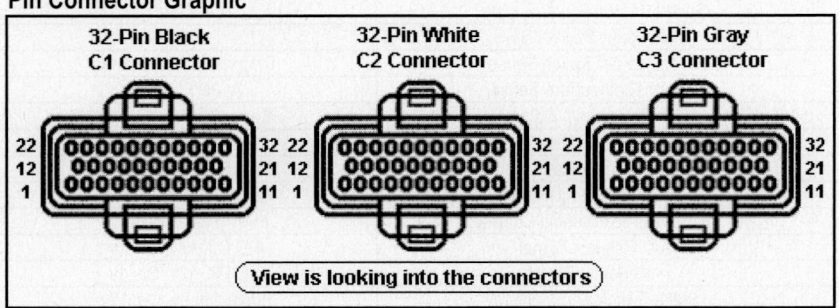

1998-2001 Pickup 5.2L V8 MFI (All) VIN Y 'C2' White Connector

PCM Pin #	Wire Color	Circuit Description (32 Pin)	Value at Hot Idle
1	VT	Transmission Temperature Sensor Signal	At 200ºF: 2.40v
2	VT/TN	Injector 7 Driver	1-4 ms
4	WT/DB	Injector 1 Driver	1-4 ms
5	YL/WT	Injector 3 Driver	1-4 ms
6	GY	Injector 5 Driver	1-4 ms
8	VT/WT	Governor Pressure Solenoid	PWM Signal: 0-12-0v
10	DG	Generator Field Control	Digital Signals: 0-12-0v
11	OR/BK	TCC Solenoid Control	At Cruise w/TCC On: 1v
12	BR/DB	Injector 6 Driver	1-4 ms
13	GY/LB	Injector 8 Driver	1-4 ms
15	TN	Injector 2 Driver	1-4 ms
16	LB/BR	Injector 4 Driver	1-4 ms
21	BR	3-4 Shift Solenoid Control	Cruise w/solenoid On: 1v
23	GY/OR	Engine Oil Pressure Sensor	1.6v at 24 psi
25	DB/BK	OSS Sensor (-) Signal	Moving: AC pulse signals
27	WT/OR	Vehicle Speed Signal	Digital Signal
28	LG/BK	OSS Sensor (+) Signal	Moving: AC pulse signals
29	LG/WT	Governor Pressure Sensor	0.58v
30	PK	Transmission Relay Control	Relay Off: 12v, On: 1v
31	OR	5-Volt Supply	4.9-5.1v

1998-2001 Pickup 5.2L V8 MFI VIN Y (All) 'C3' Gray Connector

PCM Pin #	Wire Color	Circuit Description (32 Pin)	Value at Hot Idle
1	DB/OR	A/C Clutch Relay Control	Relay Off: 12v, On: 1v
3	DB/YL	ASD Relay Control	Relay Off: 12v, On: 1v
4	TN/RD	Speed Control Vacuum Solenoid	Vacuum Increasing: 1v
5	LG/RD	Speed Control Vent Solenoid	Vacuum Decreasing: 1v
6 ('98)	LG/OR	Overdrive Off Lamp Control	Lamp Off: 12v, On: 1v
8	BR/VT	HO2S-11 Heater Control (Cal)	PWM Signal: 0-12-0v
9	DG/PK	HO2S-12 Heater Relay (Cal)	Relay Off: 12v, On: 1v
10	WT/DG	LDP Solenoid Control	PWM Signal: 0-12-0v
11	YL/RD	Speed Control Power Supply	12-14v
12	DG/OR	ASD Relay Power (B+)	12-14v
13	OR/WT	Overdrive Off Switch Sense	Switch Off: 12v, On: 1v
14	OR	LDP Switch Sense Signal	LDP Switch Closed: 0v
15	PK/YL	Battery Temperature Sensor	At 86ºF: 1.96v
19	BR/WT	Fuel Pump Relay Control	Relay Off: 12v, On: 1v
20	PK/WT	EVAP Purge Solenoid Control	PWM Signal: 0-12-0v
22	BR	A/C Switch Signal	A/C Off: 12v, On: 1v
23	LG/WT	A/C Select Signal	A/C Off: 12v, On: 1v
24	WT/PK	Brake Switch Sense Signal	Brake Off: 0v, On: 12v
25	DB	Generator Field Source	12-14v
26	DB/WT	Fuel Level Sensor Signal	Full: 0.56v, 1/2 full: 2.5v
27	PK/DB	SCI Transmit Signal	0v
28	WT/BK	CCD Bus (-) Signal	<0.050v
29	DG	SCI Receive Signal	5v
30	VT/BR	CCD Bus (+) Signal	Digital Signals: 0-5-0v
32	RD/LG	Speed Control Set Switch Signal	S/C & Set Switch On: 3.8v

Standard Colors and Abbreviations

Abbreviation	Color	Abbreviation	Color	Abbreviation	Color
BK	Black	GY	Gray	RD	Red
BL	Blue	GN	Green	TN	Tan
BR	Brown	LG	Light Green	VT	Violet
DB	Dark Blue	OR	Orange	WT	White
DG	Dark Green	PK	Pink	YL	Yellow

1990-91 Pickup 5.9L V8 TBI VIN 5, VIN Z (All) 60 Pin Connector

PCM Pin #	Wire Color	Circuit Description (60 Pin)	Value at Hot Idle
1	DG/RD	MAP Sensor Signal	1.5-1.6v
2	TN	ECT Sensor Signal	At 180°F: 2.80v
3	RD	Keep Alive Power (B+)	12-14v
4, 5	BK	Sensor Ground	<0.050v
6	PK/WT	5-Volt Supply	4.9-5.1v
7	OR	8-Volt Supply	7.9-8.1v
8	DG/BK	ASD Relay Power (B+)	12-14v
9	DB	Ignition Switch Power (B+)	12-14v
10, 13-14,	---	Not Used	---
11, 12	LB/RD	Power Ground	<0.1v
15	TN	Injector 2 Driver	1-4 ms
16	WT/DB	Injector 1 Driver	1-4 ms
19	BK/GY	Coil Driver Control	5°, 55 mph: 8° dwell
20	DG	Alternator Field Control	Digital Signals: 0-12-0v
21	BK/RD	Throttle Body Temperature	At 100°F: 2.51v
17-18, 26, 31	---	Not Used	---
22	OR/DB	TP Sensor Signal	0.6-1.0v
24	GY	Distributor Reference Signal	Digital Signals: 0-5-0v
25, 45	PK, LG	SCI Transmit, SCI Receive	0v, 5v
27	BR	A/C Clutch Signal	A/C Off: 12v, On: 1v
28	PK	Closed Throttle Switch	Switch Off: 12v, On: 1v
29	WT/PK	Brake Switch Sense Signal	Brake Off: 0v, On: 12v
30	BR/YL	PNP Switch Sense Signal	In P/N: 0v, Others: 5v
32	BK/PK	MIL (lamp) Control	MIL Off: 12v, On: 1v
33	TN/RD	Speed Control Vacuum Solenoid	Vacuum Increasing: 1v
34	DB/OR	A/C WOT Relay Control	Relay Off: 12v, On: 1v
35	GY/YL	EGR Solenoid Control	12v, at 55 mph: 1v
36	BK/OR	Air Switching Solenoid	Solenoid Off: 12v, On: 1v
37-39, 46	---	Not Used	---
40	BR	AIS Motor Control	Pulse Signals
41	BK/DG	HO2S-11 (B1 S1) Signal	0.1-1.1v
42-43	---	Not Used	---
44	OR/WT	Overdrive Lockout Switch	Switch Off: 12v, On: 1v
47	WT/OR	Vehicle Speed Signal	Digital Signal
48	BR/RD	Speed Control Set Switch Signal	S/C & Set Switch On: 3.8v
49	YL/RD	Speed Control On/Off Switch	S/C On: 1v, Off: 12v
50	WT/LG	Speed Control Resume Switch	S/C On: 1v, Off: 12v
51	DB/YL	ASD Relay Control	Relay Off: 12v, On: 1v
52	PK/BK	EVAP Purge Solenoid Control	Solenoid Off: 12v, On: 1v
53	LG/RD	Speed Control Vent Solenoid	Vacuum Decreasing: 1v
54	OR/BK	Part Throttle Unlock Solenoid	Solenoid Off: 12v, On: 1v
55	OR/LG	Overdrive Lockout Solenoid	Solenoid Off: 12v, On: 1v
56	GY/PK	Maintenance Indicator Lamp	Lamp Off: 12v, On: 1v
57-59	---	Not Used	---
60	GY/RD	AIS Motor Control	Pulse Signals

Pin Connector Graphic

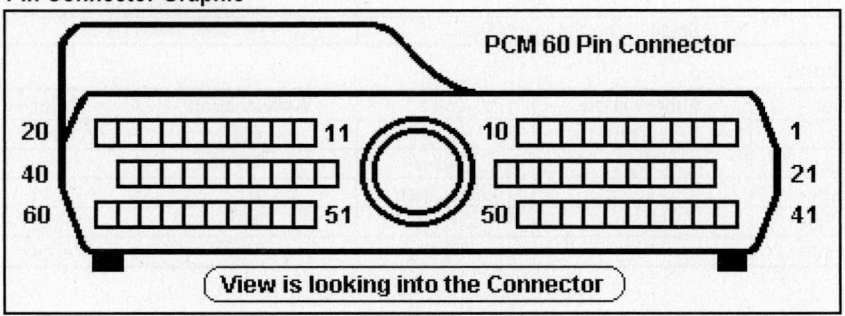

PCM 60 Pin Connector

View is looking into the Connector

1992 Pickup 5.9L V8 TBI VIN 5, VIN Z (All) 60 Pin Connector

PCM Pin #	Wire Color	Circuit Description (60 Pin)	Value at Hot Idle
1	DG/RD	MAP Sensor Signal	1.5-1.6v
2	TN/BK	ECT Sensor Signal	At 180ºF: 2.80v
3	RD	Keep Alive Power (B+)	12-14v
4	BK	Sensor Ground	<0.050v
5	BK	Sensor Ground	<0.050v
6	PK/WT	5-Volt Supply	4.9-5.1v
7	OR	8-Volt Supply	7.9-8.1v
8	DG/BK	ASD Relay Power (B+)	12-14v
9	DB	Ignition Switch Power (B+)	12-14v
10	---	Not Used	---
11	BK/TN	Power Ground	<0.1v
12	BK/TN	Power Ground	<0.1v
13-14	---	Not Used	---
15	TN	Injector 2 Driver	1-4 ms
16	WT/DB	Injector 1 Driver	1-4 ms
17-18	---	Not Used	---
19	GY	Coil Driver Control	5º, 55 mph: 8º dwell
20	DG	Alternator Field Control	Digital Signals: 0-12-0v
21	BK/RD	Throttle Body Temperature	At 100ºF: 2.51v
23	---	Not Used	---
22	OR/DB	TP Sensor Signal	0.6-1.0v
24	GY	Distributor Reference Signal	Digital Signals: 0-5-0v
25	PK	SCI Transmit	0v
26	---	Not Used	---
27	BR	A/C Clutch Signal	A/C Off: 12v, On: 1v
28	PK	Closed Throttle Switch	Switch Off: 12v, On: 1v
29	WT/PK	Brake Switch Sense Signal	Brake Off: 0v, On: 12v
30	BR/YL	PNP Switch Sense Signal	In P/N: 0v, Others: 5v
31	---	Not Used	---
32	BK/PK	MIL (lamp) Control	MIL Off: 12v, On: 1v
33	TN/RD	Speed Control Vacuum Solenoid	Vacuum Increasing: 1v
34	DB/OR	A/C WOT Relay Control	Relay Off: 12v, On: 1v
35	GY/YL	EGR Solenoid Control	12v, at 55 mph: 1v
36	BK/OR	Air Switching Solenoid	Solenoid Off: 12v, On: 1v
37-39	---	Not Used	---
40	BR/WT	AIS Motor Control	Pulse Signals
41	BK/DG	HO2S-11 (B1 S1) Signal	0.1-1.1v
42-43	---	Not Used	---
44	OR/WT	Overdrive Lockout Switch	Switch Off: 12v, On: 1v
45	LG	SCI Receive	5v
46	---	Not Used	---
47	WT/OR	Vehicle Speed Signal	Digital Signal
48	BR/RD	Speed Control Set Switch Signal	S/C & Set Switch On: 3.8v
49	YL/RD	Speed Control On/Off Switch	S/C On: 1v, Off: 12v
50	WT/LG	Speed Control Resume Switch	S/C On: 1v, Off: 12v
51	DB/YL	ASD Relay Control	Relay Off: 12v, On: 1v
52	PK/BK	EVAP Purge Solenoid Control	Solenoid Off: 12v, On: 1v
53	LG/RD	Speed Control Vent Solenoid	Vacuum Decreasing: 1v
54	OR/BK	Part Throttle Unlock Solenoid	Solenoid Off: 12v, On: 1v
55	BR	Overdrive Lockout Solenoid	Solenoid Off: 12v, On: 1v
56	GY/PK	Maintenance Indicator Lamp	Lamp Off: 12v, On: 1v
57-59	---	Not Used	---
60	GY/RD	AIS Motor Control	Pulse Signals

1993 Pickup 5.9L V8 MFI VIN 5, VIN Z (All) 60 Pin Connector

PCM Pin #	Wire Color	Circuit Description (60 Pin)	Value at Hot Idle
1	DG/RD	MAP Sensor Signal	1.5-1.6v
2	TN/BK	ECT Sensor Signal	At 180°F: 2.80v
3	RD	Keep Alive Power (B+)	12-14v
4	BK/LB	Sensor Ground	<0.050v
5	BK/WT	Sensor Ground	<0.050v
6	PK/WT	5-Volt Supply	4.9-5.1v
7	OR	8-Volt Supply	7.9-8.1v
8	---	Not Used	---
9	DB	Ignition Switch Power (B+)	12-14v
10	OR/WT	Overdrive Override Switch	Switch Off: 12v, On: 1v
11-12	BK/TN	Power Ground	<0.1v
13	LB/BR	Injector 4 Driver	1-4 ms
14	YL/WT	Injector 3 Driver	1-4 ms
15	T	Injector 2 Driver	1-4 ms
16	WT/DB	Injector 1 Driver	1-4 ms
17	DB/TN	Injector 7 Driver	1-4 ms
18	RD/YL	Injector 8 Driver	1-4 ms
19	GY	Coil Driver Control	5°, 55 mph: 8° dwell
20	DG	Alternator Field Control	Digital Signals: 0-12-0v
21	BK/RD	IAT Sensor Signal	At 100°F: 1.83v
22	OR/DB	TP Sensor Signal	0.6-1.0v
23	TN/WT	HO2S-12 (B1 S2) Signal	0.1-1.1v
24	GY/BK	Distributor Reference Signal	Digital Signals: 0-5-0v
25	PK	SCI Transmit	0v
26	---	Not Used	---
27	BR	A/C Clutch Signal	A/C Off: 12v, On: 1v
28, 31	---	Not Used	---
29	WT/PK	Brake Switch Sense Signal	Brake Off: 0v, On: 12v
30	BR/YL	PNP Switch Sense Signal	In P/N: 0v, Others: 5v
32	BK/PK	MIL (lamp) Control	MIL Off: 12v, On: 1v
33	TN/RD	Speed Control Vacuum Solenoid	Vacuum Increasing: 1v
34	DB/OR	A/C WOT Relay Control	Relay Off: 12v, On: 1v
35	GY/YL	EGR Solenoid Control	12v, at 55 mph: 1v
37	BK/OR	Overdrive Lamp Control	Lamp Off: 12v, On: 1v
38	PK/BK	Injector 5 Driver	1-4 ms
39	GY/RD	AIS Motor Control	Pulse Signals
40	BR/WT	AIS Motor Control	Pulse Signals
41	BK/DG	HO2S-11 (B1 S1) Signal	0.1-1.1v
42, 43	---	Not Used	---
44	GY	Distributor Sync Signal	Digital Signals: 0-5-0v
45	LG	SCI Receive	5v
46	---	Not Used	---
47	WT/OR	Vehicle Speed Signal	Digital Signal
48	BR/RD	Speed Control Set Switch Signal	S/C & Set Switch On: 3.8v
49	YL/RD	Speed Control On/Off Switch	S/C On: 1v, Off: 12v
50	WT/LG	Speed Control Resume Switch	S/C On: 1v, Off: 12v
51	DB/YL	ASD Relay Control	Relay Off: 12v, On: 1v
52	PK/BK	EVAP Purge Solenoid Control	Solenoid Off: 12v, On: 1v
53	LG/RD	Speed Control Vent Solenoid	Vacuum Decreasing: 1v
54	OR/BK	A/T: Overdrive Solenoid	Solenoid Off: 12v, On: 1v
54	OR/BK	M/T: Upshift Lamp Control	Lamp Off: 12v, On: 1v
55	OR/LG	Overdrive Lockout Solenoid	Solenoid Off: 12v, On: 1v
56	GY/PK	Maintenance Indicator Lamp	Lamp Off: 12v, On: 1v
57	DG/OR	ASD Relay Power (B+)	12-14v
58	LG/BK	Injector 6 Driver	1-4 ms
59	PK/BK	AIS Motor Control	Pulse Signals
60	YL/BK	AIS Motor Control	Pulse Signals

1994-95 Pickup 5.9L V8 MFI VIN 5, VIN Z (All) 60 Pin Connector

PCM Pin #	Wire Color	Circuit Description (60 Pin)	Value at Hot Idle
1	DG/RD	MAP Sensor Signal	1.5-1.6v
2	TN/BK	ECT Sensor Signal	At 180°F: 2.80v
3	RD/WT	Keep Alive Power (B+)	12-14v
4	BK/LB	Sensor Ground	<0.050v
5, 8	---	Not Used	---
6	PK/WT	5-Volt Supply	4.9-5.1v
7	OR	8-Volt Supply	7.9-8.1v
9	LG/BK	Ignition Switch Power (B+)	12-14v
10	OR/WT	Overdrive Override Switch	Switch Off: 12v, On: 1v
11-12	BK/TN	Power Ground	<0.1v
13	LB/BR	Injector 4 Driver	1-4 ms
14	YL/WT	Injector 3 Driver	1-4 ms
15	TN	Injector 2 Driver	1-4 ms
16	WT/DB	Injector 1 Driver	1-4 ms
17	DB/TN	Injector 7 Driver	1-4 ms
18	RD/YL	Injector 8 Driver	1-4 ms
19	BK/GY	Coil Driver Control	5°, 55 mph: 8° dwell
20	DG	Alternator Field Control	Digital Signals: 0-12-0v
21	BK/RD	IAT Sensor Signal	At 100°F: 1.83v
22	OR/DB	TP Sensor Signal	0.6-1.0v
23	TN/WT	HO2S-12 (B1 S2) Signal	0.1-1.1v
24	GY/BK	Distributor Reference Signal	Digital Signals: 0-5-0v
25	PK	SCI Transmit	0v
26	---	Not Used	---
27	BR	A/C Damped Pressure Switch	A/C Off: 12v, On: 1v
29	WT/PK	Brake Switch Sense Signal	Brake Off: 0v, On: 12v
30	BK/WT	PNP Switch Sense Signal	In P/N: 0v, Others: 5v
31	PK/BK	Transmission Temperature Lamp Control	Lamp Off: 12v, On: 1v
32	BK/PK	MIL (lamp) Control	MIL Off: 12v, On: 1v
33	TN/RD	Speed Control Vacuum Solenoid	Vacuum Increasing: 1v
34	DB/OR	A/C WOT Relay Control	Relay Off: 12v, On: 1v
35	GY/YL	EGR Solenoid Control	12v, at 55 mph: 1v
37	LG/OR	Overdrive Lamp Control	Lamp Off: 12v, On: 1v
38	GY	Injector 5 Driver	1-4 ms
39	GY/RD	AIS Motor Control	Pulse Signals
40	BR/WT	AIS Motor Control	Pulse Signals
41	BK/DG	HO2S-11 (B1 S1) Signal	0.1-1.1v
42	PK	Transmission Temperature Sensor Signal	At 200°F: 2.40v
43	GY/LB	Tachometer Signal	Pulse Signals
44	TN/YL	Distributor Sync Signal	Digital Signals: 0-5-0v
45	LG	SCI Receive	5v
46	---	Not Used	---
47	WT/OR	Vehicle Speed Signal	Digital Signal
48	BR/RD	Speed Control Set Switch Signal	S/C & Set Switch On: 3.8v
49	YL/RD	Speed Control On/Off Switch	S/C On: 1v, Off: 12v
50	WT/LG	Speed Control Resume Switch	S/C On: 1v, Off: 12v
51	DB/YL	ASD Relay Control	Relay Off: 12v, On: 1v
52	PK/BK	EVAP Purge Solenoid Control	Solenoid Off: 12v, On: 1v
53	LG/RD	Speed Control Vent Solenoid	Vacuum Decreasing: 1v
54	OR/BK	A/T: Overdrive Solenoid	Solenoid Off: 12v, On: 1v
54	OR/BK	M/T: Upshift Lamp Control	Lamp Off: 12v, On: 1v
55	BR	Overdrive Lockout Solenoid	Solenoid Off: 12v, On: 1v
56	GY/PK	Maintenance Indicator Lamp	Lamp Off: 12v, On: 1v
57	DG/OR	ASD Relay Power (B+)	12-14v
58	BR/DB	Injector 6 Driver	1-4 ms
59	PK/BK	AIS Motor Control	Pulse Signals
60	YL/BK	AIS Motor Control	Pulse Signals

1996-97 Pickup 5.9L V8 MFI VIN 5, VIN Z 'A' Black Connector

PCM Pin #	Wire Color	Circuit Description (32 Pin)	Value at Hot Idle
1	---	Not Used	---
2	LG/BK	Ignition Switch Power (B+)	12-14v
3	---	Not Used	---
4	BK/LB	Sensor Ground	<0.050v
5	---	Not Used	---
6	BK/WT	PNP Switch Sense Signal	In P/N: 0v, Others: 5v
7	BK/GY	Coil 1 Driver Control	5°, 55 mph: 8° dwell
8	GY/BK	CKP Sensor Signal	Digital Signals: 0-5-0v
9	---	Not Used	---
10	YL/BK	IAC 2 Driver Control	Pulse Signals
11	BR/WT	IAC 3 Driver Control	Pulse Signals
12-14	---	Not Used	---
15	BK/RD	IAT Sensor Signal	At 100°F: 1.83v
16	TN/BK	ECT Sensor Signal	At 180°F: 2.80v
17	PK/WT	5-Volt Supply	4.9-5.1v
18	TN/YL	CMP Sensor Signal	Digital Signals: 0-5-0v
19	GY/RD	IAC 1 Driver Control	Pulse Signals
20	PK/BK	IAC 4 Driver Control	Pulse Signals
21	---	Not Used	---
22	RD/WT	Keep Alive Power (B+)	12-14v
23	OR/DB	TP Sensor Signal	0.6-1.0v
24	TN/WT	HO2S-11 (B1 S1) Signal	0.1-1.1v
25	OR/BK	L/D A/T: HO2S-12 Signal	0.1-1.1v
26	BK/DG	H/D A/T: HO2S-12 Signal	0.1-1.1v
27	DG/RD	MAP Sensor Signal	1.5-1.6v
28-30	---	Not Used	---
31	BK/TN	Power Ground	<0.1v
32	BK/TN	Power Ground	<0.1v

Pin Connector Graphic

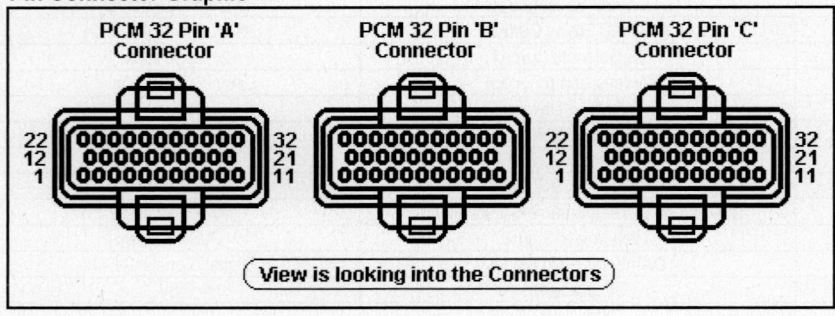

PCM 32 Pin 'A' Connector

PCM 32 Pin 'B' Connector

PCM 32 Pin 'C' Connector

View is looking into the Connectors

1996-97 Pickup 5.9L V8 MFI VIN 5, VIN Z 'B' White Connector

PCM Pin #	Wire Color	Circuit Description (32 Pin)	Value at Hot Idle
1	VT	Transmission Temperature Sensor Signal	At 200°F: 2.40v
2	PK/TN	Injector 7 Driver	1-4 ms
4	WT/DB	Injector 1 Driver	1-4 ms
5	YL/WT	Injector 3 Driver	1-4 ms
6	GY	Injector 5 Driver	1-4 ms
8	PK/WT	Governor Pressure Solenoid	PWM Signal: 0-12-0v
10	DG	Generator Field Control	Digital Signals: 0-12-0v
11	OR/BK	TCC Solenoid Control	At Cruise w/TCC On: 1v
12	BR/DB	Injector 6 Driver	1-4 ms
13	GY/LB	Injector 8 Driver	1-4 ms
15	TN	Injector 2 Driver	1-4 ms
16	LB/BR	Injector 4 Driver	1-4 ms
21	BR	3-4 Shift Solenoid Control	Solenoid Off: 12v, On: 1v
25	DB/BK	OSS Sensor (-) Signal	Moving: AC pulse signals
27	WT/OR	Vehicle Speed Signal	Digital Signal
28	LG/BK	OSS Sensor (+) Signal	Moving: AC pulse signals
29	LG/WT	Governor Pressure Sensor	0.58v
30	PK	Transmission Relay Control	Relay Off: 12v, On: 1v
31	OR	5-Volt Supply	4.9-5.1v
32	GY/YL	EGR Solenoid Control	12v, at 55 mph: 1v

1996-97 Pickup 5.9L V8 MFI VIN 5, VIN Z 'C' Gray Connector

PCM Pin #	Wire Color	Circuit Description (32 Pin)	Value at Hot Idle
1	DB/OR	A/C Clutch Relay Control	Relay Off: 12v, On: 1v
2	---	Not Used	---
3	DB/YL	ASD Relay Control	Relay Off: 12v, On: 1v
4	TN/RD	Speed Control Vacuum Solenoid	Vacuum Increasing: 1v
5	LG/RD	Speed Control Vent Solenoid	Vacuum Decreasing: 1v
6	LG/OR	Overdrive Off Lamp Control	Lamp Off: 12v, On: 1v
7	PK/BK	Transmission Temperature Lamp Control	Lamp Off: 12v, On: 1v
8-10	---	Not Used	---
11	YL/RD	Speed Control Power Supply	12-14v
12	DG/OR	ASD Relay Power (B+)	12-14v
13	OR/WT	Overdrive Off Switch Sense	Switch Off: 12v, On: 1v
14	---	Not Used	---
15	PK/YL	Battery Temperature Sensor	At 86°F: 1.96v
16	TN/YL	Generator Lamp Control	Lamp Off: 12v, On: 1v
17	BK/PK	MIL (lamp) Control	MIL Off: 12v, On: 1v
18	GY/PK	Maintenance Indicator Lamp	Lamp Off: 12v, On: 1v
19	BR/WT	Fuel Pump Relay Control	Relay Off: 12v, On: 1v
20	PK	EVAP Purge Solenoid Control	PWM Signal: 0-12-0v
21	---	Not Used	---
22	BR	A/C Request Signal	A/C Off: 12v, On: 1v
23	LG	A/C Select Signal	A/C Off: 12v, On: 1v
24	WT/PK	Brake Switch Sense Signal	Brake Off: 0v, On: 12v
21, 25,	---	Not Used	---
26	DB/WT	Fuel Level Sensor Signal	Full: 0.56v, 1/2 full: 2.5v
27	PK	SCI Transmit	0v
28	WT/BK	CCD Bus (-) Signal	<0.050v
29	DG	SCI Receive	5v
30	VT/BR	CCD Bus (+) Signal	Digital Signals: 0-5-0v
31	GY/LB	Tachometer Signal	Pulse Signals
32	RD/LG	Speed Control Set Switch Signal	S/C & Set Switch On: 3.8v

Standard Colors and Abbreviations

Abbreviation	Color	Abbreviation	Color	Abbreviation	Color
BK	Black	GY	Gray	RD	Red
BL	Blue	GN	Green	TN	Tan
BR	Brown	LG	Light Green	VT	Violet
DB	Dark Blue	OR	Orange	WT	White
DG	Dark Green	PK	Pink	YL	Yellow

1998-2002 Pickup 5.9L V8 VIN 5, VIN Z 'C1' Black Connector

PCM Pin #	Wire Color	Circuit Description (32 Pin)	Value at Hot Idle
1	---	Not Used	---
2	LG/BK	Ignition Switch Power (B+)	12-14v
3	---	Not Used	---
4	BK/LB	Sensor Ground	<0.050v
5	---	Not Used	---
6	BK/WT	PNP Switch Sense Signal	In P/N: 0v, Others: 5v
7	BK/GY	Coil 1 Driver Control	5°, 55 mph: 8° dwell
8	GY/BK	CKP Sensor Signal	Digital Signals: 0-5-0v
9	---	Not Used	---
10	YL/BK	IAC 2 Driver Control	Pulse Signals
11	BR/WT	IAC 3 Driver Control	Pulse Signals
12, 14	---	Not Used	---
13	OR	Power Takeoff Switch Sense	Switch Off: 12v, On: 1v
15	BK/RD	IAT Sensor Signal	At 100°F: 1.83v
16	TN/BK	ECT Sensor Signal	At 180°F: 2.80v
17	VT/WT	5-Volt Supply	4.9-5.1v
18	TN/YL	CMP Sensor Signal	Digital Signals: 0-5-0v
19	GY/RD	IAC 1 Driver Control	Pulse Signals
20	VT/BK	IAC 4 Driver Control	Pulse Signals
21	---	Not Used	---
22	RD/WT	Keep Alive Power (B+)	12-14v
23	OR/DB	TP Sensor Signal	0.6-1.0v
24	TN/WT	HO2S-11 (B1 S1) Signal	0.1-1.1v
24 (HD)	BK/DG	HO2S-11 (B1 S1) Signal	0.1-1.1v
25	OR/BK	L/D Trans. - HO2S-12	0.1-1.1v
25 ('00)	OR/BK	M/D A/T: HO2S-13 (Cal)	0.1-1.1v
26	BK/DG	H/D A/T: HO2S-12	0.1-1.1v
26 ('00-'01)	LG/RD	HO2S-21 (B2 S1) Signal (Cal)	0.1-1.1v
26 ('00)	LG/RD	M/D A/T: HO2S-12 (Cal)	0.1-1.1v
27	DG/RD	MAP Sensor Signal	1.5-1.6v
28-30	---	Not Used	---
29	OR/TN	HO2S-22 (B2 S2) Signal	0.1-1.1v
31	BK/TN	Power Ground	<0.1v
32	BK/TN	Power Ground	<0.1v

Pin Connector Graphic

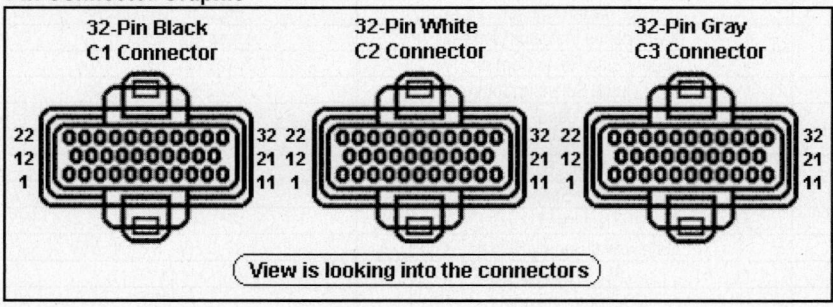

1998-2002 Pickup 5.9L V8 VIN 5, VIN Z White 'C2' Connector

PCM Pin #	Wire Color	Circuit Description (32 Pin)	Value at Hot Idle
1	VT	Transmission Temperature Sensor Signal	At 200°F: 2.40v
2	VT/TN	Injector 7 Driver	1-4 ms
4	WT/DB	Injector 1 Driver	1-4 ms
5	YL/WT	Injector 3 Driver	1-4 ms
6	GY	Injector 5 Driver	1-4 ms
8	VT/WT	Governor Pressure Solenoid	PWM Signal: 0-12-0v
10	DG	Generator Field Control	Digital Signals: 0-12-0v
11	OR/BK	TCC Solenoid Control	At Cruise w/TCC On: 1v
12	BR/DB	Injector 6 Driver	1-4 ms
13	GY/LB	Injector 8 Driver	1-4 ms
15	TN	Injector 2 Driver	1-4 ms
16	LB/BR	Injector 4 Driver	1-4 ms
21	BR	3-4 Shift Solenoid Control	Cruise w/solenoid On: 1v
23	GY/OR	Engine Oil Pressure Sensor	1.6v at 24 psi
25	DB/BK	OSS Sensor (-) Signal	Moving: AC pulse signals
27	WT/OR	Vehicle Speed Signal	Digital Signal
28	LG/BK	OSS Sensor (+) Signal	Moving: AC pulse signals
29	LG/WT	Governor Pressure Sensor	0.58v
30	PK	Transmission Relay Control	Relay Off: 12v, On: 1v
31	OR	5-Volt Supply	4.9-5.1v

1998-2002 Pickup 5.9L V8 VIN 5, VIN Z 'C3' Gray Connector

PCM Pin #	Wire Color	Circuit Description (32 Pin)	Value at Hot Idle
1	DB/OR	A/C Clutch Relay Control	Relay Off: 12v, On: 1v
3	DB/YL	ASD Relay Control	Relay Off: 12v, On: 1v
4	TN/RD	Speed Control Vacuum Solenoid	Vacuum Increasing: 1v
5	LG/RD	Speed Control Vent Solenoid	Vacuum Decreasing: 1v
6 ('98)	LG/OR	Overdrive Off Lamp Control	Lamp Off: 12v, On: 1v
8	BR/VT	HO2S-11 Heater Control (Cal)	Heater Off: 12v, On: 1v
9	DG/PK	HO2S-12 Heater Relay (Cal)	Relay Off: 12v, On: 1v
10	WT/DG	LDP Solenoid Control	PWM Signal: 0-12-0v
11	YL/RD	Speed Control Power Supply	12-14v
12	DG/OR	ASD Relay Power (B+)	12-14v
13	OR/WT	Overdrive Off Switch Sense	Switch Off: 12v, On: 1v
14	OR	LDP Switch Sense Signal	LDP Switch Closed: 0v
15	PK/YL	Battery Temperature Sensor	At 86°F: 1.96v
19	BR/WT	Fuel Pump Relay Control	Relay Off: 12v, On: 1v
20	PK/WT	EVAP Purge Solenoid Control	PWM Signal: 0-12-0v
22	BR	A/C Switch Signal	A/C Off: 12v, On: 1v
23	LG/WT	A/C Select Signal	A/C Off: 12v, On: 1v
24	WT/PK	Brake Switch Sense Signal	Brake Off: 0v, On: 12v
25	DB	Generator Field Source	12-14v
26	DB/WT	Fuel Level Sensor Signal	Full: 0.56v, 1/2 full: 2.5v
27, 29	PK, LG	SCI Transmit, SCI Receive	5v
28	WT/BK	CCD Bus (-) Signal	<0.050v
30	VT/BR	CCD Bus (+) Signal	Digital Signals: 0-5-0v
32	RD/LG	Speed Control Set Switch Signal	S/C & Set Switch On: 3.8v

Standard Colors and Abbreviations

Abbreviation	Color	Abbreviation	Color	Abbreviation	Color
BK	Black	GY	Gray	RD	Red
BL	Blue	GN	Green	TN	Tan
BR	Brown	LG	Light Green	VT	Violet
DB	Dark Blue	OR	Orange	WT	White
DG	Dark Green	PK	Pink	YL	Yellow

2003 Pickup 5.9L V8 VIN Z (All) 'C1' Black Connector

PCM Pin #	Wire Color	Circuit Description (32 Pin)	Value at Hot Idle
1	---	Not Used	---
2	LG/BK	Ignition Switch Power (B+)	12-14v
3	---	Not Used	---
4	BK/LB	Sensor Ground	<0.050v
5	---	Not Used	---
6	BR/YL	PNP Switch Sense Signal	In P/N: 0v, Others: 5v
7	BK/GY	Coil 1 Driver Control	5°, 55 mph: 8° dwell
8	GY/BK	CKP Sensor Signal	Digital Signals: 0-5-0v
9	---	Not Used	---
10	GY/RD	IAC 3 Driver Control	Pulse Signals
11	YL/BK	IAC 2 Driver Control	Pulse Signals
12	---	Not Used	---
13	OR	Power Takeoff Switch Sense	Switch Off: 12v, On: 1v
14	BR/WT	Manual Transmission Transfer Case	Switch Off: 12v, On: 1v
15	BR/RD	IAT Sensor Signal	At 100°F: 1.83v
16	TN/BK	ECT Sensor Signal	At 180°F: 2.80v
17	OR	5-Volt Supply	4.9-5.1v
18	TN/YL	CMP Sensor Signal	Digital Signals: 0-5-0v
19	VT/BK	IAC 4 Driver Control	Pulse Signals
20	BR/WT	IAC 1 Driver Control	Pulse Signals
21	---	Not Used	---
22	RD/WT	Keep Alive Power (B+)	12-14v
23	OR/DB	TP Sensor Signal	0.6-1.0v
24	BK/DG	HO2S-11 (B1 S1) Signal	0.1-1.1v
25	TN/WT	HO2S-12 (B1 S2) Signal	0.1-1.1v
26	BK/DG	H/D A/T: HO2S-12	0.1-1.1v
26	LG/RD	HO2S-21 (B2 S1) Signal (California)	0.1-1.1v
27	OR/RD	MAP Sensor Signal	1.5-1.6v
28, 30	---	Not Used	---
29	TN/WT	HO2S-22 (B2 S2) Signal	0.1-1.1v
31	BK/TN	Power Ground	<0.1v
32	BK/TN	Power Ground	<0.1v

Pin Connector Graphic

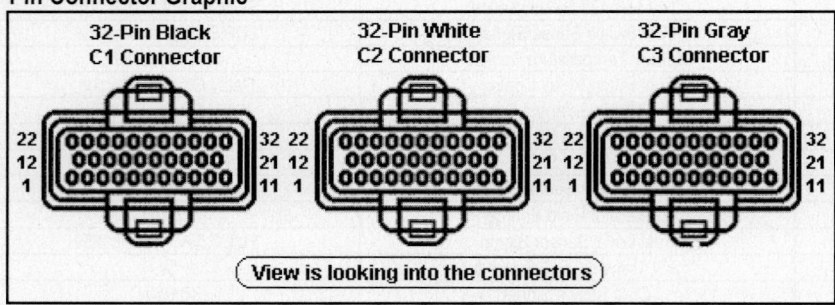

1998-2001 Pickup 5.9L V8 VIN 5, VIN Z White 'C2' Connector

PCM Pin #	Wire Color	Circuit Description (32 Pin)	Value at Hot Idle
1	VT	Transmission Temperature Sensor Signal	At 200°F: 2.40v
2	VT/TN	Injector 7 Driver	1-4 ms
3, 7, 9, 14	---	Not Used	---
4	WT/DB	Injector 1 Driver	1-4 ms
5	YL/WT	Injector 3 Driver	1-4 ms
6	GY	Injector 5 Driver	1-4 ms
8	VT/WT	Governor Pressure Solenoid	PWM Signal: 0-12-0v
10	DG	Generator Field Control	Digital Signals: 0-12-0v
11	OR/BK	TCC Solenoid Control	At Cruise w/TCC On: 1v
12	BR/DB	Injector 6 Driver	1-4 ms
13	GY/LB	Injector 8 Driver	1-4 ms
15	TN	Injector 2 Driver	1-4 ms
16	LB/BR	Injector 4 Driver	1-4 ms
17	DB/YL	Condenser Fan Relay Control	Relay Off: 12v, On: 1v
19	DB	A/C Pressure Sensor	A/C On: 0.451-4.850v
22, 24, 26, 32	---	Not Used	---
21	BR	3-4 Shift Solenoid Control	Cruise w/solenoid On: 1v
23	GY/YL	Engine Oil Pressure Sensor	1.6v at 24 psi
25	DB/BK	OSS Sensor (-) Signal	Moving: AC pulse signals
27	WT/OR	Vehicle Speed Signal	Digital Signal
28	LG/WT	OSS Sensor (+) Signal	Moving: AC pulse signals
29	LG/RD	Governor Pressure Sensor	0.58v
30	PK	Transmission Relay Control	Relay Off: 12v, On: 1v
31	VT/WT	5-Volt Supply	4.9-5.1v

1998-2001 Pickup 5.9L V8 VIN 5, VIN Z 'C3' Gray Connector

PCM Pin #	Wire Color	Circuit Description (32 Pin)	Value at Hot Idle
1	DB/OR	A/C Clutch Relay Control	Relay Off: 12v, On: 1v
2, 6-7	---	Not Used	---
3	DB/YL	ASD Relay Control	Relay Off: 12v, On: 1v
4	TN/RD	Speed Control Vacuum Solenoid	Vacuum Increasing: 1v
5	LG/RD	Speed Control Vent Solenoid	Vacuum Decreasing: 1v
8	BR/VT	HO2S-11 Heater Control	Heater Off: 12v, On: 1v
9	DB/OR	HO2S-12 Heater Relay (Downstream)	Relay Off: 12v, On: 1v
10	WT/DG	LDP Solenoid Pump Switch Sense	LDP Switch Closed: 0v
11	YL/RD	Speed Control Power Supply	12-14v
12	DG/OR	ASD Relay Power (B+)	12-14v
13	YL/DG	Overdrive Off Switch Sense	Switch Off: 12v, On: 1v
14	OR	Leak Detection Pump Solenoid Control	PWM Signal: 0-12-0v
15	PK/YL	Battery Temperature Sensor	At 86°F: 1.96v
16	BR/WT	HO2S-21 (B2 S1) Heater Control	Heater Off: 12v, On: 1v
19	BR	Fuel Pump Relay Control	Relay Off: 12v, On: 1v
20	PK/BK	EVAP Purge Solenoid Control	PWM Signal: 0-12-0v
17-18, 21-23, 31	---	Not Used	---
24	WT/PK	Brake Switch Sense Signal	Brake Off: 0v, On: 12v
25	WT/DB	Generator Field Source	12-14v
26	LB/BK	Fuel Level Sensor Signal	Full: 0.56v, 1/2 full: 2.5v
27	PK	SCI Transmit	0v
28	LG	SCI Receive	5v
30	VT/BR	PCI Data Bus (J1850)	Digital Signals: 0-7-0v
32	RD/LG	Speed Control Set Switch Signal	S/C & Set Switch On: 3.8v

Standard Colors and Abbreviations

Abbreviation	Color	Abbreviation	Color	Abbreviation	Color
BK	Black	GY	Gray	RD	Red
BL	Blue	GN	Green	TN	Tan
BR	Brown	LG	Light Green	VT	Violet
DB	Dark Blue	OR	Orange	WT	White
DG	Dark Green	PK	Pink	YL	Yellow

1994-95 Pickup 8.0L V10 MFI VIN W (All) 60 Pin Connector

PCM Pin #	Wire Color	Circuit Description (60 Pin)	Value at Hot Idle
1	DG/RD	MAP Sensor Signal	1.5-1.6v
2	TN/BK	ECT Sensor Signal	At 180°F: 2.80v
3	RD/WT	Keep Alive Power (B+)	12-14v
4	BK/LB	Sensor Ground	<0.050v
5	---	Not Used	---
6	PK/WT	5-Volt Supply	4.9-5.1v
7	OR	8-Volt Supply	7.9-8.1v
8	---	Not Used	---
9	LG/BK	Ignition Switch Power (B+)	12-14v
10	OR/WT	Overdrive Override Switch	Switch Off: 12v, On: 1v
11	BK/TN	Power Ground	<0.1v
12	BK/TN	Power Ground	<0.1v
13	LB/BR	Injector 5 & 8 Driver	1-4 ms
14	YL/WT	Injector 3 & 6 Driver	1-4 ms
15	TN	Injector 4 & 9 Driver	1-4 ms
16	WT/DB	Injector 1 & 10 Driver	1-4 ms
17	DB/TN	Coil Driver Control	5°, 55 mph: 8° dwell
18	RD/YL	Coil Driver Control	5°, 55 mph: 8° dwell
19	BK/GY	Coil Driver Control	5°, 55 mph: 8° dwell
20	DG	Alternator Field Control	Digital Signals: 0-12-0v
21	BK/RD	IAT Sensor Signal	At 100°F: 1.83v
22	OR/DB	TP Sensor Signal	0.6-1.0v
23	TN/WT	HO2S-11 (B1 S1) Signal	0.1-1.1v
24	GY/BK	Distributor Reference Signal	Digital Signals: 0-5-0v
25	PK	SCI Transmit	0v
26	---	Not Used	---
27	BR	A/C Damped Pressure Switch	A/C Off: 12v, On: 1v
28	---	Not Used	---
29	WT/PK	Brake Switch Sense Signal	Brake Off: 0v, On: 12v
30	BK/WT	Starter Relay Control	Relay Off: 12v, On: 1v
31	LG/OR	Overdrive Lamp Control	Lamp Off: 12v, On: 1v
32	BK/PK	MIL (lamp) Control	MIL Off: 12v, On: 1v
33	TN/RD	Speed Control Vacuum Solenoid	Vacuum Increasing: 1v
34	DB/OR	A/C Clutch Relay Control	Relay Off: 12v, On: 1v
35	GY/YL	EGR Solenoid Control	12v, at 55 mph: 1v
36	PK/BK	Transmission Temperature Lamp Control	Lamp Off: 12v, On: 1v
37	WT/BK	Coil Driver Control	5°, 55 mph: 8° dwell
38	BR/OR	Coil Driver Control	5°, 55 mph: 8° dwell
39	GY/RD	IAC Driver Control	Pulse Signals
40	BR/WT	IAC Driver Control	Pulse Signals

Standard Colors and Abbreviations

Abbreviation	Color	Abbreviation	Color	Abbreviation	Color
BK	Black	GY	Gray	RD	Red
BL	Blue	GN	Green	TN	Tan
BR	Brown	LG	Light Green	VT	Violet
DB	Dark Blue	OR	Orange	WT	White
DG	Dark Green	PK	Pink	YL	Yellow

1994-95 Pickup 8.0L V10 MFI VIN W (All) 60 Pin Connector

PCM Pin #	Wire Color	Circuit Description (60 Pin)	Value at Hot Idle
41	BK/DG	HO2S-12 (B1 S2) Signal	0.1-1.1v
42	VT	Transmission Temperature Sensor Signal	At 200°F: 2.40v
43	GY/LB	Tachometer Signal	Pulse Signals
44	TN/YL	CMP Sensor Signal	Digital Signals: 0-5-0v
45	LG	SCI Receive	5v
47	WT/OR	Vehicle Speed Signal	Digital Signal
48	BR/RD	Speed Control Set Switch Signal	S/C & Set Switch On: 3.8v
49	YL/RD	Speed Control On/Off Switch	S/C On: 1v, Off: 12v
50	WT/LG	Speed Control Resume Switch	S/C On: 1v, Off: 12v
51	DB/YL	ASD Relay Control	Relay Off: 12v, On: 1v
52	PK/BK	EVAP Purge Solenoid Control	Solenoid Off: 12v, On: 1v
53	LG/RD	Speed Control Vent Solenoid	Vacuum Decreasing: 1v
54	OR/BK	A/T: Overdrive Solenoid	Solenoid Off: 12v, On: 1v
54	OR/BK	M/T: SIL (lamp) Control	Lamp Off: 12v, On: 1v
55	BR	Overdrive Lockout Solenoid	Solenoid Off: 12v, On: 1v
56	GY/PK	Maintenance Indicator Lamp	Lamp Off: 12v, On: 1v
57	DG/OR	ASD Relay Power (B+)	12-14v
58	LG/BK	Injector 2 & 7 Driver	1-4 ms
59	PK/BK	IAC Driver Control	Pulse Signals
60	YL/BK	IAC Driver Control	Pulse Signals

Pin Connector Graphic

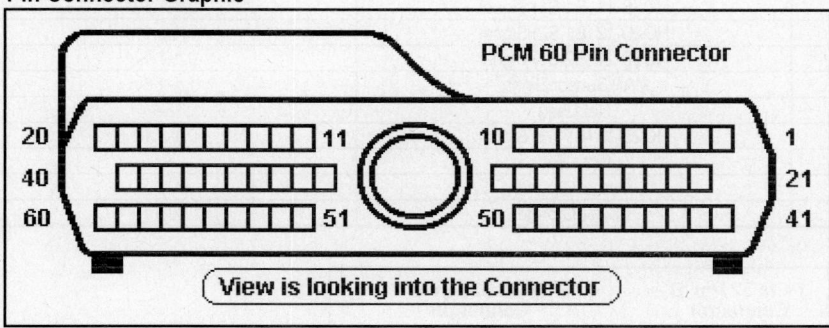

Standard Colors and Abbreviations

Abbreviation	Color	Abbreviation	Color	Abbreviation	Color
BK	Black	GY	Gray	RD	Red
BL	Blue	GN	Green	TN	Tan
BR	Brown	LG	Light Green	VT	Violet
DB	Dark Blue	OR	Orange	WT	White
DG	Dark Green	PK	Pink	YL	Yellow

1996-97 Pickup 8.0L V10 MFI VIN W (All) 'A' Black Connector

PCM Pin #	Wire Color	Circuit Description (32 Pin)	Value at Hot Idle
1	YL/GY	Coil 4 Driver Control	5°, 55 mph: 8° dwell
2	LG/BK	Ignition Switch Power (B+)	12-14v
3	RD/BK	Coil 3 Driver Control	5°, 55 mph: 8° dwell
4	BK/LB	Sensor Ground	<0.050v
5	DG/GY	Coil 5 Driver Control	5°, 55 mph: 8° dwell
6	BK/WT	PNP Switch Sense Signal	In P/N: 0v, Others: 5v
7	BK/GY	Coil 1 Driver Control	5°, 55 mph: 8° dwell
8	GY/BK	CKP Sensor Signal	Digital Signals: 0-5-0v
9	DB/WT	Coil 2 Driver Control	5°, 55 mph: 8° dwell
10	YL/BK	IAC 2 Driver Control	Pulse Signals
11	BR/WT	IAC 3 Driver Control	Pulse Signals
12-14	---	Not Used	---
15	BK/RD	IAT Sensor Signal	At 100°F: 1.83v
16	TN/BK	ECT Sensor Signal	At 180°F: 2.80v
17	PK/WT	5-Volt Supply	4.9-5.1v
18	TN/YL	CMP Sensor Signal	Digital Signals: 0-5-0v
19	GY/RD	IAC 1 Driver Control	Pulse Signals
20	PK/BK	IAC 4 Driver Control	Pulse Signals
21	---	Not Used	---
22	RD/WT	Keep Alive Power (B+)	12-14v
23	OR/DB	TP Sensor Signal	0.6-1.0v
24	TN/WT	HO2S-11 (B1 S1) Signal	0.1-1.1v
25	OR/BK	HO2S-12 (B1 S2) Signal	0.1-1.1v
26	BK/DG	HO2S-21 (B2 S1) Signal	0.1-1.1v
27	DG/RD	MAP Sensor Signal	1.5-1.6v
28	---	Not Used	---
29	TN/RD	HO2S-13 (B1 S3) Signal	0.1-1.1v
30	---	Not Used	---
31	BK/TN	Power Ground	<0.1v
32	BK/TN	Power Ground	<0.1v

Pin Connector Graphic

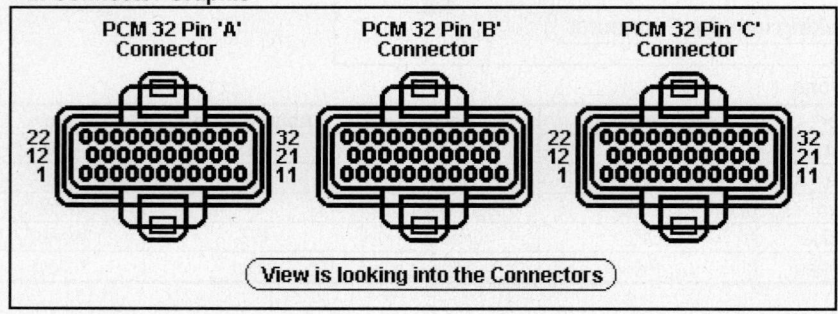

PCM 32 Pin 'A' Connector   PCM 32 Pin 'B' Connector   PCM 32 Pin 'C' Connector

View is looking into the Connectors

1996-97 Pickup 8.0L V10 MFI VIN W (All) 'B' White Connector

PCM Pin #	Wire Color	Circuit Description (32 Pin)	Value at Hot Idle
1	VT	Transmission Temperature Sensor Signal	At 200°F: 2.40v
2	PK/TN	Injector 7 Driver	1-4 ms
3	RD/BK	Injector 9 Driver	1-4 ms
4	WT/DB	Injector 1 Driver	1-4 ms
5	YL/WT	Injector 3 Driver	1-4 ms
6	GY	Injector 5 Driver	1-4 ms
8	PK/WT	Governor Pressure Solenoid	PWM Signal: 0-12-0v
10	DG	Generator Field Control	Digital Signals: 0-12-0v
11	OR/BK	TCC Solenoid Control	At Cruise w/TCC On: 1v
12	BR/DB	Injector 6 Driver	1-4 ms
13	GY/LB	Injector 8 Driver	1-4 ms
14	WT/DB	Injector 10 Driver	1-4 ms
15	TN	Injector 2 Driver	1-4 ms
16	LB/BR	Injector 4 Driver	1-4 ms
21	BR	3-4 Shift Solenoid Control	Solenoid Off: 12v, On: 1v
25	DB/BK	OSS Sensor (-) Signal	Moving: AC pulse signals
27	WT/OR	Vehicle Speed Signal	Digital Signal
28	LG/BK	OSS Sensor (+) Signal	Moving: AC pulse signals
29	LG/WT	Governor Pressure Sensor	0.58v
30	PK	Transmission Relay Control	Relay Off: 12v, On: 1v
31	OR	5-Volt Supply	4.9-5.1v
32	GY/YL	EGR Solenoid Control	12v, at 55 mph: 1v

1996-97 Pickup 8.0L V10 MFI VIN W (All) 'C' Gray Connector

PCM Pin #	Wire Color	Circuit Description (32 Pin)	Value at Hot Idle
1	DB/OR	A/C Clutch Relay Control	Relay Off: 12v, On: 1v
3	DB/YL	ASD Relay Control	Relay Off: 12v, On: 1v
4	TN/RD	Speed Control Vacuum Solenoid	Vacuum Increasing: 1v
5	LG/RD	Speed Control Vent Solenoid	Vacuum Decreasing: 1v
6	LG/OR	Overdrive Off Lamp Control	Lamp Off: 12v, On: 1v
7	PK/BK	Transmission Temperature Lamp Control	Lamp Off: 12v, On: 1v
11	YL/RD	Speed Control Power Supply	12-14v
12	DG/OR	ASD Relay Power (B+)	12-14v
13	OR/WT	Overdrive Off Switch Sense	Switch Off: 12v, On: 1v
15	PK/YL	Battery Temperature Sensor	At 86°F: 1.96v
16	TN/YL	Generator Lamp Control	Lamp Off: 12v, On: 1v
17	BK/PK	MIL (lamp) Control	MIL Off: 12v, On: 1v
18	GY/PK	Maintenance Indicator Lamp	Lamp Off: 12v, On: 1v
19	BR/WT	Fuel Pump Relay Control	Relay Off: 12v, On: 1v
20	PK	EVAP Purge Solenoid Control	PWM Signal: 0-12-0v
22	BR	A/C Request Signal	A/C Off: 12v, On: 1v
23	LG	A/C Select Signal	A/C Off: 12v, On: 1v
24	WT/PK	Brake Switch Sense Signal	Brake Off: 0v, On: 12v
26	DB/WT	Fuel Level Sensor Signal	Full: 0.56v, 1/2 full: 2.5v
27	PK/DB	SCI Transmit Signal	0v
29	DG	SCI Transmit Signal	5v
31	GY/LB	Tachometer Signal	Pulse Signals
32	RD/LG	Speed Control Set Switch Signal	S/C & Set Switch On: 3.8v

1998-2002 Pickup 8.0L V10 VIN W (All) 'C1' Black Connector

PCM Pin #	Wire Color	Circuit Description (32 Pin)	Value at Hot Idle
1	YL/GY	Coil 4 Driver Control	5°, 55 mph: 8° dwell
2	LG/BK	Ignition Switch Output (Run-Start)	12-14v
3	RD/BK	Coil 3 Driver Control	5°, 55 mph: 8° dwell
4	BK/LB	Sensor Ground	<0.1v
5	DG/GY	Coil 5 Driver Control	5°, 55 mph: 8° dwell
6	BK/WT	PNP Switch Sense Signal	In P/N: 0v, Others: 5v
7	BK/GY	Coil 1 Driver Control	5°, 55 mph: 8° dwell
8	GY/BK	CKP Sensor Signal	Digital Signals: 0-5-0v
9	DB/WT	Coil 2 Driver Control	5°, 55 mph: 8° dwell
10	YL/BK	IAC 2 Driver Control	Pulse Signals
11	BR/WT	IAC 3 Driver Control	Pulse Signals
12	---	Not Used	---
13	OR	Power Takeoff Switch Sense	Switch Off: 12v, On: 1v
14	---	Not Used	---
15	BK/RD	IAT Sensor Signal	At 100°F: 1.83v
16	TN/BK	ECT Sensor Signal	At 180°F: 2.80v
17	VT/WT	5-Volt Supply	4.9-5.1v
18	TN/YL	CMP Sensor Signal	Digital Signals: 0-5-0v
19	GY/RD	IAC 1 Driver Control	Pulse Signals
20	VT/BK	IAC 4 Driver Control	Pulse Signals
21	---	Not Used	---
22	RD/WT	Keep Alive Power (B+)	12-14v
23	OR/DB	TP Sensor Signal	0.6-1.0v
24	BK/DG	HO2S-11 (B1 S1) Signal	0.1-1.1v
25	OR/BK	HO2S-13 (B1 S3) Signal	0.1-1.1v
26	LG/RD	HO2S-12 (B1 S2) Signal	0.1-1.1v
27	DG/RD	MAP Sensor Signal	1.5-1.6v
28	---	Not Used	---
29	TN/RD	HO2S-21 (B2 S1) Signal	0.1-1.1v
29 ('00-'01)	TN/WT	HO2S-21 (B2 S1) Signal	0.1-1.1v
30	---	Not Used	---
31	BK/TN	Power Ground	<0.1v
32	BK/TN	Power Ground	<0.1v

Pin Connector Graphic

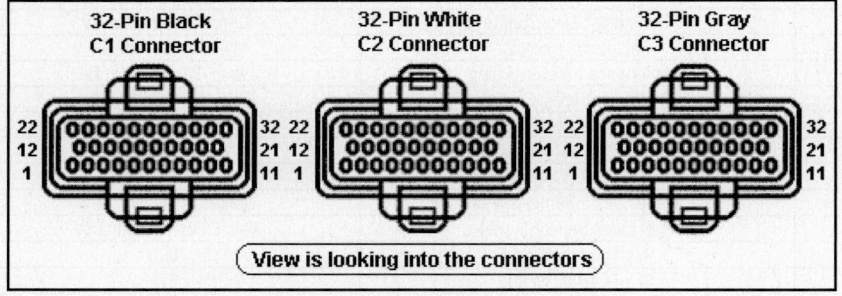

32-Pin Black C1 Connector   32-Pin White C2 Connector   32-Pin Gray C3 Connector

View is looking into the connectors

1998-2002 Pickup 8.0L V10 VIN W (All) 'C2' White Connector

PCM Pin #	Wire Color	Circuit Description (32 Pin)	Value at Hot Idle
1	VT	Transmission Temperature Sensor Signal	At 200°F: 2.40v
2	VT/TN	Injector 7 Driver	1-4 ms
3	TN/BK	Injector 9 Driver	1-4 ms
4	WT/DB	Injector 1 Driver	1-4 ms
5	YL/WT	Injector 3 Driver	1-4 ms
6	GY	Injector 5 Driver	1-4 ms
7	---	Not Used	---
8	VT/WT	Governor Pressure Solenoid	PWM Signal: 0-12-0v
9	---	Not Used	---
10	DG	Generator Field Control	Digital Signals: 0-12-0v
11	OR/BK	TCC Solenoid Control	At Cruise w/TCC On: 1v
12	BR/DB	Injector 6 Driver	1-4 ms
13	GY/LB	Injector 8 Driver	1-4 ms
14	WT	Injector 10 Driver	1-4 ms
15	TN	Injector 2 Driver	1-4 ms
16	LB/BR	Injector 4 Driver	1-4 ms
17-20	---	Not Used	---
21	BR	3-4 Shift Solenoid Control	Solenoid Off: 12v, On: 1v
22	---	Not Used	---
23	GY/OR	Engine Oil Pressure Sensor	1.6v at 24 psi
24	---	Not Used	---
25	DB/BK	OSS Sensor (-) Signal	Moving: AC pulse signals
26	---	Not Used	---
27	WT/OR	Vehicle Speed Signal	Digital Signal
28	LG/BK	OSS Sensor (+) Signal	Moving: AC pulse signals
29	LG/WT	Governor Pressure Sensor	0.58v
30	PK	Transmission Relay Control	Relay Off: 12v, On: 1v
31	OR	5-Volt Supply	4.9-5.1v
32	---	Not Used	---

Pin Connector Graphic

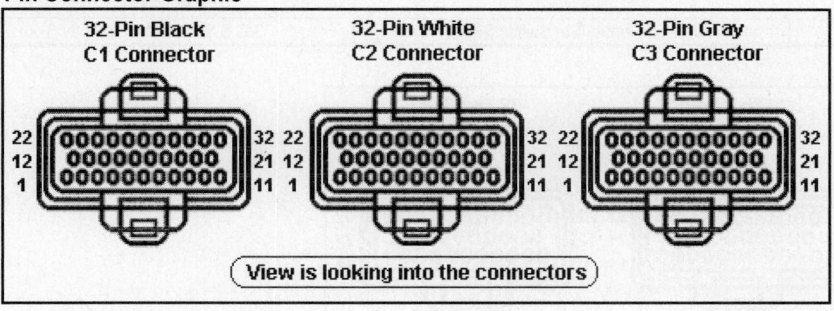

32-Pin Black
C1 Connector

32-Pin White
C2 Connector

32-Pin Gray
C3 Connector

View is looking into the connectors

1998-2002 Pickup 8.0L V10 VIN W (All) 'C3' Gray Connector

PCM Pin #	Wire Color	Circuit Description (32 Pin)	Value at Hot Idle
1	DB/OR	A/C Clutch Relay Control	Relay Off: 12v, On: 1v
2	---	Not Used	---
3	DB/YL	ASD Relay Control	Relay Off: 12v, On: 1v
4	TN/RD	Speed Control Vacuum Solenoid	Vacuum Increasing: 1v
5	LG/RD	Speed Control Vent Solenoid	Vacuum Decreasing: 1v
6 ('98)	LG/OR	Overdrive Off Lamp Control	Lamp Off: 12v, On: 1v
7	---	Not Used	
8	BR/VT	HO2S-11 Heater Control (Cal)	Heater Off: 12v, On: 1v
9	DG/PK	HO2S-12 Heater Relay (Cal)	Relay Off: 12v, On: 1v
10	WT/OR	LDP Solenoid Control	PWM Signal: 0-12-0v
10 ('00-'02)	WT/DG	LDP Solenoid Control	PWM Signal: 0-12-0v
11	YL/RD	Speed Control Power Supply	12-14v
12	DG/OR	ASD Relay Power (B+)	12-14v
13	OR/WT	Overdrive Off Switch Sense	Switch Off: 12v, On: 1v
14	OR	LDP Switch Sense Signal	LDP Switch Closed: 0v
15	PK/YL	Battery Temperature Sensor	At 86°F: 1.96v
16	BR/WT	HO2S-21 Heater Control	Heater Off: 12v, On: 1v
17-18	---	Not Used	---
19	BR/WT	Fuel Pump Relay Control	Relay Off: 12v, On: 1v
20	PK/WT	EVAP Purge Solenoid Control	PWM Signal: 0-12-0v
21	---	Not Used	---
22	BR	A/C Switch Signal	A/C Off: 12v, On: 1v
23	LG/WT	A/C Select Signal	A/C Off: 12v, On: 1v
24	WT/PK	Brake Switch Sense Signal	Brake Off: 0v, On: 12v
25	DB	Generator Field Source	12-14v
26	DB/WT	Fuel Level Sensor Signal	Full: 0.56v, 1/2 full: 2.5v
27	PK/DB	SCI Transmit Signal	0v
28	WT/BK	CCD Bus (-) Signal	<0.050v
29	DG	SCI Receive Signal	5v
30	VT/BR	CCD Bus (+) Signal	Digital Signals: 0-5-0v
31	---	Not Used	---
32	RD/LG	Speed Control Set Switch Signal	S/C & Set Switch On: 3.8v

Pin Connector Graphic

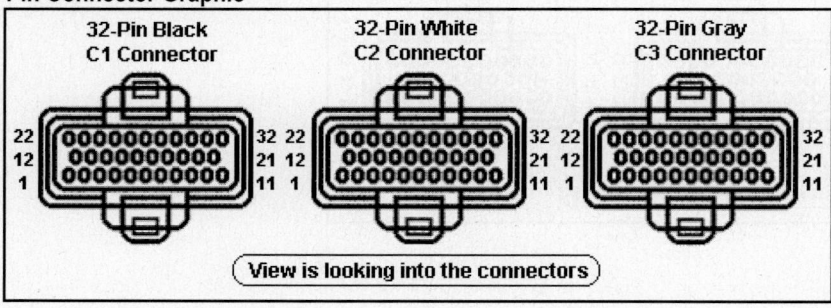

32-Pin Black
C1 Connector

32-Pin White
C2 Connector

32-Pin Gray
C3 Connector

View is looking into the connectors

2003 Pickup 8.0L V10 VIN W (All) 'C1' Black Connector

PCM Pin #	Wire Color	Circuit Description (32 Pin)	Value at Hot Idle
1	YL/GY	Coil 4 Driver Control	5°, 55 mph: 8° dwell
2	LG/BK	Ignition Switch Output (Run-Start)	12-14v
3	RD/BK	Coil 3 Driver Control	5°, 55 mph: 8° dwell
4	BK/LB	Sensor Ground	<0.1v
5	DG/GY	Coil 5 Driver Control	5°, 55 mph: 8° dwell
6	BR/YL	PNP Switch Sense Signal	In P/N: 0v, Others: 5v
7	BK/GY	Coil 1 Driver Control	5°, 55 mph: 8° dwell
8	GY/BK	CKP Sensor Signal	Digital Signals: 0-5-0v
9	DB/TN	Coil 2 Driver Control	5°, 55 mph: 8° dwell
10	GY/RD	IAC 3 Driver Control	Pulse Signals
11	YL/BK	IAC 2 Driver Control	Pulse Signals
12-14	---	Not Used	---
15	BK/RD	IAT Sensor Signal	At 100°F: 1.83v
16	TN/BK	ECT Sensor Signal	At 180°F: 2.80v
17	OR	5-Volt Supply	4.9-5.1v
18	TN/YL	CMP Sensor Signal	Digital Signals: 0-5-0v
19	VT/BK	IAC 4 Driver Control	Pulse Signals
20	BR/WT	IAC 1 Driver Control	Pulse Signals
21	---	Not Used	---
22	RD/WT	Keep Alive Power (B+)	12-14v
23	OR/DB	TP Sensor Signal	0.6-1.0v
24	BK/DG	HO2S-11 (B1 S1) Signal	0.1-1.1v
25	TN/WT	HO2S-12 (B1 S2) Signal	0.1-1.1v
26	LG/RD	HO2S-21 (B2 S1) Signal (California)	0.1-1.1v
27	OR/RD	MAP Sensor Signal	1.5-1.6v
28	---	Not Used	---
29	TN/WT	HO2S-22 (B2 S2) Signal (California)	0.1-1.1v
30	---	Not Used	---
31	BK/TN	Power Ground	<0.1v
32	BK/TN	Power Ground	<0.1v

Pin Connector Graphic

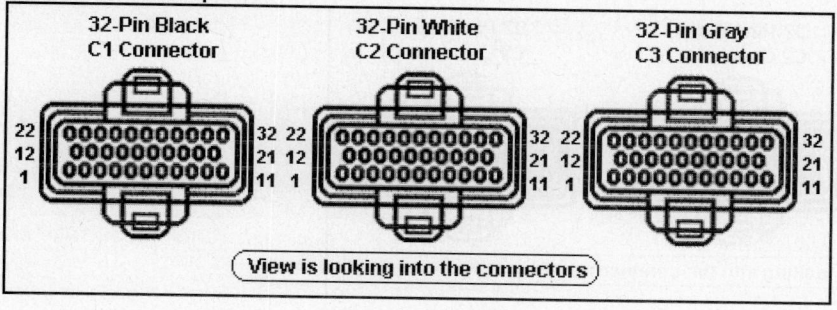

32-Pin Black C1 Connector   32-Pin White C2 Connector   32-Pin Gray C3 Connector

View is looking into the connectors

2003 Pickup 8.0L V10 VIN W (All) 'C2' White Connector

PCM Pin #	Wire Color	Circuit Description (32 Pin)	Value at Hot Idle
1	VT	Transmission Temperature Sensor Signal	At 200°F: 2.40v
2	VT/TN	Injector 7 Driver	1-4 ms
3	TN	Injector 9 Driver	1-4 ms
4	WT/DB	Injector 1 Driver	1-4 ms
5	YL/WT	Injector 3 Driver	1-4 ms
6	GY	Injector 5 Driver	1-4 ms
7	---	Not Used	---
8	VT/WT	Governor Pressure Solenoid	PWM Signal: 0-12-0v
9	---	Not Used	---
10	DG	Generator Field Control	Digital Signals: 0-12-0v
11	OR/BK	TCC Solenoid Control	At Cruise w/TCC On: 1v
12	BR/DB	Injector 6 Driver	1-4 ms
13	GY/LB	Injector 8 Driver	1-4 ms
14	WT/DB	Injector 10 Driver	1-4 ms
15	TN	Injector 2 Driver	1-4 ms
16	LB/BR	Injector 4 Driver	1-4 ms
17	DB/YL	Condenser Fan Relay Control	Relay Off: 12v, On: 1v
18-20	---	Not Used	---
21	LB/BR	3-4 Shift Solenoid Control	Solenoid Off: 12v, On: 1v
22	---	Not Used	---
23	GY/YL	Oil Pressure Sensor	1.6v at 24 psi
24	---	Not Used	---
25	DB/BK	OSS Sensor (-) Signal	Moving: AC pulse signals
26	---	Not Used	---
27	WT/OR	Vehicle Speed Signal	Digital Signals
28	LG/WT	OSS Sensor (+) Signal	Moving: AC pulse signals
29	LG/RD	Governor Pressure Sensor	0.58v
30	PK	Transmission Relay Control	Relay Off: 12v, On: 1v
31	WT	5-Volt Supply	4.9-5.1v
32	---	Not Used	---

Pin Connector Graphic

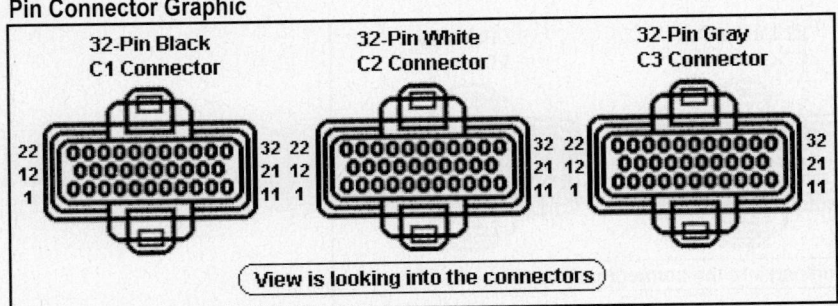

32-Pin Black C1 Connector   32-Pin White C2 Connector   32-Pin Gray C3 Connector

View is looking into the connectors

2003 Pickup 8.0L V10 VIN W (All) 'C3' Gray Connector

PCM Pin #	Wire Color	Circuit Description (32 Pin)	Value at Hot Idle
1	OR	A/C Clutch Relay Control	Relay Off: 12v, On: 1v
2	---	Not Used	---
3	DB/YL	ASD Relay Control	Relay Off: 12v, On: 1v
4	TN/RD	Speed Control Vacuum Solenoid	Vacuum Increasing: 1v
5	LG/RD	Speed Control Vent Solenoid	Vacuum Decreasing: 1v
6-7	---	Not Used	
8	BR/VT	HO2S-11 Heater Control (California)	Heater Off: 12v, On: 1v
9	DB/OR	HO2S-12 Heater Relay (California)	Relay Off: 12v, On: 1v
10	WT/OR	LDP Solenoid Control	PWM Signal: 0-12-0v
11	YL/RD	Speed Control Power Supply	12-14v
12	DG/OR	ASD Relay Output (B+)	12-14v
13	---	Not Used	---
14	OR	LDP Switch Sense Signal	LDP Switch Closed: 0v
15	PK/YL	Battery Temperature Sensor	At 86°F: 1.96v
16	BR/WT	HO2S-21 Heater Control (Except California)	Heater Off: 12v, On: 1v
17-18	---	Not Used	
19	BR	Fuel Pump Relay Control	Relay Off: 12v, On: 1v
20	PK/BK	EVAP Purge Solenoid Control	PWM Signal: 0-12-0v
21-23	---	Not Used	
24	WT/PK	Brake Switch Sense	Brake Off: 0v, On: 12v
25	WT/OR	Generator Field Source	12-14v
26	LB/BK	Fuel Level Sensor Signal	Full: 0.56v, 1/2 full: 2.5v
27	PK	SCI Transmit	0v
28	---	Not Used	---
29	LG	SCI Receive	5v
30	VT/BR	PCI Data Bus (J1850)	Digital Signals: 0-7-0v
31	---	Not Used	---
32	RD/LG	Speed Control Set Switch Signal	S/C & Set Switch On: 3.8v

Pin Connector Graphic

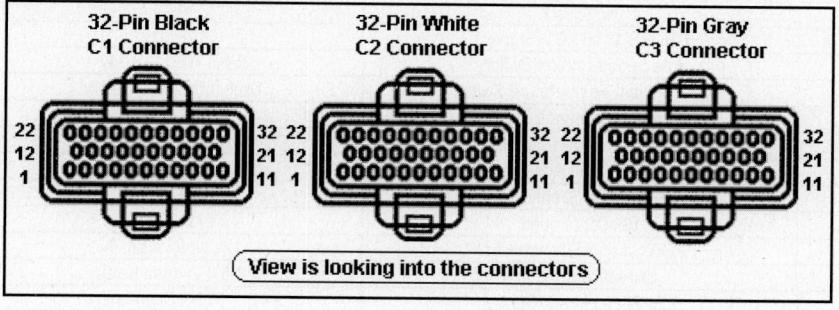

32-Pin Black C1 Connector   32-Pin White C2 Connector   32-Pin Gray C3 Connector

View is looking into the connectors

1990-92 Ram & Ram 50 2.4L I4 MFI VIN W (All) 10 Pin Connector

PCM Pin #	Wire Color	Circuit Description (10-Pin)	Value at Hot Idle
101	BK	Power Ground	<0.1v
102	RD	Ignition Switch Power (B+)	12-14v
103	BK/YL	Keep Alive Power (B+)	12-14v
104	BK/YL	Inhibitor Switch Signal	In P/N: 0v, Others: 5v
104	BK	M/T: Sensor Ground	<0.050v
105, 110	---	Not Used	---
106	BK	Power Ground	<0.1v
107	RD	Ignition Switch Power (B+)	12-14v
108	BK/RD	Starter Relay Control	KOEC: 9-11v
109	BK/BL	Fuel Pump Relay Control	F/P On: 1v, Off: 12v
PCM Pin #	**Wire Color**	**Circuit Description (18-Pin)**	**Value at Hot Idle**
51	YL/BL	Injector 1 Control Driver	1-4 ms
52	YL/BK	Injector 2 Control Driver	1-4 ms
53	BR/GN	EGR Control Solenoid (Calif.)	Solenoid Off: 12v, On: 1v
54	WT	Ignition Coil Control	5°, 55 mph: 8° dwell
55, 57	---	Not Used	---
56	WT/RD	MPI Control Relay	Relay Off: 12v, On: 1v
58	BL/YL	Idle Speed Motor Control	Pulse Signals
59	GN/BK	Idle Speed Motor Control	Pulse Signals
60	BL/GN	Injector 3 Control Driver	1-4 ms
61	LG/WT	Injector 4 Control Driver	1-4 ms
62	GN/RD	EVAP Purge Solenoid Control	Solenoid Off: 12v, On: 1v
63	RD/GN	Airflow Sensor	2.2-3.2v
64	LG/RD	MIL (lamp) Control	MIL Off: 12v, On: 1v
65	BL	A/C Clutch Relay Control	A/C Off: 12v, On: 1v
66-68	---	Not Used	---
PCM Pin #	**Wire Color**	**Circuit Description (24 Pin)**	**Value at Hot Idle**
1	YL	Data Link Connector	0v
2	GN/WTT	Data Link Connector	0v
3, 5, 9, 11	---	Not Used	---
4	WT	HO2S-11 (B1 S1) Signal	0.1-1.1v
6	YL/RD	Idle Speed Control Switch	0-6v, WOT: 12-14v
7	GN/RD	A/C Coolant Temperature Switch	Switch Off: 12v, On: 1v
8	RD/BL	Airflow Sensor VREF	4.9-5.1v
10	WT/BK	Air Temperature Sensor	At 100°F: 2.51v
12	BK/BL	Ignition Timing Adjustment	4.0-5.5v
13	GN/YL	5-Volt Supply	4.9-5.1v
14	BK	Sensor Ground	<0.050v
15	BR/WT	EGR Solenoid Control (Cal)	12v, at 55 mph: 1v
16	PK	BARO Read Sensor	3.5-4.2v (sea level)
17	BK/RD	Idle Speed Motor Control	Pulse Signals
18	YL/WT	MIL (lamp) Control	MIL Off: 12v, On: 1v
19	GN/WT	TP Sensor Signal	0.6-1.0v
20	YL/GN	ECT Sensor Signal	At 180°F: 2.80v
21	BR/GN	Distributor Reference Signal	Digital Signals: 0-5-0v
22	BR/RD	TDC Signal Output	0.2-2.0v (pulses)
23	GN/BL	5-Volt Supply	4.9-5.1v
24	BK	Sensor Ground	<0.050v

Pin Connector Graphic

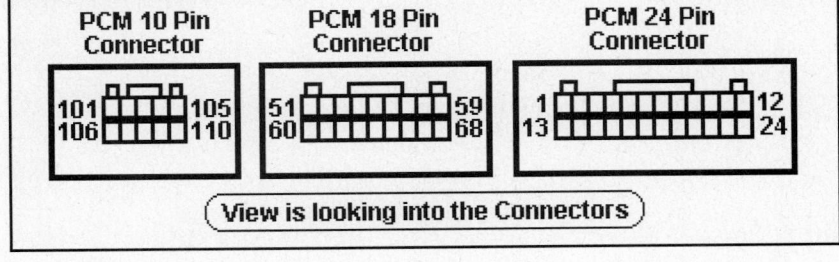

View is looking into the Connectors

1993 Ram & Ram 50 2.4L I4 MFI VIN G (California) 16, 22 & 26 Pin Connector

PCM Pin #	Wire Color	Circuit Description (16 Pin)	Value at Hot Idle
101-103	---	Not Used	---
104	WT/YL	Ignition Timing Adjustment	4.0-5.5v
105	BLK	Oxygen Sensor Ground	<0.050v
106	LG/RD	MIL (lamp) Control	MIL Off: 12v, On: 1v
107-111, 114, 116	---	Not Used	---
112	YL	Data Link Connector	0v
113	GN/WT	Data Link Connector	0v
115	GN/RD	A/C Coolant Temperature Switch	Switch Off: 12v, On: 1v
PCM Pin #	Wire Color	Circuit Description (22 Pin)	Value at Hot Idle
51	BK/YL	Starter Relay Control	KOEC: 9-11v
52	RD/BK	BARO Read Sensor	3.5-4.2v (sea level)
53	BK/BL	EGR Solenoid Control	12v, at 55 mph: 1v
54, 57-59, 62	---	Not Used	---
55	W	HO2S-12 (B1 S2) Signal	0.1-1.1v
56	W	HO2S-11 (B1 S1) Signal	0.1-1.1v
60	BK/YL	Keep Alive Power (B+)	12-14v
61	GN/BL	5-Volt Supply	4.9-5.1v
63	YL/GN	5-Volt Supply	4.9-5.1v
64	GN/WT	TP Sensor Signal	0.6-1.0v
65	PK	Airflow Sensor	2.2-3.2v
66	YL/WT	Vehicle Speed Signal	Digital Signal
67	YL/RD	Closed Throttle Switch	0-6v, WOT: 12-14v
68	BR/RD	TDC Signal Output	0.2-2.0v (pulses)
69	BR/G	Distributor Reference Signal	Digital Signals: 0-5-0v
70	WT/BK	Airflow Sensor VREF	4.9-5.1v
71	BK/YL	Inhibitor Switch Signal	Switch Off: 12v, On: 1v
71	BK	M/T: Sensor Ground	<0.050v
72	BK	Sensor Ground	<0.050v
PCM Pin #	Wire Color	Circuit Description (26 Pin)	Value at Hot Idle
1	YL/BL	Injector 1 Control Driver	1-4 ms
2	LG	Injector 3 Control Driver	1-4 ms
3, 7, 11, 16	---	Not Used	---
4	BL/YL	Idle Speed Motor Control	Pulse Signals
5	WT	Idle Speed Motor Control	Pulse Signals
6	BR	EGR Control Solenoid	Solenoid Off: 12v, On: 1v
8	BK/BL	ECM Control Relay	Relay Off: 12v, On: 1v
9	BR/YL	EVAP Purge Solenoid Control	Solenoid Off: 12v, On: 1v
10	WT	Ignition Coil Control	5°, 55 mph: 8° dwell
12	RD	Ignition Switch Power (B+)	12-14v
13, 26	BK	Sensor Ground	<0.050v
14	YL/BK	Injector 2 Control Driver	1-4 ms
15	LG/WT	Injector 4 Control Driver	1-4 ms
17	GN/BK	Idle Speed Motor Control	Pulse Signals
18	GN	Idle Speed Motor Control	Pulse Signals
19	BL	Air Temperature Sensor	At 100°F: 2.51v
20-21, 23-24	---	Not Used	---
22	BL	A/C Clutch Relay Control	A/C Off: 12v, On: 1v
25	RD	Ignition Switch Power (B+)	12-14v

Pin Connector Graphic

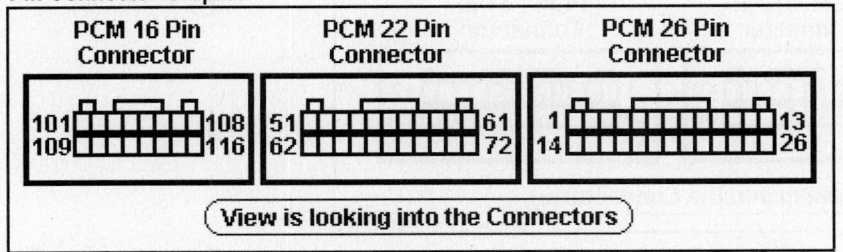

PCM 16 Pin Connector — 101 108 / 109 116
PCM 22 Pin Connector — 51 61 / 62 72
PCM 26 Pin Connector — 1 13 / 14 26

View is looking into the Connectors

1993 Ram & Ram 50 2.4L I4 MFI VIN G (Federal) 10, 18 & 24 Pin Connector

PCM Pin #	Wire Color	Circuit Description (10 Pin)	Value at Hot Idle
101	BK	Power Ground	<0.1v
102	RD	Ignition Switch Power (B+)	12-14v
103	BK/YL	Keep Alive Power (B+)	12-14v
104	BK/YL	Inhibitor Switch Signal	In P/N: 0v, Others: 5v
104	BK	M/T: Sensor Ground	<0.050v
105, 110	---	Not Used	---
106	BK	Power Ground	<0.1v
107	RD	Ignition Switch Power (B+)	12-14v
108	BK/RD	Starter Relay Control	KOEC: 9-11v
109	BK/BL	Fuel Pump Relay Control	Relay Off: 12v, On: 1v
PCM Pin #	**Wire Color**	**Circuit Description (18 Pin)**	**Value at Hot Idle**
51	YL/BL	Injector 1 Control Driver	1-4 ms
52	YL/BK	Injector 2 Control Driver	1-4 ms
53	BR/GN	EGR Control Solenoid (Calif.)	Solenoid Off: 12v, On: 1v
54	WT	Ignition Coil Control	5°, 55 mph: 8° dwell
55, 57	---	Not Used	---
56	WT/RD	ECM Control Relay	Relay Off: 12v, On: 1v
58	BL/YL	Idle Speed Motor Control	Pulse Signals
59	GN/BK	Idle Speed Motor Control	Pulse Signals
60	BL/GN	Injector 3 Control Driver	1-4 ms
61	LG/WT	Injector 4 Control Driver	1-4 ms
63	RD/GN	Airflow Sensor	2.2-3.2v
64	LG/RD	MIL (lamp) Control	MIL Off: 12v, On: 1v
65	BL	A/C Clutch Relay Control	A/C Off: 12v, On: 1v
66-68	---	Not Used	---
PCM Pin #	**Wire Color**	**Circuit Description (24 Pin)**	**Value at Hot Idle**
1	YL	Data Link Connector	0v
2	GN/WT	Data Link Connector	0v
3, 5, 9, 11	---	Not Used	---
4	WT	HO2S-11 (B1 S1) Signal	0.1-1.1v
6	YL/RD	Idle Speed Control Switch	0-6v, WOT: 12-14v
7	GN/RD	A/C Coolant Temperature Switch	Switch Off: 12v, On: 1v
8	RD/BL	Airflow Sensor VREF	4.9-5.1v
10	WT/BK	Airflow Sensor	2.2-3.2v
12	BK/BL	Ignition Timing Adjustment	4.0-5.5v
13	GN/YL	5-Volt Supply	4.9-5.1v
14	BK	Sensor Ground	<0.050v
15	BR/WT	EGR Solenoid Control (Cal)	12v, at 55 mph: 1v
16	PK	BARO Read Sensor	3.5-4.2v (sea level)
17	BK/RD	Idle Speed Motor Control	Pulse Signals
18	YL/WT	MIL (lamp) Control	MIL Off: 12v, On: 1v
19	GN/WT	TP Sensor Signal	0.6-1.0v
20	YL/GN	ECT Sensor Signal	At 180°F: 2.80v
21	BR/GN	Distributor Reference Signal	Digital Signals: 0-5-0v
22	BR/RD	TDC Signal Output	0.2-2.0v (pulses)
23	GN/BL	5-Volt Supply	4.9-5.1v
24	BK	Sensor Ground	<0.050v

Pin Connector Graphic

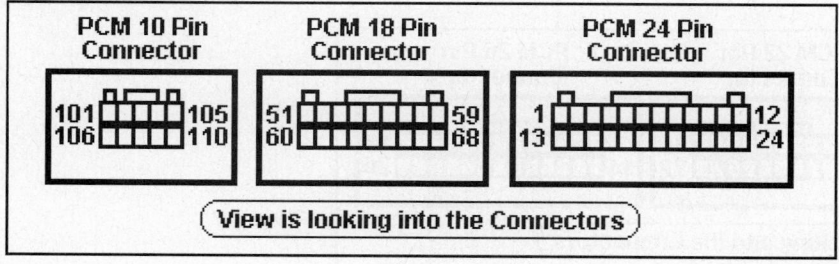

PCM 10 Pin Connector — 101 105 / 106 110

PCM 18 Pin Connector — 51 59 / 60 68

PCM 24 Pin Connector — 1 12 / 13 24

(View is looking into the Connectors)

1990-93 Ram & Ram 50 3.0L V6 MFI VIN S (All) 10 Pin Connector

PCM Pin #	Wire Color	Circuit Description (10-Pin)	Value at Hot Idle
101	BK	Power Ground	<0.1v
102, 107	RD	Ignition Switch Power (B+)	12-14v
103	BK/YL	Keep Alive Power (B+)	12-14v
104	BK/YL	Inhibitor Switch Signal	In P/N: 0v, Others: 5v
104	BK	M/T: Sensor Ground	<0.050v
105	YL/WT	Injector 5 Control Driver	1-4 ms
106	BK	Power Ground	<0.1v
108	BK/RD	Starter Relay Control	KOEC: 9-11v
109	YL/GN	Injector 6 Control Driver	1-4 ms
110	BK/WT	MPI Control Relay	Relay Off: 12v, On: 1v

1990-93 Ram & Ram 50 3.0L V6 MFI VIN S (All) 18 Pin Connector

PCM Pin #	Wire Color	Circuit Description (18 Pin)	Value at Hot Idle
51, 52	YL, YL/BK	Injector 1, Injector 2 Control Driver	1-4 ms
53	LG/Y	EGR Control Solenoid (California)	Solenoid Off: 12v, On: 1v
54	WT	Ignition Coil Control	5°, 55 mph: 8° dwell
55, 62	---	Not Used	---
56	WT/RD	MPI Control Relay	Relay Off: 12v, On: 1v
57	BR/BL	EVAP Purge Solenoid Control	Solenoid Off: 12v, On: 1v
58, 59	GN, GN/BK	Idle Speed Motor Control	Pulse Signals
60	BL/GN	Injector 3 Control Driver	1-4 ms
61	LG/WT	Injector 4 Control Driver	1-4 ms
63	BL/GN	MPI Control Relay	Relay Off: 12v, On: 1v
64	LG/RD	MIL (lamp) Control	MIL Off: 12v, On: 1v
65	BL	A/C Clutch Relay Control	A/C Off: 12v, On: 1v
66	BL/GN	MPI Control Relay	Relay Off: 12v, On: 1v
67, 68	GN, GN/BL	Idle Speed Motor Control	Pulse Signals

1990-93 Ram & Ram 50 3.0L V6 MFI VIN S (All) 24 Pin Connector

PCM Pin #	Wire Color	Circuit Description (24 Pin)	Value at Hot Idle
1, 2	YL, GN/WT	Data Link Connector	0v
3, 6, 9, 11	---	Not Used	---
4	WT	HO2S-11 (B1 S1) Signal	0.1-1.1v
5	BL/WT	PSP Switch Sense Signal	Straight: 0v, Turning: 5v
7	GN/RD	A/C Damped Pressure Switch	A/C Off: 12v, On: 1v
8	RD/BL	Airflow Sensor VREF	4.9-5.1v
10	BL/YL	Airflow Sensor Signal	2.2-3.2v
12	WT/YL	Ignition Timing Adjustment	4.0-5.5v
13	BK/BL	Fuel Pump Relay Control	F/P On: 1v, Off: 12v
14	YL/RD	Idle Speed Control Switch	0-6v, WOT: 12-14v
15	LG/BK	EGR Solenoid Control (California)	12v, at 55 mph: 1v
16	PK	BARO Read Sensor	3.5-4.2v (sea level)
17, 24	BK	Sensor Ground	<0.050v
18	YL/WT	MIL (lamp) Control	MIL Off: 12v, On: 1v
19	GN/WT	TP Sensor Signal	0.6-1.0v
20	YL/GN	ECT Sensor Signal	At 180°F: 2.80v
21	BL/WT	Distributor Reference Signal	Digital Signals: 0-5-0v
22	BL/RD	TDC Signal Output	0.2-2.0v (pulses)
23	GN/YL	Air Temperature Sensor	At 100°F: 2.51v

Pin Connector Graphic

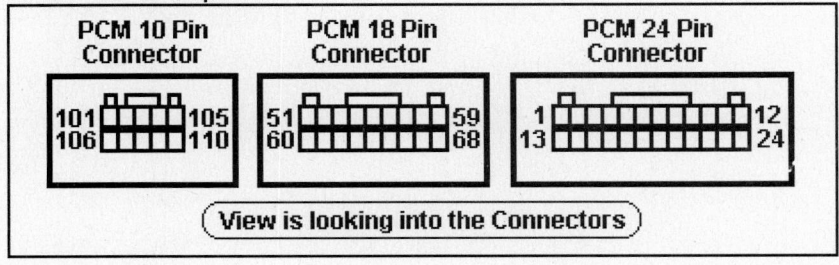

1990-91 Ramcharger 5.2L V8 TBI VIN Y (All) 60 Pin Connector

PCM Pin #	Wire Color	Circuit Description (60 Pin)	Value at Hot Idle
1	DG/RD	MAP Sensor Signal	1.5-1.6v
2	TN	ECT Sensor Signal	At 180°F: 2.80v
3	RD	Keep Alive Power (B+)	12-14v
4	BK/LB	Sensor Ground	<0.050v
5	BK	Sensor Ground	<0.050v
6	PK/WT	5-Volt Supply	4.9-5.1v
7	OR	8-Volt Supply	7.9-8.1v
8	DG/BK	ASD Relay Power (B+)	12-14v
9	DB	Ignition Switch Power (B+)	12-14v
10	---	Not Used	---
11	LB/RD	Power Ground	<0.1v
12	LB/RD	Power Ground	<0.1v
13-14	---	Not Used	---
15	TN	Injector 2 Driver	1-4 ms
16	WT/DB	Injector 1 Driver	1-4 ms
17-18	---	Not Used	---
19	BK/YL	Coil Driver Control	5°, 55 mph: 8° dwell
20	DG	Alternator Field Control	Digital Signals: 0-12-0v
21	BK/RD	Throttle Body Temperature	At 100°F: 2.51v
22	OR/DB	TP Sensor Signal	0.6-1.0v
24	GY	Distributor Reference Signal	Digital Signals: 0-5-0v
25	PK	SCI Transmit	0v
26	---	Not Used	---
27	BR	A/C Clutch Signal	A/C Off: 12v, On: 1v
28	PK	Closed Throttle Switch	Switch Off: 12v, On: 1v
29	WT/PK	Brake Switch Sense Signal	Brake Off: 0v, On: 12v
30	BR/YL	PNP Switch Sense Signal	In P/N: 0v, Others: 5v
31	---	Not Used	---
32	BK/PK	MIL (lamp) Control	MIL Off: 12v, On: 1v
33	TN/RD	Speed Control Vacuum Solenoid	Vacuum Increasing: 1v
34	DB/OR	A/C WOT Relay Control	Relay Off: 12v, On: 1v
35	GY/YL	EGR Solenoid Control	12v, at 55 mph: 1v
36	BK/OR	Air Switching Solenoid	Solenoid Off: 12v, On: 1v
37-39	---	Not Used	---
40	BR	AIS Motor Control	Pulse Signals

1990-91 Ramcharger 5.2L V8 TBI VIN Y (All) 60 Pin Connector

PCM Pin #	Wire Color	Circuit Description (60 Pin)	Value at Hot Idle
41	BK/DG	HO2S-11 (B1 S1) Signal	0.1-1.1v
42-43	---	Not Used	---
44	OR/WT	Overdrive Lockout Switch	Switch Off: 12v, On: 1v
45	LG	SCI Receive	5v
46	---	Not Used	---
47	WT/OR	Vehicle Speed Signal	Digital Signal
48	BR/RD	Speed Control Set Switch Signal	S/C & Set Switch On: 3.8v
49	YL/RD	Speed Control On/Off Switch	S/C On: 1v, Off: 12v
50	WT/LG	Speed Control Resume Switch	S/C On: 1v, Off: 12v
51	DB/YL	ASD Relay Control	Relay Off: 12v, On: 1v
52	PK/BK	EVAP Purge Solenoid Control	Solenoid Off: 12v, On: 1v
53	LG/RD	Speed Control Vent Solenoid	Vacuum Decreasing: 1v
54	OR/BK	Part Throttle Unlock Solenoid	Solenoid Off: 12v, On: 1v
55	OR/LG	Overdrive Lockout Solenoid	Solenoid Off: 12v, On: 1v
56	GY/PK	Maintenance Indicator Lamp	Lamp Off: 12v, On: 1v
57-59	---	Not Used	---
60	GY/RD	AIS Motor Control	Pulse Signals

Pin Connector Graphic

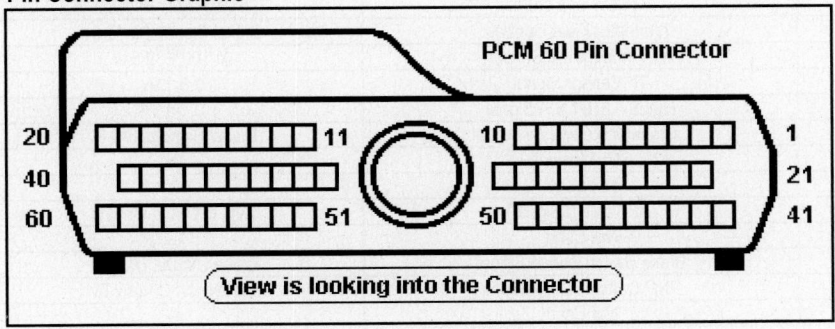

PCM 60 Pin Connector

20 11   10   1
40   21
60 51   50   41

View is looking into the Connector

1992-93 Ramcharger 5.2L V8 MFI VIN Y (All) 60 Pin Connector

PCM Pin #	Wire Color	Circuit Description (60 Pin)	Value at Hot Idle
1	DG/RD	MAP Sensor Signal	1.5-1.6v
2	TN/BK	ECT Sensor Signal	At 180°F: 2.80v
3	RD	Keep Alive Power (B+)	12-14v
4	BK/LB	Sensor Ground	<0.050v
5	BK/WT	Sensor Ground	<0.050v
6	PK/WT	5-Volt Supply	4.9-5.1v
7	OR	8-Volt Supply	7.9-8.1v
8	---	Not Used	---
9	DB	Ignition Switch Power (B+)	12-14v
10	OR/WT	Overdrive Override Switch	Switch Off: 12v, On: 1v
11	BK/TN	Power Ground	<0.1v
12	BK/TN	Power Ground	<0.1v
13	LG/BR	Injector 4 Driver	1-4 ms
14	YL/WT	Injector 3 Driver	1-4 ms
15	TN	Injector 2 Driver	1-4 ms
16	WT/DB	Injector 1 Driver	1-4 ms
17	DB/TN	Injector 7 Driver	1-4 ms
18	RD/YL	Injector 8 Driver	1-4 ms
19	GY	Coil Driver Control	5°, 55 mph: 8° dwell
20	DG	Alternator Field Control	Digital Signals: 0-12-0v
21	BK/RD	Air Temperature Sensor	At 100°F: 2.51v
22	OR/DB	TP Sensor Signal	0.6-1.0v
23 ('93)	TN/WT	HO2S-12 (B1 S2) Signal	0.1-1.1v
24	GY/BK	Distributor Reference Signal	Digital Signals: 0-5-0v
25	PK	SCI Transmit	0v
26	---	Not Used	---
27	BR	A/C Clutch Signal	A/C Off: 12v, On: 1v
28	---	Not Used	---
29	WT/PK	Brake Switch Sense Signal	Brake Off: 0v, On: 12v
30	BR/YL	PNP Switch Sense Signal	In P/N: 0v, Others: 5v
31	---	Not Used	---
32	BK/PK	MIL (lamp) Control	MIL Off: 12v, On: 1v
33	TN/RD	Speed Control Vacuum Solenoid	Vacuum Increasing: 1v
34	DB/OR	A/C WOT Relay Control	Relay Off: 12v, On: 1v
35	GY/YL	EGR Solenoid Control	12v, at 55 mph: 1v
36	---	Not used	---
37	BK/OR	Overdrive Lamp Control	Lamp Off: 12v, On: 1v
38	PK/BK	Injector 5 Driver	1-4 ms
39	GY/RD	AIS Motor Control	Pulse Signals
40	BR/WT	AIS Motor Control	Pulse Signals

Standard Colors and Abbreviations

Abbreviation	Color	Abbreviation	Color	Abbreviation	Color
BK	Black	GY	Gray	RD	Red
BL	Blue	GN	Green	TN	Tan
BR	Brown	LG	Light Green	VT	Violet
DB	Dark Blue	OR	Orange	WT	White
DG	Dark Green	PK	Pink	YL	Yellow

1992-93 Ramcharger 5.2L V8 MFI VIN Y (All) 60 Pin Connector

PCM Pin #	Wire Color	Circuit Description (60 Pin)	Value at Hot Idle
41	BK/DG	HO2S-11 (B1 S1) Signal	0.1-1.1v
42-43	---	Not Used	---
44	GY	Distributor Sync Signal	Digital Signals: 0-5-0v
45	LG	SCI Receive	5v
46	---	Not Used	---
47	WT/OR	Vehicle Speed Signal	Digital Signal
48	BR/RD	Speed Control Set Switch Signal	S/C & Set Switch On: 3.8v
49	YL/RD	Speed Control On/Off Switch	S/C On: 1v, Off: 12v
50	WT/LG	Speed Control Resume Switch	S/C On: 1v, Off: 12v
51	DB/YL	ASD Relay Control	Relay Off: 12v, On: 1v
52	PK/BK	EVAP Purge Solenoid Control	Solenoid Off: 12v, On: 1v
53	LG/RD	Speed Control Vent Solenoid	Vacuum Decreasing: 1v
54	OR/BK	Part Throttle Unlock Solenoid	Solenoid Off: 12v, On: 1v
55	OR/LG	Overdrive Lockout Solenoid	Solenoid Off: 12v, On: 1v
56	GY/PK	Maintenance Indicator Lamp	Lamp Off: 12v, On: 1v
57	DG/OR	ASD Relay Power (B+)	12-14v
58	LG/BK	Injector 6 Driver	1-4 ms
59	PK/BK	AIS Motor Control	Pulse Signals
60	YL/BK	AIS Motor Control	Pulse Signals

Pin Connector Graphic

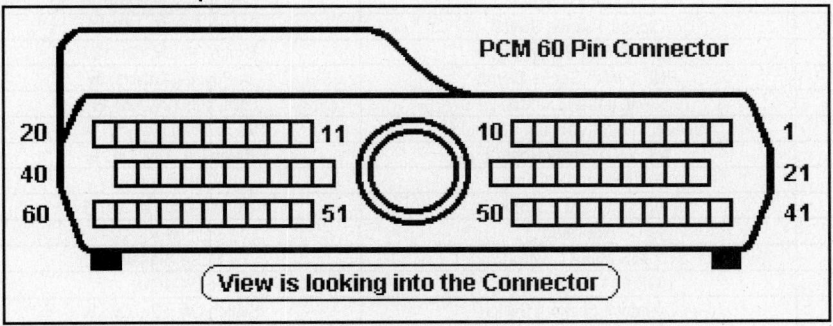

1990-91 Ramcharger 5.9L V8 TBI VIN Z (All) 60 Pin Connector

PCM Pin #	Wire Color	Circuit Description (60 Pin)	Value at Hot Idle
1	DG/RD	MAP Sensor Signal	1.5-1.6v
2	TN	ECT Sensor Signal	At 180ºF: 2.80v
3	RD	Keep Alive Power (B+)	12-14v
4	BK/LB	Sensor Ground	<0.050v
5	BK	Sensor Ground	<0.050v
6	PK/WT	5-Volt Supply	4.9-5.1v
7	OR	8-Volt Supply	7.9-8.1v
8	DG/BK	ASD Relay Power (B+)	12-14v
9	DB	Ignition Switch Power (B+)	12-14v
10	---	Not Used	---
11-12	LB/RD	Power Ground	<0.1v
13-14, 23, 26	---	Not Used	---
15	TN	Injector 2 Driver	1-4 ms
16	WT/DB	Injector 1 Driver	1-4 ms
19	BK/YL	Coil Driver Control	5º, 55 mph: 8º dwell
20	DG	Alternator Field Control	Digital Signals: 0-12-0v
21	BK/RD	Throttle Body Temperature	At 100ºF: 2.51v
22	OR/DB	TP Sensor Signal	0.6-1.0v
24	GY	Distributor Reference Signal	Digital Signals: 0-5-0v
25	PK	SCI Transmit Signal	0v
27	BR	A/C Clutch Signal	A/C Off: 12v, On: 1v
28	PK	Closed Throttle Switch	Switch Off: 12v, On: 1v
29	WT/PK	Brake Switch Sense Signal	Brake Off: 0v, On: 12v
30	BR/YL	PNP Switch Sense Signal	In P/N: 0v, Others: 5v
32	BK/PK	MIL (lamp) Control	MIL Off: 12v, On: 1v
33	TN/RD	Speed Control Vacuum Solenoid	Vacuum Increasing: 1v
34	DB/OR	A/C WOT Relay Control	Relay Off: 12v, On: 1v
35	GY/YL	EGR Solenoid Control	12v, at 55 mph: 1v
36	BK/OR	Air Switching Solenoid	Solenoid Off: 12v, On: 1v
37-39, 42-43, 46	---	Not Used	---
40	BR	AIS Motor Control	Pulse Signals
41	BK/DG	HO2S-11 (B1 S1) Signal	0.1-1.1v
44	OR/WT	Overdrive Lockout Switch	Switch Off: 12v, On: 1v
45	LG	SCI Receive Signal	5v
47	WT/OR	Vehicle Speed Signal	Digital Signal
48	BR/RD	Speed Control Set Switch Signal	S/C & Set Switch On: 3.8v
49	YL/RD	Speed Control On/Off Switch	S/C On: 1v, Off: 12v
50	WT/LG	Speed Control Resume Switch	S/C On: 1v, Off: 12v
51	DB/YL	ASD Relay Control	Relay Off: 12v, On: 1v
52	PK	EVAP Purge Solenoid Control	Solenoid Off: 12v, On: 1v
53	LG/RD	Speed Control Vent Solenoid	Vacuum Decreasing: 1v
54	OR/BK	Part Throttle Unlock Solenoid	Solenoid Off: 12v, On: 1v
55	OR/DB	Overdrive Lockout Solenoid	Solenoid Off: 12v, On: 1v
56	GY/PK	Maintenance Indicator Lamp	Lamp Off: 12v, On: 1v
57-59	---	Not Used	---
60	GY/RD	AIS Motor Control	Pulse Signals

Pin Connector Graphic

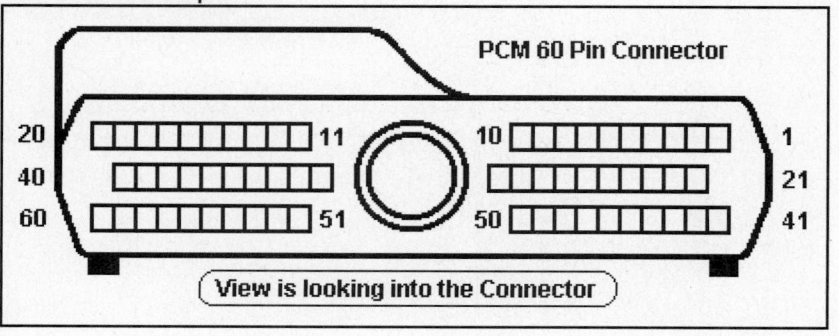

PCM 60 Pin Connector

View is looking into the Connector

1992 Ramcharger 5.9L V8 TBI VIN Z (All) 60 Pin Connector

PCM Pin #	Wire Color	Circuit Description (60 Pin)	Value at Hot Idle
1	DG/RD	MAP Sensor Signal	1.5-1.6v
2	TN/BK	ECT Sensor Signal	At 180°F: 2.80v
3	RD	Keep Alive Power (B+)	12-14v
4	BK/LB	Sensor Ground	<0.050v
5	BK	Sensor Ground	<0.050v
6	PK/WT	5-Volt Supply	4.9-5.1v
7	OR	8-Volt Supply	7.9-8.1v
8	DG/BK	ASD Relay Power (B+)	12-14v
9	DB	Ignition Switch Power (B+)	12-14v
10, 13-14, 23, 26	---	Not Used	---
11-12	LB/RD	Power Ground	<0.1v
15	TN	Injector 2 Driver	1-4 ms
16	WT/DB	Injector 1 Driver	1-4 ms
19	BK/YL	Coil Driver Control	5°, 55 mph: 8° dwell
20	DG	Alternator Field Control	Digital Signals: 0-12-0v
21	BK/RD	Throttle Body Temperature	At 100°F: 2.51v
22	OR/DB	TP Sensor Signal	0.6-1.0v
24	GY/BK	Distributor Reference Signal	Digital Signals: 0-5-0v
25	PK	SCI Transmit Signal	0v
27	BR	A/C Clutch Signal	A/C Off: 12v, On: 1v
28	PK	Closed Throttle Switch	Switch Off: 12v, On: 1v
29	WT/PK	Brake Switch Sense Signal	Brake Off: 0v, On: 12v
30	BR/YL	PNP Switch Sense Signal	In P/N: 0v, Others: 5v
32	BK/PK	MIL (lamp) Control	MIL Off: 12v, On: 1v
33	TN/RD	Speed Control Vacuum Solenoid	Vacuum Increasing: 1v
34	DB/OR	A/C WOT Relay Control	Relay Off: 12v, On: 1v
35	GY/YL	EGR Solenoid Control	12v, at 55 mph: 1v
36	BK/OR	Air Switching Solenoid	Solenoid Off: 12v, On: 1v
37-39, 42-44, 46	---	Not Used	---
40	BR	AIS Motor Control	Pulse Signals
41	BK/DG	HO2S-11 (B1 S1) Signal	0.1-1.1v
44	OR/WT	Overdrive Lockout Switch	Switch Off: 12v, On: 1v
45	LG	SCI Receive Signal	5v
47	WT/OR	Vehicle Speed Signal	Digital Signal
48	BR/RD	Speed Control Set Switch Signal	S/C & Set Switch On: 3.8v
49	YL/RD	Speed Control On/Off Switch	S/C On: 1v, Off: 12v
50	WT/LG	Speed Control Resume Switch	S/C On: 1v, Off: 12v
51	DB/YL	ASD Relay Control	Relay Off: 12v, On: 1v
52	PK	EVAP Purge Solenoid Control	Solenoid Off: 12v, On: 1v
53	LG/RD	Speed Control Vent Solenoid	Vacuum Decreasing: 1v
54	OR/BK	Part Throttle Unlock Solenoid	Solenoid Off: 12v, On: 1v
55	OR/LG	Overdrive Lockout Solenoid	Solenoid Off: 12v, On: 1v
56	GY/PK	Maintenance Indicator Lamp	Lamp Off: 12v, On: 1v
57-59	---	Not Used	---
60	GY/RD	AIS Motor Control	Pulse Signals

Pin Connector Graphic

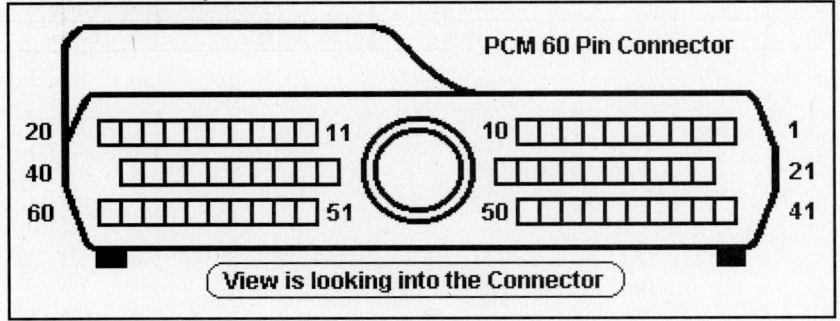

1993 Ramcharger 5.9L V8 MFI VIN Z (A/T) 60 Pin Connector

PCM Pin #	Wire Color	Circuit Description (60 Pin)	Value at Hot Idle
1	DG/RD	MAP Sensor Signal	1.5-1.6v
2	TN/BK	ECT Sensor Signal	At 180ºF: 2.80v
3	RD	Keep Alive Power (B+)	12-14v
4	BK/LB	Sensor Ground	<0.050v
5	BK/WT	Sensor Ground	<0.050v
6	PK/WT	5-Volt Supply	4.9-5.1v
7	OR	8-Volt Supply	7.9-8.1v
8	---	Not Used	---
9	DB	Ignition Switch Power (B+)	12-14v
10	OR/WT	Overdrive Override Switch	Switch Off: 12v, On: 1v
11	BK/TN	Power Ground	<0.1v
12	BK/TN	Power Ground	<0.1v
13	LB/BR	Injector 4 Driver	1-4 ms
14	YL/WT	Injector 3 Driver	1-4 ms
15	TN	Injector 2 Driver	1-4 ms
16	WT/DB	Injector 1 Driver	1-4 ms
17	DB/TN	Injector 7 Driver	1-4 ms
18	RD/YL	Injector 8 Driver	1-4 ms
19	GY	Coil Driver Control	5º, 55 mph: 8º dwell
20	DG	Alternator Field Control	Digital Signals: 0-12-0v
21	BK/RD	Throttle Body Temperature	At 100ºF: 2.51v
22	OR/DB	TP Sensor Signal	0.6-1.0v
23	TN/WT	HO2S-12 (B1 S2) Signal	0.1-1.1v
24	GY/BK	Distributor Reference Signal	Digital Signals: 0-5-0v
25	PK	SCI Transmit	0v
26	---	Not Used	---
27	BR	A/C Clutch Signal	A/C Off: 12v, On: 1v
28	---	Not Used	---
29	WT/PK	Brake Switch Sense Signal	Brake Off: 0v, On: 12v
30	BR/YL	PNP Switch Sense Signal	In P/N: 0v, Others: 5v
31	---	Not Used	---
32	BK/PK	MIL (lamp) Control	MIL Off: 12v, On: 1v
33	TN/RD	Speed Control Vacuum Solenoid	Vacuum Increasing: 1v
34	DB/OR	A/C WOT Relay Control	Relay Off: 12v, On: 1v
35	GY/YL	EGR Solenoid Control	12v, at 55 mph: 1v
36	---	Not Used	---
37	BK/OR	Overdrive Lamp Control	Lamp Off: 12v, On: 1v
38	PK/BK	Injector 5 Driver	1-4 ms
39	GY/RD	AIS Motor Control	Pulse Signals
40	BR/WT	AIS Motor Control	Pulse Signals

Standard Colors and Abbreviations

Abbreviation	Color	Abbreviation	Color	Abbreviation	Color
BK	Black	GY	Gray	RD	Red
BL	Blue	GN	Green	TN	Tan
BR	Brown	LG	Light Green	VT	Violet
DB	Dark Blue	OR	Orange	WT	White
DG	Dark Green	PK	Pink	YL	Yellow

1993 Ramcharger 5.9L V8 MFI VIN Z (A/T) 60 Pin Connector

PCM Pin #	Wire Color	Circuit Description (60 Pin)	Value at Hot Idle
41	BK/DG	HO2S-11 (B1 S1) Signal	0.1-1.1v
42-43	---	Not Used	---
44	GY	Distributor Sync Signal	Digital Signals: 0-5-0v
45	LG	SCI Receive	5v
46	---	Not Used	---
47	WT/OR	Vehicle Speed Signal	Digital Signal
48	BR/RD	Speed Control Set Switch Signal	S/C & Set Switch On: 3.8v
49	YL/RD	Speed Control On/Off Switch	S/C On: 1v, Off: 12v
50	WT/LG	Speed Control Resume Switch	S/C On: 1v, Off: 12v
51	DB/YL	ASD Relay Control	Relay Off: 12v, On: 1v
52	PK/BK	EVAP Purge Solenoid Control	Solenoid Off: 12v, On: 1v
53	LG/RD	Speed Control Vent Solenoid	Vacuum Decreasing: 1v
54	OR/BK	Part Throttle Unlock Solenoid	Solenoid Off: 12v, On: 1v
55	OR/LG	Overdrive Lockout Solenoid	Solenoid Off: 12v, On: 1v
56	GY/PK	Maintenance Indicator Lamp	Lamp Off: 12v, On: 1v
57	DG/OR	ASD Relay Power (B+)	12-14v
58	LG/BK	Injector 6 Driver	1-4 ms
59	PK/BK	AIS Motor Control	Pulse Signals
60	YL/BK	AIS Motor Control	Pulse Signals

Pin Connector Graphic

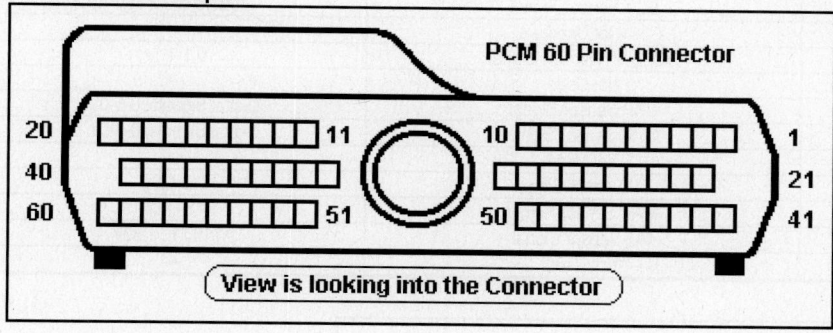

1996 Caravan 2.4L I4 DOHC VIN B (A/T) 'C1' Black Connector

PCM Pin #	Wire Color	Circuit Description (40 Pin)	Value at Hot Idle
1-2	---	Not Used	---
3	DB/TN	Coil 2 Driver Control	5°, 55 mph: 8° dwell
4	DG	Generator Field Driver	Digital Signals: 0-12-0v
5	YL/RD	Speed Control Power Supply	12-14v
6	DG/OR	ASD Relay Power (B+)	12-14v
7	YL/WT	Injector 3 Driver	1-4 ms
8	TN	Starter Relay Control	KOEC: 9-11v
9	---	Not Used	---
10	BK/TN	Power Ground	<0.1v
11	GY	Coil 1 Driver Control	5°, 55 mph: 8° dwell
12	---	Not Used	---
13	WT/DB	Injector 1 Driver	1-4 ms
14-15	---	Not Used	---
16	LB/BR	Injector 4 Driver	1-4 ms
17	TN	Injector 2 Driver	1-4 ms
18-19	---	Not Used	---
20	WT/BK	Ignition Switch Power (B+)	12-14v
21-23	---	Not Used	---
24	DB/LG	Knock Sensor Signal	0.080v AC
25	---	Not Used	---
26	TN/BK	ECT Sensor Signal	At 180°F: 2.80v
27-29	---	Not Used	---
30	BK/DG	HO2S-11 (B1 S1) Signal	0.1-1.1v
31	---	Not Used	---
32	GY/BK	CKP Sensor Signal	Digital Signals: 0-5-0v
33	TN/YL	CMP Sensor Signal	Digital Signals: 0-5-0v
34	---	Not Used	---
35	OR/DB	TP Sensor Signal	0.6-1.0v
36	DG/RD	MAP Sensor Signal	1.5-1.6v
37	BK/RD	IAT Sensor Signal	At 100°F: 1.83v
38-40	---	Not Used	---

Pin Connector Graphic

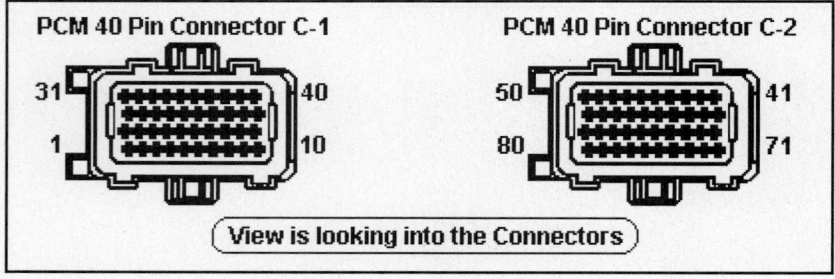

PCM 40 Pin Connector C-1

31 — 40

1 — 10

PCM 40 Pin Connector C-2

50 — 41

80 — 71

View is looking into the Connectors

1996 Caravan 2.4L i4 DOHC VIN B (A/T) 'C2' White Connector

PCM Pin #	Wire Color	Circuit Description (40 Pin)	Value at Hot Idle
41	RD/LG	Speed Control Set Switch Signal	S/C & Set Switch On: 3.8v
42	DB	A/C Pressure Switch Signal	A/C On: 0.451-4.850v
43	BK/LB	Sensor Ground	<0.050v
44	OR	8-Volt Supply	7.9-8.1v
45	DB	A/C Switch Sense Signal	A/C Off: 12v, On: 1v
46	RD/WT	Keep Alive Power (B+)	12-14v
47	---	Not Used	---
48	BR/WT	IAC 1 Driver Control	Pulse Signals
49	YL/BK	IAC 2 Driver Control	Pulse Signals
50	BK/TN	Power Ground	<0.1v
51	TN/WT	HO2S-12 (B1 S2) Signal	0.1-1.1v
52-55	---	Not Used	---
56	OR/BK	Lockup Torque Converter	At Cruise w/TCC On: 1v
57	GY/RD	IAC 3 Driver Control	Pulse Signals
58	PK/BK	IAC 4 Driver Control	Pulse Signals
59	PK/BR	CCD Bus (+)	Digital Signals: 0-5-0v
60	WT/BK	CCD Bus (-)	<0.050v
61	PK/WT	5-Volt Supply	4.9-5.1v
62	WT/PK	Brake Switch Sense Signal	Brake Off: 0v, On: 12v
63	YL/DG	Torque Management Request	Digital Signals
64	DB/OR	A/C Clutch Relay Control	A/C Off: 12v, On: 1v
65	PK	SCI Transmit	0v
66	WT/OR	Vehicle Speed Signal	Digital Signal
67	DB/YL	ASD Relay Control	Relay Off: 12v, On: 1v
68	PK/BK	EVAP Purge Solenoid Control	PWM Signal: 0-12-0v
69-72	---	Not Used	---
73	LG/DG	Radiator Fan Control Relay	Relay Off: 12v, On: 1v
74	BR	Fuel Pump Relay Control	F/P On: 1v, Off: 12v
75	LG	SCI Receive	5v
76	BK/WT	PNP Switch Sense (EATX)	In P/N: 0v, Others: 5v
77	---	Not Used	---
78	TN/RD	Speed Control Vacuum Solenoid	Vacuum Increasing: 1v
79	---	Not Used	---
80	LG/RD	Speed Control Vent Solenoid	Vacuum Decreasing: 1v

Pin Connector Graphic

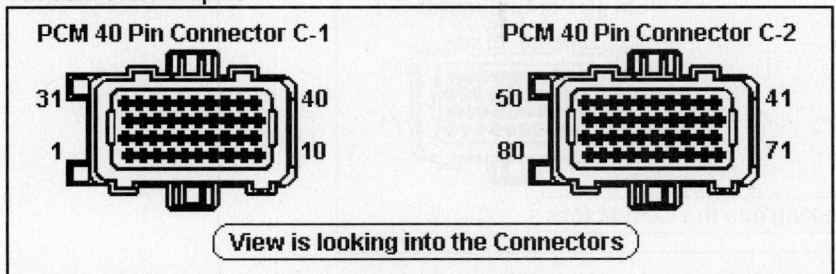

PCM 40 Pin Connector C-1

PCM 40 Pin Connector C-2

31   40   50   41

1   10   80   71

View is looking into the Connectors

1997-98 Caravan 2.4L I4 MFI VIN B (A/T) 'C1' Black Connector

PCM Pin #	Wire Color	Circuit Description (40 Pin)	Value at Hot Idle
1-2	---	Not Used	---
3	DB/TN	Coil 2 Driver Control	5°, 55 mph: 8° dwell
4	DG	Generator Field Driver	Digital Signals: 0-12-0v
5	YL/RD	Speed Control Power Supply	12-14v
6	DG/OR	ASD Relay Power (B+)	12-14v
7	YL/WT	Injector 3 Driver	1-4 ms
8	DB/OR	Starter Relay Control	KOEC: 9-11v
9	---	Not Used	---
10	BK/TN	Power Ground	<0.1v
11	GY	Coil 1 Driver Control	5°, 55 mph: 8° dwell
12	---	Not Used	---
13	WT/DB	Injector 1 Driver	1-4 ms
14-15	---	Not Used	---
16	LB/BR	Injector 4 Driver	1-4 ms
17	T	Injector 2 Driver	1-4 ms
18-19	---	Not Used	---
20	WT/BK	Ignition Switch Power (B+)	12-14v
21	---	Not Used	---
22	BK/PK	MIL (lamp) Control	MIL Off: 12v, On: 1v
23	---	Not Used	---
24	DB/LG	Knock Sensor Signal	0.080v AC
25	---	Not Used	---
26	TN/BK	ECT Sensor Signal	At 180°F: 2.80v
27-29	---	Not Used	---
30	BK/DG	HO2S-11 (B1 S1) Signal	0.1-1.1v
31	---	Not Used	---
32	GY/BK	CKP Sensor Signal	Digital Signals: 0-5-0v
33	TN/YL	CMP Sensor Signal	Digital Signals: 0-5-0v
34	---	Not Used	---
35	OR/DB	TP Sensor Signal	0.6-1.0v
36	DG/RD	MAP Sensor Signal	1.5-1.6v
37	BK/RD	IAT Sensor Signal	At 100°F: 1.83v
38-40	---	Not Used	---

Pin Connector Graphic

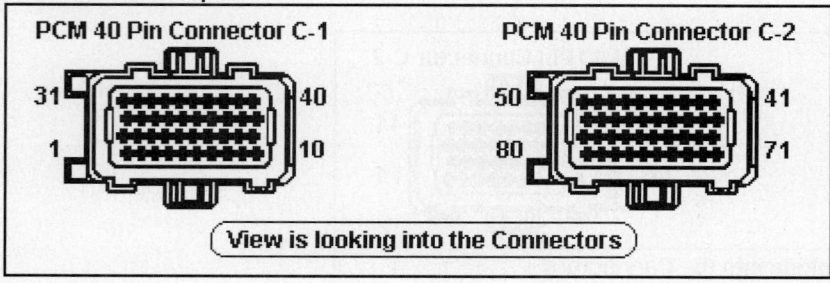

1997-98 Caravan 2.4L I4 MFI VIN B (A/T) 'C2' White Connector

PCM Pin #	Wire Color	Circuit Description (40 Pin)	Value at Hot Idle
41	RD/LG	Speed Control Set Switch Signal	S/C & Set Switch On: 3.8v
42	DB	A/C Pressure Switch Signal	A/C On: 0.451-4.850v
43	BK/LB	Sensor Ground	<0.050v
44	OR	8-Volt Supply	7.9-8.1v
45	DG	A/C Switch Signal	A/C Off: 12v, On: 1v
46	RD/WT	Keep Alive Power (B+)	12-14v
47	---	Not Used	---
48	BR/WT	IAC 3 Driver Control	Pulse Signals
49	YL/BK	IAC 2 Driver Control	Pulse Signals
50	BK/TN	Power Ground	<0.1v
51	TN/WT	HO2S-12 (B1 S2) Signal	0.1-1.1v
52-55	---	Not Used	---
56	OR/BK	Lockup Torque Converter	At Cruise w/TCC On: 1v
57	GY/RD	IAC 1 Driver Control	Pulse Signals
58	PK/BK	IAC 4 Driver Control	Pulse Signals
59	PK/BR	CCD Bus (+)	Digital Signals: 0-5-0v
60	WT/BK	CCD Bus (-)	<0.050v
61	PK/WT	5-Volt Supply	4.9-5.1v
62	WT/PK	Brake Switch Sense Signal	Brake Off: 0v, On: 12v
63	YL/DG	Torque Management Request	Digital Signals
64	DB/OR	A/C Clutch Relay Control	A/C Off: 12v, On: 1v
65, 75	PK, LG	SCI Transmit, SCI Receive	5v
66	WT/OR	Vehicle Speed Signal	Digital Signal
67	DB/YL	ASD Relay Control	Relay Off: 12v, On: 1v
68	PK/BK	EVAP Purge Solenoid Control	PWM Signal: 0-12-0v
69-71	---	Not Used	---
72	YL/BK	LDP Switch Sense Signal	LDP Switch Closed: 0v
73	LG/DB	Radiator Fan Control Relay	Relay Off: 12v, On: 1v
74	BR	Fuel Pump Relay Control	F/P On: 1v, Off: 12v
76	BK/WT	PNP Switch Sense (EATX)	In P/N: 0v, Others: 5v
77	WT/DG	LDP Solenoid Control	PWM Signal: 0-12-0v
78	TN/RD	Speed Control Vacuum Solenoid	Vacuum Increasing: 1v
79	---	Not Used	---
80	LG/RD	Speed Control Vent Solenoid	Vacuum Decreasing: 1v

Pin Connector Graphic

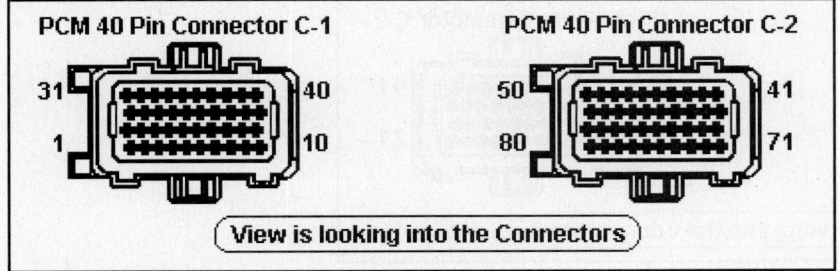

1999-2000 Caravan 2.4L I4 MFI VIN B (A/T) 'C1' Black Connector

PCM Pin #	Wire Color	Circuit Description (40 Pin)	Value at Hot Idle
1-2	---	Not Used	---
3	DB/TN	Coil 2 Driver Control	5°, 55 mph: 8° dwell
4	---	Not Used	---
5	YL/RD	Speed Control Power Supply	12-14v
6	DG/OR	ASD Relay Power (B+)	12-14v
7	YL/WT	Injector 3 Driver	1-4 ms
8	DG	Generator Field Driver	Digital Signals: 0-12-0v
9	---	Not Used	---
10	BK/TN	Power Ground	<0.1v
11	GY/RD	Coil 1 Driver Control	5°, 55 mph: 8° dwell
12	---	Not Used	---
13	WT/DB	Injector 1 Driver	1-4 ms
14-15	---	Not Used	---
16	LB/BR	Injector 4 Driver	1-4 ms
17	TN/WT	Injector 2 Driver	1-4 ms
18-19	---	Not Used	---
20	WT/BK	Ignition Switch Power (B+)	12-14v
21	---	Not Used	---
22	BK/PK	MIL (lamp) Control	MIL Off: 12v, On: 1v
23-24	---	Not Used	---
25	DB/LG	Knock Sensor Signal	0.080v AC
26	TN/BK	ECT Sensor Signal	At 180°F: 2.80v
27	BK/OR	HO2S Ground	<0.050v
28-29	---	Not Used	---
30	BK/DG	HO2S-11 (B1 S1) Signal	0.1-1.1v
31	TN	Smart Start Relay Control	KOEC: 9-11v
32	GY/BK	CKP Sensor Signal	Digital Signals: 0-5-0v
33	TN/YL	CMP Sensor Signal	Digital Signals: 0-5-0v
34, 39	---	Not Used	---
35	OR/DB	TP Sensor Signal	0.6-1.0v
36	DG/RD	MAP Sensor Signal	1.5-1.6v
37	BK/RD	IAT Sensor Signal	At 100°F: 1.83v
38	DG/LB	A/C Switch Sense Signal	A/C Off: 12v, On: 1v
40	GY/YL	EGR Solenoid Control	12v, at 55 mph: 1v

Pin Connector Graphic

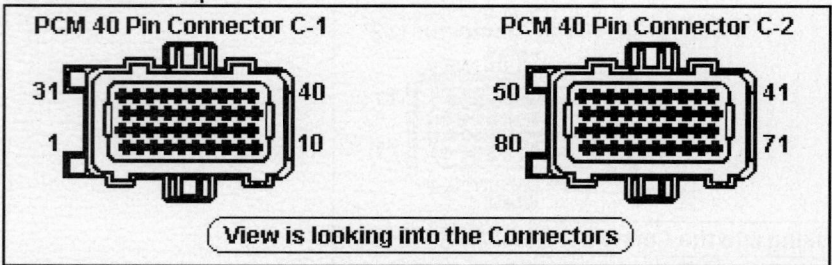

1999-2000 Caravan 2.4L I4 MFI VIN B (A/T) 'C2' White Connector

PCM Pin #	Wire Color	Circuit Description (40 Pin)	Value at Hot Idle
41	RD/LG	Speed Control Set Switch Signal	S/C & Set Switch On: 3.8v
42	DB	A/C Pressure Switch Signal	A/C On: 0.451-4.850v
43	BK/LB	Sensor Ground	<0.050v
44	OR	8-Volt Supply	7.9-8.1v
45	---	Not Used	---
46	RD/WT	Keep Alive Power (B+)	12-14v
47	---	Not Used	---
48	BR/WT	IAC 3 Driver Control	Pulse Signals
49	YL/BK	IAC 2 Driver Control	Pulse Signals
50	BK/TN	Power Ground	<0.1v
51	TN/WT	HO2S-12 (B1 S2) Signal	0.1-1.1v
52-55	---	Not Used	---
56	TN/RD	Speed Control Vacuum Solenoid	Vacuum Increasing: 1v
57	GY/RD	IAC 1 Driver Control	Pulse Signals
58	PK/BK	IAC 4 Driver Control	Pulse Signals
59	PK/BR	CCD Bus (+)	Digital Signals: 0-5-0v
60	WT/BK	CCD Bus (-)	<0.050v
61	PK/WT	5-Volt Supply	4.9-5.1v
62	WT/PK	Brake Switch Sense Signal	Brake Off: 0v, On: 12v
63	YL/DG	Torque Management Request	Digital Signals
64	DB/OR	A/C Clutch Relay Control	Relay Off: 12v, On: 1v
65	PK	SCI Transmit	0v
66	WT/OR	Vehicle Speed Signal	Digital Signal
67	DB/YL	ASD Relay Control	Relay Off: 12v, On: 1v
68	PK/BK	EVAP Purge Solenoid Control	PWM Signal: 0-12-0v
69	---	Not Used	---
70	P/R	EVAP Purge Solenoid Sense	0-1v
71	---	Not Used	---
72	YL/BK	LDP Switch Sense Signal	LDP Switch Closed: 0v
73	LG/DB	Radiator Fan Control Relay	Relay Off: 12v, On: 1v
74	BR	Fuel Pump Relay Control	Relay Off: 12v, On: 1v
75	LG	SCI Receive	5v
76	BK/WT	PNP Switch Sense (EATX)	In P/N: 0v, Others: 5v
77	WT/DG	LDP Solenoid Control	PWM Signal: 0-12-0v
78	OR/BK	Lockup Torque Converter	At Cruise w/TCC On: 1v
79	---	Not Used	---
80	LG/RD	Speed Control Vent Solenoid	Vacuum Decreasing: 1v

Pin Connector Graphic

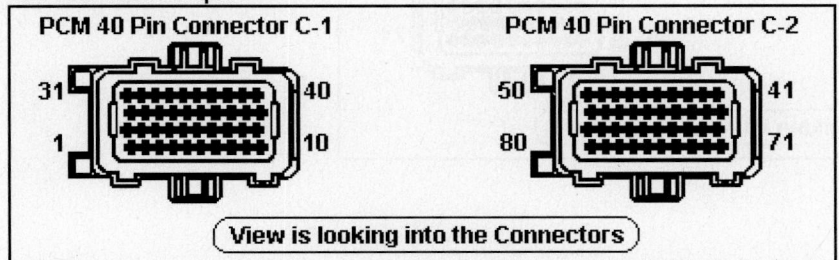

PCM 40 Pin Connector C-1

PCM 40 Pin Connector C-2

View is looking into the Connectors

2001-02 Caravan 2.4L I4 MFI VIN B (A/T) 'C1' Black Connector

PCM Pin #	Wire Color	Circuit Description (40 Pin)	Value at Hot Idle
1-2	---	Not Used	---
3	DB/TN	Coil 2 Driver Control	5°, 55 mph: 8° dwell
4	---	Not Used	---
5	VT/YL	S/C On/Off Switch Sense	12-14v
6	BR/WT	ASD Relay Power (B+)	12-14v
7	BR/LB	Injector 3 Driver	1-4 ms
8	BR/GY	Generator Field Driver	Digital Signals: 0-12-0v
9	---	Not Used	---
10	BK/BR	Power Ground	<0.1v
11	DB/GN	Coil 1 Driver Control	5°, 55 mph: 8° dwell
12	VT/GY	Engine Oil Pressure Sensor	1.6v at 24 psi
13	BR/YL	Injector 1 Driver	1-4 ms
14-15	---	Not Used	---
16	BR/TN	Injector 4 Driver	1-4 ms
17	BR/DB	Injector 2 Driver	1-4 ms
18	BR/LG	HO2S-11 (B1 S1) Heater	PWM Signal: 0-12-0v
19	---	Not Used	---
20	PK/GY	Ignition Switch Power (B+)	12-14v
21-24	---	Not Used	---
25	DB/YL	Knock Sensor Signal	0.080v AC
26	VT/OR	ECT Sensor Signal	At 180°F: 2.80v
27	BR/LG	HO2S Ground	<0.050v
28-29	---	Not Used	---
30	DB/LB	HO2S-11 (B1 S1) Signal	0.1-1.1v
31	DG/OR	Double Start Override	KOEC: 9-11v
32	BR/LB	CKP Sensor Signal	Digital Signals: 0-5-0v
33	DB/GY	CMP Sensor Signal	Digital Signals: 0-5-0v
34	---	Not Used	---
35	BR/OR	TP Sensor Signal	0.6-1.0v
36	VT/BR	MAP Sensor Signal	1.5-1.6v
37	DB/LG	IAT Sensor Signal	At 100°F: 1.83v
38-39	---	Not Used	---
40	DB/VT	EGR Solenoid Control	12v, at 55 mph: 1v

Pin Connector Graphic

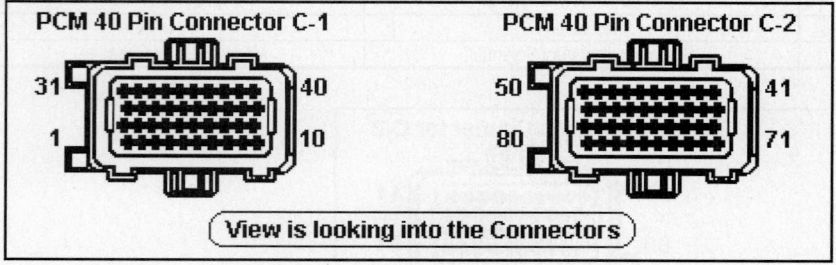

PCM 40 Pin Connector C-1

PCM 40 Pin Connector C-2

(View is looking into the Connectors)

2001-02 Caravan 2.4L I4 MFI VIN B (A/T) 'C2' White Connector

PCM Pin #	Wire Color	Circuit Description (40 Pin)	Value at Hot Idle
41	VT	Speed Control Set Switch Signal	S/C & Set Switch On: 3.8v
42	LB/BR	A/C Pressure Switch Signal	A/C On: 0.451-4.850v
43	DB/DG	Sensor Ground	<0.050v
44	BR/PK	8-Volt Supply	7.9-8.1v
45	---	Not Used	---
46	OR/RD	Keep Alive Power (B+)	12-14v
47	---	Not Used	---
48	BR/LG	IAC 3 Driver Control	Pulse Signals
49	VT/LG	IAC 2 Driver Control	Pulse Signals
50	BK/DG	Power Ground	<0.1v
51	DB/YL	HO2S-12 (B1 S2) Signal	0.1-1.1v
52-55	---	Not Used	---
56	VT/YL	Speed Control Vacuum Solenoid	Vacuum Increasing: 1v
57	VT/DG	IAC 1 Driver Control	Pulse Signals
58	BR/DG	IAC 4 Driver Control	Pulse Signals
59	WT/VT	PCI Data Bus (J1850)	Digital Signals: 0-7-0v
60	---	Not Used	---
61	PK/YL	5-Volt Supply	4.9-5.1v
62	DG/WT	Brake Switch Sense Signal	Brake Off: 0v, On: 12v
63	---	Not Used	---
64	LB/OR	A/C Clutch Relay Control	Relay Off: 12v, On: 1v
65	WT/BR	SCI Transmit Signal	0v
66	DB/OR	Vehicle Speed Signal	Digital Signal
67	BR/WT	ASD Relay Control	Relay Off: 12v, On: 1v
68	DB/WT	EVAP Purge Solenoid Control	PWM Signal: 0-12-0v
69, 71	---	Not Used	---
70	DB/BR	EVAP Purge Solenoid Sense	0-1v
72	VT/WT	LDP Switch Sense Signal	LDP Switch Closed: 0v
73	BR/VT	Radiator Fan Control Relay	Relay Off: 12v, On: 1v
74	BR	Fuel Pump Relay Control	Relay Off: 12v, On: 1v
75	WT/LG	SCI Receive Signal	5v
76	YL/DB	PNP Switch Sense (EATX)	In P/N: 0v, Others: 5v
77	VT/LB	LDP Solenoid Control	PWM Signal: 0-12-0v
78	DB/WT	A/T: Lockup Torque Converter	At Cruise w/TCC On: 1v
79	---	Not Used	---
80	VT/OR	Speed Control Vent Solenoid	Vacuum Decreasing: 1v

Pin Connector Graphic

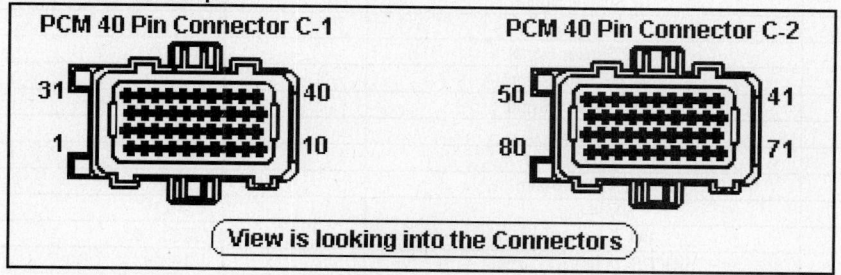

CHRYSLER PIN TABLES

2003 Caravan 2.4L I4 DOHC MFI VIN B (All) 'C1' Black Connector

PCM Pin #	Wire Color	Circuit Description (38 Pin)	Value at Hot Idle
1-8	---	Not Used	---
9	BK/BR	Power Ground	<0.1v
10	---	Not Used	---
11	PK/GY	Ignition Switch Output (Run-Start)	12-14v
12	PK/OR	Ignition Switch Output (Run-Start)	12-14v
13	DB/OR	Vehicle Speed Signal	Digital Signal
14	GY/BK	Brake Fluid Level Switch Signal	Switch Open: 12v, Closed: 0v
15-17	---	Not Used	---
18	BK/DG	Power Ground	<0.1v
19	---	Not Used	---
20	VT/GY	Oil Pressure Signal	1.6v at 24 psi
21-24	---	Not Used	---
25	WT/LG	SCI Receive (PCM) - DLC Pin 12	5v
26	WT/BR	Flash Program Enable	5v
27-28	---	Not Used	---
29	RD/WT	Battery Power (B+)	12-14v
30	YL	Ignition Switch Output (Start)	8-11v (cranking)
31	DB/YL	HO2S-12 (B1 S2) Signal	0.1-1.1v
32	DB/DG	Oxygen Sensor Return (Downstream)	<0.1v
33-35	---	Not Used	---
36	WT/BR	SCI Transmit (PCM) - DLC Pin 7	0v
37	DG/YL	SCI Transmit (TCM)	0v
38	WT/VT	PCI Data Bus (J1850) - DLC Pin 2	Digital Signals: 0-7-0v

2003 Caravan 2.4L I4 DOHC MFI VIN B (All) 'C2' Gray Connector

PCM Pin #	Wire Color	Circuit Description (38 Pin)	Value at Hot Idle
1-8	---	Not Used	---
9	DB/TN	Coil 2 Driver	5°, at 55 mph: 8° dwell
10	DB/DG	Coil 1 Driver	5°, at 55 mph: 8° dwell
11	BR/TN	Injector 4 Driver	1.0-4.0 ms
12	BR/LB	Injector 3 Driver	1.0-4.0 ms
13	BR/DB	Injector 2 Driver	1.0-4.0 ms
14	BR/YL	Injector 1 Driver	1.0-4.0 ms
15-17	---	Not Used	---
18	BR/LG	HO2S-11 (B1 S1) Heater Control	Heater Off: 12v, On: 1v
19	BR/GY	Generator Field Driver	Digital Signal: 0-12-0v
20	VT/OR	ECT Sensor Signal	At 180°F: 2.80v
21	BR/OR	TP Sensor Signal	0.6-1.0v
22	---	Not Used	---
23	VT/BR	MAP Sensor Signal	1.5-1.7v
24	BR/LG	Knock Sensor Return	<0.050v
25	DB/YL	Knock Sensor Signal	0.080v AC
26	---	Not Used	---
27	DB/DG	Sensor Ground	<0.1v
28	VR/VT	IAC Motor Return	12-14v
29	PK/YL	5-Volt Supply	4.9-5.1v
30	DB/LG	IAT Sensor Signal	At 100°F: 1.83v
31	DB/LB	HO2S-11 (B1 S1) Signal	0.1-1.1v
32	BR/DG	Oxygen Sensor Return (Upstream)	<0.1v
33	---	Not Used	---
34	DB/GY	CMP Sensor Signal	Digital Signal: 0-5-0v
35	BR/LB	CKP Sensor Signal	Digital Signal: 0-5-0v
36-37	---	Not Used	---
38	VT/GY	IAC Motor Driver	DC pulse signals: 0.8-11v

2003 Caravan 2.4L I4 DOHC MFI VIN B (All) 'C3' White Connector

PCM Pin #	Wire Color	Circuit Description (38 Pin)	Value at Hot Idle
1-2, 4	---	Not Used	---
3	BR/WT	ASD Relay Control	Relay Off: 12v, On: 1v
5	VT/OR	Speed Control Vent Solenoid	Vacuum Decreasing: 1v
6	BR/VT	Radiator Fan Relay Control	Relay Off: 12v, On: 1v
7	VT/YL	Speed Control Power Supply	12-14v
8	VT/LG	Natural Vacuum Leak Detection Pump	Solenoid Off: 12v, On: 1v
9	BR/WT	HO2S-12 (B1 S2) Heater Control	---
10	---	Not Used	---
11	LB/OR	A/C Clutch Relay Control	Relay Off: 12v, On: 1v
12	VT/YL	Speed Control Vacuum Solenoid	Vacuum Increasing: 1v
13-18, 21-22	---	Not Used	---
19	BR/WT	Automatic Shutdown Relay Output	12-14v
20	DB/WT	EVAP Purge Solenoid Control	PWM Signal: 0-12-0v
23	DG/WT	Brake Switch Sense	Brake Off: 0v, On: 12v
24-27	---	Not Used	---
28	BR/WT	Automatic Shutdown Relay Output	12-14v
29	DB/BR	EVAP Purge Sense	0-1v
30	---	Not Used	---
31	LB/BR	A/C Pressure Sensor	A/C On: 0.45-4.85v
32	DB/YL	Battery Temperature Sensor	At 100°F: 1.83v
33, 36	---	Not Used	---
34	VT	Speed Control Set Switch Signal	S/C & Set Switch On: 3.8v
35	VT/WT	Natural Vacuum Leak Detection Switch Sense	0.1v
36	---	Not Used	---
37	BR	Fuel Pump Relay Control	Relay Off: 12v, On: 1v
38	DG/OR	Starter Relay Control	KOEC: 9-11v

2003 PT Cruiser 2.4L I4 DOHC MFI VIN B (All) 'C4' Green Connector

PCM Pin #	Wire Color	Circuit Description (38 Pin)	Value at Hot Idle
1	YL/GY	Overdrive Solenoid Control	Solenoid Off: 12v, On: 1v
2	YL/LB	Underdrive Solenoid Control	Solenoid Off: 12v, On: 1v
3-5	---	Not Used	---
6	YL/DB	2-4 Solenoid Control	Solenoid Off: 12v, On: 1v
7-9	---	Not Used	---
10	DG/WT	Low/Reverse Solenoid Control	Solenoid Off: 12v, On: 1v
11, 14	---	Not Used	---
12, 13	BK/LG	Power Ground	<0.050v
15	DG/LB	TRS T1 Sense	<0.050v
16	DG/DB	TRS T3 Sense	<0.050v
17, 20-21	---	Not Used	---
18, 38	YL/BR	Transmission Control Relay Output	Relay Off: 12v, On: 1v
19, 28	YL/OR	Transmission Control Relay Output	Relay Off: 12v, On: 1v
23-26	---	Not Used	---
27	DG/GY	TRS T41 Sense	<0.050v
29	YL/TN	Low/Reverse Pressure Switch Sense	12-14v
30	YL/DG	2-4 Pressure Switch Sense	In Low/Reverse: 2-4v
31, 36	---	Not Used	---
32	DG/BR	Output Speed Sensor Signal	In 2-4 Position: 2-4v
33	DG/WT	Input Speed Sensor Signal	Moving: AC voltage
34	DG/VT	Speed Sensor Ground	Moving: AC voltage
35	DG/OR	Transmission Temperature Sensor Signal	<0.050v
37	DG/YL	TRS T42 Sense	In PRNL: 0v, Others 5v

Pin Connector Graphic

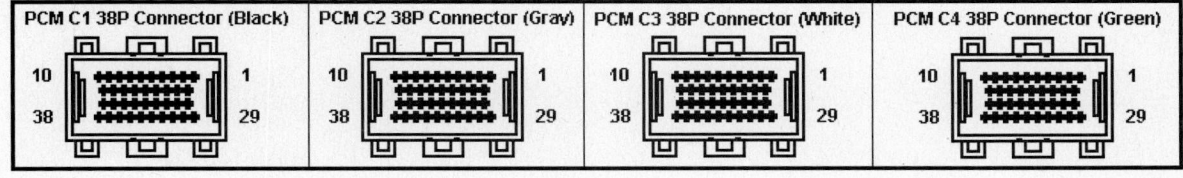

1990 Caravan 2.5L I4 SOHC Turbo MFI VIN J 60 Pin Connector

PCM Pin #	Wire Color	Circuit Description (60 Pin)	Value at Hot Idle
1	DG/RD	MAP Sensor Signal	1.5-1.6v
2	TN/WT	ECT Sensor Signal	At 180°F: 2.80v
3	RD/WT	Keep Alive Power (B+)	12-14v
4	BK/LB	Sensor Ground	<0.050v
5	BK/WT	Sensor Ground	<0.050v
6	PK/WT	5-Volt Supply	4.9-5.1v
7	OR	8-Volt Supply	7.9-8.1v
8	DG/BK	Auto Shutdown Relay Input	12-14v
9	DB	Ignition Switch Power (B+)	12-14v
10	---	Not Used	---
11	BK/TN	Power Ground	<0.1v
12	BK/TN	Power Ground	<0.1v
13	LB/BR	Injector 4 Driver	1-4 ms
14	YL/WT	Injector 3 Driver	1-4 ms
15	TN	Injector 2 Driver	1-4 ms
16	WT/DB	Injector 1 Driver	1-4 ms
17-18, 21	---	Not Used	---
19	BK/GY	Coil 1 Driver Control	5°, 55 mph: 8° dwell
20	DG	Alternator Field Control	Digital Signals: 0-12-0v
22	OR/DB	TP Sensor Signal	0.6-1.0v
23, 28	---	Not Used	---
24	GY/BK	CKP Sensor Signal	Digital Signals: 0-5-0v
25	PK	SCI Transmit	0v
26	PK/BR	CCD Bus (+)	Digital Signals: 0-5-0v
27	BR	A/C Clutch Signal	A/C Off: 12v, On: 1v
29	WT/PK	Brake Switch Sense Signal	Brake Off: 0v, On: 12v
30	BR/LB	PNP Switch Sense Signal	In P/N: 0v, Others: 5v
31	DB/PK	Radiator Fan Control Relay	Relay Off: 12v, On: 1v
32	BK/PK	MIL (lamp) Control	MIL Off: 12v, On: 1v
33	TN/RD	Speed Control Vacuum Solenoid	Vacuum Increasing: 1v
34	DB/OR	A/C WOT Relay Control	Relay Off: 12v, On: 1v
35	---	Not Used	---
36	LG/BK	Wastegate Solenoid Control	Solenoid Off: 12v, On: 1v
37-38	---	Not Used	---
39	GY/RD	AIS Motor 3 Control	Pulse Signals
40	BR/WT	AIS Motor 1 Control	Pulse Signals

Standard Colors and Abbreviations

Abbreviation	Color	Abbreviation	Color	Abbreviation	Color
BK	Black	GY	Gray	RD	Red
BL	Blue	GN	Green	TN	Tan
BR	Brown	LG	Light Green	VT	Violet
DB	Dark Blue	OR	Orange	WT	White
DG	Dark Green	PK	Pink	YL	Yellow

1990 Caravan 2.5L I4 SOHC Turbo MFI VIN J 60 Pin Connector

PCM Pin #	Wire Color	Circuit Description (60 Pin)	Value at Hot Idle
41	BK/DG	HO2S-11 (B1 S1) Signal	0.1-1.1v
42	BK/LG	Knock Sensor Signal	0.080v AC
43	GY/LB	Tachometer Signal	Pulse Signals
44	TN/YL	Fuel Monitor Output Signal	F/P On: 1v, Off: 12v
45	LG	SCI Receive	5v
46	WT/BK	CCD Bus (-)	<0.050v
47	WT/OR	Vehicle Speed Signal	Digital Signal
48	BR/RD	Speed Control Set Switch Signal	S/C & Set Switch On: 3.8v
49	YL/RD	Speed Control On/Off Switch	S/C On: 1v, Off: 12v
50	WT/LG	Speed Control Resume Switch	S/C On: 1v, Off: 12v
51	DB/YL	ASD Relay Control	Relay Off: 12v, On: 1v
52	PK/BK	EVAP Purge Solenoid Control	Solenoid Off: 12v, On: 1v
53	LG/RD	Speed Control Vent Solenoid	Vacuum Decreasing: 1v
54	OR/BK	Shift Indicator Lamp Control	SIL On: 1v, SIL Off: 12v
55	LB	BARO Read Solenoid	Solenoid Off: 12v, On: 1v
56	GY/PK	Maintenance Indicator Lamp	Lamp Off: 12v, On: 1v
57-58	---	Not Used	---
59	PK/BK	AIS Motor 4 Control	Pulse Signals
60	YL/BK	AIS Motor 2 Control	Pulse Signals

Pin Connector Graphic

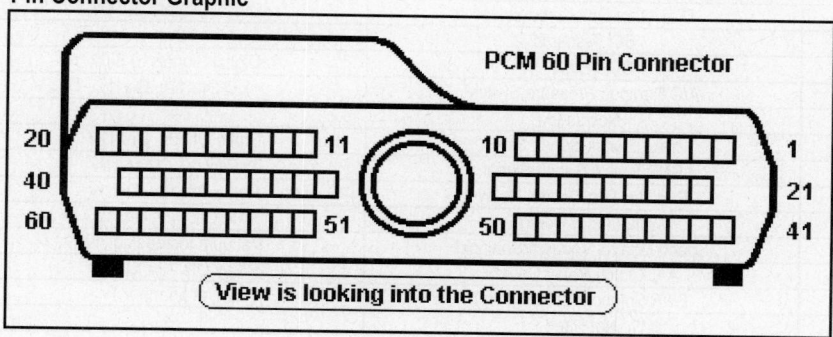

1990 Caravan 2.5L I4 SOHC TBI VIN K (All) 60 Pin Connector

PCM Pin #	Wire Color	Circuit Description (60 Pin)	Value at Hot Idle
1	DG/RD	MAP Sensor Signal	1.5-1.6v
2	TN/WT	ECT Sensor Signal	At 180°F: 2.80v
3	RD/WT	Keep Alive Power (B+)	12-14v
4	BK/LB	Sensor Ground	<0.050v
5	BK/WT	Sensor Ground	<0.050v
6	PK/WT	5-Volt Supply	4.9-5.1v
7	OR	9-Volt Supply	8.9-9.1v
8	WT	Ignition Switch (Start) Output	12-14v
9	DB	Ignition Switch Power (B+)	12-14v
10	---	Not Used	---
11	LB/RD	Power Ground	<0.1v
12	LB/RD	Power Ground	<0.1v
13-15	---	Not Used	---
16	WT/DB	Injector Control Driver	1-4 ms
17-18	---	Not Used	---
19	BK/GY	Coil Driver Control	5°, 55 mph: 8° dwell
20	DG	Alternator Field Control	Digital Signals: 0-12-0v
21	BK/RD	Charge Temperature Sensor	At 100°F: 2.51v
22	OR/DB	TP Sensor Signal	0.6-1.0v
23	---	Not Used	---
24	GY/BK	Distributor Reference Signal	Digital Signals: 0-5-0v
25	PK	SCI Transmit	0v
26	PK/BR	CCD Bus (+)	Digital Signals: 0-5-0v
27	BR	A/C Damped Pressure Switch	A/C Off: 12v, On: 1v
28	---	Not Used	---
29	WT/PK	Brake Switch Sense Signal	Brake Off: 0v, On: 12v
30	BR/YL	PNP Switch Sense Signal	In P/N: 0v, Others: 5v
31	DB/PK	Radiator Fan Control Relay	Relay Off: 12v, On: 1v
32	BK/PK	MIL (lamp) Control	MIL Off: 12v, On: 1v
33	TN/RD	Speed Control Vacuum Solenoid	Vacuum Increasing: 1v
34	DB/OR	A/C Clutch Relay Control	Relay Off: 12v, On: 1v
35	GY/YL	EGR Solenoid Control (Cal)	12v, at 55 mph: 1v
36-38	---	Not Used	---
39	GY/RD	AIS Motor 3 Control	Pulse Signals
40	BR	AIS Motor 1 Control	Pulse Signals

Standard Colors and Abbreviations

Abbreviation	Color	Abbreviation	Color	Abbreviation	Color
BK	Black	GY	Gray	RD	Red
BL	Blue	GN	Green	TN	Tan
BR	Brown	LG	Light Green	VT	Violet
DB	Dark Blue	OR	Orange	WT	White
DG	Dark Green	PK	Pink	YL	Yellow

1990 Caravan 2.5L I4 SOHC TBI VIN K (All) 60 Pin Connector

PCM Pin #	Wire Color	Circuit Description (60 Pin)	Value at Hot Idle
41	BK/DG	HO2S-11 (B1 S1) Signal	0.1-1.1v
42	---	Not Used	---
43	GY/LB	Tachometer Signal	Pulse Signals
44	---	Not Used	---
45	LG	SCI Receive	5v
46	WT/BK	CCD Bus (-)	<0.050v
47	WT/OR	Vehicle Speed Signal	Digital Signal
48	BR/RD	Speed Control Set Switch Signal	S/C & Set Switch On: 3.8v
49	YL/RD	Speed Control On/Off Switch	S/C On: 1v, Off: 12v
50	WT/LG	Speed Control Resume Switch	S/C On: 1v, Off: 12v
51	DB/YL	ASD Relay Power (B+)	Relay Off: 12v, On: 1v
52	PK/BK	EVAP Purge Solenoid Control	Solenoid Off: 12v, On: 1v
53	LG/RD	Speed Control Vent Solenoid	Vacuum Decreasing: 1v
54	OR/BK	Lockup Torque Converter	At Cruise w/TCC On: 1v
55	---	Not Used	---
56	GY/PK	Maintenance Indicator Lamp	Lamp Off: 12v, On: 1v
57-58	---	Not Used	---
59	PK/BK	AIS Motor 4 Control	Pulse Signals
60	YL/BK	AIS Motor 2 Control	Pulse Signals

Pin Connector Graphic

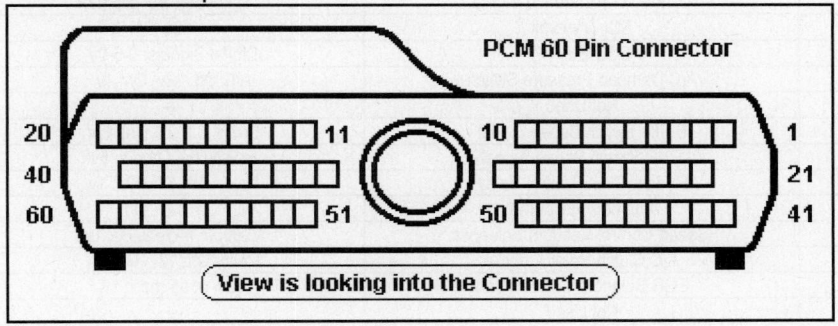

1991-92 Caravan 2.5L I4 SOHC TBI VIN K (All) 60 Pin Connector

PCM Pin #	Wire Color	Circuit Description (60 Pin)	Value at Hot Idle
1	DG/RD	MAP Sensor Signal	1.5-1.6v
2	TN/BK	ECT Sensor Signal	At 180°F: 2.80v
3	RD/WT	Keep Alive Power (B+)	12-14v
4	BK/LB	Sensor Ground	<0.050v
5	BK/WT	Sensor Ground	<0.050v
6	PK/WT	5-Volt Supply	4.9-5.1v
7	OR	9-Volt Supply	8.9-9.1v
8	WT	Ignition Switch (Start) Output	12-14v
9	DB	Ignition Switch Power (B+)	12-14v
10	---	Not Used	---
11	BK/TN	Power Ground	<0.1v
12	BK/TN	Power Ground	<0.1v
13-15	---	Not Used	---
16	WT/DB	Injector Control Driver	1-4 ms
17-18	---	Not Used	---
19	GY	Coil Driver Control	5°, 55 mph: 8° dwell
20	DG	Alternator Field Control	Digital Signals: 0-12-0v
21	BK/RD	Charge Temperature Sensor	At 100°F: 2.51v
22	OR/DB	TP Sensor Signal	0.6-1.0v
23 ('92)	RD/LG	Speed Control Set Switch Signal	S/C & Set Switch On: 3.8v
24	GY/BK	Distributor Reference Signal	Digital Signals: 0-5-0v
25	PK	SCI Transmit	0v
26	PK/BR	CCD Bus (+)	Digital Signals: 0-5-0v
27	BR	A/C Damped Pressure Switch	A/C Off: 12v, On: 1v
28	---	Not Used	---
29	WT/PK	Brake Switch Sense Signal	Brake Off: 0v, On: 12v
30	BR/YL	PNP Switch Sense Signal	In P/N: 0v, Others: 5v
31	DB/PK	Radiator Fan Control Relay	Relay Off: 12v, On: 1v
32	BK/PK	MIL (lamp) Control	MIL Off: 12v, On: 1v
33	TN/RD	Speed Control Vacuum Solenoid	Vacuum Increasing: 1v
34	DB/OR	A/C Clutch Relay Control	Relay Off: 12v, On: 1v
35	GY/YL	EGR Solenoid Control (Cal)	12v, at 55 mph: 1v
36-38	---	Not Used	---
39	GY/RD	AIS Motor 3 Control	Pulse Signals
40	BR/WT	AIS Motor 1 Control	Pulse Signals

Standard Colors and Abbreviations

Abbreviation	Color	Abbreviation	Color	Abbreviation	Color
BK	Black	GY	Gray	RD	Red
BL	Blue	GN	Green	TN	Tan
BR	Brown	LG	Light Green	VT	Violet
DB	Dark Blue	OR	Orange	WT	White
DG	Dark Green	PK	Pink	YL	Yellow

1991-92 Caravan 2.5L I4 SOHC TBI VIN K (All) 60 Pin Connector

PCM Pin #	Wire Color	Circuit Description (60 Pin)	Value at Hot Idle
41	BK/DG	HO2S-11 (B1 S1) Signal	0.1-1.1v
42	---	Not Used	---
43	GY/LB	Tachometer Signal	Pulse Signals
44	---	Not Used	---
45	LG	SCI Receive	5v
46	WT/BK	CCD Bus (-)	<0.050v
47	WT/OR	Vehicle Speed Signal	Digital Signal
48	BR/RD	Speed Control Set Switch Signal	S/C & Set Switch On: 3.8v
49	YL/RD	Speed Control On/Off Switch	S/C On: 1v, Off: 12v
50	WT/LG	Speed Control Resume Switch	S/C On: 1v, Off: 12v
51	DB/YL	ASD Relay Power (B+)	Relay Off: 12v, On: 1v
52	PK/BK	EVAP Purge Solenoid Control	Solenoid Off: 12v, On: 1v
53	LG/RD	Speed Control Vent Solenoid	Vacuum Decreasing: 1v
54	OR/BK	Lockup Torque Converter	At Cruise w/TCC On: 1v
55	---	Not Used	---
56	GY/PK	Maintenance Indicator Lamp	Lamp Off: 12v, On: 1v
57-58	---	Not Used	---
59	PK/BK	AIS Motor 4 Control	Pulse Signals
60	YL/BK	AIS Motor 2 Control	Pulse Signals

Pin Connector Graphic

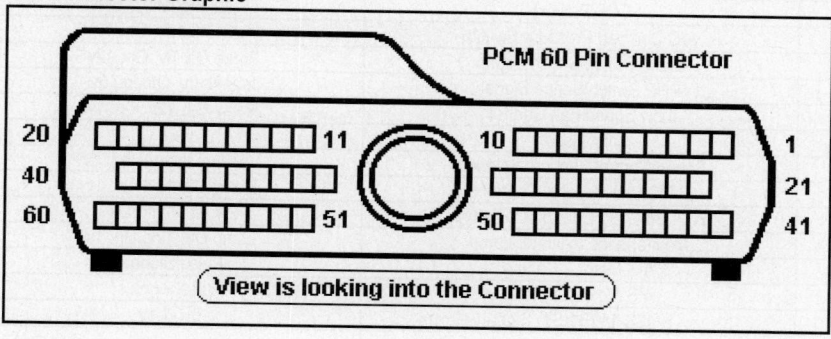

1993 Caravan 2.5L I4 SOHC TBI VIN K (All) 60 Pin Connector

PCM Pin #	Wire Color	Circuit Description (60 Pin)	Value at Hot Idle
1	DG/RD	MAP Sensor Signal	1.5-1.6v
2	TN/BK	ECT Sensor Signal	At 180°F: 2.80v
3	RD/WT	Keep Alive Power (B+)	12-14v
4	BK/LB	Sensor Ground	<0.050v
5	BK/WT	Sensor Ground	<0.050v
6	PK/WT	5-Volt Supply	4.9-5.1v
7	OR	8-Volt Supply	7.9-8.1v
8	WT	Ignition Switch (Start) Output	12-14v
9	DB	Ignition Switch Power (B+)	12-14v
10, 13-15	---	Not Used	---
11	BK/TN	Power Ground	<0.1v
12	BK/TN	Power Ground	<0.1v
16	WT/DB	Injector Control Driver	1-4 ms
17-18, 21, 28	---	Not Used	---
19	BK/GY	Coil Driver Control	5°, 55 mph: 8° dwell
20	DG	Alternator Field Control	Digital Signals: 0-12-0v
22	OR/DB	TP Sensor Signal	0.6-1.0v
23	RD/LG	Speed Control Set Switch Signal	S/C & Set Switch On: 3.8v
24	GY/BK	Distributor Reference Signal	Digital Signals: 0-5-0v
25	PK	SCI Transmit	0v
26	PK/BR	CCD Bus (+)	Digital Signals: 0-5-0v
27	BR	A/C Damped Pressure Switch	A/C Off: 12v, On: 1v
29	WT/PK	Brake Switch Sense Signal	Brake Off: 0v, On: 12v
30	BR/YL	PNP Switch Sense Signal	In P/N: 0v, Others: 5v
31	DB/PK	Low Speed Fan Control	Relay Off: 12v, On: 1v
32, 37-38	---	Not Used	---
33	TN/RD	Speed Control Vacuum Solenoid	Vacuum Increasing: 1v
34	DB/OR	A/C Clutch Relay Control	Relay Off: 12v, On: 1v
35	GY/YL	EGR Solenoid Control (California)	12v, at 55 mph: 1v
36	GY/PK	Maintenance Indicator Lamp	Lamp Off: 12v, On: 1v
39	GY/RD	AIS Motor 3 Control	Pulse Signals
40	BR/WT	AIS Motor 1 Control	Pulse Signals
4	BK/DG	HO2S-11 (B1 S1) Signal	0.1-1.1v
42-44, 48-50, 56	---	Not Used	---
45	LG	SCI Receive	5v
46	WT/BK	CCD Bus (-)	<0.050v
47	WT/OR	Vehicle Speed Signal	Digital Signal
51	DB/YL	ASD Relay Power (B+)	Relay Off: 12v, On: 1v
52	PK/BK	EVAP Purge Solenoid Control	Solenoid Off: 12v, On: 1v
53	LG/RD	Speed Control Vent Solenoid	Vacuum Decreasing: 1v
54	OR/BK	Lockup Torque Converter	At Cruise w/TCC On: 1v
55	YL	High Speed Fan Control	Relay Off: 12v, On: 1v
57	DG/OR	ASD Relay Control	Relay Off: 12v, On: 1v
58	---	Not Used	---
59	PK/BK	AIS Motor 4 Control	Pulse Signals
60	YL/BK	AIS Motor 2 Control	Pulse Signals

Pin Connector Graphic

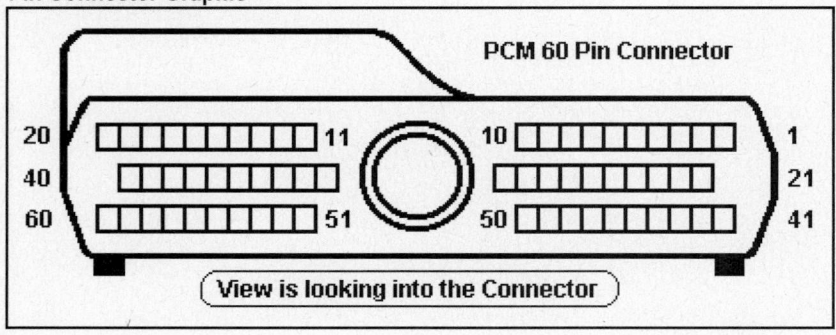

PCM 60 Pin Connector

View is looking into the Connector

1994-95 Caravan 2.5L I4 SOHC TBI VIN K (All) 60 Pin Connector

PCM Pin #	Wire Color	Circuit Description (60 Pin)	Value at Hot Idle
1	DG/RD	MAP Sensor Signal	1.5-1.6v
2	TN/BK	ECT Sensor Signal	At 180ºF: 2.80v
3	RD/WT	Keep Alive Power (B+)	12-14v
4	BK/LB	Sensor Ground	<0.050v
5	BK/WT	Sensor Ground	<0.050v
6	PK/WT	5-Volt Supply	4.9-5.1v
7	OR	8-Volt Supply	7.9-8.1v
8	WT	Ignition Switch (Start) Output	12-14v
9	DB/WT	Ignition Switch Power (B+)	12-14v
10, 13-15	---	Not Used	---
11	BK/TN	Power Ground	<0.1v
12	BK/TN	Power Ground	<0.1v
16	LB	Injector Control Driver	1-4 ms
17-18, 21, 28	---	Not Used	---
19	BK/GY	Coil Driver Control	5º, 55 mph: 8º dwell
20	DG	Alternator Field Control	Digital Signals: 0-12-0v
22	DB	TP Sensor Signal	0.6-1.0v
23	RD	Speed Control Set Switch Signal	S/C & Set Switch On: 3.8v
24	GY/BK	Distributor Reference Signal	Digital Signals: 0-5-0v
25	PK	SCI Transmit	0v
26	PK/BR	CCD Bus (+)	Digital Signals: 0-5-0v
27	BR	A/C Damped Pressure Switch	A/C Off: 12v, On: 1v
29	WT/PK	Brake Switch Sense Signal	Brake Off: 0v, On: 12v
30	BR/YL	PNP Switch Sense Signal	In P/N: 0v, Others: 5v
31	DB/PK	Low Speed Fan Control	Relay Off: 12v, On: 1v
32	---	Not Used	---
33	TN/RD	Speed Control Vacuum Solenoid	Vacuum Increasing: 1v
34	DB/OR	A/C Clutch Relay Control	Relay Off: 12v, On: 1v
35-36	---	Not Used	---
39	GY/RD	AIS Motor 3 Control	Pulse Signals
40	BR/WT	AIS Motor 1 Control	Pulse Signals
41	DG/WT	HO2S-11 (B1 S1) Signal	0.1-1.1v
42-44	---	Not Used	---
45	LG	SCI Receive	5v
46	WT/BK	CCD Bus (-)	<0.050v
47	WT/OR	Vehicle Speed Signal	Digital Signal
48-50	---	Not Used	---
51	DB/YL	ASD Relay Power (B+)	Relay Off: 12v, On: 1v
52	PK/BK	EVAP Purge Solenoid Control	Solenoid Off: 12v, On: 1v
53	LG/RD	Speed Control Vent Solenoid	Vacuum Decreasing: 1v
54	OR/BK	Lockup Torque Converter	At Cruise w/TCC On: 1v
55	YL	High Speed Fan Control	Relay Off: 12v, On: 1v
56, 58	---	Not Used	---
57	DG/OR	ASD Relay Control	Relay Off: 12v, On: 1v
59	PK/BK	AIS Motor 4 Control	Pulse Signals
60	YL/BK	AIS Motor 2 Control	Pulse Signals

Pin Connector Graphic

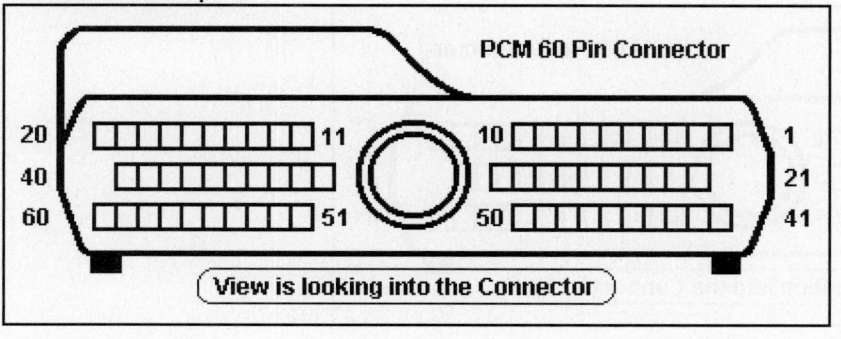

1990-91 Caravan 3.0L V6 SOHC MFI VIN 3 (A/T) 60 Pin Connector

PCM Pin #	Wire Color	Circuit Description (60 Pin)	Value at Hot Idle
1	DG/RD	MAP Sensor Signal	1.5-1.6v
2	TN/WT	ECT Sensor Signal	At 180°F: 2.80v
3	RD/WT	Keep Alive Power (B+)	12-14v
4	BK/LB	Sensor Ground	<0.050v
5	BK/WT	Sensor Ground	<0.050v
6	PK/WT	5-Volt Supply	4.9-5.1v
7	OR	9-Volt Supply	8.9-9.1v
8	DG/BK	Ignition Switch (Start) Output	12-14v
9	DB	Ignition Switch Power (B+)	12-14v
10, 13, 21	---	Not Used	---
11, 12	BK/TN	Power Ground	<0.1v
14	YL/WT	Injector 3 Driver Control	1-4 ms
15	TN	Injector 2 Driver Control	1-4 ms
16	WT/DB	Injector 1 Driver Control	1-4 ms
19	BK/GY	Coil Driver Control	5°, 55 mph: 8° dwell
20	DG	Alternator Field Control	Digital Signals: 0-12-0v
22	OR/DB	TP Sensor Signal	0.6-1.0v
23, 28	---	Not Used	---
24	GY/BK	Distributor Reference Signal	Digital Signals: 0-5-0v
25, 45	PK, LG	SCI Transmit, SCI Receive	0v, 5v
26	PK/BR	CCD Bus (+)	Digital Signals: 0-5-0v
27	BR	A/C Switch Signal	A/C Off: 12v, On: 1v
29	WT/PK	Brake Switch Sense Signal	Brake Off: 0v, On: 12v
30	BR/YL	PNP Switch Sense Signal	In P/N: 0v, Others: 5v
31	DB/PK	Radiator Fan Control Relay	Relay Off: 12v, On: 1v
32	BK/PK	MIL (lamp) Control	MIL Off: 12v, On: 1v
33	TN/RD	Speed Control Vacuum Solenoid	Vacuum Increasing: 1v
34	DB/OR	A/C WOT Relay Control	Relay Off: 12v, On: 1v
35-38, 42	---	Not Used	---
39	GY/RD	AIS Motor 3 Control	Pulse Signals
40	OR/WT	AIS Motor 1 Control	Pulse Signals
41	BK/DG	HO2S-11 (B1 S1) Signal	0.1-1.1v
43	GY/LB	Tachometer Signal	Pulse Signals
44	TN/YL	Distributor Sync Signal	Digital Signals: 0-5-0v
46	WT/BK	CCD Bus (-)	<0.050v
47	WT/OR	Vehicle Speed Signal	Digital Signal
48	BR/RD	Speed Control Set Switch Signal	S/C & Set Switch On: 3.8v
49	YL/RD	Speed Control On/Off Switch	S/C On: 1v, Off: 12v
50	WT/LG	Speed Control Resume Switch	S/C On: 1v, Off: 12v
51	DB/YL	ASD Relay Control	Relay Off: 12v, On: 1v
52	PK/BK	EVAP Purge Solenoid Control	Solenoid Off: 12v, On: 1v
53	LG/RD	Speed Control Vent Solenoid	Vacuum Decreasing: 1v
56	GY/PK	Maintenance Indicator Lamp	Lamp Off: 12v, On: 1v
57-58	---	Not Used	---
59	PK/BK	AIS Motor 4 Control	Pulse Signals
60	YL/BK	AIS Motor 2 Control	Pulse Signals

Pin Connector Graphic

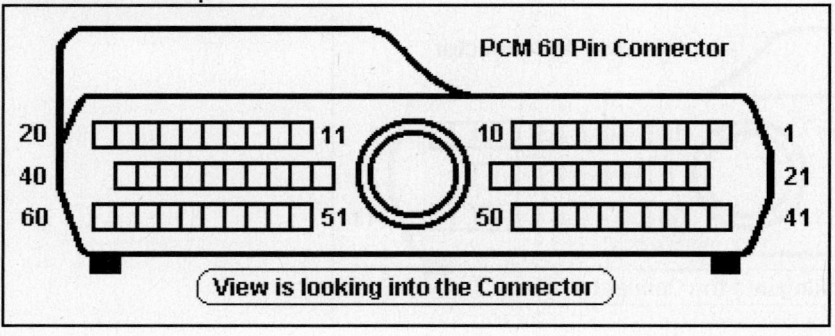

1992 Caravan 3.0L V6 SOHC MFI VIN 3 (A/T) 60 Pin Connector

PCM Pin #	Wire Color	Circuit Description (60 Pin)	Value at Hot Idle
1	DG/RD	MAP Sensor Signal	1.5-1.6v
2	TN/BK	ECT Sensor Signal	At 180ºF: 2.80v
3	RD/WT	Keep Alive Power (B+)	12-14v
4, 5	BK, BK/WT	Sensor Ground	<0.050v
6	PK/WT	5-Volt Supply	4.9-5.1v
7	OR	9-Volt Supply	8.9-9.1v
8, 10, 17-18, 21	---	Not Used	---
9	DB	Ignition Switch Power (B+)	12-14v
11, 12	BK/TN	Power Ground	<0.1v
13	LB/BR	Injector 4 Driver Control	1-4 ms
14	YL/WT	Injector 3 Driver Control	1-4 ms
15	TN	Injector 2 Driver Control	1-4 ms
16	WT/DB	Injector 1 Driver Control	1-4 ms
19	GY	Coil Driver Control	5º, 55 mph: 8º dwell
20	DG	Alternator Field Control	Digital Signals: 0-12-0v
22	OR/DB	TP Sensor Signal	0.6-1.0v
23	RD/LG	Speed Control Set Switch Signal	S/C & Set Switch On: 3.8v
24	GY/BK	Distributor Reference Signal	Digital Signals: 0-5-0v
25	PK	SCI Transmit	0v
26	PK/BR	CCD Bus (+)	Digital Signals: 0-5-0v
27	BR	A/C Damped Pressure Switch	A/C Off: 12v, On: 1v
28, 35-36	---	Not Used	---
29	WT/PK	Brake Switch Sense Signal	Brake Off: 0v, On: 12v
30	BR/YL	PNP Switch Sense Signal	In P/N: 0v, Others: 5v
31	DB/PK	Radiator Fan Control Relay	Relay Off: 12v, On: 1v
32	YL	High Speed Fan Control	Relay Off: 12v, On: 1v
33	TN/RD	Speed Control Vacuum Solenoid	Vacuum Increasing: 1v
34	DB/OR	A/C Clutch Relay Control	Relay Off: 12v, On: 1v
37, 42-43	---	Not Used	---
38	GY	Injector 5 Driver Control	1-4 ms
39	GY/RD	AIS Motor Control (open)	Pulse Signals
40	BR/WT	AIS Motor Control (close)	Pulse Signals
41	BK/DG	HO2S-11 (B1 S1) Signal	0.1-1.1v
44	TN/YL	Distributor Sync Signal	Digital Signals: 0-5-0v
45	LG	SCI Receive	5v
46	WT/BK	CCD Bus (-)	<0.050v
47	WT/OR	Vehicle Speed Signal	Digital Signal
48-50, 54-55	---	Not Used	---
51	DB/YL	ASD Relay Control	Relay Off: 12v, On: 1v
52	PK/BK	EVAP Purge Solenoid Control	Solenoid Off: 12v, On: 1v
53	LG/RD	Speed Control Vent Solenoid	Vacuum Decreasing: 1v
56	GY/PK	Maintenance Indicator Lamp	Lamp Off: 12v, On: 1v
57	DG/OR	ASD Relay Power (B+)	12-14v
58	BR/DB	Injector 6 Driver Control	1-4 ms
59	PK/BK	AIS Motor Control (close)	Pulse Signals
60	YL/BK	AIS Motor Control (close)	Pulse Signals

Pin Connector Graphic

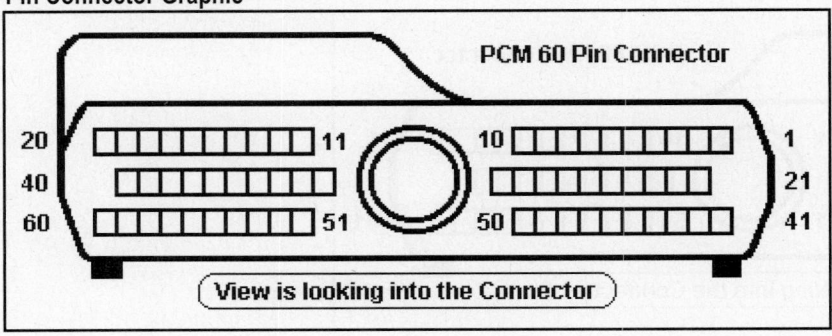

1993 Caravan 3.0L V6 SOHC MFI VIN 3 (A/T) 60 Pin Connector

PCM Pin #	Wire Color	Circuit Description (60 Pin)	Value at Hot Idle
1	DG/RD	MAP Sensor Signal	1.5-1.6v
2	TN/BK	ECT Sensor Signal	At 180°F: 2.80v
3	RD/WT	Keep Alive Power (B+)	12-14v
4, 5	BK, BK/WT	Sensor Ground	<0.050v
6	PK/WT	5-Volt Supply	4.9-5.1v
7	OR	9-Volt Supply	8.9-9.1v
8, 10, 17-18, 21	---	Not Used	---
9	DB	Ignition Switch Power (B+)	12-14v
11, 12	BK/TN	Power Ground	<0.1v
13	LB/BR	Injector 4 Driver Control	1-4 ms
14	YL/WT	Injector 3 Driver Control	1-4 ms
15	TN	Injector 2 Driver Control	1-4 ms
16	WT/DB	Injector 1 Driver Control	1-4 ms
19	BK/GY	Coil Driver Control	5°, 55 mph: 8° dwell
20	DG	Alternator Field Control	Digital Signals: 0-12-0v
22	OR/DB	TP Sensor Signal	0.6-1.0v
23	RD/LG	Speed Control Set Switch Signal	S/C & Set Switch On: 3.8v
24	GY/BK	Distributor Reference Signal	Digital Signals: 0-5-0v
25, 45	PK, LG	SCI Transmit, SCI Receive	0v, 5v
26, 46	PK, WT/BK	CCD Bus (+), CCD Bus (-)	Digital Signals: 0-5-0v, <0.050v
27	BR	A/C Damped Pressure Switch	A/C Off: 12v, On: 1v
28	---	Not Used	---
29	WT/PK	Brake Switch Sense Signal	Brake Off: 0v, On: 12v
30	BR/YL	PNP Switch Sense Signal	In P/N: 0v, Others: 5v
31	DB/PK	Radiator Fan Control Relay	Relay Off: 12v, On: 1v
32	YL	High Speed Fan Control	Relay Off: 12v, On: 1v
33	TN/RD	Speed Control Vacuum Solenoid	Vacuum Increasing: 1v
34	DB/OR	A/C Clutch Relay Control	Relay Off: 12v, On: 1v
35	GY/YL	EGR Solenoid Control (Cal)	12v, at 55 mph: 1v
36	GY/PK	Maintenance Indicator Lamp	Lamp Off: 12v, On: 1v
37, 42-43	---	Not Used	---
38	GY	Injector 5 Driver Control	1-4 ms
39	GY/RD	AIS Motor Control (open)	Pulse Signals
40	BR/WT	AIS Motor Control (close)	Pulse Signals
41	BK/DG	HO2S-11 (B1 S1) Signal	0.1-1.1v
44	TN/YL	Distributor Sync Signal	Digital Signals: 0-5-0v
47	WT/OR	Vehicle Speed Signal	Digital Signal
48-50, 55	---	Not Used	---
51	DB/YL	ASD Relay Control	Relay Off: 12v, On: 1v
52	PK/BK	EVAP Purge Solenoid Control	Solenoid Off: 12v, On: 1v
53	LG/RD	Speed Control Vent Solenoid	Vacuum Decreasing: 1v
54	OR/BK	Lockup Torque Converter	At Cruise w/TCC On: 1v
57	DG/OR	ASD Relay Power (B+)	12-14v
58	BR/DB	Injector 6 Driver Control	1-4 ms
59	PK/BK	AIS Motor Control (close)	Pulse Signals
60	YL/BK	AIS Motor Control (close)	Pulse Signals

Pin Connector Graphic

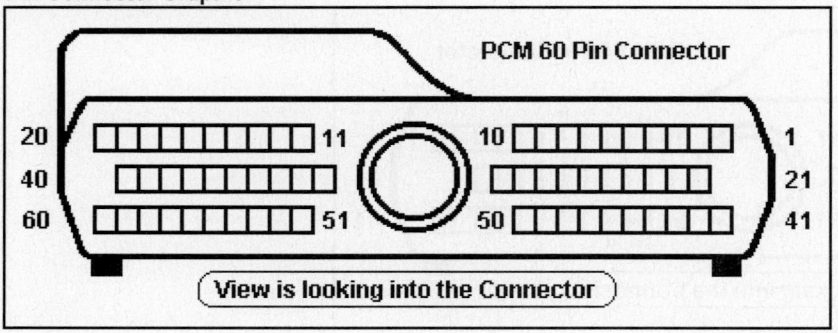

PCM 60 Pin Connector

View is looking into the Connector

1994-95 Caravan 3.0L V6 SOHC MFI VIN 3 (A/T) 60 Pin Connector

PCM Pin #	Wire Color	Circuit Description (60 Pin)	Value at Hot Idle
1	DG/RD	MAP Sensor Signal	1.5-1.6v
2	TN/BK	ECT Sensor Signal	At 180°F: 2.80v
3	RD/WT	Keep Alive Power (B+)	12-14v
4	BK/LB	Sensor Ground	<0.050v
5	BK/WT	Sensor Ground	<0.050v
6	PK/WT	5-Volt Supply	4.9-5.1v
7	OR	8-Volt Supply	7.9-8.1v
9	DB/WT	Ignition Switch Power (B+)	12-14v
10	---	Not Used	---
11	BK/TN	Power Ground	<0.1v
12	BK/TN	Power Ground	<0.1v
13	LB/BR	Injector 4 Driver Control	1-4 ms
14	YL/WT	Injector 3 Driver Control	1-4 ms
15	TN	Injector 2 Driver Control	1-4 ms
16	LB	Injector 1 Driver Control	1-4 ms
17-18	---	Not Used	
19	BK/GY	Coil Driver Control	5°, 55 mph: 8° dwell
20	DG	Alternator Field Control	Digital Signals: 0-12-0v
21	---	Not Used	---
22	DB	TP Sensor Signal	0.6-1.0v
23	RD	Speed Control Set Switch Signal	S/C & Set Switch On: 3.8v
24	GY/P	Distributor Reference Signal	Digital Signals: 0-5-0v
25	PK	SCI Transmit	0v
26	PK/BR	CCD Bus (+)	Digital Signals: 0-5-0v
27	BR	A/C Damped Pressure Switch	A/C Off: 12v, On: 1v
28	---	Not Used	---
29	WT/PK	Brake Switch Sense Signal	Brake Off: 0v, On: 12v
30	BK/OR	PNP Switch Sense Signal	In P/N: 0v, Others: 5v
31	DB/PK	Low Speed Fan Control	Relay Off: 12v, On: 1v
32	---	Not Used	---
33	TN/RD	Speed Control Vacuum Solenoid	Vacuum Increasing: 1v
34	DB/OR	A/C Clutch Relay Control	Relay Off: 12v, On: 1v
35	GY/YL	EGR Solenoid Control	Solenoid Off: 12v, On: 1v
36-37	---	Not Used	---
38	GY	Injector 5 Driver Control	1-4 ms
39	GY/RD	AIS Motor Control (open)	Pulse Signals
40	BR/WT	AIS Motor Control (close)	Pulse Signals

Standard Colors and Abbreviations

Abbreviation	Color	Abbreviation	Color	Abbreviation	Color
BK	Black	GY	Gray	RD	Red
BL	Blue	GN	Green	TN	Tan
BR	Brown	LG	Light Green	VT	Violet
DB	Dark Blue	OR	Orange	WT	White
DG	Dark Green	PK	Pink	YL	Yellow

1994-95 Caravan 3.0L V6 SOHC MFI VIN 3 (A/T) 60 Pin Connector

PCM Pin #	Wire Color	Circuit Description (60 Pin)	Value at Hot Idle
41	DG/WT	HO2S-11 (B1 S1) Signal	0.1-1.1v
42-43	---	Not Used	---
44	TN/YL	Distributor Sync Signal	Digital Signals: 0-5-0v
45	LG	SCI Receive	5v
46	WT/BK	CCD Bus (-)	<0.050v
47	WT/OR	Vehicle Speed Signal	Digital Signal
48-50	---	Not Used	---
51	DB/YL	ASD Relay Control	Relay Off: 12v, On: 1v
52	PK/BK	EVAP Purge Solenoid Control	Solenoid Off: 12v, On: 1v
53	LG/RD	Speed Control Vent Solenoid	Vacuum Decreasing: 1v
54	OR/BK	Lockup Torque Converter	At Cruise w/TCC On: 1v
55	YL	High Speed Fan Control	Relay Off: 12v, On: 1v
56	---	Not Used	---
57	DG/OR	ASD Relay Power (B+)	12-14v
58	BR/DB	Injector 6 Driver Control	1-4 ms
59	PK/BK	AIS Motor Control (close)	Pulse Signals
60	YL/BK	AIS Motor Control (close)	Pulse Signals

Pin Connector Graphic

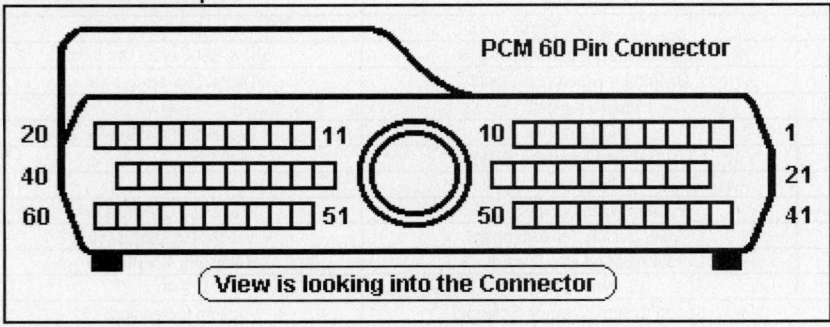

PCM 60 Pin Connector

20 · 11 · 10 · 1
40 · · · 21
60 · 51 · 50 · 41

View is looking into the Connector

1996 Caravan 3.0L V6 SOHC MFI VIN 3 (A/T) 'C1' Black Connector

PCM Pin #	Wire Color	Circuit Description (40 Pin)	Value at Hot Idle
1-3	---	Not Used	---
4	DG	Generator Field Driver	Digital Signals: 0-12-0v
5	YL/RD	Speed Control Power Supply	12-14v
6	DG/OR	ASD Relay Power (B+)	12-14v
7	YL/WT	Injector 3 Driver	1-4 ms
8	TN	Starter Relay Control	KOEC: 9-11v
9	---	Not Used	---
10	BK/TN	Power Ground	<0.1v
11	GY	Coil 1 Driver Control	5°, 55 mph: 8° dwell
12	---	Not Used	---
13	WT/DB	Injector 1 Driver	1-4 ms
14	BR/DB	Injector 6 Driver	1-4 ms
15	GY	Injector 5 Driver	1-4 ms
16	LB/BR	Injector 4 Driver	1-4 ms
17	TN	Injector 2 Driver	1-4 ms
18-19	---	Not Used	---
20	WT/BK	Ignition Switch Power (B+)	12-14v
21-25	---	Not Used	---
26	TN/BK	ECT Sensor Signal	At 180°F: 2.80v
27-29	---	Not Used	---
30	BK/DG	HO2S-11 (B1 S1) Signal	0.1-1.1v
31	---	Not Used	---
32	GY/BK	CKP Sensor Signal	Digital Signals: 0-5-0v
33	TN/YL	CMP Sensor Signal	Digital Signals: 0-5-0v
34	---	Not Used	---
35	OR/DB	TP Sensor Signal	0.6-1.0v
36	DG/RD	MAP Sensor Signal	1.5-1.6v
37-39	---	Not Used	---
40	GY/YL	EGR Solenoid Control	12v, at 55 mph: 1v

Pin Connector Graphic

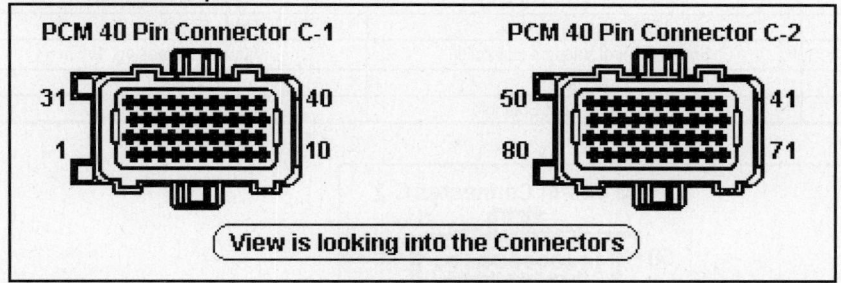

1996 Caravan 3.0L V6 SOHC MFI VIN 3 (A/T) 'C2' White Connector

PCM Pin #	Wire Color	Circuit Description (40 Pin)	Value at Hot Idle
41	RD/LG	Speed Control Set Switch Signal	S/C & Set Switch On: 3.8v
42	DB	A/C Pressure Switch Signal	A/C On: 0.451-4.850v
43	BK/LB	Sensor Ground	<0.050v
44	OR	8-Volt Supply	7.9-8.1v
45	DB	A/C Switch Sense Signal	A/C Off: 12v, On: 1v
46	RD/WT	Keep Alive Power (B+)	12-14v
47	---	Not Used	---
48	BR/WT	IAC 1 Driver Control	Pulse Signals
49	YL/BK	IAC 2 Driver Control	Pulse Signals
50	BK/TN	Power Ground	<0.1v
51	TN/WT	HO2S-12 (B1 S2) Signal	0.1-1.1v
52-55	---	Not Used	---
56	OR/BK	Lockup Torque Converter	At Cruise w/TCC On: 1v
57	GY/RD	IAC 3 Driver Control	Pulse Signals
58	PK/BK	IAC 4 Driver Control	Pulse Signals
59	PK/BR	CCD Bus (+)	Digital Signals: 0-5-0v
60	WT/BK	CCD Bus (-)	<0.050v
61	PK/WT	5-Volt Supply	4.9-5.1v
62	WT/PK	Brake Switch Sense Signal	Brake Off: 0v, On: 12v
63	YL/DG	Torque Management Request	Digital Signals
64	DB/OR	A/C Clutch Relay Control	Relay Off: 12v, On: 1v
65	PK	SCI Transmit Signal	0v
66	WT/OR	Vehicle Speed Signal	Digital Signal
67	DB/YL	ASD Relay Control	Relay Off: 12v, On: 1v
68	PK/BK	EVAP Purge Solenoid Control	PWM Signal: 0-12-0v
69-71	---	Not Used	---
72	YL/BK	LDP Switch Sense Signal	LDP Switch Closed: 0v
73	LG/DG	Radiator Fan Control Relay	Relay Off: 12v, On: 1v
74	BR	Fuel Pump Relay Control	Relay Off: 12v, On: 1v
75	LG	SCI Receive Signal	5v
76	BK/WT	PNP Switch Sense (EATX)	In P/N: 0v, Others: 5v
77	WT/DG	LDP Solenoid Control	PWM Signal: 0-12-0v
78	TN/RD	Speed Control Vacuum Solenoid	Vacuum Increasing: 1v
79	---	Not Used	---
80	LG/RD	Speed Control Vent Solenoid	Vacuum Decreasing: 1v

Pin Connector Graphic

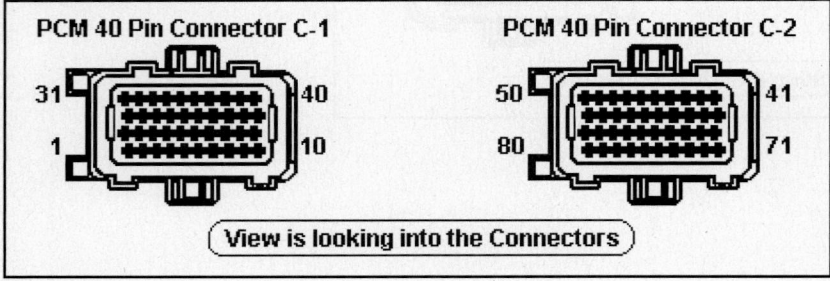

1997-98 Caravan 3.0L V6 SOHC MFI VIN 3 'C1' Black Connector

PCM Pin #	Wire Color	Circuit Description (40 Pin)	Value at Hot Idle
1-3	---	Not Used	---
4	DG	Generator Field Driver	Digital Signals: 0-12-0v
5	YL/RD	Speed Control Power Supply	12-14v
6	DG/OR	ASD Relay Power (B+)	12-14v
7	YL/WT	Injector 3 Driver	1-4 ms
8	TN	Starter Relay Control	KOEC: 9-11v
9	---	Not Used	---
10	BK/TN	Power Ground	<0.1v
11	GY	Coil 1 Driver Control	5°, 55 mph: 8° dwell
12	---	Not Used	---
13	WT/DB	Injector 1 Driver	1-4 ms
14	BR/DB	Injector 6 Driver	1-4 ms
15	GY	Injector 5 Driver	1-4 ms
16	LB/BR	Injector 4 Driver	1-4 ms
17	TN	Injector 2 Driver	1-4 ms
18-19	---	Not Used	---
20	WT/BK	Ignition Switch Power (B+)	12-14v
21	---	Not Used	---
22	BK/PK	MIL (lamp) Control	MIL Off: 12v, On: 1v
23-25	---	Not Used	---
26	TN/BK	ECT Sensor Signal	At 180°F: 2.80v
27-29	---	Not Used	---
30	BK/DG	HO2S-11 (B1 S1) Signal	0.1-1.1v
31	---	Not Used	---
32	GY/BK	CKP Sensor Signal	Digital Signals: 0-5-0v
33	TN/YL	CMP Sensor Signal	Digital Signals: 0-5-0v
34	---	Not Used	---
35	OR/DB	TP Sensor Signal	0.6-1.0v
36	DG/RD	MAP Sensor Signal	1.5-1.6v
37-39	---	Not Used	---
40	GY/YL	EGR Solenoid Control	12v, at 55 mph: 1v

Pin Connector Graphic

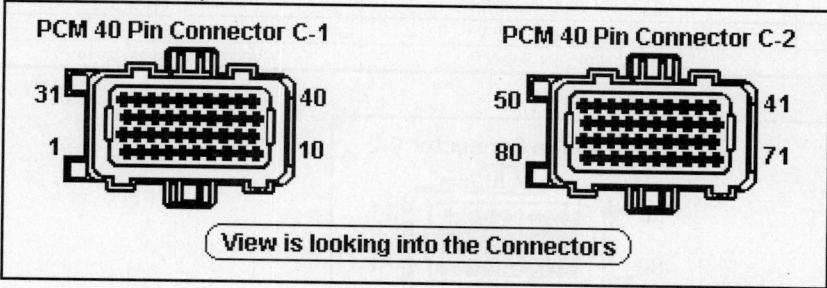

PCM 40 Pin Connector C-1

PCM 40 Pin Connector C-2

View is looking into the Connectors

1997-98 Caravan 3.0L V6 SOHC MFI VIN 3 'C2' White Connector

PCM Pin #	Wire Color	Circuit Description (40 Pin)	Value at Hot Idle
41	RD/LG	Speed Control Set Switch Signal	S/C & Set Switch On: 3.8v
42	DB	A/C Pressure Switch Signal	A/C On: 0.451-4.850v
43	BK/LB	Sensor Ground	<0.050v
44	OR	8-Volt Supply	7.9-8.1v
45	DG	A/C Switch Sense Signal	A/C Off: 12v, On: 1v
46	RD/WT	Keep Alive Power (B+)	12-14v
47	---	Not Used	---
48	BR/WT	IAC 3 Driver Control	Pulse Signals
49	YL/BK	IAC 2 Driver Control	Pulse Signals
50	BK/TN	Power Ground	<0.1v
51	TN/WT	HO2S-12 (B1 S2) Signal	0.1-1.1v
52-55	---	Not Used	---
56	OR/BK	Lockup Torque Converter	At Cruise w/TCC On: 1v
57	GY/RD	IAC 1 Driver Control	Pulse Signals
58	PK/BK	IAC 4 Driver Control	Pulse Signals
59	PK/BR	CCD Bus (+)	Digital Signals: 0-5-0v
60	WT/BK	CCD Bus (-)	<0.050v
61	PK/WT	5-Volt Supply	4.9-5.1v
62	WT/PK	Brake Switch Sense Signal	Brake Off: 0v, On: 12v
63	YL/DG	Torque Management Request	Digital Signals
64	DB/OR	A/C Clutch Relay Control	Relay Off: 12v, On: 1v
65	PK	SCI Transmit Signal	0v
66	WT/OR	Vehicle Speed Signal	Digital Signal
67	DB/YL	ASD Relay Control	Relay Off: 12v, On: 1v
68	PK/BK	EVAP Purge Solenoid Control	PWM Signal: 0-12-0v
69-71	---	Not Used	---
72	YL/BK	LDP Switch Sense Signal	LDP Switch Closed: 0v
73	LG/DB	Radiator Fan Control Relay	Relay Off: 12v, On: 1v
74	BR	Fuel Pump Relay Control	Relay Off: 12v, On: 1v
75	LG	SCI Receive Signal	5v
76	BK/WT	PNP Switch Sense (EATX)	In P/N: 0v, Others: 5v
77	WT/DG	LDP Solenoid Control	PWM Signal: 0-12-0v
78	TN/RD	Speed Control Vacuum Solenoid	Vacuum Increasing: 1v
79	---	Not Used	---
80	LG/RD	Speed Control Vent Solenoid	Vacuum Decreasing: 1v

Pin Connector Graphic

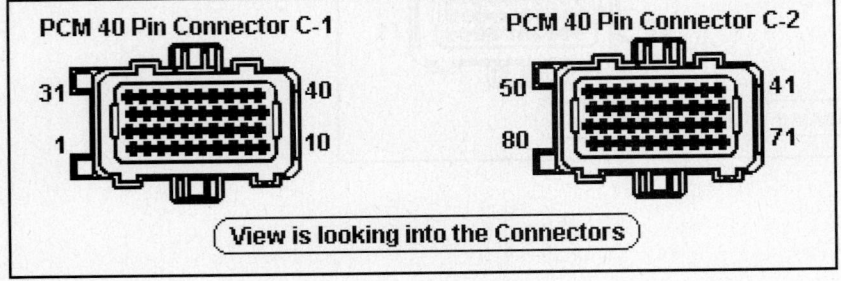

PCM 40 Pin Connector C-1

31 | 40
1 | 10

PCM 40 Pin Connector C-2

50 | 41
80 | 71

View is looking into the Connectors

1999-2000 Caravan 3.0L V6 SOHC MFI VIN 3 'C1' Black Connector

PCM Pin #	Wire Color	Circuit Description (40 Pin)	Value at Hot Idle
1	---	Not Used	---
2	RD/YL	Coil 3 Driver Control	5°, 55 mph: 8° dwell
3	DB/TN	Coil 2 Driver Control	5°, 55 mph: 8° dwell
4 ('99)	DG	Generator Field Driver	Digital Signals: 0-12-0v
5	YL/RD	Speed Control Power Supply	12-14v
6	DG/OR	ASD Relay Power (B+)	12-14v
7	YL/WT	Injector 3 Driver	1-4 ms
8 ('99)	TN	Smart Start Relay Control	KOEC: 9-11v
8	DG	Generator Field Driver	Digital Signals: 0-12-0v
9, 12	---	Not Used	---
10	BK/TN	Power Ground	<0.1v
11	GY/RD	Coil 1 Driver Control	5°, 55 mph: 8° dwell
13	WT/DB	Injector 1 Driver	1-4 ms
14	BR/DB	Injector 6 Driver	1-4 ms
15	GY	Injector 5 Driver	1-4 ms
16	LB/BR	Injector 4 Driver	1-4 ms
17	TN/WT	Injector 2 Driver	1-4 ms
18-19	---	Not Used	---
20	WT/BK	Ignition Switch Power (B+)	12-14v
21	---	Not Used	---
22	BK/PK	MIL (lamp) Control	MIL Off: 12v, On: 1v
23-24	---	Not Used	---
25	DB/LG	Knock Sensor Signal	0.080v AC
26	TN/BK	ECT Sensor Signal	At 180°F: 2.80v
27	BK/OR	HO2S Ground	<0.050v
28-29	---	Not Used	---
30	BK/DG	HO2S-11 (B1 S1) Signal	0.1-1.1v
31	---	Not Used	---
32	GY/BK	CKP Sensor Signal	Digital Signals: 0-5-0v
33	TN/YL	CMP Sensor Signal	Digital Signals: 0-5-0v
34	---	Not Used	---
35	OR/DB	TP Sensor Signal	0.6-1.0v
36	DG/RD	MAP Sensor Signal	1.5-1.6v
37-39	---	Not Used	---
40	GY/YL	EGR Solenoid Control	12v, at 55 mph: 1v

Pin Connector Graphic

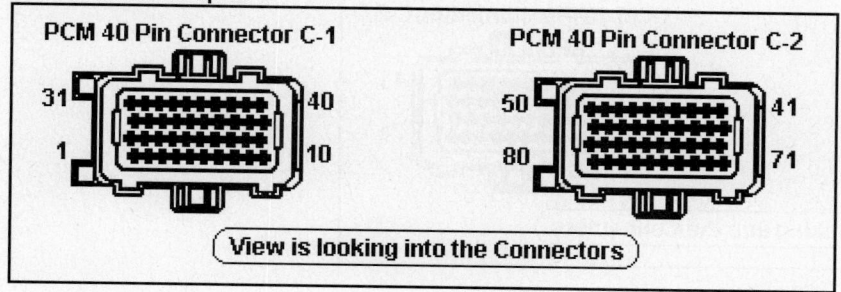

PCM 40 Pin Connector C-1

31   40
1   10

PCM 40 Pin Connector C-2

50   41
80   71

View is looking into the Connectors

1999-2000 Caravan 3.0L V6 SOHC MFI VIN 3 'C2' White Connector

PCM Pin #	Wire Color	Circuit Description (40 Pin)	Value at Hot Idle
41	RD/LG	Speed Control Set Switch Signal	S/C & Set Switch On: 3.8v
42	DB	A/C Pressure Switch Signal	A/C On: 0.451-4.850v
43	BK/LB	Sensor Ground	<0.050v
44	OR	8-Volt Supply	7.9-8.1v
45 ('99)	DG/LB	A/C Switch Sense Signal	A/C Off: 12v, On: 1v
46	RD/WT	Keep Alive Power (B+)	12-14v
47	---	Not Used	---
48	BR/WT	IAC 3 Driver Control	Pulse Signals
49	YL/BK	IAC 2 Driver Control	Pulse Signals
50	BK/TN	Power Ground	<0.1v
51	TN/WT	HO2S-12 (B1 S2) Signal	0.1-1.1v
52-56	---	Not Used	---
57	GY/RD	IAC 1 Driver Control	Pulse Signals
58	PK/BK	IAC 4 Driver Control	Pulse Signals
59	PK/BR	CCD Bus (+)	Digital Signals: 0-5-0v
60	WT/BK	CCD Bus (-)	<0.050v
61	PK/WT	5-Volt Supply	4.9-5.1v
62	WT/PK	Brake Switch Sense Signal	Brake Off: 0v, On: 12v
63	YL/DG	Torque Management Request	Digital Signals
64	DB/OR	A/C Clutch Relay Control	Relay Off: 12v, On: 1v
65	PK	SCI Transmit Signal	0v
66	WT/OR	Vehicle Speed Signal	Digital Signal
67	DB/YL	ASD Relay Control	Relay Off: 12v, On: 1v
68	PK/BK	EVAP Purge Solenoid Control	PWM Signal: 0-12-0v
69-71	---	Not Used	---
72 ('99)	YL/BK	LDP Switch Sense Signal	LDP Switch Closed: 0v
73	LG/DB	Radiator Fan Control Relay	Relay Off: 12v, On: 1v
74	BR	Fuel Pump Relay Control	Relay Off: 12v, On: 1v
75	LG	SCI Receive Signal	5v
76	BR/YL	PNP Switch Sense (EATX)	In P/N: 0v, Others: 5v
77 ('99)	WT/DG	LDP Solenoid Control	PWM Signal: 0-12-0v
78	TN/RD	Speed Control Vacuum Solenoid	Vacuum Increasing: 1v
79	---	Not Used	---
80	LG/RD	Speed Control Vent Solenoid	Vacuum Decreasing: 1v

Pin Connector Graphic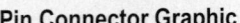

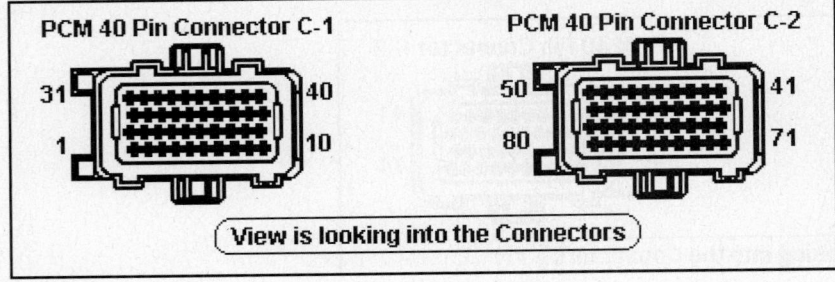

2002 Caravan 3.0L V6 OHC Flexible Fuel VIN 3 'C1' Black Connector

PCM Pin #	Wire Color	Circuit Description (40 Pin)	Value at Hot Idle
1	---	Not Used	---
2	DB/OR	Coil 3 Driver Control	5°, 55 mph: 8° dwell
3	DB/TN	Coil 2 Driver Control	5°, 55 mph: 8° dwell
4	---	Not Used	---
5	VT/YL	Speed Control Power Supply	12-14v
6	BR/WT	ASD Relay Power (B+)	12-14v
7	BR/LB	Injector 3 Driver	1-4 ms
8	BR/GY	Generator Field Driver	Digital Signals: 0-12-0v
9	---	Not Used	---
10	BK/BR	Power Ground	<0.1v
11	DB/DG	Coil 1 Driver Control	5°, 55 mph: 8° dwell
12	VT/GY	Engine Oil Pressure Sensor	1.6v at 24 psi
13	BR/YL	Injector 1 Driver	1-4 ms
14	BR/VT	Injector 6 Driver	1-4 ms
15	BR/OR	Injector 5 Driver	1-4 ms
16	BR/TN	Injector 4 Driver	1-4 ms
17	BR/DB	Injector 2 Driver	1-4 ms
18	BR/LG	HO2S-11 (B1 S1) Heater Control	Heater Off: 12v, On: 1v
19	---	Not Used	---
20	PK/GY	Ignition Switch Power (B+)	12-14v
21-24	---	Not Used	---
25	DB/YL	Knock Sensor Signal	0.080v AC
26	VT/OR	ECT Sensor Signal	At 180°F: 2.80v
27	BR/DG	Oxygen Sensor Ground (Both)	<0.050v
28-29	---	Not Used	---
30	DB/LB	HO2S-11 (B1 S1) Signal	0.1-1.1v
31	DG/OR	Starter Motor Relay Control	Relay Off: 12v, On: 1v
32	BR/LB	CKP Sensor Signal	Digital Signals: 0-5-0v
33	DB/GY	CMP Sensor Signal	Digital Signals: 0-5-0v
34	---	Not Used	---
35	BR/OR	TP Sensor Signal	0.6-1.0v
36	VT/BR	MAP Sensor Signal	1.5-1.6v
37	DB/LG	Intake Air Temperature Sensor	At 100°F: 1.83v
38-39	---	Not Used	---
40	DB/VT	EGR Solenoid Control	12v, at 55 mph: 1v

Pin Connector Graphic

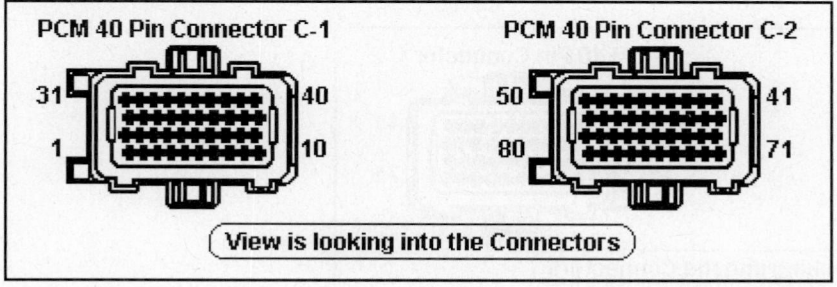

PCM 40 Pin Connector C-1

31　40
1　10

PCM 40 Pin Connector C-2

50　41
80　71

View is looking into the Connectors

2002 Caravan 3.3L V6 OHC Flexible Fuel VIN 3 'C2' White Connector

PCM Pin #	Wire Color	Circuit Description (40 Pin)	Value at Hot Idle
41	VT	Speed Control Switch Signal	S/C & Set Switch On: 3.8v
42	LB/BR	A/C Pressure Sensor	A/C On: 0.451-4.850v
43	DB/DG	Sensor Ground	<0.050v
44	BR/PK	8-Volt Supply	7.9-8.1v
45	---	Not Used	---
46	OR/RD	Keep Alive Power (B+)	12-14v
47-48	---	Not Used	
49	VT/DG	IAC 1 Driver Control	Pulse Signals
50	BK/DG	Power Ground	<0.1v
51	DB/YL	HO2S-12 (B1 S2) Signal	0.1-1.1v
52-55	---	Not Used	---
56	VT/YL	Speed Control Vacuum Solenoid	Vacuum Increasing: 1v
57	VT/LG	IAC 2 Driver Control	Pulse Signals
58	---	Not Used	---
59	WT/VT	PCI Data Bus (J1850)	Digital Signals: 0-7-0v
60	---	Not Used	---
61	PK/YL	5-Volt Supply	4.9-5.1v
62	DG/WT	Secondary Brake Switch Sense	Brake Off: 0v, On: 12v
63	DG/LG	Torque Management Request	Digital Signals
64	LB/OR	A/C Clutch Relay Control	Relay Off: 12v, On: 1v
65	WT/BR	SCI Transmit	0v
66	DB/OR	Vehicle Speed Signal	Digital Signals
67	BR/WT	ASD Relay Control	Relay Off: 12v, On: 1v
68	DB/WT	EVAP Purge Solenoid Control	PWM Signal: 0-12-0v
69	---	Not Used	---
70	DB/BR	EVAP Purge Sense	<0.1v
71	---	Not Used	---
72	VT/WT	LDP Switch Sense Signal	LDP Switch Closed: 0v
73	BR/VT	Radiator Fan Control Relay	Relay Off: 12v, On: 1v
74	BR	Fuel Pump Relay Control	Relay Off: 12v, On: 1v
75	WT/LG	SCI Receive	5v
76	YL/DB	PNP Switch Sense (TRS T41)	In P/N: 0v, Others: 5v
77	VT/LB	LDP Solenoid Control	PWM Signal: 0-12-0v
78	DB/WT	Torque Converter Clutch Solenoid Control	Solenoid Off: 12v, On: PWM 0-12-0v
79	---	Not Used	---
80	VT/OR	Speed Control Vent Solenoid	Vacuum Decreasing: 1v

Pin Connector Graphic

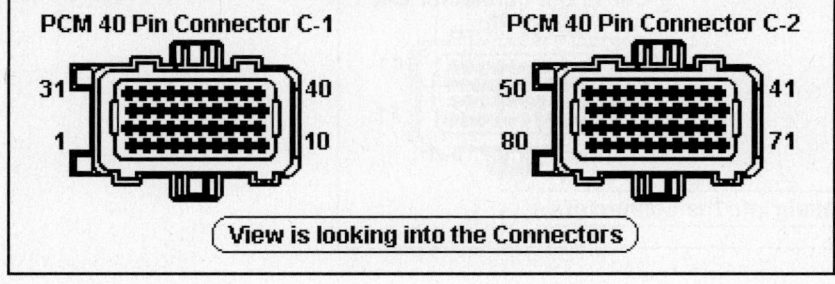

2003 Caravan 3.3L V6 OHC MFI VIN 3 'C1' Black Connector

PCM Pin #	Wire Color	Circuit Description (38 Pin)	Value at Hot Idle
1-8, 10	---	Not Used	---
9	BK/BR	Power Ground	<0.1v
11	PK/GY	Ignition Switch Output (Run-Start)	12-14v
12	PK/WT	Ignition Switch Output (Off-Run-Start)	12-14v
13-17, 19	---	Not Used	---
18	BK/DG	Power Ground	<0.1v
20	VT/GY	Oil Pressure Sensor	A/C On: 0.45-4.85v
21, 23-24	---	Not Used	---
22	VT/LG	Ambient Air Temperature Sensor	At 100ºF: 1.83v
25	WT/LG	SCI Receive (PCM)	5v
26	WT/BR	Flash Program Enable	5v
27-28	---	Not Used	---
29	OR/RD	Battery Power (B+)	12-14v
30	YL	Ignition Switch Output (Start)	8-11v (cranking)
31	DB/YL	HO2S-12 (B1 S2) Signal	0.1-1.1v
32	DB/DG	Oxygen Sensor Return (down)	<0.050v
33-35	---	Not Used	---
36	WT/BR	SCI Transmit (PCM)	0v
37	DG/YL	SCI Transmit (TCM)	0v
38	WT/VT	PCI Data Bus (J1850)	Digital Signals: 0-7-0v

2003 Caravan 3.3L V6 OHC MFI VIN 3 'C2' Gray Connector

PCM Pin #	Wire Color	Circuit Description (38 Pin)	Value at Hot Idle
1-8	---	Not Used	---
9	DB/TN	Coil 2 Driver	5º, at 55 mph: 8º dwell
10	DB/DG	Coil 1 Driver	5º, at 55 mph: 8º dwell
11	BR/TN	Injector 4 Driver Control	1.0-4.0 ms
12	BR/LB	Injector 3 Driver Control	1.0-4.0 ms
13	BR/DB	Injector 2 Driver Control	1.0-4.0 ms
14	BR/YL	Injector 1 Driver Control	1.0-4.0 ms
15-17	---	Not Used	---
18	BR/LG	HO2S-12 (B1 S2) Heater Control	Heater Off: 12v, On: 1v
19	BR/GY	Generator Field Driver	Digital Signal: 0-12-0v
20	VT/OR	ECT Sensor Signal	At 180ºF: 2.80v
21	BR/OR	TP Sensor Signal	0.6-1.0v
22	---	Not Used	---
23	VT/BR	MAP Sensor Signal	1.5-1.7v
24	BR/LG	Knock Sensor Return	<0.050v
25	DB/YL	Knock Sensor Signal	0.080v AC
26	---	Not Used	---
27	DB/DG	Sensor Ground	<0.1v
28	BR/VT	IAC Motor Sense	12-14v
29	PK/YL	5-Volt Supply	4.9-5.1v
30	DB/LG	IAT Sensor Signal	At 100ºF: 1.83v
31	DB/LB	HO2S-11 (B1 S1) Signal	0.1-1.1v
32	BR/DG	HO2S-11 (B1 S1) Ground (Upstream)	<0.1v
33	---	Not Used	---
34	DB/GY	CMP Sensor Signal	Digital Signal: 0-5-0v
35	BR/LB	CKP Sensor Signal	Digital Signal: 0-5-0v
36-37	---	Not Used	---
38	VT/GY	IAC Motor Driver	DC pulse signals: 0.8-11v

Pin Connector Graphic

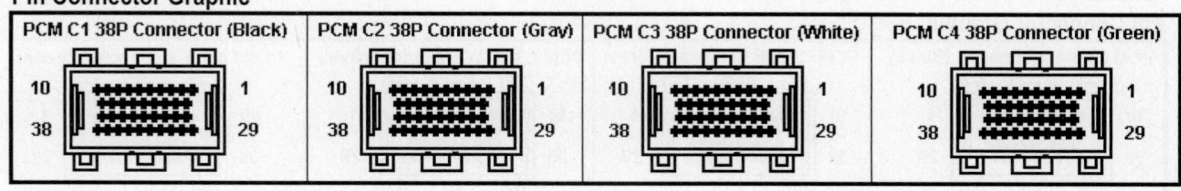

2003 Caravan 3.3L V6 OHC MFI VIN 3 'C3' White Connector

PCM Pin #	Wire Color	Circuit Description (38 Pin)	Value at Hot Idle
1-2, 4	---	Not Used	---
3	BR/WT	ASD Relay Control	Relay Off: 12v, On: 1v
5	VT/OR	Speed Control Vent Solenoid	Vacuum Decreasing: 1v
6	BR/VT	Radiator Fan Relay Control	Relay Off: 12v, On: 1v
7	VT/YL	Speed Control Power Supply	12-14v
8	WT/DG	Natural Vacuum Leak Detection Solenoid	Solenoid Off: 12v, On: 1v
9	BR/WT	HO2S-12 (B1 S2) Heater Control	Heater On: 1v, Off: 12v
10	---	Not Used	---
11	LB/OR	A/C Clutch Relay Control	Relay Off: 12v, On: 1v
12	VT/YL	Speed Control Vacuum Solenoid	Vacuum Increasing: 1v
13-18	---	Not Used	---
18, 19	BR/WT	Automatic Shutdown Relay Output	12-14v
20	DB/WT	EVAP Purge Solenoid Control	PWM Signal: 0-12-0v
21-22	---	Not Used	---
23	DG/WT	Brake Switch Sense	Brake Off: 0v, On: 12v
24-27	---	Not Used	---
28	BR/WT	Automatic Shutdown Relay Output	12-14v
29	DB/BR	EVAP Purge Solenoid Sense	0-1v
30-31, 33, 36	---	Not Used	---
31	LB/BR	A/C Pressure Sensor	A/C On: 0.451-4.850v
32	DB/YL	Battery Temperature Sensor	At 100ºF: 1.83v
34	VT	Speed Control Set Switch Signal	S/C & Set Switch On: 3.8v
35	VT/WT	Natural Vacuum Leak Detection Switch Sense	0.1v
37	BR	Fuel Pump Relay Control	Relay Off: 12v, On: 1v
38	DG/OR	Starter Relay Control	KOEC: 9-11v

2003 Caravan 3.3L V6 OHC MFI VIN 3 'C4' Green Connector

PCM Pin #	Wire Color	Circuit Description (38 Pin)	Value at Hot Idle
1	YL/GY	Overdrive Solenoid Control	Solenoid Off: 12v, On: 1v
2	YL/LB	Underdrive Solenoid Control	Solenoid Off: 12v, On: 1v
3-5	---	Not Used	---
6	YL/LB	2-4 Solenoid Control	Solenoid Off: 12v, On: 1v
7-9, 11	---	Not Used	---
10	DG/WT	Low/Reverse Solenoid Control	Solenoid Off: 12v, On: 1v
12, 13	BK/LG	Power Ground	<0.050v
14	---	Not Used	---
15	DG/LB	TRS T1 Sense	<0.050v
16	DG/DB	TRS T3 Sense	<0.050v
17, 20-21	---	Not Used	---
18	YL/BR	Transmission Control Relay Control	Relay Off: 12v, On: 1v
19	YL/OR, 38	Transmission Control Relay Output	Relay Off: 12v, On: 1v
23-26	---	Not Used	---
27	DG/GY	TRS T41 Sense	<0.050v
28, 38	YL/OR	Transmission Control Relay Output	Relay Off: 12v, On: 1v
29	YL/TN	Low/Reverse Pressure Switch Sense	12-14v
30	YL/DG	2-4 Pressure Switch Sense	In Low/Reverse: 2-4v
31, 36	---	Not Used	---
32	DG/BR	Output Speed Sensor Signal	In 2-4 Position: 2-4v
33	DG/WT	Input Speed Sensor Signal	Moving: AC voltage
34	DG/VT	Speed Sensor Ground	Moving: AC voltage
35	DG/OR	Transmission Temperature Sensor Signal	<0.050v
37	DG/YL	TRS T42 Sense	In PRNL: 0v, Others 5v

Pin Connector Graphic

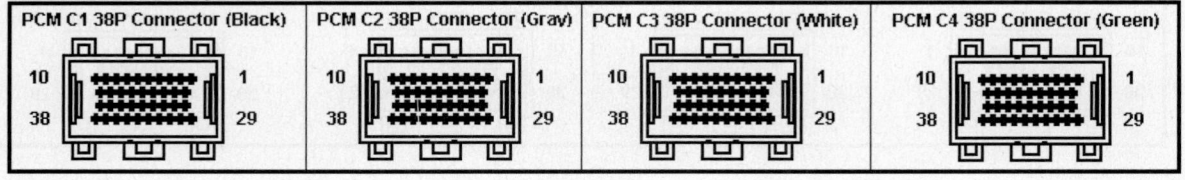

1994-95 Caravan 3.3L V6 CNG VIN J (A/T) 60 Pin Connector

PCM Pin #	Wire Color	Circuit Description (60 Pin)	Value at Hot Idle
1	DG/RD	MAP Sensor Signal	1.5-1.6v
2	TN/BK	ECT Sensor Signal	At 180°F: 2.80v
3	RD/WT	Keep Alive Power (B+)	12-14v
4, 5	BK/LB, BK	Sensor Ground	<0.050v
6	PK/WT	5-Volt Supply	4.9-5.1v
7	OR	8-Volt Supply	7.9-8.1v
8, 10	---	Not Used	---
9	DB/WT	Ignition Switch Power (B+)	12-14v
11, 12	BK/TN	Power Ground	<0.1v
13	PK or LB/BR	Injector 4 Driver Control	1-4 ms
14	PK or YL/WT	Injector 3 Driver Control	1-4 ms
15	PK/DB or Tan	Injector 2 Driver Control	1-4 ms
16	PK or LB	Injector 1 Driver Control	1-4 ms
17	DB/TN	Coil 2 Driver Control	5°, 55 mph: 8° dwell
18	DB/GY	Coil 3 Driver Control	5°, 55 mph: 8° dwell
19	BK/GY	Coil 1 Driver Control	5°, 55 mph: 8° dwell
20	DG	Alternator Field Control	Digital Signals: 0-12-0v
21	TN/PK	Fuel Temperature Sensor	0.5-4.5v
22	DB	TP Sensor Signal	0.6-1.0v
23	RD	Speed Control Set Switch Signal	S/C & Set Switch On: 3.8v
24	GY/P	CKP Sensor Signal	Digital Signals: 0-5-0v
25	PK, LG	SCI Transmit, SCI Receive	0v, 5v
26, 46	PK, WT/BK	CCD Bus (+), CCD Bus (-) Signal	Digital Signals: 0-5-0v, <0.050v
27	BR	A/C Switch Sense Signal	A/C Off: 12v, On: 1v
28, 32, 35-37, 43	---	Not Used	
29	WT/PK	Brake Switch Sense Signal	Brake Off: 0v, On: 12v
30	BK/OR	PNP Switch Sense Signal	In P/N: 0v, Others: 5v
31	DB/PK	Low Speed Fan Control	Relay Off: 12v, On: 1v
33	TN/RD	Speed Control Vacuum Solenoid	Vacuum Increasing: 1v
34	DB/OR	A/C Clutch Relay Control	Relay Off: 12v, On: 1v
38	PK or GY	Injector 5 Driver Control	1-4 ms
39	GY/RD	AIS Motor 3 Control	Pulse Signals
40	BR/WT	AIS Motor 1 Control	Pulse Signals
41	DG/WT	HO2S-11 (B1 S1) Signal	0.1-1.1v
42	DB/LB	Fuel Low Pressure Sensor	0.5-4.5v
44	TN/YL	CMP Sensor Signal	Digital Signals: 0-5-0v
47	WT/OR	Vehicle Speed Signal	Digital Signals
48-50, 52, 54, 56	---	Not Used	---
51	DB/YL	ASD Relay Control	Relay Off: 12v, On: 1v
53	LG/RD	Speed Control Vent Solenoid	Vacuum Decreasing: 1v
55	YL	High Speed Fan Control	Relay Off: 12v, On: 1v
57	DG/OR	ASD Relay Power (B+)	12-14v
58	PK/TN	Injector 6 Driver Control	1-4 ms
58 ('95)	BR/DB	Injector 6 Driver Control	1-4 ms
59	PK/BK	AIS Motor 4 Control	Pulse Signals
60	YL/BK	AIS Motor 2 Control	Pulse Signals

Pin Connector Graphic

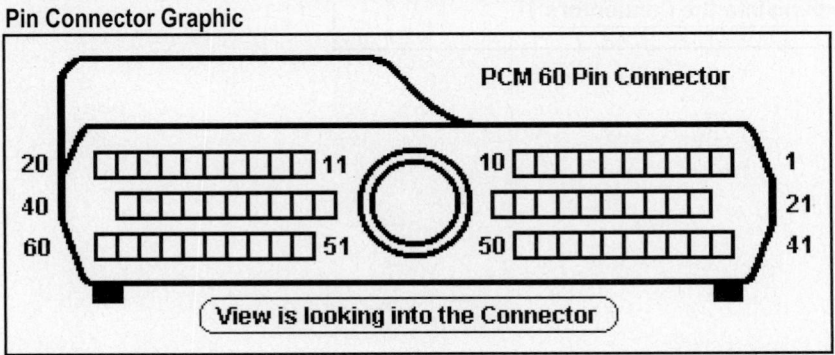

1998-2000 Caravan 3.3L V6 F/F VIN G (A/T) 'C1' Black Connector

PCM Pin #	Wire Color	Circuit Description (40 Pin)	Value at Hot Idle
1	---	Not Used	---
2	RD/YL	Coil 3 Driver (Cyl 3 & 6)	5°, 55 mph: 8° dwell
3	DB/TN	Coil 2 Driver (Cyl 2 & 5)	5°, 55 mph: 8° dwell
4	---	Not Used	
5	YL/RD	Speed Control Power Supply	12-14v
6	DG/OR	ASD Relay Power (B+)	12-14v
7	YL/WT	Injector 3 Driver	1-4 ms
8	DG	Generator Field Driver	Digital Signals: 0-12-0v
9	---	Not Used	---
10	BK/TN	Power Ground	<0.1v
11	GY/RD	Coil 1 Driver (Cyl 1 & 4)	5°, 55 mph: 8° dwell
12	---	Not Used	---
13	WT/DB	Injector 1 Driver	1-4 ms
14	BR/DB	Injector 6 Driver	1-4 ms
15	GY	Injector 5 Driver	1-4 ms
16	LB/BR	Injector 4 Driver	1-4 ms
17	TN/WT	Injector 2 Driver	1-4 ms
18-19, 21	---	Not Used	---
20	WT/BK	Ignition Switch Power (B+)	12-14v
22	BK/PK	MIL (lamp) Control	MIL Off: 12v, On: 1v
23-24	---	Not Used	
25	DB/LG	Knock Sensor Signal	0.080v AC
26	TN/BK	ECT Sensor Signal	At 180°F: 2.80v
27	BK/OR	Oxygen Sensor Ground	<0.050v
28-29	---	Not Used	---
30	BK/DG	HO2S-11 (B1 S1) Signal	0.1-1.1v
31	TN	Smart Start Relay Control	KOEC: 9-11v
32	GY/BK	CKP Sensor Signal	Digital Signals: 0-5-0v
33	TN/YL	CMP Sensor Signal	Digital Signals: 0-5-0v
34, 37	---	Not Used	---
35	OR/DB	TP Sensor Signal	0.6-1.0v
36	DG/RD	MAP Sensor Signal	1.5-1.6v
38 ('99)	DG	A/C Switch Sense Signal	A/C Off: 12v, On: 1v
38	DG/LB	A/C Switch Sense Signal	A/C Off: 12v, On: 1v
39	---	Not Used	
40	GY/YL	EGR Solenoid Control	12v, at 55 mph: 1v

Pin Connector Graphic

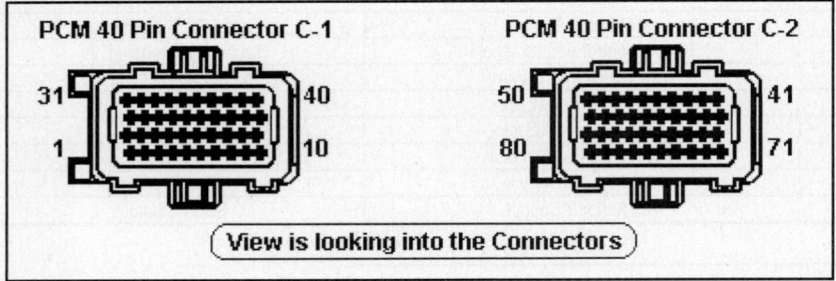

PCM 40 Pin Connector C-1

31 ... 40
1 ... 10

PCM 40 Pin Connector C-2

50 ... 41
80 ... 71

View is looking into the Connectors

1998-2000 Caravan 3.3L V6 F/F VIN G (A/T) 'C2' White Connector

PCM Pin #	Wire Color	Circuit Description (40 Pin)	Value at Hot Idle
41	RD/LG	Speed Control Set Switch Signal	S/C & Set Switch On: 3.8v
42	DB	A/C Pressure Sensor	A/C On: 0.451-4.850v
43	BK/LB	Sensor Ground	<0.050v
44	OR	8-Volt Supply	7.9-8.1v
45, 47	---	Not Used	---
46	RD/WT	Keep Alive Power (B+)	12-14v
48	BR/WT	IAC 3 Driver Control	Pulse Signals
49	YL/BK	IAC 2 Driver Control	Pulse Signals
50	BK/TN	Power Ground	<0.1v
51	TN/WT	HO2S-12 (B1 S2) Signal	0.1-1.1v
52-53	---	Not Used	---
54	YL/WT	Flex Fuel Sensor Signal	Digital Signals
56	TN/RD	Speed Control Vacuum Solenoid	Vacuum Increasing: 1v
57	GY/RD	IAC 1 Driver Control	Pulse Signals
58	PK/BK	IAC 4 Driver Control	Pulse Signals
59	PK/BR	CCD Bus (+)	Digital Signals: 0-5-0v
60	WT/BK	CCD Bus (-)	<0.050v
61	PK/WT	5-Volt Supply	4.9-5.1v
62	WT/PK	Brake Switch Sense Signal	Brake Off: 0v, On: 12v
63	YL/DG	Torque Management Request	Digital Signals
64	DB/OR	A/C Clutch Relay Control	Relay Off: 12v, On: 1v
65	PK	SCI Transmit Signal	0v
66	WT/OR	Vehicle Speed Signal	Digital Signal
67	DB/YL	ASD Relay Control	Relay Off: 12v, On: 1v
68	PK/BK	EVAP Purge Solenoid Control	PWM Signal: 0-12-0v
69	---	Not Used	---
70	P/R	EVAP Purge PWM Control	PWM Signal: 0-12-0v
71	---	Not Used	---
72	DB/WT	LDP Switch Sense Signal	LDP Switch Closed: 0v
73	LG/DB	Radiator Fan Control Relay	Relay Off: 12v, On: 1v
74	BR	Fuel Pump Relay Control	Relay Off: 12v, On: 1v
75	LG	SCI Receive Signal	5v
76	BK/WT	PNP Switch Sense (EATX)	In P/N: 0v, Others: 5v
77	WT/DG	LDP Solenoid Control	PWM Signal: 0-12-0v
78-79	---	Not Used	---
80	LG/RD	Speed Control Vent Solenoid	Vacuum Decreasing: 1v

Pin Connector Graphic

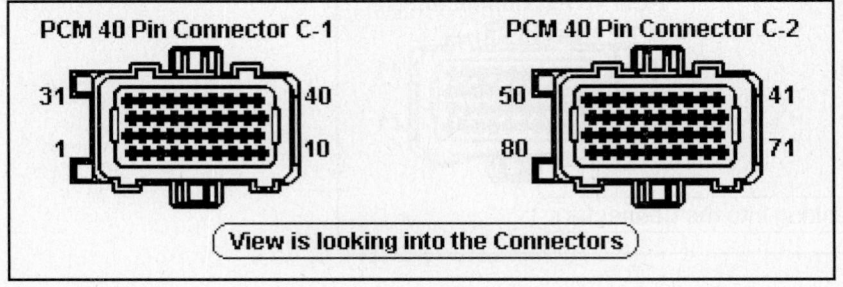

PCM 40 Pin Connector C-1

31   40

1   10

PCM 40 Pin Connector C-2

50   41

80   71

(View is looking into the Connectors)

2001 Caravan 3.3L V6 Flexible Fuel VIN G (A/T) 'C1' Black Connector

PCM Pin #	Wire Color	Circuit Description (40 Pin)	Value at Hot Idle
1	---	Not Used	---
2	DB/OR	Coil 3 Driver Control	5°, 55 mph: 8° dwell
3	DB/TN	Coil 2 Driver Control	5°, 55 mph: 8° dwell
4	---	Not Used	---
5	VT/YL	S/C On/Off Switch Sense	12-14v
6	BR/WT	ASD Relay Power (B+)	12-14v
7	BR/LB	Injector 3 Driver	1-4 ms
8	BR/GY	Generator Field Driver	Digital Signals: 0-12-0v
9	---	Not Used	---
10	BR/BR	Power Ground	<0.1v
11	DB/DG	Coil 1 Driver Control	5°, 55 mph: 8° dwell
12	VT/GY	Engine Oil Pressure Sensor	1.6v at 24 psi
13	BR/YL	Injector 1 Driver Control	1-4 ms
14	BR/VT	Injector 6 Driver Control	1-4 ms
15	BR/OR	Injector 5 Driver Control	1-4 ms
16	BR/TN	Injector 4 Driver Control	1-4 ms
17	BR/LB	Injector 2 Driver Control	1-4 ms
18	BR/LG	HO2S-11 (B1 S1) Heater	PWM Signal: 0-12-0v
19	---	Not Used	---
20	PK/GY	Ignition Switch Power (B+)	12-14v
21-24	---	Not Used	---
25	DB/YL	Knock Sensor Signal	0.080v AC
26	VT/OR	ECT Sensor Signal	At 180°F: 2.80v
27	BR/DG	Oxygen Sensor Ground	<0.050v
28-29	---	Not Used	---
30	DB/LB	HO2S-11 (B1 S1) Signal	0.1-1.1v
31	DG/OR	Double Start Override Signal	KOEC: 9-11v
32	BR/LB	CKP Sensor Signal	Digital Signals: 0-5-0v
33	DB/GY	CMP Sensor Signal	Digital Signals: 0-5-0v
34	---	Not Used	---
35	BR/OR	TP Sensor Signal	0.6-1.0v
36	VT/BR	MAP Sensor Signal	1.5-1.6v
37	DB/LG	Intake Air Temp. Signal	At 100°F: 1.83v
38-39	---	Not Used	---
40	DB/VT	EGR Solenoid Control	12v, at 55 mph: 1v

Pin Connector Graphic

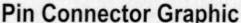

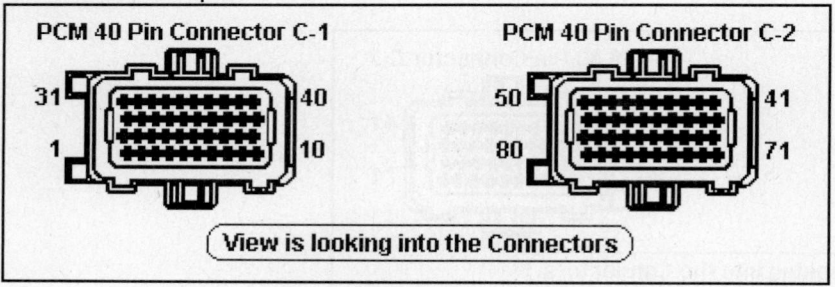

2001 Caravan 3.3L V6 Flexible Fuel VIN G (A/T) 'C2' White Connector

PCM Pin #	Wire Color	Circuit Description (40 Pin)	Value at Hot Idle
41	VT	Speed Control Set Switch Signal	S/C & Set Switch On: 3.8v
42	LB/BR	A/C Pressure Switch Signal	A/C On: 0.451-4.850v
43	DB/DG	Sensor Ground	<0.050v
44	BR/PK	8-Volt Supply	7.9-8.1v
45	---	Not Used	---
46	OR/RD	Keep Alive Power (B+)	12-14v
47-48	---	Not Used	---
49	VT/DG	IAC 1 Driver Control	Pulse Signals
50	BK/DG	Power Ground	<0.1v
51	DB/YL	HO2S-12 (B1 S2) Signal	0.1-1.1v
52-53	---	Not Used	---
54	YL/WT	Flex Fuel Sensor Signal	Digital Signals
56	VT/YL	Speed Control Vacuum Solenoid	Vacuum Increasing: 1v
57	VT/LG	IAC 2 Driver Control	Pulse Signals
58	---	Not Used	---
59	WT/VT	PCI Data Bus (J1850)	Digital Signals: 0-7-0v
60	---	Not Used	---
61	PK/YL	5-Volt Supply	4.9-5.1v
62	DG/WT	Brake Switch Sense Signal	Brake Off: 0v, On: 12v
63	DG/LG	Torque Management Request	Digital Signals
64	LB/OR	A/C Clutch Relay Control	Relay Off: 12v, On: 1v
65	WT/BR	SCI Transmit Signal	0v
66	DB/OR	Vehicle Speed Signal	Digital Signal
67	BR/WT	ASD Relay Control	Relay Off: 12v, On: 1v
68	DB/WT	EVAP Purge Solenoid Control	PWM Signal: 0-12-0v
69	---	Not Used	---
70	DB/BR	EVAP Purge Solenoid Sense	0-1v
71	---	Not Used	---
72	VT/WT	LDP Switch Sense Signal	LDP Switch Closed: 0v
73	BR/VT	Radiator Fan Control Relay	Relay Off: 12v, On: 1v
74	BR	Fuel Pump Relay Control	Relay Off: 12v, On: 1v
75	WT/LG	SCI Receive Signal	5v
76	YL/LB	TRS T41 Signal	In P/N: 0v, Others: 5v
77	VT/LB	LDP Solenoid Control	PWM Signal: 0-12-0v
78-79	---	Not Used	---
80	LG/RD	Speed Control Vent Solenoid	Vacuum Decreasing: 1v

Pin Connector Graphic

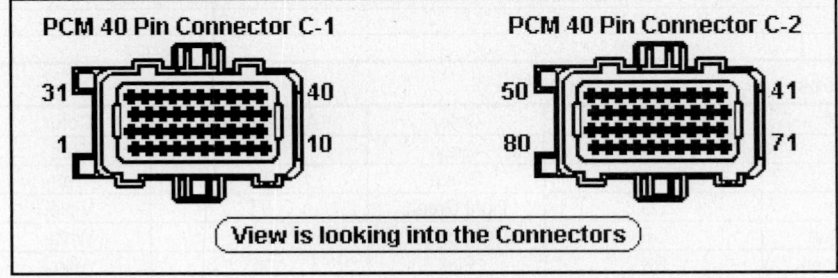

1990-91 Caravan 3.3L V6 MFI VIN R (A/T) 60 Pin Connector

PCM Pin #	Wire Color	Circuit Description (60 Pin)	Value at Hot Idle
1	DG/RD	MAP Sensor Signal	1.5-1.6v
2	TN/WT	ECT Sensor Signal	At 180°F: 2.80v
3	RD/WT	Keep Alive Power (B+)	12-14v
4	BK/LB	Sensor Ground	<0.050v
5	BK/WT	Sensor Ground	<0.050v
6	PK/WT	5-Volt Supply	4.9-5.1v
7	OR	9-Volt Supply	8.9-9.1v
8	DG/BK	Ignition Switch (Start) Output	12-14v
9	DB	Ignition Switch Power (B+)	12-14v
10	---	Not Used	---
11	LB/RD	Power Ground	<0.1v
12	LB/RD	Power Ground	<0.1v
13	---	Not Used	---
14	YL/WT	Injector 3 Driver Control	1-4 ms
15	TN	Injector 2 Driver Control	1-4 ms
16	WT/DB	Injector 1 Driver Control	1-4 ms
17	DB/BK	Coil 2 Driver Control	5°, 55 mph: 8° dwell
18	DB/GY	Coil 3 Driver Control	5°, 55 mph: 8° dwell
19	BK/GY	Coil 1 Driver Control	5°, 55 mph: 8° dwell
20	DG	Alternator Field Control	Digital Signals: 0-12-0v
21	BK/RD	Charge Temperature Sensor	At 100°F: 2.51v
22	OR/DB	TP Sensor Signal	0.6-1.0v
23	---	Not Used	---
24	GY/BK	CKP Sensor Signal	Digital Signals: 0-5-0v
25	PK	SCI Transmit	0v
26	PK/BK	CCD Bus (+)	Digital Signals: 0-5-0v
27	BR	A/C Clutch Signal	A/C Off: 12v, On: 1v
28	---	Not Used	---
29	WT/PK	Brake Switch Sense Signal	Brake Off: 0v, On: 12v
30	BR/YL	PNP Switch Sense Signal	In P/N: 0v, Others: 5v
31 ('90)	OR/PK	Radiator Fan Control Relay	Relay Off: 12v, On: 1v
31	DB/PK	Radiator Fan Control Relay	Relay Off: 12v, On: 1v
32	BK/PK	MIL (lamp) Control	MIL Off: 12v, On: 1v
33	TN/RD	Speed Control Vacuum Solenoid	Vacuum Increasing: 1v
34	DB/OR	A/C Clutch Relay Control	Relay Off: 12v, On: 1v
35	GY/YL	EGR Solenoid Control (Cal)	12v, at 55 mph: 1v
36-38	---	Not Used	---
39	GY/RD	AIS Motor 3 Control	Pulse Signals
40 ('90)	OR/WT	AIS Motor 1 Control	Pulse Signals
40	BR/WT	AIS Motor 1 Control	Pulse Signals

Standard Colors and Abbreviations

Abbreviation	Color	Abbreviation	Color	Abbreviation	Color
BK	Black	GY	Gray	RD	Red
BL	Blue	GN	Green	TN	Tan
BR	Brown	LG	Light Green	VT	Violet
DB	Dark Blue	OR	Orange	WT	White
DG	Dark Green	PK	Pink	YL	Yellow

1990-91 Caravan 3.3L V6 MFI VIN R (A/T) 60 Pin Connector

PCM Pin #	Wire Color	Circuit Description (60 Pin)	Value at Hot Idle
41	BK/DG	HO2S-11 (B1 S1) Signal	0.1-1.1v
42	BK/LG	Knock Sensor Signal	0.080v AC
43 ('90)	GY/LB	Tachometer Signal	Pulse Signals
44	TN/YL	CMP Sensor Signal	Digital Signals: 0-5-0v
45	LG	SCI Receive	5v
46	WT/BK	CCD Bus (-)	<0.050v
47	WT/OR	Vehicle Speed Signal	Digital Signal
48	BR/RD	Speed Control Set Switch Signal	S/C & Set Switch On: 3.8v
49	YL/RD	Speed Control On/Off Switch	S/C On: 1v, Off: 12v
50	WT/LG	Speed Control Resume Switch	S/C On: 1v, Off: 12v
51	DB/YL	ASD Relay Control	Relay Off: 12v, On: 1v
52	PK/BK	EVAP Purge Solenoid Control	Solenoid Off: 12v, On: 1v
53	LG/RD	Speed Control Vent Solenoid	Vacuum Decreasing: 1v
54-55	---	Not Used	---
56	GY/PK	Maintenance Indicator Lamp	Lamp Off: 12v, On: 1v
57-58	---	Not Used	---
59	PK/BK	AIS Motor 4 Control	Pulse Signals
60	YL/BK	AIS Motor 2 Control	Pulse Signals

Pin Connector Graphic

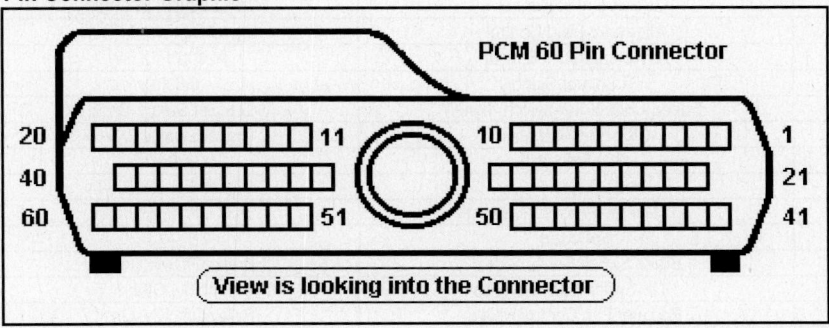

1992-93 Caravan 3.3L V6 MFI VIN R (A/T) 60 Pin Connector

PCM Pin #	Wire Color	Circuit Description (60 Pin)	Value at Hot Idle
1	DG/RD	MAP Sensor Signal	1.5-1.6v
2	TN/BK	ECT Sensor Signal	At 180°F: 2.80v
3	RD/WT	Keep Alive Power (B+)	12-14v
4	BK/LB	Sensor Ground	<0.050v
5	BK/WT	Sensor Ground	<0.050v
6	PK/WT	5-Volt Supply	4.9-5.1v
7	OR	9-Volt Supply	8.9-9.1v
8	---	Not Used	---
9	DB	Ignition Switch Power (B+)	12-14v
10	---	Not Used	---
11	BK/TN	Power Ground	<0.1v
12	BK/TN	Power Ground	<0.1v
13	LB/BR	Injector 4 Driver Control	1-4 ms
14	YL/WT	Injector 3 Driver Control	1-4 ms
15	TN	Injector 2 Driver Control	1-4 ms
16	WT/DB	Injector 1 Driver Control	1-4 ms
17	DB/YL	Coil 2 Driver Control	5°, 55 mph: 8° dwell
18	RD/YL	Coil 3 Driver Control	5°, 55 mph: 8° dwell
19	GY	Coil 1 Driver Control	5°, 55 mph: 8° dwell
19 ('93)	BK/GY	Coil 1 Driver Control	5°, 55 mph: 8° dwell
20	DG	Alternator Field Control	Digital Signals: 0-12-0v
21 ('92)	BK/RD	Air Temperature Sensor	At 100°F: 2.51v
22	OR/DB	TP Sensor Signal	0.6-1.0v
23	RD/LG	Speed Control Set Switch Signal	S/C & Set Switch On: 3.8v
24	GY/BK	CKP Sensor Signal	Digital Signals: 0-5-0v
25	PK	SCI Transmit	0v
26	PK/BR	CCD Bus (+)	Digital Signals: 0-5-0v
27	BR	A/C Damped Pressure Switch	A/C Off: 12v, On: 1v
28, 32	---	Not Used	---
29	WT/PK	Brake Switch Sense Signal	Brake Off: 0v, On: 12v
30	BR/YL	PNP Switch Sense Signal	In P/N: 0v, Others: 5v
31 ('92)	DB/PK	Radiator Fan Control Relay	Relay Off: 12v, On: 1v
31	DB/PK	Low Speed Fan Control	Relay Off: 12v, On: 1v
33	TN/RD	Speed Control Vacuum Solenoid	Vacuum Increasing: 1v
34	DB/OR	A/C Clutch Relay Control	Relay Off: 12v, On: 1v
35	GY/YL	EGR Solenoid Control	Solenoid Off: 12v, On: 1v
36-37	---	Not Used	---
38	GY	Injector 5 Driver Control	1-4 ms
39	GY/RD	AIS Motor Control (open)	Pulse Signals
40	BR/WT	AIS Motor Control (close)	Pulse Signals

Standard Colors and Abbreviations

Abbreviation	Color	Abbreviation	Color	Abbreviation	Color
BK	Black	GY	Gray	RD	Red
BL	Blue	GN	Green	TN	Tan
BR	Brown	LG	Light Green	VT	Violet
DB	Dark Blue	OR	Orange	WT	White
DG	Dark Green	PK	Pink	YL	Yellow

1992-93 Caravan 3.3L V6 MFI VIN R (A/T) 60 Pin Connector

PCM Pin #	Wire Color	Circuit Description (60 Pin)	Value at Hot Idle
41	BK/DG	HO2S-11 (B1 S1) Signal	0.1-1.1v
42-43	---	Not Used	---
44	TN/YL	CMP Sensor Signal	Digital Signals: 0-5-0v
45	LG	SCI Receive	5v
46	WT/BK	CCD Bus (-)	<0.050v
47	WT/OR	Vehicle Speed Signal	Digital Signal
48-50	---	Not Used	---
51	DB/YL	ASD Relay Control	Relay Off: 12v, On: 1v
52	PK/BK	EVAP Purge Solenoid Control	Solenoid Off: 12v, On: 1v
53	LG/RD	Speed Control Vent Solenoid	Vacuum Decreasing: 1v
54	---	Not Used	---
55	YL	High Speed Fan Control	Relay Off: 12v, On: 1v
56 ('92)	GY/PK	Maintenance Indicator Lamp	Lamp Off: 12v, On: 1v
57	DG/OR	ASD Relay Power (B+)	12-14v
58	BR/DB	Injector 6 Driver Control	1-4 ms
59	PK/BK	AIS Motor Control (close)	Pulse Signals
60	YL/BK	AIS Motor Control (close)	Pulse Signals

Pin Connector Graphic

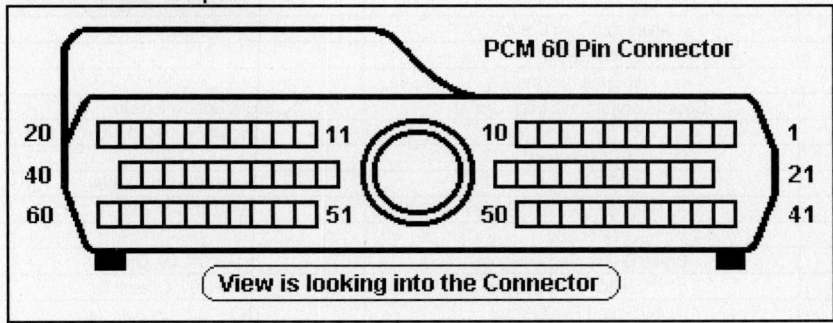

1994-95 Caravan 3.3L V6 MFI VIN R (A/T) 60 Pin Connector

PCM Pin #	Wire Color	Circuit Description (60 Pin)	Value at Hot Idle
1	DG/RD	MAP Sensor Signal	1.5-1.6v
2	TN/BK	ECT Sensor Signal	At 180ºF: 2.80v
3	RD/WT	Keep Alive Power (B+)	12-14v
4	BK/LB	Sensor Ground	<0.050v
5	BK/WT	Sensor Ground	<0.050v
6	PK/WT	5-Volt Supply	4.9-5.1v
7	OR	8-Volt Supply	7.9-8.1v
8	---	Not Used	---
9	DB/WT	Ignition Switch Power (B+)	12-14v
10	---	Not Used	---
11	BK/TN	Power Ground	<0.1v
12	BK/TN	Power Ground	<0.1v
13	LB/BR	Injector 4 Driver Control	1-4 ms
14	YL/WT	Injector 3 Driver Control	1-4 ms
15	TN	Injector 2 Driver Control	1-4 ms
16	LB	Injector 1 Driver Control	1-4 ms
17	DB/YL	Coil 2 Driver Control	5º, 55 mph: 8º dwell
18	DB/GY	Coil 3 Driver Control	5º, 55 mph: 8º dwell
19	BK/GY	Coil 1 Driver Control	5º, 55 mph: 8º dwell
20	DG	Alternator Field Control	Digital Signals: 0-12-0v
21	---	Not Used	---
22	DB	TP Sensor Signal	0.6-1.0v
23	RD	Speed Control Set Switch Signal	S/C & Set Switch On: 3.8v
24	GY/PK	CKP Sensor Signal	Digital Signals: 0-5-0v
25	PK	SCI Transmit	0v
26	PK/BR	CCD Bus (+)	Digital Signals: 0-5-0v
27	BR	A/C Damped Pressure Switch	A/C Off: 12v, On: 1v
28	---	Not Used	---
29	WT/PK	Brake Switch Sense Signal	Brake Off: 0v, On: 12v
30	BK/OR	PNP Switch Sense Signal	In P/N: 0v, Others: 5v
31	DB/PK	Low Speed Fan Control	Relay Off: 12v, On: 1v
32	---	Not Used	---
33	TN/RD	Speed Control Vacuum Solenoid	Vacuum Increasing: 1v
34	DB/OR	A/C Clutch Relay Control	Relay Off: 12v, On: 1v
35	GY/YL	EGR Solenoid Control	Solenoid Off: 12v, On: 1v
36-37	---	Not Used	---
38	GY	Injector 5 Driver Control	1-4 ms
39	GY/RD	AIS Motor Control (open)	Pulse Signals
40	BR/WT	AIS Motor Control (close)	Pulse Signals

Standard Colors and Abbreviations

Abbreviation	Color	Abbreviation	Color	Abbreviation	Color
BK	Black	GY	Gray	RD	Red
BL	Blue	GN	Green	TN	Tan
BR	Brown	LG	Light Green	VT	Violet
DB	Dark Blue	OR	Orange	WT	White
DG	Dark Green	PK	Pink	YL	Yellow

1994-95 Caravan 3.3L V6 MFI VIN R (A/T) 60 Pin Connector

PCM Pin #	Wire Color	Circuit Description (60 Pin)	Value at Hot Idle
41	DG/WT	HO2S-11 (B1 S1) Signal	0.1-1.1v
42-43	---	Not Used	---
44	TN/YL	CMP Sensor Signal	Digital Signals: 0-5-0v
45	LG	SCI Receive	5v
46	WT/BK	CCD Bus (-)	<0.050v
47	WT/OR	Vehicle Speed Signal	Digital Signal
48-50	---	Not Used	---
51	DB/YL	ASD Relay Control	Relay Off: 12v, On: 1v
52	PK/BK	EVAP Purge Solenoid Control	Solenoid Off: 12v, On: 1v
53	LG/RD	Speed Control Vent Solenoid	Vacuum Decreasing: 1v
54	---	Not Used	---
55	YL	High Speed Fan Control	Relay Off: 12v, On: 1v
56	---	Not Used	---
57	DG/OR	ASD Relay Power (B+)	12-14v
58	BR/DB	Injector 6 Driver Control	1-4 ms
59	PK/BK	AIS Motor Control (close)	Pulse Signals
60	YL/BK	AIS Motor Control (close)	Pulse Signals

Pin Connector Graphic

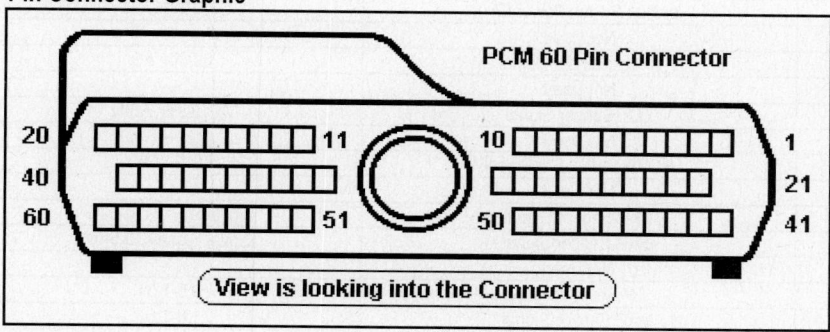

1996 Caravan 3.3L V6 MFI VIN R (A/T) 'C1' Black Connector

PCM Pin #	Wire Color	Circuit Description (40 Pin)	Value at Hot Idle
1	---	Not Used	---
2	RD/YL	Coil 3 Driver (Cyl 3 & 6)	5°, 55 mph: 8° dwell
3	DB/TN	Coil 2 Driver (Cyl 2 & 5)	5°, 55 mph: 8° dwell
4	DG	Generator Field Driver	Digital Signals: 0-12-0v
5	YL/RD	Speed Control Power Supply	12-14v
6	DG/OR	ASD Relay Power (B+)	12-14v
7	YL/WT	Injector 3 Driver	1-4 ms
8	TN	Smart Start Relay Control	KOEC: 9-11v
9	---	Not Used	---
10	BK/TN	Power Ground	<0.1v
11	GY	Coil 1 Driver (Cyl 1 & 4)	5°, 55 mph: 8° dwell
12	---	Not Used	
13	WT/DB	Injector 1 Driver	1-4 ms
14	BR/DB	Injector 6 Driver	1-4 ms
15	GY	Injector 5 Driver	1-4 ms
16	LB/BR	Injector 4 Driver	1-4 ms
17	TN	Injector 2 Driver	1-4 ms
18-19	---	Not Used	---
20	WT/BK	Ignition Switch Power (B+)	12-14v
21-23	---	Not Used	---
24	DB/LG	Knock Sensor Signal	0.080v AC
25	---	Not Used	---
26	TN/BK	ECT Sensor Signal	At 180°F: 2.80v
27-29	---	Not Used	---
30	BK/DG	HO2S-11 (B1 S1) Signal	0.1-1.1v
31	---	Not Used	---
32	GY/BK	CKP Sensor Signal	Digital Signals: 0-5-0v
33	TN/YL	CMP Sensor Signal	Digital Signals: 0-5-0v
34	---	Not Used	---
35	OR/DB	TP Sensor Signal	0.6-1.0v
36	DG/RD	MAP Sensor Signal	1.5-1.6v
37-39	---	Not Used	---
40	GY/YL	EGR Solenoid Control	12v, at 55 mph: 1v

Pin Connector Graphic

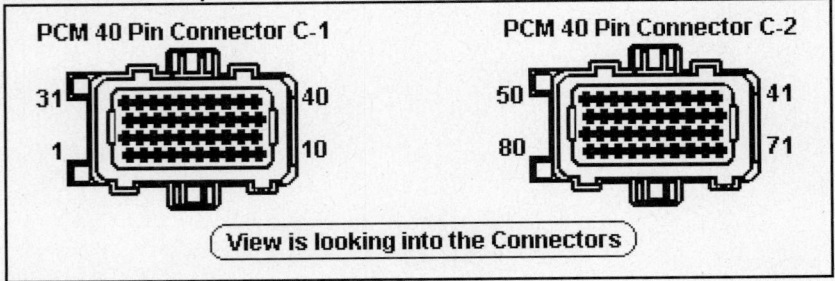

PCM 40 Pin Connector C-1

PCM 40 Pin Connector C-2

View is looking into the Connectors

1996 Caravan 3.3L V6 MFI VIN R (A/T) 'C2' White Connector

PCM Pin #	Wire Color	Circuit Description (40 Pin)	Value at Hot Idle
41	RD/LG	Speed Control Set Switch Signal	S/C & Set Switch On: 3.8v
42	DB	A/C Pressure Switch Signal	A/C On: 0.451-4.850v
43	BK/LB	Sensor Ground	<0.050v
44	OR	8-Volt Supply	7.9-8.1v
45	DB	A/C Switch Sense Signal	A/C Off: 12v, On: 1v
46	RD/WT	Keep Alive Power (B+)	12-14v
47	---	Not Used	---
48	BR/WT	IAC 3 Driver Control	Pulse Signals
49	YL/BK	IAC 2 Driver Control	Pulse Signals
50	BK/TN	Power Ground	<0.1v
52-56	---	Not Used	---
51	TN/WT	HO2S-12 (B1 S2) Signal	0.1-1.1v
57	GY/RD	IAC 1 Driver Control	Pulse Signals
58	PK/BK	IAC 4 Driver Control	Pulse Signals
59	PK/BR	CCD Bus (+)	Digital Signals: 0-5-0v
60	WT/BK	CCD Bus (-)	<0.050v
61	PK/WT	5-Volt Supply	4.9-5.1v
62	WT/PK	Brake Switch Sense Signal	Brake Off: 0v, On: 12v
63	YL/DG	Torque Management Request	Digital Signals
64	DB/OR	A/C Clutch Relay Control	Relay Off: 12v, On: 1v
65	PK	SCI Transmit Signal	0v
66	WT/OR	Vehicle Speed Signal	Digital Signal
67	DB/YL	ASD Relay Control	Relay Off: 12v, On: 1v
68	PK/BK	EVAP Purge Solenoid Control	PWM Signal: 0-12-0v
69-71	---	Not Used	---
72	YL/BK	LDP Switch Sense Signal	LDP Switch Closed: 0v
73	LG/DG	Radiator Fan Control Relay	Relay Off: 12v, On: 1v
74	BR	Fuel Pump Relay Control	Relay Off: 12v, On: 1v
75	LG	SCI Receive Signal	5v
76	BK/WT	PNP Switch Sense (EATX)	In P/N: 0v, Others: 5v
77	WT/DG	LDP Solenoid Control	PWM Signal: 0-12-0v
78	TN/RD	Speed Control Vacuum Solenoid	Vacuum Increasing: 1v
79	---	Not Used	---
80	LG/RD	Speed Control Vent Solenoid	Vacuum Decreasing: 1v

Pin Connector Graphic

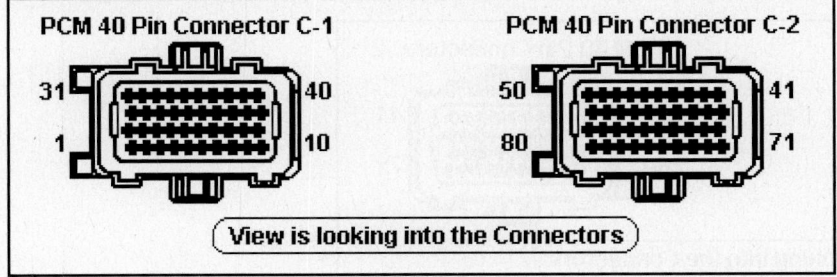

1997-98 Caravan 3.3L V6 MFI VIN R (A/T) 'C1' Black Connector

PCM Pin #	Wire Color	Circuit Description (40 Pin)	Value at Hot Idle
1	---	Not Used	---
2	RD/YL	Coil 3 Driver (Cyl 3 & 6)	5°, 55 mph: 8° dwell
3	DB/TN	Coil 2 Driver (Cyl 2 & 5)	5°, 55 mph: 8° dwell
4	DG	Generator Field Driver	Digital Signals: 0-12-0v
5	YL/RD	Speed Control Power Supply	12-14v
6	DG/OR	ASD Relay Power (B+)	12-14v
7	YL/WT	Injector 3 Driver	1-4 ms
8	TN	Smart Start Relay Control	KOEC: 9-11v
9	---	Not Used	---
10	BK/TN	Power Ground	<0.1v
11	GY/RD	Coil 1 Driver (Cyl 1 & 4)	5°, 55 mph: 8° dwell
12	---	Not Used	---
13	WT/DB	Injector 1 Driver	1-4 ms
14	BR/DB	Injector 6 Driver	1-4 ms
15	GY	Injector 5 Driver	1-4 ms
16	LB/BR	Injector 4 Driver	1-4 ms
17	TN	Injector 2 Driver	1-4 ms
18-19	---	Not Used	---
20	WT/BK	Ignition Switch Power (B+)	12-14v
21	---	Not Used	---
22	BK/PK	MIL (lamp) Control	MIL Off: 12v, On: 1v
23	---	Not Used	---
24	DB/LG	Knock Sensor Signal	0.080v AC
25	---	Not Used	---
26	TN/BK	ECT Sensor Signal	At 180°F: 2.80v
27-29	---	Not Used	---
30	BK/DG	HO2S-11 (B1 S1) Signal	0.1-1.1v
31	---	Not Used	---
32	GY/BK	CKP Sensor Signal	Digital Signals: 0-5-0v
33	TN/YL	CMP Sensor Signal	Digital Signals: 0-5-0v
34	---	Not Used	---
35	OR/DB	TP Sensor Signal	0.6-1.0v
36	DG/RD	MAP Sensor Signal	1.5-1.6v
37-39	---	Not Used	---
40	GY/YL	EGR Solenoid Control	12v, at 55 mph: 1v

Pin Connector Graphic

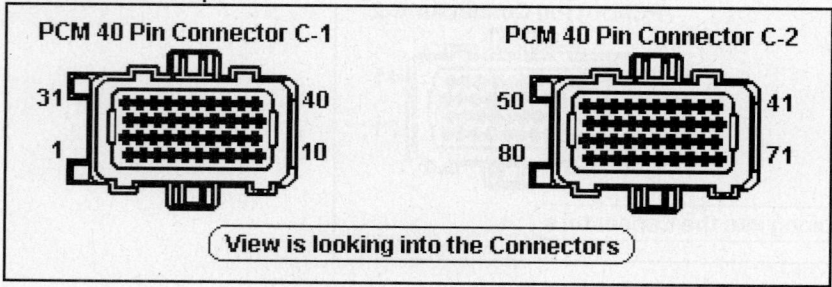

PCM 40 Pin Connector C-1

PCM 40 Pin Connector C-2

View is looking into the Connectors

1997-98 Caravan 3.3L V6 MFI VIN R (A/T) 'C2' White Connector

PCM Pin #	Wire Color	Circuit Description (40 Pin)	Value at Hot Idle
41	RD/LG	Speed Control Set Switch Signal	S/C & Set Switch On: 3.8v
42	DB	A/C Pressure Switch Signal	A/C On: 0.451-4.850v
43	BK/LB	Sensor Ground	<0.050v
44	OR	8-Volt Supply	7.9-8.1v
45	DG	A/C Switch Sense Signal	A/C Off: 12v, On: 1v
46	RD/WT	Keep Alive Power (B+)	12-14v
47	---	Not Used	---
48	BR/WT	IAC 3 Driver Control	Pulse Signals
49	YL/BK	IAC 2 Driver Control	Pulse Signals
50	BK/TN	Power Ground	<0.1v
51	TN/WT	HO2S-12 (B1 S2) Signal	0.1-1.1v
52-56	---	Not Used	---
57	GY/RD	IAC 1 Driver Control	Pulse Signals
58	PK/BK	IAC 4 Driver Control	Pulse Signals
59	PK/BR	CCD Bus (+)	Digital Signals: 0-5-0v
60	WT/BK	CCD Bus (-)	<0.050v
61	PK/WT	5-Volt Supply	4.9-5.1v
62	WT/PK	Brake Switch Sense Signal	Brake Off: 0v, On: 12v
63	YL/DG	Torque Management Request	Digital Signals
64	DB/OR	A/C Clutch Relay Control	Relay Off: 12v, On: 1v
65	PK	SCI Transmit Signal	0v
66	WT/OR	Vehicle Speed Signal	Digital Signal
67	DB/YL	ASD Relay Control	Relay Off: 12v, On: 1v
68	PK/BK	EVAP Purge Solenoid Control	PWM Signal: 0-12-0v
69-71	---	Not Used	---
72	DB/WT	LDP Switch Sense Signal	LDP Switch Closed: 0v
73	LG/DG	Radiator Fan Control Relay	Relay Off: 12v, On: 1v
74	BR	Fuel Pump Relay Control	Relay Off: 12v, On: 1v
75	LG	SCI Receive Signal	5v
76	BR/YL	PNP Switch Sense (EATX)	In P/N: 0v, Others: 5v
77	WT/DG	LDP Solenoid Control	PWM Signal: 0-12-0v
78	TN/RD	Speed Control Vacuum Solenoid	Vacuum Increasing: 1v
79	---	Not Used	---
80	LG/RD	Speed Control Vent Solenoid	Vacuum Decreasing: 1v

Pin Connector Graphic

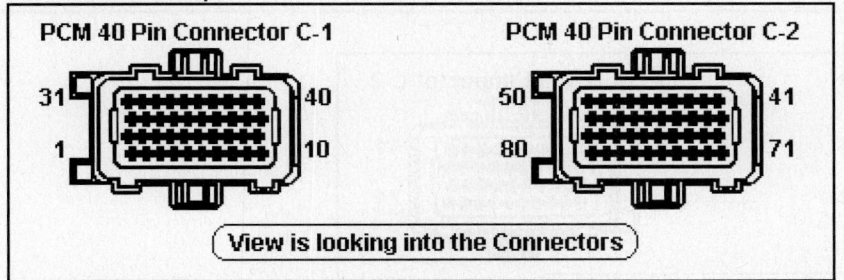

1999-2000 Caravan 3.3L V6 MFI VIN R (A/T) 'C1' Black Connector

PCM Pin #	Wire Color	Circuit Description (40 Pin)	Value at Hot Idle
1	---	Not Used	---
2	RD/YL	Coil 3 Driver (Cyl 3 & 6)	5°, 55 mph: 8° dwell
3	DB/TN	Coil 2 Driver (Cyl 2 & 5)	5°, 55 mph: 8° dwell
4	---	Not Used	---
5	YL/RD	Speed Control Power Supply	12-14v
6	DG/OR	ASD Relay Power (B+)	12-14v
7	YL/WT	Injector 3 Driver	1-4 ms
8	DG	Generator Field Driver	Digital Signals: 0-12-0v
9, 12	---	Not Used	---
10	BK/TN	Power Ground	<0.1v
11	GY/RD	Coil 1 Driver (Cyl 1 & 4)	5°, 55 mph: 8° dwell
13	WT/DB	Injector 1 Driver	1-4 ms
14	BR/DB	Injector 6 Driver	1-4 ms
15	GY	Injector 5 Driver	1-4 ms
16	LB/BR	Injector 4 Driver	1-4 ms
17	TN/WT	Injector 2 Driver	1-4 ms
18-19	---	Not Used	---
20	WT/BK	Ignition Switch Power (B+)	12-14v
21	---	Not Used	---
22	BK/PK	MIL (lamp) Control	MIL Off: 12v, On: 1v
23-24	---	Not Used	---
25	DB/LG	Knock Sensor Signal	0.080v AC
26	TN/BK	ECT Sensor Signal	At 180°F: 2.80v
27	BK/OR	HO2S Ground	<0.050v
28-29	---	Not Used	---
30	BK/DG	HO2S-11 (B1 S1) Signal	0.1-1.1v
31	TN	Smart Start Relay Control	KOEC: 9-11v
32	GY/BK	CKP Sensor Signal	Digital Signals: 0-5-0v
33	TN/YL	CMP Sensor Signal	Digital Signals: 0-5-0v
34, 37	---	Not Used	---
35	OR/DB	TP Sensor Signal	0.6-1.0v
36	DG/RD	MAP Sensor Signal	1.5-1.6v
38 ('99)	DG	A/C Switch Sense Signal	A/C Off: 12v, On: 1v
39	---	Not Used	---
38	DG/LB	A/C Switch Sense Signal	A/C Off: 12v, On: 1v
40	GY/YL	EGR Solenoid Control	12v, at 55 mph: 1v

Pin Connector Graphic

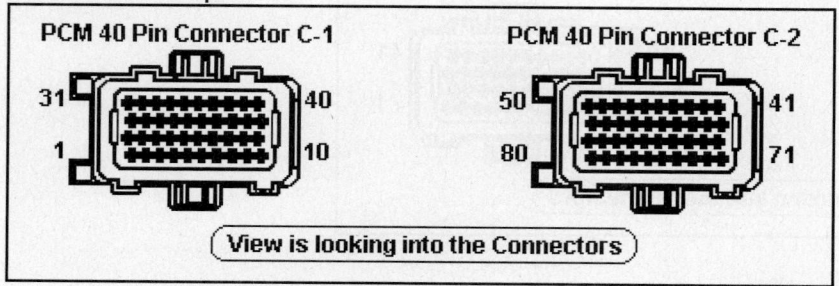

PCM 40 Pin Connector C-1

PCM 40 Pin Connector C-2

View is looking into the Connectors

1999-2000 Caravan 3.3L V6 MFI VIN R (A/T) 'C2' White Connector

PCM Pin #	Wire Color	Circuit Description (40 Pin)	Value at Hot Idle
41	RD/LG	Speed Control Set Switch Signal	S/C & Set Switch On: 3.8v
42	DB	A/C Pressure Sensor Signal	A/C On: 0.45-4.85v
43	BK/LB	Sensor Ground	<0.050v
44	OR	8-Volt Supply	7.9-8.1v
45	---	Not Used	---
46	RD/WT	Keep Alive Power (B+)	12-14v
47	---	Not Used	---
48	BR/WT	IAC 3 Driver Control	Pulse Signals
49	YL/BK	IAC 2 Driver Control	Pulse Signals
50	BK/TN	Power Ground	<0.1v
51	TN/WT	HO2S-12 (B1 S2) Signal	0.1-1.1v
52-55	---	Not Used	---
56	TN/RD	Speed Control Vacuum Solenoid	Vacuum Increasing: 1v
57	GY/RD	IAC 1 Driver Control	Pulse Signals
58	PK/BK	IAC 4 Driver Control	Pulse Signals
59	PK/BR	CCD Bus (+)	Digital Signals: 0-5-0v
60	WT/BK	CCD Bus (-)	<0.050v
61	PK/WT	5-Volt Supply	4.9-5.1v
62	WT/PK	Brake Switch Sense Signal	Brake Off: 0v, On: 12v
63	YL/DG	Torque Management Request	Digital Signals
64	DB/OR	A/C Clutch Relay Control	Relay Off: 12v, On: 1v
65	PK	SCI Transmit Signal	0v
66	WT/OR	Vehicle Speed Signal	Digital Signal
67	DB/YL	ASD Relay Control	Relay Off: 12v, On: 1v
68	PK/BK	EVAP Purge Solenoid Control	PWM Signal: 0-12-0v
69	---	Not Used	---
70	PK/RD	EVAP Purge Solenoid Sense	0-1v
71	---	Not Used	---
72	DB/WT	LDP Switch Sense Signal	LDP Switch Closed: 0v
73	LG/DB	Radiator Fan Control Relay	Relay Off: 12v, On: 1v
74	BR	Fuel Pump Relay Control	Relay Off: 12v, On: 1v
75	LG	SCI Receive Signal	5v
76	BR/YL	PNP Switch Sense (EATX)	In P/N: 0v, Others: 5v
77	WT/DG	LDP Solenoid Control	PWM Signal: 0-12-0v
78-79	---	Not Used	---
80	LG/RD	Speed Control Vent Solenoid	Vacuum Decreasing: 1v

Pin Connector Graphic

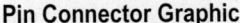

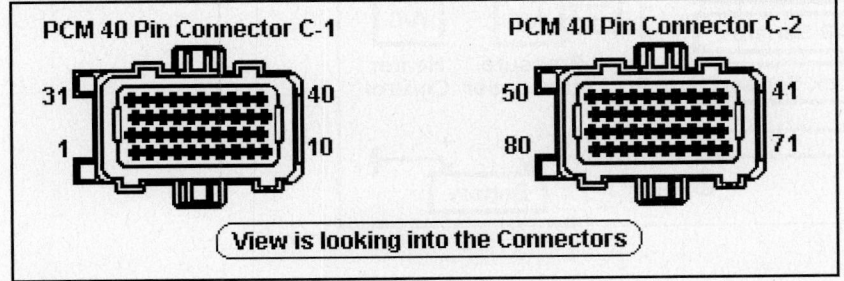

1999 Caravan 3.3L V6 MFI VIN R (A/T) Wiring Diagram (Part 1)

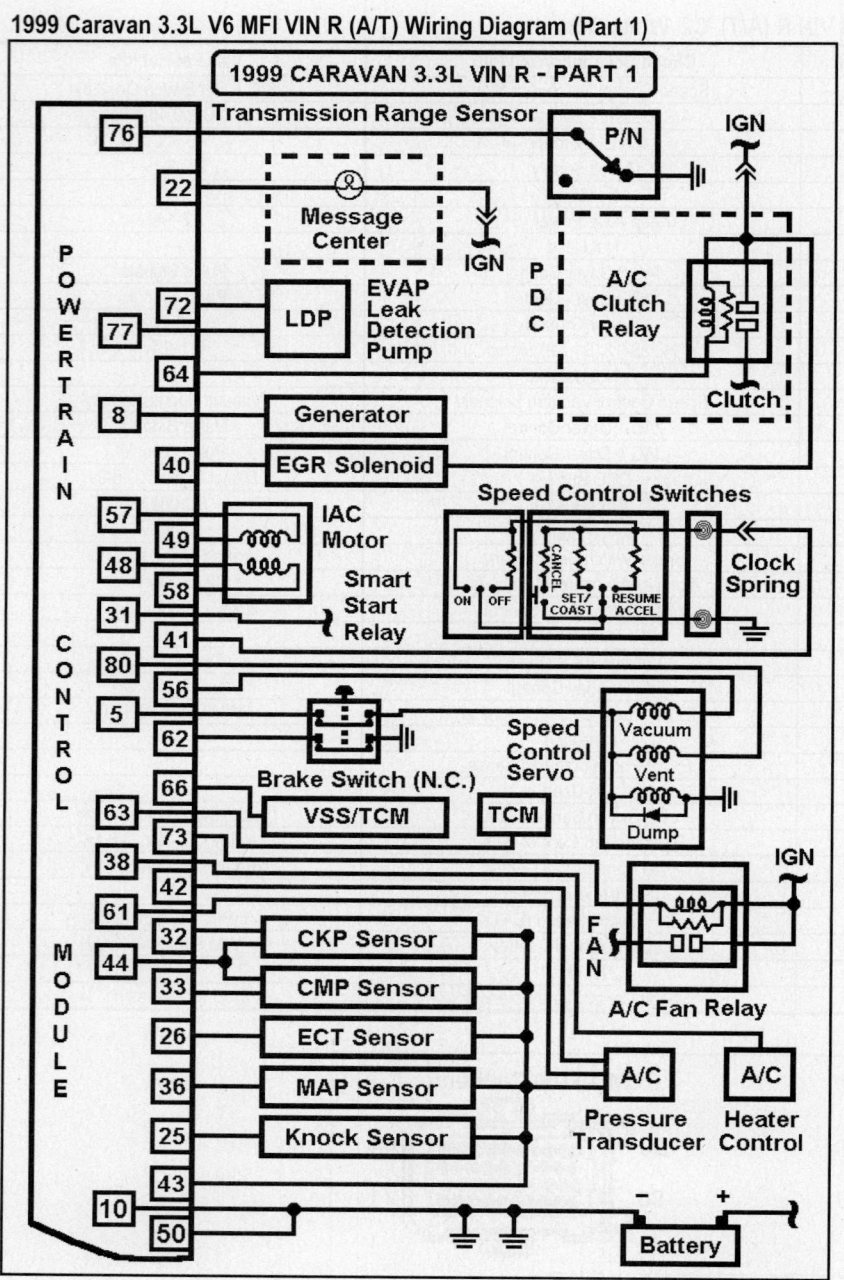

1999 Caravan 3.3L V6 MFI VIN R (A/T) Wiring Diagram (Part 2)

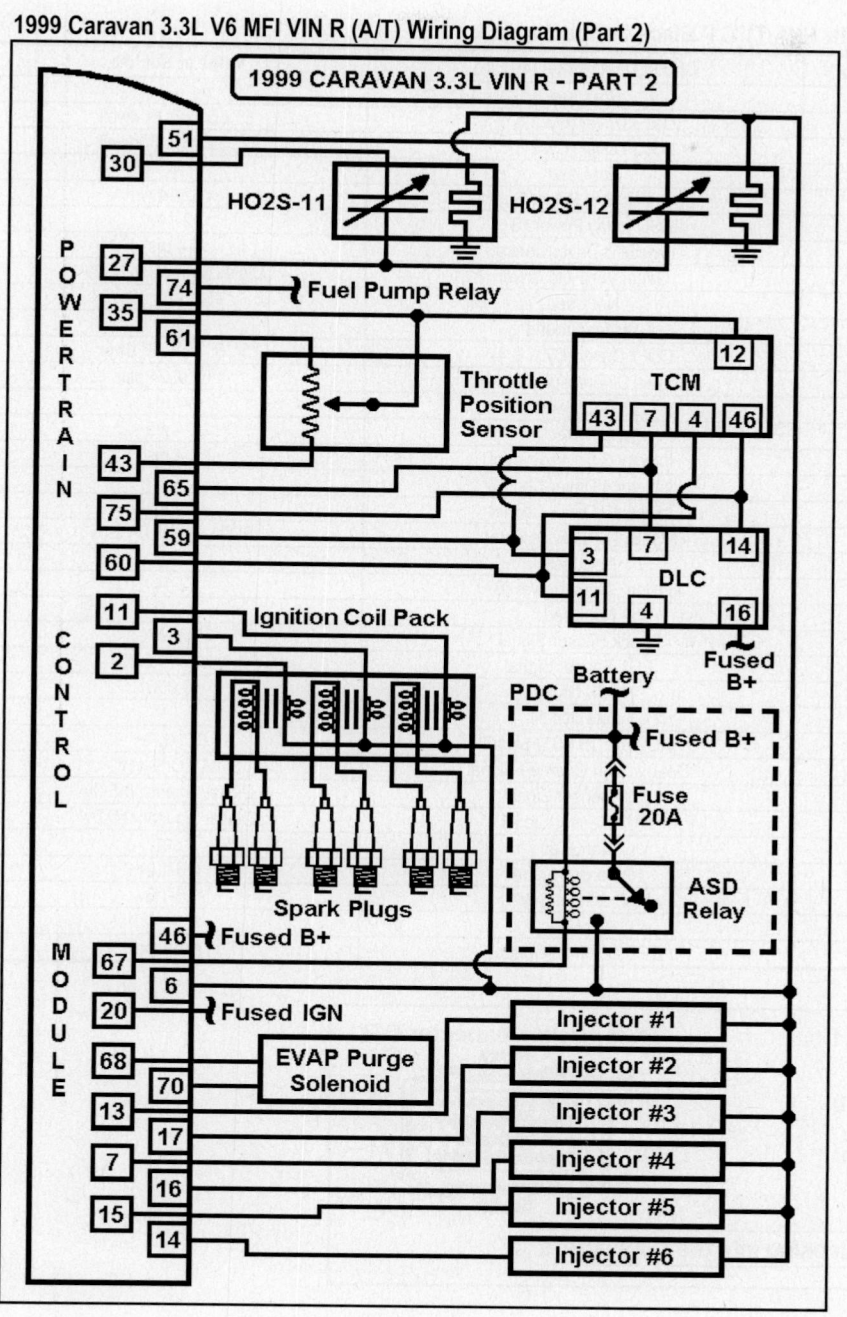

2001-02 Caravan 3.3L V6 MFI VIN R (A/T) 'C1' Black Connector

PCM Pin #	Wire Color	Circuit Description (40 Pin)	Value at Hot Idle
1	---	Not Used	---
2	DB/OR	Coil 3 Driver (Cyl 3 & 6)	5°, 55 mph: 8° dwell
3	DB/TN	Coil 2 Driver (Cyl 2 & 5)	5°, 55 mph: 8° dwell
4	---	Not Used	---
5	VT/YL	Speed Control Power Supply	12-14v
6	BR/WT	ASD Relay Power (B+)	12-14v
7	BR/LB	Injector 3 Driver Control	1-4 ms
8	BR/GY	Generator Field Driver	Digital Signals: 0-12-0v
9	---	Not Used	---
10	BK/BR	Power Ground	<0.1v
11	DB/DG	Coil 1 Driver (Cyl 1 & 4)	5°, 55 mph: 8° dwell
12	VT/GY	Engine Oil Pressure Sensor	1.6v at 24 psi
13	BR/YL	Injector 1 Driver Control	1-4 ms
14	BR/VT	Injector 6 Driver Control	1-4 ms
15	BR/OR	Injector 5 Driver Control	1-4 ms
16	BR/TN	Injector 4 Driver Control	1-4 ms
17	BR/DB	Injector 2 Driver Control	1-4 ms
18	BR/LG	HO2S-11 (B1 S1) Heater	PWM Signal: 0-12-0v
19	---	Not Used	---
20	PK/GY	Ignition Switch Power (B+)	12-14v
21-24	---	Not Used	---
25	DB/YL	Knock Sensor Signal	0.080v AC
26	VT/OR	ECT Sensor Signal	At 180°F: 2.80v
27	BR/DG	HO2S Ground	<0.050v
28-29	---	Not Used	---
30	DB/LG	HO2S-11 (B1 S1) Signal	0.1-1.1v
31	DG/OR	Double Start Override Signal	KOEC: 9-11v
32	BR/LB	CKP Sensor Signal	Digital Signals: 0-5-0v
33	DB/GY	CMP Sensor Signal	Digital Signals: 0-5-0v
34	---	Not Used	---
35	BR/OR	TP Sensor Signal	0.6-1.0v
36	VT/BR	MAP Sensor Signal	1.5-1.6v
38-39	---	Not Used	---
40	DB/VT	EGR Solenoid Control	12v, at 55 mph: 1v

Pin Connector Graphic

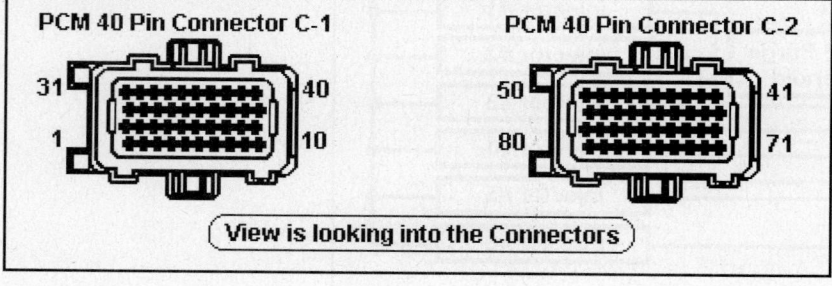

2001-02 Caravan 3.3L V6 MFI VIN R (A/T) 'C2' White Connector

PCM Pin #	Wire Color	Circuit Description (40 Pin)	Value at Hot Idle
41	VT	Speed Control Set Switch Signal	S/C & Set Switch On: 3.8v
42	LB/BR	A/C Pressure Sensor Signal	A/C On: 0.45-4.85v
43	DB/DG	Sensor Ground	<0.050v
44	BR/PK	8-Volt Supply	7.9-8.1v
45	---	Not Used	---
46	OR/RD	Keep Alive Power (B+)	12-14v
47-48	---	Not Used	---
49	VT/DG	IAC 1 Driver Control	Pulse Signals
50	BK/DG	Power Ground	<0.1v
51	DB/YL	HO2S-12 (B1 S2) Signal	0.1-1.1v
52-55	---	Not Used	---
56	VT/YL	Speed Control Vacuum Solenoid	Vacuum Increasing: 1v
57	VT/LG	IAC 2 Driver Control	Pulse Signals
58	---	Not Used	---
59	WT/VT	PCI Data Bus (J1850)	Digital Signals: 0-7-0v
60	---	Not Used	---
61	PK/YL	5-Volt Supply	4.9-5.1v
62	DG/WT	Brake Switch Sense Signal	Brake Off: 0v, On: 12v
63	DG/LG	Torque Management Request	Digital Signals
64	LB/OR	A/C Clutch Relay Control	Relay Off: 12v, On: 1v
65	WT/BR	SCI Transmit Signal	0v
66	DB/OR	Vehicle Speed Signal	Digital Signal
67	BR/WT	ASD Relay Control	Relay Off: 12v, On: 1v
68	DB/WT	EVAP Purge Solenoid Control	PWM Signal: 0-12-0v
69	---	Not Used	---
70	DB/BR	EVAP Purge Solenoid Sense	0-1v
71	---	Not Used	---
72	VT/WT	LDP Switch Sense Signal	LDP Switch Closed: 0v
73	BR/VT	Radiator Fan Control Relay	Relay Off: 12v, On: 1v
74	BR	Fuel Pump Relay Control	Relay Off: 12v, On: 1v
75	WT/LG	SCI Receive Signal	5v
76	YL/LB	TRS T41 Signal	In P/N: 0v, Others: 5v
77	VT/LB	LDP Solenoid Control	PWM Signal: 0-12-0v
78-79	---	Not Used	---
80	VT/OR	Speed Control Vent Solenoid	Vacuum Decreasing: 1v

Pin Connector Graphic

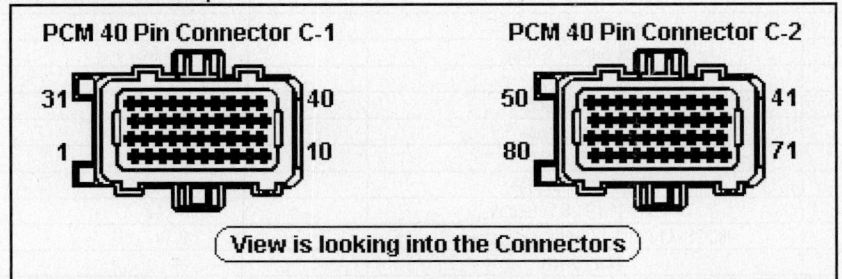

PCM 40 Pin Connector C-1

31   40

1   10

PCM 40 Pin Connector C-2

50   41

80   71

View is looking into the Connectors

2003 Caravan 3.3L V6 OHV MFI VIN R 'C1' Black Connector

PCM Pin #	Wire Color	Circuit Description (38 Pin)	Value at Hot Idle
1-8, 10	---	Not Used	---
9	BK/BR	Power Ground	<0.1v
11	PK/GY	Ignition Switch Output (Run-Start)	12-14v
12	PK/WT	Ignition Switch Output (Off-Run-Start)	12-14v
13-17, 19	---	Not Used	---
18	BK/DG	Power Ground	<0.1v
20	VT/GY	Oil Pressure Sensor	A/C On: 0.45-4.85v
21, 23-24	---	Not Used	---
22	VT/LG	Ambient Air Temperature Sensor	At 100ºF: 1.83v
25	WT/LG	SCI Receive (PCM)	5v
26	WT/BR	Flash Program Enable	5v
27-28	---	Not Used	---
29	OR/RD	Battery Power (B+)	12-14v
30	YL	Ignition Switch Output (Start)	8-11v (cranking)
31	DB/YL	HO2S-12 (B1 S2) Signal	0.1-1.1v
32	DB/DG	Oxygen Sensor Return (down)	<0.050v
33-35	---	Not Used	---
36	WT/BR	SCI Transmit (PCM)	0v
37	DG/YL	SCI Transmit (TCM)	0v
38	WT/VT	PCI Data Bus (J1850)	Digital Signals: 0-7-0v

2003 Caravan 3.3L V6 OHV MFI VIN R 'C2' Gray Connector

PCM Pin #	Wire Color	Circuit Description (38 Pin)	Value at Hot Idle
1-8	---	Not Used	---
9	DB/TN	Coil 2 Driver	5º, at 55 mph: 8º dwell
10	DB/DG	Coil 1 Driver	5º, at 55 mph: 8º dwell
11	BR/TN	Injector 4 Driver Control	1.0-4.0 ms
12	BR/LB	Injector 3 Driver Control	1.0-4.0 ms
13	BR/DB	Injector 2 Driver Control	1.0-4.0 ms
14	BR/YL	Injector 1 Driver Control	1.0-4.0 ms
15-17	---	Not Used	---
18	BR/LG	HO2S-12 (B1 S2) Heater Control	Heater Off: 12v, On: 1v
19	BR/GY	Generator Field Driver	Digital Signal: 0-12-0v
20	VT/OR	ECT Sensor Signal	At 180ºF: 2.80v
21	BR/OR	TP Sensor Signal	0.6-1.0v
22	---	Not Used	---
23	VT/BR	MAP Sensor Signal	1.5-1.7v
24	BR/LG	Knock Sensor Return	<0.050v
25	DB/YL	Knock Sensor Signal	0.080v AC
26	---	Not Used	---
27	DB/DG	Sensor Ground	<0.1v
28	BR/VT	IAC Motor Sense	12-14v
29	PK/YL	5-Volt Supply	4.9-5.1v
30	DB/LG	IAT Sensor Signal	At 100ºF: 1.83v
31	DB/LG	HO2S-11 (B1 S1) Signal	0.1-1.1v
32	BR/DG	HO2S-11 (B1 S1) Ground (Upstream)	<0.1v
33	---	Not Used	---
34	DB/GY	CMP Sensor Signal	Digital Signal: 0-5-0v
35	BR/LB	CKP Sensor Signal	Digital Signal: 0-5-0v
36-37	---	Not Used	---
38	GY/RD	IAC Motor Driver	DC pulse signals: 0.8-11v

Pin Connector Graphic

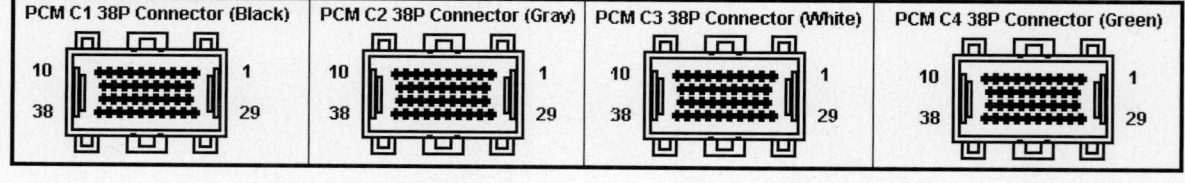

PCM C1 38P Connector (Black) PCM C2 38P Connector (Gray) PCM C3 38P Connector (White) PCM C4 38P Connector (Green)

2003 Caravan 3.3L V6 OHV MFI VIN R 'C3' White Connector

PCM Pin #	Wire Color	Circuit Description (38 Pin)	Value at Hot Idle
1-2, 4	---	Not Used	---
3	BR/WT	ASD Relay Control	Relay Off: 12v, On: 1v
5	VT/OR	Speed Control Vent Solenoid	Vacuum Decreasing: 1v
6	BR/VT	Radiator Fan Relay Control	Relay Off: 12v, On: 1v
7	VT/YL	Speed Control Power Supply	12-14v
8	WT/DG	Natural Vacuum Leak Detection Solenoid	Solenoid Off: 12v, On: 1v
9	BR/WT	HO2S-12 (B1 S2) Heater Control	Heater On: 1v, Off: 12v
10	---	Not Used	---
11	LB/OR	A/C Clutch Relay Control	Relay Off: 12v, On: 1v
12	VT/YL	Speed Control Vacuum Solenoid	Vacuum Increasing: 1v
13-18	---	Not Used	---
19	BR/WT	Automatic Shutdown Relay Output	12-14v
20	DB/WT	EVAP Purge Solenoid Control	PWM Signal: 0-12-0v
21-22	---	Not Used	---
23	DG/WT	Brake Switch Sense	Brake Off: 0v, On: 12v
24-27	---	Not Used	---
28	BR/WT	Automatic Shutdown Relay Output	12-14v
29	DB/BR	EVAP Purge Solenoid Sense	0-1v
30, 33, 36	---	Not Used	---
31	LB/BR	A/C Pressure Sensor	A/C On: 0.451-4.850v
32	DB/YL	Battery Temperature Sensor	At 100°F: 1.83v
34	VT	Speed Control Set Switch Signal	S/C & Set Switch On: 3.8v
35	VT/WT	Natural Vacuum Leak Detection Switch Sense	0.1v
37	BR	Fuel Pump Relay Control	Relay Off: 12v, On: 1v
38	DG/OR	Starter Relay Control	KOEC: 9-11v

2003 Caravan 3.3L V6 OHV MFI VIN R 'C4' Green Connector

PCM Pin #	Wire Color	Circuit Description (38 Pin)	Value at Hot Idle
1	YL/GY	Overdrive Solenoid Control	Solenoid Off: 12v, On: 1v
2	YL/LB	Underdrive Solenoid Control	Solenoid Off: 12v, On: 1v
3-5	---	Not Used	---
6	YL/DB	2-4 Solenoid Control	Solenoid Off: 12v, On: 1v
7-9, 11	---	Not Used	---
10	DG/WT	Low/Reverse Solenoid Control	Solenoid Off: 12v, On: 1v
12, 13	BK/LG	Power Ground	<0.050v
14	---	Not Used	---
15	DG/LB	TRS T1 Sense	<0.050v
16	DG/DB	TRS T3 Sense	<0.050v
17, 20-21	---	Not Used	---
18	YL/BR	Transmission Control Relay Control	Relay Off: 12v, On: 1v
19, 38	YL/OR	Transmission Control Relay Output	Relay Off: 12v, On: 1v
23-26	---	Not Used	---
27	DG/GY	TRS T41 Sense	<0.050v
28, 38	YL/OR	Transmission Control Relay Output	Relay Off: 12v, On: 1v
29	YL/TN	Low/Reverse Pressure Switch Sense	12-14v
30	YL/DG	2-4 Pressure Switch Sense	In Low/Reverse: 2-4v
31, 36	---	Not Used	---
32	DG/BR	Output Speed Sensor Signal	In 2-4 Position: 2-4v
33	DG/WT	Input Speed Sensor Signal	Moving: AC voltage
34	DG/VT	Speed Sensor Ground	Moving: AC voltage
35	DG/OR	Transmission Temperature Sensor Signal	<0.050v
37	DG/YL	TRS T42 Sense	In PRNL: 0v, Others 5v

Pin Connector Graphic

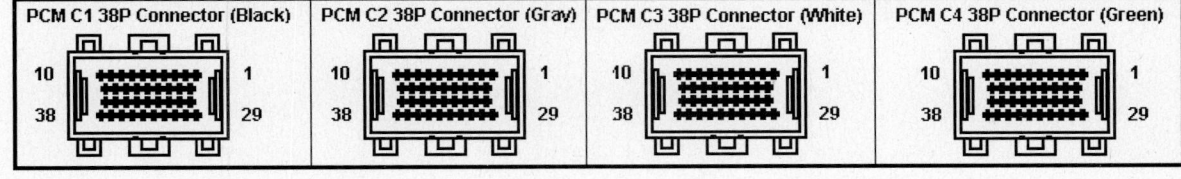

1996 Caravan 3.8L V6 MFI VIN L (A/T) 'C1' Black Connector

PCM Pin #	Wire Color	Circuit Description (40 Pin)	Value at Hot Idle
1	---	Not Used	---
2	RD/YL	Coil 3 Driver (Cyl 3 & 6)	5°, 55 mph: 8° dwell
3	DB/TN	Coil 2 Driver (Cyl 2 & 5)	5°, 55 mph: 8° dwell
4	DG	Generator Field Driver	Digital Signals: 0-12-0v
5	YL/RD	Speed Control Power Supply	12-14v
6	DG/OR	ASD Relay Power (B+)	12-14v
7	YL/WT	Injector 3 Driver Control	1-4 ms
8	TN	Smart Start Relay Control	KOEC: 9-11v
9	---	Not Used	---
10	BK/TN	Power Ground	<0.1v
11	GY	Coil 1 Driver (Cyl 1 & 4)	5°, 55 mph: 8° dwell
12	---	Not Used	---
13	WT/DB	Injector 1 Driver Control	1-4 ms
14	BR/DB	Injector 6 Driver Control	1-4 ms
15	GY	Injector 5 Driver Control	1-4 ms
16	LB/BR	Injector 4 Driver Control	1-4 ms
17	TN	Injector 2 Driver Control	1-4 ms
18-19	---	Not Used	---
20	WT/BK	Ignition Switch Power (B+)	12-14v
21-23	---	Not Used	---
24	DB/LG	Knock Sensor Signal	0.080v AC
25	---	Not Used	---
26	TN/BK	ECT Sensor Signal	At 180°F: 2.80v
27-29	---	Not Used	---
30	BK/DG	HO2S-11 (B1 S1) Signal	0.1-1.1v
31	---	Not Used	---
32	GY/BK	CKP Sensor Signal	Digital Signals: 0-5-0v
33	TN/YL	CMP Sensor Signal	Digital Signals: 0-5-0v
34	---	Not Used	---
35	OR/DB	TP Sensor Signal	0.6-1.0v
36	DG/RD	MAP Sensor Signal	1.5-1.6v
37-39	---	Not Used	---
40	GY/YL	EGR Solenoid Control	12v, at 55 mph: 1v

Pin Connector Graphic

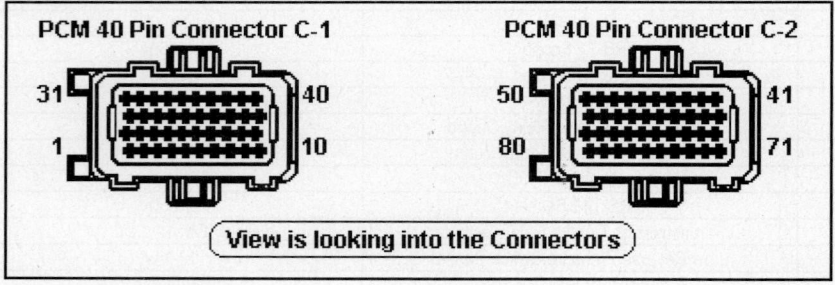

PCM 40 Pin Connector C-1

PCM 40 Pin Connector C-2

View is looking into the Connectors

1996 Caravan 3.8L V6 MFI VIN L (A/T) 'C2' White Connector

PCM Pin #	Wire Color	Circuit Description (40 Pin)	Value at Hot Idle
41	RD/LG	Speed Control Set Switch Signal	S/C & Set Switch On: 3.8v
42	DB	A/C Pressure Switch Signal	A/C On: 0.451-4.850v
43	BK/LB	Sensor Ground	<0.050v
44	OR	8-Volt Supply	7.9-8.1v
45	DB	A/C Switch Sense Signal	A/C Off: 12v, On: 1v
46	RD/WT	Keep Alive Power (B+)	12-14v
47	---	Not Used	---
48	BR/WT	IAC 1 Driver Control	Pulse Signals
49	YL/BK	IAC 2 Driver Control	Pulse Signals
50	BK/TN	Power Ground	<0.1v
51	TN/WT	HO2S-12 (B1 S2) Signal	0.1-1.1v
52-56	---	Not Used	---
57	GY/RD	IAC 3 Driver Control	Pulse Signals
58	PK/BK	IAC 4 Driver Control	Pulse Signals
59	PK/BR	CCD Bus (+)	Digital Signals: 0-5-0v
60	WT/BK	CCD Bus (-)	<0.050v
61	PK/WT	5-Volt Supply	4.9-5.1v
62	WT/PK	Brake Switch Sense Signal	Brake Off: 0v, On: 12v
63	YL/DG	Torque Management Request	Digital Signals
64	DB/OR	A/C Clutch Relay Control	Relay Off: 12v, On: 1v
65	PK	SCI Transmit Signal	0v
66	WT/OR	Vehicle Speed Signal	Digital Signal
67	DB/YL	ASD Relay Control	Relay Off: 12v, On: 1v
68	PK/BK	EVAP Purge Solenoid Control	PWM Signal: 0-12-0v
69-71	---	Not Used	---
72	YL/BK	LDP Switch Sense Signal	LDP Switch Closed: 0v
73	LG/DG	Radiator Fan Control Relay	Relay Off: 12v, On: 1v
74	BR	Fuel Pump Relay Control	Relay Off: 12v, On: 1v
75	LG	SCI Receive Signal	5v
76	BK/WT	PNP Switch Sense (EATX)	In P/N: 0v, Others: 5v
77	WT/DG	LDP Solenoid Control	PWM Signal: 0-12-0v
78	TN/RD	Speed Control Vacuum Solenoid	Vacuum Increasing: 1v
79	---	Not Used	---
80	LG/RD	Speed Control Vent Solenoid	Vacuum Decreasing: 1v

Pin Connector Graphic

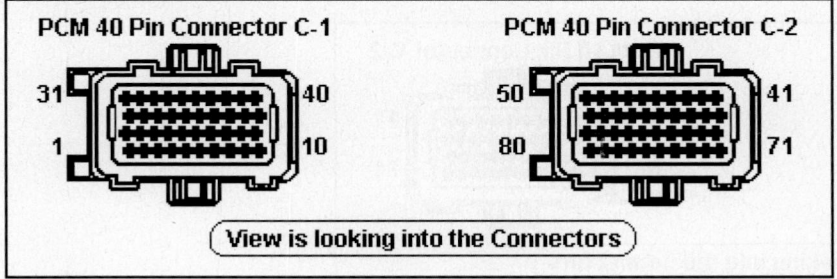

1997-98 Caravan 3.8L V6 MFI VIN L (A/T) 'C1' Black Connector

PCM Pin #	Wire Color	Circuit Description (40 Pin)	Value at Hot Idle
1	---	Not Used	---
2	RD/YL	Coil 3 Driver (Cyl 3 & 6)	5°, 55 mph: 8° dwell
3	DB/TN	Coil 2 Driver (Cyl 2 & 5)	5°, 55 mph: 8° dwell
4	DG	Generator Field Driver	Digital Signals: 0-12-0v
5	YL/RD	Speed Control Power Supply	12-14v
6	DG/OR	ASD Relay Power (B+)	12-14v
7	YL/WT	Injector 3 Driver	1-4 ms
8	TN	Smart Start Relay Control	KOEC: 9-11v
9	---	Not Used	---
10	BK/TN	Power Ground	<0.1v
11	GY/RD	Coil 1 Driver (Cyl 1 & 4)	5°, 55 mph: 8° dwell
12	---	Not Used	---
13	WT/DB	Injector 1 Driver	1-4 ms
14	BR/DB	Injector 6 Driver	1-4 ms
15	GY	Injector 5 Driver	1-4 ms
16	LB/BR	Injector 4 Driver	1-4 ms
17	TN	Injector 2 Driver	1-4 ms
18-19	---	Not Used	---
20	WT/BK	Ignition Switch Power (B+)	12-14v
21	---	Not Used	---
22	BK/PK	MIL (lamp) Control	MIL Off: 12v, On: 1v
23	---	Not Used	---
24	DB/LG	Knock Sensor Signal	0.080v AC
25	---	Not Used	---
26	TN/BK	ECT Sensor Signal	At 180°F: 2.80v
27-29	---	Not Used	---
30	BK/DG	HO2S-11 (B1 S1) Signal	0.1-1.1v
31	---	Not Used	---
32	GY/BK	CKP Sensor Signal	Digital Signals: 0-5-0v
33	TN/YL	CMP Sensor Signal	Digital Signals: 0-5-0v
34	---	Not Used	---
35	OR/DB	TP Sensor Signal	0.6-1.0v
36	DG/RD	MAP Sensor Signal	1.5-1.6v
37-39	---	Not Used	---
40	GY/YL	EGR Solenoid Control	12v, at 55 mph: 1v

Pin Connector Graphic

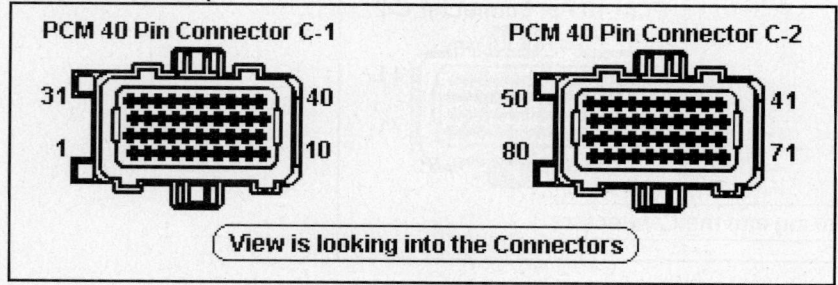

PCM 40 Pin Connector C-1

PCM 40 Pin Connector C-2

View is looking into the Connectors

1997-98 Caravan 3.8L V6 MFI VIN L (A/T) 'C2' White Connector

PCM Pin #	Wire Color	Circuit Description (40 Pin)	Value at Hot Idle
41	RD/LG	Speed Control Set Switch Signal	S/C & Set Switch On: 3.8v
42	DB	A/C Pressure Switch Signal	A/C On: 0.451-4.850v
43	BK/LB	Sensor Ground	<0.050v
44	OR	8-Volt Supply	7.9-8.1v
45	DG	A/C Switch Sense Signal	A/C Off: 12v, On: 1v
46	RD/WT	Keep Alive Power (B+)	12-14v
47	---	Not Used	---
48	BR/WT	IAC 3 Driver Control	Pulse Signals
49	YL/BK	IAC 2 Driver Control	Pulse Signals
50	BK/TN	Power Ground	<0.1v
51	TN/WT	HO2S-12 (B1 S2) Signal	0.1-1.1v
52-56	---	Not Used	---
57	GY/RD	IAC 1 Driver Control	Pulse Signals
58	PK/BK	IAC 4 Driver Control	Pulse Signals
59	PK/BR	CCD Bus (+)	Digital Signals: 0-5-0v
60	WT/BK	CCD Bus (-)	<0.050v
61	PK/WT	5-Volt Supply	4.9-5.1v
62	WT/PK	Brake Switch Sense Signal	Brake Off: 0v, On: 12v
63	YL/DG	Torque Management Request	Digital Signals
64	DB/OR	A/C Clutch Relay Control	Relay Off: 12v, On: 1v
65	PK	SCI Transmit Signal	0v
66	WT/OR	Vehicle Speed Signal	Digital Signal
67	DB/YL	ASD Relay Control	Relay Off: 12v, On: 1v
68	PK/BK	EVAP Purge Solenoid Control	PWM Signal: 0-12-0v
69-71	---	Not Used	---
72	DB/WT	LDP Switch Sense Signal	LDP Switch Closed: 0v
73	LG/DG	Radiator Fan Control Relay	Relay Off: 12v, On: 1v
74	BR	Fuel Pump Relay Control	Relay Off: 12v, On: 1v
75	LG	SCI Receive Signal	5v
76	BR/YL	PNP Switch Sense (EATX)	In P/N: 0v, Others: 5v
77	WT/DG	LDP Solenoid Control	PWM Signal: 0-12-0v
78	TN/RD	Speed Control Vacuum Solenoid	Vacuum Increasing: 1v
79	---	Not Used	---
80	LG/RD	Speed Control Vent Solenoid	Vacuum Decreasing: 1v

Pin Connector Graphic

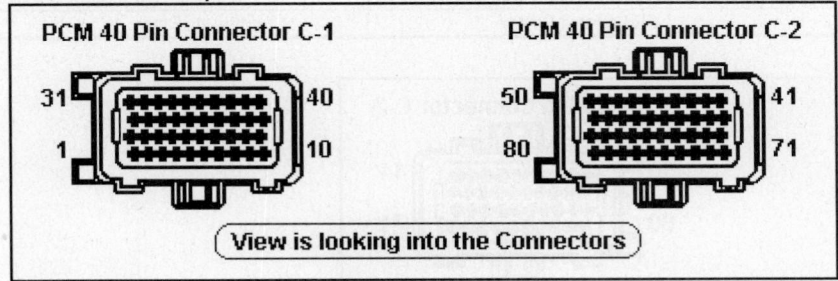

PCM 40 Pin Connector C-1

31   40
1   10

PCM 40 Pin Connector C-2

50   41
80   71

View is looking into the Connectors

1999 Caravan 3.8L V6 OHV MFI VIN L (A/T) 'C1' Black Connector

PCM Pin #	Wire Color	Circuit Description (40 Pin)	Value at Hot Idle
1	---	Not Used	---
2	RD/YL	Coil 3 Driver (Cyl 3 & 6)	5°, 55 mph: 8° dwell
3	DB/TN	Coil 2 Driver (Cyl 2 & 5)	5°, 55 mph: 8° dwell
4	---	Not Used	---
5	YL/RD	Speed Control Power Supply	12-14v
6	DG/OR	ASD Relay Power (B+)	12-14v
7	YL/WT	Injector 3 Driver	1-4 ms
8	DG	Generator Field Driver	Digital Signals: 0-12-0v
9	---	Not Used	---
10	BK/TN	Power Ground	<0.1v
11	GY/RD	Coil 1 Driver (Cyl 1 & 4)	5°, 55 mph: 8° dwell
12	---	Not Used	---
13	WT/DB	Injector 1 Driver	1-4 ms
14	BR/DB	Injector 6 Driver	1-4 ms
15	GY	Injector 5 Driver	1-4 ms
16	LB/BR	Injector 4 Driver	1-4 ms
17	TN/WT	Injector 2 Driver	1-4 ms
18-19	---	Not Used	---
20	WT/BK	Ignition Switch Power (B+)	12-14v
21	---	Not Used	---
22	BK/PK	MIL (lamp) Control	MIL Off: 12v, On: 1v
23-24	---	Not Used	---
25	DB/LG	Knock Sensor Signal	0.080v AC
26	TN/BK	ECT Sensor Signal	At 180°F: 2.80v
27	BK/OR	HO2S Ground	<0.050v
28-29	---	Not Used	---
30	BK/DG	HO2S-11 (B1 S1) Signal	0.1-1.1v
31	TN	Smart Start Relay Control	KOEC: 9-11v
32	GY/BK	CKP Sensor Signal	Digital Signals: 0-5-0v
33	TN/YL	CMP Sensor Signal	Digital Signals: 0-5-0v
34	---	Not Used	---
35	OR/DB	TP Sensor Signal	0.6-1.0v
36	DG/RD	MAP Sensor Signal	1.5-1.6v
37	---	Not Used	---
38	DG/LB	A/C Switch Sense Signal	A/C Off: 12v, On: 1v
39	---	Not Used	---
40	GY/YL	EGR Solenoid Control	12v, at 55 mph: 1v

Pin Connector Graphic

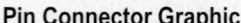

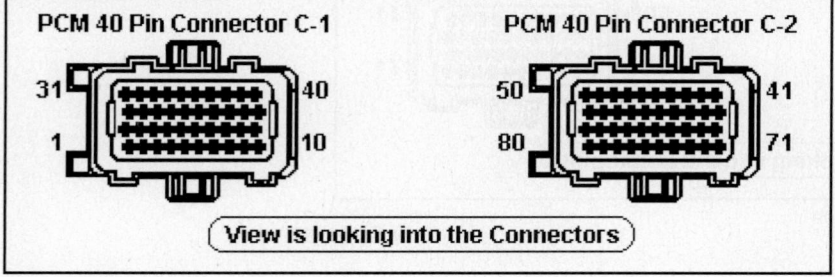

PCM 40 Pin Connector C-1

PCM 40 Pin Connector C-2

31   40   50   41

1   10   80   71

(View is looking into the Connectors)

1999 Caravan 3.8L V6 OHV MFI VIN L (A/T) 'C2' White Connector

PCM Pin #	Wire Color	Circuit Description (40 Pin)	Value at Hot Idle
41	RD/LG	Speed Control Set Switch Signal	S/C & Set Switch On: 3.8v
42	DB	A/C Pressure Switch Signal	A/C On: 0.451-4.850v
43	BK/LB	Sensor Ground	<0.050v
44	OR	8-Volt Supply	7.9-8.1v
45	---	Not Used	---
46	RD/WT	Keep Alive Power (B+)	12-14v
47	---	Not Used	---
48	BR/WT	IAC 3 Driver Control	Pulse Signals
49	YL/BK	IAC 2 Driver Control	Pulse Signals
50	BK/TN	Power Ground	<0.1v
51	TN/WT	HO2S-12 (B1 S2) Signal	0.1-1.1v
52-55	---	Not Used	---
56	TN/RD	Speed Control Vacuum Solenoid	Vacuum Increasing: 1v
57	GY/RD	IAC 1 Driver Control	Pulse Signals
58	PK/BK	IAC 4 Driver Control	Pulse Signals
59	PK/BR	CCD Bus (+)	Digital Signals: 0-5-0v
60	WT/BK	CCD Bus (-)	<0.050v
61	PK/WT	5-Volt Supply	4.9-5.1v
62	WT/PK	Brake Switch Sense Signal	Brake Off: 0v, On: 12v
63	YL/DG	Torque Management Request	Digital Signals
64	DB/OR	A/C Clutch Relay Control	Relay Off: 12v, On: 1v
65	PK	SCI Transmit Signal	0v
66	WT/OR	Vehicle Speed Signal	Digital Signal
67	DB/YL	ASD Relay Control	Relay Off: 12v, On: 1v
68	PK/BK	EVAP Purge Solenoid Control	PWM Signal: 0-12-0v
69	---	Not Used	---
70	PK/RD	EVAP Purge Solenoid Sense	0-1v
71	---	Not Used	---
72	DB/WT	LDP Switch Sense Signal	LDP Switch Closed: 0v
73	LG/DG	Radiator Fan Control Relay	Relay Off: 12v, On: 1v
74	BR	Fuel Pump Relay Control	Relay Off: 12v, On: 1v
75	LG	SCI Receive Signal	5v
76	BR/YL	PNP Switch Sense (EATX)	In P/N: 0v, Others: 5v
77	WT/DG	LDP Solenoid Control	PWM Signal: 0-12-0v
78-79	---	Not Used	---
80	LG/RD	Speed Control Vent Solenoid	Vacuum Decreasing: 1v

Pin Connector Graphic

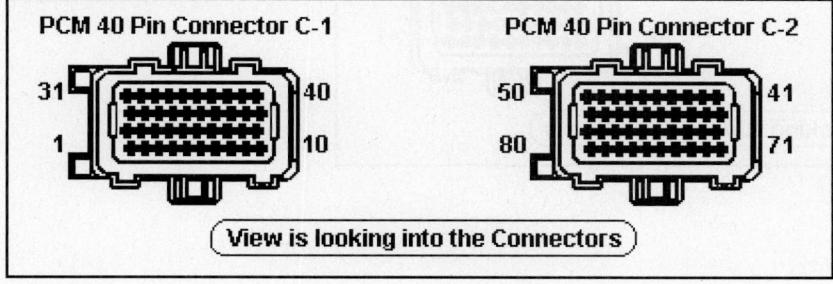

PCM 40 Pin Connector C-1

PCM 40 Pin Connector C-2

View is looking into the Connectors

1996 Grand Caravan 2.4L I4 MFI VIN B 'C1' Black Connector

PCM Pin #	Wire Color	Circuit Description (40 Pin)	Value at Hot Idle
1-2	---	Not Used	---
3	DB/TN	Coil 2 Driver Control	5°, 55 mph: 8° dwell
4	DG	Generator Field Driver	Digital Signals: 0-12-0v
5	YL/RD	Speed Control Power Supply	12-14v
6	DG/OR	ASD Relay Power (B+)	12-14v
7	YL/WT	Injector 3 Driver	1-4 ms
8	DB/OR	Starter Relay Control	KOEC: 9-11v
9	---	Not Used	---
10	BK/TN	Power Ground	<0.1v
11	GY	Coil 1 Driver Control	5°, 55 mph: 8° dwell
12	---	Not Used	---
13	WT/DB	Injector 1 Driver	1-4 ms
14-15	---	Not Used	---
16	LB/BR	Injector 4 Driver	1-4 ms
17	TN	Injector 2 Driver	1-4 ms
18-19	---	Not Used	---
20	WT/BK	Ignition Switch Power (B+)	12-14v
21-23	---	Not Used	---
24	DB/LG	Knock Sensor Signal	0.080v AC
25	---	Not Used	---
26	TN/BK	ECT Sensor Signal	At 180°F: 2.80v
27-29	---	Not Used	---
30	BK/DG	HO2S-11 (B1 S1) Signal	0.1-1.1v
31	---	Not Used	---
32	GY/BK	CKP Sensor Signal	Digital Signals: 0-5-0v
33	TN/YL	CMP Sensor Signal	Digital Signals: 0-5-0v
34	---	Not Used	---
35	OR/DB	TP Sensor Signal	0.6-1.0v
36	DG/RD	MAP Sensor Signal	1.5-1.6v
37	BK/RD	IAT Sensor Signal	At 100°F: 1.83v
38-39	---	Not Used	---
40	GY/YL	EGR Solenoid Control	12v, at 55 mph: 1v

Pin Connector Graphic

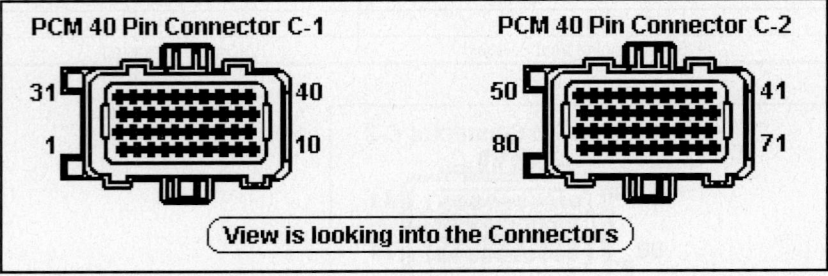

PCM 40 Pin Connector C-1

31   40
1   10

PCM 40 Pin Connector C-2

50   41
80   71

View is looking into the Connectors

1996 Grand Caravan 2.4L I4 MFI VIN B 'C2' White Connector

PCM Pin #	Wire Color	Circuit Description (40 Pin)	Value at Hot Idle
41	RD/LG	Speed Control Set Switch Signal	S/C & Set Switch On: 3.8v
42	DB	A/C Pressure Switch Signal	A/C On: 0.451-4.850v
43	BK/LB	Sensor Ground	<0.050v
44	OR	8-Volt Supply	7.9-8.1v
45	DG	A/C Switch Sense Signal	A/C Off: 12v, On: 1v
46	RD/WT	Keep Alive Power (B+)	12-14v
47	---	Not Used	---
48	BR/WT	IAC 3 Driver Control	Pulse Signals
49	YL/BK	IAC 2 Driver Control	Pulse Signals
50	BK/TN	Power Ground	<0.1v
51	TN/WT	HO2S-12 (B1 S2) Signal	0.1-1.1v
52-55	---	Not Used	---
56	OR/BK	Lockup Torque Converter	At Cruise w/TCC On: 1v
57	GY/RD	IAC 1 Driver Control	Pulse Signals
58	PK/BK	IAC 4 Driver Control	Pulse Signals
59	PK/BR	CCD Bus (+)	Digital Signals: 0-5-0v
60	WT/BK	CCD Bus (-)	<0.050v
61	PK/WT	5-Volt Supply	4.9-5.1v
62	WT/PK	Brake Switch Sense Signal	Brake Off: 0v, On: 12v
63	YL/DG	Torque Management Request	Digital Signals
64	DB/OR	A/C Clutch Relay Control	A/C Off: 12v, On: 1v
65	PK	SCI Transmit	0v
66	WT/OR	Vehicle Speed Signal	Digital Signal
67	DB/YL	ASD Relay Control	Relay Off: 12v, On: 1v
68	PK/BK	EVAP Purge Solenoid Control	PWM Signal: 0-12-0v
69-72	---	Not Used	---
73	LG	Radiator Fan Control Relay	Relay Off: 12v, On: 1v
74	BR	Fuel Pump Relay Control	F/P On: 1v, Off: 12v
75	LG	SCI Receive	5v
76	BK/WT	PNP Switch Sense (EATX)	In P/N: 0v, Others: 5v
77	---	Not Used	---
78	TN/RD	Speed Control Vacuum Solenoid	Vacuum Increasing: 1v
79	---	Not Used	---
80	LG/RD	Speed Control Vent Solenoid	Vacuum Decreasing: 1v

Pin Connector Graphic

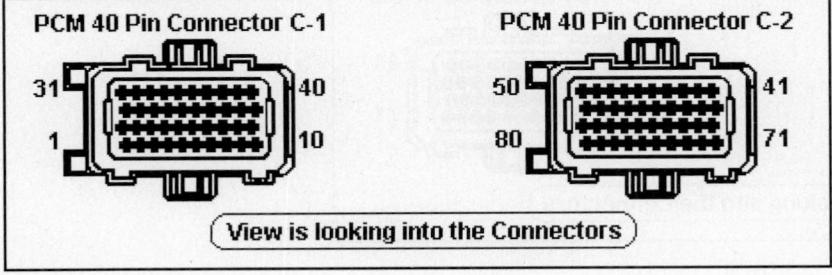

PCM 40 Pin Connector C-1   PCM 40 Pin Connector C-2

31 ... 40   50 ... 41

1 ... 10   80 ... 71

View is looking into the Connectors

1997 Grand Caravan 2.4L I4 MFI VIN B 'C1' Black Connector

PCM Pin #	Wire Color	Circuit Description (40 Pin)	Value at Hot Idle
1-2	---	Not Used	---
3	DB/TN	Coil 2 Driver Control	5°, 55 mph: 8° dwell
4	DG	Generator Field Driver	Digital Signals: 0-12-0v
5	YL/RD	Speed Control Power Supply	12-14v
6	DG/OR	ASD Relay Power (B+)	12-14v
7	YL/WT	Injector 3 Driver	1-4 ms
8	DB/OR	Starter Relay Control	KOEC: 9-11v
9	---	Not Used	---
10	BK/TN	Power Ground	<0.1v
11	GY	Coil 1 Driver Control	5°, 55 mph: 8° dwell
12	---	Not Used	---
13	WT/DB	Injector 1 Driver	1-4 ms
14-15	---	Not Used	---
16	LB/BR	Injector 4 Driver	1-4 ms
17	TN	Injector 2 Driver	1-4 ms
18-19	---	Not Used	---
20	WT/BK	Ignition Switch Power (B+)	12-14v
21	---	Not Used	---
22	BK/PK	MIL (lamp) Control	MIL Off: 12v, On: 1v
23	---	Not Used	---
24	DB/LG	Knock Sensor Signal	0.080v AC
25	---	Not Used	---
26	TN/BK	ECT Sensor Signal	At 180°F: 2.80v
27-29	---	Not Used	---
30	BK/DG	HO2S-11 (B1 S1) Signal	0.1-1.1v
31	---	Not Used	---
32	GY/BK	CKP Sensor Signal	Digital Signals: 0-5-0v
33	TN/YL	CMP Sensor Signal	Digital Signals: 0-5-0v
34	---	Not Used	---
35	OR/DB	TP Sensor Signal	0.6-1.0v
36	DG/RD	MAP Sensor Signal	1.5-1.6v
37	BK/RD	IAT Sensor Signal	At 100°F: 1.83v
38-40	---	Not Used	---

Pin Connector Graphic

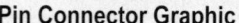

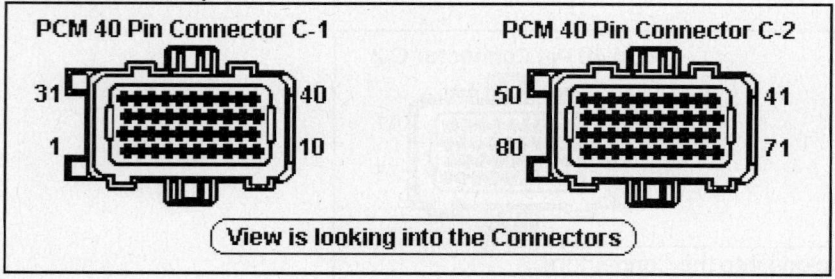

PCM 40 Pin Connector C-1

31   40
1   10

PCM 40 Pin Connector C-2

50   41
80   71

View is looking into the Connectors

1997 Grand Caravan 2.4L I4 MFI VIN B 'C2' White Connector

PCM Pin #	Wire Color	Circuit Description (40 Pin)	Value at Hot Idle
41	RD/LG	Speed Control Set Switch Signal	S/C & Set Switch On: 3.8v
42	DB	A/C Pressure Switch Signal	A/C On: 0.451-4.850v
43	BK/LB	Sensor Ground	<0.050v
44	OR	8-Volt Supply	7.9-8.1v
45	DG	A/C Switch Signal	A/C Off: 12v, On: 1v
46	RD/WT	Keep Alive Power (B+)	12-14v
47	---	Not Used	---
48	BR/WT	IAC 3 Driver Control	Pulse Signals
49	YL/BK	IAC 2 Driver Control	Pulse Signals
50	BK/TN	Power Ground	<0.1v
51	TN/WT	HO2S-12 (B1 S2) Signal	0.1-1.1v
52-55	---	Not Used	---
56	OR/BK	Lockup Torque Converter	At Cruise w/TCC On: 1v
57	GY/RD	IAC 1 Driver Control	Pulse Signals
58	PK/BK	IAC 4 Driver Control	Pulse Signals
59	PK/BR	CCD Bus (+)	Digital Signals: 0-5-0v
60	WT/BK	CCD Bus (-)	<0.050v
61	PK/WT	5-Volt Supply	4.9-5.1v
62	WT/PK	Brake Switch Sense Signal	Brake Off: 0v, On: 12v
63	YL/DG	Torque Management Request	Digital Signals
64	DB/OR	A/C Clutch Relay Control	A/C Off: 12v, On: 1v
65	PK	SCI Transmit Signal	0v
66	WT/OR	Vehicle Speed Signal	Digital Signal
67	DB/YL	ASD Relay Control	Relay Off: 12v, On: 1v
68	PK/BK	EVAP Purge Solenoid Control	PWM Signal: 0-12-0v
69-71	---	Not Used	---
72	YL/BK	LDP Switch Sense Signal	LDP Switch Closed: 0v
73	LG/DB	Radiator Fan Control Relay	Relay Off: 12v, On: 1v
74	BR	Fuel Pump Relay Control	F/P On: 1v, Off: 12v
75	LG	SCI Receive Signal	5v
76	BK/WT	PNP Switch Sense (EATX)	In P/N: 0v, Others: 5v
77	WT/DG	LDP Solenoid Control	PWM Signal: 0-12-0v
78	TN/RD	Speed Control Vacuum Solenoid	Vacuum Increasing: 1v
79	---	Not Used	---
80	LG/RD	Speed Control Vent Solenoid	Vacuum Decreasing: 1v

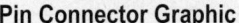

Pin Connector Graphic

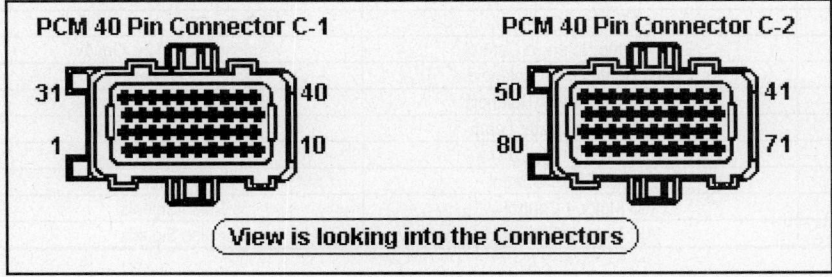

PCM 40 Pin Connector C-1

31   40
1   10

PCM 40 Pin Connector C-2

50   41
80   71

View is looking into the Connectors

1992 Grand Caravan 3.0L V6 MFI VIN 3 (A/T) 60 Pin Connector

PCM Pin #	Wire Color	Circuit Description (60 Pin)	Value at Hot Idle
1	DG/RD	MAP Sensor Signal	1.5-1.6v
2	TN/BK	ECT Sensor Signal	At 180°F: 2.80v
3	RD/WT	Keep Alive Power (B+)	12-14v
4, 5	BK, BK/WT	Sensor Ground	<0.050v
6	PK/WT	5-Volt Supply	4.9-5.1v
7	OR	8-Volt Supply	7.9-8.1v
8, 10, 17-18	---	Not Used	---
9	DB	Ignition Switch Power (B+)	12-14v
11, 12	BK/TN	Power Ground	<0.1v
13	LB/BR	Injector 4 Driver Control	1-4 ms
14	YL/WT	Injector 3 Driver Control	1-4 ms
15	TN	Injector 2 Driver Control	1-4 ms
16	WT/DB	Injector 1 Driver Control	1-4 ms
19	BK/GY	Coil 1 Driver Control	5°, 55 mph: 8° dwell
20	DG	Generator Field Control	Digital Signals: 0-12-0v
21, 28, 32	---	Not Used	---
22	OR/DB	TP Sensor Signal	0.6-1.0v
23	RD/LG	Speed Control Set Switch Signal	S/C & Set Switch On: 3.8v
24	GY/BK	Distributor Reference Signal	Digital Signals: 0-5-0v
25, 45	PK, LG	SCI Transmit, SCI Receive	0v, 5v
26	PK/BR	CCD Bus (+)	Digital Signals: 0-5-0v
27	BR	A/C Damped Pressure Switch	A/C Off: 12v, On: 1v
29	WT/PK	Brake Switch Sense Signal	Brake Off: 0v, On: 12v
30	BR/YL	PNP Switch Sense Signal	In P/N: 0v, Others: 5v
31	DB/PK	Radiator Fan Control Relay	Relay Off: 12v, On: 1v
33	TN/RD	Speed Control Vacuum Solenoid	Vacuum Increasing: 1v
34	DB/OR	A/C Clutch Relay Control	Relay Off: 12v, On: 1v
35	GY/YL	EGR Solenoid Control (California)	12v, at 55 mph: 1v
36-37, 42-43	---	Not Used	---
38	GY	Injector 5 Driver Control	1-4 ms
39	GY/RD	AIS Motor 3 Control	Pulse Signals
40	BR/WT	AIS Motor 1 Control	Pulse Signals
41	BK/DG	HO2S-11 (B1 S1) Signal	0.1-1.1v
44	TN/YL	Distributor Sync Signal	Digital Signals: 0-5-0v
46	WT/BK	CCD Bus (-)	<0.050v
47	WT/OR	Vehicle Speed Signal	Digital Signal
48-50, 55	---	Not Used	---
51	DG/YL	ASD Relay Control	Relay Off: 12v, On: 1v
52	PK/BK	EVAP Purge Solenoid Control	Solenoid Off: 12v, On: 1v
53	LG/RD	Speed Control Vent Solenoid	Vacuum Decreasing: 1v
54	OR/BK	Part Throttle Unlock Solenoid	Solenoid Off: 12v, On: 1v
56	GY/PK	Maintenance Indicator Lamp	Lamp Off: 12v, On: 1v
57	DG/OR	ASD Relay Power (B+)	12-14v
58	BR/DB	Injector 5 Driver Control	1-4 ms
59	PK/BK	AIS Motor 4 Control	Pulse Signals
60	YL/BK	AIS Motor 2 Control	Pulse Signals

Pin Connector Graphic

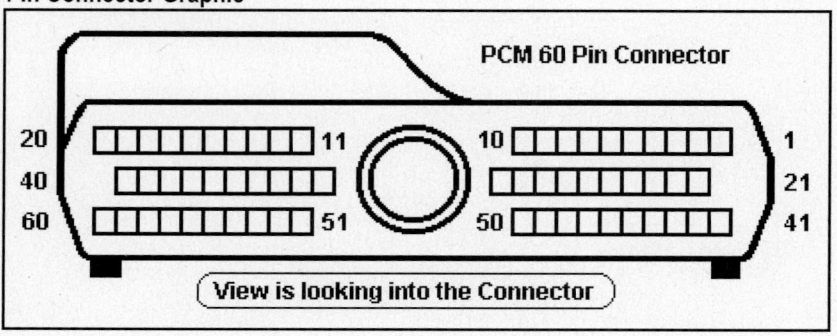

1993 Grand Caravan 3.0L V6 MFI VIN 3 (A/T) 60 Pin Connector

PCM Pin #	Wire Color	Circuit Description (60 Pin)	Value at Hot Idle
1	DG/RD	MAP Sensor Signal	1.5-1.6v
2	TN/BK	ECT Sensor Signal	At 180°F: 2.80v
3	RD/WT	Keep Alive Power (B+)	12-14v
4	BK, BK/WT	Sensor Ground	<0.050v
6	PK/WT	5-Volt Supply	4.9-5.1v
7	OR	8-Volt Supply	7.9-8.1v
8, 17-18, 21, 28	---	Not Used	---
9	DB	Ignition Switch Power (B+)	12-14v
10	---	Not Used	---
11, 12	BK/TN	Power Ground	<0.1v
13	LB/BR	Injector 4 Driver Control	1-4 ms
14	YL/WT	Injector 3 Driver Control	1-4 ms
15	TN	Injector 2 Driver Control	1-4 ms
16	WT/DB	Injector 1 Driver Control	1-4 ms
19	BK/GY	Coil 1 Driver Control	5°, 55 mph: 8° dwell
20	DG	Generator Field Control	Digital Signals: 0-12-0v
22	OR/DB	TP Sensor Signal	0.6-1.0v
23	RD/LG	Speed Control Set Switch Signal	S/C & Set Switch On: 3.8v
24	GY/BK	Distributor Reference Signal	Digital Signals: 0-5-0v
25, 45	PK, LG	SCI Transmit, SCI Receive	0v, 5v
26, 46	PK, WT/BK	CCD Bus (+), CCD Bus (-)	Digital Signals: 0-5-0v, <0.050v
27	BR	A/C Damped Pressure Switch	A/C Off: 12v, On: 1v
29	WT/PK	Brake Switch Sense Signal	Brake Off: 0v, On: 12v
30	BR/YL	PNP Switch Sense Signal	In P/N: 0v, Others: 5v
31	DB/PK	Low Speed Fan Control	Relay Off: 12v, On: 1v
32	YL	High Speed Fan Control	Relay Off: 12v, On: 1v
33	TN/RD	Speed Control Vacuum Solenoid	Vacuum Increasing: 1v
34	DB/OR	A/C Clutch Relay Control	Relay Off: 12v, On: 1v
35	GY/YL	EGR Solenoid Control (California)	12v, at 55 mph: 1v
36	GY/PK	Maintenance Indicator Lamp	Lamp Off: 12v, On: 1v
37, 42-43	---	Not Used	---
38	GY	Injector 5 Driver Control	1-4 ms
39	GY/RD	AIS Motor 3 Control	Pulse Signals
40	BR/WT	AIS Motor 1 Control	Pulse Signals
41	BK/DG	HO2S-11 (B1 S1) Signal	0.1-1.1v
44	TN/YL	Distributor Sync Signal	Digital Signals: 0-5-0v
47	WT/OR	Vehicle Speed Signal	Digital Signal
48-50, 55-56	---	Not Used	---
51	DG/Y	ASD Relay Control	Relay Off: 12v, On: 1v
52	PK/BK	EVAP Purge Solenoid Control	Solenoid Off: 12v, On: 1v
53	LG/RD	Speed Control Vent Solenoid	Vacuum Decreasing: 1v
54	OR/BK	Part Throttle Unlock Solenoid	Solenoid Off: 12v, On: 1v
57	DG/OR	ASD Relay Power (B+)	12-14v
58	BR/DB	Injector 5 Driver Control	1-4 ms
59	PK/BK	AIS Motor 4 Control	Pulse Signals
60	YL/BK	AIS Motor 2 Control	Pulse Signals

Pin Connector Graphic

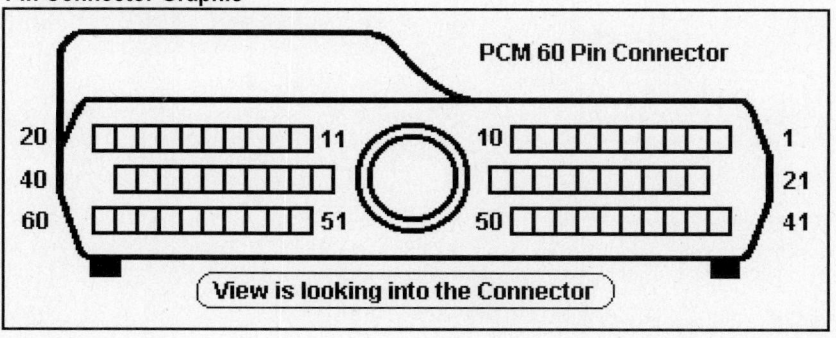

PCM 60 Pin Connector

View is looking into the Connector

1994-95 Grand Caravan 3.0L V6 MFI VIN 3 (A/T) 60 Pin Connector

PCM Pin #	Wire Color	Circuit Description (60 Pin)	Value at Hot Idle
1	DG/RD	MAP Sensor Signal	1.5-1.6v
2	TN/BK	ECT Sensor Signal	At 180ºF: 2.80v
3	RD/WT	Keep Alive Power (B+)	12-14v
4	BK/LB	Sensor Ground	<0.050v
5	BK/WT	Sensor Ground	<0.050v
6	PK/WT	5-Volt Supply	4.9-5.1v
7	OR	8-Volt Supply	7.9-8.1v
8	---	Not Used	---
9	DB/WT	Ignition Switch Power (B+)	12-14v
10	---	Not Used	---
11	BK/TN	Power Ground	<0.1v
12	BK/TN	Power Ground	<0.1v
13	LB/BR	Injector 4 Driver Control	1-4 ms
14	YL/WT	Injector 3 Driver Control	1-4 ms
15	TN	Injector 2 Driver Control	1-4 ms
16	LB	Injector 1 Driver Control	1-4 ms
17-18	---	Not Used	---
19	BK/GY	Coil Driver Control	5º, 55 mph: 8º dwell
20	DG	Alternator Field Control	Digital Signals: 0-12-0v
21	---	Not Used	---
22	DB	TP Sensor Signal	0.6-1.0v
23	RD	Speed Control Set Switch Signal	S/C & Set Switch On: 3.8v
24	GY/P	Distributor Reference Signal	Digital Signals: 0-5-0v
25	PK	SCI Transmit	0v
26	PK/BR	CCD Bus (+)	Digital Signals: 0-5-0v
27	BR	A/C Damped Pressure Switch	A/C Off: 12v, On: 1v
28	---	Not Used	---
29	WT/PK	Brake Switch Sense Signal	Brake Off: 0v, On: 12v
30	BK/OR	PNP Switch Sense Signal	In P/N: 0v, Others: 5v
31	DB/PK	Low Speed Fan Control	Relay Off: 12v, On: 1v
32	---	Not Used	---
33	TN/RD	Speed Control Vacuum Solenoid	Vacuum Increasing: 1v
34	DB/OR	A/C Clutch Relay Control	Relay Off: 12v, On: 1v
35	GY/YL	EGR Solenoid Control	Solenoid Off: 12v, On: 1v
36-37	---	Not Used	---
38	GY	Injector 5 Driver Control	1-4 ms
39	GY/RD	AIS Motor Control (open)	Pulse Signals
40	BR/WT	AIS Motor Control (close)	Pulse Signals

Standard Colors and Abbreviations

Abbreviation	Color	Abbreviation	Color	Abbreviation	Color
BK	Black	GY	Gray	RD	Red
BL	Blue	GN	Green	TN	Tan
BR	Brown	LG	Light Green	VT	Violet
DB	Dark Blue	OR	Orange	WT	White
DG	Dark Green	PK	Pink	YL	Yellow

1994-95 Grand Caravan 3.0L V6 MFI VIN 3 (A/T) 60 Pin Connector

PCM Pin #	Wire Color	Circuit Description (60 Pin)	Value at Hot Idle
41	DG/WT	HO2S-11 (B1 S1) Signal	0.1-1.1v
42-43	---	Not Used	---
44	TN/YL	Distributor Sync Signal	Digital Signals: 0-5-0v
45	LG	SCI Receive	5v
46	WT/BK	CCD Bus (-)	<0.050v
47	WT/OR	Vehicle Speed Signal	Digital Signal
48-50	---	Not Used	---
51	DB/YL	ASD Relay Control	Relay Off: 12v, On: 1v
52	PK/BK	EVAP Purge Solenoid Control	Solenoid Off: 12v, On: 1v
53	LG/RD	Speed Control Vent Solenoid	Vacuum Decreasing: 1v
54	OR/BK	Lockup Torque Converter	At Cruise w/TCC On: 1v
55	YL	High Speed Fan Control	Relay Off: 12v, On: 1v
56	---	Not Used	---
57	DG/OR	ASD Relay Power (B+)	12-14v
58	BR/DB	Injector 6 Driver Control	1-4 ms
59	PK/BK	AIS Motor Control (close)	Pulse Signals
60	YL/BK	AIS Motor Control (close)	Pulse Signals

Pin Connector Graphic

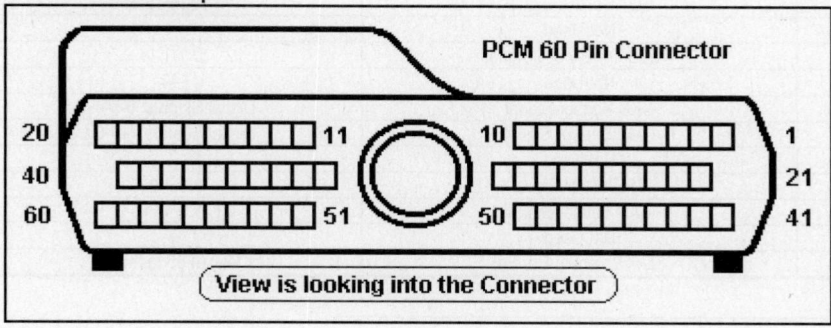

1996 Grand Caravan 3.0L V6 MFI VIN 3 (A/T) 'C1' Black Connector

PCM Pin #	Wire Color	Circuit Description (40 Pin)	Value at Hot Idle
1-3	---	Not Used	---
4	DG	Generator Field Driver	Digital Signals: 0-12-0v
5	YL/RD	Speed Control Power Supply	12-14v
6	DG/OR	ASD Relay Power (B+)	12-14v
7	YL/WT	Injector 3 Driver	1-4 ms
8	TN	Starter Relay Control	KOEC: 9-11v
9	---	Not Used	---
10	BK/TN	Power Ground	<0.1v
11	GY	Coil 1 Driver Control	5°, 55 mph: 8° dwell
12	---	Not Used	---
13	WT/DB	Injector 1 Driver	1-4 ms
14	BR/DB	Injector 6 Driver	1-4 ms
15	GY	Injector 5 Driver	1-4 ms
16	LB/BR	Injector 4 Driver	1-4 ms
17	TN	Injector 2 Driver	1-4 ms
18-19	---	Not Used	---
20	WT/BK	Ignition Switch Power (B+)	12-14v
21-25	---	Not Used	---
26	TN/BK	ECT Sensor Signal	At 180°F: 2.80v
27-29	---	Not Used	---
30	BK/DG	HO2S-11 (B1 S1) Signal	0.1-1.1v
31	---	Not Used	---
32	GY/BK	CKP Sensor Signal	Digital Signals: 0-5-0v
33	TN/YL	CMP Sensor Signal	Digital Signals: 0-5-0v
34	---	Not Used	---
35	OR/DB	TP Sensor Signal	0.6-1.0v
36	DG/RD	MAP Sensor Signal	1.5-1.6v
37-39	---	Not Used	---
40	GY/YL	EGR Solenoid Control	12v, at 55 mph: 1v

Pin Connector Graphic

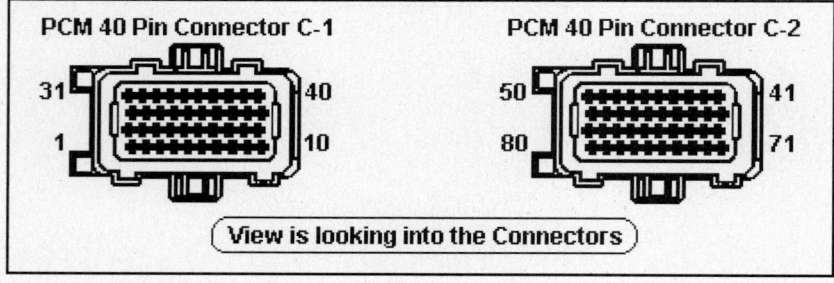

1996 Grand Caravan 3.0L V6 MFI VIN 3 (A/T) 'C2' White Connector

PCM Pin #	Wire Color	Circuit Description (40 Pin)	Value at Hot Idle
41	RD/LG	Speed Control Set Switch Signal	S/C & Set Switch On: 3.8v
42	DB	A/C Pressure Switch Signal	A/C On: 0.451-4.850v
43	BK/LB	Sensor Ground	<0.050v
44	OR	8-Volt Supply	7.9-8.1v
45	DB	A/C Switch Sense Signal	A/C Off: 12v, On: 1v
46	RD/WT	Keep Alive Power (B+)	12-14v
47	---	Not Used	---
48	BR/WT	IAC 1 Driver Control	Pulse Signals
49	YL/BK	IAC 2 Driver Control	Pulse Signals
50	BK/TN	Power Ground	<0.1v
51	TN/WT	HO2S-12 (B1 S2) Signal	0.1-1.1v
52-55	---	Not Used	---
56	OR/BK	Lockup Torque Converter	At Cruise w/TCC On: 1v
57	GY/RD	IAC 3 Driver Control	Pulse Signals
58	PK/BK	IAC 4 Driver Control	Pulse Signals
59	PK/BR	CCD Bus (+)	Digital Signals: 0-5-0v
60	WT/BK	CCD Bus (-)	<0.050v
61	PK/WT	5-Volt Supply	4.9-5.1v
62	WT/PK	Brake Switch Sense Signal	Brake Off: 0v, On: 12v
63	YL/DG	Torque Management Request	Digital Signals
64	DB/OR	A/C Clutch Relay Control	Relay Off: 12v, On: 1v
65	PK	SCI Transmit Signal	0v
66	WT/OR	Vehicle Speed Signal	Digital Signal
67	DB/YL	ASD Relay Control	Relay Off: 12v, On: 1v
68	PK/BK	EVAP Purge Solenoid Control	PWM Signal: 0-12-0v
69-71	---	Not Used	---
72	YL/BK	LDP Switch Sense Signal	LDP Switch Closed: 0v
73	LG/DG	Radiator Fan Control Relay	Relay Off: 12v, On: 1v
74	BR	Fuel Pump Relay Control	Relay Off: 12v, On: 1v
75	LG	SCI Receive Signal	5v
76	BK/WT	PNP Switch Sense (EATX)	In P/N: 0v, Others: 5v
77	WT/DG	LDP Solenoid Control	PWM Signal: 0-12-0v
78	TN/RD	Speed Control Vacuum Solenoid	Vacuum Increasing: 1v
79	---	Not Used	---
80	LG/RD	Speed Control Vent Solenoid	Vacuum Decreasing: 1v

Pin Connector Graphic

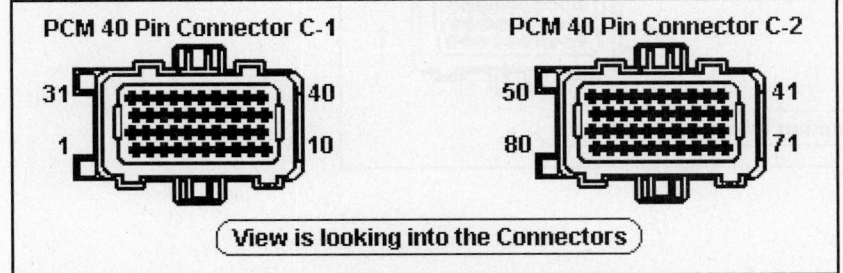

PCM 40 Pin Connector C-1
31 — 40
1 — 10

PCM 40 Pin Connector C-2
50 — 41
80 — 71

View is looking into the Connectors

1997-98 Grand Caravan 3.0L V6 MFI VIN 3 'C1' Black Connector

PCM Pin #	Wire Color	Circuit Description (40 Pin)	Value at Hot Idle
1-3	---	Not Used	---
4	DG	Generator Field Driver	Digital Signals: 0-12-0v
5	YL/RD	Speed Control Power Supply	12-14v
6	DG/OR	ASD Relay Power (B+)	12-14v
7	YL/WT	Injector 3 Driver	1-4 ms
8	TN	Starter Relay Control	KOEC: 9-11v
9	---	Not Used	---
10	BK/TN	Power Ground	<0.1v
11	GY	Coil 1 Driver Control	5°, 55 mph: 8° dwell
12	---	Not Used	---
13	WT/DB	Injector 1 Driver	1-4 ms
14	BR/DB	Injector 6 Driver	1-4 ms
15	GY	Injector 5 Driver	1-4 ms
16	LB/BR	Injector 4 Driver	1-4 ms
17	TN	Injector 2 Driver	1-4 ms
18-19	---	Not Used	---
20	WT/BK	Ignition Switch Power (B+)	12-14v
21	---	Not Used	---
22	BK/PK	MIL (lamp) Control	MIL Off: 12v, On: 1v
23-25	---	Not Used	---
26	TN/BK	ECT Sensor Signal	At 180°F: 2.80v
27-29	---	Not Used	---
30	BK/DG	HO2S-11 (B1 S1) Signal	0.1-1.1v
31	---	Not Used	---
32	GY/BK	CKP Sensor Signal	Digital Signals: 0-5-0v
33	TN/YL	CMP Sensor Signal	Digital Signals: 0-5-0v
34	---	Not Used	---
35	OR/DB	TP Sensor Signal	0.6-1.0v
36	DG/RD	MAP Sensor Signal	1.5-1.6v
37-39	---	Not Used	---
40	GY/YL	EGR Solenoid Control	12v, at 55 mph: 1v

Pin Connector Graphic

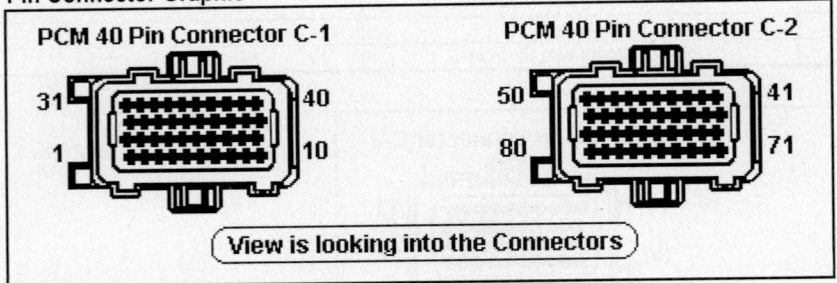

PCM 40 Pin Connector C-1

31   40
1   10

PCM 40 Pin Connector C-2

50   41
80   71

View is looking into the Connectors

1997-98 Grand Caravan 3.0L V6 MFI VIN 3 'C2' White Connector

PCM Pin #	Wire Color	Circuit Description (40 Pin)	Value at Hot Idle
41	RD/LG	Speed Control Set Switch Signal	S/C & Set Switch On: 3.8v
42	DB	A/C Pressure Switch Signal	A/C On: 0.451-4.850v
43	BK/LB	Sensor Ground	<0.050v
44	OR	8-Volt Supply	7.9-8.1v
45	DG	A/C Switch Sense Signal	A/C Off: 12v, On: 1v
46	RD/WT	Keep Alive Power (B+)	12-14v
47	---	Not Used	---
48	BR/WT	IAC 3 Driver Control	Pulse Signals
49	YL/BK	IAC 2 Driver Control	Pulse Signals
50	BK/TN	Power Ground	<0.1v
51	TN/WT	HO2S-12 (B1 S2) Signal	0.1-1.1v
52-55	---	Not Used	---
56	OR/BK	Lockup Torque Converter	At Cruise w/TCC On: 1v
57	GY/RD	IAC 1 Driver Control	Pulse Signals
58	PK/BK	IAC 4 Driver Control	Pulse Signals
59	PK/BR	CCD Bus (+)	Digital Signals: 0-5-0v
60	WT/BK	CCD Bus (-)	<0.050v
61	PK/WT	5-Volt Supply	4.9-5.1v
62	WT/PK	Brake Switch Sense Signal	Brake Off: 0v, On: 12v
63	YL/DG	Torque Management Request	Digital Signals
64	DB/OR	A/C Clutch Relay Control	Relay Off: 12v, On: 1v
65	PK	SCI Transmit Signal	0v
66	WT/OR	Vehicle Speed Signal	Digital Signal
67	DB/YL	ASD Relay Control	Relay Off: 12v, On: 1v
68	PK/BK	EVAP Purge Solenoid Control	PWM Signal: 0-12-0v
69-71	---	Not Used	---
72	YL/BK	LDP Switch Sense Signal	LDP Switch Closed: 0v
73	LG/DB	Radiator Fan Control Relay	Relay Off: 12v, On: 1v
74	BR	Fuel Pump Relay Control	Relay Off: 12v, On: 1v
75	LG	SCI Receive Signal	5v
76	BK/WT	PNP Switch Sense (EATX)	In P/N: 0v, Others: 5v
77	WT/DG	LDP Solenoid Control	PWM Signal: 0-12-0v
78	TN/RD	Speed Control Vacuum Solenoid	Vacuum Increasing: 1v
79	---	Not Used	---
80	LG/RD	Speed Control Vent Solenoid	Vacuum Decreasing: 1v

Pin Connector Graphic

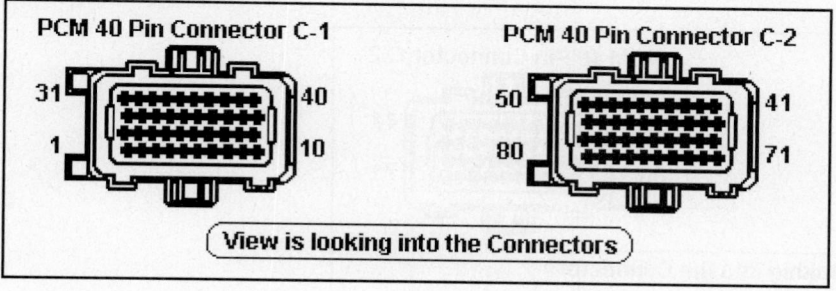

PCM 40 Pin Connector C-1

PCM 40 Pin Connector C-2

View is looking into the Connectors

1999-2000 Grand Caravan 3.0L V6 MFI VIN 3 'C1' Black Connector

PCM Pin #	Wire Color	Circuit Description (40 Pin)	Value at Hot Idle
1	---	Not Used	---
2	RD/YL	Coil 3 Driver Control	5°, 55 mph: 8° dwell
3	DB/TN	Coil 2 Driver Control	5°, 55 mph: 8° dwell
4 ('99)	DG	Generator Field Driver	Digital Signals: 0-12-0v
5	YL/RD	Speed Control Power Supply	12-14v
6	DG/OR	ASD Relay Power (B+)	12-14v
7	YL/WT	Injector 3 Driver	1-4 ms
8	TN	Smart Start Relay Control	KOEC: 9-11v
8 ('00)	DG	Generator Field Driver	Digital Signals: 0-12-0v
9	---	Not Used	---
10	BK/TN	Power Ground	<0.1v
11	GY/RD	Coil 1 Driver Control	5°, 55 mph: 8° dwell
12	---	Not Used	---
13	WT/DB	Injector 1 Driver	1-4 ms
14	BR/DB	Injector 6 Driver	1-4 ms
15	GY	Injector 5 Driver	1-4 ms
16	LB/BR	Injector 4 Driver	1-4 ms
17	TN/WT	Injector 2 Driver	1-4 ms
18-19	---	Not Used	---
20	WT/BK	Ignition Switch Power (B+)	12-14v
21	---	Not Used	---
22	BK/PK	MIL (lamp) Control	MIL Off: 12v, On: 1v
23-24	---	---	---
25	DB/LG	Knock Sensor Signal	0.080v AC
26	TN/BK	ECT Sensor Signal	At 180°F: 2.80v
27	BK/OR	HO2S Ground	<0.050v
28-29	---	---	---
30	BK/DG	HO2S-11 (B1 S1) Signal	0.1-1.1v
31	---	Not Used	---
32	GY/BK	CKP Sensor Signal	Digital Signals: 0-5-0v
33	TN/YL	CMP Sensor Signal	Digital Signals: 0-5-0v
34	---	Not Used	---
35	OR/DB	TP Sensor Signal	0.6-1.0v
36	DG/RD	MAP Sensor Signal	1.5-1.6v
37-39	---	Not Used	---
40	GY/YL	EGR Solenoid Control	12v, at 55 mph: 1v

Pin Connector Graphic

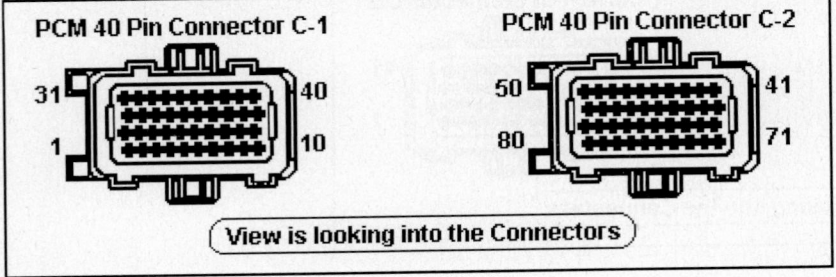

PCM 40 Pin Connector C-1

PCM 40 Pin Connector C-2

View is looking into the Connectors

1999-2000 Grand Caravan 3.0L V6 MFI VIN 3 'C2' White Connector

PCM Pin #	Wire Color	Circuit Description (40 Pin)	Value at Hot Idle
41	RD/LG	Speed Control Set Switch Signal	S/C & Set Switch On: 3.8v
42	DB	A/C Pressure Switch Signal	A/C On: 0.451-4.850v
43	BK/LB	Sensor Ground	<0.050v
44	OR	8-Volt Supply	7.9-8.1v
45 ('99)	DG/LB	A/C Switch Sense Signal	A/C Off: 12v, On: 1v
46	RD/WT	Keep Alive Power (B+)	12-14v
47	---	Not Used	---
48	BR/WT	IAC 3 Driver Control	Pulse Signals
49	YL/BK	IAC 2 Driver Control	Pulse Signals
50	BK/TN	Power Ground	<0.1v
51	TN/WT	HO2S-12 (B1 S2) Signal	0.1-1.1v
52-56	---	Not Used	---
57	GY/RD	IAC 1 Driver Control	Pulse Signals
58	PK/BK	IAC 4 Driver Control	Pulse Signals
59	PK/BR	CCD Bus (+)	Digital Signals: 0-5-0v
60	WT/BK	CCD Bus (-)	<0.050v
61	PK/WT	5-Volt Supply	4.9-5.1v
62	WT/PK	Brake Switch Sense Signal	Brake Off: 0v, On: 12v
63	YL/DG	Torque Management Request	Digital Signals
64	DB/OR	A/C Clutch Relay Control	Relay Off: 12v, On: 1v
65	PK	SCI Transmit Signal	0v
66	WT/OR	Vehicle Speed Signal	Digital Signal
67	DB/YL	ASD Relay Control	Relay Off: 12v, On: 1v
68	PK/BK	EVAP Purge Solenoid Control	PWM Signal: 0-12-0v
69-71	---	Not Used	---
72 ('99)	YL/BK	LDP Switch Sense Signal	LDP Switch Closed: 0v
73	LG/DB	Radiator Fan Control Relay	Relay Off: 12v, On: 1v
74	BR	Fuel Pump Relay Control	Relay Off: 12v, On: 1v
75	LG	SCI Receive Signal	5v
76	BR/YL	PNP Switch Sense (EATX)	In P/N: 0v, Others: 5v
77 ('99)	WT/DG	LDP Solenoid Control	PWM Signal: 0-12-0v
78	TN/RD	Speed Control Vacuum Solenoid	Vacuum Increasing: 1v
79	---	Not Used	---
80	LG/RD	Speed Control Vent Solenoid	Vacuum Decreasing: 1v

Pin Connector Graphic

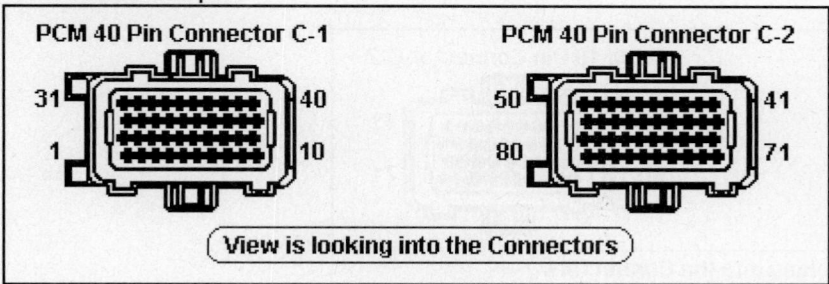

PCM 40 Pin Connector C-1

31   40
1   10

PCM 40 Pin Connector C-2

50   41
80   71

View is looking into the Connectors

2002 Grand Caravan 3.0L V6 OHV Flexible Fuel VIN 3 'C1' Black Connector

PCM Pin #	Wire Color	Circuit Description (40 Pin)	Value at Hot Idle
1	---	Not Used	---
2	DB/OR	Coil 3 Driver Control	5°, 55 mph: 8° dwell
3	DB/TN	Coil 2 Driver Control	5°, 55 mph: 8° dwell
4	---	Not Used	---
5	VT/YL	Speed Control Power Supply	12-14v
6	BR/WT	ASD Relay Power (B+)	12-14v
7	BR/LB	Injector 3 Driver	1-4 ms
8	BR/GY	Generator Field Driver	Digital Signals: 0-12-0v
9	---	Not Used	---
10	BK/BR	Power Ground	<0.1v
11	DB/DG	Coil 1 Driver Control	5°, 55 mph: 8° dwell
12	VT/GY	Engine Oil Pressure Sensor	1.6v at 24 psi
13	BR/YL	Injector 1 Driver	1-4 ms
14	BR/VT	Injector 6 Driver	1-4 ms
15	BR/OR	Injector 5 Driver	1-4 ms
16	BR/TN	Injector 4 Driver	1-4 ms
17	BR/DB	Injector 2 Driver	1-4 ms
18	BR/LG	HO2S-11 (B1 S1) Heater Control	Heater Off: 12v, On: 1v
19	---	Not Used	---
20	PK/GY	Ignition Switch Power (B+)	12-14v
21-24	---	Not Used	---
25	DB/YL	Knock Sensor Signal	0.080v AC
26	VT/OR	ECT Sensor Signal	At 180°F: 2.80v
27	BR/DG	Oxygen Sensor Ground (Both)	<0.050v
28-29	---	Not Used	---
30	DB/LB	HO2S-11 (B1 S1) Signal	0.1-1.1v
31	DG/OR	Starter Motor Relay Control	Relay Off: 12v, On: 1v
32	BR/LB	CKP Sensor Signal	Digital Signals: 0-5-0v
33	DB/GY	CMP Sensor Signal	Digital Signals: 0-5-0v
34	---	Not Used	---
35	BR/OR	TP Sensor Signal	0.6-1.0v
36	VT/BR	MAP Sensor Signal	1.5-1.6v
37	DB/LG	Intake Air Temperature Sensor	At 100°F: 1.83v
38-39	---	Not Used	---
40	DB/VT	EGR Solenoid Control	12v, at 55 mph: 1v

Pin Connector Graphic

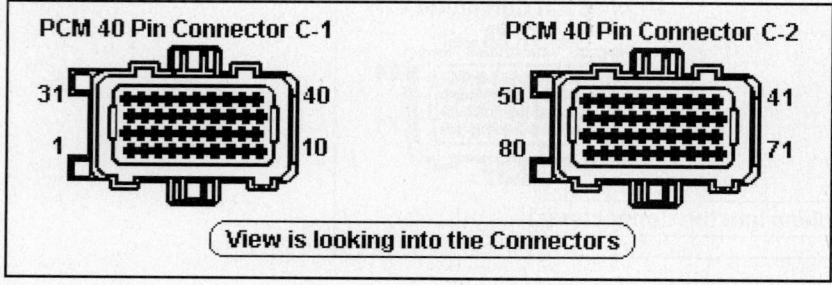

PCM 40 Pin Connector C-1

31 — 40
1 — 10

PCM 40 Pin Connector C-2

50 — 41
80 — 71

View is looking into the Connectors

2002 Grand Caravan 3.3L V6 OHV Flexible Fuel VIN 3 'C2' White Connector

PCM Pin #	Wire Color	Circuit Description (40 Pin)	Value at Hot Idle
41	VT	Speed Control Switch Signal	S/C & Set Switch On: 3.8v
42	LB/BR	A/C Pressure Sensor	A/C On: 0.451-4.850v
43	DB/DG	Sensor Ground	<0.050v
44	BR/PK	8-Volt Supply	7.9-8.1v
45	---	Not Used	---
46	OR/RD	Keep Alive Power (B+)	12-14v
47-48	---	Not Used	---
49	VT/DG	IAC 1 Driver Control	Pulse Signals
50	BK/DG	Power Ground	<0.1v
51	DB/YL	HO2S-12 (B1 S2) Signal	0.1-1.1v
52-55	---	Not Used	---
56	VT/YL	Speed Control Vacuum Solenoid	Vacuum Increasing: 1v
57	VT/LG	IAC 2 Driver Control	Pulse Signals
58	---	Not Used	---
59	WT/VT	PCI Data Bus (J1850)	Digital Signals: 0-7-0v
60	---	Not Used	---
61	PK/YL	5-Volt Supply	4.9-5.1v
62	DG/WT	Secondary Brake Switch Sense	Brake Off: 0v, On: 12v
63	DG/LG	Torque Management Request	Digital Signals
64	LB/OR	A/C Clutch Relay Control	Relay Off: 12v, On: 1v
65	WT/BR	SCI Transmit	0v
66	DB/OR	Vehicle Speed Signal	Digital Signals
67	BR/WT	ASD Relay Control	Relay Off: 12v, On: 1v
68	DB/WT	EVAP Purge Solenoid Control	PWM Signal: 0-12-0v
69	---	Not Used	---
70	DB/BR	EVAP Purge Sense	<0.1v
71	---	Not Used	---
72	VT/WT	LDP Switch Sense Signal	LDP Switch Closed: 0v
73	BR/VT	Radiator Fan Control Relay	Relay Off: 12v, On: 1v
74	BR	Fuel Pump Relay Control	Relay Off: 12v, On: 1v
75	WT/LG	SCI Receive	5v
76	YL/DB	PNP Switch Sense (TRS T41)	In P/N: 0v, Others: 5v
77	VT/LB	LDP Solenoid Control	PWM Signal: 0-12-0v
78	DB/WT	Torque Converter Clutch Solenoid Control	Solenoid Off: 12v, On: PWM 0-12-0v
79	---	Not Used	---
80	VT/OR	Speed Control Vent Solenoid	Vacuum Decreasing: 1v

Pin Connector Graphic

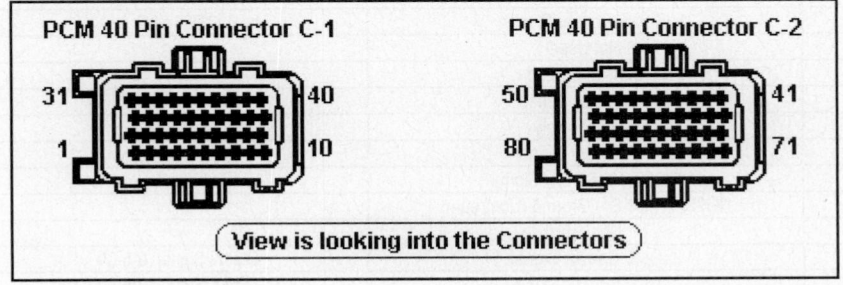

PCM 40 Pin Connector C-1   PCM 40 Pin Connector C-2

View is looking into the Connectors

2003 Grand Caravan 3.3L V6 OHV Flexible Fuel VIN 3 'C1' Black Connector

PCM Pin #	Wire Color	Circuit Description (38 Pin)	Value at Hot Idle
1-8, 10	---	Not Used	---
9	BK/BR	Power Ground	<0.1v
11	PK/GY	Ignition Switch Output (Run-Start)	12-14v
12	PK/WT	Ignition Switch Output (Off-Run-Start)	12-14v
13-17, 19	---	Not Used	---
18	BK/DG	Power Ground	<0.1v
20	VT/GY	Oil Pressure Sensor	A/C On: 0.45-4.85v
21, 23-24	---	Not Used	---
22	VT/LG	Ambient Air Temperature Sensor	At 100°F: 1.83v
25	WT/LG	SCI Receive (PCM)	5v
26	WT/BR	Flash Program Enable	5v
27-28	---	Not Used	---
29	OR/RD	Battery Power (B+)	12-14v
30	YL	Ignition Switch Output (Start)	8-11v (cranking)
31	DB/YL	HO2S-12 (B1 S2) Signal	0.1-1.1v
32	DB/DG	Oxygen Sensor Return (down)	<0.050v
33-35	---	Not Used	---
36	WT/BR	SCI Transmit (PCM)	0v
37	DG/YL	SCI Transmit (TCM)	0v
38	WT/VT	PCI Data Bus (J1850)	Digital Signals: 0-7-0v

2003 Grand Caravan 3.3L V6 OHV Flexible Fuel VIN 3 'C2' Gray Connector

PCM Pin #	Wire Color	Circuit Description (38 Pin)	Value at Hot Idle
1-8	---	Not Used	---
9	DB/TN	Coil 2 Driver	5°, at 55 mph: 8° dwell
10	DB/DG	Coil 1 Driver	5°, at 55 mph: 8° dwell
11	BR/TN	Injector 4 Driver Control	1.0-4.0 ms
12	BR/LB	Injector 3 Driver Control	1.0-4.0 ms
13	BR/DB	Injector 2 Driver Control	1.0-4.0 ms
14	BR/YL	Injector 1 Driver Control	1.0-4.0 ms
15-17	---	Not Used	---
18	BR/LG	HO2S-12 (B1 S2) Heater Control	Heater Off: 12v, On: 1v
19	BR/GY	Generator Field Driver	Digital Signal: 0-12-0v
20	VT/OR	ECT Sensor Signal	At 180°F: 2.80v
21	BR/OR	TP Sensor Signal	0.6-1.0v
22	---	Not Used	---
23	VT/BR	MAP Sensor Signal	1.5-1.7v
24	BR/LG	Knock Sensor Return	<0.050v
25	DB/YL	Knock Sensor Signal	0.080v AC
26	---	Not Used	---
27	DB/DG	Sensor Ground	<0.1v
28	BR/VT	IAC Motor Sense	12-14v
29	PK/YL	5-Volt Supply	4.9-5.1v
30	DB/LG	IAT Sensor Signal	At 100°F: 1.83v
31	DB/LB	HO2S-11 (B1 S1) Signal	0.1-1.1v
32	BR/DG	HO2S-11 (B1 S1) Ground (Upstream)	<0.1v
33	---	Not Used	---
34	DB/GY	CMP Sensor Signal	Digital Signal: 0-5-0v
35	BR/LB	CKP Sensor Signal	Digital Signal: 0-5-0v
36-37	---	Not Used	---
38	VT/GY	IAC Motor Driver	DC pulse signals: 0.8-11v

Pin Connector Graphic

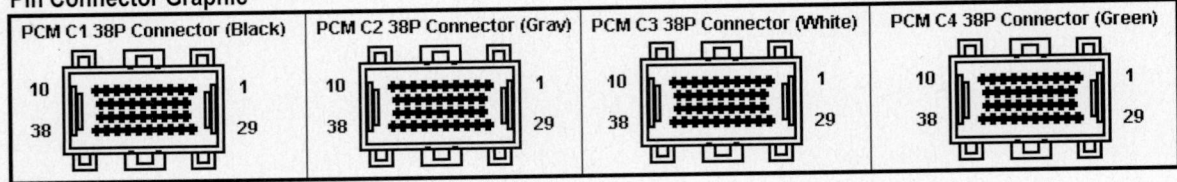

| PCM C1 38P Connector (Black) | PCM C2 38P Connector (Gray) | PCM C3 38P Connector (White) | PCM C4 38P Connector (Green) |

2003 Grand Caravan 3.3L V6 OHV Flexible Fuel VIN 3 'C3' White Connector

PCM Pin #	Wire Color	Circuit Description (38 Pin)	Value at Hot Idle
1-2, 4	---	Not Used	---
3	BR/WT	ASD Relay Control	Relay Off: 12v, On: 1v
5	VT/OR	Speed Control Vent Solenoid	Vacuum Decreasing: 1v
6	BR/VT	Radiator Fan Relay Control	Relay Off: 12v, On: 1v
7	VT/YL	Speed Control Power Supply	12-14v
8	WT/DG	Natural Vacuum Leak Detection Solenoid	Solenoid Off: 12v, On: 1v
9	BR/WT	HO2S-12 (B1 S2) Heater Control	Heater On: 1v, Off: 12v
10	---	Not Used	
11	LB/OR	A/C Clutch Relay Control	Relay Off: 12v, On: 1v
12	VT/YL	Speed Control Vacuum Solenoid	Vacuum Increasing: 1v
13-18	---	Not Used	---
18, 19	BR/WT	Automatic Shutdown Relay Output	12-14v
20	DB/WT	EVAP Purge Solenoid Control	PWM Signal: 0-12-0v
21-22	---	Not Used	---
23	DG/WT	Brake Switch Sense	Brake Off: 0v, On: 12v
24-27	---	Not Used	---
28	BR/WT	Automatic Shutdown Relay Output	12-14v
29	DB/BR	EVAP Purge Solenoid Sense	0-1v
30-31, 33, 36	---	Not Used	---
31	LB/BR	A/C Pressure Sensor	A/C On: 0.451-4.850v
32	DB/YL	Battery Temperature Sensor	At 100°F: 1.83v
34	VT	Speed Control Set Switch Signal	S/C & Set Switch On: 3.8v
35	VT/WT	Natural Vacuum Leak Detection Switch Sense	0.1v
37	BR	Fuel Pump Relay Control	Relay Off: 12v, On: 1v
38	DG/OR	Starter Relay Control	KOEC: 9-11v

2003 Grand Caravan 3.3L V6 OHV Flexible Fuel VIN 3 'C4' Green Connector

PCM Pin #	Wire Color	Circuit Description (38 Pin)	Value at Hot Idle
1	YL/GY	Overdrive Solenoid Control	Solenoid Off: 12v, On: 1v
2	YL/LB	Underdrive Solenoid Control	Solenoid Off: 12v, On: 1v
3-5	---	Not Used	---
6	YL/LB	2-4 Solenoid Control	Solenoid Off: 12v, On: 1v
7-9, 11	---	Not Used	---
10	DG/WT	Low/Reverse Solenoid Control	Solenoid Off: 12v, On: 1v
12, 13	BK/LG	Power Ground	<0.050v
14	---	Not Used	---
15	DG/LB	TRS T1 Sense	<0.050v
16	DG/DB	TRS T3 Sense	<0.050v
17, 20-21	---	Not Used	---
18	YL/BR	Transmission Control Relay Control	Relay Off: 12v, On: 1v
19	YL/OR, 38	Transmission Control Relay Output	Relay Off: 12v, On: 1v
23-26	---	Not Used	---
27	DG/GY	TRS T41 Sense	<0.050v
28, 38	YL/OR	Transmission Control Relay Output	Relay Off: 12v, On: 1v
29	YL/TN	Low/Reverse Pressure Switch Sense	12-14v
30	YL/DG	2-4 Pressure Switch Sense	In Low/Reverse: 2-4v
31, 36	---	Not Used	---
32	DG/BR	Output Speed Sensor Signal	In 2-4 Position: 2-4v
33	DG/WT	Input Speed Sensor Signal	Moving: AC voltage
34	DG/VT	Speed Sensor Ground	Moving: AC voltage
35	DG/OR	Transmission Temperature Sensor Signal	<0.050v
37	DG/YL	TRS T42 Sense	In PRNL: 0v, Others 5v

Pin Connector Graphic

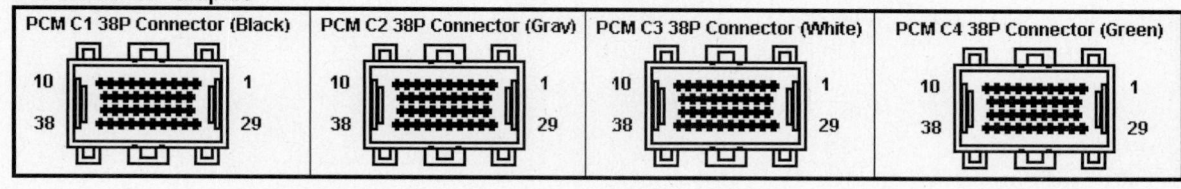

1998-2000 Grand Caravan 3.3L V6 Flexible Fuel VIN G 'C1' Black Connector

PCM Pin #	Wire Color	Circuit Description (40 Pin)	Value at Hot Idle
1	---	Not Used	---
2	RD/YL	Coil 3 Driver (Cyl 3 & 6)	5°, 55 mph: 8° dwell
3	DB/TN	Coil 2 Driver (Cyl 2 & 5)	5°, 55 mph: 8° dwell
4	---	Not Used	---
5	YL/RD	Speed Control Power Supply	12-14v
6	DG/OR	ASD Relay Power (B+)	12-14v
7	YL/WT	Injector 3 Driver	1-4 ms
8	DG	Generator Field Driver	Digital Signals: 0-12-0v
9	---	Not Used	---
10	BK/TN	Power Ground	<0.1v
11	GY/RD	Coil 1 Driver (Cyl 1 & 4)	5°, 55 mph: 8° dwell
12	---	Not Used	---
13	WT/DB	Injector 1 Driver	1-4 ms
14	BR/DB	Injector 6 Driver	1-4 ms
15	GY	Injector 5 Driver	1-4 ms
16	LB/BR	Injector 4 Driver	1-4 ms
17	TN/WT	Injector 2 Driver	1-4 ms
18-19	---	Not Used	---
20	WT/BK	Ignition Switch Power (B+)	12-14v
21	---	Not Used	---
22	BK/PK	MIL (lamp) Control	MIL Off: 12v, On: 1v
23-24	---	Not Used	---
25	DB/LG	Knock Sensor Signal	0.080v AC
26	TN/BK	ECT Sensor Signal	At 180°F: 2.80v
27	BK/OR	Oxygen Sensor Ground	<0.050v
28-29	---	Not Used	---
30	BK/DG	HO2S-11 (B1 S1) Signal	0.1-1.1v
31	TN	Smart Start Relay Control	KOEC: 9-11v
32	GY/BK	CKP Sensor Signal	Digital Signals: 0-5-0v
33	TN/YL	CMP Sensor Signal	Digital Signals: 0-5-0v
34	---	Not Used	---
35	OR/DB	TP Sensor Signal	0.6-1.0v
36	DG/RD	MAP Sensor Signal	1.5-1.6v
37	---	Not Used	---
38 ('99)	DG	A/C Switch Sense Signal	A/C Off: 12v, On: 1v
38	DG/LB	A/C Switch Sense Signal	A/C Off: 12v, On: 1v
39	---	Not Used	---
40	GY/YL	EGR Solenoid Control	12v, at 55 mph: 1v

Pin Connector Graphic

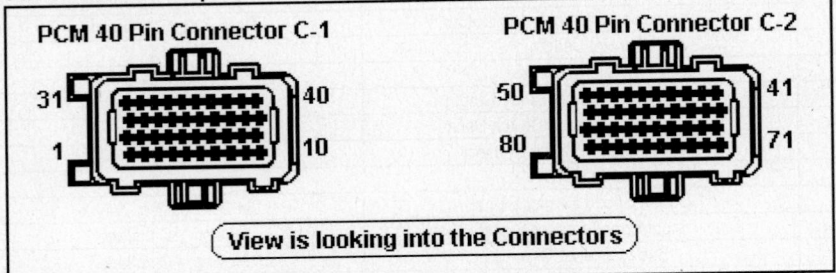

PCM 40 Pin Connector C-1

PCM 40 Pin Connector C-2

31 ... 40
1 ... 10

50 ... 41
80 ... 71

(View is looking into the Connectors)

1998-2000 Grand Caravan 3.3L V6 Flexible Fuel VIN G 'C2' White Connector

PCM Pin #	Wire Color	Circuit Description (40 Pin)	Value at Hot Idle
41	RD/LG	Speed Control Set Switch Signal	S/C & Set Switch On: 3.8v
42	DB	A/C Pressure Switch Signal	A/C On: 0.451-4.850v
43	BK/LB	Sensor Ground	<0.050v
44	OR	8-Volt Supply	7.9-8.1v
45	---	Not Used	---
46	RD/WT	Keep Alive Power (B+)	12-14v
47	---	Not Used	---
48	BR/WT	IAC 3 Driver Control	Pulse Signals
49	YL/BK	IAC 2 Driver Control	Pulse Signals
50	BK/TN	Power Ground	<0.1v
51	TN/WT	HO2S-12 (B1 S2) Signal	0.1-1.1v
52-53	---	Not Used	---
54	YL/WT	Flex Fuel Sensor Signal	Digital Signals
56	TN/RD	Speed Control Vacuum Solenoid	Vacuum Increasing: 1v
57	GY/RD	IAC 1 Driver Control	Pulse Signals
58	PK/BK	IAC 4 Driver Control	Pulse Signals
59	PK/BR	CCD Bus (+)	Digital Signals: 0-5-0v
60	WT/BK	CCD Bus (-)	<0.050v
61	PK/WT	5-Volt Supply	4.9-5.1v
62	WT/PK	Brake Switch Sense Signal	Brake Off: 0v, On: 12v
63	YL/DG	Torque Management Request	Digital Signals
64	DB/OR	A/C Clutch Relay Control	Relay Off: 12v, On: 1v
65	PK	SCI Transmit Signal	0v
66	WT/OR	Vehicle Speed Signal	Digital Signal
67	DB/YL	ASD Relay Control	Relay Off: 12v, On: 1v
68	PK/BK	EVAP Purge Solenoid Control	PWM Signal: 0-12-0v
69	---	Not Used	---
70	P/R	EVAP Purge Solenoid Sense	0-1v
71	---	Not Used	---
72	DB/WT	LDP Switch Sense Signal	LDP Switch Closed: 0v
73	LG/DB	Radiator Fan Control Relay	Relay Off: 12v, On: 1v
74	BR	Fuel Pump Relay Control	Relay Off: 12v, On: 1v
75	LG	SCI Receive Signal	5v
76	BK/WT	PNP Switch Sense (EATX)	In P/N: 0v, Others: 5v
77	WT/DG	LDP Solenoid Control	PWM Signal: 0-12-0v
78-79	---	Not Used	---
80	LG/RD	Speed Control Vent Solenoid	Vacuum Decreasing: 1v

Pin Connector Graphic

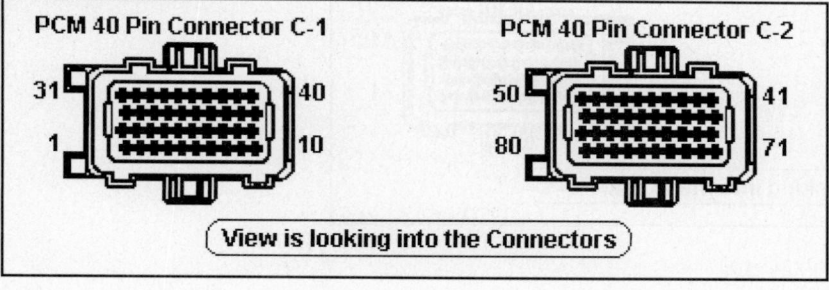

2001 Grand Caravan 3.3L V6 Flexible Fuel VIN G (A/T) 'C1' Black Connector

PCM Pin #	Wire Color	Circuit Description (40 Pin)	Value at Hot Idle
1	---	Not Used	---
2	DB/OR	Coil 3 Driver Control	5°, 55 mph: 8° dwell
3	DB/TN	Coil 2 Driver Control	5°, 55 mph: 8° dwell
4	---	Not Used	---
5	VT/YL	S/C On/Off Switch Sense	12-14v
6	BR/WT	ASD Relay Power (B+)	12-14v
7	BR/LB	Injector 3 Driver	1-4 ms
8	BR/GY	Generator Field Driver	Digital Signals: 0-12-0v
9	---	Not Used	---
10	BR/BR	Power Ground	<0.1v
11	DB/DG	Coil 1 Driver Control	5°, 55 mph: 8° dwell
12	VT/GY	Engine Oil Pressure Sensor	1.6v at 24 psi
13	BR/YL	Injector 1 Driver Control	1-4 ms
14	BR/VT	Injector 6 Driver Control	1-4 ms
15	BR/OR	Injector 5 Driver Control	1-4 ms
16	BR/TN	Injector 4 Driver Control	1-4 ms
17	BR/LB	Injector 2 Driver Control	1-4 ms
18	BR/LG	HO2S-11 (B1 S1) Heater	PWM Signal: 0-12-0v
19	---	Not Used	---
20	PK/GY	Ignition Switch Power (B+)	12-14v
21-24	---	Not Used	---
25	DB/YL	Knock Sensor Signal	0.080v AC
26	VT/OR	ECT Sensor Signal	At 180°F: 2.80v
27	BR/DG	Oxygen Sensor Ground	<0.050v
28-29	---	Not Used	---
30	DB/LB	HO2S-11 (B1 S1) Signal	0.1-1.1v
31	DG/OR	Double Start Override Signal	KOEC: 9-11v
32	BR/LB	CKP Sensor Signal	Digital Signals: 0-5-0v
33	DB/GY	CMP Sensor Signal	Digital Signals: 0-5-0v
34	---	Not Used	---
35	BR/OR	TP Sensor Signal	0.6-1.0v
36	VT/BR	MAP Sensor Signal	1.5-1.6v
37	DB/LG	Intake Air Temp. Signal	At 100°F: 1.83v
38-39	---	Not Used	---
40	DB/VT	EGR Solenoid Control	12v, at 55 mph: 1v

Pin Connector Graphic

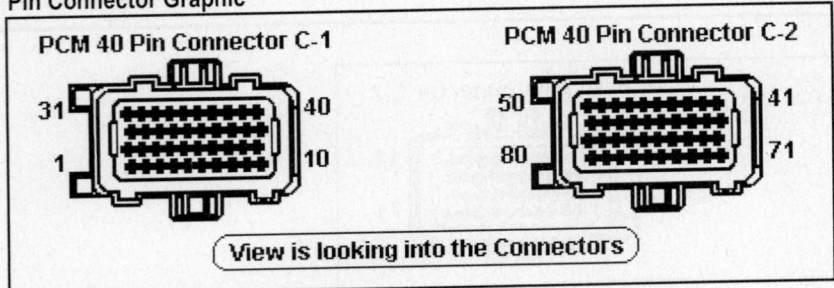

PCM 40 Pin Connector C-1

PCM 40 Pin Connector C-2

View is looking into the Connectors

2001 Grand Caravan 3.3L V6 Flexible Fuel VIN G (A/T) 'C2' White Connector

PCM Pin #	Wire Color	Circuit Description (40 Pin)	Value at Hot Idle
41	VT	Speed Control Set Switch Signal	S/C & Set Switch On: 3.8v
42	LB/BR	A/C Pressure Switch Signal	A/C On: 0.451-4.850v
43	DB/DG	Sensor Ground	<0.050v
44	BR/PK	8-Volt Supply	7.9-8.1v
45	---	Not Used	---
46	OR/RD	Keep Alive Power (B+)	12-14v
47-48	---	Not Used	---
49	VT/DG	IAC 1 Driver Control	Pulse Signals
50	BK/DG	Power Ground	<0.1v
51	DB/YL	HO2S-12 (B1 S2) Signal	0.1-1.1v
52-53	---	Not Used	---
54	YL/WT	Flex Fuel Sensor Signal	Digital Signals
56	VT/YL	Speed Control Vacuum Solenoid	Vacuum Increasing: 1v
57	VT/LG	IAC 2 Driver Control	Pulse Signals
58	---	Not Used	---
59	WT/VT	PCI Data Bus (J1850)	Digital Signals: 0-7-0v
60	---	Not Used	---
61	PK/YL	5-Volt Supply	4.9-5.1v
62	DG/WT	Brake Switch Sense Signal	Brake Off: 0v, On: 12v
63	DG/LG	Torque Management Request	Digital Signals
64	LB/OR	A/C Clutch Relay Control	Relay Off: 12v, On: 1v
65	WT/BR	SCI Transmit Signal	0v
66	DB/OR	Vehicle Speed Signal	Digital Signal
67	BR/WT	ASD Relay Control	Relay Off: 12v, On: 1v
68	DB/WT	EVAP Purge Solenoid Control	PWM Signal: 0-12-0v
69	---	Not Used	---
70	DB/BR	EVAP Purge Solenoid Sense	0-1v
71	---	Not Used	---
72	VT/WT	LDP Switch Sense Signal	LDP Switch Closed: 0v
73	BR/VT	Radiator Fan Control Relay	Relay Off: 12v, On: 1v
74	BR	Fuel Pump Relay Control	Relay Off: 12v, On: 1v
75	WT/LG	SCI Receive Signal	5v
76	YL/LB	TRS T41 Signal	In P/N: 0v, Others: 5v
77	VT/LB	LDP Solenoid Control	PWM Signal: 0-12-0v
78-79	---	Not Used	---
80	LG/RD	Speed Control Vent Solenoid	Vacuum Decreasing: 1v

Pin Connector Graphic

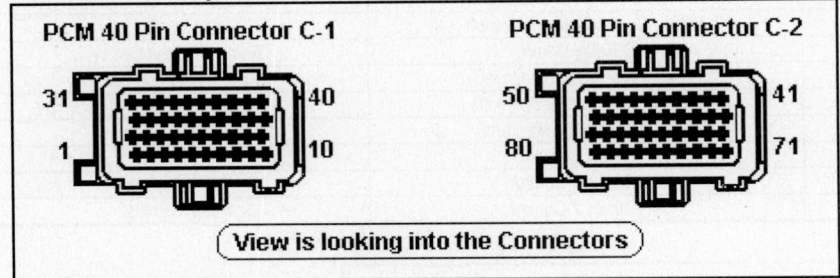

1990 Grand Caravan 3.3L V6 MFI VIN R (A/T) 60 Pin Connector

PCM Pin #	Wire Color	Circuit Description (60 Pin)	Value at Hot Idle
1	DG/RD	MAP Sensor Signal	1.5-1.6v
2	TN/WT	ECT Sensor Signal	At 180°F: 2.80v
3	RD/WT	Keep Alive Power (B+)	12-14v
4, 5	BK, BK	Sensor Ground	<0.050v
6	PK/WT	5-Volt Supply	4.9-5.1v
7	OR	8-Volt Supply	7.9-8.1v
8	DG/BK	Ignition Switch (Start) Output	12-14v
9	DB	Auto Shutdown Relay Output	12-14v
10, 13, 23, 28	---	Not Used	---
11, 12	LB/RD	Power Ground	<0.1v
14	YL/WT	Injector 3 Driver Control	1-4 ms
15	TN	Injector 2 Driver Control	1-4 ms
16	WT/DB	Injector 1 Driver Control	1-4 ms
17	DB/BK	Coil 2 Driver Control	5°, 55 mph: 8° dwell
18	DB/GY	Coil 3 Driver Control	5°, 55 mph: 8° dwell
19	BK/GY	Coil 1 Driver Control	5°, 55 mph: 8° dwell
20	DG	Alternator Field Control	Digital Signals: 0-12-0v
21	BK/RD	Charge Temperature Sensor	At 100°F: 2.51v
22	OR/DB	TP Sensor Signal	0.6-1.0v
24	GY/BK	Distributor Reference Signal	Digital Signals: 0-5-0v
25, 45	PK, LG	SCI Transmit, SCI Receive	0v, 6=5v
26, 46	BK, WT/BK	CCD Bus (+), CCD Bus (-)	Digital Signals: 0-5-0v, <0.050v
27	BR	A/C Clutch Signal	A/C Off: 12v, On: 1v
29	WT/PK	Brake Switch Sense Signal	Brake Off: 0v, On: 12v
30	BR/YL	PNP Switch Sense Signal	In P/N: 0v, Others: 5v
31	DB/PK	Radiator Fan Control Relay	Relay Off: 12v, On: 1v
32	BK/PK	MIL (lamp) Control	MIL Off: 12v, On: 1v
33	TN/RD	Speed Control Vacuum Solenoid	Vacuum Increasing: 1v
34	DB/OR	A/C Clutch Relay Control	Relay Off: 12v, On: 1v
35	GY/YL	EGR Solenoid Control (California)	Solenoid Off: 12v, On: 1v
36-38, 43, 54-55	---	Not Used	---
39, 40	GY, OR/WT	AIS Motor 3, AIS Motor 1 Control	Pulse Signals
41	BK/DG	HO2S-11 (B1 S1) Signal	0.1-1.1v
42	BK/LG	Knock Sensor Signal	0.080v AC
44	TN/YL	Distributor Sync Signal	Digital Signals: 0-5-0v
47	WT/OR	Vehicle Speed Signal	Digital Signal
48	BR/RD	Speed Control Set Switch Signal	S/C & Set Switch On: 3.8v
49	YL/RD	Speed Control On/Off Switch	S/C On: 1v, Off: 12v
50	WT/LG	Speed Control Resume Switch	S/C On: 1v, Off: 12v
51	DB/YL	ASD Relay Control	Relay Off: 12v, On: 1v
52	PK/BK	EVAP Purge Solenoid Control	Solenoid Off: 12v, On: 1v
53	LG/RD	Speed Control Vent Solenoid	Vacuum Decreasing: 1v
56	GY/PK	Maintenance Indicator Lamp	Lamp Off: 12v, On: 1v
57-58	---	Not Used	---
59	PK/BK	AIS Motor 4 Control	Pulse Signals
60	YL/BK	AIS Motor 2 Control	Pulse Signals

Pin Connector Graphic

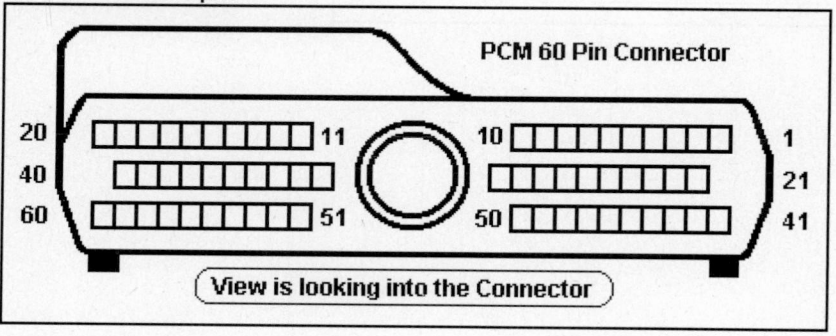

1994-95 Grand Caravan 3.3L V6 MFI VIN R (A/T) 60 Pin Connector

PCM Pin #	Wire Color	Circuit Description (60 Pin)	Value at Hot Idle
1	DG/RD	MAP Sensor Signal	1.5-1.6v
2	TN/BK	ECT Sensor Signal	At 180°F: 2.80v
3	RD/WT	Keep Alive Power (B+)	12-14v
4, 5	BK, BK/WT	Sensor Ground	<0.050v
6	PK/WT	5-Volt Supply	4.9-5.1v
7	OR	8-Volt Supply	7.9-8.1v
8, 10, 21	---	Not Used	---
9	DB/WT	Ignition Switch Power (B+)	12-14v
11, 12	BK/TN	Power Ground	<0.1v
13	LB/BR	Injector 4 Driver Control	1-4 ms
14	YL/WT	Injector 3 Driver Control	1-4 ms
15	TN	Injector 2 Driver Control	1-4 ms
16	LB	Injector 1 Driver Control	1-4 ms
17	DB/YL	Coil 2 Driver Control	5°, 55 mph: 8° dwell
18	DB/GY	Coil 3 Driver Control	5°, 55 mph: 8° dwell
19	BK/GY	Coil 1 Driver Control	5°, 55 mph: 8° dwell
20	DG	Alternator Field Control	Digital Signals: 0-12-0v
22	DB	TP Sensor Signal	0.6-1.0v
23	RD/LG	Speed Control Set Switch Signal	S/C & Set Switch On: 3.8v
24	GY/P	CKP Sensor Signal	Digital Signals: 0-5-0v
25, 45	PK, LG	SCI Transmit, SCI Receive	0v, 6=5v
26, 46	BK, WT/BK	CCD Bus (+), CCD Bus (-)	Digital Signals: 0-5-0v, <0.050v
27	BR	A/C Damped Pressure Switch	A/C Off: 12v, On: 1v
28, 32, 36-37	---	Not Used	---
29	WT/PK	Brake Switch Sense Signal	Brake Off: 0v, On: 12v
30	BK/OR	PNP Switch Sense Signal	In P/N: 0v, Others: 5v
31	DB/PK	Low Speed Fan Control	Relay Off: 12v, On: 1v
33	TN/RD	Speed Control Vacuum Solenoid	Vacuum Increasing: 1v
34	DB/OR	A/C Clutch Relay Control	Relay Off: 12v, On: 1v
35	GY/YL	EGR Solenoid Control	Solenoid Off: 12v, On: 1v
38	GY	Injector 5 Driver Control	1-4 ms
39	GY/RD	AIS Motor Control (open)	Pulse Signals
40	BR/WT	AIS Motor Control (close)	Pulse Signals
41	DG/WT	HO2S-11 (B1 S1) Signal	0.1-1.1v
42-43, 48-50, 54	---	Not Used	---
44	TN/YL	CMP Sensor Signal	Digital Signals: 0-5-0v
47	WT/OR	Vehicle Speed Signal	Digital Signal
51	DB/YL	ASD Relay Control	Relay Off: 12v, On: 1v
52	PK/BK	EVAP Purge Solenoid Control	PWM Signal: 0-12-0v
53	LG/RD	Speed Control Vent Solenoid	Vacuum Decreasing: 1v
55	YL	High Speed Fan Control	Relay Off: 12v, On: 1v
56	GY/PK	Maintenance Indicator Lamp	Lamp Off: 12v, On: 1v
57	DG/OR	ASD Relay Power (B+)	12-14v
58	BR/DB	Injector 6 Driver Control	1-4 ms
59	PK/BK	AIS Motor Control (close)	Pulse Signals
60	YL/BK	AIS Motor Control (close)	Pulse Signals

Pin Connector Graphic

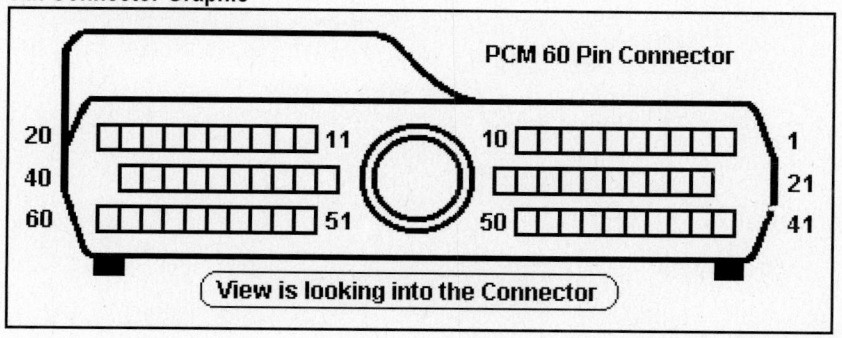

PCM 60 Pin Connector

20 — 11   10 — 1
40   21
60 — 51   50 — 41

View is looking into the Connector

1996 Grand Caravan 3.3L V6 MFI VIN R (A/T) 'C1' Black Connector

PCM Pin #	Wire Color	Circuit Description (40 Pin)	Value at Hot Idle
1	---	Not Used	---
2	RD/YL	Coil 3 Driver (Cyl 3 & 6)	5°, 55 mph: 8° dwell
3	DB/TN	Coil 2 Driver (Cyl 2 & 5)	5°, 55 mph: 8° dwell
4	DG	Generator Field Driver	Digital Signals: 0-12-0v
5	YL/RD	Speed Control Power Supply	12-14v
6	DG/OR	ASD Relay Power (B+)	12-14v
7	YL/WT	Injector 3 Driver	1-4 ms
8	TN	Starter Relay Control	KOEC: 9-11v
9	---	Not Used	---
10	BK/TN	Power Ground	<0.1v
11	GY	Coil 1 Driver (Cyl 1 & 4)	5°, 55 mph: 8° dwell
12	---	Not Used	---
13	WT/DB	Injector 1 Driver	1-4 ms
14	BR/DB	Injector 6 Driver	1-4 ms
15	GY	Injector 5 Driver	1-4 ms
16	LB/BR	Injector 4 Driver	1-4 ms
17	TN	Injector 2 Driver	1-4 ms
18-19	---	Not Used	---
20	WT/BK	Ignition Switch Power (B+)	12-14v
21-23	---	Not Used	---
24	DB/LG	Knock Sensor Signal	0.080v AC
25	---	Not Used	---
26	TN/BK	ECT Sensor Signal	At 180°F: 2.80v
27-29	---	Not Used	---
30	BK/DG	HO2S-11 (B1 S1) Signal	0.1-1.1v
31	---	Not Used	---
32	GY/BK	CKP Sensor Signal	Digital Signals: 0-5-0v
33	TN/YL	CMP Sensor Signal	Digital Signals: 0-5-0v
34	---	Not Used	---
35	OR/DB	TP Sensor Signal	0.6-1.0v
36	DG/RD	MAP Sensor Signal	1.5-1.6v
37-39	---	Not Used	---
40	GY/YL	EGR Solenoid Control	12v, at 55 mph: 1v

Pin Connector Graphic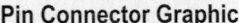

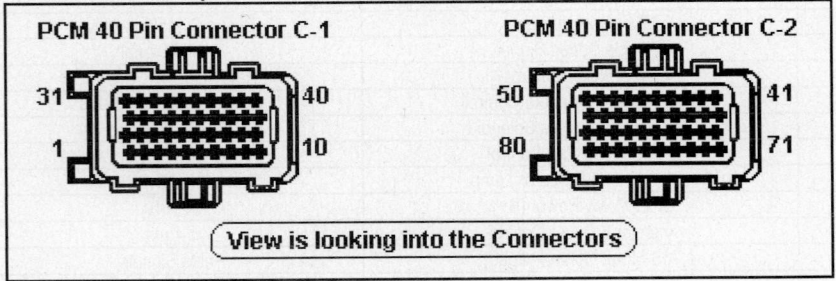

PCM 40 Pin Connector C-1

PCM 40 Pin Connector C-2

View is looking into the Connectors

1996 Grand Caravan 3.3L V6 MFI VIN R (A/T) 'C2' White Connector

PCM Pin #	Wire Color	Circuit Description (40 Pin)	Value at Hot Idle
41	RD/LG	Speed Control Set Switch Signal	S/C & Set Switch On: 3.8v
42	DB	A/C Pressure Switch Signal	A/C On: 0.451-4.850v
43	BK/LB	Sensor Ground	<0.050v
44	OR	8-Volt Supply	7.9-8.1v
45	DB	A/C Switch Sense Signal	A/C Off: 12v, On: 1v
46	RD/WT	Keep Alive Power (B+)	12-14v
47	----	Not Used	---
48	BR/WT	IAC 3 Driver Control	Pulse Signals
49	YL/BK	IAC 2 Driver Control	Pulse Signals
50	BK/TN	Power Ground	<0.1v
51	TN/WT	HO2S-12 (B1 S2) Signal	0.1-1.1v
52-56	---	Not Used	---
57	GY/RD	IAC 1 Driver Control	Pulse Signals
58	PK/BK	IAC 4 Driver Control	Pulse Signals
59	PK/BR	CCD Bus (+)	Digital Signals: 0-5-0v
60	WT/BK	CCD Bus (-)	<0.050v
61	PK/WT	5-Volt Supply	4.9-5.1v
62	WT/PK	Brake Switch Sense Signal	Brake Off: 0v, On: 12v
63	YL/DG	Torque Management Request	Digital Signals
64	DB/OR	A/C Clutch Relay Control	Relay Off: 12v, On: 1v
65, 75	PK, LG	SCI Transmit, SCI Receive	0v, 5v
66	WT/OR	Vehicle Speed Signal	Digital Signal
67	DB/YL	ASD Relay Control	Relay Off: 12v, On: 1v
68	PK/BK	EVAP Purge Solenoid Control	PWM Signal: 0-12-0v
69-71	---	Not Used	---
72	YL/BK	LDP Switch Sense Signal	LDP Switch Closed: 0v
73	LG/DG	Radiator Fan Control Relay	Relay Off: 12v, On: 1v
74	BR	Fuel Pump Relay Control	Relay Off: 12v, On: 1v
76	BK/WT	PNP Switch Sense (EATX)	In P/N: 0v, Others: 5v
77	WT/DG	LDP Solenoid Control	PWM Signal: 0-12-0v
78	TN/RD	Speed Control Vacuum Solenoid	Vacuum Increasing: 1v
79	---	Not Used	---
80	LG/RD	Speed Control Vent Solenoid	Vacuum Decreasing: 1v

Pin Connector Graphic

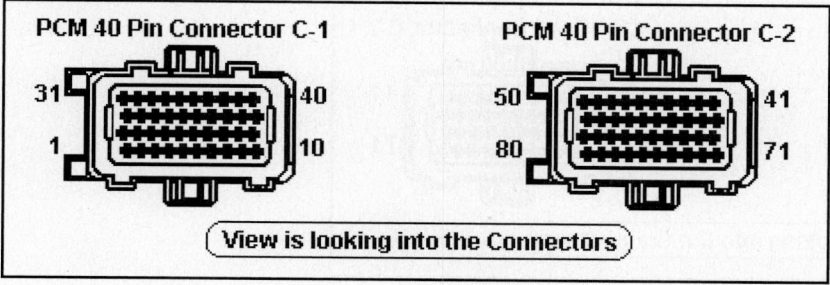

1997-98 Grand Caravan 3.3L V6 MFI VIN R 'C1' Black Connector

PCM Pin #	Wire Color	Circuit Description (40 Pin)	Value at Hot Idle
1	---	Not Used	---
2	RD/YL	Coil 3 Driver (Cyl 3 & 6)	5°, 55 mph: 8° dwell
3	DB/TN	Coil 2 Driver (Cyl 2 & 5)	5°, 55 mph: 8° dwell
4	DG	Generator Field Driver	Digital Signals: 0-12-0v
5	YL/RD	Speed Control Power Supply	12-14v
6	DG/OR	ASD Relay Power (B+)	12-14v
7	YL/WT	Injector 3 Driver	1-4 ms
8	TN	Starter Relay Control	KOEC: 9-11v
9, 12	---	Not Used	---
10	BK/TN	Power Ground	<0.1v
11	GY/RD	Coil 1 Driver (Cyl 1 & 4)	5°, 55 mph: 8° dwell
13	WT/DB	Injector 1 Driver	1-4 ms
14	BR/DB	Injector 6 Driver	1-4 ms
15	GY	Injector 5 Driver	1-4 ms
16	LB/BR	Injector 4 Driver	1-4 ms
17	TN	Injector 2 Driver	1-4 ms
18-19	---	Not Used	---
20	WT/BK	Ignition Switch Power (B+)	12-14v
21, 23	---	Not Used	---
22	BK/PK	MIL (lamp) Control	MIL Off: 12v, On: 1v
24	DB/LG	Knock Sensor Signal	0.080v AC
25	---	Not Used	---
26	TN/BK	ECT Sensor Signal	At 180°F: 2.80v
27-29	---	Not Used	---
30	BK/DG	HO2S-11 (B1 S1) Signal	0.1-1.1v
31	---	Not Used	---
32	GY/BK	CKP Sensor Signal	Digital Signals: 0-5-0v
33	TN/YL	CMP Sensor Signal	Digital Signals: 0-5-0v
34	---	Not Used	---
35	OR/DB	TP Sensor Signal	0.6-1.0v
36	DG/RD	MAP Sensor Signal	1.5-1.6v
37-39	---	Not Used	---
40	GY/YL	EGR Solenoid Control	12v, at 55 mph: 1v

Pin Connector Graphic

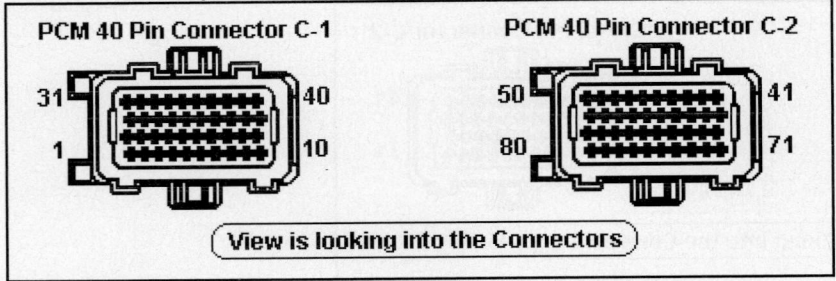

PCM 40 Pin Connector C-1

31   40

1   10

PCM 40 Pin Connector C-2

50   41

80   71

(View is looking into the Connectors)

1997-98 Grand Caravan 3.3L V6 MFI VIN R 'C2' White Connector

PCM Pin #	Wire Color	Circuit Description (40 Pin)	Value at Hot Idle
41	RD/LG	Speed Control Set Switch Signal	S/C & Set Switch On: 3.8v
42	DB	A/C Pressure Switch Signal	A/C On: 0.451-4.850v
43	BK/LB	Sensor Ground	<0.050v
44	OR	8-Volt Supply	7.9-8.1v
45	DG	A/C Switch Sense Signal	A/C Off: 12v, On: 1v
46	RD/WT	Keep Alive Power (B+)	12-14v
47	---	Not Used	---
48	BR/WT	IAC 3 Driver Control	Pulse Signals
49	YL/BK	IAC 2 Driver Control	Pulse Signals
50	BK/TN	Power Ground	<0.1v
51	TN/WT	HO2S-12 (B1 S2) Signal	0.1-1.1v
52-56	---	Not Used	---
57	GY/RD	IAC 1 Driver Control	Pulse Signals
58	PK/BK	IAC 4 Driver Control	Pulse Signals
59	PK/BR	CCD Bus (+)	Digital Signals: 0-5-0v
60	WT/BK	CCD Bus (-)	<0.050v
61	PK/WT	5-Volt Supply	4.9-5.1v
62	WT/PK	Brake Switch Sense Signal	Brake Off: 0v, On: 12v
63	YL/DG	Torque Management Request	Digital Signals
64	DB/OR	A/C Clutch Relay Control	Relay Off: 12v, On: 1v
65, 75	PK, LG	SCI Transmit, SCI Receive	0v, 5v
66	WT/OR	Vehicle Speed Signal	Digital Signal
67	DB/YL	ASD Relay Control	Relay Off: 12v, On: 1v
68	PK/BK	EVAP Purge Solenoid Control	PWM Signal: 0-12-0v
69-71	---	Not Used	---
72	DB/WT	LDP Switch Sense Signal	LDP Switch Closed: 0v
73	LG/DG	Radiator Fan Control Relay	Relay Off: 12v, On: 1v
74	BR	Fuel Pump Relay Control	Relay Off: 12v, On: 1v
76	BR/YL	PNP Switch Sense (EATX)	In P/N: 0v, Others: 5v
77	WT/DG	LDP Solenoid Control	PWM Signal: 0-12-0v
78	TN/RD	Speed Control Vacuum Solenoid	Vacuum Increasing: 1v
79	---	Not Used	---
80	LG/RD	Speed Control Vent Solenoid	Vacuum Decreasing: 1v

Pin Connector Graphic

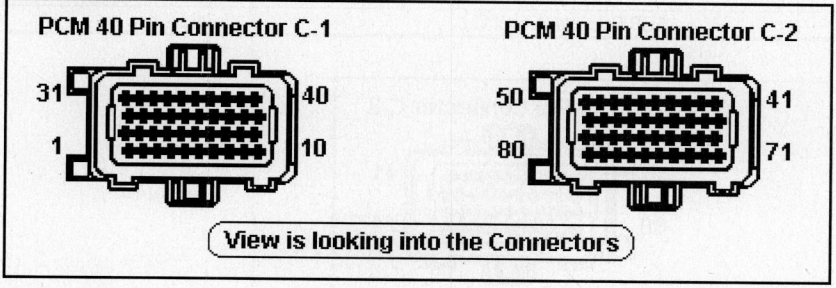

1999-2000 Grand Caravan 3.3L V6 MFI VIN R 'C1' Black Connector

PCM Pin #	W/Color	Circuit Description (40 Pin)	Value at Hot Idle
1	---	Not Used	---
2	RD/YL	Coil 3 Driver (Cyl 3 & 6)	5°, 55 mph: 8° dwell
3	DB/TN	Coil 2 Driver (Cyl 2 & 5)	5°, 55 mph: 8° dwell
4	DG	Generator Field Driver	Digital Signals: 0-12-0v
5	YL/RD	Speed Control Set Switch Signal	S/C & Set Switch On: 3.8v
6	DG/OR	ASD Relay Power (B+)	12-14v
7	YL/WT	Injector 3 Driver	1-4 ms
8	DG	Generator Field Driver	Digital Signals: 0-12-0v
9	---	Not Used	---
10	BK/TN	Power Ground	<0.1v
11	GY/RD	Coil 1 Driver (Cyl 1 & 4)	5°, 55 mph: 8° dwell
12	---	Not Used	---
13	WT/DB	Injector 1 Driver	1-4 ms
14	BR/DB	Injector 6 Driver	1-4 ms
15	GY	Injector 5 Driver	1-4 ms
16	LB/BR	Injector 4 Driver	1-4 ms
17	TN/WT	Injector 2 Driver	1-4 ms
18-19	---	Not Used	---
20	WT/BK	Ignition Switch Power (B+)	12-14v
21	---	Not Used	---
22	BK/PK	MIL (lamp) Control	MIL Off: 12v, On: 1v
23-24	---	Not Used	---
25	DB/LG	Knock Sensor Signal	0.080v AC
26	TN/BK	ECT Sensor Signal	At 180°F: 2.80v
27	BK/OR	HO2S Ground	<0.050v
28-29	---	Not Used	---
30	BK/DG	HO2S-11 (B1 S1) Signal	0.1-1.1v
31	TN	Smart Start Relay Control	KOEC: 9-11v
32	GY/BK	CKP Sensor Signal	Digital Signals: 0-5-0v
33	TN/YL	CMP Sensor Signal	Digital Signals: 0-5-0v
34, 37	---	Not Used	---
35	OR/DB	TP Sensor Signal	0.6-1.0v
36	DG/RD	MAP Sensor Signal	1.5-1.6v
38	DG/LB	A/C Switch Sense Signal	A/C Off: 12v, On: 1v
39	---	Not Used	---
40	GY/YL	EGR Solenoid Control	12v, at 55 mph: 1v

Pin Connector Graphic

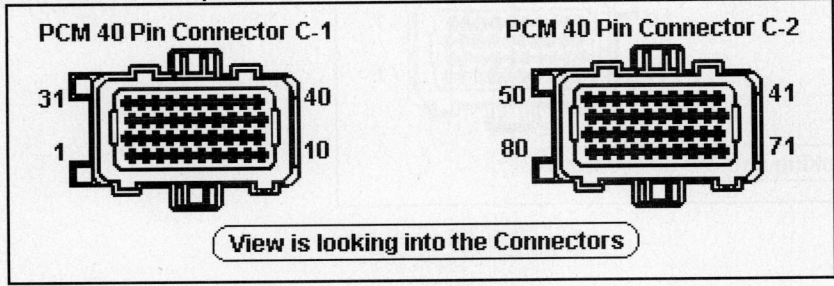

PCM 40 Pin Connector C-1

31 — 40
1 — 10

PCM 40 Pin Connector C-2

50 — 41
80 — 71

(View is looking into the Connectors)

1999-2000 Grand Caravan 3.3L V6 MFI VIN R 'C2' White Connector

PCM Pin #	W/Color	Circuit Description (40 Pin)	Value at Hot Idle
41	RD/LG	Speed Control Set Switch Signal	S/C & Set Switch On: 3.8v
42	DB	A/C Pressure Switch Signal	A/C On: 0.451-4.850v
43	BK/LB	Sensor Ground	<0.050v
44	OR	8-Volt Supply	7.9-8.1v
45	---	Not Used	---
46	RD/WT	Keep Alive Power (B+)	12-14v
47	---	Not Used	---
48	BR/WT	IAC 3 Driver Control	Pulse Signals
49	YL/BK	IAC 2 Driver Control	Pulse Signals
50	BK/TN	Power Ground	<0.1v
51	TN/WT	HO2S-12 (B1 S2) Signal	0.1-1.1v
52-55	---	Not Used	---
56	TN/RD	Speed Control Vacuum Solenoid	Vacuum Increasing: 1v
57	GY/RD	IAC 1 Driver Control	Pulse Signals
58	PK/BK	IAC 4 Driver Control	Pulse Signals
59	PK/BR	CCD Bus (+)	Digital Signals: 0-5-0v
60	WT/BK	CCD Bus (-)	<0.050v
61	PK/WT	5-Volt Supply	4.9-5.1v
62	WT/PK	Brake Switch Sense Signal	Brake Off: 0v, On: 12v
63	YL/DG	Torque Management Request	Digital Signals
64	DB/OR	A/C Clutch Relay Control	Relay Off: 12v, On: 1v
65	PK	SCI Transmit Signal	0v
66	WT/OR	Vehicle Speed Signal	Digital Signal
67	DB/YL	ASD Relay Control	Relay Off: 12v, On: 1v
68	PK/BK	EVAP Purge Solenoid Control	PWM Signal: 0-12-0v
69	---	Not Used	---
70	P/R	EVAP Purge Solenoid Sense	0-1v
71	---	Not Used	---
72	DB/WT	LDP Switch Sense Signal	LDP Switch Closed: 0v
73	LG/DB	Radiator Fan Control Relay	Relay Off: 12v, On: 1v
74	BR	Fuel Pump Relay Control	Relay Off: 12v, On: 1v
75	LG	SCI Receive Signal	5v
76	BK/YL	PNP Switch Sense (EATX)	In P/N: 0v, Others: 5v
77	WT/DG	LDP Solenoid Control	PWM Signal: 0-12-0v
78-79	---	Not Used	---
80	LG/RD	Speed Control Vent Solenoid	Vacuum Decreasing: 1v

Pin Connector Graphic

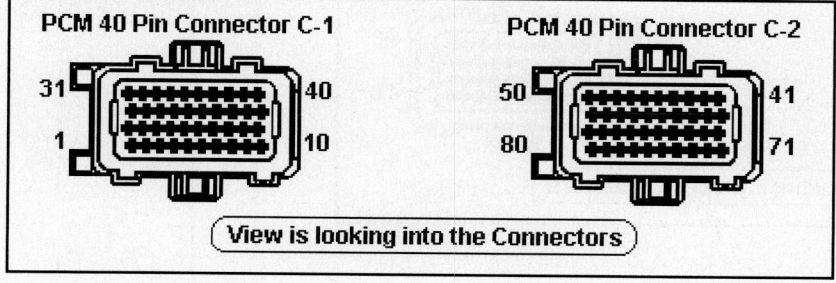

PCM 40 Pin Connector C-1

31   40
1   10

PCM 40 Pin Connector C-2

50   41
80   71

View is looking into the Connectors

2001-02 Grand Caravan 3.3L V6 OHV MFI VIN R (A/T) 'C1' Black Connector

PCM Pin #	Wire Color	Circuit Description (40 Pin)	Value at Hot Idle
1	---	Not Used	---
2	DB/OR	Coil 3 Driver (Cyl 3 & 6)	5°, 55 mph: 8° dwell
3	DB/TN	Coil 2 Driver (Cyl 2 & 5)	5°, 55 mph: 8° dwell
4	---	Not Used	---
5	VT/YL	Speed Control Power Supply	12-14v
6	BR/WT	ASD Relay Power (B+)	12-14v
7	BR/LB	Injector 3 Driver Control	1-4 ms
8	BR/GY	Generator Field Driver	Digital Signals: 0-12-0v
9	---	Not Used	---
10	BK/BR	Power Ground	<0.1v
11	DB/DG	Coil 1 Driver (Cyl 1 & 4)	5°, 55 mph: 8° dwell
12	VT/GY	Engine Oil Pressure Sensor	1.6v at 24 psi
13	BR/YL	Injector 1 Driver Control	1-4 ms
14	BR/VT	Injector 6 Driver Control	1-4 ms
15	BR/OR	Injector 5 Driver Control	1-4 ms
16	BR/TN	Injector 4 Driver Control	1-4 ms
17	BR/DB	Injector 2 Driver Control	1-4 ms
18	BR/LG	HO2S-11 (B1 S1) Heater	PWM Signal: 0-12-0v
19	---	Not Used	---
20	PK/GY	Ignition Switch Power (B+)	12-14v
21-24	---	Not Used	---
25	DB/YL	Knock Sensor Signal	0.080v AC
26	VT/OR	ECT Sensor Signal	At 180°F: 2.80v
27	BR/DG	HO2S Ground	<0.050v
28-29	---	Not Used	---
30	DB/LG	HO2S-11 (B1 S1) Signal	0.1-1.1v
31	DG/OR	Double Start Override Signal	KOEC: 9-11v
32	BR/LB	CKP Sensor Signal	Digital Signals: 0-5-0v
33	DB/GY	CMP Sensor Signal	Digital Signals: 0-5-0v
34	---	Not Used	---
35	BR/OR	TP Sensor Signal	0.6-1.0v
36	VT/BR	MAP Sensor Signal	1.5-1.6v
38-39	---	Not Used	---
40	DB/VT	EGR Solenoid Control	12v, at 55 mph: 1v

Pin Connector Graphic
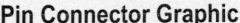

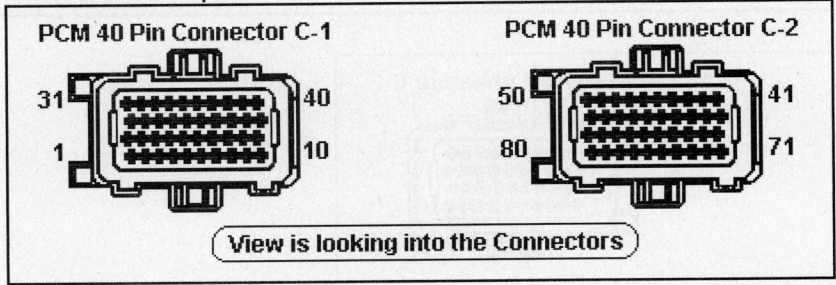

2001-02 Grand Caravan 3.3L V6 OHV MFI VIN R (A/T) 'C2' White Connector

PCM Pin #	Wire Color	Circuit Description (40 Pin)	Value at Hot Idle
41	VT	Speed Control Set Switch Signal	S/C & Set Switch On: 3.8v
42	LB/BR	A/C Pressure Sensor Signal	A/C On: 0.45-4.85v
43	DB/DG	Sensor Ground	<0.050v
44	BR/PK	8-Volt Supply	7.9-8.1v
45	---	Not Used	---
46	OR/RD	Keep Alive Power (B+)	12-14v
47-48	---	Not Used	---
49	VT/DG	IAC 1 Driver Control	Pulse Signals
50	BK/DG	Power Ground	<0.1v
51	DB/YL	HO2S-12 (B1 S2) Signal	0.1-1.1v
52-55	---	Not Used	---
56	VT/YL	Speed Control Vacuum Solenoid	Vacuum Increasing: 1v
57	VT/LG	IAC 2 Driver Control	Pulse Signals
58	---	Not Used	---
59	WT/VT	PCI Data Bus (J1850)	Digital Signals: 0-7-0v
60	---	Not Used	---
61	PK/YL	5-Volt Supply	4.9-5.1v
62	DG/WT	Brake Switch Sense Signal	Brake Off: 0v, On: 12v
63	DG/LG	Torque Management Request	Digital Signals
64	LB/OR	A/C Clutch Relay Control	Relay Off: 12v, On: 1v
65	WT/BR	SCI Transmit Signal	0v
66	DB/OR	Vehicle Speed Signal	Digital Signal
67	BR/WT	ASD Relay Control	Relay Off: 12v, On: 1v
68	DB/WT	EVAP Purge Solenoid Control	PWM Signal: 0-12-0v
69	---	Not Used	---
70	DB/BR	EVAP Purge Solenoid Sense	0-1v
71	---	Not Used	---
72	VT/WT	LDP Switch Sense Signal	LDP Switch Closed: 0v
73	BR/VT	Radiator Fan Control Relay	Relay Off: 12v, On: 1v
74	BR	Fuel Pump Relay Control	Relay Off: 12v, On: 1v
75	WT/LG	SCI Receive Signal	5v
76	YL/LB	TRS T41 Signal	In P/N: 0v, Others: 5v
77	VT/LB	LDP Solenoid Control	PWM Signal: 0-12-0v
78-79	---	Not Used	---
80	VT/OR	Speed Control Vent Solenoid	Vacuum Decreasing: 1v

Pin Connector Graphic

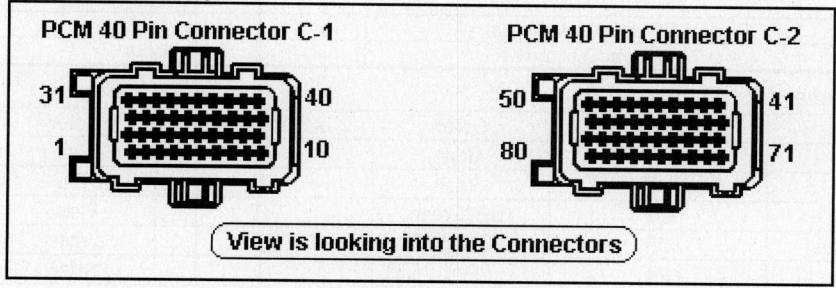

1994-95 Grand Caravan 3.8L V6 MFI VIN L (A/T) 60 Pin Connector

PCM Pin #	Wire Color	Circuit Description (60 Pin)	Value at Hot Idle
1	DG/RD	MAP Sensor Signal	1.5-1.6v
2	TN/BK	ECT Sensor Signal	At 180°F: 2.80v
3	RD/WT	Keep Alive Power (B+)	12-14v
4	BK/LB	Sensor Ground	<0.050v
5	BK/WT	Sensor Ground	<0.050v
6	PK/WT	5-Volt Supply	4.9-5.1v
7	OR	8-Volt Supply	7.9-8.1v
8	---	Not Used	---
9	DB/WT	Ignition Switch Power (B+)	12-14v
10	---	Not Used	---
11	BK/TN	Power Ground	<0.1v
12	BK/TN	Power Ground	<0.1v
13	LB/BR	Injector 4 Driver Control	1-4 ms
14	YL/WT	Injector 3 Driver Control	1-4 ms
15	TN	Injector 2 Driver Control	1-4 ms
16	LB	Injector 1 Driver Control	1-4 ms
17	DB/YL	Coil 2 Driver Control	5°, 55 mph: 8° dwell
18	DB/GY	Coil 3 Driver Control	5°, 55 mph: 8° dwell
19	BK/GY	Coil 1 Driver Control	5°, 55 mph: 8° dwell
20	DG	Alternator Field Control	Digital Signals: 0-12-0v
21	---	Not Used	---
22	DB	TP Sensor Signal	0.6-1.0v
23	R	Speed Control Set Switch Signal	S/C & Set Switch On: 3.8v
24	GY/P	CKP Sensor Signal	Digital Signals: 0-5-0v
25	PK	SCI Transmit Signal	0v
26	PK/BR	CCD Bus (+)	Digital Signals: 0-5-0v
27	BR	A/C Damped Pressure Switch	A/C Off: 12v, On: 1v
28	---	Not Used	---
29	WT/PK	Brake Switch Sense Signal	Brake Off: 0v, On: 12v
30	BK/OR	PNP Switch Sense Signal	In P/N: 0v, Others: 5v
31	DB/PK	Low Speed Fan Control	Relay Off: 12v, On: 1v
32	---	Not Used	---
33	TN/RD	Speed Control Vacuum Solenoid	Vacuum Increasing: 1v
34	DB/OR	A/C Clutch Relay Control	Relay Off: 12v, On: 1v
35	GY/YL	EGR Solenoid Control	Solenoid Off: 12v, On: 1v
36-37	---	Not Used	---
38	GY	Injector 5 Driver Control	1-4 ms
39	GY/RD	AIS Motor Control (open)	Pulse Signals
40	BR/WT	AIS Motor Control (close)	Pulse Signals

Standard Colors and Abbreviations

Abbreviation	Color	Abbreviation	Color	Abbreviation	Color
BK	Black	GY	Gray	RD	Red
BL	Blue	GN	Green	TN	Tan
BR	Brown	LG	Light Green	VT	Violet
DB	Dark Blue	OR	Orange	WT	White
DG	Dark Green	PK	Pink	YL	Yellow

1994-95 Grand Caravan 3.8L V6 MFI VIN L (A/T) 60 Pin Connector

PCM Pin #	Wire Color	Circuit Description (60 Pin)	Value at Hot Idle
41	DG/WT	HO2S-11 (B1 S1) Signal	0.1-1.1v
42-43	---	Not Used	---
44	TN/YL	CMP Sensor Signal	Digital Signals: 0-5-0v
45	LG	SCI Receive Signal	5v
46	WT/BK	CCD Bus (-)	<0.050v
47	WT/OR	Vehicle Speed Signal	Digital Signal
48-50	---	Not Used	---
51	DB/YL	ASD Relay Control	Relay Off: 12v, On: 1v
52	PK/BK	EVAP Purge Solenoid Control	PWM Signal: 0-12-0v
53	LG/RD	Speed Control Vent Solenoid	Vacuum Decreasing: 1v
54	---	Not Used	---
55	YL	High Speed Fan Control	Relay Off: 12v, On: 1v
56	---	Not Used	---
57	DG/OR	ASD Relay Power (B+)	12-14v
58	BR/DB	Injector 6 Driver Control	1-4 ms
59	PK/BK	AIS Motor Control (close)	Pulse Signals
60	YL/BK	AIS Motor Control (close)	Pulse Signals

Pin Connector Graphic

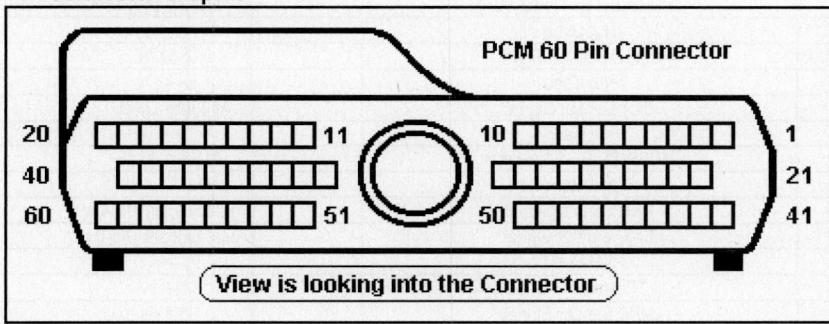

1996 Grand Caravan 3.8L V6 MFI VIN L (A/T) 'C1' Black Connector

PCM Pin #	Wire Color	Circuit Description (40 Pin)	Value at Hot Idle
1	---	Not Used	---
2	RD/YL	Coil 3 Driver (Cyl 3 & 6)	5°, 55 mph: 8° dwell
3	DB/TN	Coil 2 Driver (Cyl 2 & 5)	5°, 55 mph: 8° dwell
4	DG	Generator Field Driver	Digital Signals: 0-12-0v
5	YL/RD	Speed Control Power Supply	12-14v
6	DG/OR	ASD Relay Power (B+)	12-14v
7	YL/WT	Injector 3 Driver	1-4 ms
8	TN	Smart Start Relay Control	KOEC: 9-11v
10	BK/TN	Power Ground	<0.1v
9	---	Not Used	---
11	GY	Coil 1 Driver (Cyl 1 & 4)	5°, 55 mph: 8° dwell
12	---	Not Used	---
13	WT/DB	Injector 1 Driver	1-4 ms
14	BR/DB	Injector 6 Driver	1-4 ms
15	GY	Injector 5 Driver	1-4 ms
16	LB/BR	Injector 4 Driver	1-4 ms
17	TN	Injector 2 Driver	1-4 ms
18-19	---	Not Used	---
20	WT/BK	Ignition Switch Power (B+)	12-14v
21-23	---	Not Used	---
24	DB/LG	Knock Sensor Signal	0.080v AC
25	---	Not Used	---
26	TN/BK	ECT Sensor Signal	At 180°F: 2.80v
27-29	---	Not Used	---
30	BK/DG	HO2S-11 (B1 S1) Signal	0.1-1.1v
31	---	Not Used	---
32	GY/BK	CKP Sensor Signal	Digital Signals: 0-5-0v
33	TN/YL	CMP Sensor Signal	Digital Signals: 0-5-0v
34	---	Not Used	---
35	OR/DB	TP Sensor Signal	0.6-1.0v
36	DG/RD	MAP Sensor Signal	1.5-1.6v
37-39	---	Not Used	---
40	GY/YL	EGR Solenoid Control	12v, at 55 mph: 1v

Pin Connector Graphic

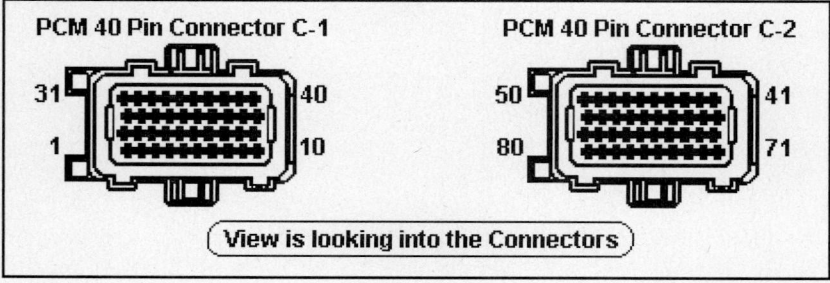

1996 Grand Caravan 3.8L V6 MFI VIN L (A/T) 'C2' White Connector

PCM Pin #	Wire Color	Circuit Description (40 Pin)	Value at Hot Idle
41	RD/LG	Speed Control Set Switch Signal	S/C & Set Switch On: 3.8v
42	DB	A/C Pressure Switch Signal	A/C On: 0.451-4.850v
43	BK/LB	Sensor Ground	<0.050v
44	OR	8-Volt Supply	7.9-8.1v
45	DB	A/C Switch Sense Signal	A/C Off: 12v, On: 1v
46	RD/WT	Keep Alive Power (B+)	12-14v
47	---	Not Used	---
48	BR/WT	IAC 1 Driver Control	Pulse Signals
49	YL/BK	IAC 2 Driver Control	Pulse Signals
50	BK/TN	Power Ground	<0.1v
51	TN/WT	HO2S-12 (B1 S2) Signal	0.1-1.1v
52-56	---	Not Used	---
57	GY/RD	IAC 3 Driver Control	Pulse Signals
58	PK/BK	IAC 4 Driver Control	Pulse Signals
59	PK/BR	CCD Bus (+)	Digital Signals: 0-5-0v
60	WT/BK	CCD Bus (-)	<0.050v
61	PK/WT	5-Volt Supply	4.9-5.1v
62	WT/PK	Brake Switch Sense Signal	Brake Off: 0v, On: 12v
63	YL/DG	Torque Management Request	Digital Signals
64	DB/OR	A/C Clutch Relay Control	Relay Off: 12v, On: 1v
65	PK	SCI Transmit Signal	0v
66	WT/OR	Vehicle Speed Signal	Digital Signal
67	DB/YL	ASD Relay Control	Relay Off: 12v, On: 1v
68	PK/BK	EVAP Purge Solenoid Control	PWM Signal: 0-12-0v
69-71	---	Not Used	---
72	YL/BK	LDP Switch Sense Signal	LDP Switch Closed: 0v
73	LG/DG	Radiator Fan Control Relay	Relay Off: 12v, On: 1v
74	BR	Fuel Pump Relay Control	Relay Off: 12v, On: 1v
75	LG	SCI Receive Signal	5v
76	BK/WT	PNP Switch Sense (EATX)	In P/N: 0v, Others: 5v
77	WT/DG	LDP Solenoid Control	PWM Signal: 0-12-0v
78	TN/RD	Speed Control Vacuum Solenoid	Vacuum Increasing: 1v
79	---	Not Used	---
80	LG/RD	Speed Control Vent Solenoid	Vacuum Decreasing: 1v

Pin Connector Graphic

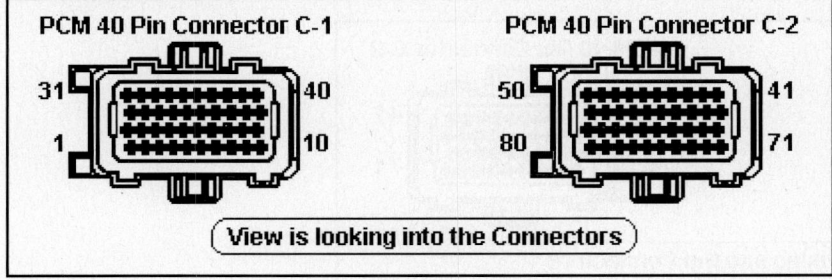

PCM 40 Pin Connector C-1

PCM 40 Pin Connector C-2

31 ... 40

1 ... 10

50 ... 41

80 ... 71

View is looking into the Connectors

1997-98 Grand Caravan 3.8L V6 MFI VIN L 'C1' Black Connector

PCM Pin #	Wire Color	Circuit Description (40 Pin)	Value at Hot Idle
1	---	Not Used	---
2	RD/YL	Coil 3 Driver (Cyl 3 & 6)	5°, 55 mph: 8° dwell
3	DB/TN	Coil 2 Driver (Cyl 2 & 5)	5°, 55 mph: 8° dwell
4	DG	Generator Field Driver	Digital Signals: 0-12-0v
5	YL/RD	Speed Control Power Supply	12-14v
6	DG/OR	ASD Relay Power (B+)	12-14v
7	YL/WT	Injector 3 Driver	1-4 ms
8	TN	Smart Start Relay Control	KOEC: 9-11v
9	---	Not Used	---
10	BK/TN	Power Ground	<0.1v
11	GY/RD	Coil 1 Driver (Cyl 1 & 4)	5°, 55 mph: 8° dwell
12	---	Not Used	---
13	WT/DB	Injector 1 Driver	1-4 ms
14	BR/DB	Injector 6 Driver	1-4 ms
15	GY	Injector 5 Driver	1-4 ms
16	LB/BR	Injector 4 Driver	1-4 ms
17	TN	Injector 2 Driver	1-4 ms
18-19	---	Not Used	---
20	WT/BK	Ignition Switch Power (B+)	12-14v
21	---	Not Used	---
22	BK/PK	MIL (lamp) Control	MIL Off: 12v, On: 1v
23	---	Not Used	---
24	DB/LG	Knock Sensor Signal	0.080v AC
25	---	Not Used	---
26	TN/BK	ECT Sensor Signal	At 180°F: 2.80v
27-29	---	Not Used	---
30	BK/DG	HO2S-11 (B1 S1) Signal	0.1-1.1v
31	---	Not Used	---
32	GY/BK	CKP Sensor Signal	Digital Signals: 0-5-0v
33	TN/YL	CMP Sensor Signal	Digital Signals: 0-5-0v
34	---	Not Used	---
35	OR/DB	TP Sensor Signal	0.6-1.0v
36	DG/RD	MAP Sensor Signal	1.5-1.6v
37-39	---	Not Used	---
40	GY/YL	EGR Solenoid Control	12v, at 55 mph: 1v

Pin Connector Graphic

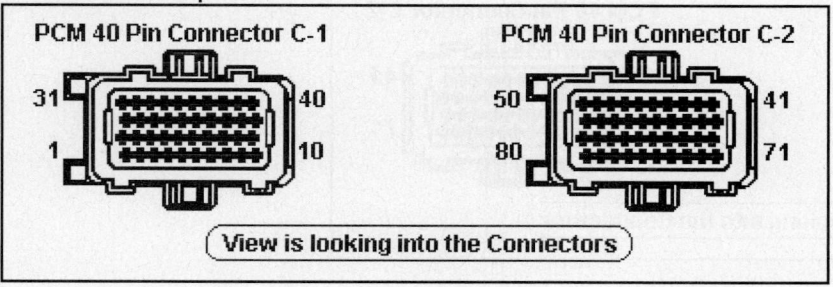

1997-98 Grand Caravan 3.8L V6 MFI VIN L 'C2' White Connector

PCM Pin #	Wire Color	Circuit Description (40 Pin)	Value at Hot Idle
41	RD/LG	Speed Control Set Switch Signal	S/C & Set Switch On: 3.8v
42	DB	A/C Pressure Switch Signal	A/C On: 0.451-4.850v
43	BK/LB	Sensor Ground	<0.050v
44	OR	8-Volt Supply	7.9-8.1v
45	DG	A/C Switch Sense Signal	A/C Off: 12v, On: 1v
46	RD/WT	Keep Alive Power (B+)	12-14v
47	---	Not Used	---
48	BR/WT	IAC 3 Driver Control	Pulse Signals
49	YL/BK	IAC 2 Driver Control	Pulse Signals
50	BK/TN	Power Ground	<0.1v
51	TN/WT	HO2S-12 (B1 S2) Signal	0.1-1.1v
52-56	---	Not Used	---
57	GY/RD	IAC 1 Driver Control	Pulse Signals
58	PK/BK	IAC 4 Driver Control	Pulse Signals
59	PK/BR	CCD Bus (+)	Digital Signals: 0-5-0v
60	WT/BK	CCD Bus (-)	<0.050v
61	PK/WT	5-Volt Supply	4.9-5.1v
62	WT/PK	Brake Switch Sense Signal	Brake Off: 0v, On: 12v
63	YL/DG	Torque Management Request	Digital Signals
64	DB/OR	A/C Clutch Relay Control	Relay Off: 12v, On: 1v
65	PK	SCI Transmit Signal	0v
66	WT/OR	Vehicle Speed Signal	Digital Signal
67	DB/YL	ASD Relay Control	Relay Off: 12v, On: 1v
68	PK/BK	EVAP Purge Solenoid Control	PWM Signal: 0-12-0v
69-71	---	Not Used	---
72	DB/WT	LDP Switch Sense Signal	LDP Switch Closed: 0v
73	LG/DG	Radiator Fan Control Relay	Relay Off: 12v, On: 1v
74	BR	Fuel Pump Relay Control	Relay Off: 12v, On: 1v
75	LG	SCI Receive Signal	5v
76	BR/YL	PNP Switch Sense (EATX)	In P/N: 0v, Others: 5v
77	WT/DG	LDP Solenoid Control	PWM Signal: 0-12-0v
78	TN/RD	Speed Control Vacuum Solenoid	Vacuum Increasing: 1v
79	---	Not Used	---
80	LG/RD	Speed Control Vent Solenoid	Vacuum Decreasing: 1v

Pin Connector Graphic

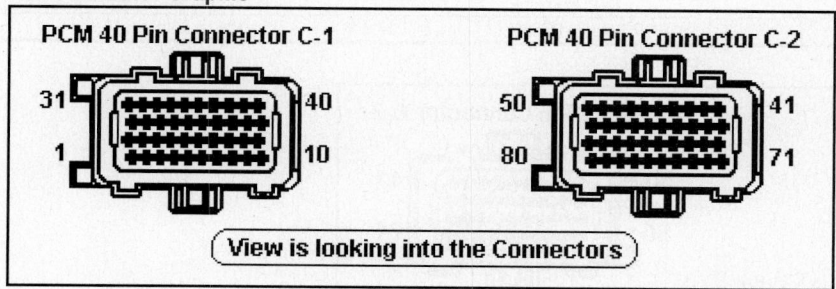

1999-2000 Grand Caravan 3.8L V6 MFI VIN L 'C1' Black Connector

PCM Pin #	Wire Color	Circuit Description (40 Pin)	Value at Hot Idle
1	---	Not Used	---
2	RD/YL	Coil 3 Driver (Cyl 3 & 6)	5º, 55 mph: 8º dwell
3	DB/TN	Coil 2 Driver (Cyl 2 & 5)	5º, 55 mph: 8º dwell
4	---	Not Used	---
5	YL/RD	Speed Control Set Switch Signal	S/C & Set Switch On: 3.8v
6	DG/OR	ASD Relay Power (B+)	12-14v
7	YL/WT	Injector 3 Driver	1-4 ms
8	DG	Generator Field Driver	Digital Signals: 0-12-0v
9	---	Not Used	---
10	BK/TN	Power Ground	<0.1v
11	GY/RD	Coil 1 Driver (Cyl 1 & 4)	5º, 55 mph: 8º dwell
12	---	Not Used	---
13	WT/DB	Injector 1 Driver	1-4 ms
14	BR/DB	Injector 6 Driver	1-4 ms
15	GY	Injector 5 Driver	1-4 ms
16	LB/BR	Injector 4 Driver	1-4 ms
17	TN/WT	Injector 2 Driver	1-4 ms
18-19	---	Not Used	---
20	WT/BK	Ignition Switch Power (B+)	12-14v
21	---	Not Used	---
22	BK/PK	MIL (lamp) Control	MIL Off: 12v, On: 1v
23-24	---	Not Used	---
25	DB/LG	Knock Sensor Signal	0.080v AC
26	TN/BK	ECT Sensor Signal	At 180ºF: 2.80v
27	BK/OR	HO2S Ground	<0.050v
28-29	---	Not Used	---
30	BK/DG	HO2S-11 (B1 S1) Signal	0.1-1.1v
31	TN	Smart Start Relay Control	KOEC: 9-11v
32	GY/BK	CKP Sensor Signal	Digital Signals: 0-5-0v
33	TN/YL	CMP Sensor Signal	Digital Signals: 0-5-0v
34	---	Not Used	---
35	OR/DB	TP Sensor Signal	0.6-1.0v
36	DG/RD	MAP Sensor Signal	1.5-1.6v
37	---	Not Used	---
38	DG/LB	A/C Switch Sense Signal	A/C Off: 12v, On: 1v
39	---	Not Used	---
40	GY/YL	EGR Solenoid Control	12v, at 55 mph: 1v

Pin Connector Graphic

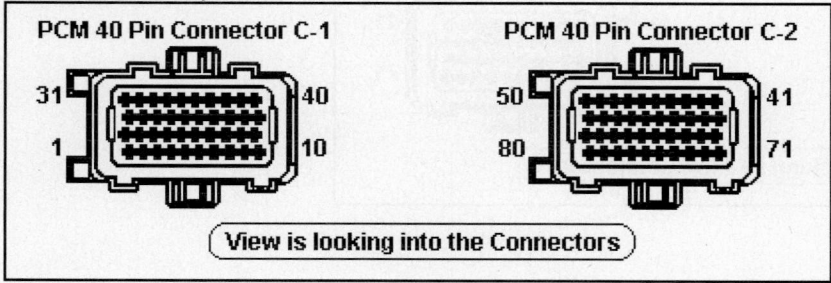

1999-2000 Grand Caravan 3.8L V6 MFI VIN L 'C2' White Connector

PCM Pin #	Wire Color	Circuit Description (40 Pin)	Value at Hot Idle
41	RD/LG	Speed Control Set Switch Signal	S/C & Set Switch On: 3.8v
42	DB	A/C Pressure Switch Signal	A/C On: 0.451-4.850v
43	BK/LB	Sensor Ground	<0.050v
44	OR	8-Volt Supply	7.9-8.1v
45	---	Not Used	---
46	RD/WT	Keep Alive Power (B+)	12-14v
47	---	Not Used	---
48	BR/WT	IAC 3 Driver Control	Pulse Signals
49	YL/BK	IAC 2 Driver Control	Pulse Signals
50	BK/TN	Power Ground	<0.1v
51	TN/WT	HO2S-12 (B1 S2) Signal	0.1-1.1v
52-55	---	Not Used	---
56	TN/RD	Speed Control Vacuum Solenoid	Vacuum Increasing: 1v
57	GY/RD	IAC 1 Driver Control	Pulse Signals
58	PK/BK	IAC 4 Driver Control	Pulse Signals
59	PK/BR	CCD Bus (+)	Digital Signals: 0-5-0v
60	WT/BK	CCD Bus (-)	<0.050v
61	PK/WT	5-Volt Supply	4.9-5.1v
62	WT/PK	Brake Switch Sense Signal	Brake Off: 0v, On: 12v
63	YL/DG	Torque Management Request	Digital Signals
64	DB/OR	A/C Clutch Relay Control	Relay Off: 12v, On: 1v
65	PK	SCI Transmit Signal	0v
66	WT/OR	Vehicle Speed Signal	Digital Signal
67	DB/YL	ASD Relay Control	Relay Off: 12v, On: 1v
68	PK/BK	EVAP Purge Solenoid Control	PWM Signal: 0-12-0v
69	---	Not Used	---
70	PK/RD	EVAP Purge Solenoid Sense	0-1v
71	---	Not Used	---
72	DB/WT	LDP Switch Sense Signal	LDP Switch Closed: 0v
73	LG/DB	Radiator Fan Control Relay	Relay Off: 12v, On: 1v
74	BR	Fuel Pump Relay Control	Relay Off: 12v, On: 1v
75	LG	SCI Receive Data	5v
76	BR/YL	PNP Switch Sense (EATX)	In P/N: 0v, Others: 5v
77	WT/DG	LDP Solenoid Control	PWM Signal: 0-12-0v
78-79	---	Not Used	---
80	LG/RD	Speed Control Vent Solenoid	Vacuum Decreasing: 1v

Pin Connector Graphic

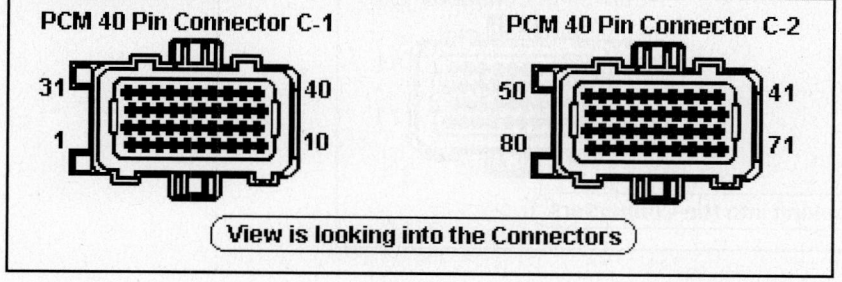

PCM 40 Pin Connector C-1   PCM 40 Pin Connector C-2

31 ... 40   50 ... 41

1 ... 10   80 ... 71

View is looking into the Connectors

2001-02 Grand Caravan 3.8L V6 MFI VIN L (A/T) 'C1' Black Connector

PCM Pin #	Wire Color	Circuit Description (40 Pin)	Value at Hot Idle
1	---	Not Used	---
2	DB/OR	Coil 3 Driver Control	5°, 55 mph: 8° dwell
3	DB/TN	Coil 2 Driver Control	5°, 55 mph: 8° dwell
4	---	Not Used	---
5	VT/YL	S/C On/Off Switch Sense	12-14v
6	BR/WT	ASD Relay Power (B+)	12-14v
7	BR/LB	Injector 3 Driver	1-4 ms
8	BR/GY	Generator Field Driver	Digital Signals: 0-12-0v
9	---	Not Used	---
10	BK/BR	Power Ground	<0.1v
11	DB/GY	Coil 1 Driver Control	5°, 55 mph: 8° dwell
12	VT/GY	Engine Oil Pressure Sensor	1.6v at 24 psi
13	BR/YL	Injector 1 Driver Control	1-4 ms
14	BR/VT	Injector 6 Driver Control	1-4 ms
15	BR/OR	Injector 5 Driver Control	1-4 ms
16	BR/TN	Injector 4 Driver Control	1-4 ms
17	BR/DB	Injector 2 Driver Control	1-4 ms
18	BR/LG	HO2S-11 (B1 S1) Heater	PWM Signal: 0-12-0v
19	---	Not Used	---
20	PK/GY	Ignition Switch Power (B+)	12-14v
21-24	---	Not Used	---
25	DB/YL	Knock Sensor Signal	0.080v AC
26	VT/OR	ECT Sensor Signal	At 180°F: 2.80v
27	BR/LG	HO2S Ground	<0.050v
28-29	---	Not Used	---
30	DB/LB	HO2S-11 (B1 S1) Signal	0.1-1.1v
31	DG/OR	Double Start Override Signal	KOEC: 9-11v
32	BR/LB	CKP Sensor Signal	Digital Signals: 0-5-0v
33	DB/GY	CMP Sensor Signal	Digital Signals: 0-5-0v
34	---	Not Used	---
35	DB/OR	TP Sensor Signal	0.6-1.0v
36	VT/BR	MAP Sensor Signal	1.5-1.6v
37	DB/LG	IAT Sensor Signal	At 100°F: 1.83v
38-39	---	Not Used	---
40	DB/VT	EGR Solenoid Control	12v, at 55 mph: 1v

Pin Connector Graphic

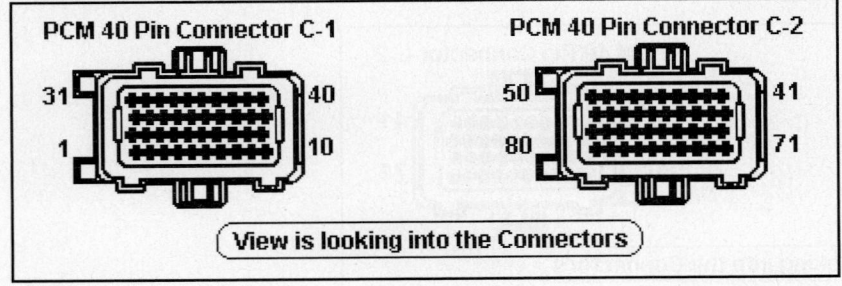

PCM 40 Pin Connector C-1

31   40
1   10

PCM 40 Pin Connector C-2

50   41
80   71

(**View is looking into the Connectors**)

2001-02 Grand Caravan 3.8L V6 MFI VIN L (A/T) 'C2' White Connector

PCM Pin #	Wire Color	Circuit Description (40 Pin)	Value at Hot Idle
41	VT	Speed Control Set Switch Signal	S/C & Set Switch On: 3.8v
42	LB/BR	A/C Pressure Switch Signal	A/C On: 0.451-4.850v
43	DB/DG	Sensor Ground	<0.050v
44	BR/PK	8-Volt Supply	7.9-8.1v
45	---	Not Used	---
46	OR/RD	Keep Alive Power (B+)	12-14v
47-48	---	Not Used	---
49	VT/DG	IAC 1 Driver Control	Pulse Signals
50	BK/DG	Power Ground	<0.1v
51	DB/YL	HO2S-12 (B1 S2) Signal	0.1-1.1v
52-55	---	Not Used	---
56	VT/YL	Speed Control Vacuum Solenoid	Vacuum Increasing: 1v
57	VT/LG	IAC 2 Driver Control	Pulse Signals
58	---	Not Used	---
59	WT/VT	PCI Data Bus (J1850)	Digital Signals: 0-7-0v
60	---	Not Used	---
61	PK/YL	5-Volt Supply	4.9-5.1v
62	DG/WT	Brake Switch Sense Signal	Brake Off: 0v, On: 12v
63	DG/LG	Torque Management Request	Digital Signals
64	LB/OR	A/C Clutch Relay Control	Relay Off: 12v, On: 1v
65	WT/BR	SCI Transmit Signal	0v
66	DB/OR	Vehicle Speed Signal	Digital Signal
67	BR/WT	ASD Relay Control	Relay Off: 12v, On: 1v
68	DB/WT	EVAP Purge Solenoid Control	PWM Signal: 0-12-0v
69	---	Not Used	---
70	DB/BR	EVAP Purge Solenoid Sense	0-1v
71	---	Not Used	---
72	VT/WT	LDP Switch Sense Signal	LDP Switch Closed: 0v
73	BR/VT	Radiator Fan Control Relay	Relay Off: 12v, On: 1v
74	BR	Fuel Pump Relay Control	Relay Off: 12v, On: 1v
75	WT/LG	SCI Receive Signal	5v
76	YL/DB	TRS TR1 Signal	In P/N: 0v, Others: 5v
77	VT/LB	LDP Solenoid Control	PWM Signal: 0-12-0v
78-79	---	Not Used	---
80	VT/OR	Speed Control Vent Solenoid	Vacuum Decreasing: 1v

Pin Connector Graphic

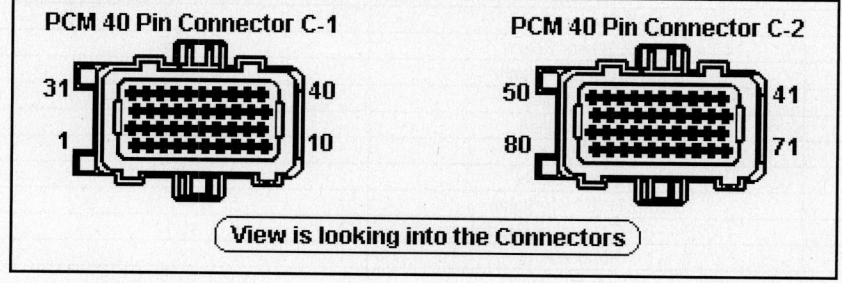

2003 Grand Caravan 3.3L V6 OHV MFI VIN L 'C1' Black Connector

PCM Pin #	Wire Color	Circuit Description (38 Pin)	Value at Hot Idle
1-8, 10	---	Not Used	---
9	BK/BR	Power Ground	<0.1v
11	PK/GY	Ignition Switch Output (Run-Start)	12-14v
12	PK/WT	Ignition Switch Output (Off-Run-Start)	12-14v
13-17, 19	---	Not Used	---
18	BK/DG	Power Ground	<0.1v
20	VT/GY	Oil Pressure Sensor	A/C On: 0.45-4.85v
21	---	Not Used	---
22	VT/LG	Ambient Air Temperature Sensor	At 100°F: 1.83v
23-24	---	Not Used	---
25	WT/LG	SCI Receive (PCM)	5v
26	WT/BR	Flash Program Enable	5v
27-28	---	Not Used	---
29	OR/RD	Battery Power (B+)	12-14v
30	YL	Ignition Switch Output (Start)	8-11v (cranking)
31	DB/YL	HO2S-12 (B1 S2) Signal	0.1-1.1v
32	DB/DG	Oxygen Sensor Return (down)	<0.050v
33-35	---	Not Used	---
36	WT/BR	SCI Transmit (PCM)	0v
37	DG/YL	SCI Transmit (TCM)	0v
38	WT/VT	PCI Data Bus (J1850)	Digital Signals: 0-7-0v

2003 Grand Caravan 3.3L V6 OHV MFI VIN R 'C2' Gray Connector

PCM Pin #	Wire Color	Circuit Description (38 Pin)	Value at Hot Idle
1-8	---	Not Used	---
9	DB/TN	Coil 2 Driver	5°, at 55 mph: 8° dwell
10	DB/DG	Coil 1 Driver	5°, at 55 mph: 8° dwell
11	BR/TN	Injector 4 Driver Control	1.0-4.0 ms
12	BR/LB	Injector 3 Driver Control	1.0-4.0 ms
13	BR/DB	Injector 2 Driver Control	1.0-4.0 ms
14	BR/YL	Injector 1 Driver Control	1.0-4.0 ms
15-17	---	Not Used	---
18	BR/LG	HO2S-12 (B1 S2) Heater Control	Heater Off: 12v, On: 1v
19	BR/GY	Generator Field Driver	Digital Signal: 0-12-0v
20	VT/OR	ECT Sensor Signal	At 180°F: 2.80v
21	BR/OR	TP Sensor Signal	0.6-1.0v
22, 26	---	Not Used	---
23	VT/BR	MAP Sensor Signal	1.5-1.7v
24	BR/LG	Knock Sensor Return	<0.050v
25	DB/YL	Knock Sensor Signal	0.080v AC
27	DB/DG	Sensor Ground	<0.1v
28	BR/VT	IAC Motor Sense	12-14v
29	PK/YL	5-Volt Supply	4.9-5.1v
30	DB/LG	IAT Sensor Signal	At 100°F: 1.83v
31	DB/LB	HO2S-11 (B1 S1) Signal	0.1-1.1v
32	BR/DG	HO2S-11 (B1 S1) Ground (Upstream)	<0.1v
33	---	Not Used	---
34	DB/GY	CMP Sensor Signal	Digital Signal: 0-5-0v
35	BR/LB	CKP Sensor Signal	Digital Signal: 0-5-0v
36-37	---	Not Used	---
38	VT/GY	IAC Motor Driver	DC pulse signals: 0.8-11v

Pin Connector Graphic

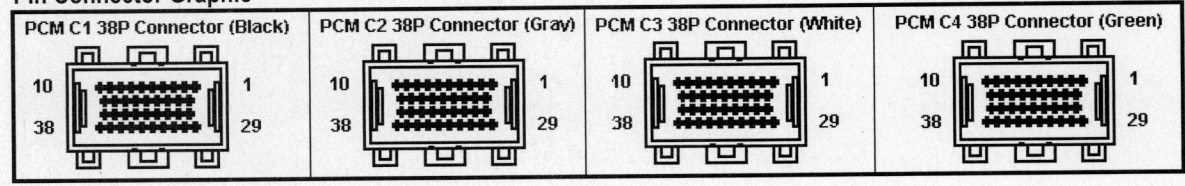

2003 Grand Caravan 3.3L V6 OHV MFI VIN R 'C3' White Connector

PCM Pin #	Wire Color	Circuit Description (38 Pin)	Value at Hot Idle
1-2, 4	---	Not Used	---
3	BR/WT	ASD Relay Control	Relay Off: 12v, On: 1v
5	VT/OR	Speed Control Vent Solenoid	Vacuum Decreasing: 1v
6	BR/VT	Radiator Fan Relay Control	Relay Off: 12v, On: 1v
7	VT/YL	Speed Control Power Supply	12-14v
8	WT/DG	Natural Vacuum Leak Detection Solenoid	Solenoid Off: 12v, On: 1v
9	BR/WT	HO2S-12 (B1 S2) Heater Control	Heater On: 1v, Off: 12v
10	---	Not Used	---
11	LB/OR	A/C Clutch Relay Control	Relay Off: 12v, On: 1v
12	VT/YL	Speed Control Vacuum Solenoid	Vacuum Increasing: 1v
13-18	---	Not Used	---
19	BR/WT	Automatic Shutdown Relay Output	12-14v
20	DB/WT	EVAP Purge Solenoid Control	PWM Signal: 0-12-0v
21-22	---	Not Used	---
23	DG/WT	Brake Switch Sense	Brake Off: 0v, On: 12v
24-27	---	Not Used	---
28	BR/WT	Automatic Shutdown Relay Output	12-14v
29	DB/BR	EVAP Purge Solenoid Sense	0-1v
30, 33, 36	---	Not Used	---
31	LB/BR	A/C Pressure Sensor	A/C On: 0.451-4.850v
32	DB/YL	Battery Temperature Sensor	At 100°F: 1.83v
34	VT	Speed Control Set Switch Signal	S/C & Set Switch On: 3.8v
35	VT/WT	Natural Vacuum Leak Detection Switch Sense	0.1v
37	BR	Fuel Pump Relay Control	Relay Off: 12v, On: 1v
38	DG/OR	Starter Relay Control	KOEC: 9-11v

2003 Grand Caravan 3.3L V6 OHV MFI VIN R 'C4' Green Connector

PCM Pin #	Wire Color	Circuit Description (38 Pin)	Value at Hot Idle
1	YL/GY	Overdrive Solenoid Control	Solenoid Off: 12v, On: 1v
2	YL/LB	Underdrive Solenoid Control	Solenoid Off: 12v, On: 1v
3-5	---	Not Used	---
6	YL/DB	2-4 Solenoid Control	Solenoid Off: 12v, On: 1v
7-9, 11, 14	---	Not Used	---
10	DG/WT	Low/Reverse Solenoid Control	Solenoid Off: 12v, On: 1v
12, 13	BK/LG	Power Ground	<0.050v
15	DG/LB	TRS T1 Sense	<0.050v
16	DG/DB	TRS T3 Sense	<0.050v
17, 20-21	---	Not Used	---
18	YL/BR	Transmission Control Relay Control	Relay Off: 12v, On: 1v
19, 38	YL/OR	Transmission Control Relay Output	Relay Off: 12v, On: 1v
22	DG/TN	Overdrive Pressure Switch Sense	In Overdrive: 2-4v
23-26	---	Not Used	---
27	DG/GY	TRS T41 Sense	<0.050v
28, 38	YL/OR	Transmission Control Relay Output	Relay Off: 12v, On: 1v
29	YL/TN	Low/Reverse Pressure Switch Sense	12-14v
30	YL/DG	2-4 Pressure Switch Sense	In Low/Reverse: 2-4v
31, 36	---	Not Used	---
32	DG/BR	Output Speed Sensor Signal	In 2-4 Position: 2-4v
33	DG/WT	Input Speed Sensor Signal	Moving: AC voltage
34	DG/VT	Speed Sensor Ground	Moving: AC voltage
35	DG/OR	Transmission Temperature Sensor Signal	<0.050v
37	DG/YL	TRS T42 Sense	In PRNL: 0v, Others 5v

Pin Connector Graphic

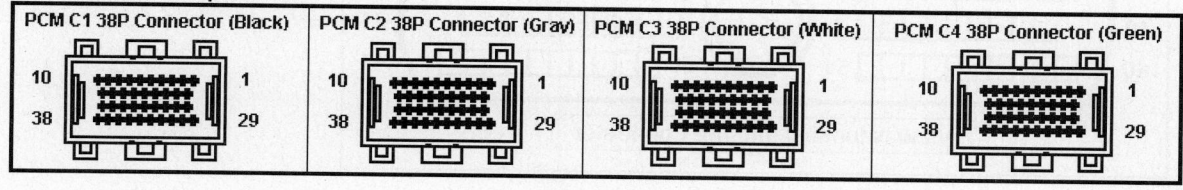

1990-91 Ram Van 3.9L V6 TBI VIN X (All) 60 Pin Connector

PCM Pin #	Wire Color	Circuit Description (60 Pin)	Value at Hot Idle
1	DG/RD	MAP Sensor Signal	1.5-1.6v
2	TN	ECT Sensor Signal	At 180°F: 2.80v
3	RD	Keep Alive Power (B+)	12-14v
4, 5	BK, BK/WT	Sensor Ground	<0.050v
6	PK/WT	5-Volt Supply	4.9-5.1v
7	OR	8-Volt Supply	7.9-8.1v
8	DG/BK	ASD Relay Power (B+)	12-14v
9	DB	Ignition Switch Power (B+)	12-14v
10, 13-14	---	Not Used	---
11, 12	LB/RD	Power Ground	<0.1v
15	TN	Injector 2 Driver	1-4 ms
16	WT/DB	Injector 1 Driver	1-4 ms
17-18	---	Not Used	---
19	BK/GY	Coil Driver Control	5°, 55 mph: 8° dwell
20	DG	Alternator Field Control	Digital Signals: 0-12-0v
21, 23	---	Not Used	---
22	OR/DB	TP Sensor Signal	0.6-1.0v
24	GY	Distributor Reference Signal	Digital Signals: 0-5-0v
25, 45	PK, LG	SCI Transmit, SCI Receive	0v, 5v
26, 31	---	Not Used	---
27	BR	A/C Request Signal	A/C Off: 12v, On: 1v
28	PK	Closed Throttle Switch	Switch Off: 12v, On: 1v
29	WT/PK	Brake Switch Sense Signal	Brake Off: 0v, On: 12v
30	BR/YL	PNP Switch Sense Signal	In P/N: 0v, Others: 5v
32	BK/PK	MIL (lamp) Control	MIL Off: 12v, On: 1v
33	TN/RD	Speed Control Vacuum Solenoid	Vacuum Increasing: 1v
34	DB/OR	A/C WOT Relay Control	Relay Off: 12v, On: 1v
35	GY/YL	EGR Solenoid Control	12v, at 55 mph: 1v
36	BK/OR	Air Switching Solenoid	Solenoid Off: 12v, On: 1v
37-39	---	Not Used	---
40	BR	AIS Motor Control	Pulse Signals
41	BK/DG	HO2S-11 (B1 S1) Signal	0.1-1.1v
42-43, 46	---	Not Used	---
44	OR/WT	Overdrive Lockout Switch	Switch Off: 12v, On: 1v
47	WT/OR	Vehicle Speed Signal	Digital Signal
48	BR/RD	Speed Control Set Switch Signal	S/C & Set Switch On: 3.8v
49	YL/RD	Speed Control On/Off Switch	S/C On: 1v, Off: 12v
50	WT/LG	Speed Control Resume Switch	S/C On: 1v, Off: 12v
51	DB/YL	ASD Relay Control	Relay Off: 12v, On: 1v
52	PK/BK	EVAP Purge Solenoid Control	Solenoid Off: 12v, On: 1v
53	LG/RD	Speed Control Vent Solenoid	Vacuum Decreasing: 1v
54	OR/BK	Part Throttle Unlock Solenoid	Solenoid Off: 12v, On: 1v
55	OR/LG	Overdrive Lockout Solenoid	Solenoid Off: 12v, On: 1v
56	GY/PK	Maintenance Indicator Lamp	Lamp Off: 12v, On: 1v
57-59	---	Not Used	---
60	GY/RD	AIS Motor Control	Pulse Signals

Pin Connector Graphic

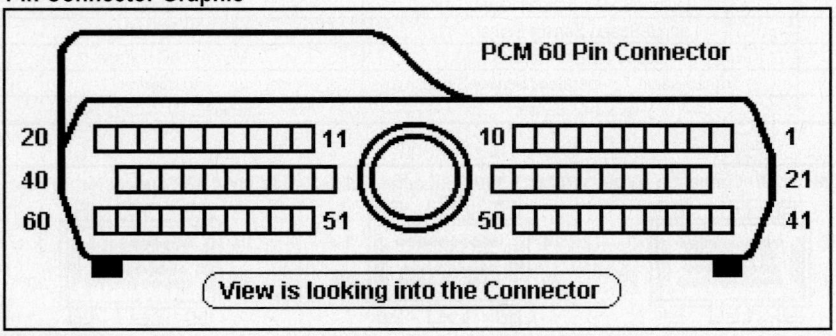

PCM 60 Pin Connector

View is looking into the Connector

1992-95 Ram Van 3.9L V6 MFI VIN X (All) 60 Pin Connector

PCM Pin #	Wire Color	Circuit Description (60 Pin)	Value at Hot Idle
1	DG/RD	MAP Sensor Signal	1.5-1.6v
2	TN/BK	ECT Sensor Signal	At 180ºF: 2.80v
3 ('92-'93)	RD	Keep Alive Power (B+)	12-14v
3	RD/WT	Keep Alive Power (B+)	12-14v
4	BK/LB	Sensor Ground	<0.050v
5 ('92-'93)	BK/WT	Sensor Ground	<0.050v
6	PK/WT	5-Volt Supply	4.9-5.1v
7	OR	8-Volt Supply	7.9-8.1v
8	---	Not Used	---
9 ('92-'93)	DB	Ignition Switch Power (B+)	12-14v
9	LG/BK	Ignition Switch Power (B+)	12-14v
10	OR/WT	Overdrive Override Switch	Switch Off: 12v, On: 1v
11	BK/TN	Power Ground	<0.1v
12	BK/TN	Power Ground	<0.1v
13	LB/BR	Injector 4 Driver	1-4 ms
14	YL/WT	Injector 3 Driver	1-4 ms
15	TN	Injector 2 Driver	1-4 ms
16 ('92-'93)	WT/DG	Injector 1 Driver	1-4 ms
16	WT/DB	Injector 1 Driver	1-4 ms
17-18, 23	---	Not Used	---
18	---	Not Used	---
19 ('92-'93)	GY	Coil Driver Control	5º, 55 mph: 8º dwell
19	BK/GY	Coil Driver Control	5º, 55 mph: 8º dwell
20	DG/W	Alternator Field Control	Digital Signals: 0-12-0v
21 ('94-'95)	BK/RD	Air Temperature Sensor	At 100ºF: 2.51v
22	OR/DB	TP Sensor Signal	0.6-1.0v
24	GY/BK	Distributor Reference Signal	Digital Signals: 0-5-0v
25	PK	SCI Transmit	0v
26	---	Not Used	---
27	BR	A/C Damped Pressure Switch	A/C Off: 12v, On: 1v
28	---	Not Used	---
29	WT/PK	Brake Switch Sense Signal	Brake Off: 0v, On: 12v
30 ('92-'93)	BR/YL	PNP Switch Sense Signal	In P/N: 0v, Others: 5v
30	BR/WT	PNP Switch Sense Signal	In P/N: 0v, Others: 5v
31	---	Not Used	---
32	BK/PK	MIL (lamp) Control	MIL Off: 12v, On: 1v
33	TN/RD	Speed Control Vacuum Solenoid	Vacuum Increasing: 1v
34	DB/OR	A/C Clutch Relay Control	Relay Off: 12v, On: 1v
35	GY/YL	EGR Solenoid Control	12v, at 55 mph: 1v
36	---	Not Used	---
37 ('92-'93)	BK/OR	Overdrive Lamp Control	Lamp Off: 12v, On: 1v
37	LG/OR	Overdrive Lamp Control	Lamp Off: 12v, On: 1v
38	PK/BK	Injector 5 Driver	1-4 ms
39	GY/RD	AIS Motor 4 Control	Pulse Signals
40	BR/WT	AIS Motor 2 Control	Pulse Signals

Standard Colors and Abbreviations

Abbreviation	Color	Abbreviation	Color	Abbreviation	Color
BK	Black	GY	Gray	RD	Red
BL	Blue	GN	Green	TN	Tan
BR	Brown	LG	Light Green	VT	Violet
DB	Dark Blue	OR	Orange	WT	White
DG	Dark Green	PK	Pink	YL	Yellow

1992-95 Ram Van 3.9L V6 MFI VIN X (All) 60 Pin Connector

PCM Pin #	Wire Color	Circuit Description (60 Pin)	Value at Hot Idle
41	BK/DG	HO2S-11 (B1 S1) Signal	0.1-1.1v
42-43	---	Not Used	---
44	TN/YL	CMP Sensor Signal	Digital Signals: 0-5-0v
45	LG	SCI Receive	5v
46	---	Not Used	---
47	WT/OR	Vehicle Speed Signal	Digital Signal
48	BR/RD	Speed Control Set Switch Signal	S/C & Set Switch On: 3.8v
49	YL/RD	Speed Control On/Off Switch	S/C On: 1v, Off: 12v
50	WT/LG	Speed Control Resume Switch	S/C On: 1v, Off: 12v
51	DB/YL	ASD Relay Control	Relay Off: 12v, On: 1v
52	PK/BK	EVAP Purge Solenoid Control	Solenoid Off: 12v, On: 1v
53	LG/RD	Speed Control Vent Solenoid	Vacuum Decreasing: 1v
54	OR/BK	A/T: Overdrive Solenoid	Solenoid Off: 12v, On: 1v
54	OR/BK	M/T: SIL (lamp) Control	Lamp Off: 12v, On: 1v
55	OR/WT	Overdrive Lockout Solenoid	Solenoid Off: 12v, On: 1v
56	GY/PK	Maintenance Indicator Lamp	Lamp Off: 12v, On: 1v
57	DG/OR	ASD Relay Power (B+)	12-14v
58	LG/BK	Injector 6 Driver	1-4 ms
59	PK/BK	AIS Motor 1 Control	Pulse Signals
60	YL/BK	AIS Motor 3 Control	Pulse Signals

Pin Connector Graphic

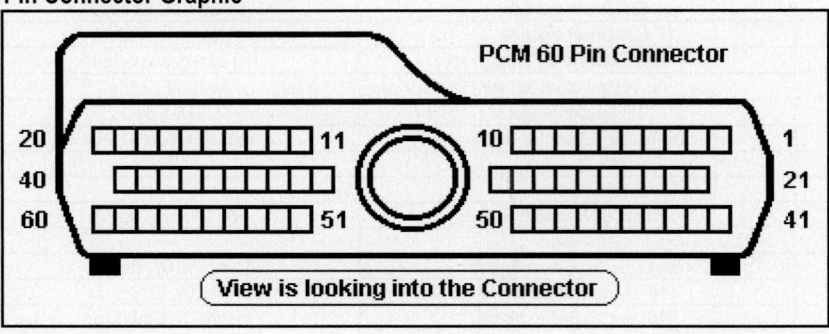

1996-97 Ram Van 3.9L V6 MFI VIN X (A/T) 'A' Black Connector

PCM Pin #	Wire Color	Circuit Description (32 Pin)	Value at Hot Idle
1	---	Not Used	---
2	LG/BK	Ignition Switch Power (B+)	12-14v
3	---	Not Used	---
4	BK/LB	Sensor Ground	<0.050v
5	---	Not Used	---
6	BR/OR	PNP Switch Sense Signal	In P/N: 0v, Others: 5v
7	GY	Coil 1 Driver Control	5°, 55 mph: 8° dwell
8	GY/BK	CKP Sensor Signal	Digital Signals: 0-5-0v
9	---	Not Used	---
10	YL/BK	IAC 2 Driver Control	Pulse Signals
11	BR/WT	IAC 3 Driver Control	Pulse Signals
12-14	---	Not Used	---
15	BK/RD	IAT Sensor Signal	At 100°F: 1.83v
16	TN/BK	ECT Sensor Signal	At 180°F: 2.80v
17	PK/WT	5-Volt Supply	4.9-5.1v
18	TN/YL	CMP Sensor Signal	Digital Signals: 0-5-0v
19	GY/RD	IAC 1 Driver Control	Pulse Signals
20	PK/BK	IAC 4 Driver Control	Pulse Signals
21	---	Not Used	---
22	RD/WT	Keep Alive Power (B+)	12-14v
23	OR/DB	TP Sensor Signal	0.6-1.0v
24	BK/LG	HO2S-11 (B1 S1) Signal	0.1-1.1v
25	TN/WT	HO2S-12 (B1 S2) Signal	0.1-1.1v
26	---	Not Used	---
27	DG/RD	MAP Sensor Signal	1.5-1.6v
28-30	---	Not Used	---
31	BK/TN	Power Ground	<0.1v
32	BK/TN	Power Ground	<0.1v

Pin Connector Graphic

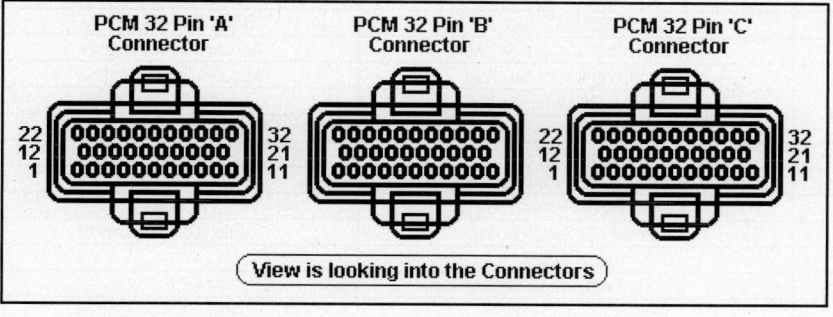

1996-97 Ram Van 3.9L V6 MFI VIN X (A/T) 'B' White Connector

PCM Pin #	Wire Color	Circuit Description (32 Pin)	Value at Hot Idle
1	GY/BK	Transmission Temperature Sensor Signal	At 200°F: 2.40v
2-3	---	Not Used	---
4	WT/DB	Injector 1 Driver	1-4 ms
5	YL/WT	Injector 3 Driver	1-4 ms
6	GY	Injector 5 Driver	1-4 ms
7	---	Not Used	---
8	PK	Governor Pressure Solenoid	PWM Signal: 0-12-0v
9	---	Not Used	---
10	DG/WT	Generator Field Control	Digital Signals: 0-12-0v
11	OR/BK	TCC Solenoid Control	At Cruise w/TCC On: 1v
12	BR/DB	Injector 6 Driver	1-4 ms
13-14	---	Not Used	---
15	TN	Injector 2 Driver	1-4 ms
16	LB/BR	Injector 4 Driver	1-4 ms
17-20	---	Not Used	---
21	OR/WT	3-4 Shift Solenoid Control	Solenoid Off: 12v, On: 1v
22-24	---	Not Used	---
25	DB	OSS Sensor (-)	Moving: AC pulse signals
26	---	Not Used	---
27	WT/OR	Vehicle Speed Signal	Digital Signal
28	LG	OSS Sensor (+)	Moving: AC pulse signals
29	LG/RD	Governor Pressure Sensor	0.58v
30	PK/LB	Transmission Relay Control	Relay Off: 12v, On: 1v
31	OR	5-Volt Supply	4.9-5.1v
32	---	Not Used	---

1996-97 Ram Van 3.9L V6 MFI VIN X (A/T) 'C' Gray Connector

PCM Pin #	Wire Color	Circuit Description (32 Pin)	Value at Hot Idle
1	DB/OR	A/C Clutch Relay Control	Relay Off: 12v, On: 1v
2	---	Not Used	---
3	DB/YL	ASD Relay Control	Relay Off: 12v, On: 1v
4	TN/RD	Speed Control Vacuum Solenoid	Vacuum Increasing: 1v
5	LG/RD	Speed Control Vent Solenoid	Vacuum Decreasing: 1v
6	LG/OR	Overdrive Off Lamp Control	Lamp Off: 12v, On: 1v
7	PK/BK	Transmission Temperature Lamp Control	Lamp Off: 12v, On: 1v
8-10	---	Not Used	---
11	YL/RD	Speed Control Power Supply	12-14v
12	DG/OR	ASD Relay Power (B+)	12-14v
13	OR/WT	Overdrive Off Switch Sense	Switch Off: 12v, On: 1v
14	---	Not Used	---
15	PK/YL	Battery Temperature Sensor	At 86°F: 1.96v
16	TN/BK	Generator Lamp Control	Lamp Off: 12v, On: 1v
17	BK/PK	MIL (lamp) Control	MIL Off: 12v, On: 1v
18	---	Not Used	---
19	BR	Fuel Pump Relay Control	Relay Off: 12v, On: 1v
20	PK/BK	EVAP Purge Solenoid Control	PWM Signal: 0-12-0v
21	---	Not Used	---
22	BR/WT	A/C Request Signal	A/C Off: 12v, On: 1v
23	LG	A/C Select Signal	A/C Off: 12v, On: 1v
24	WT/PK	Brake Switch Sense Signal	Brake Off: 0v, On: 12v
25	---	Not Used	---
26	LB/BK	Fuel Level Sensor Signal	Full: 0.56v, 1/2 full: 2.5v
27	PK	SCI Transmit Signal	0v
28	---	Not Used	---
29	LG	SCI Receive Signal	5v
30-31	---	Not Used	---
32	RD/LG	Speed Control Set Switch Signal	S/C & Set Switch On: 3.8v

1998 Ram Van 3.9L V6 MFI VIN X 'A' Black Connector

PCM Pin #	Wire Color	Circuit Description (32 Pin)	Value at Hot Idle
1	---	Not Used	---
2	LG/BK	Ignition Switch Power (B+)	12-14v
3	---	Not Used	---
4	BK/LB	Sensor Ground	<0.050v
5	---	Not Used	---
6	BR/YL	PNP Switch Sense Signal	In P/N: 0v, Others: 5v
7	GY	Coil 1 Driver Control	5°, 55 mph: 8° dwell
8	GY/BK	CKP Sensor Signal	Digital Signals: 0-5-0v
9	---	Not Used	---
10	YL/BK	IAC 2 Driver Control	Pulse Signals
11	BR/WT	IAC 3 Driver Control	Pulse Signals
12-14	---	Not Used	---
15	BK/RD	IAT Sensor Signal	At 100°F: 1.83v
16	TN/BK	ECT Sensor Signal	At 180°F: 2.80v
17	OR	5-Volt Supply	4.9-5.1v
18	TN/YL	CMP Sensor Signal	Digital Signals: 0-5-0v
19	GY/RD	IAC 1 Driver Control	Pulse Signals
20	VT/BK	IAC 4 Driver Control	Pulse Signals
21	---	Not Used	---
22	RD/WT	Battery Power (B+)	12-14v
23	OR/DB	TP Sensor Signal	0.6-1.0v
24	BK/DG	HO2S-12 (B1 S2) Signal	0.1-1.1v
25	TN/WT	HO2S-11 (B1 S1) Signal	0.1-1.1v
26	---	Not Used	---
27	DG/RD	MAP Sensor Signal	1.5-1.6v
28-30	---	Not Used	---
31	BK/TN	Power Ground	<0.1v
32	BK/TN	Power Ground	<0.1v

Pin Connector Graphic

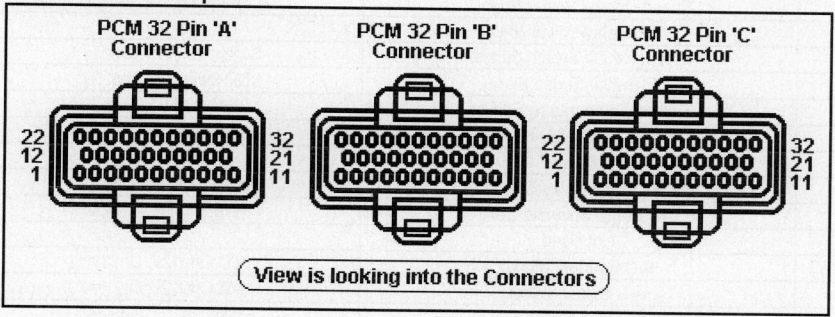

1998-2003 Ram Van 3.9L V6 MFI VIN X 'B' White Connector

PCM Pin #	Wire Color	Circuit Description (32 Pin)	Value at Hot Idle
1-3	---	Not Used	---
4	WT/DB	Injector 1 Driver Control	1-4 ms
5	YL/WT	Injector 3 Driver Control	1-4 ms
6	GY	Injector 5 Driver Control	1-4 ms
7-9	---	Not Used	---
10	DG/WT	Generator Field Control	Digital Signals: 0-12-0v
11	OR/BK	TCC Solenoid Control	At Cruise w/TCC On: 1v
12	BR/DB	Injector 6 Driver Control	1-4 ms
13-14	---	Not Used	---
15	TN	Injector 2 Driver Control	1-4 ms
16	LB/BR	Injector 4 Driver Control	1-4 ms
17-22	---	Not Used	---
23	GY/YL	Engine Oil Pressure Sensor	1.6v at 24 psi
25-26	---	Not Used	---
27	WT/OR	Vehicle Speed Signal	Digital Signal
28-32	---	Not Used	---

1998 Van 3.9L V6 MFI VIN X (A/T) 'C' Gray Connector

PCM Pin #	Wire Color	Circuit Description (32 Pin)	Value at Hot Idle
1	DB/OR	A/C Clutch Relay Control	Relay Off: 12v, On: 1v
2	---	Not Used	---
3	DB/YL	ASD Relay Control	Relay Off: 12v, On: 1v
4	TN/RD	Speed Control Vacuum Solenoid	Vacuum Increasing: 1v
5	LG/RD	Speed Control Vent Solenoid	Vacuum Decreasing: 1v
6	LG/OR	Overdrive Off Lamp Control	Lamp Off: 12v, On: 1v
7	---	Not Used	---
8	OR/RD	HO2S-11 (B1 S1) Heater	PWM Signal: 0-12-0v
9	---	Not Used	---
10	WT/DG	LDP Solenoid Control	PWM Signal: 0-12-0v
11	YL/RD	Speed Control Power Supply	12-14v
12	DG/OR	ASD Relay Power (B+)	12-14v
13	OR/WT	Overdrive Off Switch Sense	Switch Off: 12v, On: 1v
14	OR	LDP Switch Sense Signal	LDP Switch Closed: 0v
15	PK/YL	Battery Temperature Sensor	At 86°F: 1.96v
16	VT/TN	HO2S-21 (B2 S1) Heater	PWM Signal: 0-12-0v
17-18	---	Not Used	---
19	BR	Fuel Pump Relay Control	Relay Off: 12v, On: 1v
20	PK/BK	EVAP Purge Solenoid Control	PWM Signal: 0-12-0v
21, 31	---	Not Used	---
22	DB	A/C Request Signal	A/C Off: 12v, On: 1v
23	LG/GY	A/C Select Signal	A/C Off: 12v, On: 1v
24	WT/PK	Brake Switch Sense Signal	Brake Off: 0v, On: 12v
25	OR/DG	Generator Field Source	12-14v
25 ('00-'03)	WT/DB	Generator Field Source	12-14v
26	LB/BK	Fuel Level Sensor Signal	Full: 0.56v, 1/2 full: 2.5v
27	PK	SCI Transmit Signal	0v
28	WT	CCD Bus (-) Signal	<0.050v
29	LG	SCI Receive Signal	5v
30	PK/BR	CCD Bus (+) Signal	Digital Signals: 0-5-0v
32	WT/LG	Speed Control Set Switch Signal	S/C & Set Switch On: 3.8v

1999-2003 Ram Van 3.9L V6 MFI VIN X 'A' Black Connector

PCM Pin #	Wire Color	Circuit Description (32 Pin)	Value at Hot Idle
1	---	Not Used	---
2	LG/BK	Ignition Switch Power (B+)	12-14v
3	---	Not Used	---
4	BK/LB	Sensor Ground	<0.050v
5	---	Not Used	---
6	BR/YL	PNP Switch Sense Signal	In P/N: 0v, Others: 5v
7	GY	Coil 1 Driver Control	5°, 55 mph: 8° dwell
8	GY/BK	CKP Sensor Signal	Digital Signals: 0-5-0v
9	---	Not Used	---
10	YL/BK	IAC 2 Driver Control	Pulse Signals
11	BR/WT	IAC 3 Driver Control	Pulse Signals
12-14	---	Not Used	---
15	BK/RD	IAT Sensor Signal	At 100°F: 1.83v
16	TN/BK	ECT Sensor Signal	At 180°F: 2.80v
17	OR	5-Volt Supply	4.9-5.1v
18	TN/YL	CMP Sensor Signal	Digital Signals: 0-5-0v
19	GY/RD	IAC 1 Driver Control	Pulse Signals
20	VT/BK	IAC 4 Driver Control	Pulse Signals
21	---	Not Used	---
22	RD/WT	Battery Power (B+)	12-14v
22 ('01-'03)	DG/BK	Battery Power (B+)	12-14v
23	OR/DB	TP Sensor Signal	0.6-1.0v
24	BK/DG	HO2S-12 (B1 S2) Signal	0.1-1.1v
25	TN/WT	HO2S-11 (B1 S1) Signal	0.1-1.1v
26	---	Not Used	---
27	DG/RD	MAP Sensor Signal	1.5-1.6v
28-30	---	Not Used	---
31	BK/TN	Power Ground	<0.1v
32	BK/TN	Power Ground	<0.1v

Pin Connector Graphic

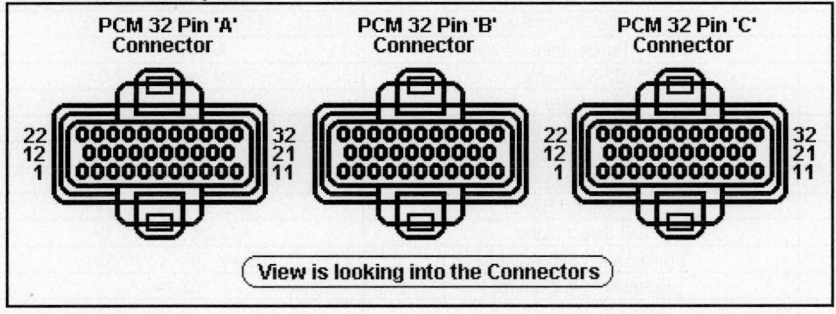

1999-2003 Ram Van 3.9L V6 MFI VIN X 'B' White Connector

PCM Pin #	Wire Color	Circuit Description (32 Pin)	Value at Hot Idle
1-3	---	Not Used	---
4	WT/DB	Injector 1 Driver Control	1-4 ms
5	YL/WT	Injector 3 Driver Control	1-4 ms
6	GY	Injector 5 Driver Control	1-4 ms
7-9	---	Not Used	---
10	DG/WT	Generator Field Control	Digital Signals: 0-12-0v
11	OR/BK	TCC Solenoid Control	At Cruise w/TCC On: 1v
12	BR/DB	Injector 6 Driver Control	1-4 ms
13-14	---	Not Used	---
15	TN	Injector 2 Driver Control	1-4 ms
16	LB/BR	Injector 4 Driver Control	1-4 ms
17-22	---	Not Used	---
23	GY/YL	Engine Oil Pressure Sensor	1.6v at 24 psi
25-26	---	Not Used	---
27	WT/OR	Vehicle Speed Signal	Digital Signal
28-32	---	Not Used	---

1998-2003 Van 3.9L V6 MFI VIN X (A/T) 'C' Gray Connector

PCM Pin #	Wire Color	Circuit Description (32 Pin)	Value at Hot Idle
1	DB/OR	A/C Clutch Relay Control	Relay Off: 12v, On: 1v
2	---	Not Used	---
3	DB/YL	ASD Relay Control	Relay Off: 12v, On: 1v
4	TN/RD	Speed Control Vacuum Solenoid	Vacuum Increasing: 1v
5	LG/RD	Speed Control Vent Solenoid	Vacuum Decreasing: 1v
6-7	---	Not Used	---
8	OR/RD	HO2S-11 (B1 S1) Heater	PWM Signal: 0-12-0v
9	---	Not Used	---
10	WT/DG	LDP Solenoid Control	PWM Signal: 0-12-0v
11	YL/RD	Speed Control Power Supply	12-14v
12	DG/OR	ASD Relay Power (B+)	12-14v
13	OR/WT	Overdrive Off Switch Sense	Switch Off: 12v, On: 1v
14	OR	LDP Switch Sense Signal	LDP Switch Closed: 0v
15	PK/YL	Battery Temperature Sensor	At 86°F: 1.96v
16	VT/TN	HO2S-21 (B2 S1) Heater Control	PWM Signal: 0-12-0v
17-18	---	Not Used	---
19	BR	Fuel Pump Relay Control	Relay Off: 12v, On: 1v
20	PK/BK	EVAP Purge Solenoid Control	PWM Signal: 0-12-0v
21, 31	---	Not Used	---
22	DB	A/C Request Signal	A/C Off: 12v, On: 1v
23	LG/GY	A/C Select Signal	A/C Off: 12v, On: 1v
24	WT/PK	Brake Switch Sense Signal	Brake Off: 0v, On: 12v
25	OR/DG	Generator Field Source	12-14v
25 ('00-'03)	WT/DB	Generator Field Source	12-14v
26	LB/BK	Fuel Level Sensor Signal	Full: 0.56v, 1/2 full: 2.5v
27	PK	SCI Transmit Signal	0v
28	WT	CCD Bus (-) Signal	<0.050v
29	LG	SCI Receive Signal	5v
30	PK/BR	CCD Bus (+) Signal	Digital Signals: 0-5-0v
32	WT/LG	Speed Control Set Switch Signal	S/C & Set Switch On: 3.8v

1993-95 Ram Van 5.2L V8 CNG VIN T (A/T) 60 Pin Connector

PCM Pin #	Wire Color	Circuit Description (60 Pin)	Value at Hot Idle
1	DG/RD	MAP Sensor Signal	1.5-1.6v
2	TN/BK	ECT Sensor Signal	At 180°F: 2.80v
3	RD	Keep Alive Power (B+)	12-14v
4	BK/LB	Sensor Ground	<0.050v
5	BK/WT	Sensor Ground	<0.050v
6	PK/WT	5-Volt Supply	4.9-5.1v
7	OR	8-Volt Supply	7.9-8.1v
8	---	Not Used	---
9	DB	Ignition Switch Power (B+)	12-14v
10	OR/WT	Overdrive Override Switch	Switch Off: 12v, On: 1v
11	BK/TN	Power Ground	<0.1v
12	BK/TN	Power Ground	<0.1v
13	W/LB	Injector 4 Driver	1-4 ms
14	WT/YL	Injector 3 Driver	1-4 ms
15	WT	Injector 2 Driver	1-4 ms
16	WT/BR	Injector 1 Driver	1-4 ms
17	WT/TN	Injector 7 Driver	1-4 ms
18	W/GY	Injector 8 Driver	1-4 ms
19	GY	Coil Driver Control	5°, 55 mph: 8° dwell
20	DG/WT	Alternator Field Control	Digital Signals: 0-12-0v
21	BK/RD	Throttle Body Temperature	At 100°F: 2.51v
22	OR/DB	TP Sensor Signal	0.6-1.0v
23	DG/OR	Fuel Low Pressure Sensor	0.5-4.5v
24	GY/BK	Distributor Reference Signal	Digital Signals: 0-5-0v
25	PK	SCI Transmit	0v
26	---	Not Used	---
27	BR	A/C Clutch Signal	A/C Off: 12v, On: 1v
28	---	Not Used	---
29	WT/PK	Brake Switch Sense Signal	Brake Off: 0v, On: 12v
30	BR/YL	PNP Switch Sense Signal	In P/N: 0v, Others: 5v
31	---	Not Used	---
32	BK/PK	MIL (lamp) Control	MIL Off: 12v, On: 1v
33	TN/RD	Speed Control Vacuum Solenoid	Vacuum Increasing: 1v
34	DB/OR	A/C WOT Relay Control	Relay Off: 12v, On: 1v
35	GY/YL	EGR Solenoid Control	12v, at 55 mph: 1v
36	---	Not Used	---
37	BK/OR	Overdrive Lamp Control	Lamp Off: 12v, On: 1v
38	WT/BK	Injector 5 Driver	1-4 ms
39	GY/RD	AIS Motor Control	Pulse Signals
40	BR/WT	AIS Motor Control	Pulse Signals

Standard Colors and Abbreviations

Abbreviation	Color	Abbreviation	Color	Abbreviation	Color
BK	Black	GY	Gray	RD	Red
BL	Blue	GN	Green	TN	Tan
BR	Brown	LG	Light Green	VT	Violet
DB	Dark Blue	OR	Orange	WT	White
DG	Dark Green	PK	Pink	YL	Yellow

1993-95 Ram Van 5.2L V8 CNG VIN T (A/T) 60 Pin Connector

PCM Pin #	Wire Color	Circuit Description (60 Pin)	Value at Hot Idle
41	BK/DG	HO2S-11 (B1 S1) Signal	0.1-1.1v
42	TN/PK	Fuel Temperature Sensor	0.5-4.5v
43	---	Not Used	---
44	TN/YL	Distributor Sync Signal	Digital Signals: 0-5-0v
45	LG	SCI Receive	5v
46	---	Not Used	---
47	WT/OR	Vehicle Speed Signal	Digital Signal
48	BR/RD	Speed Control Set Switch Signal	S/C & Set Switch On: 3.8v
49	YL/RD	Speed Control On/Off Switch	S/C On: 1v, Off: 12v
50	WT/LG	Speed Control Resume Switch	S/C On: 1v, Off: 12v
51	DB/YL	ASD Relay Control	Relay Off: 12v, On: 1v
52	---	Not Used	---
53	LG/RD	Speed Control Vent Solenoid	Vacuum Decreasing: 1v
54	OR/BK	Part Throttle Unlock Solenoid	Solenoid Off: 12v, On: 1v
55	OR/WT	Overdrive Lockout Solenoid	Solenoid Off: 12v, On: 1v
56	GY/PK	Maintenance Indicator Lamp	Lamp Off: 12v, On: 1v
57	DG/OR	ASD Relay Power (B+)	12-14v
58	WT/RD	Injector 6 Driver	1-4 ms
59	PK/BK	AIS Motor Control	Pulse Signals
60	YL/BK	AIS Motor Control	Pulse Signals

Pin Connector Graphic

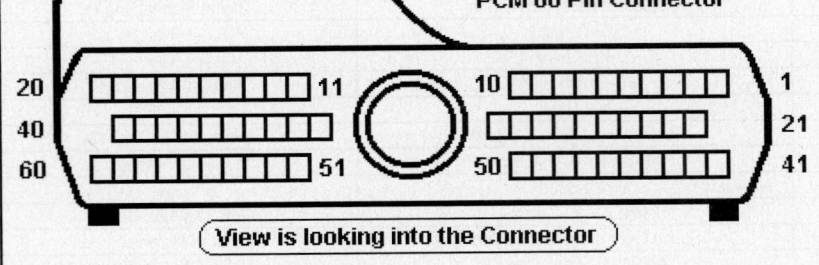

1996 Ram Van 5.2L V8 MFI CNG VIN T 'A' Black Connector

PCM Pin #	Wire Color	Circuit Description (32 Pin)	Value at Hot Idle
1	---	Not Used	---
2	LG/BK	Ignition Switch Power (B+)	12-14v
3	---	Not Used	
4	BK/LB	Sensor Ground	<0.050v
5	---	Not Used	---
6	BK/WT	PNP Switch Sense Signal	In P/N: 0v, Others: 5v
7	BK/GY	Coil 1 Driver Control	5°, 55 mph: 8° dwell
8	GY/BK	CKP Sensor Signal	Digital Signals: 0-5-0v
9	---	Not Used	---
10	YL/BK	IAC 2 Driver Control	Pulse Signals
11	BR/WT	IAC 1 Driver Control	Pulse Signals
12-14	---	Not Used	
15	BK/RD	IAT Sensor Signal	At 100°F: 1.83v
16	TN/BK	ECT Sensor Signal	At 180°F: 2.80v
17	VT/WT	5-Volt Supply	4.9-5.1v
18	TN/YL	CMP Sensor Signal	Digital Signals: 0-5-0v
19	GY/RD	IAC 3 Driver Control	Pulse Signals
20	VT/BK	IAC 4 Driver Control	Pulse Signals
21	---	Not Used	---
22	RD/WT	Keep Alive Power (B+)	12-14v
23	OR/DB	TP Sensor Signal	0.6-1.0v
24	TN/WT	HO2S-11 (B1 S1) Signal	0.1-1.1v
25	OR/BK	HO2S-12 (B1 S2) Signal	0.1-1.1v
26	---	Not Used	---
27	DG/RD	MAP Sensor Signal	1.5-1.6v
28	DG/LB	Fuel Pressure Sensor Signal	4.5v at 150 psi
29-30	---	Not Used	---
31	BK/TN	Power Ground	<0.1v
32	BK/TN	Power Ground	<0.1v

Pin Connector Graphic

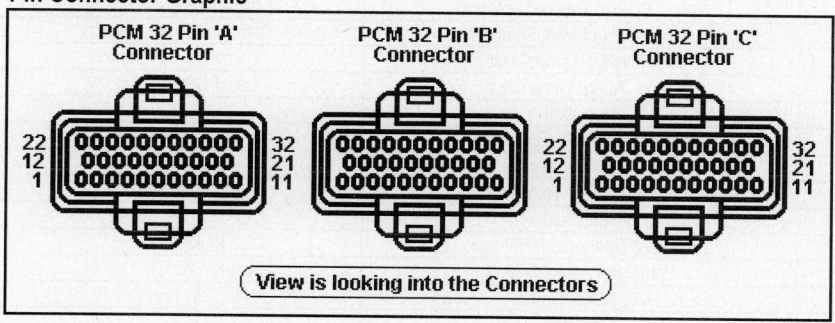

PCM 32 Pin 'A' Connector   PCM 32 Pin 'B' Connector   PCM 32 Pin 'C' Connector

View is looking into the Connectors

1996 Ram Van 5.2L V8 MFI CNG VIN T 'B' White Connector

PCM Pin #	Wire Color	Circuit Description (32 Pin)	Value at Hot Idle
1	VT	Transmission Temperature Sensor Signal	At 200ºF: 2.40v
2	VT/TN	Injector 7 Driver	1-4 ms
3, 7, 9, 11	---	Not Used	---
4	WT/DB	Injector 1 Driver	1-4 ms
5	YL/WT	Injector 3 Driver	1-4 ms
6	PK/BK	Injector 5 Driver	1-4 ms
8	VT/WT	Governor Pressure Solenoid	PWM Signal: 0-12-0v
10	DG	Generator Field Control	Digital Signals: 0-12-0v
12	BR/DB	Injector 6 Driver	1-4 ms
13	GY/LB	Injector 8 Driver	1-4 ms
14	---	Not Used	---
15	TN	Injector 2 Driver	1-4 ms
16	LB/BR	Injector 4 Driver	1-4 ms
17-20	---	Not Used	---
21	BR	Overdrive Solenoid Control	Solenoid Off: 12v, On: 1v
22	TN/PK	Fuel Temperature Sensor	0.5-4.5v
23-24	---	Not Used	---
25	DB, LG	OSS Sensor (-), (+) Signals	Moving: AC pulse signals
26	---	Not Used	---
27	WT/OR	Vehicle Speed Signal	Digital Signal
29	LG/WT	Governor Pressure Sensor	0.58v
30	PK	Transmission Relay Control	Relay Off: 12v, On: 1v
31	OR	5-Volt Supply	4.9-5.1v
32	GY/YL	EGR Solenoid Control	12v, at 55 mph: 1v

1996 Ram Van 5.2L V8 MFI CNG VIN T 'C' Gray Connector

PCM Pin #	Wire Color	Circuit Description (32 Pin)	Value at Hot Idle
1	DB/OR	A/C Clutch Relay Control	Relay Off: 12v, On: 1v
2	---	Not Used	---
3	DB/YL	ASD Relay Control	Relay Off: 12v, On: 1v
4	TN/RD	Speed Control Vacuum Solenoid	Vacuum Increasing: 1v
5	LG/RD	Speed Control Vent Solenoid	Vacuum Decreasing: 1v
6	LG/OR	Overdrive Off Lamp Control	Lamp Off: 12v, On: 1v
7	PK/BK	Trans. Temp. Lamp Driver	Lamp Off: 12v, On: 1v
8-10	---	Not Used	---
11	YL/RD	Speed Control Power Supply	12-14v
12	DG/OR	ASD Relay Power (B+)	12-14v
13	OR/WT	Overdrive Off Switch Sense	Switch Off: 12v, On: 1v
14	---	Not Used	---
15	PK/YL	Battery Temperature Sensor	At 86ºF: 1.96v
16	TN/YL	Generator Lamp Control	Lamp Off: 12v, On: 1v
17	BK/PK	MIL (lamp) Control	MIL Off: 12v, On: 1v
18	---	Not Used	---
19	BR/WT	High Pressure Fuel Shutoff Relay	Relay Off: 12v, On: 1v
20-21	---	Not Used	---
22	BR	A/C Request Signal	A/C Off: 12v, On: 1v
23	LG	A/C Select Signal	A/C Off: 12v, On: 1v
24	WT/PK	Brake Switch Sense Signal	Brake Off: 0v, On: 12v
25	---	Not Used	---
26	DB/WT	Low Fuel Sense Signal	0.5-4.5v
27, 29	PK, DG	SCI Transmit, Receive Signal	0v, 5v
28, 30	---	Not Used	---
31	GY/LB	Tachometer Signal	Pulse Signals
32	RD/LG	Speed Control Set Switch Signal	S/C & Set Switch On: 3.8v

1997 Ram Van 5.2L V8 MFI CNG VIN T 'A' Black Connector

PCM Pin #	Wire Color	Circuit Description (32 Pin)	Value at Hot Idle
1	---	Not Used	---
2	LG/BK	Ignition Switch Power (B+)	12-14v
3	---	Not Used	---
4	BK/LB	Sensor Ground	<0.050v
5	---	Not Used	---
6	BR/OR	PNP Switch Sense Signal	In P/N: 0v, Others: 5v
7	GY	Coil 1 Driver Control	5°, 55 mph: 8° dwell
8	GY/BK	CKP Sensor Signal	Digital Signals: 0-5-0v
9	---	Not Used	---
10	YL/BK	IAC 2 Driver Control	Pulse Signals
11	BR/WT	IAC 1 Driver Control	Pulse Signals
12-13	---	Not Used	---
14	DB	Fuel Level Sensor Signal	Varies: 2-10v
15	BK/RD	IAT Sensor Signal	At 100°F: 1.83v
16	TN/BK	ECT Sensor Signal	At 180°F: 2.80v
17	VT/WT	5-Volt Supply	4.9-5.1v
18	TN/YL	CMP Sensor Signal	Digital Signals: 0-5-0v
19	GY/RD	IAC 3 Driver Control	Pulse Signals
20	VT/BK	IAC 4 Driver Control	Pulse Signals
21	---	Not Used	---
22	RD/WT	Keep Alive Power (B+)	12-14v
23	OR/DB	TP Sensor Signal	0.6-1.0v
24	LG/RD	HO2S-11 (B1 S1) Signal	0.1-1.1v
25	PK/WT	HO2S-12 (B1 S2) Signal	0.1-1.1v
26	---	Not Used	---
27	DG/RD	MAP Sensor Signal	1.5-1.6v
28	DG/LB	Fuel Pressure Sensor Signal	4.5v at 150 psi
29-30	---	Not Used	---
31	BK/TN	Power Ground	<0.1v
32	BK/TN	Power Ground	<0.1v

Pin Connector Graphic

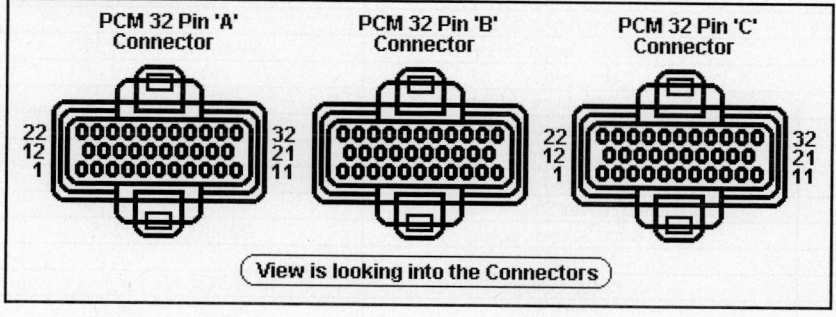

1997 Ram Van 5.2L V8 MFI CNG VIN T 'B' White Connector

PCM Pin #	Wire Color	Circuit Description (32 Pin)	Value at Hot Idle
1	GY/BK	Transmission Temperature Sensor Signal	At 200°F: 2.40v
2	VT	Injector 7 Driver	1-4 ms
3, 7, 9	---	Not Used	---
4	WT/DB	Injector 1 Driver	1-4 ms
5	YL/WT	Injector 3 Driver	1-4 ms
6	GY	Injector 5 Driver	1-4 ms
8	PK	Variable Force Solenoid	PWM Signal: 0-12-0v
10	DG/WT	Generator Field Control	Digital Signals: 0-12-0v
11	OR/BK	M/T: Shift Indicator Lamp	Lamp Off: 12v, On: 1v
12	BR/DB	Injector 6 Driver	1-4 ms
13	GY/LB	Injector 8 Driver	1-4 ms
14, 17-20	---	Not Used	---
15	TN	Injector 2 Driver	1-4 ms
16	LB/BR	Injector 4 Driver	1-4 ms
21	OR/WT	Overdrive Solenoid Control	Solenoid Off: 12v, On: 1v
22	TN/PK	Fuel Temperature Sensor	0.5-4.5v
23-24, 26	---	Not Used	---
25, 28	DB, LG	OSS Sensor (-), (+) Signals	Moving: AC pulse signals
27	WT/OR	Vehicle Speed Signal	Digital Signal
29	LG/RD	Governor Pressure Sensor	0.58v
30	VT/LB	Transmission Relay Control	Relay Off: 12v, On: 1v
31	OR	5-Volt Supply	4.9-5.1v

1997 Ram Van 5.2L V8 MFI CNG VIN T 'C' Gray Connector

PCM Pin #	Wire Color	Circuit Description (32 Pin)	Value at Hot Idle
1	DB/OR	A/C Clutch Relay Control	Relay Off: 12v, On: 1v
2, 7-10	---	Not Used	---
3	DB/YL	ASD Relay Control	Relay Off: 12v, On: 1v
4	TN/RD	Speed Control Vacuum Solenoid	Vacuum Increasing: 1v
5	LG/RD	Speed Control Vent Solenoid	Vacuum Decreasing: 1v
6	LG/OR	Overdrive Off Lamp Control	Lamp Off: 12v, On: 1v
11	YL/RD	Speed Control Power Supply	12-14v
12	DG/OR	ASD Relay Power (B+)	12-14v
13	OR/WT	Overdrive Off Switch Sense	Switch Off: 12v, On: 1v
14	---	Not Used	---
15	PK/YL	Battery Temperature Sensor	At 86°F: 1.96v
16	TN/BK	Generator Lamp Control	Lamp Off: 12v, On: 1v
17	BK/PK	MIL (lamp) Control	MIL Off: 12v, On: 1v
18	---	Not Used	---
19	BR	High Pressure Fuel Shutoff Relay	Relay Off: 12v, On: 1v
20	PK/BK	EVAP Purge Solenoid Control	PWM Signal: 0-12-0v
21	---	Not Used	---
22	BR/WT	A/C Request Signal	A/C Off: 12v, On: 1v
23	LG, DB	A/C Select Signal	A/C Off: 12v, On: 1v
24	WT/PK	Brake Switch Sense Signal	Brake Off: 0v, On: 12v
25	OR/DG	Generator Field Source (B+)	12-14v
26	---	Not Used	---
27	PK	SCI Transmit	0v
28	---	Not Used	---
29	LG	SCI Receive	5v
30	---	Not Used	---
31	GY/LB	Tachometer Signal	Pulse Signals
32	RD, WT	Speed Control Set Switch Signal	S/C & Set Switch On: 3.8v

1998-2000 Ram Van 5.2L V8 OHV Propane VIN 2 'A' Black Connector

PCM Pin #	Wire Color	Circuit Description (32 Pin)	Value at Hot Idle
1	---	Not Used	---
2	LG/BK	Ignition Switch Power (B+)	12-14v
3	---	Not Used	---
4	BK/LB	Sensor Ground	<0.050v
5	---	Not Used	---
6	BR/YL	PNP Switch Sense Signal	In P/N: 0v, Others: 5v
7	GY	Coil 1 Driver Control	5°, 55 mph: 8° dwell
8	GY/BK	CKP Sensor Signal	Digital Signals: 0-5-0v
9	---	Not Used	---
10	YL/BK	IAC 2 Driver Control	Pulse Signals
11	BR/WT	IAC 1 Driver Control	Pulse Signals
12-14	---	Not Used	---
15	BK/RD	IAT Sensor Signal	At 100°F: 1.83v
16	TN/BK	ECT Sensor Signal	At 180°F: 2.80v
17	OR	5-Volt Supply	4.9-5.1v
18	TN/YL	CMP Sensor Signal	Digital Signals: 0-5-0v
19	GY/RD	IAC 3 Driver Control	Pulse Signals
20	VT/BK	IAC 4 Driver Control	Pulse Signals
21	---	Not Used	---
22	RD/WT	Keep Alive Power (B+)	12-14v
22 ('01)	RD/WT	Keep Alive Power (B+)	12-14v
23	OR/DB	TP Sensor Signal	0.6-1.0v
24	BK/DG	HO2S-11 (B1 S1) Signal	0.1-1.1v
25	TN/WT	HO2S-12 (B1 S2) Signal	0.1-1.1v
26	---	Not Used	---
27	DG/RD	MAP Sensor Signal	1.5-1.6v
28	DG/OR	Fuel Pressure Sensor	0.5-4.5v
29-30	---	Not Used	---
31	BK/TN	Power Ground	<0.1v
32	BK/TN	Power Ground	<0.1v

Pin Connector Graphic

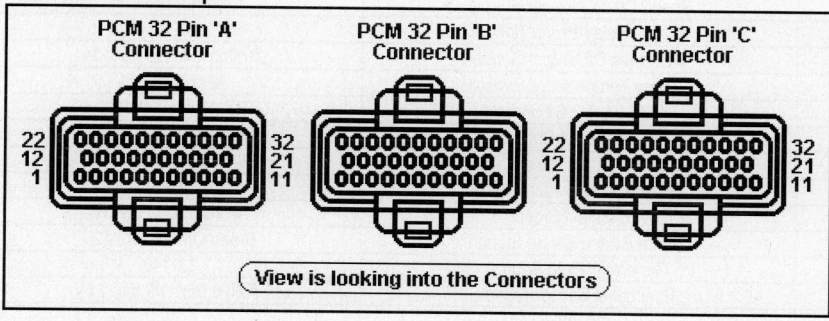

PCM 32 Pin 'A' Connector   PCM 32 Pin 'B' Connector   PCM 32 Pin 'C' Connector

View is looking into the Connectors

1998-2000 Ram Van 5.2L V8 OHV Propane VIN 2 'B' White Connector

PCM Pin #	Wire Color	Circuit Description (32 Pin)	Value at Hot Idle
1	GY/BK	Transmission Temperature Sensor Signal	At 200ºF: 2.40v
2	VT	Injector 7 Driver	1-4 ms
4	WT/DB	Injector 1 Driver	1-4 ms
5	YL/WT	Injector 3 Driver	1-4 ms
6	GY	Injector 5 Driver	1-4 ms
8	VT/WT	Governor Pressure Solenoid	PWM Signal: 0-12-0v
10	DG/WT	Generator Field Control	Digital Signals: 0-12-0v
11	OR/BK	TCC Solenoid Control	At Cruise w/TCC On: 1v
12	BR/DB	Injector 6 Driver	1-4 ms
13	GY/LB	Injector 8 Driver	1-4 ms
15	TN	Injector 2 Driver	1-4 ms
16	LB/BR	Injector 4 Driver	1-4 ms
17-20, 24	---	Not Used	---
21	BR	3-4 Shift Solenoid Control	Solenoid Off: 12v, On: 1v
22	TN/PK	Fuel Temperature Sensor	0.5-4.5v
23	GY/YL	Engine Oil Pressure Sensor	1.6v at 24 psi
25, 28	DB, LG	OSS Sensor (-), (+) Signals	Moving: AC pulse signals
27	WT/OR	Vehicle Speed Signal	Digital Signal
29	LG/RD	Governor Pressure Sensor	0.58v
30	PK	Transmission Relay Control	Relay Off: 12v, On: 1v
31	VT/WT	5-Volt Supply	4.9-5.1v

1998-2000 Ram Van 5.2L V8 OHV Propane VIN 2 'C' Gray Connector

PCM Pin #	Wire Color	Circuit Description (32 Pin)	Value at Hot Idle
1	DB/OR	A/C Clutch Relay Control	Relay Off: 12v, On: 1v
3	DB/YL	ASD Relay Control	Relay Off: 12v, On: 1v
4	TN/RD	Speed Control Vacuum Solenoid	Vacuum Increasing: 1v
5	LG/RD	Speed Control Vent Solenoid	Vacuum Decreasing: 1v
8	OR/RD	HO2S-11 (B1 S1) Heater	PWM Signal: 0-12-0v
9	BK/OR	HO2S-11 Heater Relay (Cal)	Heater Off: 12v, On: 1v
10	WT/DG	LDP Solenoid Control	PWM Signal: 0-12-0v
11	YL/RD	Speed Control Power Supply	12-14v
12	DG/OR	ASD Relay Power (B+)	12-14v
13	OR/WT	Overdrive Off Switch Sense	Switch Off: 12v, On: 1v
15	PK/YL	Battery Temperature Sensor	At 86ºF: 1.96v
16	VT/TN	HO2S-21 (B2 S1) Heater	PWM Signal: 0-12-0v
19	BR	High Pressure Fuel Shutoff Relay	Relay Off: 12v, On: 1v
20	PK/BK	EVAP Purge Solenoid Control	PWM Signal: 0-12-0v
22	DB	A/C Pressure Switch Signal	A/C On: 0.451-4.850v
23	LG/GY	A/C Select Signal	A/C Off: 12v, On: 1v
24	WT/PK	Brake Switch Sense Signal	Brake Off: 0v, On: 12v
25	WT/DB	Generator Field Source	12-14v
26	DB	Fuel Level Sensor Signal	Full: 0.56v, 1/2 full: 2.5v
27	PK	SCI Transmit Signal	0v
28	WT/BK	CCD Bus (-) Signal	<0.050v
29	LG	SCI Receive Signal	5v
30	VT/BR	CCD Bus (+) Signal	Digital Signals: 0-5-0v
32	WT/LG	Speed Control Set Switch Signal	S/C & Set Switch On: 3.8v

Standard Colors and Abbreviations

Abbreviation	Color	Abbreviation	Color	Abbreviation	Color
BK	Black	GY	Gray	RD	Red
BL	Blue	GN	Green	TN	Tan
BR	Brown	LG	Light Green	VT	Violet
DB	Dark Blue	OR	Orange	WT	White
DG	Dark Green	PK	Pink	YL	Yellow

1999-2003 Ram Van 5.2L V8 OHV CNG VIN T 'A' Black Connector

PCM Pin #	Wire Color	Circuit Description (32 Pin)	Value at Hot Idle
1	---	Not Used	---
2	LG/BK	Ignition Switch Power (B+)	12-14v
3	---	Not Used	---
4	BK/LB	Sensor Ground	<0.050v
5	---	Not Used	---
6	BR/YL	PNP Switch Sense Signal	In P/N: 0v, Others: 5v
7	GY	Coil 1 Driver Control	5°, 55 mph: 8° dwell
8	GY/BK	CKP Sensor Signal	Digital Signals: 0-5-0v
9	---	Not Used	---
10	YL/BK	IAC 2 Driver Control	Pulse Signals
11	BR/WT	IAC 3 Driver Control	Pulse Signals
12-14	---	Not Used	---
15	BK/RD	IAT Sensor Signal	At 100°F: 1.83v
16	TN/BK	ECT Sensor Signal	At 180°F: 2.80v
17	OR	5-Volt Supply	4.9-5.1v
18	TN/YL	CMP Sensor Signal	Digital Signals: 0-5-0v
19	GY/RD	IAC 1 Driver Control	Pulse Signals
20	VT/BK	IAC 4 Driver Control	Pulse Signals
21	---	Not Used	---
22	RD/WT	Keep Alive Power (B+)	12-14v
23	OR/DB	TP Sensor Signal	0.6-1.0v
24	BK/DG	HO2S-11 (B1 S1) Signal	0.1-1.1v
25	TN/WT	HO2S-12 (B1 S2) Signal	0.1-1.1v
26	---	Not Used	---
27	DG/RD	MAP Sensor Signal	1.5-1.6v
28	DG/OR	Fuel Pressure Sensor	0.5-4.5v
29-30	---	Not Used	---
31	BK/TN	Power Ground	<0.1v
32	BK/TN	Power Ground	<0.1v

Pin Connector Graphic

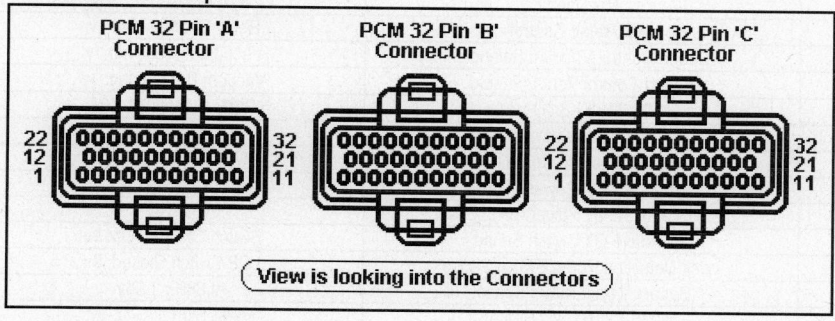

PCM 32 Pin 'A' Connector
PCM 32 Pin 'B' Connector
PCM 32 Pin 'C' Connector

View is looking into the Connectors

1999-2003 Ram Van 5.2L V8 OHV CNG VIN T 'B' White Connector

PCM Pin #	Wire Color	Circuit Description (32 Pin)	Value at Hot Idle
1	GY/BK	Transmission Temperature Sensor Signal	At 200°F: 2.40v
2	VT	Injector 7 Driver	1-4 ms
3, 7	---	Not Used	---
4	WT/DB	Injector 1 Driver	1-4 ms
5	YL/WT	Injector 3 Driver	1-4 ms
6	GY	Injector 5 Driver	1-4 ms
8	VT/BK	Governor Pressure Solenoid	PWM Signal: 0-12-0v
9, 14	---	Not Used	---
10	DG/WT	Generator Field Control	Digital Signals: 0-12-0v
11	OR/BK	TCC Solenoid Control	At Cruise w/TCC On: 1v
12	BR/DB	Injector 6 Driver	1-4 ms
13	GY/LB	Injector 8 Driver	1-4 ms
15	TN	Injector 2 Driver	1-4 ms
16	LB/BR	Injector 4 Driver	1-4 ms
17-20	---	Not Used	---
21	BR	3-4 Shift Solenoid Control	Solenoid Off: 12v, On: 1v
22	TN/PK	Fuel Temperature Sensor	0.5-4.5v
23	GY/YL	Engine Oil Pressure Sensor	1.6v at 24 psi
24	---	Not Used	---
25	DB/BK	OSS Sensor (-) Signals	Moving: AC pulse signals
27	WT/OR	Vehicle Speed Signal	Digital Signals
28	LG/WT	OSS Sensor (+) Signals	Moving: AC pulse signals
29	LG/RD	Governor Pressure Sensor	0.58v
30	PK	Transmission Relay Control	Relay Off: 12v, On: 1v
31	VT/WT	5-Volt Supply	4.9-5.1v
32	---	Not Used	---

1999-2003 Ram Van 5.2L V8 OHV CNG VIN T 'C' Gray Connector

PCM Pin #	Wire Color	Circuit Description (32 Pin)	Value at Hot Idle
1	DB/OR	A/C Clutch Relay Control	Relay Off: 12v, On: 1v
2	---	Not Used	---
3	DB/YL	ASD Relay Control	Relay Off: 12v, On: 1v
4	TN/RD	Speed Control Vacuum Solenoid	Vacuum Increasing: 1v
5	LG/RD	Speed Control Vent Solenoid	Vacuum Decreasing: 1v
8	OR/RD	HO2S-11 (B1 S1) Heater	PWM Signal: 0-12-0v
9	---	Not Used	---
10	WT/DG	LDP Solenoid Control	PWM Signal: 0-12-0v
11	YL/RD	Speed Control Power Supply	12-14v
12	DG/OR	ASD Relay Power (B+)	12-14v
13	OR/WT	Overdrive Off Switch Sense	Switch Off: 12v, On: 1v
14	OR	Leak Detection Pump Switch Sense	LDP Switch Closed: 0v
15	PK/YL	Battery Temperature Sensor	At 86°F: 1.96v
16	VT/TN	HO2S-12 (B1 S2) Heater	PWM Signal: 0-12-0v
17-18	---	Not Used	---
19	BR	High Pressure Fuel Shutoff Relay	Relay Off: 12v, On: 1v
20-21	---	Not Used	---
22	DB	A/C Pressure Switch Signal	A/C On: 0.451-4.850v
23	LG/GY	A/C Switch Sense	A/C Off: 12v, On: 1v
24	WT/PK	Brake Switch Sense	Brake Off: 0v, On: 12v
25	WT/DB	Generator Field Source	12-14v
26	LB/BK	Fuel Level Sensor Signal	Full: 0.56v, 1/2 full: 2.5v
27	PK	SCI Transmit	0v
28	WT/BK	CCD Bus (-) Signal	<0.050v
29	LG	SCI Receive	5v
30	VT/BR	CCD Bus (+) Signal	Digital Signals: 0-5-0v
31	---	Not Used	---
32	WT/LG	Speed Control Set Switch Signal	S/C & Set Switch On: 3.8v

1990-91 Ram Van 5.2L V8 TBI VIN Y (A/T) 60 Pin Connector

PCM Pin #	Wire Color	Circuit Description (60 Pin)	Value at Hot Idle
1	DG/RD	MAP Sensor Signal	1.5-1.6v
2	TN	ECT Sensor Signal	At 180°F: 2.80v
3	RD	Keep Alive Power (B+)	12-14v
4, 5	BK, BK/WT	Sensor Ground	<0.050v
6	PK/WT	5-Volt Supply	4.9-5.1v
7	OR	8-Volt Supply	7.9-8.1v
8	DG/BK	ASD Relay Power (B+)	12-14v
9	DB	Ignition Switch Power (B+)	12-14v
10, 13-14	---	Not Used	---
11, 12	LB/RD	Power Ground	<0.1v
15	TN	Injector 2 Driver	1-4 ms
16	WT/DB	Injector 1 Driver	1-4 ms
17-18, 23, 26	---	Not Used	---
19	BK/GY	Coil Driver Control	5°, 55 mph: 8° dwell
20	DG	Alternator Field Control	Digital Signals: 0-12-0v
21	BK/RD	Throttle Body Temperature	At 100°F: 2.51v
22	OR/DB	TP Sensor Signal	0.6-1.0v
24	GY	Distributor Reference Signal	Digital Signals: 0-5-0v
25	PK	SCI Transmit	0v
27	BR	A/C Clutch Signal	A/C Off: 12v, On: 1v
28	PK	Closed Throttle Switch	Switch Off: 12v, On: 1v
29	WT/PK	Brake Switch Sense Signal	Brake Off: 0v, On: 12v
30	BR/YL	PNP Switch Sense Signal	In P/N: 0v, Others: 5v
31, 37-39	---	Not Used	---
32	GY/PK	MIL (lamp) Control	MIL Off: 12v, On: 1v
33	TN/RD	Speed Control Vacuum Solenoid	Vacuum Increasing: 1v
34	DB/OR	A/C WOT Relay Control	Relay Off: 12v, On: 1v
35	GY/YL	EGR Solenoid Control	12v, at 55 mph: 1v
36	BK/OR	Air Switching Solenoid	Solenoid Off: 12v, On: 1v
40	BR	AIS Motor Control	Pulse Signals
41	BK/DG	HO2S-11 (B1 S1) Signal	0.1-1.1v
42-43, 46	---	Not Used	---
44	OR/WT	Overdrive Lockout Switch	Switch Off: 12v, On: 1v
45	LG	SCI Receive	5v
47	WT/OR	Vehicle Speed Signal	Digital Signal
48	BR/RD	Speed Control Set Switch Signal	S/C & Set Switch On: 3.8v
49	YL/RD	Speed Control On/Off Switch	S/C On: 1v, Off: 12v
50	WT/LG	Speed Control Resume Switch	S/C On: 1v, Off: 12v
51	DB/YL	ASD Relay Control	Relay Off: 12v, On: 1v
52	PK	EVAP Purge Solenoid Control	Solenoid Off: 12v, On: 1v
53	LG/RD	Speed Control Vent Solenoid	Vacuum Decreasing: 1v
54	OR/BK	Part Throttle Unlock Solenoid	Solenoid Off: 12v, On: 1v
55	OR/DB	Overdrive Lockout Solenoid	Solenoid Off: 12v, On: 1v
56	GY/PK	Maintenance Indicator Lamp	Lamp Off: 12v, On: 1v
57-59	---	Not Used	---
60	GY/RD	AIS Motor Control	Pulse Signals

Pin Connector Graphic

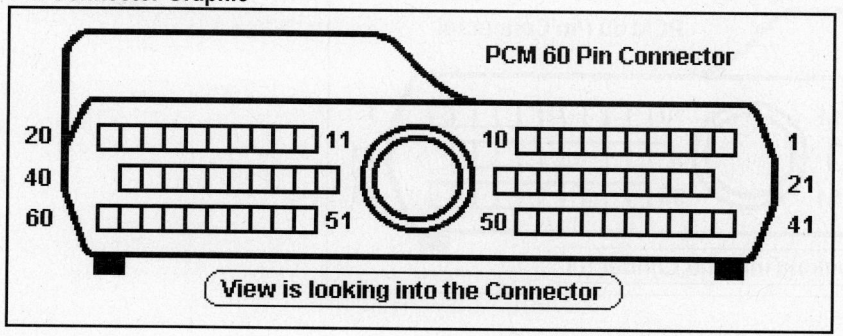

PCM 60 Pin Connector

View is looking into the Connector

1992-93 Ram Van 5.2L V8 MFI VIN Y (A/T) 60 Pin Connector

PCM Pin #	Wire Color	Circuit Description (60 Pin)	Value at Hot Idle
1	DG/RD	MAP Sensor Signal	1.5-1.6v
2	TN/BK	ECT Sensor Signal	At 180°F: 2.80v
3	RD	Battery Power (B+)	12-14v
4, 5	BK, BK/WT	Sensor Ground	<0.050v
6, 7	PK/WT, OR	5-Volt Supply, 8-Volt Supply	4.9-5.1v, 7.9-8.1v
8, 23, 26, 31, 36	---	Not Used	---
9	DB	Ignition Switch Power (B+)	12-14v
10	OR/WT	Overdrive Override Switch	Switch Off: 12v, On: 1v
11, 12	BK/TN	Power Ground	<0.1v
13	LG/BR	Injector 4 Driver	1-4 ms
14	YL/WT	Injector 3 Driver	1-4 ms
15	TN	Injector 2 Driver	1-4 ms
16	WT/DB	Injector 1 Driver	1-4 ms
17	DB/T	Injector 7 Driver	1-4 ms
18	RD/YL	Injector 8 Driver	1-4 ms
19	GY	Coil Driver Control	5°, 55 mph: 8° dwell
20	DG	Alternator Field Control	Digital Signals: 0-12-0v
21	BK/RD	Air Temperature Sensor	At 100°F: 2.51v
22	OR/DB	TP Sensor Signal	0.6-1.0v
24	GY/BK	Distributor Reference Signal	Digital Signals: 0-5-0v
25, 45	PK, LG	SCI Transmit, SCI Receive	0v, 5v
27	BR	A/C Clutch Signal	A/C Off: 12v, On: 1v
29	WT/PK	Brake Switch Sense Signal	Brake Off: 0v, On: 12v
30	BR/YL	PNP Switch Sense Signal	In P/N: 0v, Others: 5v
32	BK/PK	MIL (lamp) Control	MIL Off: 12v, On: 1v
33	TN/RD	Speed Control Vacuum Solenoid	Vacuum Increasing: 1v
34	DB/OR	A/C WOT Relay Control	Relay Off: 12v, On: 1v
35	GY/YL	EGR Solenoid Control	12v, at 55 mph: 1v
37	BK/OR	Overdrive Lamp Control	Lamp Off: 12v, On: 1v
38	PK/BK	Injector 5 Driver	1-4 ms
39, 40	GY, BR/WT	AIS Motor Control	Pulse Signals
41	BK/DG	HO2S-11 (B1 S1) Signal	0.1-1.1v
42-43, 46	---	Not Used	---
44	TN/YL	Distributor Sync Signal	Digital Signals: 0-5-0v
47	WT/OR	Vehicle Speed Signal	Digital Signal
48	BR/RD	Speed Control Set Switch Signal	S/C & Set Switch On: 3.8v
49	YL/RD	Speed Control On/Off Switch	S/C On: 1v, Off: 12v
50	WT/LG	Speed Control Resume Switch	S/C On: 1v, Off: 12v
51	DB/YL	ASD Relay Control	Relay Off: 12v, On: 1v
52	PK/BK	EVAP Purge Solenoid Control	Solenoid Off: 12v, On: 1v
53	LG/RD	Speed Control Vent Solenoid	Vacuum Decreasing: 1v
54	OR/BK	Part Throttle Unlock Solenoid	Solenoid Off: 12v, On: 1v
55	OR/WT	Overdrive Lockout Solenoid	Solenoid Off: 12v, On: 1v
56	GY/PK	Maintenance Indicator Lamp	Lamp Off: 12v, On: 1v
57	DG/OR	ASD Relay Power (B+)	12-14v
58	LG/BK	Injector 6 Driver	1-4 ms
59, 60	PK, YL/BK	AIS Motor Control	Pulse Signals

Pin Connector Graphic

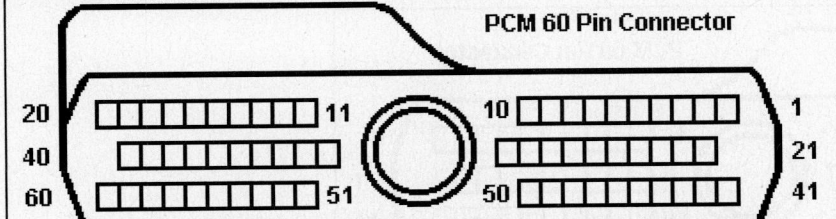

PCM 60 Pin Connector

View is looking into the Connector

1994-95 Ram Van 5.2L V8 MFI VIN Y (A/T) 60 Pin Connector

PCM Pin #	Wire Color	Circuit Description (60 Pin)	Value at Hot Idle
1	DG/RD	MAP Sensor Signal	1.5-1.6v
2	TN/BK	ECT Sensor Signal	At 180°F: 2.80v
3	RD	Keep Alive Power (B+)	12-14v
4	BK/LB	Sensor Ground	<0.050v
5 ('94)	BK/WT	Sensor Ground	<0.050v
6	PK/WT	5-Volt Supply	4.9-5.1v
7	OR	8-Volt Supply	7.9-8.1v
8	---	Not Used	---
9	DB	Ignition Switch Power (B+)	12-14v
10	OR/WT	Overdrive Override Switch	Switch Off: 12v, On: 1v
11	BK/TN	Power Ground	<0.1v
12	BK/TN	Power Ground	<0.1v
13	LB/BR	Injector 4 Driver	1-4 ms
14	YL/WT	Injector 3 Driver	1-4 ms
15	TN	Injector 2 Driver	1-4 ms
16	WT/DB	Injector 1 Driver	1-4 ms
17	DB/TN	Injector 7 Driver	1-4 ms
18	DB/GY	Injector 8 Driver	1-4 ms
19	GY	Coil Driver Control	5°, 55 mph: 8° dwell
20	DG/WT	Alternator Field Control	Digital Signals: 0-12-0v
21	BK/RD	IAT Sensor Signal	At 100°F: 1.83v
22	OR/DB	TP Sensor Signal	0.6-1.0v
23	---	Not Used	---
24	GY/BK	Distributor Reference Signal	Digital Signals: 0-5-0v
25	PK	SCI Transmit	0v
26	---	Not Used	---
27	BR	A/C Cycling Switch	A/C Off: 12v, On: 1v
28	---	Not Used	---
29	WT/PK	Brake Switch Sense Signal	Brake Off: 0v, On: 12v
30	BR/O	PNP Switch Sense Signal	In P/N: 0v, Others: 5v
31	---	Not Used	---
32	BK/PK	MIL (lamp) Control	MIL Off: 12v, On: 1v
33	TN/RD	Speed Control Vacuum Solenoid	Vacuum Increasing: 1v
34	DB/OR	A/C WOT Relay Control	Relay Off: 12v, On: 1v
35	GY/YL	EGR Solenoid Control	12v, at 55 mph: 1v
36	---	Not Used	---
37	BK/OR	Overdrive Lamp Control	Lamp Off: 12v, On: 1v
38	PK/BK	Injector 5 Driver	1-4 ms
39	GY/RD	AIS Motor Control	Pulse Signals
40	BR/WT	AIS Motor Control	Pulse Signals

Standard Colors and Abbreviations

Abbreviation	Color	Abbreviation	Color	Abbreviation	Color
BK	Black	GY	Gray	RD	Red
BL	Blue	GN	Green	TN	Tan
BR	Brown	LG	Light Green	VT	Violet
DB	Dark Blue	OR	Orange	WT	White
DG	Dark Green	PK	Pink	YL	Yellow

1994-95 Ram Van 5.2L V8 MFI VIN Y (A/T) 60 Pin Connector

PCM Pin #	Wire Color	Circuit Description (60 Pin)	Value at Hot Idle
41	BK/DG	HO2S-11 (B1 S1) Signal	0.1-1.1v
42-43	---	Not Used	---
44	TN/YL	CMP Sensor Signal	Digital Signals: 0-5-0v
45	LG	SCI Receive	5v
46	---	Not Used	---
47	WT/OR	Vehicle Speed Signal	Digital Signal
48	BR/RD	Speed Control Set Switch Signal	S/C & Set Switch On: 3.8v
49	YL/RD	Speed Control On/Off Switch	S/C On: 1v, Off: 12v
50	WT/LG	Speed Control Resume Switch	S/C On: 1v, Off: 12v
51	DB/YL	ASD Relay Control	Relay Off: 12v, On: 1v
52	PK/BK	EVAP Purge Solenoid Control	Solenoid Off: 12v, On: 1v
53	LG/RD	Speed Control Vent Solenoid	Vacuum Decreasing: 1v
54	OR/BK	A/T: Overdrive Solenoid	Solenoid Off: 12v, On: 1v
55	OR/WT	Overdrive Lockout Solenoid	Solenoid Off: 12v, On: 1v
56	---	Not Used	---
57	DG/OR	ASD Relay Power (B+)	12-14v
58	LG/BK	Injector 6 Driver	1-4 ms
59	PK/BK	AIS Motor Control	Pulse Signals
60	YL/BK	AIS Motor Control	Pulse Signals

Pin Connector Graphic

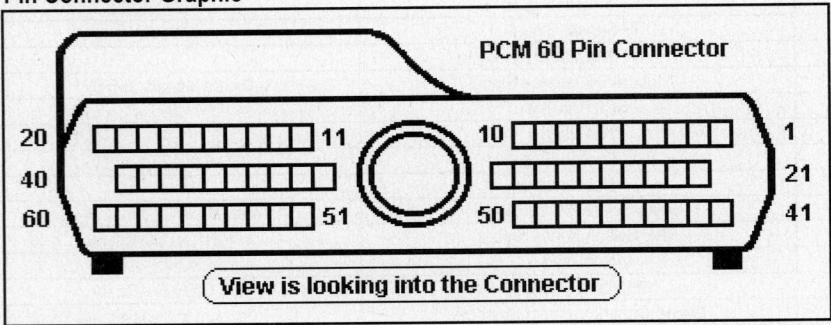

1996-97 Ram Van 5.2L V8 MFI VIN Y (A/T) 'A' Black Connector

PCM Pin #	Wire Color	Circuit Description (32 Pin)	Value at Hot Idle
1	---	Not Used	---
2	LG/BK	Ignition Switch Power (B+)	12-14v
3	---	Not Used	---
4	BK/LB	Sensor Ground	<0.050v
5	---	Not Used	---
6	BR/OR	PNP Switch Sense Signal	In P/N: 0v, Others: 5v
7	GY	Coil 1 Driver Control	5°, 55 mph: 8° dwell
8	GY/BK	CKP Sensor Signal	Digital Signals: 0-5-0v
9	---	Not Used	---
10	YL/BK	IAC 2 Driver Control	Pulse Signals
11	BR/WT	IAC 1 Driver Control	Pulse Signals
12-14	---	Not Used	---
15	BK/RD	IAT Sensor Signal	At 100°F: 1.83v
16	TN/BK	ECT Sensor Signal	At 180°F: 2.80v
17	PK/WT	5-Volt Supply	4.9-5.1v
18	TN/YL	CMP Sensor Signal	Digital Signals: 0-5-0v
19	GY/RD	IAC 3 Driver Control	Pulse Signals
20	PK/BK	IAC 4 Driver Control	Pulse Signals
21	---	Not Used	---
22	RD/WT	Keep Alive Power (B+)	12-14v
23	OR/DB	TP Sensor Signal	0.6-1.0v
24	BK/DG	HO2S-11 (B1 S1) Signal	0.1-1.1v
25	TN/WT	HO2S-12 (B1 S2) Signal	0.1-1.1v
27	DG/RD	MAP Sensor Signal	1.5-1.6v
28-30	---	Not Used	---
31-32	BK/TN	Power Ground	<0.1v

Pin Connector Graphic

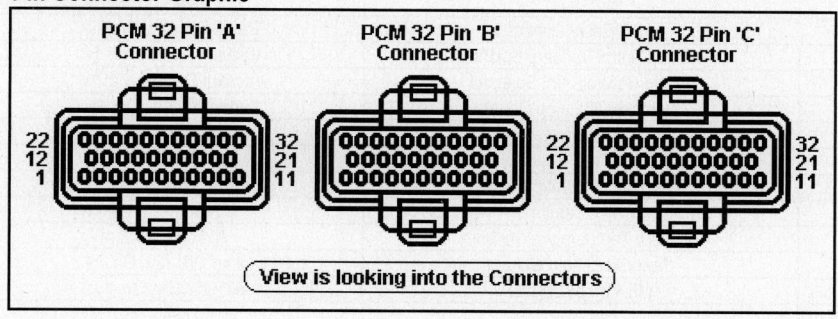

1996-97 Ram Van 5.2L V8 MFI VIN Y (A/T) 'B' White Connector

PCM Pin #	Wire Color	Circuit Description (32 Pin)	Value at Hot Idle
1	GY/BK	Transmission Temperature Sensor Signal	At 200°F: 2.40v
2	PK	Injector 7 Driver	1-4 ms
3	---	Not Used	---
4	WT/DB	Injector 1 Driver	1-4 ms
5	YL/WT	Injector 3 Driver	1-4 ms
6	GY	Injector 5 Driver	1-4 ms
7	---	Not Used	---
8	PK/WT	Governor Pressure Solenoid	PWM Signal: 0-12-0v
9, 14	---	Not Used	---
10	DG/WT	Generator Field Control	Digital Signals: 0-12-0v
11	OR/BK	TCC Solenoid Control	At Cruise w/TCC On: 1v
12	BR/DB	Injector 6 Driver	1-4 ms
13	GY/LB	Injector 8 Driver	1-4 ms
15	TN	Injector 2 Driver	1-4 ms
16	LB/BR	Injector 4 Driver	1-4 ms
21	OR/WT	3-4 Shift Solenoid Control	Solenoid Off: 12v, On: 1v
25	DB/BK	OSS Sensor (-) Signal	Moving: AC pulse signals
27	WT/OR	Vehicle Speed Signal	Digital Signal
28	LG/WT	OSS Sensor (+) Signal	Moving: AC pulse signals
29	LG/RD	Governor Pressure Sensor	0.58v
30	PK/LB	Transmission Relay Control	Relay Off: 12v, On: 1v
31	OR	5-Volt Supply	4.9-5.1v
32 ('96)	GY/YL	EGR Solenoid Control	12v, at 55 mph: 1v

1996-97 Ram Van 5.2L V8 MFI VIN Y (A/T) 'C' Gray Connector

PCM Pin #	Wire Color	Circuit Description (32 Pin)	Value at Hot Idle
1	DB/OR	A/C Clutch Relay Control	Relay Off: 12v, On: 1v
2, 7-10	---	Not Used	---
3	DB/YL	ASD Relay Control	Relay Off: 12v, On: 1v
4	TN/RD	Speed Control Vacuum Solenoid	Vacuum Increasing: 1v
5	LG/RD	Speed Control Vent Solenoid	Vacuum Decreasing: 1v
6	LG/OR	Overdrive Off Lamp Control	Lamp Off: 12v, On: 1v
11	YL/RD	Speed Control Power Supply	12-14v
12	DG/OR	ASD Relay Power (B+)	12-14v
13	OR/WT	Overdrive Off Switch Sense	Switch Off: 12v, On: 1v
14	---	Not Used	---
15	PK/YL	Battery Temperature Sensor	At 86°F: 1.96v
16	TN/BK	Generator Lamp Control	Lamp Off: 12v, On: 1v
17	BK/PK	MIL (lamp) Control	MIL Off: 12v, On: 1v
18	---	Not Used	---
19	BR	Fuel Pump Relay Control	Relay Off: 12v, On: 1v
20	PK/BK	EVAP Purge Solenoid Control	PWM Signal: 0-12-0v
21, 25, 28	---	Not Used	---
22	BR/WT	A/C Request Signal	A/C Off: 12v, On: 1v
23	LG	A/C Select Signal	A/C Off: 12v, On: 1v
24	WT/PK	Brake Switch Sense Signal	Brake Off: 0v, On: 12v
26	DB/WT	Fuel Level Sensor Signal	Full: 0.56v, 1/2 full: 2.5v
27	PK	SCI Transmit Signal	0v
29	LG/WT	SCI Receive Signal	5v
30-31	---	Not Used	---
32	RD/LG	Speed Control Set Switch Signal	S/C & Set Switch On: 3.8v

Standard Colors and Abbreviations

Abbreviation	Color	Abbreviation	Color	Abbreviation	Color
BK	Black	GY	Gray	RD	Red
BL	Blue	GN	Green	TN	Tan
BR	Brown	LG	Light Green	VT	Violet
DB	Dark Blue	OR	Orange	WT	White
DG	Dark Green	PK	Pink	YL	Yellow

1998-2000 Ram Van 5.2L V8 MFI VIN Y 'C1' Black Connector

PCM Pin #	Wire Color	Circuit Description (32 Pin)	Value at Hot Idle
1, 3, 5, 9	---	Not Used	---
2	LG/BK	Ignition Switch Power (B+)	12-14v
4	BK/LB	Sensor Ground	<0.050v
6	BR/YL	PNP Switch Sense Signal	In P/N: 0v, Others: 5v
7	GY	Coil 1 Driver Control	5°, 55 mph: 8° dwell
8	GY/BK	CKP Sensor Signal	Digital Signals: 0-5-0v
10	YL/BK	IAC 2 Driver Control	Pulse Signals
11	BR/WT	IAC 1 Driver Control	Pulse Signals
14, 21	---	Not Used	---
15	BK/RD	IAT Sensor Signal	At 100°F: 1.83v
16	TN/BK	ECT Sensor Signal	At 180°F: 2.80v
17	OR	5-Volt Supply	4.9-5.1v
18	TN/YL	CMP Sensor Signal	Digital Signals: 0-5-0v
19	GY/RD	IAC 3 Driver Control	Pulse Signals
20	VT/BK	IAC 4 Driver Control	Pulse Signals
22	RD/WT	Keep Alive Power (B+)	12-14v
23	OR/DB	TP Sensor Signal	0.6-1.0v
24	TN/WT	HO2S-11 (B1 S1) Signal	0.1-1.1v
25	BK/DG	HO2S-12 (B1 S2) Signal	0.1-1.1v
26	---	Not Used	---
27	DG/RD	MAP Sensor Signal	1.5-1.6v
28-30	---	Not Used	---
31-32	BK/TN	Power Ground	<0.1v

1998-2000 Ram Van 5.2L V8 MFI VIN Y 'C2' White Connector

PCM Pin #	Wire Color	Circuit Description (32 Pin)	Value at Hot Idle
1	GY/BK	Transmission Temperature Sensor Signal	At 200°F: 2.40v
2	VT	Injector 7 Driver	1-4 ms
3, 7, 9, 14	---	Not Used	---
4	WT/DB	Injector 1 Driver	1-4 ms
5	YL/WT	Injector 3 Driver	1-4 ms
6	GY	Injector 5 Driver	1-4 ms
8	PK/WT	Governor Pressure Solenoid	PWM Signal: 0-12-0v
10	DG/WT	Generator Field Control	Digital Signals: 0-12-0v
11	OR/BK	TCC Solenoid Control	At Cruise w/TCC On: 1v
12	BR/DB	Injector 6 Driver	1-4 ms
13	GY/LB	Injector 8 Driver	1-4 ms
15	TN	Injector 2 Driver	1-4 ms
16	LB/BR	Injector 4 Driver	1-4 ms
17-20	---	Not Used	---
21	OR/WT	Overdrive Solenoid Control	Cruise w/solenoid On: 1v
22	---	Not Used	---
24, 26	---	Not Used	---
23	GY	Engine Oil Pressure Sensor	1.6v at 24 psi
25	DB/BK	OSS Sensor (-) Signal	Moving: AC pulse signals
27	WT/OR	Vehicle Speed Signal	Digital Signal
28	LG/WT	OSS Sensor (+) Signal	Moving: AC pulse signals
29	LG/RD	Governor Pressure Sensor	0.58v
30	PK/LB	Transmission Relay Control	Relay Off: 12v, On: 1v
31	PK/WT	5-Volt Supply	4.9-5.1v
32	---	Not Used	---

Standard Colors and Abbreviations

Abbreviation	Color	Abbreviation	Color	Abbreviation	Color
BK	Black	GY	Gray	RD	Red
BL	Blue	GN	Green	TN	Tan
BR	Brown	LG	Light Green	VT	Violet
DB	Dark Blue	OR	Orange	WT	White
DG	Dark Green	PK	Pink	YL	Yellow

1998-2000 Ram Van 5.2L V8 MFI VIN Y 'C3' Gray Connector

PCM Pin #	Wire Color	Circuit Description (32 Pin)	Value at Hot Idle
1	DB/OR	A/C Clutch Relay Control	Relay Off: 12v, On: 1v
2	---	Not Used	---
3	DB/YL	ASD Relay Control	Relay Off: 12v, On: 1v
4	TN/RD	Speed Control Vacuum Solenoid	Vacuum Increasing: 1v
5	LG/RD	Speed Control Vent Solenoid	Vacuum Decreasing: 1v
6 ('98)	LG/OR	Overdrive Off Lamp Control	Lamp Off: 12v, On: 1v
7-9	---	Not Used	---
10	WT/DG	LDP Solenoid Control	PWM Signal: 0-12-0v
11	YL/RD	Speed Control Power Supply	12-14v
12	DG/OR	ASD Relay Power (B+)	12-14v
13	OR/WT	Overdrive Off Switch Sense	Switch Off: 12v, On: 1v
14	OR	Leak Detection Pump Switch Sense	LDP Switch Closed: 0v
15	PK/YL	Battery Temperature Sensor	At 86°F: 1.96v
16-18	---	Not Used	---
19	BR	Fuel Pump Relay Control	Relay Off: 12v, On: 1v
20	PK/BK	EVAP Purge Solenoid Control	PWM Signal: 0-12-0v
21	---	Not Used	---
22	DB	A/C Pressure Switch Signal	A/C On: 0.451-4.850v
23	LG/GY	A/C Select Signal	A/C Off: 12v, On: 1v
24	WT/PK	Brake Switch Sense Signal	Brake Off: 0v, On: 12v
25	O/DG	Generator Field Source	12-14v
25 ('00)	WT/DB	Generator Field Source	12-14v
26	LB/BK	Fuel Level Sensor Signal	Full: 0.56v, 1/2 full: 2.5v
27	PK	SCI Transmit	0v
28	WT/BK	CCD Bus (-)	<0.050v
29	LG/WT	SCI Receive	5v
30	PK/BR	CCD Bus (+)	Digital Signals: 0-5-0v
31	---	Not Used	---
32	WT/LG	Speed Control Set Switch Signal	S/C & Set Switch On: 3.8v

Pin Connector Graphic

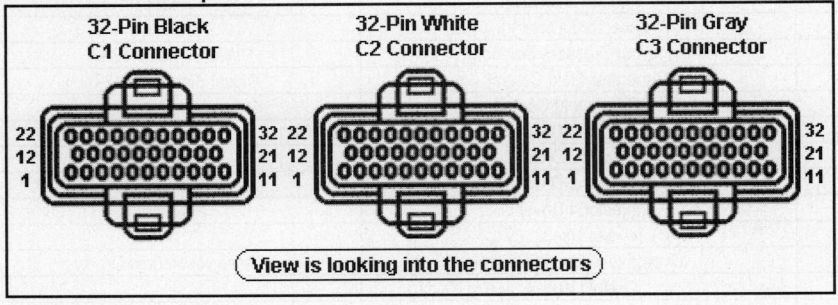

32-Pin Black C1 Connector · 32-Pin White C2 Connector · 32-Pin Gray C3 Connector

View is looking into the connectors

2001-03 Ram Van 5.2L V8 OHV MFI VIN Y (A/T) 'C1' Black Connector

PCM Pin #	Wire Color	Circuit Description (32 Pin)	Value at Hot Idle
1	---	Not Used	---
2	LG/BK	Ignition Switch Power (B+)	12-14v
3	---	Not Used	---
4	BK/LB	Sensor Ground	<0.050v
5	---	Not Used	---
6	BR/YL	PNP Switch Sense Signal	In P/N: 0v, Others: 5v
7	GY	Coil 1 Driver Control	5°, 55 mph: 8° dwell
8	GY/BK	CKP Sensor Signal	Digital Signals: 0-5-0v
9	---	Not Used	---
10	YL/BK	IAC 2 Driver Control	Pulse Signals
11	BR/WT	IAC 1 Driver Control	Pulse Signals
14	---	Not Used	---
15	BK/RD	IAT Sensor Signal	At 100°F: 1.83v
16	TN/BK	ECT Sensor Signal	At 180°F: 2.80v
17	OR	5-Volt Supply	4.9-5.1v
18	TN/YL	CMP Sensor Signal	Digital Signals: 0-5-0v
19	GY/RD	IAC 3 Driver Control	Pulse Signals
20	VT/BK	IAC 4 Driver Control	Pulse Signals
21	---	Not Used	---
22	DG/BK	Keep Alive Power (B+)	12-14v
23	OR/DB	TP Sensor Signal	0.6-1.0v
24	BK/DG	HO2S-11 (B1 S1) Signal	0.1-1.1v
25	TN/WT	HO2S-12 (B1 S2) Signal	0.1-1.1v
26	---	Not Used	---
27	DG/RD	MAP Sensor Signal	1.5-1.6v
28-30	---	Not Used	---
31-32	BK/TN	Power Ground	<0.1v

2001 Ram Van 5.2L V8 OHV MFI VIN Y (A/T) White 'C2' Connector

PCM Pin #	Wire Color	Circuit Description (32 Pin)	Value at Hot Idle
1	GY/BK	Transmission Temperature Sensor Signal	At 200°F: 2.40v
2	VT	Injector 7 Driver	1-4 ms
3	---	Not Used	---
4	WT/DB	Injector 1 Driver	1-4 ms
5	YL/WT	Injector 3 Driver	1-4 ms
6	GY	Injector 5 Driver	1-4 ms
7	---	Not Used	---
8	VT/BK	Governor Pressure Solenoid	PWM Signal: 0-12-0v
9	---	Not Used	---
10	DG/WT	Generator Field Control	Digital Signals: 0-12-0v
11	OR/BK	TCC Solenoid Control	At Cruise w/TCC On: 1v
12	BR/DB	Injector 6 Driver	1-4 ms
13	GY/LB	Injector 8 Driver	1-4 ms
14	---	Not Used	---
15	TN	Injector 2 Driver	1-4 ms
16	LB/BR	Injector 4 Driver	1-4 ms
17-20	---	Not Used	---
21	BR	3-4 Shift Solenoid Control	Cruise w/solenoid On: 1v
22	---	Not Used	---
23	GY/YL	Oil Pressure Sensor	1.6v at 24 psi
24	---	Not Used	---
25	DB/BK	OSS Sensor (-) Signal	Moving: AC pulse signals
26	---	Not Used	---
27	WT/OR	Vehicle Speed Signal	Digital Signal
28	LG/WT	OSS Sensor (+) Signal	Moving: AC pulse signals
29	LG/RD	Governor Pressure Sensor	0.58v
30	PK	Transmission Relay Control	Relay Off: 12v, On: 1v
31	VT/WT	5-Volt Supply	4.9-5.1v
32	---	Not Used	---

2001-03 Ram Van 5.2L V8 OHV MFI VIN Y (A/T) Gray 'C1' Connector

PCM Pin #	Wire Color	Circuit Description (32 Pin)	Value at Hot Idle
1	DB/OR	A/C Clutch Relay Control	Relay Off: 12v, On: 1v
2	---	Not Used	---
3	DB/YL	ASD Relay Control	Relay Off: 12v, On: 1v
4	TN/RD	Speed Control Vacuum Solenoid	Vacuum Increasing: 1v
5	LG/RD	Speed Control Vent Solenoid	Vacuum Decreasing: 1v
6-7	---	Not Used	---
8	OR/RD	HO2S-11 (B1 S1) Heater	PWM Signal: 0-12-0v
9	---	Not Used	---
10	WT/DG	LDP Solenoid Control	PWM Signal: 0-12-0v
11	YL/RD	Speed Control Power Supply	12-14v
12	DG/OR	ASD Relay Power (B+)	12-14v
13	OR/WT	Overdrive Off Switch Sense	Switch Off: 12v, On: 1v
14	OR	LDP Switch Sense Signal	LDP Switch Closed: 0v
15	PK/YL	Battery Temperature Sensor	At 86°F: 1.96v
16	VT/TN	HO2S-21 (B2 S1) Heater	PWM Signal: 0-12-0v
17-18	---	Not Used	---
19	BR	Fuel Pump Relay Control	Relay Off: 12v, On: 1v
20	PK/BK	EVAP Purge Solenoid Control	PWM Signal: 0-12-0v
21	---	Not Used	---
22	DB	A/C Pressure Switch Signal	A/C On: 0.451-4.850v
23	LG/GY	A/C Select Signal	A/C Off: 12v, On: 1v
24	WT/PK	Brake Switch Sense Signal	Brake Off: 0v, On: 12v
25	WT/DB	Generator Field Source	12-14v
26	LB/BK	Fuel Level Sensor Signal	Full: 0.56v, 1/2 full: 2.5v
27	PK	SCI Transmit	0v
28	WT/BK	CCD Bus (-)	<0.050v
29	LG	SCI Receive	5v
30	VT/BR	CCD Bus (+)	Digital Signals: 0-5-0v
31	---	Not Used	---
32	WT/LG	Speed Control Set Switch Signal	S/C & Set Switch On: 3.8v

Pin Connector Graphic

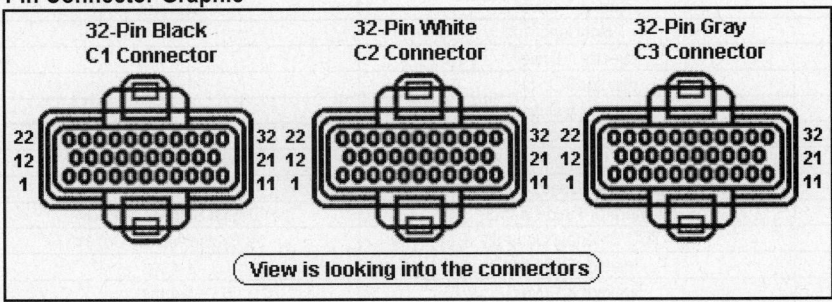

1990-92 Ram Van 5.9L V8 TBI VIN Z (A/T) 60 Pin Connector

PCM Pin #	Wire Color	Circuit Description (60 Pin)	Value at Hot Idle
1	DG/RD	MAP Sensor Signal	1.5-1.6v
2	TN	ECT Sensor Signal	At 180°F: 2.80v
2 ('92)	TN/WT	ECT Sensor Signal	At 180°F: 2.80v
3	RD	Keep Alive Power (B+)	12-14v
4	BK/LB	Sensor Ground	<0.050v
5	BK/WT	Sensor Ground	<0.050v
6	PK/WT	5-Volt Supply	4.9-5.1v
7	OR	8-Volt Supply	7.9-8.1v
8	DG/BK	ASD Relay Power (B+)	12-14v
9	DB	Ignition Switch Power (B+)	12-14v
10, 13-14	---	Not Used	---
11	LB/RD	Power Ground	<0.1v
12	LB/RD	Power Ground	<0.1v
15	TN	Injector 2 Driver	1-4 ms
16	WT/DB	Injector 1 Driver	1-4 ms
17-18, 23	---	Not Used	---
19	BK/GY	Coil Driver Control	5°, 55 mph: 8° dwell
20	DG	Alternator Field Control	Digital Signals: 0-12-0v
21	BK/RD	Throttle Body Temperature	At 100°F: 2.51v
22	OR/DB	TP Sensor Signal	0.6-1.0v
24	GY	Distributor Reference Signal	Digital Signals: 0-5-0v
24 ('92)	GY/BK	Distributor Reference Signal	Digital Signals: 0-5-0v
25	PK	SCI Transmit	0v
26, 31	---	Not Used	---
27	BR	A/C Clutch Signal	A/C Off: 12v, On: 1v
28	PK	Closed Throttle Switch	Switch Off: 12v, On: 1v
29	WT/PK	Brake Switch Sense Signal	Brake Off: 0v, On: 12v
30	BR/YL	PNP Switch Sense Signal	In P/N: 0v, Others: 5v
32	GY/PK	MIL (lamp) Control	MIL Off: 12v, On: 1v
33	TN/RD	Speed Control Vacuum Solenoid	Vacuum Increasing: 1v
34	DB/OR	A/C WOT Relay Control	Relay Off: 12v, On: 1v
35	GY/YL	EGR Solenoid Control	12v, at 55 mph: 1v
36	BK/OR	Air Switching Solenoid	Solenoid Off: 12v, On: 1v
37-39	---	Not Used	---
40	BR	AIS Motor Control	Pulse Signals
40 ('92)	BR/WT	AIS Motor Control	Pulse Signals

Standard Colors and Abbreviations

Abbreviation	Color	Abbreviation	Color	Abbreviation	Color
BK	Black	GY	Gray	RD	Red
BL	Blue	GN	Green	TN	Tan
BR	Brown	LG	Light Green	VT	Violet
DB	Dark Blue	OR	Orange	WT	White
DG	Dark Green	PK	Pink	YL	Yellow

1990-92 Ram Van 5.9L V8 TBI VIN Z (A/T) 60 Pin Connector

PCM Pin #	Wire Color	Circuit Description (60 Pin)	Value at Hot Idle
41	BK/DG	HO2S-11 (B1 S1) Signal	0.1-1.1v
42-43	---	Not Used	
44	OR/WT	Overdrive Lockout Switch	Switch Off: 12v, On: 1v
45	LG	SCI Receive	5v
46	---	Not Used	---
47	WT/OR	Vehicle Speed Signal	Digital Signal
48	BR/RD	Speed Control Set Switch Signal	S/C & Set Switch On: 3.8v
49	YL/RD	Speed Control On/Off Switch	S/C On: 1v, Off: 12v
50	WT/LG	Speed Control Resume Switch	S/C On: 1v, Off: 12v
51	DB/YL	ASD Relay Control	Relay Off: 12v, On: 1v
52	PK	EVAP Purge Solenoid Control	Solenoid Off: 12v, On: 1v
53	LG/RD	Speed Control Vent Solenoid	Vacuum Decreasing: 1v
54	OR/BK	Part Throttle Unlock Solenoid	Solenoid Off: 12v, On: 1v
55	OR/DB	Overdrive Lockout Solenoid	Solenoid Off: 12v, On: 1v
55 ('92)	OR/LG	Overdrive Lockout Solenoid	Solenoid Off: 12v, On: 1v
56	GY/PK	Maintenance Indicator Lamp	Lamp Off: 12v, On: 1v
57-59	---	Not Used	---
60	GY/RD	AIS Motor Control	Pulse Signals

Pin Connector Graphic

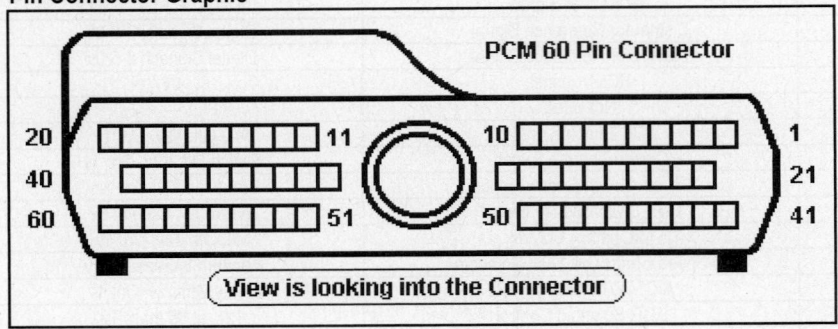

PCM 60 Pin Connector

View is looking into the Connector

1993-95 Ram Van 5.9L V8 MFI VIN Z (A/T) 60 Pin Connector

PCM Pin #	Wire Color	Circuit Description (60 Pin)	Value at Hot Idle
1	DG/RD	MAP Sensor Signal	1.5-1.6v
2	TN/BK	ECT Sensor Signal	At 180ºF: 2.80v
3	RD	Keep Alive Power (B+)	12-14v
4	BK/LB	Sensor Ground	<0.050v
5 ('93-'94)	BK/WT	Sensor Ground	<0.050v
6	PK/WT	5-Volt Supply	4.9-5.1v
7	OR	8-Volt Supply	7.9-8.1v
8	---	Not Used	---
9	DB	Ignition Switch Power (B+)	12-14v
10	OR/WT	Overdrive Override Switch	Switch Off: 12v, On: 1v
11	BK/TN	Power Ground	<0.1v
12	BK/TN	Power Ground	<0.1v
13	LB/BR	Injector 4 Driver	1-4 ms
14	YL/WT	Injector 3 Driver	1-4 ms
15	TN	Injector 2 Driver	1-4 ms
16	WT/DB	Injector 1 Driver	1-4 ms
17	DB/TN	Injector 7 Driver	1-4 ms
18	DB/GY	Injector 8 Driver	1-4 ms
19 ('93)	GY	Coil Driver Control	5º, 55 mph: 8º dwell
19	BK/GY	Coil Driver Control	5º, 55 mph: 8º dwell
20	DG/WT	Alternator Field Control	Digital Signals: 0-12-0v
21	BK/RD	IAT Sensor Signal	At 100ºF: 1.83v
22	OR/DB	TP Sensor Signal	0.6-1.0v
23	---	Not Used	---
24	GY/BK	Distributor Reference Signal	Digital Signals: 0-5-0v
25	PK	SCI Transmit	0v
26, 28	---	Not Used	---
27 ('93)	BR	A/C Clutch Signal	A/C Off: 12v, On: 1v
27	BR	A/C Damped Pressure Switch	A/C Off: 12v, On: 1v
29	WT/PK	Brake Switch Sense Signal	Brake Off: 0v, On: 12v
30	BR/O	PNP Switch Sense Signal	In P/N: 0v, Others: 5v
31	---	Not Used	---
32	BK/PK	MIL (lamp) Control	MIL Off: 12v, On: 1v
33	TN/RD	Speed Control Vacuum Solenoid	Vacuum Increasing: 1v
34	DB/OR	A/C WOT Relay Control	Relay Off: 12v, On: 1v
35	GY/YL	EGR Solenoid Control	12v, at 55 mph: 1v
36	---	Not Used	---
37	BK/OR	Overdrive Lamp Control	Lamp Off: 12v, On: 1v
38	PK/BK	Injector 5 Driver	1-4 ms
39	GY/RD	AIS Motor Control	Pulse Signals
40	BR/WT	AIS Motor Control	Pulse Signals

Standard Colors and Abbreviations

Abbreviation	Color	Abbreviation	Color	Abbreviation	Color
BK	Black	GY	Gray	RD	Red
BL	Blue	GN	Green	TN	Tan
BR	Brown	LG	Light Green	VT	Violet
DB	Dark Blue	OR	Orange	WT	White
DG	Dark Green	PK	Pink	YL	Yellow

1993-95 Ram Van 5.9L V8 MFI VIN Z (A/T) 60 Pin Connector

PCM Pin #	Wire Color	Circuit Description (60 Pin)	Value at Hot Idle
41	BK/DG	HO2S-11 (B1 S1) Signal	0.1-1.1v
42-43	---	Not Used	---
44	TN/YL	CMP Sensor Signal	Digital Signals: 0-5-0v
45	LG	SCI Receive	5v
46	---	Not Used	---
47	WT/OR	Vehicle Speed Signal	Digital Signal
48	BR/RD	Speed Control Set Switch Signal	S/C & Set Switch On: 3.8v
49	YL/RD	Speed Control On/Off Switch	S/C On: 1v, Off: 12v
50	WT/LG	Speed Control Resume Switch	S/C On: 1v, Off: 12v
51	DB/YL	ASD Relay Control	Relay Off: 12v, On: 1v
52	PK/BK	EVAP Purge Solenoid Control	Solenoid Off: 12v, On: 1v
53	LG/RD	Speed Control Vent Solenoid	Vacuum Decreasing: 1v
54	OR/BK	Part Throttle Unlock Solenoid	Solenoid Off: 12v, On: 1v
55	OR/WT	Overdrive Lockout Solenoid	Solenoid Off: 12v, On: 1v
56	GY/PK	Maintenance Indicator Lamp	Lamp Off: 12v, On: 1v
57	DG/OR	ASD Relay Power (B+)	12-14v
58	LG/BK	Injector 6 Driver	1-4 ms
59	PK/BK	AIS Motor Control	Pulse Signals
60	YL/BK	AIS Motor Control	Pulse Signals

Pin Connector Graphic

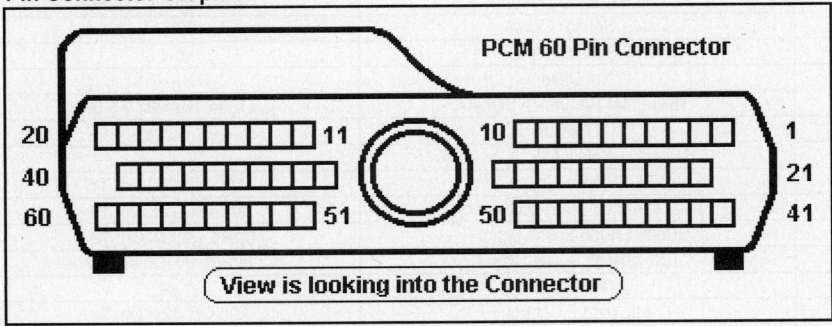

Standard Colors and Abbreviations

Abbreviation	Color	Abbreviation	Color	Abbreviation	Color
BK	Black	GY	Gray	RD	Red
BL	Blue	GN	Green	TN	Tan
BR	Brown	LG	Light Green	VT	Violet
DB	Dark Blue	OR	Orange	WT	White
DG	Dark Green	PK	Pink	YL	Yellow

1996-97 Ram Van 5.9L V8 MFI VIN Z (A/T) 'A' Black Connector

PCM Pin #	Wire Color	Circuit Description (32 Pin)	Value at Hot Idle
2	LG/BK	Ignition Switch Power (B+)	12-14v
4	BK/LB	Sensor Ground	<0.050v
6	BR/O	PNP Switch Sense Signal	In P/N: 0v, Others: 5v
7	GY	Coil 1 Driver Control	5°, 55 mph: 8° dwell
8	GY/BK	CKP Sensor Signal	Digital Signals: 0-5-0v
10	YL/BK	IAC 2 Driver Control	Pulse Signals
11	BR/WT	IAC 1 Driver Control	Pulse Signals
15	BK/RD	IAT Sensor Signal	At 100°F: 1.83v
16	TN/BK	ECT Sensor Signal	At 180°F: 2.80v
17	PK/WT	5-Volt Supply	4.9-5.1v
18	TN/YL	CMP Sensor Signal	Digital Signals: 0-5-0v
19	GY/RD	IAC 3 Driver Control	Pulse Signals
20	PK/BK	IAC 4 Driver Control	Pulse Signals
22	RD/WT	Keep Alive Power (B+)	12-14v
23	OR/DB	TP Sensor Signal	0.6-1.0v
24	BK/DG	HO2S-11 (B1 S1) Signal	0.1-1.1v
25	TN/WT	HO2S-12 (B1 S2) Signal	0.1-1.1v
27	DG/RD	MAP Sensor Signal	1.5-1.6v
31-32	BK/TN	Power Ground	<0.1v

1996-97 Ram Van 5.9L V8 MFI VIN Z (A/T) 'B' White Connector

PCM Pin #	Wire Color	Circuit Description (32 Pin)	Value at Hot Idle
1	P	Transmission Temperature Sensor Signal	At 200°F: 2.40v
2	PK/TN	Injector 7 Driver	1-4 ms
4	WT/DB	Injector 1 Driver	1-4 ms
5	YL/WT	Injector 3 Driver	1-4 ms
6	GY	Injector 5 Driver	1-4 ms
8	PK/WT	Governor Pressure Solenoid	PWM Signal: 0-12-0v
10	DG	Generator Field Control	Digital Signals: 0-12-0v
11	OR/BK	TCC Solenoid Control	At Cruise w/TCC On: 1v
12	BR/DB	Injector 6 Driver	1-4 ms
13	GY/LB	Injector 8 Driver	1-4 ms
15	T	Injector 2 Driver	1-4 ms
16	LB/BR	Injector 4 Driver	1-4 ms
21	BR	3-4 Shift Solenoid Control	Solenoid Off: 12v, On: 1v
25	DB/BK	OSS Sensor (-) Signal	Moving: AC pulse signals
27	WT/OR	Vehicle Speed Signal	Digital Signal
28	LG/WT	OSS Sensor (+) Signal	Moving: AC pulse signals
29	LG/RD	Governor Pressure Sensor	0.58v
30	PK/LB	Transmission Relay Control	Relay Off: 12v, On: 1v
31	O	5-Volt Supply	4.9-5.1v
32	GY/YL	EGR Solenoid Control	12v, at 55 mph: 1v

Standard Colors and Abbreviations

Abbreviation	Color	Abbreviation	Color	Abbreviation	Color
BK	Black	GY	Gray	RD	Red
BL	Blue	GN	Green	TN	Tan
BR	Brown	LG	Light Green	VT	Violet
DB	Dark Blue	OR	Orange	WT	White
DG	Dark Green	PK	Pink	YL	Yellow

1996-97 Ram Van 5.9L V8 MFI VIN Z (A/T) 'C' Gray Connector

PCM Pin #	Wire Color	Circuit Description (32 Pin)	Value at Hot Idle
1	DB/OR	A/C Clutch Relay Control	Relay Off: 12v, On: 1v
2	---	Not Used	---
3	DB/YL	ASD Relay Control	Relay Off: 12v, On: 1v
4	TN/RD	Speed Control Vacuum Solenoid	Vacuum Increasing: 1v
5	LG/RD	Speed Control Vent Solenoid	Vacuum Decreasing: 1v
6	LG/OR	Overdrive Off Lamp Control	Lamp Off: 12v, On: 1v
7-10	---	Not Used	---
11	YL/RD	Speed Control Power Supply	12-14v
12	DG/OR	ASD Relay Power (B+)	12-14v
13	OR/WT	Overdrive Off Switch Sense	Switch Off: 12v, On: 1v
14	---	Not Used	---
15	PK/YL	Battery Temperature Sensor	At 86°F: 1.96v
16	TN/YL	Generator Lamp Control	Lamp Off: 12v, On: 1v
17	BK/PK	MIL (lamp) Control	MIL Off: 12v, On: 1v
18	---	Not Used	---
19	BR	Fuel Pump Relay Control	Relay Off: 12v, On: 1v
20	PK/BK	EVAP Purge Solenoid Control	PWM Signal: 0-12-0v
21	---	Not Used	---
22	BR/WT	A/C Request Signal	A/C Off: 12v, On: 1v
23	LG	A/C Select Signal	A/C Off: 12v, On: 1v
24	WT/PK	Brake Switch Sense Signal	Brake Off: 0v, On: 12v
25	---	Not Used	---
26	LB/BK	Fuel Level Sensor Signal	Full: 0.56v, 1/2 full: 2.5v
27	PK	SCI Transmit	0v
28	---	Not Used	---
29	LG/WT	SCI Receive	5v
30-31	---	Not Used	---
32	WT/LG	Speed Control Set Switch Signal	S/C & Set Switch On: 3.8v

Pin Connector Graphic

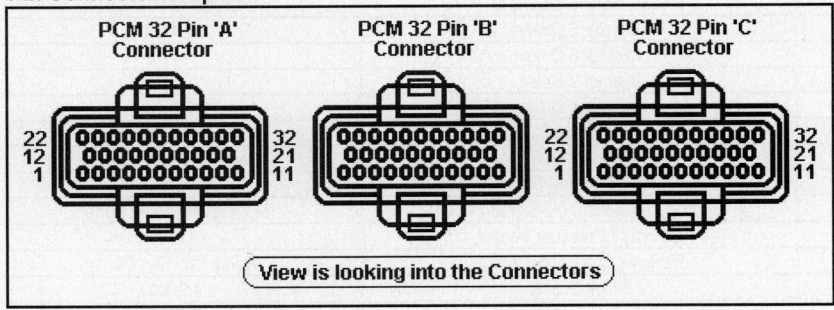

PCM 32 Pin 'A' Connector   PCM 32 Pin 'B' Connector   PCM 32 Pin 'C' Connector

View is looking into the Connectors

1998-2000 Ram Van 5.9L V8 MFI VIN Z 'C1' Black Connector

PCM Pin #	Wire Color	Circuit Description (32 Pin)	Value at Hot Idle
2	LG/BK	Ignition Switch Power (B+)	12-14v
4	BK/LB	Sensor Ground	<0.050v
6	BK/WT	PNP Switch Sense Signal	In P/N: 0v, Others: 5v
6 ('00)	BR/YL	PNP Switch Sense Signal	In P/N: 0v, Others: 5v
7	GY	Coil 1 Driver Control	5°, 55 mph: 8° dwell
8	GY/BK	CKP Sensor Signal	Digital Signals: 0-5-0v
10	YL/BK	IAC 2 Driver Control	Pulse Signals
11	BR/WT	IAC 1 Driver Control	Pulse Signals
15	BK/RD	IAT Sensor Signal	At 100°F: 1.83v
16	TN/BK	ECT Sensor Signal	At 180°F: 2.80v
17	OR	5-Volt Supply	4.9-5.1v
18	TN/YL	CMP Sensor Signal	Digital Signals: 0-5-0v
19	GY/RD	IAC 3 Driver Control	Pulse Signals
20	PK/BK	IAC 4 Driver Control	Pulse Signals
22	RD/WT	Keep Alive Power (B+)	12-14v
23	OR/DB	TP Sensor Signal	0.6-1.0v
24	TN/WT	HO2S-11 (B1 S1) Signal	0.1-1.1v
25	BK/DG	HO2S-12 (B1 S2) Signal	0.1-1.1v
27	DG/RD	MAP Sensor Signal	1.5-1.6v
31-32	BK/TN	Power Ground	<0.1v

1998-2000 Ram Van 5.9L V8 MFI VIN Z White 'C2' Connector

PCM Pin #	Wire Color	Circuit Description (32 Pin)	Value at Hot Idle
1	GY/BK	Transmission Temperature Sensor Signal	At 200°F: 2.40v
2	PK	Injector 7 Driver	1-4 ms
4	WT/DB	Injector 1 Driver	1-4 ms
5	YL/WT	Injector 3 Driver	1-4 ms
6	GY	Injector 5 Driver	1-4 ms
8	PK/WT	Governor Pressure Solenoid	PWM Signal: 0-12-0v
10	DG/WT	Generator Field Control	Digital Signals: 0-12-0v
11	OR/BK	TCC Solenoid Control	At Cruise w/TCC On: 1v
12	BR/DB	Injector 6 Driver	1-4 ms
13	GY/LB	Injector 8 Driver	1-4 ms
15	TN	Injector 2 Driver	1-4 ms
16	LB/BR	Injector 4 Driver	1-4 ms
21	OR/WT	Overdrive Solenoid Control	Cruise w/solenoid On: 1v
21 ('00)	BR	Overdrive Solenoid Control	Cruise w/solenoid On: 1v
23	GY	Engine Oil Pressure Sensor	1.6v at 24 psi
25	DB/BK	OSS Sensor (-) Signal	Moving: AC pulse signals
27	WT/OR	Vehicle Speed Signal	Digital Signal
28	LG/WT	OSS Sensor (+) Signal	Moving: AC pulse signals
29	LG/RD	Governor Pressure Sensor	0.58v
30	LB/PK	Transmission Relay Control	Relay Off: 12v, On: 1v
31	PK/WT	5-Volt Supply	4.9-5.1v

Standard Colors and Abbreviations

Abbreviation	Color	Abbreviation	Color	Abbreviation	Color
BK	Black	GY	Gray	RD	Red
BL	Blue	GN	Green	TN	Tan
BR	Brown	LG	Light Green	VT	Violet
DB	Dark Blue	OR	Orange	WT	White
DG	Dark Green	PK	Pink	YL	Yellow

1998-2000 Ram Van 5.9L V8 MFI VIN Z 'C3' Gray Connector

PCM Pin #	Wire Color	Circuit Description (32 Pin)	Value at Hot Idle
1	DB/OR	A/C Clutch Relay Control	Relay Off: 12v, On: 1v
2	---	Not Used	
3	DB/YL	ASD Relay Control	Relay Off: 12v, On: 1v
4	TN/RD	Speed Control Vacuum Solenoid	Vacuum Increasing: 1v
5	LG/RD	Speed Control Vent Solenoid	Vacuum Decreasing: 1v
6 ('98)	LG/OR	Overdrive Off Lamp Control	Lamp Off: 12v, On: 1v
7-9	---	Not Used	---
10	WT/DG	LDP Solenoid Control	PWM Signal: 0-12-0v
11	YL/RD	Speed Control Power Supply	12-14v
12	DG/OR	ASD Relay Power (B+)	12-14v
13	OR/WT	Overdrive Off Switch Sense	Switch Off: 12v, On: 1v
14	OR	LDP Switch Sense Signal	LDP Switch Closed: 0v
15	PK/YL	Battery Temperature Sensor	At 86°F: 1.96v
16-18	---	Not Used	---
19	BR	Fuel Pump Relay Control	Relay Off: 12v, On: 1v
20	PK/W	EVAP Purge Solenoid Control	PWM Signal: 0-12-0v
20 ('00)	PK/BK	EVAP Purge Solenoid Control	PWM Signal: 0-12-0v
21	---	Not Used	---
22	DB	A/C Pressure Switch Signal	A/C On: 0.451-4.850v
23	LG/GY	A/C Select Signal	A/C Off: 12v, On: 1v
24	WT/PK	Brake Switch Sense Signal	Brake Off: 0v, On: 12v
25	OR/DB	Generator Field Source	12-14v
25 ('00)	WT/DB	Generator Field Source	12-14v
26	LB/BK	Fuel Level Sensor Signal	Full: 0.56v, 1/2 full: 2.5v
27	PK	SCI Transmit	0v
28	WT/BK	CCD Bus (-)	<0.050v
29	LG/WT	SCI Receive	5v
30	PK/BR	CCD Bus (+)	Digital Signals: 0-5-0v
31	---	Not Used	---
32	WT/LG	Speed Control Set Switch Signal	S/C & Set Switch On: 3.8v

Pin Connector Graphic

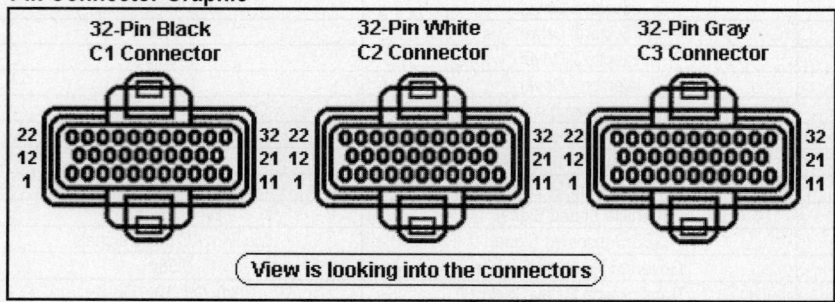

32-Pin Black C1 Connector   32-Pin White C2 Connector   32-Pin Gray C3 Connector

View is looking into the connectors

2001-03 Ram Van 5.9L V8 OHV MFI VIN Z (A/T) 'C1' Black Connector

PCM Pin #	Wire Color	Circuit Description (32 Pin)	Value at Hot Idle
1	---	Not Used	---
2	LG/BK	Ignition Switch Power (B+)	12-14v
3	---	Not Used	---
4	BK/LB	Sensor Ground	<0.050v
5	---	Not Used	---
6	BR/YL	PNP Switch Sense Signal	In P/N: 0v, Others: 5v
7	GY	Coil 1 Driver Control	5°, 55 mph: 8° dwell
8	GY/BK	CKP Sensor Signal	Digital Signals: 0-5-0v
9	---	Not Used	---
10	YL/BK	IAC 2 Driver Control	Pulse Signals
11	BR/WT	IAC 1 Driver Control	Pulse Signals
12-14	---	Not Used	---
15	BK/RD	IAT Sensor Signal	At 100°F: 1.83v
16	TN/BK	ECT Sensor Signal	At 180°F: 2.80v
17	OR	5-Volt Supply	4.9-5.1v
18	TN/YL	CMP Sensor Signal	Digital Signals: 0-5-0v
19	GY/RD	IAC 3 Driver Control	Pulse Signals
20	VT/BK	IAC 4 Driver Control	Pulse Signals
21	---	Not Used	---
22	DG/BK	Keep Alive Power (B+)	12-14v
23	OR/DB	TP Sensor Signal	0.6-1.0v
24	DB/DG	HO2S-11 (B1 S1) Signal	0.1-1.1v
25	TN/WT	HO2S-12 (B1 S2) Signal	0.1-1.1v
26	---	Not Used	---
27	DG/RD	MAP Sensor Signal	1.5-1.6v
28-32	---	Not Used	---

2001-03 Ram Van 5.9L V8 OHV MFI VIN Z (A/T) White 'C2' Connector

PCM Pin #	Wire Color	Circuit Description (32 Pin)	Value at Hot Idle
1	GY/BK	Transmission Temperature Sensor Signal	At 200°F: 2.40v
2	VT	Injector 7 Driver	1-4 ms
3	---	Not Used	---
4	WT/DB	Injector 1 Driver	1-4 ms
5	YL/WT	Injector 3 Driver	1-4 ms
6	GY	Injector 5 Driver	1-4 ms
7	---	Not Used	---
8	VT/BK	Governor Pressure Solenoid	PWM Signal: 0-12-0v
9	---	Not Used	---
10	DG/WT	Generator Field Control	Digital Signals: 0-12-0v
11	OR/BK	TCC Solenoid Control	At Cruise w/TCC On: 1v
12	BR/DB	Injector 6 Driver	1-4 ms
13	GY/LB	Injector 8 Driver	1-4 ms
14	---	Not Used	---
15	TN	Injector 2 Driver	1-4 ms
16	LB/BR	Injector 4 Driver	1-4 ms
17-20	---	Not Used	---
21	BR	3-4 Shift Solenoid Control	Cruise w/solenoid On: 1v
22	---	Not Used	---
23	GY/YL	Engine Oil Pressure Sensor	1.6v at 24 psi
24	---	Not Used	---
25	DB	OSS Sensor (-) Signal	Moving: AC pulse signals
26	---	Not Used	---
27	WT/OR	Vehicle Speed Signal	Digital Signal
28	LG/RD	OSS Sensor (+) Signal	Moving: AC pulse signals
29	LG/RD	Governor Pressure Sensor	0.58v
30	PK	Transmission Relay Control	Relay Off: 12v, On: 1v
31	VT/WT	5-Volt Supply	4.9-5.1v
32	---	Not Used	---

2001-03 Ram Van 5.9L V8 OHV MFI VIN Z (A/T) 'C3' Gray Connector

PCM Pin #	Wire Color	Circuit Description (32 Pin)	Value at Hot Idle
1	DB/OR	A/C Clutch Relay Control	Relay Off: 12v, On: 1v
2	---	Not Used	---
3	DB/YL	ASD Relay Control	Relay Off: 12v, On: 1v
4	TN/RD	Speed Control Vacuum Solenoid	Vacuum Increasing: 1v
5	LG/RD	Speed Control Vent Solenoid	Vacuum Decreasing: 1v
6-7	---	Not Used	---
8	OR/RD	HO2S-11 (B1 S1) Heater	PWM Signal: 0-12-0v
9	---	Not Used	---
10	WT/DG	LDP Solenoid Control	PWM Signal: 0-12-0v
11	YL/RD	Speed Control Power Supply	12-14v
12	DG/OR	ASD Relay Power (B+)	12-14v
13	OR/WT	Overdrive Off Switch Sense	Switch Off: 12v, On: 1v
14	OR	LDP Switch Sense Signal	LDP Switch Closed: 0v
15	PK/YL	Battery Temperature Sensor	At 86°F: 1.96v
16	VT/TN	HO2S-21 (B2 S1) Heater	PWM Signal: 0-12-0v
17-18	---	Not Used	---
19	BR	Fuel Pump Relay Control	Relay Off: 12v, On: 1v
20	PK/BK	EVAP Purge Solenoid Control	PWM Signal: 0-12-0v
21	---	Not Used	---
22	DB	A/C Pressure Switch Signal	A/C On: 0.451-4.850v
23	LG/GY	A/C Select Signal	A/C Off: 12v, On: 1v
24	WT/PK	Brake Switch Sense Signal	Brake Off: 0v, On: 12v
25	WT/DB	Generator Field Source	12-14v
26	LB/BK	Fuel Level Sensor Signal	Full: 0.56v, 1/2 full: 2.5v
27	PK	SCI Transmit	0v
28	WT/BK	CCD Bus (-)	<0.050v
29	LG	SCI Receive	5v
30	VT/BR	CCD Bus (+)	Digital Signals: 0-5-0v
31	---	Not Used	---
32	WT/LG	Speed Control Set Switch Signal	S/C & Set Switch On: 3.8v

Pin Connector Graphic

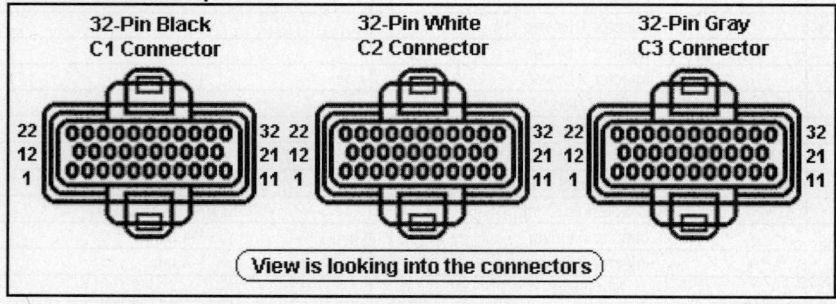

| 32-Pin Black C1 Connector | 32-Pin White C2 Connector | 32-Pin Gray C3 Connector |

View is looking into the connectors

SPORT UTILITY VEHICLE CONTENTS

SUV Pin Tables

Introduction

This section of Chrysler Diagnostic Manual contains Pin Tables for Dodge and Jeep SUV's from 1990-2003. It can be used to assist in repair of both Code and No Code faults related to the PCM.

Vehicle Coverage

- 1990-2001 Jeep Cherokee SUV Applications
- 1990-92 Jeep Comanche SUV Applications
- 1999-2003 Dodge Durango SUV Applications
- 1993-2003 Jeep Grand Cherokee SUV Applications
- 1990 Jeep Liberty SUV Applications
- 1990-2003 Jeep Wrangler SUV Applications
- 1990 Jeep Wagoneer SUV Applications

How to Use This Section

This section can be used to look up the location of a particular pin, a wire color or a "known good" value of a PCM circuit. To locate the PCM information for a particular vehicle, find the model, correct engine size (with VIN Code) and finally the year of the vehicle.

For example, to look up the PCM terminals for a 1998 Cherokee 2.5L VIN P, go to Contents Page 1 and find the text string shown below:

Then turn to Page 13-25 to find the following PCM related information.

1998-2000 Cherokee 2.5L I4 MFI VIN P 'A' Black Connector

PCM Pin #	Wire Color	Circuit Description (32 Pin)	Value at Hot Idle
15	BK/RD	IAT Sensor Signal	At 100ºF: 1.83v
16	TN/BK	ECT Sensor Signal	At 180ºF: 2.80v
17	OR	5-Volt Supply	4.9-5.1v
18	TN/YL	CMP Sensor Signal	Digital Signals: 0-5-0v
19	GY/RD	IAC 1 Driver Control	Pulse Signals
20	PK/BK	IAC 4 Driver Control	Pulse Signals

In this example, the Camshaft Position (CMP) signal circuit is connected to Pin 18 of the PCM (32 Pin) connector with a TN/YL wire on 1998-2000 models. The Hot Idle values shown here (0-5-0v) are the nominal values with the engine at idle speed in Park or Neutral position.

The acronym (All) in the Title at the top of the table indicates that information in this table is for all 1990 Cherokee models for either a Manual Transmission (M/T) or an Automatic Transmission (A/T).

DODGE PIN TABLES

1998-99 Durango 3.9L V6 MFI VIN X 'A' Black Connector

PCM Pin #	Wire Color	Circuit Description (32 Pin)	Value at Hot Idle
1, 3, 5, 9	---	Not Used	---
2	LG/BK	Ignition Switch Power (B+)	12-14v
4	BK/LB	Sensor Ground	<0.1v
6	BK/WT	PNP Switch Sense Signal	In P/N: 0v, Others: 5v
7	BK/GY	Coil 1 Driver Control	5°, 55 mph: 8° dwell
8	GY/BK	CKP Sensor Signal	Digital Signals: 0-5-0v
10	YL/BK	IAC 2 Driver Control	Pulse Signals
11	BR/WT	IAC 3 Driver Control	Pulse Signals
12-14, 21	---	Not Used	---
15	BK/RD	IAT Sensor Signal	At 100°F: 1.83v
16	TN/BK	ECT Sensor Signal	At 180°F: 2.80v
17	PK/WT	5-Volt Supply	4.9-5.1v
18	TN/YL	CMP Sensor Signal	Digital Signals: 0-5-0v
19	GY/RD	IAC 1 Driver Control	Pulse Signals
20	PK/BK	IAC 4 Driver Control	Pulse Signals
22	RD/WT	Fused Battery Power (B+)	12-14v
23	OR/DB	TP Sensor Signal	0.6-1.0v
24	TN/WT	HO2S-11 (B1 S1) Signal	0.1-1.1v
25	PK/WT	HO2S-12 (B1 S2) Signal	0.1-1.1v
27	DG/RD	MAP Sensor Signal	1.5-1.6v
26, 28-30	---	Not Used	---
31, 32	BK/TN	Power Ground	<0.1v

1998-99 Durango 3.9L V6 MFI VIN X 'B' White Connector

PCM Pin #	Wire Color	Circuit Description (32 Pin)	Value at Hot Idle
1	GY/BK	Transmission Temperature Sensor	At 200°F: 2.40v
2-3, 7, 9	---	Not Used	---
4	WT/DB	Injector 1 Driver	1-4 ms
5	YL/WT	Injector 3 Driver	1-4 ms
6	GY	Injector 5 Driver	1-4 ms
7, 9, 13-14	---	Not Used	---
8	PK/WT	Governor Pressure Solenoid	PWM Signal: 0-12-0v
10	DG/WT	Generator Field Control	Digital Signals: 0-12-0v
11	OR/BK	TCC Solenoid Control	At Cruise w/TCC On: <1v
12	BR/DB	Injector 6 Driver	1-4 ms
15	TN	Injector 2 Driver	1-4 ms
16	LB/BR	Injector 4 Driver	1-4 ms
17-20, 22	---	Not Used	---
21	BR	Overdrive Solenoid Control	Cruise w/solenoid on: <1v
21 ('99)	BR	3-4 Shift Solenoid Control	Solenoid Off: 12v, On: 1v
23	GY/YL	Engine Oil Pressure Sensor	1.6v at 24 psi
24, 26, 32	---	Not Used	---
25	DB	OSS Sensor (-) Signal	Moving: AC pulse signals
27	WT/OR	Vehicle Speed Signal	Digital Signal
28	LG	OSS Sensor (+) Signal	Moving: AC pulse signals
29	LG/RD	Governor Pressure Signal	0.58v
30	PK	Transmission Relay Control	Relay Off: 12v, On: 1v
31	OR	5-Volt Supply	4.9-5.1v

Standard Colors and Abbreviations

Abbreviation	Color	Abbreviation	Color	Abbreviation	Color
BK	Black	GY	Gray	RD	Red
BL	Blue	GN	Green	TN	Tan
BR	Brown	LG	Light Green	VT	Violet
DB	Dark Blue	OR	Orange	WT	White
DG	Dark Green	PK	Pink	YL	Yellow

1998-99 Durango 3.9L V6 MFI VIN X 'C' Gray Connector

PCM Pin #	Wire Color	Circuit Description (32 Pin)	Value at Hot Idle
1	DB/OR	A/C Clutch Relay Control	Relay Off: 12v, On: 1v
2, 7-9	---	Not Used	---
3	DB/YL	ASD Relay Control	Relay Off: 12v, On: 1v
4	TN/RD	Speed Control Vacuum Solenoid	Vacuum Increasing: 1v
5	LG/RD	Speed Control Vent Solenoid	Vacuum Decreasing: 1v
6 ('98)	LG/OR	Overdrive 'Off' Lamp Control	O/D On: 1v, O/D Off: 12v
10	WT/DG	LDP Solenoid Control	PWM Signal: 0-12-0v
11	YL/RD	Speed Control Power Supply	12-14v
12	DG/OR	ASD Relay Output	12-14v
13	OR/WT	Overdrive 'Off' Switch Sense	Switch On: 1v, Off: 12v
14	OR	LDP Switch Sense Signal	Closed: 0v, Open: 12v
15	PK/YL	Battery Temperature Sensor	At 86°F: 1.96v
16-18	---	Not Used	---
19	DB/WT	Fuel Pump Relay Control	Relay Off: 12v, On: 1v
20	VT/BK	EVAP Purge Solenoid Control	PWM Signal: 0-12-0v
21	---	Not Used	---
22	BR	A/C Request Signal	A/C Off: 12v, On: 1v
23	LG/WT	A/C Select Signal	A/C Off: 12v, On: 1v
24	WT/PK	Brake Switch Sense Signal	Brake Off: 0v, On: 12v
25	WT/PK	Generator Field Source	12-14v
26	DB	Fuel Level Sensor Signal	Full: 0.56v, 1/2 full: 2.5v
27	PK	SCI Transmit	0v
28	WT/PK	CCD Bus (-)	<0.050v
29	PK/WT	SCI Receive	5v
30	PK/BR	CCD Bus (+)	Digital Signals: 0-5-0v
31	---	Not Used	---
32	BR/YL	Speed Control Switch Signal	S/C & Set Switch On: 3.8v

Pin Connector Graphic

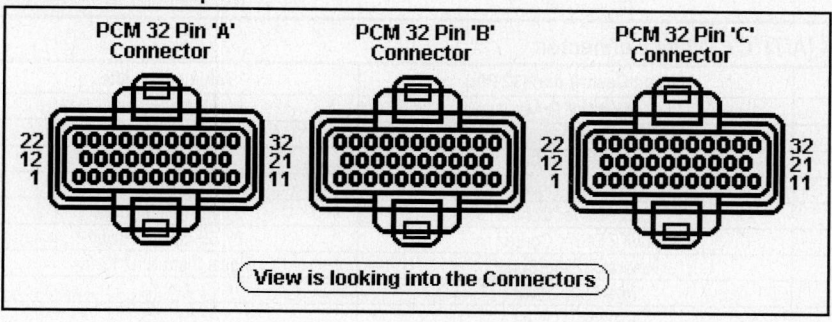

2000 Durango 4.7L V8 MFI VIN N (A/T) 'C1' Black Connector

PCM Pin #	Wire Color	Circuit Description (32 Pin)	Value at Hot Idle
1	TN/OR	Coil 3 Driver Control	5º, 55 mph: 8º dwell
2	LG/BK	Ignition Switch Power (B+)	12-14v
3	TN/LG	Coil 4 Driver Control	5º, 55 mph: 8º dwell
4	BK/LB	Sensor Ground	<0.1v
5	TN/LB	Coil 6 Driver Control	5º, 55 mph: 8º dwell
6	BK/WT	PNP Switch Sense Signal	In P/N: 0v, Others: 5v
7	BK/GY	Coil 1 Driver Control	5º, 55 mph: 8º dwell
8	GY/BK	CKP Sensor Signal	Digital Signals: 0-5-0v
9	RD/YL	Coil 8 Driver Control	5º, 55 mph: 8º dwell
10	YL/BK	IAC 2 Driver Control	Pulse Signals
11	BR/WT	IAC 3 Driver Control	Pulse Signals
12	DB/OR	PSP Switch Sense Signal	Straight: 0v, Turning: 5v
13-14	---	Not Used	---
15	BK/RD	IAT Sensor Signal	At 100ºF: 1.83v
16	TN/BK	ECT Sensor Signal	At 180ºF: 2.80v
17	PK/WT	5-Volt Supply	4.9-5.1v
18	TN/YL	CMP Sensor Signal	Digital Signals: 0-5-0v
19	GY/RD	IAC 1 Driver Control	Pulse Signals
20	PK/BK	IAC 4 Driver Control	Pulse Signals
21	TN/RD	Coil 5 Driver Control	5º, 55 mph: 8º dwell
22	RD/WT	Fused Battery Power (B+)	12-14v
23	OR/DB	TP Sensor Signal	0.6-1.0v
24	LG/RD	HO2S-11 (B1 S1) Signal	0.1-1.1v
25	TN/WT	HO2S-12 (B1 S2) Signal	0.1-1.1v
26	OR/TN	HO2S-21 (B2 S1) Signal (Cal)	0.1-1.1v
27	DG/RD	MAP Sensor Signal	1.5-1.6v
28, 30	---	Not Used	---
29	PK/WT	HO2S-22 (B2 S2) Signal (Cal)	0.1-1.1v
31, 32	BK/TN	Power Ground	<0.1v

2000 Durango 4.7L V8 MFI VIN N (A/T) 'C2' White Connector

PCM Pin #	Wire Color	Circuit Description (32 Pin)	Value at Hot Idle
2	VT	Injector 7 Driver	1-4 ms
4	WT/DB	Injector 1 Driver	1-4 ms
5	YL/WT	Injector 3 Driver	1-4 ms
6	GY	Injector 5 Driver	1-4 ms
7	DB/TN	Coil 7 Driver Control	5º, 55 mph: 8º dwell
9	TN/PK	Coil 2 Driver Control	5º, 55 mph: 8º dwell
10	DG	Generator Field Control	Digital Signals: 0-12-0v
12	BR/DB	Injector 6 Driver	1-4 ms
13	GY/LB	Injector 8 Driver	1-4 ms
15	TN	Injector 2 Driver	1-4 ms
16	LB/BR	Injector 4 Driver	1-4 ms
17	DB/PK	Radiator Fan Relay Control	Relay Off: 12v, On: 1v
23	GY/YL	Engine Oil Pressure Sensor	1.6v at 24 psi
27	WT/OR	Vehicle Speed Signal	Digital Signal

Standard Colors and Abbreviations

Abbreviation	Color	Abbreviation	Color	Abbreviation	Color
BK	Black	GY	Gray	RD	Red
BL	Blue	GN	Green	TN	Tan
BR	Brown	LG	Light Green	VT	Violet
DB	Dark Blue	OR	Orange	WT	White
DG	Dark Green	PK	Pink	YL	Yellow

2000 Durango 4.7L V8 MFI VIN N (A/T) 'C3' 32 Pin Black Connector

PCM Pin #	Wire Color	Circuit Description (32 Pin)	Value at Hot Idle
1	DB/OR	A/C Clutch Relay Control	Relay Off: 12v, On: 1v
3	DB/YL	ASD Relay Control	Relay Off: 12v, On: 1v
4	TN/RD	Speed Control Vacuum Solenoid	Vacuum Increasing: 1v
5	LG/RD	Speed Control Vent Solenoid	Vacuum Decreasing: 1v
8	DG/BK	HO2S-11 Heater Relay (Cal)	Relay Off: 12v, On: 1v
9	OR/RD	HO2S-12 Heater Relay (Cal)	Relay Off: 12v, On: 1v
10	WT/DG	LDP Solenoid Control	PWM Signal: 0-12-0v
11	YL/RD	Speed Control Power Supply	12-14v
12	DG/OR	ASD Relay Output	12-14v
13	YL/DG	Torque Management Relay	Relay Off: 12v, On: 1v
14	OR	LDP Switch Sense Signal	Closed: 0v, Open: 12v
15	PK/YL	Battery Temperature Sensor	At 86°F: 1.96v
19	DB/WT	Fuel Pump Relay Control	Relay Off: 12v, On: 1v
20	PK/BK	EVAP Purge Solenoid Control	PWM Signal: 0-12-0v
22	BR	A/C Switch Signal	A/C Off: 12v, On: 1v
23	LG/WT	A/C Select Signal	A/C Off: 12v, On: 1v
24	WT/PK	Brake Switch Sense Signal	Brake Off: 0v, On: 12v
25	WL/DB	Generator Field Source	12-14v
26	DB	Fuel Level Sensor Signal	Full: 0.56v, 1/2 full: 2.5v
27	PK/DB	SCI Transmit	0v
28	WT/BK	CCD Bus (-)	<0.050v
29	PK/WT	SCI Receive	5v
30	PT/BR	CCD Bus (+)	Digital Signals: 0-5-0v
32	RD/LG	Speed Control Switch Signal	S/C & Set Switch On: 3.8v

Pin Connector Graphic

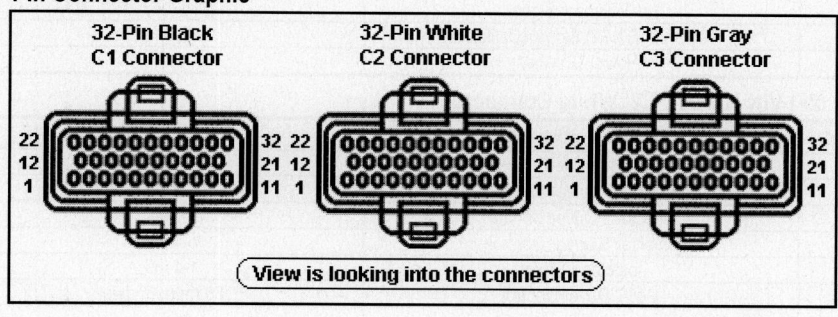

2001-02 Durango 4.7L V8 SOHC MFI VIN N (A/T) 'C1' Black Connector

PCM Pin #	Wire Color	Circuit Description (32 Pin)	Value at Hot Idle
1	TN/OR	Coil 3 Driver Control	5°, 55 mph: 8° dwell
2	LG/BK	Ignition Switch Power (B+)	12-14v
3	TN/LG	Coil 4 Driver Control	5°, 55 mph: 8° dwell
4	BK/LB	Sensor Ground	<0.1v
5	TN/LB	Coil 6 Driver Control	5°, 55 mph: 8° dwell
6	BK/WT	PNP Switch Sense Signal	In P/N: 0v, Others: 5v
7	BK/GY	Coil 1 Driver Control	5°, 55 mph: 8° dwell
8	GY/BK	CKP Sensor Signal	Digital Signals: 0-5-0v
9	LB/RD	Coil 8 Driver Control	5°, 55 mph: 8° dwell
10	YL/BK	IAC 2 Driver Control	Pulse Signals
11	BR/WT	IAC 3 Driver Control	Pulse Signals
12	DB/OR	PSP Switch Sense Signal	Straight: 0v, Turning: 5v
13-14, 28, 30	---	Not Used	---
15	BK/RD	IAT Sensor Signal	At 100°F: 1.83v
16	TN/BK	ECT Sensor Signal	At 180°F: 2.80v
17	OR	5-Volt Supply	4.9-5.1v
18	TN/YL	CMP Sensor Signal	Digital Signals: 0-5-0v
19	GY/RD	IAC 1 Driver Control	Pulse Signals
20	VT/BK	IAC 4 Driver Control	Pulse Signals
21	TN/DG	Coil 5 Driver Control	5°, 55 mph: 8° dwell
22	RD/WT	Fused Battery Power (B+)	12-14v
23	OR/DB	TP Sensor Signal	0.6-1.0v
24	BK/DG	HO2S-11 (B1 S1) Signal	0.1-1.1v
25	TN/WT	HO2S-12 (B1 S2) Signal	0.1-1.1v
26	LG/RD	HO2S-21 (B2 S1) Signal (California)	0.1-1.1v
27	DG/RD	MAP Sensor Signal	1.5-1.6v
29	TN/WT	HO2S-22 (B2 S2) Signal (California)	0.1-1.1v
31, 32	BK/TN	Power Ground	<0.1v

2001-02 Durango 4.7L V8 SOHC MFI VIN N (A/T) 'C2' White Connector

PCM Pin #	Wire Color	Circuit Description (32 Pin)	Value at Hot Idle
1, 3	---	Not Used	---
2	VT	Injector 7 Driver Control	1-4 ms
4	WT/DB	Injector 1 Driver Control	1-4 ms
5	YL/WT	Injector 3 Driver Control	1-4 ms
6	GY	Injector 5 Driver	1-4 ms
7	DB/TN	Coil 7 Driver Control	5°, 55 mph: 8° dwell
8, 11	---	Not Used	---
9	TN/PK	Coil 2 Driver Control	5°, 55 mph: 8° dwell
10	DG	Generator Field Control	Digital Signals: 0-12-0v
12	BR/DB	Injector 6 Driver Control	1-4 ms
13	GY/LB	Injector 8 Driver Control	1-4 ms
14	---	Not Used	---
15	TN	Injector 2 Driver Control	1-4 ms
16	LB/BR	Injector 4 Driver Control	1-4 ms
17	DB/PK	Radiator Fan Relay Control	Relay Off: 12v, On: 1v
18-20, 24-26	---	Not Used	---
23	GY/YL	Engine Oil Pressure Sensor	1.6v at 24 psi
27	WT/OR	Vehicle Speed Signal	Digital Signals
28-30, 32	---	Not Used	---
31	VT/WT	5-Volt Supply	4.9-5.1v

Standard Colors and Abbreviations

Abbreviation	Color	Abbreviation	Color	Abbreviation	Color
BK	Black	GY	Gray	RD	Red
BL	Blue	GN	Green	TN	Tan
BR	Brown	LG	Light Green	VT	Violet
DB	Dark Blue	OR	Orange	WT	White
DG	Dark Green	PK	Pink	YL	Yellow

2001-02 Durango 4.7L V8 SOHC MFI VIN N (A/T) 'C3' 32 Pin Black Connector

PCM Pin #	Wire Color	Circuit Description (32 Pin)	Value at Hot Idle
1	DB/OR	A/C Clutch Relay Control	Relay Off: 12v, On: 1v
2	---	Not Used	---
3	DB/YL	ASD Relay Control	Relay Off: 12v, On: 1v
4	TN/RD	Speed Control Vacuum Solenoid	Vacuum Increasing: 1v
5	LG/RD	Speed Control Vent Solenoid	Vacuum Decreasing: 1v
6-7	---	Not Used	
8	VT/WT	HO2S-11 Heater Relay (California)	Relay Off: 12v, On: 1v
9	DG/BK	HO2S-12 Heater Relay (California)	Relay Off: 12v, On: 1v
10	WT/DG	LDP Solenoid Control	PWM Signal: 0-12-0v
11	YL/RD	Speed Control Power Supply	12-14v
12	DG/OR	ASD Relay Output	12-14v
13	YL/DG	Torque Management Relay	Relay Off: 12v, On: 1v
14	OR	LDP Switch Sense Signal	Closed: 0v, Open: 12v
15	PK/YL	Battery Temperature Sensor	At 86°F: 1.96v
16	VT/OR	HO2S-12 Heater Control (California)	Digital Signals: 0-12-0v
17-18	---	Not Used	---
19	BR	Fuel Pump Relay Control	Relay Off: 12v, On: 1v
20	PK/BK	EVAP Purge Solenoid Control	PWM Signal: 0-12-0v
21	---	Not Used	
22	BR	A/C Switch Signal	A/C Off: 12v, On: 1v
23	---	Not Used	---
24	WT/PK	Brake Switch Sense Signal	Brake Off: 0v, On: 12v
25	WT/DB	Generator Field Source	12-14v
26	DB/WT	Fuel Pump Relay Control	Relay Off: 12v, On: 1v
27	PK	SCI Transmit	0v
28	---	Not Used	---
29	LG	SCI Receive	5v
30	VT/TN	PCI Data Bus (J1850)	Digital Signals: 0-7-0v
31	---	Not Used	---
32	RD/LG	Speed Control Switch Signal	S/C & Set Switch On: 3.8v

Pin Connector Graphic

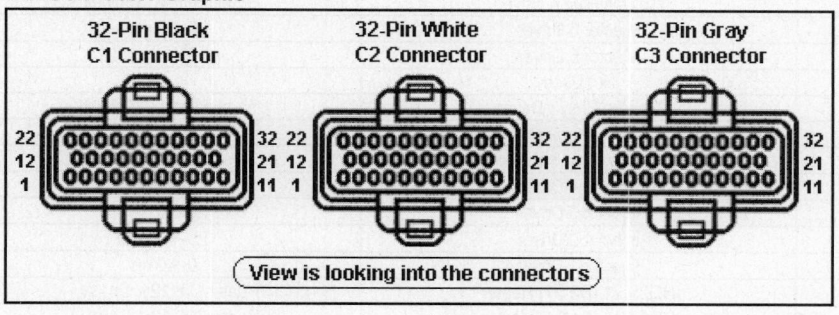

32-Pin Black C1 Connector   32-Pin White C2 Connector   32-Pin Gray C3 Connector

View is looking into the connectors

2003 Durango 4.7L V8 SOHC MFI VIN N (All) 'C1' 38 Pin Black Connector

PCM Pin #	Wire Color	Circuit Description (38 Pin)	Value at Hot Idle
1	LB/RD	Coil On Plug 8 Driver	5°, at 55 mph: 8° dwell
2, 6-8, 10	---	Not Used	---
3	BR	Coil On Plug 7 Driver	5°, at 55 mph: 8° dwell
4	GY/LG	Injector 8 Driver	1.0-4.0 ms
5	VT	Injector 7 Driver	1.0-4.0 ms
9, 18	BK/TN	Power Ground	<0.050v
11	LG/BK	Ignition Switch Output (Run-Start)	12-14v
12	RD/WT	Ignition Switch Output (Off-Run-Start)	12-14v
13	WT/OR	Vehicle Speed Signal	Digital Signals
14-17, 19	---	Not Used	---
20	GY/YL	Oil Pressure Sensor	1.6v at 24 psi
21	LB/BR	A/C Pressure Sensor	A/C On: 0.45-4.85v
22	VT/OR	Ambient Temperature Sensor	At 86°F: 1.96v
23-24, 28	---	Not Used	---
25	LG	SCI Receive (PCM)	5v
26	PK/LB	SCI Receive (PCM)	5v
27	VT/WT	5-Volt Supply	4.9-5.1v
29	RD/WT	Fused Power (B+)	12-14v
30	RD/YL	Fused Ignition Output (Start)	9-11v
31	TN/WT	HO2S-12 (B1 S2) Signal	0.1-1.1v
32	BR/DG	HO2S Return (Downstream)	<0.050v
33	TN/WT	HO2S-22 (B1 S2) Signal (California)	0.1-1.1v
34-35	---	Not Used	---
36	PK	SCI Transmit (PCM)	0v
37	WT/DG	SCI Transmit (PCM)	0v
38	WT/VT	PCI Data Bus (J1850)	Digital Signals: 0-7-0v

2003 Durango 4.7L V8 SOHC MFI VIN N (All) 'C2' 38 Pin Gray Connector

PCM Pin #	Wire Color	Circuit Description (38 Pin)	Value at Hot Idle
1	TN/LB	Coil On Plug 6 Driver	5°, at 55 mph: 8° dwell
2	TN/DG	Coil On Plug 5 Driver	5°, at 55 mph: 8° dwell
3	TN/LG	Coil On Plug 4 Driver	5°, at 55 mph: 8° dwell
4	BR/DB	Injector 6 Driver	1.0-4.0 ms
5	GY	Injector 5 Driver	1.0-4.0 ms
6, 8, 15-16	---	Not Used	---
7	TN/OR	Coil On Plug 3 Driver	5°, at 55 mph: 8° dwell
9	TN/PK	Coil On Plug 2 Driver	5°, at 55 mph: 8° dwell
10	TN/RD	Coil On Plug 1 Driver	5°, at 55 mph: 8° dwell
11	LB/BR	Injector 4 Driver	1.0-4.0 ms
12	YL/WT	Injector 3 Driver	1.0-4.0 ms
13	TN	Injector 2 Driver	1.0-4.0 ms
14	WT/DG	Injector 1 Driver	1.0-4.0 ms
17	BR/VT	HO2S-21 (B2 S1) Heater	Heater Off: 12v, On: 1v
18	BR/OR	HO2S-11 (B1 S1) Heater	Heater Off: 12v, On: 1v
19	DG	Generator Field Driver	Digital Signals: 0-12-0v
20	VT/OR	ECT Sensor Signal	At 180°F: 2.80v
21	BR/OR	TP Sensor 1 Signal	0.6-1.0v
22, 24-26	---	Not Used	---
23	VT/BR	MAP Sensor Signal	1.5-1.7v
27	BK/LB	Sensor Ground	<0.050v
28	YL/BK	IAC Motor Return	12-14v
29	OR	5-Volt Supply	4.9-5.1v
30	DB/LG	IAT Sensor Signal	At 100°F: 1.83v
31	BK/DG	HO2S-11 (B1 S1) Signal	0.1-1.1v
32	DB/DG	HO2S Return (Upstream)	<0.050v
33	LG/RD	HO2S-21 (B2 S1) Signal (California)	0.1-1.1v
34	DB/GY	CMP Sensor Signal	Digital Signal: 0-5-0v
35	DB/WT	CKP Sensor Signal	Digital Signal: 0-5-0v
36-37	---	Not Used	---
38	GY/RD	IAC Motor Driver Control	DC pulses: 0.8-11v

2003 Durango 4.7L V8 SOHC MFI VIN N (All) 'C3' 38 Pin White Connector

PCM Pin #	Wire Color	Circuit Description (38 Pin)	Value at Hot Idle
1-2, 4, 13-18, 22	---	Not Used	---
3	DB/YL	ASD Relay Control	Relay Off: 12v, On: 1v
5	LG/RD	Speed Control Vent Solenoid	Vacuum Decreasing: 1v
6	DB/PK	Low Speed Fan Relay	Relay Off: 12v, On: 1v
7	VT/YL	Speed Control Power Supply	12-14v
8	WT/DG	Natural Vacuum Leak Detection Solenoid	Solenoid Off: 12v, On: 1v
9	BR/WT	HO2S-12 (B1 S2) Heater Control	Heater Off: 12v, On: 1v
10	BR/GY	HO2S-22 (B2 S2) Heater (California)	Heater Off: 12v, On: 1v
11	LB/OR	A/C Clutch Relay Control	Relay Off: 12v, On: 1v
12	VT/DG	Speed Control Vacuum Solenoid	Vacuum Increasing: 1v
19, 28	DG/OR	ASD Relay Output	12-14v
20	PK/BK	EVAP Purge Solenoid Control	PWM Signal: 0-12-0v
21	BK/WT	Park Neutral Switch Sense (T41)	In P/N: 0v, Others: 5v
24-27, 31, 36	---	Not Used	---
23	WT/PK	Brake Switch Signal	Brake Off: 0v, On: 12v
29	DB/BR	EVAP Purge Solenoid Sense	<0.1v
30	DB/OR	Power Steering Pressure Switch	Straight: 0v, Turning: 5v
32	PK/YL	Battery Temperature Sensor	At 86°F: 1.96v
33	DB/WT	Fuel Level Signal	Full: 0.56v, 1/2 full: 2.5v
34	VT/TN	Speed Control Set Switch Signal	S/C & Set Switch On: 3.8v
35	OR	Natural Vacuum Leak Detection Switch	Open: 12v, Closed: 0v
37	BR	Fuel Pump Relay Control	Relay Off: 12v, On: 1v
38	DG/OR	Starter Relay Control	Relay Off: 12v, On: 1v

2003 Durango 4.7L V8 SOHC MFI VIN N (A/T Only) 'C4' 38 Pin Green Connector

1	YL/GY	Overdrive Solenoid Control	Solenoid Off: 12v, On: 1v
2	DG/WT	4C Solenoid Control	Solenoid Off: 12v, On: 1v
3, 5, 7, 9	---	Not Used	---
4	YL/DB	MS Solenoid Control	Solenoid Off: 12v, On: 1v
6	WT/DB	2C Solenoid Control	Solenoid Off: 12v, On: 1v
8	YL/LB	Underdrive Solenoid Control	Solenoid Off: 12v, On: 1v
10	LG	Low/Reverse Solenoid control	Solenoid Off: 12v, On: 1v
11	VT/LG	Pressure Control Solenoid	PWM Signals (0-12-0v)
12-14	BK/RD	Power Ground	<0.1v
15	DG/LB	TRS T1 Sense	In NOL: 0v, Others: 5v
16	DG/DB	TRS T3 Sense	In P3L: 0v, Others: 5v
17	OR/WT	Overdrive Off Switch Sense	In Overdrive: 2-4v
18	YL/BR	Transmission Control Relay Control	Relay Off: 12v, On: 1v
19, 28, 38	RD	Transmission Control Relay Output	12-14v
20	DB	4C Pressure Switch Sense	In 4th Position: 2-4v
21	GY	Underdrive Pressure Switch Sense	In Underdrive Position: 2-4v
22	YL/LG	Overdrive Pressure Switch Sense	In Overdrive Position: 2-4v
23-25, 27, 36	---	Not Used	---
26	PK/OR	TRS T2 Sense	In P/N: 0v, Others: 5v
29	YL/TN	Low/Reverse Pressure Switch	In Low/Reverse: 2-4v
30	YL/DG	2C Pressure Switch	In 2-4 Position: 2-4v
31	VT/TN	Line Pressure Switch	In 2-4 Position: 2-4v
32	LG/WT	Output Speed Sensor	Moving: AC voltage
33	RD/BK	Input Speed Sensor	Moving: AC voltage
34	DB/BK	Speed Sensor Ground	<0.050v
35	DG/OR	Transmission Temperature Sensor	3.2-3.4v at 104°F
37	DG/YL	TRS T42 Sense	In PRNL: 0v, Others 5v

Pin Connector Graphic

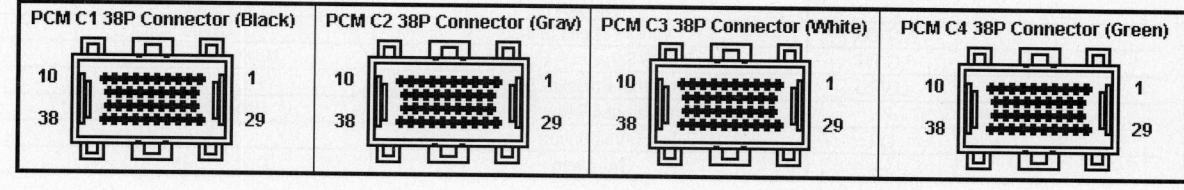

1998-2000 Durango 5.2L V8 MFI VIN Y 'C1' Black Connector

PCM Pin #	Wire Color	Circuit Description (32 Pin)	Value at Hot Idle
1	---	Not Used	---
2	LG/BK	Ignition Switch Power (B+)	12-14v
3	---	Not Used	---
4	BK/LB	Sensor Ground	<0.1v
5	---	Not Used	---
6	BK/WT	PNP Switch Sense Signal	In P/N: 0v, Others: 5v
7	BK/GY	Coil 1 Driver Control	5°, 55 mph: 8° dwell
8	GY/BK	CKP Sensor Signal	Digital Signals: 0-5-0v
9	---	Not Used	---
10	YL/BK	IAC 2 Driver Control	Pulse Signals
11	BR/WT	IAC 3 Driver Control	Pulse Signals
12-14	---	Not Used	---
15	BK/RD	IAT Sensor Signal	At 100°F: 1.83v
16	TN/BK	ECT Sensor Signal	At 180°F: 2.80v
17	PK/WT	5-Volt Supply	4.9-5.1v
18	TN/YL	CMP Sensor Signal	Digital Signals: 0-5-0v
19	GY/RD	IAC 1 Driver Control	Pulse Signals
20	VT/BK	IAC 4 Driver Control	Pulse Signals
21	---	Not Used	---
22	RD/WT	Fused Battery Power (B+)	12-14v
23	OR/DB	TP Sensor Signal	0.6-1.0v
24	TN/WT	HO2S-11 (B1 S1) Signal	0.1-1.1v
25	ON/BK	HO2S-12 (B1 S2) Signal	0.1-1.1v
26	---	Not Used	---
27	DG/RD	MAP Sensor Signal	1.5-1.6v
31, 32	BK/TN	Power Ground	<0.1v

1998-2000 Durango 5.2L V8 MFI VIN Y 'C2' White Connector

PCM Pin #	Wire Color	Circuit Description (32 Pin)	Value at Hot Idle
1	PK or GY	Transmission Temperature Sensor	At 200°F: 2.40v
2	PK/WT	Injector 7 Driver	1-4 ms
3	---	Not Used	---
4	WT/DB	Injector 1 Driver	1-4 ms
5	YL/WT	Injector 3 Driver	1-4 ms
6	GY	Injector 5 Driver	1-4 ms
7	---	Not Used	---
8 (2000)	PK	Variable Force Solenoid	PWM Signal: 0-12-0v
8 ('98-'99)	PK/WT	Governor Pressure Signal	0.58v
9	---	Not Used	---
10	DG	Generator Field Control	Digital Signals: 0-12-0v
11	OR/BK	TCC Solenoid Control	At Cruise w/TCC On: <1v
12	BR/DB	Injector 6 Driver	1-4 ms
13	GY/LB	Injector 8 Driver	1-4 ms
14	---	Not Used	---
15	TN	Injector 2 Driver	1-4 ms
16	LB/BR	Injector 4 Driver	1-4 ms
17 (2000)	DB/PK	Radiator Fan Relay Control	Relay Off: 12v, On: 1v
18-20	---	Not Used	---
21	BR	Overdrive Solenoid Control	Cruise w/solenoid on: <1v
22	---	Not Used	---
23	GY/YL	Engine Oil Pressure Sensor	1.6v at 24 psi
24	---	Not Used	---
25	DB/BK	OSS Sensor (-) Signal	Moving: AC pulse signals
26	---	Not Used	---
27	WT/OR	Vehicle Speed Signal	Digital Signal
28	LG/WT	OSS Sensor (+) Signal	Moving: AC pulse signals
29 (2000)	LG/RD	Governor Pressure Signal	0.58v
30	PK	Transmission Relay Control	Relay Off: 12v, On: 1v
31	OR	5-Volt Supply	4.9-5.1v
32	---	Not Used	---

1998-2000 Durango 5.2L V8 MFI VIN Y 'C3' Gray Connector

PCM Pin #	Wire Color	Circuit Description (32 Pin)	Value at Hot Idle
1	DB/OR	A/C Clutch Relay Control	Relay Off: 12v, On: 1v
2 ('98-'99)	DB/PK	Radiator Fan Control Relay	Relay Off: 12v, On: 1v
3	DB/YL	ASD Relay Control	Relay Off: 12v, On: 1v
4	TN/RD	Speed Control Vacuum Solenoid	Vacuum Increasing: 1v
5	LG/RD	Speed Control Vent Solenoid	Vacuum Decreasing: 1v
8	DG/BK	HO2S-11 HTR Control (CAL)	Relay Off: 12v, On: 1v
9	OR/RD	HO2S-12 HTR Control (CAL)	Relay Off: 12v, On: 1v
10 ('98-'99)	YL/DG	LDP Solenoid Control	PWM Signal: 0-12-0v
10	WT/DG	LDP Solenoid Control	PWM Signal: 0-12-0v
11	YL/RD	Speed Control Power Supply	12-14v
12	DG/OR	ASD Relay Output	12-14v
13	OR/WT	Overdrive 'Off' Switch Sense	Switch On: 1v, Off: 12v
14 ('98-'99)	YL/WT	LDP Switch Sense Signal	Closed: 0v, Open: 12v
14	OR	LDP Switch Sense Signal	Closed: 0v, Open: 12v
15	PK/YL	Battery Temperature Sensor	At 86°F: 1.96v
19 ('98-'99)	LB/O	Fuel Pump Relay Control	Relay Off: 12v, On: 1v
19	DB/WT	Fuel Pump Relay Control	Relay Off: 12v, On: 1v
20	PK/BK	EVAP Purge Solenoid Control	PWM Signal: 0-12-0v
22	BR	A/C Switch Signal	A/C Off: 12v, On: 1v
23	LG/WT	A/C Select Signal	A/C Off: 12v, On: 1v
24	WT/PK	Brake Switch Sense Signal	Brake Off: 0v, On: 12v
25 ('98-'99)	DG/BK	Generator Field Source	12-14v
25	WT/DB	Generator Field Source	12-14v
26	DB	Fuel Level Sensor Signal	Full: 0.56v, 1/2 full: 2.5v
27 ('98-'99)	PK	SCI Transmit	0v
27	PK/DB	SCI Transmit	0v
28	WT/BK	CCD Bus (-)	<0.050v
29 ('98-'99)	LG	SCI Receive	5v
29	PK/WT	SCI Receive	5v
30	PK/BR	CCD Bus (+)	Digital Signals: 0-5-0v
32	RD/LG	Speed Control Switch Signal	S/C & Set Switch On: 3.8v

Pin Connector Graphic

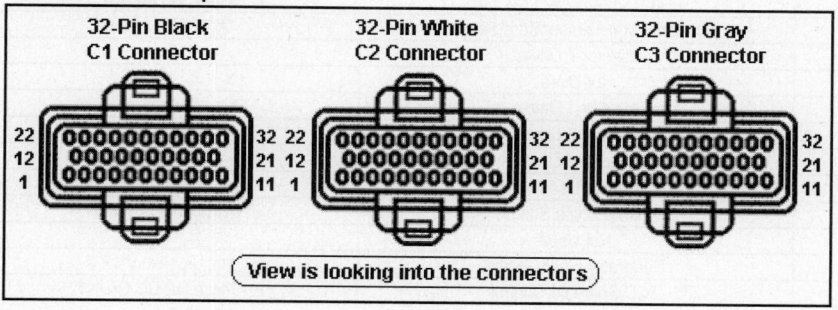

1998-2000 Durango 5.9L V8 MFI VIN Z 'A' Black Connector

PCM Pin #	Wire Color	Circuit Description (32 Pin)	Value at Hot Idle
1	---	Not Used	---
2	LG/BK	Ignition Switch Power (B+)	12-14v
3	---	Not Used	---
4	BK/LB	Sensor Ground	<0.1v
5	---	Not Used	---
6	BK/WT	PNP Switch Sense Signal	In P/N: 0v, Others: 5v
7	BK/GY	Coil 1 Driver Control	5°, 55 mph: 8° dwell
8	GY/BK	CKP Sensor Signal	Digital Signals: 0-5-0v
9	---	Not Used	---
10	YL/BK	IAC 2 Driver Control	Pulse Signals
11	BR/WT	IAC 3 Driver Control	Pulse Signals
12-14	---	Not Used	---
15	BK/RD	IAT Sensor Signal	At 100°F: 1.83v
16	TN/BK	ECT Sensor Signal	At 180°F: 2.80v
17	PK/WT	5-Volt Supply	4.9-5.1v
18	TN/YL	CMP Sensor Signal	Digital Signals: 0-5-0v
19	GY/RD	IAC 1 Driver Control	Pulse Signals
20	PK/BK	IAC 4 Driver Control	Pulse Signals
21	---	Not Used	---
22	RD/WT	Fused Battery Power (B+)	12-14v
23	OR/DB	TP Sensor Signal	0.6-1.0v
24	TN/WT	HO2S-11 (B1 S1) Signal	0.1-1.1v
25 ('98-'99)	OR/BK	HO2S-13 (Bank 1 Sensor 3)	0.1-1.1v
25	OR/BK	HO2S-12 (B1 S2) Signal	0.1-1.1v
26 ('98-'99)	OR/TN	HO2S-21 (B2 S1) Signal	0.1-1.1v
27	DG/RD	MAP Sensor Signal	1.5-1.6v
28	---	Not Used	---
29 ('98-'99)	TN/WT	HO2S-12 (B1 S2) Signal	0.1-1.1v
30	---	Not Used	---
31, 32	BK/TN	Power Ground	<0.1v

1998-2000 Durango 5.9L V8 MFI VIN Z 'B' White Connector

PCM Pin #	Wire Color	Circuit Description (32 Pin)	Value at Hot Idle
1	GY/BK	Transmission Temperature Sensor	At 200°F: 2.40v
2	PK	Injector 7 Driver	1-4 ms
3	---	Not Used	---
4	WT/DB	Injector 1 Driver	1-4 ms
5	YL/WT	Injector 3 Driver	1-4 ms
6	GY	Injector 5 Driver	1-4 ms
7	---	Not Used	---
8	PK/WT	Variable Force Solenoid	PWM Signal: 0-12-0v
9	---	Not Used	---
10	DG	Generator Field Control	Digital Signals: 0-12-0v
11	OR/BK	TCC Solenoid Control	At Cruise w/TCC On: <1v
12	BR/DB	Injector 6 Driver	1-4 ms
13	GY/LB	Injector 8 Driver	1-4 ms
15	TN	Injector 2 Driver	1-4 ms
16	LB/BR	Injector 4 Driver	1-4 ms
17 (2000)	DB/PK	Radiator Fan Control Relay	Relay Off: 12v, On: 1v
21	BR	3-4 Shift Solenoid Control	Cruise w/solenoid on: <1v
23	GY/YL	Engine Oil Pressure Sensor	1.6v at 24 psi
25	DB	OSS Sensor (-) Signal	Moving: AC pulse signals
27	WT/OR	Vehicle Speed Signal	Digital Signal
29	LG/RD	Governor Pressure Signal	0.58v
28	LG	OSS Sensor (+) Signal	Moving: AC pulse signals
30	PK	Transmission Relay Control	Relay Off: 12v, On: 1v
31	OR	5-Volt Supply	4.9-5.1v

1998-2000 Durango 5.9L V8 MFI VIN Z 'C' Gray Connector

PCM Pin #	Wire Color	Circuit Description (32 Pin)	Value at Hot Idle
1	DB/OR	A/C Clutch Relay Control	Relay Off: 12v, On: 1v
2	---	Not Used	---
3	DB/YL	ASD Relay Control	Relay Off: 12v, On: 1v
4	TN/RD	Speed Control Vacuum Solenoid	Vacuum Increasing: 1v
5	LG/RD	Speed Control Vent Solenoid	Vacuum Decreasing: 1v
6-7	---	Not Used	---
8	DG/BK	HO2S-11 HTR Control (California)	Relay Off: 12v, On: 1v
9	OR/RD	HO2S-12 HTR Control (California)	Relay Off: 12v, On: 1v
10 ('98-'99)	YL/DG	LDP Solenoid Control	PWM Signal: 0-12-0v
10	WT/DG	LDP Solenoid Control	PWM Signal: 0-12-0v
11	YL/RD	Speed Control Power Supply	12-14v
12	DG/OR	ASD Relay Output	12-14v
13	OR/WT	Overdrive 'Off' Switch Sense	Switch On: 1v, Off: 12v
14	OR	LDP Switch Sense Signal	Closed: 0v, Open: 12v
15	PK/YL	Battery Temperature Sensor	At 86°F: 1.96v
16-18, 21	---	Not Used	---
19	DB/WT	Fuel Pump Relay Control	Relay Off: 12v, On: 1v
20	PK/BK	EVAP Purge Solenoid Control	PWM Signal: 0-12-0v
22	BR	A/C Switch Signal	A/C Off: 12v, On: 1v
23	LG/WT	A/C Select Signal	A/C Off: 12v, On: 1v
24	WT/PK	Brake Switch Sense Signal	Brake Off: 0v, On: 12v
25	WT/DB	Generator Field Source	12-14v
26	DB	Fuel Level Sensor Signal	Full: 0.56v, 1/2 full: 2.5v
27	PK	SCI Transmit	0v
28	WT	CCD Bus (-)	<0.050v
29	PK/WT	SCI Receive	5v
30	PK/BR	CCD Bus (+)	Digital Signal: 0-5-0v
31	---	Not Used	---
32	RD/LG	Speed Control Switch Signal	S/C & Set Switch On: 3.8v

Pin Connector Graphic

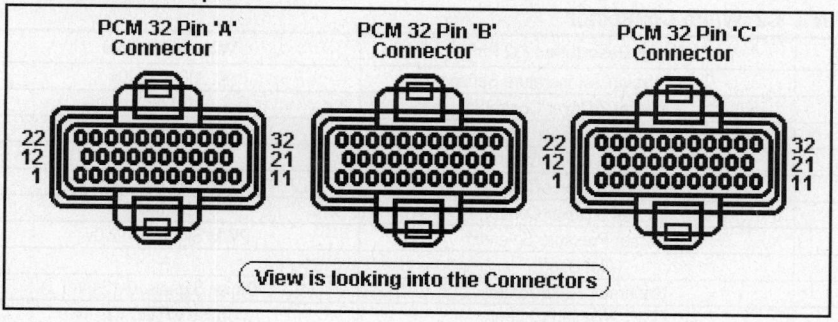

View is looking into the Connectors

2001-03 Durango 5.9L V8 MFI VIN Z 'C1' Black Connector

PCM Pin #	Wire Color	Circuit Description (32 Pin)	Value at Hot Idle
1	---	Not Used	---
2	LG/BK	Ignition Switch Power (Run-Start)	12-14v
3	---	Not Used	---
4	BK/LB	Sensor Ground	<0.1v
5	---	Not Used	---
6	BK/WT	PNP Switch Sense Signal	In P/N: 0v, Others: 5v
7	BK/GY	Coil 1 Driver Control	5°, 55 mph: 8° dwell
8	GY/BK	CKP Sensor Signal	Digital Signals: 0-5-0v
9	---	Not Used	---
10	YL/BK	IAC 2 Driver Control	Pulse Signals
11	BR/WT	IAC 3 Driver Control	Pulse Signals
12-14	---	Not Used	---
15	BK/RD	IAT Sensor Signal	At 100°F: 1.83v
16	TN/BK	ECT Sensor Signal	At 180°F: 2.80v
17	OR	5-Volt Supply	4.9-5.1v
18	TN/YL	CMP Sensor Signal	Digital Signals: 0-5-0v
19	GY/RD	IAC 1 Driver Control	Pulse Signals
20	VT/BK	IAC 4 Driver Control	Pulse Signals
21	---	Not Used	---
22	RD/WT	Fused Battery Power (B+)	12-14v
23	OR/DB	TP Sensor Signal	0.6-1.0v
24	BK/DG	HO2S-11 (B1 S1) Signal	0.1-1.1v
25	TN/WT	HO2S-12 (B1 S2) Signal (California)	0.1-1.1v
26	LG/RD	HO2S-21 (B2 S1) Signal	0.1-1.1v
27	DG/RD	MAP Sensor Signal	1.5-1.6v
28	---	Not Used	---
29	TN/WT	HO2S-22 (B2 S2) Signal (California)	0.1-1.1v
30	---	Not Used	---
31	BK/TN	Power Ground	<0.1v
32	BK/TN	Power Ground	<0.1v

2001-03 Durango 5.9L V8 MFI VIN Z 'C2' White Connector

PCM Pin #	Wire Color	Circuit Description (32 Pin)	Value at Hot Idle
1	GY/BK	Transmission Temperature Sensor	At 200°F: 2.40v
2	VT	Injector 7 Driver Control	1-4 ms
3, 7	---	Not Used	---
4	WT/DB	Injector 1 Driver Control	1-4 ms
5	YL/WT	Injector 3 Driver Control	1-4 ms
6	GY	Injector 5 Driver Control	1-4 ms
8	VT/WT	Governor Pressure Solenoid	PWM Signal: 0-12-0v
9, 14	---	Not Used	---
10	DB	Generator Field Control	Digital Signals: 0-12-0v
11	OR/BK	TCC Solenoid Control	At Cruise w/TCC On: 1v
12	BR/DB	Injector 6 Driver Control	1-4 ms
13	GY/LB	Injector 8 Driver Control	1-4 ms
15	TN	Injector 2 Driver Control	1-4 ms
16	LB/BR	Injector 4 Driver Control	1-4 ms
17	DB/PK	Radiator Fan Control Relay	Relay Off: 12v, On: 1v
18, 22	---	Not Used	---
19	DB	AC Pressure Sensor Signal	0.90v at 79 psi
21	BR	Overdrive Solenoid Control	Cruise w/solenoid on: <1v
23	GY/YL	Engine Oil Pressure Sensor	1.6v at 24 psi
24, 26	---	Not Used	---
25	DB/BK	OSS Sensor (-) Signal	Moving: AC pulse signals
27	WT/OR	Vehicle Speed Sensor	Digital Signals
28	LG/WT	OSS Sensor (+)	Moving: AC pulse signals
29	LG/RD	Governor Pressure Signal	0.58v
30	PK	Transmission Relay Control	Relay Off: 12v, On: 1v
31	VT/WT	5-Volt Supply	4.9-5.1v
32	---	Not Used	---

2001-03 Durango 5.9L V8 MFI VIN Z 'C3' Gray Connector

PCM Pin #	Wire Color	Circuit Description (32 Pin)	Value at Hot Idle
1	DB/OR	A/C Clutch Relay Control	Relay Off: 12v, On: 1v
2	---	Not Used	---
3	DB/YL	ASD Relay Control	Relay Off: 12v, On: 1v
4	TN/RD	Speed Control Vacuum Solenoid	Vacuum Increasing: 1v
5	LG/RD	Speed Control Vent Solenoid	Vacuum Decreasing: 1v
6-7	---	Not Used	---
8	VT/WT	HO2S-11 (B1 S1) Heater Control	Relay Off: 12v, On: 1v
9	DG/BK	HO2S-12 (B1 S2) Heater Control	Relay Off: 12v, On: 1v
10	WT/DG	LDP Solenoid Control	PWM Signal: 0-12-0v
11	YL/RD	Speed Control Power Supply	12-14v
12	DG/OR	ASD Relay Output	12-14v
13	OR/WT	Overdrive 'Off' Switch Sense	Switch On: 1v, Off: 12v
14	OR	LDP Switch Sense Signal	Closed: 0v, Open: 12v
15	PK/YL	Battery Temperature Sensor	At 86°F: 1.96v
16	VT/OR	HO2S-21 (B2 S1) Heater Control	Heater Off: 12v, On: PWM 0-12-0v
17-18	---	Not Used	---
19	BR	Fuel Pump Relay Control	Relay Off: 12v, On: 1v
20	PK/BK	EVAP Purge Solenoid Control	PWM Signal: 0-12-0v
21	---	Not Used	---
22	BR	A/C Switch Signal	A/C Off: 12v, On: 1v
23	---	Not Used	---
24	WT/PK	Brake Switch Sense	Brake Off: 0v, On: 12v
25	WT/DB	Generator Field Source	12-14v
26	DB/WT	Fuel Level Sensor Signal	Full: 0.56v, 1/2 full: 2.5v
27	PK	SCI Transmit	0v
28	---	Not Used	---
29	LG	SCI Receive	5v
30	VT/YL	PCI Data Bus (J1850)	Digital Signals: 0-7-0v
31	---	Not Used	---
32	RD/LG	Speed Control Switch Signal	S/C & Set Switch On: 3.8v

Pin Connector Graphic

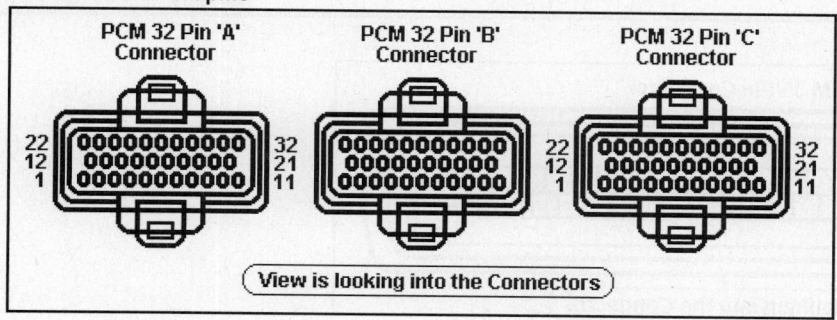

PCM 32 Pin 'A' Connector PCM 32 Pin 'B' Connector PCM 32 Pin 'C' Connector

View is looking into the Connectors

1990 Cherokee 2.5L I4 TBI VIN E (All) 35 Pin Connector

PCM Pin #	Wire Color	Circuit Description (35 Pin)	Value at Hot Idle
1-2	BK	Power Ground	<0.1v
3	YL	Ignition Switch Run Output	12-14v
4	RD	Fused Battery Power (B+)	12-14v
5	GN	EGR Solenoid Control	12v, at 55 mph: 1v
6	OR	Fuel Pump Relay Control	Relay Off: 12v, On: 1v
7	BK	Latch Relay B+ Control	Relay Off: 12v, On: 1v
8	GY/BK	PSP Switch Sense Signal	Straight: 0v, Turning: 5v
8	BL/OR	PSP Switch Sense Signal	Straight: 0v, Turning: 5v
10	BK	Sensor Ground	<0.1v
11	RD	CKP Sensor (+) Signal	Digital Signals: 0-5-0v
12	BK	PNP Switch Sense Signal	In P/N: 0v, Others: 5v
13	BR	TP Sensor Ground	<0.1v
14	TN	Air Temperature Sensor	At 100°F: 2.51v
15	TN	ECT Sensor Signal	At 180°F: 2.80v
16	RD	MAP Sensor VREF	4.9-5.1v
17	BK	Map Sensor Ground	<0.1v
18	PK	Shift Indicator Lamp Control	Lamp Off: 12v, On: 1v
19	PK	B+ Latch Relay Output	Relay Off: 12v, On: 1v
21	LB	Injector Control Driver	1-4 ms
22	BL	A/C Clutch Relay Control	A/C Off: 12v, On: 1v
23	BR	Idle Speed Motor (-) Retract	Pulse Signals
24	LG	Idle Speed Motor (+) Extend	Pulse Signals
25	GY	Closed Throttle Sensor	Open: 12v, Closed: 1v
27	OR	Ignition Coil Control	5°, 55 mph: 8° dwell
28	WT	CKP Sensor (-) Signal	Digital Signals: 0-5-0v
29	DG	Starter Input Signal	Cranking: 9-11v
30	LG	A/C Select Switch Sense	A/C Off: 12v, On: 1v
31	YL/GN	TP Sensor Signal	0.6-1.0v
32	BR	ECT Sensor Ground	<0.1v
33	PK	MAP Sensor Signal	1.5-1.6v
34	TN	A/C Request Switch Sense	A/C Off: 12v, On: 1v
35	GY	HO2S-11 (B1 S1) Signal	0.1-1.1v

Pin Connector Graphic

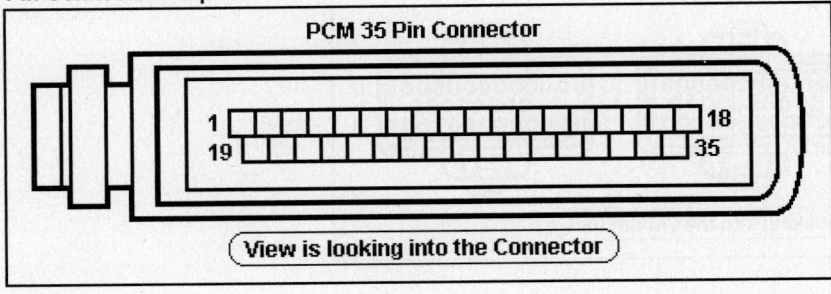

1991-92 Cherokee 2.5L I4 MFI VIN P (M/T) 60 Pin Connector

PCM Pin #	Wire Color	Circuit Description (60 Pin)	Value at Hot Idle
1	DG/RD	MAP Sensor Signal	1.5-1.6v
2	TN/BK	ECT Sensor Signal	At 180ºF: 2.80v
3	RD	Fused Battery Power (B+)	12-14v
4	BK/LB	Sensor Ground	<0.1v
5	BK/WT	Sensor Ground	<0.1v
6	PK/WT	5-Volt Supply	4.9-5.1v
7	OR	8-Volt Supply	7.9-8.1v
8 ('91)	BR	Starter Input Signal	Cranking: 9-11v
9	DB	Ignition Switch Power (B+)	12-14v
10	DB/OR	PSP Switch Sense Signal	Straight: 0v, Turning: 5v
11	BK/TN	Power Ground	<0.1v
12	BK/TN	Power Ground	<0.1v
13	LB/BR	Injector 4 Driver	1-4 ms
14	YL/WT	Injector 3 Driver	1-4 ms
15	TN	Injector 2 Driver	1-4 ms
16	WT/DB	Injector 1 Driver	1-4 ms
17-18	---	Not Used	---
19	GY	Coil 1 Driver Control	5º, 55 mph: 8º dwell
20	DG	Alternator Field Control	Digital Signals: 0-12-0v
21	BK/RD	Air Temperature Sensor	At 100ºF: 2.51v
22	OR/DB	TP Sensor Signal	0.6-1.0v
23	---	Not Used	---
24	GY/BK	Distributor Reference Pickup	Digital Signals: 0-5-0v
25	PK	SCI Transmit	0v
26	---	Not Used	---
27	LB	A/C Request Switch Sense	A/C Off: 12v, On: 1v
28	LG	A/C Select Switch Sense	A/C Off: 12v, On: 1v
29	WT/PK	Brake Switch Sense Signal	Brake Off: 0v, On: 12v
30	---	Not Used	---
31 ('92)	DB/PK	Radiator Fan Control Relay	Relay Off: 12v, On: 1v
32	BK/PK	MIL (lamp) Control	MIL Off: 12v, On: 1v
33	TN/RD	Speed Control Vacuum Solenoid	Vacuum Increasing: 1v
34	DB/OR	A/C Clutch Relay Control	Relay Off: 12v, On: 1v
35	---	Not Used	---
36	DG/YL	Alternator Lamp Control	Lamp Off: 12v, On: 1v
37	RD/DB	Ballast Bypass Relay Control	Relay Off: 12v, On: 1v
38	---	Not Used	---
39	GY/RD	AIS Motor 'D' Control	Pulse Signals
40	BR/WT	AIS Motor 'B' Control	Pulse Signals

Standard Colors and Abbreviations

Abbreviation	Color	Abbreviation	Color	Abbreviation	Color
BK	Black	GY	Gray	RD	Red
BL	Blue	GN	Green	TN	Tan
BR	Brown	LG	Light Green	VT	Violet
DB	Dark Blue	OR	Orange	WT	White
DG	Dark Green	PK	Pink	YL	Yellow

1991-92 Cherokee 2.5L I4 MFI VIN P (M/T) 60 Pin Connector

PCM Pin #	Wire Color	Circuit Description (60 Pin)	Value at Hot Idle
41	BK/DG	HO2S-11 (B1 S1) Signal	0.1-1.1v
42	---	Not Used	---
43	GY/LB	Tachometer Signal	Pulse Signals
44	TN/YL	Distributor Sync Pickup	Digital Signals: 0-5-0v
45	LG	SCI Receive	5v
46	---	Not Used	---
47	WT/OR	Vehicle Speed Signal	Digital Signal
48	BR/RD	Speed Control Set Switch	S/C & Set Switch On: 3.8v
49	YL/RD	Speed Control On/Off Switch	S/C Off: 12v, On: 1v
50	WT/LG	Speed Control Resume Switch	S/C Off: 12v, On: 1v
51	DB/YL	ASD Relay Control	Relay Off: 12v, On: 1v
52	---	Not Used	---
53	LG/RD	Speed Control Vent Solenoid	Vacuum Decreasing: 1v
54	OR/BK	Shift Indicator Lamp Control	Lamp Off: 12v, On: 1v
55	---	Not Used	---
56	GY/PK	Maintenance Indicator Lamp	Lamp Off: 12v, On: 1v
57	DG/OR	ASD Relay Output	12-14v
58	---	Not Used	---
59	PK/BK	AIS Motor 'D' Control	Pulse Signals
60	YL/BK	AIS Motor 'C' Control	Pulse Signals

Pin Connector Graphic

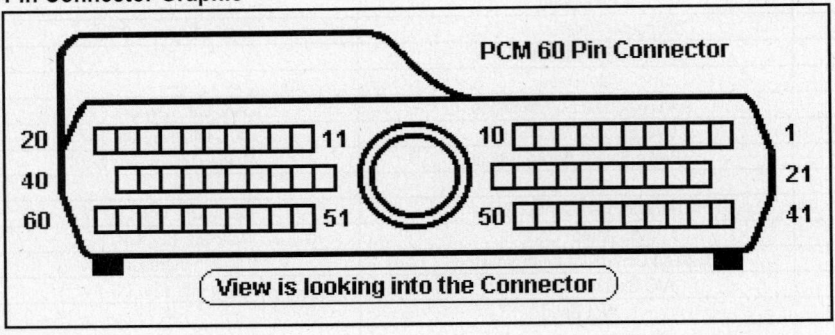

1993-95 Cherokee 2.5L I4 MFI VIN P (All) 60 Pin Connector

PCM Pin #	Wire Color	Circuit Description (60 Pin)	Value at Hot Idle
1	DG/RD	MAP Sensor Signal	1.5-1.6v
2	TN/BK	ECT Sensor Signal	At 180°F: 2.80v
3	RD	Fused Battery Power (B+)	12-14v
4	BK/LB	Sensor Ground	<0.1v
5	BK/WT	Sensor Ground	<0.1v
6	PK/WT	5-Volt Supply	4.9-5.1v
7	OR	8-Volt Supply	7.9-8.1v
8	---	Not Used	---
9	DB	Ignition Switch Power (B+)	12-14v
10	DB/WT	PSP Switch Sense Signal	Straight: 0v, Turning: 5v
10 ('94-'95)	PK	PSP Switch Sense Signal	Straight: 0v, Turning: 5v
11	BK/TN	Power Ground	<0.1v
12	BK/TN	Power Ground	<0.1v
13	LB/BR	Injector 4 Driver	1-4 ms
14	YL/WT	Injector 3 Driver	1-4 ms
15	TN	Injector 2 Driver	1-4 ms
16	WT/DB	Injector 1 Driver	1-4 ms
17-18	---	Not Used	---
19	GY	Coil 1 Driver Control	5°, 55 mph: 8° dwell
20	DG	Alternator Field Control	Digital Signals: 0-12-0v
21	BK/RD	IAT Sensor Signal	At 100°F: 1.83v
22	OR/DB	TP Sensor Signal	0.6-1.0v
23	---	Not Used	---
24	GY/BK	CKP Sensor Signal	Digital Signals: 0-5-0v
25	PK	SCI Transmit	0v
26	---	Not Used	---
27	LB	A/C Request Switch Sense	A/C Off: 12v, On: 1v
28	LG	A/C Select Switch Sense	A/C Off: 12v, On: 1v
29	WT/PK	Brake Switch Signal	Brake Off: 0v, On: 12v
30	BR/YL	PNP Switch Sense Signal	In P/N: 0v, Others: 5v
31	---	Not Used	---
32	BK/PK	MIL (lamp) Control	MIL Off: 12v, On: 1v
33	TN/RD	Speed Control Vacuum Solenoid	Vacuum Increasing: 1v
34	DB/OR	A/C Clutch Relay Control	Relay Off: 12v, On: 1v
35	---	Not Used	---
36	DG/YL	Alternator Lamp Control	Lamp Off: 12v, On: 1v
37 ('93)	RD/DB	Ballast Bypass Relay Control	Relay Off: 12v, On: 1v
38	---	Not Used	---
39	GY/RD	AIS Motor 1 Circuit Control	Pulse Signals
40	BR/WT	AIS Motor 3 Circuit Control	Pulse Signals

Standard Colors and Abbreviations

Abbreviation	Color	Abbreviation	Color	Abbreviation	Color
BK	Black	GY	Gray	RD	Red
BL	Blue	GN	Green	TN	Tan
BR	Brown	LG	Light Green	VT	Violet
DB	Dark Blue	OR	Orange	WT	White
DG	Dark Green	PK	Pink	YL	Yellow

1993-95 Cherokee 2.5L I4 MFI VIN P (All) 60 Pin Connector

PCM Pin #	Wire Color	Circuit Description (60 Pin)	Value at Hot Idle
41	BK/DG	HO2S-11 (B1 S1) Signal	0.1-1.1v
42	---	Not Used	---
43	GY/LB	Tachometer Signal	Pulse Signals
44	TN/YL	CMP Sensor Signal	Digital Signals: 0-5-0v
45	LG	SCI Receive	5v
46	---	Not Used	---
47	WT/OR	Vehicle Speed Signal	Digital Signals
48	BR/RD	Speed Control Set Switch	S/C & Set Switch On: 3.8v
49	YL/RD	Speed Control On/Off Switch	S/C Off: 12v, On: 1v
50	WT/LG	Speed Control Resume Switch	S/C Off: 12v, On: 1v
51	DB/YL	ASD Relay Control	Relay Off: 12v, On: 1v
52	---	Not Used	---
53	LG/RD	Speed Control Vent Solenoid	Vacuum Decreasing: 1v
54	OR/BK	A/T: Lockup Torque Converter	TCC on at Cruise: 1v
54	OR/BK	M/T: SIL (lamp) Control	Lamp Off: 12v, On: 1v
55	---	Not Used	---
56 ('93)	GY/PK	Maintenance Indicator Lamp	Lamp Off: 12v, On: 1v
57	DG/OR	ASD Relay Output	12-14v
58	---	Not Used	---
59	PK/BK	AIS Motor 4 Circuit Control	Pulse Signals
60	YL/BK	AIS Motor 2 Circuit Control	Pulse Signals

Pin Connector Graphic

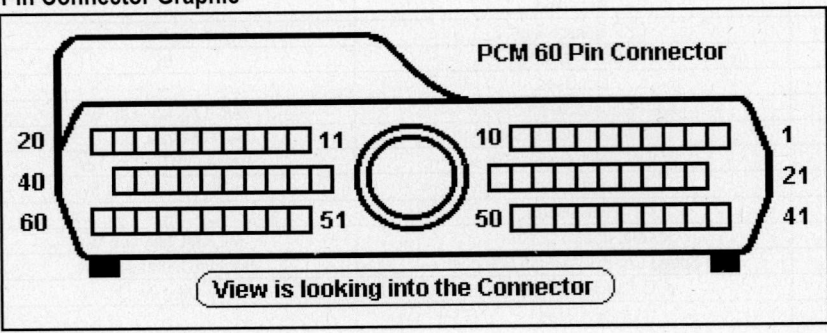

1996 Cherokee 2.5L I4 MFI VIN P (All) 'A' Black Connector

PCM Pin #	Wire Color	Circuit Description (32 Pin)	Value at Hot Idle
1	---	Not Used	---
2	DB	Ignition Switch Power (B+)	12-14v
3	---	Not Used	---
4	BK/LB	Sensor Ground	<0.1v
5	---	Not Used	---
6	BR/YL	PNP Switch Sense Signal	In P/N: 0v, Others: 5v
6	BK	M/T: Sensor Ground	<0.1v
7	GY	Coil 1 Driver Control	5°, 55 mph: 8° dwell
8	GY/BK	CKP Sensor Signal	Digital Signals: 0-5-0v
9	---	Not Used	---
10	YL/BK	IAC 2 Driver Control	Pulse Signals
11	BR/WT	IAC 3 Driver Control	Pulse Signals
12	DB/OR	PSP Switch Sense Signal	Straight: 0v, Turning: 5v
13-14	---	Not Used	---
15	BK/RD	IAT Sensor Signal	At 100ºF: 1.83v
16	TN/BK	ECT Sensor Signal	At 180ºF: 2.80v
17	PK/WT	5-Volt Supply	4.9-5.1v
18	TN/YL	CMP Sensor Signal	Digital Signals: 0-5-0v
19	GY/RD	IAC 1 Driver Control	Pulse Signals
20	PK/BK	IAC 4 Driver Control	Pulse Signals
21	---	Not Used	---
22	RD	Fused Battery Power (B+)	12-14v
23	OR/DB	TP Sensor Signal	0.6-1.0v
24	BK/DG	HO2S-11 (B1 S1) Signal	0.1-1.1v
25	TN/WT	HO2S-12 (B1 S2) Signal	0.1-1.1v
26	---	Not Used	---
27	DG/RD	MAP Sensor Signal	1.5-1.6v
28-30	---	Not Used	---
31-32	BK/TN	Power Ground	<0.1v

1996 Cherokee 2.5L I4 MFI VIN P (All) 'B' White Connector

PCM Pin #	Wire Color	Circuit Description (32 Pin)	Value at Hot Idle
1-3	---	Not Used	---
4	WT/DB	Injector 1 Driver	1-4 ms
5	YL/WT	Injector 3 Driver	1-4 ms
6-9	---	Not Used	---
10	DG	Generator Field Control	Digital Signals: 0-12-0v
11	OR/BK	A/T: TCC Solenoid Control	TCC on at Cruise: 1v
11	OR/BK	M/T: SIL (lamp) Control	Lamp Off: 12v, On: 1v
12-14	---	Not Used	---
15	TN	Injector 2 Driver	1-4 ms
16	LB/BR	Injector 4 Driver	1-4 ms
17-26	---	Not Used	---
27	WT/OR	Vehicle Speed Signal	Digital Signal
28-30	---	Not Used	---
31	RD/YL	5-Volt Supply	4.9-5.1v
32	---	Not Used	---

Standard Colors and Abbreviations

Abbreviation	Color	Abbreviation	Color	Abbreviation	Color
BK	Black	GY	Gray	RD	Red
BL	Blue	GN	Green	TN	Tan
BR	Brown	LG	Light Green	VT	Violet
DB	Dark Blue	OR	Orange	WT	White
DG	Dark Green	PK	Pink	YL	Yellow

1996 Cherokee 2.5L I4 MFI VIN P (All) 'C' Gray Connector

PCM Pin #	Wire Color	Circuit Description (32 Pin)	Value at Hot Idle
1	DB/OR	A/C Clutch Relay Control	Relay Off: 12v, On: 1v
2	---	Not Used	---
3	DB/YL	ASD Relay Control	Relay Off: 12v, On: 1v
4	TN/RD	Speed Control Vacuum Solenoid	Vacuum Increasing: 1v
5	LG/RD	Speed Control Vent Solenoid	Vacuum Decreasing: 1v
6-10	---	Not Used	---
11	YL/RD	Speed Control Power Supply	12-14v
12	DG/OR	ASD Relay Output	12-14v
13-14	---	Not Used	---
15	PK/YL	Battery Temperature Sensor	At 86°F: 1.96v
16	PK/BK	Generator Lamp Control	Lamp Off: 12v, On: 1v
17	BK/PK	MIL (lamp) Control	MIL Off: 12v, On: 1v
18	---	Not Used	---
19	BR	Fuel Pump Relay Control	Relay Off: 12v, On: 1v
20	PK/BK	EVAP Purge Solenoid Control	PWM Signal: 0-12-0v
21	---	Not Used	---
22	DB/OR	A/C Request Signal	A/C Off: 12v, On: 1v
23	LG	A/C Select Signal	A/C Off: 12v, On: 1v
24	WT/PK	Brake Switch Sense Signal	Brake Off: 0v, On: 12v
25	---	Not Used	---
26	DB	Fuel Level Sensor Signal	Full: 0.5v, 1/2 full: 2.5v
27	PK	SCI Transmit	0v
28	WT/BK	CCD Bus (-)	<0.050v
29	LG	SCI Receive	5v
30	PK/BR	CCD Bus (+)	Digital Signals: 0-5-0v
31	GY/LB	Tachometer Signal	Pulse Signals
32	WT/LG	Speed Control Switch Signal	S/C & Set Switch On: 3.8v

Pin Connector Graphic

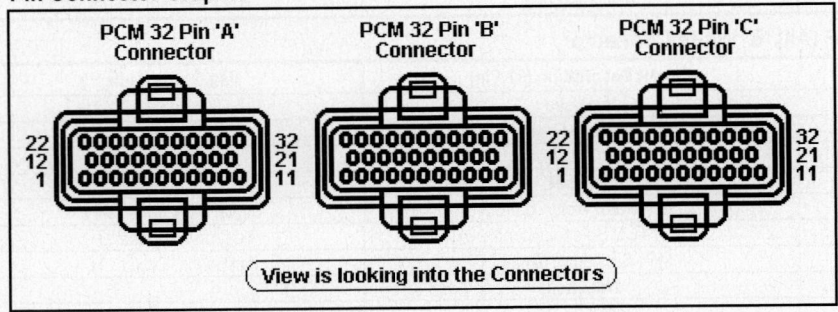

1997 Cherokee 2.5L I4 MFI VIN P (All) 'A' Black Connector

PCM Pin #	Wire Color	Circuit Description (32 Pin)	Value at Hot Idle
1	---	Not Used	---
2	DB/WT	Ignition Switch Power (B+)	12-14v
3	---	Not Used	---
4	BR/YL	Sensor Ground	<0.1v
5	---	Not Used	---
6	BK/WT	A/T: PNP Switch Signal	In P/N: 0v, Others: 5v
6	BK	M/T: Sensor Ground	<0.1v
7	GY	Coil 1 Driver Control	5°, 55 mph: 8° dwell
8	GY/BK	CKP Sensor Signal	Digital Signals: 0-5-0v
9	---	Not Used	---
10	YL/BK	IAC 2 Driver Control	Pulse Signals
11	BR/WT	IAC 3 Driver Control	Pulse Signals
12	DB/BR	PSP Switch Sense Signal	Straight: 0v, Turning: 5v
13-14	---	Not Used	---
15	BK/RD	IAT Sensor Signal	At 100°F: 1.83v
16	TN/BK	ECT Sensor Signal	At 180°F: 2.80v
17	OR	5-Volt Supply	4.9-5.1v
18	TN/YL	CMP Sensor Signal	Digital Signals: 0-5-0v
19	GY/RD	IAC 1 Driver Control	Pulse Signals
20	PK/BK	IAC 4 Driver Control	Pulse Signals
21	---	Not Used	---
22	DG/BK	Fused Battery Power (B+)	12-14v
23	OR/DB	TP Sensor Signal	0.6-1.0v
24	BK/DG	HO2S-11 (B1 S1) Signal	0.1-1.1v
25	TN/BK	HO2S-12 (B1 S2) Signal	0.1-1.1v
26	---	Not Used	---
27	DG/RD	MAP Sensor Signal	1.5-1.6v
28-30	---	Not Used	---
31	BK/TN	Power Ground	<0.1v
32	BK/TN	Power Ground	<0.1v

1997 Cherokee 2.5L I4 MFI VIN P (All) 'B' White Connector

PCM Pin #	Wire Color	Circuit Description (32 Pin)	Value at Hot Idle
1-3	---	Not Used	---
4	WT/DB	Injector 1 Driver	1-4 ms
5	YL/WT	Injector 3 Driver	1-4 ms
6-9	---	Not Used	---
10	DG	Generator Field Control	Digital Signals: 0-12-0v
11-14	---	Not Used	---
15	T	Injector 2 Driver	1-4 ms
16	LB/BR	Injector 4 Driver	1-4 ms
17-22	---	Not Used	---
23	GY/YL	Engine Oil Pressure Sensor	1.6v at 24 psi
24-26	---	Not Used	---
27	WT/OR	Vehicle Speed Signal	Digital Signals
28-30	---	Not Used	---
31	PK/OR	5-Volt Supply	4.9-5.1v
32	---	Not Used	---

Standard Colors and Abbreviations

Abbreviation	Color	Abbreviation	Color	Abbreviation	Color
BK	Black	GY	Gray	RD	Red
BL	Blue	GN	Green	TN	Tan
BR	Brown	LG	Light Green	VT	Violet
DB	Dark Blue	OR	Orange	WT	White
DG	Dark Green	PK	Pink	YL	Yellow

1997 Cherokee 2.5L I4 MFI VIN P (All) 'C' Gray Connector

PCM Pin #	Wire Color	Circuit Description (32 Pin)	Value at Hot Idle
1	DB/OR	A/C Clutch Relay Control	Relay Off: 12v, On: 1v
2	---	Not Used	---
3	DB/YL	ASD Relay Control	Relay Off: 12v, On: 1v
4	TN/RD	Speed Control Vacuum Solenoid	Vacuum Increasing: 1v
5	LG/RD	Speed Control Vent Solenoid	Vacuum Decreasing: 1v
6-10	---	Not Used	---
11	YL/RD	Speed Control Power Supply	12-14v
12	DG/OR	ASD Relay Output	12-14v
13-14	---	Not Used	---
15	PK/YL	Battery Temperature Sensor	At 86°F: 1.96v
16-18	---	Not Used	---
19	BR	Fuel Pump Relay Control	Relay Off: 12v, On: 1v
20	PK/BK	EVAP Purge Solenoid Control	PWM Signal: 0-12-0v
21	---	Not Used	---
22	DB/WT	A/C Request Signal	A/C Off: 12v, On: 1v
23	LG	A/C Select Signal	A/C Off: 12v, On: 1v
24	WT/PK	Brake Switch Sense Signal	Brake Off: 0v, On: 12v
25	---	Not Used	---
26	DB/LG	Fuel Level Sensor Signal	Full: 0.5v, 1/2 full: 2.5v
27	PK	SCI Transmit	0v
28	WT/BK	CCD Bus (-)	<0.050v
29	LG	SCI Receive	5v
30	PK/BR	CCD Bus (+)	Digital Signals: 0-5-0v
31	---	Not Used	---
32	RD/LG	Speed Control Switch Signal	S/C & Set Switch On: 3.8v

Pin Connector Graphic

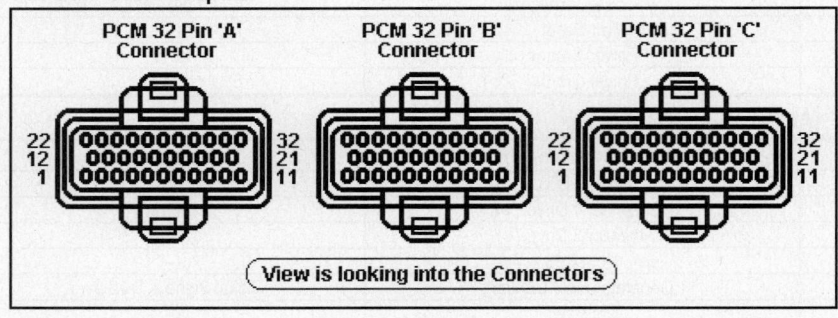

1998-2000 Cherokee 2.5L I4 MFI VIN P 'A' Black Connector

PCM Pin #	Wire Color	Circuit Description (32 Pin)	Value at Hot Idle
1	---	Not Used	---
2	DB/WT	Ignition Switch Power (B+)	12-14v
3	---	Not Used	---
4	BR/YL	Sensor Ground	<0.1v
5	---	Not Used	---
6	BK/WT	A/T: PNP Switch Signal	In P/N: 0v, Others: 5v
6	BK	M/T: Sensor Ground	<0.1v
7	GY	Coil 1 Driver Control	5°, 55 mph: 8° dwell
8	GY/BK	CKP Sensor Signal	Digital Signals: 0-5-0v
9	---	Not Used	---
10	YL/BK	IAC 2 Driver Control	Pulse Signals
11	BR/WT	IAC 3 Driver Control	Pulse Signals
12	DB/BR	PSP Switch Sense Signal	Straight: 0v, Turning: 5v
13-14	---	Not Used	---
15	BK/RD	IAT Sensor Signal	At 100°F: 1.83v
16	TN/BK	ECT Sensor Signal	At 180°F: 2.80v
17	OR	5-Volt Supply	4.9-5.1v
18	TN/YL	CMP Sensor Signal	Digital Signals: 0-5-0v
19	GY/RD	IAC 1 Driver Control	Pulse Signals
20	PK/BK	IAC 4 Driver Control	Pulse Signals
21	---	Not Used	---
22	DG/BK	Fused Battery Power (B+)	12-14v
23	OR/DB	TP Sensor Signal	0.6-1.0v
24	BK/DG	HO2S-11 (B1 S1) Signal	0.1-1.1v
25 ('98)	TN/BK	HO2S-12 (B1 S2) Signal	0.1-1.1v
25	TN/WT	HO2S-12 (B1 S2) Signal	0.1-1.1v
26	---	Not Used	---
27	DG/RD	MAP Sensor Signal	1.5-1.6v
28-30	---	Not Used	---
31-32	BK/TN	Power Ground	<0.1v

1998-2000 Cherokee 2.5L I4 MFI VIN P 'B' White Connector

PCM Pin #	Wire Color	Circuit Description (32 Pin)	Value at Hot Idle
1-3	---	Not Used	---
4	WT/DB	Injector 1 Driver	1-4 ms
5	YL/WT	Injector 3 Driver	1-4 ms
6-9	---	Not Used	---
10	DG	Generator Field Control	Digital Signals: 0-12-0v
11-12	---	Not Used	---
13	OR/BK	TCC Solenoid Control	TCC on at Cruise: 1v
14	---	Not Used	---
15	TN	Injector 2 Driver	1-4 ms
16	LB/BR	Injector 4 Driver	1-4 ms
17-22	---	Not Used	---
23	GY/YL	Engine Oil Pressure Sensor	1.6v at 24 psi
24-26	---	Not Used	---
27	WT/OR	Vehicle Speed Signal	Digital Signal
28-30	---	Not Used	---
31	PK/OR	5-Volt Supply	4.9-5.1v
32	---	Not Used	---

Standard Colors and Abbreviations

Abbreviation	Color	Abbreviation	Color	Abbreviation	Color
BK	Black	GY	Gray	RD	Red
BL	Blue	GN	Green	TN	Tan
BR	Brown	LG	Light Green	VT	Violet
DB	Dark Blue	OR	Orange	WT	White
DG	Dark Green	PK	Pink	YL	Yellow

1998-2000 Cherokee 2.5L I4 MFI VIN P 'C' Gray Connector

PCM Pin #	Wire Color	Circuit Description (32 Pin)	Value at Hot Idle
1	DB/OR	A/C Clutch Relay Control	Relay Off: 12v, On: 1v
2	DB/PK	Radiator Fan Control Relay	Relay Off: 12v, On: 1v
3	DB/YL	ASD Relay Control	Relay Off: 12v, On: 1v
4	TN/RD	Speed Control Vacuum Solenoid	Vacuum Increasing: 1v
5	LG/RD	Speed Control Vent Solenoid	Vacuum Decreasing: 1v
6-9	---	Not Used	---
10	WT/DG	LDP Solenoid Control	PWM Signal: 0-12-0v
11	YL/RD	Speed Control Power Supply	12-14v
12	DG/OR	ASD Relay Output	12-14v
13	---	Not Used	---
14	WT/OR	LDP Switch Sense Signal	Closed: 0v, Open: 12v
15	PK/YL	Battery Temperature Sensor	At 86°F: 1.96v
16-18	---	Not Used	---
19	BR	Fuel Pump Relay Control	Relay Off: 12v, On: 1v
20	PK/BK	EVAP Purge Solenoid Control	PWM Signal: 0-12-0v
21	---	Not Used	
22	DB/WT	A/C High Pressure Switch	Switch open: 12v, closed: 0v
23	LG	A/C Select Signal	A/C Off: 12v, On: 1v
24	WT/PK	Brake Switch Sense Signal	Brake Off: 0v, On: 12v
25	DG/OR	Generator Field Source	12-14v
26	DB/LG	Fuel Level Sensor Signal	Full: 0.5v, 1/2 full: 2.5v
27	PK	SCI Transmit	0v
28	WT/BK	CCD Bus (-)	<0.050v
29	LG/BK	SCI Receive	5v
30	PK/BR	CCD Bus (+)	Digital Signals: 0-5-0v
31	---	Not Used	---
32	RD/LG	Speed Control Switch Signal	S/C & Set Switch On: 3.8v

Pin Connector Graphic

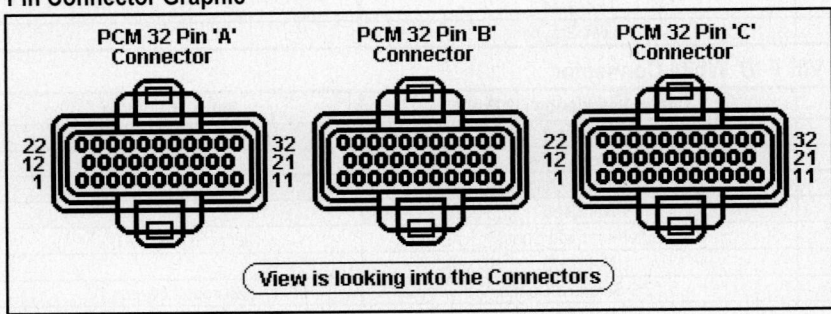

PCM 32 Pin 'A' Connector   PCM 32 Pin 'B' Connector   PCM 32 Pin 'C' Connector

View is looking into the Connectors

2000 Cherokee 2.5L I4 MFI VIN P Wiring Schematic (Part 1)

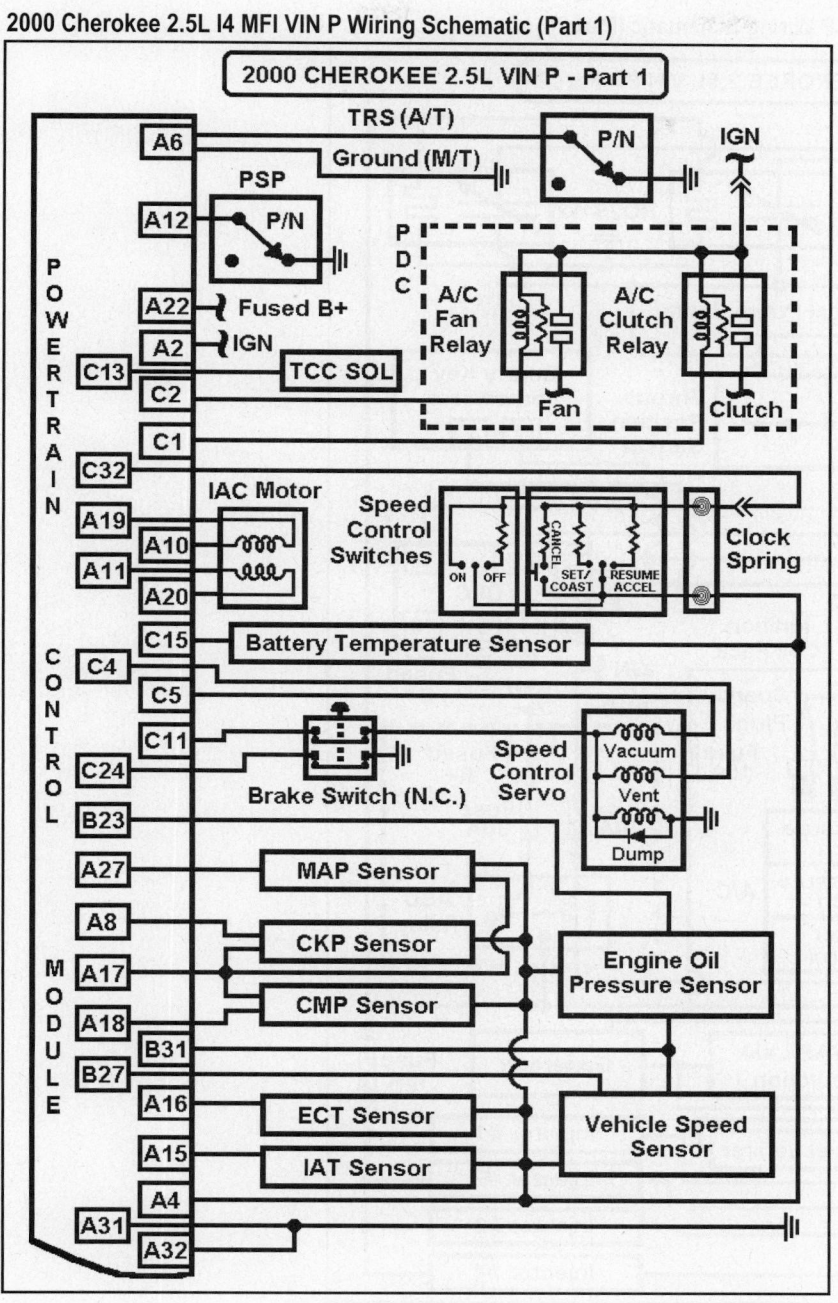

2000 Cherokee 2.5L I4 MFI VIN P Wiring Schematic (Part 2)

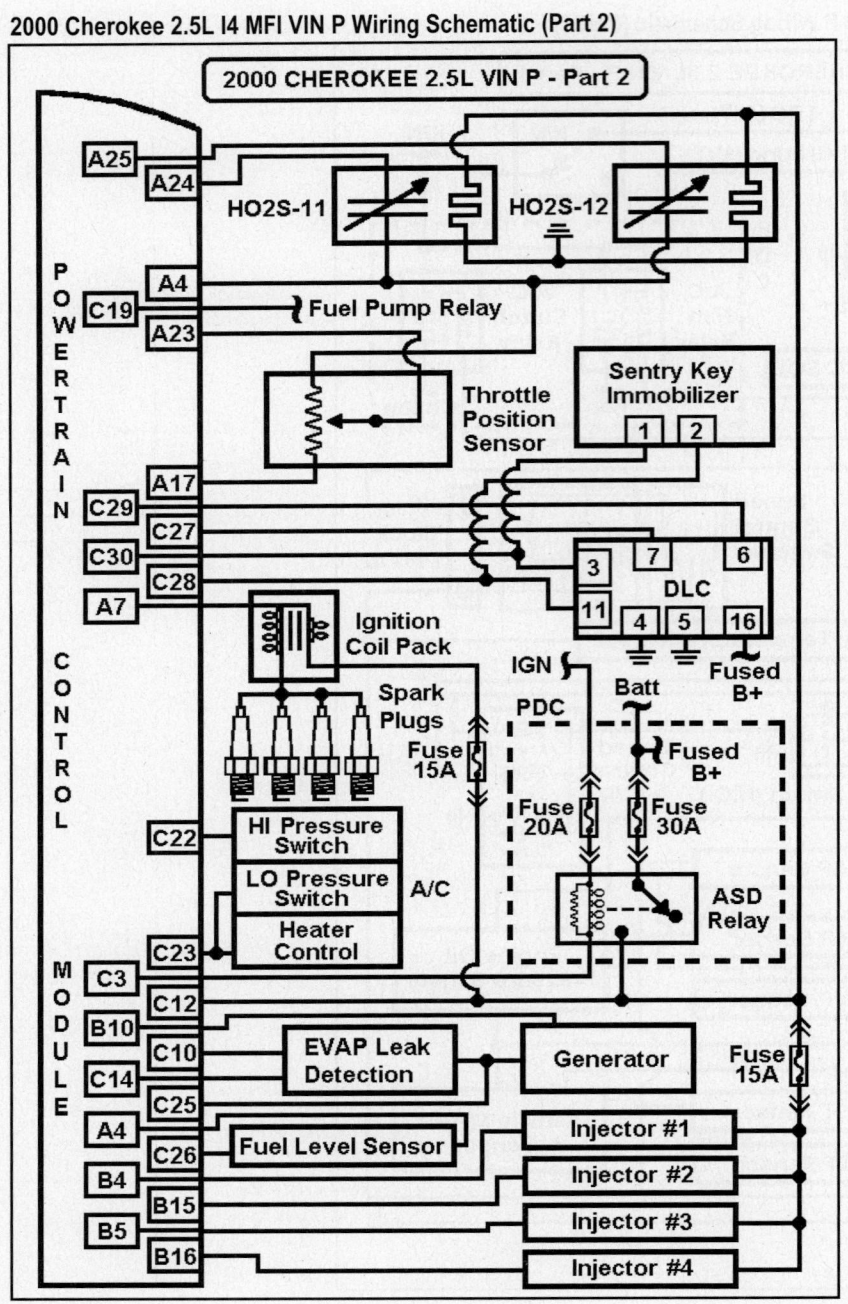

1990 Cherokee 4.0L V6 MFI VIN L Row 'A' 24 Pin Connector

PCM Pin #	Wire Color	Circuit Description (24 Pin)	Value at Hot Idle
1	TN	Injector 3 Control Driver	1-4 ms
2	BR	Injector 6 Control Driver	1-4 ms
3	LG	Injector 2 Control Driver	1-4 ms
4	YL	Injector 4 Control Driver	1-4 ms
5	OR	Fuel Pump Relay Control	Relay Off: 12v, On: 1v
6	---	Not Used	---
7	GN/YL	HO2S Relay Control	Relay Off: 12v, On: 1v
8	PK	Shift Indicator Lamp Control	Lamp Off: 12v, On: 1v
9	BK	Latch Relay B+ Control	Relay Off: 12v, On: 1v
10	GN	EGR Solenoid Control	12v, at 55 mph: 1v
11	---	Not Used	---
12	BL	A/C Clutch Relay Control	Relay Off: 12v, On: 1v

1990 Cherokee 4.0L V6 MFI VIN L Row 'B' 24 Pin Connector

PCM Pin #	Wire Color	Circuit Description (24 Pin)	Value at Hot Idle
1	LB	Injector 1 Control Driver	1-4 ms
2	WT	Injector 5 Control Driver	1-4 ms
3	RD/YL	Idle Speed Motor 'A' Control	Pulse Signals
4	BL/YL	Idle Speed Motor 'D' Control	Pulse Signals
5	GN/BK	Idle Speed Motor 'C' Control	Pulse Signals
6	PK/BK	Idle Speed Motor 'B' Control	Pulse Signals
7	RD	Battery Power (B+)	12-14v
8	YL	Ignition Switch Run Output	12-14v
9	---	Not Used	---
10	PK	B+ Latch Relay Output	Relay Off: 12v, On: 1v
11	BK	Power Ground	<0.1v
12	BK	Power Ground	<0.1v

Pin Connector Graphic

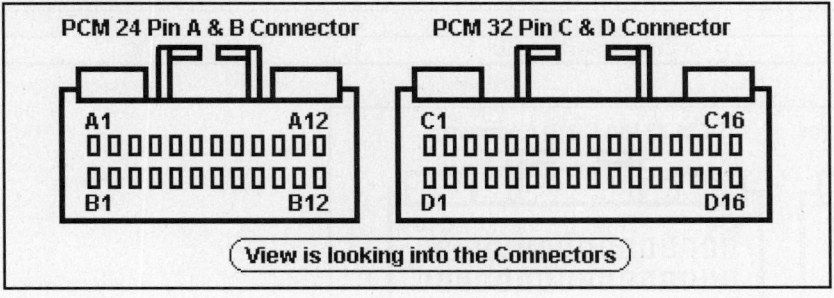

PCM 24 Pin A & B Connector   PCM 32 Pin C & D Connector

A1   A12   C1   C16
B1   B12   D1   D16

View is looking into the Connectors

1990 Cherokee 4.0L V6 MFI VIN L Row 'C' 32 Pin Connector

PCM Pin #	Wire Color	Circuit Description (32 Pin)	Value at Hot Idle
1	PK	CKP Sensor (+) Signal	Digital Signals: 0-5-0v
2	TN	A/C Request Switch Sense	A/C Off: 12v, On: 1v
3	GN	Starter Input Signal	Cranking: 9-11v
4	BK	PNP Switch Sense Signal	In P/N: 0v, Others: 5v
5	GY	Distributor Sync Pickup (+)	Digital Signals: 0-5-0v
6	PK	MAP Sensor Signal	1.5-1.6v
7	YL/GN	TP Sensor Signal	0.6-1.0v
8	TN	Air Temperature Sensor	At 100°F: 2.51v
9	---	Not Used	
10	TN	ECT Sensor Signal	At 180°F: 2.80v
11	OR	Fuel Pump Relay Output	Relay Off: 12v, On: 1v
12	BR/PK	SCI Transmit	0v
13	---	Not Used	---
14	RD	MAP Sensor VREF	4.9-5.1v
15	LB	TP Sensor VREF	4.9-5.1v
16	BL	Distributor Sync Pickup (-)	Digital Signals: 0-5-0v

1990 Cherokee 4.0L V6 MFI VIN L Row 'D' 32 Pin Connector

PCM Pin #	Wire Color	Circuit Description (32 Pin)	Value at Hot Idle
1	WT	CKP Sensor (-) Signal	Digital Signals: 0-5-0v
2	LG	A/C Select Switch Sense	A/C Off: 12v, On: 1v
3	BR	Sensor Ground	<0.1v
4-7	---	Not Used	---
8	YL	Knock Sensor (-) Signal	<0.050v
9	GY	HO2S-11 (B1 S1) Signal	0.1-1.1v
10	OR	Fuel Pump Relay Control	Relay Off: 12v, On: 1v
11	BK/WT	SCI Receive	5v
12	---	Not Used	---
13	YL	Ignition Coil Control	5°, 55 mph: 8° dwell
14-15	---	Not Used	---
16	PK	Knock Sensor Signal	0.080v AC

Pin Connector Graphic

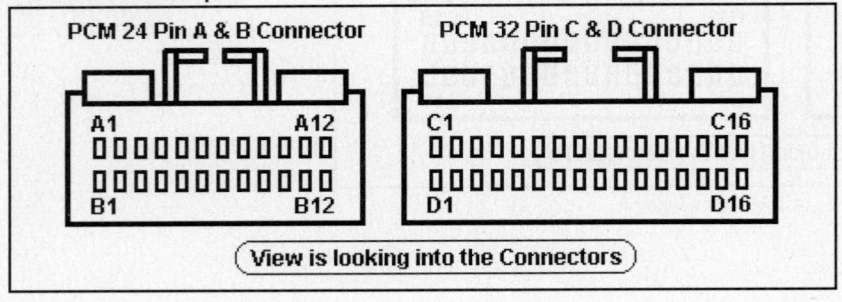

PCM 24 Pin A & B Connector   PCM 32 Pin C & D Connector

A1   A12   C1   C16

B1   B12   D1   D16

View is looking into the Connectors

1991-92 Cherokee 4.0L V6 MFI VIN S (All) 60 Pin Connector

PCM Pin #	Wire Color	Circuit Description (60 Pin)	Value at Hot Idle
1	DG/RD	MAP Sensor Signal	1.5-1.6v
2	TN/BK	ECT Sensor Signal	At 180ºF: 2.80v
3	RD	Fused Battery Power (B+)	12-14v
4	BK/LB	Sensor Ground	<0.1v
5	BK	Sensor Ground	<0.1v
6	PK	5-Volt Supply	4.9-5.1v
7	OR	8-Volt Supply	7.9-8.1v
8	BR	Starter Input Signal	Cranking: 9-11v
9	DB	Ignition Switch Power (B+)	12-14v
10	---	Not Used	---
11	BK/TN	Power Ground	<0.1v
12	BK/TN	Power Ground	<0.1v
13	LB/BR	Injector 4 Driver	1-4 ms
14	YL/WT	Injector 3 Driver	1-4 ms
15	TN	Injector 2 Driver	1-4 ms
16	WT/DB	Injector 1 Driver	1-4 ms
17-18	---	---	---
19	GY	Coil 1 Driver Control	5º, 55 mph: 8º dwell
20	DG	Alternator Field Control	Digital Signals: 0-12-0v
21	BK/RD	Air Temperature Sensor	At 100ºF: 2.51v
22	OR/DB	TP Sensor Signal	0.6-1.0v
23	---	Not Used	---
24	GY/BK	Distributor Reference Pickup	Digital Signals: 0-5-0v
25	PK	SCI Transmit	0v
26	PK/BR	CCD Bus (+)	Digital Signals: 0-5-0v
27	LB	A/C Request Switch Sense	A/C Off: 12v, On: 1v
28	LG	A/C Select Switch Sense	A/C Off: 12v, On: 1v
29	WT/PK	Brake Switch Sense Signal	Brake Off: 0v, On: 12v
30	BR/YL	PNP Switch Sense Signal	In P/N: 0v, Others: 5v
31	DB/PK	Radiator Fan Control Relay	Relay Off: 12v, On: 1v
32	BK/PK	MIL (lamp) Control	MIL Off: 12v, On: 1v
33	TN/RD	Speed Control Vacuum Solenoid	Vacuum Increasing: 1v
34	DB/OR	A/C Clutch Relay Control	Relay Off: 12v, On: 1v
35	---	Not Used	---
36	DG/YL	Alternator Lamp Control	Lamp Off: 12v, On: 1v
37	RD/DB	Ballast Bypass Relay Control	Relay Off: 12v, On: 1v
38	PK/BK	Injector 5 Driver	1-4 ms
39	GY/RD	AIS Motor 'D' Control	Pulse Signals
40	BR/WT	AIS Motor 'B' Control	Pulse Signals

Standard Colors and Abbreviations

Abbreviation	Color	Abbreviation	Color	Abbreviation	Color
BK	Black	GY	Gray	RD	Red
BL	Blue	GN	Green	TN	Tan
BR	Brown	LG	Light Green	VT	Violet
DB	Dark Blue	OR	Orange	WT	White
DG	Dark Green	PK	Pink	YL	Yellow

1991-92 Cherokee 4.0L V6 MFI VIN S (All) 60 Pin Connector

PCM Pin #	Wire Color	Circuit Description (60 Pin)	Value at Hot Idle
41	BK/DG	HO2S-11 (B1 S1) Signal	0.1-1.1v
42	---	Not Used	---
43	GY/LB	Tachometer Signal	Pulse Signals
44	TN/YL	Distributor Sync Pickup	Digital Signals: 0-5-0v
45	LG	SCI Receive	5v
46	WT	CCD Bus (-)	<0.050v
47	WT/OR	Vehicle Speed Signal	Digital Signals
48	BR/RD	Speed Control Set Switch	S/C & Set Switch On: 3.8v
49	YL/RD	Speed Control On/Off Switch	S/C Off: 12v, On: 1v
50	WT/LG	Speed Control Resume Switch	S/C Off: 12v, On: 1v
51	DB/YL	ASD Relay Control	Relay Off: 12v, On: 1v
52	---	Not Used	---
53	LG/RD	Speed Control Vent Solenoid	Vacuum Decreasing: 1v
54	OR/BK	A/T: Lockup Torque Converter	TCC on at Cruise: 1v
54	OR/BK	M/T: SIL (lamp) Control	Lamp Off: 12v, On: 1v
55	---	Not Used	---
56	GY/PK	Maintenance Indicator Lamp	Lamp Off: 12v, On: 1v
57	DG/OR	ASD Relay Output	12-14v
58	LG/BK	Injector 6 Driver	1-4 ms
59	PK/BK	AIS Motor 'D' Control	Pulse Signals
60	YL/BK	AIS Motor 'C' Control	Pulse Signals

Pin Connector Graphic

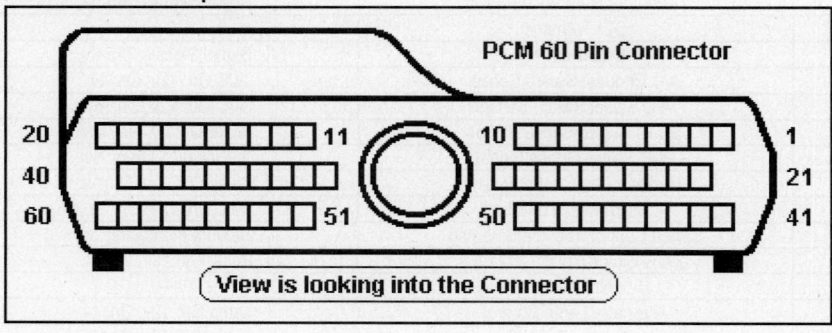

1993-95 Cherokee 4.0L V6 MFI VIN S (All) 60 Pin Connector

PCM Pin #	Wire Color	Circuit Description (60 Pin)	Value at Hot Idle
1	DG/RD	MAP Sensor Signal	1.5-1.6v
2	TN/BK	ECT Sensor Signal	At 180°F: 2.80v
3	RD	Fused Battery Power (B+)	12-14v
4	BK/LB	Sensor Ground	<0.1v
5	BK/WT	Sensor Ground	<0.1v
6	PK/WT	5-Volt Supply	4.9-5.1v
7	OR	8-Volt Supply	7.9-8.1v
8	---	Not Used	---
9	DB	Ignition Switch Power (B+)	12-14v
10	PK	Extended Idle Switch (Police)	Switch Off: 12v, On: 1v
11	BK/TN	Power Ground	<0.1v
12	BK/TN	Power Ground	<0.1v
13	LB/BR	Injector 4 Driver	1-4 ms
14	YL/WT	Injector 3 Driver	1-4 ms
15	TN	Injector 2 Driver	1-4 ms
16	WT/DB	Injector 1 Driver	1-4 ms
17-18	---	Not Used	---
19	GY	Coil 1 Driver Control	5°, 55 mph: 8° dwell
20	DG	Alternator Field Control	Digital Signals: 0-12-0v
21	BK/RD	Air Temperature Sensor	At 100°F: 2.51v
22	OR/DB	TP Sensor Signal	0.6-1.0v
23	---	Not Used	---
24	GY/BK	Distributor Reference Pickup	Digital Signals: 0-5-0v
25	PK	SCI Transmit	0v
26	PK/BR	CCD Bus (+)	Digital Signals: 0-5-0v
27	LB	A/C Request Switch Sense	A/C Off: 12v, On: 1v
28	LG	A/C Select Switch Sense	A/C Off: 12v, On: 1v
29	WT/PK	Brake Switch Sense Signal	Brake Off: 0v, On: 12v
30	BR/YL	PNP Switch Sense Signal	In P/N: 0v, Others: 5v
31	DB/PK	Radiator Fan Control Relay	Relay Off: 12v, On: 1v
32	BK/PK	MIL (lamp) Control	MIL Off: 12v, On: 1v
33	TN/RD	Speed Control Vacuum Solenoid	Vacuum Increasing: 1v
34	DB/OR	A/C Clutch Relay Control	Relay Off: 12v, On: 1v
35	---	Not Used	---
36	DG/YL	Alternator Lamp Control	Lamp Off: 12v, On: 1v
37 ('93)	RD/DB	Ballast Bypass Relay Control	Relay Off: 12v, On: 1v
38	PK/BK	Injector 5 Driver	1-4 ms
39	GY/RD	AIS Motor 1 Circuit Control	Pulse Signals
40	BR/WT	AIS Motor 3 Circuit Control	Pulse Signals

Standard Colors and Abbreviations

Abbreviation	Color	Abbreviation	Color	Abbreviation	Color
BK	Black	GY	Gray	RD	Red
BL	Blue	GN	Green	TN	Tan
BR	Brown	LG	Light Green	VT	Violet
DB	Dark Blue	OR	Orange	WT	White
DG	Dark Green	PK	Pink	YL	Yellow

1993-95 Cherokee 4.0L V6 MFI VIN S (All) 60 Pin Connector

PCM Pin #	Wire Color	Circuit Description (60 Pin)	Value at Hot Idle
41	BK/DG	HO2S-11 (B1 S1) Signal	0.1-1.1v
42	---	Not Used	---
43	GY/LB	Tachometer Signal	Pulse Signals
44	TN/YL	Distributor Sync Pickup	Digital Signals: 0-5-0v
45	LG	SCI Receive	5v
46	WT/BK	CCD Bus (-)	<0.050v
47	WT/OR	Vehicle Speed Signal	Digital Signal
48	BR/RD	Speed Control Set Switch	S/C & Set Switch On: 3.8v
49	YL/RD	Speed Control On/Off Switch	S/C Off: 12v, On: 1v
50	WT/LG	Speed Control Resume Switch	S/C Off: 12v, On: 1v
51	DB/YL	ASD Relay Control	Relay Off: 12v, On: 1v
52	---	Not Used	---
53	LG/RD	Speed Control Vent Solenoid	Vacuum Decreasing: 1v
54	OR/BK	A/T: Lockup Torque Converter	TCC on at Cruise: 1v
54	OR/BK	M/T: SIL (lamp) Control	Lamp Off: 12v, On: 1v
55	---	Not Used	---
56 ('93)	GY/PK	Maintenance Indicator Lamp	Lamp Off: 12v, On: 1v
57	DG/OR	ASD Relay Output	12-14v
58	LG/BK	Injector 6 Driver	1-4 ms
59	PK/BK	AIS Motor 4 Circuit Control	Pulse Signals
60	YL/BK	AIS Motor 2 Circuit Control	Pulse Signals

Pin Connector Graphic

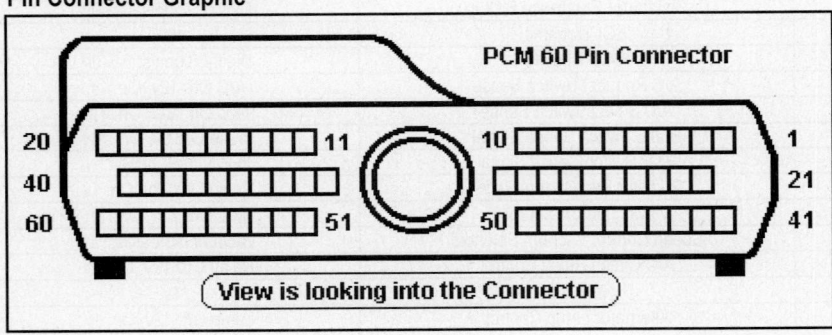

1996 Cherokee 4.0L V6 MFI VIN S (All) 'A' Black Connector

PCM Pin #	Wire Color	Circuit Description (32 Pin)	Value at Hot Idle
1	---	Not Used	---
2	DB	Ignition Switch Power (B+)	12-14v
3	---	Not Used	---
4	BK/LB	A/T: Analog Sensor Ground	<0.1v
5	---	Not Used	---
6	BR/YL	PNP Switch Sense Signal	In P/N: 0v, Others: 5v
6	BK	M/T: Analog Sensor Ground	<0.1v
7	GY	Coil 1 Driver Control	5º, 55 mph: 8º dwell
8	GY/BK	CKP Sensor Signal	Digital Signals: 0-5-0v
9	---	Not Used	---
10	YL/BK	IAC 2 Driver Control	Pulse Signals
11	BR/WT	IAC 3 Driver Control	Pulse Signals
12	DB/OR	Extended Idle Switch (Police Special)	Switch Off: 12v, On: 1v
13-14	---	Not Used	---
15	BK/RD	IAT Sensor Signal	At 100ºF: 1.83v
16	TN/BK	ECT Sensor Signal	At 180ºF: 2.80v
17	PK/WT	5-Volt Supply	4.9-5.1v
18	TN/YL	CMP Sensor Signal	Digital Signals: 0-5-0v
19	GY/RD	IAC 1 Driver Control	Pulse Signals
20	PK/BK	IAC 4 Driver Control	Pulse Signals
21	---	Not used	---
22	RD	Fused Battery Power (B+)	12-14v
23	OR/DB	TP Sensor Signal	0.6-1.0v
24	BK/DG	HO2S-11 (B1 S1) Signal	0.1-1.1v
25	TN/WT	HO2S-12 (B1 S2) Signal	0.1-1.1v
26	---	Not Used	---
27	DG/RD	MAP Sensor Signal	1.5-1.6v
28-30	---	Not Used	---
31-32	BK/TN	Power Ground	<0.1v

1996 Cherokee 4.0L V6 MFI VIN S (All) 'B' White Connector

PCM Pin #	Wire Color	Circuit Description (32 Pin)	Value at Hot Idle
1-3	---	Not Used	---
4	WT/DB	Injector 1 Driver	1-4 ms
5	YL/WT	Injector 3 Driver	1-4 ms
6	GY	Injector 5 Driver	1-4 ms
7-9	---	Not Used	---
10	DG	Generator Field Control	Digital Signals: 0-12-0v
11	OR/BK	A/T: TCC Solenoid Control	TCC on at Cruise: 1v
11	OR/BK	M/T: Upshift (lamp) Control	Lamp Off: 12v, On: 1v
12	BR/DB	Injector 6 Driver	1-4 ms
13-14	---	Not Used	---
15	TN	Injector 2 Driver	1-4 ms
16	LB/BR	Injector 4 Driver	1-4 ms
17-26	---	Not Used	---
27	WT/OR	Vehicle Speed Signal	Digital Signal
28-30	---	Not Used	---
31	OR	5-Volt Supply	4.9-5.1v
32	---	Not Used	---

Standard Colors and Abbreviations

Abbreviation	Color	Abbreviation	Color	Abbreviation	Color
BK	Black	GY	Gray	RD	Red
BL	Blue	GN	Green	TN	Tan
BR	Brown	LG	Light Green	VT	Violet
DB	Dark Blue	OR	Orange	WT	White
DG	Dark Green	PK	Pink	YL	Yellow

1996 Cherokee 4.0L V6 MFI VIN S (All) 'C' Gray Connector

PCM Pin #	Wire Color	Circuit Description (32 Pin)	Value at Hot Idle
1	DB/OR	A/C Clutch Relay Control	Relay Off: 12v, On: 1v
2	DB/PK	Radiator Fan Control Relay	Relay Off: 12v, On: 1v
3	DB/YL	ASD Relay Control	Relay Off: 12v, On: 1v
4	TN/RD	Speed Control Vacuum Solenoid	Vacuum Increasing: 1v
5	LG/RD	Speed Control Vent Solenoid	Vacuum Decreasing: 1v
6-10	---	Not Used	---
11	YL/RD	Speed Control Power Supply	12-14v
12	DG/OR	ASD Relay Output	12-14v
13-14	---	Not Used	---
15	PK/YL	Battery Temperature Sensor	At 86°F: 1.96v
16	PK/BK	Generator Lamp Control	Lamp Off: 12v, On: 1v
17	BK/PK	MIL (lamp) Control	MIL Off: 12v, On: 1v
18	---	Not Used	---
19	BR	Fuel Pump Relay Control	Relay Off: 12v, On: 1v
20	PK/BK	EVAP Purge Solenoid Control	PWM Signal: 0-12-0v
21	---	Not Used	---
22	DB/OR	A/C Request Signal	A/C Off: 12v, On: 1v
23	LG	A/C Select Signal	A/C Off: 12v, On: 1v
24	WT/PK	Brake Switch Sense Signal	Brake Off: 0v, On: 12v
25	---	Not Used	---
26	DB	Fuel Level Sensor Signal	Full: 0.5v, 1/2 full: 2.5v
27	PK	SCI Transmit	0v
28	WT/BK	CCD Bus (-)	<0.050v
29	LG	SCI Receive	5v
30	PK/BR	CCD Bus (+)	Digital Signals: 0-5-0v
31	GY/LB	Tachometer Signal	Pulse Signals
32	WT/LG	Speed Control Switch Signal	S/C & Set Switch On: 3.8v

Pin Connector Graphic

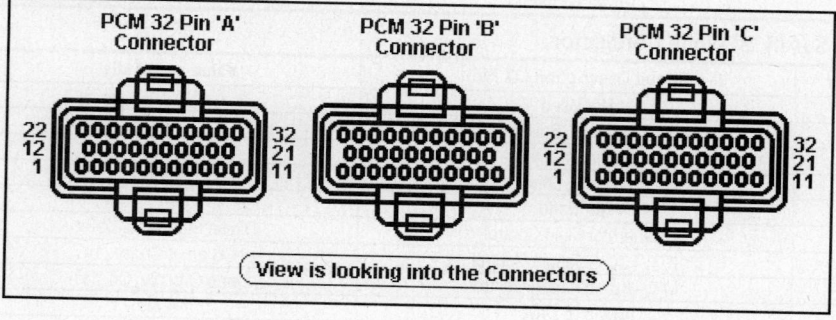

PCM 32 Pin 'A' Connector PCM 32 Pin 'B' Connector PCM 32 Pin 'C' Connector

View is looking into the Connectors

1997 Cherokee 4.0L V6 MFI VIN S (All) 'A' Black Connector

PCM Pin #	Wire Color	Circuit Description (32 Pin)	Value at Hot Idle
1, 3, 5	---	Not Used	---
2	DB/WT	Ignition Switch Power (B+)	12-14v
4	BR/YL	A/T: Analog Sensor Ground	<0.1v
6	BK/WT	PNP Switch Sense Signal	In P/N: 0v, Others: 5v
6	BK	M/T: Analog Sensor Ground	<0.1v
7	GY	Coil 1 Driver Control	5°, 55 mph: 8° dwell
8	GY/BK	CKP Sensor Signal	Digital Signals: 0-5-0v
9	---	Not Used	---
10	YL/BK	IAC 2 Driver Control	Pulse Signals
11	BR/WT	IAC 3 Driver Control	Pulse Signals
12	GY	Extended Idle Switch (Police)	Switch Off: 12v, On: 1v
13-14	---	Not Used	---
15	BK/RD	IAT Sensor Signal	At 100°F: 1.83v
16	TN/BK	ECT Sensor Signal	At 180°F: 2.80v
17	OR	5-Volt Supply	4.9-5.1v
18	TN/YL	CMP Sensor Signal	Digital Signals: 0-5-0v
19	GY/RD	IAC 1 Driver Control	Pulse Signals
20	PK/BK	IAC 4 Driver Control	Pulse Signals
21	---	Not Used	---
22	DG/BK	Fused Battery Power (B+)	12-14v
23	OR/DB	TP Sensor Signal	0.6-1.0v
24	BK/DG	HO2S-11 (B1 S1) Signal	0.1-1.1v
25	TN/BK	HO2S-12 (B1 S2) Signal	0.1-1.1v
26	---	Not Used	---
27	DG/RD	MAP Sensor Signal	1.5-1.6v
28-30	---	Not Used	---
31	BK/TN	Power Ground	<0.1v
32	BK/TN	Power Ground	<0.1v

1997 Cherokee 4.0L V6 MFI VIN S (All) 'B' White Connector

PCM Pin #	Wire Color	Circuit Description (32 Pin)	Value at Hot Idle
1-3	---	Not Used	---
4	WT/DB	Injector 1 Driver	1-4 ms
5	YL/WT	Injector 3 Driver	1-4 ms
6	PK/BK	Injector 5 Driver	1-4 ms
7-9	---	Not Used	---
10	DG	Generator Field Control	Digital Signals: 0-12-0v
11-14	---	Not Used	---
15	TN	Injector 2 Driver	1-4 ms
16	LB/BR	Injector 4 Driver	1-4 ms
17-22	---	Not Used	---
23	GY/YL	Engine Oil Pressure Sensor	1.6v at 24 psi
24-26	---	Not Used	---
27	WT/OR	Vehicle Speed Signal	Digital Signal
28-30	---	Not Used	---
31	PK/OR	5-Volt Supply	4.9-5.1v
32	---	Not Used	---

Standard Colors and Abbreviations

Abbreviation	Color	Abbreviation	Color	Abbreviation	Color
BK	Black	GY	Gray	RD	Red
BL	Blue	GN	Green	TN	Tan
BR	Brown	LG	Light Green	VT	Violet
DB	Dark Blue	OR	Orange	WT	White
DG	Dark Green	PK	Pink	YL	Yellow

1997 Cherokee 4.0L V6 MFI VIN S (All) 'C' Gray Connector

PCM Pin #	Wire Color	Circuit Description (32 Pin)	Value at Hot Idle
1	DB/OR	A/C Clutch Relay Control	Relay Off: 12v, On: 1v
2	---	Not Used	---
3	DB/YL	ASD Relay Control	Relay Off: 12v, On: 1v
4	TN/RD	Speed Control Vacuum Solenoid	Vacuum Increasing: 1v
5	LG/RD	Speed Control Vent Solenoid	Vacuum Decreasing: 1v
6-10	---	Not Used	---
11	YL/RD	Speed Control Power Supply	12-14v
12	DG/OR	ASD Relay Output	12-14v
13-14	---	Not Used	---
15	PK/YL	Battery Temperature Sensor	At 86°F: 1.96v
16-18	---	Not Used	---
19	BR	Fuel Pump Relay Control	Relay Off: 12v, On: 1v
20	PK/BK	EVAP Purge Solenoid Control	PWM Signal: 0-12-0v
21	---	Not Used	---
22	DB/WT	A/C Request Signal	A/C Off: 12v, On: 1v
23	LG	A/C Select Signal	A/C Off: 12v, On: 1v
24	WT/PK	Brake Switch Sense Signal	Brake Off: 0v, On: 12v
25	---	Not Used	---
26	DB/LG	Fuel Level Sensor Signal	Full: 0.5v, 1/2 full: 2.5v
27	PK	SCI Transmit	0v
28	WT/BK	CCD Bus (-)	<0.050v
29	LG	SCI Receive	5v
30	PK/BR	CCD Bus (+)	Digital Signals: 0-5-0v
31	---	Not Used	---
32	RD/LG	Speed Control Switch Signal	S/C & Set Switch On: 3.8v

Pin Connector Graphic

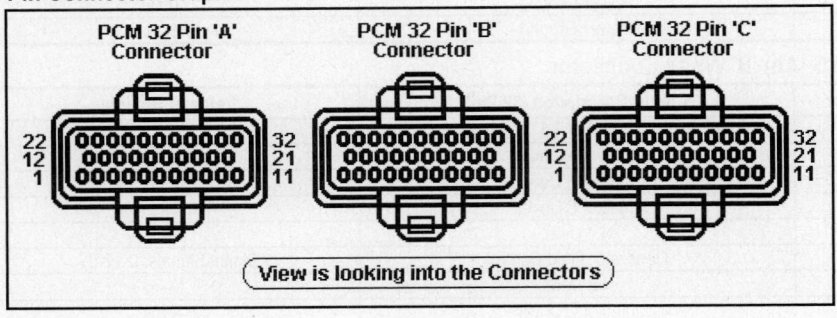

1998-99 Cherokee 4.0L V6 VIN S (All) 'A' Black Connector

PCM Pin #	Wire Color	Circuit Description (32 Pin)	Value at Hot Idle
1	---	Not Used	---
2	DB/WT	Ignition Switch Power (B+)	12-14v
3	---	Not Used	---
4	BR/YL	A/T: Analog Sensor Ground	<0.050v
5	---	Not Used	---
6	BK/WT	PNP Switch Sense Signal	In P/N: 0v, Others: 5v
6	BK	M/T: Analog Sensor Ground	<0.050v
7	GY	Coil 1 Driver Control	5°, 55 mph: 8° dwell
8	GY/BK	CKP Sensor Signal	Digital Signals: 0-5-0v
9	---	Not Used	---
10	YL/BK	IAC 2 Driver Control	Pulse Signals
11	BR/WT	IAC 3 Driver Control	Pulse Signals
12	GY	Extended Idle Switch (Police)	Switch Off: 12v, On: 1v
13-14	---	Not Used	---
15	BK/RD	IAT Sensor Signal	At 100°F: 1.83v
16	TN/BK	ECT Sensor Signal	At 180°F: 2.80v
17	OR	5-Volt Supply	4.9-5.1v
18	TN/YL	CMP Sensor Signal	Digital Signals: 0-5-0v
19	GY/RD	IAC 1 Driver Control	Pulse Signals
20	PK/BK	IAC 4 Driver Control	Pulse Signals
21	---	Not Used	---
22	DG/BK	Fused Battery Power (B+)	12-14v
23	OR/DB	TP Sensor Signal	0.6-1.0v
24	BK/DG	HO2S-11 (B1 S1) Signal	0.1-1.1v
25	TN/WT	HO2S-12 (B1 S2) Signal	0.1-1.1v
27	DG/RD	MAP Sensor Signal	1.5-1.6v
28-30	---	Not Used	---
31	BK/TN	Power Ground	<0.1v
32	BK/TN	Power Ground	<0.1v

1998-99 Cherokee 4.0L V6 VIN S (All) 'B' White Connector

PCM Pin #	Wire Color	Circuit Description (32 Pin)	Value at Hot Idle
1-3	---	Not Used	---
4	WT/DB	Injector 1 Driver	1-4 ms
5	YL/WT	Injector 3 Driver	1-4 ms
6	PK/BK	Injector 5 Driver	1-4 ms
7-9	---	Not Used	---
10	DG	Generator Field Control	Digital Signals: 0-12-0v
11 ('98)	OR/BK	TCC Solenoid Control	TCC on at Cruise: 1v
12	LG/BK	Injector 6 Driver	1-4 ms
13-14	---	Not Used	---
15	TN	Injector 2 Driver	1-4 ms
16	LB/BR	Injector 4 Driver	1-4 ms
17-22	---	Not Used	---
23	GY/YL	Engine Oil Pressure Sensor	1.6v at 24 psi
24-26	---	Not Used	---
27	WT/OR	Vehicle Speed Signal	Digital Signal
28-30	---	Not Used	---
31	PK/OR	5-Volt Supply	4.9-5.1v
32	---	Not Used	---

Standard Colors and Abbreviations

Abbreviation	Color	Abbreviation	Color	Abbreviation	Color
BK	Black	GY	Gray	RD	Red
BL	Blue	GN	Green	TN	Tan
BR	Brown	LG	Light Green	VT	Violet
DB	Dark Blue	OR	Orange	WT	White
DG	Dark Green	PK	Pink	YL	Yellow

1998-99 Cherokee 4.0L V6 VIN S (All) 'C' Gray Connector

PCM Pin #	Wire Color	Circuit Description (32 Pin)	Value at Hot Idle
1	DB/OR	A/C Clutch Relay Control	Relay Off: 12v, On: 1v
2	DB/PK	Radiator Fan Control Relay	Relay Off: 12v, On: 1v
3	DB/YL	ASD Relay Control	Relay Off: 12v, On: 1v
4	TN/RD	Speed Control Vacuum Solenoid	Vacuum Increasing: 1v
5	LG/RD	Speed Control Vent Solenoid	Vacuum Decreasing: 1v
6-9	---	Not Used	---
10	WT/DG	LDP Solenoid Control	PWM Signal: 0-12-0v
11	YL/RD	Speed Control Power Supply	12-14v
12	DG/OR	ASD Relay Output	12-14v
13	---	Not Used	
14 ('98)	WT/OR	LDP Switch Sense Signal	Closed: 0v, Open: 12v
14	OR	Battery Temperature Sensor	At 86°F: 1.96v
15 ('98)	PK/YL	Battery Temperature Sensor	At 86°F: 1.96v
15	PK/YL	LDP Switch Sense Signal	Closed: 0v, Open: 12v
16-18	---	Not Used	---
19	BR	Fuel Pump Relay Control	Relay Off: 12v, On: 1v
20	PK/BK	EVAP Purge Solenoid Control	PWM Signal: 0-12-0v
21	---	Not Used	---
22	DB/WT	A/C Request Signal	A/C Off: 12v, On: 1v
23	LG	A/C Select Signal	A/C Off: 12v, On: 1v
24	WT/PK	Brake Switch Sense Signal	Brake Off: 0v, On: 12v
25	DG/OR	Generator Field Source	12-14v
26	DB/LG	Fuel Level Sensor Signal	Full: 0.5v, 1/2 full: 2.5v
27	PK	SCI Transmit	0v
28	WT/BK	CCD Bus (-)	<0.050v
29	LG/BK	SCI Receive	5v
30	PK/BR	CCD Bus (+)	Digital Signals: 0-5-0v
31	---	Not Used	---
32	RD/LG	Speed Control Switch Signal	S/C & Set Switch On: 3.8v

Pin Connector Graphic

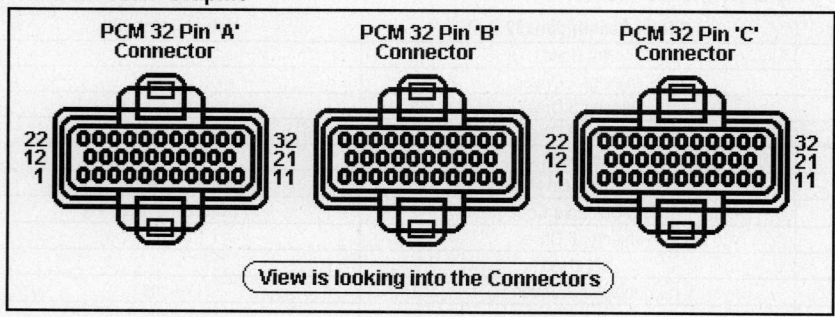

2000-01 Cherokee 4.0L V6 VIN S (All) 'A' Black Connector

PCM Pin #	Wire Color	Circuit Description (32 Pin)	Value at Hot Idle
1	RD/YL	Ignition Coil 3 Driver	5°, 55 mph: 8° dwell
2	DB/WT	Ignition Switch Power (B+)	12-14v
3	---	Not Used	---
4	BR/WT	Sensor Ground	<0.050v
5	---	Not Used	---
6	BK/WT	PNP Switch Sense Signal	In P/N: 0v, Others: 5v
6	GY	Ignition Coil 1 Driver	5°, 55 mph: 8° dwell
8	GY/BK	CKP Sensor Signal	Digital Signals: 0-5-0v
9	---	Not Used	---
10	YL/BK	IAC 2 Driver Control	Pulse Signals
11	BR/WT	IAC 3 Driver Control	Pulse Signals
12	GY	Extended Idle Switch (California Police)	Switch Off: 12v, On: 1v
13-14	---	Not Used	---
15	BK/RD	IAT Sensor Signal	At 100°F: 1.83v
16	TN/BK	ECT Sensor Signal	At 180°F: 2.80v
17	OR	5-Volt Supply	4.9-5.1v
18	TN/YL	CMP Sensor Signal	Digital Signals: 0-5-0v
19	GY/RD	IAC 1 Driver Control	Pulse Signals
20	VT/BK	IAC 4 Driver Control	Pulse Signals
21	---	Not Used	---
22	DG/BK	Fused Battery Power (B+)	12-14v
23	OR/DB	TP Sensor Signal	0.6-1.0v
24	BK/DG	HO2S-11 (B1 S1) Signal	0.1-1.1v
25	TN/WT	HO2S-12 (B1 S2) Signal	0.1-1.1v
26	LG/RD	HO2S-21 (B2 S1) Signal	0.1-1.1v
27	DG/RD	MAP Sensor Signal	1.5-1.6v
28-30	---	Not Used	---
31	BK/TN	Power Ground	<0.1v
32	BK/TN	Power Ground	<0.1v

2000-01 Cherokee 4.0L V6 VIN S (All) 'B' White Connector

PCM Pin #	Wire Color	Circuit Description (32 Pin)	Value at Hot Idle
1-3	---	Not Used	---
4	WT/DB	Injector 1 Driver Control	1-4 ms
5	YL/WT	Injector 3 Driver Control	1-4 ms
6	PK/BK	Injector 5 Driver Control	1-4 ms
7-8	---	Not Used	---
9	DB/TN	Ignition Coil 2 Driver	5°, 55 mph: 8° dwell
10	DG	Generator Field Control	Digital Signals: 0-12-0v
11	---	Not Used	---
12	LG/BK	Injector 6 Driver Control	1-4 ms
13-14	---	Not Used	---
15	T	Injector 2 Driver Control	1-4 ms
16	LB/BR	Injector 4 Driver Control	1-4 ms
17-22	---	Not Used	---
23	GY/YL	Engine Oil Pressure Sensor	1.6v at 24 psi
24-26	---	Not Used	---
27	WT/OR	Vehicle Speed Signal	Digital Signal
28-30	---	Not Used	---
31	VT/OR	5-Volt Supply	4.9-5.1v
32	---	Not Used	---

Standard Colors and Abbreviations

Abbreviation	Color	Abbreviation	Color	Abbreviation	Color
BK	Black	GY	Gray	RD	Red
BL	Blue	GN	Green	TN	Tan
BR	Brown	LG	Light Green	VT	Violet
DB	Dark Blue	OR	Orange	WT	White
DG	Dark Green	PK	Pink	YL	Yellow

2000-01 Cherokee 4.0L V6 VIN S (All) 'C' Gray Connector

PCM Pin #	Wire Color	Circuit Description (32 Pin)	Value at Hot Idle
1	DB/OR	A/C Clutch Relay Control	Relay Off: 12v, On: 1v
2	DB/PK	Radiator Fan Control Relay	Relay Off: 12v, On: 1v
3	DB/YL	ASD Relay Control	Relay Off: 12v, On: 1v
4	TN/RD	Speed Control Vacuum Solenoid	Vacuum Increasing: 1v
5	LG/RD	Speed Control Vent Solenoid	Vacuum Decreasing: 1v
6-7	---	Not Used	---
8	DB/OR	HO2S-11 Heater Relay	Relay Off: 12v, On: 1v
9	BR/VT	HO2S-12 Heater Relay	Relay Off: 12v, On: 1v
10	WT/DG	LDP Solenoid Control	PWM Signal: 0-12-0v
11	YL/RD	Speed Control Power Supply	12-14v
12	DG/WT	ASD Relay Output	12-14v
13	---	Not Used	---
14	WT/OR	LDP Switch Sense Signal	Closed: 0v, Open: 12v
15	PK/YL	Battery Temperature Sensor	At 86°F: 1.96v
16-18	---	Not Used	---
19	BR	Fuel Pump Relay Control	Relay Off: 12v, On: 1v
20	PK/BK	EVAP Purge Solenoid Control	PWM Signal: 0-12-0v
21	---	Not Used	---
22	DB/WT	A/C Request Signal	A/C Off: 12v, On: 1v
23	LG	A/C Select Signal	A/C Off: 12v, On: 1v
24	WT/PK	Brake Switch Sense Signal	Brake Off: 0v, On: 12v
25	DG/OR	Generator Field Source	12-14v
26	DB/LG	Fuel Level Sensor Signal	Full: 0.5v, 1/2 full: 2.5v
27	PK	SCI Transmit	0v
28	WT/BK	CCD Bus (-)	<0.050v
29	LG/BK	SCI Receive	5v
30	VT/BR	CCD Bus (+)	Digital Signals: 0-5-0v
31	---	Not Used	---
32	RD/LG	Speed Control Switch Signal	S/C & Set Switch On: 3.8v

Pin Connector Graphic

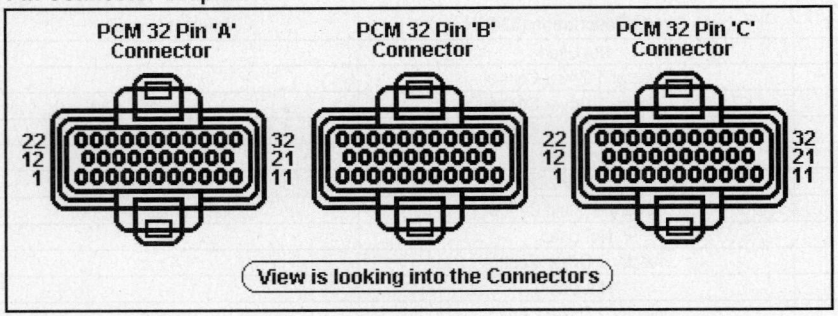

1990 Comanche 2.5L I4 TBI VIN E (All) 35 Pin Connector

PCM Pin #	Wire Color	Circuit Description (35 Pin)	Value at Hot Idle
1	BK	Power Ground	<0.1v
2	BK	Power Ground	<0.1v
3	YL	Ignition Switch Run Output	12-14v
4	RD	Fused Battery Power (B+)	12-14v
5	GN	EGR Solenoid Control	12v, at 55 mph: 1v
6	OR	Fuel Pump Relay Control	Relay Off: 12v, On: 1v
7	BK	Latch Relay B+ Control	Relay Off: 12v, On: 1v
8	GY/BK	PSP Switch Sense Signal	Straight: 0v, Turning: 5v
8	BL/OR	PSP Switch Sense Signal	Straight: 0v, Turning: 5v
10	BK	Sensor Ground	<0.1v
11	RD	CKP Sensor (+) Signal	Digital Signals: 0-5-0v
12	BK	PNP Switch Sense Signal	In P/N: 0v, Others: 5v
13	BR	TP Sensor Ground	<0.1v
14	TN	Air Temperature Sensor	At 100°F: 2.51v
15	TN	ECT Sensor Signal	At 180°F: 2.80v
16	RD	MAP Sensor VREF	4.9-5.1v
17	BK	Map Sensor Ground	<0.1v
18	PK	Shift Indicator Lamp Control	Lamp Off: 12v, On: 1v
19	PK	B+ Latch Relay Output	Relay Off: 12v, On: 1v
21	LB	Injector Control Driver	1-4 ms
22	BL	A/C Clutch Relay Control	A/C Off: 12v, On: 1v
23	BR	Idle Speed Motor (-) Retract	Pulse Signals
24	LG	Idle Speed Motor (+) Extend	Pulse Signals
25	GY	Closed Throttle Sensor	Open: 12v, Closed: 1v
27	OR	Ignition Coil Control	5°, 55 mph: 8° dwell
28	WT	CKP Sensor (-) Signal	Digital Signals: 0-5-0v
29	GN	Starter Input Signal	Cranking: 9-11v
30	LG	A/C Select Switch Sense	A/C Off: 12v, On: 1v
31	YL/GN	TP Sensor Signal	0.6-1.0v
32	BR	ECT Sensor Ground	<0.1v
33	PK	MAP Sensor Signal	1.5-1.6v
34	TN	A/C Request Switch Sense	A/C Off: 12v, On: 1v
35	GY	HO2S-11 (B1 S1) Signal	0.1-1.1v

Pin Connector Graphic

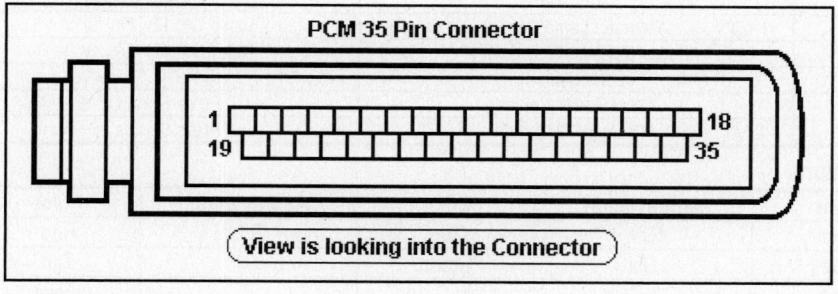

PCM 35 Pin Connector

1   18
19   35

View is looking into the Connector

1991-92 Comanche 2.5L I4 MFI VIN P (All) 60 Pin Connector

PCM Pin #	Wire Color	Circuit Description (60 Pin)	Value at Hot Idle
1	DG/RD	MAP Sensor Signal	1.5-1.6v
2	TN/BK	ECT Sensor Signal	At 180°F: 2.80v
3	RD	Fused Battery Power (B+)	12-14v
4	BK/LB	Sensor Ground	<0.1v
5	BK/WT	Sensor Ground	<0.1v
6	PK/WT	5-Volt Supply	4.9-5.1v
7	OR	8-Volt Supply	7.9-8.1v
8 ('91)	BR	Starter Input Signal	Cranking: 9-11v
9	DB	Ignition Switch Power (B+)	12-14v
10	DB/OR	PSP Switch Sense Signal	Straight: 0v, Turning: 5v
10 ('92)	DB/WT	PSP Switch Sense Signal	Straight: 0v, Turning: 5v
11	BK/TN	Power Ground	<0.1v
12	BK/TN	Power Ground	<0.1v
13	LB/BR	Injector 4 Driver	1-4 ms
14	YL/WT	Injector 3 Driver	1-4 ms
15	TN	Injector 2 Driver	1-4 ms
16	WT/DB	Injector 1 Driver	1-4 ms
17-18	---	Not Used	---
19	GY	Coil 1 Driver Control	5°, 55 mph: 8° dwell
20	DG	Alternator Field Control	Digital Signals: 0-12-0v
21	BK/RD	Air Temperature Sensor	At 100°F: 2.51v
22	OR/DB	TP Sensor Signal	0.6-1.0v
23	---	Not Used	---
24	GY/BK	Distributor Reference Pickup	Digital Signals: 0-5-0v
25	PK	SCI Transmit	0v
26	---	Not Used	---
27	LB	A/C Request Switch Sense	A/C Off: 12v, On: 1v
28	LG	A/C Select Switch Sense	A/C Off: 12v, On: 1v
29	WT/PK	Brake Switch Sense Signal	Brake Off: 0v, On: 12v
30	---	Not Used	---
31 ('92)	DB/PK	Radiator Fan Control Relay	Relay Off: 12v, On: 1v
32	BK/PK	MIL (lamp) Control	MIL Off: 12v, On: 1v
33	TN/RD	Speed Control Vacuum Solenoid	Vacuum Increasing: 1v
34	DB/OR	A/C Clutch Relay Control	Relay Off: 12v, On: 1v
35	---	Not Used	---
36	DG/YL	Alternator Lamp Control	Lamp Off: 12v, On: 1v
37	RD/DB	Ballast Bypass Relay Control	Relay Off: 12v, On: 1v
38	---	Not Used	---
39	GY/RD	AIS Motor 'D' Control	Pulse Signals
40	BR/WT	AIS Motor 'B' Control	Pulse Signals

Standard Colors and Abbreviations

Abbreviation	Color	Abbreviation	Color	Abbreviation	Color
BK	Black	GY	Gray	RD	Red
BL	Blue	GN	Green	TN	Tan
BR	Brown	LG	Light Green	VT	Violet
DB	Dark Blue	OR	Orange	WT	White
DG	Dark Green	PK	Pink	YL	Yellow

1991-92 Comanche 2.5L I4 MFI VIN P (All) 60 Pin Connector

PCM Pin #	Wire Color	Circuit Description (60 Pin)	Value at Hot Idle
41	BK/DG	HO2S-11 (B1 S1) Signal	0.1-1.1v
42	---	Not Used	---
43	GY/LB	Tachometer Signal	Pulse Signals
44	TN/YL	Distributor Sync Pickup	Digital Signals: 0-5-0v
45	LG	SCI Receive	5v
46	---	Not Used	---
47	WT/OR	Vehicle Speed Signal	Digital Signal
48	BR/RD	Speed Control Set Switch	S/C & Set Switch On: 3.8v
49	YL/RD	Speed Control On/Off Switch	S/C Off: 12v, On: 1v
50	WT/LG	Speed Control Resume Switch	S/C Off: 12v, On: 1v
51	DB/YL	ASD Relay Control	Relay Off: 12v, On: 1v
52	---	Not Used	---
53	LG/RD	Speed Control Vent Solenoid	Vacuum Decreasing: 1v
54	OR/BK	Shift Indicator Lamp Control	Lamp Off: 12v, On: 1v
55	---	Not Used	---
56	GY/PK	Maintenance Indicator Lamp	Lamp Off: 12v, On: 1v
57	DG/OR	ASD Relay Output	12-14v
58	---	Not Used	---
59	PK/BK	AIS Motor 'D' Control	Pulse Signals
60	YL/BK	AIS Motor 'C' Control	Pulse Signals

Pin Connector Graphic

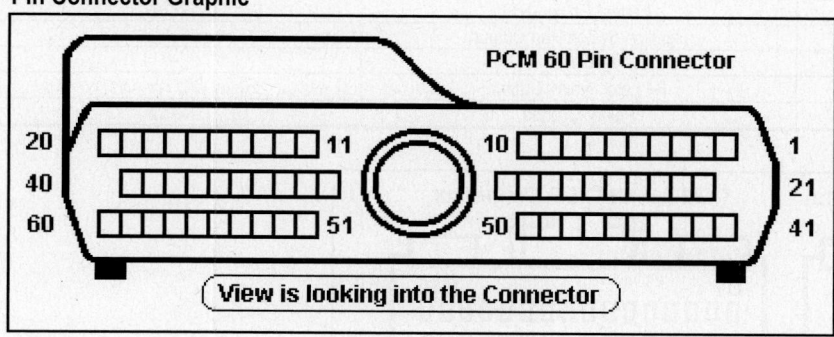

1990 Comanche 4.0L V6 MFI VIN L Row 'A' 24 Pin Connector

PCM Pin #	Wire Color	Circuit Description (24 Pin)	Value at Hot Idle
1	TN	Injector 3 Control Driver	1-4 ms
2	BR	Injector 6 Control Driver	1-4 ms
3	LG	Injector 2 Control Driver	1-4 ms
4	YL	Injector 4 Control Driver	1-4 ms
5	OR	Fuel Pump Relay Control	Relay Off: 12v, On: 1v
6	---	Not Used	---
7	GN/YL	HO2S Relay Control	Relay Off: 12v, On: 1v
8	PK	M/T: SIL (lamp) Control	Lamp Off: 12v, On: 1v
9	BK	Latch Relay B+ Control	Relay Off: 12v, On: 1v
10	GN	EGR Solenoid Control	12v, at 55 mph: 1v
11	---	Not Used	---
12	BL	A/C Clutch Relay Control	Relay Off: 12v, On: 1v

1990 Comanche 4.0L V6 MFI VIN L Row 'B' 24 Pin Connector

PCM Pin #	Wire Color	Circuit Description (24 Pin)	Value at Hot Idle
1	LB	Injector 1 Control Driver	1-4 ms
2	WT	Injector 5 Control Driver	1-4 ms
3	RD/YL	Idle Speed Motor 'A' Control	Pulse Signals
4	BL/YL	Idle Speed Motor 'D' Control	Pulse Signals
5	GN/BK	Idle Speed Motor 'C' Control	Pulse Signals
6	PK/BK	Idle Speed Motor 'B' Control	Pulse Signals
7	RD	Battery Power (B+)	12-14v
8	YL	Ignition Switch Run Output	12-14v
9	---	Not Used	---
10	PK	B+ Latch Relay Output	Relay Off: 12v, On: 1v
11-12	BK	Power Ground	<0.1v

Pin Connector Graphic

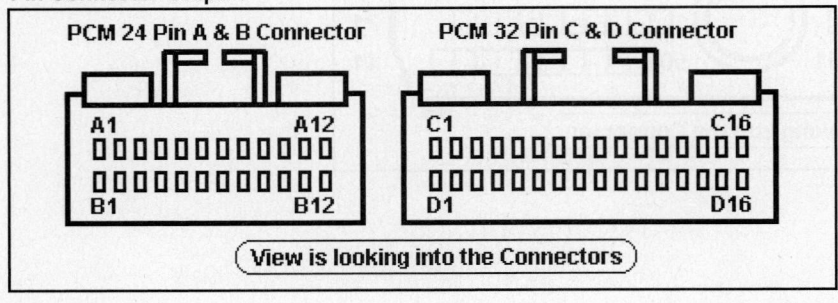

1990 Comanche 4.0L V6 MFI VIN L Row 'C' 32 Pin Connector

PCM Pin #	Wire Color	Circuit Description (32 Pin)	Value at Hot Idle
1	PK	CKP Sensor (+) Signal	Digital Signals: 0-5-0v
2	TN	A/C Request Switch Sense	A/C Off: 12v, On: 1v
3	GN	Starter Input Signal	Cranking: 9-11v
4	BK	PNP Switch Sense Signal	In P/N: 0v, Others: 5v
5	GY	Distributor Sync Pickup (+)	Digital Signals: 0-5-0v
6	PK	MAP Sensor Signal	1.5-1.6v
7	YL/GN	TP Sensor Signal	0.6-1.0v
8	TN	Air Temperature Sensor	At 100ºF: 2.51v
9	---	Not Used	---
10	TN	ECT Sensor Signal	At 180ºF: 2.80v
11	OO	Fuel Pump Relay Output	Relay Off: 12v, On: 1v
12	BR/PK	SCI Transmit	0v
13	---	Not Used	---
14	RD	MAP Sensor VREF	4.9-5.1v
15	LB	TP Sensor VREF	4.9-5.1v
16	BL	Distributor Sync Pickup (-)	Digital Signals: 0-5-0v

1990 Comanche 4.0L V6 MFI VIN L Row 'D' 32 Pin Connector

PCM Pin #	Wire Color	Circuit Description (32 Pin)	Value at Hot Idle
1	WT	CKP Sensor (-) Signal	Digital Signals: 0-5-0v
2	LG	A/C Select Switch Sense	A/C Off: 12v, On: 1v
3	BR	Sensor Ground	<0.1v
4-7	---	Not Used	---
8	YL	Knock Sensor (-) Signal	<0.050v
9	GY	HO2S-11 (B1 S1) Signal	0.1-1.1v
10	OR	Fuel Pump Relay Control	Relay Off: 12v, On: 1v
11	BK/WT	SCI Receive	5v
12	---	Not Used	---
13	YL	Ignition Coil Control	5º, 55 mph: 8º dwell
14-15	---	Not Used	---
16	PK	Knock Sensor Signal	0.080v AC

Pin Connector Graphic

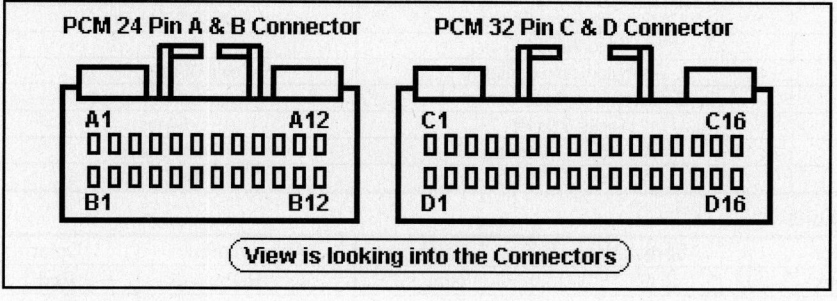

1991-92 Comanche 4.0L V6 MFI VIN S (All) 60 Pin Connector

PCM Pin #	Wire Color	Circuit Description (60 Pin)	Value at Hot Idle
1	DG/RD	MAP Sensor Signal	1.5-1.6v
2	TN/BK	ECT Sensor Signal	At 180°F: 2.80v
3	RD	Fused Battery Power (B+)	12-14v
4	BK/LB	Sensor Ground	<0.1v
5	BK	Sensor Ground	<0.1v
5 ('92)	BK/WT	Sensor Ground	<0.1v
6	PK	5-Volt Supply	4.9-5.1v
7	OR	8-Volt Supply	7.9-8.1v
8 ('91)	BR	Starter Input Signal	Cranking: 9-11v
9	DB	Ignition Switch Power (B+)	12-14v
10	---	Not Used	---
11	BK/TN	Power Ground	<0.1v
12	BK/TN	Power Ground	<0.1v
13	LB/BR	Injector 4 Driver	1-4 ms
14	YL/WT	Injector 3 Driver	1-4 ms
15	TN	Injector 2 Driver	1-4 ms
16	WT/DB	Injector 1 Driver	1-4 ms
17-18	---	Not Used	---
19	GY	Coil 1 Driver Control	5°, 55 mph: 8° dwell
20	DG	Alternator Field Control	Digital Signals: 0-12-0v
21	BK/RD	Air Temperature Sensor	At 100°F: 2.51v
22	OR/DB	TP Sensor Signal	0.6-1.0v
23	---	Not Used	---
24	GY/BK	Distributor Reference Pickup	Digital Signals: 0-5-0v
25	PK	SCI Transmit	0v
26	PK/BR	CCD Bus (+)	Digital Signals: 0-5-0v
27	LB	A/C Request Switch Sense	A/C Off: 12v, On: 1v
28	LG	A/C Select Switch Sense	A/C Off: 12v, On: 1v
29	WT/PK	Brake Switch Sense Signal	Brake Off: 0v, On: 12v
30	BR/YL	PNP Switch Sense Signal	In P/N: 0v, Others: 5v
31	DB/PK	Radiator Fan Control Relay	Relay Off: 12v, On: 1v
32	BK/PK	MIL (lamp) Control	MIL Off: 12v, On: 1v
33	TN/RD	Speed Control Vacuum Solenoid	Vacuum Increasing: 1v
34	DB/OR	A/C Clutch Relay Control	Relay Off: 12v, On: 1v
35	---	Not Used	---
36	DG/YL	Alternator Lamp Control	Lamp Off: 12v, On: 1v
37	RD/DB	Ballast Bypass Relay Control	Relay Off: 12v, On: 1v
38	PK/BK	Injector 5 Driver	1-4 ms
39	GY/RD	AIS Motor 'D' Control	Pulse Signals
40	BR/WT	AIS Motor 'B' Control	Pulse Signals

Standard Colors and Abbreviations

Abbreviation	Color	Abbreviation	Color	Abbreviation	Color
BK	Black	GY	Gray	RD	Red
BL	Blue	GN	Green	TN	Tan
BR	Brown	LG	Light Green	VT	Violet
DB	Dark Blue	OR	Orange	WT	White
DG	Dark Green	PK	Pink	YL	Yellow

1991-92 Comanche 4.0L V6 MFI VIN S (All) 60 Pin Connector

PCM Pin #	Wire Color	Circuit Description (60 Pin)	Value at Hot Idle
41	BK/DG	HO2S-11 (B1 S1) Signal	0.1-1.1v
42	---	Not Used	---
43	GY/LB	Tachometer Signal	Pulse Signals
44	TN/YL	Distributor Sync Pickup	Digital Signals: 0-5-0v
45	LG	SCI Receive	5v
46	WT	CCD Bus (-)	<0.050v
46 ('92)	WT/BK	CCD Bus (-)	<0.050v
47	WT/OR	Vehicle Speed Signal	Digital Signal
48	BR/RD	Speed Control Set Switch	S/C & Set Switch On: 3.8v
49	YL/RD	Speed Control On/Off Switch	S/C Off: 12v, On: 1v
50	WT/LG	Speed Control Resume Switch	S/C Off: 12v, On: 1v
51	DB/YL	ASD Relay Control	Relay Off: 12v, On: 1v
52	---	Not Used	---
53	LG/RD	Speed Control Vent Solenoid	Vacuum Decreasing: 1v
54	OR/BK	A/T: Lockup Torque Converter	TCC on at Cruise: 1v
54	OR/BK	M/T: SIL (lamp) Control	Lamp Off: 12v, On: 1v
55	---	Not Used	---
56	GY/PK	Maintenance Indicator Lamp	Lamp Off: 12v, On: 1v
57	DG/OR	ASD Relay Output	12-14v
58	LG/BK	Injector 6 Driver	1-4 ms
59	PK/BK	AIS Motor 'D' Control	Pulse Signals
60 ('91)	Y	AIS Motor 'C' Control	Pulse Signals
60	YL/BK	AIS Motor 'C' Control	Pulse Signals

Pin Connector Graphic

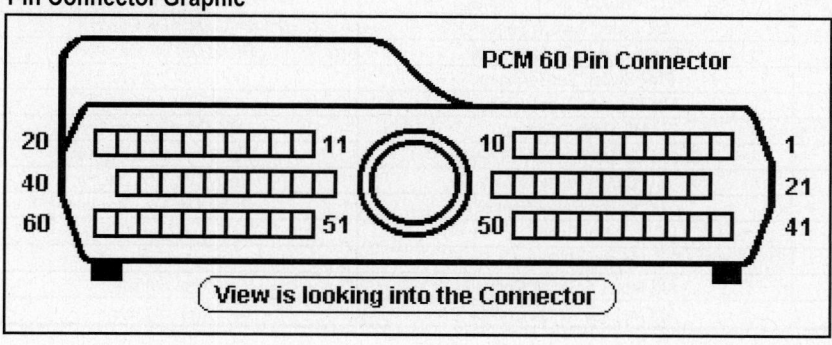

1993-95 Grand Cherokee 4.0L V6 VIN S (All) 60 Pin Connector

PCM Pin #	Wire Color	Circuit Description (60 Pin)	Value at Hot Idle
1	RD/WT	MAP Sensor Signal	1.5-1.6v
2	TN/BK	ECT Sensor Signal	At 180°F: 2.80v
3	RD	Fused Battery Power (B+)	12-14v
4	BK/LB	Sensor Ground	<0.1v
5	BK/TN	Sensor Ground	<0.1v
6	PK/WT	5-Volt Supply	4.9-5.1v
7	WT/BK	8-Volt Supply	7.9-8.1v
8	---	Not Used	---
9	LB/RD	Ignition Switch Power (B+)	12-14v
10	---	Not Used	---
11	BK/TN	Power Ground	<0.1v
12	BK/TN	Power Ground	<0.1v
13	LB/BR	Injector 4 Driver	1-4 ms
14	YL/WT	Injector 3 Driver	1-4 ms
15	TN	Injector 2 Driver	1-4 ms
16	W/LB	Injector 1 Driver	1-4 ms
17-18	---	Not Used	---
19	GY/WT	Coil 1 Driver Control	5°, 55 mph: 8° dwell
20	DG	Alternator Field Control	Digital Signals: 0-12-0v
21	BK/RD	IAT Sensor Signal	At 100°F: 2.51v
22	OR/DB	TP Sensor Signal	0.6-1.0v
23, 31	---	Not Used	---
24	RD/LG	CKP Sensor Signal	Digital Signals: 0-5-0v
25	BK	SCI Transmit	0v
26	PK/BR	CCD Bus (+)	Digital Signals: 0-5-0v
27	DB/OR	A/C Pressure Switch Signal	A/C Off: 12v, On: 1v
28	LG	A/C Select Switch Sense	A/C Off: 12v, On: 1v
29	BR	Brake Switch Sense Signal	Brake Off: 0v, On: 12v
30	BR/WT	PNP Switch Sense Signal	In P/N: 0v, Others: 5v
30 ('94-'95)	BK/WT	PNP Switch Sense Signal	In P/N: 0v, Others: 5v
32	BK/PK	MIL (lamp) Control	MIL Off: 12v, On: 1v
33	TN/RD	Speed Control Vacuum Solenoid	Vacuum Increasing: 1v
34	DB/OR	A/C Clutch Relay Control	Relay Off: 12v, On: 1v
34 ('94-'95)	DB/RD	A/C Clutch Relay Control	Relay Off: 12v, On: 1v
35-37	---	Not Used	---
38	GY	Injector 5 Driver	1-4 ms
39	YL/BK	AIS Motor 4 Circuit Control	Pulse Signals
40	BR/WT	AIS Motor 2 Circuit Control	Pulse Signals

Standard Colors and Abbreviations

Abbreviation	Color	Abbreviation	Color	Abbreviation	Color
BK	Black	GY	Gray	RD	Red
BL	Blue	GN	Green	TN	Tan
BR	Brown	LG	Light Green	VT	Violet
DB	Dark Blue	OR	Orange	WT	White
DG	Dark Green	PK	Pink	YL	Yellow

1993-95 Grand Cherokee 4.0L V6 VIN S (All) 60 Pin Connector

PCM Pin #	Wire Color	Circuit Description (60 Pin)	Value at Hot Idle
41	BK/OR	HO2S-11 (B1 S1) Signal	0.1-1.1v
42	---	Not Used	---
43	GY/LB	Tachometer Signal	Pulse Signals
44	GY/BK	CMP Sensor Signal	Digital Signals: 0-5-0v
45	BK/YL	SCI Receive	5v
46	WT/GY	CCD Bus (-)	<0.050v
47	WT/OR	Vehicle Speed Signal	Digital Signals
48	BR/RD	Speed Control Set Switch	S/C & Set Switch On: 3.8v
49	YL/RD	Speed Control On/Off Switch	S/C Off: 12v, On: 1v
50	WT/LG	Speed Control Resume Switch	S/C Off: 12v, On: 1v
51	PK	Fuel Pump Relay Control	Relay Off: 12v, On: 1v
52	---	Not Used	---
53	LG/RD	Speed Control Vent Solenoid	Vacuum Decreasing: 1v
54	OR/BK	M/T: SIL (lamp) Control	Lamp Off: 12v, On: 1v
55	---	Not Used	---
56 ('93)	GY/PK	Maintenance Indicator Lamp	Lamp Off: 12v, On: 1v
57	DG/BK	ASD Relay Output	12-14v
58	BR/YL	Injector 6 Driver	1-4 ms
59	GY/RD	AIS Motor 1 Circuit Control	Pulse Signals
60	PK/BK	AIS Motor 3 Circuit Control	Pulse Signals

Pin Connector Graphic

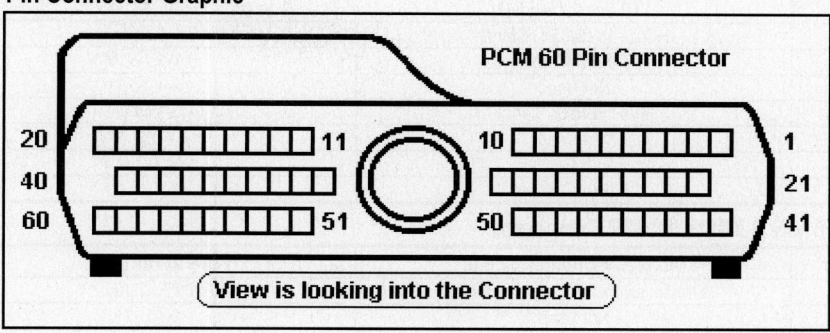

1996-97 Grand Cherokee 4.0L V6 VIN S Black 'A' 32P Connector

PCM Pin #	Wire Color	Circuit Description (32 Pin)	Value at Hot Idle
1	---	Not Used	---
2	OR	Ignition Switch Power (B+)	12-14v
3	---	Not Used	---
4	BK/LB	Sensor Ground	<0.1v
5	---	Not Used	---
6	BK/WT	PNP Switch Sense Signal	In P/N: 0v, Others: 5v
7	GY/WT	Coil 1 Driver Control	5°, 55 mph: 8° dwell
8	RD/LG	CKP Sensor Signal	Digital Signals: 0-5-0v
9	---	Not Used	---
10	PK/BK	IAC 4 Driver Control	Pulse Signals
11	BR/WT	IAC 3 Driver Control	Pulse Signals
13	---	Not Used	---
14 ('96)	YL/RD	Fused Battery Power (B+)	12-14v
15	BK/RD	IAT Sensor Signal	At 100°F: 1.83v
16	TN/BK	ECT Sensor Signal	At 180°F: 2.80v
17	WT/BK	5-Volt Supply	4.9-5.1v
18	GY/BK	CMP Sensor Signal	Digital Signals: 0-5-0v
19	YL/BK	IAC 2 Driver Control	Pulse Signals
20	GY/RD	IAC 1 Driver Control	Pulse Signals
21	---	Not Used	---
22	RD/YL	Fused Battery Power (B+)	12-14v
23	OR/DB	TP Sensor Signal	0.6-1.0v
24	BK/OR	HO2S-11 (B1 S1) Signal	0.1-1.1v
25	BK/PK	HO2S-12 (B1 S2) Signal	0.1-1.1v
26	---	Not Used	---
27	RD/WT	MAP Sensor Signal	1.5-1.6v
28-30	---	Not Used	---
31	BK/TN	Power Ground	<0.1v
32	BK/TN	Power Ground	<0.1v

1996-97 Grand Cherokee 4.0L V6 VIN S White 'B' 32P Connector

PCM Pin #	Wire Color	Circuit Description (32 Pin)	Value at Hot Idle
1	PK	Transmission Temperature Sensor	At 200°F: 2.40v
2-3	---	Not Used	---
4	WT/DB	Injector 1 Driver	1-4 ms
5	YL/WT	Injector 3 Driver	1-4 ms
6	GY	Injector 5 Driver	1-4 ms
7	---	Not Used	---
8	PK	Governor Pressure Solenoid	PWM Signal: 0-12-0v
10	DG	Generator Field Control	Digital Signals: 0-12-0v
11	DG/LB	TCC Solenoid Control	TCC on at Cruise: 1v
12	BR/YL	Injector 6 Driver	1-4 ms
13-14	---	Not Used	---
15	TN	Injector 2 Driver	1-4 ms
16	LB/BR	Injector 4 Driver	1-4 ms
17-20	---	Not Used	---
21	BR	Overdrive Solenoid Control	Solenoid on at Cruise: 1v
22	---	Not Used	---
23	GY/WT	Engine Oil Pressure Sensor	1.6v at 24 psi
24	---	Not Used	---
25	DB	OSS Sensor (-) Signal	AC pulse signals
26	---	Not Used	---
27	WT/OR	Vehicle Speed Signal	Digital Signal
28	LG	OSS Sensor (+) Signal	AC pulse signals
29	LG	Governor Pressure Signal	0.58v
30	BR/OR	Transmission Relay Control	Relay Off: 12v, On: 1v
31	PK/WT	5-Volt Supply	4.9-5.1v

1996-97 Grand Cherokee 4.0L V6 VIN S Gray 'C' 32P Connector

PCM Pin #	Wire Color	Circuit Description (32 Pin)	Value at Hot Idle
1	DB/RD	A/C Clutch Relay Control	Relay Off: 12v, On: 1v
2	---	Not Used	---
3	PK/WT	ASD Relay Control	Relay Off: 12v, On: 1v
4	TN/RD	Speed Control Vacuum Solenoid	Vacuum Increasing: 1v
5	LG/RD	Speed Control Vent Solenoid	Vacuum Decreasing: 1v
6	BR/YL	O/D Indicator (lamp) Control	O/D On: 1v, Off: 12v
7-9	---	Not Used	---
10	DG/RD	LDP Solenoid Control	PWM Signal: 0-12-0v
11	YL/RD	Speed Control Power Supply	12-14v
12	DG/OR	ASD Relay Output	12-14v
13	OR	Overdrive 'Off' Switch Sense	Switch On: 1v, Off: 12v
14	PK/RD	LDP Switch Sense Signal	Closed: 0v, Open: 12v
15	PK/YL	Battery Temperature Sensor	At 86°F: 1.96v
16-18	---	Not Used	---
19	DB	Fuel Pump Relay Control	Relay Off: 12v, On: 1v
20	PK/BK	EVAP Purge Solenoid Control	PWM Signal: 0-12-0v
21	---	Not Used	---
22	DB/BK	A/C Request Signal	A/C Off: 12v, On: 1v
23	---	Not Used	---
24	BR	Brake Switch Sense Signal	Brake Off: 0v, On: 12v
25	---	Not Used	---
26	LB/BK	Fuel Level Sensor Signal	Full: 0.5v, 1/2 full: 2.5v
27	BK/PK	SCI Transmit	0v
28	WT/BK	CCD Bus (-)	<0.050v
29	BK/WT	SCI Receive	5v
30	PK/BR	CCD Bus (+)	Digital Signals: 0-5-0v
31	---	Not Used	---
32	PK	Speed Control Switch Signal	S/C & Set Switch On: 3.8v

Pin Connector Graphic

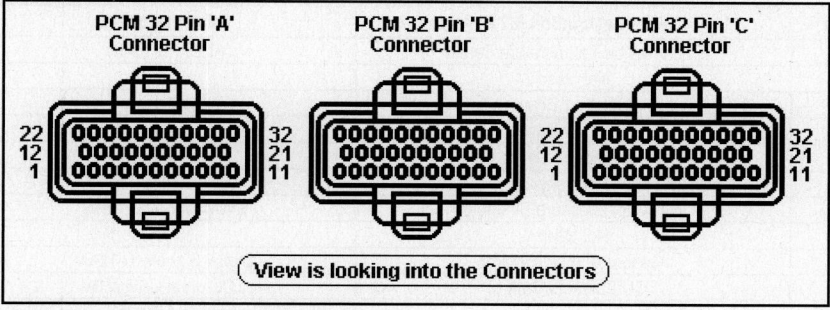

Standard Colors and Abbreviations

Abbreviation	Color	Abbreviation	Color	Abbreviation	Color
BK	Black	GY	Gray	RD	Red
BL	Blue	GN	Green	TN	Tan
BR	Brown	LG	Light Green	VT	Violet
DB	Dark Blue	OR	Orange	WT	White
DG	Dark Green	PK	Pink	YL	Yellow

1998 Grand Cherokee 4.0L V6 VIN S 'A' Black Connector

PCM Pin #	Wire Color	Circuit Description (32 Pin)	Value at Hot Idle
1	---	Not Used	---
2	OR	Ignition Switch Power (B+)	12-14v
3	---	Not Used	---
4	BK/LB	Sensor Ground	<0.1v
5	---	Not Used	---
6	BK/WT	PNP Switch Sense Signal	In P/N: 0v, Others: 5v
7	GY/WT	Coil 1 Driver Control	5°, 55 mph: 8° dwell
8	RD/LG	CKP Sensor Signal	Digital Signals: 0-5-0v
9	---	Not Used	---
10	PK/BK	IAC 4 Driver Control	Pulse Signals
11	BR/WT	IAC 3 Driver Control	Pulse Signals
12-14	---	Not Used	---
15	BK/RD	IAT Sensor Signal	At 100°F: 1.83v
16	TN/BK	ECT Sensor Signal	At 180°F: 2.80v
17	WT/BK	5-Volt Supply	4.9-5.1v
18	GY/BK	CMP Sensor Signal	Digital Signals: 0-5-0v
19	YL/BK	IAC 2 Driver Control	Pulse Signals
20	GY/RD	IAC 1 Driver Control	Pulse Signals
21	---	Not Used	---
22	RD/YL	Fused Battery Power (B+)	12-14v
23	OR/DB	TP Sensor Signal	0.6-1.0v
24	BK/OR	HO2S-11 (B1 S1) Signal	0.1-1.1v
25	BK/PK	HO2S-12 (B1 S2) Signal	0.1-1.1v
26	---	Not Used	---
27	RD/WT	MAP Sensor Signal	1.5-1.6v
28-30	---	Not Used	---
31	BK/TN	Power Ground	<0.1v
32	BK/TN	Power Ground	<0.1v

1998 Grand Cherokee 4.0L V6 VIN S 'B' White Connector

PCM Pin #	Wire Color	Circuit Description (32 Pin)	Value at Hot Idle
1	PK	Transmission Temperature Sensor	At 200°F: 2.40v
2-3	---	Not Used	---
4	WT/DB	Injector 1 Driver	1-4 ms
5	YL/WT	Injector 3 Driver	1-4 ms
6	GY	Injector 5 Driver	1-4 ms
7	---	Not Used	---
8	PK	Governor Pressure Solenoid	PWM Signal: 0-12-0v
9	---	Not Used	---
10	DG	Generator Field Control	Digital Signals: 0-12-0v
11	DG/LB	TCC Solenoid Control	TCC on at Cruise: 1v
12	BR/YL	Injector 6 Driver	1-4 ms
13-14	---	Not Used	---
15	TN	Injector 2 Driver	1-4 ms
16	LB/BR	Injector 4 Driver	1-4 ms
17-20	---	Not Used	---
21	BR	Overdrive Solenoid Control	Solenoid on at Cruise: 1v
22	---	Not Used	---
23	GY/WT	Engine Oil Pressure Sensor	1.6v at 24 psi
24	---	Not Used	---
25	DB/BK	OSS Sensor Signal (-)	AC pulse signals
26	---	Not Used	---
27	WT/OR	Vehicle Speed Signal	Digital Signal
28	LG/WT	OSS Sensor Signal (+)	AC pulse signals
29	LG	Governor Pressure Signal	0.58v
30	BR/OR	Transmission Relay Control	Relay Off: 12v, On: 1v
31	PK/OR	5-Volt Supply	4.9-5.1v
32	---	Not Used	---

1998 Grand Cherokee 4.0L V6 VIN S 'C' Gray Connector

PCM Pin #	Wire Color	Circuit Description (32 Pin)	Value at Hot Idle
1	DB/RD	A/C Clutch Relay Control	Relay Off: 12v, On: 1v
2	---	Not Used	---
3	PK/WT	ASD Relay Control	Relay Off: 12v, On: 1v
4	TN/RD	Speed Control Vacuum Solenoid	Vacuum Increasing: 1v
5	LG/RD	Speed Control Vent Solenoid	Vacuum Decreasing: 1v
6	BR/YL	O/D Indicator (lamp) Control	O/D On: 1v, Off: 12v
7-9	---	Not Used	---
10	DG/RD	LDP Solenoid Control	PWM Signal: 0-12-0v
11	YL/RD	Speed Control Power Supply	12-14v
12	DG/OR	ASD Relay Output	12-14v
13	OR	Overdrive 'Off' Switch Sense	Switch On: 1v, Off: 12v
14	PK/RD	LDP Switch Sense Signal	Closed: 0v, Open: 12v
15	RD/YL	Battery Temperature Sensor	At 86ºF: 1.96v
16-18	---	Not Used	---
19	DB	Fuel Pump Relay Control	Relay Off: 12v, On: 1v
20	PK/BK	EVAP Purge Solenoid Control	PWM Signal: 0-12-0v
21	---	Not Used	---
22	DB/BK	A/C Request Signal	A/C Off: 12v, On: 1v
23	---	Not Used	---
24	BR	Brake Switch Sense Signal	Brake Off: 0v, On: 12v
25	DG/PK	Generator Field Source	12-14v
26	LB/BK	Fuel Level Sensor Signal	Full: 0.5v, 1/2 full: 2.5v
27	BK/PK	SCI Transmit	0v
28	WT/BK	CCD Bus (-)	<0.050v
29	BK/WT	SCI Receive	5v
30	PK/BR	CCD Bus (+)	Digital Signals: 0-5-0v
31	---	Not Used	---
32	PK	Speed Control Switch Signal	S/C & Set Switch On: 3.8v

Pin Connector Graphic

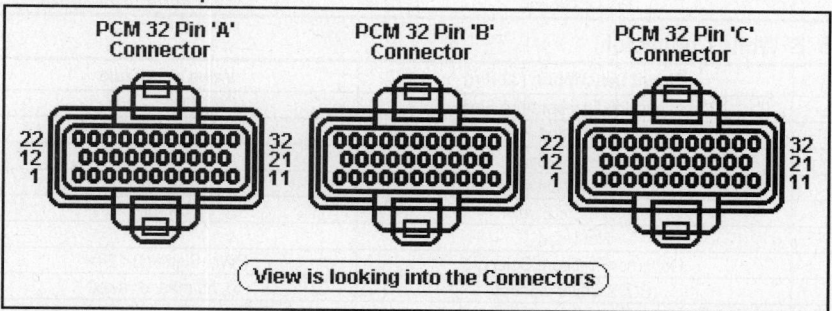

Standard Colors and Abbreviations

Abbreviation	Color	Abbreviation	Color	Abbreviation	Color
BK	Black	GY	Gray	RD	Red
BL	Blue	GN	Green	TN	Tan
BR	Brown	LG	Light Green	VT	Violet
DB	Dark Blue	OR	Orange	WT	White
DG	Dark Green	PK	Pink	YL	Yellow

1999 Grand Cherokee 4.0L VIN S 'A' Black Connector

PCM Pin #	Wire Color	Circuit Description (32 Pin)	Value at Hot Idle
1	TN/OR	Ignition Coil 3 Driver Control	5°, 55 mph: 8° dwell
2	OR/DB	Ignition Switch Power (B+)	12-14v
3	---	Not Used	---
4	BK/LB	Sensor Ground	<0.050v
5	---	Not Used	---
6	BR/YL	PNP Switch Sense Signal	In P/N: 0v, Others: 5v
7	TN/RD	Ignition Coil 1 Driver Control	5°, 55 mph: 8° dwell
8	GY/BK	CKP Sensor Signal	Digital Signals: 0-5-0v
9	---	Not Used	---
10	YL/BK	IAC 2 Driver Control	Pulse Signals
11	BR/WT	IAC 3 Driver Control	Pulse Signals
12-14	---	Not Used	---
15	BK/RD	IAT Sensor Signal	At 100°F: 1.83v
16	TN/BK	ECT Sensor Signal	At 180°F: 2.80v
17	OR	5-Volt Supply	4.9-5.1v
18	TN/YL	CMP Sensor Signal	Digital Signals: 0-5-0v
19	GY/RD	IAC 1 Driver Control	Pulse Signals
20	VT/BK	IAC 4 Driver Control	Pulse Signals
21	---	Not Used	---
22	RD/BK	Fused Battery Power (B+)	12-14v
23	OR/RD	TP Sensor Signal	0.6-1.0v
24	BK/DG	HO2S-11 (B1 S1) Signal	0.1-1.1v
25	BK/DG	HO2S-12 (B1 S2) Signal	0.1-1.1v
26	TN/WT	HO2S-21 (B2 S1) Signal (Cal)	0.1-1.1v
27	DG/OR	MAP Sensor Signal	1.5-1.6v
28	---	Not Used	---
29	PK/WT	HO2S-22 (B2 S2) Signal (Cal)	0.1-1.1v
30	---	Not Used	---
31	BK/WT	Power Ground	<0.1v
32	BK/TN	Power Ground	<0.1v

2000 Grand Cherokee 4.0L VIN S 'B' White Connector

PCM Pin #	Wire Color	Circuit Description (32 Pin)	Value at Hot Idle
1	VT	Transmission Temperature Sensor	At 200°F: 2.40v
2-3	---	Not Used	---
4	WT/DB	Injector 1 Driver	1-4 ms
5	YL/WT	Injector 3 Driver	1-4 ms
6	GY	Injector 5 Driver	1-4 ms
7	---	Not Used	---
8	PK	Governor Pressure Solenoid	PWM Signal: 0-12-0v
9	TN/PK	Coil 2 Driver Control	5°, 55 mph: 8° dwell
10	DG	Generator Field Control	Digital Signals: 0-12-0v
11	DG/LB	TCC Solenoid Control	TCC on at Cruise: 1v
12	BR/DB	Injector 6 Driver	1-4 ms
13-14	---	Not Used	---
15	TN	Injector 2 Driver	1-4 ms
16	LB/BR	Injector 4 Driver	1-4 ms
17	DB/PK	Radiator Fan Control Relay	Relay Off: 12v, On: 1v
18-20	---	Not Used	---
21	BR/WT	3-4 Shift Solenoid Control	Solenoid on at Cruise: 1v
22, 24	---	Not Used	---
23	GY/YL	Engine Oil Pressure Sensor	1.6v at 24 psi
25	DB/BK	OSS Sensor Signal (-)	AC pulse signals
26	---	Not Used	---
27	WT/OR	Vehicle Speed Signal	Digital Signal
28	LG/WT	OSS Sensor Signal (+)	AC pulse signals
29	LG/RD	Governor Pressure Signal	0.58v
30	PK/YL	Transmission Relay Control	Relay Off: 12v, On: 1v
31	PK/OR	5-Volt Supply	4.9-5.1v
32	---	Not Used	---

1999 Grand Cherokee 4.0L VIN S 'C' Gray Connector

PCM Pin #	Wire Color	Circuit Description (32 Pin)	Value at Hot Idle
1	DB/OR	A/C Clutch Relay Control	Relay Off: 12v, On: 1v
2	---	Not Used	---
3	DB/YL	ASD Relay Control	Relay Off: 12v, On: 1v
4	TN/RD	Speed Control Vacuum Solenoid	Vacuum Increasing: 1v
5	LG/RD	Speed Control Vent Solenoid	Vacuum Decreasing: 1v
6-7	---	Not Used	---
8	GY	HO2S-12-22 Relay Signal	Relay Off: 12v, On: 1v
9	LB	HO2S-12-21 Relay Signal	Relay Off: 12v, On: 1v
10	WT/DG	LDP Solenoid Control	PWM Signal: 0-12-0v
11	OR/DG	Speed Control Power Supply	12-14v
12	DG/LG	ASD Relay Output	12-14v
13	OR/YL	Torque Management Relay	Relay Off: 12v, On: 1v
14	OR/PK	LDP Switch Sense Signal	Closed: 0v, Open: 12v
15	PK/LG	Battery Temperature Sensor	At 86°F: 1.96v
16-18	---	Not Used	---
19	BR	Fuel Pump Relay Control	Relay Off: 12v, On: 1v
20	PK/BK	EVAP Purge Solenoid Control	PWM Signal: 0-12-0v
21	---	Not Used	---
22	DB	A/C Pressure Signal	0.90v at 79 psi
23	---	Not Used	---
24	WT/PK	Brake Switch Sense Signal	Brake Off: 0v, On: 12v
25	LB/RD	Generator Field Source	12-14v
26	LB/YL	Fuel Level Sensor Signal	Full: 0.5v, 1/2 full: 2.5v
27	PK	SCI Transmit	0v
28	WT/BK	CCD Bus (-)	<0.050v
29	LG/DG	SCI Receive	5v
30	YL/PK	CCD Bus (+)	Digital Signals: 0-5-0v
31	---	Not Used	---
32	RD/LG	Speed Control Switch Signal	S/C & Set Switch On: 3.8v

Pin Connector Graphic

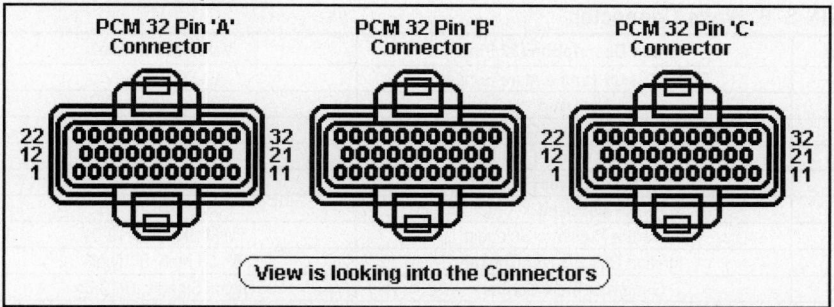

Standard Colors and Abbreviations

Abbreviation	Color	Abbreviation	Color	Abbreviation	Color
BK	Black	GY	Gray	RD	Red
BL	Blue	GN	Green	TN	Tan
BR	Brown	LG	Light Green	VT	Violet
DB	Dark Blue	OR	Orange	WT	White
DG	Dark Green	PK	Pink	YL	Yellow

2000-01 Grand Cherokee 4.0L VIN S 'A' Black Connector

PCM Pin #	Wire Color	Circuit Description (32 Pin)	Value at Hot Idle
1	TN/OR	Ignition Coil 3 Driver Control	5°, 55 mph: 8° dwell
2	OR/DB	Ignition Switch Power (B+)	12-14v
3	---	Not Used	---
4	BK/LB	Sensor Ground	<0.050v
5	---	Not Used	---
6	BR/YL	PNP Switch Sense Signal	In P/N: 0v, Others: 5v
7	TN/RD	Ignition Coil 1 Driver Control	5°, 55 mph: 8° dwell
8	GY/BK	CKP Sensor Signal	Digital Signals: 0-5-0v
9	---	Not Used	---
10	YL/BK	IAC 2 Driver Control	Pulse Signals
11	BR/WT	IAC 3 Driver Control	Pulse Signals
12-14	---	Not Used	---
15	BK/RD	IAT Sensor Signal	At 100°F: 1.83v
16	TN/BK	ECT Sensor Signal	At 180°F: 2.80v
17	OR	5-Volt Supply	4.9-5.1v
18	TN/YL	CMP Sensor Signal	Digital Signals: 0-5-0v
19	GY/RD	IAC 1 Driver Control	Pulse Signals
20	VT/BK	IAC 4 Driver Control	Pulse Signals
21	---	Not Used	---
22	RD/BK	Fused Battery Power (B+)	12-14v
23	OR/RD	TP Sensor Signal	0.6-1.0v
24	LG/RD	HO2S-11 (B1 S1) Signal	0.1-1.1v
25	BK/DG	HO2S-12 (B1 S2) Signal	0.1-1.1v
26	TN/WT	HO2S-21 (B2 S1) Signal	0.1-1.1v
27	DG/RD	MAP Sensor Signal	1.5-1.6v
28	---	Not Used	---
29	PK/WT	HO2S-22 (B2 S2) Signal	0.1-1.1v
30	---	Not Used	---
31	BK/WT	Power Ground	<0.1v
32	BK/TN	Power Ground	<0.1v

2000-01 Grand Cherokee 4.0L VIN S 'B' White Connector

PCM Pin #	Wire Color	Circuit Description (32 Pin)	Value at Hot Idle
1	VT	Transmission Temperature Sensor	At 200°F: 2.40v
2-3	---	Not Used	---
4	WT/DB	Injector 1 Driver Control	1-4 ms
5	YL/WT	Injector 3 Driver Control	1-4 ms
6	GY	Injector 5 Driver Control	1-4 ms
7	---	Not Used	---
8	PK	Governor Pressure Solenoid	PWM Signal: 0-12-0v
9	TN/PK	Ignition Coil 2 Driver Control	5°, 55 mph: 8° dwell
10	DG	Generator Field Control	Digital Signals: 0-12-0v
11	DG/LB	TCC Solenoid Control	TCC on at Cruise: 1v
12	BR/DB	Injector 6 Driver Control	1-4 ms
13-14	---	Not Used	---
15	TN	Injector 2 Driver Control	1-4 ms
16	LB/BR	Injector 4 Driver Control	1-4 ms
17	DB/PK	Radiator Fan Control Relay	Relay Off: 12v, On: 1v
18, 20, 22	---	Not Used	---
19	DB	AC Pressure Sensor Signal	0.90v at 79 psi
21	BR/WT	3-4 Shift Solenoid Control	Solenoid on at Cruise: 1v
23	GY/YL	Engine Oil Pressure Sensor	1.6v at 24 psi
24, 26	---	Not Used	---
25	DB/BK	OSS Sensor Signal (-)	AC pulse signals
27	WT/OR	Vehicle Speed Signal	Digital Signal
28	LG/WT	OSS Sensor Signal (+)	AC pulse signals
29	LG/RD	Governor Pressure Signal	0.58v
30	PK/YL	Transmission Relay Control	Relay Off: 12v, On: 1v
31	VT/BK	5-Volt Supply	4.9-5.1v
32	---	Not Used	---

2000-01 Grand Cherokee 4.0L VIN S 'C' Gray Connector

PCM Pin #	Wire Color	Circuit Description (32 Pin)	Value at Hot Idle
1	DB/OR	A/C Clutch Relay Control	Relay Off: 12v, On: 1v
2	---	Not Used	---
3	DB/YL	ASD Relay Control	Relay Off: 12v, On: 1v
4	TN/RD	Speed Control Vacuum Solenoid	Vacuum Increasing: 1v
5	LG/RD	Speed Control Vent Solenoid	Vacuum Decreasing: 1v
6-7	---	Not Used	---
8	GY	HO2S-11 Heater Relay	Relay Off: 12v, On: 1v
9	LB	HO2S-12 Heater Relay	Relay Off: 12v, On: 1v
10	WT/DG	LDP Solenoid Control	PWM Signal: 0-12-0v
11	OR/DG	Speed Control Power Supply	12-14v
12	DG/LG	ASD Relay Output	12-14v
13	OR/YL	Torque Management Request	Digital Signals
14	OR/PK	LDP Switch Sense Signal	Closed: 0v, Open: 12v
15	VT/LG	Battery Temperature Sensor	At 86°F: 1.96v
16	DB/WT	HO2S-21 Heater Relay	Relay Off: 12v, On: 1v
17-18	---	Not Used	---
19	BR	Fuel Pump Relay Control	Relay Off: 12v, On: 1v
20	PK/BK	EVAP Purge Solenoid Control	PWM Signal: 0-12-0v
21	---	Not Used	---
22	DB/YL	A/C Switch Sense	A/C Off: 12v, On: 1v
23	---	Not Used	---
24	WT/PK	Brake Switch Sense Signal	Brake Off: 0v, On: 12v
25	LB/RD	Generator Field Source	12-14v
26	LB/YL	Fuel Level Sensor Signal	Full: 0.5v, 1/2 full: 2.5v
27	PK	SCI Transmit	0v
28	WT/BK	CCD Bus (-)	<0.050v
28 ('01)	---	Not Used	---
29	LG/DG	SCI Receive	5v
30	YL/PK	CCD Bus (+)	Digital Signals: 0-5-0v
30 ('01)	YL/VT	PCI Data Bus (J1850)	Digital Signals: 0-7-0v
31	---	Not Used	---
32	RD/LG	Speed Control Switch Signal	S/C & Set Switch On: 3.8v

Pin Connector Graphic

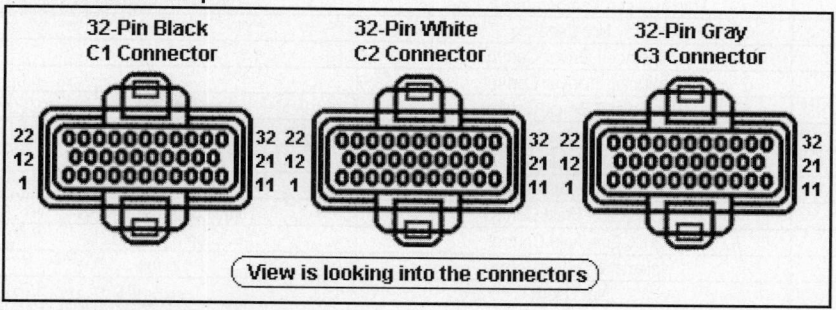

Standard Colors and Abbreviations

Abbreviation	Color	Abbreviation	Color	Abbreviation	Color
BK	Black	GY	Gray	RD	Red
BL	Blue	GN	Green	TN	Tan
BR	Brown	LG	Light Green	VT	Violet
DB	Dark Blue	OR	Orange	WT	White
DG	Dark Green	PK	Pink	YL	Yellow

2002-03 Grand Cherokee 4.0L VIN S 'A' Black Connector

PCM Pin #	Wire Color	Circuit Description (32 Pin)	Value at Hot Idle
1	TN/OR	Ignition Coil 3 Driver Control	5°, 55 mph: 8° dwell
2	OR/DB	Ignition Switch Power (B+)	12-14v
3, 5	---	Not Used	---
4	BK/LB	Sensor Ground	<0.050v
6	BK/WT	Park Neutral Position Switch	In P/N: 0v, Others: 5v
7	TN/RD	Ignition Coil 1 Driver Control	5°, 55 mph: 8° dwell
8	GY/BK	CKP Sensor Signal	Digital Signals: 0-5-0v
9	---	Not Used	---
10	YL/BK	IAC 2 Driver Control	Pulse Signals
11	BR/WT	IAC 3 Driver Control	Pulse Signals
12-13	---	Not Used	---
14	LG/BK	Transfer Case Position Switch	Switch Open: 12v, Closed: 0v
15	BK/RD	IAT Sensor Signal	At 100°F: 1.83v
16	TN/BK	ECT Sensor Signal	At 180°F: 2.80v
17	OR	5-Volt Supply	4.9-5.1v
18	TN/YL	CMP Sensor Signal	Digital Signals: 0-5-0v
19	GY/RD	IAC 1 Driver Control	Pulse Signals
20	VT/BK	IAC 4 Driver Control	Pulse Signals
21	---	Not Used	---
22	RD/BK	Fused Battery Power (B+)	12-14v
23	OR/RD	TP Sensor Signal	0.6-1.0v
24	BK/DG	HO2S-11 (B1 S1) Signal	0.1-1.1v
25	TN/WT	HO2S-12 (B1 S2) Signal	0.1-1.1v
26	LG/RD	HO2S-21 (B2 S1) Signal	0.1-1.1v
27	DG/RD	MAP Sensor Signal	1.5-1.6v
28	---	Not Used	---
29	TN/WT	HO2S-22 (B2 S2) Signal	0.1-1.1v
30	---	Not Used	---
31	BK/WT	Power Ground	<0.1v
32	BK/TN	Power Ground	<0.1v

2002-3 Grand Cherokee 4.0L VIN S 'B' White Connector

PCM Pin #	Wire Color	Circuit Description (32 Pin)	Value at Hot Idle
1	VT	Transmission Temperature Sensor	At 200°F: 2.40v
2-3	---	Not Used	---
4	WT/DB	Injector 1 Driver Control	1-4 ms
5	YL/WT	Injector 3 Driver Control	1-4 ms
6	GY	Injector 5 Driver Control	1-4 ms
7	---	Not Used	---
8	PK	Governor Pressure Solenoid	PWM Signal: 0-12-0v
9	TN/PK	Ignition Coil 2 Driver Control	5°, 55 mph: 8° dwell
10	DG	Generator Field Control	Digital Signals: 0-12-0v
11	LB	TCC Solenoid Control	TCC on at Cruise: 1v
12	BR/DB	Injector 6 Driver Control	1-4 ms
13-14	---	Not Used	---
15	TN	Injector 2 Driver Control	1-4 ms
16	LB/BR	Injector 4 Driver Control	1-4 ms
17	LG	Radiator Fan Motor Relay	Relay Off: 12v, On: 1v
18, 20, 22	---	Not Used	---
19	DB	AC Pressure Sensor Signal	0.90v at 79 psi
21	BR	3-4 Shift Solenoid Control	Solenoid on at Cruise: 1v
23	GY/YL	Engine Oil Pressure Sensor	1.6v at 24 psi
24, 26	---	Not Used	---
25	DB/BK	Speed Sensor Ground	AC pulse signals
27	DG/YL	Vehicle Speed Signal	Digital Signals
28	LG/WT	Output Speed Sensor Signal (+)	AC pulse signals
29	LG/RD	Governor Pressure Signal	0.58v
30	PK/YL	Transmission Relay Control	Relay Off: 12v, On: 1v
31	VT/BK	5-Volt Supply	4.9-5.1v
32	---	Not Used	---

2002-03 Grand Cherokee 4.0L VIN S 'C' Gray Connector

PCM Pin #	Wire Color	Circuit Description (32 Pin)	Value at Hot Idle
1	DB/OR	A/C Clutch Relay Control	Relay Off: 12v, On: 1v
2	---	Not Used	---
3	DB/YL	ASD Relay Control	Relay Off: 12v, On: 1v
4	TN/RD	Speed Control Vacuum Solenoid	Vacuum Increasing: 1v
5	LG/RD	Speed Control Vent Solenoid	Vacuum Decreasing: 1v
6-7	---	Not Used	---
8	BR/OR	HO2S-11 (B1 S1) Heater Control	Heater Off: 12v, On: 1v
9	RD/YL	Downstream Heater Control Relay	Relay Off: 12v, On: 1v
10	WT/DG	LDP Solenoid Control	PWM Signal: 0-12-0v
11	OR/DG	Speed Control Power Supply	12-14v
12	DG/LG	ASD Relay Output	12-14v
13	OR/WT	Overdrive Off Switch Sense	Switch Off: 12v, On: 1v
14	OR/PK	LDP Switch Sense Signal	Closed: 0v, Open: 12v
15	VT/LG	Battery Temperature Sensor	At 86°F: 1.96v
16	BR/WT	HO2S-12 (B1 S2) Heater Control	Heater Off: 12v, On: 1v
17-18	---	Not Used	---
19	BR	Fuel Pump Relay Control	Relay Off: 12v, On: 1v
20	PK/BK	EVAP Purge Solenoid Control	PWM Signal: 0-12-0v
21-23	---	Not Used	---
24	WT/PK	Brake Switch Sense Signal	Brake Off: 0v, On: 12v
25	WT/DB	Generator Field Source	12-14v
26	LB/YL	Fuel Level Sensor Signal	Full: 0.5v, 1/2 full: 2.5v
27	PK	SCI Transmit	0v
28	---	Not Used	---
29	LG	SCI Receive	5v
30	WT/YL	PCI Data Bus (J1850)	Digital Signals: 0-7-0v
31	---	Not Used	---
32	DG/LG	Speed Control Switch Signal	S/C & Set Switch On: 3.8v

Pin Connector Graphic

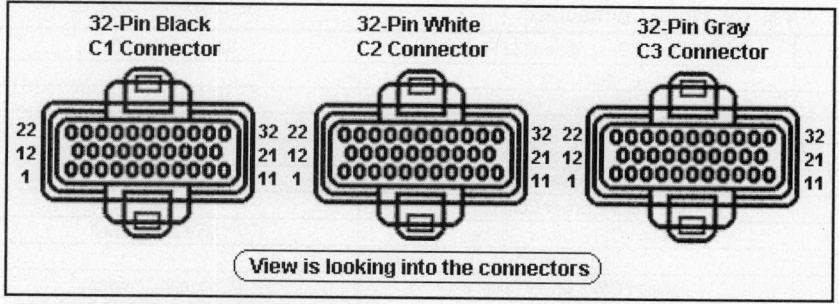

32-Pin Black C1 Connector   **32-Pin White C2 Connector**   **32-Pin Gray C3 Connector**

View is looking into the connectors

Standard Colors and Abbreviations

Abbreviation	Color	Abbreviation	Color	Abbreviation	Color
BK	Black	GY	Gray	RD	Red
BL	Blue	GN	Green	TN	Tan
BR	Brown	LG	Light Green	VT	Violet
DB	Dark Blue	OR	Orange	WT	White
DG	Dark Green	PK	Pink	YL	Yellow

1999-2000 Grand Cherokee 4.7L VIN N 'C1' Black Connector

PCM Pin #	Wire Color	Circuit Description (32 Pin)	Value at Hot Idle
1	TN/OR	Ignition Coil 3 Driver Control	7°, 55 mph: 9° dwell
2	OR/DB	Ignition Switch Power (B+)	12-14v
3	TN/LG	Ignition Coil 4 Driver Control	7°, 55 mph: 9° dwell
4	BK/LB	Sensor Ground	<0.1v
5	TN/LB	Ignition Coil 6 Driver Control	7°, 55 mph: 9° dwell
6	BR/YL	PNP Switch Sense Signal	In P/N: 0v, Others: 5v
7	TN/RD	Ignition Coil 1 Driver Control	7°, 55 mph: 9° dwell
8	GY/BK	CKP Sensor Signal	Digital Signals: 0-5-0v
9	LB/RD	Ignition Coil 8 Driver Control	7°, 55 mph: 9° dwell
10	YL/BK	IAC 2 Driver Control	Pulse Signals
11	BR/WT	IAC 3 Driver Control	Pulse Signals
12-14	---	---	---
15	BK/RD	IAT Sensor Signal	At 100°F: 1.83v
16	TN/BK	ECT Sensor Signal	At 180°F: 2.80v
17	OR	5-Volt Supply	4.9-5.1v
18	TN/YL	CMP Sensor Signal	Digital Signals: 0-5-0v
19	GY/BK	IAC 1 Driver Control	Pulse Signals
20	VT/BK	IAC 4 Driver Control	Pulse Signals
21	TN/DG	Ignition Coil 5 Driver Control	7°, 55 mph: 9° dwell
22	RD/BK	Fused Battery Power (B+)	12-14v
23	OR/RD	TP Sensor Signal	0.6-1.0v
24	LG/RD	HO2S-11 (B1 S1) Signal	0.1-1.1v
25	BK/DG	HO2S-12 (B1 S2) Signal	0.1-1.1v
26	TN/WT	HO2S-21 (B2 S1) Signal	0.1-1.1v
27	DG/RD	MAP Sensor Signal	1.5-1.6v
28, 30	---	Not Used	---
29	PK/WT	HO2S-22 (B2 S2) Signal	0.1-1.1v
31	BK/WT	Power Ground	<0.1v
32	BK/TN	Power Ground	<0.1v

1999-2000 Grand Cherokee 4.7L VIN N 'C2' White Connector

PCM Pin #	Wire Color	Circuit Description (32 Pin)	Value at Hot Idle
1	---	Not Used	---
2	DB/TN	Injector 7 Driver Control	1-4 ms
3	---	Not Used	---
4	WT/DB	Injector 1 Driver Control	1-4 ms
5	YL/WT	Injector 3 Driver Control	1-4 ms
6	GY	Injector 5 Driver Control	1-4 ms
7	BR	Ignition Coil 7 Driver Control	7°, 55 mph: 9° dwell
8	---	Not Used	---
9	TN/PK	Ignition Coil 2 Driver Control	7°, 55 mph: 9° dwell
10	DG	Generator Field Control	Digital Signals: 0-12-0v
11	---	Not Used	---
12	BR/DB	Injector 6 Driver Control	1-4 ms
13	DB/GY	Injector 8 Driver Control	1-4 ms
14	---	Not Used	---
15	TN	Injector 2 Driver Control	1-4 ms
16	LB/BR	Injector 4 Driver Control	1-4 ms
17	DB/RD	Radiator Fan Control Relay	Relay Off: 12v, On: 1v
18, 20	---	Not Used	---
19	DB	A/C Pressure Signal	0.90v at 79 psi
21	BR/WT	3-4 Shift Solenoid Control	Solenoid on at Cruise: 1v
22	---	Not Used	---
23	GY/YL	Engine Oil Pressure Sensor	1.6v at 24 psi
24-26	---	Not Used	---
27	WT/OR	Vehicle Speed Signal	Digital Signal
28-30	---	Not Used	---
31	VT/BK	5-Volt Supply	4.9-5.1v
32	---	Not Used	---

1999-2000 Grand Cherokee 4.7L VIN N 'C3' Gray Connector

PCM Pin #	Wire Color	Circuit Description (32 Pin)	Value at Hot Idle
1	DB/OR	A/C Clutch Relay Control	Relay Off: 12v, On: 1v
2	---	Not Used	---
3	DB/YL	ASD Relay Control	Relay Off: 12v, On: 1v
4	TN/RD	Speed Control Vacuum Solenoid	Vacuum Increasing: 1v
5	LG/RD	Speed Control Vent Solenoid	Vacuum Decreasing: 1v
6-7	---	Not Used	---
8	GY	HO2S-12-22 Relay Signal	Relay Off: 12v, On: 1v
9	LB	HO2S-12-21 Relay Signal	Relay Off: 12v, On: 1v
10	WT/DG	LDP Solenoid Control	PWM Signal: 0-12-0v
11	OR/DG	Speed Control Power Supply	12-14v
12	DG/LG	ASD Relay Output	12-14v
13	OR/YL	Torque Management Relay	Relay Off: 12v, On: 1v
14	OR/PK	LDP Switch Sense Signal	Closed: 0v, Open: 12v
15	PK/LG	Battery Temperature Sensor	At 86°F: 1.96v
16-18	---	Not Used	---
19	BR	Fuel Pump Relay Control	Relay Off: 12v, On: 1v
20	PK/BK	EVAP Purge Solenoid Control	PWM Signal: 0-12-0v
21	---	Not Used	---
22	DB	A/C Pressure Switch Signal	A/C On: 0.451-4.850v
23	---	Not Used	---
24	WT/PK	Brake Switch Sense Signal	Brake Off: 0v, On: 12v
25	LB/RD	Generator Field Source	12-14v
26	LB/YL	Fuel Level Sensor Signal	Full: 0.5v, 1/2 full: 2.5v
27	PK	SCI Transmit	0v
28	WT/BK	CCD Bus (-)	<0.050v
29	LG/DG	SCI Receive	5v
30	YL/PK	CCD Bus (+)	Digital Signals: 0-5-0v
31	---	Not Used	---
32	RD/LG	Speed Control Switch Signal	S/C & Set Switch On: 3.8v

Pin Connector Graphic

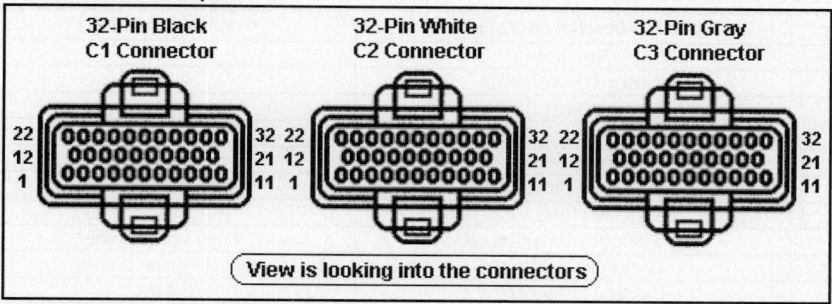

Standard Colors and Abbreviations

Abbreviation	Color	Abbreviation	Color	Abbreviation	Color
BK	Black	GY	Gray	RD	Red
BL	Blue	GN	Green	TN	Tan
BR	Brown	LG	Light Green	VT	Violet
DB	Dark Blue	OR	Orange	WT	White
DG	Dark Green	PK	Pink	YL	Yellow

2001 Grand Cherokee 4.7L VIN N 'C1' Black Connector

PCM Pin #	Wire Color	Circuit Description (32 Pin)	Value at Hot Idle
1	TN/OR	Ignition Coil 3 Driver Control	7°, 55 mph: 9° dwell
2	OR/DB	Ignition Switch Power (B+)	12-14v
3	TN/LG	Ignition Coil 4 Driver Control	7°, 55 mph: 9° dwell
4	BK/LB	Sensor Ground	<0.1v
5	TN/LB	Ignition Coil 6 Driver Control	7°, 55 mph: 9° dwell
6	BR/YL	PNP Switch Sense Signal	In P/N: 0v, Others: 5v
7	TN/RD	Ignition Coil 1 Driver Control	7°, 55 mph: 9° dwell
8	GY/BK	CKP Sensor Signal	Digital Signals: 0-5-0v
9	LB/RD	Ignition Coil 8 Driver Control	7°, 55 mph: 9° dwell
10	YL/BK	IAC 2 Driver Control	Pulse Signals
11	BR/WT	IAC 3 Driver Control	Pulse Signals
12-14	---	---	---
15	BK/RD	IAT Sensor Signal	At 100°F: 1.83v
16	TN/BK	ECT Sensor Signal	At 180°F: 2.80v
17	OR	5-Volt Supply	4.9-5.1v
18	TN/YL	CMP Sensor Signal	Digital Signals: 0-5-0v
19	GY/BK	IAC 1 Driver Control	Pulse Signals
20	VT/BK	IAC 4 Driver Control	Pulse Signals
21	TN/DG	Ignition Coil 5 Driver Control	7°, 55 mph: 9° dwell
22	RD/BK	Fused Battery Power (B+)	12-14v
23	OR/RD	TP Sensor Signal	0.6-1.0v
24	LG/RD	HO2S-11 (B1 S1) Signal	0.1-1.1v
25	BK/DG	HO2S-12 (B1 S2) Signal	0.1-1.1v
26	TN/WT	HO2S-21 (B2 S1) Signal	0.1-1.1v
27	DG/RD	MAP Sensor Signal	1.5-1.6v
28, 30	---	Not Used	---
29	PK/WT	HO2S-22 (B2 S2) Signal	0.1-1.1v
31	BK/WT	Power Ground	<0.1v
32	BK/TN	Power Ground	<0.1v

2001 Grand Cherokee 4.7L VIN N 'C2' White Connector

PCM Pin #	Wire Color	Circuit Description (32 Pin)	Value at Hot Idle
1, 3	---	Not Used	---
2	DB/TN	Injector 7 Driver Control	1-4 ms
4	WT/DB	Injector 1 Driver Control	1-4 ms
5	YL/WT	Injector 3 Driver Control	1-4 ms
6	GY	Injector 5 Driver Control	1-4 ms
7	BR	Ignition Coil 7 Driver Control	7°, 55 mph: 9° dwell
8, 11	---	Not Used	---
9	TN/PK	Ignition Coil 2 Driver Control	7°, 55 mph: 9° dwell
10	DG	Generator Field Control	Digital Signals: 0-12-0v
12	BR/DB	Injector 6 Driver Control	1-4 ms
13	DB/GY	Injector 8 Driver Control	1-4 ms
14	---	Not Used	---
15	TN	Injector 2 Driver Control	1-4 ms
16	LB/BR	Injector 4 Driver Control	1-4 ms
17	DB/RD	Radiator Fan Control Relay	Relay Off: 12v, On: 1v
18, 20	---	Not Used	---
19	DB	A/C Pressure Signal	0.90v at 79 psi
21	BR/WT	3-4 Shift Solenoid Control	Solenoid on at Cruise: 1v
22, 24	---	Not Used	---
23	GY/YL	Engine Oil Pressure Sensor	1.6v at 24 psi
25	DB/BK	Output Speed Sensor (N)	AC pulse signals
26	---	Not Used	---
27	WT/OR	Vehicle Speed Signal	Digital Signal
28	LG/WT	OSS Sensor Signal (P)	AC pulse signals
29	LG/RD	Governor Pressure Signal	0.58v
30	PK/YL	Transmission Control Relay	Relay Off: 12v, On: 1v
31	VT/BK	5-Volt Supply	4.9-5.1v
32	---	Not Used	---

2001 Grand Cherokee 4.7L VIN N 'C3' Gray Connector

PCM Pin #	Wire Color	Circuit Description (32 Pin)	Value at Hot Idle
1	DB/OR	A/C Clutch Relay Control	Relay Off: 12v, On: 1v
2	---	Not Used	---
3	DB/YL	ASD Relay Control	Relay Off: 12v, On: 1v
4	TN/RD	Speed Control Vacuum Solenoid	Vacuum Increasing: 1v
5	LG/RD	Speed Control Vent Solenoid	Vacuum Decreasing: 1v
6-7	---	Not Used	---
8	GY	HO2S-12-22 Relay Signal	Relay Off: 12v, On: 1v
9	LB	HO2S-12-21 Relay Signal	Relay Off: 12v, On: 1v
10	WT/DG	LDP Solenoid Control	PWM Signal: 0-12-0v
11	OR/DG	Speed Control Power Supply	12-14v
12	DG/LG	ASD Relay Output	12-14v
13	OR/YL	Torque Management Relay	Relay Off: 12v, On: 1v
14	OR/PK	LDP Switch Sense Signal	Closed: 0v, Open: 12v
15	PK/LG	Battery Temperature Sensor	At 86°F: 1.96v
16-18	---	Not Used	---
19	BR	Fuel Pump Relay Control	Relay Off: 12v, On: 1v
20	PK/BK	EVAP Purge Solenoid Control	PWM Signal: 0-12-0v
21	---	Not Used	---
22	DB	A/C Pressure Switch Signal	A/C On: 0.451-4.850v
23	---	Not Used	---
24	WT/PK	Brake Switch Sense Signal	Brake Off: 0v, On: 12v
25	LB/RD	Generator Field Source	12-14v
26	LB/YL	Fuel Level Sensor Signal	Full: 0.5v, 1/2 full: 2.5v
27	PK	SCI Transmit	0v
28	WT/BK	CCD Bus (-)	<0.050v
29	LG/DG	SCI Receive	5v
30	YL/PK	CCD Bus (+)	Digital Signals: 0-5-0v
31	---	Not Used	---
32	RD/LG	Speed Control Switch Signal	S/C & Set Switch On: 3.8v

Pin Connector Graphic

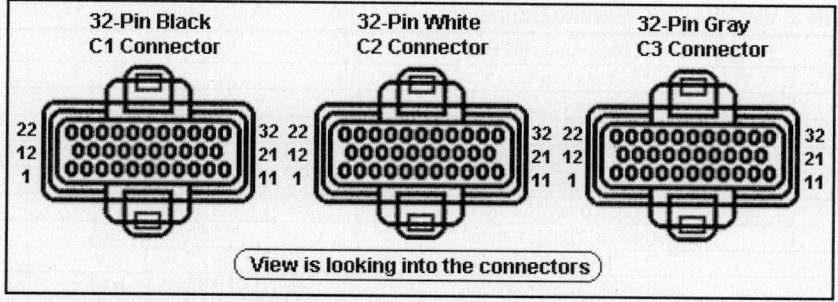

32-Pin Black C1 Connector   **32-Pin White C2 Connector**   **32-Pin Gray C3 Connector**

View is looking into the connectors

Standard Colors and Abbreviations

Abbreviation	Color	Abbreviation	Color	Abbreviation	Color
BK	Black	GY	Gray	RD	Red
BL	Blue	GN	Green	TN	Tan
BR	Brown	LG	Light Green	VT	Violet
DB	Dark Blue	OR	Orange	WT	White
DG	Dark Green	PK	Pink	YL	Yellow

2002-03 Grand Cherokee 4.7L VIN J, VIN N (A/T) 'C1' Black Connector

PCM Pin #	Wire Color	Circuit Description (32 Pin)	Value at Hot Idle
1	TN/OR	Ignition Coil 3 Driver Control	7°, 55 mph: 9° dwell
2	OR/DB	Ignition Switch Power (B+)	12-14v
3	TN/LG	Ignition Coil 4 Driver Control	7°, 55 mph: 9° dwell
4	BK/LB	Sensor Ground	<0.1v
5	TN/LB	Ignition Coil 6 Driver Control	7°, 55 mph: 9° dwell
6	BK/WT	PNP Switch Sense Signal	In P/N: 0v, Others: 5v
7	TN/RD	Ignition Coil 1 Driver Control	7°, 55 mph: 9° dwell
8	GY/BK	CKP Sensor Signal	Digital Signals: 0-5-0v
9	LB/RD	Ignition Coil 8 Driver Control	7°, 55 mph: 9° dwell
10	YL/BK	IAC 2 Driver Control	Pulse Signals
11	BR/WT	IAC 3 Driver Control	Pulse Signals
12-13	---	---	---
14	LG/BK	Transfer Case Position Switch	Switch Open: 12v, Closed: 0v
15	BK/RD	IAT Sensor Signal	At 100°F: 1.83v
16	TN/BK	ECT Sensor Signal	At 180°F: 2.80v
17	OR	5-Volt Supply	4.9-5.1v
18	TN/YL	CMP Sensor Signal	Digital Signals: 0-5-0v
19	GY/BK	IAC 1 Driver Control	Pulse Signals
20	VT/BK	IAC 4 Driver Control	Pulse Signals
21	TN/DG	Ignition Coil 5 Driver Control	7°, 55 mph: 9° dwell
22	RD/BK	Fused Battery Power (B+)	12-14v
23	OR/RD	TP Sensor Signal	0.6-1.0v
24	LG/RD	HO2S-11 (B1 S1) Signal	0.1-1.1v
25	BK/DG	HO2S-12 (B1 S2) Signal	0.1-1.1v
26	LG/RD	HO2S-21 (B2 S1) Signal	0.1-1.1v
27	DG/RD	MAP Sensor Signal	1.5-1.6v
28, 30	---	Not Used	---
29	PK/WT	HO2S-22 (B2 S2) Signal	0.1-1.1v
31	BK/WT	Power Ground	<0.1v
32	BK/TN	Power Ground	<0.1v

2002-03 Grand Cherokee 4.7L VIN J, VIN N (A/T) 'C2' White Connector

PCM Pin #	Wire Color	Circuit Description (32 Pin)	Value at Hot Idle
1	---	Not Used	---
2	VT	Injector 7 Driver Control	1-4 ms
3	---	Not Used	---
4	WT/DB	Injector 1 Driver Control	1-4 ms
5	YL/WT	Injector 3 Driver Control	1-4 ms
6	GY	Injector 5 Driver Control	1-4 ms
7	BR	Ignition Coil 7 Driver Control	7°, 55 mph: 9° dwell
8	---	Not Used	---
9	TN/PK	Ignition Coil 2 Driver Control	7°, 55 mph: 9° dwell
10	DG	Generator Field Control	Digital Signals: 0-12-0v
11	---	Not Used	---
12	BR/DB	Injector 6 Driver Control	1-4 ms
13	GY/LB	Injector 8 Driver Control	1-4 ms
14	---	Not Used	---
15	TN	Injector 2 Driver Control	1-4 ms
16	LB/BR	Injector 4 Driver Control	1-4 ms
17	LG	Radiator Fan Control Relay	Relay Off: 12v, On: 1v
18	---	Not Used	---
19	DB	A/C Pressure Signal	0.90v at 79 psi
20-22	---	Not Used	---
23	GY/YL	Engine Oil Pressure Sensor	1.6v at 24 psi
24-26	---	Not Used	---
27	DG/YL	Vehicle Speed Signal	Digital Signal
28-30	---	Not Used	---
31	VT/BK	5-Volt Supply	4.9-5.1v
32	---	Not Used	---

2002-03 Grand Cherokee 4.7L VIN J, VIN N (A/T) 'C3' Gray Connector

PCM Pin #	Wire Color	Circuit Description (32 Pin)	Value at Hot Idle
1	DB/OR	A/C Clutch Relay Control	Relay Off: 12v, On: 1v
2	---	Not Used	---
3	DB/YL	ASD Relay Control	Relay Off: 12v, On: 1v
4	TN/RD	Speed Control Vacuum Solenoid	Vacuum Increasing: 1v
5	LG/RD	Speed Control Vent Solenoid	Vacuum Decreasing: 1v
6	---	Not Used	---
7	DB/LG	Knock Sensor 1 Signal	0.080v AC
8	BR/OR	HO2S-11(B1 S1) Heater Control	Heater Off: 12v, On: 1v
9	RD/YL	Downstream Heater Relay Control	Relay Off: 12v, On: 1v
10	WT/DG	LDP Solenoid Control	PWM Signal: 0-12-0v
11	OR/DG	Speed Control Power Supply	12-14v
12	DG/LG	ASD Relay Output	12-14v
13	YL/DG	Torque Management Relay	Relay Off: 12v, On: 1v
14	OR/PK	LDP Switch Sense Signal	Closed: 0v, Open: 12v
15	VT/LG	Battery Temperature Sensor	At 86°F: 1.96v
16	BR/WT	HO2S-12 (B1 S1) Heater Control	Heater Off: 12v, On: 1v
17	---	Not Used	---
18	GY/BK	Knock Sensor 2 Signal	0.080v AC
19	BR	Fuel Pump Relay Control	Relay Off: 12v, On: 1v
20	PK/BK	EVAP Purge Solenoid Control	PWM Signal: 0-12-0v
21-23	---	Not Used	---
24	WT/PK	Brake Switch Sense	Brake Off: 0v, On: 12v
25	WT/DB	Generator Field Source	12-14v
26	LB/YL	Fuel Level Sensor Signal	Full: 0.5v, 1/2 full: 2.5v
27	PK	SCI Transmit	0v
28	---	Not Used	---
29	LG/DG	SCI Receive	5v
30	VT/YL	PCI Bus Signal (J1850)	Digital Signals: 0-7-0v
31	---	Not Used	---
32	RD/LG	Speed Control Switch Signal	S/C & Set Switch On: 3.8v

Pin Connector Graphic

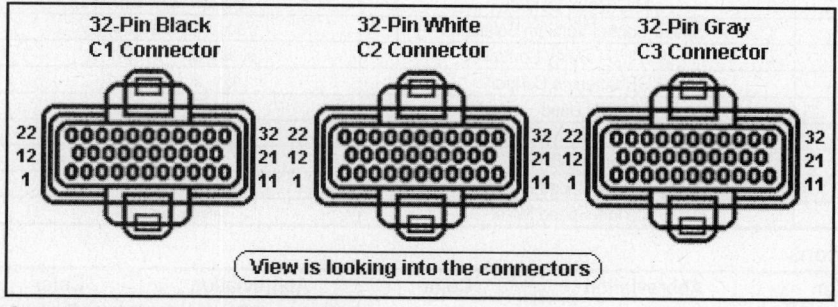

32-Pin Black C1 Connector   32-Pin White C2 Connector   32-Pin Gray C3 Connector

View is looking into the connectors

Standard Colors and Abbreviations

Abbreviation	Color	Abbreviation	Color	Abbreviation	Color
BK	Black	GY	Gray	RD	Red
BL	Blue	GN	Green	TN	Tan
BR	Brown	LG	Light Green	VT	Violet
DB	Dark Blue	OR	Orange	WT	White
DG	Dark Green	PK	Pink	YL	Yellow

1993 Grand Cherokee 5.2L V8 MFI VIN Y (A/T) 60 Pin Connector

PCM Pin #	Wire Color	Circuit Description (60 Pin)	Value at Hot Idle
1	RD/WT	MAP Sensor Signal	1.5-1.6v
2	TN/BK	ECT Sensor Signal	At 180°F: 2.80v
3	PK/BK	Fused Battery Power (B+)	12-14v
4	BK/LB	Sensor Ground	<0.1v
5	BK/TN	Sensor Ground	<0.1v
6	PK/WT	5-Volt Supply	4.9-5.1v
7	WT/BK	8-Volt Supply	7.9-8.1v
8	LG/BK	Starter Input Signal	Cranking: 9-11v
9	LB/RD	Ignition Switch Power (B+)	12-14v
10	OR/BK	Overdrive Override Switch	Switch On: 1v, Off: 12v
11	BK/TN	Power Ground	<0.1v
12	BK/TN	Power Ground	<0.1v
13	LB/BR	Injector 4 Driver	1-4 ms
14	YL/WT	Injector 3 Driver	1-4 ms
15	TN	Injector 2 Driver	1-4 ms
16	WT/DB	Injector 1 Driver	1-4 ms
17	DB/WT	Injector 7 Driver	1-4 ms
18	DB/YL	Injector 8 Driver	1-4 ms
19	GY/WT	Coil Driver (Dwell)	7°, 55 mph: 9° dwell
20	DG	Alternator Field Control	Digital Signals: 0-12-0v
21	BK/RD	Air Temperature Sensor	At 100°F: 2.51v
22	OR/DB	TP Sensor Signal	0.6-1.0v
23	---	Not Used	---
24	RD/LG	Distributor Reference Pickup	Digital Signals: 0-5-0v
25	BK	SCI Transmit	0v
26	PK/BR	CCD Bus (+)	Digital Signals: 0-5-0v
27	DB/OR	A/C Damped Pressure Switch	A/C Off: 12v, On: 1v
28	LG	A/C Request Signal	A/C Off: 12v, On: 1v
29	BR	Brake Switch Sense Signal	Brake Off: 0v, On: 12v
30	BK/WT	PNP Switch Sense Signal	In P/N: 0v, Others: 5v
31	---	Not Used	---
32	BK/PK	MIL (lamp) Control	MIL Off: 12v, On: 1v
33	TN/RD	Speed Control Vacuum Solenoid	Vacuum Increasing: 1v
34	DB/RD	A/C WOT Relay Control	Relay Off: 12v, On: 1v
35	GY/YL	EGR Solenoid Control	12v, at 55 mph: 1v
36	---	Not Used	---
37	PK/OR	Overdrive (lamp) Control	O/D On: 1v, O/D Off: 12v
38	PK/BK	Injector 5 Driver	1-4 ms
39	YL/BK	Auto Idle Speed Motor	Pulse Signals
40	BR/WT	Auto Idle Speed Motor	Pulse Signals

Standard Colors and Abbreviations

Abbreviation	Color	Abbreviation	Color	Abbreviation	Color
BK	Black	GY	Gray	RD	Red
BL	Blue	GN	Green	TN	Tan
BR	Brown	LG	Light Green	VT	Violet
DB	Dark Blue	OR	Orange	WT	White
DG	Dark Green	PK	Pink	YL	Yellow

1993 Grand Cherokee 5.2L V8 MFI VIN Y (A/T) 60 Pin Connector

PCM Pin #	Wire Color	Circuit Description (60 Pin)	Value at Hot Idle
41	BK/OR	HO2S-11 (B1 S1) Signal	0.1-1.1v
42	---	Not Used	---
43	GY/LB	Tachometer Signal	Pulse Signals
44	GY/BK	Distributor Sync Pickup	Digital Signals: 0-5-0v
45	BK/YL	SCI Receive	5v
46	WT/GY	CCD Bus (-)	<0.050v
47	WT/OR	Vehicle Speed Signal	Digital Signal
48	BR/RD	Speed Control Set Switch	S/C & Set Switch On: 3.8v
49	YL/RD	Speed Control On/Off Switch	S/C Off: 12v, On: 1v
50	WT/LG	Speed Control Resume Switch	S/C Off: 12v, On: 1v
51	PK	Fuel Pump Relay Control	Relay Off: 12v, On: 1v
52	PK/BK	EVAP Purge Solenoid Control	Solenoid Off: 12v, On: 1v
53	LG/RD	Speed Control Vent Solenoid	Vacuum Decreasing: 1v
54	PK/YL	EMCC Solenoid	Solenoid Off: 12v, On: 1v
55	BR/LG	Overdrive Lockout Solenoid	Solenoid Off: 12v, On: 1v
56	GY/PK	Maintenance Indicator Lamp	Lamp Off: 12v, On: 1v
57	DG/BK	ASD Relay Output	12-14v
58	BR/YL	Injector 6 Driver	1-4 ms
59	GY/RD	Auto Idle Speed Motor	Pulse Signals
60	PK/BK	Auto Idle Speed Motor	Pulse Signals

Pin Connector Graphic

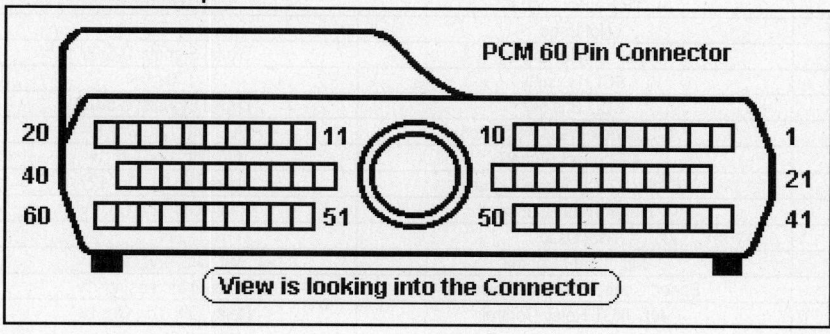

1994-95 Grand Cherokee 5.2L V8 MFI VIN Y 60 Pin Connector

PCM Pin #	Wire Color	Circuit Description (60 Pin)	Value at Hot Idle
1	RD/WT	MAP Sensor Signal	1.5-1.6v
2	TN/BK	ECT Sensor Signal	At 180ºF: 2.80v
3	RD	Fused Battery Power (B+)	12-14v
4	BK/LB	Sensor Ground	<0.1v
5	BK/TN	Power Ground	<0.1v
6	PK/WT	5-Volt Supply	4.9-5.1v
7	WT/BK	8-Volt Supply	7.9-8.1v
8	---	Not Used	---
9	LB	Ignition Switch Power (B+)	12-14v
10	OR/BK	Overdrive Override Switch	Switch On: 1v, Off: 12v
11	BK/TN	Power Ground	<0.1v
12	BK/TN	Power Ground	<0.1v
13	LB/BR	Injector 4 Driver	1-4 ms
14	YL/WT	Injector 3 Driver	1-4 ms
15	TN	Injector 2 Driver	1-4 ms
16	WT/DB	Injector 1 Driver	1-4 ms
17	DB/WT	Injector 7 Driver	1-4 ms
18	DB/YL	Injector 8 Driver	1-4 ms
19	GY/WT	Coil Driver (Dwell)	7º, 55 mph: 9º dwell
20	DG	Alternator Field Control	Digital Signals: 0-12-0v
21	BK/RD	IAT Sensor Signal	At 100ºF: 1.83v
22	OR/DB	TP Sensor Signal	0.6-1.0v
23	---	Not Used	---
24	RD/LG	CKP Sensor Signal	Digital Signals: 0-5-0v
25	BK	SCI Transmit	0v
26	PK/BR	CCD Bus (+)	Digital Signals: 0-5-0v
27	DB/OR	A/C Damped Pressure Switch	A/C Off: 12v, On: 1v
28	LG	A/C Select Signal	A/C Off: 12v, On: 1v
29	BR	Brake Switch Sense Signal	Brake Off: 0v, On: 12v
30	BK/WT	Starter Relay Control	Relay Off: 12v, On: 1v
31	---	Not Used	---
32	BK/PK	MIL (lamp) Control	MIL Off: 12v, On: 1v
33	TN/RD	Speed Control Vacuum Solenoid	Vacuum Increasing: 1v
34	DB/RD	A/C WOT Relay Control	Relay Off: 12v, On: 1v
35	GY/YL	EGR Solenoid Control	12v, at 55 mph: 1v
36	---	Not Used	---
37	PK/OR	Overdrive (lamp) Control	O/D On: 1v, O/D Off: 12v
38	GY	Injector 5 Driver	1-4 ms
39	YL/BK	AIS Motor 4 Circuit Control	Pulse Signals
40	BR/WT	AIS Motor 2 Circuit Control	Pulse Signals

Standard Colors and Abbreviations

Abbreviation	Color	Abbreviation	Color	Abbreviation	Color
BK	Black	GY	Gray	RD	Red
BL	Blue	GN	Green	TN	Tan
BR	Brown	LG	Light Green	VT	Violet
DB	Dark Blue	OR	Orange	WT	White
DG	Dark Green	PK	Pink	YL	Yellow

1994-95 Grand Cherokee 5.2L V8 MFI VIN Y 60 Pin Connector

PCM Pin #	Wire Color	Circuit Description (60 Pin)	Value at Hot Idle
41	BK/OR	HO2S-11 (B1 S1) Signal	0.1-1.1v
42	---	Not Used	---
43	GY/LB	Tachometer Signal	Pulse Signals
44	GY/BK	CMP Sensor Signal	Digital Signals: 0-5-0v
45	BK/YL	SCI Receive	5v
47	WT/OR	Vehicle Speed Signal	Digital Signal
48	BR/RD	Speed Control Set Switch	S/C & Set Switch On: 3.8v
49	YL/RD	Speed Control On/Off Switch	S/C Off: 12v, On: 1v
50	WT/LG	Speed Control Resume Switch	S/C Off: 12v, On: 1v
51	PK	Fuel Pump Relay Control	Relay Off: 12v, On: 1v
52	PK/BK	EVAP Purge Solenoid Control	Solenoid Off: 12v, On: 1v
53	LG/RD	Speed Control Vent Solenoid	Vacuum Decreasing: 1v
54	PK/YL	TCC Solenoid Control	TCC on at Cruise: 1v
55	BR/LG	Overdrive Lockout Solenoid	Solenoid Off: 12v, On: 1v
56	---	Not Used	---
57	DG/BK	ASD Relay Output	12-14v
58	BR/YL	Injector 6 Driver	1-4 ms
59	PK/BK	AIS Motor 1 Circuit Control	Pulse Signals
60	YL/BK	AIS Motor 3 Circuit Control	Pulse Signals

Pin Connector Graphic

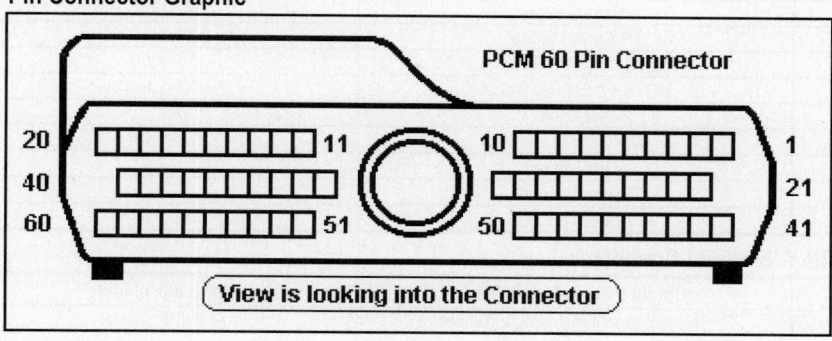

1996-98 Grand Cherokee 5.2L VIN Y 'A' Black Connector

PCM Pin #	Wire Color	Circuit Description (32 Pin)	Value at Hot Idle
1	---	Not Used	---
2	OR	Ignition Switch Power (B+)	12-14v
3	---	Not Used	---
4	BK/LB	Sensor Ground	<0.1v
5	---	Not Used	---
6	BK/WT	PNP Switch Sense Signal	In P/N: 0v, Others: 5v
7	GY/WT	Coil 1 Driver Control	7°, 55 mph: 9° dwell
8	RD/LG	CKP Sensor Signal	Digital Signals: 0-5-0v
9	---	Not Used	---
10	PK/BK	IAC 4 Driver Control	Pulse Signals
11	BR/WT	IAC 3 Driver Control	Pulse Signals
12-13	---	Not Used	---
14	YL/RD	Fused Battery Power (B+)	12-14v
15	BK/RD	IAT Sensor Signal	At 100°F: 1.83v
16	TN/BK	ECT Sensor Signal	At 180°F: 2.80v
17	WT/BK	5-Volt Supply	4.9-5.1v
18	GY/BK	CMP Sensor Signal	Digital Signals: 0-5-0v
19	YL/RD	IAC 2 Driver Control	Pulse Signals
20	GY/RD	IAC 1 Driver Control	Pulse Signals
21	---	Not Used	---
22	RD/YL	Fused Battery Power (B+)	12-14v
23	OR/DB	TP Sensor Signal	0.6-1.0v
24	BK/OR	HO2S-11 (B1 S1) Signal	0.1-1.1v
25	BK/PK	HO2S-12 (B1 S2) Signal	0.1-1.1v
26	---	Not Used	---
27	RD/WT	MAP Sensor Signal	1.5-1.6v
28-30	---	Not Used	---
31	BK/TN	Power Ground	<0.1v
32	BK/TN	Power Ground	<0.1v

1996-98 Grand Cherokee 5.2L VIN Y 'B' White Connector

PCM Pin #	Wire Color	Circuit Description (32 Pin)	Value at Hot Idle
1	PK	Transmission Temperature Sensor	At 200°F: 2.40v
2	DB/WT	Injector 7 Driver	1-4 ms
3	---	Not Used	---
4	WT/DB	Injector 1 Driver	1-4 ms
5	YL/WT	Injector 3 Driver	1-4 ms
6	GY	Injector 5 Driver	1-4 ms
7	---	Not Used	---
8	PK	Governor Pressure Solenoid	PWM Signal: 0-12-0v
10	DG	Generator Field Control	Digital Signals: 0-12-0v
11	DG/LB	TCC Solenoid Control	TCC on at Cruise: 1v
12	BR/YL	Injector 6 Driver	1-4 ms
13	DB/YL	Injector 8 Driver	1-4 ms
14	---	Not Used	---
15	TN	Injector 2 Driver	1-4 ms
16	LB/BR	Injector 4 Driver	1-4 ms
17-20	---	Not Used	---
21	BR	Overdrive Solenoid Control	Solenoid Off: 12v, On: 1v
22, 24	---	Not Used	---
23	GY/WT	Engine Oil Pressure Sensor	1.6v at 24 psi
25	DB/BK	OSS Sensor Signal (-)	AC pulse signals
26	---	Not Used	---
27	WT/OR	Vehicle Speed Signal	Digital Signal
28	LG/WT	OSS Sensor Signal (+)	AC pulse signals
29	LG	Governor Pressure Signal	0.58v
30	BR/OR	Transmission Relay Control	Relay Off: 12v, On: 1v
31	PK/WT	5-Volt Supply	4.9-5.1v
32	---	Not Used	---

1996-98 Grand Cherokee 5.2L VIN Y 'C' Gray Connector

PCM Pin #	Wire Color	Circuit Description (32 Pin)	Value at Hot Idle
1	DB/RD	A/C Clutch Relay Control	Relay Off: 12v, On: 1v
2	---	Not Used	
3	PK/WT	ASD Relay Control	Relay Off: 12v, On: 1v
4	TN/RD	Speed Control Vacuum Solenoid	Vacuum Increasing: 1v
5	LG/RD	Speed Control Vent Solenoid	Vacuum Decreasing: 1v
6	BR/YL	Overdrive 'Off' (lamp) Control	O/D On: 1v, O/D Off: 12v
7-9	---	Not Used	---
10	DG/RD	LDP Solenoid Control	PWM Signal: 0-12-0v
11	YL/RD	Speed Control Power Supply	12-14v
12	DG/OR	ASD Relay Output	12-14v
13	OR	Overdrive 'Off' Switch Sense	Switch On: 1v, Off: 12v
14	PK/RD	LDP Switch Sense Signal	Closed: 0v, Open: 12v
15	RD/YL	Battery Temperature Sensor	At 86°F: 1.96v
16-18	---	Not Used	---
19	DB	Fuel Pump Relay Control	Relay Off: 12v, On: 1v
20	PK/BK	EVAP Purge Solenoid Control	PWM Signal: 0-12-0v
21	---	Not Used	---
22	DB/BK	A/C Damped Pressure Switch	A/C Off: 12v, On: 1v
23	---	Not Used	---
24	BR	Brake Switch Sense Signal	Brake Off: 0v, On: 12v
25 ('98)	DG/PK	Generator Field Source	12-14v
26	LB/BK	Fuel Level Sensor Signal	Full: 0.5v, 1/2 full: 2.5v
27	BK/PK	SCI Transmit	0v
28	WT/BK	CCD Bus (+)	Digital Signals: 0-5-0v
29	BK/WT	SCI Receive	5v
30	PK/BR	CCD Bus (-)	<0.050v
31	---	Not Used	---
32	RD/LG	Speed Control Switch Signal	S/C & Set Switch On: 3.8v
32 ('98)	PK	Speed Control Switch Signal	S/C & Set Switch On: 3.8v

Pin Connector Graphic

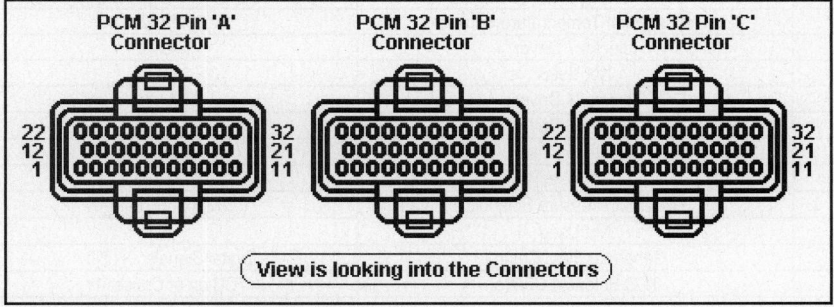

Standard Colors and Abbreviations

Abbreviation	Color	Abbreviation	Color	Abbreviation	Color
BK	Black	GY	Gray	RD	Red
BL	Blue	GN	Green	TN	Tan
BR	Brown	LG	Light Green	VT	Violet
DB	Dark Blue	OR	Orange	WT	White
DG	Dark Green	PK	Pink	YL	Yellow

1998 Grand Cherokee 5.9L V8 VIN Z 'A' Black Connector

PCM Pin #	Wire Color	Circuit Description (32 Pin)	Value at Hot Idle
1	---	Not Used	---
2	OR	Ignition Switch Power (B+)	12-14v
3	---	Not Used	---
4	BK/LB	Sensor Ground	<0.1v
5	---	Not Used	---
6	BK/WT	PNP Switch Sense Signal	In P/N: 0v, Others: 5v
7	GY/WT	Coil 1 Driver Control	7°, 55 mph: 9° dwell
8	RD/LG	CKP Sensor Signal	Digital Signals: 0-5-0v
9	---	Not Used	---
10	PK/BK	IAC 4 Driver Control	Pulse Signals
11	BR/WT	IAC 3 Driver Control	Pulse Signals
12-14	---	Not Used	---
15	BK/RD	IAT Sensor Signal	At 100°F: 1.83v
16	TN/BK	ECT Sensor Signal	At 180°F: 2.80v
17	WT/BK	5-Volt Supply	4.9-5.1v
18	GY/BK	CMP Sensor Signal	Digital Signals: 0-5-0v
19	YL/BK	IAC 2 Driver Control	Pulse Signals
20	GY/RD	IAC 1 Driver Control	Pulse Signals
21	---	Not Used	---
22	RD/YL	Fused Battery Power (B+)	12-14v
23	OR/DB	TP Sensor Signal	0.6-1.0v
24	BK/OR	HO2S-11 (B1 S1) Signal	0.1-1.1v
25	BK/PK	HO2S-12 (B1 S2) Signal	0.1-1.1v
26	---	Not Used	---
27	RD/WT	MAP Sensor Signal	1.5-1.6v
28-30	---	Not Used	---
31	BK/TN	Power Ground	<0.1v
32	BK/TN	Power Ground	<0.1v

1998 Grand Cherokee 5.9L V8 VIN Z 'B' White Connector

PCM Pin #	Wire Color	Circuit Description (32 Pin)	Value at Hot Idle
1	PK	Transmission Temperature Sensor	At 200°F: 2.40v
2	DB/WT	Injector 7 Driver	1-4 ms
3	---	Not Used	---
4	WT/DB	Injector 1 Driver	1-4 ms
5	YL/WT	Injector 3 Driver	1-4 ms
6	GY	Injector 5 Driver	1-4 ms
7	---	Not Used	---
8	PK	Governor Pressure Solenoid	PWM Signal: 0-12-0v
9	---	Not Used	---
10	DG	Generator Field Control	Digital Signals: 0-12-0v
11	DG/LB	TCC Solenoid Control	TCC on at Cruise: 1v
12	BR/YL	Injector 6 Driver	1-4 ms
13	DB/YL	Injector 8 Driver	1-4 ms
14	---	Not Used	---
15	TN	Injector 2 Driver	1-4 ms
16	LB/BR	Injector 4 Driver	1-4 ms
17-20	---	Not Used	---
21	BR	Overdrive Solenoid Control	Solenoid on at Cruise: 1v
22	---	Not Used	---
23	GY/WT	Engine Oil Pressure Sensor	1.6v at 24 psi
24, 26	---	Not Used	---
25	DB/BK	OSS Sensor Signal (-)	AC pulse signals
27	WT/OR	Vehicle Speed Signal	Digital Signal
28	LG/WT	OSS Sensor Signal (+)	AC pulse signals
29	LG	Governor Pressure Signal	0.58v
30	BR/OR	Transmission Relay Control	Relay Off: 12v, On: 1v
31	PK/WT	5-Volt Supply	4.9-5.1v
32	---	Not Used	---

1998 Grand Cherokee 5.9L V8 VIN Z 'C' Gray Connector

PCM Pin #	Wire Color	Circuit Description (32 Pin)	Value at Hot Idle
1	DB/RD	A/C Clutch Relay Control	Relay Off: 12v, On: 1v
2	---	Not Used	---
3	PK/WT	ASD Relay Control	Relay Off: 12v, On: 1v
4	TN/RD	Speed Control Vacuum Solenoid	Vacuum Increasing: 1v
5	LG/RD	Speed Control Vent Solenoid	Vacuum Decreasing: 1v
6	BR/YL	Overdrive 'Off' (lamp) Control	O/D On: 1v, O/D Off: 12v
7-9	---	Not Used	---
10	DG/RD	LDP Solenoid Control	PWM Signal: 0-12-0v
11	YL/RD	Speed Control Power Supply	12-14v
12	DG/OR	ASD Relay Output	12-14v
13	OR	Overdrive 'Off' Switch Sense	Switch On: 1v, Off: 12v
14	PK/RD	LDP Switch Sense Signal	Closed: 0v, Open: 12v
15	RD/YL	Battery Temperature Sensor	At 86°F: 1.96v
16-18	---	Not Used	---
19	DB	Fuel Pump Relay Control	Relay Off: 12v, On: 1v
20	PK/BK	EVAP Purge Solenoid Control	PWM Signal: 0-12-0v
21	---	Not Used	---
22	DB/BK	A/C Switch Sense	A/C Off: 12v, On: 1v
23	---	Not Used	---
24	BR	Brake Switch Sense Signal	Brake Off: 0v, On: 12v
25	DG/PK	Generator Field Source	12-14v
26	LB/BK	Fuel Level Sensor Signal	Full: 0.5v, 1/2 full: 2.5v
27	BK/PK	SCI Transmit	0v
28	WT/BK	CCD Bus (-)	<0.050v
29	BK/WT	SCI Receive	5v
30	PK/BR	CCD Bus (+)	Digital Signals: 0-5-0v
31	---	Not Used	---
32	PK	Speed Control Switch Signal	S/C & Set Switch On: 3.8v

Pin Connector Graphic

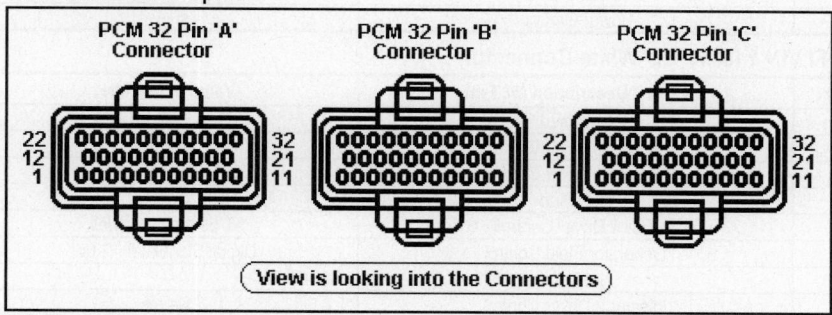

PCM 32 Pin 'A' Connector
PCM 32 Pin 'B' Connector
PCM 32 Pin 'C' Connector

View is looking into the Connectors

Standard Colors and Abbreviations

Abbreviation	Color	Abbreviation	Color	Abbreviation	Color
BK	Black	GY	Gray	RD	Red
BL	Blue	GN	Green	TN	Tan
BR	Brown	LG	Light Green	VT	Violet
DB	Dark Blue	OR	Orange	WT	White
DG	Dark Green	PK	Pink	YL	Yellow

2002-03 Liberty 2.4L I4 DOHC MFI VIN 1 (M/T) 'C1' Black Connector

PCM Pin #	Wire Color	Circuit Description (32 Pin)	Value at Hot Idle
1	---	Not Used	---
2	DB/WT	Ignition Switch Power (Start-Run)	12-14v
3	---	Not Used	---
4	BK/LB	Sensor Ground	<0.050v
5-6	---	Not Used	---
7	BK/GY	Coil 1 Driver Control	5°, 55 mph: 8° dwell
8	GY/BK	CKP Sensor Signal	Digital Signals: 0-5-0v
9	---	Not Used	---
10	YL/BK	IAC 2 Driver Control	Pulse Signals
11	BR/WT	IAC 1 Driver Control	Pulse Signals
12	DB/OR	PSP Switch Sense Signal	Straight: 0v, Turning: 5v
13	YL/RD	Clutch Interlock Relay Control	Clutch Out: 12v, In: 0v
14	BR/WT	Transfer Case Position Switch	Switch Open: 12v, Closed: 0v
15	BK/RD	IAT Sensor Signal	At 100°F: 1.83v
16	TN/BK	ECT Sensor Signal	At 180°F: 2.80v
17	OR	5-Volt Supply	4.9-5.1v
18	TN/YL	CMP Sensor Signal	Digital Signals: 0-5-0v
19	GY/RD	IAC 3 Driver Control	DC Pulse Signals
20	VT/BK	IAC 4 Driver Control	DC Pulse Signals
21	---	Not Used	---
22	RD/WT	Battery Power (B+)	12-14v
23	OR/DB	TP Sensor Signal	0.6-1.0v
24	BK/DG	HO2S-11 (B1 S1) Signal	0.1-1.1v
25	TN/WT	HO2S-12 (B1 S2) Signal	0.1-1.1v
26	---	Not Used	---
27	DG/RD	MAP Sensor Signal	1.5-1.6v
28-30	---	Not Used	---
31	BK/DB	Power Ground	<0.1v
32	BK/DB	Power Ground	<0.1v

2002-03 Liberty 2.4L I4 DOHC MFI VIN 1 (M/T) 'C2' White Connector

PCM Pin #	Wire Color	Circuit Description (32 Pin)	Value at Hot Idle
1-3	---	Not Used	---
4	WT/DB	Injector 1 Driver Control	1-4 ms
5	YL/WT	Injector 3 Driver Control	1-4 ms
6-8	---	Not Used	---
9	DB/TN	Coil 2 Driver Control	5°, 55 mph: 8° dwell
10	DG	Generator Field Control	Digital Signals: 0-12-0v
11-14	---	Not Used	---
15	TN	Injector 2 Driver Control	1-4 ms
16	LB/BR	Injector 4 Driver Control	1-4 ms
17	LG	High Speed Radiator Fan Relay	Relay Off: 12v, On: 1v
18	---	Not Used	---
19	DB	A/C Pressure Sensor	A/C On: 0.451-4.850v
20-22	---	Not Used	---
23	GY/YL	Oil Pressure Sensor	1.6v at 24 psi
24-30	---	Not Used	---
31	VT/WT	5-Volt Supply	4.9-5.1v
32	---	Not Used	---

Standard Colors and Abbreviations

Abbreviation	Color	Abbreviation	Color	Abbreviation	Color
BK	Black	GY	Gray	RD	Red
BL	Blue	GN	Green	TN	Tan
BR	Brown	LG	Light Green	VT	Violet
DB	Dark Blue	OR	Orange	WT	White
DG	Dark Green	PK	Pink	YL	Yellow

2002-03 Liberty 2.4L I4 DOHC MFI VIN 1 (M/T) 'C3' Gray Connector

PCM Pin #	Wire Color	Circuit Description (32 Pin)	Value at Hot Idle
1	DG	A/C Clutch Relay Control	Relay Off: 12v, On: 1v
2	---	Not Used	---
3	DB/YL	ASD Relay Control	Relay Off: 12v, On: 1v
4	TN/RD	Speed Control Vacuum Solenoid	Vacuum Increasing: 1v
5	LG/RD	Speed Control Vent Solenoid	Vacuum Decreasing: 1v
6	TN	Clutch Switch Override Relay Control	Relay Off: 12v, On: 1v
7	---	Not Used	---
8	BR/OR	HO2S-11 (B1 S1) Heater Control	Heater Off: 12v, On: 1v
9	RD/YL	Oxygen Sensor Downstream Relay Control	Relay Off: 12v, On: 1v
10	WT/DG	LDP Solenoid Control	PWM Signal: 0-12-0v
11	YL/RD	Speed Control Power Supply	12-14v
12	OR/DG	ASD Relay Output	12-14v
13	YL/DG	Torque Management Request	Digital Signals
14	OR	LDP Switch Sense Signal	Closed: 0v, Open: 12v
15	PK/YL	Battery Temperature Sensor	At 86°F: 1.96v
16	BR/WT	HO2S-12 (B1 S2) Heater Control	Heater Off: 12v, On: 1v
17	DG/YL	Vehicle Speed Signal	Digital Signals
18	---	Not Used	---
19	BR	Fuel Pump Relay Control	Relay Off: 12v, On: 1v
20	PK/BK	EVAP Purge Solenoid Control	PWM Signal: 0-12-0v
21	---	Not Used	---
22	DB/OR	A/C Switch Sense	A/C Off: 12v, On: 1v
23	---	Not Used	---
24	WT/PK	Brake Switch Sense	Brake Off: 0v, On: 12v
25	WT/DB	Generator Field Source	12-14v
26	DB/WT	Fuel Level Sensor Signal	Full: 0.5v, 1/2 full: 2.5v
27	PK	SCI Transmit	0v
28	---	Not Used	---
29	LG	SCI Receive	5v
30	VT/YL	PCI Bus Signal (J1850)	Digital Signals: 0-7-0v
31	---	Not Used	---
32	RD/LG	Speed Control Switch Signal	S/C & Set Switch On: 3.8v

Pin Connector Graphic

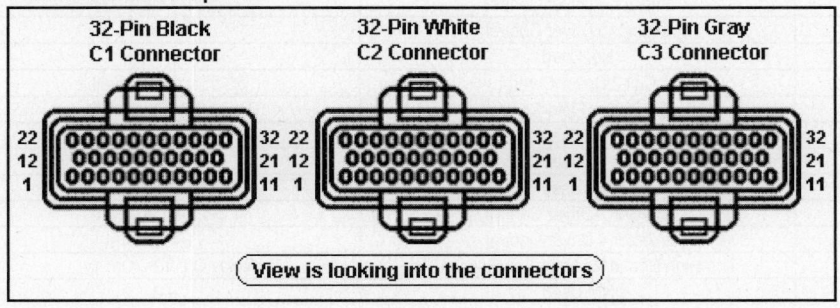

2002-03 Liberty 3.7L V6 SOHC MFI VIN K (All) 'C1' Black Connector

PCM Pin #	Wire Color	Circuit Description (32 Pin)	Value at Hot Idle
1	---	Not Used	---
2	DB/WT	Ignition Switch Power (Start-Run)	12-14v
3	---	Not Used	---
4	BK/LB	Sensor Ground	<0.050v
5-6	---	Not Used	---
7	BK/GY	Coil 1 Driver Control	5°, 55 mph: 8° dwell
8	GY/BK	CKP Sensor Signal	Digital Signals: 0-5-0v
9	---	Not Used	---
10	YL/BK	IAC 2 Driver Control	Pulse Signals
11	BR/WT	IAC 1 Driver Control	Pulse Signals
12	DB/OR	PSP Switch Sense Signal	Straight: 0v, Turning: 5v
13	YL/RD	Clutch Interlock Relay Control	Clutch Out: 12v, In: 0v
14	BR/WT	Transfer Case Position Switch	Switch Open: 12v, Closed: 0v
15	BK/RD	IAT Sensor Signal	At 100°F: 1.83v
16	TN/BK	ECT Sensor Signal	At 180°F: 2.80v
17	OR	5-Volt Supply	4.9-5.1v
18	TN/YL	CMP Sensor Signal	Digital Signals: 0-5-0v
19	GY/RD	IAC 3 Driver Control	DC Pulse Signals
20	VT/BK	IAC 4 Driver Control	DC Pulse Signals
21, 26	---	Not Used	---
22	RD/WT	Battery Power (B+)	12-14v
23	OR/DB	TP Sensor Signal	0.6-1.0v
24	BK/DG	HO2S-11 (B1 S1) Signal	0.1-1.1v
25	TN/WT	HO2S-12 (B1 S2) Signal	0.1-1.1v
27	DG/RD	MAP Sensor Signal	1.5-1.6v
28-30	---	Not Used	---
31	BK/DB	Power Ground	<0.1v
32	BK/DB	Power Ground	<0.1v

2002-03 Liberty 3.7L V6 SOHC MFI VIN K (All) 'C2' White Connector

PCM Pin #	Wire Color	Circuit Description (32 Pin)	Value at Hot Idle
1-3	---	Not Used	---
4	WT/DB	Injector 1 Driver Control	1-4 ms
5	YL/WT	Injector 3 Driver Control	1-4 ms
6	GY	Injector 5 Driver Control	1-4 ms
7-8	---	Not Used	---
9	DB/TN	Coil 2 Driver Control	5°, 55 mph: 8° dwell
10	DG	Generator Field Control	Digital Signals: 0-12-0v
11	---	Not Used	---
12	BR/DB	Injector 6 Driver Control	1-4 ms
13-14	---	Not Used	---
15	TN	Injector 2 Driver Control	1-4 ms
16	LB/BR	Injector 4 Driver Control	1-4 ms
17	LG	High Speed Radiator Fan Relay	Relay Off: 12v, On: 1v
18	---	Not Used	---
19	DB	A/C Pressure Sensor	A/C On: 0.451-4.850v
20-22	---	Not Used	---
23	GY/YL	Oil Pressure Sensor	1.6v at 24 psi
24-30	---	Not Used	---
31	VT/WT	5-Volt Supply	4.9-5.1v
32	---	Not Used	---

Standard Colors and Abbreviations

Abbreviation	Color	Abbreviation	Color	Abbreviation	Color
BK	Black	GY	Gray	RD	Red
BL	Blue	GN	Green	TN	Tan
BR	Brown	LG	Light Green	VT	Violet
DB	Dark Blue	OR	Orange	WT	White
DG	Dark Green	PK	Pink	YL	Yellow

2002-03 Liberty 3.7L V6 SOHC MFI VIN K (All) 'C3' Gray Connector

PCM Pin #	Wire Color	Circuit Description (32 Pin)	Value at Hot Idle
1	DG	A/C Clutch Relay Control	Relay Off: 12v, On: 1v
2	---	Not Used	---
3	DB/YL	ASD Relay Control	Relay Off: 12v, On: 1v
4	TN/RD	Speed Control Vacuum Solenoid	Vacuum Increasing: 1v
5	LG/RD	Speed Control Vent Solenoid	Vacuum Decreasing: 1v
6	TN	Clutch Switch Override Relay Control	Relay Off: 12v, On: 1v
7	---	Not Used	---
8	BR/OR	HO2S-11 (B1 S1) Heater Control	Heater Off: 12v, On: 1v
9	RD/YL	Oxygen Sensor Downstream Relay Control	Relay Off: 12v, On: 1v
10	WT/DG	LDP Solenoid Control	PWM Signal: 0-12-0v
11	YL/RD	Speed Control Power Supply	12-14v
12	OR/DG	ASD Relay Output	12-14v
13	YL/DG	Torque Management Request	Digital Signals
14	OR	LDP Switch Sense Signal	Closed: 0v, Open: 12v
15	PK/YL	Battery Temperature Sensor	At 86°F: 1.96v
16	BR/WT	HO2S-12 (B1 S2) Heater Control	Heater Off: 12v, On: 1v
17	DG/YL	Vehicle Speed Signal	Digital Signals
18	---	Not Used	---
19	BR	Fuel Pump Relay Control	Relay Off: 12v, On: 1v
20	PK/BK	EVAP Purge Solenoid Control	PWM Signal: 0-12-0v
21	---	Not Used	---
22	DB/OR	A/C Switch Sense	A/C Off: 12v, On: 1v
23	---	Not Used	---
24	WT/PK	Brake Switch Sense	Brake Off: 0v, On: 12v
25	WT/DB	Generator Field Source	12-14v
26	DB/WT	Fuel Level Sensor Signal	Full: 0.5v, 1/2 full: 2.5v
27	PK	SCI Transmit	0v
28	---	Not Used	---
29	LG	SCI Receive	5v
30	VT/YL	PCI Bus Signal (J1850)	Digital Signals: 0-7-0v
31	---	Not Used	---
32	RD/LG	Speed Control Switch Signal	S/C & Set Switch On: 3.8v

Pin Connector Graphic

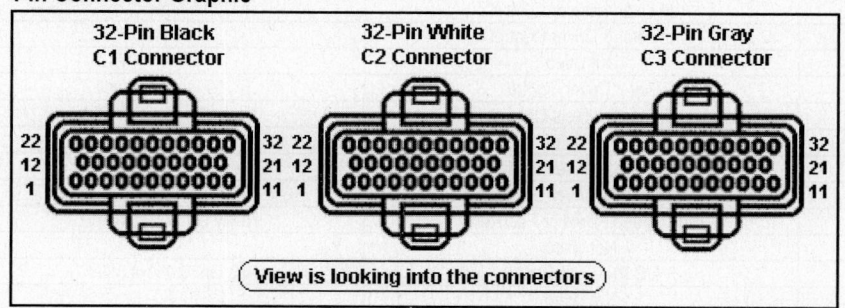

2003 Wrangler 2.4L I4 DOHC MFI VIN 1 (All) 'C1' Black Connector

PCM Pin #	Wire Color	Circuit Description (32 Pin)	Value at Hot Idle
1, 3	---	Not Used	---
2	DB/WT	Ignition Switch Power (Start-Run)	12-14v
4	BK/LB	Sensor Ground	<0.050v
5	---	Not Used	---
6	BK/WT	PNP Switch Sense	In P/N: 0v, Others: 5v
7	BK/GY	Coil 1 Driver Control	5°, 55 mph: 8° dwell
8	GY/BK	CKP Sensor Signal	Digital Signals: 0-5-0v
9	---	Not Used	---
10	YL/BK	IAC 2 Driver Control	Pulse Signals
11	BR/WT	IAC 1 Driver Control	Pulse Signals
12	DB/OR	PSP Switch Sense Signal	Straight: 0v, Turning: 5v
13	YL/RD	Ignition Switch Power (Start)	Cranking: 9-11v
14	---	Not Used	---
15	BK/RD	IAT Sensor Signal	At 100°F: 1.83v
16	TN/BK	ECT Sensor Signal	At 180°F: 2.80v
17	OR	5-Volt Supply	4.9-5.1v
18	TN/YL	CMP Sensor Signal	Digital Signals: 0-5-0v
19	GY/RD	IAC 3 Driver Control	DC Pulse Signals
20	VT/BK	IAC 4 Driver Control	DC Pulse Signals
21	---	Not Used	---
22	RD/WT	Battery Power (B+)	12-14v
23	OR/DB	TP Sensor Signal	0.6-1.0v
24	BK/DG	HO2S-11 (B1 S1) Signal	0.1-1.1v
25	TN/WT	HO2S-12 (B1 S2) Signal	0.1-1.1v
26	---	Not Used	---
27	DG/RD	MAP Sensor Signal	1.5-1.6v
28-30	---	Not Used	---
31	BK/TN	Power Ground	<0.1v
32	BK/TN	Power Ground	<0.1v

2003 Wrangler 2.4L I4 DOHC MFI VIN 1 (All) 'C2' White Connector

PCM Pin #	Wire Color	Circuit Description (32 Pin)	Value at Hot Idle
1-3	---	Not Used	---
4	WT/DB	Injector 1 Driver Control	1-4 ms
5	YL/WT	Injector 3 Driver Control	1-4 ms
6-8	---	Not Used	---
9	DB/TN	Coil 2 Driver Control	5°, 55 mph: 8° dwell
10	DG	Generator Field Control	Digital Signals: 0-12-0v
11-14	---	Not Used	---
15	TN	Injector 2 Driver Control	1-4 ms
16	LB/BR	Injector 4 Driver Control	1-4 ms
17	LG	High Speed Radiator Fan Relay	Relay Off: 12v, On: 1v
18	---	Not Used	---
19	DB	A/C Pressure Sensor	A/C On: 0.451-4.850v
20-22	---	Not Used	---
23	GY/YL	Oil Pressure Sensor	1.6v at 24 psi
24-26	---	Not Used	---
27	WT/OR	Vehicle Speed Signal	Digital Signals
28-30	---	Not Used	---
31	VT/WT	5-Volt Supply	4.9-5.1v
32	---	Not Used	---

Standard Colors and Abbreviations

Abbreviation	Color	Abbreviation	Color	Abbreviation	Color
BK	Black	GY	Gray	RD	Red
BL	Blue	GN	Green	TN	Tan
BR	Brown	LG	Light Green	VT	Violet
DB	Dark Blue	OR	Orange	WT	White
DG	Dark Green	PK	Pink	YL	Yellow

2003 Wrangler 2.4L I4 DOHC MFI VIN I (All) 'C3' Gray Connector

PCM Pin #	Wire Color	Circuit Description (32 Pin)	Value at Hot Idle
1	OR	A/C Clutch Relay Control	Relay Off: 12v, On: 1v
2	---	Not Used	---
3	DB/YL	ASD Relay Control	Relay Off: 12v, On: 1v
4	TN/RD	Speed Control Vacuum Solenoid	Vacuum Increasing: 1v
5	LG/RD	Speed Control Vent Solenoid	Vacuum Decreasing: 1v
6-9	---	Not Used	---
10	WT/DG	LDP Solenoid Control	PWM Signal: 0-12-0v
11	YL/RD	Speed Control Power Supply	12-14v
12	DG/PK	ASD Relay Output	12-14v
13	YL/DG	Torque Management Request	Digital Signals
14	OR	LDP Switch Sense Signal	Closed: 0v, Open: 12v
15	PK/YL	Battery Temperature Sensor	At 86°F: 1.96v
16-18	---	Not Used	---
19	BR	Fuel Pump Relay Control	Relay Off: 12v, On: 1v
20	PK/BK	EVAP Purge Solenoid Control	PWM Signal: 0-12-0v
21	---	Not Used	---
22	DB/OR	A/C Switch Signal	A/C Off: 12v, On: 1v
23	LG	A/C Select Signal	A/C Off: 12v, On: 1v
24	WT/PK	Brake Switch Sense Signal	Brake Off: 0v, On: 12v
25	WT/DB	Generator Field Source	12-14v
26	DB/LG	Fuel Level Sensor Signal	Full: 0.5v, 1/2 full: 2.5v
27	PK	SCI Transmit	0v
28	---	Not Used	---
29	LG/WT	SCI Receive	5v
30	VT/YL	PCI Bus Signal (J1850)	Digital Signals: 0-7-0v
31	---	Not Used	---
32	RD/LG	Speed Control Switch Signal	S/C & Set Switch On: 3.8v

Pin Connector Graphic

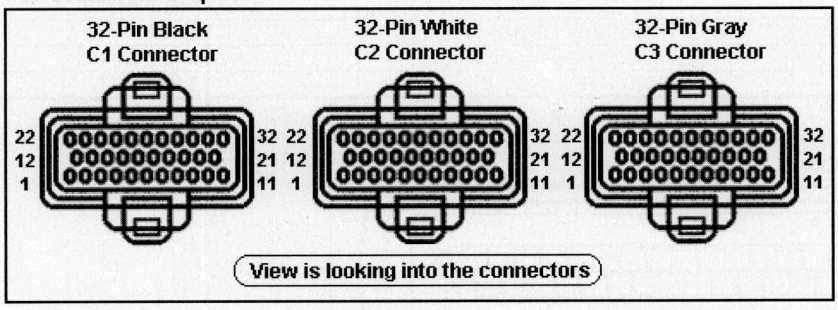

1990 Wrangler 2.5L I4 TBI VIN E (M/T) 35 Pin Connector

PCM Pin #	Wire Color	Circuit Description (35 Pin)	Value at Hot Idle
1	BK	Power Ground	<0.1v
2	BK	Power Ground	<0.1v
3	YL	Ignition Switch Run Output	12-14v
4	RD	Fused Battery Power (B+)	12-14v
5	BL	EGR Solenoid Control	12v, at 55 mph: 1v
6	OR	Fuel Pump Relay Control	Relay Off: 12v, On: 1v
7	BK	Latch Relay B+ Control	Relay Off: 12v, On: 1v
8	GY/BK	WOT Relay Control	Relay Off: 12v, On: 1v
10	BK	Sensor Ground	<0.1v
11	RD	CKP Sensor (+) Signal	Digital Signals: 0-5-0v
12	BK	PNP Switch Sense Signal	In P/N: 0v, Others: 5v
13	BR	TP Sensor Ground	<0.1v
14	TN	Air Temperature Sensor	At 100°F: 2.51v
15	TN	ECT Sensor Signal	At 180°F: 2.80v
16	RD	MAP Sensor VREF	4.9-5.1v
17	BK	Map Sensor Ground	<0.1v
18	PK	Shift Indicator Lamp Control	Lamp Off: 12v, On: 1v
19	PK	B+ Latch Relay Output	Relay Off: 12v, On: 1v
21	LB	Injector Control Driver	1-4 ms
22	BL	A/C Clutch Relay Control	A/C Off: 12v, On: 1v
23	BR	Idle Speed Motor (-) Retract	Pulse Signals
24	LG	Idle Speed Motor (+) Extend	Pulse Signals
25	GY	Closed Throttle Sensor	Open: 12v, Closed: 1v
27	OR	Ignition Coil Control	5°, 55 mph: 8° dwell
28	WT	CKP Sensor (-) Signal	Digital Signals: 0-5-0v
29	GN	Starter Input Signal	Cranking: 9-11v
30	LG	A/C Select Switch Sense	A/C Off: 12v, On: 1v
31	YL/GN	TP Sensor Signal	0.6-1.0v
32	BR	ECT Sensor Ground	<0.1v
33	PK	MAP Sensor Signal	1.5-1.6v
34	LB	A/C Request Switch Sense	A/C Off: 12v, On: 1v
35	OR	HO2S-11 (B1 S1) Signal	0.1-1.1v

Pin Connector Graphic

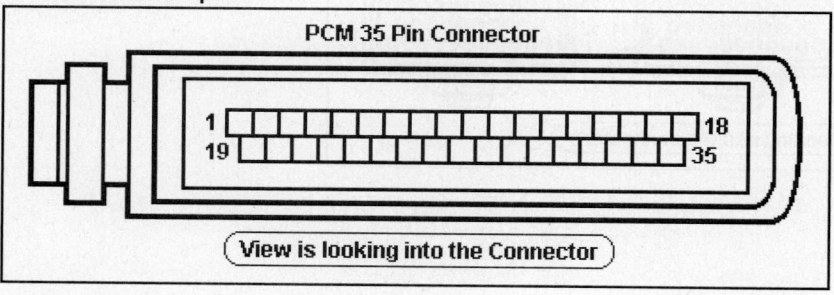

PCM 35 Pin Connector

1 ... 18
19 ... 35

View is looking into the Connector

1991-92 Wrangler 2.5L I4 MFI VIN P (All) 60 Pin Connector

PCM Pin #	Wire Color	Circuit Description (60 Pin)	Value at Hot Idle
1	RD/WT	MAP Sensor Signal	1.5-1.6v
2	TN	ECT Sensor Signal	At 180°F: 2.80v
3	PK/YL	Fused Battery Power (B+)	12-14v
4	BR/RD	Sensor Ground	<0.1v
5	BK/YL	Sensor Ground	<0.1v
6	BR/YL	5-Volt Supply	4.9-5.1v
7	WT/BK	8-Volt Supply	7.9-8.1v
8 ('91)	DG/LG	Starter Input Signal	Cranking: 9-11v
9	YL/WT	Ignition Switch Power (B+)	12-14v
10	---	Not Used	---
11, 12	BK	Power Ground	<0.1v
13	YL	Injector 4 Driver	1-4 ms
14	TN/YL	Injector 3 Driver	1-4 ms
15	LG	Injector 2 Driver	1-4 ms
16	LB	Injector 1 Driver	1-4 ms
17-18, 23, 26	---	Not Used	---
19	YL/BK	Coil 1 Driver Control	5°, 55 mph: 8° dwell
20	TN/BK	Alternator Field Control	Digital Signals: 0-12-0v
21	TN/DB	Air Temperature Sensor	At 100°F: 2.51v
22	YL/DG	TP Sensor Signal	0.6-1.0v
24	RD/DG	Distributor Reference Pickup	Digital Signals: 0-5-0v
25	BK/RD	SCI Transmit	0v
27	LB/WT	A/C Request Switch Sense	A/C Off: 12v, On: 1v
28	LB/RD	A/C Select Switch Sense	A/C Off: 12v, On: 1v
29	LB/YL	Brake Switch Sense Signal	Brake Off: 0v, On: 12v
30	BK/WT	PNP Switch Sense Signal	In P/N: 0v, Others: 5v
31, 33, 35-38	---	Not Used	---
32	DG/RD	MIL (lamp) Control	MIL Off: 12v, On: 1v
34	DB/WT	A/C Clutch Relay Control	Relay Off: 12v, On: 1v
39	DG/BK	AIS Motor 'A' Control	Pulse Signals
40	RD/YL	AIS Motor 'B' Control	Pulse Signals
41	GY	HO2S-11 (B1 S1) Signal	0.1-1.1v
42, 46	---	Not Used	---
43	OR	Tachometer Signal	Pulse Signals
44	GY/BK	Distributor Sync Pickup	Digital Signals: 0-5-0v
45	BK/PK	SCI Receive	5v
47	DB	Vehicle Speed Signal	Digital Signal
48-50	---	Not Used	---
51	OR/DG	ASD Relay Control	Relay Off: 12v, On: 1v
52-53, 55	---	Not Used	---
54	BR	Shift Indicator Lamp Control	Lamp Off: 12v, On: 1v
56	DG/OR	Maintenance Indicator Lamp	Lamp Off: 12v, On: 1v
57	DG/BK	ASD Relay Output	12-14v
58	---	Not Used	---
59	DB/YL	AIS Motor 'D' Control	Pulse Signals
60	PK/BK	AIS Motor 'C' Control	Pulse Signals

Pin Connector Graphic

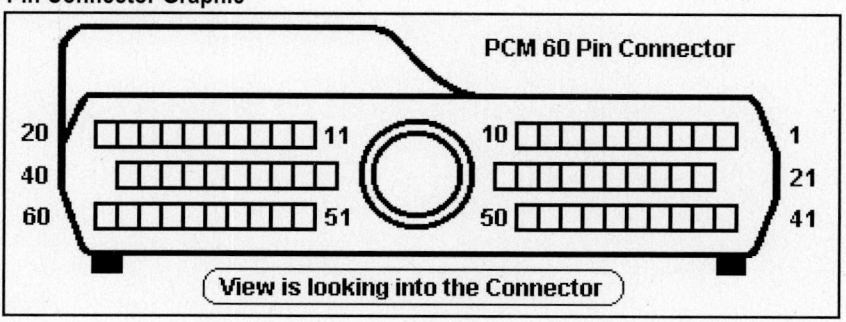

1993-95 Wrangler 2.5L I4 MFI VIN P (All) 60 Pin Connector

PCM Pin #	Wire Color	Circuit Description (60 Pin)	Value at Hot Idle
1	DG/RD	MAP Sensor Signal	1.5-1.6v
2	TN/BK	ECT Sensor Signal	At 180°F: 2.80v
3	RD/WT	Fused Battery Power (B+)	12-14v
4	BK/LB	Sensor Ground	<0.1v
5	BK/WT	Sensor Ground	<0.1v
6	PK/WT	5-Volt Supply	4.9-5.1v
7	OR	8-Volt Supply	7.9-8.1v
8	---	Not Used	---
9	DB/WT	Ignition Switch Power (B+)	12-14v
10	---	Not Used	---
11	BK	Power Ground	<0.1v
12	BK	Power Ground	<0.1v
13	LB/BR	Injector 4 Driver	1-4 ms
14	YL/WT	Injector 3 Driver	1-4 ms
15	TN	Injector 2 Driver	1-4 ms
16	WT/DB	Injector 1 Driver	1-4 ms
17-18	---	Not Used	---
19	GY	Coil 1 Driver Control	5°, 55 mph: 8° dwell
20	DG	Alternator Field Control	Digital Signals: 0-12-0v
21	BK/RD	IAT Sensor Signal	At 100°F: 1.83v
22	OR/DB	TP Sensor Signal	0.6-1.0v
23	---	Not Used	---
24	GY/BK	CKP Sensor Signal	Digital Signals: 0-5-0v
25	BK	SCI Transmit	0v
26	---	Not Used	---
27	LB	A/C Request Switch Sense	A/C Off: 12v, On: 1v
28	BR	A/C Select Switch Sense	A/C Off: 12v, On: 1v
29	WT/PK	Brake Switch Sense Signal	Brake Off: 0v, On: 12v
30	BR/YL	PNP Switch Sense Signal	In P/N: 0v, Others: 5v
31	---	Not Used	---
32	BK/PK	MIL (lamp) Control	MIL Off: 12v, On: 1v
33	---	Not Used	---
34	DB/OR	A/C Clutch Relay Control	Relay Off: 12v, On: 1v
35-38	---	Not Used	---
39	GY/RD	AIS Motor 1 Circuit Control	Pulse Signals
40	BR/WT	AIS Motor 3 Circuit Control	Pulse Signals

Standard Colors and Abbreviations

Abbreviation	Color	Abbreviation	Color	Abbreviation	Color
BK	Black	GY	Gray	RD	Red
BL	Blue	GN	Green	TN	Tan
BR	Brown	LG	Light Green	VT	Violet
DB	Dark Blue	OR	Orange	WT	White
DG	Dark Green	PK	Pink	YL	Yellow

1993-95 Wrangler 2.5L I4 MFI VIN P (All) 60 Pin Connector

PCM Pin #	Wire Color	Circuit Description (60 Pin)	Value at Hot Idle
41	BK/DG	HO2S-11 (B1 S1) Signal	0.1-1.1v
42	---	Not Used	---
43	GY/LB	Tachometer Signal	Pulse Signals
44	TN/YL	CMP Sensor Signal	Digital Signals: 0-5-0v
45	LG	SCI Receive	5v
46	---	Not Used	---
47	WT/OR	Vehicle Speed Signal	Digital Signal
48-50	---	Not Used	---
51	DB/YL	ASD Relay Control	Relay Off: 12v, On: 1v
52-53	---	Not Used	---
54	OR/BK	Shift Indicator Lamp Control	Lamp Off: 12v, On: 1v
55	---	Not Used	---
56 ('93)	GY/PK	Maintenance Indicator Lamp	Lamp Off: 12v, On: 1v
57	DG/OR	ASD Relay Output	12-14v
58	---	Not Used	---
59	PK/BK	AIS Motor 4 Circuit Control	Pulse Signals
60	YL/BK	AIS Motor 2 Circuit Control	Pulse Signals

Pin Connector Graphic

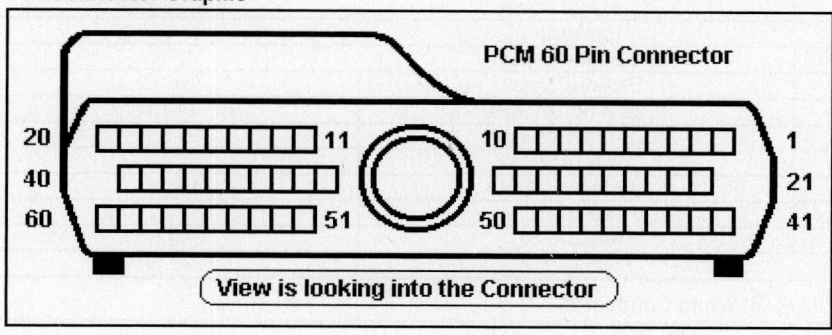

1997 Wrangler 2.5L I4 MFI VIN P (All) 'A' Black Connector

PCM Pin #	Wire Color	Circuit Description (32 Pin)	Value at Hot Idle
1	---	Not Used	---
2	RD/LG	Ignition Switch Power (B+)	12-14v
3	---	Not Used	---
4	BR/YL	Sensor Ground	<0.1v
5	---	Not Used	---
6	BK/LB	PNP Switch Sense Signal	In P/N: 0v, Others: 5v
7	GY	Coil 1 Driver Control	5°, 55 mph: 8° dwell
8	GY/BK	CKP Sensor Signal	Digital Signals: 0-5-0v
9	---	Not Used	---
10	YL/BK	IAC 2 Driver Control	Pulse Signals
11	BR/WT	IAC 3 Driver Control	Pulse Signals
12	DB/BR	PSP Switch Sense Signal	Straight: 0v, Turning: 5v
13-14	---	Not Used	---
15	BK/RD	IAT Sensor Signal	At 100°F: 1.83v
16	TN/BK	ECT Sensor Signal	At 180°F: 2.80v
17	OR	5-Volt Supply	4.9-5.1v
18	TN/YL	CMP Sensor Signal	Digital Signals: 0-5-0v
19	GY/RD	IAC 1 Driver Control	Pulse Signals
20	PK/BK	IAC 4 Driver Control	Pulse Signals
21	---	Not Used	---
22	RD/WT	Fused Battery Power (B+)	12-14v
23	OR/DB	TP Sensor Signal	0.6-1.0v
24	BK/DG	HO2S-11 (B1 S1) Signal	0.1-1.1v
25	TN/WT	HO2S-12 (B1 S2) Signal	0.1-1.1v
26	---	Not Used	---
27	DG/RD	MAP Sensor Signal	1.5-1.6v
28-30	---	Not Used	---
31	BK/TN	Power Ground	<0.1v
32	BK/TN	Power Ground	<0.1v

1997 Wrangler 2.5L I4 MFI VIN P (All) 'B' White Connector

PCM Pin #	Wire Color	Circuit Description (32 Pin)	Value at Hot Idle
1-3	---	Not Used	---
4	WT/DB	Injector 1 Driver	1-4 ms
5	YL/WT	Injector 3 Driver	1-4 ms
6-9	---	Not Used	---
10	DG	Generator Field Control	Digital Signals: 0-12-0v
11	OR/LG	TCC Solenoid Control (ATX)	TCC on at Cruise: 1v
12-14	---	Not Used	---
15	TN	Injector 2 Driver	1-4 ms
16	LB/BR	Injector 4 Driver	1-4 ms
17-22	---	Not Used	---
23	GY/YL	Engine Oil Pressure Sensor	1.6v at 24 psi
24-26	---	Not Used	---
27	WT/OR	Vehicle Speed Signal	Digital Signal
28-30	---	Not Used	---
31	PK/OR	5-Volt Supply	4.9-5.1v
32	---	Not Used	---

Standard Colors and Abbreviations

Abbreviation	Color	Abbreviation	Color	Abbreviation	Color
BK	Black	GY	Gray	RD	Red
BL	Blue	GN	Green	TN	Tan
BR	Brown	LG	Light Green	VT	Violet
DB	Dark Blue	OR	Orange	WT	White
DG	Dark Green	PK	Pink	YL	Yellow

1997 Wrangler 2.5L I4 MFI VIN P (All) 'C' Gray Connector

PCM Pin #	Wire Color	Circuit Description (32 Pin)	Value at Hot Idle
1	DB/OR	A/C Clutch Relay Control	Relay Off: 12v, On: 1v
2	---	Not Used	
3	DB/YL	ASD Relay Control	Relay Off: 12v, On: 1v
4-11	---	Not Used	---
12	DG/OR	ASD Relay Output	12-14v
13-14	---	Not Used	---
15	PK/YL	Battery Temperature Sensor	At 86°F: 1.96v
16-18	---	Not Used	---
19	BR	Fuel Pump Relay Control	Relay Off: 12v, On: 1v
20	PK/BK	EVAP Purge Solenoid Control	PWM Signal: 0-12-0v
21	---	Not Used	---
22	DB/WT	A/C Request Signal	A/C Off: 12v, On: 1v
23	LG/DG	A/C Select Signal	A/C Off: 12v, On: 1v
24	WT/PK	Brake Switch Sense Signal	Brake Off: 0v, On: 12v
25	DG/OR	Generator Field Source	12-14v
26	DB/LG	Fuel Level Sensor Signal	Full: 0.5v, 1/2 full: 2.5v
27	PK	SCI Transmit	0v
28	WT/BK	CCD Bus (-)	<0.050v
29	LG	SCI Receive	5v
30	PK/BR	CCD Bus (+)	Digital Signals: 0-5-0v
31-32	---	---	---

Pin Connector Graphic

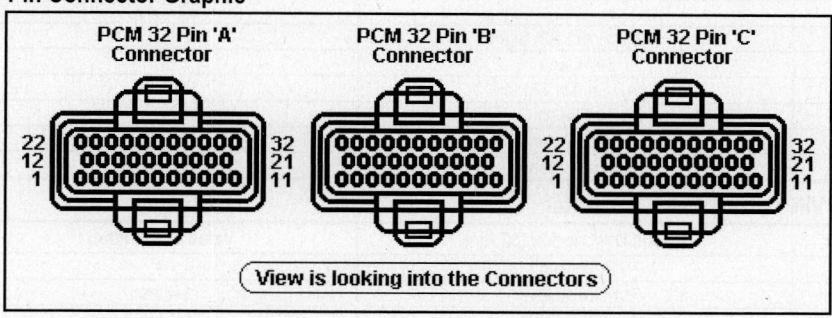

1998-2000 Wrangler 2.5L I4 MFI VIN P 'C1' Black Connector

PCM Pin #	Wire Color	Circuit Description (32 Pin)	Value at Hot Idle
1, 3	---	Not Used	---
2 ('98)	RD/LG	Ignition Switch Power (B+)	12-14v
2	DB	Ignition Switch Power (B+)	12-14v
4	BR/YL	Sensor Ground	<0.1v
5	---	Not Used	---
6	BR/LB	PNP Switch Sense Signal	In P/N: 0v, Others: 5v
7	GY	Coil 1 Driver Control	5°, 55 mph: 8° dwell
8	GY/BK	CKP Sensor Signal	Digital Signals: 0-5-0v
9	---	---	---
10	YL/BK	IAC 2 Driver Control	Pulse Signals
11	BR/WT	IAC 3 Driver Control	Pulse Signals
12	DB/BR	PSP Switch Sense Signal	Straight: 0v, Turning: 5v
13-14	---	Not Used	---
15	BK/RD	IAT Sensor Signal	At 100°F: 1.83v
16	TN/BK	ECT Sensor Signal	At 180°F: 2.80v
17	OR	5-Volt Supply	4.9-5.1v
18	TN/YL	CMP Sensor Signal	Digital Signals: 0-5-0v
19	GY/RD	IAC 1 Driver Control	Pulse Signals
20	PK/BK	IAC 4 Driver Control	Pulse Signals
21	---	Not Used	---
22	RD/WT	Fused Battery Power (B+)	12-14v
23	OR/DB	TP Sensor Signal	0.6-1.0v
24	BK/DG	HO2S-11 (B1 S1) Signal	0.1-1.1v
25	TN/WT	HO2S-12 (B1 S2) Signal	0.1-1.1v
26	---	Not Used	---
27	DG/RD	MAP Sensor Signal	1.5-1.6v
28-30	---	Not Used	---
31	BK/TN	Power Ground	<0.1v
32	BK/TN	Power Ground	<0.1v

1998-2000 Wrangler 2.5L I4 MFI VIN P 'C2' White Connector

PCM Pin #	Wire Color	Circuit Description (32 Pin)	Value at Hot Idle
1-3	---	Not Used	---
4	WT/DB	Injector 1 Driver	1-4 ms
5	YL/WT	Injector 3 Driver	1-4 ms
6-9	---	Not Used	---
10	DG	Generator Field Control	Digital Signals: 0-12-0v
11	OR/LG	TCC Solenoid Control (ATX)	TCC on at Cruise: 1v
12-14	---	Not Used	---
15	TN	Injector 2 Driver	1-4 ms
16	LB/BR	Injector 4 Driver	1-4 ms
17-22	---	Not Used	---
23	GY/YL	Engine Oil Pressure Sensor	1.6v at 24 psi
24-26	---	Not Used	---
27	WT/OR	Vehicle Speed Signal	Digital Signal
28-30	---	Not Used	---
31	PK/OR	5-Volt Supply	4.9-5.1v
32	---	Not Used	---

Standard Colors and Abbreviations

Abbreviation	Color	Abbreviation	Color	Abbreviation	Color
BK	Black	GY	Gray	RD	Red
BL	Blue	GN	Green	TN	Tan
BR	Brown	LG	Light Green	VT	Violet
DB	Dark Blue	OR	Orange	WT	White
DG	Dark Green	PK	Pink	YL	Yellow

1998-2000 Wrangler 2.5L I4 MFI VIN P 'C3' Gray Connector

PCM Pin #	Wire Color	Circuit Description (32 Pin)	Value at Hot Idle
1	DB/OR	A/C Clutch Relay Control	Relay Off: 12v, On: 1v
2	---	Not Used	---
3	DB/YL	ASD Relay Control	Relay Off: 12v, On: 1v
4	TN/RD	Speed Control Vacuum Solenoid	Vacuum Increasing: 1v
5	LG/RD	Speed Control Vent Solenoid	Vacuum Decreasing: 1v
6-9	---	Not Used	---
10	WT/DG	LDP Solenoid Control	PWM Signal: 0-12-0v
11	YL/RD	Speed Control Power Supply	12-14v
12	DG/PK	ASD Relay Output	12-14v
13	---	Not Used	---
14	WT/OR	LDP Switch Sense Signal	Closed: 0v, Open: 12v
15	PK/YL	Battery Temperature Sensor	At 86°F: 1.96v
16-18	---	Not Used	---
19	BR	Fuel Pump Relay Control	Relay Off: 12v, On: 1v
20	PK/BK	EVAP Purge Solenoid Control	PWM Signal: 0-12-0v
21	---	Not Used	---
22	DB/WT	A/C Switch Signal	A/C Off: 12v, On: 1v
23	LG	A/C Select Signal	A/C Off: 12v, On: 1v
24	WT/PK	Brake Switch Sense Signal	Brake Off: 0v, On: 12v
25	DG/OR	Generator Field Source	12-14v
26	DB/LG	Fuel Level Sensor Signal	Full: 0.5v, 1/2 full: 2.5v
27	PK	SCI Transmit	0v
28	WT/BK	CCD Bus (-)	<0.050v
29	LG	SCI Receive	5v
30	PK/BR	CCD Bus (+)	Digital Signals: 0-5-0v
31	---	Not Used	---
32	RD/LG	Speed Control Switch Signal	S/C & Set Switch On: 3.8v

Pin Connector Graphic

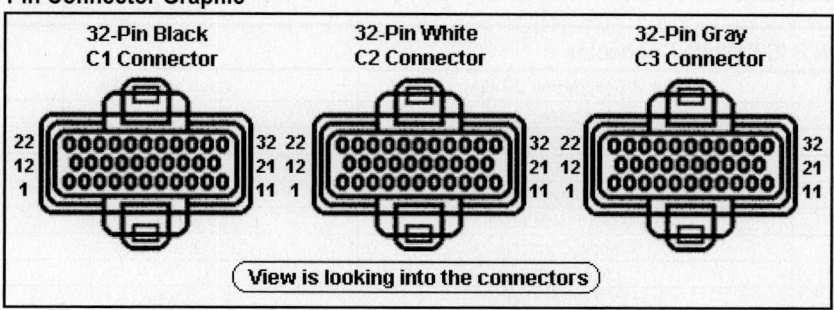

2001-02 Wrangler 2.5L I4 MFI VIN P 'C1' Black Connector

PCM Pin #	Wire Color	Circuit Description (32 Pin)	Value at Hot Idle
1	---	Not Used	---
2	DB	Ignition Switch Power (B+)	12-14v
3	---	Not Used	---
4	BK/LB	Sensor Ground	<0.050v
5	---	Not Used	---
6	BR/LB	PNP Switch Sense Signal	In P/N: 0v, Others: 5v
7	GY	Coil 1 Driver Control	5°, 55 mph: 8° dwell
8	GY/BK	CKP Sensor Signal	Digital Signals: 0-5-0v
9	---	Not Used	---
10	YL/BK	IAC 2 Driver Control	Pulse Signals
11	BR/WT	IAC 3 Driver Control	Pulse Signals
12	DB/BR	PSP Switch Sense Signal	Straight: 0v, Turning: 5v
13-14	---	Not Used	---
15	BK/RD	IAT Sensor Signal	At 100°F: 1.83v
16	TN/BK	ECT Sensor Signal	At 180°F: 2.80v
17	OR	5-Volt Supply	4.9-5.1v
18	TN/YL	CMP Sensor Signal	Digital Signals: 0-5-0v
19	GY/RD	IAC 1 Driver Control	Pulse Signals
20	VT/BK	IAC 4 Driver Control	Pulse Signals
21	---	Not Used	---
22	RD/WT	Fused Battery Power (B+)	12-14v
23	OR/DB	TP Sensor Signal	0.6-1.0v
24	BK/DG	HO2S-11 (B1 S1) Signal	0.1-1.1v
25	TN/WT	HO2S-12 (B1 S2) Signal	0.1-1.1v
26	---	Not Used	---
27	DG/RD	MAP Sensor Signal	1.5-1.6v
28-30	---	Not Used	---
31	BK/TN	Power Ground	<0.1v
32	BK/TN	Power Ground	<0.1v

2001-02 Wrangler 2.5L I4 MFI VIN P 'C2' White Connector

PCM Pin #	Wire Color	Circuit Description (32 Pin)	Value at Hot Idle
1-3	---	Not Used	---
4	WT/DB	Injector 1 Driver Control	1-4 ms
5	YL/WT	Injector 3 Driver Control	1-4 ms
6-9	---	Not Used	---
10	DG	Generator Field Control	Digital Signals: 0-12-0v
11	OR/LG	TCC Solenoid Control (ATX)	TCC on at Cruise: 1v
12-14	---	Not Used	---
15	TN	Injector 2 Driver Control	1-4 ms
16	LB/BR	Injector 4 Driver Control	1-4 ms
17-22	---	Not Used	---
23	GY/YL	Engine Oil Pressure Sensor	1.6v at 24 psi
24-26	---	Not Used	---
27	WT/OR	Vehicle Speed Signal	Digital Signal
28-30	---	Not Used	---
31	VT/OR	5-Volt Supply	4.9-5.1v
32	---	Not Used	---

Standard Colors and Abbreviations

Abbreviation	Color	Abbreviation	Color	Abbreviation	Color
BK	Black	GY	Gray	RD	Red
BL	Blue	GN	Green	TN	Tan
BR	Brown	LG	Light Green	VT	Violet
DB	Dark Blue	OR	Orange	WT	White
DG	Dark Green	PK	Pink	YL	Yellow

2001-02 Wrangler 2.5L I4 MFI VIN P 'C3' Gray Connector

PCM Pin #	Wire Color	Circuit Description (32 Pin)	Value at Hot Idle
1	DB/OR	A/C Clutch Relay Control	Relay Off: 12v, On: 1v
2	---	Not Used	---
3	DB/YL	ASD Relay Control	Relay Off: 12v, On: 1v
4	TN/RD	Speed Control Vacuum Solenoid	Vacuum Increasing: 1v
5	LG/RD	Speed Control Vent Solenoid	Vacuum Decreasing: 1v
6-9	---	Not Used	---
10	WT/DG	LDP Solenoid Control	PWM Signal: 0-12-0v
11	YL/RD	Speed Control Power Supply	12-14v
12	DG/PK	ASD Relay Output	12-14v
13	---	Not Used	---
14	OR	LDP Switch Sense Signal	Closed: 0v, Open: 12v
15	PK/YL	Battery Temperature Sensor	At 86°F: 1.96v
16-18	---	Not Used	---
19	BR	Fuel Pump Relay Control	Relay Off: 12v, On: 1v
20	PK/BK	EVAP Purge Solenoid Control	PWM Signal: 0-12-0v
21	---	Not Used	---
22	DB/OR	A/C Switch Signal	A/C Off: 12v, On: 1v
23	LG	A/C Select Signal	A/C Off: 12v, On: 1v
24	WT/PK	Brake Switch Sense Signal	Brake Off: 0v, On: 12v
25	WT/DB	Generator Field Source	12-14v
26	DB/LG	Fuel Level Sensor Signal	Full: 0.5v, 1/2 full: 2.5v
27	PK	SCI Transmit	0v
28	---	Not Used	---
29	LG	SCI Receive	5v
30	VT/YL	PCI Bus Signal (J1850)	Digital Signals: 0-7-0v
31	---	Not Used	---
32	RD/LG	Speed Control Switch Signal	S/C & Set Switch On: 3.8v

Pin Connector Graphic

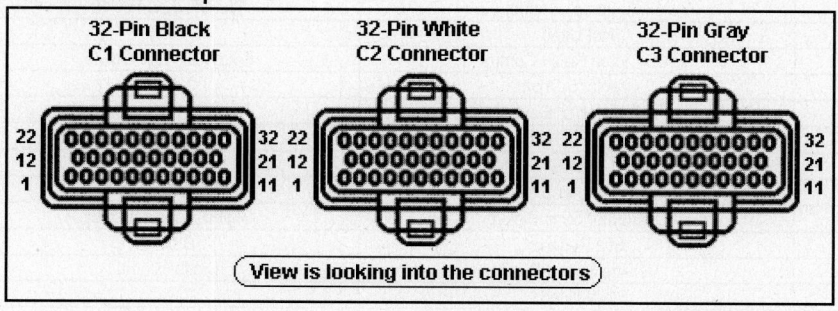

1991-92 Wrangler 4.0L V6 MFI VIN S (All) 60 Pin Connector

PCM Pin #	Wire Color	Circuit Description (60 Pin)	Value at Hot Idle
1	RD/WT	MAP Sensor Signal	1.5-1.6v
2	TN	ECT Sensor Signal	At 180°F: 2.80v
3	PK/YL	Fused Battery Power (B+)	12-14v
4	BR/RD	Sensor Ground	<0.1v
5	BK/YL	Sensor Ground	<0.1v
6	BR/YL	5-Volt Supply	4.9-5.1v
7	WT/BK	8-Volt Supply	7.9-8.1v
8 ('91)	DG/LG	Starter Input Signal	Cranking: 9-11v
9	YL	Ignition Switch Power (B+)	12-14v
9 ('91)	W/Y	Ignition Switch Power (B+)	12-14v
10, 17-18, 23	---	Not Used	---
11	BK	Power Ground	<0.1v
12	BK	Power Ground	<0.1v
13	YL	Injector 4 Driver	1-4 ms
14	TN/YL	Injector 3 Driver	1-4 ms
15	LG	Injector 2 Driver	1-4 ms
16	LB	Injector 1 Driver	1-4 ms
19	YL/BK	Coil 1 Driver Control	5°, 55 mph: 8° dwell
20	TN/BK	Alternator Field Control	Digital Signals: 0-12-0v
21	TN/DB	Air Temperature Sensor	At 100°F: 2.51v
22	YL/DG	TP Sensor Signal	0.6-1.0v
24	RD/DG	Distributor Reference Pickup	Digital Signals: 0-5-0v
25	BK	SCI Transmit	0v
26, 31, 33	---	Not Used	---
27	LB/WT	A/C Request Switch Sense	A/C Off: 12v, On: 1v
28	LB/RD	A/C Select Switch Sense	A/C Off: 12v, On: 1v
29	LB/YL	Brake Switch Sense Signal	Brake Off: 0v, On: 12v
30	BK/WT	PNP Switch Sense Signal	In P/N: 0v, Others: 5v
32	DG/RD	MIL (lamp) Control	MIL Off: 12v, On: 1v
35-37, 42, 46	---	Not Used	---
34	DB/WT	A/C Clutch Relay Control	Relay Off: 12v, On: 1v
38	WT	Injector 5 Driver	1-4 ms
39	DG/BK	AIS Motor 'D' Control	Pulse Signals
40	RD/YL	AIS Motor 'B' Control	Pulse Signals
41	GY	HO2S-11 (B1 S1) Signal	0.1-1.1v
43	OR	Tachometer Signal	Pulse Signals
44	GY/BK	Distributor Sync Pickup	Digital Signals: 0-5-0v
45	BK	SCI Receive	5v
47	DB	Vehicle Speed Signal	Digital Signal
48-50, 52-53, 55	---	Not Used	---
51	OR/DG	ASD Relay Control	Relay Off: 12v, On: 1v
54	BR	M/T: SIL (lamp) Control	Lamp Off: 12v, On: 1v
56	DG/OR	Maintenance Indicator Lamp	Lamp Off: 12v, On: 1v
57	DG/BK	ASD Relay Output	12-14v
58	BR/DG	Injector 6 Driver	1-4 ms
59	DB/YL	AIS Motor 'D' Control	Pulse Signals
60	PK/BK	AIS Motor 'C' Control	Pulse Signals

Pin Connector Graphic

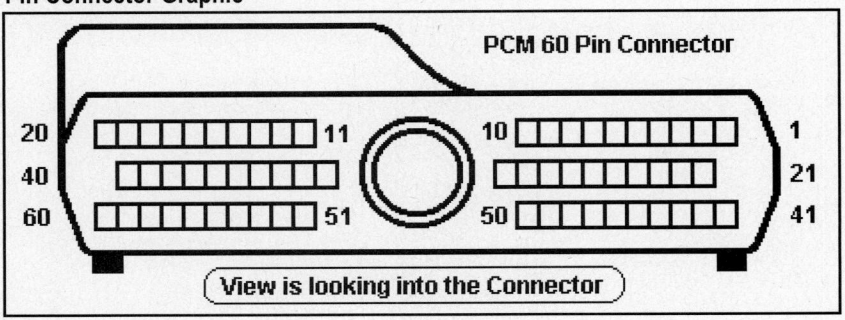

1993-95 Wrangler 4.0L V6 MFI VIN S (All) 60 Pin Connector

PCM Pin #	Wire Color	Circuit Description (60 Pin)	Value at Hot Idle
1	DG/RD	MAP Sensor Signal	1.5-1.6v
2	TN/BK	ECT Sensor Signal	At 180°F: 2.80v
3	RD/WT	Fused Battery Power (B+)	12-14v
4	BK/LB	Sensor Ground	<0.1v
5	BK/WT	Sensor Ground	<0.1v
6	PK/WT	5-Volt Supply	4.9-5.1v
7	OR	8-Volt Supply	7.9-8.1v
8	---	Not Used	---
9	DB/WT	Ignition Switch Power (B+)	12-14v
10	---	Not Used	---
11	BK	Power Ground	<0.1v
12	BK	Power Ground	<0.1v
13	LB/BR	Injector 4 Driver	1-4 ms
14	YL/WT	Injector 3 Driver	1-4 ms
15	TN	Injector 2 Driver	1-4 ms
16	WT/DB	Injector 1 Driver	1-4 ms
17-18	---	Not Used	---
19	GY	Coil 1 Driver Control	5°, 55 mph: 8° dwell
20	DG	Alternator Field Control	Digital Signals: 0-12-0v
21	BK/RD	Air Temperature Sensor	At 100°F: 2.51v
22	OR/DB	TP Sensor Signal	0.6-1.0v
23	---	Not Used	---
24	GY/BK	Distributor Reference Pickup	Digital Signals: 0-5-0v
25	BK	SCI Transmit	0v
26	---	Not Used	---
27	LB	A/C Request Switch Sense	A/C Off: 12v, On: 1v
28	BR	A/C Select Switch Sense	A/C Off: 12v, On: 1v
29	WT/PK	Brake Switch Sense Signal	Brake Off: 0v, On: 12v
30	BR/YL	PNP Switch Sense Signal	In P/N: 0v, Others: 5v
31	---	Not Used	---
32	BK/PK	MIL (lamp) Control	MIL Off: 12v, On: 1v
33	---	Not Used	---
34	DB/OR	A/C Clutch Relay Control	Relay Off: 12v, On: 1v
35-37	---	Not Used	---
38	PK/BK	Injector 5 Driver	1-4 ms
39	GY/RD	AIS Motor 1 Circuit Control	Pulse Signals
40	BR/WT	AIS Motor 3 Circuit Control	Pulse Signals

Standard Colors and Abbreviations

Abbreviation	Color	Abbreviation	Color	Abbreviation	Color
BK	Black	GY	Gray	RD	Red
BL	Blue	GN	Green	TN	Tan
BR	Brown	LG	Light Green	VT	Violet
DB	Dark Blue	OR	Orange	WT	White
DG	Dark Green	PK	Pink	YL	Yellow

1993-95 Wrangler 4.0L V6 MFI VIN S (All) 60 Pin Connector

PCM Pin #	Wire Color	Circuit Description (60 Pin)	Value at Hot Idle
41	BK/DG	HO2S-11 (B1 S1) Signal	0.1-1.1v
42	---	Not Used	---
43	GY/LB	Tachometer Signal	Pulse Signals
44	TN/YL	Distributor Sync Pickup	Digital Signals: 0-5-0v
45	LG	SCI Receive	5v
46	---	Not Used	---
47	WT/OR	Vehicle Speed Signal	Digital Signal
48-50	---	Not Used	---
51	DB/YL	ASD Relay Control	Relay Off: 12v, On: 1v
52-53	---	Not Used	---
54	OR/BK	M/T: SIL (lamp) Control	Lamp Off: 12v, On: 1v
55	---	Not Used	---
56 ('93)	GY/PK	Maintenance Indicator Lamp	Lamp Off: 12v, On: 1v
57	DG/OR	ASD Relay Output	12-14v
58	LG/BK	Injector 6 Driver	1-4 ms
59	PK/BK	AIS Motor 4 Circuit Control	Pulse Signals
60	YL/BK	AIS Motor 2 Circuit Control	Pulse Signals

Pin Connector Graphic

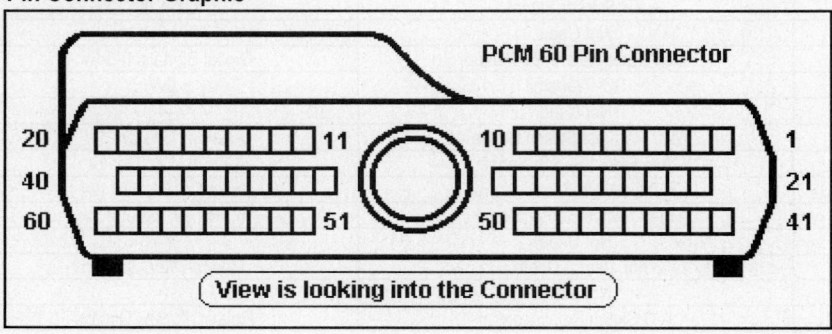

1997 Wrangler 4.0L V6 MFI VIN S 'A' Black Connector

PCM Pin #	Wire Color	Circuit Description (32 Pin)	Value at Hot Idle
1, 3	---	Not Used	---
2	RD/LG	Ignition Switch Power (B+)	12-14v
4	BR/YL	Sensor Ground	<0.1v
5, 9	---	Not Used	---
6	BR/LB	PNP Switch Sense Signal	In P/N: 0v, Others: 5v
7	GY	Coil 1 Driver Control	5°, 55 mph: 8° dwell
8	GY/BK	CKP Sensor Signal	Digital Signals: 0-5-0v
10	YL/BK	IAC 2 Driver Control	Pulse Signals
11	BR/WT	IAC 3 Driver Control	Pulse Signals
12-14	---	Not Used	---
15	BK/RD	IAT Sensor Signal	At 100°F: 1.83v
16	TN/BK	ECT Sensor Signal	At 180°F: 2.80v
17	OR	5-Volt Supply	4.9-5.1v
18	TN/YL	CMP Sensor Signal	Digital Signals: 0-5-0v
19	GY/RD	IAC 1 Driver Control	Pulse Signals
20	PK/BK	IAC 4 Driver Control	Pulse Signals
21	---	Not Used	---
22	RD/WT	Fused Battery Power (B+)	12-14v
23	OR/DB	TP Sensor Signal	0.6-1.0v
24	BK/DG	HO2S-11 (B1 S1) Signal	0.1-1.1v
25	TN/BK	HO2S-12 (B1 S2) Signal	0.1-1.1v
26	---	Not Used	---
27	DG/RD	MAP Sensor Signal	1.5-1.6v
28-30	---	Not Used	---
31	BK/TN	Power Ground	<0.1v
32	BK/TN	Power Ground	<0.1v

1997 Wrangler 4.0L V6 MFI VIN S 'B' White Connector

PCM Pin #	Wire Color	Circuit Description (32 Pin)	Value at Hot Idle
1-3	---	Not Used	---
4	WT/DB	Injector 1 Driver	1-4 ms
5	YL/WT	Injector 3 Driver	1-4 ms
6	PK/BK	Injector 5 Driver	1-4 ms
7-9	---	Not Used	---
10	DG	Generator Field Control	Digital Signals: 0-12-0v
11	OR/LG	TCC Solenoid Control	TCC on at Cruise: 1v
12	LG/BK	Injector 6 Driver	1-4 ms
13-14	---	Not Used	---
15	TN	Injector 2 Driver	1-4 ms
16	LB/BR	Injector 4 Driver	1-4 ms
17-22	---	Not Used	---
23	GY/YL	Engine Oil Pressure Sensor	1.6v at 24 psi
24-26	---	Not Used	---
27	WT/OR	Vehicle Speed Signal	Digital Signal
28-30	---	Not Used	---
31	PK/OR	5-Volt Supply	4.9-5.1v
32	---	Not Used	---

Standard Colors and Abbreviations

Abbreviation	Color	Abbreviation	Color	Abbreviation	Color
BK	Black	GY	Gray	RD	Red
BL	Blue	GN	Green	TN	Tan
BR	Brown	LG	Light Green	VT	Violet
DB	Dark Blue	OR	Orange	WT	White
DG	Dark Green	PK	Pink	YL	Yellow

1997 Wrangler 4.0L V6 MFI VIN S 'C' Gray Connector

PCM Pin #	Wire Color	Circuit Description (32 Pin)	Value at Hot Idle
1	DB/OR	A/C Clutch Relay Control	Relay Off: 12v, On: 1v
2	---	Not Used	---
3	DB/YL	ASD Relay Control	Relay Off: 12v, On: 1v
4-11	---	Not Used	---
12	DG/OR	ASD Relay Output	12-14v
13-14	---	Not Used	---
15	PK/YL	Battery Temperature Sensor	At 86°F: 1.96v
16-18	---	Not Used	---
19	BR	Fuel Pump Relay Control	Relay Off: 12v, On: 1v
20	PK/BK	EVAP Purge Solenoid Control	PWM Signal: 0-12-0v
21	---	Not Used	---
22	DB/WT	A/C Request Signal	A/C Off: 12v, On: 1v
23	LG/DG	A/C Select Signal	A/C Off: 12v, On: 1v
24	WT/PK	Brake Switch Sense Signal	Brake Off: 0v, On: 12v
25	DG/OR	Generator Field Source	12-14v
26	DB/LG	Fuel Level Sensor Signal	Full: 0.5v, 1/2 full: 2.5v
27	PK	SCI Transmit	0v
28	WT/BK	CCD Bus (-)	<0.050v
29	LG	SCI Receive	5v
30	PK/BR	CCD Bus (+)	Digital Signals: 0-5-0v
31-32	---	Not Used	---

Pin Connector Graphic

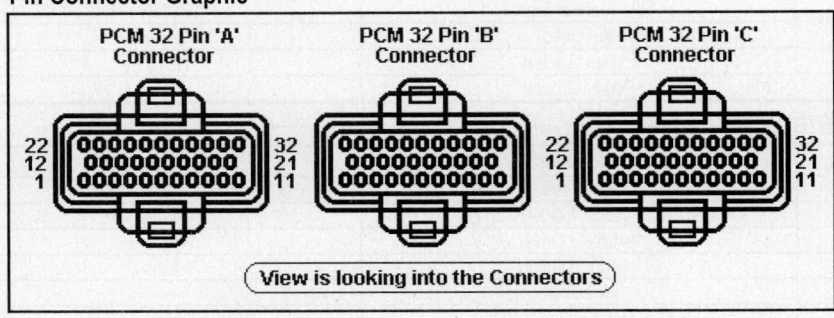

1998-2000 Wrangler 4.0L V6 VIN S (All) 'C1' Black Connector

PCM Pin #	Wire Color	Circuit Description (32 Pin)	Value at Hot Idle
1, 3	---	Not Used	---
2 ('98)	RD/LG	Ignition Switch Power (B+)	12-14v
2	DB	Ignition Switch Power (B+)	12-14v
4	BR/YL	Sensor Ground	<0.1v
5	---	Not Used	---
6	BR/LB	PNP Switch Sense Signal	In P/N: 0v, Others: 5v
7	GY	Coil 1 Driver Control	5°, 55 mph: 8° dwell
8	GY/BK	CKP Sensor Signal	Digital Signals: 0-5-0v
9	---	Not Used	---
10	YL/BK	IAC 2 Driver Control	Pulse Signals
11	BR/WT	IAC 3 Driver Control	Pulse Signals
12-14	---	Not Used	---
15	BK/RD	IAT Sensor Signal	At 100°F: 1.83v
16	TN/BK	ECT Sensor Signal	At 180°F: 2.80v
17	OR	5-Volt Supply	4.9-5.1v
18	TN/YL	CMP Sensor Signal	Digital Signals: 0-5-0v
19	GY/RD	IAC 1 Driver Control	Pulse Signals
20	PK/BK	IAC 4 Driver Control	Pulse Signals
21	---	Not Used	---
22	RD/WT	Fused Battery Power (B+)	12-14v
23	OR/DB	TP Sensor Signal	0.6-1.0v
24	BK/DG	HO2S-11 (B1 S1) Signal	0.1-1.1v
25	TN/WT	HO2S-12 (B1 S2) Signal	0.1-1.1v
26	---	Not Used	---
27	DG/RD	MAP Sensor Signal	1.5-1.6v
28-30	---	Not Used	---
31, 32	BK/TN	Power Ground	<0.1v

1998-2000 Wrangler 4.0L V6 VIN S (All) 'C2' White Connector

PCM Pin #	Wire Color	Circuit Description (32 Pin)	Value at Hot Idle
1-3	---	Not Used	---
4	WT/DB	Injector 1 Driver	1-4 ms
5	YL/WT	Injector 3 Driver	1-4 ms
6	PK/BK	Injector 5 Driver	1-4 ms
7-9	---	Not Used	---
10	DG	Generator Field Control	Digital Signals: 0-12-0v
11	OR/LG	TCC Solenoid Control (ATX)	TCC on at Cruise: 1v
12	LG/BK	Injector 6 Driver	1-4 ms
13-14	---	Not Used	---
15	TN	Injector 2 Driver	1-4 ms
16	LB/BR	Injector 4 Driver	1-4 ms
17-22	---	Not Used	---
23	GY/YL	Engine Oil Pressure Sensor	1.6v at 24 psi
24-26	---	Not Used	---
27	WT/OR	Vehicle Speed Signal	Digital Signal
28-30	---	Not Used	---
31	PK/OR	5-Volt Supply	4.9-5.1v
32	---	Not Used	---

Standard Colors and Abbreviations

Abbreviation	Color	Abbreviation	Color	Abbreviation	Color
BK	Black	GY	Gray	RD	Red
BL	Blue	GN	Green	TN	Tan
BR	Brown	LG	Light Green	VT	Violet
DB	Dark Blue	OR	Orange	WT	White
DG	Dark Green	PK	Pink	YL	Yellow

1998-2000 Wrangler 4.0L V6 VIN S (All) 'C3' Gray Connector

PCM Pin #	Wire Color	Circuit Description (32 Pin)	Value at Hot Idle
1	DB/OR	A/C Clutch Relay Control	Relay Off: 12v, On: 1v
2	---	Not Used	---
3	DB/YL	ASD Relay Control	Relay Off: 12v, On: 1v
4	TN/RD	Speed Control Vacuum Solenoid	Vacuum Increasing: 1v
5	LG/RD	Speed Control Vent Solenoid	Vacuum Decreasing: 1v
6-9	---	Not Used	---
10	WT/DG	LDP Solenoid Control	PWM Signal: 0-12-0v
11	YL/RD	Speed Control Power Supply	12-14v
12	DG/PK	ASD Relay Output	12-14v
13	---	Not Used	---
14	WT/OR	LDP Switch Sense Signal	Closed: 0v, Open: 12v
15	PK/YL	Battery Temperature Sensor	At 86°F: 1.96v
16-18	---	Not Used	---
19	BR	Fuel Pump Relay Control	Relay Off: 12v, On: 1v
20	PK/BK	EVAP Purge Solenoid Control	PWM Signal: 0-12-0v
21	---	Not Used	---
22	DB/WT	A/C Switch Signal	A/C Off: 12v, On: 1v
23	LG	A/C Select Signal	A/C Off: 12v, On: 1v
24	WT/PK	Brake Switch Sense Signal	Brake Off: 0v, On: 12v
25	DG/OR	Generator Field Source	12-14v
26	DB/LG	Fuel Level Sensor Signal	Full: 0.5v, 1/2 full: 2.5v
27	PK	SCI Transmit	0v
28	WT/BK	CCD Bus (-)	<0.050v
29	LG	SCI Receive	5v
30	PK/BR	CCD Bus (+)	Digital Signals: 0-5-0v
31	---	Not Used	---
32	RD/LG	Speed Control Switch Signal	S/C & Set Switch On: 3.8v

Pin Connector Graphic

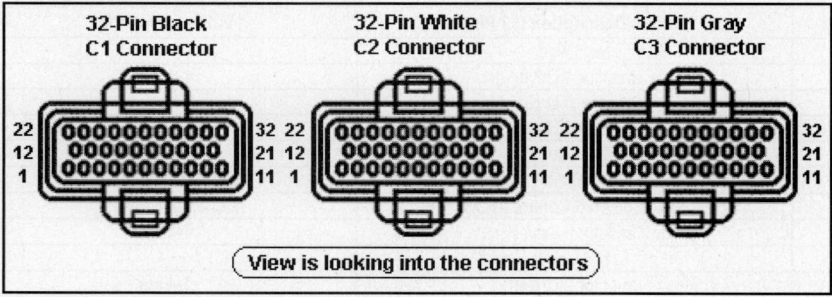

2001 Wrangler 4.0L V6 VIN S (All) 'A' Black Connector

PCM Pin #	Wire Color	Circuit Description (32 Pin)	Value at Hot Idle
1	RD/YL	Ignition Coil 3 Driver Control	5°, 55 mph: 8° dwell
2	DB	Ignition Switch Power (B+)	12-14v
3	---	Not Used	---
4	BK/LB	Sensor Ground	<0.050v
5	---	Not Used	---
6	BR/LB	PNP Switch Sense Signal	In P/N: 0v, Others: 5v
7	GY	Ignition Coil 1 Driver Control	5°, 55 mph: 8° dwell
8	GY/BK	CKP Sensor Signal	Digital Signals: 0-5-0v
9	---	Not Used	---
10	YL/BK	IAC 2 Driver Control	Pulse Signals
11	BR/WT	IAC 3 Driver Control	Pulse Signals
12-14	---	Not Used	---
15	BK/RD	IAT Sensor Signal	At 100°F: 1.83v
16	TN/BK	ECT Sensor Signal	At 180°F: 2.80v
17	OR	5-Volt Supply	4.9-5.1v
18	TN/YL	CMP Sensor Signal	Digital Signals: 0-5-0v
19	GY/RD	IAC 1 Driver Control	Pulse Signals
20	VT/BK	IAC 4 Driver Control	Pulse Signals
21	---	Not Used	---
22	RD/WT	Fused Battery Power (B+)	12-14v
23	OR/DB	TP Sensor Signal	0.6-1.0v
24	BK/DG	HO2S-11 (B1 S1) Signal	0.1-1.1v
25	TN/WT	HO2S-12 (B1 S2) Signal	0.1-1.1v
26	LG/RD	HO2S-21 (B2 S1) Signal (California)	0.1-1.1v
27	DG/RD	MAP Sensor Signal	1.5-1.6v
28	---	Not Used	---
29	TN	HO2S-22 (B2 S2) Signal (California)	0.1-1.1v
30	---	Not Used	---
31-32	BK/TN	Power Ground	<0.1v

2001 Wrangler 4.0L V6 VIN S (All) 'B' White Connector

PCM Pin #	Wire Color	Circuit Description (32 Pin)	Value at Hot Idle
1-3	---	Not Used	---
4	WT/DB	Injector 1 Driver Control	1-4 ms
5	YL/WT	Injector 3 Driver Control	1-4 ms
6	PK/BK	Injector 5 Driver Control	1-4 ms
7-8	---	Not Used	---
9	DB/TN	Injector 2 Driver Control	1-4 ms
10	DG	Generator Field Control	Digital Signals: 0-12-0v
11	OR/LG	TCC Solenoid Control (ATX)	TCC on at Cruise: 1v
12	LG/BK	Injector 6 Driver Control	1-4 ms
13-14	---	Not Used	---
15	TN	Injector 2 Driver Control	1-4 ms
16	LB/BR	Injector 4 Driver Control	1-4 ms
17-22	---	Not Used	---
23	GY/YL	Engine Oil Pressure Sensor	1.6v at 24 psi
24-26	---	Not Used	---
27	WT/OR	Vehicle Speed Signal	Digital Signal
28-30	---	Not Used	---
31	VT/OR	5-Volt Supply	4.9-5.1v
32	---	Not Used	---

Standard Colors and Abbreviations

Abbreviation	Color	Abbreviation	Color	Abbreviation	Color
BK	Black	GY	Gray	RD	Red
BL	Blue	GN	Green	TN	Tan
BR	Brown	LG	Light Green	VT	Violet
DB	Dark Blue	OR	Orange	WT	White
DG	Dark Green	PK	Pink	YL	Yellow

2001 Wrangler 4.0L V6 MFI VIN S (All) 'C' Gray Connector

PCM Pin #	Wire Color	Circuit Description (32 Pin)	Value at Hot Idle
1	DB/OR	A/C Clutch Relay Control	Relay Off: 12v, On: 1v
2	---	Not Used	
3	DB/YL	ASD Relay Control	Relay Off: 12v, On: 1v
4	TN/RD	Speed Control Vacuum Solenoid	Vacuum Increasing: 1v
5	LG/RD	Speed Control Vent Solenoid	Vacuum Decreasing: 1v
6-7	---	Not Used	---
8	BR/OR	HO2S-11 Heater Relay	Relay Off: 12v, On: 1v
9	RD/YL	HO2S-12 Heater Relay	Relay Off: 12v, On: 1v
10	WT/DG	LDP Solenoid Control	PWM Signal: 0-12-0v
11	YL/RD	Speed Control Power Supply	12-14v
12	DG/PK	ASD Relay Output	12-14v
13	---	Not Used	---
14	OR	LDP Switch Sense Signal	Closed: 0v, Open: 12v
15	PK/YL	Battery Temperature Sensor	At 86°F: 1.96v
16	BR/WT	HO2S-21 Heater Control	Relay Off: 12v, On: 1v
17-18	---	Not Used	
19	BR	Fuel Pump Relay Control	Relay Off: 12v, On: 1v
20	PK/BK	EVAP Purge Solenoid Control	PWM Signal: 0-12-0v
21	---	Not Used	---
22	DB/OR	A/C Switch Signal	A/C Off: 12v, On: 1v
23	LG	A/C Select Signal	A/C Off: 12v, On: 1v
24	WT/PK	Brake Switch Sense Signal	Brake Off: 0v, On: 12v
25	WT/DB	Generator Field Source	12-14v
26	DB/LG	Fuel Level Sensor Signal	Full: 0.5v, 1/2 full: 2.5v
27	PK	SCI Transmit	0v
28	---	Not Used	---
29	LG	SCI Receive	5v
30	VT/YL	PCI Bus Signal (J1850)	Digital Signals: 0-7-0v
31	---	Not Used	---
32	RD/LG	Speed Control Switch Signal	S/C & Set Switch On: 3.8v

Pin Connector Graphic

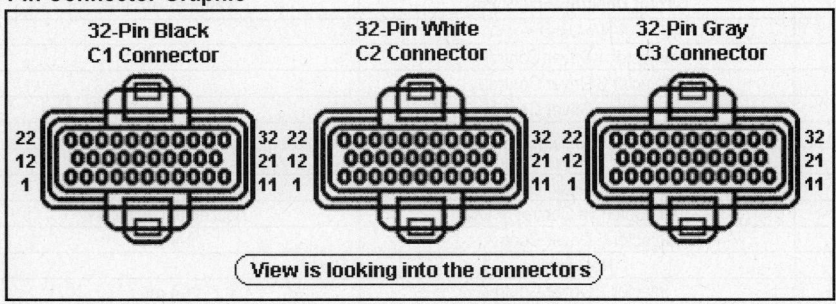

2002-03 Wrangler 4.0L V6 VIN S (All) 'A' Black Connector

PCM Pin #	Wire Color	Circuit Description (32 Pin)	Value at Hot Idle
1	RD/YL	Ignition Coil 3 Driver Control	5º, 55 mph: 8º dwell
2	DB/WT	Ignition Switch Power (Start-Run)	12-14v
3, 5	---	Not Used	---
4	BK/LB	Sensor Ground	<0.050v
6	BK/WT	PNP Switch Sense Signal	In P/N: 0v, Others: 5v
7	BR/GY	Ignition Coil 1 Driver Control	5º, 55 mph: 8º dwell
8	GY/BK	CKP Sensor Signal	Digital Signals: 0-5-0v
9	---	Not Used	---
10	YL/BK	IAC 2 Driver Control	DC Pulse Signals
11	BR/WT	IAC 3 Driver Control	DC Pulse Signals
12, 14	---	Not Used	---
13	YL/RD	Ignition Switch Power (Start)	12-14v
15	BK/RD	IAT Sensor Signal	At 100ºF: 1.83v
16	TN/BK	ECT Sensor Signal	At 180ºF: 2.80v
17	OR	5-Volt Supply	4.9-5.1v
18	TN/YL	CMP Sensor Signal	Digital Signals: 0-5-0v
19	GY/RD	IAC 1 Driver Control	DC Pulse Signals
20	VT/BK	IAC 4 Driver Control	DC Pulse Signals
21	---	Not Used	---
22	RD/WT	Battery Power (B+)	12-14v
23	OR/DB	TP Sensor Signal	0.6-1.0v
24	BK/DG	HO2S-11 (B1 S1) Signal	0.1-1.1v
25	TN/WT	HO2S-12 (B1 S2) Signal	0.1-1.1v
26	LG/RD	HO2S-21 (B2 S1) Signal	0.1-1.1v
27	DG/RD	MAP Sensor Signal	1.5-1.6v
28	---	Not Used	---
29	TN/WT	HO2S-22 (B2 S2) Signal	0.1-1.1v
30	---	Not Used	---
31-32	BK/TN	Power Ground	<0.1v

2002-03 Wrangler 4.0L V6 VIN S (All) 'B' White Connector

PCM Pin #	Wire Color	Circuit Description (32 Pin)	Value at Hot Idle
1-3	---	Not Used	---
4	WT/DB	Injector 1 Driver Control	1-4 ms
5	YL/WT	Injector 3 Driver Control	1-4 ms
6	GY	Injector 5 Driver Control	1-4 ms
7-8	---	Not Used	---
9	DB/TN	Injector 2 Driver Control	1-4 ms
10	DG	Generator Field Control	Digital Signals: 0-12-0v
11	---	Not Used	---
12	BR/DB	Injector 6 Driver Control	1-4 ms
13-14	---	Not Used	---
15	TN	Injector 2 Driver Control	1-4 ms
16	LB/BR	Injector 4 Driver Control	1-4 ms
17-22	---	Not Used	---
23	GY/YL	Oil Pressure Sensor	1.6v at 24 psi
24-26	---	Not Used	---
27	WT/OR	Vehicle Speed Signal	Digital Signals
28-30	---	Not Used	---
31	VT/WT	5-Volt Supply	4.9-5.1v
32	---	Not Used	---

Standard Colors and Abbreviations

Abbreviation	Color	Abbreviation	Color	Abbreviation	Color
BK	Black	GY	Gray	RD	Red
BL	Blue	GN	Green	TN	Tan
BR	Brown	LG	Light Green	VT	Violet
DB	Dark Blue	OR	Orange	WT	White
DG	Dark Green	PK	Pink	YL	Yellow

2002-03 Wrangler 4.0L V6 MFI VIN S (All) 'C' Gray Connector

PCM Pin #	Wire Color	Circuit Description (32 Pin)	Value at Hot Idle
1	DB/OR	A/C Clutch Relay Control	Relay Off: 12v, On: 1v
2	---	Not Used	---
3	DB/YL	ASD Relay Control	Relay Off: 12v, On: 1v
4	TN/RD	Speed Control Vacuum Solenoid	Vacuum Increasing: 1v
5	LG/RD	Speed Control Vent Solenoid	Vacuum Decreasing: 1v
6-7	---	Not Used	---
8	BR/OR	Oxygen Sensor Upstream Relay Control	Relay Off: 12v, On: 1v
9	RD/YL	HO2S-12 Heater Control	Heater Off: 12v, On: 1v
10	WT/DG	LDP Solenoid Control	PWM Signal: 0-12-0v
11	YL/RD	Speed Control Power Supply	12-14v
12	DG/PK	ASD Relay Output	12-14v
13	YL/DG	Torque Management Request Signal	Digital Signals
14	OR	LDP Switch Sense Signal	Closed: 0v, Open: 12v
15	PK/YL	Battery Temperature Sensor	At 86°F: 1.96v
16	BR/WT	HO2S-21 Heater Control	Relay Off: 12v, On: 1v
17-18	---	Not Used	---
19	BR	Fuel Pump Relay Control	Relay Off: 12v, On: 1v
20	PK/BK	EVAP Purge Solenoid Control	PWM Signal: 0-12-0v
21	---	Not Used	---
22	DB/OR	A/C Switch Sense	A/C Off: 12v, On: 1v
23	LG	A/C Select Signal	A/C Off: 12v, On: 1v
24	WT/PK	Brake Switch Sense	Brake Off: 0v, On: 12v
25	WT/DB	Generator Field Source	12-14v
26	DB/LG	Fuel Level Sensor Signal	Full: 0.5v, 1/2 full: 2.5v
27	PK	SCI Transmit	0v
28	---	Not Used	---
29	LG/WT	SCI Receive	5v
30	VT/YL	PCI Bus Signal (J1850)	Digital Signals: 0-7-0v
31	---	Not Used	---
32	RD/LB	Speed Control Switch Signal	S/C & Set Switch On: 3.8v

Pin Connector Graphic

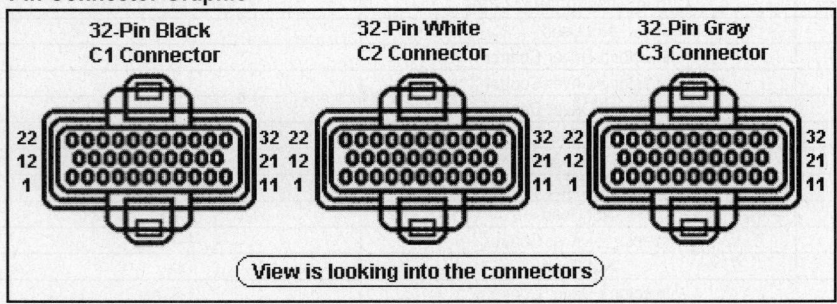

1990 Wagoneer 4.0L V6 MFI VIN L (All) Row A

PCM Pin #	Wire Color	Circuit Description (24 Pin)	Value at Hot Idle
1	TN	Injector 3 Control Driver	1-4 ms
2	BR	Injector 6 Control Driver	1-4 ms
3	LG	Injector 2 Control Driver	1-4 ms
4	YL	Injector 4 Control Driver	1-4 ms
5	OR	Fuel Pump Relay Control	Relay Off: 12v, On: 1v
6	---	Not Used	---
7	GN/YL	HO2S Relay Control	Relay Off: 12v, On: 1v
8	PK	Shift Indicator Lamp Control	Lamp Off: 12v, On: 1v
9	BK	Latch Relay B+ Control	Relay Off: 12v, On: 1v
10	GN	EGR Solenoid Control	12v, at 55 mph: 1v
11	---	Not Used	---
12	BL	A/C Clutch Relay Control	Relay Off: 12v, On: 1v

1990 Wagoneer 4.0L V6 MFI VIN L (All) Row B

PCM Pin #	Wire Color	Circuit Description (24 Pin)	Value at Hot Idle
1	LB	Injector 1 Control Driver	1-4 ms
2	WT	Injector 5 Control Driver	1-4 ms
3	RD/YL	Idle Speed Motor 'A' Control	Pulse Signals
4	BL/YL	Idle Speed Motor 'D' Control	Pulse Signals
5	GN/BK	Idle Speed Motor 'C' Control	Pulse Signals
6	PK/BK	Idle Speed Motor 'B' Control	Pulse Signals
7	RD	Battery Power (B+)	12-14v
8	YL	Ignition Switch Run Output	12-14v
9	---	Not Used	---
10	PK	B+ Latch Relay Output	Relay Off: 12v, On: 1v
11-12	BK	Power Ground	<0.1v

Pin Connector Graphic

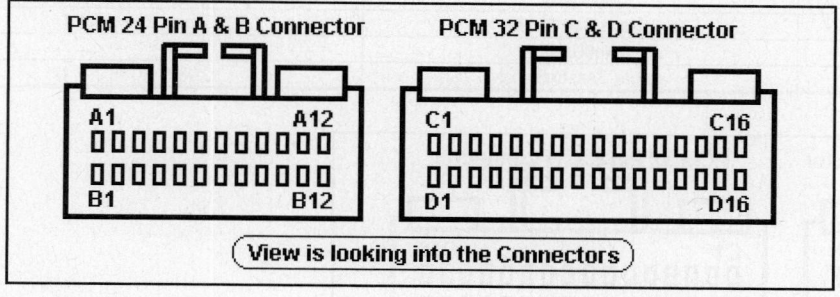

1990 Wagoneer 4.0L V6 MFI VIN L Row 'C' 32 Pin Connector

PCM Pin #	Wire Color	Circuit Description (32 Pin)	Value at Hot Idle
1	PK	CKP Sensor (+) Signal	Digital Signals: 0-5-0v
2	TN	A/C Request Switch Sense	A/C Off: 12v, On: 1v
3	GN	Starter Input Signal	Cranking: 9-11v
4	BK	PNP Switch Sense Signal	In P/N: 0v, Others: 5v
5	GY	Distributor Sync Pickup (+)	Digital Signals: 0-5-0v
6	PK	MAP Sensor Signal	1.5-1.6v
7	YL/GN	TP Sensor Signal	0.6-1.0v
8	TN	Air Temperature Sensor	At 100°F: 2.51v
9	---	Not Used	---
10	TN	ECT Sensor Signal	At 180°F: 2.80v
11	OR	Fuel Pump Relay Output	Relay Off: 12v, On: 1v
12	BR/PK	SCI Transmit	0v
13	---	Not Used	---
14	RD	MAP Sensor VREF	4.9-5.1v
15	LB	TP Sensor VREF	4.9-5.1v
16	BL	Distributor Sync Pickup (-)	Digital Signals: 0-5-0v

1990 Wagoneer 4.0L V6 MFI VIN L (All) Row 'D' 32 Pin Connector

PCM Pin #	Wire Color	Circuit Description (32 Pin)	Value at Hot Idle
1	WT	CKP Sensor (-) Signal	Digital Signals: 0-5-0v
2	LG	A/C Select Switch Sense	A/C Off: 12v, On: 1v
3	BR	Sensor Ground	<0.1v
4-7	---	Not Used	---
8	YL	Knock Sensor (-) Signal	<0.050v
9	GY	HO2S-11 (B1 S1) Signal	0.1-1.1v
10	OR	Fuel Pump Relay Control	Relay Off: 12v, On: 1v
11	BK/WT	SCI Receive	5v
12	---	Not Used	---
13	YL	Ignition Coil Control	5°, 55 mph: 8° dwell
14-15	---	Not Used	---
16	PK	Knock Sensor Signal	0.080v AC

Pin Connector Graphic

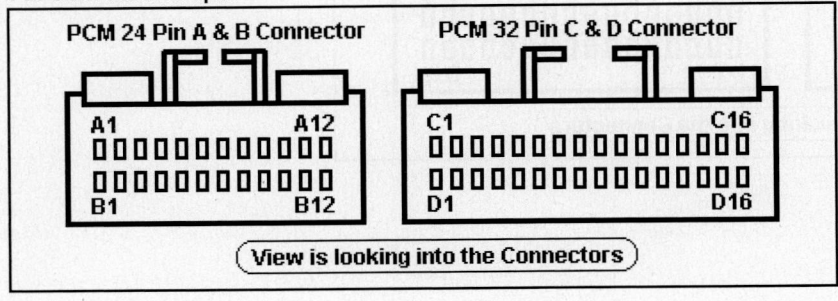

PCM 24 Pin A & B Connector

PCM 32 Pin C & D Connector

A1 A12
B1 B12

C1 C16
D1 D16

View is looking into the Connectors

Manual ISBN 1-4018-7412-6/Part No. 27412

With the *Chilton® 2005 Labor Guide*, professional technicians gain access to labor times for vehicle brands and models that conform to current Automotive Aftermarket Industry Association standards. Thousands of labor times for 1981 through 2005 domestic and imported vehicles reflect technicians' use of aftermarket tools and training. Updates based on technical hotline input, Original Equipment Manufacturer (OEM) warranty times, and technical editor evaluation include more diagnostic labor times than ever before. Labor operations have been rewritten to conform to the most recent industry standards. Prior model coverage has been re-evaluated by experts to ensure accuracy. Chilton labor times are accepted by insurance and extended warranty companies.

Labor Guide Manual Benefits:

- 2,500 pages of Chilton labor times
- each OEM is arranged alphabetically by section for easy reference
- improved indexing means easier access to today's repair industry standards

Hardcover manual is 8 7/8" x 11", ©2005

Labor Guide CD-ROM Benefits:

- easy-to-use software to create and print professional-quality estimates and invoices
- three user-defined levels of labor rates correspond to different types of job scenarios, for "real-world" application
- functions as a database of aftermarket labor times for monitoring warranty and insurance claims
- software keeps track of customers and prior estimates for time-saving recall
- customizable application allows service writers to add labor operations and times, and parts companies to add labor times to existing parts ordering systems

CD-ROM ISBN 1-4018-7818-0/Part No. 27818

Previous Year Editions

Chilton 2004 Labor Guide Manual, **ISBN 1-4018-4356-5/Part No. 24356**

Chilton 2004 Labor Guide CD-ROM, **ISBN 1-4018-4357-3/Part No. 24357**

For the most up-to-date service and repair information anywhere, look no further than the newly updated *Chilton® 2005 Mechanical Service Manuals – Annual Editions*! Still the lowest-priced professional repair manuals on the market, this series of manufacturer-based books now features an easier-to-handle, two-volume Asian Manual set. Increased model coverage over the 2004 editions is supported by more illustrations in each section, making fast, accurate repairs and reassembly easier than ever before. With modernized content, it's no wonder that more professionals trust Chilton Professional Manuals for their mechanical service and repair needs.

Mechanical Service Manual Benefits:

- all books are grouped by manufacturer to make accessing information simple
- step-by-step procedures from drive train to chassis and related components help yield fast accurate results
- comprehensive, technically-detailed content is organized by model and system, and is supported by exploded-view illustrations, diagrams, and specification charts for added clarity
- most mechanical systems are included, such as engines, suspensions, steering components, and more
- special tools are described and clearly illustrated so that performing repairs is as easy and quick as possible

Chilton 2005 Ford Mechanical Service Manual
      ISBN 1-4018-6719-7/Part No. 26719
Chilton 2005 General Motors Mechanical Service Manual
      ISBN 1-4018-7146-1/Part No. 27146
Chilton 2005 Chrysler Mechanical Service Manual
      ISBN 1-4018-6718-9/Part No. 26718
Chilton 2005 Asian Mechanical Service Manual (Complete Set of 2 manuals)
      ISBN 1-4018-7180-1/Part No.
Chilton 2005 Asian Mechanical Service Manual, Acura - Mazda
      ISBN 1-4018-6716-2/Part No. 26716
Chilton 2005 Asian Mechanical Service Manual, Mitsubishi - Toyota
      ISBN 1-4018-6717-0/Part No. 26717
Chilton 2005 European Mechanical Service Manual
      ISBN 1-4018-6720-0/Part No. 26720

Manuals are 8 1/2" x 11", ©2005

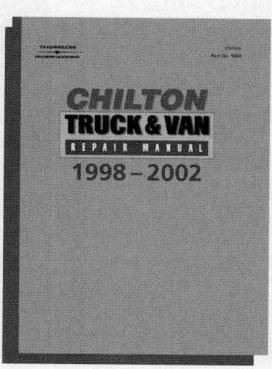

The *Chilton® Perennial Editions* contain repair and maintenance information for popular mechanical systems that may not be available elsewhere. They offer a wide range of repair information on cars, trucks, vans, and SUVs dating back to the early 1960s, and as current as 2002. Information for 1993 and later model years includes scheduled maintenance interval charts.

Benefits:

• covers the most common vehicle models found in the repair aftermarket today

• gain quick understanding of systems using exploded-view illustrations, diagrams, and charts

• simplify tough jobs with easy-to-follow removal and installation instructions for heater core and other components

• obtain complete coverage of repair procedures from drive train to chassis and associated components

Auto Repair Manual, 1998-2002, 1,426 pages
ISBN 0-8019-9362-8/Part No. 9362
Auto Repair Manual, 1993-1997, 2,064 pages
ISBN 0-8019-7919-6/Part No. 7919
Auto Repair Manual, 1988-1992, 1,284 pages
ISBN 0-8019-7906-4/Part No. 7906
Auto Repair Manual, 1980-1987, 1,344 pages
ISBN 0-8019-7670-7/Part No. 7670

Import Car Repair Manual, 1998-2002, 1,792 pps
ISBN 0-8019-9363-6/Part No. 9363
Import Car Repair Manual, 1993-1997, 2,080 pps
ISBN 0-8019-7920-X/Part No. 7920
Import Car Repair Manual, 1988-1992, 1,632 pages
ISBN 0-8019-7907-2/Part No. 7907
Import Car Repair Manual, 1980-1987, 1,488 pages
ISBN 0-8019-7672-3/Part No. 7672

Truck & Van Repair Manual, 1998-2002, 1,408 pages
ISBN 0-8019-9364-4/Part No. 9364
Truck & Van Repair Manual, 1993-1997, 2,096 pages
ISBN 0-8019-7921-8/Part No. 7921
Truck & Van Repair Manual, 1991-1995, 1,664 pages
ISBN 0-8019-7911-0/Part No. 7911
Truck & Van Repair Manual, 1986-1990, 1,536 pages
ISBN 0-8019-7902-1/Part No. 7902
Truck & Van Repair Manual, 1979-1986, 1,440 pages
ISBN 0-8019-7655-3/Part No. 7655

SUV Repair Manual, 1998-2002, 1,292 pages
ISBN 0-8019-9365-2/Part No. 9365

Hardcover manuals are 8 1/2" x 11".

Chilton Collector's Editions - *Reference Manuals for Vintage Vehicles*
Auto Repair Manual, 1964-1971, ISBN 0-8019-5974-8/Part No. 5974,
Truck & Van Repair Manual, 1961-1971, ISBN 0-8019-6198-X/Part No. 6198
Truck & Van Repair Manual, 1971-1978, ISBN 0-8019-7012-1/Part No. 7012

ASE Test Preparation Series

Thomson Delmar Learning
ISBN 1-4018-5182-7
Part No. 25182

(Complete Set: A1-A8, L1, P2 X1, C1)
Thomson Delmar Learning has developed comprehensive ASE Test Preparation Manuals to help automotive technicians increase their success on these certification programs. The material covers the topics one might find during the test process. The booklets include many review questions and answers, as well as detailed descriptions of the repairs involved. Designed to look like the actual test, participants will feel more comfortable with practice, which will translate into greater success in taking the actual tests. The design of the Delmar Learning product also includes helpful test taking hints and student preparation ideas designed to enhance success.

BENEFITS
- The history of the ASE
- Test-taking strategies
- Tasks lists and overview
- Sample test questions
- ASE-style exams
- Explanations to the answers (right and wrong)
- Glossary of terms

(A1) Automotive Engine Repair, 2E

1-4018-2040-9
Part No. 22040

General Engine Diagnosis, Cylinder Head and Valve Train Diagnosis and Repair, Engine Block Diagnosis and Repair, Lubrication and Cooling Systems Diagnosis and Repair, and Fuel, Electrical, Ignition and Exhaust Systems Inspection and Service.

(A2) Automotive Transmissions and Transaxles, 2E

1-4018-2041-7
Part No. 22041

General Transmission/ Transaxle Diagnosis (Mechanical/Hydraulic Systems and Electronic Systems), Transmission/Transaxle Maintenance and Adjustment, In-Vehicle Transmission/Transaxle Repair, Off-Vehicle Transmission/Transaxle Repair.

(A3) Automotive Manual Drive Trains and Axles, 2E

1-4018-2042-5
Part No. 22042

Clutch Diagnosis and Repair, Transmission Diagnosis and Repair, Transaxle Diagnosis and Repair, Drive Shaft/Half Shaft and Universal Joint/Constant Velocity (CV) Joint Diagnosis and Repair (Front and Rear Wheel Drive), Rear Axle Diagnosis and Repair, Four Wheel Drive/All Wheel Drive Component Diagnosis and Repair.

(A4) Automotive Suspension and Steering, 2E

1-4018-2043-3
Part No. 22043

Steering Systems Diagnosis and Repair (Steering Columns and Manual Steering Gears, Power Assisted Steering Units, Steering Linkage), Suspension Systems Diagnosis and Repair (Front Suspensions, Rear Suspensions, Miscellaneous Services), Wheel Alignment Diagnosis, Adjustment and Repair, and Wheel and Tire Diagnosis and Repair.

(A5) Automotive Brakes, 2E

1-4018-2044-1
Part No. 22044

Hydraulic System Diagnosis and Repair, Drum Brake Diagnosis and Repair, Disc Brake Diagnosis and Repair, Power Assist Units Diagnosis and Repair, Miscellaneous Systems Diagnosis and Repair, Antilock Brake Systems (ABS) Diagnosis and Repair.

(A6) Automotive Electrical-Electronic Systems, 2E

1-4018-2045-X
Part No. 22045

General Electrical/Electronic Systems Diagnosis, Battery Diagnosis and Service, Starting Systems Diagnosis and Repair, Charging Systems Diagnosis and Repair, Lighting Systems Diagnosis and Repair, Gauges, Warning Devices and Driver Information Systems Diagnosis and Repair, Horn and Wiper/Washer Diagnosis and Repair.

(A7) Automotive Heating and Air Conditioning, 2E

1-4018-2046-8
Part No. 22046

The manual for A7 includes the following topics: A/C System Diagnosis and Repair, Refrigeration System Component Diagnosis and Repair, Heating and Engine Cooling Systems Diagnosis and Repair, Operating Systems and Related Controls Diagnosis and Repair, Refrigerant Recovery, Recycling, Handling and Retrofit.

(A8) Automotive Engine Performance, 2E

1-4018-2047-6
Part No. 22047

The manual for A8 includes the following topics: General Engine Diagnosis, Ignition System Diagnosis and Repair, Fuel, Air Induction, and Exhaust Systems Diagnosis and Repair, Emissions Control Systems Diagnosis and Repair (Including OBDII), Computerized Engine controls Diagnosis and Repair (Including OBDII), Engine Electrical Systems diagnosis and Repair.

(L1) Automotive Advance Engine Performance, 2E

1-4018-2049-2
Part No. 22049

The manual for L1 includes the following topics: General Powertrain Diagnosis, Computerized Powertrain Controls Diagnosis (Including OBDII), Ignition System Diagnosis, Fuel Systems and Air Induction Systems Diagnosis, Emission Control Systems Diagnosis, I/M Failure Diagnosis.

(P2) Automobile Parts Specialist, 2E

1-4018-2048-4
Part No. 22048

The manual for P2 includes the following topics: General Operations, Customer Relations and Sales Skills, Vehicle Systems Knowledge, Vehicle Identification, Cataloging Skills, Inventory Management, Merchandising.

(X1) Exhaust Systems

1-4018-2050-6
Part No. 22050

Exhaust Systems includes the following topics: Exhaust Systems Inspection and Repair, Emissions Systems Diagnosis, Exhaust System Fabrication, Exhaust System Installation, Exhaust System Repair Regulations.

(C1) Service Consultant

See next page for details

ASE Test Preparation Series in Español!

Thomson Delmar Learning
ISBN 1-4018-1530-8

(Complete Set: A1-A8, L1, P2, X1)
Now available in Español – the first of its kind for Spanish-speaking technicians! This comprehensive package of ASE test preparation booklets are intended for any Spanish-speaking automotive technician who is preparing to take an ASE examination. The series includes questions that relate to each competency required for certification by ASE. In addition to a multitude of questions, the reason why each answer is right or wrong is explained, along with task lists and overview, test-taking strategies, and more.

(A1) Reparación de Motores, 2A Edición
1-4018-1014-4/Part No. 21014

(A2) Transmisión Automática/ Eje de Transmision Automática, 2A Edición
1-4018-1015-2/Part No. 21015

(A3) Tren de y Mando Ejes Manuales, 2A Edición
1-4018-1016-0/Part No. 21016

(A4) Suspensión y Dirección, 2A Edición
1-4018-1017-9/Part No. 21017

(A5) Frenos, 2A Edición
1-4018-1018-7/Part No. 21018

(A6) Sistemas Eléctricos/ Electrónicos, 2A Edición
1-4018-1019-5/Part No. 21019

(A7) Calefacción y Aire Acondicionado, 2A Edición
1-4018-1020-9/Part No. 21020

(A8) Funcionamiento de Motores, 2A Edición
1-4018-1021-7/Part No. 21021

(L1) Especialista en el Funciommiato Avansado de Motores, 2A Edición
1-4018-1022-5/Part No. 21022

(P2) Especialista en Partes de Automovil, 2A Edición
1-4018-1023-3/Part No. 21023

(X1) Sistemas de Escape, 2A Edición
1-4018-1024-1/Part No. 21024

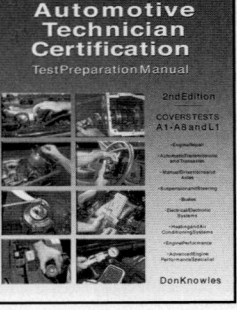

Automotive ASE Preparation Video Series

Thomson Delmar Learning

ISBN 0-7668-3168-X *(Complete Set of 12 Tapes)*
ISBN 0-7668-8042-7 *(Complete Set of 3 CD-ROMs)*

Thomson Delmar Learning's Automotive ASE Test Prep Videos present test takers with a review of the A1-A8, L1, and P2 tests prior to taking the exam. Each tape summarizes key topics and key task areas through live action and animation. Actual technicians, authentic automotive shops, and late-model vehicles are featured for an up-to-date look and feel. Safety is emphasized throughout each tape. An overview tape introduces test takers to the ASE testing style.

BENEFITS OF THE VIDEO SERIES

- lively, easy to follow videos emphasize safety throughout
- covers major task areas and topics for each of the ASE exams
- accompanying Instructor's Guide helps users comprehend and retain information presented

Complete Set of 12 Tapes (with Instructor's Guide), ©2001

Tape 1: Overview of ASE, 0-7668-2484-5
Tape 2: A1 Engine Repair, 0-7668-2485-3
Tape 3: A2 Automatic Transmission, 0-7668-2498-5
Tape 4: A3 Manual Transmission, 0-7668-2499-3
Tape 5: A4 Steering and Suspension, 0-7668-2500-0
Tape 6: A5 Automotive Brakes, 0-7668-2501-9
Tape 7: A6 Electricity/Electronics, 0-7668-2493-4
Tape 8: A7 Air Conditioning, 0-7668-2486-1
Tape 9: A8 Engine Performance, 0-7668-2494-2
Tape 10: P2 Parts Specialist, 0-7668-2487-X
Tape 11: L1 Advanced Engine Performance (Part 1), 0-7668-2491-8
Tape 12: L1 Advanced Engine Performance (Part 2), 0-7668-2492-6

BUNDLES

Bundle 1: Specialty Topics (Set of 4 Tapes) includes Overview of ASE, A1 Engine Repair, A7 Air Conditioning, and P2 Parts Specialist, 0-7668-2483-7
Bundle 2: Engine Performance/Electronics (Set of 4 Tapes) includes L1 Part 1, L1 Part 2, A6 Electricity/ Electronics, and A8 Engine Performance, 0-7668-2490-X
Bundle 3: Undercar (Set of 4 Tapes) includes A2 Automatic Transmissions, A3 Manual Transmissions, A4 Steering and Suspension, and A5 Automotive Brakes, 0-7668-2497-7

CD-ROM COURSEWARE

Based on the ASE Test Prep Series, the CD-ROMs offer the following in addition to the video content:

- Gradebook
- Pre-test/Post-test
- Ability to modify
- Video Glossary
- Variety of question types
- Remediation
- Video File Server compatible

CD-ROM 1: Specialty Topics CD-ROM includes Overview of ASE, A1 Engine Repair, A7 Air Conditioning, and P2 Parts Specialist, 0-7668-2489-6
CD-ROM 2: Engine Performance/Electronics CD-ROM includes L1 Part 1, L1 Part 2, A6 Electricity/ Electronics, and A8 Engine Performance, 0-7668-2496-9
CD-ROM 3: Undercar CD-ROM includes A2 Automatic Transmissions, A3 Manual Transmissions, A4 Steering and Suspension, and A5 Automotive Brakes, 0-7668-2503-5

The ASE "Passing Lane" Package

Thomson Delmar Learning

ISBN 0-7668-4338-6
(Complete Set: A1-A8, L1, P2)

The most comprehensive test preparation for Automotive Tests A1-A8, L1, and P2. Combining the most thorough ASE Test Preparation books with the latest in ASE videos, this package provides a program of self-study for the automotive ASE Tests.

EACH BOOK IN THE SERIES BENEFITS:

- test-taking strategies
- tasks lists and overview
- sample test questions
- ASE-style exams
- explanations to the answers
- glossary of terms

EACH VIDEO IN THE SERIES BENEFITS:

- lively, easy to follow videos emphasize safety throughout
- covers major task areas and topics for each of the ASE exams
- accompanying Activity Sheets help comprehend and retain information

(A1) Automotive Engine Repair Book/Video, 0-7668-4181-2
(A2) Automotive Transmissions and Transaxles Book/Video, 0-7668-4182-0
(A3) Automotive Manual Drive Trains and Axles Book/Video, 0-7668-4183-9
(A4) Automotive Suspension and Steering Book/Video, 0-7668-4184-7
(A5) Automotive Brakes Book/Video, 0-7668-4185-5
(A6) Automotive Electrical-Electronics Systems Book/Video, 0-7668-4186-3
(A7) Automotive Heating and Air Conditioning Book/Video, 0-7668-4187-1
(A8) Automotive Engine Performance Book/Video, 0-7668-4188-X
(L1) Automotive Advanced Engine Performance Book/Video, 0-7668-4189-8
(P2) Automobile Parts Specialist Book/Video, 0-7668-4190-1

Automotive Technician Certification Test Preparation Manual, 2E

Don Knowles

ISBN 0-7668-1948-5/ Part No. 11948

The second edition of Certified ASE Master Technician Don Knowles' popular ASE test preparation book adds coverage of the L1 Advanced Engine Performance test to its coverage of automotive tests A1 through A8. All nine tests covered in this book reflect year 2000 task lists, including the updated composite vehicle in the L1 test. This revised edition contains at least one practice question for every ASE task in the tests. Also included is the updated and expanded coverage of electronic automatic transmissions, electronically controlled automatic transmissions, electronically controlled 4 wheel drive and steering, ABS systems, wiring diagrams, and repairing electronic components.

BENEFITS

- a new section has been added on computer-controlled automatic transmissions and transaxles including those used in OBD II vehicles
- new information has been included on electronically-controlled 4WD systems and ABS systems
- the chapter on Electrical/Electronic Systems has been expanded to include information on reading wiring diagrams and inspecting, testing, and repairing electronic components
- a complete chapter has been added to prepare technicians for the Advanced Engine Performance (L1) test

CONTENTS

Engine Repair Automatic Transmission/ Transaxle. Manual Drive Train and Axles. Suspension and Steering. Brakes. Electrical/Electronic Systems. Heating, Ventilation, and Air Conditioning Systems. Engine Performance. Advanced Engine Performance.

788 pp, 8½≤ x 11≤, softcover, ©2001

This pioneering eight-book series offers automotive repair shop owners and those wanting to be shop owners the necessary business and customer service skills to run a successful automotive service facility.

The series covers three main topical areas: personnel management, business management, and sales and marketing. Each book provides a framework to help technicians make consistent, high-quality, and productive service a part of every day shop operations. According to the author, "Great performance coupled with increased customer loyalty, trust, and operational excellence will almost always result in increased profits."

Automotive Service Management Series Benefits:

- real-world approach reflects author's experience as a fourth generation technician, a repair & service company owner, and an automotive industry trainer
- all-inclusive coverage spans from designing an automotive repair facility floor plan through financial management techniques, customer/staff relations, and more
- length of each book makes it easy to incorporate this series into workshops, seminars, and training/education courses
- information is available "as is" or for customization

Total Customer Relationship Management
ISBN 1-4018-2657-1/Part No. 22657
From Intent to Implementation
ISBN 1-4018-2658-X/Part No. 22658
Operational Excellence
ISBN 1-4018-2659-8/Part No. 22659
Building a Team
ISBN 1-4018-2660-1/Part No. 22660
The High Performance Shop
ISBN 1-4018-2661-X/Part No. 22661
Safety Communications
ISBN 1-4018-2662-8/Part No. 22662
Managing Dollars with Sense
ISBN 1-4018-2663-6/Part No. 22663
Operations Management
ISBN 1-4018-2665-2/Part No. 22665
Entire Set of 8 Books
ISBN 1-4018-2499-4/Part No. 2499

Softcover manuals are 8 1/2" x 11", ©2003

ABOUT THE AUTHOR

Mitch Schneider is a fourth generation mechanic/technician and is a frequent speaker at major conventions and meetings of automotive industry trade organizations. Schneider is also an award-winning journalist and is a regular contributor and senior contributing editor for *Motor Age* magazine. He provides commentary on the evolving relationship between service dealers, jobbers, warehouse directors and manufacturers.

Schneider has also appeared on the TNN cable show "Truckin' USA" where he hosted the "Tech Tips" segment. In addition to operating the award-winning Schneider's Automotive for 22 years in Simi Valley, CA, he is also the president and founder of Schneider's Future-Tech, a service company specializing in conducting management seminars for automotive service dealers, jobbers, warehouse distribution companies, and manufacturers.

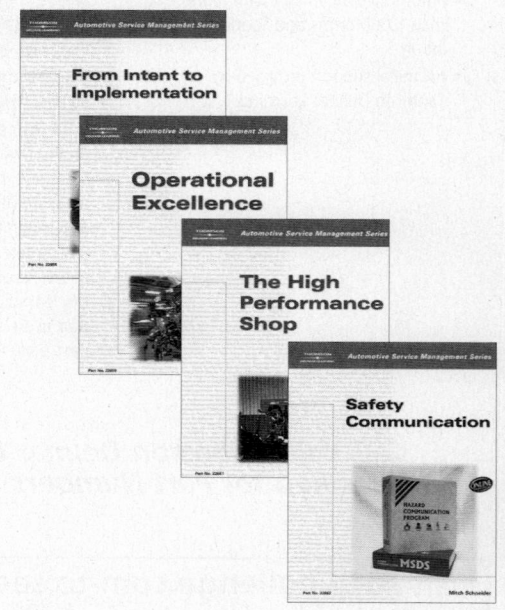

Delmar's Automotive Dictionary
David W. South & Boyce Dwiggins
ISBN 0-8273-7405-4

This handy, ready-reference dictionary provides the automotive engineer, technician, mechanic, student, enthusiast or layperson with a single source for the most up-to-date definitions available of technical, professional and informal terminology used in today's automotive world. It is descriptive and covers the wide scope of terms pertinent to the automotive field. With multiple definitions and aids, and proper pronunciation of terms, this dictionary is a must for all!

BENEFITS

- over 3000 terms comprehensively covering more than 100 subject areas
- enhanced by a list of acronyms and abbreviations
- up-to-date definitions of today's automotive terminology
- aids for proper pronunciation
- each term has multiple definitions

281 pp, 6≤ x 9≤, softcover, ©1997

Practical Problems in Mathematics for Automotive Technicians, 5E
George Morre, Todd Sformo & Larry Sformo
ISBN 0-8273-7944-7

By showing how to apply math solutions to everyday problems, this all-in-one math reference transforms the "remove it and replace it" mechanic into a complete automotive technician. The book builds from math basics to cover more complex topics--not to mention such workplace issues as invoices and scale reading of test meters. Each easy-to-read chapter features step-by-step instructions, diagrams, charts and examples to make the problem-solving process a snap.

256 pp, 7⅛≤ x 9¼≤, softcover, ©1998
Instructor's Manual   **0-8273-7945-5**

Math for the Automotive Trade, 3E
John C. Peterson & William deKryger
ISBN 0-8273-6712-0

Math for Automotive Trades, 3E provides excellent examples and problems that reflect technological requirements of workers in automotive technology. The text has three parts: review of basic mathematics skills, math applications to specific automotive situations, and an examination of measurement aspects beginning with angle and linear measurements and ending with an extensive look at measurement tools used in the automotive trade.

345 pp, 8½≤ x 11≤, softcover, ©1995
Instructor's Manual   **0-8273-6713-9**